ALGEBRA

ARITHMETIC OPERATIONS

$$a(b + c) = ab + ac \qquad \frac{a}{b} + \frac{c}{d} = \frac{ad + bc}{bd}$$

$$\frac{a + c}{b} = \frac{a}{b} + \frac{c}{b} \qquad \frac{\frac{a}{b}}{\frac{c}{d}} = \frac{a}{b} \times \frac{d}{c} = \frac{ad}{bc}$$

EXPONENTS AND RADICALS

$$x^m x^n = x^{m+n} \qquad \frac{x^m}{x^n} = x^{m-n}$$

$$(x^m)^n = x^{mn} \qquad x^{-n} = \frac{1}{x^n}$$

$$(xy)^n = x^n y^n \qquad \left(\frac{x}{y}\right)^n = \frac{x^n}{y^n}$$

$$x^{1/n} = \sqrt[n]{x} \qquad x^{m/n} = \sqrt[n]{x^m} = (\sqrt[n]{x})^m$$

$$\sqrt[n]{xy} = \sqrt[n]{x}\,\sqrt[n]{y} \qquad \sqrt[n]{\frac{x}{y}} = \frac{\sqrt[n]{x}}{\sqrt[n]{y}}$$

$$\sqrt[m]{\sqrt[n]{x}} = \sqrt[n]{\sqrt[m]{x}} = \sqrt[mn]{x}$$

FACTORING SPECIAL POLYNOMIALS

$$x^2 - y^2 = (x + y)(x - y)$$

$$x^3 + y^3 = (x + y)(x^2 - xy + y^2)$$

$$x^3 - y^3 = (x - y)(x^2 + xy + y^2)$$

BINOMIAL THEOREM

$$(x + y)^2 = x^2 + 2xy + y^2 \qquad (x - y)^2 = x^2 - 2xy + y^2$$

$$(x + y)^3 = x^3 + 3x^2y + 3xy^2 + y^3$$

$$(x - y)^3 = x^3 - 3x^2y + 3xy^2 - y^3$$

$$(x + y)^n = x^n + nx^{n-1}y + \frac{n(n-1)}{2}x^{n-2}y^2 + \cdots + \binom{n}{k}x^{n-k}y^k + \cdots + nxy^{n-1} + y^n$$

where $\displaystyle\binom{n}{k} = \frac{k(k-1)\cdots(k-n+1)}{1\cdot 2\cdot 3\cdot\cdots\cdot n}$

QUADRATIC FORMULA

If $ax^2 + bx + c = 0$, then

$$x = \frac{-b \pm \sqrt{b^2 - 4ac}}{2a}$$

INEQUALITIES AND ABSOLUTE VALUE

If $a < b$ and $b < c$, then $a < c$.

If $a < b$, then $a + c < b + c$.

If $a < b$ and $c > 0$, then $ca < cb$.

If $a < b$ and $c < 0$, then $ca > cb$.

If $a > 0$, then

$|x| = a$ means $x = a$ or $x = -a$

$|x| < a$ means $-a < x < a$

$|x| > a$ means $x > a$ or $x < -a$

GEOMETRY

GEOMETRIC FORMULAS

Formulas for area A, circumference C, and volume V:

Triangle

$A = \frac{1}{2}bh$

$= \frac{1}{2}ab\sin\theta$

Circle

$A = \pi r^2$

$C = 2\pi r$

Sector of Circle

$A = \frac{1}{2}r^2\theta$

$s = r\theta$ (θ in radians)

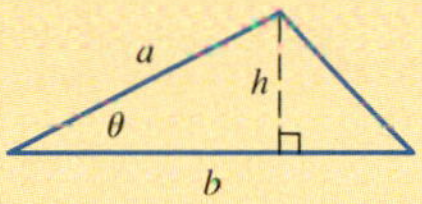

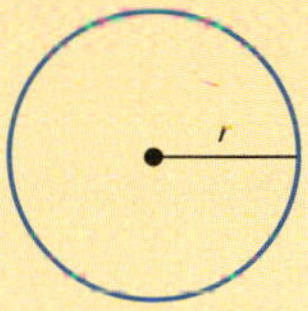

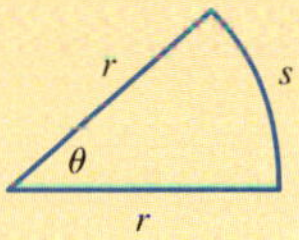

Sphere

$V = \frac{4}{3}\pi r^3$

$A = 4\pi r^2$

Cylinder

$V = \pi r^2 h$

Cone

$V = \frac{1}{3}\pi r^2 h$

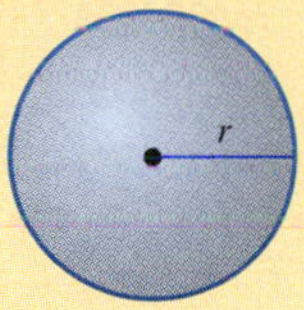

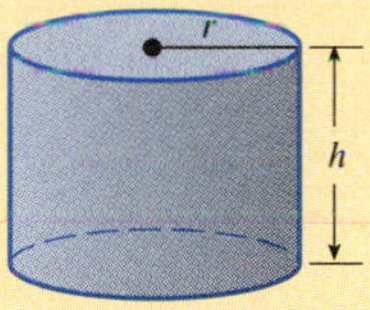

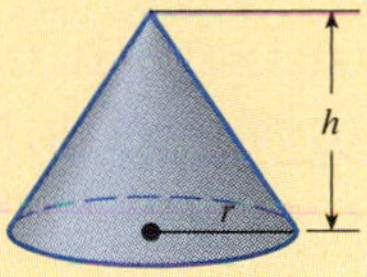

DISTANCE AND MIDPOINT FORMULAS

Distance between $P_1(x_1, y_1)$ and $P_2(x_2, y_2)$:

$$d = \sqrt{(x_2 - x_1)^2 + (y_2 - y_1)^2}$$

Midpoint of $\overline{P_1P_2}$: $\left(\frac{x_1 + x_2}{2}, \frac{y_1 + y_2}{2}\right)$

LINES

Slope of line through $P_1(x_1, y_1)$ and $P_2(x_2, y_2)$:

$$m = \frac{y_2 - y_1}{x_2 - x_1}$$

Point-slope equation of line through $P_1(x_1, y_1)$ with slope m:

$$y - y_1 = m(x - x_1)$$

Slope-intercept equation of line with slope m and y-intercept b:

$$y = mx + b$$

CIRCLES

Equation of the circle with center (h, k) and radius r:

$$(x - h)^2 + (y - k)^2 = r^2$$

TRIGONOMETRY

ANGLE MEASUREMENT

π radians $= 180°$

$1° = \dfrac{\pi}{180}$ rad $\qquad$ 1 rad $= \dfrac{180°}{\pi}$

$s = r\theta$

(θ in radians)

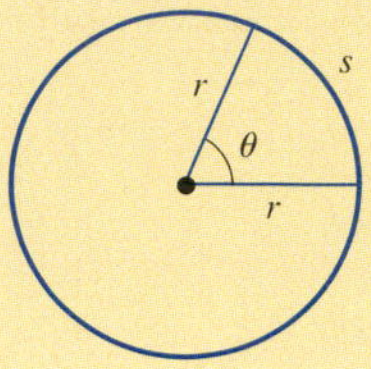

RIGHT ANGLE TRIGONOMETRY

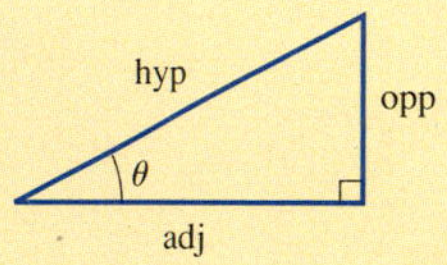

$$\sin\theta = \frac{\text{opp}}{\text{hyp}} \qquad \csc\theta = \frac{\text{hyp}}{\text{opp}}$$

$$\cos\theta = \frac{\text{adj}}{\text{hyp}} \qquad \sec\theta = \frac{\text{hyp}}{\text{adj}}$$

$$\tan\theta = \frac{\text{opp}}{\text{adj}} \qquad \cot\theta = \frac{\text{adj}}{\text{opp}}$$

TRIGONOMETRIC FUNCTIONS

$$\sin\theta = \frac{y}{r} \qquad \csc\theta = \frac{r}{y}$$

$$\cos\theta = \frac{x}{r} \qquad \sec\theta = \frac{r}{x}$$

$$\tan\theta = \frac{y}{x} \qquad \cot\theta = \frac{x}{y}$$

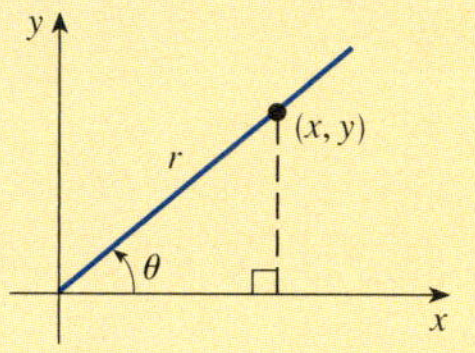

GRAPHS OF THE TRIGONOMETRIC FUNCTIONS

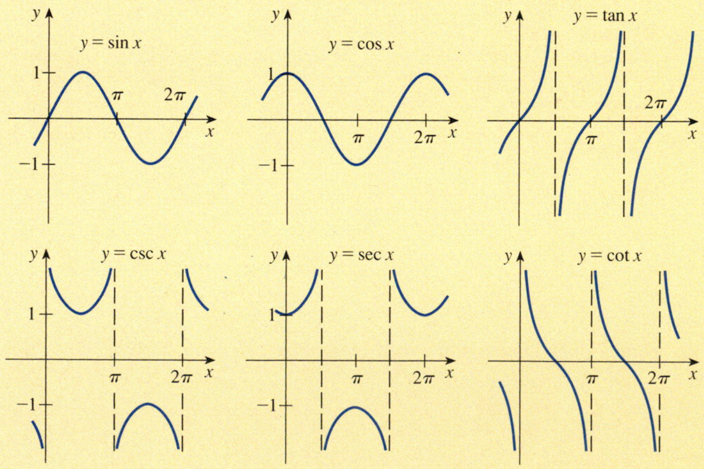

TRIGONOMETRIC FUNCTIONS OF IMPORTANT ANGLES

θ	radians	$\sin\theta$	$\cos\theta$	$\tan\theta$
0°	0	0	1	0
30°	$\pi/6$	1/2	$\sqrt{3}/2$	$\sqrt{3}/3$
45°	$\pi/4$	$\sqrt{2}/2$	$\sqrt{2}/2$	1
60°	$\pi/3$	$\sqrt{3}/2$	1/2	$\sqrt{3}$
90°	$\pi/2$	1	0	—

FUNDAMENTAL IDENTITIES

$$\csc\theta = \frac{1}{\sin\theta} \qquad \sec\theta = \frac{1}{\cos\theta}$$

$$\tan\theta = \frac{\sin\theta}{\cos\theta} \qquad \cot\theta = \frac{\cos\theta}{\sin\theta}$$

$$\cot\theta = \frac{1}{\tan\theta} \qquad \sin^2\theta + \cos^2\theta = 1$$

$$1 + \tan^2\theta = \sec^2\theta \qquad 1 + \cot^2\theta = \csc^2\theta$$

$$\sin(-\theta) = -\sin\theta \qquad \cos(-\theta) = \cos\theta$$

$$\tan(-\theta) = -\tan\theta \qquad \sin\left(\frac{\pi}{2} - \theta\right) = \cos\theta$$

$$\cos\left(\frac{\pi}{2} - \theta\right) = \sin\theta \qquad \tan\left(\frac{\pi}{2} - \theta\right) = \cot\theta$$

THE LAW OF SINES

$$\frac{\sin A}{a} = \frac{\sin B}{b} = \frac{\sin C}{c}$$

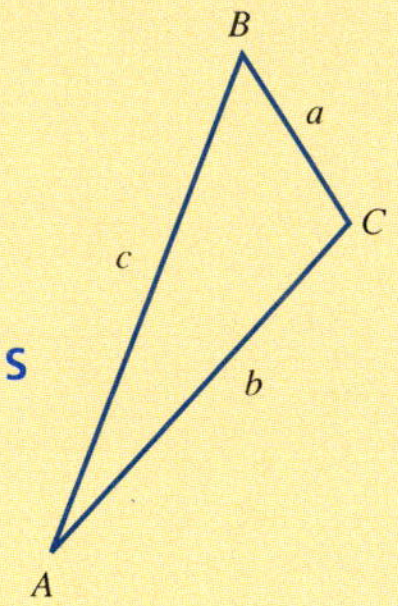

THE LAW OF COSINES

$$a^2 = b^2 + c^2 - 2bc\cos A$$

$$b^2 = a^2 + c^2 - 2ac\cos B$$

$$c^2 = a^2 + b^2 - 2ab\cos C$$

ADDITION AND SUBTRACTION FORMULAS

$$\sin(x + y) = \sin x\cos y + \cos x\sin y$$

$$\sin(x - y) = \sin x\cos y - \cos x\sin y$$

$$\cos(x + y) = \cos x\cos y - \sin x\sin y$$

$$\cos(x - y) = \cos x\cos y + \sin x\sin y$$

$$\tan(x + y) = \frac{\tan x + \tan y}{1 - \tan x\tan y}$$

$$\tan(x - y) = \frac{\tan x - \tan y}{1 + \tan x\tan y}$$

DOUBLE-ANGLE FORMULAS

$$\sin 2x = 2\sin x\cos x$$

$$\cos 2x = \cos^2 x - \sin^2 x = 2\cos^2 x - 1 = 1 - 2\sin^2 x$$

$$\tan 2x = \frac{2\tan x}{1 - \tan^2 x}$$

HALF-ANGLE FORMULAS

$$\sin^2 x = \frac{1 - \cos 2x}{2} \qquad \cos^2 x = \frac{1 + \cos 2x}{2}$$

CALCULUS
EARLY VECTORS

CALCULUS
EARLY VECTORS

JAMES STEWART
McMaster University
and
University of Toronto

with the assistance of members of the
Texas A&M University Department of Mathematics

Arthur Belmonte
Albert Boggess
Dante DeBlassie
Bruce Lowe
Jeffrey Morgan
Kirby Smith
Philip Yasskin

BROOKS/COLE
CENGAGE Learning™

Australia • Brazil • Japan • Korea • Mexico • Singapore • Spain • United Kingdom • United States

BROOKS/COLE
CENGAGE Learning™

Calculus: Early Vectors
James Stewart

GWO A GARY W. OSTEDT BOOK

Publisher: Gary W. Ostedt

Marketing Team: Caroline Croley, Christine Davis, and Debra Johnston

Editorial Associate: Carol Benedict

Production Editor: Jamie Sue Brooks

Production Service: *TECH*-arts

Manuscript Editor: Kathi Townes, *TECH*-arts

Electronic File Editor: Stephanie Kuhns, *TECH*-arts

Interior Design: *TECH*-arts

Cover Design: Vernon T. Boes

Cover Sculpture: Christian Haase

Cover Photo: Ed Young

Illustrations: Brian Betsill, *TECH*-arts

Typesetting: Sandy Senter, The Beacon Group

CREDITS: page 75, Courtesy of CN Tower, Toronto, Canada; page 340, Manfred Kage / Peter Arnold, Inc.

Library of Congress Control Number: 98025748

Softcover ISBN-13: 978-0-534-34941-7
Softcover ISBN-10: 0-534-34941-2
Hardcover ISBN-13: 978-0-534-49348-6
Hardcover ISBN-10: 0-534-49348-3

Brooks/Cole
10 Davis Drive
Belmont, CA 94002-3098
USA

Cengage Learning is a leading provider of customized learning solutions with office locations around the globe, including Singapore, the United Kingdom, Australia, Mexico, Brazil, and Japan. Locate your local office at: **www.cengage.com/global**

Cengage Learning products are represented in Canada by Nelson Education, Ltd.

To learn more about Brooks/Cole, visit **www.cengage.com/brookscole**

Purchase any of our products at your local college store or at our preferred online store **www.ichapters.com**

Printed in the United States of America
10 11 12 13 14 15 11 10 09

PREFACE

The idea for this book came from the Mathematics Department of Texas A&M University. A large number of their students take Engineering and Physics courses concurrently with calculus and these courses require the concept of vectors to explain velocity, force, and so on. The traditional calculus curriculum postpones the study of vectors until the third semester; consequently, many students learn vector concepts from Physics or Engineering departments during their freshman year without any reinforcement from Mathematics departments.

Al Boggess and his colleagues at Texas A&M University asked me to write a version of my calculus text that would introduce vectors at a very early stage and integrate them throughout. I thought that idea made a lot of sense, but I was involved in another writing project at the time and so I suggested that they write the first draft themselves. They drew upon material from my book *Calculus, Early Transcendentals, Third Edition* (henceforth abbreviated as ET) and rearranged it to fit their curriculum, providing new exposition and new exercises where needed. They class-tested the new version for two years. I then took part in the writing, revising the manuscript in collaboration with them. The result is this Preliminary Edition of *Calculus: Early Vectors.*

Although the impetus for this book came from Texas A&M, a number of other universities have expressed interest in it as well. Before we publish the first edition of this book, I would like to have input from other potential users with respect to both broad structure and fine details. If you have any comments or suggestions, particularly on the best way of integrating vectors into first-semester calculus, please send them to me in care of my publisher, Brooks/Cole Publishing Company, 511 Forest Lodge Road, Pacific Grove CA 93950, or e-mail to info@brookscole.com.

VECTORS

The goal of *Calculus: Early Vectors* is to introduce vectors early in the first semester, in a manner that does not dramatically change the rest of the calculus curriculum. Vectors are introduced in two stages. Chapter 1 introduces two-dimensional vectors along with basic operations, including the dot product, and two-dimensional vector functions. Three-dimensional vectors (along with three-dimensional geometry and cross products) are treated in Chapter 11 (covered at the end of the second semester or beginning of the third). Our experience has shown that a full treatment of three-dimensional vectors typically overwhelms the average first-semester calculus student. A thorough treatment in Chapter 1 would also postpone discussion of other calculus topics so that the resulting course would no longer resemble the traditional first-semester course.

- Chapter 1 on vectors is relatively short. It introduces two-dimensional vectors along with basic vector operations, including sums, scalar multiplication, dot product, and projection. Applications to problems involving force and work are discussed in detail. In fact, the definition of the dot product is motivated by the need to compute the work done by a force that does not point in the direction of motion.

- Vector functions and parametric curves are introduced in Chapter 1 in the context of projectiles. Limits and rates of change of such functions are discussed in Chapter 2, immediately following the development of the scalar case.
- Derivatives of vector functions, tangent vectors, and tangents to parametric curves are treated in Chapter 3. (See Sections 3.7 and 3.9.)
- Chapter 4 (on exponential functions and logarithms) is almost identical to the corresponding chapter of ET.
- Chapter 5 (applications of differentiation) contains the streamlined approach to graphing that I have used in *Calculus: Concepts and Contexts.* In addition to a more reformed approach that takes advantage of technology (where appropriate), this streamlined approach allows the typical first-semester calculus course to cover topics through Chapter 6 (integration), corresponding roughly to the topics of most traditional first-semester courses.
- Chapters 7–9 (techniques of integration and their applications) are largely unchanged from ET.
- Chapter 10 (series) uses the more streamlined approach contained in *Calculus: Concepts and Contexts.*
- Three-dimensional vectors and geometry are treated in Chapter 11. Here the basic concepts developed in Chapter 1 are reviewed quickly and generalized to three dimensions. We view this repetition as giving the important reinforcement this concept needs because it is so heavily used in Physics and Engineering. Cross products are then introduced with the notion of torque serving as the motivation. A student who has transferred from a more traditional first-semester calculus course that does not contain vectors will be introduced to all the vector concepts from Chapter 1 in the three-dimensional setting covered in Chapter 11.
- Chapters 12–14 (partial derivatives, multiple integrals, and vector calculus) are largely unchanged from ET with the following exception. Polar and spherical coordinates are now introduced in Chapter 13 (multiple integrals) immediately before they are needed for integrals in polar coordinates and spherical coordinates.

TECHNOLOGY

For many years I have experimented with calculus laboratories for my own students, first with graphing software for computers, then with graphing calculators, and finally with computer algebra systems. Those of us who have watched our students use these machines know how enlivening such experiences can be. We have seen from the expressions on their faces how these devices can engage our students' attention and make them active learners.

Despite my enthusiasm for technology, I think there are potential dangers for misusing it. When I first started using technology, I tended to use it too much, but then I started to see where it is appropriate and where it is not. Many topics in calculus can be explained with chalk and blackboard (and reinforced with pencil and paper exercises) more simply, more quickly, and more clearly than with technology. Other topics cry out for the use of machines. What is important is the *appropriate* use of technology, which can be characterized as involving the *interaction* between technology and calculus. In short, technology is not a panacea, but, when used appropriately, it can be a powerful stimulus to learning.

This textbook can be used either with or without technology and I use two special symbols to indicate clearly when a particular type of machine is required. The icon indicates an example or exercise that requires the use of either a graphing calculator or a computer with graphing software. (Section 3 in Review and Preview discusses the use of these graphing devices and some of the pitfalls that can arise. Section 5.4 is a good example of what I mean by the interaction between technology and calculus.) The symbol CAS is reserved for problems in which the full resources of a computer algebra system (like Derive, Maple, or Mathematica) are required. In all cases we assume that the student knows how to use the machine—we rarely give explicit commands.

Some of the exercises designated by or CAS require considerable time for their completion. Instructors should therefore consult the solutions manual to determine the complexity of a problem before assigning it. Some of those problems explore the shape of a family of curves depending on one or more parameters. Other such projects involve technology in very different ways. See, for instance, pages 651 (logistic sequences) and 524.

VISUALIZATION

One of the themes of the calculus reform movement is the Rule of Four: Topics should be presented verbally, numerically, graphically, and symbolically, wherever possible. See pages 150 and 588 for examples of how the Rule of Four comes into play. You will also see that I include substantial work with tabular functions and numerical estimates of sums of series.

Many examples and exercises promote visual thinking. Given the graph of a function, I think it is important for a student to be able to sketch the graph of its derivative (page 152) and also to sketch the graph of an antiderivative (page 348) in a qualitative manner. See pages 194, 208, 232, 344, 419, 428, 507, 657, 696, 747, 872, and 891 for other examples of exercises that test students' visual understanding.

There are hundreds of new computer-generated figures that illustrate examples. These are not just pretty pictures—they constantly remind students of the geometric meaning behind the result of a calculation. I have also tried to provide more visual insight into formulas and their proofs (see, for instance, pages 162 and 290).

PROBLEM SOLVING

My educational philosophy was strongly influenced by attending the lectures of George Polya and Gabor Szego when I was a student at Stanford University. Both Polya and Szego consistently introduced a topic by relating it to something concrete or familiar. Wherever practical, I have introduced topics with an intuitive geometrical or physical description and attempted to tie mathematical concepts to the students' experience.

I found Polya's lectures on problem solving very inspirational and his books *How to Solve It, Mathematical Discovery,* and *Mathematics and Plausible Reasoning* have become the core text material for a mathematical problem-solving course that I have instituted and taught at McMaster University. I have adapted these problem-solving strategies to the study of calculus both explicitly, by outlining strategies, and implicitly, by illustration and example.

Students usually have difficulties in situations that involve no single well-defined procedure for obtaining the answer. I think nobody has improved very much on Polya's four-stage problem-solving strategy and, accordingly, I have included in this edition a version of Polya's strategy in Section 4 of Review and Preview, together with several examples and exercises involving precalculus material. I have also rewritten the solutions to certain examples in a more patient manner to make the problem-solving principles more apparent. (See, for instance, Example 1 on page 215.)

The classic calculus situations where problem-solving skills are especially important are related rates problems, maximum and minimum problems, and techniques of integration. In these and other situations I have adapted Polya's strategies to the matter at hand. See, in particular, Section 8.6 (Strategy for Integration).

I include what I call *Problems Plus* after odd-numbered chapters. These are problems that go beyond the usual exercises in one way or another and require a higher level of problem-solving ability. The very fact that they do not occur in the context of any particular chapter makes them a little more challenging. For instance, a problem that occurs after Chapter 10 need not have anything to do with Chapter 10. I particularly value problems in which a student has to combine methods from two or three different chapters. The examples in the Problems Plus sections serve not as solutions to imitate (there are no problems like them), but rather as examples of how to tackle a challenging calculus problem. (See the example on page 360.) Many of the problems have a geometric flavor (see Problems 9, 10, 18, 27 after Chapter 3 and Problem 14 after Chapter 10). I have been testing these Problems Plus on my own students by putting them on assignments, tests, and exams. Because of their challenging nature I grade these problems in a different way. Here I reward a student significantly for ideas toward a solution and for recognizing which problem-solving principles are relevant. My aim is to teach my students to be unafraid to tackle a problem the likes of which they have never seen before.

REAL WORLD APPLICATIONS

I have included substantial applied problems that I believe will capture the attention of students. See, for instance, Problem 10 on page 421 (investigating the shape of a can), Problem 7 on page 523 (positioning a shortstop to make the best relay to home plate), and Problem 9 (choosing a seat in a movie theater) and Problem 10 (explaining the formation and location of rainbows) on page 524. These are all extended problems that would make good projects. They happen to be located in the *Applications Plus* sections, which occur after even-numbered chapters (starting with Chapter 4) and are a counterpart to the Problems Plus. (Again the idea is often to combine ideas and techniques from different parts of the book.) But there are many good applied problems in the ordinary sections of the book as well. (See, for instance, Exercise 56 on page 340 and Exercise 32 on page 233).

THE TEACHING PACKAGE

Calculus: Early Vectors, Preliminary Edition, is supported by ancillaries designed to enhance student understanding and to facilitate creative instruction. The *Student Solutions Manual* (by Jeffery A. Cole, *Anoka-Ramsey Community College,* James Stewart, *McMaster University,* Daniel Anderson, *University of Iowa,* and Daniel Drucker, *Wayne State University*) includes detailed solutions to all odd-numbered text exercises. The *Complete Solutions Manual,* by the same team of authors, provides detailed solutions to all text exercises.

In addition, *Test Items* and *Computerized Testing* are available, offering multiple-choice and short-answer questions. Computerized testing allows instructors to insert their own questions and to customize the questions provided, with some algorithmically generated. Full-color *Transparencies* are available of over 100 of the more complex text diagrams.

ACKNOWLEDGMENTS

I am grateful to the following reviewers for sharing their knowledge and judgment with me. I have learned something from each of them.

Loren Argabright, *Drexel University*

Mark Bridger, *Northeastern University*

Philip Crooke, *Vanderbilt University*

Harvey Keynes, *University of Minnesota*

Marshall Leitman, *Case Western Reserve University*

Douglas B. Meade, *University of South Carolina*

Fabio A. Milner, *Purdue University*

Hiram Paley, *University of Illinois at Urbana–Champaign*

Rick Smith, *University of Florida*

Robert L. Wilson, *University of Wisconsin–Madison*

In addition, I thank Jeff Cole for his checking the accuracy of the manuscript and careful proofreading of the text, Kathi Townes and the staff of TECH·arts for their production coordination and interior illustration, Sandy Senter of The Beacon Group for typesetting the text, Christian Haase for the cover sculpture, Ed Young for the cover photograph, and the following Brooks/Cole staff: Jamie Sue Brooks, production editor; Vernon T. Boes, art director; Caroline Croley, Christine Davis, and Debra Johnston, marketing team; Beth Wilbur, project development editor, Linda Row, assistant editor, and Carol Benedict, editorial associate. They have all done an outstanding job.

Finally, I thank Al Boggess and his colleagues at Texas A&M University, not only for suggesting the idea of a calculus text with early vectors, but also for working with me to make it a reality. My publisher, Gary W. Ostedt, also deserves special thanks for arranging this collaboration and for giving us the benefit of his extensive experience and keen editorial insight.

JAMES STEWART

TO THE STUDENT

Reading a calculus textbook is different from reading a newspaper or a novel, or even a physics book. Don't be discouraged if you have to read a passage more than once in order to understand it. You should have pencil and paper at hand to make a calculation or sketch a diagram.

Some students start by trying their homework problems and only read the text if they get stuck on an exercise. I suggest that a far better plan is to read and understand a section of the text before attempting the exercises. In particular, you should study the definitions to see the exact meanings of the terms.

Part of the aim of this course is to train you to think logically. Learn to write the solutions of the exercises in a connected step-by-step fashion with explanatory words and symbols—not just a string of disconnected equations or formulas.

The answers to the odd-numbered exercises appear at the back of the book, in Appendix L. There are often several different forms in which to express an answer, so if your answer differs from mine, don't immediately assume you are wrong. There may be an algebraic or trigonometric identity that connects the answers. For example, if the answer given in the back of the book is $\sqrt{2} - 1$ and you obtain $1/(1 + \sqrt{2})$, then you are right and rationalizing the denominator will show that the expressions are equivalent.

The symbol indicates an example or exercise that requires the use of either a graphing calculator or a computer with graphing software. (Section 3 in Review and Preview discusses the use of these graphing devices and some of the pitfalls that you may encounter.) The icon CAS is reserved for problems in which the full resources of a computer algebra system (like Derive, Maple, or Mathematica) are required. You will also encounter the symbol ⊘, which warns you against committing an error. I have placed this symbol in the margin in situations where I have observed that a large proportion of my students tend to make the same mistake.

Calculus is an exciting subject; I hope you find it both useful and interesting in its own right.

A NOTE ON LOGIC

In understanding the theorems it is important to know the meaning of certain logical terms and symbols. If P and Q are mathematical statements, then $P \Rightarrow Q$ is read as "P implies Q" and means the same as "If P is true, then Q is true." The *converse* of a theorem of the form $P \Rightarrow Q$ is the statement $Q \Rightarrow P$. (The converse of a theorem may or may not be true. For example, the converse of the statement "If it rains, then I take my umbrella" is "If I take my umbrella, then it rains.") The symbol $\Longleftrightarrow$ indicates that two statements are equivalent. Thus $P \Longleftrightarrow Q$ means that both $P \Rightarrow Q$ and $Q \Rightarrow P$. The phrase "if and only if" is also used in this situation. Thus "P is true if and only if Q is true" means the same as $P \Longleftrightarrow Q$. The *contrapositive* of a theorem $P \Rightarrow Q$ is the statement that $\sim Q \Rightarrow \sim P$, where $\sim P$ means *not* P. So the contrapositive says "If Q is false, then P is false." Unlike converses, the contrapositive of a theorem is always true.

CONTENTS

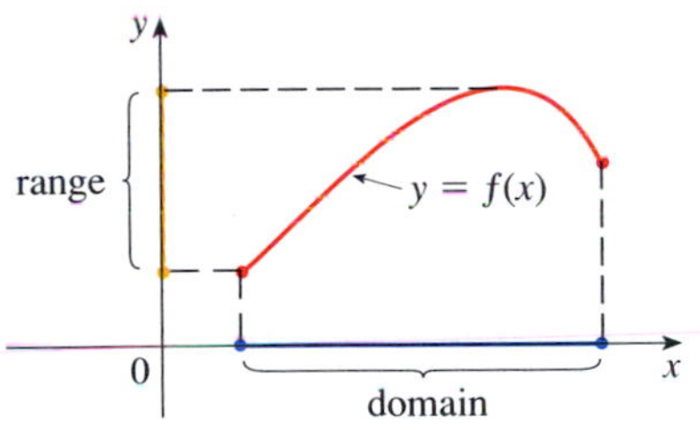

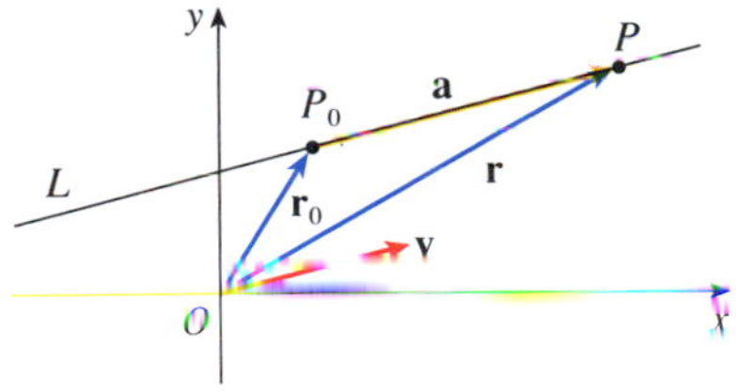

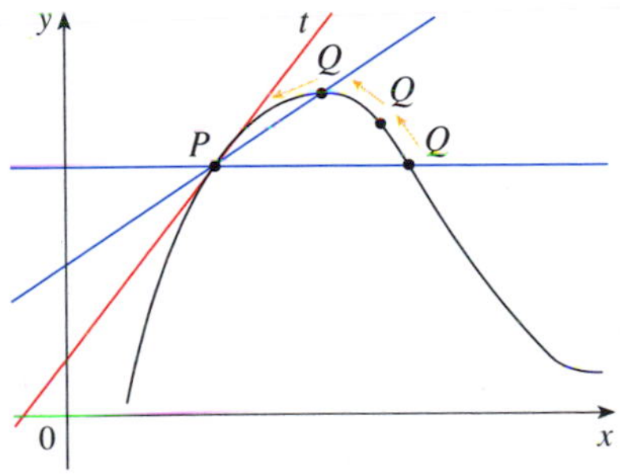

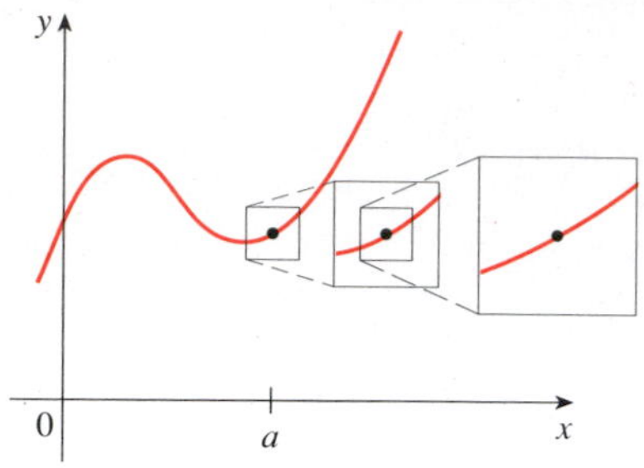
y
0
a
x

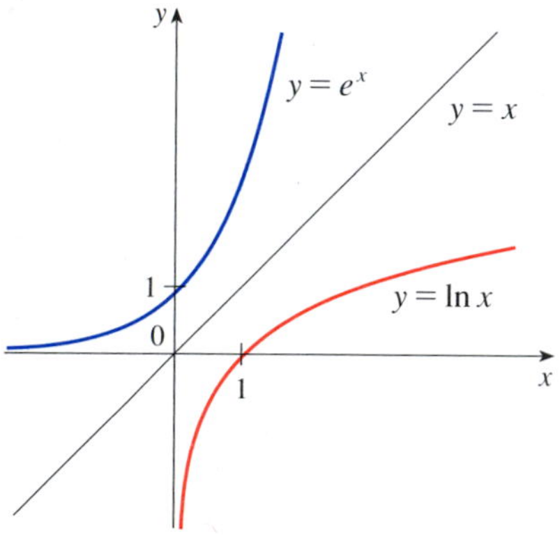
y
$y = e^x$
$y = x$
1
0
$y = \ln x$
1
x

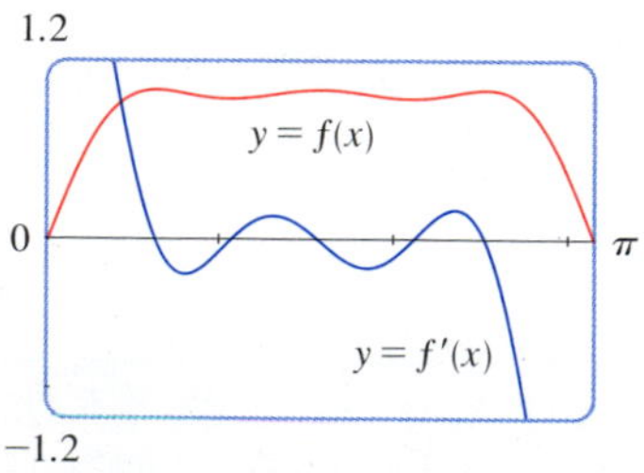
1.2
$y = f(x)$
0
π
$y = f'(x)$
−1.2

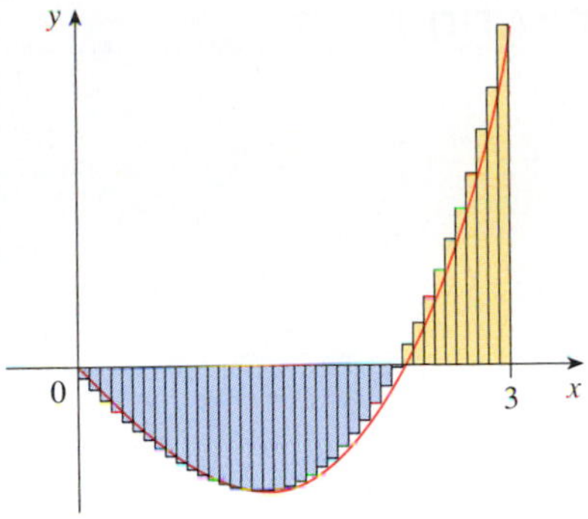
y
0
3
x

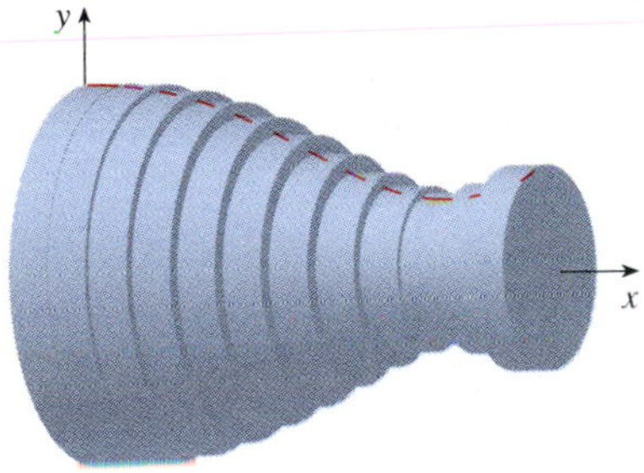
y
x

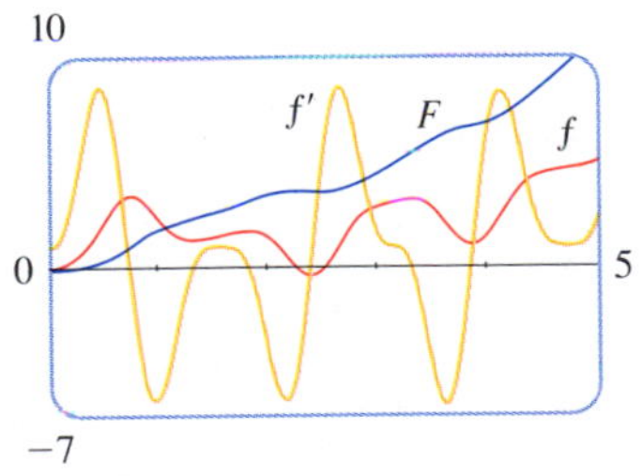
10
f′
F
f
0
5
−7

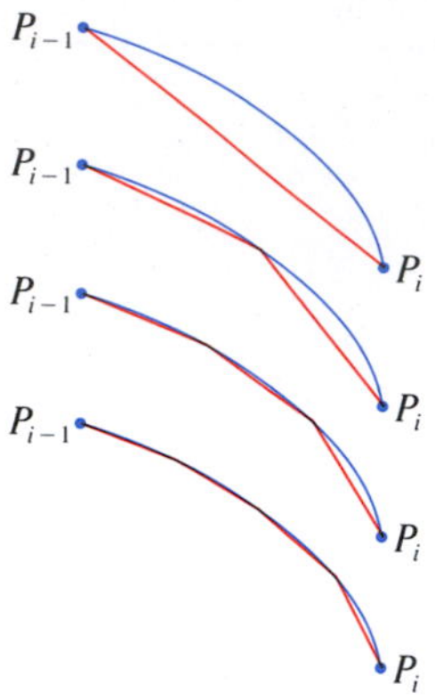
P_{i-1}
P_i

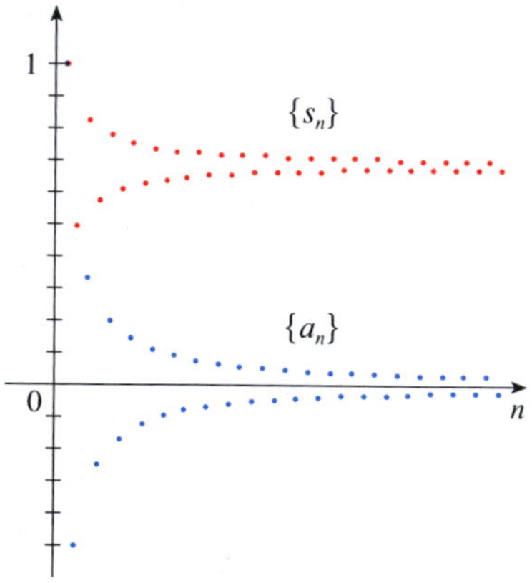
1
0
$\{s_n\}$
$\{a_n\}$
n

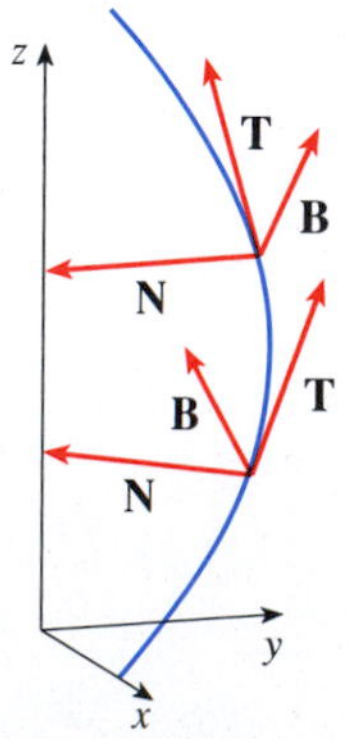
z
T
B
N
B
T
N
y
x

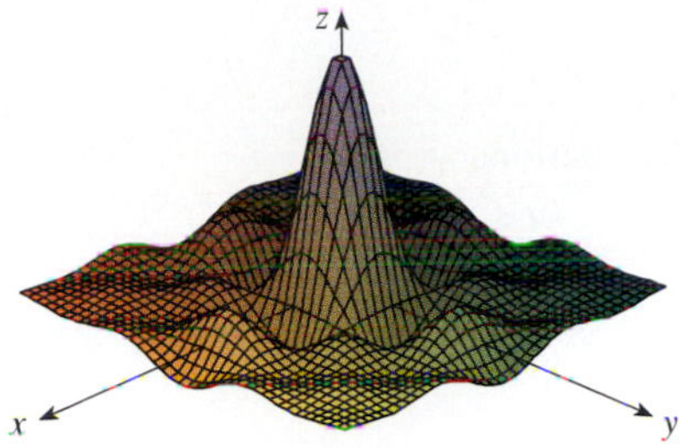
z
x
y

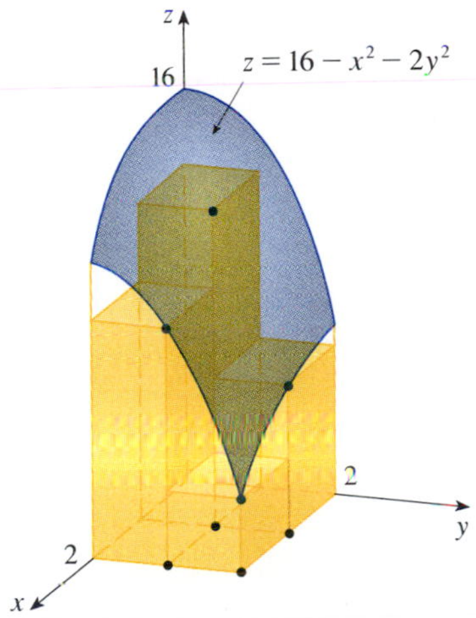
z
16
$z = 16 - x^2 - 2y^2$
2
2
y
x

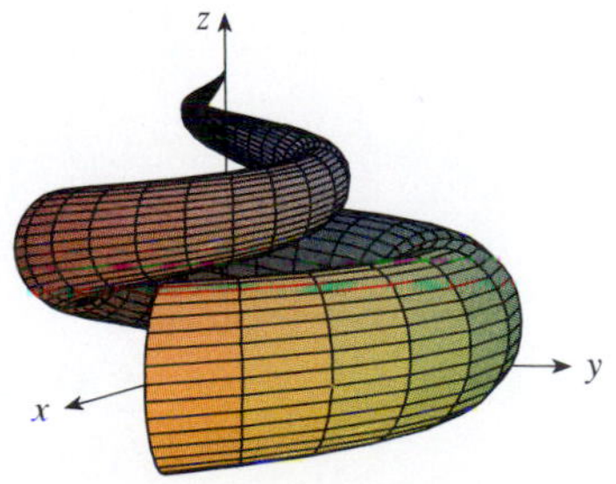
z
x
y

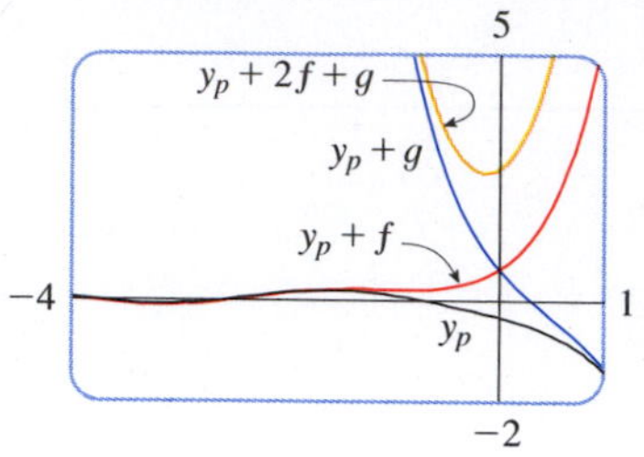

15 SECOND-ORDER DIFFERENTIAL EQUATIONS 944

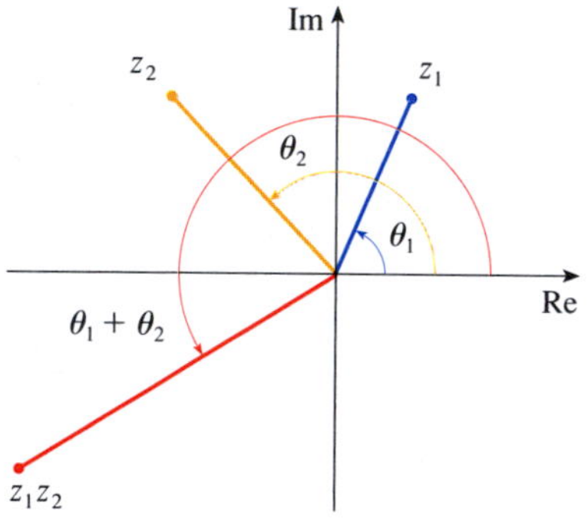

APPENDIXES A1

INDEX A125

CALCULUS
EARLY VECTORS

REVIEW AND PREVIEW

In most sciences one generation tears down what another has built and what one has established another undoes. In mathematics alone each generation builds a new story to the old structure.

HERMANN HANKEL

The fundamental objects that we deal with in calculus are functions. So we first review the basic ideas concerning functions, their graphs, and ways of combining them. We list the main types of functions that occur in calculus and its applications and we review the procedures for shifting, stretching, and reflecting their graphs. We then discuss the use of graphing calculators and graphing software for computers. This chapter also contains a discussion of the principles of problem solving that will be useful throughout the book. In the last section we give a preview of some of the principal ideas of calculus. Although it is not absolutely necessary to read this last section, it does provide an overview of the subject and a brief look at some of the reasons for studying calculus.

1 FUNCTIONS AND THEIR GRAPHS

The area A of a circle depends on the radius r of the circle. The rule that connects r and A is given by the equation $A = \pi r^2$. With each positive number r there is associated one value of A, and we say that A is a *function* of r.

The number N of bacteria in a culture depends on the time t. If the culture starts with 5000 bacteria and the population doubles every hour, then after t hours the number of bacteria will be $N = (5000)2^t$. This is the rule that connects t and N. For each value of t there is a corresponding value of N, and we say that N is a function of t.

The cost C of mailing a first-class letter depends on the weight w of the letter. Although there is no single neat formula that connects w and C, the post office has a rule for determining C when w is known.

Each of these examples describes a rule whereby, given a number (r, t, or w), another number (A, N, or C) is assigned. In each case we say that the second number is a function of the first number.

A **function** f is a rule that assigns to each element x in a set A exactly one element, called $f(x)$, in a set B.

We usually consider functions for which the sets A and B are sets of real numbers. The set A is called the **domain** of the function. The number $f(x)$ is the **value of f at x** and is read "f of x." The **range** of f is the set of all possible values of $f(x)$ as x varies throughout the domain, that is, $\{f(x) \mid x \in A\}$.

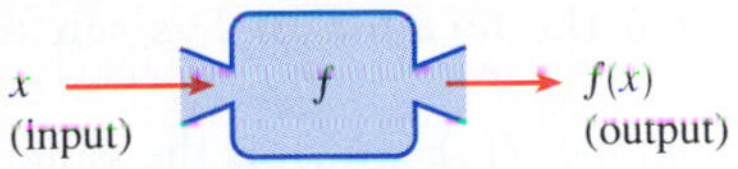

FIGURE 1
Machine diagram for a function f

It is helpful to think of a function as a **machine** (see Figure 1). If x is in the domain of the function f, then when x enters the machine, it is accepted as an input and the machine produces an output $f(x)$ according to the rule of the function. Thus, we can think of the domain as the set of all possible inputs and the range as the set of all possible outputs.

The preprogrammed functions in a calculator are good examples of a function as a machine. For example, the $\sqrt{x}$ key on your calculator is such a function. First you input x into the display. Then you press the key labeled $\sqrt{x}$. If $x < 0$, then x is not in the domain of this function; that is, x is not an acceptable input, and the calculator will indicate an error. If $x \geqslant 0$, then an approximation to $\sqrt{x}$ will appear in the display. Thus, the $\sqrt{x}$ key on your calculator is not quite the same as the exact mathematical function f defined by $f(x) = \sqrt{x}$. (See Appendix G.)

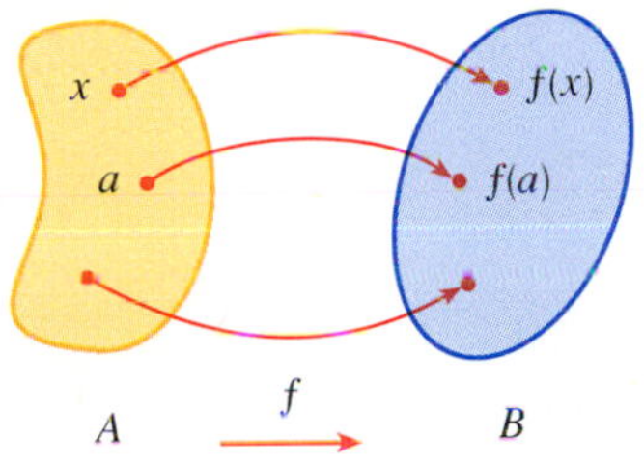

FIGURE 2
Arrow diagram for f

Another way to picture a function is by an **arrow diagram** as in Figure 2. Each arrow connects an element of A to an element of B. The arrow indicates that $f(x)$ is associated with x, $f(a)$ is associated with a, and so on.

EXAMPLE 1 The squaring function assigns to each real number x its square x^2. It is defined by the equation

$$f(x) = x^2$$

The values of f are found by substituting for x in this equation. For example,

$$f(3) = 3^2 = 9 \qquad f(-2) = (-2)^2 = 4$$

The domain of f is the set $\mathbb{R}$ of all real numbers. The range of f consists of all values of $f(x)$, that is, all numbers of the form x^2. But $x^2 \geqslant 0$ for all numbers x, and any nonnegative number c is a square since $c = (\sqrt{c}\,)^2 = f(\sqrt{c}\,)$. Therefore, the range of f is $\{y \mid y \geqslant 0\} = [0, \infty)$. The machine diagram and arrow diagram for this function are shown in Figure 3.

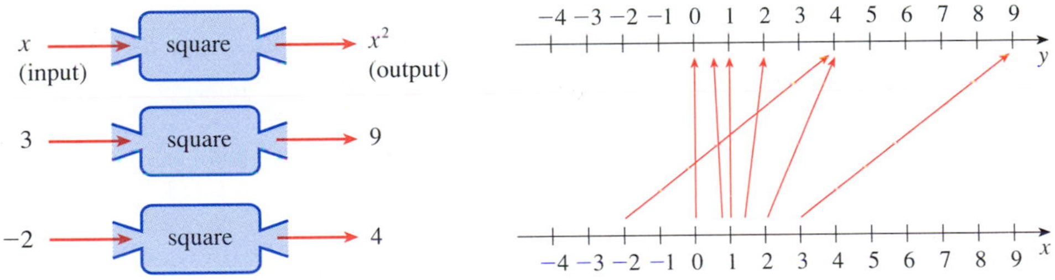

FIGURE 3 (a) Machine diagram (b) Arrow diagram ■

EXAMPLE 2 If we define a function g by

$$g(x) = x^2 \qquad 0 \leqslant x \leqslant 3$$

See Appendix A for interval notation and diagrams.

then the domain of g is given as the closed interval $[0, 3]$. This is different from the function f given in Example 1 because in considering g we are restricting our attention to those values of x between 0 and 3. The range of g is

$$\{x^2 \mid 0 \leqslant x \leqslant 3\} = \{y \mid 0 \leqslant y \leqslant 9\} = [0, 9]$$ ■

In Examples 1 and 2 the domain of the function was given explicitly. But **if a function is given by a formula and the domain is not stated explicitly, the convention**

is that the domain is the set of all numbers for which the formula makes sense and defines a real number.

We should distinguish between a function f and the number $f(x)$, which is the value of f at x. Nonetheless, it is common to abbreviate an expression such as

$$\text{“the function } f \text{ defined by } f(x) = x^2 + x\text{”}$$

to

$$\text{“the function } f(x) = x^2 + x\text{”}$$

EXAMPLE 3 Find the domain of the function $f(x) = \dfrac{1}{x^2 - x}$.

SOLUTION Since

$$f(x) = \frac{1}{x^2 - x} = \frac{1}{x(x-1)}$$

and division by 0 is not allowed, we see that $f(x)$ is not defined when $x = 0$ or $x = 1$. Thus, the domain of f is

$$\{x \mid x \neq 0, x \neq 1\}$$

which could also be written in interval notation as

$$(-\infty, 0) \cup (0, 1) \cup (1, \infty)$$ ■

EXAMPLE 4 Find the domain of $h(x) = \sqrt{2 - x - x^2}$.

SOLUTION Since the square root of a negative number is not defined (as a real number), the domain of h consists of all values of x such that

$$2 - x - x^2 \geqslant 0$$

For a review of solving inequalities, see Appendix A.

We solve this inequality by factoring. Since $2 - x - x^2 = (2 + x)(1 - x)$, the product changes sign when $x = -2$ or $x = 1$ as indicated in the following chart:

Interval	$2 + x$	$1 - x$	$(2 + x)(1 - x)$
$x \leqslant -2$	$-$	$+$	$-$
$-2 \leqslant x \leqslant 1$	$+$	$+$	$+$
$x \geqslant 1$	$+$	$-$	$-$

Therefore, the domain of h is

$$\{x \mid -2 \leqslant x \leqslant 1\} = [-2, 1]$$ ■

A symbol that represents an arbitrary number in the *domain* of a function f is called an **independent variable**. A symbol that represents a number in the *range* of f is called a **dependent variable**. For example, the squaring function of Example 1 could be defined by saying that each number x is assigned the number y by the rule $y = x^2$. Then x is the independent variable and y is the dependent variable. In the example on bacteria at the beginning of this section, t is the independent variable, N is the dependent variable, and they are connected by the equation $N = (5000)2^t$. In general, a function describes how one quantity depends on another. For instance, we say that population is a function of time and that pressure is a function of temperature.

EXAMPLE 5 A rectangular storage container with an open top has a volume of 10 m^3. The length of its base is twice its width. Material for the base costs \$10 per square meter; material for the sides costs \$6 per square meter. Express the cost of materials as a function of the width of the base.

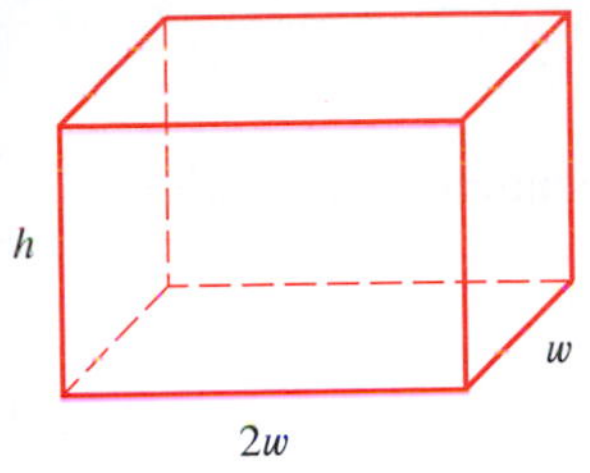

FIGURE 4

SOLUTION We draw a diagram as in Figure 4 and introduce notation by letting w and $2w$ be the width and length of the base, respectively, and h be the height.

The area of the base is $(2w)w = 2w^2$, so the cost, in dollars, of the material for the base is $10(2w^2)$. Two of the sides have area wh and the other two have area $2wh$, so the cost of the material for the sides is $6[2(wh) + 2(2wh)]$. The total cost is therefore

$$C = 10(2w^2) + 6[2(wh) + 2(2wh)] = 20w^2 + 36wh$$

To express C as a function of w alone, we need to eliminate h and we do so by using the fact that the volume is 10 m^3. Thus

$$w(2w)h = 10$$

which gives
$$h = \frac{10}{2w^2} = \frac{5}{w^2}$$

In setting up applied functions as in Example 5, it may be useful to review the principles of problem solving as discussed in Section 4, particularly Step 1: Understanding the problem.

Substituting this into the expression for C, we have

$$C = 20w^2 + 36w\left(\frac{5}{w^2}\right) = 20w^2 + \frac{180}{w}$$

Therefore, the equation

$$C(w) = 20w^2 + \frac{180}{w} \qquad w > 0$$

expresses C as a function of w. ■

GRAPHS OF FUNCTIONS

We have seen how to picture functions using machine diagrams and arrow diagrams. A third method for visualizing a function is its graph. If f is a function with domain A, its **graph** is the set of ordered pairs

$$\{(x, f(x)) \mid x \in A\}$$

In other words, the graph of f consists of all points (x, y) in the coordinate plane such that $y = f(x)$ and x is in the domain of f.

The graph of a function f gives us a useful picture of the behavior or "life history" of a function. Since the y-coordinate of any point (x, y) on the graph is $y = f(x)$, we can read the value of $f(x)$ from the graph as being the height of the graph above the point x (see Figure 5). The graph of f also allows us to picture the domain and range of f on the x-axis and y-axis as in Figure 6.

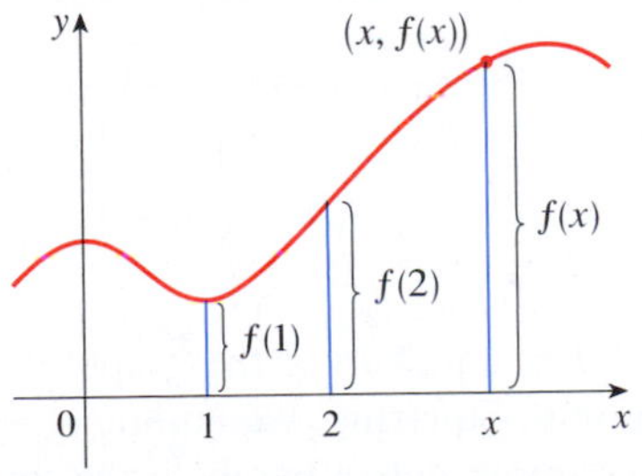

FIGURE 5

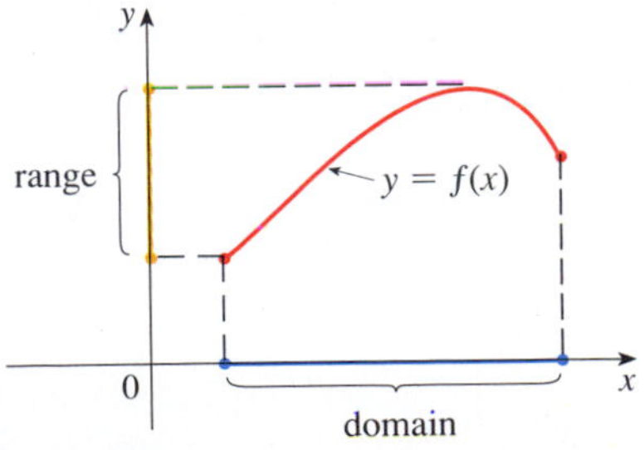

FIGURE 6

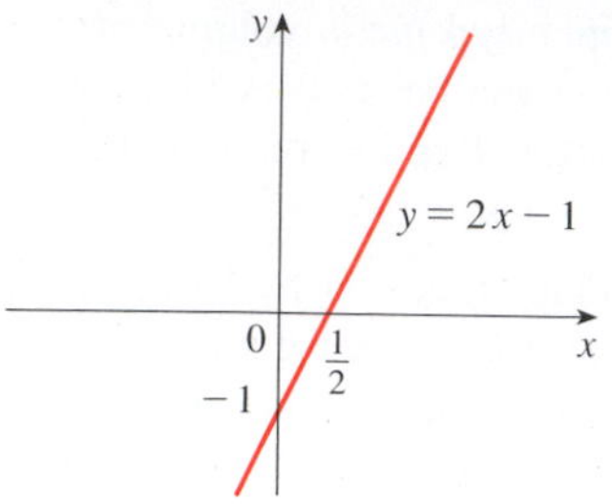

FIGURE 7

EXAMPLE 6 Sketch the graph of the function $f(x) = 2x - 1$.

SOLUTION The equation of the graph is $y = 2x - 1$, and we recognize this as being the equation of a line with slope 2 and y-intercept -1. This enables us to sketch the graph of f in Figure 7. ■

EXAMPLE 7 Sketch the graph of $f(x) = x^3$.

SOLUTION We list some functional values and the corresponding points on the graph in the following table:

x	0	$\frac{1}{2}$	1	2	$-\frac{1}{2}$	-1	-2
$f(x) = x^3$	0	$\frac{1}{8}$	1	8	$-\frac{1}{8}$	-1	-8
(x, x^3)	$(0, 0)$	$(\frac{1}{2}, \frac{1}{8})$	$(1, 1)$	$(2, 8)$	$(-\frac{1}{2}, -\frac{1}{8})$	$(-1, -1)$	$(-2, -8)$

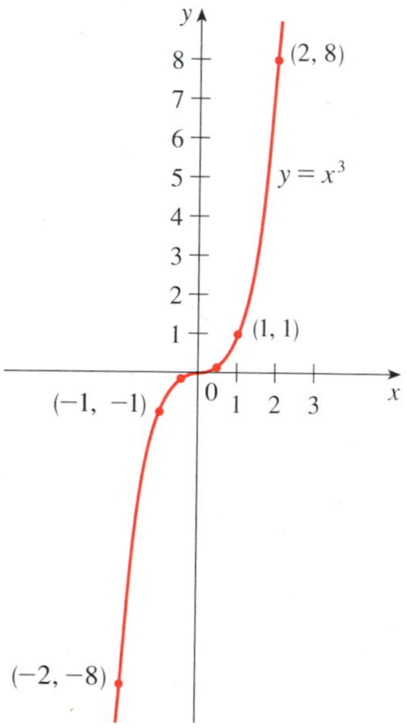

FIGURE 8

Then we plot these points and join them by a smooth curve to obtain the graph shown in Figure 8. At the present stage we cannot be absolutely certain that the graph is exactly as shown. But the use of a graphing calculator or computer to plot a much larger number of points supports the sketch we have made, and we will later develop calculus techniques that confirm the picture. ■

EXAMPLE 8 Sketch the graph of $f(x) = x^2$.

SOLUTION We could plot points and join them to produce the graph shown in Figure 9. Or we could simply draw the graph by recognizing that $y = x^2$ is the equation of a parabola that opens upward and has its vertex at the origin. (See Appendix C.)

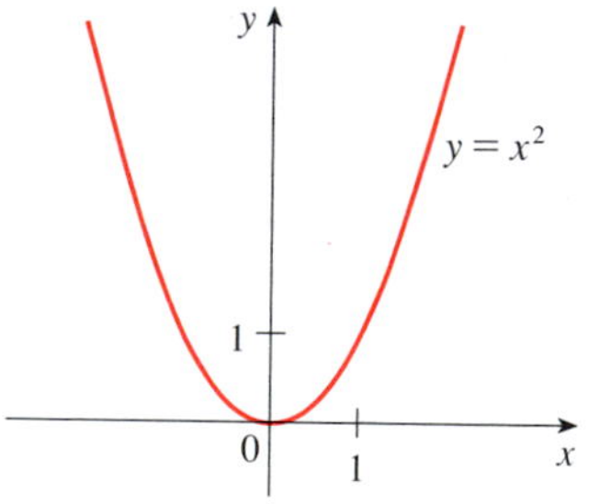

FIGURE 9

■

EXAMPLE 9 Sketch the graph of $f(x) = \sqrt{x + 2}$.

SOLUTION 1 We first observe that $\sqrt{x + 2}$ is defined when $x + 2 \geqslant 0$, so the domain of f is $\{x \mid x \geqslant -2\} = [-2, \infty)$. Then we plot the points given by the following table and use them to produce the sketch in Figure 10.

x	-2	-1	0	1	2	3
$f(x) = \sqrt{x + 2}$	0	1	$\sqrt{2}$	$\sqrt{3}$	2	$\sqrt{5}$

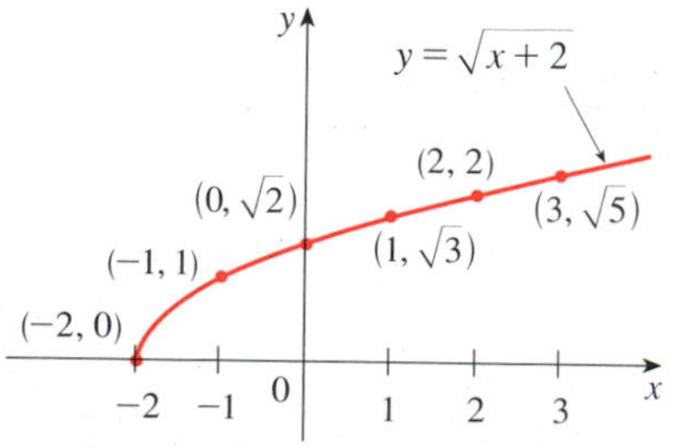

FIGURE 10

SOLUTION 2 The equation of the graph of f is $y = \sqrt{x + 2}$ and we note that this implies that $y \geqslant 0$. Squaring, we obtain $y^2 = x + 2$, or $x = y^2 - 2$, $y \geqslant 0$. The equation $x = y^2$ represents a parabola that opens to the right and has its vertex at the origin. If we shift it two units to the left, we get the parabola $x = y^2 - 2$ shown in

Figure 11(a). But since $y \geqslant 0$, the graph of $y = \sqrt{x+2}$ is just the top half of this parabola [Figure 11(b)]. [The bottom half of the parabola is the graph of the function $y = -\sqrt{x+2}$ shown in Figure 11(c).]

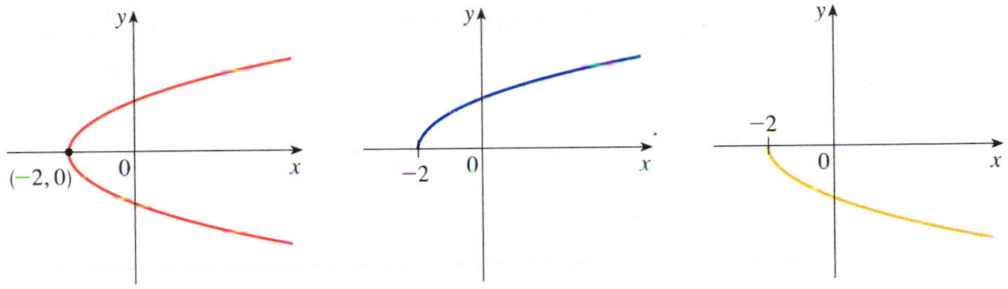

FIGURE 11 (a) $x = y^2 - 2$ (b) $y = \sqrt{x+2}$ (c) $y = -\sqrt{x+2}$ ■

The graph of a function is a curve in the xy-plane. But the question arises: Which curves in the xy-plane are graphs of functions? This is answered by the following test.

THE VERTICAL LINE TEST A curve in the xy-plane is the graph of a function of x if and only if no vertical line intersects the curve more than once.

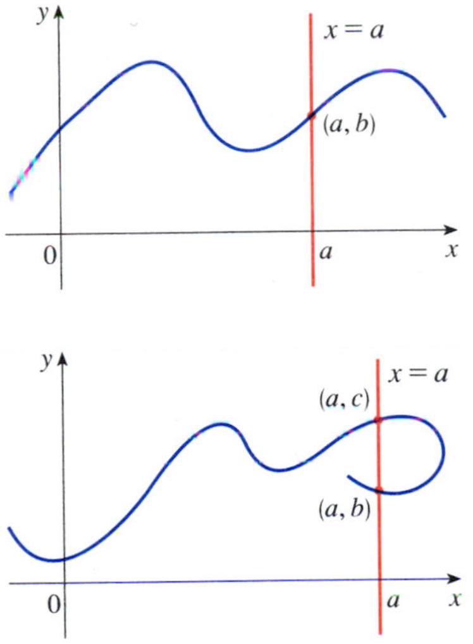

FIGURE 12

The reason for the truth of the Vertical Line Test can be seen in Figure 12. If each vertical line $x = a$ intersects a curve only once, at (a, b), then exactly one functional value is defined by $f(a) = b$. But if a line $x = a$ intersects the curve twice, at (a, b) and (a, c), then the curve cannot represent a function because a function cannot assign two different values to a.

For example, the parabola $x = y^2 - 2$ shown in Figure 11(a) is not the graph of a function of x because, as you can see, there are vertical lines that intersect the parabola twice. The parabola, however, does contain the graphs of *two* functions of x; the upper and lower halves of the parabola are the graphs of the functions $f(x) = \sqrt{x+2}$ and $g(x) = -\sqrt{x+2}$ [Figures 11(b) and (c)]. We observe that if we reverse the roles of x and y, then the equation $x = h(y) = y^2 - 2$ does define x as a function of y (with y as the independent variable and x as the dependent variable) and the parabola now appears as the graph of the function h.

The functions we have looked at so far have been defined by means of simple formulas. But many functions are not given by such formulas. Here are some examples: the cost of mailing a first-class letter as a function of its weight, the population of New York City as a function of time, and the cost of a taxi ride as a function of distance. The following examples give further illustrations.

EXAMPLE 10 When you turn on a hot water faucet, the temperature T of the water depends on how long the water has been running. Draw a rough graph of T as a function of the time t that has elapsed since the faucet was turned on.

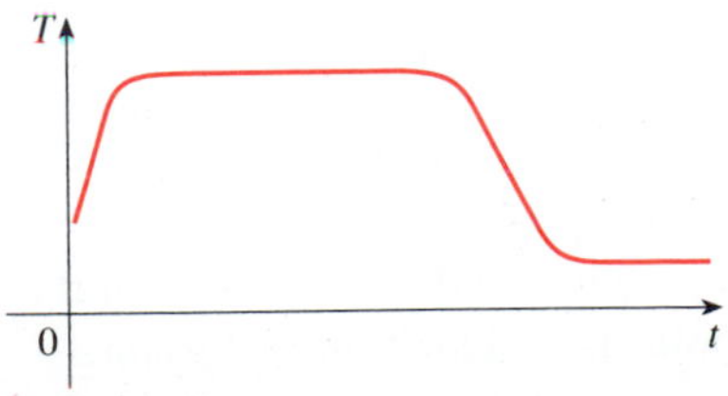

FIGURE 13

SOLUTION The initial temperature of the running water is close to room temperature because of the water that has been in the pipes. When the water from the hot water tank starts coming out, T increases quickly. In the next phase, T is constant at the temperature of the water in the tank. When the tank is drained, T decreases to the temperature of the water supply. This enables us to make the rough sketch of T as a function of t in Figure 13. ■

A more accurate graph of the function in Example 10 could be obtained by using a thermometer to measure the temperature of the water at 10-second intervals. In general, scientists collect experimental data and use them to sketch the graphs of functions. The next example illustrates this procedure.

t	$C(t)$
0	0.0800
2	0.0570
4	0.0408
6	0.0295
8	0.0210

EXAMPLE 11 The data shown in the margin come from an experiment on the lactonization of hydroxyvaleric acid at 25 °C. They give the concentration $C(t)$ of this acid (in moles per liter) after t minutes. Use these data to draw an approximation to the graph of the concentration function. Then use this graph to estimate the concentration after 5 minutes.

SOLUTION We plot the five points corresponding to the data from the table and draw a smooth curve through them as in Figure 14. Then we use the graph to estimate that the concentration after 5 min is

$$C(5) \approx 0.035 \text{ mole/liter}$$

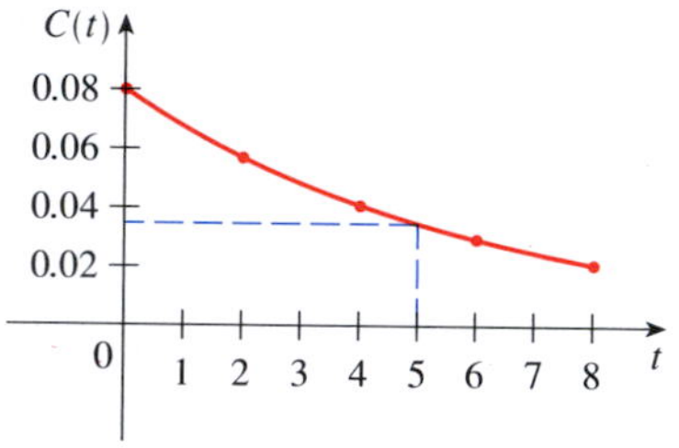

FIGURE 14

■

PIECEWISE DEFINED FUNCTIONS

The functions in the following four examples are defined by different formulas in different parts of their domains.

EXAMPLE 12 A function f is defined by

$$f(x) = \begin{cases} 1 - x & \text{if } x \leqslant 1 \\ x^2 & \text{if } x > 1 \end{cases}$$

Evaluate $f(0)$, $f(1)$, and $f(2)$ and sketch the graph.

SOLUTION Remember that a function is a rule. For this particular function the rule is the following: First look at the value of the input x. If it happens that $x \leqslant 1$, then the value of $f(x)$ is $1 - x$. On the other hand, if $x > 1$, then the value of $f(x)$ is x^2.

$$\text{Since } 0 \leqslant 1, \text{ we have } f(0) = 1 - 0 = 1.$$

$$\text{Since } 1 \leqslant 1, \text{ we have } f(1) = 1 - 1 = 0.$$

$$\text{Since } 2 > 1, \text{ we have } f(2) = 2^2 = 4.$$

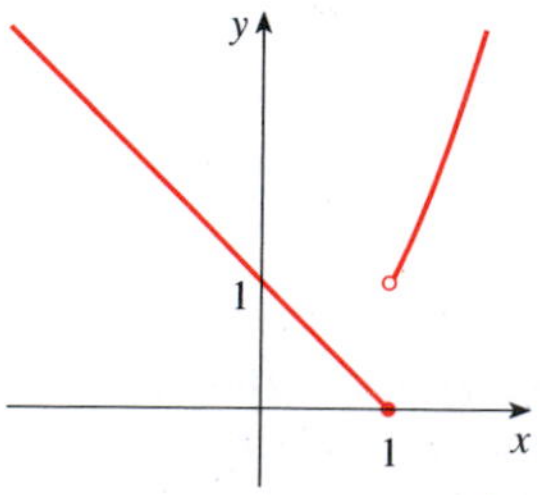

FIGURE 15

How do we draw the graph of f? We observe that if $x \leqslant 1$, then $f(x) = 1 - x$, so the part of the graph of f that lies to the left of the vertical line $x = 1$ must coincide with the line $y = 1 - x$, which has slope -1 and y-intercept 1. If $x > 1$, then $f(x) = x^2$, so the part of the graph of f that lies to the right of the line $x = 1$ must coincide with the graph of $y = x^2$, which is a parabola. This enables us to sketch the graph in Figure 15. The solid dot indicates that the point is included on the graph; the open dot indicates that the point is excluded from the graph. ■

For a more extensive review of absolute values, see Appendix A.

The next example of a piecewise defined function is the absolute value function. Recall that the **absolute value** of a number a, denoted by $|a|$, is the distance from a to 0 on the real number line. Distances are always positive or 0, so we have

$$|a| \geqslant 0 \qquad \text{for every number } a$$

For example,

$$|3| = 3 \qquad |-3| = 3 \qquad |0| = 0 \qquad |\sqrt{2} - 1| = \sqrt{2} - 1 \qquad |3 - \pi| = \pi - 3$$

In general, we have

$$|a| = a \qquad \text{if } a \geqslant 0$$
$$|a| = -a \qquad \text{if } a < 0$$

(Remember that if a is negative, then $-a$ is positive.)

EXAMPLE 13 Sketch the graph of the absolute value function $f(x) = |x|$.

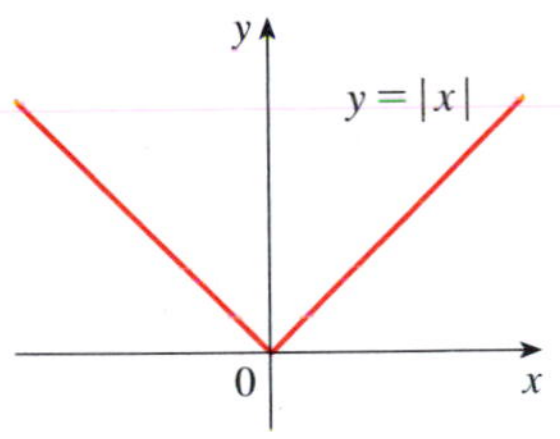

FIGURE 16

SOLUTION From the preceding discussion we know that

$$|x| = \begin{cases} x & \text{if } x \geqslant 0 \\ -x & \text{if } x < 0 \end{cases}$$

Using the same method as in Example 12, we see that the graph of f coincides with the line $y = x$ to the right of the y-axis and coincides with the line $y = -x$ to the left of the y-axis (see Figure 16). ■

EXAMPLE 14 Find a formula for the function f graphed in Figure 17.

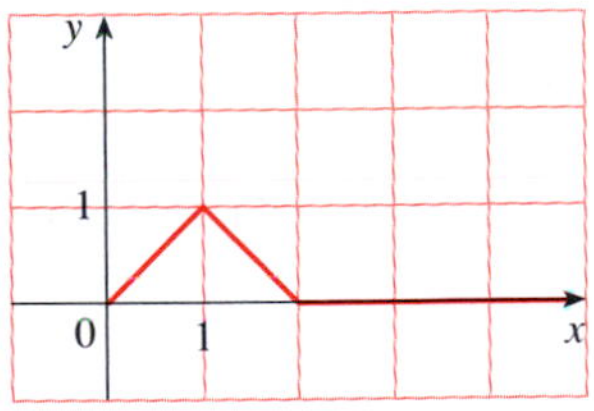

FIGURE 17

SOLUTION The line through $(0, 0)$ and $(1, 1)$ has slope $m = 1$ and y-intercept $b = 0$, so its equation is $y = x$. Thus, for the part of the graph of f that joins $(0, 0)$ to $(1, 1)$, we have

$$f(x) = x \qquad \text{if } 0 \leqslant x \leqslant 1$$

The line through $(1, 1)$ and $(2, 0)$ has slope $m = -1$, so its point-slope form is

$$y - 0 = (-1)(x - 2) \qquad \text{or} \qquad y = 2 - x$$

So we have $$f(x) = 2 - x \qquad \text{if } 1 < x \leqslant 2$$

We also see that the graph of f coincides with the x-axis for $x > 2$. Putting this information together, we have the following three-piece formula for f:

$$f(x) = \begin{cases} x & \text{if } 0 \leqslant x \leqslant 1 \\ 2 - x & \text{if } 1 < x \leqslant 2 \\ 0 & \text{if } x > 2 \end{cases}$$ ■

EXAMPLE 15 The cost of a long-distance daytime phone call from Toronto to Rome on a calling card is \$6.60 for the initial three-minute period and \$1.65 for each additional minute (or part of a minute). Draw the graph of the cost C (in dollars) of the phone call as a function of the time t (in minutes).

SOLUTION Let $C(t)$ be the cost for t minutes. Since $t > 0$, the domain of the function is $(0, \infty)$. From the given information, we have

$$
\begin{aligned}
C(t) &= 6.60 && \text{if } 0 < t \leqslant 3 \\
C(t) &= 6.60 + 1.65 = 8.25 && \text{if } 3 < t \leqslant 4 \\
C(t) &= 6.60 + 2(1.65) = 9.90 && \text{if } 4 < t \leqslant 5 \\
C(t) &= 6.60 + 3(1.65) = 11.55 && \text{if } 5 < t \leqslant 6
\end{aligned}
$$

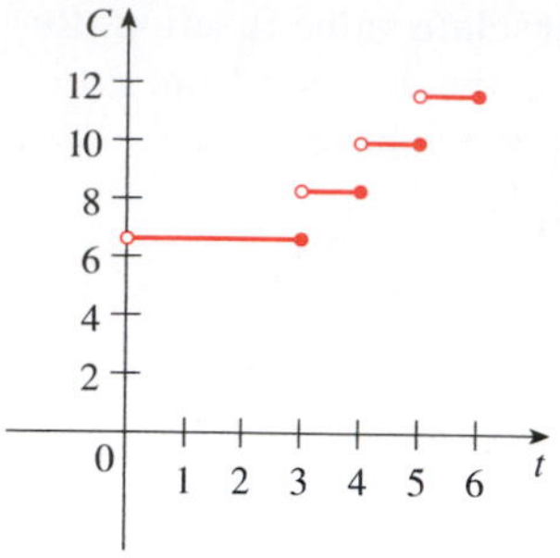

FIGURE 18

and so on. The graph is shown in Figure 18. You can see why functions similar to this one are called **step functions**—they jump from one value to the next. Such functions will be studied in Chapter 2. ■

SYMMETRY

If a function f satisfies $f(-x) = f(x)$ for every number x in its domain, then f is called an **even function**. For instance, the function $f(x) = x^2$ is even because

$$f(-x) = (-x)^2 = x^2 = f(x)$$

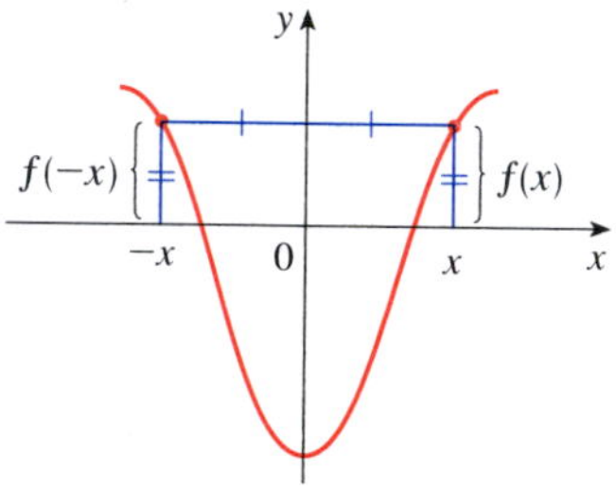

FIGURE 19
An even function

The geometric significance of an even function is that its graph is symmetric with respect to the y-axis (see Figure 19). This means that if we have plotted the graph of f for $x \geqslant 0$, we obtain the entire graph simply by reflecting about the y-axis.

If f satisfies $f(-x) = -f(x)$ for every number x in its domain, then f is called an **odd function**. For example, the function $f(x) = x^3$ is odd because

$$f(-x) = (-x)^3 = -x^3 = -f(x)$$

The graph of an odd function is symmetric about the origin (see Figure 20). If we already have the graph of f for $x \geqslant 0$, we can obtain the entire graph by rotating through $180°$ about the origin. For instance, in Example 7 we need only have plotted the graph of $y = x^3$ for $x \geqslant 0$ and then rotated that part about the origin.

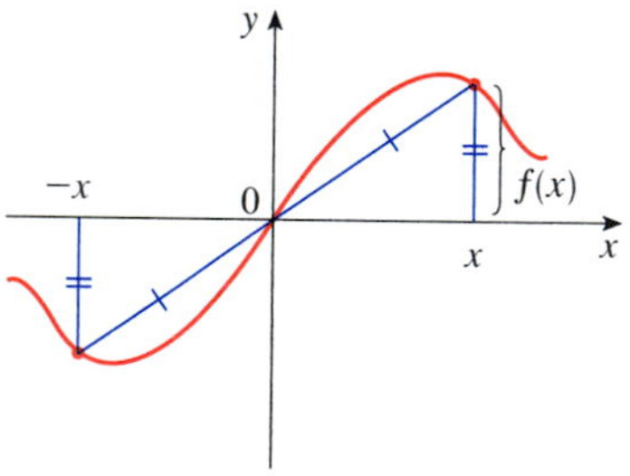

FIGURE 20
An odd function

EXAMPLE 16 Determine whether each of the following functions is even, odd, or neither even nor odd:

(a) $f(x) = x^5 + x$ (b) $g(x) = 1 - x^4$ (c) $h(x) = 2x - x^2$

SOLUTION

(a)
$$
\begin{aligned}
f(-x) &= (-x)^5 + (-x) = (-1)^5 x^5 + (-x) \\
&= -x^5 - x = -(x^5 + x) \\
&= -f(x)
\end{aligned}
$$

Therefore, f is an odd function.

(b)
$$g(-x) = 1 - (-x)^4 = 1 - x^4 = g(x)$$

So g is even.

(c)
$$h(-x) = 2(-x) - (-x)^2 = -2x - x^2$$

Since $h(-x) \neq h(x)$ and $h(-x) \neq -h(x)$, we conclude that h is neither even nor odd. ■

The graphs of the functions in Example 16 are shown in Figure 21. The graph of f was drawn by plotting points for $x \geqslant 0$ and rotating about the origin. The graph of g was drawn by plotting points for $x \geqslant 0$ and reflecting about the y-axis. Notice that the graph of h is symmetric neither about the y-axis nor about the origin.

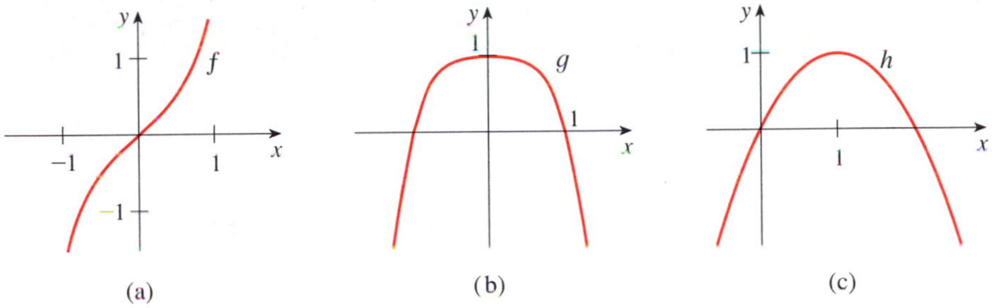

FIGURE 21

COMBINATIONS OF FUNCTIONS

Two functions f and g can be combined to form new functions $f + g$, $f - g$, fg, and f/g in a manner similar to the way we add, subtract, multiply, and divide real numbers.

If we define the sum $f + g$ by the equation

(1) $$(f + g)(x) = f(x) + g(x)$$

then the right side of Equation 1 makes sense if both $f(x)$ and $g(x)$ are defined, that is, if x belongs to the domain of f and also to the domain of g. If the domain of f is A and the domain of g is B, then the domain of $f + g$ is the intersection of these domains, that is, $A \cap B$.

Notice that the $+$ sign on the left side of Equation 1 stands for the operation of addition of *functions*, but the $+$ sign on the right side of the equation stands for addition of the *numbers* $f(x)$ and $g(x)$.

Similarly, we can define the difference $f - g$ and the product fg, and their domains are also $A \cap B$. But in defining the quotient f/g we must remember not to divide by 0.

ALGEBRA OF FUNCTIONS Let f and g be functions with domains A and B. Then the functions $f + g$, $f - g$, fg, and f/g are defined as follows:

$$(f + g)(x) = f(x) + g(x) \qquad \text{domain} = A \cap B$$

$$(f - g)(x) = f(x) - g(x) \qquad \text{domain} = A \cap B$$

$$(fg)(x) = f(x)g(x) \qquad \text{domain} = A \cap B$$

$$\left(\frac{f}{g}\right)(x) = \frac{f(x)}{g(x)} \qquad \text{domain} = \{x \in A \cap B \mid g(x) \neq 0\}$$

EXAMPLE 17 If $f(x) = \sqrt{x}$ and $g(x) = \sqrt{4 - x^2}$, find the functions $f + g$, $f - g$, fg, and f/g.

Another way to solve $4 - x^2 \geqslant 0$:
$(2 - x)(2 + x) \geqslant 0$

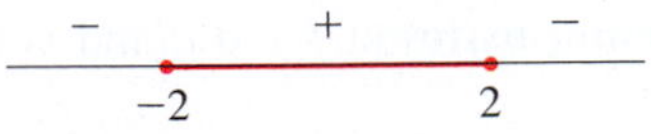

SOLUTION The domain of $f(x) = \sqrt{x}$ is $[0, \infty)$. The domain of $g(x) = \sqrt{4 - x^2}$ consists of all numbers x such that $4 - x^2 \geqslant 0$, that is, $x^2 \leqslant 4$. Taking square roots of both sides, we get $|x| \leqslant 2$, or $-2 \leqslant x \leqslant 2$, so the domain of g is the interval $[-2, 2]$. The intersection of the domains of f and g is

$$[0, \infty) \cap [-2, 2] = [0, 2]$$

Thus, according to the definitions, we have

$$(f + g)(x) = \sqrt{x} + \sqrt{4 - x^2} \qquad 0 \leqslant x \leqslant 2$$

$$(f - g)(x) = \sqrt{x} - \sqrt{4 - x^2} \qquad 0 \leqslant x \leqslant 2$$

$$(fg)(x) = \sqrt{x}\,\sqrt{4 - x^2} = \sqrt{4x - x^3} \qquad 0 \leqslant x \leqslant 2$$

$$\left(\frac{f}{g}\right)(x) = \frac{\sqrt{x}}{\sqrt{4 - x^2}} = \sqrt{\frac{x}{4 - x^2}} \qquad 0 \leqslant x < 2$$

Notice that the domain of f/g is the interval $[0, 2)$ because we must exclude the points where $g(x) = 0$, that is, $x = \pm 2$. ■

The graph of the function $f + g$ is obtained from the graphs of f and g by **graphical addition**. This means that we add corresponding y-coordinates as in Figure 22. Figure 23 shows the result of using this procedure to graph the function $f + g$ from Example 17.

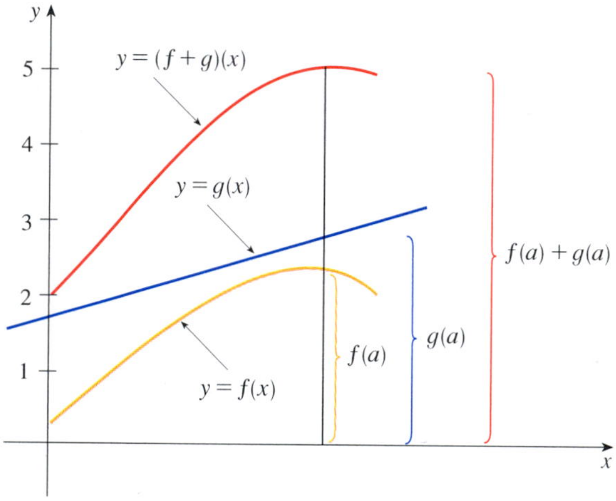

FIGURE 22

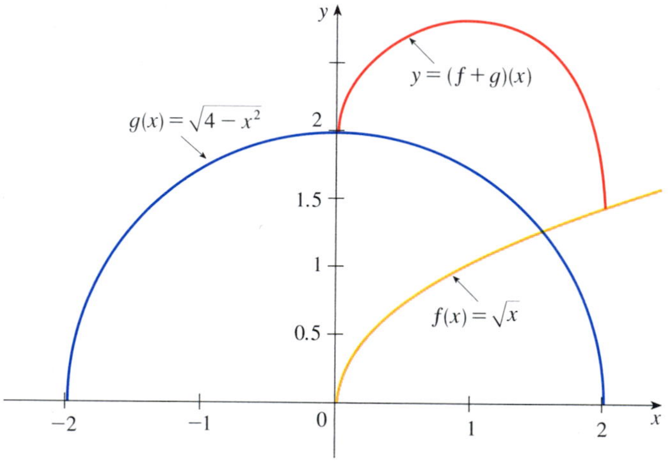

FIGURE 23

COMPOSITION OF FUNCTIONS

There is another way of combining two functions to get a new function. For example, suppose that $y = f(u) = \sqrt{u}$ and $u = g(x) = x^2 + 1$. Since y is a function of u and u is, in turn, a function of x, it follows that y is ultimately a function of x. We compute this by substitution:

$$y = f(u) = f(g(x)) = f(x^2 + 1) = \sqrt{x^2 + 1}$$

The procedure is called **composition** because the new function is *composed* of the two given functions f and g.

In general, given any two functions f and g, we start with a number x in the domain of g and find its image $g(x)$. If this number $g(x)$ is in the domain of f, then we can calculate the value of $f(g(x))$. The result is a new function $h(x) = f(g(x))$ obtained by substituting g into f. It is called the **composition** (or **composite**) of f and g and is denoted by $f \circ g$ ("f circle g").

DEFINITION Given two functions f and g, the **composite function** $f \circ g$ (also called the **composition** of f and g) is defined by

$$(f \circ g)(x) = f(g(x))$$

The domain of $f \circ g$ is the set of all x in the domain of g such that $g(x)$ is in the domain of f. In other words, $(f \circ g)(x)$ is defined whenever both $g(x)$ and $f(g(x))$ are defined. The best way to picture $f \circ g$ is by a machine diagram (Figure 24) or an arrow diagram (Figure 25).

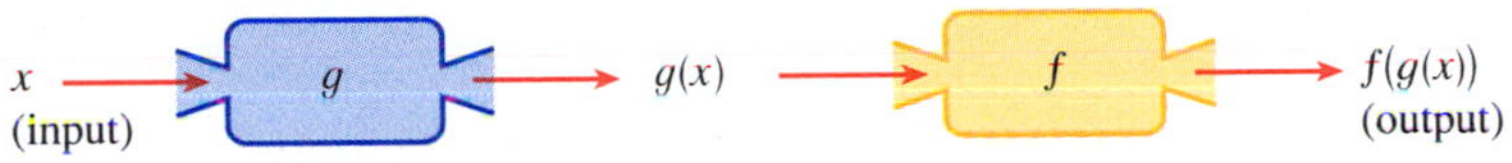

FIGURE 24
The $f \circ g$ machine is composed of the g machine (first) and then the f machine.

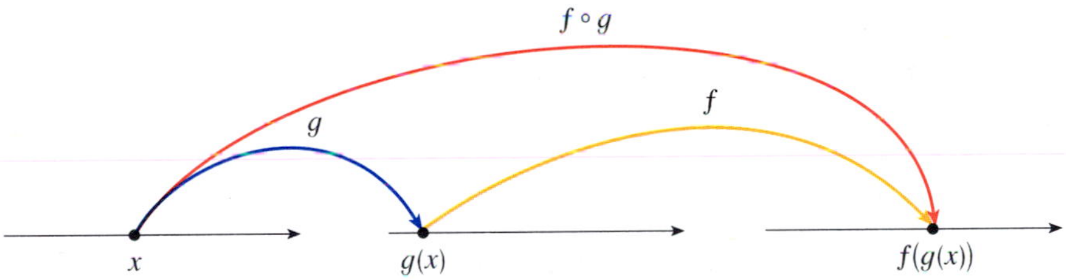

FIGURE 25
Arrow diagram for $f \circ g$

EXAMPLE 18 If $f(x) = x^2$ and $g(x) = x - 3$, find the composite functions $f \circ g$ and $g \circ f$ and state their domains.

SOLUTION We have

$$(f \circ g)(x) = f(g(x)) = f(x - 3) = (x - 3)^2$$
$$(g \circ f)(x) = g(f(x)) = g(x^2) = x^2 - 3$$

The domains of both $f \circ g$ and $g \circ f$ are $\mathbb{R}$ (the set of all real numbers). ■

NOTE: You can see from Example 18 that, in general, $f \circ g \neq g \circ f$. Remember, the notation $f \circ g$ means that the function g is applied first and then f is applied second. In Example 18, $f \circ g$ is the function that *first* subtracts 3 and *then* squares; $g \circ f$ is the function that *first* squares and *then* subtracts 3.

EXAMPLE 19 If $f(x) = \sqrt{x}$ and $g(x) = \sqrt{2 - x}$, find each function and its domain.

(a) $f \circ g$ (b) $g \circ f$ (c) $f \circ f$ (d) $g \circ g$

SOLUTION

(a) $$(f \circ g)(x) = f(g(x)) = f(\sqrt{2 - x}) = \sqrt{\sqrt{2 - x}} = \sqrt[4]{2 - x}$$

The domain of $f \circ g$ is $\{x \mid 2 - x \geqslant 0\} = \{x \mid x \leqslant 2\} = (-\infty, 2]$.

(b) $$(g \circ f)(x) = g(f(x)) = g(\sqrt{x}) = \sqrt{2 - \sqrt{x}}$$

For $\sqrt{x}$ to be defined we must have $x \geqslant 0$. For $\sqrt{2 - \sqrt{x}}$ to be defined we must have $2 - \sqrt{x} \geqslant 0$, that is, $\sqrt{x} \leqslant 2$, or $x \leqslant 4$. Thus, we have $0 \leqslant x \leqslant 4$, so the domain of $g \circ f$ is the closed interval $[0, 4]$.

If $0 \leqslant a \leqslant b$, then $a^2 \leqslant b^2$.

(c) $$(f \circ f)(x) = f(f(x)) = f(\sqrt{x}) = \sqrt{\sqrt{x}} = \sqrt[4]{x}$$

The domain of $f \circ f$ is $[0, \infty)$.

(d) $$(g \circ g)(x) = g(g(x)) = g(\sqrt{2 - x}) = \sqrt{2 - \sqrt{2 - x}}$$

This expression is defined when $2 - x \geqslant 0$, that is, $x \leqslant 2$, and $2 - \sqrt{2 - x} \geqslant 0$. This latter inequality is equivalent to $\sqrt{2 - x} \leqslant 2$, or $2 - x \leqslant 4$, that is, $x \geqslant -2$. Thus $-2 \leqslant x \leqslant 2$, so the domain of $g \circ g$ is the closed interval $[-2, 2]$. ■

It is possible to take the composition of three or more functions. For instance, the composite function $f \circ g \circ h$ is found by first applying h, then g, and then f as follows:

$$(f \circ g \circ h)(x) = f(g(h(x)))$$

EXAMPLE 20 Find $f \circ g \circ h$ if $f(x) = x/(x + 1)$, $g(x) = x^{10}$, and $h(x) = x + 3$.

SOLUTION

$$(f \circ g \circ h)(x) = f(g(h(x))) = f(g(x + 3)) = f((x + 3)^{10}) = \frac{(x + 3)^{10}}{(x + 3)^{10} + 1}$$ ■

So far we have used composition to build complicated functions from simpler ones. But in calculus it is sometimes useful to be able to decompose a complicated function into simpler ones, as in the following example.

EXAMPLE 21 Given $F(x) = \sqrt[4]{x + 9}$, find functions f and g such that $F = f \circ g$.

SOLUTION Since the formula for F says to first add 9 and then take the fourth root, we let

$$g(x) = x + 9 \qquad \text{and} \qquad f(x) = \sqrt[4]{x}$$

Then

$$(f \circ g)(x) = f(g(x)) = f(x + 9) = \sqrt[4]{x + 9} = F(x)$$ ■

INCREASING AND DECREASING FUNCTIONS

The graph shown in Figure 26 rises from A to B, falls from B to C, and rises again from C to D. The function f is said to be increasing on the interval $[a, b]$, decreasing on $[b, c]$, and increasing again on $[c, d]$. Notice that if x_1 and x_2 are any two numbers between a and b with $x_1 < x_2$, then $f(x_1) < f(x_2)$. We use this as the defining property of an increasing function.

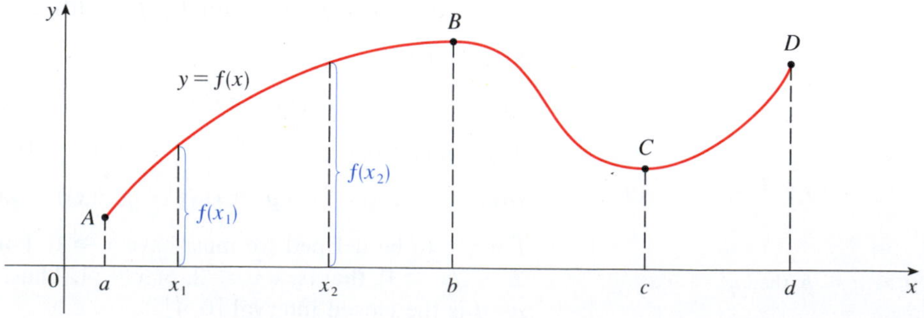

FIGURE 26

A function f is called **increasing** on an interval I if

$$f(x_1) < f(x_2) \quad \text{whenever } x_1 < x_2 \text{ in } I$$

It is called **decreasing** on I if

$$f(x_1) > f(x_2) \quad \text{whenever } x_1 < x_2 \text{ in } I$$

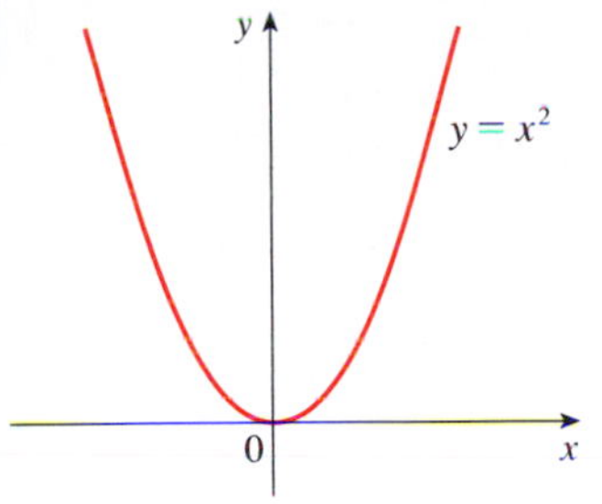

FIGURE 27

In the definition of an increasing function it is important to realize that the inequality $f(x_1) < f(x_2)$ must be satisfied for *every* pair of numbers x_1 and x_2 in I with $x_1 < x_2$.

You can see from Figure 27 that the function $f(x) = x^2$ is decreasing on the interval $(-\infty, 0]$ and increasing on the interval $[0, \infty)$.

EXERCISES 1

1. If $f(x) = 2x^2 + 3x - 4$, find $f(0)$, $f(2)$, $f(\sqrt{2})$, $f(1 + \sqrt{2})$, $f(-x)$, $f(x + 1)$, $2f(x)$, and $f(2x)$.

2. If $g(x) = x^3 + 2x^2 - 3$, find $g(0)$, $g(3)$, $g(-x)$, and $g(1 + h)$.

3–4 ■ Find $f(2 + h)$, $f(x + h)$, and $\dfrac{f(x + h) - f(x)}{h}$, where $h \neq 0$.

3. $f(x) = x - x^2$

4. $f(x) = \dfrac{x}{x + 1}$

5–6 ■ Draw a machine diagram, an arrow diagram, and a graph for the given function.

5. $f(x) = \sqrt{x}, \quad 0 \leqslant x \leqslant 4$

6. $f(x) = 2/x, \quad 1 \leqslant x \leqslant 4$

7. The domain of f is $A = \{1, 2, 3, 4, 5, 6\}$ and $f(1) = 2$, $f(2) = 1$, $f(3) = 0$, $f(4) = 1$, $f(5) = 2$, and $f(6) = 4$. What is the range of f? Draw an arrow diagram and a graph for f.

8–14 ■ Find the domain and range of the function.

8. $f(x) = 2x + 7, \quad -1 \leqslant x \leqslant 6$

9. $f(x) = 6 - 4x, \quad -2 \leqslant x \leqslant 3$

10. $g(x) = \dfrac{2}{3x - 5}$

11. $h(x) = \sqrt{2x - 5}$

12. $h(x) = \sqrt[4]{7 - 3x}$

13. $F(x) = \sqrt{1 - x^2}$

14. $F(x) = 1 - \sqrt{x}$

15–22 ■ Find the domain of the function.

15. $f(x) = \dfrac{x + 2}{x^2 - 1}$

16. $f(x) = \dfrac{x^4}{x^2 + x - 6}$

17. $g(x) = \sqrt[4]{x^2 - 6x}$

18. $g(x) = \sqrt{x^2 - 2x - 8}$

19. $\phi(x) = \sqrt{\dfrac{x}{\pi - x}}$

20. $\phi(x) = \sqrt{\dfrac{x^2 - 2x}{x - 1}}$

21. $f(t) = \sqrt[3]{t - 1}$

22. $f(t) = \sqrt{t^2 + 1}$

23–50 ■ Find the domain and sketch the graph of the function.

23. $f(x) = 3 - 2x$

24. $f(x) = \dfrac{x + 3}{2}, \quad -2 \leqslant x \leqslant 2$

25. $f(x) = x^2 + 2x - 1$

26. $f(x) = -x^2 + 6x - 7$

27. $g(x) = \sqrt{-x}$

28. $g(x) = \sqrt{6 - 2x}$

29. $h(x) = \sqrt{4 - x^2}$

30. $h(x) = \sqrt{x^2 - 4}$

31. $F(x) = \dfrac{1}{x}$

32. $F(x) = \dfrac{2}{x + 4}$

33. $G(x) = |x| + x$

34. $G(x) = |x| - x$

35. $H(x) = |2x|$

36. $H(x) = |2x - 3|$

37. $f(x) = x/|x|$

38. $f(x) = |x^2 - 1|$

39. $f(x) = \dfrac{x^2 - 1}{x - 1}$

40. $f(x) = \dfrac{x^2 + 5x + 6}{x + 2}$

41. $f(x) = \begin{cases} 0 & \text{if } x < 2 \\ 1 & \text{if } x \geqslant 2 \end{cases}$

42. $f(x) = \begin{cases} -1 & \text{if } x < -1 \\ 1 & \text{if } -1 \leqslant x \leqslant 1 \\ -1 & \text{if } x > 1 \end{cases}$

43. $f(x) = \begin{cases} x & \text{if } x \leqslant 0 \\ x + 1 & \text{if } x > 0 \end{cases}$

44. $f(x) = \begin{cases} 2x + 3 & \text{if } x < -1 \\ 3 - x & \text{if } x \geqslant -1 \end{cases}$

45. $f(x) = \begin{cases} -1 & \text{if } x < -1 \\ x & \text{if } -1 \leqslant x \leqslant 1 \\ 1 & \text{if } x > 1 \end{cases}$

46. $f(x) = \begin{cases} |x| & \text{if } |x| \leqslant 1 \\ 1 & \text{if } |x| > 1 \end{cases}$

47. $f(x) = \begin{cases} x + 2 & \text{if } x \leqslant -1 \\ x^2 & \text{if } x > -1 \end{cases}$

48. $f(x) = \begin{cases} 1 - x^2 & \text{if } x \leqslant 2 \\ 2x - 7 & \text{if } x > 2 \end{cases}$

49. $f(x) = \begin{cases} -1 & \text{if } x \leqslant -1 \\ 3x + 2 & \text{if } |x| < 1 \\ 7 - 2x & \text{if } x \geqslant 1 \end{cases}$

50. $f(x) = \begin{cases} \sqrt{-x} & \text{if } x < 0 \\ x & \text{if } 0 \leqslant x \leqslant 2 \\ \sqrt{x - 2} & \text{if } x > 2 \end{cases}$

51–54 ■ State whether the curve is the graph of a function of x. If it is, state the domain and range of the function.

51.

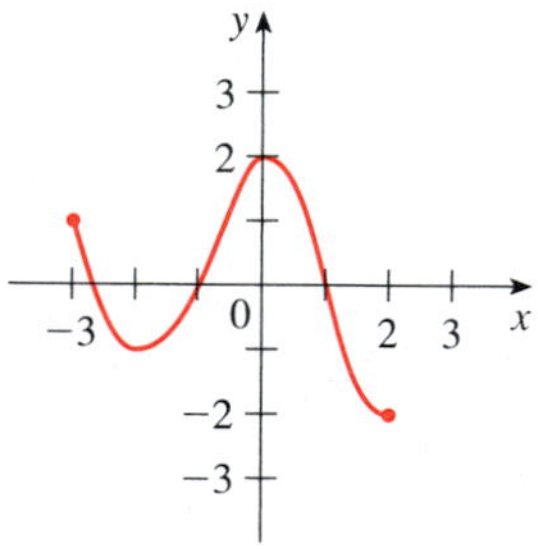

52.

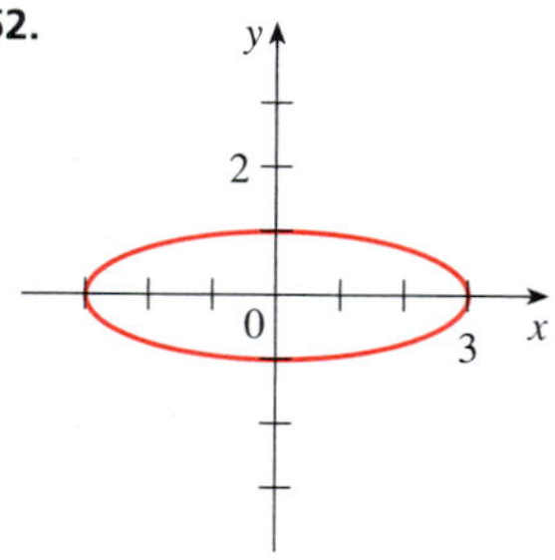

53.

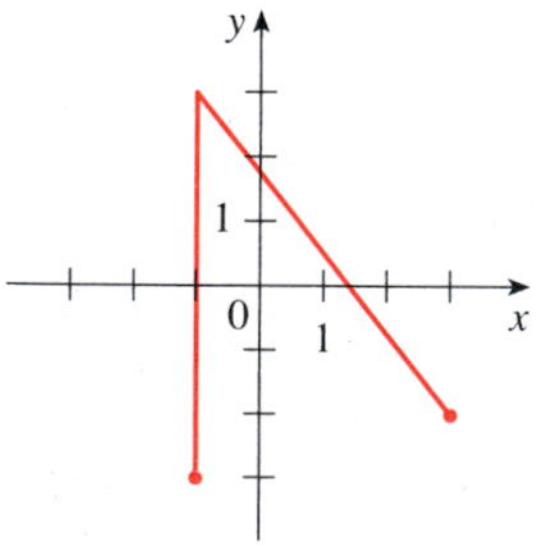

54.

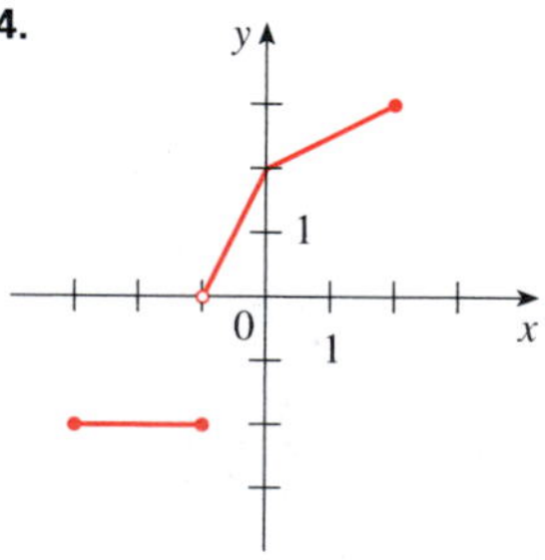

55–60 ■ Find a function whose graph is the given curve.

55. The line segment joining the points $(-2, 1)$ and $(4, -6)$

56. The line segment joining the points $(-3, -2)$ and $(6, 3)$

57. The bottom half of the parabola $x + (y - 1)^2 = 0$

58. The top half of the circle $(x - 1)^2 + y^2 = 1$

59.

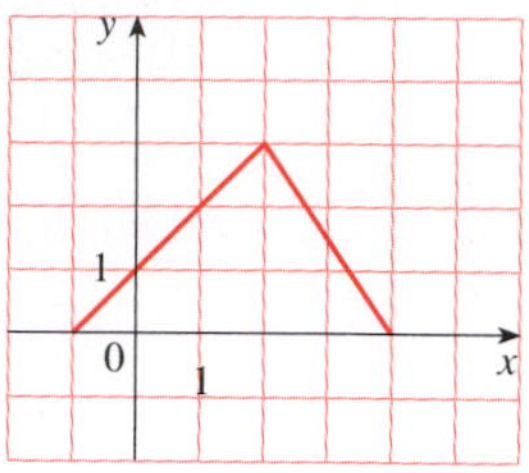

60.

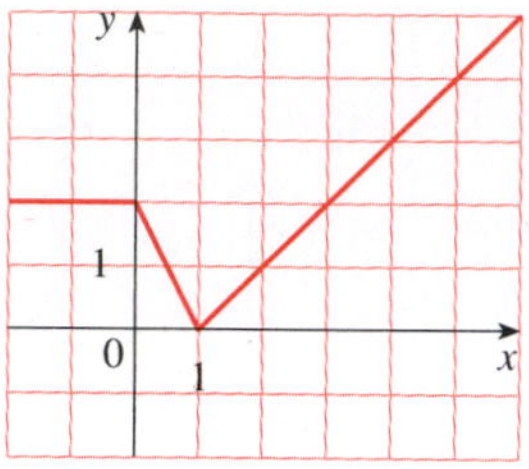

61–65 ■ Find a formula for the described function and state its domain.

61. A rectangle has perimeter 20 m. Express the area of the rectangle as a function of the length of one of its sides.

62. A rectangle has area 16 m^2. Express the perimeter of the rectangle as a function of the length of one of its sides.

63. Express the area of an equilateral triangle as a function of the length of a side.

64. Express the surface area of a cube as a function of its volume.

65. An open rectangular box with volume 2 m^3 has a square base. Express the surface area of the box as a function of the length of a side of the base.

66. A Norman window has the shape of a rectangle surmounted by a semicircle. If the perimeter of the window is 30 ft, express the area A of the window as a function of the width x of the window.

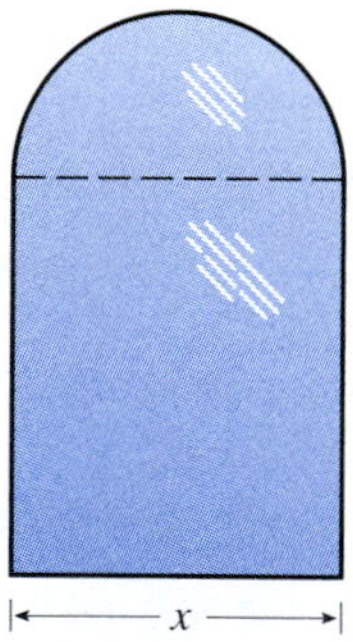

67. A box with an open top is to be constructed from a rectangular piece of cardboard with dimensions 12 in. by 20 in. by cutting out equal squares of side x at each corner and then folding up the sides as in the figure. Express the volume V of the box as a function of x.

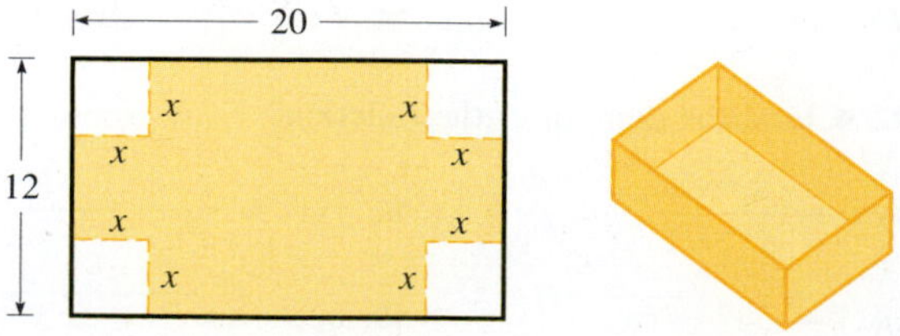

68. A taxi company charges two dollars for the first mile (or part of a mile) and 20 cents for each succeeding tenth of a mile (or part). Express the cost C (in dollars) of a ride as a function of the distance x traveled (in miles) for $0 < x < 2$, and sketch the graph of this function.

69. (a) As dry air moves upward, it expands and cools. If the ground temperature is 20 °C and the temperature at a height of 1 km is 10 °C, express the temperature T (in °C) as a function of the height h (in kilometers), assuming the function is linear.
(b) Draw the graph of the function in part (a). What does the slope represent?
(c) What is the temperature at a height of 2.5 km?

70. The monthly cost of driving a car depends on the number of miles driven. Lynn found that in May it cost her \$380 to drive 480 mi and in June it cost her \$460 to drive 800 mi.
(a) Express the monthly cost C as a function of the distance driven d, assuming that a linear function gives a suitable model.
(b) Use part (a) to predict the cost of driving 1500 miles per month.
(c) Draw the graph of the function. What does the slope of the line represent?
(d) What does the y-intercept of the graph represent?
(e) Why is a linear function a suitable model in this situation?

71. You put some ice cubes in a glass, fill the glass with cold water, and then let the glass sit on a table. Sketch a rough graph of the temperature of the water as a function of the elapsed time.

72. Sketch a rough graph of the number of hours of daylight as a function of the time of year.

73. Sketch a rough graph of the outdoor temperature as a function of time during a typical spring day.

74. You place a frozen pie in an oven and bake it for an hour. Then you take it out and let it cool before eating it. Sketch a rough graph of the temperature of the pie as a function of time.

75. Temperature readings T (in °F) were recorded every two hours from midnight to noon on a day in May. The time t was measured in hours from midnight.

t	0	2	4	6	8	10	12
T	45	42	40	40	44	50	60

(a) Use the readings to sketch the graph of T as a function of t.
(b) Use the graph to estimate the temperature at 11 A.M.

76. The population P (in thousands) of a city from 1970 to 1980 is shown in the table.

t	1970	1972	1974	1976	1978	1980
P	71	73	78	87	102	123

(a) Draw a graph of P as a function of time.
(b) Use the graph to estimate the population in 1979.

77–82 ■ Determine whether f is even, odd, or neither. If f is even or odd, use symmetry to sketch its graph.

77. $f(x) = x^{-2}$ **78.** $f(x) = x^{-3}$

79. $f(x) = x^2 + x$ **80.** $f(x) = x^4 - 4x^2$

81. $f(x) = x^3 - x$ **82.** $f(x) = 3x^3 + 2x^2 + 1$

83–84 ■ Find $f + g$, $f - g$, fg, and f/g and state their domains.

83. $f(x) = x^3 + 2x^2$, $g(x) = 3x^2 - 1$

84. $f(x) = \sqrt{1 + x}$, $g(x) = \sqrt{1 - x}$

85–86 ■ Use the graphs of f and g and the method of graphical addition to sketch the graph of $f + g$.

85. $f(x) = x$, $g(x) = 1/x$ **86.** $f(x) = x^3$, $g(x) = -x^2$

87–94 ■ Find the functions $f \circ g$, $g \circ f$, $f \circ f$, and $g \circ g$ and their domains.

87. $f(x) = 2x^2 - x$, $g(x) = 3x + 2$

88. $f(x) = \sqrt{x - 1}$, $g(x) = x^2$

89. $f(x) = 1/x$, $g(x) = x^3 + 2x$

90. $f(x) = \dfrac{1}{x - 1}$, $g(x) = \dfrac{x - 1}{x + 1}$

91. $f(x) = \sqrt[3]{x}$, $g(x) = 1 - \sqrt{x}$

92. $f(x) = \sqrt{x^2 - 1}$, $g(x) = \sqrt{1 - x}$

93. $f(x) = \dfrac{x + 2}{2x + 1}$, $g(x) = \dfrac{x}{x - 2}$

94. $f(x) = \dfrac{1}{\sqrt{x}}$, $g(x) = x^2 - 4x$

95–98 ■ Find $f \circ g \circ h$.

95. $f(x) = x - 1$, $g(x) = \sqrt{x}$, $h(x) = x - 1$

96. $f(x) = \dfrac{1}{x}$, $g(x) = x^3$, $h(x) = x^2 + 2$

97. $f(x) = x^4 + 1$, $g(x) = x - 5$, $h(x) = \sqrt{x}$

98. $f(x) = \sqrt{x}$, $g(x) = \dfrac{x}{x - 1}$, $h(x) = \sqrt[3]{x}$

99–102 ■ Express the function in the form $f \circ g$.

99. $F(x) = (x - 9)^5$

100. $F(x) = \sqrt{x} + 1$

101. $G(x) = \dfrac{x^2}{x^2 + 4}$

102. $G(x) = \dfrac{1}{x + 3}$

103–104 ■ Express the function in the form $f \circ g \circ h$.

103. $H(x) = \dfrac{1}{x^2 + 1}$

104. $H(x) = \sqrt[3]{\sqrt{x} - 1}$

105. A stone is dropped into a lake, creating a circular ripple that travels outward at a speed of 60 cm/s. Express the area of this circle as a function of time t (in seconds).

106. A spherical balloon is being inflated. If the radius of the balloon is increasing at a rate of 1 cm/s, express the volume of the balloon as a function of time t (in seconds).

107. If $f(x) = 3x + 5$ and $h(x) = 3x^2 + 3x + 2$, find a function g such that $f \circ g = h$.

108. If $f(x) = x + 4$ and $h(x) = 4x - 1$, find a function g such that $g \circ f = h$.

109. Let $f(x) = 1/x$ and $g(x) = x$. How does $f \circ f$ differ from g?

2 TYPES OF FUNCTIONS; SHIFTING AND SCALING

In solving calculus problems you will find that it is helpful to be familiar with the graphs of some commonly occurring functions. We classify various types of functions as follows.

CONSTANT FUNCTIONS The constant function $f(x) = c$ has domain $\mathbb{R}$ and its range consists of the single number c. Its graph is a horizontal line and is illustrated in Figure 1 for $c = 2$.

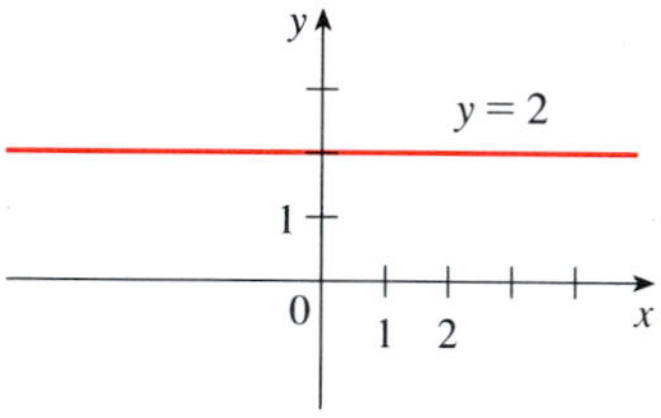

FIGURE 1

POWER FUNCTIONS A function of the form $f(x) = x^a$, where a is a constant, is called a **power function**. We consider several cases.

(a) $a = n$, a positive integer.

The graphs of $f(x) = x^n$ for $n = 1, 2, 3, 4$, and 5 are shown in Figure 2. We already know the shape of the graphs of $y = x$ (a line through the origin with slope 1), $y = x^2$ (a parabola), and $y = x^3$ (Example 7 in Section 1).

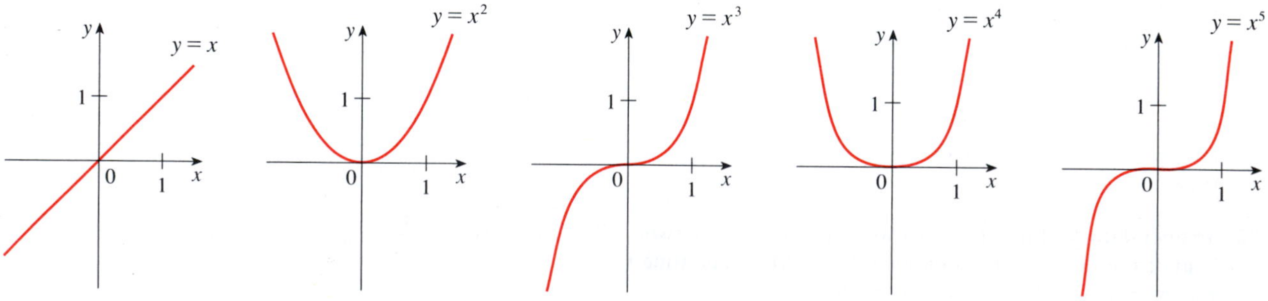

FIGURE 2 Graphs of $f(x) = x^n$ for $n = 1, 2, 3, 4, 5$

The general shape of the graph of $f(x) = x^n$ depends on whether n is even or odd. If n is even, then $f(x) = x^n$ is an even function and its graph is similar to the parabola $y = x^2$. If n is odd, then $f(x) = x^n$ is an odd function and its graph is similar to that of $y = x^3$. Notice from Figure 3, however, that as n increases, the graph of $y = x^n$ becomes flatter near 0 and steeper when $|x| \geqslant 1$. (If x is small, then x^2 is smaller, x^3 is even smaller, x^4 is smaller still, and so on.)

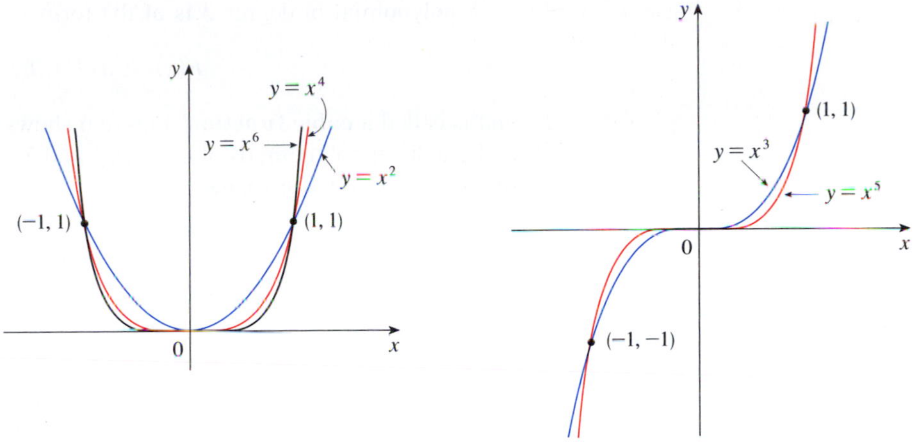

FIGURE 3

(b) $a = -1$.
The graph of the reciprocal function $f(x) = x^{-1} = 1/x$ is shown in Figure 4. Its graph has the equation $y = 1/x$ or $xy = 1$. This is an equilateral hyperbola with the coordinate axes as its asymptotes.

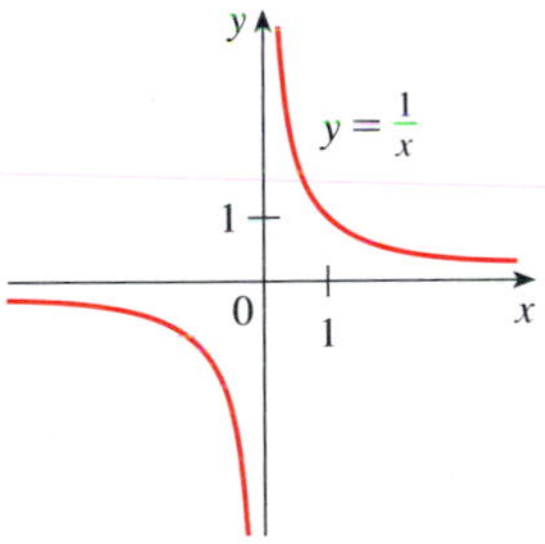

FIGURE 4

(c) $a = 1/n$, n a positive integer.
The function $f(x) = x^{1/n} = \sqrt[n]{x}$ is a **root function**. For $n = 2$ it is the square root function $f(x) = \sqrt{x}$ whose domain is $[0, \infty)$ and whose graph is the upper half of the parabola $x = y^2$ [see Figure 5(a)]. For other even values of n, the graph of $y = \sqrt[n]{x}$ is similar to that of $y = \sqrt{x}$. For $n = 3$ we have the cube root function $f(x) = \sqrt[3]{x}$ whose domain is $\mathbb{R}$ (recall that every real number has a cube root) and whose graph is shown in Figure 5(b). The graph of $y = \sqrt[n]{x}$ for n odd ($n > 3$) is similar to that of $y = \sqrt[3]{x}$.

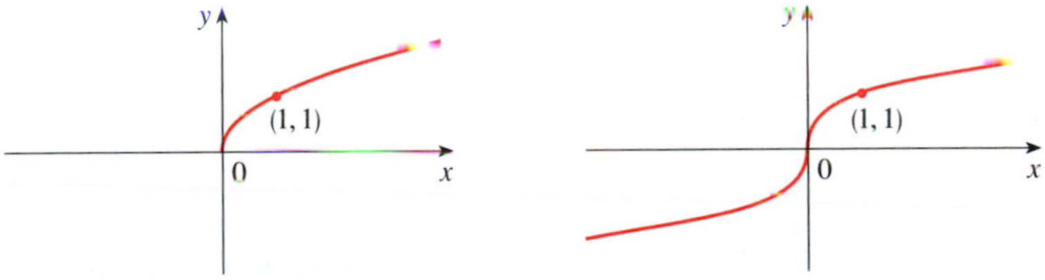

FIGURE 5
Graphs of root functions

(a) $f(x) = \sqrt{x}$ (b) $f(x) = \sqrt[3]{x}$

POLYNOMIALS A function P is called a **polynomial** if

$$P(x) = a_n x^n + a_{n-1} x^{n-1} + \cdots + a_2 x^2 + a_1 x + a_0$$

where n is a nonnegative integer and the numbers $a_0, a_1, a_2, \ldots, a_n$ are constants called the **coefficients** of the polynomial. The domain of any polynomial is $\mathbb{R} = (-\infty, \infty)$. If the leading coefficient $a_n \neq 0$, then the **degree** of the polynomial is n. For example, the function

$$P(x) = 2x^6 - x^4 + \tfrac{2}{5}x^3 + \sqrt{2}$$

is a polynomial of degree 6.

A polynomial of degree 1 is of the form $P(x) = ax + b$ and is called a **linear function** because its graph is the line $y = ax + b$ (slope a, y-intercept b).

A polynomial of degree 2 is of the form $P(x) = ax^2 + bx + c$ and is called a **quadratic function**. The graph of a quadratic function is always a parabola obtained by shifting the parabola $y = ax^2$ (see Example 3).

A polynomial of degree 3 is of the form

$$P(x) = ax^3 + bx^2 + cx + d$$

and is called a **cubic function**. Figure 6 shows the graph of a cubic function in part (a) and graphs of polynomials of degrees 4 and 5 in parts (b) and (c). We will see later why the graphs have these shapes.

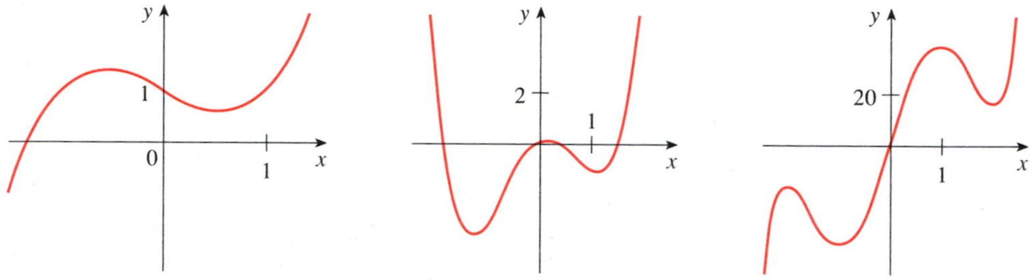

FIGURE 6

(a) $y = x^3 - x + 1$ (b) $y = x^4 - 3x^2 + x$ (c) $y = 3x^5 - 25x^3 + 60x$

Polynomials are commonly used to model various quantities that occur in the natural and social sciences. For instance, in Section 3.3 we will explain why economists often use a polynomial $P(x)$ to represent the cost of producing x units of a commodity.

RATIONAL FUNCTIONS A **rational function** f is a ratio of two polynomials:

$$f(x) = \frac{P(x)}{Q(x)}$$

where P and Q are polynomials. The domain consists of all values of x such that $Q(x) \neq 0$. For example, the function

$$f(x) = \frac{2x^4 - x^2 + 1}{x^2 - 4}$$

is a rational function with domain $\{x \mid x \neq \pm 2\}$. Its graph is shown in Figure 7.

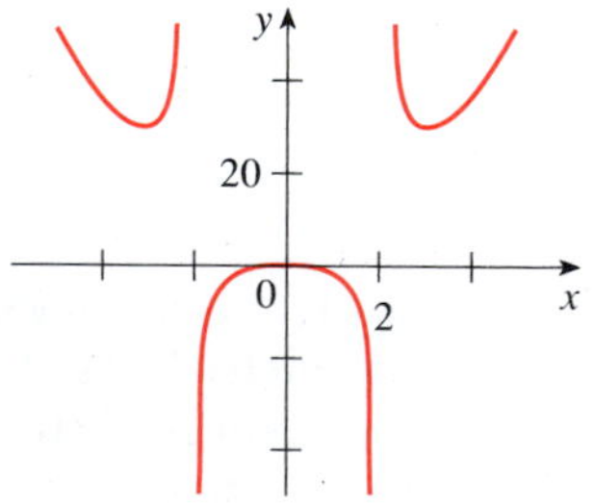

FIGURE 7

$f(x) = \dfrac{2x^4 - x^2 + 1}{x^2 - 4}$

ALGEBRAIC FUNCTIONS A function f is called an **algebraic function** if it can be constructed using algebraic operations (addition, subtraction, multiplication, division, and taking roots) starting with polynomials. Any rational function is automatically an algebraic function. Here are two more examples:

$$f(x) = \sqrt{x^2 + 1} \qquad g(x) = \frac{x^4 - 16x^2}{x + \sqrt{x}} + (x - 2)\sqrt[3]{x + 1}$$

When we sketch algebraic functions in Chapter 5 we will see that their graphs can assume a variety of shapes. Figure 8 illustrates some of the possibilities.

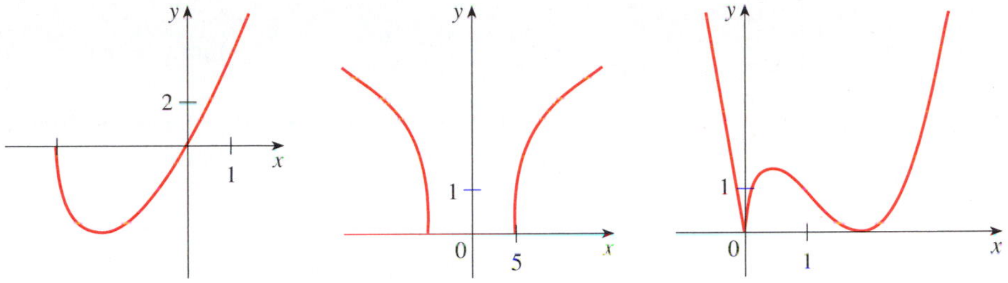

FIGURE 8

(a) $f(x) = x\sqrt{x+3}$ (b) $g(x) = \sqrt[4]{x^2 - 25}$ (c) $h(x) = x^{2/3}(x-2)^2$

TRIGONOMETRIC FUNCTIONS Trigonometry and the trigonometric functions are reviewed on the front endpaper and in Appendix D. In calculus the convention is that radian measure is always used (except when otherwise indicated). For example, when we use the function $f(x) = \sin x$, it is understood that $\sin x$ means the sine of the angle whose radian measure is x. Thus, the graphs of the sine and cosine functions are as shown in Figure 9.

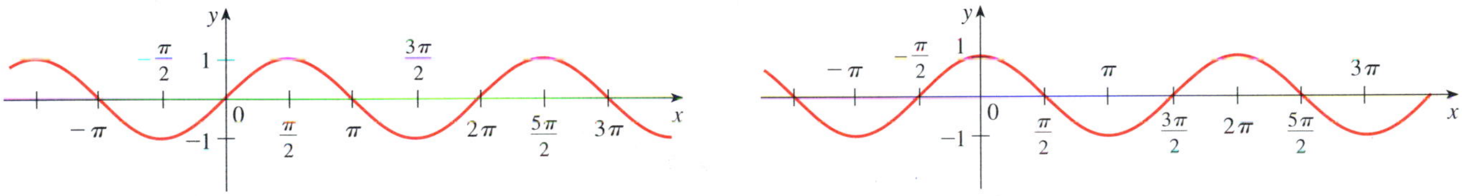

(a) $f(x) = \sin x$ (b) $g(x) = \cos x$

FIGURE 9

Notice that for both the sine and cosine functions the domain is $(-\infty, \infty)$ and the range is the closed interval $[-1, 1]$. Thus, for all values of x, we have

$$-1 \leqslant \sin x \leqslant 1 \qquad -1 \leqslant \cos x \leqslant 1$$

Also, the zeros of the sine function occur at the integer multiples of π; that is,

$$\sin x = 0 \quad \text{when} \quad x = n\pi \quad n \text{ an integer}$$

An important property of the sine and cosine functions is that they are periodic functions and have period 2π. This means that, for all values of x,

$$\sin(x + 2\pi) = \sin x \qquad \cos(x + 2\pi) = \cos x$$

The periodic nature of these functions makes them suitable for modeling periodic phenomena such as tides, vibrating springs, and sound waves.

The tangent function is related to the sine and cosine functions by the equation

$$\tan x = \frac{\sin x}{\cos x}$$

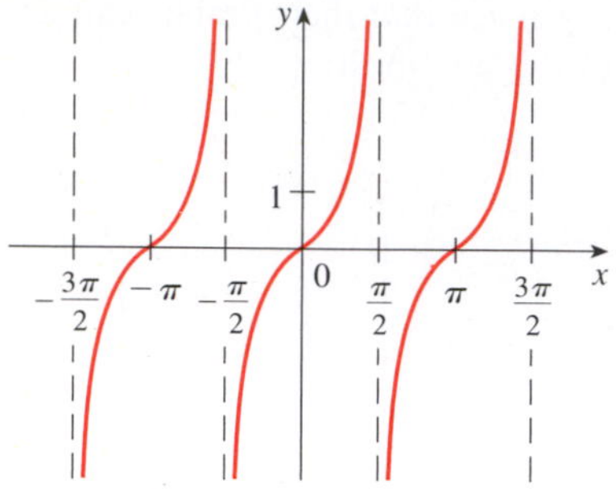

FIGURE 10
$y = \tan x$

and its graph is shown in Figure 10. It is undefined when $\cos x = 0$, that is, when $x = \pm\pi/2, \pm 3\pi/2, \ldots$. Its range is $(-\infty, \infty)$. Notice that the tangent function has period π:

$$\tan(x + \pi) = \tan x \qquad \text{for all } x$$

The remaining three trigonometric functions (cosecant, secant, and cotangent) are the reciprocals of the sine, cosine, and tangent functions. Their graphs are shown in Appendix D.

EXPONENTIAL FUNCTIONS These are the functions of the form $f(x) = a^x$, where the base a is a positive constant. The graphs of $y = 2^x$ and $y = (0.5)^x$ are shown in Figure 11. In both cases the domain is $(-\infty, \infty)$ and the range is $(0, \infty)$.

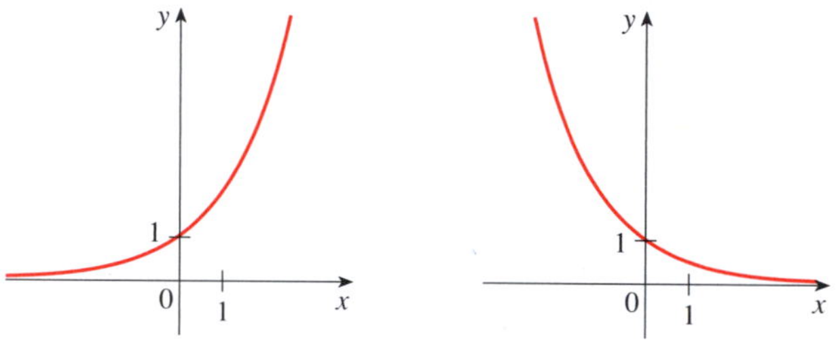

FIGURE 11

(a) $y = 2^x$ (b) $y = (0.5)^x$

Exponential functions will be studied in detail in Chapter 4 and we will see that they are useful for modeling population growth if $a > 1$ and radioactive decay if $a < 1$.

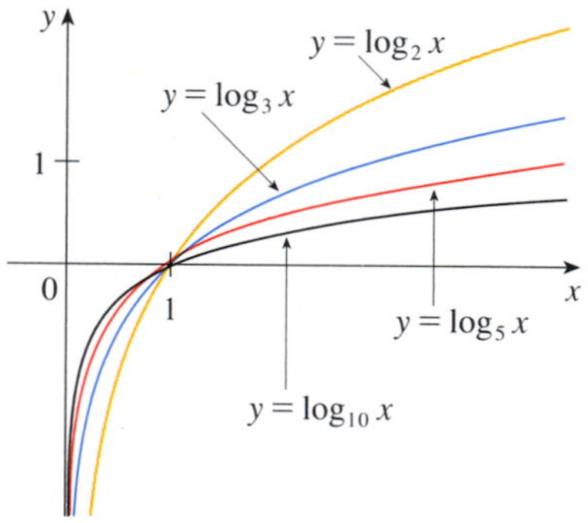

FIGURE 12

LOGARITHMIC FUNCTIONS These are the functions $f(x) = \log_a x$, where the base a is a positive constant. They are the inverse functions of the exponential functions and will also be studied in Chapter 4. Figure 12 shows the graphs of four logarithmic functions with various bases. In each case the domain is $(0, \infty)$, the range is $(-\infty, \infty)$, and the function increases slowly when $x > 1$.

TRANSCENDENTAL FUNCTIONS These are functions that are not algebraic. The set of transcendental functions includes the trigonometric, inverse trigonometric, exponential, and logarithmic functions, but it also includes a vast number of other functions that have never been named. In Chapter 10 we will study transcendental functions that are defined as sums of infinite series.

EXAMPLE 1 Classify the following functions as one of the types of functions that we have discussed.

(a) $f(x) = 5^x$ (b) $g(x) = x^5$

(c) $h(x) = \dfrac{1 + x}{1 - \sqrt{x}}$ (d) $u(t) = 1 - t + 5t^4$

SOLUTION

(a) $f(x) = 5^x$ is an exponential function. (The x is the exponent.)
(b) $g(x) = x^5$ is a power function. (The x is the base.)

(c) $h(x) = \dfrac{1 + x}{1 - \sqrt{x}}$ is an algebraic function.

(d) $u(t) = 1 - t + 5t^4$ is a polynomial of degree 4. ■

TRANSFORMATIONS OF FUNCTIONS

By applying certain transformations to the graph of a given function we can obtain the graphs of certain related functions and thereby reduce the amount of work in graphing. Let's first consider **translations**. By adding the constant function $g(x) = c > 0$ to a given function f by graphical addition, we see that the graph of $y = f(x) + c$ is just the graph of $y = f(x)$ shifted upward a distance of c units. Likewise, if $g(x) = f(x - c)$, where $c > 0$, then the value of g at x is the same as the value of f at $x - c$ (c units to the left of x). Therefore, the graph of $y = f(x - c)$ is just the graph of $y = f(x)$ shifted c units to the right (see Figure 13).

VERTICAL AND HORIZONTAL SHIFTS Suppose $c > 0$. To obtain the graph of

$y = f(x) + c$, shift the graph of $y = f(x)$ a distance c units upward

$y = f(x) - c$, shift the graph of $y = f(x)$ a distance c units downward

$y = f(x - c)$, shift the graph of $y = f(x)$ a distance c units to the right

$y = f(x + c)$, shift the graph of $y = f(x)$ a distance c units to the left

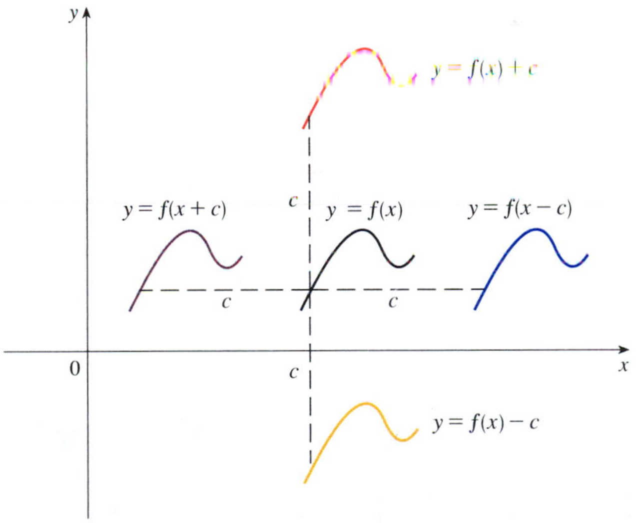

FIGURE 13
Translating the graph of f

Now let's consider the **stretching** and **reflecting** transformations. By multiplying the given function f by the constant function $g(x) = c$, where $c > 1$, we see that the graph of $y = cf(x)$ is the graph of $y = f(x)$ stretched by a factor of c in the vertical direction. The graph of $y = -f(x)$ is the graph of $y = f(x)$ reflected about the x-axis because the point (x, y) is replaced by the point $(x, -y)$. (See Figure 14 and the following chart, where the results of other stretching, compressing, and reflecting transformations are also given.)

VERTICAL AND HORIZONTAL STRETCHING AND REFLECTING Suppose $c > 1$. To obtain the graph of

$y = cf(x)$, stretch the graph of $y = f(x)$ vertically by a factor of c

$y = (1/c)f(x)$, compress the graph of $y = f(x)$ vertically by a factor of c

$y = f(cx)$, compress the graph of $y = f(x)$ horizontally by a factor of c

$y = f(x/c)$, stretch the graph of $y = f(x)$ horizontally by a factor of c

$y = -f(x)$, reflect the graph of $y = f(x)$ about the x-axis

$y = f(-x)$, reflect the graph of $y = f(x)$ about the y-axis

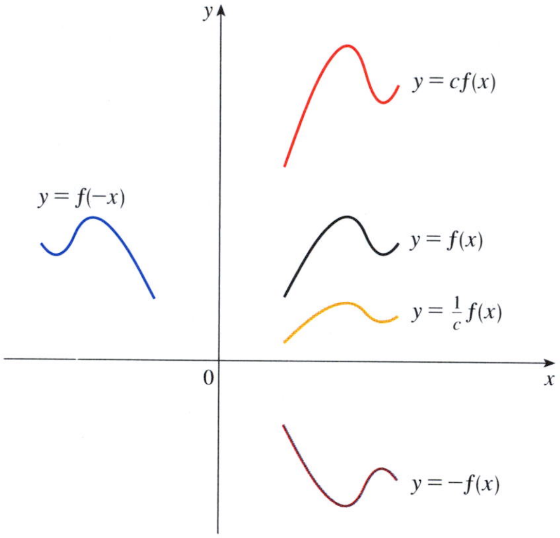

FIGURE 14
Stretching and reflecting the graph of f

Figure 15 illustrates these stretching transformations when applied to the cosine function with $c = 2$.

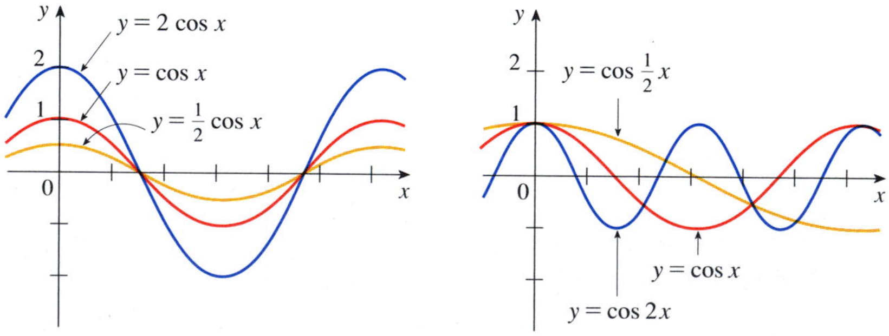

FIGURE 15

EXAMPLE 2 Given the graph of $y = \sqrt{x}$, use transformations to graph $y = \sqrt{x} - 2$, $y = \sqrt{x - 2}$, $y = -\sqrt{x}$, $y = 2\sqrt{x}$, and $y = \sqrt{-x}$.

SOLUTION The graph of the square root function $y = \sqrt{x}$, obtained from Figure 5, is shown in Figure 16(a). In the other parts of the figure we sketch $y = \sqrt{x} - 2$ by shifting 2 units downward, $y = \sqrt{x - 2}$ by shifting 2 units to the right, $y = -\sqrt{x}$

by reflecting about the x-axis, $y = 2\sqrt{x}$ by stretching vertically by a factor of 2, and $y = \sqrt{-x}$ by reflecting about the y-axis.

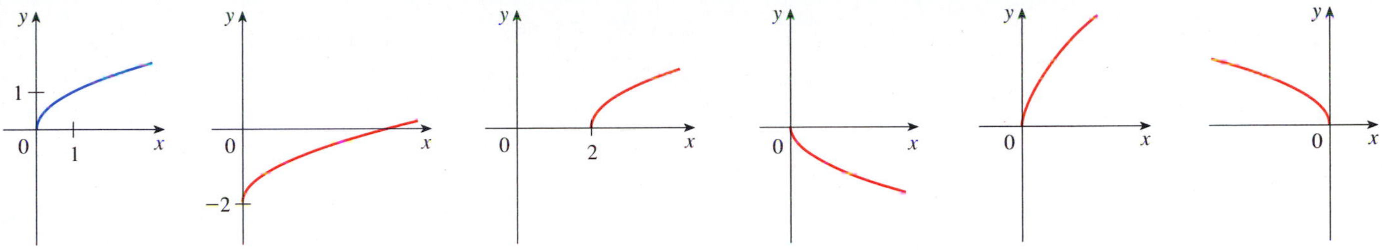

(a) $y = \sqrt{x}$ (b) $y = \sqrt{x} - 2$ (c) $y = \sqrt{x-2}$ (d) $y = -\sqrt{x}$ (e) $y = 2\sqrt{x}$ (f) $y = \sqrt{-x}$

FIGURE 16

EXAMPLE 3 Sketch the graph of the function $f(x) = x^2 + 6x + 10$.

SOLUTION Completing the square, we write the equation of the graph as

$$y = x^2 + 6x + 10 = (x + 3)^2 + 1$$

This means we obtain the desired graph by starting with the parabola $y = x^2$ and shifting 3 units to the left and then 1 unit upward (see Figure 17).

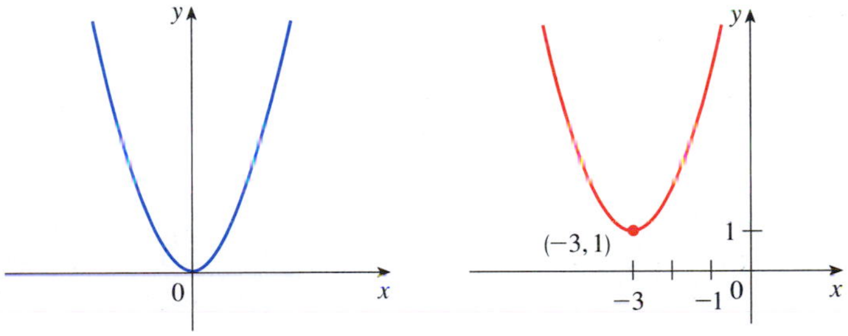

FIGURE 17 (a) $y = x^2$ (b) $y = (x + 3)^2 + 1$

EXAMPLE 4 Sketch the graphs of the functions:
(a) $y = \sin 2x$ (b) $y = 1 - \sin x$

SOLUTION
(a) We obtain the graph of $y = \sin 2x$ from that of $y = \sin x$ by compressing horizontally by a factor of 2 (see Figures 18 and 19).

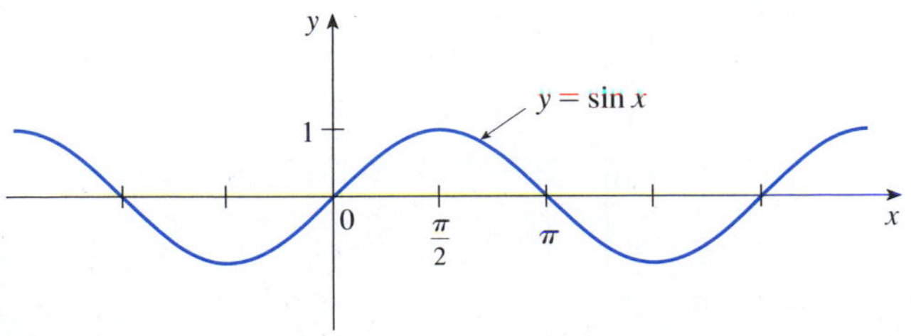

FIGURE 18

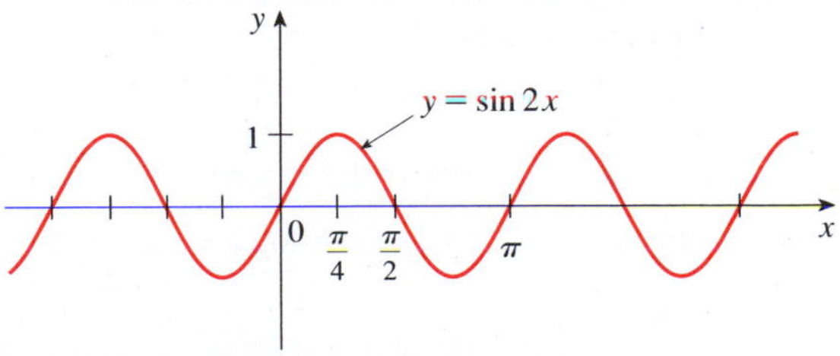

FIGURE 19

(b) To obtain the graph of $y = 1 - \sin x$, we again start with $y = \sin x$. We reflect about the x-axis to get the graph of $y = -\sin x$ and then we shift 1 unit upward to get $y = 1 - \sin x$ (see Figure 20).

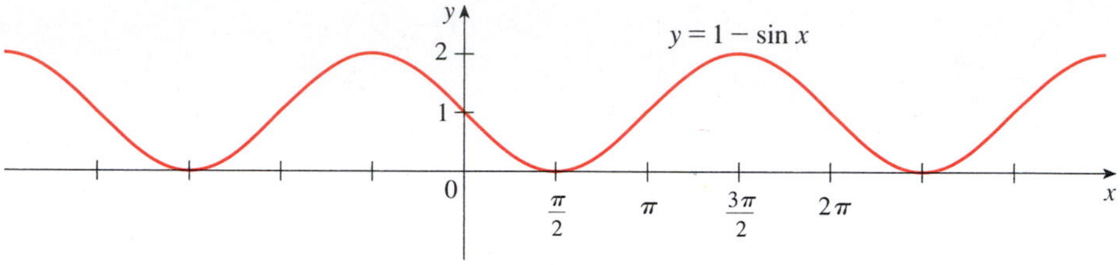

FIGURE 20 ■

Another transformation of some interest is taking the absolute value of a function. If $y = |f(x)|$, then according to the definition of absolute value, $y = f(x)$ when $f(x) \geq 0$ and $y = -f(x)$ when $f(x) < 0$. This tells us how to get the graph of $y = |f(x)|$ from the graph of $y = f(x)$: The part of the graph that lies above the x-axis remains the same; the part that lies below the x-axis is reflected about the x-axis.

EXAMPLE 5 Sketch the graph of the function $y = |x^2 - 1|$.

SOLUTION We first graph the parabola $y = x^2 - 1$ in Figure 21(a) by shifting the parabola $y = x^2$ downward 1 unit. We see that the graph lies below the x-axis when $-1 < x < 1$, so we reflect that part of the graph about the x-axis to obtain the graph of $y = |x^2 - 1|$ in Figure 21(b).

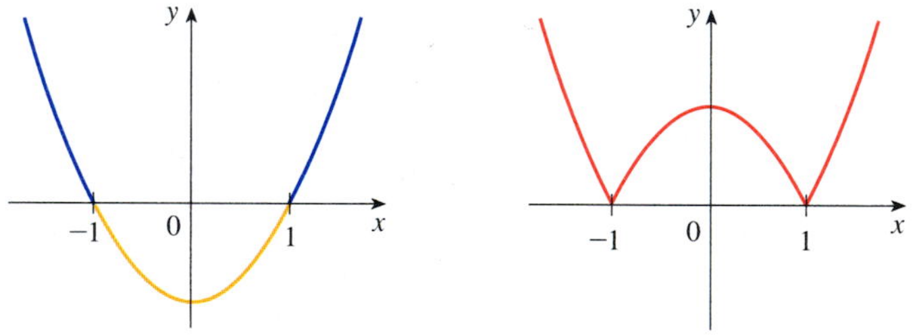

FIGURE 21 (a) $y = x^2 - 1$ (b) $y = |x^2 - 1|$ ■

EXERCISES 2

1–2 ■ Classify each function as a power function, root function, polynomial (state its degree), rational function, algebraic function, trigonometric function, exponential function, or logarithmic function.

1. (a) $f(x) = \sqrt[5]{x}$ (b) $g(x) = \sqrt{1 - x^2}$
(c) $h(x) = x^9 + x^4$ (d) $r(x) = \dfrac{x^2 + 1}{x^3 + x}$
(e) $s(x) = \tan 2x$ (f) $t(x) = \log_{10} x$

2. (a) $y = \dfrac{x - 6}{x + 6}$ (b) $y = x + \dfrac{x^2}{\sqrt{x - 1}}$
(c) $y = 10^x$ (d) $y = x^{10}$
(e) $y = 2t^6 + t^4 - \pi$ (f) $y = \cos\theta + \sin\theta$

3. The graph of f is given.

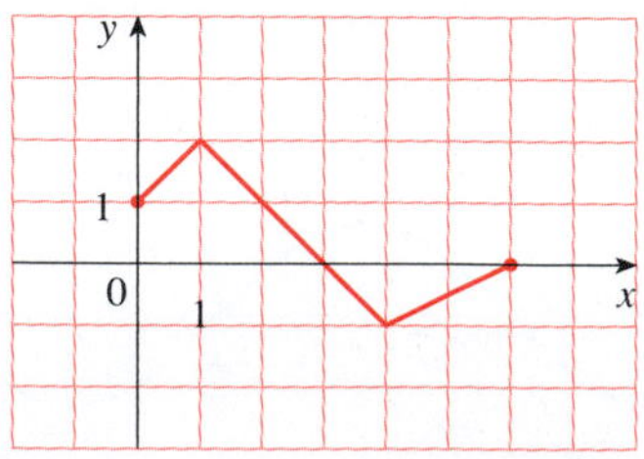

Use it to graph the following functions:
(a) $y = f(2x)$ (b) $y = f(\tfrac{1}{2}x)$
(c) $y = f(-x)$ (d) $y = -f(-x)$

4. The graph of f is given. Draw the graphs of the following functions:
(a) $y = f(x + 4)$ (b) $y = f(x) + 4$
(c) $y = 2f(x)$ (d) $y = -\frac{1}{2}f(x) + 3$

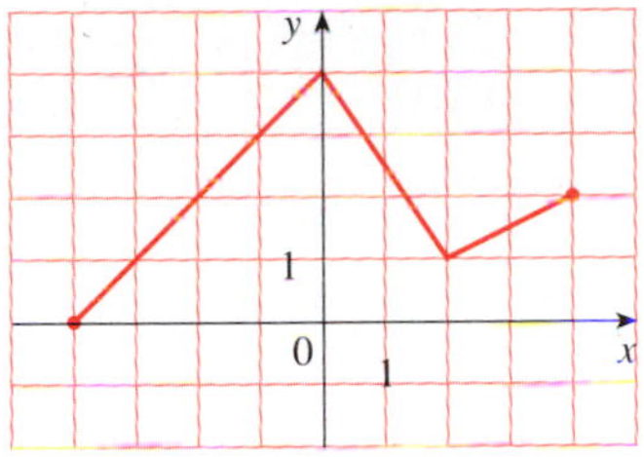

5–26 ■ Graph each function, not by plotting points, but by starting with the graph of one of the standard functions given in this section, and then applying the appropriate transformations.

5. $y = -1/x$

6. $y = -x^3$

7. $y = 2 \sin x$

8. $y = 1 + \sqrt{x}$

9. $y = (x - 1)^3 + 2$

10. $y = 2 - \cos x$

11. $y = \tan 2x$

12. $y = \sqrt[3]{x + 2}$

13. $y = \cos(x/2)$

14. $y = x^2 + x + 1$

15. $y = \dfrac{1}{x - 3}$

16. $y = -2 \sin \pi x$

17. $y = \dfrac{1}{3} \sin\left(x - \dfrac{\pi}{6}\right)$

18. $y = 2 + \dfrac{1}{x + 1}$

19. $y = 1 + 2x - x^2$

20. $y = \frac{1}{2}\sqrt{x + 4} - 3$

21. $y = 2 - \sqrt{x + 1}$

22. $y = 1 - (x - 8)^6$

23. $y = |x^2 - 2x|$

24. $y = |\cos x|$

25. $y = ||x| - 1|$

26. $y = |||x| - 2| - 1|$

27. (a) How is the graph of $y = f(|x|)$ related to the graph of f?
(b) Sketch the graph of $y = \sin|x|$.

28. Sketch the graph of $y = \sqrt{|x|}$.

29. The graph of f is given. Sketch the graph of $y = 1/f(x)$.

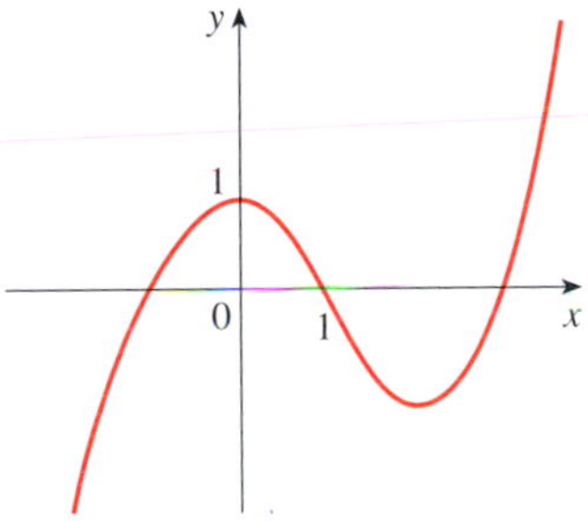

3 GRAPHING CALCULATORS AND COMPUTERS

In this section we assume that you have access to a graphing calculator or a computer with graphing software. We will see that the use of such a device enables us to graph more complicated functions and to solve more complex problems than would otherwise be possible. We also point out some of the pitfalls that can occur with these machines.

Graphing calculators and computers can give very accurate graphs of functions. But we will see in Chapter 5 that only through the use of calculus can we be sure that we have uncovered all the interesting aspects of a graph.

A graphing calculator or computer displays a rectangular portion of the graph of a function in a **display window** or **viewing screen**, which we refer to as a **viewing rectangle**. The default screen often gives an incomplete or misleading picture, so it is important to choose the viewing rectangle with care. If we choose the x-values to range from a minimum value of *Xmin* $= a$ to a maximum value of *Xmax* $= b$ and the y-values to range from a minimum of *Ymin* $= c$ to a maximum of *Ymax* $= d$, then the portion of the graph lies in the rectangle

$$[a, b] \times [c, d] = \{(x, y) \mid a \leqslant x \leqslant b, c \leqslant y \leqslant d\}$$

shown in Figure 1. We refer to this rectangle as the $[a, b]$ *by* $[c, d]$ *viewing rectangle*.

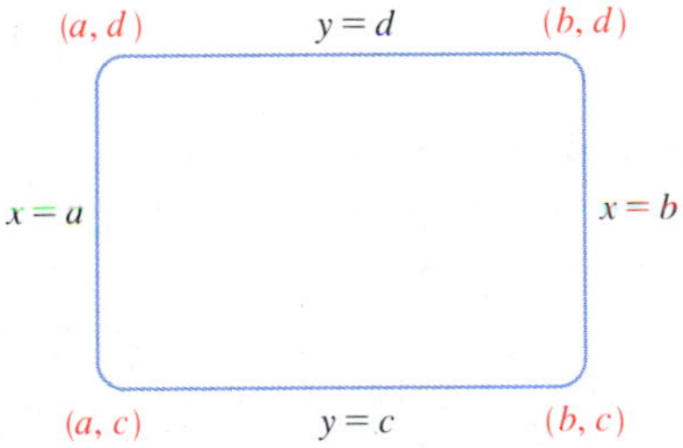

FIGURE 1
The viewing rectangle $[a, b]$ by $[c, d]$

The machine draws the graph of a function f much as you would. It plots points of the form $(x, f(x))$ for a certain number of equally spaced values of x between a and b. If an x-value is not in the domain of f, or if $f(x)$ lies outside the viewing rectangle, it moves on to the next x-value. The machine connects each point to the preceding plotted point to form a representation of the graph of f.

EXAMPLE 1 Draw the graph of the function $f(x) = x^2 + 3$ in each of the following viewing rectangles.

(a) $[-2, 2]$ by $[-2, 2]$ (b) $[-4, 4]$ by $[-4, 4]$

(c) $[-10, 10]$ by $[-5, 30]$ (d) $[-50, 50]$ by $[-100, 1000]$

SOLUTION For part (a) we select the range by setting *Xmin* $= -2$, *Xmax* $= 2$, *Ymin* $= -2$, and *Ymax* $= 2$. The resulting graph is shown in Figure 2(a). The display window is blank! A moment's thought provides the explanation: Notice that $x^2 \geqslant 0$ for all x, so $x^2 + 3 \geqslant 3$ for all x. Thus, the range of the function $f(x) = x^2 + 3$ is $[3, \infty)$. This means that the graph of f lies entirely outside the viewing rectangle $[-2, 2]$ by $[-2, 2]$.

The graphs for the viewing rectangles in parts (b), (c), and (d) are also shown in Figure 2. Observe that we get a more complete picture in parts (c) and (d), but in part (d) it is not clear that the y-intercept is 3.

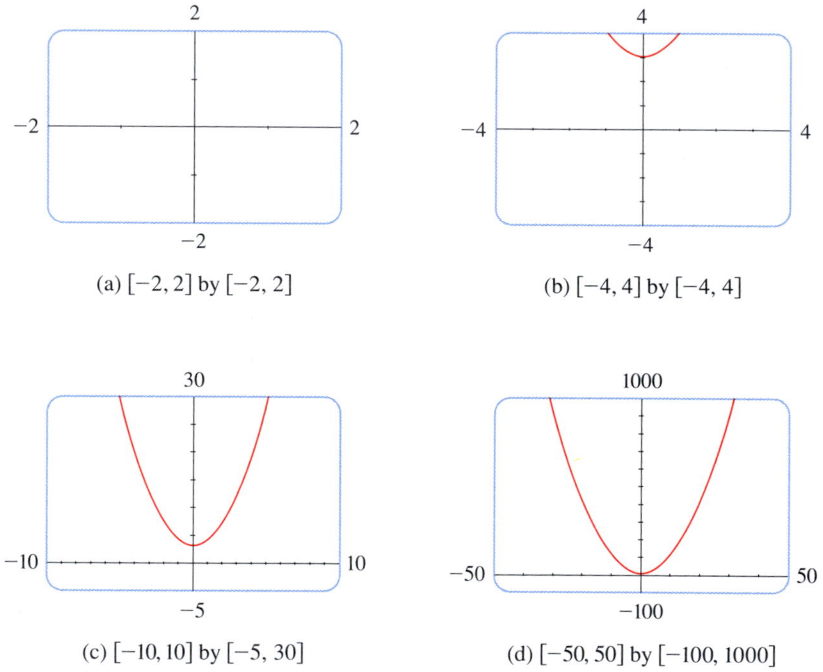

FIGURE 2
Graphs of $f(x) = x^2 + 3$

■

We see from Example 1 that the choice of a viewing rectangle can make a big difference in the appearance of a graph. Sometimes it is necessary to change to a larger viewing rectangle to obtain a more complete picture, a more global view, of the graph. In the next example we see that knowledge of the domain and range of a function sometimes provides us with enough information to select a good viewing rectangle.

EXAMPLE 2 Determine an appropriate viewing rectangle for the function $f(x) = \sqrt{8 - 2x^2}$ and use it to graph f.

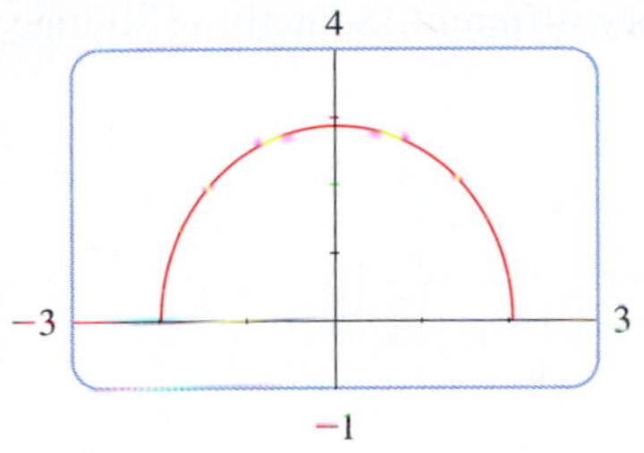

FIGURE 3

SOLUTION The expression for $f(x)$ is defined when

$$8 - 2x^2 \geqslant 0 \iff 2x^2 \leqslant 8 \iff x^2 \leqslant 4 \iff |x| \leqslant 2 \iff -2 \leqslant x \leqslant 2$$

Therefore, the domain of f is the interval $[-2, 2]$. Also,

$$0 \leqslant \sqrt{8 - 2x^2} \leqslant \sqrt{8} = 2\sqrt{2} \approx 2.83$$

so the range of f is the interval $[0, 2\sqrt{2}\,]$.

We choose the viewing rectangle so that the x-interval is somewhat larger than the domain and the y-interval is larger than the range. Taking the viewing rectangle to be $[-3, 3]$ by $[-1, 4]$, we get the graph shown in Figure 3. ■

EXAMPLE 3 Graph the function $y = x^3 - 49x$.

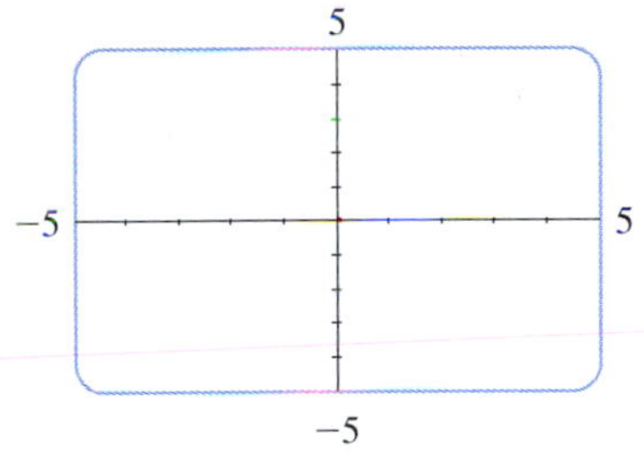

FIGURE 4

SOLUTION Here the domain is $\mathbb{R}$, the set of all real numbers. That doesn't help us choose a viewing rectangle. Let's experiment. If we start with the viewing rectangle $[-5, 5]$ by $[-5, 5]$, we get the graph in Figure 4. On most calculators the screen appears to be blank, but it is not quite blank because the point $(0, 0)$ has been plotted. It turns out that for all the other x-values that the calculator chooses between -5 and 5, the values of $f(x)$ are greater than 5 or less than -5, so the corresponding points on the graph lie outside the viewing rectangle.

If we use the zoom-out feature of a graphing calculator to change the viewing rectangle to $[-10, 10]$ by $[-10, 10]$, we get the picture shown in Figure 5(a). The graph appears to consist of vertical lines, but we know that can't be correct. If we look carefully while the graph is being drawn, we see that the graph leaves the screen and reappears during the graphing process. This indicates that we need to see more in the vertical direction, so we change the viewing rectangle to $[-10, 10]$ by $[-100, 100]$. The resulting graph is shown in Figure 5(b). It still doesn't quite reveal all the main features of the function, so we try $[-10, 10]$ by $[-200, 200]$ in Figure 5(c). Now we are more confident that we have arrived at an appropriate viewing rectangle. In Chapter 5, we will be able to see that the graph shown in Figure 5(c) does indeed reveal all the main features of the function.

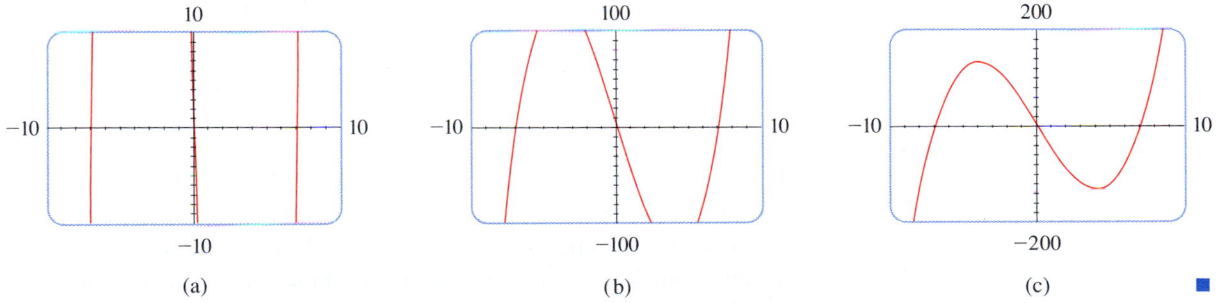

FIGURE 5
$f(x) = x^3 - 49x$

■

EXAMPLE 4 Graph the function $f(x) = \sin 50x$ in an appropriate viewing rectangle.

SOLUTION Figure 6(a) shows the graph of f produced by a graphing calculator using the viewing rectangle $[-12, 12]$ by $[-1.5, 1.5]$. At first glance the graph appears to be reasonable. But if we change the viewing rectangle to the ones shown in the following parts of Figure 6, the graphs look very different. Something strange is happening.

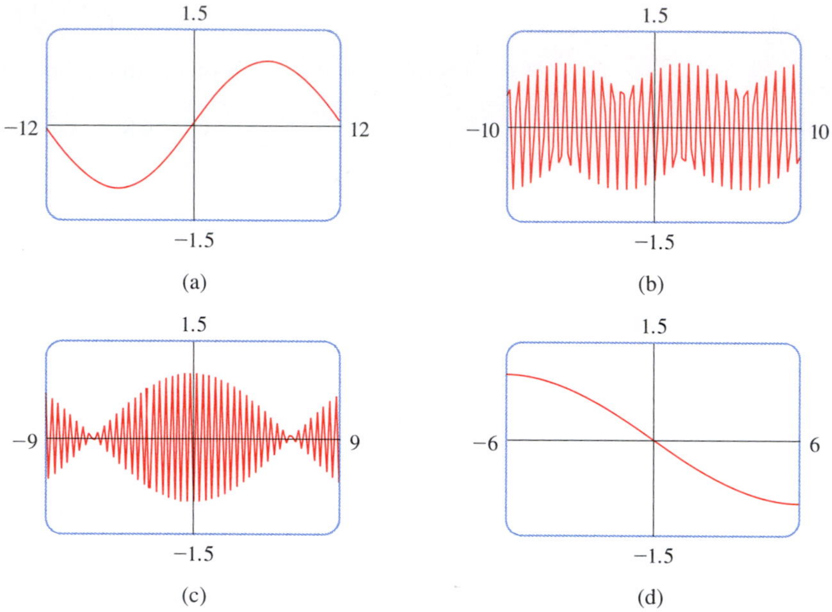

FIGURE 6
Graphs of $f(x) = \sin 50x$ in four viewing rectangles

In order to explain the big differences in appearance of these graphs and to find an appropriate viewing rectangle, we need to find the period of the function $y = \sin 50x$. We know that the function $y = \sin x$ has period 2π, so the period of $y = \sin 50x$ is

$$\frac{2\pi}{50} = \frac{\pi}{25} \approx 0.126$$

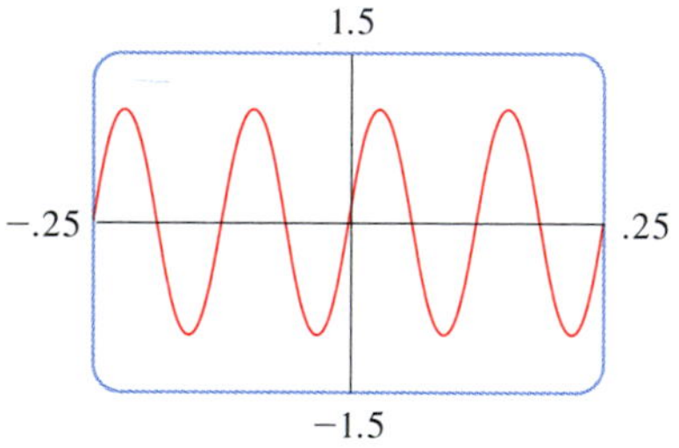

FIGURE 7
$f(x) = \sin 50x$

This suggests that we should deal only with small values of x in order to show just a few oscillations of the graph. If we choose the viewing rectangle $[-0.25, 0.25]$ by $[-1.5, 1.5]$, we get the graph shown in Figure 7.

Now we see what went wrong in Figure 6. The oscillations of $y = \sin 50x$ are so rapid that when the calculator plots points and joins them, it misses most of the maximum and minimum points and therefore gives a very misleading impression of the graph. ■

We have seen that the use of an inappropriate viewing rectangle can give a misleading impression of the graph of a function. In Examples 1 and 3 we solved the problem by changing to a larger viewing rectangle. In Example 4 we had to make the viewing rectangle smaller. In the next example we look at a function for which there is no single viewing rectangle that reveals the true shape of the graph.

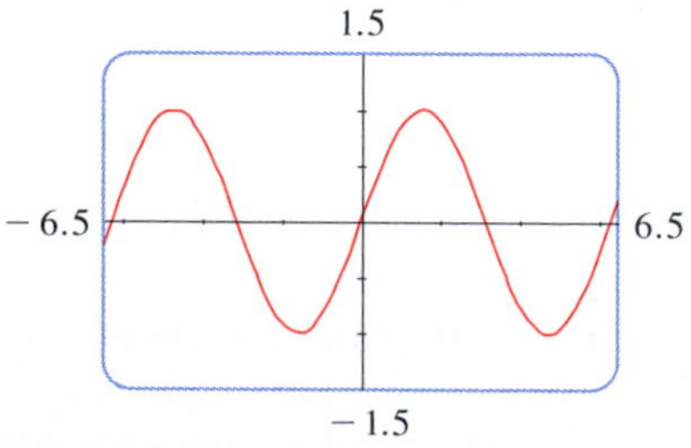

FIGURE 8

EXAMPLE 5 Graph the function $f(x) = \sin x + \frac{1}{100}\cos 100x$.

SOLUTION Figure 8 shows the graph of f produced by a graphing calculator with viewing rectangle $[-6.5, 6.5]$ by $[-1.5, 1.5]$. It looks much like the graph of $y = \sin x$, but perhaps with some bumps attached. If we zoom in to the viewing rectangle $[-0.1, 0.1]$ by $[-0.1, 0.1]$, we can see much more clearly the shape of these

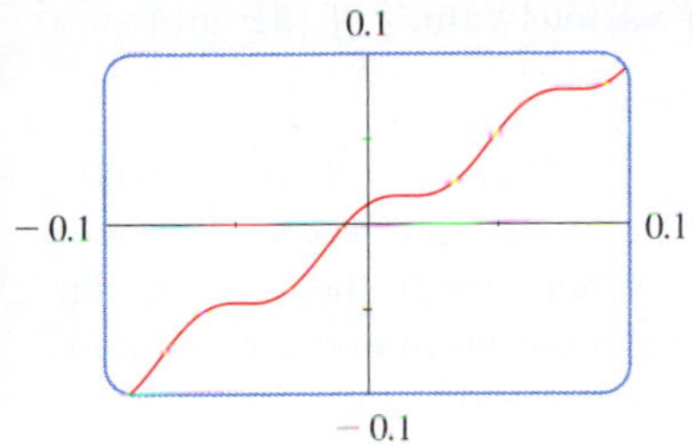

FIGURE 9

bumps in Figure 9. The reason for this behavior is that the second term, $\frac{1}{100}\cos 100x$, is very small in comparison with the first term, $\sin x$. Thus, we really need two graphs to see the true nature of this function. ■

EXAMPLE 6 Draw the graph of the function $y = \dfrac{1}{1 - x}$.

SOLUTION Figure 10(a) shows the graph produced by a graphing calculator with viewing rectangle $[-9, 9]$ by $[-9, 9]$. In connecting successive points on the graph, the calculator produced a steep line segment from the top to the bottom of the screen. That line segment is not truly part of the graph. Notice that the domain of the function $y = 1/(1 - x)$ is $\{x \mid x \neq 1\}$. We can eliminate the extraneous near-vertical line by experimenting with a change of scale. When we change to the smaller viewing rectangle $[-5, 5]$ by $[-5, 5]$, we obtain the much better graph in Figure 10(b).

Another way to avoid the extraneous line is to change the graphing mode on the calculator so that the dots are not connected.

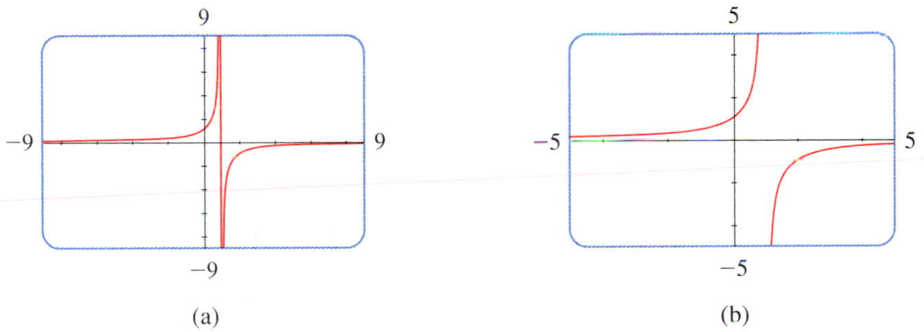

FIGURE 10
$y = \dfrac{1}{1 - x}$

■

EXAMPLE 7 Graph the function $y = \sqrt[3]{x}$.

SOLUTION Some graphing devices display the graph shown in Figure 11, whereas others produce a graph like that in Figure 12. We know from Section 2 (Figure 5) that the graph in Figure 12 is correct, so what happened in Figure 11? The explanation is that, in some machines, $x^{1/3}$ is computed as $e^{(1/3)\ln x}$ and $\ln x$ is not defined for $x < 0$, so only the right half of the graph is produced.

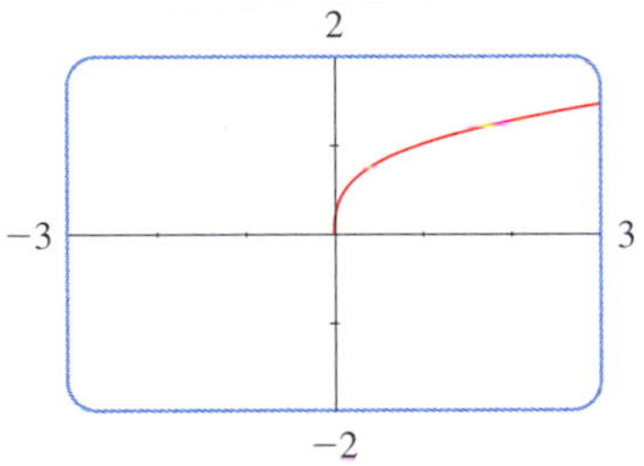

FIGURE 11

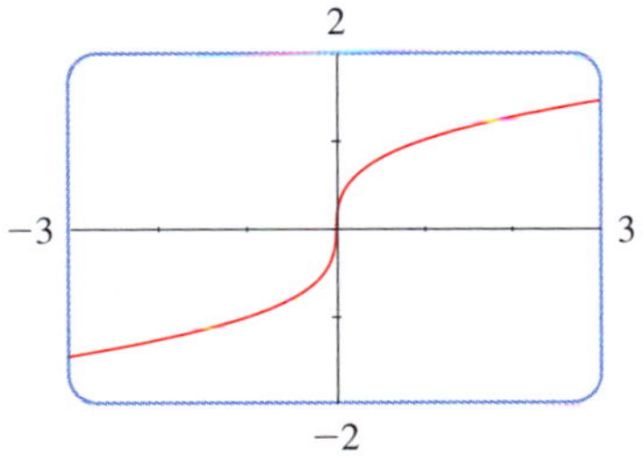

FIGURE 12

You should experiment with your own machine to see which of these two graphs is produced. If you get the graph in Figure 11, you can obtain the correct picture by graphing the function

$$f(x) = \frac{x}{|x|} \cdot |x|^{1/3}$$

Notice that this function is equal to $\sqrt[3]{x}$ (except when $x = 0$). ■

EXAMPLE 8 Graph the function $y = x^3 + cx$ for various values of the number c. How does the graph change when c is changed?

SOLUTION Figure 13 shows the graphs of $y = x^3 + cx$ for $c = 2, 1, 0, -1$, and -2. We see that, for positive values of c, the graph increases from left to right with no maximum or minimum points (peaks or valleys). When $c = 0$, the curve is flat at the origin. When c is negative, the curve has a maximum point and a minimum point. As c decreases, the maximum point becomes higher and the minimum point lower.

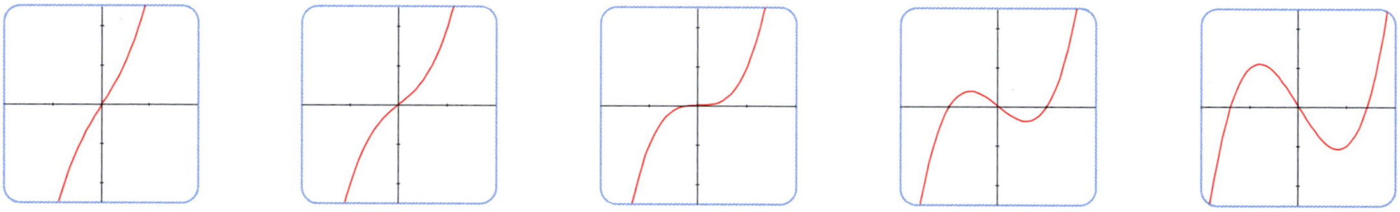

(a) $y = x^3 + 2x$ (b) $y = x^3 + x$ (c) $y = x^3$ (d) $y = x^3 - x$ (e) $y = x^3 - 2x$

FIGURE 13
Several members of the family of functions $y = x^3 + cx$, all graphed in the viewing rectangle $[-2, 2]$ by $[-2.5, 2.5]$

■

EXAMPLE 9 Find the solution of the equation $\cos x = x$ correct to two decimal places.

SOLUTION The solutions of the equation $\cos x = x$ are the x-coordinates of the points of intersection of the curves $y = \cos x$ and $y = x$. From Figure 14(a) we see that there is only one solution and it lies between 0 and 1. Zooming in to the viewing rectangle $[0, 1]$ by $[0, 1]$, we see from Figure 14(b) that the root lies between 0.7 and 0.8. So we zoom in further to the viewing rectangle $[0.7, 0.8]$ by $[0.7, 0.8]$ in Figure 14(c). By moving the cursor to the intersection point of the two curves, or by inspection and the fact that the x-scale is 0.01, we see that the root of the equation is about 0.74.

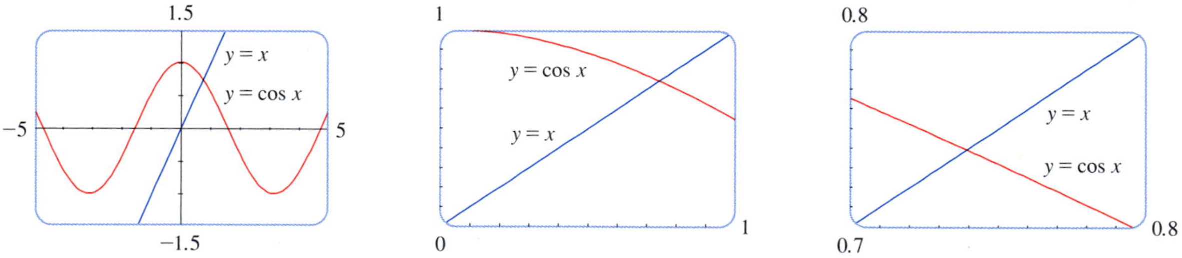

FIGURE 14
Locating the roots of $\cos x = x$

(a) $[-5, 5]$ by $[-1.5, 1.5]$, x-scale = 1

(b) $[0, 1]$ by $[0, 1]$, x-scale = 0.1

(c) $[0.7, 0.8]$ by $[0.7, 0.8]$, x-scale = 0.01

■

EXERCISES 3

1. Use a graphing calculator or computer to determine which of the given viewing rectangles produces the most appropriate graph of the function $f(x) = x^4 + 2$.
(a) $[-2, 2]$ by $[-2, 2]$ (b) $[0, 4]$ by $[0, 4]$
(c) $[-4, 4]$ by $[-4, 4]$ (d) $[-8, 8]$ by $[-4, 40]$
(e) $[-40, 40]$ by $[-80, 800]$

2. Use a graphing calculator or computer to determine which of the given viewing rectangles produces the most appropriate graph of the function $f(x) = x^2 + 7x + 6$.
(a) $[-5, 5]$ by $[-5, 5]$ (b) $[0, 10]$ by $[-20, 100]$
(c) $[-15, 8]$ by $[-20, 100]$
(d) $[-10, 3]$ by $[-100, 20]$

3. Use a graphing calculator or computer to determine which of the given viewing rectangles produces the most appropriate graph of the function $f(x) = 10 + 25x - x^3$.
(a) $[-4, 4]$ by $[-4, 4]$ (b) $[-10, 10]$ by $[-10, 10]$
(c) $[-20, 20]$ by $[-100, 100]$
(d) $[-100, 100]$ by $[-200, 200]$

4. Use a graphing calculator or computer to determine which of the given viewing rectangles produces the most appropriate graph of the function $f(x) = \sqrt{8x - x^2}$.
(a) $[-4, 4]$ by $[-4, 4]$ (b) $[-5, 5]$ by $[0, 100]$
(c) $[-10, 10]$ by $[-10, 40]$ (d) $[-2, 10]$ by $[-2, 6]$

5–22 ■ Determine an appropriate viewing rectangle for the given function and use it to draw the graph.

5. $f(x) = 4 + 6x - x^2$
6. $f(x) = 0.3x^2 + 1.7x - 3$
7. $f(x) = \sqrt[4]{256 - x^2}$
8. $f(x) = \sqrt{12x - 17}$
9. $f(x) = 0.01x^3 - x^2 + 5$
10. $f(x) = x(x + 6)(x - 9)$
11. $y = \dfrac{1}{x^2 + 25}$
12. $y = \dfrac{x}{x^2 + 25}$
13. $y = x^4 - 4x^3$
14. $y = x^3 + \dfrac{1}{x}$
15. $y = \dfrac{2x - 1}{x + 3}$
16. $y = 2x - |x^2 - 5|$
17. $f(x) = \cos 100x$
18. $f(x) = 3 \sin 120x$
19. $f(x) = \sin(x/40)$
20. $y = \tan 25x$
21. $y = 3^{\cos(x^2)}$
22. $y = x^2 + 0.02 \sin 50x$

23. Graph the ellipse $4x^2 + 2y^2 = 1$ by graphing the functions whose graphs are the upper and lower halves of the ellipse.

24. Graph the hyperbola $y^2 - 9x^2 = 1$ by graphing the functions whose graphs are the upper and lower halves of the hyperbola.

25. Draw the graph of the function

$$f(x) = \begin{cases} x^3 - 2x + 1 & \text{if } x \leq 1 \\ \sqrt[3]{x - 1} & \text{if } x > 1 \end{cases}$$

This can be done by using a graphing device to graph the functions $g(x) = x^3 - 2x + 1$ and $h(x) = \sqrt[3]{x - 1}$ and then drawing the graph of f (by hand) by combining the graph of g to the left of $x = 1$ with the graph of h to the right of $x = 1$. Alternatively, on many graphing calculators it can be produced using the logical functions in the calculator. For example, on the TI-81 the following equation gives the desired graph:

$$y = (x < 1)(x^3 - 2x + 1) + (x > 1)\sqrt[3]{x - 1}$$

26. Draw the graph of the function

$$f(x) = \begin{cases} \sin x & \text{if } x < 0 \\ \sqrt[3]{2x - x^2} & \text{if } 0 \leq x \leq 2 \\ x - 2 & \text{if } x > 2 \end{cases}$$

27–30 ■ Find all solutions of the equation correct to two decimal places.

27. $3x^3 + x^2 + x - 2 = 0$

28. $x^4 + 8x + 16 = 2x^3 + 8x^2$

29. $2 \sin x = x$
30. $2 \cos x = 2 - x$

31. Use graphs to determine which of the functions $f(x) = 10x^2$ and $g(x) = x^3/10$ is eventually larger (that is, larger when x is very large).

32. Use graphs to determine which of the functions $f(x) = x^4 - 100x^3$ and $g(x) = x^3$ is eventually larger.

33. (a) Compare the rates of growth of the functions $f(x) = 2^x$ and $g(x) = x^5$ by drawing the graphs of both functions in the following viewing rectangles:
(i) $[0, 5]$ by $[0, 20]$ (ii) $[0, 25]$ by $[0, 10^7]$
(iii) $[0, 50]$ by $[0, 10^8]$
(b) Find the solutions of the equation $2^x = x^5$ correct to one decimal place.

34. (a) Compare the rates of growth of the functions $f(x) = 3^x$ and $g(x) = x^4$ by drawing the graphs of both functions in the following viewing rectangles:
(i) $[-4, 4]$ by $[0, 20]$ (ii) $[0, 10]$ by $[0, 5000]$
(iii) $[0, 20]$ by $[0, 10^5]$
(b) Find the solutions of the equation $3^x = x^4$ correct to two decimal places.

35. For what values of x is it true that $|\sin x - x| < 0.1$?

36. Graph the polynomials $P(x) = 3x^5 - 5x^3 + 2x$ and $Q(x) = 3x^5$ on the same screen, first using the viewing rectangle $[-2, 2]$ by $[-2, 2]$ and then changing to $[-10, 10]$ by $[-10{,}000, 10{,}000]$. What do you observe from these graphs?

37. (a) Graph the root functions $y = \sqrt{x}$, $y = \sqrt[4]{x}$, and $y = \sqrt[6]{x}$ on the same screen using the viewing rectangle $[-1, 4]$ by $[-1, 3]$.
(b) Graph the root functions $y = x$, $y = \sqrt[3]{x}$, and $y = \sqrt[5]{x}$ on the same screen using the viewing rectangle $[-3, 3]$ by $[-2, 2]$. (See Example 7.)
(c) Graph the root functions $y = \sqrt{x}$, $y = \sqrt[3]{x}$, $y = \sqrt[4]{x}$, and $y = \sqrt[5]{x}$ on the same screen using the viewing rectangle $[-1, 3]$ by $[-1, 2]$.
(d) What conclusions can you make from these graphs?

38. (a) Graph the functions $y = 1/x$ and $y = 1/x^3$ on the same screen using the viewing rectangle $[-3, 3]$ by $[-3, 3]$.
(b) Graph the functions $y = 1/x^2$ and $y = 1/x^4$ on the same screen using the same viewing rectangle as in part (a).
(c) Graph all of the functions in parts (a) and (b) on the same screen using the viewing rectangle $[-1, 3]$ by $[-1, 3]$.
(d) What conclusions can you make from these graphs?

39. Graph the function $f(x) = x^4 + cx^2 + x$ for several values of c. How does the graph change when c changes?

40. Graph the function $f(x) = \sqrt{1 + cx^2}$ for various values of c. Describe how changing the value of c affects the graph.

41. Graph the function $y = x^n 2^{-x}$, $x \geqslant 0$, for $n = 1, 2, 3, 4, 5$, and 6. How does the graph change as n increases?

42. The curves with equations

$$y = \frac{|x|}{\sqrt{c - x^2}}$$

are called **bullet-nose curves**. Graph some of these curves to see why. What happens as c increases?

43. What happens to the graph of the equation $y^2 = cx^3 + x^2$ as c varies?

44. This exercise explores the effect of the inner function g on a composite function $y = f(g(x))$.

(a) Graph the function $y = \sin(\sqrt{x})$ using the viewing rectangle $[0, 400]$ by $[-1.5, 1.5]$. How does this graph differ from the graph of the sine function?

(b) Graph the function $y = \sin(x^2)$ using the viewing rectangle $[-5, 5]$ by $[-1.5, 1.5]$. How does this graph differ from the graph of the sine function?

4 PRINCIPLES OF PROBLEM SOLVING

There are no hard and fast rules that will ensure success in solving problems. However, it is possible to outline some general steps in the problem-solving process and to give some principles that may be useful in the solution of certain problems. These steps and principles are just common sense made explicit. They have been adapted from George Polya's book *How To Solve It*.

STEP 1. UNDERSTAND THE PROBLEM

The first step is to read the problem and make sure that you understand it clearly. Ask yourself the following questions:

What is the unknown?

What are the given quantities?

What are the given conditions?

For many problems it is useful to

draw a diagram

and identify the given and required quantities on the diagram.

Usually it is necessary to

introduce suitable notation

In choosing symbols for the unknown quantities we often use letters such as a, b, c, m, n, x, and y, but in some cases it helps to use initials as suggestive symbols, for instance, V for volume or t for time.

STEP 2. THINK OF A PLAN

Find a connection between the given information and the unknown that will enable you to calculate the unknown. It often helps to ask yourself explicitly: "How can I relate the given to the unknown?" If you do not see a connection immediately, the following ideas may be helpful in devising a plan.

TRY TO RECOGNIZE SOMETHING FAMILIAR Relate the given situation to previous knowledge. Look at the unknown and try to recall a more familiar problem that has a similar unknown.

TRY TO RECOGNIZE PATTERNS Some problems are solved by recognizing that some kind of pattern is occurring. The pattern could be geometric, or numerical, or algebraic. If you can see regularity or repetition in a problem, you might be able to guess what the continuing pattern is and then prove it.

USE ANALOGY Try to think of an analogous problem, that is, a similar problem, a related problem, but one that is easier than the original problem. If you can solve the similar, simpler problem, then it might give you the clues you need to solve the original, more difficult problem. For instance, if a problem involves very large numbers, you could first try a similar problem with smaller numbers. Or if the problem involves three-dimensional geometry, you could look for a similar problem in two-dimensional geometry. Or if the problem you start with is a general one, you could first try a special case.

INTRODUCE SOMETHING EXTRA It may sometimes be necessary to introduce something new, an auxiliary aid, to help make the connection between the given and the unknown. For instance, in a problem where a diagram is useful the auxiliary aid could be a new line drawn in a diagram. In a more algebraic problem it could be a new unknown that is related to the original unknown.

TAKE CASES We may sometimes have to split a problem into several cases and give a different argument for each of the cases. For instance, we often have to use this strategy in dealing with absolute value.

WORK BACKWARD Sometimes it is useful to imagine that your problem is solved and work backward, step by step, until you arrive at the given data. Then you may be able to reverse your steps and thereby construct a solution to the original problem. This procedure is commonly used in solving equations. For instance, in solving the equation $3x - 5 = 7$, we suppose that x is a number that satisfies $3x - 5 = 7$ and work backward. We add 5 to each side of the equation and then divide each side by 3 to get $x = 4$. Since each of these steps can be reversed, we have solved the problem.

ESTABLISH SUBGOALS In a complex problem it is often useful to set subgoals (in which the desired situation is only partially fulfilled). If we can first reach these subgoals, then we may be able to build on them to reach our final goal.

INDIRECT REASONING Sometimes it is appropriate to attack a problem indirectly. In using proof by contradiction to prove that P implies Q we assume that P is true and Q is false and try to see why this cannot happen. Somehow we have to use this information and arrive at a contradiction to what we absolutely know is true.

MATHEMATICAL INDUCTION In proving statements that involve a positive integer n, it is frequently helpful to use the Principle of Mathematical Induction, which is discussed in Appendix E.

STEP 3. CARRY OUT THE PLAN

In Step 2 a plan was devised. In carrying out that plan we have to check each stage of the plan and write the details that prove that each stage is correct.

STEP 4. LOOK BACK

Having completed our solution, it is wise to look back over it, partly to see if we have made errors in the solution and partly to see if we can think of an easier way to solve the problem. Another reason for looking back is that it will familiarize us with the method of solution and this may be useful for solving a future problem. Descartes said, "Every problem that I solved became a rule which served afterwards to solve other problems."

These principles of problem solving are illustrated in the following examples. Before you look at the solutions, try to solve these problems yourself, referring to these Principles of Problem Solving if you get stuck. You may find it useful to refer to this section from time to time as you solve the exercises in the remaining chapters of this book.

EXAMPLE 1 A driver sets out on a journey. For the first half of the distance she drives at the leisurely pace of 30 mi/h; she drives the second half at 60 mi/h. What is her average speed on this trip?

PRELIMINARY THOUGHTS It is tempting to take the average of the speeds and say that the average speed for the entire trip is

$$\frac{30 + 60}{2} = 45 \text{ mi/h}$$

But is this simple-minded approach really correct?

Try a special case

Let's look at an easily calculated special case. Suppose that the total distance traveled is 120 mi. Since the first 60 mi is traveled at 30 mi/h, it takes 2 h. The second 60 mi is traveled at 60 mi/h, so it takes 1 h. Thus, the total time is $2 + 1 = 3$ h and the average speed is

$$\frac{120}{3} = 40 \text{ mi/h}$$

So our guess of 45 mi/h is wrong.

Understand the problem

SOLUTION We need to look more carefully at the meaning of average speed. It is defined as

$$\text{average speed} = \frac{\text{distance traveled}}{\text{time elapsed}}$$

Introduce notation

Let d be the distance traveled on each half of the trip. Let t_1 and t_2 be the times taken for the first and second halves of the trip. Now we are in a position to write down the information that we are given.

State what is given

For the first half of the trip we have

$$30 = \frac{d}{t_1} \tag{1}$$

and for the second half we have

$$60 = \frac{d}{t_2} \tag{2}$$

Identify the unknown

Let's now identify the quantity we are asked to find:

$$\text{average speed for entire trip} = \frac{\text{total distance}}{\text{total time}} = \frac{2d}{t_1 + t_2}$$

Connect the given with the unknown

To calculate this quantity we need to know t_1 and t_2, so we solve Equations 1 and 2 for these times:

$$t_1 = \frac{d}{30} \qquad t_2 = \frac{d}{60}$$

Now we have the ingredients needed to calculate the desired quantity:

$$\text{average speed} = \frac{2d}{t_1 + t_2} = \frac{2d}{\dfrac{d}{30} + \dfrac{d}{60}}$$

$$= \frac{60(2d)}{60\left(\dfrac{d}{30} + \dfrac{d}{60}\right)} \qquad \text{(Multiply numerator and denominator by 60)}$$

$$= \frac{120d}{2d + d} = \frac{120d}{3d} = 40$$

The average speed for the entire trip is 40 mi/h. ■

As the next example illustrates, it is often necessary to use the problem-solving principle of *taking cases* when dealing with absolute values.

EXAMPLE 2 Solve the inequality $|x - 3| + |x + 2| < 11$.

SOLUTION Recall the definition of absolute value:

$$|x| = \begin{cases} x & \text{if } x \geqslant 0 \\ -x & \text{if } x < 0 \end{cases}$$

It follows that

$$|x - 3| = \begin{cases} x - 3 & \text{if } x - 3 \geqslant 0 \\ -(x - 3) & \text{if } x - 3 < 0 \end{cases}$$

$$= \begin{cases} x - 3 & \text{if } x \geqslant 3 \\ -x + 3 & \text{if } x < 3 \end{cases}$$

Similarly

$$|x + 2| = \begin{cases} x + 2 & \text{if } x + 2 \geqslant 0 \\ -(x + 2) & \text{if } x + 2 < 0 \end{cases}$$

$$= \begin{cases} x + 2 & \text{if } x \geqslant -2 \\ -x - 2 & \text{if } x < -2 \end{cases}$$

Take cases

These expressions show that we must consider three cases:

$$x < -2 \qquad -2 \leqslant x < 3 \qquad x \geqslant 3$$

CASE I If $x < -2$, we have

$$|x - 3| + |x + 2| < 11$$

$$-x + 3 - x - 2 < 11$$

$$-2x < 10$$

$$x > -5$$

CASE II If $-2 \leqslant x < 3$, the given inequality becomes

$$-x + 3 + x + 2 < 11$$

$$5 < 11 \qquad \text{(always true)}$$

CASE III If $x \geq 3$, the inequality becomes

$$x - 3 + x + 2 < 11$$
$$2x < 12$$
$$x < 6$$

Combining cases I, II, and III, we see that the inequality is satisfied when $-5 < x < 6$. So the solution is the interval $(-5, 6)$. ■

EXAMPLE 3 Express the hypotenuse h of a right triangle in terms of its area A and its perimeter P.

Understand the problem

SOLUTION Let us first sort out the information by identifying the unknown quantity and the data:

$$\text{Unknown: } h$$

$$\text{Given quantities: } A, P$$

Draw a diagram

Connect the given with the unknown

Introduce something extra

It helps to draw a diagram and we do so in Figure 1.

In order to connect the given quantities to the unknown, we introduce two extra variables a and b, which are the lengths of the other two sides of the triangle. This enables us to express the given condition, which is that the triangle is right-angled, by the Pythagorean Theorem:

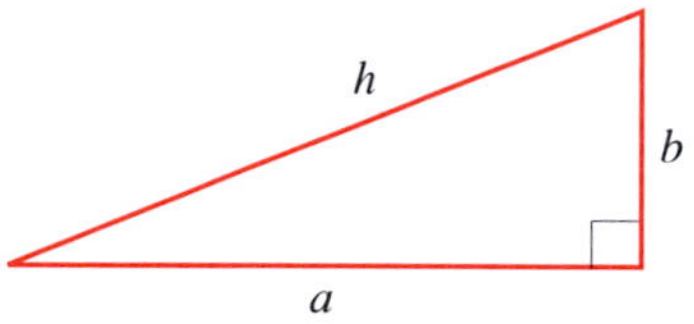

FIGURE 1

$$h^2 = a^2 + b^2$$

The other connections among the variables come by writing expressions for the area and perimeter:

$$A = \tfrac{1}{2}ab \qquad P = a + b + h$$

Since A and P are given, notice that we now have three equations in the three unknowns a, b, and h:

$$h^2 = a^2 + b^2 \tag{3}$$

$$A = \tfrac{1}{2}ab \tag{4}$$

$$P = a + b + h \tag{5}$$

Relate to the familiar

Although we have the correct number of equations, they are not easy to solve in a straightforward fashion. But if we use the problem-solving strategy of trying to recognize something familiar, then we can solve these equations by an easier method. Look at the right sides of Equations 3, 4, and 5. Do these expressions remind you of anything familiar? Notice that they contain the ingredients of a familiar formula:

$$(a + b)^2 = a^2 + 2ab + b^2$$

Using this idea, we express $(a + b)^2$ in two ways. From Equations 3 and 4 we have

$$(a + b)^2 = (a^2 + b^2) + 2ab = h^2 + 4A$$

From Equation 5 we have

$$(a + b)^2 = (P - h)^2 = P^2 - 2Ph + h^2$$

Thus
$$h^2 + 4A = P^2 - 2Ph + h^2$$
$$2Ph = P^2 - 4A$$
$$h = \frac{P^2 - 4A}{2P}$$

This is the required expression. ■

EXAMPLE 4 Find the final digit in the number 3^{459}.

Analogy

SOLUTION First notice that 3^{459} is a very large number—far too large for a calculator. Therefore, we attack this problem by first looking at analogous problems. A similar, but simpler, problem would be to find the final digit in 3^9 or 3^{59}. In fact, let's start with the exponents 1, 2, 3, . . . and see what happens.

Number	Final Digit
3^1	3
3^2	9
3^3	7
3^4	1
3^5	3
3^6	9
3^7	7
3^8	1

Pattern

By now you can see a pattern. The final digits occur in a cycle with length 4: 3, 9, 7, 1, 3, 9, 7, 1, 3, 9, 7, 1, ... Which number occurs in the 459th position? If we divide 459 by 4, the remainder is 3. So the final digit is the third number in the cycle, namely, 7.

The final digit in the number 3^{459} is 7. ■

EXAMPLE 5 Show that $\sqrt{2}$ is an irrational number.

Indirect reasoning

SOLUTION In this situation it is necessary to use proof by contradiction. Suppose that $\sqrt{2}$ is a rational number. This means that

$$\sqrt{2} = \frac{m}{n} \qquad (6)$$

where m and n are integers and $n \neq 0$. In fact, we can assume that m and n have no common factor; that is, any common factors have already been divided out.

Squaring both sides of Equation 6, we have

$$2 = \frac{m^2}{n^2} \qquad \text{or} \qquad m^2 = 2n^2 \qquad (7)$$

This equation shows that m^2 is an even number. It follows that m is an even number. [If m were odd, then $m = 2k + 1$ for some integer k, so $m^2 = (2k + 1)^2 = 4k^2 + 4k + 1$, which is 1 more than an even number and therefore odd.] But, since m is even, it must be of the form $m = 2k$, where k is an integer. Putting this expression into Equation 7, we have

$$(2k)^2 = 2n^2 \qquad 4k^2 = 2n^2 \qquad n^2 = 2k^2$$

This shows that n^2 is even and so n is even.

We have shown that both m and n are even. But this contradicts our assumption that m and n have no common factor.

We have arrived at a contradiction, so we conclude that our hypothesis that $\sqrt{2}$ is rational is false. Thus, $\sqrt{2}$ is irrational. ■

EXERCISES 4

1–2 ■ Solve the inequality.

1. $|x + 1| + |x + 4| \leqslant 5$ **2.** $|x - 1| - |x - 3| \geqslant 5$

3–4 ■ Solve the equation.

3. $|2x - 1| - |x + 5| = 3$ **4.** $||2x + 1| + 5| = 10$

5. Find the final digit in the number 947^{362}.

6. How many digits does the number $8^{15} \cdot 5^{37}$ have?

7. If $f_0(x) = x^2$ and $f_{n+1}(x) = f_0(f_n(x))$ for $n = 0, 1, 2, \ldots$, find a formula for $f_n(x)$.

8. If $f_0(x) = \dfrac{1}{2 - x}$ and $f_{n+1} = f_0 \circ f_n$ for $n = 0, 1, 2, \ldots$, find $f_{100}(3)$.

9–10 ■ Sketch the graph of the function.

9. $f(x) = |x^2 - 4|x| + 3|$ **10.** $g(x) = |x^2 - 1| - |x^2 - 4|$

11. Use a calculator to find the value of the expression

$$\sqrt{3 + 2\sqrt{2}} - \sqrt{3 - 2\sqrt{2}}$$

The number looks very simple. Show that the calculated value is correct.

12. Use a calculator to evaluate

$$\frac{\sqrt{2} + \sqrt{6}}{\sqrt{2 + \sqrt{3}}}$$

Show that the calculated value is correct.

13. Draw the graph of the equation $|x| + |y| = 1 + |xy|$.

14. Draw the graph of the equation

$$x^2y - y^3 - 5x^2 + 5y^2 = 0$$

without making a table of values.

15. Sketch the region in the plane consisting of all points (x, y) such that $|x| + |y| \leqslant 1$.

16. Sketch the region in the plane consisting of all points (x, y) such that

$$|x - y| + |x| - |y| \leqslant 2$$

17. (a) The radius of the earth is about 3960 mi. What length of ribbon would you need to wrap around the earth at the equator?

(b) How much more ribbon would you need if you raised the ribbon one foot above the earth?

18. Two runners start running laps at the same time from the same starting position. George takes 50 s to run a lap; Sue takes 30 s to run a lap. When will the runners next be even with each other?

19. In a right triangle, the hypotenuse has length 5 cm and another side has length 3 cm. What is the length of the altitude that is perpendicular to the base?

20. The perimeter of a right triangle is 60 cm and the altitude perpendicular to the hypotenuse is 12 cm. Find the lengths of the three sides.

21. Prove that $\sqrt{3}$ is irrational.

22. The sum of two numbers is 4 and their product is 1. Find the sum of their cubes.

23. Player A has a higher batting average than Player B for the first half of the baseball season. Player A also has a higher batting average than Player B for the second half of the season. Prove, or disprove, that Player A has a higher batting average than Player B for the entire season.

24. Prove that at any party there are two people who know the same number of people. Assume that if A knows B, then B knows A. Assume also that everyone knows himself or herself. [*Hint*: Use indirect reasoning.]

25. A car with tires that have a radius of 15 in. was driven on a trip and the odometer indicated that the distance traveled was 400 mi. Two weeks later, with snow tires installed, the odometer indicated that the distance for the return trip, over the same route, was 390 mi. Find the radius of the snow tires.

26. A spoonful of cream is taken from a cup of cream and put into a cup of coffee and stirred. Then a spoonful of this mixture is taken and put into the cup of cream. Is there now more cream in the coffee cup, or more coffee in the cup of cream?

5 A PREVIEW OF CALCULUS

Calculus is fundamentally different from the mathematics that you have studied previously. Calculus is less static and more dynamic. It is concerned with change and motion; it deals with quantities that approach other quantities. For that reason it may be useful to have an overview of the subject before beginning its intensive study. In this section we give a glimpse of some of the main ideas of calculus by showing how limits arise when we attempt to solve a variety of problems.

THE AREA PROBLEM

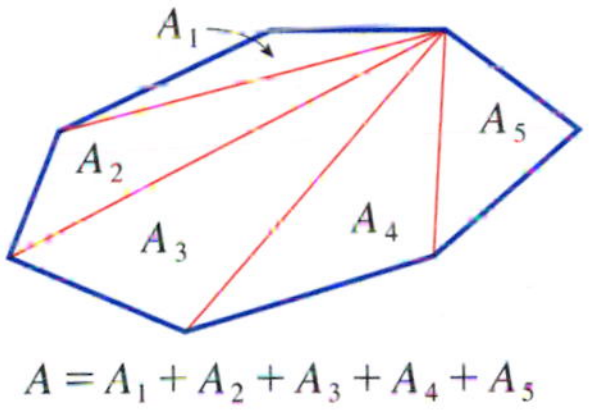

FIGURE 1

The origins of calculus go back at least 2500 years to the ancient Greeks, who found areas using the "method of exhaustion." They knew how to find the area A of any polygon by dividing it into triangles as in Figure 1 and adding the areas of these triangles.

It is a much more difficult problem to find the area of a curved figure. The Greek method of exhaustion was to inscribe polygons in the figure and circumscribe polygons about the figure and then let the number of sides of the polygons increase. Figure 2 illustrates this process for the special case of a circle with inscribed regular polygons.

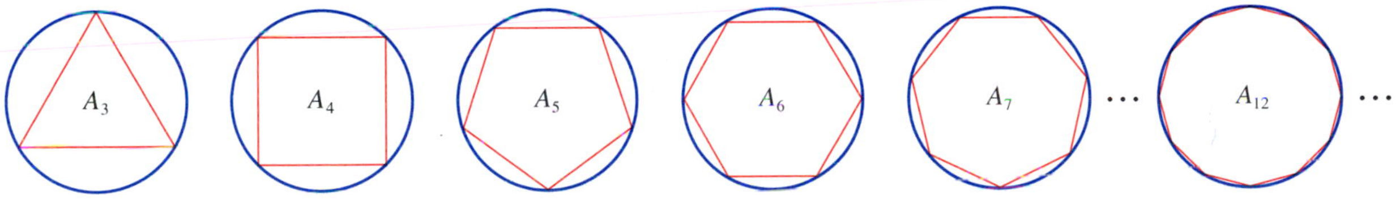

FIGURE 2

Let A_n be the area of the inscribed polygon with n sides. As n increases, it appears that A_n becomes closer and closer to the area of the circle. We say that the area of the circle is the *limit* of the areas of the inscribed polygons, and we write

$$A = \lim_{n \to \infty} A_n$$

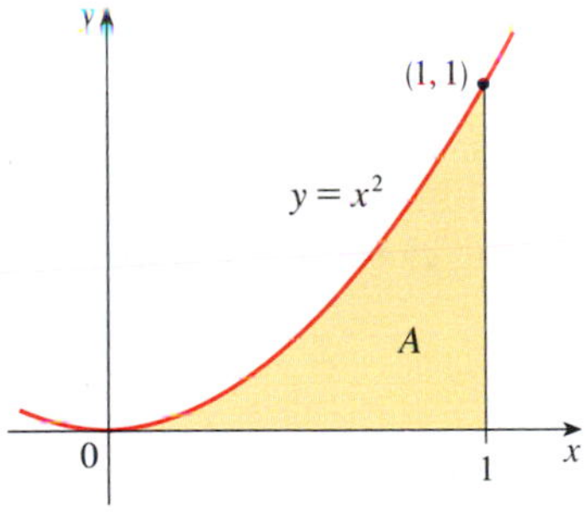

FIGURE 3

The Greeks themselves did not use limits explicitly. However, by indirect reasoning, Eudoxus (fifth century B.C.) used exhaustion to prove the familiar formula for the area of a circle: $A = \pi r^2$.

We will use a similar idea in Chapter 6 to find areas of regions of the type shown in Figure 3. We will approximate the desired area A by areas of rectangles (as in Figure 4), let the width of the rectangles decrease, and then calculate A as the limit of these sums of areas of rectangles.

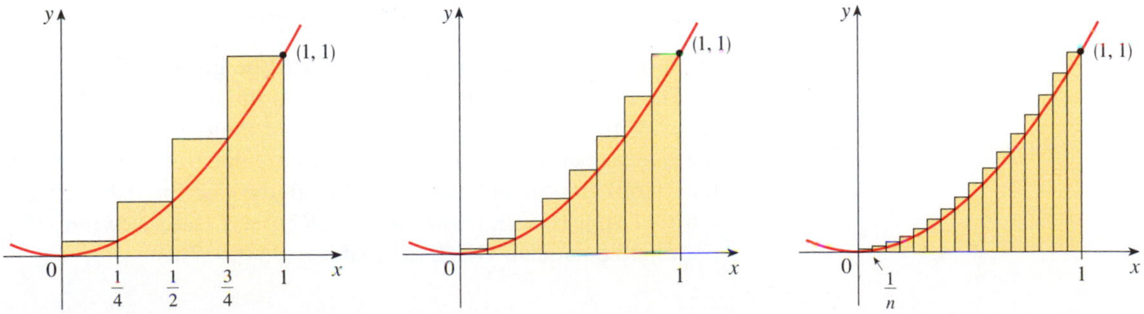

FIGURE 4

The area problem is the central problem in the branch of calculus called *integral calculus*. The techniques that we will develop in Chapter 6 for finding areas will also enable us to compute the volume of a solid, the length of a curve, the force of water against a dam, the mass and center of gravity of a rod, and the work done in pumping water out of a tank.

THE TANGENT PROBLEM

Consider the problem of trying to find the equation of the tangent line t to a curve with equation $y = f(x)$ at a given point P. (We will give a precise definition of a tangent line in Chapter 2. For now you can think of it as a line that touches the curve at P as in Figure 5.) Since we know that the point P lies on the tangent line, we can find the equation of t if we know its slope m. The problem is that we need two points to compute the slope and we know only one point, P, on t. To get around the problem we first find an approximation to m by taking a nearby point Q on the curve and computing the slope m_{PQ} of the secant line PQ. From Figure 6 we see that

(1)
$$m_{PQ} = \frac{f(x) - f(a)}{x - a}$$

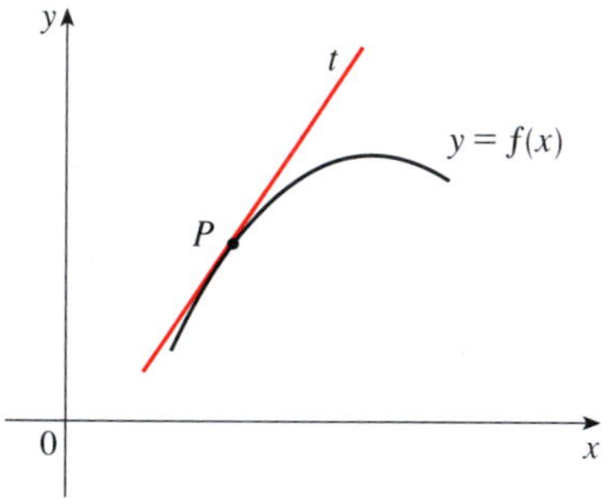

FIGURE 5
The tangent line at P

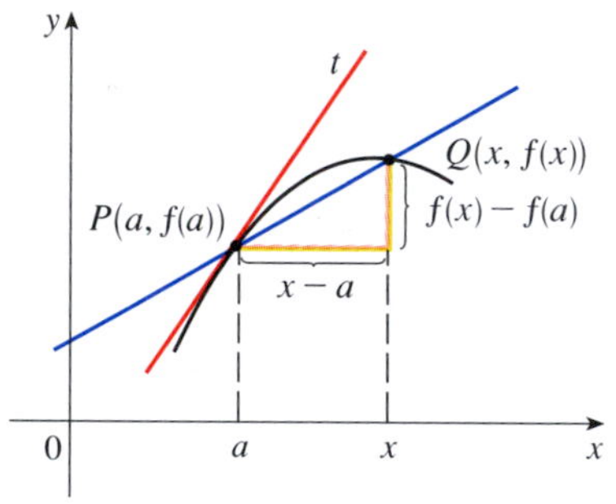

FIGURE 6
The secant line PQ

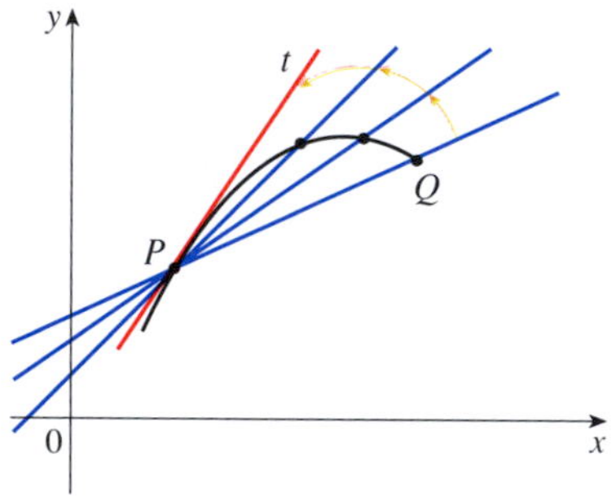

FIGURE 7
Secant lines approaching the tangent line

Now imagine that Q moves along the curve toward P as in Figure 7. You can see that the secant line rotates and approaches the tangent line as its limiting position. This means that the slope m_{PQ} of the secant line becomes closer and closer to the slope m of the tangent line. We write

$$m = \lim_{Q \to P} m_{PQ}$$

and we say that m is the limit of m_{PQ} as Q approaches P along the curve. Since x approaches a as Q approaches P, we could also use Equation 1 to write

(2)
$$m = \lim_{x \to a} \frac{f(x) - f(a)}{x - a}$$

Specific examples of this procedure will be given in Chapter 2.

The tangent problem has given rise to the branch of calculus called *differential calculus*, which was not invented until more than 2000 years after integral calculus. The main ideas behind differential calculus are due to the French mathematician Pierre Fermat (1601–1665) and were developed by the English mathematicians John Wallis (1616–1703), Isaac Barrow (1630–1677), and Isaac Newton (1642–1723) and the German mathematician Gottfried Leibniz (1648–1716).

The two branches of calculus and their chief problems, the area problem and the tangent problem, appear to be very different, but it turns out that there is a very close connection between them. The tangent problem and the area problem are inverse problems in a sense that will be described in Chapter 6.

VELOCITY

When we look at the speedometer of a car and read that the car is traveling at 48 mi/h, what does that information indicate to us? We know that if the velocity remains constant, then after an hour we will have traveled 48 mi. But if the velocity of the car varies, what does it mean to say that the velocity at a given instant is 48 mi/h?

In order to analyze this question, let us analyze the motion of a car that travels along a straight road and assume that we can measure the distance traveled by the car (in feet) at 1-second intervals as in the following chart:

t = time elapsed (s)	0	1	2	3	4	5
d = distance (ft)	0	2	10	25	43	78

As a first step toward finding the velocity after 2 seconds have elapsed, let's find the average velocity during the time interval $2 \leqslant t \leqslant 4$:

$$\begin{aligned} \text{average velocity} &= \frac{\text{distance traveled}}{\text{time elapsed}} \\ &= \frac{43 - 10}{4 - 2} \\ &= 16.5 \text{ ft/s} \end{aligned}$$

Similarly, the average velocity in the time interval $2 \leqslant t \leqslant 3$ is

$$\text{average velocity} = \frac{25 - 10}{3 - 2} = 15 \text{ ft/s}$$

We have the feeling that the velocity at the instant $t = 2$ cannot be much different from the average velocity during a short time interval starting at $t = 2$. So let's imagine that the distance traveled has been measured at 0.1-second time intervals as in the following chart:

t	2.0	2.1	2.2	2.3	2.4	2.5
d	10.00	11.02	12.16	13.45	14.96	16.80

Then we can compute, for instance, the average velocity over the time interval [2, 2.5]:

$$\text{average velocity} = \frac{16.80 - 10.00}{2.5 - 2} = 13.6 \text{ ft/s}$$

The results of such calculations are shown in the following chart:

Time interval	[2, 3]	[2, 2.5]	[2, 2.4]	[2, 2.3]	[2, 2.2]	[2, 2.1]
Average velocity (ft/s)	15.0	13.6	12.4	11.5	10.8	10.2

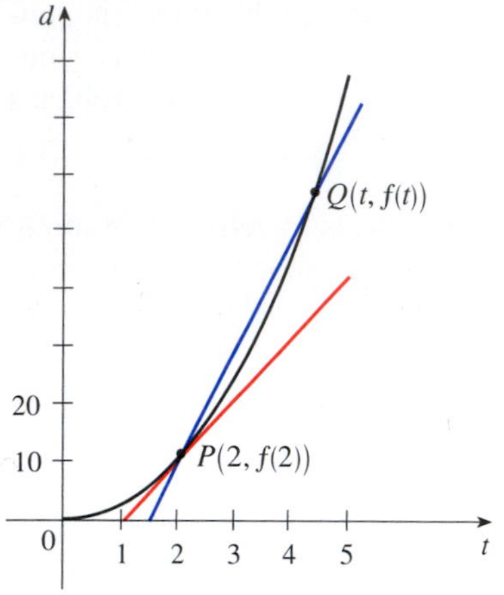

FIGURE 8

The average velocities over successively smaller intervals appear to be getting closer to a number near 10, and so we expect that the velocity at exactly $t = 2$ is about 10 ft/s. In Chapter 2 we will define the instantaneous velocity of a moving object as the limiting value of the average velocities over smaller and smaller time intervals.

In Figure 8 we show a graphical representation of the motion of the car by plotting the distance traveled as a function of time. If we write $d = f(t)$, then $f(t)$ is the number of feet traveled after t seconds. The average velocity in the time interval $[2, t]$ is

$$\text{average velocity} = \frac{\text{distance traveled}}{\text{time elapsed}} = \frac{f(t) - f(2)}{t - 2}$$

which is the same as the slope of the secant line PQ in Figure 8. The velocity v when $t = 2$ is the limiting value of this average velocity as t approaches 2; that is,

$$v = \lim_{t \to 2} \frac{f(t) - f(2)}{t - 2}$$

and we recognize from Equation 2 that this is the same as the slope of the tangent line to the curve at P.

Thus, when we solve the tangent problem in differential calculus, we are also solving problems concerning velocities. The same techniques also enable us to solve problems involving rates of change in all of the natural and social sciences.

THE LIMIT OF A SEQUENCE

In the fifth century B.C. the Greek philosopher Zeno of Elea posed four problems, now known as *Zeno's paradoxes*, that were intended to challenge some of the ideas concerning space and time that were held in his day. Zeno's second paradox concerns a race between the Greek hero Achilles and a tortoise that has been given a head start. Zeno argued as follows that Achilles could never pass the tortoise: Suppose that Achilles starts at position a_1 and the tortoise starts at position t_1 (see Figure 9).

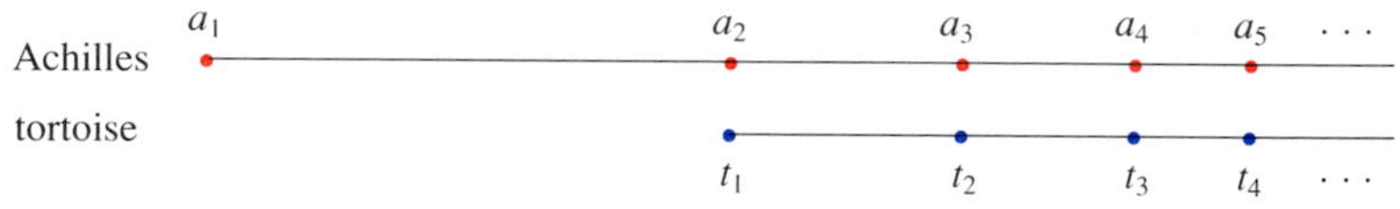

FIGURE 9

When Achilles reaches the point $a_2 = t_1$, the tortoise is farther ahead at position t_2. When Achilles reaches $a_3 = t_2$, the tortoise is at t_3. This process continues indefinitely and so it appears that the tortoise will always be ahead! But this defies common sense.

One way of explaining this paradox is with the idea of a *sequence*. The successive positions of Achilles $(a_1, a_2, a_3, \ldots)$ or the successive positions of the tortoise $(t_1, t_2, t_3, \ldots)$ form what is known as a sequence.

In general, a sequence $\{a_n\}$ is a set of numbers written in a definite order. For instance, the sequence

$$\{1, \tfrac{1}{2}, \tfrac{1}{3}, \tfrac{1}{4}, \tfrac{1}{5}, \ldots\}$$

can be described by giving the following formula for the nth term:

$$a_n = \frac{1}{n}$$

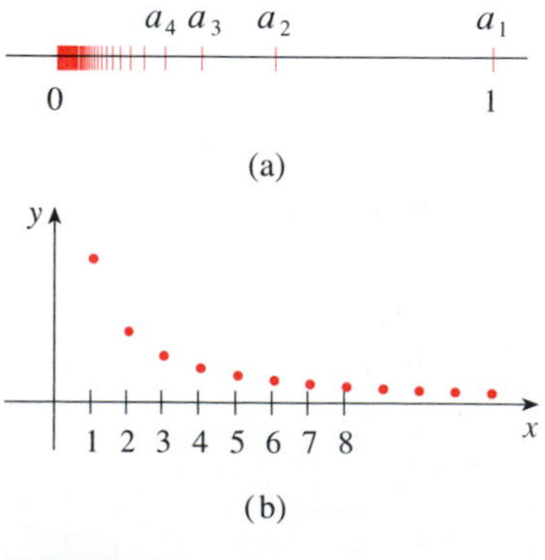

FIGURE 10

We can visualize this sequence by plotting its terms on a number line as in Figure 10(a) or by drawing its graph as in Figure 10(b). Observe from either picture that the terms of the sequence $a_n = 1/n$ are becoming closer and closer to 0 as n increases. In fact we can make the terms as small as we please by making n large enough. We say that the limit of the sequence is 0, and we indicate this by writing

$$\lim_{n\to\infty} \frac{1}{n} = 0$$

In general, the notation

$$\lim_{n\to\infty} a_n = L$$

is used if the terms a_n approach the number L as n becomes large. This means that the numbers a_n can be made as close as we like to the number L by taking n sufficiently large.

The concept of the limit of a sequence occurs whenever we use the decimal representation of a real number. For instance, if

$$a_1 = 3.1$$
$$a_2 = 3.14$$
$$a_3 = 3.141$$
$$a_4 = 3.1415$$
$$a_5 = 3.14159$$
$$a_6 = 3.141592$$
$$a_7 = 3.1415926$$
$$\vdots$$

then $$\lim_{n\to\infty} a_n = \pi$$

The terms in this sequence are rational approximations to π.

Let's return to Zeno's paradox. The successive positions of Achilles and the tortoise form sequences $\{a_n\}$ and $\{t_n\}$, where $a_n < t_n$ for all n. But using the more precise treatment of sequences given in Chapter 10, it can be shown that both sequences have the same limit:

$$\lim_{n\to\infty} a_n = p = \lim_{n\to\infty} t_n$$

It is precisely at this point p that Achilles overtakes the tortoise.

THE SUM OF A SERIES

Another of Zeno's paradoxes, as passed on to us by Aristotle, is the following: "A man standing in a room cannot walk to the wall. In order to do so, he would first have to go half the distance, then half the remaining distance, and then again half of what still remains. This process can always be continued and can never be ended." (See Figure 11.)

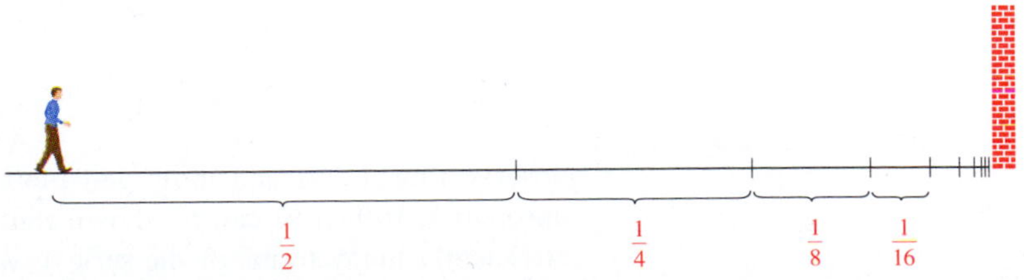

FIGURE 11

Of course, we know that the man can actually reach the wall, so this suggests that perhaps the total distance can be expressed as the sum of infinitely many smaller distances as follows:

$$1 = \frac{1}{2} + \frac{1}{4} + \frac{1}{8} + \frac{1}{16} + \cdots + \frac{1}{2^n} + \cdots \tag{3}$$

Zeno was arguing that it does not make sense to add infinitely many numbers together. But there are other situations in which we implicitly use infinite sums. For instance, in decimal notation, the symbol $0.\overline{3} = 0.3333\ldots$ means

$$\frac{3}{10} + \frac{3}{100} + \frac{3}{1000} + \frac{3}{10{,}000} + \cdots$$

and so, in some sense, it must be true that

$$\frac{3}{10} + \frac{3}{100} + \frac{3}{1000} + \frac{3}{10{,}000} + \cdots = \frac{1}{3}$$

More generally, if d_n denotes the nth digit in the decimal representation of a number, then

$$0.d_1d_2d_3d_4\ldots = \frac{d_1}{10} + \frac{d_2}{10^2} + \frac{d_3}{10^3} + \cdots + \frac{d_n}{10^n} + \cdots$$

Therefore, some infinite sums, or infinite series as they are called, have a meaning. But we must define carefully what the sum of an infinite series is.

Returning to the series in Equation 3, we denote by s_n the sum of the first n terms of the series. Thus

$$\begin{aligned}
s_1 &= \tfrac{1}{2} = 0.5 \\
s_2 &= \tfrac{1}{2} + \tfrac{1}{4} = 0.75 \\
s_3 &= \tfrac{1}{2} + \tfrac{1}{4} + \tfrac{1}{8} = 0.875 \\
s_4 &= \tfrac{1}{2} + \tfrac{1}{4} + \tfrac{1}{8} + \tfrac{1}{16} = 0.9375 \\
s_5 &= \tfrac{1}{2} + \tfrac{1}{4} + \tfrac{1}{8} + \tfrac{1}{16} + \tfrac{1}{32} = 0.96875 \\
s_6 &= \tfrac{1}{2} + \tfrac{1}{4} + \tfrac{1}{8} + \tfrac{1}{16} + \tfrac{1}{32} + \tfrac{1}{64} = 0.984375 \\
s_7 &= \tfrac{1}{2} + \tfrac{1}{4} + \tfrac{1}{8} + \tfrac{1}{16} + \tfrac{1}{32} + \tfrac{1}{64} + \tfrac{1}{128} = 0.9921875 \\
&\vdots \\
s_{10} &= \tfrac{1}{2} + \tfrac{1}{4} + \cdots + \tfrac{1}{1024} \approx 0.99902344 \\
&\vdots \\
s_{16} &= \frac{1}{2} + \frac{1}{4} + \cdots + \frac{1}{2^{16}} \approx 0.99998474
\end{aligned}$$

Observe that as we add more and more terms, the partial sums become closer and closer to 1. In fact, it can be shown that by taking n large enough (that is, by adding sufficiently many terms of the series), we can make the partial sum s_n as close as we please to the number 1. It therefore seems reasonable to say that the sum of the infinite series is 1 and to write

$$\frac{1}{2} + \frac{1}{4} + \frac{1}{8} + \cdots + \frac{1}{2^n} + \cdots = 1$$

In other words, the reason the sum of the series is 1 is that

$$\lim_{n \to \infty} s_n = 1$$

In Chapter 10 we will make these ideas precise and we will use Newton's idea of combining infinite series with differential and integral calculus.

SUMMARY

We have seen that the concept of a limit arises in trying to find the area of a region, the slope of a tangent to a curve, the velocity of a car, or the sum of an infinite series. In each case the common theme is the calculation of a quantity as the limit of other, easily calculated, quantities. It is this basic idea of a limit that sets calculus apart from other areas of mathematics. In fact, we could define calculus as the part of mathematics that deals with limits.

Sir Isaac Newton invented his version of calculus in order to explain the motion of the planets around the Sun. Today calculus is used not only in calculating the orbits of satellites and spacecraft and in the study of astronomy, nuclear physics, electricity, thermodynamics, acoustics, design of machines, chemical reactions, growth of organisms, weather prediction, and the calculation of life insurance premiums, but also in such everyday concerns as fencing off a field so as to enclose the maximum area or computing the most economical speed for driving a car. Some of these uses of calculus will be explored throughout this book.

1

INTRODUCTION TO VECTORS AND VECTOR FUNCTIONS

> The notion of a vector . . . entered mathematics quietly.
>
> MORRIS KLINE

The concept of a vector is one that is vital to many applications of mathematics. For example, the velocity of a moving mass is a vector quantity. The force that moves a mass is also a vector quantity. Velocity and force have both magnitude and direction, which are the two defining properties of a vector. This chapter defines two-dimensional vectors along with their basic operations of addition, scalar multiplication, and dot product. Vector-valued functions can be used to describe the motion of objects through space. These are discussed at the end of this chapter.

1.1 VECTORS

The term **vector** is used by scientists to indicate a quantity (such as velocity or force) that has both magnitude and direction. A vector is often represented by an arrow or a directed line segment. The length of the arrow represents the *magnitude* of the vector and the arrow points in the *direction* of the vector. For instance, Figure 1 shows a particle moving along a path in the plane and its velocity vector **v** at a specific location of the particle. Here the length of the arrow represents the speed of the particle and it points in the direction that the particle is moving.

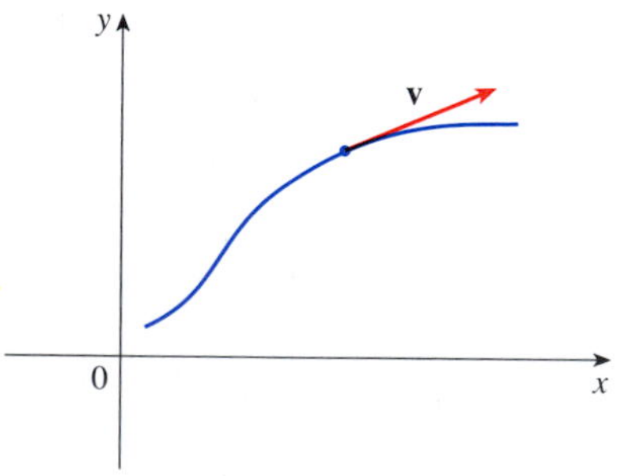

FIGURE 1

The velocity vector of a particle

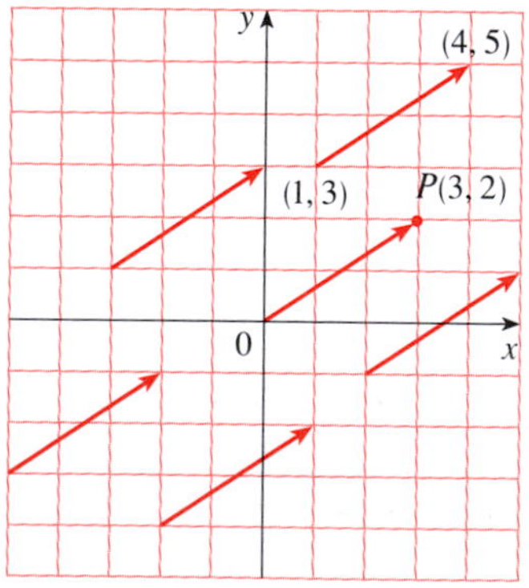

FIGURE 2

Representations of the vector $\mathbf{v} = \langle 3, 2 \rangle$

In Figure 2 notice that all of the arrows are equivalent in the sense that they have the same length and point in the same direction even though they are in different locations. Each of these directed line segments has the property that the terminal point is reached from the initial point by a displacement of three units to the right and two upward. We regard each of the directed line segments as equivalent representations of a single entity called a **vector**. In other words, we can regard a vector **v** as a set of equivalent directed line segments. These line segments are characterized by the numbers 3 and 2, and we symbolize this situation by writing $\mathbf{v} = \langle 3, 2 \rangle$. Thus, a two-dimensional vector can be thought of as an ordered pair of real numbers. We use the notation $\langle a, b \rangle$ for the ordered pair that refers to a vector so as not to confuse it with the ordered pair (a, b) that refers to a point in the plane. A vector can be indicated by printing a letter in boldface (**v**) or by putting an arrow above the letter ($\vec{v}$).

(1) DEFINITION A **two-dimensional vector** is an ordered pair $\mathbf{a} = \langle a_1, a_2 \rangle$ of real numbers. The numbers a_1 and a_2 are called the **components** of **a**.

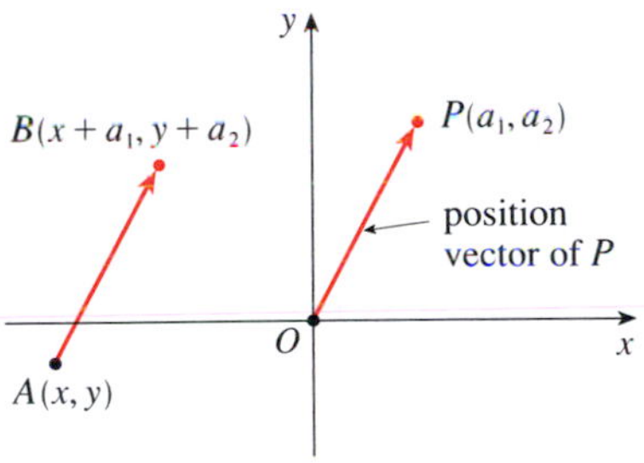

FIGURE 3
Representations of $\mathbf{a} = \langle a_1, a_2 \rangle$

A **representation** of the vector $\mathbf{a} = \langle a_1, a_2 \rangle$ is a directed line segment $\overrightarrow{AB}$ from any point $A(x, y)$ to the point $B(x + a_1, y + a_2)$. A particular representation of **a** is the directed line segment $\overrightarrow{OP}$ from the origin to the point $P(a_1, a_2)$, and $\langle a_1, a_2 \rangle$ is called the **position vector** of the point $P(a_1, a_2)$ (see Figure 3). Observe that if $\mathbf{a} = \langle a_1, a_2 \rangle$ is a vector that has the representation $\overrightarrow{AB}$, where the initial point is $A(x_1, y_1)$ and the terminal point is $B(x_2, y_2)$, then we must have $x_1 + a_1 = x_2$ and $y_1 + a_2 = y_2$, and so $a_1 = x_2 - x_1$ and $a_2 = y_2 - y_1$. Thus, we have the following:

(2) Given the points $A(x_1, y_1)$ and $B(x_2, y_2)$, the vector **a** with representation $\overrightarrow{AB}$ is

$$\mathbf{a} = \langle x_2 - x_1, y_2 - y_1 \rangle$$

EXAMPLE 1 Find the vector represented by the directed line segment with initial point $A(2, -3)$ and terminal point $B(-2, 1)$.

SOLUTION By (2), the vector corresponding to $\overrightarrow{AB}$ is

$$\mathbf{a} = \langle -2 - 2, 1 - (-3) \rangle = \langle -4, 4 \rangle$$ ■

The **magnitude** or **length** of the vector **v** is the length of any of its representations and is denoted by the symbol $|\mathbf{v}|$ or $\|\mathbf{v}\|$. By using the distance formula to compute the length of a segment OP, we obtain the following:

(3) The length of the vector $\mathbf{a} = \langle a_1, a_2 \rangle$ is

$$|\mathbf{a}| = \sqrt{a_1^2 + a_2^2}$$

The length of the vector $\overrightarrow{AB}$ from $A(x_1, y_1)$ to $B(x_2, y_2)$ is

$$|\overrightarrow{AB}| = \sqrt{(x_2 - x_1)^2 + (y_2 - y_1)^2}$$

The only vector with length 0 is the **zero vector** $\mathbf{0} = \langle 0, 0 \rangle$. This vector is also the only vector with no specific direction.

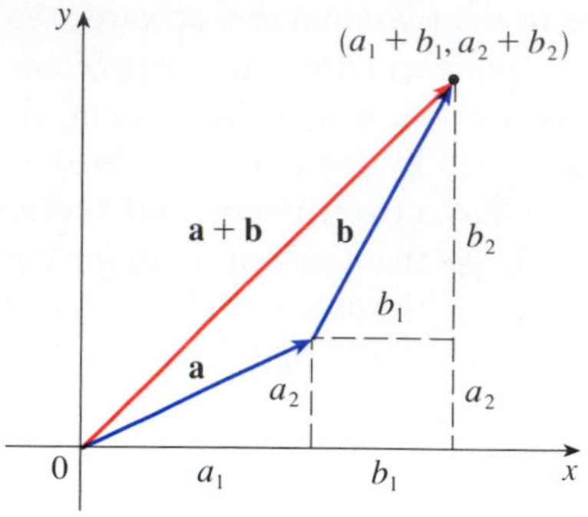

FIGURE 4
Triangle Law

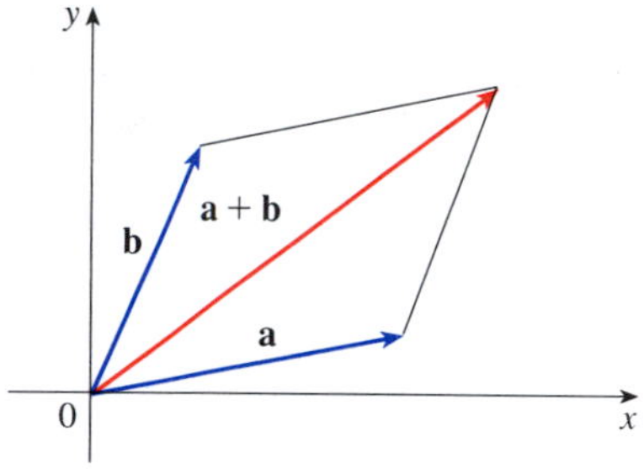

FIGURE 5
Parallelogram Law

According to the following definition, we add vectors by adding the corresponding components of two vectors.

> **(4) VECTOR ADDITION** If $\mathbf{a} = \langle a_1, a_2 \rangle$ and $\mathbf{b} = \langle b_1, b_2 \rangle$, then the vector $\mathbf{a} + \mathbf{b}$ is defined by
>
> $$\mathbf{a} + \mathbf{b} = \langle a_1 + b_1, a_2 + b_2 \rangle$$

Definition 4 is illustrated geometrically in Figure 4. You can see why the definition of vector addition is sometimes called the **Triangle Law**. Alternatively, another interpretation of vector addition is shown in Figure 5 and is called the **Parallelogram Law**.

It's possible to multiply a vector by a real number c. (In this context we call the real number c a **scalar** to distinguish it from a vector.) For instance, we want $2\mathbf{a}$ to be the same vector as $\mathbf{a} + \mathbf{a}$, so

$$2\langle a_1, a_2 \rangle = \langle a_1, a_2 \rangle + \langle a_1, a_2 \rangle = \langle 2a_1, 2a_2 \rangle$$

In general, we multiply a vector by a scalar by multiplying each component by that scalar.

> **(5) MULTIPLICATION OF A VECTOR BY A SCALAR** If c is a scalar and $\mathbf{a} = \langle a_1, a_2 \rangle$, then the vector $c\mathbf{a}$ is defined by
>
> $$c\mathbf{a} = \langle ca_1, ca_2 \rangle$$

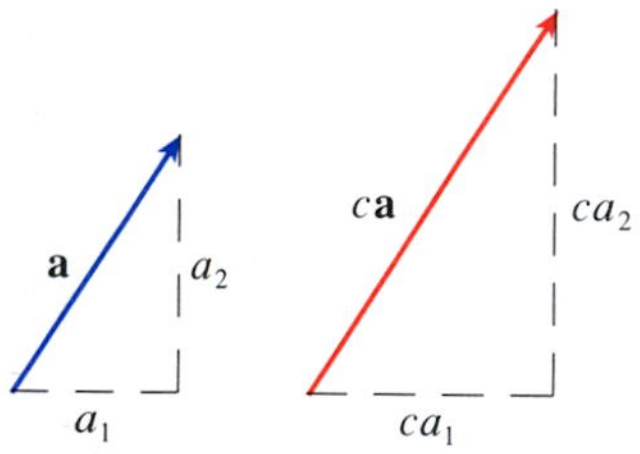

FIGURE 6

Definition 5 is illustrated by Figure 6.

Let's see how the scalar multiple $c\mathbf{a}$ compares with the original vector $\mathbf{a}$. If $\mathbf{a} = \langle a_1, a_2 \rangle$, then

$$|c\mathbf{a}| = \sqrt{(ca_1)^2 + (ca_2)^2} = \sqrt{c^2(a_1^2 + a_2^2)}$$
$$= \sqrt{c^2}\,\sqrt{a_1^2 + a_2^2} = |c|\,|\mathbf{a}|$$

so the length of $c\mathbf{a}$ is $|c|$ times the length of $\mathbf{a}$.

If $a_1 \neq 0$, we can talk about the slope of $\mathbf{a}$ as being a_2/a_1. But, if $c \neq 0$, then the slope of $c\mathbf{a}$ is $ca_2/ca_1 = a_2/a_1$, the same as the slope of $\mathbf{a}$. If $c > 0$, then a_1 and ca_1 have the same sign. Also, a_2 and ca_2 have the same sign. This means that $\mathbf{a}$ and $c\mathbf{a}$ have the same direction. On the other hand, if $c < 0$, then a_1 and ca_1 have opposite signs, as do a_2 and ca_2, so $\mathbf{a}$ and $c\mathbf{a}$ have opposite directions. In particular, the vector $-\mathbf{a} = (-1)\mathbf{a}$ has the same length as $\mathbf{a}$ but points in the opposite direction. Illustrations of representations are shown in Figure 7.

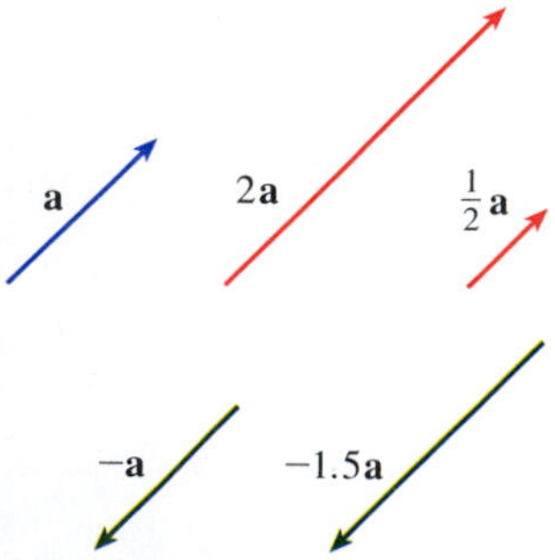

FIGURE 7
Scalar multiples of **a**

Two vectors $\mathbf{a}$ and $\mathbf{b}$ are called **parallel** if $\mathbf{b} = c\mathbf{a}$ for some scalar c.

By the **difference** $\mathbf{a} - \mathbf{b}$ of two vectors, we mean

$$\mathbf{a} - \mathbf{b} = \mathbf{a} + (-\mathbf{b})$$

so if $\mathbf{a} = \langle a_1, a_2 \rangle$ and $\mathbf{b} = \langle b_1, b_2 \rangle$, then

$$\mathbf{a} - \mathbf{b} = \langle a_1 - b_1, a_2 - b_2 \rangle$$

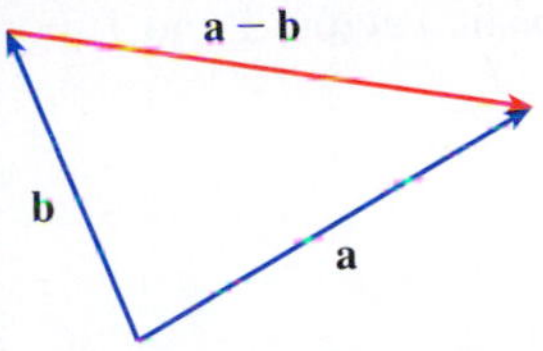

FIGURE 8
Geometric interpretation of the difference of two vectors

Since $(\mathbf{a} - \mathbf{b}) + \mathbf{b} = \mathbf{a}$, the vector $\mathbf{a} - \mathbf{b}$, when added to $\mathbf{b}$, gives $\mathbf{a}$. This is illustrated in Figure 8 by means of the Triangle Law.

EXAMPLE 2 If $\mathbf{a} = \langle 4, 3\rangle$ and $\mathbf{b} = \langle -2, 1\rangle$, find $|\mathbf{a}|$ and the vectors $\mathbf{a} + \mathbf{b}$, $\mathbf{a} - \mathbf{b}$, $3\mathbf{b}$, and $2\mathbf{a} + 5\mathbf{b}$.

SOLUTION
$$|\mathbf{a}| = \sqrt{4^2 + 3^2} = \sqrt{25} = 5$$

$$\mathbf{a} + \mathbf{b} = \langle 4, 3\rangle + \langle -2, 1\rangle$$
$$= \langle 4 - 2, 3 + 1\rangle = \langle 2, 4\rangle$$

$$\mathbf{a} - \mathbf{b} = \langle 4, 3\rangle - \langle -2, 1\rangle$$
$$= \langle 4 - (-2), 3 - 1\rangle = \langle 6, 2\rangle$$

$$3\mathbf{b} = 3\langle -2, 1\rangle = \langle 3(-2), 3(1)\rangle = \langle -6, 3\rangle$$

$$2\mathbf{a} + 5\mathbf{b} = 2\langle 4, 3\rangle + 5\langle -2, 1\rangle$$
$$= \langle 8, 6\rangle + \langle -10, 5\rangle = \langle -2, 11\rangle$$ ■

(6) PROPERTIES OF VECTORS If $\mathbf{a}$, $\mathbf{b}$, and $\mathbf{c}$ are vectors and c and d are scalars, then

1. $\mathbf{a} + \mathbf{b} = \mathbf{b} + \mathbf{a}$
2. $\mathbf{a} + (\mathbf{b} + \mathbf{c}) = (\mathbf{a} + \mathbf{b}) + \mathbf{c}$
3. $\mathbf{a} + \mathbf{0} = \mathbf{a}$
4. $\mathbf{a} + (-\mathbf{a}) = \mathbf{0}$
5. $c(\mathbf{a} + \mathbf{b}) = c\mathbf{a} + c\mathbf{b}$
6. $(c + d)\mathbf{a} = c\mathbf{a} + d\mathbf{a}$
7. $(cd)\mathbf{a} = c(d\mathbf{a})$
8. $1\mathbf{a} = \mathbf{a}$

The eight properties of vectors in Theorem 6 can be readily verified using Definitions 4 and 5. For instance, here is the verification of Property 1:

$$\mathbf{a} + \mathbf{b} = \langle a_1, a_2\rangle + \langle b_1, b_2\rangle = \langle a_1 + b_1, a_2 + b_2\rangle$$
$$= \langle b_1 + a_1, b_2 + a_2\rangle = \langle b_1, b_2\rangle + \langle a_1, a_2\rangle$$
$$= \mathbf{b} + \mathbf{a}$$

The remaining proofs are left as exercises.

Two vectors play a special role. Let

$$\mathbf{i} = \langle 1, 0\rangle \qquad \mathbf{j} = \langle 0, 1\rangle$$

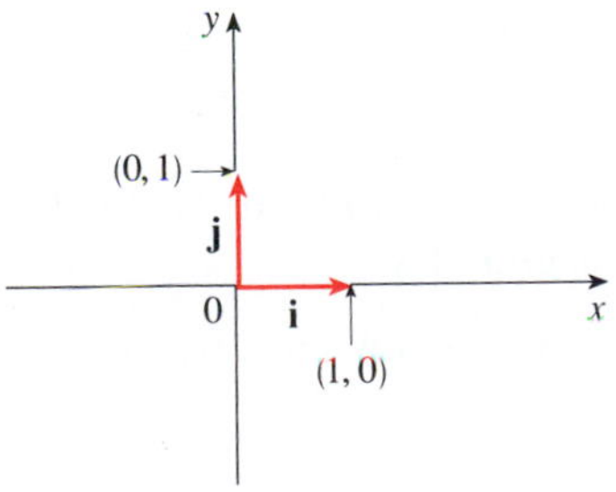

FIGURE 9
Standard basis vectors

Then $\mathbf{i}$ and $\mathbf{j}$ are vectors that have length 1 and point in the directions of the positive x- and y-axes (see Figure 9). We refer to these as **standard basis vectors**.

If $\mathbf{a} = \langle a_1, a_2\rangle$, then we can write

$$\mathbf{a} = \langle a_1, a_2\rangle = \langle a_1, 0\rangle + \langle 0, a_2\rangle$$
$$= a_1\langle 1, 0\rangle + a_2\langle 0, 1\rangle$$

(7) $$\mathbf{a} = a_1\mathbf{i} + a_2\mathbf{j}$$

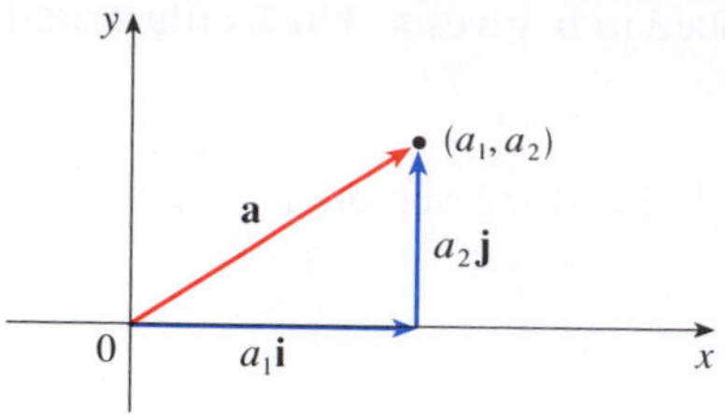

FIGURE 10
$\mathbf{a} = a_1\mathbf{i} + a_2\mathbf{j}$

Thus, any vector can be expressed in terms of the standard basis vectors **i** and **j**. For instance,

$$\langle 1, -2 \rangle = \mathbf{i} - 2\mathbf{j}$$

See Figure 10 for the geometric interpretation of Equation 7.

EXAMPLE 3 If $\mathbf{a} = \mathbf{i} + 2\mathbf{j}$ and $\mathbf{b} = 4\mathbf{i} + 7\mathbf{j}$, express the vector $2\mathbf{a} + 3\mathbf{b}$ in terms of **i** and **j**.

SOLUTION Using Properties 1, 2, 5, 6, and 7 of Theorem 6, we have

$$\begin{aligned} 2\mathbf{a} + 3\mathbf{b} &= 2(\mathbf{i} + 2\mathbf{j}) + 3(4\mathbf{i} + 7\mathbf{j}) \\ &= 2\mathbf{i} + 4\mathbf{j} + 12\mathbf{i} + 21\mathbf{j} = 14\mathbf{i} + 25\mathbf{j} \end{aligned}$$ ■

A **unit vector** is a vector whose length is 1. For instance, **i** and **j** are unit vectors. In general, if $\mathbf{a} \neq \mathbf{0}$, then the unit vector that has the same direction as **a** is

$$\textbf{(8)} \qquad \mathbf{u} = \frac{1}{|\mathbf{a}|}\mathbf{a} = \frac{\mathbf{a}}{|\mathbf{a}|}$$

In order to verify this, we let $c = 1/|\mathbf{a}|$. Then $\mathbf{u} = c\mathbf{a}$ and c is a positive scalar, so **u** has the same direction as **a**. Also

$$|\mathbf{u}| = |c\mathbf{a}| = |c||\mathbf{a}| = \frac{1}{|\mathbf{a}|}|\mathbf{a}| = 1$$

EXAMPLE 4 Find the unit vector in the direction of the vector $3\mathbf{i} - 4\mathbf{j}$.

SOLUTION The given vector has length

$$|3\mathbf{i} - 4\mathbf{j}| = \sqrt{3^2 + (-4)^2} = \sqrt{25} = 5$$

so, by Equation 8, the unit vector with the same direction is

$$\tfrac{1}{5}(3\mathbf{i} - 4\mathbf{j}) = \tfrac{3}{5}\mathbf{i} - \tfrac{4}{5}\mathbf{j}$$ ■

We conclude this section by considering one of the many applications of vectors in physics and engineering. A force is represented by a vector because it has both a magnitude (measured in pounds or newtons, N) and a direction. If several forces are acting on an object, the **resultant force** experienced by the object is the vector sum of these forces.

EXAMPLE 5 A 100-lb weight hangs from two wires as shown in Figure 11. Find the tensions (forces) $\mathbf{T}_1$ and $\mathbf{T}_2$ in both wires and their magnitudes.

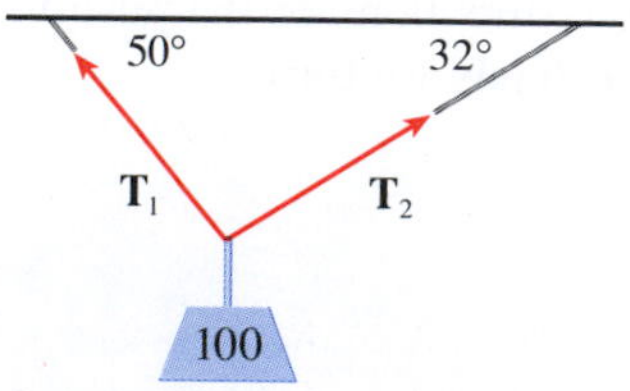

FIGURE 11

SOLUTION We first express $\mathbf{T}_1$ and $\mathbf{T}_2$ in terms of their horizontal and vertical components. From Figure 12 and the Triangle Law we see that

$$\textbf{(9)} \qquad \mathbf{T}_1 = -|\mathbf{T}_1|\cos 50°\,\mathbf{i} + |\mathbf{T}_1|\sin 50°\,\mathbf{j}$$

$$\textbf{(10)} \qquad \mathbf{T}_2 = |\mathbf{T}_2|\cos 32°\,\mathbf{i} + |\mathbf{T}_2|\sin 32°\,\mathbf{j}$$

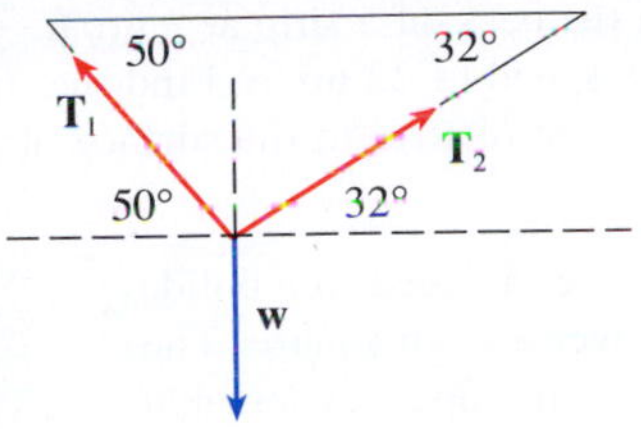

FIGURE 12

The resultant $\mathbf{T}_1 + \mathbf{T}_2$ of the tensions counterbalances the weight $\mathbf{w}$ and so we must have

$$\mathbf{T}_1 + \mathbf{T}_2 = -\mathbf{w} = 100\,\mathbf{j}$$

Thus

$$(-|\mathbf{T}_1|\cos 50^\circ + |\mathbf{T}_2|\cos 32^\circ)\,\mathbf{i} + (|\mathbf{T}_1|\sin 50^\circ + |\mathbf{T}_2|\sin 32^\circ)\,\mathbf{j} = 100\,\mathbf{j}$$

Equating components, we get

$$-|\mathbf{T}_1|\cos 50^\circ + |\mathbf{T}_2|\cos 32^\circ = 0$$

$$|\mathbf{T}_1|\sin 50^\circ + |\mathbf{T}_2|\sin 32^\circ = 100$$

Solving the first of these equations for $|\mathbf{T}_2|$ and substituting into the second, we get

$$|\mathbf{T}_1|\sin 50^\circ + \frac{|\mathbf{T}_1|\cos 50^\circ}{\cos 32^\circ}\sin 32^\circ = 100$$

So the magnitudes of the tensions are

$$|\mathbf{T}_1| = \frac{100}{\sin 50^\circ + \tan 32^\circ \cos 50^\circ} \approx 85.64 \text{ lb}$$

and

$$|\mathbf{T}_2| = \frac{|\mathbf{T}_1|\cos 50^\circ}{\cos 32^\circ} \approx 64.91 \text{ lb}$$

Substituting these values in (9) and (10), we obtain the tension vectors

$$\mathbf{T}_1 \approx -55.05\,\mathbf{i} + 65.60\,\mathbf{j} \qquad \mathbf{T}_2 \approx 55.05\,\mathbf{i} + 34.40\,\mathbf{j}$$ ■

EXERCISES 1.1

1–4 ■ Find a vector $\mathbf{a}$ with representation given by the directed line segment $\overrightarrow{AB}$. Draw $\overrightarrow{AB}$ and the equivalent representation starting at the origin.

1. $A(1, 3)$, $B(4, 4)$

2. $A(-3, 4)$, $B(-1, 0)$

3. $A(3, -1)$, $B(3, -3)$

4. $A(4, -1)$, $B(1, 2)$

5–8 ■ Find the sum of the given vectors and illustrate geometrically.

5. $\langle 2, 3\rangle$, $\langle 3, -4\rangle$

6. $\langle -1, 2\rangle$, $\langle 5, 3\rangle$

7. $\langle 1, 0\rangle$, $\langle 0, 1\rangle$

8. $\langle 3, 2\rangle$, $\langle 1, -3\rangle$

9–16 ■ Find $|\mathbf{a}|$, $\mathbf{a} + \mathbf{b}$, $\mathbf{a} - \mathbf{b}$, $2\mathbf{a}$, and $3\mathbf{a} + 4\mathbf{b}$.

9. $\mathbf{a} = \langle 5, -12\rangle$, $\mathbf{b} = \langle -2, 8\rangle$

10. $\mathbf{a} = \langle -1, 2\rangle$, $\mathbf{b} = \langle 4, 3\rangle$

11. $\mathbf{a} = \langle 2, -3\rangle$, $\mathbf{b} = \langle 1, 4\rangle$

12. $\mathbf{a} = \langle 2, -1\rangle$, $\mathbf{b} = \langle 6, 7\rangle$

13. $\mathbf{a} = \mathbf{i} - \mathbf{j}$, $\mathbf{b} = \mathbf{i} + \mathbf{j}$

14. $\mathbf{a} = 2\,\mathbf{i} + 3\,\mathbf{j}$, $\mathbf{b} = 3\,\mathbf{i} - 2\,\mathbf{j}$

15. $\mathbf{a} = \mathbf{i} + \mathbf{j}$, $\mathbf{b} = 2\,\mathbf{i} - \mathbf{j}$

16. $\mathbf{a} = 6\,\mathbf{i}$, $\mathbf{b} = \mathbf{i} - 2\,\mathbf{j}$

17–20 ■ Find a unit vector that has the same direction as the given vector.

17. $\langle 1, 2\rangle$

18. $\langle 3, -5\rangle$

19. $\mathbf{i} + \mathbf{j}$

20. $2\,\mathbf{i} - 4\,\mathbf{j}$

21–22 ■ Express $\mathbf{i}$ and $\mathbf{j}$ in terms of $\mathbf{a}$ and $\mathbf{b}$.

21. $\mathbf{a} = 2\,\mathbf{i} + 3\,\mathbf{j}$, $\mathbf{b} = \mathbf{i} - \mathbf{j}$

22. $\mathbf{a} = \mathbf{i} - 2\,\mathbf{j}$, $\mathbf{b} = 3\,\mathbf{i} + \mathbf{j}$

23. If A, B, and C are the vertices of a triangle, find $\overrightarrow{AB} + \overrightarrow{BC} + \overrightarrow{CA}$.

24. Let C be the point on the line segment AB that is twice as far from B as it is from A. If $\mathbf{a} = \overrightarrow{OA}$, $\mathbf{b} = \overrightarrow{OB}$, and $\mathbf{c} = \overrightarrow{OC}$, show that $\mathbf{c} = \frac{2}{3}\,\mathbf{a} + \frac{1}{3}\,\mathbf{b}$.

25. (a) Draw the vectors $\mathbf{a} = \langle 3, 2 \rangle$, $\mathbf{b} = \langle 2, -1 \rangle$, and $\mathbf{c} = \langle 7, 1 \rangle$.
(b) Show, by means of a sketch, that there are scalars s and t such that $\mathbf{c} = s\mathbf{a} + t\mathbf{b}$.
(c) Use the sketch to estimate the values of s and t.
(d) Find the exact values of s and t.

26. Suppose that $\mathbf{a}$ and $\mathbf{b}$ are nonzero vectors that are not parallel and $\mathbf{c}$ is any vector in the plane determined by $\mathbf{a}$ and $\mathbf{b}$. Give a geometric argument to show that $\mathbf{c}$ can be written as $\mathbf{c} = s\mathbf{a} + t\mathbf{b}$ for suitable scalars s and t. Then give an argument using components.

27. Two forces $\mathbf{F}_1$ and $\mathbf{F}_2$ with magnitudes 10 lb and 12 lb act on an object at a point P as shown in the figure. Find the resultant force $\mathbf{F}$ acting at P as well as its magnitude and its direction. (Indicate the direction by finding the angle θ shown in the figure.)

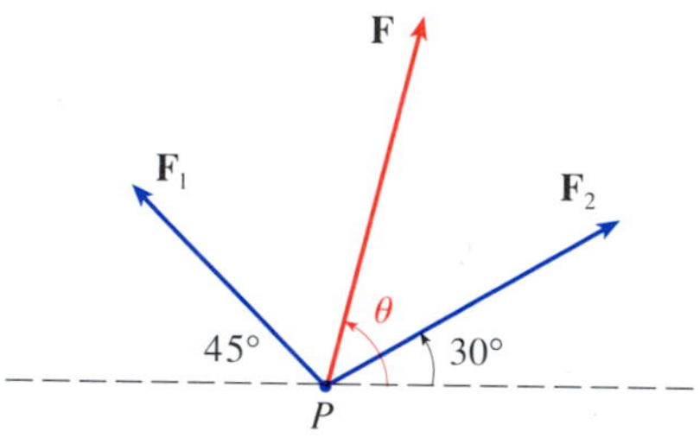

28. Velocities have both direction and magnitude and thus are vectors. The magnitude of a velocity vector is called *speed.* Suppose that a wind is blowing from the direction N45°W at a speed of 50 km/h. (This means that the direction from which the wind blows is 45° west of the northerly direction.) A pilot is steering a plane in the direction N60°E at an airspeed (speed in still air) of 250 km/h. The *true course,* or *track,* of the plane is the direction of the resultant of the velocity vectors of the plane and the wind. The *ground speed* of the plane is the magnitude of the resultant. Find the true course and the ground speed of the plane.

29. A woman walks due west on the deck of a ship at 3 mi/h. The ship is moving north at a speed of 22 mi/h. Find the speed and direction of the woman relative to the surface of the water.

30. Ropes 3 m and 5 m in length are fastened to a holiday decoration that is suspended over a town square. The decoration has a mass of 5 kg. The ropes, fastened at different heights, make angles of 52° and 40° with the horizontal. Find the tension in each wire and the magnitude of each tension. [*Note:* Gravity exerts a downward force of $5(9.8) = 49$ N on a 5-kg mass.]

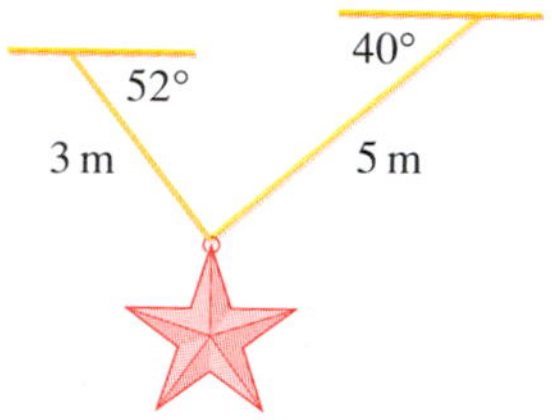

31. If $\mathbf{r} = \langle x, y \rangle$ and $\mathbf{r}_0 = \langle x_0, y_0 \rangle$, describe the set of all points (x, y) such that $|\mathbf{r} - \mathbf{r}_0| = 1$.

32. If $\mathbf{r} = \langle x, y \rangle$, $\mathbf{r}_1 = \langle x_1, y_1 \rangle$, and $\mathbf{r}_2 = \langle x_2, y_2 \rangle$, describe the set of all points (x, y) such that $|\mathbf{r} - \mathbf{r}_1| + |\mathbf{r} - \mathbf{r}_2| = k$, where $k > |\mathbf{r}_1 - \mathbf{r}_2|$.

33. Prove Property 2 of Theorem 6.

34. Prove Property 5 of Theorem 6.

35. Prove Property 6 of Theorem 6.

36. Use vectors to prove that the line joining the midpoints of two sides of a triangle is parallel to the third side and half its length.

37. A quadrilateral has one pair of opposite sides parallel and of equal length. Use vectors to prove that the other pair of opposite sides is parallel and of equal length.

1.2 THE DOT PRODUCT

So far we have added two vectors and multiplied a vector by a scalar. The question arises: Is it possible to multiply two vectors so that their product is a useful quantity? One such product is the dot product, which we consider in this section. Another is the cross product, which is discussed in Chapter 11.

WORK AND THE DOT PRODUCT

An example of a situation in physics and engineering where we need to combine two vectors occurs in calculating the work done by a force. The work done by a constant force F in moving an object through a distance d is $W = Fd$, but this applies only when the force is directed along the line of motion of the object. Suppose, however, that the

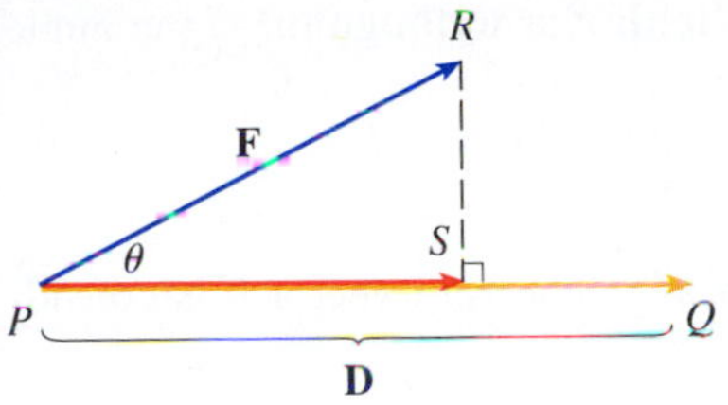

FIGURE 1

constant force is a vector $\mathbf{F} = \overrightarrow{PR}$ pointing in some other direction, as in Figure 1. If the force moves the object from P to Q, then the **displacement vector** is $\mathbf{D} = \overrightarrow{PQ}$. So here we have two vectors: the force $\mathbf{F}$ and the displacement $\mathbf{D}$. The **work** done by $\mathbf{F}$ is defined as the magnitude of the displacement, $|\mathbf{D}|$, multiplied by the magnitude of the applied force in the direction of the motion, which, from Figure 1, is

$$|\overrightarrow{PS}| = |\mathbf{F}|\cos\theta$$

So the work done by $\mathbf{F}$ is defined to be

$$W = |\mathbf{D}|(|\mathbf{F}|\cos\theta) = |\mathbf{F}||\mathbf{D}|\cos\theta \tag{1}$$

Notice that work is a scalar quantity; it has no direction. But its value depends on the angle between the force and displacement vectors.

We use the expression in Equation 1 to define the dot product of two vectors even when they don't represent force or displacement.

(2) DEFINITION The **dot product** of two nonzero vectors **a** and **b** is the number

$$\mathbf{a} \cdot \mathbf{b} = |\mathbf{a}||\mathbf{b}|\cos\theta$$

where θ is the angle between **a** and **b**, $0 \leq \theta \leq \pi$. If either **a** or **b** is **0**, we define $\mathbf{a} \cdot \mathbf{b} = 0$.

This product is called the **dot product** because of the dot in the notation $\mathbf{a} \cdot \mathbf{b}$. The result of computing $\mathbf{a} \cdot \mathbf{b}$ is not a vector. It is a real number, that is, a scalar. For this reason, the dot product is sometimes called the **scalar product**.

In the example of finding the work done by a force $\mathbf{F}$ in moving an object through a displacement $\mathbf{D} = \overrightarrow{PQ}$ by calculating $\mathbf{F} \cdot \mathbf{D} = |\mathbf{F}||\mathbf{D}|\cos\theta$, it makes no sense for the angle θ between $\mathbf{F}$ and $\mathbf{D}$ to be $\pi/2$ or larger because movement from P to Q couldn't take place. We make no such restriction in our general definition of $\mathbf{a} \cdot \mathbf{b}$, however, and allow θ to be any angle from 0 to π.

EXAMPLE 1 If the vectors **a** and **b** have lengths 4 and 6, and the angle between them is $\pi/3$, find $\mathbf{a} \cdot \mathbf{b}$.

SOLUTION According to the definition,

$$\mathbf{a} \cdot \mathbf{b} = |\mathbf{a}||\mathbf{b}|\cos(\pi/3) = 4 \cdot 6 \cdot \tfrac{1}{2} = 12$$ ■

EXAMPLE 2 A crate is hauled 8 m up a ramp under a constant force of 20 N applied at an angle of 25° to the ramp. Find the work done.

SOLUTION If $\mathbf{F}$ and $\mathbf{D}$ are the force and displacement vectors, as pictured in Figure 2, then the work done is

$$\begin{aligned} W &= \mathbf{F} \cdot \mathbf{D} = |\mathbf{F}||\mathbf{D}|\cos 25^\circ \\ &= (20)(8)\cos 25^\circ \approx 145 \text{ N·m} \end{aligned}$$

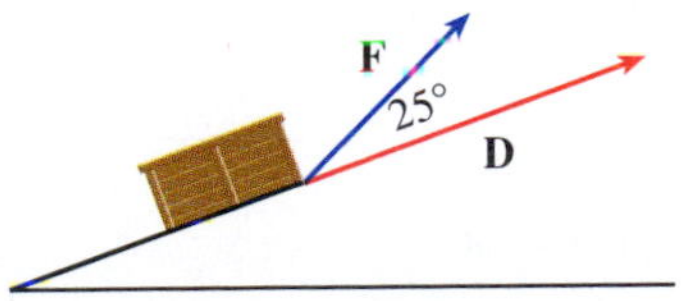

FIGURE 2

In the SI metric system the unit for work is a newton-meter, which is called a joule (J). So we can write $W \approx 145$ J. ■

Two nonzero vectors $\mathbf{a}$ and $\mathbf{b}$ are called **perpendicular** or **orthogonal** if the angle between them is $\theta = \pi/2$. For such vectors we have

$$\mathbf{a} \cdot \mathbf{b} = |\mathbf{a}||\mathbf{b}|\cos(\pi/2) = 0$$

and conversely if $\mathbf{a} \cdot \mathbf{b} = 0$, then $\cos\theta = 0$, so $\theta = \pi/2$. The zero vector $\mathbf{0}$ is considered to be perpendicular to all vectors. Therefore

(3) Two vectors $\mathbf{a}$ and $\mathbf{b}$ are orthogonal if and only if $\mathbf{a} \cdot \mathbf{b} = 0$.

Because $\cos\theta > 0$ if $0 \leqslant \theta < \pi/2$ and $\cos\theta < 0$ if $\pi/2 < \theta \leqslant \pi$, we see that $\mathbf{a} \cdot \mathbf{b}$ is positive for $\theta < \pi/2$ and negative for $\theta > \pi/2$. We can think of $\mathbf{a} \cdot \mathbf{b}$ as measuring the extent to which $\mathbf{a}$ and $\mathbf{b}$ point in the same direction. The dot product $\mathbf{a} \cdot \mathbf{b}$ is positive if $\mathbf{a}$ and $\mathbf{b}$ point in the same general direction, 0 if they are perpendicular, and negative if they point in generally opposite directions. In the extreme case where $\mathbf{a}$ and $\mathbf{b}$ point in exactly the same direction we have $\theta = 0$, so $\cos\theta = 1$ and

$$\mathbf{a} \cdot \mathbf{b} = |\mathbf{a}||\mathbf{b}|$$

If $\mathbf{a}$ and $\mathbf{b}$ point in exactly opposite directions, then $\theta = \pi$ and so $\cos\theta = -1$ and $\mathbf{a} \cdot \mathbf{b} = -|\mathbf{a}||\mathbf{b}|$.

THE DOT PRODUCT IN COMPONENT FORM

Suppose we are given two vectors in component form:

$$\mathbf{a} = \langle a_1, a_2 \rangle \qquad \mathbf{b} = \langle b_1, b_2 \rangle$$

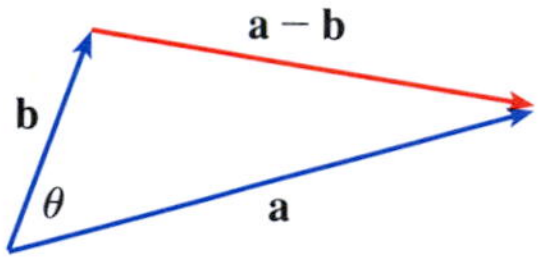

FIGURE 3

We want to find a convenient expression for $\mathbf{a} \cdot \mathbf{b}$ in terms of these components. If we apply the Law of Cosines to the triangle in Figure 3, we get

$$\begin{aligned} |\mathbf{a} - \mathbf{b}|^2 &= |\mathbf{a}|^2 + |\mathbf{b}|^2 - 2|\mathbf{a}||\mathbf{b}|\cos\theta \\ &= |\mathbf{a}|^2 + |\mathbf{b}|^2 - 2\,\mathbf{a} \cdot \mathbf{b} \end{aligned}$$

Solving for the dot product, we obtain

$$\begin{aligned} \mathbf{a} \cdot \mathbf{b} &= \tfrac{1}{2}(|\mathbf{a}|^2 + |\mathbf{b}|^2 - |\mathbf{a} - \mathbf{b}|^2) \\ &= \tfrac{1}{2}[a_1^2 + a_2^2 + b_1^2 + b_2^2 - (a_1 - b_1)^2 - (a_2 - b_2)^2] \\ &= a_1b_1 + a_2b_2 \end{aligned}$$

(4) The dot product of $\mathbf{a} = \langle a_1, a_2 \rangle$ and $\mathbf{b} = \langle b_1, b_2 \rangle$ is

$$\mathbf{a} \cdot \mathbf{b} = a_1b_1 + a_2b_2$$

EXAMPLE 3

$$\langle 2, 4 \rangle \cdot \langle 3, -1 \rangle = 2(3) + 4(-1) = 2$$

$$\langle -1, 7 \rangle \cdot \langle 6, 2 \rangle = (-1)(6) + 7(2) = 8$$

$$(\mathbf{i} + 2\mathbf{j}) \cdot (3\mathbf{i} - \mathbf{j}) = 1(3) + 2(-1) = 1$$

■

EXAMPLE 4 A force is given by a vector $\mathbf{F} = 3\mathbf{i} + 4\mathbf{j}$ and moves a particle from the point $P(2, 1)$ to the point $Q(4, 6)$. Find the work done.

SOLUTION The displacement vector is $\mathbf{D} = \overrightarrow{PQ} = \langle 2, 5 \rangle$ and so the work done is

$$\begin{aligned} W &= \mathbf{F} \cdot \mathbf{D} = \langle 3, 4 \rangle \cdot \langle 2, 5 \rangle \\ &= 6 + 20 = 26 \end{aligned}$$

If the unit of length is meters and the magnitude of the force is measured in newtons, then the work done is 26 J. ■

The dot product obeys many of the laws that hold for ordinary products of real numbers. These are stated in the following theorem.

PROPERTIES OF THE DOT PRODUCT If $\mathbf{a}$, $\mathbf{b}$, and $\mathbf{c}$ are vectors and c is a scalar, then

1. $\mathbf{a} \cdot \mathbf{a} = |\mathbf{a}|^2$
2. $\mathbf{a} \cdot \mathbf{b} = \mathbf{b} \cdot \mathbf{a}$
3. $\mathbf{a} \cdot (\mathbf{b} + \mathbf{c}) = \mathbf{a} \cdot \mathbf{b} + \mathbf{a} \cdot \mathbf{c}$
4. $(c\mathbf{a}) \cdot \mathbf{b} = c(\mathbf{a} \cdot \mathbf{b}) = \mathbf{a} \cdot (c\mathbf{b})$
5. $\mathbf{0} \cdot \mathbf{a} = 0$

Properties 1, 2, and 5 are immediate consequences of the definition of a dot product. Property 3 is best proved using components:

$$\begin{aligned} \mathbf{a} \cdot (\mathbf{b} + \mathbf{c}) &= \langle a_1, a_2 \rangle \cdot \langle b_1 + c_1, b_2 + c_2 \rangle \\ &= a_1(b_1 + c_1) + a_2(b_2 + c_2) \\ &= a_1b_1 + a_1c_1 + a_2b_2 + a_2c_2 \\ &= (a_1b_1 + a_2b_2) + (a_1c_1 + a_2c_2) \\ &= \mathbf{a} \cdot \mathbf{b} + \mathbf{a} \cdot \mathbf{c} \end{aligned}$$

The remaining proofs are left as exercises.

EXAMPLE 5 Find the angle between the vectors $\mathbf{a} = \langle 2, 2 \rangle$ and $\mathbf{b} = \langle 5, -3 \rangle$.

SOLUTION Since

$$|\mathbf{a}| = \sqrt{2^2 + 2^2} = \sqrt{8} \qquad \text{and} \qquad |\mathbf{b}| = \sqrt{5^2 + (-3)^2} = \sqrt{34}$$

and since

$$\mathbf{a} \cdot \mathbf{b} = 2(5) + 2(-3) = 4$$

we have, from the definition of the dot product,

$$\cos\theta = \frac{\mathbf{a} \cdot \mathbf{b}}{|\mathbf{a}||\mathbf{b}|} = \frac{4}{\sqrt{8}\sqrt{34}}$$

So the angle between $\mathbf{a}$ and $\mathbf{b}$ is

$$\theta = \cos^{-1}\left(\frac{4}{\sqrt{8}\sqrt{34}}\right) \approx 1.33 \text{ rad} \quad (\text{or } 76^\circ)$$ ■

EXAMPLE 6 Show that $2\,\mathbf{i} + 3\,\mathbf{j}$ is perpendicular to $3\,\mathbf{i} - 2\,\mathbf{j}$.

SOLUTION Since

$$(2\,\mathbf{i} + 3\,\mathbf{j}) \cdot (3\,\mathbf{i} - 2\,\mathbf{j}) = 2(3) + 3(-2) = 0$$

these vectors are perpendicular by (3). ■

PROJECTIONS

Figure 4 shows representations $\overrightarrow{PQ}$ and $\overrightarrow{PR}$ of two vectors $\mathbf{a}$ and $\mathbf{b}$ with the same initial point P. If S is the foot of the perpendicular from R to the line containing $\overrightarrow{PQ}$, then the vector with representation $\overrightarrow{PS}$ is called the **vector projection** of $\mathbf{b}$ onto $\mathbf{a}$ and is denoted by $\text{proj}_{\mathbf{a}}\,\mathbf{b}$. The **scalar projection** of $\mathbf{b}$ onto $\mathbf{a}$ (also called the **component of b along a**) is defined to be the magnitude of the vector projection, which is the number $|\mathbf{b}|\cos\theta$, where θ is the angle between $\mathbf{a}$ and $\mathbf{b}$ (see Figure 5).

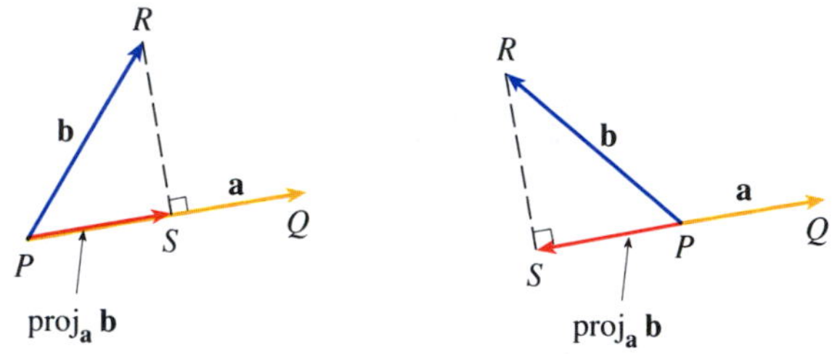

FIGURE 4
Vector projections

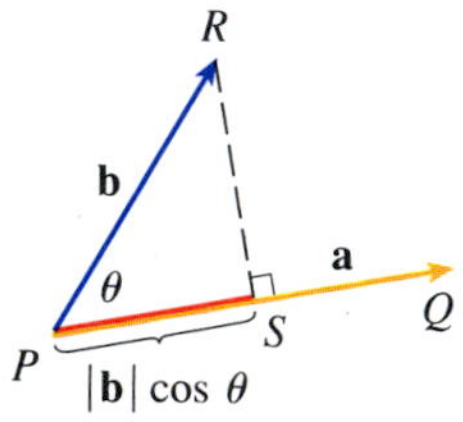

FIGURE 5
Scalar projection

This is denoted by $\text{comp}_{\mathbf{a}}\,\mathbf{b}$. Observe that it is negative if $\pi/2 < \theta \leqslant \pi$. (Note that we used the component of the force $\mathbf{F}$ along the displacement $\mathbf{D}$, $\text{comp}_{\mathbf{D}}\,\mathbf{F}$, at the beginning of this section.) The equation

$$\mathbf{a} \cdot \mathbf{b} = |\mathbf{a}|\,|\mathbf{b}|\cos\theta = |\mathbf{a}|\,(|\mathbf{b}|\cos\theta)$$

shows that the dot product of $\mathbf{a}$ and $\mathbf{b}$ can be interpreted as the length of $\mathbf{a}$ times the scalar projection of $\mathbf{b}$ onto $\mathbf{a}$. Since

$$|\mathbf{b}|\cos\theta = \frac{\mathbf{a} \cdot \mathbf{b}}{|\mathbf{a}|} = \frac{\mathbf{a}}{|\mathbf{a}|} \cdot \mathbf{b}$$

the component of $\mathbf{b}$ along $\mathbf{a}$ can be computed by taking the dot product of $\mathbf{b}$ with the unit vector in the direction of $\mathbf{a}$. To summarize:

Scalar projection of **b** onto **a**:	$\text{comp}_{\mathbf{a}}\,\mathbf{b} = \dfrac{\mathbf{a} \cdot \mathbf{b}}{|\mathbf{a}|}$
Vector projection of **b** onto **a**:	$\text{proj}_{\mathbf{a}}\,\mathbf{b} = \left(\dfrac{\mathbf{a} \cdot \mathbf{b}}{|\mathbf{a}|}\right)\dfrac{\mathbf{a}}{|\mathbf{a}|} = \dfrac{\mathbf{a} \cdot \mathbf{b}}{|\mathbf{a}|^2}\,\mathbf{a}$

EXAMPLE 7 Find the scalar projection and vector projection of $\mathbf{b} = \langle 1, 2\rangle$ onto $\mathbf{a} = \langle -2, 3\rangle$.

SOLUTION Since $|\mathbf{a}| = \sqrt{(-2)^2 + 3^2} = \sqrt{13}$, the scalar projection of $\mathbf{b}$ onto $\mathbf{a}$ is

$$\text{comp}_{\mathbf{a}}\,\mathbf{b} = \frac{\mathbf{a} \cdot \mathbf{b}}{|\mathbf{a}|} = \frac{(-2)(1) + 3(2)}{\sqrt{13}} = \frac{4}{\sqrt{13}}$$

The vector projection is this scalar projection times the unit vector in the direction of **a**:

$$\text{proj}_{\mathbf{a}}\mathbf{b} = \frac{4}{\sqrt{13}}\frac{\mathbf{a}}{|\mathbf{a}|} = \frac{4}{\sqrt{13}}\frac{\langle -2, 3\rangle}{\sqrt{13}} = \left\langle -\frac{8}{13}, \frac{12}{13}\right\rangle$$ ■

EXAMPLE 8 A force **F** has magnitude 300 N and acts in the direction N30°E (30° east of north). Find the vector projection of **F** in the direction N60°E.

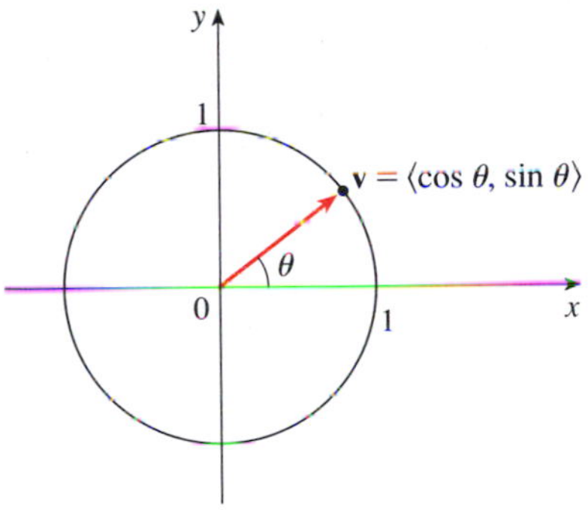

FIGURE 6

SOLUTION A general unit vector is $\mathbf{v} = \langle \cos\theta, \sin\theta\rangle$, where θ is the angle from the positive x-axis to **v** (see Figure 6). So the unit vector in the direction N60°E (for which $\theta = 30°$) is

$$\mathbf{a} = \langle \cos 30°, \sin 30°\rangle = \left\langle \frac{\sqrt{3}}{2}, \frac{1}{2}\right\rangle$$

The magnitude of the force vector **F** is 300 N and the angle for this vector is $\theta = 60°$, so

$$\mathbf{F} = 300\langle \cos 60°, \sin 60°\rangle = \langle 150, 150\sqrt{3}\,\rangle$$

The vector projection in the direction N60°E is

$$\text{proj}_{\mathbf{a}}\mathbf{F} = \frac{\mathbf{a}\cdot\mathbf{F}}{|\mathbf{a}|^2}\mathbf{a} = (\mathbf{a}\cdot\mathbf{F})\mathbf{a} \qquad (\text{since } |\mathbf{a}| = 1)$$

We have

$$\mathbf{a}\cdot\mathbf{F} = \left\langle \frac{\sqrt{3}}{2}, \frac{1}{2}\right\rangle \cdot \langle 150, 150\sqrt{3}\,\rangle$$

$$= \frac{\sqrt{3}}{2}\cdot 150 + \frac{1}{2}\cdot 150\sqrt{3} = 150\sqrt{3}$$

So

$$\text{proj}_{\mathbf{a}}\mathbf{F} = (\mathbf{a}\cdot\mathbf{F})\mathbf{a} = 150\sqrt{3}\left\langle \frac{\sqrt{3}}{2}, \frac{1}{2}\right\rangle$$

$$= \langle 225, 75\sqrt{3}\,\rangle$$ ■

ORTHOGONAL COMPLEMENTS

Given a vector **a** in the plane, it is sometimes useful to construct a vector that is perpendicular to **a** and has the same length as **a**. A simple sketch will show that if **a** is a nonzero vector, then there are two vectors that are pendicular to **a** and have the same length as **a**. One such vector is denoted by $\mathbf{a}^{\perp}$.

> **(5) DEFINITION** Given the nonzero vector $\mathbf{a} = \langle a_1, a_2\rangle$, the **orthogonal complement** of **a** is the vector $\mathbf{a}^{\perp} = \langle -a_2, a_1\rangle$.

It is easy to check that $\mathbf{a}^{\perp}$ is perpendicular to **a** and has the same length as **a** as follows:

$$\mathbf{a}^{\perp}\cdot\mathbf{a} = \langle -a_2, a_1\rangle \cdot \langle a_1, a_2\rangle = -a_2a_1 + a_1a_2 = 0$$

and $$|\mathbf{a}^{\perp}| = \sqrt{(-a_2)^2 + (a_1)^2} = \sqrt{a_1^2 + a_2^2} = |\mathbf{a}|$$

Orthogonal complements are sometimes used to help compute distances.

EXAMPLE 9 Find the distance from the point $(3, 1)$ to the line $x + y = 1$.

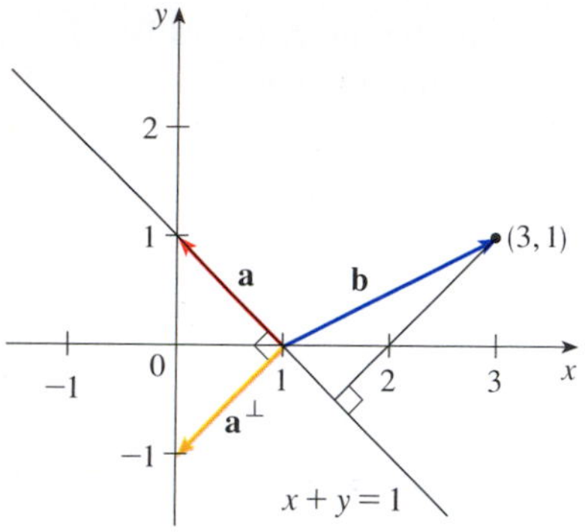

FIGURE 7

SOLUTION Note that the points $(0, 1)$ and $(1, 0)$ lie on the line $x + y = 1$ (see Figure 7). Consequently, the displacement vector

$$\mathbf{a} = \langle 0, 1\rangle - \langle 1, 0\rangle = \langle -1, 1\rangle$$

is parallel to this line, and as a result

$$\mathbf{a}^{\perp} = \langle -1, -1\rangle$$

is perpendicular to this line.

Thus, if we form the displacement vector

$$\mathbf{b} = \langle 3, 1\rangle - \langle 1, 0\rangle = \langle 2, 1\rangle$$

then $|\text{comp}_{\mathbf{a}^{\perp}} \mathbf{b}|$ gives the distance from the point $(3, 1)$ to the line $x + y = 1$. Since

$$\text{comp}_{\mathbf{a}^{\perp}} \mathbf{b} = \frac{\mathbf{a}^{\perp} \cdot \mathbf{b}}{|\mathbf{a}^{\perp}|} = \frac{(-1)(2) + (-1)(1)}{\sqrt{(-1)^2 + (-1)^2}}$$

$$= \frac{-3}{\sqrt{2}}$$

the distance is $|-3/\sqrt{2}| = 3/\sqrt{2}$. ■

EXERCISES 1.2

1. Which of the following expressions are meaningful? Which are meaningless? Explain.
(a) $(\mathbf{a} \cdot \mathbf{b}) \cdot \mathbf{c}$
(b) $(\mathbf{a} \cdot \mathbf{b})\mathbf{c}$
(c) $|\mathbf{a}|(\mathbf{b} \cdot \mathbf{c})$
(d) $\mathbf{a} \cdot (\mathbf{b} + \mathbf{c})$
(e) $\mathbf{a} \cdot \mathbf{b} + \mathbf{c}$
(f) $|\mathbf{a}| \cdot (\mathbf{b} + \mathbf{c})$

2. Find the dot product of two vectors if their lengths are 6 and $\frac{1}{3}$ and the angle between them is $\pi/4$.

3–8 ■ Find $\mathbf{a} \cdot \mathbf{b}$.

3. $|\mathbf{a}| = 2$, $|\mathbf{b}| = 3$, the angle between $\mathbf{a}$ and $\mathbf{b}$ is $\pi/3$

4. $\mathbf{a} = \langle -2, -8\rangle$, $\mathbf{b} = \langle 6, -4\rangle$

5. $\mathbf{a} = \langle 4, 7\rangle$, $\mathbf{b} = \langle -2, 1\rangle$

6. $\mathbf{a} = \langle -1, -2\rangle$, $\mathbf{b} = \langle 2, 8\rangle$

7. $\mathbf{a} = 2\mathbf{i} + 3\mathbf{j}$, $\mathbf{b} = \mathbf{i} - 3\mathbf{j}$

8. $\mathbf{a} = \mathbf{i} - \mathbf{j}$, $\mathbf{b} = \mathbf{i} + 2\mathbf{j}$

9–10 ■ If $\mathbf{u}$ is a unit vector, find $\mathbf{u} \cdot \mathbf{v}$ and $\mathbf{u} \cdot \mathbf{w}$.

9.

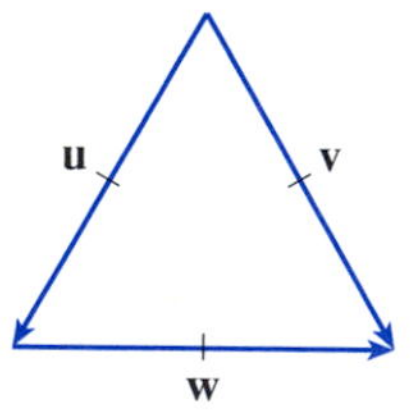

10.

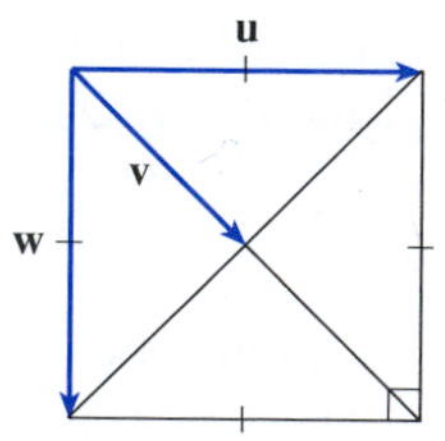

11. (a) Show that $\mathbf{i} \cdot \mathbf{j} = 0$.
(b) Show that $\mathbf{i} \cdot \mathbf{i} = \mathbf{j} \cdot \mathbf{j} = 1$.

12. A street vendor sells a hamburgers and b hot dogs on a given day. He charges \$2 for a hamburger and \$1.50 for a hot dog. If $\mathbf{A} = \langle a, b \rangle$ and $\mathbf{P} = \langle 2, 1.5 \rangle$, what is the meaning of the dot product $\mathbf{A} \cdot \mathbf{P}$?

13–18 ■ Find the angle between the vectors. (First find an exact expression and then approximate to the nearest degree.)

13. $\mathbf{a} = \langle 1, 2 \rangle, \quad \mathbf{b} = \langle 3, 4 \rangle$

14. $\mathbf{a} = \langle 6, 0 \rangle, \quad \mathbf{b} = \langle 5, 3 \rangle$

15. $\mathbf{a} = \langle 1, 2 \rangle, \quad \mathbf{b} = \langle 12, -5 \rangle$

16. $\mathbf{a} = \langle 3, 1 \rangle, \quad \mathbf{b} = \langle 2, 4 \rangle$

17. $\mathbf{a} = 6\mathbf{i} - 2\mathbf{j}, \quad \mathbf{b} = \mathbf{i} + \mathbf{j}$

18. $\mathbf{a} = \mathbf{i} + \mathbf{j}, \quad \mathbf{b} = 2\mathbf{j}$

19–20 ■ Find, correct to the nearest degree, the three angles of the triangle with the given vertices.

19. $A(1, 2), \quad B(6, 1), \quad C(-1, -2)$

20. $P(0, 1), \quad Q(2, 1), \quad R(5, 4)$

21–26 ■ Determine whether the given vectors are orthogonal, parallel, or neither.

21. $\mathbf{a} = \langle 2, -4 \rangle, \quad \mathbf{b} = \langle -1, 2 \rangle$

22. $\mathbf{a} = \langle 2, 1 \rangle, \quad \mathbf{b} = \langle 4, 2 \rangle$

23. $\mathbf{a} = \langle 2, 8 \rangle, \quad \mathbf{b} = \langle -1, 2 \rangle$

24. $\mathbf{a} = \langle -1, 5 \rangle, \quad \mathbf{b} = \langle 4, 2 \rangle$

25. $\mathbf{a} = 3\mathbf{i} + \mathbf{j}, \quad \mathbf{b} = -3\mathbf{i} + 9\mathbf{j}$

26. $\mathbf{a} = 2\mathbf{i} + 6\mathbf{j}, \quad \mathbf{b} = -3\mathbf{i} + \mathbf{j}$

27–30 ■ Find the values of x such that the given vectors are orthogonal.

27. $x\mathbf{i} - 2\mathbf{j}, \quad x\mathbf{i} + 8\mathbf{j}$

28. $x\mathbf{i} + 2x\mathbf{j}, \quad x\mathbf{i} - 2\mathbf{j}$

29. $\langle x, 1 \rangle, \quad \langle 4, x \rangle$

30. $\langle x, x \rangle, \quad \langle 1, x \rangle$

31. Find a unit vector that is orthogonal to $\mathbf{i} + 3\mathbf{j}$.

32. For what values of c is the angle between the vectors $\langle 1, 1 \rangle$ and $\langle 1, c \rangle$ equal to $60°$?

33–38 ■ Find the scalar and vector projections of **b** onto **a**.

33. $\mathbf{a} = \langle 2, 3 \rangle, \quad \mathbf{b} = \langle 4, 1 \rangle$

34. $\mathbf{a} = \langle 3, -1 \rangle, \quad \mathbf{b} = \langle 2, 3 \rangle$

35. $\mathbf{a} = \langle 4, 2 \rangle, \quad \mathbf{b} = \langle 1, 1 \rangle$

36. $\mathbf{a} = \langle -1, -2 \rangle, \quad \mathbf{b} = \langle 3, 3 \rangle$

37. $\mathbf{a} = \mathbf{i}, \quad \mathbf{b} = \mathbf{i} - \mathbf{j}$

38. $\mathbf{a} = 2\mathbf{i} - 3\mathbf{j}, \quad \mathbf{b} = \mathbf{i} + 6\mathbf{j}$

39–42 ■ Find $\text{proj}_{\mathbf{a}^\perp} \mathbf{b}$.

39. $\mathbf{a} = \langle 2, 1 \rangle, \quad \mathbf{b} = \langle 1, 1 \rangle$

40. $\mathbf{a} = \langle 2, 1 \rangle, \quad \mathbf{b} = \langle 1, -1 \rangle$

41. $\mathbf{a} = \langle -2, 3 \rangle, \quad \mathbf{b} = \langle 0, 2 \rangle$

42. $\mathbf{a} = \langle -3, 2 \rangle, \quad \mathbf{b} = \langle 1, 2 \rangle$

43. Find the distance from the point $(3, 7)$ to the line $y = 2x$.

44. Find the distance from the point $(0, 3)$ to the line $y = -3x$.

45. Find the distance from the point $(-2, 3)$ to the line $3x - 4y + 5 = 0$.

46. Find the distance from the point $(0, 4)$ to the line $2x + 5y = -3$.

47. Find the distance between the parallel lines $y = 2x + 3$ and $y - 2x = 9$.

48. Find the distance between the parallel lines $y = -4x + 3$ and $y = -4x + 12$.

49. Show that if **a** and **b** are nonzero vectors, then the vector

$$\text{orth}_{\mathbf{a}} \mathbf{b} = \mathbf{b} - \text{proj}_{\mathbf{a}} \mathbf{b}$$

is orthogonal to **a**. This vector is referred to as an **orthogonal projection** of **b** onto **a**. How is the vector $\text{orth}_{\mathbf{a}} \mathbf{b}$ related to the vector $\text{proj}_{\mathbf{a}^\perp} \mathbf{b}$?

50. For the vectors in Exercise 34, find $\text{orth}_{\mathbf{a}} \mathbf{b}$ and illustrate by drawing the vectors **a**, **b**, $\text{proj}_{\mathbf{a}} \mathbf{b}$, and $\text{orth}_{\mathbf{a}} \mathbf{b}$.

51. If $\mathbf{a} = \langle 3, -1 \rangle$, find a vector **b** such that $\text{comp}_{\mathbf{a}} \mathbf{b} = 2$.

52. Suppose that **a** and **b** are nonzero vectors.
(a) Under what circumstances is $\text{comp}_{\mathbf{a}} \mathbf{b} = \text{comp}_{\mathbf{b}} \mathbf{a}$?
(b) Under what circumstances is $\text{proj}_{\mathbf{a}} \mathbf{b} = \text{proj}_{\mathbf{b}} \mathbf{a}$?

53. A constant force with vector representation $\mathbf{F} = 10\mathbf{i} + 18\mathbf{j}$ moves an object along a straight line from the point $(2, 3)$ to the point $(4, 9)$. Find the work done if the distance is measured in meters and the magnitude of the force is measured in newtons.

54. Find the work done by a force of 20 lb acting in the direction N50°W in moving an object 4 ft due west.

55. A woman exerts a horizontal force of 25 lb on a crate as she pushes it up a ramp that is 10 ft long and inclined at an angle of 20° above the horizontal. Find the work done on the box.

56. A wagon is pulled a distance of 100 m along a horizontal path by a constant force of 50 N. The handle of the wagon is at an angle of 30° above the horizontal. How much work is done?

57. (a) Use a scalar projection to show that the distance from a point $P_1(x_1, y_1)$ to the line with equation $ax + by + c = 0$ is

$$\frac{|ax_1 + by_1 + c|}{\sqrt{a^2 + b^2}}$$

(b) Use this formula to find the distance from the point $(-2, 3)$ to the line $3x - 4y + 5 = 0$.

58. If $\mathbf{r} = \langle x, y \rangle$, $\mathbf{a} = \langle a_1, a_2 \rangle$, and $\mathbf{b} = \langle b_1, b_2 \rangle$, show that the vector equation $(\mathbf{r} - \mathbf{a}) \cdot (\mathbf{r} - \mathbf{b}) = 0$ represents a circle, and find its center and radius.

59. Prove Properties 2 and 5 of the dot product.

60. Prove Property 4 of the dot product.

61. Use the definition of the dot product to prove the Cauchy-Schwarz Inequality:

$$|\mathbf{a} \cdot \mathbf{b}| \leq |\mathbf{a}||\mathbf{b}|$$

62. The Triangle Inequality for vectors is $|\mathbf{a} + \mathbf{b}| \leq |\mathbf{a}| + |\mathbf{b}|$.

(a) Give a geometric interpretation of the Triangle Inequality.

(b) Use the Cauchy-Schwarz Inequality from Exercise 61 to prove the Triangle Inequality. [*Hint:* Use the fact that $|\mathbf{a} + \mathbf{b}|^2 = (\mathbf{a} + \mathbf{b}) \cdot (\mathbf{a} + \mathbf{b})$ and use Property 3 of the dot product.]

63. The Parallelogram Law states that

$$|\mathbf{a} + \mathbf{b}|^2 + |\mathbf{a} - \mathbf{b}|^2 = 2|\mathbf{a}|^2 + 2|\mathbf{b}|^2$$

(a) Give a geometric interpretation of the Parallelogram Law.

(b) Prove the Parallelogram Law. (See the hint in Exercise 62.)

1.3 VECTOR FUNCTIONS

PARAMETRIC CURVES

In Review and Preview we discussed functions of the form $y = f(x)$ and their graphs. For some applications a more convenient way to present a curve in the plane is by describing its x- and y-coordinates as functions of some other variable, say t. Curves of this type lead us to a discussion of parametric curves and vector functions. To motivate this discussion, we consider the following example.

EXAMPLE 1 Suppose a cannonball is fired from a cannon with a muzzle velocity of 1000 ft/s. Suppose that the cannon is fired at an angle of $\pi/3$ radians, or 60°, as measured from the horizontal (see Figure 1).

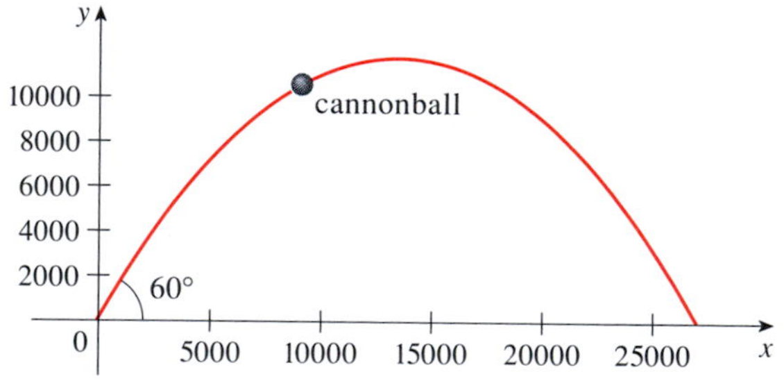

FIGURE 1

The position (x, y) of the cannonball at time t seconds after the cannon is fired is given by $(x, y) = (x(t), y(t))$, where

$$x(t) = 500t \qquad y(t) = 500\sqrt{3}\,t - 16t^2$$

We have assumed that the cannon is located at the origin and we have ignored air resistance.

A detailed derivation of this formula for the position of the cannonball will be given later in the text. Here we give a rough idea of its derivation. At time $t = 0$ the velocity vector for the cannonball has length 1000 and points in the direction of the line that makes an angle of $\pi/3$ radians with the positive x-axis. The horizontal component of this velocity vector is $1000 \cos(\pi/3) = 500$ (feet per second). Therefore, after t seconds have elapsed the horizontal position of the cannonball will have moved a distance of $500t$ feet, and hence the formula $x(t) = 500t$. Similarly, the vertical component of the velocity at time $t = 0$ is $1000 \sin(\pi/3) = 500\sqrt{3}$, which leads to the term $500\sqrt{3}\,t$ in the expression for $y(t)$. The term $-16t^2$ is present because of gravity, which gradually pulls the cannonball back to Earth.

The following table gives positions of the cannonball at various values of t:

t (in seconds)	Position of cannonball (coordinates in feet)
0	(0, 0)
1	(500, 850)
10	(5000, 7060)
20	(10000, 10921)
30	(15000, 11581)
40	(20000, 9041)
50	(25000, 3301)
54.13	(27063, 0)

Note that at $t = 54.13$ s the cannonball has returned to Earth (its y-coordinate is zero). The corresponding x-coordinate is 27,063 ft, which represents the horizontal distance traveled. ■

A curve of the type $x = x(t)$, $y = y(t)$, as in the example of the cannonball, is called a **parametric curve** and the variable t is called a **parameter**. In some cases the parameter t has physical significance, such as time in Example 1. In other cases the parameter has no physical meaning.

VECTOR FUNCTIONS

For each value of the parameter t we may view the point $(x(t), y(t))$ on a parametric curve as the endpoint of the vector

$$\mathbf{r}(t) = \langle x(t), y(t) \rangle$$

or

$$\mathbf{r}(t) = x(t)\,\mathbf{i} + y(t)\,\mathbf{j}$$

which begins at the origin and ends at the point $(x(t), y(t))$. When viewed this way, $\mathbf{r}(t)$ is called the *position vector* for the point with coordinates $(x(t), y(t))$. Figure 2 shows a position vector for the function in Example 1.

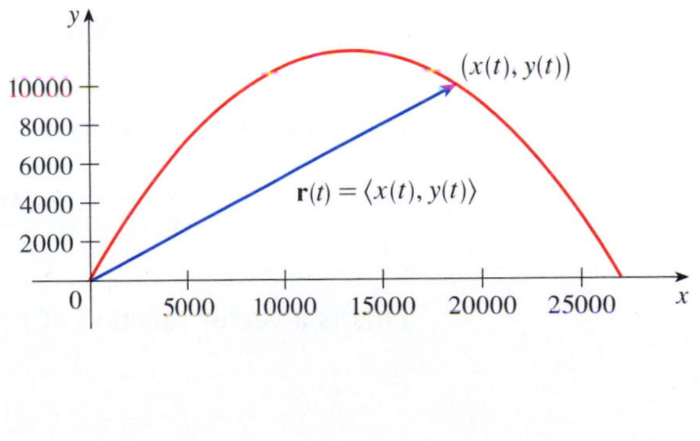

FIGURE 2
$x(t) = 500t$, $y(t) = 500\sqrt{3}t - 16t^2$

A function such as **r**, whose range is a set of vectors, is called a **vector function** (or **vector-valued function**) of t.

EXAMPLE 2 The vector function that represents the position vector of the cannonball in Example 1 is

$$\mathbf{r}(t) = \langle 500t, 500\sqrt{3}\,t - 16t^2 \rangle$$

■

EXAMPLE 3 Sketch the curve defined by the vector function

$$\mathbf{r}(t) = (t^2 - 2t)\,\mathbf{i} + (t + 1)\,\mathbf{j}$$

SOLUTION The endpoint of the vector $\mathbf{r}(t)$ is (x, y) where $x = t^2 - 2t$ and $y = t + 1$. Each value of t gives a point on the curve, as shown in the following table. For instance, if $t = 0$, then $x = 0$, $y = 1$ and so the corresponding point is $(0, 1)$. In Figure 3, we plot the points (x, y) determined by several values of the parameter and join them to produce a curve.

t	x	y
-2	8	-1
-1	3	0
0	0	1
1	-1	2
2	0	3
3	3	4
4	8	5

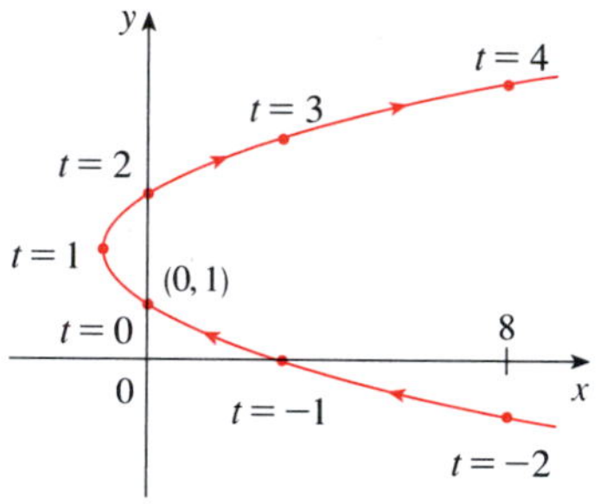

FIGURE 3

■

EXAMPLE 4 The position of a satellite of mass m at time t is $(2t, t^2)$. A planet of mass M is located at $(0, 1)$. Find the force of gravity at time t that is exerted by the planet on the satellite.

SOLUTION The gravitational force exerted on a mass m by another mass M is a vector that points from m to M and has magnitude GmM/r^2, where G is the gravitational constant and r is the distance from m to M. Since the satellite is located at $(2t, t^2)$ and the planet is located at $(0, 1)$, the distance between them is

$$r = \sqrt{(0 - 2t)^2 + (1 - t^2)^2} = t^2 + 1$$

A unit vector that points from the satellite to the planet is

$$\mathbf{R}(t) = \frac{\langle 0, 1 \rangle - \langle 2t, t^2 \rangle}{r} = \frac{\langle -2t, 1 - t^2 \rangle}{r}$$

Thus, the force of gravity at time t is

$$\mathbf{F}(t) = \frac{GmM}{r^2}\mathbf{R}(t) = GmM\left\langle \frac{-2t}{(t^2 + 1)^3}, \frac{1 - t^2}{(t^2 + 1)^3} \right\rangle$$

This is a vector function of t. ■

Parametric equations are often used to describe curves that are difficult or impossible to describe in the form $y = f(x)$. Consider the following example, where our parameter is θ instead of t.

EXAMPLE 5 The curve

$$x = a\cos\theta \qquad y = a\sin\theta$$

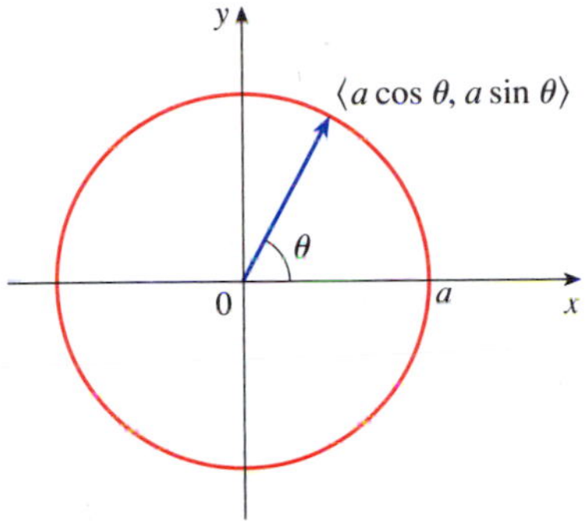

FIGURE 4
$x = a\cos\theta,\ y = a\sin\theta$

describes a circle of radius a. In fact, these equations can be transformed into

$$x^2 + y^2 = a^2(\cos^2\theta + \sin^2\theta) = a^2$$

which is the Cartesian form of the equation of a circle that is centered at the origin and has radius a. Figure 4 shows the graph of this equation.

In this case, the parameter θ represents the angle formed by the position vector $\langle a\cos\theta, a\sin\theta\rangle$ and the positive x-axis. As θ varies from 0 to 2π, the point $(a\cos\theta, a\sin\theta)$ traces out the circle in a counterclockwise direction. A circle cannot be described by an equation of the form $y = f(x)$ because vertical lines intersect the circle at more than one point. ■

It is important to realize that a curve has many parametric representations. For example, the circle in Example 5 can also be parametrized by

$$x = a\cos 2t \qquad y = a\sin 2t$$

In this parametrization the point $(x, y) = (a\cos 2t, a\sin 2t)$ traces the same circle as before. However, the point (x, y) makes a complete revolution around the circle as t varies from 0 to π, which is twice as fast as in our original parametrization.

Any curve of the form $y = f(x)$ can be put into parametric form by letting $x = t$ and then $y = f(t)$. The parametrization $x = t^3$, $y = f(t^3)$ is a different parametrization that describes the same curve $y = f(x)$. The parametrization $x = t^2$, $y = f(t^2)$ describes that portion of the graph of $y = f(x)$ with $x \geqslant 0$. For example, the parametrizations $x = t$, $y = t^2$ and $x = t^3$, $y = t^6$ both describe the parabola $y = x^2$. The parametrization $x = t^2$, $y = t^4$ describes the right half of this parabola: $y = x^2$ with $x \geqslant 0$.

VECTOR EQUATION OF A LINE

A line is determined by a point P_0 and a direction. The direction of a line in the plane is determined by specifying a slope or by specifying a vector that is parallel to the line. The latter case leads to the vector equation of a line. Let $\mathbf{v}$ be a vector parallel to a line L. Let P be an arbitrary point on L and let $\mathbf{r}_0$ and $\mathbf{r}$ be the position vectors of P_0 and P (so $\mathbf{r}_0 = \overrightarrow{OP_0}$ and $\mathbf{r} = \overrightarrow{OP}$ in Figure 5). If $\mathbf{a}$ is the vector with representation $\overrightarrow{P_0P}$, as in Figure 5, then the Triangle Law gives $\mathbf{r} = \mathbf{r}_0 + \mathbf{a}$. But since $\mathbf{a}$ and $\mathbf{v}$ are parallel, there is a scalar t such that $\mathbf{a} = t\mathbf{v}$. Thus, we can write the equation of L as follows.

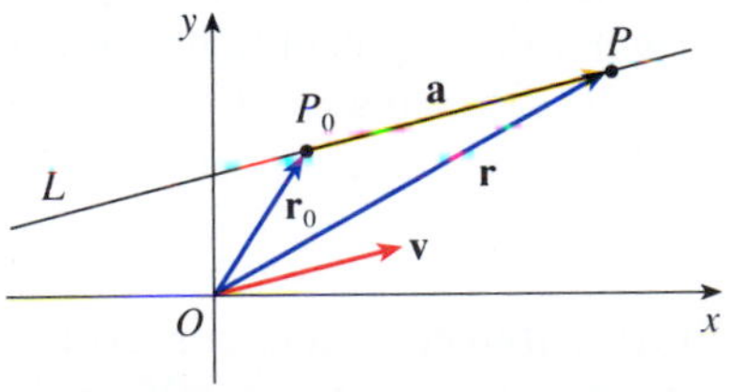

FIGURE 5

(1) VECTOR EQUATION OF A LINE

$$\mathbf{r}(t) = \mathbf{r}_0 + t\mathbf{v}$$

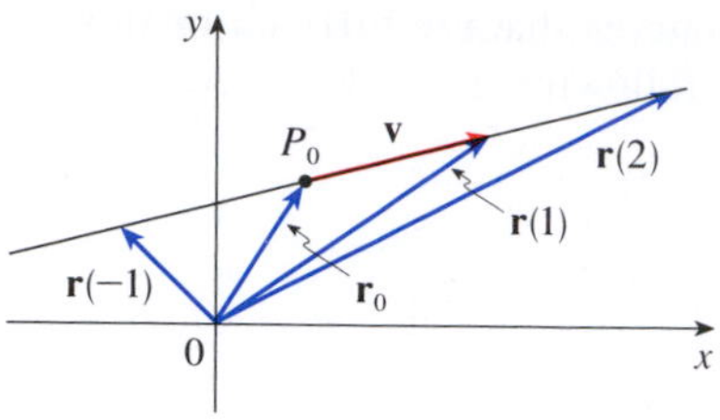

FIGURE 6

As Figure 6 shows, the endpoints of the vectors $\mathbf{r}(0) = \mathbf{r}_0$, $\mathbf{r}(1) = \mathbf{r}_0 + \mathbf{v}$, $\mathbf{r}(2) = \mathbf{r}_0 + 2\mathbf{v}$, and $\mathbf{r}(-1) = \mathbf{r}_0 - \mathbf{v}$ all lie along the line that passes through the point P_0 and is parallel to $\mathbf{v}$. When $t > 0$ the endpoint of $\mathbf{r}(t)$ lies on the ray that points in the direction of $\mathbf{v}$, whereas if $t < 0$ the endpoint of $\mathbf{r}(t)$ lies on the ray that points in the direction opposite that of $\mathbf{v}$.

If $\mathbf{r}_0 = \langle x_0, y_0 \rangle$ and $\mathbf{v} = \langle a, b \rangle$, then the vector equation of a line can be written parametrically as follows:

$$\begin{aligned} \langle x(t), y(t) \rangle &= \mathbf{r}(t) = \mathbf{r}_0 + t\mathbf{v} \\ &= \langle x_0, y_0 \rangle + t\langle a, b \rangle = \langle x_0 + at, y_0 + bt \rangle \end{aligned}$$

By equating the components, we have the following parametric equations of a line.

(2) Parametric equations of the line that passes through the point $P_0(x_0, y_0)$ and is parallel to the vector $\langle a, b \rangle$ are given by

$$x(t) = x_0 + at \qquad y(t) = y_0 + bt$$

Note that if $\mathbf{v}$ is parallel to the line, then any multiple of $\mathbf{v}$ is also parallel to the line and can be used to obtain a vector equation or parametrized equations of a line.

EXAMPLE 6 Find parametric equations of the line that passes through the points $(3, -2)$ and $(5, 7)$.

SOLUTION First we find a vector that is parallel to the line. One such vector is found by taking the difference between the two given points (vectors) that lie on the line, that is,

$$\mathbf{v} = \langle 5, 7 \rangle - \langle 3, -2 \rangle = \langle 2, 9 \rangle$$

Any nonzero multiple of $\mathbf{v}$ can also be used. We can use any point lying on the line, for example $(3, -2)$, for the point P_0. The vector equation of the line becomes

$$\begin{aligned} \mathbf{r}(t) &= \mathbf{r}_0 + t\mathbf{v} \\ &= \langle 3, -2 \rangle + t\langle 2, 9 \rangle \end{aligned}$$

or, equivalently, $$\mathbf{r}(t) = (3 + 2t)\,\mathbf{i} + (-2 + 9t)\,\mathbf{j}$$

Parametric equations of this line are

$$x(t) = 3 + 2t \qquad y(t) = -2 + 9t$$ ■

NOTE: In Example 6, if we restrict the parameter t to $0 \leqslant t \leqslant 1$, then the parametric equations $x(t) = 3 + 2t$, $y(t) = -2 + 9t$ represent a line segment that starts at $(3, -2)$ and ends at $(5, 7)$. Different parametrizations of the same line can be obtained by choosing P_0 to be $(5, 7)$ or any other point on the line. Likewise, we can use any multiple of $\mathbf{v}$, such as $\langle 4, 18 \rangle$, $\langle 1, 4.5 \rangle$, or $\langle -6, -27 \rangle$.

CONVERTING BETWEEN CARTESIAN EQUATIONS AND PARAMETRIZED EQUATIONS

In some situations we may find it useful to eliminate the parameter t and convert the parametric equations of a curve to an equation involving only x and y; this form is called a *Cartesian equation*. In Example 5 we converted the equations of a circle to a Cartesian equation. For the case of a line, the elimination of t is even easier.

EXAMPLE 7 Identify the Cartesian equation of the parametrized line given in Example 6.

SOLUTION Parametric equations of the line are given by

$$x = 3 + 2t \qquad y = -2 + 9t$$

Solving the first equation for t gives $t = (x - 3)/2$. Substituting this expression into the second equation gives

$$y = -2 + 9\left(\frac{x-3}{2}\right)$$

$$2y = -4 + 9x - 27$$

or

$$9x - 2y = 31$$

Note that both of the original points, $(3, -2)$ and $(5, 7)$, satisfy this equation. ■

EXAMPLE 8 Identify the curve traced by the cannonball in Example 1.

SOLUTION In Example 1 we found that the position of the cannonball at time t is given by the parametric equations

$$x = 500t \qquad y = 500t\sqrt{3} - 16t^2$$

Solving the first equation for t in terms of x, we obtain $t = x/500$. Substituting $t = x/500$ into the expression for y, we obtain

$$y = 500\frac{x}{500}\sqrt{3} - 16\left(\frac{x}{500}\right)^2$$

$$= \sqrt{3}\,x - \frac{16x^2}{250{,}000}$$

So the path of the cannonball is a parabola that opens downward. ■

EXAMPLE 9 Identify the Cartesian equation of the curve of Example 3, shown in Figure 3.

SOLUTION Parametric equations of the curve in Example 3 are $x = t^2 - 2t$, $y = t + 1$. From the second equation, we obtain $t = y - 1$. Substituting into the first equation gives

$$x = (y - 1)^2 - 2(y - 1) = y^2 - 4y + 3$$

So the Cartesian equation for this curve is $x = y^2 - 4y + 3$, which we recognize as the equation of a parabola. ■

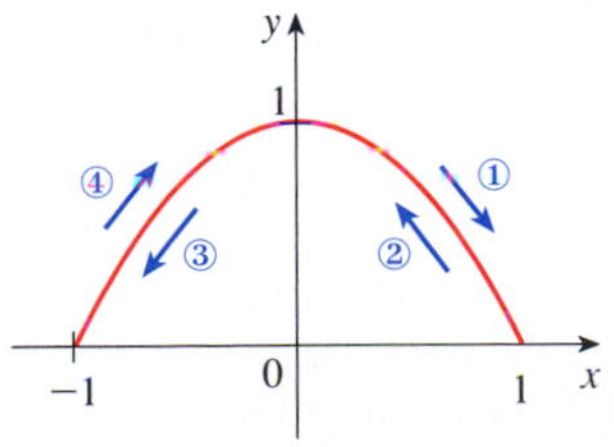

FIGURE 7
$y = 1 - x^2, \ -1 \leqslant x \leqslant 1$

EXAMPLE 10 Identify the curve given in parametric form by

$$x = \sin t \qquad y = \cos^2 t \qquad 0 \leqslant t \leqslant 2\pi$$

SOLUTION We use a trigonometric identity to express y in terms of x:

$$y = \cos^2 t = 1 - \sin^2 t = 1 - x^2$$

As t varies from 0 to 2π, $x = \sin t$ varies from 0 to 1, then to -1, and then back to 0. Therefore, the curve is the portion of the parabola $y = 1 - x^2$ with $-1 \leqslant x \leqslant 1$ (see Figure 7). ■

EXERCISES 1.3

1–13 ■
(a) Sketch the curve represented by the parametric equations.
(b) Eliminate the parameter to find the Cartesian equation of the curve.

1. $x = 1 - t, \quad y = 2 + 3t$

2. $x = 2t - 1, \quad y = 2 - t, \quad -3 \leq t \leq 3$

3. $x = 3t^2, \quad y = 2 + 5t, \quad 0 \leq t \leq 2$

4. $x = 2t - 1, \quad y = t^2 - 1$

5. $x = \sqrt{t}, \quad y = 1 - t$

6. $x = \sin\theta, \quad y = \cos\theta, \quad 0 \leq \theta \leq \pi$

7. $x = 3\cos\theta, \quad y = 2\sin\theta, \quad 0 \leq \theta \leq 2\pi$

8. $x = \sin^2\theta, \quad y = \cos^2\theta$

9. $x = \sec\theta, \quad y = \tan\theta, \quad -\pi/2 < \theta < \pi/2$

10. $x = \cos^2 t, \quad y = \cos^4 t$

11. $x = \cos t, \quad y = \cos 2t$

12. $x = \dfrac{1 - t^2}{1 + t^2}, \quad y = \dfrac{2t}{1 + t^2}$

13. $x = \dfrac{1 - t}{1 + t}, \quad y = t^2$

14–19 ■ Describe the motion of a particle with position (x, y) or $\mathbf{r}(t)$ as t varies in the given interval.

14. $x = \cos \pi t, \quad y = \sin \pi t, \quad 1 \leq t \leq 2$

15. $x = 2 + \cos t, \quad y = 3 + \sin t, \quad 0 \leq t \leq 2\pi$

16. $\mathbf{r}(t) = (8t - 3)\,\mathbf{i} + (2 - t)\,\mathbf{j}, \quad 0 \leq t \leq 1$

17. $\mathbf{r}(t) = (\cos^2 t)\,\mathbf{i} + (\cos t)\,\mathbf{j}, \quad 0 \leq t \leq 4\pi$

18. $\mathbf{r}(t) = \langle 2\sin t, 3\cos t\rangle, \quad 0 \leq t \leq 2\pi$

19. $\mathbf{r}(t) = \langle \sin t, \csc t\rangle, \quad \pi/6 \leq t \leq 1$

20–23 ■ Sketch the curve represented by the parametric equations.

20. $x = 3(t^2 - 3), \quad y = t^3 - 3t$

21. $x = t\cos t, \quad y = t\sin t$

22. $x = \tan\theta + \sin\theta, \quad y = \cos\theta$

23. $x = \dfrac{3t}{1 + t^3}, \quad y = \dfrac{3t^2}{1 + t^3}$

24. Show that the parametric equations

$$x = x_1 + (x_2 - x_1)t$$
$$y = y_1 + (y_2 - y_1)t$$

where $0 \leq t \leq 1$, describe the line segment that joins the points $P_1(x_1, y_1)$ and $P_2(x_2, y_2)$.

25–27 ■ Find (a) a vector equation and (b) parametric equations for the line passing through the given pair of points.

25. $(1, 3)$ and $(-2, 7)$

26. $(-3, 4)$ and $(2, 8)$

27. $(4, -1)$ and $(-2, 5)$

28–30 ■ Find (a) a vector equation, (b) parametric equations, and (c) a Cartesian equation for the line that passes through the given point P and is parallel to the specified vector $\mathbf{a}$.

28. $P(1, 3), \quad \mathbf{a} = \langle 1, -2\rangle$

29. $P(-4, 5), \quad \mathbf{a} = \langle -2, 6\rangle$

30. $P(2, 5), \quad \mathbf{a} = \langle 3, 0\rangle$

31–33 ■ Determine whether the lines L_1 and L_2 given by the vector equations are parallel, perpendicular, or neither. If the lines are not parallel, find their point of intersection.

31. L_1: $\mathbf{r}(t) = (2 + 3t)\,\mathbf{i} + (-1 + 4t)\,\mathbf{j}$
L_2: $\mathbf{r}(t) = (1 + 4t)\,\mathbf{i} + (2 - 2t)\,\mathbf{j}$

32. L_1: $\mathbf{r}(t) = (-4 + 2t)\,\mathbf{i} + (5 + t)\,\mathbf{j}$
L_2: $\mathbf{r}(t) = (2 + 3t)\,\mathbf{i} + (4 - 6t)\,\mathbf{j}$

33. L_1: $\mathbf{r}(t) = (2 - t)\,\mathbf{i} + (-3 + 5t)\,\mathbf{j}$
L_2: $\mathbf{r}(t) = (8 + 10t)\,\mathbf{i} + (2 + 2t)\,\mathbf{j}$

34. This exercise generalizes Example 1. If a projectile is fired with an initial velocity of v_0 feet per second at an angle α above the horizontal and air resistance is assumed to be negligible, then its position after t seconds is given by the parametric equations

$$x = (v_0 \cos\alpha)t$$
$$y = (v_0 \sin\alpha)t - \tfrac{1}{2}gt^2$$

where g is the acceleration due to gravity (32 ft/s^2).
(a) If a gun is fired with $\alpha = 30°$ and $v_0 = 1500$ ft/s, when will the bullet hit the ground? How far from the gun will it hit the ground?
(b) What is the maximum height reached by the bullet?
(c) Show that the path is parabolic by eliminating the parameter t.

35. An object is moving in the xy-plane and its position after t seconds is $\mathbf{r}(t) = (t^2 + 2t)\,\mathbf{i} + (t + 4)\,\mathbf{j}$.
(a) Find the position of the object at time $t = 2$.
(b) At what time is the object at the point $(15, 7)$?
(c) Does the object pass through the point $(20, 9)$?
(d) Find an equation in x and y whose graph is the path of the object.

36. An object is moving in the xy-plane and its position after t seconds is $\mathbf{r}(t) = \langle t - 3, t^2 - 2t \rangle$.
(a) Find the position of the object at time $t = 5$.
(b) At what time is the object at the point $(1, 8)$?
(c) Does the object pass through the point $(3, 20)$?
(d) Find an equation in x and y whose graph is the path of the object.

37. Investigate the family of curves defined by the parametric equations $x = t^2$, $y = t^3 - ct$. How does the shape change as c increases? Illustrate by graphing several members of the family.

38. The **swallowtail catastrophe curves** are defined by the parametric equations $x = 2ct - 4t^3$, $y = -ct^2 + 3t^4$. Graph several of these curves. What features do the curves have in common? How do they change when c increases?

39. The curves with equations $x = a \sin nt$, $y = b \cos t$ are called **Lissajous figures**. Investigate how these curves vary when a, b, and n vary. (Take n to be a positive integer.)

40. Match the parametric equations with the graphs at the right (labeled I–VI). Give reasons for your choices. (Do not use a graphing device.)
(a) $x = t^3 - 2t, \quad y = t^2 - t$
(b) $x = t^3 - 1, \quad y = 2 - t^2$
(c) $x = \sin 3t, \quad y = \sin 4t$
(d) $x = t + \sin 2t, \quad y = t + \sin 3t$
(e) $x = \sin(t + \sin t), \quad y = \cos(t + \cos t)$
(f) $x = \cos t, \quad y = \sin(t + \sin 5t)$

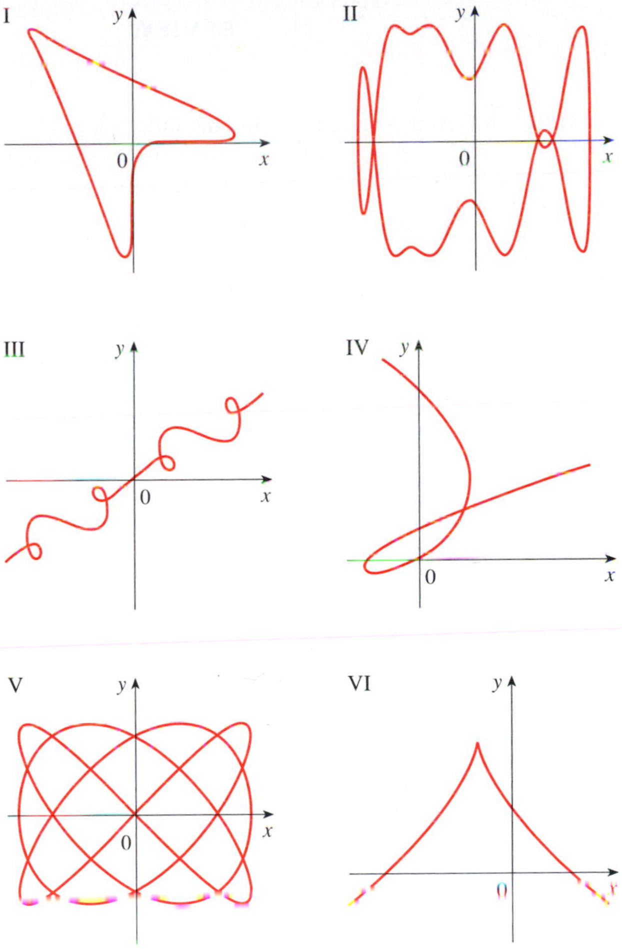

1 REVIEW

KEY TOPICS ■ Define, state, or discuss the following.

1. Addition of vectors
2. Multiplication of a vector by a scalar
3. Length of a vector
4. Parallel vectors
5. Properties of vectors
6. Standard basis vectors
7. Unit vector
8. Displacement vector
9. Dot product and its properties
10. Scalar product
11. Angle between two vectors
12. Work done by a constant force over a direction vector
13. Orthogonal vectors
14. Vector and scalar projections
15. Orthogonal complement
16. Parametric equations of a curve
17. Vector function
18. Vector equation of a line
19. Converting between Cartesian equations and parametrized equations

EXERCISES

1–2 ■ Find a vector $\mathbf{a}$ with representation given by the directed line segment $\overrightarrow{AB}$. Draw (a) $\overrightarrow{AB}$, (b) an equivalent representation for $\mathbf{a}$ starting at the origin, and (c) an equivalent representation starting at the point $P(1, 2)$.

1. $A(2, 1)$, $B(3, -1)$

2. $A(-1, 0)$, $B(-2, -2)$

3–6 ■ Find the sum of the given vectors and illustrate geometrically.

3. $2\mathbf{i} - 3\mathbf{j}$, $-4\mathbf{i} + \mathbf{j}$

4. $3\mathbf{i} + 5\mathbf{j}$, $2\mathbf{i} - \mathbf{j}$

5. $\langle 4, 3 \rangle$, $\langle -1, 2 \rangle$

6. $\langle -1, 6 \rangle$, $\langle 3, 1 \rangle$

7–10 ■ In each of the following situations, $\mathbf{a}$ is a vector that is represented by the directed line segment $\overrightarrow{AB}$.

7. Given $\mathbf{a} = -2\mathbf{i} + 3\mathbf{j}$ and $A(1, 0)$, find B.

8. Given $\mathbf{a} = \langle 4, -2 \rangle$ and $A(1, 2)$, find B.

9. Given $\mathbf{a} = \langle -3, 4 \rangle$ and $B(1, 5)$, find A.

10. Given $\mathbf{a} = -4\mathbf{i} - 5\mathbf{j}$ and $B(-2, 3)$, find A.

11–14 ■ Find (a) the length of $\mathbf{a}$, (b) a unit vector having the same direction as $\mathbf{a}$, and (c) a unit vector that is orthogonal to $\mathbf{a}$. Then find (d) $-4\mathbf{a} + 3\mathbf{b}$ and (e) scalars s and t such that $\mathbf{c} = s\mathbf{a} + t\mathbf{b}$.

11. $\mathbf{a} = \langle 3, -4 \rangle$, $\mathbf{b} = \langle 1, 3 \rangle$, $\mathbf{c} = \langle 2, 1 \rangle$

12. $\mathbf{a} = \langle -2, 3 \rangle$, $\mathbf{b} = \langle 6, 1 \rangle$, $\mathbf{c} = \langle 2, 0 \rangle$

13. $\mathbf{a} = \mathbf{i} - 7\mathbf{j}$, $\mathbf{b} = 2\mathbf{i} + 3\mathbf{j}$, $\mathbf{c} = 4\mathbf{j}$

14. $\mathbf{a} = 2\mathbf{i} + 5\mathbf{j}$, $\mathbf{b} = -\mathbf{i} + 3\mathbf{j}$, $\mathbf{c} = \mathbf{i} + \mathbf{j}$

15–20 ■ Find (a) $\mathbf{a} \cdot \mathbf{b}$, (b) the angle between $\mathbf{a}$ and $\mathbf{b}$, (c) the scalar projection of $\mathbf{b}$ onto $\mathbf{a}$, and (d) the vector projection of $\mathbf{b}$ onto $\mathbf{a}$.

15. $\mathbf{a} = \langle 1, 2 \rangle$, $\mathbf{b} = \langle -2, 3 \rangle$

16. $\mathbf{a} = \langle -2, 1 \rangle$, $\mathbf{b} = \langle 1, 3 \rangle$

17. $\mathbf{a} = -3\mathbf{i} + 4\mathbf{j}$, $\mathbf{b} = -2\mathbf{j}$

18. $\mathbf{a} = 2\mathbf{i} - \mathbf{j}$, $\mathbf{b} = 3\mathbf{i}$

19. $\mathbf{a} = \langle 7, -6 \rangle$, $\mathbf{b} = \langle 6, 7 \rangle$

20. $\mathbf{a} = \langle 4, 6 \rangle$, $\mathbf{b} = \langle -3, 2 \rangle$

21. Find the distance from the point $(2, 3)$ to the line $2x - y = 7$.

22. Find the distance from the point $(-1, 2)$ to the line $y = 4x + 3$.

23. Find the distance between the lines $2x + 4y = 7$ and $2x + 4y = 9$.

24. Find the distance between the lines $y = 4x + 3$ and $y = 4x - 5$.

25. A constant force $\mathbf{F} = 5\mathbf{i} + 6\mathbf{j}$ moves an object along the straight line from the point $(-1, 2)$ to the point $(2, 3)$. Find the work done by $\mathbf{F}$.

26. Find the work done by the constant force $\mathbf{F} = 3\mathbf{i} + 7\mathbf{j}$ in moving an object from $(0, 0)$ to $(2, 4)$ along the line $y = 2x$.

27–30 ■
(a) Sketch the curve represented by the parametric equations.
(b) Eliminate the parameter to find a Cartesian equation for the curve.

27. $x = 1 - t^2, \quad y = 1 - t, \quad -1 \leqslant t \leqslant 1$

28. $x = t^2 + 1, \quad y = t^2 - 1$

29. $x = 1 + \sin t, \quad y = 2 + \cos t$

30. $x = 1 + \cos t, \quad y = 1 + \sin^2 t$

31–34 ■ Find (a) a vector equation, (b) parametric equations, and (c) a Cartesian equation for the given line.

31. The line containing the points $(1, 2)$ and $(-3, 4)$

32. The line containing the points $(-1, 1)$ and $(2, 5)$

33. The line containing the point $(2, 3)$ and parallel to $\mathbf{i} - 5\mathbf{j}$

34. The line containing the point $(2, -1)$ and parallel to $2\mathbf{i} + 3\mathbf{j}$

35. A particle is moving in the xy-plane and its position after t seconds is $\mathbf{r}(t) = (t^2 + t)\,\mathbf{i} + (t - 4)\,\mathbf{j}$.
(a) Find the position of the particle at time $t = 2$.
(b) At what time is the particle at the point $(12, -1)$?
(c) Does the particle pass through the point $(4, 8)$?

36. A particle is moving in the xy-plane and its position (x, y) at time t is given by $x = 3t + 1$, $y = t^2 - t$.
(a) Find the position of the particle at time $t = 3$.
(b) At what time is the particle at the point $(16, 20)$?
(c) Does the particle pass through the point $(7, 3)$?

37. The graph of $2x - 3y = 7$ is a line in the xy-plane. What is the relationship between the vector $2\mathbf{i} - 3\mathbf{j}$ and the line? In general, how is the vector $a\mathbf{i} + b\mathbf{j}$ related to the line $ax + by = c$?

2

LIMITS AND RATES OF CHANGE

■ The calculus was the first achievement of modern mathematics and it is difficult to overestimate its importance. I think it defines more unequivocally than anything else the inception of modern mathematics; and the system of mathematical analysis, which is its logical development, still constitutes the greatest technical advance in exact thinking.

JOHN VON NEUMANN

In A Preview of Calculus (Section 5 in Review and Preview) we saw how the idea of a limit underlies the various branches of calculus. Thus, it is appropriate to begin our study of calculus by investigating limits and their properties.

2.1 THE TANGENT AND VELOCITY PROBLEMS

In this section we see how limits arise when we attempt to find a tangent to a curve or the velocity of an object.

THE TANGENT PROBLEM

The word *tangent* is derived from the Latin word *tangens*, which means "touching." Thus, a tangent to a curve is a line that touches the curve. How can this idea be made precise?

For a circle we could simply follow Euclid and say that a tangent is a line that intersects the circle once and only once as in Figure 1(a). For more complicated curves this definition is inadequate. Figure 1(b) shows two lines l and t passing through a point P on a curve C. The line l intersects C only once, but it certainly doesn't look like what we think of as a tangent. The line t, on the other hand, looks like a tangent but it intersects C twice.

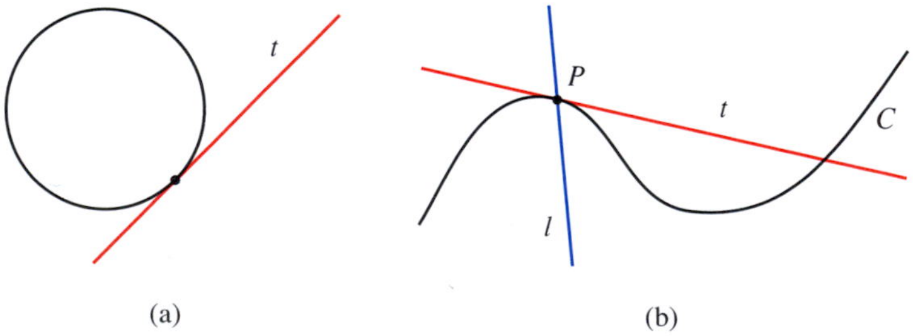

FIGURE 1

To be specific, let's look at the problem of trying to find a tangent line t to the parabola $y = x^2$ in the following example.

EXAMPLE 1 Find an equation of the tangent line to the parabola $y = x^2$ at the point $P(1, 1)$.

SOLUTION We will be able to find the equation of the tangent line t as soon as we know its slope m. The difficulty is that we know only one point, P, on t, whereas

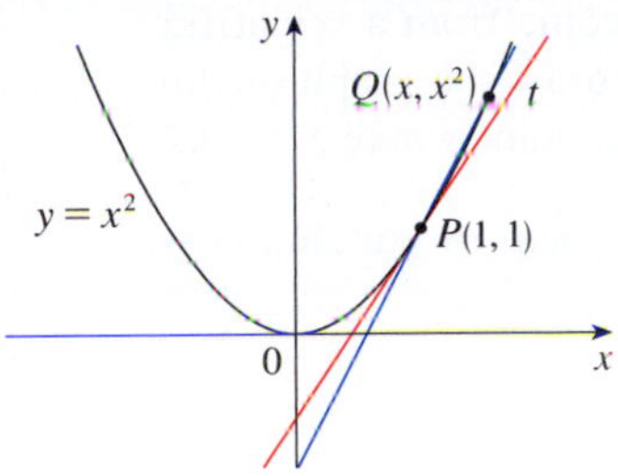

FIGURE 2

we need two points to compute the slope. But observe that we can compute an approximation to m by choosing a nearby point $Q(x, x^2)$ on the parabola (as in Figure 2) and computing the slope m_{PQ} of the secant line PQ.

We choose $x \neq 1$ so that $Q \neq P$. Then

$$m_{PQ} = \frac{x^2 - 1}{x - 1}$$

For instance, for the point $Q(1.5, 2.25)$ we have

$$m_{PQ} = \frac{2.25 - 1}{1.5 - 1} = \frac{1.25}{0.5} = 2.5$$

x	m_{PQ}
2	3
1.5	2.5
1.1	2.1
1.01	2.01
1.001	2.001

x	m_{PQ}
0	1
0.5	1.5
0.9	1.9
0.99	1.99
0.999	1.999

The tables in the margin show the values of m_{PQ} for several values of x close to 1. The closer Q is to P, the closer x is to 1 and, it appears from the tables, the closer m_{PQ} is to 2. This suggests that the slope of the tangent line t should be $m = 2$.

We say that the slope of the tangent line is the *limit* of the slopes of the secant lines, and we express this symbolically by writing

$$\lim_{Q \to P} m_{PQ} = m$$

and

$$\lim_{x \to 1} \frac{x^2 - 1}{x - 1} = 2$$

Assuming that the slope of the tangent line is indeed 2, we use the point-slope form of the equation of a line to write the equation of the tangent line through $(1, 1)$ as

$$y - 1 = 2(x - 1) \qquad \text{or} \qquad y = 2x - 1$$

Figure 3 illustrates the limiting process that occurs in this example. As Q approaches P along the parabola, the corresponding secant lines rotate about P and approach the tangent line.

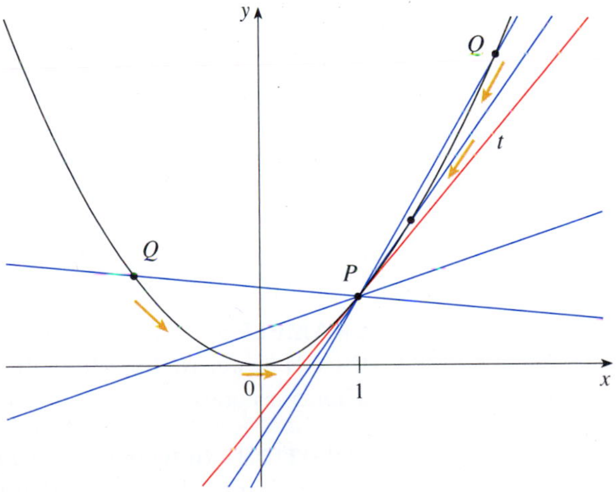

FIGURE 3

■

Many functions that occur in science are not described by an explicit equation; they are defined by experimental data. The next example shows how to estimate the slope of the tangent line to the graph of such a function.

x	y
0.0	3.86
0.1	3.71
0.2	3.40
0.3	3.02
0.4	2.35
0.5	1.46

EXAMPLE 2 Suppose the data in the margin, which come from a scientific experiment, define y as a function of x. Use the data to draw the graph of this function and estimate the slope of the tangent line at the point where $x = 0.2$.

SOLUTION In Figure 4 we plot the given data and use them to sketch a curve that approximates the graph of the function.

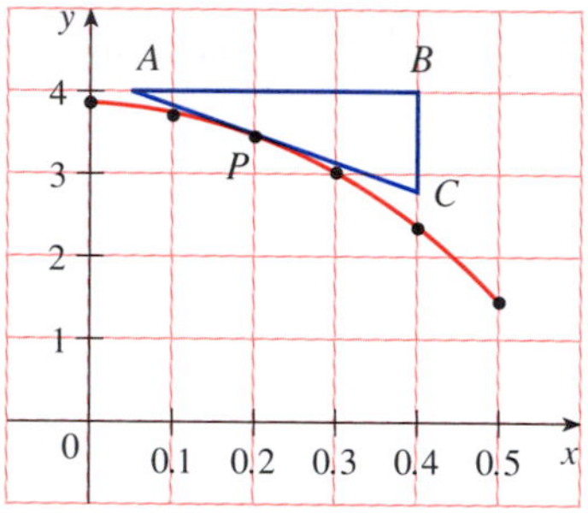

FIGURE 4

Given the points $P(0.2, 3.40)$ and $Q(0.0, 3.86)$ on the graph, we find that the slope of the secant line PQ is

$$m_{PQ} = \frac{3.86 - 3.40}{0.0 - 0.2} = -2.3$$

Q	m_{PQ}
(0.0, 3.86)	−2.3
(0.1, 3.71)	−3.1
(0.3, 3.02)	−3.8
(0.4, 2.35)	−5.3
(0.5, 1.46)	−6.5

The table in the margin shows the results of similar calculations for the slopes of other secant lines. From this table we would expect the slope of the tangent line to lie somewhere between -3.1 and -3.8.

We now draw an approximation to the tangent line at P and measure the sides of the triangle ABC. This gives an estimate of the slope of the tangent line as

$$-\frac{|BC|}{|AB|} \approx -\frac{1.2}{0.35} \approx -3.4$$

■

THE VELOCITY PROBLEM: LINEAR MOTION

If you watch the speedometer of a car while traveling in city traffic, you see that the needle does not stay still for very long; that is, the velocity of the car is not constant. We assume from watching the speedometer that the car has a definite velocity at each moment, but how is the "instantaneous" velocity defined? Let's investigate the example of a falling ball.

EXAMPLE 3 Suppose that a ball is dropped from the upper observation deck of the CN Tower in Toronto, 450 m above the ground. Find the velocity of the ball after 5 seconds.

SOLUTION In trying to solve this problem we use the fact, discovered by Galileo almost four centuries ago, that the distance fallen by any freely falling body is proportional to the square of the time it has been falling. (This neglects air resistance.) If the distance fallen after t seconds is denoted by $s(t)$ and measured in meters, then Galileo's law is expressed by the equation

$$s(t) = 4.9t^2$$

Courtesy of CN Tower.

The CN Tower in Toronto is currently the highest freestanding building in the world.

The difficulty in finding the velocity after 5 s is that we are dealing with a single instant of time ($t = 5$) so no time interval is involved. However, we can approximate the desired quantity by computing the average velocity over the brief time interval of a tenth of a second from $t = 5$ to $t = 5.1$:

$$\begin{aligned} \text{average velocity} &= \frac{\text{distance traveled}}{\text{time elapsed}} \\ &= \frac{s(5.1) - s(5)}{5.1 - 5} \\ &= \frac{4.9(5.1)^2 - 4.9(5)^2}{0.1} = 49.49 \text{ m/s} \end{aligned}$$

The following table shows the results of similar calculations of the average velocity over successively smaller time periods.

Time interval	Average velocity (m/s)
$5 \leqslant t \leqslant 6$	53.9
$5 \leqslant t \leqslant 5.1$	49.49
$5 \leqslant t \leqslant 5.05$	49.245
$5 \leqslant t \leqslant 5.01$	49.049
$5 \leqslant t \leqslant 5.001$	49.0049

It appears that as we shorten the time period, the average velocity is becoming closer to 49 m/s. The **instantaneous velocity** when $t = 5$ is defined to be the limiting value of these average velocities over shorter and shorter time periods that start at $t = 5$. Thus, the (instantaneous) velocity after 5 s is

$$v = 49 \text{ m/s}$$

■

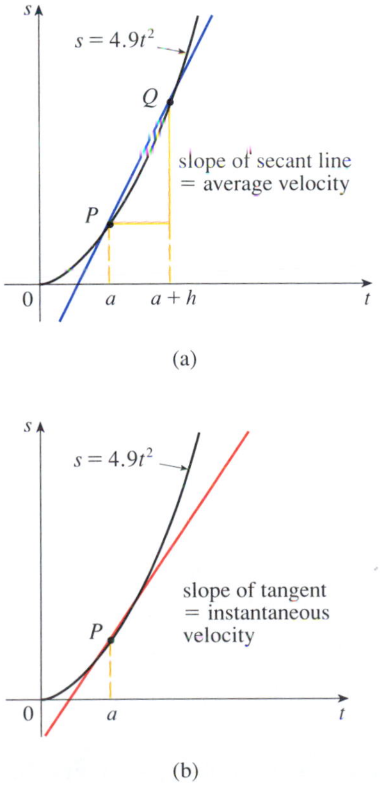

FIGURE 5

You may have the feeling that the calculations used in solving this problem are very similar to those used earlier in this section to find tangents. In fact, there is a close connection between the tangent problem and the problem of finding velocities. If we draw the graph of the distance function of the ball (as in Figure 5) and we consider the points

$$P(a, 4.9a^2) \qquad \text{and} \qquad Q(a + h, 4.9(a + h)^2)$$

on the graph, then the slope of the secant line PQ is

$$m_{PQ} = \frac{4.9(a + h)^2 - 4.9a^2}{(a + h) - a}$$

which is the same as the average velocity over the time interval $[a, a + h]$. Therefore, the velocity at time $t = a$ (the limit of these average velocities as h approaches 0) must be equal to the slope of the tangent line at P (the limit of the slopes of the secant lines).

THE VELOCITY PROBLEM: CURVILINEAR MOTION

Example 3 dealt with a ball that is dropped from a height of 450 m and falls straight down to the ground. So the path of the ball is a line, and we have linear motion. The instantaneous velocity of the ball after 5 s is 49 m/s. This is in fact a vector quantity with the direction of the motion of the ball understood to be downward toward the ground.

More realistic motion problems deal with objects that do not move in a straight line—a ball being thrown off a tall building, a projectile being fired from a gun, and so on. The paths of such objects are curves, and we have curvilinear motion. The position of the object at time t may be described by a vector function $\mathbf{r}(t)$, such as in the following example.

EXAMPLE 4 Suppose that an artillery shell is fired from a gun and its position $\mathbf{r}(t)$ at time t is

$$\mathbf{r}(t) = (433t)\,\mathbf{i} + (250t - 4.9t^2)\,\mathbf{j}$$

where distance is measured in meters and t in seconds. Find the instantaneous velocity of the shell after 40 s.

SOLUTION The instantaneous velocity (rate of change of position with respect to time) is the limit of the average velocities over successively smaller periods of time. Here the average velocity between times $t = 40$ and $t = 40 + h$ is given by

$$\begin{aligned}\text{average velocity} &= \frac{\text{change in position}}{\text{time elapsed}}\\ &= \frac{\mathbf{r}(40 + h) - \mathbf{r}(40)}{h}\\ &= \frac{1}{h}[\mathbf{r}(40 + h) - \mathbf{r}(40)]\end{aligned}$$

Notice that the last expression is a scalar multiple of a vector difference.

Time interval (in seconds)	h (in seconds)	$\frac{1}{h}[\mathbf{r}(40 + h) - \mathbf{r}(40)]$ (components in m/s)
$40 \leqslant t \leqslant 40.1$	0.1	$433\,\mathbf{i} - 142.49\,\mathbf{j}$
$40 \leqslant t \leqslant 40.01$	0.01	$433\,\mathbf{i} - 142.049\,\mathbf{j}$
$40 \leqslant t \leqslant 40.001$	0.001	$433\,\mathbf{i} - 142.0049\,\mathbf{j}$
$40 \leqslant t \leqslant 40.0001$	0.0001	$433\,\mathbf{i} - 142.00049\,\mathbf{j}$
$39.9 \leqslant t \leqslant 40$	−0.1	$433\,\mathbf{i} - 141.51\,\mathbf{j}$
$39.99 \leqslant t \leqslant 40$	−0.01	$433\,\mathbf{i} - 141.951\,\mathbf{j}$
$39.999 \leqslant t \leqslant 40$	−0.001	$433\,\mathbf{i} - 141.9951\,\mathbf{j}$
$39.9999 \leqslant t \leqslant 40$	−0.0001	$433\,\mathbf{i} - 141.99951\,\mathbf{j}$

From the table, it appears that the instantaneous velocity at time $t = 40$ s is $\mathbf{v} = 433\,\mathbf{i} - 142\,\mathbf{j}$. The units of each component are meters per second. ■

THE TANGENT PROBLEM: VECTOR VERSION

In Example 1 we found an equation of the tangent line to a curve at a given point. In a similar way we can find a vector tangent to a curve at a given point. This is illustrated in the next example.

EXAMPLE 5 Find a vector tangent to the parabola

$$\mathbf{r}(t) = \langle x(t), y(t) \rangle = \langle t, t^2 \rangle$$

at the point $P(1, 1)$. Use this vector to find parametric equations for the tangent line to the parabola at P.

The arrow in Figure 6 represents the secant vector $\mathbf{r}(t) - \mathbf{r}(1)$.

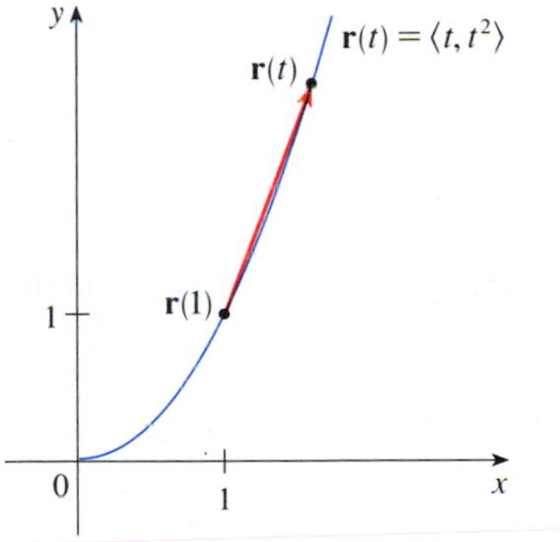

FIGURE 6

SOLUTION We first note that the point $(1, 1)$ corresponds to $t = 1$. To find a tangent vector at $(1, 1)$, we compute secant vectors given by

$$\frac{\mathbf{r}(t) - \mathbf{r}(1)}{t - 1} = \frac{1}{t - 1}[\mathbf{r}(t) - \mathbf{r}(1)]$$

for values of t closer and closer to 1, as shown in the following table. This difference quotient is analogous to the average rate of change of position with respect to time.

t	$\frac{1}{t-1}[\mathbf{r}(t) - \mathbf{r}(1)]$
1.1	$\langle 1, 2.1 \rangle$
1.01	$\langle 1, 2.01 \rangle$
1.001	$\langle 1, 2.001 \rangle$
1.0001	$\langle 1, 2.0001 \rangle$
0.9	$\langle 1, 1.9 \rangle$
0.99	$\langle 1, 1.99 \rangle$
0.999	$\langle 1, 1.999 \rangle$
0.9999	$\langle 1, 1.9999 \rangle$

These values suggest that the tangent vector is

$$\mathbf{v} = \lim_{t \to 1} \frac{\mathbf{r}(t) - \mathbf{r}(1)}{t - 1} = \langle 1, 2 \rangle$$

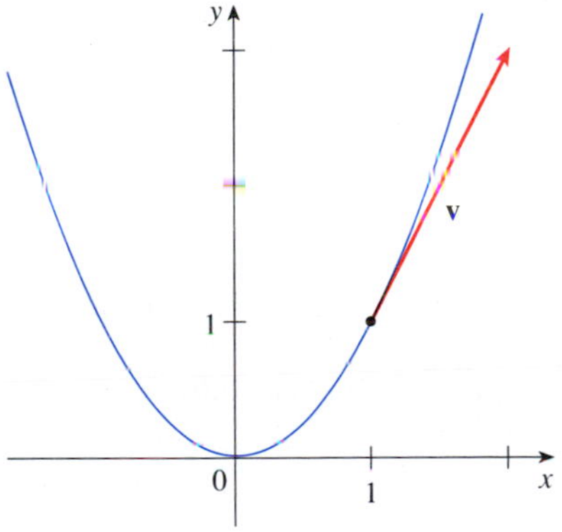

FIGURE 7

This vector is illustrated in Figure 7.

Using $\mathbf{v}$, we have the following vector equation of the tangent line to $\mathbf{r}(t)$ at $P(1, 1)$:

$$\begin{aligned} \mathbf{L}(t) &= \mathbf{r}(1) + t\mathbf{v} \\ &= \langle 1, 1 \rangle + t\langle 1, 2 \rangle \\ &= \langle 1 + t, 1 + 2t \rangle \end{aligned}$$

Parametric equations for the tangent line are then

$$x = 1 + t \qquad y = 1 + 2t$$

Eliminating the parameter gives $y = 1 + 2(x - 1)$, or $y = 2x - 1$, the same tangent line as we obtained in Example 1. ■

The examples in this section show that in order to solve tangent and velocity problems we must be able to find limits. After studying methods for computing limits in the next five sections, we will return to the problems of finding tangents and velocities in Section 2.7.

EXERCISES 2.1

1. The experimental data in the table define y as a function of x.

x	0	1	2	3	4	5
y	2.6	2.0	1.1	1.3	2.1	3.5

(a) If P is the point $(3, 1.3)$, find the slopes of the secant lines PQ when Q is the point on the graph with $x = 0$, 1, 2, 4, and 5.

(b) Use a graph of the function to estimate the slope of the tangent line at P.

2. The point $P(1, 3)$ lies on the curve $y = 1 + x + x^2$.

(a) If Q is the point $(x, 1 + x + x^2)$, find the slope of the secant line PQ for the following values of x:
(i) 2 (ii) 1.5 (iii) 1.1 (iv) 1.01
(v) 1.001 (vi) 0 (vii) 0.5 (viii) 0.9
(ix) 0.99 (x) 0.999

(b) Using the results of part (a), guess the value of the slope of the tangent line to the curve at $P(1, 3)$.

(c) Using the slope from part (b), find an equation of the tangent line to the curve at $P(1, 3)$.

3. The point $P(2, 3)$ lies on the curve $\mathbf{r}(t) = \langle t, t^2 - 1\rangle$ for $t = 2$.

(a) If Q is the point $(t, t^2 - 1)$, find the secant vector $\dfrac{1}{t-2}\overrightarrow{PQ}$ for each of the following values of t:
(i) 3, 2.5, 2.1, 2.01, 2.001
(ii) 1, 1.5, 1.9, 1.99, 1.999

(b) Using the results of part (a), guess the resulting tangent vector to the curve at $P(2, 3)$.

(c) Using the vector in part (b), write a vector equation for the tangent line to the curve at $P(2, 3)$.

4. The point $P(1, 2)$ lies on the curve $\mathbf{r}(t) = (t^2)\,\mathbf{i} + (2t)\,\mathbf{j}$ for $t = 1$.

(a) If Q is the point $(t^2, 2t)$, find the secant vector $\dfrac{1}{t-1}\overrightarrow{PQ}$ for each of the following values of t:
(i) 2, 1.5, 1.1, 1.01, 1.001
(ii) 0, 0.5, 0.9, 0.99, 0.999

(b) Using the results of part (a), guess the resulting tangent vector to the curve at $P(1, 2)$.

(c) Using the vector in part (b), write a vector equation for the tangent line to the curve at $P(1, 2)$.

5. A ball is projected into the air. The ball's position at time t is $\mathbf{r}(t) = (3t)\,\mathbf{i} + (45t - 16t^2)\,\mathbf{j}$, where distance is measured in feet and time in seconds.

(a) Find the average velocity of the ball for the time period that begins when $t = 2$ and lasts
(i) 0.5 s (ii) 0.1 s
(iii) 0.05 s (iv) 0.01 s

(b) Find the instantaneous velocity of the ball when $t = 2$.

6. The location at time t of an object moving in the xy-plane is given by

$$\mathbf{r}(t) = \langle t^2, t^3 + 1\rangle$$

where distance is measured in meters and time in seconds. So at time $t = 1$ s, the object is located at the point $(1, 2)$.

(a) Find the average velocity of the object for the time period that begins when $t = 1$ and lasts
(i) 0.5 s (ii) 0.1 s
(iii) 0.05 s (iv) 0.01 s

(b) Find the instantaneous velocity of the ball when it is at the point $(1, 2)$.

7. The point $P(4, 2)$ lies on the curve $y = \sqrt{x}$.

(a) If Q is the point $(x, \sqrt{x})$, use your calculator to find the slope of the secant line PQ (correct to six decimal places) for the following values of x:
(i) 5 (ii) 4.5 (iii) 4.1 (iv) 4.01
(v) 4.001 (vi) 3 (vii) 3.5 (viii) 3.9
(ix) 3.99 (x) 3.999

(b) Using the results of part (a), guess the value of the slope of the tangent line to the curve at $P(4, 2)$.

(c) Using the slope from part (b), find an equation of the tangent line to the curve at $P(4, 2)$.

8. The point $P(0.5, 2)$ lies on the curve $y = 1/x$.

(a) If Q is the point $(x, 1/x)$, use your calculator to find the slope of the secant line PQ (correct to six decimal places) for the following values of x:
(i) 2 (ii) 1 (iii) 0.9 (iv) 0.8
(v) 0.7 (vi) 0.6 (vii) 0.55 (viii) 0.51
(ix) 0.45 (x) 0.49

(b) Using the results of part (a), guess the value of the slope of the tangent line to the curve at $P(0.5, 2)$.

(c) Using the slope from part (b), find an equation of the tangent line to the curve at $P(0.5, 2)$.

(d) Sketch the curve, two of the secant lines, and the tangent line.

9. If a ball is thrown into the air with a velocity of 40 ft/s, its height in feet after t seconds is given by $y = 40t - 16t^2$.

(a) Find the average velocity for the time period beginning when $t = 2$ and lasting
(i) 0.5 s (ii) 0.1 s
(iii) 0.05 s (iv) 0.01 s

(b) Find the instantaneous velocity when $t = 2$.

10. If an arrow is shot upward on the moon with a velocity of 58 m/s, its height in meters after t seconds is given by $h = 58t - 0.83t^2$.

(a) Find the average velocity over the given time intervals:
(i) $[1, 2]$ (ii) $[1, 1.5]$ (iii) $[1, 1.1]$
(iv) $[1, 1.01]$ (v) $[1, 1.001]$

(b) Find the instantaneous velocity after 1 s.

11. The displacement (in feet) of a particle moving in a straight line is given by $s = t^3/6$, where t is measured in seconds.
(a) Find the average velocity over the following time periods:
(i) $[1, 3]$ (ii) $[1, 2]$
(iii) $[1, 1.5]$ (iv) $[1, 1.1]$
(b) Find the instantaneous velocity when $t = 1$.
(c) Draw the graph of s as a function of t and draw the secant lines whose slopes are the average velocities found in part (a).
(d) Draw the tangent line whose slope is the instantaneous velocity from part (b).

12. The position of a car is given by the values in the table.

t (seconds)	0	1	2	3	4	5
s (feet)	0	10	32	70	119	178

(a) Find the average velocity for the time period beginning when $t = 2$ and lasting
(i) 3 s (ii) 2 s (iii) 1 s
(b) Use the graph of s as a function of t to estimate the instantaneous velocity when $t = 2$.

2.2 THE LIMIT OF A FUNCTION

Having seen in the preceding section how limits arise when we want to find the tangent to a curve or the velocity of an object, we now turn our attention to limits in general and methods for computing them.

Let's investigate the behavior of the function f defined by $f(x) = x^2 - x + 2$ for values of x near 2. The following table gives values of $f(x)$ for values of x close to 2 but not equal to 2.

x	$f(x)$	x	$f(x)$
1.0	2.000000	3.0	8.000000
1.5	2.750000	2.5	5.750000
1.8	3.440000	2.2	4.640000
1.9	3.710000	2.1	4.310000
1.95	3.852500	2.05	4.152500
1.99	3.970100	2.01	4.030100
1.995	3.985025	2.005	4.015025
1.999	3.997001	2.001	4.003001

From the table and the graph of f (a parabola) shown in Figure 1 we see that when x is close to 2 (on either side of 2), $f(x)$ is close to 4. In fact, it appears that we can make

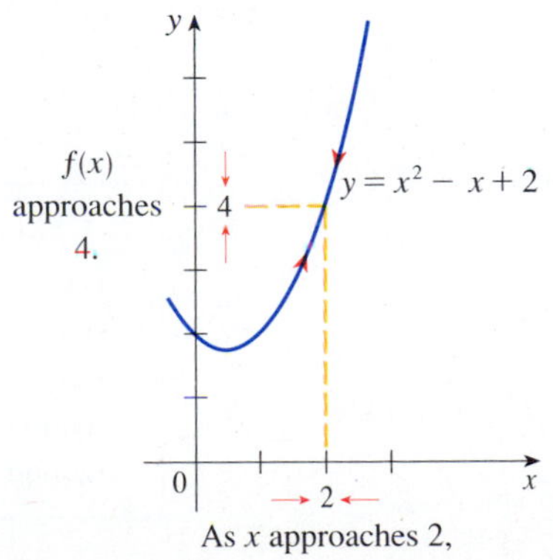

FIGURE 1

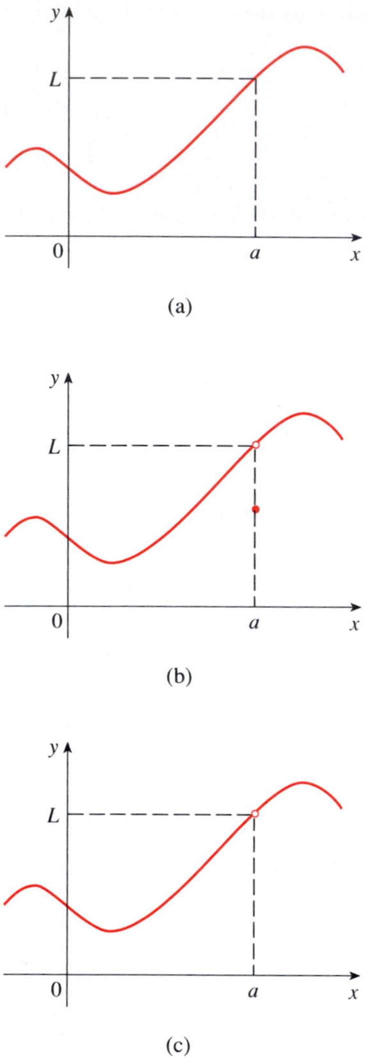

FIGURE 2

$\lim_{x \to a} f(x) = L$ in all three cases

the values of $f(x)$ as close as we like to 4 by taking x sufficiently close to 2. We express this by saying "the limit of the function $f(x) = x^2 - x + 2$ as x approaches 2 is equal to 4." The notation for this is

$$\lim_{x \to 2} (x^2 - x + 2) = 4$$

In general, we use the following notation.

(1) DEFINITION We write

$$\lim_{x \to a} f(x) = L$$

and say "the limit of $f(x)$, as x approaches a, equals L"

if we can make the values of $f(x)$ arbitrarily close to L (as close to L as we like) by taking x to be sufficiently close to a but not equal to a.

Roughly speaking, this says that the values of $f(x)$ get closer and closer to the number L as x gets closer and closer to the number a (from either side of a) but $x \neq a$. A more precise definition will be given in Section 2.4.

An alternative notation for

$$\lim_{x \to a} f(x) = L$$

is

$$f(x) \to L \qquad \text{as} \qquad x \to a$$

which is usually read "$f(x)$ approaches L as x approaches a."

Notice the phrase "but $x \neq a$" in the definition of limit. This means that in finding the limit of $f(x)$ as x approaches a, we never consider $x = a$. In fact, $f(x)$ need not even be defined when $x = a$. The only thing that matters is how f is defined *near* a.

Figure 2 shows the graphs of three functions. Note that in part (c), $f(a)$ is not defined and in part (b), $f(a) \neq L$. But in each case, regardless of what happens at a, $\lim_{x \to a} f(x) = L$.

EXAMPLE 1 Guess the value of $\lim_{x \to 1} \dfrac{x - 1}{x^2 - 1}$.

SOLUTION Notice that the function $f(x) = (x - 1)/(x^2 - 1)$ is not defined when $x = 1$, but that doesn't matter because the definition of $\lim_{x \to a} f(x)$ says that we consider values of x that are close to a but not equal to a. The following table gives values of $f(x)$ (correct to six decimal places) for values of x that approach 1 (but are not equal to 1).

$x < 1$	$f(x)$
0.5	0.666667
0.9	0.526316
0.99	0.502513
0.999	0.500250
0.9999	0.500025

$x > 1$	$f(x)$
1.5	0.400000
1.1	0.476190
1.01	0.497512
1.001	0.499750
1.0001	0.499975

On the basis of the values in the table, we make the guess that

$$\lim_{x \to 1} \frac{x - 1}{x^2 - 1} = 0.5$$ ■

Example 1 is illustrated by the graph of f in Figure 3. Now let's change f slightly by giving it the value 2 when $x = 1$ and calling the resulting function g:

$$g(x) = \begin{cases} \dfrac{x - 1}{x^2 - 1} & \text{if } x \neq 1 \\ 2 & \text{if } x = 1 \end{cases}$$

This new function g still has the same limit as x approaches 1 (see Figure 4).

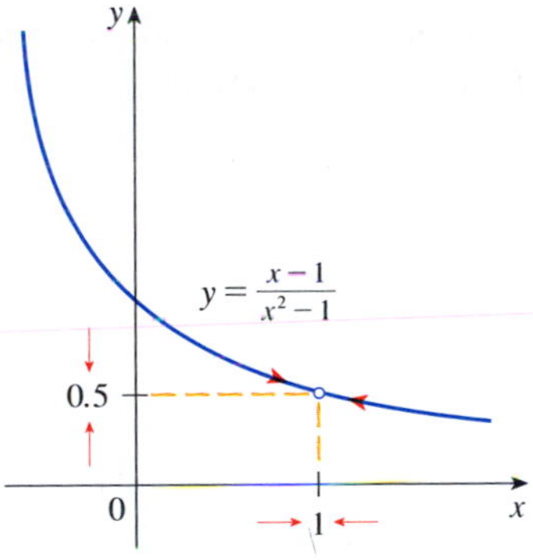

FIGURE 3

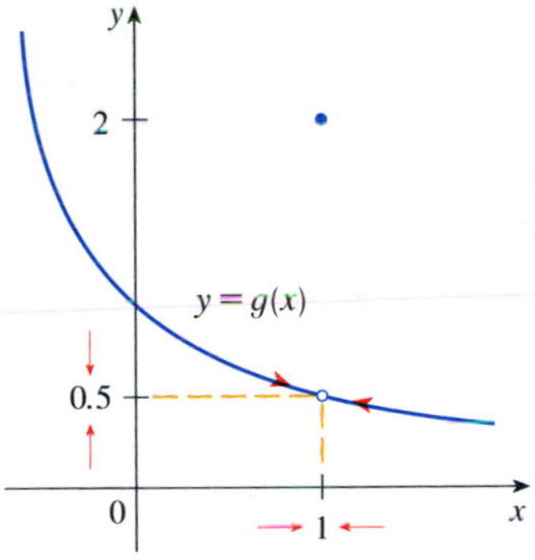

FIGURE 4

EXAMPLE 2 Find $\displaystyle\lim_{t \to 0} \frac{\sqrt{t^2 + 9} - 3}{t^2}$.

SOLUTION The following table lists values of the function for several values of t near 0.

t	$\dfrac{\sqrt{t^2 + 9} - 3}{t^2}$
±1.0	0.16228
±0.5	0.16553
±0.1	0.16662
±0.05	0.16666
±0.01	0.16667

As t approaches 0, the values of the function seem to approach 0.1666666... and so we guess that

$$\lim_{t \to 0} \frac{\sqrt{t^2 + 9} - 3}{t^2} = \frac{1}{6}$$ ■

t	$\dfrac{\sqrt{t^2 + 9} - 3}{t^2}$
±0.0005	0.16800
±0.0001	0.20000
±0.00005	0.00000
±0.00001	0.00000

In Example 2 what would have happened if we had taken even smaller values of t? The table in the margin displays the results from one calculator; you can see that something strange seems to be happening.

If you try these calculations on your own calculator you might get different values, but eventually you will get the value 0 if you make t sufficiently small. Does this mean that the answer is really 0 instead of $\frac{1}{6}$? No, the value of the limit is $\frac{1}{6}$, as we will show in the next section. The problem is that the calculator gave false values because $\sqrt{t^2 + 9}$ is very close to 3 when t is small. For a further explanation, see Appendix G (Lies My Calculator and Computer Told Me) and, in particular, the section called The Perils of Subtraction.

Something similar happens when we try to graph the function

$$f(t) = \frac{\sqrt{t^2 + 9} - 3}{t^2}$$

of Example 2 on a graphing calculator or computer. Parts (a) and (b) of Figure 5 show quite accurate graphs of f and when we use the trace mode (if available), we can estimate easily that the limit is about $\frac{1}{6}$. But if we zoom in too far, as in parts (c) and (d), then we get inaccurate graphs, again because of problems with subtraction.

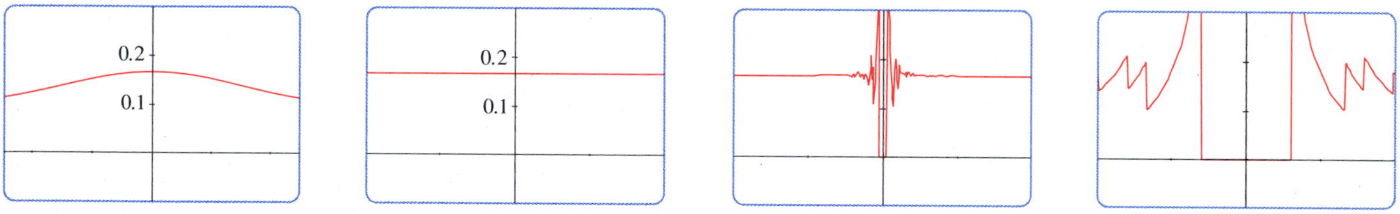

(a) $[-5, 5]$ by $[-0.1, 0.3]$ (b) $[-0.1, 0.1]$ by $[-0.1, 0.3]$ (c) $[-10^{-6}, 10^{-6}]$ by $[-0.1, 0.3]$ (d) $[-10^{-7}, 10^{-7}]$ by $[-0.1, 0.3]$

FIGURE 5

EXAMPLE 3 Find $\lim_{x \to 0} \frac{\sin x}{x}$.

SOLUTION Again the function $f(x) = (\sin x)/x$ is not defined when $x = 0$. Using a calculator (and remembering that, if $x \in \mathbb{R}$, $\sin x$ means the sine of the angle whose *radian* measure is x), we construct the accompanying table of values correct to eight decimal places. From the table and Figure 6 (drawn with the aid of the table) we guess that

$$\lim_{x \to 0} \frac{\sin x}{x} = 1$$

This guess is in fact correct, as will be proved in Chapter 3 using a geometric argument. ■

x	$\frac{\sin x}{x}$
± 1.0	0.84147098
± 0.5	0.95885108
± 0.4	0.97354586
± 0.3	0.98506736
± 0.2	0.99334665
± 0.1	0.99833417
± 0.05	0.99958339
± 0.01	0.99998333
± 0.005	0.99999583
± 0.001	0.99999983

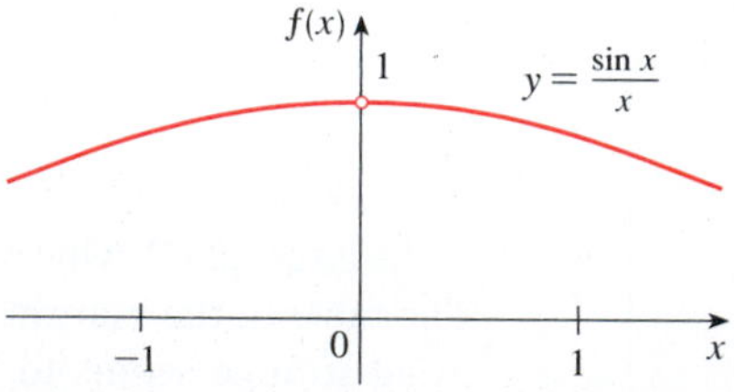

FIGURE 6

COMPUTER ALGEBRA SYSTEMS

Computer algebra systems (CAS) have commands that compute limits. Because of the types of pitfalls demonstrated in Examples 2, 4, and 5, they do not find limits by numerical experimentation. Instead, they use more sophisticated techniques such as computing infinite series. If you have access to a CAS, use the limit command to compute the limits in the examples of this section and to check your answers in the exercises of this chapter.

EXAMPLE 4 Find $\lim_{x \to 0} \sin \dfrac{\pi}{x}$.

SOLUTION Once again the function $f(x) = \sin(\pi/x)$ is undefined at 0. Evaluating the function for some small values of x, we get

$$f(1) = \sin \pi = 0 \qquad f(\tfrac{1}{2}) = \sin 2\pi = 0$$

$$f(\tfrac{1}{3}) = \sin 3\pi = 0 \qquad f(\tfrac{1}{4}) = \sin 4\pi = 0$$

$$f(0.1) = \sin 10\pi = 0 \qquad f(0.01) = \sin 100\pi = 0$$

Similarly, $f(0.001) = f(0.0001) = 0$. On the basis of this information we might be tempted to guess that

$$\lim_{x \to 0} \sin \frac{\pi}{x} = 0$$

but this time our guess is wrong. Note that although $f(1/n) = \sin n\pi = 0$ for any integer n, it is also true that $f(x) = 1$ for infinitely many values of x that approach 0. [In fact, $\sin(\pi/x) = 1$ when

$$\frac{\pi}{x} = \frac{\pi}{2} + 2n\pi$$

and, solving for x, we get $x = 2/(4n + 1)$.] The graph of f is given in Figure 7.

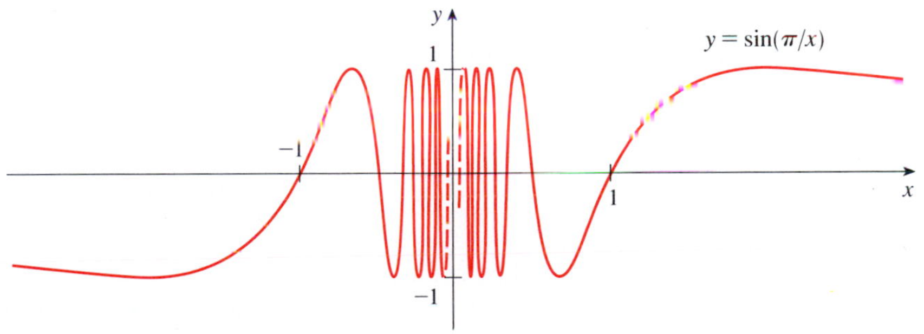

FIGURE 7

The broken lines indicate that the values of $\sin(\pi/x)$ oscillate between 1 and -1 infinitely often as x approaches 0. (See Exercise 31.) Since the values of $f(x)$ do not approach a fixed number as x approaches 0,

$$\lim_{x \to 0} \sin \frac{\pi}{x} \text{ does not exist}$$

■

x	$x^3 + \dfrac{\cos 5x}{10{,}000}$
1	1.000028
0.5	0.124920
0.1	0.001088
0.05	0.000222
0.01	0.000101

EXAMPLE 5 Find $\lim_{x \to 0} \left(x^3 + \dfrac{\cos 5x}{10{,}000} \right)$.

SOLUTION As before, we construct a table of values. From the table in the margin it appears that

$$\lim_{x \to 0} \left(x^3 + \frac{\cos 5x}{10{,}000} \right) = 0$$

x	$x^3 + \dfrac{\cos 5x}{10{,}000}$
0.005	0.00010009
0.001	0.00010000

But if we persevere with smaller values of x, the table in the margin suggests that

$$\lim_{x \to 0} \left(x^3 + \frac{\cos 5x}{10{,}000} \right) = 0.000100 = \frac{1}{10{,}000}$$

Later we will see that $\lim_{x \to 0} \cos 5x = 1$ and then it follows that the limit is 0.0001. ■

Examples 4 and 5 illustrate some of the pitfalls in guessing the value of a limit. It is easy to guess the wrong value if we use inappropriate values of x, but it is difficult to know when to stop calculating values. And, as the discussion after Example 2 shows, sometimes calculators and computers give the wrong values. In the next two sections, however, we develop foolproof methods for calculating limits.

EXAMPLE 6 The Heaviside function H is defined by

$$H(t) = \begin{cases} 0 & \text{if } t < 0 \\ 1 & \text{if } t \geqslant 0 \end{cases}$$

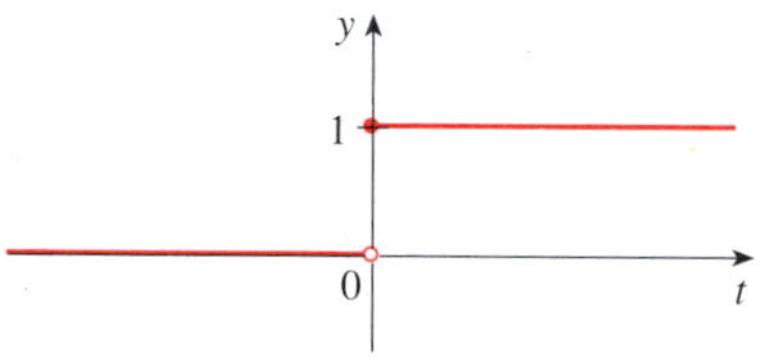

FIGURE 8

[This function is named after the electrical engineer Oliver Heaviside (1850–1925) and can be used to describe an electric current that is switched on at time $t = 0$.] Its graph is shown in Figure 8.

As t approaches 0 from the left, $H(t)$ approaches 0. As t approaches 0 from the right, $H(t)$ approaches 1. There is no single number that $H(t)$ approaches as t approaches 0. Therefore $\lim_{t \to 0} H(t)$ does not exist. ■

ONE-SIDED LIMITS

We noticed in Example 6 that $H(t)$ approaches 0 as t approaches 0 from the left and $H(t)$ approaches 1 as t approaches 0 from the right. We indicate this situation symbolically by writing

$$\lim_{t \to 0^-} H(t) = 0 \qquad \text{and} \qquad \lim_{t \to 0^+} H(t) = 1$$

The symbol "$t \to 0^-$" indicates that we consider only values of t that are less than 0. Likewise, "$t \to 0^+$" indicates that we consider only values of t that are greater than 0.

(2) DEFINITION We write

$$\lim_{x \to a^-} f(x) = L$$

and say the **left-hand limit of $f(x)$ as x approaches a** [or the **limit of $f(x)$ as x approaches a from the left**] is equal to L if we can make the values of $f(x)$ arbitrarily close to L by taking x to be sufficiently close to a and x less than a.

Notice that Definition 2 differs from Definition 1 only in that we require x to be less than a. Similarly, if we require that x be greater than a, we get "the **right-hand limit of $f(x)$ as x approaches a** is equal to L" and we write

$$\lim_{x \to a^+} f(x) = L$$

Thus, the symbol "$x \to a^+$" means that we consider only $x > a$. These definitions are illustrated in Figure 9.

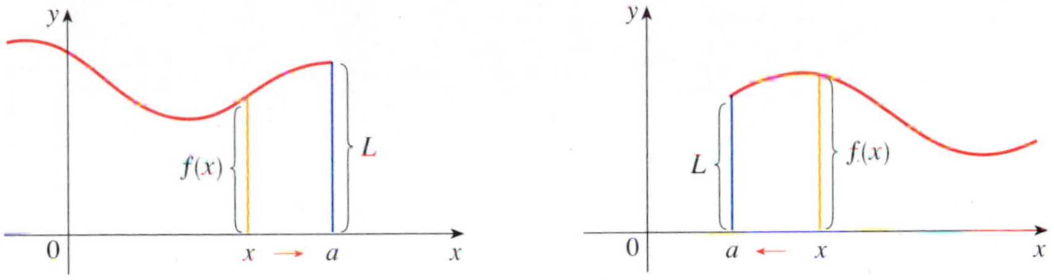

FIGURE 9

(a) $\lim_{x \to a^-} f(x) = L$ (b) $\lim_{x \to a^+} f(x) = L$

By comparing Definition 1 with the definitions of one-sided limits, we see that the following is true.

(3) $\lim_{x \to a} f(x) = L$ if and only if $\lim_{x \to a^-} f(x) = L$ and $\lim_{x \to a^+} f(x) = L$

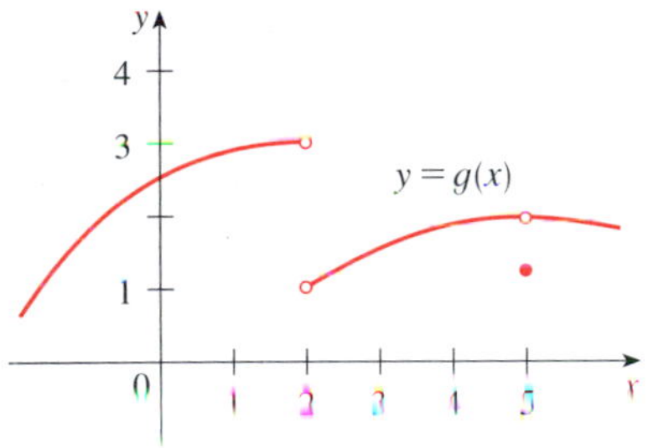

FIGURE 10

EXAMPLE 7 The graph of a function g is shown in Figure 10. Use it to state the values (if they exist) of the following:

(a) $\lim_{x \to 2^-} g(x)$ (b) $\lim_{x \to 2^+} g(x)$

(c) $\lim_{x \to 2} g(x)$ (d) $\lim_{x \to 5^-} g(x)$

(e) $\lim_{x \to 5^+} g(x)$ (f) $\lim_{x \to 5} g(x)$

SOLUTION From the graph we see that

$$\text{(a)} \lim_{x \to 2^-} g(x) = 3 \quad \text{and} \quad \text{(b)} \lim_{x \to 2^+} g(x) = 1$$

(c) Since the left and right limits are different, we conclude from (3) that $\lim_{x \to 2} g(x)$ does not exist.

The graph also shows that

$$\text{(d)} \lim_{x \to 5^-} g(x) = 2 \quad \text{and} \quad \text{(e)} \lim_{x \to 5^+} g(x) = 2$$

(f) This time the left and right limits are the same and so, by (3), we have

$$\lim_{x \to 5} g(x) = 2$$

Despite this fact, notice that $g(5) \neq 2$. ■

INFINITE LIMITS

EXAMPLE 8 Find $\lim_{x \to 0} \dfrac{1}{x^2}$ if it exists.

SOLUTION As x becomes close to 0, x^2 also becomes close to 0, and $1/x^2$ becomes very large. (See the following table.) In fact, it appears from the graph of the function $f(x) = 1/x^2$ shown in Figure 11 that the values of $f(x)$ can be made arbitrarily large by taking x close enough to 0. Thus, the values of $f(x)$ do not approach a number, so $\lim_{x\to 0} (1/x^2)$ does not exist.

x	$\frac{1}{x^2}$
± 1	1
± 0.5	4
± 0.2	25
± 0.1	100
± 0.05	400
± 0.01	10,000
± 0.001	1,000,000

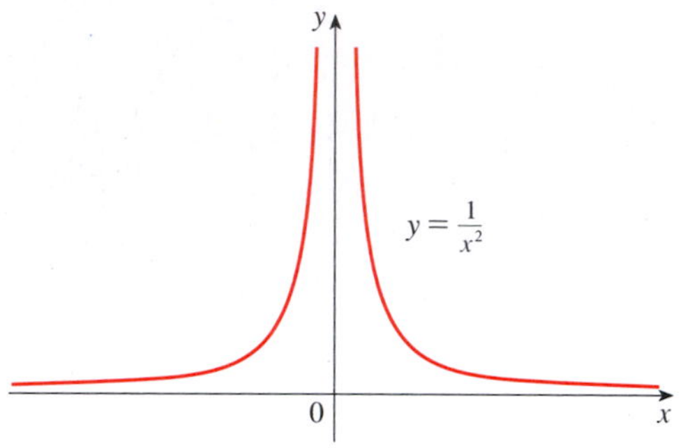

FIGURE 11

To indicate the kind of behavior exhibited in Example 8 we use the notation

$$\lim_{x\to 0} \frac{1}{x^2} = \infty$$

This doesn't mean that we are regarding ∞ as a number. Nor does it mean that the limit exists. It simply expresses the particular way in which the limit does not exist: $1/x^2$ can be made as large as we like by taking x close enough to 0.

In general, we write symbolically

$$\lim_{x\to a} f(x) = \infty$$

to indicate that the values of $f(x)$ become larger and larger (or "increase without bound") as x becomes closer and closer to a.

(4) DEFINITION Let f be a function defined on both sides of a, except possibly at a itself. Then

$$\lim_{x\to a} f(x) = \infty$$

means that the values of $f(x)$ can be made arbitrarily large (as large as we please) by taking x sufficiently close to a (but not equal to a).

Another notation for $\lim_{x\to a} f(x) = \infty$ is

$$f(x) \to \infty \qquad \text{as} \qquad x \to a$$

Again the symbol ∞ is not a number, but the expression $\lim_{x\to a} f(x) = \infty$ is often read as

"the limit of $f(x)$, as x approaches a, is infinity"

or "$f(x)$ becomes infinite as x approaches a"

or "$f(x)$ increases without bound as x approaches a"

This definition is illustrated graphically in Figure 12.

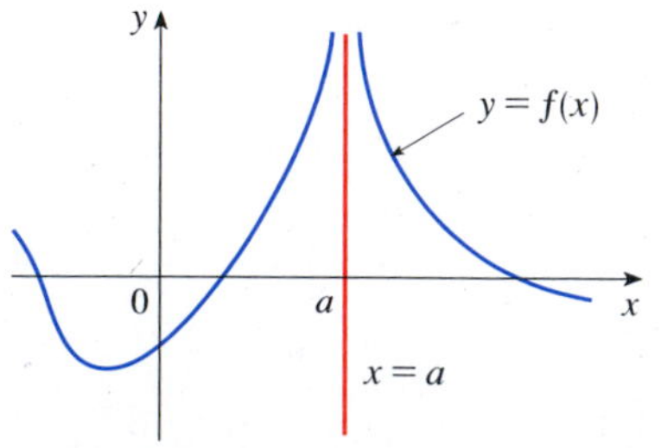

FIGURE 12
$\lim_{x\to a} f(x) = \infty$

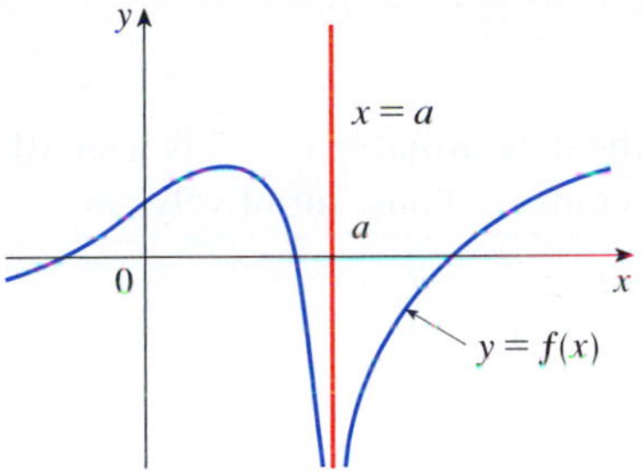

FIGURE 13

$\lim_{x \to a} f(x) = -\infty$

A similar sort of limit, for functions that become large negative as x gets close to a, is defined in Definition 5 and is illustrated in Figure 13.

(5) DEFINITION Let f be defined on both sides of a, except possibly at a itself. Then

$$\lim_{x \to a} f(x) = -\infty$$

means that the values of $f(x)$ can be made arbitrarily large negative by taking x sufficiently close to a (but not equal to a).

The symbol $\lim_{x \to a} f(x) = -\infty$ can be read as "the limit of $f(x)$, as x approaches a, is negative infinity" or "$f(x)$ decreases without bound as x approaches a." As an example we have

$$\lim_{x \to 0} \left(-\frac{1}{x^2} \right) = -\infty$$

Similar definitions can be given for the one-sided infinite limits

$$\lim_{x \to a^-} f(x) = \infty \qquad \lim_{x \to a^+} f(x) = \infty$$

$$\lim_{x \to a^-} f(x) = -\infty \qquad \lim_{x \to a^+} f(x) = -\infty$$

remembering that "$x \to a^-$" means that we consider only values of x that are less than a, and similarly "$x \to a^+$" means that we consider only $x > a$. Illustrations of these four cases are given in Figure 14.

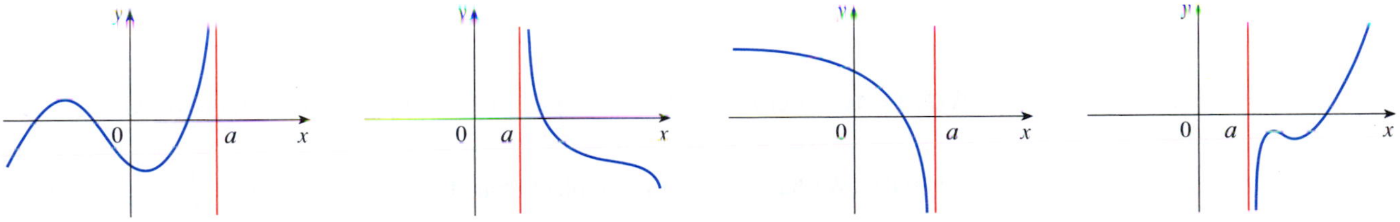

(a) $\lim_{x \to a^-} f(x) = \infty$ (b) $\lim_{x \to a^+} f(x) = \infty$ (c) $\lim_{x \to a^-} f(x) = -\infty$ (d) $\lim_{x \to a^+} f(x) = -\infty$

FIGURE 14

(6) DEFINITION The line $x = a$ is called a **vertical asymptote** of the curve $y = f(x)$ if at least one of the following statements is true:

$$\lim_{x \to a} f(x) = \infty \qquad \lim_{x \to a^-} f(x) = \infty \qquad \lim_{x \to a^+} f(x) = \infty$$

$$\lim_{x \to a} f(x) = -\infty \qquad \lim_{x \to a^-} f(x) = -\infty \qquad \lim_{x \to a^+} f(x) = -\infty$$

For instance, the y-axis is a vertical asymptote of the curve $y = 1/x^2$ because $\lim_{x \to 0} (1/x^2) = \infty$. In Figure 14 the line $x = a$ is a vertical asymptote in each of the four cases shown.

EXAMPLE 9 Find $\lim_{x \to 3^+} \frac{2}{x - 3}$ and $\lim_{x \to 3^-} \frac{2}{x - 3}$.

SOLUTION If x is close to 3 but larger than 3, then the denominator $x - 3$ is a small positive number and so $2/(x - 3)$ is a large positive number. Thus, intuitively, we see that

$$\lim_{x \to 3^+} \frac{2}{x - 3} = \infty$$

Likewise, if x is close to 3 but smaller than 3, then $x - 3$ is a small negative number and so $2/(x - 3)$ is a numerically large negative number. Thus

$$\lim_{x \to 3^-} \frac{2}{x - 3} = -\infty$$

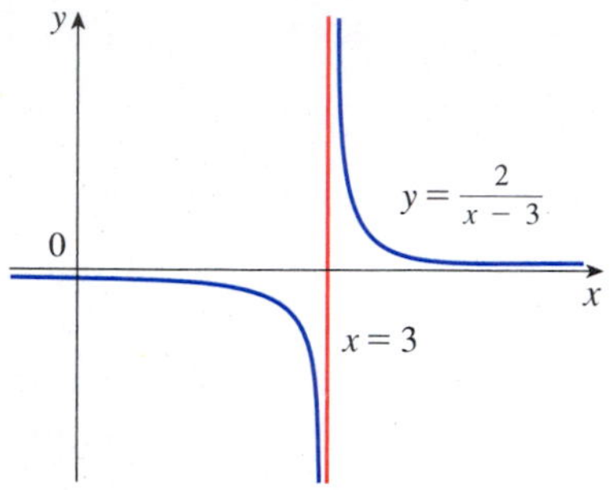

FIGURE 15

The graph of the curve $y = 2/(x - 3)$ is given in Figure 15. The line $x = 3$ is a vertical asymptote. ■

LIMITS OF VECTOR FUNCTIONS

Limits of a vector function $\mathbf{r}(t) = \langle f(t), g(t) \rangle$ essentially involve limits of the two component functions f and g. We use the following notation.

Two vectors, **a** and **b**, are close if $|\mathbf{a} - \mathbf{b}|$ is small or, equivalently, each component of $\mathbf{a} - \mathbf{b}$ is small.

(7) DEFINITION We write

$$\lim_{t \to a} \mathbf{r}(t) = \mathbf{b}$$

and say "the limit of $\mathbf{r}(t)$, as t approaches a, equals $\mathbf{b}$"

if we can make the vectors $\mathbf{r}(t)$ arbitrarily close to $\mathbf{b}$ (as close to $\mathbf{b}$ as we like) by taking t to be sufficiently close to a but not equal to a.

EXAMPLE 10 Let $\mathbf{r}(t) = \left(\frac{t - 1}{t^2 - 1} \right) \mathbf{i} + (t^2 + 1)\,\mathbf{j}$. Find $\lim_{t \to 1} \mathbf{r}(t)$ if it exists.

SOLUTION We have seen in Example 1 that if t is close to 1, then $\frac{t - 1}{t^2 - 1}$ is close to $\frac{1}{2}$. Also, if t is close to 1, then $t^2 + 1$ is close to 2. This implies that if t is close to 1, then $\left(\frac{t - 1}{t^2 - 1} \right) \mathbf{i} + (t^2 + 1)\,\mathbf{j}$ is close to $\frac{1}{2}\mathbf{i} + 2\mathbf{j}$, and we guess that

$$\lim_{t \to 1} \mathbf{r}(t) = \left[\lim_{t \to 1} \frac{t - 1}{t^2 - 1} \right] \mathbf{i} + \left[\lim_{t \to 1} (t^2 + 1) \right] \mathbf{j} = \tfrac{1}{2}\mathbf{i} + 2\mathbf{j}$$ ■

Example 10 illustrates that to find the limit of a vector function we take the limit of each component.

If $\mathbf{r}(t) = \langle f(t), g(t) \rangle$, then

$$\lim_{t \to a} \mathbf{r}(t) = \left\langle \lim_{t \to a} f(t), \lim_{t \to a} g(t) \right\rangle$$

provided the limits of the component functions exist.

EXERCISES 2.2

1. For the function f whose graph is given, state the value of the given quantity, if it exists.
   (a) $\lim_{x\to 1} f(x)$ (b) $\lim_{x\to 3^-} f(x)$ (c) $\lim_{x\to 3^+} f(x)$
   (d) $\lim_{x\to 3} f(x)$ (e) $f(3)$ (f) $\lim_{x\to -2^-} f(x)$
   (g) $\lim_{x\to -2^+} f(x)$ (h) $\lim_{x\to -2} f(x)$ (i) $f(-2)$

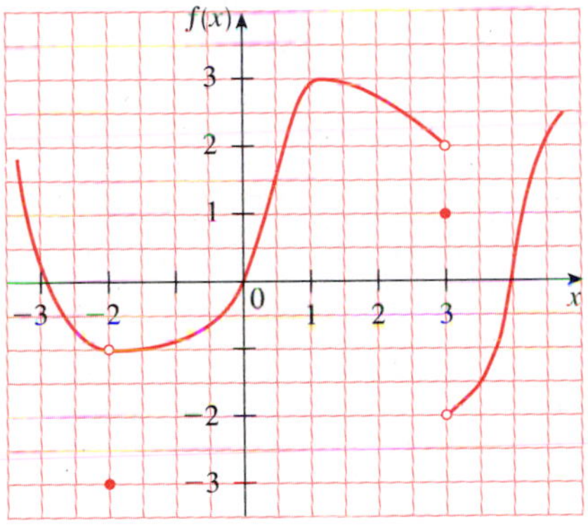

2. For the function g whose graph is given, state the value of the given quantity, if it exists.
   (a) $\lim_{x\to -2^-} g(x)$ (b) $\lim_{x\to -2^+} g(x)$ (c) $\lim_{x\to -2} g(x)$
   (d) $g(-2)$ (e) $\lim_{x\to 2^-} g(x)$ (f) $\lim_{x\to 2^+} g(x)$
   (g) $\lim_{x\to 2} g(x)$ (h) $g(2)$ (i) $\lim_{x\to 4^+} g(x)$
   (j) $\lim_{x\to 4^-} g(x)$ (k) $g(0)$ (l) $\lim_{x\to 0} g(x)$

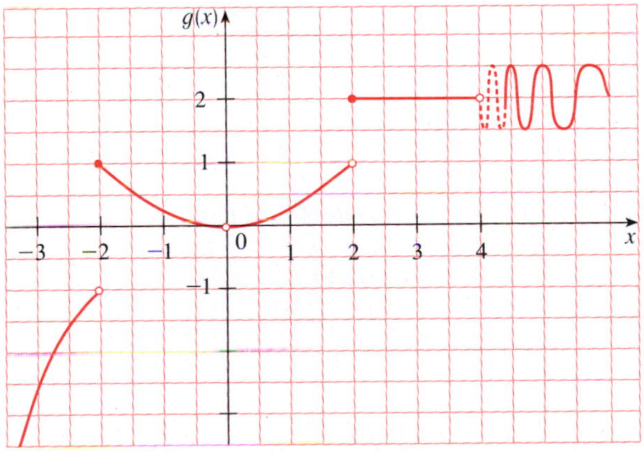

3. State the value of the limit, if it exists, from the given graph.
   (a) $\lim_{x\to 3} f(x)$ (b) $\lim_{x\to 1} f(x)$ (c) $\lim_{x\to -3} f(x)$
   (d) $\lim_{x\to 2^-} f(x)$ (e) $\lim_{x\to 2^+} f(x)$ (f) $\lim_{x\to 2} f(x)$

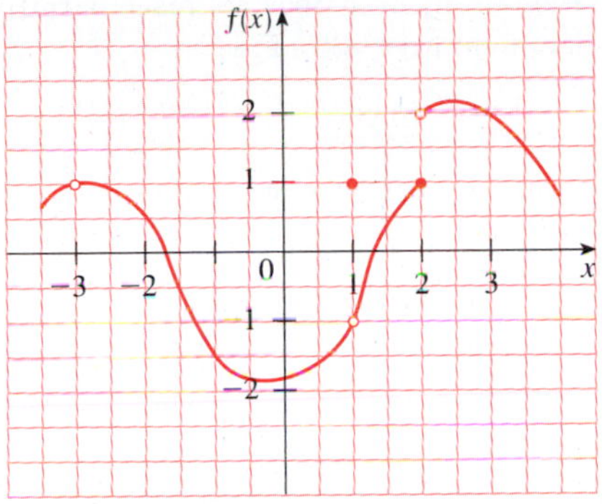

4. State the value of the limit, if it exists, from the given graph.
   (a) $\lim_{x\to 1} g(x)$ (b) $\lim_{x\to 0} g(x)$ (c) $\lim_{x\to 2} g(x)$
   (d) $\lim_{x\to -2} g(x)$ (e) $\lim_{x\to -1^-} g(x)$ (f) $\lim_{x\to -1} g(x)$

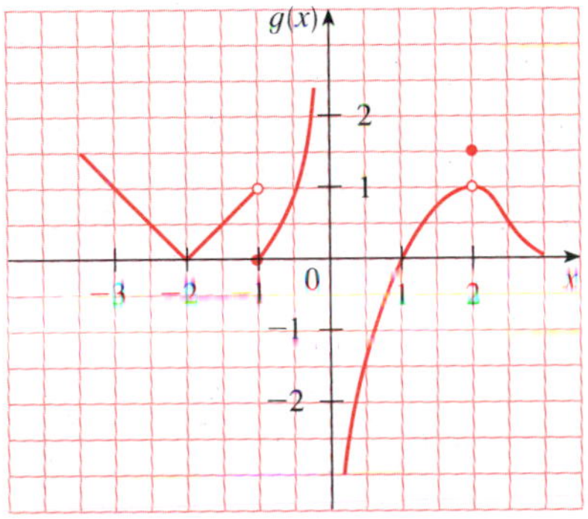

5. For the function f whose graph is shown, state the following.
   (a) $\lim_{x\to 3} f(x)$ (b) $\lim_{x\to 7} f(x)$ (c) $\lim_{x\to -4} f(x)$
   (d) $\lim_{x\to -9^-} f(x)$ (e) $\lim_{x\to -9^+} f(x)$
   (f) The equations of the vertical asymptotes

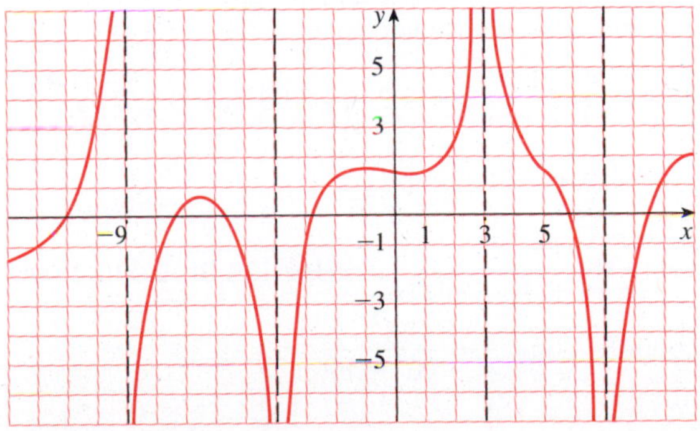

6. A patient receives a 150-mg injection of a drug every 4 hours. The graph shows the amount $f(t)$ of the drug in the bloodstream after t hours. (Later we will be able to compute dosages and time intervals to ensure that the concentration of the drug does not reach harmful levels. See Problem 6 on page 720.) Find

$$\lim_{t \to 12^-} f(t) \qquad \text{and} \qquad \lim_{t \to 12^+} f(t)$$

and explain the significance of these one-sided limits.

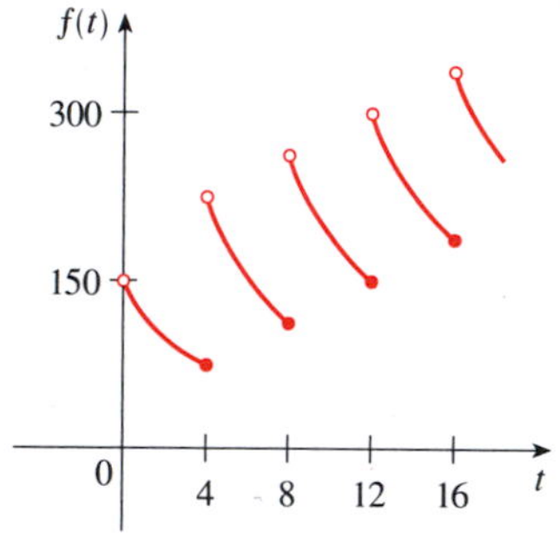

7. (a) Sketch the graph of the function

$$f(x) = \begin{cases} 1 - x & \text{if } x \leq 0 \\ 1 & \text{if } x > 0 \end{cases}$$

(b) Use the graph from part (a) to state the value of each of the following limits, if it exists.

(i) $\lim_{x \to 0^-} f(x)$ (ii) $\lim_{x \to 0^+} f(x)$ (iii) $\lim_{x \to 0} f(x)$

8. (a) Sketch the graph of the function

$$g(x) = \begin{cases} 2 - x & \text{if } x < -1 \\ x & \text{if } -1 \leq x < 1 \\ 4 & \text{if } x = 1 \\ 4 - x & \text{if } x > 1 \end{cases}$$

(b) Use the graph from part (a) to state the value of each of the following limits, if it exists.

(i) $\lim_{x \to -1^-} g(x)$ (ii) $\lim_{x \to -1^+} g(x)$ (iii) $\lim_{x \to -1} g(x)$

(iv) $\lim_{x \to 1^-} g(x)$ (v) $\lim_{x \to 1^+} g(x)$ (vi) $\lim_{x \to 1} g(x)$

9–16 ■ Evaluate the function at the given numbers (correct to six decimal places). Use the results to guess the value of the limit, or state that it does not exist.

9. $g(x) = \dfrac{x - 1}{x^3 - 1}$; $x = 0.2, 0.4, 0.6, 0.8, 0.9, 0.99, 1.8, 1.6,$ $1.4, 1.2, 1.1, 1.01$; $\lim_{x \to 1} \dfrac{x - 1}{x^3 - 1}$

10. $g(x) = \dfrac{1 - x^2}{x^2 + 3x - 10}$; $x = 3, 2.1, 2.01, 2.001, 2.0001,$ 2.00001; $\lim_{x \to 2^+} \dfrac{1 - x^2}{x^2 + 3x - 10}$

11. $F(x) = \dfrac{(1/\sqrt{x}) - \frac{1}{5}}{x - 25}$; $x = 26, 25.5, 25.1, 25.05, 25.01,$ $24, 24.5, 24.9, 24.95, 24.99$; $\lim_{x \to 25} \dfrac{(1/\sqrt{x}) - \frac{1}{5}}{x - 25}$

12. $F(t) = \dfrac{\sqrt[3]{t} - 1}{\sqrt{t} - 1}$; $t = 1.5, 1.2, 1.1, 1.01, 1.001$; $\lim_{t \to 1} \dfrac{\sqrt[3]{t} - 1}{\sqrt{t} - 1}$

13. $f(x) = \dfrac{1 - \cos x}{x^2}$; $x = 1, 0.5, 0.4, 0.3, 0.2, 0.1, 0.05, 0.01$; $\lim_{x \to 0} \dfrac{1 - \cos x}{x^2}$

14. $g(x) = \dfrac{\cos x - 1}{\sin x}$; $x = 1, 0.5, 0.4, 0.3, 0.2, 0.1, 0.05,$ 0.01; $\lim_{x \to 0} \dfrac{\cos x - 1}{\sin x}$

15. $\mathbf{r}(t) = (2t - 1)\,\mathbf{i} + \left(\dfrac{t^2 - 1}{t - 1}\right)\mathbf{j}$;
$t = 0.2, 0.4, 0.6, 0.8, 0.9, 0.99$;
$t = 1.8, 1.6, 1.4, 1.2, 1.1, 1.01$; $\lim_{t \to 1} \mathbf{r}(t)$

16. $\mathbf{r}(t) = \left\langle \dfrac{t^2 - 4t + 4}{t - 2}, \dfrac{t^2 - 4}{2t - 4} \right\rangle$;
$t = 1.2, 1.4, 1.6, 1.8, 1.9, 1.99$;
$t = 2.8, 2.6, 2.4, 2.2, 2.1, 2.01$; $\lim_{t \to 2} \mathbf{r}(t)$

17–22 ■ Determine the infinite limit.

17. $\lim_{x \to 5^+} \dfrac{6}{x - 5}$ **18.** $\lim_{x \to 5^-} \dfrac{6}{x - 5}$ **19.** $\lim_{x \to 3} \dfrac{1}{(x - 3)^8}$

20. $\lim_{x \to 0} \dfrac{x - 1}{x^2(x + 2)}$ **21.** $\lim_{x \to -2^+} \dfrac{x - 1}{x^2(x + 2)}$ **22.** $\lim_{x \to \pi^-} \csc x$

23. Determine $\lim_{x \to 1^-} \dfrac{1}{x^3 - 1}$ and $\lim_{x \to 1^+} \dfrac{1}{x^3 - 1}$

(a) by evaluating $f(x) = 1/(x^3 - 1)$ for values of x that approach 1 from the left and from the right,

(b) by reasoning as in Example 9,

(c) from a graph of f.

24. (a) Find the vertical asymptotes of the function

$$y = \frac{x}{x^2 - x - 2}$$

(b) Confirm your answer to part (a) by graphing the function.

25. The existence of the limit

$$\lim_{x \to 0} (1 + x)^{1/x}$$

will be established in Chapter 10. Estimate the value of the limit to five decimal places by first evaluating the function for $x = 1, 0.1, 0.01, 0.001, 0.0001, 0.00001, 0.000001, 0.0000001, 0.00000001$, and 0.000000001.

26. The slope of the tangent line to the graph of the exponential function $y = 2^x$ at the point $(0, 1)$ is $\lim_{x\to 0} (2^x - 1)/x$. Estimate the slope to three decimal places by first evaluating $(2^x - 1)/x$ for $x = 0.5, 0.1, 0.05, 0.01, 0.005, 0.001, 0.0005$, and 0.0001.

27. (a) Evaluate the function $f(x) = x^2 - (2^x/1000)$ for $x = 1, 0.8, 0.6, 0.4, 0.2, 0.1$, and 0.05, and guess the value of

$$\lim_{x\to 0}\left(x^2 - \frac{2^x}{1000}\right)$$

(b) Evaluate $f(x)$ for $x = 0.04, 0.02, 0.01, 0.005, 0.003$, and 0.001. Guess again.

28. (a) Evaluate $h(x) = (\tan x - x)/x^3$ for $x = 1, 0.5, 0.1, 0.05, 0.01$, and 0.005.

(b) Guess the value of $\lim_{x\to 0} \dfrac{\tan x - x}{x^3}$.

(c) Evaluate $h(x)$ for successively smaller values of x until you finally reach 0 values for $h(x)$. Are you still confident that your guess in part (b) is correct? In Section 4.8 a method for evaluating the limit will be explained. See Appendix G for an explanation of what went wrong with the calculator computations.

29. By graphing the function $f(x) = (\tan 4x)/x$ and zooming in toward the point where the graph crosses the y-axis, estimate the value of $\lim_{x\to 0} f(x)$.

30. Estimate the value of

$$\lim_{x\to 0} \frac{6^x - 2^x}{x}$$

by graphing the function $y = (6^x - 2^x)/x$. State your answer to two decimal places.

31. Graph the function $f(x) = \sin(\pi/x)$ of Example 4 in the viewing rectangle $[-1, 1]$ by $[-1, 1]$. Then zoom in toward the origin several times. Comment on the behavior of this function.

32. Graph the function h of Exercise 28 in the viewing rectangle $[-1, 1]$ by $[0, 1]$. Then zoom in toward the point where the graph crosses the y-axis to estimate the limit of $h(x)$ as x approaches 0. Continue to zoom in until you observe distortions in the graph of h. Compare with the results of Exercise 28.

2.3 CALCULATING LIMITS USING THE LIMIT LAWS

In Section 2.2 we used calculators and graphs to guess the values of limits, but we saw that such methods don't always lead to the correct answer. In this section we use the following properties of limits, called the *Limit Laws*, to calculate limits.

LIMIT LAWS Suppose that c is a constant and the limits

$$\lim_{x\to a} f(x) \quad\text{and}\quad \lim_{x\to a} g(x)$$

exist. Then

1. $\lim_{x\to a} [f(x) + g(x)] = \lim_{x\to a} f(x) + \lim_{x\to a} g(x)$

2. $\lim_{x\to a} [f(x) - g(x)] = \lim_{x\to a} f(x) - \lim_{x\to a} g(x)$

3. $\lim_{x\to a} [cf(x)] = c \lim_{x\to a} f(x)$

4. $\lim_{x\to a} [f(x)g(x)] = \lim_{x\to a} f(x) \cdot \lim_{x\to a} g(x)$

5. $\lim_{x\to a} \dfrac{f(x)}{g(x)} = \dfrac{\lim_{x\to a} f(x)}{\lim_{x\to a} g(x)}$ if $\lim_{x\to a} g(x) \neq 0$

The analogues of these laws also hold when f and g are vector functions (except for Law 5). If f and g are vector-valued, then the product in Law 4 should be interpreted as a dot product.

The five limit laws can be stated verbally as follows:

Sum Law
1. The limit of a sum is the sum of the limits.

Difference Law
2. The limit of a difference is the difference of the limits.

Constant Multiple Law
3. The limit of a constant times a function is the constant times the limit of the function.

Product Law
4. The limit of a product is the product of the limits.

Quotient Law
5. The limit of a quotient is the quotient of the limits (provided that the limit of the denominator is not 0).

It's easy to believe that these properties are true. For instance, if $f(x)$ is close to L and $g(x)$ is close to M, it's reasonable to conclude that $f(x) + g(x)$ is close to $L + M$. This gives us an intuitive basis for believing that Law 1 is true. In Section 2.4 we give a precise definition of a limit and use it to prove this law. The proofs of the remaining laws are given in Appendix F.

If we use the Product Law repeatedly with $g(x) = f(x)$, we obtain the following law.

Power Law

6. $\lim_{x \to a} [f(x)]^n = \left[\lim_{x \to a} f(x)\right]^n$ where n is a positive integer

In applying these six limit laws we need to use two special limits:

7. $\lim_{x \to a} c = c$ **8.** $\lim_{x \to a} x = a$

These limits are obvious from an intuitive point of view (state them in words or draw graphs of $y = c$ and $y = x$), but proofs based on the precise definition are requested in the exercises for Section 2.4.

If we now put $f(x) = x$ in Law 6 and use Law 8, we get another useful special limit.

9. $\lim_{x \to a} x^n = a^n$ where n is a positive integer

A similar limit holds for roots as follows. (For square roots the proof is outlined in Exercise 31 in Section 2.4.)

10. $\lim_{x \to a} \sqrt[n]{x} = \sqrt[n]{a}$ where n is a positive integer
(If n is even, we assume that $a > 0$.)

More generally, we have the following law, which is proved as a consequence of Law 10 in Section 2.5.

Root Law

11. $\lim_{x \to a} \sqrt[n]{f(x)} = \sqrt[n]{\lim_{x \to a} f(x)}$ where n is a positive integer

(If n is even, we assume that $\lim_{x \to a} f(x) > 0$.)

EXAMPLE 1 Find $\lim_{x \to 5} (2x^2 - 3x + 4)$ and justify each step.

SOLUTION

$$\begin{aligned}
\lim_{x \to 5} (2x^2 - 3x + 4) &= \lim_{x \to 5} (2x^2) - \lim_{x \to 5} (3x) + \lim_{x \to 5} 4 && \text{(by Laws 2 and 1)} \\
&= 2 \lim_{x \to 5} x^2 - 3 \lim_{x \to 5} x + \lim_{x \to 5} 4 && \text{(by 3)} \\
&= 2(5^2) - 3(5) + 4 && \text{(by 9, 8, and 7)} \\
&= 39
\end{aligned}$$

■

EXAMPLE 2 Find $\lim_{x \to -2} \dfrac{x^3 + 2x^2 - 1}{5 - 3x}$ and justify each step.

SOLUTION We start by using Law 5, but its use is only fully justified at the final stage when we see that the limits of the numerator and denominator exist and the limit of the denominator is not 0.

$$\begin{aligned}
\lim_{x \to -2} \frac{x^3 + 2x^2 - 1}{5 - 3x} &= \frac{\lim_{x \to -2} (x^3 + 2x^2 - 1)}{\lim_{x \to -2} (5 - 3x)} && \text{(by Law 5)} \\
&= \frac{\lim_{x \to -2} x^3 + 2 \lim_{x \to -2} x^2 - \lim_{x \to -2} 1}{\lim_{x \to -2} 5 - 3 \lim_{x \to -2} x} && \text{(by 1, 2, and 3)} \\
&= \frac{(-2)^3 + 2(-2)^2 - 1}{5 - 3(-2)} && \text{(by 9, 8, and 7)} \\
&= -\frac{1}{11}
\end{aligned}$$

■

EXAMPLE 3 Calculate $\lim_{x \to 1} \left[\sqrt[5]{x^2 - x} + (x^3 + x)^9\right]$ and justify each step.

SOLUTION

$$\begin{aligned}
&\lim_{x \to 1} \left[\sqrt[5]{x^2 - x} + (x^3 + x)^9\right] \\
&\quad = \lim_{x \to 1} \sqrt[5]{x^2 - x} + \lim_{x \to 1} (x^3 + x)^9 && \text{(by Law 1)} \\
&\quad = \sqrt[5]{\lim_{x \to 1} (x^2 - x)} + \left[\lim_{x \to 1} (x^3 + x)\right]^9 && \text{(by 11 and 6)} \\
&\quad = \sqrt[5]{\lim_{x \to 1} x^2 - \lim_{x \to 1} x} + \left[\lim_{x \to 1} x^3 + \lim_{x \to 1} x\right]^9 && \text{(by 2 and 1)} \\
&\quad = \sqrt[5]{1^2 - 1} + [1^3 + 1]^9 && \text{(by 9 and 8)} \\
&\quad = 2^9 = 512
\end{aligned}$$

■

NEWTON AND LIMITS

Isaac Newton was born on Christmas Day in 1642, the year of Galileo's death. When he entered Cambridge University in 1661 Newton didn't know much mathematics, but he learned quickly by reading Euclid and Descartes and by attending the lectures of Isaac Barrow. Cambridge was closed because of the plague in 1665 and 1666, and Newton returned home to reflect on what he had learned. Those two years were amazingly productive for at that time he made four of his major discoveries: (1) his representation of functions as sums of infinite series, including the binomial theorem, (2) his work on differential and integral calculus, (3) his laws of motion and law of universal gravitation, and (4) his prism experiments on the nature of light and color. Because of a fear of controversy and criticism, he was reluctant to publish his discoveries and it wasn't until 1687, at the urging of the astronomer Halley, that Newton published *Principia Mathematica*. In this work, the greatest scientific treatise ever written, Newton set forth his version of calculus and used it to investigate mechanics, fluid dynamics, and wave motion, and to explain the motion of planets and comets.

The beginnings of calculus are found in the calculations of areas and volumes by ancient Greek scholars such as Eudoxus and Archimedes. Although aspects of the idea of a limit are implicit in their "method of exhaustion," Eudoxus and Archimedes never explicitly formulated the concept of a limit. Likewise, mathematicians such as Cavalieri, Fermat, and Barrow, the immediate precursors of Newton in the development of calculus, did not actually use limits. It was Isaac Newton who was the first to talk explicitly about limits. He explained that the main idea behind limits is that quantities "approach nearer than by any given difference." Newton stated that the limit was the basic concept in calculus, but it was left to later mathematicians like Cauchy (see page 106) to clarify his ideas about limits.

NOTE: If we let $f(x) = 2x^2 - 3x + 4$, then $f(5) = 39$. In other words, we would have gotten the correct answer in Example 1 by substituting 5 for x. Similarly, direct substitution provides the correct answers in Examples 2 and 3. The functions in Examples 1 and 2 are a polynomial and a rational function, respectively, and similar use of the Limit Laws proves that direct substitution always works for such functions (see Exercises 73 and 74). We state this fact as follows.

> If f is a polynomial or a rational function and a is in the domain of f, then
>
> $$\lim_{x \to a} f(x) = f(a)$$

Functions with this direct substitution property are called *continuous at a* and will be studied in Section 2.5. However, not all limits can be evaluated by direct substitution, as the following examples show.

EXAMPLE 4 Find $\displaystyle \lim_{x \to 1} \frac{x^2 - 1}{x - 1}$.

SOLUTION Let $f(x) = (x^2 - 1)/(x - 1)$. We cannot find the limit by substituting $x = 1$ because $f(1)$ is not defined. Nor can we apply the Quotient Law because the limit of the denominator is 0. Instead we need to do some preliminary algebra. We factor the numerator as a difference of squares:

$$\frac{x^2 - 1}{x - 1} = \frac{(x - 1)(x + 1)}{x - 1}$$

The numerator and denominator have a common factor of $x - 1$. When we take the limit as x approaches 1, we have $x \neq 1$ and so $x - 1 \neq 0$. Therefore, we can cancel the common factor and compute the limit as follows:

$$\begin{aligned} \lim_{x \to 1} \frac{x^2 - 1}{x - 1} &= \lim_{x \to 1} \frac{(x - 1)(x + 1)}{x - 1} \\ &= \lim_{x \to 1} (x + 1) \\ &= 1 + 1 = 2 \end{aligned}$$

The limit in this example arose in Section 2.1 when we were trying to find the tangent to the parabola $y = x^2$ at the point (1, 1). ■

EXAMPLE 5 Find $\lim_{x \to 1} g(x)$ where

$$g(x) = \begin{cases} x + 1 & \text{if } x \neq 1 \\ \pi & \text{if } x = 1 \end{cases}$$

SOLUTION Here g is defined at $x = 1$ and $g(1) = \pi$, but the value of a limit as x approaches 1 does not depend on the value of the function at 1. Since $g(x) = x + 1$ for $x \neq 1$, we have

$$\lim_{x \to 1} g(x) = \lim_{x \to 1} (x + 1) = 2$$ ■

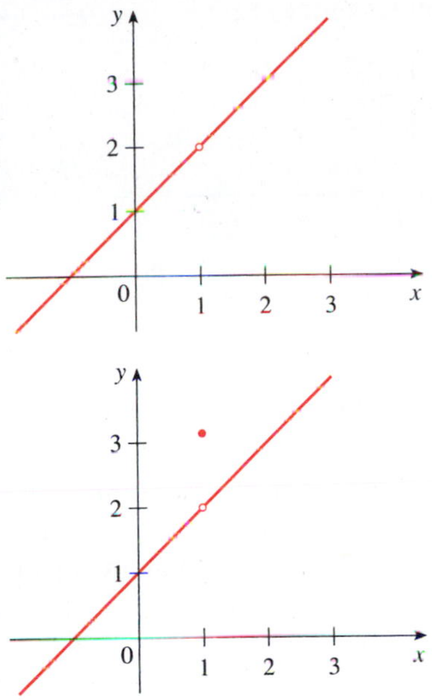

FIGURE 1

The graphs of function f (from Example 3) and g from Example 4.

Note that the values of the functions in Examples 4 and 5 are identical except when $x = 1$ (see Figure 1) and so they have the same limit as x approaches 1.

EXAMPLE 6 If

$$\mathbf{r}(t) = \left\langle \frac{t-1}{t^2-1}, t^2+1 \right\rangle$$

find $\lim_{t\to 1} \mathbf{r}(t)$.

SOLUTION As in Example 4, we have

$$\lim_{t\to 1} \frac{t-1}{t^2-1} = \lim_{t\to 1} \frac{t-1}{(t-1)(t+1)}$$
$$= \lim_{t\to 1} \frac{1}{t+1}$$
$$= \frac{1}{2}$$

Also, from the Limit Laws, $\lim_{t\to 1} (t^2+1) = 2$. So,

$$\lim_{t\to 1} \mathbf{r}(t) = \left\langle \lim_{t\to 1} \frac{t-1}{t^2-1}, \lim_{t\to 1} (t^2+1) \right\rangle$$
$$= \left\langle \tfrac{1}{2}, 2 \right\rangle$$

This agrees with our guess in Example 10 of Section 2.2. ■

EXAMPLE 7 Evaluate $\displaystyle\lim_{h\to 0} \frac{(3+h)^2 - 9}{h}$.

SOLUTION If we define

$$F(h) = \frac{(3+h)^2 - 9}{h}$$

then, as in Example 4, we cannot compute $\lim_{h\to 0} F(h)$ by letting $h = 0$ since $F(0)$ is undefined. But if we simplify $F(h)$ algebraically, we find

$$F(h) = \frac{(9 + 6h + h^2) - 9}{h} = \frac{6h + h^2}{h} = 6 + h$$

(Recall that we consider only $h \neq 0$ when letting h approach 0.) Thus

$$\lim_{h\to 0} \frac{(3+h)^2 - 9}{h} = \lim_{h\to 0} (6 + h) = 6$$ ■

EXAMPLE 8 Find $\displaystyle\lim_{t\to 0} \frac{\sqrt{t^2+9} - 3}{t^2}$.

SOLUTION We cannot apply the Quotient Law immediately, since the limit of the denominator is 0. Here the preliminary algebra consists of rationalizing the numerator:

$$\lim_{t \to 0} \frac{\sqrt{t^2 + 9} - 3}{t^2} = \lim_{t \to 0} \frac{\sqrt{t^2 + 9} - 3}{t^2} \cdot \frac{\sqrt{t^2 + 9} + 3}{\sqrt{t^2 + 9} + 3}$$

$$= \lim_{t \to 0} \frac{(t^2 + 9) - 9}{t^2(\sqrt{t^2 + 9} + 3)} = \lim_{t \to 0} \frac{t^2}{t^2(\sqrt{t^2 + 9} + 3)}$$

$$= \lim_{t \to 0} \frac{1}{\sqrt{t^2 + 9} + 3} = \frac{1}{\sqrt{\lim_{t \to 0} (t^2 + 9)} + 3} = \frac{1}{3 + 3} = \frac{1}{6}$$

This calculation confirms the guess that we made in Example 2 in Section 2.2. ■

Some limits are best calculated by first finding the left- and right-hand limits. The following theorem is a reminder of what we discovered in Section 2.2. It says that a two-sided limit exists if and only if both of the one-sided limits exist and are equal. (For the proof, see Exercise 34 in Section 2.4.)

(1) THEOREM $\lim_{x \to a} f(x) = L$ if and only if $\lim_{x \to a^-} f(x) = L = \lim_{x \to a^+} f(x)$

When computing one-sided limits we use the fact that the Limit Laws also hold for one-sided limits.

EXAMPLE 9 Show that $\lim_{x \to 0} |x| = 0$.

SOLUTION Recall that

$$|x| = \begin{cases} x & \text{if } x \geqslant 0 \\ -x & \text{if } x < 0 \end{cases}$$

Since $|x| = x$ for $x > 0$, we have

$$\lim_{x \to 0^+} |x| = \lim_{x \to 0^+} x = 0$$

For $x < 0$ we have $|x| = -x$ and so

$$\lim_{x \to 0^-} |x| = \lim_{x \to 0^-} (-x) = 0$$

Therefore, by Theorem 1,

$$\lim_{x \to 0} |x| = 0$$

■

EXAMPLE 10 Prove that $\lim_{x \to 0} \frac{|x|}{x}$ does not exist.

SOLUTION

$$\lim_{x \to 0^+} \frac{|x|}{x} = \lim_{x \to 0^+} \frac{x}{x} = \lim_{x \to 0^+} 1 = 1$$

$$\lim_{x \to 0^-} \frac{|x|}{x} = \lim_{x \to 0^-} \frac{-x}{x} = \lim_{x \to 0^-} (-1) = -1$$

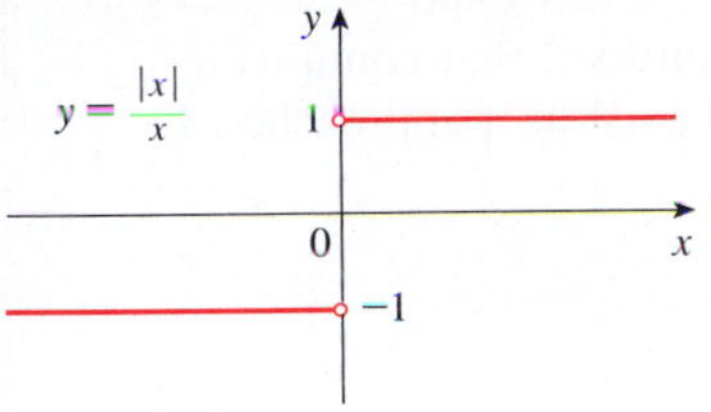

FIGURE 2

Since the right- and left-hand limits are different, it follows from Theorem 1 that $\lim_{x\to 0} |x|/x$ does not exist. The graph of the function $f(x) = |x|/x$ is shown in Figure 2. ■

EXAMPLE 11 If

$$f(x) = \begin{cases} \sqrt{x-4} & \text{if } x > 4 \\ 8 - 2x & \text{if } x < 4 \end{cases}$$

determine whether $\lim_{x\to 4} f(x)$ exists.

It is shown in Example 3 in Section 2.4 that $\lim_{x\to 0^+} \sqrt{x} = 0$.

SOLUTION Since $f(x) = \sqrt{x-4}$ for $x > 4$, we have

$$\lim_{x\to 4^+} f(x) = \lim_{x\to 4^+} \sqrt{x-4} = \sqrt{4-4} = 0$$

Since $f(x) = 8 - 2x$ for $x < 4$, we have

$$\lim_{x\to 4^-} f(x) = \lim_{x\to 4^-} (8 - 2x) = 8 - 2 \cdot 4 = 0$$

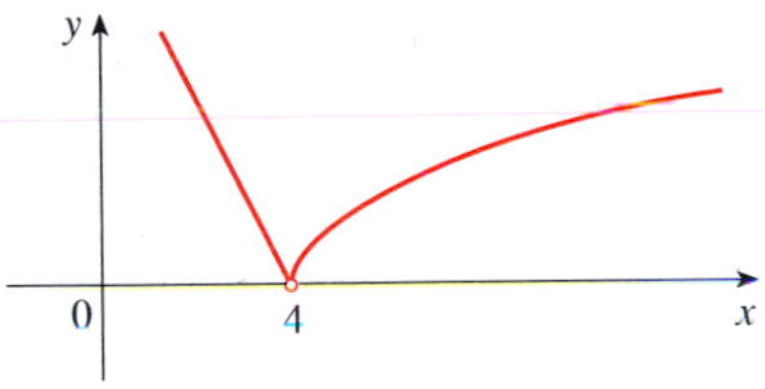

FIGURE 3

The right- and left-hand limits are equal. Thus, the limit exists and

$$\lim_{x\to 4} f(x) = 0$$

The graph of f is shown in Figure 3. ■

EXAMPLE 12 The *greatest integer function* is defined by $[\![x]\!]$ = the largest integer that is less than or equal to x. (For instance, $[\![4]\!] = 4$, $[\![4.8]\!] = 4$, $[\![\pi]\!] = 3$, $[\![\sqrt{2}]\!] = 1$, $[\![-\tfrac{1}{2}]\!] = -1$.) Show that $\lim_{x\to 3} [\![x]\!]$ does not exist.

Other notations for $[\![x]\!]$ are $[x]$ and $\lfloor x \rfloor$.

SOLUTION The graph of the greatest integer function is shown in Figure 4. Since $[\![x]\!] = 3$ for $3 \leqslant x < 4$, we have

$$\lim_{x\to 3^+} [\![x]\!] = \lim_{x\to 3^+} 3 = 3$$

Since $[\![x]\!] = 2$ for $2 \leqslant x < 3$, we have

$$\lim_{x\to 3^-} [\![x]\!] = \lim_{x\to 3^-} 2 = 2$$

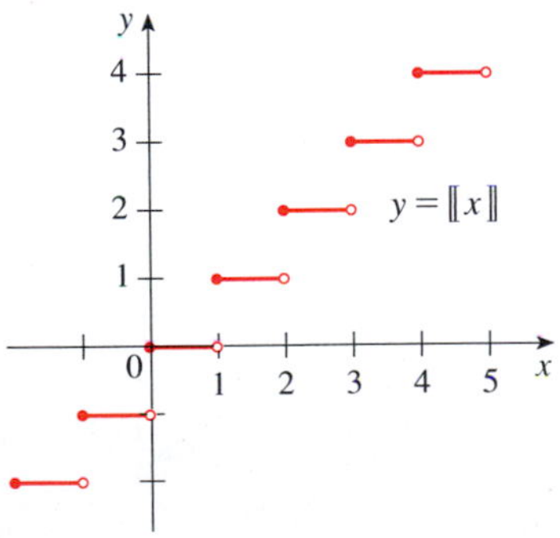

FIGURE 4

Greatest integer function

Because these one-sided limits are not equal, $\lim_{x\to 3} [\![x]\!]$ does not exist by Theorem 1. ■

NOTE: If you have a calculator with an "integer part" key INT, note that $\text{INT}(x) = [\![x]\!]$ for $x \geqslant 0$, but be warned that in some calculators $\text{INT}(x) = -[\![-x]\!]$ for $x < 0$. For instance, these calculators would compute $\text{INT}(-3.6) = -3$, whereas $[\![-3.6]\!] = -4$.

The next two theorems give two additional properties of limits. Their proofs can be found in Appendix F.

(2) THEOREM If $f(x) \leqslant g(x)$ for all x in an open interval that contains a (except possibly at a) and the limits of f and g both exist as x approaches a, then

$$\lim_{x \to a} f(x) \leqslant \lim_{x \to a} g(x)$$

(3) THE SQUEEZE THEOREM If $f(x) \leqslant g(x) \leqslant h(x)$ for all x in an open interval that contains a (except possibly at a) and

$$\lim_{x \to a} f(x) = \lim_{x \to a} h(x) = L$$

then

$$\lim_{x \to a} g(x) = L$$

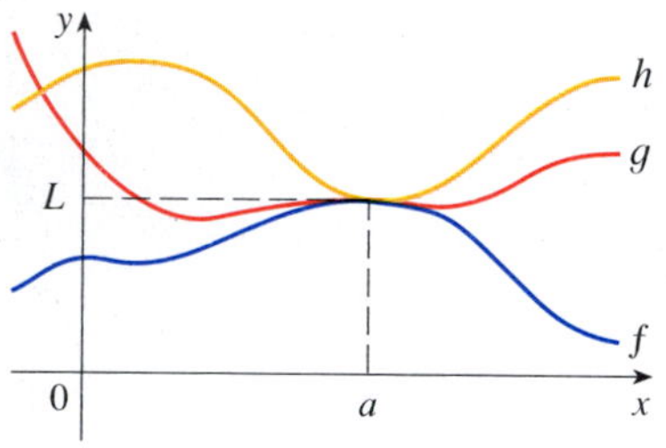

FIGURE 5

The Squeeze Theorem, sometimes called the Sandwich Theorem or the Pinching Theorem, is illustrated by Figure 5. It says that if $g(x)$ is squeezed between $f(x)$ and $h(x)$ near a, and if f and h have the same limit L at a, then g is forced to have the same limit L at a.

EXAMPLE 13 Show that $\lim_{x \to 0} x \sin \dfrac{1}{x} = 0$.

SOLUTION First note that we *cannot* use

$$\lim_{x \to 0} x \sin \frac{1}{x} = \lim_{x \to 0} x \cdot \lim_{x \to 0} \sin \frac{1}{x}$$

because $\lim_{x \to 0} \sin(1/x)$ does not exist. (See Example 4 in Section 2.2.) However, since

$$-1 \leqslant \sin \frac{1}{x} \leqslant 1$$

we have, as illustrated by Figure 6,

$$-|x| \leqslant x \sin \frac{1}{x} \leqslant |x|$$

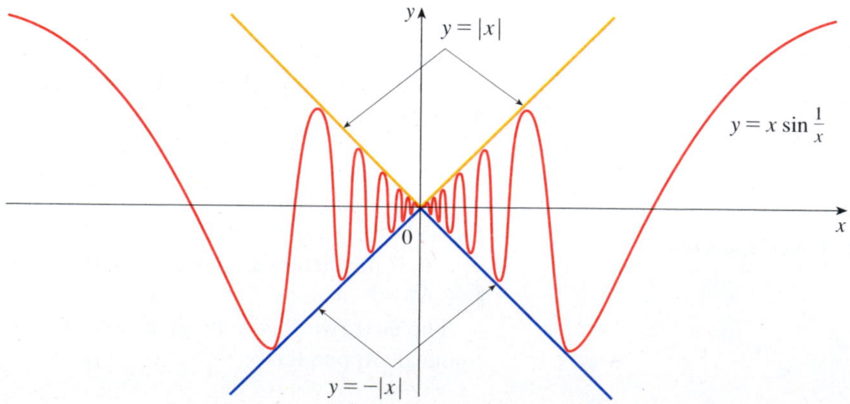

FIGURE 6

We know that

$$\lim_{x\to 0} |x| = 0 \qquad \text{and} \qquad \lim_{x\to 0} -|x| = 0$$

from Example 9. Taking $f(x) = -|x|$, $g(x) = x\sin(1/x)$, and $h(x) = |x|$ in the Squeeze Theorem, we obtain

$$\lim_{x\to 0} x\sin\frac{1}{x} = 0$$

■

EXERCISES 2.3

1–12 ■ Evaluate the given limit and justify each step by indicating the appropriate Limit Law(s).

1. $\lim_{x\to 4} (5x^2 - 2x + 3)$

2. $\lim_{x\to -3} (x^3 + 2x^2 + 6)$

3. $\lim_{x\to 2} (x^2 + 1)(x^2 + 4x)$

4. $\lim_{x\to -2} (x^2 + x + 1)^5$

5. $\lim_{x\to -1} \dfrac{x - 2}{x^2 + 4x - 3}$

6. $\lim_{t\to -2} \dfrac{t^3 - t^2 - t + 10}{t^2 + 3t + 2}$

7. $\lim_{x\to -1} \sqrt{x^3 + 2x + 7}$

8. $\lim_{x\to 64} (\sqrt[3]{x} + 3\sqrt{x})$

9. $\lim_{t\to -2} (t + 1)^9(t^2 - 1)$

10. $\lim_{r\to 3} (r^4 - 7r + 4)^{2/3}$

11. $\lim_{w\to -2} \sqrt[3]{\dfrac{4w + 3w^3}{3w + 10}}$

12. $\lim_{y\to 3} \dfrac{3(8y^2 - 1)}{2y^2(y - 1)^4}$

13. Given that

$$\lim_{x\to a} f(x) = -3 \qquad \lim_{x\to a} g(x) = 0 \qquad \lim_{x\to a} h(x) = 8$$

find the limits that exist.

(a) $\lim_{x\to a} [f(x) + h(x)]$

(b) $\lim_{x\to a} [f(x)]^2$

(c) $\lim_{x\to a} \sqrt[3]{h(x)}$

(d) $\lim_{x\to a} \dfrac{1}{f(x)}$

(e) $\lim_{x\to a} \dfrac{f(x)}{h(x)}$

(f) $\lim_{x\to a} \dfrac{g(x)}{f(x)}$

(g) $\lim_{x\to a} \dfrac{f(x)}{g(x)}$

(h) $\lim_{x\to a} \dfrac{2f(x)}{h(x) - f(x)}$

14. (a) What is wrong with the equation $\dfrac{x^2 + x - 6}{x - 2} = x + 3$?

(b) In view of part (a), explain why the equation

$$\lim_{x\to 2} \frac{x^2 + x - 6}{x - 2} = \lim_{x\to 2} (x + 3)$$

is correct.

15–38 ■ Evaluate each limit, if it exists.

15. $\lim_{x\to -3} \dfrac{x^2 - x + 12}{x + 3}$

16. $\lim_{x\to -3} \dfrac{x^2 - x - 12}{x + 3}$

17. $\lim_{x\to -1} \dfrac{x^2 - x - 2}{x + 1}$

18. $\lim_{x\to 1} \dfrac{x^2 - x - 2}{x + 1}$

19. $\lim_{t\to 1} \dfrac{t^3 - t}{t^2 - 1}$

20. $\lim_{x\to -1} \dfrac{x^2 - x - 3}{x + 1}$

21. $\lim_{h\to 0} \dfrac{(h - 5)^2 - 25}{h}$

22. $\lim_{x\to 1} \dfrac{x^3 - 1}{x^2 - 1}$

23. $\lim_{h\to 0} \dfrac{(1 + h)^4 - 1}{h}$

24. $\lim_{h\to 0} \dfrac{(2 + h)^3 - 8}{h}$

25. $\lim_{x\to -2} \dfrac{x + 2}{x^2 - x - 6}$

26. $\lim_{x\to 1} \dfrac{x^2 + x - 2}{x^2 - 3x + 2}$

27. $\lim_{t\to 9} \dfrac{9 - t}{3 - \sqrt{t}}$

28. $\lim_{t\to 2} \dfrac{t^2 + t - 6}{t^2 - 4}$

29. $\lim_{t\to 0} \dfrac{\sqrt{2 - t} - \sqrt{2}}{t}$

30. $\lim_{x\to 2} \dfrac{x^4 - 16}{x - 2}$

31. $\lim_{x\to 9} \dfrac{x^2 - 81}{\sqrt{x} - 3}$

32. $\lim_{x\to 1} \left[\dfrac{1}{x - 1} - \dfrac{2}{x^2 - 1}\right]$

33. $\lim_{t\to 0} \left[\dfrac{1}{t\sqrt{1 + t}} - \dfrac{1}{t}\right]$

34. $\lim_{h\to 0} \dfrac{(3 + h)^{-1} - 3^{-1}}{h}$

35. $\lim_{x\to 0} \dfrac{x}{\sqrt{1 + 3x} - 1}$

36. $\lim_{x\to 2} \dfrac{\frac{1}{x} - \frac{1}{2}}{x - 2}$

37. $\lim_{x\to 2} \dfrac{x - \sqrt{3x - 2}}{x^2 - 4}$

38. $\lim_{x\to 1} \dfrac{\sqrt{x} - x^2}{1 - \sqrt{x}}$

39. $\lim_{t\to 1} \mathbf{r}(t)$, $\mathbf{r}(t) = (2t - 3)\,\mathbf{i} + \dfrac{t^2 - t}{t - 1}\mathbf{j}$

40. $\lim_{t\to 3} \mathbf{r}(t)$, $\mathbf{r}(t) = \dfrac{t^2 - 9}{t^2 - 3}\mathbf{i} + \sqrt{t^2 - 5}\,\mathbf{j}$

41. $\lim_{t\to 2} \mathbf{r}(t)$, $\mathbf{r}(t) = \left\langle \dfrac{4 - t}{2 - \sqrt{t}}, \dfrac{t^2 - 4}{t - 2} \right\rangle$

42. $\lim_{t\to 1} \mathbf{r}(t)$, $\mathbf{r}(t) = \left\langle \dfrac{t^3 - 1}{t^2 - 1}, \dfrac{t^2 - 2t + 1}{t - 1} \right\rangle$

43. The position of a moving particle at time t, $0 \le t < 2$, is given by $\mathbf{r}(t) = \left(\dfrac{t^2 - 4}{t^2 - t - 2}\right)\mathbf{i} + (2t)\,\mathbf{j}$. What is the anticipated position of the particle at time $t = 2$?

44. The position of a moving particle at time t, $0 \leqslant t < 5$, is given by $\mathbf{r}(t) = \left\langle \dfrac{2t - 10}{t - 5}, \dfrac{t - 5}{t^2 - 4t - 5} \right\rangle$. What is the anticipated position of the particle at time $t = 5$?

45. Use the Squeeze Theorem to show that $\lim_{x \to 0} x^2 \cos 20\pi x = 0$. Illustrate by graphing the functions $f(x) = -x^2$, $g(x) = x^2 \cos 20\pi x$, and $h(x) = x^2$ on the same screen.

46. Use the Squeeze Theorem to show that

$$\lim_{x \to 0} \sqrt{x^3 + x^2} \sin \frac{\pi}{x} = 0$$

Illustrate by graphing the functions f, g, and h (in the notation of the Squeeze Theorem) on the same screen.

47. If $1 \leqslant f(x) \leqslant x^2 + 2x + 2$ for all x, find $\lim_{x \to -1} f(x)$.

48. If $3x \leqslant f(x) \leqslant x^3 + 2$ for $0 \leqslant x \leqslant 2$, evaluate $\lim_{x \to 1} f(x)$.

49. Prove that $\lim\limits_{x \to 0} x^2 \sin \dfrac{1}{x} = 0$.

50. Prove that $\lim\limits_{x \to 0^+} \sqrt{x} \cos^4 x = 0$.

51–64 ▪ Find the limit if it exists. If the limit does not exist, explain why. The symbol ⟦ ⟧ denotes the greatest integer function, defined in Example 12.

51. $\lim\limits_{x \to 4^-} \sqrt{16 - x^2}$

52. $\lim\limits_{x \to -1.5^+} (\sqrt{3 + 2x} + x)$

53. $\lim\limits_{x \to -4} |x + 4|$

54. $\lim\limits_{x \to -4^-} \dfrac{|x + 4|}{x + 4}$

55. $\lim\limits_{x \to 2} \dfrac{|x - 2|}{x - 2}$

56. $\lim\limits_{x \to 1.5} \dfrac{2x^2 - 3x}{|2x - 3|}$

57. $\lim\limits_{x \to -2^+} [\![x]\!]$

58. $\lim\limits_{x \to -2} [\![x]\!]$

59. $\lim\limits_{x \to -2.4} [\![x]\!]$

60. $\lim\limits_{x \to 8^+} (\sqrt{x - 8} + [\![x + 1]\!])$

61. $\lim\limits_{x \to 1^+} \sqrt{x^2 + x - 2}$

62. $\lim\limits_{x \to -2^-} \sqrt{x^2 + x - 2}$

63. $\lim\limits_{x \to 0^-} \left(\dfrac{1}{x} - \dfrac{1}{|x|} \right)$

64. $\lim\limits_{x \to 0^+} \left(\dfrac{1}{x} - \dfrac{1}{|x|} \right)$

65. The *signum* (or sign) *function*, denoted by sgn, is defined by

$$\operatorname{sgn} x = \begin{cases} -1 & \text{if } x < 0 \\ 0 & \text{if } x = 0 \\ 1 & \text{if } x > 0 \end{cases}$$

(a) Sketch the graph of this function.
(b) Find each of the following limits or explain why it does not exist.
(i) $\lim\limits_{x \to 0^+} \operatorname{sgn} x$ (ii) $\lim\limits_{x \to 0^-} \operatorname{sgn} x$
(iii) $\lim\limits_{x \to 0} \operatorname{sgn} x$ (iv) $\lim\limits_{x \to 0} |\operatorname{sgn} x|$

66. Let

$$f(x) = \begin{cases} x^2 - 2x + 2 & \text{if } x < 1 \\ 3 - x & \text{if } x \geqslant 1 \end{cases}$$

(a) Find $\lim_{x \to 1^-} f(x)$ and $\lim_{x \to 1^+} f(x)$.
(b) Does $\lim_{x \to 1} f(x)$ exist?
(c) Sketch the graph of f.

67. Let

$$g(x) = \begin{cases} -x^3 & \text{if } x < -1 \\ (x + 2)^2 & \text{if } x > -1 \end{cases}$$

(a) Find $\lim_{x \to -1^-} g(x)$ and $\lim_{x \to -1^+} g(x)$.
(b) Does $\lim_{x \to -1} g(x)$ exist?
(c) Sketch the graph of g.

68. Let

$$h(x) = \begin{cases} x & \text{if } x < 0 \\ x^2 & \text{if } 0 < x \leqslant 2 \\ 8 - x & \text{if } x > 2 \end{cases}$$

(a) Evaluate each of the following limits if it exists.
(i) $\lim\limits_{x \to 0^+} h(x)$ (ii) $\lim\limits_{x \to 0} h(x)$ (iii) $\lim\limits_{x \to 1} h(x)$
(iv) $\lim\limits_{x \to 2^-} h(x)$ (v) $\lim\limits_{x \to 2^+} h(x)$ (vi) $\lim\limits_{x \to 2} h(x)$
(b) Sketch the graph of h.

69. (a) If n is an integer, evaluate
(i) $\lim\limits_{x \to n^-} [\![x]\!]$ (ii) $\lim\limits_{x \to n^+} [\![x]\!]$
(b) For what values of a does $\lim\limits_{x \to a} [\![x]\!]$ exist?

70. Let $f(x) = x - [\![x]\!]$.
(a) Sketch the graph of f.
(b) If n is an integer, evaluate
(i) $\lim\limits_{x \to n^-} f(x)$ (ii) $\lim\limits_{x \to n^+} f(x)$
(c) For what values of a does $\lim_{x \to a} f(x)$ exist?
(d) If you have a calculator with a "fractional part" key FRAC, explain how FRAC is related to f. (See the note after Example 12.) Sketch the graphs of INT and FRAC.

71. Let $F(x) = \dfrac{x^2 - 1}{|x - 1|}$.
(a) Find
(i) $\lim\limits_{x \to 1^+} F(x)$ (ii) $\lim\limits_{x \to 1^-} F(x)$
(b) Does $\lim_{x \to 1} F(x)$ exist?
(c) Sketch the graph of F.

72. Let $g(x) = [\![x/2]\!]$.
(a) Sketch the graph of g.
(b) Evaluate each of the following limits if it exists.
(i) $\lim\limits_{x \to 1^+} g(x)$ (ii) $\lim\limits_{x \to 1^-} g(x)$ (iii) $\lim\limits_{x \to 1} g(x)$
(iv) $\lim\limits_{x \to 2^+} g(x)$ (v) $\lim\limits_{x \to 2^-} g(x)$ (vi) $\lim\limits_{x \to 2} g(x)$
(c) For what values of a does $\lim_{x \to a} g(x)$ exist?

73. If p is a polynomial, show that $\lim_{x \to a} p(x) = p(a)$.

74. If r is a rational function, use Exercise 73 to show that $\lim_{x \to a} r(x) = r(a)$ for every number a in the domain of r.

75. If

$$f(x) = \begin{cases} x^2 & \text{if } x \text{ is rational} \\ 0 & \text{if } x \text{ is irrational} \end{cases}$$

prove that $\lim_{x \to 0} f(x) = 0$.

76. Show by means of an example that $\lim_{x \to a} [f(x) + g(x)]$ may exist even though neither $\lim_{x \to a} f(x)$ nor $\lim_{x \to a} g(x)$ exists.

77. Show by means of an example that $\lim_{x \to a} [f(x)g(x)]$ may exist even though neither $\lim_{x \to a} f(x)$ nor $\lim_{x \to a} g(x)$ exists.

78. Evaluate $\displaystyle\lim_{x \to 2} \frac{\sqrt{6 - x} - 2}{\sqrt{3 - x} - 1}$.

79. Evaluate $\displaystyle\lim_{x \to 0} \frac{\sqrt[3]{1 + cx} - 1}{x}$.

80. Evaluate $\displaystyle\lim_{x \to 1} \frac{\sqrt[3]{x} - 1}{\sqrt{x} - 1}$.

81. Is there a number a such that

$$\lim_{x \to -2} \frac{3x^2 + ax + a + 3}{x^2 + x - 2}$$

exists? If so, find the value of a and the value of the limit.

82. Find numbers a and b such that

$$\lim_{x \to 0} \frac{\sqrt{ax + b} - 2}{x} = 1$$

83. Use the Squeeze Theorem to show that

$$\lim_{x \to 0} x^2 \left[\!\left[\frac{1}{4x^2} \right]\!\right] = \frac{1}{4}$$

84. The figure shows a fixed circle C_1 with equation

$$(x - 1)^2 + y^2 = 1$$

and a shrinking circle C_2 with radius r and center the origin. P is the point $(0, r)$, Q is the upper point of intersection of the two circles, and R is the point of intersection of the line PQ and the x-axis. What happens to R as C_2 shrinks, that is, as $r \to 0^+$?

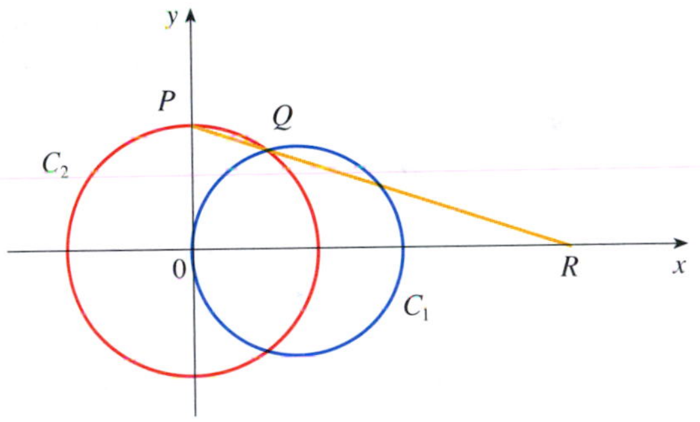

2.4 THE PRECISE DEFINITION OF A LIMIT

The intuitive definition of a limit given in Section 2.2 is inadequate for some purposes because such phrases as "x is close to 2" and "$f(x)$ gets closer and closer to L" are vague. In order to be able to prove conclusively that

$$\lim_{x \to 0} \left(x^3 + \frac{\cos 5x}{10{,}000} \right) = 0.0001 \qquad \text{or} \qquad \lim_{x \to 0} \frac{\sin x}{x} = 1$$

we must make the definition of a limit precise.

To motivate the precise definition of a limit, let's consider the function

$$f(x) = \begin{cases} 2x - 1 & \text{if } x \neq 3 \\ 6 & \text{if } x = 3 \end{cases}$$

Intuitively it is clear that when x is close to 3 but $x \neq 3$, then $f(x)$ is close to 5, and so $\lim_{x \to 3} f(x) = 5$.

To obtain more detailed information about how $f(x)$ varies when x is close to 3, let's ask the following question:

How close to 3 does x have to be so that $f(x)$ differs from 5 by less than 0.1?

It is traditional to use the Greek letter δ (delta) in this situation.

The distance from x to 3 is $|x - 3|$ and the distance from $f(x)$ to 5 is $|f(x) - 5|$, so our problem is to find a number δ such that

$$|f(x) - 5| < 0.1 \quad \text{if} \quad |x - 3| < \delta \quad \text{but } x \neq 3$$

If $|x - 3| > 0$, then $x \neq 3$, so an equivalent formulation of our problem is to find a number δ such that

$$|f(x) - 5| < 0.1 \quad \text{if} \quad 0 < |x - 3| < \delta$$

Notice that if $0 < |x - 3| < (0.1)/2 = 0.05$, then

$$|f(x) - 5| = |(2x - 1) - 5| = |2x - 6| = 2|x - 3| < 0.1$$

that is, $$|f(x) - 5| < 0.1 \quad \text{if} \quad 0 < |x - 3| < 0.05$$

Thus, an answer to the problem is given by $\delta = 0.05$; that is, if x is within a distance of 0.05 from 3, then $f(x)$ will be within a distance of 0.1 from 5.

If we change the number 0.1 in our problem to the smaller number 0.01, then by using the same method we find that $f(x)$ will differ from 5 by less than 0.01 provided that x differs from 3 by less than $(0.01)/2 = 0.005$:

$$|f(x) - 5| < 0.01 \quad \text{if} \quad 0 < |x - 3| < 0.005$$

Similarly, $$|f(x) - 5| < 0.001 \quad \text{if} \quad 0 < |x - 3| < 0.0005$$

If, instead of tolerating an error of 0.1 or 0.01 or 0.001, we want accuracy to within a tolerance of an arbitrary positive number ε (the Greek letter epsilon), then we find as before that

(1) $$|f(x) - 5| < \varepsilon \quad \text{if} \quad 0 < |x - 3| < \delta = \frac{\varepsilon}{2}$$

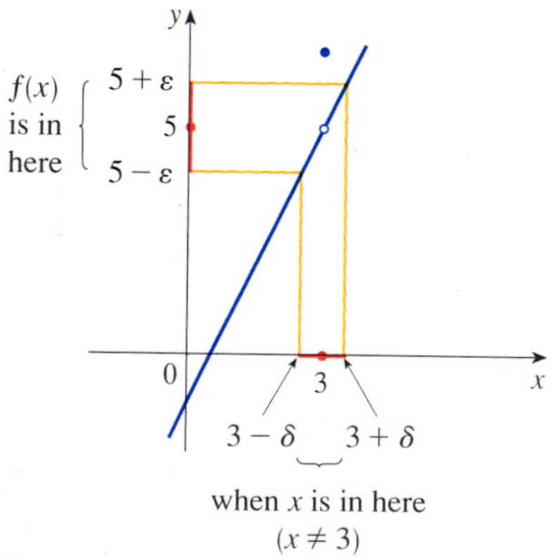

FIGURE 1

This is a precise way of saying that $f(x)$ is close to 5 when x is close to 3 because (1) says that we can make the values of $f(x)$ within an arbitrary distance ε from 5 by taking the values of x within a distance $\varepsilon/2$ from 3 (but $x \neq 3$).

Note that (1) can be rewritten as

$$5 - \varepsilon < f(x) < 5 + \varepsilon \quad \text{whenever} \quad 3 - \delta < x < 3 + \delta \quad (x \neq 3)$$

and this is illustrated in Figure 1. By taking the values of x ($\neq 3$) to lie in the interval $(3 - \delta, 3 + \delta)$ we can make the values of $f(x)$ lie in the interval $(5 - \varepsilon, 5 + \varepsilon)$.

Using (1) as a model, we give a precise definition of a limit.

(2) DEFINITION Let f be a function defined on some open interval that contains the number a, except possibly at a itself. Then we say that the **limit of $f(x)$ as x approaches a is L**, and we write

$$\lim_{x \to a} f(x) = L$$

if for every number $\varepsilon > 0$ there is a corresponding number $\delta > 0$ such that

$$|f(x) - L| < \varepsilon \quad \text{whenever} \quad 0 < |x - a| < \delta$$

Another way of writing the last line of this definition is

$$\text{if} \quad 0 < |x - a| < \delta \quad \text{then} \quad |f(x) - L| < \varepsilon$$

Another notation for $\lim_{x \to a} f(x) = L$ is

$$f(x) \to L \quad \text{as} \quad x \to a$$

Since $|x - a|$ is the distance from x to a and $|f(x) - L|$ is the distance from $f(x)$ to L, and since ε can be arbitrarily small, the definition of a limit can be expressed in words as follows:

> $\lim_{x \to a} f(x) = L$ means that the distance between $f(x)$ and L can be made arbitrarily small by taking the distance from x to a sufficiently small (but not 0).

Alternatively,

> $\lim_{x \to a} f(x) = L$ means that the values of $f(x)$ can be made as close as we please to L by taking x close enough to a (but not equal to a).

We can also reformulate Definition 2 in terms of intervals by observing that the inequality $|x - a| < \delta$ is equivalent to $-\delta < x - a < \delta$, which in turn can be written as $a - \delta < x < a + \delta$. Also, $0 < |x - a|$ is true if and only if $x - a \neq 0$, that is, $x \neq a$. Similarly, if f is a real-valued function, then the inequality $|f(x) - L| < \varepsilon$ is equivalent to the inequalities $L - \varepsilon < f(x) < L + \varepsilon$. Therefore, in terms of intervals, Definition 2 can be stated as follows:

> $\lim_{x \to a} f(x) = L$ means that for every $\varepsilon > 0$ (no matter how small ε is) we can find $\delta > 0$ such that if x lies in the open interval $(a - \delta, a + \delta)$ and $x \neq a$, then $f(x)$ lies in the open interval $(L - \varepsilon, L + \varepsilon)$.

We interpret this statement geometrically by representing a function by an arrow diagram as in Figure 2, where f maps a subset of $\mathbb{R}$ onto another subset of $\mathbb{R}$.

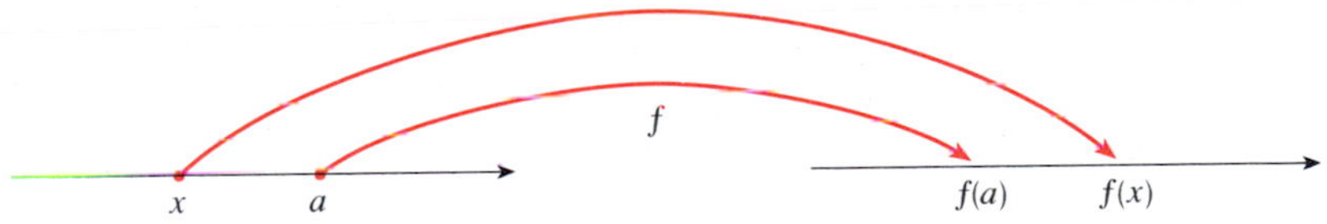

FIGURE 2

The definition of limit says that if any small interval $(L - \varepsilon, L + \varepsilon)$ is given around L, then we can find an interval $(a - \delta, a + \delta)$ around a such that f maps all the points in $(a - \delta, a + \delta)$ (except possibly a) into the interval $(L - \varepsilon, L + \varepsilon)$ (see Figure 3).

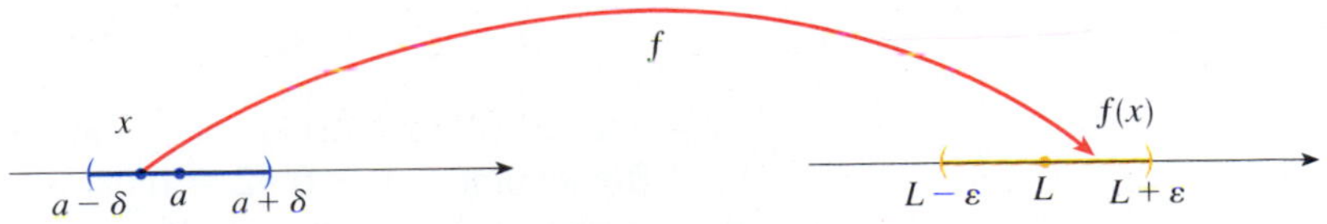

FIGURE 3

Another geometric interpretation of limits can be given in terms of the graph of a function. If $\varepsilon > 0$ is given, then we draw the horizontal lines $y = L + \varepsilon$ and

$y = L - \varepsilon$ and the graph of f (see Figure 4). If $\lim_{x \to a} f(x) = L$, then we can find a number $\delta > 0$ such that if we restrict x to lie in the interval $(a - \delta, a + \delta)$ and take $x \neq a$, then the curve $y = f(x)$ lies between the lines $y = L - \varepsilon$ and $y = L + \varepsilon$ (see Figure 5). You can see that if such a δ has been found, then any smaller δ will also work.

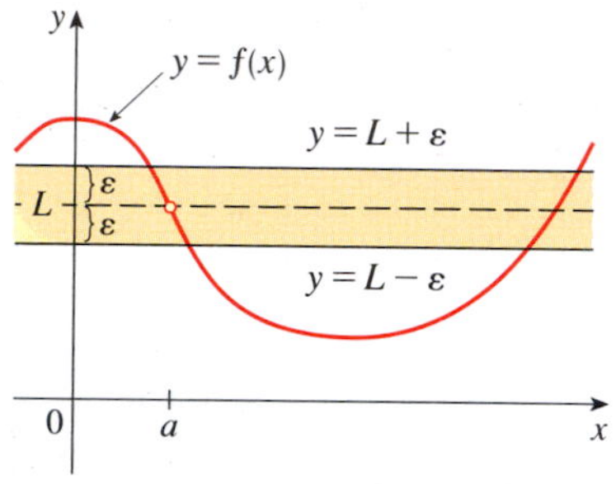

FIGURE 4

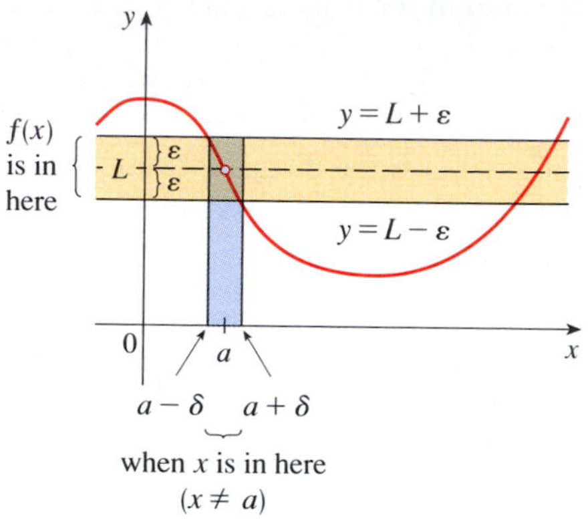

FIGURE 5

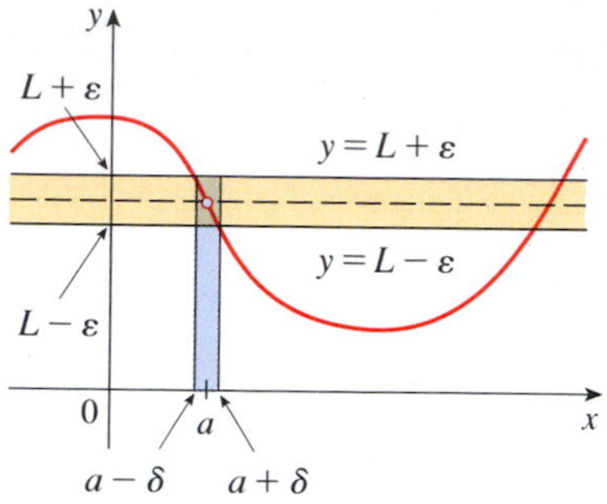

FIGURE 6

It is important to realize that the process illustrated in Figures 4 and 5 must work for *every* positive number ε no matter how small it is chosen. Figure 6 shows that if a smaller ε is chosen, then a smaller δ may be required.

EXAMPLE 1 Use a graph to find a number δ such that

$$|(x^3 - 5x + 6) - 2| < 0.2 \qquad \text{whenever} \qquad |x - 1| < \delta$$

In other words, find a number δ that corresponds to $\varepsilon = 0.2$ in the definition of a limit for the function $f(x) = x^3 - 5x + 6$ with $a = 1$ and $L = 2$.

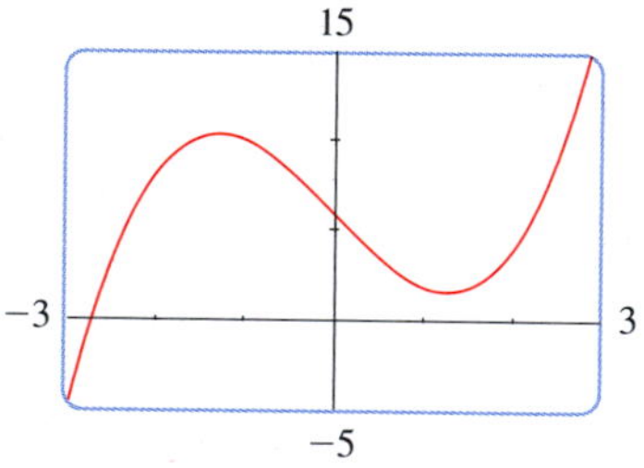

FIGURE 7

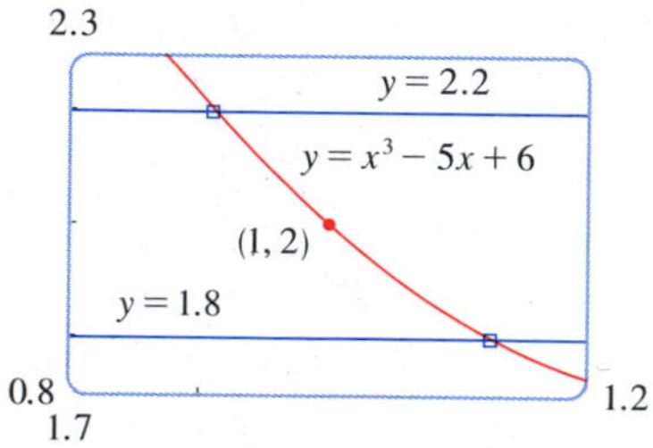

FIGURE 8

SOLUTION A graph of f is shown in Figure 7; we are interested in the region near the point $(1, 2)$. Notice that we can rewrite the inequality

$$|(x^3 - 5x + 6) - 2| < 0.2$$

as

$$1.8 < x^3 - 5x + 6 < 2.2$$

So we need to determine the values of x for which the curve $y = x^3 - 5x + 6$ lies between the horizontal lines $y = 1.8$ and $y = 2.2$. Therefore, we graph the curves $y = x^3 - 5x + 6$, $y = 1.8$, and $y = 2.2$ near the point $(1, 2)$ in Figure 8. Then we use the cursor to estimate that the x-coordinate of the point of intersection of the line $y = 2.2$ and the curve $y = x^3 - 5x + 6$ is about 0.911. Similarly, $y = x^3 - 5x + 6$ intersects the line $y = 1.8$ when $x \approx 1.124$. So, rounding to be safe, we can say that

$$1.8 < x^3 - 5x + 6 < 2.2 \qquad \text{whenever} \qquad 0.92 < x < 1.12$$

This interval $(0.92, 1.12)$ is not symmetric about $x = 1$. The distance from $x = 1$ to the left endpoint is $1 - 0.92 = 0.08$ and the distance to the right endpoint is 0.12. We can choose δ to be the smaller of these numbers, that is, $\delta = 0.08$. Then we can rewrite our inequalities in terms of distances as follows:

$$|(x^3 - 5x + 6) - 2| < 0.2 \qquad \text{whenever} \qquad |x - 1| < 0.08$$

This just says that by keeping x within 0.08 of 1, we are able to keep $f(x)$ within 0.2 of 2.

Although we chose $\delta = 0.08$, any smaller positive value of δ would also have worked. ■

The graphical procedure in Example 1 gives an illustration of the definition for $\varepsilon = 0.2$, but it does not *prove* that the limit is equal to 2. A proof has to provide a δ for *every* ε.

In proving limit statements it may be helpful to think of the definition of limit as a challenge. First it challenges you with a number ε. Then you must be able to produce a suitable δ. You have to be able to do this for *every* $\varepsilon > 0$, not just a particular ε.

Imagine a contest between two people, A and B, and imagine yourself to be B. Person A stipulates that the fixed number L should be approximated by the values of $f(x)$ to within a degree of accuracy ε (say, 0.01). Person B then responds by finding a number δ such that $|f(x) - L| < \varepsilon$ whenever $0 < |x - a| < \delta$. Then A may become more exacting and challenge B with a smaller value of ε (say, 0.0001). Again B has to respond by finding a corresponding δ. Usually the smaller the value of ε, the smaller the corresponding value of δ must be. If B always wins, no matter how small A makes ε, then $\lim_{x\to a} f(x) = L$.

EXAMPLE 2 Prove that $\displaystyle\lim_{x\to 3} (4x - 5) = 7$.

SOLUTION

1. *Preliminary analysis of the problem (guessing a value for δ).* Let ε be a given positive number. We want to find a number δ such that

$$|(4x - 5) - 7| < \varepsilon \qquad \text{whenever} \qquad 0 < |x - 3| < \delta$$

But $|(4x - 5) - 7| = |4x - 12| = |4(x - 3)| = 4|x - 3|$. Therefore, we want

$$4|x - 3| < \varepsilon \qquad \text{whenever} \qquad 0 < |x - 3| < \delta$$

that is,

$$|x - 3| < \frac{\varepsilon}{4} \qquad \text{whenever} \qquad 0 < |x - 3| < \delta$$

This suggests that we should choose $\delta = \varepsilon/4$.

2. *Proof (showing that the δ works).* Given $\varepsilon > 0$, choose $\delta = \varepsilon/4$. If $0 < |x - 3| < \delta$, then

$$|(4x - 5) - 7| = |4x - 12| = 4|x - 3| < 4\delta = 4\left(\frac{\varepsilon}{4}\right) = \varepsilon$$

Thus

$$|(4x - 5) - 7| < \varepsilon \qquad \text{whenever} \qquad 0 < |x - 3| < \delta$$

Therefore, by the definition of a limit,

$$\lim_{x\to 3} (4x - 5) = 7$$

This example is illustrated by Figure 9. ■

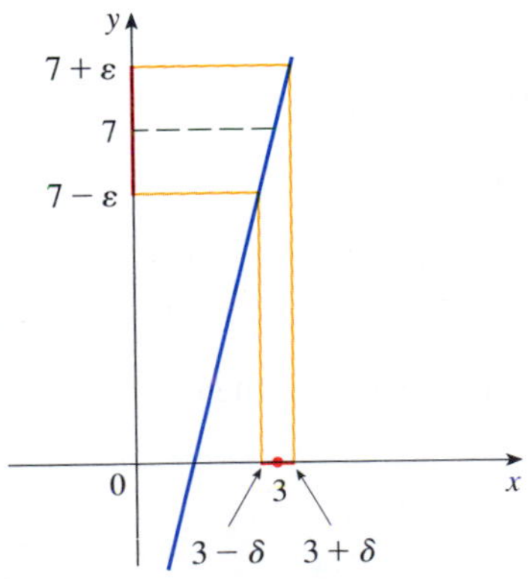

FIGURE 9

Note that in the solution of Example 2 there were two stages—guessing and proving. We made a preliminary analysis that enabled us to guess a value for δ. But then in the second stage we had to go back and prove in a careful, logical fashion that we had made a correct guess. This procedure is typical of much of mathematics. Sometimes it is necessary to first make an intelligent guess about the answer to a problem and then prove that the guess is correct.

The intuitive definitions of one-sided limits that were given in Section 2.2 can be precisely reformulated as follows.

(3) DEFINITION OF LEFT-HAND LIMIT

$$\lim_{x \to a^-} f(x) = L$$

if for every number $\varepsilon > 0$ there is a corresponding number $\delta > 0$ such that

$$|f(x) - L| < \varepsilon \qquad \text{whenever} \qquad a - \delta < x < a$$

(4) DEFINITION OF RIGHT-HAND LIMIT

$$\lim_{x \to a^+} f(x) = L$$

if for every number $\varepsilon > 0$ there is a corresponding number $\delta > 0$ such that

$$|f(x) - L| < \varepsilon \qquad \text{whenever} \qquad a < x < a + \delta$$

Notice that Definition 3 is the same as Definition 2 except that x is restricted to lie in the *left* half $(a - \delta, a)$ of the interval $(a - \delta, a + \delta)$. In Definition 4, x is restricted to lie in the *right* half $(a, a + \delta)$ of the interval $(a - \delta, a + \delta)$.

EXAMPLE 3 Use Definition 4 to prove that $\lim_{x \to 0^+} \sqrt{x} = 0$.

SOLUTION

1. *Guessing a value for δ.* Let ε be a given positive number. Here $a = 0$ and $L = 0$, so we want to find a number δ such that

$$|\sqrt{x} - 0| < \varepsilon \qquad \text{whenever} \qquad 0 < x < \delta$$

that is,

$$\sqrt{x} < \varepsilon \qquad \text{whenever} \qquad 0 < x < \delta$$

or, squaring both sides of the inequality $\sqrt{x} < \varepsilon$, we get

$$x < \varepsilon^2 \qquad \text{whenever} \qquad 0 < x < \delta$$

This suggests that we should choose $\delta = \varepsilon^2$.

2. *Showing that δ works.* Given $\varepsilon > 0$, let $\delta = \varepsilon^2$. If $0 < x < \delta$, then

$$\sqrt{x} < \sqrt{\delta} = \sqrt{\varepsilon^2} = \varepsilon$$

so

$$|\sqrt{x} - 0| < \varepsilon$$

According to Definition 4, this shows that $\lim_{x \to 0^+} \sqrt{x} = 0$. ■

CAUCHY AND LIMITS

After the invention of calculus in the 17th century, there followed a period of free development of the subject in the 18th century. Mathematicians like the Bernoulli brothers and Euler were eager to exploit the power of calculus and boldly explored the consequences of this new and wonderful mathematical theory without worrying too much about whether their proofs were completely correct.

The 19th century, by contrast, was the Age of Rigor in mathematics. There was a movement to go back to the foundations of the subject—to provide careful definitions and rigorous proofs. At the forefront of this movement was the French mathematician Augustin-Louis Cauchy (1789–1857), who started out as a military engineer before becoming a mathematics professor in Paris. Cauchy took Newton's idea of a limit, which was kept alive in the 18th century by the French mathematician Jean d'Alembert, and made it more precise. His definition of a limit reads as follows: "When the successive values attributed to a variable approach indefinitely a fixed value so as to end by differing from it by as little as one wishes, this last is called the *limit* of all the others." But when Cauchy used this definition in examples and proofs, he often employed delta-epsilon inequalities similar to the ones in this section. A typical Cauchy proof starts with: "Designate by δ and ε two very small numbers; . . ." He used ε because of the correspondence between epsilon and the French word *erreur*. Later, the German mathematician Karl Weierstrass (1815–1897) stated the definition of a limit exactly as in our Definition 2.

EXAMPLE 4 Prove that $\lim_{x \to 3} x^2 = 9$.

SOLUTION

1. *Guessing a value for δ.* Let $\varepsilon > 0$ be given. We have to find a number $\delta > 0$ such that

$$|x^2 - 9| < \varepsilon \quad \text{whenever} \quad 0 < |x - 3| < \delta$$

To connect $|x^2 - 9|$ with $|x - 3|$ we write $|x^2 - 9| = |(x + 3)(x - 3)|$. Then we want

$$|x + 3||x - 3| < \varepsilon \quad \text{whenever} \quad 0 < |x - 3| < \delta$$

Notice that if we can find a positive constant C such that $|x + 3| < C$, then

$$|x + 3||x - 3| < C|x - 3|$$

and we can make $C|x - 3| < \varepsilon$ by taking $|x - 3| < \varepsilon/C = \delta$.

We can find such a number C if we restrict x to lie in some interval centered at 3. In fact, since we are interested only in values of x that are close to 3, it is reasonable to assume that x is within a distance 1 from 3, that is, $|x - 3| < 1$. Then $2 < x < 4$, so $5 < x + 3 < 7$. Thus, we have $|x + 3| < 7$, and so $C = 7$ is a suitable choice for the constant.

But now there are two restrictions on $|x - 3|$, namely,

$$|x - 3| < 1 \quad \text{and} \quad |x - 3| < \frac{\varepsilon}{C} = \frac{\varepsilon}{7}$$

To make sure that both of these inequalities are satisfied, we take δ to be the smaller of the two numbers 1 and $\varepsilon/7$. The notation for this is $\delta = \min\{1, \varepsilon/7\}$.

2. *Showing that δ works.* Given $\varepsilon > 0$, let $\delta = \min\{1, \varepsilon/7\}$. If $0 < |x - 3| < \delta$, then $|x - 3| < 1 \Rightarrow 2 < x < 4 \Rightarrow |x + 3| < 7$ (as in part 1). We also have $|x - 3| < \varepsilon/7$, so

$$|x^2 - 9| = |x + 3||x - 3| < 7 \cdot \frac{\varepsilon}{7} = \varepsilon$$

This shows that $\lim_{x \to 3} x^2 = 9$. ■

As Example 4 shows, it's not always easy to prove that limit statements are true using the ε, δ definition. In fact, for a more complicated function such as

$$f(x) = \frac{6x^2 - 8x + 9}{2x^2 - 1}$$

a proof would require a great deal of ingenuity. Fortunately this is not necessary because the Limit Laws stated in Section 2.3 can be proved using Definition 2, and then the limits of complicated functions can be found rigorously from the Limit Laws without resorting to the definition directly.

For instance, we prove the Sum Law: If $\lim_{x \to a} f(x) = L$ and $\lim_{x \to a} g(x) = M$ both exist, then

$$\lim_{x \to a} [f(x) + g(x)] = L + M$$

The remaining Limit Laws are proved in the exercises and in Appendix F.

PROOF OF THE SUM LAW Let $\varepsilon > 0$ be given. We must find $\delta > 0$ such that

$$|f(x) + g(x) - (L + M)| < \varepsilon \quad \text{whenever} \quad 0 < |x - a| < \delta$$

Triangle Inequality:
$$|a + b| \leq |a| + |b|$$
(See Appendix A.)

Using the Triangle Inequality we can write

$$\textbf{(5)} \qquad \begin{aligned} |f(x) + g(x) - (L + M)| &= |(f(x) - L) + (g(x) - M)| \\ &\leq |f(x) - L| + |g(x) - M| \end{aligned}$$

We make $|f(x) + g(x) - (L + M)|$ less than ε by making each of the terms $|f(x) - L|$ and $|g(x) - M|$ less than $\varepsilon/2$.

Since $\varepsilon/2 > 0$ and $\lim_{x \to a} f(x) = L$, there exists a number $\delta_1 > 0$ such that

$$|f(x) - L| < \frac{\varepsilon}{2} \quad \text{whenever} \quad 0 < |x - a| < \delta_1$$

Similarly, since $\lim_{x \to a} g(x) = M$, there exists a number $\delta_2 > 0$ such that

$$|g(x) - M| < \frac{\varepsilon}{2} \quad \text{whenever} \quad 0 < |x - a| < \delta_2$$

Let $\delta = \min\{\delta_1, \delta_2\}$. Notice that

$$\text{if} \quad 0 < |x - a| < \delta \quad \text{then} \quad 0 < |x - a| < \delta_1 \quad \text{and} \quad 0 < |x - a| < \delta_2$$

and so $$|f(x) - L| < \frac{\varepsilon}{2} \quad \text{and} \quad |g(x) - M| < \frac{\varepsilon}{2}$$

Therefore, by (5),

$$\begin{aligned} |f(x) + g(x) - (L + M)| &\leq |f(x) - L| + |g(x) - M| \\ &< \frac{\varepsilon}{2} + \frac{\varepsilon}{2} = \varepsilon \end{aligned}$$

To summarize,

$$|f(x) + g(x) - (L + M)| < \varepsilon \quad \text{whenever} \quad 0 < |x - a| < \delta$$

Thus, by the definition of a limit,

$$\lim_{x \to a} [f(x) + g(x)] = L + M$$

□

INFINITE LIMITS

Infinite limits can also be defined in a precise way. The following is a precise version of Definition 4 in Section 2.2.

(6) DEFINITION Let f be a function defined on some open interval that contains the number a, except possibly at a itself. Then

$$\lim_{x \to a} f(x) = \infty$$

means that for every positive number M there is a corresponding number $\delta > 0$ such that

$$f(x) > M \quad \text{whenever} \quad 0 < |x - a| < \delta$$

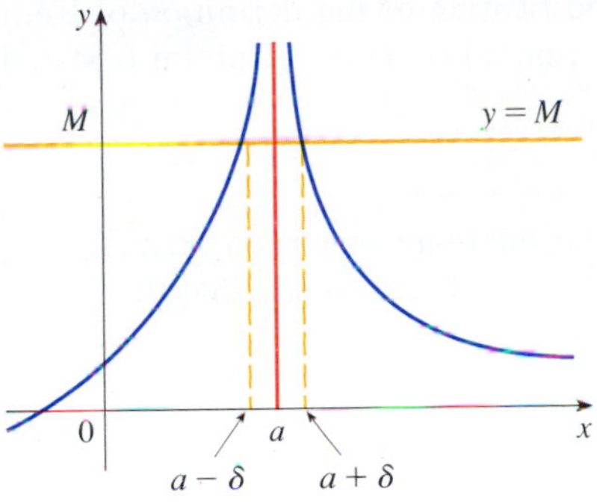

FIGURE 10

This says that the value of $f(x)$ can be made arbitrarily large (larger than any given number M) by taking x close enough to a (within a distance δ, where δ depends on M). A geometric illustration is shown in Figure 10.

Given any horizontal line $y = M$, we can find a number $\delta > 0$ such that if we restrict x to lie in the interval $(a - \delta, a + \delta)$, then the curve $y = f(x)$ lies above the line $y = M$. You can see that if a larger M is chosen, then a smaller δ may be required.

EXAMPLE 5 Use Definition 6 to prove that $\lim_{x \to 0} \dfrac{1}{x^2} = \infty$.

SOLUTION

1. *Guessing a value for* δ. Given $M > 0$, we want to find $\delta > 0$ such that

$$\frac{1}{x^2} > M \qquad \text{whenever} \qquad 0 < |x - 0| < \delta$$

that is,
$$x^2 < \frac{1}{M} \qquad \text{whenever} \qquad 0 < |x| < \delta$$

or
$$|x| < \frac{1}{\sqrt{M}} \qquad \text{whenever} \qquad 0 < |x| < \delta$$

This suggests that we should take $\delta = 1/\sqrt{M}$.

2. *Showing that this* δ *works.* If $M > 0$ is given, let $\delta = 1/\sqrt{M}$. If $0 < |x - 0| < \delta$, then

$$|x| < \delta \quad \Rightarrow \quad x^2 < \delta^2$$
$$\Rightarrow \quad \frac{1}{x^2} > \frac{1}{\delta^2} = M$$

Thus
$$\frac{1}{x^2} > M \qquad \text{whenever} \qquad 0 < |x - 0| < \delta$$

Therefore, by Definition 6,

$$\lim_{x \to 0} \frac{1}{x^2} = \infty$$

■

Similarly, the following is a precise version of Definition 5 in Section 2.2. It is illustrated by Figure 11.

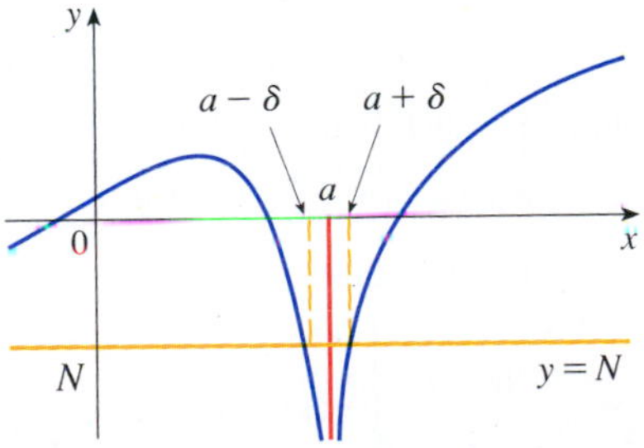

FIGURE 11

(7) DEFINITION Let f be a function defined on some open interval that contains the number a, except possibly at a itself. Then

$$\lim_{x \to a} f(x) = -\infty$$

means that for every negative number N there is a corresponding number $\delta > 0$ such that

$$f(x) < N \qquad \text{whenever} \qquad 0 < |x - a| < \delta$$

LIMITS OF VECTOR FUNCTIONS

We conclude this section by mentioning that a modification of the definition of limit given in Definition 2 applies in the case of vector functions. This definition is stated precisely as follows.

(8) DEFINITION Let $\mathbf{r}(t)$ be a vector function defined on some open interval that contains the number a, except possibly at a itself. Then we say that the **limit of $\mathbf{r}(t)$ as t approaches a is $\mathbf{L}$**, and we write

$$\lim_{t\to a} \mathbf{r}(t) = \mathbf{L}$$

if for every number $\varepsilon > 0$ there is a corresponding number $\delta > 0$ such that

$$|\mathbf{r}(t) - \mathbf{L}| < \varepsilon \qquad \text{whenever} \qquad 0 < |t - a| < \delta$$

Keep in mind that $\mathbf{r}(t)$ and $\mathbf{L}$ are vectors and that $|\mathbf{r}(t) - \mathbf{L}|$ should be interpreted as the length of the vector $\mathbf{r}(t) - \mathbf{L}$. Thus, $\lim_{t\to a} \mathbf{r}(t) = \mathbf{L}$ means that the vector $\mathbf{r}(t)$ can be made as close as we please to the vector $\mathbf{L}$ by taking t close enough to a (but not equal to a). If $\mathbf{r}(t) = \langle f(t), g(t)\rangle$ and $\mathbf{L} = \langle l_1, l_2\rangle$, then $\lim_{t\to a} \mathbf{r}(t) = \mathbf{L}$ is equivalent to $\lim_{t\to a} f(t) = l_1$ and $\lim_{t\to a} g(t) = l_2$ as mentioned at the end of Section 2.2. (Exercise 43 asks for a rigorous proof of this fact.)

EXERCISES 2.4

1. How close to 3 do we have to take x so that $6x + 1$ is within a distance of (a) 0.1 and (b) 0.01 from 19?

2. How close to 2 do we have to take x so that $8x - 5$ is within a distance of (a) 0.01, (b) 0.001, and (c) 0.0001 from 11?

3. Use the given graph of $f(x) = 1/x$ to find a number δ such that

$$\left|\frac{1}{x} - 0.5\right| < 0.2 \qquad \text{whenever} \qquad |x - 2| < \delta$$

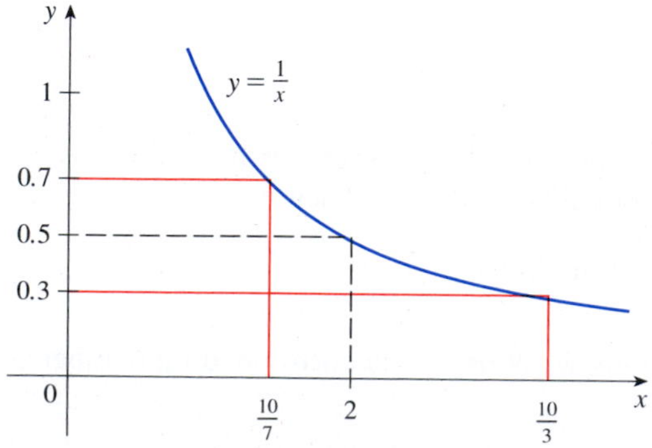

4. Use the given graph of $f(x) = x^2$ to find a number δ such that

$$|x^2 - 1| < \tfrac{1}{2} \qquad \text{whenever} \qquad |x - 1| < \delta$$

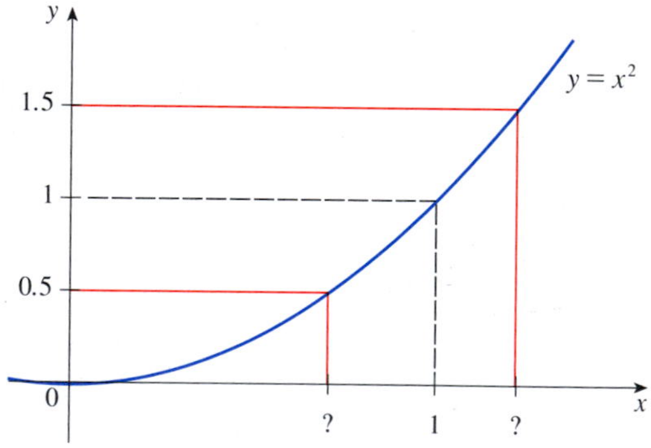

5. Use a graph to find a number δ such that

$$|\sqrt{4x + 1} - 3| < 0.5 \qquad \text{whenever} \qquad |x - 2| < \delta$$

6. Use a graph to find a number δ such that

$$|\sin x - \tfrac{1}{2}| < 0.1 \qquad \text{whenever} \qquad \left|x - \frac{\pi}{6}\right| < \delta$$

7. For the limit $\lim_{x \to 1} (4 + x - 3x^3) = 2$

illustrate the definition by finding values of δ that correspond to $\varepsilon = 1$ and $\varepsilon = 0.1$.

8. For the limit $\lim_{x \to 2} \dfrac{4x + 1}{3x - 4} = 4.5$

illustrate the definition by finding values of δ that correspond to $\varepsilon = 0.5$ and $\varepsilon = 0.1$.

9. Use a graph to find a number δ such that

$$\frac{x}{(x^2 + 1)(x - 1)^2} > 100 \quad \text{whenever} \quad 0 < |x - 1| < \delta$$

10. For the limit $\lim_{x \to 0} \cot^2 x = \infty$

illustrate the definition by finding values of δ that correspond to (a) $M = 100$ and (b) $M = 1000$.

11–14 ■ Prove each statement using the ε, δ definition of limit and illustrate with a diagram like Figure 9.

11. $\lim_{x \to 2} (3x - 2) = 4$

12. $\lim_{x \to 4} (5 - 2x) = -3$

13. $\lim_{x \to -1} (5x + 8) = 3$

14. $\lim_{x \to -1} (3 - 4x) = 7$

15–28 ■ Prove each statement using the ε, δ definition of limit.

15. $\lim_{x \to 2} \dfrac{x}{7} = \dfrac{2}{7}$

16. $\lim_{x \to 4} \left(\dfrac{x}{3} + 1\right) = \dfrac{7}{3}$

17. $\lim_{x \to -5} \left(4 - \dfrac{3x}{5}\right) = 7$

18. $\lim_{x \to 2} \dfrac{x^2 + x - 6}{x - 2} = 5$

19. $\lim_{x \to a} x = a$

20. $\lim_{x \to a} c = c$

21. $\lim_{x \to 0} x^2 = 0$

22. $\lim_{x \to 0} x^3 = 0$

23. $\lim_{x \to 0} |x| = 0$

24. $\lim_{x \to 9^-} \sqrt[4]{9 - x} = 0$

25. $\lim_{x \to 2} (x^2 - 4x + 5) = 1$

26. $\lim_{x \to 3} (x^2 + x - 4) = 8$

27. $\lim_{x \to -2} (x^2 - 1) = 3$

28. $\lim_{x \to 2} x^3 = 8$

29. Verify that another possible choice of δ for showing that $\lim_{x \to 3} x^2 = 9$ in Example 4 is $\delta = \min\{2, \varepsilon/8\}$.

30. Prove that $\lim_{x \to 2} \dfrac{1}{x} = \dfrac{1}{2}$.

31. Prove that $\lim_{x \to a} \sqrt{x} = \sqrt{a}$ if $a > 0$.

$$\left[\textit{Hint}: \text{Use } \left|\sqrt{x} - \sqrt{a}\right| = \frac{|x - a|}{\sqrt{x} + \sqrt{a}}. \right]$$

32. If H is the Heaviside function defined in Example 6 in Section 2.2, prove, using Definition 2, that $\lim_{t \to 0} H(t)$ does not exist. [*Hint:* Use an indirect proof: Suppose that the limit is L. Take $\varepsilon = \frac{1}{2}$ in the definition of a limit and try to arrive at a contradiction.]

33. If the function f is defined by

$$f(x) = \begin{cases} 0 & \text{if } x \text{ is rational} \\ 1 & \text{if } x \text{ is irrational} \end{cases}$$

prove that $\lim_{x \to 0} f(x)$ does not exist.

34. By comparing Definitions 2, 3, and 4, prove Theorem 1 in Section 2.3.

35. How close to -3 do we have to take x so that

$$\frac{1}{(x + 3)^4} > 10{,}000$$

36. Prove, using Definition 6, that $\lim_{x \to -3} \dfrac{1}{(x + 3)^4} = \infty$.

37. Prove that $\lim_{x \to -1^-} \dfrac{5}{(x + 1)^3} = -\infty$.

38. Suppose that $\lim_{x \to a} f(x) = \infty$ and $\lim_{x \to a} g(x) = c$, where c is a real number. Prove each statement.

(a) $\lim_{x \to a} [f(x) + g(x)] = \infty$

(b) $\lim_{x \to a} [f(x)g(x)] = \infty$ if $c > 0$

(c) $\lim_{x \to a} [f(x)g(x)] = -\infty$ if $c < 0$

39. Complete the following in analogy with Definition 4: Let $\mathbf{r}$ be a vector function defined on some open interval that contains the number a, except possibly at a itself. Then $\lim_{t \to a^+} \mathbf{r}(t) = \mathbf{v}$ if ________.

40. Complete the following in analogy with Definition 3: Let $\mathbf{r}$ be a vector function defined on some open interval that contains the number a, except possibly at a itself. Then $\lim_{t \to a^-} \mathbf{r}(t) = \mathbf{v}$ if ________.

41. Let $\mathbf{r}(t) = (2t + 1)\,\mathbf{i} + (t + 2)\,\mathbf{j}$. For each of the following cases, how close to 3 do we have to make t so that $\mathbf{r}(t)$ is within a distance ε from $7\mathbf{i} + 5\mathbf{j}$?

(a) $\varepsilon = 0.01$ (b) $\varepsilon = 0.001$ (c) $\varepsilon = 0.0001$

42. Let $\mathbf{r}(t) = \langle 3t - 4, 4t + 1 \rangle$. For each of the following cases, how close to 2 do we have to make t so that $\mathbf{r}(t)$ is within a distance ε from the vector $\langle 2, 9 \rangle$?

(a) $\varepsilon = 0.01$ (b) $\varepsilon = 0.001$ (c) $\varepsilon = 0.0001$

43. Show that if $\mathbf{r}(t) = \langle f(t), g(t) \rangle$ and $\mathbf{L} = \langle l_1, l_2 \rangle$, then $\lim_{t \to a} \mathbf{r}(t) = \mathbf{L}$ is equivalent to $\lim_{t \to a} f(t) = l_1$ and $\lim_{t \to a} g(t) = l_2$.

2.5 CONTINUITY

We noticed in Section 2.3 that the limit of a function as x approaches a can often be found simply by calculating the value of the function at a. Functions with this property are called *continuous at a*. We will see that the mathematical definition of continuity corresponds closely with the meaning of the word *continuity* in everyday language. (A continuous process is one that takes place gradually, without interruption or abrupt change.)

(1) DEFINITION A function f is **continuous at a number *a*** if

$$\lim_{x \to a} f(x) = f(a)$$

If f is not continuous at a, we say f is **discontinuous at *a***, or f has a **discontinuity** at a. Notice that Definition 1 implicitly requires three things if f is continuous at a:

1. $f(a)$ is defined (that is, a is in the domain of f).
2. $\lim_{x \to a} f(x)$ exists (so f must be defined on an open interval that contains a).
3. $\lim_{x \to a} f(x) = f(a)$.

Intuitively, f is continuous at a if $f(x)$ gets closer and closer to $f(a)$ as x gets closer and closer to a. Thus, a continuous function f has the property that a small change in x produces only a small change in $f(x)$. In fact, the change in $f(x)$ can be kept as small as we please by keeping the change in x sufficiently small.

Physical phenomena are usually continuous. For instance, the displacement or velocity of a vehicle varies continuously with time, as does a person's height. But discontinuities do occur in such situations as electric currents. [See Example 6 in Section 2.2, where the Heaviside function is discontinuous at 0 because $\lim_{t \to 0} H(t)$ does not exist.]

Geometrically, you can think of a function that is continuous at every number in an interval as a function whose graph has no break in it. The graph can be drawn without removing your pen from the paper.

EXAMPLE 1 $f(x) = \dfrac{x^2 - x - 2}{x - 2}$

This function is discontinuous at $a = 2$, since $f(2)$ is not defined. ■

EXAMPLE 2 $f(x) = \begin{cases} \dfrac{1}{x^2} & \text{if } x \neq 0 \\ 1 & \text{if } x = 0 \end{cases}$

This function is discontinuous at $a = 0$. Here $f(0) = 1$ is defined but

$$\lim_{x \to 0} f(x) = \lim_{x \to 0} \frac{1}{x^2}$$

does not exist. (See Example 8 in Section 2.2.) ■

EXAMPLE 3 $f(x) = \begin{cases} \dfrac{x^2 - x - 2}{x - 2} & \text{if } x \neq 2 \\ 1 & \text{if } x = 2 \end{cases}$

This function has a discontinuity at $a = 2$. Here $f(2) = 1$ is defined and

$$\lim_{x \to 2} f(x) = \lim_{x \to 2} \frac{x^2 - x - 2}{x - 2} = \lim_{x \to 2} \frac{(x - 2)(x + 1)}{x - 2} = \lim_{x \to 2} (x + 1) = 3$$

exists. But

$$\lim_{x \to 2} f(x) \neq f(2)$$

so f is not continuous at 2. ■

EXAMPLE 4 The greatest integer function $f(x) = [\![x]\!]$ has discontinuities at all of the integers because $\lim_{x \to n} [\![x]\!]$ does not exist if n is an integer. (See Example 12 and Exercise 69 in Section 2.3.) ■

Figure 1 shows the graphs of the functions in Examples 1–4. In each case the graph cannot be drawn without lifting the pen from the paper because a hole or break or jump occurs in the graph. The kind of discontinuity illustrated in parts (a) and (c) is called **removable** because we could remove the discontinuity by redefining f at 2. [The function $g(x) = x + 1$ is continuous.] The discontinuity in part (b) is called an **infinite discontinuity**. The discontinuities in part (d) are called **jump discontinuities** because the function "jumps" from one value to another.

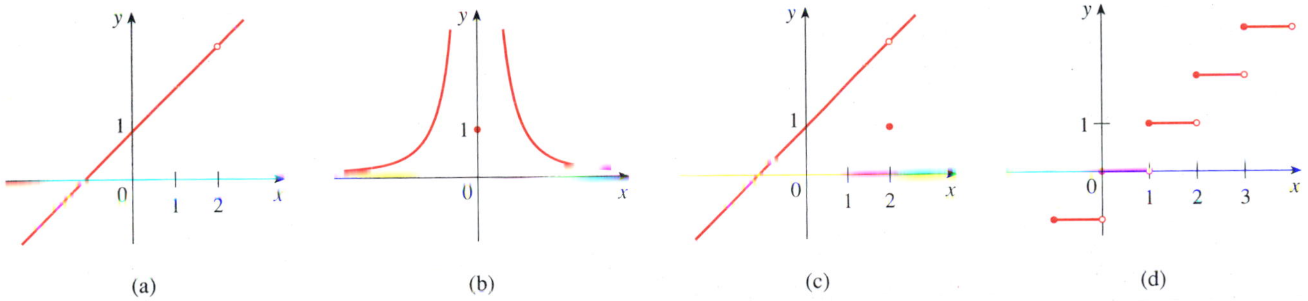

FIGURE 1

(2) DEFINITION A function f is **continuous from the right at a number a** if

$$\lim_{x \to a^+} f(x) = f(a)$$

and f is **continuous from the left at a** if

$$\lim_{x \to a^-} f(x) = f(a)$$

EXAMPLE 5 At each integer n, the function $f(x) = [\![x]\!]$ [see Figure 1(d)] is continuous from the right but discontinuous from the left because

$$\lim_{x \to n^+} f(x) = \lim_{x \to n^+} [\![x]\!] = n = f(n)$$

but

$$\lim_{x \to n^-} f(x) = \lim_{x \to n^-} [\![x]\!] = n - 1 \neq f(n)$$ ■

(3) DEFINITION A function f is **continuous on an interval** if it is continuous at every number in the interval. (At an endpoint of the interval we understand *continuous* to mean *continuous from the right* or *continuous from the left*.)

EXAMPLE 6 Show that the function $f(x) = 1 - \sqrt{1 - x^2}$ is continuous on the interval $[-1, 1]$.

SOLUTION If $-1 < a < 1$, then using the Limit Laws, we have

$$\begin{aligned}\lim_{x \to a} f(x) &= \lim_{x \to a} (1 - \sqrt{1 - x^2}) \\ &= 1 - \lim_{x \to a} \sqrt{1 - x^2} && \text{(by Laws 2 and 7)} \\ &= 1 - \sqrt{\lim_{x \to a} (1 - x^2)} && \text{(by 11)} \\ &= 1 - \sqrt{1 - a^2} && \text{(by 2, 7, and 9)} \\ &= f(a)\end{aligned}$$

Thus, by Definition 1, f is continuous at a if $-1 < a < 1$. We must also calculate the right-hand limit at -1 and the left-hand limit at 1.

$$\begin{aligned}\lim_{x \to -1^+} f(x) &= \lim_{x \to -1^+} (1 - \sqrt{1 - x^2}) \\ &= 1 - \sqrt{\lim_{x \to -1^+} (1 - x^2)} && \text{(as above)} \\ &= 1 - \sqrt{1 - 1^2} \\ &= 1 = f(-1)\end{aligned}$$

So f is continuous from the right at -1. Similarly,

$$\begin{aligned}\lim_{x \to 1^-} f(x) &= \lim_{x \to 1^-} (1 - \sqrt{1 - x^2}) \\ &= 1 - \sqrt{\lim_{x \to 1^-} (1 - x^2)} = 1 - 0 = 1 = f(1)\end{aligned}$$

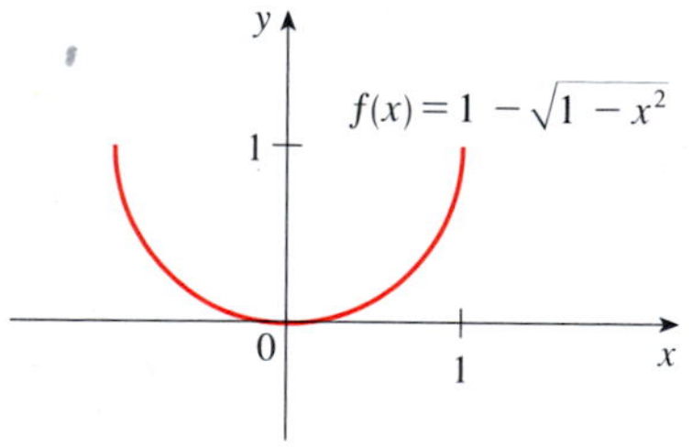

FIGURE 2

So f is continuous from the left at 1. Therefore, according to Definition 3, f is continuous on $[-1, 1]$.

The graph of f is sketched in Figure 2. It is the lower half of the circle $x^2 + (y - 1)^2 = 1$. ■

Instead of always using Definitions 1, 2, and 3 to verify the continuity of a function as we did in Example 6, it is often convenient to use the next theorem, which shows how to build up complicated continuous functions from simple ones.

(4) THEOREM If f and g are continuous at a and c is a constant, then the following functions are also continuous at a:

1. $f + g$ **2.** $f - g$ **3.** cf **4.** fg **5.** $\dfrac{f}{g}$ if $g(a) \neq 0$

PROOF Each of the five parts of this theorem follows from the corresponding Limit Law in Section 2.3. For instance, we give the proof of part 1. Since f and g are continuous at a, we have

$$\lim_{x \to a} f(x) = f(a) \qquad \text{and} \qquad \lim_{x \to a} g(x) = g(a)$$

Therefore

$$\begin{aligned} \lim_{x \to a} (f + g)(x) &= \lim_{x \to a} [f(x) + g(x)] \\ &= \lim_{x \to a} f(x) + \lim_{x \to a} g(x) \qquad \text{(by Law 1)} \\ &= f(a) + g(a) \\ &= (f + g)(a) \end{aligned}$$

This shows that $f + g$ is continuous at a. □

It follows from Theorem 4 and Definition 3 that if f and g are continuous on an interval, then so are the functions $f + g, f - g, cf, fg$, and (if g is never 0) f/g. The following theorem was stated in Section 2.3.

(5) THEOREM
(a) Any polynomial is continuous everywhere; that is, it is continuous on $\mathbb{R} = (-\infty, \infty)$.
(b) Any rational function is continuous wherever it is defined; that is, it is continuous on its domain.

PROOF
(a) A polynomial is a function of the form

$$P(x) = c_n x^n + c_{n-1} x^{n-1} + \cdots + c_1 x + c_0$$

where $c_0, c_1, \ldots, c_n$ are constants. We know that

$$\lim_{x \to a} c_0 = c_0 \qquad \text{(by Law 7)}$$

and

$$\lim_{x \to a} x^m = a^m \qquad m = 1, 2, \ldots, n \qquad \text{(by 9)}$$

This equation is precisely the statement that the function $f(x) = x^m$ is a continuous function. Thus, by part 3 of Theorem 4, the function $g(x) = cx^m$ is continuous. Since P is a sum of functions of this form and a constant function, it follows from part 1 of Theorem 4 that P is continuous.

(b) A rational function is a function of the form

$$f(x) = \frac{P(x)}{Q(x)}$$

where P and Q are polynomials. The domain of f is $D = \{x \in \mathbb{R} \mid Q(x) \neq 0\}$. We know from part (a) that P and Q are continuous everywhere. Thus, by part 5 of Theorem 4, f is continuous at every number in D. □

As an illustration of Theorem 5, observe that the volume of a sphere varies continuously with its radius because the formula

$$V(r) = \tfrac{4}{3}\pi r^3$$

shows that V is a polynomial function of r. Likewise, if a ball is thrown vertically into the air with a velocity of 50 ft/s, then the height of the ball in feet after t seconds is given by the formula $h = 50t - 16t^2$. Again this is a polynomial function, so the height is a continuous function of the elapsed time.

Knowledge of which functions are continuous enables us to evaluate some limits very quickly, as the following example shows. Compare it with Example 2 in Section 2.3.

EXAMPLE 7 Find $\displaystyle\lim_{x \to -2} \frac{x^3 + 2x^2 - 1}{5 - 3x}$.

SOLUTION The function

$$f(x) = \frac{x^3 + 2x^2 - 1}{5 - 3x}$$

is rational, so by Theorem 5 it is continuous on its domain. Therefore

$$\lim_{x \to -2} \frac{x^3 + 2x^2 - 1}{5 - 3x} = \lim_{x \to -2} f(x) = f(-2)$$

$$= \frac{(-2)^3 + 2(-2)^2 - 1}{5 - 3(-2)} = -\frac{1}{11}$$

■

The following theorem is a consequence of Law 10 of limits.

(6) THEOREM If n is a positive even integer, then $f(x) = \sqrt[n]{x}$ is continuous on $[0, \infty)$. If n is a positive odd integer, then f is continuous on $(-\infty, \infty)$.

EXAMPLE 8 On what intervals is each function continuous?

(a) $f(x) = x^{100} - 2x^{37} + 75$

(b) $g(x) = \dfrac{x^2 + 2x + 17}{x^2 - 1}$

(c) $h(x) = \sqrt{x} + \dfrac{x + 1}{x - 1} - \dfrac{x + 1}{x^2 + 1}$

SOLUTION

(a) f is a polynomial, so it is continuous on $(-\infty, \infty)$ by Theorem 5(a).

(b) g is a rational function, so by Theorem 5(b), it is continuous on its domain, which is $D = \{x \mid x^2 - 1 \neq 0\} = \{x \mid x \neq \pm 1\}$. Thus, g is continuous on the intervals $(-\infty, -1)$, $(-1, 1)$, and $(1, \infty)$.

(c) We can write $h(x) = F(x) + G(x) - H(x)$, where

$$F(x) = \sqrt{x} \qquad G(x) = \frac{x + 1}{x - 1} \qquad H(x) = \frac{x + 1}{x^2 + 1}$$

F is continuous on $[0, \infty)$ by Theorem 6. G is a rational function, so it is continuous everywhere except when $x - 1 = 0$, that is, $x = 1$. H is also a rational function, but its denominator is never 0, so H is continuous everywhere. Thus, by parts 1 and 2 of Theorem 4, h is continuous on the intervals $[0, 1)$ and $(1, \infty)$. ■

Another way of combining continuous functions f and g to get a new continuous function is to form the composite function $f \circ g$. This fact is a consequence of the following theorem.

(7) THEOREM If f is continuous at b and $\lim_{x \to a} g(x) = b$, then

$$\lim_{x \to a} f(g(x)) = f(b) = f\left(\lim_{x \to a} g(x)\right)$$

Intuitively this theorem is reasonable because if x is close to a, then $g(x)$ is close to b, and since f is continuous at b, if $g(x)$ is close to b, then $f(g(x))$ is close to $f(b)$. A proof of Theorem 7 is given in Appendix F.

Let's now apply Theorem 7 in the special case where $f(x) = \sqrt[n]{x}$, with n being a positive integer. Then

$$f(g(x)) = \sqrt[n]{g(x)}$$

and

$$f\left(\lim_{x \to a} g(x)\right) = \sqrt[n]{\lim_{x \to a} g(x)}$$

If we put these expressions into Theorem 7 we get

$$\lim_{x \to a} \sqrt[n]{g(x)} = \sqrt[n]{\lim_{x \to a} g(x)}$$

and so Limit Law 11 has now been proved. (We assume that the roots exist.)

(8) THEOREM If g is continuous at a and f is continuous at $g(a)$, then $(f \circ g)(x) = f(g(x))$ is continuous at a.

This theorem is often expressed informally by saying "a continuous function of a continuous function is a continuous function."

PROOF Since g is continuous at a, we have

$$\lim_{x \to a} g(x) = g(a)$$

Since f is continuous at $b = g(a)$, we can apply Theorem 7 to obtain

$$\lim_{x \to a} f(g(x)) = f(g(a))$$

which is precisely the statement that the function $h(x) = f(g(x))$ is continuous at a; that is, $f \circ g$ is continuous at a. □

EXAMPLE 9 Where are the following functions continuous?

$$\text{(a)}\quad h(x) = |x| \qquad\qquad \text{(b)}\quad F(x) = \frac{1}{\sqrt{x^2+7}-4}$$

SOLUTION

(a) Since $|x| = \sqrt{x^2}$ for all x, we have $h(x) = f(g(x))$, where

$$g(x) = x^2 \qquad \text{and} \qquad f(x) = \sqrt{x}$$

Now g is continuous on $\mathbb{R}$ since it is a polynomial and f is continuous on $[0, \infty)$ by Theorem 6. Thus, $h = f \circ g$ is continuous on $\mathbb{R}$ by Theorem 8.

(b) Notice that F can be broken up as the composition of four continuous functions:

$$F = f \circ g \circ h \circ k \qquad \text{or} \qquad F(x) = f(g(h(k(x))))$$

$$\text{where}\quad f(x) = \frac{1}{x} \qquad g(x) = x - 4 \qquad h(x) = \sqrt{x} \qquad k(x) = x^2 + 7$$

We know that each of these functions is continuous on its domain (by Theorems 5 and 6), so by Theorem 8, F is continuous on its domain, which is

$$\{x \in \mathbb{R} \mid \sqrt{x^2+7} \neq 4\} = \{x \mid x \neq \pm 3\} = (-\infty, -3) \cup (-3, 3) \cup (3, \infty)$$ ■

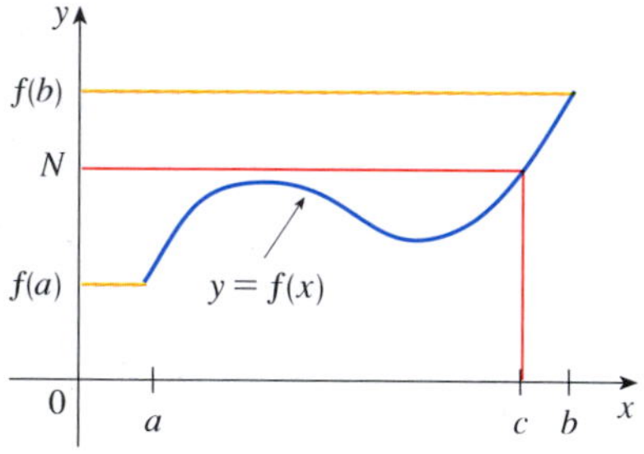

(a)

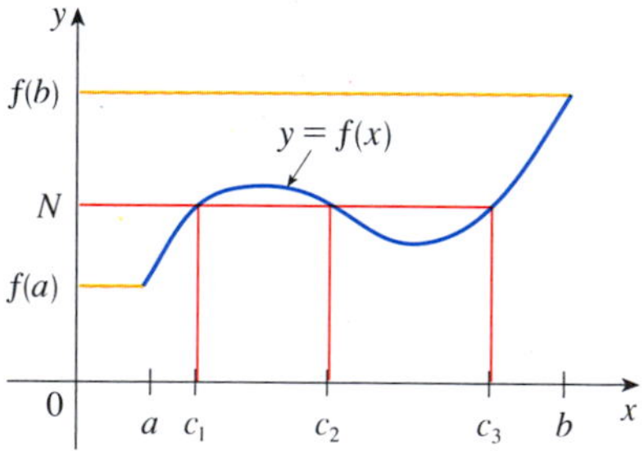

(b)

FIGURE 3

An important property of continuous functions is expressed by the following theorem, whose proof is found in more advanced books on calculus.

> **(9) THE INTERMEDIATE VALUE THEOREM** Suppose that f is continuous on the closed interval $[a, b]$ and let N be any number strictly between $f(a)$ and $f(b)$. Then there exists a number c in (a, b) such that $f(c) = N$.

The Intermediate Value Theorem states that a continuous function takes on every intermediate value between the function values $f(a)$ and $f(b)$. It is illustrated by Figure 3. Note that the value N can be taken on once [as in part (a)] or more than once [as in part (b)].

If we think of a continuous function as a function whose graph has no hole or break, then it is easy to believe that the Intermediate Value Theorem is true. In geometric terms it says that if any horizontal line $y = N$ is given between $y = f(a)$ and $y = f(b)$ as in Figure 4, then the graph of f cannot jump over the line. It must intersect $y = N$ somewhere.

It is important that the function f in Theorem 9 be continuous. The Intermediate Value Theorem is not true in general for discontinuous functions (see Exercise 41).

One use of the Intermediate Value Theorem is in locating roots of equations as in the following example.

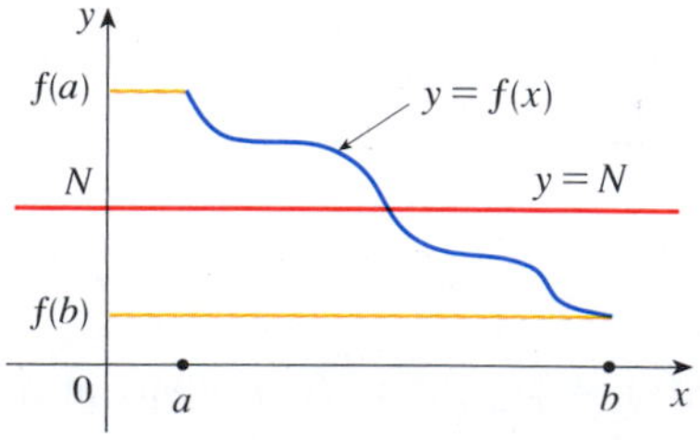

FIGURE 4

EXAMPLE 10 Show that there is a root of the equation

$$4x^3 - 6x^2 + 3x - 2 = 0$$

between 1 and 2.

SOLUTION Let $f(x) = 4x^3 - 6x^2 + 3x - 2$. We are looking for a solution of the given equation, that is, a number c between 1 and 2 such that $f(c) = 0$. Therefore, we take $a = 1$, $b = 2$, and $N = 0$ in Theorem 9. We have

$$f(1) = 4 - 6 + 3 - 2 = -1 < 0 \quad \text{and} \quad f(2) = 32 - 24 + 6 - 2 = 12 > 0$$

Thus $f(1) < 0 < f(2)$, that is, $N = 0$ is a number between $f(1)$ and $f(2)$. Now f is continuous since it is a polynomial, so the Intermediate Value Theorem says there is a number c between 1 and 2 such that $f(c) = 0$. In other words, the equation $4x^3 - 6x^2 + 3x - 2 = 0$ has a root c in the interval $(1, 2)$.

In fact, we can locate a root more precisely by using the Intermediate Value Theorem again. Since

$$f(1.2) = -0.128 < 0 \quad \text{and} \quad f(1.3) = 0.548 > 0$$

a root must lie between 1.2 and 1.3. A calculator gives, by trial and error,

$$f(1.22) = -0.007008 < 0 \quad \text{and} \quad f(1.23) = 0.056068 > 0$$

so a root lies in the interval $(1.22, 1.23)$. ■

We can use a graphing calculator or computer to illustrate the use of the Intermediate Value Theorem in Example 10. Figure 5 shows the graph of f in the viewing rectangle $[-1, 3]$ by $[-3, 3]$ and you can see the graph crossing the x-axis between 1 and 2. Figure 6 shows the result of zooming in to the viewing rectangle $[1.2, 1.3]$ by $[-0.2, 0.2]$.

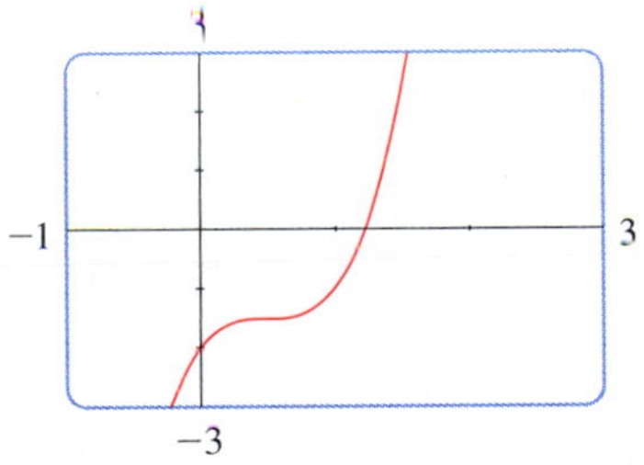

FIGURE 5

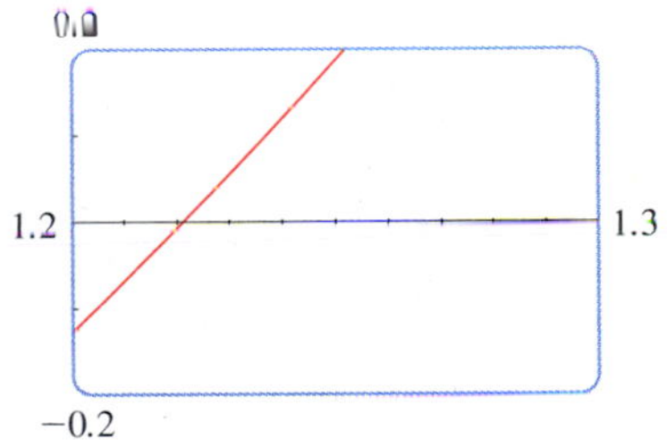

FIGURE 6

In fact, the Intermediate Value Theorem plays a role in the very way these graphing devices work. A computer calculates a finite number of points on the graph and turns on the pixels that contain these calculated points. It assumes that the function is continuous and takes on all the intermediate values between two consecutive points. The computer therefore connects the pixels by turning on the intermediate pixels.

CONTINUITY OF VECTOR FUNCTIONS

We conclude this section by mentioning that if $\mathbf{r}$ is a vector function, then the definition of continuity of $\mathbf{r}$ at $t = a$ is the same as that for real-valued functions. That is, a vector function $\mathbf{r}(t)$ is continuous at $t = a$ if $\lim_{t \to a} \mathbf{r}(t) = \mathbf{r}(a)$. If $\mathbf{r}(t) = \langle f(t), g(t) \rangle$, then the continuity of $\mathbf{r}(t)$ at $t = a$ is equivalent to the continuity of both of the component functions f and g at $t = a$ (see Exercise 61).

EXERCISES 2.5

1. (a) From the graph of f, state the numbers at which f is discontinuous.
   (b) For each of the numbers stated in part (a), state whether f is continuous from the right, or from the left, or neither.

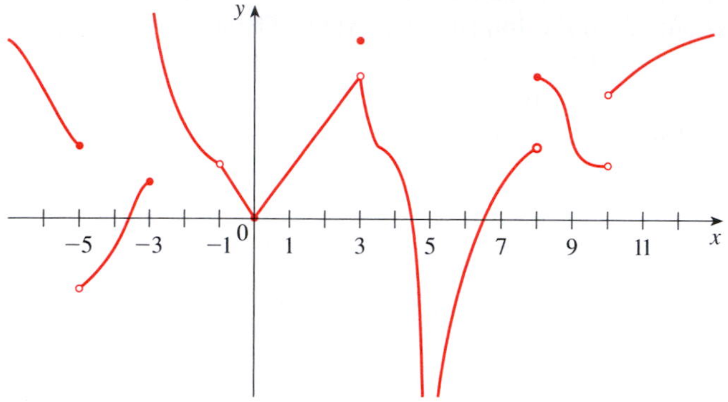

2. From the graph of g, state the intervals on which g is continuous.

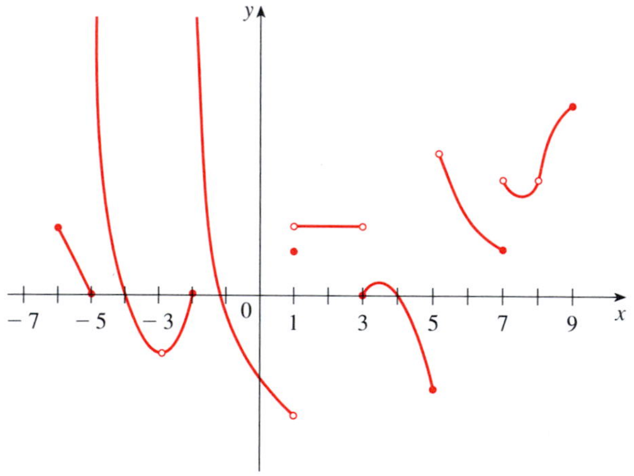

3–7 ■ Use the definition of continuity and the properties of limits to show that the function is continuous at the given number.

3. $f(x) = x^4 - 5x^3 + 6, \quad a = 3$

4. $f(x) = x^2 + (x - 1)^9, \quad a = 2$

5. $f(x) = 1 + \sqrt{x^2 - 9}, \quad a = 5$

6. $g(x) = \dfrac{x + 1}{2x^2 - 1}, \quad a = 4$

7. $g(t) = \dfrac{\sqrt[3]{t}}{(t + 1)^4}, \quad a = -8$

8–11 ■ Use the definition of continuity and the properties of limits to show that each function is continuous on the given interval.

8. $f(x) = x + \sqrt{x - 1}, \quad [1, \infty)$

9. $f(x) = x\sqrt{16 - x^2}, \quad [-4, 4]$

10. $F(x) = \dfrac{x + 1}{x - 3}, \quad (-\infty, 3)$

11. $f(x) = (x^2 - 1)^8, \quad (-\infty, \infty)$

12–18 ■ Explain why each function is discontinuous at the given point. Sketch the graph of the function.

12. $f(x) = \dfrac{x^2 - 1}{x + 1} \qquad a = -1$

13. $f(x) = -\dfrac{1}{(x - 1)^2} \qquad a = 1$

14. $f(x) = \begin{cases} \dfrac{x^2 - 1}{x + 1} & \text{if } x \neq -1 \\ 6 & \text{if } x = -1 \end{cases} \qquad a = -1$

15. $f(x) = \begin{cases} -\dfrac{1}{(x - 1)^2} & \text{if } x \neq 1 \\ 0 & \text{if } x = 1 \end{cases} \qquad a = 1$

16. $f(x) = \begin{cases} \dfrac{x^2 - 2x - 8}{x - 4} & \text{if } x \neq 4 \\ 3 & \text{if } x = 4 \end{cases} \qquad a = 4$

17. $f(x) = \begin{cases} x^2 - 2 & \text{if } x \neq -3 \\ 5 & \text{if } x = -3 \end{cases} \qquad a = -3$

18. $f(x) = \begin{cases} 1 - x & \text{if } x \leq 2 \\ x^2 - 2x & \text{if } x > 2 \end{cases} \qquad a = 2$

19–27 ■ Use Theorems 4, 5, 6, and 8 to show that each function is continuous on its domain. State the domain.

19. $f(x) = (x + 1)(x^3 + 8x + 9)$

20. $G(x) = \dfrac{x^4 + 17}{6x^2 + x - 1}$

21. $H(x) = \dfrac{1}{\sqrt{x + 1}}$

22. $f(t) = 2t + \sqrt{25 - t^2}$

23. $h(x) = \sqrt[5]{x - 1}\,(x^2 - 2)$

24. $g(t) = \dfrac{1}{t + \sqrt{t^2 - 4}}$

25. $F(t) = (t^2 + t + 1)^{3/2}$

26. $H(x) = \sqrt{\dfrac{x - 2}{5 + x}}$

27. $L(x) = |x^3 - x|$

28. (a) Use Theorems 4, 5, 6, and 8 to show that the function $y = \sqrt[5]{x^2 - x} + (x^3 + x)^9$ is continuous on $(-\infty, \infty)$.
   (b) Use part (a) to give an easier method for calculating the limit in Example 3 in Section 2.3.

29. Let

$$f(x) = \begin{cases} x - 1 & \text{for } x < 3 \\ 5 - x & \text{for } x \geq 3 \end{cases}$$

Show that f is continuous on $(-\infty, \infty)$.

30. Let

$$g(x) = \begin{cases} x & \text{if } x < 0 \\ x^2 & \text{if } 0 \leq x \leq 1 \\ x^3 & \text{if } x > 1 \end{cases}$$

Show that g is continuous on $(-\infty, \infty)$.

31–35 ■ Find the points at which f is discontinuous. At which of these points is f continuous from the right, from the left, or neither? Sketch the graph of f.

31. $f(x) = \begin{cases} (x-1)^3 & \text{if } x < 0 \\ (x+1)^3 & \text{if } x \geq 0 \end{cases}$

32. $f(x) = \begin{cases} 2x + 1 & \text{if } x \leq -1 \\ 3x & \text{if } -1 < x < 1 \\ 2x - 1 & \text{if } x \geq 1 \end{cases}$

33. $f(x) = \begin{cases} 1/x & \text{if } x < -1 \\ x & \text{if } -1 \leq x \leq 1 \\ 1/x^2 & \text{if } x > 1 \end{cases}$

34. $f(x) = \begin{cases} \sqrt{-x} & \text{if } x < 0 \\ 1 & \text{if } 0 < x \leq 1 \\ \sqrt{x} & \text{if } x > 1 \end{cases}$

35. $f(x) = [\![2x]\!]$

36. If your monthly salary is now \$3200 and you are guaranteed a 3% raise every 6 months, then your monthly salary is given by

$$S(t) = 3200(1.03)^{[\![t/6]\!]}$$

where t is measured in months. Sketch a graph of your salary function for $0 \leq t \leq 24$ and discuss its continuity.

37. For what value of the constant c is the function f continuous on $(-\infty, \infty)$?

$$f(x) = \begin{cases} cx + 1 & \text{if } x \leq 3 \\ cx^2 - 1 & \text{if } x > 3 \end{cases}$$

38. Find the constant c that makes g continuous on $(-\infty, \infty)$.

$$g(x) = \begin{cases} x^2 - c^2 & \text{if } x < 4 \\ cx + 20 & \text{if } x \geq 4 \end{cases}$$

39. Find the values of c and d that make h continuous on $\mathbb{R}$.

$$h(x) = \begin{cases} 2x & \text{if } x < 1 \\ cx^2 + d & \text{if } 1 \leq x \leq 2 \\ 4x & \text{if } x > 2 \end{cases}$$

40. Which of the following functions f has a removable discontinuity at a? If the discontinuity is removable, find a function g that agrees with f for $x \neq a$ and is continuous on $\mathbb{R}$.

(a) $f(x) = \dfrac{x^2 - 2x - 8}{x + 2}, \quad a = -2$

(b) $f(x) = \dfrac{x - 7}{|x - 7|}, \quad a = 7$

(c) $f(x) = \dfrac{x^3 + 64}{x + 4}, \quad a = -4$

(d) $f(x) = \dfrac{3 - \sqrt{x}}{9 - x}, \quad a = 9$

41. Let

$$f(x) = \begin{cases} 1 - x^2 & \text{if } 0 \leq x \leq 1 \\ 1 + \dfrac{x}{2} & \text{if } 1 < x \leq 2 \end{cases}$$

(a) Show that f is not continuous on $[0, 2]$.
(b) Show that f does not take on all values between $f(0)$ and $f(2)$.

42. Use the Intermediate Value Theorem to prove that there is a positive number c such that $c^2 = 2$. (This proves the existence of the number $\sqrt{2}$.)

43. If $f(x) = x^3 - x^2 + x$, show that there is a number c such that $f(c) = 10$.

44. If $g(x) = x^5 - 2x^3 + x^2 + 2$, show that there is a number c such that $g(c) = -1$.

45–48 ■ Use the Intermediate Value Theorem to show that there is a root of the given equation in the given interval.

45. $x^3 - 3x + 1 = 0, \quad (0, 1)$

46. $x^5 - 2x^4 - x - 3 = 0, \quad (2, 3)$

47. $x^3 + 2x = x^2 + 1, \quad (0, 1)$

48. $x^2 = \sqrt{x + 1}, \quad (1, 2)$

49–50 ■ (a) Prove that the equation has at least one real root. (b) Use your calculator to find an interval of length 0.01 that contains a root.

49. $x^3 - x + 1 = 0$

50. $x^5 - x^2 + 2x + 3 = 0$

51–52 ■ (a) Prove that the equation has at least one real root. (b) Use your graphing device to find the root correct to three decimal places.

51. $x^5 - x^2 - 4 = 0$

52. $\sqrt{x - 5} = \dfrac{1}{x + 3}$

53. Prove that f is continuous at a if and only if

$$\lim_{h \to 0} f(a + h) = f(a)$$

54. (a) Prove Theorem 4, part 3.
(b) Prove Theorem 4, part 5.

55. For what values of x is f continuous?

$$f(x) = \begin{cases} 0 & \text{if } x \text{ is rational} \\ 1 & \text{if } x \text{ is irrational} \end{cases}$$

56. For what values of x is g continuous?

$$g(x) = \begin{cases} 0 & \text{if } x \text{ is rational} \\ x & \text{if } x \text{ is irrational} \end{cases}$$

57. Is there a number that is exactly 1 more than its cube?

58. (a) Prove that if f is a continuous function on an interval, then so is $|f|$.
(b) Is the converse of this statement also true? If so, prove it. If not, find a counterexample.

59. A Tibetan monk leaves the monastery at 7:00 A.M. and takes his usual path to the top of the mountain, arriving at 7:00 P.M. The following morning, he starts at 7:00 A.M. at the top and takes the same path back, arriving at the monastery at 7:00 P.M. Use the Intermediate Value Theorem to show that there is a point on the path that the monk will cross at exactly the same time of day on both days.

60. A **fixed point** of a function f is a number c in its domain such that $f(c) = c$. (The function doesn't move c; it stays fixed.)
(a) Sketch the graph of a continuous function with domain $[0, 1]$ whose range also lies in $[0, 1]$. Locate a fixed point of f.
(b) Try to draw the graph of a continuous function with domain $[0, 1]$ and range in $[0, 1]$ that does *not* have a fixed point. What is the obstacle?
(c) Use the Intermediate Value Theorem to prove that any continuous function with domain $[0, 1]$ and range in $[0, 1]$ must have a fixed point.

61. Prove that the vector function $\mathbf{r}(t) = \langle f(t), g(t) \rangle$ is continuous at $t = a$ if and only if each of the component functions $f(t)$ and $g(t)$ is continuous at $t = a$.

2.6 LIMITS AT INFINITY; HORIZONTAL ASYMPTOTES

In Sections 2.2 and 2.4 we investigated infinite limits and vertical asymptotes. There we let x approach a number and the result was that the values of y became arbitrarily large (positive or negative). In this section we let x become arbitrarily large (positive or negative) and see what happens to y.

Let's begin by investigating the behavior of the function f defined by

$$f(x) = \frac{x^2 - 1}{x^2 + 1}$$

as x becomes large. The accompanying table gives values of this function correct to six decimal places, and the graph of f has been drawn by a computer in Figure 1.

x	$f(x)$
0	−1
±1	0
±2	0.600000
±3	0.800000
±4	0.882353
±5	0.923077
±10	0.980198
±50	0.999200
±100	0.999800
±1000	0.999998

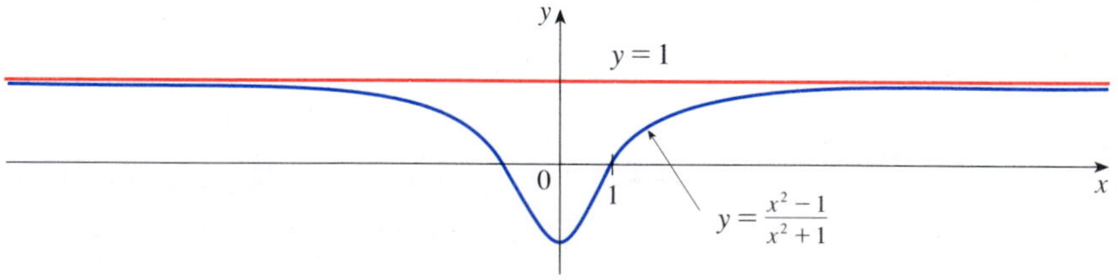

FIGURE 1

As x grows larger and larger you can see that the values of $f(x)$ get closer and closer to 1. In fact, it seems that we can make the values of $f(x)$ as close as we like to 1 by taking x sufficiently large. This situation is expressed symbolically by writing

$$\lim_{x \to \infty} \frac{x^2 - 1}{x^2 + 1} = 1$$

In general we use the symbolism

$$\lim_{x \to \infty} f(x) = L$$

to indicate that the values of $f(x)$ become closer and closer to L as x becomes larger and larger.

> **(1) DEFINITION** Let f be a function defined on some interval (a, ∞). Then
>
> $$\lim_{x \to \infty} f(x) = L$$
>
> means that the values of $f(x)$ can be made arbitrarily close to L by taking x sufficiently large.

Another notation for $\lim_{x \to \infty} f(x) = L$ is

$$f(x) \to L \qquad \text{as } x \to \infty$$

The symbol ∞ does not represent a number. Nonetheless, the expression $\lim_{x \to \infty} f(x) = L$ is often read as

"the limit of $f(x)$, as x approaches infinity, is L"

or "the limit of $f(x)$, as x becomes infinite, is L"

or "the limit of $f(x)$, as x increases without bound, is L"

The meaning of such phrases is given by Definition 1. A more precise definition, similar to the ε, δ definition of Section 2.4, is given at the end of this section.

Geometric illustrations of Definition 1 are shown in Figure 2. Notice that there are many ways for the graph of f to approach the line $y = L$ (which is called a *horizontal asymptote*).

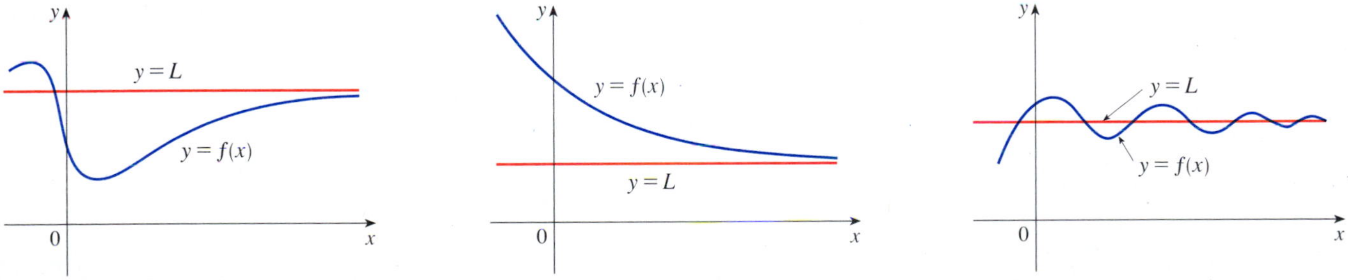

FIGURE 2
Examples illustrating $\lim_{x \to \infty} f(x) = L$

Referring back to Figure 1, we see that for numerically large negative values of x, the values of $f(x)$ are close to 1. By letting x decrease through negative values without bound, we can make $f(x)$ as close as we like to 1. This is expressed by writing

$$\lim_{x \to -\infty} \frac{x^2 - 1}{x^2 + 1} = 1$$

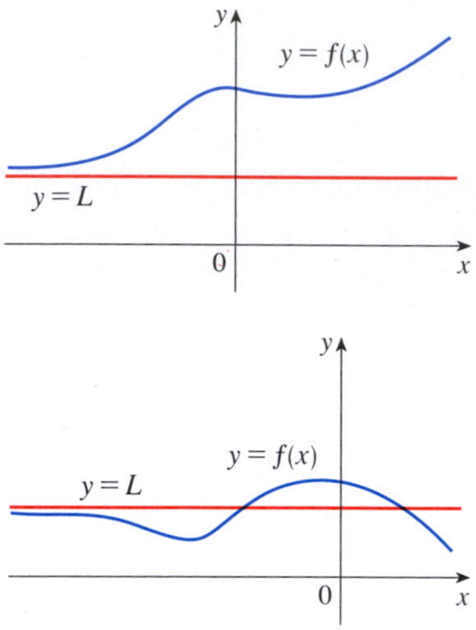

FIGURE 3
Examples illustrating $\lim_{x\to-\infty} f(x) = L$

The general definition is as follows.

(2) DEFINITION Let f be a function defined on some interval $(-\infty, a)$. Then

$$\lim_{x\to-\infty} f(x) = L$$

means that the values of $f(x)$ can be made arbitrarily close to L by taking x sufficiently large negative.

Again, the symbol $-\infty$ does not represent a number, but the expression $\lim_{x\to-\infty} f(x) = L$ is often read as

"the limit of $f(x)$, as x approaches negative infinity, is L"

Definition 2 is illustrated in Figure 3.

(3) DEFINITION The line $y = L$ is called a **horizontal asymptote** of the curve $y = f(x)$ if either

$$\lim_{x\to\infty} f(x) = L \qquad \text{or} \qquad \lim_{x\to-\infty} f(x) = L$$

For instance, the curve illustrated in Figure 1 has the line $y = 1$ as a horizontal asymptote because

$$\lim_{x\to\infty} \frac{x^2 - 1}{x^2 + 1} = 1$$

The curve $y = f(x)$ sketched in Figure 4 has both $y = -1$ and $y = 2$ as horizontal asymptotes because

$$\lim_{x\to\infty} f(x) = -1 \qquad \text{and} \qquad \lim_{x\to-\infty} f(x) = 2$$

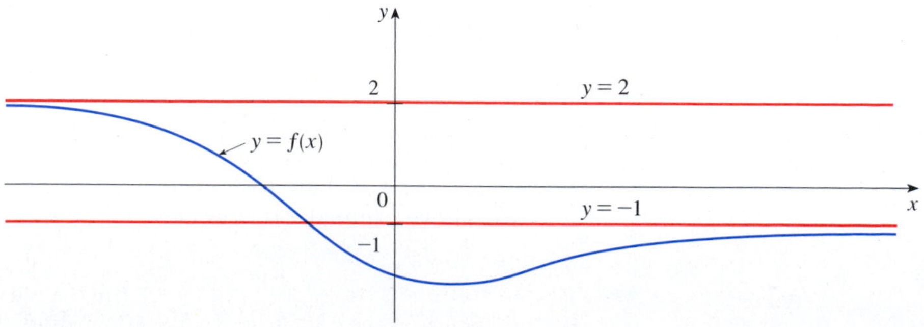

FIGURE 4

EXAMPLE 1 Find $\lim_{x\to\infty} \frac{1}{x}$ and $\lim_{x\to-\infty} \frac{1}{x}$.

SOLUTION Observe that when x is large, $1/x$ is small. For instance,

$$\frac{1}{100} = 0.01$$

$$\frac{1}{10{,}000} = 0.0001$$

$$\frac{1}{1{,}000{,}000} = 0.000001$$

In fact, by taking x large enough, we can make $1/x$ as close to 0 as we please. Therefore, according to Definition 1, we have

$$\lim_{x\to\infty} \frac{1}{x} = 0$$

Similar reasoning shows that when x is large negative, $1/x$ is small negative, so we also have

$$\lim_{x\to-\infty} \frac{1}{x} = 0$$

It follows that the line $y = 0$ (the x-axis) is a horizontal asymptote of the curve $y = 1/x$. (This is an equilateral hyperbola; see Figure 5.) ■

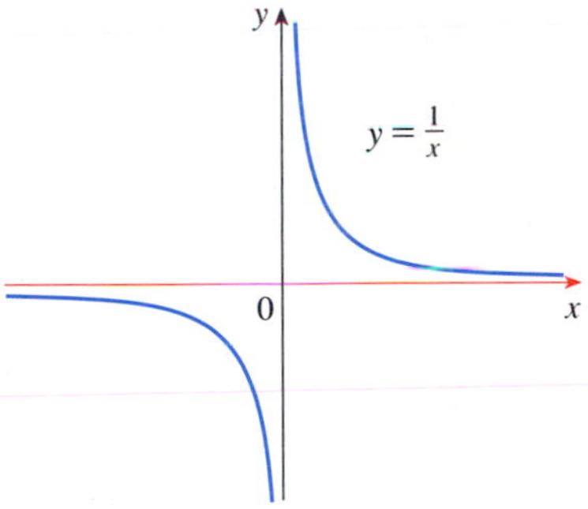

FIGURE 5

$\lim_{x\to\infty} \frac{1}{x} = 0, \quad \lim_{x\to-\infty} \frac{1}{x} = 0$

Most of the Limit Laws that were given in Section 2.3 also hold for limits at infinity. *It can be proved that the Limit Laws listed in Section 2.3 (with the exception of Laws 8, 9, and 10) are also valid if "$x \to a$" is replaced by "$x \to \infty$" or "$x \to -\infty$."* In particular, if we combine Laws 6 and 11 with the results of Example 1 we obtain the following important rule for calculating limits.

(4) THEOREM If $r > 0$ is a rational number, then

$$\lim_{x\to\infty} \frac{1}{x^r} = 0$$

If $r > 0$ is a rational number such that x^r is defined for all x, then

$$\lim_{x\to-\infty} \frac{1}{x^r} = 0$$

EXAMPLE 2 Evaluate

$$\lim_{x\to\infty} \frac{3x^2 - x - 2}{5x^2 + 4x + 1}$$

and indicate which properties of limits are used at each stage.

SOLUTION To evaluate the limit at infinity of a rational function, we first divide both the numerator and denominator by the highest power of x that occurs. (We may assume that $x \neq 0$ since we are interested only in large values of x.) In this case the highest power of x is x^2, so we have

$$\lim_{x\to\infty} \frac{3x^2 - x - 2}{5x^2 + 4x + 1} = \lim_{x\to\infty} \frac{3 - \dfrac{1}{x} - \dfrac{2}{x^2}}{5 + \dfrac{4}{x} + \dfrac{1}{x^2}}$$

$$= \frac{\lim_{x\to\infty}\left(3 - \dfrac{1}{x} - \dfrac{2}{x^2}\right)}{\lim_{x\to\infty}\left(5 + \dfrac{4}{x} + \dfrac{1}{x^2}\right)} \qquad \text{(by Law 5)}$$

$$= \frac{\lim_{x\to\infty} 3 - \lim_{x\to\infty} \dfrac{1}{x} - 2\lim_{x\to\infty} \dfrac{1}{x^2}}{\lim_{x\to\infty} 5 + 4\lim_{x\to\infty} \dfrac{1}{x} + \lim_{x\to\infty} \dfrac{1}{x^2}} \qquad \text{(by 1, 2, and 3)}$$

$$= \frac{3 - 0 - 0}{5 + 0 + 0} \qquad \text{(by 7 and Theorem 4)}$$

$$= \frac{3}{5}$$

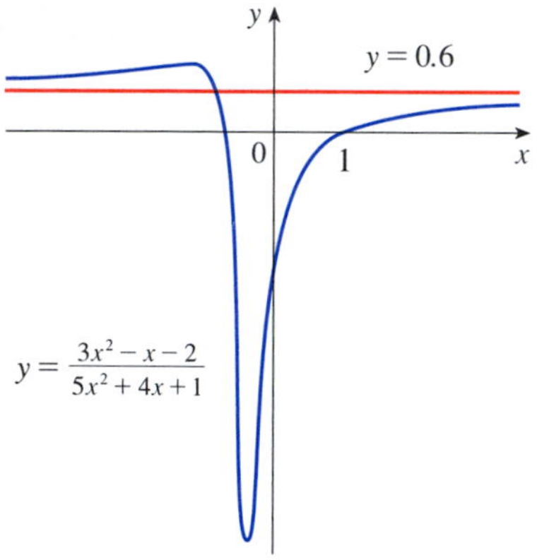

FIGURE 6

A similar calculation shows that the limit as $x \to -\infty$ is also $\frac{3}{5}$. Figure 6 illustrates the results of these calculations by showing how the graph of the given rational function approaches the horizontal asymptote $y = \frac{3}{5}$. ■

EXAMPLE 3 Find the horizontal and vertical asymptotes of the graph of the function

$$f(x) = \frac{\sqrt{2x^2 + 1}}{3x - 5}$$

SOLUTION Dividing both numerator and denominator by x and using the properties of limits, we have

$$\lim_{x\to\infty} \frac{\sqrt{2x^2 + 1}}{3x - 5} = \lim_{x\to\infty} \frac{\sqrt{2 + \dfrac{1}{x^2}}}{3 - \dfrac{5}{x}} \qquad \text{(since } \sqrt{x^2} = x \text{ for } x > 0\text{)}$$

$$= \frac{\lim_{x\to\infty} \sqrt{2 + \dfrac{1}{x^2}}}{\lim_{x\to\infty}\left(3 - \dfrac{5}{x}\right)} = \frac{\sqrt{\lim_{x\to\infty} 2 + \lim_{x\to\infty} \dfrac{1}{x^2}}}{\lim_{x\to\infty} 3 - 5\lim_{x\to\infty} \dfrac{1}{x}}$$

$$= \frac{\sqrt{2 + 0}}{3 - 5 \cdot 0} = \frac{\sqrt{2}}{3}$$

Therefore, the line $y = \sqrt{2}/3$ is a horizontal asymptote of the graph of f.

In computing the limit as $x \to -\infty$, we must remember that for $x < 0$, we have $\sqrt{x^2} = |x| = -x$, so when we divide the numerator by x, when $x < 0$, we get

$$\frac{1}{x}\sqrt{2x^2+1} = -\frac{1}{\sqrt{x^2}}\sqrt{2x^2+1} = -\sqrt{2+\frac{1}{x^2}}$$

Therefore

$$\lim_{x\to-\infty}\frac{\sqrt{2x^2+1}}{3x-5} = \lim_{x\to-\infty}\frac{-\sqrt{2+\dfrac{1}{x^2}}}{3-\dfrac{5}{x}}$$

$$= \frac{-\sqrt{2+\lim\limits_{x\to-\infty}\dfrac{1}{x^2}}}{3-5\lim\limits_{x\to-\infty}\dfrac{1}{x}} = -\frac{\sqrt{2}}{3}$$

Thus, the line $y = -\sqrt{2}/3$ is also a horizontal asymptote.

The vertical asymptote is likely to occur when the denominator, $3x - 5$, is 0, that is, when $x = \frac{5}{3}$. If x is close to $\frac{5}{3}$ and $x > \frac{5}{3}$, then the denominator is close to 0 and $3x - 5$ is positive. The numerator $\sqrt{2x^2+1}$ is always positive, so $f(x)$ is positive. Therefore

$$\lim_{x\to(5/3)^+}\frac{\sqrt{2x^2+1}}{3x-5} = \infty$$

If x is close to $\frac{5}{3}$ but $x < \frac{5}{3}$, then $3x - 5 < 0$ and so $f(x)$ is large negative. Thus

$$\lim_{x\to(5/3)^-}\frac{\sqrt{2x^2+1}}{3x-5} = -\infty$$

The vertical asymptote is $x = \frac{5}{3}$. All three asymptotes are shown in Figure 7. ■

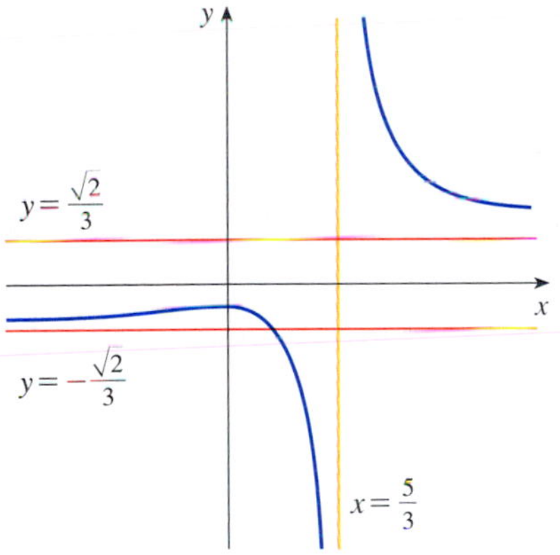

FIGURE 7
$y = \dfrac{\sqrt{2x^2+1}}{3x-5}$

EXAMPLE 4 Compute $\lim\limits_{x\to\infty}(\sqrt{x^2+1} - x)$.

SOLUTION We first multiply numerator and denominator by the conjugate radical:

$$\lim_{x\to\infty}(\sqrt{x^2+1} - x) = \lim_{x\to\infty}(\sqrt{x^2+1} - x)\,\frac{\sqrt{x^2+1}+x}{\sqrt{x^2+1}+x}$$

$$= \lim_{x\to\infty}\frac{(x^2+1)-x^2}{\sqrt{x^2+1}+x} = \lim_{x\to\infty}\frac{1}{\sqrt{x^2+1}+x}$$

The Squeeze Theorem could be used to show that this limit is 0. But an easier method is to divide numerator and denominator by x. Doing this and using the Limit Laws, we obtain

$$\lim_{x\to\infty}(\sqrt{x^2+1} - x) = \lim_{x\to\infty}\frac{\dfrac{1}{x}}{\sqrt{1+\dfrac{1}{x^2}}+1}$$

$$= \frac{0}{\sqrt{1+0}+1} = 0$$

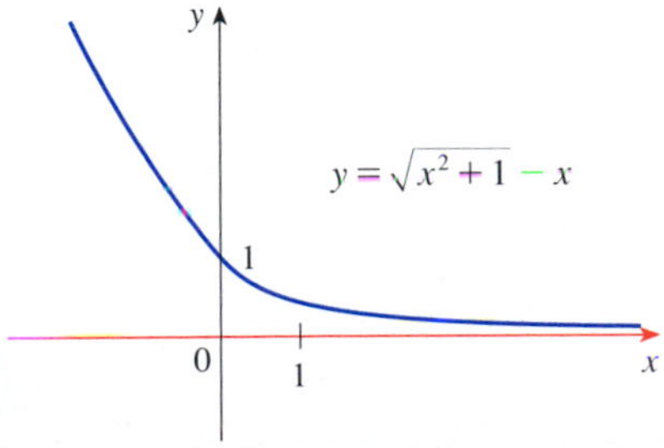

FIGURE 8

Figure 8 illustrates this result. ■

EXAMPLE 5 Evaluate (a) $\lim_{x\to\infty} \sin x$ and (b) $\lim_{x\to\infty} \sin\dfrac{1}{x}$.

SOLUTION

(a) As x increases, the values of $\sin x$ oscillate between 1 and -1 infinitely often. Thus, $\lim_{x\to\infty} \sin x$ does not exist.

The problem-solving strategy for part (b) is *Introduce Something Extra* (see Section 4 in Review and Preview). Here, the something extra, the auxiliary aid, is the new variable t.

(b) If we let $t = 1/x$, then $t \to 0^+$ as $x \to \infty$, so we have

$$\lim_{x\to\infty} \sin\frac{1}{x} = \lim_{t\to 0^+} \sin t = 0$$

(See Exercise 55.) ■

INFINITE LIMITS AT INFINITY

The notation

$$\lim_{x\to\infty} f(x) = \infty$$

is used to indicate that the values of $f(x)$ become large as x becomes large. Similar meanings are attached to the following symbols:

$$\lim_{x\to-\infty} f(x) = \infty \qquad \lim_{x\to\infty} f(x) = -\infty \qquad \lim_{x\to-\infty} f(x) = -\infty$$

EXAMPLE 6 Find $\lim_{x\to\infty} x^3$ and $\lim_{x\to-\infty} x^3$.

SOLUTION When x becomes large, x^3 also becomes large. For instance,

$$10^3 = 1000 \qquad 100^3 = 1{,}000{,}000 \qquad 1000^3 = 1{,}000{,}000{,}000$$

In fact we can make x^3 as big as we like by taking x large enough. Therefore, we can write

$$\lim_{x\to\infty} x^3 = \infty$$

Similarly, when x is large negative, so is x^3. Thus

$$\lim_{x\to-\infty} x^3 = -\infty$$

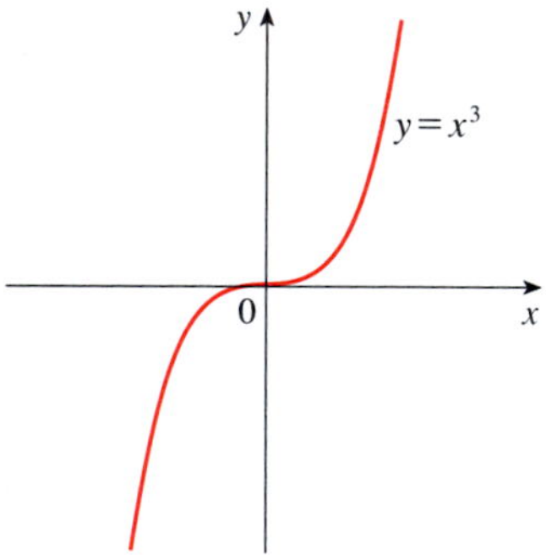

FIGURE 9
$\lim_{x\to\infty} x^3 = \infty$, $\lim_{x\to-\infty} x^3 = -\infty$

These limit statements can also be seen from the graph of $y = x^3$ in Figure 9. ■

EXAMPLE 7 Find $\lim_{x\to\infty} (x^2 - x)$.

SOLUTION Note that we *cannot* write

$$\begin{aligned}\lim_{x\to\infty} (x^2 - x) &= \lim_{x\to\infty} x^2 - \lim_{x\to\infty} x \\ &= \infty - \infty\end{aligned}$$

The Limit Laws cannot be applied to infinite limits because ∞ is not a number

($\infty - \infty$ cannot be defined). However, we can write

$$\lim_{x\to\infty} (x^2 - x) = \lim_{x\to\infty} x(x-1) = \infty$$

because both x and $x - 1$ become arbitrarily large. ■

EXAMPLE 8 Find $\lim_{x\to\infty} \dfrac{x^2 + x}{3 - x}$.

SOLUTION 1 As in Example 2 we can evaluate the limit at infinity of a rational function by dividing numerator and denominator by the highest power of x that occurs:

$$\lim_{x\to\infty} \frac{x^2+x}{3-x} = \lim_{x\to\infty} \frac{1 + \dfrac{1}{x}}{\dfrac{3}{x^2} - \dfrac{1}{x}} = -\infty$$

because $1 + 1/x \to 1$ and $3/x^2 - 1/x \to 0$ through negative values as $x \to \infty$. (Note that $3/x^2 - 1/x < 0$ for $x > 3$.)

SOLUTION 2 It is also possible to find the given limit by dividing numerator and denominator by x instead of by x^2:

Notice that x is the highest power that occurs in the *denominator.*

$$\lim_{x\to\infty} \frac{x^2+x}{3-x} = \lim_{x\to\infty} \frac{x+1}{\dfrac{3}{x} - 1} = -\infty$$

because $x + 1 \to \infty$ and $3/x - 1 \to -1$ as $x \to \infty$. ■

The next example shows that by using infinite limits at infinity, together with intercepts, we can get a rough idea of the graph of a polynomial even without computing derivatives.

EXAMPLE 9 Sketch the graph of $y = (x-2)^4(x+1)^3(x-1)$ by finding its intercepts and its limits as $x \to \infty$ and as $x \to -\infty$.

SOLUTION The y-intercept is $f(0) = (-2)^4(1)^3(-1) = -16$ and the x-intercepts are found by setting $y = 0$: $x = 2, -1, 1$. Notice that since $(x-2)^4$ is positive, the function does not change sign at 2; thus the graph does not cross the x-axis at 2. The graph crosses the axis at -1 and 1.

When x is large, all three factors are large, so

$$\lim_{x\to\infty} (x-2)^4(x+1)^3(x-1) = \infty$$

When x is large negative, the first factor is large positive and the second and third factors are both large negative, so

$$\lim_{x\to-\infty} (x-2)^4(x+1)^3(x-1) = \infty$$

Combining this information, we give a rough sketch of the graph in Figure 10. ■

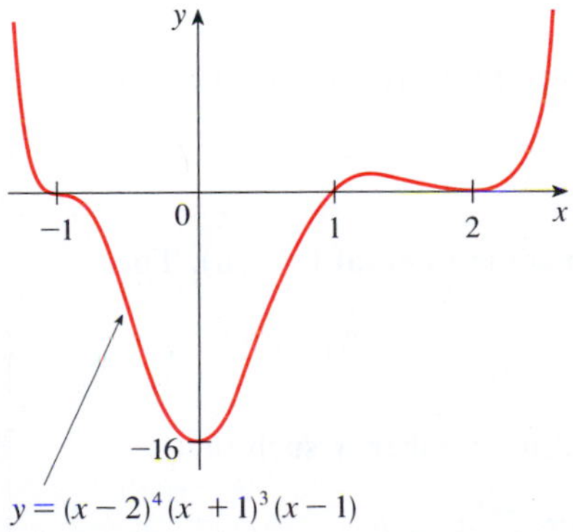

FIGURE 10

PRECISE DEFINITIONS

Definition 1 can be stated precisely as follows:

(5) DEFINITION Let f be a function defined on some interval (a, ∞). Then

$$\lim_{x \to \infty} f(x) = L$$

means that for every $\varepsilon > 0$ there is a corresponding number N such that

$$|f(x) - L| < \varepsilon \qquad \text{whenever} \qquad x > N$$

In words, this says that the values of $f(x)$ can be made arbitrarily close to L (within a distance ε, where ε is any positive number) by taking x sufficiently large (larger than N, where N depends on ε). Graphically it says that by choosing x large enough (larger than some number N) we can make the graph of f lie between the given horizontal lines $y = L - \varepsilon$ and $y = L + \varepsilon$ as in Figure 11. This must be true no matter how small we choose ε. Figure 12 shows that if a smaller value of ε is chosen, then a larger value of N may be required.

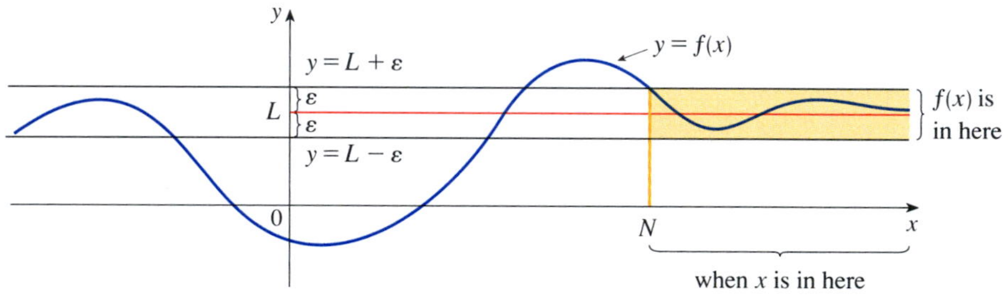

FIGURE 11
$\lim_{x \to \infty} f(x) = L$

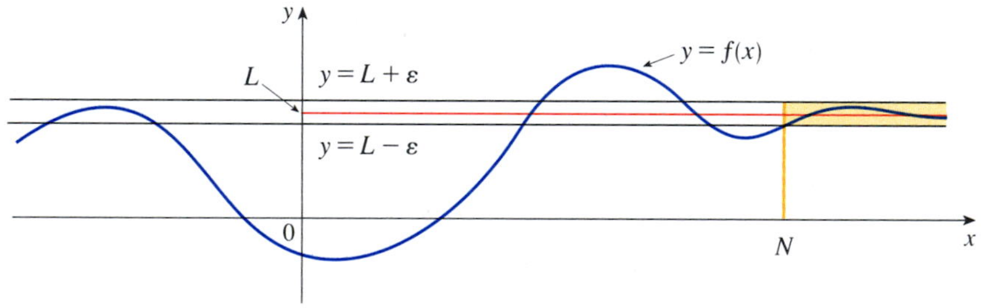

FIGURE 12
$\lim_{x \to \infty} f(x) = L$

Similarly, a precise version of Definition 2 is given by Definition 6, which is illustrated in Figure 13.

(6) DEFINITION Let f be a function defined on some interval $(-\infty, a)$. Then

$$\lim_{x \to -\infty} f(x) = L$$

means that for every $\varepsilon > 0$ there is a corresponding number N such that

$$|f(x) - L| < \varepsilon \qquad \text{whenever} \qquad x < N$$

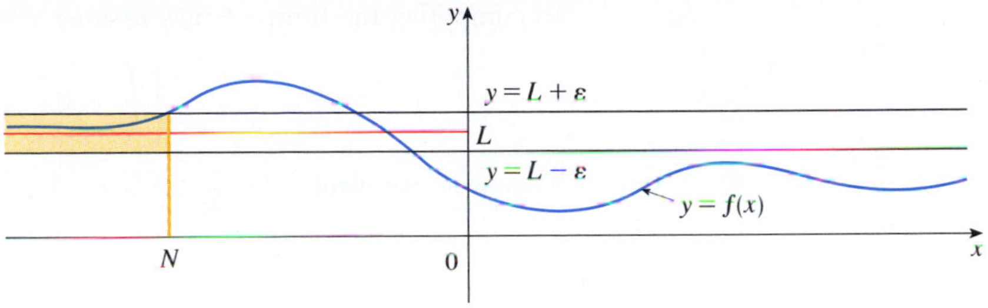

FIGURE 13
$\lim_{x \to -\infty} f(x) = L$

In Example 2 we calculated that

$$\lim_{x \to \infty} \frac{3x^2 - x - 2}{5x^2 + 4x + 1} = \frac{3}{5}$$

In the next example we use a graphing device to relate this statement to Definition 5 with $L = \frac{3}{5}$ and $\varepsilon = 0.1$.

EXAMPLE 10 Use a graph to find a number N such that

$$\left| \frac{3x^2 - x - 2}{5x^2 + 4x + 1} - 0.6 \right| < 0.1 \quad \text{whenever} \quad x > N$$

SOLUTION We rewrite the given inequality as

$$0.5 < \frac{3x^2 - x - 2}{5x^2 + 4x + 1} < 0.7$$

We need to determine the values of x for which the given curve lies between the horizontal lines $y = 0.5$ and $y = 0.7$. So we graph the curve and these lines in Figure 14. Then we use the cursor to estimate that the curve crosses the line $y = 0.5$ when $x \approx 6.7$. To the right of this number the curve stays between the lines $y = 0.5$ and $y = 0.7$. Rounding to be safe, we can say that

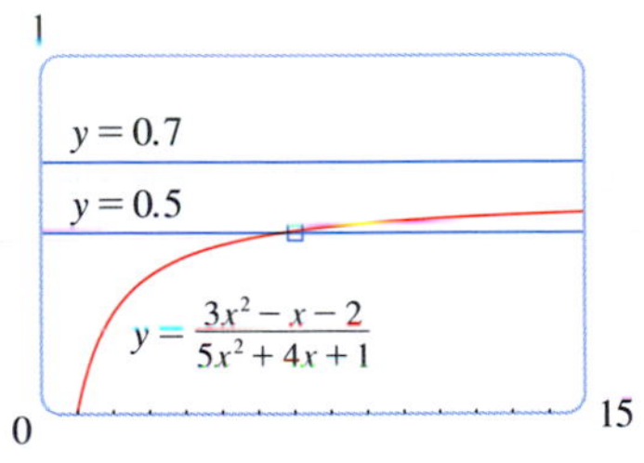

FIGURE 14

$$\left| \frac{3x^2 - x - 2}{5x^2 + 4x + 1} - 0.6 \right| < 0.1 \quad \text{whenever} \quad x > 7$$

In other words, for $\varepsilon = 0.1$ we can choose $N = 7$ (or any larger number) in Definition 5. ■

EXAMPLE 11 Use Definition 5 to prove that $\lim_{x \to \infty} \frac{1}{x} = 0$.

SOLUTION

1. *Preliminary analysis of the problem: guessing a value for N.* Given $\varepsilon > 0$, we want to find N such that

$$\left| \frac{1}{x} - 0 \right| < \varepsilon \quad \text{whenever} \quad x > N$$

In computing the limit we may assume $x > 0$, in which case

$$\left|\frac{1}{x} - 0\right| = \left|\frac{1}{x}\right| = \frac{1}{x}$$

Therefore, we want

$$\frac{1}{x} < \varepsilon \qquad \text{whenever} \qquad x > N$$

that is,

$$x > \frac{1}{\varepsilon} \qquad \text{whenever} \qquad x > N$$

This suggests that we should take $N = 1/\varepsilon$.

2. *Proof* (*showing that N works*). Given $\varepsilon > 0$, we choose $N = 1/\varepsilon$. Let $x > N$. Then

$$\left|\frac{1}{x} - 0\right| = \frac{1}{|x|} = \frac{1}{x} < \frac{1}{N} = \varepsilon$$

Thus

$$\left|\frac{1}{x} - 0\right| < \varepsilon \qquad \text{whenever} \qquad x > N$$

Therefore, by Definition 5,

$$\lim_{x\to\infty} \frac{1}{x} = 0$$

Figure 15 illustrates the proof by showing some values of ε and the corresponding values of N.

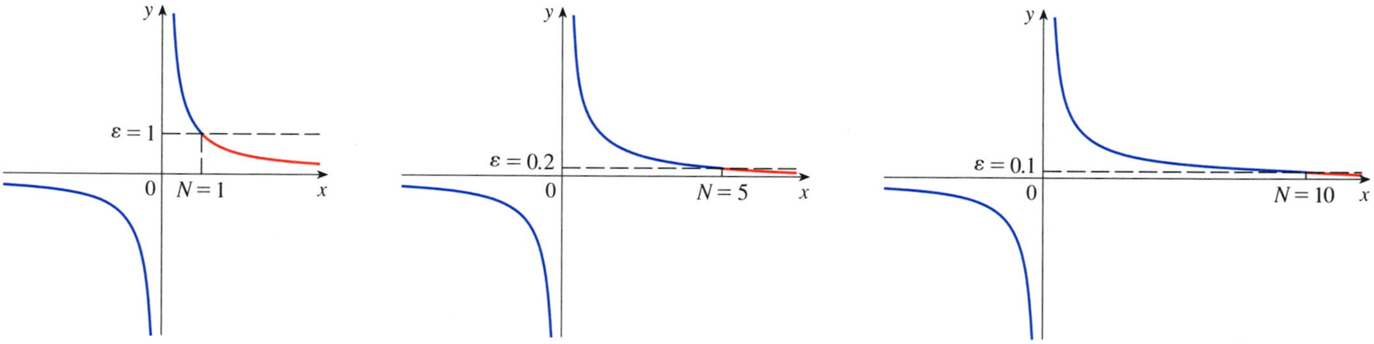

FIGURE 15

Finally we note that an infinite limit at infinity can be defined as follows. The geometric illustration is given in Figure 16.

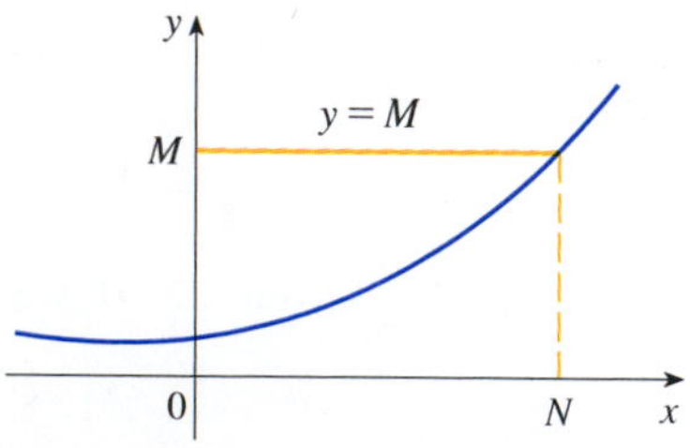

FIGURE 16
$\lim_{x\to\infty} f(x) = \infty$

(7) DEFINITION Let f be a function defined on some interval (a, ∞). Then

$$\lim_{x\to\infty} f(x) = \infty$$

means that for every positive number M there is a corresponding number $N > 0$ such that

$$f(x) > M \qquad \text{whenever} \qquad x > N$$

EXERCISES 2.6

1–8 ■ Evaluate the limit and justify each step by indicating the appropriate properties of limits.

1. $\displaystyle\lim_{x\to\infty} \frac{1}{x\sqrt{x}}$

2. $\displaystyle\lim_{x\to\infty} \frac{5+2x}{3-x}$

3. $\displaystyle\lim_{x\to\infty} \frac{x+4}{x^2-2x+5}$

4. $\displaystyle\lim_{t\to\infty} \frac{7t^3+4t}{2t^3-t^2+3}$

5. $\displaystyle\lim_{x\to-\infty} \frac{(1-x)(2+x)}{(1+2x)(2-3x)}$

6. $\displaystyle\lim_{x\to\infty} \sqrt{\frac{2x^2-1}{x+8x^2}}$

7. $\displaystyle\lim_{x\to\infty} \frac{1}{3+\sqrt{x}}$

8. $\displaystyle\lim_{x\to\infty} \frac{\sin^2 x}{x^2}$

9–30 ■ Find the limit.

9. $\displaystyle\lim_{r\to\infty} \frac{r^4-r^2+1}{r^5+r^3-r}$

10. $\displaystyle\lim_{t\to-\infty} \frac{6t^2+5t}{(1-t)(2t-3)}$

11. $\displaystyle\lim_{x\to\infty} \frac{\sqrt{1+4x^2}}{4+x}$

12. $\displaystyle\lim_{x\to-\infty} \frac{\sqrt{x^2+4x}}{4x+1}$

13. $\displaystyle\lim_{x\to\infty} \frac{1-\sqrt{x}}{1+\sqrt{x}}$

14. $\displaystyle\lim_{x\to\infty} \left(\sqrt{x^2+3x+1}-x\right)$

15. $\displaystyle\lim_{x\to\infty} \left(\sqrt{x^2+1}-\sqrt{x^2-1}\right)$

16. $\displaystyle\lim_{x\to-\infty} \left(x+\sqrt{x^2+2x}\right)$

17. $\displaystyle\lim_{x\to\infty} \left(\sqrt{1+x}-\sqrt{x}\right)$

18. $\displaystyle\lim_{x\to\infty} \left(\sqrt[3]{1+x}-\sqrt[3]{x}\right)$

19. $\displaystyle\lim_{x\to-\infty} \left(\sqrt{x^2+x+1}+x\right)$

20. $\displaystyle\lim_{x\to\infty} \cos x$

21. $\displaystyle\lim_{x\to\infty} \sqrt{x}$

22. $\displaystyle\lim_{x\to-\infty} \sqrt[3]{x}$

23. $\displaystyle\lim_{x\to\infty} \left(x-\sqrt{x}\right)$

24. $\displaystyle\lim_{x\to\infty} \left(x+\sqrt{x}\right)$

25. $\displaystyle\lim_{x\to-\infty} (x^3-5x^2)$

26. $\displaystyle\lim_{x\to\infty} (x^2-x^4)$

27. $\displaystyle\lim_{x\to\infty} \frac{x^7-1}{x^6+1}$

28. $\displaystyle\lim_{x\to\infty} \frac{x^3-1}{x^4+1}$

29. $\displaystyle\lim_{x\to\infty} \frac{\sqrt{x}+3}{x+3}$

30. $\displaystyle\lim_{x\to\infty} \frac{x}{\sqrt{x}-1}$

31. Guess the value of the limit

$$\lim_{x\to\infty} \frac{x^2}{2^x}$$

by evaluating the function $f(x) = x^2/2^x$ for $x = 0, 1, 2, 3, 4, 5, 6, 7, 8, 9, 10, 20, 50$, and 100.

32. Guess the value of the limit

$$\lim_{x\to\infty} x^2 \sin\frac{5}{x^2}$$

by evaluating $f(x) = x^2 \sin(5/x^2)$ for $x = 1, 2, 3, 4, 5, 6, 7, 8, 9, 10, 20, 50$, and 100. Then confirm your guess by evaluating this limit exactly.

33–38 ■ Find the horizontal and vertical asymptotes of each curve. Check your work by graphing the curve and estimating the asymptotes.

33. $y = \dfrac{x}{x+4}$

34. $y = \dfrac{x^2+4}{x^2-1}$

35. $y = \dfrac{x^3}{x^2+3x-10}$

36. $y = \dfrac{x^3+1}{x^3+x}$

37. $h(x) = \dfrac{x}{\sqrt[4]{x^4+1}}$

38. $F(x) = \dfrac{x-9}{\sqrt{4x^2+3x+2}}$

39–42 ■ Find the limits as $x \to \infty$ and $x \to -\infty$. Use this information, together with intercepts, to give a rough sketch of the graph as in Example 9.

39. $y = x^2(x-2)(1-x)$

40. $y = (2+x)^3(1-x)(3-x)$

41. $y = (x+4)^5(x-3)^4$

42. $y = (1-x)(x-3)^2(x-5)^2$

43. Let P and Q be polynomials. Find

$$\lim_{x\to\infty} \frac{P(x)}{Q(x)}$$

if the degree of P is (a) less than the degree of Q and (b) greater than the degree of Q.

44. Make a rough sketch of the curve $y = x^n$ (n an integer) for the following five cases:

(i) $n = 0$ (ii) $n > 0$, n odd
(iii) $n > 0$, n even (iv) $n < 0$, n odd
(v) $n < 0$, n even

Then use these sketches to find the following limits.

(a) $\displaystyle\lim_{x\to 0^+} x^n$ (b) $\displaystyle\lim_{x\to 0^-} x^n$
(c) $\displaystyle\lim_{x\to\infty} x^n$ (d) $\displaystyle\lim_{x\to-\infty} x^n$

45. Find $\lim_{x \to \infty} f(x)$ if

$$\frac{4x - 1}{x} < f(x) < \frac{4x^2 + 3x}{x^2}$$

for all $x > 5$.

46. (a) A tank contains 5000 L of pure water. Brine that contains 30 g of salt per liter of water is pumped into the tank at a rate of 25 L/min. Show that the concentration of salt after t minutes (in grams per liter) is

$$C(t) = \frac{30t}{200 + t}$$

(b) What happens to the concentration as $t \to \infty$?

47. Use a graph to find a number N such that

$$\left| \frac{6x^2 + 5x - 3}{2x^2 - 1} - 3 \right| < 0.2 \qquad \text{whenever} \qquad x > N$$

48. For the limit

$$\lim_{x \to \infty} \frac{\sqrt{4x^2 + 1}}{x + 1} = 2$$

illustrate Definition 5 by finding values of N that correspond to $\varepsilon = 0.5$ and $\varepsilon = 0.1$.

49. For the limit

$$\lim_{x \to -\infty} \frac{\sqrt{4x^2 + 1}}{x + 1} = -2$$

illustrate Definition 6 by finding values of N that correspond to $\varepsilon = 0.5$ and $\varepsilon = 0.1$.

50. For the limit

$$\lim_{x \to \infty} \frac{2x + 1}{\sqrt{x + 1}} = \infty$$

illustrate Definition 7 by finding a value of N that corresponds to $M = 100$.

51. (a) How large do we have to take x so that $1/x^2 < 0.0001$?
(b) Taking $r = 2$ in Theorem 4, we have the statement

$$\lim_{x \to \infty} \frac{1}{x^2} = 0$$

Prove this directly using Definition 5.

52. (a) How large do we have to take x so that $1/\sqrt{x} < 0.0001$?
(b) Taking $r = \frac{1}{2}$ in Theorem 4, we have the statement

$$\lim_{x \to \infty} \frac{1}{\sqrt{x}} = 0$$

Prove this directly using Definition 5.

53. Use Definition 6 to prove that $\lim_{x \to -\infty} \dfrac{1}{x} = 0$.

54. Prove, using Definition 7, that $\lim_{x \to \infty} x^3 = \infty$.

55. Prove that

$$\lim_{x \to \infty} f(x) = \lim_{t \to 0^+} f\left(\frac{1}{t}\right)$$

and

$$\lim_{x \to -\infty} f(x) = \lim_{t \to 0^-} f\left(\frac{1}{t}\right)$$

if these limits exist.

2.7 TANGENTS, VELOCITIES, AND OTHER RATES OF CHANGE

In Section 2.1 we guessed the values of slopes of tangent lines and velocities on the basis of numerical evidence. Now that we have defined limits and have learned techniques for computing them, we return to the tangent and velocity problems with the ability to calculate slopes of tangents, velocities, and other rates of change.

TANGENTS

If a curve C has equation $y = f(x)$ and we want to find the tangent to C at the point $P(a, f(a))$, then we consider a nearby point $Q(x, f(x))$, where $x \neq a$, and compute the slope of the secant line PQ:

$$m_{PQ} = \frac{f(x) - f(a)}{x - a}$$

Then we let Q approach P along the curve C by letting x approach a. If m_{PQ} approaches a number m, then we define the *tangent* t to be the line through P with slope m. (This amounts to saying that the tangent line is the limiting position of the secant line PQ as Q approaches P. See Figure 1.)

(1) DEFINITION The **tangent line** to the curve $y = f(x)$ at the point $P(a, f(a))$ is the line through P with slope

$$m = \lim_{x \to a} \frac{f(x) - f(a)}{x - a}$$

provided that this limit exists.

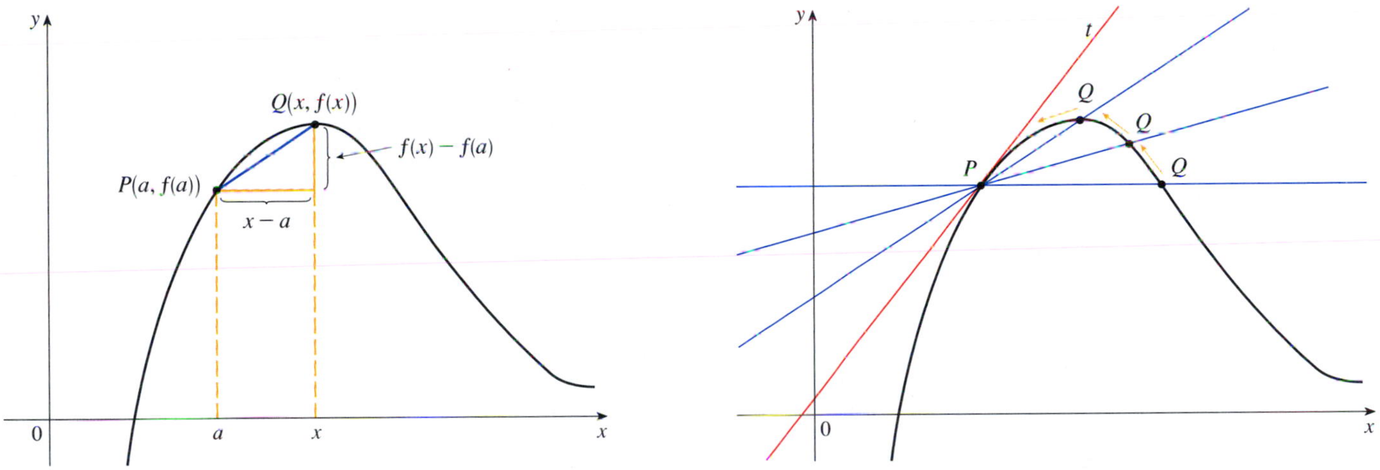

(a) (b)

FIGURE 1

In our first example we confirm the guess we made in Example 1 in Section 2.1.

EXAMPLE 1 Find the equation of the tangent line to the parabola $y = x^2$ at the point $P(1, 1)$.

SOLUTION Here we have $a = 1$ and $f(x) = x^2$, so the slope is

$$m = \lim_{x \to 1} \frac{f(x) - f(1)}{x - 1} = \lim_{x \to 1} \frac{x^2 - 1}{x - 1}$$

$$= \lim_{x \to 1} \frac{(x - 1)(x + 1)}{x - 1}$$

$$= \lim_{x \to 1} (x + 1) = 1 + 1 = 2$$

Point-slope form for a line through the point (x_1, y_1) with slope m:
$y - y_1 = m(x - x_1)$

Using the point-slope form of the equation of a line, we find that the equation of the tangent line at $(1, 1)$ is

$$y - 1 = 2(x - 1) \qquad \text{or} \qquad y = 2x - 1$$

■

There is another expression for the slope of a tangent line that is sometimes easier to use. Let

$$h = x - a$$

Then
$$x = a + h$$

so the slope of the secant line PQ is

$$m_{PQ} = \frac{f(a + h) - f(a)}{h}$$

(See Figure 2 where the case $h > 0$ is illustrated and Q is to the right of P. If $h < 0$, however, Q would be to the left of P.)

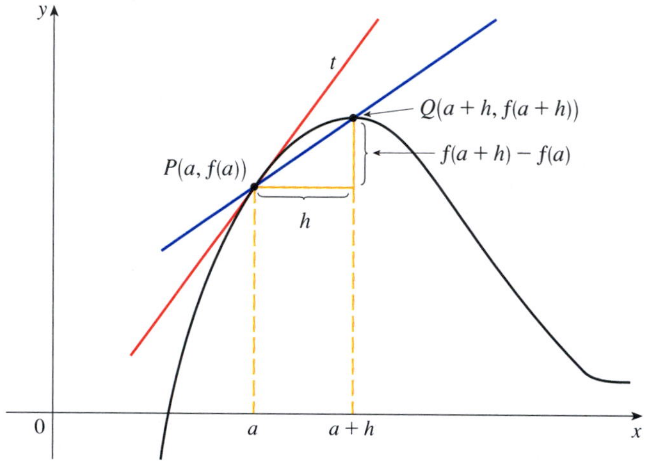

FIGURE 2

Notice that as x approaches a, h approaches 0 and so the expression for the slope of the tangent line in Definition 1 becomes

(2)
$$m = \lim_{h \to 0} \frac{f(a + h) - f(a)}{h}$$

EXAMPLE 2 Find the equation of the tangent line to the hyperbola $y = 3/x$ at the point $(3, 1)$.

SOLUTION Let $f(x) = 3/x$. Then the slope of the tangent at $(3, 1)$ is

$$\begin{aligned} m &= \lim_{h \to 0} \frac{f(3 + h) - f(3)}{h} \\ &= \lim_{h \to 0} \frac{\dfrac{3}{3 + h} - 1}{h} = \lim_{h \to 0} \frac{\dfrac{3 - (3 + h)}{3 + h}}{h} \\ &= \lim_{h \to 0} \frac{-h}{h(3 + h)} = \lim_{h \to 0} \left(-\frac{1}{3 + h}\right) \\ &= -\frac{1}{3} \end{aligned}$$

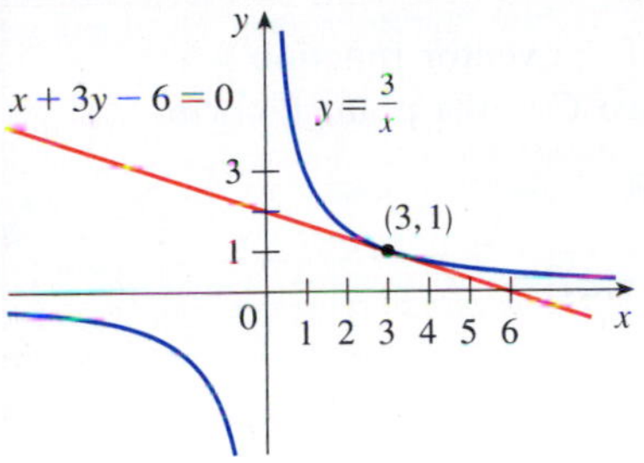

FIGURE 3

Therefore, the equation of the tangent at the point (3, 1) is

$$y - 1 = -\tfrac{1}{3}(x - 3)$$

which simplifies to $$x + 3y - 6 = 0$$

The hyperbola and its tangent are shown in Figure 3. ■

EXAMPLE 3 Find the slopes of the tangent lines to the graph of the function $f(x) = \sqrt{2x + 3}$ at the points $(\frac{1}{2}, 2)$, $(3, 3)$, and $(5, \sqrt{13})$.

SOLUTION Since three slopes are requested, it is efficient to start by finding the slope at the general point $(a, \sqrt{2a + 3})$:

$$m = \lim_{h \to 0} \frac{f(a + h) - f(a)}{h} = \lim_{h \to 0} \frac{\sqrt{2(a + h) + 3} - \sqrt{2a + 3}}{h}$$

Rationalize the numerator

$$= \lim_{h \to 0} \frac{\sqrt{2(a + h) + 3} - \sqrt{2a + 3}}{h} \cdot \frac{\sqrt{2(a + h) + 3} + \sqrt{2a + 3}}{\sqrt{2(a + h) + 3} + \sqrt{2a + 3}}$$

$$= \lim_{h \to 0} \frac{(2a + 2h + 3) - (2a + 3)}{h(\sqrt{2a + 2h + 3} + \sqrt{2a + 3})} = \lim_{h \to 0} \frac{2h}{h(\sqrt{2a + 2h + 3} + \sqrt{2a + 3})}$$

Continuous function of h

$$= \lim_{h \to 0} \frac{2}{\sqrt{2a + 2h + 3} + \sqrt{2a + 3}} = \frac{2}{\sqrt{2a + 3} + \sqrt{2a + 3}} = \frac{1}{\sqrt{2a + 3}}$$

At the point $(\frac{1}{2}, 2)$, $a = \frac{1}{2}$, so the slope of the tangent is

$$m = \frac{1}{\sqrt{2(\frac{1}{2}) + 3}} = \frac{1}{2}$$

At (3, 3), we have

$$m = \frac{1}{\sqrt{2(3) + 3}} = \frac{1}{3}$$

and at $(5, \sqrt{13})$,

$$m = \frac{1}{\sqrt{2(5) + 3}} = \frac{1}{\sqrt{13}}$$ ■

TANGENT VECTORS

In Example 5 in Section 2.1 we found a tangent vector to the parabola traced by the vector function $\mathbf{r}(t) = \langle t, t^2 \rangle$ at the point $P(1, 1)$ corresponding to $t = 1$ by computing the limit of secant vectors $[\mathbf{r}(t) - \mathbf{r}(1)]/(t - 1)$ as $t \to 1$. This procedure can be generalized to find a tangent vector to a curve traced by a function $\mathbf{r}(t) = \langle x(t), y(t) \rangle$ at the point P corresponding to the vector $\mathbf{r}(a) = \langle x(a), y(a) \rangle$. We consider a nearby point Q corresponding to $\mathbf{r}(t) = \langle x(t), y(t) \rangle$, where $t \neq a$, and compute the secant vector

$$\mathbf{v}_{PQ} = \frac{\mathbf{r}(t) - \mathbf{r}(a)}{t - a} = \frac{1}{t - a}[\mathbf{r}(t) - \mathbf{r}(a)]$$

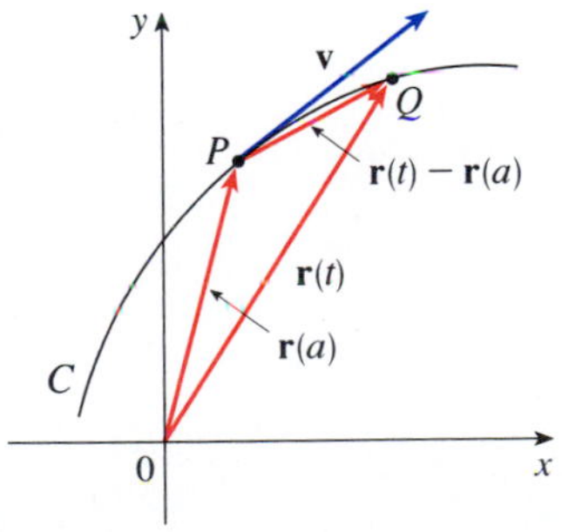

FIGURE 4

In Figure 4, let Q approach P along the curve by letting t approach a. The limit of $\mathbf{v}_{PQ}$ as Q approaches P (if it exists) is defined to be the tangent vector.

(3) DEFINITION Let the curve C be the graph of the vector function $\mathbf{r}(t) = \langle x(t), y(t)\rangle$. The **tangent vector** to the curve C at the point P corresponding to $\mathbf{r}(a)$ is given by

$$\mathbf{v} = \lim_{t \to a} \frac{1}{t - a}[\mathbf{r}(t) - \mathbf{r}(a)]$$

provided that this limit exists.

An alternative formulation of the tangent vector is given by the following.

(4) DEFINITION

$$\mathbf{v} = \lim_{h \to 0} \frac{1}{h}[\mathbf{r}(a + h) - \mathbf{r}(a)]$$

In our next example we confirm our guess made in Example 5 in Section 2.1.

EXAMPLE 4 Find the tangent vector and parametric equations for the line tangent to the parabola

$$\mathbf{r}(t) = \langle t, t^2\rangle$$

at the point $P(1, 1)$.

SOLUTION The point P is obtained by taking $t = 1$, so in Definition 3 we have $a = 1$. A vector that is tangent to the parabola at P is given by

$$\begin{aligned}
\mathbf{v} &= \lim_{t \to 1} \frac{1}{t - 1}[\mathbf{r}(t) - \mathbf{r}(1)] \\
&= \lim_{t \to 1} \frac{1}{t - 1}[\langle t, t^2\rangle - \langle 1, 1\rangle] \\
&= \lim_{t \to 1} \left\langle \frac{t - 1}{t - 1}, \frac{t^2 - 1}{t - 1} \right\rangle \\
&= \lim_{t \to 1} \langle 1, t + 1\rangle \\
&= \langle 1, 2\rangle \quad \text{or} \quad \mathbf{i} + 2\mathbf{j}
\end{aligned}$$

The tangent line to C at $P(1, 1)$ is given by

$$\begin{aligned}
\mathbf{L}(t) &= \langle 1, 1\rangle + t\langle 1, 2\rangle \\
&= \langle 1 + t, 1 + 2t\rangle
\end{aligned}$$

So parametric equations for the tangent line are

$$x = 1 + t \qquad y = 1 + 2t$$

■

VELOCITIES: LINEAR MOTION

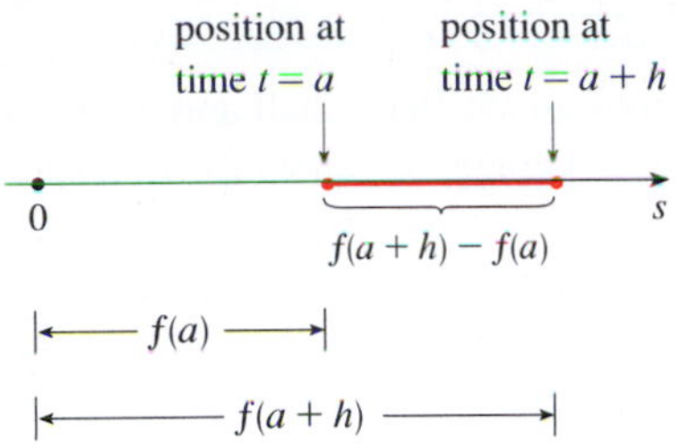

FIGURE 5

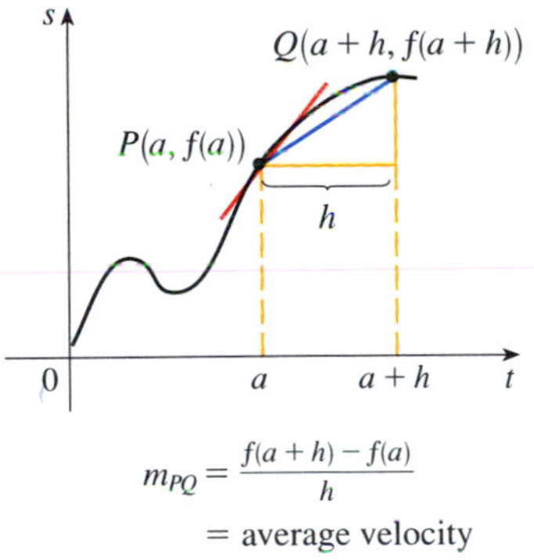

FIGURE 6

In Section 2.1 we investigated the motion of a ball dropped from the CN Tower and defined its velocity to be the limiting value of average velocities over shorter and shorter time periods.

In general, suppose an object moves along a straight line according to an equation of motion $s = f(t)$, where s is the displacement (directed distance) of the object from the origin at time t. The function f that describes the motion is called the **position function** of the object. In the time interval from $t = a$ to $t = a + h$ the change in position is $f(a + h) - f(a)$ (see Figure 5). The average velocity over this time interval is

$$\text{average velocity} = \frac{\text{displacement}}{\text{time}} = \frac{f(a + h) - f(a)}{h}$$

which is the same as the slope of the secant line PQ in Figure 6.

Now suppose we compute the average velocities over shorter and shorter time intervals $[a, a + h]$. In other words, we let h approach 0. As in the example of the falling ball, we define the **velocity** (or **instantaneous velocity**) $v(a)$ at time $t = a$ to be the limit of these average velocities:

(5)
$$v(a) = \lim_{h \to 0} \frac{f(a + h) - f(a)}{h}$$

This means that the velocity at time $t = a$ is equal to the slope of the tangent line at P. (Compare Equations 2 and 5.)

Now that we know how to compute limits, let's reconsider the problem of the falling ball.

Recall from Section 2.1: The distance (in meters) fallen after t seconds is $4.9t^2$.

EXAMPLE 5 Suppose that a ball is dropped from the upper observation deck of the CN Tower, 450 m above the ground.

(a) What is the velocity of the ball after 5 seconds?

(b) How fast is the ball traveling when it hits the ground?

SOLUTION We first use the equation of motion $s = f(t) = 4.9t^2$ to find the velocity $v(a)$ after a seconds:

$$v(a) = \lim_{h \to 0} \frac{f(a + h) - f(a)}{h} = \lim_{h \to 0} \frac{4.9(a + h)^2 - 4.9a^2}{h}$$

$$= \lim_{h \to 0} \frac{4.9(a^2 + 2ah + h^2 - a^2)}{h} = \lim_{h \to 0} \frac{4.9(2ah + h^2)}{h}$$

$$= \lim_{h \to 0} 4.9(2a + h) = 9.8a$$

(a) The velocity after 5 s is $v(5) = (9.8)(5) = 49$ m/s.

(b) Since the observation deck is 450 m above the ground, the ball will hit the ground at the time t_1 when $s(t_1) = 450$, that is,

$$4.9t_1^2 = 450$$

This gives

$$t_1^2 = \frac{450}{4.9} \quad \text{and} \quad t_1 = \sqrt{\frac{450}{4.9}} \approx 9.6 \text{ s}$$

The velocity of the ball as it hits the ground is therefore

$$v(t_1) = 9.8t_1 = 9.8\sqrt{\frac{450}{4.9}} \approx 94 \text{ m/s}$$

■

VELOCITIES: CURVILINEAR MOTION

In Example 4 in Section 2.1 we investigated the motion of an artillery shell and guessed its instantaneous velocity to be the limiting value of the average velocities over shorter and shorter time periods.

In general, suppose an object moves in the xy-plane in such a way that its position at time t is given by the position vector $\mathbf{r}(t)$. (So the path of the object is the graph of the vector function $\mathbf{r}$. See Figure 7.) In the time interval from $t = a$ to $t = a + h$, the change in position of the object is $\mathbf{r}(a + h) - \mathbf{r}(a)$. The average velocity over this time interval is

$$\text{average velocity} = \frac{\mathbf{r}(a + h) - \mathbf{r}(a)}{h}$$

$$= \frac{1}{h}[\mathbf{r}(a + h) - \mathbf{r}(a)]$$

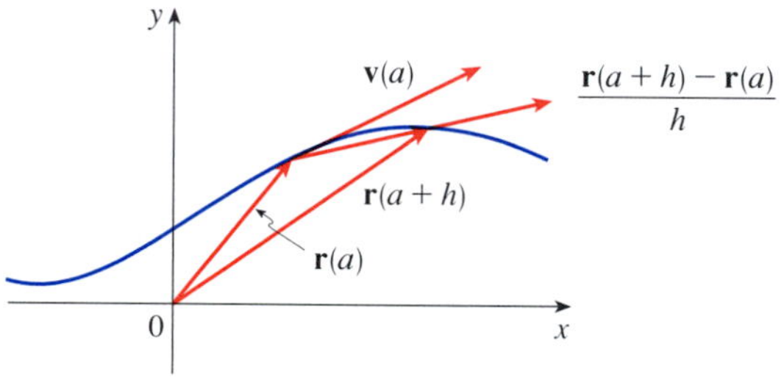

FIGURE 7

We compute the average velocities over shorter and shorter time intervals $[a, a + h]$ by letting h approach 0. We define the **instantaneous velocity** $\mathbf{v}(a)$ at time $t = a$ to be the limit of these average velocities:

(6) $$\mathbf{v}(a) = \lim_{t \to a} \frac{1}{t - a}[\mathbf{r}(t) - \mathbf{r}(a)] = \lim_{h \to 0} \frac{1}{h}[\mathbf{r}(a + h) - \mathbf{r}(a)]$$

This means that the velocity at time $t = a$ is the same as the tangent vector to the graph of $\mathbf{r}$ *at the point corresponding to* $\mathbf{r}(a)$. (Compare Equations 6 and 2.)

We now reconsider the problem of the motion of the artillery shell from Example 4 of Section 2.1.

EXAMPLE 6 An artillery shell is fired from a gun and its position $\mathbf{r}(t)$ in meters at time t seconds is given by

$$\mathbf{r}(t) = (433t)\,\mathbf{i} + (250t - 4.9t^2)\,\mathbf{j}$$

Find the instantaneous velocity of the shell after 40 seconds.

SOLUTION We use Formula 6 to find the instantaneous velocity $\mathbf{v}(a)$ after a seconds:

$$\begin{aligned}
\mathbf{v}(a) &= \lim_{h\to 0}\frac{1}{h}[\mathbf{r}(a+h)-\mathbf{r}(a)] \\
&= \lim_{h\to 0}\frac{1}{h}\{([433(a+h)]\,\mathbf{i}+[250(a+h)-4.9(a+h)^2]\,\mathbf{j})-[(433a)\,\mathbf{i}+(250a-4.9a^2)\,\mathbf{j}]\} \\
&= \lim_{h\to 0}\frac{1}{h}([433(a+h)-433a]\,\mathbf{i}+[(250(a+h)-4.9(a+h)^2)-(250a-4.9a^2)]\,\mathbf{j}) \\
&= \lim_{h\to 0}\left[\left(\frac{433h}{h}\right)\mathbf{i}+\left(\frac{250h-9.8ah-4.9h^2}{h}\right)\mathbf{j}\right] \\
&= \lim_{h\to 0}[433\,\mathbf{i}+(250-9.8a-4.9h)\,\mathbf{j}] \\
&= 433\,\mathbf{i}+(250-9.8a)\,\mathbf{j}
\end{aligned}$$

Therefore, the instantaneous velocity after 40 s is

$$\mathbf{v}(40) = 433\,\mathbf{i}+[250-(9.8)(40)]\,\mathbf{j} = 433\,\mathbf{i}-142\,\mathbf{j}$$

where the units of each component are meters per second. ■

The **speed** of a particle moving along a curve is defined to be the magnitude of the velocity vector. So in Example 6 the speed of the artillery shell at time $t = 40$ is

$$|\mathbf{v}(40)| = \sqrt{433^2+(-142)^2} \approx 455.7 \text{ m/s}$$

OTHER RATES OF CHANGE

Suppose y is a quantity that depends on another quantity x. Thus, y is a function of x and we write $y = f(x)$. If x changes from x_1 to x_2, then the change in x (also called the **increment** of x) is

$$\Delta x = x_2 - x_1$$

and the corresponding change in y is

$$\Delta y = f(x_2) - f(x_1)$$

The difference quotient

$$\frac{\Delta y}{\Delta x} = \frac{f(x_2)-f(x_1)}{x_2-x_1}$$

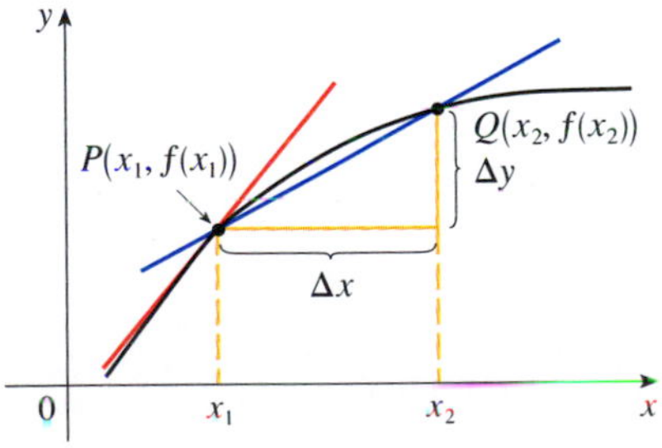

average rate of change $= m_{PQ}$
instantaneous rate of change = slope of tangent at P

FIGURE 8

is called the **average rate of change of y with respect to x** over the interval $[x_1, x_2]$ and can be interpreted as the slope of the secant line PQ in Figure 8. By analogy with velocity, we consider the average rate of change over smaller and smaller intervals by letting x_2 approach x_1 and therefore letting Δx approach 0. The limit of these average rates of change is called the **(instantaneous) rate of change of y with respect to x** at

$x = x_1$, which is interpreted as the slope of the tangent to the curve $y = f(x)$ at $P(x_1, f(x_1))$:

(7)

$$\text{instantaneous rate of change} = \lim_{\Delta x \to 0} \frac{\Delta y}{\Delta x} = \lim_{x_2 \to x_1} \frac{f(x_2) - f(x_1)}{x_2 - x_1}$$

EXAMPLE 7 Temperature readings T (in degrees Celsius) were recorded every hour starting at midnight on a day in April in Whitefish, Montana. The time x is measured in hours from midnight. The data are given in the following table.

x (h)	0	1	2	3	4	5	6	7	8	9	10	11	12
T (°C)	6.5	6.1	5.6	4.9	4.2	4.0	4.0	4.8	6.1	8.3	10.0	12.1	14.3

x (h)	13	14	15	16	17	18	19	20	21	22	23	24
T (°C)	16.0	17.3	18.2	18.8	17.6	16.0	14.1	11.5	10.2	9.0	7.9	7.0

(a) Find the average rate of change of temperature with respect to time

(i) from noon to 3 P.M.
(ii) from noon to 2 P.M.
(iii) from noon to 1 P.M.

(b) Estimate the instantaneous rate of change at noon.

SOLUTION

(a) (i) From noon to 3 P.M. the temperature changes from 14.3 °C to 18.2 °C, so

$$\Delta T = T(15) - T(12) = 18.2 - 14.3 = 3.9\,°\text{C}$$

while the change in time is $\Delta x = 15 - 12 = 3$ h. Therefore, the average rate of change of temperature with respect to time is

$$\frac{\Delta T}{\Delta x} = \frac{3.9}{3} = 1.3\,°\text{C/h}$$

(ii) From noon to 2 P.M. the average rate of change is

$$\frac{\Delta T}{\Delta x} = \frac{T(14) - T(12)}{14 - 12} = \frac{17.3 - 14.3}{2} = 1.5\,°\text{C/h}$$

(iii) From noon to 1 P.M. the average rate of change is

$$\frac{\Delta T}{\Delta x} = \frac{T(13) - T(12)}{13 - 12} = \frac{16.0 - 14.3}{1} = 1.7\,°\text{C/h}$$

(b) We plot the given data in Figure 9 and use them to sketch a smooth curve that approximates the graph of the temperature function. Then we draw the tangent at the point P where $x = 12$ and, after measuring the sides of triangle ABC, we estimate that the slope of the tangent line is

$$\frac{|BC|}{|AC|} = \frac{10.3}{5.5} \approx 1.9$$

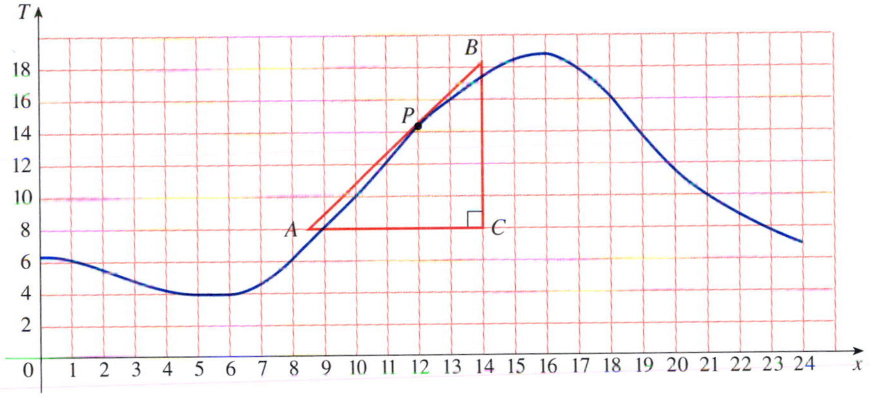

FIGURE 9

Therefore, the instantaneous rate of change of temperature with respect to time at noon is about 1.9 °C/h. ■

The velocity of a particle is the rate of change of displacement with respect to time. Physicists are interested in other rates of change as well—for instance, the rate of change of work with respect to time (which is called power). Chemists who study a chemical reaction are interested in the rate of change in the concentration of a reactant with respect to time (called the rate of reaction). A steel manufacturer is interested in the rate of change of the cost of producing x tons of steel per day with respect to x (called the marginal cost). A biologist is interested in the rate of change of the population of a colony of bacteria with respect to time. In fact, the computation of rates of change is important in all of the natural sciences, in engineering, and even in the social sciences. Further examples will be given in Section 3.3.

All these rates of change can be interpreted as slopes of tangents. This gives added significance to the solution of the tangent problem. Whenever we solve a problem involving tangent lines, we are not just solving a problem in geometry. We are also implicitly solving a great variety of problems involving rates of change in science and engineering.

EXERCISES 2.7

1. (a) Find the slope of the tangent line to the parabola $y = x^2 + 2x$ at the point $(-3, 3)$
 (i) using Definition 1
 (ii) using Equation 2
(b) Find the equation of the tangent line in part (a).
(c) Graph the parabola and the tangent line.

2. (a) Find the slope of the tangent line to the curve $y = x^3$ at the point $(-1, -1)$
 (i) using Definition 1
 (ii) using Equation 2
(b) Find the equation of the tangent line in part (a).
(c) Graph the curve and the tangent line.

3–6 ■ Find the equation of the tangent line to the curve at the given point.

3. $y = 1 - 2x - 3x^2$, $(-2, -7)$

4. $y = 1/\sqrt{x}$, $(1, 1)$

5. $y = 1/x^2$, $(-2, \frac{1}{4})$

6. $y = x/(1 - x)$, $(0, 0)$

7. (a) Find the slope of the tangent to the curve $y = 2/(x + 3)$ at the point where $x = a$.
(b) Find the slopes of the tangent lines at the points whose x-coordinates are (i) -1, (ii) 0, and (iii) 1.

8. (a) Find the slope of the tangent to the parabola $y = 1 + x + x^2$ at the point where $x = a$.
(b) Find the slopes of the tangent lines at the points whose x-coordinates are (i) -1, (ii) $-\frac{1}{2}$, and (iii) 1.
(c) Graph the curve and the three tangents.

9. (a) Find the slope of the tangent to the curve $y = x^3 - 4x + 1$ at the point where $x = a$.
(b) Find the equations of the tangent lines at the points $(1, -2)$ and $(2, 1)$.
(c) Graph the curve and both tangents.

10. (a) Find the slope of the tangent to the curve $y = 1/\sqrt{5 - 2x}$ at the point where $x = a$.
(b) Find the equations of the tangent lines at the points $(2, 1)$ and $(-2, \frac{1}{3})$.
(c) Graph the curve and both tangents.

11. If a ball is thrown into the air with a velocity of 40 ft/s, its height (in feet) after t seconds is given by $y = 40t - 16t^2$. Find the velocity when $t = 2$.

12. If an arrow is shot upward on the moon with a velocity of 58 m/s, its height (in meters) after t seconds is given by

$$h = 58t - 0.83t^2$$

(a) Find the velocity of the arrow after 1 s.
(b) Find the velocity of the arrow when $t = a$.
(c) When will the arrow hit the moon?
(d) With what velocity will the arrow hit the moon?

13. The displacement (in meters) of a particle moving in a straight line is given by the equation of motion

$$s = 4t^3 + 6t + 2$$

where t is measured in seconds. Find the velocity of the particle at times $t = a$, $t = 1$, $t = 2$, and $t = 3$.

14. The displacement (in meters) of a particle moving in a straight line is given by $s = t^2 - 8t + 18$, where t is measured in seconds.
(a) Find the average velocities over the following time intervals:
(i) $[3, 4]$ (ii) $[3.5, 4]$
(iii) $[4, 5]$ (iv) $[4, 4.5]$
(b) Find the instantaneous velocity when $t = 4$.
(c) Draw the graph of s as a function of t and draw the secant lines whose slopes are the average velocities in part (a) and the tangent line whose slope is the instantaneous velocity in part (b).

15. The graph shows the position function of a car. Use the shape of the graph to explain your answers to the following questions.
(a) What was the initial velocity of the car?
(b) Was the car going faster at B or at C?
(c) Was the car slowing down or speeding up at A, B, and C?
(d) What happened between D and E?

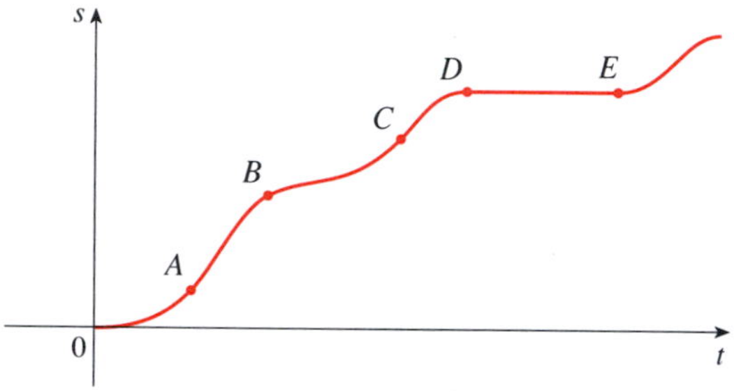

16. A roast turkey is taken from an oven when its temperature has reached 185 °F and is placed on a table in a room where the temperature is 75 °F. The graph shows how the temperature of the turkey decreases and eventually approaches room temperature. (In Section 4.5 we will be able to use Newton's Law of Cooling to find an equation for T as a function of time.) By measuring the slope of the tangent, estimate the rate of change of the temperature after an hour.

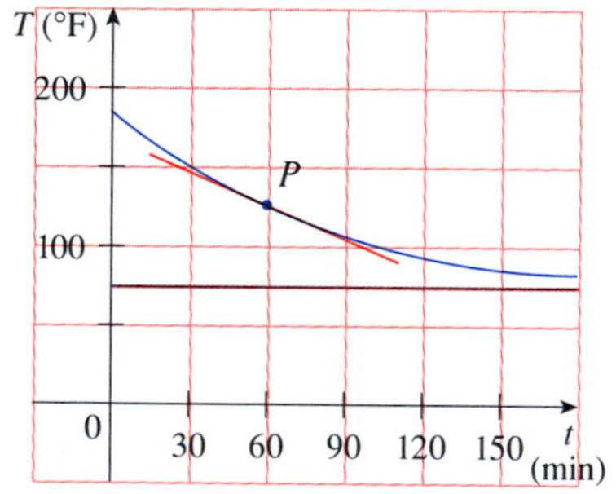

17. (a) Use the data in Example 7 to find the average rate of change of temperature with respect to time
(i) from 8 P.M. to 11 P.M.
(ii) from 8 P.M. to 10 P.M.
(iii) from 8 P.M. to 9 P.M.
(b) Estimate the instantaneous rate of change of T with respect to x at 8 P.M. by measuring the slope of a tangent.

18. The population P (in thousands) of a city from 1990 to 1996 is given in the following table.

(a) Find the average rate of growth
(i) from 1992 to 1996 (ii) from 1992 to 1995
(iii) from 1992 to 1994 (iv) from 1992 to 1993

(b) Estimate the instantaneous rate of growth in 1992 by measuring the slope of a tangent.

Year	P (in 1000s)
1990	105
1991	110
1992	117
1993	126
1994	137
1995	150
1996	164

19. The cost (in dollars) of producing x units of a certain commodity is $C(x) = 5000 + 10x + 0.05x^2$.
(a) Find the average rate of change of C with respect to x when the production level is changed
(i) from $x = 100$ to $x = 105$
(ii) from $x = 100$ to $x = 101$
(b) Find the instantaneous rate of change of C with respect to x when $x = 100$. (This is called the *marginal cost*. Its significance will be explained in Section 3.3.)

20. If a cylindrical tank holds 100,000 gallons of water, which can be drained from the bottom of the tank in 1 h, then Torricelli's Law gives the volume V of water remaining in the tank after t minutes as

$$V(t) = 100{,}000\left(1 - \frac{t}{60}\right)^2 \qquad 0 \leqslant t \leqslant 60$$

Find the rate at which the water is flowing out of the tank (the instantaneous rate of change of V with respect to t) after 20 min.

21. Let $\mathbf{r}(t) = (t^2 + t)\,\mathbf{i} + \dfrac{2}{t}\,\mathbf{j}$.
(a) Find a vector tangent to the curve given by the graph of $\mathbf{r}(t)$ at the point where $t = 2$.
(b) Find parametric equations for the tangent line to the curve at the point where $t = 2$.

22. Let $\mathbf{r}(t) = (t^2 - t - 2)\,\mathbf{i} + 4t^2\,\mathbf{j}$.
(a) Find a vector tangent to the curve given by the graph of $\mathbf{r}(t)$ at the point $(-2, 4)$.
(b) Find parametric equations for the tangent line to the curve at the point $(-2, 4)$.

23. (Compare with Exercise 11.) If a ball is thrown into the air with a velocity of $10\,\mathbf{i} + 30\,\mathbf{j}$ ft/s, its position after t seconds is given by

$$\mathbf{r}(t) = 10t\,\mathbf{i} + (30t - 16t^2)\,\mathbf{j}$$

(a) Find the velocity of the ball when $t = 1$.
(b) What is the speed of the ball when $t = 1$?

24. (Compare with Exercise 12.) If an arrow is shot upward on the moon with a velocity of $15\,\mathbf{i} + 40\,\mathbf{j}$ m/s, its position after t seconds is given by

$$\mathbf{r}(t) = 15t\,\mathbf{i} + (40t - 0.83t^2)\,\mathbf{j}$$

(a) Find the velocity of the arrow after 1 s.
(b) Find the velocity of the arrow when $t = a$.
(c) When will the arrow hit the moon?
(d) With what velocity will the arrow hit the moon?
(e) With what speed will the arrow hit the moon?

2 REVIEW

KEY TOPICS ■ Define, state, or discuss the following.

1. Limit of $f(x)$ as x approaches a
2. One-sided limits
3. Infinite limits
4. Limit Laws
5. The Squeeze Theorem
6. Continuous at a number a
7. Discontinuities of a function
8. Continuous on an interval
9. Intermediate Value Theorem
10. Limit at infinity
11. Horizontal asymptote
12. Infinite limit at infinity
13. Tangent line
14. Tangent vector
15. Velocity
16. Average and instantaneous rates of change

EXERCISES

1–12 ■ Determine whether the statement is true or false.

1. $\lim_{x\to 4}\left(\dfrac{2x}{x-4}-\dfrac{8}{x-4}\right)=\lim_{x\to 4}\dfrac{2x}{x-4}-\lim_{x\to 4}\dfrac{8}{x-4}$

2. $\lim_{x\to 1}\dfrac{x^2+6x-7}{x^2+5x-6}=\dfrac{\lim_{x\to 1}(x^2+6x-7)}{\lim_{x\to 1}(x^2+5x-6)}$

3. $\lim_{x\to 1}\dfrac{x-3}{x^2+2x-4}=\dfrac{\lim_{x\to 1}(x-3)}{\lim_{x\to 1}(x^2+2x-4)}$

4. If $\lim_{x\to 5} f(x)=2$ and $\lim_{x\to 5} g(x)=0$, then $\lim_{x\to 5}[f(x)/g(x)]$ does not exist.

5. If $f(x)>1$ for all x and $\lim_{x\to 0} f(x)$ exists, then $\lim_{x\to 0} f(x)>1$.

6. Let f be a function such that $\lim_{x\to 4} f(x)=7$. If $3.9<x<4.1$, then $6.9<f(x)<7.1$.

7. Let f be a function such that $\lim_{x\to 0} f(x)=6$. Then there exists a number δ such that if $0<|x|<\delta$, then $|f(x)-6|<1$.

8. If $\lim_{x\to 6} f(x)g(x)$ exists, then the limit must be $f(6)g(6)$.

9. If p is a polynomial, then $\lim_{x\to b} p(x)=p(b)$.

10. If $f(1)>0$ and $f(3)<0$, then there exists a number c between 1 and 3 such that $f(c)=0$.

11. If f is continuous at 5 and $f(5)=2$ and $f(4)=3$, then $\lim_{x\to 2} f(4x^2-11)=2$.

12. If f is continuous on $[-1,1]$ and $f(-1)=4$ and $f(1)=3$, then there exists a number r such that $|r|<1$ and $f(r)=\pi$.

13–34 ■ Find the limit.

13. $\lim_{x\to 4}\sqrt{x+\sqrt{x}}$

14. $\lim_{x\to 0^-}\sqrt{-x}$

15. $\lim_{t\to -1}\dfrac{t+1}{t^3-t}$

16. $\lim_{t\to 4}\dfrac{t-4}{t^2-3t-4}$

17. $\lim_{h\to 0}\dfrac{(1+h)^2-1}{h}$

18. $\lim_{h\to 0}\dfrac{(1+h)^{-2}-1}{h}$

19. $\lim_{x\to -1}\dfrac{x^2-x-2}{x^2+3x-2}$

20. $\lim_{x\to -1}\dfrac{x^2-x-2}{x^2+3x+2}$

21. $\lim_{t\to 6}\dfrac{17}{(t-6)^2}$

22. $\lim_{x\to -6^+}\dfrac{x}{x+6}$

23. $\lim_{s\to 16}\dfrac{4-\sqrt{s}}{s-16}$

24. $\lim_{v\to 2}\dfrac{v^2+2v-8}{v^4-16}$

25. $\lim_{x\to 8^-}\dfrac{|x-8|}{x-8}$

26. $\lim_{x\to 9^+}\left(\sqrt{x-9}+[\![x+1]\!]\right)$

27. $\lim_{x\to 0}\dfrac{1-\sqrt{1-x^2}}{x}$

28. $\lim_{x\to 2}\dfrac{\sqrt{x+2}-\sqrt{2x}}{x^2-2x}$

29. $\lim_{x\to\infty}\dfrac{1+2x-x^2}{1-x+2x^2}$

30. $\lim_{x\to 3^+}\dfrac{\sqrt{x^2-9}}{2x-6}$

31. $\lim_{x\to\infty}\dfrac{\sqrt{x^2-9}}{2x-6}$

32. $\lim_{x\to -\infty}\dfrac{\cos^2 x}{x^2}$

33. $\lim_{x\to\infty}\left(\sqrt[3]{x}-\dfrac{x}{3}\right)$

34. $\lim_{x\to\infty}\left(\sqrt{x^2+x+1}-\sqrt{x^2-x}\right)$

35–38 ■ Find the limit.

35. $\lim_{t\to 3}\left(\dfrac{2t^2-18}{t-3}\mathbf{i}+t^2\mathbf{j}\right)$

36. $\lim_{t\to 4}\left(\dfrac{t^2}{t+4}\mathbf{i}+\dfrac{t-4}{t^2-16}\mathbf{j}\right)$

37. $\lim_{t\to 1}\left\langle\dfrac{1-\sqrt{t}}{1-t},\, t^3-4t\right\rangle$

38. $\lim_{t\to 2}\left\langle 4t-3,\, \dfrac{t^2+t-6}{t^2-t-2}\right\rangle$

39–42 ■ Prove each statement using the precise definition of a limit.

39. $\lim_{x\to 5}(7x-27)=8$

40. $\lim_{x\to 0}\sqrt[3]{x}=0$

41. $\lim_{x\to 2}(x^2-3x)=-2$

42. $\lim_{x\to 4^+}\dfrac{2}{\sqrt{x-4}}=\infty$

43. If $2x-1\leqslant f(x)\leqslant x^2$ for $0<x<3$, find $\lim_{x\to 1} f(x)$.

44. Prove that $\lim_{x\to 0} x^2\cos(1/x^2)=0$.

45. Let

$$f(x) = \begin{cases} \sqrt{-x} & \text{if } x < 0 \\ 3 - x & \text{if } 0 \leq x < 3 \\ (x-3)^2 & \text{if } x > 3 \end{cases}$$

(a) Evaluate each limit, if it exists.

(i) $\lim_{x \to 0^+} f(x)$ (ii) $\lim_{x \to 0^-} f(x)$ (iii) $\lim_{x \to 0} f(x)$

(iv) $\lim_{x \to 3^-} f(x)$ (v) $\lim_{x \to 3^+} f(x)$ (vi) $\lim_{x \to 3} f(x)$

(b) Where is f discontinuous?

(c) Sketch the graph of f.

46. Let

$$g(x) = \begin{cases} 2x - x^2 & \text{if } 0 \leq x \leq 2 \\ 2 - x & \text{if } 2 < x \leq 3 \\ x - 4 & \text{if } 3 < x < 4 \\ \pi & \text{if } x \geq 4 \end{cases}$$

(a) For each of the numbers 2, 3, and 4, discover whether g is continuous from the left, continuous from the right, or continuous at the number.

(b) Sketch the graph of g.

47–48 ■ Show that the function is continuous on its domain. State the domain.

47. $f(x) = \dfrac{x+1}{x^2 + x + 1}$ **48.** $g(x) = \dfrac{\sqrt{x^2 - 9}}{x^2 - 2}$

49–50 ■ Use the Intermediate Value Theorem to show that there is a root of the equation in the given interval.

49. $2x^3 + x^2 + 2 = 0$, $(-2, -1)$

50. $x^4 + 1 = 1/x$, $(0.5, 1)$

51. (a) Find the slope of the tangent line to the curve $y = 9 - 2x^2$ at the point $(2, 1)$.

(b) Find the equation of this tangent line.

52. Find the equations of the tangent lines to the curve $y = 2/(1 - 3x)$ at the points with x-coordinates 0 and -1.

53. Let $\mathbf{r}(t) = (3t^2 - 4t)\,\mathbf{i} + (5t + 2)\,\mathbf{j}$.

(a) Find a vector tangent to the curve given by the graph of $\mathbf{r}(t)$ at the point where $t = 2$.

(b) Find a vector equation for the tangent line to the curve at the point where $t = 2$.

(c) Find parametric equations for the tangent line to the curve at the point where $t = 2$.

54. Let $\mathbf{r}(t) = \langle 1 - 4t, 2t - 3t^2 \rangle$.

(a) Find a vector tangent to the curve given by the graph of $\mathbf{r}(t)$ at the point where $t = 3$.

(b) Find a vector equation for the tangent line to the curve at the point where $t = 3$.

(c) Find parametric equations for the tangent line to the curve at the point where $t = 3$.

55. The position at time t of an object moving in the xy-plane is given by $\mathbf{r}(t) = \langle t^2 + 3t, 4t^2 \rangle$, where distance is measured in feet and time t in seconds.

(a) Find the average velocity of the object over the following time periods.

(i) $[2, 4]$ (ii) $[2, 3]$

(iii) $[2, 2.5]$ (iv) $[2, 2.1]$

(b) Find the instantaneous velocity of the object when $t = 2$.

56. A ball is tossed into the air. Its position at time t is given by $\mathbf{r}(t) = 5t\,\mathbf{i} + (100t - 16t^2)\,\mathbf{j}$, where distance is measured in feet and time t in seconds.

(a) Find the velocity of the ball when $t = 2$.

(b) Find the speed of the ball when $t = 2$.

(c) How high does the ball go?

(d) When will the ball hit the ground?

57. The displacement (in meters) of an object moving in a straight line is given by $s = 1 + 2t + t^2/4$, where t is measured in seconds.

(a) Find the average velocity over the following time periods.

(i) $[1, 3]$ (ii) $[1, 2]$

(iii) $[1, 1.5]$ (iv) $[1, 1.1]$

(b) Find the instantaneous velocity when $t = 1$.

58. According to Boyle's Law, if the temperature of a confined gas is held fixed, then the product of the pressure P and the volume V is a constant. Suppose that, for a certain gas, $PV = 800$, where P is measured in pounds per square inch and V is measured in cubic inches.

(a) Find the average rate of change of P as V increases from 200 in^3 to 250 in^3.

(b) Express V as a function of P and show that the instantaneous rate of change of V with respect to P is proportional to the inverse square of P.

59. Use a graph to find a number δ such that

$$\left| \frac{x+1}{x-1} - 3 \right| < 0.2 \quad \text{whenever} \quad |x - 2| < \delta$$

60. Graph the curve $y = (x + 1)/(x - 1)$ and the tangent lines to this curve at the points $(2, 3)$ and $(-1, 0)$.

61. Suppose that $|f(x)| \leq g(x)$ for all x, where $\lim_{x \to a} g(x) = 0$. Find $\lim_{x \to a} f(x)$.

62. Let $f(x) = [\![x]\!] + [\![-x]\!]$.

(a) For what values of a does $\lim_{x \to a} f(x)$ exist?

(b) At what numbers is f discontinuous?

63. If

$$\lim_{x \to a} [f(x) + g(x)] = 2 \quad \text{and} \quad \lim_{x \to a} [f(x) - g(x)] = 1$$

find $\lim_{x \to a} f(x)g(x)$.

3

DERIVATIVES

Mathematics compares the most diverse phenomena and discovers the secret analogies that unite them.

JOSEPH FOURIER

In this chapter we begin our study of differential calculus, which is concerned with how one quantity changes in relation to another quantity. The central concept of differential calculus is the *derivative*. After learning how to calculate derivatives, we use them to solve problems involving rates of change.

3.1 DERIVATIVES

In Section 2.7 we defined the slope of the tangent to a curve with equation $y = f(x)$ at the point where $x = a$ to be

$$m = \lim_{h \to 0} \frac{f(a + h) - f(a)}{h} \tag{1}$$

We also saw that the velocity of an object with position function $s = f(t)$ at time $t = a$ is

$$v(a) = \lim_{h \to 0} \frac{f(a + h) - f(a)}{h}$$

In fact, limits of the form

$$\lim_{h \to 0} \frac{f(a + h) - f(a)}{h}$$

arise whenever we calculate a rate of change in any of the sciences or engineering, such as a rate of reaction in chemistry or a marginal cost in economics. Since this type of limit occurs so widely, it is given a special name and notation.

$f'(a)$ is read "f prime of a."

(2) DEFINITION The **derivative of a function f at a number a**, denoted by $f'(a)$, is

$$f'(a) = \lim_{h \to 0} \frac{f(a + h) - f(a)}{h}$$

if this limit exists.

If we write $x = a + h$, then $h = x - a$ and h approaches 0 if and only if x approaches a. Therefore, an equivalent way of stating the definition of the derivative, as we saw in finding tangent lines, is

(3)
$$f'(a) = \lim_{x \to a} \frac{f(x) - f(a)}{x - a}$$

EXAMPLE 1 Find the derivative of the function $f(x) = x^2 - 8x + 9$ at the number a.

SOLUTION From Definition 2 we have

$$\begin{aligned} f'(a) &= \lim_{h \to 0} \frac{f(a + h) - f(a)}{h} \\ &= \lim_{h \to 0} \frac{[(a + h)^2 - 8(a + h) + 9] - [a^2 - 8a + 9]}{h} \\ &= \lim_{h \to 0} \frac{a^2 + 2ah + h^2 - 8a - 8h + 9 - a^2 + 8a - 9}{h} \\ &= \lim_{h \to 0} \frac{2ah + h^2 - 8h}{h} = \lim_{h \to 0} (2a + h - 8) \\ &= 2a - 8 \end{aligned}$$

INTERPRETATIONS OF THE DERIVATIVE

THE INTERPRETATION OF THE DERIVATIVE AS THE SLOPE OF A TANGENT In Section 2.7 we defined the tangent line to the curve $y = f(x)$ at the point $P(a, f(a))$ to be the line that passes through P and has slope m given by Equation 1. Since, by Definition 2, this is the same as the derivative $f'(a)$, we can now say that *the tangent line to $y = f(x)$ at $(a, f(a))$ is the line through $(a, f(a))$ whose slope is equal to $f'(a)$, the derivative of f at a.* Thus, the geometric interpretation of a derivative [as defined by either (2) or (3)] is as shown in Figure 1.

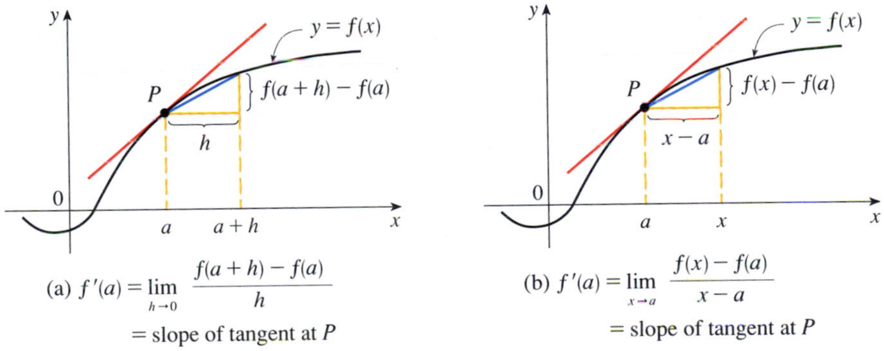

FIGURE 1
Geometric interpretation of the derivative

Using the point-slope form of the equation of a line, we have the following:

(4) If $f'(a)$ exists, then an equation of the tangent line to the curve $y = f(x)$ at the point $(a, f(a))$ is
$$y - f(a) = f'(a)(x - a)$$

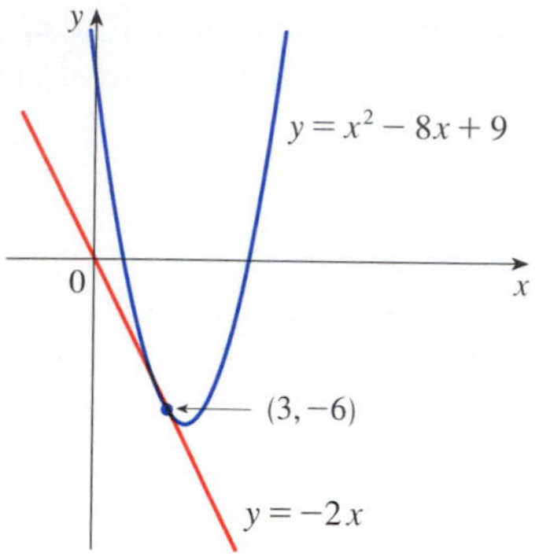

FIGURE 2

EXAMPLE 2 Find an equation of the tangent line to the parabola $y = x^2 - 8x + 9$ at the point $(3, -6)$.

SOLUTION From Example 1 we know that the derivative of $f(x) = x^2 - 8x + 9$ at the number a is $f'(a) = 2a - 8$. Therefore, the slope of the tangent line at $(3, -6)$ is $f'(3) = 2(3) - 8 = -2$. Thus, the equation of the tangent line, shown in Figure 2, is

$$y - (-6) = (-2)(x - 3) \qquad \text{or} \qquad y = -2x$$

■

EXAMPLE 3 Let $f(x) = 2^x$. Estimate the value of $f'(0)$ in two ways:
(a) By using Definition 2 and taking successively smaller values of h.
(b) By interpreting $f'(0)$ as the slope of a tangent and using a graphing calculator to zoom in on the graph of $y = 2^x$.

SOLUTION
(a) From Definition 2 we have

$$f'(0) = \lim_{h \to 0} \frac{f(h) - f(0)}{h} = \lim_{h \to 0} \frac{2^h - 1}{h}$$

h	$\dfrac{2^h - 1}{h}$
0.1	0.718
0.01	0.696
0.001	0.693
0.0001	0.693
−0.1	0.670
−0.01	0.691
−0.001	0.693
−0.0001	0.693

Since we are not yet able to evaluate this limit exactly, we use a calculator to approximate the values of $(2^h - 1)/h$. From the numerical evidence in the table we see that as h approaches 0, these values appear to approach a number near 0.69. So our estimate is

$$f'(0) \approx 0.69$$

(b) In Figure 3 we graph the curve $y = 2^x$ and zoom in toward the point $(0, 1)$. We see that the closer we get to $(0, 1)$, the more the curve looks like a straight line. In fact in Figure 3(c) the curve is practically indistinguishable from its tangent line at $(0, 1)$. Since the x-scale and the y-scale are both 0.01, we estimate that the slope of this line is

$$\frac{0.14}{0.20} = 0.7$$

So our estimate of the derivative is $f'(0) \approx 0.7$. In Chapter 4 we will show that, correct to six decimal places, $f'(0) \approx 0.693147$.

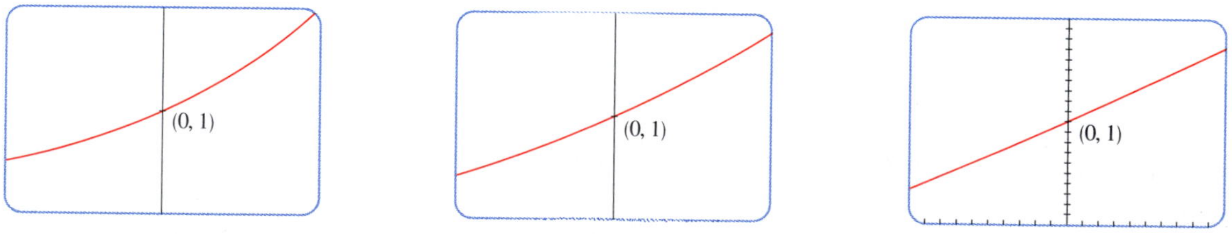

(a) $[-1, 1]$ by $[0, 2]$ (b) $[-0.5, 0.5]$ by $[0.5, 1.5]$ (c) $[-0.1, 0.1]$ by $[0.9, 1.1]$ ■

FIGURE 3
Zooming in on the graph of $y = 2^x$ near $(0, 1)$

THE INTERPRETATION OF THE DERIVATIVE AS A RATE OF CHANGE In Section 2.7 we defined the instantaneous rate of change of $y = f(x)$ with respect to x at $x = x_1$ as the limit of the average rate of change over smaller and smaller intervals. If the interval is $[x_1, x_2]$, then the change in x is $\Delta x = x_2 - x_1$, the corresponding change in y is

$$\Delta y = f(x_2) - f(x_1)$$

and

(5) $$\text{instantaneous rate of change} = \lim_{\Delta x \to 0} \frac{\Delta y}{\Delta x} = \lim_{x_2 \to x_1} \frac{f(x_2) - f(x_1)}{x_2 - x_1}$$

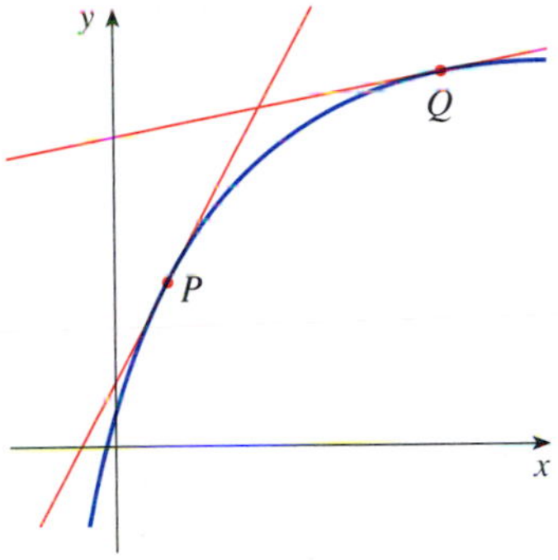

FIGURE 4
The y-values are changing rapidly at P and slowly at Q.

From Equation 3 we recognize this limit as being the derivative of f at x_1, that is, $f'(x_1)$.

This gives a second interpretation of the *derivative $f'(a)$ as the instantaneous rate of change of $y = f(x)$ with respect to x when $x = a$.* The connection with the first interpretation is that if we sketch the curve $y = f(x)$, then the instantaneous rate of change is the slope of the tangent to this curve at the point where $x = a$. This means that when the derivative is large (and therefore the curve is steep, as at the point P in Figure 4), the y-values change rapidly. When the derivative is small, the curve is relatively flat and the y-values change slowly.

In particular, if $s = f(t)$ is the position function of a particle that moves along a straight line, then $f'(a)$ is the rate of change of the displacement s with respect to the time t. In other words, *$f'(a)$ is the velocity of the particle at time $t = a$.* (See Section 2.7.) The *speed* of the particle is the absolute value of the velocity, that is, $|f'(a)|$.

EXAMPLE 4 The position of a particle is given by the equation of motion $s = f(t) = t^2 - 8t + 9$, where t is measured in seconds and s in meters. Find the velocity and the speed after 2 seconds.

SOLUTION Again from Example 1 we know that the derivative of the position function is $f'(t) = 2t - 8$. Thus, the velocity after 2 s is $f'(2) = 2(2) - 8 = -4$ m/s, and the speed is $|f'(2)| = |-4| = 4$ m/s. ■

THE DERIVATIVE AS A FUNCTION

If we replace a by x in Definition 2, we obtain

(6) $$f'(x) = \lim_{h \to 0} \frac{f(x + h) - f(x)}{h}$$

Given a function f, we associate with it a new function f', called the **derivative of f**, defined by Equation 6. We know that the value of f' at x, $f'(x)$, can be interpreted geometrically as the slope of the tangent line to the graph of f at the point $(x, f(x))$.

The function f' is called the derivative of f because it has been "derived" from f by the limiting operation in Equation 6. The domain of f' is the set $\{x \mid f'(x) \text{ exists}\}$ and may be smaller than the domain of f.

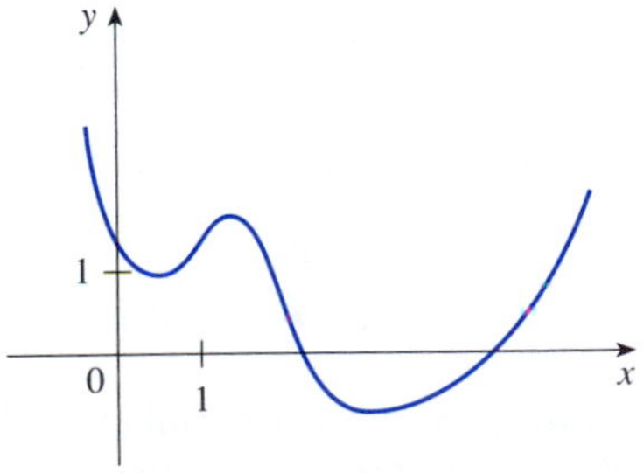

FIGURE 5

EXAMPLE 5 The graph of a function f is given in Figure 5. Use it to sketch the graph of the derivative f'.

SOLUTION We can estimate the value of the derivative at any value of x by drawing the tangent at the point $(x, f(x))$ and estimating its slope. For instance, for

$x = 5$ we draw the tangent at P and estimate its slope to be about $\frac{3}{2}$, so $f'(5) \approx 1.5$. This allows us to plot the point $P'(5, 1.5)$ on the graph of f' directly beneath P. Repeating this procedure at several points, we get the graph shown in Figure 6.

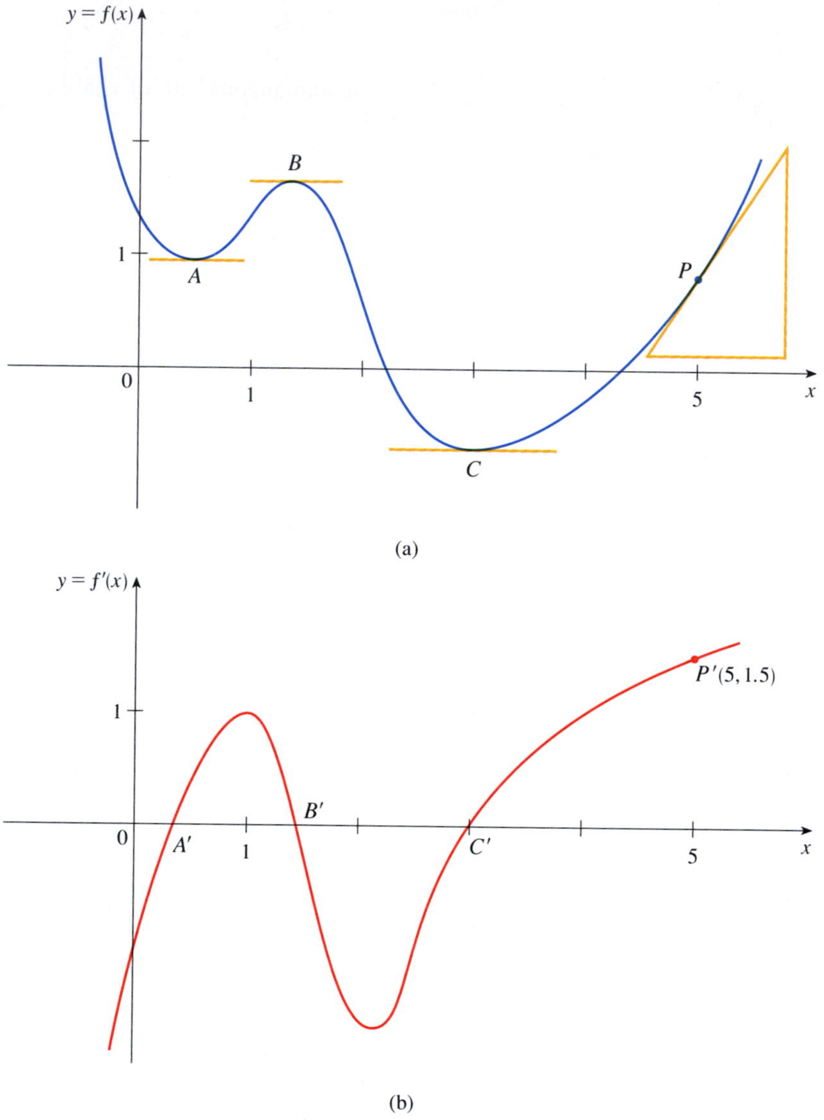

FIGURE 6

Notice that the tangents at A, B, and C are horizontal, so the derivative is 0 there and the graph of f' crosses the x-axis at A', B', and C'. Between A and B the tangents have positive slope, so $f'(x)$ is positive there. But between B and C the tangents have negative slope, so $f'(x)$ is negative there. ■

EXAMPLE 6 If $f(x) = \sqrt{x - 1}$, find the derivative of f. State the domain of f'.

SOLUTION When using Equation 6 to compute a derivative, we must remember that the variable is h and that x is temporarily regarded as a constant during the calculation of the limit.

$$f'(x) = \lim_{h \to 0} \frac{f(x+h) - f(x)}{h}$$

$$= \lim_{h \to 0} \frac{\sqrt{x+h-1} - \sqrt{x-1}}{h}$$

$$= \lim_{h \to 0} \frac{\sqrt{x+h-1} - \sqrt{x-1}}{h} \cdot \frac{\sqrt{x+h-1} + \sqrt{x-1}}{\sqrt{x+h-1} + \sqrt{x-1}}$$

$$= \lim_{h \to 0} \frac{(x+h-1) - (x-1)}{h\left(\sqrt{x+h-1} + \sqrt{x-1}\right)}$$

$$= \lim_{h \to 0} \frac{1}{\sqrt{x+h-1} + \sqrt{x-1}}$$

$$= \frac{1}{\sqrt{x-1} + \sqrt{x-1}} = \frac{1}{2\sqrt{x-1}}$$

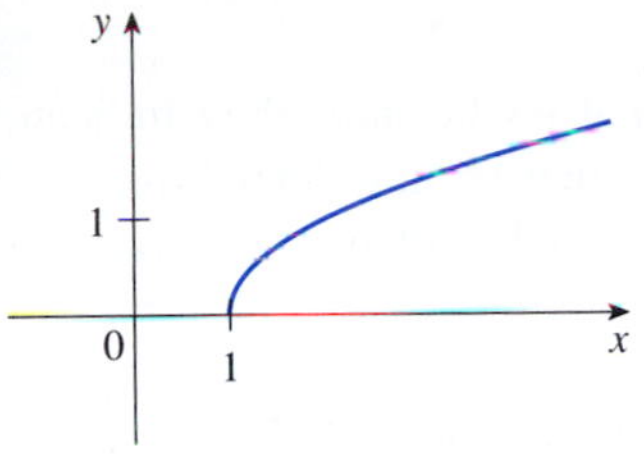

(a) $y = \sqrt{x-1}$

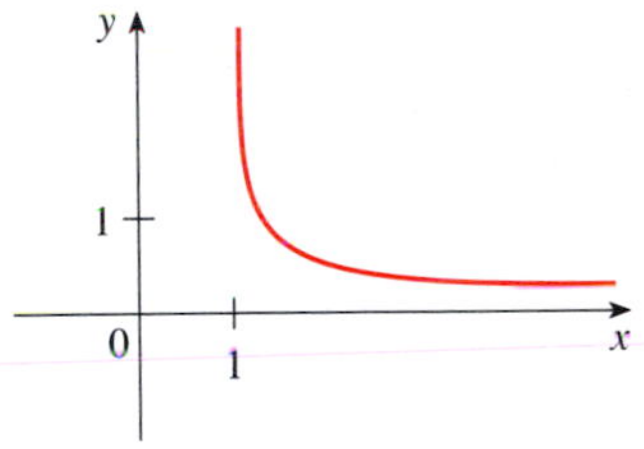

(b) $y' = \dfrac{1}{2\sqrt{x-1}}$

FIGURE 7

We see that $f'(x)$ exists if $x > 1$, so the domain of f' is $(1, \infty)$. This is smaller than the domain of f, which is $[1, \infty)$. ■

Let's check to see that the result of Example 6 is reasonable. When x is close to 1, $\sqrt{x-1}$ is close to 0, so $f'(x) = 1/(2\sqrt{x-1})$ is very large and this corresponds to the steep tangent lines near $(1, 0)$ in Figure 7(a). When x is large, $f'(x)$ is very small and this corresponds to the almost horizontal tangent lines at the far right of the graph. Figure 7(b) shows the graph of the derivative $y' = f'(x)$. Notice the relationship between the graphs of f and f'.

EXAMPLE 7 Find f' if $f(x) = \dfrac{1-x}{2+x}$.

SOLUTION

$$f'(x) = \lim_{h \to 0} \frac{f(x+h) - f(x)}{h}$$

$$= \lim_{h \to 0} \frac{\dfrac{1-(x+h)}{2+(x+h)} - \dfrac{1-x}{2+x}}{h}$$

$$\frac{\dfrac{a}{b} - \dfrac{c}{d}}{e} = \frac{ad - bc}{bd} \cdot \frac{1}{e}$$

$$= \lim_{h \to 0} \frac{(1-x-h)(2+x) - (1-x)(2+x+h)}{h(2+x+h)(2+x)}$$

$$= \lim_{h \to 0} \frac{(2-x-2h-x^2-xh) - (2-x+h-x^2-xh)}{h(2+x+h)(2+x)}$$

$$= \lim_{h \to 0} \frac{-3h}{h(2+x+h)(2+x)}$$

$$= \lim_{h \to 0} \frac{-3}{(2+x+h)(2+x)} = -\frac{3}{(2+x)^2}$$ ■

OTHER NOTATIONS

If we use the traditional notation $y = f(x)$ to indicate that the independent variable is x and the dependent variable is y, then some common alternative notations for the derivative are as follows:

$$f'(x) = y' = \frac{dy}{dx} = \frac{df}{dx} = \frac{d}{dx} f(x) = Df(x) = D_x f(x)$$

The symbols D and d/dx are called **differentiation operators** because they indicate the operation of **differentiation**, which is the process of calculating a derivative.

The symbol dy/dx, which was introduced by Leibniz, should not be regarded as a ratio (for the time being); it is simply a synonym for $f'(x)$. Nonetheless, it is a very useful and suggestive notation, especially when used in conjunction with increment notation. Referring to Equation 5, we can rewrite the definition of derivative in Leibniz notation in the form

$$\frac{dy}{dx} = \lim_{\Delta x \to 0} \frac{\Delta y}{\Delta x}$$

If we want to indicate the value of a derivative dy/dx in Leibniz notation at a specific number a, we use the notation

$$\left.\frac{dy}{dx}\right|_{x=a} \qquad \text{or} \qquad \left.\frac{dy}{dx}\right]_{x=a}$$

which is a synonym for $f'(a)$.

Gottfried Wilhelm Leibniz was born in Leipzig in 1646 and studied law, theology, philosophy, and mathematics at the university there, graduating with a bachelor's degree at age 17. After earning his doctorate in law at age 20, Leibniz entered the diplomatic service and spent most of his life traveling to the capitals of Europe on political missions. In particular, he worked to avert a French military threat against Germany and attempted to reconcile the Catholic and Protestant churches.

His serious study of mathematics did not begin until 1672 while he was on a diplomatic mission in Paris. There he built a calculating machine and met scientists, like Huygens, who directed his attention to the latest developments in mathematics and science. Leibniz sought to develop a symbolic logic and system of notation that would simplify logical reasoning. In particular, the version of calculus that he published in 1684 established the notation and the rules for finding derivatives that we use today.

Unfortunately, a dreadful priority dispute arose in the 1690s between the followers of Newton and those of Leibniz as to who had invented calculus first. Leibniz was even accused of plagiarism by members of the Royal Society in England. The truth is that each man invented calculus independently. Newton arrived at his version of calculus first but, because of his fear of controversy, did not publish it immediately. So Leibniz's 1684 account of calculus was the first to be published.

(7) DEFINITION A function f is **differentiable at a** if $f'(a)$ exists. It is **differentiable on an open interval** (a, b) [or (a, ∞) or $(-\infty, a)$ or $(-\infty, \infty)$] if it is differentiable at every number in the interval.

EXAMPLE 8 Where is the function $f(x) = |x|$ differentiable?

SOLUTION If $x > 0$, then $|x| = x$ and we can choose h small enough that $x + h > 0$ and hence $|x + h| = x + h$. Therefore, for $x > 0$ we have

$$f'(x) = \lim_{h \to 0} \frac{|x + h| - |x|}{h}$$

$$= \lim_{h \to 0} \frac{(x + h) - x}{h} = \lim_{h \to 0} \frac{h}{h} = \lim_{h \to 0} 1 = 1$$

and so f is differentiable for any $x > 0$.

Similarly, for $x < 0$ we have $|x| = -x$ and h can be chosen small enough that $x + h < 0$ and so $|x + h| = -(x + h)$. Therefore, for $x < 0$,

$$f'(x) = \lim_{h \to 0} \frac{|x + h| - |x|}{h}$$

$$= \lim_{h \to 0} \frac{-(x + h) - (-x)}{h} = \lim_{h \to 0} \frac{-h}{h} = \lim_{h \to 0} (-1) = -1$$

and so f is differentiable for any $x < 0$.

For $x = 0$ we have to investigate

$$f'(0) = \lim_{h\to 0} \frac{f(0+h) - f(0)}{h}$$

$$= \lim_{h\to 0} \frac{|0+h| - |0|}{h} \qquad \text{(if it exists)}$$

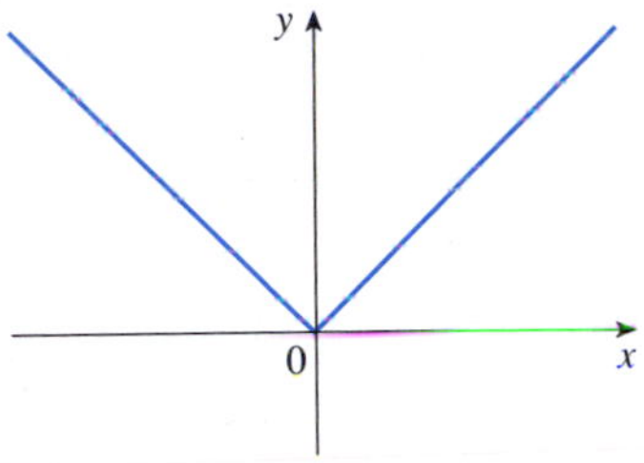

(a) $y = f(x) = |x|$

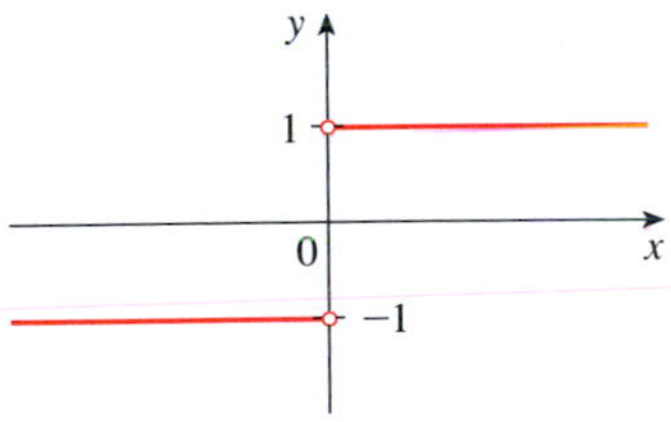

(b) $y = f'(x)$

FIGURE 8

Let's compute the left and right limits separately:

$$\lim_{h\to 0^+} \frac{|0+h| - |0|}{h} = \lim_{h\to 0^+} \frac{|h|}{h} = \lim_{h\to 0^+} \frac{h}{h} = \lim_{h\to 0^+} 1 = 1$$

and $$\lim_{h\to 0^-} \frac{|0+h| - |0|}{h} = \lim_{h\to 0^-} \frac{|h|}{h} = \lim_{h\to 0^-} \frac{-h}{h} = -1$$

Since these limits are different, $f'(0)$ does not exist. Thus, f is differentiable at all x except 0.

A formula for f' is given by

$$f'(x) = \begin{cases} 1 & \text{if } x > 0 \\ -1 & \text{if } x < 0 \end{cases}$$

and its graph is shown in Figure 8(b). The fact that $f'(0)$ does not exist is reflected geometrically in the fact that the curve $y = |x|$ does not have a tangent line at $(0, 0)$. [See Figure 8(a).] ■

(8) THEOREM If f is differentiable at a, then f is continuous at a.

PROOF To prove that f is continuous at a, we have to show that $\lim_{x\to a} f(x) = f(a)$. We do this by showing that the difference $f(x) - f(a)$ approaches 0.

For $x \neq a$ we can divide and multiply by $x - a$:

$$f(x) - f(a) = \frac{f(x) - f(a)}{x - a}(x - a)$$

We did this in order to involve the difference quotient. [We somehow have to use the fact that f is differentiable at a, that is, $f'(a)$ exists.] Thus, we can use the Product Law and Equation 3 to write

$$\lim_{x\to a} [f(x) - f(a)] = \lim_{x\to a} \frac{f(x) - f(a)}{x - a}(x - a)$$

$$= \lim_{x\to a} \frac{f(x) - f(a)}{x - a} \lim_{x\to a} (x - a)$$

$$= f'(a) \cdot 0 = 0$$

Therefore

$$\begin{aligned}\lim_{x \to a} f(x) &= \lim_{x \to a} [f(a) + (f(x) - f(a))] \\ &= \lim_{x \to a} f(a) + \lim_{x \to a} [f(x) - f(a)] \\ &= f(a) + 0 = f(a)\end{aligned}$$

and so f is continuous at a. □

NOTE: The converse of Theorem 8 is false; that is, there are functions that are continuous but not differentiable. For instance, the function $f(x) = |x|$ is continuous at 0 because

$$\lim_{x \to 0} f(x) = \lim_{x \to 0} |x| = 0 = f(0)$$

(See Example 9 in Section 2.3.) But in Example 8 we showed that f is not differentiable at 0.

HOW CAN A FUNCTION FAIL TO BE DIFFERENTIABLE?

In Example 8 we saw that the function $y = |x|$ is not differentiable at 0. In general, if the graph of a function f has a "corner" or "kink" in it, then the graph of f has no tangent at that point and f is not differentiable there. [In trying to compute $f'(a)$, we find that the left and right limits are different.]

Theorem 8 gives another way for a function not to have a derivative. It says that if f is not continuous at a, then f is not differentiable at a. So at any discontinuity (for instance, a jump discontinuity) f fails to be differentiable.

A third possibility is that the curve has a **vertical tangent line** when $x = a$, that is, f is continuous at a and

$$\lim_{x \to a} |f'(x)| = \infty$$

This means that the tangent lines become steeper and steeper as $x \to a$. Figure 9 shows one way that this can happen; Figure 10(c) shows another. Figure 10 illustrates the three possibilities that we have discussed.

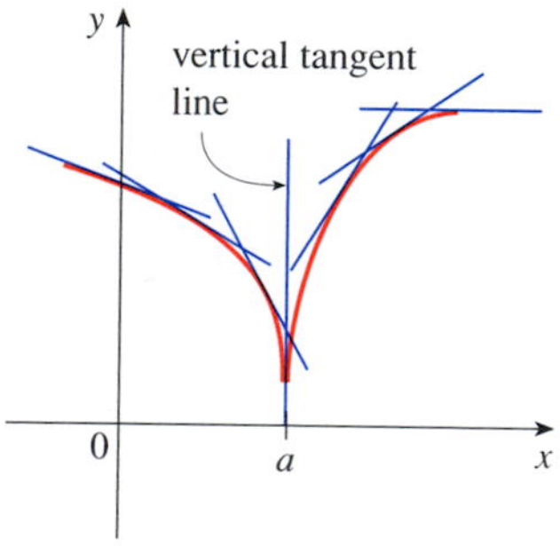

FIGURE 9

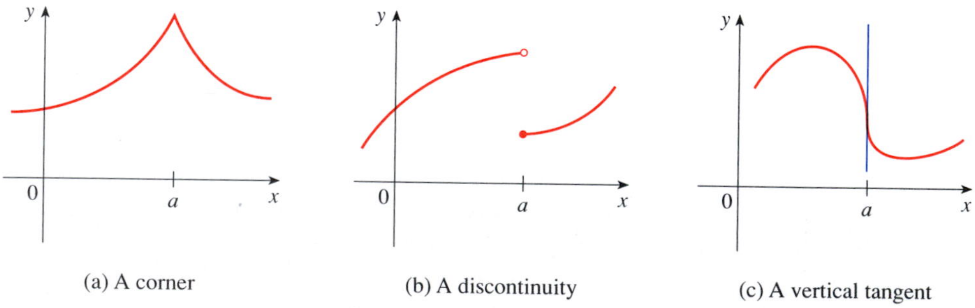

FIGURE 10
Three ways for f not to be differentiable at a

A graphing calculator or computer provides another way of looking at differentiability. If f is differentiable at a, then when we zoom in toward the point $(a, f(a))$ the graph straightens out and appears more and more like a line. (See Figure 11. We saw a specific example of this in Figure 3.) But no matter how much we zoom in toward a point like the ones in Figures 9 and 10(a), we cannot eliminate the sharp point or corner. (See Figure 12.)

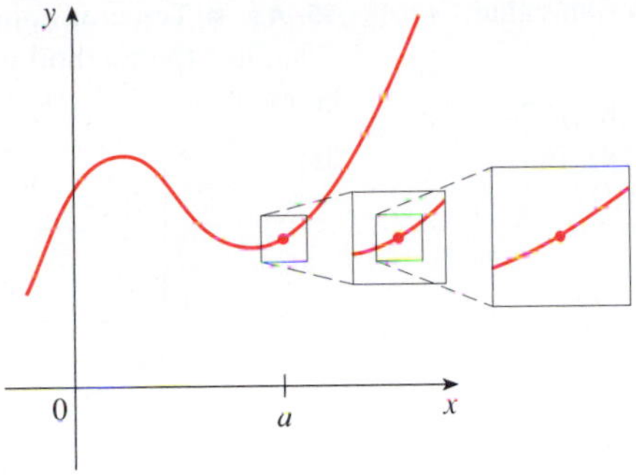

FIGURE 11
f is differentiable at a.

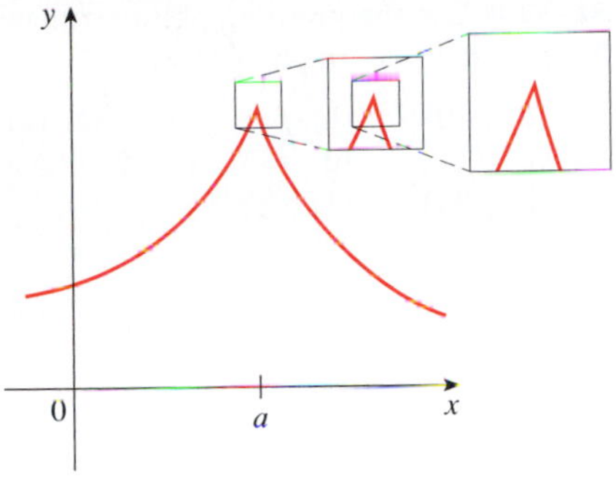

FIGURE 12
f is not differentiable at a.

EXERCISES 3.1

1. If $f(x) = 3x^2 - 5x$, find $f'(2)$ and use it to find the equation of the tangent line to the parabola $y = 3x^2 - 5x$ at the point $(2, 2)$.

2. If $g(x) = 1 - x^3$, find $g'(0)$ and use it to find the equation of the tangent line to the curve $y = 1 - x^3$ at the point $(0, 1)$.

3. (a) If $F(x) = x^3 - 5x + 1$, find $F'(1)$ and use it to find the equation of the tangent line to the curve $y = x^3 - 5x + 1$ at the point $(1, -3)$.
(b) Illustrate part (a) by graphing the curve and the tangent line on the same screen.

4. (a) If $G(x) = x/(1 + 2x)$, find $G'(a)$ and use it to find the equation of the tangent line to the curve $y = x/(1 + 2x)$ at the point $\left(-\frac{1}{4}, -\frac{1}{2}\right)$.
(b) Illustrate part (a) by graphing the curve and the tangent line on the same screen.

5–6 ■ A particle moves along a straight line with equation of motion $s = f(t)$, where s is measured in meters and t in seconds. Find the velocity when $t = 2$.

5. $f(t) = t^2 - 6t - 5$

6. $f(t) = 2t^3 - t + 1$

7–12 ■ Find $f'(a)$.

7. $f(x) = 1 + x - 2x^2$

8. $f(x) = x^3 + 3x$

9. $f(x) = \dfrac{x}{2x - 1}$

10. $f(x) = \dfrac{x}{x^2 - 1}$

11. $f(x) = \dfrac{2}{\sqrt{3 - x}}$

12. $f(x) = \sqrt{x - 1}$

13–18 ■ Each limit represents the derivative of some function f at some number a. State f and a in each case.

13. $\displaystyle\lim_{h\to 0} \frac{\sqrt{1 + h} - 1}{h}$

14. $\displaystyle\lim_{h\to 0} \frac{(2 + h)^3 - 8}{h}$

15. $\displaystyle\lim_{x\to 1} \frac{x^9 - 1}{x - 1}$

16. $\displaystyle\lim_{x\to 3\pi} \frac{\cos x + 1}{x - 3\pi}$

17. $\displaystyle\lim_{t\to 0} \frac{\sin\left(\frac{\pi}{2} + t\right) - 1}{t}$

18. $\displaystyle\lim_{x\to 0} \frac{3^x - 1}{x}$

19–28 ■ Find the derivative of the given function using the definition of derivative. State the domain of the function and the domain of its derivative.

19. $f(x) = 5x + 3$

20. $f(x) = 18$

21. $f(x) = x^3 - x^2 + 2x$

22. $f(x) = \sqrt{6 - x}$

23. $g(x) = \sqrt{1 + 2x}$

24. $f(x) = \dfrac{x + 1}{x - 1}$

25. $G(x) = \dfrac{4 - 3x}{2 + x}$

26. $g(x) = \dfrac{1}{x^2}$

27. $f(x) = x^4$

28. $F(x) = \dfrac{1}{\sqrt{x - 1}}$

29. Compute the derivatives of the functions $f(x) = x$, $f(x) = x^2$, and $f(x) = x^3$ and observe the result of Exercise 27. On the basis of these results, guess the derivative of $f(x) = x^n$ for n a positive integer. Test your guess by computing the derivative of $f(x) = x^5$.

30. (a) If $f(t) = 6/(1 + t^2)$, find $f'(t)$.
(b) Check to see that your answer to part (a) is reasonable by comparing the graphs of f and f'.

31. (a) If $f(x) = x - (2/x)$, find $f'(x)$.
(b) Check to see that your answer to part (a) is reasonable by comparing the graphs of f and f'.

32–33 ■ Use the given graph to estimate each function value. Then sketch the graph of f'.

32. (a) $f'(0)$ (b) $f'(1)$ (c) $f'(2)$ (d) $f'(3)$ (e) $f'(4)$ (f) $f'(5)$

33. (a) $f'(1)$ (b) $f'(2)$ (c) $f'(3)$ (d) $f'(4)$

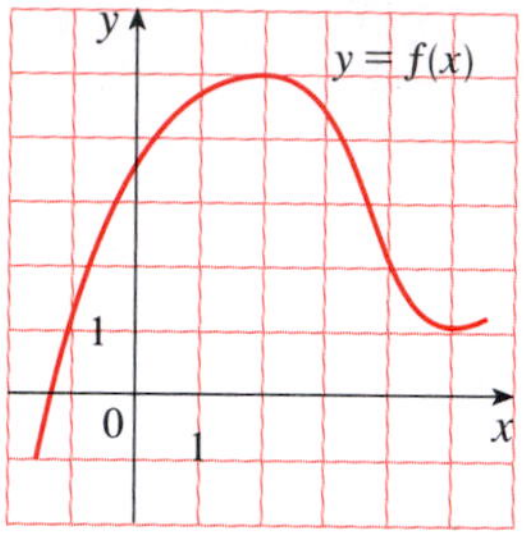

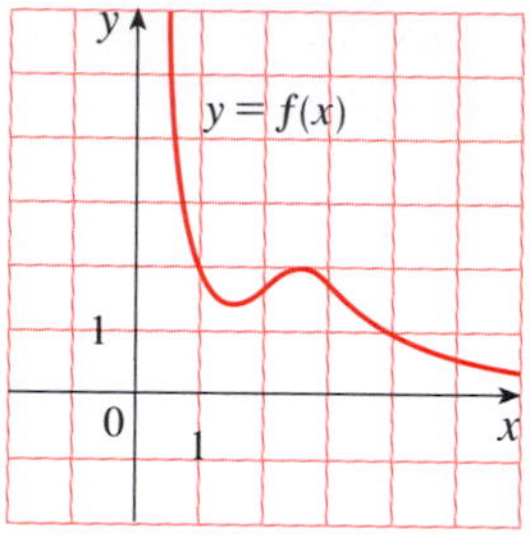

34. Match the graph of each function in (a)–(d) with the graph of its derivative in I–IV. Give reasons for your choices.

(a)

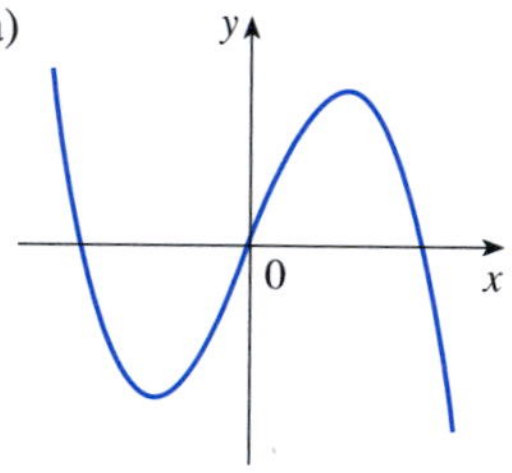

(b)

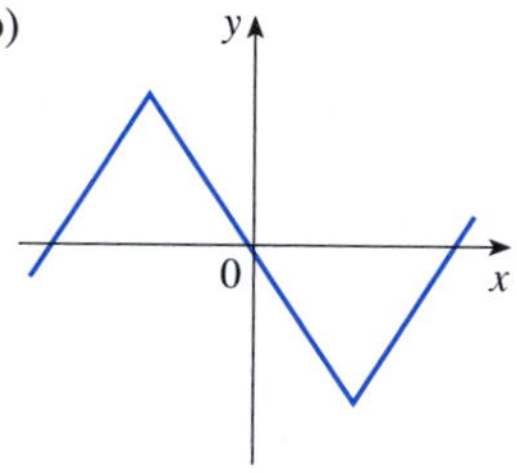

(c)

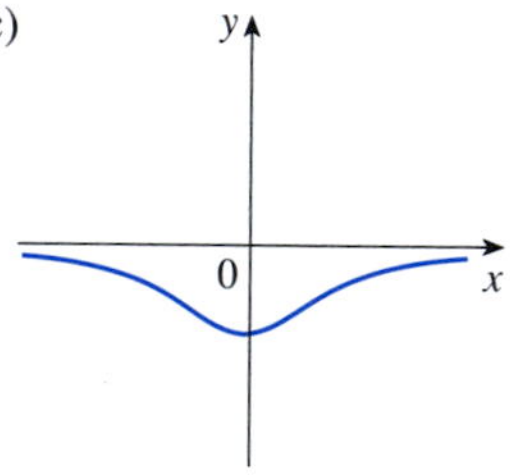

(d)

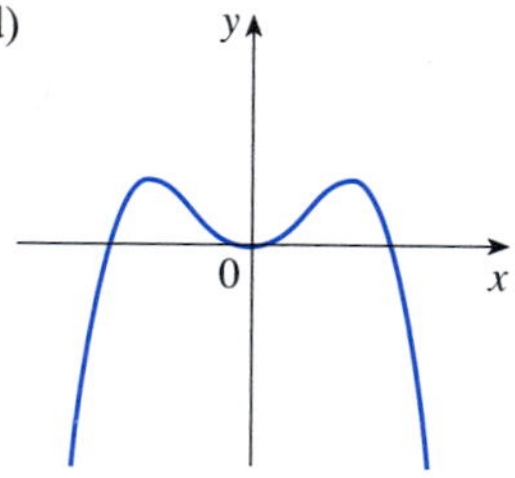

I

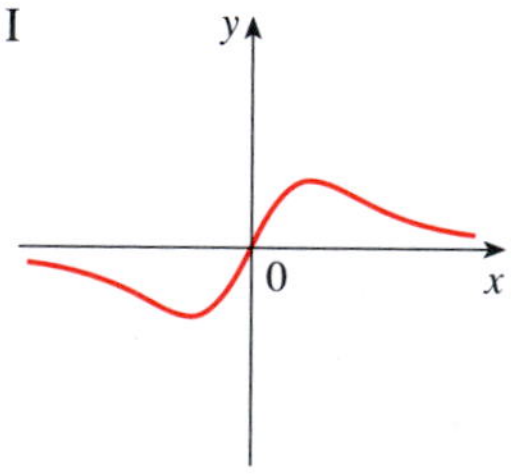

II

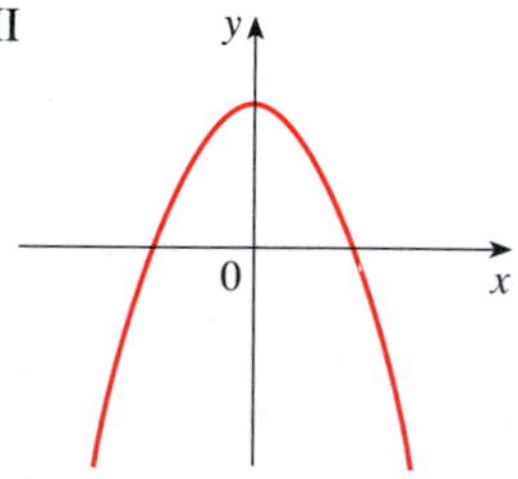

III

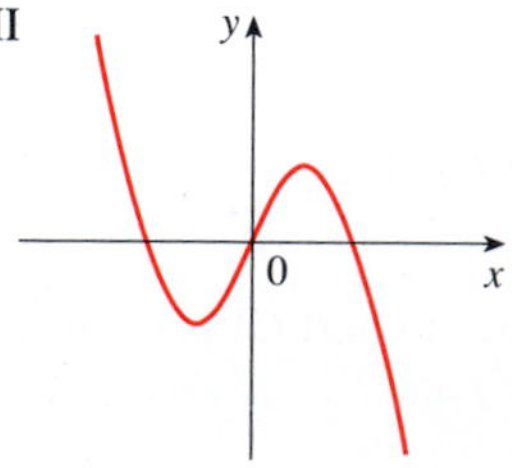

IV

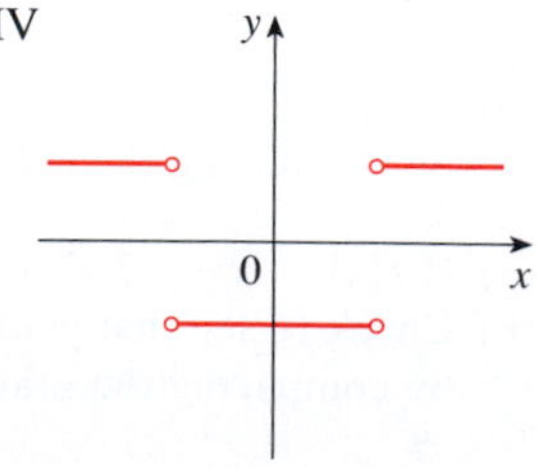

35–43 ■ Trace or copy the graph of the given function f. Then use the method of Example 5 to sketch the graph of f' below it.

35.

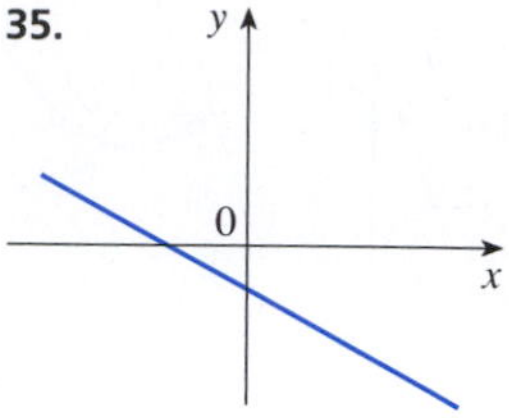

36.

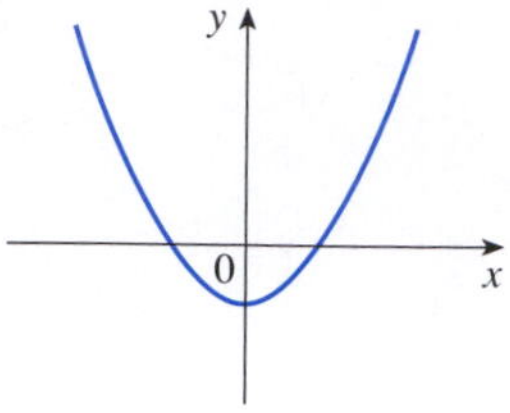

37.

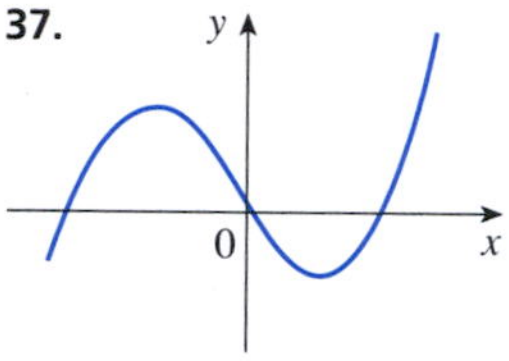

38.

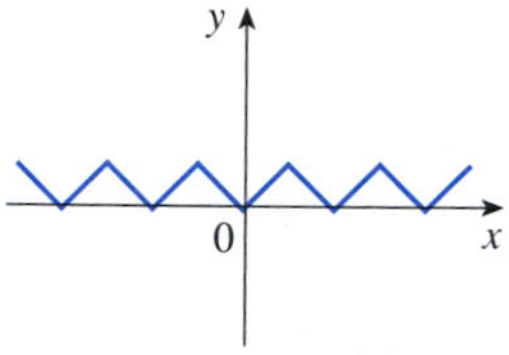

39.

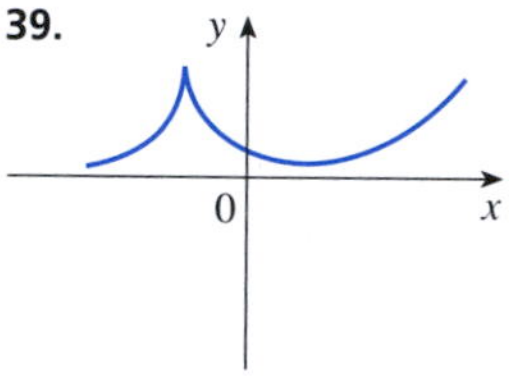

40.

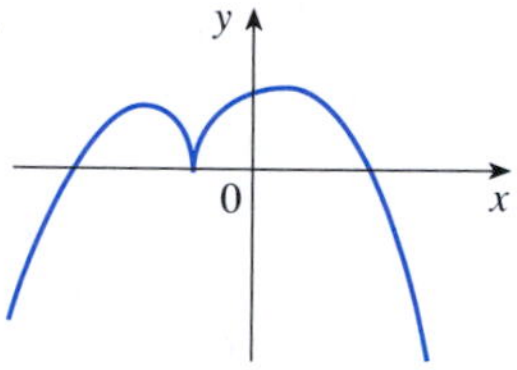

41.

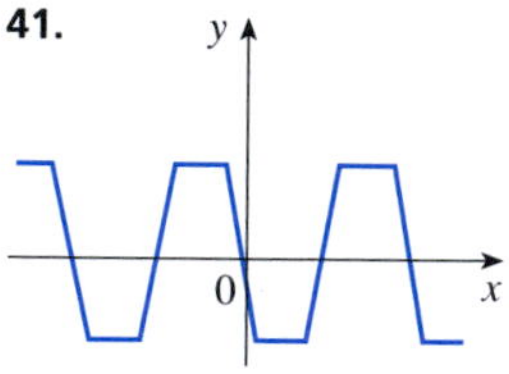

42.

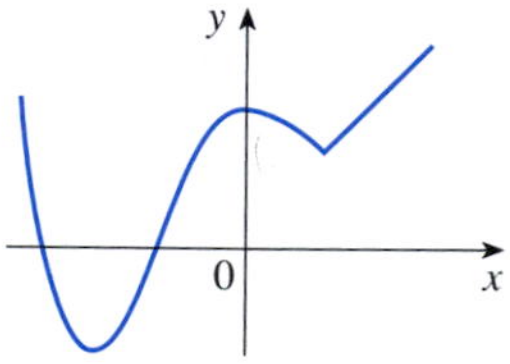

43.

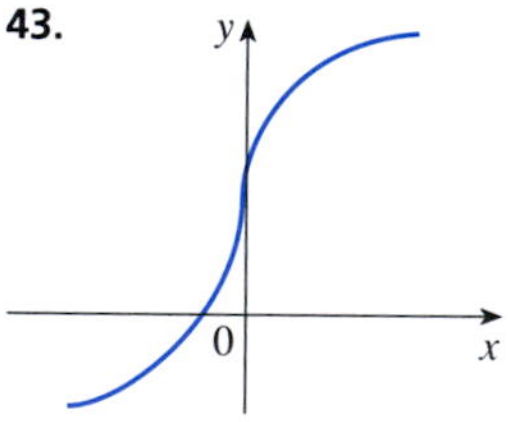

44. Make a careful sketch of the graph of the sine function and below it sketch the graph of its derivative in the same manner as in Exercises 35–43. Can you guess what the derivative of the sine function is from its graph?

45. Let $f(x) = 3^x$. Estimate the value of $f'(1)$ in two ways:
(a) By using Definition 2 and taking successively smaller values of h.
(b) By zooming in on the graph of $y = 3^x$ and estimating the slope.

46. Let $g(x) = \tan x$. Estimate the value of $g'(\pi/4)$ in two ways:
(a) By using Definition 2 and taking successively smaller values of h.
(b) By zooming in on the graph of $y = \tan x$ and estimating the slope.

47. A function f is given by the data in the table. Find approximate values for $f'(x)$ when $x = 0.1, 0.2, 0.3, 0.4, 0.5, 0.6$, and 0.7.

x	0.0	0.1	0.2	0.3	0.4	0.5	0.6	0.7	0.8
$f(x)$	5.0	4.1	4.0	4.6	5.5	6.2	6.5	6.1	4.7

48. A function g is given by the data in the table. Find approximate values for $g'(x)$ when $x = 2, 4, 6, 8, 10, 12$, and 14. Then sketch the graph of g'.

x	0	2	4	6	8	10	12	14	16
$g(x)$	1.8	4.7	6.3	6.8	3.9	2.5	2.0	1.8	1.7

49. Let $f(x) = \sqrt[3]{x}$.
(a) If $a \neq 0$, use Equation 3 to find $f'(a)$.
(b) Show that $f'(0)$ does not exist.
(c) Show that $y = \sqrt[3]{x}$ has a vertical tangent line at $(0, 0)$. (Recall the shape of the graph of f. See Figure 5 in Section 2 of Review and Preview.)

50. (a) If $g(x) = x^{2/3}$, show that $g'(0)$ does not exist.
(b) If $a \neq 0$, find $g'(a)$.
(c) Show that $y = x^{2/3}$ has a vertical tangent line at $(0, 0)$.
(d) Illustrate part (c) by graphing $y = x^{2/3}$.

51. The graph of f is given. State, with reasons, the numbers at which f is not differentiable.

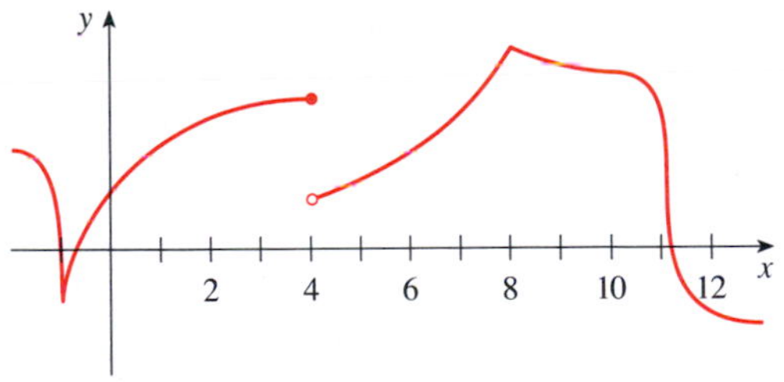

52. The graph of g is given.
(a) At what numbers is g discontinuous?
(b) At what numbers is g not differentiable?

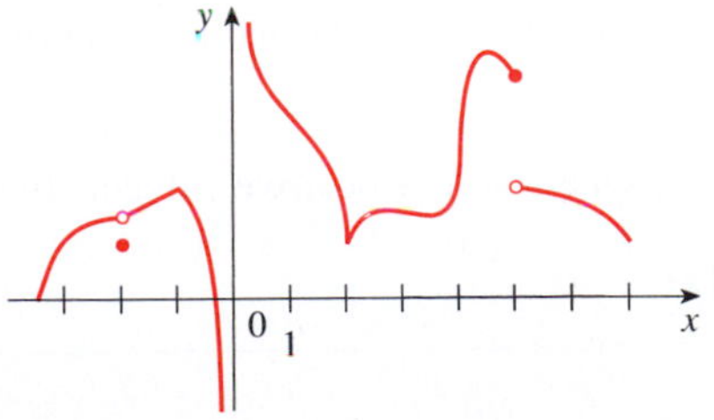

53. Show that the function $f(x) = |x - 6|$ is not differentiable at 6. Find a formula for f' and sketch its graph.

54. Where is the greatest integer function $f(x) = [\![x]\!]$ not differentiable? Find a formula for f' and sketch its graph.

55. (a) Sketch the graph of the function $f(x) = x|x|$.
(b) For what values of x is f differentiable?
(c) Find a formula for f'.

56. Where (and why) is the following function discontinuous? Where is it not differentiable? Sketch its graph.

$$g(x) = \begin{cases} \dfrac{x^3 - x}{x^2 + x} & \text{if } x < 1 \quad (x \neq 0) \\ 0 & \text{if } x = 0 \\ 1 - x & \text{if } x \geq 1 \end{cases}$$

57. The **left-hand** and **right-hand derivatives** of f at a are defined by

$$f'_-(a) = \lim_{h \to 0^-} \frac{f(a + h) - f(a)}{h}$$

and $$f'_+(a) = \lim_{h \to 0^+} \frac{f(a + h) - f(a)}{h}$$

if these limits exist. Then $f'(a)$ exists if and only if these one-sided derivatives exist and are equal.
(a) Find $f'_-(0.6)$ and $f'_+(0.6)$ for the function

$$f(x) = |5x - 3|$$

(b) Show that $f'(0.6)$ does not exist.

58. (a) Use the definitions in Exercise 57 to compute $f'_-(4)$ and $f'_+(4)$ for the function

$$f(x) = \begin{cases} 0 & \text{if } x \leq 0 \\ 5 - x & \text{if } 0 < x < 4 \\ \dfrac{1}{5 - x} & \text{if } x \geq 4 \end{cases}$$

(b) Sketch the graph of f.
(c) Where is f discontinuous?
(d) Where is f not differentiable?

59–60 ■ Determine whether or not $f'(0)$ exists.

59. $f(x) = \begin{cases} x \sin \dfrac{1}{x} & \text{if } x \neq 0 \\ 0 & \text{if } x = 0 \end{cases}$

60. $f(x) = \begin{cases} x^2 \sin \dfrac{1}{x} & \text{if } x \neq 0 \\ 0 & \text{if } x = 0 \end{cases}$

61. Recall that a function f is called *even* if $f(-x) = f(x)$ for all x in its domain and *odd* if $f(-x) = -f(x)$ for all such x. Prove each of the following.
(a) The derivative of an even function is an odd function.
(b) The derivative of an odd function is an even function.

62. When you turn on a hot-water faucet, the temperature T of the water depends on how long the water has been running.
(a) Sketch a possible graph of T as a function of the time t that has elapsed since the faucet was turned on.
(b) Describe how the rate of change of T with respect to t varies as t increases.
(c) Sketch a graph of the derivative of T.

63. Let ℓ be the tangent line to the parabola $y = x^2$ at the point $(1, 1)$. The *angle of inclination* of ℓ is the angle ϕ that ℓ makes with the positive direction of the x-axis. Calculate ϕ correct to the nearest degree.

64. Suppose f is a function that satisfies the equation

$$f(x + y) = f(x) + f(y) + x^2y + xy^2$$

for all real numbers x and y. Suppose also that

$$\lim_{x \to 0} \frac{f(x)}{x} = 1$$

(a) Find $f(0)$. (b) Find $f'(0)$. (c) Find $f'(x)$.

3.2 DIFFERENTIATION FORMULAS

If it were always necessary to compute derivatives directly from the definition, as we did in the preceding section, such computations would be tedious and the evaluation of some limits would require ingenuity. Fortunately, several rules have been developed for finding derivatives without having to use the definition directly. These formulas greatly simplify the task of differentiation.

(1) THEOREM If f is a constant function, $f(x) = c$, then $f'(x) = 0$.

This result is geometrically evident because the graph of a constant function is a horizontal line, which has slope 0, but a formal proof is also easy.

PROOF

$$f'(x) = \lim_{h \to 0} \frac{f(x + h) - f(x)}{h} = \lim_{h \to 0} \frac{c - c}{h}$$

$$= \lim_{h \to 0} 0 = 0$$

□

In Leibniz notation, Theorem 1 can be written as

$$\frac{d}{dx} c = 0$$

The next theorem gives a formula for differentiating the power function $f(x) = x^n$. In the preceding section (Exercises 27 and 29) we found that $D(x) = 1$, $D(x^2) = 2x$, $D(x^3) = 3x^2$, $D(x^4) = 4x^3$, and so it seems reasonable to make the guess that $D(x^n) = nx^{n-1}$. We give two proofs of this fact; the second uses the Binomial Theorem.

(2) THE POWER RULE If $f(x) = x^n$, where n is a positive integer, then

$$f'(x) = nx^{n-1}$$

FIRST PROOF The formula

$$x^n - a^n = (x - a)(x^{n-1} + x^{n-2}a + \cdots + xa^{n-2} + a^{n-1})$$

can be verified simply by multiplying out the right-hand side (or by summing the second factor as a geometric series). Thus, by using Equation 3.1.3 for $f'(a)$ and then using the equation above, we get

$$\begin{aligned} f'(a) &= \lim_{x \to a} \frac{f(x) - f(a)}{x - a} = \lim_{x \to a} \frac{x^n - a^n}{x - a} \\ &= \lim_{x \to a} (x^{n-1} + x^{n-2}a + \cdots + xa^{n-2} + a^{n-1}) \\ &= a^{n-1} + a^{n-2}a + \cdots + aa^{n-2} + a^{n-1} \\ &= na^{n-1} \end{aligned}$$

SECOND PROOF

$$f'(x) = \lim_{h \to 0} \frac{f(x + h) - f(x)}{h} = \lim_{h \to 0} \frac{(x + h)^n - x^n}{h}$$

The Binomial Theorem is given on the front endpaper.

Expanding $(x + h)^n$ by the Binomial Theorem, we get

$$\begin{aligned} f'(x) &= \lim_{h \to 0} \frac{\left[x^n + nx^{n-1}h + \dfrac{n(n-1)}{2} x^{n-2}h^2 + \cdots + nxh^{n-1} + h^n\right] - x^n}{h} \\ &= \lim_{h \to 0} \frac{nx^{n-1}h + \dfrac{n(n-1)}{2} x^{n-2}h^2 + \cdots + nxh^{n-1} + h^n}{h} \\ &= \lim_{h \to 0} \left[nx^{n-1} + \frac{n(n-1)}{2} x^{n-2}h + \cdots + nxh^{n-2} + h^{n-1}\right] \\ &= nx^{n-1} \end{aligned}$$

because every term except the first has h as a factor and therefore approaches 0. □

The Power Rule can be written in Leibniz notation as

$$\frac{d}{dx}(x^n) = nx^{n-1}$$

We illustrate the Power Rule using various notations in Example 1.

EXAMPLE 1

(a) If $f(x) = x^6$, then $f'(x) = 6x^5$.

(b) If $y = x^{1000}$, then $y' = 1000x^{999}$.

(c) If $y = t^4$, then $\dfrac{dy}{dt} = 4t^3$.

(d) $\dfrac{d}{dr}(r^3) = 3r^2$

(e) $D_u(u^m) = mu^{m-1}$ ■

The next differentiation formulas tell us that *the derivative of a constant times a function is the constant times the derivative of the function*, and *the derivative of a sum (or difference) of functions is the sum (or difference) of the derivatives* (assuming these derivatives exist).

(3) THEOREM Suppose c is a constant and $f'(x)$ and $g'(x)$ exist.

(a) If $F(x) = cf(x)$, then $F'(x)$ exists and $F'(x) = cf'(x)$.

(b) If $G(x) = f(x) + g(x)$, then $G'(x)$ exists and $G'(x) = f'(x) + g'(x)$.

(c) If $H(x) = f(x) - g(x)$, then $H'(x)$ exists and $H'(x) = f'(x) - g'(x)$.

In short:

(a) $(cf)' = cf'$ (b) $(f + g)' = f' + g'$ (c) $(f - g)' = f' - g'$

GEOMETRIC INTERPRETATION OF FORMULA (a):

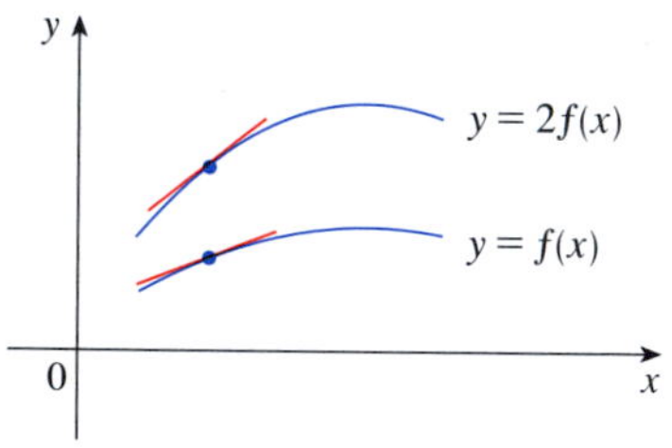

Multiplying by $c = 2$ stretches the graph vertically by a factor of 2. All the rises have been doubled but the runs stay the same. So the slopes are doubled, too.

PROOF

(a)
$$\begin{aligned} F'(x) &= \lim_{h\to 0} \frac{F(x+h) - F(x)}{h} = \lim_{h\to 0} \frac{cf(x+h) - cf(x)}{h} \\ &= \lim_{h\to 0} c\left[\frac{f(x+h) - f(x)}{h}\right] \\ &= c \lim_{h\to 0} \frac{f(x+h) - f(x)}{h} \qquad \text{(by Law 3 of limits)} \\ &= cf'(x) \end{aligned}$$

(b)
$$\begin{aligned} G'(x) &= \lim_{h\to 0} \frac{G(x+h) - G(x)}{h} \\ &= \lim_{h\to 0} \frac{[f(x+h) + g(x+h)] - [f(x) + g(x)]}{h} \\ &= \lim_{h\to 0} \left[\frac{f(x+h) - f(x)}{h} + \frac{g(x+h) - g(x)}{h}\right] \\ &= \lim_{h\to 0} \frac{f(x+h) - f(x)}{h} + \lim_{h\to 0} \frac{g(x+h) - g(x)}{h} \qquad \text{(by Law 1)} \\ &= f'(x) + g'(x) \end{aligned}$$

Part (c) can be proved similarly using Law 2 of limits from Section 2.3. □

Using Leibniz notation, the results of Theorem 3 can be summarized as follows.

(a) $\dfrac{d}{dx}(cf) = c\dfrac{df}{dx}$ (b) $\dfrac{d}{dx}(f + g) = \dfrac{df}{dx} + \dfrac{dg}{dx}$

(c) $\dfrac{d}{dx}(f - g) = \dfrac{df}{dx} - \dfrac{dg}{dx}$

The result of Theorem 3(b) can be extended to the sum of any number of functions. For instance, using this theorem twice, we get

$$(f + g + h)' = [(f + g) + h]' = (f + g)' + h' = f' + g' + h'$$

Theorem 3 can be combined with the Power Rule to differentiate any polynomial, as the following examples demonstrate.

EXAMPLE 2

$$\frac{d}{dx}(x^8 + 12x^5 - 4x^4 + 10x^3 - 6x + 5)$$
$$= \frac{d}{dx}(x^8) + 12\frac{d}{dx}(x^5) - 4\frac{d}{dx}(x^4) + 10\frac{d}{dx}(x^3) - 6\frac{d}{dx}(x) + \frac{d}{dx}(5)$$
$$= 8x^7 + 12(5x^4) - 4(4x^3) + 10(3x^2) - 6(1) + 0$$
$$= 8x^7 + 60x^4 - 16x^3 + 30x^2 - 6$$

■

EXAMPLE 3 If $f(x) = x^4 - x^3 + x^2 - x + 1$, find an equation of the tangent to the graph of f at the point $(1, 1)$.

SOLUTION The slope is $f'(1)$, which we calculate as follows:

$$f'(x) = 4x^3 - 3x^2 + 2x - 1$$
$$f'(1) = 4 - 3 + 2 - 1 = 2$$

Therefore, an equation of the tangent at $(1, 1)$ is

$$y - 1 = 2(x - 1)$$

or

$$2x - y - 1 = 0$$

■

Next we need a formula for the derivative of a product of two functions. By analogy with Theorem 3(b) and (c), one might be tempted to guess, as Leibniz did three centuries ago, that the derivative of a product is the product of the derivatives. We can see, however, that this guess is wrong by looking at a particular example. Let $f(x) = x$ and $g(x) = x^2$. Then the Power Rule gives $f'(x) = 1$ and $g'(x) = 2x$. But $(fg)(x) = x^3$, so $(fg)'(x) = 3x^2$. Thus $(fg)' \neq f'g'$. The correct formula was discovered by Leibniz (soon after his false start) and is called the Product Rule.

THE PRODUCT RULE If $F(x) = f(x)g(x)$ and $f'(x)$ and $g'(x)$ both exist, then

$$F'(x) = f(x)g'(x) + g(x)f'(x)$$

In short:

$$(fg)' = fg' + gf'$$

PROOF

$$F'(x) = \lim_{h \to 0} \frac{F(x + h) - F(x)}{h}$$
$$= \lim_{h \to 0} \frac{f(x + h)g(x + h) - f(x)g(x)}{h}$$

In order to evaluate this limit, we would like to separate the functions f and g as in the proof of Theorem 3(b). We can achieve this separation by adding and subtracting the term $f(x + h)g(x)$ in the numerator:

$$F'(x) = \lim_{h \to 0} \frac{f(x + h)g(x + h) - f(x + h)g(x) + f(x + h)g(x) - f(x)g(x)}{h}$$

$$= \lim_{h \to 0} \left[f(x + h)\frac{g(x + h) - g(x)}{h} + g(x)\frac{f(x + h) - f(x)}{h} \right]$$

$$= \lim_{h \to 0} f(x + h) \cdot \lim_{h \to 0} \frac{g(x + h) - g(x)}{h} + \lim_{h \to 0} g(x) \cdot \lim_{h \to 0} \frac{f(x + h) - f(x)}{h}$$

$$= f(x)g'(x) + g(x)f'(x)$$

Note that $\lim_{h \to 0} g(x) = g(x)$ because $g(x)$ is a constant with respect to the variable h. Also, since f is differentiable at x, it is continuous at x by Theorem 3.1.8, and so $\lim_{h \to 0} f(x + h) = f(x)$. (See Exercise 53 in Section 2.5.) □

When expressed in Leibniz notation, the Product Rule becomes

$$\frac{d}{dx}(fg) = f\frac{dg}{dx} + g\frac{df}{dx}$$

In words, this says that *the derivative of a product of two functions is the first function times the derivative of the second function plus the second function times the derivative of the first function.*

EXAMPLE 4 Find $F'(x)$ if $F(x) = (6x^3)(7x^4)$.

SOLUTION By the Product Rule, we have

$$F'(x) = (6x^3)\frac{d}{dx}(7x^4) + (7x^4)\frac{d}{dx}(6x^3)$$

$$= (6x^3)(28x^3) + (7x^4)(18x^2)$$

$$= 168x^6 + 126x^6 = 294x^6$$ ■

Notice that we could verify the answer to Example 4 directly by first multiplying the factors:

$$F(x) = (6x^3)(7x^4) = 42x^7 \quad \Rightarrow \quad F'(x) = 42(7x^6) = 294x^6$$

But later we will meet functions, such as $y = x^2 \sin x$, for which the Product Rule is the only possible method.

THE QUOTIENT RULE If $F(x) = f(x)/g(x)$ and both $f'(x)$ and $g'(x)$ exist, then $F'(x)$ exists and

$$F'(x) = \frac{g(x)f'(x) - f(x)g'(x)}{[g(x)]^2}$$

In short: $$\left(\frac{f}{g}\right)' = \frac{gf' - fg'}{g^2}$$

PROOF

$$F'(x) = \lim_{h\to 0} \frac{F(x+h) - F(x)}{h} = \lim_{h\to 0} \frac{\dfrac{f(x+h)}{g(x+h)} - \dfrac{f(x)}{g(x)}}{h}$$

$$= \lim_{h\to 0} \frac{f(x+h)g(x) - f(x)g(x+h)}{hg(x+h)g(x)}$$

We can separate f and g in this expression by adding and subtracting the term $f(x)g(x)$ in the numerator:

$$F'(x) = \lim_{h\to 0} \frac{f(x+h)g(x) - f(x)g(x) + f(x)g(x) - f(x)g(x+h)}{hg(x+h)g(x)}$$

$$= \lim_{h\to 0} \frac{g(x)\dfrac{f(x+h) - f(x)}{h} - f(x)\dfrac{g(x+h) - g(x)}{h}}{g(x+h)g(x)}$$

$$= \frac{\lim_{h\to 0} g(x) \cdot \lim_{h\to 0} \dfrac{f(x+h) - f(x)}{h} - \lim_{h\to 0} f(x) \cdot \lim_{h\to 0} \dfrac{g(x+h) - g(x)}{h}}{\lim_{h\to 0} g(x+h) \cdot \lim_{h\to 0} g(x)}$$

$$= \frac{g(x)f'(x) - f(x)g'(x)}{[g(x)]^2}$$

Again g is continuous by Theorem 3.1.8, so $\lim_{h\to 0} g(x+h) = g(x)$. □

The Quotient Rule, in Leibniz notation, becomes

$$\frac{d}{dx}\left(\frac{f(x)}{g(x)}\right) = \frac{g(x)\dfrac{d}{dx}f(x) - f(x)\dfrac{d}{dx}g(x)}{[g(x)]^2}$$

In words, this says that the *derivative of a quotient is the denominator times the derivative of the numerator minus the numerator times the derivative of the denominator, all divided by the square of the denominator.*

The theorems of this section show that any polynomial is differentiable on $\mathbb{R}$ and any rational function is differentiable on its domain. Furthermore, the Quotient Rule and the other differentiation formulas enable us to compute the derivative of any rational function, as the next example illustrates.

We can use a graphing device to check that the answer to Example 5 is plausible. Figure 1 shows the graphs of the function of Example 5 and its derivative. Notice that when y grows rapidly (near -2), y' is large. And when y grows slowly, y' is near 0.

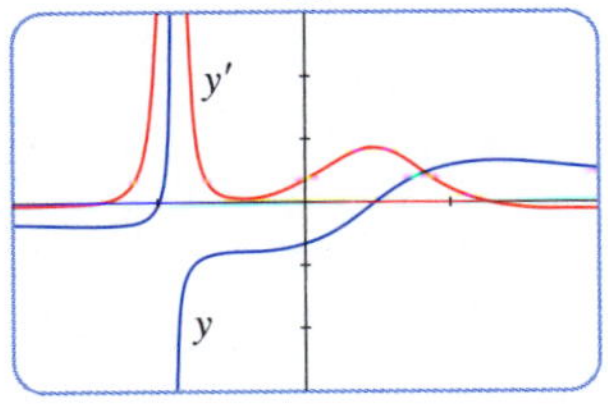

FIGURE 1

EXAMPLE 5 Let $y = \dfrac{x^2 + x - 2}{x^3 + 6}$.

Then

$$y' = \frac{(x^3+6)\,D(x^2+x-2) - (x^2+x-2)\,D(x^3+6)}{(x^3+6)^2}$$

$$= \frac{(x^3+6)(2x+1) - (x^2+x-2)(3x^2)}{(x^3+6)^2}$$

$$= \frac{(2x^4 + x^3 + 12x + 6) - (3x^4 + 3x^3 - 6x^2)}{(x^3+6)^2}$$

$$= \frac{-x^4 - 2x^3 + 6x^2 + 12x + 6}{(x^3+6)^2}$$ ■

The Quotient Rule can also be used to extend the Power Rule to the case where the exponent is a negative integer.

(4) THEOREM If $f(x) = x^{-n}$, where n is a positive integer, then

$$f'(x) = -nx^{-n-1}$$

PROOF

$$\begin{aligned} f'(x) &= \frac{d}{dx}(x^{-n}) = \frac{d}{dx}\left(\frac{1}{x^n}\right) \\ &= \frac{x^n D(1) - 1 \cdot D(x^n)}{(x^n)^2} \\ &= \frac{x^n \cdot 0 - 1 \cdot nx^{n-1}}{x^{2n}} \\ &= \frac{-nx^{n-1}}{x^{2n}} = -nx^{n-1-2n} = -nx^{-n-1} \end{aligned}$$

□

EXAMPLE 6

(a) If $y = \dfrac{1}{x}$, then $\dfrac{dy}{dx} = \dfrac{d}{dx}(x^{-1}) = -x^{-2} = -\dfrac{1}{x^2}$.

(b) $\dfrac{d}{dt}\left(\dfrac{6}{t^3}\right) = 6\dfrac{d}{dt}(t^{-3}) = 6(-3)t^{-4} = -\dfrac{18}{t^4}$ ■

By (2) and (4) the Power Rule holds if the exponent n is a positive or negative integer. If $n = 0$, then $x^0 = 1$, which we know has a derivative of 0. Thus, the Power Rule holds for any integer n. In fact, it also holds for *any real number* n, as we will prove in Chapter 4. (A proof for rational values of n is indicated in Exercise 34 in Section 3.6.) In the meantime we state the general version and use it in the examples and exercises.

(5) THE POWER RULE (GENERAL VERSION) Let n be any real number. If $f(x) = x^n$, then $f'(x) = nx^{n-1}$.

Or, in Leibniz notation, $\dfrac{d}{dx}(x^n) = nx^{n-1}$.

EXAMPLE 7

(a) If $f(x) = x^\pi$, then $f'(x) = \pi x^{\pi-1}$.

(b) $\dfrac{d}{dx}\sqrt{x} = \dfrac{d}{dx}(x^{1/2}) = \frac{1}{2}x^{(1/2)-1} = \frac{1}{2}x^{-1/2} = \dfrac{1}{2\sqrt{x}}$

(c) Let

$$y = \frac{1}{\sqrt[3]{x^2}}$$

Then

$$\begin{aligned} \frac{dy}{dx} &= \frac{d}{dx}(x^{-2/3}) = -\tfrac{2}{3}x^{-(2/3)-1} \\ &= -\tfrac{2}{3}x^{-5/3} \end{aligned}$$

■

EXAMPLE 8 Differentiate the function $f(t) = \sqrt{t}(1 - t)$.

SOLUTION 1 Using the Product Rule, we have

$$\begin{aligned} f'(t) &= \sqrt{t}\,\frac{d}{dt}(1 - t) + (1 - t)\frac{d}{dt}\sqrt{t} \\ &= \sqrt{t}(-1) + (1 - t)\cdot\tfrac{1}{2}t^{-1/2} \\ &= -\sqrt{t} + \frac{1 - t}{2\sqrt{t}} = \frac{1 - 3t}{2\sqrt{t}} \end{aligned}$$

SOLUTION 2 If we first use the laws of exponents, then we can proceed directly without using the Product Rule.

$$\begin{aligned} f(t) &= \sqrt{t} - t\sqrt{t} = t^{1/2} - t^{3/2} \\ f'(t) &= \tfrac{1}{2}t^{-1/2} - \tfrac{3}{2}t^{1/2} \end{aligned}$$

which is equivalent to the answer given in Solution 1. ■

EXAMPLE 9 At what points on the hyperbola $xy = 12$ is the tangent line parallel to the line $3x + y = 0$?

SOLUTION Since $xy = 12$ can be written as $y = 12/x$, we have

$$\frac{dy}{dx} = 12\frac{d}{dx}(x^{-1}) = 12(-x^{-2}) = -\frac{12}{x^2}$$

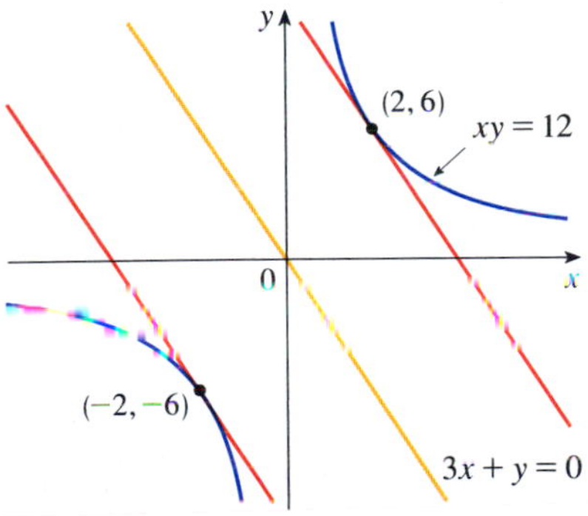

FIGURE 2

Let the x-coordinate of one of the points in question be a. Then the slope of the tangent line at that point is $-12/a^2$. This tangent line will be parallel to the line $3x + y = 0$, or $y = -3x$, if it has the same slope, that is, -3. Equating slopes, we get

$$-\frac{12}{a^2} = -3 \qquad \text{or} \qquad a^2 = 4 \qquad \text{or} \qquad a = \pm 2$$

Therefore, the required points are $(2, 6)$ and $(-2, -6)$. The hyperbola and the tangents are shown in Figure 2. ■

We summarize the differentiation formulas we have learned so far in the following table.

TABLE OF DIFFERENTIATION FORMULAS

$(cf)' = cf'$	$(f + g)' = f' + g'$
$(f - g)' = f' - g'$	$(fg)' = fg' + gf'$
$\left(\dfrac{f}{g}\right)' = \dfrac{gf' - fg'}{g^2}$	$\dfrac{d}{dx}c = 0$
$\dfrac{d}{dx}(x^n) = nx^{n-1}$	

EXERCISES 3.2

1–34 ■ Differentiate each function.

1. $f(x) = x^2 - 10x + 100$
2. $g(x) = x^{100} + 50x + 1$
3. $V(r) = \frac{4}{3}\pi r^3$
4. $s(t) = t^8 + 6t^7 - 18t^2 + 2t$
5. $F(x) = (16x)^3$
6. $G(y) = (y^2 + 1)(2y - 7)$
7. $Y(t) = 6t^{-9}$
8. $R(x) = \dfrac{\sqrt{10}}{x^7}$
9. $g(x) = x^2 + \dfrac{1}{x^2}$
10. $f(t) = \sqrt{t} - \dfrac{1}{\sqrt{t}}$
11. $h(x) = \dfrac{x + 2}{x - 1}$
12. $f(u) = \dfrac{1 - u^2}{1 + u^2}$
13. $G(s) = (s^2 + s + 1)(s^2 + 2)$
14. $H(t) = \sqrt[3]{t}\,(t + 2)$
15. $y = \dfrac{x^2 + 4x + 3}{\sqrt{x}}$
16. $y = \dfrac{\sqrt{x} - 1}{\sqrt{x} + 1}$
17. $y = \sqrt{5x}$
18. $y = x^{4/3} - x^{2/3}$
19. $y = \dfrac{1}{x^4 + x^2 + 1}$
20. $y = x^2 + x + x^{-1} + x^{-2}$
21. $y = ax^2 + bx + c$
22. $y = A + \dfrac{B}{x} + \dfrac{C}{x^2}$
23. $y = \dfrac{3t - 7}{t^2 + 5t - 4}$
24. $y = \dfrac{4t + 5}{2 - 3t}$
25. $y = x + \sqrt[5]{x^2}$
26. $y = x^4 - \sqrt[4]{x}$
27. $u = x^{\sqrt{2}}$
28. $u = \sqrt[3]{t^2} + 2\sqrt{t^3}$
29. $v = x\sqrt{x} + \dfrac{1}{x^2\sqrt{x}}$
30. $v = \dfrac{6}{\sqrt[3]{t^5}}$
31. $f(x) = \dfrac{x}{x + \dfrac{c}{x}}$
32. $f(x) = \dfrac{ax + b}{cx + d}$
33. $f(x) = \dfrac{x^5}{x^3 - 2}$
34. $s = \sqrt{t}\,(t^3 - \sqrt{t} + 1)$

35. The general polynomial of degree n has the form

$$P(x) = a_n x^n + a_{n-1}x^{n-1} + \cdots + a_2x^2 + a_1x + a_0$$

where $a_n \neq 0$. Find the derivative of P.

36–39 ■ Find the equation of the tangent line to the given curve at the specified point.

36. $y = \dfrac{x}{x - 3}$, $(6, 2)$
37. $y = x + \dfrac{4}{x}$, $(2, 4)$
38. $y = x^{5/2}$, $(4, 32)$
39. $y = x + \sqrt{x}$, $(1, 2)$

40. (a) The curve $y = x/(1 + x^2)$ is called a **serpentine**. Find an equation of the tangent line to this curve at the point $(3, 0.3)$.
(b) Illustrate part (a) by graphing the curve and the tangent line on the same screen.

41. (a) The curve $y = 1/(1 + x^2)$ is called a **witch of Agnesi**. Find an equation of the tangent line to this curve at the point $(-1, \frac{1}{2})$.
(b) Illustrate part (a) by graphing the curve and the tangent line on the same screen.

42. (a) If $f(x) = x/(x^2 - 1)$, find $f'(x)$.
(b) Check to see that your answer to part (a) is reasonable by comparing the graphs of f and f'.

43. (a) If $f(x) = 3x^{15} - 5x^3 + 3$, find $f'(x)$.
(b) Check to see that your answer to part (a) is reasonable by comparing the graphs of f and f'.

44. Find the equations of the tangent lines to the curve $y = (x - 1)/(x + 1)$ that are parallel to the line $x - 2y = 1$.

45. At what point on the curve $y = x\sqrt{x}$ is the tangent line parallel to the line $3x - y + 6 = 0$?

46. For what values of x does the graph of $f(x) = 2x^3 - 3x^2 - 6x + 87$ have a horizontal tangent?

47. Find the points on the curve $y = x^3 - x^2 - x + 1$ where the tangent is horizontal.

48. Draw a diagram to show that there are two tangent lines to the parabola $y = x^2$ that pass through the point $(0, -4)$. Find the coordinates of the points where these tangent lines intersect the parabola.

49. How many tangent lines to the curve $y = x/(x + 1)$ pass through the point $(1, 2)$? At which points do these tangent lines touch the curve?

50. Find the equations of both lines through the point $(2, -3)$ that are tangent to the parabola $y = x^2 + x$.

51. Show that the curve $y = 6x^3 + 5x - 3$ has no tangent line with slope 4.

52. A manufacturer of cartridges for stereo systems has designed a stylus with parabolic cross-section as shown in the figure. The equation of the parabola is $y = 16x^2$, where

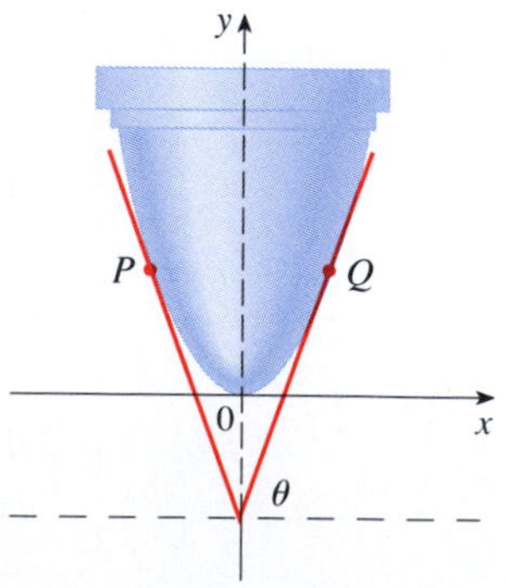

x and y are measured in millimeters. If the stylus sits in a record groove whose sides make an angle of θ with the horizontal direction, where $\tan\theta = 1.75$, find the points of contact P and Q of the stylus with the groove.

53–56 ■ Find the equation of the normal line to the curve at the given point. (The **normal line** to a curve C at a point P is, by definition, the line that passes through P and is perpendicular to the tangent line to C at P.) Also sketch the curve and the normal line in Exercises 53–55.

53. $y = 1 - x^2$, $(2, -3)$ **54.** $y = \dfrac{1}{x-1}$, $(2, 1)$

55. $y = \sqrt[3]{x}$, $(-8, -2)$

56. $y = f(x)$, $(a, f(a))$

57. At what point on the curve $y = x^4$ does the normal line have slope 16?

58. Where does the normal line to the parabola $y = x - x^2$ at the point $(1, 0)$ intersect the parabola a second time? Illustrate with a sketch.

59. Suppose that $f(5) = 1$, $f'(5) = 6$, $g(5) = -3$, and $g'(5) = 2$. Find the values of (a) $(fg)'(5)$, (b) $(f/g)'(5)$, and (c) $(g/f)'(5)$.

60. If $f(3) = 4$, $g(3) = 2$, $f'(3) = -6$, and $g'(3) = 5$, find the following numbers:
(a) $(f + g)'(3)$ (b) $(fg)'(3)$
(c) $(f/g)'(3)$ (d) $\left(\dfrac{f}{f-g}\right)'(3)$

61. If f and g are the functions whose graphs are shown, let $u(x) = f(x)g(x)$ and $v(x) = f(x)/g(x)$.
(a) Find $u'(1)$. (b) Find $v'(5)$.

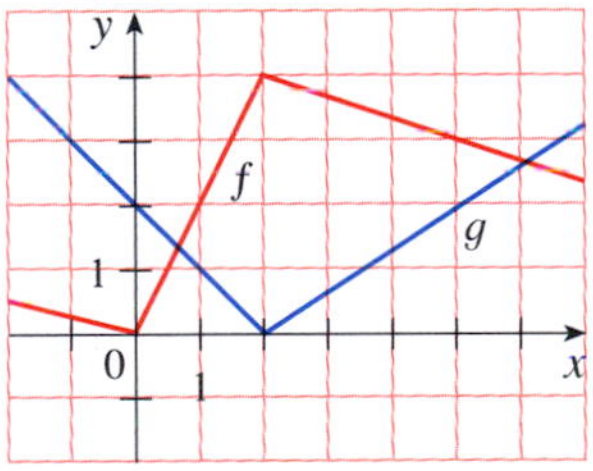

62. If f is a differentiable function, find an expression for the derivative of each of the following functions.
(a) $y = x^2 f(x)$ (b) $y = \dfrac{f(x)}{x^2}$
(c) $y = \dfrac{x^2}{f(x)}$ (d) $y = \dfrac{1 + xf(x)}{\sqrt{x}}$

63. (a) Use the Product Rule twice to prove that if f, g, and h are differentiable, then

$$(fgh)' = f'gh + fg'h + fgh'$$

(b) Taking $f = g = h$ in part (a), show that

$$\frac{d}{dx}[f(x)]^3 = 3[f(x)]^2 f'(x)$$

64–66 ■ Use Exercise 63 to differentiate each function.

64. $y = (x + 5)(x^2 + 7)(x - 3)$

65. $y = \sqrt{x}(x^4 + x + 1)(2x - 3)$

66. $y = (x^4 + 3x^3 + 17x + 82)^3$

67. Let

$$f(x) = \begin{cases} 2 - x & \text{if } x \le 1 \\ x^2 - 2x + 2 & \text{if } x > 1 \end{cases}$$

Is f differentiable at 1? Sketch the graphs of f and f'.

68. At what numbers is the following function g differentiable?

$$g(x) = \begin{cases} -1 - 2x & \text{if } x < -1 \\ x^2 & \text{if } -1 \le x \le 1 \\ x & \text{if } x > 1 \end{cases}$$

Give a formula for g' and sketch the graphs of g and g'.

69. (a) For what values of x is the function $f(x) = |x^2 - 9|$ differentiable? Find a formula for f'.
(b) Sketch the graphs of f and f'.

70. Where is the function $h(x) = |x - 1| + |x + 2|$ differentiable? Give a formula for h' and sketch the graphs of h and h'.

71. For what values of a and b is the line $2x + y = b$ tangent to the parabola $y = ax^2$ when $x = 2$?

72. Let

$$f(x) = \begin{cases} x^2 & \text{if } x \le 2 \\ mx + b & \text{if } x > 2 \end{cases}$$

Find the values of m and b that make f differentiable everywhere.

73. An easy proof of the Quotient Rule can be given if we make the prior assumption that $F'(x)$ exists, where $F = f/g$. Write $f = Fg$; then differentiate using the Product Rule and solve the resulting equation for F'.

74. A tangent line is drawn to the hyperbola $xy = c$ at a point P.
(a) Show that the midpoint of the line segment cut off this tangent line by the coordinate axes is P.
(b) Show that the triangle formed by the tangent line and the coordinate axes always has the same area, no matter where P is located on the hyperbola.

75. Evaluate $\displaystyle\lim_{x \to 1} \frac{x^{1000} - 1}{x - 1}$.

76. Draw a diagram showing two perpendicular lines that intersect on the y-axis and are both tangent to the parabola $y = x^2$. Where do these lines intersect?

3.3 RATES OF CHANGE IN THE NATURAL AND SOCIAL SCIENCES

Recall from Section 3.1 that if $y = f(x)$, then the derivative dy/dx can be interpreted as the rate of change of y with respect to x. In this section we examine some of the applications of this idea to physics, chemistry, biology, economics, and other sciences.

Let's recall from Section 2.7 the basic idea behind rates of change. If x changes from x_1 to x_2, then the change in x is

$$\Delta x = x_2 - x_1$$

and the corresponding change in y is

$$\Delta y = f(x_2) - f(x_1)$$

The difference quotient

$$\frac{\Delta y}{\Delta x} = \frac{f(x_2) - f(x_1)}{x_2 - x_1}$$

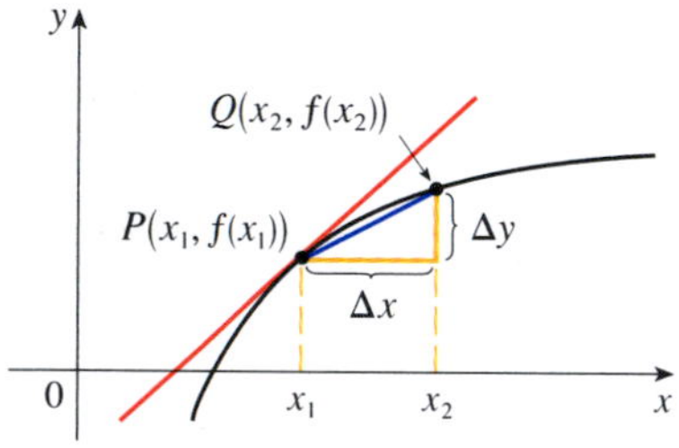

m_{PQ} = average rate of change
$m = f'(x_1)$ = instantaneous rate of change

FIGURE 1

is the **average rate of change of y with respect to x** over the interval $[x_1, x_2]$ and can be interpreted as the slope of the secant line PQ in Figure 1. Its limit as $\Delta x \to 0$ is the derivative $f'(x_1)$, which can therefore be interpreted as the instantaneous rate of change of y with respect to x or the slope of the tangent line at $P(x_1, f(x_1))$. Using Leibniz notation, we write the process in the form

$$\frac{dy}{dx} = \lim_{\Delta x \to 0} \frac{\Delta y}{\Delta x}$$

Whenever the function $y = f(x)$ has a specific interpretation in one of the sciences, its derivative will have a specific interpretation as a rate of change. We now look at some of these interpretations in the natural and social sciences.

PHYSICS

If $s = f(t)$ is the position function of a particle that is moving in a straight line, then $\Delta s/\Delta t$ represents the average velocity over a time period Δt, and $v = ds/dt$ represents the instantaneous **velocity** (the rate of change of displacement with respect to time). This was discussed in Sections 2.7 and 3.1, but now that we know the differentiation formulas, we are able to solve velocity problems more easily.

EXAMPLE 1 The position of a particle is given by the equation

$$s = f(t) = t^3 - 6t^2 + 9t$$

where t is measured in seconds and s in meters.

(a) Find the velocity at time t.
(b) What is the velocity after 2 s? After 4 s?
(c) When is the particle at rest?
(d) When is the particle moving in the positive direction?
(e) Draw a diagram to represent the motion of the particle.
(f) Find the total distance traveled by the particle during the first five seconds.

SOLUTION

(a) The velocity function is the derivative of the position function.

$$s = f(t) = t^3 - 6t^2 + 9t$$

$$v(t) = \frac{ds}{dt} = 3t^2 - 12t + 9$$

(b) The velocity after 2 s means the instantaneous velocity when $t = 2$, that is,

$$v(2) = \left.\frac{ds}{dt}\right|_{t=2} = 3(2)^2 - 12(2) + 9 = -3 \text{ m/s}$$

The velocity after 4 s is

$$v(4) = 3(4)^2 - 12(4) + 9 = 9 \text{ m/s}$$

(c) The particle is at rest when $v(t) = 0$, that is,

$$3t^2 - 12t + 9 = 3(t^2 - 4t + 3) = 3(t - 1)(t - 3) = 0$$

and this is true when $t = 1$ or $t = 3$. Thus, the particle is at rest after 1 s and after 3 s.

(d) The particle moves in the positive direction when $v(t) > 0$, that is,

$$3t^2 - 12t + 9 = 3(t - 1)(t - 3) > 0$$

This inequality is true when both factors are positive ($t > 3$) or when both factors are negative ($t < 1$). Thus, the particle moves in the positive direction in the time intervals $t < 1$ and $t > 3$. It moves in the negative direction when $1 < t < 3$.

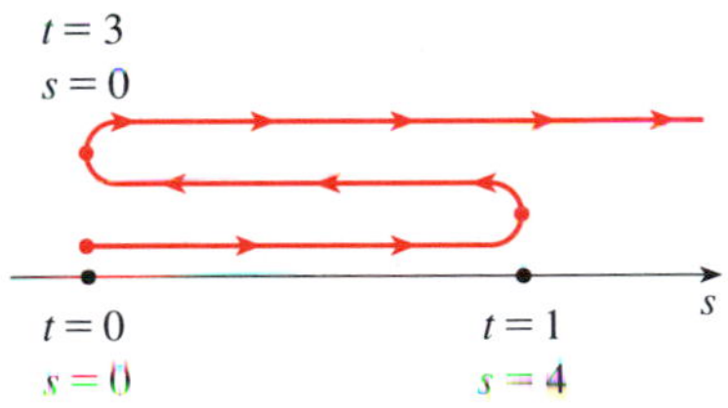

FIGURE 2

(e) The motion of the particle is illustrated schematically in Figure 2.

(f) The distance traveled in the first second is

$$|f(1) - f(0)| = |4 - 0| = 4 \text{ m}$$

From $t = 1$ to $t = 3$ the distance traveled is

$$|f(3) - f(1)| = |0 - 4| = 4 \text{ m}$$

From $t = 3$ to $t = 5$ the distance traveled is

$$|f(5) - f(3)| = |20 - 0| = 20 \text{ m}$$

The total distance is $4 + 4 + 20 = 28$ m. ■

EXAMPLE 2 If a rod or piece of wire is *homogeneous,* then its linear density is uniform and is defined as the mass per unit length ($\rho = m/l$) and measured in kilograms per meter. Suppose, however, that the rod is not homogeneous but that its mass measured from its left end to a point x is $m = f(x)$ as shown in Figure 3.

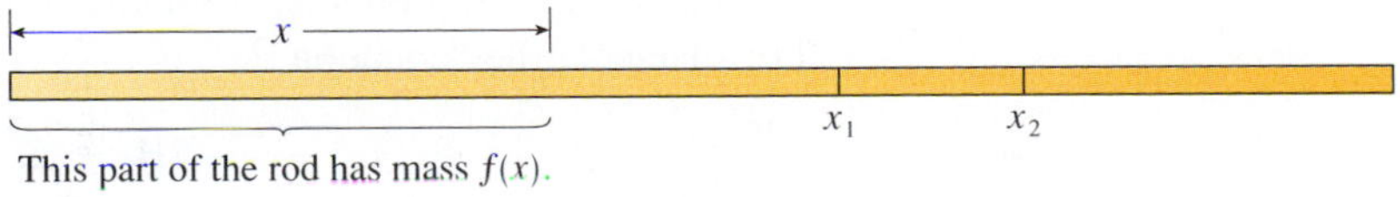

FIGURE 3

The mass of the part of the rod that lies between $x = x_1$ and $x = x_2$ is given by $\Delta m = f(x_2) - f(x_1)$, so the average density of that part of the rod is

$$\text{average density} = \frac{\Delta m}{\Delta x} = \frac{f(x_2) - f(x_1)}{x_2 - x_1}$$

If we now let $\Delta x \to 0$ (that is, $x_2 \to x_1$), we are computing the average density over a smaller and smaller interval. The **linear density** ρ at x_1 is the limit of these average densities as $\Delta x \to 0$; that is, the linear density is the rate of change of mass with respect to length. Symbolically,

$$\rho = \lim_{\Delta x \to 0} \frac{\Delta m}{\Delta x} = \frac{dm}{dx}$$

Thus, the linear density of the rod is the derivative of mass with respect to length.

For instance, if $m = f(x) = \sqrt{x}$, where x is measured in meters and m in kilograms, then the average density of the part of the rod given by $1 \leqslant x \leqslant 1.2$ is

$$\frac{\Delta m}{\Delta x} = \frac{f(1.2) - f(1)}{1.2 - 1} = \frac{\sqrt{1.2} - 1}{0.2} \approx 0.48 \text{ kg/m}$$

while the density right at $x = 1$ is

$$\rho = \left.\frac{dm}{dx}\right|_{x=1} = \left.\frac{1}{2\sqrt{x}}\right|_{x=1} = 0.50 \text{ kg/m}$$ ■

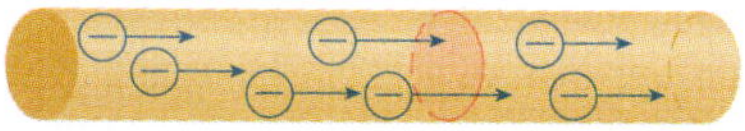

FIGURE 4

EXAMPLE 3 A current exists whenever electric charges move. Figure 4 shows part of a wire and electrons moving through a shaded plane surface. If ΔQ is the net charge that passes through this surface during a time period Δt, then the average current during this time interval is defined as

$$\text{average current} = \frac{\Delta Q}{\Delta t} = \frac{Q_2 - Q_1}{t_2 - t_1}$$

If we take the limit of this average current over smaller and smaller time intervals, we get what is called the **current** I at a given time t_1:

$$I = \lim_{\Delta t \to 0} \frac{\Delta Q}{\Delta t} = \frac{dQ}{dt}$$

Thus, the current is the rate at which charge flows through a surface. ■

Velocity, density, and current are not the only rates of change that are important in physics. Others include power (the rate at which work is done), the rate of heat flow, temperature gradient (the rate of change of temperature with respect to position), and the rate of decay of a radioactive substance in nuclear physics.

CHEMISTRY

EXAMPLE 4 A chemical reaction results in the formation of one or more substances (called products) from one or more starting materials (called reactants). For instance, the "equation"

$$2H_2 + O_2 \longrightarrow 2H_2O$$

indicates that two molecules of hydrogen and one molecule of oxygen form two molecules of water. Let's consider the reaction

$$A + B \longrightarrow C$$

where A and B are the reactants and C is the product. The **concentration** of a reactant A is the number of moles (6.022×10^{23} molecules) per liter and is denoted by [A]. The concentration varies during a reaction, so [A], [B], and [C] are all functions of time (t). The average rate of reaction of the product C over a time interval $t_1 \leq t \leq t_2$ is

$$\frac{\Delta[\mathrm{C}]}{\Delta t} = \frac{[\mathrm{C}](t_2) - [\mathrm{C}](t_1)}{t_2 - t_1}$$

But chemists are more interested in the **instantaneous rate of reaction**, which is obtained by taking the limit of the average rate of reaction as the time interval Δt approaches 0:

$$\text{rate of reaction} = \lim_{\Delta t \to 0} \frac{\Delta[\mathrm{C}]}{\Delta t} = \frac{d[\mathrm{C}]}{dt}$$

Since the concentration of the product increases as the reaction proceeds, the derivative $d[\mathrm{C}]/dt$ will be positive. (You can see intuitively that the slope of the tangent to the graph of an increasing function is positive.) Thus, the rate of reaction of C is positive. The concentrations of the reactants, however, decrease during the reaction, so, to make the rates of reaction of A and B positive numbers, we put minus signs in front of the derivatives $d[\mathrm{A}]/dt$ and $d[\mathrm{B}]/dt$. Since [A] and [B] each decrease at the same rate that [C] increases, we have

$$\text{rate of reaction} = \frac{d[\mathrm{C}]}{dt} = -\frac{d[\mathrm{A}]}{dt} = -\frac{d[\mathrm{B}]}{dt}$$

More generally, it turns out that for a reaction of the form

$$a\mathrm{A} + b\mathrm{B} \longrightarrow c\mathrm{C} + d\mathrm{D}$$

we have

$$-\frac{1}{a}\frac{d[\mathrm{A}]}{dt} = -\frac{1}{b}\frac{d[\mathrm{B}]}{dt} = \frac{1}{c}\frac{d[\mathrm{C}]}{dt} = \frac{1}{d}\frac{d[\mathrm{D}]}{dt}$$

The rate of reaction can be determined by graphical methods (see Exercise 20). In some cases we can use the rate of reaction to find explicit formulas for the concentrations as functions of time (see Exercise 7 in Section 4.5 and Exercise 31 in Section 9.1). ■

EXAMPLE 5 One of the quantities of interest in thermodynamics is *compressibility*. If a given substance is kept at a constant temperature, then its volume V depends on its pressure P. We can consider the rate of change of volume with respect to pressure—namely, the derivative dV/dP. As P increases, V decreases, so $dV/dP < 0$. The compressibility is defined by introducing a minus sign and dividing this derivative by the volume V:

$$\text{isothermal compressibility} = \beta = -\frac{1}{V}\frac{dV}{dP}$$

Thus, β measures how fast, per unit volume, the volume of a substance decreases as the pressure on it increases at constant temperature.

For instance, the volume V (in cubic meters) of a sample of air at 25 °C was found to be related to the pressure P (in kilopascals) by the equation

$$V = \frac{5.3}{P}$$

The rate of change of V with respect to P when $P = 50$ kPa is

$$\left.\frac{dV}{dP}\right|_{P=50} = -\left.\frac{5.3}{P^2}\right|_{P=50}$$
$$= -\frac{5.3}{2500} = -0.00212 \text{ m}^3/\text{kPa}$$

The compressibility at that pressure is

$$\beta = -\left.\frac{1}{V}\frac{dV}{dP}\right|_{P=50} = \frac{0.00212}{\frac{5.3}{50}} = 0.02 \ (\text{m}^3/\text{kPa})/\text{m}^3$$

■

BIOLOGY

EXAMPLE 6 Let $n = f(t)$ be the number of individuals in an animal or plant population at time t. The change in the population size between the times $t = t_1$ and $t = t_2$ is $\Delta n = f(t_2) - f(t_1)$, and so the average rate of growth during the time period $t_1 \leqslant t \leqslant t_2$ is

$$\text{average rate of growth} = \frac{\Delta n}{\Delta t} = \frac{f(t_2) - f(t_1)}{t_2 - t_1}$$

The **instantaneous rate of growth** is obtained from this average rate of growth by letting the time period Δt approach 0:

$$\text{growth rate} = \lim_{\Delta t \to 0} \frac{\Delta n}{\Delta t} = \frac{dn}{dt}$$

Strictly speaking, this is not quite accurate because the actual graph of a population function $n = f(t)$ would be a step function that is discontinuous whenever a birth or death occurs and therefore not differentiable. However, for a large animal or plant population, we can replace the graph by a smooth approximating curve as in Figure 5.

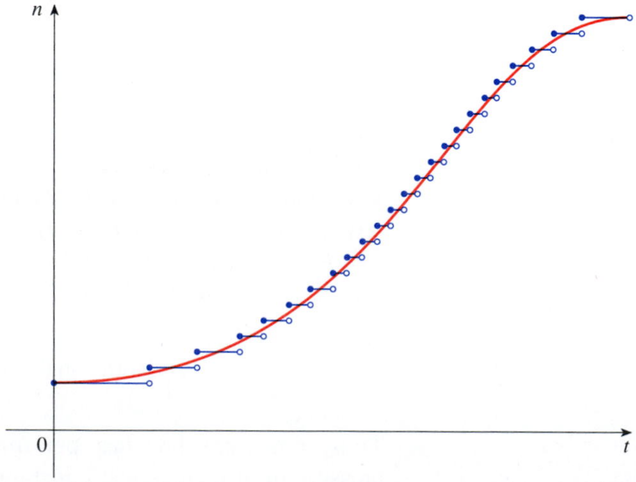

FIGURE 5
A smooth curve approximating a growth function

To be more specific, consider a population of bacteria in a homogeneous nutrient medium. Suppose that by sampling the population at certain intervals it is determined that the population doubles every hour. If the initial population is n_0 and the time t is measured in hours, then

$$f(1) = 2f(0) = 2n_0$$

$$f(2) = 2f(1) = 2^2 n_0$$

$$f(3) = 2f(2) = 2^3 n_0$$

and, in general,

$$f(t) = 2^t n_0$$

The population function is $n = n_0 2^t$.

This is an example of an exponential function. In Chapter 4 we will discuss exponential functions in general and at that time we will be able to compute their derivatives and thereby determine the rate of growth of the bacteria population. ■

EXAMPLE 7 When we consider the flow of blood through a blood vessel, such as a vein or artery, we can take the shape of the blood vessel to be a cylindrical tube with radius R and length l as illustrated in Figure 6.

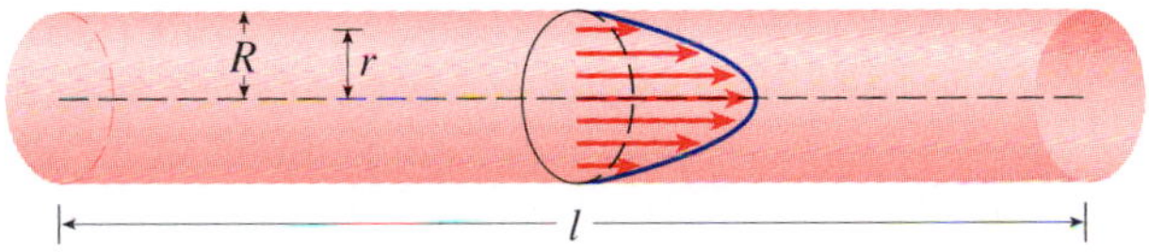

FIGURE 6
Blood flow in an artery

Because of friction at the walls of the tube, the velocity v of the blood is greatest along the central axis of the tube and decreases as the distance r from the axis increases until v becomes 0 at the wall. The relationship between v and r is given by the **law of laminar flow** discovered by the French physician Poiseuille in 1840. This states that

$$v = \frac{P}{4\eta l}(R^2 - r^2) \tag{1}$$

where η is the viscosity of the blood and P is the pressure difference between the ends of the tube. If P and l are constant, then v is a function of r with domain $[0, R]$. [For more detailed information see D. A. McDonald, *Blood Flow in Arteries* (London: Arnold, 1960).]

The average rate of change of the velocity as we move from $r = r_1$ outward to $r = r_2$ is

$$\frac{\Delta v}{\Delta r} = \frac{v(r_2) - v(r_1)}{r_2 - r_1}$$

and if we let $\Delta r \to 0$, we obtain the instantaneous rate of change of velocity with respect to r:

$$\text{velocity gradient} = \lim_{\Delta r \to 0} \frac{\Delta v}{\Delta r} = \frac{dv}{dr}$$

Using Equation 1, we obtain

$$\frac{dv}{dr} = \frac{P}{4\eta l}(0 - 2r) = -\frac{Pr}{2\eta l}$$

In a typical human artery we can take $\eta = 0.027$, $R = 0.008$ cm, $l = 2$ cm, and $P = 4000$ dynes/cm^2, which gives

$$\begin{aligned} v &= \frac{4000}{4(0.027)2}(0.000064 - r^2) \\ &\approx 1.85 \times 10^4(6.4 \times 10^{-5} - r^2) \end{aligned}$$

At $r = 0.002$ cm the blood is flowing at a speed of

$$\begin{aligned} v(0.002) &\approx 1.85 \times 10^4(64 \times 10^{-6} - 4 \times 10^{-6}) \\ &= 1.11 \text{ cm/s} \end{aligned}$$

and the velocity gradient at that point is

$$\left.\frac{dv}{dr}\right|_{r=0.002} = -\frac{4000(0.002)}{2(0.027)2} \approx -74 \text{ (cm/s)/cm}$$

■

ECONOMICS

EXAMPLE 8 Suppose $C(x)$ is the total cost that a company incurs in producing x units of a certain commodity. The function C is called a **cost function**. If the number of items produced is increased from x_1 to x_2, the additional cost is $\Delta C = C(x_2) - C(x_1)$, and the average rate of change of the cost is

$$\frac{\Delta C}{\Delta x} = \frac{C(x_2) - C(x_1)}{x_2 - x_1} = \frac{C(x_1 + \Delta x) - C(x_1)}{\Delta x}$$

The limit of this quantity as $\Delta x \to 0$, that is, the instantaneous rate of change of cost with respect to the number of items produced, is called the **marginal cost** by economists:

$$\text{marginal cost} = \lim_{\Delta x \to 0} \frac{\Delta C}{\Delta x} = \frac{dC}{dx}$$

[Since x can usually take on only integer values, it may not make literal sense to let Δx approach 0, but we can always replace $C(x)$ by a smooth approximating function as in Example 6.]

Taking $\Delta x = 1$ and n large (so that Δx is small compared to n), we have

$$C'(n) \approx C(n + 1) - C(n)$$

Thus, the marginal cost of producing n units is approximately equal to the cost of producing one more unit [the $(n + 1)$st unit].

It is often appropriate to represent a total cost function by a polynomial

$$C(x) = a + bx + cx^2 + dx^3$$

where a represents the overhead cost (rent, heat, maintenance) and the other terms represent the cost of raw materials, labor, and so on. (The cost of raw materials may be proportional to x, but labor costs might depend partly on higher powers of x because of overtime costs and inefficiencies involved in large-scale operations.)

For instance, suppose a company has estimated that the cost (in dollars) of producing x items is

$$C(x) = 10{,}000 + 5x + 0.01x^2$$

Then the marginal cost function is

$$C'(x) = 5 + 0.02x$$

The marginal cost at the production level of 500 items is

$$C'(500) = 5 + (0.02)500 = \$15/\text{item}$$

This gives the rate at which costs are increasing with respect to the production level when $x = 500$.

The cost of producing the 501st item is

$$\begin{aligned} C(501) - C(500) &= [10{,}000 + 5(501) + 0.01(501)^2] \\ &\quad - [10{,}000 + 5(500) + 0.01(500)^2] \\ &= \$15.01 \end{aligned}$$

Notice that $C'(500) \approx C(501) - C(500)$. ■

Economists also study marginal demand, marginal revenue, and marginal profit, which are the derivatives of the demand, revenue, and profit functions. These will be considered in Chapter 5 after we have developed techniques for finding the maximum and minimum values of functions.

OTHER SCIENCES

Rates of change occur in all the sciences. A geologist is interested in knowing the rate at which an intruded body of molten rock cools by conduction of heat into surrounding rocks. An engineer wants to know the rate at which water flows into or out of a reservoir. An urban geographer is interested in the rate of change of the population density in a city as the distance from the city center increases. A meteorologist is concerned with the rate of change of atmospheric pressure with respect to height. (See Exercise 17 in Section 4.5.)

In psychology, those interested in learning theory study the so-called learning curve, which graphs the performance $P(t)$ of someone learning a skill as a function of the training time t. Of particular interest is the rate at which performance improves as time passes, that is, dP/dt.

In sociology, differential calculus is used in analyzing the spread of rumors (or innovations or fads or fashions). If $p(t)$ denotes the proportion of a population that knows a rumor by time t, then the derivative dp/dt represents the rate of spread of the rumor. (See Exercise 55 in Section 4.1.)

SUMMARY

Velocity, density, current, power, and temperature gradient in physics, rate of reaction and compressibility in chemistry, rate of growth and blood velocity gradient in biology, marginal cost and marginal profit in economics, rate of heat flow in geology, rate of improvement of performance in psychology, rate of spread of a rumor in sociology—these are all special cases of a single mathematical concept, the derivative.

This is an illustration of the fact that part of the power of mathematics lies in its abstractness. A single abstract mathematical concept (such as the derivative) can have different interpretations in each of the sciences. When we develop the properties of the mathematical concept once and for all, we can then turn around and apply these results to all of the sciences. This is much more efficient than developing properties of special concepts in each separate science. The French mathematician Joseph Fourier (1768–1830) put it succinctly: "Mathematics compares the most diverse phenomena and discovers the secret analogies that unite them."

EXERCISES 3.3

1–6 ■ A particle moves according to a law of motion $s = f(t)$, $t \geq 0$, where t is measured in seconds and s in feet.
(a) Find the velocity at time t.
(b) What is the velocity after 2 s?
(c) When is the particle at rest?
(d) When is the particle moving in the positive direction?
(e) Find the total distance traveled during the first 4 s.
(f) Draw a diagram like Figure 2 to illustrate the motion of the particle.

1. $f(t) = t^2 - 6t + 9$

2. $f(t) = 4t^3 - 9t^2 + 6t + 2$

3. $f(t) = 2t^3 - 9t^2 + 12t + 1$

4. $f(t) = t^4 - 4t + 1$

5. $s = \dfrac{t}{t^2 + 1}$

6. $s = \sqrt{t}\,(5 - 5t + 2t^2)$

7. The position function of a particle is given by $s = t^3 - 4.5t^2 - 7t$, $t \geq 0$. When does the particle reach a velocity of 5 m/s?

8. If a ball is thrown vertically upward with a velocity of 80 ft/s, then its height after t seconds is $s = 80t - 16t^2$.
(a) What is the maximum height reached by the ball?
(b) What is the velocity of the ball when it is 96 ft above the ground on its way up? On its way down?

9. (a) Find the average rate of change of the volume of a cube with respect to its edge length x as x changes from
(i) 5 to 6 (ii) 5 to 5.1 (iii) 5 to 5.01
(b) Find the instantaneous rate of change when $x = 5$.
(c) Show that the rate of change of the volume of a cube with respect to its edge length (at any x) is equal to half the surface area of the cube.

10. (a) Find the average rate of change of the area of a circle with respect to its radius r as r changes from
(i) 2 to 3 (ii) 2 to 2.5 (iii) 2 to 2.1
(b) Find the instantaneous rate of change when $r = 2$.
(c) Show that the rate of change of the area of a circle with respect to its radius (at any r) is equal to the circumference of the circle.

11. A stone is dropped into a lake, creating a circular ripple that travels outward at a speed of 60 cm/s. Find the rate at which the area within the circle is increasing after (a) 1 s, (b) 3 s, and (c) 5 s.

12. (a) The volume of a growing spherical cell is $V = \frac{4}{3}\pi r^3$, where the radius r is measured in micrometers ($1\ \mu m = 10^{-6}$ m). Find the average rate of change of V with respect to r when r changes from
(i) 5 to 8 μm (ii) 5 to 6 μm (iii) 5 to 5.1 μm
(b) Find the instantaneous rate of change of V with respect to r when $r = 5\ \mu$m.

13. A spherical balloon is being inflated. Find the rate of increase of the surface area ($S = 4\pi r^2$) with respect to the radius r when r is (a) 1 ft, (b) 2 ft, and (c) 3 ft.

14. Show that the rate of change of the volume of a sphere with respect to its radius is equal to its surface area.

15. The mass of the part of a metal rod that lies between its left end and a point x meters to the right is $3x^2$ kg. Find the linear density (see Example 2) when x is (a) 1 m, (b) 2 m, and (c) 3 m.

16. If a tank holds 5000 gallons of water, which drains from the bottom of the tank in 40 min, then Torricelli's Law gives the volume V of water remaining in the tank after t minutes as

$$V = 5000\left(1 - \frac{t}{40}\right)^2 \qquad 0 \leq t \leq 40$$

Find the rate at which water is draining from the tank after (a) 5 min, (b) 10 min, and (c) 20 min.

17. The quantity of charge Q in coulombs (C) that has passed through a surface up to time t (measured in seconds) is given by $Q(t) = t^3 - 2t^2 + 6t + 2$. Find the current when (a) $t = 0.5$ s and (b) $t = 1$ s. [See Example 3. The unit of current is an ampere (1 A = 1 C/s).]

18. Newton's Law of Gravitation says that the magnitude F of the force exerted by a body of mass m on a body of mass M is

$$F = \frac{GmM}{r^2}$$

where G is the gravitational constant and r is the distance between the bodies. If the bodies are moving, find the rate of change of F with respect to r.

19. Boyle's Law states that when a sample of gas is compressed at a constant temperature, the product of the pressure and the volume remains constant: $PV = C$.
(a) Find the rate of change of volume with respect to pressure.
(b) Prove that the isothermal compressibility (see Example 5) is given by $\beta = 1/P$.

20. The data in the following table concern the lactonization of hydroxyvaleric acid at 25 °C. They give the concentration $C(t)$ of this acid in moles per liter after t minutes.

t	0	2	4	6	8
$C(t)$	0.0800	0.0570	0.0408	0.0295	0.0210

(a) Find the average rate of reaction for the following time intervals:
(i) $2 \leq t \leq 6$ (ii) $2 \leq t \leq 4$ (iii) $0 \leq t \leq 2$
(b) Plot the points from the table and draw a smooth curve through them as an approximation to the graph of the concentration function. Then draw the tangent at $t = 2$ and use it to estimate the instantaneous rate of reaction when $t = 2$.

21. If, in Example 4, one molecule of the product C is formed from one molecule of the reactant A and one molecule of the reactant B, and the initial concentrations of A and B have a common value $[A] = [B] = a$ moles/L, then $[C] = a^2kt/(akt + 1)$, where k is a constant.
(a) Find the rate of reaction at time t.
(b) Show that if $x = [C]$, then

$$\frac{dx}{dt} = k(a - x)^2$$

22. If f is the focal length of a convex lens and an object is placed at a distance p from the lens, then its image will be at a distance q from the lens, where f, p, and q are related by the *lens equation*

$$\frac{1}{f} = \frac{1}{p} + \frac{1}{q}$$

Find the rate of change of p with respect to q.

23. The mass of glucose in a metabolic experiment decreased according to the equation $m = 5 - (0.02)t^2$, where t is measured in hours. Find the rate of change of the amount of glucose after 1 h.

24. The population of a slowly growing bacterial colony after t hours is given by $n = 100 + 24t + 2t^2$. Find the growth rate after 2 h.

25. Use the law of laminar flow (see Example 7) to find the velocity gradient at a point 0.005 cm from the central axis of a blood vessel with radius 0.01 cm, length 3 cm, pressure difference 3000 dynes/cm^2, and viscosity $\eta = 0.027$.

26. If R denotes the reaction of the body to some stimulus of strength x, the *sensitivity* S is defined to be the rate of change of the reaction with respect to x. A particular example is that when the brightness x of a light source is increased, the eye reacts by decreasing the area R of the pupil. The experimental formula

$$R = \frac{40 + 24x^{0.4}}{1 + 4x^{0.4}}$$

describes the dependence of R on x when R is measured in square millimeters and x is measured in appropriate units of brightness.
(a) Find the sensitivity.
(b) Illustrate part (a) by graphing both R and S as functions of x. Comment on the values of R and S at low levels of brightness. Is this what you would expect?

27–30 ■ A cost function is given for a certain commodity. Find the marginal cost function. Then compare the marginal cost at the production level of 100 units with the cost of producing the 101st item.

27. $C(x) = 420 + 1.5x + 0.002x^2$

28. $C(x) = 1200 + \dfrac{x}{10} + \dfrac{x^2}{10{,}000}$

29. $C(x) = 2000 + 3x + 0.01x^2 + 0.0002x^3$

30. $C(x) = 2500 + 2\sqrt{x}$

3.4 DERIVATIVES OF TRIGONOMETRIC FUNCTIONS

A review of the trigonometric functions is given in Appendix D.

Before starting this section, you might need to review the trigonometric functions. In particular it is important to remember that when we talk about the function f defined for all real numbers x by

$$f(x) = \sin x$$

it is understood that $\sin x$ means the sine of the angle whose *radian* measure is x. A similar convention holds for the other trigonometric functions cos, tan, csc, sec, and cot.

In order to compute the derivatives of these functions, we first have to evaluate some trigonometric limits.

(1) THEOREM $$\lim_{\theta \to 0} \sin\theta = 0$$

If we let $f(\theta) = \sin\theta$, then $f(0) = \sin 0 = 0$ and so Theorem 1 tells us that $\lim_{\theta \to 0} f(\theta) = f(0)$. This is precisely the statement that the sine function is continuous at 0.

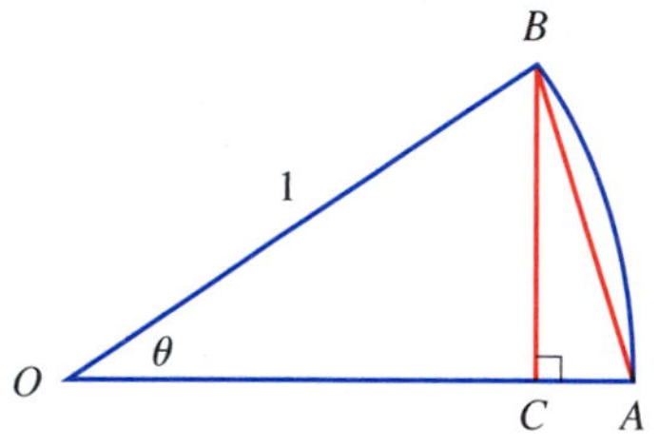

FIGURE 1

PROOF In calculating $\lim_{\theta \to 0^+} \sin\theta$, we may assume that $0 < \theta < \pi/2$. Figure 1 shows a sector of a circle with center O, central angle θ, and radius 1. BC is drawn perpendicular to OA. By the definition of radian measure, we have arc $AB = \theta$. Also $|BC| = |OB| \sin\theta = \sin\theta$. From the diagram we see that

$$|BC| < |AB| < \text{arc } AB$$

Therefore

$$0 < \sin\theta < \theta \tag{2}$$

Since we know that $\lim_{\theta \to 0^+} 0 = 0$ and $\lim_{\theta \to 0^+} \theta = 0$, it follows from the Squeeze Theorem that

$$\lim_{\theta \to 0^+} \sin\theta = 0$$

If $-\pi/2 < \theta < 0$, then $0 < -\theta < \pi/2$, so by (2) we have

$$0 < \sin(-\theta) < -\theta$$

or

$$0 < -\sin\theta < -\theta$$

which implies

$$\theta < \sin\theta < 0$$

This inequality, together with the fact that $\lim_{\theta \to 0^-} \theta = 0$ and the Squeeze Theorem, shows that $\lim_{\theta \to 0^-} \sin\theta = 0$. Thus

$$\lim_{\theta \to 0} \sin\theta = 0$$ □

(3) COROLLARY $$\lim_{\theta \to 0} \cos\theta = 1$$

Again, by the definition of continuity and the fact that $\cos 0 = 1$, Corollary 3 says that the cosine function is continuous at 0.

PROOF Using $\sin^2\theta + \cos^2\theta = 1$ together with $\cos\theta \geqslant 0$ for $-\pi/2 \leqslant \theta \leqslant \pi/2$, we have $\cos\theta = \sqrt{1 - \sin^2\theta}$ for those values of θ. Thus, using the properties of limits, we have

$$\begin{aligned}\lim_{\theta \to 0} \cos\theta &= \lim_{\theta \to 0} \sqrt{1 - \sin^2\theta} \\ &= \sqrt{\lim_{\theta \to 0} (1 - \sin^2\theta)}\end{aligned}$$

$$= \sqrt{1 - 0} \qquad \text{(by Theorem 1)}$$

$$= 1$$ □

In Example 3 in Section 2.2 we made the guess that $\lim_{x \to 0} (\sin x)/x = 1$. We are now in a position to prove that the guess is correct.

(4) THEOREM $$\lim_{\theta \to 0} \frac{\sin \theta}{\theta} = 1$$

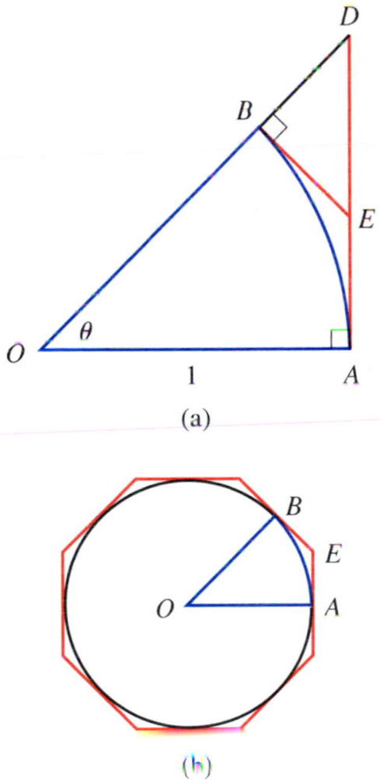

FIGURE 2

PROOF First suppose that $0 < \theta < \pi/2$. Again, Figure 2(a) shows a sector of a circle with center O, central angle θ, and radius 1. Let the tangents at A and B intersect at E. You can see from Figure 2(b) that the circumference of a circle is smaller than the length of a circumscribed polygon, so arc $AB < |AE| + |EB|$. Thus

$$\begin{aligned} \theta = \text{arc } AB &< |AE| + |EB| \\ &< |AE| + |ED| \\ &= |AD| = |OA| \tan \theta \\ &= \tan \theta \end{aligned}$$

(In Appendix F the inequality $\theta \leqslant \tan \theta$ is proved directly from the definition of the length of an arc without resorting to geometric intuition as we did above.) Therefore, we have

$$\theta < \frac{\sin \theta}{\cos \theta}$$

or $$\cos \theta < \frac{\sin \theta}{\theta}$$

From the proof of Theorem 1 we have $\sin \theta < \theta$, so

(5) $$\cos \theta < \frac{\sin \theta}{\theta} < 1$$

If $-\pi/2 < \theta < 0$, then $0 < -\theta < \pi/2$, so (5) gives

(6) $$\cos(-\theta) < \frac{\sin(-\theta)}{-\theta} < 1$$

But $\cos(-\theta) = \cos \theta$ and $\sin(-\theta) = -\sin \theta$, so (6), when simplified, becomes (5). Thus (5) holds whenever $\theta \in (-\pi/2, \pi/2)$ and $\theta \neq 0$. Also, we know that $\lim_{\theta \to 0} 1 = 1$ and $\lim_{\theta \to 0} \cos \theta = 1$ by Corollary 3. Hence, by the Squeeze Theorem,

$$\lim_{\theta \to 0} \frac{\sin \theta}{\theta} = 1$$ □

(7) COROLLARY $$\lim_{\theta \to 0} \frac{\cos \theta - 1}{\theta} = 0$$

PROOF To put the function in a form in which we can use the limits we know, we multiply numerator and denominator by $\cos\theta + 1$:

$$\lim_{\theta\to 0}\frac{\cos\theta - 1}{\theta} = \lim_{\theta\to 0}\left[\frac{\cos\theta - 1}{\theta}\cdot\frac{\cos\theta + 1}{\cos\theta + 1}\right] = \lim_{\theta\to 0}\frac{\cos^2\theta - 1}{\theta(\cos\theta + 1)}$$

$$= \lim_{\theta\to 0}\frac{-\sin^2\theta}{\theta(\cos\theta + 1)} = -\lim_{\theta\to 0}\frac{\sin\theta}{\theta}\cdot\frac{\sin\theta}{\cos\theta + 1}$$

$$= -\lim_{\theta\to 0}\frac{\sin\theta}{\theta}\cdot\lim_{\theta\to 0}\frac{\sin\theta}{\cos\theta + 1} = -1\cdot\left(\frac{0}{1+1}\right) = 0$$

Here we have used Theorem 4 together with Theorem 1 and Corollary 3. □

Another method for proving Corollary 7 is outlined in Exercise 41.

EXAMPLE 1 Find $\lim_{x\to 0}\dfrac{\sin 7x}{4x}$.

SOLUTION In order to apply Theorem 4, we first rewrite the function as follows:

Note that $\sin 7x \neq 7\sin x$.

$$\frac{\sin 7x}{4x} = \frac{7}{4}\left(\frac{\sin 7x}{7x}\right)$$

Notice that as $x \to 0$, we have $7x \to 0$, and so, by Theorem 4 with $\theta = 7x$,

$$\lim_{x\to 0}\frac{\sin 7x}{7x} = \lim_{7x\to 0}\frac{\sin(7x)}{7x} = 1$$

Thus

$$\lim_{x\to 0}\frac{\sin 7x}{4x} = \lim_{x\to 0}\frac{7}{4}\left(\frac{\sin 7x}{7x}\right)$$

$$= \frac{7}{4}\lim_{x\to 0}\frac{\sin 7x}{7x} = \frac{7}{4}\cdot 1 = \frac{7}{4}$$ ■

Calculate $\lim_{x\to 0} x\cot x$.

$$\lim_{x\to 0} x\cot x = \lim_{x\to 0} x\,\frac{\cos x}{\sin x}$$

$$= \lim_{x\to 0}\frac{\cos x}{\dfrac{\sin x}{x}} = \frac{\lim_{x\to 0}\cos x}{\lim_{x\to 0}\dfrac{\sin x}{x}}$$

$$= \tfrac{1}{1} \qquad \text{(by Corollary 3 and Theorem 4)}$$

$$= 1$$ ■

DERIVATIVES

If you sketch the graph of the function $f(x) = \sin x$ and use the interpretation of $f'(x)$ as the slope of the tangent to the sine curve in order to sketch the graph of f' (see Exercise 44 in Section 3.1), then it looks as if the graph of f' may be the same as the cosine curve (see Figure 3). This is confirmed in the next theorem by using the definition of derivative to calculate $f'(x)$.

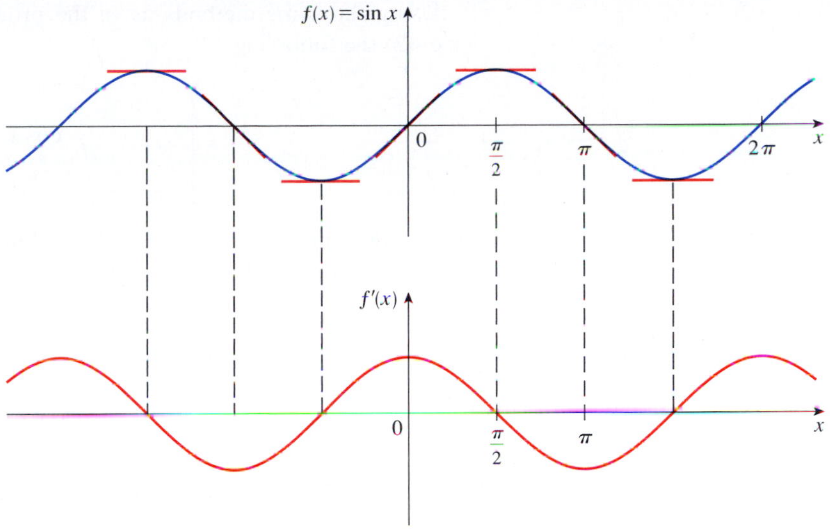

FIGURE 3

(8)

$$\frac{d}{dx} \sin x = \cos x$$

PROOF If $f(x) = \sin x$, then

$$\begin{aligned} f'(x) &= \lim_{h \to 0} \frac{f(x+h) - f(x)}{h} \\ &= \lim_{h \to 0} \frac{\sin(x+h) - \sin x}{h} \\ &= \lim_{h \to 0} \frac{\sin x \cos h + \cos x \sin h - \sin x}{h} \\ &= \lim_{h \to 0} \left[\sin x \left(\frac{\cos h - 1}{h} \right) + \cos x \left(\frac{\sin h}{h} \right) \right] \\ &= \lim_{h \to 0} \sin x \cdot \lim_{h \to 0} \frac{\cos h - 1}{h} + \lim_{h \to 0} \cos x \cdot \lim_{h \to 0} \frac{\sin h}{h} \\ &= \sin x \cdot 0 + \cos x \cdot 1 \quad \text{(by Corollary 7 and Theorem 4)} \\ &= \cos x \end{aligned}$$

□

We've used the addition formula for sine. See Appendix D.

Figure 4 shows the graphs of the function of Example 3 and its derivative. Notice that $y' = 0$ whenever y has a horizontal tangent.

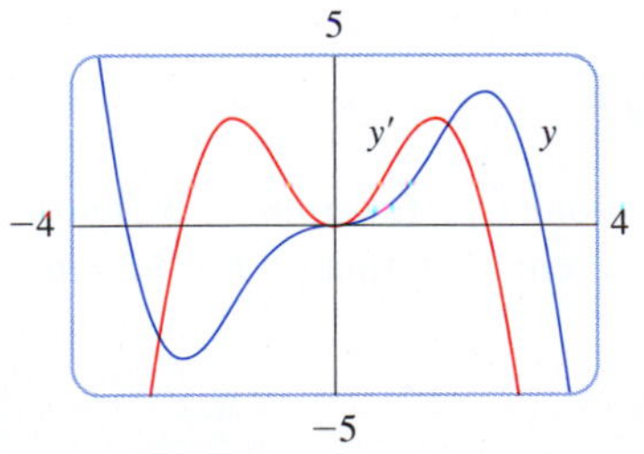

FIGURE 4

EXAMPLE 3 Differentiate $y = x^2 \sin x$.

SOLUTION Using the Product Rule and Theorem 8, we have

$$\begin{aligned} \frac{dy}{dx} &= x^2 \frac{d}{dx} \sin x + \sin x \frac{d}{dx} (x^2) \\ &= x^2 \cos x + 2x \sin x \end{aligned}$$

■

Using the same methods as in the proof of Theorem 8, one can prove (see Exercise 42) the following.

(9)
$$\frac{d}{dx}\cos x = -\sin x$$

The tangent function can also be differentiated from first principles, but it is easier to use the Quotient Rule together with Equations 8 and 9:

$$\begin{aligned}\frac{d}{dx}\tan x &= \frac{d}{dx}\left(\frac{\sin x}{\cos x}\right)\\ &= \frac{\cos x\, D\sin x - \sin x\, D\cos x}{\cos^2 x}\\ &= \frac{\cos x \cdot \cos x - \sin x\,(-\sin x)}{\cos^2 x}\\ &= \frac{\cos^2 x + \sin^2 x}{\cos^2 x}\\ &= \frac{1}{\cos^2 x} = \sec^2 x\end{aligned}$$

(10)
$$\frac{d}{dx}\tan x = \sec^2 x$$

The derivatives of the remaining trigonometric functions, csc, sec, and cot, can also be found easily using the Quotient Rule (see Exercises 17–19). We collect all the differentiation formulas for trigonometric functions in the following table.

When you memorize this table it is helpful to notice that the minus signs go with the derivatives of the "cofunctions," that is, cosine, cosecant, and cotangent.

TABLE OF DERIVATIVES OF TRIGONOMETRIC FUNCTIONS

$\frac{d}{dx}(\sin x) = \cos x$	$\frac{d}{dx}(\csc x) = -\csc x \cot x$
$\frac{d}{dx}(\cos x) = -\sin x$	$\frac{d}{dx}(\sec x) = \sec x \tan x$
$\frac{d}{dx}(\tan x) = \sec^2 x$	$\frac{d}{dx}(\cot x) = -\csc^2 x$

These differentiation formulas show that each trigonometric function is differentiable at every number in its domain. It follows from Theorem 3.1.8 that *all the trigonometric functions are continuous on their domains.*

In particular, this means that, for all values of a,

$$\lim_{x\to a}\sin x = \sin a \qquad \text{and} \qquad \lim_{x\to a}\cos x = \cos a$$

EXAMPLE 4 Differentiate $f(x) = \dfrac{\sec x}{1 + \tan x}$.

SOLUTION The Quotient Rule gives

$$f'(x) = \frac{(1 + \tan x)\, D \sec x - \sec x\, D(1 + \tan x)}{(1 + \tan x)^2}$$

$$= \frac{(1 + \tan x) \sec x \tan x - \sec x \cdot \sec^2 x}{(1 + \tan x)^2}$$

$$= \frac{\sec x\, [\tan x + \tan^2 x - \sec^2 x]}{(1 + \tan x)^2}$$

$$= \frac{\sec x\, (\tan x - 1)}{(1 + \tan x)^2}$$

In simplifying the answer we have used the identity $\tan^2 x + 1 = \sec^2 x$. ■

EXERCISES 3.4

1–16 ■ Find each limit.

1. $\lim_{x \to 0} (x^2 + \cos x)$

2. $\lim_{x \to 0} \cos(\sin x)$

3. $\lim_{x \to \pi/3} (\sin x - \cos x)$

4. $\lim_{x \to \pi} x^2 \sec x$

5. $\lim_{x \to \pi/4} \dfrac{\sin x}{3x}$

6. $\lim_{x \to 0} \dfrac{\sin x}{3x}$

7. $\lim_{t \to 0} \dfrac{\sin 5t}{t}$

8. $\lim_{t \to 0} \dfrac{\sin 8t}{\sin 9t}$

9. $\lim_{\theta \to 0} \dfrac{\sin(\cos \theta)}{\sec \theta}$

10. $\lim_{\theta \to 0} \dfrac{\cos \theta - 1}{\sin \theta}$

11. $\lim_{x \to \pi/4} \dfrac{\tan x}{4x}$

12. $\lim_{x \to 0} \dfrac{\tan x}{4x}$

13. $\lim_{\theta \to 0} \dfrac{\sin^2 \theta}{\theta}$

14. $\lim_{h \to 0} \dfrac{\sin 5h}{\tan 3h}$

15. $\lim_{x \to 0} \dfrac{\tan 3x}{3 \tan 2x}$

16. $\lim_{t \to 0} \dfrac{\sin^2 3t}{t^2}$

17. Prove that $\dfrac{d}{dx}(\csc x) = -\csc x \cot x$.

18. Prove that $\dfrac{d}{dx}(\sec x) = \sec x \tan x$.

19. Prove that $\dfrac{d}{dx}(\cot x) = -\csc^2 x$.

20–31 ■ Find $\dfrac{dy}{dx}$.

20. $y = \cos x - 2 \tan x$

21. $y = \sin x + \cos x$

22. $y = x \csc x$

23. $y = \csc x \cot x$

24. $y = \dfrac{\sin x}{1 + \cos x}$

25. $y = \dfrac{\tan x}{x}$

26. $y = \dfrac{\tan x - 1}{\sec x}$

27. $y = \dfrac{x}{\sin x + \cos x}$

28. $y = 2x(\sqrt{x} - \cot x)$

29. $y = x^{-3} \sin x \tan x$

30. $y = x \sin x \cos x$

31. $y = \dfrac{x^2 \tan x}{\sec x}$

32–34 ■ Find the equation of the tangent line to the given curve at the specified point.

32. $y = 2 \sin x, \quad (\pi/6, 1)$

33. $y = \tan x, \quad (\pi/4, 1)$

34. $y = \sec x - 2 \cos x, \quad (\pi/3, 1)$

35. (a) Find an equation of the tangent line to the curve $y = x \cos x$ at the point $(\pi, -\pi)$.
(b) Illustrate part (a) by graphing the curve and the tangent line on the same screen.

36. (a) If $f(x) = 2x + \cot x$, find $f'(x)$.
(b) Check to see that your answer to part (a) is reasonable by graphing both f and f' for $0 < x < \pi$.

37. For what values of x does the graph of $f(x) = x + 2 \sin x$ have a horizontal tangent?

38. Find the points on the curve $y = (\cos x)/(2 + \sin x)$ at which the tangent is horizontal.

39. A ladder 10 ft long rests against a vertical wall. Let θ be the angle between the top of the ladder and the wall and let

x be the distance from the bottom of the ladder to the wall. If the bottom of the ladder slides away from the wall, how fast does x change with respect to θ when $\theta = \pi/3$?

40. An object with weight W is dragged along a horizontal plane by a force acting along a rope attached to the object. If the rope makes an angle θ with the plane, then the magnitude of the force is

$$F = \frac{\mu W}{\mu \sin\theta + \cos\theta}$$

where μ is a constant called the *coefficient of friction.*
(a) Find the rate of change of F with respect to θ.
(b) When is this rate of change equal to 0?
(c) If $W = 50$ lb and $\mu = 0.6$, draw the graph of F as a function of θ and use it to locate the value of θ for which $dF/d\theta = 0$. Is the value consistent with your answer to part (b)?

41. If we put $\theta = 2x$ in the identity $\cos 2x = 1 - 2\sin^2 x$, it becomes $\cos\theta = 1 - 2\sin^2(\theta/2)$. Use this identity to give an alternative proof of Corollary 7.

42. Prove, using the definition of derivative, that if $f(x) = \cos x$, then $f'(x) = -\sin x$.

43–54 ■ Find each limit.

43. $\displaystyle\lim_{x\to 0} \frac{\cot 2x}{\csc x}$

44. $\displaystyle\lim_{x\to 0} \frac{1 - \cos x}{2x^2}$

45. $\displaystyle\lim_{x\to \pi} \frac{\tan x}{\sin 2x}$

46. $\displaystyle\lim_{x\to \pi/4} \frac{\sin x - \cos x}{\cos 2x}$

47. $\displaystyle\lim_{\theta\to 0} \frac{\sin\theta}{\theta + \tan\theta}$

48. $\displaystyle\lim_{x\to 1} \frac{\sin(x - 1)}{x^2 + x - 2}$

49. $\displaystyle\lim_{x\to 0} \frac{\cos x \sin x - \tan x}{x^2 \sin x}$

50. $\displaystyle\lim_{y\to 0} \left(\lim_{x\to 0} \frac{\cos x \sin y}{x - y} \right)$

51. $\displaystyle\lim_{x\to 0} \frac{\sin(\sin x)}{\sin x}$

52. $\displaystyle\lim_{x\to 0} \frac{\sin(\sin x)}{x}$

53. $\displaystyle\lim_{x\to\infty} \cos\frac{1}{x}$

54. $\displaystyle\lim_{x\to\infty} \left(x - x\cos\frac{1}{x} \right)$

55. Differentiate each trigonometric identity to obtain a new (or familiar) identity.

(a) $\tan x = \dfrac{\sin x}{\cos x}$ (b) $\sec x = \dfrac{1}{\cos x}$

(c) $\sin x + \cos x = \dfrac{1 + \cot x}{\csc x}$

56. A semicircle with diameter PQ sits on an isosceles triangle PQR to form a region shaped like an ice cream cone, as shown in the figure. If $A(\theta)$ is the area of the semicircle and $B(\theta)$ is the area of the triangle, find

$$\lim_{\theta\to 0^+} \frac{A(\theta)}{B(\theta)}$$

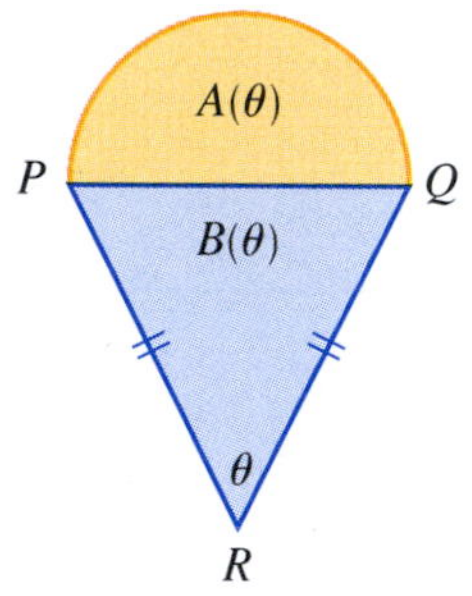

57. The figure shows a circular arc of length s and a chord of length d, both subtended by a central angle θ. Find

$$\lim_{\theta\to 0^+} \frac{s}{d}$$

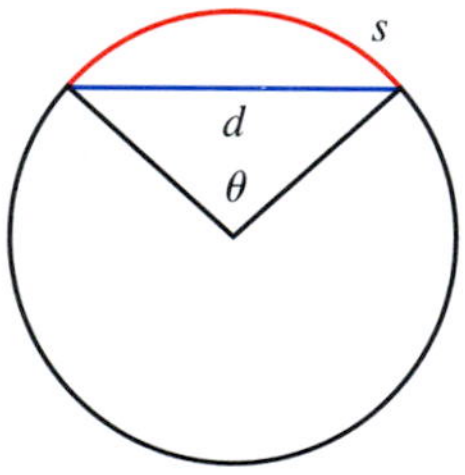

3.5 THE CHAIN RULE

Suppose you were asked to differentiate the function

$$F(x) = \sqrt{x^2 + 1}$$

The differentiation formulas you learned in the preceding section do not enable you to calculate $F'(x)$.

Observe that F is a composite function. If we let $y = f(u) = \sqrt{u}$ and let $u = g(x) = x^2 + 1$, then we can write $y = F(x) = f(g(x))$, that is, $F = f \circ g$. (See Section 1 in Review and Preview for a review of composite functions.) We know how to differentiate both f and g, so it would be useful to have a rule that tells us how to find the derivative of $F = f \circ g$ in terms of the derivatives of f and g.

It turns out that the derivative of the composite function $f \circ g$ is the product of the derivatives of f and g. This fact is one of the most important of the differentiation rules and is called the *Chain Rule*. It seems plausible if we interpret derivatives as rates of change. Regard du/dx as the rate of change of u with respect to x, dy/du as the rate of change of y with respect to u, and dy/dx as the rate of change of y with respect to x. If u changes twice as fast as x and y changes three times as fast as u, then it seems reasonable that y changes six times as fast as x, and so we expect that

$$\frac{dy}{dx} = \frac{dy}{du}\frac{du}{dx}$$

THE CHAIN RULE If the derivatives $g'(x)$ and $f'(g(x))$ both exist, and $F = f \circ g$ is the composite function defined by $F(x) = f(g(x))$, then $F'(x)$ exists and is given by the product

$$F'(x) = f'(g(x))g'(x)$$

In Leibniz notation, if $y = f(u)$ and $u = g(x)$ are both differentiable functions, then

$$\frac{dy}{dx} = \frac{dy}{du}\frac{du}{dx}$$

COMMENTS ON THE PROOF OF THE CHAIN RULE Let Δu be the change in u corresponding to a change of Δx in x, that is,

$$\Delta u = g(x + \Delta x) - g(x)$$

Then the corresponding change in y is

$$\Delta y = f(u + \Delta u) - f(u)$$

It is tempting to write

$$\begin{aligned} \frac{dy}{dx} &= \lim_{\Delta x \to 0} \frac{\Delta y}{\Delta x} \\ &= \lim_{\Delta x \to 0} \frac{\Delta y}{\Delta u} \cdot \frac{\Delta u}{\Delta x} \qquad (1) \\ &= \lim_{\Delta x \to 0} \frac{\Delta y}{\Delta u} \cdot \lim_{\Delta x \to 0} \frac{\Delta u}{\Delta x} \\ &= \lim_{\Delta u \to 0} \frac{\Delta y}{\Delta u} \cdot \lim_{\Delta x \to 0} \frac{\Delta u}{\Delta x} \\ &= \frac{dy}{du}\frac{du}{dx} \end{aligned}$$

(Note that $\Delta u \to 0$ as $\Delta x \to 0$ since g is continuous.)

The only flaw in this reasoning is that in (1) it might happen that $\Delta u = 0$ (even when $\Delta x \neq 0$) and, of course, we can't divide by 0. At the end of this section we present a complete proof of the Chain Rule by showing how to argue in the case where $\Delta u = 0$. □

The Chain Rule can be written either in the prime notation

(2) $$(f \circ g)'(x) = f'(g(x))g'(x)$$

or, if $y = f(u)$ and $u = g(x)$, in Leibniz notation:

(3) $$\frac{dy}{dx} = \frac{dy}{du}\frac{du}{dx}$$

Equation 3 is easy to remember because if dy/du and du/dx were quotients, then we could cancel du. Remember, however, that du has not been defined and du/dx should not be thought of as an actual quotient.

EXAMPLE 1 Find $F'(x)$ if $F(x) = \sqrt{x^2 + 1}$.

SOLUTION 1 (using Equation 2): At the beginning of this section we expressed F as $F(x) = (f \circ g)(x) = f(g(x))$ where $f(u) = \sqrt{u}$ and $g(x) = x^2 + 1$. Since

$$f'(u) = \tfrac{1}{2}u^{-1/2} = \frac{1}{2\sqrt{u}} \qquad \text{and} \qquad g'(x) = 2x$$

we have

$$\begin{aligned} F'(x) &= f'(g(x))g'(x) \\ &= \frac{1}{2\sqrt{x^2+1}} \cdot 2x = \frac{x}{\sqrt{x^2+1}} \end{aligned}$$

SOLUTION 2 (using Equation 3): If we let $u = x^2 + 1$ and $y = \sqrt{u}$, then

$$\begin{aligned} F'(x) &= \frac{dy}{du}\frac{du}{dx} = \frac{1}{2\sqrt{u}}(2x) \\ &= \frac{1}{2\sqrt{x^2+1}}(2x) = \frac{x}{\sqrt{x^2+1}} \end{aligned}$$ ■

When using Formula 3 we should bear in mind that dy/dx refers to the derivative of y when y is considered as a function of x (called the *derivative of y with respect to x*), whereas dy/du refers to the derivative of y when considered as a function of u (the derivative of y with respect to u). For instance, in Example 1, y can be considered as a function of x $\left(y = \sqrt{x^2 + 1}\right)$ and also as a function of u $\left(y = \sqrt{u}\right)$. Note that

$$\frac{dy}{dx} = F'(x) = \frac{x}{\sqrt{x^2+1}} \qquad \text{whereas} \qquad \frac{dy}{du} = f'(u) = \frac{1}{2\sqrt{u}}$$

EXAMPLE 2 If $y = u^3 + u^2 + 1$, where $u = 2x^2 - 1$, find

$$\left.\frac{dy}{dx}\right|_{x=2}$$

SOLUTION 1 Using the Chain Rule, we have

$$\frac{dy}{dx} = \frac{dy}{du}\frac{du}{dx} = (3u^2 + 2u)(4x)$$

When $x = 2$, $u = 2(2)^2 - 1 = 7$, so

$$\left.\frac{dy}{dx}\right|_{x=2} = (3 \cdot 7^2 + 2 \cdot 7)(4 \cdot 2) = 161 \cdot 8 = 1288$$

SOLUTION 2 We could solve this problem without using the Chain Rule by expressing y explicitly as a function of x:

$$\begin{aligned} y = u^3 + u^2 + 1 &= (2x^2 - 1)^3 + (2x^2 - 1)^2 + 1 \\ &= (8x^6 - 12x^4 + 6x^2 - 1) + (4x^4 - 4x^2 + 1) + 1 \\ &= 8x^6 - 8x^4 + 2x^2 + 1 \end{aligned}$$

Therefore

$$\frac{dy}{dx} = 48x^5 - 32x^3 + 4x$$

$$\left.\frac{dy}{dx}\right|_{x=2} = 48 \cdot 2^5 - 32 \cdot 2^3 + 4 \cdot 2 = 1288$$

Solution 1 is clearly preferable. ■

Let's make explicit the special case of the Chain Rule where the outer function f is a power function. If $y = [g(x)]^n$, then we can write $y = f(u) = u^n$ where $u = g(x)$. By using the Chain Rule and then the Power Rule, we get

$$\frac{dy}{dx} = \frac{dy}{du}\frac{du}{dx} = nu^{n-1}\frac{du}{dx} = n[g(x)]^{n-1}g'(x)$$

(4) THE POWER RULE COMBINED WITH THE CHAIN RULE If n is any real number and $u = g(x)$ is differentiable, then

$$\frac{d}{dx}(u^n) = nu^{n-1}\frac{du}{dx}$$

Alternatively,

$$\frac{d}{dx}[g(x)]^n = n[g(x)]^{n-1} \cdot g'(x)$$

Notice that the derivative in Example 1 could be calculated by taking $n = \frac{1}{2}$ in (4).

EXAMPLE 3 Differentiate $y = (x^3 - 1)^{100}$.

SOLUTION Taking $u = g(x) = x^3 - 1$ and $n = 100$ in (4), we have

$$\begin{aligned} \frac{dy}{dx} = \frac{d}{dx}(x^3 - 1)^{100} &= 100(x^3 - 1)^{99}\frac{d}{dx}(x^3 - 1) \\ &= 100(x^3 - 1)^{99} \cdot 3x^2 \\ &= 300x^2(x^3 - 1)^{99} \end{aligned}$$

■

Computer algebra systems have commands that differentiate functions. Although we have developed several differentiation formulas in this chapter, it is useful to be able to check your work if the function involved is complicated. If you have access to a computer algebra system, use the differentiation command to find the derivatives in the examples of this section and to check your answers in the exercises.

EXAMPLE 4 Find $f'(x)$ if $f(x) = \dfrac{1}{\sqrt[3]{x^2 + x + 1}}$.

SOLUTION First rewrite f: $f(x) = (x^2 + x + 1)^{-1/3}$. Thus

$$
\begin{aligned}
f'(x) &= -\tfrac{1}{3}(x^2 + x + 1)^{-4/3} \frac{d}{dx}(x^2 + x + 1) \\
&= -\tfrac{1}{3}(x^2 + x + 1)^{-4/3}(2x + 1)
\end{aligned}
$$

■

EXAMPLE 5 Find the derivative of the function

$$g(t) = \left(\frac{t - 2}{2t + 1}\right)^9$$

SOLUTION Combining the Power Rule, Chain Rule, and Quotient Rule, we get

$$
\begin{aligned}
g'(t) &= 9\left(\frac{t - 2}{2t + 1}\right)^8 \frac{d}{dt}\left(\frac{t - 2}{2t + 1}\right) \\
&= 9\left(\frac{t - 2}{2t + 1}\right)^8 \frac{(2t + 1) \cdot 1 - 2(t - 2)}{(2t + 1)^2} \\
&= \frac{45(t - 2)^8}{(2t + 1)^{10}}
\end{aligned}
$$

■

EXAMPLE 6 Differentiate $y = (2x + 1)^5(x^3 - x + 1)^4$.

The graphs of the functions y and y' in Example 6 are shown in Figure 1. Notice that y' is large when y increases rapidly and $y' = 0$ when y has a horizontal tangent. So our answer appears to be reasonable.

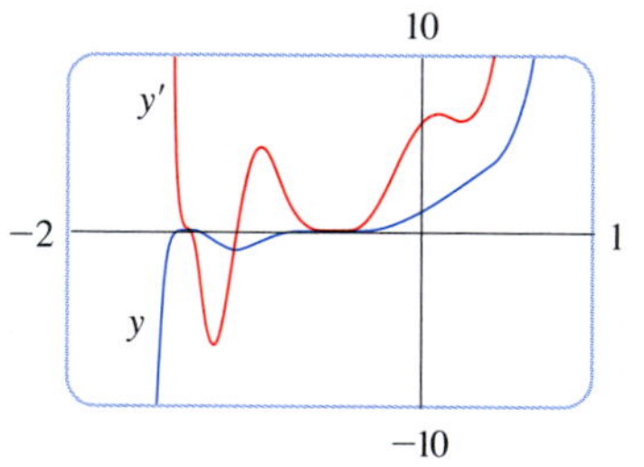

FIGURE 1

SOLUTION In this example we must use the Product Rule before using the Chain Rule:

$$
\begin{aligned}
\frac{dy}{dx} &= (2x + 1)^5 \frac{d}{dx}(x^3 - x + 1)^4 + (x^3 - x + 1)^4 \frac{d}{dx}(2x + 1)^5 \\
&= (2x + 1)^5 \cdot 4(x^3 - x + 1)^3 \frac{d}{dx}(x^3 - x + 1) \\
&\quad + (x^3 - x + 1)^4 \cdot 5(2x + 1)^4 \frac{d}{dx}(2x + 1) \\
&= 4(2x + 1)^5(x^3 - x + 1)^3(3x^2 - 1) + 5(x^3 - x + 1)^4(2x + 1)^4 \cdot 2
\end{aligned}
$$

By using common factors, we could write the answer as

$$\frac{dy}{dx} = 2(2x + 1)^4(x^3 - x + 1)^3(17x^3 + 6x^2 - 9x + 3)$$

■

NOTE: In using the Chain Rule we work from the outside to the inside. Formula 2 says that *we differentiate the outer function f [at the inner function g(x)] and then we multiply by the derivative of the inner function.*

$$\frac{d}{dx} \underbrace{f}_{\substack{\text{outer}\\\text{function}}} \underbrace{(g(x))}_{\substack{\text{evaluated}\\\text{at inner}\\\text{function}}} = \underbrace{f'}_{\substack{\text{derivative}\\\text{of outer}\\\text{function}}} \underbrace{(g(x))}_{\substack{\text{evaluated}\\\text{at inner}\\\text{function}}} \cdot \underbrace{g'(x)}_{\substack{\text{derivative}\\\text{of inner}\\\text{function}}}$$

EXAMPLE 7 Differentiate (a) $y = \sin(x^2)$ and (b) $y = \sin^2 x$.

SOLUTION

(a) If $y = \sin(x^2)$, then the outer function is the sine function and the inner function is the squaring function, so the Chain Rule gives

$$\frac{dy}{dx} = \frac{d}{dx} \underbrace{\sin}_{\substack{\text{outer}\\\text{function}}} \underbrace{(x^2)}_{\substack{\text{evaluated}\\\text{at inner}\\\text{function}}} = \underbrace{\cos}_{\substack{\text{derivative}\\\text{of outer}\\\text{function}}} \underbrace{(x^2)}_{\substack{\text{evaluated}\\\text{at inner}\\\text{function}}} \cdot \underbrace{2x}_{\substack{\text{derivative}\\\text{of inner}\\\text{function}}}$$

$$= 2x\cos(x^2)$$

(b) Note that $\sin^2 x = (\sin x)^2$. Here the outer function is the squaring function and the inner function is the sine function. So

$$\frac{dy}{dx} = \frac{d}{dx} \underbrace{(\sin x)^2}_{\substack{\text{inner}\\\text{function}}} = \underbrace{2}_{\substack{\text{derivative}\\\text{of outer}\\\text{function}}} \cdot \underbrace{(\sin x)}_{\substack{\text{evaluated}\\\text{at inner}\\\text{function}}} \cdot \underbrace{\cos x}_{\substack{\text{derivative}\\\text{of inner}\\\text{function}}}$$

The answer can be left as $2\sin x\cos x$ or written as $\sin 2x$ (by a trigonometric identity known as the double-angle formula). ■

In Example 7(a) we combined the Chain Rule with the rule for differentiating the sine function. In general, if $y = \sin u$, where u is a differentiable function of x, then, by the Chain Rule,

$$\frac{dy}{dx} = \frac{dy}{du}\frac{du}{dx} = \cos u \frac{du}{dx}$$

Thus

$$\frac{d}{dx}(\sin u) = \cos u \frac{du}{dx}$$

In a similar fashion, all of the formulas for differentiating trigonometric functions can be combined with the Chain Rule.

The reason for the name "Chain Rule" becomes clear when we make a longer chain by adding another link. Suppose that $y = f(u)$, $u = g(x)$, and $x = h(t)$, where f, g, and h are differentiable functions. Then to compute the derivative of y with respect to t, we use the Chain Rule twice:

$$\frac{dy}{dt} = \frac{dy}{dx}\frac{dx}{dt} = \frac{dy}{du}\frac{du}{dx}\frac{dx}{dt}$$

EXAMPLE 8 If $f(x) = \sin(\cos(\tan x))$, then

$$f'(x) = \cos(\cos(\tan x)) \frac{d}{dx} \cos(\tan x)$$

$$= \cos(\cos(\tan x))[-\sin(\tan x)] \frac{d}{dx}(\tan x)$$

$$= -\cos(\cos(\tan x)) \sin(\tan x) \sec^2 x$$

Notice that the Chain Rule has been used twice. ■

We use the problem-solving strategy of *Taking cases.* See Section 4 in Review and Preview.

PROOF OF THE CHAIN RULE We use the notation introduced on page 187 and we consider two cases.

CASE I $du/dx \neq 0$

In this case $\Delta u \neq 0$ if Δx is sufficiently small (otherwise du/dx would be 0). But then we can divide and multiply by Δu in Equation 1, so the proof of the Chain Rule given there is correct.

Here we consider two subcases within Case II.

CASE II $du/dx = 0$

In this case $\Delta u = 0$ for some values of Δx and $\Delta u \neq 0$ for other values of Δx. If $\Delta x \to 0$ through values such that $\Delta u \neq 0$, then we can write, as before,

$$\frac{\Delta y}{\Delta x} = \frac{\Delta y}{\Delta u} \cdot \frac{\Delta u}{\Delta x} \quad \to \quad \frac{dy}{du} \cdot \frac{du}{dx} = \frac{dy}{du} \cdot 0 = 0$$

If $\Delta x \to 0$ through values such that $\Delta u = 0$, we have

$$\Delta y = f(u + \Delta u) - f(u) = f(u) - f(u) = 0$$

and so
$$\frac{\Delta y}{\Delta x} = 0$$

In both cases, $\Delta y/\Delta x \to 0$, so

$$\frac{dy}{dx} = 0 = \frac{dy}{du}\frac{du}{dx}$$

The Chain Rule is verified because both sides are 0 in this case. □

EXERCISES 3.5

1–4 ■ Find dy/dx and $dy/dx]_{x=1}$ in two ways: (a) using the Chain Rule and (b) without using the Chain Rule, as in Example 2.

1. $y = u^2, \quad u = x^2 + 2x + 3$

2. $y = u^2 - 2u + 3, \quad u = 5 - 6x$

3. $y = u^3, \quad u = x + (1/x)$

4. $y = u - u^2, \quad u = \sqrt{x} + \sqrt[3]{x}$

5–48 ■ Find the derivative of the function.

5. $F(x) = (x^2 + 4x + 6)^5$

6. $F(x) = (x^3 - 5x)^4$

7. $G(x) = (3x - 2)^{10}(5x^2 - x + 1)^{12}$

8. $g(t) = (6t^2 + 5)^3(t^3 - 7)^4$

9. $f(t) = (2t^2 - 6t + 1)^{-8}$

10. $f(t) = \dfrac{1}{(t^2 - 2t - 5)^4}$

11. $g(x) = \sqrt{x^2 - 7x}$

12. $k(x) = \sqrt[3]{1 + \sqrt{x}}$

13. $h(t) = \left(t - \dfrac{1}{t}\right)^{3/2}$

14. $F(s) = \sqrt{s^3 + 1}\,(s^2 + 1)^4$

15. $F(y) = \left(\dfrac{y - 6}{y + 7}\right)^3$

16. $s(t) = \sqrt[4]{\dfrac{t^3 + 1}{t^3 - 1}}$

17. $f(z) = \dfrac{1}{\sqrt[5]{2z - 1}}$

18. $f(x) = \dfrac{x}{\sqrt{7 - 3x}}$

19. $y = (2x - 5)^4(8x^2 - 5)^{-3}$

20. $y = (x^2 + 1)\sqrt[3]{x^2 + 2}$

21. $y = \tan 3x$

22. $y = 4 \sec 5x$

23. $y = \cos(x^3)$

24. $y = \cos^3 x$

25. $y = (1 + \cos^2 x)^6$

26. $y = \tan(x^2) + \tan^2 x$

27. $y = \cos(\tan x)$

28. $y = \sin(\sin x)$

29. $y = \sec^2 2x - \tan^2 2x$

30. $y = \sqrt{1 + 2\tan x}$

31. $y = \csc(x/3)$

32. $y = \cot \sqrt[3]{1 + x^2}$

33. $y = \sin^3 x + \cos^3 x$

34. $y = \sin^2(\cos 4x)$

35. $y = \sin \dfrac{1}{x}$

36. $y = \dfrac{\sin^2 x}{\cos x}$

37. $y = \dfrac{1 + \sin 2x}{1 - \sin 2x}$

38. $y = x \sin \dfrac{1}{x}$

39. $y = \tan^2(x^3)$

40. $y = (\sin \sqrt{x^2 + 1}\,)^{\sqrt{2}}$

41. $y = \cos^2(\cos x) + \sin^2(\cos x)$

42. $y = \sin(\sin(\sin x))$

43. $y = \sqrt{x + \sqrt{x}}$

44. $y = \sqrt{x + \sqrt{x + \sqrt{x}}}$

45. $f(x) = [x^3 + (2x - 1)^3]^3$

46. $g(t) = \sqrt[4]{(1 - 3t)^4 + t^4}$

47. $y = \sin(\tan \sqrt{\sin x}\,)$

48. $y = \sqrt{\cos(\sin^2 x)}$

49–52 ■ Find the equation of the tangent line to the curve at the given point.

49. $y = (x^3 - x^2 + x - 1)^{10}, \quad (1, 0)$

50. $y = \sqrt{x + (1/x)}, \quad (1, \sqrt{2}\,)$

51. $y = \dfrac{8}{\sqrt{4 + 3x}}, \quad (4, 2)$

52. $y = \sin x + \cos 2x, \quad (\pi/6, 1)$

53. (a) Find an equation of the tangent line to the curve $y = \tan(\pi x^2/4)$ at the point $(1, 1)$.
(b) Illustrate part (a) by graphing the curve and the tangent line on the same screen.

54. (a) The curve $y = |x|/\sqrt{2 - x^2}$ is called a **bullet-nose curve.** Find an equation of the tangent line to this curve at the point $(1, 1)$.
(b) Illustrate part (a) by graphing the curve and the tangent line on the same screen.

55. (a) If $f(x) = \sqrt{1 - x^2}/x$, find $f'(x)$.
(b) Check to see that your answer to part (a) is reasonable by comparing the graphs of f and f'.

56. (a) If $f(x) = 1/(\cos^2 \pi x + 9 \sin^2 \pi x)$, find $f'(x)$.
(b) Check to see that your answer to part (a) is reasonable by comparing the graphs of f and f'.

57. Find all points on the graph of the function $f(x) = 2 \sin x + \sin^2 x$ at which the tangent line is horizontal.

58. Find the x-coordinates of all points on the curve $y = \sin 2x - 2 \sin x$ at which the tangent line is horizontal.

59. Suppose that $F(x) = f(g(x))$ and $g(3) = 6$, $g'(3) = 4$, $f'(3) = 2$, and $f'(6) = 7$. Find $F'(3)$.

60. Suppose that $w = u \circ v$ and $u(0) = 1$, $v(0) = 2$, $u'(0) = 3$, $u'(2) = 4$, $v'(0) = 5$, and $v'(2) = 6$. Find $w'(0)$.

61. The displacement of a particle on a vibrating string is given by the equation

$$s(t) = 10 + \tfrac{1}{4}\sin(10\pi t)$$

where s is measured in centimeters and t in seconds. Find the velocity of the particle after t seconds.

62. If the equation of motion of a particle is given by $s = A \cos(\omega t + \delta)$, the particle is said to undergo *simple harmonic motion.*
(a) Find the velocity of the particle at time t.
(b) When is the velocity 0?

63. A Cepheid variable star is a star whose brightness alternately increases and decreases. The most easily visible such star is Delta Cephei, for which the interval between times of maximum brightness is 5.4 days. The average brightness of this star is 4.0 and its brightness changes by ± 0.35. In view of these data, the brightness of Delta Cephei at time t, where t is measured in days, has been modeled by the function

$$B(t) = 4.0 + 0.35 \sin(2\pi t/5.4)$$

(a) Find the rate of change of the brightness after t days.
(b) Find, correct to two decimal places, the rate of increase after one day.

64. The frequency of vibrations of a vibrating violin string is given by

$$f = \frac{1}{2L}\sqrt{\frac{T}{\rho}}$$

where L is the length of the string, T is its tension, and ρ is its linear density. [See Chapter 11 in D. E. Hall, *Musical Acoustics,* 2d ed. (Pacific Grove, CA: Brooks/Cole, 1991).] Find the rate of change of the frequency with respect to
(a) the length (when T and ρ are constant),
(b) the tension (when L and ρ are constant), and
(c) the linear density (when L and T are constant).

65. Let h be differentiable on $[0, \infty)$ and define G by $G(x) = h(\sqrt{x}\,)$.
(a) Where is G differentiable?
(b) Find an expression for $G'(x)$.

66. Suppose f is differentiable on $\mathbb{R}$ and α is a real number. Let $F(x) = f(x^\alpha)$ and $G(x) = [f(x)]^\alpha$. Find expressions for (a) $F'(x)$ and (b) $G'(x)$.

67. Suppose f is differentiable on $\mathbb{R}$. Let $F(x) = f(\cos x)$ and $G(x) = \cos(f(x))$. Find expressions for (a) $F'(x)$ and (b) $G'(x)$.

68. If $g(t) = [f(\sin t)]^2$, where f is a differentiable function, find $g'(t)$.

69. If $g(x) = f(b + mx) + f(b - mx)$, where f is differentiable at b, find $g'(0)$.

70. Suppose $y = f(x)$ is a curve that always lies above the x-axis and never has a horizontal tangent, where f is differentiable everywhere. For what value of y is the rate of change of y^5 with respect to x eighty times the rate of change of y with respect to x?

71. Use the Chain Rule to give easier proofs of Exercise 61 in Section 3.1.

72. Use the Chain Rule and the Product Rule to give an alternative proof of the Quotient Rule. [*Hint:* Write $f(x)/g(x) = f(x)[g(x)]^{-1}$.]

73. If n is a positive integer, prove that

$$\frac{d}{dx}(\sin^n x \cos nx) = n \sin^{n-1} x \cos(n+1)x$$

74. Find a formula for $D(\cos^n x \cos nx)$ that is similar to the one in Exercise 73.

75–77 ■ Find the derivative of each function. [*Hint:* $|x| = \sqrt{x^2}$.]

75. $f(x) = |x|$

76. $g(x) = |x|/x$

77. $h(x) = x|2x - 1|$

78. (a) Sketch the graph of the function $f(x) = |\sin x|$.
(b) At what points is f not differentiable?
(c) Give a formula for f' and sketch its graph.
(d) Do the same for $g(x) = \sin|x|$.

79. Use the Chain Rule to show that if θ is measured in degrees, then

$$\frac{d}{d\theta}(\sin\theta) = \frac{\pi}{180}\cos\theta$$

(This gives one reason for the convention that radian measure is always used when dealing with trigonometric functions in calculus: the differentiation formulas would not be as simple if we used degree measure.)

80. If f and g are the functions whose graphs are shown, let $u(x) = f(g(x))$, $v(x) = g(f(x))$, and $w(x) = g(g(x))$. Find each derivative, if it exists.
(a) $u'(1)$ (b) $v'(1)$ (c) $w'(1)$

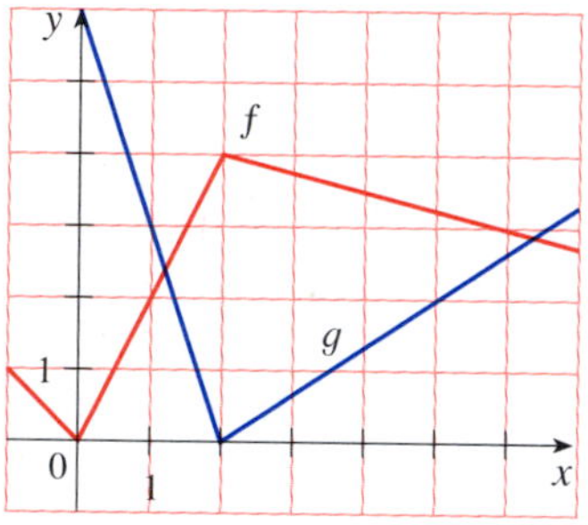

3.6 IMPLICIT DIFFERENTIATION

The functions that we have met so far can be described by expressing one variable explicitly in terms of another variable—for example,

$$y = \sqrt{x^3 + 1} \quad \text{or} \quad y = x\sin x$$

or, in general, $y = f(x)$. Some functions, however, are defined implicitly by a relation between x and y such as

(1) $$x^2 + y^2 = 25$$

or

(2) $$x^3 + y^3 = 6xy$$

Sometimes it may be possible to solve such an equation for y as an explicit function (or several functions) of x. For instance, if we solve Equation 1 for y, we get $y = \pm\sqrt{25 - x^2}$, so two functions determined by the implicit Equation 1 are $f(x) = \sqrt{25 - x^2}$ and $g(x) = -\sqrt{25 - x^2}$. The graphs of f and g are the upper and lower semicircles of the circle $x^2 + y^2 = 25$ (see Figure 1).

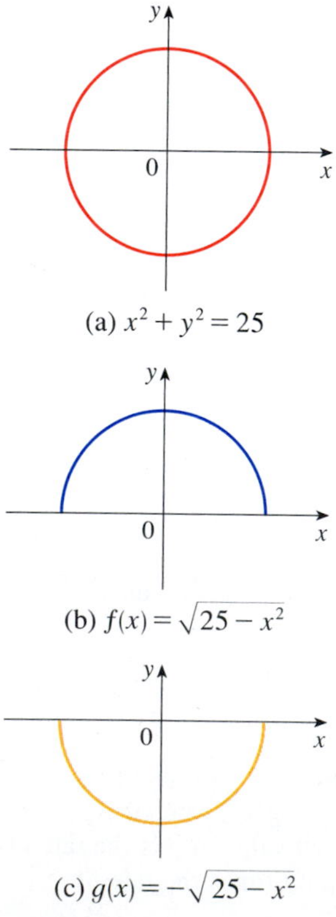

FIGURE 1

It would be very difficult (though possible) to solve Equation 2 for y explicitly as a function of x. Nonetheless, it is the equation of a curve called the *folium of Descartes* shown in Figure 2 and it implicitly defines y as several functions of x. The graphs of three such functions are shown in Figure 3. When we say that f is a function defined implicitly by Equation 2, we mean that the equation

$$x^3 + [f(x)]^3 = 6xf(x)$$

is true for all values of x in the domain of f.

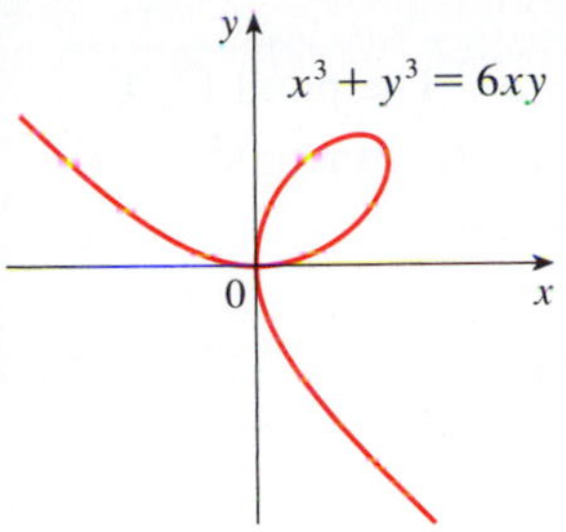

FIGURE 2
The folium of Descartes

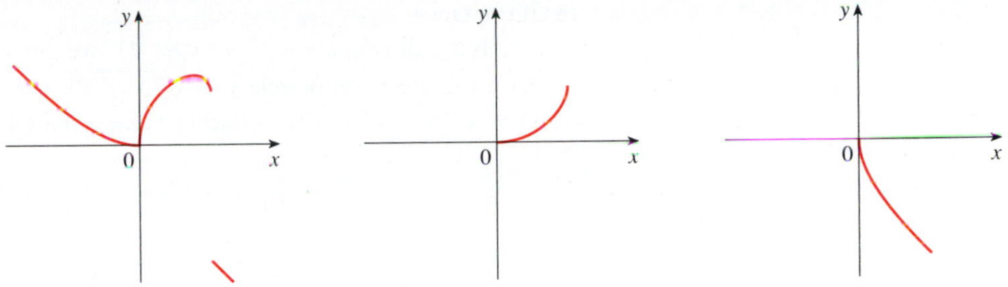

FIGURE 3
Graphs of three functions defined by the folium of Descartes

Fortunately it is not necessary to solve an equation for y in terms of x in order to find the derivative of y. Instead we can use the method of **implicit differentiation**. This consists of differentiating both sides of the relation with respect to x and then solving the resulting equation for y'. In the examples and exercises of this section it is always assumed that the given equation determines y implicitly as a differentiable function of x so that the method of implicit differentiation can be applied.

EXAMPLE 1

(a) If $x^2 + y^2 = 25$, find $\dfrac{dy}{dx}$.

(b) Find the equation of the tangent to the circle $x^2 + y^2 = 25$ at the point $(3, 4)$.

SOLUTION 1

(a) Differentiate both sides of the equation $x^2 + y^2 = 25$:

$$\frac{d}{dx}(x^2 + y^2) = \frac{d}{dx}(25)$$

$$\frac{d}{dx}(x^2) + \frac{d}{dx}(y^2) = 0$$

Remembering that y is a function of x and using the Chain Rule, we have

$$\frac{d}{dx}(y^2) = 2y\frac{dy}{dx}$$

Thus
$$2x + 2y\frac{dy}{dx} = 0$$

Now we solve this equation for dy/dx:

$$\frac{dy}{dx} = -\frac{x}{y}$$

(b) At the point $(3, 4)$ we have $x = 3$ and $y = 4$, so

$$\frac{dy}{dx} = -\frac{3}{4}$$

An equation of the tangent to the circle at $(3, 4)$ is therefore

$$y - 4 = -\tfrac{3}{4}(x - 3)$$

or
$$3x + 4y = 25$$

SOLUTION 2

(b) Solving the equation $x^2 + y^2 = 25$, we get $y = \pm\sqrt{25 - x^2}$. The point $(3, 4)$ lies on the upper semicircle $y = \sqrt{25 - x^2}$ and so we consider the function $f(x) = \sqrt{25 - x^2}$. Differentiating f using the Chain Rule, we have

$$\begin{aligned} f'(x) &= \tfrac{1}{2}(25 - x^2)^{-1/2} \frac{d}{dx}(25 - x^2) \\ &= \tfrac{1}{2}(25 - x^2)^{-1/2}(-2x) \\ &= -\frac{x}{\sqrt{25 - x^2}} \\ f'(3) &= -\frac{3}{\sqrt{25 - 3^2}} = -\frac{3}{4} \end{aligned}$$

and, as in Solution 1, the equation of the tangent is $3x + 4y = 25$. ■

NOTE 1: Example 1 illustrates that even when it is possible to solve an equation explicitly for y in terms of x, it may be easier to use implicit differentiation.

NOTE 2: The expression $dy/dx = -x/y$ gives the derivative in terms of both x and y. It is correct no matter which function y is determined by the given equation. For instance, for $y = f(x) = \sqrt{25 - x^2}$ we have

$$\frac{dy}{dx} = -\frac{x}{y} = -\frac{x}{\sqrt{25 - x^2}}$$

whereas for $y = g(x) = -\sqrt{25 - x^2}$ we have

$$\frac{dy}{dx} = -\frac{x}{y} = -\frac{x}{-\sqrt{25 - x^2}} = \frac{x}{\sqrt{25 - x^2}}$$

EXAMPLE 2

(a) Find y' if $x^3 + y^3 = 6xy$.

(b) Find the tangent to the folium of Descartes $x^3 + y^3 = 6xy$ at the point $(3, 3)$.

SOLUTION

(a) Differentiating both sides of $x^3 + y^3 = 6xy$ with respect to x, regarding y as a function of x, and using the Chain Rule on the y^3 term and the Product Rule on the $6xy$ term, we get

$$3x^2 + 3y^2y' = 6y + 6xy'$$

or

$$x^2 + y^2y' = 2y + 2xy'$$

We now solve for y':

$$(y^2 - 2x)y' = 2y - x^2$$

or

$$y' = \frac{2y - x^2}{y^2 - 2x}$$

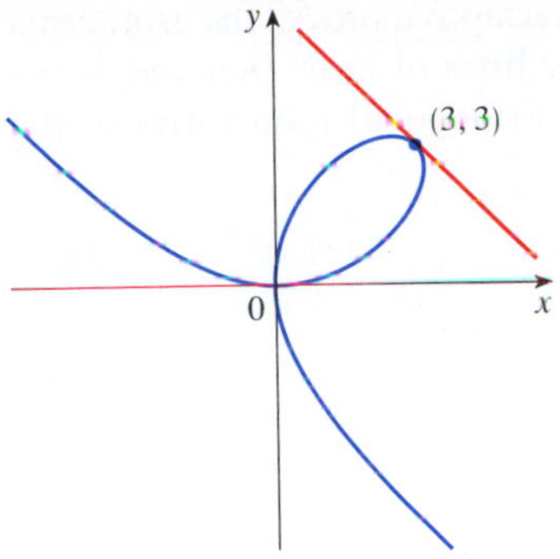

FIGURE 4

(b) When $x = y = 3$,

$$y' = \frac{2 \cdot 3 - 3^2}{3^2 - 2 \cdot 3} = -1$$

and a glance at Figure 4 confirms that this is a reasonable value for the slope at (3, 3). So an equation of the tangent to the folium at (3, 3) is

$$y - 3 = -1(x - 3) \qquad \text{or} \qquad x + y = 6$$

■

NOTE: There is a formula for the three roots of a cubic equation that is like the quadratic formula but much more complicated. If this formula is used to solve the equation $x^3 + y^3 = 6xy$ for y in terms of x, then three functions determined by the equation are

$$y = f(x) = \sqrt[3]{-\frac{x^3}{2} + \sqrt{\frac{x^6}{4} - 8x^3}} + \sqrt[3]{-\frac{x^3}{2} - \sqrt{\frac{x^6}{4} - 8x^3}}$$

and

$$y = \frac{1}{2}\left[-f(x) \pm \sqrt{-3}\left(\sqrt[3]{-\frac{x^3}{2} + \sqrt{\frac{x^6}{4} - 8x^3}} - \sqrt[3]{-\frac{x^3}{2} - \sqrt{\frac{x^6}{4} - 8x^3}}\right)\right]$$

The Norwegian mathematician Niels Abel proved in 1824 that no general formula can be given for the roots of a fifth-degree equation. Later the French mathematician Evariste Galois proved that it is impossible to find a general formula for the roots of an nth-degree equation (in terms of algebraic operations on the coefficients) if n is any integer larger than 4.

You can see that the method of implicit differentiation saves a lot of work in cases such as this. Moreover, implicit differentiation works just as easily for equations such as

$$y^5 + 3x^2y^2 + 5x^4 = 12$$

which are *impossible* to solve for y in terms of x.

EXAMPLE 3 Find y' if $\sin(x + y) = y^2 \cos x$.

SOLUTION Differentiating implicitly with respect to x and remembering that y is a function of x, we get

$$\cos(x + y) \cdot (1 + y') = 2yy'\cos x + y^2(-\sin x)$$

(Note that we have used the Chain Rule on the left side and the Product Rule and Chain Rule on the right side.) If we collect the terms that involve y', we get

$$\cos(x + y) + y^2 \sin x = (2y \cos x)y' - \cos(x + y) \cdot y'$$

So

$$y' = \frac{y^2 \sin x + \cos(x + y)}{2y \cos x - \cos(x + y)}$$

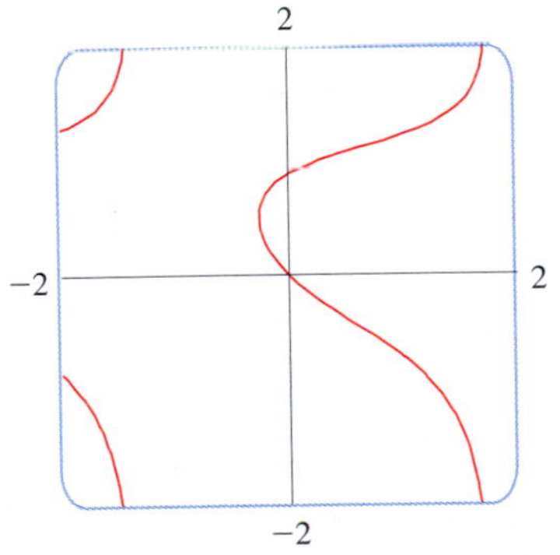

FIGURE 5

Figure 5 shows part of the curve $\sin(x + y) = y^2 \cos x$. As a check on our calculation, notice that $y' = -1$ when $x = y = 0$ and it appears from the graph that the slope is approximately -1 at the origin. ■

Two curves are called **orthogonal** if at each point of intersection their tangent lines are perpendicular. In the next example we use implicit differentiation to show that two families of curves are **orthogonal trajectories** of each other; that is, every curve in one family is orthogonal to every curve in the other family. Orthogonal families arise in several areas of physics. For example, the lines of force in an electrostatic field are orthogonal to the lines of constant potential. In thermodynamics, the isotherms (curves of equal temperature) are orthogonal to the flow lines of heat. In aerodynamics, the streamlines (curves of direction of airflow) are orthogonal trajectories of the velocity-equipotential curves.

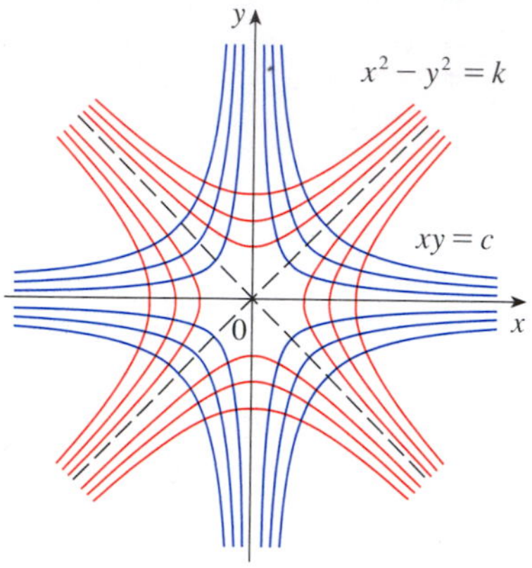

FIGURE 6

EXAMPLE 4 The equation

(3) $$xy = c \qquad c \neq 0$$

represents a family of hyperbolas. (Different values of the constant c give different hyperbolas. See Figure 6.) The equation

(4) $$x^2 - y^2 = k \qquad k \neq 0$$

represents another family of hyperbolas with asymptotes $y = \pm x$. Show that every curve in the family (3) is orthogonal to every curve in the family (4); that is, the families are orthogonal trajectories of each other.

SOLUTION Implicit differentiation of Equation 3 gives

(5) $$y + x\frac{dy}{dx} = 0 \qquad \text{so} \qquad \frac{dy}{dx} = -\frac{y}{x}$$

Implicit differentiation of Equation 4 gives

(6) $$2x - 2y\frac{dy}{dx} = 0 \qquad \text{so} \qquad \frac{dy}{dx} = \frac{x}{y}$$

From (5) and (6) we see that at any point of intersection of curves from each family, the slopes of the tangents are negative reciprocals of each other. Therefore, the curves intersect at right angles. ■

EXERCISES 3.6

1–4 ■
(a) Find y' by implicit differentiation.
(b) Solve the equation explicitly for y and differentiate to get y' in terms of x.
(c) Check that your solutions to parts (a) and (b) are consistent by substituting the expression for y into your solution for part (a).

1. $x^2 + 3x + xy = 5$

2. $\dfrac{x^2}{2} + \dfrac{y^2}{4} = 1$

3. $2y^2 + xy = x^2 + 3$

4. $\sqrt{x} + \sqrt{y} = 4$

5–16 ■ Find dy/dx by implicit differentiation.

5. $x^2 - xy + y^3 = 8$

6. $\sqrt{xy} - 2x = \sqrt{y}$

7. $2y^2 + \sqrt[3]{xy} = 3x^2 + 17$

8. $y^5 + 3x^2y^2 + 5x^4 = 12$

9. $x^4 + y^4 = 16$

10. $\sqrt{x + y} + \sqrt{xy} = 6$

11. $\dfrac{y}{x - y} = x^2 + 1$

12. $x\sqrt{1 + y} + y\sqrt{1 + 2x} = 2x$

13. $\cos(x - y) = y \sin x$

14. $x \sin y + \cos 2y = \cos y$

15. $xy = \cot(xy)$

16. $x \cos y + y \cos x = 1$

17–18 ■ Regard y as the independent variable and x as the dependent variable, and use implicit differentiation to find dx/dy.

17. $y^4 + x^2y^2 + yx^4 = y + 1$

18. $(x^2 + y^2)^2 = ax^2y$

19. If $x[f(x)]^3 + xf(x) = 6$ and $f(3) = 1$, find $f'(3)$.

20. If $[g(x)]^2 + 12x = x^2g(x)$ and $g(4) = 12$, find $g'(4)$.

21–26 ■ Find an equation of the tangent line to the curve at the given point.

21. $\dfrac{x^2}{16} - \dfrac{y^2}{9} = 1, \quad (-5, \frac{9}{4})$ (hyperbola)

22. $\dfrac{x^2}{9} + \dfrac{y^2}{36} = 1, \quad (-1, 4\sqrt{2})$ (ellipse)

23. $y^2 = x^3(2 - x)$
$(1, 1)$
(piriform)

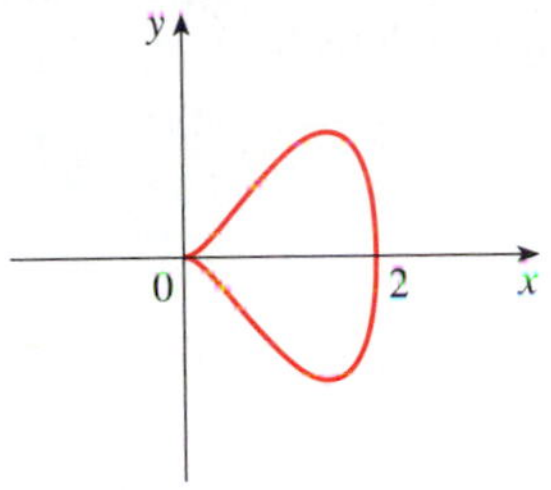

24. $x^{2/3} + y^{2/3} = 4$
$(-3\sqrt{3}, 1)$
(astroid)

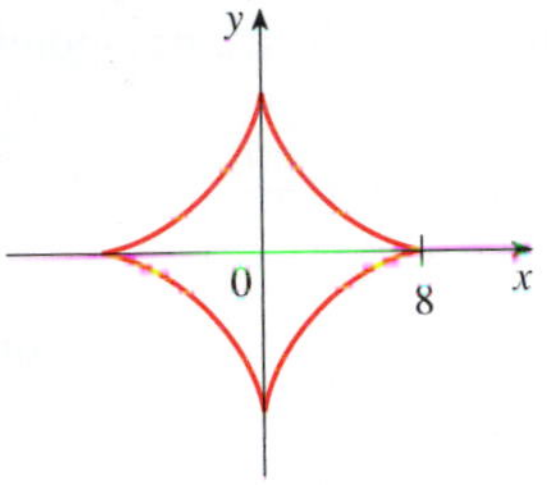

25. $2(x^2 + y^2)^2 = 25(x^2 - y^2)$
$(3, 1)$
(lemniscate)

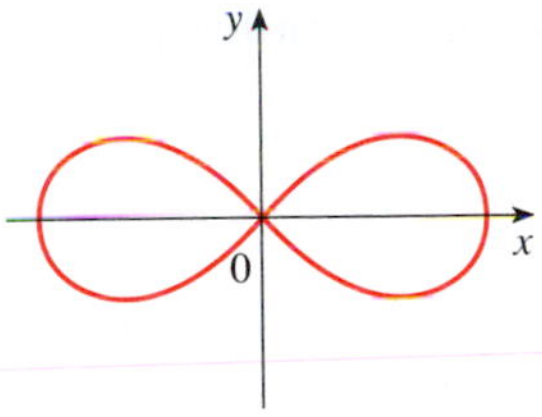

26. $x^2y^2 = (y + 1)^2(4 - y^2)$
$(0, -2)$
(conchoid of Nicomedes)

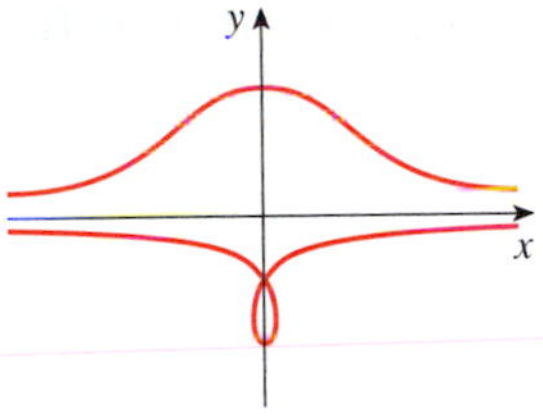

27. (a) The curve with equation $y^2 = 5x^4 - x^2$ is called a **kampyle of Eudoxus**. Find an equation of the tangent line to this curve at the point $(1, 2)$.
(b) Illustrate part (a) by graphing the curve and the tangent line on a common screen. (If your graphing device will graph implicitly defined curves, then use that capability. If not, you can still graph this curve by graphing its upper and lower halves separately.)

28. (a) The curve with equation $y^2 = x^3 + 3x^2$ is called **Tschirnhauser's cubic**. Find an equation of the tangent line to this curve at the point $(1, -2)$.
(b) At what points does this curve have a horizontal tangent?
(c) Illustrate parts (a) and (b) by graphing the curve and the tangent lines on a common screen.

29. Find the points on the lemniscate in Exercise 25 where the tangent is horizontal.

30. Show by implicit differentiation that the tangent to the ellipse

$$\frac{x^2}{a^2} + \frac{y^2}{b^2} = 1$$

at the point (x_0, y_0) is

$$\frac{x_0 x}{a^2} + \frac{y_0 y}{b^2} = 1$$

31. Find the equation of the tangent line to the hyperbola

$$\frac{x^2}{a^2} - \frac{y^2}{b^2} = 1$$

at the point (x_0, y_0).

32. Show that the sum of the x- and y-intercepts of any tangent line to the curve $\sqrt{x} + \sqrt{y} = \sqrt{c}$ is equal to c.

33. Show, using implicit differentiation, that any tangent line at a point P to a circle with center O is perpendicular to the radius OP.

34. The Power Rule (3.2.5) can be proved using implicit differentiation for the case where n is a rational number, $n = p/q$, and $y = f(x) = x^n$ is assumed beforehand to be a differentiable function. If $y = x^{p/q}$, then $y^q = x^p$. Use implicit differentiation to show that

$$y' = \frac{p}{q}x^{(p/q)-1}$$

35–36 ■ Show that the given curves are orthogonal.

35. $2x^2 + y^2 = 3, \quad x = y^2$

36. $x^2 - y^2 = 5, \quad 4x^2 + 9y^2 = 72$

37–40 ■ Show that the given families of curves are orthogonal trajectories of each other. Sketch both families of curves on the same axes.

37. $x^2 + y^2 = r^2, \quad ax + by = 0$

38. $x^2 + y^2 = ax, \quad x^2 + y^2 = by$

39. $y = cx^2, \quad x^2 + 2y^2 = k$

40. $y = ax^3, \quad x^2 + 3y^2 = b$

41. The equation $x^2 - xy + y^2 = 3$ represents a "rotated ellipse," that is, an ellipse whose axes are not parallel to the coordinate axes. Find the points at which this ellipse crosses the x-axis and show that the tangent lines at these points are parallel.

42. (a) Where does the normal line to the ellipse $x^2 - xy + y^2 = 3$ at the point $(-1, 1)$ intersect the ellipse a second time?
(b) Illustrate part (a) by graphing the ellipse and the normal line.

43. Find all points on the curve $x^2y^2 + xy = 2$ where the slope of the tangent line is -1.

44. Find the equations of both the tangent lines to the ellipse $x^2 + 4y^2 = 36$ that pass through the point $(12, 3)$.

45. The figure shows a lamp located three units to the right of the y-axis and a shadow created by the elliptical region $x^2 + 4y^2 \leq 5$. If the point $(-5, 0)$ is on the edge of the shadow, how far above the x-axis is the lamp located?

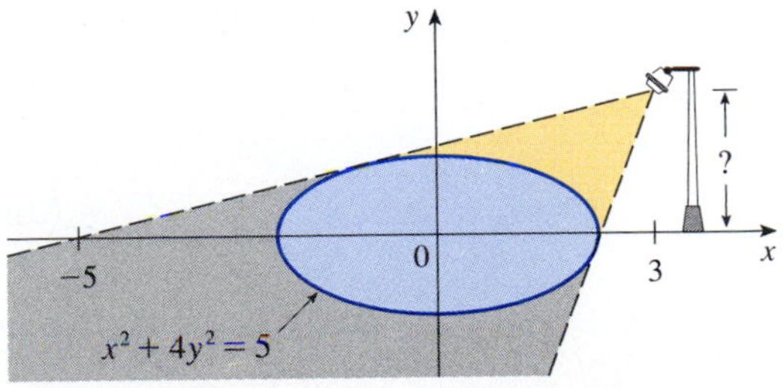

3.7 DERIVATIVES OF VECTOR FUNCTIONS

In Section 2.7 we defined the tangent vector to a curve defined by a vector function $\mathbf{r}(t) = \langle x(t), y(t) \rangle$ at the point $(x(a), y(a))$ corresponding to $t = a$ to be

(1) $$\mathbf{v} = \lim_{h \to 0} \frac{\mathbf{r}(a + h) - \mathbf{r}(a)}{h}$$

We also used limits of this form for the calculation of the velocity vector of a projectile whose position at time t is given by the vector function $\mathbf{r}(t)$.

As with the case of a scalar function (a function that is real-valued rather than vector-valued), we define the limit in (1) to be the derivative of the vector function $\mathbf{r}$ at $t = a$.

(2) DEFINITION The **derivative of a vector function r at a number *a***, denoted by $\mathbf{r}'(a)$, is

$$\mathbf{r}'(a) = \lim_{h \to 0} \frac{\mathbf{r}(a + h) - \mathbf{r}(a)}{h}$$

if this limit exists.

If we write $t = a + h$, then $h = t - a$ and h approaches 0 if and only if t approaches a. Therefore, an equivalent way of stating the derivative is

(3) $$\mathbf{r}'(a) = \lim_{t \to a} \frac{\mathbf{r}(t) - \mathbf{r}(a)}{t - a}$$

It should be emphasized that the numerators $\mathbf{r}(a + h) - \mathbf{r}(a)$ and $\mathbf{r}(t) - \mathbf{r}(a)$ are vectors. Therefore, the derivative of a vector function at $t = a$ is a vector rather than a scalar. Comparing with Definition 2.7.3, we see that the derivative, $\mathbf{r}'(a)$, is a tangent vector to the curve traced by $\mathbf{r}(t)$ at $t = a$.

We now show that the derivative of a vector function $\mathbf{r}(t) = \langle x(t), y(t) \rangle$ involves computing the derivative of each of the component functions $x(t)$ and $y(t)$. This follows from the following computation.

$$\begin{aligned}
\mathbf{r}'(t) &= \lim_{h \to 0} \frac{\mathbf{r}(t + h) - \mathbf{r}(t)}{h} && \text{[definition of } \mathbf{r}'(t)\text{]} \\
&= \lim_{h \to 0} \frac{\langle x(t + h), y(t + h) \rangle - \langle x(t), y(t) \rangle}{h} \\
&= \lim_{h \to 0} \left\langle \frac{x(t + h) - x(t)}{h}, \frac{y(t + h) - y(t)}{h} \right\rangle \\
&= \left\langle \lim_{h \to 0} \frac{x(t + h) - x(t)}{h}, \lim_{h \to 0} \frac{y(t + h) - y(t)}{h} \right\rangle \\
&= \langle x'(t), y'(t) \rangle && \text{[definition of } x'(t), y'(t)\text{]}
\end{aligned}$$

If either $x'(t)$ or $y'(t)$ does not exist, then the derivative of the vector function $\mathbf{r}(t)$ does not exist either. We summarize this discussion as follows.

(4) If $\mathbf{r}(t) = \langle x(t), y(t)\rangle$ is a vector function, then

$$\mathbf{r}'(t) = \langle x'(t), y'(t)\rangle = x'(t)\,\mathbf{i} + y'(t)\,\mathbf{j}$$

if both $x'(t)$ and $y'(t)$ exist.

EXAMPLE 1 Find the derivative of the vector function

$$\mathbf{r}(t) = \langle t^2 + 2t, t^3 - 4t\rangle$$

Find a tangent vector of unit length to the curve traced by $\mathbf{r}(t)$ at $t = 1$.

SOLUTION Taking the derivative of each component of $\mathbf{r}$ gives

$$\mathbf{r}'(t) = \langle 2t + 2, 3t^2 - 4\rangle$$

This is the derivative of $\mathbf{r}$ at any value of t. Letting $t = 1$ gives $\mathbf{r}'(1) = \langle 4, -1\rangle$, which is a tangent vector to the curve traced by $\mathbf{r}(t)$ at $t = 1$ (see Figure 1). Dividing by the length of this vector gives a tangent vector, $(1/\sqrt{17}\,)\langle 4, -1\rangle$, of unit length.

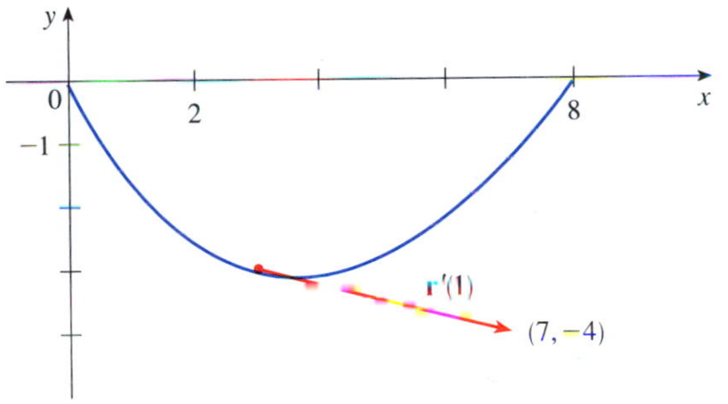

FIGURE 1

VELOCITY AND SPEED

As discussed in Section 2.7, the instantaneous velocity of a projectile is computed by (2), which is the limit of the average velocity of the projectile over shorter and shorter time intervals. Therefore, if $\mathbf{r}(t)$ represents the position of a projectile at time t, then its derivative, $\mathbf{r}'(t)$, represents its instantaneous velocity at time t. The *instantaneous speed* of the particle is defined to be the length of its velocity. These ideas are summarized as follows.

(5) DEFINITION If $\mathbf{r}(t) = \langle x(t), y(t)\rangle$ is a vector function representing the position of a particle at time t, then

instantaneous velocity at time $t = \mathbf{r}'(t) = \langle x'(t), y'(t)\rangle$

instantaneous speed at time $t = |\mathbf{r}'(t)| = \sqrt{[x'(t)]^2 + [y'(t)]^2}$

It is important to emphasize that the velocity is a vector quantity (with both a length and a direction), but speed is a scalar quantity (speed equals the length of the velocity vector, without any associated direction).

EXAMPLE 2 Consider a cannonball that is fired from a cannon with a muzzle velocity of 1000 ft/s, inclined at an angle of 60° from the horizontal. (See Example 1 in Section 1.3.) The position of the cannonball t seconds after the cannon is fired is given by

$$x = 500t \qquad \text{and} \qquad y = 500\sqrt{3}\,t - 16t^2$$

Compute the instantaneous velocity and speed of the cannonball at $t = 35$. Compare the instantaneous velocity at $t = 35$ to the average velocity over the time interval $35 \leq t \leq 35.5$.

SOLUTION The associated vector function that describes the position of the cannonball at time t is

$$r(t) = \langle 500t, 500\sqrt{3}\,t - 16t^2 \rangle$$

The velocity at time t is given by

$$\text{velocity} = \mathbf{r}'(t) = \langle 500, 500\sqrt{3} - 32t \rangle$$

So the instantaneous velocity at $t = 35$ is

$$\begin{aligned} \mathbf{r}'(35) &= \langle 500, 500\sqrt{3} - 32(35) \rangle \\ &\approx \langle 500, -253.97 \rangle \end{aligned}$$

The units for each component are feet per second. The speed is the length of this vector:

$$\begin{aligned} \text{speed} = |\mathbf{r}'(35)| &\approx \sqrt{500^2 + (-253.97)^2} \\ &\approx 560.81 \text{ ft/s} \end{aligned}$$

By comparison, the average velocity over the time interval $35 \leq t \leq 35.5$ is given by

$$\begin{aligned} \text{average velocity} &= \frac{\mathbf{r}(35.5) - \mathbf{r}(35)}{35.5 - 35} \\ &= \left\langle \frac{x(35.5) - x(35)}{0.5}, \frac{y(35.5) - y(35)}{0.5} \right\rangle \\ &\approx \langle 500, -261.97 \rangle \end{aligned}$$

in units of feet per second. ■

Note that the horizontal component of the cannonball in Example 2 is $x = 500t$, which changes at a constant rate of 500 ft/s. This is why the horizontal component of the instantaneous velocity is the same as the horizontal component of the average velocity. On the other hand, the vertical component is $y = 500\sqrt{3}\,t - 16t^2$, which does not change at a constant rate (its derivative is $y' = 500\sqrt{3} - 32t$, which varies with t). Therefore, the vertical component of the instantaneous velocity is different from the vertical component of the average velocity.

EXERCISES 3.7

1–6 ■ Sketch the curve with the given vector equation. Indicate with an arrow the direction in which t increases.

1. $\mathbf{r}(t) = \langle t, 2t \rangle$

2. $\mathbf{r}(t) = \langle t^2, t \rangle$

3. $\mathbf{r}(t) = \sin t\, \mathbf{i} + \cos t\, \mathbf{j}$

4. $\mathbf{r}(t) = \mathbf{i} + \cos t\, \mathbf{j}$

5. $\mathbf{r}(t) = \langle t^3 + 1, 2t \rangle$

6. $\mathbf{r}(t) = \langle t^2, t^3 \rangle$

7–10 ■ Find the domain and derivative of the vector function.

7. $\mathbf{r}(t) = \langle t^2, t \rangle$

8. $\mathbf{r}(t) = t \sin t\, \mathbf{i} + t \cos t\, \mathbf{j}$

9. $\mathbf{r}(t) = \langle t^2 - 4, \sqrt{t - 4} \rangle$

10. $\mathbf{r}(t) = \left\langle \dfrac{t^2}{t + 1}, \dfrac{t}{t^2 - 1} \right\rangle$

11–14 ■ Find a tangent vector of unit length at the point with the given value of the parameter t.

11. $\mathbf{r}(t) = \langle 2t, 3t^2 \rangle, \quad t = 1$

12. $\mathbf{r}(t) = 2 \sin t\, \mathbf{i} + 3 \cos t\, \mathbf{j}, \quad t = \pi/6$

13. $\mathbf{r}(t) = (1 + t^3)\, \mathbf{i} + t^2 \mathbf{j}, \quad t = 1$

14. $\mathbf{r}(t) = \langle t \cos t, t \sin t \rangle, \quad t = \pi/4$

15–18 ■ The vector function $\mathbf{r}(t)$ represents the position of a particle at time t. Find the velocity and speed at the given value of t.

15. $\mathbf{r}(t) = \langle t, 25t - 5t^2 \rangle, \quad t = 1$

16. $\mathbf{r}(t) = \langle 4 \cos t, 3 \sin t \rangle, \quad t = \pi/3$

17. $\mathbf{r}(t) = (t^4 + 7t)\, \mathbf{i} + (t^2 + 2t)\, \mathbf{j}, \quad t = 1$

18. $\mathbf{r}(t) = \sqrt{t^2 + 5}\, \mathbf{i} + t\, \mathbf{j}, \quad t = 2$

19. The curves traced by $\mathbf{r}_1(t) = \langle t, t^2 \rangle$ and $\mathbf{r}_2(t) = \langle \sin t, \sin 2t \rangle$ intersect at the origin. Find their angle of intersection correct to the nearest degree.

20. At what point do the curves traced by $\mathbf{r}_1(t) = \langle 1 - t, 3 + t^2 \rangle$ and $\mathbf{r}_2(s) = \langle s - 2, s^2 \rangle$ intersect? Find their angle of intersection correct to the nearest degree.

21. If $\mathbf{u}$ and $\mathbf{v}$ are vector functions of t, show that

$$\frac{d}{dt}[\mathbf{u}(t) \cdot \mathbf{v}(t)] = \mathbf{u}(t) \cdot \mathbf{v}'(t) + \mathbf{v}(t) \cdot \mathbf{u}'(t)$$

22. If $\mathbf{r}(t) \neq \mathbf{0}$, show that $\dfrac{d}{dt}\,|\mathbf{r}(t)| = \dfrac{1}{|\mathbf{r}(t)|}\,\mathbf{r}(t) \cdot \mathbf{r}'(t)$.
[*Hint:* Differentiate the equation $|\mathbf{r}(t)|^2 = \mathbf{r}(t) \cdot \mathbf{r}(t)$.]

23. If a curve has the property that the position vector $\mathbf{r}(t)$ is always perpendicular to the tangent vector $\mathbf{r}'(t)$, show that the curve lies on a circle with center the origin.
[*Hint:* Differentiate the equation $|\mathbf{r}(t)|^2 = \mathbf{r}(t) \cdot \mathbf{r}(t)$.]

3.8 HIGHER DERIVATIVES

If f is a differentiable function, then its derivative f' is also a function, so f' may have a derivative of its own, denoted by $(f')' = f''$. This new function f'' is called the **second derivative** of f because it is the derivative of the derivative of f. Thus

$$f''(x) = \frac{d}{dx}(f'(x)) = \frac{d}{dx}\left(\frac{d}{dx} f(x)\right)$$

For instance, if $f(x) = x^8$, then $f'(x) = 8x^7$, so

$$f''(x) = \frac{d}{dx}(f'(x)) = \frac{d}{dx}(8x^7) = 56x^6$$

NOTATION: If $y = f(x)$, then

$$y'' = f''(x) = \frac{d}{dx}\left(\frac{dy}{dx}\right) = \frac{d^2y}{dx^2} = D^2 f(x) = D_x^2 f(x)$$

The symbol D^2 indicates that the operation of differentiation is performed twice.

Similarly the *third derivative* f''' is the derivative of the second derivative: $f''' = (f'')'$. If $y = f(x)$, then alternative notations for the third derivative are

$$y''' = f'''(x) = \frac{d}{dx}\left(\frac{d^2y}{dx^2}\right) = \frac{d^3y}{dx^3} = D^3f(x) = D_x^3f(x)$$

The process can be continued. The fourth derivative f'''' is usually denoted by $f^{(4)}$. In general the nth derivative of f is denoted by $f^{(n)}$ and is obtained from f by differentiating n times. If $y = f(x)$, we write

$$y^{(n)} = f^{(n)}(x) = \frac{d^ny}{dx^n} = D^nf(x) = D_x^nf(x)$$

EXAMPLE 1 If $y = x^3 - 6x^2 - 5x + 3$

then

$$y' = 3x^2 - 12x - 5$$

$$y'' = 6x - 12$$

$$y''' = 6$$

$$y^{(4)} = 0$$

and in fact $y^{(n)} = 0$ for all $n \geqslant 4$. ■

EXAMPLE 2 If $f(x) = \dfrac{1}{x}$, find $f^{(n)}(x)$.

Figure 1 shows the graph of the function $f(x) = 1/x$ of Example 2 together with the graphs of its first two derivatives. Check to see that these three graphs are consistent with the geometric interpretations: $f'(x)$ is the slope of the tangent to $y = f(x)$ at $(x, f(x))$ and $f''(x)$ is the slope of the tangent to $y = f'(x)$.

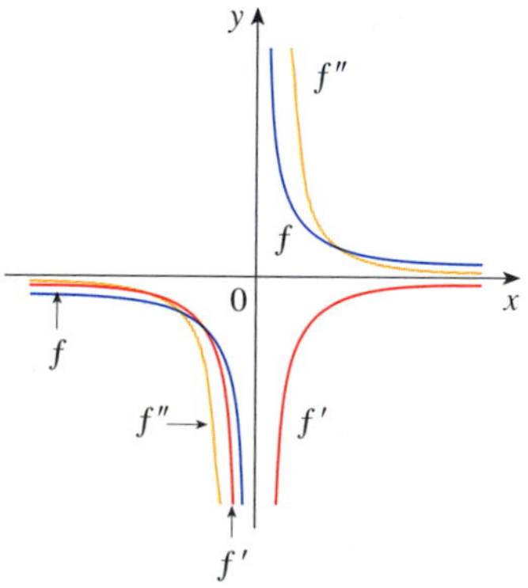

FIGURE 1

SOLUTION

$$f(x) = \frac{1}{x} = x^{-1}$$

$$f'(x) = -x^{-2} = \frac{-1}{x^2}$$

$$f''(x) = (-2)(-1)x^{-3} = \frac{2}{x^3}$$

$$f'''(x) = -3 \cdot 2 \cdot 1 \cdot x^{-4}$$

$$f^{(4)}(x) = 4 \cdot 3 \cdot 2 \cdot 1 \cdot x^{-5}$$

$$f^{(5)}(x) = -5 \cdot 4 \cdot 3 \cdot 2 \cdot 1 \cdot x^{-6} = -5!\,x^{-6}$$

$$\vdots$$

$$f^{(n)}(x) = (-1)^n n(n-1)(n-2)\cdots 2 \cdot 1 \cdot x^{-(n+1)}$$

or

$$f^{(n)}(x) = \frac{(-1)^n n!}{x^{n+1}}$$

Here we have used the factorial symbol $n!$ for the product of the first n positive integers.

$$n! = 1 \cdot 2 \cdot 3 \cdots (n-1) \cdot n$$

■

EXAMPLE 3 Find $D^{27}\cos x$.

SOLUTION The first few derivatives of $\cos x$ are as follows:

Look for a pattern.

$$D\cos x = -\sin x$$

$$D^2\cos x = -\cos x$$

$$D^3\cos x = \sin x$$

$$D^4\cos x = \cos x$$

$$D^5\cos x = -\sin x$$

We see that the successive derivatives occur in a cycle of length 4 and, in particular, $D^n\cos x = \cos x$ whenever n is a multiple of 4. Therefore

$$D^{24}\cos x = \cos x$$

and, differentiating three more times, we have

$$D^{27}\cos x = \sin x$$ ■

ACCELERATION

If $s = s(t)$ is the position function of an object that moves in a straight line, we know that its first derivative represents the velocity $v(t)$ of the object as a function of time:

$$v(t) = s'(t) = \frac{ds}{dt}$$

The instantaneous rate of change of velocity with respect to time is called the **acceleration** $a(t)$ of the object. Thus, the acceleration function is the derivative of the velocity function and is therefore the second derivative of the position function:

$$a(t) = v'(t) = s''(t)$$

or, in Leibniz notation,

$$a = \frac{dv}{dt} = \frac{d^2s}{dt^2}$$

EXAMPLE 4 The equation of motion of a particle is $s = 2t^3 - 5t^2 + 3t + 4$, where s is measured in centimeters and t in seconds. Find the acceleration as a function of time. What is the acceleration after 2 s?

SOLUTION The velocity and acceleration are

$$v(t) = \frac{ds}{dt} = 6t^2 - 10t + 3$$

$$a(t) = \frac{dv}{dt} = 12t - 10$$

The units for acceleration are centimeters per second per second, written as cm/s^2.

The acceleration after 2 s is $a(2) = 14\ cm/s^2$. ■

Acceleration for vector motion is computed in a similar manner. If $\mathbf{r}(t) = \langle x(t), y(t)\rangle$ represents the position of an object, then $\mathbf{r}'(t) = \langle x'(t), y'(t)\rangle$ represents velocity. So

$$\text{acceleration} = \mathbf{a}(t) = \mathbf{r}''(t) = \langle x''(t), y''(t)\rangle$$

Note that acceleration is a vector quantity, as is velocity. Writing this equality in terms of the basis vectors $\mathbf{i} = \langle 1, 0\rangle$ and $\mathbf{j} = \langle 0, 1\rangle$, we have

$$\mathbf{r}''(t) = x''(t)\,\mathbf{i} + y''(t)\,\mathbf{j}$$

EXAMPLE 5 Compute the acceleration for the cannonball in Example 2 of Section 3.7 at $t = 35$.

SOLUTION The velocity of the cannonball at any time t was computed to be

$$\mathbf{r}'(t) = \langle 500, 500\sqrt{3} - 32t\rangle$$

Differentiating this one more time gives the acceleration, $\mathbf{r}''(t)$:

$$\mathbf{r}''(t) = \langle 0, -32\rangle$$

where each component is given in feet per second per second. In this case, the acceleration is independent of time t. In particular, $\langle 0, -32\rangle$ is the acceleration of the cannonball at $t = 35$.

Note that the acceleration is a constant vector that points downward with a magnitude of 32, which is the acceleration constant due to Earth's gravity. Since we are ignoring air resistance, the only force acting on the cannonball (after it leaves the cannon) is the force of gravity. Therefore, our computation of the acceleration agrees with Newton's law: *force* = *mass* × *acceleration*, which implies that the acceleration should be a constant vector that points downward with a magnitude equal to that of gravity. ■

EXAMPLE 6 A satellite at a height 400 km above Earth completes an orbit about every 94 minutes. Find the equation of motion for the satellite. Also find the velocity and acceleration vectors and the satellite's speed when the satellite is directly overhead. (The radius of Earth is approximately 6600 km.)

SOLUTION First we fix a coordinate system with the center of Earth at the origin. The path of the satellite is approximately a circle centered at the origin and having radius $R = 7000$ km. Let's suppose that when the satellite is overhead, its position is $(R, 0)$, as shown in Figure 2. (This choice is arbitrary and is made for convenience.)

The circular motion of the satellite is given by

$$\mathbf{r}(t) = \langle R\cos\omega t, R\sin\omega t\rangle$$

where t represents time (in hours) and ω is a constant (called the *angular velocity*). The value of ω is determined as follows: When $\omega t = 2\pi$, the satellite will have traveled once around Earth. Since 94 min $= 1.5\overline{6}$ h, we have

$$\omega(1.5\overline{6}) = 2\pi$$

or

$$\omega = 4.011 \text{ rad/h}$$

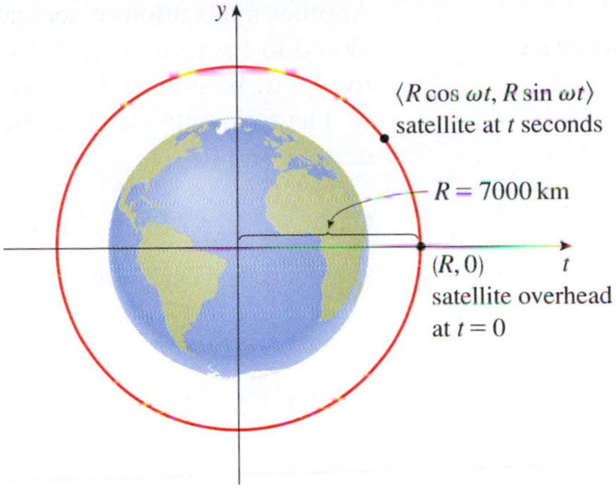

FIGURE 2

To find the velocity and acceleration, we differentiate the expression for $\mathbf{r}(t)$ to obtain

$$\text{velocity} = \mathbf{r}'(t) = \langle -R\omega \sin \omega t, R\omega \cos \omega t \rangle$$

$$\text{acceleration} = \mathbf{r}''(t) = \langle -R\omega^2 \cos \omega t, -R\omega^2 \sin \omega t \rangle$$

Since we have arranged our coordinate system so that the satellite is overhead at $(R, 0)$, which corresponds to $t = 0$, we can substitute $t = 0$ into the expressions for the velocity and acceleration vectors to obtain

$$\begin{aligned} \text{velocity} &= \mathbf{r}'(0) = \langle 0, R\omega \rangle \\ &\approx \langle 0, 2.81 \times 10^4 \rangle \\ \text{acceleration} &= \mathbf{r}''(0) = \langle -R\omega^2, 0 \rangle \\ &\approx \langle -1.13 \times 10^5, 0 \rangle \end{aligned}$$

where the velocity components are measured in km/h and the acceleration components are measured in km/h^2. The speed of the satellite is equal to the magnitude of the velocity vector, or 2.81×10^4 km/h. Note that the velocity vector points straight upward (tangential to the path of the satellite) and the acceleration vector points directly in toward the center of Earth (see Figure 3). Since the only force acting on the satellite is gravity, its acceleration should (and does) point in the direction of the pull of gravity. Converting the units of the acceleration vector to metric units, the magnitude of the acceleration vector becomes 8.7 m/s^2. Notice that the acceleration due to Earth's gravity at a height of 400 km is somewhat less than the value of 9.8 m/s^2 at Earth's surface.

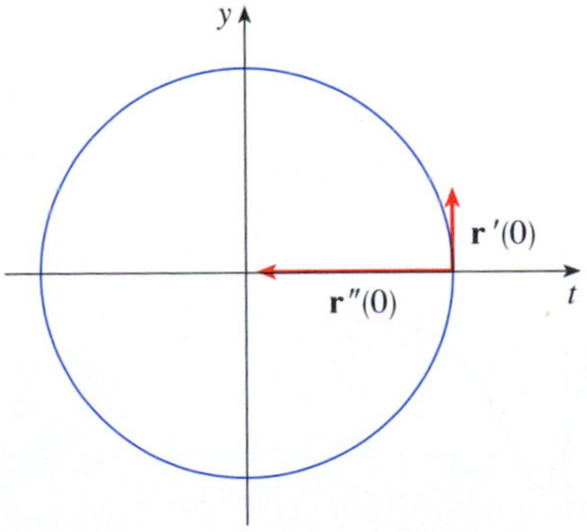

FIGURE 3
The velocity vector $\mathbf{r}'(0)$ and the acceleration vector $\mathbf{r}''(0)$

■

IMPLICIT SECOND DERIVATIVES

Another application of second derivatives occurs in curve sketching. (This idea is explored in Exercise 62 and developed further in Section 5.3.) Second and higher derivatives will be used in Chapter 10 to represent functions as sums of infinite series.

The following example shows how to find the second derivative of a function that is defined implicitly.

EXAMPLE 7 Find y'' if $x^4 + y^4 = 16$.

SOLUTION Differentiating the equation implicitly with respect to x, we get

$$4x^3 + 4y^3y' = 0$$

Solving for y' gives

$$y' = -\frac{x^3}{y^3} \tag{1}$$

To find y'' we differentiate this expression for y' using the Quotient Rule and remembering that y is a function of x:

$$y'' = \frac{d}{dx}\left(-\frac{x^3}{y^3}\right) = -\frac{y^3 D(x^3) - x^3 D(y^3)}{(y^3)^2}$$

$$= -\frac{y^3 \cdot 3x^2 - x^3(3y^2y')}{y^6}$$

If we now substitute Equation 1 into this expression we get

$$y'' = -\frac{3x^2y^3 - 3x^3y^2\left(\dfrac{-x^3}{y^3}\right)}{y^6}$$

$$= -\frac{3(x^2y^4 + x^6)}{y^7} = -\frac{3x^2(y^4 + x^4)}{y^7}$$

But the values of x and y must satisfy the original equation $x^4 + y^4 = 16$. So the answer simplifies to

$$y'' = -\frac{3x^2(16)}{y^7} = -48\frac{x^2}{y^7}$$

■

Figure 4 shows the graph of the curve $x^4 + y^4 = 16$ of Example 7. Notice that it is a stretched and flattened version of the circle $x^2 + y^2 = 4$. For this reason it is sometimes called a "fat circle." It starts out very steep on the left but quickly becomes very flat. This can be seen from the expression $y' = -x^3/y^3 = -(x/y)^3$.

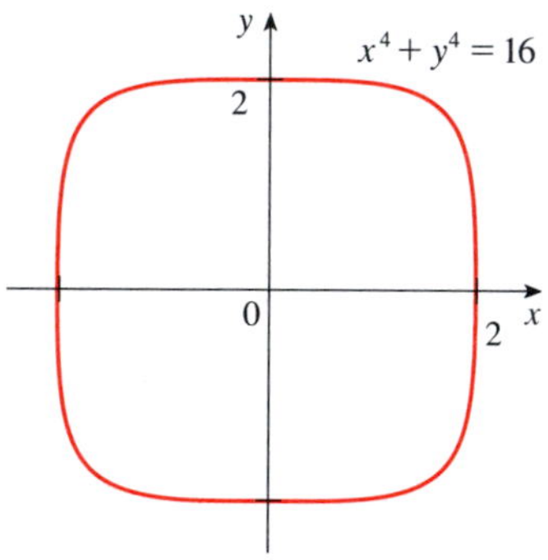

FIGURE 4

EXERCISES 3.8

1. The figure shows the graphs of f, f', and f''. Identify each curve, and explain your choices.

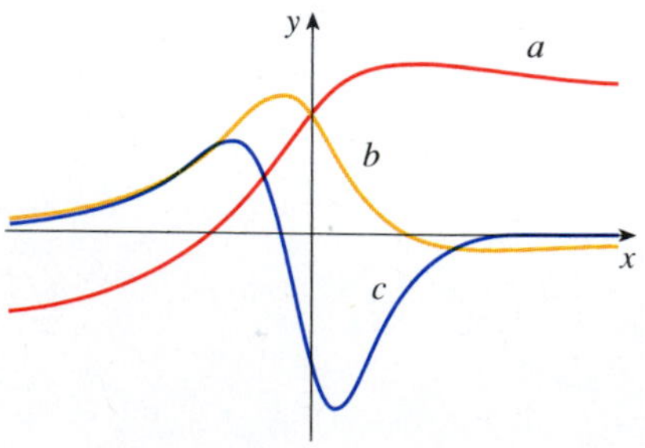

2. The figure shows the graphs of three functions. One is the position function of a car, one is the velocity of the car,

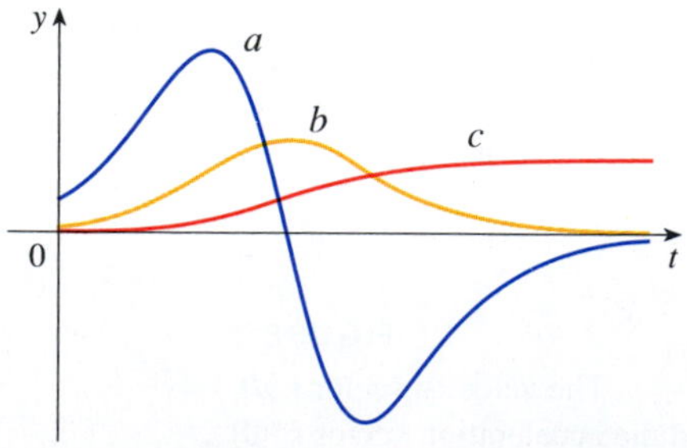

and one is its acceleration. Identify each curve, and explain your choices.

3–14 ■ Find the first and second derivatives of each function.

3. $f(x) = x^4 - 3x^3 + 16x$

4. $f(t) = t^{10} - 2t^7 + t^4 - 6t + 8$

5. $h(x) = \sqrt{x^2 + 1}$

6. $G(r) = \sqrt{r} + \sqrt[3]{r}$

7. $F(s) = (3s + 5)^8$

8. $g(u) = \dfrac{1}{\sqrt{1 - u}}$

9. $y = \dfrac{x}{1 - x}$

10. $y = x^n$

11. $y = (1 - x^2)^{3/4}$

12. $y = \dfrac{x^2}{x + 1}$

13. $H(t) = \tan^3(2t - 1)$

14. $g(s) = s^2 \cos s$

15. (a) If $f(x) = 2\cos x + \sin^2 x$, find $f'(x)$ and $f''(x)$.
(b) Check to see that your answers to part (a) are reasonable by comparing the graphs of f, f', and f''.

16. (a) If $f(x) = x/(x^2 + 1)$, find $f'(x)$ and $f''(x)$.
(b) Check to see that your answers to part (a) are reasonable by comparing the graphs of f, f', and f''.

17–18 ■ Find y'''.

17. $y = \sqrt{5t - 1}$

18. $y = \dfrac{1 - x}{1 + x}$

19. If $f(x) = (2 - 3x)^{-1/2}$, find $f(0)$, $f'(0)$, $f''(0)$, and $f'''(0)$.

20. If $g(t) = (2 - t^2)^6$, find $g(0)$, $g'(0)$, $g''(0)$, and $g'''(0)$.

21. If $f(\theta) = \cot \theta$, find $f'''(\pi/6)$.

22. If $g(x) = \sec x$, find $g^{(4)}(\pi/4)$.

23–26 ■ Find y'' by implicit differentiation.

23. $x^3 + y^3 = 1$

24. $\sqrt{x} + \sqrt{y} = 1$

25. $x^2 + 6xy + y^2 = 8$

26. $\dfrac{x^2}{a^2} - \dfrac{y^2}{b^2} = 1$

27. Find the first 73 derivatives of

$$f(x) = x - x^2 + x^3 - x^4 + x^5 - x^6$$

28–31 ■ Find a formula for $f^{(n)}(x)$.

28. $f(x) = \sqrt{x}$

29. $f(x) = x^n$

30. $f(x) = \dfrac{1}{(1 - x)^2}$

31. $f(x) = \dfrac{1}{3x^3}$

32–34 ■ Find the given derivative by finding the first few derivatives and observing the pattern that occurs.

32. $D^{99} \sin x$

33. $D^{50} \cos 2x$

34. $D^{35}\, x \sin x$

35–38 ■ The equation of motion is given for a particle, where s is in meters and t is in seconds. Find (a) the velocity and acceleration as functions of t, (b) the acceleration after 1 s, and (c) the acceleration at the instants when the velocity is 0.

35. $s = t^3 - 3t$

36. $s = t^2 - t + 1$

37. $s = At^2 + Bt + C$

38. $s = 2t^3 - 7t^2 + 4t + 1$

39–40 ■ An equation of motion is given, where s is in meters and t in seconds. Find (a) the times at which the acceleration is 0 and (b) the displacement and velocity at these times.

39. $s = t^4 - 4t^3 + 2$

40. $s = 2t^3 - 9t^2$

41–44 ■
(a) Sketch the curve traced by the vector equation.
(b) Find $\mathbf{r}'(t)$ and $\mathbf{r}''(t)$.
(c) Sketch the position vector $\mathbf{r}(t)$, the tangent vector $\mathbf{r}'(t)$, and $\mathbf{r}''(t)$ for the given value of t.

41. $\mathbf{r}(t) = \langle \cos t, \sin t \rangle, \quad t = \pi/4$

42. $\mathbf{r}(t) = t^3\,\mathbf{i} + t^2\,\mathbf{j}, \quad t = 1$

43. $\mathbf{r}(t) = (1 + t)\,\mathbf{i} + t^2\,\mathbf{j}, \quad t = 1$

44. $\mathbf{r}(t) = \langle 2\cos t, 3\sin t \rangle, \quad t = \pi/3$

45–48 ■ Find the acceleration at the given value of t.

45. $\mathbf{r}(t) = \langle t, 25t - 5t^2 \rangle, \quad t = 1$

46. $\mathbf{r}(t) = \langle 4\cos t, 3\sin t \rangle, \quad t = \pi/3$

47. $\mathbf{r}(t) = (t^4 + 7t)\,\mathbf{i} + (t^2 + 2t)\,\mathbf{j}, \quad t = 1$

48. $\mathbf{r}(t) = \sqrt{t^2 + 5}\,\mathbf{i} + t\,\mathbf{j}, \quad t = 2$

49. A satellite completes one orbit of Earth along the equator at an altitude of 1000 km every 1 h 48 min. Find the velocity, speed, and acceleration of the satellite at any given time. [*Note:* Earth's radius is approximately 6600 km.]

50. A satellite completes one orbit of Earth along the equator at an altitude of 2000 km every 2 h 8 min. Find the velocity, speed, and acceleration of the satellite at any given time. [*Note:* Earth's radius is approximately 6600 km.]

51. Compute the velocity, speed, and acceleration of the cannonball in Example 2 of Section 3.7 at the time the ball hits the ground.

52. Compute the velocity, speed, and acceleration of the cannonball in Exercise 51 when the ball reaches its maximum height.

53. A mass attached to a vertical spring has position function given by $y(t) = A \sin \omega t$, where A is the amplitude of its oscillations and ω is a constant.
(a) Find the velocity and acceleration as functions of time.
(b) Show that the acceleration is proportional to the displacement y.
(c) Show that the speed is a maximum when the acceleration is 0.

54. An object moves in such a way that its velocity v is never 0 and is related to its displacement s by the equation $v = \sqrt{2gs + c}$ where c and g are constants. Show that the acceleration is constant.

55. Find a second-degree polynomial P such that $P(2) = 5$, $P'(2) = 3$, and $P''(2) = 2$.

56. Find a third-degree polynomial Q such that $Q(1) = 1$, $Q'(1) = 3$, $Q''(1) = 6$, and $Q'''(1) = 12$.

57. If P is a polynomial of degree n, show that $P^{(m)}(x) = 0$ for $m > n$.

58. If $f(x) = |x^2 - x|$, find f' and f''. What are their domains? Sketch the graphs of all three functions.

59–61 ■ The function g is a twice differentiable function. Find f'' in terms of g, g', and g''.

59. $f(x) = xg(x^2)$

60. $f(x) = \dfrac{g(x)}{x}$

61. $f(x) = g(\sqrt{x})$

62. If $f(x) = 3x^5 - 10x^3 + 5$, graph both f and f''. On what intervals is $f''(x) > 0$? On those intervals, how is the graph of f related to its tangent lines? What about the intervals where $f''(x) < 0$?

63. (a) Compute the first few derivatives of the function $f(x) = 1/(x^2 + x)$ until you see that the computations are becoming algebraically unmanageable.

(b) Use the identity

$$\frac{1}{x(x+1)} = \frac{1}{x} - \frac{1}{x+1}$$

to compute the derivatives much more easily. Then find an expression for $f^{(n)}(x)$. This method of splitting up a fraction in terms of simpler fractions, called *partial fractions,* will be pursued further in Section 8.4.

64. (a) If $F(x) = f(x)g(x)$, where f and g have derivatives of all orders, show that

$$F'' = f''g + 2f'g' + fg''$$

(b) Find similar formulas for F''' and $F^{(4)}$.

(c) Guess a formula for $F^{(n)}$.

65. If $y = f(u)$ and $u = g(x)$, where f and g are twice differentiable functions, show that

$$\frac{d^2y}{dx^2} = \frac{d^2y}{du^2}\left(\frac{du}{dx}\right)^2 + \frac{dy}{du}\frac{d^2u}{dx^2}$$

66. If $y = f(u)$ and $u = g(x)$, where f and g possess third derivatives, find a formula for d^3y/dx^3 similar to the one given in Exercise 65.

3.9 SLOPES AND TANGENTS OF PARAMETRIC CURVES

Suppose that a curve C is given by the parametric equations

$$x = x(t) \qquad y = y(t)$$

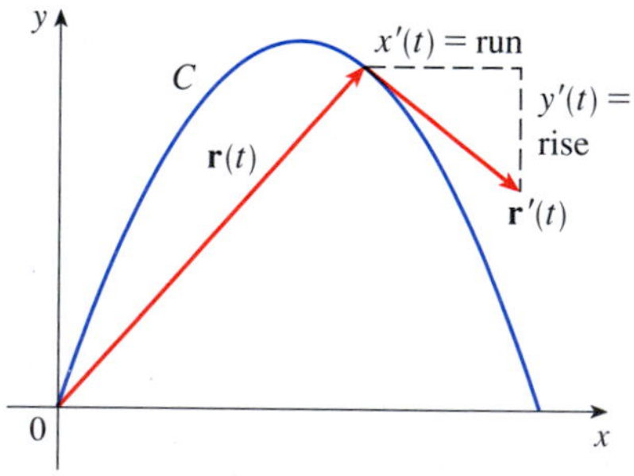

FIGURE 1

In Section 3.7 the point $(x(t), y(t))$ was viewed as the endpoint of the position vector $\mathbf{r}(t) = \langle x(t), y(t)\rangle$. The derivative $\mathbf{r}'(t) = \langle x'(t), y'(t)\rangle$ is a vector that is tangent to the curve C as shown in Figure 1.

The slope of the tangent vector and, hence, the slope of the tangent line is given by

$$\text{slope} = \frac{\text{rise}}{\text{run}} = \frac{y'(t)}{x'(t)}$$

Another way to see this is by using the Chain Rule

$$\frac{dy}{dt} = \frac{dy}{dx}\frac{dx}{dt}$$

which can be written as

$$\frac{dy}{dx} = \frac{\dfrac{dy}{dt}}{\dfrac{dx}{dt}} \tag{1}$$

A convenient way to remember this is by cancelling the dt's.

EXAMPLE 1 Show that the parametric equations

$$x = 3\cos t \qquad y = 2\sin t \qquad 0 \leqslant t \leqslant 2\pi$$

parametrize an ellipse. Find equations of the tangent and normal lines to this ellipse at the point $(\frac{3}{2}, \sqrt{3})$.

SOLUTION To show that the curve is an ellipse, we convert the parametric equations $x = 3\cos t$, $y = 2\sin t$ to rectangular form:

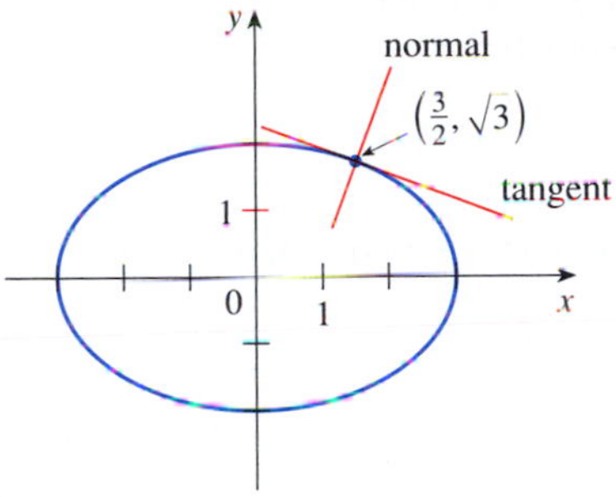

FIGURE 2

$$\left(\frac{x}{3}\right)^2 + \left(\frac{y}{2}\right)^2 = \cos^2 t + \sin^2 t = 1$$

The equation $(x/3)^2 + (y/2)^2 = 1$ is the standard form of the equation of an ellipse with x-intercepts ± 3 and y-intercepts ± 2 (see Figure 2).

To find the tangent line at the given point, we must first find the value of the parameter t that corresponds to the given point $(\frac{3}{2}, \sqrt{3})$. Equating $3\cos t$ with $3/2$ and $2\sin t$ with $\sqrt{3}$, we see that $\cos t = \frac{1}{2}$ and $\sin t = \sqrt{3}/2$, so $t = \pi/3$. Differentiating the parametric equations, we find

$$\frac{dy}{dx} = \frac{dy/dt}{dx/dt} = \frac{2\cos t}{-3\sin t}$$

Letting $t = \pi/3$, we obtain

$$\text{slope} = \frac{2}{-3\sqrt{3}}$$

An equation of the tangent line through $(\frac{3}{2}, \sqrt{3})$ is

$$y - \sqrt{3} = \frac{2}{-3\sqrt{3}}\left(x - \frac{3}{2}\right)$$

The slope of the normal line is $3\sqrt{3}/2$ (the negative reciprocal of the slope of the tangent line). So an equation of the normal line is

$$y - \sqrt{3} = \frac{3\sqrt{3}}{2}\left(x - \frac{3}{2}\right)$$

■

EXAMPLE 2 A curve C is defined by the parametric equations

$$x = t^2 \qquad y = t^3 - 3t$$

(a) Show that C has two tangents at the point $(3, 0)$ and find their equations.
(b) Find the point(s) on C where the tangent is horizontal or vertical.

SOLUTION
(a) The first task is to find the values of t that correspond to the point $(3, 0)$. The equation $3 = x = t^2$ corresponds to $t = \sqrt{3}$ and $t = -\sqrt{3}$. These two values of t also satisfy the equation $0 = y = t^3 - 3t = t(t^2 - 3)$. The two different values of t indicate that C crosses itself at $(3, 0)$ and probably has two different tangents there.

Since

$$\frac{dy}{dx} = \frac{\dfrac{dy}{dt}}{\dfrac{dx}{dt}} = \frac{3t^2 - 3}{2t}$$

the slope of the tangent line when $t = \sqrt{3}$ is $6/(2\sqrt{3}) = \sqrt{3}$ and the slope when $t = -\sqrt{3}$ is $-\sqrt{3}$. Therefore, we obtain the two tangent lines

$$y = \sqrt{3}\,(x - 3) \qquad \text{and} \qquad y = -\sqrt{3}\,(x - 3)$$

(b) The curve C has a vertical tangent when the slope dy/dx is undefined. Since

$$\frac{dy}{dx} = \frac{\dfrac{dy}{dt}}{\dfrac{dx}{dt}}$$

a vertical tangent occurs when $dx/dt = 0$ and $dy/dt \neq 0$. Now $dx/dt = 2t = 0$ at $t = 0$, which corresponds to the point $(0, 0)$ on C. A horizontal tangent occurs when $dy/dx = 0$ or, in other words, when $dy/dt = 0$ and $dx/dt \neq 0$. For the curve C, we have $dy/dt = 3t^2 - 3 = 0$, at $t = 1$ or -1. The corresponding points on C are $(1, -2)$ and $(1, 2)$. The curve C and all the tangents are shown in Figure 3.

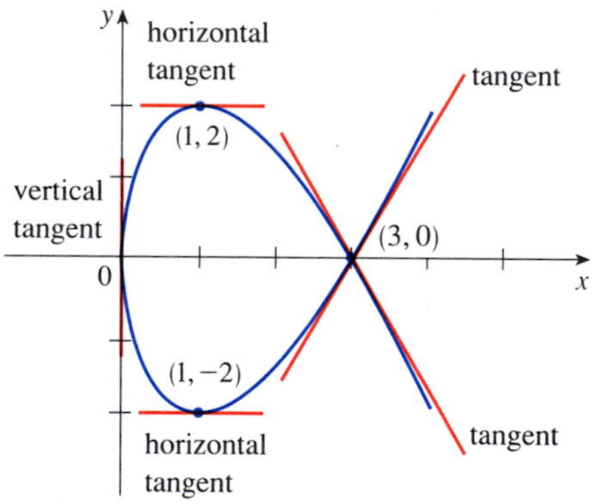

FIGURE 3

■

EXAMPLE 3 The curve traced out by a point P on the circumference of a circle as the circle rolls along a straight line is called a **cycloid**. Suppose the radius of the circle is r and the point P starts at the origin.
(a) Find parametric equations of the cycloid.
(b) For $r = 1$, find an equation of the tangent line to the cycloid at the point corresponding to $\theta = \pi/3$.

SOLUTION

(a) We choose the parameter θ, which we define as the angle formed by the radial line from P to the center of the circle and the vertical line through the center. When the circle has rotated through θ radians, the distance it has rolled from the origin is

$$|OT| = \text{arc } PT = r\theta$$

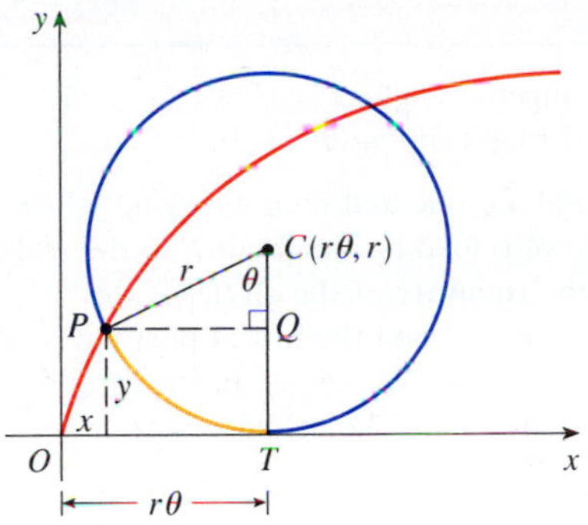

FIGURE 4

and so the center of the circle is $C(r\theta, r)$. Let the coordinates of P be (x, y). Then from Figure 4 we see that

$$x = |OT| - |PQ| = r\theta - r\sin\theta = r(\theta - \sin\theta)$$

$$y = |TC| - |QC| = r - r\cos\theta = r(1 - \cos\theta)$$

Therefore, parametric equations of the cycloid are

$$x = r(\theta - \sin\theta) \qquad y = r(1 - \cos\theta) \qquad \theta \in \mathbb{R}$$

One arch of the cycloid comes from one rotation of the circle and so is described by $0 \leqslant \theta \leqslant 2\pi$.

Although it is possible to eliminate the parameter θ, the resulting Cartesian equation $y = f(x)$ is very complicated and not as convenient to work with as the parametric equations.

(b) When $r = 1$, the equations of the cycloid are

$$x = \theta - \sin\theta \qquad y = 1 - \cos\theta$$

The slope of the tangent line at an arbitrary point on the cycloid is

$$\frac{dy}{dx} = \frac{\dfrac{dy}{d\theta}}{\dfrac{dx}{d\theta}} = \frac{\sin\theta}{1 - \cos\theta}$$

When $\theta = \pi/3$ the slope of the tangent is

$$\text{slope} = \frac{\sin(\pi/3)}{1 - \cos(\pi/3)} = \sqrt{3}$$

The point on the cycloid corresponding to $\theta = \pi/3$ is

$$x = \left(\frac{\pi}{3} - \sin\frac{\pi}{3}\right) = \frac{\pi}{3} - \frac{\sqrt{3}}{2} \qquad y = \left(1 - \cos\frac{\pi}{3}\right) = \frac{1}{2}$$

Therefore, the tangent line is

$$y - \frac{1}{2} = \sqrt{3}\left(x - \frac{\pi}{3} + \frac{\sqrt{3}}{2}\right) \qquad \text{or} \qquad y = \sqrt{3}\,x + \left(2 - \frac{\pi}{\sqrt{3}}\right)$$

Figure 5 shows the curve C and tangent at $\theta = \pi/3$.

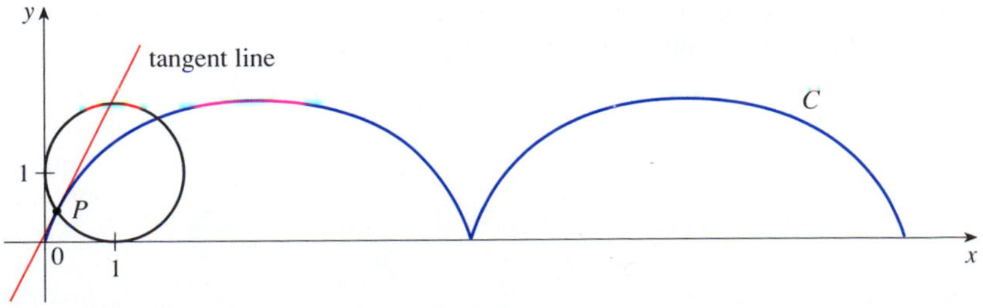

FIGURE 5

■

EXERCISES 3.9

1–6 ■ Find an equation of the tangent to the curve at the point corresponding to the given value of the parameter.

1. $x = t^2 + t,\ y = t^2 - t;\quad t = 0$

2. $x = 1 - t^3,\ y = t^2 - 3t + 1;\quad t = 1$

3. $x = t^2 + t,\ y = \sqrt{t};\quad t = 4$

4. $x = 2\sin\theta,\ y = 3\cos\theta;\quad \theta = \pi/4$

5. $\mathbf{r}(t) = t\sin t\,\mathbf{i} + t\cos t\,\mathbf{j};\quad t = \pi$

6. $\mathbf{r}(t) = \cos t\,\mathbf{i} + \tan t\,\mathbf{j};\quad t = \pi/4$

7–10 ■ Find an equation of the tangent to the curve at the given point or endpoint of the vector by two methods: (a) without eliminating the parameter and (b) by first eliminating the parameter.

7. $x = 1 - t,\ y = 1 - t^2;\quad (1, 1)$

8. $x = 2t + 3,\ y = t^2 + 2t;\quad (5, 3)$

9. $\mathbf{r}(t) = 5\cos t\,\mathbf{i} + 5\sin t\,\mathbf{j};\quad 3\mathbf{i} + 4\mathbf{j}$

10. $\mathbf{r}(t) = t^3\mathbf{i} + t^2\mathbf{j};\quad \mathbf{i} + \mathbf{j}$

11–16 ■ Find the point(s) on the curve where the tangent is horizontal or vertical.

11. $x = 1 - 2\cos t,\ y = 2 + 3\sin t$

12. $x = t^3 - 3t^2,\ y = t^3 - 3t$

13. $x = t(t^2 - 3),\ y = 3(t^2 - 3)$

14. $x = \sin 2t,\ y = \sin t$

15. $x = \dfrac{3t}{1 + t^3},\ y = \dfrac{3t^2}{1 + t^3}$

16. $x = a(\cos\theta - \cos^2\theta),\ y = a(\sin\theta - \sin\theta\cos\theta)$

17. Show that the curve $x = \cos t,\ y = \sin t\cos t$ has two tangents at $(0, 0)$ and find their equations. Sketch the curve.

18. At what point does the curve

$$x = 1 - 2\cos^2 t \qquad y = (\tan t)(1 - 2\cos^2 t)$$

cross itself? Find equations of both tangents at that point.

19. At what point does the curve $x = t(t^2 - 3),\ y = 3(t^2 - 3)$ cross itself? Find equations of both tangents at that point.

20. Show that the cycloid $x = r(\theta - \sin\theta),\ y = r(1 - \cos\theta)$ has a vertical tangent at the origin by showing that $\lim_{\theta\to 0} |dy/dx| \to \infty$.

21. At what points on the curve $x = t^3 + 4t,\ y = 6t^2$ is the tangent parallel to the line with equations $x = -7t,\ y = 12t - 5$?

22. Find equations of the tangents to the curve $x = 3t^2 + 1,\ y = 2t^3 + 1$ that pass through the point $(4, 3)$.

23. A string is wound around a circle and then unwound while being held taut. The curve traced by the point P at the end of the string is called the **involute** of the circle. If the circle has radius r and center O and the initial position of P is $(r, 0)$, and if the parameter θ is chosen as in the figure, show that parametric equations of the involute are

$$x = r(\cos\theta + \theta\sin\theta) \qquad y = r(\sin\theta - \theta\cos\theta)$$

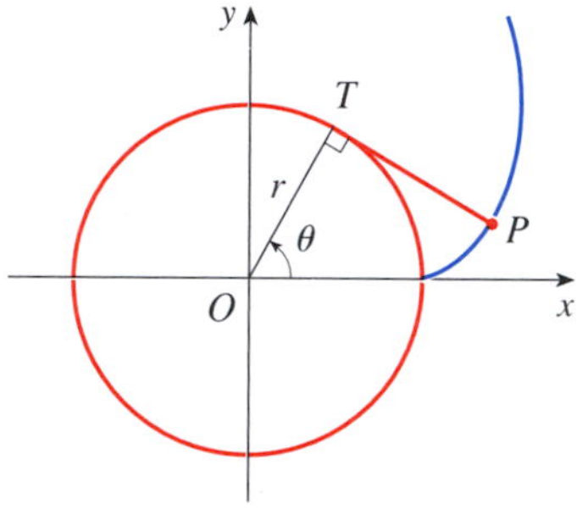

24. Let P be a point at a distance d from the center of a circle of radius r. The curve traced out by P as the circle rolls along a straight line is called a **trochoid**. (Think of the motion of a point on a spoke of a bicycle wheel.) The cycloid is the special case of a trochoid with $d = r$. Using the same parameter θ as for the cycloid and assuming the line is the x-axis and $\theta = 0$ when P is at one of its lowest points, show that the parametric equations of the trochoid are

$$x = r\theta - d\sin\theta$$
$$y = r - d\cos\theta$$

Sketch the trochoid for the cases $d < r$ and $d > r$.

25. (a) A circle C of radius b rolls on the inside of a larger circle with center O and radius a. The curve traced out by a fixed point P on C is called a **hypocycloid**. If the initial position of P is $(a, 0)$ and the parameter θ is chosen as in the figure, show that parametric equations of the hypocycloid are

$$x = (a - b)\cos\theta + b\cos\left(\frac{a - b}{b}\theta\right)$$

$$y = (a - b)\sin\theta - b\sin\left(\frac{a - b}{b}\theta\right)$$

(b) If $b = a/4$, the hypocycloid is called a **hypocycloid of four cusps**, or an **astroid**. Show that in this case the parametric equations reduce to

$$x = a\cos^3\theta \qquad y = a\sin^3\theta$$

and sketch the curve.

(c) Graph the hypocycloid for several other values of a/b to show the various possibilities that can occur.

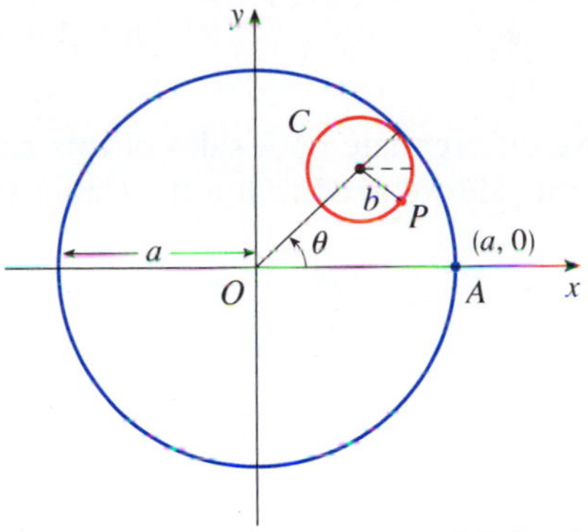

26. If the circle C of Exercise 25 rolls on the *outside* of the larger circle, then the curve traced out by P is called an **epicycloid**. (The special case in which $a = b$ is called a **cardioid**, because it is shaped like a heart.)

(a) Find parametric equations for the epicycloid.

(b) Graph the epicycloid for several values of a/b to show the various possibilities that can occur.

27. (a) Find the slope of the tangent line to the trochoid $x = r\theta - d\sin\theta$, $y = r - d\cos\theta$ in terms of θ. (See Exercise 24.)

(b) Show that if $d < r$, then the trochoid does not have a vertical tangent.

3.10 RELATED RATES

In a related rates problem the idea is to compute the rate of change of one quantity in terms of the rate of change of another quantity (which may be more easily measured). The procedure is to find an equation that relates the two quantities and then use the Chain Rule to differentiate both sides with respect to time.

EXAMPLE 1 Air is being pumped into a spherical balloon so that its volume increases at a rate of 100 cm^3/s. How fast is the radius of the balloon increasing when the diameter is 50 cm?

According to the Principles of Problem Solving discussed in Section 4 of Review and Preview, the first step is to *Understand the problem*. This includes reading the problem carefully, identifying the given and the unknown, and introducing suitable notation.

SOLUTION We start by identifying two things:

the *given information:*

the rate of increase of the volume of air is 100 cm^3/s

and the *unknown*:

the rate of increase of the radius when the diameter is 50 cm

In order to express these quantities mathematically we introduce some suggestive *notation*:

Let V be the volume of the balloon and let r be its radius.

The key thing to remember is that rates of change are derivatives. In this problem, the volume and the radius are both functions of the time t. The rate of increase of the volume with respect to time is the derivative dV/dt and the rate of increase of the radius is dr/dt. We can therefore restate the given and the unknown as follows:

Given: $\dfrac{dV}{dt} = 100\ cm^3/s$

Unknown: $\dfrac{dr}{dt}$ when $r = 25$ cm

The second stage of problem solving is to *Think of a plan* for connecting the given and the unknown.

In order to connect dV/dt and dr/dt we first relate V and r by the formula for the volume of a sphere:

$$V = \frac{4}{3}\pi r^3$$

In order to use the given information, we differentiate both sides of this equation with respect to t. To differentiate the right side we need to use the Chain Rule:

$$\frac{dV}{dt} = \frac{dV}{dr}\frac{dr}{dt} = 4\pi r^2 \frac{dr}{dt}$$

Now we solve for the unknown quantity:

$$\frac{dr}{dt} = \frac{1}{4\pi r^2}\frac{dV}{dt}$$

If we put $r = 25$ and $dV/dt = 100$ in this equation, we obtain

$$\frac{dr}{dt} = \frac{1}{4\pi(25)^2}100 = \frac{1}{25\pi}$$

The radius of the balloon is increasing at the rate of $1/(25\pi)$ cm/s. ■

EXAMPLE 2 A ladder 10 ft long rests against a vertical wall. If the bottom of the ladder slides away from the wall at a rate of 1 ft/s, how fast is the top of the ladder sliding down the wall when the bottom of the ladder is 6 ft from the wall?

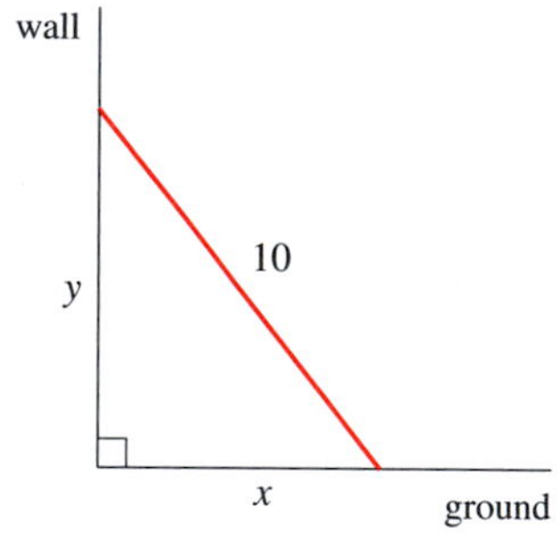

FIGURE 1

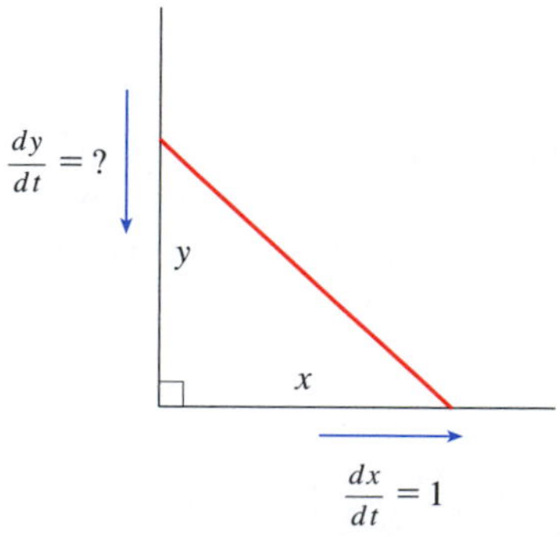

FIGURE 2

SOLUTION We first draw a diagram and label it as in Figure 1. Let x feet be the distance from the bottom of the ladder to the wall and y feet the distance from the top of the ladder to the ground. Note that x and y are both functions of t (time).

We are given that $dx/dt = 1$ ft/s and we are asked to find dy/dt when $x = 6$ ft (see Figure 2). In this question, the relationship between x and y is given by the Pythagorean Theorem:

$$x^2 + y^2 = 100$$

Differentiating each side with respect to t using the Chain Rule, we have

$$2x\frac{dx}{dt} + 2y\frac{dy}{dt} = 0$$

and solving this equation for the desired rate, we obtain

$$\frac{dy}{dt} = -\frac{x}{y}\frac{dx}{dt}$$

When $x = 6$, the Pythagorean Theorem gives $y = 8$ and so, substituting these values and $dx/dt = 1$, we have

$$\frac{dy}{dt} = -\frac{6}{8}(1) = -\frac{3}{4} \text{ ft/s}$$ ■

EXAMPLE 3 A water tank has the shape of an inverted circular cone with base radius 2 m and height 4 m. If water is being pumped into the tank at a rate of 2 m^3/min, find the rate at which the water level is rising when the water is 3 m deep.

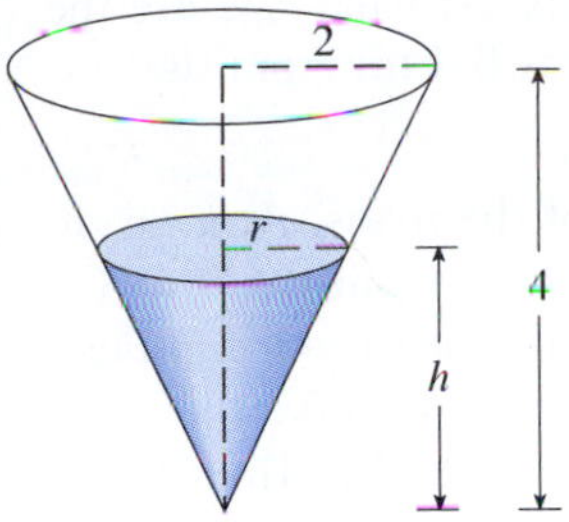

FIGURE 3

SOLUTION We first sketch the cone and label it as in Figure 3. Let V, r, and h be the volume of the water, the radius of the surface, and the height at time t, where t is measured in minutes.

We are given that $dV/dt = 2$ m^3/min and we are asked to find dh/dt when h is 3 m. The quantities V and h are related by the equation

$$V = \tfrac{1}{3}\pi r^2 h$$

but it is very useful to express V as a function of h alone. In order to eliminate r we use the similar triangles in Figure 3 to write

$$\frac{r}{h} = \frac{2}{4} \qquad r = \frac{h}{2}$$

and the expression for V becomes

$$V = \frac{1}{3}\pi\left(\frac{h}{2}\right)^2 h = \frac{\pi}{12}h^3$$

Now we can differentiate each side with respect to t:

$$\frac{dV}{dt} = \frac{\pi}{4}h^2\frac{dh}{dt}$$

so

$$\frac{dh}{dt} = \frac{4}{\pi h^2}\frac{dV}{dt}$$

Substituting $h = 3$ m and $dV/dt = 2$ m^3/min, we have

$$\frac{dh}{dt} = \frac{4}{\pi(3)^2}\cdot 2 = \frac{8}{9\pi} \approx 0.28 \text{ m/min}$$

■

Look back: What have we learned from Examples 1–3 that will help us solve future problems?

STRATEGY It is useful to recall some of the problem-solving principles from Section 4 of Review and Preview and adapt them to related rates in light of our experience in Examples 1–3:

1. Read the problem carefully.
2. Draw a diagram if possible.
3. Introduce notation. Assign symbols to all quantities that are functions of time.
4. Express the given information and the required rate in terms of derivatives.
5. Write an equation that relates the various quantities of the problem. If necessary, use the geometry of the situation to eliminate one of the variables by substitution (as in Example 3).
6. Use the Chain Rule to differentiate both sides of the equation with respect to t.
7. Substitute the given information into the resulting equation and solve for the unknown rate.

WARNING: A common error is to substitute the given numerical information (for quantities that vary with time) too early. This should be done only *after* the differentiation. (Step 7 follows Step 6.) For instance, in Example 3 we dealt with general values of h until we finally substituted $h = 3$ at the last stage. (If we had put $h = 3$ earlier, we would have gotten $dV/dt = 0$, which is clearly wrong.)

The following examples are further illustrations of the strategy.

EXAMPLE 4 Car A is traveling west at 50 mi/h and car B is traveling north at 60 mi/h. Both are headed for the intersection of the two roads. At what rate are the cars approaching each other when car A is 0.3 mi and car B is 0.4 mi from the intersection?

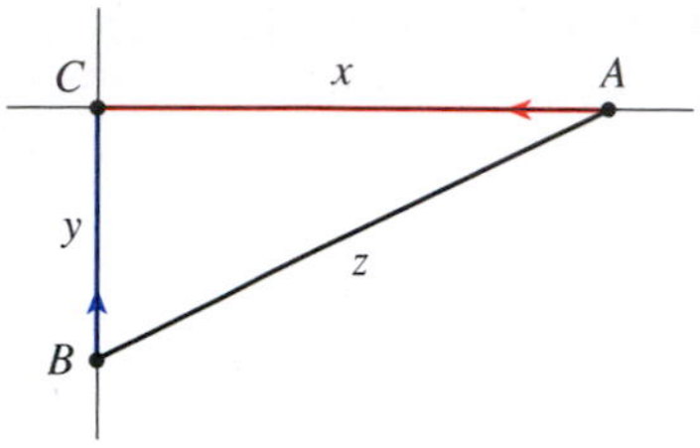

FIGURE 4

SOLUTION We draw Figure 4 where C is the intersection of the roads. At a given time t, let x be the distance from car A to C, let y be the distance from car B to C, and let z be the distance between the cars, where x, y, and z are measured in miles.

We are given that $dx/dt = -50$ mi/h and $dy/dt = -60$ mi/h. (We take the derivatives to be negative since x and y are decreasing.) We are asked to find dz/dt. The equation that relates x, y, and z is given by the Pythagorean Theorem:

$$z^2 = x^2 + y^2$$

Differentiating each side with respect to t, we have

$$2z\frac{dz}{dt} = 2x\frac{dx}{dt} + 2y\frac{dy}{dt}$$

$$\frac{dz}{dt} = \frac{1}{z}\left(x\frac{dx}{dt} + y\frac{dy}{dt}\right)$$

When $x = 0.3$ mi and $y = 0.4$ mi, the Pythagorean Theorem gives $z = 0.5$ mi, so

$$\frac{dz}{dt} = \frac{1}{0.5}[0.3(-50) + 0.4(-60)]$$

$$= -78 \text{ mi/h}$$

The cars are approaching each other at a rate of 78 mi/h. ■

EXAMPLE 5 A man walks along a straight path at a speed of 4 ft/s. A searchlight is located on the ground 20 ft from the path and is kept focused on the man. At what rate is the searchlight rotating when the man is 15 ft from the point on the path closest to the searchlight?

FIGURE 5

SOLUTION We draw Figure 5 and let x be the distance from the point on the path closest to the searchlight to the man. We let θ be the angle between the beam of the searchlight and the perpendicular to the path.

We are given that $dx/dt = 4$ ft/s and are asked to find $d\theta/dt$ when $x = 15$. The equation that relates x and θ can be written from Figure 5:

$$\frac{x}{20} = \tan\theta \qquad x = 20\tan\theta$$

Differentiating each side with respect to t, we get

$$\frac{dx}{dt} = 20\sec^2\theta\,\frac{d\theta}{dt}$$

so $$\frac{d\theta}{dt} = \tfrac{1}{20}\cos^2\theta\,\frac{dx}{dt} = \tfrac{1}{20}\cos^2\theta(4) = \tfrac{1}{5}\cos^2\theta$$

When $x = 15$, the length of the beam is 25 (by the Pythagorean Theorem), so

$\cos\theta = \frac{4}{5}$ and

$$\frac{d\theta}{dt} = \frac{1}{5}\left(\frac{4}{5}\right)^2 = \frac{16}{125} = 0.128$$

The searchlight is rotating at a rate of 0.128 rad/s. ■

EXERCISES 3.10

1. If V is the volume of a cube with edge length x, find dV/dt in terms of dx/dt.

2. If A is the area of a circle with radius r, find dA/dt in terms of dr/dt.

3. If $xy = 1$ and $dx/dt = 4$, find dy/dt when $x = 2$.

4. If $x^2 + 3xy + y^2 = 1$ and $dy/dt = 2$, find dx/dt when $y = 1$.

5. A spherical snowball is melting in such a way that its volume is decreasing at a rate of 1 cm^3/min. At what rate is the diameter decreasing when the diameter is 10 cm?

6. If a snowball melts so that its surface area decreases at a rate of 1 cm^2/min, find the rate at which the diameter decreases when the diameter is 10 cm.

7. A street light is at the top of a 15-ft-tall pole. A man 6 ft tall walks away from the pole with a speed of 5 ft/s along a straight path.
(a) How fast is the tip of his shadow moving when he is 40 ft from the pole?
(b) How fast is his shadow lengthening at that point?

8. A spotlight on the ground shines on a wall 12 m away. If a man 2 m tall walks from the spotlight toward the building at a speed of 1.6 m/s, how fast is his shadow on the building decreasing when he is 4 m from the building?

9. A plane flying horizontally at an altitude of 1 mi and a speed of 500 mi/h passes directly over a radar station. Find the rate at which the distance from the plane to the station is increasing when it is 2 mi away from the station.

10. A baseball diamond is a square with side 90 ft. A batter hits the ball and runs toward first base with a speed of 24 ft/s.
(a) At what rate is his distance from second base decreasing when he is halfway to first base?
(b) At what rate is his distance from third base increasing at the same moment?

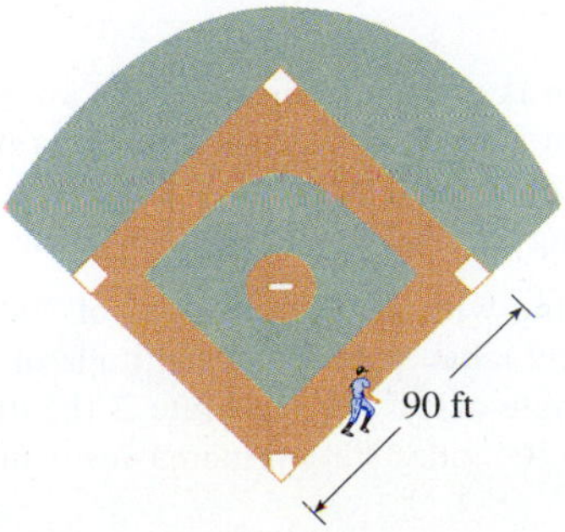

11. Two cars start moving from the same point. One travels south at 60 mi/h and the other travels west at 25 mi/h. At what rate is the distance between the cars increasing two hours later?

12. At noon, ship A is 150 km west of ship B. Ship A is sailing east at 35 km/h and ship B is sailing north at 25 km/h. How fast is the distance between the ships changing at 4:00 P.M.?

13. At noon, ship A is 100 km west of ship B. Ship A is sailing south at 35 km/h and ship B is sailing north at 25 km/h. How fast is the distance between the ships changing at 4:00 P.M.?

14. A man starts walking north at 4 ft/s from a point P. Five minutes later a woman starts walking south at 5 ft/s from a point 500 ft due east of P. At what rate are the people moving apart 15 min after the woman starts walking?

15. The altitude of a triangle is increasing at a rate of 1 cm/min while the area of the triangle is increasing at a rate of 2 cm^2/min. At what rate is the base of the triangle changing when the altitude is 10 cm and the area is 100 cm^2?

16. A boat is pulled into a dock by a rope attached to the bow of the boat and passing through a pulley on the dock that is 1 m higher than the bow of the boat. If the rope is pulled in at a rate of 1 m/s, how fast is the boat approaching the dock when it is 8 m from the dock?

17. Water is leaking out of an inverted conical tank at a rate of 10,000 cm^3/min at the same time that water is being pumped into the tank at a constant rate. The tank has height 6 m and the diameter at the top is 4 m. If the water level is rising at a rate of 20 cm/min when the height of the water is 2 m, find the rate at which water is being pumped into the tank.

18. A trough is 10 ft long and its ends have the shape of isosceles triangles that are 3 ft across at the top and have a height of 1 ft. If the trough is filled with water at a rate of 12 ft^3/min, how fast is the water level rising when the water is 6 inches deep?

19. A water trough is 10 m long and a cross-section has the shape of an isosceles trapezoid that is 30 cm wide at the bottom, 80 cm wide at the top, and has height 50 cm. If the trough is being filled with water at the rate of 0.2 m^3/min, how fast is the water level rising when the water is 30 cm deep?

20. A swimming pool is 20 ft wide, 40 ft long, 3 ft deep at the shallow end, and 9 ft deep at its deepest point. A cross-section is shown in the figure. If the pool is being filled at a rate of 0.8 ft^3/min, how fast is the water level rising when the depth at the deepest point is 5 ft?

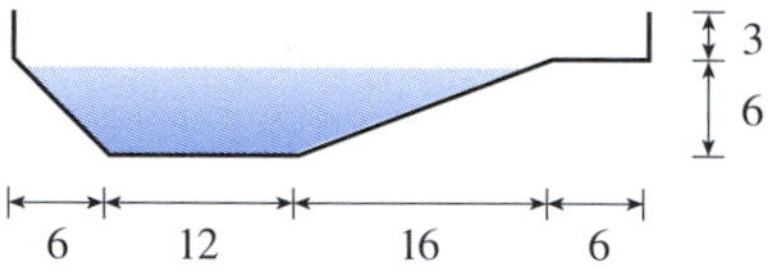

21. Gravel is being dumped from a conveyor belt at a rate of 30 ft^3/min and its coarseness is such that it forms a pile in the shape of a cone whose base diameter and height are always equal. How fast is the height of the pile increasing when the pile is 10 ft high?

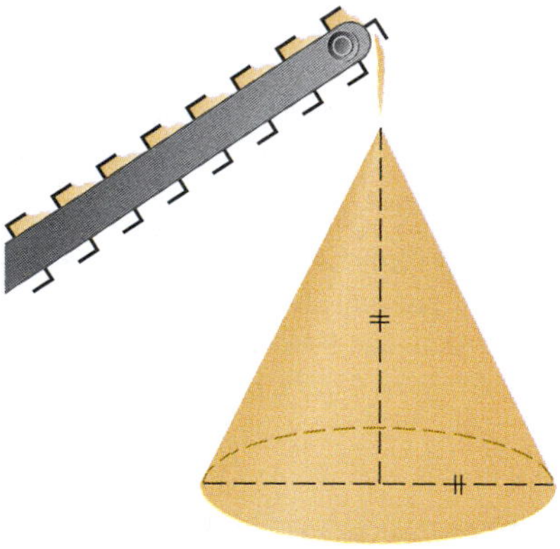

22. A kite 100 ft above the ground moves horizontally at a speed of 8 ft/s. At what rate is the angle between the string and the horizontal decreasing when 200 ft of string have been let out?

23. Two sides of a triangle are 4 m and 5 m in length and the angle between them is increasing at a rate of 0.06 rad/s. Find the rate at which the area of the triangle is increasing when the angle between the sides of fixed length is $\pi/3$.

24. Two sides of a triangle have lengths 12 m and 15 m. The angle between them is increasing at a rate of $2°$/min. How fast is the length of the third side increasing when the angle between the sides of fixed length is 60°?

25. Boyle's Law states that when a sample of gas is compressed at a constant temperature, the pressure P and volume V satisfy the equation $PV = C$, where C is a constant. Suppose that at a certain instant the volume is 600 cm^3, the pressure is 150 kPa, and the pressure is increasing at a rate of 20 kPa/min. At what rate is the volume decreasing at this instant?

26. When air expands adiabatically (without gaining or losing heat), its pressure P and volume V are related by the equation $PV^{1.4} = C$, where C is a constant. Suppose that at a certain instant the volume is 400 cm^3 and the pressure is 80 kPa and is decreasing at a rate of 10 kPa/min. At what rate is the volume increasing at this instant?

27. A television camera is positioned 4000 ft from the base of a rocket launching pad. A rocket rises vertically and its speed is 600 ft/s when it has risen 3000 ft.
(a) How fast is the distance from the television camera to the rocket changing at that moment?
(b) If the television camera is always kept focused on the rocket, how fast is the camera's angle of elevation changing at that same moment?

28. Two carts, A and B, are connected by a rope 39 ft long that passes over a pulley P. The point Q is on the floor 12 ft directly beneath P and between the carts. Cart A is being pulled away from Q at a speed of 2 ft/s. How fast is cart B moving toward Q at the instant when cart A is 5 ft from Q?

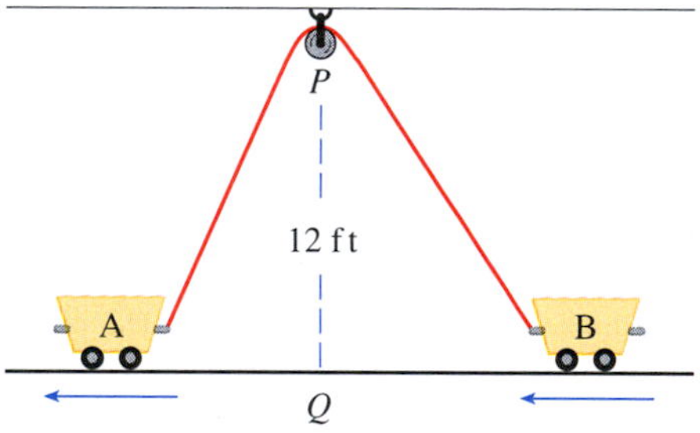

29. A ladder 10 ft long rests against a vertical wall. If the bottom of the ladder slides away from the wall at a speed of 2 ft/s, how fast is the angle between the top of the ladder and the wall changing when the angle is $\pi/4$ radians?

30. A lighthouse is on a small island 3 km away from the nearest point P on a straight shoreline and its light makes four revolutions per minute. How fast is the beam of light moving along the shoreline when it is 1 km from P?

31. A plane flying with a constant speed of 300 km/h passes over a ground radar station at an altitude of 1 km and climbs at angle of 30°. At what rate is the distance from the plane to the radar station increasing 1 min later?

32. Two people start from the same point. One walks east at 3 mi/h and the other walks northeast at 2 mi/h. How fast is the distance between the people changing after 15 min?

33. A runner runs around a circular track of radius 100 m at a constant speed of 7 m/s. The runner's friend is standing at a distance 200 m from the center of the track. How fast is the distance between the friends changing when the distance between them is 200 m?

34. The minute hand on a watch is 8 mm long and the hour hand is 4 mm long. How fast is the distance between the tips of the hands changing at one o'clock?

3.11 DIFFERENTIALS; LINEAR AND QUADRATIC APPROXIMATIONS

We have used the Leibniz notation dy/dx to denote the derivative of y with respect to x, but we have regarded it as a single entity and not a ratio. In this section we give the quantities dy and dx separate meanings in such a way that their ratio is equal to the derivative. We also see that these quantities, called *differentials*, are useful in finding approximate values of functions.

(1) DEFINITION Let $y = f(x)$, where f is a differentiable function. Then the **differential** dx is an independent variable; that is, dx can be given the value of any real number. The **differential** dy is then defined in terms of dx by the equation

$$dy = f'(x)\,dx$$

NOTE 1: The differentials dx and dy are both variables, but dx is an independent variable, whereas dy is a dependent variable—it depends on the values of x and dx. If dx is given a specific value and x is taken to be some specific number in the domain of f, then the numerical value of dy is determined.

NOTE 2: If $dx \neq 0$, we can divide both sides of the equation in Definition 1 by dx to obtain

$$\frac{dy}{dx} = f'(x)$$

We have seen similar equations before, but now the left side can genuinely be interpreted as a ratio of differentials.

EXAMPLE 1

(a) Find dy if $y = x^3 + 2x^2$.

(b) Find the value of dy when $x = 2$ and $dx = 0.1$.

SOLUTION

(a) If $f(x) = x^3 + 2x^2$, then $f'(x) = 3x^2 + 4x$, so

$$dy = (3x^2 + 4x)\,dx$$

(b) Substituting $x = 2$ and $dx = 0.1$ in the expression for dy, we have

$$dy = (3 \cdot 2^2 + 4 \cdot 2)0.1 = 2$$

■

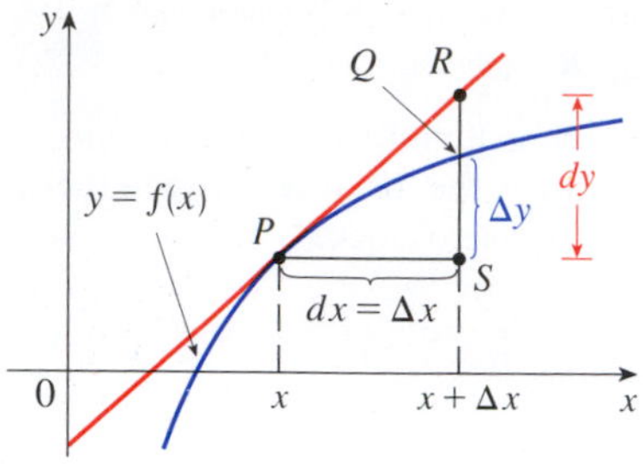

FIGURE 1

The geometric meaning of differentials is shown in Figure 1. Let $P(x, f(x))$ and $Q(x + \Delta x, f(x + \Delta x))$ be points on the graph of f and set $dx = \Delta x$. The corresponding change in y is

$$\Delta y = f(x + \Delta x) - f(x)$$

The slope of the tangent line PR is the derivative $f'(x)$. Thus, the directed distance from S to R is $f'(x)\,dx = dy$. Therefore, dy represents the amount that the tangent line rises or falls, whereas Δy represents the amount that the curve $y = f(x)$ rises or falls when x changes by an amount dx.

Since

$$\frac{dy}{dx} = \lim_{\Delta x \to 0} \frac{\Delta y}{\Delta x}$$

we have

(2) $$\frac{\Delta y}{\Delta x} \approx \frac{dy}{dx}$$

when Δx is small. (Geometrically, this says that the slope of the secant line PQ is very close to the slope of the tangent line at P when Δx is small.) If we take $dx = \Delta x$, then (2) becomes

(3) $$\Delta y \approx dy$$

which says that if Δx is small, then the actual change in y is approximately equal to the differential dy. (Again this is geometrically evident in the case illustrated by Figure 1.)

The approximation given by (3) can be used in computing approximate values of functions. Suppose that $f(a)$ is a known number and an approximate value is to be calculated for $f(a + \Delta x)$ where Δx is small. Since

$$f(a + \Delta x) = f(a) + \Delta y$$

(3) gives

(4) $$f(a + \Delta x) \approx f(a) + dy$$

EXAMPLE 2 Compare the values of Δy and dy if $y = f(x) = x^3 + x^2 - 2x + 1$ and x changes (a) from 2 to 2.05 and (b) from 2 to 2.01.

Figure 2 shows the function in Example 2 and a comparison of dy and Δy when $a = 2$. The viewing rectangle is $[1.8, 2.5]$ by $[6, 18]$.

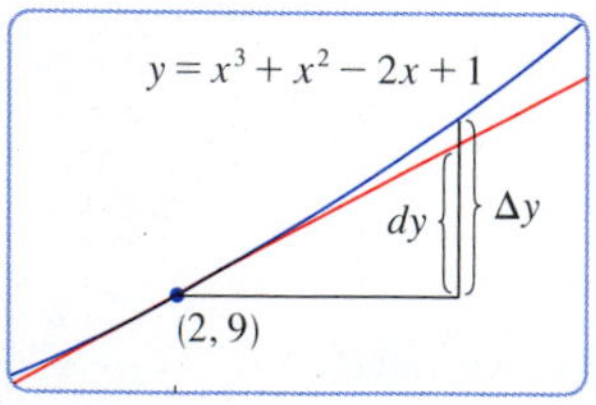

FIGURE 2

SOLUTION

(a) We have

$$f(2) = 2^3 + 2^2 - 2(2) + 1 = 9$$

$$f(2.05) = (2.05)^3 + (2.05)^2 - 2(2.05) + 1 = 9.717625$$

$$\Delta y = f(2.05) - f(2) = 0.717625$$

In general, $$dy = f'(x)\,dx = (3x^2 + 2x - 2)\,dx$$

When $x = 2$ and $dx = \Delta x = 0.05$, this becomes

$$dy = [3(2)^2 + 2(2) - 2]0.05 = 0.7$$

(b)
$$f(2.01) = (2.01)^3 + (2.01)^2 - 2(2.01) + 1 = 9.140701$$
$$\Delta y = f(2.01) - f(2) = 0.140701$$

When $dx = \Delta x = 0.01$,

$$dy = [3(2)^2 + 2(2) - 2]0.01 = 0.14$$

■

Notice that the approximation $\Delta y \approx dy$ becomes better as Δx becomes smaller in Example 2. Notice also that dy was easier to compute than Δy. For more complicated functions it may be impossible to compute Δy exactly. In such cases the approximation by differentials is especially useful.

EXAMPLE 3 Use differentials to find an approximate value for $\sqrt[3]{65}$.

SOLUTION Let $y = f(x) = \sqrt[3]{x} = x^{1/3}$. Then

$$dy = \tfrac{1}{3}x^{-2/3}\,dx$$

Since $f(64) = 4$, we take $a = 64$ and $dx = \Delta x = 1$. This gives

$$dy = \tfrac{1}{3}(64)^{-2/3}(1) = \frac{1}{3 \cdot 16} = \frac{1}{48}$$

Therefore, (4) gives

$$\sqrt[3]{65} = f(64 + 1) \approx f(64) + dy$$
$$= 4 + \tfrac{1}{48} \approx 4.021$$

■

NOTE: The actual value of $\sqrt[3]{65}$ is 4.0207257.... Thus, the approximation by differentials in Example 3 is accurate to three decimal places even when $\Delta x = 1$.

EXAMPLE 4 The radius of a sphere was measured and found to be 21 cm with a possible error in measurement of at most 0.05 cm. What is the maximum error in using this value of the radius to compute the volume of the sphere?

SOLUTION If the radius of the sphere is r, then its volume is $V = \tfrac{4}{3}\pi r^3$. If the error in the measured value of r is denoted by $dr = \Delta r$, then the corresponding error in the calculated value of V is ΔV, which can be approximated by the differential

$$dV = 4\pi r^2\,dr$$

When $r = 21$ and $dr = 0.05$, this becomes

$$dV = 4\pi(21)^2 0.05 \approx 277$$

The maximum error in the calculated volume is about 277 cm^3. ■

NOTE: Although the possible error in Example 4 may appear to be rather large, a better picture of the error is given by the **relative error**, which is computed by dividing the error by the total volume:

$$\frac{\Delta V}{V} \approx \frac{dV}{V} \approx \frac{277}{38{,}792} \approx 0.00714$$

Thus, a relative error of $dr/r = 0.05/21 \approx 0.0024$ in the radius produces a relative error of about 0.007 in the volume. The errors could also be expressed as **percentage errors** of 0.24% in the radius and 0.7% in the volume.

LINEAR APPROXIMATIONS

The equation of the tangent line to the curve $y = f(x)$ at $(a, f(a))$ is

$$y = f(a) + f'(a)(x - a)$$

and the right side of this equation is just $f(a) + f'(a)\,dx$, or $f(a) + dy$, so when we use the approximation in (4) we are, in effect, using the tangent line at $P(a, f(a))$ as an approximation to the curve $y = f(x)$ when x is near a. For this reason, the approximation

$$f(x) \approx f(a) + f'(a)(x - a) \tag{5}$$

is called the **linear approximation** or **tangent line approximation** of f at a, and the function

$$L(x) = f(a) + f'(a)(x - a) \tag{6}$$

(whose graph is the tangent line) is called the **linearization** of f at a.

EXAMPLE 5 Find the linearization of the function $f(x) = \sqrt{x + 3}$ at $a = 1$ and use it to approximate the numbers $\sqrt{3.98}$ and $\sqrt{4.05}$.

SOLUTION The derivative of $f(x) = (x + 3)^{1/2}$ is

$$f'(x) = \tfrac{1}{2}(x + 3)^{-1/2} = \frac{1}{2\sqrt{x + 3}}$$

and so we have $f(1) = 2$ and $f'(1) = \frac{1}{4}$. Putting these values into Equation 6, we see that the linearization is

$$L(x) = f(1) + f'(1)(x - 1) = 2 + \tfrac{1}{4}(x - 1) = \frac{7}{4} + \frac{x}{4}$$

The corresponding linear approximation (5) is

$$\sqrt{x + 3} \approx \frac{7}{4} + \frac{x}{4}$$

In particular, we have:

$$\sqrt{3.98} \approx \tfrac{7}{4} + \tfrac{0.98}{4} = 1.995 \qquad \text{and} \qquad \sqrt{4.05} \approx \tfrac{7}{4} + \tfrac{1.05}{4} = 2.0125$$ ■

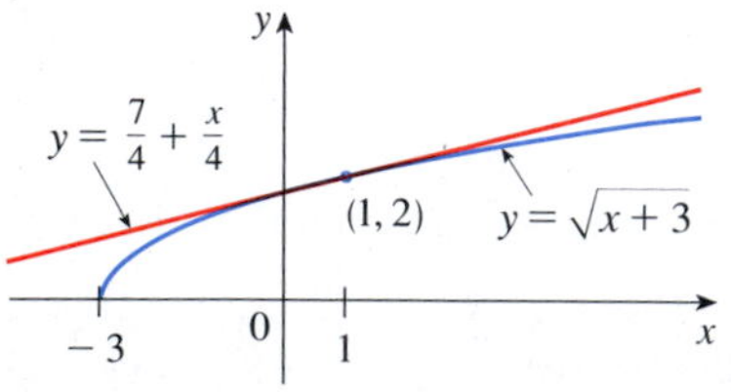

FIGURE 3

The linear approximation in Example 5 is illustrated in Figure 3. You can see that, indeed, the tangent line approximation is a good approximation to the given function when x is near 1. Of course, a calculator could give us approximations for $\sqrt{3.98}$ and $\sqrt{4.05}$, but the linear approximation gives an approximation over an entire interval.

How good is the approximation that we obtained in Example 5? The next example shows that, if we have access to a graphing calculator or computer, we can determine an interval throughout which a linear approximation provides a specified accuracy.

EXAMPLE 6 For what values of x is the linear approximation

$$\sqrt{x+3} \approx \frac{7}{4} + \frac{x}{4}$$

accurate to within 0.5? What about accuracy to within 0.1?

SOLUTION Accuracy to within 0.5 means that the functions should differ by less than 0.5:

$$\left| \sqrt{x+3} - \left(\frac{7}{4} + \frac{x}{4} \right) \right| < 0.5$$

Equivalently, we could write

$$\sqrt{x+3} - 0.5 < \frac{7}{4} + \frac{x}{4} < \sqrt{x+3} + 0.5$$

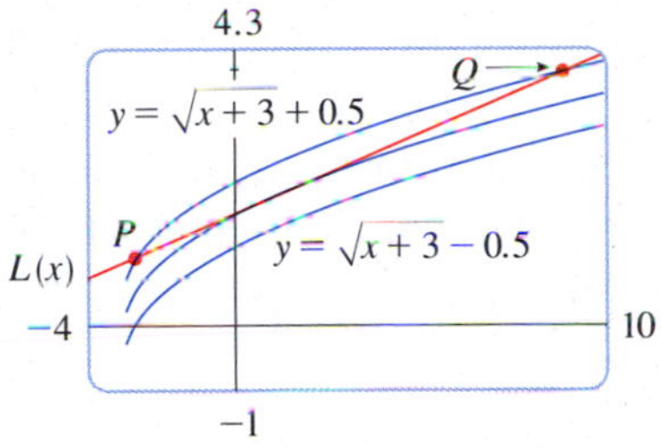

FIGURE 4

This says that the linear approximation should lie between the curves obtained by shifting the curve $y = \sqrt{x+3}$ upward and downward by an amount 0.5. Figure 4 shows the tangent line $y = (7 + x)/4$ intersecting the upper curve $y = \sqrt{x+3} + 0.5$ at P and Q. Zooming in and using the cursor, we estimate that the x-coordinate of P is about -2.65 and the x-coordinate of Q is about 8.66. Thus, we see from the graph that the approximation

$$\sqrt{x+3} \approx \frac{7}{4} + \frac{x}{4}$$

is accurate to within 0.5 when $-2.6 < x < 8.6$. (We have rounded to be safe.)

Similarly, from Figure 5 we see that the approximation is accurate to within 0.1 when $-1.1 < x < 4.0$. ■

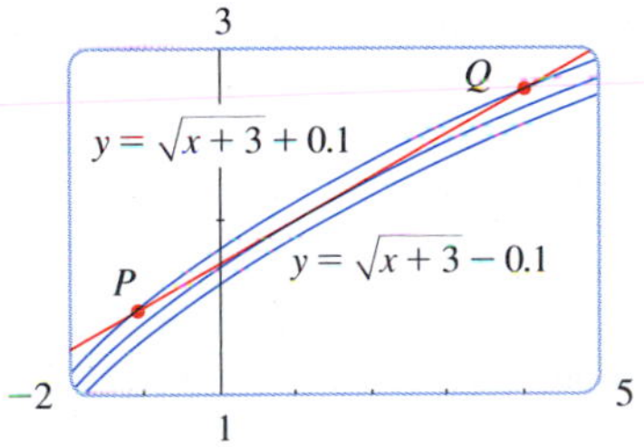

FIGURE 5

QUADRATIC APPROXIMATIONS

The tangent line approximation $L(x)$ is the best first-degree (linear) approximation to $f(x)$ near $x = a$ because $f(x)$ and $L(x)$ have the same rate of change (derivative) at a. For a better approximation than a linear one, let's try a second-degree (quadratic) approximation $P(x)$. In other words, we approximate a curve by a parabola instead of by a straight line. To make sure that the approximation is a good one, we stipulate the following:

(7) $P(a) = f(a)$ (P and f should have the same value at a.)

(8) $P'(a) = f'(a)$ (P and f should have the same rate of change at a.)

(9) $P''(a) = f''(a)$ (The slopes of P and f should change at the same rate.)

EXAMPLE 7 Find the quadratic approximation to the function $f(x) = \cos x$ near 0.

SOLUTION If we let A, B, and C be the coefficients of the quadratic function, then

$$P(x) = A + Bx + Cx^2 \qquad f(x) = \cos x$$

$$P'(x) = B + 2Cx \qquad f'(x) = -\sin x$$

$$P''(x) = 2C \qquad f''(x) = -\cos x$$

So, taking $a = 0$, our three conditions (7), (8), and (9) become

$$P(0) = f(0): \qquad A = \cos 0 = 1$$

$$P'(0) = f'(0): \qquad B = -\sin 0 = 0$$

$$P''(0) = f''(0): \qquad 2C = -\cos 0 = -1 \quad \Rightarrow \quad C = -\tfrac{1}{2}$$

The desired quadratic function is $P(x) = 1 - \frac{1}{2}x^2$, so the quadratic approximation is

$$\cos x \approx 1 - \tfrac{1}{2}x^2$$

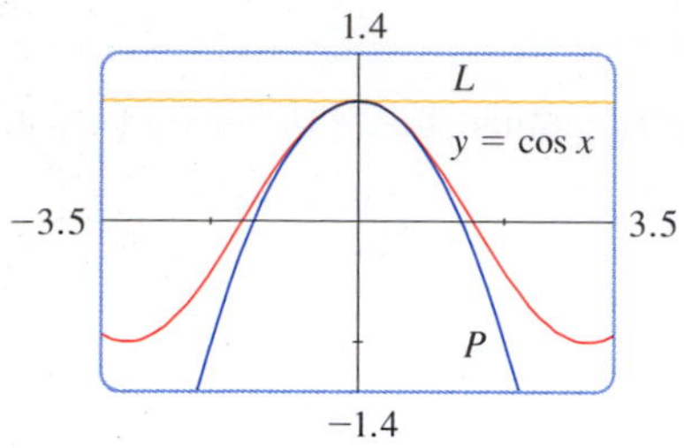

FIGURE 6

Figure 6 shows a graph of the cosine function together with its linear approximation $L(x) = 1$ and quadratic approximation $P(x) = 1 - \frac{1}{2}x^2$ near 0. You can see that the quadratic approximation is much better than the linear one.

EXAMPLE 8 For what values of x is the quadratic approximation

$$\cos x \approx 1 - \tfrac{1}{2}x^2$$

accurate to within 0.1?

SOLUTION As in Example 6, accuracy to within 0.1 means that

$$\left|\cos x - \left(1 - \tfrac{1}{2}x^2\right)\right| < 0.1$$

or

$$\cos x - 0.1 < 1 - \tfrac{1}{2}x^2 < \cos x + 0.1$$

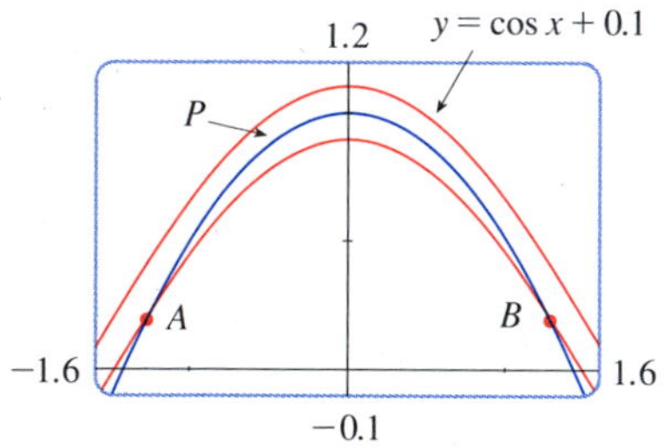

FIGURE 7

From Figure 7 we see that this is true between A and B. Zooming in and using the cursor, we find that the x-coordinates of B and A are about ± 1.26. Rounding to be safe, we see that the approximation

$$\cos x \approx 1 - \tfrac{1}{2}x^2$$

is accurate to within 0.1 when $-1.2 < x < 1.2$.

In general, if we want to approximate a function f by a quadratic function P near a number a, it's best to write P in the form

$$P(x) = A + B(x - a) + C(x - a)^2$$

Then
$$P'(x) = B + 2C(x - a)$$

and
$$P''(x) = 2C$$

Applying the conditions (7), (8), and (9), we get

$$P(a) = f(a): \qquad A = f(a)$$

$$P'(a) = f'(a): \qquad B = f'(a)$$

$$P''(a) = f''(a): \qquad 2C = f''(a) \quad \Rightarrow \quad C = \tfrac{1}{2}f''(a)$$

Therefore, the desired quadratic function is

$$P(x) = f(a) + f'(a)(x - a) + \frac{f''(a)}{2}(x - a)^2$$

(10) The quadratic approximation to $f(x)$ near a is

$$f(x) \approx f(a) + f'(a)(x - a) + \frac{f''(a)}{2}(x - a)^2$$

Notice that the first two terms on the right side of (10) give the linear approximation. Higher-degree approximations are discussed in Exercise 55 and in Chapter 10.

EXAMPLE 9 Find the quadratic approximation to $f(x) = \sqrt{x + 3}$ near $a = 1$.

SOLUTION From Example 5 we have $f(1) = 2$, $f'(1) = \frac{1}{4}$, and

$$f'(x) = \tfrac{1}{2}(x + 3)^{-1/2}$$

So

$$f''(x) = -\tfrac{1}{4}(x + 3)^{-3/2} \quad \Rightarrow \quad f''(1) = -\tfrac{1}{32}$$

From (10), the quadratic approximation is

$$\sqrt{x + 3} \approx f(1) + f'(1)(x - 1) + \frac{f''(1)}{2}(x - 1)^2$$

$$= 2 + \tfrac{1}{4}(x - 1) - \tfrac{1}{64}(x - 1)^2$$

■

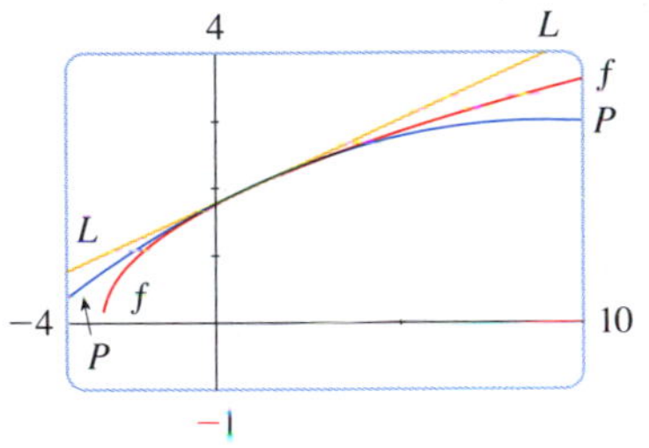

FIGURE 8

Figure 8 shows the function $f(x) = \sqrt{x + 3}$ together with its linear approximation $L(x)$ and its quadratic approximation $P(x)$. You can see that $P(x)$ is a better approximation than $L(x)$ and this is borne out by the numerical values in the following chart.

	from $L(x)$	actual value	from $P(x)$
$\sqrt{3.98}$	1.995	1.99499373...	1.99499375
$\sqrt{4.05}$	2.0125	2.01246118...	2.01246094
$\sqrt{4.2}$	2.05	2.04939015...	2.04937500

EXERCISES 3.11

1–6 ■ Find the differential of the function.

1. $y = x^5$

2. $y = \sqrt[4]{x}$

3. $y = \sqrt{x^4 + x^2 + 1}$

4. $y = \dfrac{x - 2}{2x + 3}$

5. $y = \sin 2x$

6. $y = x \tan x$

7–12 ■ (a) Find the differential dy and (b) evaluate dy for the given values of x and dx.

7. $y = 1 - x^2, \quad x = 5, dx = \frac{1}{2}$

8. $y = x^4 - 3x^3 + x - 1, \quad x = 2, dx = 0.1$

9. $y = (x^2 + 5)^3, \quad x = 1, dx = 0.05$

10. $y = \sqrt{1 - x}, \quad x = 0, dx = 0.02$

11. $y = \cos x, \quad x = \pi/6, dx = 0.05$

12. $y = \sin x, \quad x = \pi/6, dx = -0.1$

13–16 ■ Compute Δy and dy for the given values of x and $dx = \Delta x$. Then sketch a diagram like Figure 1 showing the line segments with lengths dx, dy, and Δy.

13. $y = x^2, \quad x = 1, \Delta x = 0.5$

14. $y = \sqrt{x}, \quad x = 1, \Delta x = 1$

15. $y = 6 - x^2, \quad x = -2, \Delta x = 0.4$

16. $y = 16/x, \quad x = 4, \Delta x = -1$

17–18 ■ Compute Δy, dy, and $\Delta y - dy$ for the given value of x and for each of the following values of $dx = \Delta x$: 1, 0.5, 0.1, and 0.01.

17. $y = 2x^3 + 3x - 4, \quad x = 3$

18. $y = x^4 + x^2 + 1, \quad x = 1$

19–24 ■ Use differentials to find an approximate value for the given number.

19. $\sqrt{36.1}$

20. $\sqrt[3]{1.02} + \sqrt[4]{1.02}$

21. $\frac{1}{10.1}$

22. $(1.97)^6$

23. $\sin 59°$

24. $\cos 31.5°$

25. The edge of a cube was found to be 30 cm with a possible error in measurement of 0.1 cm. Use differentials to estimate the maximum possible error in computing (a) the volume of the cube and (b) the surface area of the cube.

26. The radius of a circular disk is given as 24 cm with a maximum error in measurement of 0.2 cm.
(a) Use differentials to estimate the maximum error in the calculated area of the disk.
(b) What is the relative error?

27. The circumference of a sphere was measured to be 84 cm with a possible error of 0.5 cm.
(a) Use differentials to estimate the maximum error in the calculated surface area.
(b) What is the relative error?

28. Do Exercise 27 with "surface area" replaced by "volume."

29. (a) Use differentials to find a formula for the approximate volume of a thin cylindrical shell with height h, inner radius r, and thickness Δr.
(b) What is the error involved in using the formula from part (a)?

30. Use differentials to estimate the amount of paint needed to apply a coat of paint 0.05 cm thick to a hemispherical dome with diameter 50 m.

31–34 ■ Find the linearization $L(x)$ of the function at a.

31. $f(x) = x^3, \quad a = 1$

32. $f(x) = 1/\sqrt{2 + x}, \quad a = 0$

33. $f(x) = 1/x, \quad a = 4$

34. $f(x) = \sqrt[3]{x}, \quad a = -8$

35–38 ■ Verify the linear approximation at $a = 0$.

35. $\sqrt{1 + x} \approx 1 + \frac{1}{2}x$

36. $\sin x \approx x$

37. $1/(1 + 2x)^4 \approx 1 - 8x$

38. $1/\sqrt{4 - x} \approx \frac{1}{2} + \frac{1}{16}x$

39. Find the linear approximation of the function $f(x) = \sqrt{1 - x}$ at $a = 0$ and use it to approximate the numbers $\sqrt{0.9}$ and $\sqrt{0.99}$. Illustrate with a sketch like Figure 3.

40. Find the linear approximation of the function $g(x) = \sqrt[3]{1 + x}$ at $a = 0$ and use it to approximate the numbers $\sqrt[3]{0.95}$ and $\sqrt[3]{1.1}$. Illustrate with a sketch like Figure 3.

41–44 ■ Determine the values of x for which the linear approximation is accurate to within 0.1.

41. $\sqrt{1 + x} \approx 1 + \frac{1}{2}x$

42. $\sin x \approx x$

43. $1/(1 + 2x)^4 \approx 1 - 8x$

44. $1/\sqrt{4 - x} \approx \frac{1}{2} + \frac{1}{16}x$

45–48 ■ Find the quadratic approximation to f near a.

45. $f(x) = 1/x, \quad a = 4$

46. $f(x) = \sqrt[3]{x}, \quad a = -8$

47. $f(x) = \sec x, \quad a = 0$

48. $f(x) = \sin x, \quad a = \pi/6$

49–50 ■ Find the linear and quadratic approximations to f near a. Illustrate by graphing f and both approximations.

49. $f(x) = \sqrt{x}, \quad a = 1$

50. $f(x) = \tan x, \quad a = \pi/4$

51. (a) Find the linear and quadratic approximations to $f(x) = \cos x$ near $\pi/6$.
(b) Illustrate part (a) by graphing f and both approximations.
(c) Determine the values of x for which the linear approximation is accurate to within 0.1.
(d) Determine the values of x for which the quadratic approximation is accurate to within 0.1.

52. (a) Find the linear and quadratic approximations to $f(x) = 1/(1 + x^2)$ near 1.
(b) Illustrate part (a) by graphing f and both approximations.
(c) Determine the values of x for which the linear approximation is accurate to within 0.1.
(d) Determine the values of x for which the quadratic approximation is accurate to within 0.1.
(e) Compare the exact values of $f(x)$ for $x = 0.9, 1.1, 1.2$, and 1.3 with the linear and quadratic approximations for these values of x.

53. Establish the following rules for working with differentials (where c denotes a constant and u and v are functions of x).
(a) $dc = 0$
(b) $d(cu) = c\,du$
(c) $d(u + v) = du + dv$
(d) $d(uv) = u\,dv + v\,du$
(e) $d\left(\dfrac{u}{v}\right) = \dfrac{v\,du - u\,dv}{v^2}$
(f) $d(x^n) = nx^{n-1}\,dx$

54. Let f be a function such that $f(1) = 2$ and whose derivative is known to be $f'(x) = \sqrt{x^3 + 1}$. [You are not given a formula for $f(x)$. Don't try to guess one—you won't succeed.]
(a) Use a linear approximation to estimate the value of $f(1.1)$.
(b) Do you think the true value of $f(1.1)$ is less than or greater than your estimate? Why?
(c) Use a quadratic approximation to estimate the value of $f(1.1)$.

55. Instead of being satisfied with a linear or quadratic approximation to $f(x)$ near 0, suppose that we want an approximation by an nth-degree polynomial

$$P(x) = a_0 + a_1 x + a_2 x^2 + \cdots + a_n x^n$$

If we insist that P and all of its first n derivatives should agree with those of f at 0, show that

$$a_k = \frac{f^{(k)}(0)}{k!} \qquad \text{for } k = 1, 2, \ldots, n$$

(The resulting polynomial P is called a **Taylor polynomial of degree n** and is discussed in Chapter 10.) Find the third-degree Taylor polynomial for $\sin x$ near 0.

3.12 NEWTON'S METHOD

Many problems in science, engineering, and mathematics lead to the problem of finding the roots of an equation of the form $f(x) = 0$ where f is a differentiable function. For a quadratic equation $ax^2 + bx + c = 0$ there is a well-known formula for the roots. For third- and fourth-degree equations there are also formulas for the roots but they are extremely complicated. If f is a polynomial of degree 5 or higher, there is no such formula (see the note on page 197). Likewise there is no formula that will enable us to find the exact roots of a transcendental equation such as $\cos x = x$. However, there are methods that give *approximations* to the roots of such equations.

One such method is called **Newton's method** or the **Newton-Raphson method**. The idea behind it is shown in Figure 1, where the root that we are trying to find is labeled r. We start with a first approximation x_1, which is obtained by guessing or from a rough sketch of the graph of f. Consider the tangent line L to the curve $y = f(x)$ at the point $(x_1, f(x_1))$ and look at the x-intercept of L, labeled x_2. If x_1 is close to r, it appears that x_2 is even closer to r and we use it as the second approximation to r. To find a formula for x_2 in terms of x_1 we use the fact that the slope of L is $f'(x_1)$, so its equation is

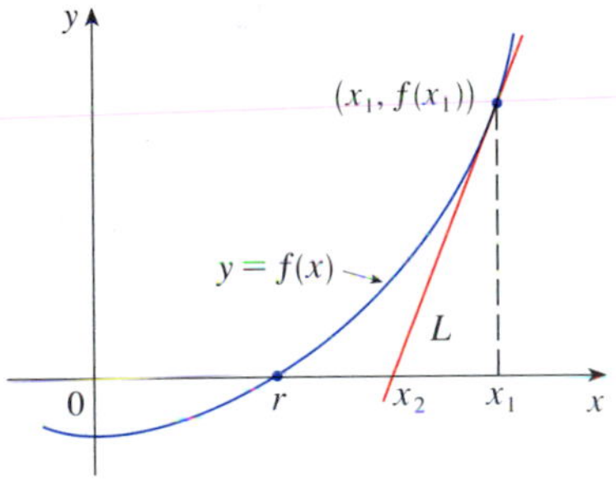

FIGURE 1

$$y - f(x_1) = f'(x_1)(x - x_1)$$

Since the x-intercept of L is x_2, we set $y = 0$ and obtain

$$0 - f(x_1) = f'(x_1)(x_2 - x_1)$$

If $f'(x_1) \neq 0$, we can solve this equation for x_2:

$$x_2 = x_1 - \frac{f(x_1)}{f'(x_1)}$$

Next we repeat this procedure with x_1 replaced by x_2, using the tangent line at $(x_2, f(x_2))$. This gives a third approximation:

$$x_3 = x_2 - \frac{f(x_2)}{f'(x_2)}$$

If we keep repeating this process we obtain a sequence of approximations $x_1, x_2, x_3, x_4, \ldots$ as shown in Figure 2. In general, if the nth approximation is x_n and $f'(x_n) \neq 0$, then the next approximation is given by

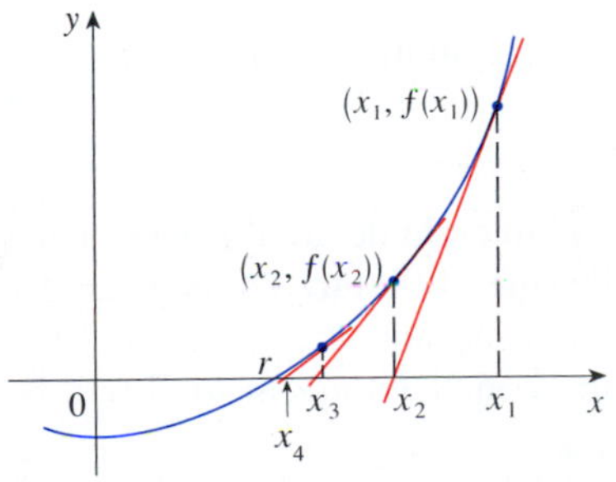

FIGURE 2

(1) $$x_{n+1} = x_n - \frac{f(x_n)}{f'(x_n)}$$

See Section 10.1 for a discussion of general sequences.

If the numbers x_n become closer and closer to r as n becomes large, then we say that the sequence converges to r and we write

$$\lim_{n\to\infty} x_n = r$$

Although the sequence of successive approximations converges to the desired root for functions of the type illustrated in Figure 2, in certain circumstances the sequence may not converge. For example, consider the situation shown in Figure 3. You can see that x_2 is a worse approximation than x_1. This is likely to be the case when $f'(x_1)$ is close to 0. It might even happen that an approximation (such as x_3 in Figure 3) falls outside the domain of f. Then Newton's method fails and a better initial approximation x_1 should be chosen. See Exercises 25–28 for specific examples in which Newton's method works very slowly or does not work at all.

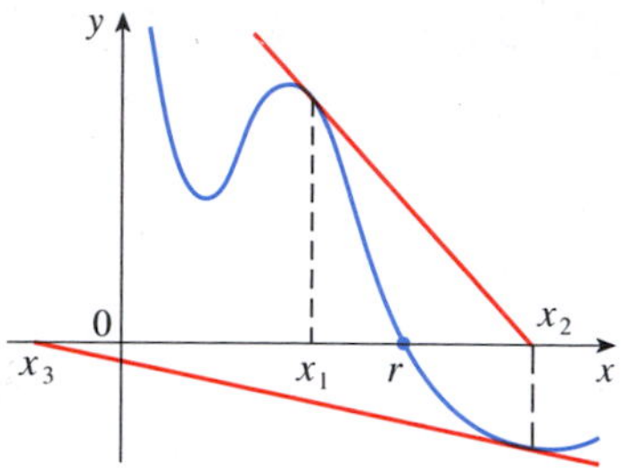

FIGURE 3

EXAMPLE 1 Starting with $x_1 = 2$, find the third approximation x_3 to the root of the equation $x^3 - 2x - 5 = 0$.

SOLUTION We apply Newton's method with

$$f(x) = x^3 - 2x - 5 \qquad \text{and} \qquad f'(x) = 3x^2 - 2$$

Newton himself used this equation to illustrate his method and he chose $x_1 = 2$ after some experimentation because $f(1) = -6$, $f(2) = -1$, and $f(3) = 16$. Equation 1 becomes

$$x_{n+1} = x_n - \frac{x_n^3 - 2x_n - 5}{3x_n^2 - 2}$$

With $n = 1$ we have

$$x_2 = x_1 - \frac{x_1^3 - 2x_1 - 5}{3x_1^2 - 2}$$

$$= 2 - \frac{2^3 - 2(2) - 5}{3(2)^2 - 2} = 2.1$$

Then with $n = 2$ we obtain

$$x_3 = x_2 - \frac{x_2^3 - 2x_2 - 5}{3x_2^2 - 2}$$

$$= 2.1 - \frac{(2.1)^3 - 2(2.1) - 5}{3(2.1)^2 - 2} \approx 2.0946$$

It turns out that this third approximation $x_3 \approx 2.0946$ is accurate to four decimal places. ■

Suppose that we want to achieve a given accuracy, say to eight decimal places, using Newton's method. How do we know when to stop? The rule of thumb that is generally used is that we can stop when successive approximations x_n and x_{n+1} agree to eight decimal places. (A precise statement concerning accuracy in Newton's method will be given as Exercise 29 in Section 10.9.)

Notice that the procedure in going from n to $n + 1$ is the same for all values of n. (It is called an *iterative* process.) This means that Newton's method is particularly convenient for use with a programmable calculator or a computer.

EXAMPLE 2 Use Newton's method to find $\sqrt[6]{2}$ correct to eight decimal places.

SOLUTION First we observe that finding $\sqrt[6]{2}$ is equivalent to finding the positive root of the equation

$$x^6 - 2 = 0$$

so we take $f(x) = x^6 - 2$. Then $f'(x) = 6x^5$ and Formula 1 (Newton's method) becomes

$$x_{n+1} = x_n - \frac{x_n^6 - 2}{6x_n^5}$$

If we choose $x_1 = 1$ as the initial approximation, then we obtain

$$x_2 \approx 1.16666667$$
$$x_3 \approx 1.12644368$$
$$x_4 \approx 1.12249707$$
$$x_5 \approx 1.12246205$$
$$x_6 \approx 1.12246205$$

Since x_5 and x_6 agree to eight decimal places, we conclude that

$$\sqrt[6]{2} \approx 1.12246205$$

to eight decimal places. ■

EXAMPLE 3 Find, correct to six decimal places, the root of the equation $\cos x = x$.

SOLUTION We first rewrite the equation in standard form:

$$\cos x - x = 0$$

Therefore, we let $f(x) = \cos x - x$. Then $f'(x) = -\sin x - 1$, so Formula 1 becomes

$$x_{n+1} = x_n - \frac{\cos x_n - x_n}{-\sin x_n - 1} = x_n + \frac{\cos x_n - x_n}{\sin x_n + 1}$$

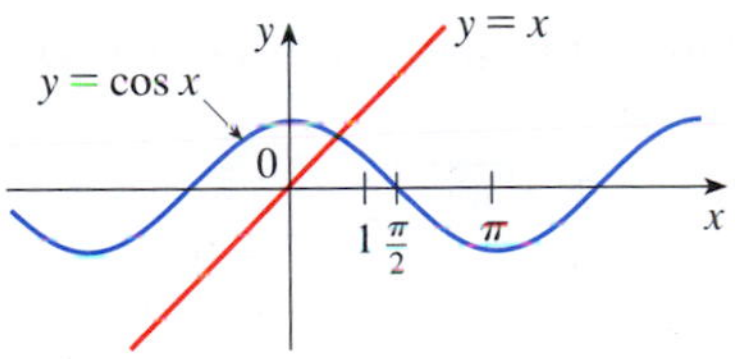

FIGURE 4

In order to guess a suitable value for x_1 we sketch the graphs of $y = \cos x$ and $y = x$ in Figure 4. It appears that they intersect at a point whose x-coordinate is somewhat less than 1, so let's take $x_1 = 1$ as a convenient first approximation. Then

$$x_2 \approx 0.75036387$$
$$x_3 \approx 0.73911289$$
$$x_4 \approx 0.73908513$$
$$x_5 \approx 0.73908513$$

Since x_4 and x_5 agree to six decimal places (eight, in fact), we conclude that the root of the equation, correct to six decimal places, is 0.739085. ■

If you have a graphing calculator or computer, it can be used to provide an initial approximation x_1 for Newton's method. In fact, you might wonder why we bother at all with Newton's method if a graphing device is available. Isn't it easier to zoom in repeatedly and find the roots as we did in Section 3 of Review and Preview? If only one or two decimal places of accuracy are required, then indeed Newton's method is inappropriate and a graphing device suffices. But if six or eight decimal places are required, then repeated zooming becomes tiresome; it is usually faster and more efficient to use Newton's method. The next example shows how to use both methods in tandem—the graphing device to get started and Newton's method to finish.

EXAMPLE 4 Find all roots of the equation $x^4 - 5x^3 + 4x^2 - x + 13 = 0$ correct to eight decimal places.

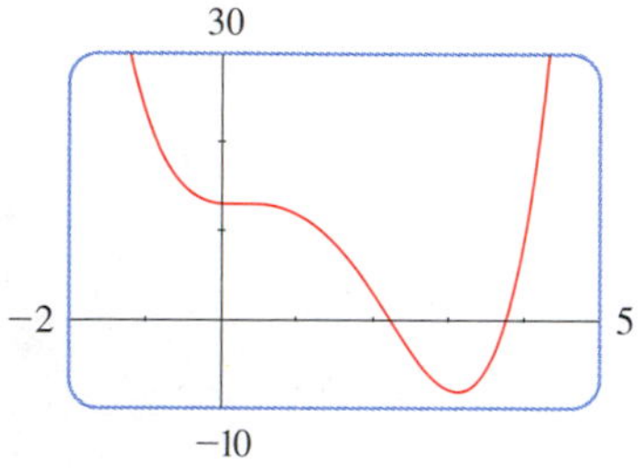

FIGURE 5

SOLUTION Figure 5 shows a graph of $f(x) = x^4 - 5x^3 + 4x^2 - x + 13$ and we see that the roots are near 2.2 and 3.8. The formula for Newton's method is

$$x_{n+1} = x_n - \frac{x_n^4 - 5x_n^3 + 4x_n^2 - x_n + 13}{4x_n^3 - 15x_n^2 + 8x_n - 1}$$

Using Newton's method with the initial approximations from the graph, we get

$$\begin{array}{ll} x_1 = 2.2 & x_1 = 3.8 \\ x_2 \approx 2.22577566 & x_2 \approx 3.76551041 \\ x_3 \approx 2.22578253 & x_3 \approx 3.76419061 \\ x_4 \approx 2.22578253 & x_4 \approx 3.76418872 \\ & x_5 \approx 3.76418872 \end{array}$$

The roots of the given equation, correct to eight decimal places, are 2.22578253 and 3.76418872. ■

EXERCISES 3.12

1. The figure shows the graph of a function f. Suppose that Newton's method is used to approximate the root r of the equation $f(x) = 0$ with initial approximation $x_1 = 1$. Draw the tangent lines that are used to find x_2 and x_3, and estimate the numerical values of x_2 and x_3.

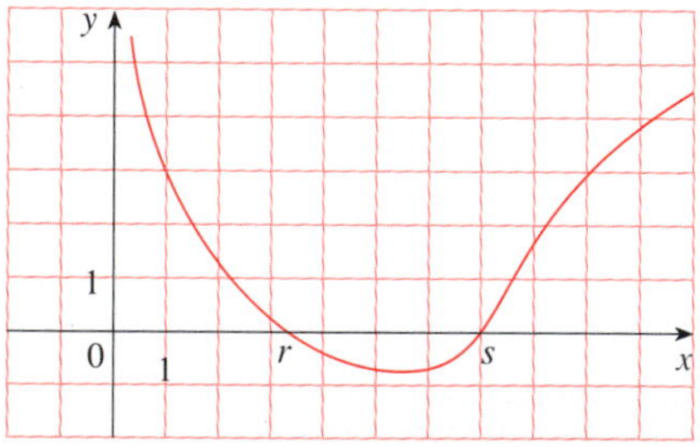

2. Follow the instructions for Exercise 1 but use $x_1 = 9$ as the starting approximation for finding the root s.

3–6 ■ Use Newton's method with the specified initial approximation x_1 to find x_3, the third approximation to the root of the given equation. (Give your answer to four decimal places.)

3. $x^3 + x + 1 = 0, \quad x_1 = -1$
4. $x^3 + x^2 + 2 = 0, \quad x_1 = -2$
5. $x^5 - 10 = 0, \quad x_1 = 1.5$
6. $x^7 - 100 = 0, \quad x_1 = 2$

7–8 ■ Use Newton's method to approximate the given number correct to eight decimal places.

7. $\sqrt[4]{22}$
8. $\sqrt[10]{100}$

9–12 ■ Use Newton's method to approximate the indicated root of the equation correct to six decimal places.

9. The root of $x^3 - 2x - 1 = 0$ in the interval $[1, 2]$
10. The root of $x^4 + x^3 - 22x^2 - 2x + 41 = 0$ in the interval $[1, 2]$
11. The positive root of $2 \sin x = x$
12. The root of $\tan x = x$ in the interval $(\pi/2, 3\pi/2)$

13–18 ■ Use Newton's method to find all roots of the equation correct to six decimal places.

13. $x^3 = 4x - 1$
14. $x^5 = 5x - 2$
15. $x^4 = 1 + x - x^2$
16. $(x - 2)^4 = x/2$
17. $2 \cos x = 2 - x$
18. $\sin \pi x = x$

19–22 ■ Use Newton's method to find all the roots of the equation correct to eight decimal places. Start by drawing a graph to find initial approximations.

19. $x^4 + 3x^3 - x - 10 = 0$

20. $x^9 - x^6 + 2x^4 + 5x - 14 = 0$

21. $\sqrt{x^2 - x + 1} = 2\sin \pi x$

22. $\cos(x^2 + 1) = x^3$

23. (a) Apply Newton's method to the equation $x^2 - a = 0$ to derive the following square-root algorithm (used by the ancient Babylonians to compute $\sqrt{a}$):

$$x_{n+1} = \frac{1}{2}\left(x_n + \frac{a}{x_n}\right)$$

(b) Use part (a) to compute $\sqrt{1000}$ correct to six decimal places.

24. (a) Apply Newton's method to the equation $1/x - a = 0$ to derive the following reciprocal algorithm:

$$x_{n+1} = 2x_n - ax_n^2$$

(This algorithm enables a computer to find reciprocals without actually dividing.)

(b) Use part (a) to compute $1/1.6984$ correct to six decimal places.

25. Explain why Newton's method does not work for finding the root of the equation $x^3 - 3x + 6 = 0$ if the initial approximation is chosen to be $x_1 = 1$.

26. (a) Use Newton's method with $x_1 = 1$ to find the root of the equation $x^3 - x = 1$ correct to six decimal places.

(b) Solve the equation in part (a) using $x_1 = 0.6$ as the initial approximation.

(c) Solve the equation in part (a) using $x_1 = 0.57$. Sketch the graph of $f(x) = x^3 - x - 1$ to show why x_2 is such a poor approximation. [*Hint*: You definitely need a programmable calculator for this problem.]

27. Show that Newton's method fails when applied to the equation $\sqrt[3]{x} = 0$ with any initial approximation $x_1 \neq 0$.

28. If

$$f(x) = \begin{cases} \sqrt{x} & \text{if } x \geq 0 \\ -\sqrt{-x} & \text{if } x < 0 \end{cases}$$

then the root of the equation $f(x) = 0$ is $x = 0$. Explain why Newton's method fails to find the root no matter which initial approximation $x_1 \neq 0$ is used. Illustrate your explanation with a sketch.

29. A grain silo consists of a cylindrical main section, with height 30 ft, and a hemispherical roof. In order to achieve a total volume of 15,000 ft^3 (including the part inside the roof section), what would the radius of the silo have to be?

30. In the figure, the length of the chord AB is 4 cm and the length of the arc AB is 5 cm. Find the central angle θ, in radians, correct to four decimal places. Then give the answer to the nearest degree.

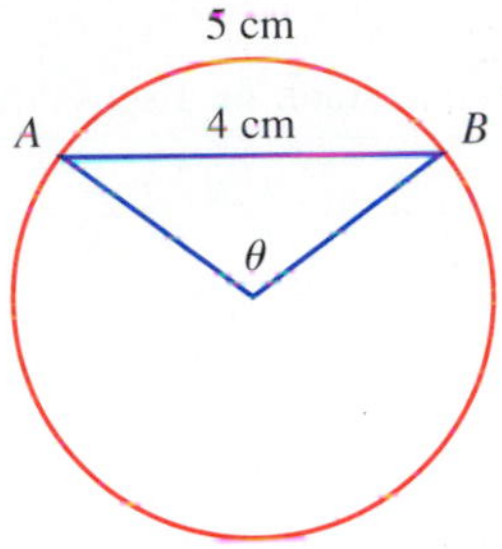

31. A car dealer sells a new car for \$18,000. He also offers to sell the same car for payments of \$375 per month for five years. What monthly interest rate is this dealer charging?

To solve this problem you will need to use the formula for the present value A of an annuity consisting of n equal payments of size R with interest rate i per time period:

$$A = \frac{R}{i}[1 - (1 + i)^{-n}]$$

Replacing i by x, show that

$$48x(1 + x)^{60} - (1 + x)^{60} + 1 = 0$$

Use Newton's method to solve this equation.

32. The figure shows the Sun located at the origin and Earth at the point $(1, 0)$. (The unit here is the distance between the centers of Earth and the Sun, called an *astronomical unit*: 1 AU $\approx 1.496 \times 10^8$ km.) There are five locations L_1, L_2, L_3, L_4, and L_5 in this plane of rotation of Earth about the Sun where a satellite remains motionless with respect to Earth because the gravitational attractions of Earth and the Sun acting on the satellite balance each other. These locations are called *libration points*. (A solar research satellite has been placed at one of these libration points.) If m_1 is the mass of the Sun, m_2 is the mass of Earth, and $r = m_2/(m_1 + m_2)$, it turns out that the x-coordinate of L_1 is the unique root of the fifth-degree equation

$$p(x) = x^5 - (2 + r)x^4 + (1 + 2r)x^3 - (1 - r)x^2 + 2(1 - r)x + r - 1 = 0$$

and the x-coordinate of L_2 is the root of the equation

$$p(x) - 2rx^2 = 0$$

Using the value $r \approx 3.04042 \times 10^{-6}$, find the locations of the libration points (a) L_1 and (b) L_2.

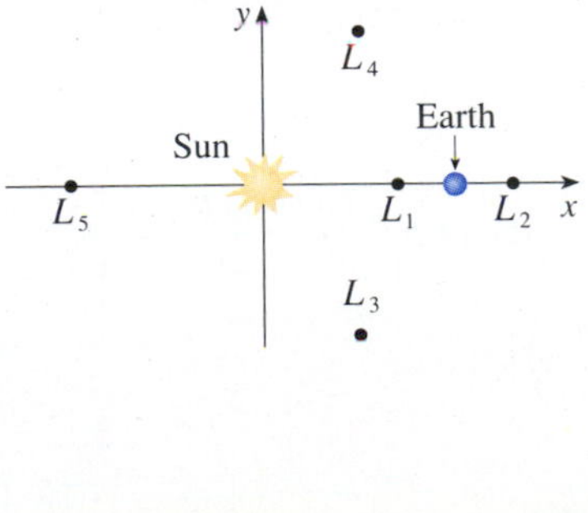

3 REVIEW

KEY TOPICS ■ Define, state, or discuss the following.

1. Derivative of a function
2. Interpretations of the derivative
3. Equation of the tangent line to $y = f(x)$ at $(a, f(a))$
4. Differentiable function
5. Relation between differentiability and continuity
6. Power Rule
7. Product Rule
8. Quotient Rule
9. Derivatives of trigonometric functions
10. Chain Rule
11. Implicit differentiation
12. Orthogonal curves
13. Second derivative; higher derivatives
14. Position function, velocity, acceleration
15. Average and instantaneous rates of change
16. Derivatives of vector functions
17. Tangent vector
18. Slopes and tangents of parametric curves
19. Related rates
20. Differentials
21. Linear approximation
22. Linearization of f at a
23. Quadratic approximation
24. Newton's method

EXERCISES

1–12 ■ Determine whether the statement is true or false.

1. If f is continuous at a, then f is differentiable at a.
2. If f and g are differentiable, then
$$\frac{d}{dx}[f(x) + g(x)] = f'(x) + g'(x).$$
3. If f and g are differentiable, then
$$\frac{d}{dx}[f(x)g(x)] = f'(x)g'(x).$$
4. If f and g are differentiable, then
$$\frac{d}{dx}[f(g(x))] = f'(g(x))g'(x).$$
5. If f is differentiable, then $\dfrac{d}{dx}\sqrt{f(x)} = \dfrac{f'(x)}{2\sqrt{f(x)}}$.
6. If f is differentiable, then $\dfrac{d}{dx}f(\sqrt{x}) = \dfrac{f'(x)}{2\sqrt{x}}$.
7. $\dfrac{d}{dx}|x^2 + x| = |2x + 1|$
8. If $f'(r)$ exists, then $\lim_{x \to r} f(x) = f(r)$.
9. If $g(x) = x^5$, then $\lim_{x \to 2} \dfrac{g(x) - g(2)}{x - 2} = 80$.
10. $\dfrac{d^2y}{dx^2} = \left(\dfrac{dy}{dx}\right)^2$
11. An equation of the tangent line to the parabola $y = x^2$ at $(-2, 4)$ is $y - 4 = 2x(x + 2)$.
12. $D(\tan^2 x) = D(\sec^2 x)$

13–16 ■ Find $f'(x)$ from first principles, that is, directly from the definition of a derivative.

13. $f(x) = x^3 + 5x + 4$
14. $f(x) = \dfrac{4 - x}{3 + x}$
15. $f(x) = \sqrt{3 - 5x}$
16. $f(x) = x \sin x$

17–40 ■ Calculate y'.

17. $y = (x + 2)^8(x + 3)^6$
18. $y = \sqrt[3]{x} + \dfrac{1}{\sqrt[3]{x}}$
19. $y = \dfrac{x}{\sqrt{9 - 4x}}$
20. $y = \left(x + \dfrac{1}{x^2}\right)^{\sqrt{7}}$
21. $x^2y^3 + 3y^2 = x - 4y$
22. $y = (1 - x^{-1})^{-1}$
23. $y = \sqrt{x\sqrt{x\sqrt{x}}}$
24. $y = -2/\sqrt[4]{x^3}$
25. $y = x/(8 - 3x)$
26. $y\sqrt{x - 1} + x\sqrt{y - 1} = xy$
27. $y = \sqrt[5]{x \tan x}$
28. $y = \sin(\cos x)$
29. $x^2 = y(y + 1)$
30. $y = 1/\sqrt[3]{x + \sqrt{x}}$
31. $y = \dfrac{(x - 1)(x - 4)}{(x - 2)(x - 3)}$
32. $y = \sqrt{\sin\sqrt{x}}$

33. $y = \tan \sqrt{1 - x}$

34. $y = \dfrac{1}{\sin(x - \sin x)}$

35. $y = \sin(\tan \sqrt{1 + x^3})$

36. $y = \dfrac{(x + \lambda)^4}{x^4 + \lambda^4}$

37. $y = \cot(3x^2 + 5)$

38. $y = \dfrac{\sin mx}{x}$

39. $y = \cos^2(\tan x)$

40. $x \tan y = y - 1$

41. If $f(x) = 1/(2x - 1)^5$, find $f''(0)$.

42. If $g(t) = \csc 2t$, find $g'''(-\pi/8)$.

43. Find y'' if $x^6 + y^6 = 1$.

44. Find $f^{(n)}(x)$ if $f(x) = 1/(2 - x)$.

45–46 ■ Find the limit.

45. $\lim_{x \to 0} \dfrac{\sec x}{1 - \sin x}$

46. $\lim_{t \to 0} \dfrac{t^3}{\tan^3 2t}$

47–50 ■ Find the equation of the tangent to the curve at the given point.

47. $y = \dfrac{x}{x^2 - 2}$, $(2, 1)$

48. $\sqrt{x} + \sqrt{y} = 3$, $(4, 1)$

49. $y = \tan x$, $(\pi/3, \sqrt{3})$

50. $y = x\sqrt{1 + x^2}$, $(1, \sqrt{2})$

51. At what points on the curve $y = \sin x + \cos x$, $0 \leqslant x \leqslant 2\pi$, is the tangent line horizontal?

52. Find the points on the ellipse $x^2 + 2y^2 = 1$ where the tangent line has slope 1.

53. If $f(x) = (x - a)(x - b)(x - c)$, show that

$$\frac{f'(x)}{f(x)} = \frac{1}{x - a} + \frac{1}{x - b} + \frac{1}{x - c}$$

54. (a) By differentiating the double-angle formula

$$\cos 2x = \cos^2 x - \sin^2 x$$

obtain the double-angle formula for the sine function.

(b) By differentiating the addition formula

$$\sin(x + a) = \sin x \cos a + \cos x \sin a$$

obtain the addition formula for the cosine function.

55. Suppose that $h(x) = f(x)g(x)$ and $F(x) = f(g(x))$, where $f(2) = 3$, $g(2) = 5$, $g'(2) = 4$, $f'(2) = -2$, and $f'(5) = 11$. Find (a) $h'(2)$ and (b) $F'(2)$.

56. (a) Graph the function $f(x) = x - 2 \sin x$ in the viewing rectangle $[0, 8]$ by $[-1, 8]$.
(b) On which interval is the average rate of change larger: $[1, 2]$ or $[2, 3]$?
(c) At which value of x is the instantaneous rate of change larger: $x = 2$ or $x = 5$?
(d) Check your visual estimates in part (c) by computing $f'(x)$ and comparing the numerical values of $f'(2)$ and $f'(5)$.

57. The figure shows the graphs of f, f', and f''. Identify each curve, and explain your choices.

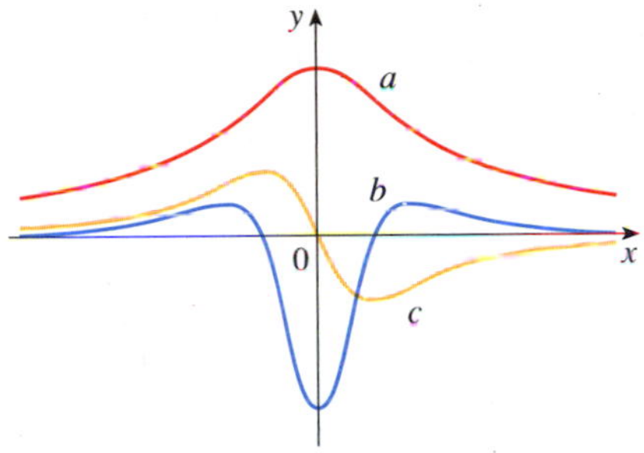

58. The graph of a function f is given here. Trace or copy it and use slopes of tangents to sketch the graph of f' beneath it.

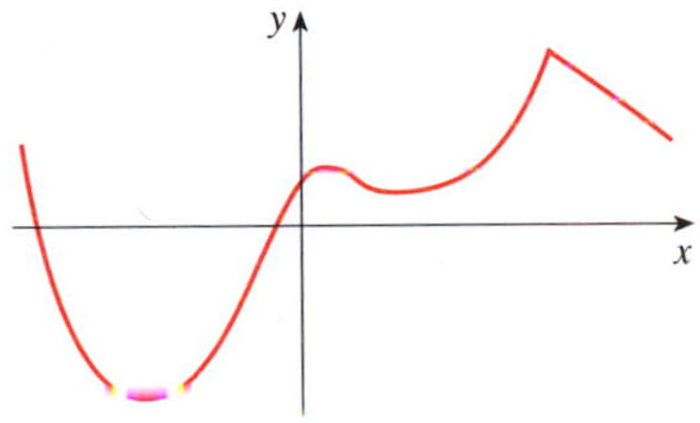

59. (a) If $f(x) = x\sqrt{5 - x}$, find $f'(x)$.
(b) Find the equations of the tangent lines to the curve $y = x\sqrt{5 - x}$ at the points $(1, 2)$ and $(4, 4)$.
(c) Illustrate part (b) by graphing the curve and tangent lines.
(d) Check to see that your answer to part (a) is reasonable by comparing the graphs of f and f'.

60. (a) If $f(x) = 4x - \tan x$, $-\pi/2 < x < \pi/2$, find f' and f''.
(b) Check to see that your answers to part (a) are reasonable by comparing the graphs of f, f', and f''.

61–66 ■ Find f' in terms of g'.

61. $f(x) = x^2 g(x)$

62. $f(x) = g(x^2)$

63. $f(x) = [g(x)]^2$

64. $f(x) = x^a g(x^b)$

65. $f(x) = g(g(x))$

66. $f(x) = g(\tan \sqrt{x})$

67–69 ■ Find h' in terms of f' and g'.

67. $h(x) = \dfrac{f(x)g(x)}{f(x) + g(x)}$

68. $h(x) = \sqrt{\dfrac{f(x)}{g(x)}}$

69. $h(x) = f(g(\sin 4x))$

70. A particle moves along a horizontal line so that its coordinate at time t is $x = \sqrt{b^2 + c^2t^2}$, $t \geq 0$, where b and c are positive constants.
(a) Find the velocity and acceleration functions.
(b) Show that the particle always moves in the positive direction.

71. A particle moves on a vertical line so that its coordinate at time t is $y = t^3 - 12t + 3$, $t \geq 0$.
(a) Find the velocity and acceleration functions.
(b) When is the particle moving upward and when is it moving downward?
(c) Find the distance that the particle travels in the time interval $0 \leq t \leq 3$.

72. The volume of a right circular cone is $V = \pi r^2h/3$, where r is the radius of the base and h is the height.
(a) Find the rate of change of the volume with respect to the height if the radius is constant.
(b) Find the rate of change of the volume with respect to the radius if the height is constant.

73–76 ■
(a) Sketch the curve described by the vector equation.
(b) Find $\mathbf{r}'(t)$ and $\mathbf{r}''(t)$.
(c) Sketch the position vector $\mathbf{r}(t)$, the tangent vector $\mathbf{r}'(t)$, and the vector $\mathbf{r}''(t)$ for the given value of t.

73. $\mathbf{r}(t) = \langle \sin 2t, \cos 2t \rangle, \quad t = \pi/4$

74. $\mathbf{r}(t) = (t^3 + 2t)\,\mathbf{i} + (4t - 5)\,\mathbf{j}, \quad t = 2$

75. $\mathbf{r}(t) = (4t^2 - 6t)\,\mathbf{i} + t^3\mathbf{j}, \quad t = 2$

76. $\mathbf{r}(t) = \langle -\sin t, -\cos t \rangle, \quad t = \pi/4$

77–80 ■ The vector function $\mathbf{r}(t)$ represents the position of a particle at time t. Find the position, velocity, acceleration, and speed of the particle at the given time.

77. $\mathbf{r}(t) = (t^2 - 4t)\,\mathbf{i} + (t^3 - 5t)\,\mathbf{j}, \quad t = 1$

78. $\mathbf{r}(t) = \langle 6\sin t, 3\cos t \rangle, \quad t = \pi/4$

79. $\mathbf{r}(t) = \langle 3\cos 2t, -4\sin 2t \rangle, \quad t = \pi/8$

80. $\mathbf{r}(t) = (t^3 - 4t^2 + 2)\,\mathbf{i} + (2t^2 - 3t)\,\mathbf{j}, \quad t = 2$

81–84 ■ Find an equation of the line tangent to the curve at the point corresponding to the given value of the parameter.

81. $x = t^2 + 2t, \ y = t^3 - t; \quad t = 1$

82. $x = t^2 + 1, \ y = t^2 - 1; \quad t = 2$

83. $\mathbf{r}(t) = t\cos t\,\mathbf{i} + t\sin t\,\mathbf{j}; \quad t = \pi/4$

84. $\mathbf{r}(t) = (t^6 + t^3)\,\mathbf{i} + (t^4 + t^2)\,\mathbf{j}; \quad t = 1$

85. The mass of part of a wire is $x(1 + \sqrt{x})$ kilograms, where x is measured in meters from one end of the wire. Find the linear density of the wire when $x = 4$ m.

86. The cost, in dollars, of manufacturing x units of a given item is $950 + 12x + 0.01x^2$. Find the marginal cost at a production level of 200 units. Compare this with the cost of producing the 201st unit.

87. The volume of a cube is increasing at a rate of 10 cm^3/min. How fast is the surface area increasing when the length of an edge is 30 cm?

88. A paper cup has the shape of a cone with height 10 cm and radius 3 cm (at the top). If water is poured into the cup at a rate of 2 cm^3/s, how fast is the water level rising when the water is 5 cm deep?

89. A balloon is rising at a constant speed of 5 ft/s. A boy is cycling along a straight road at a speed of 15 ft/s. When he passes under the balloon it is 45 ft above him. How fast is the distance between the boy and the balloon increasing 3 s later?

90. A waterskier skis over the ramp shown in the figure at a speed of 30 ft/s. How fast is she rising as she leaves the ramp?

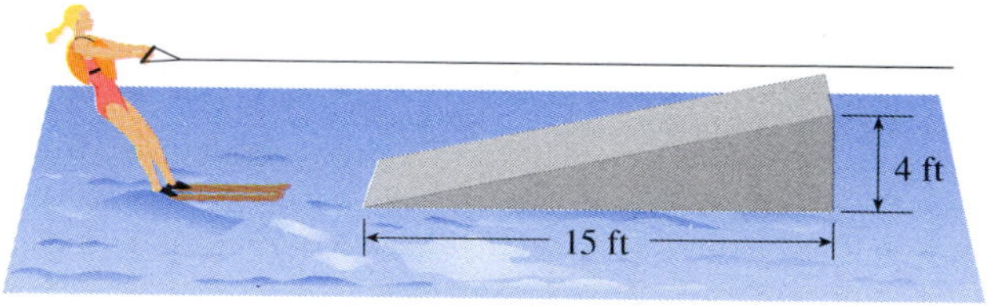

91. The angle of elevation of the Sun is decreasing at a rate of 0.25 rad/h. How fast is the shadow cast by a 400-ft-tall building increasing when the angle of elevation of the Sun is $\pi/6$?

92. Find dy if $y = (4 - x^2)^{3/2}$.

93. Evaluate dy if $y = x^3 - 2x^2 + 1$, $x = 2$, and $dx = 0.2$.

94. Use differentials to approximate $8 + \sqrt{143.6}$.

95. Find the linearization of $f(x) = \sqrt[3]{1 + 3x}$ at $a = 0$. State the corresponding linear approximation and use it to give an approximate value for $\sqrt[3]{1.03}$.

96. A window has the shape of a square surmounted by a semicircle. The base of the window is measured as having width 60 cm with a possible error in measurement of 0.1 cm. Use differentials to estimate the maximum error possible in computing the area of the window.

97. Determine the values of x for which the linear approximation given in Exercise 95 is accurate to within 0.1.

98. (a) Find the linear and quadratic approximations to $f(x) = \sqrt{25 - x^2}$ near 3.
(b) Illustrate part (a) by graphing f and both approximations.
(c) For what values of x is the linear approximation accurate to within 0.1?
(d) For what values of x is the quadratic approximation accurate to within 0.1?

99. Use Newton's method to find the root of the equation $x^4 + x - 1 = 0$ in the interval $[0, 1]$ correct to six decimal places.

100. Use Newton's method to find all roots of the equation $6 \cos x = x$ correct to six decimal places.

101. Find the coordinates (correct to four decimal places) of the point on the curve $y = x^6 + 2x^2 - 8x + 3$ at which the tangent is horizontal.

102–104 ■ Express the limit as a derivative and evaluate.

102. $\lim_{x \to 1} \dfrac{x^{17} - 1}{x - 1}$

103. $\lim_{h \to 0} \dfrac{(2 + h)^6 - 64}{h}$

104. $\lim_{\theta \to \pi/3} \dfrac{\cos \theta - 0.5}{\theta - \pi/3}$

105. Suppose f is a differentiable function such that $f(1) = 1$, $f(2) = 2$, $f'(1) = 1$, $f'(2) = 2$, and $f'(3) = 3$. If $g(x) = f(x^3 + f(x^2 + f(x)))$, evaluate $g'(1)$.

106. Suppose f is a differentiable function such that $f(g(x)) = x$ and $f'(x) = 1 + [f(x)]^2$. Show that $g'(x) = 1/(1 + x^2)$.

107. Evaluate $\lim_{x \to 0} \dfrac{\sqrt{1 + \tan x} - \sqrt{1 + \sin x}}{x^3}$.

108. Draw a diagram to show that there are two lines that are tangent to both of the parabolas $y = -1 - x^2$ and $y = 1 + x^2$. Find the coordinates of the points at which these tangents touch the parabolas.

109. Suppose f is a function with the property that $|f(x)| \leqslant x^2$ for all x. Show that $f(0) = 0$. Then show that $f'(0) = 0$.

110. Show that the length of the portion of any tangent line to the astroid $x^{2/3} + y^{2/3} = a^{2/3}$ cut off by the coordinate axes is constant.

PROBLEMS PLUS

The problems in this collection don't necessarily have anything to do with Chapter 3. They are meant to test your understanding of all of the material presented so far in this book. Some of them will challenge your problem-solving skills, so it is a good idea to refer back to the discussion of the principles of problem solving in Section 4 of Review and Preview. Many of the problems will require a considerable amount of time to think through, so don't be discouraged if you can't solve them right away.

Before you look at the following example, cover up the solution and try to solve it yourself first.

EXAMPLE 1 Let P be a point on the curve $y = x^3$ and suppose the tangent line at P intersects the curve again at Q. Prove that the slope at Q is four times the slope at P.

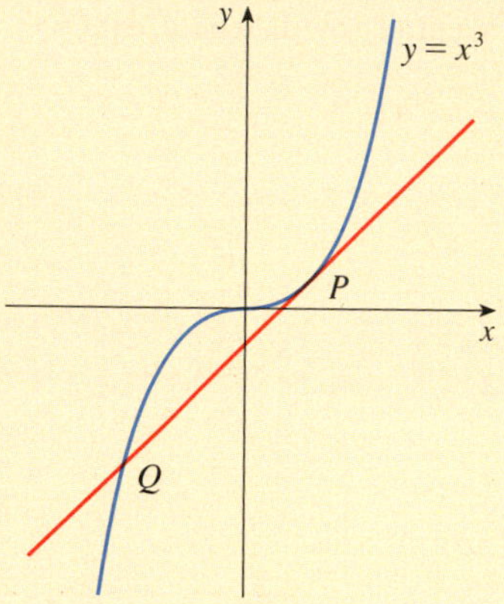

FIGURE 1

SOLUTION The first step is to understand the problem; this includes drawing a diagram, introducing notation, and identifying the given and the unknown. So we sketch the curve $y = x^3$ in Figure 1, pick a point P, and draw the tangent line at P intersecting again at Q. It certainly appears that the curve is steeper at Q, so what we are asked to prove is indeed plausible.

The notation we need is for the x-coordinate of P; let it be a. (Don't let it be x; this notation might lead to incorrect equations later.) The given information is that P lies on the curve. This means that the y-coordinate of P must be a^3, so P is the point (a, a^3). Since $y' = 3x^2$, the slope at P is $3a^2$.

The unknown is the slope at Q. To find it we have to know the x-coordinate of Q, which we can find by solving the equations of the curve and the tangent line at P. Using the point-slope form, we get the equation of the tangent:

$$y - a^3 = 3a^2(x - a) \qquad \text{or} \qquad y = 3a^2x - 2a^3$$

So to find Q we need to solve the equations

$$y = x^3 \qquad \text{and} \qquad y = 3a^2x - 2a^3$$

This gives

$$x^3 = 3a^2x - 2a^3$$

(1)
$$x^3 - 3a^2x + 2a^3 = 0$$

It's not easy to solve this cubic equation, but we get a head start if we think about the meaning of its roots. The solutions of Equation 1 are the x-coordinates of the points of intersection of the tangent line and the curve. We already know the x-coordinates of one of those points: $x = a$. So $x - a$ must be a factor of the left side of Equation 1. Dividing $x - a$ into $x^3 - 3a^2x + 2a^3$ by long division, we get the following factorization:

$$(x - a)(x^2 + ax - 2a^2) = 0$$

(2)
$$(x - a)(x - a)(x + 2a) = 0$$

Thus, the x-coordinate of Q must be $x = -2a$. Then the slope at P is

$$3(-2a)^2 = 12a^2 = 4(3a^2)$$

Therefore, the slope at Q is four times the slope at P.

LOOK BACK:

- Is there an easier way?
- Maybe the method of solution will help us to solve future problems.

In looking over our solution we might notice that we could save some work. A slightly shorter method is suggested by Equation 2. It could be written as

$$(x - a)^2(x + 2a) = 0$$

In retrospect, we might have anticipated that $(x - a)^2$ is a factor because the line is tangent to the curve at P, so a has to be a *double* root. Then we wouldn't have had to use long division. ■

PROBLEMS

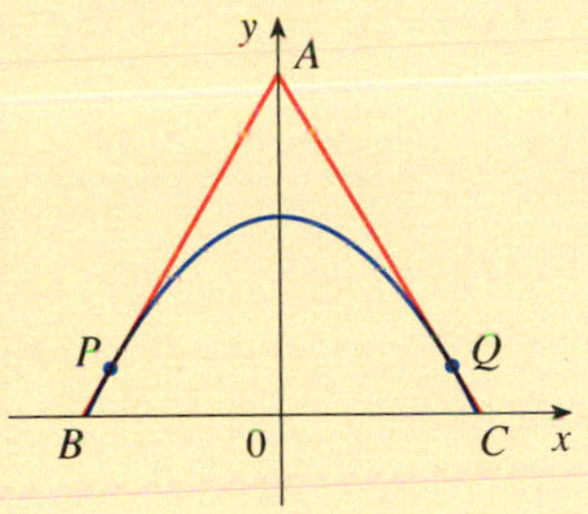

FIGURE FOR PROBLEM 1

1. Find points P and Q on the parabola $y = 1 - x^2$ so that the triangle ABC formed by the x-axis and the tangent lines at P and Q is an equilateral triangle.

2. (a) Find the sum of the finite geometric series $1 + x + x^2 + \cdots + x^n$.
   (b) Find a formula for the sum $1 + 2x + 3x^2 + \cdots + nx^{n-1}$.

3. Solve the inequality $1 + x + x^2 + \cdots + x^{100} \geqslant 0$.

4. (a) Solve the inequality $|x + 1| + |x - 2| < 7$.
   (b) Sketch the graph of the function $f(x) = |x + 1| + |x - 2|$.
   (c) Where is f continuous?
   (d) Where is f differentiable?
   (e) Sketch the graph of $g(x) = |x| + |x + 1| + |x - 1|$.
   (f) Where is g differentiable?

5. Evaluate $\displaystyle\lim_{x \to 0} \frac{|2x - 1| - |2x + 1|}{x}$.

6. If f is differentiable at a, where $a > 0$, evaluate the following limit in terms of $f'(a)$:

$$\lim_{x \to a} \frac{f(x) - f(a)}{\sqrt{x} - \sqrt{a}}$$

7. Prove that $\dfrac{d^n}{dx^n}(\sin^4 x + \cos^4 x) = 4^{n-1}\cos(4x + n\pi/2)$.

8. Find the nth derivative of the function $f(x) = x^n/(1 - x)$.

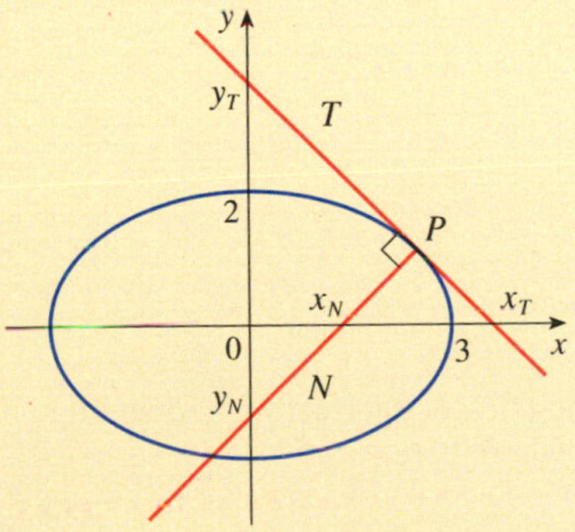

FIGURE FOR PROBLEM 9

9. Let T and N be the tangent and normal lines to the ellipse $x^2/9 + y^2/4 = 1$ at any point P on the ellipse in the first quadrant. Let x_T and y_T be the x- and y-intercepts of T and x_N and y_N be the intercepts of N. As P moves along the ellipse in the first quadrant (but not on the axes), what values can x_T, y_T, x_N, and y_N take on? First try to guess the answers just by looking at the figure. Then use calculus to solve the problem and see how good your intuition is.

10. If the ellipse of Problem 9 is replaced by a more general ellipse $x^2/a^2 + y^2/b^2 = 1$, where $a > b > 0$, what values can x_N and y_N take on? Express your answers in terms of a, b, c, and e, where $c^2 = a^2 - b^2$, and $e = c/a$ is the eccentricity of the ellipse. Interpret your results geometrically.

11. (a) Find the domain of the function $f(x) = \sqrt{1 - \sqrt{2 - \sqrt{3 - x}}}$.
    (b) Find $f'(x)$.

12. (a) Sketch the graph of the function $g(x) = |x^2 - 4| - |x^2 - 9|$.
    (b) Find $g'(x)$.

13. (a) Sketch the graph of the function $h(x) = |x^2 - 6|x| + 8|$.
(b) At what values of x is h not differentiable?

14. If $[\![x]\!]$ denotes the greatest integer function, sketch the regions in the plane defined by the following equations.
(a) $[\![x]\!]^2 + [\![y]\!]^2 = 1$ (b) $[\![x]\!]^2 - [\![y]\!]^2 = 3$ (c) $[\![x + y]\!]^2 = 1$ (d) $[\![x]\!] + [\![y]\!] = 1$

15. Tangent lines T_1 and T_2 are drawn at two points P_1 and P_2 on the parabola $y = x^2$ and they intersect at a point P. Another tangent line T is drawn at a point between P_1 and P_2; it intersects T_1 at Q_1 and T_2 at Q_2. Show that

$$\frac{|PQ_1|}{|PP_1|} + \frac{|PQ_2|}{|PP_2|} = 1$$

16. Evaluate $\lim_{x \to 0} \frac{\sin(3 + x)^2 - \sin 9}{x}$.

17. Suppose that f is differentiable at 0 and

$$\lim_{x \to 0} \frac{f(x)}{x} = 4 \qquad \lim_{x \to 0} \frac{g(x)}{x} = 2$$

(a) Find $f(0)$. (b) Find $f'(0)$. (c) Find $\lim_{x \to 0} \frac{g(x)}{f(x)}$.

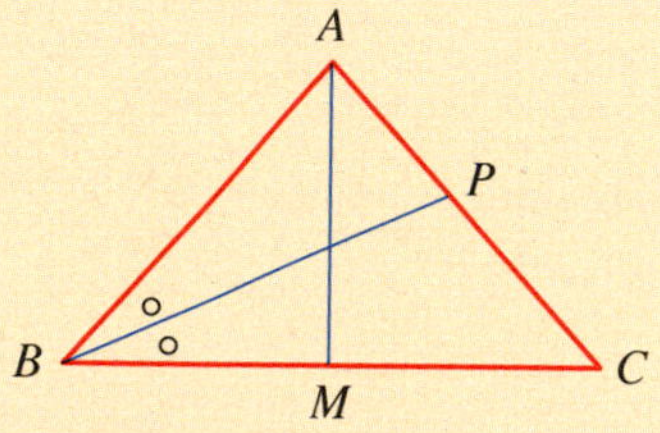

FIGURE FOR PROBLEM 18

18. (a) The figure shows an isosceles triangle ABC with $\angle B = \angle C$. The bisector of angle B intersects the side AC at the point P. Suppose that the base BC remains fixed but the altitude $|AM|$ of the triangle approaches 0, so A approaches the midpoint M of BC. What happens to P during this process? Does it have a limiting position? If so, find it.
(b) Try to sketch the path traced out by P during this process. Then find the equation of this curve and use the equation to sketch the curve.

19. Evaluate $\lim_{x \to 0} \frac{\sin(a + 2x) - 2\sin(a + x) + \sin a}{x^2}$.

20. (a) Draw a diagram and interpret the quotient

$$\frac{f(x + h) - f(x - h)}{2h}$$

as the slope of a secant line.
(b) If f is differentiable at x, show that

$$\lim_{h \to 0} \frac{f(x + h) - f(x - h)}{2h} = f'(x)$$

[*Hint*: In the proof of the Product Rule we used the trick of subtracting and adding the same quantity in a numerator. A similar trick works here.]
(c) Show that it is possible for the limit

$$\lim_{h \to 0} \frac{f(x + h) - f(x - h)}{2h}$$

to exist but for $f'(x)$ not to exist. [*Hint*: Consider $f(x) = |x|$.]

21. (a) Use the identity for $\tan(x - y)$ (see Equation 14b in Appendix D) to show that if two lines L_1 and L_2 intersect at an angle α, then

$$\tan\alpha = \frac{m_2 - m_1}{1 + m_1 m_2}$$

where m_1 and m_2 are the slopes of L_1 and L_2, respectively.

(b) The **angle between the curves** C_1 and C_2 at a point of intersection P is defined to be the angle between the tangent lines to C_1 and C_2 at P (if these tangent lines exist). Use part (a) to find, correct to the nearest degree, the angle between each pair of curves at each point of intersection.

(i) $y = x^2$ and $y = (x - 2)^2$

(ii) $x^2 - y^2 = 3$ and $x^2 - 4x + y^2 + 3 = 0$

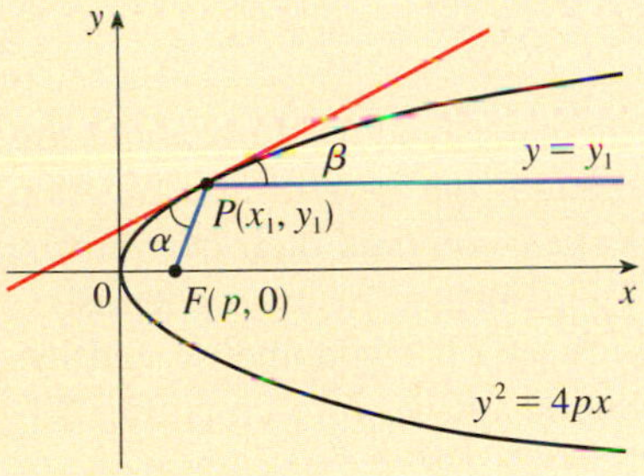

FIGURE FOR PROBLEM 22

22. Let $P(x_1, y_1)$ be a point on the parabola $y^2 = 4px$ with focus $F(p, 0)$. Let α be the angle between the parabola and the line segment FP and let β be the angle between the horizontal line $y = y_1$ and the parabola as in the figure. Prove that $\alpha = \beta$. (Thus, by a principle of geometrical optics, light from a source placed at F will be reflected along a line parallel to the x-axis. This explains why paraboloids, the surfaces obtained by rotating parabolas about their axes, are used as the shape of some automobile headlights and mirrors for telescopes.)

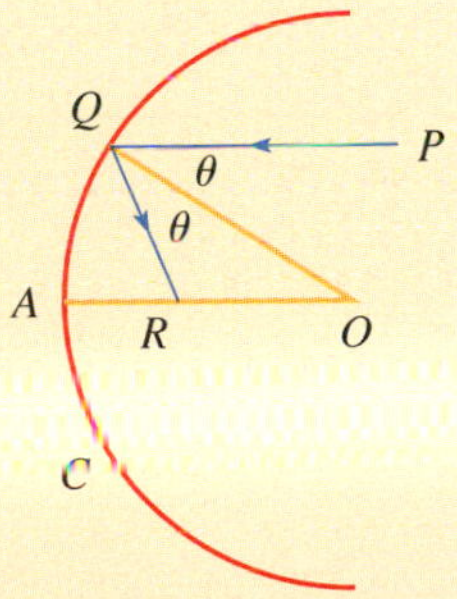

FIGURE FOR PROBLEM 23

23. Suppose that we replace the parabolic mirror of Problem 22 by a spherical mirror. Although the mirror has no focus, we can show the existence of an *approximate* focus. In the figure, C is a semicircle with center O. A ray of light coming in toward the mirror parallel to the axis along the line PQ will be reflected to the point R on the axis so that $\angle PQO = \angle OQR$ (the angle of incidence is equal to the angle of reflection). What happens to the point R as P is taken closer and closer to the axis?

24. Given an ellipse $x^2/a^2 + y^2/b^2 = 1$, where $a \neq b$, find the equation of the set of all points from which there are two tangents to the curve whose slopes are (a) reciprocals and (b) negative reciprocals.

25. Find the two points on the curve $y = x^4 - 2x^2 - x$ that have a common tangent line.

26. Suppose that three points on the parabola $y = x^2$ have the property that their normal lines intersect at a common point. Show that the sum of their x-coordinates is 0.

27. A *lattice point* in the plane is a point with integer coordinates. Suppose that circles with radius r are drawn using all lattice points as centers. Find the smallest value of r such that any line with slope $\frac{2}{5}$ intersects some of these circles.

4

INVERSE FUNCTIONS: EXPONENTIAL, LOGARITHMIC, AND INVERSE TRIGONOMETRIC FUNCTIONS

■ Each problem that I solved became a rule which served afterwards to solve other problems.

RENÉ DESCARTES

Two of the most important functions that occur in mathematics and its applications are the exponential function $f(x) = a^x$ and its inverse function, the logarithmic function $g(x) = \log_a x$. In this chapter we investigate their properties, compute their derivatives, and use them to describe exponential growth and decay in such sciences as chemistry, physics, biology, and economics. We also study the inverses of trigonometric and hyperbolic functions. Finally, we look at a method (l'Hospital's Rule) for computing difficult limits and apply it to sketching curves.

4.1 EXPONENTIAL FUNCTIONS AND THEIR DERIVATIVES

An **exponential function** is a function of the form

$$f(x) = a^x$$

where a is a positive constant. It is defined in five stages:

1. If $x = n$, a positive integer, then

$$a^n = \underbrace{a \cdot a \cdot \cdots \cdot a}_{n \text{ factors}}$$

2. If $x = 0$, then $a^0 = 1$.

3. If $x = -n$, n a positive integer, then

$$a^{-n} = \frac{1}{a^n}$$

4. If x is a rational number, $x = p/q$, where p and q are integers and $q > 0$, then

$$a^x = a^{p/q} = \sqrt[q]{a^p}$$

5. If x is an irrational number, we wish to define a^x so as to fill in the holes of the graph of the function $y = a^x$, where x is rational. In other words, we want to make $f(x) = a^x$, $x \in \mathbb{R}$, a continuous function. Since any irrational number

can be approximated as closely as we like by a rational number, we define

(1)
$$a^x = \lim_{r \to x} a^r \qquad r \text{ rational}$$

It can be shown that this definition uniquely specifies a^x and makes the function $f(x) = a^x$ continuous.

To illustrate Equation 1 let us take $x = \sqrt{2}$. The number $\sqrt{2}$ is an irrational number that has a nonrepeating decimal representation:

$$\sqrt{2} = 1.414213562373095\ldots$$

This means that $\sqrt{2}$ is the limit of the sequence of rational numbers

$$1, \quad 1.4, \quad 1.41, \quad 1.414, \quad 1.4142, \quad 1.41421, \quad 1.414213, \quad \ldots$$

According to Equation 1, $a^{\sqrt{2}}$ is the limit of the sequence

$$a^1, \quad a^{1.4}, \quad a^{1.41}, \quad a^{1.414}, \quad a^{1.4142}, \quad a^{1.41421}, \quad a^{1.414213}, \quad \ldots$$

and each term of this sequence has been defined in stage 4. In fact, if $a > 1$, then $a^{\sqrt{2}}$ is the unique number that satisfies all of the following inequalities:

$$a^1 < a^{\sqrt{2}} < a^2$$
$$a^{1.4} < a^{\sqrt{2}} < a^{1.5}$$
$$a^{1.41} < a^{\sqrt{2}} < a^{1.42}$$
$$a^{1.414} < a^{\sqrt{2}} < a^{1.415}$$
$$a^{1.4142} < a^{\sqrt{2}} < a^{1.4143}$$
$$\vdots$$

For example, if $a = 2$, then this approximation process allows us to compute $2^{\sqrt{2}}$ correct to six decimal places:

$$2^{\sqrt{2}} \approx 2.665144$$

Figures 1 and 2 illustrate the definition of the exponential function with base $a = 2$. Figure 1 is a representation of the graph of $y = 2^x$ up to stage 4, that is, for x rational.

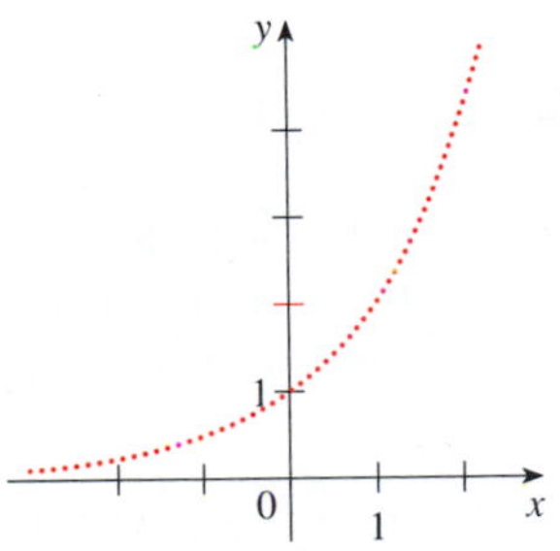

FIGURE 1
Representation of $y = 2^x$, x rational

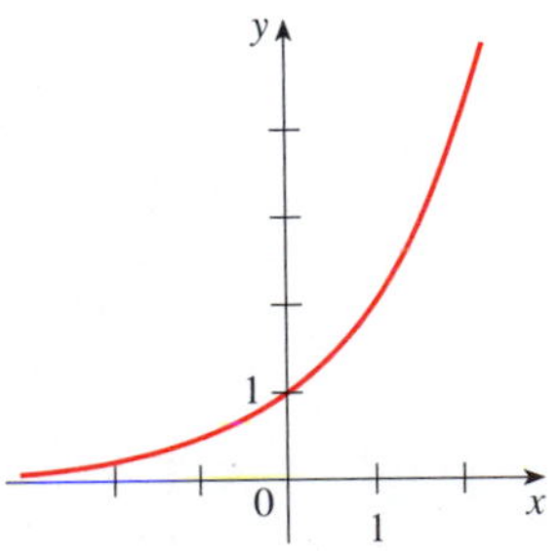

FIGURE 2
$y = 2^x$, x real

Figure 2 shows how stage 5 completes the picture by filling in all the holes to give the graph of the continuous function $y = 2^x$, $x \in \mathbb{R}$.

The graphs of the function $y = a^x$ are shown in Figure 3 for various values of the base a. Notice that all of these graphs pass through the same point $(0, 1)$ because $a^0 = 1$ for $a \neq 0$. Notice also that as the base a gets larger, the exponential function grows more rapidly (for $x > 0$).

Figure 4 shows how the exponential function $y = 2^x$ compares with the power function $y = x^2$. The graphs intersect three times, but ultimately the exponential curve $y = 2^x$ grows far more rapidly than the parabola $y = x^2$ (see also Figure 5).

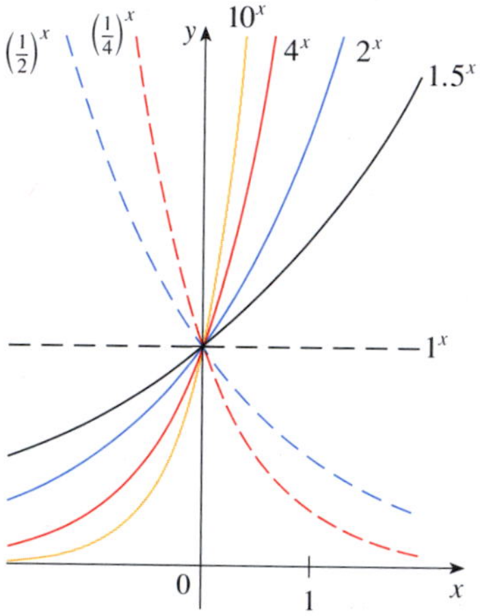

FIGURE 3

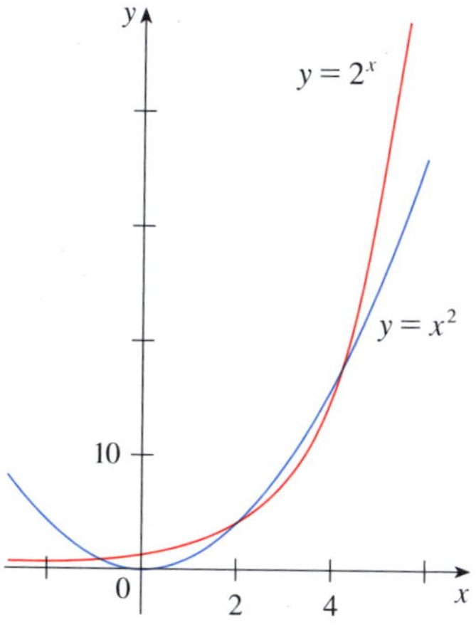

FIGURE 4

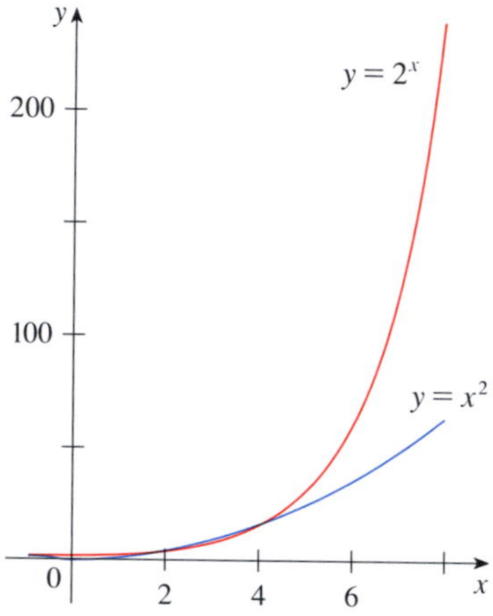

FIGURE 5

A function is called *increasing* where its graph rises and *decreasing* where its graph falls. See Section 5.3 for the precise definitions.

You can see from Figure 3 that there are basically three kinds of exponential functions $y = a^x$. If $0 < a < 1$, the exponential function decreases; if $a = 1$, it is a constant; and if $a > 1$, it increases. These three cases are illustrated in Figure 6. Since $(1/a)^x = 1/a^x = a^{-x}$, the graph of $y = (1/a)^x$ is just the reflection of the graph of $y = a^x$ about the y-axis.

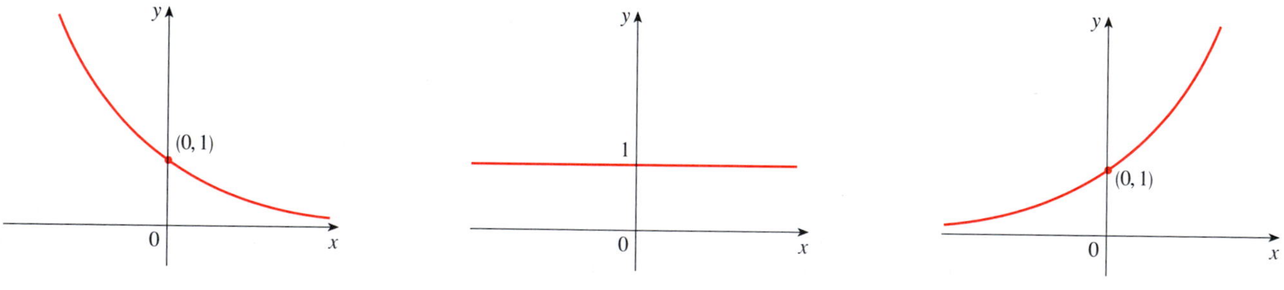

(a) $y = a^x$, $0 < a < 1$ (b) $y = 1^x$ (c) $y = a^x$, $a > 1$

FIGURE 6

The properties of the exponential function are summarized in the following theorem.

> **(2) THEOREM** If $a > 0$ and $a \neq 1$, then $f(x) = a^x$ is a continuous function with domain $\mathbb{R}$ and range $(0, \infty)$. In particular, $a^x > 0$ for all x. If $0 < a < 1$, $f(x) = a^x$ is a decreasing function; if $a > 1$, f is an increasing function. If $a, b > 0$ and $x, y \in \mathbb{R}$, then
>
> 1. $a^{x+y} = a^x a^y$ 2. $a^{x-y} = \dfrac{a^x}{a^y}$ 3. $(a^x)^y = a^{xy}$ 4. $(ab)^x = a^x b^x$

The reason for the importance of the exponential function lies in properties 1–4, which are called the **Laws of Exponents**. If x and y are rational numbers, then these laws are well known from elementary algebra. For arbitrary real numbers x and y these laws can be deduced from the special case where the exponents are rational by using Equation 1.

The following limits can be read from the graphs shown in Figure 6 or proved from the definition of a limit at infinity. (See Exercise 85 in Section 4.3.)

(3) If $a > 1$, then $\lim_{x\to\infty} a^x = \infty$ and $\lim_{x\to-\infty} a^x = 0$

If $0 < a < 1$, then $\lim_{x\to\infty} a^x = 0$ and $\lim_{x\to-\infty} a^x = \infty$

In particular, if $a \neq 1$, then the x-axis is a horizontal asymptote of the graph of the exponential function $y = a^x$.

EXAMPLE 1

(a) Find $\lim_{x\to\infty} (2^{-x} - 1)$.

(b) Sketch the graph of the function $y = 2^{-x} - 1$.

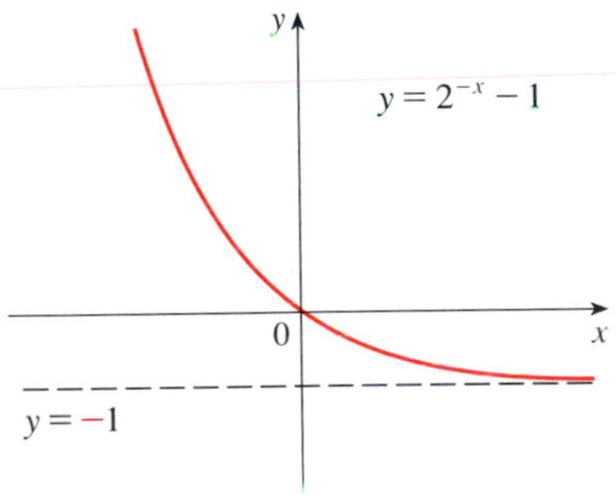

FIGURE 7

SOLUTION

(a)
$$\lim_{x\to\infty} (2^{-x} - 1) = \lim_{x\to\infty} \left[\left(\tfrac{1}{2}\right)^x - 1\right]$$
$$= 0 - 1 \qquad [\text{by (3) with } a = \tfrac{1}{2} < 1]$$
$$= -1$$

(b) We write $y = (\frac{1}{2})^x - 1$ as above. The graph of $y = (\frac{1}{2})^x$ is shown in Figure 3, so we shift it down one unit to obtain the graph of $y = (\frac{1}{2})^x - 1$ shown in Figure 7. (For a review of shifting graphs, see Section 2 in Review and Preview.) Part (a) shows that the line $y = -1$ is a horizontal asymptote. ■

DERIVATIVES OF EXPONENTIAL FUNCTIONS

Let's try to compute the derivative of the exponential function $f(x) = a^x$ using the definition of a derivative:

$$f'(x) = \lim_{h\to0} \frac{f(x+h) - f(x)}{h} = \lim_{h\to0} \frac{a^{x+h} - a^x}{h}$$
$$= \lim_{h\to0} \frac{a^x a^h - a^x}{h} = \lim_{h\to0} \frac{a^x(a^h - 1)}{h}$$
$$f'(x) = a^x \lim_{h\to0} \frac{a^h - 1}{h}$$

Notice that the limit is the value of the derivative of f at 0, that is,

$$\lim_{h\to0} \frac{a^h - 1}{h} = f'(0)$$

Therefore, we have shown that if the exponential function $f(x) = a^x$ is differentiable at 0, then it is differentiable everywhere and

$$f'(x) = f'(0)a^x \qquad \textbf{(4)}$$

This equation says that *the rate of change of any exponential function is proportional to the function itself.* (The slope is proportional to the height.)

h	$\dfrac{2^h - 1}{h}$	$\dfrac{3^h - 1}{h}$
0.1	0.7177	1.1612
0.01	0.6956	1.1047
0.001	0.6934	1.0992
0.0001	0.6932	1.0987

Numerical evidence for the existence of $f'(0)$ is given in the accompanying table for the cases $a = 2$ and $a = 3$. (Values are stated correct to four decimal places.) It appears that the limits exist and

$$\text{for } a = 2, \quad f'(0) = \lim_{h \to 0} \frac{2^h - 1}{h} \approx 0.69$$

$$\text{for } a = 3, \quad f'(0) = \lim_{h \to 0} \frac{3^h - 1}{h} \approx 1.10$$

In fact, it follows from the results of Section 6.6 that the limits exist and, correct to six decimal places, the values are

(5) $$\left.\frac{d}{dx} 2^x\right|_{x=0} \approx 0.693147 \qquad \left.\frac{d}{dx} 3^x\right|_{x=0} \approx 1.098612$$

Thus, from Equation 4, we have

$$\frac{d}{dx}(2^x) \approx (0.69)2^x \qquad \frac{d}{dx}(3^x) \approx (1.10)3^x$$

Of all possible choices for the base a in Equation 4, the simplest differentiation formula occurs when $f'(0) = 1$. In view of the estimates of $f'(0)$ for $a = 2$ and $a = 3$, it seems reasonable that there is a number a between 2 and 3 for which $f'(0) = 1$. It is traditional to denote this value by the letter e.* Thus

(6) $$e \text{ is the number such that } \lim_{h \to 0} \frac{e^h - 1}{h} = 1$$

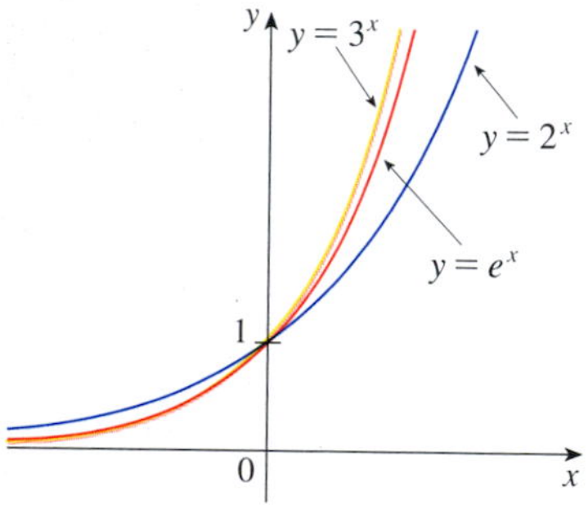

FIGURE 8

Geometrically, this means that of all the possible exponential functions $y = a^x$, the function $f(x) = e^x$ is the one whose tangent line at $(0, 1)$ has a slope $f'(0)$ that is exactly 1. (See Figures 8 and 9.)

If we put $a = e$ and therefore $f'(0) = 1$ in Equation 4, it becomes the following important differentiation formula:

(7) $$\frac{d}{dx} e^x = e^x$$

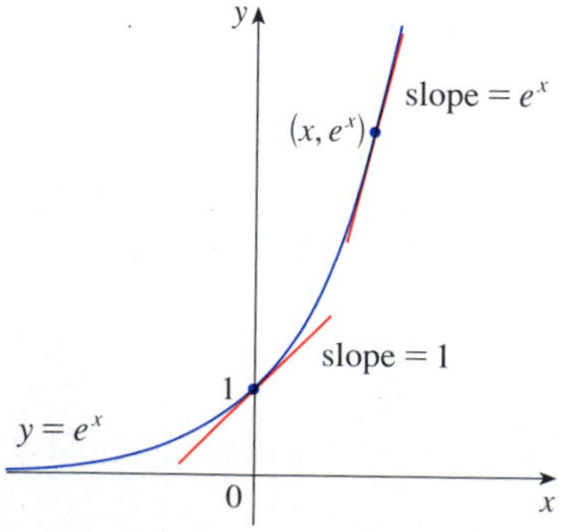

FIGURE 9

Thus, the exponential function $f(x) = e^x$ has the property that it is its own derivative. The geometrical significance of this fact is that the slope of a tangent line to the curve $y = e^x$ is equal to the y-coordinate of the point (see Figure 9).

EXAMPLE 2 Differentiate the function $y = e^{\tan x}$.

SOLUTION To use the Chain Rule, we let $u = \tan x$. Then we have $y = e^u$, so

$$\frac{dy}{dx} = \frac{dy}{du}\frac{du}{dx} = e^u \frac{du}{dx} = e^{\tan x} \sec^2 x$$ ■

*The notation e for this number was chosen by the Swiss mathematician Leonhard Euler in 1727, probably because it is the first letter of the word *exponential*.

In general, if we combine Formula 7 with the Chain Rule, as in Example 2, we get

$$\frac{d}{dx} e^u = e^u \frac{du}{dx} \tag{8}$$

EXAMPLE 3 Find y' if $y = e^{-4x} \sin 5x$.

SOLUTION Using Formula 8 and the Product Rule, we have

$$y' = e^{-4x}(-4)\sin 5x + e^{-4x}(\cos 5x)(5) = e^{-4x}(5\cos 5x - 4\sin 5x)$$ ■

We have seen that e is a number that lies somewhere between 2 and 3, but we can use Equation 4 to estimate the numerical value of e more accurately. Let $e = 2^c$. Then $e^x = 2^{cx}$. If $f(x) = 2^x$, then from Equation 4 we have $f'(x) = k2^x$, where the value of k is $f'(0) \approx 0.693147$. Thus, by the Chain Rule,

$$e^x = \frac{d}{dx} e^x = \frac{d}{dx} 2^{cx} = k2^{cx} \frac{d}{dx}(cx) = ck2^{cx}$$

Putting $x = 0$, we have $1 = ck$, so $c = 1/k$ and

$$e = 2^{1/k} \approx 2^{1/0.693147} \approx 2.71828$$

It can be shown that the approximate value to 20 decimal places is

$$e \approx 2.71828182845904523536$$

The decimal expansion of e is nonrepeating because e is an irrational number.

EXAMPLE 4 In Example 6 in Section 3.3 we considered a population of bacteria cells in a homogeneous nutrient medium. We showed that if the population doubles every hour, then the population after t hours is

$$n = n_0 2^t$$

where n_0 is the initial population. Now we can use (4) and (5) to compute the growth rate:

The rate of growth is proportional to the size of the population.

$$\frac{dn}{dt} \approx n_0(0.693147)2^t$$

For instance, if the initial population is $n_0 = 1000$ cells, then the growth rate after 2 h is

$$\left.\frac{dn}{dt}\right|_{t=2} \approx (1000)(0.693147)2^t\big|_{t=2}$$

$$= (4000)(0.693147) \approx 2773 \text{ cells/h}$$ ■

The exponential function $f(x) = e^x$ is one of the most frequently occurring functions in calculus and its applications, so it is important to be familiar with its graph

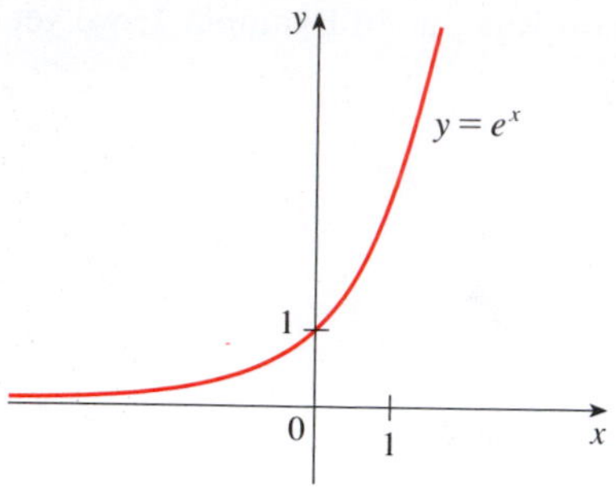

FIGURE 10

(Figure 10) and properties. We summarize these properties as follows, using the fact that this function is just a special case of the exponential functions considered in Theorem 2 but with base $a = e > 1$.

(9) PROPERTIES OF THE EXPONENTIAL FUNCTION The exponential function $f(x) = e^x$ is an increasing continuous function with domain $\mathbb{R}$ and range $(0, \infty)$. Thus, $e^x > 0$ for all x. Also

$$\lim_{x \to -\infty} e^x = 0 \qquad \lim_{x \to \infty} e^x = \infty$$

So the x-axis is a horizontal asymptote of $f(x) = e^x$.

EXAMPLE 5 Find $\displaystyle\lim_{x \to \infty} \frac{e^{2x}}{e^{2x} + 1}$.

SOLUTION We divide numerator and denominator by e^{2x}:

$$\lim_{x \to \infty} \frac{e^{2x}}{e^{2x} + 1} = \lim_{x \to \infty} \frac{1}{1 + e^{-2x}} = \frac{1}{1 + \lim_{x \to \infty} e^{-2x}}$$

$$= \frac{1}{1 + 0} = 1$$

We have used the fact that $t = -2x \to -\infty$ as $x \to \infty$ and so

$$\lim_{x \to \infty} e^{-2x} = \lim_{t \to -\infty} e^t = 0$$

■

EXAMPLE 6

(a) Find $\lim_{x \to 0^-} e^{1/x}$.

(b) Find the horizontal and vertical asymptotes of $y = e^{1/x}$.

SOLUTION

(a) As $x \to 0^-$, $1/x \to -\infty$, so by (9),

$$\lim_{x \to 0^-} e^{1/x} = 0$$

(b) Since $1/x \to 0$ as $x \to \pm\infty$, we have

$$\lim_{x \to \pm\infty} e^{1/x} = 1$$

so $y = 1$ is a horizontal asymptote. Since $1/x \to \infty$ as $x \to 0^+$, we have

$$\lim_{x \to 0^+} e^{1/x} = \infty$$

so $x = 0$ is a vertical asymptote. ■

EXERCISES 4.1

1–4 ■ Graph the given functions on a common screen. How are these graphs related?

1. $y = 2^x, \quad y = e^x, \quad y = 5^x, \quad y = 20^x$

2. $y = e^x, \quad y = e^{-x}, \quad y = 8^x, \quad y = 8^{-x}$

3. $y = 3^x, \quad y = 10^x, \quad y = \left(\frac{1}{3}\right)^x, \quad y = \left(\frac{1}{10}\right)^x$

4. $y = 0.9^x, \quad y = 0.6^x, \quad y = 0.3^x, \quad y = 0.1^x$

5–12 ■ Make a rough sketch of the graph of each function. Do not use a calculator. Just use the graphs given in Figures 3, 6, and 8 and, if necessary, the transformations of Section 2 of Review and Preview.

5. $y = 2^x + 1$

6. $y = 2^{x+1}$

7. $y = 3^{-x}$

8. $y = -3^x$

9. $y = -3^{-x}$

10. $y = 2^{|x|}$

11. $y = 3 - e^x$

12. $y = 2 + 5(1 - e^{-x})$

13–22 ■ Find each limit.

13. $\lim_{x\to\infty} (1.1)^x$

14. $\lim_{x\to-\infty} (1.1)^x$

15. $\lim_{x\to-\infty} (\pi/4)^x$

16. $\lim_{x\to\infty} (2\pi/7)^x$

17. $\lim_{x\to\infty} \dfrac{e^{3x} - e^{-3x}}{e^{3x} + e^{-3x}}$

18. $\lim_{x\to-\infty} \dfrac{e^{3x} - e^{-3x}}{e^{3x} + e^{-3x}}$

19. $\lim_{x\to 1^-} e^{2/(x-1)}$

20. $\lim_{x\to 1^+} e^{2/(x-1)}$

21. $\lim_{x\to(\pi/2)^-} \dfrac{2}{1 + e^{\tan x}}$

22. $\lim_{x\to 0^-} \dfrac{2}{1 + e^{\cot x}}$

23. Show that if the graphs of $f(x) = x^2$ and $g(x) = 2^x$ are drawn on a coordinate grid where the unit of measurement is 1 inch, then at a distance 2 ft to the right of the origin the height of the graph of f is 48 ft but the height of the graph of g is about 265 mi.

24. Compare the rates of growth of the functions $f(x) = x^5$ and $g(x) = 5^x$ by graphing both functions in several viewing rectangles. Find all points of intersection of the graphs correct to one decimal place.

25. (a) Use a calculator to evaluate the quantity $(4^h - 1)/h$ for $h = 0.1, 0.01, 0.001$, and 0.0001. What does the quantity represent?

(b) Estimate the value of the limit

$$\lim_{h\to 0} \frac{4^h - 1}{h}$$

correct to two decimal places.

(c) What does the limit in part (b) represent?

26. Use a calculator to estimate the values of the limits to two decimal places. What is the significance of these limits?

(a) $\lim_{h\to 0} \dfrac{2.7^h - 1}{h}$

(b) $\lim_{h\to 0} \dfrac{2.8^h - 1}{h}$

27–42 ■ Differentiate each function.

27. $f(x) = e^{\sqrt{x}}$

28. $f(x) = xe^{-x^2}$

29. $y = xe^{2x}$

30. $g(x) = e^{-5x}\cos 3x$

31. $h(t) = \sqrt{1 - e^t}$

32. $h(\theta) = e^{\sin 5\theta}$

33. $y = e^{x\cos x}$

34. $y = \dfrac{e^{-x^2}}{x}$

35. $y = e^{-1/x}$

36. $y = e^{x+e^x}$

37. $y = \tan(e^{3x-2})$

38. $y = \sqrt[3]{2x + e^{3x}}$

39. $y = \dfrac{e^{3x}}{1 + e^x}$

40. $y = \dfrac{e^x + e^{-x}}{e^x - e^{-x}}$

41. $y = x^e$

42. $y = \sec(e^{\tan x^2})$

43–44 ■ Find an equation of the tangent line to the curve at the given point.

43. $y = e^{-x}\sin x$, $(\pi, 0)$

44. $y = x^2e^{-x}$, $(1, 1/e)$

45. Find y' if $\cos(x - y) = xe^x$.

46. Find an equation of the tangent line to the curve $2e^{xy} = x + y$ at the point $(0, 2)$.

47. Show that the function $y = e^{2x} + e^{-3x}$ satisfies the differential equation $y'' + y' - 6y = 0$.

48. Show that the function $y = Ae^{-x} + Bxe^{-x}$ satisfies the differential equation $y'' + 2y' + y = 0$.

49. For what values of r does the function $y = e^{rx}$ satisfy the equation $y'' + 5y' - 6y = 0$?

50. Find the values of λ for which $y = e^{\lambda x}$ satisfies the equation $y + y' = y''$.

51. If $f(x) = e^{-2x}$, find $f^{(8)}(x)$.

52. Find the thousandth derivative of $f(x) = xe^{-x}$.

53. (a) Use the Intermediate Value Theorem to show that there is a root of the equation $e^x + x = 0$.

(b) Use Newton's method to find the root of the equation in part (a) correct to six decimal places.

54. Use a graph to find an initial approximation (to one decimal place) to the root of the equation $e^{-x^2} = x^3 + x - 3$. Then use Newton's method to find the root correct to six decimal places.

55. Under certain circumstances a rumor spreads according to the equation

$$p(t) = \frac{1}{1 + ae^{-kt}}$$

where $p(t)$ is the proportion of the population that knows the rumor at time t and a and k are positive constants. [In Section 9.1 we will see that this is a reasonable equation for $p(t)$.]

(a) Find $\lim_{t\to\infty} p(t)$.

(b) Find the rate of spread of the rumor.

(c) Graph p for the case $a = 10$, $k = 0.5$ with t measured in hours. Use the graph to estimate how long it will take for 80% of the population to hear the rumor.

56. (a) Compare the rates of growth of $f(x) = x^{10}$ and $g(x) = e^x$ by graphing both f and g in several viewing rectangles. When does the graph of g finally surpass the graph of f?

(b) Find a viewing rectangle that shows how the function $h(x) = e^x/x^{10}$ behaves for large x.

(c) Find a number N such that

$$\frac{e^x}{x^{10}} > 10^{10} \qquad \text{whenever } x > N$$

57. Evaluate $\displaystyle\lim_{x \to \pi} \frac{e^{\sin x} - 1}{x - \pi}$.

58. Plot the parametric curve $x = e^t$, $y = e^{-t}$ for $-1 \leqslant t \leqslant 1$. Find an equation of the tangent line to this curve at the point corresponding to $t = 0$.

59. Plot the parametric curve $x = e^{t/3} \cos t$, $y = e^{t/3} \sin t$ for $-\pi \leqslant t \leqslant \pi$. Identify the shape of the graph. [*Hint:* $x = \cos t$, $y = \sin t$ lies on the unit circle.]

60. Find an equation of the tangent line to the curve in Exercise 59 at the point corresponding to $t = \pi/4$.

61. Find dy/dx if $x = e^{-t}$ and $y = te^{2t}$.

62. Find dy/dx if $x = te^{-t}$ and $y = e^t$.

4.2 INVERSE FUNCTIONS

The common theme that links the functions of this chapter is that they occur as pairs of inverse functions. The only functions that possess inverse functions are one-to-one functions, so we start by reviewing that concept.

Let's compare the functions f and g whose arrow diagrams are shown in Figure 1.

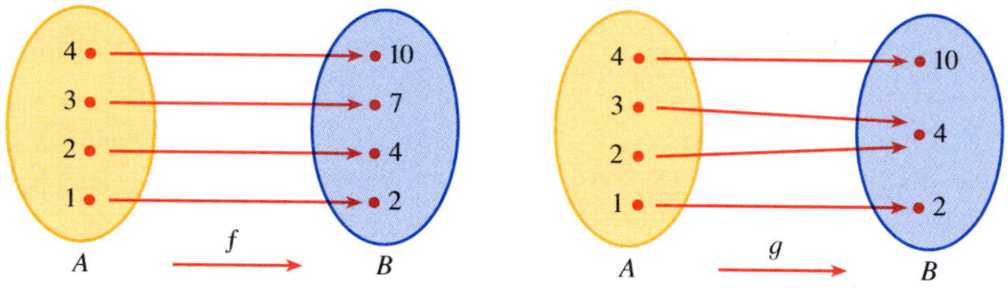

FIGURE 1

Note that f never takes on the same value twice (any two numbers in A have different images), whereas g does take on the same value twice (both 2 and 3 have the same image, 4). In symbols,

$$g(2) = g(3)$$

but

$$f(x_1) \neq f(x_2) \qquad \text{whenever } x_1 \neq x_2$$

Functions that have this latter property are called *one-to-one functions.*

> **(1) DEFINITION** A function f with domain A is called a **one-to-one function** if no two elements of A have the same image; that is,
>
> $$f(x_1) \neq f(x_2) \qquad \text{whenever } x_1 \neq x_2$$

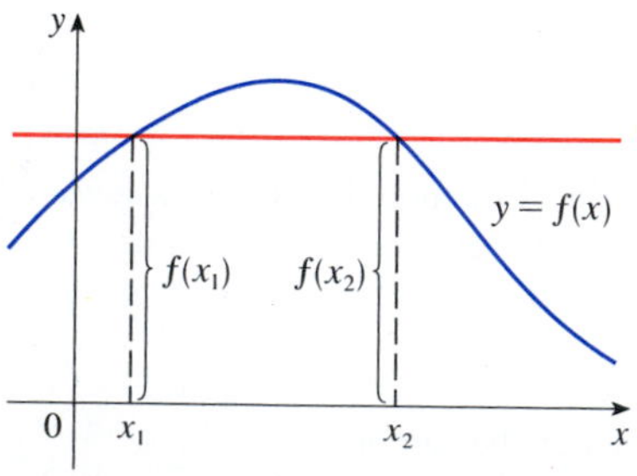

FIGURE 2
This function is not one-to-one because $f(x_1) = f(x_2)$.

If a horizontal line intersects the graph of f in more than one point, then we see from Figure 2 that there are numbers x_1 and x_2 such that $f(x_1) = f(x_2)$. This means that f is not one-to-one. Therefore, we have the following geometric method for determining whether a function is one-to-one.

> **HORIZONTAL LINE TEST** A function is one-to-one if and only if no horizontal line intersects its graph more than once.

EXAMPLE 1 Is the function $f(x) = x^3$ one-to-one?

SOLUTION 1 If $x_1 \neq x_2$, then $x_1^3 \neq x_2^3$ (two different numbers cannot have the same cube). Therefore, by Definition 1, $f(x) = x^3$ is one-to-one.

SOLUTION 2 From Figure 3 we see that no horizontal line intersects the graph of $f(x) = x^3$ more than once. Therefore, by the Horizontal Line Test, f is one-to-one. ■

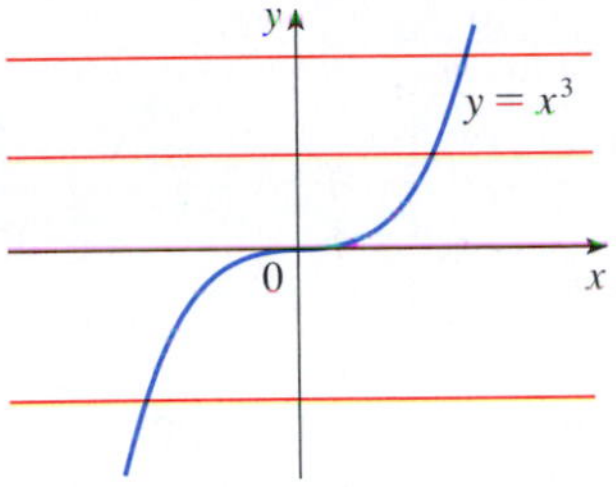

FIGURE 3
$f(x) = x^3$ is one-to-one.

EXAMPLE 2 Is the function $g(x) = x^2$ one-to-one?

SOLUTION 1 This function is not one-to-one because, for instance,

$$g(1) = 1 = g(-1)$$

and so 1 and -1 have the same image.

SOLUTION 2 From Figure 4 we see that there are horizontal lines that intersect the graph of g more than once. Therefore, by the Horizontal Line Test, g is not one-to-one. ■

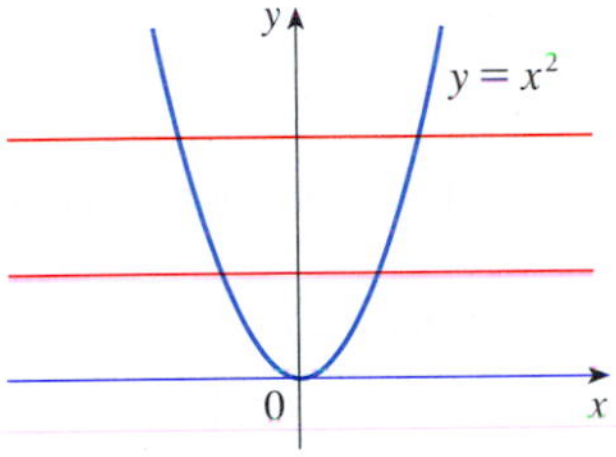

FIGURE 4
$g(x) = x^2$ is not one-to-one.

One-to-one functions are important because they are precisely the functions that possess inverse functions according to the following definition.

(2) DEFINITION Let f be a one-to-one function with domain A and range B. Then its **inverse function** f^{-1} has domain B and range A and is defined by

$$f^{-1}(y) = x \iff f(x) = y$$

for any y in B.

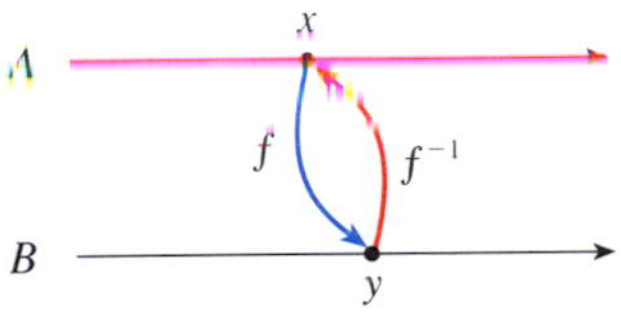

FIGURE 5

This definition says that if f maps x into y, then f^{-1} maps y back into x. (If f were not one-to-one, then f^{-1} would not be uniquely defined.) The arrow diagram in Figure 5 indicates that f^{-1} reverses the effect of f. Note that

$$\text{domain of } f^{-1} = \text{range of } f$$

$$\text{range of } f^{-1} = \text{domain of } f$$

For example, the inverse function of $f(x) = x^3$ is $f^{-1}(x) = x^{1/3}$ because if $y = x^3$, then

$$f^{-1}(y) = f^{-1}(x^3) = (x^3)^{1/3} = x$$

Caution: Do not mistake the -1 in f^{-1} for an exponent. Thus

$$f^{-1}(x) \quad \text{does } \textit{not} \text{ mean} \quad \frac{1}{f(x)}$$

The reciprocal $1/f(x)$ could, however, be written as $[f(x)]^{-1}$.

The letter x is traditionally used as the independent variable, so when we concentrate on f^{-1} rather than on f, we usually reverse the roles of x and y in Definition 2 and write

(3)
$$f^{-1}(x) = y \iff f(y) = x$$

By substituting for y in Definition 2 and substituting for x in (3), we get the following **cancellation equations**:

(4)

$$f^{-1}(f(x)) = x \quad \text{for every } x \text{ in } A$$

$$f(f^{-1}(x)) = x \quad \text{for every } x \text{ in } B$$

The first cancellation equation says that if we start with x, apply f, and then apply f^{-1}, we arrive back at x, where we started. Thus, f^{-1} undoes what f does. The second equation says that f undoes what f^{-1} does.

For example, if $f(x) = x^3$, then $f^{-1}(x) = x^{1/3}$ and the cancellation equations become

$$f^{-1}(f(x)) = (x^3)^{1/3} = x$$

$$f(f^{-1}(x)) = (x^{1/3})^3 = x$$

These equations simply say that the cube function and the cube root function cancel each other.

Let's now see how to compute inverse functions. If we have a function $y = f(x)$ and are able to solve this equation for x in terms of y, then according to Definition 2 we must have $x = f^{-1}(y)$. If we then interchange x and y, we have $y = f^{-1}(x)$, which is the desired equation.

(5) HOW TO FIND THE INVERSE FUNCTION OF A ONE-TO-ONE FUNCTION f

Step 1. Write $y = f(x)$.
Step 2. Solve this equation for x in terms of y (if possible).
Step 3. Interchange x and y. The resulting equation is $y = f^{-1}(x)$.

EXAMPLE 3 Find the inverse function of $f(x) = x^3 + 2$.

SOLUTION According to (5) we first write

$$y = x^3 + 2$$

Then we solve this equation for x:

$$x^3 = y - 2$$

$$x = \sqrt[3]{y - 2}$$

Finally, we interchange x and y:

$$y = \sqrt[3]{x - 2}$$

Therefore, the inverse function is $f^{-1}(x) = \sqrt[3]{x - 2}$. ■

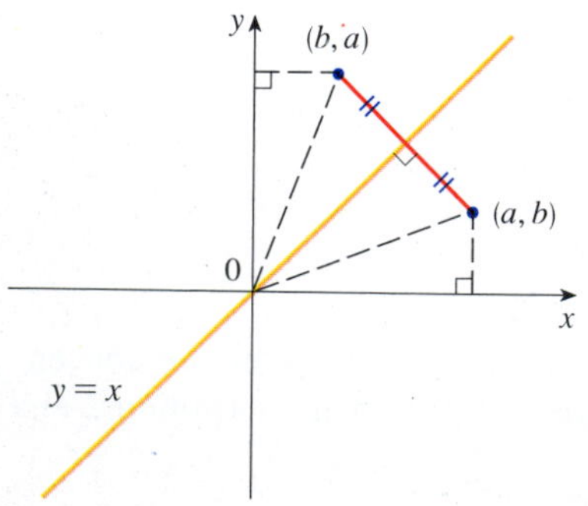

FIGURE 6

The principle of interchanging x and y to find the inverse function also gives us the method for obtaining the graph of f^{-1} from the graph of f. Since $f(a) = b$ if and only if $f^{-1}(b) = a$, the point (a, b) is on the graph of f if and only if the point (b, a) is on the graph of f^{-1}. But we get the point (b, a) from (a, b) by reflecting about the line $y = x$ (see Figure 6).

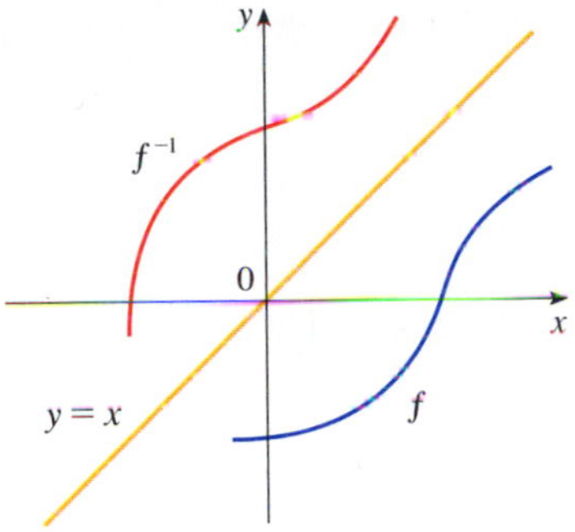

FIGURE 7

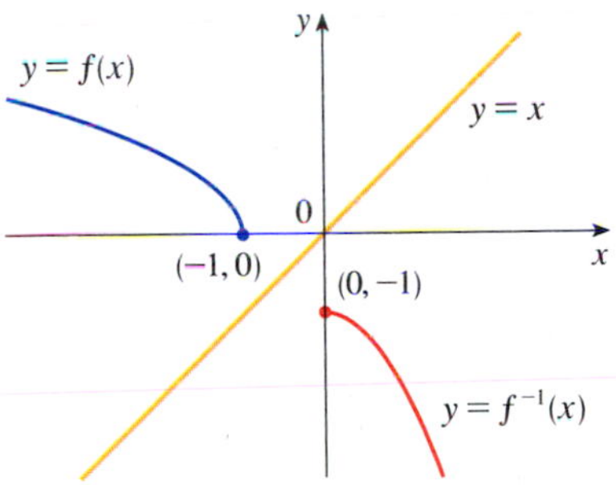

FIGURE 8

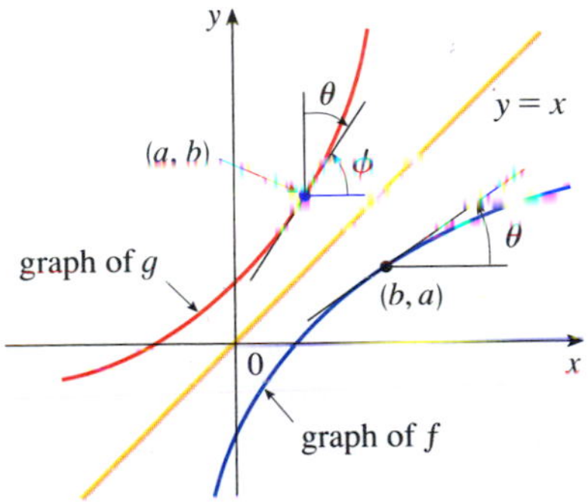

FIGURE 9

Therefore, as illustrated by Figure 7:

> The graph of f^{-1} is obtained by reflecting the graph of f about the line $y = x$.

EXAMPLE 4 Sketch the graphs of $f(x) = \sqrt{-1 - x}$ and its inverse function using the same coordinate axes.

SOLUTION First we sketch the curve $y = \sqrt{-1 - x}$ (the top half of the parabola $y^2 = -1 - x$ or $x = -y^2 - 1$) and then we reflect about the line $y = x$ to get the graph of f^{-1} (see Figure 8). ■

Now let's look at inverse functions from the point of view of calculus. Suppose that f is both one-to-one and continuous. We think of a continuous function as one whose graph has no break in it. (It consists of just one piece.) Since the graph of f^{-1} is obtained from the graph of f by reflecting about the line $y = x$, the graph of f^{-1} has no break in it either (see Figure 7). Thus, we would expect that f^{-1} is also a continuous function.

This geometrical argument does not prove the following theorem but at least it makes the theorem plausible. A proof can be found in Appendix F.

> **(6) THEOREM** If f is a one-to-one continuous function defined on an interval, then its inverse function f^{-1} is also continuous.

Now suppose that f is a one-to-one differentiable function. Geometrically, we can think of a differentiable function as one whose graph has no corner or kink in it. We get the graph of f^{-1} by reflecting the graph of f about the line $y = x$, so the graph of f^{-1} has no corner or kink in it either. We therefore expect that f^{-1} is also differentiable (except where its tangents are vertical). In fact, we can predict the value of the derivative of f^{-1} at a given point by a geometric argument. In Figure 9 the graphs of f and its inverse $g = f^{-1}$ are shown. If $f(b) = a$, then $g(a) = f^{-1}(a) = b$ and $g'(a)$ is the slope of the tangent to the graph of g at (a, b), which is $\tan\phi$. Likewise, $f'(b) = \tan\theta$. From Figure 9 we see that $\theta + \phi = \pi/2$, so

$$g'(a) = \tan\phi = \tan\left(\frac{\pi}{2} - \theta\right) = \frac{1}{\tan\theta} = \frac{1}{f'(b)}$$

that is,
$$g'(a) = \frac{1}{f'(g(a))}$$

> **(7) THEOREM** If f is a one-to-one differentiable function with inverse function $g = f^{-1}$ and $f'(g(a)) \neq 0$, then the inverse function is differentiable at a and
>
> $$g'(a) = \frac{1}{f'(g(a))}$$

PROOF Write the definition of derivative as in Equation 3.1.3:

$$g'(a) = \lim_{x \to a} \frac{g(x) - g(a)}{x - a}$$

By (3) we have

$$g(x) = y \iff f(y) = x$$

and

$$g(a) = b \iff f(b) = a$$

Since f is differentiable, it is continuous, so $g = f^{-1}$ is continuous by Theorem 6. Thus, if $x \to a$, then $g(x) \to g(a)$, that is, $y \to b$. Therefore

$$\begin{aligned} g'(a) &= \lim_{x \to a} \frac{g(x) - g(a)}{x - a} = \lim_{y \to b} \frac{y - b}{f(y) - f(b)} \\ &= \lim_{y \to b} \frac{1}{\dfrac{f(y) - f(b)}{y - b}} = \frac{1}{\lim\limits_{y \to b} \dfrac{f(y) - f(b)}{y - b}} \\ &= \frac{1}{f'(b)} = \frac{1}{f'(g(a))} \end{aligned}$$

□

NOTE 1: Replacing a by the general number x in the formula of Theorem 7, we get

(8) $$g'(x) = \frac{1}{f'(g(x))}$$

If we write $y = g(x)$, then $f(y) = x$, so Equation 8, when expressed in Leibniz notation, becomes

$$\frac{dy}{dx} = \frac{1}{\dfrac{dx}{dy}}$$

NOTE 2: If it is known in advance that f^{-1} is differentiable, then its derivative can be computed more easily than in the proof of Theorem 7 by using implicit differentiation. If $y = f^{-1}(x)$, then $f(y) = x$. Differentiating the equation $f(y) = x$ implicitly with respect to x, remembering that y is a function of x, and using the Chain Rule, we get

$$f'(y)\frac{dy}{dx} = 1$$

Therefore

$$\frac{dy}{dx} = \frac{1}{f'(y)} = \frac{1}{\dfrac{dx}{dy}}$$

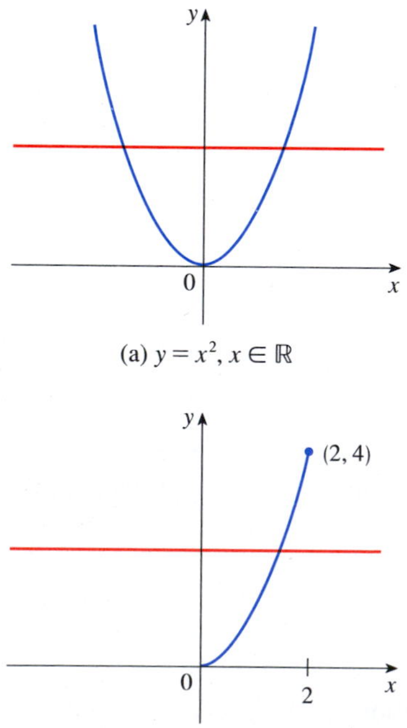

(a) $y = x^2, x \in \mathbb{R}$

(b) $f(x) = x^2, 0 \leqslant x \leqslant 2$

FIGURE 10

EXAMPLE 5 Although the function $y = x^2$, $x \in \mathbb{R}$, is not one-to-one and therefore does not have an inverse function, we can turn it into a one-to-one function by restricting its domain. For instance, the function $f(x) = x^2$, $0 \leqslant x \leqslant 2$, is one-to-one (by the Horizontal Line Test) and has domain $[0, 2]$ and range $[0, 4]$ (see Figure 10). Thus, f has an inverse function $g = f^{-1}$ with domain $[0, 4]$ and range $[0, 2]$.

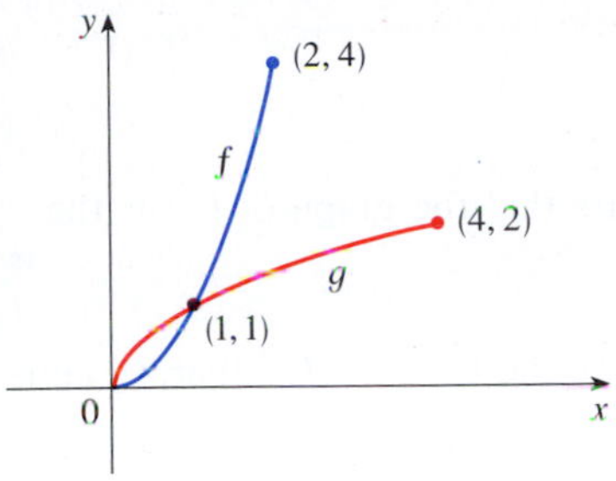

FIGURE 11

Without computing a formula for g' we can still calculate $g'(1)$. Since $f(1) = 1$, we have $g(1) = 1$. Also $f'(x) = 2x$. So by Theorem 7 we have

$$g'(1) = \frac{1}{f'(g(1))} = \frac{1}{f'(1)} = \frac{1}{2}$$

In this case it is easy to find g explicitly. In fact, $g(x) = \sqrt{x}$, $0 \leqslant x \leqslant 4$. [In general we could use the method given by (5).] Then $g'(x) = 1/(2\sqrt{x})$, so $g'(1) = \frac{1}{2}$, which agrees with the preceding computation. The functions f and g are graphed in Figure 11. ■

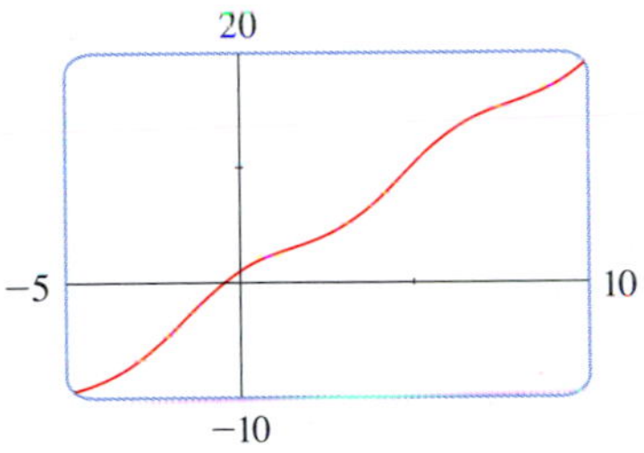

FIGURE 12

EXAMPLE 6 If $f(x) = 2x + \cos x$ and $g = f^{-1}$, find $g'(1)$.

SOLUTION Notice from the graph of f in Figure 12 that f is one-to-one by the Horizontal Line Test.

To use Theorem 7 we need to know $g(1)$ and we can find it by inspection:

$$f(0) = 1 \quad \Rightarrow \quad g(1) = 0$$

Therefore
$$g'(1) = \frac{1}{f'(g(1))} = \frac{1}{f'(0)} = \frac{1}{2 - \sin 0} = \frac{1}{2}$$
■

Many graphing devices won't plot the inverse of a given function directly, but we can obtain the desired graph by using the parametric graphing capability of such a device. As discussed in Section 1.3, parametric curves are defined by a pair of equations $x = x(t)$, $y = y(t)$ and each value of the number t (called a *parameter*) determines a point (x, y) on the curve. If we start with an ordinary curve $y = f(x)$, where f is a one-to-one function, then we can represent it as a parametric curve by writing the parametric equations

$$x = t \qquad y = f(t)$$

We know that the graph of the inverse function is obtained by interchanging the x- and y-coordinates of the points on the graph of f. Therefore, parametric equations for the graph of f^{-1} are

$$x = f(t) \qquad y = t$$

EXAMPLE 7 Show that the function $f(x) = \sqrt{x^3 + x^2 + x + 1}$ is one-to-one and graph both f and f^{-1}.

SOLUTION We plot the graph in Figure 13 and observe that f is one-to-one by the Horizontal Line Test.

To graph f and f^{-1} on the same screen we use parametric graphs. Parametric equations for the graph of f are

$$x = t \qquad y = \sqrt{t^3 + t^2 + t + 1}$$

and parametric equations for the graph of f^{-1} are

$$x = \sqrt{t^3 + t^2 + t + 1} \qquad y = t$$

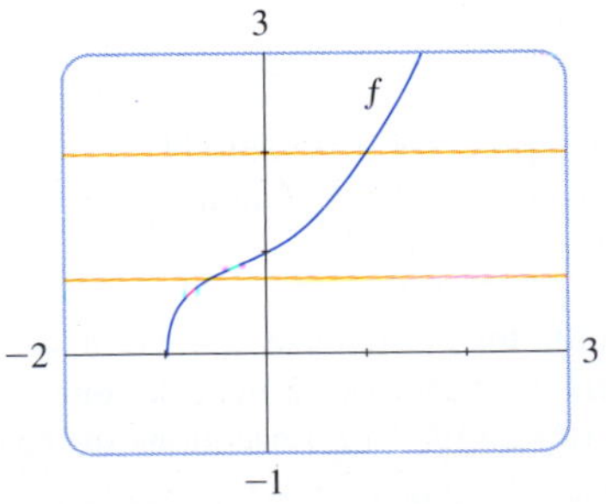

FIGURE 13

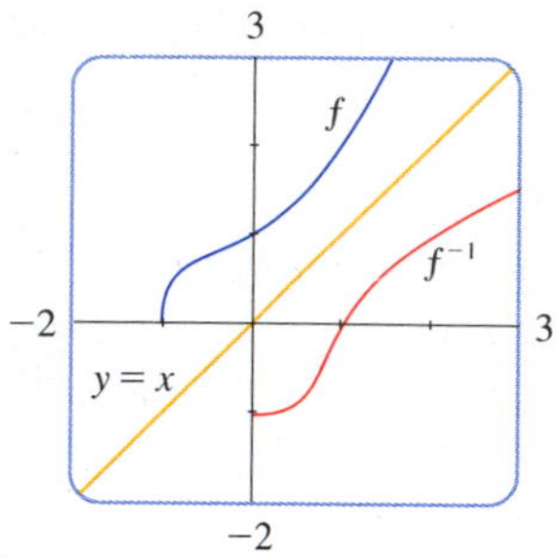

FIGURE 14

Let's also plot the line $y = x$:

$$x = t \qquad y = t$$

Figure 14 shows all three graphs and, indeed, it appears that the graph of f^{-1} is the reflection of the graph of f in the line $y = x$. ■

In Exercise 38 we show how to graph the derivative of the function f^{-1} that we considered in Example 7.

EXERCISES 4.2

1–6 ■ The graph of a function f is shown. Determine whether f is one-to-one.

1.

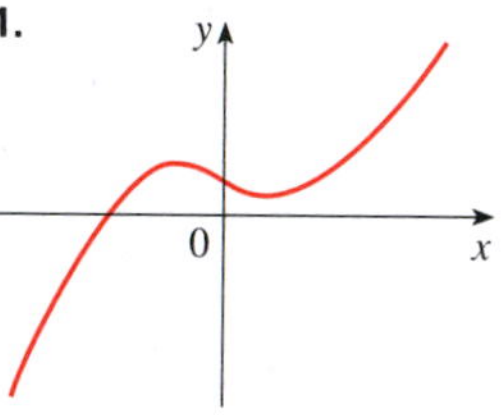

2.

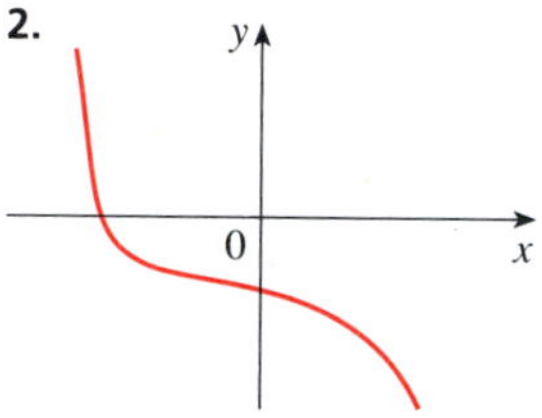

3.

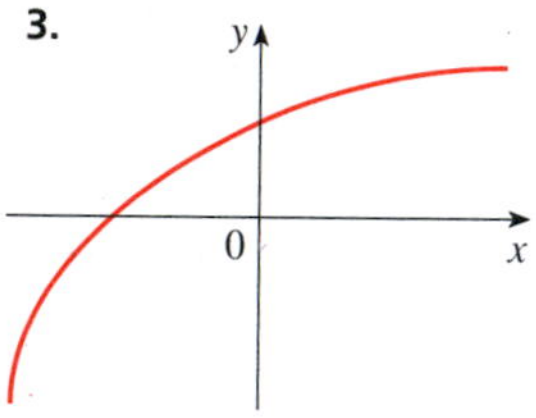

4.

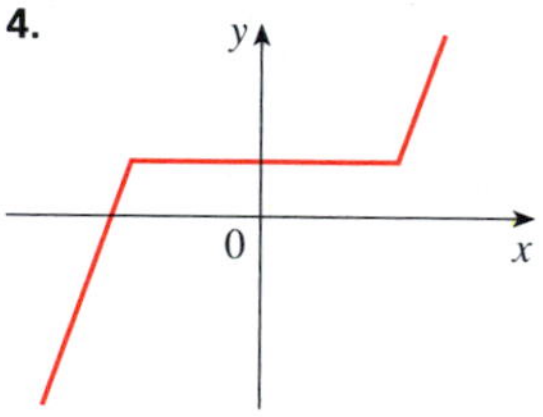

5.

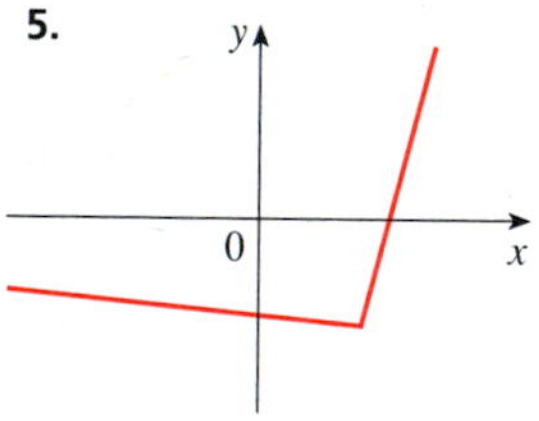

6.

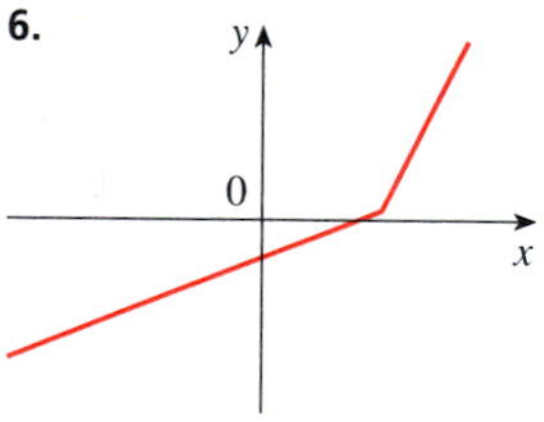

7–12 ■ Determine whether the function is one-to-one.

7. $f(x) = 7x - 3$

8. $f(x) = x^2 - 2x + 5$

9. $g(x) = \sqrt{x}$

10. $g(x) = |x|$

11. $h(x) = x^4 + 5$

12. $h(x) = x^4 + 5, \quad 0 \leqslant x \leqslant 2$

13–18 ■ Show that f is one-to-one and find its inverse function.

13. $f(x) = 4x + 7$

14. $f(x) = \dfrac{x - 2}{x + 2}$

15. $f(x) = \dfrac{1 + 3x}{5 - 2x}$

16. $f(x) = 5 - 4x^3$

17. $f(x) = \sqrt{2 + 5x}$

18. $f(x) = x^2 + x, \quad x \geqslant -\frac{1}{2}$

19–24 ■

(a) Show that f is one-to-one.
(b) Use Theorem 7 to find $g'(a)$, where $g = f^{-1}$.
(c) Calculate $g(x)$ and state the domain and range of g.
(d) Calculate $g'(a)$ from the formula in part (c) and check that it agrees with the result of part (b).
(e) Sketch the graphs of f and g on the same axes.

19. $f(x) = 2x + 1, \quad a = 3$

20. $f(x) = 6 - x, \quad a = 2$

21. $f(x) = x^3, \quad a = 8$

22. $f(x) = \sqrt{x - 2}, \quad a = 2$

23. $f(x) = 9 - x^2, \quad 0 \leqslant x \leqslant 3, \quad a = 8$

24. $f(x) = 1/(x - 1), \quad x > 1, \quad a = 2$

25–30 ■ Find $g'(a)$, where g is the inverse function of the given function.

25. $f(x) = x^3 + x + 1, \quad a = 1$

26. $f(x) = x^5 - x^3 + 2x, \quad a = 2$

27. $f(x) = 3 + x^2 + \tan(\pi x/2), \quad -1 < x < 1, \quad a = 3$

28. $f(x) = \sqrt{x^3 + x^2 + x + 1}, \quad a = 2$

29. $f(x) = e^x, \quad a = 1$

30. $f(x) = 3 + x + e^x, \quad a = 4$

31. Suppose g is the inverse function of f and $f(4) = 5$, $f'(4) = \frac{2}{3}$. Find $g'(5)$.

32. Suppose g is the inverse function of a differentiable function f and let $G(x) = 1/g(x)$. If $f(3) = 2$ and $f'(3) = \frac{1}{9}$, find $G'(2)$.

33–34 ■ Find an explicit formula for f^{-1} and use it to graph f^{-1}, f, and the line $y = x$ on the same screen. To check your work, see whether the graphs of f and f^{-1} are reflections in the line.

33. $f(x) = 1 - 2/x^2, \quad x > 0$

34. $f(x) = \sqrt{x^2 + 2x}, \quad x > 0$

35–36 ■ Show that f is one-to-one. Then graph f, f^{-1}, and $y = x$ on the same screen using parametric graphs.

35. $f(x) = \sqrt{x^2 + 1} - x$

36. $f(x) = x + \sin x$

37. The equation $\sqrt[5]{x} - \sqrt[5]{y} = y$ defines y implicitly as a one-to-one function f of x.
(a) Find an explicit expression for $f^{-1}(x)$.
(b) Graph f using a parametric graph.

38. If a one-to-one function f is given, we can use Theorem 7 to graph the derivative of f^{-1}. According to that theorem, the points on the graph of $(f^{-1})'$ are of the form $(x, 1/f'(f^{-1}(x)))$. If we change to the variable $t = f^{-1}(x)$, then $x = f(t)$ and so we can write the points as $(f(t), 1/f'(t))$. This means that the parametric equations

$$x = f(t) \qquad y = 1/f'(t)$$

can be used to graph $(f^{-1})'$. Use this method to graph the derivative of the function f^{-1} considered in Example 7. Check your work by comparing with the graph of f^{-1} in Figure 14.

39. If n is a positive integer, then $y = \sqrt[n]{x}$ has domain $\mathbb{R}$ if n is odd and domain $[0, \infty)$ if n is even. It is the inverse function of the power function. Use the method of Note 2 to find dy/dx.

40. Show that $h(x) = \sin x$, $x \in \mathbb{R}$, is not one-to-one, but its restriction $f(x) = \sin x$, $-\pi/2 \leqslant x \leqslant \pi/2$, is one-to-one. Compute the derivative of $f^{-1} = \sin^{-1}$ by the method of Note 2.

41. If f is a one-to-one, twice differentiable function with inverse function g, show that

$$g''(x) = -\frac{f''(g(x))}{[f'(g(x))]^3}$$

4.3 LOGARITHMIC FUNCTIONS

If $a > 0$ and $a \neq 1$, the exponential function $f(x) = a^x$ is either increasing or decreasing and so it is one-to-one. It therefore has an inverse function f^{-1}, which is called the **logarithmic function with base *a*** and is denoted by $\log_a$. If we use the formulation of an inverse function given by (4.2.3),

$$f^{-1}(x) = y \iff f(y) = x$$

then we have

(1)
$$\log_a x = y \iff a^y = x$$

Thus, if $x > 0$, $\log_a x$ is the exponent to which the base a must be raised to give x.

EXAMPLE 1 Evaluate (a) $\log_3 81$, (b) $\log_{25} 5$, and (c) $\log_{10} 0.001$.

SOLUTION
(a) $\log_3 81 = 4$ because $3^4 = 81$
(b) $\log_{25} 5 = \frac{1}{2}$ because $25^{1/2} = 5$
(c) $\log_{10} 0.001 = -3$ because $10^{-3} = 0.001$ ■

The cancellation equations (4.2.4), when applied to $f(x) = a^x$ and $f^{-1}(x) = \log_a x$, become

(2)
$$\log_a(a^x) = x \quad \text{for every } x \in \mathbb{R}$$
$$a^{\log_a x} = x \quad \text{for every } x > 0$$

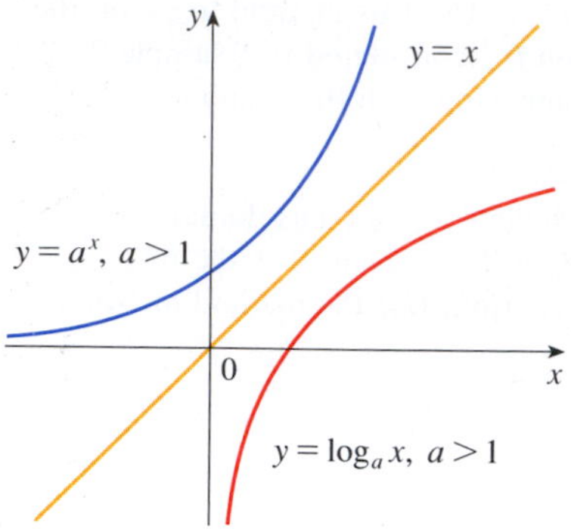

FIGURE 1

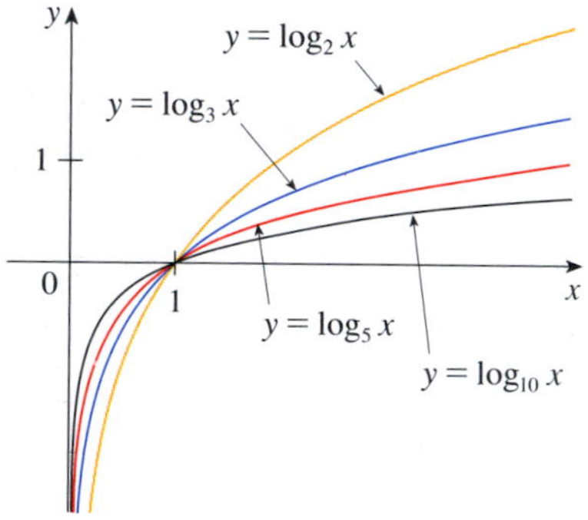

FIGURE 2

The logarithmic function $\log_a$ has domain $(0, \infty)$ and range $\mathbb{R}$ and is continuous since it is the inverse of a continuous function, namely, the exponential function. Its graph is the reflection of the graph of $y = a^x$ about the line $y = x$.

Figure 1 shows the case where $a > 1$. (The most important logarithmic functions have base $a > 1$.) The fact that $y = a^x$ is a very rapidly increasing function for $x > 0$ is reflected in the fact that $y = \log_a x$ is a very slowly increasing function for $x > 1$.

Figure 2 shows the graphs of $y = \log_a x$ with various values of the base a. Since $\log_a 1 = 0$, the graphs of all logarithmic functions pass through the point $(1, 0)$.

The following theorem summarizes the properties of logarithmic functions.

(3) THEOREM If $a > 1$, the function $f(x) = \log_a x$ is a one-to-one, continuous, increasing function with domain $(0, \infty)$ and range $\mathbb{R}$. If $x, y > 0$, then

1. $\log_a(xy) = \log_a x + \log_a y$ **2.** $\log_a\left(\dfrac{x}{y}\right) = \log_a x - \log_a y$

3. $\log_a(x^y) = y \log_a x$

Properties 1, 2, and 3 follow from the corresponding properties of exponential functions given in Section 4.1.

EXAMPLE 2 Use the properties of logarithms in Theorem 3 to evaluate the following: (a) $\log_4 2 + \log_4 32$ and (b) $\log_2 80 - \log_2 5$.

SOLUTION

(a) Using Property 1 in Theorem 3, we have

$$\log_4 2 + \log_4 32 = \log_4(2 \cdot 32) = \log_4 64 = 3$$

since $4^3 = 64$.

(b) Using Property 2 we have

$$\log_2 80 - \log_2 5 = \log_2\left(\tfrac{80}{5}\right) = \log_2 16 = 4$$

since $2^4 = 16$. ■

The limits of exponential functions given in Section 4.1 are reflected in the following limits of logarithmic functions. (Compare with Figure 1.)

(4) If $a > 1$, then

$$\lim_{x \to \infty} \log_a x = \infty \quad \text{and} \quad \lim_{x \to 0^+} \log_a x = -\infty$$

In particular, the y-axis is a vertical asymptote of the curve $y = \log_a x$.

EXAMPLE 3 Find $\displaystyle\lim_{x \to 0} \log_{10}(\tan^2 x)$.

SOLUTION As $x \to 0$, we know that $t = \tan^2 x \to \tan^2 0 = 0$ and the values of t are positive. So by (4) with $a = 10 > 1$, we have

$$\lim_{x \to 0} \log_{10}(\tan^2 x) = \lim_{t \to 0^+} \log_{10} t = -\infty$$

■

NATURAL LOGARITHMS

Of all possible bases a for logarithms, we will see in the next section that the most convenient choice of a base is the number e, which was defined in Section 4.1. The logarithm with base e is called the **natural logarithm** and has a special notation:

$$\log_e x = \ln x$$

NOTATION FOR LOGARITHMS

Most textbooks in calculus and the sciences, as well as calculators, use the notation $\ln x$ for the natural logarithm and $\log x$ for the "common logarithm," $\log_{10} x$. In the more advanced mathematical and scientific literature and in computer languages, however, the notation $\log x$ usually denotes the natural logarithm.

If we put $a = e$ and $\log_e = \ln$ in (1) and (2), then the defining properties of the natural logarithm function become

(5)
$$\ln x = y \iff e^y = x$$

(6)
$$\ln(e^x) = x \qquad x \in \mathbb{R}$$
$$e^{\ln x} = x \qquad x > 0$$

In particular, if we set $x = 1$, we get

$$\ln e = 1$$

EXAMPLE 4 Find x if $\ln x = 5$.

SOLUTION 1 From (5) we see that

$$\ln x = 5 \qquad \text{means} \qquad e^5 = x$$

Therefore, $x = e^5$.

(If you have trouble working with the "ln" notation, just replace it by $\log_e$. Then the equation becomes $\log_e x = 5$; so by the definition of logarithm, $e^5 = x$.)

SOLUTION 2 Start with the equation

$$\ln x = 5$$

and apply the exponential function to both sides of the equation:

$$e^{\ln x} = e^5$$

But the second cancellation equation in (6) says that $e^{\ln x} = x$. Therefore, $x = e^5$. ■

EXAMPLE 5 Solve the equation $e^{5-3x} = 10$.

SOLUTION We take natural logarithms of both sides of the equation and use (6):

$$\ln(e^{5-3x}) = \ln 10$$
$$5 - 3x = \ln 10$$
$$3x = 5 - \ln 10$$
$$x = \tfrac{1}{3}(5 - \ln 10)$$

Since the natural logarithm is found on scientific calculators, we can approximate the solution to four decimal places: $x \approx 0.8991$. ■

EXAMPLE 6 Express $\ln a + \frac{1}{2}\ln b$ as a single logarithm.

SOLUTION Using Properties 3 and 1 of logarithms, we have

$$\begin{aligned}\ln a + \tfrac{1}{2}\ln b &= \ln a + \ln b^{1/2}\\ &= \ln a + \ln\sqrt{b}\\ &= \ln(a\sqrt{b})\end{aligned}$$

■

The following formula shows that logarithms with any base can be expressed in terms of the natural logarithm.

(7) For any positive number a ($a \neq 1$), we have

$$\log_a x = \frac{\ln x}{\ln a}$$

PROOF Let $y = \log_a x$. Then, from (1), we have $a^y = x$. Taking natural logarithms of both sides of this equation, we get $y\ln a = \ln x$. Therefore,

$$y = \frac{\ln x}{\ln a}$$

□

Scientific calculators have a key for natural logarithms, so Formula 7 enables us to use a calculator to compute a logarithm with any base (as in the next example). Similarly, Formula 7 allows us to graph any logarithmic function on a graphing calculator or computer (see Exercises 22–24).

EXAMPLE 7 Evaluate $\log_8 5$ correct to six decimal places.

SOLUTION Formula 7 gives

$$\log_8 5 = \frac{\ln 5}{\ln 8} \approx 0.773976$$

■

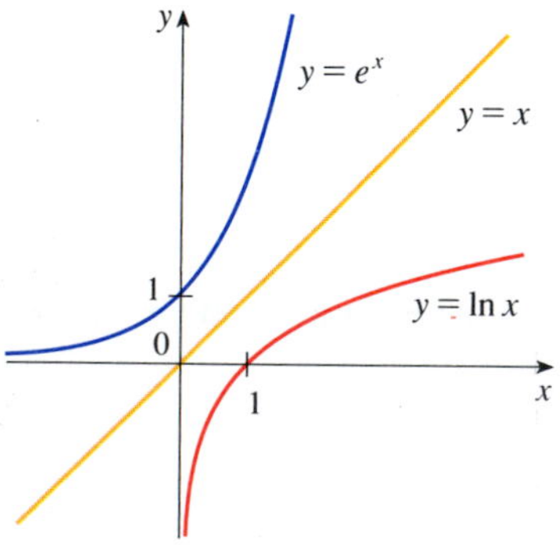

FIGURE 3

The graphs of the exponential function $y = e^x$ and its inverse function, the natural logarithm function, are shown in Figure 3.

In common with all other logarithmic functions with base greater than 1, the natural logarithm is a continuous, increasing function defined on $(0, \infty)$ and the y-axis is a vertical asymptote.

If we put $a = e$ in (4), then we have the following limits:

(8)
$$\lim_{x\to\infty} \ln x = \infty \qquad \lim_{x\to 0^+} \ln x = -\infty$$

EXAMPLE 8 Sketch the graph of the function $y = \ln(x - 2) - 1$.

SOLUTION We start with the graph of $y = \ln x$ as given in Figure 3. Using the transformations of Section 2 in Review and Preview, we shift it two units to the right to get the graph of $y = \ln(x - 2)$ and then we shift it one unit downward to get the graph of $y = \ln(x - 2) - 1$ (see Figure 4). Notice that the line $x = 2$ is a vertical asymptote since

$$\lim_{x \to 2^+} [\ln(x - 2) - 1] = -\infty$$

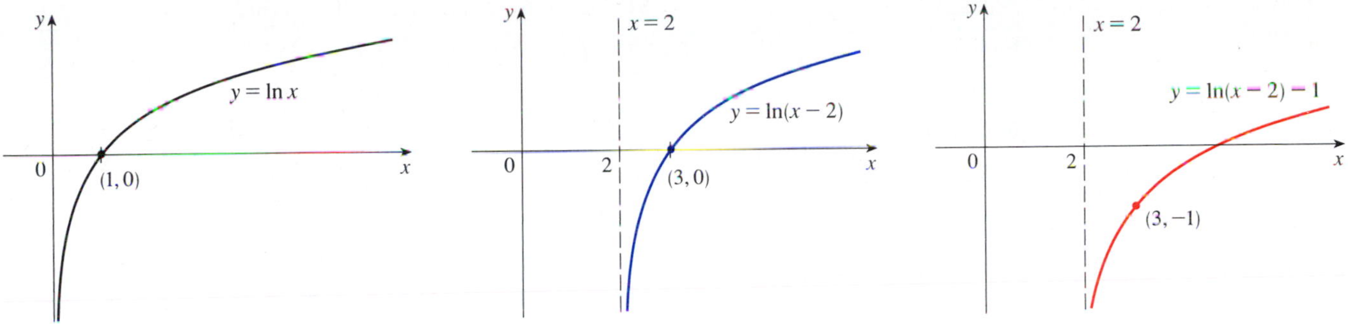

FIGURE 4

■

We have seen that $\ln x \to \infty$ as $x \to \infty$. But this happens *very* slowly. In fact, $\ln x$ grows more slowly than any power of x. To illustrate this fact, we compare approximate values of the functions $y = \ln x$ and $y = x^{1/2} = \sqrt{x}$ in the following table and we graph them in Figures 5 and 6.

x	1	2	5	10	50	100	500	1000	10,000	100,000
$\ln x$	0	0.69	1.61	2.30	3.91	4.6	6.2	6.9	9.2	11.5
$\sqrt{x}$	1	1.41	2.24	3.17	7.07	10.0	22.4	31.6	100	316
$\dfrac{\ln x}{\sqrt{x}}$	0	0.49	0.72	0.73	0.55	0.46	0.28	0.22	0.09	0.04

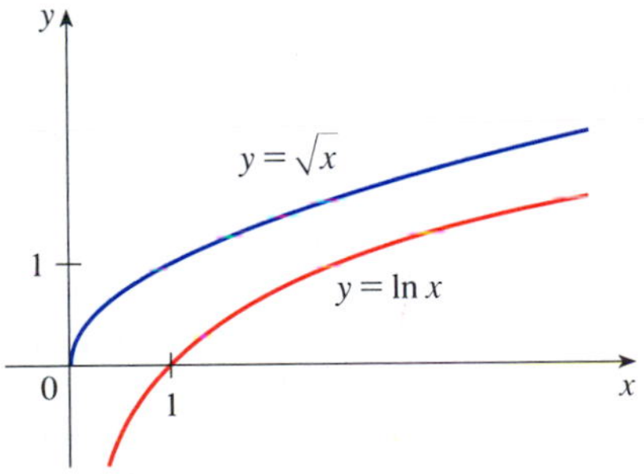

FIGURE 5

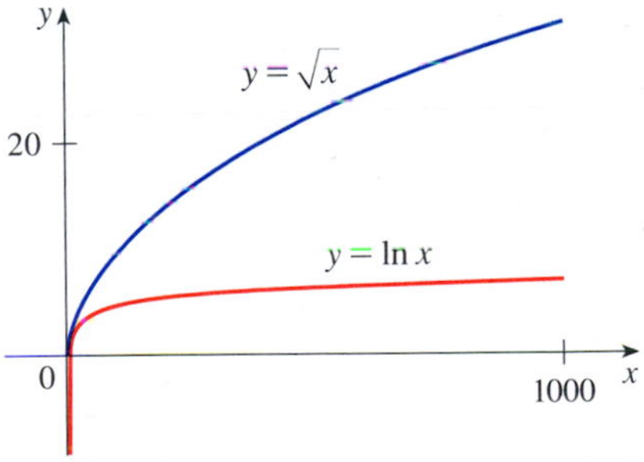

FIGURE 6

You can see that initially the graphs of $y = \sqrt{x}$ and $y = \ln x$ grow at comparable rates, but eventually the root function far surpasses the logarithm. In fact, we will be able to show in Section 4.8 that

$$\lim_{x \to \infty} \frac{\ln x}{x^p} = 0$$

for any power $p > 0$. So for large x, the values of $\ln x$ are very small compared with x^p. (See Exercise 86.)

EXERCISES 4.3

1–14 ■ Evaluate each expression.

1. $\log_2 64$

2. $\log_6 \frac{1}{36}$

3. $\log_8 2$

4. $\log_8 4$

5. $\log_3 \frac{1}{27}$

6. $e^{\ln 6}$

7. $\ln e^{\sqrt{2}}$

8. $\log_3 3^{\sqrt{5}}$

9. $\log_{10} 1.25 + \log_{10} 80$

10. $\log_3 108 - \log_3 4$

11. $\log_8 6 - \log_8 3 + \log_8 4$

12. $\log_5 10 + \log_5 20 - 3\log_5 2$

13. $2^{(\log_2 3 + \log_2 5)}$

14. $e^{3\ln 2}$

15–20 ■ Express the given quantity as a single logarithm.

15. $\log_5 a + \log_5 b - \log_5 c$

16. $\log_2 x + 5\log_2(x + 1) + \frac{1}{2}\log_2(x - 1)$

17. $2\ln 4 - \ln 2$

18. $\ln 10 + \frac{1}{2}\ln 9$

19. $\frac{1}{3}\ln x - 4\ln(2x + 3)$

20. $\ln x + a\ln y - b\ln z$

21. Use Formula 7 to evaluate each logarithm correct to six decimal places.
(a) $\log_2 5$ (b) $\log_5 26.05$
(c) $\log_3 e$ (d) $\log_{0.7} 14$

22–24 ■ Use Formula 7 to graph the given functions on a common screen. How are these graphs related?

22. $y = \log_2 x,\quad y = \log_4 x,\quad y = \log_6 x,\quad y = \log_8 x$

23. $y = \log_{1.5} x,\quad y = \ln x,\quad y = \log_{10} x,\quad y = \log_{50} x$

24. $y = \ln x,\quad y = \log_{10} x,\quad y = e^x,\quad y = 10^x$

25–34 ■ Make a rough sketch of the graph of each function. Do not use a calculator. Just use the graphs given in Figures 1, 2, and 3 and, if necessary, the transformations of Section 2 in Review and Preview.

25. $y = \log_{10}(x + 5)$

26. $y = 1 + \log_5(x - 1)$

27. $y = -\ln x$

28. $y = \ln(-x)$

29. $y = -\ln(-x)$

30. $y = \ln|x|$

31. $y = \ln(x^2)$

32. $y = \ln(1/x)$

33. $y = \ln(x + 3)$

34. $y = \ln|x + 3|$

35–54 ■ Solve each equation for x.

35. $\log_2 x = 3$

36. $2^{x-5} = 3$

37. $e^x = 16$

38. $\ln x = -1$

39. $\ln(2x - 1) = 3$

40. $e^{3x-4} = 2$

41. $3^{x+2} = m$

42. $5^{\log_5(2x)} = 6$

43. $\ln x = \ln 5 + \ln 8$

44. $\ln x^2 = 2\ln 4 - 4\ln 2$

45. $\ln(e^{2x-1}) = 5$

46. $\ln x + \ln(x - 1) = 1$

47. $\ln(\ln x) = 1$

48. $e^{e^x} = 10$

49. $2^{3^x} = 5$

50. $\log_2(\log_3(\log_4 x)) = C$

51. $\ln(x + 6) + \ln(x - 3) = \ln 5 + \ln 2$

52. $\ln\left(\dfrac{x - 2}{x - 1}\right) = 1 + \ln\left(\dfrac{x - 3}{x - 1}\right)$

53. $e^{ax} = Ce^{bx}$, where $a \neq b$

54. $7e^x - e^{2x} = 12$

55–58 ■ Find the solution of the equation correct to four decimal places.

55. $\ln(x - 5) = 3$

56. $e^{5x-1} = 12$

57. $e^{2-3x} = 20$

58. $2^{-x} = 5$

59. Suppose that the graph of $y = \log_2 x$ is drawn on a coordinate grid where the unit of measurement is an inch. How many miles to the right of the origin do we have to move before the height of the curve reaches 3 ft?

60. The velocity of a particle that moves in a straight line under the influence of viscous forces is $v(t) = ce^{-kt}$ where c and k are positive constants.
(a) Show that the acceleration is proportional to the velocity.
(b) Explain the significance of the number c.
(c) At what time is the velocity equal to half the initial velocity?

61. The geologist C. F. Richter defined the magnitude of an earthquake to be $\log_{10}(I/S)$ where I is the intensity of the quake (measured by the amplitude of a seismograph 100 km from the earthquake) and S is the intensity of a "standard" earthquake (where the amplitude is only 1 micron $= 10^{-4}$ cm). The 1989 Loma Prieta earthquake that shook San Francisco had a magnitude of 7.1 on the Richter scale. The 1906 San Francisco earthquake was 16 times as intense. What was its magnitude on the Richter scale?

62. A sound so faint that it can just be heard has intensity $I_0 = 10^{-12}$ watt/m^2 at a frequency of 1000 hertz (Hz). The loudness, in decibels (dB), of a sound with intensity I is then defined to be $L = 10\log_{10}(I/I_0)$. Amplified rock music is measured at 120 dB. The noise from a motor-driven lawn mower is measured at 106 dB. Find the ratio of the intensity of the rock music to that of the mower.

63–70 ■ Find each limit.

63. $\lim_{x \to 5^+} \ln(x - 5)$

64. $\lim_{x \to 0^+} \log_{10}(4x)$

65. $\lim_{x \to \infty} \log_2(x^2 - x)$

66. $\lim_{x \to 0^+} \ln(\sin x)$

67. $\lim_{x \to (\pi/2)^-} \log_{10}(\cos x)$

68. $\lim_{x \to \infty} \dfrac{\ln x}{1 + \ln x}$

69. $\lim_{x\to\infty} \ln(1 + e^{-x^2})$

70. $\lim_{x\to\infty} [\ln(2 + x) - \ln(1 + x)]$

71–74 ■ Find the domain and range of each function.

71. $f(x) = \log_{10}(1 - x)$

72. $g(x) = \ln(4 - x^2)$

73. $F(t) = \sqrt{t}\,\ln(t^2 - 1)$

74. $G(t) = \ln(t^3 - t)$

75–80 ■ Find the inverse function of each function.

75. $y = \ln(x + 3)$

76. $y = 2^{10^x}$

77. $y = e^{\sqrt{x}}$

78. $y = (\ln x)^2, \quad x \geqslant 1$

79. $y = \dfrac{10^x}{10^x + 1}$

80. $y = \dfrac{1 + e^x}{1 - e^x}$

81. (a) Show that the function $f(x) = \ln(x + \sqrt{x^2 + 1}\,)$ is an odd function.
(b) Find the inverse function of f.

82. Find an equation of the tangent to the curve $y = e^{-x}$ that is perpendicular to the line $2x - y = 8$.

83. Without using a calculator, determine which of the numbers $\log_{10} 99$ or $\log_9 82$ is larger.

84. Any function of the form $f(x) = [g(x)]^{h(x)}$, where $g(x) > 0$, can be analyzed as a power of e by writing $g(x) = e^{\ln g(x)}$ so that $f(x) = e^{h(x)\ln g(x)}$. Using this device, calculate:
(a) $\lim_{x\to\infty} x^{\ln x}$
(b) $\lim_{x\to 0^+} x^{-\ln x}$
(c) $\lim_{x\to 0^+} x^{1/x}$
(d) $\lim_{x\to\infty} (\ln 2x)^{-\ln x}$

85. Let $a > 1$. Prove, using Definitions 2.6.6 and 2.6.7, that
(a) $\lim_{x\to-\infty} a^x = 0$
(b) $\lim_{x\to\infty} a^x = \infty$

86. (a) Compare the rates of growth of $f(x) = x^{0.1}$ and $g(x) = \ln x$ by graphing both f and g in several viewing rectangles. When does the graph of f finally surpass the graph of g?
(b) Graph the function $h(x) = (\ln x)/x^{0.1}$ in a viewing rectangle that displays the behavior of the function as $x \to \infty$.
(c) Find a number N such that

$$\frac{\ln x}{x^{0.1}} < 0.1 \quad \text{whenever} \quad x > N$$

87. Solve the inequality $\ln(x^2 - 2x - 2) \leqslant 0$.

88. A **prime number** is a positive integer that has no factors other than 1 and itself. The first few primes are 2, 3, 5, 7, 11, 13, 17, ... We denote by $\pi(n)$ the number of primes that are less than or equal to n. For instance, $\pi(15) = 6$ because there are six primes smaller than 15.
(a) Calculate the numbers $\pi(25)$ and $\pi(100)$.
[*Hint:* To find $\pi(100)$, first compile a list of the primes up to 100 using the *sieve of Eratosthenes:* Write the numbers from 2 to 100 and cross out all multiples of 2. Then cross out all multiples of 3. The next remaining number is 5, so cross out all remaining multiples of it, and so on.]
(b) By inspecting tables of prime numbers and tables of logarithms, the great mathematician Gauss made the guess in 1792 (when he was 15) that the number of primes up to n is approximately $n/\ln n$ when n is large. More precisely, he conjectured that

$$\lim_{n\to\infty} \frac{\pi(n)}{n/\ln n} = 1$$

This was finally proved, a hundred years later, by Hadamard and de la Vallée Poussin and is called the **Prime Number Theorem**. Provide evidence for the truth of this theorem by computing the ratio of $\pi(n)$ to $n/\ln n$ for $n = 100$, 1000, 10^4, 10^5, 10^6, and 10^7. Use the following data: $\pi(1000) = 168$, $\pi(10^4) = 1229$, $\pi(10^5) = 9592$, $\pi(10^6) = 78{,}498$, $\pi(10^7) = 664{,}579$.
(c) Use the Prime Number Theorem to estimate the number of primes up to a billion.

4.4 DERIVATIVES OF LOGARITHMIC FUNCTIONS

In this section we find the derivatives of the logarithmic functions $y = \log_a x$ and the exponential functions $y = a^x$. We start with the natural logarithmic function $y = \ln x$. We know that it is differentiable because it is the inverse of the differentiable function $y = e^x$.

$$\frac{d}{dx}(\ln x) = \frac{1}{x} \tag{1}$$

PROOF Let $y = \ln x$. Then

$$e^y = x$$

Differentiating this equation implicitly with respect to x, we get

$$e^y \frac{dy}{dx} = 1$$

and so

$$\frac{dy}{dx} = \frac{1}{e^y} = \frac{1}{x}$$

□

EXAMPLE 1 Differentiate $y = \ln(x^3 + 1)$.

SOLUTION To use the Chain Rule we let $u = x^3 + 1$. Then $y = \ln u$, so

$$\frac{dy}{dx} = \frac{dy}{du}\frac{du}{dx} = \frac{1}{u}\frac{du}{dx} = \frac{1}{x^3+1}(3x^2) = \frac{3x^2}{x^3+1}$$

■

In general, if we combine Formula 1 with the Chain Rule as in Example 1, we get

(2) $$\frac{d}{dx}\ln u = \frac{1}{u}\frac{du}{dx} \qquad \text{or} \qquad \frac{d}{dx}\ln g(x) = \frac{g'(x)}{g(x)}$$

EXAMPLE 2 Find $\frac{d}{dx}\ln(\sin x)$.

SOLUTION Using (2), we have

$$\frac{d}{dx}\ln(\sin x) = \frac{1}{\sin x}\frac{d}{dx}\sin x = \frac{1}{\sin x}\cos x = \cot x$$

■

EXAMPLE 3 Differentiate $f(x) = \sqrt{\ln x}$.

SOLUTION This time the logarithm is the inner function, so the Chain Rule gives

$$f'(x) = \tfrac{1}{2}(\ln x)^{-1/2}\frac{d}{dx}(\ln x) = \frac{1}{2\sqrt{\ln x}}\cdot\frac{1}{x} = \frac{1}{2x\sqrt{\ln x}}$$

■

Figure 1 shows the graph of the function f of Example 4 together with the graph of its derivative. It gives a visual check on our calculation. Notice that $f'(x)$ is large negative when f is rapidly decreasing and $f'(x) = 0$ when f has a minimum.

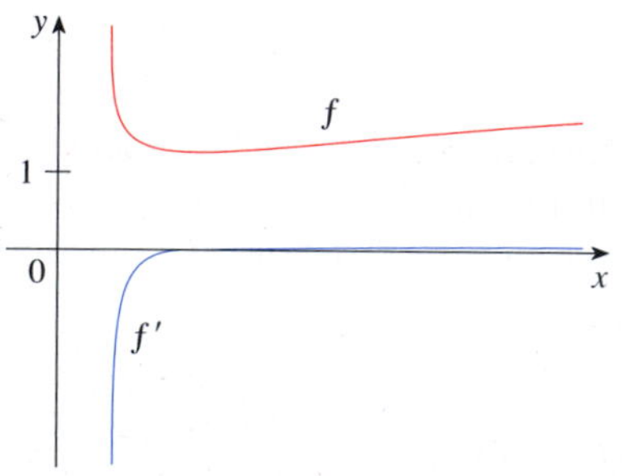

FIGURE 1

EXAMPLE 4 Find $\frac{d}{dx}\ln\frac{x+1}{\sqrt{x-2}}$.

SOLUTION 1

$$\frac{d}{dx}\ln\frac{x+1}{\sqrt{x-2}} = \frac{1}{\dfrac{x+1}{\sqrt{x-2}}}\frac{d}{dx}\frac{x+1}{\sqrt{x-2}}$$

$$= \frac{\sqrt{x-2}}{x+1}\,\frac{1\cdot\sqrt{x-2} - (x+1)\left(\frac{1}{2}\right)(x-2)^{-1/2}}{x-2}$$

$$= \frac{x-2-\frac{1}{2}(x+1)}{(x+1)(x-2)} = \frac{x-5}{2(x+1)(x-2)}$$

SOLUTION 2 If we first simplify the given function using the laws of logarithms, then the differentiation becomes easier:

$$\frac{d}{dx} \ln \frac{x+1}{\sqrt{x-2}} = \frac{d}{dx} [\ln(x+1) - \tfrac{1}{2} \ln(x-2)]$$

$$= \frac{1}{x+1} - \frac{1}{2}\left(\frac{1}{x-2}\right)$$

(This answer can be left as written, but if we used a common denominator we would see that it gives the same answer as in Solution 1.) ■

EXAMPLE 5 Find $f'(x)$ if $f(x) = \ln|x|$.

SOLUTION Since

$$f(x) = \begin{cases} \ln x & \text{if } x > 0 \\ \ln(-x) & \text{if } x < 0 \end{cases}$$

it follows that

$$f'(x) = \begin{cases} \dfrac{1}{x} & \text{if } x > 0 \\ \dfrac{1}{-x}(-1) = \dfrac{1}{x} & \text{if } x < 0 \end{cases}$$

Thus, $f'(x) = 1/x$ for all $x \neq 0$. ■

The result of this example is worth remembering:

(3) $$\frac{d}{dx} \ln|x| = \frac{1}{x}$$

GENERAL LOGARITHMIC AND EXPONENTIAL FUNCTIONS

Formula 7 in Section 4.3 expresses a logarithmic function with base a in terms of the natural logarithmic function:

$$\log_a x = \frac{\ln x}{\ln a}$$

Since $\ln a$ is a constant, we can differentiate as follows:

$$\frac{d}{dx} \log_a x = \frac{d}{dx} \frac{\ln x}{\ln a} = \frac{1}{\ln a} \frac{d}{dx} \ln x = \frac{1}{x \ln a}$$

(4) $$\frac{d}{dx} \log_a x = \frac{1}{x \ln a}$$

EXAMPLE 6

$$\frac{d}{dx} \log_{10}(2 + \sin x) = \frac{1}{(2 + \sin x)\ln 10} \frac{d}{dx}(2 + \sin x) = \frac{\cos x}{(2 + \sin x)\ln 10}$$ ■

From Formula 4 we see one of the main reasons that natural logarithms (logarithms with base e) are used in calculus: The differentiation formula is simplest when $a = e$ because $\ln e = 1$.

EXPONENTIAL FUNCTIONS WITH BASE a In Section 4.1 we showed that the derivative of the general exponential function $f(x) = a^x$, $a > 0$, is a constant multiple of itself:

$$f'(x) = f'(0)a^x \qquad \text{where} \qquad f'(0) = \lim_{h \to 0} \frac{a^h - 1}{h}$$

We are now in a position to show that the value of the constant is $f'(0) = \ln a$.

(5) $$\frac{d}{dx} a^x = a^x \ln a$$

PROOF We use the fact that $e^{\ln a} = a$:

$$\frac{d}{dx} a^x = \frac{d}{dx} (e^{\ln a})^x = \frac{d}{dx} e^{(\ln a)x}$$

$$= e^{(\ln a)x} \frac{d}{dx} (\ln a)x$$

$$= (e^{\ln a})^x (\ln a) = a^x \ln a$$ □

In Example 6 in Section 3.3 we considered a population of bacteria cells that doubles every hour and saw that the population after t hours is $n = n_0 2^t$, where n_0 is the initial population. Formula 5 enables us to find the growth rate:

$$\frac{dn}{dt} = n_0 2^t \ln 2$$

EXAMPLE 7 Combining Formula 5 with the Chain Rule, we have

$$\frac{d}{dx} 10^{x^2} = 10^{x^2} (\ln 10) \frac{d}{dx} x^2 = (2 \ln 10) x 10^{x^2}$$ ■

LOGARITHMIC DIFFERENTIATION

The calculation of derivatives of complicated functions involving products, quotients, or powers can often be simplified by taking logarithms. The method used in the following example is called **logarithmic differentiation**.

EXAMPLE 8 Differentiate

$$y = \frac{x^{3/4} \sqrt{x^2 + 1}}{(3x + 2)^5}$$

SOLUTION We take logarithms of both sides of the equation:

$$\ln y = \tfrac{3}{4} \ln x + \tfrac{1}{2} \ln(x^2 + 1) - 5 \ln(3x + 2)$$

Differentiating implicitly with respect to x gives

$$\frac{1}{y}\frac{dy}{dx} = \frac{3}{4}\cdot\frac{1}{x} + \frac{1}{2}\cdot\frac{2x}{x^2+1} - 5\cdot\frac{3}{3x+2}$$

Solving for dy/dx, we get

$$\frac{dy}{dx} = y\left(\frac{3}{4x} + \frac{x}{x^2+1} - \frac{15}{3x+2}\right)$$

$$= \frac{x^{3/4}\sqrt{x^2+1}}{(3x+2)^5}\left(\frac{3}{4x} + \frac{x}{x^2+1} - \frac{15}{3x+2}\right)$$ ■

STEPS IN LOGARITHMIC DIFFERENTIATION

1. Take logarithms of both sides of an equation $y = f(x)$.
2. Differentiate implicitly with respect to x.
3. Solve the resulting equation for y'.

If $f(x) < 0$ for some values of x, then $\ln f(x)$ is not defined, but we can write $|y| = |f(x)|$ and use Equation 3. We illustrate this procedure by proving the general version of the Power Rule, as promised in Section 3.2.

POWER RULE If n is any real number and $f(x) = x^n$, then

$$f'(x) = nx^{n-1}$$

PROOF Let $y = x^n$ and use logarithmic differentiation:

If $x = 0$, we can show that $f'(0) = 0$ for $n > 1$ directly from the definition of a derivative.

$$\ln|y| = \ln|x|^n = n\ln|x| \qquad x \neq 0$$

Therefore

$$\frac{y'}{y} = \frac{n}{x}$$

Hence

$$y' = n\frac{y}{x} = n\frac{x^n}{x} = nx^{n-1}$$ □

You should distinguish carefully between the Power Rule ($Dx^n = nx^{n-1}$), where the base is variable and the exponent is constant, and the rule for differentiating exponential functions ($Da^x = a^x \ln a$), where the base is constant and the exponent is variable. In general there are four cases for exponents and bases:

1. $\dfrac{d}{dx}(a^b) = 0$ (a and b are constants)
2. $\dfrac{d}{dx}[f(x)]^b = b[f(x)]^{b-1}f'(x)$
3. $\dfrac{d}{dx}[a^{g(x)}] = a^{g(x)}(\ln a)g'(x)$
4. To find $(d/dx)[f(x)]^{g(x)}$, logarithmic differentiation can be used, as in the next example.

EXAMPLE 9 Differentiate $y = x^{\sqrt{x}}$.

Figure 2 illustrates Example 9 by showing the graphs of $f(x) = x^{\sqrt{x}}$ and its derivative.

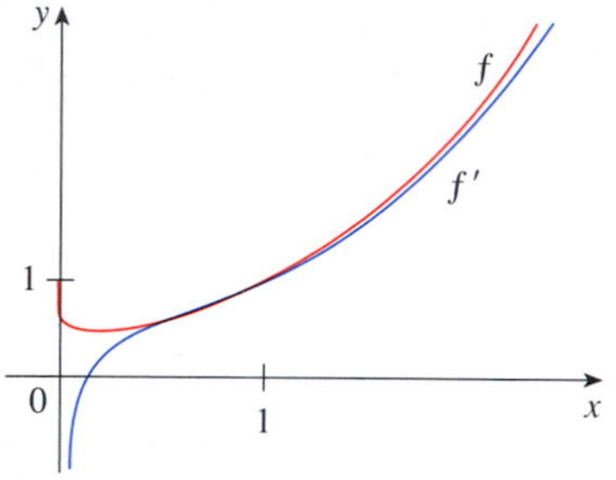

FIGURE 2

SOLUTION 1 Using logarithmic differentiation, we have

$$\ln y = \ln x^{\sqrt{x}} = \sqrt{x}\ln x$$

$$\frac{y'}{y} = \frac{1}{2\sqrt{x}}\ln x + \sqrt{x}\cdot\frac{1}{x}$$

$$y' = y\left(\frac{\ln x}{2\sqrt{x}} + \frac{1}{\sqrt{x}}\right) = x^{\sqrt{x}}\left(\frac{\ln x + 2}{2\sqrt{x}}\right)$$

SOLUTION 2 Another method is to write $x^{\sqrt{x}} = (e^{\ln x})^{\sqrt{x}}$:

$$\frac{d}{dx}\left(x^{\sqrt{x}}\right) = \frac{d}{dx}\left(e^{\sqrt{x}\ln x}\right)$$

$$= e^{\sqrt{x}\ln x}\frac{d}{dx}\left(\sqrt{x}\ln x\right)$$

$$= x^{\sqrt{x}}\left(\frac{\ln x + 2}{2\sqrt{x}}\right) \qquad \text{(as above)}$$

■

THE NUMBER e AS A LIMIT

We have shown that if $f(x) = \ln x$, then $f'(x) = 1/x$. Thus, $f'(1) = 1$. We now use this fact to express the number e as a limit.

From the definition of a derivative as a limit, we have

$$f'(1) = \lim_{h\to 0}\frac{f(1+h) - f(1)}{h} = \lim_{x\to 0}\frac{f(1+x) - f(1)}{x}$$

$$= \lim_{x\to 0}\frac{\ln(1+x) - \ln 1}{x} = \lim_{x\to 0}\frac{1}{x}\ln(1+x)$$

$$= \lim_{x\to 0}\ln(1+x)^{1/x} = \ln\left[\lim_{x\to 0}(1+x)^{1/x}\right] \qquad \text{(since ln is continuous)}$$

Because $f'(1) = 1$, we have

$$\ln\left[\lim_{x\to 0}(1+x)^{1/x}\right] = 1$$

Therefore

x	$(1 + x)^{1/x}$
0.1	2.59374246
0.01	2.70481383
0.001	2.71692393
0.0001	2.71814593
0.00001	2.71826824
0.000001	2.71828047
0.0000001	2.71828169
0.00000001	2.71828182

(6)
$$\lim_{x\to 0}(1+x)^{1/x} = e$$

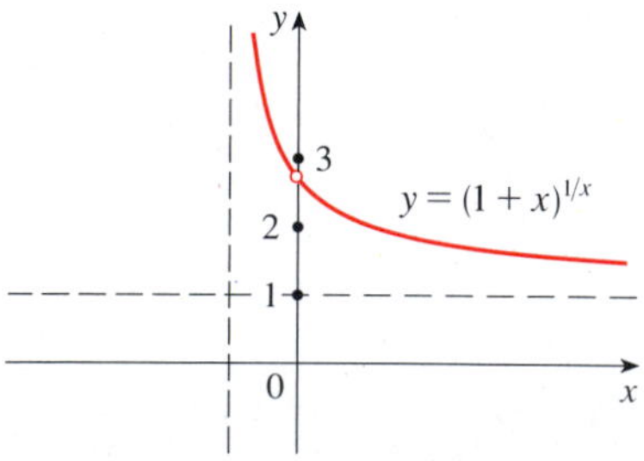

FIGURE 3

Formula 6 is illustrated by the graph of the function $y = (1 + x)^{1/x}$ in Figure 3 and a table of values for small values of x.

If we put $n = 1/x$ in Formula 6, then $n \to \infty$ as $x \to 0^+$ and so an alternative expression for e is

(7)
$$e = \lim_{n\to\infty}\left(1 + \frac{1}{n}\right)^n$$

EXERCISES 4.4

1–6 ■ Find f' and state the domains of f and f'.

1. $f(x) = \ln(x + 1)$

2. $f(x) = \cos(\ln x)$

3. $f(x) = x^2 \ln(1 - x^2)$

4. $f(x) = \ln \ln \ln x$

5. $f(x) = \log_3(x^2 - 4)$

6. $f(x) = \sqrt{3 - 2^x}$

7–10 ■ Find y' and y''.

7. $y = x \ln x$

8. $y = \ln(ax)$

9. $y = \log_{10} x$

10. $y = \ln(\sec x + \tan x)$

11–44 ■ Differentiate each function.

11. $f(x) = \sqrt{x}\ln x$

12. $f(x) = \log_{10}\left(\dfrac{x}{x-1}\right)$

13. $g(x) = \ln\dfrac{a - x}{a + x}$

14. $h(x) = \ln(x + \sqrt{x^2 - 1})$

15. $F(x) = \ln\sqrt{x}$

16. $G(x) = \sqrt{\ln x}$

17. $f(t) = \log_2(t^4 - t^2 + 1)$

18. $h(y) = \ln(y^3 \sin y)$

19. $g(u) = \dfrac{1 - \ln u}{1 + \ln u}$

20. $G(u) = \ln\sqrt{\dfrac{3u + 2}{3u - 2}}$

21. $y = (\ln \sin x)^3$

22. $y = \ln(x + \ln x)$

23. $y = \dfrac{\ln x}{1 + x^2}$

24. $y = \ln(x\sqrt{1 - x^2}\sin x)$

25. $y = \ln|x^3 - x^2|$

26. $y = \ln|\tan 2x|$

27. $F(x) = e^x \ln x$

28. $G(x) = 5^{\tan x}$

29. $f(t) = \pi^{-t}$

30. $g(x) = 1.6^x + x^{1.6}$

31. $h(t) = t^3 - 3^t$

32. $y = 2^{3^x}$

33. $y = \ln(e^{-x} + xe^{-x})$

34. $y = x^x$

35. $y = x^{\sin x}$

36. $y = (\sin x)^x$

37. $y = x^{e^x}$

38. $y = x^{1/x}$

39. $y = (\ln x)^x$

40. $y = x^{\ln x}$

41. $y = x^{1/\ln x}$

42. $y = (\sin x)^{\cos x}$

43. $y = \cos(x^{\sqrt{x}})$

44. $y = x^{x^x}$

45. If $f(x) = \dfrac{x}{\ln x}$, find $f'(e)$.

46. If $f(x) = x^2 \ln x$, find $f'(1)$.

47–48 ■ Find $f'(x)$. Check that your answer is reasonable by comparing the graphs of f and f'.

47. $f(x) = \sin x + \ln x$

48. $f(x) = x^{\cos x}$

49–52 ■ Find an equation of the tangent line to the curve at the given point.

49. $y = \ln \ln x, \quad (e, 0)$

50. $y = 10^x, \quad (1, 10)$

51. $x = \ln t, \quad y = te^t, \quad (0, e)$

$\left[\textit{Hint:} \text{ Recall that } dy/dx = \dfrac{dy/dt}{dx/dt}.\right]$

52. $x = 1 + t^2, \quad y = t \ln t, \quad (2, 0)$

53. Find y' if $y = \ln(x^2 + y^2)$.

54. Find y' if $x^y = y^x$.

55. Find a formula for $f^{(n)}(x)$ if $f(x) = \ln(x - 1)$.

56. Find $\dfrac{d^9}{dx^9}(x^8 \ln x)$.

57–58 ■ Use a graph to estimate the roots of the equation. Then use these estimates as the initial approximations in Newton's method to find the roots correct to six decimal places.

57. $\ln x = e^{-x}$

58. $\ln(4 - x^2) = x$

59–64 ■ Use logarithmic differentiation to find the derivative of each function.

59. $y = (3x - 7)^4(8x^2 - 1)^3$

60. $y = x^{2/5}(x^2 + 8)^4 e^{x^2 + x}$

61. $y = \dfrac{(x + 1)^4(x - 5)^3}{(x - 3)^8}$

62. $y = \sqrt{\dfrac{x^2 + 1}{x + 1}}$

63. $y = \dfrac{e^x\sqrt{x^5 + 2}}{(x + 1)^4(x^2 + 3)^2}$

64. $y = \dfrac{(x^3 + 1)^4 \sin^2 x}{\sqrt[3]{x}}$

65. If g is the inverse function of $f(x) = 2x + \ln x$, find $g'(2)$.

66. If $f(x) = e^x + \ln x$ and $h(x) = f^{-1}(x)$, find $h'(e)$.

67. Use the definition of derivative to prove that

$$\lim_{x \to 0} \frac{\ln(1 + x)}{x} = 1$$

68. (a) Find the linear and quadratic approximations to $f(x) = \ln x$ near 1.

(b) Illustrate part (a) by graphing f and both approximations.

(c) For what values of x is the linear approximation accurate to within 0.1?

(d) For what values of x is the quadratic approximation accurate to within 0.1?

69. Show that $\lim_{n \to \infty}\left(1 + \dfrac{x}{n}\right)^n = e^x$ for any $x > 0$.

4.5 EXPONENTIAL GROWTH AND DECAY

In many natural phenomena, quantities grow or decay at a rate proportional to their size. For instance, if $y = f(t)$ is the number of individuals in a population of animals or bacteria at time t, then it seems reasonable to expect that the rate of growth $f'(t)$ is proportional to the population $f(t)$; that is, $f'(t) = kf(t)$ for some constant k. Indeed, under ideal conditions (unlimited environment, adequate nutrition, immunity to disease) the mathematical model given by the equation $f'(t) = kf(t)$ predicts what actually happens fairly accurately. Another example occurs in nuclear physics where the mass of a radioactive substance decays at a rate proportional to the mass. In chemistry, the rate of a unimolecular first-order reaction is proportional to the concentration of the substance. In finance, the value of a savings account with continuously compounded interest increases at a rate proportional to that value.

In general, if $y(t)$ is the value of a quantity y at time t and if the rate of change of y with respect to t is proportional to its size $y(t)$ at any time, then

(1) $$\frac{dy}{dt} = ky$$

where k is a constant. Equation 1 is sometimes called the **law of natural growth** (if $k > 0$) or the **law of natural decay** (if $k < 0$). It is called a **differential equation** because it involves an unknown function y and its derivative dy/dt. (This and other differential equations will be studied in Section 9.1.)

It is not hard to think of a solution of Equation 1. This equation asks us to find a function whose derivative is a constant multiple of itself. We have met such functions in this chapter. Any exponential function of the form $y(t) = Ce^{kt}$, where C is a constant, satisfies

$$y'(t) = Cke^{kt} = k(Ce^{kt}) = ky(t)$$

Conversely, we can show that *any* function that satisfies $dy/dt = ky$ must be of the form $y = Ce^{kt}$.

To see that this is true we assume that $y > 0$ and write the equation $dy/dt = ky$ in the form

$$\frac{1}{y}\frac{dy}{dt} = k$$

By the Chain Rule, this is the same as

$$\frac{d}{dt}\ln y = \frac{d}{dt}kt$$

In Chapter 5 we will show that two functions with the same derivative must differ by a constant. So

$$\ln y = kt + C$$

To solve for y, we apply the exponential function to each side of this equation:

$$y = e^{\ln y} = e^{kt+C} = e^{kt}e^{C}$$

Writing $e^C = A$, we have

$$y(t) = Ae^{kt}$$

To see the significance of the constant A, we observe that

$$y(0) = Ae^{k \cdot 0} = A$$

Therefore, A is the initial value of the function and we have proved the following theorem.

(2) THEOREM The only solutions of the differential equation $dy/dt = ky$ are the exponential functions

$$y(t) = y(0)e^{kt}$$

EXAMPLE 1 A bacteria culture starts with 1000 bacteria, and after 2 h the population is 2500 bacteria. Assuming that the culture grows at a rate proportional to its size, find the population after 6 h.

SOLUTION Let $y(t)$ be the number of bacteria after t hours. Then $y(0) = 1000$ and $y(2) = 2500$. Since we are assuming $dy/dt = ky$, Theorem 2 gives

$$y(t) = y(0)e^{kt} = 1000e^{kt}$$

$$y(2) = 1000e^{2k} = 2500$$

Therefore $\qquad e^{2k} = 2.5 \qquad$ and $\qquad 2k = \ln 2.5$

Substituting the value of $k = \frac{1}{2}\ln 2.5$ back into the expression for $y(t)$, we have

$$y(t) = 1000e^{\ln 2.5(t/2)} \tag{3}$$

Since $e^{\ln 2.5} = 2.5$, an alternative expression for Equation 3 is

$$y(t) = 1000(2.5)^{t/2}$$

and so $\qquad y(6) = 1000(2.5)^3 = 15{,}625$ ■

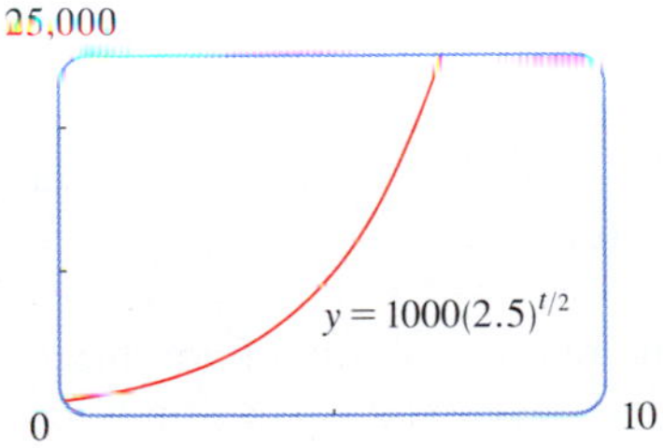

FIGURE 1

Figure 1 shows a graph of the size of the bacteria population in Example 1. The exponential model has also been applied to predict the size of animal populations and even human populations (see Exercises 5 and 6). Other mathematical models of population growth, which take into account limits to growth, will be studied in Section 9.1.

EXAMPLE 2 The *half-life* of radium-226 ($^{226}_{88}Ra$) is 1590 years. This means that the rate of decay is proportional to the amount present, and half of any given quantity will disintegrate in 1590 years.

(a) A sample of radium-226 has a mass of 100 mg. Find a formula for the mass of $^{226}_{88}Ra$ that remains after t years.

(b) Find the mass after 1000 years correct to the nearest milligram.

(c) When will the mass be reduced to 30 mg?

SOLUTION

(a) Let $y(t)$ be the mass of radium-226 (in milligrams) that remains after t years. Then $dy/dt = ky$ and $y(0) = 100$, so Theorem 2 gives

$$y(t) = y(0)e^{kt} = 100e^{kt}$$

In order to determine the value of k, we use the fact that $y(1590) = 50$. Thus

$$100e^{1590k} = 50 \qquad \text{so} \qquad e^{1590k} = \tfrac{1}{2}$$

and
$$1590k = \ln \tfrac{1}{2} = -\ln 2$$

$$k = -\frac{\ln 2}{1590}$$

Therefore
$$y(t) = 100e^{-(\ln 2/1590)t}$$

As in Example 1 we could use the fact that $e^{\ln 2} = 2$ to write the expression for $y(t)$ in the alternative form

$$y(t) = 100 \times 2^{-t/1590}$$

(b) The mass after 1000 years is

$$y(1000) = 100e^{-(\ln 2/1590)1000} \approx 65 \text{ mg}$$

(c) We want to find the value of t such that $y(t) = 30$, that is,

$$100e^{-(\ln 2/1590)t} = 30$$

or
$$e^{-(\ln 2/1590)t} = 0.3$$

We solve this equation for t by taking the natural logarithm of both sides:

$$-\frac{\ln 2}{1590}t = \ln 0.3$$

Thus
$$t = -1590\frac{\ln 0.3}{\ln 2} \approx 2762 \text{ years}$$ ■

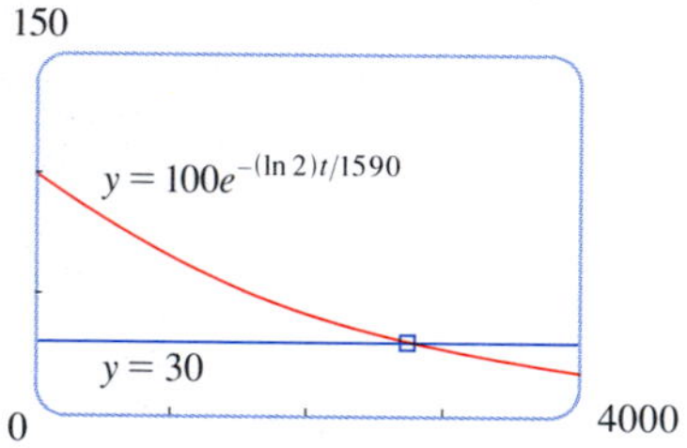

FIGURE 2

As a check on our work in Example 2, we use a graphing device to draw the graph of $y(t)$ in Figure 2 together with the horizontal line $y = 30$. These curves intersect when $t \approx 2800$ and this agrees with the answer to part (c).

EXAMPLE 3 Newton's Law of Cooling states that the rate of cooling of an object is proportional to the temperature difference between the object and its surroundings, provided that this difference is not too large. Suppose the object takes 40 min to cool from 30 °C to 24 °C in a room that is kept at 20 °C.
(a) What was the temperature of the object 15 min after it was 30 °C?
(b) How long will it take the object to cool down to 21 °C?

SOLUTION
(a) Let $y(t)$ be the temperature t minutes after it was 30 °C. Then Newton's Law of Cooling states that

$$\frac{dy}{dt} = k(y - 20)$$

This differential equation is not quite the same as Equation 1, so we introduce a new function $u(t) = y(t) - 20$. Then $du/dt = dy/dt$, so

$$\frac{du}{dt} = ku$$

Therefore, by Theorem 2, we have

$$u(t) = u(0)e^{kt} = 10e^{kt}$$

We are given that $y(40) = 24$, so $u(40) = 4$ and

$$10e^{40k} = 4 \qquad e^{40k} = 0.4$$

Taking logarithms, we have

$$k = \frac{\ln 0.4}{40}$$

Thus $\qquad u(t) = 10e^{(\ln 0.4)t/40}$

(4) $\qquad y(t) = 20 + 10e^{(\ln 0.4)t/40}$

$$y(15) = 20 + 10e^{(\ln 0.4)15/40} \approx 27.1\,°\text{C}$$

(b) We have $y(t) = 21$ when

$$20 + 10e^{(\ln 0.4)t/40} = 21$$

$$e^{(\ln 0.4)t/40} = \tfrac{1}{10}$$

$$\frac{(\ln 0.4)t}{40} = \ln \tfrac{1}{10} = -\ln 10$$

$$t = -40\frac{\ln 10}{\ln 0.4} \approx 100.5$$

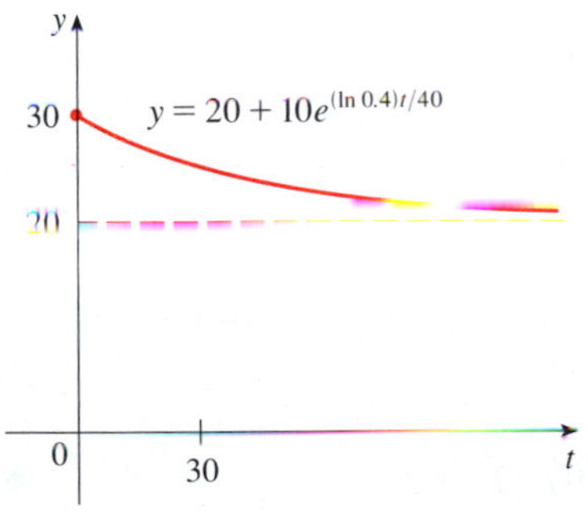

FIGURE 3

The object will cool down to 21°C after about 1 h 41 min. ■

Notice that in Example 3, $\ln(0.4) < 0$ since $0.4 < 1$. Thus, from Equation 4, we have

$$\lim_{t\to\infty} y(t) = 20 + \lim_{t\to\infty} 10e^{(\ln 0.4)t/40}$$

$$= 20 + 0 = 20$$

which is to be expected. The graph of the temperature function is shown in Figure 3.

EXAMPLE 4 If \$1000 is invested at 6% interest, compounded annually, then after 1 year the investment is worth \$1000(1.06) = \$1060; after 2 years it is worth \$[1000(1.06)]1.06 = \$1123.60; and after t years it is worth $\$1000(1.06)^t$. In general, if an amount A_0 is invested at an interest rate i ($i = 0.06$ in this example), then after t years it is worth $A_0(1 + i)^t$. Usually, however, interest is compounded more frequently, say n times a year. Then in each compounding period the interest rate is i/n and there are nt compounding periods in t years, so the value of the investment is

$$A_0\left(1 + \frac{i}{n}\right)^{nt}$$

For instance, after 3 years at 6% interest a \$1000 investment will be worth

$$\$1000(1.06)^3 = \$1191.02 \quad \text{with annual compounding}$$

$$\$1000(1.03)^6 = \$1194.05 \quad \text{with semiannual compounding}$$

$$\$1000(1.015)^{12} = \$1195.62 \quad \text{with quarterly compounding}$$

$$\$1000(1.005)^{36} = \$1196.68 \quad \text{with monthly compounding}$$

$$\$1000\left(1 + \frac{0.06}{365}\right)^{365\cdot 3} = \$1197.20 \quad \text{with daily compounding}$$

You can see that the interest paid increases as the number of compounding periods (n) increases. If we let $n \to \infty$, then we will be compounding the interest continuously and the value of the investment will be

$$\begin{aligned} A(t) &= \lim_{n\to\infty} A_0\left(1 + \frac{i}{n}\right)^{nt} = \lim_{n\to\infty} A_0\left[\left(1 + \frac{i}{n}\right)^{n/i}\right]^{it} \\ &= A_0\left[\lim_{n\to\infty}\left(1 + \frac{i}{n}\right)^{n/i}\right]^{it} \\ &= A_0\left[\lim_{m\to\infty}\left(1 + \frac{1}{m}\right)^{m}\right]^{it} && \text{(where } m = n/i\text{)} \\ &= A_0 e^{it} && \text{[by (4.4.7)]} \end{aligned}$$

This formula for continuous compounding $[A(t) = A_0e^{it}]$ should be compared with the expression for exponential growth in Theorem 2: $y(t) = y(0)e^{kt}$. When differentiated, it becomes

$$\frac{dA}{dt} = iA_0e^{it} = iA(t)$$

which says that, with continuous compounding of interest, the rate of increase of an investment is proportional to its size.

Returning to the example of \$1000 invested for 3 years at 6% interest, we see that with continuous compounding of interest the value of the investment will be

$$\begin{aligned} A(3) &= \$1000e^{(0.06)3} \\ &= \$1000e^{0.18} = \$1197.22 \end{aligned}$$

■

EXERCISES 4.5

1–4 ■ Assume that the population grows at a rate proportional to its size. If you have a graphing device, use a graph to check your answer to part (c).

1. A common inhabitant of human intestines is the bacterium *Escherichia coli.* A cell of this bacterium in a nutrient broth medium divides into two cells every 20 minutes. The initial population of a culture is 100 cells.
(a) Find an expression for the number of cells after t hours.
(b) Find the number of cells after 10 hours.
(c) When will the population reach 10,000 cells?

2. A bacteria culture starts with 4000 bacteria and the population triples every half-hour.
(a) Find an expression for the number of bacteria after t hours.
(b) Find the number of bacteria after 20 min.
(c) When will the population reach 20,000?

3. A bacteria culture starts with 500 bacteria and after 3 hours there are 8000 bacteria.
(a) Find an expression for the number of bacteria after t hours.
(b) Find the number of bacteria after 4 hours.
(c) When will the population reach 30,000?

4. The count in a bacteria culture was 400 after 2 hours and 25,600 after 6 hours.
(a) What was the initial population of the culture?
(b) Find an expression for the population after t hours.
(c) In what period of time does the population double?
(d) When will the population reach 100,000?

5. The table gives estimates of the world population, in millions, over two centuries:

Year	1750	1800	1850	1900	1950
Population	728	906	1171	1608	2517

(a) Use the exponential model and the population figures for 1750 and 1800 to predict the world population in 1900 and 1950. Compare with the actual figures.
(b) Use the exponential model and the population figures for 1850 and 1900 to predict the world population in 1950. Compare with the actual population.
(c) Use the exponential model and the population figures for 1900 and 1950 to predict the world population in 1992. Compare with the actual 1992 population of 5.4 billion and try to explain the discrepancy.

6. The table gives the population of the United States, in millions, for the years 1900–1990.

Year	Population
1900	76
1910	92
1920	106
1930	123
1940	131
1950	150
1960	179
1970	203
1980	227
1990	250

(a) Use the exponential model and the census figures for 1900 and 1910 to predict the population in 1990. Compare with the actual figure and try to explain the discrepancy.
(b) Use the exponential model and the census figures for 1970 and 1980 to predict the population in 1990. Compare with the actual population. Then use this model to predict the population in the years 2000 and 2010.
(c) Draw a graph showing both of the exponential functions in parts (a) and (b) together with a plot of the actual population. Are these models reasonable ones?

7. Experiments show that if the chemical reaction

$$N_2O_5 \rightarrow 2NO_2 + \tfrac{1}{2}O_2$$

takes place at 45 °C, the rate of reaction of dinitrogen pentoxide is proportional to its concentration as follows:

$$-\frac{d[N_2O_5]}{dt} = 0.0005[N_2O_5]$$

(See Example 4 in Section 3.3.)
(a) Find an expression for the concentration $[N_2O_5]$ after t seconds if the initial concentration is C.
(b) How long will the reaction take to reduce the concentration of N_2O_5 to 90% of its original value?

8. Polonium-210 has a half-life of 140 days (see Example 2).
(a) If a sample has a mass of 200 mg, find a formula for the mass that remains after t days.
(b) Find the mass after 100 days.
(c) When will the mass be reduced to 10 mg?
(d) Sketch the graph of the mass function.

9. Polonium-214 has a very short half-life of 1.4×10^{-4} s.
(a) If a sample has a mass of 50 mg, find a formula for the mass that remains after t seconds.
(b) Find the mass that remains after a hundredth of a second.
(c) How long would it take for the mass to decay to 40 mg?

10. After 3 days a sample of radon-222 decayed to 58% of its original amount.
(a) What is the half-life of radon-222?
(b) How long would it take the sample to decay to 10% of its original amount?

11. Scientists can determine the age of ancient objects by a method called *radiocarbon dating*. The bombardment of the upper atmosphere by cosmic rays converts nitrogen to a radioactive isotope of carbon, ^{14}C, with a half-life of about 5730 years. Vegetation absorbs carbon dioxide through the atmosphere and animal life assimilates ^{14}C through food chains. When a plant or animal dies it stops replacing its carbon and the amount of ^{14}C begins to decrease through radioactive decay. Therefore, the level of radioactivity must also decay exponentially. A parchment fragment was discovered that had about 74% as much ^{14}C radioactivity as does plant material on earth today. Estimate the age of the parchment.

12. A curve passes through the point $(0, 5)$ and has the property that the slope of the curve at every point P is twice the y-coordinate of P. What is the equation of the curve?

13. A certain object cools at a rate (in °C/min) equal to one-tenth of the difference between its temperature and that of the surrounding air. If a room is kept at 21 °C and the temperature of the object is 33 °C, find an expression for the temperature of the object t minutes later.

14. A thermometer is taken from a room where the temperature is 20 °C to the outdoors, where the temperature is 5 °C. After one minute the thermometer reads 12 °C. Use

Newton's Law of Cooling to answer the following questions.
(a) What will the reading on the thermometer be after one more minute?
(b) When will the thermometer read 6 °C?

15. A roast turkey is taken from an oven when its temperature has reached 185 °F and is placed on a table in a room where the temperature is 75 °F.
(a) If the temperature of the turkey is 150 °F after half an hour, what is the temperature after 45 min?
(b) When will the turkey have cooled to 100 °F?

16. On a hot day a thermometer is taken outside from an air-conditioned room where the temperature is 21 °C. After one minute it reads 27 °C and after 2 minutes it reads 30 °C.
(a) What is the outdoor temperature?
(b) Sketch the graph of the temperature function.

17. The rate of change of atmospheric pressure P with respect to altitude h is proportional to P, provided that the temperature is constant. At 15 °C the pressure is 101.3 kPa at sea level and 87.14 kPa at $h = 1000$ m.
(a) What is the pressure at an altitude of 3000 m?
(b) What is the pressure at the top of Mount McKinley, at an altitude of 6187 m?

18. If \$500 is borrowed at 14% interest, find the amounts due at the end of 2 years if the interest is compounded (a) annually, (b) quarterly, (c) monthly, (d) daily, (e) hourly, and (f) continuously.

19. If \$3000 is invested at 5% interest, find the value of the investment at the end of 5 years if the interest is compounded (a) annually, (b) semiannually, (c) monthly, (d) weekly, (e) daily, and (f) continuously.

20. How long will it take an investment to double in value if the interest rate is 6% compounded continuously?

21. A tank contains 1500 L of brine with a concentration of 0.3 kg of salt per liter. In order to dilute the solution, pure water is run into the tank at a rate of 20 L/min and the resulting solution, which is stirred continuously, runs out at the same rate.
(a) How many kilograms of salt will remain after half an hour?
(b) When will the concentration be reduced to 0.2 kg of salt per liter?

22. Do Exercise 21 if, instead of pure water, brine with a concentration of 0.1 kg of salt per liter is used.

4.6 INVERSE TRIGONOMETRIC FUNCTIONS

In this section we apply the ideas of Section 4.2 to find the derivatives of the so-called inverse trigonometric functions. We have a slight difficulty in this task: Because the trigonometric functions are not one-to-one, they do not have inverse functions. The difficulty is overcome by restricting the domains of these functions so that they become one-to-one.

You can see from Figure 1 that the sine function $y = \sin x$ is not one-to-one (use the Horizontal Line Test). But the function $f(x) = \sin x$, $-\pi/2 \leqslant x \leqslant \pi/2$ (see Figure 2), *is* one-to-one. The inverse function of this restricted sine function f exists and is denoted by $\sin^{-1}$ or arcsin. It is called the **inverse sine function** or the **arcsine function**.

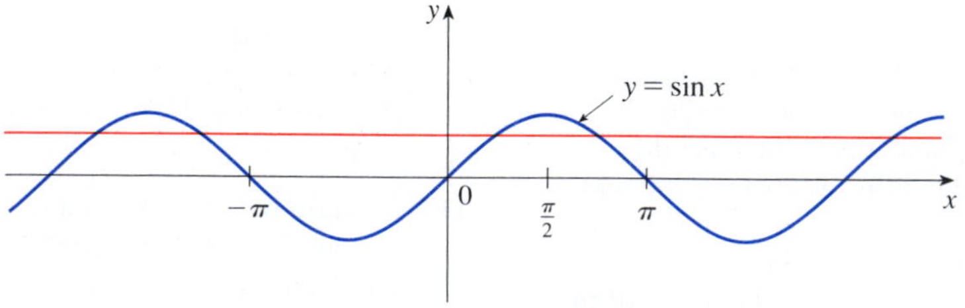

FIGURE 1

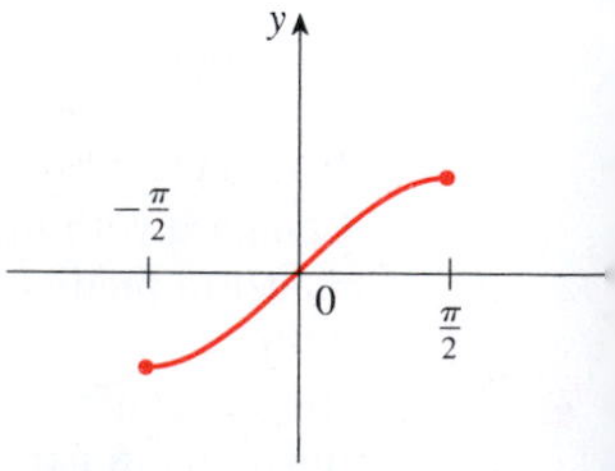

FIGURE 2

Since the definition of an inverse function says that

$$f^{-1}(x) = y \iff f(y) = x$$

we have

(1) $$\sin^{-1}x = y \iff \sin y = x \quad \text{and} \quad -\frac{\pi}{2} \leqslant y \leqslant \frac{\pi}{2}$$

$\sin^{-1}x \neq \dfrac{1}{\sin x}$

Thus, if $-1 \leqslant x \leqslant 1$, $\sin^{-1}x$ is the number between $-\pi/2$ and $\pi/2$ whose sine is x.

EXAMPLE 1 Evaluate (a) $\sin^{-1}\frac{1}{2}$ and (b) $\tan(\arcsin\frac{1}{3})$.

SOLUTION

(a) We have

$$\sin^{-1}\tfrac{1}{2} = \frac{\pi}{6}$$

because $\sin(\pi/6) = \frac{1}{2}$ and $\pi/6$ lies between $-\pi/2$ and $\pi/2$.

(b) Let $\theta = \arcsin\frac{1}{3}$. Then we can draw a right triangle with angle θ as in Figure 3 and deduce from the Pythagorean Theorem that the third side has length $\sqrt{9-1} = 2\sqrt{2}$. This enables us to read from the triangle that

$$\tan(\arcsin\tfrac{1}{3}) = \tan\theta = \frac{1}{2\sqrt{2}}$$ ■

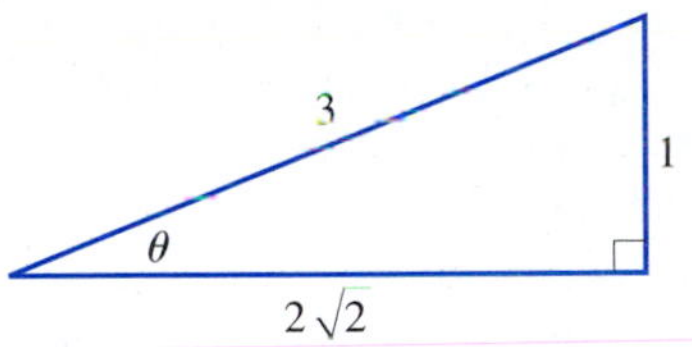

FIGURE 3

The cancellation equations for inverse functions become, in this case,

(2)
$$\sin^{-1}(\sin x) = x \quad \text{for } -\frac{\pi}{2} \leqslant x \leqslant \frac{\pi}{2}$$
$$\sin(\sin^{-1}x) = x \quad \text{for } -1 \leqslant x \leqslant 1$$

We must be careful when using the first cancellation equation because it is valid only when x lies in the interval $[-\pi/2, \pi/2]$. The following example shows how to proceed when x lies outside this interval.

EXAMPLE 2 Evaluate:

(a) $\sin(\sin^{-1}0.6)$ (b) $\sin^{-1}\left(\sin\dfrac{\pi}{12}\right)$ (c) $\sin^{-1}\left(\sin\dfrac{2\pi}{3}\right)$

SOLUTION

(a) Since 0.6 lies between -1 and 1, the second cancellation equation in (2) gives

$$\sin(\sin^{-1}0.6) = 0.6$$

(b) Since $\pi/12$ lies between $-\pi/2$ and $\pi/2$, the first cancellation equation gives

$$\sin^{-1}\left(\sin\frac{\pi}{12}\right) = \frac{\pi}{12}$$

(c) Since $2\pi/3$ does not lie in the interval $[-\pi/2, \pi/2]$, we cannot use the cancellation equation. Instead we note that $\sin(2\pi/3) = \sqrt{3}/2$ and $\sin^{-1}(\sqrt{3}/2) = \pi/3$ because $\pi/3$ lies between $-\pi/2$ and $\pi/2$. Therefore

$$\sin^{-1}\left(\sin\frac{2\pi}{3}\right) = \sin^{-1}\left(\frac{\sqrt{3}}{2}\right) = \frac{\pi}{3}$$ ■

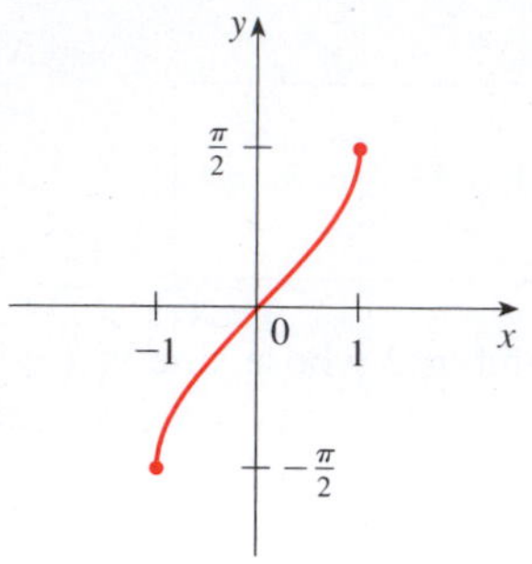

FIGURE 4
$y = \sin^{-1}x$

The inverse sine function, $\sin^{-1}$, has domain $[-1, 1]$ and range $[-\pi/2, \pi/2]$, and its graph, shown in Figure 4, is obtained from that of the restricted sine function (Figure 2) by reflection about the line $y = x$.

We know that the sine function f is continuous, so the inverse sine function is also continuous. We also know from Section 3.4 that the sine function is differentiable, so the inverse sine function is also differentiable. We could calculate the derivative of $\sin^{-1}$ by the formula in Theorem 4.2.7, but since we know that $\sin^{-1}$ is differentiable, we can just as easily calculate it by implicit differentiation as follows.

Let $y = \sin^{-1}x$. Then $\sin y = x$ and $-\pi/2 \leqslant y \leqslant \pi/2$. Differentiating $\sin y = x$ implicitly with respect to x, we obtain

$$\cos y \frac{dy}{dx} = 1$$

and

$$\frac{dy}{dx} = \frac{1}{\cos y}$$

Now $\cos y \geqslant 0$ since $-\pi/2 \leqslant y \leqslant \pi/2$, so

$$\cos y = \sqrt{1 - \sin^2 y} = \sqrt{1 - x^2}$$

Therefore

$$\frac{dy}{dx} = \frac{1}{\cos y} = \frac{1}{\sqrt{1 - x^2}}$$

(3)
$$\frac{d}{dx}(\sin^{-1}x) = \frac{1}{\sqrt{1 - x^2}} \qquad -1 < x < 1$$

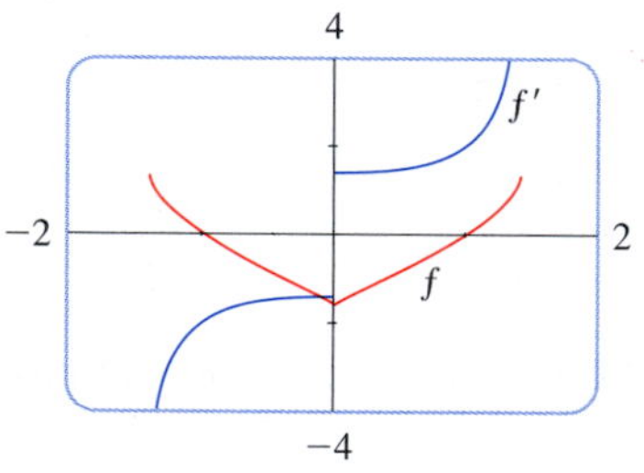

FIGURE 5

The graphs of the function f of Example 3 and its derivative are shown in Figure 5. Notice that f is not differentiable at 0 and this is consistent with the fact that the graph of f' makes a sudden jump at $x = 0$.

EXAMPLE 3 If $f(x) = \sin^{-1}(x^2 - 1)$, find (a) the domain of f, (b) $f'(x)$, and (c) the domain of f'.

SOLUTION

(a) Since the domain of the inverse sine function is $[-1, 1]$, the domain of f is

$$\{x \mid -1 \leqslant x^2 - 1 \leqslant 1\} = \{x \mid 0 \leqslant x^2 \leqslant 2\}$$
$$= \{x \mid |x| \leqslant \sqrt{2}\} = [-\sqrt{2}, \sqrt{2}]$$

(b) Combining Formula 3 with the Chain Rule, we have

$$f'(x) = \frac{1}{\sqrt{1 - (x^2 - 1)^2}} \frac{d}{dx}(x^2 - 1)$$
$$= \frac{1}{\sqrt{1 - (x^4 - 2x^2 + 1)}} 2x = \frac{2x}{\sqrt{2x^2 - x^4}}$$

(c) The domain of f' is

$$\{x \mid -1 < x^2 - 1 < 1\} = \{x \mid 0 < x^2 < 2\}$$
$$= \{x \mid 0 < |x| < \sqrt{2}\} = (-\sqrt{2}, 0) \cup (0, \sqrt{2})$$ ■

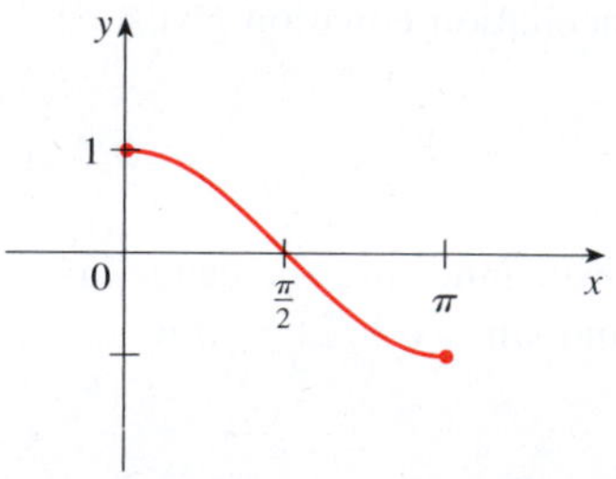

FIGURE 6
$y = \cos x, 0 \leqslant x \leqslant \pi$

The **inverse cosine function** is handled similarly. The restricted cosine function $f(x) = \cos x$, $0 \leqslant x \leqslant \pi$, is one-to-one (see Figure 6) and so it has an inverse function denoted by $\cos^{-1}$ or arccos.

(4)
$$\cos^{-1}x = y \iff \cos y = x \text{ and } 0 \leqslant y \leqslant \pi$$

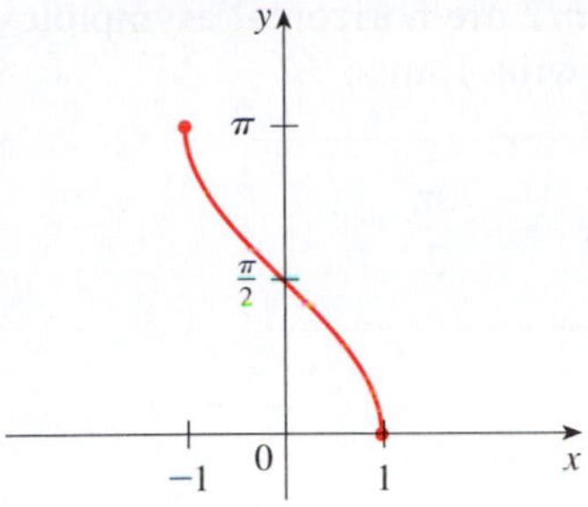

FIGURE 7
$y = \cos^{-1}x$

The cancellation equations are

(5)
$$\cos^{-1}(\cos x) = x \quad \text{for } 0 \leqslant x \leqslant \pi$$
$$\cos(\cos^{-1}x) = x \quad \text{for } -1 \leqslant x \leqslant 1$$

The inverse cosine function, $\cos^{-1}$, has domain $[-1, 1]$ and range $[0, \pi]$ and is a continuous function whose graph is shown in Figure 7. Its derivative is given by

(6)
$$\frac{d}{dx}(\cos^{-1}x) = -\frac{1}{\sqrt{1 - x^2}} \qquad -1 < x < 1$$

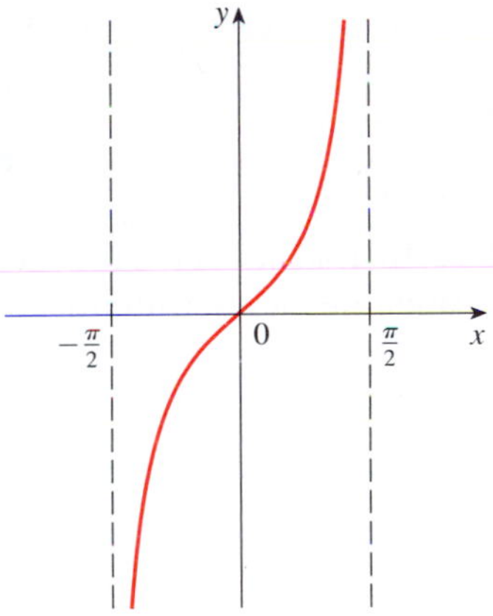

FIGURE 8
$y = \tan x, -\frac{\pi}{2} < x < \frac{\pi}{2}$

Formula 6 can be proved by the same method as for Formula 3 and is left as Exercise 25.

The tangent function can be made one-to-one by restricting it to the interval $(-\pi/2, \pi/2)$. Thus, the **inverse tangent function** is defined as the inverse of the function $f(x) = \tan x$, $-\pi/2 < x < \pi/2$ (see Figure 8). It is denoted by $\tan^{-1}$ or arctan.

(7)
$$\tan^{-1}x = y \iff \tan y = x \quad \text{and} \quad -\frac{\pi}{2} < y < \frac{\pi}{2}$$

EXAMPLE 4 Simplify the expression $\cos(\tan^{-1}x)$.

SOLUTION 1 Let $y = \tan^{-1}x$. Then $\tan y = x$ and $-\pi/2 < y < \pi/2$. We want to find $\cos y$ but, since $\tan y$ is known, it is easier to find $\sec y$ first:

$$\sec^2 y = 1 + \tan^2 y = 1 + x^2$$

$$\sec y = \sqrt{1 + x^2} \qquad (\text{since } \sec y > 0 \text{ for } -\pi/2 < y < \pi/2)$$

Thus $$\cos(\tan^{-1}x) = \cos y = \frac{1}{\sec y} = \frac{1}{\sqrt{1 + x^2}}$$

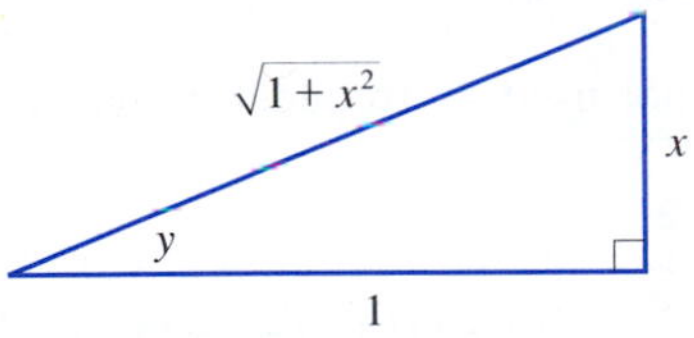

FIGURE 9

SOLUTION 2 Instead of using trigonometric identities as in Solution 1, it is perhaps easier to use a diagram. If $y = \tan^{-1}x$, then $\tan y = x$, and we can read from Figure 9 (which illustrates the case $y > 0$) that

$$\cos(\tan^{-1}x) = \cos y = \frac{1}{\sqrt{1 + x^2}}$$ ■

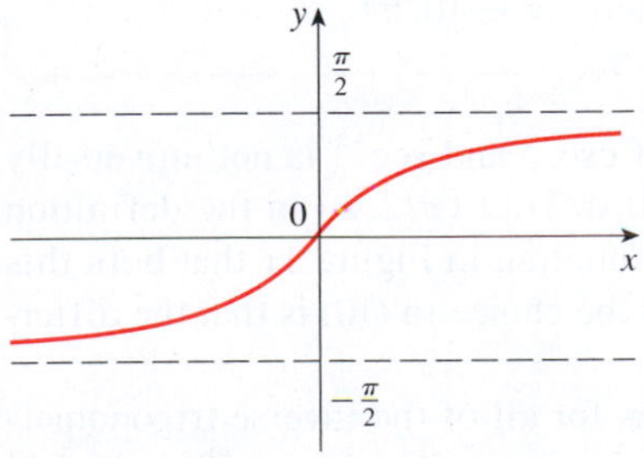

FIGURE 10
$y = \tan^{-1}x = \arctan x$

The inverse tangent function, $\tan^{-1} = \arctan$, has domain $\mathbb{R}$ and range $(-\pi/2, \pi/2)$. Its graph is shown in Figure 10.

We know that

$$\lim_{x \to (\pi/2)^-} \tan x = \infty \quad \text{and} \quad \lim_{x \to -(\pi/2)^+} \tan x = -\infty$$

and so the lines $x = \pm\pi/2$ are vertical asymptotes of the graph of tan. Since the graph of $\tan^{-1}$ is obtained by reflecting the graph of the restricted tangent function about the line $y = x$, it follows that the lines $y = \pi/2$ and $y = -\pi/2$ are horizontal asymptotes of the graph of $\tan^{-1}$. This fact is expressed by the following limits:

(8) $$\lim_{x\to\infty} \tan^{-1}x = \frac{\pi}{2} \qquad \lim_{x\to-\infty} \tan^{-1}x = -\frac{\pi}{2}$$

EXAMPLE 5 Evaluate $\lim_{x\to2^+} \arctan\left(\frac{1}{x-2}\right)$.

SOLUTION Since

$$\frac{1}{x-2} \to \infty \qquad \text{as } x \to 2^+$$

(8) gives

$$\lim_{x\to2^+} \arctan\left(\frac{1}{x-2}\right) = \frac{\pi}{2}$$ ■

Since tan is differentiable, $\tan^{-1}$ is also differentiable. To find its derivative, let $y = \tan^{-1}x$. Then $\tan y = x$. Differentiating this latter equation implicitly with respect to x, we have

$$\sec^2 y \frac{dy}{dx} = 1$$

and so $$\frac{dy}{dx} = \frac{1}{\sec^2 y} = \frac{1}{1 + \tan^2 y} = \frac{1}{1 + x^2}$$

(9) $$\frac{d}{dx}(\tan^{-1}x) = \frac{1}{1 + x^2}$$

The remaining inverse trigonometric functions are not used as frequently and are summarized here.

(10)
$$y = \csc^{-1}x \ (|x| \geqslant 1) \iff \csc y = x \ \text{ and } \ y \in (0, \pi/2] \cup (\pi, 3\pi/2]$$
$$y = \sec^{-1}x \ (|x| \geqslant 1) \iff \sec y = x \ \text{ and } \ y \in [0, \pi/2) \cup [\pi, 3\pi/2)$$
$$y = \cot^{-1}x \ (x \in \mathbb{R}) \iff \cot y = x \ \text{ and } \ y \in (0, \pi)$$

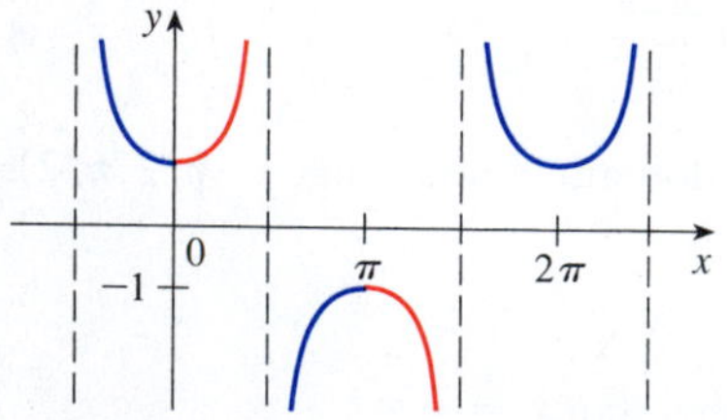

FIGURE 11
$y = \sec x$

The choice of intervals for y in the definitions of $\csc^{-1}$ and $\sec^{-1}$ is not universally agreed upon. For instance, some authors use $y \in [0, \pi/2) \cup (\pi/2, \pi]$ in the definition of $\sec^{-1}$. [You can see from the graph of the secant function in Figure 11 that both this choice and the one in (10) will work.] The reason for the choice in (10) is that the differentiation formulas are simpler (see Exercise 71).

We collect in Table 11 the differentiation formulas for all of the inverse trigonometric functions. The proofs of the formulas for the derivatives of $\csc^{-1}$, $\sec^{-1}$, and $\cot^{-1}$ are left as Exercises 27–29.

(11) TABLE OF DERIVATIVES OF INVERSE TRIGONOMETRIC FUNCTIONS

$$\frac{d}{dx}(\sin^{-1}x) = \frac{1}{\sqrt{1-x^2}} \qquad \frac{d}{dx}(\csc^{-1}x) = -\frac{1}{x\sqrt{x^2-1}}$$

$$\frac{d}{dx}(\cos^{-1}x) = -\frac{1}{\sqrt{1-x^2}} \qquad \frac{d}{dx}(\sec^{-1}x) = \frac{1}{x\sqrt{x^2-1}}$$

$$\frac{d}{dx}(\tan^{-1}x) = \frac{1}{1+x^2} \qquad \frac{d}{dx}(\cot^{-1}x) = -\frac{1}{1+x^2}$$

Each of these formulas can be combined with the Chain Rule. For instance, if u is a differentiable function of x, then

$$\frac{d}{dx}(\sin^{-1}u) = \frac{1}{\sqrt{1-u^2}}\frac{du}{dx} \quad \text{and} \quad \frac{d}{dx}(\tan^{-1}u) = \frac{1}{1+u^2}\frac{du}{dx}$$

EXAMPLE 6 Differentiate (a) $y = \dfrac{1}{\sin^{-1}x}$ and (b) $f(x) = x\tan^{-1}\sqrt{x}$.

SOLUTION

(a)
$$\frac{dy}{dx} = \frac{d}{dx}(\sin^{-1}x)^{-1} = -(\sin^{-1}x)^{-2}\frac{d}{dx}(\sin^{-1}x)$$

$$= -\frac{1}{(\sin^{-1}x)^2\sqrt{1-x^2}}$$

(b)
$$f'(x) = \tan^{-1}\sqrt{x} + x\frac{1}{1+(\sqrt{x})^2}\frac{1}{2}x^{-1/2}$$

$$= \tan^{-1}\sqrt{x} + \frac{\sqrt{x}}{2(1+x)}$$

■

EXERCISES 4.6

1–18 ■ Find the exact value of the expression.

1. $\cos^{-1}(-1)$

2. $\sin^{-1}(0.5)$

3. $\tan^{-1}\sqrt{3}$

4. $\arctan(-1)$

5. $\csc^{-1}\sqrt{2}$

6. $\arcsin 1$

7. $\cot^{-1}(-\sqrt{3})$

8. $\sec^{-1}2$

9. $\sin(\sin^{-1}0.7)$

10. $\sin^{-1}(\sin 1)$

11. $\tan^{-1}\left(\tan\dfrac{4\pi}{3}\right)$

12. $\tan(\cos^{-1}0.5)$

13. $\sin(\cos^{-1}\frac{4}{5})$

14. $\sec(\arctan 2)$

15. $\arcsin\left(\sin\dfrac{5\pi}{4}\right)$

16. $\sin(2\sin^{-1}\frac{3}{5})$

17. $\cos(2\sin^{-1}\frac{5}{13})$

18. $\sin[\sin^{-1}\frac{1}{3} + \sin^{-1}\frac{2}{3}]$

19. Prove that $\cos(\sin^{-1}x) = \sqrt{1-x^2}$ for $-1 \leqslant x \leqslant 1$.

20–22 ■ Simplify each expression as in Exercise 19.

20. $\tan(\sin^{-1}x)$

21. $\sin(\tan^{-1}x)$

22. $\sin(2\cos^{-1}x)$

23–24 ■ Graph the given functions on the same screen. How are these graphs related?

23. $y = \sin x,\ -\pi/2 \leqslant x \leqslant \pi/2;\quad y = \sin^{-1}x;\quad y = x$

24. $y = \tan x,\ -\pi/2 < x < \pi/2;\quad y = \tan^{-1}x;\quad y = x$

25. Prove Formula 6 for the derivative of $\cos^{-1}$ by the same method as for Formula 3.

26. (a) Prove that $\sin^{-1}x + \cos^{-1}x = \pi/2$.
(b) Use part (a) to prove Formula 6.

27. Prove that $D\cot^{-1}x = -\dfrac{1}{1+x^2}$.

28. Prove that $D\sec^{-1}x = \dfrac{1}{x\sqrt{x^2-1}}$, $|x| > 1$.

29. Prove that $D\csc^{-1}x = -\dfrac{1}{x\sqrt{x^2-1}}$, $|x| > 1$.

30–49 ■ Find the derivative of the function. Simplify where possible.

30. $f(x) = \sin^{-1}(2x - 1)$

31. $g(x) = \tan^{-1}(x^3)$

32. $y = (\sin^{-1}x)^2$

33. $y = \sin^{-1}(x^2)$

34. $h(x) = (\arcsin x)\ln x$

35. $H(x) = (1 + x^2)\arctan x$

36. $f(t) = (\cos^{-1}t)/t$

37. $g(t) = \sin^{-1}(4/t)$

38. $F(t) = \sqrt{1-t^2} + \sin^{-1}t$

39. $G(t) = \cos^{-1}\sqrt{2t-1}$

40. $y = \tan^{-1}\left(\dfrac{x}{a}\right) + \ln\sqrt{\dfrac{x-a}{x+a}}$

41. $y = \sec^{-1}\sqrt{1+x^2}$

42. $y = x\cos^{-1}x - \sqrt{1-x^2}$

43. $y = \tan^{-1}(\sin x)$

44. $y = \sin^{-1}\left(\dfrac{\cos x}{1+\sin x}\right)$

45. $y = (\tan^{-1}x)^{-1}$

46. $y = \tan^{-1}(x - \sqrt{1+x^2})$

47. $y = x^2\cot^{-1}(3x)$

48. $y = x\sin x\csc^{-1}x$

49. $y = \arccos\left(\dfrac{b + a\cos x}{a + b\cos x}\right),\quad 0 \leqslant x \leqslant \pi,\ a > b > 0$

50–55 ■ Find the derivative of each function. State the domains of the function and its derivative.

50. $f(x) = \cos^{-1}(\sin^{-1}x)$

51. $g(x) = \sin^{-1}(3x + 1)$

52. $F(x) = \sqrt{\sin^{-1}(2/x)}$

53. $S(x) = \sin^{-1}(\tan^{-1}x)$

54. $R(t) = \arcsin(2^t)$

55. $U(t) = 2^{\arctan t}$

56. If $f(x) = x\tan^{-1}x$, find $f'(1)$.

57. If $g(x) = x\sin^{-1}(x/4) + \sqrt{16-x^2}$, find $g'(2)$.

58. If $h(x) = (3\tan^{-1}x)^4$, find $h'(3)$.

59–60 ■ Find $f'(x)$. Check that your answer is reasonable by comparing the graphs of f and f'.

59. $f(x) = e^x - x^2\arctan x$

60. $f(x) = x\arcsin(1 - x^2)$

61–64 ■ Find each limit.

61. $\displaystyle\lim_{x\to -1^+} \sin^{-1}x$

62. $\displaystyle\lim_{x\to\infty} \sin^{-1}\left(\frac{x+1}{2x+1}\right)$

63. $\displaystyle\lim_{x\to\infty} \tan^{-1}(x^2)$

64. $\displaystyle\lim_{x\to\infty} \tan^{-1}(x - x^2)$

65. A ladder 10 ft long leans against a vertical wall. If the bottom of the ladder slides away from the base of the wall at a speed of 2 ft/s, how fast is the angle between the ladder and the wall changing when the bottom of the ladder is 6 ft from the base of the wall?

66. A lighthouse is on a small island 3 km away from the nearest point P on a straight shoreline and its light makes four revolutions per minute. How fast is the beam of light moving along the shoreline when it is 1 km from P?

67. Sketch a graph of $y = \sec^{-1}x$.

68. Prove that, for $xy \neq 1$,

$$\arctan x + \arctan y = \arctan\frac{x+y}{1-xy}$$

if the left side lies between $-\pi/2$ and $\pi/2$.

69. Use the result of Exercise 68 to prove the following:
(a) $\arctan\frac{1}{2} + \arctan\frac{1}{3} = \pi/4$
(b) $2\arctan\frac{1}{3} + \arctan\frac{1}{7} = \pi/4$

70. (a) Sketch a graph of the function $f(x) = \sin(\sin^{-1}x)$.
(b) Sketch a graph of the function $g(x) = \sin^{-1}(\sin x)$, $x \in \mathbb{R}$.
(c) Show that $g'(x) = \dfrac{\cos x}{|\cos x|}$.
(d) Sketch a graph of $h(x) = \cos^{-1}(\sin x)$, $x \in \mathbb{R}$, and find its derivative.

71. Some authors define $y = \sec^{-1}x \iff \sec y = x$ and $y \in [0, \pi/2) \cup (\pi/2, \pi]$. Show that if this definition is adopted, then, instead of the formula given in Exercise 28, we have

$$\frac{d}{dx}(\sec^{-1}x) = \frac{1}{|x|\sqrt{x^2-1}} \qquad |x| > 1$$

72. Let $f(x) = x\arctan(1/x)$ if $x \neq 0$ and $f(0) = 0$.
(a) Is f continuous at 0?
(b) Is f differentiable at 0?

4.7 HYPERBOLIC FUNCTIONS

Certain combinations of the exponential functions e^x and e^{-x} arise so frequently in mathematics and its applications that they deserve to be given special names. In many ways they are analogous to the trigonometric functions, and they have the same relationship to the hyperbola that the trigonometric functions have to the circle. For this reason they are collectively called **hyperbolic functions** and individually called **hyperbolic sine**, **hyperbolic cosine**, and so on.

DEFINITION OF THE HYPERBOLIC FUNCTIONS

$$\sinh x = \frac{e^x - e^{-x}}{2} \qquad \operatorname{csch} x = \frac{1}{\sinh x}$$

$$\cosh x = \frac{e^x + e^{-x}}{2} \qquad \operatorname{sech} x = \frac{1}{\cosh x}$$

$$\tanh x = \frac{\sinh x}{\cosh x} \qquad \coth x = \frac{\cosh x}{\sinh x}$$

The graphs of hyperbolic sine and cosine can be sketched using graphical addition as in Figures 1 and 2.

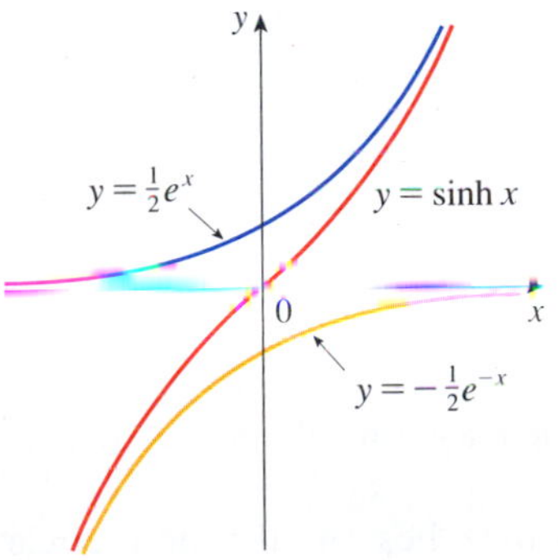

FIGURE 1
$y = \sinh x = \frac{1}{2}e^x - \frac{1}{2}e^{-x}$

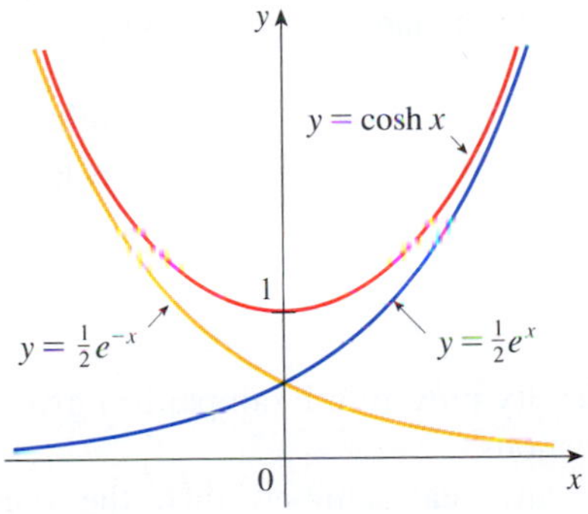

FIGURE 2
$y = \cosh x = \frac{1}{2}e^x + \frac{1}{2}e^{-x}$

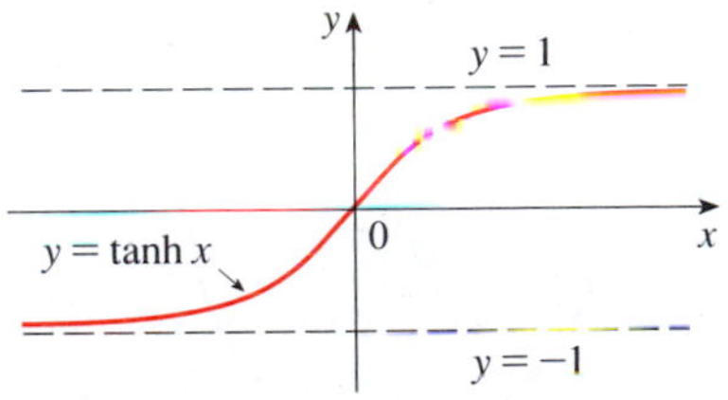

FIGURE 3

Note that sinh has domain $\mathbb{R}$ and range $\mathbb{R}$, while cosh has domain $\mathbb{R}$ and range $[1, \infty)$. The graph of tanh is shown in Figure 3. It has the horizontal asymptotes $y = \pm 1$. (See Exercise 23.)

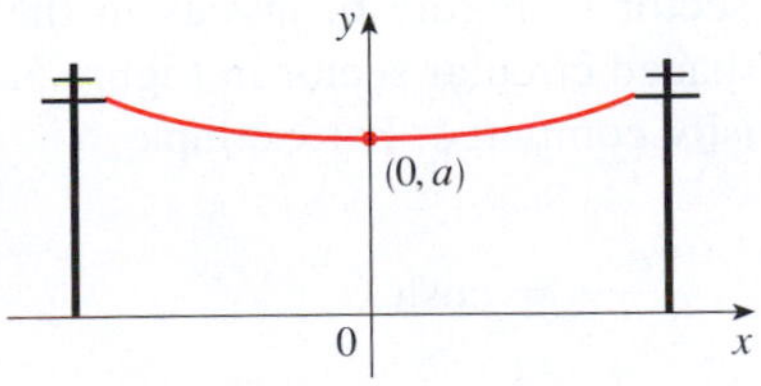

FIGURE 4
The catenary $y = C + a\cosh(x/a)$

Some of the mathematical uses of hyperbolic functions will be seen in Chapter 8. Applications to science and engineering occur whenever an entity such as light, velocity, electricity, or radioactivity is gradually absorbed or extinguished, for the decay can be represented by hyperbolic functions. The most famous application is the use of hyperbolic cosine to describe the shape of a hanging wire. It can be proved that if a heavy flexible cable (such as a telephone or power line) is suspended between two points at the same height, then it takes the shape of a curve with equation $y = C + a\cosh(x/a)$ called a *catenary* (see Figure 4). (The Latin word *catena* means "chain.")

The hyperbolic functions satisfy a number of identities that are analogues of well-known trigonometric identities. We list some of them here and leave most of the proofs to the exercises.

HYPERBOLIC IDENTITIES

$$\sinh(-x) = -\sinh x \qquad \cosh(-x) = \cosh x$$

$$\cosh^2 x - \sinh^2 x = 1 \qquad 1 - \tanh^2 x = \operatorname{sech}^2 x$$

$$\sinh(x + y) = \sinh x \cosh y + \cosh x \sinh y$$

$$\cosh(x + y) = \cosh x \cosh y + \sinh x \sinh y$$

EXAMPLE 1 Prove (a) $\cosh^2 x - \sinh^2 x = 1$ and (b) $1 - \tanh^2 x = \operatorname{sech}^2 x$.

SOLUTION

(a)
$$\cosh^2 x - \sinh^2 x = \left(\frac{e^x + e^{-x}}{2}\right)^2 - \left(\frac{e^x - e^{-x}}{2}\right)^2$$
$$= \frac{e^{2x} + 2 + e^{-2x}}{4} - \frac{e^{2x} - 2 + e^{-2x}}{4}$$
$$= \tfrac{4}{4} = 1$$

(b) We start with the identity proved in part (a):

$$\cosh^2 x - \sinh^2 x = 1$$

If we divide both sides by $\cosh^2 x$, we get

$$1 - \frac{\sinh^2 x}{\cosh^2 x} = \frac{1}{\cosh^2 x}$$

or

$$1 - \tanh^2 x = \operatorname{sech}^2 x$$ ■

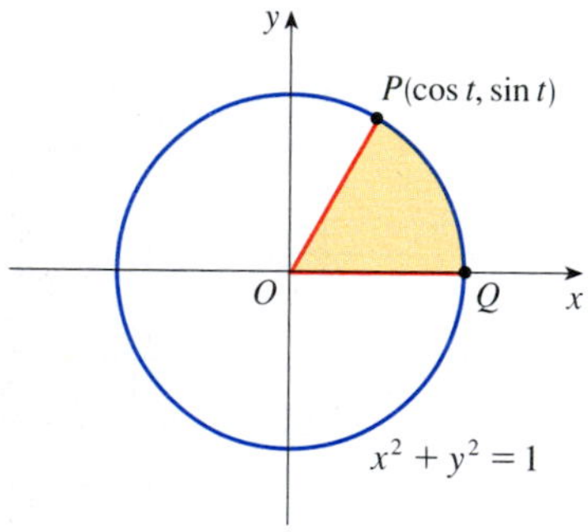

FIGURE 5

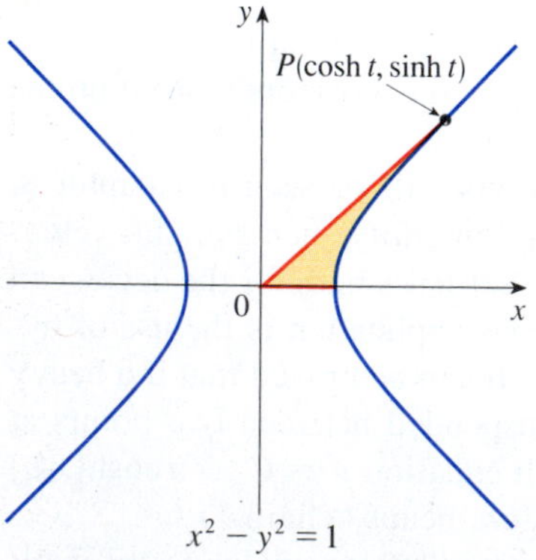

FIGURE 6

The identity proved in Example 1(a) gives a clue to the reason for the name "hyperbolic" functions.

If t is any real number, then the point $P(\cos t, \sin t)$ lies on the unit circle $x^2 + y^2 = 1$ because $\cos^2 t + \sin^2 t = 1$. In fact, t can be interpreted as the radian measure of $\angle POQ$ in Figure 5. For this reason the trigonometric functions are sometimes called *circular* functions.

Likewise, if t is any real number, then the point $P(\cosh t, \sinh t)$ lies on the right branch of the hyperbola $x^2 - y^2 = 1$ because $\cosh^2 t - \sinh^2 t = 1$ and $\cosh t \geqslant 1$. This time, t does not represent the measure of an angle. However, it turns out that t represents twice the area of the shaded hyperbolic sector in Figure 6, just as in the trigonometric case t represents twice the area of the shaded circular sector in Figure 5.

The derivatives of the hyperbolic functions are easily computed. For example,

$$\frac{d}{dx}(\sinh x) = \frac{d}{dx}\left(\frac{e^x - e^{-x}}{2}\right) = \frac{e^x + e^{-x}}{2} = \cosh x$$

We list the differentiation formulas for the hyperbolic functions as Table 1. The remaining proofs are left as exercises. Note the analogy with the differentiation formulas for trigonometric functions, but beware that the signs are sometimes different.

(1) TABLE OF DERIVATIVES OF HYPERBOLIC FUNCTIONS

$$\frac{d}{dx}\sinh x = \cosh x \qquad \frac{d}{dx}\operatorname{csch} x = -\operatorname{csch} x \coth x$$

$$\frac{d}{dx}\cosh x = \sinh x \qquad \frac{d}{dx}\operatorname{sech} x = -\operatorname{sech} x \tanh x$$

$$\frac{d}{dx}\tanh x = \operatorname{sech}^2 x \qquad \frac{d}{dx}\coth x = -\operatorname{csch}^2 x$$

EXAMPLE 2 Any of these differentiation rules can be combined with the Chain Rule. For instance,

$$\frac{d}{dx}(\cosh\sqrt{x}) = \sinh\sqrt{x}\cdot\frac{d}{dx}\sqrt{x} = \frac{\sinh\sqrt{x}}{2\sqrt{x}}$$ ■

INVERSE HYPERBOLIC FUNCTIONS

You can see from Figures 1 and 3 that sinh and tanh are one-to-one functions and so they have inverse functions denoted by $\sinh^{-1}$ and $\tanh^{-1}$. Figure 2 shows that cosh is not one-to-one, but when restricted to the domain $[0, \infty)$ it becomes one-to-one. The inverse hyperbolic cosine function is defined as the inverse of this restricted function.

(2)

$$y = \sinh^{-1}x \iff \sinh y = x$$
$$y = \cosh^{-1}x \iff \cosh y = x \text{ and } y \geqslant 0$$
$$y = \tanh^{-1}x \iff \tanh y = x$$

The remaining inverse hyperbolic functions are defined similarly (see Exercise 28).

We can sketch the graphs of $\sinh^{-1}$, $\cosh^{-1}$, and $\tanh^{-1}$ in Figures 7, 8, and 9 by using Figures 1, 2, and 3.

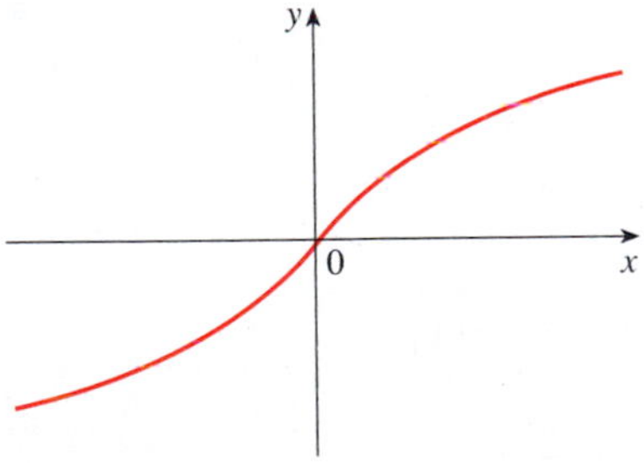

FIGURE 7
$y = \sinh^{-1}x$
domain $= \mathbb{R}$
range $= \mathbb{R}$

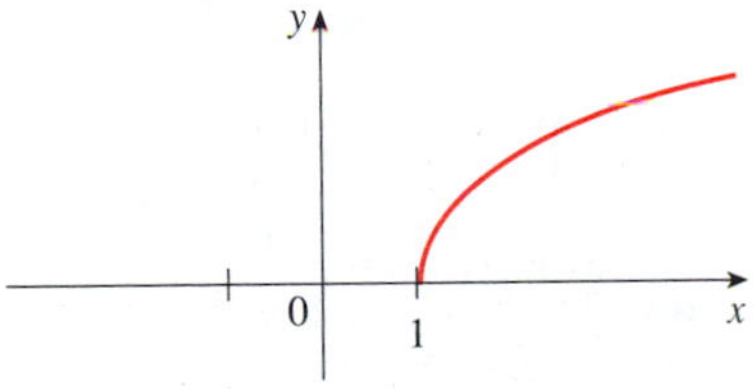

FIGURE 8
$y = \cosh^{-1}x$
domain $= [1, \infty)$
range $= [0, \infty)$

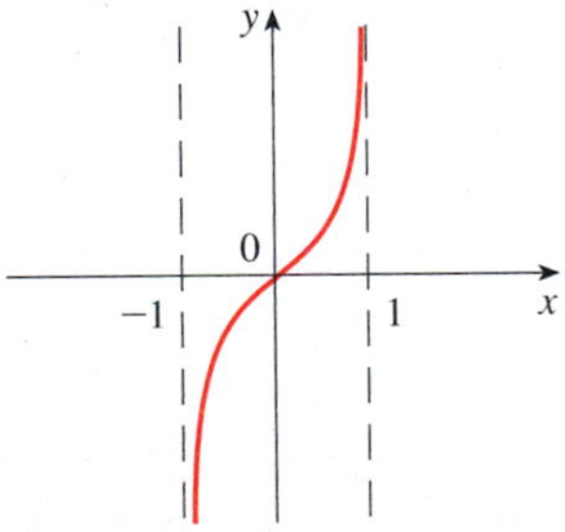

FIGURE 9
$y = \tanh^{-1}x$
domain $= (-1, 1)$
range $= \mathbb{R}$

Since the hyperbolic functions are defined in terms of exponential functions, it is not surprising to learn that the inverse hyperbolic functions can be expressed in terms of logarithms. In particular, we have:

(3)	$\sinh^{-1}x = \ln(x + \sqrt{x^2 + 1})$	$x \in \mathbb{R}$
(4)	$\cosh^{-1}x = \ln(x + \sqrt{x^2 - 1})$	$x \geqslant 1$
(5)	$\tanh^{-1}x = \frac{1}{2}\ln\left(\frac{1 + x}{1 - x}\right)$	$-1 < x < 1$

EXAMPLE 3 Show that $\sinh^{-1}x = \ln(x + \sqrt{x^2 + 1})$.

SOLUTION Let $y = \sinh^{-1}x$. Then

$$x = \sinh y = \frac{e^y - e^{-y}}{2}$$

so

$$e^y - 2x - e^{-y} = 0$$

or, multiplying by e^y,

$$e^{2y} - 2xe^y - 1 = 0$$

This is really a quadratic equation in e^y:

$$(e^y)^2 - 2x(e^y) - 1 = 0$$

Solving by the quadratic formula, we get

$$e^y = \frac{2x \pm \sqrt{4x^2 + 4}}{2} = x \pm \sqrt{x^2 + 1}$$

Note that $e^y > 0$, but $x - \sqrt{x^2 + 1} < 0$ (because $x < \sqrt{x^2 + 1}$). Thus, the minus sign is inadmissible and we have

$$e^y = x + \sqrt{x^2 + 1}$$

Therefore

$$y = \ln(e^y) = \ln(x + \sqrt{x^2 + 1})$$

(See Exercise 25 for another method.) ■

(6) TABLE OF DERIVATIVES OF INVERSE HYPERBOLIC FUNCTIONS

$$\frac{d}{dx}\sinh^{-1}x = \frac{1}{\sqrt{1 + x^2}} \qquad \frac{d}{dx}\operatorname{csch}^{-1}x = -\frac{1}{|x|\sqrt{x^2 + 1}}$$

$$\frac{d}{dx}\cosh^{-1}x = \frac{1}{\sqrt{x^2 - 1}} \qquad \frac{d}{dx}\operatorname{sech}^{-1}x = -\frac{1}{x\sqrt{1 - x^2}}$$

$$\frac{d}{dx}\tanh^{-1}x = \frac{1}{1 - x^2} \qquad \frac{d}{dx}\coth^{-1}x = \frac{1}{1 - x^2}$$

The inverse hyperbolic functions are all differentiable because the hyperbolic functions are differentiable. The formulas in Table 6 can be proved either by the method for inverse functions or by differentiating Formulas 3, 4, and 5.

EXAMPLE 4 Prove that $\dfrac{d}{dx}\sinh^{-1}x = \dfrac{1}{\sqrt{1+x^2}}$.

SOLUTION 1 Let $y = \sinh^{-1}x$. Then $\sinh y = x$. If we differentiate this equation implicitly with respect to x, we get

$$\cosh y \frac{dy}{dx} = 1$$

Since $\cosh^2 y - \sinh^2 y = 1$ and $\cosh y \geqslant 0$, we have $\cosh y = \sqrt{1 + \sinh^2 y}$, so

$$\frac{dy}{dx} = \frac{1}{\cosh y} = \frac{1}{\sqrt{1+\sinh^2 y}} = \frac{1}{\sqrt{1+x^2}}$$

SOLUTION 2 From Equation 3 (proved in Example 3), we have

$$\begin{aligned}\frac{d}{dx}\sinh^{-1}x &= \frac{d}{dx}\ln\left(x+\sqrt{x^2+1}\right)\\ &= \frac{1}{x+\sqrt{x^2+1}}\frac{d}{dx}\left(x+\sqrt{x^2+1}\right)\\ &= \frac{1}{x+\sqrt{x^2+1}}\left(1+\frac{x}{\sqrt{x^2+1}}\right)\\ &= \frac{\sqrt{x^2+1}+x}{\left(x+\sqrt{x^2+1}\right)\sqrt{x^2+1}}\\ &= \frac{1}{\sqrt{x^2+1}}\end{aligned}$$

■

EXAMPLE 5 Find $\dfrac{d}{dx}\tanh^{-1}(\sin x)$.

SOLUTION Using Table 6 and the Chain Rule, we have

$$\begin{aligned}\frac{d}{dx}\tanh^{-1}(\sin x) &= \frac{1}{1-(\sin x)^2}\frac{d}{dx}(\sin x)\\ &= \frac{1}{1-\sin^2 x}\cos x\\ &= \frac{\cos x}{\cos^2 x} = \sec x\end{aligned}$$

■

EXERCISES 4.7

1–6 ■ Find the numerical value of each expression.

1. (a) $\sinh 0$ (b) $\cosh 0$
2. (a) $\tanh 0$ (b) $\tanh 1$
3. (a) $\sinh(\ln 2)$ (b) $\sinh 2$
4. (a) $\cosh 3$ (b) $\cosh(\ln 3)$
5. (a) $\operatorname{sech} 0$ (b) $\cosh^{-1} 1$
6. (a) $\sinh 1$ (b) $\sinh^{-1} 1$

7–19 ■ Prove each identity.

7. $\sinh(-x) = -\sinh x$
(This shows that sinh is an odd function.)
8. $\cosh(-x) = \cosh x$
(This shows that cosh is an even function.)
9. $\cosh x + \sinh x = e^x$
10. $\cosh x - \sinh x = e^{-x}$
11. $\sinh(x+y) = \sinh x \cosh y + \cosh x \sinh y$

12. $\cosh(x + y) = \cosh x \cosh y + \sinh x \sinh y$

13. $\coth^2 x - 1 = \operatorname{csch}^2 x$

14. $\tanh(x + y) = \dfrac{\tanh x + \tanh y}{1 + \tanh x \tanh y}$

15. $\sinh 2x = 2 \sinh x \cosh x$

16. $\cosh 2x = \cosh^2 x + \sinh^2 x$

17. $\tanh(\ln x) = \dfrac{x^2 - 1}{x^2 + 1}$ **18.** $\dfrac{1 + \tanh x}{1 - \tanh x} = e^{2x}$

19. $(\cosh x + \sinh x)^n = \cosh nx + \sinh nx$
(n any real number)

20. If $\sinh x = \frac{3}{4}$, find the values of the other hyperbolic functions at x.

21. If $\tanh x = \frac{4}{5}$, find the values of the other hyperbolic functions at x.

22. (a) Use the graphs of sinh, cosh, and tanh in Figures 1–3 to draw the graphs of csch, sech, and coth.
(b) Check the graphs that you sketched in part (a) by using a graphing device to produce them.

23. Use the definitions of the hyperbolic functions to find the following limits:
(a) $\lim_{x \to \infty} \tanh x$ (b) $\lim_{x \to -\infty} \tanh x$
(c) $\lim_{x \to \infty} \sinh x$ (d) $\lim_{x \to -\infty} \sinh x$
(e) $\lim_{x \to \infty} \operatorname{sech} x$ (f) $\lim_{x \to \infty} \coth x$
(g) $\lim_{x \to 0^+} \coth x$ (h) $\lim_{x \to 0^-} \coth x$
(i) $\lim_{x \to -\infty} \operatorname{csch} x$

24. Prove the formulas given in Table 1 for the derivatives of the following functions: (a) cosh, (b) tanh, (c) csch, (d) sech, and (e) coth.

25. Give an alternative solution to Example 3 by letting $y = \sinh^{-1}x$ and then using Exercise 9 and Example 1(a) with x replaced by y.

26. Prove Equation 4.

27. Prove Equation 5 using (a) the method of Example 3 and (b) Exercise 18 with x replaced by y.

28. For each of the following functions (i) give a definition like those in (2), (ii) sketch the graph, and (iii) find a formula similar to Equation 3.
(a) csch^{-1} (b) sech^{-1} (c) $\coth^{-1}$

29. Prove the formulas given in Table 6 for the derivatives of the following functions.
(a) $\cosh^{-1}$ (b) $\tanh^{-1}$ (c) csch^{-1}
(d) sech^{-1} (e) $\coth^{-1}$

30–47 ■ Find the derivative of each function.

30. $f(x) = e^x \sinh x$ **31.** $f(x) = \tanh 3x$

32. $g(x) = \cosh^4 x$ **33.** $h(x) = \cosh(x^4)$

34. $F(x) = e^{\coth 2x}$ **35.** $G(x) = x^2 \operatorname{sech} x$

36. $f(t) = \ln(\sinh t)$ **37.** $H(t) = \tanh(e^t)$

38. $y = \cos(\sinh x)$ **39.** $y = x^{\cosh x}$

40. $y = e^{\tanh x} \cosh(\cosh x)$ **41.** $y = \cosh^{-1}(x^2)$

42. $y = \sqrt{x} \sinh^{-1} \sqrt{x}$

43. $y = x \ln(\operatorname{sech} 4x)$

44. $y = x \tanh^{-1}x + \ln \sqrt{1 - x^2}$

45. $y = x \sinh^{-1}(x/3) - \sqrt{9 + x^2}$

46. $y = \operatorname{sech}^{-1} \sqrt{1 - x^2}$, $x > 0$

47. $y = \coth^{-1} \sqrt{x^2 + 1}$

48–49 ■ Eliminate t from the parametric equations and describe the given curve as an equation in x and y.

48. $x = \sinh t$, $y = \cosh t$

49. $x = 4 \cosh t$, $y = 3 \sinh t$

50–51 ■ Compute a tangent vector of unit length to the curve at the given point.

50. $x = \sinh t$, $y = \cosh t$, $(0, 1)$

51. $x = 4 \cosh t$, $y = 3 \sinh t$, $(4, 0)$

52. Using principles from physics it can be shown that when a cable is hung between two poles, it takes the shape of a curve $y = f(x)$ that satisfies the differential equation

$$\frac{d^2y}{dx^2} = \frac{\rho g}{T} \sqrt{1 + \left(\frac{dy}{dx}\right)^2}$$

where ρ is the linear density of the cable, g is the acceleration due to gravity, and T is the tension in the cable at its lowest point. Verify that the function

$$y = f(x) = \frac{T}{\rho g} \cosh\left(\frac{\rho g x}{T}\right)$$

is a solution of this differential equation.

53. (a) Show that any function of the form

$$y = A \sinh mx + B \cosh mx$$

satisfies the differential equation $y'' = m^2 y$.
(b) Find $y = y(x)$ such that $y'' = 9y$, $y(0) = -4$, and $y'(0) = 6$.

54. Evaluate $\lim_{x \to \infty} \dfrac{\sinh x}{e^x}$.

55. At what point of the curve $y = \cosh x$ does the tangent have slope 1?

56. If $x = \ln(\sec \theta + \tan \theta)$, show that $\sec \theta = \cosh x$.

4.8 INDETERMINATE FORMS AND L'HOSPITAL'S RULE

Suppose we are trying to sketch the graph of the function

$$F(x) = \frac{2^x - 1}{x}$$

Although F is not defined when $x = 0$, we need to know how F behaves *near* 0. In particular, we would like to know the value of the limit

(1) $$\lim_{x \to 0} \frac{2^x - 1}{x}$$

But we cannot apply Law 5 of limits (the limit of a quotient is the quotient of the limits) to (1) because the limit of the denominator is 0. In fact, although the limit in (1) exists, its value is not obvious because both numerator and denominator approach 0 and $\frac{0}{0}$ is not defined.

In general, if we have a limit of the form

$$\lim_{x \to a} \frac{f(x)}{g(x)}$$

where both $f(x) \to 0$ and $g(x) \to 0$ as $x \to a$, then this limit may or may not exist and is called an **indeterminate form of type $\frac{0}{0}$**. We met some limits of this type in Chapter 2. For rational functions, we can cancel common factors:

$$\lim_{x \to 1} \frac{x^2 - x}{x^2 - 1} = \lim_{x \to 1} \frac{x(x - 1)}{(x + 1)(x - 1)} = \lim_{x \to 1} \frac{x}{x + 1} = \frac{1}{2}$$

We used a geometric argument to show that

$$\lim_{x \to 0} \frac{\sin x}{x} = 1$$

But these methods do not work for limits such as (1), so in this section we introduce a systematic method, known as l'Hospital's Rule, for the evaluation of indeterminate forms.

Another situation in which a limit is not obvious occurs when we try to sketch the curve $y = (\ln x)/x$. In searching for horizontal asymptotes or other aspects of the curve for large values of x, we need to evaluate the limit

(2) $$\lim_{x \to \infty} \frac{\ln x}{x}$$

It is not obvious how to evaluate this limit because both numerator and denominator become large as $x \to \infty$. There is a struggle between numerator and denominator. If the numerator wins, the limit will be ∞; if the denominator wins, the answer will be 0. Or there may be some compromise, in which case the answer may be some finite positive number.

In general, if we have a limit of the form

$$\lim_{x \to a} \frac{f(x)}{g(x)}$$

where both $f(x) \to \infty$ (or $-\infty$) and $g(x) \to \infty$ (or $-\infty$), then the limit may or may not exist and is called an **indeterminate form of type ∞/∞**. We saw in Section 2.6 that this type of limit can be evaluated for certain functions, including rational functions, by dividing numerator and denominator by the highest power of x that occurs. For instance,

$$\lim_{x\to\infty} \frac{x^2 - 1}{2x^2 + 1} = \lim_{x\to\infty} \frac{1 - \dfrac{1}{x^2}}{2 + \dfrac{1}{x^2}} = \frac{1 - 0}{2 + 0} = \frac{1}{2}$$

This method does not work for limits such as (2), but l'Hospital's Rule also applies to this type of indeterminate form.

L'Hospital's Rule is named after a French nobleman, the Marquis de l'Hospital (1661–1704), but was discovered by a Swiss mathematician, John Bernoulli (1667–1748).

(3) L'HOSPITAL'S RULE Suppose f and g are differentiable and $g'(x) \neq 0$ on an open interval I that contains a (except possibly at a). Suppose that

$$\lim_{x\to a} f(x) = 0 \quad \text{and} \quad \lim_{x\to a} g(x) = 0$$

or that

$$\lim_{x\to a} f(x) = \pm\infty \quad \text{and} \quad \lim_{x\to a} g(x) = \pm\infty$$

(In other words, we have an indeterminate form of type $\frac{0}{0}$ or ∞/∞.) Then

$$\lim_{x\to a} \frac{f(x)}{g(x)} = \lim_{x\to a} \frac{f'(x)}{g'(x)}$$

if the limit on the right side exists (or is ∞ or $-\infty$).

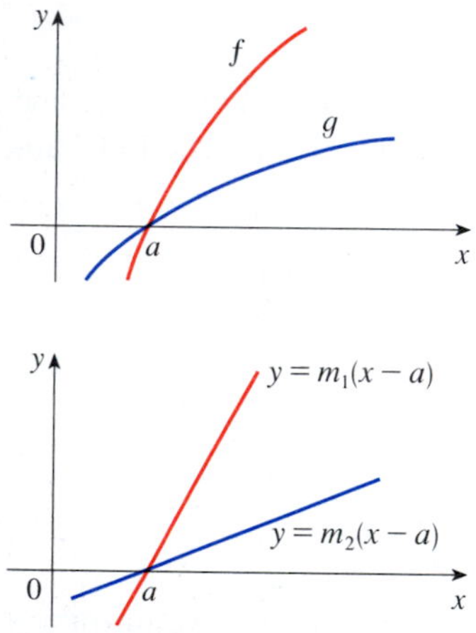

FIGURE 1

Figure 1 suggests visually why l'Hospital's Rule might be true. The first graph shows two differentiable functions f and g, each of which approaches 0 as $x \to a$. If we were to zoom in toward the point $(a, 0)$, the graphs would start to look almost linear. But if the functions were actually linear, as in the second graph, then their ratio would be

$$\frac{m_1(x - a)}{m_2(x - a)} = \frac{m_1}{m_2}$$

which is the ratio of their derivatives. This suggests that

$$\lim_{x\to a} \frac{f(x)}{g(x)} = \lim_{x\to a} \frac{f'(x)}{g'(x)}$$

NOTE 1: L'Hospital's Rule says that the limit of a quotient of functions is equal to the limit of the quotient of their derivatives, provided that the conditions stated in (3) are satisfied. It is especially important to verify the conditions regarding the limits of f and g before using l'Hospital's Rule.

NOTE 2: L'Hospital's Rule is also valid for one-sided limits and for limits at infinity or negative infinity; that is, in (3) "$x \to a$" can be replaced by any of the following symbols: $x \to a^+$, $x \to a^-$, $x \to \infty$, $x \to -\infty$.

NOTE 3: For the special case in which $f(a) = g(a) = 0$, f' and g' are continuous, and $g'(a) \neq 0$, it is easy to see why l'Hospital's Rule is true. In fact, using the alternate form of the definition of a derivative, we have

$$\lim_{x\to a} \frac{f(x)}{g(x)} = \lim_{x\to a} \frac{f(x) - f(a)}{g(x) - g(a)}$$

$$= \lim_{x\to a} \frac{\dfrac{f(x) - f(a)}{x - a}}{\dfrac{g(x) - g(a)}{x - a}} = \frac{\displaystyle\lim_{x\to a} \frac{f(x) - f(a)}{x - a}}{\displaystyle\lim_{x\to a} \frac{g(x) - g(a)}{x - a}}$$

$$= \frac{f'(a)}{g'(a)} = \lim_{x\to a} \frac{f'(x)}{g'(x)}$$

The general version of l'Hospital's Rule is more difficult; its proof can be found in more advanced books.

EXAMPLE 1 Find $\displaystyle\lim_{x\to 0}\frac{2^x - 1}{x}$.

SOLUTION Since $\lim_{x\to 0}(2^x - 1) = 0$ and $\lim_{x\to 0} x = 0$, we can apply l'Hospital's Rule:

$$\lim_{x\to 0}\frac{2^x - 1}{x} = \lim_{x\to 0}\frac{\dfrac{d}{dx}(2^x - 1)}{\dfrac{d}{dx}(x)} = \lim_{x\to 0}\frac{2^x \ln 2}{1} = \ln 2$$

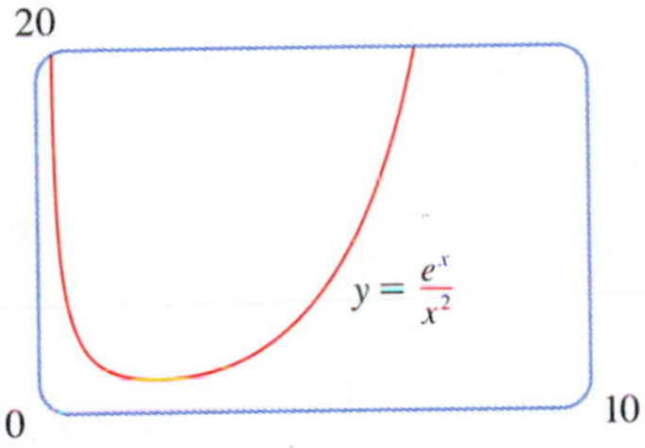

FIGURE 2

The graph of the function of Example 2 is shown in Figure 2. We have noticed previously that exponential functions grow far more rapidly than power functions, so the result of Example 2 is not unexpected. See also Exercise 77.

EXAMPLE 2 Calculate $\displaystyle\lim_{x\to\infty}\frac{e^x}{x^2}$.

SOLUTION We have $\lim_{x\to\infty} e^x = \infty$ and $\lim_{x\to\infty} x^2 = \infty$, so l'Hospital's Rule gives

$$\lim_{x\to\infty}\frac{e^x}{x^2} = \lim_{x\to\infty}\frac{e^x}{2x}$$

Since $e^x \to \infty$ and $2x \to \infty$ as $x \to \infty$, the limit on the right side is also indeterminate, but a second application of l'Hospital's Rule gives

$$\lim_{x\to\infty}\frac{e^x}{x^2} = \lim_{x\to\infty}\frac{e^x}{2x} = \lim_{x\to\infty}\frac{e^x}{2} = \infty$$

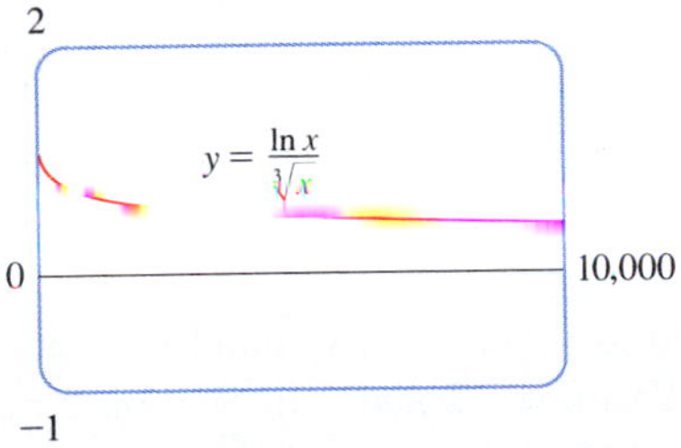

FIGURE 3

The graph of the function of Example 3 is shown in Figure 3. We have discussed previously the slow growth of logarithms, so it is not surprising that this ratio approaches 0 as $x \to \infty$. See also Exercise 78.

EXAMPLE 3 Calculate $\displaystyle\lim_{x\to\infty}\frac{\ln x}{\sqrt[3]{x}}$.

SOLUTION Since $\ln x \to \infty$ and $\sqrt[3]{x} \to \infty$ as $x \to \infty$, l'Hospital's Rule applies:

$$\lim_{x\to\infty}\frac{\ln x}{\sqrt[3]{x}} = \lim_{x\to\infty}\frac{\dfrac{1}{x}}{\frac{1}{3}x^{-2/3}}$$

Notice that the limit on the right side is now indeterminate of type $\frac{0}{0}$. But instead of applying l'Hospital's Rule a second time as we did in Example 2, we simplify the expression and see that a second application is unnecessary:

$$\lim_{x\to\infty}\frac{\ln x}{\sqrt[3]{x}} = \lim_{x\to\infty}\frac{\dfrac{1}{x}}{\frac{1}{3}x^{-2/3}} = \lim_{x\to\infty}\frac{3}{\sqrt[3]{x}} = 0$$

EXAMPLE 4 Find $\displaystyle\lim_{x\to 0}\frac{\tan x - x}{x^3}$. (See Exercise 28 in Section 2.2.)

SOLUTION Noting that both $\tan x - x \to 0$ and $x^3 \to 0$ as $x \to 0$, we use l'Hospital's Rule:

$$\lim_{x\to 0}\frac{\tan x - x}{x^3} = \lim_{x\to 0}\frac{\sec^2 x - 1}{3x^2}$$

The graph in Figure 4 gives visual confirmation of the result of Example 4. If we were to zoom in too far, however, we would get an inaccurate graph because $\tan x$ is close to x when x is small. See Exercise 32 in Section 2.2.

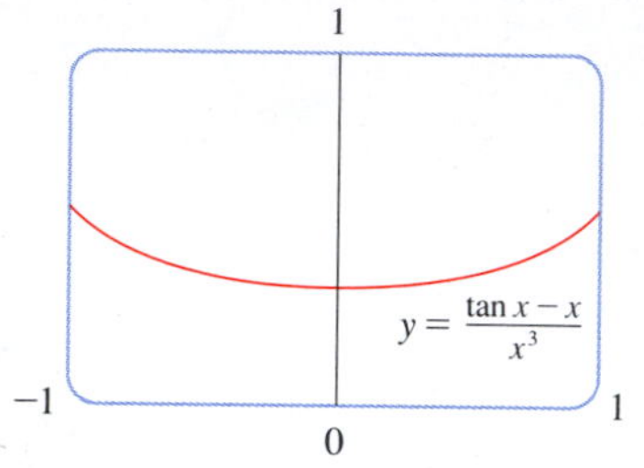

FIGURE 4

Since the limit on the right side is still indeterminate of type $\frac{0}{0}$, we apply l'Hospital's Rule again:

$$\lim_{x \to 0} \frac{\sec^2 x - 1}{3x^2} = \lim_{x \to 0} \frac{2 \sec^2 x \tan x}{6x}$$

Again both numerator and denominator approach 0, so a third application of l'Hospital's Rule is necessary. Putting together all three steps, we get

$$\lim_{x \to 0} \frac{\tan x - x}{x^3} = \lim_{x \to 0} \frac{\sec^2 x - 1}{3x^2} = \lim_{x \to 0} \frac{2 \sec^2 x \tan x}{6x}$$

$$= \lim_{x \to 0} \frac{4 \sec^2 x \tan^2 x + 2 \sec^4 x}{6} = \frac{2}{6} = \frac{1}{3}$$ ■

EXAMPLE 5 Find $\displaystyle\lim_{x \to \pi^-} \frac{\sin x}{1 - \cos x}$.

SOLUTION If we blindly attempted to use l'Hospital's Rule, we would get

$$\lim_{x \to \pi^-} \frac{\sin x}{1 - \cos x} = \lim_{x \to \pi^-} \frac{\cos x}{\sin x} = -\infty$$

This is *wrong!* Although the numerator $\sin x \to 0$ as $x \to \pi^-$, notice that the denominator $(1 - \cos x)$ does not approach 0, so l'Hospital's Rule cannot be applied here.

The required limit is, in fact, easy to find because the function is continuous and the denominator is nonzero at π:

$$\lim_{x \to \pi^-} \frac{\sin x}{1 - \cos x} = \frac{\sin \pi}{1 - \cos \pi} = \frac{0}{1 - (-1)} = 0$$ ■

Example 5 shows what can go wrong if you use l'Hospital's Rule without thinking. Other limits *can* be found using l'Hospital's Rule but are more easily found by other methods. (See Examples 4, 7, and 8 in Section 2.3, Examples 2 and 3 in Section 2.6, Examples 1 and 2 in Section 3.4, and the discussion at the beginning of this section.) So when evaluating any limit, you should consider other methods before using l'Hospital's Rule.

INDETERMINATE PRODUCTS

If $\lim_{x \to a} f(x) = 0$ and $\lim_{x \to a} g(x) = \infty$ (or $-\infty$), then it is not clear what the value of $\lim_{x \to a} f(x)g(x)$, if any, will be. There is a struggle between f and g. If f wins, the answer will be 0; if g wins, the answer will be ∞ (or $-\infty$). Or there may be a compromise where the answer is a finite nonzero number. This kind of limit is called an **indeterminate form of type $0 \cdot \infty$**. We can deal with it by writing the product fg as a quotient:

$$fg = \frac{f}{1/g} \qquad \text{or} \qquad fg = \frac{g}{1/f}$$

This converts the given limit into an indeterminate form of type $\frac{0}{0}$ or ∞/∞ so that we can use l'Hospital's Rule.

EXAMPLE 6 Evaluate $\displaystyle\lim_{x \to 0^+} x \ln x$.

SOLUTION The given limit is indeterminate because $x \to 0^+$ while $\ln x \to -\infty$. Writing $x = 1/(1/x)$, we have $1/x \to \infty$ as $x \to 0^+$, so l'Hospital's Rule gives

$$\lim_{x \to 0^+} x \ln x = \lim_{x \to 0^+} \frac{\ln x}{\dfrac{1}{x}} = \lim_{x \to 0^+} \frac{\dfrac{1}{x}}{-\dfrac{1}{x^2}}$$

$$= \lim_{x \to 0^+} (-x) = 0$$

■

INDETERMINATE DIFFERENCES

If $\lim_{x \to a} f(x) = \infty$ and $\lim_{x \to a} g(x) = \infty$, then the limit

$$\lim_{x \to a} [f(x) - g(x)]$$

is called an **indeterminate form of type $\infty - \infty$**. Again there is a contest between f and g. Will the answer be ∞ (f wins) or will it be $-\infty$ (g wins) or will they compromise on a finite number? To find out, we try to convert the difference into a quotient (for instance, by using a common denominator or rationalization, or factoring out a common factor) so that we have an indeterminate form of type $\frac{0}{0}$ or ∞/∞.

EXAMPLE 7 Compute $\displaystyle\lim_{x \to (\pi/2)^-} (\sec x - \tan x)$.

SOLUTION First notice that $\sec x \to \infty$ and $\tan x \to \infty$ as $x \to (\pi/2)^-$, so the limit is indeterminate. Here we use a common denominator:

$$\lim_{x \to (\pi/2)^-} (\sec x - \tan x) = \lim_{x \to (\pi/2)^-} \left(\frac{1}{\cos x} - \frac{\sin x}{\cos x} \right)$$

$$= \lim_{x \to (\pi/2)^-} \frac{1 - \sin x}{\cos x} = \lim_{x \to (\pi/2)^-} \frac{-\cos x}{-\sin x} = 0$$

Note that the use of l'Hospital's Rule is justified because $1 - \sin x \to 0$ and $\cos x \to 0$ as $x \to (\pi/2)^-$. ■

INDETERMINATE POWERS

Several indeterminate forms arise from the limit

$$\lim_{x \to a} [f(x)]^{g(x)}$$

1. $\displaystyle\lim_{x \to a} f(x) = 0$ and $\displaystyle\lim_{x \to a} g(x) = 0$ type 0^0
2. $\displaystyle\lim_{x \to a} f(x) = \infty$ and $\displaystyle\lim_{x \to a} g(x) = 0$ type ∞^0
3. $\displaystyle\lim_{x \to a} f(x) = 1$ and $\displaystyle\lim_{x \to a} g(x) = \pm\infty$ type 1^∞

Each of these three cases can be treated either by taking the natural logarithm:

$$\text{let} \quad y = [f(x)]^{g(x)}, \quad \text{then} \quad \ln y = g(x) \ln f(x)$$

or by writing the function as an exponential:

$$[f(x)]^{g(x)} = e^{g(x) \ln f(x)}$$

(Recall that both of these methods were used in differentiating such functions.) In either method we are led to the indeterminate product $g(x) \ln f(x)$, which is of type $0 \cdot \infty$.

EXAMPLE 8 Calculate $\lim_{x\to 0^+} (1 + \sin 4x)^{\cot x}$.

SOLUTION First notice that as $x \to 0^+$, we have $1 + \sin 4x \to 1$ and $\cot x \to \infty$, so the given limit is indeterminate. Let

$$y = (1 + \sin 4x)^{\cot x}$$

Then $$\ln y = \ln[(1 + \sin 4x)^{\cot x}] = \cot x \ln(1 + \sin 4x)$$

so l'Hospital's Rule gives

$$\lim_{x\to 0^+} \ln y = \lim_{x\to 0^+} \frac{\ln(1 + \sin 4x)}{\tan x}$$

$$= \lim_{x\to 0^+} \frac{\dfrac{4\cos 4x}{1 + \sin 4x}}{\sec^2 x} = 4$$

So far we have computed the limit of $\ln y$, but what we want is the limit of y. To find this we use the fact that $y = e^{\ln y}$:

$$\lim_{x\to 0^+} (1 + \sin 4x)^{\cot x} = \lim_{x\to 0^+} y$$

$$= \lim_{x\to 0^+} e^{\ln y}$$

$$= e^4$$

■

EXAMPLE 9 Find $\lim_{x\to 0^+} x^x$.

SOLUTION Notice that this limit is indeterminate since $0^x = 0$ for any $x > 0$ but $x^0 = 1$ for any $x \neq 0$. We could proceed as in Example 8 or by writing the function as an exponential:

$$x^x = (e^{\ln x})^x = e^{x\ln x}$$

Now l'Hospital's Rule gives

$$\lim_{x\to 0^+} x\ln x = \lim_{x\to 0^+} \frac{\ln x}{\dfrac{1}{x}}$$

$$= \lim_{x\to 0^+} \frac{\dfrac{1}{x}}{\dfrac{-1}{x^2}}$$

$$= \lim_{x\to 0^+} (-x) = 0$$

Therefore $$\lim_{x\to 0^+} x^x = \lim_{x\to 0^+} e^{x\ln x} = e^0 = 1$$

■

EXERCISES 4.8

1–68 ■ Find the limit.

1. $\lim_{x\to 2} \frac{x-2}{x^2-4}$

2. $\lim_{x\to 1} \frac{x^2+3x-4}{x-1}$

3. $\lim_{x\to -1} \frac{x^6-1}{x^4-1}$

4. $\lim_{x\to 1} \frac{x^a-1}{x^b-1}$

5. $\lim_{x\to 0} \frac{e^x-1}{\sin x}$

6. $\lim_{x\to 1} \frac{\ln x}{x-1}$

7. $\lim_{x\to 0} \frac{\sin x}{x^3}$

8. $\lim_{x\to \pi} \frac{\tan x}{x}$

9. $\lim_{x\to 0} \frac{\tan x}{x+\sin x}$

10. $\lim_{x\to 3\pi/2} \frac{\cos x}{x-(3\pi/2)}$

11. $\lim_{x\to \infty} \frac{\ln x}{x}$

12. $\lim_{x\to 0^+} \frac{\ln x}{\sqrt{x}}$

13. $\lim_{x\to \infty} \frac{e^x}{x^3}$

14. $\lim_{x\to \infty} \frac{(\ln x)^3}{x^2}$

15. $\lim_{x\to a} \frac{\sqrt[3]{x}-\sqrt[3]{a}}{x-a}$, $a \neq 0$

16. $\lim_{x\to 0} \frac{6^x-2^x}{x}$

17. $\lim_{x\to 0} \frac{e^x-1-x}{x^2}$

18. $\lim_{x\to 0} \frac{e^x-1-x-(x^2/2)}{x^3}$

19. $\lim_{x\to 0} \frac{\sin x}{e^x}$

20. $\lim_{x\to 0} \frac{\sin^2 x}{\tan(x^2)}$

21. $\lim_{x\to 0} \frac{1-\cos x}{x^2}$

22. $\lim_{x\to 0} \frac{\sin x - x}{x^3}$

23. $\lim_{x\to 2^-} \frac{\ln x}{\sqrt{2-x}}$

24. $\lim_{x\to 0} \frac{\sin x}{\sinh x}$

25. $\lim_{x\to \infty} \frac{\ln \ln x}{\sqrt{x}}$

26. $\lim_{x\to \infty} \frac{\ln(1+e^x)}{5x}$

27. $\lim_{x\to 0} \frac{\tan^{-1}(2x)}{3x}$

28. $\lim_{x\to 0} \frac{x}{\sin^{-1}(3x)}$

29. $\lim_{x\to 0} \frac{\tan \alpha x}{x}$

30. $\lim_{x\to 0} \frac{\sin mx}{\sin nx}$

31. $\lim_{x\to 0} \frac{\tan 2x}{\tanh 3x}$

32. $\lim_{x\to 0} \frac{\sin^{10} x}{\sin(x^{10})}$

33. $\lim_{x\to 0} \frac{x+\sin 3x}{x-\sin 3x}$

34. $\lim_{x\to 0} \frac{2x-\sin^{-1}x}{2x+\cos^{-1}x}$

35. $\lim_{x\to 0} \frac{e^{4x}-1}{\cos x}$

36. $\lim_{x\to 0} \frac{2x-\sin^{-1}x}{2x+\tan^{-1}x}$

37. $\lim_{x\to 0} \frac{\tan x-\sin x}{x^3}$

38. $\lim_{x\to 0} \frac{\cos mx-\cos nx}{x^2}$

39. $\lim_{x\to 0^+} \sqrt{x}\ln x$

40. $\lim_{x\to -\infty} xe^x$

41. $\lim_{x\to \infty} e^{-x}\ln x$

42. $\lim_{x\to(\pi/2)^-} \sec 7x \cos 3x$

43. $\lim_{x\to \infty} x^3 e^{-x^2}$

44. $\lim_{x\to 0^+} \sqrt{x}\sec x$

45. $\lim_{x\to \pi} (x-\pi)\cot x$

46. $\lim_{x\to 1^+} (x-1)\tan(\pi x/2)$

47. $\lim_{x\to 0} \left(\frac{1}{x^4}-\frac{1}{x^2}\right)$

48. $\lim_{x\to 0} (\csc x - \cot x)$

49. $\lim_{x\to 0} \left(\frac{1}{x}-\csc x\right)$

50. $\lim_{x\to 1} \left(\frac{1}{\ln x}-\frac{1}{x-1}\right)$

51. $\lim_{x\to \infty} \left(x-\sqrt{x^2-1}\right)$

52. $\lim_{x\to \infty} \left(\sqrt{x^2+x+1}-\sqrt{x^2-x}\right)$

53. $\lim_{x\to \infty} \left(\frac{x^3}{x^2-1}-\frac{x^3}{x^2+1}\right)$

54. $\lim_{x\to \infty} (xe^{1/x}-x)$

55. $\lim_{x\to 0^+} x^{\sin x}$

56. $\lim_{x\to 0^+} (\sin x)^{\tan x}$

57. $\lim_{x\to 0} (1-2x)^{1/x}$

58. $\lim_{x\to \infty} \left(1+\frac{a}{x}\right)^{bx}$

59. $\lim_{x\to \infty} \left(1+\frac{3}{x}+\frac{5}{x^2}\right)^x$

60. $\lim_{x\to \infty} \left(1+\frac{1}{x^2}\right)^x$

61. $\lim_{x\to \infty} x^{1/x}$

62. $\lim_{x\to \infty} (e^x+x)^{1/x}$

63. $\lim_{x\to 0^+} (\cot x)^{\sin x}$

64. $\lim_{x\to \infty} \left(1+\frac{1}{x}\right)^{x^2}$

65. $\lim_{x\to \infty} \left(\frac{x}{x+1}\right)^x$

66. $\lim_{x\to 0} (\cos 3x)^{5/x}$

67. $\lim_{x\to 0^+} (-\ln x)^x$

68. $\lim_{x\to \infty} \left(\frac{2x-3}{2x+5}\right)^{2x+1}$

69–70 ■ Use a graph to estimate the value of the limit. Then use l'Hospital's Rule to find the exact value.

69. $\lim_{x\to \infty} x\,[\ln(x+5)-\ln x]$

70. $\lim_{x\to \pi/4} (\tan x)^{\tan 2x}$

71–72 ■ Illustrate l'Hospital's Rule by graphing both $f(x)/g(x)$ and $f'(x)/g'(x)$ near $x = 0$ to see that these ratios have the same limit as $x \to 0$. Also, calculate the exact value of the limit.

71. $f(x) = e^x - 1$, $g(x) = x^3 + 4x$

72. $f(x) = 2x\sin x$, $g(x) = \sec x - 1$

73. The first appearance in print of l'Hospital's Rule was in the book *Analyse des infiniment petits* published by the Marquis de l'Hospital in 1696. This was the first calculus *textbook* ever published and the example that the Marquis

used in that book to illustrate his rule was to find the limit of the function

$$y = \frac{\sqrt{2a^3x - x^4} - a\sqrt[3]{aax}}{a - \sqrt[4]{ax^3}}$$

as x approaches a. (At that time it was common to write aa instead of a^2.) Solve this problem.

74. The figure shows a sector of a circle with central angle θ. Let $A(\theta)$ be the area of the segment between the chord PR and the arc PR. Let $B(\theta)$ be the area of the triangle PQR. Find $\lim_{\theta \to 0^+} A(\theta)/B(\theta)$.

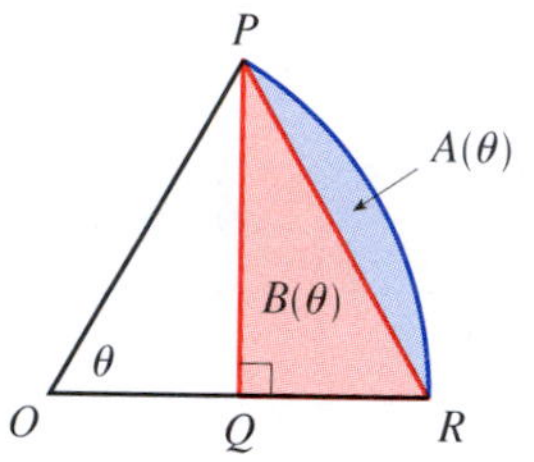

75. If f' is continuous, use l'Hospital's Rule to show that

$$\lim_{h \to 0} \frac{f(x + h) - f(x - h)}{2h} = f'(x)$$

(A proof without the use of l'Hospital's Rule is outlined in Problems Plus 20 on page 240.)

76. If f'' is continuous, show that

$$\lim_{h \to 0} \frac{f(x + h) - 2f(x) + f(x - h)}{h^2} = f''(x)$$

77. Prove that

$$\lim_{x \to \infty} \frac{e^x}{x^n} = \infty$$

for any integer n. This shows that the exponential function approaches infinity faster than any power of x.

78. Prove that

$$\lim_{x \to \infty} \frac{\ln x}{x^p} = 0$$

for any number $p > 0$. This shows that the logarithmic function approaches ∞ more slowly than any power of x.

79. Let

$$f(x) = \begin{cases} e^{-1/x^2} & \text{if } x \neq 0 \\ 0 & \text{if } x = 0 \end{cases}$$

(a) Use the definition of derivative to compute $f'(0)$.

(b) Show that f has derivatives of all orders that are defined on $\mathbb{R}$. [*Hint:* First show by induction that there is a polynomial $p_n(x)$ and a nonnegative integer k_n such that $f^{(n)}(x) = p_n(x)f(x)/x^{k_n}$ for $x \neq 0$.]

80. Let

$$f(x) = \begin{cases} |x|^x & \text{if } x \neq 0 \\ 1 & \text{if } x = 0 \end{cases}$$

(a) Show that f is continuous at 0.

(b) Investigate graphically whether f is differentiable at 0 by zooming in several times toward the point (0, 1) on the graph of f.

(c) Show that f is not differentiable at 0. How can you reconcile this fact with the appearance of the graphs in part (b)?

4 REVIEW

KEY TOPICS ■ Define, state, or discuss the following.

1. Exponential functions
2. Graphs of exponential functions
3. Properties of exponential functions
4. Limits of exponential functions
5. One-to-one function
6. Horizontal Line Test
7. Inverse function
8. Cancellation equations
9. Procedure for finding an inverse function
10. Graph of an inverse function
11. Continuity and differentiability of an inverse function
12. Formula for the derivative of an inverse function
13. Logarithmic function with base a
14. Graphs of logarithmic functions
15. Properties of logarithms
16. Limits of logarithmic functions
17. The number e
18. Natural logarithm

19. Derivatives of logarithmic and exponential functions
20. Logarithmic differentiation
21. Law of natural growth or decay
22. Inverse trigonometric functions
23. Derivatives of inverse trigonometric functions
24. Hyperbolic functions
25. Hyperbolic identities
26. Derivatives of hyperbolic functions
27. Inverse hyperbolic functions
28. L'Hospital's Rule
29. Indeterminate quotients, products, differences, and powers

EXERCISES

1–16 ■ Determine whether each statement is true or false.

1. The function $f(x) = \cos x$, $-\pi/2 \leqslant x \leqslant \pi/2$, is one-to-one.
2. $\tan^{-1}(-1) = 3\pi/4$
3. If $0 < a < b$, then $\ln a < \ln b$.
4. $\pi^{\sqrt{5}} = e^{\sqrt{5}\ln \pi}$
5. You can always divide by e^x.
6. If $a > 0$ and $b > 0$, then $\ln(a + b) = \ln a + \ln b$.
7. If $x > 0$, then $(\ln x)^6 = 6 \ln x$.
8. $\dfrac{d}{dx} 10^x = x10^{x-1}$
9. $\dfrac{d}{dx} \ln 10 = \dfrac{1}{10}$
10. The inverse function of $y = e^{3x}$ is $y = \frac{1}{3} \ln x$.
11. $\cos^{-1}x = \dfrac{1}{\cos x}$
12. $\tan^{-1}x = \dfrac{\sin^{-1}x}{\cos^{-1}x}$
13. $\cosh x \geqslant 1$ for all x
14. $\dfrac{d}{dx} \ln|x| = \dfrac{1}{|x|}$
15. $\dfrac{d}{dx} \log_8 x = \dfrac{1}{(3\ln 2)x}$
16. $\displaystyle\lim_{x\to\pi^-} \frac{\tan x}{1 - \cos x} = -\infty$

17–25 ■ Sketch a rough graph of the function without using a calculator.

17. $y = e^x$
18. $y = e^{-x}$
19. $y = -e^{-x}$
20. $y = 1 + 7^x$
21. $y = \ln x$
22. $y = \ln(x - 1)$
23. $y = 2 - \ln x$
24. $y = e^x \cos x$
25. $y = \tan^{-1}x$

26. Let $a > 1$. For large values of x, which of the functions $y = x^a$, $y = a^x$, and $y = \log_a x$ has the largest values and which has the smallest values?

27–34 ■ Solve each equation for x.

27. $e^x = 5$
28. $\ln x = 2$
29. $\log_{10}(e^x) = 1$
30. $e^{e^x} = 2$
31. $\ln(x^\pi) = 2$
32. $\ln(x + 1) - \ln x = 1$
33. $\tan x = 4$
34. $\sin^{-1}x = 1$

35–61 ■ Calculate y'.

35. $y = \log_{10}(x^2 - x)$
36. $y = \sqrt{2}^{\,x}$
37. $y = \dfrac{\sqrt{x + 1}\,(2 - x)^5}{(x + 3)^7}$
38. $y = \ln(\csc 5x)$
39. $y = e^{cx}(c \sin x - \cos x)$
40. $y = \sin^{-1}(e^x)$
41. $y = \ln(\sec^2 x)$
42. $y = \ln(x^2 e^x)$
43. $y = xe^{-1/x}$
44. $y = \ln|\csc 3x + \cot 3x|$
45. $y = (\cos^{-1}x)^{\sin^{-1}x}$
46. $y = x^r e^{sx}$
47. $y = e^{e^x}$
48. $y = 5^{x \tan x}$
49. $y = \ln\left(\dfrac{1}{x}\right) + \dfrac{1}{\ln x}$
50. $xe^y = y - 1$
51. $y = 7^{\sqrt{2x}}$
52. $y = e^{\cos x} + \cos(e^x)$
53. $y = \ln(\cosh 3x)$
54. $y = \ln\left|\dfrac{x^2 - 4}{2x + 5}\right|$
55. $y = \cosh^{-1}(\sinh x)$
56. $y = x \tanh^{-1}\sqrt{x}$
57. $y = \ln \sin x - \frac{1}{2}\sin^2 x$
58. $y = (c/x)^x$
59. $y = \sin^{-1}\left(\dfrac{x - 1}{x + 1}\right)$
60. $y = \arctan(\arcsin \sqrt{x}\,)$
61. $y = \ln \sqrt[4]{\dfrac{x^2 + x + 1}{x^2 - x + 1}} + \dfrac{1}{2\sqrt{3}}\left[\tan^{-1}\left(\dfrac{2x + 1}{\sqrt{3}}\right) + \tan^{-1}\left(\dfrac{2x - 1}{\sqrt{3}}\right)\right]$

62. If

$$y = \frac{x}{\sqrt{a^2 - 1}} - \frac{2}{\sqrt{a^2 - 1}} \arctan \frac{\sin x}{a + \sqrt{a^2 - 1} + \cos x}$$

show that $y' = \dfrac{1}{a + \cos x}$.

63–64 ■ Find $f^{(n)}(x)$.

63. $f(x) = 2^x$

64. $f(x) = \ln(2x)$

65. Use mathematical induction to show that if $f(x) = xe^x$, then $f^{(n)}(x) = (x + n)e^x$.

66. Find y' if $y = x + \arctan y$.

67–68 ■ Find an equation of the tangent to the curve at the given point.

67. $y = \ln(e^x + e^{2x})$, $(0, \ln 2)$

68. $y = x \ln x$, (e, e)

69. Show that $f(x) = \sqrt{x} - \sqrt{x - 1}$ is one-to-one. Then graph f, f^{-1}, and $y = x$ on the same screen using parametric graphs.

70. If $f(x) = xe^{\sin x}$, find $f'(x)$. Graph f and f' on the same screen and comment.

71. At what point on the curve $y = [\ln(x + 4)]^2$ is the tangent horizontal?

72. Find an equation of the tangent to the curve $y = e^x$ that is parallel to the line $x - 4y = 1$.

73. Find an equation of the tangent to the curve $y = e^x$ that passes through the origin.

74. The function $C(t) = K(e^{-at} - e^{-bt})$, where a, b, and K are positive constants and $b > a$, is used to model the concentration at time t of a drug injected into the bloodstream.
(a) Show that $\lim_{t\to\infty} C(t) = 0$.
(b) Find $C'(t)$, the rate at which the drug is cleared from circulation.
(c) When is this rate equal to 0?

75–96 ■ Evaluate each limit.

75. $\lim_{x\to-\infty} 10^{-x}$

76. $\lim_{x\to\infty} \dfrac{4^x}{4^x - 1}$

77. $\lim_{x\to0^+} \ln(\tan x)$

78. $\lim_{x\to-\infty} e^{x/2} \cos x$

79. $\lim_{x\to-4^+} e^{1/(x+4)}$

80. $\lim_{x\to-1^+} e^{\tanh^{-1}x}$

81. $\lim_{x\to\infty} \dfrac{e^x}{e^{2x} + e^{-x}}$

82. $\lim_{x\to0} (1 + x)^{2/x}$

83. $\lim_{x\to1} \cos^{-1}\left(\dfrac{x}{x + 1}\right)$

84. $\lim_{x\to-\infty} \tan^{-1}(x^4)$

85. $\lim_{x\to\pi} \dfrac{\sin x}{x^2 - \pi^2}$

86. $\lim_{x\to0} \dfrac{e^{ax} - e^{bx}}{x}$

87. $\lim_{x\to\infty} \dfrac{\ln(\ln x)}{\ln x}$

88. $\lim_{x\to0} \dfrac{1 + \sin x - \cos x}{1 - \sin x - \cos x}$

89. $\lim_{x\to0} \dfrac{\ln(1 - x) + x + \dfrac{x^2}{2}}{x^3}$

90. $\lim_{x\to\pi/2} \left(\dfrac{\pi}{2} - x\right) \tan x$

91. $\lim_{x\to0^+} (\sin x)(\ln x)^2$

92. $\lim_{x\to0} (\csc^2 x - x^{-2})$

93. $\lim_{x\to1} (\ln x)^{\sin x}$

94. $\lim_{x\to1} x^{1/(1-x)}$

95. $\lim_{x\to0^+} \dfrac{\sqrt[3]{x} - 1}{\sqrt[4]{x} - 1}$

96. $\lim_{x\to\infty} \tan^{-1}\left(\dfrac{\sqrt{x}}{\ln x}\right)$

97. A bacteria culture starts with 1000 bacteria and the growth rate is proportional to the number of bacteria. After 2 h the population is 9000.
(a) Find an expression for the number of bacteria after t hours.
(b) Find the number of bacteria after 3 h.
(c) In what period of time does the number of bacteria double?

98. An isotope of strontium, Sr^{90}, has a half-life of 25 years.
(a) Find the mass of Sr^{90} that remains from a sample of 18 mg after t years.
(b) How long would it take for the mass to decay to 2 mg?

99. An amount of \$10,000 is invested at an interest rate of 6% per year. Find the value of the investment at the end of 4 years if the interest is compounded (a) annually, (b) semiannually, (c) quarterly, (d) monthly, (e) daily, and (f) continuously.

100. A cup of coffee has a temperature of 200 °F and is in a room that has a temperature of 70 °F. After 10 min the temperature of the coffee is 150 °F.
(a) What is the temperature of the coffee after 15 min?
(b) When will the coffee have cooled to 100 °F?

101. Let $C(t)$ be the concentration of a drug in the bloodstream. As the body eliminates the drug, $C(t)$ decreases at a rate that is proportional to the amount of the drug that is present at the time. Thus $C'(t) = -kC(t)$, where k is a positive number called the *elimination constant* of the drug.
(a) If C_0 is the concentration at time $t = 0$, find the concentration at time t.
(b) If the body eliminates half the drug in 30 h, how long does it take to eliminate 90% of the drug?

102. (a) Show that there is exactly one root of the equation $\ln x = 3 - x$ and that it lies between 2 and e.
(b) Find the root of the equation in part (a) correct to four decimal places.

103. An equation of motion of the form $s = Ae^{-ct}\cos(\omega t + \delta)$ represents damped oscillation of an object. Find the velocity and acceleration of the object.

104–109 ■ Find f' in terms of g'.

104. $f(x) = g(e^x)$

105. $f(x) = e^{g(x)}$

106. $f(x) = g(\ln x)$

107. $f(x) = \ln|g(x)|$

108. $f(x) = xe^{g(\sqrt{x})}$

109. $f(x) = \ln g(e^x)$

110. If $f(x) = x + x^2 + e^x$ and $g(x) = f^{-1}(x)$, find $g'(1)$.

111. If g is the inverse function of $f(x) = \ln x + \tan^{-1}x$, find $g'(\pi/4)$.

112. Show that

$$\cos\{\arctan[\sin(\text{arccot}\, x)]\} = \sqrt{\frac{x^2 + 1}{x^2 + 2}}$$

113. Show that $\sin^{-1}(\tanh x) = \tan^{-1}(\sinh x)$.

APPLICATIONS PLUS

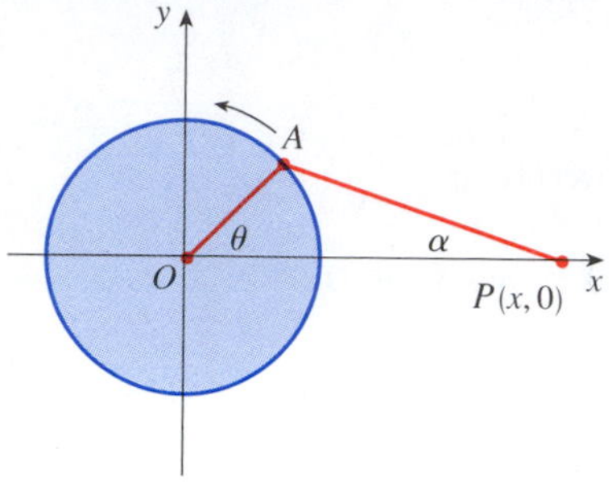

FIGURE FOR PROBLEM 1

1. The figure shows a rotating wheel with radius 40 cm and a connecting rod AP with length 1.2 m. The pin P slides back and forth along the x-axis as the wheel rotates counterclockwise at a rate of 360 revolutions per minute.
   (a) Find the angular velocity of the connecting rod, $d\alpha/dt$, in radians per second, when $\theta = \pi/3$.
   (b) Express the distance $x = |OP|$ in terms of θ.
   (c) Find an expression for the velocity of the pin P in terms of θ.

2. If we start from 0° longitude and proceed in the counterclockwise direction, we can let $T(x)$ denote the temperature at the point x at any given time. Assume that T is a continuous function of x.
   (a) Use the function T to show that at any fixed time there are at least two diametrically opposite points on the equator that have exactly the same temperature.
   (b) Does the result in (a) hold for points lying on any circle on the earth's surface?
   (c) Does the result in (a) hold for barometric pressure and for altitude above sea level?

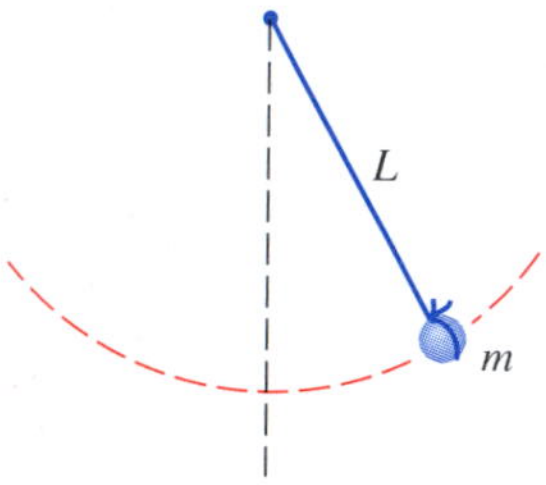

FIGURE FOR PROBLEM 3

3. A simple pendulum consists of a mass m swinging at the end of a (massless) rod or string of length L. The time T for one complete swing of the pendulum is known as the period and, if the pendulum swings through only a small angle, it is given by $T = 2\pi\sqrt{L/g}$, where g is the acceleration due to gravity.
   (a) Show that a small change dL in the length produces a change in the period dT satisfying

$$\frac{dT}{T} = \frac{dL}{2L}$$

   (b) Suppose that a pendulum clock loses 15 seconds per hour. How should the length of the pendulum be adjusted?
   (c) The preceding formula for the period can be used to measure the acceleration due to gravity. Assume that the error in measuring the length L is negligible, and express the error dg in the acceleration of gravity in terms of the error dT in measuring the period.

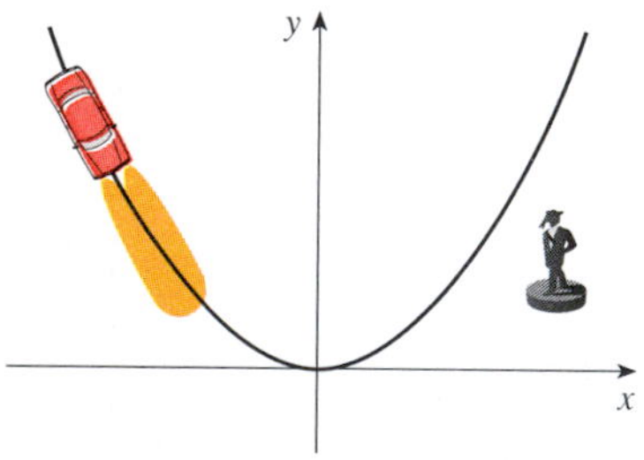

FIGURE FOR PROBLEM 4

4. A car is traveling at night along a highway shaped like a parabola with its vertex at the origin. The car starts at a point 100 m west and 100 m north of the origin and travels in an easterly direction. There is a statue located 100 m east and 50 m north of the origin. At what point on the highway will the car's headlights illuminate the statue?

5. Assume that a snowball melts so that its volume decreases at a rate proportional to its surface area. If it takes 3 h for the snowball to decrease to half its original volume, how much longer will it take for the snowball to melt completely?

6. A glucose solution is administered intravenously into the bloodstream at a constant rate r. As the glucose is added, it is converted into other substances and removed from the bloodstream at a rate that is proportional to the concentration at that time. The mathe-

matical model for the concentration $C = C(t)$ of the glucose solution in the bloodstream is

$$\frac{dC}{dt} = r - kC = k\left(\frac{r}{k} - C\right)$$

where k is a positive constant.

(a) Suppose that the concentration at time $t = 0$ is C_0. Determine the concentration at any time t by solving the differential equation. [*Hint:* Consider the function $u(t) = (r/k) - C(t)$.]

(b) Assuming that $C_0 < r/k$, find $\lim_{t\to\infty} C(t)$ and interpret this result.

7. The annual sales of a new company are expected to grow at a rate that is proportional to the difference between the sales at time t and an upper limit of \$25 million. The sales are \$0 initially and are projected to be \$3.5 million at the end of three years.

(a) Determine a differential equation that expresses the rate of growth of sales for the company.

(b) Determine the annual sales of the company at any time t by solving the differential equation in part (a).

(c) What will the annual sales be at the end of the eighth year?

(d) How long will it take for the annual sales to be \$12.5 million?

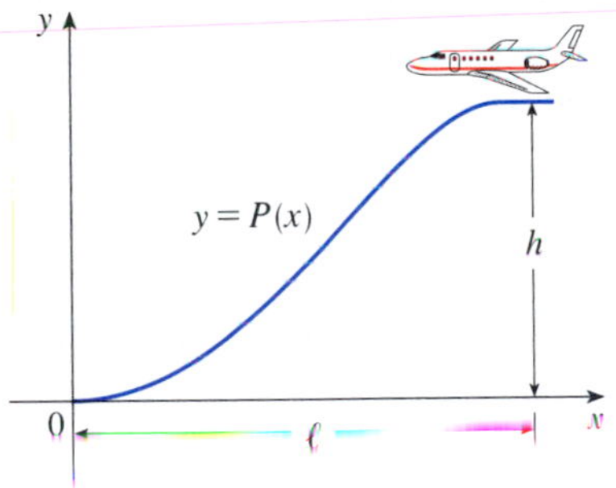

FIGURE FOR PROBLEM 8

8. An approach path for an aircraft landing is shown in the figure and satisfies the following conditions:

(i) The cruising altitude is h when descent starts at a horizontal distance ℓ from touchdown at the origin.

(ii) The pilot must maintain a constant horizontal speed v throughout descent.

(iii) The absolute value of the vertical acceleration should not exceed a constant k (which is much less than the acceleration due to gravity).

(a) Find a cubic polynomial $P(x) = ax^3 + bx^2 + cx + d$ that satisfies condition (i) by imposing suitable conditions on $P(x)$ and $P'(x)$ at the start of descent and at touchdown.

(b) Use conditions (ii) and (iii) to show that

$$\frac{6hv^2}{\ell^2} \leq k$$

(c) Suppose that an airline decides not to allow vertical acceleration of a plane to exceed $k = 860 \text{ mi/h}^2$. If the cruising altitude of a plane is 35,000 ft and the speed is 300 mi/h, how far away from the airport should the pilot start descent?

5

APPLICATIONS OF DIFFERENTIATION

The greatest mathematicians, as Archimedes, Newton, and Gauss, always united theory and applications in equal measure.

FELIX KLEIN

For since the fabric of the universe is most perfect and the work of a most wise Creator, nothing at all takes place in the universe in which some rule of maximum or minimum does not appear.

LEONHARD EULER

We have already investigated some of the applications of derivatives, but now that we know the differentiation rules we are in a better position to pursue the applications of differentiation in greater depth. In particular, we learn how to find maximum and minimum values of functions. This is a useful skill because many practical problems require us to minimize a cost or maximize an area or somehow find the best possible outcome of a situation.

5.1 WHAT DOES f' SAY ABOUT f?

Many of the applications of calculus depend on our ability to deduce facts about a function f from information concerning its derivative f'. Because $f'(x)$ represents the slope of the curve $y = f(x)$ at the point $(x, f(x))$, it tells us the direction in which the curve proceeds at each point. So it is reasonable to expect that information about $f'(x)$ will provide us with information about $f(x)$.

In particular, to see how the derivative of f can tell us where a function is increasing or decreasing, look at Figure 1. (Increasing functions and decreasing functions were defined in Section 1 of Review and Preview.) Between A and B and between C and D, the tangent lines have positive slope and so $f'(x) > 0$. Between B and C, the tangent lines have negative slope and so $f'(x) < 0$. Thus, it appears that f increases when $f'(x)$ is positive and decreases when $f'(x)$ is negative.

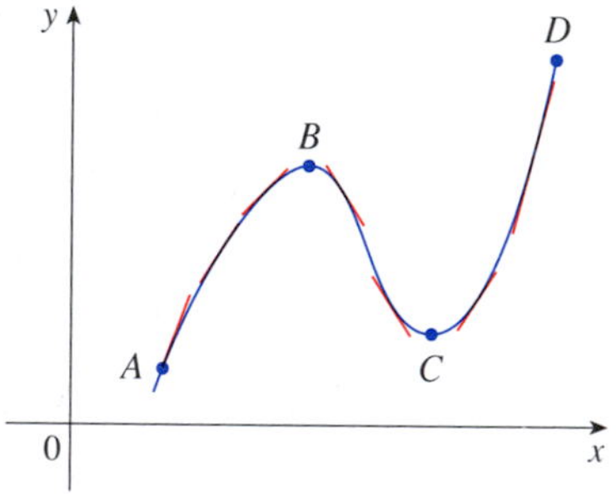

FIGURE 1

It turns out, as we will see in Section 5.3, that what we observed for the function graphed in Figure 1 is always true. We state the general result as follows.

> If $f'(x) > 0$ on an interval, then f is increasing on that interval.
>
> If $f'(x) < 0$ on an interval, then f is decreasing on that interval.

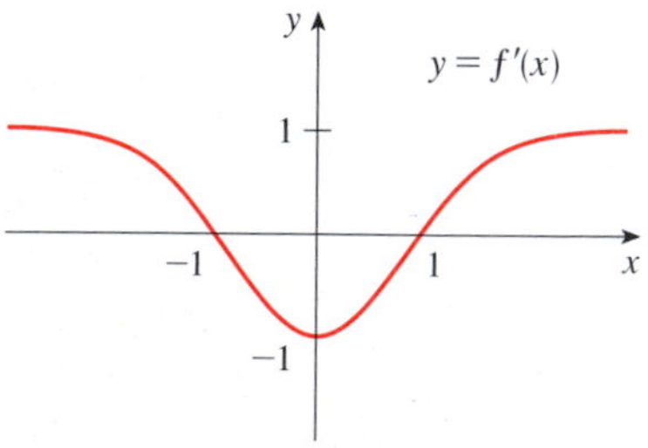

FIGURE 2

EXAMPLE 1

(a) If it is known that the graph of the derivative f' of a function is as shown in Figure 2, what can we say about f?

(b) If it is known that $f(0) = 0$, sketch a possible graph of f.

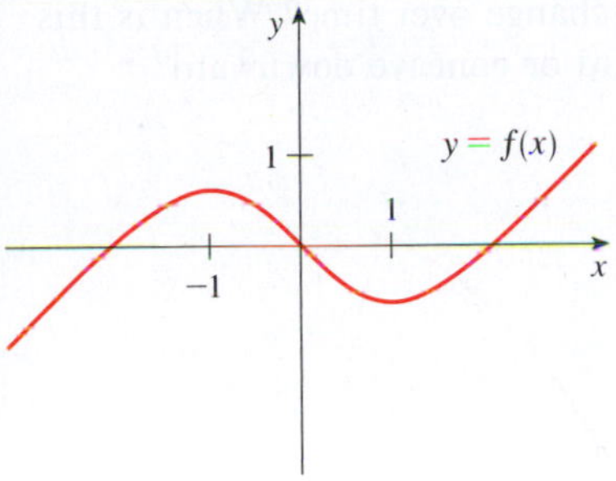

FIGURE 3

SOLUTION

(a) We observe from Figure 2 that $f'(x)$ is negative when $-1 < x < 1$, so the original function f must be decreasing on the interval $(-1, 1)$. Similarly, $f'(x)$ is positive for $x < -1$ and for $x > 1$, so f is increasing on the intervals $(-\infty, -1)$ and $(1, \infty)$. Also note that, since $f'(-1) = 0$ and $f'(1) = 0$, the graph of f has horizontal tangents when $x = \pm 1$.

(b) We use the information from part (a), and the fact that the graph passes through the origin, to sketch a possible graph of f in Figure 3. Notice that $f'(0) = -1$, so we have drawn the curve $y = f(x)$ passing through the origin with a slope of -1. Notice also that $f'(x) \to 1$ as $x \to \pm\infty$ (from Figure 2). So the slope of the curve $y = f(x)$ approaches 1 as x becomes large (positive or negative). That is why we have drawn the graph of f in Figure 3 progressively straighter as $x \to \pm\infty$. ■

We say that the function f in Example 1 has a **local maximum** at -1 because near $x = -1$ the values of $f(x)$ are at least as big as the neighboring values. Note that $f'(x)$ is positive to the left of -1 and negative just to the right of -1. Similarly, f has a **local minimum** at 1, where the derivative changes from negative to positive. In Section 5.5 we will develop these observations into a general method for finding optimal values of functions.

WHAT DOES f'' SAY ABOUT f?

Let's see how the sign of $f''(x)$ affects the appearance of the graph of f. Since $f'' = (f')'$, we know that if $f''(x)$ is positive, then f' is an increasing function. This says that the slopes of the tangent lines of the curve $y = f(x)$ increase from left to right. Figure 4 shows the graph of such a function. The slope of this curve becomes progressively larger as x increases and we observe that, as a consequence, the curve bends upward. Such a curve is called **concave upward**.

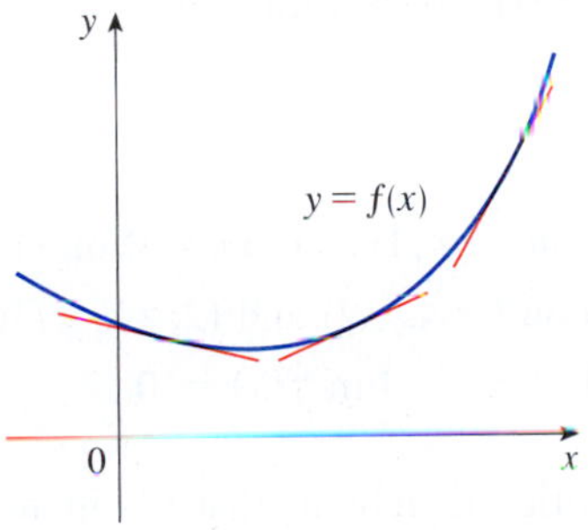

FIGURE 4
Since $f''(x) > 0$, the slopes increase and f is concave upward.

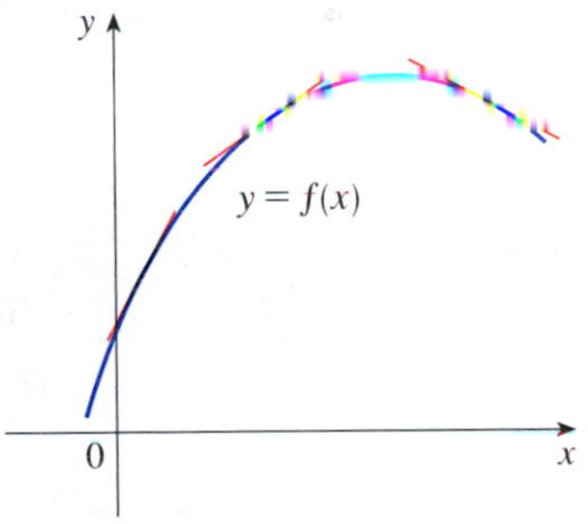

FIGURE 5
Since $f''(x) < 0$, the slopes decrease and f is concave downward.

In Figure 5, however, $f''(x)$ is negative, which means that f' is decreasing. Thus, the slopes of f decrease from left to right and the curve bends downward. This curve is called **concave downward**. We summarize our discussion as follows. (Concavity is discussed in greater detail in Section 5.3.)

> If $f''(x) > 0$ on an interval, then f is concave upward on that interval.
>
> If $f''(x) < 0$ on an interval, then f is concave downward on that interval.

EXAMPLE 2 Figure 6 shows a population graph for Cyprian honeybees raised in an apiary. How does the rate of population increase change over time? When is this rate highest? Over what intervals is P concave upward or concave downward?

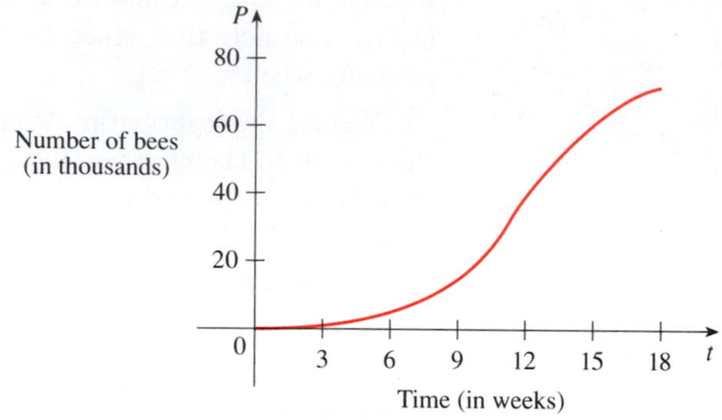

FIGURE 6

SOLUTION By looking at the slope of the curve as t increases, we see that the rate of increase of the population is initially very small, then gets larger until it reaches a maximum at about $t = 12$ weeks, and decreases as the population begins to level off. As the population approaches its maximum value of about 75,000 (called the *carrying capacity*), the rate of increase, $P'(t)$, approaches 0. The curve is concave upward on $(0, 12)$ and concave downward on $(12, 18)$. ■

In Example 2, the population curve changed from concave upward to concave downward at approximately the point $(12, 38{,}000)$. This point is called an *inflection point* of the curve. The significance of this point is that the rate of population increase has its maximum value there. In general, a point where a curve changes its direction of concavity is called an **inflection point**.

EXAMPLE 3 Sketch a possible graph of a function f that satisfies the following conditions:

(i) $f'(x) > 0$ on $(-\infty, 1)$, $f'(x) < 0$ on $(1, \infty)$

(ii) $f''(x) > 0$ on $(-\infty, -2)$ and $(2, \infty)$, $f''(x) < 0$ on $(-2, 2)$

(iii) $\lim_{x \to -\infty} f(x) = -2$, $\lim_{x \to \infty} f(x) = 0$

SOLUTION Condition (i) tells us that f is increasing on $(-\infty, 1)$ and decreasing on $(1, \infty)$. Condition (ii) says that f is concave upward on $(-\infty, -2)$ and $(2, \infty)$, and concave downward on $(-2, 2)$. From condition (iii) we know that the graph of f has two horizontal asymptotes: $y = -2$ and $y = 0$.

We first draw the horizontal asymptote $y = -2$ as a dashed line. (See Figure 7.) We then draw the graph of f approaching this asymptote at the far left, increasing to

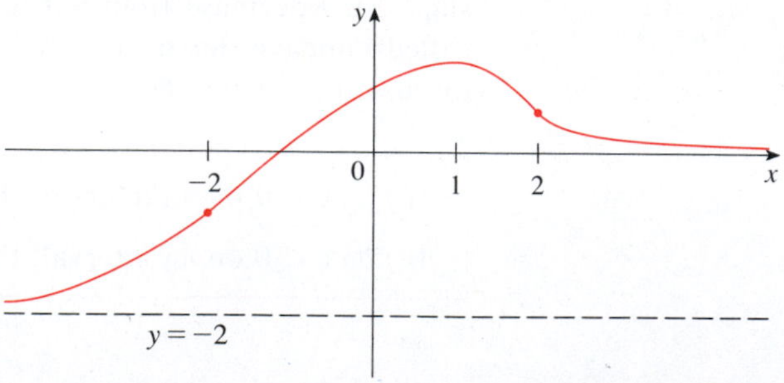

FIGURE 7

its maximum point at $x = 1$ and decreasing toward the x-axis as $x \to \infty$. We also make sure that the graph has inflection points when $x = -2$ and 2. Notice that the curve bends upward for $x < -2$ and $x > 2$, and bends downward when x is between -2 and 2. ■

EXERCISES 5.1

1–2 ■ The graph of the *derivative* f' of a function f is shown.

(a) On what intervals is f increasing or decreasing?

(b) At what values of x does f have a local maximum or minimum?

(c) If it is known that $f(0) = 0$, sketch a possible graph of f.

1.

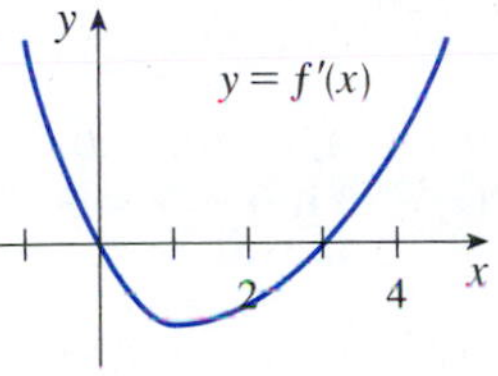

2.

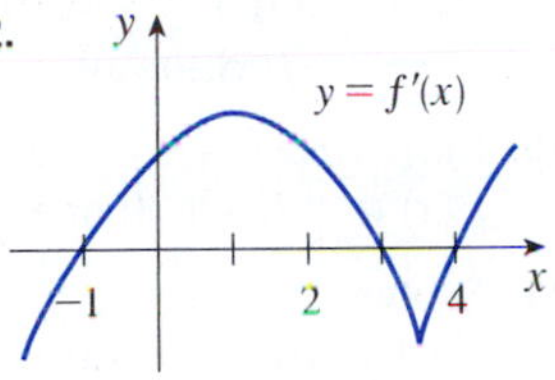

3. Use the given graph of f to estimate the intervals on which the derivative f' is increasing or decreasing.

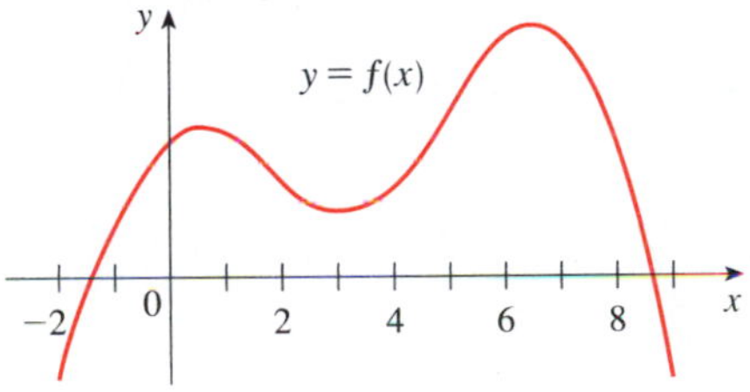

4. (a) Sketch a curve whose slope is always positive and increasing.

(b) Sketch a curve whose slope is always positive and decreasing.

(c) Give equations for curves with these properties.

5. The president announces that the national deficit is increasing, but at a decreasing rate. Interpret this statement in terms of a function and its derivatives.

6. A graph of a population of yeast cells in a new laboratory culture as a function of time is shown.

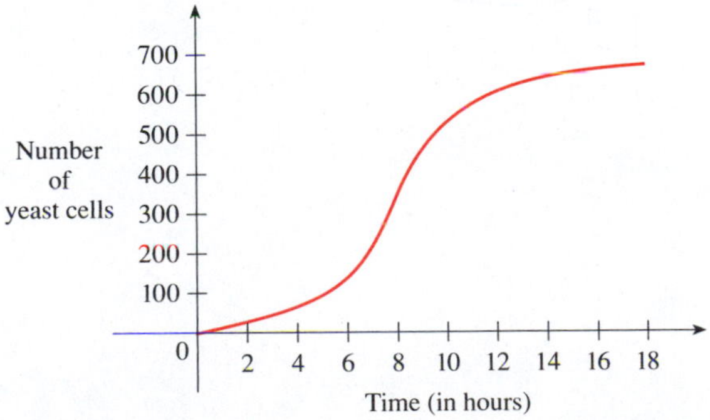

(a) Describe how the rate of population increase varies.

(b) When is this rate highest?

(c) On what intervals is the population function concave upward or downward?

(d) Estimate the coordinates of the inflection point.

7. The table gives population densities for ring-necked pheasants (in number of pheasants per acre) on Pelee Island, Ontario.

(a) Describe how the rate of change of population varies.

(b) Estimate the inflection points of the graph. What is the significance of these points?

t	1927	1930	1932	1934	1936	1938	1940
$P(t)$	0.1	0.6	2.5	4.6	4.8	3.5	3.0

8. A particle is moving along a horizontal straight line. The graph of its position function (the distance to the right of a fixed point as a function of time) is shown.

(a) When is the particle moving toward the right and when is it moving toward the left?

(b) When does the particle have positive acceleration and when does it have negative acceleration?

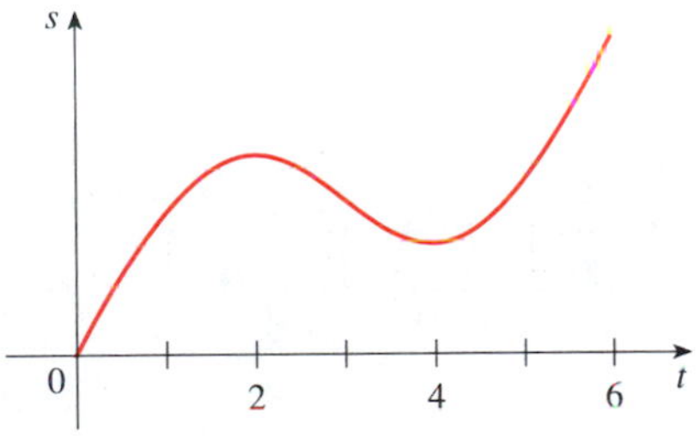

9. Let $K(t)$ be a measure of the knowledge you gain by studying for a test for t hours. Which do you think is larger, $K(8) - K(7)$ or $K(3) - K(2)$? Is the graph of K concave upward or concave downward? Why?

10. Coffee is being poured into the mug shown in the figure at a constant rate (measured in volume per unit time). Sketch

a rough graph of the depth of the coffee in the mug as a function of time. Account for the shape of the graph in terms of concavity. What is the significance of the inflection point?

11–12 ■ The graph of the derivative f' of a function f is shown.

(a) On what intervals is f increasing or decreasing?

(b) At what values of x does f have a local maximum or minimum?

(c) On what intervals is f concave upward or downward?

(d) State the x-coordinates of the points of inflection.

(e) Assuming that f is continuous and $f(0) = 0$, sketch a graph of f.

11.

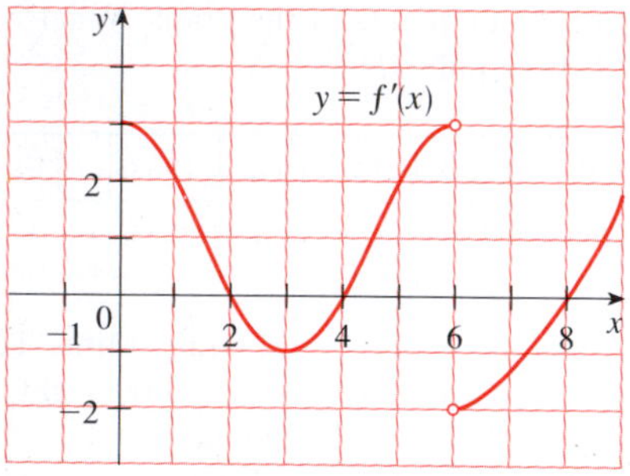

12.

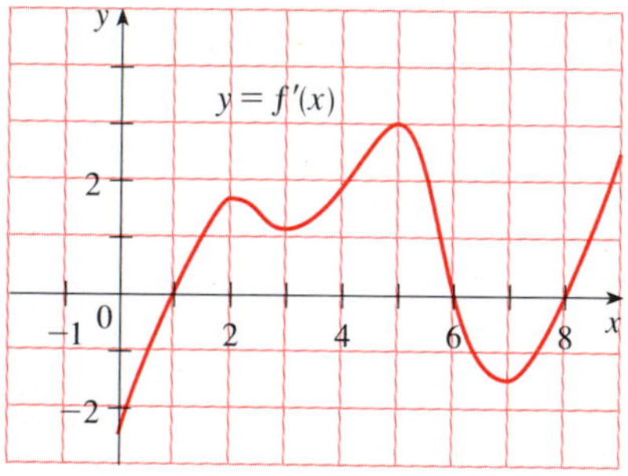

13. Sketch the graph of a function whose first and second derivatives are always negative.

14. Sketch the graph of a function whose first derivative is always negative and whose second derivative is always positive.

15–18 ■ Sketch the graph of a function that satisfies the given conditions.

15. $f'(-1) = f'(1) = 0$, $f'(x) < 0$ if $|x| < 1$, $f'(x) > 0$ if $|x| > 1$, $f(-1) = 4$, $f(1) = 0$, $f''(x) < 0$ if $x < 0$, $f''(x) > 0$ if $x > 0$

16. $f'(-1) = 0$, $f'(1)$ does not exist, $f'(x) < 0$ if $|x| < 1$, $f'(x) > 0$ if $|x| > 1$, $f(-1) = 4$, $f(1) = 0$, $f''(x) < 0$ if $x \neq 1$

17. $f'(2) = 0$, $f(2) = -1$, $f(0) = 0$, $f'(x) < 0$ if $0 < x < 2$, $f'(x) > 0$ if $x > 2$, $f''(x) < 0$ if $0 \leqslant x < 1$ or if $x > 4$, $f''(x) > 0$ if $1 < x < 4$, $\lim_{x\to\infty} f(x) = 1$, $f(-x) = f(x)$ for all x

18. $\lim_{x\to 3} f(x) = -\infty$, $f''(x) < 0$ if $x \neq 3$, $f'(0) = 0$, $f'(x) > 0$ if $x < 0$ or $x > 3$, $f'(x) < 0$ if $0 < x < 3$

19. Suppose $f'(x) = xe^{-x^2}$.

(a) On what interval is f increasing? On what interval is f decreasing?

(b) Does f have a maximum or minimum value?

20. If $f'(x) = e^{-x^2}$, what can you say about f?

21–22 ■ The graph of a function f is shown. Which graph has f as its derivative?

21.

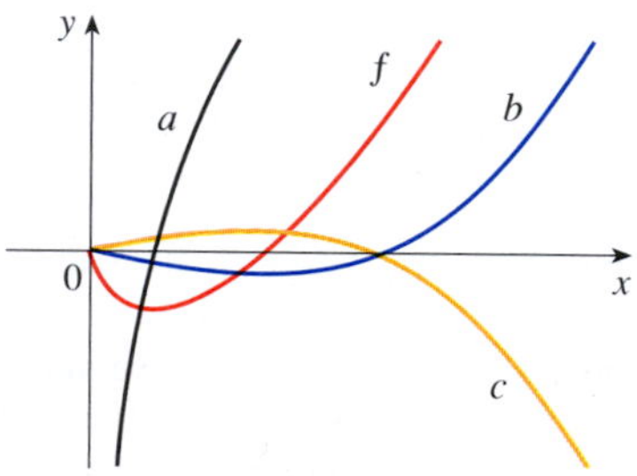

22.

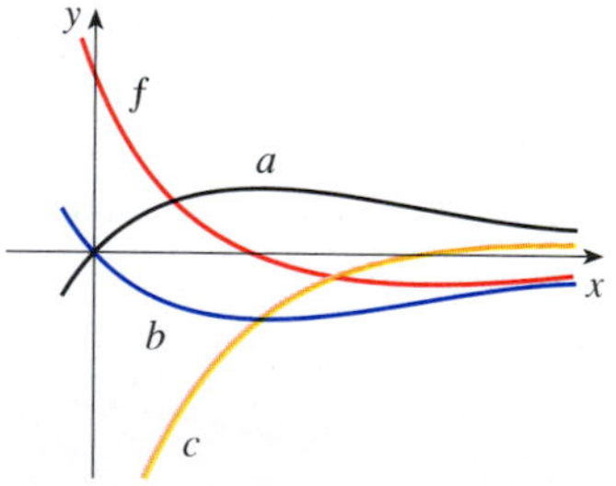

5.2 MAXIMUM AND MINIMUM VALUES

Some of the most important applications of differential calculus are *optimization problems*, in which we are required to find the optimal (best) way of doing something. Here are examples of such problems that we will solve in this chapter:

- What is the shape of a can that minimizes manufacturing costs?
- What is the maximum acceleration of a space shuttle? (This is an important question to the astronauts who have to withstand the effects of acceleration.)
- What is the radius of a contracted windpipe that expels air most rapidly during a cough?
- At what angle should blood vessels branch so as to minimize the energy expended by the heart in pumping blood?

These problems can be reduced to finding the maximum or minimum values of a function. Let's first explain exactly what we mean by maximum and minimum values.

(1) DEFINITION A function f has an **absolute maximum** (or **global maximum**) at c if $f(c) \geqslant f(x)$ for all x in D, where D is the domain of f. The number $f(c)$ is called the **maximum value** of f on D. Similarly, f has an **absolute minimum** at c if $f(c) \leqslant f(x)$ for all x in D and the number $f(c)$ is called the **minimum value** of f on D. The maximum and minimum values of f are called the **extreme values** of f.

Figure 1 shows the graph of a function f with absolute maximum at d and absolute minimum at a. Note that $(d, f(d))$ is the highest point on the graph and $(a, f(a))$ is the lowest point.

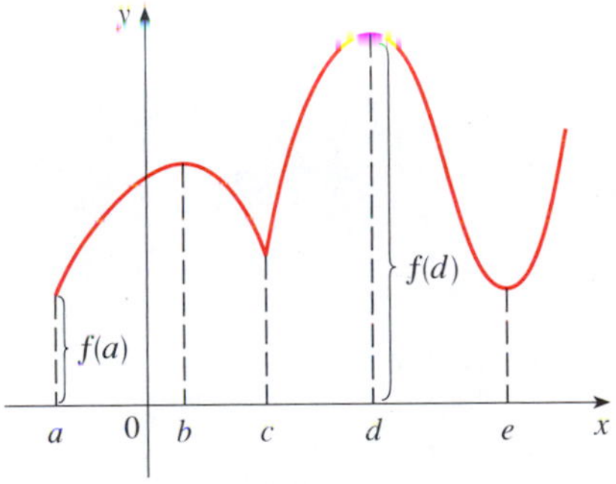

FIGURE 1
Minimum value $f(a)$, maximum value $f(d)$

In Figure 1, if we consider only values of x near b [for instance, if we restrict our attention to the interval (a, c)], then $f(b)$ is the largest of those values of $f(x)$ and is called a *local maximum value* of f. Likewise, $f(c)$ is called a *local minimum value* of f because $f(c) \leqslant f(x)$ for x near c [in the interval (b, d), for instance]. The function f also has a local minimum at e. In general, we have the following definition.

(2) DEFINITION A function f has a **local maximum** (or **relative maximum**) at c if $f(c) \geqslant f(x)$ when x is near c. [This means that $f(c) \geqslant f(x)$ for all x in some open interval containing c.] Similarly, f has a **local minimum** at c if $f(c) \leqslant f(x)$ when x is near c.

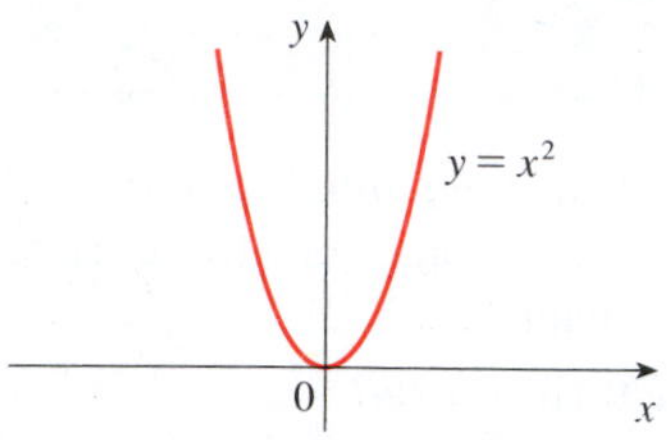

FIGURE 2
Minimum value 0, no maximum

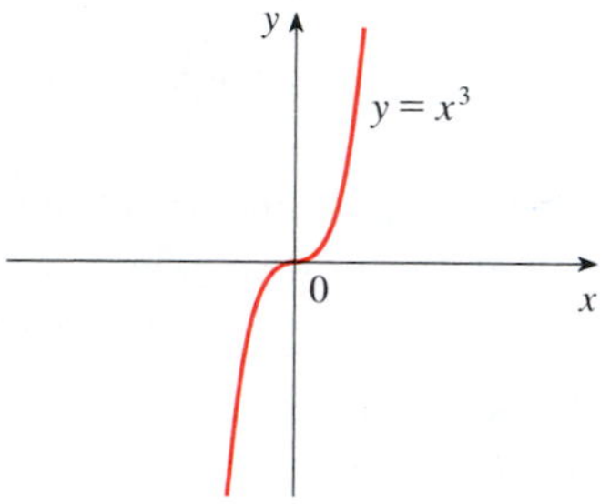

FIGURE 3
No minimum, no maximum

EXAMPLE 1 The function $f(x) = \cos x$ takes on its (local and absolute) maximum value of 1 infinitely many times, since $\cos 2n\pi = 1$ for any integer n and $-1 \leqslant \cos x \leqslant 1$ for all x. Likewise, $\cos(2n + 1)\pi = -1$ is its minimum value, where n is any integer. ■

EXAMPLE 2 If $f(x) = x^2$, then $f(x) \geqslant f(0)$ because $x^2 \geqslant 0$ for all x. Therefore, $f(0) = 0$ is the absolute (and local) minimum value of f. This corresponds to the fact that the origin is the lowest point on the parabola $y = x^2$ (see Figure 2). However, there is no highest point on the parabola and so this function has no maximum value. ■

EXAMPLE 3 From the graph of the function $f(x) = x^3$, shown in Figure 3, we see that this function has neither an absolute maximum value nor an absolute minimum value. In fact, it has no local extreme values either. ■

EXAMPLE 4 The graph of the function

$$f(x) = 3x^4 - 16x^3 + 18x^2 \qquad -1 \leqslant x \leqslant 4$$

is shown in Figure 4. You can see that $f(1) = 5$ is a local maximum, whereas the absolute maximum is $f(-1) = 37$. [This absolute maximum is not a local maximum because it occurs at an endpoint.] Also, $f(0) = 0$ is a local minimum and $f(3) = -27$ is both a local and an absolute minimum.

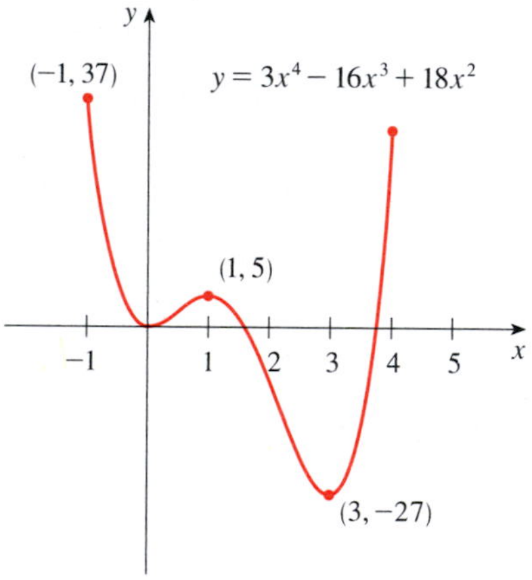

FIGURE 4

■

We have seen that some functions have extreme values, whereas others do not. The following theorem gives conditions under which a function is guaranteed to possess extreme values.

> **(3) THE EXTREME VALUE THEOREM** If f is continuous on a closed interval $[a, b]$, then f attains an absolute maximum value $f(c)$ and an absolute minimum value $f(d)$ at some numbers c and d in $[a, b]$.

The Extreme Value Theorem is illustrated in Figure 5. Note that an extreme value can be taken on more than once. Although the Extreme Value Theorem is intuitively very plausible, it is difficult to prove and so we omit the proof.

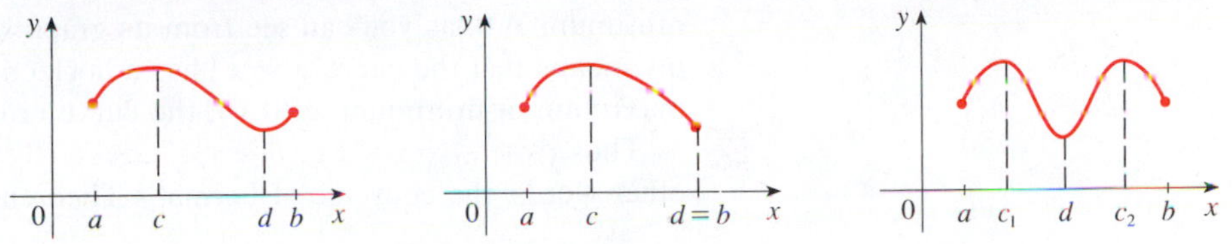

FIGURE 5

Figures 6 and 7 show that a function need not possess extreme values if either hypothesis (continuity or closed interval) is omitted from the Extreme Value Theorem.

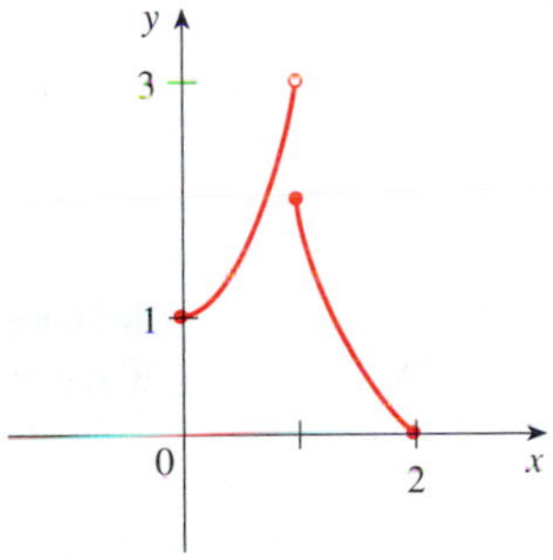

FIGURE 6
This function has minimum value $f(2) = 0$, but no maximum value.

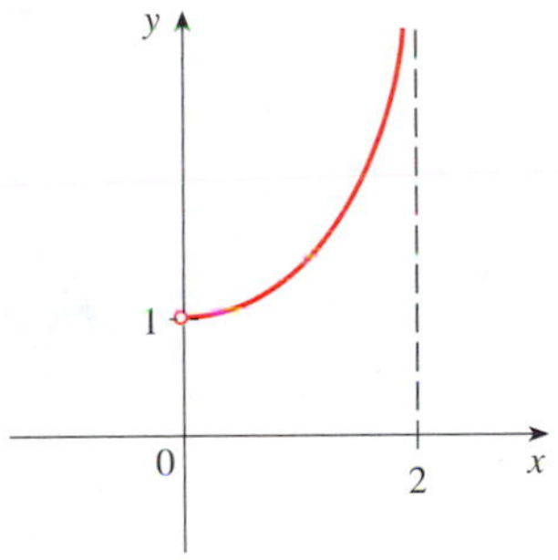

FIGURE 7
This continuous function has no maximum or minimum.

The function f whose graph is shown in Figure 6 is defined on the closed interval $[0, 2]$ but has no maximum value. (Notice that the range of f is $[0, 3)$. The function takes on values arbitrarily close to 3, but never actually attains the value 3.) This does not contradict the Extreme Value Theorem because f is not continuous. [Nonetheless, a discontinuous function *could* have maximum and minimum values. See Exercise 13(b).]

The function g shown in Figure 7 is continuous on the open interval $(0, 2)$ but has neither a maximum nor a minimum value. [The range of g is $(1, \infty)$. The function takes on arbitrarily large values.] This does not contradict the Extreme Value Theorem because the interval $(0, 2)$ is not closed.

The Extreme Value Theorem says that a continuous function on a closed interval has a maximum value and a minimum value, but it does not tell us how to find these extreme values. We start by looking for local extreme values.

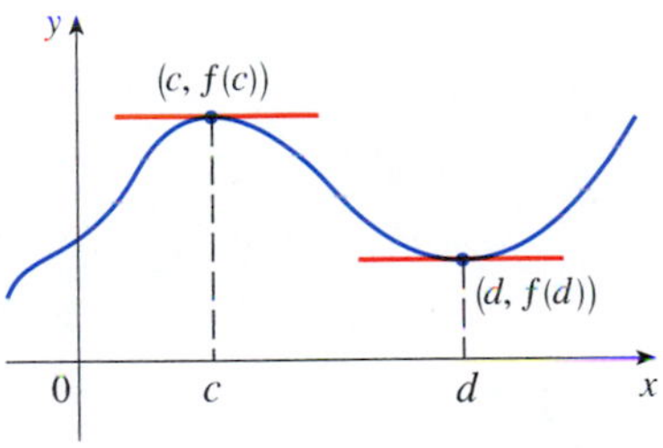

FIGURE 8

Figure 8 shows the graph of a function f with a local maximum at c and a local minimum at d. It appears that at the maximum and minimum points the tangent line is horizontal and therefore has slope 0. We know that the derivative is the slope of the tangent line, so it appears that $f'(c) = 0$ and $f'(d) = 0$. The following theorem says that this is always true for differentiable functions.

Fermat's Theorem is named after Pierre Fermat (1601–1665), a French lawyer who took up mathematics as a hobby. Despite his amateur status, Fermat was one of the two inventors of analytic geometry (Descartes was the other). His methods for finding tangents to curves and maximum and minimum values (before the invention of limits and derivatives) made him a forerunner of Newton in the creation of differential calculus.

(4) FERMAT'S THEOREM If f has a local maximum or minimum at c, and if $f'(c)$ exists, then $f'(c) = 0$.

Our intuition suggests that Fermat's Theorem is true. A rigorous proof, using the definition of a derivative, is given in Appendix F.

Although Fermat's Theorem is very useful, we have to guard against reading too much into it. If $f(x) = x^3$, then $f'(x) = 3x^2$, so $f'(0) = 0$. But f has no maximum or minimum at 0, as you can see from its graph in Figure 9. The fact that $f'(0) = 0$ simply means that the curve $y = x^3$ has a horizontal tangent at $(0, 0)$. Instead of having a maximum or minimum at $(0, 0)$, the curve crosses its horizontal tangent there.

Thus, when $f'(c) = 0$, f doesn't necessarily have a maximum or minimum at c. (In other words, the converse of Fermat's Theorem is false in general.)

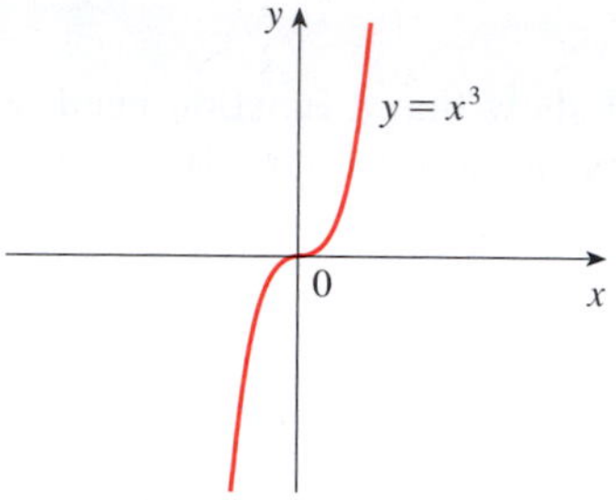

FIGURE 9

If $f(x) = x^3$, then $f'(0) = 0$ but f has no minimum or maximum.

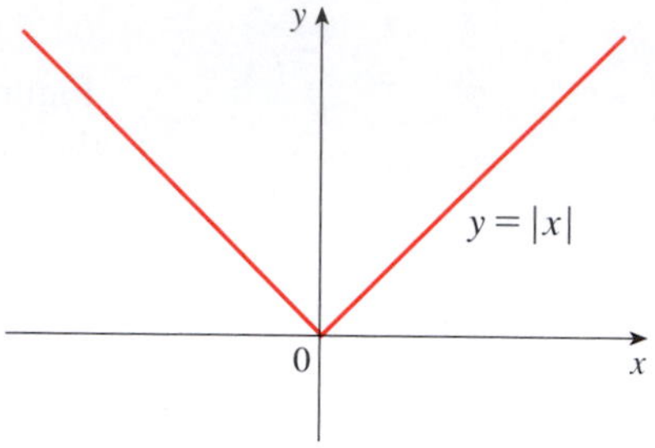

FIGURE 10

If $f(x) = |x|$, then $f(0) = 0$ is a minimum value, but $f'(0)$ does not exist.

We should bear in mind that there may be an extreme value where $f'(c)$ does not exist. For instance, the function $f(x) = |x|$ has its (local and absolute) minimum value at 0 (see Figure 10), but that value cannot be found by setting $f'(x) = 0$ because, as was shown in Example 8 in Section 3.1, $f'(0)$ does not exist.

Fermat's Theorem does suggest that we should at least *start* looking for extreme values of f at the numbers c where $f'(c) = 0$ or where $f'(c)$ does not exist. Such numbers are given a special name.

(5) DEFINITION A **critical number** of a function f is a number c in the domain of f such that either $f'(c) = 0$ or $f'(c)$ does not exist.

Figure 11 shows a graph of the function f in Example 5. It supports our answer because there is a horizontal tangent when $x = 1.5$ and a vertical tangent when $x = 0$.

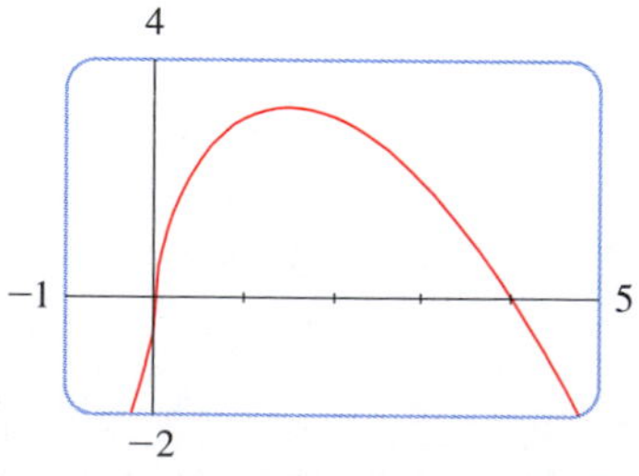

FIGURE 11

EXAMPLE 5 Find the critical numbers of $f(x) = x^{3/5}(4 - x)$.

SOLUTION The Product Rule gives

$$f'(x) = \tfrac{3}{5}x^{-2/5}(4 - x) + x^{3/5}(-1)$$
$$= \frac{3(4 - x) - 5x}{5x^{2/5}} = \frac{12 - 8x}{5x^{2/5}}$$

[The same result could be obtained by first writing $f(x) = 4x^{3/5} - x^{8/5}$.] Therefore, $f'(x) = 0$ if $12 - 8x = 0$, that is, $x = \frac{3}{2}$, and $f'(x)$ does not exist when $x = 0$. Thus, the critical numbers are $\frac{3}{2}$ and 0. ■

In terms of critical numbers, Fermat's Theorem can be rephrased as follows (compare Definition 5 with Theorem 4):

(6) If f has a local extremum at c, then c is a critical number of f.

To find an absolute maximum or minimum of a continuous function on a closed interval, we note that either it is a local extremum [in which case it occurs at a critical number by (6)] or it occurs at an endpoint of the interval. Thus, the following three-step procedure always works.

(7) THE CLOSED INTERVAL METHOD To find the *absolute* maximum and minimum values of a continuous function f on a closed interval $[a, b]$:

1. Find the values of f at the critical numbers of f in (a, b).
2. Find the values of f at the endpoints of the interval.
3. The largest of the values from steps 1 and 2 is the absolute maximum value; the smallest of these values is the absolute minimum value.

We can estimate maximum and minimum values very easily using a graphing calculator or a computer with graphing software. But, as Example 6 shows, calculus is needed to find the *exact* values.

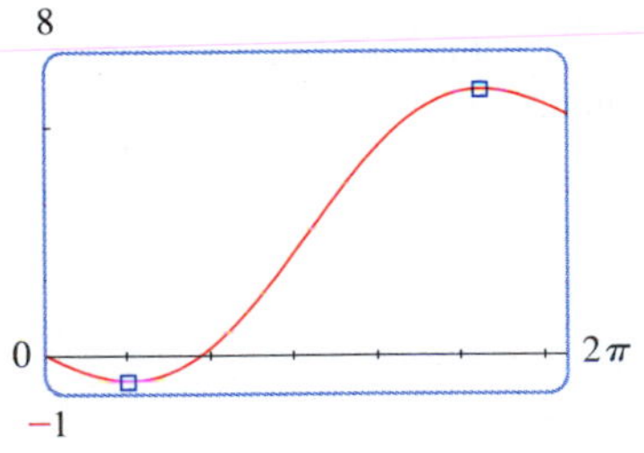

FIGURE 12

EXAMPLE 6

(a) Use a graphing device to estimate the absolute minimum and maximum values of the function $f(x) = x - 2\sin x, 0 \leqslant x \leqslant 2\pi$.

(b) Use calculus to find the exact minimum and maximum values.

SOLUTION

(a) Figure 12 shows a graph of f in the viewing rectangle $[0, 2\pi]$ by $[-1, 8]$. By moving the cursor close to the maximum point, we see that the y-coordinates do not change very much in the vicinity of the maximum. The absolute maximum value is about 6.97 and it occurs when $x \approx 5.2$. Similarly, by moving the cursor close to the minimum point, we see that the absolute minimum value is about -0.68 and it occurs when $x \approx 1.0$. It is possible to get more accurate estimates by zooming in toward the maximum and minimum points, but instead let's use calculus.

(b) The function $f(x) = x - 2\sin x$ is continuous on $[0, 2\pi]$. Since $f'(x) = 1 - 2\cos x$, we have $f'(x) = 0$ when $\cos x = \frac{1}{2}$ and this occurs when $x = \pi/3$ or $5\pi/3$. The values of f at these critical points are

$$f(\pi/3) = \frac{\pi}{3} - 2\sin\frac{\pi}{3} = \frac{\pi}{3} - \sqrt{3} \approx -0.684853$$

and

$$f(5\pi/3) = \frac{5\pi}{3} - 2\sin\frac{5\pi}{3} = \frac{5\pi}{3} + \sqrt{3} \approx 6.968039$$

The values of f at the endpoints are

$$f(0) = 0 \qquad \text{and} \qquad f(2\pi) = 2\pi \approx 6.28$$

Comparing these four numbers and using procedure (7), we see that the absolute minimum value is $f(\pi/3) = \pi/3 - \sqrt{3}$ and the absolute maximum value is $f(5\pi/3) = 5\pi/3 + \sqrt{3}$. The values from part (a) serve as a check on our work. ■

EXAMPLE 7 A model for the food-price index (the price of a representative "basket" of foods) between 1984 and 1994 is given by the function

$$I(t) = 0.00009045t^5 + 0.001438t^4 - 0.06561t^3 + 0.4598t^2 - 0.6270t + 99.33$$

where t is measured in years since midyear 1984, so $0 \leqslant t \leqslant 10$, and $I(t)$ is measured in 1987 dollars and scaled such that $I(3) = 100$. Estimate the times when food was cheapest and most expensive during the period 1984–1994.

Figure 13 shows the graph of the food-price index function I of Example 7. This model is based on the data points shown.

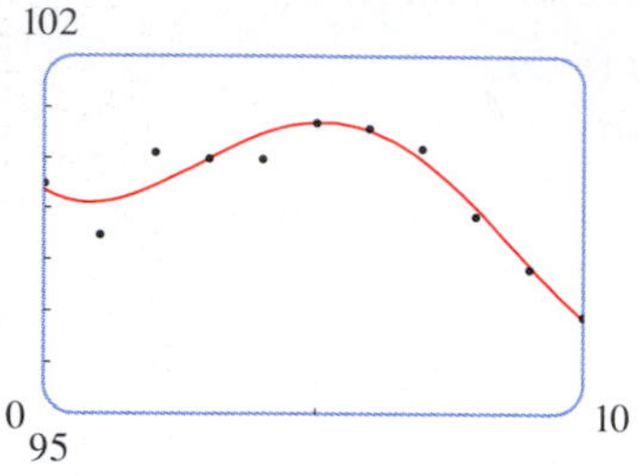

FIGURE 13

SOLUTION We apply the Closed Interval Method (7) to the continuous function I on $[0, 10]$. Its derivative is

$$I'(t) = 0.00045225t^4 + 0.005752t^3 - 0.19683t^2 + 0.9196t - 0.6270$$

Since I' exists for all t, the only critical numbers of I occur when $I'(t) = 0$. We use a root-finder on a computer algebra system (or a graphing device) to find that $I'(t) = 0$ when $t \approx -29.7186$, 0.8231, 5.1309, or 11.0459, but only the second and third roots lie in the interval $[0, 10]$. The values of I at these critical numbers are

$$I(0.8231) \approx 99.09 \qquad \text{and} \qquad I(5.1309) \approx 100.67$$

The values of I at the endpoints of the interval are

$$I(0) = 99.33 \qquad I(10) \approx 96.86$$

Comparing these four numbers, we see that food was most expensive at $t \approx 5.1309$ (corresponding roughly to August, 1989) and cheapest at $t = 10$ (midyear 1994). ■

EXERCISES 5.2

1. Explain the difference between an absolute minimum and a local minimum.

2. Suppose f is a continuous function defined on a closed interval $[a, b]$.
   (a) What theorem guarantees the existence of an absolute maximum value and an absolute minimum value for f?
   (b) What steps would you take to find those maximum and minimum values?

3–4 ■ For each of the numbers a, b, c, d, e, r, s, and t, state whether the function whose graph is shown has an absolute maximum or minimum, a local maximum or minimum, or neither a maximum nor a minimum.

3.

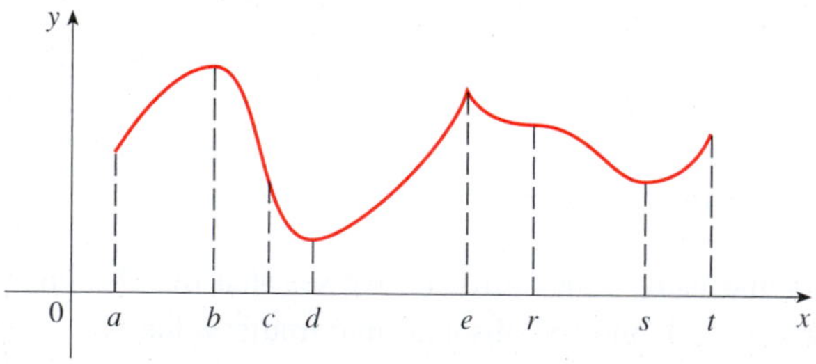

4.

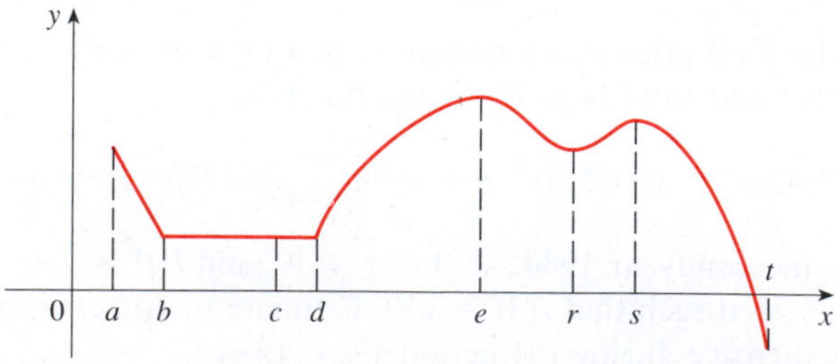

5–6 ■ Use the graph to state the absolute and local maximum and minimum values of the function.

5.

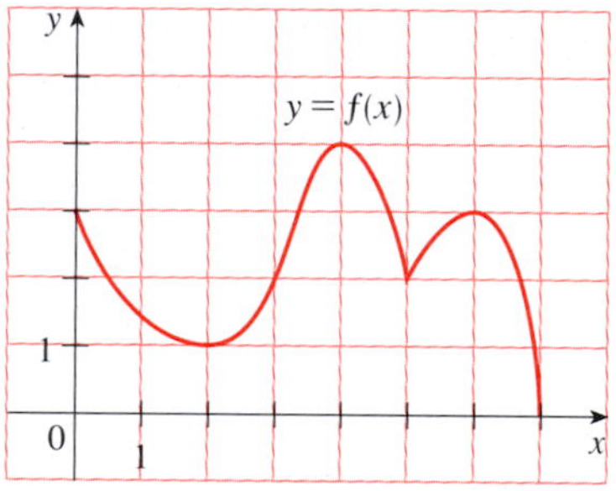

6.

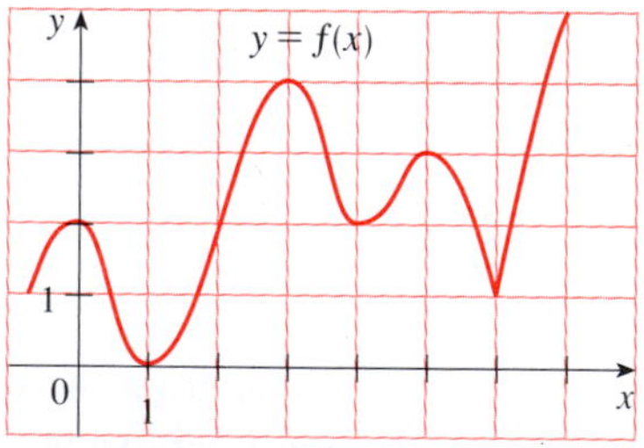

7–10 ■ Sketch the graph of a function f that is continuous on $[0, 3]$ and has the given properties.

7. Absolute maximum at 0, absolute minimum at 3, local minimum at 1, local maximum at 2

8. Absolute maximum at 1, absolute minimum at 2

9. 2 is a critical number, but f has no local maximum or minimum

10. Absolute minimum at 0, absolute maximum at 2, local maxima at 1 and 2, local minimum at 1.5

11. (a) Sketch the graph of a function that has a local maximum at 2 and is differentiable at 2.
(b) Sketch the graph of a function that has a local maximum at 2 and is continuous but not differentiable at 2.
(c) Sketch the graph of a function that has a local maximum at 2 and is not continuous at 2.

12. (a) Sketch the graph of a function on $[-1, 2]$ that has an absolute maximum but no local maximum.
(b) Sketch the graph of a function on $[-1, 2]$ that has a local maximum but no absolute maximum.

13. (a) Sketch the graph of a function on $[-1, 2]$ that has an absolute maximum but no absolute minimum.
(b) Sketch the graph of a function on $[-1, 2]$ that is discontinuous but has both an absolute maximum and an absolute minimum.

14. (a) Sketch the graph of a function that has two local maxima, one local minimum, and no absolute minimum.
(b) Sketch the graph of a function that has three local minima, two local maxima, and seven critical numbers.

15–24 ■ Find the absolute and local maximum and minimum values of f. Begin by sketching its graph by hand. (Use the graphs and transformations of Section 2 in Review and Preview.)

15. $f(x) = 1 + 2x, \quad x \geqslant -1$

16. $f(x) = 1 - x^2, \quad 0 < x \leqslant 1$

17. $f(x) = 1 - x^2, \quad -2 \leqslant x \leqslant 1$

18. $f(t) = 1/t, \quad 0 < t < 1$

19. $f(\theta) = \sin\theta, \quad -2\pi \leqslant \theta \leqslant 2\pi$

20. $f(\theta) = \tan\theta, \quad -\pi/4 \leqslant \theta < \pi/2$

21. $f(x) = x^5$

22. $f(x) = 2 - x^4$

23. $f(x) = 1 - e^{-x}, \quad x \geqslant 0$

24. $f(x) = \begin{cases} x^2 & \text{if } -1 \leqslant x < 0 \\ 2 - x^2 & \text{if } 0 \leqslant x \leqslant 1 \end{cases}$

25–36 ■ Find the critical numbers of the function.

25. $f(x) = 4x^3 - 9x^2 - 12x + 3$

26. $f(t) = t^3 + 6t^2 + 3t - 1$

27. $s(t) = t^4 + 4t^3 + 2t^2$

28. $g(x) = |x + 1|$

29. $f(r) = \dfrac{r}{r^2 + 1}$

30. $f(z) = \dfrac{z + 1}{z^2 + z + 1}$

31. $F(x) = x^{4/5}(x - 4)^2$

32. $G(x) = \sqrt[3]{x^2 - x}$

33. $f(\theta) = \sin^2(2\theta)$

34. $g(\theta) = \theta + \sin\theta$

35. $f(x) = x \ln x$

36. $f(x) = xe^{2x}$

37–46 ■ Find the absolute maximum and absolute minimum values of f on the given interval.

37. $f(x) = x^2 - 2x + 2, \quad [0, 3]$

38 $f(x) = x^3 - 12x + 1, \quad [-3, 5]$

39. $f(x) = 3x^5 - 5x^3 - 1, \quad [-2, 2]$

40. $f(x) = \sqrt{9 - x^2}, \quad [-1, 2]$

41. $f(x) = x^2 + 2/x, \quad [\frac{1}{2}, 2]$

42. $f(x) = \dfrac{x}{x + 1}, \quad [1, 2]$

43. $f(x) = \sin x + \cos x, \quad [0, \pi/3]$

44. $f(x) = x - 2\cos x, \quad [-\pi, \pi]$

45. $f(x) = xe^{-x}, \quad [0, 2]$

46. $f(x) = (\ln x)/x, \quad [1, 3]$

47–50 ■
(a) Use a graph to estimate the absolute maximum and minimum values of the function to two decimal places.
(b) Use calculus to find the exact maximum and minimum values.

47. $f(x) = x^3 - 8x + 1, \quad -3 \leqslant x \leqslant 3$

48. $f(x) = e^{x^3 - x}, \quad -1 \leqslant x \leqslant 0$

49. $f(x) = x\sqrt{x - x^2}$

50. $f(x) = (\cos x)/(2 + \sin x), \quad 0 \leqslant x \leqslant 2\pi$

51. Between 0 °C and 30 °C, the volume V (in cubic centimeters) of 1 kg of water at a temperature T is given approximately by the formula

$$V = 999.87 - 0.06426T + 0.0085043T^2 - 0.0000679T^3$$

Find the temperature at which water has its maximum density.

52. An object with weight W is dragged along a horizontal plane by a force acting along a rope attached to the object. If the rope makes an angle θ with the plane, then the magnitude of the force is

$$F = \frac{\mu W}{\mu \sin\theta + \cos\theta}$$

where μ is a positive constant called the *coefficient of friction* and where $0 \leqslant \theta \leqslant \pi/2$. Show that F is minimized when $\tan\theta = \mu$.

53. The Hubble Space Telescope was deployed on April 24, 1990, by the space shuttle *Discovery*. A model for the velocity of the shuttle during this mission, from liftoff at $t = 0$ s until the solid rocket boosters were jettisoned at $t = 126$ s, is given by

$$v(t) = 0.001302t^3 - 0.09029t^2 + 23.61t - 3.083$$

(in feet per second). Using this model, estimate the

absolute maximum and minimum values of the *acceleration* of the shuttle between liftoff and the jettisoning of the boosters.

CAS **54.** On May 7, 1992, the space shuttle *Endeavour* was launched on mission STS-49, the purpose of which was to install a new perigee kick motor in an Intelsat communications satellite. The following table gives the velocity data for the shuttle between liftoff and the jettisoning of the solid rocket boosters.

Event	Time (s)	Velocity (ft/s)
Launch	0	0
Begin roll maneuver	10	185
End roll maneuver	15	319
Throttle to 89%	20	447
Throttle to 67%	32	742
Throttle to 104%	59	1325
Maximum dynamic pressure	62	1445
Solid rocket booster separation	125	4151

(a) Use a computer algebra system or a graphing calculator with a statistics package to find the cubic polynomial that best models the velocity of the shuttle for the time interval $t \in [0, 125]$. Then graph this polynomial.

(b) Find a model for the acceleration of the shuttle and use it to estimate the maximum and minimum values of the acceleration during the first 125 s.

55. When a foreign object lodged in the trachea (windpipe) forces a person to cough, the diaphragm thrusts upward causing an increase in pressure in the lungs. This is accompanied by a contraction of the trachea, making a narrower channel for the expelled air to flow through. For a given amount of air to escape in a fixed time, it must move faster through the narrower channel than the wider one. The greater the velocity of the airstream, the greater the force on the foreign object. X rays show that the radius of the circular tracheal tube contracts to about two-thirds of its normal radius during a cough. According to a mathematical model of coughing, the velocity v of the airstream is related to the radius r of the trachea by the equation

$$v(r) = k(r_0 - r)r^2 \qquad \tfrac{1}{2}r_0 \leqslant r \leqslant r_0$$

where k is a constant and r_0 is the normal radius of the trachea. The restriction on r is due to the fact that the tracheal wall stiffens under pressure and a contraction greater than $r_0/2$ is prevented (otherwise the person would suffocate).

(a) Determine the value of r in the interval $[r_0/2, r_0]$ at which v has an absolute maximum. How does this compare with experimental evidence?

(b) What is the absolute maximum value of v on the interval?

(c) Sketch the graph of v on the interval $[0, r_0]$.

56. A cubic function is a polynomial of degree 3; that is, it has the form $f(x) = ax^3 + bx^2 + cx + d$, where $a \neq 0$.

(a) Show that a cubic function can have two, one, or no critical number(s). Give examples and sketches to illustrate the three possibilities.

(b) How many local extreme values can a cubic function have?

5.3 DERIVATIVES AND THE SHAPES OF CURVES

In Section 5.1 we discussed how the signs of the first and second derivatives $f'(x)$ and $f''(x)$ influence the shape of the graph of f. Here we revisit those facts, giving an indication of why they are true and using them, together with the differentation formulas of Chapter 3, to explain the shapes of graphs.

We start with a fact, known as the Mean Value Theorem, that will be useful not only for present purposes but also for explaining why some of the other basic results of calculus are true.

THE MEAN VALUE THEOREM If f is a differentiable function on the interval $[a, b]$, then there exists a number c between a and b such that

(1) $$f'(c) = \frac{f(b) - f(a)}{b - a}$$

or, equivalently,

(2) $$f(b) - f(a) = f'(c)(b - a)$$

The Mean Value Theorem was first formulated by Joseph-Louis Lagrange (1736–1813), born in Italy of a French father and an Italian mother. He was a child prodigy and became a professor in Turin at the tender age of 19. Lagrange made great contributions to number theory, theory of functions, theory of equations, and analytical and celestial mechanics. In particular, he applied calculus to the analysis of the stability of the solar system. At the invitation of Frederick the Great, he succeeded Euler at the Berlin Academy and, when Frederick died, Lagrange accepted King Louis XVI's invitation to Paris, where he was given apartments in the Louvre. He was a kind and quiet man, though, living only for science.

We can see that this theorem is reasonable by interpreting it geometrically. Figures 1 and 2 show the points $A(a, f(a))$ and $B(b, f(b))$ on the graphs of two differentiable functions.

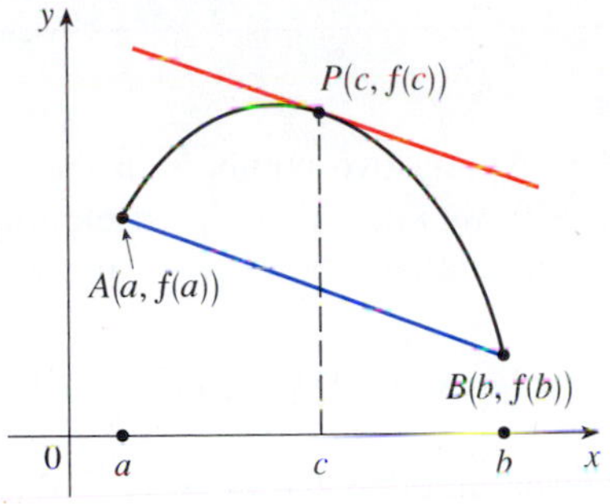

FIGURE 1

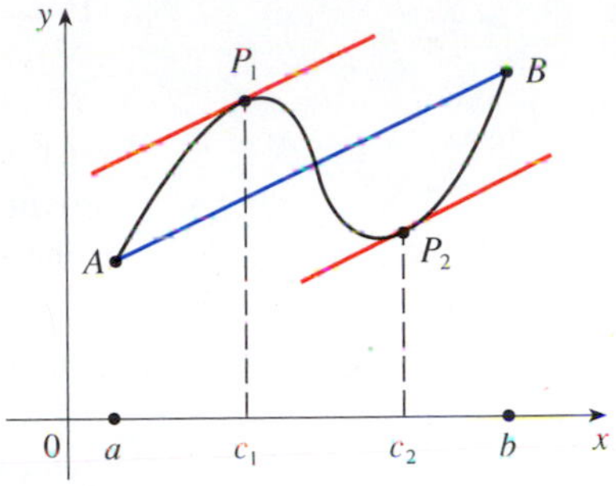

FIGURE 2

The slope of the secant line AB is

$$m_{AB} = \frac{f(b) - f(a)}{b - a}$$

which is the same expression as on the right side of Equation 1. Since $f'(c)$ is the slope of the tangent line at the point $(c, f(c))$, the Mean Value Theorem, in the form given by Equation 1, says that there is at least one point $P(c, f(c))$ on the graph where the slope of the tangent line is the same as the slope of the secant line AB. In other words, there is a point P where the tangent line is parallel to the secant line AB. It seems clear that there is one such point P in Figure 1 and two such points P_1 and P_2 in Figure 2. Because our intuition tells us that the Mean Value Theorem is true, we take it as the starting point for the development of the main facts of calculus. (When calculus is developed from first principles, however, the Mean Value Theorem is proved as a consequence of the axioms that define the real number system.)

EXAMPLE 1 If an object moves in a straight line with position function $s = f(t)$, then the average velocity between $t = a$ and $t = b$ is

$$\frac{f(b) - f(a)}{b - a}$$

and the velocity at $t = c$ is $f'(c)$. Thus, the Mean Value Theorem tells us that at some time $t = c$ between a and b the instantaneous velocity $f'(c)$ is equal to that average velocity. For instance, if a car traveled 180 km in 2 h, then the speedometer must have read 90 km/h at least once. ■

The main significance of the Mean Value Theorem is that it enables us to obtain information about a function from information about its derivative. Our immediate use of this principle is to prove the basic facts concerning increasing and decreasing functions. (See Exercises 43 and 44 for another use.)

INCREASING AND DECREASING FUNCTIONS

In Section 1 of Review and Preview we defined increasing functions and decreasing functions and in Section 5.1 we observed from graphs that a function with a positive derivative is increasing. We now deduce this fact from the Mean Value Theorem.

Let's abbreviate the name of this test to the I/D Test.

INCREASING / DECREASING TEST

(a) If $f'(x) > 0$ on an interval, then f is increasing on that interval.

(b) If $f'(x) < 0$ on an interval, then f is decreasing on that interval.

PROOF

(a) Let x_1 and x_2 be any two numbers in the interval with $x_1 < x_2$. Because we are given that $f'(x) > 0$, we know that f is differentiable on $[x_1, x_2]$. So, by the Mean Value Theorem there is a number c between x_1 and x_2 such that

$$f(x_2) - f(x_1) = f'(c)(x_2 - x_1) \qquad (3)$$

Now $f'(c) > 0$ by assumption and $x_2 - x_1 > 0$ because $x_1 < x_2$. Thus, the right side of Equation 3 is positive, and so

$$f(x_2) - f(x_1) > 0 \qquad \text{or} \qquad f(x_1) < f(x_2)$$

This shows that f is increasing.

Part (b) is proved similarly. □

EXAMPLE 2 Find where the function $f(x) = 3x^4 - 4x^3 - 12x^2 + 5$ is increasing and where it is decreasing.

SOLUTION $$f'(x) = 12x^3 - 12x^2 - 24x = 12x(x - 2)(x + 1)$$

To use the I/D Test we have to know where $f'(x) > 0$ and where $f'(x) < 0$. This depends on the signs of the three factors of $f'(x)$, namely, $12x$, $x - 2$, and $x + 1$. We divide the real line into intervals whose endpoints are the critical numbers $-1, 0$, and 2 and arrange our work in a chart. A plus sign indicates that the given expression is positive, and a minus sign indicates that it is negative. The last column of the chart gives the conclusion based on the I/D Test. For instance, $f'(x) < 0$ for $0 < x < 2$, so f is decreasing on $(0, 2)$. (It would also be true to say that f is decreasing on the closed interval $[0, 2]$.)

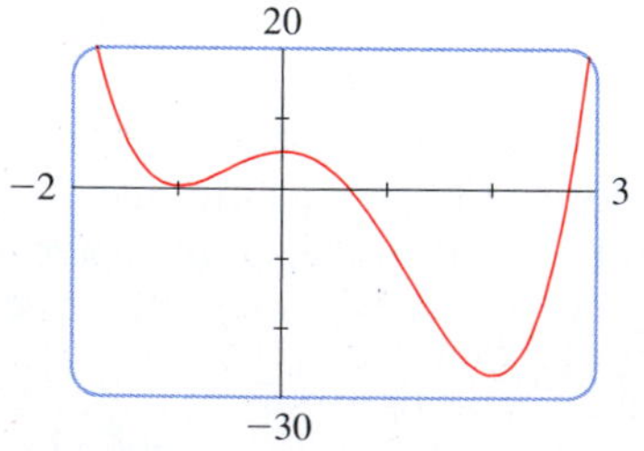

FIGURE 3

Interval	$12x$	$x - 2$	$x + 1$	$f'(x)$	f
$x < -1$	−	−	−	−	decreasing on $(-\infty, -1)$
$-1 < x < 0$	−	−	+	+	increasing on $(-1, 0)$
$0 < x < 2$	+	−	+	−	decreasing on $(0, 2)$
$x > 2$	+	+	+	+	increasing on $(2, \infty)$

The graph of f shown in Figure 3 confirms the information in the chart. ■

Recall from Section 5.2 that if f has a local maximum or minimum at c, then c must be a critical number of f (by Fermat's Theorem), but not every critical number gives rise to a maximum or a minimum. We therefore need a test that will tell us whether or not f has a local maximum or minimum at a critical number.

You can see from Figure 4(a) that $f(c)$ is a local maximum value of f because f increases for $x < c$ and decreases for $x > c$. Or, in terms of derivatives, $f'(x) > 0$ for $x < c$ and $f'(x) < 0$ for $x > c$. In other words, the sign of $f'(x)$ changes from positive to negative at c. This observation is the basis of the following test.

THE FIRST DERIVATIVE TEST Suppose that c is a critical number of a continuous function f.

(a) If f' changes from positive to negative at c, then f has a local maximum at c.

(b) If f' changes from negative to positive at c, then f has a local minimum at c.

(c) If f' does not change sign at c (that is, f' is positive on both sides of c or negative on both sides), then f has no local maximum or minimum at c.

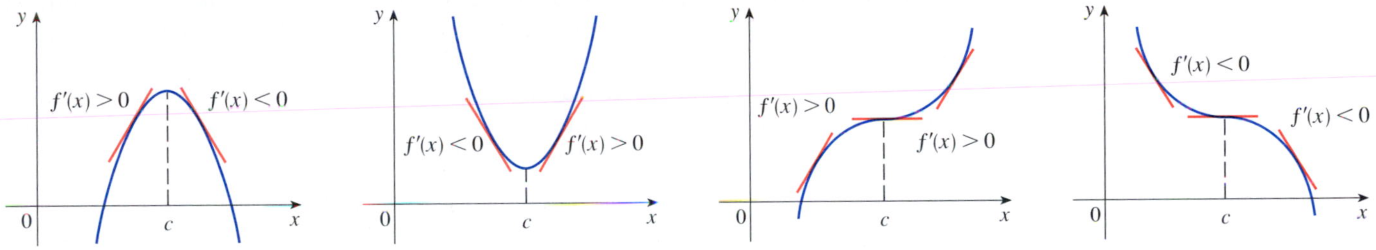

(a) Local maximum (b) Local minimum (c) No maximum or minimum (d) No maximum or minimum

FIGURE 4

The First Derivative Test is a consequence of the I/D Test. In part (a), for instance, since the sign of $f'(x)$ changes from positive to negative at c, f is increasing to the left of c and decreasing to the right of c. It follows that f has a local maximum at c.

It is easy to remember the First Derivative Test by visualizing diagrams such as those in Figure 4.

EXAMPLE 3 Find the local minimum and maximum values of the function f in Example 2.

SOLUTION From the chart in the solution to Example 2 we see that $f'(x)$ changes from negative to positive at -1, so $f(-1) = 0$ is a local minimum value by the First Derivative Test. Similarly, f' changes from negative to positive at 2, so $f(2) = -27$ is also a local minimum value. As previously noted, $f(0) = 5$ is a local maximum value because $f'(x)$ changes from positive to negative at 0. ■

CONCAVITY

Let's recall the definition of concavity from Section 5.1.

A function (or its graph) is called **concave upward** on an interval I if f' is an increasing function on I. It is called **concave downward** on I if f' is decreasing on I.

Notice in Figure 5 that the slopes of the tangent lines increase from left to right on the interval (a, b), so f' is increasing and f is concave upward (abbreviated CU) on (a, b). [It can be proved that this is equivalent to saying that the graph of f lies above all of its tangent lines on (a, b).] Similarly, the slopes of the tangent lines decrease from left to right on (b, c), so f' is decreasing and f is concave downward (CD) on (b, c).

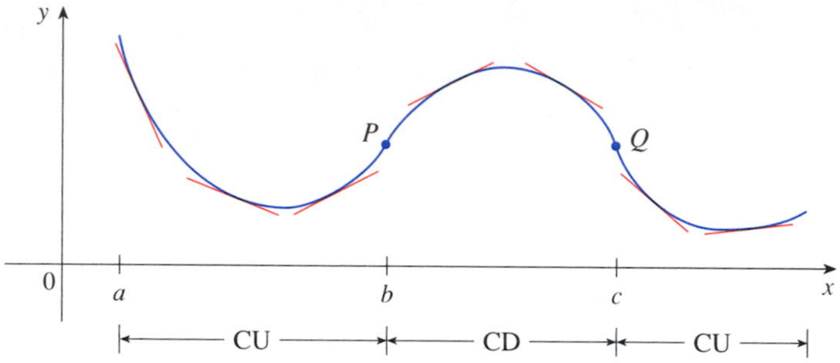

FIGURE 5

A point where a curve changes its direction of concavity is called an **inflection point**. The curve in Figure 5 changes from concave upward to concave downward at P and from concave downward to concave upward at Q, so both P and Q are inflection points.

Because $f'' = (f')'$, we know that if $f''(x)$ is positive, then f' is an increasing function and so f is concave upward. Similarly, if $f''(x)$ is negative, then f' is decreasing and f is concave downward. Thus, we have the following test for concavity.

CONCAVITY TEST

(a) If $f''(x) > 0$ for all x in I, then the graph of f is concave upward on I.

(b) If $f''(x) < 0$ for all x in I, then the graph of f is concave downward on I.

In view of the Concavity Test, there is a point of inflection at any point where the second derivative changes sign. A consequence of the Concavity Test is the following test for maximum and minimum values.

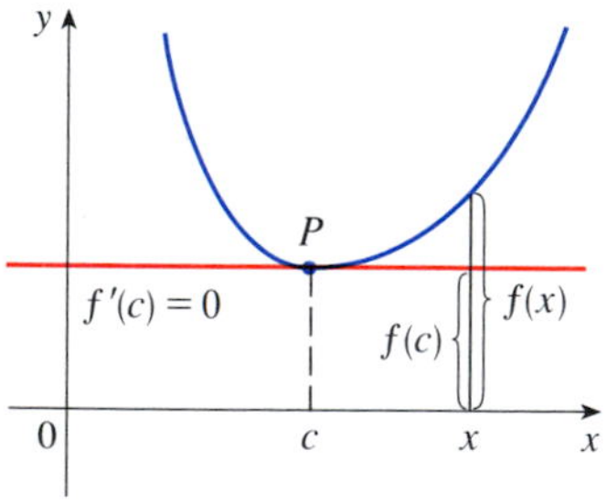

FIGURE 6
$f''(c) > 0$, concave upward

THE SECOND DERIVATIVE TEST Suppose f'' is continuous near c.

(a) If $f'(c) = 0$ and $f''(c) > 0$, then f has a local minimum at c.

(b) If $f'(c) = 0$ and $f''(c) < 0$, then f has a local maximum at c.

For instance, part (a) is true because $f''(x) > 0$ near c and so f is concave upward near c. This means that the graph of f lies *above* its horizontal tangent at c and so f has a local minimum at c. (See Figure 6.)

EXAMPLE 4 Discuss the curve $y = x^4 - 4x^3$ with respect to concavity, points of inflection, and local maxima and minima. Use this information to sketch the curve.

SOLUTION If $f(x) = x^4 - 4x^3$, then

$$f'(x) = 4x^3 - 12x^2 = 4x^2(x - 3)$$

$$f''(x) = 12x^2 - 24x = 12x(x - 2)$$

To find the critical numbers we set $f'(x) = 0$ and obtain $x = 0$ and $x = 3$. To use the Second Derivative Test we evaluate f'' at these critical numbers:

$$f''(0) = 0 \qquad f''(3) = 36 > 0$$

Since $f'(3) = 0$ and $f''(3) > 0$, $f(3) = -27$ is a local minimum. Since $f''(0) = 0$, the Second Derivative Test gives no information about the critical number 0. But since $f'(x) < 0$ for $x < 0$ and also for $0 < x < 3$, the First Derivative Test tells us that f does not have a local maximum or minimum at 0. [In fact, the expression for $f'(x)$ shows that f decreases to the left of 3 and increases to the right of 3.]

Since $f''(x) = 0$ when $x = 0$ or 2, we divide the real line into intervals with these numbers as endpoints and complete the following chart.

Interval	$f''(x) = 12x(x - 2)$	Concavity
$(-\infty, 0)$	+	upward
$(0, 2)$	−	downward
$(2, \infty)$	+	upward

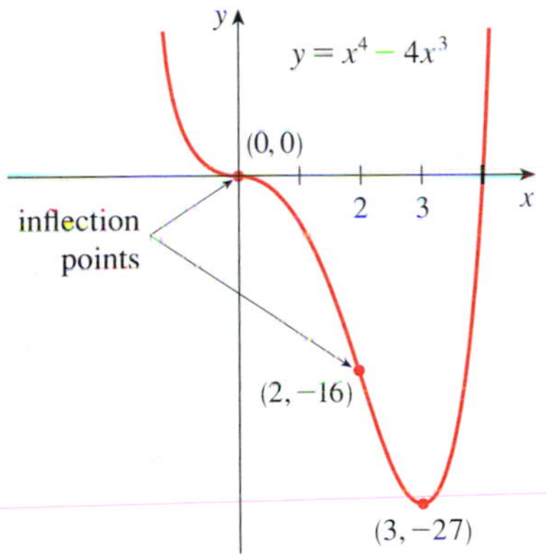

FIGURE 7

The point $(0, 0)$ is an inflection point since the curve changes from concave upward to concave downward there. Also $(2, -16)$ is an inflection point since the curve changes from concave downward to concave upward there.

Using the local minimum, the intervals of concavity, and the inflection points, we sketch the curve in Figure 7. ■

NOTE: The Second Derivative Test is inconclusive when $f''(c) = 0$. In other words, at such a point there might be a maximum, there might be a minimum, or there might be neither (as in Example 4). This test also fails when $f''(c)$ does not exist. In such cases the First Derivative Test must be used. In fact, even when both tests apply, the First Derivative Test is often the easier one to use.

EXAMPLE 5 Sketch the graph of the function $f(x) = x^{2/3}(6 - x)^{1/3}$.

SOLUTION Calculation of the first two derivatives gives

Use the differentiation rules to check these calculations.

$$f'(x) = \frac{4 - x}{x^{1/3}(6 - x)^{2/3}} \qquad f''(x) = \frac{-8}{x^{4/3}(6 - x)^{5/3}}$$

Since $f'(x) = 0$ when $x = 4$ and $f'(x)$ does not exist when $x = 0$ or $x = 6$, the critical numbers are 0, 4, and 6.

Interval	$4 - x$	$x^{1/3}$	$(6 - x)^{2/3}$	$f'(x)$	f
$x < 0$	+	−	+	−	decreasing on $(-\infty, 0)$
$0 < x < 4$	+	+	+	+	increasing on $(0, 4)$
$4 < x < 6$	−	+	+	−	decreasing on $(4, 6)$
$x > 6$	−	+	+	−	decreasing on $(6, \infty)$

To find the local extreme values we use the First Derivative Test. Since f' changes from negative to positive at 0, $f(0) = 0$ is a local minimum. Since f' changes from positive to negative at 4, $f(4) = 2^{5/3}$ is a local maximum. The sign of f' does not change at 6, so there is no minimum or maximum there. (The Second Derivative Test could be used at 4 but not at 0 or 6 since f'' does not exist there.)

Looking at the expression for $f''(x)$ and noting that $x^{4/3} \geq 0$ for all x, we have $f''(x) < 0$ for $x < 0$ and for $0 < x < 6$ and $f''(x) > 0$ for $x > 6$. So f is concave downward on $(-\infty, 0)$ and $(0, 6)$ and concave upward on $(6, \infty)$, and the only inflection point is $(6, 0)$. The graph is sketched in Figure 8. Note that the curve has vertical tangents at $(0, 0)$ and $(6, 0)$ because $|f'(x)| \to \infty$ as $x \to 0$ and as $x \to 6$.

Try reproducing the graph in Figure 8 with a graphing calculator or computer. Some machines produce the complete graph, some produce only the portion to the right of the y-axis, and some produce only the portion between $x = 0$ and $x = 6$. For an explanation and cure, see Example 7 in Section 3 of Review and Preview. An equivalent expression that gives the correct graph is

$$y = (x^2)^{1/3} \cdot \frac{6 - x}{|6 - x|} |6 - x|^{1/3}$$

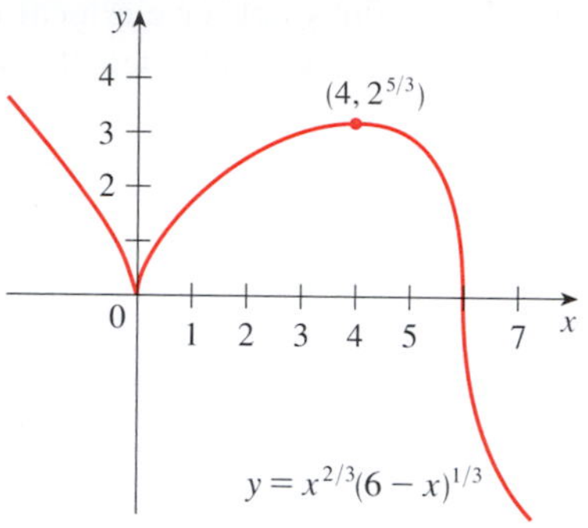

FIGURE 8

■

EXAMPLE 6 Use the first and second derivatives of $f(x) = e^{1/x}$, together with asymptotes, to sketch its graph.

SOLUTION Notice that the domain of f is $\{x \mid x \neq 0\}$, so we check for vertical asymptotes by computing the left and right limits as $x \to 0$. As $x \to 0^+$, we know that $t = 1/x \to \infty$, so

$$\lim_{x \to 0^+} e^{1/x} = \lim_{t \to \infty} e^t = \infty$$

and this shows that $x = 0$ is a vertical asymptote. As $x \to 0^-$, we have $t = 1/x \to -\infty$, so

$$\lim_{x \to 0^-} e^{1/x} = \lim_{t \to -\infty} e^t = 0$$

As $x \to \pm\infty$, we have $1/x \to 0$ and so

$$\lim_{x \to \pm\infty} e^{1/x} = e^0 = 1$$

This shows that $y = 1$ is a horizontal asymptote.

Now let's compute the derivative. The Chain Rule gives

$$f'(x) = -\frac{e^{1/x}}{x^2}$$

Since $e^{1/x} > 0$ and $x^2 > 0$ for all $x \neq 0$, we have $f'(x) < 0$ for all $x \neq 0$. Thus, f is decreasing on $(-\infty, 0)$ and on $(0, \infty)$. There is no critical number, so the function has no maximum or minimum. The second derivative is

$$f''(x) = -\frac{x^2 e^{1/x}(-1/x^2) - e^{1/x}(2x)}{x^4}$$

$$= \frac{e^{1/x}(2x + 1)}{x^4}$$

Since $e^{1/x} > 0$ and $x^4 > 0$, we have $f''(x) > 0$ when $x > -\frac{1}{2}$ $(x \neq 0)$ and $f''(x) < 0$ when $x < -\frac{1}{2}$. So the curve is concave downward on $(-\infty, -\frac{1}{2})$ and concave upward on $(-\frac{1}{2}, 0)$ and on $(0, \infty)$. The inflection point is $(-\frac{1}{2}, e^{-2})$.

To sketch the graph of f we first draw the horizontal asymptote $y = 1$ (as a dashed line), together with the parts of the curve near the asymptotes in a preliminary sketch [Figure 9(a)]. These parts reflect the information concerning limits and the fact that f is decreasing on both $(-\infty, 0)$ and $(0, \infty)$. Notice that we have indicated that $f(x) \to 0$ as $x \to 0^-$ even though $f(0)$ does not exist. In Figure 9(b) we finish the sketch by incorporating the information concerning concavity and the inflection point. In Figure 9(c) we check our work with a graphing device.

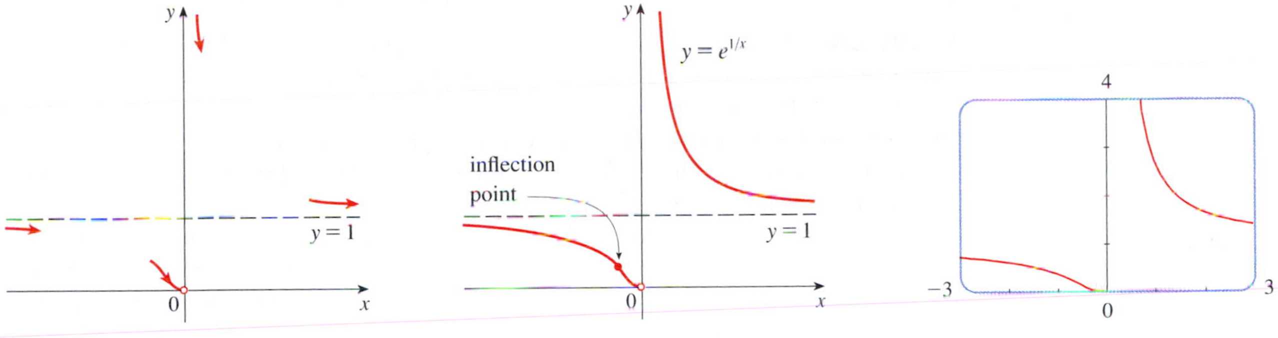

(a) Preliminary sketch (b) Finished sketch (c) Computer confirmation

FIGURE 9

■

EXAMPLE 7 A population of honeybees raised in an apiary started with 50 bees at time $t = 0$ and was modeled by the function

$$P(t) = \frac{75{,}200}{1 + 1503e^{-0.5932t}}$$

where t is the time in weeks, $0 \leqslant t \leqslant 25$. Use a graph to estimate the time at which the bee population was growing fastest. Then use derivatives to give a more accurate estimate.

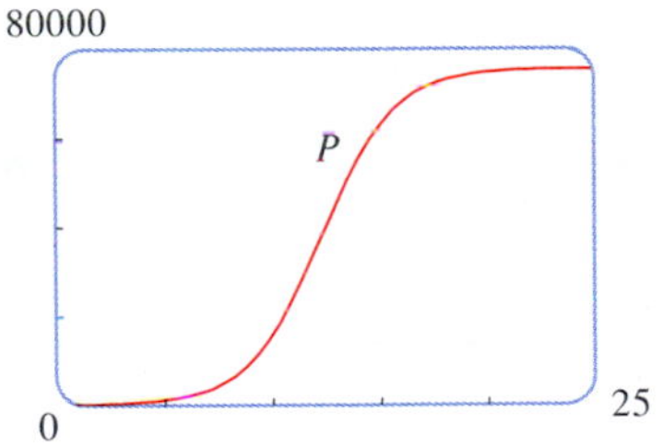

FIGURE 10

SOLUTION The population grows fastest when the population curve $y = P(t)$ has the steepest tangent line. From the graph of P in Figure 10, we estimate that the steepest tangent occurs when $t \approx 12$, so the bee population was growing most rapidly after about 12 weeks.

For a better estimate we calculate the derivative $P'(t)$, which is the rate of increase of the bee population:

$$P'(t) = \frac{67{,}046{,}785.92e^{-0.5932t}}{(1 + 1503e^{-0.5932t})^2}$$

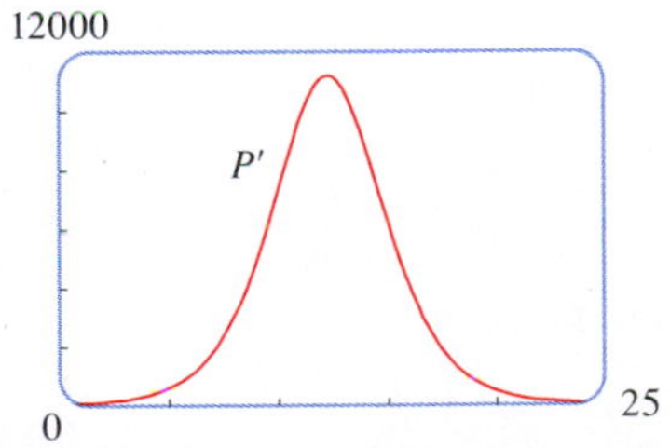

FIGURE 11

We graph P' in Figure 11 and observe that P' has its maximum value when $t \approx 12.3$.

To get a still better estimate we note that f' has its maximum value when f' changes from increasing to decreasing. This happens when f changes from concave upward to concave downward, so we ask a CAS to compute the second derivative:

$$P''(t) \approx \frac{119555093144e^{-1.1864t}}{(1 + 1503e^{-0.5932t})^3} - \frac{39772153e^{-0.5932t}}{(1 + 1503e^{-0.5932t})^2}$$

We could plot this function to see where it changes from positive to negative, but instead let's have the CAS solve the equation $P''(t) = 0$. It gives the answer $t \approx 12.3318$. ■

Our final example is concerned with *families* of functions. This means that the functions in the family are related to each other by a formula that contains one or more arbitrary constants. Each value of the constant gives rise to a member of the family and the idea is to see how the graph of the function changes as the constant changes.

EXAMPLE 8 Investigate the family of functions given by $f(x) = cx + \sin x$. What features do the members of this family have in common? How do they differ?

SOLUTION The derivative is $f'(x) = c + \cos x$. If $c > 1$, then $f'(x) > 0$ for all x (since $\cos x \geqslant -1$), so f is always increasing. If $c = 1$, then $f'(x) = 0$ when x is an odd multiple of π, but f just has horizontal tangents there and is still an increasing function. Similarly, if $c \leqslant -1$, then f is always decreasing. If $-1 < c < 1$, then the equation $c + \cos x = 0$ has infinitely many solutions $[x = 2n\pi \pm \cos^{-1}(-c)]$ and f has infinitely many minima and maxima.

The second derivative is $f''(x) = -\sin x$, which is negative when $0 < x < \pi$ and, in general, when $2n\pi < x < (2n + 1)\pi$, where n is any integer. Thus, *all* members of the family are concave downward on $(0, \pi)$, $(2\pi, 3\pi)$, . . . and concave upward on $(\pi, 2\pi)$, $(3\pi, 4\pi)$, ... This is illustrated by several members of the family in Figure 12.

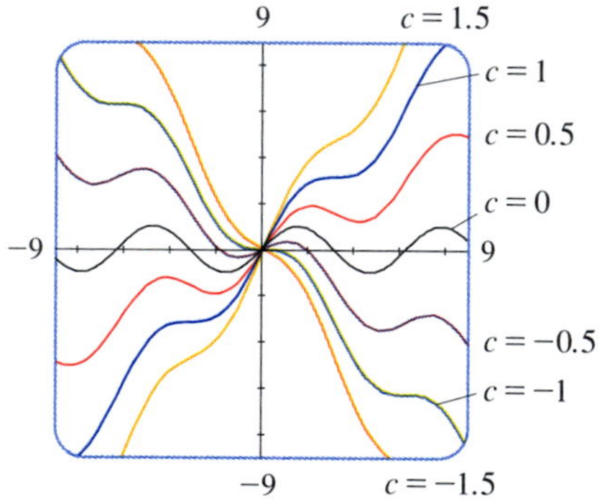

FIGURE 12 ■

EXERCISES 5.3

1. Use the graph of f to estimate the values of c that satisfy the conclusion of the Mean Value Theorem for the interval $[0, 8]$.

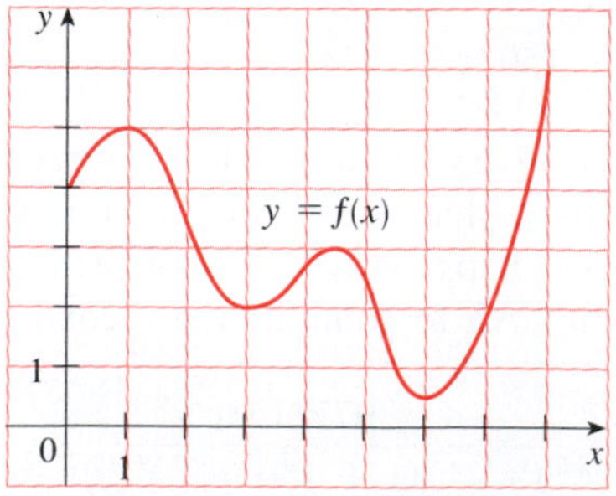

2. From the given graph of g, state
   (a) the largest open intervals on which g is concave upward,

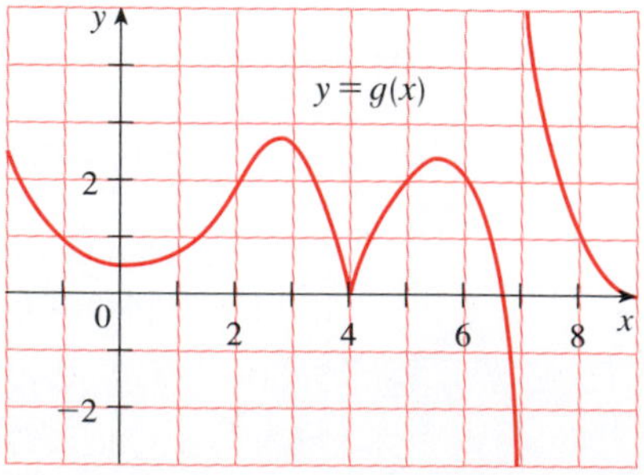

(b) the largest open intervals on which g is concave downward, and
(c) the coordinates of the points of inflection.

3. (a) How do you determine where a function is increasing or decreasing?
(b) How do you determine where a curve is concave upward or concave downward?
(c) How do you locate inflection points?

4. (a) State the First Derivative Test.
(b) State the Second Derivative Test. Under what circumstances is it inconclusive? What do you do if it fails?

5. The graph of the second derivative f'' of a function f is shown. State the x-coordinates of the inflection points of f. Give reasons for your answers.

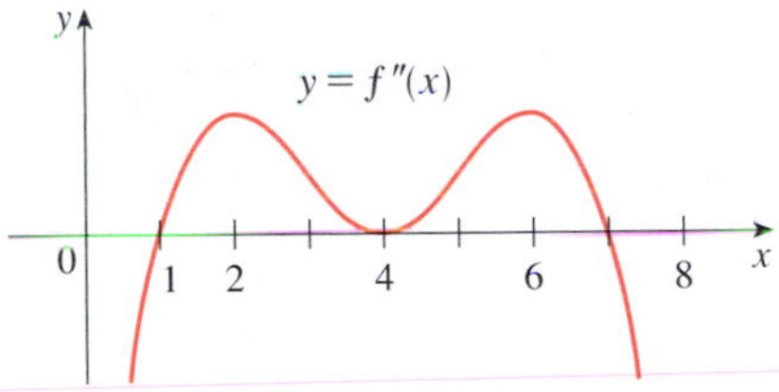

6. The graph of the first derivative f' of a function f is shown.
(a) On what intervals is f increasing? Explain.
(b) At what values of x does f have a local maximum or minimum? Explain.
(c) On what intervals is f concave upward or concave downward? Explain.
(d) What are the x-coordinates of the inflection points of f? Why?

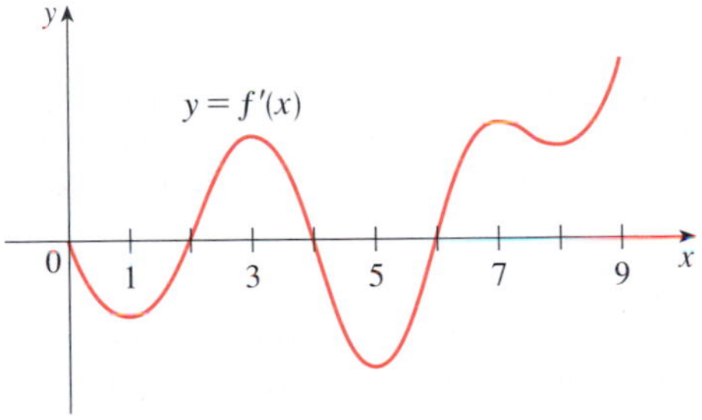

7–12 ■
(a) Find the intervals on which f is increasing or decreasing.
(b) Find the local maximum and minimum values of f.
(c) Find the intervals of concavity and the inflection points.

7. $f(x) = x^6 + 192x + 17$

8. $f(x) = 2\sin x + \sin^2 x$

9. $y = xe^x$

10. $y = x^2 e^x$

11. $y = (\ln x)/\sqrt{x}$

12. $y = x \ln x$

13–20 ■
(a) Find the intervals on which f is increasing or decreasing.
(b) Find the local maximum and minimum values of f.
(c) Find the intervals of concavity and the inflection points.
(d) Use the information from parts (a), (b), and (c) to sketch the graph of f. Check your work with a graphing device.

13. $f(x) = 1 - 3x + 5x^2 - x^3$

14. $f(x) = x^4 - 6x^2$

15. $f(x) = (x^2 - 1)^3$

16. $f(x) = x\sqrt{x + 1}$

17. $f(x) = x^{1/3}(x + 3)^{2/3}$

18. $f(x) = 2x + \cot x, \quad 0 < x < \pi$

19. $f(x) = 2\cos x + \sin^2 x$

20. $f(x) = \ln(1 + x^2)$

21–26 ■
(a) Find the vertical and horizontal asymptotes.
(b) Find the intervals of increase or decrease.
(c) Find the local maximum and minimum values.
(d) Find the intervals of concavity and the inflection points.
(e) Use the information from parts (a)–(d) to sketch the graph of f.

21. $f(x) = \dfrac{1 + x^2}{1 - x^2}$

22. $f(x) = \dfrac{x}{(x - 1)^2}$

23. $f(x) = \sqrt{x^2 + 1} - x$

24. $f(x) = x\tan x, \quad -\pi/2 < x < \pi/2$

25. $f(x) = e^{-1/(x+1)}$

26. $f(x) = \ln(\tan^2 x)$

27–28 ■
(a) Use a graph of f to give a rough estimate of the intervals of concavity and the coordinates of the points of inflection.
(b) Use a graph of f'' to give better estimates.

27. $f(x) = 3x^5 - 40x^3 + 30x^2$

28. $f(x) = 2\cos x + \sin 2x, \quad 0 \leqslant x \leqslant 2\pi$

29–30 ■
(a) Use a graph of f to estimate the maximum and minimum values. Then find the exact values.
(b) Estimate the value of x at which f increases most rapidly. Then find the exact value.

29. $f(x) = \dfrac{x + 1}{\sqrt{x^2 + 1}}$

30. $f(x) = x^2 e^{-x}$

31. For the period from 1980 to 1994, the percentage of households in the United States with at least one VCR has been modeled by the function

$$V(t) = \frac{75}{1 + 74e^{-0.6t}}$$

where the time t is measured in years since midyear 1980, so $0 \leqslant t \leqslant 14$. Use a graph to estimate the time at which the number of VCRs was increasing most rapidly. Then use derivatives to give a more accurate estimate.

32. The family of bell-shaped curves

$$y = \frac{1}{\sigma\sqrt{2\pi}} e^{-(x-\mu)^2/(2\sigma^2)}$$

occurs in probability and statistics, where it is called the *normal density function*. The constant μ is called the *mean* and the positive constant σ is called the *standard deviation*. For simplicity, let's scale the function so as to remove the factor $1/(\sigma\sqrt{2\pi})$ and let's analyze the special case where $\mu = 0$. So we study the function

$$f(x) = e^{-x^2/(2\sigma^2)}$$

(a) Find the asymptote, maximum value, and inflection points of f.
(b) What role does σ play in the shape of the curve?
(c) Illustrate by graphing four members of this family on the same screen.

CAS **33–34** ■ Estimate the intervals of concavity to one decimal place by using a computer algebra system to compute and graph f''.

33. $f(x) = \dfrac{x^3 - 10x + 5}{\sqrt{x^2 + 4}}$

34. $f(x) = \dfrac{(x + 1)^3(x^2 + 5)}{(x^3 + 1)(x^2 + 4)}$

35. Find a cubic function $f(x) = ax^3 + bx^2 + cx + d$ that has a local maximum value of 3 at -2 and a local minimum value of 0 at 1.

36. For what values of the numbers a and b does the function

$$f(x) = axe^{bx^2}$$

have the maximum value $f(2) = 1$?

37–40 ■ Assume that all of the functions are twice differentiable.

37. If f and g are concave upward on I, show that $f + g$ is concave upward on I.

38. If f is positive and concave upward on I, show that the function $g(x) = [f(x)]^2$ is concave upward on I.

39. If f and g are positive increasing concave upward functions on I, show that the product function fg is concave upward on I.

40. Suppose f and g are both concave upward on $(-\infty, \infty)$. Under what condition on f will the composite function $h(x) = f(g(x))$ be concave upward?

41. Show that $\tan x > x$ for $0 < x < \pi/2$. [*Hint:* Show that $f(x) = \tan x - x$ is increasing on $(0, \pi/2)$.]

42. (a) Show that $e^x \geqslant 1 + x$ for $x \geqslant 0$.
(b) Deduce that $e^x \geqslant 1 + x + \frac{1}{2}x^2$ for $x \geqslant 0$.
(c) Use mathematical induction to prove that for $x \geqslant 0$ and any positive integer n,

$$e^x \geqslant 1 + x + \frac{x^2}{2!} + \cdots + \frac{x^n}{n!}$$

43. Suppose that $f(0) = -3$ and $f'(x) \leqslant 5$ for all values of x. The inequality gives a restriction on the rate of growth of f, which then imposes a restriction on the possible values of f. Use the Mean Value Theorem to determine how large $f(4)$ can possibly be.

44. Suppose that $1 \leqslant f'(x) \leqslant 4$ for all x in $[2, 5]$. Show that $3 \leqslant f(5) - f(2) \leqslant 12$.

45. Two runners start a race at the same time and finish in a tie. Prove that at some time during the race they have the same velocity. [*Hint:* Consider $f(t) = g(t) - h(t)$ where g and h are the position functions of the two runners.]

46. At 2:00 P.M. a car's speedometer reads 30 mi/h. At 2:10 P.M. it reads 50 mi/h. Show that at some time between 2:00 and 2:10 the acceleration is exactly 120 mi/h^2.

47. Show that a cubic function (a third-degree polynomial) always has exactly one point of inflection. If its graph has three x-intercepts x_1, x_2, and x_3, show that the x-coordinate of the inflection point is $(x_1 + x_2 + x_3)/3$.

48. For what values of c does the polynomial $P(x) = x^4 + cx^3 + x^2$ have two inflection points? One inflection point? None? Illustrate by graphing P for several values of c. How does the graph change as c decreases?

5.4 GRAPHING WITH CALCULUS *AND* CALCULATORS

If you have not already read Section 3 of Review and Preview, you should do so now. In particular, it explains how to avoid some of the pitfalls of using graphing devices by choosing appropriate viewing rectangles.

When we graph with technology, the theme is the *interaction* between calculus and graphing devices. We start with a graph produced by a graphing calculator or computer and then we refine it. We use calculus to make sure that we reveal all the important aspects of the curve.

EXAMPLE 1 Graph the polynomial $f(x) = 2x^6 + 3x^5 + 3x^3 - 2x^2$. Use the graphs of f' and f'' to estimate all maximum and minimum points and intervals of concavity.

SOLUTION If we specify a domain but not a range, many graphing devices will deduce a suitable range from the values computed. Figure 1 shows the plot from one such device if we specify that $-5 \leqslant x \leqslant 5$. Although this viewing rectangle is useful for showing that the asymptotic behavior (or end behavior) is the same as for $y = 2x^6$, it is obviously hiding some finer detail. So we change to the viewing rectangle $[-3, 2]$ by $[-50, 100]$ shown in Figure 2.

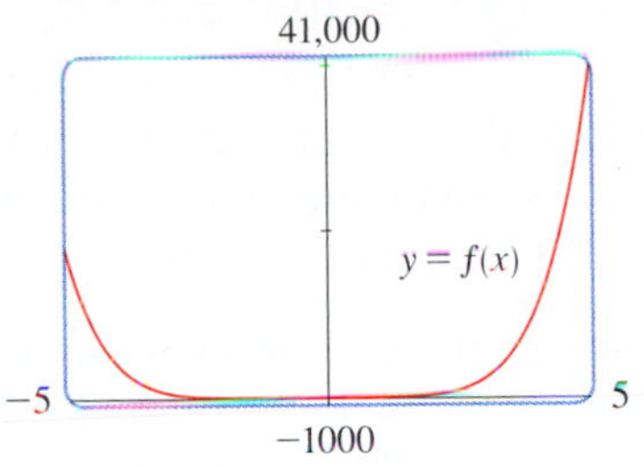

FIGURE 1

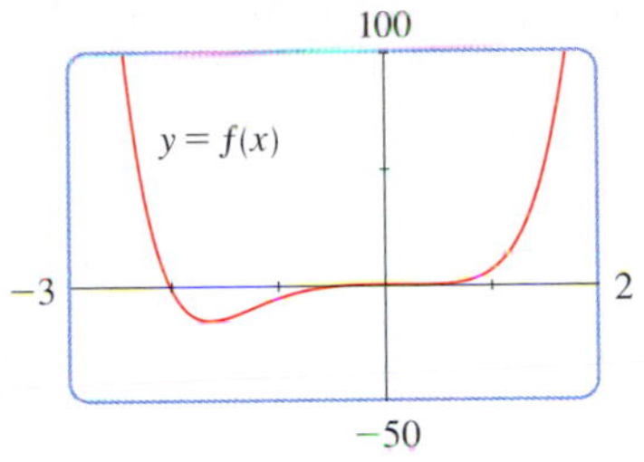

FIGURE 2

From this graph it appears that there is an absolute minimum value of about -15.33 when $x \approx -1.62$ (by using the cursor) and f is decreasing on $(-\infty, -1.62)$ and increasing on $(-1.62, \infty)$. Also there appears to be a horizontal tangent at the origin and inflection points when $x = 0$ and when x is somewhere between -2 and -1.

Now let's try to confirm these impressions using calculus. We differentiate and get

$$f'(x) = 12x^5 + 15x^4 + 9x^2 - 4x \qquad f''(x) = 60x^4 + 60x^3 + 18x - 4$$

When we graph f' in Figure 3 we see that $f'(x)$ changes from negative to positive when $x \approx -1.62$; this confirms (by the First Derivative Test) the minimum value that we found earlier. But, perhaps to our surprise, we also notice that $f'(x)$ changes from positive to negative when $x = 0$ and from negative to positive when $x \approx 0.35$. This means that f has a local maximum at 0 and a local minimum when $x \approx 0.35$, but these were hidden in Figure 2. Indeed, if we now zoom in toward the origin in Figure 4, we see what we missed before: a local maximum value of 0 when $x = 0$ and a local minimum value of about -0.1 when $x \approx 0.35$.

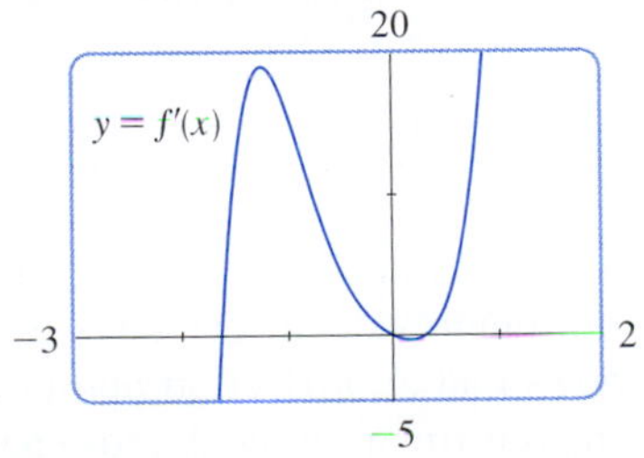

FIGURE 3

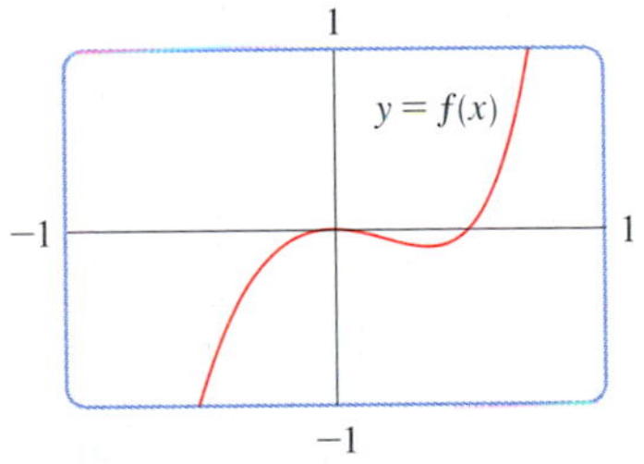

FIGURE 4

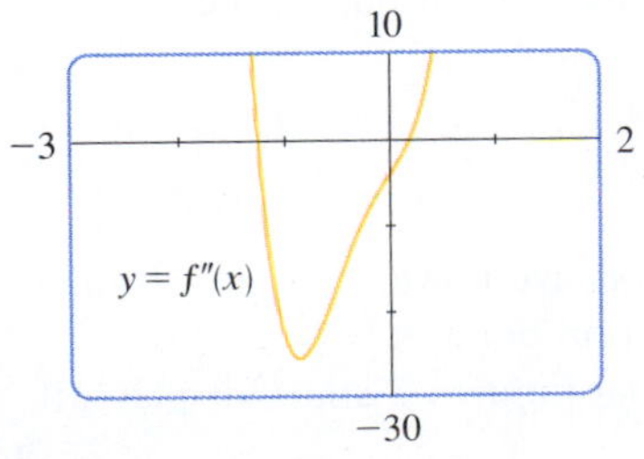

FIGURE 5

What about concavity and inflection points? From Figures 2 and 4 there appear to be inflection points when x is a little to the left of -1 and when x is a little to the right of 0. But it is difficult to determine inflection points from the graph of f, so we graph the second derivative f'' in Figure 5. We see that f'' changes from positive to negative when $x \approx -1.23$ and from negative to positive when $x \approx 0.19$. So, correct to two decimal places, f is concave upward on $(-\infty, -1.23)$ and $(0.19, \infty)$ and concave downward on $(-1.23, 0.19)$. The inflection points are $(-1.23, -10.18)$ and $(0.19, -0.05)$.

We have discovered that no single graph reveals all the important features of this polynomial. But Figures 2 and 4, when taken together, do provide an accurate picture. ■

EXAMPLE 2 Draw the graph of the function

$$f(x) = \frac{x^2 + 7x + 3}{x^2}$$

in a viewing rectangle that contains all the important features of the function. Estimate the maximum and minimum values and the intervals of concavity. Then use calculus to find these quantities exactly.

SOLUTION Figure 6, produced by a computer with automatic scaling, is a disaster. Some graphing calculators use $[-10, 10]$ by $[-10, 10]$ as the default viewing rectangle, so let's try it. We get the graph shown in Figure 7; it's a major improvement.

The y-axis appears to be a vertical asymptote and indeed it is because

$$\lim_{x \to 0} \frac{x^2 + 7x + 3}{x^2} = \infty$$

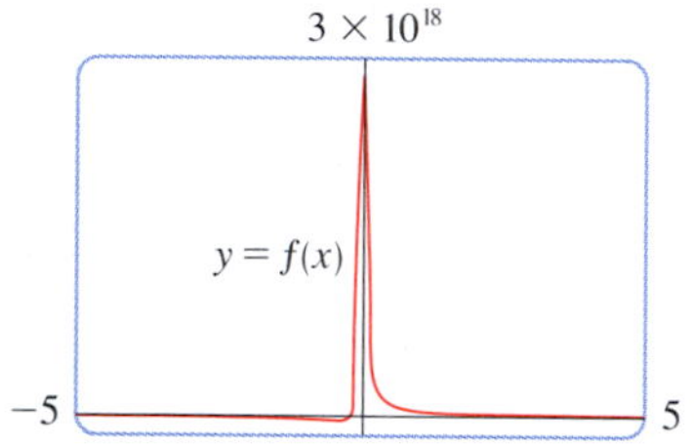

FIGURE 6

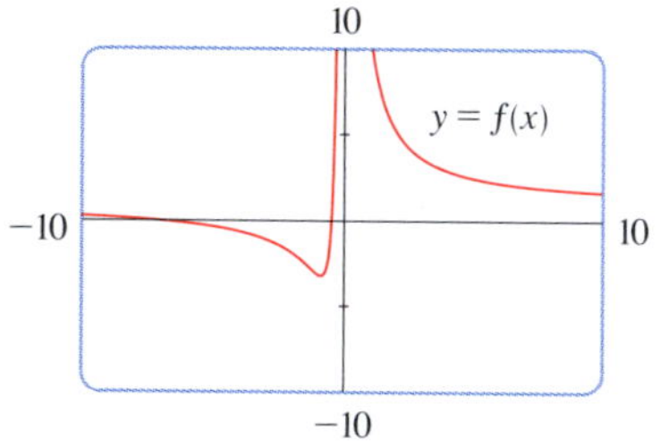

FIGURE 7

Figure 7 also allows us to estimate the x-intercepts: about -0.5 and -6.5. The exact values are obtained by using the quadratic formula to solve the equation $x^2 + 7x + 3 = 0$; we get $x = (-7 \pm \sqrt{37})/2$.

To get a better look at horizontal asymptotes we change to the viewing rectangle $[-20, 20]$ by $[-5, 10]$ in Figure 8. It appears that $y = 1$ is the horizontal asymptote and this is easily confirmed:

$$\lim_{x \to \pm\infty} \frac{x^2 + 7x + 3}{x^2} = \lim_{x \to \pm\infty} \left(1 + \frac{7}{x} + \frac{3}{x^2}\right) = 1$$

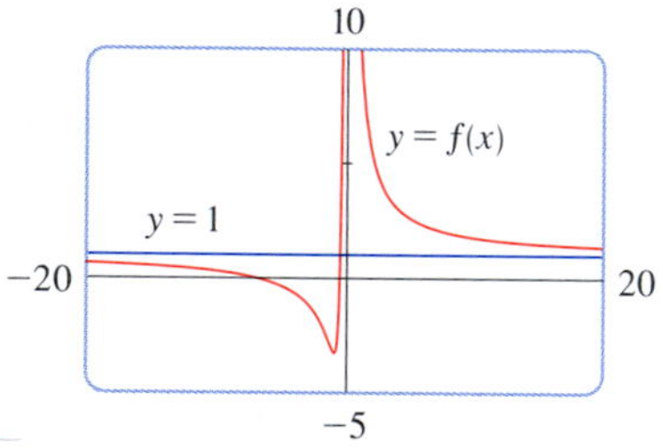

FIGURE 8

To estimate the minimum value we zoom in to the viewing rectangle $[-3, 0]$ by $[-4, 2]$ in Figure 9. The cursor indicates that the absolute minimum value is about -3.1 when $x \approx -0.9$ and we see that the function decreases on $(-\infty, -0.9)$ and $(0, \infty)$ and increases on $(-0.9, 0)$. The exact values are obtained by differentiating:

$$f'(x) = -\frac{7}{x^2} - \frac{6}{x^3} = -\frac{7x + 6}{x^3}$$

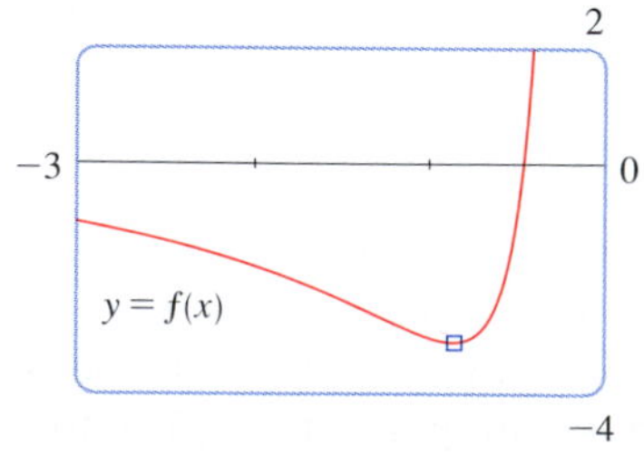

FIGURE 9

This shows that $f'(x) > 0$ when $-\frac{6}{7} < x < 0$ and $f'(x) < 0$ when $x < -\frac{6}{7}$ and when $x > 0$. The exact minimum value is $f(-\frac{6}{7}) = -\frac{37}{12} \approx -3.08$.

Figure 9 also shows that an inflection point occurs somewhere between $x = -1$ and $x = -2$. We could estimate it much more accurately using the graph of the second derivative, but in this case it is just as easy to find exact values. Since

$$f''(x) = \frac{14}{x^3} + \frac{18}{x^4} = 2\,\frac{7x + 9}{x^4}$$

we see that $f''(x) > 0$ when $x > -\frac{9}{7}$ $(x \neq 0)$. So f is concave upward on $(-\frac{9}{7}, 0)$ and $(0, \infty)$ and concave downward on $(-\infty, -\frac{9}{7})$. The inflection point is $(-\frac{9}{7}, -\frac{71}{27})$.

The analysis using the first two derivatives shows that Figures 7 and 8 display all the major aspects of the curve. ■

EXAMPLE 3 Graph the function $f(x) = \dfrac{x^2(x+1)^3}{(x-2)^2(x-4)^4}$.

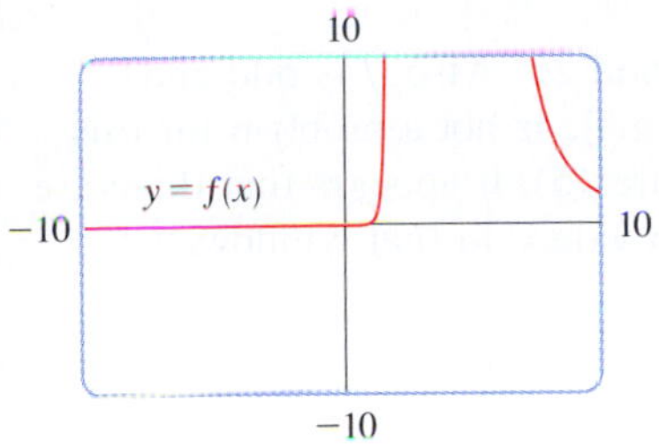

FIGURE 10

SOLUTION Drawing on our experience with a rational function in Example 2, let's start by graphing f in the viewing rectangle $[-10, 10]$ by $[-10, 10]$. From Figure 10 we have the feeling that we are going to have to zoom in to see some finer detail and also to zoom out to see the larger picture. But, as a guide to intelligent zooming, let's first take a close look at the expression for $f(x)$. Because of the factors $(x-2)^2$ and $(x-4)^4$ in the denominator we expect $x = 2$ and $x = 4$ to be the vertical asymptotes. Indeed

$$\lim_{x\to 2} \frac{x^2(x+1)^3}{(x-2)^2(x-4)^4} = \infty \qquad \text{and} \qquad \lim_{x\to 4} \frac{x^2(x+1)^3}{(x-2)^2(x-4)^4} = \infty$$

To find the horizontal asymptotes we divide numerator and denominator by x^6:

$$\frac{x^2(x+1)^3}{(x-2)^2(x-4)^4} = \frac{\dfrac{1}{x}\left(1+\dfrac{1}{x}\right)^3}{\left(1-\dfrac{2}{x}\right)^2\left(1-\dfrac{4}{x}\right)^4} \to 0 \quad \text{as} \quad x \to \pm\infty$$

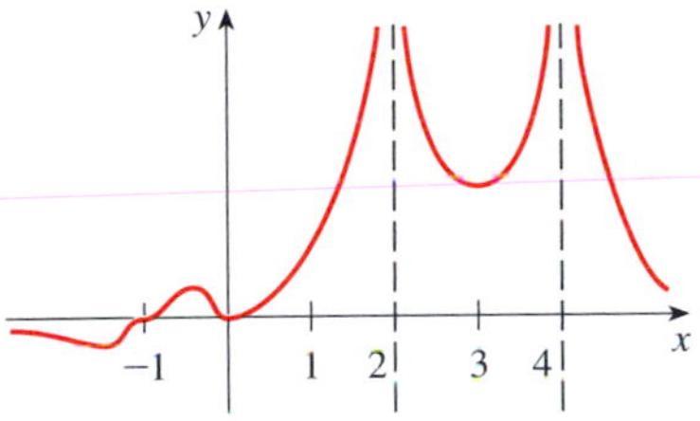

FIGURE 11

so the x-axis is the horizontal asymptote.

It's also very useful to consider the behavior of the graph near the x-intercepts. Since x^2 is positive, $f(x)$ does not change sign at 0 and so its graph doesn't cross the x-axis at 0. But, because of the factor $(x+1)^3$, the graph does cross the x-axis at -1 and has a horizontal tangent there. Putting all this information together, but without using derivatives, we see that the curve has to look something like the one in Figure 11.

Now that we know what to look for, we zoom in (several times) to produce the graphs in Figures 12 and 13 and zoom out (several times) to get Figure 14.

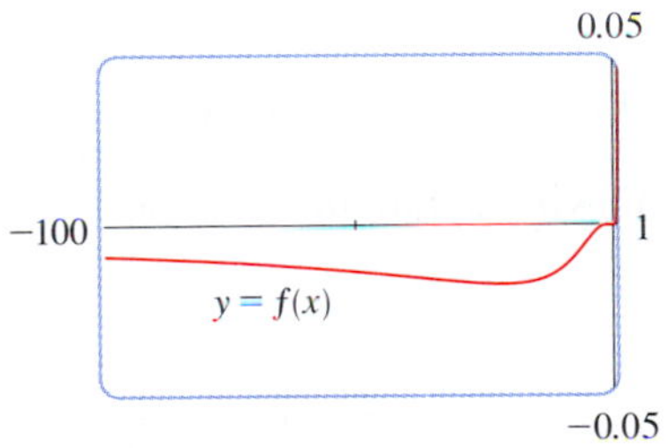

FIGURE 12

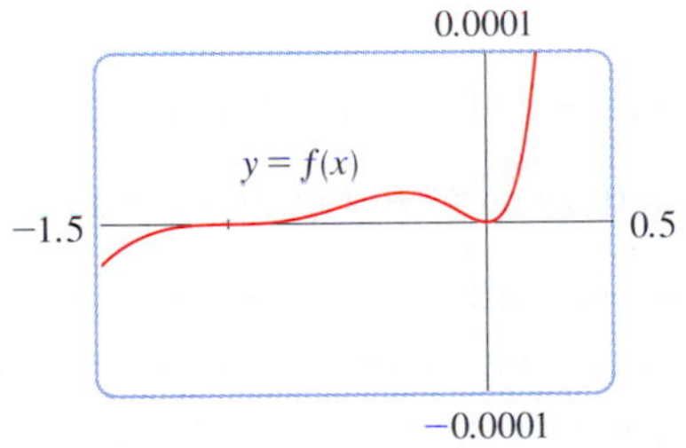

FIGURE 13

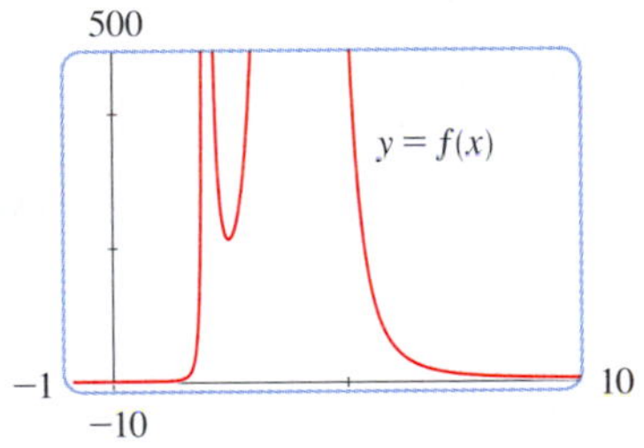

FIGURE 14

We can read from these graphs that the absolute minimum is about -0.02 and occurs when $x \approx -20$. There is also a local maximum ≈ 0.00002 when $x \approx -0.3$ and a local minimum ≈ 211 when $x \approx 2.5$. These graphs also show three inflection points near -35, -5, and -1 and two between -1 and 0. To estimate the inflection points closely we would need to graph f'', but to compute f'' by hand is an unreasonable chore. If you have a computer algebra system, then it is easy (see Exercise 13).

We have seen that, for this particular function, *three* graphs (Figures 12, 13, and 14) are necessary to convey all the useful information. The only way to display all these features of the function on a single graph is to draw it by hand. Despite the exaggerations and distortions, Figure 11 does manage to summarize the essential nature of the function. ■

The family of functions

$$f(x) = \sin(x + \sin cx)$$

where c is a constant, occurs in applications to frequency modulation (FM) synthesis. A sine wave is modulated by a wave with a different frequency ($\sin cx$). The case where $c = 2$ is studied in Example 4. Exercise 15 explores another special case.

EXAMPLE 4 Graph the function $f(x) = \sin(x + \sin 2x)$. For $0 \leqslant x \leqslant \pi$, locate all maximum and minimum values, intervals of increase and decrease, and inflection points correct to one decimal place.

SOLUTION We first note that f is periodic with period 2π. Also, f is odd and $|f(x)| \leqslant 1$ for all x. So the choice of a viewing rectangle is not a problem for this function: we start with $[0, \pi]$ by $[-1.1, 1.1]$ (see Figure 15). It appears that there are three local maximum values and two local minimum values in that window.

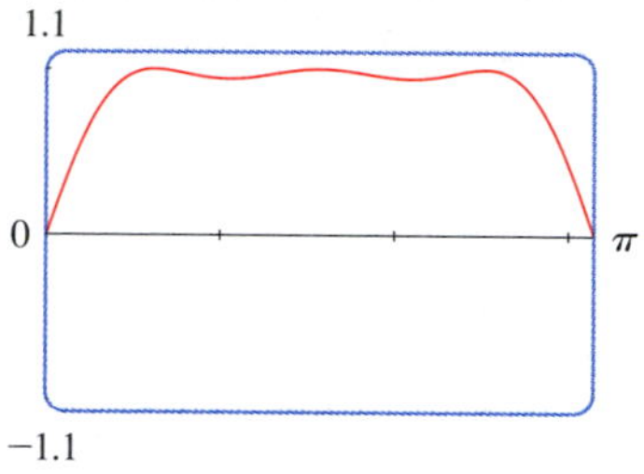

FIGURE 15

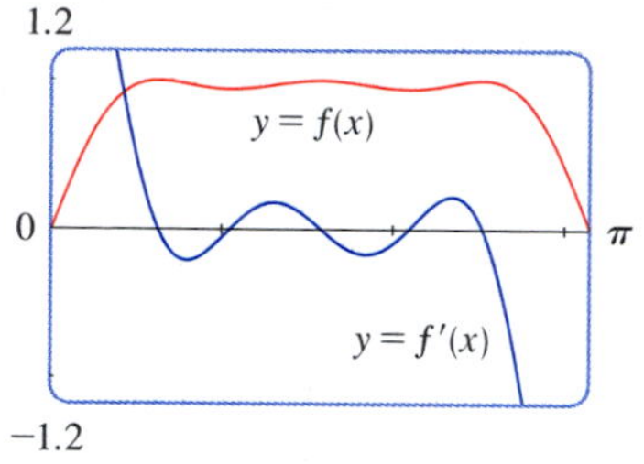

FIGURE 16

To confirm this and locate them more accurately, we calculate that

$$f'(x) = \cos(x + \sin 2x) \cdot (1 + 2\cos 2x)$$

and graph both f and f' in Figure 16. Using zoom-in and the First Derivative Test, we find the following values to one decimal place.

Intervals of increase:	$(0, 0.6), (1.0, 1.6), (2.1, 2.5)$
Intervals of decrease:	$(0.6, 1.0), (1.6, 2.1), (2.5, \pi)$
Local maximum values:	$f(0.6) \approx 1, f(1.6) \approx 1, f(2.5) \approx 1$
Local minimum values:	$f(1.0) \approx 0.94, f(2.1) \approx 0.94$

The second derivative is

$$f''(x) = -(1 + 2\cos 2x)^2 \sin(x + \sin 2x) - 4\sin 2x \cos(x + \sin 2x)$$

Graphing both f and f'' in Figure 17, we obtain the following approximate values:

Concave upward on:	$(0.8, 1.3), (1.8, 2.3)$
Concave downward on:	$(0, 0.8), (1.3, 1.8), (2.3, \pi)$
Inflection points:	$(0, 0), (0.8, 0.975), (1.3, 0.97), (1.8, 0.97), (2.3, 0.975)$

Having checked that Figure 15 does indeed represent f accurately for $0 \leqslant x \leqslant \pi$, we can state that the extended graph in Figure 18 represents f accurately for $-2\pi \leqslant x \leqslant 2\pi$.

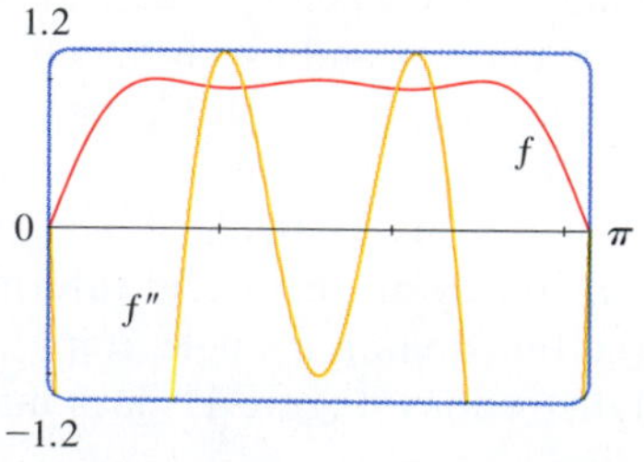

FIGURE 17

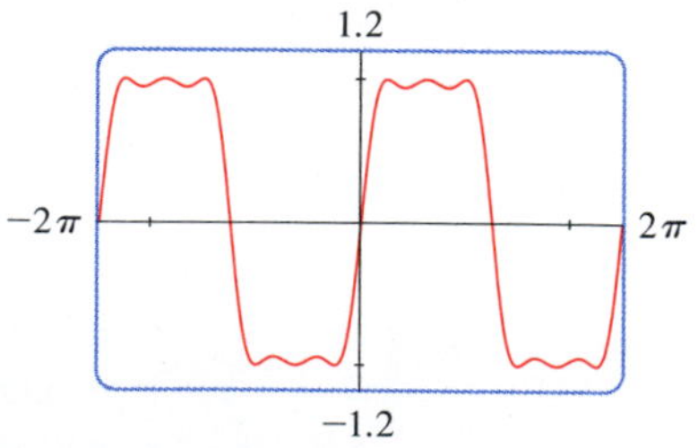

FIGURE 18 ■

EXAMPLE 5 How does the graph of $f(x) = 1/(x^2 + 2x + c)$ vary as c varies?

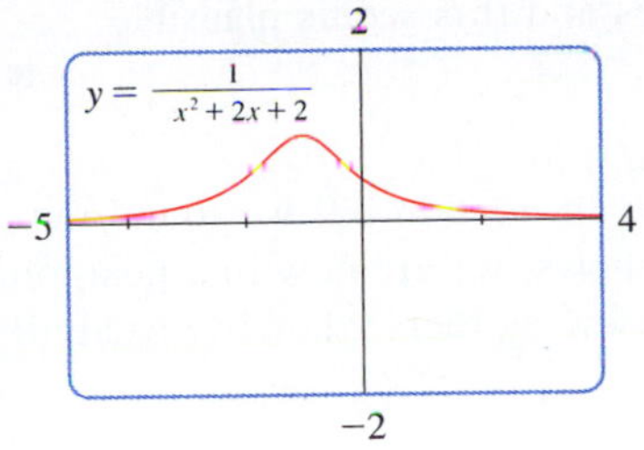

FIGURE 19
$c = 2$

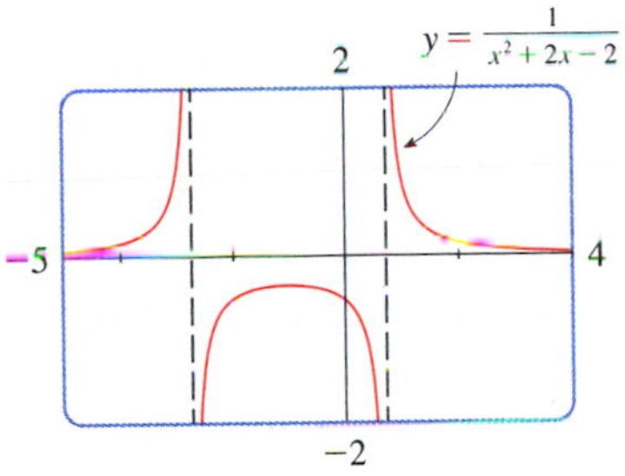

FIGURE 20
$c = -2$

SOLUTION The graphs in Figures 19 and 20 (the special cases $c = 2$ and $c = -2$) show two very different-looking curves. Before drawing any more graphs, let's see what members of this family have in common. Since

$$\lim_{x \to \pm\infty} \frac{1}{x^2 + 2x + c} = 0$$

for any value of c, they all have the x-axis as a horizontal asymptote. A vertical asymptote will occur when $x^2 + 2x + c = 0$. Solving this quadratic equation, we get $x = -1 \pm \sqrt{1 - c}$. When $c > 1$, there is no vertical asymptote (as in Figure 19). When $c = 1$ the graph has a single vertical asymptote $x = -1$ because

$$\lim_{x \to -1} \frac{1}{x^2 + 2x + 1} = \lim_{x \to -1} \frac{1}{(x + 1)^2} = \infty$$

When $c < 1$ there are two vertical asymptotes: $x = -1 + \sqrt{1 - c}$ and $x = -1 - \sqrt{1 - c}$ (as in Figure 20).

Now we compute the derivative:

$$f'(x) = -\frac{2x + 2}{(x^2 + 2x + c)^2}$$

This shows that $f'(x) = 0$ when $x = -1$ (if $c \neq 1$), $f'(x) > 0$ when $x < -1$, and $f'(x) < 0$ when $x > -1$. For $c \geqslant 1$ this means that f increases on $(-\infty, -1)$ and decreases on $(-1, \infty)$. For $c > 1$, there is an absolute maximum value $f(-1) = 1/(c - 1)$. For $c < 1$, $f(-1) = 1/(c - 1)$ is a local maximum value and the intervals of increase and decrease are interrupted at the vertical asymptotes.

Figure 21 is a "slide show" displaying five members of the family, all graphed in the viewing rectangle $[-5, 4]$ by $[-2, 2]$.

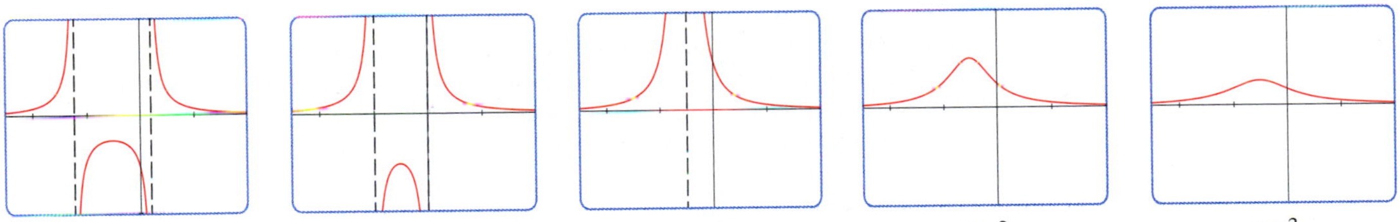

$c = -1$ $c = 0$ $c = 1$ $c = 2$ $c = 3$

FIGURE 21
The family of functions
$f(x) = \frac{1}{x^2 + 2x + c}$

As predicted, $c = 1$ is the value at which a transition takes place from two vertical asymptotes to one, and then to none. As c increases from 1, we see that the maximum point becomes lower; this is explained by the fact that $1/(c - 1) \to 0$ as $c \to \infty$. As c decreases from 1, the vertical asymptotes become more widely separated because the distance between them is $2\sqrt{1 - c}$, which becomes large as $c \to -\infty$. Again, the maximum point approaches the x-axis because $1/(c - 1) \to 0$ as $c \to -\infty$.

There is clearly no inflection point when $c \leqslant 1$. For $c > 1$ we calculate that

$$f''(x) = \frac{2(3x^2 + 6x + 4 - c)}{(x^2 + 2x + c)^3}$$

and deduce that inflection points occur when $x = -1 \pm \sqrt{3(c - 1)}/3$. So the inflection points become more spread out as c increases and this seems plausible from the last two parts of Figure 21. ■

In Section 1.3 we introduced parametric curves and in Section 3.9 we found tangents to parametric curves. But, as our final example shows, we are now in a position to use calculus to ensure that a parameter interval or a viewing rectangle will reveal all the important aspects of a curve.

EXAMPLE 6 Graph the curve with parametric equations

$$x(t) = t^2 + t + 1 \qquad y(t) = 3t^4 - 8t^3 - 18t^2 + 25$$

in a viewing rectangle that displays the important features of the curve. Find the coordinates of the interesting points on the curve.

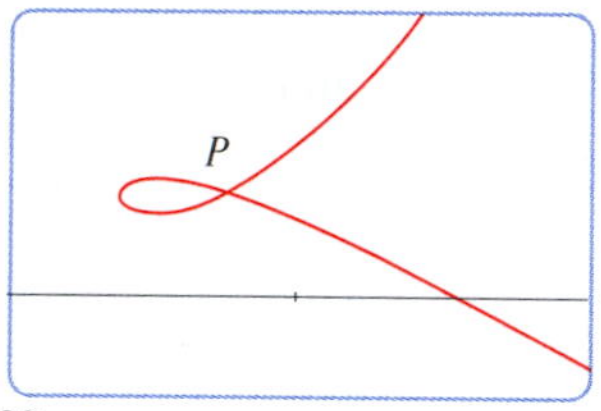

FIGURE 22

SOLUTION Figure 22 shows the graph of this curve in the viewing rectangle $[0, 4]$ by $[-20, 60]$. Zooming in toward the point P where the curve intersects itself, we estimate that the coordinates of P are $(1.50, 22.25)$. We estimate the highest point on the loop has coordinates $(1, 25)$, the lowest point $(1, 18)$, and the leftmost point $(0.75, 21.7)$. To be sure that we have discovered all the interesting aspects of the curve, however, we need to use calculus. From Equation 3.9.1, we have

$$\frac{dy}{dx} = \frac{dy/dt}{dx/dt} = \frac{12t^3 - 24t^2 - 36t}{2t + 1}$$

The vertical tangent occurs when $dx/dt = 2t + 1 = 0$, that is, $t = -\frac{1}{2}$. So the exact coordinates of the leftmost point of the loop are $x(-\frac{1}{2}) = 0.75$ and $y(-\frac{1}{2}) = 21.6875$. Also,

$$\frac{dy}{dt} = 12t(t^2 - 2t - 3) = 12t(t + 1)(t - 3)$$

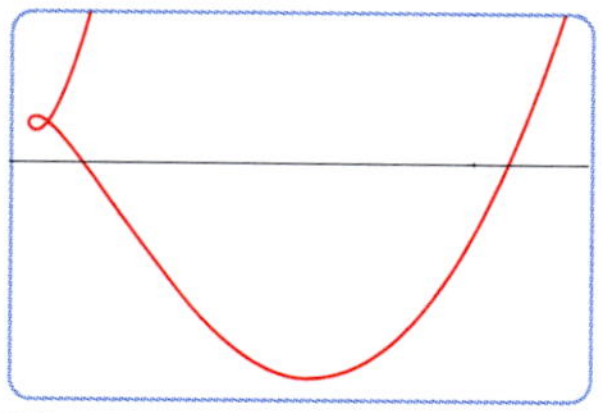

FIGURE 23

and so horizontal tangents occur when $t = 0, -1$, and 3. The bottom of the loop corresponds to $t = -1$ and, indeed, its coordinates are $x(-1) = 1$ and $y(-1) = 18$. Similarly, the coordinates of the top of the loop are exactly what we estimated: $x(0) = 1$ and $y(0) = 25$. But what about the parameter value $t = 3$? The corresponding point on the curve has coordinates $x(3) = 13$ and $y(3) = -110$. Figure 23 shows the graph of the curve in the viewing rectangle $[0, 25]$ by $[-120, 80]$. This shows that the point $(13, -110)$ is the lowest point on the curve. We can now be confident that there are no hidden maximum or minimum points. ■

EXERCISES 5.4

1–6 ■ Produce graphs of f that reveal all the important aspects of the curve. In particular, you should use graphs of f' and f'' to estimate the intervals of increase and decrease, extreme values, intervals of concavity, and inflection points.

1. $f(x) = 4x^4 - 7x^2 + 4x + 6$

2. $f(x) = 8x^5 + 45x^4 + 80x^3 + 90x^2 + 200x$

3. $f(x) = \sqrt[3]{x^2 - 3x - 5}$

4. $f(x) = \dfrac{x^4 + x^3 - 2x^2 + 2}{x^2 + x - 2}$

5. $f(x) = x^2 \sin x, \; -7 \leqslant x \leqslant 7$

6. $f(x) = \sin x + \frac{1}{3}\sin 3x$

7–8 ■ Produce graphs of f that reveal all the important aspects of the curve. Estimate the intervals of increase and decrease, extreme values, intervals of concavity, and inflection points, and use calculus to find these quantities exactly.

7. $f(x) = 8x^3 - 3x^2 - 10$

8. $f(x) = x\sqrt{9 - x^2}$

9–10 ▪ Produce a graph of f that shows all the important aspects of the curve. Estimate the local maximum and minimum values and then use calculus to find these values exactly. Use a graph of f'' to estimate the inflection points.

9. $f(x) = e^{x^3 - x}$ **10.** $f(x) = e^{\cos x}$

11–12 ▪ Sketch the graph by hand using asymptotes and intercepts, but not derivatives. Then use your sketch as a guide to producing graphs (with a graphing device) that display the major features of the curve. Use these graphs to estimate the maximum and minimum values.

11. $f(x) = \dfrac{(x+4)(x-3)^2}{x^4(x-1)}$

12. $f(x) = \dfrac{10x(x-1)^4}{(x-2)^3(x+1)^2}$

CAS **13.** If f is the function considered in Example 3, use a computer algebra system to calculate f' and then graph it to confirm that all the maximum and minimum values are as given in the example. Calculate f'' and use it to estimate the intervals of concavity and inflection points.

CAS **14.** If f is the function of Exercise 12, find f' and f'' and use their graphs to estimate the intervals of increase and decrease and the concavity of f.

15. In Example 4 we considered a member of the family of functions $f(x) = \sin(x + \sin cx)$ that occur in FM synthesis. Here we investigate the function with $c = 3$. Start by graphing f in the viewing rectangle $[0, \pi]$ by $[-1.2, 1.2]$. How many local maximum points do you see? The graph has more than are visible to the naked eye. To discover the hidden maximum and minimum points you will need to examine the graph of f' very carefully. In fact, it helps to look at the graph of f'' at the same time. Find all the maximum and minimum values and inflection points. Then graph f in the viewing rectangle $[-2\pi, 2\pi]$ by $[-1.2, 1.2]$ and comment on symmetry.

16. Use a graph to estimate the coordinates of the leftmost point on the curve

$$x = t^4 - t^2 \qquad y = t + \ln t$$

Then use calculus to find the exact coordinates.

17–18 ▪ Graph the curve in a viewing rectangle that displays all the important aspects of the curve. At what points does the curve have vertical or horizontal tangents?

17. $x = t^4 - 2t^3 - 2t^2, \quad y = t^3 - t$

18. $x = t^4 + 4t^3 - 8t^2, \quad y = 2t^2 - t$

19. Investigate the family of curves given by the parametric equations $x = t^3 - ct$, $y = t^2$. In particular, determine the values of c for which there is a loop and find the point where the curve intersects itself. What happens to the loop as c increases? Find the coordinates of the leftmost and rightmost points of the loop.

20. The family of functions $f(t) = C(e^{-at} - e^{-bt})$, where a, b, and C are positive numbers and $b > a$, has been used to model the concentration of a drug injected into the blood at time $t = 0$. Graph several members of this family. What do they have in common? For fixed values of C and a, discover graphically what happens as b increases. Then use calculus to prove what you have discovered.

21–25 ▪ Describe how the graph of f varies as c varies. Graph several members of the family to illustrate the trends that you discover. In particular, you should investigate how maximum and minimum points and inflection points move when c changes. You should also identify any transitional values of c at which the basic shape of the curve changes.

21. $f(x) = \dfrac{cx}{1 + c^2x^2}$ **22.** $f(x) = \ln(x^2 + c)$

23. $f(x) = e^{-c/x^2}$ **24.** $f(x) = \dfrac{1}{(1 - x^2)^2 + cx^2}$

25. $f(x) = x^4 + cx^2$

26. Investigate the family of curves given by the equation $f(x) = x^4 + cx^2 + x$. Start by determining the transitional value of c at which the number of inflection points changes. Then graph several members of the family to see what shapes are possible. There is another transitional value of c at which the number of critical numbers changes. Try to discover it graphically. Then prove what you have discovered.

5.5 APPLIED MAXIMUM AND MINIMUM PROBLEMS

The methods we have learned in this chapter for finding extreme values have practical applications in many areas of life. A businessperson wants to minimize costs and maximize profits. Fermat's Principle in optics states that light follows the path that takes the least time. In this section and the next we solve such problems as maximizing areas, volumes, and profits and minimizing distances, times, and costs.

In solving such practical problems the greatest challenge is often to convert the word problem into a maximum-minimum problem by setting up the function that is to be maximized or minimized. Let's recall the problem-solving principles discussed in Section 4 of Review and Preview and adapt them to this situation:

STEPS IN SOLVING APPLIED MAXIMUM AND MINIMUM PROBLEMS

1. UNDERSTAND THE PROBLEM. The first step is to read the problem carefully until it is clearly understood. Ask yourself: What is the unknown? What are the given quantities? What are the given conditions?

2. DRAW A DIAGRAM. In most problems it is useful to draw a diagram and identify the given and required quantities on the diagram.

3. INTRODUCE NOTATION. Assign a symbol to the quantity that is to be maximized or minimized (let us call it Q for now). Also select symbols $(a, b, c, \ldots, x, y)$ for other unknown quantities and label the diagram with these symbols. It may help to use initials as suggestive symbols—for example, A for area, h for height, t for time.

4. Express Q in terms of some of the other symbols from step 3.

5. If Q has been expressed as a function of more than one variable in step 4, use the given information to find relationships (in the form of equations) among these variables. Then use these equations to eliminate all but one of the variables in the expression for Q. Thus, Q will be given as a function of *one* variable x, say, $Q = f(x)$. Write the domain of this function.

6. Use the methods of Sections 5.2 and 5.3 to find the *absolute* maximum or minimum value of f. In particular, if the domain of f is a closed interval, then the Closed Interval Method (5.2.7) can be used.

EXAMPLE 1 A farmer has 2400 ft of fencing and wants to fence off a rectangular field that borders a straight river. He needs no fence along the river. What are the dimensions of the field that has the largest area?

Understand the problem
Analogy: Try special cases
Draw diagrams

SOLUTION In order to get a feeling for what is happening in this problem let's experiment with some special cases. Figure 1 (not to scale) shows three possible ways of laying out the 2400 ft of fencing. We see that when we try shallow, wide fields or deep, narrow fields, we get relatively small areas. It seems plausible that there is some intermediate configuration that produces the largest area.

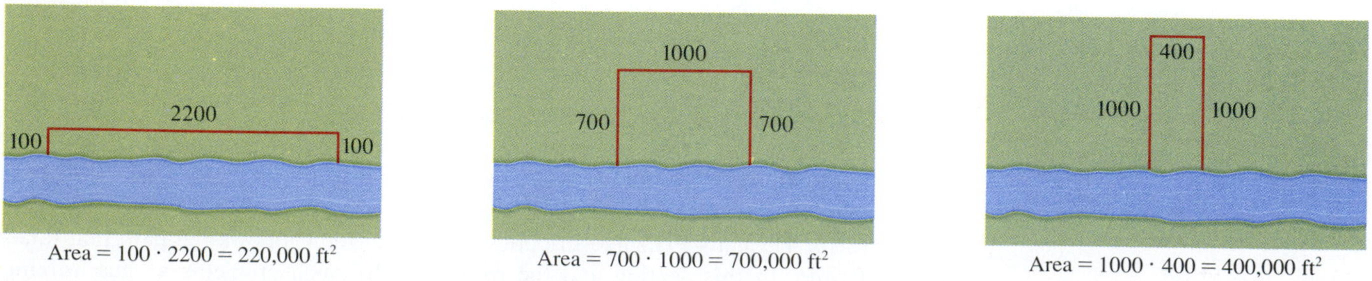

FIGURE 1

Introduce notation

Figure 2 illustrates the general case. We wish to maximize the area A of the rectangle. Let x and y be the width and length of the rectangle (in feet). Then we express A in terms of x and y:

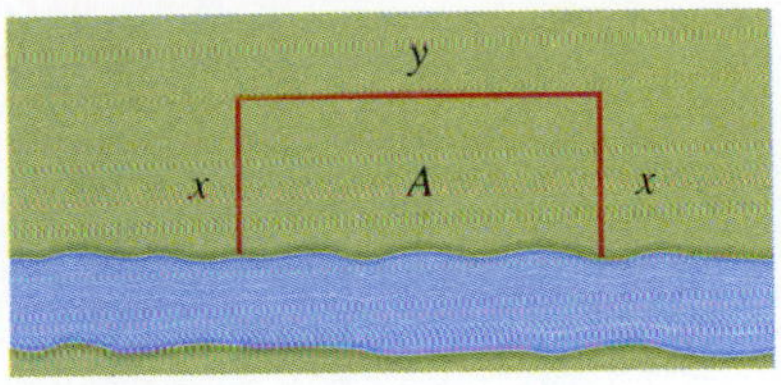

FIGURE 2

$$A = xy$$

We want to express A as a function of just one variable, so we eliminate y by expressing it in terms of x. To do this we use the given information that the total length of the fencing is 2400 ft. Thus

$$2x + y = 2400$$

From this equation we have $y = 2400 - 2x$, which gives

$$A = x(2400 - 2x) = 2400x - 2x^2$$

Note that $x \geqslant 0$ and $x \leqslant 1200$ (otherwise $A < 0$). So the function that we wish to maximize is

$$A(x) = 2400x - 2x^2 \qquad 0 \leqslant x \leqslant 1200$$

The derivative is $A'(x) = 2400 - 4x$, so to find the critical numbers we solve the equation

$$2400 - 4x = 0$$

which gives $x = 600$. The maximum value of A must occur either at this critical number or at an endpoint of the interval. Since $A(0) = 0$, $A(600) = 720,000$, and $A(1200) = 0$, the Closed Interval Method (5.2.7) gives the maximum value as $A(600) = 720,000$.

[Alternatively, we could have observed that $A''(x) = -4 < 0$ for all x, so A is always concave downward and the local maximum at $x = 600$ must be an absolute maximum.]

Thus, the rectangular field should be 600 ft deep and 1200 ft wide. ■

EXAMPLE 2 A can is to be made to hold 1 L of oil. Find the dimensions that will minimize the cost of the metal to manufacture the can.

SOLUTION Draw the diagram as in Figure 3 where r is the radius and h the height (in centimeters). In order to minimize the cost of the metal, we minimize the total surface area of the cylinder (top, bottom, and sides), which is

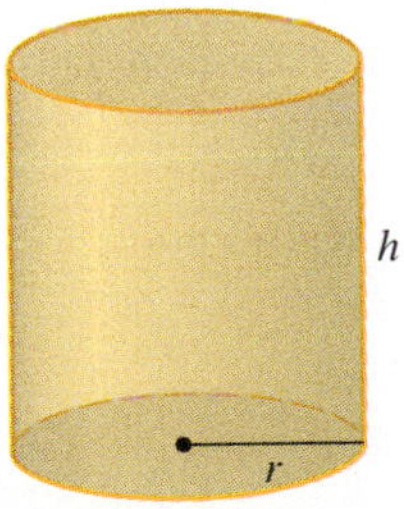

FIGURE 3

$$A = 2\pi r^2 + 2\pi rh$$

To eliminate h we use the fact that the volume is given as 1 L, which we take to be 1000 cm^3. Thus

$$\pi r^2 h = 1000$$

which gives $h = 1000/(\pi r^2)$. Substitution of this into the expression for A gives

$$A = 2\pi r^2 + 2\pi r\left(\frac{1000}{\pi r^2}\right) = 2\pi r^2 + \frac{2000}{r}$$

Therefore, the function that we want to minimize is

$$A(r) = 2\pi r^2 + \frac{2000}{r} \qquad r > 0$$

To find the critical numbers, we differentiate:

$$A'(r) = 4\pi r - \frac{2000}{r^2} = \frac{4(\pi r^3 - 500)}{r^2}$$

Then $A'(r) = 0$ when $\pi r^3 = 500$, so the only critical number is $r = \sqrt[3]{500/\pi}$.

Since the domain of A is $(0, \infty)$, we cannot use the argument of Example 1 concerning endpoint extrema. But we can observe that $A'(r) < 0$ for $r < \sqrt[3]{500/\pi}$ and $A'(r) > 0$ for $r > \sqrt[3]{500/\pi}$, so A is decreasing for *all* r to the left of the critical number and increasing for *all* r to the right. Thus, $r = \sqrt[3]{500/\pi}$ must give rise to an *absolute* minimum.

[Alternatively, we could argue that $A(r) \to \infty$ as $r \to 0^+$ and $A(r) \to \infty$ as $r \to \infty$, so there must be a minimum value of $A(r)$, which must occur at the critical number. See Figure 4.]

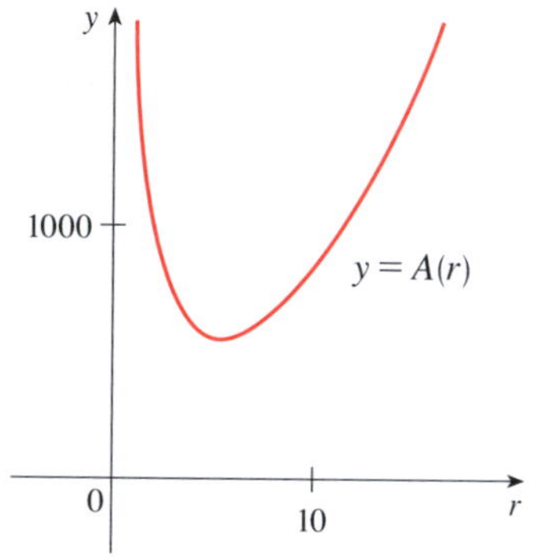

FIGURE 4

The value of h corresponding to $r = \sqrt[3]{500/\pi}$ is

$$h = \frac{1000}{\pi r^2} = \frac{1000}{\pi(500/\pi)^{2/3}} = 2\sqrt[3]{\frac{500}{\pi}} = 2r$$

Thus, to minimize the cost of the can, the radius should be $\sqrt[3]{500/\pi}$ cm and the height should be equal to twice the radius, namely, the diameter. ■

NOTE: The argument used in Example 2 to justify the absolute minimum is a variant of the First Derivative Test (which applies only to local extrema) and is stated here for future reference.

FIRST DERIVATIVE TEST FOR ABSOLUTE EXTREMA Suppose that c is a critical number of a continuous function f defined on an interval.

(a) If $f'(x) > 0$ for all $x < c$ and $f'(x) < 0$ for all $x > c$, then $f(c)$ is the absolute maximum value of f.

(b) If $f'(x) < 0$ for all $x < c$ and $f'(x) > 0$ for all $x > c$, then $f(c)$ is the absolute minimum value of f.

EXAMPLE 3 Find the point on the parabola $y^2 = 2x$ that is closest to the point $(1, 4)$.

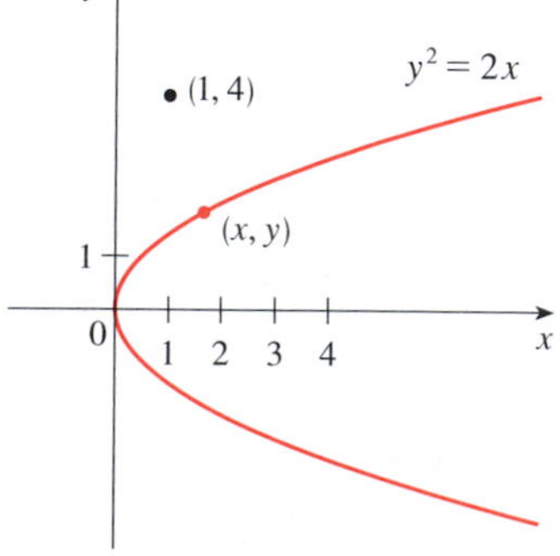

FIGURE 5

SOLUTION The distance between the point $(1, 4)$ and the point (x, y) is

$$d = \sqrt{(x - 1)^2 + (y - 4)^2}$$

(See Figure 5.) But if (x, y) lies on the parabola, then $x = y^2/2$, so the expression for d becomes

$$d = \sqrt{\left(\frac{y^2}{2} - 1\right)^2 + (y - 4)^2}$$

(Alternatively, we could have substituted $y = \sqrt{2x}$ to get d in terms of x alone.) Instead of minimizing d, we minimize its square:

$$d^2 = f(y) = \left(\frac{y^2}{2} - 1\right)^2 + (y - 4)^2$$

(You should convince yourself that the minimum of d occurs at the same point as the minimum of d^2, but d^2 is easier to work with.) Differentiating, we obtain

$$f'(y) = 2\left(\frac{y^2}{2} - 1\right)y + 2(y - 4) = y^3 - 8$$

so $f'(y) = 0$ when $y = 2$. Observe that $f'(y) < 0$ when $y < 2$ and $f'(y) > 0$ when $y > 2$, so by the First Derivative Test for Absolute Extrema, the absolute minimum occurs when $y = 2$. (Or we could simply say that because of the geometric nature of the problem, it is obvious that there is a closest point but not a farthest point.) The corresponding value of x is $x = y^2/2 = 2$. Thus, the point on $y^2 = 2x$ closest to $(1, 4)$ is $(2, 2)$. ■

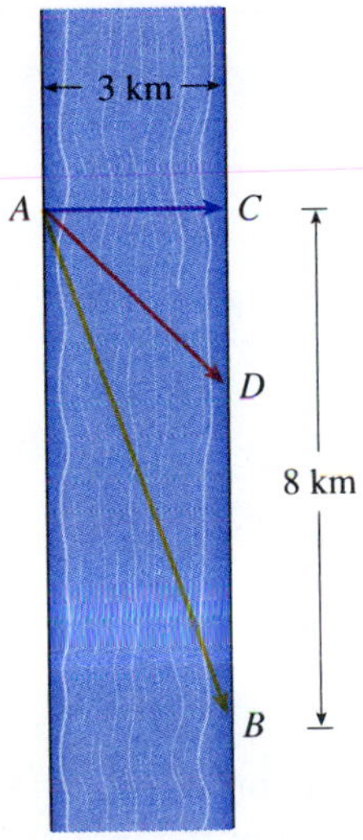

FIGURE 6

EXAMPLE 4 A man is at point A on a bank of a straight river, 3 km wide, and wants to reach point B, 8 km downstream on the opposite bank, as quickly as possible (see Figure 6). He could row his boat directly across the river to point C and then run to B, or he could row directly to B, or he could row to some point D between C and B and then run to B. If he can row at 6 km/h and run at 8 km/h, where should he land to reach B as soon as possible?

SOLUTION Let x be the distance from C to D. Then the running distance is $|DB| = 8 - x$ and the Pythagorean Theorem gives the rowing distance as $|AD| = \sqrt{x^2 + 9}$. We assume the speed of the water is 0 km/h and use the equation

$$\text{time} = \frac{\text{distance}}{\text{rate}}$$

Then the rowing time is $\sqrt{x^2 + 9}/6$ and the running time is $(8 - x)/8$, so the total time T as a function of x is

$$T(x) = \frac{\sqrt{x^2 + 9}}{6} + \frac{8 - x}{8}$$

The domain of this function T is $[0, 8]$. Notice that if $x = 0$ he rows to C and if $x = 8$ he rows directly to B. The derivative of T is

$$T'(x) = \frac{x}{6\sqrt{x^2 + 9}} - \frac{1}{8}$$

Thus, using the fact that $x \geqslant 0$, we have

$$T'(x) = 0 \iff \frac{x}{6\sqrt{x^2 + 9}} = \frac{1}{8} \iff 4x = 3\sqrt{x^2 + 9}$$

$$\iff 16x^2 = 9(x^2 + 9) \iff 7x^2 = 81$$

$$\iff x = \frac{9}{\sqrt{7}}$$

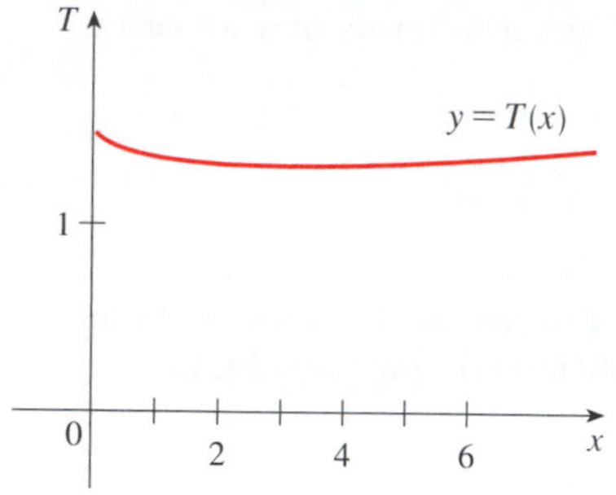

FIGURE 7

The only critical number is $x = 9/\sqrt{7}$. To see whether the minimum occurs at this critical number or at an endpoint of the domain $[0, 8]$, we evaluate T at all three points:

$$T(0) = 1.5 \qquad T\left(\frac{9}{\sqrt{7}}\right) = 1 + \frac{\sqrt{7}}{8} \approx 1.33 \qquad T(8) = \frac{\sqrt{73}}{6} \approx 1.42$$

Since the smallest of these values of T occurs when $x = 9/\sqrt{7}$, the absolute minimum value of T must occur there. Figure 7 illustrates this calculation by showing the graph of T.

Thus, the man should land the boat at a point $9/\sqrt{7}$ km ($\approx$3.4 km) downstream from his starting point. ■

EXAMPLE 5 Find the area of the largest rectangle that can be inscribed in a semicircle of radius r.

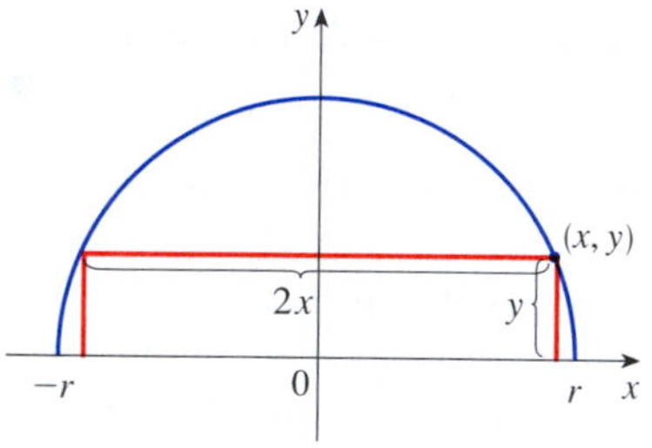

FIGURE 8

SOLUTION 1 Let's take the semicircle to be the upper half of the circle $x^2 + y^2 = r^2$ with center the origin. Then the word *inscribed* means that the rectangle has two vertices on the semicircle and two vertices on the x-axis as shown in Figure 8.

Let (x, y) be the vertex that lies in the first quadrant. Then the rectangle has sides of lengths $2x$ and y, so its area is

$$A = 2xy$$

To eliminate y we use the fact that (x, y) lies on the circle $x^2 + y^2 = r^2$ and so $y = \sqrt{r^2 - x^2}$. Thus

$$A = 2x\sqrt{r^2 - x^2}$$

The domain of this function is $0 \leqslant x \leqslant r$. Its derivative is

$$A' = 2\sqrt{r^2 - x^2} - \frac{2x^2}{\sqrt{r^2 - x^2}} = \frac{2(r^2 - 2x^2)}{\sqrt{r^2 - x^2}}$$

which is 0 when $2x^2 = r^2$, that is, $x = r/\sqrt{2}$ (since $x \geqslant 0$). This value of x gives a maximum value of A since $A(0) = 0$ and $A(r) = 0$. Therefore, the area of the largest inscribed rectangle is

$$A\left(\frac{r}{\sqrt{2}}\right) = 2\frac{r}{\sqrt{2}}\sqrt{r^2 - \frac{r^2}{2}} = r^2$$

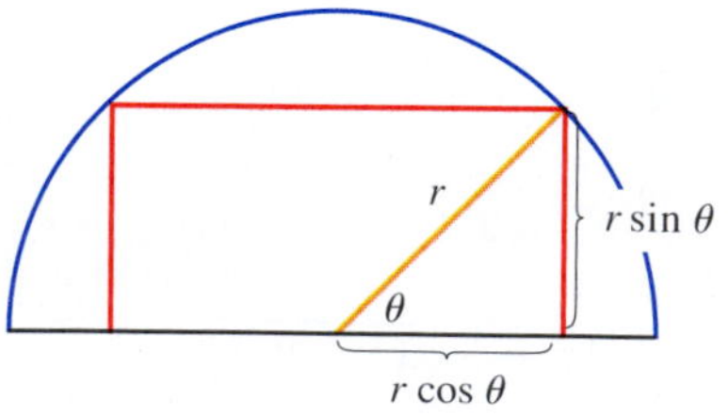

FIGURE 9

SOLUTION 2 A simpler solution is possible if we think of using an angle as a variable. Let θ be the angle shown in Figure 9. Then the area of the rectangle is

$$\begin{aligned} A(\theta) &= (2r\cos\theta)(r\sin\theta) \\ &= r^2(2\sin\theta\cos\theta) \\ &= r^2\sin 2\theta \end{aligned}$$

We know that $\sin 2\theta$ has a maximum value of 1 and it occurs when $2\theta = \pi/2$. So $A(\theta)$ has a maximum value of r^2 and it occurs when $\theta = \pi/4$.

Notice that this trigonometric solution does not involve differentiation. In fact, we didn't need to use calculus at all. ■

EXERCISES 5.5

1. Find two numbers whose sum is 100 and whose product is a maximum.

2. Find two numbers whose difference is 100 and whose product is a minimum.

3. Find two positive numbers whose product is 100 and whose sum is a minimum.

4. Show that of all the rectangles with a given area, the one with smallest perimeter is a square.

5. Show that of all the rectangles with a given perimeter, the one with greatest area is a square.

6. A farmer wants to fence an area of 1.5 million square feet in a rectangular field and then divide it in half with a fence parallel to one of the sides of the rectangle. How can he do this so as to minimize the cost of the fence?

7. A farmer with 750 ft of fencing wants to enclose a rectangular area and then divide it into four pens with fencing parallel to one side of the rectangle. What is the largest possible total area of the four pens?

8. A box with a square base and open top must have a volume of 32,000 cm^3. Find the dimensions of the box that minimize the amount of material used.

9. If 1200 cm^2 of material is available to make a box with a square base and an open top, find the largest possible volume of the box.

10. A rectangular storage container with an open top is to have a volume of 10 m^3. The length of its base is twice the width. Material for the base costs \$10 per square meter. Material for the sides costs \$6 per square meter. Find the cost of materials for the cheapest such container.

11. Do Exercise 10 assuming the container has a lid that is made from the same material as the sides.

12. A box with an open top is to be constructed from a square piece of cardboard, 3 ft wide, by cutting out a square from each of the four corners and bending up the sides. Find the largest volume that such a box can have.

13. Find the point on the line $y = 2x - 3$ that is closest to the origin.

14. Find the point on the line $2x + 3y + 5 = 0$ that is closest to the point $(-1, -2)$.

15. Find the points on the hyperbola $y^2 - x^2 = 4$ that are closest to the point $(2, 0)$.

16. Find the point on the parabola $x + y^2 = 0$ that is closest to the point $(0, -3)$.

17. A particle moves in the xy-plane in such a way that its position at time t is $\mathbf{r}(t) = (t - \sin t)\,\mathbf{i} + (1 - \cos t)\,\mathbf{j}$. Find the particle's maximum speed.

18. A particle moves in the xy-plane in such a way that its position at time t is $\mathbf{r}(t) = (1 + 3\cos t)\,\mathbf{i} + 2\sin t\,\mathbf{j}$. Find the particle's maximum speed.

19. Find the dimensions of the rectangle of largest area that can be inscribed in a circle of radius r.

20. Find the area of the largest rectangle that can be inscribed in the ellipse $x^2/a^2 + y^2/b^2 = 1$.

21. Find the dimensions of the rectangle of largest area that can be inscribed in an equilateral triangle of side L if one side of the rectangle lies on the base of the triangle.

22. Find the dimensions of the rectangle of largest area that has its base on the x-axis and its other two vertices above the x-axis and lying on the parabola $y = 8 - x^2$.

23. Find the dimensions of the isosceles triangle of largest area that can be inscribed in a circle of radius r.

24. Find the area of the largest rectangle that can be inscribed in a right triangle with legs of lengths 3 cm and 4 cm if two sides of the rectangle lie along the legs.

25. A right circular cylinder is inscribed in a sphere of radius r. Find the largest possible volume of such a cylinder.

26. A right circular cylinder is inscribed in a cone with height h and base radius r. Find the largest possible volume of such a cylinder.

27. A right circular cylinder is inscribed in a sphere of radius r. Find the largest possible surface area of such a cylinder.

28. A Norman window has the shape of a rectangle surmounted by a semicircle. (Thus, the diameter of the semicircle is equal to the width of the rectangle.) If the perimeter of the window is 30 ft, find the dimensions of the window so that the greatest possible amount of light is admitted.

29. The top and bottom margins of a poster are each 6 cm and the side margins are each 4 cm. If the area of printed material on the poster is fixed at 384 cm^2, find the dimensions of the poster with the smallest area.

30. A poster is to have an area of 180 in^2 with 1-inch margins at the bottom and sides and a 2-inch margin at the top. What dimensions will give the largest printed area?

31. A piece of wire 10 m long is cut into two pieces. One piece is bent into a square and the other is bent into an equilateral triangle. How should the wire be cut so that the total area enclosed is (a) a maximum? (b) A minimum?

32. Answer Exercise 31 if one piece is bent into a square and the other into a circle.

33. A cylindrical can without a top is made to contain V cm^3 of liquid. Find the dimensions that will minimize the cost of the metal to make the can.

34. A fence 8 ft tall runs parallel to a tall building at a distance of 4 ft from the building. What is the length of the shortest ladder that will reach from the ground over the fence to the wall of the building?

35. A conical drinking cup is made from a circular piece of paper of radius R by cutting out a sector and joining the edges CA and CB. Find the maximum capacity of such a cup.

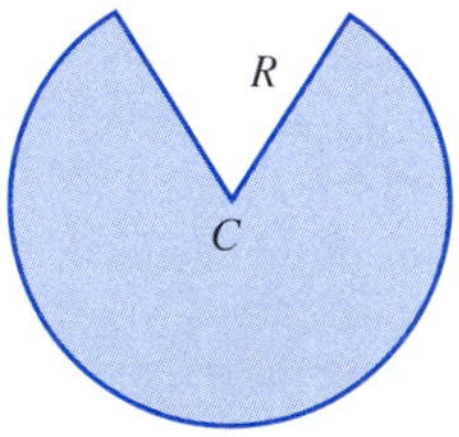

36. For a fish swimming at a speed v relative to the water, the energy expenditure per unit time is proportional to v^3. It is believed that migrating fish try to minimize the total energy required to swim a fixed distance. If the fish are swimming against a current u ($u < v$), then the time required to swim a distance L is $L/(v - u)$ and the total energy E required to swim the distance is given by

$$E(v) = av^3 \cdot \frac{L}{v - u}$$

where a is the proportionality constant.

(a) Determine the value of v that minimizes E.
(b) Sketch the graph of E.

Note: This result has been verified experimentally; migrating fish swim against a current at a speed 50% greater than the current speed.

37. In a beehive, each cell is a regular hexagonal prism, open at one end with a trihedral angle at the other end. It is believed that bees form their cells in such a way as to minimize the surface area for a given volume, thus using the least amount of wax in cell construction. Examination of these cells has shown that the measure of the apex angle θ is amazingly consistent. Based on the geometry of the cell, it can be shown that the surface area S is given by

$$S = 6sh - \tfrac{3}{2}s^2 \cot\theta + (3s^2\sqrt{3}/2)\csc\theta$$

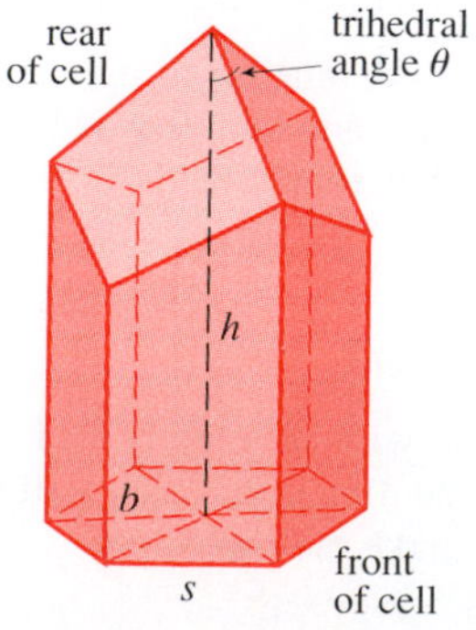

where s, the length of the sides of the hexagon, and h, the height, are constants.

(a) Calculate $dS/d\theta$.
(b) What angle should the bees prefer?
(c) Determine the minimum surface area of the cell (in terms of s and h).

Note: Actual measurements of the angle θ in beehives have been made, and the measures of these angles seldom differ from the calculated value by more than 2°.

38. A boat leaves a dock at 2:00 P.M. and travels due south at a speed of 20 km/h. Another boat has been heading due east at 15 km/h and reaches the same dock at 3:00 P.M. At what time were the two boats closest together?

39. Solve the problem in Example 4 if the river is 5 km wide and point B is only 5 km downstream from A.

40. A woman at a point A on the shore of a circular lake with radius 2 mi wants to be at the point C diametrically opposite A on the other side of the lake in the shortest possible time. She can walk at the rate of 4 mi/h and row a boat at 2 mi/h. At what angle θ to the diameter should she row?

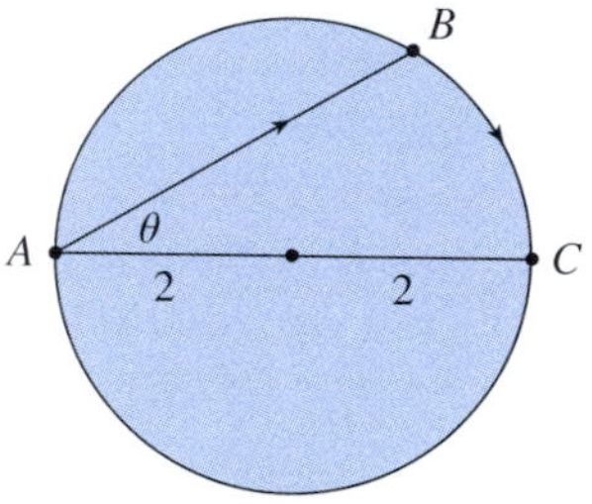

41. The illumination of an object by a light source is directly proportional to the strength of the source and inversely proportional to the square of the distance from the source. If two light sources, one three times as strong as the other, are placed 10 ft apart, where should an object be placed on the line between the sources so as to receive the least illumination?

42. Find an equation of the line through the point (3, 5) that cuts off the least area from the first quadrant.

43. Show that of all the isosceles triangles with a given perimeter, the one with the greatest area is equilateral.

44. The frame for a kite is to be made from six pieces of wood. The four exterior pieces have been cut with the

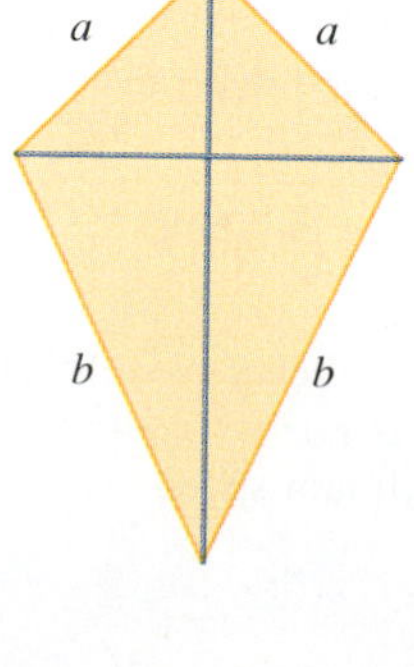

lengths indicated in the figure. To maximize the area of the kite, how long should the diagonal pieces be?

45. Let v_1 be the velocity of light in air and v_2 the velocity of light in water. According to Fermat's Principle, a ray of light will travel from a point A in the air to a point B in the water by a path ACB that minimizes the time taken. Show that

$$\frac{\sin\theta_1}{\sin\theta_2} = \frac{v_1}{v_2}$$

where θ_1 (the angle of incidence) and θ_2 (the angle of refraction) are as shown in the figure. This equation is known as Snell's Law.

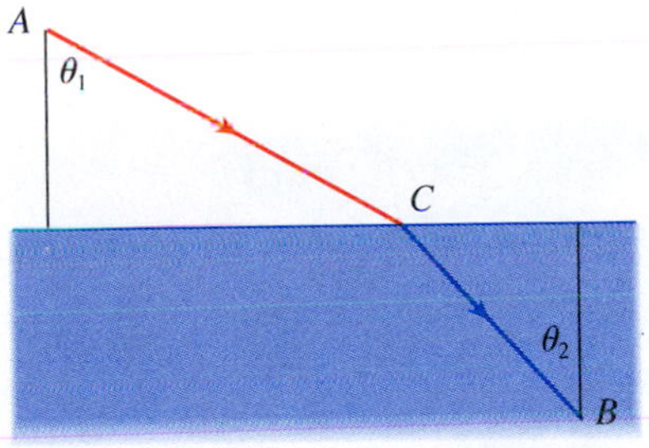

46. Two vertical poles PQ and ST are secured by a rope PRS going from the top of the first pole to a point R on the ground between the poles and then to the top of the second pole as in the figure. Show that the shortest length of such a rope occurs when $\theta_1 = \theta_2$.

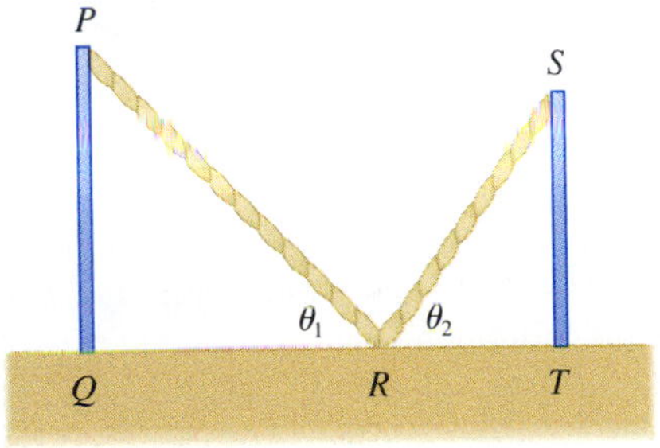

47. The upper left-hand corner of a piece of paper 8 in. wide by 12 in. long is folded over to the right-hand edge as in the figure. How would you fold it so as to minimize the length of the fold? In other words, how would you choose x to minimize y?

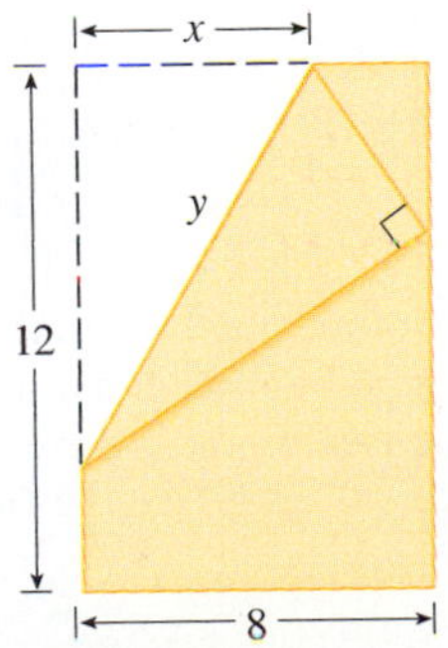

48. A steel pipe is being carried down a hallway 9 ft wide. At the end of the hall there is a right-angled turn into a narrower hallway 6 ft wide. What is the length of the longest pipe that can be carried horizontally around the corner?

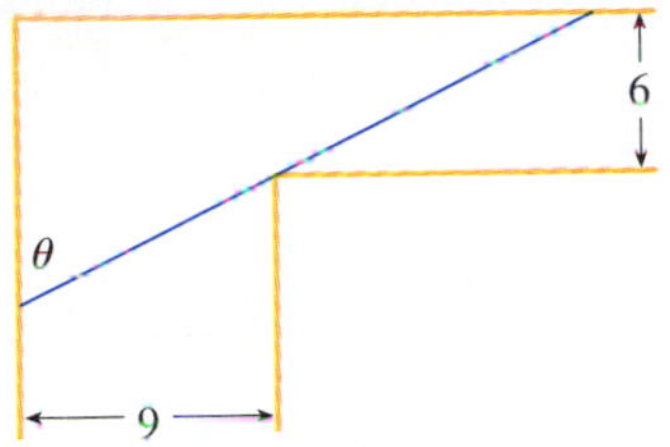

49. An observer stands at a point P, 1 unit away from a track. Two runners start at the point S in the figure and run along the track. One runner runs three times as fast as the other. Find the maximum value of the observer's angle of sight θ between the runners. [*Hint:* Maximize $\tan\theta$.]

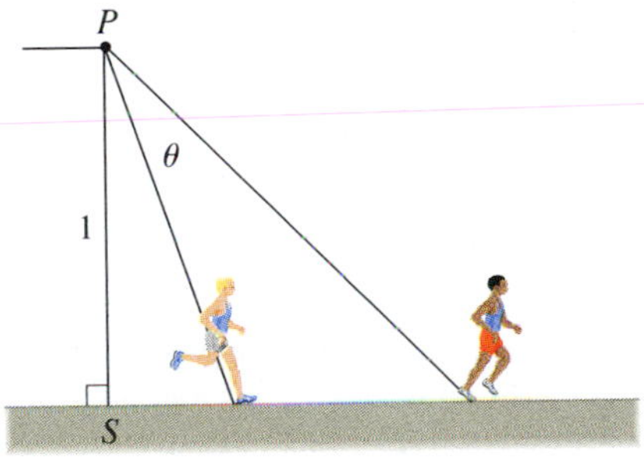

50. A rain gutter is to be constructed from a metal sheet of width 30 cm by bending up one-third of the sheet on each side through an angle θ. How should θ be chosen so that the gutter will carry the maximum amount of water?

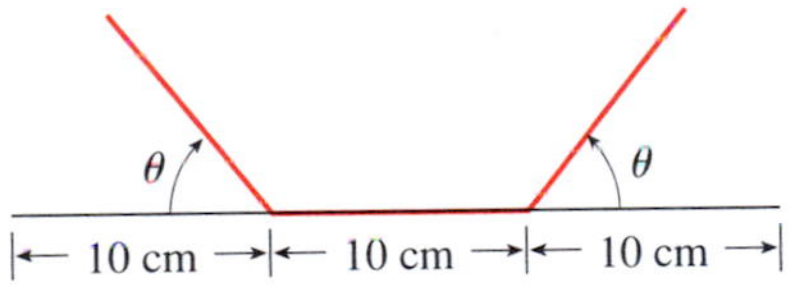

51. Where should the point P be chosen on the line segment AB so as to maximize the angle θ?

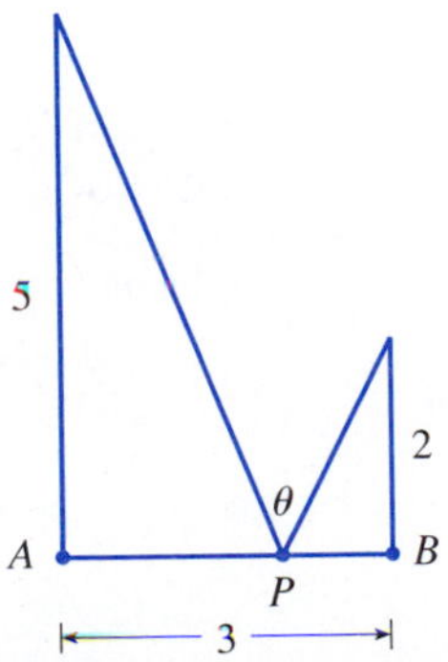

52. A painting in an art gallery has height h and is hung so that its lower edge is a distance d above the eye of an observer (as in the figure). How far from the wall should the observer stand to get the best view? (In other words, where should the observer stand so as to maximize the angle θ subtended at his eye by the painting?)

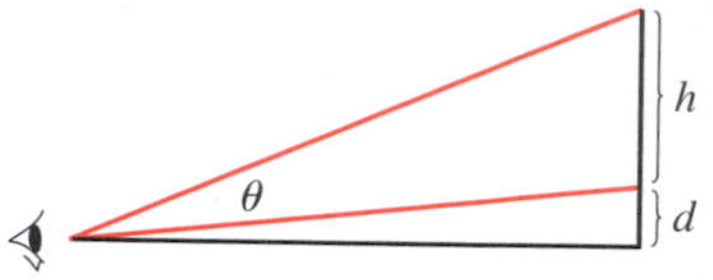

53. Find the maximum area of a rectangle that can be circumscribed about a given rectangle with length L and width W.

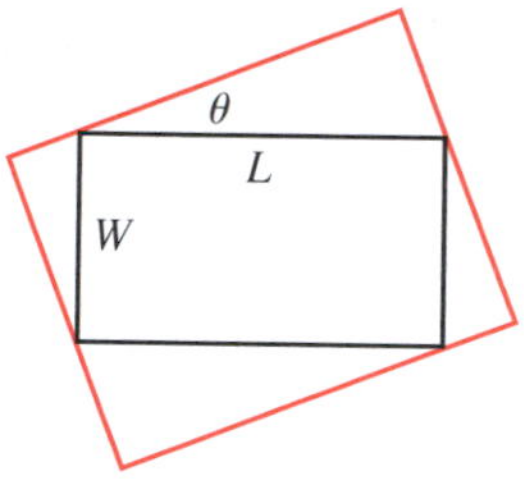

54. The blood vascular system consists of blood vessels (arteries, arterioles, capillaries, and veins) that convey blood from the heart to the organs and back to the heart. This system should work so as to minimize the energy expended by the heart in pumping the blood. In particular, this energy is reduced when the resistance of the blood is lowered. One of Poiseuille's Laws gives the resistance R of the blood as

$$R = C\frac{L}{r^4}$$

where L is the length of the blood vessel, r is the radius, and C is a positive constant determined by the viscosity of the blood. (Poiseuille established this law experimentally but it also follows from Equation 9.7.4.) The figure shows a main blood vessel with radius r_1 branching at an angle θ into a smaller vessel with radius r_2.

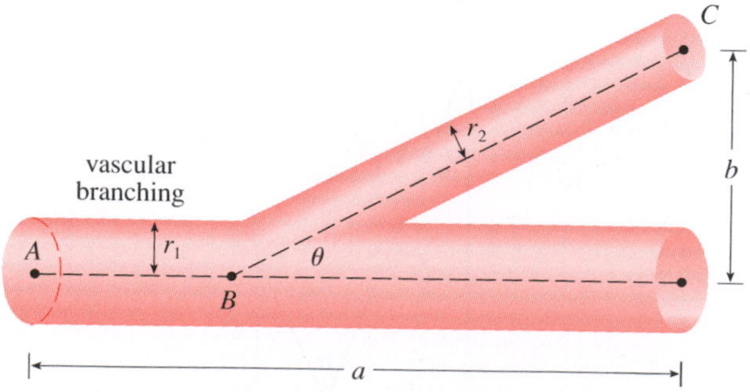

(a) Use Poiseuille's Law to show that the total resistance of the blood along the path ABC is

$$R = C\left(\frac{a - b\cot\theta}{r_1^4} + \frac{b\csc\theta}{r_2^4}\right)$$

where a and b are the distances shown in the figure.

(b) Prove that this resistance is minimized when

$$\cos\theta = \frac{r_2^4}{r_1^4}$$

(c) Find the optimal branching angle (correct to the nearest degree) when the radius of the smaller blood vessel is two-thirds the radius of the larger vessel.

55. A point P needs to be located somewhere on the line AD so that the total length L of cables linking P to the points A, B, and C is minimized (see the figure). Express L as a function of $x = |AP|$ and use the graphs of L and dL/dx to estimate the minimum value.

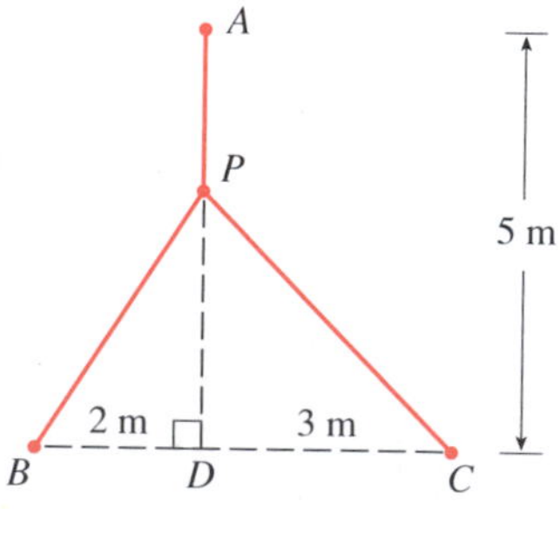

56. Two light sources of identical strength are placed 10 m apart. An object is to be placed at a point P on a line ℓ parallel to the line joining the light sources and at a distance of d meters from it (see the figure). We want to locate P on ℓ so that the intensity of illumination is minimized. We need to use the fact that the intensity of illumination for a single source is directly proportional to the strength of the source and inversely proportional to the square of the distance from the source.

(a) Find an expression for the intensity $I(x)$ at the point P.

(b) If $d = 5$ m, use graphs of $I(x)$ and $I'(x)$ to show that the intensity is minimized when $x = 5$ m, that is, when P is at the midpoint of ℓ.

(c) If $d = 10$ m, show that the intensity (perhaps surprisingly) is *not* minimized at the midpoint.

(d) Somewhere between $d = 5$ m and $d = 10$ m there is a transitional value of d at which the point of minimal illumination abruptly changes. Estimate this value of d.

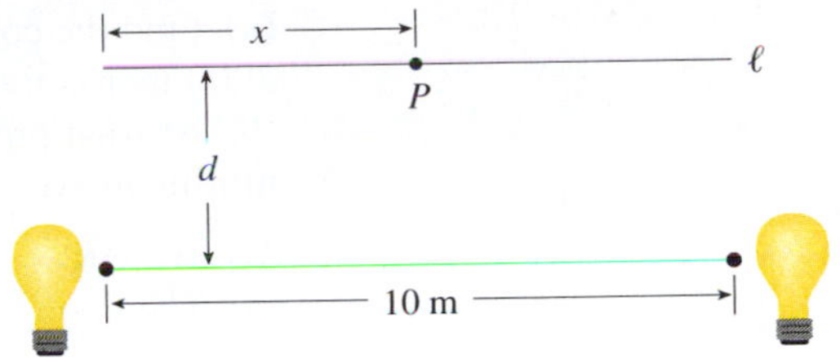

5.6 APPLICATIONS TO ECONOMICS

In Section 3.3 we introduced the idea of marginal cost. Recall that if $C(x)$, the **cost function**, is the cost of producing x units of a certain product, then the **marginal cost** is the rate of change of C with respect to x. In other words, the marginal cost function is the derivative, $C'(x)$, of the cost function.

The graph of a typical cost function is shown in Figure 1. The marginal cost $C'(x)$ is the slope of the tangent to the cost curve at $(x, C(x))$. Notice that the cost curve is initially concave downward (the marginal cost is decreasing) because of economies of scale (more efficient use of the fixed costs of production). But eventually there is an inflection point and the cost curve becomes concave upward (the marginal cost is increasing) perhaps because of overtime costs or the inefficiencies of a large-scale operation.

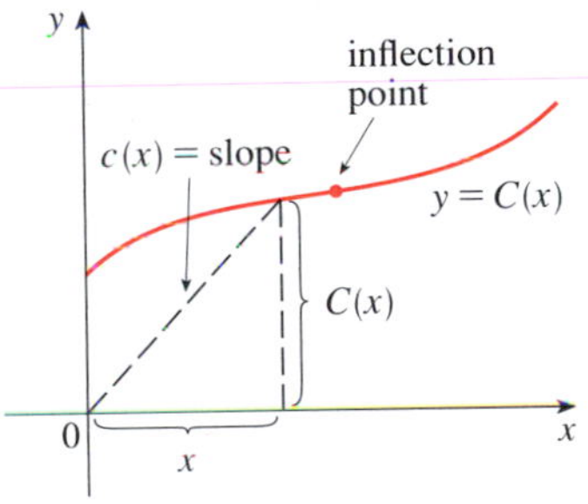

FIGURE 1
Cost function

The **average cost function**

$$c(x) = \frac{C(x)}{x} \tag{1}$$

represents the cost per unit when x units are produced. We sketch a typical average cost function in Figure 2 by noting that $C(x)/x$ is the slope of the line that joins the origin to the point $(x, C(x))$ in Figure 1. It appears that there will be an absolute minimum. To find it we locate the critical point of c by using the Quotient Rule to differentiate Equation 1:

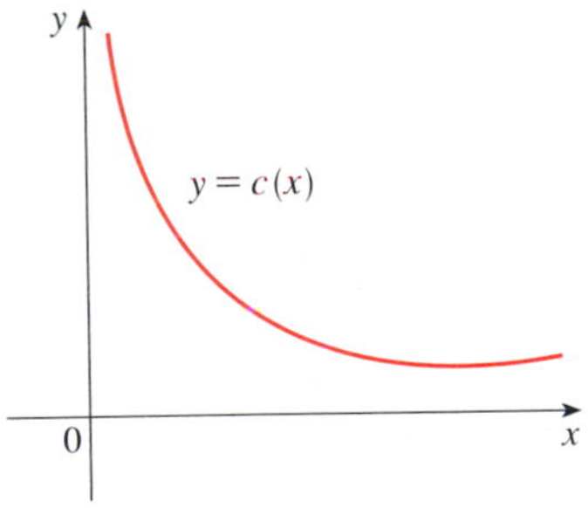

FIGURE 2
Average cost function

$$c'(x) = \frac{xC'(x) - C(x)}{x^2}$$

Now $c'(x) = 0$ when $xC'(x) - C(x) = 0$ and this gives

$$C'(x) = \frac{C(x)}{x} = c(x)$$

Therefore:

> If the average cost is a minimum, then
>
> marginal cost = average cost

EXAMPLE 1 A company estimates that the cost (in dollars) of producing x items is $C(x) = 2600 + 2x + 0.001x^2$. (See Example 8 in Section 3.3 for an explanation of why it's reasonable to model a cost function by a polynomial.)
(a) Find the cost, average cost, and marginal cost of producing 1000 items, 2000 items, and 3000 items.
(b) At what production level will the average cost be lowest, and what is this minimum average cost?

SOLUTION
(a) The average cost function is

$$c(x) = \frac{C(x)}{x} = \frac{2600}{x} + 2 + 0.001x$$

The marginal cost function is

$$C'(x) = 2 + 0.002x$$

x	$C(x)$	$c(x)$	$C'(x)$
1000	5,600.00	5.60	4.00
2000	10,600.00	5.30	6.00
3000	17,600.00	5.87	8.00

We use these expressions to fill in the accompanying table giving the cost, average cost, and marginal cost (in dollars, or dollars per item, rounded to the nearest cent).
(b) To minimize the average cost we must have

$$\text{marginal cost} = \text{average cost}$$

$$C'(x) = c(x)$$

$$2 + 0.002x = \frac{2600}{x} + 2 + 0.001x$$

This equation simplifies to

$$0.001x = \frac{2600}{x}$$

so

$$x^2 = \frac{2600}{0.001} = 2{,}600{,}000$$

and

$$x = \sqrt{2{,}600{,}000} \approx 1612$$

To see that this production level actually gives a minimum we note that $c''(x) = 5200/x^3 > 0$, so c is concave upward on its entire domain. The minimum average cost is

$$c(1612) = \frac{2600}{1612} + 2 + 0.001(1612) = \$5.22/\text{item}$$

■

Now let's consider marketing. Let $p(x)$ be the price per unit that the company can charge if it sells x units. Then p is called the **demand function** (or **price function**) and we would expect it to be a decreasing function of x. If x units are sold and the price per unit is $p(x)$, then the total revenue is

$$R(x) = xp(x)$$

and R is called the **revenue function** (or **sales function**). The derivative R' of the revenue function is called the **marginal revenue function** and is the rate of change of revenue with respect to the number of units sold.

If x units are sold, then the total profit is

$$P(x) = R(x) - C(x)$$

and P is called the **profit function**. The **marginal profit function** is P', the derivative of the profit function. In order to maximize profit we look for the critical numbers of P, that is, the numbers where the marginal profit is 0. But if

$$P'(x) = R'(x) - C'(x) = 0$$

then $$R'(x) = C'(x)$$

Therefore:

> If the profit is a maximum, then
>
> marginal revenue = marginal cost

To ensure that this condition gives a maximum we could use the Second Derivative Test. Note that

$$P''(x) = R''(x) - C''(x) < 0$$

when $$R''(x) < C''(x)$$

and this condition says that the rate of increase of marginal revenue is less than the rate of increase of marginal cost. Thus, the profit will be a maximum when

$$R'(x) = C'(x) \qquad \text{and} \qquad R''(x) < C''(x)$$

EXAMPLE 2 Determine the production level that will maximize the profit for a company with cost and demand functions

$$C(x) = 3800 + 5x - \frac{x^2}{1000} \qquad p(x) = 50 - \frac{x}{100}$$

SOLUTION The revenue function is

$$R(x) = xp(x) = 50x - \frac{x^2}{100}$$

so the marginal revenue function is

$$R'(x) = 50 - \frac{x}{50}$$

and the marginal cost function is

$$C'(x) = 5 - \frac{x}{500}$$

Thus, marginal revenue is equal to marginal cost when

$$50 - \frac{x}{50} = 5 - \frac{x}{500}$$

Solving, we get $$x = 2500$$

To check that this gives a maximum we compute the second derivatives:

$$R''(x) = -\tfrac{1}{50} \qquad C''(x) = -\tfrac{1}{500}$$

Thus, $R''(x) < C''(x)$ for all values of x. Therefore, a production level of 2500 units will maximize the profit. ■

EXAMPLE 3 A store has been selling 200 compact disc players a week at \$350 each. A market survey indicates that for each \$10 rebate offered to the buyers, the number of sets sold will increase by 20 a week. Find the demand function and the revenue function. How large a rebate should the store offer to maximize its revenue?

SOLUTION If x is the number of CD players sold per week, then the weekly increase in sales is $x - 200$. For each increase of 20 players sold, the price is decreased by \$10. So for each additional player sold the decrease in price will be $\frac{1}{20} \times 10$ and the demand function is

$$p(x) = 350 - \tfrac{10}{20}(x - 200) = 450 - \tfrac{1}{2}x$$

The revenue function is

$$R(x) = xp(x) = 450x - \tfrac{1}{2}x^2$$

Since $R'(x) = 450 - x$, we see that $R'(x) = 0$ when $x = 450$. This value of x gives an absolute maximum by the First Derivative Test (or simply by observing that the graph of R is a parabola that opens downward). The corresponding price is

$$p(450) = 450 - \tfrac{1}{2}(450) = 225$$

and the rebate is $350 - 225 = 125$. Therefore, to maximize revenue the store should offer a rebate of \$125. ■

EXERCISES 5.6

1. A manufacturer keeps precise records of the cost $C(x)$ of producing x items and produces the graph of the cost function shown in the figure.
(a) Explain why $C(0) > 0$.
(b) What is the significance of the inflection point?
(c) Use the graph of C to sketch the graph of the marginal cost function.

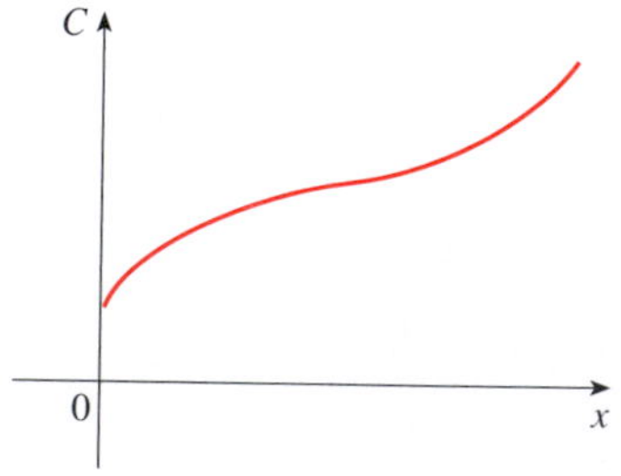

2. The figure shows graphs of the cost and revenue functions reported by a manufacturer.
(a) Identify on the graph the value of x for which the profit is maximized.
(b) Sketch a graph of the profit function.
(c) Sketch a graph of the marginal profit function.

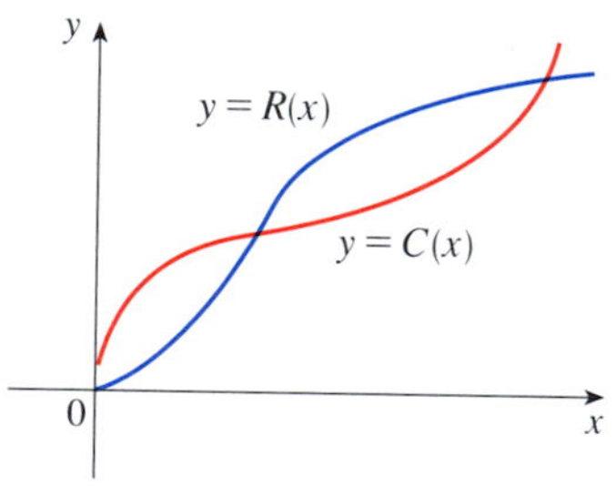

3–8 ■ For each cost function, find (a) the cost, average cost, and marginal cost of producing 1000 units; (b) the production level that will minimize the average cost; and (c) the minimum average cost.

3. $C(x) = 10{,}000 + 25x + x^2$

4. $C(x) = 1600 + 8x + 0.01x^2$

5. $C(x) = 45 + \dfrac{x}{2} + \dfrac{x^2}{560}$

6. $C(x) = 2000 + 10x + 0.001x^3$

7. $C(x) = 2\sqrt{x} + x^2/8000$

8. $C(x) = 1000 + 96x + 2x^{3/2}$

9–14 ■ For the given cost and demand functions, find the production level that will maximize profit.

9. $C(x) = 680 + 4x + 0.01x^2$, $p(x) = 12$

10. $C(x) = 680 + 4x + 0.01x^2$, $p(x) = 12 - x/500$

11. $C(x) = 1200 + 25x - 0.0001x^2$, $p(x) = 55 - x/1000$

12. $C(x) = 900 + 110x - 0.1x^2 + 0.02x^3$, $p(x) = 260 - 0.1x$

13. $C(x) = 1450 + 36x - x^2 + 0.001x^3$, $p(x) = 60 - 0.01x$

14. $C(x) = 10{,}000 + 28x - 0.01x^2 + 0.002x^3$, $p(x) = 90 - 0.02x$

15–16 ■ Find the production level at which the marginal cost function starts to increase.

15. $C(x) = 0.001x^3 - 0.3x^2 + 6x + 900$

16. $C(x) = 0.0002x^3 - 0.25x^2 + 4x + 1500$

17. A baseball team plays in a stadium that holds 55,000 spectators. With ticket prices at \$10, the average attendance has been 27,000. When ticket prices were lowered to \$8, the average attendance rose to 33,000.
(a) Find the demand function, assuming that it is linear.
(b) How should ticket prices be set to maximize revenue?

18. During the summer months Terry makes and sells necklaces on the beach. Last summer he sold the necklaces for \$10 each and his sales averaged 20 per day. When he increased the price by \$1, he found that he lost two sales per day.
(a) Find the demand function, assuming that it is linear.
(b) If the material for each necklace costs Terry \$6, what should the selling price be to maximize his profit?

19. A manufacturer has been selling 1000 television sets a week at \$450 each. A market survey indicates that for each \$10 rebate offered to the buyer, the number of sets sold will increase by 100 per week.
(a) Find the demand function.
(b) How large a rebate should the company offer the buyer in order to maximize its revenue?
(c) If its weekly cost function is $C(x) = 68{,}000 + 150x$, how should it set the size of the rebate in order to maximize its profit?

20. The manager of a 100-unit apartment complex knows from experience that all units will be occupied if the rent is \$400 per month. A market survey suggests that, on the average, one additional unit will remain vacant for each \$5 increase in rent. What rent should the manager charge to maximize revenue?

5.7 ANTIDERIVATIVES

We know how to solve the derivative problem: given a function, find its derivative. But many problems in mathematics and its applications require us to solve the inverse of the derivative problem: given a function f, find a function F whose derivative is f. If such a function F exists, it is called an *antiderivative* of f.

(1) DEFINITION A function F is called an **antiderivative** of f on an interval I if $F'(x) = f(x)$ for all x in I.

For instance, let $f(x) = x^2$. It is not difficult to discover an antiderivative of f if we keep the Power Rule in mind. In fact, if $F(x) = \frac{1}{3}x^3$, then $F'(x) = x^2 = f(x)$. But the function $G(x) = \frac{1}{3}x^3 + 100$ also satisfies $G'(x) = x^2$. Therefore, both F and G are antiderivatives of f. Indeed, any function of the form $H(x) = \frac{1}{3}x^3 + C$, where C is a constant, is an antiderivative of f. The question arises: Are there any others?

The following theorem says that f has no other antiderivative. A proof of Theorem 2, using the Mean Value Theorem, is outlined in Exercise 85.

(2) THEOREM If F is an antiderivative of f on an interval I, then the most general antiderivative of f on I is

$$F(x) + C$$

where C is an arbitrary constant.

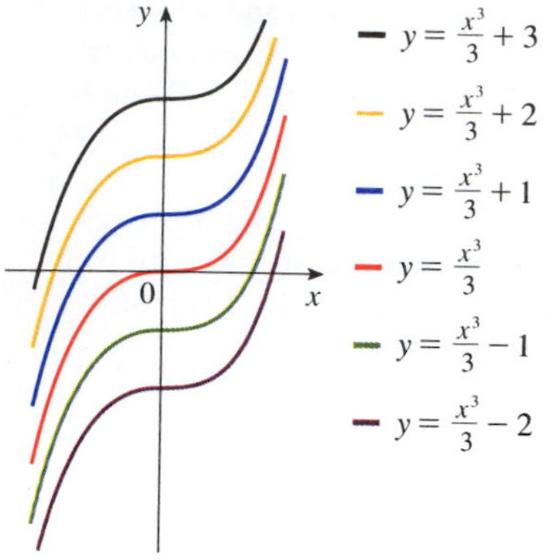

FIGURE 1
The antiderivatives of $f(x)=x^2$

Going back to the function $f(x) = x^2$, we see that the general antiderivative of f is $x^3/3 + C$. By assigning specific values to the constant C we obtain a family of functions whose graphs are vertical translates of one another (see Figure 1).

EXAMPLE 1 Find the most general antiderivative of each of the following functions:
(a) $f(x) = \sin x$
(b) $f(x) = x^n, \quad n \geqslant 0$
(c) $f(x) = x^{-3}$
(d) $f(x) = 1/x$

SOLUTION
(a) If $F(x) = -\cos x$, then $F'(x) = \sin x$, so an antiderivative of sine is $-$cosine. By Theorem 2, the most general antiderivative is $G(x) = -\cos x + C$.

(b) $$\frac{d}{dx}\left(\frac{x^{n+1}}{n+1}\right) = \frac{(n+1)x^n}{n+1} = x^n$$

Thus, the general antiderivative of $f(x) = x^n$ is

$$F(x) = \frac{x^{n+1}}{n+1} + C$$

This is valid for $n \geqslant 0$ since then $f(x) = x^n$ is defined on the interval $(-\infty, \infty)$.

(c) If we put $n = -3$ in part (b) we get the particular antiderivative

$$F(x) = \frac{x^{-2}}{-2}$$

by the same calculation. But notice that $f(x) = x^{-3}$ is not defined at $x = 0$. Thus, Theorem 2 tells us only that the general antiderivative of f is $x^{-2}/(-2) + C$ on any interval that does not contain 0. So the general antiderivative of $f(x) = 1/x^3$ is

$$F(x) = \begin{cases} -\dfrac{1}{2x^2} + C_1 & \text{if } x > 0 \\[2ex] -\dfrac{1}{2x^2} + C_2 & \text{if } x < 0 \end{cases}$$

(d) Formula 4.4.3 shows that a particular antiderivative of $1/x$ is $\ln|x|$. But since f is not defined where $x = 0$, Theorem 2 tells us that the most general antiderivative is

$$F(x) = \begin{cases} \ln x + C_1 & \text{if } x > 0 \\ \ln(-x) + C_2 & \text{if } x < 0 \end{cases}$$

■

As in Example 1, every differentiation formula, when read from right to left, gives rise to an antidifferentiation formula. In Table 3 we list some particular antiderivatives. Each formula in the table is true because the derivative of the function in the right column appears in the left column. In particular, the first formula says that the antiderivative of a constant times a function is the constant times the antiderivative of the function. The second formula says that the antiderivative of a sum is the sum of the antiderivatives. (We use the notation $F' = f$, $G' = g$.)

(3) TABLE OF ANTIDIFFERENTIATION FORMULAS

Function	Particular antiderivative
$cf(x)$	$cF(x)$
$f(x) + g(x)$	$F(x) + G(x)$
$x^n \ (n \neq -1)$	$\dfrac{x^{n+1}}{n+1}$
$1/x$	$\ln\|x\|$
e^x	e^x
$\cos x$	$\sin x$
$\sin x$	$-\cos x$
$\sec^2 x$	$\tan x$
$\dfrac{1}{\sqrt{1-x^2}}$	$\sin^{-1}x$
$\dfrac{1}{1+x^2}$	$\tan^{-1}x$

To obtain the most general antiderivative from the particular ones in Table 3 we have to add a constant (or constants) as in Example 1.

EXAMPLE 2 Find all functions g such that

$$g'(x) = 4\sin x - 3x^5 + 6\sqrt[4]{x^3}$$

SOLUTION We want to find an antiderivative of

$$f(x) = g'(x) = 4\sin x - 3x^5 + 6x^{3/4}$$

Using the formulas in Table 3 together with Theorem 2, we obtain

$$g(x) = 4(-\cos x) - 3\frac{x^6}{6} + 6\frac{x^{7/4}}{\frac{7}{4}} + C$$

$$= -4\cos x - \frac{x^6}{2} + \frac{24}{7}x^{7/4} + C$$

■

In applications of calculus it is very common to have a situation as in Example 2, where it is required to find a function, given knowledge about its derivatives. An equation that involves the derivatives of a function is called a **differential equation**. These will be studied in some detail in Section 9.1 and Chapter 15, but for the present we can solve some elementary differential equations. The general solution of a differential equation involves an arbitrary constant (or constants) as in Example 2. However, there may be some extra conditions given that will determine the constants and therefore uniquely specify the solution.

EXAMPLE 3 Find f if $f'(x) = x\sqrt{x} + 5(1 + x^2)^{-1}$ and $f(0) = 2$.

SOLUTION The general antiderivative of

$$f'(x) = x^{3/2} + \frac{5}{1 + x^2}$$

is
$$f(x) = \frac{x^{5/2}}{\frac{5}{2}} + 5\tan^{-1}x + C = \tfrac{2}{5}x^{5/2} + 5\tan^{-1}x + C$$

To determine C we use the fact that $f(0) = 2$:

$$f(0) = 0 + 5\tan^{-1}0 + C = 2$$

Thus, we have $C = 2$, so the particular solution is

$$f(x) = \tfrac{2}{5}x^{5/2} + 5\tan^{-1}x + 2$$ ■

EXAMPLE 4 Find f if $f''(x) = 12x^2 + 6x - 4$, $f(0) = 4$, and $f(1) = 1$.

SOLUTION The general antiderivative of $f''(x) = 12x^2 + 6x - 4$ is

$$f'(x) = 12\frac{x^3}{3} + 6\frac{x^2}{2} - 4x + C = 4x^3 + 3x^2 - 4x + C$$

Using the antidifferentiation rules once more, we find that

$$f(x) = 4\frac{x^4}{4} + 3\frac{x^3}{3} - 4\frac{x^2}{2} + Cx + D = x^4 + x^3 - 2x^2 + Cx + D$$

To determine C and D we use the given conditions that $f(0) = 4$ and $f(1) = 1$. Since $f(0) = 0 + D = 4$, we have $D = 4$. Since

$$f(1) = 1 + 1 - 2 + C + 4 = 1$$

we have $C = -3$. Therefore, the required function is

$$f(x) = x^4 + x^3 - 2x^2 - 3x + 4$$ ■

THE GEOMETRY OF ANTIDERIVATIVES

If we are given the graph of a function f, it seems reasonable that we should be able to sketch the graph of an antiderivative F. Suppose, for instance, that we are given that $F(0) = 1$. Then we have a place to start, the point $(0, 1)$, and the direction in which we move our pencil is given at each stage by the derivative $F'(x) = f(x)$. In the next example we use the principles of this chapter to show how to graph F even when we don't have a formula for f. This would be the case, for instance, when $f(x)$ is determined by experimental data.

EXAMPLE 5 The graph of a function f is given in Figure 2. Make a rough sketch of an antiderivative F, given that $F(0) = 2$.

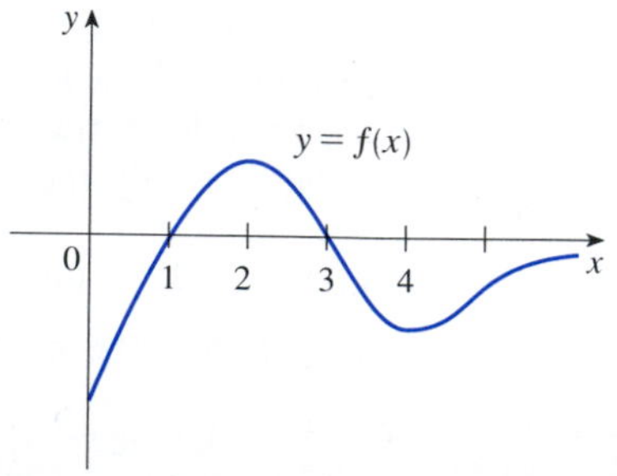

FIGURE 2

SOLUTION We are guided by the fact that the slope of $y = F(x)$ is $f(x)$. We start at the point $(0, 2)$ and draw F as an initially decreasing function since $f(x)$ is negative

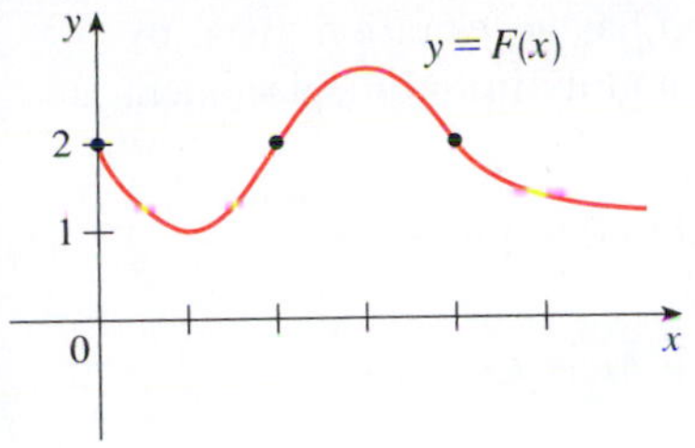

FIGURE 3

when $0 < x < 1$. Notice that $f(1) = f(3) = 0$, so F has horizontal tangents when $x = 1$ and $x = 3$. For $1 < x < 3$, $f(x)$ is positive and so F is increasing. We see that F has a local minimum when $x = 1$ and a local maximum when $x = 3$. For $x > 3$, $f(x)$ is negative and so F is decreasing on $(3, \infty)$. Since $f(x) \to 0$ as $x \to \infty$, the graph of F becomes flatter as $x \to \infty$. Also notice that $F''(x) = f'(x)$ changes from positive to negative at $x = 2$ and from negative to positive at $x = 4$, so F has inflection points when $x = 2$ and $x = 4$. We use this information to sketch the graph of the antiderivative in Figure 3. ■

EXAMPLE 6 If $f(x) = \sqrt{1 + x^3} - x$, sketch the graph of the antiderivative F that satisfies the initial condition $F(-1) = 0$.

SOLUTION We could try all day to think of a formula for an antiderivative of f and still be unsuccessful. A second possibility would be to draw the graph of f first and then use it to graph F as in Example 5. That would work, but instead let's create a more accurate graph by using what is called a **direction field**.

Since $f(0) = 1$, the graph of F has slope 1 when $x = 0$. So we draw several short tangent segments with slope 1, all centered at $x = 0$. We do the same for several other values of x and the result is shown in Figure 4. It is called a direction field because each segment indicates the direction in which the curve $y = F(x)$ proceeds at that point.

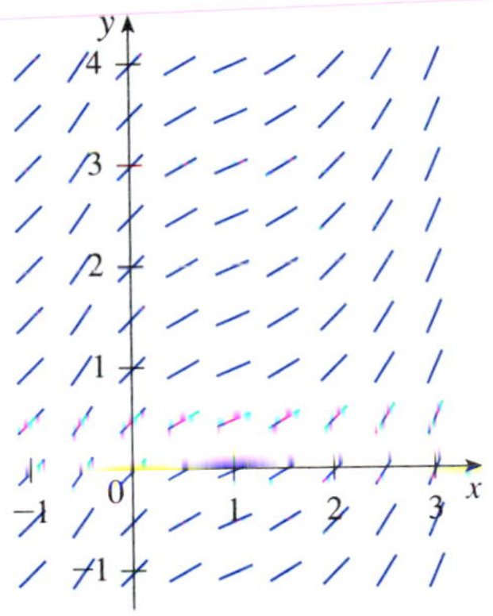

FIGURE 4
A direction field for $f(x) = \sqrt{1 + x^3} - x$. The slope of the line segments above $x = a$ is $f(a)$.

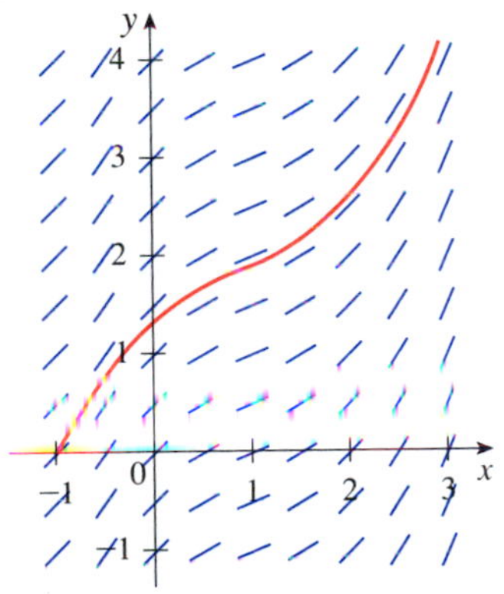

FIGURE 5
The graph of an antiderivative follows the direction field.

Now we use the direction field to sketch the graph of F. Because of the initial condition $F(-1) = 0$, we start at the point $(-1, 0)$ and draw the graph so that it follows the directions of the tangent segments. The result is pictured in Figure 5. Any other antiderivative would be obtained by shifting the graph of F upward or downward. ■

RECTILINEAR MOTION

Antidifferentiation is particularly useful in analyzing the motion of an object moving in a straight line. Recall that if the object has position function $s = f(t)$, then the velocity function is $v(t) = s'(t)$. This means that the position function is an antiderivative of the velocity function. Likewise, the acceleration function is $a(t) = v'(t)$, so the velocity function is an antiderivative of the acceleration. If the acceleration and the initial values $s(0)$ and $v(0)$ are known, then the position function can be found by antidifferentiating twice.

EXAMPLE 7 A particle moves in a straight line and has acceleration given by $a(t) = 6t + 4$. Its initial velocity is $v(0) = -6$ cm/s and its initial displacement is $s(0) = 9$ cm. Find its position function $s(t)$.

SOLUTION Since $v'(t) = a(t) = 6t + 4$, antidifferentiation gives

$$v(t) = 6\frac{t^2}{2} + 4t + C = 3t^2 + 4t + C$$

Note that $v(0) = C$. But we are given that $v(0) = -6$, so $C = -6$ and

$$v(t) = 3t^2 + 4t - 6$$

Since $v(t) = s'(t)$, s is the antiderivative of v:

$$s(t) = 3\frac{t^3}{3} + 4\frac{t^2}{2} - 6t + D = t^3 + 2t^2 - 6t + D$$

This gives $s(0) = D$. We are given that $s(0) = 9$, so $D = 9$ and the required position function is

$$s(t) = t^3 + 2t^2 - 6t + 9$$ ■

An object near Earth's surface is subject to a gravitational force that produces a downward acceleration denoted by g. For motion close to Earth we may assume that g is constant, its value being about 9.8 m/s^2 (or 32 ft/s^2).

EXAMPLE 8 A ball is thrown upward with a speed of 48 ft/s from the edge of a cliff 432 ft above the ground. Find its height above the ground t seconds later. When does it reach its maximum height? When does it hit the ground?

SOLUTION The motion is vertical and we choose the positive direction to be upward. At time t the distance above the ground is $s(t)$ and the velocity $v(t)$ is decreasing. Therefore, the acceleration must be negative and we have

$$a(t) = \frac{dv}{dt} = -32$$

Taking antiderivatives, we have

$$v(t) = -32t + C$$

To determine C we use the given information that $v(0) = 48$. This gives $48 = 0 + C$, so

$$v(t) = -32t + 48$$

The maximum height is reached when $v(t) = 0$, that is, after 1.5 s. Since $s'(t) = v(t)$, we antidifferentiate again and obtain

$$s(t) = -16t^2 + 48t + D$$

Using the fact that $s(0) = 432$, we have $432 = 0 + D$, and so

$$s(t) = -16t^2 + 48t + 432$$

The expression for $s(t)$ is valid until the ball hits the ground. This happens when $s(t) = 0$, that is, when

$$-16t^2 + 48t + 432 = 0$$

or, equivalently,

$$t^2 - 3t - 27 = 0$$

Using the quadratic formula to solve this equation, we get

$$t = \frac{3 \pm 3\sqrt{13}}{2}$$

We reject the solution with the minus sign since it gives a negative value for t. Therefore, the ball hits the ground after $3(1 + \sqrt{13})/2 \approx 6.9$ s. ■

ANTIDERIVATIVES OF VECTOR FUNCTIONS

The definition of antiderivative extends to the case of vector functions.

(4) DEFINITION A vector function $\mathbf{R}(t) = X(t)\,\mathbf{i} + Y(t)\,\mathbf{j}$ is called an **antiderivative** of $\mathbf{r}(t) = x(t)\,\mathbf{i} + y(t)\,\mathbf{j}$ on an interval I if $\mathbf{R}'(t) = \mathbf{r}(t)$ for all t in I; that is, $X'(t) = x(t)$ and $Y'(t) = y(t)$.

In other words, to find an antiderivative of $\mathbf{r}$, find an antiderivative of each of its components. By using Theorem 2 and knowing one particular antiderivative of $\mathbf{r}$ on I, we can give the most general antiderivative as follows.

(5) THEOREM If $\mathbf{R}$ is an antiderivative of $\mathbf{r}$ on an interval I, then the most general antiderivative of $\mathbf{r}$ on I is

$$\mathbf{R}(t) + \mathbf{C}$$

where $\mathbf{C}$ is an arbitrary constant vector.

EXAMPLE 9 Find the most general antiderivative of $\mathbf{r}(t) = \mathbf{i} + 3t\,\mathbf{j}$.

SOLUTION The $\mathbf{i}$ component of $\mathbf{r}$ is 1, and an antiderivative of 1 is t. The $\mathbf{j}$ component of $\mathbf{r}$ is $3t$, which has $\frac{3}{2}t^2$ as an antiderivative. Thus, the most general antiderivative of $\mathbf{r}$ is

$$t\,\mathbf{i} + \tfrac{3}{2}t^2\,\mathbf{j} + \mathbf{C}$$ ■

Recall that if $\mathbf{r}(t)$ is a vector function of t that represents the position of an object at time t, then the velocity at time t is

$$\mathbf{v}(t) = \mathbf{r}'(t)$$

and the acceleration at time t is

$$\mathbf{a}(t) = \mathbf{v}'(t) = \mathbf{r}''(t)$$

Thus, the position function is an antiderivative of the velocity, and the velocity is an antiderivative of the acceleration.

EXAMPLE 10 A moving particle starts at an initial position $\mathbf{r}(0) = \mathbf{i}$ with initial velocity $\mathbf{v}(0) = \mathbf{i} - \mathbf{j}$. Its acceleration is $\mathbf{a}(t) = 4t\,\mathbf{i} + 6t\,\mathbf{j}$. Find its velocity and position at time t.

SOLUTION Since $\mathbf{v}(t)$ is an antiderivative of $\mathbf{a}(t) = 4t\,\mathbf{i} + 6t\,\mathbf{j}$, we have

$$\mathbf{v}(t) = 2t^2\mathbf{i} + 3t^2\mathbf{j} + \mathbf{C}$$

To determine the value of the constant vector $\mathbf{C}$, we use the fact that $\mathbf{v}(0) = \mathbf{i} - \mathbf{j}$. The preceding equation gives $\mathbf{v}(0) = \mathbf{C}$, so $\mathbf{C} = \mathbf{i} - \mathbf{j}$ and

$$\begin{aligned}\mathbf{v}(t) &= 2t^2\mathbf{i} + 3t^2\mathbf{j} + \mathbf{i} - \mathbf{j}\\ &= (2t^2 + 1)\,\mathbf{i} + (3t^2 - 1)\,\mathbf{j}\end{aligned}$$

Since $\mathbf{r}(t)$ is an antiderivative of $\mathbf{v}(t)$, we have

$$\mathbf{r}(t) = \left(\tfrac{2}{3}t^3 + t\right)\mathbf{i} + (t^3 - t)\,\mathbf{j} + \mathbf{D}$$

Putting $t = 0$, we find that $\mathbf{D} = \mathbf{r}(0) = \mathbf{i}$, so

$$\mathbf{r}(t) = \left(\tfrac{2}{3}t^3 + t + 1\right)\mathbf{i} + (t^3 - t)\,\mathbf{j}$$

■

If the force that acts on a particle is known, then the acceleration can be found from **Newton's Second Law of Motion**. The vector version of this law states that if, at any time t, a force $\mathbf{F}(t)$ acts on an object of mass m producing an acceleration $\mathbf{a}(t)$, then

$$\mathbf{F}(t) = m\mathbf{a}(t)$$

EXAMPLE 11 A projectile is fired with angle of elevation α and initial velocity $\mathbf{v}_0$ (see Figure 6). Assuming that air resistance is negligible and the only external force is due to gravity, find the position function $\mathbf{r}(t)$ of the projectile. What value of α maximizes the range (the horizontal distance traveled)?

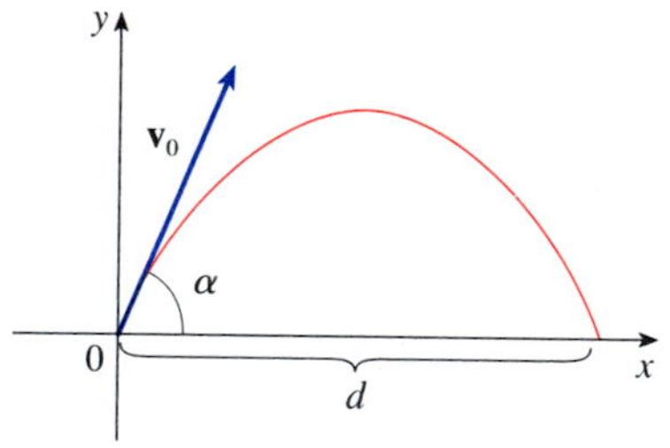

FIGURE 6

SOLUTION We set up the axes so that the projectile starts at the origin. Since the force due to gravity acts downward, we have

$$\mathbf{F} = m\mathbf{a} = -mg\,\mathbf{j}$$

where $g = |\mathbf{a}| \approx 9.8\ \text{m/s}^2$. Thus

$$\mathbf{a} = -g\,\mathbf{j}$$

Since $\mathbf{v}'(t) = \mathbf{a}$, we have

$$\mathbf{v}(t) = -gt\,\mathbf{j} + \mathbf{C}$$

where $\mathbf{C} = \mathbf{v}(0) = \mathbf{v}_0$. Therefore

$$\mathbf{r}'(t) = \mathbf{v}(t) = -gt\,\mathbf{j} + \mathbf{v}_0$$

Antidifferentiating again, we obtain

$$\mathbf{r}(t) = -\tfrac{1}{2}gt^2\mathbf{j} + t\mathbf{v}_0 + \mathbf{D}$$

But $\mathbf{D} = \mathbf{r}(0) = \mathbf{0}$, so the position vector of the projectile is given by

(6) $$\mathbf{r}(t) = -\tfrac{1}{2}gt^2\mathbf{j} + t\mathbf{v}_0$$

If we write $|\mathbf{v}_0| = v_0$ (the initial speed of the projectile), then

$$\mathbf{v}_0 = v_0\cos\alpha\,\mathbf{i} + v_0\sin\alpha\,\mathbf{j}$$

and Equation 6 becomes

$$\mathbf{r}(t) = (v_0\cos\alpha)t\,\mathbf{i} + [(v_0\sin\alpha)t - \tfrac{1}{2}gt^2]\,\mathbf{j}$$

The parametric equations of the trajectory are therefore

(7) $$x = (v_0\cos\alpha)t \qquad y = (v_0\sin\alpha)t - \tfrac{1}{2}gt^2$$

The horizontal distance d is the value of x when $y = 0$. Setting $y = 0$, we obtain $t = 0$ (when the projectile is fired) or $t = (2v_0\sin\alpha)/g$ (when the projectile lands). This last value of t then gives

$$d = x = (v_0\cos\alpha)\frac{2v_0\sin\alpha}{g} = \frac{v_0^2\sin 2\alpha}{g}$$

Clearly, d has its maximum value when $\sin 2\alpha = 1$, that is, $\alpha = \pi/4$. ■

EXERCISES 5.7

1–16 ■ Find the most general antiderivative of the function. (Check your answer by differentiation.)

1. $f(x) = 12x^2 + 6x - 5$

2. $f(x) = x^3 - 4x^2 + 17$

3. $f(x) = 6x^9 - 4x^7 + 3x^2 + 1$

4. $f(x) = x^{99} - 2x^{49} - 1$

5. $f(x) = \sqrt{x} + \sqrt[3]{x}$

6. $f(x) = \sqrt[3]{x^2} - \sqrt{x^3}$

7. $f(x) = 6/x^5$

8. $f(x) = \dfrac{3}{x^2} - \dfrac{5}{x^4}$

9. $g(t) = (t^3 + 2t^2)/\sqrt{t}$

10. $f(x) = x^{2/3} + 2x^{-1/3}$

11. $h(x) = \sin x - 2\cos x$

12. $f(t) = \sin t - 2\sqrt{t}$

13. $f(t) = \sec^2 t + t^2$

14. $f(\theta) = \theta + \sec\theta\tan\theta$

15. $f(x) = 2x + 5(1 - x^2)^{-1/2}$

16. $f(x) = \dfrac{x^2 + x + 1}{x}$

17–40 ■ Find $f(x)$.

17. $f''(x) = x^2 + x^3$

18. $f''(x) = 60x^4 - 45x^2$

19. $f''(x) = 1$

20. $f''(x) = \sin x$

21. $f'''(x) = 24x$

22. $f'''(x) = \sqrt{x}$

23. $f'(x) = 4x + 3,\quad f(0) = -9$

24. $f'(x) = 12x^2 - 24x + 1,\quad f(1) = -2$

25. $f'(x) = 3\sqrt{x} - 1/\sqrt{x},\quad f(1) = 2$

26. $f'(x) = 1 + 1/x^2,\ x > 0,\quad f(1) = 1$

27. $f'(x) = 3\cos x + 5\sin x,\quad f(0) = 4$

28. $f'(x) = 3x^{-2},\quad f(1) = f(-1) = 0$

29. $f'(x) = 2/x,\quad x < 0,\quad f(-1) = 7$

30. $f'(x) = 4 - 3(1 + x^2)^{-1},\quad f(1) = 0$

31. $f''(x) = x,\quad f(0) = -3,\quad f'(0) = 2$

32. $f''(x) = 20x^3 - 10, \quad f(1) = 1, \quad f'(1) = -5$

33. $f''(x) = x^2 + 3\cos x, \quad f(0) = 2, \quad f'(0) = 3$

34. $f''(x) = x + \sqrt{x}, \quad f(1) = 1, \quad f'(1) = 2$

35. $f''(x) = 6x + 6, \quad f(0) = 4, \quad f(1) = 3$

36. $f''(x) = 12x^2 - 6x + 2, \quad f(0) = 1, \quad f(2) = 11$

37. $f''(x) = 1/x^3, \quad x > 0, \quad f(1) = 0, \quad f(2) = 0$

38. $f''(x) = 3e^x + 5\sin x, \quad f(0) = 1, \quad f'(0) = 2$

39. $f''(x) = x^{-2}, \quad x > 0, \quad f(1) = 0, \quad f(2) = 0$

40. $f'''(x) = \sin x, \quad f(0) = 1, \quad f'(0) = 1, \quad f''(0) = 1$

41. Given that the graph of f passes through the point $(1, 6)$ and that the slope of its tangent line at $(x, f(x))$ is $2x + 1$, find $f(2)$.

42. Find a function f such that $f'(x) = x^3$ and the line $x + y = 0$ is tangent to the graph of f.

43–44 ■ The graph of a function f is shown. Which graph is an antiderivative of f and why?

43.

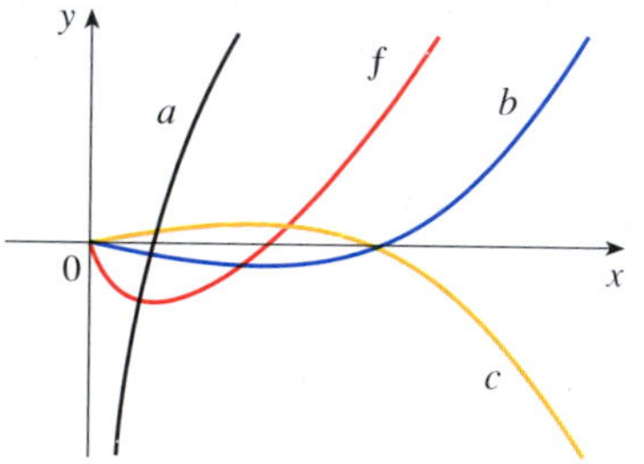

44.

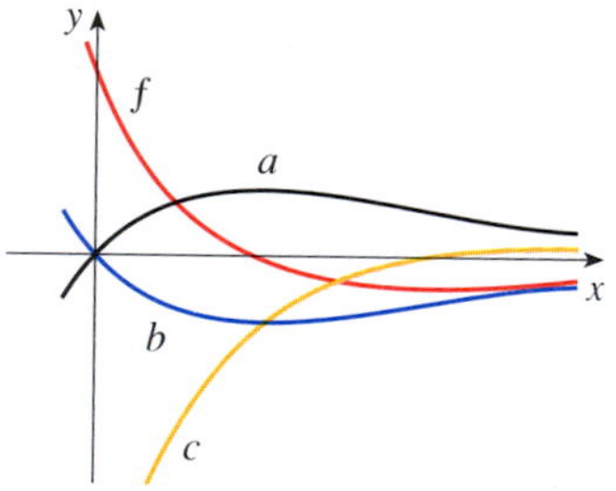

45. The graph of a function is shown in the figure. Make a rough sketch of an antiderivative F, given that $F(0) = 0$.

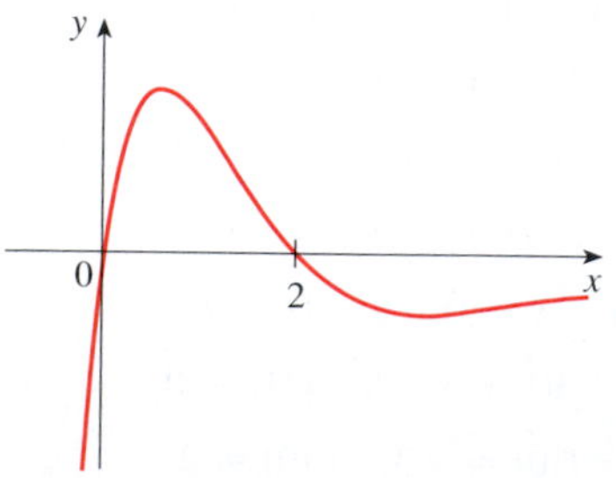

46. The graph of the velocity function of a car is shown in the figure. Sketch the graph of the position function.

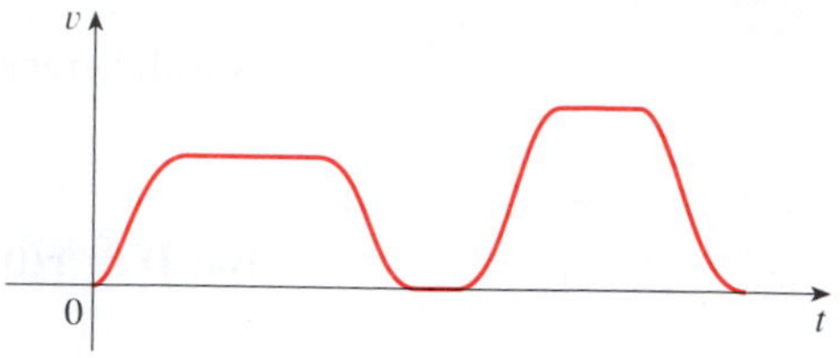

47–48 ■ Draw a graph of f and use it to make a rough sketch of the antiderivative that passes through the origin.

47. $f(x) = \sin(x^2), \quad 0 \leqslant x \leqslant 4$

48. $f(x) = 1/(x^4 + 1)$

49–50 ■ A direction field is given for a function. Use it to draw the antiderivative F that satisfies $F(0) = -2$.

49.

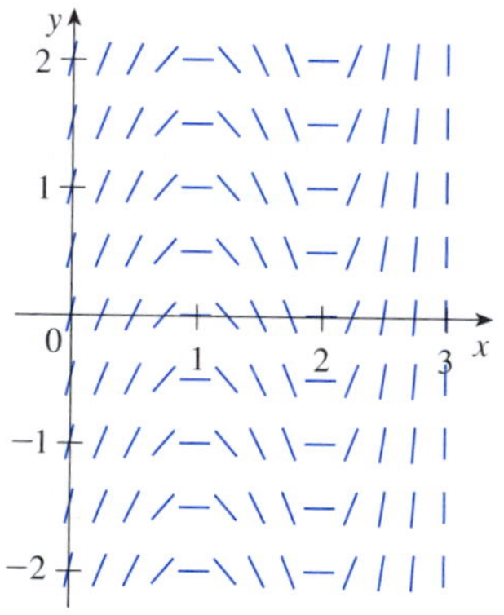

50.

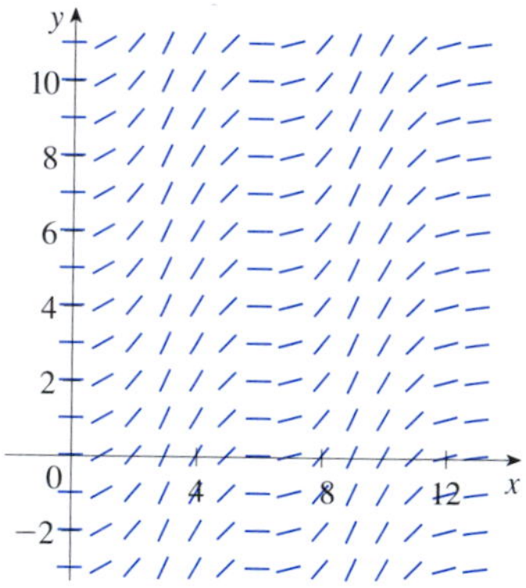

51–52 ■ Use a direction field to graph the antiderivative that satisfies $F(0) = 0$.

51. $f(x) = \dfrac{\sin x}{x}, \quad 0 < x < 2\pi$

52. $f(x) = x\tan x, \quad -\pi/2 < x < \pi/2$

53. A function is defined by the following experimental data. Use a direction field to sketch the graph of its antiderivative if the initial condition is $F(0) = 0$.

x	0	0.2	0.4	0.6	0.8	1.0	1.2	1.4	1.6
$f(x)$	0	0.2	0.5	0.8	1.0	0.6	0.2	0	−0.1

54. (a) Draw a direction field for the function $f(x) = 1/x^2$ and use it to sketch several members of the family of antiderivatives.

(b) Compute the general antiderivative explicitly and sketch several particular antiderivatives. Compare with your sketch in part (a).

55–60 ■ A particle is moving with the given data. Find the position of the particle.

55. $v(t) = 3 - 2t$, $s(0) = 4$

56. $v(t) = 3\sqrt{t}$, $s(1) = 5$

57. $a(t) = 3t + 8$, $s(0) = 1$, $v(0) = -2$

58. $a(t) = \cos t + \sin t$, $s(0) = 0$, $v(0) = 5$

59. $a(t) = t^2 - t$, $s(0) = 0$, $s(6) = 12$

60. $a(t) = 10 + 3t - 3t^2$, $s(0) = 0$, $s(2) = 10$

61. A stone is dropped from the upper observation deck (the Space Deck) of the CN Tower, 450 m above the ground.
(a) Find the distance of the stone above ground level at time t.
(b) How long does it take the stone to reach the ground?
(c) With what velocity does it strike the ground?

62. Answer Exercise 61 if the stone is thrown downward with a speed of 5 m/s.

63. Answer Exercise 61 if the stone is thrown upward with a speed of 5 m/s.

64. Show that for motion in a straight line with constant acceleration a, initial velocity v_0, and initial displacement s_0, the displacement after time t is

$$s = \tfrac{1}{2}at^2 + v_0 t + s_0$$

65. An object is projected upward with initial velocity v_0 meters per second from a point s_0 meters above the ground. Show that

$$[v(t)]^2 = v_0^2 - 19.6[s(t) - s_0]$$

66. Two balls are thrown upward from the edge of the cliff in Example 8. The first is thrown with a speed of 48 ft/s and the second is thrown 1 s later with a speed of 24 ft/s. Do the balls ever pass each other?

67. A company estimates that the marginal cost (in dollars per item) of producing x items is $1.92 - 0.002x$. If the cost of producing one item is \$562, find the cost of producing 100 items.

68. The linear density of a rod of length 1 m is given by $\rho(x) = 1/\sqrt{x}$ in grams per centimeter, where x is measured in centimeters from one end of the rod. Find the mass of the rod.

69. Since raindrops grow as they fall, their surface area increases and therefore the resistance to their falling increases. A raindrop has an initial downward velocity of 10 m/s and its downward acceleration is

$$a = \begin{cases} 9 - 0.9t & \text{if } 0 \leqslant t \leqslant 10 \\ 0 & \text{if } t > 10 \end{cases}$$

If the raindrop is initially 500 m above the ground, how long does it take to fall?

70. A car is traveling at 50 mi/h when the brakes are fully applied, producing a constant deceleration of 40 ft/s². What is the distance covered before the car comes to a stop?

71. What constant acceleration is required to increase the speed of a car from 30 mi/h to 50 mi/h in 5 s?

72. A car braked with a constant deceleration of 40 ft/s², producing skid marks measuring 160 ft before coming to a stop. How fast was the car traveling when the brakes were first applied?

73. A stone was dropped off a cliff and hit the ground with a speed of 120 ft/s. What is the height of the cliff?

74–75 ■ Find the vector functions that describe the velocity and position of a particle that has an acceleration, initial velocity, and initial position as specified.

74. $\mathbf{a}(t) = 2\,\mathbf{j}$, $\mathbf{v}(0) = \mathbf{i} - \mathbf{j}$, $\mathbf{r}(0) = \mathbf{0}$

75. $\mathbf{a}(t) = -10\,\mathbf{j}$, $\mathbf{v}(0) = \mathbf{i} + \mathbf{j}$, $\mathbf{r}(0) = \mathbf{0}$

76–77 ■
(a) Find the vector functions that describe the velocity and position of a particle that has an acceleration, initial velocity, and initial position as specified.
(b) Use a calculator or computer to graph the path of the particle.

76. $\mathbf{a}(t) = 2\,\mathbf{i} + t\,\mathbf{j}$, $\mathbf{v}(0) = \mathbf{i} - \mathbf{j}$, $\mathbf{r}(0) = \mathbf{i} + \mathbf{j}$

77. $\mathbf{a}(t) = t^2\mathbf{i} + \cos 2t\,\mathbf{j}$, $\mathbf{v}(0) = \mathbf{j}$, $\mathbf{r}(0) = \mathbf{i}$

78. A force with magnitude 20 N acts in the positive y-direction on an object with mass 4 kg. The object starts at the origin with initial velocity $\mathbf{v}(0) = \mathbf{i} - \mathbf{j}$. Find its position function and its speed at time t.

79. A projectile is fired with an initial speed of 500 m/s and angle of elevation 30°. Find (a) the range of the projectile, (b) the maximum height reached, and (c) the speed at impact.

80. Rework Exercise 79 if the projectile is fired from a position of 200 m above the ground.

81. A ball is thrown at an angle of 45° to the ground. If the ball lands 90 m away, what was its initial speed?

82. A gun has muzzle speed 120 m/s. What angle of elevation should be used to hit an object 500 m away?

83. Show that the trajectory of a projectile is a parabola by eliminating the parameter t from the parametric equations

$$x = (v_0 \cos\alpha)t \qquad y = (v_0 \sin\alpha)t - \tfrac{1}{2}gt^2$$

84. A rocket burning its onboard fuel while moving through space has velocity $\mathbf{v}(t)$ and mass $m(t)$ at time t. If exhaust gases escape with velocity $\mathbf{v}_e$ relative to the rocket, it can be deduced from Newton's Second Law of Motion that

$$m\frac{d\mathbf{v}}{dt} = \frac{dm}{dt}\mathbf{v}_e$$

(a) Show that $\mathbf{v}(t) = \mathbf{v}(0) - \ln \dfrac{m(0)}{m(t)}\mathbf{v}_e$.

(b) For the rocket to accelerate in a straight line from rest to twice the speed of its own exhaust gases, what fraction of its initial mass would the rocket have to burn as fuel?

85. To prove Theorem 2, let F and G be any two antiderivatives of f on I and let $H = G - F$.

(a) If x_1 and x_2 are any two numbers in I with $x_1 < x_2$, apply the Mean Value Theorem on the interval $[x_1, x_2]$ to show that $H(x_1) = H(x_2)$. Why does this show that H is a constant function?

(b) Deduce Theorem 2 from the result of part (a).

5 REVIEW

KEY TOPICS ■ Define, state, or discuss the following.

1. Absolute maximum and minimum values
2. Local maximum and minimum values
3. Extreme Value Theorem
4. Fermat's Theorem
5. Critical number
6. Closed Interval Method
7. Mean Value Theorem
8. Increasing/Decreasing Test
9. First Derivative Test
10. Concave upward; concave downward
11. Test for Concavity
12. Point of inflection
13. Second Derivative Test
14. First Derivative Test for Absolute Extrema
15. Antiderivative
16. Antiderivative of a vector function

EXERCISES

1–16 ■ Determine whether the statement is true or false.

1. If $f'(c) = 0$, then f has a local maximum or minimum at c.
2. If f has an absolute minimum value at c, then $f'(c) = 0$.
3. If f is continuous on (a, b), then f attains an absolute maximum value $f(c)$ and an absolute minimum value $f(d)$ at some numbers c and d in (a, b).
4. If f is continuous on $[-1, 1]$ and differentiable on $(-1, 1)$ and $f(-1) = f(1)$, then there is a number c such that $|c| < 1$ and $f'(c) = 0$.
5. If f is continuous on $[1, 6]$ and $f'(x) < 0$ for $x \in (1, 6)$, then f is decreasing on $[1, 6]$.
6. If $f''(c) = 0$, then $(c, f(c))$ is an inflection point of the curve $y = f(x)$.
7. If $f'(x) = g'(x)$ for $0 < x < 1$, then $f(x) = g(x)$ for $0 < x < 1$.
8. There exists a function f such that $f(1) = -2$, $f(3) = 0$, and $f'(x) > 1$ for all x.
9. There exists a function f such that $f(x) > 0$, $f'(x) < 0$, and $f''(x) > 0$ for all x.
10. There exists a function f such that $f(x) < 0$, $f'(x) < 0$, and $f''(x) > 0$ for all x.
11. If f and g are increasing on an interval I, then $f + g$ is increasing on I.
12. If f and g are increasing on an interval I, then $f - g$ is increasing on I.
13. If f and g are increasing on an interval I, then fg is increasing on I.
14. If f and g are positive increasing functions on an interval I, then fg is increasing on I.
15. If f is increasing and $f(x) > 0$ on I, then $g(x) = 1/f(x)$ is decreasing on I.
16. The most general antiderivative of $f(x) = x^{-2}$ is $F(x) = (-1/x) + C$.

17. The graph of the derivative f' of a function f is given.
(a) On what intervals is f increasing or decreasing?
(b) At what values of x does f have a local maximum or minimum?
(c) Where is f concave upward or downward?
(d) If $f(0) = 0$, sketch a possible graph of f.

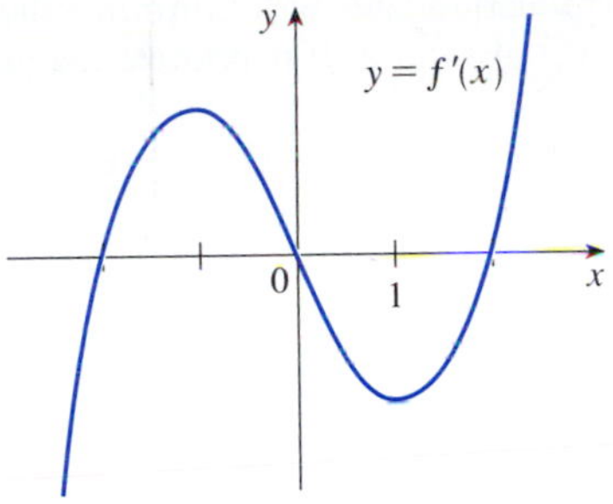

18. The figure shows the graph of the derivative f' of a function f.
(a) Sketch the graph of f''.
(b) Sketch a possible graph of f.

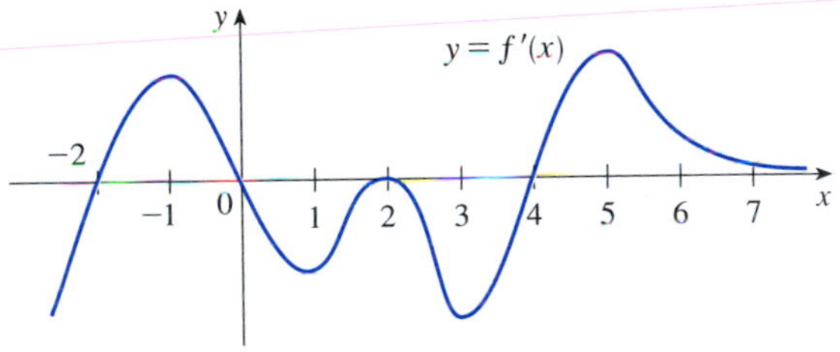

19–24 ■ Find the local and absolute extreme values of the function on the given interval.

19. $f(x) = x^3 - 12x + 5, \quad [-5, 3]$

20. $f(x) = 3x^5 - 25x^3 + 60x, \quad [-1, 3]$

21. $f(x) = \dfrac{x - 2}{x + 2}, \quad [0, 4]$

22. $f(x) = \sqrt{x^2 + 4x + 8}, \quad [-3, 0]$

23. $f(x) = x - \sqrt{2}\sin x, \quad [0, \pi]$

24. $f(x) = 2x + 2\cos x - 4\sin x - \cos 2x, \quad [0, \pi]$

25–38 ■
(a) Find the vertical and horizontal asymptotes, if any.
(b) Find the intervals of increase or decrease.
(c) Find the local maximum and minimum values.
(d) Find the intervals of concavity and the inflection points.
(e) Use the information from parts (a)–(d) to sketch the graph of f. Check your work with a graphing device.

25. $y = 1 + x + x^3$

26. $y = 3x^4 - 4x^3 - 12x^2 + 2$

27. $y = \dfrac{1}{x(x - 3)^2}$

28. $y = \dfrac{1}{x^2 - x - 6}$

29. $y = x\sqrt{5 - x}$

30. $y = \dfrac{1}{x} + \dfrac{1}{x + 1}$

31. $y = \sqrt{1 - x^2}/x$

32. $y = x + \sqrt{1 - x}$

33. $y = \sqrt{x} - \sqrt[3]{x}$

34. $y = 4x - \tan x, \quad -\pi/2 < x < \pi/2$

35. $y = \sin^{-1}(1/x)$

36. $y = e^{2x - x^2}$

37. $y = e^x + e^{-3x}$

38. $y = \ln(x^2 - 1)$

39–42 ■ Produce graphs of f that reveal all the important aspects of the curve. Use graphs of f' and f'' to estimate the intervals of increase and decrease, extreme values, intervals of concavity, and inflection points. In Exercise 39 use calculus to find these quantities exactly.

39. $f(x) = \dfrac{x^2 - 1}{x^3}$

40. $f(x) = \dfrac{\sqrt[3]{x}}{1 - x}$

41. $f(x) = 3x^6 - 5x^5 + x^4 - 5x^3 - 2x^2 + 2$

42. $f(x) = \sin x \cos^2 x, \quad 0 \leqslant x \leqslant 2\pi$

43. Graph $f(x) = e^{-1/x^2}$ in a viewing rectangle that shows all the main aspects of this function. Estimate the inflection points. Then use calculus to find them exactly.

44. Investigate the family of functions $f(x) = cxe^{-cx^2}$. What happens to the maximum and minimum points and inflection points as c changes? Illustrate your conclusions by graphing several members of the family.

CAS **45.** If $f(x) = \arctan(\cos(3\arcsin x))$, use the graphs of f, f', and f'' to estimate the x-coordinates of the maximum and minimum points and inflection points of f.

CAS **46.** If $f(x) = \ln(2x + x\sin x)$, use the graphs of f, f', and f'' to estimate the intervals of increase and the inflection points of f on the interval $(0, 15]$.

47. Show that the equation $x^{101} + x^{51} + x - 1 = 0$ has exactly one real root.

48. Suppose that f is continuous on $[0, 4]$, $f(0) = 1$, and $2 \leqslant f'(x) \leqslant 5$ for all x in $(0, 4)$. Show that $9 \leqslant f(4) \leqslant 21$.

49. By applying the Mean Value Theorem to the function $f(x) = x^{1/5}$ on the interval $[32, 33]$, show that

$$2 < \sqrt[5]{33} < 2.0125$$

50. A number x_0 is called a **fixed point** of a function f if $f(x_0) = x_0$. Show that if $f'(x) < 1$ for all x, then f has at most one fixed point.

51–52 ■ Sketch the graph of a function that satisfies the given conditions.

51. $f(0) = 0,\ f'(-2) = f'(1) = f'(9) = 0,$
$\lim_{x\to\infty} f(x) = 0,\ \lim_{x\to 6} f(x) = -\infty,$
$f'(x) < 0$ on $(-\infty, -2), (1, 6)$, and $(9, \infty)$,
$f'(x) > 0$ on $(-2, 1)$ and $(6, 9)$,
$f''(x) > 0$ on $(-\infty, 0)$ and $(12, \infty)$,
$f''(x) < 0$ on $(0, 6)$ and $(6, 12)$

52. $f(0) = 0$, f is continuous and even, $f'(x) = 2x$ if $0 < x < 1$, $f'(x) = -1$ if $1 < x < 3$, $f'(x) = 1$ if $x > 3$

53. For what values of the constants a and b is $(1, 6)$ a point of inflection of the curve $y = x^3 + ax^2 + bx + 1$?

54. Let $g(x) = f(x^2)$, where f is twice differentiable for all x, $f'(x) > 0$ for all $x \neq 0$, and f is concave downward on $(-\infty, 0)$ and concave upward on $(0, \infty)$.
(a) At what numbers does g have an extreme value?
(b) Discuss the concavity of g.

55. Show that the shortest distance from the point (x_1, y_1) to the straight line $Ax + By + C = 0$ is

$$\frac{|Ax_1 + By_1 + C|}{\sqrt{A^2 + B^2}}$$

56. Find the point on the hyperbola $xy = 8$ that is closest to the point $(3, 0)$.

57. Find the smallest possible area of an isosceles triangle that is circumscribed about a circle of radius r.

58. Find the volume of the largest circular cone that can be inscribed in a sphere of radius r.

59. In ΔABC, D lies on AB, $CD \perp AB$, $|AD| = |BD| = 4$ cm, and $|CD| = 5$ cm. Where should a point P be chosen on CD so that the sum $|PA| + |PB| + |PC|$ is a minimum?

60. Do Exercise 59 when $|CD| = 2$ cm.

61. The velocity of a wave of length L in deep water is

$$v = K\sqrt{\frac{L}{C} + \frac{C}{L}}$$

where K and C are known positive constants. What is the length of the wave that gives the minimum velocity?

62. A metal storage tank with volume V is to be constructed in the shape of a right circular cylinder surmounted by a hemisphere. What dimensions will require the least amount of metal?

63. A hockey team plays in an arena with a seating capacity of 15,000 spectators. With ticket prices at \$12, average attendance at a game has been 11,000. A market survey indicates that for each dollar that ticket prices are lowered, the average attendance will increase by 1000. How should the owners of the team set ticket prices to maximize their revenue from ticket sales?

64. A manufacturer determines that the cost of making x units of a commodity is

$$C(x) = 250{,}000 + 0.84x + 0.0002x^2$$

and the demand function is

$$p(x) = 10 - 0.05x$$

(a) What production level will minimize the average cost?
(b) What level of sales will maximize the profit?

65–72 ■ Find $f(x)$.

65. $f'(x) = x - \sqrt[4]{x}$

66. $f'(x) = 2/\sqrt{x^5}$

67. $f'(x) = \dfrac{1 + x}{\sqrt{x}},\ f(1) = 0$

68. $f'(x) = 1 + 2\sin x - \cos x,\ f(0) = 3$

69. $f''(x) = x^3 + x,\ f(0) = -1,\ f'(0) = 1$

70. $f''(x) = x^4 - 4x^2 + 3x - 2,\ f(0) = 0,\ f(1) = 1$

71. $f'(x) = 2/(1 + x^2),\ f(0) = -1$

72. $f'(x) = e^x - (1/x),\ f(1) = 0$

73–76 ■ Find the vector function $\mathbf{r}(t)$ that gives the position of a particle at time t having the given acceleration, initial velocity, and initial position.

73. $\mathbf{a}(t) = \mathbf{i} + 2\mathbf{j},\quad \mathbf{v}(0) = \mathbf{j},\quad \mathbf{r}(0) = -\mathbf{i} + \mathbf{j}$

74. $\mathbf{a}(t) = -\mathbf{i} + 4\mathbf{j},\quad \mathbf{v}(0) = 2\mathbf{i} + 3\mathbf{j},\quad \mathbf{r}(0) = \mathbf{i} + 2\mathbf{j}$

75. $\mathbf{a}(t) = 2t\,\mathbf{i} + 3\mathbf{j},\quad \mathbf{v}(0) = \mathbf{i} - \mathbf{j},$ initial position $(1, 2)$

76. $\mathbf{a}(t) = 2\mathbf{i} + 3t\,\mathbf{j},\quad \mathbf{v}(0) = -2\mathbf{i} + \mathbf{j},$ initial position $(2, -2)$

77. At time $t = 0$ a particle is located at the point $(1, 2)$. It travels in a straight line to the point $(4, 1)$, has speed 2 at $(1, 2)$, and constant acceleration $3\mathbf{i} - \mathbf{j}$. Find $\mathbf{r}(t)$, the position of the object at time t.

78. At time $t = 0$ a particle is located at the point $(2, 5)$. It travels in a straight line to the point $(-1, 2)$, has speed 2 at $(2, 5)$, and constant acceleration $-\mathbf{i} - \mathbf{j}$. Find $\mathbf{r}(t)$, the position of the object at time t.

79. Use a graphing device to draw a graph of the function $f(x) = x^2\sin(x^2)$, $0 \leqslant x \leqslant \pi$, and use that graph to sketch the antiderivative F of f that satisfies the initial condition $F(0) = 0$.

80. Investigate the family of curves given by

$$f(x) = x^4 + x^3 + cx^2$$

In particular you should determine the transitional value of c at which the number of critical numbers changes and the transitional value at which the number of inflection points changes. Illustrate the various possible shapes with graphs.

81. A canister is dropped from a helicopter 500 m above the ground. Its parachute does not open, but the canister has been designed to withstand an impact velocity of 100 m/s. Will it burst or not?

82. In an automobile race along a straight road, car A passed car B twice. Prove that at some time during the race their accelerations were equal.

83. A rectangular beam will be cut from a cylindrical log of radius 10 inches.

(a) Show that the beam of maximal cross-sectional area is a square.

(b) Four rectangular planks will be cut from the four sections of the log that remain after cutting the square beam. Determine the dimensions of the planks that will have maximal cross-sectional area.

(c) Suppose that the strength of a rectangular beam is proportional to the product of its width and the square of its depth. Find the dimensions of the strongest beam that can be cut from the cylindrical log.

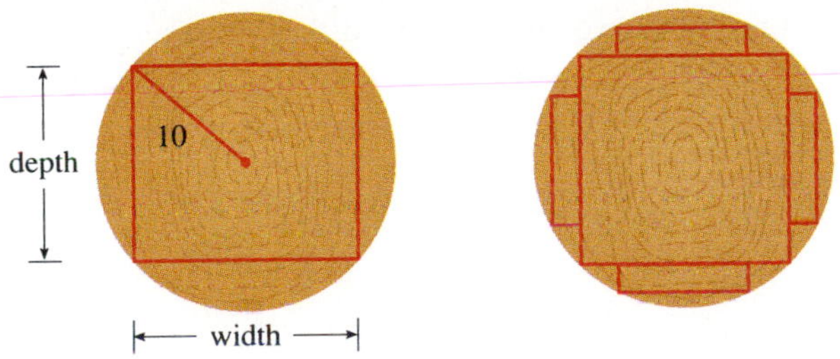

84. If a projectile is fired with an initial velocity v at an angle of inclination θ from the horizontal, then its trajectory, neglecting air resistance, is the parabola

$$y = (\tan\theta)x - \frac{g}{2v^2\cos^2\theta}x^2 \qquad 0 \leqslant \theta \leqslant \frac{\pi}{2}$$

(a) Suppose the projectile is fired from the base of a plane that is inclined at an angle α, $\alpha > 0$, from the horizontal, as shown in the figure. Show that the range of the projectile, measured up the slope, is given by

$$R(\theta) = \frac{2v^2\cos\theta\sin(\theta - \alpha)}{g\cos^2\alpha}$$

(b) Determine θ so that R is a maximum.

(c) Suppose the plane is at an angle α *below* the horizontal. Determine the range R in this case, and determine the angle at which the projectile should be fired to maximize R.

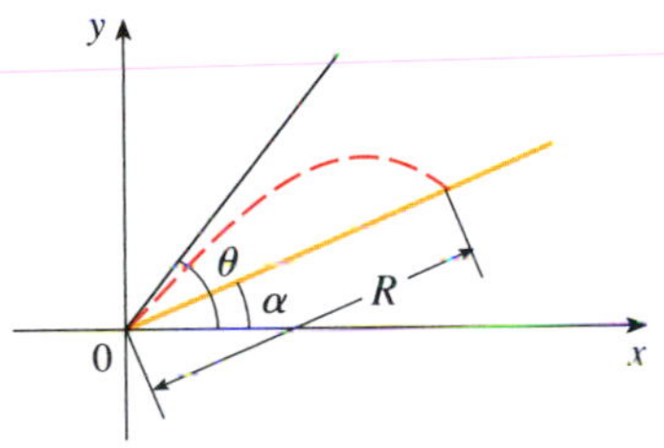

PROBLEMS PLUS

Cover up the solution to the example and try it yourself.

EXAMPLE For what values of c does the equation $\ln x = cx^2$ have exactly one solution?

SOLUTION One of the most important principles of problem solving is to draw a diagram, even if the problem as stated doesn't explicitly mention a geometric situation. Our present problem can be reformulated geometrically as follows: For what values of c does the curve $y = \ln x$ intersect the curve $y = cx^2$ in exactly one point?

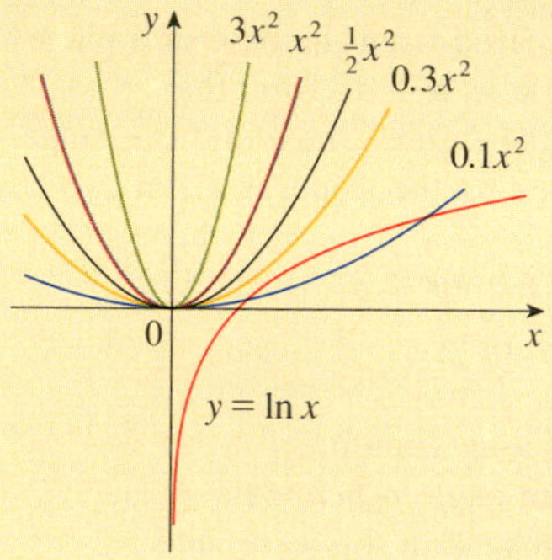

FIGURE 1

Let's start by graphing $y = \ln x$ and $y = cx^2$ for various values of c. We know that, for $c \neq 0$, $y = cx^2$ is a parabola that opens upward if $c > 0$ and downward if $c < 0$. Figure 1 shows the parabolas $y = cx^2$ for several positive values of c. Most of them don't intersect $y = \ln x$ at all and one intersects twice. We have the feeling that there must be a value of c (somewhere between 0.1 and 0.3) for which the curves intersect exactly once, as in Figure 2.

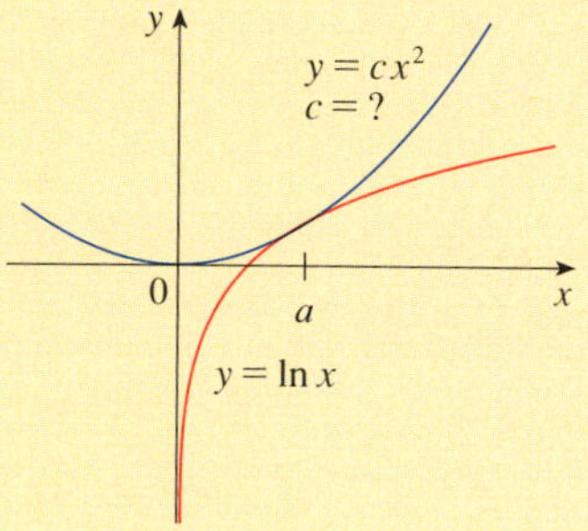

FIGURE 2

To find that particular value of c, we let a be the x-coordinate of the single point of intersection. In other words, $\ln a = ca^2$, so a is the unique solution of the given equation. We see from Figure 2 that the curves just touch, so they have a common tangent line when $x = a$. That means the curves $y = \ln x$ and $y = cx^2$ have the same slope when $x = a$. Therefore

$$\frac{1}{a} = 2ca$$

Solving the equations $\ln a = ca^2$ and $1/a = 2ca$, we get

$$\ln a = ca^2 = c \cdot \frac{1}{2c} = \frac{1}{2}$$

Thus, $a = e^{1/2}$ and

$$c = \frac{\ln a}{a^2} = \frac{\ln e^{1/2}}{e} = \frac{1}{2e}$$

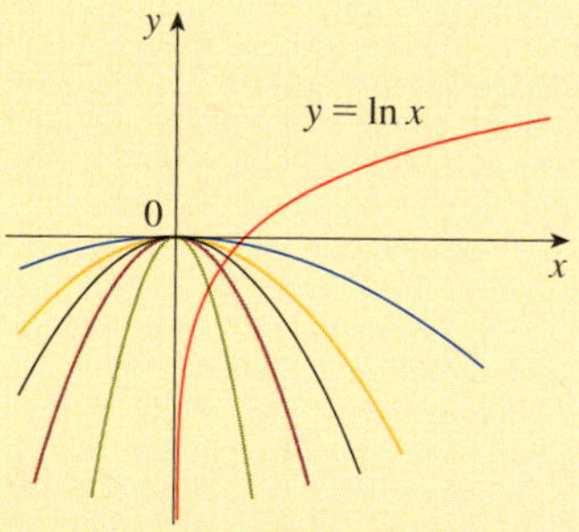

FIGURE 3

For negative values of c we have the situation illustrated in Figure 3: All parabolas $y = cx^2$ with negative values of c intersect $y = \ln x$ exactly once. And let's not forget about $c = 0$: The curve $y = 0x^2 = 0$ is just the x-axis, which intersects $y = \ln x$ exactly once.

To summarize, the required values of c are $c = 1/(2e)$ and $c \leq 0$. ■

PROBLEMS

1. If a rectangle has its base on the x-axis and two vertices on the curve $y = e^{-x^2}$, show that the rectangle has the largest possible area when the two vertices are at the points of inflection of the curve.

2. Prove that $\log_2 5$ is an irrational number.

3. If x, y, and z are positive numbers, prove that

$$\frac{(x^2 + 1)(y^2 + 1)(z^2 + 1)}{xyz} \geqslant 8$$

4. Find all functions f such that $f''' = f''$.

5. Find the highest and lowest points on the curve $x^2 + xy + y^2 = 12$.

6. An arc PQ of a circle subtends a central angle θ as in the figure. Let $A(\theta)$ be the area between the chord PQ and the arc PQ. Let $B(\theta)$ be the area between the tangent lines PR, QR, and the arc. Find

$$\lim_{\theta \to 0^+} \frac{A(\theta)}{B(\theta)}$$

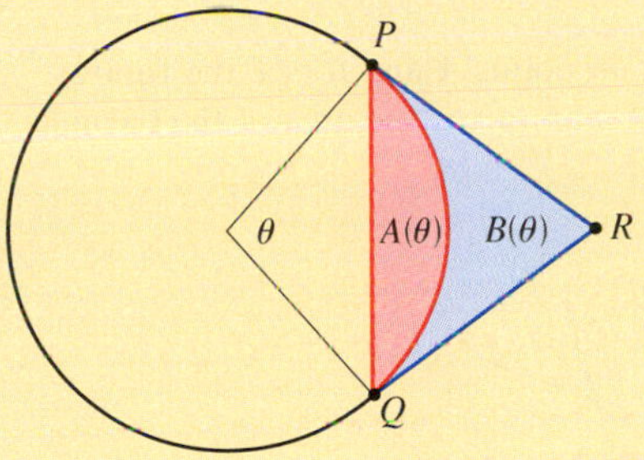

FIGURE FOR PROBLEM 6

7. Show that

$$\frac{d^n}{dx^n} e^{ax} \sin bx = r^n e^{ax} \sin(bx + n\theta)$$

where $r^2 = a^2 + b^2$ and $\theta = \tan^{-1}(b/a)$.

8. Find a function f such that $f'(-1) = \frac{1}{2}$, $f'(0) = 0$, and $f''(x) > 0$ for all x, or prove that such a function cannot exist.

9. Find the absolute maximum value of the function $f(x) = \dfrac{1}{1 + |x|} + \dfrac{1}{1 + |x - 2|}$.

10. Find the point where the curves $y = x^3 - 3x + 4$ and $y = 3(x^2 - x)$ are tangent to each other, that is, have a common tangent line. Illustrate by sketching both curves and the common tangent.

11. The figure shows a circle with radius 1 inscribed in the parabola $y = x^2$. Find the center of the circle.

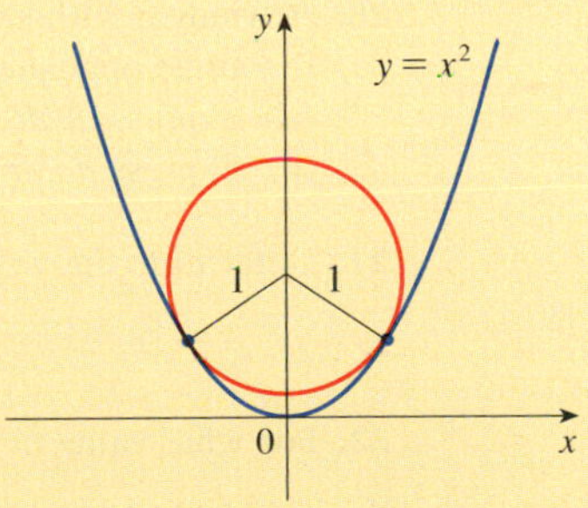

FIGURE FOR PROBLEM 11

12. Sketch the region in the plane consisting of all points (x, y) such that

$$2xy \leqslant |x - y| \leqslant x^2 + y^2$$

13. Show that, for $x > 0$,

$$\frac{x}{1 + x^2} < \tan^{-1}x < x$$

14. Sketch the graph of a function f such that $f'(x) < 0$ for all x, $f''(x) > 0$ for $|x| > 1$, $f''(x) < 0$ for $|x| < 1$, and $\lim_{x \to \pm\infty} [f(x) + x] = 0$.

15. Show that $f(x) = [\![x]\!] + \sqrt{x - [\![x]\!]}$ is continuous and increasing on $(-\infty, \infty)$.

16. Draw the graphs of the functions $f(x) = (-1)^{[\![1/x]\!]}$ and $g(x) = x(-1)^{[\![1/x]\!]}$ for $\frac{1}{6} \leqslant |x| \leqslant 2$. Is it possible to define $g(0)$ in such a way that g becomes continuous at 0?

17. The line $y = mx + b$ intersects the parabola $y = x^2$ in points A and B (see the figure). Find the point P on the arc AOB of the parabola that maximizes the area of the triangle PAB.

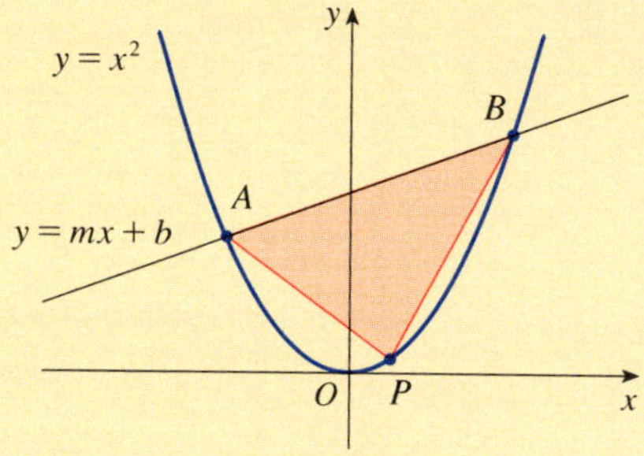

FIGURE FOR PROBLEM 17

18. $ABCD$ is a square piece of paper with sides of length 1 m. A quarter-circle is drawn from B to D with center A. The piece of paper is folded along EF, with E on AB and F on AD, so that A falls on the quarter-circle. Determine the maximum and minimum areas that the triangle AEF could have.

19. Find $\displaystyle\lim_{x \to \infty} \frac{x}{[\![x]\!]}$.

20. A triangle with sides a, b, and c varies with time t, but its area never changes. Let θ be the angle opposite the side of length a and suppose θ always remains acute.
(a) Express $d\theta/dt$ in terms of b, c, θ, db/dt, and dc/dt.
(b) Express da/dt in terms of the quantities in part (a).

21. Determine the values of the number a for which function f has no critical number:

$$f(x) = (a^2 + a - 6)\cos 2x + (a - 2)x + \cos 1$$

22. For what value of a is it true that

$$\lim_{x \to \infty} \left(\frac{x + a}{x - a}\right)^x = e$$

23. Let ABC be a triangle with $\angle BAC = 120^\circ$ and $|AB| \cdot |AC| = 1$.
(a) Express the length of the angle bisector AD in terms of $x = |AB|$.
(b) Find the largest possible value of $|AD|$.

24. (a) Let ABC be a triangle with right angle A and hypotenuse $a = |BC|$ (see the figure). If the inscribed circle touches the hypotenuse at D, show that

$$|CD| = \tfrac{1}{2}(|BC| + |AC| - |AB|)$$

(b) If $\theta = \frac{1}{2}\angle C$, express the radius r of the inscribed circle in terms of a and θ.
(c) If a is fixed and θ varies, find the maximum value of r.

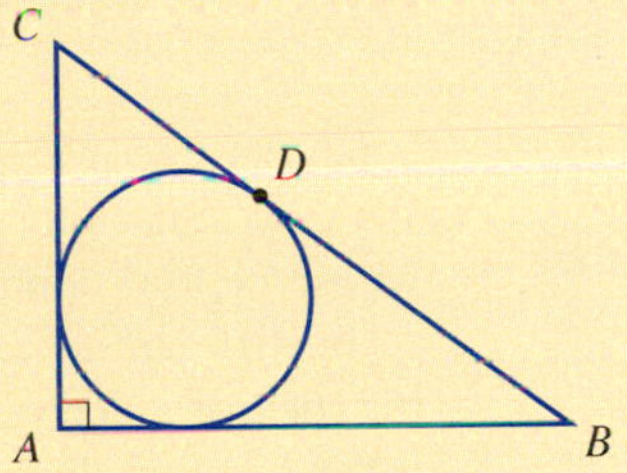

FIGURE FOR PROBLEM 24

25. Evaluate $\lim_{n\to\infty} \left(\frac{1}{\sqrt{n}\sqrt{n+1}} + \frac{1}{\sqrt{n}\sqrt{n+2}} + \cdots + \frac{1}{\sqrt{n}\sqrt{n+n}} \right)$.

26. For which positive numbers a is it true that $a^x \geqslant 1 + x$ for all x?

27. For which positive numbers a does the curve $y = a^x$ intersect the line $y = x$?

28. How many points of intersection do the curves $y = a^x$ and $y = \log_a x$ have? First, experiment by graphing these curves on the same screen for various values of a. Then try to prove what you have discovered.

29. Prove that $\cosh(\sinh x) < \sinh(\cosh x)$ for all x.

30. Suppose that the roots of a fourth-degree polynomial are consecutive terms of an arithmetic sequence; that is, they are equally spaced. Draw a diagram to make it seem plausible that the roots of its derivative are also consecutive terms of an arithmetic sequence. Then prove it.

31. Let $f(x) = a_1 \sin x + a_2 \sin 2x + \cdots + a_n \sin nx$, where $a_1, a_2, \ldots, a_n$ are real numbers and n is a positive integer. If it is given that $|f(x)| \leqslant |\sin x|$ for all x, show that

$$|a_1 + 2a_2 + \cdots + na_n| \leqslant 1$$

6

INTEGRALS

The experimental verification of a theory concerning any natural phenomenon generally rests on the result of an integration.

J. W. MELLOR

Now is a good time to read (or reread) A Preview of Calculus, which is Section 5 in Review and Preview. It discusses the unifying ideas of calculus and helps put in perspective where we have been and where we are going.

Chapters 2–5 were concerned with the branch of calculus called differential calculus. We now turn our attention to another branch of calculus, called integral calculus, whose central concept is the definite integral.

In differential calculus, the tangent problem led us to formulate, in terms of limits, the idea of a derivative, which later turned out to be applicable, through velocities and other rates of change, to a variety of applied problems.

In integral calculus, the area problem leads us to formulate, again in terms of limits, the idea of an integral, which will later be used to find volumes, lengths of curves, work, and forces, and to solve problems in chemistry, physics, biology, and economics.

There is a connection between the two branches of calculus. It is called the Fundamental Theorem of Calculus and we see in this chapter that it greatly simplifies the solution of many problems.

6.1 SIGMA NOTATION

In finding areas and evaluating integrals we often encounter sums with many terms. A convenient way of writing such sums uses the Greek letter Σ (capital sigma, corresponding to our letter S) and is called **sigma notation**.

This tells us to end with $i = n$.

This tells us to add. $\sum_{i=m}^{n} a_i$

This tells us to start with $i = m$.

(1) DEFINITION If $a_m, a_{m+1}, \ldots, a_n$ are real numbers and m and n are integers such that $m \leq n$, then

$$\sum_{i=m}^{n} a_i = a_m + a_{m+1} + a_{m+2} + \cdots + a_{n-1} + a_n$$

With function notation, Definition 1 can be written as

$$\sum_{i=m}^{n} f(i) = f(m) + f(m+1) + f(m+2) + \cdots + f(n-1) + f(n)$$

Thus, the symbol $\Sigma_{i=m}^{n}$ indicates a summation in which the letter i (called the **index of summation**) takes on the values $m, m+1, \ldots, n$. Other letters can also be used as the index of summation.

EXAMPLE 1

(a) $\sum_{i=1}^{4} i^2 = 1^2 + 2^2 + 3^2 + 4^2 = 30$

(b) $\sum_{i=3}^{n} i = 3 + 4 + 5 + \cdots + (n-1) + n$

(c) $\sum_{j=0}^{5} 2^j = 2^0 + 2^1 + 2^2 + 2^3 + 2^4 + 2^5 = 63$

(d) $\sum_{k=1}^{n} \frac{1}{k} = 1 + \frac{1}{2} + \frac{1}{3} + \cdots + \frac{1}{n}$

(e) $\sum_{i=1}^{3} \frac{i-1}{i^2+3} = \frac{1-1}{1^2+3} + \frac{2-1}{2^2+3} + \frac{3-1}{3^2+3} = 0 + \frac{1}{7} + \frac{1}{6} = \frac{13}{42}$

(f) $\sum_{i=1}^{4} 2 = 2 + 2 + 2 + 2 = 8$ ■

EXAMPLE 2 Write the sum $2^3 + 3^3 + \cdots + n^3$ in sigma notation.

SOLUTION There is no unique way of writing a sum in sigma notation. We could write

$$2^3 + 3^3 + \cdots + n^3 = \sum_{i=2}^{n} i^3$$

or

$$2^3 + 3^3 + \cdots + n^3 = \sum_{j=1}^{n-1} (j+1)^3$$

or

$$2^3 + 3^3 + \cdots + n^3 = \sum_{k=0}^{n-2} (k+2)^3$$ ■

The following theorem gives three simple rules for working with sigma notation.

(2) THEOREM If c is any constant (that is, it does not depend on i), then

(a) $\sum_{i=m}^{n} ca_i = c \sum_{i=m}^{n} a_i$

(b) $\sum_{i=m}^{n} (a_i + b_i) = \sum_{i=m}^{n} a_i + \sum_{i=m}^{n} b_i$

(c) $\sum_{i=m}^{n} (a_i - b_i) = \sum_{i=m}^{n} a_i - \sum_{i=m}^{n} b_i$

PROOF To see why these rules are true, all we have to do is write both sides in expanded form. Rule (a) is just the distributive property of real numbers:

$$ca_m + ca_{m+1} + \cdots + ca_n = c(a_m + a_{m+1} + \cdots + a_n)$$

Rule (b) follows from the associative and commutative properties:

$$(a_m + b_m) + (a_{m+1} + b_{m+1}) + \cdots + (a_n + b_n)$$
$$= (a_m + a_{m+1} + \cdots + a_n) + (b_m + b_{m+1} + \cdots + b_n)$$

Rule (c) is proved similarly. □

EXAMPLE 3 Find $\sum_{i=1}^{n} 1$.

SOLUTION

$$\sum_{i=1}^{n} 1 = \underbrace{1 + 1 + \cdots + 1}_{n \text{ terms}} = n$$

■

EXAMPLE 4 Prove the formula for the sum of the first n positive integers:

$$\sum_{i=1}^{n} i = 1 + 2 + 3 + \cdots + n = \frac{n(n+1)}{2}$$

SOLUTION This formula can be proved by mathematical induction (see Appendix E) or by the following method used by the German mathematician Karl Friedrich Gauss (1777–1855) when he was ten years old.

Write the sum S twice, once in the usual order and once in reverse order:

$$S = 1 + \quad 2 \quad + \quad 3 \quad + \cdots + (n-1) + n$$

$$S = n + (n-1) + (n-2) + \cdots + \quad 2 \quad + 1$$

Adding all columns vertically, we get

$$2S = (n+1) + (n+1) + (n+1) + \cdots + (n+1) + (n+1)$$

On the right side there are n terms, each of which is $n + 1$, so

$$2S = n(n+1) \qquad \text{or} \qquad S = \frac{n(n+1)}{2}$$

■

EXAMPLE 5 Prove the formula for the sum of the squares of the first n positive integers:

$$\sum_{i=1}^{n} i^2 = 1^2 + 2^2 + 3^2 + \cdots + n^2 = \frac{n(n+1)(2n+1)}{6}$$

SOLUTION 1 Let S be the desired sum. We start with the *telescoping sum* (or collapsing sum):

Most terms cancel in pairs.

$$\sum_{i=1}^{n} [(1+i)^3 - i^3] = (2^3 - 1^3) + (3^3 - 2^3) + (4^3 - 3^3) + \cdots + [(n+1)^3 - n^3]$$

$$= (n+1)^3 - 1^3 = n^3 + 3n^2 + 3n$$

On the other hand, using Theorem 2 and Examples 3 and 4, we have

$$\sum_{i=1}^{n} [(1+i)^3 - i^3] = \sum_{i=1}^{n} [3i^2 + 3i + 1] = 3\sum_{i=1}^{n} i^2 + 3\sum_{i=1}^{n} i + \sum_{i=1}^{n} 1$$

$$= 3S + 3\frac{n(n+1)}{2} + n = 3S + \tfrac{3}{2}n^2 + \tfrac{5}{2}n$$

Thus, we have

$$n^3 + 3n^2 + 3n = 3S + \tfrac{3}{2}n^2 + \tfrac{5}{2}n$$

Solving this equation for S, we obtain

$$3S = n^3 + \tfrac{3}{2}n^2 + \tfrac{1}{2}n$$

or

$$S = \frac{2n^3 + 3n^2 + n}{6} = \frac{n(n+1)(2n+1)}{6}$$

PRINCIPLE OF MATHEMATICAL INDUCTION:

Let S_n be a statement involving the positive integer n. Suppose that
1. S_1 is true.
2. If S_k is true, then S_{k+1} is true.
Then S_n is true for all positive integers n.

See Appendix E for a more thorough discussion of mathematical induction.

SOLUTION 2 Let S_n be the given formula.

1. S_1 is true because $\quad 1^2 = \dfrac{1(1+1)(2\cdot 1+1)}{6}$
2. Assume that S_k is true; that is,

$$1^2 + 2^2 + 3^2 + \cdots + k^2 = \frac{k(k+1)(2k+1)}{6}$$

Then

$$\begin{aligned}1^2 + 2^2 + 3^2 + \cdots + (k+1)^2 &= (1^2 + 2^2 + 3^2 + \cdots + k^2) + (k+1)^2 \\ &= \frac{k(k+1)(2k+1)}{6} + (k+1)^2 \\ &= (k+1)\frac{k(2k+1) + 6(k+1)}{6} \\ &= (k+1)\frac{2k^2 + 7k + 6}{6} \\ &= \frac{(k+1)(k+2)(2k+3)}{6} \\ &= \frac{(k+1)[(k+1)+1][2(k+1)+1]}{6}\end{aligned}$$

So S_{k+1} is true.

By the Principle of Mathematical Induction, S_n is true for all n. ■

We list the results of Examples 3, 4, and 5 together with similar results for cubes and fourth powers (see Exercises 37–40) as Theorem 3. These formulas are needed for finding areas in the next section.

(3) THEOREM Let c be a constant and n a positive integer. Then

(a) $\displaystyle\sum_{i=1}^{n} 1 = n$ (b) $\displaystyle\sum_{i=1}^{n} c = nc$

(c) $\displaystyle\sum_{i=1}^{n} i = \frac{n(n+1)}{2}$ (d) $\displaystyle\sum_{i=1}^{n} i^2 = \frac{n(n+1)(2n+1)}{6}$

(e) $\displaystyle\sum_{i=1}^{n} i^3 = \left[\frac{n(n+1)}{2}\right]^2$ (f) $\displaystyle\sum_{i=1}^{n} i^4 = \frac{n(n+1)(2n+1)(3n^2+3n-1)}{30}$

EXAMPLE 6 Evaluate $\sum_{i=1}^{n} i(4i^2 - 3)$.

SOLUTION Using Theorems 2 and 3, we have

$$\begin{aligned}\sum_{i=1}^{n} i(4i^2 - 3) &= \sum_{i=1}^{n} (4i^3 - 3i) = 4\sum_{i=1}^{n} i^3 - 3\sum_{i=1}^{n} i \\ &= 4\left[\frac{n(n+1)}{2}\right]^2 - 3\frac{n(n+1)}{2} \\ &= \frac{n(n+1)[2n(n+1) - 3]}{2} \\ &= \frac{n(n+1)(2n^2 + 2n - 3)}{2}\end{aligned}$$ ■

The type of calculation in Example 7 arises in the next section when we compute areas.

EXAMPLE 7 Find $\lim_{n\to\infty} \sum_{i=1}^{n} \frac{3}{n}\left[\left(\frac{i}{n}\right)^2 + 1\right]$.

SOLUTION

$$\begin{aligned}\lim_{n\to\infty} \sum_{i=1}^{n} \frac{3}{n}\left[\left(\frac{i}{n}\right)^2 + 1\right] &= \lim_{n\to\infty} \sum_{i=1}^{n} \left[\frac{3}{n^3}i^2 + \frac{3}{n}\right] \\ &= \lim_{n\to\infty} \left[\frac{3}{n^3}\sum_{i=1}^{n} i^2 + \frac{3}{n}\sum_{i=1}^{n} 1\right] \\ &= \lim_{n\to\infty} \left[\frac{3}{n^3}\frac{n(n+1)(2n+1)}{6} + \frac{3}{n}\cdot n\right] \\ &= \lim_{n\to\infty} \left[\frac{1}{2}\cdot\frac{n}{n}\cdot\left(\frac{n+1}{n}\right)\left(\frac{2n+1}{n}\right) + 3\right] \\ &= \lim_{n\to\infty} \left[\frac{1}{2}\cdot 1\left(1 + \frac{1}{n}\right)\left(2 + \frac{1}{n}\right) + 3\right] \\ &= \tfrac{1}{2}\cdot 1\cdot 1\cdot 2 + 3 = 4\end{aligned}$$ ■

EXERCISES 6.1

1–10 ■ Write the sum in expanded form.

1. $\sum_{i=1}^{5} \sqrt{i}$

2. $\sum_{i=1}^{6} \frac{1}{i+1}$

3. $\sum_{i=4}^{6} 3^i$

4. $\sum_{i=4}^{6} i^3$

5. $\sum_{k=0}^{4} \frac{2k-1}{2k+1}$

6. $\sum_{k=5}^{8} x^k$

7. $\sum_{i=1}^{n} i^{10}$

8. $\sum_{j=n}^{n+3} j^2$

9. $\sum_{j=0}^{n-1} (-1)^j$

10. $\sum_{i=1}^{n} f(x_i)\,\Delta x_i$

11–20 ■ Write the sum in sigma notation.

11. $1 + 2 + 3 + 4 + \cdots + 10$

12. $\sqrt{3} + \sqrt{4} + \sqrt{5} + \sqrt{6} + \sqrt{7}$

13. $\frac{1}{2} + \frac{2}{3} + \frac{3}{4} + \frac{4}{5} + \cdots + \frac{19}{20}$

14. $\frac{3}{7} + \frac{4}{8} + \frac{5}{9} + \frac{6}{10} + \cdots + \frac{23}{27}$

15. $2 + 4 + 6 + 8 + \cdots + 2n$

16. $1 + 3 + 5 + 7 + \cdots + (2n - 1)$

17. $1 + 2 + 4 + 8 + 16 + 32$

18. $\frac{1}{1} + \frac{1}{4} + \frac{1}{9} + \frac{1}{16} + \frac{1}{25} + \frac{1}{36}$

19. $x + x^2 + x^3 + \cdots + x^n$

20. $1 - x + x^2 - x^3 + \cdots + (-1)^n x^n$

21–36 ■ Find the value of the sum.

21. $\sum_{i=4}^{8} (3i - 2)$

22. $\sum_{i=3}^{6} i(i + 2)$

23. $\sum_{j=1}^{6} 3^{j+1}$

24. $\sum_{k=0}^{8} \cos k\pi$

25. $\sum_{n=1}^{20} (-1)^n$

26. $\sum_{i=1}^{100} 4$

27. $\sum_{i=0}^{4} (2^i + i^2)$

28. $\sum_{i=-2}^{4} 2^{3-i}$

29. $\sum_{i=1}^{n} 2i$

30. $\sum_{i=1}^{n} (2 - 5i)$

31. $\sum_{i=1}^{n} (i^2 + 3i + 4)$

32. $\sum_{i=1}^{n} (3 + 2i)^2$

33. $\sum_{i=1}^{n} (i + 1)(i + 2)$

34. $\sum_{i=1}^{n} i(i + 1)(i + 2)$

35. $\sum_{i=1}^{n} (i^3 - i - 2)$

36. $\sum_{k=1}^{n} k^2(k^2 - k + 1)$

37. Prove formula (b) of Theorem 3.

38. Prove formula (e) of Theorem 3 using mathematical induction.

39. Prove formula (e) of Theorem 3 using a method similar to that of Example 5, Solution 1 [start with $(1 + i)^4 - i^4$].

40. Prove formula (e) of Theorem 3 using the following method published by Abu Bekr Mohammed ibn Alhusain Alkarchi in about A.D. 1010. The figure shows a square $ABCD$ in which sides AB and AD have been divided into segments of lengths 1, 2, 3, . . . , n. Thus, the side of the square has length $n(n + 1)/2$ so the area is $[n(n + 1)/2]^2$. But the area is also the sum of the areas of the n "gnomons" $G_1, G_2, \ldots, G_n$ shown in the figure. Show that the area of G_i is i^3 and conclude that formula (e) is true.

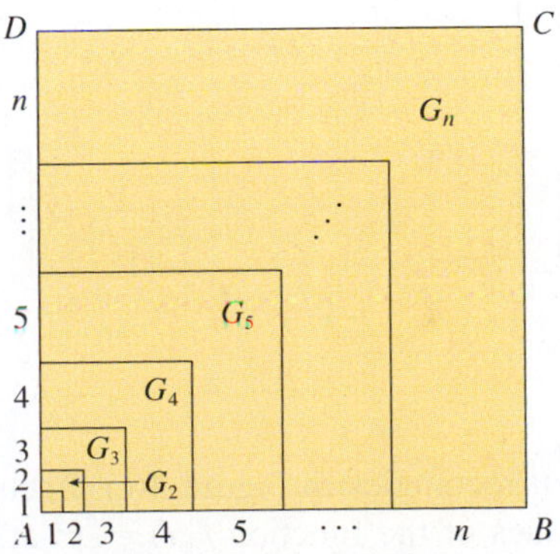

41. Evaluate each telescoping sum.

(a) $\sum_{i=1}^{n} [i^4 - (i - 1)^4]$ (b) $\sum_{i=1}^{100} (5^i - 5^{i-1})$

(c) $\sum_{i=3}^{99} \left(\frac{1}{i} - \frac{1}{i + 1}\right)$ (d) $\sum_{i=1}^{n} (a_i - a_{i-1})$

42. Prove the generalized triangle inequality

$$\left|\sum_{i=1}^{n} a_i\right| \leqslant \sum_{i=1}^{n} |a_i|$$

43–46 ■ Find each limit.

43. $\lim_{n\to\infty} \sum_{i=1}^{n} \frac{1}{n}\left(\frac{i}{n}\right)^2$

44. $\lim_{n\to\infty} \sum_{i=1}^{n} \frac{1}{n}\left[\left(\frac{i}{n}\right)^3 + 1\right]$

45. $\lim_{n\to\infty} \sum_{i=1}^{n} \frac{2}{n}\left[\left(\frac{2i}{n}\right)^3 + 5\left(\frac{2i}{n}\right)\right]$

46. $\lim_{n\to\infty} \sum_{i=1}^{n} \frac{3}{n}\left[\left(1 + \frac{3i}{n}\right)^3 - 2\left(1 + \frac{3i}{n}\right)\right]$

47. Prove the formula for the sum of a finite geometric series with first term a and common ratio r:

$$\sum_{i=1}^{n} ar^{i-1} = a + ar + ar^2 + \cdots + ar^{n-1} = \frac{a(r^n - 1)}{r - 1}$$

48. Evaluate $\sum_{i=1}^{n} \frac{3}{2^{i-1}}$.

49. Evaluate $\sum_{i=1}^{n} (2i + 2^i)$.

50. Evaluate $\sum_{i=1}^{m} \left[\sum_{j=1}^{n} (i + j)\right]$

51. Find the number n such that $\sum_{i=1}^{n} i = 78$.

52. (a) Use the product formula for $\sin x \cos y$ (see 18a in Appendix D) to show that

$$2 \sin \tfrac{1}{2}x \cos ix = \sin(i + \tfrac{1}{2})x - \sin(i - \tfrac{1}{2})x$$

(b) Use the identity in part (a) and telescoping sums to prove the formula

$$\sum_{i=1}^{n} \cos ix = \frac{\sin(n + \frac{1}{2})x - \sin \frac{1}{2}x}{2 \sin \frac{1}{2}x}$$

where x is not an integer multiple of 2π. Deduce that

$$\sum_{i=1}^{n} \cos ix = \frac{\sin \frac{1}{2}nx \cos \frac{1}{2}(n + 1)x}{\sin \frac{1}{2}x}$$

53. Use the method of Exercise 52 to prove the formula

$$\sum_{i=1}^{n} \sin ix = \frac{\sin \frac{1}{2}nx \sin \frac{1}{2}(n + 1)x}{\sin \frac{1}{2}x}$$

where x is not an integer multiple of 2π.

6.2 AREA

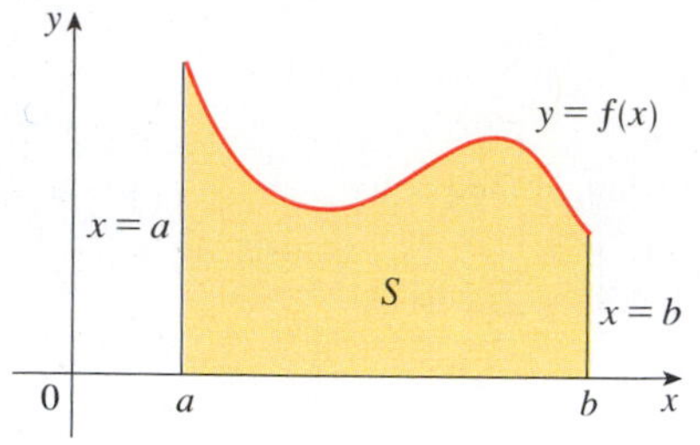

FIGURE 1
$S = \{(x, y) \mid a \leqslant x \leqslant b, 0 \leqslant y \leqslant f(x)\}$

We begin by attempting to solve the *area problem:* Find the area of the region S that lies under the curve $y = f(x)$ from a to b. This means that S, illustrated in Figure 1, is bounded by the graph of a function f [where $f(x) \geqslant 0$], the vertical lines $x = a$ and $x = b$, and the x-axis.

In trying to solve the area problem we have to ask ourselves: What is the meaning of the word *area*? This question is easy to answer for regions with straight sides. For a rectangle, the area is defined as the product of the length and the width. The area of a triangle is half the base times the height. The area of a polygon is found by dividing it into triangles (as in Figure 2) and adding the areas of the triangles.

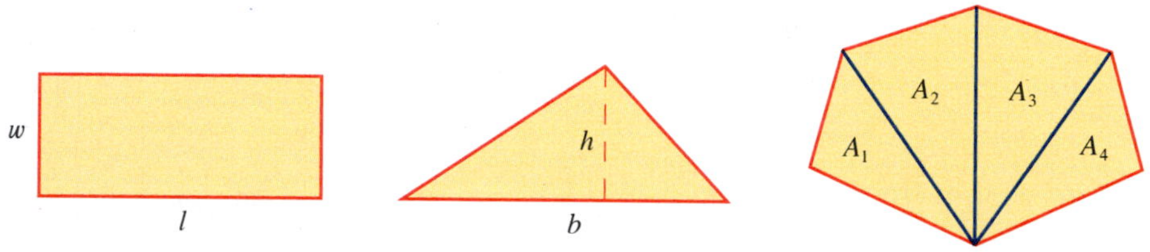

FIGURE 2 $A = lw$ $A = \frac{1}{2}bh$ $A = A_1 + A_2 + A_3 + A_4$

However, it is not so easy to find the area of a region with curved sides. We all have an intuitive idea of what the area of a region is. But part of the area problem is to make this intuitive idea precise by giving an exact definition of area.

Recall that in defining a tangent we first approximated the slope of the tangent line by slopes of secant lines and then we took the limit of these approximations. We pursue a similar idea for areas. We first approximate the region S by polygons and then we take the limit of the areas of these polygons. The following example illustrates the procedure.

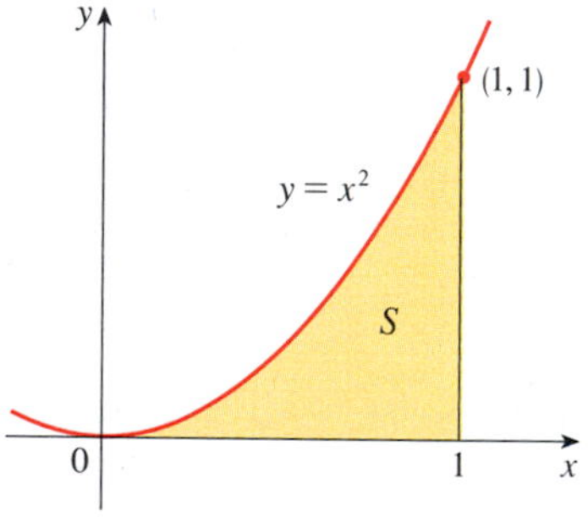

FIGURE 3

EXAMPLE 1 Let's try to find the area under the parabola $y = x^2$ from 0 to 1 (the parabolic region illustrated in Figure 3). One method of approximating the desired area is to divide the interval $[0, 1]$ into subintervals of equal length and consider the rectangles whose bases are these subintervals and whose heights are the values of the function at the right-hand endpoints of these subintervals. Figure 4 shows the approximation of the parabolic region by four, eight, and n rectangles.

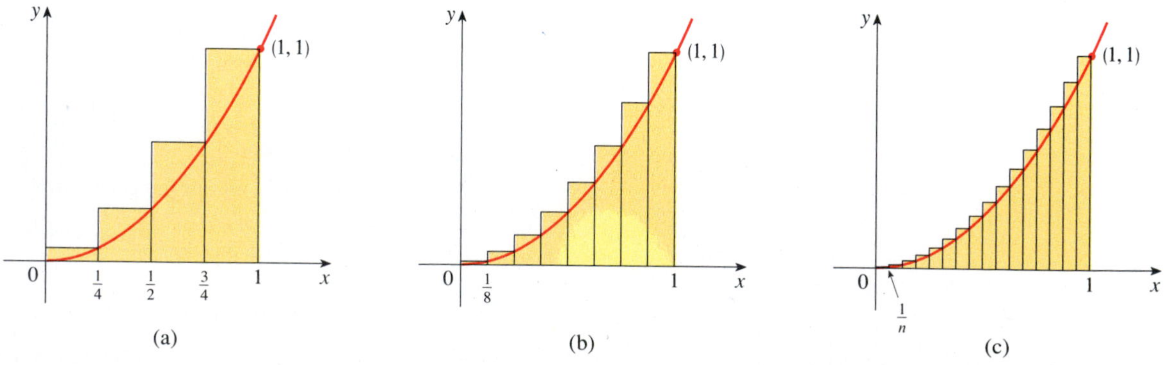

FIGURE 4

Let S_n be the sum of the areas of the n rectangles in Figure 4(c). Each rectangle has width $1/n$ and the heights are the values of the function $f(x) = x^2$ at the points

$1/n, 2/n, 3/n, \ldots, n/n$; that is, the heights are $(1/n)^2$, $(2/n)^2$, $(3/n)^2$, ..., $(n/n)^2$. Thus

$$\begin{aligned} S_n &= \frac{1}{n}\left(\frac{1}{n}\right)^2 + \frac{1}{n}\left(\frac{2}{n}\right)^2 + \frac{1}{n}\left(\frac{3}{n}\right)^2 + \cdots + \frac{1}{n}\left(\frac{n}{n}\right)^2 \\ &= \frac{1}{n}\frac{1}{n^2}(1^2 + 2^2 + 3^2 + \cdots + n^2) \\ &= \frac{1}{n^3}\sum_{i=1}^{n} i^2 \end{aligned}$$

Using the formula for the sum of the squares of the first n integers [Formula 6.1.3(d)], we can write

$$S_n = \frac{1}{n^3}\frac{n(n+1)(2n+1)}{6} = \frac{(n+1)(2n+1)}{6n^2}$$

For instance, the sum of the areas of the four shaded rectangles in Figure 4(a) is

$$S_4 = \frac{5(9)}{6(16)} = 0.46875$$

and the sum of the areas of the eight rectangles in Figure 4(b) is

$$S_8 = \frac{9(17)}{6(64)} = 0.3984375$$

n	S_n
10	0.385000
20	0.358750
30	0.350185
50	0.343400
100	0.338350
1000	0.333834

The results of similar calculations are shown in the table in the margin.

It looks as if S_n is becoming closer to $\frac{1}{3}$ as n increases. In fact,

$$\begin{aligned} \lim_{n\to\infty} S_n &= \lim_{n\to\infty} \frac{(n+1)(2n+1)}{6n^2} \\ &= \lim_{n\to\infty} \frac{1}{6}\left(\frac{n+1}{n}\right)\left(\frac{2n+1}{n}\right) \\ &= \lim_{n\to\infty} \frac{1}{6}\left(1 + \frac{1}{n}\right)\left(2 + \frac{1}{n}\right) \\ &= \tfrac{1}{6} \cdot 1 \cdot 2 = \tfrac{1}{3} \end{aligned}$$

From Figure 4 it appears that, as n increases, S_n becomes a better and better approximation to the area of the parabolic segment. Therefore, we *define* the area A to be the limit of the sums of the areas of the approximating rectangles, that is,

$$A = \lim_{n\to\infty} S_n = \tfrac{1}{3}$$ ■

In applying the idea of Example 1 to the more general region S of Figure 1, we have no need to use rectangles of equal width. We start by subdividing the interval $[a, b]$ into n smaller subintervals by choosing partition points $x_0, x_1, x_2, \ldots, x_n$ so that

$$a = x_0 < x_1 < x_2 < \cdots < x_{n-1} < x_n = b$$

Then the n subintervals are

$$[x_0, x_1], [x_1, x_2], [x_2, x_3], \ldots, [x_{n-1}, x_n]$$

This subdivision is called a **partition** of $[a, b]$ and we denote it by P. We use the notation Δx_i for the length of the ith subinterval $[x_{i-1}, x_i]$. Thus

$$\Delta x_i = x_i - x_{i-1}$$

The length of the longest subinterval is denoted by $\|P\|$ and is called the **norm** of P. Thus

$$\|P\| = \max\{\Delta x_1, \Delta x_2, \ldots, \Delta x_n\}$$

Figure 5 illustrates one possible partition of $[a, b]$.

FIGURE 5

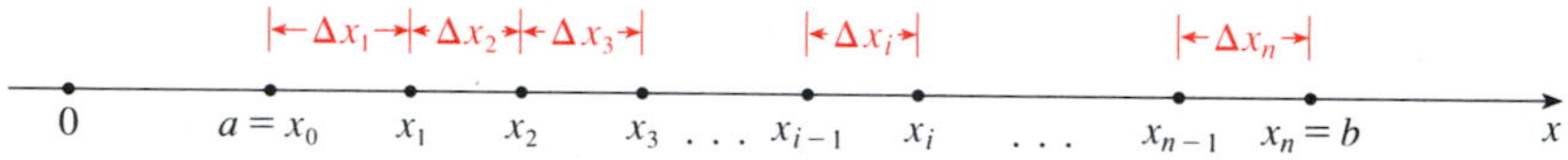

By drawing the lines $x = a, x = x_1, x = x_2, \ldots, x = b$, we use the partition P to divide the region S into strips $S_1, S_2, \ldots, S_n$ as in Figure 6. Next we approximate these strips S_i by rectangles R_i. To do this we choose a number x_i^* in each subinterval $[x_{i-1}, x_i]$ and construct a rectangle R_i with base Δx_i and height $f(x_i^*)$ as in Figure 7.

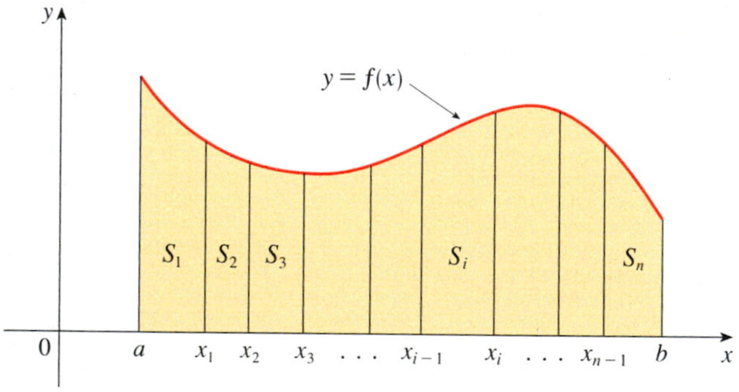

FIGURE 6

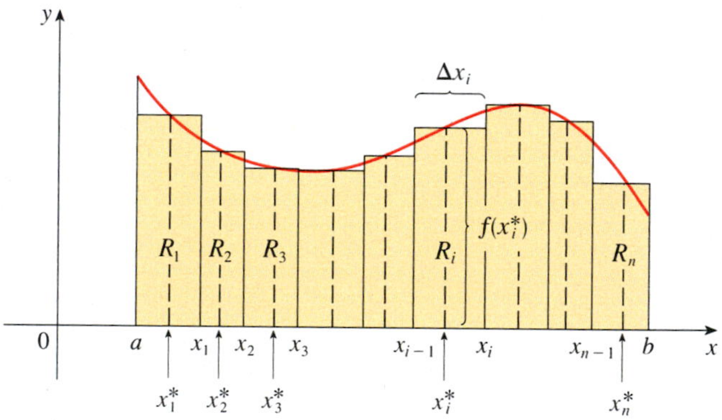

FIGURE 7

Each point x_i^* can be anywhere in its subinterval—at the right endpoint (as in Example 1) or at the left endpoint or somewhere between the endpoints. The area of the ith rectangle R_i is

$$A_i = f(x_i^*)\,\Delta x_i$$

The n rectangles $R_1, \ldots, R_n$ form a polygonal approximation to the region S. What we think of intuitively as the area of S is approximated by the sum of the areas of these rectangles, which is

(1) $$\sum_{i=1}^{n} A_i = \sum_{i=1}^{n} f(x_i^*)\,\Delta x_i = f(x_1^*)\,\Delta x_1 + \cdots + f(x_n^*)\,\Delta x_n$$

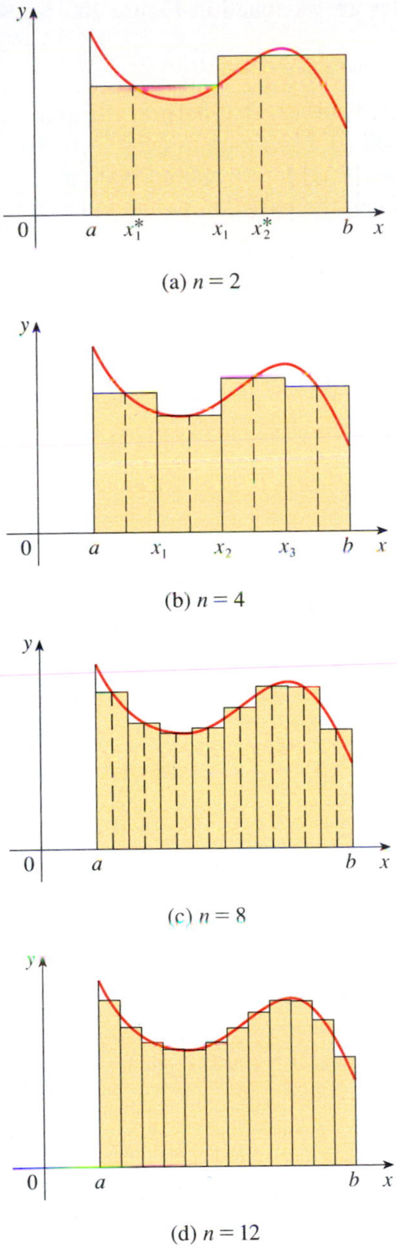

FIGURE 8

Figure 8 shows this approximation for partitions with $n = 2, 4, 8$, and 12. Notice that this approximation appears to become better and better as the strips become thinner and thinner, that is, as $\|P\| \to 0$. Therefore, we define the **area** A of the region S as the limiting value (if it exists) of the areas of the approximating polygons, that is, the limit of the sum (1) of the areas of the approximating rectangles. In symbols:

(2)
$$A = \lim_{\|P\| \to 0} \sum_{i=1}^{n} f(x_i^*) \, \Delta x_i$$

The preceding discussion and the diagrams in Figures 7 and 8 show that the definition of area in (2) corresponds to our intuitive feeling of what area ought to be.

The limit in (2) may or may not exist. It can be shown that if f is continuous, then this limit does exist; that is, the region has an area. [The precise meaning of the limit in Definition 2 is that for every $\varepsilon > 0$ there is a corresponding number $\delta > 0$ such that

$$\left| A - \sum_{i=1}^{n} f(x_i^*) \, \Delta x_i \right| < \varepsilon \qquad \text{whenever} \qquad \|P\| < \delta$$

In other words, the area can be approximated by a sum of areas of rectangles to within an arbitrary degree of accuracy (ε) by taking the norm of the partition sufficiently small.]

EXAMPLE 2

(a) If the interval $[0, 3]$ is divided into subintervals by the partition P and the set of partition points is $\{0, 0.6, 1.2, 1.6, 2, 2.5, 3\}$, find $\|P\|$.

(b) If $f(x) = x^2 - 4x + 5$ and x_i^* is chosen to be the left endpoint of the ith subinterval, find the sum of the areas of the approximating rectangles.

(c) Sketch the approximating rectangles.

SOLUTION

(a) We are given $x_0 = 0$, $x_1 = 0.6$, $x_2 = 1.2$, $x_3 = 1.6$, $x_4 = 2$, $x_5 = 2.5$, and $x_6 = 3$, so

$$\Delta x_1 = 0.6 - 0 = 0.6 \qquad \Delta x_2 = 1.2 - 0.6 = 0.6$$
$$\Delta x_3 = 1.6 - 1.2 = 0.4 \qquad \Delta x_4 = 2 - 1.6 = 0.4$$
$$\Delta x_5 = 2.5 - 2 = 0.5 \qquad \Delta x_6 = 3 - 2.5 = 0.5$$

(See Figure 9.) Therefore

$$\|P\| = \max\{0.6, 0.6, 0.4, 0.4, 0.5, 0.5\} = 0.6$$

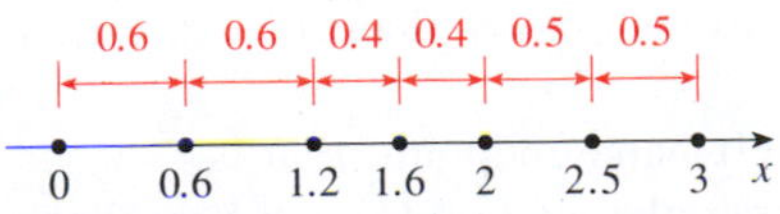

FIGURE 9

(b) Since $x_i^* = x_{i-1}$, the sum of the areas of the approximating rectangles is, by (1),

$$\begin{aligned}\sum_{i=1}^{6} f(x_i^*) \, \Delta x_i &= \sum_{i=1}^{6} f(x_{i-1}) \, \Delta x_i \\ &= f(0) \, \Delta x_1 + f(0.6) \, \Delta x_2 + f(1.2) \, \Delta x_3 + f(1.6) \, \Delta x_4 + f(2) \, \Delta x_5 \\ &\quad + f(2.5) \, \Delta x_6 \\ &= 5(0.6) + 2.96(0.6) + 1.64(0.4) + 1.16(0.4) + 1(0.5) + 1.25(0.5) \\ &= 7.021\end{aligned}$$

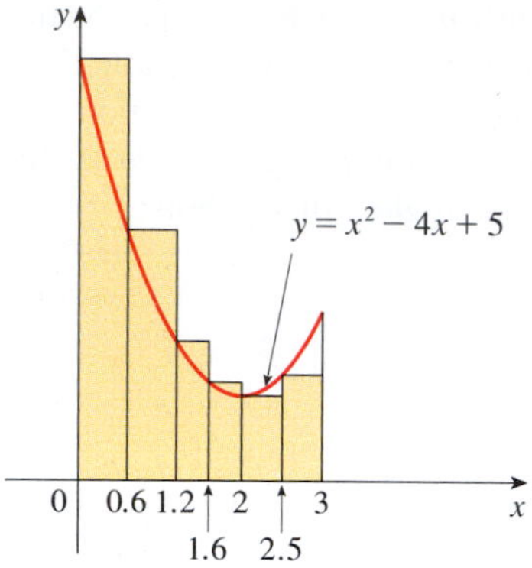

FIGURE 10

(c) The graph of f and the approximating rectangles are sketched in Figure 10. ■

EXAMPLE 3 Find the area under the parabola $y = x^2 + 1$ from 0 to 2.

SOLUTION Since $f(x) = x^2 + 1$ is continuous, the limit (2) that defines the area must exist for all possible partitions P of the interval $[0, 2]$ as long as $\|P\| \to 0$. To simplify things let's take the partition P that divides $[0, 2]$ into n subintervals of equal length. (This is called a regular partition.) Then the partition points are

$$x_0 = 0,\ x_1 = \frac{2}{n},\ x_2 = \frac{4}{n},\ \ldots,\ x_i = \frac{2i}{n},\ \ldots,\ x_n = \frac{2n}{n} = 2$$

and

$$\Delta x_1 = \Delta x_2 = \cdots = \Delta x_i = \cdots = \Delta x_n = \frac{2}{n}$$

so the norm of P is

$$\|P\| = \max\{\Delta x_i\} = \frac{2}{n}$$

The point x_i^* can be chosen to be anywhere in the ith subinterval. For the sake of definiteness, let's choose it to be the right-hand endpoint:

$$x_i^* = x_i = \frac{2i}{n}$$

Since $\|P\| = 2/n$, the condition $\|P\| \to 0$ is equivalent to $n \to \infty$. So the definition of area (2) becomes

$$A = \lim_{\|P\|\to 0} \sum_{i=1}^{n} f(x_i^*)\,\Delta x_i = \lim_{n\to\infty} \sum_{i=1}^{n} f\left(\frac{2i}{n}\right)\frac{2}{n}$$

$$= \lim_{n\to\infty} \sum_{i=1}^{n} \left[\left(\frac{2i}{n}\right)^2 + 1\right]\frac{2}{n} = \lim_{n\to\infty} \sum_{i=1}^{n} \left[\frac{8i^2}{n^3} + \frac{2}{n}\right]$$

$$= \lim_{n\to\infty} \left[\frac{8}{n^3}\sum_{i=1}^{n} i^2 + \frac{2}{n}\sum_{i=1}^{n} 1\right] \qquad \text{(by Theorem 6.1.2)}$$

$$= \lim_{n\to\infty} \left[\frac{8}{n^3}\cdot\frac{n(n+1)(2n+1)}{6} + \frac{2}{n}\cdot n\right] \qquad \text{(by Theorem 6.1.3)}$$

$$= \lim_{n\to\infty} \left[\frac{4}{3}\cdot 1\cdot\left(1 + \frac{1}{n}\right)\left(2 + \frac{1}{n}\right) + 2\right]$$

$$= \tfrac{4}{3}\cdot 1\cdot 1\cdot 2 + 2 = \tfrac{14}{3}$$

The sum in this calculation is represented by the areas of the shaded rectangles in Figure 11. Notice that in this case, with our choice of x_i^* as the right-hand endpoint and since f is increasing, $f(x_i^*)$ is the maximum value of f on $[x_{i-1}, x_i]$, so the sum R_n of the areas of the approximating rectangles is always *greater* than the exact area $A = \frac{14}{3}$.

We could just as well have chosen x_i^* to be the left-hand endpoint, that is, $x_i^* = x_{i-1} = 2(i-1)/n$. Then $f(x_i^*)$ is the minimum value of f on $[x_{i-1}, x_i]$, so the sum L_n of the areas of the approximating rectangles in Figure 12 is always *less* than A.

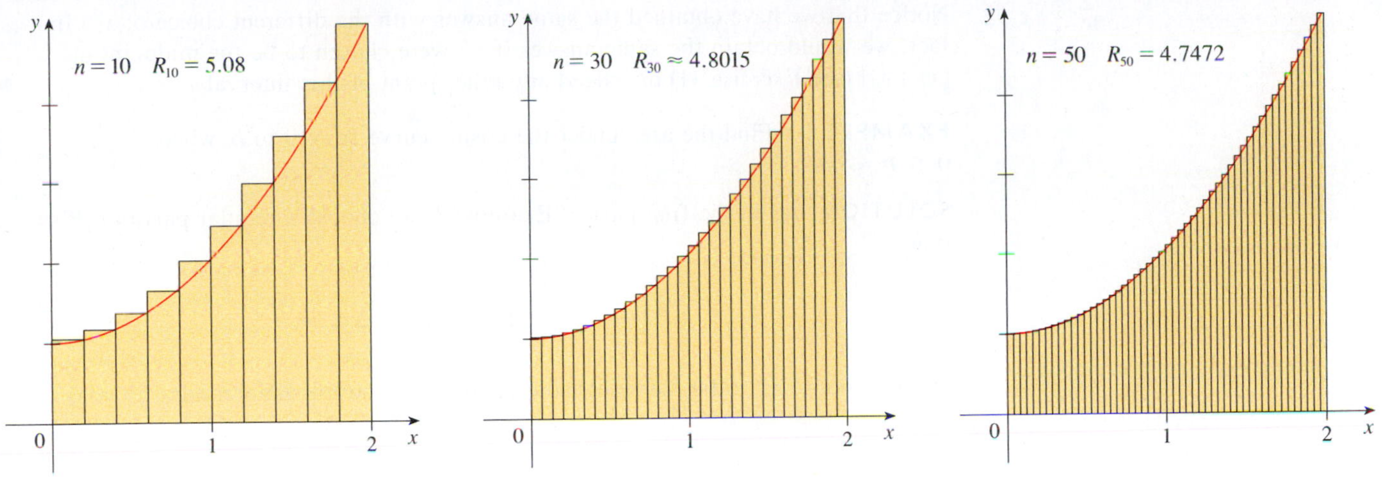

FIGURE 11 Right sums

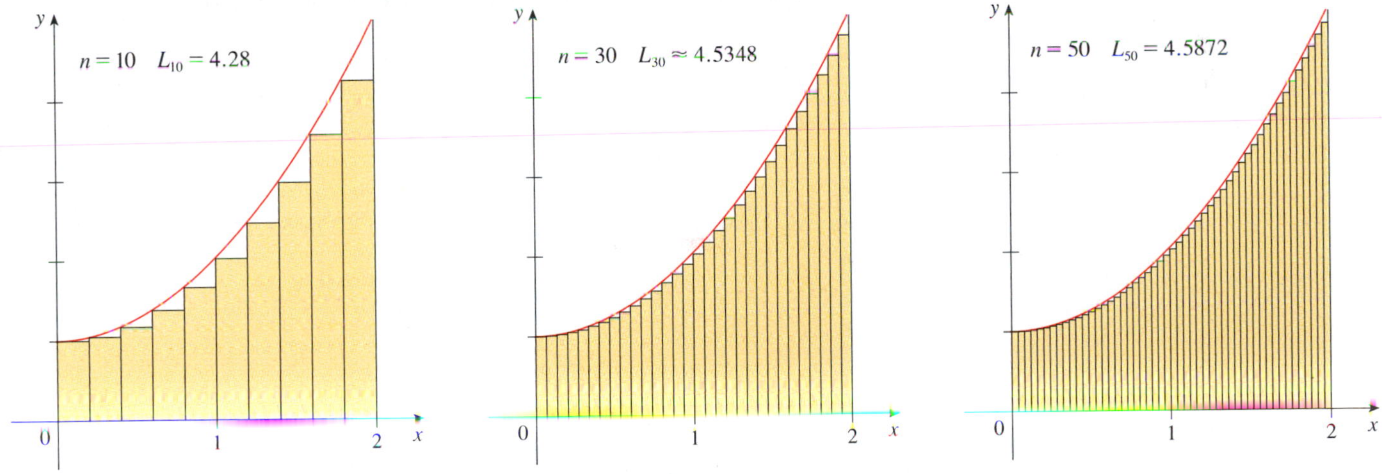

FIGURE 12 Left sums

The calculation with this choice is as follows:

$$\begin{aligned}
A &= \lim_{\|P\|\to 0} \sum_{i=1}^{n} f(x_i^*)\,\Delta x_i \\
&= \lim_{n\to\infty} \sum_{i=1}^{n} f\left(\frac{2(i-1)}{n}\right)\frac{2}{n} \\
&= \lim_{n\to\infty} \sum_{i=1}^{n} \left\{\left[\frac{2(i-1)}{n}\right]^2 + 1\right\}\frac{2}{n} \\
&= \lim_{n\to\infty} \sum_{i=1}^{n} \left[\frac{8}{n^3}(i^2 - 2i + 1) + \frac{2}{n}\right] \\
&= \lim_{n\to\infty} \left[\frac{8}{n^3}\sum_{i=1}^{n} i^2 - \frac{16}{n^3}\sum_{i=1}^{n} i + \frac{8}{n^3}\sum_{i=1}^{n} 1 + \frac{2}{n}\sum_{i=1}^{n} 1\right] \\
&= \lim_{n\to\infty} \left[\frac{8}{n^3}\,\frac{n(n+1)(2n+1)}{6} - \frac{16}{n^3}\,\frac{n(n+1)}{2} + \frac{8}{n^3}n + \frac{2}{n}n\right] \\
&= \lim_{n\to\infty} \left[\tfrac{4}{3}\cdot 1\cdot\left(1 + \frac{1}{n}\right)\left(2 + \frac{1}{n}\right) - \frac{8}{n}\left(1 + \frac{1}{n}\right) + \frac{8}{n^2} + 2\right] \\
&= \tfrac{4}{3}\cdot 1\cdot 1\cdot 2 - 0\cdot 1 + 0 + 2 = \tfrac{14}{3}
\end{aligned}$$

Notice that we have obtained the same answer with the different choice of x_i^*. In fact, we would obtain the same answer if x_i^* were chosen to be the midpoint of $[x_{i-1}, x_i]$ (see Exercise 11) or indeed any other point of this interval. ■

EXAMPLE 4 Find the area under the cosine curve from 0 to b, where $0 \leq b \leq \pi/2$.

SOLUTION As in the first part of Example 3, we choose a regular partition P so that

$$\|P\| = \Delta x_1 = \Delta x_2 = \cdots = \Delta x_n = \frac{b}{n}$$

and we choose x_i^* to be the right-hand endpoint of the ith subinterval:

$$x_i^* = x_i = \frac{ib}{n}$$

Since $\|P\| = b/n \to 0$ as $n \to \infty$, the area under the cosine curve from 0 to b is

(3) $$A = \lim_{\|P\| \to 0} \sum_{i=1}^{n} f(x_i^*)\,\Delta x_i = \lim_{n \to \infty} \sum_{i=1}^{n} \cos\left(i\frac{b}{n}\right)\frac{b}{n} = \lim_{n \to \infty} \frac{b}{n} \sum_{i=1}^{n} \cos\left(i\frac{b}{n}\right)$$

To evaluate this limit we use the formula of Exercise 52 in Section 6.1:

$$\sum_{i=1}^{n} \cos ix = \frac{\sin \frac{1}{2}nx \cos \frac{1}{2}(n+1)x}{\sin \frac{1}{2}x}$$

with $x = b/n$. Then Equation 3 becomes

(4) $$A = \lim_{n \to \infty} \frac{b}{n} \frac{\sin \frac{1}{2}b \cos\left[\dfrac{(n+1)b}{2n}\right]}{\sin \dfrac{b}{2n}}$$

Now $$\cos\left[\frac{(n+1)b}{2n}\right] = \cos\left[\left(1 + \frac{1}{n}\right)\frac{b}{2}\right] \to \cos\frac{b}{2} \qquad \text{as } n \to \infty$$

since cosine is continuous. Letting $t = b/n$ and using Theorem 3.4.4, we have

$$\lim_{n \to \infty} \frac{b}{n} \cdot \frac{1}{\sin \dfrac{b}{2n}} = \lim_{t \to 0^+} \frac{t}{\sin \dfrac{t}{2}} = \lim_{t \to 0^+} 2 \cdot \frac{\dfrac{t}{2}}{\sin \dfrac{t}{2}} = 2$$

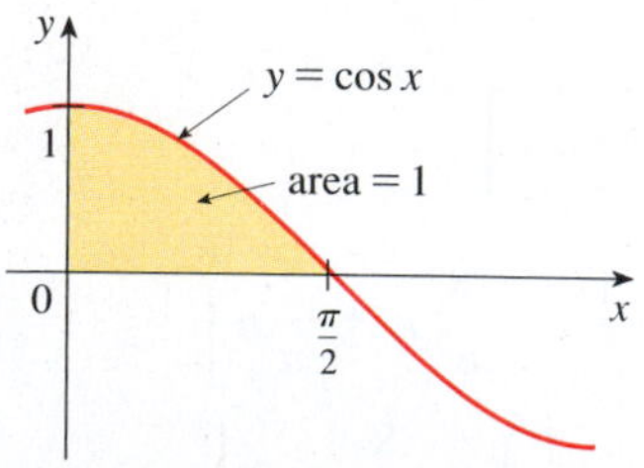

FIGURE 13

Putting these limits in Equation 4, we obtain

$$A = 2 \sin\frac{b}{2} \cos\frac{b}{2} = \sin b$$

In particular, taking $b = \pi/2$, we have proved that the area under the cosine curve from 0 to $\pi/2$ is $\sin(\pi/2) = 1$ (see Figure 13). ■

NOTE: The area calculations in Examples 3 and 4 are not easy. We will see in Section 6.4, however, that the Fundamental Theorem of Calculus gives a much easier method for computing these areas.

EXERCISES 6.2

1–8 ■ You are given a function f, an interval, partition points, and a description of the point x_i^* within the ith subinterval.
(a) Find $\|P\|$.
(b) Find the sum of the areas of the approximating rectangles, as given in (1).
(c) Sketch the graph of f and the approximating rectangles.

1. $f(x) = 16 - x^2$, $[0, 4]$, $\{0, 1, 2, 3, 4\}$, $x_i^* =$ left endpoint

2. $f(x) = 16 - x^2$, $[0, 4]$, $\{0, 1, 2, 3, 4\}$, $x_i^* =$ right endpoint

3. $f(x) = 16 - x^2$, $[0, 4]$, $\{0, 1, 2, 3, 4\}$, $x_i^* =$ midpoint

4. $f(x) = 2x + 1$, $[0, 4]$, $\{0, 0.5, 1, 2, 4\}$, $x_i^* =$ left endpoint

5. $f(x) = x^3 + 2$, $[-1, 2]$, $\{-1, -0.5, 0, 0.5, 1.0, 1.5, 2\}$, $x_i^* =$ right endpoint

6. $f(x) = 1/(x + 1)$, $[0, 2]$, $\{0, 0.5, 1.0, 1.5, 2\}$, $x_1^* = 0.25$, $x_2^* = 1$, $x_3^* = 1.25$, $x_4^* = 2$

7. $f(x) = 2 \sin x$, $[0, \pi]$, $\{0, \pi/4, \pi/2, 3\pi/4, \pi\}$, $x_1^* = \pi/6$, $x_2^* = \pi/3$, $x_3^* = 2\pi/3$, $x_4^* = 5\pi/6$

8. $f(x) = 4 \cos x$, $[0, \pi/2]$, $\{0, \pi/6, \pi/4, \pi/3, \pi/2\}$, $x_i^* =$ left endpoint

9. (a) Sketch a graph of the region that lies under the parabola $y = x^2 - 2x + 2$ from $x = 0$ to $x = 3$ and use it to make a rough visual estimate of the area of the region.
(b) Find an expression for R_n, the sum of the areas of the n approximating rectangles, taking x_i^* in (1) to be the right endpoint and using subintervals of equal length.
(c) Find the numerical values of the approximating areas R_n for $n = 6, 12$, and 24.
(d) Find the exact area of the region.

10. (a) Use a graphing device to sketch a graph of the region that lies under the curve $y = 4x - x^3$ from $x = 0$ to $x = 2$ and use it to make a rough visual estimate of the area of the region.
(b) Find an expression for R_n, the sum of the areas of the n approximating rectangles, taking x_i^* in (1) to be the right endpoint and using subintervals of equal length.
(c) Find the numerical values of the approximating areas R_n for $n = 10$, 20, and 30.
(d) Find the exact area of the region.

11. Find the area from Example 3 taking x_i^* to be the midpoint of $[x_{i-1}, x_i]$. Illustrate the approximating rectangles with a sketch.

12. Find the area under the curve $y = x^3$ from 0 to 1 using subintervals of equal length and taking x_i^* in (2) to be the (a) left endpoint, (b) right endpoint, and (c) midpoint of the ith subinterval. In each case, sketch the approximating rectangles.

13–18 ■ Use (2) to find the area under the given curve from a to b. Use equal subintervals and take x_i^* to be the right endpoint of the ith subinterval. Sketch the region.

13. $y = 2x + 1$, $a = 0, b = 5$

14. $y = x^2 + 3x - 2$, $a = 1, b = 4$

15. $y = 2x^2 - 4x + 5$, $a = -3, b = 2$

16. $y = x^3 + 2x$, $a = 0, b = 2$

17. $y = x^3 + 2x^2 + x$, $a = 0, b = 1$

18. $y = x^4 + 3x + 2$, $a = 0, b = 3$

19–20 ■ If you have a programmable calculator (or a computer), it is possible to evaluate the expression (1) for the sum of areas of approximating rectangles, even for large values of n, using looping. (On a TI use the Is> command, on a Casio use Isz, on an HP or in BASIC use a FOR-NEXT loop.) Compute the sum of the areas of approximating rectangles using equal subintervals and right endpoints for $n = 10$, 30, and 50. Then guess the value of the exact area.

19. The region under $y = \sin x$ from 0 to π

20. The region under $y = 1/x^2$ from 1 to 2

CAS **21.** Some computer algebra systems have commands that will draw approximating rectangles and evaluate the sums of their areas, at least if x_i^* is a left or right endpoint. (For instance, in Maple use leftbox, rightbox, leftsum, and rightsum.)
(a) If $f(x) = \sqrt{x}$, $1 \leqslant x \leqslant 4$, find the left and right sums for $n = 10$, 30, and 50.
(b) Illustrate by graphing the rectangles in part (a).
(c) Show that the exact area under f lies between 4.6 and 4.7.

CAS **22.** (a) If $f(x) = \sin(\sin x)$, $0 \leqslant x \leqslant \pi/2$, use the commands discussed in Exercise 21 to find the left and right sums for $n = 10$, 30, and 50.
(b) Illustrate by graphing the rectangles in part (a).
(c) Show that the exact area under f lies between 0.87 and 0.91.

23–24 ■ Determine a region whose area is equal to the given limit. Do not evaluate the limit.

23. $\displaystyle \lim_{n \to \infty} \sum_{i=1}^{n} \frac{\pi}{4n} \tan \frac{i\pi}{4n}$

24. $\displaystyle \lim_{n \to \infty} \sum_{i=1}^{n} \frac{3}{n} \sqrt{1 + \frac{3i}{n}}$

25. Find the area under the curve $y = \sin x$ from 0 to π. [*Hint:* Use equal subintervals and right endpoints, and use Exercise 53 in Section 6.1.]

26. (a) Let A_n be the area of a polygon with n equal sides inscribed in a circle with radius r. By dividing the polygon into n congruent triangles with central angle $2\pi/n$, show that

$$A_n = \tfrac{1}{2}nr^2\sin(2\pi/n)$$

(b) Show that $\lim_{n\to\infty} A_n = \pi r^2$. [*Hint:* Use Theorem 3.4.4.]

6.3 THE DEFINITE INTEGRAL

We saw in the preceding section that a limit of the form

(1)
$$\lim_{\|P\|\to 0}\sum_{i=1}^{n} f(x_i^*)\,\Delta x_i$$

arises when we compute an area. It turns out that this same type of limit occurs in a wide variety of situations even when f is not necessarily a positive function. In Chapters 7 and 9 we will see that limits of the form (1) also arise in finding lengths of curves, volumes of solids, areas of surfaces, centers of mass, fluid pressure, and work, as well as other quantities. We therefore give this type of limit a special name and notation.

(2) DEFINITION OF A DEFINITE INTEGRAL If f is a function defined on a closed interval $[a, b]$, let P be a partition of $[a, b]$ with partition points $x_0, x_1, \ldots, x_n$, where

$$a = x_0 < x_1 < x_2 < \cdots < x_n = b$$

Choose points x_i^* in $[x_{i-1}, x_i]$ and let $\Delta x_i = x_i - x_{i-1}$ and $\|P\| = \max\{\Delta x_i\}$. Then the **definite integral of f from a to b** is

$$\int_a^b f(x)\,dx = \lim_{\|P\|\to 0}\sum_{i=1}^{n} f(x_i^*)\,\Delta x_i$$

if this limit exists. If the limit does exist, then f is called **integrable** on the interval $[a, b]$.

NOTE 1: The symbol $\int$ was introduced by Leibniz and is called an **integral sign**. It is an elongated S and was chosen because an integral is a limit of sums. In the notation $\int_a^b f(x)\,dx$, $f(x)$ is called the **integrand** and a and b are called the **limits of integration**; a is the **lower limit** and b is the **upper limit**. The symbol dx has no meaning by itself; $\int_a^b f(x)\,dx$ is all one symbol. The procedure of calculating an integral is called **integration**.

NOTE 2: The definite integral $\int_a^b f(x)\,dx$ is a number; it does not depend on x. In fact, we could use any letter in place of x without changing the value of the integral:

$$\int_a^b f(x)\,dx = \int_a^b f(t)\,dt = \int_a^b f(r)\,dr$$

NOTE 3: The sum

(3)
$$\sum_{i=1}^{n} f(x_i^*)\,\Delta x_i$$

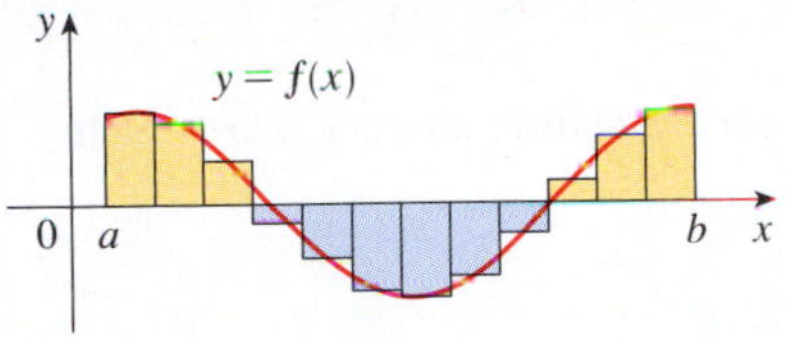

FIGURE 1

that occurs in Definition 2 is called a **Riemann sum** after the German mathematician Bernhard Riemann (1826–1866). The definite integral is sometimes called the **Riemann integral**. If f happens to be positive, then the Riemann sum can be interpreted as a sum of areas of approximating rectangles. [Compare (3) with (6.2.1).] If f takes on both positive and negative values, as in Figure 1, then the Riemann sum is the sum of the areas of the rectangles that lie above the x-axis and the *negatives* of the areas of the rectangles that lie below the x-axis (the areas of the gold rectangles *minus* the areas of the blue rectangles).

EXAMPLE 1 Let $f(x) = 1 + 5x$ and consider the partition P of the interval $[-2, 1]$ by means of the set of partition points $\{-2, -1.5, -1, -0.3, 0.2, 1\}$. In this example, $a = -2$, $b = 1$, $n = 5$, and $x_0 = -2$, $x_1 = -1.5$, $x_2 = -1$, $x_3 = -0.3$, $x_4 = 0.2$, and $x_5 = 1$. The lengths of the subintervals are

$$\Delta x_1 = -1.5 - (-2) = 0.5 \qquad \Delta x_2 = -1 - (-1.5) = 0.5$$
$$\Delta x_3 = -0.3 - (-1) = 0.7 \qquad \Delta x_4 = 0.2 - (-0.3) = 0.5$$
$$\Delta x_5 = 1 - 0.2 = 0.8$$

Thus, the norm of the partition P is

$$\|P\| = \max\{0.5, 0.5, 0.7, 0.5, 0.8\} = 0.8$$

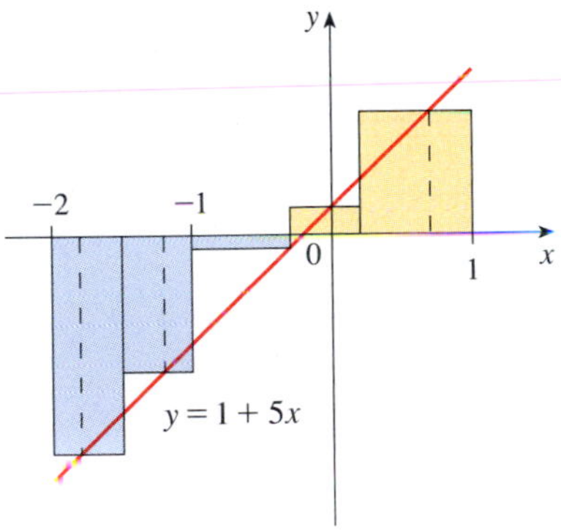

FIGURE 2

Suppose we choose $x_1^* = -1.8$, $x_2^* = -1.2$, $x_3^* = -0.3$, $x_4^* = 0$, and $x_5^* = 0.7$. Then the corresponding Riemann sum is

$$\begin{aligned}\sum_{i=1}^{5} f(x_i^*)\,\Delta x_i &= f(-1.8)\,\Delta x_1 + f(-1.2)\,\Delta x_2 + f(-0.3)\,\Delta x_3 + f(0)\,\Delta x_4 + f(0.7)\,\Delta x_5 \\ &= (-8)(0.5) + (-5)(0.5) + (-0.5)(0.7) + 1(0.5) + (4.5)(0.8) \\ &= -2.75\end{aligned}$$

Notice that, in this example, f is not a positive function and so the Riemann sum does not represent a sum of areas of rectangles. But it does represent the sum of the areas of the gold rectangles (above the x-axis) minus the sum of the areas of the blue rectangles (below the axis) in Figure 2. ■

NOTE 4: An integral need not represent an area. But for *positive* functions, an integral can be interpreted as an area. In fact, comparing Definition 2 with the definition of area (6.2.2), we see the following:

> For the special case where $f(x) \geqslant 0$,
>
> $$\int_a^b f(x)\,dx = \text{the area under the graph of } f \text{ from } a \text{ to } b$$

In general, a definite integral can be interpreted as a difference of areas:

$$\int_a^b f(x)\,dx = A_1 - A_2$$

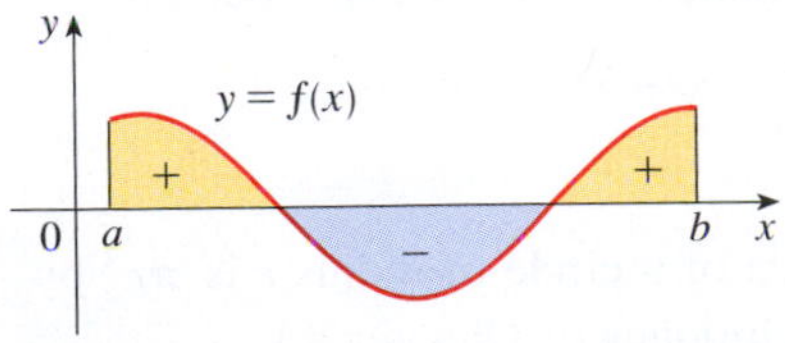

FIGURE 3

where A_1 is the area of the region above the x-axis and below the graph of f and A_2 is the area of the region below the x-axis and above the graph of f. (This seems reasonable from a comparison of Figures 1 and 3, but we will be able to see it more clearly in Section 7.1.)

Bernhard Riemann received his Ph.D. under the direction of the legendary Gauss at the University of Göttingen and remained there to teach. Gauss, who was not in the habit of praising other mathematicians, spoke of Riemann's "creative, active, truly mathematical mind and gloriously fertile originality." The definition (2) of an integral that we use is due to Riemann. He also made major contributions to the theory of functions of a complex variable, mathematical physics, number theory, and the foundations of geometry. Riemann's broad concept of space and geometry turned out to be the right setting, 50 years later, for Einstein's general relativity theory. Riemann's health was poor throughout his life, and he died of tuberculosis at the age of 39.

NOTE 5: The precise meaning of the limit that defines the integral in Definition 2 is as follows:

$\int_a^b f(x)\,dx = I$ means that for every $\varepsilon > 0$ there is a corresponding number $\delta > 0$ such that

$$\left| I - \sum_{i=1}^{n} f(x_i^*)\,\Delta x_i \right| < \varepsilon$$

for all partitions P of $[a, b]$ with $\|P\| < \delta$ and for all possible choices of x_i^* in $[x_{i-1}, x_i]$.

This means that a definite integral can be approximated to within any desired degree of accuracy by a Riemann sum.

NOTE 6: In Definition 2 we are dealing with a function f defined on an interval $[a, b]$, so we are implicitly assuming that $a < b$. But for some purposes it is useful to extend the definition of $\int_a^b f(x)\,dx$ to the case where $a > b$ or $a = b$ as follows:

$$\text{If } a > b, \text{ then } \int_a^b f(x)\,dx = -\int_b^a f(x)\,dx.$$

$$\text{If } a = b, \text{ then } \int_a^a f(x)\,dx = 0.$$

EXAMPLE 2 Express

$$\lim_{\|P\|\to 0} \sum_{i=1}^{n} [(x_i^*)^3 + x_i^* \sin x_i^*]\,\Delta x_i$$

as an integral on the interval $[0, \pi]$.

SOLUTION Comparing the given limit with the limit in Definition 2, we see that they will be identical if we choose

$$f(x) = x^3 + x \sin x$$

We are given that $a = 0$ and $b = \pi$. Therefore, by Definition 2, we have

$$\lim_{\|P\|\to 0} \sum_{i=1}^{n} [(x_i^*)^3 + x_i^* \sin x_i^*]\,\Delta x_i = \int_0^{\pi} (x^3 + x \sin x)\,dx$$ ■

EXAMPLE 3 Evaluate the following integrals by interpreting each in terms of areas.

(a) $\int_0^1 \sqrt{1 - x^2}\,dx$ (b) $\int_0^3 (x - 1)\,dx$

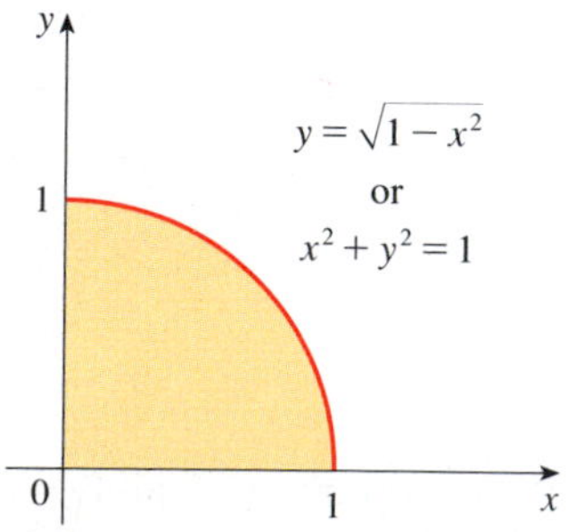

FIGURE 4

SOLUTION

(a) Since $f(x) = \sqrt{1 - x^2} \geqslant 0$ we can interpret this integral as the area under the curve $y = \sqrt{1 - x^2}$ from 0 to 1. But, since $y^2 = 1 - x^2$, we get $x^2 + y^2 = 1$, which shows that the graph of f is the quarter-circle with radius 1 in Figure 4. Therefore

$$\int_0^1 \sqrt{1 - x^2}\,dx = \tfrac{1}{4}\pi(1)^2 = \frac{\pi}{4}$$

(In Section 8.3 we will be able to *prove* that the area of a circle of radius r is πr^2 by evaluating the integral $\int_0^r \sqrt{r^2 - x^2}\,dx$ using the techniques of Chapter 8.)

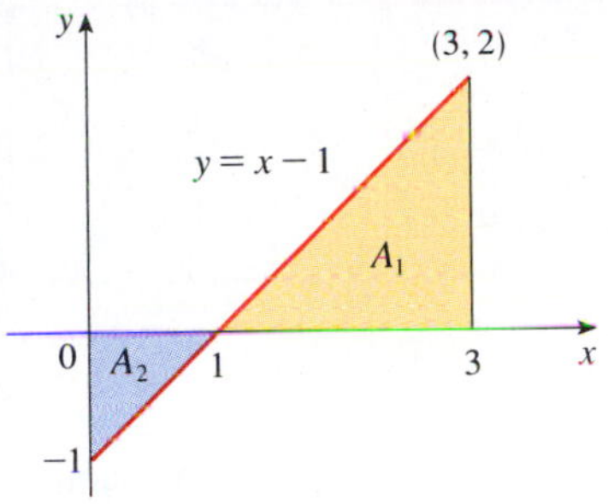

FIGURE 5

(b) The graph of $y = x - 1$ is the line with slope 1 shown in Figure 5. We compute the integral as the difference of the areas of the two triangles:

$$\int_0^3 (x - 1)\,dx = A_1 - A_2 = \tfrac{1}{2}(2 \cdot 2) - \tfrac{1}{2}(1 \cdot 1) = 1.5$$

■

The integrals in Example 3 were simple to evaluate because we were able to express them in terms of areas of simple regions, but not all integrals are that easy. In fact, the integrals of some functions don't even exist. So the question arises: Which functions are integrable? A partial answer is given by the following theorem, which is proved in courses on advanced calculus.

(4) THEOREM If f is either continuous or increasing or decreasing, on $[a, b]$, then f is integrable on $[a, b]$; that is, the definite integral $\int_a^b f(x)\,dx$ exists.

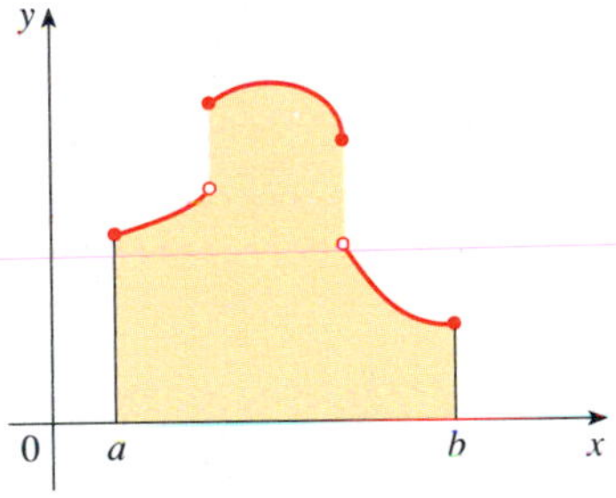

FIGURE 6
Discontinuous integrable function

If f is discontinuous at some points in $[a, b]$, then $\int_a^b f(x)\,dx$ might exist or it might not exist (see Exercises 70 and 71). If f has only a finite number of discontinuities and these are all jump discontinuities, then f is called **piecewise continuous** and it turns out that f is integrable. (See Figure 6.)

It can be shown that if f is integrable on $[a, b]$, then f must be a **bounded function** on $[a, b]$; that is, there exists a number M such that $|f(x)| \leqslant M$ for all x in $[a, b]$. Geometrically, this means that the graph of f lies between the horizontal lines $y = M$ and $y = -M$. In particular, if f has an infinite discontinuity at some point in $[a, b]$, then f is not bounded and is therefore not integrable. (See Exercise 70 and Figure 7.)

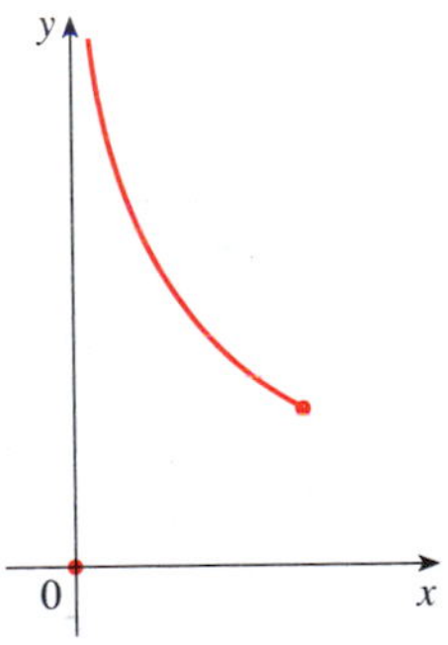

FIGURE 7
Nonintegrable function

If f is integrable on $[a, b]$, then the Riemann sums (3) must approach $\int_a^b f(x)\,dx$ as $\|P\| \to 0$ no matter how the partitions P are chosen and no matter how the points x_i^* are chosen in $[x_{i-1}, x_i]$. Therefore, if it is known beforehand that f is integrable on $[a, b]$ (for instance, if it is known that f is continuous or increasing or decreasing), then in calculating the value of an integral we are free to choose partitions P and points x_i^* in any way we like as long as $\|P\| \to 0$. For purposes of calculation, it is often convenient to take P to be a **regular partition**; that is, all the subintervals have the same length Δx. Then

$$\Delta x = \Delta x_1 = \Delta x_2 = \cdots = \Delta x_n = \frac{b - a}{n}$$

and $\qquad x_0 = a, \quad x_1 = a + \Delta x, \quad x_2 = a + 2\,\Delta x, \quad \ldots, \quad x_i = a + i\,\Delta x$

If we choose x_i^* to be the right endpoint of the ith subinterval, then

$$x_i^* = x_i = a + i\,\Delta x = a + i\,\frac{b - a}{n}$$

Since $\|P\| = \Delta x = (b - a)/n$, we have $\|P\| \to 0$ as $n \to \infty$, so Definition 2 gives

$$\int_a^b f(x)\,dx = \lim_{\|P\| \to 0} \sum_{i=1}^{n} f(x_i^*)\,\Delta x$$

$$= \lim_{n \to \infty} \sum_{i=1}^{n} f\left(a + i\,\frac{b - a}{n}\right) \frac{b - a}{n}$$

Since $(b - a)/n$ does not depend on i, Theorem 6.1.2 allows us to take it in front of the sigma sign, and we have the following formula for calculating integrals.

(5) THEOREM If f is integrable on $[a, b]$, then

$$\int_a^b f(x)\,dx = \lim_{n\to\infty} \frac{b-a}{n} \sum_{i=1}^{n} f\left(a + i\frac{b-a}{n}\right)$$

EXAMPLE 4 Evaluate $\int_0^3 (x^3 - 5x)\,dx$.

SOLUTION Here we have $f(x) = x^3 - 5x$, $a = 0$, and $b = 3$. Since f is continuous, we know it is integrable and so Theorem 5 gives

$$\int_0^3 (x^3 - 5x)\,dx = \lim_{n\to\infty} \frac{3}{n} \sum_{i=1}^{n} f\left(\frac{3i}{n}\right) = \lim_{n\to\infty} \frac{3}{n} \sum_{i=1}^{n} \left[\left(\frac{3i}{n}\right)^3 - 5\left(\frac{3i}{n}\right)\right]$$

$$= \lim_{n\to\infty} \left[\frac{81}{n^4} \sum_{i=1}^{n} i^3 - \frac{45}{n^2} \sum_{i=1}^{n} i\right]$$

$$= \lim_{n\to\infty} \left\{\frac{81}{n^4}\left[\frac{n(n+1)}{2}\right]^2 - \frac{45}{n^2}\frac{n(n+1)}{2}\right\}$$

$$= \lim_{n\to\infty} \left[\frac{81}{4}\left(1 + \frac{1}{n}\right)^2 - \frac{45}{2}\left(1 + \frac{1}{n}\right)\right]$$

$$= \tfrac{81}{4} - \tfrac{45}{2} = -\tfrac{9}{4} = -2.25$$

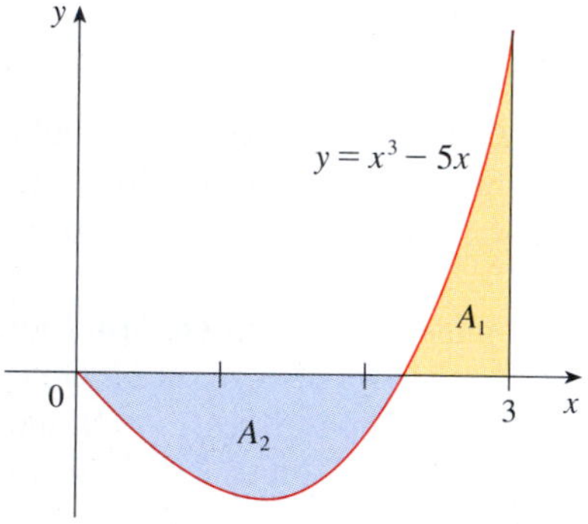

FIGURE 8

This integral cannot be interpreted as an area because f takes on both positive and negative values. But it can be interpreted as the difference of areas $A_1 - A_2$, where A_1 and A_2 are shown in Figure 8.

Figure 9 illustrates the calculation by showing the positive and negative terms in the right Riemann sum R_n for $n = 40$. The values in the table show the Riemann sums approaching the exact value of the integral, -2.25, as $n \to \infty$.

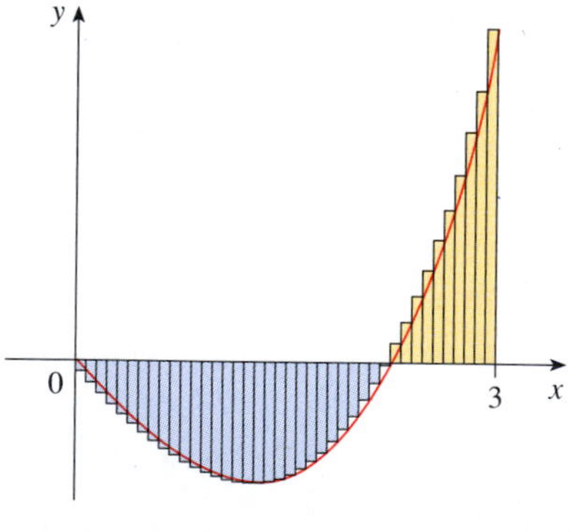

FIGURE 9

n	R_n
40	−1.7873
100	−2.0680
500	−2.2139
1000	−2.2320
5000	−2.2464

■

A much simpler method for evaluating the integral in Example 4 will be given in Section 6.4 after we have proved the Fundamental Theorem of Calculus.

THE MIDPOINT RULE

We often choose x_i^* to be the right endpoint of the ith subinterval because it is convenient for computing the limit. But if the purpose is to find an *approximation* to an integral, it is usually better to choose x_i^* to be the midpoint of the interval, which we denote by $\bar{x}_i$. Any Riemann sum is an approximation to an integral, but if we use midpoints and a regular partition we get the following approximation:

MIDPOINT RULE

$$\int_a^b f(x)\,dx \approx \sum_{i=1}^{n} f(\bar{x}_i)\,\Delta x = \Delta x[f(\bar{x}_1) + \cdots + f(\bar{x}_n)]$$

where $\Delta x = \dfrac{b-a}{n}$

and $\bar{x}_i = \frac{1}{2}(x_{i-1} + x_i) = \text{midpoint of } [x_{i-1}, x_i]$

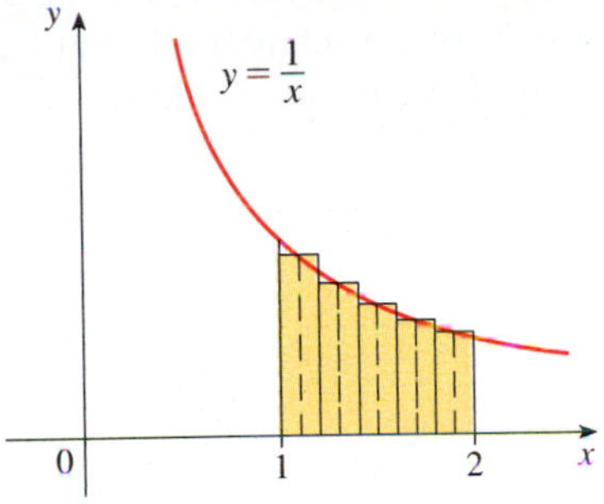

FIGURE 10

EXAMPLE 5 Use the Midpoint Rule with $n = 5$ to approximate $\int_1^2 \frac{1}{x}\,dx$.

SOLUTION The partition points are 1, 1.2, 1.4, 1.6, 1.8, and 2.0, so the midpoints of the five intervals are 1.1, 1.3, 1.5, 1.7, and 1.9. The width of the intervals is $\Delta x = (2-1)/5 = \frac{1}{5}$, so the Midpoint Rule gives

$$\int_1^2 \frac{1}{x}\,dx \approx \Delta x[f(1.1) + f(1.3) + f(1.5) + f(1.7) + f(1.9)]$$

$$= \frac{1}{5}\left(\frac{1}{1.1} + \frac{1}{1.3} + \frac{1}{1.5} + \frac{1}{1.7} + \frac{1}{1.9}\right)$$

$$\approx 0.691908$$

Since $f(x) = 1/x > 0$ for $1 \leqslant x \leqslant 2$, the integral represents an area and the approximation given by the Midpoint Rule is the sum of the areas of the rectangles shown in Figure 10. ■

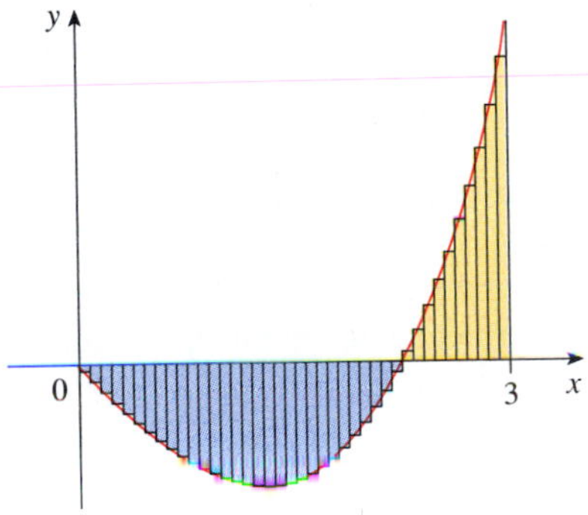

FIGURE 11
$M_{40} \approx -2.2563$

At the moment we don't know how accurate the approximation in Example 5 is, but in Section 8.8 we will learn a method for estimating the error involved in using the Midpoint Rule. At that time we will discuss other methods for approximating definite integrals.

If we apply the Midpoint Rule to the integral in Example 4, we get the picture in Figure 11. The approximation $M_{40} \approx -2.2563$ is much closer to the true value -2.25 than the right endpoint approximation, $R_{40} \approx -1.7873$ shown in Figure 9.

PROPERTIES OF THE DEFINITE INTEGRAL

We now develop some basic properties of integrals that will help us to evaluate integrals in a simple manner.

PROPERTIES OF THE INTEGRAL Suppose that all of the following integrals exist. Then

1. $\int_a^b c\,dx = c(b-a)$, where c is any constant
2. $\int_a^b [f(x) + g(x)]\,dx = \int_a^b f(x)\,dx + \int_a^b g(x)\,dx$
3. $\int_a^b cf(x)\,dx = c\int_a^b f(x)\,dx$, where c is any constant
4. $\int_a^b [f(x) - g(x)]\,dx = \int_a^b f(x)\,dx - \int_a^b g(x)\,dx$
5. $\int_a^b f(x)\,dx = \int_a^c f(x)\,dx + \int_c^b f(x)\,dx$

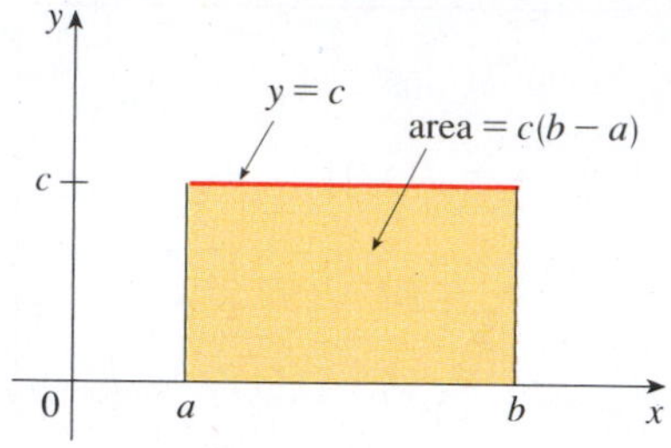

FIGURE 12

$\int_a^b c\,dx = c(b-a)$

The proof of Property 1 is requested in Exercise 15. This property says that the integral of a constant function $f(x) = c$ is the constant times the length of the interval. If $c > 0$ and $a < b$, this is to be expected because $c(b - a)$ is the area of the shaded rectangle in Figure 12.

PROOF OF PROPERTY 2 Since $\int_a^b [f(x) + g(x)]\,dx$ exists, we can compute it using a regular partition and choosing x_i^* to be the right endpoint of the ith subinterval, that is, $x_i^* = x_i$. Using the fact that the limit of a sum is the sum of the limits, we have

$$\int_a^b [f(x) + g(x)]\,dx = \lim_{n\to\infty} \sum_{i=1}^n [f(x_i) + g(x_i)]\,\Delta x$$

$$= \lim_{n\to\infty} \left[\sum_{i=1}^n f(x_i)\,\Delta x + \sum_{i=1}^n g(x_i)\,\Delta x \right] \qquad \text{(by Theorem 6.1.2)}$$

$$= \lim_{n\to\infty} \sum_{i=1}^n f(x_i)\,\Delta x + \lim_{n\to\infty} \sum_{i=1}^n g(x_i)\,\Delta x$$

$$= \int_a^b f(x)\,dx + \int_a^b g(x)\,dx$$

□

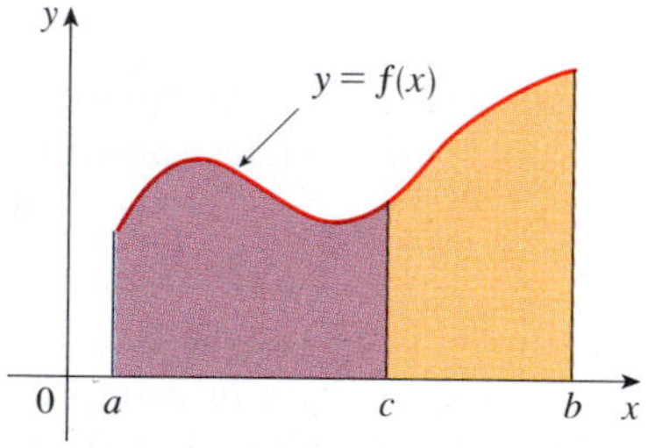

FIGURE 13

Property 2 says that the integral of a sum is the sum of the integrals. Property 3 can be proved in a similar manner (see Exercise 65) and says that the integral of a constant times a function is the constant times the integral of the function. In other words, a constant (but *only* a constant) can be taken in front of an integral sign. Property 4 is proved by writing $f - g = f + (-g)$ and using Properties 2 and 3 with $c = -1$.

Property 5 is somewhat more complicated and is proved in Appendix F, but for the case where $f(x) \geqslant 0$ and $a < c < b$, it can be seen from the geometric interpretation in Figure 13. For positive functions f, $\int_a^b f(x)\,dx$ is the total area under $y = f(x)$ from a to b, which is the sum of $\int_a^c f(x)\,dx$ (the area from a to c) and $\int_c^b f(x)\,dx$ (the area from c to b).

EXAMPLE 6 Use the properties of integrals and the results

$$\int_a^b x\,dx = \frac{b^2 - a^2}{2} \qquad\qquad \int_0^{\pi/2} \cos x\,dx = 1$$

(from Exercise 21 in this section and Example 4 in Section 6.2) to evaluate the following integrals.

(a) $\int_0^{\pi/2} (x + 3\cos x)dx$ (b) $\int_{-4}^5 |x|\,dx$

SOLUTION

(a) Using Properties 2 and 3 of integrals, we get

$$\int_0^{\pi/2} (x + 3\cos x)\,dx = \int_0^{\pi/2} x\,dx + 3\int_0^{\pi/2} \cos x\,dx = \frac{\pi^2}{8} + 3$$

(b) Since

$$|x| = \begin{cases} x & \text{if } x \geqslant 0 \\ -x & \text{if } x < 0 \end{cases}$$

we use Property 5 to split the integral at 0:

$$\int_{-4}^{5} |x|\,dx = \int_{-4}^{0} |x|\,dx + \int_{0}^{5} |x|\,dx$$

$$= \int_{-4}^{0} (-x)\,dx + \int_{0}^{5} x\,dx = -\int_{-4}^{0} x\,dx + \int_{0}^{5} x\,dx$$

$$= -\tfrac{1}{2}[0^2 - (-4)^2] + \tfrac{1}{2}[5^2 - 0^2] = 20.5$$

Notice that Properties 1–5 are true whether $a < b$, $a = b$, or $a > b$. The following properties, however, are true only if $a \leqslant b$.

ORDER PROPERTIES OF THE INTEGRAL Suppose the following integrals exist and $a \leqslant b$:

6. If $f(x) \geqslant 0$ for $a \leqslant x \leqslant b$, then $\int_a^b f(x)\,dx \geqslant 0$.

7. If $f(x) \geqslant g(x)$ for $a \leqslant x \leqslant b$, then $\int_a^b f(x)\,dx \geqslant \int_a^b g(x)\,dx$.

8. If $m \leqslant f(x) \leqslant M$ for $a \leqslant x \leqslant b$, then

$$m(b - a) \leqslant \int_a^b f(x)\,dx \leqslant M(b - a)$$

9. $\left|\int_a^b f(x)\,dx\right| \leqslant \int_a^b |f(x)|\,dx$

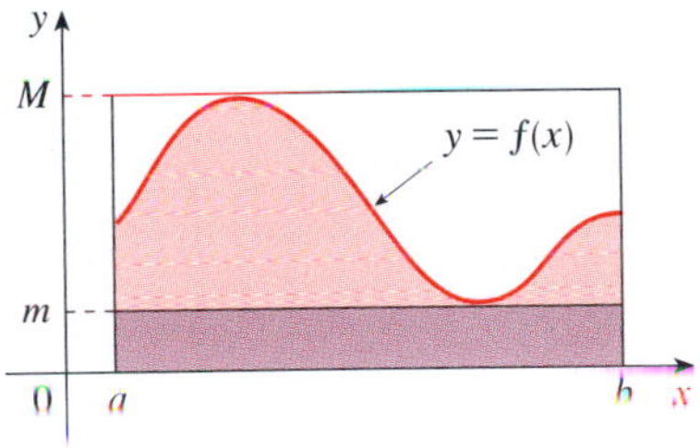

FIGURE 14

If $f(x) \geqslant 0$, then $\int_a^b f(x)\,dx$ represents the area under the graph of f, so the geometric interpretation of Property 6 is simply that areas are positive. But the property can be proved from the definition of an integral (Exercise 66). Property 7 says that a bigger function has a bigger integral. It follows from Properties 6 and 4 because $f - g \geqslant 0$.

Property 8 is illustrated by Figure 14 for the case where $f(x) \geqslant 0$. If f is continuous we could take m and M to be the absolute minimum and maximum values of f on the interval $[a, b]$. In this case Property 8 says that the area under the graph of f is greater than the area of the rectangle with height m and less than the area of the rectangle with height M.

PROOF OF PROPERTY 8 Since $m \leqslant f(x) \leqslant M$, Property 7 gives

$$\int_a^b m\,dx \leqslant \int_a^b f(x)\,dx \leqslant \int_a^b M\,dx$$

Using Property 1 to evaluate the integrals on the left- and right-hand sides, we obtain

$$m(b - a) \leqslant \int_a^b f(x)\,dx \leqslant M(b - a)$$

The proof of Property 9 is left as Exercise 67.

EXAMPLE 7 Use Property 8 to estimate the value of $\int_1^4 \sqrt{x}\,dx$.

SOLUTION Since $f(x) = \sqrt{x}$ is an increasing function, its absolute minimum on $[1, 4]$ is $m = f(1) = 1$ and its absolute maximum on $[1, 4]$ is $M = f(4) = \sqrt{4} = 2$. Thus, Property 8 gives

$$1(4 - 1) \leqslant \int_1^4 \sqrt{x}\,dx \leqslant 2(4 - 1)$$

or

$$3 \leqslant \int_1^4 \sqrt{x}\,dx \leqslant 6$$

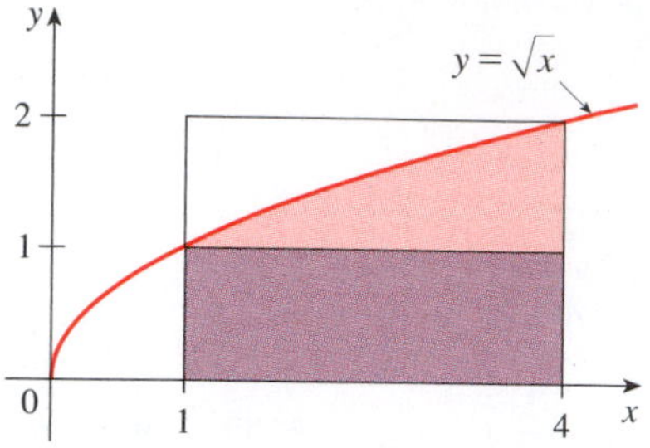

FIGURE 15

The result of Example 7 is illustrated in Figure 15. The area under $y = \sqrt{x}$ from 1 to 4 is greater than the area of the lower rectangle and less than the area of the large rectangle.

EXAMPLE 8 Show that $\int_1^4 \sqrt{1 + x^2}\, dx \geqslant 7.5$.

SOLUTION The minimum value of $f(x) = \sqrt{1 + x^2}$ on $[1, 4]$ is $m = f(1) = \sqrt{2}$, since f is increasing. Thus, Property 8 gives

$$\int_1^4 \sqrt{1 + x^2}\, dx \geqslant \sqrt{2}(4 - 1) = 3\sqrt{2} \approx 4.24$$

This result is not good enough, so instead we use Property 7. Notice that

$$1 + x^2 > x^2 \quad \Rightarrow \quad \sqrt{1 + x^2} > \sqrt{x^2} = |x|$$

Since $|x| = x$ for $x > 0$, we have $\sqrt{1 + x^2} > x$ for $1 \leqslant x \leqslant 4$. Thus, by Property 7,

$$\int_1^4 \sqrt{1 + x^2}\, dx \geqslant \int_1^4 x\, dx = \tfrac{1}{2}(4^2 - 1^2) = 7.5$$

[Here we have used the fact that $\int_a^b x\, dx = (b^2 - a^2)/2$ from Exercise 21.] ■

EXERCISES 6.3

1–6 ■ You are given a function f, an interval, partition points that define a partition P, and points x_i^* in the ith subinterval. (a) Find $\|P\|$. (b) Find the Riemann sum (3).

1. $f(x) = 7 - 2x$, $[1, 5]$, $\{1, 1.6, 2.2, 3.0, 4.2, 5\}$, $x_i^* =$ midpoint

2. $f(x) = 3x - 1$, $[-2, 2]$, $\{-2, -1.2, -0.6, 0, 0.8, 1.6, 2\}$, $x_i^* =$ midpoint

3. $f(x) = 2 - x^2$, $[-2, 2]$, $\{-2, -1.4, -1, 0, 0.8, 1.4, 2\}$, $x_i^* =$ right endpoint

4. $f(x) = x + x^2$, $[-2, 0]$, $\{-2, -1.5, -1, -0.7, -0.4, 0\}$, $x_i^* =$ left endpoint

5. $f(x) = x^3$, $[-1, 1]$, $\{-1, -0.5, 0, 0.5, 1\}$, $x_1^* = -1$, $x_2^* = -0.4$, $x_3^* = 0.2$, $x_4^* = 1$

6. $f(x) = \sin x$, $[-\pi/2, \pi]$, $\{-\pi/2, -1, 0, 1, 2, \pi\}$, $x_1^* = -1.5$, $x_2^* = -0.5$, $x_3^* = 0.5$, $x_4^* = 1.5$, $x_5^* = 3$

7. The graph of a function f is given. Estimate $\int_0^8 f(x)\, dx$ using four equal subintervals with (a) right endpoints, (b) left endpoints, and (c) midpoints.

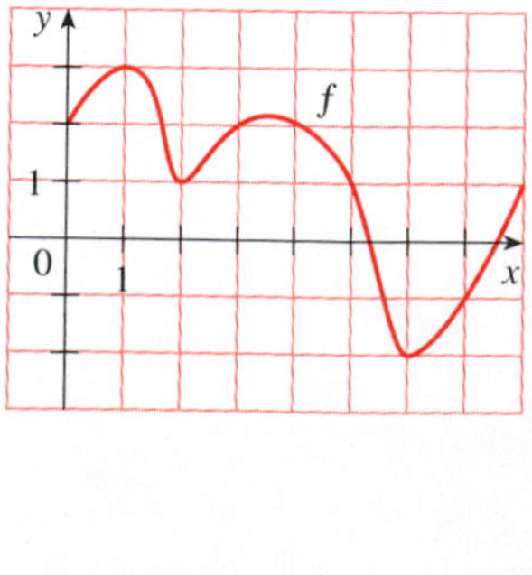

8. The table gives the values of a function obtained from an experiment. Use them to estimate $\int_0^6 f(x)\, dx$ using three equal subintervals with (a) right endpoints, (b) left endpoints, and (c) midpoints. If the function is known to be a decreasing function, can you say whether your estimates are less than or greater than the exact value of the integral?

x	0	1	2	3	4	5	6
$f(x)$	9.3	9.0	8.3	6.5	2.3	−7.6	−10.5

9–12 ■ Use the Midpoint Rule with the given value of n to approximate each integral. Round each answer to four decimal places.

9. $\int_0^5 x^3\, dx$, $n = 5$

10. $\int_1^3 \frac{1}{2x - 7}\, dx$, $n = 4$

11. $\int_1^2 \sqrt{1 + x^2}\, dx$, $n = 10$

12. $\int_0^{\pi/4} \tan x\, dx$, $n = 4$

CAS **13.** If you have a CAS that evaluates midpoint approximations and graphs the corresponding rectangles (use middlesum and middlebox commands in Maple), check the answer to Exercise 11 and illustrate with a graph. Then repeat with $n = 20$ and $n = 30$.

14. With a programmable calculator or computer (see the instructions for Exercise 19 in Section 6.2) compute the left and right Riemann sums for the function $f(x) = \sqrt{1 + x^2}$ on the interval $[1, 2]$ with $n = 100$. Explain why these estimates show that

$$1.805 < \int_1^2 \sqrt{1 + x^2}\,dx < 1.815$$

Deduce that the approximation using the Midpoint Rule with $n = 10$ in Exercise 11 is accurate to two decimal places.

15–20 ■ Use Theorem 5 to evaluate the integral.

15. $\int_a^b c\,dx$

16. $\int_{-2}^7 (6 - 2x)\,dx$

17. $\int_1^4 (x^2 - 2)\,dx$

18. $\int_1^5 (2 + 3x - x^2)\,dx$

19. $\int_0^b (x^3 + 4x)\,dx$

20. $\int_0^1 (x^3 - 5x^4)\,dx$

21. Prove that $\int_a^b x\,dx = \dfrac{b^2 - a^2}{2}$.

22. Prove that $\int_a^b x^2\,dx = \dfrac{b^3 - a^3}{3}$.

23–28 ■ Evaluate each integral by interpreting it in terms of areas.

23. $\int_1^3 (1 + 2x)\,dx$

24. $\int_{-2}^2 \sqrt{4 - x^2}\,dx$

25. $\int_{-3}^0 \left(1 + \sqrt{9 - x^2}\right)dx$

26. $\int_{-1}^3 (2 - x)\,dx$

27. $\int_{-2}^2 (1 - |x|)\,dx$

28. $\int_0^3 |3x - 5|\,dx$

29–32 ■ Express each limit as a definite integral on the given interval.

29. $\lim_{\|P\|\to 0} \sum_{i=1}^{n} [2(x_i^*)^2 - 5x_i^*]\Delta x_i$, $[0, 1]$

30. $\lim_{\|P\|\to 0} \sum_{i=1}^{n} \sqrt{x_i^*}\,\Delta x_i$, $[1, 4]$

31. $\lim_{\|P\|\to 0} \sum_{i=1}^{n} \cos x_i\,\Delta x_i$, $[0, \pi]$

32. $\lim_{\|P\|\to 0} \sum_{i=1}^{n} \dfrac{\tan x_i}{x_i}\Delta x_i$, $[2, 4]$

33–35 ■ Express the limit as a definite integral.

33. $\lim_{n\to\infty} \sum_{i=1}^{n} \dfrac{i^4}{n^5}$ [*Hint:* Consider $f(x) = x^4$.]

34. $\lim_{n\to\infty} \dfrac{1}{n} \sum_{i=1}^{n} \dfrac{1}{1 + (i/n)^2}$

35. $\lim_{n\to\infty} \sum_{i=1}^{n} \left[3\left(1 + \dfrac{2i}{n}\right)^5 - 6\right]\dfrac{2}{n}$

36. Evaluate $\int_1^1 x^2 \cos x\,dx$.

37. Given that $\int_4^9 \sqrt{x}\,dx = \frac{38}{3}$, what is $\int_9^4 \sqrt{t}\,dt$?

38. (a) Find an approximation to the integral $\int_0^4 (x^2 - 3x)\,dx$ using a Riemann sum with right endpoints and $n = 8$.
(b) Draw a diagram like Figure 1 to illustrate the approximation in part (a).
(c) Evaluate $\int_0^4 (x^2 - 3x)\,dx$.
(d) Interpret the integral in part (c) as a difference of areas and illustrate with a diagram like Figure 3.

39–44 ■ Use the properties of integrals to evaluate each integral. You may assume from Section 6.2 that

$$\int_0^b \cos x\,dx = \sin b$$

and you may use the results of Exercises 21 and 22.

39. $\int_{-4}^{-1} \sqrt{3}\,dx$

40. $\int_3^6 (4 - 7x)\,dx$

41. $\int_1^4 (2x^2 - 3x + 1)\,dx$

42. $\int_0^1 (5\cos x + 4x)\,dx$

43. $\int_{-1}^1 f(x)\,dx$ where $f(x) = \begin{cases} -2x & \text{if } -1 \leqslant x < 0 \\ 3x^2 & \text{if } 0 \leqslant x \leqslant 1 \end{cases}$

44. $\int_3^4 f(x)\,dx + \int_1^3 f(x)\,dx + \int_4^1 f(x)\,dx$

45–48 ■ Write the given sum or difference as a single integral in the form $\int_a^b f(x)\,dx$.

45. $\int_1^3 f(x)\,dx + \int_3^6 f(x)\,dx + \int_6^{12} f(x)\,dx$

46. $\int_5^8 f(x)\,dx + \int_0^5 f(x)\,dx$

47. $\int_2^{10} f(x)\,dx - \int_2^7 f(x)\,dx$

48. $\int_{-3}^5 f(x)\,dx - \int_{-3}^0 f(x)\,dx + \int_5^6 f(x)\,dx$

49–54 ■ Use the properties of integrals to verify each inequality without evaluating the integrals.

49. $\int_0^{\pi/4} \sin^3 x\,dx \leqslant \int_0^{\pi/4} \sin^2 x\,dx$

50. $\int_1^2 \sqrt{5 - x}\,dx \geqslant \int_1^2 \sqrt{x + 1}\,dx$

51. $\int_4^6 \dfrac{1}{x}\,dx \leqslant \int_4^6 \dfrac{1}{8 - x}\,dx$

52. $\dfrac{\pi}{6} \leqslant \int_{\pi/6}^{\pi/2} \sin x\,dx \leqslant \dfrac{\pi}{3}$

53. $2 \leqslant \int_{-1}^{1} \sqrt{1 + x^2}\, dx \leqslant 2\sqrt{2}$

54. $\left| \int_{0}^{2\pi} f(x) \sin 2x\, dx \right| \leqslant \int_{0}^{2\pi} | f(x) |\, dx$

55–60 ■ Use Property 8 to estimate the value of the integral.

55. $\int_{1}^{2} \frac{1}{x}\, dx$

56. $\int_{0}^{2} \sqrt{x^3 + 1}\, dx$

57. $\int_{-3}^{0} (x^2 + 2x)\, dx$

58. $\int_{\pi/4}^{\pi/3} \cos x\, dx$

59. $\int_{-1}^{1} \sqrt{1 + x^4}\, dx$

60. $\int_{\pi/4}^{3\pi/4} \sin^2 x\, dx$

61–64 ■ Use properties of integrals, together with Exercises 21 and 22, to prove the inequality.

61. $\int_{1}^{3} \sqrt{x^4 + 1}\, dx \geqslant 26/3$

62. $\int_{2}^{5} \sqrt{x^2 - 1}\, dx \leqslant 10.5$

63. $\int_{0}^{\pi/2} x \sin x\, dx \leqslant \pi^2/8$

64. $\left| \int_{0}^{\pi} x^2 \cos x\, dx \right| \leqslant \pi^3/3$

65. Prove Property 3 of integrals.

66. Prove Property 6 of integrals.

67. Prove Property 9 of integrals.
[*Hint:* $-| f(x) | \leqslant f(x) \leqslant | f(x) |$.]

68. Suppose that f is continuous on $[a, b]$ and $f(x) > 0$ for all x in $[a, b]$. Prove that $\int_{a}^{b} f(x)\, dx > 0$. [*Hint:* Use the Extreme Value Theorem and Property 8.]

69. Which of the following functions are integrable on the interval $[0, 2]$?
(a) $f(x) = x^2 \sin x$
(b) $f(x) = \sec x$
(c) $f(x) = \begin{cases} x + 1 & \text{if } 0 \leqslant x < 1 \\ 2 - x & \text{if } 1 \leqslant x \leqslant 2 \end{cases}$
(d) $f(x) = \begin{cases} (x - 1)^{-2} & \text{if } x \neq 1 \\ 1 & \text{if } x = 1 \end{cases}$

70. Let

$$f(x) = \begin{cases} \dfrac{1}{x} & \text{if } 0 < x \leqslant 1 \\ 0 & \text{if } x = 0 \end{cases}$$

(a) Show that f is not continuous on $[0, 1]$.
(b) Show that f is unbounded on $[0, 1]$.
(c) Show that $\int_{0}^{1} f(x)\, dx$ does not exist, that is, f is not integrable on $[0, 1]$. [*Hint:* Show that the first term in the Riemann sum, $f(x_1^*)\, \Delta x_1$, can be made arbitrarily large.]

71. Let

$$f(x) = \begin{cases} 0 & \text{if } x \text{ is rational} \\ 1 & \text{if } x \text{ is irrational} \end{cases}$$

Show that f is bounded but not integrable on $[a, b]$. [*Hint:* Show that, no matter how small $\| P \|$ is, some Riemann sums are 0 whereas others are equal to $b - a$.]

72. Evaluate $\int_{0}^{1} e^x\, dx$. [*Hint:* Use Theorem 6.3.5, sum a geometric series, and then use l'Hospital's Rule.]

73. Find $\int_{1}^{2} x^{-2}\, dx$. [*Hint:* Use a regular partition but choose x_i^* to be the geometric mean of x_{i-1} and x_i ($x_i^* = \sqrt{x_{i-1} x_i}$) and use the identity

$$\frac{1}{m(m + 1)} = \frac{1}{m} - \frac{1}{m + 1}$$

74. (a) Draw the graph of the function $f(x) = \cos(x^2)$ in the viewing rectangle $[0, 2]$ by $[-1, 1]$.
(b) If we define a new function g by $g(x) = \int_{0}^{x} \cos(t^2)\, dt$, then $g(x)$ is the area under the graph of f from 0 to x [until $f(x)$ becomes negative, at which point $g(x)$ becomes a difference of areas]. Use the graph of f from part (a) to estimate the value of $g(x)$ when $x = 0, 0.2, 0.4, 0.6, \ldots$ up to $x = 2$. At what value of x does $g(x)$ start to decrease?
(c) Use the information from part (a) to sketch a rough graph of g.
(d) Sketch a more accurate graph of g by using your calculator or computer to estimate $g(0.2), g(0.4), \ldots$. (Use the integration command, if available, or the Midpoint Rule.)
(e) Use your graph of g from part (d) to sketch the graph of g' using the interpretation of $g'(x)$ as the slope of a tangent line. How does the graph of g' compare with the graph of f?

6.4 THE FUNDAMENTAL THEOREM OF CALCULUS

The Fundamental Theorem of Calculus is appropriately named because it establishes a connection between the two branches of calculus: differential calculus and integral calculus. Differential calculus arose from the tangent problem, whereas integral calculus arose from a seemingly unrelated problem, the area problem. Newton's teacher at Cambridge, Isaac Barrow (1630–1677), discovered that these two problems are actually closely related. In fact, he realized that differentiation and integration are inverse processes. The Fundamental Theorem of Calculus gives the precise inverse relationship between the derivative and the integral. It was Newton and Leibniz who exploited this relationship and used it to develop calculus into a systematic mathematical method. In particular, they saw that the Fundamental Theorem enabled them to compute areas and integrals very easily without having to compute them as limits of sums as we did in Sections 6.2 and 6.3.

In order to motivate the Fundamental Theorem, let f be a continuous function on $[a, b]$ and define a new function g by

(1) $$g(x) = \int_a^x f(t)\,dt$$

Observe that g depends only on x, which appears as the variable upper limit in the integral. If x is a fixed number, then the integral $\int_a^x f(t)\,dt$ is a definite number. If we then let x vary, the number $\int_a^x f(t)\,dt$ also varies and defines a function of x denoted by $g(x)$. For instance, if we take $f(t) = t$ and $a = 0$, then, using Exercise 21 in Section 6.3, we have

$$g(x) = \int_0^x t\,dt = \frac{x^2}{2}$$

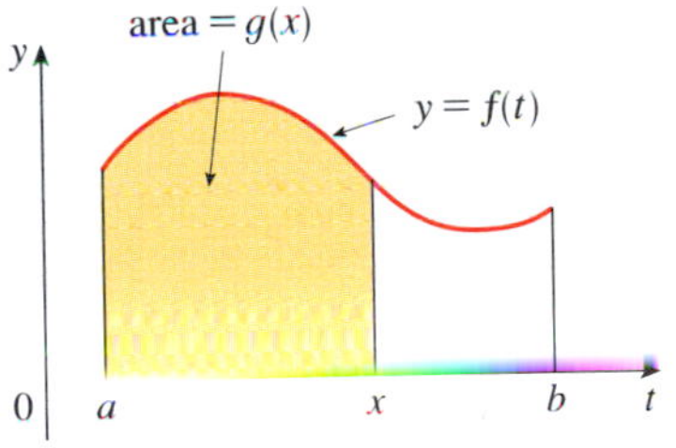

FIGURE 1

Notice that $g'(x) = x$, that is, $g' = f$. In other words, if g is defined as the integral of f by Equation 1, then g turns out to be an antiderivative of f, at least in this case.

To see why this might be generally true we consider any continuous function f with $f(x) \geqslant 0$. Then $g(x) = \int_a^x f(t)\,dt$ can be interpreted as the area under the graph of f from a to x, where x can vary from a to b. (Think of g as the "area so far" function; see Figure 1.)

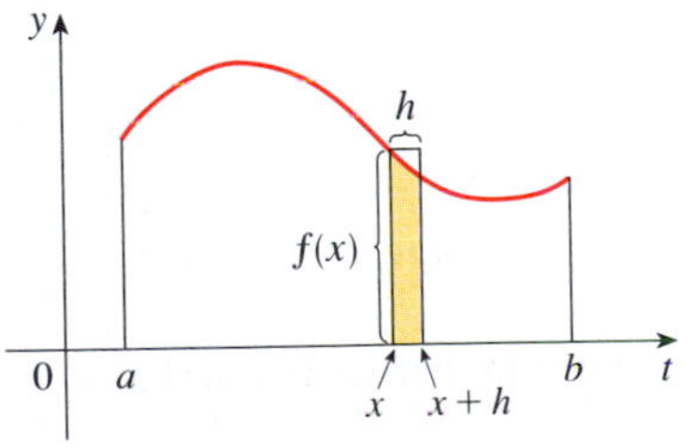

FIGURE 2

In order to compute $g'(x)$ from the definition of derivative we first observe that, for $h > 0$, $g(x + h) - g(x)$ is obtained by subtracting areas, so it is the area under the graph of f from x to $x + h$ (the gold area in Figure 2). For small h you can see from the figure that this area is approximately equal to the area of the rectangle with height $f(x)$ and width h:

$$g(x + h) - g(x) \approx hf(x) \qquad \frac{g(x + h) - g(x)}{h} \approx f(x)$$

Intuitively, we therefore expect that

$$g'(x) = \lim_{h \to 0} \frac{g(x + h) - g(x)}{h} = f(x)$$

The fact that this is true, even when f is not necessarily positive, is the first part of the Fundamental Theorem of Calculus.

(2) THE FUNDAMENTAL THEOREM OF CALCULUS, PART 1 If f is continuous on $[a, b]$, then the function g defined by

$$g(x) = \int_a^x f(t)\,dt \qquad a \leqslant x \leqslant b$$

is continuous on $[a, b]$ and differentiable on (a, b), and $g'(x) = f(x)$.

PROOF If x and $x + h$ are in (a, b), then

$$\begin{aligned} g(x+h) - g(x) &= \int_a^{x+h} f(t)\,dt - \int_a^x f(t)\,dt \\ &= \left(\int_a^x f(t)\,dt + \int_x^{x+h} f(t)\,dt\right) - \int_a^x f(t)\,dt \qquad \text{(by Property 5)} \\ &= \int_x^{x+h} f(t)\,dt \end{aligned}$$

and so, for $h \neq 0$,

(3) $$\frac{g(x+h) - g(x)}{h} = \frac{1}{h}\int_x^{x+h} f(t)\,dt$$

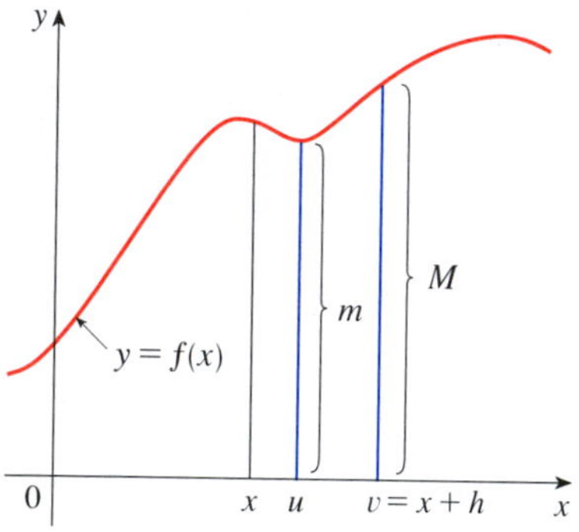

FIGURE 3

For now let's assume that $h > 0$. Since f is continuous on $[x, x + h]$, the Extreme Value Theorem says that there are numbers u and v in $[x, x + h]$ such that $f(u) = m$ and $f(v) = M$, where m and M are the absolute minimum and maximum values of f on $[x, x + h]$ (see Figure 3).

By Property 8 of integrals, we have

$$mh \leqslant \int_x^{x+h} f(t)\,dt \leqslant Mh$$

that is,

$$f(u)h \leqslant \int_x^{x+h} f(t)\,dt \leqslant f(v)h$$

Since $h > 0$, we can divide this inequality by h:

$$f(u) \leqslant \frac{1}{h}\int_x^{x+h} f(t)\,dt \leqslant f(v)$$

Now we use Equation 3 to replace the middle part of this inequality:

(4) $$f(u) \leqslant \frac{g(x+h) - g(x)}{h} \leqslant f(v)$$

Inequality 4 can be proved in a similar manner for the case where $h < 0$ (see Exercise 97).

Now we let $h \to 0$. Then $u \to x$ and $v \to x$, since u and v lie between x and $x + h$. Therefore

$$\lim_{h\to 0} f(u) = \lim_{u\to x} f(u) = f(x) \qquad \lim_{h\to 0} f(v) = \lim_{v\to x} f(v) = f(x)$$

because f is continuous at x. We conclude, from (4) and the Squeeze Theorem, that

(5) $$g'(x) = \lim_{h\to 0} \frac{g(x+h) - g(x)}{h} = f(x)$$

If $x = a$ or b, then Equation 5 can be interpreted as a one-sided limit. Then Theorem 3.1.8 (modified for one-sided limits) shows that g is continuous on $[a, b]$. □

Using Leibniz notation for derivatives, we can write Theorem 2 as

(6) $$\frac{d}{dx}\int_a^x f(t)\,dt = f(x)$$

when f is continuous. Roughly speaking, Equation 6 says that if we first integrate f and then differentiate the result, we get back to the original function f.

EXAMPLE 1 Find the derivative of the function $g(x) = \int_0^x \sqrt{1 + t^2}\,dt$.

SOLUTION Since $f(t) = \sqrt{1 + t^2}$ is continuous, Part 1 of the Fundamental Theorem of Calculus gives

$$g'(x) = \sqrt{1 + x^2}$$ ■

EXAMPLE 2 Although a formula of the form $g(x) = \int_a^x f(t)\,dt$ may seem like a strange way of defining a function, books on physics, chemistry, and statistics are full of such functions. For instance the **Fresnel function**

$$S(x) = \int_0^x \sin(\pi t^2/2)\,dt$$

is named after the French physicist Augustin Fresnel (1788–1827) who is famous for his works in optics. This function first appeared in Fresnel's theory of the diffraction of light waves, but more recently it has been applied to the design of highways.

Part 1 of the Fundamental Theorem tells us how to differentiate the Fresnel function:

$$S'(x) = \sin(\pi x^2/2)$$

This means that we can apply all the methods of differential calculus to analyze S (see Exercise 91).

Figure 4 shows the graphs of $f(x) = \sin(\pi x^2/2)$ and the Fresnel function $S(x) = \int_0^x f(t)\,dt$. A computer was used to graph S by computing the value of this integral for many values of x. It does indeed look as if $S(x)$ is the area under the graph of f from 0 to x [until $x \approx 1.4$ when $S(x)$ becomes a difference of areas]. Figure 5 shows a larger part of the graph of S.

If we now start with the graph of S in Figure 4 and think about what its derivative should look like, it seems reasonable that $S'(x) = f(x)$. [For instance, S is increasing when $f(x) > 0$ and decreasing when $f(x) < 0$.] So this gives a visual confirmation of Part 1 of the Fundamental Theorem of Calculus. ■

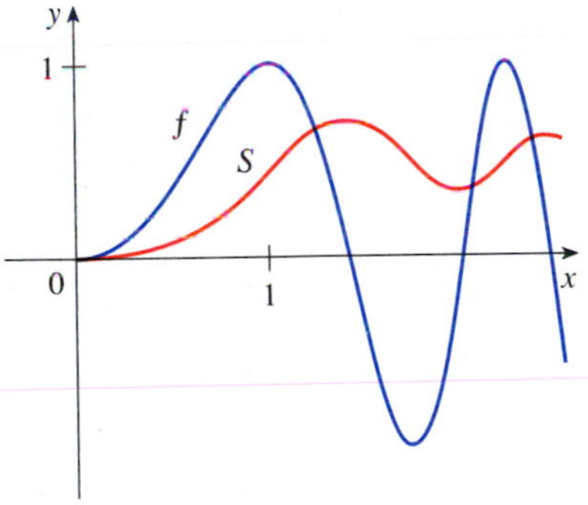

FIGURE 4
$f(x) = \sin(\pi x^2/2)$
$S(x) = \int_0^x \sin(\pi t^2/2)\,dt$

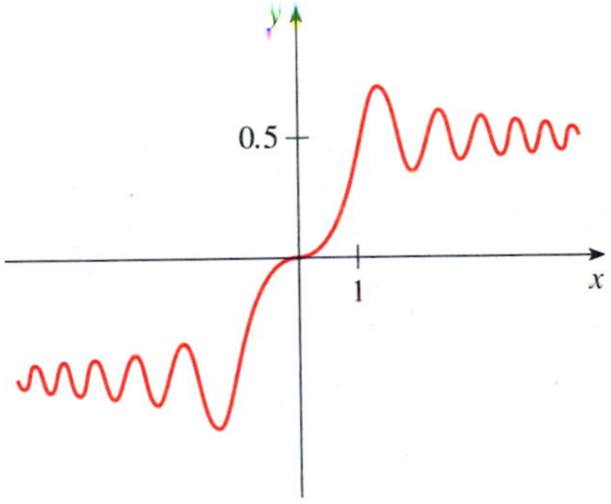

FIGURE 5
The Fresnel function
$S(x) = \int_0^x \sin(\pi t^2/2)\,dt$

EXAMPLE 3 Find $\displaystyle\frac{d}{dx}\int_1^{x^4} \sec t\,dt$.

SOLUTION Here we have to be careful to use the Chain Rule in conjunction with Part 1 of the Fundamental Theorem. Let $u = x^4$. Then

$$\frac{d}{dx}\int_1^{x^4} \sec t\,dt = \frac{d}{dx}\int_1^u \sec t\,dt$$

$$= \frac{d}{du}\left[\int_1^u \sec t\,dt\right]\frac{du}{dx} \qquad \text{(by the Chain Rule)}$$

$$= \sec u\,\frac{du}{dx} \qquad \text{(by Theorem 2)}$$

$$= \sec(x^4) \cdot 4x^3$$ ■

In Section 6.3 we computed integrals from the definition as a limit of Riemann sums and we saw that this procedure is sometimes long and difficult. The second part of the Fundamental Theorem of Calculus, which follows easily from the first part, provides us with a much simpler method for the evaluation of integrals.

(7) THE FUNDAMENTAL THEOREM OF CALCULUS, PART 2 If f is continuous on $[a, b]$, then

$$\int_a^b f(x)\,dx = F(b) - F(a)$$

where F is any antiderivative of f, that is, $F' = f$.

PROOF Let $g(x) = \int_a^x f(t)\,dt$. We know from Part 1 that $g'(x) = f(x)$; that is, g is an antiderivative of f. If F is any other antiderivative of f on $[a, b]$, then we know from Theorem 5.7.2 that F and g differ by a constant:

(8) $$F(x) = g(x) + C$$

for $a < x < b$. But both F and g are continuous on $[a, b]$ and so, by taking limits of both sides of Equation 8 (as $x \to a^+$ and $x \to b^-$), we see that it also holds when $x = a$ and $x = b$.

If we put $x = a$ in the formula for $g(x)$, we get

$$g(a) = \int_a^a f(t)\,dt = 0$$

So, using Equation 8 with $x = b$ and $x = a$, we have

$$F(b) - F(a) = [g(b) + C] - [g(a) + C]$$

$$= g(b) - g(a) = g(b) = \int_a^b f(t)\,dt$$

□

Part 2 of the Fundamental Theorem states that if we know an antiderivative F of f, then we can evaluate $\int_a^b f(x)\,dx$ simply by subtracting the values of F at the endpoints of the interval $[a, b]$. It is very surprising that $\int_a^b f(x)\,dx$, which was defined by a complicated procedure involving all of the values of $f(x)$ for $a \leqslant x \leqslant b$, can be found by knowing the values of $F(x)$ at only two points, a and b.

EXAMPLE 4 Evaluate the integral $\int_{-2}^{1} x^3\,dx$.

SOLUTION The function $f(x) = x^3$ is continuous on $[-2, 1]$ and we know from Section 5.7 that an antiderivative is $F(x) = \frac{1}{4}x^4$, so Part 2 of the Fundamental Theorem gives

$$\int_{-2}^{1} x^3\,dx = F(1) - F(-2) = \tfrac{1}{4}(1)^4 - \tfrac{1}{4}(-2)^4 = -\tfrac{15}{4}$$

■

We use the notation

$$F(x)\Big]_a^b = F(b) - F(a)$$

and so the equation of Theorem 7 can be written as

$$\int_a^b f(x)\,dx = F(x)\Big]_a^b \qquad \text{where} \qquad F' = f$$

Other common notations are $F(x)\,|_a^b$ and $[F(x)]_a^b$.

EXAMPLE 5 Find the area under the parabola $y = x^2 + 1$ from 0 to 2.

SOLUTION An antiderivative of $f(x) = x^2 + 1$ is $F(x) = \frac{1}{3}x^3 + x$. The required area A is found using Part 2 of the Fundamental Theorem:

In applying the Fundamental Theorem we use a particular antiderivative F of f. It is not necessary to use the most general antiderivative.

$$A = \int_0^2 (x^2 + 1)\,dx = \frac{x^3}{3} + x\Big]_0^2 = \left(\frac{2^3}{3} + 2\right) - \left(\frac{0^3}{3} + 0\right) = \frac{14}{3}$$ ■

If you compare the calculation in Example 5 with the one in Example 3 in Section 6.2, you see that the Fundamental Theorem gives a much shorter method.

EXAMPLE 6 Find the area under the cosine curve from 0 to b, where $0 \leqslant b \leqslant \pi/2$.

SOLUTION Since an antiderivative of $f(x) = \cos x$ is $F(x) = \sin x$, we have

$$A = \int_0^b \cos x\,dx = \sin x\Big]_0^b = \sin b - \sin 0 = \sin b$$ ■

A comparison of Example 6 with Example 4 in Section 6.2 reveals the power of the Fundamental Theorem of Calculus. When the French mathematician Gilles de Roberval first found the area under the sine and cosine curves in 1635, this was a very challenging problem that required a great deal of ingenuity. (Recall that the calculation in Example 4 in Section 6.2 depended on some little-known trigonometric identities.) But in the 1660s and 1670s when the Fundamental Theorem was discovered by Barrow and exploited by Newton and Leibniz, such problems became very easy, as you can see from Example 6.

NOTATION: Because of the relation given by the Fundamental Theorem between antiderivatives and integrals, the notation $\int f(x)\,dx$ is traditionally used for an antiderivative of f and is called an **indefinite integral**. Thus

(9) $$\int f(x)\,dx = F(x) \qquad \text{means} \qquad F'(x) = f(x)$$

⊘ You should distinguish carefully between definite and indefinite integrals. A definite integral $\int_a^b f(x)\,dx$ is a number, whereas an indefinite integral $\int f(x)\,dx$ is a function. The connection between them is given by Part 2 of the Fundamental Theorem. If f is continuous on $[a, b]$, then

(10) $$\int_a^b f(x)\,dx = \int f(x)\,dx\Big]_a^b$$

The effectiveness of the Fundamental Theorem depends on having a supply of antiderivatives of functions. We therefore restate the Table of Antidifferentiation Formulas from Section 5.7, together with a few others, in the notation of indefinite integrals. Any formula can be verified by differentiating the function on the right side and obtaining the integrand. For instance,

$$\int \sec^2 x\,dx = \tan x \qquad \text{since} \qquad \frac{d}{dx}(\tan x) = \sec^2 x$$

(11) TABLE OF INDEFINITE INTEGRALS

$$\int cf(x)\,dx = c\int f(x)\,dx$$

$$\int [f(x) + g(x)]\,dx = \int f(x)\,dx + \int g(x)\,dx$$

$$\int x^n\,dx = \frac{x^{n+1}}{n+1} + C \quad (n \neq -1) \qquad \int \frac{1}{x}\,dx = \ln|x| + C$$

$$\int e^x\,dx = e^x + C \qquad \int a^x\,dx = \frac{a^x}{\ln a} + C$$

$$\int \sin x\,dx = -\cos x + C \qquad \int \cos x\,dx = \sin x + C$$

$$\int \sec^2 x\,dx = \tan x + C \qquad \int \csc^2 x\,dx = -\cot x + C$$

$$\int \sec x \tan x\,dx = \sec x + C \qquad \int \csc x \cot x\,dx = -\csc x + C$$

$$\int \frac{1}{x^2+1}\,dx = \tan^{-1}x + C \qquad \int \frac{1}{\sqrt{1-x^2}}\,dx = \sin^{-1}x + C$$

Recall from Theorem 5.7.2 that the most general antiderivative *on a given interval* is obtained by adding a constant to a particular antiderivative. **We adopt the convention that when a formula for a general indefinite integral is given, it is valid only on an interval.** Thus, we write

$$\int \frac{1}{x^2}\,dx = -\frac{1}{x} + C$$

with the understanding that it is valid on the interval $(0, \infty)$ or on the interval $(-\infty, 0)$. This is true despite the fact that the general antiderivative of the function $f(x) = 1/x^2$, $x \neq 0$, is

$$F(x) = \begin{cases} -\dfrac{1}{x} + C_1 & \text{if } x < 0 \\[2ex] -\dfrac{1}{x} + C_2 & \text{if } x > 0 \end{cases}$$

EXAMPLE 7 Find the general indefinite integral

$$\int (10x^4 + 3\sec^2 x)\,dx$$

SOLUTION Using our convention and Table 11, we have

$$\int (10x^4 + 3\sec^2 x)\,dx = 10\frac{x^5}{5} + 3\tan x + C$$

Check the answer by differentiating it.

$$= 2x^5 + 3\tan x + C$$

■

EXAMPLE 8 Evaluate $\int_0^3 (x^3 - 5x)\,dx$.

SOLUTION Using Part 2 of the Fundamental Theorem and Table 11, we have

$$\int_0^3 (x^3 - 5x)\,dx = \frac{x^4}{4} - 5\frac{x^2}{2}\bigg]_0^3$$

$$= \left(\tfrac{1}{4}\cdot 3^4 - \tfrac{5}{2}\cdot 3^2\right) - \left(\tfrac{1}{4}\cdot 0^4 - \tfrac{5}{2}\cdot 0^2\right)$$

$$= \tfrac{81}{4} - \tfrac{45}{2} - 0 + 0 = -2.25$$

Compare this calculation with Example 4 in Section 6.3. ■

EXAMPLE 9 $\int_0^1 e^x\,dx = e^x\Big]_0^1 = e^1 - e^0 = e - 1$ ■

EXAMPLE 10

$$\int_1^9 \frac{2t^2 + t^2\sqrt{t} - 1}{t^2}\,dt = \int_1^9 (2 + t^{1/2} - t^{-2})\,dt$$

$$= 2t + \frac{t^{3/2}}{\frac{3}{2}} - \frac{t^{-1}}{-1}\bigg]_1^9$$

$$= 2t + \tfrac{2}{3}t^{3/2} + \frac{1}{t}\bigg]_1^9$$

$$= \left[2\cdot 9 + \tfrac{2}{3}(9)^{3/2} + \tfrac{1}{9}\right] - \left(2\cdot 1 + \tfrac{2}{3}\cdot 1^{3/2} + \tfrac{1}{1}\right)$$

$$= 18 + 18 + \tfrac{1}{9} - 2 - \tfrac{2}{3} - 1 = 32\tfrac{4}{9}$$

■

EXAMPLE 11 What is wrong with the following calculation?

$$\int_{-1}^3 \frac{1}{x^2}\,dx = \frac{x^{-1}}{-1}\bigg]_{-1}^3 = -\frac{1}{3} - 1 = -\frac{4}{3}$$

SOLUTION To start, we notice that this calculation must be wrong because $f(x) = 1/x^2 \geqslant 0$ and Property 6 of integrals says that $\int_a^b f(x)\,dx \geqslant 0$ when $f \geqslant 0$. The Fundamental Theorem of Calculus applies to continuous functions. It cannot be applied here because $f(x) = 1/x^2$ is not continuous on $[-1, 3]$. In fact, $f(0)$ is not defined, but even if it were defined [say, by $f(0) = 0$], the integral would not exist because f is unbounded on $[-1, 3]$. In fact, we know that $\lim_{x\to 0}(1/x^2) = \infty$. Thus

$$\int_{-1}^3 \frac{1}{x^2}\,dx \quad \text{does not exist}$$

■

APPLICATIONS OF THE FUNDAMENTAL THEOREM

Suppose a particle is moving along a straight line with position function $s(t)$, velocity function $v(t)$, and acceleration function $a(t)$. Since $s'(t) = v(t)$, the Fundamental Theorem of Calculus gives

$$\int_{t_1}^{t_2} v(t)\,dt = s(t_2) - s(t_1) \tag{12}$$

The right side of Equation 12 is the change of position, or *displacement*, of the particle during the time interval $[t_1, t_2]$. Thus, Equation 12 enables us to compute the displacement, by integration, if the velocity function is known. Similarly, since $v'(t) = a(t)$, the Fundamental Theorem of Calculus gives

$$\int_{t_1}^{t_2} a(t)\,dt = v(t_2) - v(t_1)$$

If we want to calculate the distance traveled during the time interval, we have to consider the intervals when $v(t) \geqslant 0$ (the particle moves to the right) and also the intervals when $v(t) \leqslant 0$ (the particle moves to the left). In both cases the distance is computed by integrating $|v(t)|$, the speed. Therefore

$$\text{total distance traveled} = \int_{t_1}^{t_2} |v(t)|\,dt \tag{13}$$

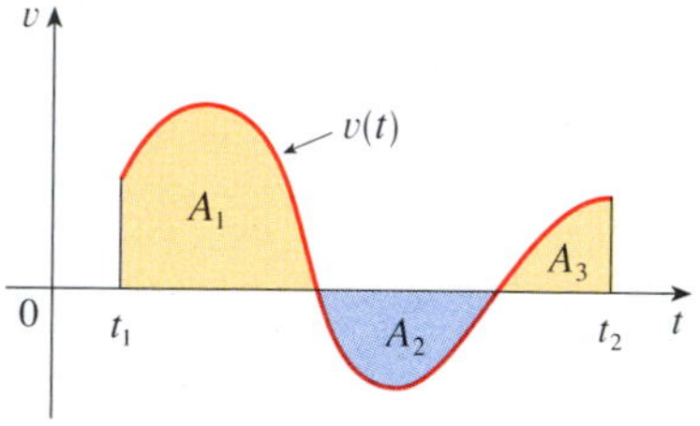

FIGURE 6

$$\text{displacement} = \int_{t_1}^{t_2} v(t)\,dt = A_1 - A_2 + A_3$$

$$\text{distance} = \int_{t_1}^{t_2} |v(t)|\,dt = A_1 + A_2 + A_3$$

Figure 6 shows how both displacement and distance traveled can be interpreted in terms of areas under a velocity curve.

EXAMPLE 12 A particle moves along a line so that its velocity at time t is $v(t) = t^2 - t - 6$ (measured in meters per second).
(a) Find the displacement of the particle during the time period $1 \leqslant t \leqslant 4$.
(b) Find the distance traveled during this time period.

SOLUTION
(a) By Equation 12, the displacement is

$$s(4) - s(1) = \int_1^4 v(t)\,dt = \int_1^4 (t^2 - t - 6)\,dt$$
$$= \left[\frac{t^3}{3} - \frac{t^2}{2} - 6t\right]_1^4 = -\frac{9}{2}$$

This means that the particle moved 4.5 m to the left.
(b) Note that $v(t) = t^2 - t - 6 = (t - 3)(t + 2)$ and so $v(t) \leqslant 0$ on the interval $[1, 3]$ and $v(t) \geqslant 0$ on $[3, 4]$. Thus, from Equation 13, the distance traveled is

$$\int_1^4 |v(t)|\,dt = \int_1^3 (-v(t))\,dt + \int_3^4 v(t)\,dt$$
$$= \int_1^3 (-t^2 + t + 6)\,dt + \int_3^4 (t^2 - t - 6)\,dt$$
$$= \left[-\frac{t^3}{3} + \frac{t^2}{2} + 6t\right]_1^3 + \left[\frac{t^3}{3} - \frac{t^2}{2} - 6t\right]_3^4$$
$$= \frac{61}{6}\text{ m}$$

■

The Fundamental Theorem of Calculus can be applied to all the rates of change in the natural and social sciences that were discussed in Section 3.3. For instance, if the mass of a rod measured from the left end to a point x is $m(x)$ and the linear density is

$\rho(x)$, then $m'(x) = \rho(x)$ (see Example 2 in Section 3.3) and so, by the Fundamental Theorem of Calculus, we have

$$m(x_2) - m(x_1) = \int_{x_1}^{x_2} \rho(x)\, dx$$

This enables us to compute the mass of a segment of the rod if the linear density is known.

If the rate of growth dn/dt of a population is known (see Example 6 in Section 3.3), then the Fundamental Theorem of Calculus allows us to compute the population at time t, in terms of the initial population n_0, as

$$n(t) = n_0 + \int_0^t \frac{dn}{dt}\, dt$$

DIFFERENTIATION AND INTEGRATION AS INVERSE PROCESSES

We end this section by bringing together the two parts of the Fundamental Theorem.

THE FUNDAMENTAL THEOREM OF CALCULUS Suppose f is continuous on $[a, b]$.

1. If $g(x) = \int_a^x f(t)\, dt$, then $g'(x) = f(x)$.
2. $\int_a^b f(x)\, dx = F(b) - F(a)$, where F is any antiderivative of f, that is, $F' = f$.

We noted that Part 1 can be rewritten as

$$\frac{d}{dx} \int_a^x f(t)\, dt = f(x)$$

which says that if f is integrated and then the result is differentiated, we arrive back at the original function f. Since $F'(x) = f(x)$, Part 2 can be rewritten as

$$\int_a^b F'(x)\, dx = F(b) - F(a)$$

This version says that if we take a function F, first differentiate it, and then integrate the result, we arrive back at the original function F, but in the form $F(b) - F(a)$. Taken together, the two parts of the Fundamental Theorem of Calculus say that differentiation and integration are inverse processes. Each undoes what the other does.

The Fundamental Theorem of Calculus is unquestionably the most important theorem in calculus and, indeed, it ranks as one of the great accomplishments of the human mind. Before it was discovered, from the time of Eudoxus and Archimedes to the time of Galileo and Fermat, problems of finding areas, volumes, and lengths of curves were so difficult that only a genius could meet the challenge. But now, armed with the systematic method that Newton and Leibniz fashioned out of the Fundamental Theorem, we will see in the chapters to come that these challenging problems are accessible to all of us.

EXERCISES 6.4

1. Let $g(x) = \int_0^x f(t)\, dt$, where f is the function whose graph is shown.

   (a) Evaluate $g(0)$, $g(1)$, $g(2)$, $g(3)$, and $g(6)$.

   (b) On what intervals is g increasing?

   (c) Where does g have a maximum value?

   (d) Sketch a rough graph of g.

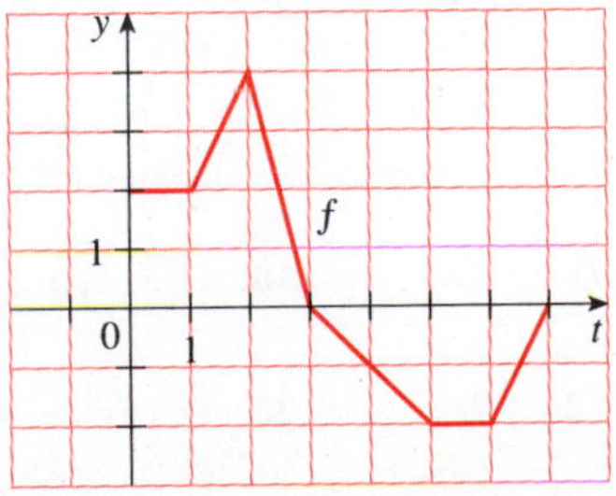

2. Let $g(x) = \int_{-3}^{x} f(t)\,dt$, where f is the function whose graph is shown.
(a) Evaluate $g(-3)$ and $g(3)$.
(b) Estimate $g(-2)$, $g(-1)$, and $g(0)$.
(c) On what interval is g increasing?
(d) Where does g have a maximum value?
(e) Sketch a rough graph of g.
(f) Use the graph in part (e) to sketch the graph of $g'(x)$. Compare with the graph of f.

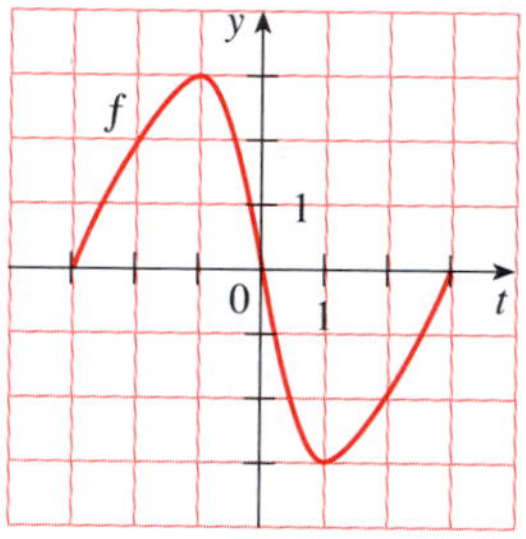

3–4 ■ Sketch the area represented by $g(x)$. Then find $g'(x)$ in two ways: (a) by using Part 1 of the Fundamental Theorem and (b) by evaluating the integral using Part 2 and then differentiating.

3. $g(x) = \int_{0}^{x} (1 + t^2)\,dt$

4. $g(x) = \int_{\pi}^{x} (2 + \cos t)\,dt$

5–16 ■ Use Part 1 of the Fundamental Theorem of Calculus to find the derivative of the given function.

5. $g(x) = \int_{1}^{x} (t^2 - 1)^{20}\,dt$

6. $g(x) = \int_{-1}^{x} \sqrt{t^3 + 1}\,dt$

7. $g(u) = \int_{\pi}^{u} \dfrac{1}{1 + t^4}\,dt$

8. $g(t) = \int_{0}^{t} \sin(x^2)\,dx$

9. $F(x) = \int_{x}^{2} \cos(t^2)\,dt$

$\left[\textit{Hint: } \int_{x}^{2} \cos(t^2)\,dt = -\int_{2}^{x} \cos(t^2)\,dt \right]$

10. $F(x) = \int_{x}^{4} (2 + \sqrt{u})^8\,du$

11. $h(x) = \int_{2}^{1/x} \sin^4 t\,dt$

12. $h(x) = \int_{1}^{\sqrt{x}} \dfrac{s^2}{s^2 + 1}\,ds$

13. $y = \int_{\tan x}^{17} \sin(t^4)\,dt$

14. $y = \int_{x^2}^{\pi} \dfrac{\sin t}{t}\,dt$

15. $y = \int_{0}^{5x+1} \dfrac{1}{u^2 - 5}\,du$

16. $y = \int_{-5}^{\sin x} t\cos(t^3)\,dt$

17–60 ■ Use Part 2 of the Fundamental Theorem of Calculus to evaluate the integral, or state that it does not exist.

17. $\int_{-2}^{4} (3x - 5)\,dx$

18. $\int_{1}^{2} x^{-2}\,dx$

19. $\int_{0}^{1} (1 - 2x - 3x^2)\,dx$

20. $\int_{1}^{2} (5x^2 - 4x + 3)\,dx$

21. $\int_{-3}^{0} (5y^4 - 6y^2 + 14)\,dy$

22. $\int_{0}^{1} (y^9 - 2y^5 + 3y)\,dy$

23. $\int_{0}^{4} \sqrt{x}\,dx$

24. $\int_{0}^{1} x^{3/7}\,dx$

25. $\int_{1}^{3} \left(\dfrac{1}{t^2} - \dfrac{1}{t^4} \right) dt$

26. $\int_{1}^{2} \dfrac{t^6 - t^2}{t^4}\,dt$

27. $\int_{1}^{2} \dfrac{x^2 + 1}{\sqrt{x}}\,dx$

28. $\int_{0}^{2} (x^3 - 1)^2\,dx$

29. $\int_{0}^{1} u(\sqrt{u} + \sqrt[3]{u})\,du$

30. $\int_{-1}^{1} \dfrac{3}{t^4}\,dt$

31. $\int_{-2}^{3} |x^2 - 1|\,dx$

32. $\int_{1}^{2} \left(x + \dfrac{1}{x} \right)^2 dx$

33. $\int_{3}^{3} \sqrt{x^5 + 2}\,dx$

34. $\int_{-1}^{2} |x - x^2|\,dx$

35. $\int_{-4}^{2} \dfrac{2}{x^6}\,dx$

36. $\int_{1}^{-1} (x - 1)(3x + 2)\,dx$

37. $\int_{1}^{4} \left(\sqrt{t} - \dfrac{2}{\sqrt{t}} \right) dt$

38. $\int_{1}^{8} \left(\sqrt[3]{r} + \dfrac{1}{\sqrt[3]{r}} \right) dr$

39. $\int_{-1}^{0} (x + 1)^3\,dx$

40. $\int_{-5}^{-2} \dfrac{x^4 - 1}{x^2 + 1}\,dx$

41. $\int_{\pi/4}^{\pi/3} \sin t\,dt$

42. $\int_{0}^{\pi/2} (\cos\theta + 2\sin\theta)\,d\theta$

43. $\int_{\pi/2}^{\pi} \sec x \tan x\,dx$

44. $\int_{\pi/3}^{\pi/2} \csc x \cot x\,dx$

45. $\int_{\pi/6}^{\pi/3} \csc^2\theta\,d\theta$

46. $\int_{\pi/4}^{\pi} \sec^2\theta\,d\theta$

47. $\int_{4}^{8} \dfrac{1}{x}\,dx$

48. $\int_{\ln 3}^{\ln 6} 8e^x\,dx$

49. $\int_{8}^{9} 2^t\,dt$

50. $\int_{-e^2}^{-e} \dfrac{3}{x}\,dx$

51. $\int_{1}^{\sqrt{3}} \dfrac{6}{1 + x^2}\,dx$

52. $\int_{0}^{0.5} \dfrac{dx}{\sqrt{1 - x^2}}$

53. $\int_{1}^{e} \dfrac{x^2 + x + 1}{x}\,dx$

54. $\int_{4}^{9} \left(\sqrt{x} + \dfrac{1}{\sqrt{x}} \right)^2 dx$

55. $\int_{0}^{1} (\sqrt[4]{x^5} + \sqrt[5]{x^4})\,dx$

56. $\int_{1}^{8} \dfrac{x - 1}{\sqrt[3]{x^2}}\,dx$

57. $\int_{-1}^{2} (x - 2|x|)\,dx$

58. $\int_{0}^{2} (x^2 - |x - 1|)\,dx$

59. $\int_{0}^{2} f(x)\,dx$ where $f(x) = \begin{cases} x^4 & \text{if } 0 \leqslant x < 1 \\ x^5 & \text{if } 1 \leqslant x \leqslant 2 \end{cases}$

60. $\int_{-\pi}^{\pi} f(x)\,dx$ where $f(x) = \begin{cases} x & \text{if } -\pi \leqslant x \leqslant 0 \\ \sin x & \text{if } 0 < x \leqslant \pi \end{cases}$

61–64 ■ Use a graph to give a rough estimate of the area of the region that lies beneath the given curve. Then find the exact area.

61. $y = \sqrt[3]{x},\ 0 \leqslant x \leqslant 27$

62. $y = x^{-4},\ 1 \leqslant x \leqslant 6$

63. $y = \sin x,\ 0 \leqslant x \leqslant \pi$

64. $y = \sec^2 x,\ 0 \leqslant x \leqslant \pi/3$

65. Use a graph to estimate the x-intercepts of the curve $y = x + x^2 - x^4$. Then use this information to estimate the area of the region that lies under the curve and above the x-axis.

66. Repeat Exercise 65 for the curve $y = 2x + 3x^4 - 2x^6$.

67–70 ■ Verify by differentiation that each formula is correct.

67. $\displaystyle\int \frac{1}{\sqrt{(a^2 - x^2)^3}}\,dx = \frac{x}{a^2\sqrt{a^2 - x^2}} + C$

68. $\displaystyle\int \frac{1}{x^2\sqrt{x^2 + a^2}}\,dx = -\frac{\sqrt{x^2 + a^2}}{a^2 x} + C$

69. $\displaystyle\int \sin^2 x\,dx = \frac{x}{2} - \frac{\sin 2x}{4} + C$

70. $\displaystyle\int x^2 \sin x\,dx = -x^2\cos x + 2\int x\cos x\,dx$

71–76 ■ Find the general indefinite integral.

71. $\displaystyle\int x\sqrt{x}\,dx$

72. $\displaystyle\int \sqrt{x}\,(x^2 - 1/x)\,dx$

73. $\displaystyle\int (2 - \sqrt{x}\,)^2\,dx$

74. $\displaystyle\int (\cos x - 2\sin x)\,dx$

75. $\displaystyle\int (2x + \sec x \tan x)\,dx$

76. $\displaystyle\int \left(x^2 + 1 + \frac{1}{x^2 + 1}\right)dx$

77–78 ■ The velocity function (in meters per second) is given for a particle moving along a line. Find (a) the displacement and (b) the distance traveled by the particle during the given time interval.

77. $v(t) = 3t - 5,\quad 0 \leqslant t \leqslant 3$

78. $v(t) = t^2 - 2t - 8,\quad 1 \leqslant t \leqslant 6$

79–80 ■ The acceleration function (in m/s^2) and the initial velocity are given for a particle moving along a line. Find (a) the velocity at time t and (b) the distance traveled during the given time interval.

79. $a(t) = t + 4,\quad v(0) = 5,\quad 0 \leqslant t \leqslant 10$

80. $a(t) = 2t + 3,\quad v(0) = -4,\quad 0 \leqslant t \leqslant 3$

81. The linear density of a rod of length 4 m is given by $\rho(x) = 9 + 2\sqrt{x}$ measured in kilograms per meter, where x is measured in meters from one end of the rod. Find the total mass of the rod.

82. An animal population is increasing at a rate of $200 + 50t$ per year (where t is measured in years). By how much does the animal population increase between the fourth and tenth years?

83. The velocity of a car was read from its speedometer at ten-second intervals and recorded in the table. Use the Midpoint Rule to estimate the distance traveled by the car.

t (s)	0	10	20	30	40	50	60	70	80	90	100
v (mi/h)	0	38	52	58	55	51	56	53	50	47	45

84. Water leaked from a tank at a rate of $r(t)$ liters per hour, where the graph of r is as shown. Express the total amount of water that leaked out during the first four hours as a definite integral. Then use the Midpoint Rule to estimate that amount.

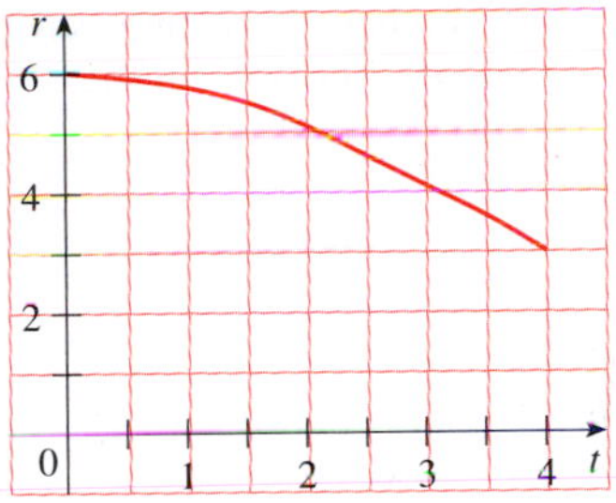

85–88 ■ Find the derivative of each function.

85. $\displaystyle g(x) = \int_{2x}^{3x} \frac{u - 1}{u + 1}\,du$

$\left[\textit{Hint: } \displaystyle\int_{2x}^{3x} f(u)\,du = \int_{2x}^{0} f(u)\,du + \int_{0}^{3x} f(u)\,du\right]$

86. $\displaystyle g(x) = \int_{\tan x}^{x^2} \frac{1}{\sqrt{2 + t^4}}\,dt$

87. $\displaystyle y = \int_{\sqrt{x}}^{x^3} \sqrt{t}\,\sin t\,dt$

88. $\displaystyle y = \int_{\cos x}^{5x} \cos(u^2)\,du$

89. If $\displaystyle F(x) = \int_1^x f(t)\,dt$, where $\displaystyle f(t) = \int_1^{t^2} \frac{\sqrt{1 + u^4}}{u}\,du$, find $F''(2)$.

90. Find the interval on which the curve

$$y = \int_0^x \frac{1}{1 + t + t^2}\,dt$$

is concave upward.

91. The Fresnel function S was defined in Example 2 and graphed in Figures 4 and 5.

(a) At what values of x does this function have local maximum values?

(b) On what intervals is the function concave upward?

(c) Use a graph to solve the following equation correct to one decimal place:

$$\int_0^x \sin(\pi t^2/2)\,dt = 0.2$$

CAS **92.** The **sine integral function**

$$\mathrm{Si}(x) = \int_0^x \frac{\sin t}{t}\, dt$$

is important in electrical engineering. [The integrand $f(t) = (\sin t)/t$ is not defined when $t = 0$ but we know that its limit is 1 when $t \to 0$. So we define $f(0) = 1$ and this makes f a continuous function everywhere.]

(a) Draw the graph of Si.

(b) At what values of x does this function have local maximum values?

(c) Find the coordinates of the first inflection point to the right of the origin.

(d) Does this function have horizontal asymptotes?

(e) Solve the following equation correct to one decimal place:

$$\int_0^x \frac{\sin t}{t}\, dt = 1$$

93–94 ■ Let $g(x) = \int_0^x f(t)\, dt$, where f is the function whose graph is shown.

(a) At what values of x do the local maximum and minimum values of g occur?

(b) Where does g attain its absolute maximum value?

(c) On what intervals is g concave downward?

(d) Sketch the graph of g.

93.

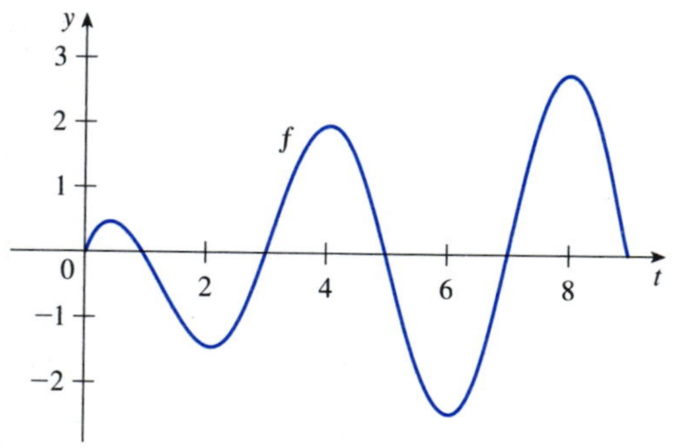

94.

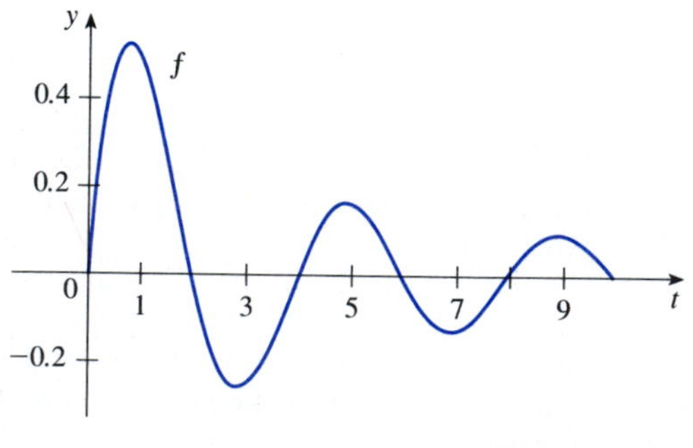

95–96 ■ Evaluate the limit by first recognizing the sum as a Riemann sum for a function defined on [0, 1].

95. $\displaystyle \lim_{n\to\infty} \sum_{i=1}^{n} \frac{i^3}{n^4}$

96. $\displaystyle \lim_{n\to\infty} \frac{1}{n}\left(\sqrt{\frac{1}{n}} + \sqrt{\frac{2}{n}} + \sqrt{\frac{3}{n}} + \cdots + \sqrt{\frac{n}{n}}\right)$

97. Justify (4) for the case where $h < 0$.

98. If f is continuous and g and h are differentiable functions, find a formula for

$$\frac{d}{dx}\int_{g(x)}^{h(x)} f(t)\, dt$$

99. (a) Show that $1 \leqslant \sqrt{1 + x^3} \leqslant 1 + x^3$ for $x \geqslant 0$.

(b) Show that $1 \leqslant \int_0^1 \sqrt{1 + x^3}\, dx \leqslant 1.25$.

100. Let

$$f(x) = \begin{cases} 0 & \text{if } x < 0 \\ x & \text{if } 0 \leqslant x \leqslant 1 \\ 2 - x & \text{if } 1 < x \leqslant 2 \\ 0 & \text{if } x > 2 \end{cases}$$

and $$g(x) = \int_0^x f(t)\, dt$$

(a) Find an expression for $g(x)$ similar to the one for $f(x)$.

(b) Sketch the graphs of f and g.

(c) Where is f differentiable? Where is g differentiable?

101. If $w'(t)$ is the rate of growth of a child in pounds per year, what does $\int_5^{10} w'(t)\, dt$ represent?

102. Find a function f such that $f(1) = 0$ and $f'(x) = 2^x/x$.

103. Find a function f and a number a such that

$$6 + \int_a^x \frac{f(t)}{t^2}\, dt = 2\sqrt{x}$$

for all $x > 0$.

104. Suppose h is a function such that $h(1) = -2$, $h'(1) = 2$, $h''(1) = 3$, $h(2) = 6$, $h'(2) = 5$, $h''(2) = 13$, and h'' is continuous everywhere. Evaluate $\int_1^2 h''(u)\, du$.

6.5 THE SUBSTITUTION RULE

The Fundamental Theorem of Calculus reduces the problem of integration to the problem of antidifferentiation. But the antidifferentiation formulas in Table 11 in Section 6.4 do not suffice to evaluate integrals such as

$$\int 2x\sqrt{1+x^2}\,dx \tag{1}$$

Differentials were defined in Section 3.11.

In such cases the task is simplified by changing from the variable x to a new variable. Suppose that we let u be the quantity under the root sign in (1), $u = 1 + x^2$. Then the differential of u is $du = 2x\,dx$. Notice that if the dx in the notation for an integral were to be interpreted as a differential, then the differential $2x\,dx$ would occur in (1) so, formally, without justifying our calculation, we could write

$$\int 2x\sqrt{1+x^2}\,dx = \int \sqrt{1+x^2}\,2x\,dx = \int \sqrt{u}\,du = \tfrac{2}{3}u^{3/2} + C \tag{2}$$
$$= \tfrac{2}{3}(x^2+1)^{3/2} + C$$

But now we could check that we have the correct answer by using the Chain Rule to differentiate the function on the right side of Equation 2:

$$\frac{d}{dx}\left[\tfrac{2}{3}(x^2+1)^{3/2} + C\right] = \tfrac{2}{3}\cdot\tfrac{3}{2}(x^2+1)^{1/2}\cdot 2x = 2x\sqrt{x^2+1}$$

In general, this method works if we have an integral of the form $\int f(g(x))g'(x)\,dx$. Observe that if $F' = f$, then

$$\int F'(g(x))g'(x)\,dx = F(g(x)) + C \tag{3}$$

because, by the Chain Rule,

$$\frac{d}{dx}[F(g(x))] = F'(g(x))g'(x)$$

If we make the "change of variable" or "substitution" $u = g(x)$, then from Equation 3 we have

$$\int F'(g(x))g'(x)\,dx = F(g(x)) + C = F(u) + C = \int F'(u)\,du$$

or, writing $F' = f$, we get

$$\int f(g(x))g'(x)\,dx = \int f(u)\,du$$

Thus, we have proved the following rule:

(4) THE SUBSTITUTION RULE If $u = g(x)$ is a differentiable function whose range is an interval I and f is continuous on I, then

$$\int f(g(x))g'(x)\,dx = \int f(u)\,du$$

Notice that the Substitution Rule for integration was proved using the Chain Rule for differentiation. Notice also that if $u = g(x)$, then $du = g'(x)\,dx$, so a way of remembering the Substitution Rule is to think of dx and du in (4) as differentials.

Thus, the Substitution Rule says: **It is permissible to operate with dx and du after integral signs as if they were differentials.**

EXAMPLE 1 Find $\int x^3 \cos(x^4 + 2)\,dx$.

SOLUTION We will make the substitution $u = x^4 + 2$ because its differential is $du = 4x^3\,dx$, which, apart from the constant factor 4, occurs in the integral. Thus, using $x^3\,dx = du/4$ and the Substitution Rule, we have

$$\begin{aligned}\int x^3 \cos(x^4 + 2)\,dx &= \int \cos u \cdot \tfrac{1}{4}\,du\\ &= \tfrac{1}{4}\int \cos u\,du\\ &= \tfrac{1}{4}\sin u + C\\ &= \tfrac{1}{4}\sin(x^4 + 2) + C\end{aligned}$$

Check the answer by differentiating it.

Notice that at the final stage we had to return to the original variable x. ■

The idea behind the Substitution Rule is to replace a relatively complicated integral by a simpler integral. This is accomplished by changing from the original variable x to a new variable u that is a function of x. Thus, in Example 1 we replaced the integral $\int x^3 \cos(x^4 + 2)\,dx$ by the simpler integral $\frac{1}{4}\int \cos u\,du$.

The main challenge in using the Substitution Rule is to think of an appropriate substitution. You should try to choose u to be some function in the integrand whose differential also occurs (except for a constant factor). This was the case in Example 1. If that is not possible, try choosing u to be some complicated part of the integrand.

EXAMPLE 2 Evaluate $\int \sqrt{3x + 4}\,dx$.

SOLUTION 1 Let $u = 3x + 4$. Then $du = 3\,dx$, so $dx = du/3$. Thus, the Substitution Rule gives

$$\begin{aligned}\int \sqrt{3x + 4}\,dx &= \int \sqrt{u}\,\frac{du}{3} = \tfrac{1}{3}\int u^{1/2}\,du\\ &= \frac{1}{3}\cdot\frac{u^{3/2}}{3/2} + C = \tfrac{2}{9}u^{3/2} + C\\ &= \tfrac{2}{9}(3x + 4)^{3/2} + C\end{aligned}$$

SOLUTION 2 Another possible substitution is $u = \sqrt{3x + 4}$. Then

$$du = \frac{3\,dx}{2\sqrt{3x + 4}} \qquad \text{and} \qquad dx = \tfrac{2}{3}\sqrt{3x + 4}\,du = \tfrac{2}{3}u\,du$$

(Or observe that $u^2 = 3x + 4$, so $2u\,du = 3\,dx$.) Therefore

$$\int \sqrt{3x + 4}\,dx = \int u \cdot \tfrac{2}{3}u\,du = \tfrac{2}{3}\int u^2\,du$$

$$= \frac{2}{3} \cdot \frac{u^3}{3} + C = \tfrac{2}{9}u^3 + C$$

$$= \tfrac{2}{9}(3x + 4)^{3/2} + C$$

■

EXAMPLE 3 Find $\displaystyle\int \frac{x}{\sqrt{1 - 4x^2}}\,dx$.

SOLUTION Let $u = 1 - 4x^2$. Then $du = -8x\,dx$, so $x\,dx = -\frac{1}{8}\,du$ and

$$\int \frac{x}{\sqrt{1 - 4x^2}}\,dx = -\tfrac{1}{8}\int \frac{du}{\sqrt{u}} = -\tfrac{1}{8}\int u^{-1/2}\,du$$

$$= -\tfrac{1}{8}(2\sqrt{u}) + C = -\tfrac{1}{4}\sqrt{1 - 4x^2} + C$$

■

The answer to Example 3 could be checked by differentiation, but instead let's check it visually. In Figure 1 we have used a computer to graph both the integrand $f(x) = x/\sqrt{1 - 4x^2}$ and its indefinite integral $g(x) = -\frac{1}{4}\sqrt{1 - 4x^2}$ (we take the case $C = 0$). Notice that $g(x)$ decreases when $f(x)$ is negative, increases when $f(x)$ is positive, and has its minimum value when $f(x) = 0$. So it seems reasonable, from the graphical evidence, that g is an antiderivative of f.

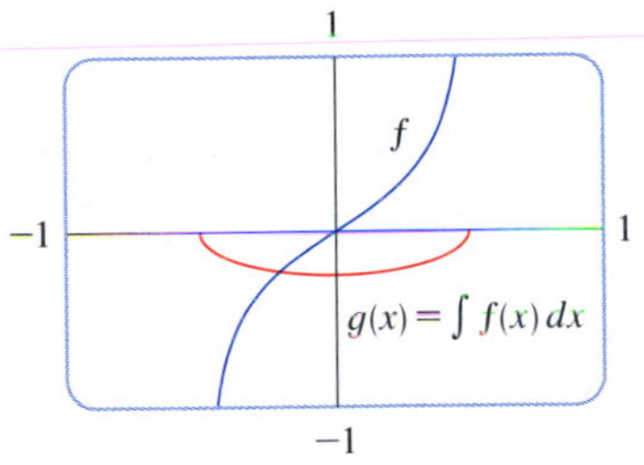

FIGURE 1

$f(x) = \dfrac{x}{\sqrt{1 - 4x^2}}$

$g(x) = \displaystyle\int f(x)\,dx = -\tfrac{1}{4}\sqrt{1 - 4x^2}$

EXAMPLE 4 Calculate $\displaystyle\int \cos 5x\,dx$.

SOLUTION If we let $u = 5x$, then $du = 5\,dx$, so $dx = \frac{1}{5}\,du$. Therefore

$$\int \cos 5x\,dx = \tfrac{1}{5}\int \cos u\,du = \tfrac{1}{5}\sin u + C = \tfrac{1}{5}\sin 5x + C$$

■

EXAMPLE 5 Find $\displaystyle\int \sqrt{1 + x^2}\,x^5\,dx$.

SOLUTION An appropriate substitution becomes more obvious if we factor x^5 as $x^4 \cdot x$. Let $u = 1 + x^2$. Then $du = 2x\,dx$, so $x\,dx = du/2$. Also $x^2 = u - 1$, so $x^4 = (u - 1)^2$:

$$\int \sqrt{1 + x^2}\,x^5\,dx = \int \sqrt{1 + x^2}\,x^4 \cdot x\,dx$$

$$= \int \sqrt{u}\,(u - 1)^2\,\frac{du}{2} = \tfrac{1}{2}\int \sqrt{u}\,(u^2 - 2u + 1)\,du$$

$$= \tfrac{1}{2}\int (u^{5/2} - 2u^{3/2} + u^{1/2})\,du$$

$$= \tfrac{1}{2}\left(\tfrac{2}{7}u^{7/2} - 2 \cdot \tfrac{2}{5}u^{5/2} + \tfrac{2}{3}u^{3/2}\right) + C$$

$$= \tfrac{1}{7}(1 + x^2)^{7/2} - \tfrac{2}{5}(1 + x^2)^{5/2} + \tfrac{1}{3}(1 + x^2)^{3/2} + C$$

■

EXAMPLE 6 Calculate $\int \tan x\, dx$.

SOLUTION First we write tangent in terms of sine and cosine:

$$\int \tan x\, dx = \int \frac{\sin x}{\cos x}\, dx$$

This suggests that we should substitute $u = \cos x$ since then $du = -\sin x\, dx$ and so $\sin x\, dx = -du$:

$$\int \tan x\, dx = \int \frac{\sin x}{\cos x}\, dx = -\int \frac{du}{u}$$

$$= -\ln|u| + C = -\ln|\cos x| + C$$

Since $-\ln|\cos x| = \ln(1/|\cos x|) = \ln|\sec x|$, the result of Example 6 can also be written as

(5)
$$\int \tan x\, dx = \ln|\sec x| + C$$

When evaluating a *definite* integral by substitution, two methods are possible. One method is to evaluate the indefinite integral first and then use the Fundamental Theorem. For instance, using the result of Example 2, we have

$$\int_0^4 \sqrt{3x + 4}\, dx = \int \sqrt{3x + 4}\, dx\Big]_0^4 = \tfrac{2}{9}(3x + 4)^{3/2}\Big]_0^4$$

$$= \tfrac{2}{9}(16)^{3/2} - \tfrac{2}{9}(4)^{3/2} = \tfrac{2}{9}(64 - 8) = \tfrac{112}{9}$$

Another method, which is usually preferable, is to change the limits of integration when the variable is changed.

(6) THE SUBSTITUTION RULE FOR DEFINITE INTEGRALS If g' is continuous on $[a, b]$ and f is continuous on the range of g, then

$$\int_a^b f(g(x))g'(x)\, dx = \int_{g(a)}^{g(b)} f(u)\, du$$

PROOF Let F be an antiderivative of f. Then, by (3), $F(g(x))$ is an antiderivative of $f(g(x))g'(x)$, so by Part 2 of the Fundamental Theorem of Calculus, we have

$$\int_a^b f(g(x))g'(x)\, dx = F(g(x))\Big]_a^b = F(g(b)) - F(g(a))$$

But, applying the Fundamental Theorem a second time, we also have

$$\int_{g(a)}^{g(b)} f(u)\, du = F(u)\Big]_{g(a)}^{g(b)} = F(g(b)) - F(g(a))$$

□

This rule says that when using a substitution in a definite integral, we must put everything in terms of the new variable u—not only x and dx but also the limits of integration. The new limits of integration are the values of u that correspond to $x = a$ and $x = b$.

The geometric interpretation of Example 7 is shown in Figure 2. The substitution $u = 3x + 4$ stretches the interval $[0, 4]$ by a factor of 3 and translates it to the right by 4 units. The Substitution Rule shows that the two areas are equal.

EXAMPLE 7 Evaluate $\int_0^4 \sqrt{3x + 4}\, dx$ using (6).

SOLUTION Using the substitution from Solution 1 of Example 2, we have $u = 3x + 4$ and $dx = du/3$. To find the new limits of integration we note that

$$\text{when } x = 0,\ u = 4 \qquad \text{and} \qquad \text{when } x = 4,\ u = 16$$

Therefore

$$\int_0^4 \sqrt{3x + 4}\, dx = \tfrac{1}{3}\int_4^{16} \sqrt{u}\, du$$

$$= \tfrac{1}{3} \cdot \tfrac{2}{3} u^{3/2} \Big]_4^{16}$$

$$= \tfrac{2}{9}(16^{3/2} - 4^{3/2}) = \tfrac{112}{9}$$

■

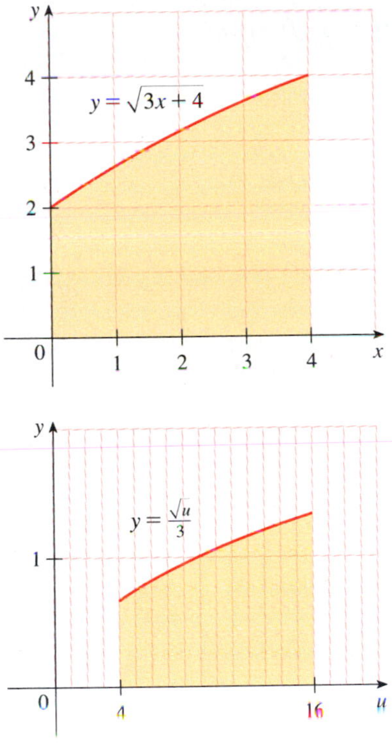

FIGURE 2

Observe that when using (6) we do not return to the variable x after integrating. We simply evaluate the expression in u between the appropriate values of u.

EXAMPLE 8 Evaluate $\displaystyle\int_1^2 \frac{dx}{(3 - 5x)^2}$.

SOLUTION Let $u = 3 - 5x$. Then $du = -5\, dx$, so $dx = -du/5$. When $x = 1$, $u = -2$ and when $x = 2$, $u = -7$.

$$\int_1^2 \frac{dx}{(3 - 5x)^2} = -\frac{1}{5}\int_{-2}^{-7} \frac{du}{u^2}$$

$$= \frac{1}{5}\left[-\frac{1}{u}\right]_{-2}^{-7} = \frac{1}{5u}\Big]_{-2}^{-7}$$

$$= \tfrac{1}{5}\left(-\tfrac{1}{7} + \tfrac{1}{2}\right) = \tfrac{1}{14}$$

■

EXAMPLE 9 Calculate $\displaystyle\int_1^e \frac{\ln x}{x}\, dx$.

SOLUTION We let $u = \ln x$ because its differential $du = dx/x$ occurs in the integral. When $x = 1$, $u = \ln 1 = 0$; when $x = e$, $u = \ln e = 1$. Thus

$$\int_1^e \frac{\ln x}{x}\, dx = \int_0^1 u\, du = \frac{u^2}{2}\Big]_0^1 = \frac{1}{2}$$

■

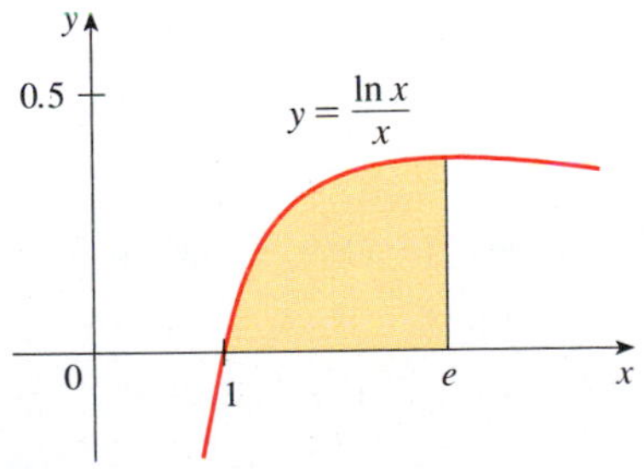

FIGURE 3

Since the function $f(x) = (\ln x)/x$ in Example 9 is positive for $x > 1$, the integral represents the area of the shaded region in Figure 3.

The next theorem uses the Substitution Rule for Definite Integrals (6) to simplify the calculation of integrals of functions that possess symmetry properties.

(7) INTEGRALS OF SYMMETRIC FUNCTIONS Suppose f is continuous on $[-a, a]$.

(a) If f is even $[f(-x) = f(x)]$, then $\int_{-a}^{a} f(x)\, dx = 2\int_0^a f(x)\, dx$.

(b) If f is odd $[f(-x) = -f(x)]$, then $\int_{-a}^{a} f(x)\, dx = 0$.

PROOF Using Property 5 of integrals, we have

$$\textbf{(8)}\quad \int_{-a}^{a} f(x)\,dx = \int_{-a}^{0} f(x)\,dx + \int_{0}^{a} f(x)\,dx = -\int_{0}^{-a} f(x)\,dx + \int_{0}^{a} f(x)\,dx$$

In the first integral on the right side we make the substitution $u = -x$. Then $du = -dx$ and when $x = -a$, $u = a$. Therefore

$$\textbf{(9)}\quad -\int_{0}^{-a} f(x)\,dx = -\int_{0}^{a} f(-u)\,(-du) = \int_{0}^{a} f(-u)\,du$$

(a) If f is even, then $f(-u) = f(u)$ so Equations 8 and 9 give

$$\int_{-a}^{a} f(x)\,dx = \int_{0}^{a} f(-u)\,du + \int_{0}^{a} f(x)\,dx$$

$$= \int_{0}^{a} f(u)\,du + \int_{0}^{a} f(x)\,dx$$

$$= 2\int_{0}^{a} f(x)\,dx$$

(b) If f is odd, then $f(-u) = -f(u)$ and so

$$\int_{-a}^{a} f(x)\,dx = \int_{0}^{a} f(-u)\,du + \int_{0}^{a} f(x)\,dx$$

$$= -\int_{0}^{a} f(u)\,du + \int_{0}^{a} f(x)\,dx = 0$$

□

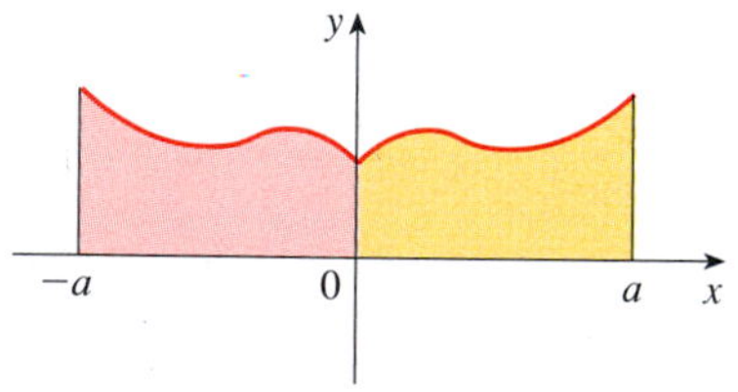

(a) f even, $\int_{-a}^{a} f(x)\,dx = 2\int_{0}^{a} f(x)\,dx$

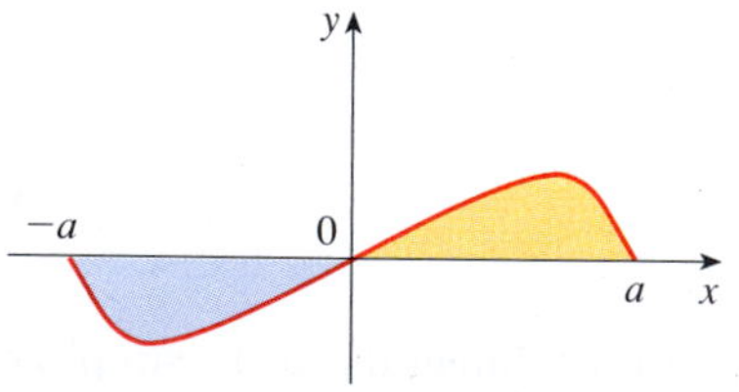

(b) f odd, $\int_{-a}^{a} f(x)\,dx = 0$

FIGURE 4

Theorem 7 is illustrated by Figure 4. For the case where f is positive and even, part (a) says that the area under $y = f(x)$ from $-a$ to a is twice the area from 0 to a because of symmetry. Recall that an integral $\int_a^b f(x)\,dx$ can be expressed as the area above the x-axis and below $y = f(x)$ minus the area below the axis and above the curve. Thus, part (b) says the integral is 0 because the areas cancel.

EXAMPLE 10 Since $f(x) = x^6 + 1$ satisfies $f(-x) = f(x)$, it is even and so

$$\int_{-2}^{2} (x^6 + 1)\,dx = 2\int_{0}^{2} (x^6 + 1)\,dx$$

$$= 2\left[\frac{x^7}{7} + x\right]_0^2 = 2\left(\tfrac{128}{7} + 2\right) = \tfrac{284}{7}$$

■

EXAMPLE 11 Since $f(x) = (\tan x)/(1 + x^2 + x^4)$ satisfies $f(-x) = -f(x)$, it is odd and so

$$\int_{-1}^{1} \frac{\tan x}{1 + x^2 + x^4}\,dx = 0$$

■

EXERCISES 6.5

1–6 ■ Evaluate the integral by making the given substitution.

1. $\int x(x^2-1)^{99}\,dx, \quad u = x^2 - 1$

2. $\int \frac{x^2}{\sqrt{2+x^3}}\,dx, \quad u = 2 + x^3$

3. $\int \sin 4x\,dx, \quad u = 4x$

4. $\int \frac{dx}{(2x+1)^2}, \quad u = 2x + 1$

5. $\int \frac{x+3}{(x^2+6x)^2}\,dx, \quad u = x^2 + 6x$

6. $\int \sec a\theta \tan a\theta\,d\theta, \quad u = a\theta$

7–48 ■ Evaluate the indefinite integral.

7. $\int (2x+1)(x^2+x+1)^3\,dx$

8. $\int x^3(1-x^4)^5\,dx$

9. $\int \sqrt{x-1}\,dx$

10. $\int \sqrt[3]{1-x}\,dx$

11. $\int x^3\sqrt{2+x^4}\,dx$

12. $\int x(x^2+1)^{3/2}\,dx$

13. $\int \frac{2}{(t+1)^6}\,dt$

14. $\int \frac{1}{(1-3t)^4}\,dt$

15. $\int (1-2y)^{1.3}\,dy$

16. $\int \sqrt[5]{3-5y}\,dy$

17. $\int \cos 2\theta\,d\theta$

18. $\int \sec^2 3\theta\,d\theta$

19. $\int \frac{x}{\sqrt[4]{x+2}}\,dx$

20. $\int \frac{x^2}{\sqrt{1-x}}\,dx$

21. $\int t\sin(t^2)\,dt$

22. $\int \frac{(1+\sqrt{x})^9}{\sqrt{x}}\,dx$

23. $\int x^3(1-x^2)^{3/2}\,dx$

24. $\int t^2\cos(1-t^3)\,dt$

25. $\int \sec x\tan x\sqrt{1+\sec x}\,dx$

26. $\int \frac{\cos\sqrt{x}}{\sqrt{x}}\,dx$

27. $\int \cos^4 x\sin x\,dx$

28. $\int \frac{ax+b}{\sqrt{ax^2+2bx+c}}\,dx$

29. $\int \sin(2x+3)\,dx$

30. $\int \cos(7-3x)\,dx$

31. $\int (\sin 3\alpha - \sin 3x)\,dx$

32. $\int \sqrt[3]{x^3+1}\,x^5\,dx$

33. $\int x^a\sqrt{b+cx^{a+1}}\,dx \quad (c \neq 0, a \neq -1)$

34. $\int \cos x\cos(\sin x)\,dx$

35. $\int \frac{dx}{2x-1}$

36. $\int \frac{x}{x^2+1}\,dx$

37. $\int \frac{(\ln x)^2}{x}\,dx$

38. $\int xe^{x^2}\,dx$

39. $\int e^x(1+e^x)^{10}\,dx$

40. $\int \frac{\tan^{-1}x}{1+x^2}\,dx$

41. $\int \frac{dx}{x\ln x}$

42. $\int e^x\sin(e^x)\,dx$

43. $\int \frac{e^x+1}{e^x}\,dx$

44. $\int \frac{e^x}{e^x+1}\,dx$

45. $\int \frac{x+1}{x^2+2x}\,dx$

46. $\int \frac{\sin x}{1+\cos^2 x}\,dx$

47. $\int \frac{1+x}{1+x^2}\,dx$

48. $\int \frac{x}{1+x^4}\,dx$

49–52 ■ Evaluate the indefinite integral. Illustrate and check that your answer is reasonable by graphing both the function and its antiderivative (take $C = 0$).

49. $\int \frac{3x-1}{(3x^2-2x+1)^4}\,dx$

50. $\int \frac{x}{\sqrt{x^2+1}}\,dx$

51. $\int \sin^3 x\cos x\,dx$

52. $\int \tan^2\theta\sec^2\theta\,d\theta$

53–74 ■ Evaluate the definite integral, if it exists.

53. $\int_0^1 (2x-1)^{100}\,dx$

54. $\int_0^{-4} \sqrt{1-2x}\,dx$

55. $\int_0^1 (x^4+x)^5(4x^3+1)\,dx$

56. $\int_2^3 \frac{3x^2-1}{(x^3-x)^2}\,dx$

57. $\int_1^2 x\sqrt{x-1}\,dx$

58. $\int_0^4 \frac{x}{\sqrt{1+2x}}\,dx$

59. $\int_0^1 \cos\pi t\,dt$

60. $\int_0^{\pi/4} \sin 4t\,dt$

61. $\int_1^4 \frac{1}{x^2}\sqrt{1+\frac{1}{x}}\,dx$

62. $\int_0^2 \frac{dx}{(2x-3)^2}$

63. $\int_0^{\pi/3} \frac{\sin\theta}{\cos^2\theta}\,d\theta$

64. $\int_{-\pi/2}^{\pi/2} \frac{x^2\sin x}{1+x^6}\,dx$

65. $\int_0^{13} \frac{dx}{\sqrt[3]{(1+2x)^2}}$

66. $\int_{-\pi/3}^{\pi/3} \sin^5\theta\,d\theta$

67. $\int_0^4 \frac{dx}{(x-2)^3}$

68. $\int_0^a x\sqrt{a^2-x^2}\,dx$

69. $\int_0^a x\sqrt{x^2+a^2}\,dx \quad (a > 0)$

70. $\int_{-a}^a x\sqrt{x^2+a^2}\,dx$

71. $\int_0^3 \frac{dx}{2x+3}$

72. $\int_0^1 t^2 2^{-t^3}\,dt$

73. $\int_e^{e^4} \frac{dx}{x\sqrt{\ln x}}$

74. $\int_0^{1/2} \frac{\sin^{-1}x}{\sqrt{1-x^2}}\,dx$

75–76 ■ Use a graph to give a rough estimate of the area of the region that lies under the given curve. Then find the exact area.

75. $y = \sqrt{2x+1}$, $0 \leq x \leq 1$

76. $y = 2\sin x - \sin 2x$, $0 \leq x \leq \pi$

77. Evaluate $\int_{-2}^{2} (x+3)\sqrt{4-x^2}\,dx$ by writing it as a sum of two integrals and interpreting one of those integrals in terms of an area.

78. Evaluate $\int_0^1 x\sqrt{1-x^4}\,dx$ by making a substitution and interpreting the resulting integral in terms of an area.

79. Breathing is cyclic and a full respiratory cycle from the beginning of inhalation to the end of exhalation takes about 5 s. The maximum rate of air flow into the lungs is about 0.5 L/s. This explains, in part, why the function $f(t) = \frac{1}{2}\sin(2\pi t/5)$ has often been used to model the rate of air flow into the lungs. Use this model to find the volume of inhaled air in the lungs at time t.

80. Alabama Instruments Company has set up a production line to manufacture a new calculator. The rate of production of these calculators after t weeks is

$$\frac{dx}{dt} = 5000\left(1 - \frac{100}{(t+10)^2}\right) \text{ calculators/week}$$

(Notice that production approaches 5000 per week as time goes on, but the initial production is lower because of the workers' unfamiliarity with the new techniques.) Find the number of calculators produced from the beginning of the third week to the end of the fourth week.

81. If f is continuous and $\int_0^4 f(x)\,dx = 10$, find $\int_0^2 f(2x)\,dx$.

82. If f is continuous and $\int_0^9 f(x)\,dx = 4$, find $\int_0^3 xf(x^2)\,dx$.

83. If f is continuous on $\mathbb{R}$, prove that

$$\int_a^b f(-x)\,dx = \int_{-b}^{-a} f(x)\,dx$$

For the case where $f(x) \geq 0$, draw a diagram to interpret this equation geometrically as an equality of areas.

84. If f is continuous on $\mathbb{R}$, prove that

$$\int_a^b f(x+c)\,dx = \int_{a+c}^{b+c} f(x)\,dx$$

For the case where $f(x) \geq 0$, draw a diagram to interpret this equation geometrically as an equality of areas.

85. If a and b are positive numbers, show that

$$\int_0^1 x^a(1-x)^b\,dx = \int_0^1 x^b(1-x)^a\,dx$$

86. Use the substitution $u = \pi - x$ to show that

$$\int_0^\pi xf(\sin x)\,dx = \frac{\pi}{2}\int_0^\pi f(\sin x)\,dx$$

87. Use Exercise 86 to evaluate the integral

$$\int_0^\pi \frac{x\sin x}{1+\cos^2 x}\,dx$$

6.6 THE LOGARITHM DEFINED AS AN INTEGRAL

In this section we use the Fundamental Theorem of Calculus to give an alternative treatment of the functions $\ln x$, e^x, a^x, $\log_a x$, and their derivatives. Instead of starting with a^x and defining $\log_a x$ as its inverse, this time we start by defining $\ln x$ as an integral and then define the exponential function as its inverse. In this section you should bear in mind that we do not use the definitions and results of Sections 4.1, 4.3, and 4.4.

THE NATURAL LOGARITHM

We first define $\ln x$ as an integral.

(1) DEFINITION The **natural logarithmic function** is the function defined by

$$\ln x = \int_1^x \frac{1}{t}\,dt \qquad x > 0$$

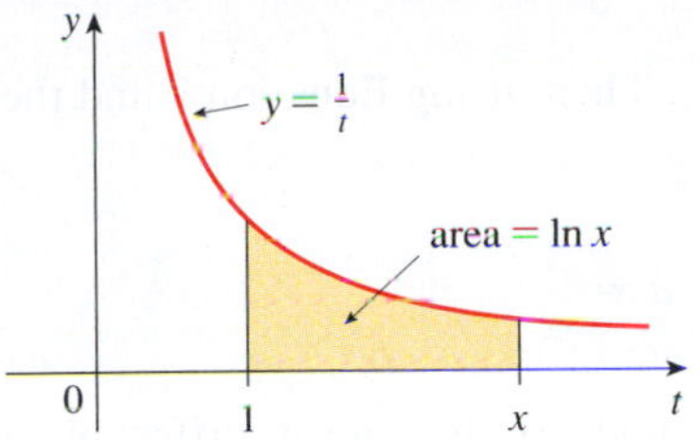

FIGURE 1

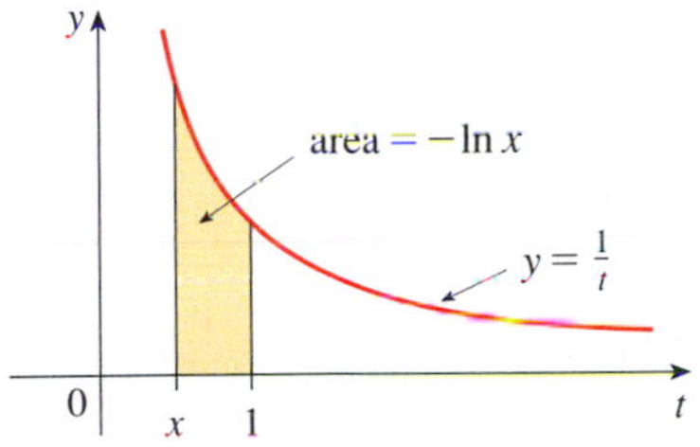

FIGURE 2

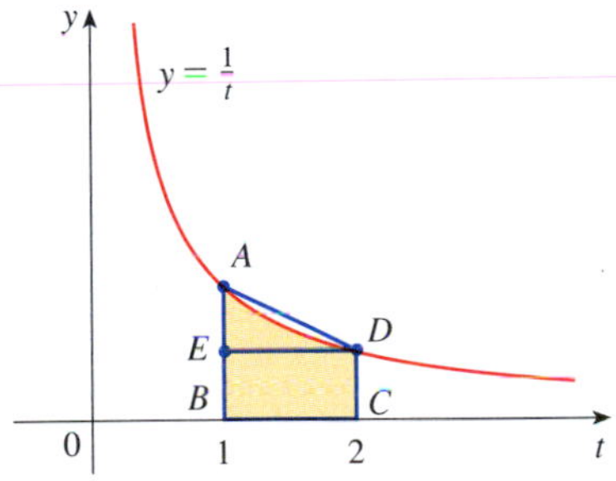

FIGURE 3

The existence of this function depends on Theorem 6.3.4, which guarantees that a continuous function is integrable. If $x > 1$, then $\ln x$ can be interpreted geometrically as the area under the hyperbola $y = 1/t$ from $t = 1$ to $t = x$ (see Figure 1). For $x = 1$, we have

$$\ln 1 = \int_1^1 \frac{1}{t}\,dt = 0$$

For $0 < x < 1$,

$$\ln x = \int_1^x \frac{1}{t}\,dt = -\int_x^1 \frac{1}{t}\,dt < 0$$

and so $\ln x$ is the negative of the area shown in Figure 2.

EXAMPLE 1

(a) By comparing areas, show that $\frac{1}{2} < \ln 2 < \frac{3}{4}$.

(b) Use the Midpoint Rule with $n = 10$ to estimate the value of $\ln 2$.

SOLUTION

(a) We can interpret $\ln 2$ as the area under the curve $y = 1/t$ from 1 to 2. From Figure 3 we see that this area is larger than the area of rectangle $BCDE$ and smaller than the area of trapezoid $ABCD$. Thus, we have

$$\frac{1}{2} \cdot 1 < \ln 2 < 1 \cdot \frac{1}{2}\left(1 + \frac{1}{2}\right)$$

$$\frac{1}{2} < \ln 2 < \frac{3}{4}$$

(b) If we use the Midpoint Rule with $f(t) = 1/t$, $n = 10$, and $\Delta t = 0.1$, we get

$$\ln 2 = \int_1^2 \frac{1}{t}\,dt \approx (0.1)[f(1.05) + f(1.15) + \cdots + f(1.95)]$$

$$= (0.1)\left(\frac{1}{1.05} + \frac{1}{1.15} + \cdots + \frac{1}{1.95}\right) \approx 0.693$$

■

Notice that the integral that defines $\ln x$ is exactly the type of integral discussed in Part 1 of the Fundamental Theorem of Calculus (see Section 6.4). In fact, using that theorem, we have

$$\frac{d}{dx}\int_1^x \frac{1}{t}\,dt = \frac{1}{x}$$

and so

(2)
$$\frac{d}{dx}\ln x = \frac{1}{x}$$

We now use this differentiation rule to prove the following properties of the logarithm function.

(3) LAWS OF LOGARITHMS If x and y are positive numbers and r is a rational number, then

1. $\ln(xy) = \ln x + \ln y$ **2.** $\ln\left(\dfrac{x}{y}\right) = \ln x - \ln y$ **3.** $\ln(x^r) = r\ln x$

PROOF

1. Let $f(x) = \ln(ax)$, where a is a positive constant. Then, using Equation 2 and the Chain Rule, we have

$$f'(x) = \frac{1}{ax}\frac{d}{dx}(ax) = \frac{1}{ax}\cdot a = \frac{1}{x}$$

Therefore, $f(x)$ and $\ln x$ have the same derivative and so they must differ by a constant:

$$\ln(ax) = \ln x + C$$

Putting $x = 1$ in this equation, we get $\ln a = \ln 1 + C = 0 + C = C$. Thus

$$\ln(ax) = \ln x + \ln a$$

If we now replace the constant a by any number y, we have

$$\ln(xy) = \ln x + \ln y$$

2. Using Law 1 with $x = 1/y$, we have

$$\ln\frac{1}{y} + \ln y = \ln\left(\frac{1}{y}\cdot y\right) = \ln 1 = 0$$

and so

$$\ln\frac{1}{y} = -\ln y$$

Using Law 1 again, we have

$$\ln\left(\frac{x}{y}\right) = \ln\left(x\cdot\frac{1}{y}\right) = \ln x + \ln\frac{1}{y} = \ln x - \ln y$$

The proof of Law 3 is left as an exercise. □

In order to graph $y = \ln x$, we first determine its limits:

(4) (a) $\displaystyle\lim_{x\to\infty}\ln x = \infty$ (b) $\displaystyle\lim_{x\to 0^+}\ln x = -\infty$

PROOF

(a) Using Law 3 with $x = 2$ and $r = n$ (where n is any positive integer), we have $\ln(2^n) = n\ln 2$. Now $\ln 2 > 0$, so this shows that $\ln(2^n) \to \infty$ as $n \to \infty$. But $\ln x$ is an increasing function since its derivative $1/x > 0$. Therefore, $\ln x \to \infty$ as $x \to \infty$.

(b) If we let $t = 1/x$, then $t \to \infty$ as $x \to 0^+$. Thus, using (a), we have

$$\lim_{x\to 0^+}\ln x = \lim_{t\to\infty}\ln\left(\frac{1}{t}\right) = \lim_{t\to\infty}(-\ln t) = -\infty$$ □

If $y = \ln x, x > 0$, then

$$\frac{dy}{dx} = \frac{1}{x} > 0 \quad\text{and}\quad \frac{d^2y}{dx^2} = -\frac{1}{x^2} < 0$$

which shows that $\ln x$ is increasing and concave downward on $(0, \infty)$. Putting this information together with (4), we draw the graph of $y = \ln x$ in Figure 4.

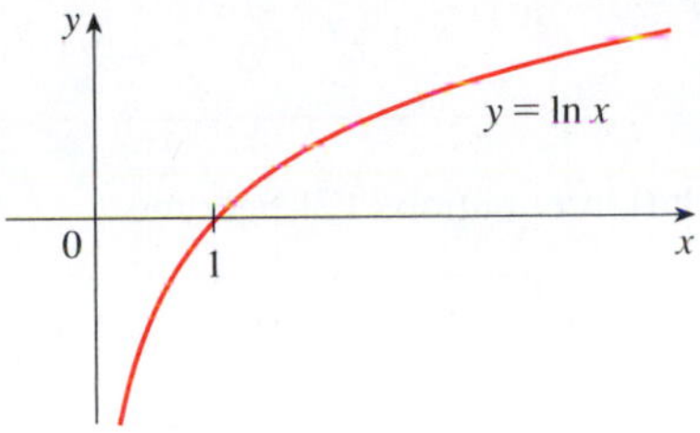

FIGURE 4

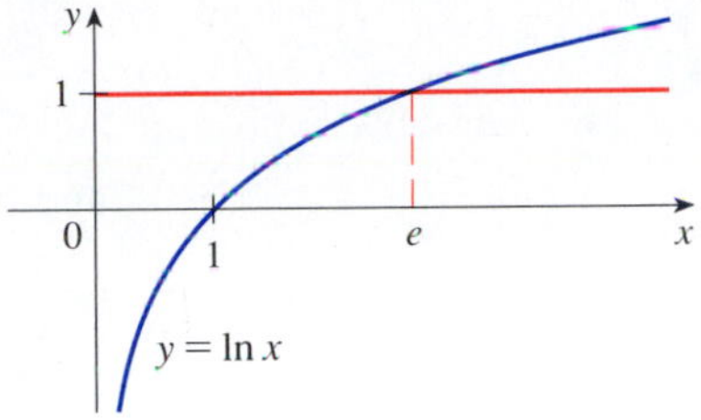

FIGURE 5

Since $\ln 1 = 0$ and $\ln x$ is an increasing continuous function that takes on arbitrarily large values, the Intermediate Value Theorem shows that there is a number where $\ln x$ takes on the value 1 (see Figure 5). This important number is denoted by e.

(5) DEFINITION e is the number such that $\ln e = 1$.

We will show (in Theorem 18) that this definition is consistent with our previous definition of e.

THE NATURAL EXPONENTIAL FUNCTION

Since ln is an increasing function, it is one-to-one and therefore has an inverse function, which we denote by exp. Thus, according to the definition of an inverse function,

$f^{-1}(x) = y \iff f(y) = x$

(6) $$\exp(x) = y \iff \ln y = x$$

and the cancellation equations are

$f^{-1}(f(x)) = x$
$f(f^{-1}(x)) = x$

(7) $$\exp(\ln x) = x \quad \text{and} \quad \ln(\exp x) = x$$

In particular, we have

$$\exp(0) = 1 \quad \text{since } \ln 1 = 0$$
$$\exp(1) = e \quad \text{since } \ln e = 1$$

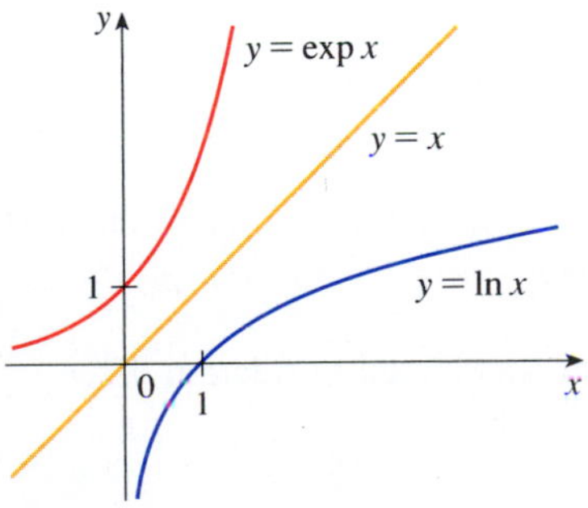

FIGURE 6

We obtain the graph of $y = \exp x$ by reflecting the graph of $y = \ln x$ about the line $y = x$ (see Figure 6). The domain of exp is the range of ln, that is, $(-\infty, \infty)$; the range of exp is the domain of ln, that is, $(0, \infty)$.

If r is any rational number, then the third law of logarithms gives

$$\ln(e^r) = r \ln e = r$$

Therefore, by (6), $$\exp(r) = e^r$$

Thus $\exp(x) = e^x$ whenever x is a rational number. This leads us to define e^x, even for irrational values of x, by the equation

$$e^x = \exp(x)$$

In other words, for the reasons given, we define e^x to be the inverse of the function $\ln x$. In this notation (6) becomes

(8)
$$e^x = y \iff \ln y = x$$

and the cancellation equations (7) become

(9)
$$e^{\ln x} = x \qquad x > 0$$

(10)
$$\ln(e^x) = x \qquad \text{for all } x$$

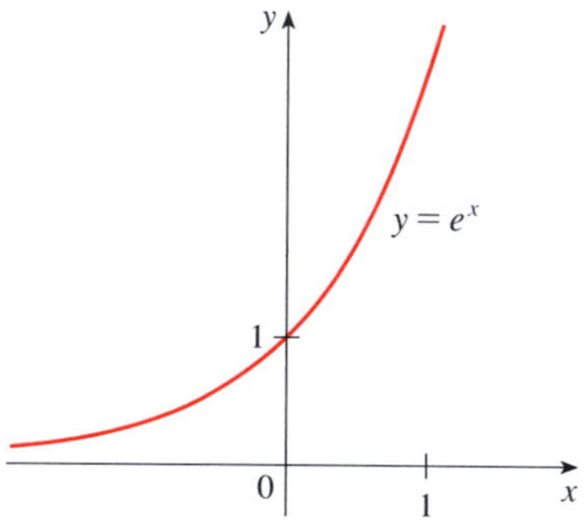

FIGURE 7

The natural exponential function $f(x) = e^x$ is one of the most frequently occurring functions in calculus and its applications, so it is important to be familiar with its graph (Figure 7) and its properties (which follow from the fact that it is the inverse of the natural logarithmic function).

PROPERTIES OF THE EXPONENTIAL FUNCTION The exponential function $f(x) = e^x$ is an increasing continuous function with domain $\mathbb{R}$ and range $(0, \infty)$. Thus, $e^x > 0$ for all x. Also

$$\lim_{x \to -\infty} e^x = 0 \qquad \lim_{x \to \infty} e^x = \infty$$

So the x-axis is a horizontal asymptote of $f(x) = e^x$.

We now verify that f has the other properties expected of an exponential function.

(11) LAWS OF EXPONENTS If x and y are real numbers and r is rational, then

1. $e^{x+y} = e^x e^y$ **2.** $e^{x-y} = \dfrac{e^x}{e^y}$ **3.** $(e^x)^r = e^{rx}$

PROOF OF LAW 1 Using the first law of logarithms and Equation 10, we have

$$\ln(e^x e^y) = \ln(e^x) + \ln(e^y) = x + y = \ln(e^{x+y})$$

Since ln is a one-to-one function, it follows that $e^x e^y = e^{x+y}$.

Laws 2 and 3 are proved similarly (see Exercises 6 and 7). As we will see in the next section, Law 3 actually holds when r is any real number. □

We now prove the differentiation formula for e^x that was stated in Section 4.1.

(12)
$$\frac{d}{dx} e^x = e^x$$

PROOF The function $y = e^x$ is differentiable because it is the inverse function of $y = \ln x$, which we know is differentiable. To find its derivative, we use the inverse function method. Let $y = e^x$. Then $\ln y = x$ and, differentiating this latter equation implicitly with respect to x, we get

$$\frac{1}{y}\frac{dy}{dx} = 1$$

$$\frac{dy}{dx} = y = e^x$$

□

GENERAL EXPONENTIAL FUNCTIONS

If $a > 0$ and r is any rational number, then by (9) and (11),

$$a^r = (e^{\ln a})^r = e^{r \ln a}$$

Therefore, even for irrational numbers x, we *define*

(13)
$$a^x = e^{x \ln a}$$

Thus, for instance,

$$2^{\sqrt{3}} = e^{\sqrt{3}\ln 2} \approx e^{1.20} \approx 3.32$$

The function $f(x) = a^x$ is called the **exponential function with base *a***. Notice that a^x is positive for all x because e^x is positive for all x. The general laws of exponents follow from Definition 13 together with the laws of exponents for e^x.

(14) LAWS OF EXPONENTS If x and y are real numbers and $a, b > 0$, then

1. $a^{x+y} = a^x a^y$ **2.** $a^{x-y} = a^x/a^y$ **3.** $(a^x)^y = a^{xy}$ **4.** $(ab)^x = a^x b^x$

PROOF

1. Using Definition 13 and the laws of exponents for e^x, we have

$$a^{x+y} = e^{(x+y)\ln a} = e^{x\ln a + y\ln a}$$
$$= e^{x\ln a}e^{y\ln a} = a^x a^y$$

3.
$$(a^x)^y = e^{y\ln(a^x)} = e^{y\ln(e^{x\ln a})}$$
$$= e^{yx\ln a} = e^{xy\ln a} = a^{xy}$$

The remaining proofs are left as exercises. □

Definition 13 also allows us to extend one of the laws of logarithms. We know that $\ln(a^r) = r\ln a$ when r is rational. But if we now let r be *any* real number we have, from Definition 13,

$$\ln a^r = \ln(e^{r\ln a}) = r\ln a$$

The differentiation formula for exponential functions is also a consequence of Definition 13:

(15)
$$\frac{d}{dx}a^x = a^x \ln a$$

PROOF

$$\frac{d}{dx} a^x = \frac{d}{dx} (e^{x \ln a}) = e^{x \ln a} \left(\frac{d}{dx} x \ln a \right)$$

$$= a^x \ln a$$ □

If $a > 1$, then $\ln a > 0$, so $D_x a^x = a^x \ln a > 0$, which shows that $y = a^x$ is increasing (see Figure 8). If $0 < a < 1$, then $\ln a < 0$ and so $y = a^x$ is decreasing (see Figure 9).

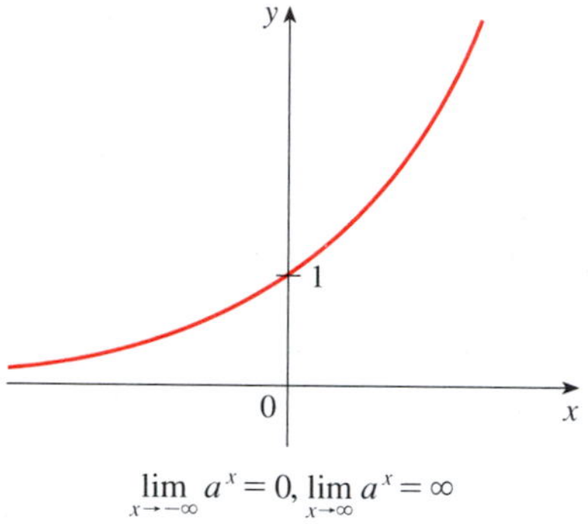

FIGURE 8
$y = a^x, a > 1$

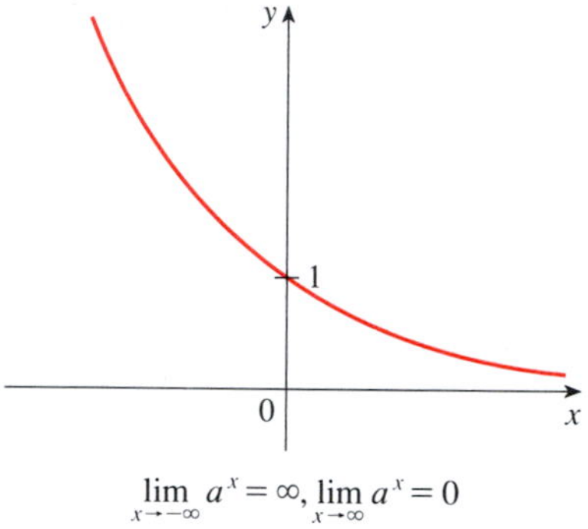

FIGURE 9
$y = a^x, 0 < a < 1$

GENERAL LOGARITHMIC FUNCTIONS

If $a > 0$ and $a \neq 1$, then $f(x) = a^x$ is a one-to-one function. Its inverse function is called the **logarithmic function with base *a*** and is denoted by $\log_a$. Thus

(16)
$$\log_a x = y \iff a^y = x$$

In particular, we see that

$$\log_e x = \ln x$$

The laws of logarithms are similar to those for the natural logarithm and can be deduced from the laws of exponents (see Exercise 10).

To differentiate $y = \log_a x$, we first recall Formula 4.3.7:

$$\log_a x = \frac{\ln x}{\ln a}$$

Since $\ln a$ is a constant, we can differentiate as follows:

$$\frac{d}{dx} \log_a x = \frac{d}{dx} \frac{\ln x}{\ln a} = \frac{1}{\ln a} \frac{d}{dx} \ln x = \frac{1}{x \ln a}$$

(17)
$$\frac{d}{dx} \log_a x = \frac{1}{x \ln a}$$

THE NUMBER e EXPRESSED AS A LIMIT

In this section we defined e as the number such that $\ln e = 1$. The next theorem shows that this is the same as the number e defined in Section 4.1 (see Equation 4.4.6).

(18) THEOREM $\lim_{x \to 0} (1 + x)^{1/x} = e$

PROOF Let $f(x) = \ln x$. Then $f'(x) = 1/x$, so $f'(1) = 1$. But, by the definition of derivative,

$$f'(1) = \lim_{h \to 0} \frac{f(1 + h) - f(1)}{h} = \lim_{x \to 0} \frac{f(1 + x) - f(1)}{x}$$

$$= \lim_{x \to 0} \frac{\ln(1 + x) - \ln 1}{x} = \lim_{x \to 0} \frac{1}{x} \ln(1 + x)$$

$$= \lim_{x \to 0} \ln(1 + x)^{1/x} = \ln\left[\lim_{x \to 0} (1 + x)^{1/x}\right] \qquad \text{(since ln is continuous)}$$

Because $f'(1) = 1$, we have

$$\ln\left[\lim_{x \to 0} (1 + x)^{1/x}\right] = 1$$

Therefore

$$\lim_{x \to 0} (1 + x)^{1/x} = e$$

□

EXERCISES 6.6

1. (a) By comparing areas, show that
$$\tfrac{1}{3} < \ln 1.5 < \tfrac{5}{12}$$
(b) Use the Midpoint Rule with $n = 10$ to estimate $\ln 1.5$.

2. Refer to Example 1.
(a) Find the equation of the tangent line to the curve $y = 1/t$ that is parallel to the secant line AD.
(b) Use part (a) to show that $\ln 2 > 0.66$.

3. By comparing areas, show that
$$\frac{1}{2} + \frac{1}{3} + \cdots + \frac{1}{n} < \ln n < 1 + \frac{1}{2} + \frac{1}{3} + \cdots + \frac{1}{n - 1}$$

4. (a) By comparing areas, show that $\ln 2 < 1 < \ln 3$.
(b) Deduce that $2 < e < 3$.

5. Prove the third law of logarithms. [*Hint:* Start by showing that both sides of the equation have the same derivative.]

6. Prove the second law of exponents for e^x [see (11)].

7. Prove the third law of exponents for e^x [see (11)].

8. Prove the second law of exponents [see (14)].

9. Prove the fourth law of exponents [see (14)].

10. Deduce the following laws of logarithms from (14):
(a) $\log_a(xy) = \log_a x + \log_a y$
(b) $\log_a(x/y) = \log_a x - \log_a y$
(c) $\log_a(x^y) = y \log_a x$

6 REVIEW

KEY TOPICS ■ Define, state, or discuss the following.

1. Sigma notation
2. Partition of $[a, b]$
3. Norm of a partition
4. Area under a curve
5. Riemann sum
6. Definite integral of f from a to b
7. Integrable function
8. Integrand; limits of integration
9. Midpoint Rule
10. Piecewise continuous function
11. Bounded function
12. Properties of the definite integral
13. Fundamental Theorem of Calculus
14. Indefinite integral
15. Substitution Rule
16. Integrals of even or odd functions

EXERCISES

1–16 ■ Determine whether the statement is true or false.

1. $\sum_{i=1}^{n} (a_i + b_i) = \sum_{i=1}^{n} a_i + \sum_{i=1}^{n} b_i$

2. $\sum_{i=1}^{n} a_i b_i = \left(\sum_{i=1}^{n} a_i\right)\left(\sum_{i=1}^{n} b_i\right)$

3. $\dfrac{d}{dx}\sum_{i=1}^{n} f_i(x) = \sum_{i=1}^{n} \dfrac{d}{dx} f_i(x)$

4. $\sum_{i=1}^{12} 3 = 36$

5. If f and g are integrable on $[a, b]$, then
$$\int_a^b [f(x) + g(x)]\,dx = \int_a^b f(x)\,dx + \int_a^b g(x)\,dx$$

6. If f and g are integrable on $[a, b]$, then
$$\int_a^b [f(x)g(x)]\,dx = \left(\int_a^b f(x)\,dx\right)\left(\int_a^b g(x)\,dx\right)$$

7. If f is integrable on $[a, b]$ and $f(x) \geqslant 0$, then
$$\int_a^b \sqrt{f(x)}\,dx = \sqrt{\int_a^b f(x)\,dx}$$

8. If f' is continuous on $[1, 3]$, then $\int_1^3 f'(v)\,dv = f(3) - f(1)$.

9. If f and g are continuous and $f(x) \geqslant g(x)$ for $a \leqslant x \leqslant b$, then
$$\int_a^b f(x)\,dx \geqslant \int_a^b g(x)\,dx$$

10. If f and g are differentiable and $f(x) \geqslant g(x)$ for $a \leqslant x \leqslant b$, then $f'(x) \geqslant g'(x)$ for $a < x < b$.

11. $\int_{-1}^{1} \left(x^5 - 6x^9 + \dfrac{\sin x}{(1 + x^4)^2}\right) dx = 0$

12. If $\int_0^1 f(x)\,dx$ exists, then f is a bounded function on $[0, 1]$.

13. $\int_{-2}^{1} \dfrac{1}{x^4}\,dx = -\frac{3}{8}$

14. $\int_0^2 (x - x^3)\,dx$ represents the area under the curve $y = x - x^3$ from 0 to 2.

15. All continuous functions have derivatives.

16. All continuous functions have antiderivatives.

17. The figure shows the graphs of f, f', and $\int_0^x f(t)\,dt$. Identify each graph, and explain your choices.

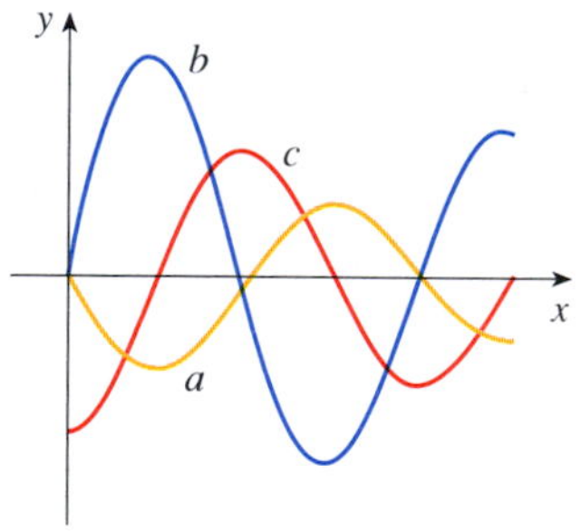

18. Evaluate:
(a) $\displaystyle\int_0^{\pi/2} \frac{d}{dx}\left(\sin\frac{x}{2}\cos\frac{x}{3}\right) dx$
(b) $\displaystyle\frac{d}{dx}\int_0^{\pi/2} \sin\frac{x}{2}\cos\frac{x}{3}\,dx$
(c) $\displaystyle\frac{d}{dx}\int_x^{\pi/2} \sin\frac{t}{2}\cos\frac{t}{3}\,dt$

19. Find the Riemann sum for the function $f(x) = 2 + (x - 2)^2$ on the interval $[0, 2]$ using a regular partition with $n = 4$ and choosing x_i^* to be the left endpoint of the ith subinterval. Sketch the graph of f and the approximating rectangles.

20. Do Exercise 19 if x_i^* is the midpoint of the ith subinterval.

21–22 ■ Evaluate each integral without using the Fundamental Theorem of Calculus.

21. $\int_2^4 (3 - 4x)\,dx$

22. $\int_0^5 (x^3 - 2x^2)\,dx$

23–58 ■ Evaluate the integral, if it exists.

23. $\int_0^5 (x^3 - 2x^2)\,dx$

24. $\int_0^b (x^3 + 4x - 1)\,dx$

25. $\int_0^1 (1 - x^9)\,dx$

26. $\int_0^1 (1 - x)^9\,dx$

27. $\int_1^8 \sqrt[3]{x}(x - 1)\,dx$

28. $\int_1^4 \dfrac{x^2 - x + 1}{\sqrt{x}}\,dx$

29. $\int_0^2 x^2(1 + 2x^3)^3\,dx$

30. $\int_0^4 x\sqrt{16 - 3x}\,dx$

31. $\int_3^{11} \dfrac{dx}{\sqrt{2x + 3}}$

32. $\int_0^2 \dfrac{x}{(x^2 - 1)^2}\,dx$

33. $\int_{-2}^{-1} \dfrac{dx}{(2x + 3)^4}$

34. $\int_{-1}^1 \dfrac{x + x^3 + x^5}{1 + x^2 + x^4}\,dx$

35. $\int \dfrac{x^4}{(2 + x^5)^6}\,dx$

36. $\int (1 - x)\sqrt{2x - x^2}\,dx$

37. $\int \sin \pi x\,dx$

38. $\int \csc^2 3t\,dt$

39. $\int \dfrac{\cos(1/t)}{t^2}\,dt$

40. $\int \sin x \sec^2(\cos x)\,dx$

41. $\int_0^{2\sqrt{3}} \dfrac{1}{x^2 + 4}\,dx$

42. $\int_1^2 \dfrac{1}{2 - 3x}\,dx$

43. $\int_2^4 \dfrac{1 + x - x^2}{x^2}\,dx$

44. $\int_0^1 e^{\pi t}\,dt$

45. $\int \dfrac{e^x}{e^x + 1}\,dx$

46. $\int_{\ln 3}^{\ln 6} 8e^x\,dx$

47. $\int \dfrac{e^{\sqrt{x}}}{\sqrt{x}}\,dx$

48. $\int \dfrac{\cos(\ln x)}{x}\,dx$

49. $\int \tan x \ln(\cos x)\,dx$

50. $\int \dfrac{x}{\sqrt{1 - x^4}}\,dx$

51. $\int \dfrac{x^3}{1 + x^4}\,dx$

52. $\int \dfrac{e^x}{(e^x + 1)\ln(e^x + 1)}\,dx$

53. $\int \dfrac{\sec\theta \tan\theta}{1 + \sec\theta}\,d\theta$

54. $\int \dfrac{1}{\sqrt{x}\,(1 + x)}\,dx$

55. $\int \cosh 3t\,dt$

56. $\int \dfrac{x^2}{2^{x^3}}\,dx$

57. $\int_0^{2\pi} |\sin x|\,dx$

58. $\int_0^8 |x^2 - 6x + 8|\,dx$

59–60 ■ Evaluate the indefinite integral. Illustrate and check that your answer is reasonable by graphing both the function and its antiderivative (take $C = 0$).

59. $\int \dfrac{\cos x}{\sqrt{1 + \sin x}}\,dx$

60. $\int \dfrac{x^3}{\sqrt{x^2 + 1}}\,dx$

61–62 ■ Use a graph to give a rough estimate of the area of the region that lies under the given curve. Then find the exact area.

61. $y = x\sqrt{x}$, $0 \leqslant x \leqslant 4$

62. $y = x^2\sqrt{1 + x^3}$, $0 \leqslant x \leqslant 2$

63–68 ■ Find the derivative of the function.

63. $F(x) = \int_1^x \sqrt{1 + t^4}\,dt$

64. $F(x) = \int_\pi^x \tan(s^2)\,ds$

65. $g(x) = \int_0^{x^3} \dfrac{t}{\sqrt{1 + t^3}}\,dt$

66. $g(x) = \int_1^{\cos x} \sqrt[3]{1 - t^2}\,dt$

67. $y = \int_{\sqrt{x}}^x \dfrac{\cos\theta}{\theta}\,d\theta$

68. $y = \int_{2x}^{3x+1} \sin(t^4)\,dt$

69–70 ■ Use Property 8 of integrals to estimate the value of the integral.

69. $\int_1^3 \sqrt{x^2 + 3}\,dx$

70. $\int_3^5 \dfrac{1}{x + 1}\,dx$

71–74 ■ Use the properties of integrals to verify each inequality.

71. $\int_0^1 x^2 \cos x\,dx \leqslant \frac{1}{3}$

72. $\int_{\pi/4}^{\pi/2} \dfrac{\sin x}{x}\,dx \leqslant \dfrac{\sqrt{2}}{2}$

73. $\int_0^1 e^x \cos x\,dx \leqslant e - 1$

74. $\int_0^1 x \sin^{-1} x\,dx \leqslant \pi/4$

75. Use the Midpoint Rule with $n = 5$ to approximate $\int_0^1 \sqrt{1 + x^3}\,dx$.

76. A particle moves along a line with velocity function $v(t) = t^2 - t$. Find (a) the displacement and (b) the distance traveled by the particle during the time interval $[0, 5]$.

77. Evaluate $\int_{-4}^4 \sqrt{16 - x^2}\,dx$ by interpreting the integral as an area.

78. Let

$$f(x) = \begin{cases} -x - 1 & \text{if } -3 \leqslant x \leqslant 0 \\ -\sqrt{1 - x^2} & \text{if } 0 \leqslant x \leqslant 1 \end{cases}$$

Evaluate $\int_{-3}^1 f(x)\,dx$ by interpreting the integral as a difference of areas.

79. Find an antiderivative F of $f(x) = x^2\sin(x^2)$ such that $F(1) = 0$.

80. The Fresnel function $S(x) = \int_0^x \sin(\pi t^2/2)\,dt$ was introduced in Section 6.4. Fresnel also used the function $C(x) = \int_0^x \cos(\pi t^2/2)\,dt$ in his theory of the diffraction of light waves.

(a) On what intervals is C increasing?

(b) On what intervals is C concave upward?

CAS (c) Use a graph to solve the following equation correct to one decimal place:

$$\int_0^x \cos\frac{\pi t^2}{2}\,dt = 0.7$$

CAS (d) Plot the graphs of C and S on the same screen. How are these graphs related?

81. Estimate the value of the number c such that the area under the curve $y = \sinh cx$ between $x = 0$ and $x = 1$ is equal to 1.

82. What is the area of the largest rectangle in the first quadrant with two sides on the axes and one vertex on the curve $y = e^{-x}$?

83. What is the area of the largest triangle in the first quadrant with two sides on the axes and the third side tangent to the curve $y = e^{-x}$?

84. Suppose that the temperature in a long thin rod placed along the x-axis is initially $C/(2a)$ if $|x| \leqslant a$ and 0 if $|x| > a$. It can be shown that if the heat diffusivity of the rod is k, then the temperature of the rod at the point x at time t is

$$T(x,t) = \frac{C}{a\sqrt{4\pi kt}} \int_0^a e^{-(x-u)^2/(4kt)}\,du$$

To find the temperature distribution that results from an initial hot spot concentrated at the origin, we need to compute

$$\lim_{a\to 0} T(x,t)$$

Use l'Hospital's Rule to find this limit.

85. Evaluate the integral $\int_0^1 \sqrt{x}\,dx$ without using the Fundamental Theorem of Calculus. [*Hint:* Use partitions with $x_i^* = x_i = i^2/n^2$.]

86. Find a function f and a value of the constant a such that

$$2\int_a^x f(t)\,dt = 2\sin x - 1$$

87. If f' is continuous on $[a, b]$, show that

$$2\int_a^b f(x)f'(x)\,dx = [f(b)]^2 - [f(a)]^2$$

88. Find $\displaystyle\lim_{h\to 0} \frac{1}{h}\int_2^{2+h} \sqrt{1+t^3}\,dt$.

89. If f is a continuous function such that

$$\int_0^x f(t)\,dt = xe^{2x} + \int_0^x e^{-t} f(t)\,dt$$

for all x, find an explicit formula for $f(x)$.

90. The figure shows two regions in the first quadrant: $A(t)$ is the area under the curve $y = \sin(x^2)$ from 0 to t, and $B(t)$ is the area of the triangle with vertices O, P, and $(t, 0)$. Find $\lim_{t\to 0^+} A(t)/B(t)$.

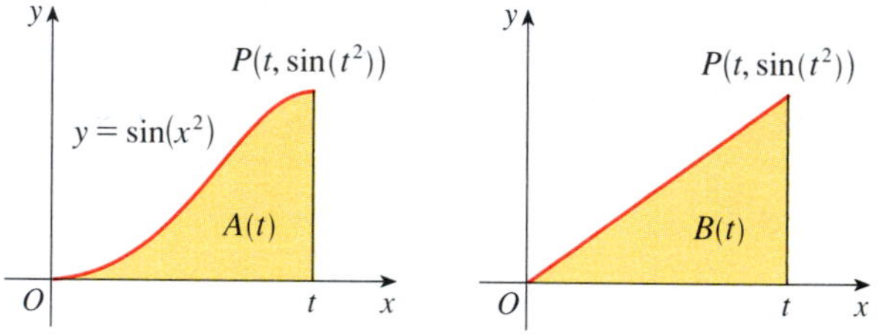

91. If f is continuous on $[0, 1]$, prove that

$$\int_0^1 f(x)\,dx = \int_0^1 f(1-x)\,dx$$

92. Evaluate

$$\lim_{n\to\infty} \frac{1}{n}\left[\left(\frac{1}{n}\right)^9 + \left(\frac{2}{n}\right)^9 + \left(\frac{3}{n}\right)^9 + \cdots + \left(\frac{n}{n}\right)^9\right]$$

APPLICATIONS PLUS

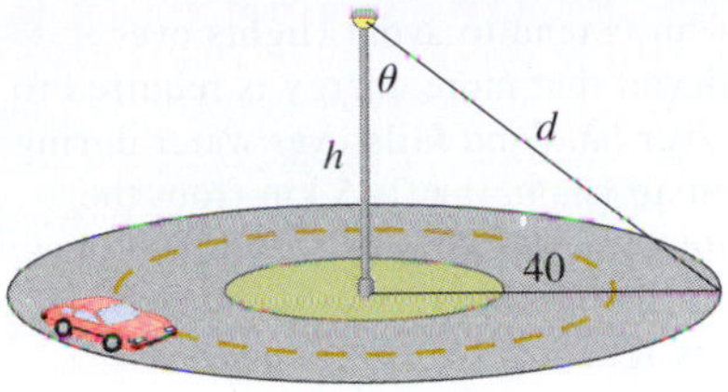

FIGURE FOR PROBLEM 1

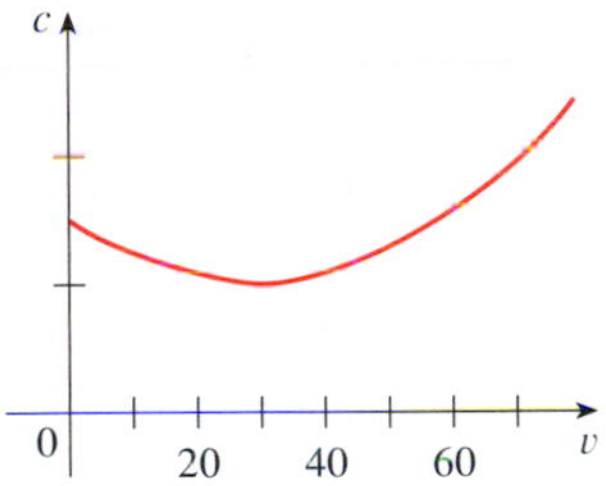

FIGURE FOR PROBLEM 2

1. A light is to be placed atop a pole of height h feet to illuminate a busy traffic circle, which has a radius of 40 ft. The intensity of illumination I at any point P on the circle is directly proportional to the cosine of the angle θ (see the figure) and inversely proportional to the square of the distance d from the source.
   (a) How tall should the light pole be to maximize I?
   (b) Suppose that the light pole is h feet tall and that a woman is walking away from the base of the pole at the rate of 4 ft/s. At what rate is the intensity of the light at the point on her back 4 ft above the ground decreasing when she reaches the traffic circle?

2. The graph shows the fuel consumption c of a car (measured in gallons per hour) as a function of the speed v of the car. At very low speeds the engine runs inefficiently, so initially c decreases as the speed increases. But at high speeds the fuel consumption increases. You can see that $c(v)$ is minimized for this car when $v \approx 30$ mi/h. However, for fuel efficiency, what must be minimized is not the consumption in gallons per hour but rather the fuel consumption in gallons *per mile*. Let's call this consumption G. Using the graph, estimate the speed at which G has its minimum value.

3. A high-speed "bullet" train accelerates and decelerates at the rate of 4 ft/s^2. Its maximum cruising speed is 90 mi/h.
   (a) What is the maximum distance the train can travel if it accelerates from rest until it reaches its cruising speed and then runs at that speed for 15 minutes?
   (b) Suppose that the train starts from rest and must come to a complete stop in 15 minutes. What is the maximum distance it can travel under these conditions?

4. A high-tech company purchases a new computing system whose initial value is V. The system will depreciate at the rate $f = f(t)$ and will accumulate maintenance costs at the rate $g = g(t)$, where t is the time measured in months. The company wants to determine the optimum time to replace the system.
   (a) Let

$$C(t) = \frac{1}{t}\int_0^t [f(s) + g(s)]\,ds$$

   Show that the critical numbers of C occur at the numbers t where $C(t) = f(t) + g(t)$.
   (b) Suppose that

$$f(t) = \begin{cases} \dfrac{V}{15} - \dfrac{V}{450}t & 0 < t \leq 30 \\ 0 & t > 30 \end{cases} \quad \text{and} \quad g(t) = \frac{Vt^2}{12{,}900} \quad t > 0$$

   Determine the length of time T for the total depreciation $D(t) = \int_0^t f(s)\,ds$ to equal the initial value V.
   (c) Determine the absolute minimum of C on $(0, T]$.
   (d) Sketch the graphs of C and $f + g$ in the same coordinate system, and verify the result in part (a) in this case.

5. A manufacturing company has a major piece of equipment that depreciates at the (continuous) rate $f = f(t)$, where t is the time measured in months since its last overhaul. Because a fixed cost A is incurred each time the machine is overhauled, the company wants to determine the optimal time T (in months) between overhauls.
   (a) Show that $\int_0^t f(s)\,ds$ represents the loss in value of the machine over the period of time t since the last overhaul.
   (b) Let $C = C(t)$ be given by

$$C(t) = \frac{1}{t}\left[A + \int_0^t f(s)\,ds\right]$$

   What does C represent and why would the company want to minimize C?
   (c) Show that C has a minimum value at the numbers $t = T$ where $C(T) = f(T)$.

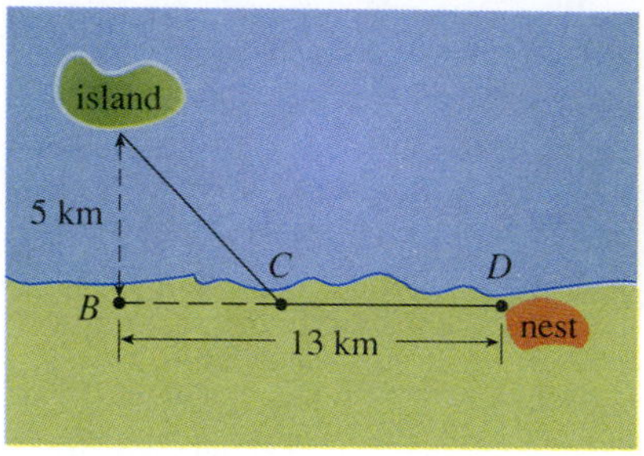

FIGURE FOR PROBLEM 6

6. Ornithologists have determined that some species of birds tend to avoid flights over large bodies of water during daylight hours. It is believed that more energy is required to fly over water than land because air generally rises over land and falls over water during the day. A bird with these tendencies is released from an island that is 5 km from the nearest point B on a straight shoreline, flies to a point C on the shoreline, and then flies along the shoreline to its nesting area D. Assume that the bird instinctively chooses a path that will minimize its energy expenditure. Points B and D are 13 km apart.
   (a) In general, if it takes 1.4 times as much energy to fly over water as land, to what point C should the bird fly in order to minimize the total energy expended in returning to its nesting area?
   (b) Let W denote the energy (in joules) required to fly over water and L the energy required to fly over land. Determine the ratio W/L corresponding to the minimum expenditure of energy.
   (c) What should the value of W/L be in order for the bird to fly directly to its nesting area D? What should the value of W/L be for the bird to fly to B and then along the shoreline to D?

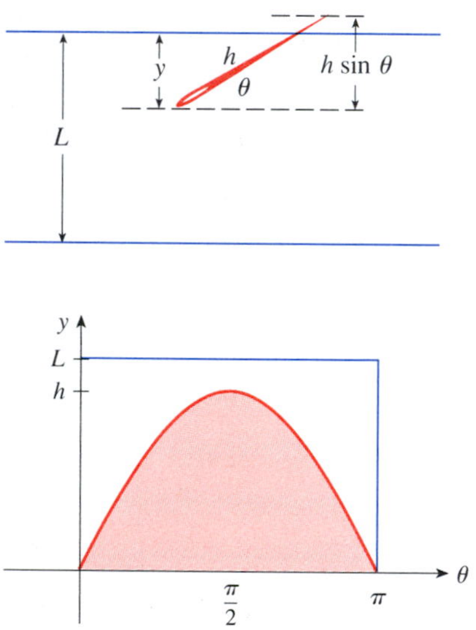

FIGURE FOR PROBLEM 7

7. In a famous 18th-century problem, known as *Buffon's needle problem,* a needle of length h is dropped onto a flat surface (for example, a table) on which parallel lines L units apart, $L \geq h$, have been drawn. The problem is to determine the probability that the needle will come to rest intersecting one of the lines. Assume that the lines run east-west, parallel to the x-axis in a rectangular coordinate system. Let y be the distance from the "southern" end of the needle to the nearest line to the north. (If the needle's southern end lies on a line, let $y = 0$. If the needle happens to lie east-west, let the "western" end be the "southern" end.) Let θ be the angle that the needle makes with a ray extending eastward from the "southern" end. Then $0 \leq y \leq L$ and $0 \leq \theta < \pi$. Note that the needle intersects one of the lines only when $y < h\sin\theta$. Now, the total set of possibilities for the needle can be identified with the rectangular region $0 \leq y \leq L$, $0 \leq \theta < \pi$, and the proportion of times that the needle intersects a line is the ratio

$$\frac{\text{area under } y = h\sin\theta}{\text{area of rectangle}}$$

This ratio is the probability that the needle intersects a line.
   (a) What is the probability that the needle will intersect a line if $h = L$?
   (b) What is the probability that the needle will intersect a line if $h = L/2$?
   (c) What is the probability that the needle will intersect a line if $h = L/5$?

8. A model rocket is fired vertically upward from rest. Its acceleration for the first three seconds is $a(t) = 60t$ at which time the fuel is exhausted and it becomes a freely "falling" body. After 17 seconds, the rocket's parachute opens, and the (downward) velocity slows linearly to -18 ft/s in 5 s. The rocket then "floats" to the ground at that rate.
   (a) Determine the position function s and the velocity function v (for all times t). Sketch the graphs of s and v.
   (b) At what time does the rocket reach its maximum height and what is that height?
   (c) At what time does the rocket land?

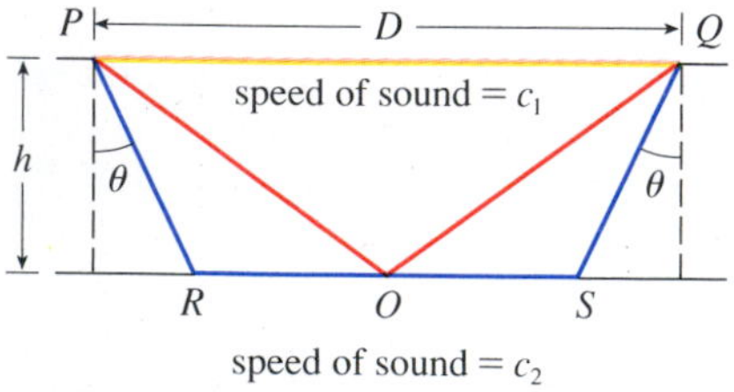

FIGURE FOR PROBLEM 9

9. The speeds of sound c_1 in an upper layer and c_2 in a lower layer of rock and the thickness h of the upper layer can be determined by seismic exploration if the speed of sound in the lower layer is greater than the speed in the upper layer. A dynamite charge is set off at a point P and the transmitted signals are recorded at a point Q, which is a distance D from P. The first signal to arrive at Q travels along the surface and takes T_1

seconds. The next signal travels from P to a point R, from R to S in the lower layer, and then to Q, taking T_2 seconds. The third signal is reflected off the lower layer at the point O and takes T_3 seconds to reach Q.

(a) Express T_1, T_2, and T_3 in terms of D, h, c_1, c_2, and θ.

(b) Show that T_2 is a minimum when $\sin\theta = c_1/c_2$.

(c) Suppose that $D = 1$ km, $T_1 = \frac{1}{4}$ s, $T_2 = \frac{1}{3}$ s, $T_3 = 3/(4\sqrt{5})$ s. Find c_1, c_2, and h.

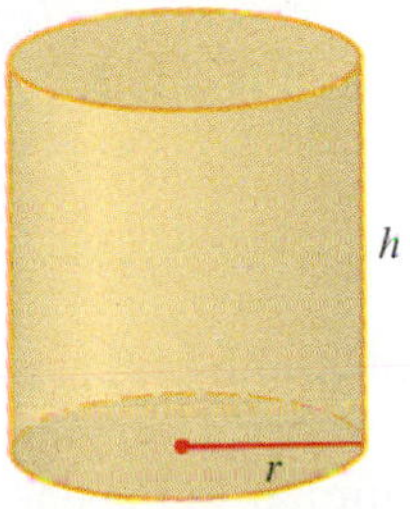

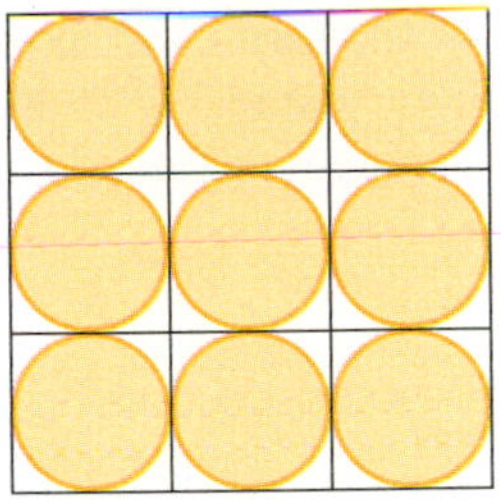

Discs cut from squares

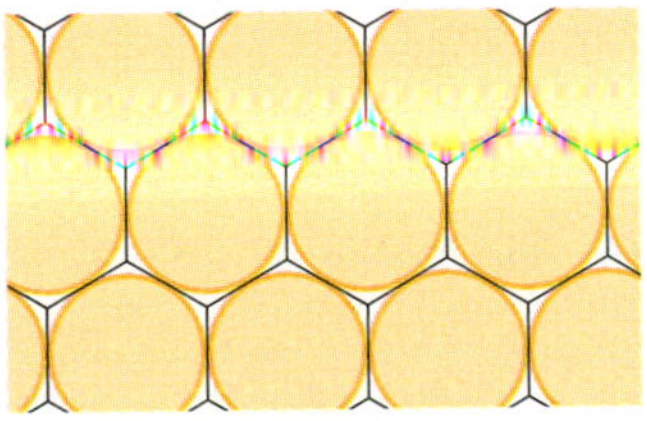

Discs cut from hexagons

FIGURE FOR PROBLEM 10

10. In this problem we investigate the most economical shape for a can. We first interpret this to mean that the volume V of a cylindrical can is given and we need to find the height h and radius r that minimize the cost of the metal to make the can (see the figure). If we disregard any waste metal in the manufacturing process, then the problem is to minimize the surface area of the cylinder. We solved this problem in Example 2 in Section 5.5 and we found that $h = 2r$, that is, the height should be the same as the diameter. But if you go to your cupboard or your supermarket with a ruler, you will discover that the height is usually greater than the diameter and the ratio h/r varies from 2 up to about 3.8. Let's see if we can explain this phenomenon.

(a) The material for the cans is cut out of sheets of metal. The cylindrical sides are formed by bending rectangles; these rectangles are cut from the sheet with little or no waste. But if the top and bottom discs are cut out of squares of side $2r$ (as in the figure), this leaves considerable waste metal, which may be recycled but has little or no value to the can makers. If this is the case, show that the amount of metal used is minimized when

$$\frac{h}{r} = \frac{8}{\pi} \approx 2.55$$

(b) A more efficient packing of the discs is obtained by dividing the metal sheet into hexagons and cutting the circular lids and bases from the hexagons (see the figure). Show that if this strategy is adopted, then

$$\frac{h}{r} = \frac{4\sqrt{3}}{\pi} \approx 2.21$$

(c) The values of h/r that we got in parts (a) and (b) are a little closer to the ones that actually occur on supermarket shelves, but they still don't account for everything. If we look more closely at some real cans, we see that the lid and the base are formed from discs with radius larger than r that are shaped over the ends. If we allow for this we would increase h/r. More significantly, in addition to the cost of the metal we need to incorporate the manufacturing of the can into the cost. Let's assume that most of the expense is incurred in joining the sides and the rims of the cans. If we cut the discs from hexagons as in part (b), then the total cost is proportional to

$$4\sqrt{3}\,r^2 + 2\pi rh + k(4\pi r + h)$$

where k is the reciprocal of the length that can be joined for the cost of one unit area of metal. Show that this expression is minimized when

$$\frac{\sqrt[3]{V}}{k} = \sqrt[3]{\frac{\pi h}{r}} \cdot \frac{2\pi - h/r}{\pi h/r - 4\sqrt{3}}$$

(d) Plot $\sqrt[3]{V}/k$ as a function of $x = h/r$ and use your graph to argue that when a can is large or joining is cheap, we should make h/r approximately 2.21 [as in part (b)]. But when the can is small or joining is costly, h/r should be substantially larger.

This shows that large cans should be almost square but small cans should be tall and thin. The next time you're in a supermarket take a look at the relative shapes of cans and you will see that this is usually the case.

7

APPLICATIONS OF INTEGRATION

The calculus is the greatest aid we have to the appreciation of physical truth in the broadest sense of the word.

W. F. OSGOOD

In this chapter we illustrate some of the applications of the definite integral by using it to compute areas between curves, volumes of solids, and the work done by a varying force. The common theme is the following general method, which is similar to the one we used to find areas under curves. We break up a quantity Q into a large number of small parts. We then approximate each small part by a quantity of the form $f(x_i^*)\,\Delta x_i$ and thus approximate Q by a Riemann sum. Then we take the limit and express Q as an integral. Finally we evaluate the integral using the Fundamental Theorem of Calculus.

7.1 AREAS BETWEEN CURVES

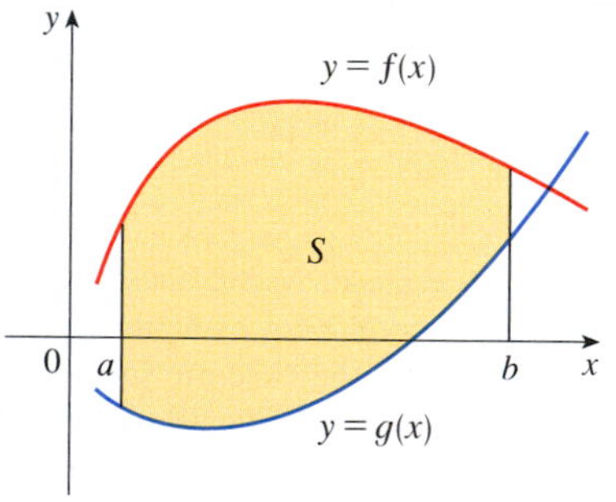

FIGURE 1
$S = \{(x, y) \mid a \leq x \leq b, g(x) \leq y \leq f(x)\}$

So far we have defined and calculated areas of regions that lie under the graphs of functions. In this section we use integrals to find areas of more general regions.

Consider the region S that lies between two curves $y = f(x)$ and $y = g(x)$ and between the vertical lines $x = a$ and $x = b$, where f and g are continuous functions and $f(x) \geq g(x)$ for all x in $[a, b]$ (see Figure 1).

Let P be a partition of $[a, b]$ by points x_i and choose points x_i^* in $[x_{i-1}, x_i]$. Let $\Delta x_i = x_i - x_{i-1}$ and $\|P\| = \max\{\Delta x_i\}$. Figure 2 shows the approximating rectangles with base Δx_i and height $f(x_i^*) - g(x_i^*)$. The Riemann sum

$$\sum_{i=1}^{n} [f(x_i^*) - g(x_i^*)]\,\Delta x_i$$

is therefore an approximation to what we intuitively think of as the area of S.

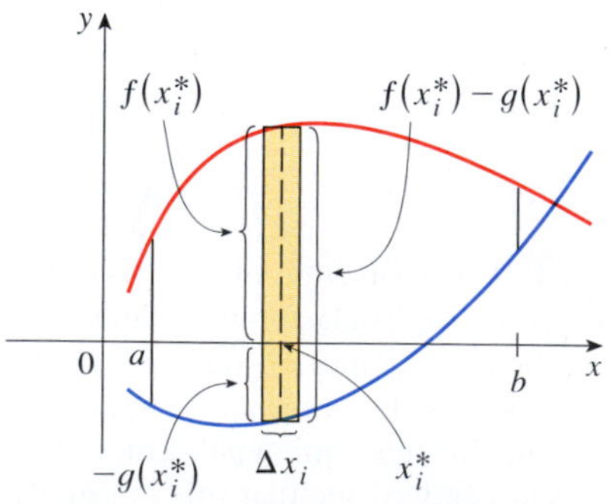

(a) Typical rectangle

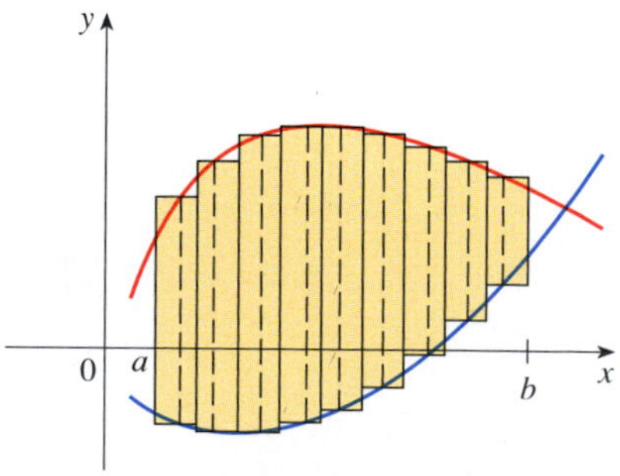

(b) Approximating rectangles

FIGURE 2

This approximation appears to become better and better as $\|P\| \to 0$. Therefore, we define the **area** A of S as the limiting value of the areas of these approximating rectangles.

(1) $$A = \lim_{\|P\|\to 0} \sum_{i=1}^{n} [f(x_i^*) - g(x_i^*)]\Delta x_i$$

We recognize the limit in (1) as a Riemann integral that exists because $f - g$ is continuous. Therefore:

(2) The area of the region bounded by the curves $y = f(x)$, $y = g(x)$, and the lines $x = a$ and $x = b$, where f and g are continuous and $f(x) \geqslant g(x)$ for all x in $[a, b]$, is

$$A = \int_a^b [f(x) - g(x)]\,dx$$

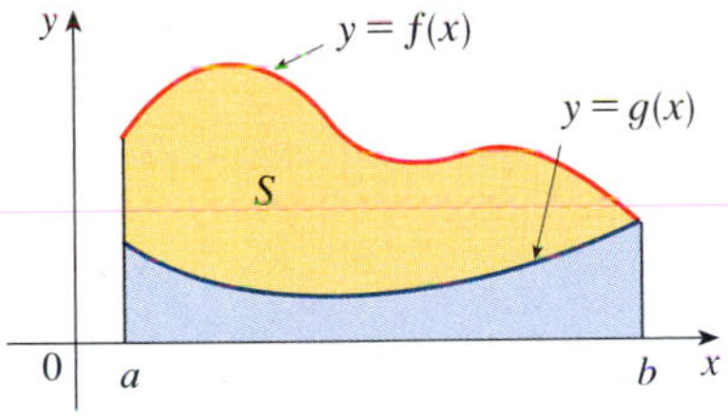

FIGURE 3

$A = \int_a^b f(x)\,dx - \int_a^b g(x)\,dx$

Notice that in the special case where $g(x) = 0$, S is the region under the graph of f and our general definition of area (1) reduces to our previous definition (6.2.2).

In the case where both f and g are positive, you can see from Figure 3 why (2) is true:

$$A = [\text{area under } y = f(x)] - [\text{area under } y = g(x)]$$
$$= \int_a^b f(x)\,dx - \int_a^b g(x)\,dx = \int_a^b [f(x) - g(x)]\,dx$$

EXAMPLE 1 Find the area of the region bounded by the parabolas $y = x^2$ and $y = 2x - x^2$.

SOLUTION We first find the points of intersection of the parabolas by solving their equations simultaneously. This gives $x^2 = 2x - x^2$, or $2x^2 = 2x$. Thus $x(x - 1) = 0$, so $x = 0$ or 1. The points of intersection are $(0, 0)$ and $(1, 1)$ and the region is shown in Figure 4.

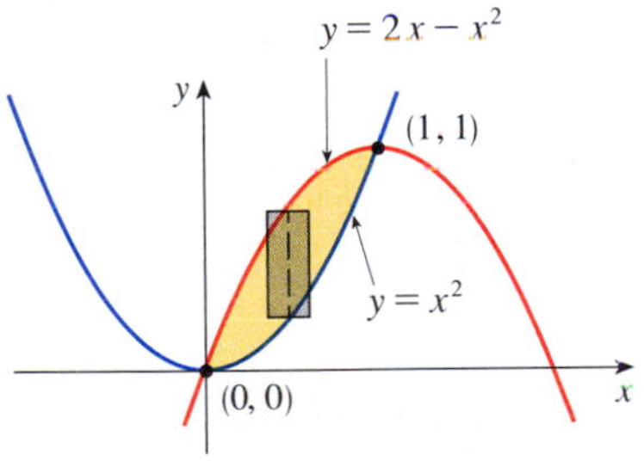

FIGURE 4

$\int_a^b (\text{upper } y - \text{lower } y)\,dx$

When using the formula for area (2) it is important to ensure that $f(x) \geqslant g(x)$ when $a \leqslant x \leqslant b$. In this case you can see from the diagram that

$$2x - x^2 \geqslant x^2 \qquad \text{for } 0 \leqslant x \leqslant 1$$

and so we choose $f(x) = 2x - x^2$, $g(x) = x^2$, $a = 0$, and $b = 1$. Then (2) gives the required area as

$$A = \int_0^1 [(2x - x^2) - x^2]\,dx$$
$$= \int_0^1 2(x - x^2)\,dx = 2\left[\frac{x^2}{2} - \frac{x^3}{3}\right]_0^1$$
$$= 2\left[\tfrac{1}{2} - \tfrac{1}{3}\right] = \tfrac{1}{3}$$ ■

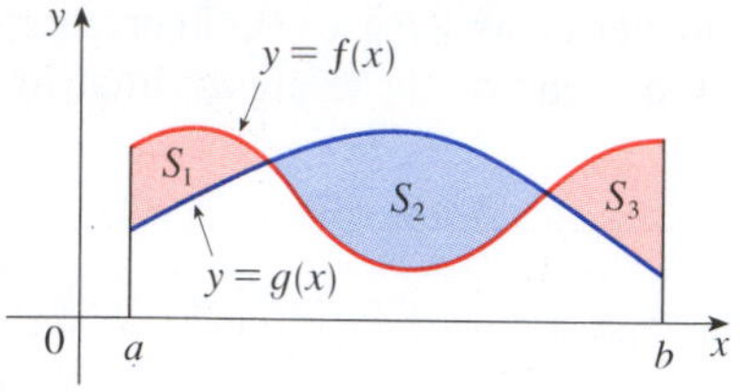

FIGURE 5

If we are asked to find the area between the curves $y = f(x)$ and $y = g(x)$ where $f(x) \geqslant g(x)$ for some values of x but $g(x) \geqslant f(x)$ for other values of x, then we split the given region S into several regions $S_1, S_2, \ldots$ with areas $A_1, A_2, \ldots$ as shown in Figure 5. We then define the area of the region S to be the sum of the areas of the smaller regions $S_1, S_2, \ldots$: $A = A_1 + A_2 + \cdots$. Since

$$|f(x) - g(x)| = \begin{cases} f(x) - g(x) & \text{when } f(x) \geqslant g(x) \\ g(x) - f(x) & \text{when } g(x) \geqslant f(x) \end{cases}$$

we have the following expression for A:

(3) The area between the curves $y = f(x)$ and $y = g(x)$ and between $x = a$ and $x = b$ is

$$A = \int_a^b |f(x) - g(x)|\,dx$$

When evaluating the integral in (3), however, we must still split it into integrals corresponding to $A_1, A_2, \ldots$.

EXAMPLE 2 Find the area of the region bounded by the curves $y = \sin x$, $y = \cos x$, $x = 0$, and $x = \pi/2$.

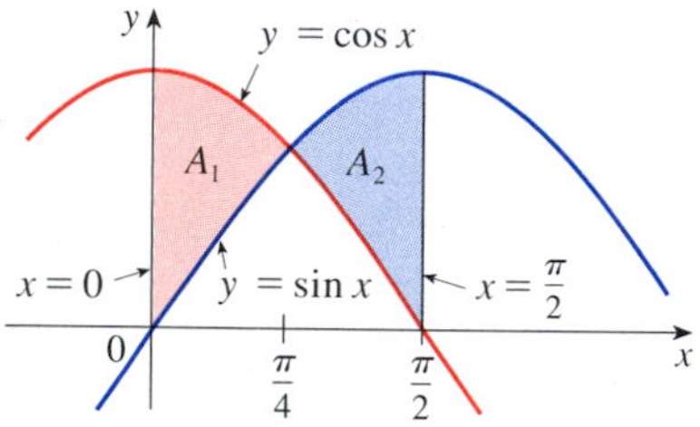

FIGURE 6

SOLUTION The points of intersection occur when $\sin x = \cos x$, that is, when $x = \pi/4$ (since $0 \leqslant x \leqslant \pi/2$). The region is sketched in Figure 6. Observe that $\cos x \geqslant \sin x$ when $0 \leqslant x \leqslant \pi/4$ but $\sin x \geqslant \cos x$ when $\pi/4 \leqslant x \leqslant \pi/2$. Therefore, the required area is

$$\begin{aligned} A &= \int_0^{\pi/2} |\cos x - \sin x|\,dx \\ &= A_1 + A_2 \\ &= \int_0^{\pi/4} (\cos x - \sin x)\,dx + \int_{\pi/4}^{\pi/2} (\sin x - \cos x)\,dx \\ &= [\sin x + \cos x]_0^{\pi/4} + [-\cos x - \sin x]_{\pi/4}^{\pi/2} \\ &= \left(\frac{1}{\sqrt{2}} + \frac{1}{\sqrt{2}} - 0 - 1\right) + \left(-0 - 1 + \frac{1}{\sqrt{2}} + \frac{1}{\sqrt{2}}\right) \\ &= 2\sqrt{2} - 2 \end{aligned}$$

In this particular example we could have saved some work by noticing that the region is symmetric about $x = \pi/4$ and so

$$A = 2A_1 = 2\int_0^{\pi/4} (\cos x - \sin x)\,dx$$

■

Sometimes it is difficult, or even impossible, to find the points of intersection of two curves exactly. As in the following example, we can use a graphing calculator or computer to find approximate values for the intersection points and then proceed as before.

EXAMPLE 3 Find the approximate area of the region bounded by the curves $y = x/\sqrt{x^2 + 1}$ and $y = x^4 - x$.

SOLUTION If we were to try to find the exact intersection points, we would have to solve the equation

$$\frac{x}{\sqrt{x^2+1}} = x^4 - x$$

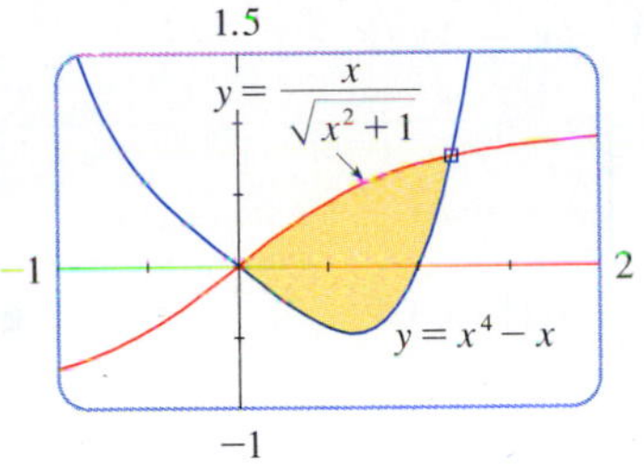

FIGURE 7

This looks like a very difficult equation to solve exactly (in fact, it's impossible), so instead we use a graphing device to draw the graphs of the two curves in Figure 7. One intersection point is the origin. We zoom in toward the other point of intersection and find that $x \approx 1.18$. (If greater accuracy is required, we could use Newton's method or a root-finder, if available on our graphing device.) Thus an approximation to the area between the curves is

$$A \approx \int_0^{1.18} \left[\frac{x}{\sqrt{x^2+1}} - (x^4 - x) \right] dx$$

To integrate the first term we use the substitution $u = x^2 + 1$. Then $du = 2x\,dx$, and when $x = 1.18$, $u \approx 2.39$. So

$$\begin{aligned} A &\approx \tfrac{1}{2}\int_1^{2.39} \frac{du}{\sqrt{u}} - \int_0^{1.18} (x^4 - x)\,dx \\ &= \sqrt{u}\,\Big]_1^{2.39} - \left[\frac{x^5}{5} - \frac{x^2}{2} \right]_0^{1.18} \\ &= \sqrt{2.39} - 1 - \frac{(1.18)^5}{5} + \frac{(1.18)^2}{2} \approx 0.785 \end{aligned}$$

■

EXAMPLE 4 Find the area of the region bounded by the line $y = x - 1$ and the parabola $y^2 = 2x + 6$.

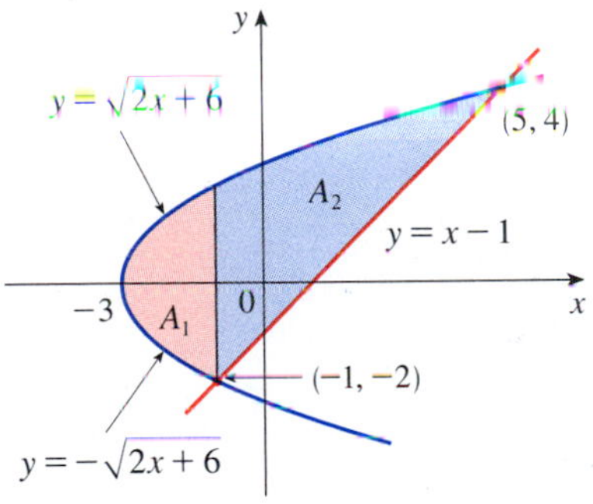

FIGURE 8

SOLUTION Solving $y^2 = 2x + 6$ and $y = x - 1$, we get $2x + 6 = (x - 1)^2$, which gives $x^2 - 4x - 5 = (x - 5)(x + 1) = 0$. Thus, $x = 5$ or -1 and the points of intersection are $(5, 4)$ and $(-1, -2)$. The region is sketched in Figure 8. Notice that just because we found the x-coordinates of the points of intersection to be -1 and 5, we cannot simply integrate a difference of functions between -1 and 5 or we miss the area labeled A_1.

In order to use Formula 2 we note that the parabola $y^2 = 2x + 6$ defines two functions given by $y = \sqrt{2x+6}$ (whose graph is the upper half of the parabola) and $y = -\sqrt{2x+6}$ (the lower half). Since the lower boundary of the region consists of part of a parabola and part of a line, we break up the area as $A = A_1 + A_2$, where the lower and upper boundaries for A_1 are given by

$$y = -\sqrt{2x+6} \qquad \text{and} \qquad y = \sqrt{2x+6}$$

and the lower and upper boundaries for A_2 are

$$y = x - 1 \qquad \text{and} \qquad y = \sqrt{2x+6}$$

Thus, the required area is

$$\begin{aligned} A &= A_1 + A_2 \\ &= \int_{-3}^{-1} \left[\sqrt{2x+6} - \left(-\sqrt{2x+6}\right)\right] dx + \int_{-1}^{5} \left[\sqrt{2x+6} - (x - 1)\right] dx \\ &= 2\int_{-3}^{-1} \sqrt{2x+6}\,dx + \int_{-1}^{5} \sqrt{2x+6}\,dx - \int_{-1}^{5} (x - 1)\,dx \end{aligned}$$

In the first two integrals we make the substitution $u = 2x + 6$ (so $du = 2\,dx$) and change the limits of integration accordingly:

$$\begin{aligned} A &= 2\int_0^4 \sqrt{u}\,\frac{du}{2} + \int_4^{16} \sqrt{u}\,\frac{du}{2} - \int_{-1}^{5} (x-1)\,dx \\ &= \left[\tfrac{2}{3}u^{3/2}\right]_0^4 + \left[\tfrac{1}{3}u^{3/2}\right]_4^{16} - \left[\frac{x^2}{2} - x\right]_{-1}^{5} \\ &= \tfrac{2}{3}(8) + \tfrac{1}{3}(64 - 8) - \left(\tfrac{25}{2} - 5 - \tfrac{1}{2} - 1\right) = 18 \end{aligned}$$

■

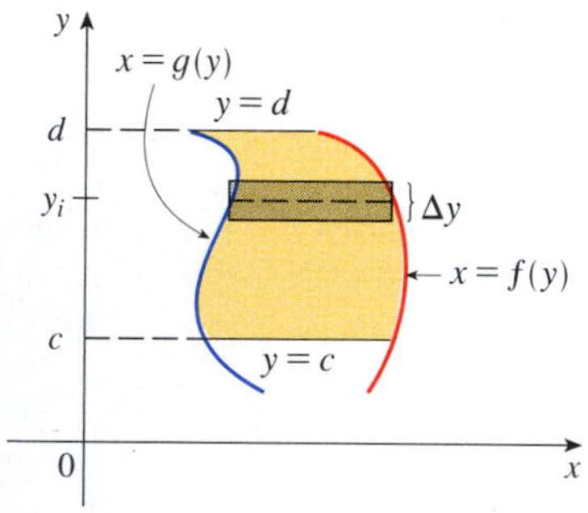

FIGURE 9

There is an easier method for solving Example 4. Instead of regarding y as a function of x, let's regard x as a function of y. In general, if a region is bounded by curves with equations $x = f(y)$, $x = g(y)$, $y = c$, and $y = d$, where f and g are continuous and $f(y) \geqslant g(y)$ for $c \leqslant y \leqslant d$ (see Figure 9), then its area is

(4)
$$A = \int_c^d [f(y) - g(y)]\,dy$$

EXAMPLE 5 Use Equation 4 to find the area in Example 4.

SOLUTION This time we do not have to split the region into two parts. Solving the equation of the parabola for x, we get $x = (y^2/2) - 3$, which represents a single function of y. Also, we write the equation of the line as $x = y + 1$. Notice that

$$y + 1 \geqslant \frac{y^2}{2} - 3 \qquad \text{for } -2 \leqslant y \leqslant 4$$

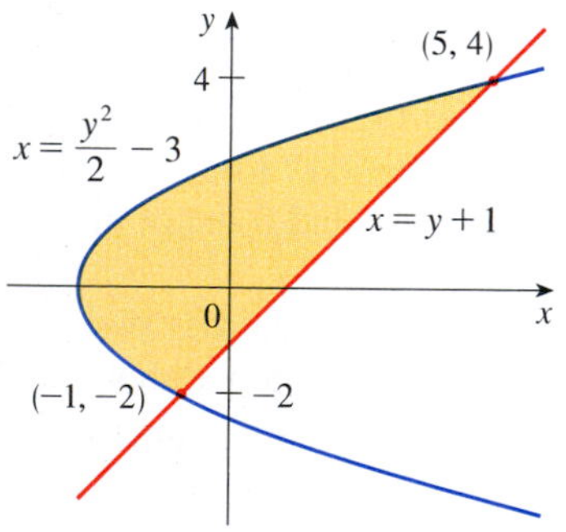

FIGURE 10

This means that the left boundary of the region is $x = (y^2/2) - 3$ and the right boundary is $x = y + 1$ (see Figure 10). We must integrate between the appropriate y-values, $y = -2$ and $y = 4$. Thus, Formula 4 gives

$$\begin{aligned} A &= \int_{-2}^{4} \left[(y+1) - \left(\frac{y^2}{2} - 3\right)\right] dy \\ &= \int_{-2}^{4} \left[-\frac{y^2}{2} + y + 4\right] dy \\ &= -\frac{1}{2}\left(\frac{y^3}{3}\right) + \frac{y^2}{2} + 4y\,\Bigg]_{-2}^{4} \\ &= -\tfrac{1}{6}(64) + 8 + 16 - \left(\tfrac{4}{3} + 2 - 8\right) = 18 \end{aligned}$$

■

Recall that the definite integral of a function can be interpreted as an area only when the function is positive. However we can use Formula 2 to prove the interpretation of an integral in terms of areas that was given in Section 6.3 as follows: For the case where $f(x) = 0$ and $g(x) < 0$, Formula 2 becomes

$$A = \int_a^b [0 - g(x)]\,dx = -\int_a^b g(x)\,dx$$

and so
$$\int_a^b g(x)\,dx = -A$$

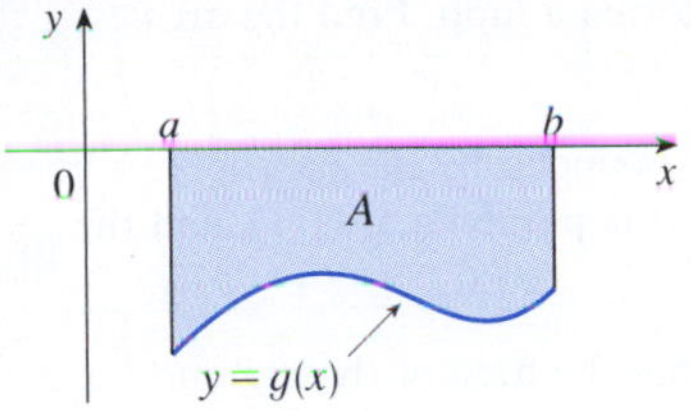

FIGURE 11

$\int_a^b g(x)\,dx = -A$

This says that the integral of a negative function can be interpreted as the *negative* of the area of the region *above* its graph between a and b (see Figure 11).

In general, if f takes on both positive and negative values, then we can interpret $\int_a^b f(x)\,dx$ as a difference of areas:

$$\int_a^b f(x)\,dx = A_1 - A_2$$

where A_1 is the area of the region between a and b that lies above the x-axis and below the graph of f, and A_2 is the area below the x-axis and above the graph of f. For instance, if, as illustrated in Figure 12, $f(x) \geqslant 0$ on $[a, c]$ and $[d, b]$ and $f(x) \leqslant 0$ on $[c, d]$, then

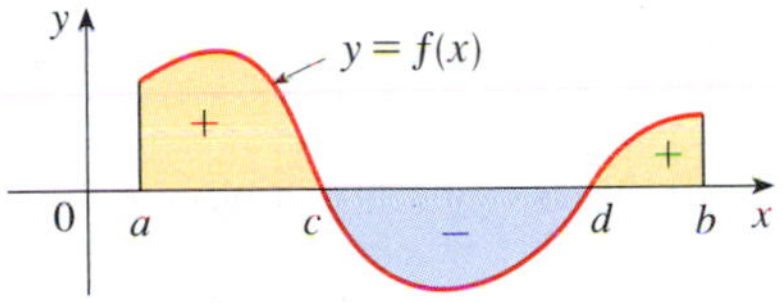

FIGURE 12

$$\begin{aligned}\int_a^b f(x)\,dx &= \int_a^c f(x)\,dx + \int_c^d f(x)\,dx + \int_d^b f(x)\,dx \\ &= \left(\int_a^c f(x)\,dx + \int_d^b f(x)\,dx\right) - \left(-\int_c^d f(x)\,dx\right) \\ &= A_1 - A_2\end{aligned}$$

EXERCISES 7.1

1–4 ■ Find the area of the shaded region.

1.

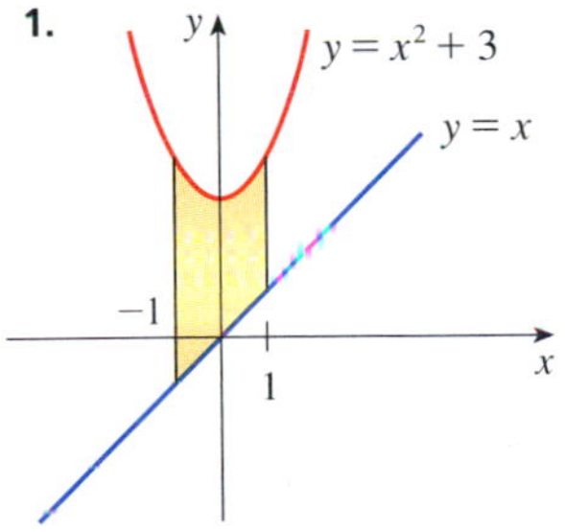

2.

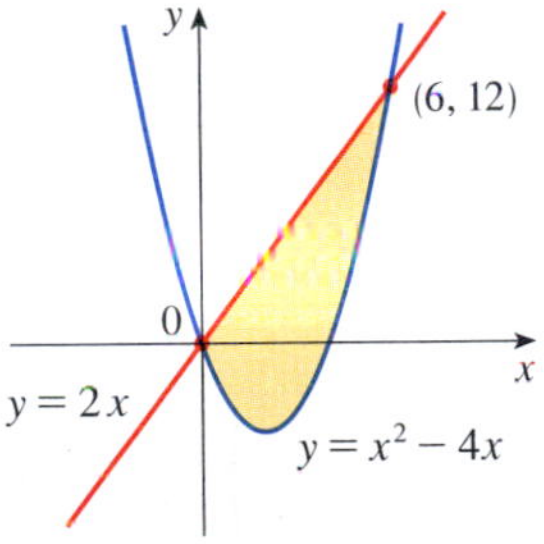

3.

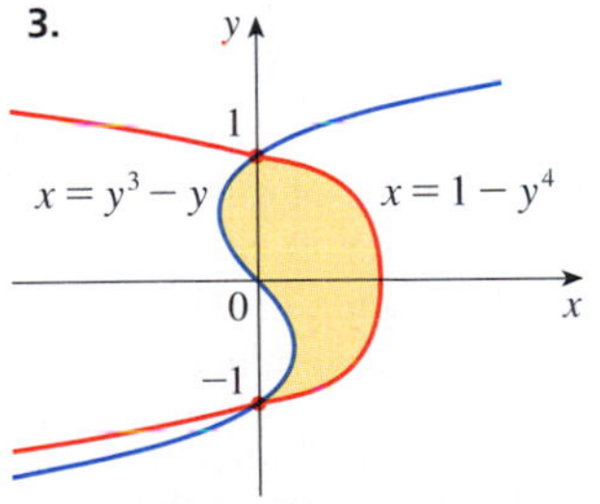

4.

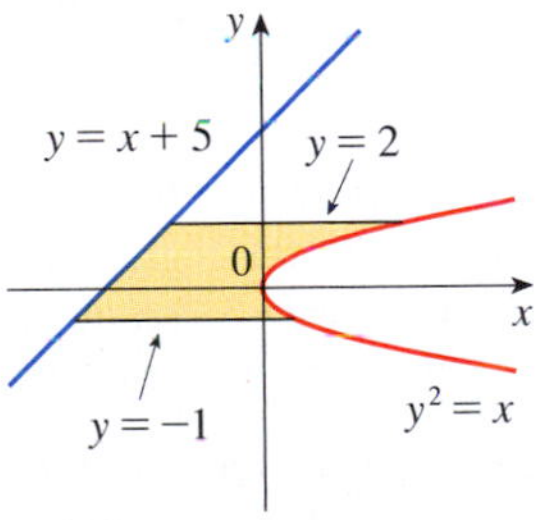

5–34 ■ Sketch the region bounded by the given curves and find the area of the region.

5. $y = x, \quad y = x^2$

6. $y = x, \quad y = x^3$

7. $y = x^2, \quad y^2 = x$

8. $y = x^2, \quad y = x^4$

9. $y = \sqrt{x}, \quad y = x/2$

10. $y = \sqrt{x - 1}, \quad x - 3y + 1 = 0$

11. $y = 4x^2, \quad y = x^2 + 3$

12. $y = x^4 - x^2, \quad y = 1 - x^2$

13. $y = x^2 + 2, \quad y = 2x + 5, \quad x = 0, \quad x = 6$

14. $y = x^2 + 1, \quad y = 3 - x^2, \quad x = -2, \quad x = 2$

15. $y^2 = x, \quad x - 2y = 3$

16. $x + y^2 = 2, \quad x + y = 0$

17. $x = 1 - y^2, \quad x = y^2 - 1$

18. $y = x^3 - 4x^2 + 3x, \quad y = x^2 - x$

19. $y = x, \quad y = \sin x, \quad x = -\pi/4, \quad x = \pi/2$

20. $y = \cos x, \quad y = \sec^2 x, \quad x = -\pi/4, \quad x = \pi/4$

21. $y = \cos x, \quad y = \sin 2x, \quad x = 0, \quad x = \pi/2$

22. $y = \sin x, \quad y = \sin 2x, \quad x = 0, \quad x = \pi/2$

23. $y = \cos x, \quad y = \sin 2x, \quad x = \pi/2, \quad x = \pi$

24. $y = \sin x, \quad y = \cos 2x, \quad x = 0, \quad x = \pi/4$

25. $y = |x|, \quad y = (x + 1)^2 - 7, \quad x = -4$

26. $y = |x - 1|, \quad y = x^2 - 3, \quad x = 0$

27. $x = 3y, \quad x + y = 0, \quad 7x + 3y = 24$

28. $y = x\sqrt{1 - x^2}, \quad y = x - x^3$

29. $y = 1/x, \quad y = 1/x^2, \quad x = 1, \quad x = 2$

30. $y = 1/x, \quad x = 0, \quad y = 1, \quad y = 2$

31. $y = x^2, \quad y = 2/(x^2 + 1)$

32. $y = 2^x, \quad y = 5^x, \quad x = -1, \quad x = 1$

33. $y = e^x, \quad y = e^{3x}, \quad x = 1$

34. $y = e^x, \quad y = e^{-x}, \quad x = -2, \quad x = 1$

35–36 ■ Find the area of the region bounded by the given curves by two methods: (a) integrating with respect to x, and (b) integrating with respect to y.

35. $4x + y^2 = 0, \quad y = 2x + 4$

36. $x + 1 = 2(y - 2)^2, \quad x + 6y = 7$

37–38 ■ Use calculus to find the area of the triangle with the given vertices.

37. $(0, 0), (1, 8), (4, 3)$

38. $(-2, 5), (0, -3), (5, 2)$

39–40 ■ Evaluate the integral and interpret it as the area of a region. Sketch the region.

39. $\int_0^2 |x^2 - x^3|\, dx$

40. $\int_0^{\pi} \left| \sin x - \frac{2}{\pi} x \right| dx$

41–42 ■ Evaluate the integral and interpret it as a difference of areas. Illustrate with a sketch like Figure 12.

41. $\int_{-1}^2 x^3\, dx$

42. $\int_{\pi/4}^{5\pi/2} \sin x\, dx$

43–44 ■ Use the Midpoint Rule with $n = 4$ to approximate the area of the region bounded by the given curves.

43. $y = \sqrt{1 + x^3}, \quad y = 1 - x, \quad x = 2$

44. $y = x \tan x, \quad y = x$

45–48 ■ Use a graph to find approximate x-coordinates of the points of intersection of the given curves. Then find (approximately) the area of the region bounded by the curves.

45. $y = x^2, \quad y = 2 \cos x$

46. $y = x^4, \quad y = 3x - x^3$

47. $y = \sqrt{x + 1}, \quad y = x^2$

48. $y = x^4 - 1, \quad y = x \sin(x^2)$

49–50 ■ Use a graph to find approximate x-coordinates of the points of intersection of the given curves. Then use the Midpoint Rule with $n = 4$ to approximate the area of the region bounded by the curves.

49. $y = 1 + 3x - 2x^2, \quad y = \sqrt{1 + x^4}$

50. $y = x^2 - x, \quad y = \sin(x^2)$

51. The curve with equation $y^2 = x^2(x + 3)$ is called **Tschirnhauser's cubic**. If you graph this curve you will see that part of the curve forms a loop. Find the area enclosed by the loop.

52. Find the area of the region bounded by the parabola $y = x^2$, the tangent line to this parabola at $(1, 1)$, and the x-axis.

53. Find the values of c such that the area of the region bounded by the parabolas $y = x^2 - c^2$ and $y = c^2 - x^2$ is 576.

54. Find a positive continuous function f such that the area under the graph of f from 0 to t is $A(t) = t^3$ for all $t > 0$.

55. Find the number b such that the line $y = b$ divides the region bounded by the curves $y = x^2$ and $y = 4$ into two regions with equal area.

56. (a) Find the number a such that the line $x = a$ bisects the area under the curve $y = 1/x^2$, $1 \leq x \leq 4$.
(b) Find the number b such that the line $y = b$ bisects the area in part (a).

57. Two cars, A and B, start side by side and accelerate from rest. The figure shows the graphs of their velocity functions.
(a) Which car is ahead after one minute?
(b) Which car is ahead after two minutes?
(c) Estimate the time at which the cars are again side by side.

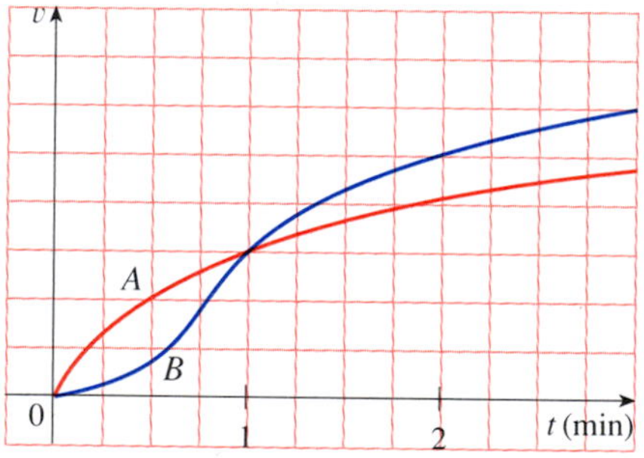

58. Suppose that $0 < c < \pi/2$. For what value of c is the area of the region enclosed by the curves $y = \cos x$, $y = \cos(x - c)$, and $x = 0$ equal to the area of the region enclosed by the curves $y = \cos(x - c)$, $x = \pi$, and $y = 0$?

59. For what values of m do the line $y = mx$ and the curve $y = x/(x^2 + 1)$ enclose a region? Find the area of the region.

7.2 VOLUME

In trying to find the volume of a solid we face the same type of problem as in finding areas. We have an intuitive idea of what volume means, but we must make this idea precise by using calculus to give an exact definition of volume.

We start with a simple type of solid called a **cylinder** (or, more precisely, a *right cylinder*). As illustrated in Figure 1(a), a cylinder is bounded by a plane region B_1, called the **base**, and a congruent region B_2 in a parallel plane. The cylinder consists of all points on line segments perpendicular to the base that join B_1 to B_2. If the area of the base is A and the height of the cylinder (the distance from B_1 to B_2) is h, then the volume V of the cylinder is defined as

$$V = Ah$$

In particular, if the base is a circle with radius r, then the cylinder is a circular cylinder with volume $V = \pi r^2 h$ [see Figure 1(b)], and if the base is a rectangle with length l and width w, then the cylinder is a rectangular box (also called a *rectangular parallelepiped*) with volume $V = lwh$ [see Figure 1(c)].

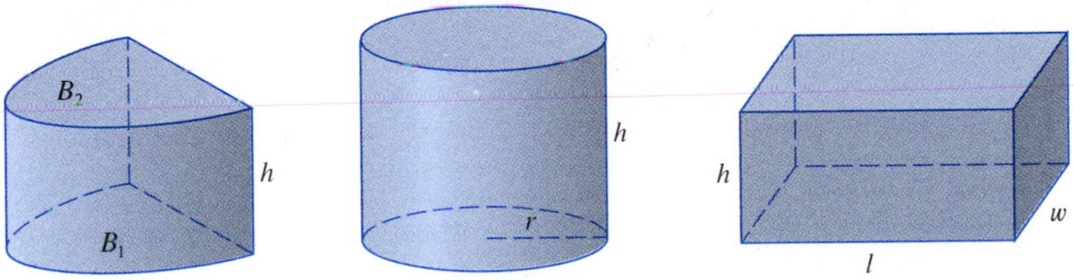

FIGURE 1

(a) Cylinder $V = Ah$

(b) Circular cylinder $V = \pi r^2 h$

(c) Rectangular box $V = lwh$

Now let S be any solid. The intersection of S with a plane is a plane region that is called a **cross-section** of S. Suppose that the area of the cross-section of S in a plane P_x perpendicular to the x-axis and passing through the point x is $A(x)$, where $a \leqslant x \leqslant b$. (See Figure 2. Think of slicing S with a knife through x and computing the area of this slice.) The cross-sectional area $A(x)$ will vary as x increases from a to b.

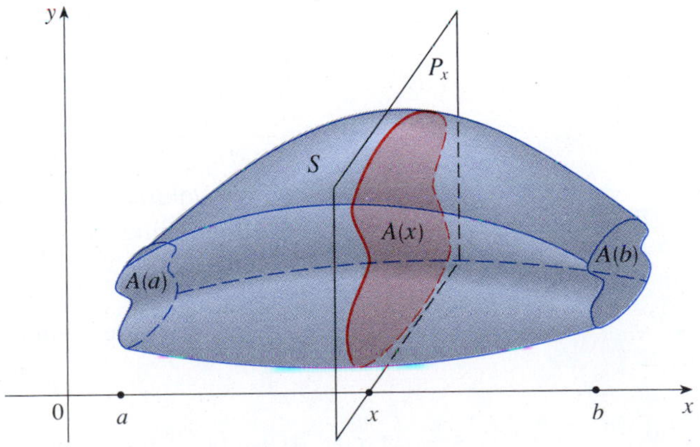

FIGURE 2

Let us consider a partition P of the interval $[a, b]$ by points x_i such that $a = x_0 < x_1 < \cdots < x_n = b$. The planes P_{x_i} will slice S into smaller "slabs." (Think of slicing a loaf of bread.) If we choose numbers x_i^* in $[x_{i-1}, x_i]$, we can approximate the ith slab S_i (the part of S that lies between the planes $P_{x_{i-1}}$ and P_{x_i}) by a cylinder with base area $A(x_i^*)$ and height $\Delta x_i = x_i - x_{i-1}$ (see Figure 3).

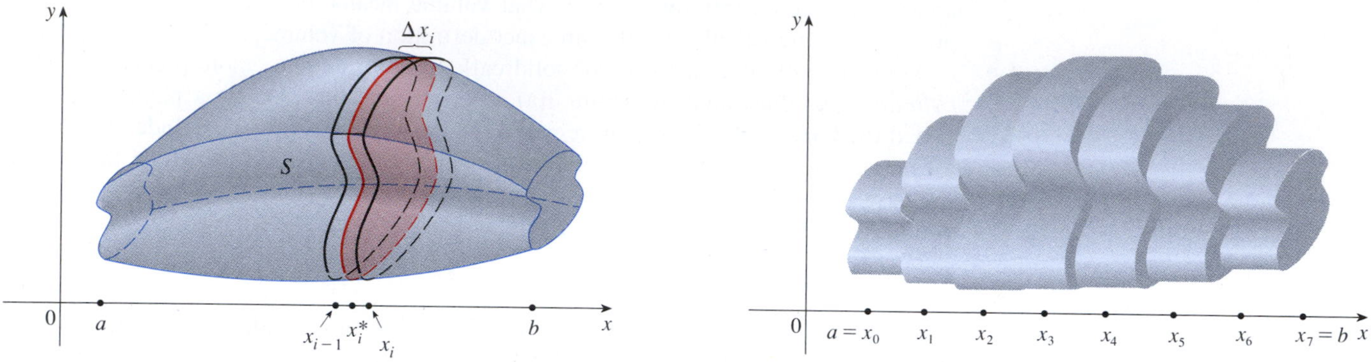

FIGURE 3

The volume of this cylinder is $A(x_i^*)\,\Delta x_i$ so an approximation to our intuitive conception of the volume of the ith slab S_i is

$$V(S_i) \approx A(x_i^*)\,\Delta x_i$$

Adding the volumes of these slabs, we get an approximation to the total volume (that is, what we think of intuitively as the volume):

$$V \approx \sum_{i=1}^{n} A(x_i^*)\,\Delta x_i$$

This approximation appears to become better and better as $\|P\| \to 0$. (Think of the slices as becoming thinner and thinner.) Therefore, we *define* the volume as the limit of these sums as $\|P\| \to 0$. But we recognize the limit of Riemann sums as a definite integral and so we have the following definition.

(1) DEFINITION OF VOLUME Let S be a solid that lies between the planes P_a and P_b. If the cross-sectional area of S in the plane P_x is $A(x)$, where A is an integrable function, then the **volume** of S is

$$V = \lim_{\|P\| \to 0} \sum_{i=1}^{n} A(x_i^*)\,\Delta x_i = \int_a^b A(x)\,dx$$

When we use the volume formula $V = \int_a^b A(x)\,dx$ it is important to remember that $A(x)$ is the area of a moving cross-section obtained by slicing through x perpendicular to the x-axis.

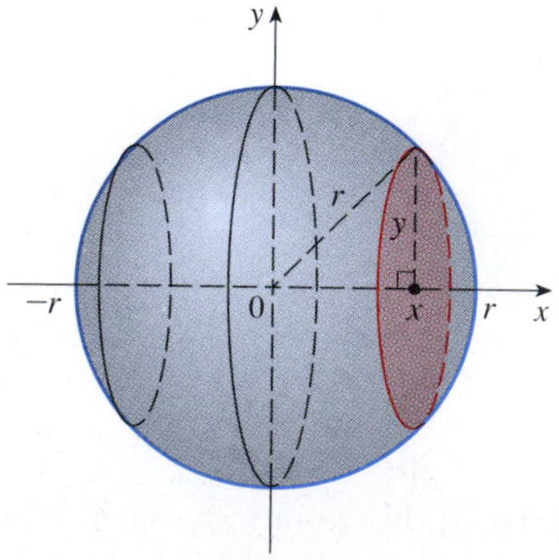

FIGURE 4

EXAMPLE 1 Show that the volume of a sphere of radius r is

$$V = \tfrac{4}{3}\pi r^3$$

SOLUTION If we place the sphere so that its center is at the origin (see Figure 4), then the plane P_x intersects the sphere in a circle whose radius (from the Pythagorean Theorem) is $y = \sqrt{r^2 - x^2}$. So the cross-sectional area is

$$A(x) = \pi y^2 = \pi(r^2 - x^2)$$

Using the formula with $a = -r$ and $b = r$, we have

$$V = \int_{-r}^{r} A(x)\,dx = \int_{-r}^{r} \pi(r^2 - x^2)\,dx$$

$$= 2\pi \int_0^r (r^2 - x^2)\,dx \qquad \text{(The integrand is even.)}$$

$$= 2\pi\left[r^2x - \frac{x^3}{3}\right]_0^r = 2\pi\left(r^3 - \frac{r^3}{3}\right)$$

$$= \tfrac{4}{3}\pi r^3$$

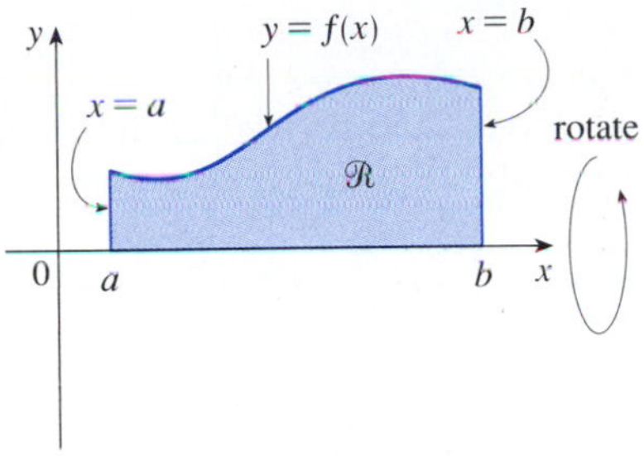

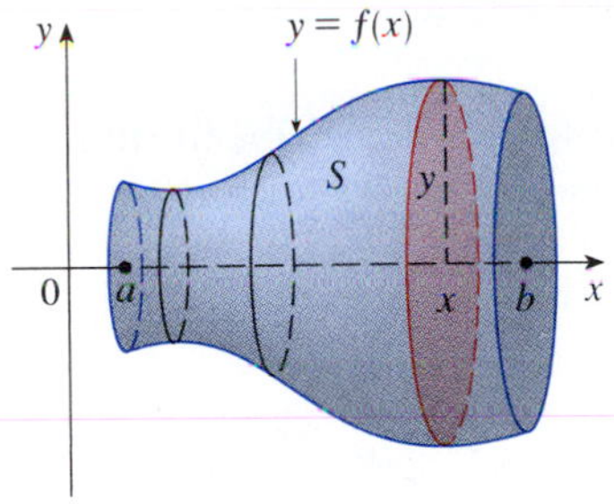

FIGURE 5

The sphere in Example 1 is an example of a **solid of revolution** since it can be obtained by revolving a circle about a diameter. In general, let S be the solid obtained by revolving the plane region $\mathcal{R}$ bounded by $y = f(x)$, $y = 0$, $x = a$, and $x = b$ about the x-axis (see Figure 5).

Because S is obtained by rotation, a cross-section through x perpendicular to the x-axis is a circular disk with radius $|y| = |f(x)|$ and so the cross-sectional area is

$$A(x) = \pi y^2 = \pi[f(x)]^2$$

Thus, using the basic volume formula $V = \int_a^b A(x)\,dx$, we have the following **formula for a volume of revolution:**

$$\textbf{(2)} \qquad V = \int_a^b \pi[f(x)]^2\,dx$$

In particular, since the sphere can be obtained by rotating the region under the semicircle

$$y = \sqrt{r^2 - x^2} \qquad -r \leqslant x \leqslant r$$

about the x-axis, Example 1 could be done using Formula 2 with $f(x) = \sqrt{r^2 - x^2}$, $a = -r$, and $b = r$.

Figure 6 illustrates the process that is represented by Formula 2. Part (a) shows the ith approximating cylinder. In this case it is a circular cylinder, or disk, and is swept out by a rectangle of width Δx_i. Part (b) shows how six such disks fit together to approximate the solid. Part (c) shows that when we increase n we usually get a better approximation to the desired volume. You can see from Figure 6 why the use of Formula 2 is called the **disk method**.

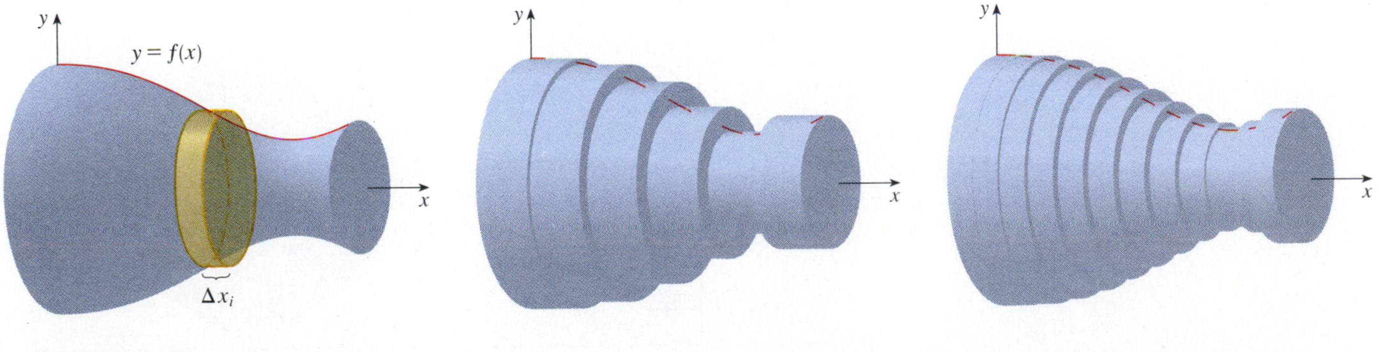

(a) ith approximating cylinder (b) $n = 6$ (c) $n = 12$

FIGURE 6

EXAMPLE 2

(a) Find the volume of the solid obtained by rotating about the x-axis the region under the curve $y = \sqrt{x}$ from 0 to 1.

(b) Illustrate the definition of volume by drawing the approximating cylinders for this solid.

SOLUTION

(a) The region is shown in Figure 7(a). Taking $a = 0$, $b = 1$, and $f(x) = \sqrt{x}$ in Formula 2, we find that the volume of the solid [illustrated in Figure 7(b)] is

$$V = \int_0^1 \pi(\sqrt{x})^2\,dx = \pi \int_0^1 x\,dx = \pi \frac{x^2}{2}\bigg]_0^1 = \frac{\pi}{2}$$

Did we get a reasonable answer in Example 2? As a check on our work, let's replace the given region by a square with base $[0, 1]$ and height 1. If we rotate this square, we get a cylinder with radius 1, height 1, and volume $\pi \cdot 1^2 \cdot 1 = \pi$. We computed that the given solid has half this volume. That seems about right.

(b) According to the definition of volume in terms of Riemann sums, we have

$$V = \lim_{\|P\|\to 0} \sum_{i=1}^{n} A(x_i^*)\,\Delta x_i = \lim_{\|P\|\to 0} \sum_{i=1}^{n} \pi(\sqrt{x_i^*})^2\,\Delta x_i$$

Taking x_i^* to be the right endpoint of $[x_{i-1}, x_i]$, we can interpret the Riemann sum as the approximating volume of the flat circular cylinders (or disks) shown in Figure 7(c).

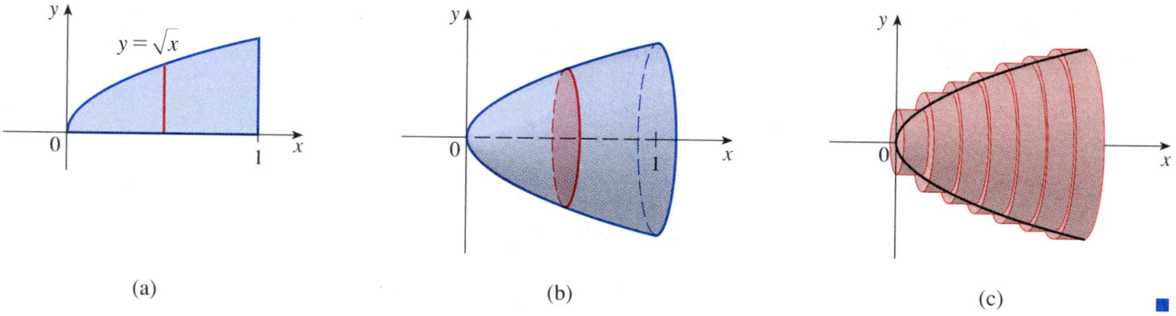

FIGURE 7

Formula 2 applies only when the axis of rotation is the x-axis. If, as illustrated in Figure 8, the region bounded by the curves $x = g(y)$, $x = 0$, $y = c$, and $y = d$ is rotated about the y-axis, then the corresponding volume of revolution is

(3)
$$V = \int_c^d \pi[g(y)]^2\,dy$$

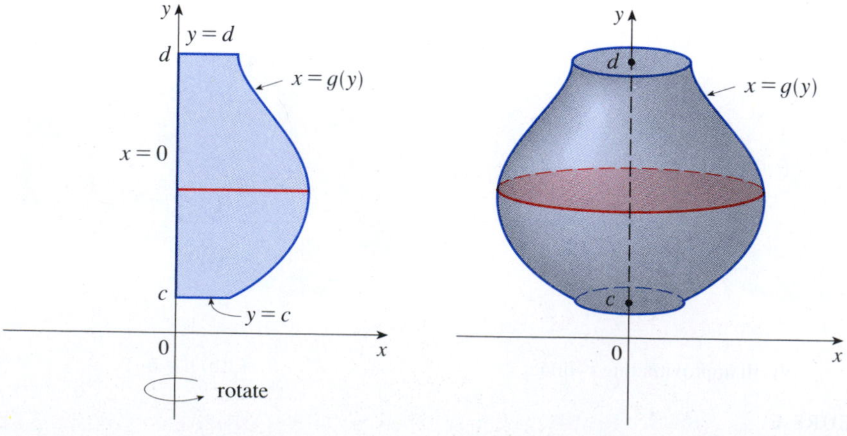

FIGURE 8

EXAMPLE 3 Find the volume of the solid obtained by rotating the region bounded by $y = x^3$, $y = 8$, and $x = 0$ around the y-axis.

SOLUTION Since the axis of rotation is the y-axis, we write the curve $y = x^3$ in the form $x = \sqrt[3]{y}$. The region and the solid with a typical disk are sketched in Figure 9. Using Formula 3 with $c = 0$, $d = 8$, and $g(y) = \sqrt[3]{y}$, we have

$$V = \int_0^8 \pi(\sqrt[3]{y})^2\,dy = \pi\int_0^8 y^{2/3}\,dy$$

$$= \pi\left[\tfrac{3}{5}y^{5/3}\right]_0^8 = \frac{96\pi}{5}$$

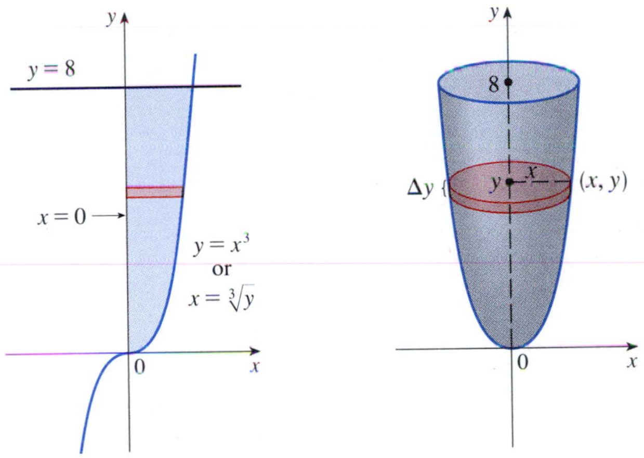

FIGURE 9

■

EXAMPLE 4 The region $\mathcal{R}$ bounded by the curves $y = x$ and $y = x^2$ is rotated about the x-axis. Find the volume of the resulting solid.

SOLUTION 1 The curves $y = x$ and $y = x^2$ intersect at the points $(0, 0)$ and $(1, 1)$. The region between them, the solid of rotation, and a cross-section perpendicular to the x-axis are shown in Figure 10. A cross-section in the plane P_x has the shape of an annulus (a ring) with inner radius x^2 and outer radius x, so the cross-sectional area is

$$A(x) = \pi x^2 - \pi(x^2)^2 = \pi(x^2 - x^4)$$

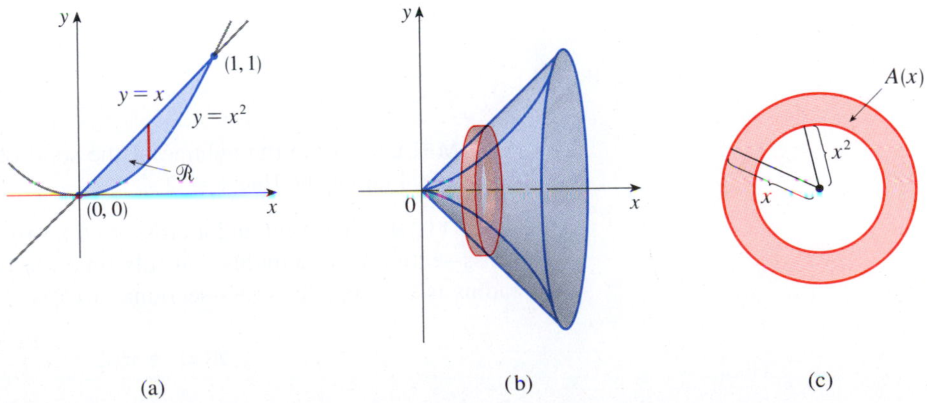

FIGURE 10

Using Definition 1, we have

$$V = \int_0^1 A(x)\,dx = \int_0^1 \pi(x^2 - x^4)\,dx$$
$$= \pi\left[\frac{x^3}{3} - \frac{x^5}{5}\right]_0^1 = \frac{2\pi}{15}$$

SOLUTION 2 Another method is to subtract volumes. If V_1 is the volume of the solid obtained by rotating the region under $y = x$ from 0 to 1 around the x-axis and V_2 is the corresponding volume for the curve $y = x^2$, then

$$V = V_1 - V_2$$

Applying Formula 2 to calculate each of V_1 and V_2, we have

$$V = \int_0^1 \pi x^2\,dx - \int_0^1 \pi(x^2)^2\,dx$$
$$= \pi\int_0^1 (x^2 - x^4)\,dx = \frac{2\pi}{15}$$ ■

In general, let S be the solid generated when the region bounded by the curves $y = f(x)$, $y = g(x)$, $x = a$, and $x = b$ [where $f(x) \geqslant g(x)$] is rotated about the x-axis (see Figure 11). Then, either by the method of Solution 1 of Example 4 (called the **washer method** because the approximating cylinders have the shape of washers) or by the method of Solution 2 (subtracting volumes), we see that the volume of S is

(4)
$$V = \pi\int_a^b \{[f(x)]^2 - [g(x)]^2\}\,dx$$

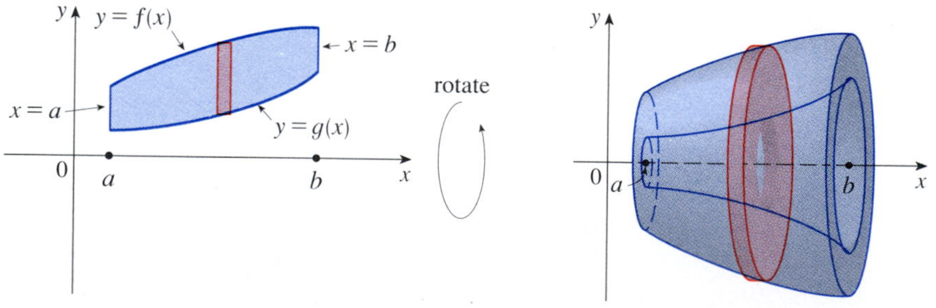

FIGURE 11

EXAMPLE 5 Find the volume of the solid obtained by rotating the region in Example 4 about the line $y = 2$.

SOLUTION The solid and a cross-section are shown in Figure 12. Again a cross-section is an annulus, but this time the inner radius is $2 - x$ and the outer radius is $2 - x^2$. The cross-sectional area is

$$A(x) = \pi(2 - x^2)^2 - \pi(2 - x)^2$$

and so the volume of S is

$$V = \int_0^1 A(x)\,dx = \pi \int_0^1 [(2 - x^2)^2 - (2 - x)^2]\,dx$$

$$= \pi \int_0^1 (x^4 - 5x^2 + 4x)\,dx$$

$$= \pi \left[\frac{x^5}{5} - 5\frac{x^3}{3} + 4\frac{x^2}{2} \right]_0^1 = \frac{8\pi}{15}$$

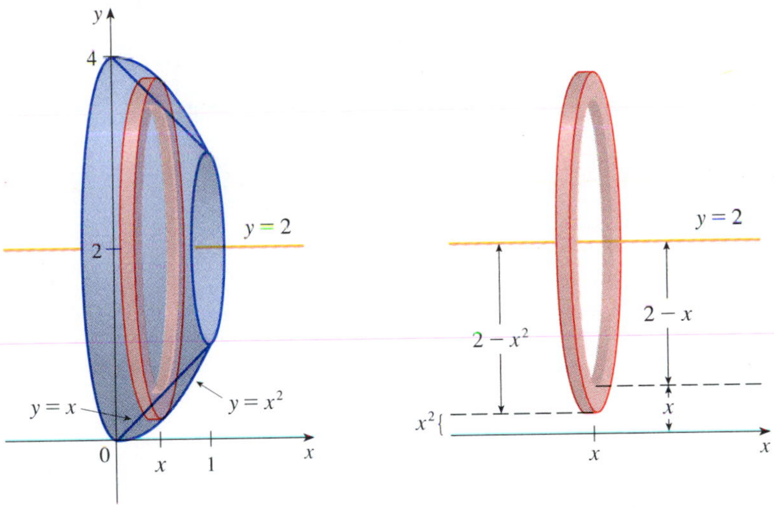

FIGURE 12

EXAMPLE 6 Find the volume of the solid obtained by rotating the region in Example 4 about the y-axis.

SOLUTION Figure 13 shows a cross-section perpendicular to the y-axis. It is an annulus with inner radius y and outer radius $\sqrt{y}$, so the cross-sectional area is

$$A(y) = \pi(\sqrt{y}\,)^2 - \pi y^2$$

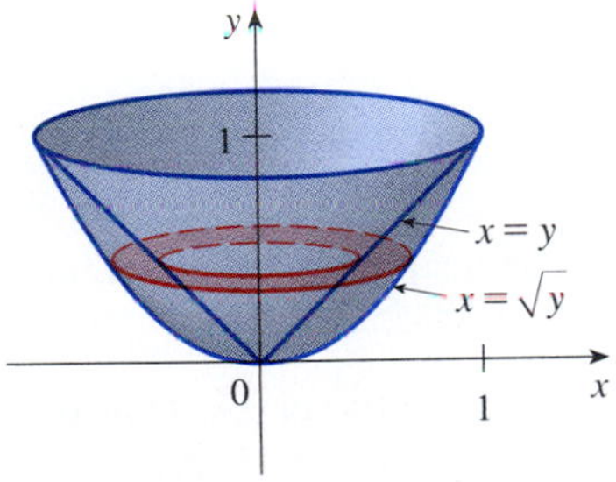

FIGURE 13

and the volume is

$$V = \int_0^1 A(y)\,dy = \pi \int_0^1 [(\sqrt{y}\,)^2 - y^2]\,dy$$

$$= \pi \int_0^1 (y - y^2)\,dy$$

$$= \pi \left[\frac{y^2}{2} - \frac{y^3}{3} \right]_0^1 = \frac{\pi}{6}$$

Another method would be to subtract volumes as in Solution 2 of Example 4. Notice that the volume in this example is larger than the volume of revolution about the x-axis in Example 4. ■

Although we have several volume formulas that are applicable in different situations, it is probably best not to memorize Formula 2, 3, or 4 but rather to remember the basic volume formula $V = \int_a^b A(x)\,dx$, which applies to any situation. We conclude this section by finding the volumes of three solids that are not solids of revolution.

EXAMPLE 7 A solid has a circular base of radius 1. Parallel cross-sections perpendicular to the base are equilateral triangles. Find the volume of the solid.

SOLUTION Let's take the circle to be $x^2 + y^2 = 1$. The solid, its base, and a typical cross-section at a distance x from the origin are shown in Figure 14.

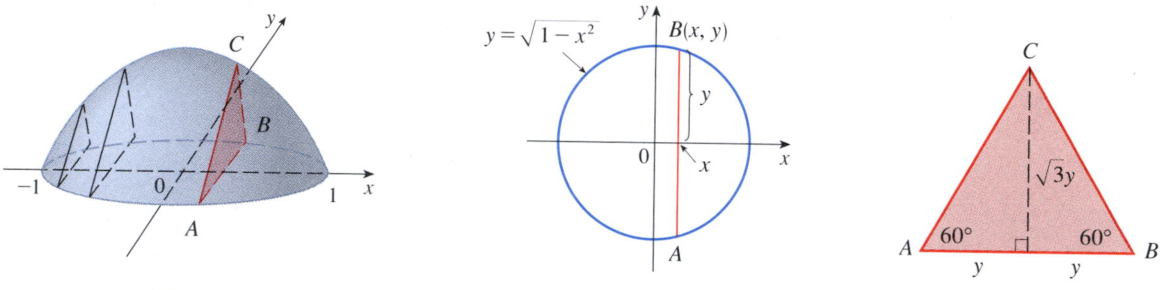

FIGURE 14 (a) The solid (b) Its base (c) A cross-section

Since B lies on the circle, we have $y = \sqrt{1 - x^2}$ and so the base of the triangle ABC is $|AB| = 2\sqrt{1 - x^2}$. Since the triangle is equilateral, we see from Figure 14(c) that its height is $\sqrt{3}\,y = \sqrt{3}\sqrt{1 - x^2}$. The cross-sectional area is therefore

$$A(x) = \tfrac{1}{2} \cdot 2\sqrt{1 - x^2} \cdot \sqrt{3}\sqrt{1 - x^2} = \sqrt{3}\,(1 - x^2)$$

and the volume of the solid is

$$V = \int_{-1}^{1} A(x)\,dx = \int_{-1}^{1} \sqrt{3}\,(1 - x^2)\,dx$$

$$= \sqrt{3}\left[x - \frac{x^3}{3}\right]_{-1}^{1} = \frac{4\sqrt{3}}{3}$$

■

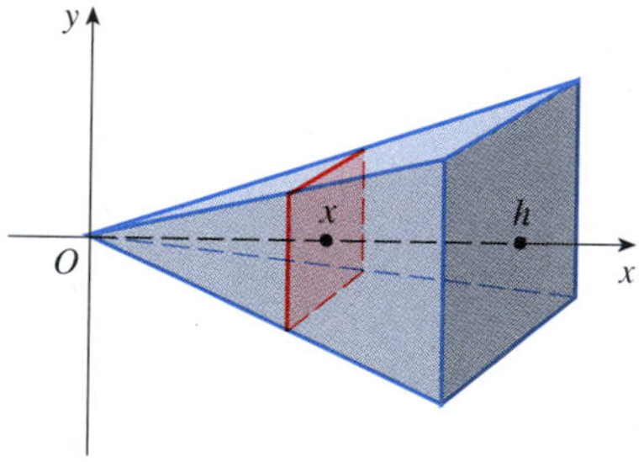

FIGURE 15

EXAMPLE 8 Find the volume of a pyramid whose base is a square with side L and whose height is h.

SOLUTION In order to use the formula in Definition 1 we place the origin O at the vertex of the pyramid and the x-axis along its central axis as in Figure 15. Any plane P_x that passes through x and is perpendicular to the x-axis intersects the pyramid in a square with side of length s, say. We can express s in terms of x by observing from the similar triangles in Figure 16 that

$$\frac{x}{h} = \frac{s/2}{L/2} = \frac{s}{L}$$

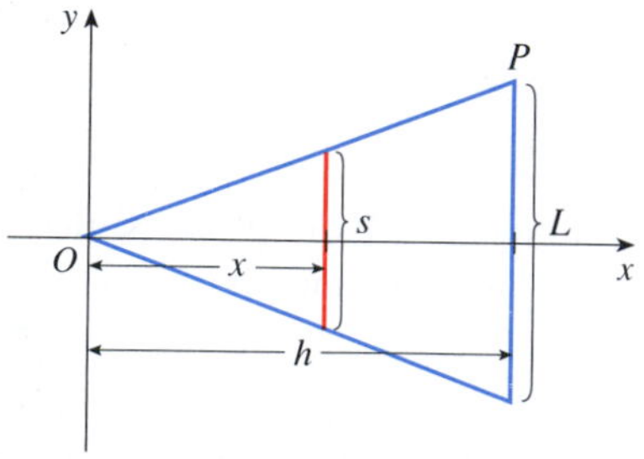

FIGURE 16

and so $s = Lx/h$. [Another method is to observe that the line OP has slope $L/(2h)$ and so its equation is $y = Lx/(2h)$.] Thus, the cross-sectional area is

$$A(x) = s^2 = \frac{L^2}{h^2}x^2$$

Taking $a = 0$ and $b = h$ in Definition 1, we see that the volume of the pyramid is

$$V = \int_0^h A(x)\,dx = \int_0^h \frac{L^2}{h^2}x^2\,dx$$

$$= \frac{L^2}{h^2}\frac{x^3}{3}\bigg]_0^h = \frac{L^2h}{3}$$

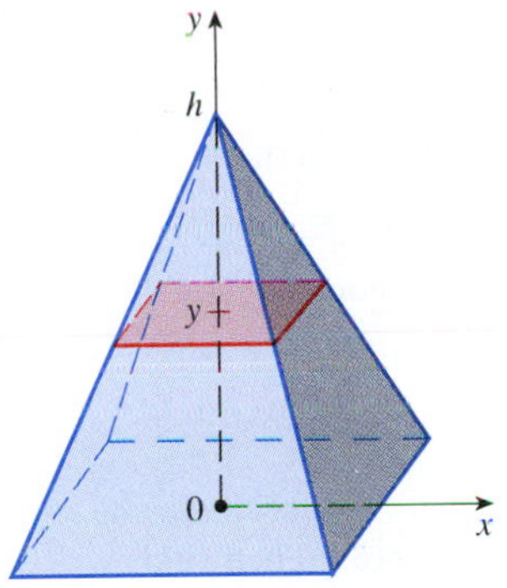

FIGURE 17

NOTE: It was not necessary to place the vertex of the pyramid at the origin in Example 8. We did so merely to make the equations simple. If, instead, we had placed the center of the base at the origin and the vertex on the positive y-axis as in Figure 17, you can verify that we would have obtained the integral

$$V = \int_0^h \frac{L^2}{h^2}(h - y)^2\,dy = \frac{L^2h}{3}$$

EXAMPLE 9 A wedge is cut out of a circular cylinder of radius 4 by two planes. One plane is perpendicular to the axis of the cylinder. The other intersects the first at an angle of 30° along a diameter of the cylinder. Find the volume of the wedge.

SOLUTION If we place the x-axis along the diameter where the planes meet, then the base of the solid is a semicircle with equation $y = \sqrt{16 - x^2}$, $-4 \leqslant x \leqslant 4$. A cross-section perpendicular to the x-axis at a distance x from the origin is a triangle ABC as shown in Figure 18 whose base is $y = \sqrt{16 - x^2}$ and whose height is $|BC| = y\tan 30° = \sqrt{16 - x^2}/\sqrt{3}$. Thus, the cross-sectional area is

$$A(x) = \tfrac{1}{2}\sqrt{16 - x^2}\cdot\frac{1}{\sqrt{3}}\sqrt{16 - x^2} = \frac{16 - x^2}{2\sqrt{3}}$$

and the volume is

$$V = \int_{-4}^4 A(x)\,dx = \int_{-4}^4 \frac{16 - x^2}{2\sqrt{3}}\,dx$$

$$= \frac{1}{\sqrt{3}}\int_0^4 (16 - x^2)\,dx = \frac{1}{\sqrt{3}}\left[16x - \frac{x^3}{3}\right]_0^4$$

$$= \frac{128}{3\sqrt{3}}$$

For another method see Exercise 67.

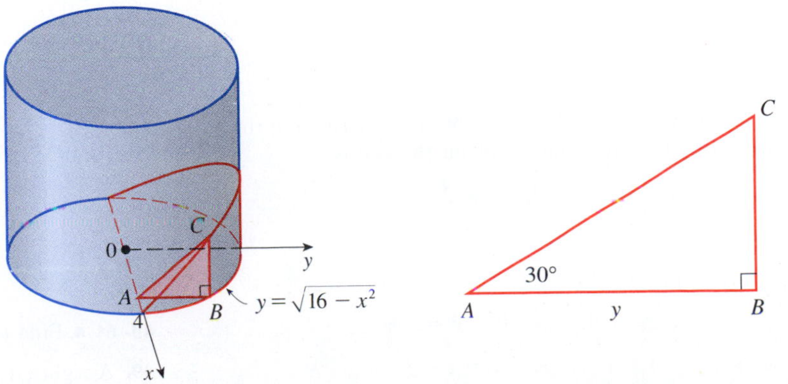

FIGURE 18

EXERCISES 7.2

1–12 ■ Find the volume of the solid obtained by rotating the region bounded by the given curves about the specified axis. Sketch the region, the solid, and a typical disk or "washer."

1. $y = x^2$, $x = 1$, $y = 0$; about the x-axis
2. $y^2 = x^3$, $x = 4$, $y = 0$; about the x-axis
3. $x + y = 1$, $x = 0$, $y = 0$; about the x-axis
4. $y = \sqrt{x - 1}$, $x = 2$, $x = 5$, $y = 0$; about the x-axis
5. $y = x^2$, $y = 4$, $x = 0$, $x = 2$; about the y-axis
6. $x = y - y^2$, $x = 0$; about the y-axis
7. $y = x^2$, $y^2 = x$; about the x-axis
8. $y = x^2 + 1$, $y = 3 - x^2$; about the x-axis
9. $y^2 = x$, $x = 2y$; about the y-axis
10. $y = 2x - x^2$, $y = 0$, $x = 0$, $x = 1$; about the y-axis
11. $y = x^4$, $y = 1$; about $y = 2$
12. $y = x$, $y = 0$, $x = 2$, $x = 4$; about $x = 1$

13–24 ■ Refer to the figure and find the volume generated by rotating the given region about the given line.

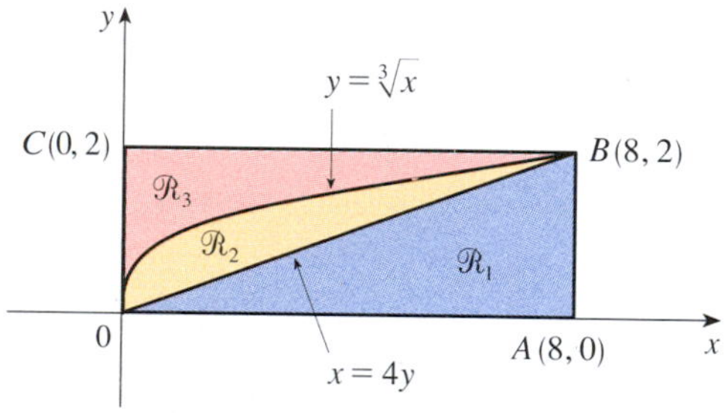

13. $\mathcal{R}_1$ about OA
14. $\mathcal{R}_1$ about OC
15. $\mathcal{R}_1$ about AB
16. $\mathcal{R}_1$ about BC
17. $\mathcal{R}_2$ about OA
18. $\mathcal{R}_2$ about OC
19. $\mathcal{R}_2$ about BC
20. $\mathcal{R}_2$ about AB
21. $\mathcal{R}_3$ about OA
22. $\mathcal{R}_3$ about OC
23. $\mathcal{R}_3$ about BC
24. $\mathcal{R}_3$ about AB

25–32 ■ Find the volume of the solid obtained by rotating the region bounded by the given curves about the x-axis.

25. $y = x^2 - 1$, $y = 0$, $x = 0$, $x = 2$
26. $y = -1/x$, $y = 0$, $x = 1$, $x = 3$
27. $y = e^x$, $y = 0$, $x = 0$, $x = 1$
28. $y = 1/\sqrt{x + 1}$, $y = 0$, $x = 0$, $x = 1$
29. $y = \sec x$, $y = 1$, $x = -1$, $x = 1$
30. $y = \cos x$, $y = \sin x$, $x = 0$, $x = \pi/4$
31. $y = |x + 2|$, $y = 0$, $x = -3$, $x = 0$
32. $y = [\![x]\!]$, $x = 1$, $x = 6$, $y = 0$

33–38 ■ Set up, but do not evaluate, an integral for the volume of the solid obtained by rotating the region bounded by the given curves about the specified line.

33. $y = \ln x$, $y = 1$, $x = 1$; about the x-axis
34. $y = \sqrt{x - 1}$, $y = 0$, $x = 5$; about the y-axis
35. $x - y = 1$, $y = (x - 4)^2 + 1$; about $y = 7$
36. $y = \cos x$, $y = 0$, $x = 0$, $x = \pi/2$; about $y = 1$
37. $y = \cos x$, $y = 0$, $x = 0$, $x = \pi/2$; about $y = -1$
38. $2x + 3y = 6$, $(y - 1)^2 = 4 - x$; about $x = -5$

39–40 ■ Use a graph to find approximate x-coordinates of the points of intersection of the given curves. Then find (approximately) the volume of the solid obtained by rotating about the x-axis the region bounded by these curves.

39. $y = x^2$, $y = \sqrt{x + 1}$
40. $y = x^4$, $y = 3x - x^3$

41–42 ■ Sketch and find the volume of the solid obtained by rotating the region under the graph of f about the x-axis.

41. $f(x) = \begin{cases} 3 & \text{if } 0 \leqslant x \leqslant 1 \\ 1 & \text{if } 1 < x < 4 \\ 3 & \text{if } 4 \leqslant x \leqslant 5 \end{cases}$

42. $f(x) = \begin{cases} \frac{1}{2} & \text{if } 0 \leqslant x < 1 \\ x^2 - 2x + 2 & \text{if } 1 \leqslant x \leqslant 2 \end{cases}$

43–48 ■ Each integral represents the volume of a solid. Describe the solid.

43. $\pi \int_0^{\pi/4} \tan^2 x \, dx$
44. $\pi \int_1^2 y^6 \, dy$
45. $\pi \int_0^1 (y - y^2) \, dy$
46. $\pi \int_0^4 [16 - (x - 2)^4] \, dx$
47. $\pi \int_0^1 [(5 - 2x^2)^2 - (5 - 2x)^2] \, dx$
48. $\pi \int_{\pi/4}^{\pi/2} [(2 + \sin x)^2 - (2 + \cos x)^2] \, dx$

49–61 ■ Find the volume of the described solid S.

49. A right circular cone with height h and base radius r

50. A frustum of a right circular cone with height h, lower base radius R, and top radius r

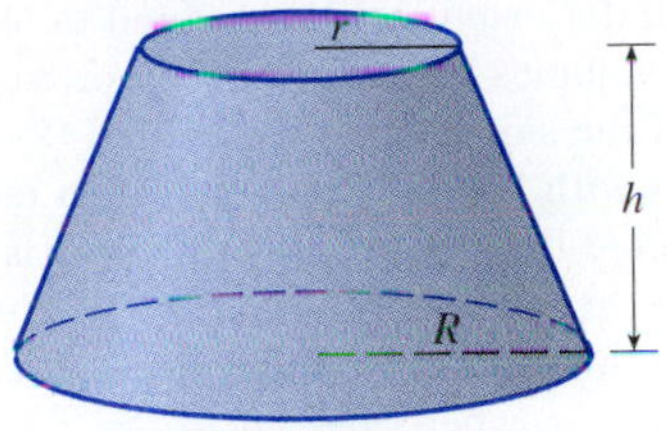

51. A cap of a sphere with radius r and height h

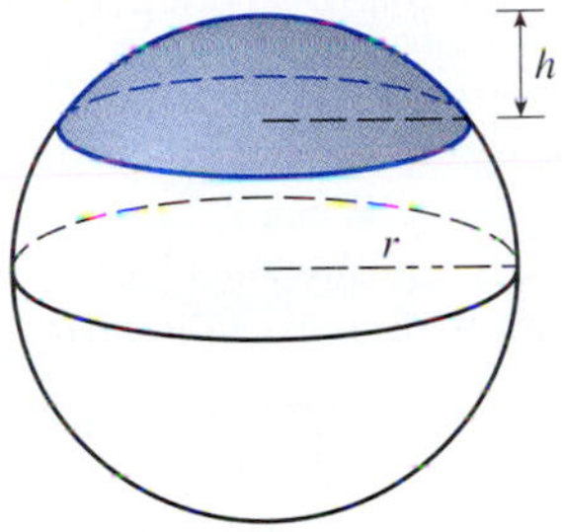

52. A frustum of a pyramid with square base of side b, square top of side a, and height h

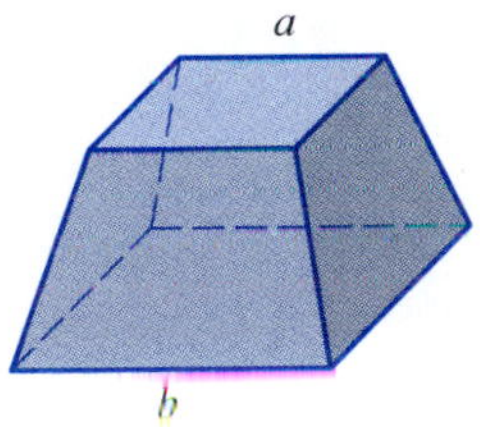

53. A pyramid with height h and rectangular base with dimensions b and $2b$

54. A pyramid with height h and base an equilateral triangle with side a (a tetrahedron)

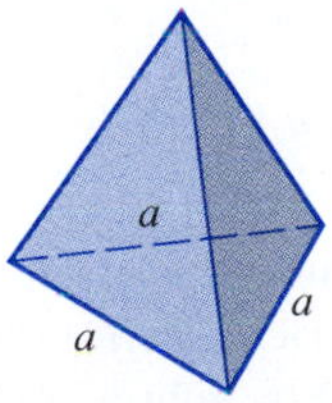

55. A tetrahedron with three mutually perpendicular faces and three mutually perpendicular edges with lengths 3 cm, 4 cm, and 5 cm

56. The base of S is a circular disk with radius r. Parallel cross-sections perpendicular to the base are squares.

57. The base of S is an elliptical region with boundary curve $9x^2 + 4y^2 = 36$. Cross-sections perpendicular to the x-axis are isosceles right triangles with hypotenuse in the base.

58. The base of S is the parabolic region $\{(x, y) \mid x^2 \leq y \leq 1\}$. Cross-sections perpendicular to the y-axis are equilateral triangles.

59. S has the same base as in Exercise 58 but cross-sections perpendicular to the y-axis are squares.

60. The base of S is the triangular region with vertices $(0, 0)$, $(2, 0)$, and $(0, 1)$. Cross-sections perpendicular to the x-axis are semicircles.

61. S has the same base as in Exercise 60 but cross-sections perpendicular to the x-axis are isosceles triangles with height equal to the base.

62. The base of S is a circular disk with radius r. Parallel cross-sections perpendicular to the base are isosceles triangles with height h and unequal side in the base.
(a) Set up an integral for the volume of S.
(b) By interpreting the integral as an area, find the volume of S.

63. (a) Set up an integral for the volume of a solid torus (the donut-shaped solid shown in the figure) with radii r and R.
(b) By interpreting the integral as an area, find the volume of the torus.

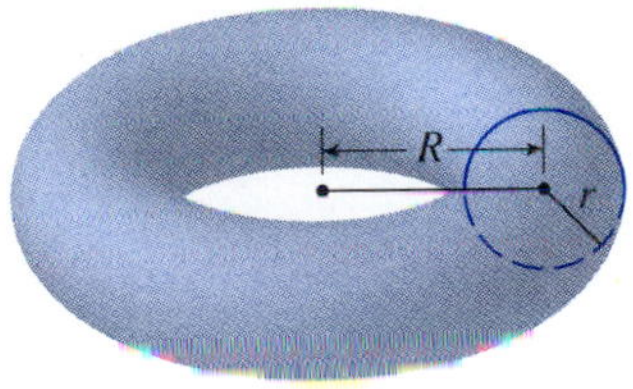

64. Find the volume common to two circular cylinders, each with radius r, if the axes of the cylinders intersect at right angles.

65. Find the volume common to two spheres, each with radius r, if the center of each sphere lies on the surface of the other sphere.

66. A bowl is shaped like a hemisphere with diameter 30 cm. A ball with diameter 10 cm is placed in the bowl and water is poured into the bowl to a depth of h centimeters. Find the volume of water in the bowl.

67. Solve Example 9 taking cross-sections to be parallel to the line of intersection of the two planes.

68. Solve Example 9 if the planes intersect at an angle of 45°.

69. A hole of radius r is bored through a cylinder of radius $R > r$ at right angles to the axis of the cylinder. Set up, but do not evaluate, an integral for the volume cut out.

70. A hole of radius r is bored through the center of a sphere of radius $R > r$. Find the volume of the remaining portion of the sphere.

71. (a) Cavalieri's Principle states that if a family of parallel planes gives equal cross-sectional areas for two solids S_1 and S_2, then the volumes of S_1 and S_2 are equal. Prove Cavalieri's Principle.

(b) Use this principle to find the volume of the oblique cylinder shown in the figure.

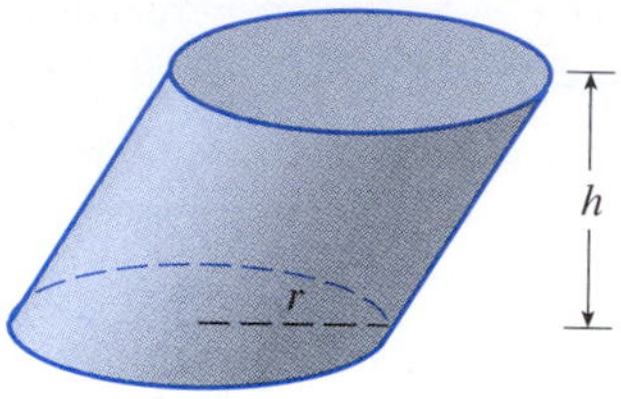

72. A log 10 m long is cut at 1-meter intervals and its cross-sectional areas A (at a distance x from the end of the log) are listed in the following table. Use the Midpoint Rule with $n = 5$ to estimate the volume of the log.

x (m)	0	1	2	3	4	5	6	7	8	9	10
A (m^2)	0.68	0.65	0.64	0.61	0.58	0.59	0.53	0.55	0.52	0.50	0.48

73. Some of the pioneers of calculus, such as Kepler and Newton, were inspired by the problem of finding the volumes of wine barrels. (In fact Kepler published a book *Stereometria doliorum* in 1715 devoted to methods for finding the volumes of barrels.) They often approximated the shape of the sides by parabolas.

(a) A barrel with height h and maximum radius R is constructed by rotating about the x-axis the parabola $y = R - cx^2$, $-h/2 \leq x \leq h/2$, where c is a positive constant. Show that the radius of each end of the barrel is $r = R - d$, where $d = ch^2/4$.

(b) Show that the volume enclosed by the barrel is

$$V = \tfrac{1}{3}\pi h\left(2R^2 + r^2 - \tfrac{2}{5}d^2\right)$$

74. Suppose that a region $\mathcal{R}$ has area A and lies above the x-axis. When $\mathcal{R}$ is rotated about the x-axis, it sweeps out a solid with volume V_1. When $\mathcal{R}$ is rotated about the line $y = -k$ (where k is a positive number), it sweeps out a solid with volume V_2. Express V_2 in terms of V_1, k, and A.

75. Water in an open bowl evaporates at a rate proportional to the area of the surface of the water. (This means that the rate of decrease of the volume is proportional to the area of the surface.) Show that the depth of the water decreases at a constant rate, regardless of the shape of the bowl.

7.3 VOLUMES BY CYLINDRICAL SHELLS

Some volume problems are very difficult to handle by the methods of the preceding section. For instance, we consider the problem of finding the volume of the solid obtained by rotating about the y-axis the region bounded by $y = x(x-1)^2$ and $y = 0$. If we were to use the "washer method," we would first have to locate the local maximum point (a, b) of $y = x(x-1)^2$ using the methods of Chapter 5. Then we would have to solve the equation $y = x(x-1)^2$ for x in terms of y to obtain the functions $x = g_1(y)$ and $x = g_2(y)$ shown in Figure 1. This step would be difficult because it involves the cubic formula. Finally we would find the volume using

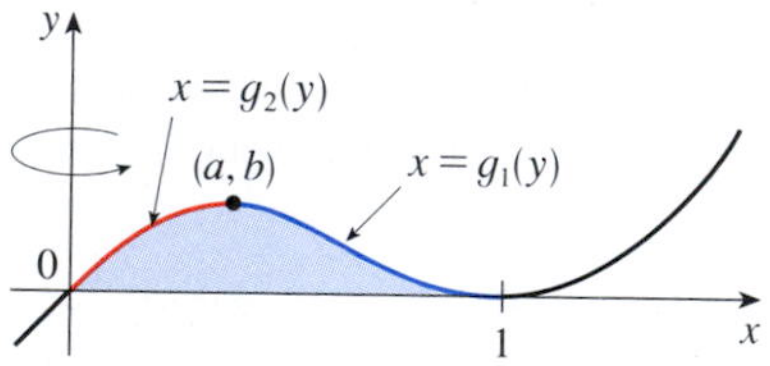

FIGURE 1

$$V = \pi \int_0^b \{[g_1(y)]^2 - [g_2(y)]^2\}\,dy$$

Fortunately, there is a method, called the **method of cylindrical shells**, that is easier to use in such a case. Figure 2 shows a cylindrical shell with inner radius r_1, outer radius r_2, and height h. Its volume V is calculated by subtracting the volume V_1 of the inner cylinder from the volume V_2 of the outer cylinder:

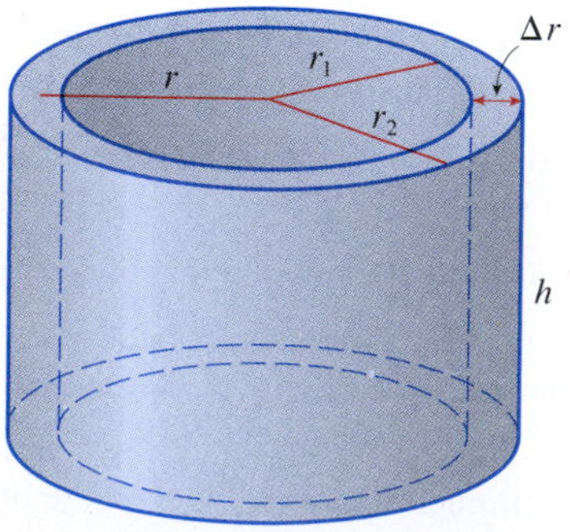

FIGURE 2

$$\begin{aligned} V &= V_2 - V_1 \\ &= \pi r_2^2 h - \pi r_1^2 h = \pi(r_2^2 - r_1^2)h \\ &= \pi(r_2 + r_1)(r_2 - r_1)h \\ &= 2\pi \frac{r_2 + r_1}{2} h(r_2 - r_1) \end{aligned}$$

If we let $\Delta r = r_2 - r_1$ (the thickness of the shell) and $r = \frac{1}{2}(r_2 + r_1)$ (the average radius of the shell), then this formula for the volume of a cylindrical shell becomes

(1)
$$V = 2\pi r h \Delta r$$

and it can be remembered as

$$V = [\text{circumference}]\,[\text{height}]\,[\text{thickness}]$$

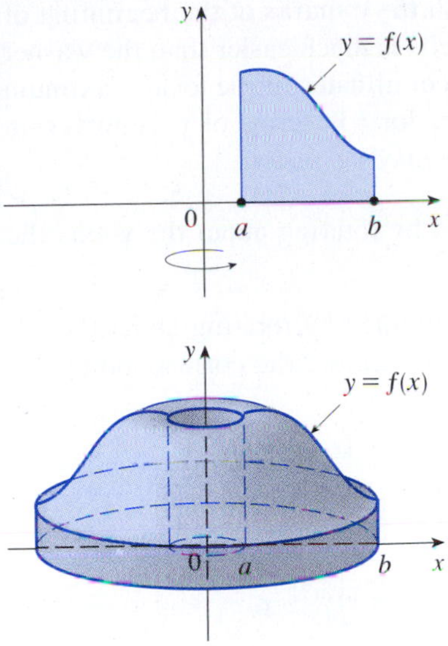

FIGURE 3

Now let S be the solid obtained by rotating about the y-axis the region bounded by $y = f(x)$ [where $f(x) \geqslant 0$], $y = 0$, $x = a$, and $x = b$, where $b > a \geqslant 0$ (see Figure 3). Let P be a partition of $[a, b]$ by points x_i with $a = x_0 < x_1 < \cdots < x_n = b$ and let x_i^* be the midpoint of $[x_{i-1}, x_i]$, that is, $x_i^* = \frac{1}{2}(x_{i-1} + x_i)$. If the rectangle with base $[x_{i-1}, x_i]$ and height $f(x_i^*)$ is rotated about the y-axis, then the result is a cylindrical shell with average radius x_i^*, height $f(x_i^*)$, and thickness $\Delta x_i = x_i - x_{i-1}$ (see Figure 4), so by Formula 1 its volume is

$$V_i = 2\pi x_i^* f(x_i^*) \Delta x_i$$

Therefore an approximation to the volume V of S is given by the sum of the volumes of these shells:

$$V \approx \sum_{i=1}^{n} V_i = \sum_{i=1}^{n} 2\pi x_i^* f(x_i^*) \Delta x_i$$

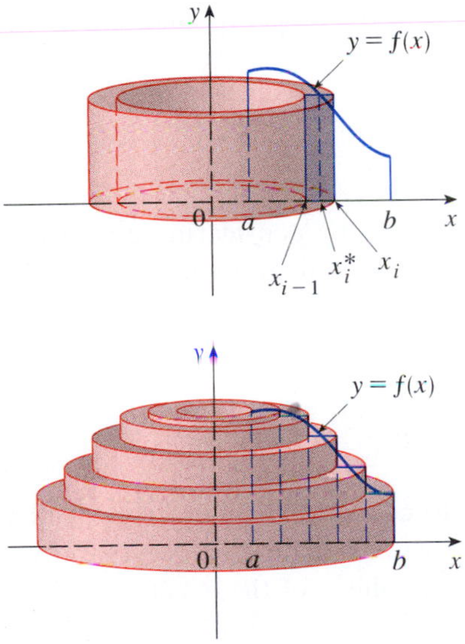

FIGURE 4

This approximation appears to become better as $\|P\| \to 0$. But, from the definition of an integral, we know that

$$\lim_{\|P\| \to 0} \sum_{i=1}^{n} 2\pi x_i^* f(x_i^*) \Delta x_i = \int_a^b 2\pi x f(x)\, dx$$

Thus, the following formula appears to be plausible.

(2) The volume of the solid in Figure 3 is

$$V = \int_a^b 2\pi x f(x)\, dx \qquad \text{where } 0 \leqslant a < b$$

The argument using cylindrical shells makes Formula 2 seem reasonable, but later we will be able to prove it (see Exercise 61 in Section 8.1).

EXAMPLE 1 Find the volume of the solid obtained by rotating about the y-axis the region bounded by $y = x(x - 1)^2$ and $y = 0$.

SOLUTION The region and solid are sketched in Figures 1 and 5. Formula 2 gives the volume of rotation as

$$V = \int_0^1 2\pi x[x(x - 1)^2]\, dx$$

$$= 2\pi \int_0^1 (x^4 - 2x^3 + x^2)\, dx$$

$$= 2\pi \left[\frac{x^5}{5} - 2\frac{x^4}{4} + \frac{x^3}{3} \right]_0^1 = \frac{\pi}{15}$$ ■

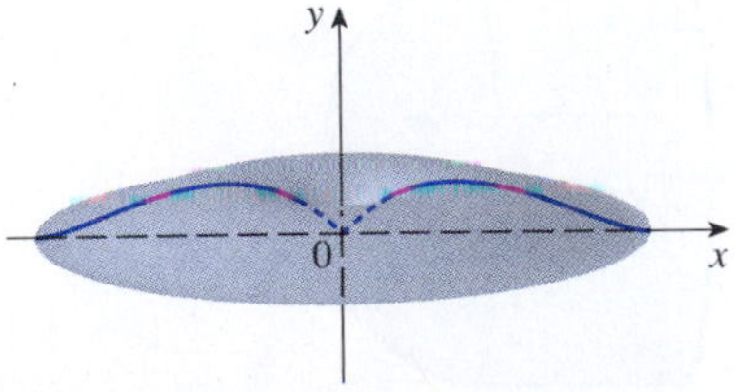

FIGURE 5

NOTE: Comparing the solution of Example 1 with the remarks at the beginning of this section, we see that the method of cylindrical shells is much easier than the washer method for this problem. We did not have to find the coordinates of the local maximum and we did not have to solve the equation of the curve for x in terms of y. However, in other examples the methods of the preceding section may be easier.

EXAMPLE 2 Find the volume of the solid obtained by rotating about the y-axis the region between $y = x$ and $y = x^2$.

SOLUTION We use Formula 2 to find the volume obtained by rotating about the y-axis the region under $y = x$ from 0 to 1 and then we subtract the corresponding volume for the region under $y = x^2$. Thus

$$V = \int_0^1 2\pi x \cdot x\,dx - \int_0^1 2\pi x \cdot x^2\,dx$$

$$= 2\pi \int_0^1 (x^2 - x^3)\,dx = 2\pi \left[\frac{x^3}{3} - \frac{x^4}{4} \right]_0^1 = \frac{\pi}{6}$$

Comparison with Example 6 in Section 7.2 shows that the washer and shell methods take about the same amount of work in this problem. ■

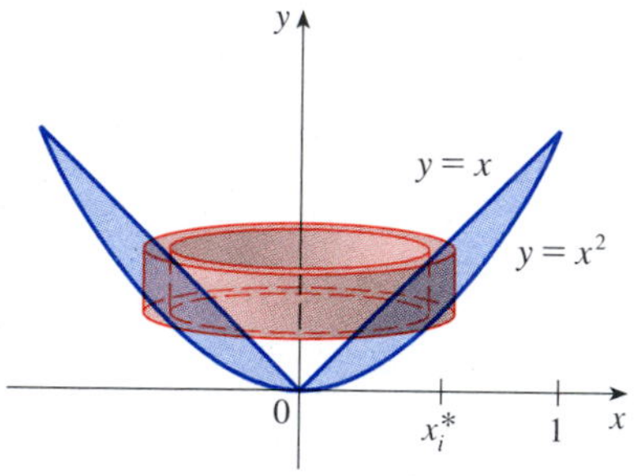

FIGURE 6

Another way of looking at Example 2 is to consider cylindrical shells of height $x_i^* - (x_i^*)^2$ as shown in Figure 6. This leads to

$$V = \int_0^1 2\pi x(x - x^2)\,dx$$

In general, by subtracting volumes as in Example 2 or by considering cylindrical shells directly, we see that the volume of the solid generated by rotating about the y-axis the region between the curves $y = f(x)$ and $y = g(x)$ from a to b [where $f(x) \geqslant g(x)$ and $0 \leqslant a < b$] is

$$V = \int_a^b 2\pi x[f(x) - g(x)]\,dx$$

The method of cylindrical shells also allows us to compute volumes of revolution about the x-axis. If we interchange the roles of x and y in Formula 2, we see that if we rotate the region in Figure 7 about the x-axis, then the volume of the resulting solid is

(3) $$V = \int_c^d 2\pi y g(y)\,dy$$

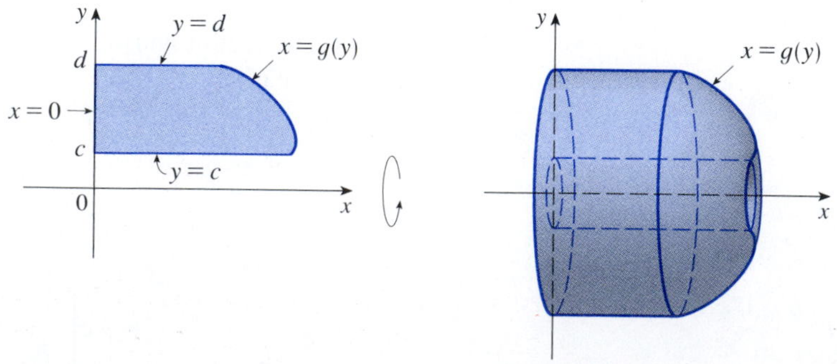

FIGURE 7

EXAMPLE 3 Use cylindrical shells to find the volume of the solid obtained by rotating about the x-axis the region under the curve $y = \sqrt{x}$ from 0 to 1.

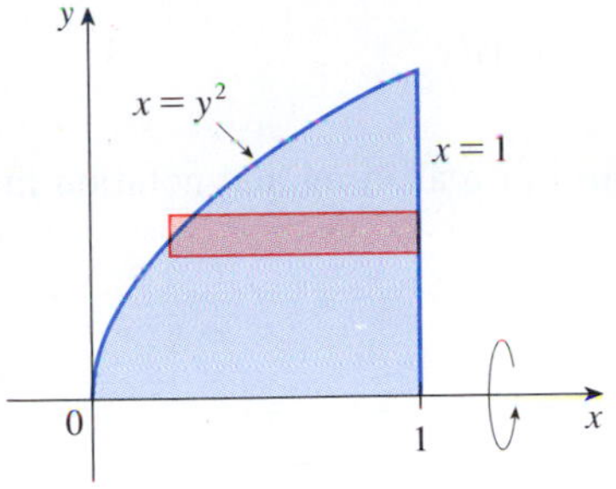

FIGURE 8

SOLUTION This problem was solved using disks in Example 2 in Section 7.2. To use shells we relabel $y = \sqrt{x}$ in the figure in that example as $x = y^2$ in Figure 8. Then by using Formula 3 and subtracting volumes, we get

$$V = \int_0^1 2\pi y \cdot 1\,dy - \int_0^1 2\pi y \cdot y^2\,dy$$

$$= 2\pi \int_0^1 (y - y^3)\,dy = 2\pi\left[\frac{y^2}{2} - \frac{y^4}{4}\right]_0^1 = \frac{\pi}{2}$$

In this problem the disk method was simpler. ■

EXAMPLE 4 Find the volume of the solid obtained by rotating the region bounded by $y = x - x^2$ and $y = 0$ about the line $x = 2$.

SOLUTION Since the axis of rotation is $x = 2$ instead of the y-axis, we cannot use Formula 2 so we go back to the basic method of deriving it using cylindrical shells. Figure 9 shows a rectangle with base $[x_{i-1}, x_i]$ rotated about $x = 2$ to form a cylindrical shell with average radius $2 - x_i^*$, height $x_i^* - (x_i^*)^2$, and thickness Δx_i, where x_i^* is the midpoint of $[x_{i-1}, x_i]$.

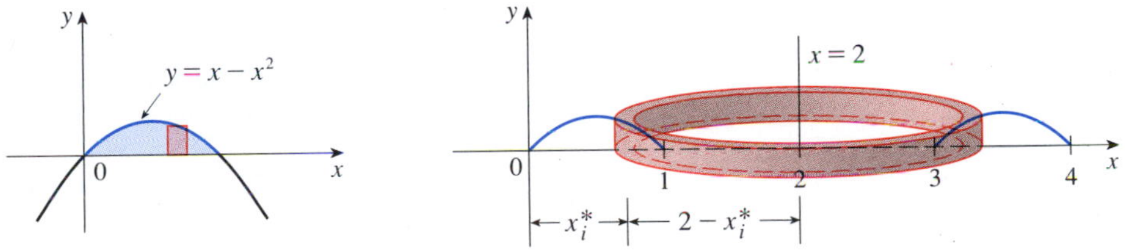

FIGURE 9

By Formula 1 the volume of this shell is

$$\underbrace{2\pi(2 - x_i^*)}_{\text{circumference}}\ \underbrace{[x_i^* - (x_i^*)^2]}_{\text{height}}\ \underbrace{\Delta x_i}_{\text{thickness}}$$

and so the volume of the given solid is

$$V = \lim_{\|P\|\to 0} \sum_{i=1}^{n} 2\pi(2 - x_i^*)[x_i^* - (x_i^*)^2]\,\Delta x_i$$

$$= \int_0^1 2\pi(2 - x)(x - x^2)\,dx$$

$$= 2\pi \int_0^1 (x^3 - 3x^2 + 2x)\,dx$$

$$= 2\pi\left[\frac{x^4}{4} - x^3 + x^2\right]_0^1 = \frac{\pi}{2}$$ ■

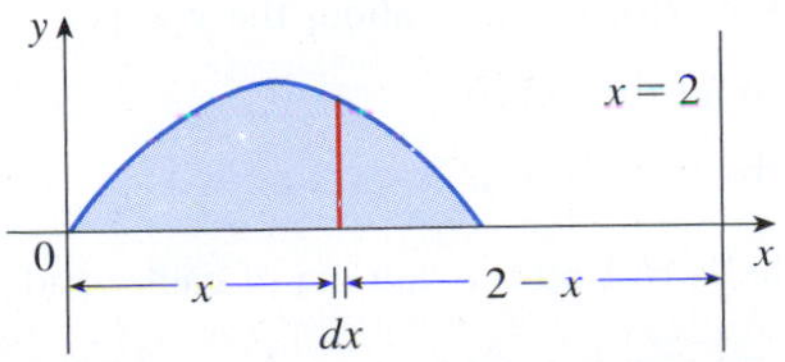

FIGURE 10

The procedure for setting up the integral in Example 4 is often abbreviated heuristically as follows: We think of a very narrow rectangle of width dx as in Figure 10. When rotated about the line $x = 2$, it produces a very thin shell of radius $2 - x$, height $x - x^2$, and thickness dx; therefore, its volume is $dV = 2\pi(2 - x)(x - x^2)\,dx$. The process of adding a large number of such volumes and taking the limit is abbreviated by writing

$$V = \int dV = \int_0^1 2\pi(2 - x)(x - x^2)\,dx$$

It is customary, particularly among scientists, to use similar reasoning and notation in all applications of integration.

EXERCISES 7.3

1–8 ■ Use the method of cylindrical shells to find the volume generated by rotating the region bounded by the given curves about the y-axis.

1. $y = x^2,\quad y = 0,\quad x = 1,\quad x = 2$

2. $y = 1/x,\quad y = 0,\quad x = 1,\quad x = 10$

3. $y = e^{-x^2},\quad y = 0,\quad x = 0,\quad x = 1$

4. $y = \sin(x^2),\quad y = 0,\quad x = 0,\quad x = \sqrt{\pi}$

5. $y = x^2,\quad y = 4,\quad x = 0$

6. $y^2 = x,\quad x = 2y$

7. $y = x^2 - x^3,\quad y = 0$

8. $y = x^2 - 6x + 10,\quad y = -x^2 + 6x - 6$

9–14 ■ Use the method of cylindrical shells to find the volume of the solid obtained by rotating the region bounded by the given curves about the x-axis.

9. $x = \sqrt[4]{y},\quad x = 0,\quad y = 16$

10. $x = y^2,\quad x = 0,\quad y = 2,\quad y = 5$

11. $y = x^2,\quad y = 9$

12. $y^2 - 6y + x = 0,\quad x = 0$

13. $y = \sqrt{x},\quad y = 0,\quad x + y = 2$

14. $y = x,\quad x = 0,\quad x + y = 2$

15–20 ■ Use the method of cylindrical shells as in Example 4 to find the volume generated by rotating the region bounded by the given curves about the specified axis. Sketch the region and a typical shell.

15. $y = \sqrt{x}, y = 0, x = 1, x = 4$; about the y-axis

16. $y = x^2, y = 0, x = -2, x = -1$; about the y-axis

17. $y = x^2, y = 0, x = 1, x = 2$; about $x = 1$

18. $y = x^2, y = 0, x = 1, x = 2$; about $x = 4$

19. $y = \sqrt{x - 1}, y = 0, x = 5$; about $y = 3$

20. $y = 4x - x^2, y = 8x - 2x^2$; about $x = -2$

21–26 ■ Set up, but do not evaluate, an integral for the volume of the solid obtained by rotating the region bounded by the given curves about the specified axis.

21. $y = \sin x, y = 0, x = 2\pi, x = 3\pi$; about the y-axis

22. $y = 1/(1 + x^2), y = 0, x = 0, x = 3$; about the y-axis

23. $x = \cos y, x = 0, y = 0, y = \pi/4$; about the x-axis

24. $y = -x^2 + 7x - 10, y = x - 2$; about the x-axis

25. $y = x^4, y = \sin(\pi x/2)$; about $x = -1$

26. $x = 4 - y^2, x = 8 - 2y^2$; about $y = 5$

27–30 ■ Each integral represents the volume of a solid. Describe the solid.

27. $\int_0^{\pi/2} 2\pi x \cos x\,dx$

28. $\int_0^9 2\pi y^{3/2}\,dy$

29. $\int_0^1 2\pi(x^3 - x^7)\,dx$

30. $\int_0^{\pi} 2\pi(4 - x)\sin^4 x\,dx$

31–32 ■ Use a graph to estimate the x-coordinates of the points of intersection of the given curves. Then use this information to estimate the volume of the solid obtained by rotating about the y-axis the region enclosed by these curves.

31. $y = 0,\quad y = x + x^2 - x^4$

32. $y = x^4,\quad y = 3x - x^3$

33–38 ■ The region bounded by the given curves is rotated about the specified axis. Find the volume of the resulting solid by any method.

33. $y = x^2 + x - 2, y = 0$; about the x-axis

34. $y = x^2 - 3x + 2, y = 0$; about the y-axis

35. $x = 1 - y^2, x = 0$; about the y-axis

36. $y = x\sqrt{1 + x^3}, y = 0, x = 0, x = 2$; about the y-axis

37. $x^2 + (y - 1)^2 = 1$; about the y-axis

38. $x^2 + (y - 1)^2 = 1$; about the x-axis

39–41 ■ Use cylindrical shells to find the volume of each solid.

39. A sphere of radius r

40. The solid torus of Exercise 63 in Section 7.2

41. A right circular cone with height h and base radius r

42. Suppose you make napkin rings by drilling holes with different diameters through two wooden balls (which also have different diameters). You discover that both napkin rings have the same height h, as shown in the figure.

(a) Guess which ring has more wood in it.

(b) Check your guess: Use cylindrical shells to compute the volume of a napkin ring created by drilling a hole with radius r through the center of a sphere of radius R and express the answer in terms of h.

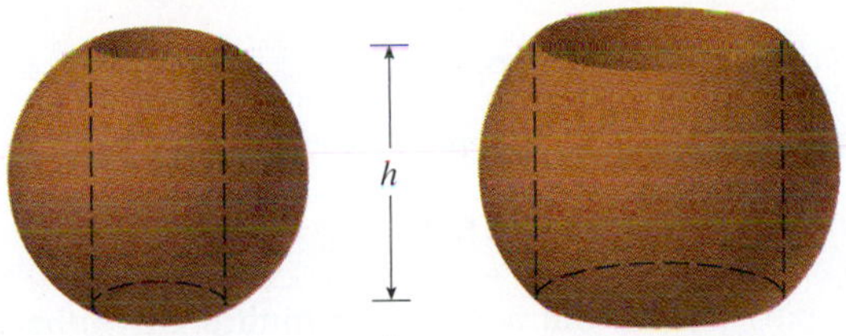

43. Formula 2 is valid only if $a \geqslant 0$. Show that if $a < b \leqslant 0$, the volume formula becomes

$$V = -\int_a^b 2\pi x f(x)\,dx$$

44. Find a formula for the volume of the solid obtained when the region under the graph of f from a to b is rotated about the vertical line $x = c$ in the cases (a) $c \leqslant a < b$ and (b) $a < b \leqslant c$.

45. Use the Midpoint Rule with $n = 4$ to estimate the volume obtained by rotating about the y-axis the region under the curve $y = \tan x$, $0 \leqslant x \leqslant \pi/4$.

46. If the region shown in the figure is rotated about the y-axis to form a solid, use the Midpoint Rule with $n = 5$ to estimate the volume of the solid.

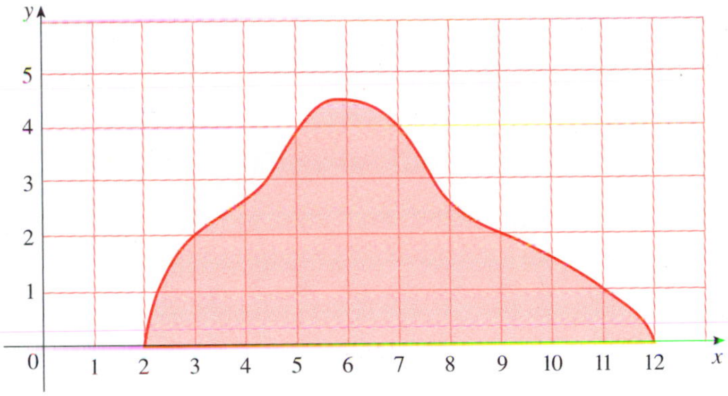

7.4 WORK

As mentioned in Section 1.2, the term **work** is used in everyday language to mean the total amount of effort required to perform a task. In physics it has a technical meaning that depends on the idea of a **force**. Intuitively you can think of a force as describing a push or pull on an object—for example, a horizontal push of a book across a table or the downward pull of Earth's gravity on a ball. In general, if an object moves along a straight line with position function $s(t)$, then the force F on the object (in the same direction) is defined by Newton's Second Law of Motion as the product of its mass m and its acceleration:

(1)
$$F = m\frac{d^2s}{dt^2}$$

In the SI metric system, the mass is measured in kilograms (kg), the displacement in meters (m), the time in seconds (s), and the force in newtons ($\text{N} = \text{kg}\cdot\text{m/s}^2$). Thus, a force of 1 N acting on a mass of 1 kg produces an acceleration of 1 m/s^2. In the British engineering system the fundamental unit is chosen to be the unit of force, which is the pound.

In the case of constant acceleration, the force F is also constant and the work done is defined to be the product of the force F and the distance d that the object moves:

(2)
$$W = Fd \qquad \text{work} = \text{force} \times \text{distance}$$

If F is measured in newtons and d in meters, then the unit for W is a newton-meter, which is called a joule (J). If F is measured in pounds and d in feet, then the unit for W is a foot-pound (ft-lb), which is about 1.36 J.

EXAMPLE 1
(a) How much work is done in lifting a 1.2-kg book off the floor to put it on a desk that is 0.7 m high? Use the fact that the acceleration due to gravity is $g = 9.8 \text{ m/s}^2$.
(b) How much work is done in lifting a 20-lb weight 6 ft off the ground?

SOLUTION
(a) The force exerted is equal and opposite to that exerted by gravity, so Equation 1 gives

$$F = mg = (1.2)(9.8) = 11.76 \text{ N}$$

and then Equation 2 gives the work done as

$$W = Fd = (11.76)(0.7) \approx 8.2 \text{ J}$$

(b) Here the force is given as $F = 20$ lb, so the work done is

$$W = Fd = 20 \cdot 6 = 120 \text{ ft-lb}$$

Notice that in part (b), unlike part (a), we did not have to multiply by g because we were given the *weight* (which is a force) and not the mass of the object. ■

Equation 2 defines work as long as the force is constant, but what happens if the force is variable? Let's suppose that the object moves along the x-axis in the positive direction from $x = a$ to $x = b$ and at each point x between a and b a force $f(x)$ acts on the object, where f is a continuous function. Let P be a partition of $[a, b]$ by points x_i $(i = 1, 2, \ldots, n)$ and let $\Delta x_i = x_i - x_{i-1}$. Choose any x_i^* in the ith subinterval $[x_{i-1}, x_i]$. Then the force at that point is $f(x_i^*)$. If $\|P\|$ is small, then Δx_i is small, and since f is continuous, the values of f do not change very much over the interval $[x_{i-1}, x_i]$. In other words, f is almost constant on the interval and so the work W_i that is done in moving the particle from x_{i-1} to x_i is approximately given by Equation 2:

$$W_i \approx f(x_i^*)\,\Delta x_i$$

Thus, we can approximate the total work by

$$W \approx \sum_{i=1}^{n} f(x_i^*)\,\Delta x_i \tag{3}$$

It seems that this approximation becomes better as we make $\|P\|$ smaller (and therefore n larger). Therefore, we define the **work done in moving the object from a to b** as the limit of this quantity as $\|P\| \to 0$. Since the right side of (3) is a Riemann sum, we recognize its limit as being a definite integral, and so

$$W = \lim_{\|P\| \to 0} \sum_{i=1}^{n} f(x_i^*)\,\Delta x_i = \int_a^b f(x)\,dx \tag{4}$$

EXAMPLE 2 When a particle is at a distance x feet from the origin, a force of $x^2 + 2x$ pounds acts on it. How much work is done in moving it from $x = 1$ to $x = 3$?

SOLUTION

$$W = \int_1^3 (x^2 + 2x)\,dx = \frac{x^3}{3} + x^2 \Big]_1^3 = \frac{50}{3}$$

The work done is $16\frac{2}{3}$ ft-lb. ■

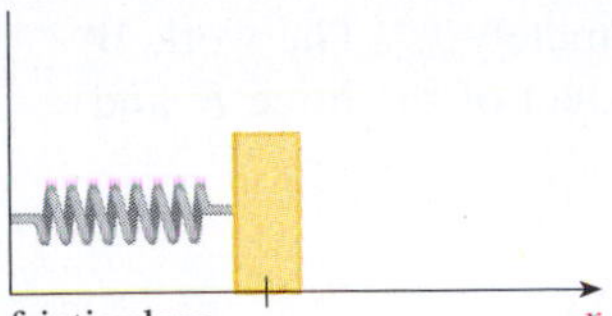

(a) Natural position of spring

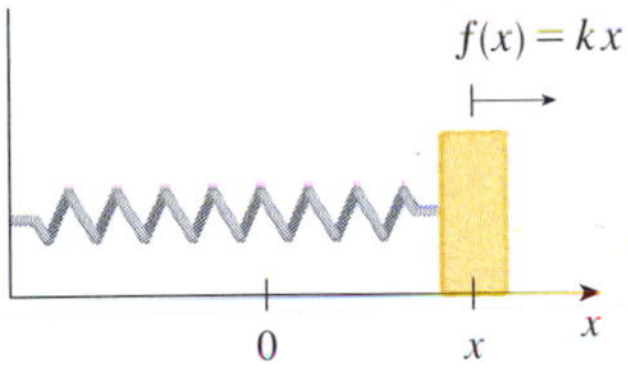

(b) Stretched position of spring

FIGURE 1
Hooke's Law

In the next example we use **Hooke's Law** from physics, which states that the force required to maintain a spring stretched x units beyond its natural length is proportional to x:

$$f(x) = kx$$

where k is a positive constant (called the **spring constant**). Hooke's Law holds provided that x is not too large (see Figure 1).

EXAMPLE 3 A force of 40 N is required to hold a spring that has been stretched from its natural length of 10 cm to a length of 15 cm. How much work is done in stretching the spring from 15 cm to 18 cm?

SOLUTION According to Hooke's Law, the force required to hold the spring stretched x meters beyond its natural length is $f(x) = kx$. When the spring is stretched from 10 cm to 15 cm, the amount stretched is 5 cm = 0.05 m. This means that $f(0.05) = 40$, so

$$0.05k = 40 \qquad k = \frac{40}{0.05} = 800$$

Thus, $f(x) = 800x$ and the work done in stretching the spring from 15 cm to 18 cm is

$$W = \int_{0.05}^{0.08} 800x\,dx = 800\frac{x^2}{2}\bigg]_{0.05}^{0.08}$$

$$= 400[(0.08)^2 - (0.05)^2] = 1.56 \text{ J}$$ ■

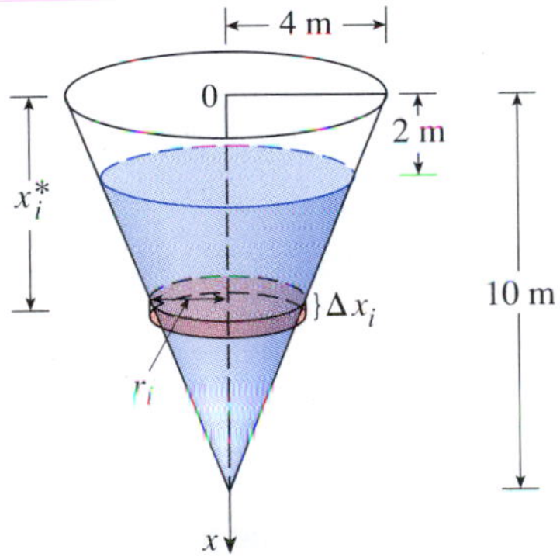

FIGURE 2

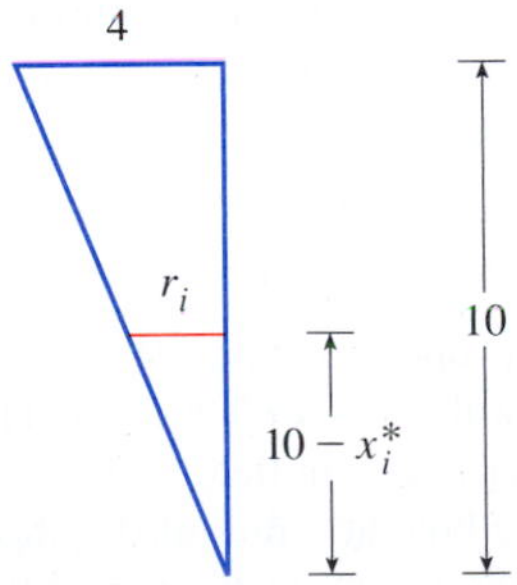

FIGURE 3

EXAMPLE 4 A tank has the shape of an inverted circular cone with height 10 m and base radius 4 m. It is filled with water to a height of 8 m. Find the work required to empty the tank by pumping all of the water to the top of the tank. (The density of water is 1000 kg/m^3.)

SOLUTION Let's measure depths from the top of the tank by introducing a vertical coordinate line as in Figure 2. The water extends from a depth of 2 m to a depth of 10 m and so we take a partition P of the interval $[2, 10]$ by points x_i with $2 = x_0 < x_1 < \cdots < x_n = 10$ and choose x_i^* in the ith subinterval. This divides the water into n layers. The ith layer is approximated by a circular cylinder with radius r_i and height Δx_i. We can compute r_i from similar triangles using Figure 3 as follows:

$$\frac{r_i}{10 - x_i^*} = \frac{4}{10} \qquad r_i = \tfrac{2}{5}(10 - x_i^*)$$

Thus, an approximation to the volume of the ith layer of water is

$$V_i \approx \pi r_i^2\,\Delta x_i = \frac{4\pi}{25}(10 - x_i^*)^2\,\Delta x_i$$

and so its mass is

$$m_i = \text{density} \times \text{volume}$$

$$\approx 1000 \cdot \frac{4\pi}{25}(10 - x_i^*)^2\,\Delta x_i = 160\pi(10 - x_i^*)^2\,\Delta x_i$$

The force required to raise this layer must overcome the force of gravity and so

$$F_i = m_i g \approx (9.8)160\pi(10 - x_i^*)^2\,\Delta x_i$$

$$\approx 1570\pi(10 - x_i^*)^2\,\Delta x_i$$

Each particle in the layer must travel a distance of approximately x_i^*. The work W_i done to raise this layer to the top is approximately the product of the force F_i and the distance x_i^*:

$$W_i \approx F_i x_i^* \approx 1570\pi x_i^*(10 - x_i^*)^2 \Delta x_i$$

To find the total work done in emptying the entire tank, we add the contributions of each of the n layers and then take the limit as $\|P\| \to 0$:

$$\begin{aligned} W &= \lim_{\|P\|\to 0} \sum_{i=1}^{n} 1570\pi x_i^*(10 - x_i^*)^2 \Delta x_i \\ &= \int_2^{10} 1570\pi x(10 - x)^2\, dx \\ &= 1570\pi \int_2^{10} (100x - 20x^2 + x^3)\, dx \\ &= 1570\pi \left[50x^2 - \frac{20x^3}{3} + \frac{x^4}{4} \right]_2^{10} \\ &= 1570\pi\left(\tfrac{2048}{3}\right) \approx 3.4 \times 10^6 \text{ J} \end{aligned}$$

■

EXERCISES 7.4

1. Find the work done in pushing a car a distance of 8 m while exerting a constant force of 900 N.

2. How much work is done by a weightlifter in raising a 60-kg barbell from the floor to a height of 2 m?

3. A particle is moved along the x-axis by a force that measures $5x^2 + 1$ pounds at a point x feet from the origin. Find the work done in moving the particle from the origin to a distance of 10 ft.

4. When a particle is at a distance x meters from the origin, a force of $\cos(\pi x/3)$ newtons acts on it. How much work is done in moving the particle from $x = 1$ to $x = 2$?

5. A force of 10 lb is required to hold a spring stretched 4 in. beyond its natural length. How much work is done in stretching it from its natural length to 6 in. beyond its natural length?

6. A spring has a natural length of 20 cm. If a 25-N force is required to keep it stretched to a length of 30 cm, how much work is required to stretch it from 20 cm to 25 cm?

7. Suppose that 2 J of work are needed to stretch a spring from its natural length of 30 cm to a length of 42 cm. How much work is needed to stretch it from 35 cm to 40 cm?

8. If the work required to stretch a spring 1 ft beyond its natural length is 12 ft-lb, how much work is needed to stretch it 9 in. beyond its natural length?

9. How far beyond its natural length will a force of 30 N keep the spring in Exercise 7 stretched?

10. If 6 J of work are needed to stretch a spring from 10 cm to 12 cm and another 10 J are needed to stretch it from 12 cm to 14 cm, what is the natural length of the spring?

11. A heavy rope, 50 ft long, weighs 0.5 lb/ft and hangs over the edge of a building 120 ft high. How much work is done in pulling the rope to the top of the building?

12. A uniform cable hanging over the edge of a tall building is 40 ft long and weighs 60 lb. How much work is required to pull 10 ft of the cable to the top?

13. A cable that weighs 2 lb/ft is used to lift 800 lb of coal up a mineshaft 500 ft deep. Find the work done.

14. A bucket that weighs 4 lb and a rope of negligible weight are used to draw water from a well that is 80 ft deep. The bucket starts with 40 lb of water and is pulled up at a rate of 2 ft/s but water leaks out of a hole at a rate of 0.2 lb/s. Find the work done in pulling the bucket to the top of the well.

15. An aquarium 2 m long, 1 m wide, and 1 m deep is full of water. Find the work needed to pump half of the water out of the aquarium.

16. A circular swimming pool has a diameter of 24 ft, the sides are 5 ft high, and the depth of the water is 4 ft. How

much work is required to pump all of the water out over the side?

17–20 ■ A tank is full of water. Find the work required to pump the water out of the outlet. In Exercises 19 and 20 use the fact that water weighs 62.5 lb/ft^3.

17.

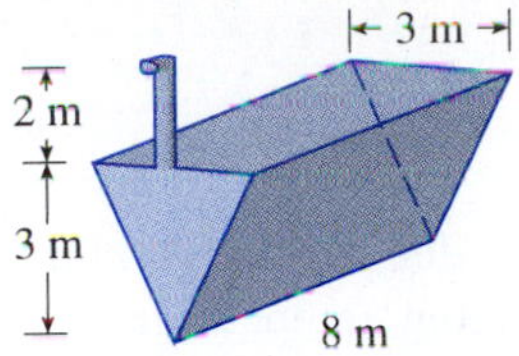

18.

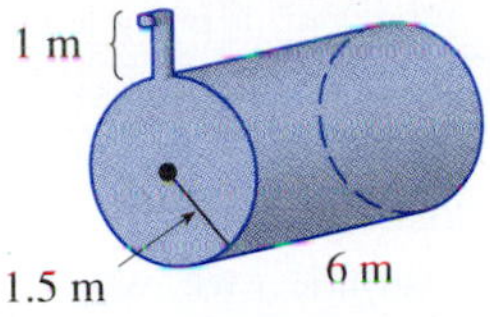

19.

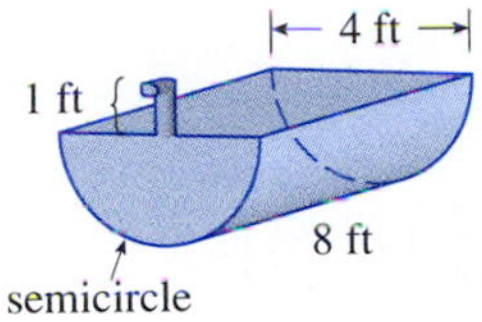

20.

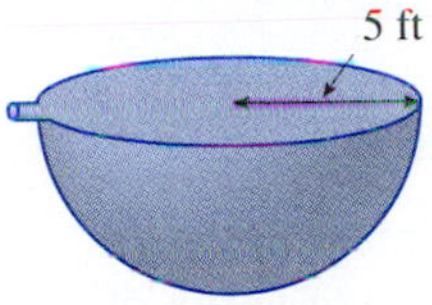

21. Suppose that for the tank in Exercise 17 the pump breaks down after 4.7×10^5 J of work has been done. What is the depth of the water remaining in the tank?

22. Solve Exercise 18 if the tank is half full of oil that has a density of 920 kg/m^3.

23. When gas expands in a cylinder with radius r, the pressure at any given time is a function of the volume: $P = P(V)$. The force exerted by the gas on the piston (see the figure) is the product of the pressure and the area: $F = \pi r^2 P$. Show that the work done by the gas when the volume expands from volume V_1 to volume V_2 is

$$W = \int_{V_1}^{V_2} P\,dV$$

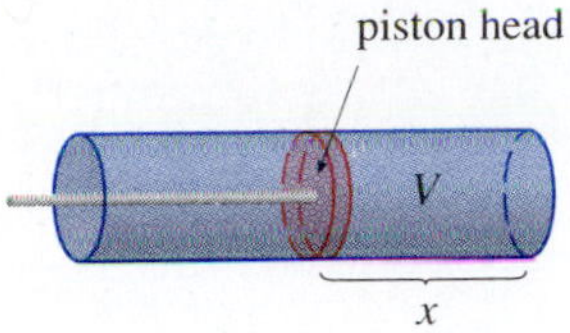

24. In a steam engine the pressure P and volume V of steam satisfy the equation $PV^{1.4} = k$, where k is a constant. (This is true for adiabatic expansion, that is, expansion in which there is no heat transfer between the cylinder and its surroundings.) Use Exercise 23 to calculate the work done by the engine during a cycle when the steam starts at a pressure of 160 lb/in^2 and a volume of 100 in^3 and expands to a volume of 800 in^3.

25. Newton's Law of Gravitation states that two bodies with masses m_1 and m_2 attract each other with a force

$$F = G\frac{m_1 m_2}{r^2}$$

where r is the distance between the bodies and G is the gravitational constant. If one of the bodies is fixed, find the work needed to move the other from $r = a$ to $r = b$.

26. Use Newton's Law of Gravitation to compute the work required to launch a 1000-kg satellite vertically to an orbit 1000 km high. You may assume that Earth's mass is 5.98×10^{24} kg and is concentrated at its center. Take the radius of Earth to be 6.37×10^6 m and $G = 6.67 \times 10^{-11}$ N·m^2/kg^2.

7.5 AVERAGE VALUE OF A FUNCTION

It's easy to calculate the average value of finitely many numbers $y_1, y_2, \ldots, y_n$:

$$y_{\text{ave}} = \frac{y_1 + y_2 + \cdots + y_n}{n}$$

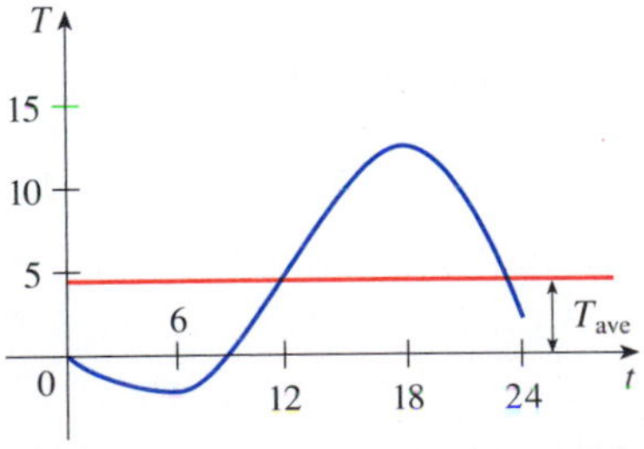

FIGURE 1

But how do we compute the average temperature during a day if infinitely many temperature readings are possible? Figure 1 shows the graph of a temperature function $T(t)$ (where t is measured in hours, T in °C) and a guess at the average temperature, T_{ave}.

In general, let us try to compute the average value of a function $y = f(x)$, $a \leqslant x \leqslant b$. We start by dividing the interval $[a, b]$ into n equal subintervals, each with length $\Delta x = (b - a)/n$. Then we choose points $x_1^*, \ldots, x_n^*$ in successive subintervals and calculate the average of the numbers $f(x_1^*), \ldots, f(x_n^*)$:

$$\frac{f(x_1^*) + \cdots + f(x_n^*)}{n}$$

(For example, if f represents a temperature function and $n = 24$, this means that we take temperature readings every hour and average them.) Since $\Delta x = (b - a)/n$, we can write $n = (b - a)/\Delta x$ and the average value becomes

$$\frac{f(x_1^*) + \cdots + f(x_n^*)}{\dfrac{b-a}{\Delta x}} = \frac{1}{b-a}\left[f(x_1^*)\,\Delta x + \cdots + f(x_n^*)\,\Delta x\right]$$

$$= \frac{1}{b-a}\sum_{i=1}^{n} f(x_i^*)\,\Delta x$$

If we let n increase, we would be computing the average value of a large number of closely spaced values. (For example, we would be averaging temperature readings taken every minute or even every second.) The limiting value is

$$\lim_{n\to\infty}\frac{1}{b-a}\sum_{i=1}^{n} f(x_i^*)\,\Delta x = \frac{1}{b-a}\int_a^b f(x)\,dx$$

by the definition of a definite integral.

Therefore, we define the **average value of f** on the interval $[a, b]$ as

(1)
$$f_{\text{ave}} = \frac{1}{b-a}\int_a^b f(x)\,dx$$

EXAMPLE 1 Find the average value of the function $f(x) = 1 + x^2$ over the interval $[-1, 2]$.

SOLUTION With $a = -1$ and $b = 2$ we have

$$f_{\text{ave}} = \frac{1}{b-a}\int_a^b f(x)\,dx = \frac{1}{2-(-1)}\int_{-1}^{2}(1 + x^2)\,dx$$

$$= \frac{1}{3}\left[x + \frac{x^3}{3}\right]_{-1}^{2} = 2$$

■

The question arises: Is there a number c at which the value of f is exactly equal to the average value of the function, that is, $f(c) = f_{\text{ave}}$? The following theorem says that this is true for continuous functions.

MEAN VALUE THEOREM FOR INTEGRALS If f is continuous on $[a, b]$, then there exists a number c in $[a, b]$ such that

$$\int_a^b f(x)\,dx = f(c)(b-a)$$

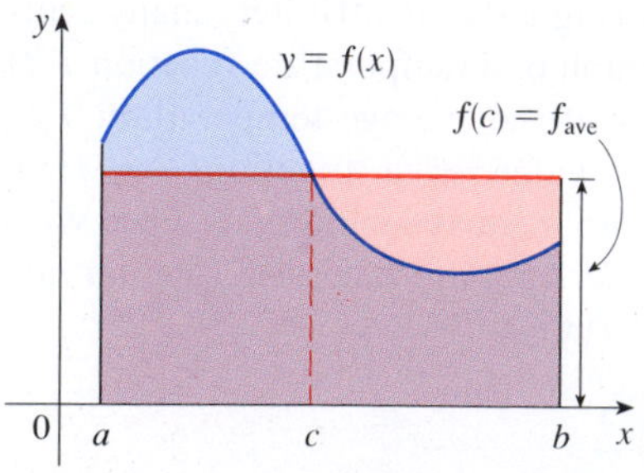

FIGURE 2

The Mean Value Theorem for Integrals is a consequence of the Mean Value Theorem for derivatives. The proof is outlined in Exercise 19.

The geometric interpretation of the Mean Value Theorem for Integrals is that, for *positive* functions f, there is a number c such that the rectangle with base $[a, b]$ and height $f(c)$ has the same area as the region under the graph of f from a to b (see Figure 2).

EXAMPLE 2 Since $f(x) = 1 + x^2$ is continuous on the interval $[-1, 2]$, the Mean Value Theorem for Integrals says there is a number c in $[-1, 2]$ such that

$$\int_{-1}^{2} (1 + x^2)\,dx = f(c)[2 - (-1)]$$

In this particular case we can find c explicitly. From Example 1 we know that

$$f(c) = f_{\text{ave}} = 2$$

Therefore $\quad 1 + c^2 = 2 \qquad c^2 = 1$

Thus, in this case there happen to be two numbers $c = \pm 1$ in the interval $[-1, 2]$ that work in the Mean Value Theorem for Integrals. ■

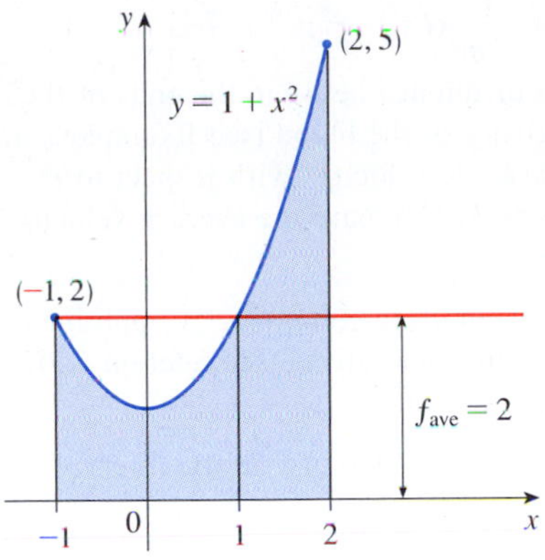

FIGURE 3

Examples 1 and 2 are illustrated by Figure 3.

EXAMPLE 3 Show that the average velocity of a car over a time interval $[t_1, t_2]$ is the same as the average of its velocities during the trip.

SOLUTION If $s(t)$ is the displacement of the car at time t, then, by definition, the average velocity of the car over the interval is

$$\frac{\Delta s}{\Delta t} = \frac{s(t_2) - s(t_1)}{t_2 - t_1}$$

On the other hand, the average value of the velocity function on the interval is

$$\begin{aligned} v_{\text{ave}} &= \frac{1}{t_2 - t_1}\int_{t_1}^{t_2} v(t)\,dt = \frac{1}{t_2 - t_1}\int_{t_1}^{t_2} s'(t)\,dt \\ &= \frac{1}{t_2 - t_1}[s(t_2) - s(t_1)] \qquad \text{(by the Fundamental Theorem)} \\ &= \frac{s(t_2) - s(t_1)}{t_2 - t_1} = \text{average velocity} \end{aligned}$$

■

EXERCISES 7.5

1–6 ■ Find the average value of f on the given interval.

1. $f(x) = x^2 - 2x, \quad [0, 3]$
2. $f(x) = \sin x, \quad [0, \pi]$
3. $f(x) = x^4, \quad [-1, 1]$
4. $f(x) = 1/x, \quad [1, 4]$
5. $f(x) = \sin^2 x \cos x, \quad [-\pi/2, \pi/4]$
6. $f(x) = \sqrt{x}, \quad [4, 9]$

7–10 ■
(a) Find the average value of f on the given interval.
(b) Find c such that $f_{\text{ave}} = f(c)$.
(c) Sketch the graph of f and a rectangle whose area is the same as the area under the graph of f.

7. $f(x) = 4 - x^2, \quad [0, 2]$
8. $f(x) = 4x - x^2, \quad [0, 3]$
9. $f(x) = x^3 - x + 1, \quad [0, 2]$
10. $f(x) = x\sin(x^2), \quad [0, \sqrt{\pi}]$
11. If f is continuous and $\int_1^3 f(x)\,dx = 8$, show that f takes on the value 4 at least once on the interval $[1, 3]$.
12. Find the numbers b such that the average value of $f(x) = 2 + 6x - 3x^2$ on the interval $[0, b]$ is equal to 3.
13. The temperature (in °F) in a certain city t hours after 9 A.M. was approximated by the function

$$T(t) = 50 + 14\sin\frac{\pi t}{12}$$

Find the average temperature during the period from 9 A.M. to 9 P.M.

14. The temperature of a metal rod, 5 m long, is $4x$ (in °C) at a distance x meters from one end of the rod. What is the average temperature of the rod?

15. The linear density in a rod 8 m long is $12/\sqrt{x+1}$ kg/m, where x is measured in meters from one end of the rod. Find the average density of the rod.

16. If a freely falling body starts from rest, then its displacement is given by $s = \frac{1}{2}gt^2$. Let the velocity after a time T be v_T. Show that if we compute the average of the velocities with respect to t we get $v_{\text{ave}} = \frac{1}{2}v_T$, but if we compute the average of the velocities with respect to s we get $v_{\text{ave}} = \frac{2}{3}v_T$.

17. Use the model given in Exercise 79 in Section 6.5 to compute the average volume of inhaled air in the lungs in one respiratory cycle.

18. The velocity v of blood that flows in a blood vessel with radius R and length l at a distance r from the central axis is

$$v(r) = \frac{P}{4\eta l}(R^2 - r^2)$$

where P is the pressure difference between the ends of the vessel and η is the viscosity of the blood (see Example 7 in Section 3.3). Find the average velocity (with respect to r) over the interval $0 \leqslant r \leqslant R$. Compare the average velocity with the maximum velocity.

19. Prove the Mean Value Theorem for Integrals by applying the Mean Value Theorem for derivatives (see Section 5.3) to the function $F(x) = \int_a^x f(t)\,dt$.

20. If $f_{\text{ave}}[a, b]$ denotes the average value of f on the interval $[a, b]$ and $a < c < b$, show that

$$f_{\text{ave}}[a, b] = \frac{c-a}{b-a} f_{\text{ave}}[a, c] + \frac{b-c}{b-a} f_{\text{ave}}[c, b]$$

7 REVIEW

KEY TOPICS ■ Define, state, or discuss the following.

1. Area between curves
2. Volume of a general solid
3. Volume of a solid of revolution: disk (or washer) method
4. Volume of a solid of revolution: shell method
5. Work
6. Average value of a function
7. Mean Value Theorem for Integrals

EXERCISES

1. Two regions $\mathcal{R}_1$ and $\mathcal{R}_2$ are shown in the figure. Express each of the following quantities as an integral.
(a) The area of $\mathcal{R}_1$
(b) The area of $\mathcal{R}_2$
(c) The volume of the solid obtained by rotating $\mathcal{R}_1$ about the x-axis
(d) The volume of the solid obtained by rotating $\mathcal{R}_1$ about the y-axis
(e) The volume of the solid obtained by rotating $\mathcal{R}_2$ about the x-axis
(f) The volume of the solid obtained by rotating $\mathcal{R}_2$ about the y-axis

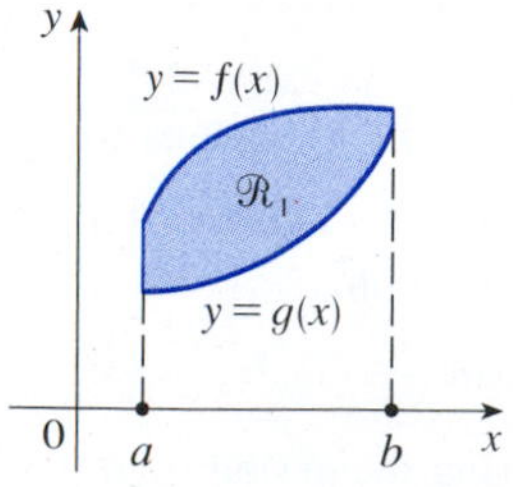

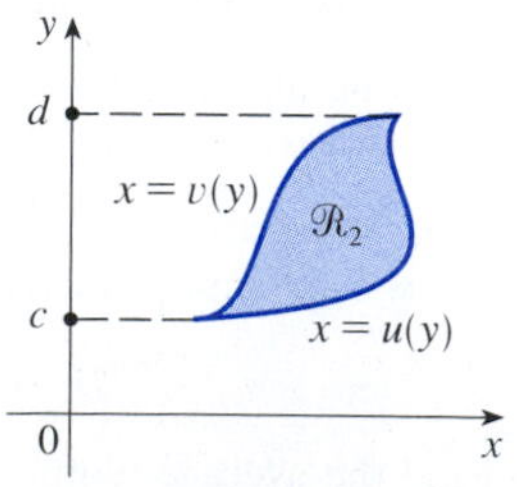

2–8 ■ Find the area of the region bounded by the given curves.

2. $y = 4 + 3x - x^2$, $y = 0$

3. $y = x^2 - 6x$, $y = 12x - 2x^2$

4. $y = x^2 - 6$, $y = 12 - x^2$, $x = -5$, $x = 5$

5. $y = x^3$, $x = y^3$

6. $x - 2y + 7 = 0$, $y^2 - 6y - x = 0$

7. $y = \sin x$, $y = -\cos x$, $x = 0$, $x = \pi$

8. $y = x^3$, $y = x^2 - 4x + 4$, $x = 0$, $x = 2$

9–13 ■ Find the volume of the solid obtained by rotating the region bounded by the given curves about the specified axis.

9. $y = \sqrt{x-1}$, $y = 0$, $x = 3$; about the x-axis

10. $y = x^3$, $y = x^2$; about the x-axis

11. $x + 3 = 4y - y^2$, $x = 0$; about the x-axis

12. $y = x^3$, $y = 8$, $x = 0$; about the y-axis

13. $x^2 - y^2 = a^2$, $x = a + h$ (where $a > 0$, $h > 0$); about the y-axis

14–16 ■ Set up, but do not evaluate, an integral for the volume of the solid obtained by rotating the region bounded by the given curves about the specified axis.

14. $y = \cos x$, $y = 0$, $x = 3\pi/2$, $x = 5\pi/2$; about the y-axis

15. $y = x^3$, $y = x^2$; about $y = 1$

16. $y = x^3$, $y = 8$, $x = 0$; about $x = 2$

17. Find the volumes of the solids obtained by rotating the region bounded by the curves $y = x$ and $y = x^2$ about the following lines:
(a) The x-axis (b) The y-axis (c) $y = 2$

18. Let $\mathcal{R}$ be the region in the first quadrant bounded by the curves $y = x^3$ and $y = 2x - x^2$. Calculate the following quantities.
(a) The area of $\mathcal{R}$
(b) The volume obtained by rotating $\mathcal{R}$ about the x-axis
(c) The volume obtained by rotating $\mathcal{R}$ about the y-axis

19. Let $\mathcal{R}$ be the region bounded by the curves $y = \tan(x^2)$, $x = 1$, and $y = 0$. Use the Midpoint Rule with $n = 4$ to estimate the following.
(a) The area of $\mathcal{R}$
(b) The volume obtained by rotating $\mathcal{R}$ about the x-axis

20. Let $\mathcal{R}$ be the region bounded by the curves $y = 1 - x^2$ and $y = x^6 - x + 1$. Estimate the following.
(a) The x-coordinates of the points of intersection of the curves
(b) The area of $\mathcal{R}$
(c) The volume generated when $\mathcal{R}$ is rotated about the x-axis
(d) The volume generated when $\mathcal{R}$ is rotated about the y-axis

21–24 ■ Each integral represents the volume of a solid. Describe the solid.

21. $\int_0^{\pi} \pi \sin^2 x\, dx$

22. $\int_0^{\pi} 2\pi x \sin x\, dx$

23. $\int_0^{2} 2\pi y(4 - y^2)\, dy$

24. $\int_0^{1} \pi[(2 - x^2)^2 - (2 - \sqrt{x})^2]\, dx$

25. The base of a solid is a circular disk with radius 3. Find the volume of the solid if parallel cross-sections perpendicular to the base are isosceles right triangles with hypotenuse lying along the base.

26. The base of a solid is the region bounded by the parabolas $y = x^2$ and $y = 2 - x^2$. Find the volume of the solid if the cross-sections perpendicular to the x-axis are squares with one side lying along the base.

27. The height of a monument is 20 m. A horizontal cross-section at a distance x meters from the top is an equilateral triangle with side $x/4$ meters. Find the volume of the monument.

28. (a) The base of a solid is a square with vertices at $(1, 0)$, $(0, 1)$, $(-1, 0)$, and $(0, -1)$. Each cross-section perpendicular to the x-axis is a semicircle. Find the volume of the solid.
(b) Show that by cutting the solid of part (a), we can rearrange it to form a cone. Thus compute its volume more simply.

29. A force of 30 N is required to maintain a spring stretched from its natural length of 12 cm to a length of 15 cm. How much work is done in stretching the spring from 12 cm to 20 cm?

30. A 1600-lb elevator is suspended by a 200-ft cable that weighs 10 lb/ft. How much work is required to raise the elevator from the basement to the third floor, a distance of 30 ft?

31. A tank full of water has the shape of a paraboloid of revolution as in the figure; that is, its shape is obtained by rotating a parabola about a vertical axis.
(a) If its height is 4 ft and the radius at the top is 4 ft, find the work required to pump the water out of the tank.
(b) After 4000 ft-lb of work has been done, what is the depth of the water remaining in the tank?

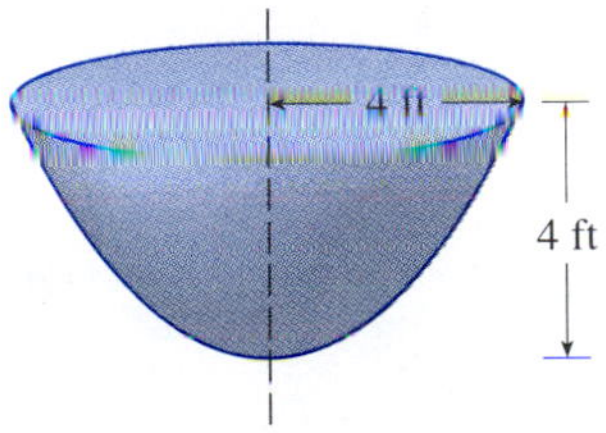

32. Find the average value of the function $f(x) = x^3$ on the interval $[2, 4]$.

33. If f is a continuous function, what is the limit as $h \to 0$ of the average value of f over the interval $[x, x + h]$?

34. Let $\mathcal{R}_1$ be the region bounded by $y = x^2$, $y = 0$, and $x = b$, where $b > 0$. Let $\mathcal{R}_2$ be the region bounded by $y = x^2$, $x = 0$, and $y = b^2$.
(a) Is there a value of b such that $\mathcal{R}_1$ and $\mathcal{R}_2$ have the same area?
(b) Is there a value of b such that $\mathcal{R}_1$ sweeps out the same volume when rotated about the x-axis and the y-axis?
(c) Is there a value of b such that $\mathcal{R}_1$ and $\mathcal{R}_2$ sweep out the same volume when rotated about the x-axis?
(d) Is there a value of b such that $\mathcal{R}_1$ and $\mathcal{R}_2$ sweep out the same volume when rotated about the y-axis?

Cover the solution to the example and try it yourself first.

EXAMPLE Evaluate $\lim_{x \to 3} \left(\frac{x}{x - 3} \int_3^x \frac{\sin t}{t}\, dt \right)$.

SOLUTION Let's start by having a preliminary look at the ingredients of the function. What happens to the first factor, $x/(x - 3)$, when x approaches 3? The numerator approaches 3 and the denominator approaches 0, so we have

$$\frac{x}{x-3} \to \infty \quad \text{as} \quad x \to 3^+ \qquad \text{and} \qquad \frac{x}{x-3} \to -\infty \quad \text{as} \quad x \to 3^-$$

The second factor approaches $\int_3^3 (\sin t)/t\, dt$, which is 0. It's not clear what happens to the function as a whole. (One factor is becoming large while the other is becoming small.) So how do we proceed?

The principles of problem solving are discussed in Section 4 of Review and Preview.

One of the principles of problem solving is: *Try to recognize something familiar.* Is there a part of the function that reminds us of something we've seen before? Well, the integral

$$\int_3^x \frac{\sin t}{t}\, dt$$

has x as its upper limit of integration and that type of integral occurs in Part 1 of the Fundamental Theorem of Calculus:

$$\frac{d}{dx} \int_a^x f(t)\, dt = f(x)$$

This suggests that differentiation might be involved.

Once we start thinking about differentiation, the denominator $(x - 3)$ reminds us of something else that should be familiar: One of the forms of the definition of the derivative in Chapter 3 is

$$F'(a) = \lim_{x \to a} \frac{F(x) - F(a)}{x - a}$$

and with $a = 3$ this becomes

$$F'(3) = \lim_{x \to 3} \frac{F(x) - F(3)}{x - 3}$$

So what is the function F in our situation? Notice that if we define

$$F(x) = \int_3^x \frac{\sin t}{t}\, dt$$

then $F(3) = 0$. What about the factor x in the numerator? That's just a red herring, so let's factor it out and put together the calculation:

$$\begin{aligned}
\lim_{x \to 3} \left(\frac{x}{x - 3} \int_3^x \frac{\sin t}{t}\, dt \right) &= \left(\lim_{x \to 3} x \right) \cdot \lim_{x \to 3} \frac{\int_3^x \frac{\sin t}{t}\, dt}{x - 3} \\
&= 3 \lim_{x \to 3} \frac{F(x) - F(3)}{x - 3} = 3F'(3) \\
&= 3 \frac{\sin 3}{3} \qquad \text{(Part 1 of the Fundamental Theorem)} \\
&= \sin 3
\end{aligned}$$

■

PROBLEMS

These problems involve all of the material of this book to the end of Chapter 7. To solve a particular problem you might have to combine a formula from Chapter 6 with a technique from Chapter 3. You might even have to look up a formula from trigonometry that you have forgotten.

1. If $x \sin \pi x = \int_0^{x^2} f(t)\,dt$, where f is a continuous function, find $f(4)$.

2. In this problem we approximate the sine function on the interval $[0, \pi]$ by three quadratic functions, each of which has the same zeros as the sine function on this interval.
   (a) Find a quadratic function f such that $f(0) = f(\pi) = 0$ and which has the same maximum value as sin on $[0, \pi]$.
   (b) Find a quadratic function g such that $g(0) = g(\pi) = 0$ and which has the same rate of change as the sine function at 0 and π.
   (c) Find a quadratic function h such that $h(0) = h(\pi) = 0$ and the area under h from 0 to π is the same as for the sine function.
   (d) Illustrate by graphing f, g, h, and the sine function in the same viewing rectangle $[0, \pi]$ by $[0, 1]$. Identify which graph belongs to each function.

3. Show that $\dfrac{1}{17} \leqslant \displaystyle\int_1^2 \frac{1}{1 + x^4}\,dx \leqslant \frac{7}{24}$.

4. Find the point on the parabola $y = 1 - x^2$ at which the tangent line cuts from the first quadrant the triangle with the smallest area.

5. Suppose the curve $y = f(x)$ passes through the origin and the point $(1, 1)$. Find the value of the integral $\int_0^1 f'(x)\,dx$.

6. (a) Graph several members of the family of functions $f(x) = (2cx - x^2)/c^3$ for $c > 0$ and look at the regions enclosed by these curves and the x-axis. Make a conjecture about how the areas of these regions are related.
   (b) Prove your conjecture in part (a).
   (c) Take another look at the graphs in part (a) and use them to sketch the curve traced out by the vertices (highest points) of the family of functions. Can you guess what kind of curve this is?
   (d) Find the equation of the curve you sketched in part (c).

7. Find a function f such that $f(1) = -1$, $f(4) = 7$, and $f'(x) > 3$ for all x, or prove that such a function cannot exist.

8. Find a continuous function f such that $\int_0^x f(t)\,dt = 3f(x) - 2$ for all x.

9. A solid is generated by rotating about the x-axis the region bounded by the x-axis, the y-axis, and the curve $y = f(x)$, $x \geqslant 0$, where f is a positive function. The volume generated by the part of the curve from $x = 0$ to $x = b$ is b^2 for all $b > 0$. Find the function f.

10. Use an integral to estimate the sum $\displaystyle\sum_{i=1}^{10000} \sqrt{i}$.

11. Suppose that a line intersects the parabola $y = x^2$ in two points A and B as shown in the figure. Let C be the point on the parabola where the tangent line is parallel to the line through A and B. Show that the area of the parabolic segment cut off from the parabola by the line is four-thirds the area of triangle ABC.

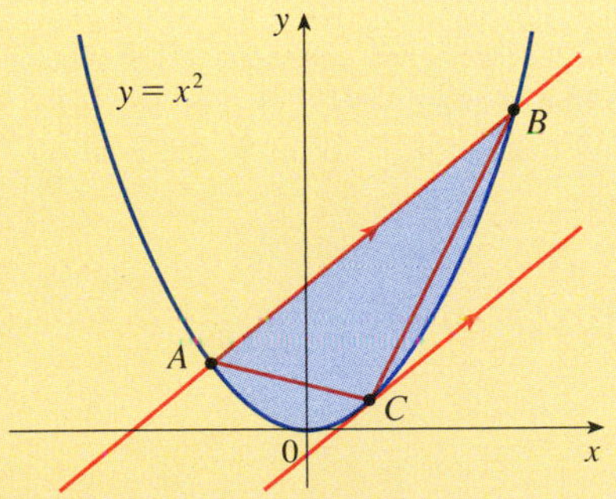

FIGURE FOR PROBLEM 11

12. If $f(x) = \displaystyle\int_0^{g(x)} \frac{1}{\sqrt{1 + t^3}}\,dt$, where $g(x) = \displaystyle\int_0^{\cos x} [1 + \sin(t^2)]\,dt$, find $f'(\pi/2)$.

13. Evaluate $\displaystyle\lim_{x \to 0} \frac{1}{x} \int_0^x (1 - \tan 2t)^{1/t}\,dt$.

14. A right circular cone with height 1 meter and base radius r is to be separated into three pieces of equal volume by cutting twice parallel to the base. At what heights should the cuts be made?

15. Find the interval $[a, b]$ for which the value of the integral $\int_a^b (2 + x - x^2)\,dx$ is a maximum.

16. Suppose f is continuous, $f(0) = 0$, $f(1) = 1$, $f'(x) > 0$, and $\int_0^1 f(x)\,dx = \frac{1}{3}$. Find the value of the integral $\int_0^1 f^{-1}(y)\,dy$.

17. Show that $f(x) = \int_1^x \sqrt{1 + t^3}\,dt$ is one-to-one and find $(f^{-1})'(0)$.

18. The figure shows a region consisting of all points inside a square that are closer to the center than to the sides of the square. Find the area of the region.

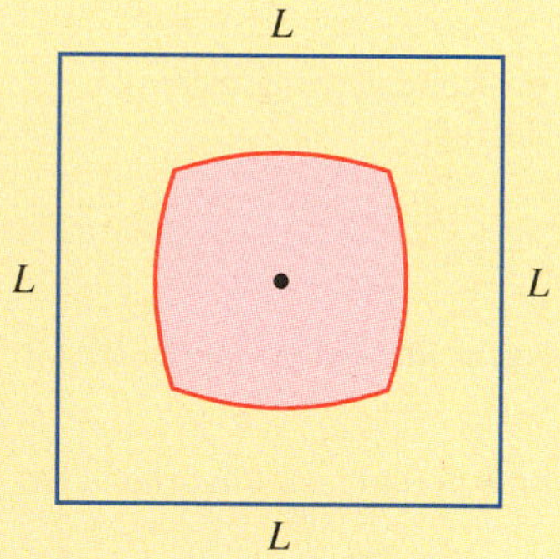

FIGURE FOR PROBLEM 18

19. Find $\dfrac{d^2}{dx^2} \displaystyle\int_0^x \left(\int_1^{\sin t} \sqrt{1 + u^4}\,du \right) dt$.

20. Sketch the set of all points (x, y) such that $|x + y| \leq e^x$.

21. Prove that if f is continuous, then $\displaystyle\int_0^x f(u)(x - u)\,du = \int_0^x \left(\int_0^u f(t)\,dt \right) du$.

22. A circular disk of radius r is used in an evaporator and is rotated in a vertical plane. If it is to be partially submerged in the liquid so as to maximize the exposed wetted area of the disk, show that the center of the disk should be positioned at a height $r/\sqrt{1 + \pi^2}$ above the surface of the liquid.

23. Find the value of c such that the limit

$$\lim_{x \to \infty} x^c e^{-2x} \int_0^x e^{2t} \sqrt{t^2 + 1}\,dt$$

exists (as a finite number) and is nonzero. Also find the value of the limit.

24. A drinking cup filled with water has the shape of a cone with height h and semivertical angle θ (see the figure). A ball is placed carefully in the cup, thereby displacing some of the water and making it overflow. What is the radius of the ball that causes the greatest volume of water to spill out of the cup?

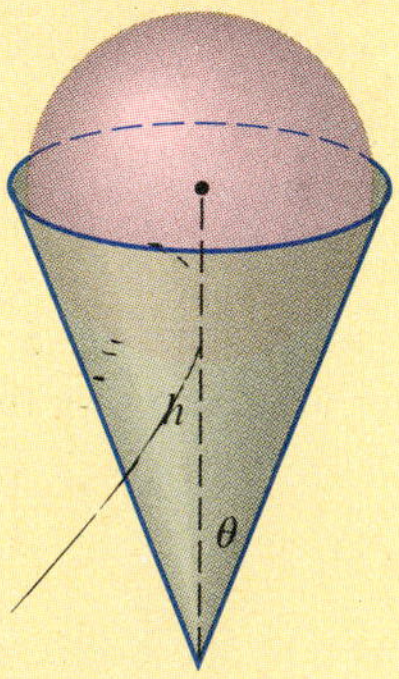

FIGURE FOR PROBLEM 24

25. If $f(x) = \int_0^x x^2 \sin(t^2)\,dt$, find $f'(x)$.

26. (a) Evaluate $\int_0^n [\![x]\!]\,dx$, where n is a positive integer.

(b) Evaluate $\int_a^b [\![x]\!]\,dx$, where a and b are real numbers with $0 \leq a < b$.

27. The figure shows a curve C with the property that, for every point P on the middle curve $y = 2x^2$, the areas A and B are equal. Find an equation for C.

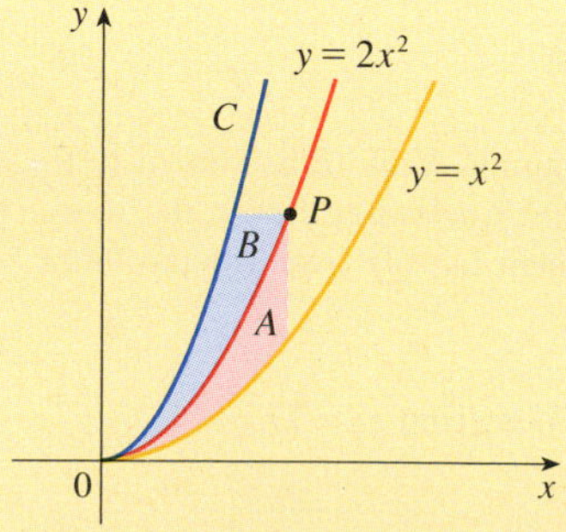

FIGURE FOR PROBLEM 27

28. If the tangent at a point P on the curve $y = x^3$ intersects the curve again at Q, let A be the area of the region bounded by the curve and the line segment PQ. Let B be the area of the region defined in the same way starting with Q instead of P. What is the relationship between A and B?

29. Show that if $a > 0$ and $1 < c < e^a$, then $\int_0^a |e^x - c|\,dx \geq (e^{a/2} - 1)^2$.

30. For any number c, we let $f(x)$ be the smaller of the two numbers $(x - c)^2$ and $(x - c - 2)^2$. Then we define

$$g(c) = \int_0^1 f(x)\,dx$$

Find the maximum and minimum values of $g(c)$ if $-2 \leq c \leq 2$.

31. Find the volume of the solid obtained by rotating the region bounded by the line $y = x$ and the parabola $y = x^2$ about the line $y = x$.

32. In Section 6.1 we found formulas for the sums of the kth powers of the first n integers when $k = 1, 2, 3$, and 4. In this problem we derive formulas for any k. These formulas were first published in 1713 by the Swiss mathematician James Bernoulli in his book *Ars Conjectandi.*

(a) The **Bernoulli polynomials** B_n are defined by $B_0(x) = 1$, $B_n'(x) = B_{n-1}(x)$, and $\int_0^1 B_n(x)\,dx = 0$ for $n = 1, 2, 3, \ldots$. Find $B_n(x)$ for $n = 1, 2, 3$, and 4.

(b) Use the Fundamental Theorem of Calculus to show that $B_n(0) = B_n(1)$ for $n \geqslant 2$.

(c) If we introduce the **Bernoulli numbers** $b_n = n!B_n(0)$, then we can write

$$B_0(x) = b_0 \qquad\qquad B_1(x) = \frac{x}{1!} + \frac{b_1}{1!}$$

$$B_2(x) = \frac{x^2}{2!} + \frac{b_1}{1!}\frac{x}{1!} + \frac{b_2}{2!} \qquad\qquad B_3(x) = \frac{x^3}{3!} + \frac{b_1}{1!}\frac{x^2}{2!} + \frac{b_2}{2!}\frac{x}{1!} + \frac{b_3}{3!}$$

and, in general,

$$B_n(x) = \frac{1}{n!}\sum_{k=0}^{n}\binom{n}{k}b_k x^{n-k} \qquad \text{where} \qquad \binom{n}{k} = \frac{n!}{k!(n-k)!}$$

[The numbers $\binom{n}{k}$ are the binomial coefficients.] Use part (b) to show that, for $n \geqslant 2$,

$$b_n = \sum_{k=0}^{n}\binom{n}{k}b_k$$

and therefore

$$b_{n-1} = -\frac{1}{n}\left[\binom{n}{0}b_0 + \binom{n}{1}b_1 + \binom{n}{2}b_2 + \cdots + \binom{n}{n-2}b_{n-2}\right]$$

This gives an efficient way of computing the Bernoulli numbers and therefore the Bernoulli polynomials.

(d) Show that $B_n(1 - x) = (-1)^n B_n(x)$ and deduce that $b_{2n+1} = 0$ for $n > 0$.

(e) Use parts (c) and (d) to calculate b_6 and b_8. Then calculate the polynomials B_5, B_6, B_7, B_8, and B_9.

(f) Graph the Bernoulli polynomials $B_1, B_2, \ldots, B_9$ for $0 \leqslant x \leqslant 1$. What pattern do you notice in the graphs?

(g) Use mathematical induction to prove that $B_{k+1}(x + 1) - B_{k+1}(x) = \dfrac{x^k}{k!}$.

(h) By putting $x = 0, 1, 2, \ldots, n$ in part (g), prove that

$$1^k + 2^k + 3^k + \cdots + n^k = k![B_{k+1}(n + 1) - B_{k+1}(0)] = k!\int_0^{n+1} B_k(x)\,dx$$

(i) Use part (h) with $k = 3$ and the formula for B_4 in part (a) to confirm the formula for the sum of the first n cubes in Section 6.1.

(j) Show that the formula in part (h) can be written symbolically as

$$1^k + 2^k + 3^k + \cdots + n^k = \frac{1}{k+1}[(n + 1 + b)^{k+1} - b^{k+1}]$$

where the expression $(n + 1 + b)^{k+1}$ is to be expanded formally using the Binomial Theorem and each power b^i is to be replaced by the Bernoulli number b_i.

(k) Use part (j) to find a formula for $1^5 + 2^5 + 3^5 + \cdots + n^5$.

8

TECHNIQUES OF INTEGRATION

> Common integration is only the memory of differentiation. The different devices by which integration is accomplished are changes, not from the known to the unknown, but from forms in which memory will not serve us to those in which it will.
>
> AUGUSTUS DE MORGAN

Because of the Fundamental Theorem of Calculus, we can integrate a function if we know an antiderivative, that is, an indefinite integral. We summarize here the most important integrals that we have learned so far.

$$\int x^n\,dx = \frac{x^{n+1}}{n+1} + C \quad (n \neq -1) \qquad \int \frac{1}{x}\,dx = \ln|x| + C$$

$$\int e^x\,dx = e^x + C \qquad \int a^x\,dx = \frac{a^x}{\ln a} + C$$

$$\int \sin x\,dx = -\cos x + C \qquad \int \cos x\,dx = \sin x + C$$

$$\int \sec^2 x\,dx = \tan x + C \qquad \int \csc^2 x\,dx = -\cot x + C$$

$$\int \sec x \tan x\,dx = \sec x + C \qquad \int \csc x \cot x\,dx = -\csc x + C$$

$$\int \sinh x\,dx = \cosh x + C \qquad \int \cosh x\,dx = \sinh x + C$$

$$\int \tan x\,dx = \ln|\sec x| + C \qquad \int \cot x\,dx = \ln|\sin x| + C$$

$$\int \frac{1}{x^2 + a^2}\,dx = \frac{1}{a}\tan^{-1}\left(\frac{x}{a}\right) + C \qquad \int \frac{1}{\sqrt{a^2 - x^2}}\,dx = \sin^{-1}\left(\frac{x}{a}\right) + C$$

In this chapter we develop techniques for using these basic integration formulas to obtain indefinite integrals of more complicated functions. We learned the most important method of integration, the Substitution Rule, in Section 6.5. The other general technique, integration by parts, is presented in Section 8.1. Then we learn methods that are special to particular classes of functions such as trigonometric functions and rational functions.

Integration is not as straightforward as differentiation; there are no rules that absolutely guarantee obtaining an indefinite integral of a function. Therefore, in Section 8.6 we discuss a strategy for integration.

8.1 INTEGRATION BY PARTS

Every differentiation rule has a corresponding integration rule. For instance, the Substitution Rule for integration corresponds to the Chain Rule for differentiation. The rule that corresponds to the Product Rule for differentiation is called the rule for integration by parts.

The Product Rule states that if f and g are differentiable functions, then

$$\frac{d}{dx}[f(x)g(x)] = f'(x)g(x) + f(x)g'(x)$$

In the notation for indefinite integrals this equation becomes

$$\int [f'(x)g(x) + f(x)g'(x)]\,dx = f(x)g(x)$$

or

$$\int f'(x)g(x)\,dx + \int f(x)g'(x)\,dx = f(x)g(x)$$

We can rearrange this latter equation as

(1)
$$\int f(x)g'(x)\,dx = f(x)g(x) - \int f'(x)g(x)\,dx$$

Formula 1 is called **the formula for integration by parts**. It is perhaps easier to remember in the following notation. Let $u = f(x)$ and $v = g(x)$. Then $du = f'(x)\,dx$ and $dv = g'(x)\,dx$, so, by the Substitution Rule, the formula for integration by parts becomes

(2)
$$\int u\,dv = uv - \int v\,du$$

EXAMPLE 1 Find $\int x \sin x\,dx$.

SOLUTION USING FORMULA 1 Suppose we choose $f(x) = x$ and $g'(x) = \sin x$. Then $f'(x) = 1$ and $g(x) = -\cos x$. (For g we can choose *any* antiderivative of g'.) Thus, using Formula 1, we have

$$\begin{aligned}\int x \sin x\,dx &= f(x)g(x) - \int f'(x)g(x)\,dx \\ &= x(-\cos x) - \int (-\cos x)\,dx \\ &= -x\cos x + \int \cos x\,dx \\ &= -x\cos x + \sin x + C\end{aligned}$$

It is wise to check the answer by differentiating it. If we do so, we get $x \sin x$, as expected.

It is helpful to use the pattern:

$u = \square \qquad dv = \square$
$du = \square \qquad v = \square$

SOLUTION USING FORMULA 2 Let

$$u = x \qquad dv = \sin x\,dx$$

Then

$$du = dx \qquad v = -\cos x$$

and so

$$\int x \sin x\,dx = \int \overset{u}{x}\,\overbrace{\sin x\,dx}^{dv} = \overset{u}{x}\,\overbrace{(-\cos x)}^{v} - \int \overbrace{(-\cos x)}^{v}\,\overbrace{dx}^{du}$$

$$= -x\cos x + \int \cos x\,dx$$

$$= -x\cos x + \sin x + C$$

■

NOTE: Our objective in using integration by parts is to obtain a simpler integral than the one we started with. Thus, in Example 1 we started with $\int x \sin x\,dx$ and expressed it in terms of the simpler integral $\int \cos x\,dx$. If we had chosen $u = \sin x$ and $dv = x\,dx$, then $du = \cos x\,dx$ and $v = x^2/2$, so integration by parts gives

$$\int x \sin x\,dx = (\sin x)\frac{x^2}{2} - \frac{1}{2}\int x^2 \cos x\,dx$$

But $\int x^2 \cos x\,dx$ is a more difficult integral than the one we started with. In general, when deciding on a choice for u and dv, we usually try to choose $u = f(x)$ to be a function that becomes simpler when differentiated (or at least not more complicated) as long as $dv = g'(x)\,dx$ can be readily integrated to give v.

EXAMPLE 2 Evaluate $\int \ln x\,dx$.

SOLUTION Here we do not have much choice for u and dv. Let

$$u = \ln x \qquad dv = dx$$

Then

$$du = \frac{1}{x}\,dx \qquad v = x$$

Integrating by parts, we get

$$\int \ln x\,dx = x\ln x - \int x\,\frac{dx}{x}$$

$$= x\ln x - \int dx$$

$$= x\ln x - x + C$$

Check the answer by differentiating it.

Integration by parts is effective in this example because the derivative of the function $f(x) = \ln x$ is simpler than f. ■

EXAMPLE 3 Find $\int x^2 e^x\,dx$.

SOLUTION Let

$$u = x^2 \qquad dv = e^x\,dx$$

Then

$$du = 2x\,dx \qquad v = e^x$$

Integration by parts gives

$$\int x^2 e^x\,dx = x^2 e^x - 2\int x e^x\,dx \tag{3}$$

The integral that we obtained, $\int xe^x\,dx$, is simpler than the original integral but is still not obvious. Therefore, we use integration by parts a second time, this time with $u = x$ and $dv = e^x\,dx$. Then $du = dx$, $v = e^x$, and

$$\int xe^x\,dx = xe^x - \int e^x\,dx$$
$$= xe^x - e^x + C$$

Putting this in Equation 3, we get

$$\int x^2e^x\,dx = x^2e^x - 2\int xe^x\,dx$$
$$= x^2e^x - 2(xe^x - e^x + C)$$
$$= x^2e^x - 2xe^x + 2e^x + C_1 \qquad \text{where } C_1 = -2C$$ ■

EXAMPLE 4 Evaluate $\int e^x \sin x\,dx$.

SOLUTION Let $u = e^x$ and $dv = \sin x\,dx$. Then $du = e^x\,dx$ and $v = -\cos x$, so integration by parts gives

$$\int e^x \sin x\,dx = -e^x\cos x + \int e^x \cos x\,dx \tag{4}$$

The integral that we have obtained, $\int e^x\cos x\,dx$, is no simpler than the original one, but at least it is no more difficult. Having had success in the preceding example integrating by parts twice, we persevere and integrate by parts again. This time we use $u = e^x$ and $dv = \cos x\,dx$. Then $du = e^x\,dx$, $v = \sin x$, and

$$\int e^x \cos x\,dx = e^x\sin x - \int e^x \sin x\,dx \tag{5}$$

At first glance, it appears as if we have accomplished nothing because we have arrived at $\int e^x \sin x\,dx$, which is where we started. However, if we put Equation 5 into Equation 4 we get

$$\int e^x \sin x\,dx = -e^x\cos x + e^x\sin x - \int e^x \sin x\,dx$$

This can be regarded as an equation to be solved for the unknown integral. Solving, we obtain

$$2\int e^x \sin x\,dx = -e^x\cos x + e^x\sin x$$

and, dividing by 2 and adding the constant of integration, we get

$$\int e^x \sin x\,dx = \tfrac{1}{2}e^x(\sin x - \cos x) + C$$ ■

Figure 1 illustrates Example 4 by showing the graphs of $f(x) = e^x \sin x$ and $F(x) = \frac{1}{2}e^x(\sin x - \cos x)$. As a visual check on our work, notice that $f(x) = 0$ when F has a maximum or minimum.

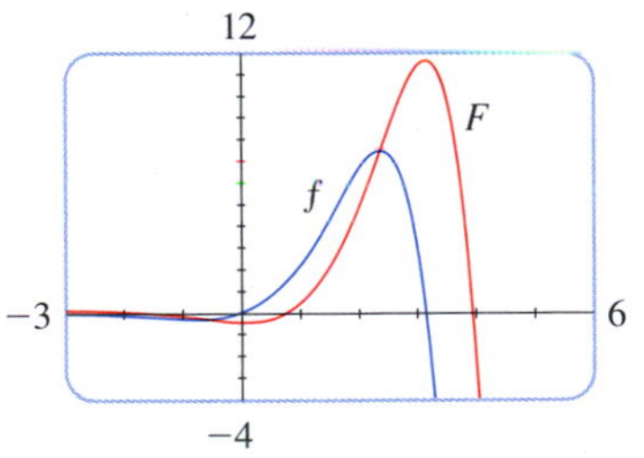

FIGURE 1

If we combine the formula for integration by parts with Part 2 of the Fundamental Theorem of Calculus, we can evaluate definite integrals by parts. Evaluating both sides of Formula 1 between a and b, assuming f' and g' are continuous, and using the Fundamental Theorem in the form of Equation 6.4.10, we obtain

$$\int_a^b f(x)g'(x)\,dx = f(x)g(x)\Big]_a^b - \int_a^b f'(x)g(x)\,dx \tag{6}$$

EXAMPLE 5 Calculate $\int_0^1 \tan^{-1}x\,dx$.

SOLUTION Let $u = \tan^{-1}x \qquad dv = dx$

Then $du = \dfrac{dx}{1 + x^2} \qquad v = x$

So Formula 6 gives

$$\int_0^1 \tan^{-1}x\,dx = x\tan^{-1}x\Big]_0^1 - \int_0^1 \frac{x}{1+x^2}\,dx$$

$$= 1 \cdot \tan^{-1}1 - 0 \cdot \tan^{-1}0 - \int_0^1 \frac{x}{1+x^2}\,dx$$

$$= \frac{\pi}{4} - \int_0^1 \frac{x}{1+x^2}\,dx$$

Since $\tan^{-1}x \geqslant 0$ for $x \geqslant 0$, the integral in Example 5 can be interpreted as the area of the region shown in Figure 2.

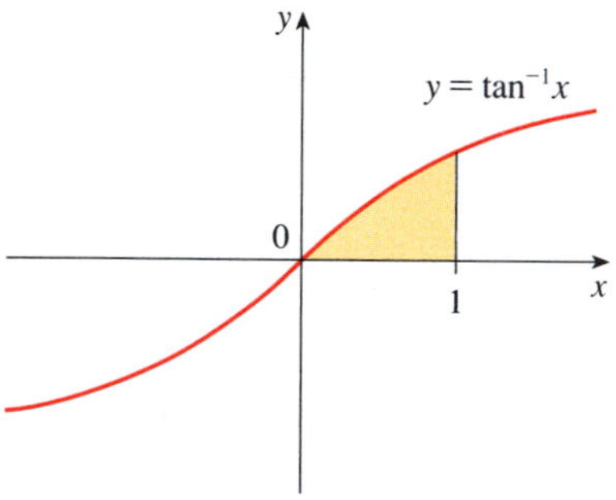

FIGURE 2

To evaluate this integral we use the substitution $t = 1 + x^2$ (since u has another meaning in this example). Then $dt = 2x\,dx$, so $x\,dx = dt/2$. When $x = 0$, $t = 1$; when $x = 1$, $t = 2$; so

$$\int_0^1 \frac{x}{1+x^2}\,dx = \tfrac{1}{2}\int_1^2 \frac{dt}{t} = \tfrac{1}{2}\ln|t|\Big]_1^2$$

$$= \tfrac{1}{2}(\ln 2 - \ln 1) = \tfrac{1}{2}\ln 2$$

Therefore $\displaystyle\int_0^1 \tan^{-1}x\,dx = \frac{\pi}{4} - \int_0^1 \frac{x}{1+x^2}\,dx = \frac{\pi}{4} - \frac{\ln 2}{2}$ ■

EXAMPLE 6 Prove the reduction formula

(7) $$\int \sin^n x\,dx = -\frac{1}{n}\cos x \sin^{n-1}x + \frac{n-1}{n}\int \sin^{n-2}x\,dx$$

where $n \geqslant 2$ is an integer.

SOLUTION Let $u = \sin^{n-1}x \qquad dv = \sin x\,dx$

Then $du = (n-1)\sin^{n-2}x\cos x\,dx \qquad v = -\cos x$

so integration by parts gives

$$\int \sin^n x\,dx = -\cos x \sin^{n-1}x + (n-1)\int \sin^{n-2}x\cos^2 x\,dx$$

Since $\cos^2 x = 1 - \sin^2 x$, we have

$$\int \sin^n x\,dx = -\cos x \sin^{n-1}x + (n-1)\int \sin^{n-2}x\,dx - (n-1)\int \sin^n x\,dx$$

As in Example 4, we solve this equation for the desired integral by taking the last term on the right side to the left side. Thus, we have

$$n\int \sin^n x\,dx = -\cos x \sin^{n-1}x + (n-1)\int \sin^{n-2}x\,dx$$

or $$\int \sin^n x\,dx = -\frac{1}{n}\cos x \sin^{n-1}x + \frac{n-1}{n}\int \sin^{n-2}x\,dx$$ ■

The reduction formula (7) is useful because by using it repeatedly we could eventually express $\int \sin^n x\,dx$ in terms of $\int \sin x\,dx$ (if n is odd) or $\int (\sin x)^0\,dx = \int dx$ (if n is even).

EXERCISES 8.1

1–30 ■ Evaluate the integral.

1. $\int xe^{2x}\,dx$

2. $\int x\cos x\,dx$

3. $\int x\sin 4x\,dx$

4. $\int x\ln x\,dx$

5. $\int x^2\cos 3x\,dx$

6. $\int x^2\sin 2x\,dx$

7. $\int (\ln x)^2\,dx$

8. $\int \sin^{-1}x\,dx$

9. $\int \theta\sin\theta\cos\theta\,d\theta$

10. $\int \theta\sec^2\theta\,d\theta$

11. $\int t^2\ln t\,dt$

12. $\int t^3e^t\,dt$

13. $\int e^{2\theta}\sin 3\theta\,d\theta$

14. $\int e^{-\theta}\cos 3\theta\,d\theta$

15. $\int y\sinh y\,dy$

16. $\int y\cosh ay\,dy$

17. $\int_0^1 te^{-t}\,dt$

18. $\int_1^4 \sqrt{t}\,\ln t\,dt$

19. $\int_0^{\pi/2} x\cos 2x\,dx$

20. $\int_0^1 x^2e^{-x}\,dx$

21. $\int_0^{1/2} \cos^{-1}x\,dx$

22. $\int_{\pi/4}^{\pi/2} x\csc^2 x\,dx$

23. $\int \cos x\ln(\sin x)\,dx$

24. $\int x^3e^{x^2}\,dx$

25. $\int (2x + 3)e^x\,dx$

26. $\int x5^x\,dx$

27. $\int \cos(\ln x)\,dx$

28. $\int x\tan^{-1}x\,dx$

29. $\int_1^4 \ln\sqrt{x}\,dx$

30. $\int \sin(\ln x)\,dx$

31–34 ■ First make a substitution and then use integration by parts to evaluate the integral.

31. $\int \sin\sqrt{x}\,dx$

32. $\int x^5\cos(x^3)\,dx$

33. $\int x^5e^{x^2}\,dx$

34. $\int_1^4 e^{\sqrt{x}}\,dx$

35–36 ■ Evaluate the indefinite integral. Illustrate, and check that your answer is reasonable, by graphing both the function and its antiderivative (take $C = 0$).

35. $\int x\cos\pi x\,dx$

36. $\int \sqrt{x}\,\ln x\,dx$

37. (a) Use the reduction formula in Example 6 to show that

$$\int \sin^2 x\,dx = \frac{x}{2} - \frac{\sin 2x}{4} + C$$

(b) Use part (a) and the reduction formula to evaluate $\int \sin^4 x\,dx$.

38. (a) Prove the reduction formula

$$\int \cos^n x\,dx = \frac{1}{n}\cos^{n-1}x\sin x + \frac{n-1}{n}\int \cos^{n-2}x\,dx$$

(b) Use part (a) to evaluate $\int \cos^2 x\,dx$.

(c) Use parts (a) and (b) to evaluate $\int \cos^4 x\,dx$.

39. (a) Use the reduction formula in Example 6 to show that

$$\int_0^{\pi/2} \sin^n x\,dx = \frac{n-1}{n}\int_0^{\pi/2} \sin^{n-2}x\,dx$$

where $n \geqslant 2$ is an integer.

(b) Use part (a) to evaluate $\int_0^{\pi/2} \sin^3 x\,dx$ and $\int_0^{\pi/2} \sin^5 x\,dx$.

(c) Use part (a) to show that, for odd powers of sin,

$$\int_0^{\pi/2} \sin^{2n+1}x\,dx = \frac{2\cdot 4\cdot 6\cdot\cdots\cdot 2n}{3\cdot 5\cdot 7\cdot\cdots\cdot(2n+1)}$$

40. Prove that, for even powers of sin,

$$\int_0^{\pi/2} \sin^{2n}x\,dx = \frac{1\cdot 3\cdot 5\cdot\cdots\cdot(2n-1)}{2\cdot 4\cdot 6\cdot\cdots\cdot 2n}\,\frac{\pi}{2}$$

41–44 ■ Use integration by parts to prove the reduction formula.

41. $\int (\ln x)^n\,dx = x(\ln x)^n - n\int (\ln x)^{n-1}\,dx$

42. $\int x^ne^x\,dx = x^ne^x - n\int x^{n-1}e^x\,dx$

43. $\int (x^2 + a^2)^n\,dx$

$$= \frac{x(x^2+a^2)^n}{2n+1} + \frac{2na^2}{2n+1}\int (x^2+a^2)^{n-1}\,dx \quad (n \neq -\tfrac{1}{2})$$

44. $\int \sec^n x\,dx = \dfrac{\tan x\sec^{n-2}x}{n-1} + \dfrac{n-2}{n-1}\int \sec^{n-2}x\,dx$ $(n \neq 1)$

45. Use Exercise 41 to find $\int (\ln x)^3\,dx$.

46. Use Exercise 42 to find $\int x^4e^x\,dx$.

47–48 ■ Find the area of the region bounded by the given curves.

47. $y = \sin^{-1}x, \quad y = 0, \quad x = 0.5$

48. $y = 5\ln x, \quad y = x\ln x$

49–50 ■ Use a graph to find approximate x-coordinates of the points of intersection of the given curves. Then find (approximately) the area of the region bounded by the curves.

49. $y = x^2, \quad y = xe^{-x/2}$

50. $y = x^2 - 5, \quad y = \ln x$

51–54 ■ Use the method of cylindrical shells to find the volume generated by rotating the region bounded by the given curves about the specified axis.

51. $y = \sin x$, $y = 0$, $x = 2\pi$, $x = 3\pi$; about the y-axis

52. $y = e^x$, $y = e^{-x}$, $x = 1$; about the y-axis

53. $y = e^{-x}$, $y = 0$, $x = -1$, $x = 0$; about $x = 1$

54. $y = e^x$, $x = 0$, $y = \pi$; about the x-axis

55. A particle that moves along a straight line has velocity $v(t) = t^2e^{-t}$ meters per second after t seconds. How far will it travel during the first t seconds?

56. If $f(0) = g(0) = 0$, show that

$$\int_0^a f(x)g''(x)\,dx = f(a)g'(a) - f'(a)g(a) + \int_0^a f''(x)g(x)\,dx$$

57. Use integration by parts to show that

$$\int f(x)\,dx = xf(x) - \int xf'(x)\,dx$$

58. If f and g are inverse functions and f' is continuous, prove that

$$\int_a^b f(x)\,dx = bf(b) - af(a) - \int_{f(a)}^{f(b)} g(y)\,dy$$

[*Hint:* Use Exercise 57 and make the substitution $y = f(x)$.]

59. Use Exercise 58 to evaluate $\int_1^e \ln x\,dx$.

60. In the case where f and g are positive functions and $b > a > 0$, draw a diagram to give a geometric interpretation of Exercise 58.

61. We arrived at Formula 7.3.2, $V = \int_a^b 2\pi x f(x)\,dx$, by using cylindrical shells, but now we can use integration by parts to deduce it from Formula 7.2.3, at least for the case where f is one-to-one and therefore has an inverse function g. Use the figure to show that

$$V = \pi b^2 d - \pi a^2 c - \int_c^d \pi[g(y)]^2\,dy$$

Make the substitution $y = f(x)$ and then use integration by parts on the resulting integral to prove that $V = \int_a^b 2\pi x f(x)\,dx$.

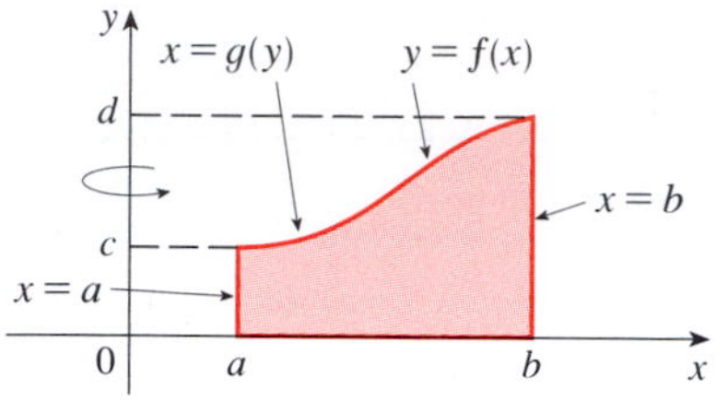

62. Let $I_n = \int_0^{\pi/2} \sin^n x\,dx$.

(a) Show that $I_{2n+2} \leqslant I_{2n+1} \leqslant I_{2n}$.

(b) Use Exercise 40 to show that

$$\frac{I_{2n+2}}{I_{2n}} = \frac{2n+1}{2n+2}$$

(c) Use parts (a) and (b) to show that

$$\frac{2n+1}{2n+2} \leqslant \frac{I_{2n+1}}{I_{2n}} \leqslant 1$$

and deduce that $\lim_{n\to\infty} I_{2n+1}/I_{2n} = 1$.

(d) Use part (c) and Exercises 39 and 40 to show that

$$\lim_{n\to\infty} \frac{2}{1}\cdot\frac{2}{3}\cdot\frac{4}{3}\cdot\frac{4}{5}\cdot\frac{6}{5}\cdot\frac{6}{7}\cdot\cdots\cdot\frac{2n}{2n-1}\cdot\frac{2n}{2n+1} = \frac{\pi}{2}$$

This formula is usually written as an infinite product:

$$\frac{\pi}{2} = \frac{2}{1}\cdot\frac{2}{3}\cdot\frac{4}{3}\cdot\frac{4}{5}\cdot\frac{6}{5}\cdot\frac{6}{7}\cdot\cdots$$

and is called the *Wallis product.*

(e) We construct rectangles as follows. Start with a square of area 1 and attach rectangles of area 1 alternately beside or on top of the previous rectangle. (See the figure.) Find the limit of the ratios of width to height of these rectangles.

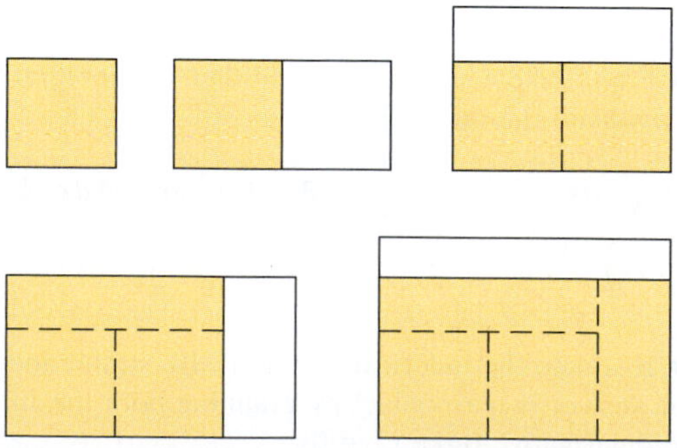

8.2 TRIGONOMETRIC INTEGRALS

In this section we use trigonometric identities to integrate certain combinations of trigonometric functions. We start with powers of sine and cosine.

EXAMPLE 1 Evaluate $\int \cos^3 x\,dx$.

SOLUTION Here the appropriate identity is $\cos^2 x = 1 - \sin^2 x$. We write

$$\cos^3 x = \cos^2 x \cdot \cos x = (1 - \sin^2 x)\cos x$$

It is useful to have the extra factor of $\cos x$ because if we make the substitution $u = \sin x$, then we have $du = \cos x\,dx$. Thus

$$\begin{aligned}\int \cos^3 x\,dx &= \int \cos^2 x \cdot \cos x\,dx = \int (1 - \sin^2 x)\cos x\,dx \\ &= \int (1 - u^2)\,du = u - \tfrac{1}{3}u^3 + C \\ &= \sin x - \tfrac{1}{3}\sin^3 x + C\end{aligned}$$

The method used in Example 1 suggests the following general strategy to be used in evaluating integrals of the form $\int \sin^m x \cos^n x\,dx$, where $m \geqslant 0$ and $n \geqslant 0$ are integers and either m or n is odd.

HOW TO EVALUATE $\int \sin^m x \cos^n x\,dx$

(a) If the power of cosine is odd ($n = 2k + 1$), save one cosine factor and use $\cos^2 x = 1 - \sin^2 x$ to express the remaining factors in terms of sine:

$$\begin{aligned}\int \sin^m x \cos^{2k+1} x\,dx &= \int \sin^m x\,(\cos^2 x)^k \cos x\,dx \\ &= \int \sin^m x\,(1 - \sin^2 x)^k \cos x\,dx\end{aligned}$$

Then substitute $u = \sin x$.

(b) If the power of sine is odd ($m = 2k + 1$), save one sine factor and use $\sin^2 x = 1 - \cos^2 x$ to express the remaining factors in terms of cosine:

$$\begin{aligned}\int \sin^{2k+1} x \cos^n x\,dx &= \int (\sin^2 x)^k \cos^n x \sin x\,dx \\ &= \int (1 - \cos^2 x)^k \cos^n x \sin x\,dx\end{aligned}$$

Then substitute $u = \cos x$.

EXAMPLE 2 Find $\int \sin^5 x \cos^2 x \, dx$.

Figure 1 shows the graphs of the integrand $\sin^5 x \cos^2 x$ in Example 2 and its indefinite integral (with $C = 0$). Which is which?

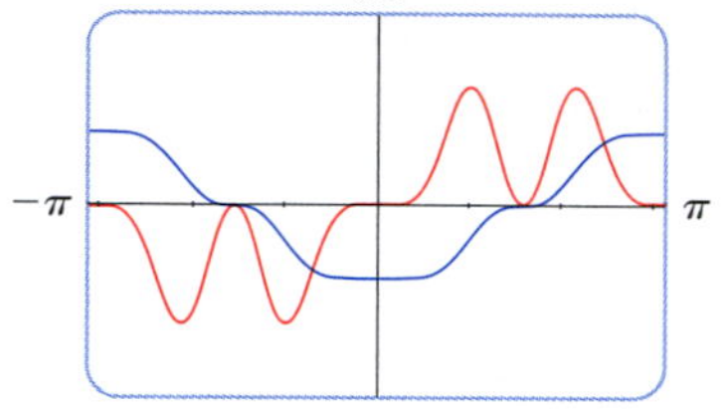

FIGURE 1

SOLUTION Here the power of sine is odd and so we proceed as in case (b), substituting $u = \cos x$:

$$\begin{aligned}
\int \sin^5 x \cos^2 x \, dx &= \int \sin^4 x \cos^2 x \sin x \, dx \\
&= \int (1 - \cos^2 x)^2 \cos^2 x \sin x \, dx \\
&= \int (1 - u^2)^2 u^2 (-du) = -\int (u^2 - 2u^4 + u^6) \, du \\
&= -\left(\frac{u^3}{3} - 2\frac{u^5}{5} + \frac{u^7}{7}\right) + C \\
&= -\tfrac{1}{3}\cos^3 x + \tfrac{2}{5}\cos^5 x - \tfrac{1}{7}\cos^7 x + C
\end{aligned}$$

The advice in (a) and (b) works if either sine or cosine has an odd exponent. In the remaining case (both m and n are even), we proceed as follows.

(c) If the powers of both sine and cosine are even, use the half-angle identities

$$\sin^2 x = \tfrac{1}{2}(1 - \cos 2x) \qquad \cos^2 x = \tfrac{1}{2}(1 + \cos 2x)$$

It is sometimes helpful to use the identity

$$\sin x \cos x = \tfrac{1}{2}\sin 2x$$

EXAMPLE 3 Evaluate $\int_0^{\pi} \sin^2 x \, dx$.

Example 3 shows that the area of the region shown in Figure 2 is $\pi/2$.

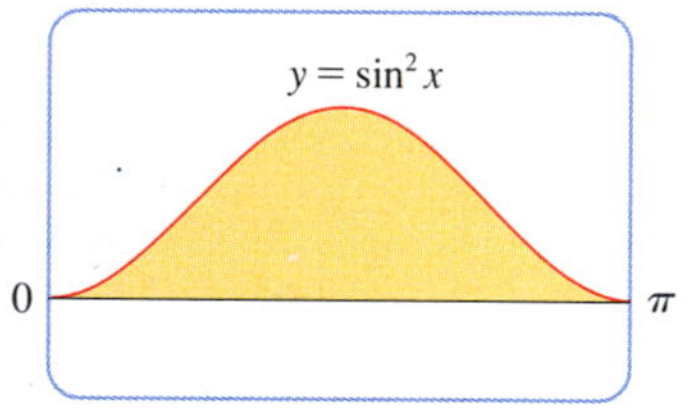

FIGURE 2

SOLUTION Here $m = 2$ and $n = 0$, so we use the half-angle formula for $\sin^2 x$:

$$\begin{aligned}
\int_0^{\pi} \sin^2 x \, dx &= \tfrac{1}{2}\int_0^{\pi} (1 - \cos 2x) \, dx = \left[\tfrac{1}{2}\left(x - \tfrac{1}{2}\sin 2x\right)\right]_0^{\pi} \\
&= \tfrac{1}{2}\left(\pi - \tfrac{1}{2}\sin 2\pi\right) - \tfrac{1}{2}\left(0 - \tfrac{1}{2}\sin 0\right) = \tfrac{1}{2}\pi
\end{aligned}$$

Notice that we mentally made the substitution $u = 2x$ when integrating $\cos 2x$. Another method for evaluating this integral was given in Exercise 37 in Section 8.1.

EXAMPLE 4 Find $\int \sin^4 x \, dx$.

SOLUTION It's possible to evaluate this integral using the reduction formula for $\int \sin^n x \, dx$ (Equation 8.1.7) together with Example 1 (as in Exercise 37 in Section 8.1), but another method is to write $\sin^4 x = (\sin^2 x)^2$ and use (c):

$$\begin{aligned}
\int \sin^4 x \, dx &= \int (\sin^2 x)^2 \, dx \\
&= \int \left(\frac{1 - \cos 2x}{2}\right)^2 dx \\
&= \tfrac{1}{4}\int (1 - 2\cos 2x + \cos^2 2x) \, dx
\end{aligned}$$

Since $\cos^2 2x$ occurs, we must use another half-angle formula

$$\cos^2 2x = \tfrac{1}{2}(1 + \cos 4x)$$

This gives

$$\int \sin^4 x\, dx = \tfrac{1}{4}\int \left[1 - 2\cos 2x + \tfrac{1}{2}(1 + \cos 4x)\right] dx$$

$$= \tfrac{1}{4}\int \left(\tfrac{3}{2} - 2\cos 2x + \tfrac{1}{2}\cos 4x\right) dx$$

$$= \tfrac{1}{4}\left(\tfrac{3}{2}x - \sin 2x + \tfrac{1}{8}\sin 4x\right) + C$$ ■

Integrals of the form $\int \tan^m x \sec^n x\, dx$ can be integrated in the following cases.

HOW TO EVALUATE $\int \tan^m x \sec^n x\, dx$

(a) If the power of secant is even ($n = 2k$), save a factor of $\sec^2 x$ and use $\sec^2 x = 1 + \tan^2 x$ to express the remaining factors in terms of $\tan x$:

$$\int \tan^m x \sec^{2k} x\, dx = \int \tan^m x (\sec^2 x)^{k-1} \sec^2 x\, dx$$

$$= \int \tan^m x (1 + \tan^2 x)^{k-1} \sec^2 x\, dx$$

Then substitute $u = \tan x$.

(b) If the power of tangent is odd ($m = 2k + 1$), save a factor of $\sec x \tan x$ and use $\tan^2 x = \sec^2 x - 1$ to express the remaining factors in terms of $\sec x$:

$$\int \tan^{2k+1} x \sec^n x\, dx = \int (\tan^2 x)^k \sec^{n-1} x \sec x \tan x\, dx$$

$$= \int (\sec^2 x - 1)^k \sec^{n-1} x \sec x \tan x\, dx$$

Then substitute $u = \sec x$.

EXAMPLE 5 Evaluate $\int \tan^6 x \sec^4 x\, dx$.

SOLUTION Since the secant has an even exponent, we factor $\sec^2 x$ from the integrand and substitute $u = \tan x$ so that $du = \sec^2 x\, dx$. The rest of the integrand is then expressed completely in terms of $\tan x$ by means of the identity $\sec^2 x = 1 + \tan^2 x$:

$$\int \tan^6 x \sec^4 x\, dx = \int \tan^6 x \sec^2 x \sec^2 x\, dx$$

$$= \int \tan^6 x (1 + \tan^2 x) \sec^2 x\, dx$$

$$= \int u^6(1 + u^2)\, du = \int (u^6 + u^8)\, du$$

$$= \frac{u^7}{7} + \frac{u^9}{9} + C$$

$$= \tfrac{1}{7}\tan^7 x + \tfrac{1}{9}\tan^9 x + C$$ ■

EXAMPLE 6 Find $\int \tan^5 x \sec^7 x\, dx$.

SOLUTION Since the power of $\tan x$ is odd, we factor $\tan x \sec x$ from the integrand and substitute $u = \sec x$ so that $du = \sec x \tan x\, dx$. Since an even power of $\tan x$ remains, we use the identity $\tan^2 x = \sec^2 x - 1$ to express the remainder of the integral completely in terms of $\sec x$:

$$\begin{aligned}
\int \tan^5 x \sec^7 x\, dx &= \int \tan^4 x \sec^6 x \sec x \tan x\, dx \\
&= \int (\sec^2 x - 1)^2 \sec^6 x \sec x \tan x\, dx \\
&= \int (u^2 - 1)^2 u^6\, du = \int (u^{10} - 2u^8 + u^6)\, du \\
&= \frac{u^{11}}{11} - 2\frac{u^9}{9} + \frac{u^7}{7} + C \\
&= \tfrac{1}{11}\sec^{11} x - \tfrac{2}{9}\sec^9 x + \tfrac{1}{7}\sec^7 x + C
\end{aligned}$$

■

If $n = 0$, only $\tan x$ occurs. Here we use $\tan^2 x = \sec^2 x - 1$ and, if necessary, the formula

$$\int \tan x\, dx = \ln|\sec x| + C$$

EXAMPLE 7 Find $\int \tan^3 x\, dx$.

SOLUTION

$$\begin{aligned}
\int \tan^3 x\, dx &= \int \tan x \tan^2 x\, dx \\
&= \int \tan x\,(\sec^2 x - 1)\, dx \\
&= \int \tan x \sec^2 x\, dx - \int \tan x\, dx \\
&= \frac{\tan^2 x}{2} - \ln|\sec x| + C
\end{aligned}$$

In the first integral we mentally substituted $u = \tan x$ so that $du = \sec^2 x\, dx$. ■

If n is odd and m is even, we express the integrand completely in terms of $\sec x$. Powers of $\sec x$ may require integration by parts.

EXAMPLE 8 Find $\int \sec x\, dx$.

SOLUTION We multiply numerator and denominator by $\sec x + \tan x$:

$$\begin{aligned}
\int \sec x\, dx &= \int \sec x\, \frac{\sec x + \tan x}{\sec x + \tan x}\, dx \\
&= \int \frac{\sec^2 x + \sec x \tan x}{\sec x + \tan x}\, dx
\end{aligned}$$

If we substitute $u = \sec x + \tan x$, then $du = (\sec x \tan x + \sec^2 x)\, dx$, so the integral becomes $\int (1/u)\,du = \ln|u| + C$. Thus, we have

(1) $$\int \sec x\,dx = \ln|\sec x + \tan x| + C$$

■

The method of Example 8 was admittedly very tricky, but we need Formula 1 for our future work.

EXAMPLE 9 Find $\int \sec^3 x\,dx$.

SOLUTION Here we integrate by parts with

$$u = \sec x \qquad dv = \sec^2 x\,dx$$
$$du = \sec x \tan x\,dx \qquad v = \tan x$$

$$\begin{aligned}\int \sec^3 x\,dx &= \sec x \tan x - \int \sec x \tan^2 x\,dx \\ &= \sec x \tan x - \int \sec x\,(\sec^2 x - 1)\,dx \\ &= \sec x \tan x - \int \sec^3 x\,dx + \int \sec x\,dx\end{aligned}$$

Using Formula 1 and solving for the required integral, we get

$$\int \sec^3 x\,dx = \tfrac{1}{2}(\sec x \tan x + \ln|\sec x + \tan x|) + C$$

■

Integrals such as the one in Example 9 may seem very special but they occur frequently in applications of integration, as we will see in Chapter 9. Integrals of the form $\int \cot^m x \csc^n x\,dx$ can be found by similar methods because of the identity $1 + \cot^2 x = \csc^2 x$.

(2) To evaluate the integrals (a) $\int \sin mx \cos nx\,dx$, (b) $\int \sin mx \sin nx\,dx$, or (c) $\int \cos mx \cos nx\,dx$, use the corresponding identity:

(a) $\sin A \cos B = \tfrac{1}{2}[\sin(A - B) + \sin(A + B)]$

(b) $\sin A \sin B = \tfrac{1}{2}[\cos(A - B) - \cos(A + B)]$

(c) $\cos A \cos B = \tfrac{1}{2}[\cos(A - B) + \cos(A + B)]$

EXAMPLE 10 Evaluate $\int \sin 4x \cos 5x\,dx$.

SOLUTION This integral could be evaluated using integration by parts, but it is easier to use the identity in Equation 2(a) as follows:

$$\begin{aligned}\int \sin 4x \cos 5x\,dx &= \int \tfrac{1}{2}[\sin(-x) + \sin 9x]\,dx \\ &= \tfrac{1}{2}\int (-\sin x + \sin 9x)\,dx \\ &= \tfrac{1}{2}(\cos x - \tfrac{1}{9}\cos 9x) + C\end{aligned}$$

■

EXERCISES 8.2

1–44 ■ Evaluate the integral.

1. $\int_0^{\pi/2} \sin^2 3x\,dx$

2. $\int_0^{\pi/2} \cos^2 x\,dx$

3. $\int \cos^4 x\,dx$

4. $\int \sin^3 x\,dx$

5. $\int \sin^3 x \cos^4 x\,dx$

6. $\int \sin^4 x \cos^3 x\,dx$

7. $\int_0^{\pi/4} \sin^4 x \cos^2 x\,dx$

8. $\int_0^{\pi/2} \sin^2 x \cos^2 x\,dx$

9. $\int (1 - \sin 2x)^2\,dx$

10. $\int \sin\left(x + \dfrac{\pi}{6}\right)\cos x\,dx$

11. $\int \cos^5 x \sin^5 x\,dx$

12. $\int \sin^6 x\,dx$

13. $\int \sin^3 x \sqrt{\cos x}\,dx$

14. $\int x \sin^3(x^2)\,dx$

15. $\int \cos^2 x \tan^3 x\,dx$

16. $\int \cot^5 x \sin^2 x\,dx$

17. $\int \dfrac{1 - \sin x}{\cos x}\,dx$

18. $\int \dfrac{dx}{1 - \sin x}$

19. $\int \tan^2 x\,dx$

20. $\int \tan^4 x\,dx$

21. $\int \sec^4 x\,dx$

22. $\int \sec^6 x\,dx$

23. $\int_0^{\pi/4} \tan^4 x \sec^2 x\,dx$

24. $\int_0^{\pi/4} \tan^2 x \sec^4 x\,dx$

25. $\int \tan x \sec^3 x\,dx$

26. $\int \tan^3 x \sec^3 x\,dx$

27. $\int \tan^5 x\,dx$

28. $\int \tan^6 x\,dx$

29. $\int_0^{\pi/3} \tan^5 x \sec x\,dx$

30. $\int_0^{\pi/3} \tan^5 x \sec^3 x\,dx$

31. $\int \dfrac{\sec^2 x}{\cot x}\,dx$

32. $\int \tan^2 x \sec x\,dx$

33. $\int_{\pi/6}^{\pi/2} \cot^2 x\,dx$

34. $\int_{\pi/4}^{\pi/2} \cot^3 x\,dx$

35. $\int \cot^4 x \csc^4 x\,dx$

36. $\int \cot^3 x \csc^4 x\,dx$

37. $\int \csc x\,dx$

38. $\int \csc^3 x\,dx$

39. $\int \sin 5x \sin 2x\,dx$

40. $\int \sin 3x \cos x\,dx$

41. $\int \cos 3x \cos 4x\,dx$

42. $\int \sin 3x \sin 6x\,dx$

43. $\int \dfrac{1 - \tan^2 x}{\sec^2 x}\,dx$

44. $\int \dfrac{\cos x + \sin x}{\sin 2x}\,dx$

45–46 ■ Evaluate the indefinite integral. Illustrate, and check that your answer is reasonable, by graphing both the integrand and its antiderivative (taking $C = 0$).

45. $\int \sin^5 x\,dx$

46. $\int \sin^4 x \cos^4 x\,dx$

47. Find the average value of the function $f(x) = \sin^2 x \cos^3 x$ on the interval $[-\pi, \pi]$.

48. Evaluate $\int \sin x \cos x\,dx$ by four methods: (a) the substitution $u = \cos x$, (b) the substitution $u = \sin x$, (c) the identity $\sin 2x = 2 \sin x \cos x$, and (d) integration by parts. Explain the different appearances of the answers.

49–50 ■ Find the area of the region bounded by the given curves.

49. $y = \sin x,\quad y = \sin^3 x,\quad x = 0,\quad x = \pi/2$

50. $y = \sin x,\quad y = 2\sin^2 x,\quad x = 0,\quad x = \pi/2$

51–52 ■ Use a graph of the integrand to guess the value of the integral. Then use the methods of this section to prove that your guess is correct.

51. $\int_0^{2\pi} \cos^3 x\,dx$

52. $\int_0^2 \sin 2\pi x \cos 5\pi x\,dx$

53–56 ■ Find the volume obtained by rotating the region bounded by the given curves about the specified axis.

53. $y = \sin x,\ x = \pi/2,\ x = \pi,\ y = 0$; about the x-axis

54. $y = \tan^2 x,\ y = 0,\ x = 0,\ x = \pi/4$; about the x-axis

55. $y = \cos x,\ y = 0,\ x = 0,\ x = \pi/2$; about $y = -1$

56. $y = \cos x,\ y = 0,\ x = 0,\ x = \pi/2$; about $y = 1$

57. A particle moves on a straight line with velocity function $v(t) = \sin \omega t \cos^2 \omega t$. Find its position function $s = f(t)$ if $f(0) = 0$.

58. Household electricity is supplied in the form of alternating current that varies from 155 V to -155 V with a frequency of 60 cycles per second (Hz). The voltage is thus given by the equation

$$E(t) = 155 \sin(120\,\pi t)$$

where t is the time in seconds. Voltmeters read the RMS (root-mean-square) voltage, which is the square root of the average value of $[E(t)]^2$ over one cycle. Calculate the RMS voltage of household current.

59–61 ■ Prove each formula, where m and n are positive integers.

59. $\int_{-\pi}^{\pi} \sin mx \cos nx\, dx = 0$

60. $\int_{-\pi}^{\pi} \sin mx \sin nx\, dx = \begin{cases} 0 & \text{if } m \neq n \\ \pi & \text{if } m = n \end{cases}$

61. $\int_{-\pi}^{\pi} \cos mx \cos nx\, dx = \begin{cases} 0 & \text{if } m \neq n \\ \pi & \text{if } m = n \end{cases}$

62. A *finite Fourier series* is given by the sum

$$f(x) = \sum_{n=1}^{N} a_n \sin nx$$

$$= a_1 \sin x + a_2 \sin 2x + \cdots + a_N \sin Nx$$

Show that the mth coefficient a_m is given by the formula

$$a_m = \frac{1}{\pi} \int_{-\pi}^{\pi} f(x) \sin mx\, dx$$

8.3 TRIGONOMETRIC SUBSTITUTION

In finding the area of a circle or an ellipse, an integral of the form $\int \sqrt{a^2 - x^2}\, dx$ arises, where $a > 0$. If it were $\int x\sqrt{a^2 - x^2}\, dx$, the substitution $u = a^2 - x^2$ would be effective but, as it stands, $\int \sqrt{a^2 - x^2}\, dx$ is more difficult. If we change the variable from x to θ by the substitution $x = a \sin\theta$, then the identity $1 - \sin^2\theta = \cos^2\theta$ allows us to get rid of the root sign because

$$\sqrt{a^2 - x^2} = \sqrt{a^2 - a^2\sin^2\theta} = \sqrt{a^2(1 - \sin^2\theta)} = \sqrt{a^2\cos^2\theta} = a|\cos\theta|$$

Notice the difference between the substitution $u = a^2 - x^2$ (in which the new variable is a function of the old one) and the substitution $x = a\sin\theta$ (the old variable is a function of the new one).

In general we can make a substitution of the form $x = g(t)$ by using the Substitution Rule in reverse. To make our calculations simpler, we assume that g has an inverse function, that is, g is one-to-one. In this case, if we replace u by x and x by t in the Substitution Rule (Equation 6.5.4), we obtain

$$\int f(x)\, dx = \int f(g(t))g'(t)\, dt$$

This kind of substitution is called *inverse substitution.*

We can make the inverse substitution $x = a\sin\theta$ provided that it defines a one-to-one function. This can be accomplished by restricting θ to lie in the interval $[-\pi/2, \pi/2]$.

In the following table we list trigonometric substitutions that are effective for the given radical expressions because of the given trigonometric identities. In each case the restriction on θ is imposed to ensure that the function that defines the substitution is one-to-one. (These are the same intervals used in Section 4.6 in defining the inverse functions.)

TABLE OF TRIGONOMETRIC SUBSTITUTIONS

Expression	Substitution	Identity
$\sqrt{a^2 - x^2}$	$x = a\sin\theta,\quad -\frac{\pi}{2} \leqslant \theta \leqslant \frac{\pi}{2}$	$1 - \sin^2\theta = \cos^2\theta$
$\sqrt{a^2 + x^2}$	$x = a\tan\theta,\quad -\frac{\pi}{2} < \theta < \frac{\pi}{2}$	$1 + \tan^2\theta = \sec^2\theta$
$\sqrt{x^2 - a^2}$	$x = a\sec\theta,\quad 0 \leqslant \theta < \frac{\pi}{2}$ or $\pi \leqslant \theta < \frac{3\pi}{2}$	$\sec^2\theta - 1 = \tan^2\theta$

EXAMPLE 1 Evaluate $\displaystyle\int \frac{\sqrt{9 - x^2}}{x^2}\,dx$.

SOLUTION Let $x = 3\sin\theta$, where $-\pi/2 \leqslant \theta \leqslant \pi/2$. Then $dx = 3\cos\theta\,d\theta$ and

$$\sqrt{9 - x^2} = \sqrt{9 - 9\sin^2\theta} = \sqrt{9\cos^2\theta} = 3|\cos\theta| = 3\cos\theta$$

(Note that $\cos\theta \geqslant 0$ because $-\pi/2 \leqslant \theta \leqslant \pi/2$.) Thus, the Inverse Substitution Rule gives

$$\begin{aligned}\int \frac{\sqrt{9 - x^2}}{x^2}\,dx &= \int \frac{3\cos\theta}{9\sin^2\theta}\,3\cos\theta\,d\theta \\ &= \int \frac{\cos^2\theta}{\sin^2\theta}\,d\theta = \int \cot^2\theta\,d\theta \\ &= \int (\csc^2\theta - 1)\,d\theta \\ &= -\cot\theta - \theta + C\end{aligned}$$

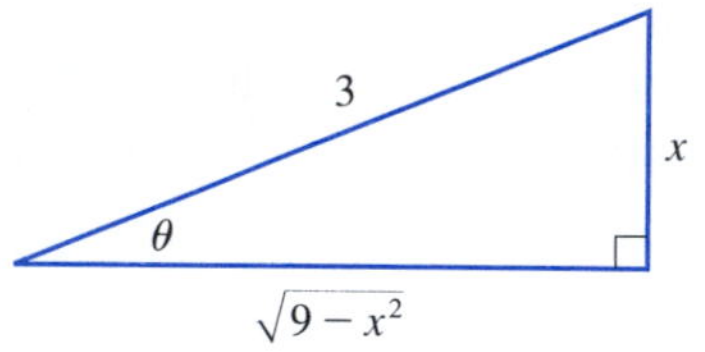

FIGURE 1

$\sin\theta = \dfrac{x}{3}$

Since this is an indefinite integral, we must return to the original variable x. This can be done either by using trigonometric identities to express $\cot\theta$ in terms of $\sin\theta = x/3$ or by drawing a diagram, as in Figure 1, where θ is interpreted as an angle of a right triangle. Since $\sin\theta = x/3$, we label the opposite side and the hypotenuse as having lengths x and 3. Then the Pythagorean Theorem gives the length of the adjacent side as $\sqrt{9 - x^2}$, so we can simply read the value of $\cot\theta$ from the figure:

$$\cot\theta = \frac{\sqrt{9 - x^2}}{x}$$

(Although $\theta > 0$ in the diagram, this expression for $\cot\theta$ is valid even when $\theta < 0$.) Since $\sin\theta = x/3$, we have $\theta = \sin^{-1}(x/3)$ and so

$$\int \frac{\sqrt{9 - x^2}}{x^2}\,dx = -\frac{\sqrt{9 - x^2}}{x} - \sin^{-1}\left(\frac{x}{3}\right) + C$$

■

EXAMPLE 2 Find the area enclosed by the ellipse

$$\frac{x^2}{a^2} + \frac{y^2}{b^2} = 1$$

SOLUTION Solving the equation of the ellipse for y, we get

$$\frac{y^2}{b^2} = 1 - \frac{x^2}{a^2} = \frac{a^2 - x^2}{a^2} \qquad \text{or} \qquad y = \pm\frac{b}{a}\sqrt{a^2 - x^2}$$

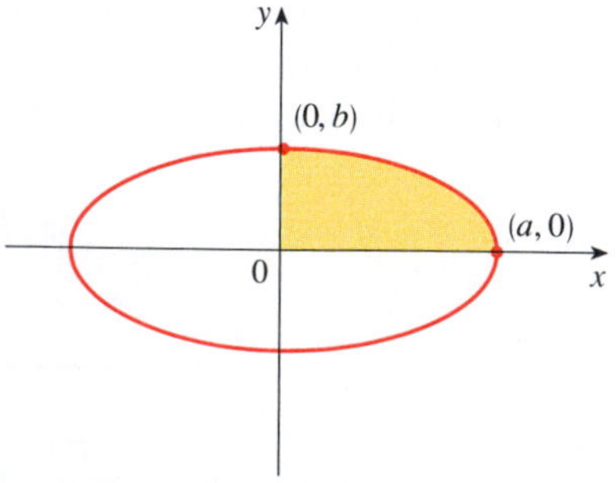

FIGURE 2

$\dfrac{x^2}{a^2} + \dfrac{y^2}{b^2} = 1$

Because the ellipse is symmetric with respect to both axes, the total area A is four times the area in the first quadrant (see Figure 2). The part of the ellipse in the first quadrant is given by the function

$$y = \frac{b}{a}\sqrt{a^2 - x^2} \qquad 0 \leqslant x \leqslant a$$

and so

$$\tfrac{1}{4}A = \int_0^a \frac{b}{a}\sqrt{a^2 - x^2}\,dx$$

To evaluate this integral we substitute $x = a\sin\theta$. Then $dx = a\cos\theta\,d\theta$. To change the limits of integration we note that when $x = 0$, $\sin\theta = 0$, so $\theta = 0$; when $x = a$, $\sin\theta = 1$, so $\theta = \pi/2$. Also

$$\sqrt{a^2 - x^2} = \sqrt{a^2 - a^2\sin^2\theta} = \sqrt{a^2\cos^2\theta} = a\,|\cos\theta| = a\cos\theta$$

since $0 \leqslant \theta \leqslant \pi/2$. Therefore

$$\begin{aligned} A &= 4\frac{b}{a}\int_0^a \sqrt{a^2 - x^2}\,dx = 4\frac{b}{a}\int_0^{\pi/2} a\cos\theta \cdot a\cos\theta\,d\theta \\ &= 4ab\int_0^{\pi/2}\cos^2\theta\,d\theta = 4ab\int_0^{\pi/2}\tfrac{1}{2}(1 + \cos 2\theta)\,d\theta \\ &= 2ab\left[\theta + \tfrac{1}{2}\sin 2\theta\right]_0^{\pi/2} = 2ab\left[\frac{\pi}{2} + 0 - 0\right] \\ &= \pi ab \end{aligned}$$

We have shown that the area of an ellipse with semiaxes a and b is πab. In particular, taking $a = b = r$, we have proved the famous formula that the area of a circle with radius r is πr^2. ■

NOTE: Since the integral in Example 2 was a definite integral, we changed the limits of integration and did not have to convert back to the original variable x.

EXAMPLE 3 Find $\displaystyle\int \frac{1}{x^2\sqrt{x^2 + 4}}\,dx$.

SOLUTION Let $x = 2\tan\theta$, $-\pi/2 < \theta < \pi/2$. Then $dx = 2\sec^2\theta\,d\theta$ and

$$\sqrt{x^2 + 4} = \sqrt{4(\tan^2\theta + 1)} = \sqrt{4\sec^2\theta} = 2|\sec\theta| = 2\sec\theta$$

Thus, we have

$$\int \frac{dx}{x^2\sqrt{x^2 + 4}} = \int \frac{2\sec^2\theta\,d\theta}{4\tan^2\theta \cdot 2\sec\theta} = \frac{1}{4}\int \frac{\sec\theta}{\tan^2\theta}\,d\theta$$

To evaluate this trigonometric integral we put everything in terms of $\sin\theta$ and $\cos\theta$:

$$\frac{\sec\theta}{\tan^2\theta} = \frac{1}{\cos\theta} \cdot \frac{\cos^2\theta}{\sin^2\theta} = \frac{\cos\theta}{\sin^2\theta}$$

Therefore, making the substitution $u = \sin\theta$, we have

$$\begin{aligned} \int \frac{dx}{x^2\sqrt{x^2 + 4}} &= \frac{1}{4}\int \frac{\cos\theta}{\sin^2\theta}\,d\theta = \frac{1}{4}\int \frac{du}{u^2} \\ &= \frac{1}{4}\left(-\frac{1}{u}\right) + C = -\frac{1}{4\sin\theta} + C \\ &= -\frac{\csc\theta}{4} + C \end{aligned}$$

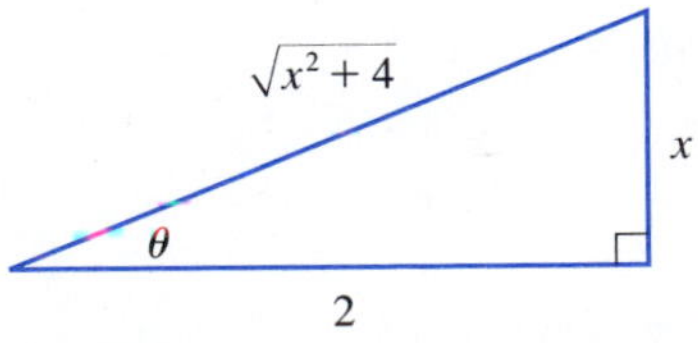

FIGURE 3

$\tan\theta = \dfrac{x}{2}$

We use Figure 3 to determine that $\csc\theta = \sqrt{x^2 + 4}/x$ and so

$$\int \frac{dx}{x^2\sqrt{x^2 + 4}} = -\frac{\sqrt{x^2 + 4}}{4x} + C$$ ■

EXAMPLE 4 Find $\int \frac{x}{\sqrt{x^2 + 4}}\,dx$.

SOLUTION It would be possible to use the trigonometric substitution $x = 2\tan\theta$ here (as in Example 3). But the direct substitution $u = x^2 + 4$ is simpler, because then $du = 2x\,dx$ and

$$\int \frac{x}{\sqrt{x^2 + 4}}\,dx = \frac{1}{2}\int \frac{du}{\sqrt{u}} = \sqrt{u} + C = \sqrt{x^2 + 4} + C$$

■

NOTE: Example 4 illustrates the fact that even when trigonometric substitutions are possible, they may not give the easiest solution. You should look for a simpler method first.

EXAMPLE 5 Evaluate $\int \frac{dx}{\sqrt{x^2 - a^2}}$, where $a > 0$.

SOLUTION 1 We let $x = a\sec\theta$ where $0 < \theta < \pi/2$ or $\pi < \theta < 3\pi/2$. Then $dx = a\sec\theta\tan\theta\,d\theta$ and

$$\sqrt{x^2 - a^2} = \sqrt{a^2(\sec^2\theta - 1)} = \sqrt{a^2\tan^2\theta} = a\,|\tan\theta| = a\tan\theta$$

Therefore
$$\int \frac{dx}{\sqrt{x^2 - a^2}} = \int \frac{a\sec\theta\tan\theta}{a\tan\theta}\,d\theta$$
$$= \int \sec\theta\,d\theta = \ln|\sec\theta + \tan\theta| + C$$

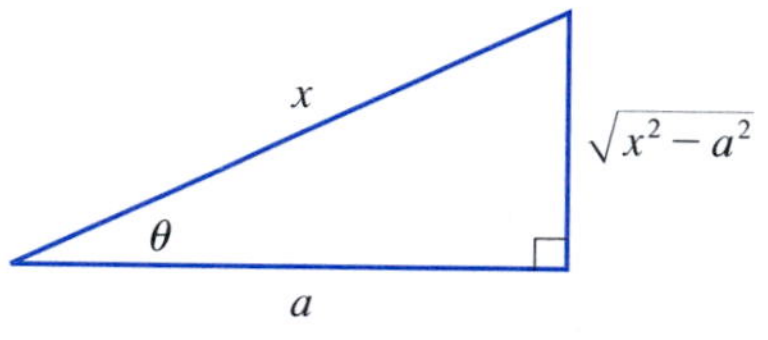

FIGURE 4

$\sec\theta = \frac{x}{a}$

The triangle in Figure 4 gives $\tan\theta = \sqrt{x^2 - a^2}/a$, so we have

$$\int \frac{dx}{\sqrt{x^2 - a^2}} = \ln\left|\frac{x}{a} + \frac{\sqrt{x^2 - a^2}}{a}\right| + C$$
$$= \ln|x + \sqrt{x^2 - a^2}| - \ln a + C$$

Writing $C_1 = C - \ln a$, we have

(1) $$\int \frac{dx}{\sqrt{x^2 - a^2}} = \ln|x + \sqrt{x^2 - a^2}| + C_1$$

SOLUTION 2 For $x > 0$ the hyperbolic substitution $x = a\cosh t$ can also be used. Using the identity $\cosh^2 y - \sinh^2 y = 1$, we have

$$\sqrt{x^2 - a^2} = \sqrt{a^2(\cosh^2 t - 1)} = \sqrt{a^2\sinh^2 t} = a\sinh t$$

Since $dx = a\sinh t\,dt$, we obtain

$$\int \frac{dx}{\sqrt{x^2 - a^2}} = \int \frac{a\sinh t\,dt}{a\sinh t} = \int dt = t + C$$

Since $\cosh t = x/a$, we have $t = \cosh^{-1}(x/a)$ and

(2) $$\int \frac{dx}{\sqrt{x^2 - a^2}} = \cosh^{-1}\left(\frac{x}{a}\right) + C$$

Although Formulas 1 and 2 look quite different, they are actually equivalent by Formula 4.7.4.

■

NOTE: As Example 5 illustrates, hyperbolic substitutions can be used in place of trigonometric substitutions and sometimes they lead to simpler answers. But we usually use trigonometric substitutions because trigonometric identities are more familiar than hyperbolic identities.

EXAMPLE 6 Find $\displaystyle\int_0^{3\sqrt{3}/2} \frac{x^3}{(4x^2+9)^{3/2}}\,dx$.

SOLUTION First we note that $(4x^2+9)^{3/2} = (\sqrt{4x^2+9})^3$ so trigonometric substitution is appropriate. Although $\sqrt{4x^2+9}$ is not quite one of the expressions in the table of trigonometric substitutions, it becomes one of them if we make the preliminary substitution $u = 2x$. When we combine this with the tangent substitution, we have $x = \frac{3}{2}\tan\theta$, which gives $dx = \frac{3}{2}\sec^2\theta\,d\theta$ and

$$\sqrt{4x^2+9} = \sqrt{9\tan^2\theta+9} = 3\sec\theta$$

When $x = 0$, $\tan\theta = 0$, so $\theta = 0$; when $x = 3\sqrt{3}/2$, $\tan\theta = \sqrt{3}$, so $\theta = \pi/3$.

$$\int_0^{3\sqrt{3}/2} \frac{x^3}{(4x^2+9)^{3/2}}\,dx = \int_0^{\pi/3} \frac{\frac{27}{8}\tan^3\theta}{27\sec^3\theta}\,\frac{3}{2}\sec^2\theta\,d\theta$$

$$= \frac{3}{16}\int_0^{\pi/3} \frac{\tan^3\theta}{\sec\theta}\,d\theta = \frac{3}{16}\int_0^{\pi/3} \frac{\sin^3\theta}{\cos^2\theta}\,d\theta$$

$$= \frac{3}{16}\int_0^{\pi/3} \frac{1-\cos^2\theta}{\cos^2\theta}\sin\theta\,d\theta$$

Now we substitute $u = \cos\theta$ so that $du = -\sin\theta\,d\theta$. When $\theta = 0$, $u = 1$; when $\theta = \pi/3$, $u = 1/2$. Therefore

$$\int_0^{3\sqrt{3}/2} \frac{x^3}{(4x^2+9)^{3/2}}\,dx = -\frac{3}{16}\int_1^{1/2} \frac{1-u^2}{u^2}\,du = \frac{3}{16}\int_1^{1/2} (1-u^{-2})\,du$$

$$= \frac{3}{16}\left[u + \frac{1}{u}\right]_1^{1/2} = \frac{3}{16}\left[\left(\tfrac{1}{2}+2\right) - (1+1)\right] = \frac{3}{32}$$ ■

EXAMPLE 7 Evaluate $\displaystyle\int \frac{x}{\sqrt{3-2x-x^2}}\,dx$.

SOLUTION We can transform the integrand into a function for which trigonometric substitution is appropriate by first completing the square under the root sign:

$$3 - 2x - x^2 = 3 - (x^2+2x) = 3 + 1 - (x^2+2x+1)$$
$$= 4 - (x+1)^2$$

This suggests that we make the substitution $u = x + 1$. Then $du = dx$ and $x = u - 1$, so

$$\int \frac{x}{\sqrt{3-2x-x^2}}\,dx = \int \frac{u-1}{\sqrt{4-u^2}}\,du$$

Figure 5 shows the graphs of the integrand in Example 7 and its indefinite integral (with $C = 0$). Which is which?

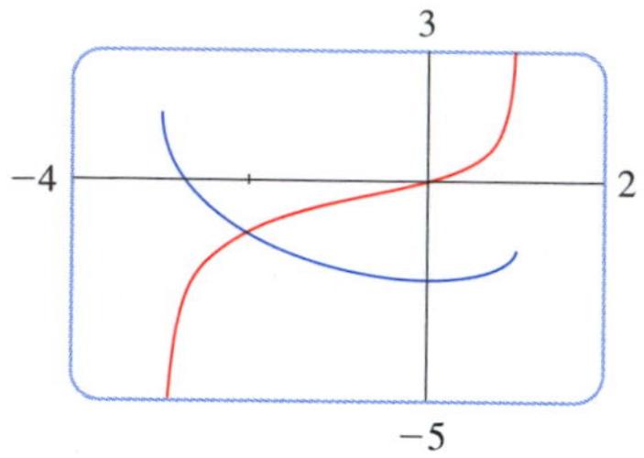

FIGURE 5

We now substitute $u = 2\sin\theta$, giving $du = 2\cos\theta\,d\theta$ and $\sqrt{4 - u^2} = 2\cos\theta$, so

$$\begin{aligned}\int \frac{x}{\sqrt{3 - 2x - x^2}}\,dx &= \int \frac{2\sin\theta - 1}{2\cos\theta}\,2\cos\theta\,d\theta \\ &= \int (2\sin\theta - 1)\,d\theta \\ &= -2\cos\theta - \theta + C \\ &= -\sqrt{4 - u^2} - \sin^{-1}\left(\frac{u}{2}\right) + C \\ &= -\sqrt{3 - 2x - x^2} - \sin^{-1}\left(\frac{x + 1}{2}\right) + C\end{aligned}$$

EXERCISES 8.3

1–28 ■ Evaluate the integral.

1. $\displaystyle\int_{1/2}^{\sqrt{3}/2} \frac{1}{x^2\sqrt{1 - x^2}}\,dx$

2. $\displaystyle\int_0^2 x^3\sqrt{4 - x^2}\,dx$

3. $\displaystyle\int \frac{x}{\sqrt{1 - x^2}}\,dx$

4. $\displaystyle\int x\sqrt{4 - x^2}\,dx$

5. $\displaystyle\int \sqrt{1 - 4x^2}\,dx$

6. $\displaystyle\int_0^2 \frac{x^3}{\sqrt{x^2 + 4}}\,dx$

7. $\displaystyle\int_0^3 \frac{dx}{\sqrt{9 + x^2}}$

8. $\displaystyle\int_0^1 \sqrt{x^2 + 1}\,dx$

9. $\displaystyle\int \frac{dx}{x^3\sqrt{x^2 - 16}}$

10. $\displaystyle\int \frac{\sqrt{x^2 - a^2}}{x^4}\,dx$

11. $\displaystyle\int \frac{\sqrt{9x^2 - 4}}{x}\,dx$

12. $\displaystyle\int \frac{dx}{x^2\sqrt{16x^2 - 9}}$

13. $\displaystyle\int \frac{x^2}{(a^2 - x^2)^{3/2}}\,dx$

14. $\displaystyle\int \frac{x^2}{\sqrt{5 - x^2}}\,dx$

15. $\displaystyle\int \frac{dx}{x\sqrt{x^2 + 3}}$

16. $\displaystyle\int \frac{x}{(x^2 + 4)^{5/2}}\,dx$

17. $\displaystyle\int_0^{2/3} x^3\sqrt{4 - 9x^2}\,dx$

18. $\displaystyle\int_0^3 x^2\sqrt{9 - x^2}\,dx$

19. $\displaystyle\int 5x\sqrt{1 + x^2}\,dx$

20. $\displaystyle\int \frac{dx}{(4x^2 - 25)^{3/2}}$

21. $\displaystyle\int \sqrt{2x - x^2}\,dx$

22. $\displaystyle\int \frac{dx}{\sqrt{x^2 + 4x + 8}}$

23. $\displaystyle\int \frac{1}{\sqrt{9x^2 + 6x - 8}}\,dx$

24. $\displaystyle\int \frac{x^2}{\sqrt{4x - x^2}}\,dx$

25. $\displaystyle\int \frac{dx}{(x^2 + 2x + 2)^2}$

26. $\displaystyle\int \frac{dx}{(5 - 4x - x^2)^{5/2}}$

27. $\displaystyle\int e^t\sqrt{9 - e^{2t}}\,dt$

28. $\displaystyle\int \sqrt{e^{2t} - 9}\,dt$

29. (a) Use trigonometric substitution to show that

$$\int \frac{dx}{\sqrt{x^2 + a^2}} = \ln\left(x + \sqrt{x^2 + a^2}\right) + C$$

(b) Use the hyperbolic substitution $x = a\sinh t$ to show that

$$\int \frac{dx}{\sqrt{x^2 + a^2}} = \sinh^{-1}\left(\frac{x}{a}\right) + C$$

These formulas are connected by Formula 4.7.3.

30. Evaluate

$$\int \frac{x^2}{(x^2 + a^2)^{3/2}}\,dx$$

(a) by trigonometric substitution and (b) by the hyperbolic substitution $x = a\sinh t$.

31. Prove the formula $A = \frac{1}{2}r^2\theta$ for the area of a sector of a circle with radius r and central angle θ. [*Hint:* Assume $0 < \theta < \pi/2$ and place the center of the circle at the origin so it has the equation $x^2 + y^2 = r^2$. Then A is the sum of the area of the triangle POQ and the area of the region PQR in the figure.]

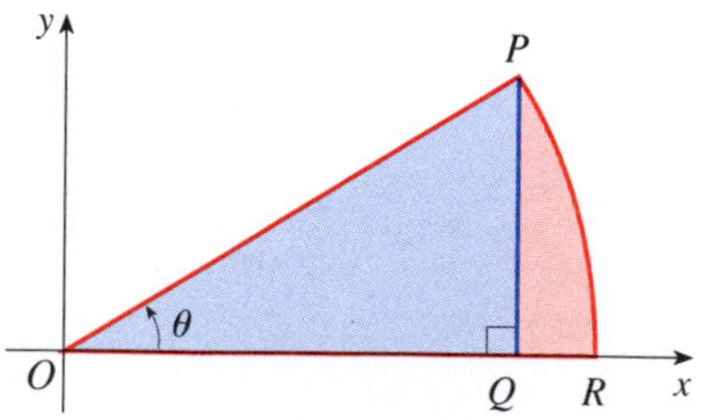

32. Find the area of the region bounded by the hyperbola $9x^2 - 4y^2 = 36$ and the line $x = 3$.

33. Use a graph to approximate the roots of the equation $x^2\sqrt{4 - x^2} = 2 - x$. Then approximate the area bounded by the curve $y = x^2\sqrt{4 - x^2}$ and the line $y = 2 - x$.

34. Evaluate the integral

$$\int \frac{dx}{x^4\sqrt{x^2-2}}$$

Graph the integrand and its indefinite integral on the same screen and check that your answer is reasonable.

35. Find the area of the crescent-shaped region (called a *lune*) bounded by arcs of circles with radii r and R. (See the figure.)

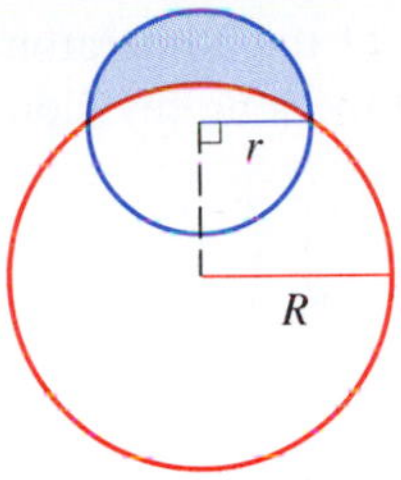

36. A water storage tank has the shape of a cylinder with diameter 10 ft. It is mounted so that the circular cross-sections are vertical. If the depth of the water is 7 ft, what percentage of the total capacity is being used?

37. A torus is generated by rotating the circle $x^2 + (y - R)^2 = r^2$ about the x-axis. Find the volume enclosed by the torus.

38. Use the result of Example 2 to find the volume enclosed by the ellipsoid

$$\frac{x^2}{a^2} + \frac{y^2}{b^2} + \frac{z^2}{c^2} = 1$$

(See Figure 1 in Section 11.5.)

8.4 INTEGRATION OF RATIONAL FUNCTIONS BY PARTIAL FRACTIONS

In this section we show how to integrate any rational function (a ratio of polynomials) by expressing it as a sum of simpler fractions, called *partial fractions*, that we already know how to integrate. To illustrate the method, observe that by taking the fractions $2/(x-1)$ and $1/(x+2)$ to a common denominator we obtain

$$\frac{2}{x-1} - \frac{1}{x+2} = \frac{2(x+2)-(x-1)}{(x-1)(x+2)} = \frac{x+5}{x^2+x-2}$$

If we now reverse the procedure we see how to integrate the function on the right side of this equation:

$$\int \frac{x+5}{x^2+x-2}\,dx = \int \left(\frac{2}{x-1} - \frac{1}{x+2}\right) dx$$
$$= 2\ln|x-1| - \ln|x+2| + C$$

To see how the method of partial fractions works in general, let's consider a rational function

$$f(x) = \frac{P(x)}{Q(x)}$$

where P and Q are polynomials. It's possible to express f as a sum of simpler fractions provided that the degree of P is less than the degree of Q. Such a rational function is called *proper*. Recall that if

$$P(x) = a_n x^n + a_{n-1}x^{n-1} + \cdots + a_1 x + a_0$$

where $a_n \neq 0$, then the degree of P is n and we write $\deg(P) = n$.

If f is improper, that is, $\deg(P) \geqslant \deg(Q)$, then we must take the preliminary step of dividing Q into P (by long division) until a remainder $R(x)$ is obtained such that $\deg(R) < \deg(Q)$. The division statement is

$$f(x) = \frac{P(x)}{Q(x)} = S(x) + \frac{R(x)}{Q(x)} \qquad \textbf{(1)}$$

where S and R are also polynomials.

As the following example illustrates, sometimes this preliminary step is all that is required.

EXAMPLE 1 Find $\displaystyle\int \frac{x^3 + x}{x - 1}\,dx$.

$$
\begin{array}{r}
x^2 + x + 2 \\
x - 1\overline{)x^3 \qquad + x} \\
\underline{x^3 - x^2} \\
x^2 + x \\
\underline{x^2 - x} \\
2x \\
\underline{2x - 2} \\
2
\end{array}
$$

SOLUTION Since the degree of the numerator is greater than the degree of the denominator, we first perform the long division. This enables us to write

$$\int \frac{x^3 + x}{x - 1}\,dx = \int \left(x^2 + x + 2 + \frac{2}{x - 1}\right) dx$$

$$= \frac{x^3}{3} + \frac{x^2}{2} + 2x + 2\ln|x - 1| + C$$

The next step is to factor the denominator $Q(x)$ as far as possible. It can be shown that any polynomial Q can be factored as a product of linear factors (of the form $ax + b$) and irreducible quadratic factors (of the form $ax^2 + bx + c$, where $b^2 - 4ac < 0$). For instance, if $Q(x) = x^4 - 16$, we could factor it as

$$Q(x) = (x^2 - 4)(x^2 + 4) = (x - 2)(x + 2)(x^2 + 4)$$

The third step is to express the proper rational function $R(x)/Q(x)$ (from Equation 1) as a sum of **partial fractions** of the form

$$\frac{A}{(ax + b)^i} \qquad \text{or} \qquad \frac{Ax + B}{(ax^2 + bx + c)^j}$$

A theorem in algebra guarantees that it is always possible to do this. We explain the details for the four cases that occur.

CASE I The denominator $Q(x)$ is a product of distinct linear factors.
This means that we can write

$$Q(x) = (a_1x + b_1)(a_2x + b_2)\cdots(a_kx + b_k)$$

where no factor is repeated. In this case the partial fraction theorem states that there exist constants $A_1, A_2, \ldots, A_k$ such that

$$\textbf{(2)} \qquad \frac{R(x)}{Q(x)} = \frac{A_1}{a_1x + b_1} + \frac{A_2}{a_2x + b_2} + \cdots + \frac{A_k}{a_kx + b_k}$$

These constants can be determined as in the following example.

EXAMPLE 2 Evaluate $\displaystyle\int \frac{x^2 + 2x - 1}{2x^3 + 3x^2 - 2x}\,dx$.

SOLUTION Since the degree of the numerator is less than the degree of the denominator, we do not need to divide. We factor the denominator as

$$2x^3 + 3x^2 - 2x = x(2x^2 + 3x - 2) = x(2x - 1)(x + 2)$$

Since the denominator has three distinct linear factors, the partial fraction decomposition of the integrand (2) has the form

(3)
$$\frac{x^2 + 2x - 1}{x(2x - 1)(x + 2)} = \frac{A}{x} + \frac{B}{2x - 1} + \frac{C}{x + 2}$$

To determine the values of A, B, and C, we multiply both sides of this equation by $x(2x - 1)(x + 2)$, obtaining

(4)
$$x^2 + 2x - 1 = A(2x - 1)(x + 2) + Bx(x + 2) + Cx(2x - 1)$$

Expanding the right side of Equation 4 and writing it in the standard form for polynomials, we get

(5)
$$x^2 + 2x - 1 = (2A + B + 2C)x^2 + (3A + 2B - C)x - 2A$$

Another method for finding A, B, and C is given in the note after this example.

The polynomials in Equation 5 are identical, so their coefficients must be equal. The coefficient of x^2 on the right side, $2A + B + 2C$, must equal the coefficient of x^2 on the left side—namely, 1. Likewise the coefficients of x are equal and the constant terms are equal. This gives the following system of equations for A, B, and C:

$$\begin{aligned} 2A + \ B + 2C &= 1 \\ 3A + 2B - \ C &= 2 \\ -2A \qquad\qquad &= -1 \end{aligned}$$

Figure 1 shows the graphs of the integrand in Example 2 and its indefinite integral (with $K = 0$). Which is which?

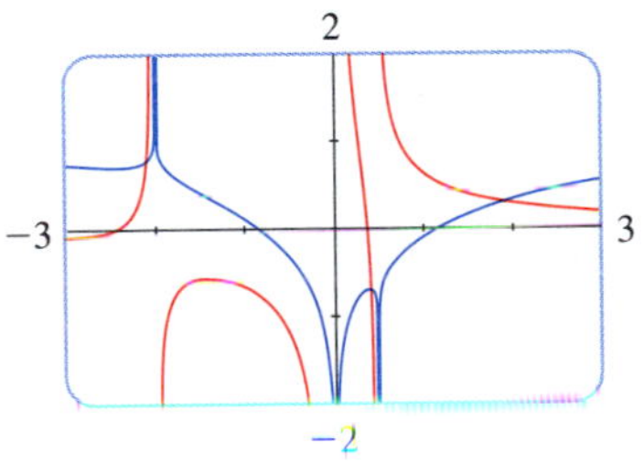

FIGURE 1

Solving, we get $A = \frac{1}{2}$, $B = \frac{1}{5}$, and $C = -\frac{1}{10}$, and so

$$\int \frac{x^2 + 2x - 1}{2x^3 + 3x^2 - 2x}\,dx = \int \left[\frac{1}{2}\,\frac{1}{x} + \frac{1}{5}\,\frac{1}{2x - 1} - \frac{1}{10}\,\frac{1}{x + 2}\right] dx$$
$$= \tfrac{1}{2}\ln|x| + \tfrac{1}{10}\ln|2x - 1| - \tfrac{1}{10}\ln|x + 2| + K$$

In integrating the middle term we have made the mental substitution $u = 2x - 1$, which gives $du = 2\,dx$ and $dx = du/2$. ■

NOTE: We can use an alternative method to find the coefficients A, B, and C in Example 2. Equation 4 is an identity; it is true for every value of x. Let's choose values of x that simplify the equation. If we put $x = 0$ in Equation 4, then the second and third terms on the right side vanish and the equation becomes $-2A = -1$ or $A = \frac{1}{2}$. Likewise, $x = \frac{1}{2}$ gives $5B/4 = \frac{1}{4}$ and $x = -2$ gives $10C = -1$, so $B = \frac{1}{5}$ and $C = -\frac{1}{10}$. (You may object that Equation 3 is not valid for $x = 0, \frac{1}{2}$, or -2, so why should Equation 4 be valid for those values? In fact, Equation 4 is true for all values of x, even $x = 0, \frac{1}{2}$, and -2. See Exercise 69 for the reason.)

EXAMPLE 3 Find $\displaystyle\int \frac{dx}{x^2 - a^2}$, where $a \neq 0$.

SOLUTION The method of partial fractions gives

$$\frac{1}{x^2 - a^2} = \frac{1}{(x - a)(x + a)} = \frac{A}{x - a} + \frac{B}{x + a}$$

and therefore

$$A(x + a) + B(x - a) = 1$$

Using the method of the preceding note, we put $x = a$ in this equation and get $A(2a) = 1$, so $A = 1/(2a)$. If we put $x = -a$, we get $B(-2a) = 1$, so $B = -1/(2a)$. Thus

$$\int \frac{dx}{x^2 - a^2} = \frac{1}{2a} \int \left[\frac{1}{x - a} - \frac{1}{x + a} \right] dx$$

$$= \frac{1}{2a}[\ln|x - a| - \ln|x + a|] + C$$

Since $\ln x - \ln y = \ln(x/y)$, we can write the integral as

(6) $$\int \frac{dx}{x^2 - a^2} = \frac{1}{2a} \ln \left| \frac{x - a}{x + a} \right| + C$$

See Exercises 58–61 for ways of using Formula 6. ■

CASE II **$Q(x)$ is a product of linear factors, some of which are repeated.** Suppose the first linear factor $(a_1x + b_1)$ is repeated r times; that is, $(a_1x + b_1)^r$ occurs in the factorization of $Q(x)$. Then instead of the single term $A_1/(a_1x + b_1)$ in Equation 2, we would use

(7) $$\frac{A_1}{a_1x + b_1} + \frac{A_2}{(a_1x + b_1)^2} + \cdots + \frac{A_r}{(a_1x + b_1)^r}$$

By way of illustration we could write

$$\frac{x^3 - x + 1}{x^2(x - 1)^3} = \frac{A}{x} + \frac{B}{x^2} + \frac{C}{x - 1} + \frac{D}{(x - 1)^2} + \frac{E}{(x - 1)^3}$$

but we prefer to work out in detail a simpler example.

EXAMPLE 4 Find $\int \frac{x^4 - 2x^2 + 4x + 1}{x^3 - x^2 - x + 1}\,dx$.

SOLUTION The first step is to divide. The result of long division is

$$\frac{x^4 - 2x^2 + 4x + 1}{x^3 - x^2 - x + 1} = x + 1 + \frac{4x}{x^3 - x^2 - x + 1}$$

The second step is to factor the denominator $Q(x) = x^3 - x^2 - x + 1$. Since $Q(1) = 0$, we know that $x - 1$ is a factor and we obtain

$$x^3 - x^2 - x + 1 = (x - 1)(x^2 - 1) = (x - 1)(x - 1)(x + 1)$$

$$= (x - 1)^2(x + 1)$$

Since the linear factor $x - 1$ occurs twice, the partial fraction decomposition is

$$\frac{4x}{(x - 1)^2(x + 1)} = \frac{A}{x - 1} + \frac{B}{(x - 1)^2} + \frac{C}{x + 1}$$

Multiplying by $(x - 1)^2(x + 1)$, we get

(8) $$4x = A(x - 1)(x + 1) + B(x + 1) + C(x - 1)^2$$

$$= (A + C)x^2 + (B - 2C)x + (-A + B + C)$$

Another method for finding the coefficients:
Put $x = 1$ in (8): $B = 2$.
Put $x = -1$: $C = -1$.
Put $x = 0$: $A = B + C = 1$.

Now we equate coefficients:

$$\begin{aligned} A \qquad + \ C &= 0 \\ B - 2C &= 4 \\ -A + B + \ C &= 0 \end{aligned}$$

Solving, we obtain $A = 1$, $B = 2$, and $C = -1$, so

$$\begin{aligned} \int \frac{x^4 - 2x^2 + 4x + 1}{x^3 - x^2 - x + 1}\,dx &= \int \left[x + 1 + \frac{1}{x - 1} + \frac{2}{(x - 1)^2} - \frac{1}{x + 1} \right] dx \\ &= \frac{x^2}{2} + x + \ln|x - 1| - \frac{2}{x - 1} - \ln|x + 1| + K \\ &= \frac{x^2}{2} + x - \frac{2}{x - 1} + \ln\left|\frac{x - 1}{x + 1}\right| + K \end{aligned}$$

CASE III **$Q(x)$ contains irreducible quadratic factors, none of which is repeated.** If $Q(x)$ has the factor $ax^2 + bx + c$, where $b^2 - 4ac < 0$, then, in addition to the partial fractions in Equations 2 and 7, the expression for $R(x)/Q(x)$ will have a term of the form

$$\frac{Ax + B}{ax^2 + bx + c} \tag{9}$$

where A and B are constants to be determined. For instance, the function given by $f(x) = x/[(x - 2)(x^2 + 1)(x^2 + 4)]$ has a partial fraction decomposition of the form

$$\frac{x}{(x - 2)(x^2 + 1)(x^2 + 4)} = \frac{A}{x - 2} + \frac{Bx + C}{x^2 + 1} + \frac{Dx + E}{x^2 + 4}$$

The term given in (9) can be integrated by completing the square and using the formula

$$\int \frac{dx}{x^2 + a^2} = \frac{1}{a}\tan^{-1}\left(\frac{x}{a}\right) + C \tag{10}$$

EXAMPLE 5 Evaluate $\displaystyle\int \frac{2x^2 - x + 4}{x^3 + 4x}\,dx$.

SOLUTION Since $x^3 + 4x = x(x^2 + 4)$ cannot be factored further, we write

$$\frac{2x^2 - x + 4}{x(x^2 + 4)} = \frac{A}{x} + \frac{Bx + C}{x^2 + 4}$$

Multiplying by $x(x^2 + 4)$, we have

$$\begin{aligned} 2x^2 - x + 4 &= A(x^2 + 4) + (Bx + C)x \\ &= (A + B)x^2 + Cx + 4A \end{aligned}$$

Equating coefficients, we obtain

$$A + B = 2 \qquad C = -1 \qquad 4A = 4$$

Thus $A = 1$, $B = 1$, and $C = -1$ and so

$$\int \frac{2x^2 - x + 4}{x^3 + 4x}\,dx = \int \left[\frac{1}{x} + \frac{x - 1}{x^2 + 4}\right] dx$$

In order to integrate the second term we split it into two parts:

$$\int \frac{x - 1}{x^2 + 4}\,dx = \int \frac{x}{x^2 + 4}\,dx - \int \frac{1}{x^2 + 4}\,dx$$

We make the substitution $u = x^2 + 4$ in the first of these integrals so that $du = 2x\,dx$. We evaluate the second integral by means of Formula 10 with $a = 2$:

$$\int \frac{2x^2 - x + 4}{x(x^2 + 4)}\,dx = \int \frac{1}{x}\,dx + \int \frac{x}{x^2 + 4}\,dx - \int \frac{1}{x^2 + 4}\,dx$$

$$= \ln|x| + \tfrac{1}{2}\ln(x^2 + 4) - \tfrac{1}{2}\tan^{-1}\left(\frac{x}{2}\right) + K$$ ■

EXAMPLE 6 Evaluate $\displaystyle\int \frac{4x^2 - 3x + 2}{4x^2 - 4x + 3}\,dx$.

SOLUTION Since the degree of the numerator is not less than the degree of the denominator, we first divide and obtain

$$\frac{4x^2 - 3x + 2}{4x^2 - 4x + 3} = 1 + \frac{x - 1}{4x^2 - 4x + 3}$$

Notice that the quadratic $4x^2 - 4x + 3$ is irreducible because its discriminant is $b^2 - 4ac = -32 < 0$. This means it cannot be factored, so we do not need to use the partial fraction technique.

To integrate the given function we complete the square in the denominator:

$$4x^2 - 4x + 3 = (2x - 1)^2 + 2$$

This suggests that we make the substitution $u = 2x - 1$. Then, $du = 2\,dx$ and $x = (u + 1)/2$, so

$$\int \frac{4x^2 - 3x + 2}{4x^2 - 4x + 3}\,dx = \int \left(1 + \frac{x - 1}{4x^2 - 4x + 3}\right) dx$$

$$= x + \tfrac{1}{2}\int \frac{\tfrac{1}{2}(u + 1) - 1}{u^2 + 2}\,du$$

$$= x + \tfrac{1}{4}\int \frac{u - 1}{u^2 + 2}\,du$$

$$= x + \tfrac{1}{4}\int \frac{u}{u^2 + 2}\,du - \tfrac{1}{4}\int \frac{1}{u^2 + 2}\,du$$

$$= x + \tfrac{1}{8}\ln(u^2 + 2) - \frac{1}{4}\cdot\frac{1}{\sqrt{2}}\tan^{-1}\left(\frac{u}{\sqrt{2}}\right) + C$$

$$= x + \tfrac{1}{8}\ln(4x^2 - 4x + 3) - \frac{1}{4\sqrt{2}}\tan^{-1}\left(\frac{2x - 1}{\sqrt{2}}\right) + C$$ ■

NOTE: Example 6 illustrates the general procedure for integrating a partial fraction of the form

$$\frac{Ax + B}{ax^2 + bx + c} \qquad \text{where } b^2 - 4ac < 0$$

We complete the square in the denominator and then make a substitution that brings the integral into the form

$$\int \frac{Cu + D}{u^2 + a^2}\,du = C\int \frac{u}{u^2 + a^2}\,du + D\int \frac{1}{u^2 + a^2}\,du$$

Then the first integral is a logarithm and the second is expressed in terms of $\tan^{-1}$.

CASE IV $Q(x)$ **contains a repeated irreducible quadratic factor.**
If $Q(x)$ has the factor $(ax^2 + bx + c)^r$, where $b^2 - 4ac < 0$, then instead of the single partial fraction (9), the sum

$$\textbf{(11)} \qquad \frac{A_1x + B_1}{ax^2 + bx + c} + \frac{A_2x + B_2}{(ax^2 + bx + c)^2} + \cdots + \frac{A_rx + B_r}{(ax^2 + bx + c)^r}$$

occurs in the partial fraction decomposition of $R(x)/Q(x)$. Each of the terms in (11) can be integrated by completing the square and making a tangent substitution.

It would be extremely tedious to work out by hand the numerical values of the coefficients in Example 7. Most computer algebra systems, however, can find the numerical values very quickly. For instance, the Maple command

convert(f, parfrac, x)

or the Mathematica command

Apart[f]

gives the following values:

$A = -1,\ B = \frac{1}{8},\ C = D = -1,$
$E = -\frac{1}{8},\ F = \frac{15}{8},\ G = H = \frac{3}{4},$
$I = -\frac{1}{2},\ J = \frac{1}{2}$

EXAMPLE 7 Write out the form of the partial fraction decomposition of the function

$$\frac{x^3 + x^2 + 1}{x(x - 1)(x^2 + x + 1)(x^2 + 1)^3}$$

SOLUTION

$$\frac{x^3 + x^2 + 1}{x(x - 1)(x^2 + x + 1)(x^2 + 1)^3}$$

$$= \frac{A}{x} + \frac{B}{x - 1} + \frac{Cx + D}{x^2 + x + 1} + \frac{Ex + F}{x^2 + 1} + \frac{Gx + H}{(x^2 + 1)^2} + \frac{Ix + J}{(x^2 + 1)^3}$$ ■

EXAMPLE 8 Evaluate $\displaystyle\int \frac{1 - 3x + 2x^2 - x^3}{x(x^2 + 1)^2}\,dx$.

SOLUTION The form of the partial fraction decomposition is

$$\frac{1 - 3x + 2x^2 - x^3}{x(x^2 + 1)^2} = \frac{A}{x} + \frac{Bx + C}{x^2 + 1} + \frac{Dx + E}{(x^2 + 1)^2}$$

Multiplying by $x(x^2 + 1)^2$, we have

$$\begin{aligned} -x^3 + 2x^2 - 3x + 1 &= A(x^2 + 1)^2 + (Bx + C)x(x^2 + 1) + (Dx + E)x \\ &= A(x^4 + 2x^2 + 1) + B(x^4 + x^2) + C(x^3 + x) + Dx^2 + Ex \\ &= (A + B)x^4 + Cx^3 + (2A + B + D)x^2 + (C + E)x + A \end{aligned}$$

If we equate coefficients, we get the system

$$A + B = 0 \qquad C = -1 \qquad 2A + B + D = 2 \qquad C + E = -3 \qquad A = 1$$

which has the solution $A = 1$, $B = -1$, $C = -1$, $D = 1$, and $E = -2$. Thus

$$\int \frac{1 - 3x + 2x^2 - x^3}{x(x^2 + 1)^2}\,dx$$

$$= \int \left(\frac{1}{x} - \frac{x + 1}{x^2 + 1} + \frac{x - 2}{(x^2 + 1)^2}\right) dx$$

$$= \int \frac{dx}{x} - \int \frac{x}{x^2 + 1}\,dx - \int \frac{dx}{x^2 + 1} + \int \frac{x\,dx}{(x^2 + 1)^2} - 2\int \frac{dx}{(x^2 + 1)^2}$$

$$= \ln|x| - \tfrac{1}{2}\ln(x^2 + 1) - \tan^{-1}x - \frac{1}{2(x^2 + 1)} - 2\int \frac{dx}{(x^2 + 1)^2}$$

To evaluate the final integral we substitute $x = \tan\theta$. Then we have $dx = \sec^2\theta\,d\theta$ and $x^2 + 1 = \sec^2\theta$, so

$$\int \frac{dx}{(x^2 + 1)^2} = \int \frac{\sec^2\theta}{\sec^4\theta}\,d\theta = \int \cos^2\theta\,d\theta$$

$$= \tfrac{1}{2}\int (1 + \cos 2\theta)\,d\theta = \frac{\theta}{2} + \frac{\sin 2\theta}{4} + K$$

$$= \frac{\theta}{2} + \frac{\sin\theta\cos\theta}{2} + K$$

$$= \tfrac{1}{2}\left(\tan^{-1}x + \frac{x}{\sqrt{x^2 + 1}} \cdot \frac{1}{\sqrt{x^2 + 1}}\right) + K \qquad \text{(from Figure 2)}$$

$$= \tfrac{1}{2}\left(\tan^{-1}x + \frac{x}{x^2 + 1}\right) + K$$

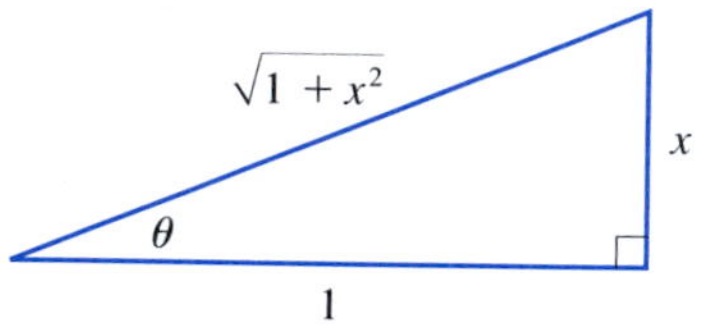

FIGURE 2

Thus

$$\int \frac{1 - 3x + 2x^2 - x^3}{x(x^2 + 1)^2}\,dx$$

$$= \ln|x| - \tfrac{1}{2}\ln(x^2 + 1) - \tan^{-1}x - \frac{1}{2(x^2 + 1)} - \tan^{-1}x - \frac{x}{x^2 + 1} + K_1$$

$$= \ln\frac{|x|}{\sqrt{x^2 + 1}} - 2\tan^{-1}x - \frac{2x + 1}{2(x^2 + 1)} + K_1 \qquad \text{where } K_1 = -2K$$

Finally, we note that sometimes partial fractions can be avoided when integrating a rational function. For instance, although the integral

$$\int \frac{x^2 + 1}{x(x^2 + 3)}\,dx$$

could be evaluated by the method of Case III, it is much easier to observe that if $u = x(x^2 + 3) = x^3 + 3x$, then $du = (3x^2 + 3)\,dx$ and so

$$\int \frac{x^2 + 1}{x(x^2 + 3)}\,dx = \tfrac{1}{3}\ln|x^3 + 3x| + C$$

EXERCISES 8.4

1–16 ■ Write out the form of the partial fraction decomposition of the function (as in Example 7). Do not determine the numerical values of the coefficients.

1. $\dfrac{1}{(x-1)(x+2)}$

2. $\dfrac{7}{2x^2+5x-12}$

3. $\dfrac{x^2+3x-4}{(2x-1)^2(2x+3)}$

4. $\dfrac{x^3-x^2}{(x-6)(5x+3)^3}$

5. $\dfrac{1}{x^4-x^3}$

6. $\dfrac{1+x+x^2}{(x+1)(x+2)^2(x+3)^3}$

7. $\dfrac{x^2+1}{x^2-1}$

8. $\dfrac{x^4+x^3-x^2-x+1}{x^3-x}$

9. $\dfrac{x^2-2}{x(x^2+2)}$

10. $\dfrac{x^3-4x^2+2}{(x^2+1)(x^2+2)}$

11. $\dfrac{x^4+x^2+1}{(x^2+1)(x^2+4)^2}$

12. $\dfrac{1+16x}{(2x-3)(x+5)^2(x^2+x+1)}$

13. $\dfrac{x^4}{(x^2+9)^3}$

14. $\dfrac{19x}{(x-1)^3(4x^2+5x+3)^2}$

15. $\dfrac{x^3+x^2+1}{x^4+x^3+2x^2}$

16. $\dfrac{1}{x^6-x^3}$

17–54 ■ Evaluate the integral.

17. $\displaystyle\int \frac{x^2}{x+1}\,dx$

18. $\displaystyle\int \frac{x}{x-5}\,dx$

19. $\displaystyle\int_2^4 \frac{4x-1}{(x-1)(x+2)}\,dx$

20. $\displaystyle\int_3^7 \frac{1}{(x+1)(x-2)}\,dx$

21. $\displaystyle\int \frac{6x-5}{2x+3}\,dx$

22. $\displaystyle\int \frac{1}{(x+a)(x+b)}\,dx$

23. $\displaystyle\int \frac{x^2+1}{x^2-x}\,dx$

24. $\displaystyle\int_0^2 \frac{x^3+x^2-12x+1}{x^2+x-12}\,dx$

25. $\displaystyle\int_0^1 \frac{2x+3}{(x+1)^2}\,dx$

26. $\displaystyle\int \frac{1}{x(x+1)(2x+3)}\,dx$

27. $\displaystyle\int_2^3 \frac{6x^2+5x-3}{x^3+2x^2-3x}\,dx$

28. $\displaystyle\int_0^1 \frac{x}{x^2+4x+4}\,dx$

29. $\displaystyle\int \frac{1}{(x-1)^2(x+4)}\,dx$

30. $\displaystyle\int \frac{x^2}{(x-3)(x+2)^2}\,dx$

31. $\displaystyle\int \frac{5x^2+3x-2}{x^3+2x^2}\,dx$

32. $\displaystyle\int \frac{18-2x-4x^2}{x^3+4x^2+x-6}\,dx$

33. $\displaystyle\int \frac{x^2+2x}{x^3+3x^2+4}\,dx$

34. $\displaystyle\int \frac{dx}{x^2(x-1)^2}$

35. $\displaystyle\int \frac{x^2}{(x+1)^3}\,dx$

36. $\displaystyle\int \frac{x^3}{(x+1)^3}\,dx$

37. $\displaystyle\int \frac{dx}{x^4-x^2}$

38. $\displaystyle\int \frac{2x^3-x}{x^4-x^2+1}\,dx$

39. $\displaystyle\int_0^1 \frac{x^3}{x^2+1}\,dx$

40. $\displaystyle\int_0^1 \frac{x-1}{x^2+2x+2}\,dx$

41. $\displaystyle\int_0^1 \frac{x}{x^2+x+1}\,dx$

42. $\displaystyle\int_{-1/2}^{1/2} \frac{4x^2+5x+7}{4x^2+4x+5}\,dx$

43. $\displaystyle\int \frac{3x^2-4x+5}{(x-1)(x^2+1)}\,dx$

44. $\displaystyle\int \frac{2x+3}{x^3+3x}\,dx$

45. $\displaystyle\int \frac{1}{x^3-1}\,dx$

46. $\displaystyle\int \frac{x^3}{x^3+1}\,dx$

47. $\displaystyle\int \frac{x^2-2x-1}{(x-1)^2(x^2+1)}\,dx$

48. $\displaystyle\int \frac{x^4}{x^4-1}\,dx$

49. $\displaystyle\int \frac{3x^3-x^2+6x-4}{(x^2+1)(x^2+2)}\,dx$

50. $\displaystyle\int \frac{x^3-2x^2+x+1}{x^4+5x^2+4}\,dx$

51. $\displaystyle\int \frac{x-3}{(x^2+2x+4)^2}\,dx$

52. $\displaystyle\int \frac{x^4+1}{x(x^2+1)^2}\,dx$

53. $\displaystyle\int \frac{(2\sin x-3)\cos x}{\sin^2 x-3\sin x+2}\,dx$

54. $\displaystyle\int \frac{\sin x\cos^2 x}{5+\cos^2 x}\,dx$

55. Use a graph of $f(x)=1/(x^2-2x-3)$ to decide whether $\int_0^2 f(x)\,dx$ is positive or negative. Use the graph to give a rough estimate of the value of the integral and then use partial fractions to find the exact value.

56. Graph both $y=1/(x^3-2x^2)$ and an antiderivative on the same screen.

57. Formula 6 can be rewritten as

$$\int \frac{dx}{a^2-x^2}=\frac{1}{2a}\ln\left|\frac{x+a}{x-a}\right|+K$$

Show, by a hyperbolic substitution or by differentiation, that an alternative formula is

$$\int \frac{dx}{a^2-x^2}=\begin{cases}\dfrac{1}{a}\tanh^{-1}\left(\dfrac{x}{a}\right)+C & \text{if } |x|<a\\[2ex] \dfrac{1}{a}\coth^{-1}\left(\dfrac{x}{a}\right)+C & \text{if } |x|>a\end{cases}$$

58–61 ■ Evaluate the integral by completing the square and using Formula 6 or the formula in Exercise 57.

58. $\displaystyle\int \frac{dx}{x^2+2x-3}$

59. $\displaystyle\int \frac{dx}{x^2-2x}$

60. $\displaystyle\int \frac{2x+1}{4x^2+12x-7}\,dx$

61. $\displaystyle\int \frac{x}{x^2+x-1}\,dx$

62–63 ■ Find the area of the region under the given curve from a to b.

62. $y = \dfrac{1}{x^2 - 6x + 8}$, $a = 5$, $b = 10$

63. $y = \dfrac{x + 1}{x - 1}$, $a = 2$, $b = 3$

64. The region under the curve $y = 1/(x^2 + 3x + 2)$ from $x = 0$ to $x = 1$ is rotated about the x-axis. Find the volume of the resulting solid.

65. If the region in Exercise 64 is rotated about the y-axis, what is the volume of the resulting solid?

66. Factor $x^4 + 1$ as a difference of squares by first adding and subtracting the same quantity. Use this factorization to evaluate $\int 1/(x^4 + 1)\,dx$.

CAS **67.** (a) Use a computer algebra system to find the partial fraction decomposition of the function

$$f(x) = \frac{4x^3 - 27x^2 + 5x - 32}{30x^5 - 13x^4 + 50x^3 - 286x^2 - 299x - 70}$$

(b) Use part (a) to find $\int f(x)\,dx$ (by hand) and compare with the result of using the CAS to integrate f directly. Comment on any discrepancy.

CAS **68.** (a) Find the partial fraction decomposition of the function

$$f(x) = \frac{12x^5 - 7x^3 - 13x^2 + 8}{100x^6 - 80x^5 + 116x^4 - 80x^3 + 41x^2 - 20x + 4}$$

(b) Use part (a) to find $\int f(x)\,dx$ and graph f and its indefinite integral on the same screen.

(c) Use the graph of f to discover the main features of the graph of $\int f(x)\,dx$.

69. Suppose that F, G, and Q are polynomials and

$$\frac{F(x)}{Q(x)} = \frac{G(x)}{Q(x)}$$

for all x except when $Q(x) = 0$. Prove that $F(x) = G(x)$ for all x. [*Hint:* Use continuity.]

70. If f is a quadratic function such that $f(0) = 1$ and

$$\int \frac{f(x)}{x^2(x + 1)^3}\,dx$$

is a rational function, find the value of $f'(0)$.

8.5 RATIONALIZING SUBSTITUTIONS

By means of appropriate substitutions, some functions can be changed into rational functions and therefore integrated by the methods of the preceding section. In particular, when an integrand contains an expression of the form $\sqrt[n]{g(x)}$, then the substitution $u = \sqrt[n]{g(x)}$ may be effective.

EXAMPLE 1 Evaluate $\displaystyle\int \frac{\sqrt{x + 4}}{x}\,dx$.

SOLUTION Let $u = \sqrt{x + 4}$. Then $u^2 = x + 4$, so $x = u^2 - 4$ and $dx = 2u\,du$. Therefore

$$\begin{aligned} \int \frac{\sqrt{x + 4}}{x}\,dx &= \int \frac{u}{u^2 - 4}\,2u\,du \\ &= 2\int \frac{u^2}{u^2 - 4}\,du \\ &= 2\int \left(1 + \frac{4}{u^2 - 4}\right) du \end{aligned}$$

We can evaluate this integral either by factoring $u^2 - 4$ as $(u - 2)(u + 2)$ and using partial fractions or by using Formula 8.4.6 with $a = 2$:

$$\int \frac{\sqrt{x + 4}}{x}\,dx = 2\int du + 8\int \frac{du}{u^2 - 4}$$

$$= 2u + 8 \cdot \frac{1}{2 \cdot 2} \ln \left| \frac{u-2}{u+2} \right| + C$$

$$= 2\sqrt{x+4} + 2\ln \left| \frac{\sqrt{x+4}-2}{\sqrt{x+4}+2} \right| + C \qquad \blacksquare$$

EXAMPLE 2 Find $\displaystyle\int \frac{dx}{\sqrt{x} - \sqrt[3]{x}}$.

SOLUTION If we were to substitute $u = \sqrt{x}$, then the square root would disappear but a cube root would remain. On the other hand, the substitution $u = \sqrt[3]{x}$ would eliminate the cube root but leave a square root. We can eliminate both roots by means of the substitution $u = \sqrt[6]{x}$. (Note that 6 is the least common multiple of 2 and 3.)

Let $u = \sqrt[6]{x}$. Then $x = u^6$, so $dx = 6u^5\,du$ and $\sqrt{x} = u^3$, $\sqrt[3]{x} = u^2$. Thus

$$\int \frac{dx}{\sqrt{x} - \sqrt[3]{x}} = \int \frac{6u^5\,du}{u^3 - u^2} = 6 \int \frac{u^3}{u-1}\,du$$

$$= 6 \int \left(u^2 + u + 1 + \frac{1}{u-1} \right) du \qquad \text{(by long division)}$$

$$= 6\left(\frac{u^3}{3} + \frac{u^2}{2} + u + \ln|u-1| \right) + C$$

$$= 2\sqrt{x} + 3\sqrt[3]{x} + 6\sqrt[6]{x} + 6\ln|\sqrt[6]{x} - 1| + C \qquad \blacksquare$$

Weierstrass was also responsible for the ε-δ definition of a limit. See the note on page 106.

The German mathematician Karl Weierstrass (1815–1897) noticed that the substitution $t = \tan(x/2)$ will convert any rational function of $\sin x$ and $\cos x$ into an ordinary rational function. Let

$$t = \tan\left(\frac{x}{2}\right) \qquad -\pi < x < \pi$$

Then
$$\cos\left(\frac{x}{2}\right) = \frac{1}{\sec\left(\frac{x}{2}\right)} = \frac{1}{\sqrt{1 + \tan^2\left(\frac{x}{2}\right)}} = \frac{1}{\sqrt{1+t^2}}$$

$$\sin\left(\frac{x}{2}\right) = \cos\left(\frac{x}{2}\right)\tan\left(\frac{x}{2}\right) = \frac{t}{\sqrt{1+t^2}}$$

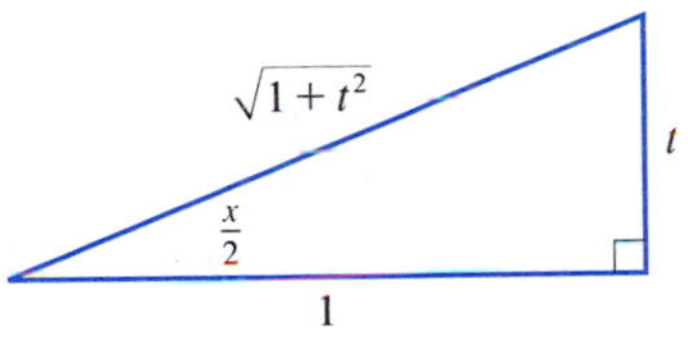

FIGURE 1
$\tan\left(\frac{x}{2}\right) = t$

These expressions can also be seen from Figure 1. Therefore

$$\sin x = 2\sin\left(\frac{x}{2}\right)\cos\left(\frac{x}{2}\right) = 2\frac{t}{\sqrt{1+t^2}} \cdot \frac{1}{\sqrt{1+t^2}} = \frac{2t}{1+t^2}$$

$$\cos x = \cos^2\left(\frac{x}{2}\right) - \sin^2\left(\frac{x}{2}\right) = \frac{1-t^2}{1+t^2}$$

Since $t = \tan(x/2)$, we have $x = 2\tan^{-1}t$, so

$$dx = \frac{2}{1+t^2}\,dt$$

Thus, if we make the Weierstrass substitution $t = \tan(x/2)$, then we have

$$\textbf{(1)} \qquad \sin x = \frac{2t}{1 + t^2} \qquad \cos x = \frac{1 - t^2}{1 + t^2} \qquad dx = \frac{2}{1 + t^2}\, dt$$

You can see from (1) that the Weierstrass substitution transforms any rational function of $\sin x$ and $\cos x$ into a rational function of t.

EXAMPLE 3 Find $\displaystyle\int \frac{1}{3\sin x - 4\cos x}\, dx$.

SOLUTION Let $t = \tan(x/2)$. Then, using the expression in (1), we have

$$\begin{aligned}
\int \frac{1}{3\sin x - 4\cos x}\, dx &= \int \frac{1}{3\left(\dfrac{2t}{1+t^2}\right) - 4\left(\dfrac{1-t^2}{1+t^2}\right)} \frac{2\, dt}{1+t^2} \\
&= 2\int \frac{dt}{3(2t) - 4(1 - t^2)} \\
&= \int \frac{dt}{2t^2 + 3t - 2} \\
&= \int \frac{dt}{(2t - 1)(t + 2)} \\
&= \int \left[\frac{2}{5}\frac{1}{2t - 1} - \frac{1}{5}\frac{1}{t + 2}\right] dt \qquad \text{(using partial fractions)} \\
&= \tfrac{1}{5}\big[\ln|2t - 1| - \ln|t + 2|\big] + C \\
&= \frac{1}{5}\ln\left|\frac{2t - 1}{t + 2}\right| + C \\
&= \frac{1}{5}\ln\left|\frac{2\tan(x/2) - 1}{\tan(x/2) + 2}\right| + C
\end{aligned}$$

■

EXERCISES 8.5

1–30 ■ Evaluate the integral.

1. $\displaystyle\int_0^1 \frac{1}{1 + \sqrt{x}}\, dx$

2. $\displaystyle\int_0^1 \frac{1}{1 + \sqrt[3]{x}}\, dx$

3. $\displaystyle\int \frac{\sqrt{x}}{x + 1}\, dx$

4. $\displaystyle\int \frac{1}{x\sqrt{x + 1}}\, dx$

5. $\displaystyle\int \frac{1}{x - \sqrt[3]{x}}\, dx$

6. $\displaystyle\int \frac{1}{x - \sqrt{x + 2}}\, dx$

7. $\displaystyle\int_5^{10} \frac{x^2}{\sqrt{x - 1}}\, dx$

8. $\displaystyle\int_1^3 \frac{\sqrt{x - 1}}{x + 1}\, dx$

9. $\displaystyle\int \frac{1}{\sqrt{1 + \sqrt{x}}}\, dx$

10. $\displaystyle\int_{1/3}^3 \frac{\sqrt{x}}{x^2 + x}\, dx$

11. $\displaystyle\int \frac{\sqrt{x} + 1}{\sqrt{x} - 1}\, dx$

12. $\displaystyle\int \frac{\sqrt[3]{x} + 1}{\sqrt[3]{x} - 1}\, dx$

13. $\displaystyle\int \frac{x^3}{\sqrt[3]{x^2 + 1}}\, dx$

14. $\displaystyle\int \frac{\sqrt{x}}{\sqrt{x} - \sqrt[3]{x}}\, dx$

15. $\displaystyle\int \frac{1}{\sqrt{x} + \sqrt[4]{x}}\, dx$

16. $\displaystyle\int \frac{1}{\sqrt[3]{x} + \sqrt[4]{x}}\, dx$

17. $\displaystyle\int \sqrt{\frac{1 - x}{x}}\, dx$

18. $\displaystyle\int \frac{\cos x}{\sin^2 x + \sin x}\, dx$

19. $\displaystyle\int \frac{e^{2x}}{e^{2x} + 3e^x + 2}\, dx$

20. $\displaystyle\int \frac{1}{\sqrt{1 + e^x}}\, dx$

21. $\displaystyle\int \sqrt{1 - e^x}\, dx$

22. $\displaystyle\int \frac{dx}{3 - 5\sin x}$

23. $\displaystyle\int_0^{\pi/2} \frac{1}{\sin x + \cos x}\, dx$

24. $\displaystyle\int_{\pi/3}^{\pi/2} \frac{1}{1 + \sin x - \cos x}\, dx$

25. $\displaystyle\int \frac{1}{3\sin x + 4\cos x}\,dx$

26. $\displaystyle\int \frac{1}{\sin x + \tan x}\,dx$

27. $\displaystyle\int \frac{1}{2\sin x + \sin 2x}\,dx$

28. $\displaystyle\int \frac{\sec x}{1 + \sin x}\,dx$

29. $\displaystyle\int \frac{dx}{a\sin x + b\cos x} \quad (b > 0)$

30. $\displaystyle\int \frac{dx}{a^2\sin^2 x + b^2\cos^2 x} \quad (a, b \neq 0)$

31. (a) Use the Weierstrass substitution $t = \tan(x/2)$ to prove the formula

$$\int \sec x\,dx = \ln\left|\frac{1 + \tan(x/2)}{1 - \tan(x/2)}\right| + C$$

(b) Show that

$$\int \sec x\,dx = \ln\left|\tan\left(\frac{\pi}{4} + \frac{x}{2}\right)\right| + C$$

by using the formula for $\tan(x + y)$ and part (a).

32. Find a formula for $\int \csc x\,dx$ similar to the one in Exercise 31(a).

33. Show that the formula in Exercise 31(a) agrees with Formula 8.2.1.

34. If $a \neq 0$, evaluate the integral

$$\int \frac{dx}{a\sin^2 x + b\sin x\cos x + c\cos^2 x}$$

[*Hint:* Make the substitution $u = \tan x$ and consider separately the cases in which $b^2 - 4ac$ is positive, zero, or negative.]

8.6 STRATEGY FOR INTEGRATION

Integration is more challenging than differentiation. In finding the derivative of a function it is obvious which differentiation formula we should apply. But it may not be obvious which technique we should use to integrate a given function.

Until now individual techniques have been applied in each section. For instance, we usually used substitution in Exercises 6.5, integration by parts in Exercises 8.1, and partial fractions in Exercises 8.4. But in this section we present a collection of miscellaneous integrals in random order and the main challenge is to recognize which technique or formula to use. No hard and fast rules can be given as to which method applies in a given situation, but we give some advice on strategy that you may find useful.

A prerequisite for strategy selection is a knowledge of the basic integration formulas. In the following table we have collected the integrals from our previous list together with several additional formulas that we have learned in this chapter. Most of them should be memorized. It is useful to know them all, but the ones marked with an asterisk need not be memorized since they are easily derived. Formula 19 can be avoided by using partial fractions, and trigonometric substitutions can be used in place of Formula 20.

TABLE OF INTEGRATION FORMULAS

Constants of integration have been omitted.

1. $\displaystyle\int x^n\,dx = \frac{x^{n+1}}{n+1} \quad (n \neq -1)$	2. $\displaystyle\int \frac{1}{x}\,dx = \ln\lvert x\rvert$
3. $\displaystyle\int e^x\,dx = e^x$	4. $\displaystyle\int a^x\,dx = \frac{a^x}{\ln a}$
5. $\displaystyle\int \sin x\,dx = -\cos x$	6. $\displaystyle\int \cos x\,dx = \sin x$
7. $\displaystyle\int \sec^2 x\,dx = \tan x$	8. $\displaystyle\int \csc^2 x\,dx = -\cot x$
9. $\displaystyle\int \sec x\tan x\,dx = \sec x$	10. $\displaystyle\int \csc x\cot x\,dx = -\csc x$

11. $\int \sec x\,dx = \ln|\sec x + \tan x|$ **12.** $\int \csc x\,dx = \ln|\csc x - \cot x|$

13. $\int \tan x\,dx = \ln|\sec x|$ **14.** $\int \cot x\,dx = \ln|\sin x|$

15. $\int \sinh x\,dx = \cosh x$ **16.** $\int \cosh x\,dx = \sinh x$

17. $\int \frac{dx}{x^2 + a^2} = \frac{1}{a}\tan^{-1}\left(\frac{x}{a}\right)$ **18.** $\int \frac{dx}{\sqrt{a^2 - x^2}} = \sin^{-1}\left(\frac{x}{a}\right)$

***19.** $\int \frac{dx}{x^2 - a^2} = \frac{1}{2a}\ln\left|\frac{x-a}{x+a}\right|$ ***20.** $\int \frac{dx}{\sqrt{x^2 \pm a^2}} = \ln\left|x + \sqrt{x^2 \pm a^2}\right|$

Once you are armed with these basic integration formulas, if you do not immediately see how to attack a given integral, you might try the following four-step strategy.

1. SIMPLIFY THE INTEGRAND IF POSSIBLE Sometimes the use of algebraic manipulation or trigonometric identities will simplify the integrand and make the method of integration obvious. Here are some examples:

$$\int \sqrt{x}(1 + \sqrt{x})\,dx = \int (\sqrt{x} + x)\,dx$$

$$\int \frac{\tan\theta}{\sec^2\theta}\,d\theta = \int \frac{\sin\theta}{\cos\theta}\cos^2\theta\,d\theta$$

$$= \int \sin\theta\cos\theta\,d\theta = \tfrac{1}{2}\int \sin 2\theta\,d\theta$$

$$\int (\sin x + \cos x)^2\,dx = \int (\sin^2 x + 2\sin x\cos x + \cos^2 x)\,dx$$

$$= \int (1 + 2\sin x\cos x)\,dx$$

2. LOOK FOR AN OBVIOUS SUBSTITUTION Try to find some function $u = g(x)$ in the integrand whose differential $du = g'(x)\,dx$ also occurs, apart from a constant factor. For instance, in the integral

$$\int \frac{x}{x^2 - 1}\,dx$$

we notice that if $u = x^2 - 1$, then $du = 2x\,dx$. Therefore, we use the substitution $u = x^2 - 1$ instead of the method of partial fractions.

3. CLASSIFY THE INTEGRAND ACCORDING TO ITS FORM If steps 1 and 2 have not led to the solution, then we take a look at the form of the integrand $f(x)$.

(a) *Trigonometric functions.* If $f(x)$ is a product of powers of $\sin x$ and $\cos x$, of $\tan x$ and $\sec x$, or of $\cot x$ and $\csc x$, then we use the substitutions recommended in Section 8.2. If f is a trigonometric function that is not of those types but is still a rational function of $\sin x$ and $\cos x$, then we use the Weierstrass substitution $t = \tan(x/2)$.

(b) *Rational functions.* If f is a rational function, we use the procedure of Section 8.4 involving partial fractions.

(c) *Integration by parts.* If $f(x)$ is a product of a power of x (or a polynomial) and a transcendental function (such as a trigonometric, exponential, or logarithmic function), then we try integration by parts, choosing u and dv according to the advice given in Section 8.1. If you look at the functions in Exercises 8.1 you will see that most of them are the type just described.

(d) *Radicals.* Particular kinds of substitutions are recommended when certain radicals appear.

(i) If $\sqrt{\pm x^2 \pm a^2}$ occurs, we use a trigonometric substitution according to the table in Section 8.3.

(ii) If $\sqrt[n]{ax + b}$ occurs, we use the rationalizing substitution $u = \sqrt[n]{ax + b}$. More generally, this sometimes works for $\sqrt[n]{g(x)}$.

4. TRY AGAIN If the first three steps have not produced the answer, remember that there are basically only two methods of integration: substitution and parts.

(a) *Try substitution.* Even if no substitution is obvious (step 2), some inspiration or ingenuity (or even desperation) may suggest an appropriate substitution.

(b) *Try parts.* Although integration by parts is used most of the time on products of the form described in step 3(c), it is sometimes effective on single functions. Looking at Section 8.1, we see that it works on $\tan^{-1}x$, $\sin^{-1}x$, and $\ln x$, and these are all inverse functions.

(c) *Manipulate the integrand.* Algebraic manipulations (perhaps rationalizing the denominator or using trigonometric identities) may be useful in transforming the integral into an easier form. These manipulations may be more substantial than in step 1 and may involve some ingenuity. Here is an example:

$$\int \frac{dx}{1 - \cos x} = \int \frac{1}{1 - \cos x} \cdot \frac{1 + \cos x}{1 + \cos x}\, dx = \int \frac{1 + \cos x}{1 - \cos^2 x}\, dx$$

$$= \int \frac{1 + \cos x}{\sin^2 x}\, dx = \int \left(\csc^2 x + \frac{\cos x}{\sin^2 x} \right) dx$$

(d) *Relate the problem to previous problems.* When you have built up some experience in integration you may be able to use a method on a given integral that is similar to a method you have already used on a previous integral. Or you may even be able to express the given integral in terms of a previous one. For instance, $\int \tan^2 x \sec x\, dx$ is a challenging integral, but if we make use of the identity $\tan^2 x = \sec^2 x - 1$, we can write

$$\int \tan^2 x \sec x\, dx = \int \sec^3 x\, dx - \int \sec x\, dx$$

and if $\int \sec^3 x\, dx$ has previously been evaluated (see Example 9 in Section 8.2), then that calculation can be used in the present problem.

(e) *Use several methods.* Sometimes two or three methods are required to evaluate an integral. The evaluation could involve several successive substitutions of different types or it might combine integration by parts with one or more substitutions.

In the following examples we indicate a method of attack but do not fully work out the integral.

EXAMPLE 1 $\displaystyle\int \frac{\tan^3 x}{\cos^3 x}\,dx$

In step 1 we rewrite the integral:

$$\int \frac{\tan^3 x}{\cos^3 x}\,dx = \int \tan^3 x \sec^3 x\,dx$$

The integral is now of the form $\int \tan^m x \sec^n x\,dx$ with m odd, so we can use the advice in Section 8.2.

Alternatively, if in step 1 we had written

$$\int \frac{\tan^3 x}{\cos^3 x}\,dx = \int \frac{\sin^3 x}{\cos^3 x}\,\frac{1}{\cos^3 x}\,dx = \int \frac{\sin^3 x}{\cos^6 x}\,dx$$

then we could have continued as follows with the substitution $u = \cos x$:

$$\int \frac{\sin^3 x}{\cos^6 x}\,dx = \int \frac{1 - \cos^2 x}{\cos^6 x}\,\sin x\,dx = \int \frac{1 - u^2}{u^6}(-du)$$

$$= \int \frac{u^2 - 1}{u^6}\,du = \int (u^{-4} - u^{-6})\,du$$ ■

EXAMPLE 2 $\displaystyle\int e^{\sqrt{x}}\,dx$

According to step 3(d)(ii) we substitute $u = \sqrt{x}$. Then $x = u^2$, so $dx = 2u\,du$ and

$$\int e^{\sqrt{x}}\,dx = 2\int ue^u\,du$$

The integrand is now a product of u and the transcendental function e^u so it can be integrated by parts. ■

EXAMPLE 3 $\displaystyle\int \frac{x^5 + 1}{x^3 - 3x^2 - 10x}\,dx$

No algebraic simplification or substitution is obvious, so steps 1 and 2 do not apply here. The integrand is a rational function so we apply the procedure of Section 8.4, remembering that the first step is to divide. ■

EXAMPLE 4 $\displaystyle\int \frac{dx}{x\sqrt{\ln x}}$

Here step 2 is all that is needed. We substitute $u = \ln x$ because its differential is $du = dx/x$, which occurs in the integral. ■

EXAMPLE 5 $\displaystyle\int \sqrt{\frac{1 - x}{1 + x}}\,dx$

Although the rationalizing substitution

$$u = \sqrt{\frac{1-x}{1+x}}$$

works here [step 3(d)(ii)], it leads to a very complicated rational function. An easier method is to do some algebraic manipulation [either as step 1 or as step 4(c)]. Multiplying numerator and denominator by $\sqrt{1-x}$, we have

$$\int \sqrt{\frac{1-x}{1+x}}\,dx = \int \frac{1-x}{\sqrt{1-x^2}}\,dx$$

$$= \int \frac{1}{\sqrt{1-x^2}}\,dx - \int \frac{x}{\sqrt{1-x^2}}\,dx$$

$$= \sin^{-1}x + \sqrt{1-x^2} + C$$

■

The question arises: Will our strategy for integration enable us to find the integral of every continuous function? In particular, can we use it to evaluate $\int e^{x^2}\,dx$? The answer is no, at least not in terms of the functions that we are familiar with.

The functions that we have been dealing with in this book are called **elementary functions**. These are the polynomials, rational functions, power functions (x^a), exponential functions (a^x), logarithmic functions, trigonometric and inverse trigonometric functions, hyperbolic and inverse hyperbolic functions, and all functions that can be obtained from these by the five operations of addition, subtraction, multiplication, division, and composition. For instance, the function

$$f(x) = \sqrt{\frac{x^2-1}{x^3+2x-1}} + \ln(\cosh x) - xe^{\sin 2x}$$

is an elementary function.

If f is an elementary function, then f' is an elementary function but $\int f(x)\,dx$ need not be an elementary function. Consider $f(x) = e^{x^2}$. Since f is continuous, its integral exists, and if we define the function F by

$$F(x) = \int_0^x e^{t^2}\,dt$$

then we know from Part 1 of the Fundamental Theorem of Calculus that

$$F'(x) = e^{x^2}$$

Thus $f(x) = e^{x^2}$ has an antiderivative F, but it has been proved that F is not an elementary function. This means that no matter how hard we try, we will never succeed in evaluating $\int e^{x^2}\,dx$ in terms of the functions we know. (In Chapter 10, however, we will see how to express $\int e^{x^2}\,dx$ as an infinite series.) The same can be said of the following integrals:

$$\int \frac{e^x}{x}\,dx \qquad \int \sin(x^2)\,dx \qquad \int \cos(e^x)\,dx$$

$$\int \sqrt{x^3+1}\,dx \qquad \int \frac{1}{\ln x}\,dx \qquad \int \frac{\sin x}{x}\,dx$$

You may be assured, though, that the integrals in the following exercises are all elementary functions.

EXERCISES 8.6

1–80 ■ Evaluate the integral.

1. $\int \frac{2x+5}{x-3}\,dx$

2. $\int e^{x+e^x}\,dx$

3. $\int \sin^2 x \cos^3 x\,dx$

4. $\int \frac{\sin x - \cos x}{\sin x + \cos x}\,dx$

5. $\int_0^{1/2} \frac{x}{\sqrt{1-x^2}}\,dx$

6. $\int_1^2 x^3 \ln x\,dx$

7. $\int \frac{\sqrt{x-2}}{x+2}\,dx$

8. $\int \frac{x}{(x+2)^2}\,dx$

9. $\int \ln(1+x^2)\,dx$

10. $\int \frac{\sqrt{1+\ln x}}{x\ln x}\,dx$

11. $\int_0^1 (1+\sqrt{x})^8\,dx$

12. $\int_0^{\pi/4} \tan^3 x \sec^4 x\,dx$

13. $\int \frac{x}{x^2-2x+2}\,dx$

14. $\int x\sin^{-1}x\,dx$

15. $\int \frac{\sqrt{9-x^2}}{x}\,dx$

16. $\int \frac{x}{x^2+3x+2}\,dx$

17. $\int x^2\cosh x\,dx$

18. $\int \frac{x^3+x+1}{x^4+2x^2+4x}\,dx$

19. $\int \frac{\cos x}{1+\sin^2 x}\,dx$

20. $\int \cos\sqrt{x}\,dx$

21. $\int_0^1 \cos \pi x \tan \pi x\,dx$

22. $\int \frac{e^{2x}}{1+e^x}\,dx$

23. $\int e^{3x}\cos 5x\,dx$

24. $\int \cos 3x \cos 5x\,dx$

25. $\int \frac{dx}{x^3+x^2+x+1}$

26. $\int x^2\ln(1+x)\,dx$

27. $\int x^5 e^{-x^3}\,dx$

28. $\int \tan^2 4x\,dx$

29. $\int \frac{1}{\sqrt{9x^2+12x-5}}\,dx$

30. $\int x^2\tan^{-1}x\,dx$

31. $\int \sqrt[3]{x}(1-\sqrt{x})\,dx$

32. $\int \frac{dx}{e^x-e^{-x}}$

33. $\int \frac{x}{x^4+2x^2+10}\,dx$

34. $\int \frac{1}{x+\sqrt[3]{x}}\,dx$

35. $\int \sin^2 x\cos^4 x\,dx$

36. $\int \frac{1}{\sqrt{5-4x-x^2}}\,dx$

37. $\int \frac{x}{1-x^2+\sqrt{1-x^2}}\,dx$

38. $\int \frac{1+\cos x}{\sin x}\,dx$

39. $\int \frac{e^x}{e^{2x}-1}\,dx$

40. $\int \frac{1}{x^3-8}\,dx$

41. $\int_{-1}^1 x^5\cosh x\,dx$

42. $\int_{\pi/4}^{\pi/3} \frac{\ln(\tan x)}{\sin x\cos x}\,dx$

43. $\int_{-3}^3 |x^3+x^2-2x|\,dx$

44. $\int_0^{\pi/4} \cos^5\theta\,d\theta$

45. $\int \cot x\ln(\sin x)\,dx$

46. $\int \frac{1+e^x}{1-e^x}\,dx$

47. $\int \frac{x}{(x^2+1)(x^2+4)}\,dx$

48. $\int \frac{dx}{4-5\sin x}$

49. $\int x\sqrt[3]{x+c}\,dx$

50. $\int e^{\sqrt[3]{x}}\,dx$

51. $\int \frac{1}{x+4+4\sqrt{x+1}}\,dx$

52. $\int \frac{x^3+1}{x^3-x^2}\,dx$

53. $\int (x^2+4x-3)\sin 2x\,dx$

54. $\int \sin x\cos(\cos x)\,dx$

55. $\int \frac{x}{\sqrt{16-x^4}}\,dx$

56. $\int \frac{x^3}{(x+1)^{10}}\,dx$

57. $\int \cot^3 2x\csc^3 2x\,dx$

58. $\int (x+\sin x)^2\,dx$

59. $\int \frac{e^{\arctan x}}{1+x^2}\,dx$

60. $\int \frac{dx}{x(x^4+1)}$

61. $\int t^3 e^{-2t}\,dt$

62. $\int \frac{\sqrt{t}}{1+\sqrt[3]{t}}\,dt$

63. $\int \sin x\sin 2x\sin 3x\,dx$

64. $\int_1^3 |\ln(x/2)|\,dx$

65. $\int \sqrt{\frac{1+x}{1-x}}\,dx$

66. $\int \frac{x\ln x}{\sqrt{x^2-1}}\,dx$

67. $\int \frac{x+a}{x^2+a^2}\,dx$

68. $\int \sqrt{1+x-x^2}\,dx$

69. $\int \frac{x^4}{x^{10}+16}\,dx$

70. $\int \frac{x+2}{x^2+x+2}\,dx$

71. $\int x\sec x\tan x\,dx$

72. $\int \frac{x}{x^4-a^4}\,dx$

73. $\int \frac{1}{\sqrt{x+1}+\sqrt{x}}\,dx$

74. $\int \frac{1}{1+2e^x-e^{-x}}\,dx$

75. $\int \frac{\arctan\sqrt{x}}{\sqrt{x}}\,dx$

76. $\int \frac{\ln(x+1)}{x^2}\,dx$

77. $\int \frac{1}{e^{3x}-e^x}\,dx$

78. $\int \frac{1+\cos^2 x}{1-\cos^2 x}\,dx$

79. $\int \frac{dx}{x\sqrt{2x-25}}$

80. $\int \frac{\sin 2x}{\sqrt{9-\cos^4 x}}\,dx$

8.7 USING TABLES OF INTEGRALS AND COMPUTER ALGEBRA SYSTEMS

In this section we describe how to use tables and computer algebra systems to integrate functions that have elementary antiderivatives. You should bear in mind, though, that even the most powerful computer algebra systems cannot find explicit formulas for the antiderivatives of functions like e^{x^2} or the other functions described at the end of Section 8.6.

TABLES OF INTEGRALS

Tables of indefinite integrals are very useful when we are confronted by an integral that is difficult to evaluate by hand and we don't have access to a computer algebra system. A relatively brief table of 120 integrals is provided in Appendix K. More extensive tables are available in the *CRC Mathematical Tables* (463 entries) or in Gradshteyn and Ryzhik's *Table of Integrals, Series and Products* (New York: Academic Press, 1979), which contains hundreds of pages of integrals. It should be remembered, however, that integrals do not often occur in exactly the form listed in a table. Usually one of the methods of this chapter, such as substitution or integration by parts, is required to transform a given integral into one of the forms in the table.

EXAMPLE 1 The region bounded by the curves $y = \arctan x$, $y = 0$, and $x = 1$ is rotated about the y-axis. Find the volume of the resulting solid.

SOLUTION Using the method of cylindrical shells, we see that the volume is

$$V = \int_0^1 2\pi x \arctan x \, dx$$

The Table of Integrals appears in Appendix K.

In the section of the Table of Integrals entitled *Inverse Trigonometric Forms* we locate Formula 92:

$$\int u \tan^{-1} u \, du = \frac{u^2 + 1}{2} \tan^{-1} u - \frac{u}{2} + C$$

Thus, the volume is

$$V = 2\pi \int_0^1 x \tan^{-1} x \, dx = 2\pi \left[\frac{x^2 + 1}{2} \tan^{-1} x - \frac{x}{2} \right]_0^1$$

$$= \pi \left[(x^2 + 1) \tan^{-1} x - x \right]_0^1 = \pi (2 \tan^{-1} 1 - 1)$$

$$= \pi [2(\pi/4) - 1] = \tfrac{1}{2}\pi^2 - \pi$$

EXAMPLE 2 Use the Table of Integrals to find $\displaystyle\int \frac{x^2}{\sqrt{5 - 4x^2}} \, dx$.

SOLUTION If we look at the section of the table entitled *Forms involving* $\sqrt{a^2 - u^2}$, we see that the closest entry is number 34:

$$\int \frac{u^2}{\sqrt{a^2 - u^2}} \, du = -\frac{u}{2}\sqrt{a^2 - u^2} + \frac{a^2}{2} \sin^{-1}\left(\frac{u}{a}\right) + C$$

This is not exactly what we have, so we make the substitution $u = 2x$:

$$\int \frac{x^2}{\sqrt{5 - 4x^2}} \, dx = \int \frac{(u/2)^2}{\sqrt{5 - u^2}} \frac{du}{2} = \frac{1}{8} \int \frac{u^2}{\sqrt{5 - u^2}} \, du$$

Then we use Formula 34 with $a^2 = 5$:

$$\int \frac{x^2}{\sqrt{5 - 4x^2}}\,dx = \frac{1}{8}\int \frac{u^2}{\sqrt{5 - u^2}}\,du = \frac{1}{8}\left[-\frac{u}{2}\sqrt{5 - u^2} + \frac{5}{2}\sin^{-1}\frac{u}{\sqrt{5}}\right] + C$$

$$= -\frac{x}{8}\sqrt{5 - 4x^2} + \frac{5}{16}\sin^{-1}\left(\frac{2x}{\sqrt{5}}\right) + C$$

■

EXAMPLE 3 Use the Table of Integrals to find $\int x^3 \sin x\,dx$.

SOLUTION We look in the section called *Trigonometric Forms* and use the reduction formula in entry 84 with $n = 3$:

$$\int x^3 \sin x\,dx = -x^3 \cos x + 3\int x^2 \cos x\,dx$$

Then we use entries 85 and 82:

$$\int x^2 \cos x\,dx = x^2 \sin x - 2\int x \sin x\,dx$$

$$= x^2 \sin x - 2(\sin x - x\cos x) + K$$

Combining these calculations, we get

$$\int x^3 \sin x\,dx = -x^3 \cos x + 3x^2 \sin x + 6x\cos x - 6\sin x + C$$

where $C = 3K$. ■

EXAMPLE 4 Use the Table of Integrals to find $\int x\sqrt{x^2 + 2x + 4}\,dx$.

SOLUTION Since the table gives forms involving $\sqrt{a^2 + x^2}$, $\sqrt{a^2 - x^2}$, and $\sqrt{x^2 - a^2}$, but not $\sqrt{ax^2 + bx + c}$, we first complete the square:

$$x^2 + 2x + 4 = (x + 1)^2 + 3$$

Therefore, we make the substitution $u = x + 1$:

$$\int x\sqrt{x^2 + 2x + 4}\,dx = \int (u - 1)\sqrt{u^2 + 3}\,du$$

$$= \int u\sqrt{u^2 + 3}\,du - \int \sqrt{u^2 + 3}\,du$$

The first integral is evaluated using the substitution $t = u^2 + 3$:

$$\int u\sqrt{u^2 + 3}\,du = \tfrac{1}{2}\int \sqrt{t}\,dt = \tfrac{1}{2}\cdot\tfrac{2}{3}t^{3/2} = \tfrac{1}{3}(u^2 + 3)^{3/2}$$

For the second integral we use Formula 21 with $a = \sqrt{3}$:

$$\int \sqrt{u^2 + 3}\,du = \frac{u}{2}\sqrt{u^2 + 3} + \tfrac{3}{2}\ln\left(u + \sqrt{u^2 + 3}\right)$$

Thus

$$\int x\sqrt{x^2 + 2x + 4}\,dx$$

$$= \tfrac{1}{3}(x^2 + 2x + 4)^{3/2} - \frac{x + 1}{2}\sqrt{x^2 + 2x + 4} - \tfrac{3}{2}\ln\left(x + 1 + \sqrt{x^2 + 2x + 4}\right) + C$$

■

COMPUTER ALGEBRA SYSTEMS

We have seen that the use of tables involves matching the form of the given integrand with the forms of the integrands in the tables. Computers are particularly good at matching patterns, so it is not surprising that computer algebra systems are able to integrate many of the functions that have elementary antiderivatives. And just as we used substitutions in conjunction with tables, a CAS can perform substitutions that transform a given integral into one that occurs in its stored formulas. That does not mean that integration by hand is an obsolete skill. We will see that a hand computation sometimes produces an indefinite integral in a form that is more convenient than a machine answer.

To begin, let's see what happens when we ask a machine to integrate the relatively simple function $y = 1/(3x - 2)$. Using the substitution $u = 3x - 2$, an easy calculation by hand gives

$$\int \frac{1}{3x - 2}\,dx = \tfrac{1}{3}\ln|3x - 2| + C$$

whereas Derive, Mathematica, and Maple all return the answer

$$\tfrac{1}{3}\ln(3x - 2)$$

The first thing to notice is that computer algebra systems omit the constant of integration. In other words, they produce a *particular* antiderivative, not the most general one. Therefore, when making use of a machine integration, we might have to add a constant. Secondly, the absolute value signs are omitted in the machine answer. That is fine if our problem is concerned only with values of x greater than $\tfrac{2}{3}$. But if we are interested in other values of x, then we need to insert the absolute value symbol.

In the next example we reconsider the integral of Example 4, but this time we ask a machine for the answer.

EXAMPLE 5 Use a computer algebra system to find $\int x\sqrt{x^2 + 2x + 4}\,dx$.

SOLUTION Maple responds with the answer

$$\tfrac{1}{3}(x^2 + 2x + 4)^{3/2} - \tfrac{1}{4}(2x + 2)\sqrt{x^2 + 2x + 4} - \tfrac{3}{2}\ln(2\sqrt{x^2 + 2x + 4} + 2x + 2)$$

Notice that this is equivalent to the answer we got in Example 4 because the third term can be rewritten as

$$-\tfrac{3}{2}\ln\left[2\left(\sqrt{x^2 + 2x + 4} + x + 1\right)\right] = -\tfrac{3}{2}\ln 2 - \tfrac{3}{2}\ln\left(x + 1 + \sqrt{x^2 + 2x + 4}\right)$$

The extra term $-\tfrac{3}{2}\ln 2$ can be absorbed into the constant of integration.

Mathematica gives the answer

$$\left(\frac{5}{6} + \frac{x}{6} + \frac{x^2}{3}\right)\sqrt{x^2 + 2x + 4} - \frac{3}{2}\,\text{arcsinh}\left(\frac{1 + x}{\sqrt{3}}\right)$$

The first term corresponds to the first two terms in the answer in Example 4. The last terms are equivalent because of the identity

This is Equation 4.7.3.

$$\text{arcsinh}\,x = \ln\left(x + \sqrt{x^2 + 1}\right)$$

Derive gives the answer

$$\tfrac{1}{6}\sqrt{x^2 + 2x + 4}\,(2x^2 + x + 5) - \tfrac{3}{2}\ln\left(\sqrt{x^2 + 2x + 4} + x + 1\right)$$

The first term is like the first term in the Mathematica answer. The second term is identical to the last term in Example 4. ■

EXAMPLE 6 Use a CAS to evaluate $\int x(x^2+5)^8\,dx$.

SOLUTION Maple and Mathematica give the same answer:

$$\tfrac{1}{18}x^{18}+\tfrac{5}{2}x^{16}+50x^{14}+\tfrac{1750}{3}x^{12}+4375x^{10}+21875x^8+\tfrac{218750}{3}x^6+156250x^4+\tfrac{390625}{2}x^2$$

It's clear that both systems must have expanded $(x^2+5)^8$ by the Binomial Theorem and then integrated each term.

If we integrate by hand instead, using the substitution $u=x^2+5$, we get

Derive also gives this answer.

$$\int x(x^2+5)^8\,dx=\tfrac{1}{18}(x^2+5)^9+C$$

For most purposes, this is a more convenient form of the answer. ■

EXAMPLE 7 Use a CAS to find $\int \sin^5x\cos^2x\,dx$.

SOLUTION In Example 2 in Section 8.2 we found that

$$\text{(1)}\qquad \int \sin^5x\cos^2x\,dx=-\tfrac{1}{3}\cos^3x+\tfrac{2}{5}\cos^5x-\tfrac{1}{7}\cos^7x+C$$

Derive and Maple report the answer

$$-\tfrac{1}{7}\sin^4x\cos^3x-\tfrac{4}{35}\sin^2x\cos^3x-\tfrac{8}{105}\cos^3x$$

whereas Mathematica produces

$$-\tfrac{5}{64}\cos x-\tfrac{1}{192}\cos 3x+\tfrac{3}{320}\cos 5x-\tfrac{1}{448}\cos 7x$$

We suspect that there are trigonometric identities which show these three answers are equivalent. Indeed, if we ask Derive, Maple, and Mathematica to simplify their expressions using trigonometric identities, they ultimately produce the same form of the answer as in Equation 1. ■

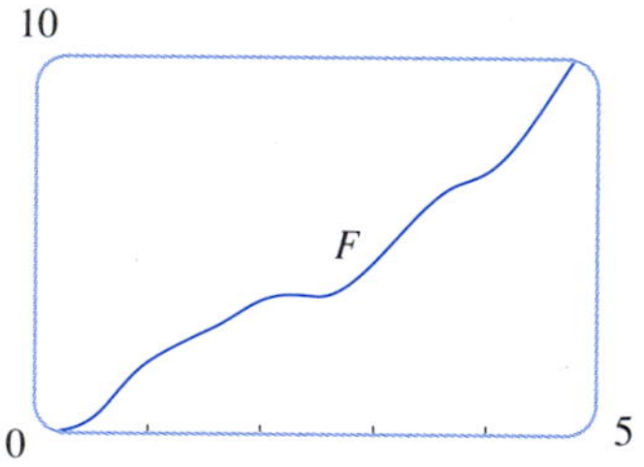

FIGURE 1

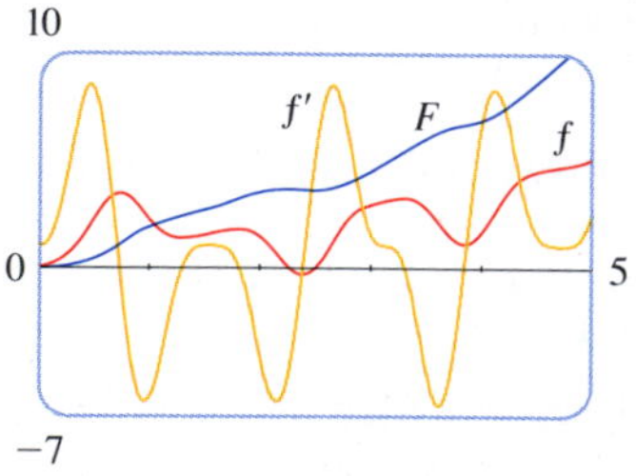

FIGURE 2

EXAMPLE 8 If $f(x)=x+60\sin^4x\cos^5x$, find the antiderivative F of f such that $F(0)=0$. Graph F for $0\leqslant x\leqslant 5$. Where does F have extreme values and inflection points?

SOLUTION The antiderivative of f produced by Maple is

$$F(x)=\tfrac{1}{2}x^2-\tfrac{20}{3}\sin^3x\cos^6x-\tfrac{20}{7}\sin x\cos^6x+\tfrac{4}{7}\cos^4x\sin x+\tfrac{16}{21}\cos^2x\sin x+\tfrac{32}{21}\sin x$$

and we note that $F(0)=0$. This expression could probably be simplified, but there is no need to do so because a computer algebra system can graph this version of F as easily as any other version. A graph of F is shown in Figure 1. To locate the extreme values of F we graph its derivative $F'=f$ in Figure 2 and observe that F has a local maximum when $x\approx 2.3$ and a local minimum when $x\approx 2.5$. The graph of $F''=f'$ in Figure 2 shows that F has inflection points when $x\approx 0.7, 1.3, 1.8, 2.4, 3.3$, and 3.9. ■

EXERCISES 8.7

1–26 ■ Use the Table of Integrals in Appendix K to evaluate the integral.

1. $\int e^{-3x}\cos 4x\,dx$
2. $\int \csc^3(x/2)\,dx$
3. $\int \frac{\sqrt{9x^2-1}}{x^2}\,dx$
4. $\int \frac{\sqrt{4-3x^2}}{x}\,dx$
5. $\int x^2e^{3x}\,dx$
6. $\int \frac{\sin x\cos x}{\sqrt{1+\sin x}}\,dx$
7. $\int x\sin^{-1}(x^2)\,dx$
8. $\int x^3\sin^{-1}(x^2)\,dx$
9. $\int e^x\operatorname{sech}(e^x)\,dx$
10. $\int x^2\cos 3x\,dx$
11. $\int \sqrt{5-4x-x^2}\,dx$
12. $\int \frac{x^5}{x^2+\sqrt{2}}\,dx$
13. $\int \sec^5 x\,dx$
14. $\int \sin^6 2x\,dx$
15. $\int \sin^2 x\cos x\ln(\sin x)\,dx$
16. $\int \frac{dx}{e^x(1+2e^x)}$
17. $\int \sqrt{2+3\cos x}\tan x\,dx$
18. $\int \frac{x}{\sqrt{x^2-4x}}\,dx$
19. $\int_0^{\pi/2}\cos^5 x\,dx$
20. $\int_0^1 x^4e^{-x}\,dx$
21. $\int \frac{x^4\,dx}{\sqrt{x^{10}-2}}$
22. $\int e^x\cos(3x+4)\,dx$
23. $\int e^x\ln(1+e^x)\,dx$
24. $\int x^2\tan^{-1}x\,dx$
25. $\int \sqrt{e^{2x}-1}\,dx$
26. $\int e^{\sin x}\sin 2x\,dx$

27. Find the volume of the solid obtained when the region under the curve $y = 1/(1+5x)^2$ from 0 to 1 is rotated about the y-axis.

28. The region under the curve $y = \tan^2 x$ from 0 to $\pi/4$ is rotated about the x-axis. Find the volume of the resulting solid.

29. Verify Formula 53 in the Table of Integrals (a) by differentiation and (b) by using the substitution $t = a + bu$.

30. Verify Formula 31 (a) by differentiation and (b) by substituting $u = a\sin\theta$.

CAS **31–38** ■ Use a computer algebra system to evaluate the integral. Compare the answer with the result of using tables. If the answers are not the same, show that they are equivalent.

31. $\int x^2\sqrt{5-x^2}\,dx$
32. $\int x^2(1+x^3)^4\,dx$
33. $\int \sin^3 x\cos^2 x\,dx$
34. $\int \tan^2 x\sec^4 x\,dx$
35. $\int x\sqrt{1+2x}\,dx$
36. $\int \sin^4 x\,dx$
37. $\int \tan^3 x\,dx$
38. $\int x^5\sqrt{x^2+1}\,dx$

CAS **39–40** ■ Use a CAS to find an antiderivative F of f such that $F(0) = 0$. Graph f and F and locate approximately the x-coordinates of the extreme points and inflection points of F.

39. $f(x) = \frac{x^2-1}{x^4+x^2+1}$

40. $f(x) = xe^{-x}\sin x,\ -5 \leqslant x \leqslant 5$

CAS **41–42** ■ Use a graphing device to draw a graph of f and use this graph to make a rough sketch, by hand, of the graph of the antiderivative F such that $F(0) = 0$. Then use a CAS to find F explicitly and graph it. Compare the machine graph with your sketch.

41. $f(x) = \sin^4 x\cos^6 x,\ 0 \leqslant x \leqslant \pi$

42. $f(x) = \frac{x^3-x}{x^6+1}$

8.8 APPROXIMATE INTEGRATION

There are two situations in which it is impossible to find the exact value of a definite integral.

The first situation arises from the fact that in order to evaluate $\int_a^b f(x)\,dx$ using the Fundamental Theorem of Calculus we need to know an antiderivative of f. Sometimes, however, it is difficult, or even impossible, to find an antiderivative (see Section 8.6). For example, it is impossible to evaluate the following integrals exactly:

$$\int_0^1 e^{x^2}\,dx \qquad \int_{-1}^1 \sqrt{1+x^3}\,dx$$

The second situation arises when the function is determined from a scientific experiment through instrument readings. There may be no formula for the function (see Example 7).

In both cases we need to find approximate values of definite integrals. We already know one such method. Recall that the definite integral is defined as a limit of Riemann sums, so any Riemann sum could be used as an approximation to the integral. In particular, let us take a partition of $[a, b]$ into n subintervals of equal length $\Delta x = (b - a)/n$. Then we have

$$\int_a^b f(x)\,dx \approx \sum_{i=1}^{n} f(x_i^*)\,\Delta x$$

where x_i^* is any point in the ith subinterval $[x_{i-1}, x_i]$ of the partition. If x_i^* is chosen to be the left endpoint of the interval, then $x_i^* = x_{i-1}$ and we have

(1) $$\int_a^b f(x)\,dx \approx L_n = \sum_{i=1}^{n} f(x_{i-1})\,\Delta x$$

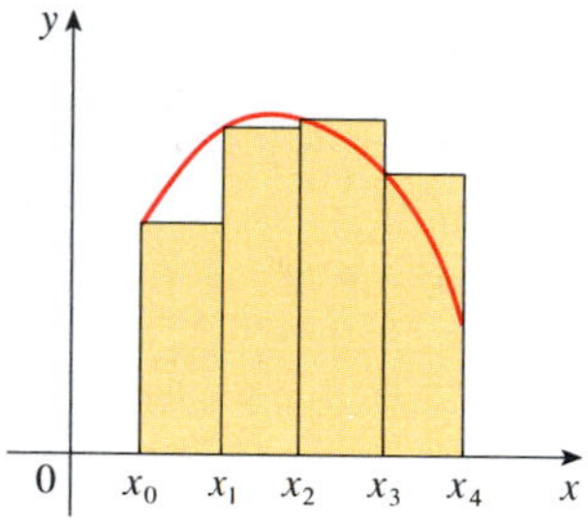

(a) Left endpoint approximation

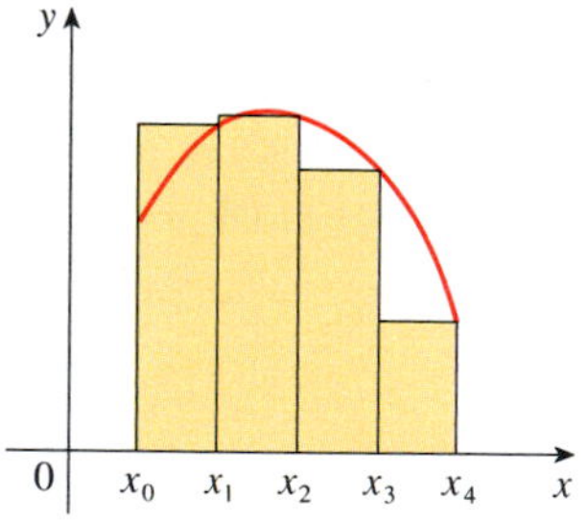

(b) Right endpoint approximation

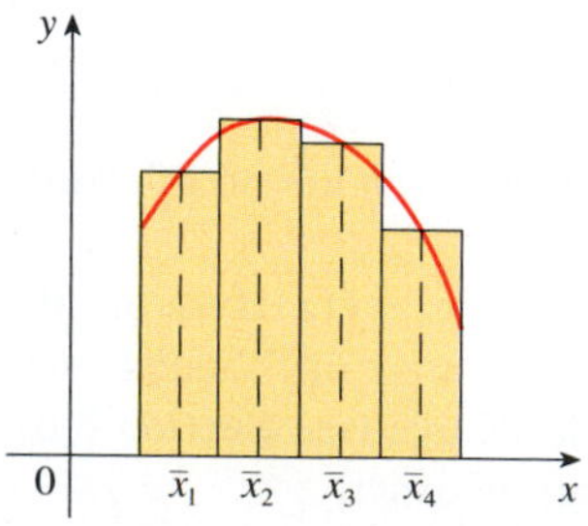

(c) Midpoint approximation

FIGURE 1

If $f(x) \geqslant 0$, then the integral represents an area and (1) represents an approximation of this area by the rectangles shown in Figure 1(a). If we choose x_i^* to be the right endpoint, then $x_i^* = x_i$ and we have

(2) $$\int_a^b f(x)\,dx \approx R_n = \sum_{i=1}^{n} f(x_i)\,\Delta x$$

[See Figure 1(b).] The approximations L_n and R_n defined by Equations 1 and 2 are called the **left endpoint approximation** and **right endpoint approximation**, respectively.

In Section 6.3 we also considered the case where x_i^* is chosen to be the midpoint $\bar{x}_i$ of the subinterval $[x_{i-1}, x_i]$. Figure 1(c) shows the midpoint approximation M_n, which appears to be better than L_n or R_n.

(3) MIDPOINT RULE

$$\int_a^b f(x)\,dx \approx M_n = \Delta x[f(\bar{x}_1) + f(\bar{x}_2) + \cdots + f(\bar{x}_n)]$$

where $$\Delta x = \frac{b - a}{n}$$

and $$\bar{x}_i = \tfrac{1}{2}(x_{i-1} + x_i) = \text{midpoint of } [x_{i-1}, x_i]$$

Another approximation results from averaging the approximations in (1) and (2):

$$\int_a^b f(x)\,dx \approx \frac{1}{2}\left[\sum_{i=1}^{n} f(x_{i-1})\,\Delta x + \sum_{i=1}^{n} f(x_i)\,\Delta x\right]$$

$$= \frac{\Delta x}{2}[(f(x_0) + f(x_1)) + (f(x_1) + f(x_2)) + \cdots + (f(x_{n-1}) + f(x_n))]$$

$$= \frac{\Delta x}{2}[f(x_0) + 2f(x_1) + 2f(x_2) + \cdots + 2f(x_{n-1}) + f(x_n)]$$

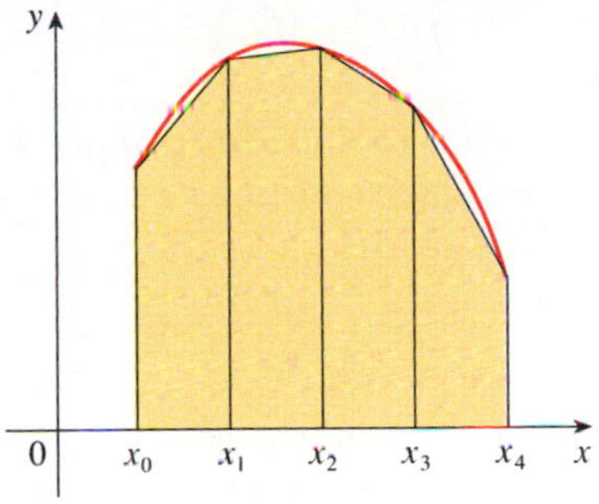

FIGURE 2
Trapezoidal approximation

(4) TRAPEZOIDAL RULE

$$\int_a^b f(x)\,dx \approx T_n = \frac{\Delta x}{2}[f(x_0) + 2f(x_1) + 2f(x_2) + \cdots + 2f(x_{n-1}) + f(x_n)]$$

where $\Delta x = (b - a)/n$ and $x_i = a + i\Delta x$.

The reason for the name Trapezoidal Rule can be seen from Figure 2, which illustrates the case $f(x) \geqslant 0$. The area of the trapezoid that lies above the ith subinterval is

$$\Delta x\left(\frac{f(x_{i-1}) + f(x_i)}{2}\right) = \frac{\Delta x}{2}[f(x_{i-1}) + f(x_i)]$$

and if we add the areas of all these trapezoids, we get the right side of (4).

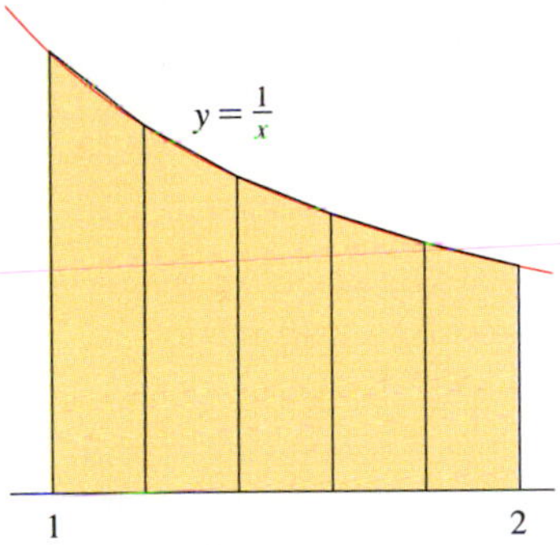

FIGURE 3

EXAMPLE 1 Use (a) the Trapezoidal Rule and (b) the Midpoint Rule with $n = 5$ to approximate the integral $\int_1^2 (1/x)\,dx$.

SOLUTION

(a) With $n = 5$, $a = 1$, and $b = 2$, we have $\Delta x = (2 - 1)/5 = 0.2$, and so the Trapezoidal Rule gives

$$\int_1^2 \frac{1}{x}\,dx \approx T_5 = \frac{0.2}{2}[f(1) + 2f(1.2) + 2f(1.4) + 2f(1.6) + 2f(1.8) + f(2)]$$

$$= 0.1\left[\frac{1}{1} + \frac{2}{1.2} + \frac{2}{1.4} + \frac{2}{1.6} + \frac{2}{1.8} + \frac{1}{2}\right]$$

$$\approx 0.695635$$

This approximation is illustrated in Figure 3.

(b) The midpoints of the five intervals are 1.1, 1.3, 1.5, 1.7, and 1.9, so the Midpoint Rule gives

$$\int_1^2 \frac{1}{x}\,dx \approx \Delta x[f(1.1) + f(1.3) + f(1.5) + f(1.7) + f(1.9)]$$

$$= \frac{1}{5}\left(\frac{1}{1.1} + \frac{1}{1.3} + \frac{1}{1.5} + \frac{1}{1.7} + \frac{1}{1.9}\right)$$

$$\approx 0.691908$$

This approximation is illustrated in Figure 4. ■

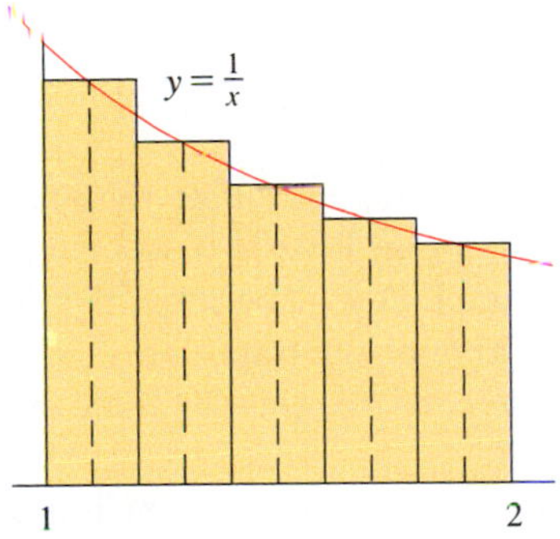

FIGURE 4

In Example 1 we deliberately chose an integral whose value can be computed explicitly so that we can see how accurate the Trapezoidal and Midpoint Rules are. By the Fundamental Theorem of Calculus,

$$\int_1^2 \frac{1}{x}\,dx = \ln x\Big]_1^2 = \ln 2 = 0.693147\ldots$$

$\int_a^b f(x)\,dx$ = approximation + error

The **error** in using an approximation is defined to be the amount that needs to be added to the approximation to make it exact. From the values in Example 1 we see that the errors in the Trapezoidal and Midpoint approximations for $n = 5$ are

$$E_T \approx -0.002488 \qquad \text{and} \qquad E_M \approx 0.001239$$

In general, we have

$$E_T = \int_a^b f(x)\,dx - T_n \qquad \text{and} \qquad E_M = \int_a^b f(x)\,dx - M_n$$

The following tables show the results of calculations similar to those in Example 1, but for $n = 5$, 10, and 20 and for the left and right endpoint approximations as well as the Trapezoidal and Midpoint Rules.

Approximations to $\int_1^2 \frac{1}{x}\,dx$

n	L_n	R_n	T_n	M_n
5	0.745635	0.645635	0.695635	0.691908
10	0.718771	0.668771	0.693771	0.692835
20	0.705803	0.680803	0.693303	0.693069

Corresponding errors

n	E_L	E_R	E_T	E_M
5	−0.052488	0.047512	−0.002488	0.001239
10	−0.025624	0.024376	−0.000624	0.000312
20	−0.012656	0.012344	−0.000156	0.000078

We can make several observations from these tables:

1. In all of the methods we get more accurate approximations when we increase the value of n. (But very large values of n result in so many arithmetic operations that we have to beware of accumulated round-off error.)
2. The errors in the left and right endpoint approximations are opposite in sign and appear to decrease by a factor of about 2 when we double the value of n.
3. The Trapezoidal and Midpoint Rules are much more accurate than the endpoint approximations.
4. The errors in the Trapezoidal and Midpoint Rules are opposite in sign and appear to decrease by a factor of about 4 when we double the value of n.
5. The size of the error in the Midpoint Rule is about half the size of the error in the Trapezoidal Rule.

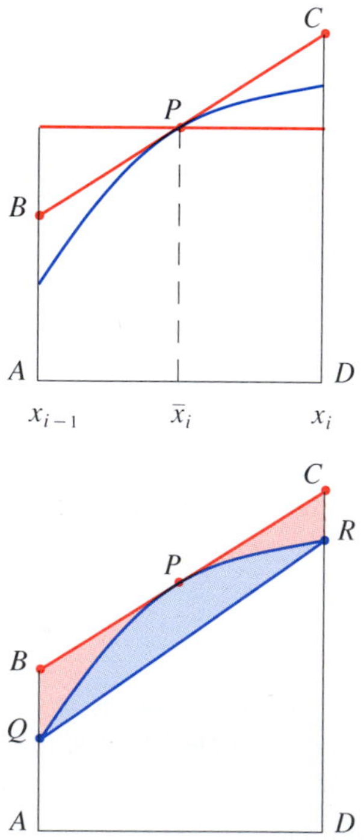

FIGURE 5

Figure 5 shows why we can expect the Midpoint Rule to be more accurate than the Trapezoidal Rule. The area of a typical rectangle in the Midpoint Rule is the same as the trapezoid $ABCD$ whose upper side is tangent to the graph at P. The area of this trapezoid is closer to the area under the graph than is the area of the trapezoid $AQRD$ used in the Trapezoidal Rule. [The midpoint error (shaded red) is smaller than the trapezoidal error (shaded blue).]

These observations are corroborated in the following error estimates, which are proved in books on numerical analysis. The fact that the estimates depend on the size of the second derivative is not surprising if you look at Figure 5, because $f''(x)$ measures how much the graph is curved.

(5) ERROR BOUNDS Suppose $|f''(x)| \leq K$ for $a \leq x \leq b$. If E_T and E_M are the errors in the Trapezoidal and Midpoint Rules, then

$$|E_T| \leq \frac{K(b-a)^3}{12n^2} \qquad \text{and} \qquad |E_M| \leq \frac{K(b-a)^3}{24n^2}$$

Let us apply this error estimate to the Trapezoidal approximation in Example 1. If $f(x) = 1/x$, then $f'(x) = -1/x^2$ and $f''(x) = 2/x^3$. Since $1 \leqslant x \leqslant 2$, we have $1/x \leqslant 1$, so

$$|f''(x)| = \left|\frac{2}{x^3}\right| \leqslant \frac{2}{1^3} = 2$$

Therefore, taking $K = 2$, $a = 1$, $b = 2$, and $n = 5$ in the error estimate (5), we see that

$$|E_T| \leqslant \frac{2(2-1)^3}{12(5)^2} \approx 0.006667$$

Comparing this error estimate of 0.006667 with the actual error of about 0.002488, we see that it can happen that the actual error is substantially less than the upper bound for the error given by (5).

EXAMPLE 2 How large should we take n in order to guarantee that the Trapezoidal and Midpoint approximations for $\int_1^2 (1/x)\,dx$ are accurate to within 0.0001?

SOLUTION We saw in the preceding calculation that $|f''(x)| \leqslant 2$ for $1 \leqslant x \leqslant 2$, so we can take $K = 2$, $a = 1$, and $b = 2$ in (5). Accuracy to within 0.0001 means that the size of the error should be less than 0.0001. Therefore, we choose n so that

$$\frac{2(1)^3}{12n^2} < 0.0001$$

Solving the inequality for n, we get

$$n^2 > \frac{2}{12(0.0001)}$$

or

$$n > \frac{1}{\sqrt{0.0006}} \approx 40.8$$

Thus, $n = 41$ will ensure the desired accuracy.

For the same accuracy with the Midpoint Rule we choose n so that

$$\frac{2(1)^3}{24n^2} < 0.0001$$

which gives

$$n > \frac{1}{\sqrt{0.0012}} \approx 29$$

■

EXAMPLE 3

(a) Use the Midpoint Rule with $n = 10$ to approximate the integral $\int_0^1 e^{x^2}\,dx$.

(b) Give an upper bound for the error involved in this approximation.

SOLUTION

(a) Since $a = 0$, $b = 1$, and $n = 10$, the Midpoint Rule gives

$$\int_0^1 e^{x^2}\,dx \approx \Delta x[f(0.05) + f(0.15) + \cdots + f(0.85) + f(0.95)]$$

$$= 0.1[e^{0.0025} + e^{0.0225} + e^{0.0625} + e^{0.1225} + e^{0.2025} + e^{0.3025} + e^{0.4225} + e^{0.5625} + e^{0.7225} + e^{0.9025}]$$

$$\approx 1.460393$$

Figure 6 illustrates this approximation.

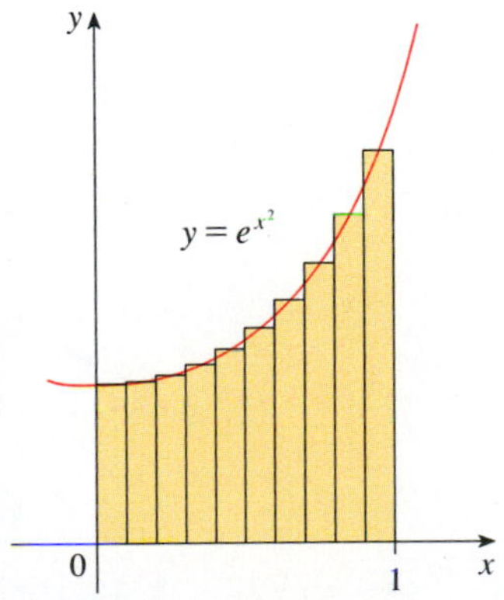

FIGURE 6

(b) Since $f(x) = e^{x^2}$, we have $f'(x) = 2xe^{x^2}$ and $f''(x) = (2 + 4x^2)e^{x^2}$. Also, since $0 \leqslant x \leqslant 1$, we have $x^2 \leqslant 1$ and so

$$0 \leqslant f''(x) = (2 + 4x^2)e^{x^2} \leqslant 6e$$

Error estimates are upper bounds for the error. They give theoretical, worst-case scenarios. The actual error in this case turns out to be about 0.0023.

Taking $K = 6e$, $a = 0$, $b = 1$, and $n = 10$ in the error estimate (5), we see that an upper bound for the error is

$$\frac{6e(1)^3}{24(10)^2} = \frac{e}{400} \approx 0.007$$

■

SIMPSON'S RULE Another rule for approximate integration results from using parabolas instead of straight line segments to approximate a curve. As before, we take a partition of $[a, b]$ into n subintervals of equal length $h = \Delta x = (b - a)/n$, but this time we assume that n is an *even* number. Then on each consecutive pair of intervals we approximate the curve $y = f(x) \geqslant 0$ by a parabola as shown in Figure 7. If $y_i = f(x_i)$, then $P_i(x_i, y_i)$ is the point on the curve lying above x_i. A typical parabola passes through three consecutive points P_i, P_{i+1}, and P_{i+2}.

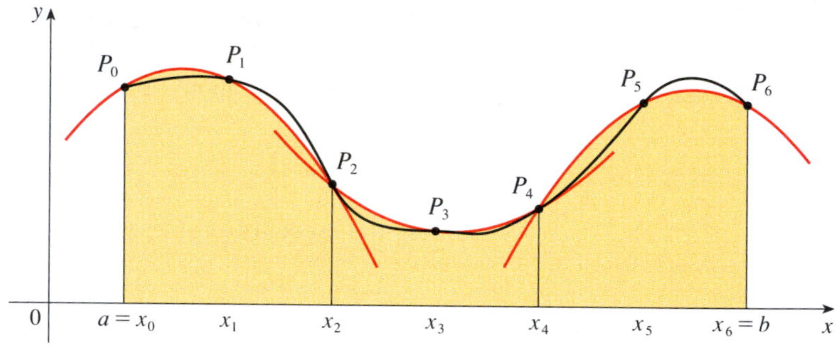

FIGURE 7

In order to simplify our calculations, we first consider the case where $x_0 = -h$, $x_1 = 0$, and $x_2 = h$. (See Figure 8.) We know that the equation of the parabola through P_0, P_1, and P_2 is of the form $y = Ax^2 + Bx + C$ and so the area under the parabola from $x = -h$ to $x = h$ is

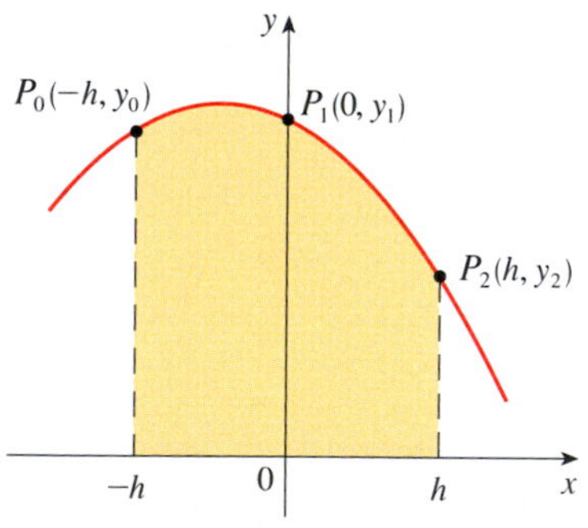

FIGURE 8

$$\int_{-h}^{h} (Ax^2 + Bx + C)\,dx = A\frac{x^3}{3} + B\frac{x^2}{2} + Cx\Big]_{-h}^{h}$$

$$= A\frac{h^3}{3} + B\frac{h^2}{2} + Ch + A\frac{h^3}{3} - B\frac{h^2}{2} + Ch$$

$$= \frac{h}{3}(2Ah^2 + 6C)$$

But, since the parabola passes through $P_0(-h, y_0)$, $P_1(0, y_1)$, and $P_2(h, y_2)$, we have

$$y_0 = A(-h)^2 + B(-h) + C = Ah^2 - Bh + C$$

$$y_1 = C$$

$$y_2 = Ah^2 + Bh + C$$

and therefore $$y_0 + 4y_1 + y_2 = 2Ah^2 + 6C$$

Thus, we can rewrite the area under the parabola as

$$\frac{h}{3}(y_0 + 4y_1 + y_2)$$

Now by shifting this parabola horizontally we do not change the area under it. This means that the area under the parabola through P_0, P_1, and P_2 from $x = x_0$ to $x = x_2$ in Figure 7 is still

$$\frac{h}{3}(y_0 + 4y_1 + y_2)$$

Similarly, the area under the parabola through P_2, P_3, and P_4 from $x = x_2$ to $x = x_4$ is

$$\frac{h}{3}(y_2 + 4y_3 + y_4)$$

If we compute the areas under all the parabolas in this manner and add the results, we get

$$\begin{aligned}\int_a^b f(x)\,dx &\approx \frac{h}{3}(y_0 + 4y_1 + y_2) + \frac{h}{3}(y_2 + 4y_3 + y_4) + \cdots + \frac{h}{3}(y_{n-2} + 4y_{n-1} + y_n)\\ &= \frac{h}{3}(y_0 + 4y_1 + 2y_2 + 4y_3 + 2y_4 + \cdots + 2y_{n-2} + 4y_{n-1} + y_n)\end{aligned}$$

Although we have derived this approximation for the case in which $f(x) \geqslant 0$, it is a reasonable approximation for any continuous function f and is called Simpson's Rule after the English mathematician Thomas Simpson (1710–1761). Note the pattern of coefficients: 1, 4, 2, 4, 2, 4, 2, . . . , 4, 2, 4, 1.

Thomas Simpson was a weaver who taught himself mathematics and went on to become one of the best English mathematicians of the 18th century. What we call Simpson's Rule was actually known to Cavalieri and Gregory in the 17th century, but Simpson popularized it in his best-selling calculus textbook, entitled *A New Treatise of Fluxions.*

(6) SIMPSON'S RULE

$$\int_a^b f(x)\,dx \approx S_n = \frac{\Delta x}{3}[f(x_0) + 4f(x_1) + 2f(x_2) + 4f(x_3) + \cdots + 2f(x_{n-2}) + 4f(x_{n-1}) + f(x_n)]$$

where n is even and $\Delta x = (b - a)/n$.

EXAMPLE 4 Use Simpson's Rule with $n = 10$ to approximate $\int_1^2 (1/x)\,dx$.

SOLUTION Putting $f(x) = 1/x$, $n = 10$, and $\Delta x = 0.1$ in (6), we obtain

$$\begin{aligned}\int_1^2 \frac{1}{x}\,dx &\approx S_{10}\\ &= \frac{\Delta x}{3}[f(1) + 4f(1.1) + 2f(1.2) + 4f(1.3) + \cdots + 2f(1.8) + 4f(1.9) + f(2)]\\ &= \frac{0.1}{3}\left[\frac{1}{1} + \frac{4}{1.1} + \frac{2}{1.2} + \frac{4}{1.3} + \frac{2}{1.4} + \frac{4}{1.5} + \frac{2}{1.6} + \frac{4}{1.7} + \frac{2}{1.8} + \frac{4}{1.9} + \frac{1}{2}\right]\\ &\approx 0.693150\end{aligned}$$

■

Notice that, in Example 4, Simpson's Rule gives us a much better approximation ($S_{10} \approx 0.693150$) to the true value of the integral ($\ln 2 \approx 0.693147\ldots$) than does the Trapezoidal Rule ($T_{10} \approx 0.693771$) or the Midpoint Rule ($M_{10} \approx 0.692835$). It turns out (see Exercise 40) that the approximations in Simpson's Rule are weighted averages of those in the Trapezoidal and Midpoint Rules:

$$S_{2n} = \tfrac{1}{3}T_n + \tfrac{2}{3}M_n$$

(Recall that E_T and E_M have opposite signs and $|E_M|$ is about half the size of $|E_T|$.)

In Exercises 27 and 28 you are asked to demonstrate, in particular cases, that the error in Simpson's Rule decreases by a factor of about 16 when n is doubled. That is consistent with the appearance of n^4 in the denominator of the following error estimate for Simpson's Rule. It is analogous to the estimates given in (5) for the Trapezoidal and Midpoint Rules but it uses the fourth derivative of f.

(7) ERROR BOUND FOR SIMPSON'S RULE Suppose that $|f^{(4)}(x)| \leqslant K$ for $a \leqslant x \leqslant b$. If E_S is the error involved in using Simpson's Rule, then

$$|E_S| \leqslant \frac{K(b-a)^5}{180n^4}$$

EXAMPLE 5 How large should we take n in order to guarantee that the Simpson's Rule approximation for $\int_1^2 (1/x)\,dx$ is accurate to within 0.0001?

SOLUTION If $f(x) = 1/x$, then $f^{(4)}(x) = 24/x^5$. Since $x \geqslant 1$, we have $1/x \leqslant 1$ and so

$$|f^{(4)}(x)| = \left|\frac{24}{x^5}\right| \leqslant 24$$

Many calculators and computer algebra systems have a built-in algorithm that computes an approximation of a definite integral. Some of these machines use Simpson's Rule; others use more sophisticated techniques such as *adaptive* numerical integration. This means that if a function fluctuates much more on a certain part of the interval than it does elsewhere, then that part gets divided into more subintervals. This strategy reduces the number of calculations required to achieve a prescribed accuracy.

Therefore, we can take $K = 24$ in (7). Thus, for an error less than 0.0001 we should choose n so that

$$\frac{24(1)^5}{180n^4} < 0.0001$$

This gives

$$n^4 > \frac{24}{180(0.0001)}$$

or

$$n > \frac{1}{\sqrt[4]{0.00075}} \approx 6.04$$

Therefore, $n = 8$ (n must be even) gives the desired accuracy. (Compare this with Example 2 where we obtained $n = 41$ for the Trapezoidal Rule and $n = 29$ for the Midpoint Rule.) ■

EXAMPLE 6

(a) Use Simpson's Rule with $n = 10$ to approximate the integral $\int_0^1 e^{x^2}\,dx$.
(b) Estimate the error involved in this approximation.

SOLUTION

(a) If $n = 10$, then $\Delta x = 0.1$ and Simpson's Rule gives

$$\int_0^1 e^{x^2}\,dx \approx \frac{\Delta x}{3}[f(0) + 4f(0.1) + 2f(0.2) + \cdots + 2f(0.8) + 4f(0.9) + f(1)]$$

$$= \frac{0.1}{3}[e^0 + 4e^{0.01} + 2e^{0.04} + 4e^{0.09} + 2e^{0.16} + 4e^{0.25} + 2e^{0.36} + 4e^{0.49} + 2e^{0.64} + 4e^{0.81} + e^1]$$

$$\approx 1.462681$$

Figure 9 illustrates the calculation in Example 6. Notice that the parabolic arcs are so close to the graph of $y = e^{x^2}$ that they are practically indistinguishable from it.

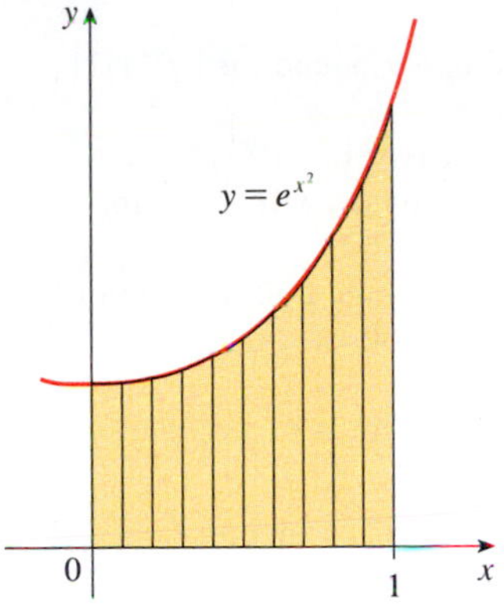

FIGURE 9

(b) The fourth derivative of $f(x) = e^{x^2}$ is

$$f^{(4)}(x) = (12 + 48x^2 + 16x^4)e^{x^2}$$

and so, since $0 \leqslant x \leqslant 1$, we have

$$0 \leqslant f^{(4)}(x) \leqslant (12 + 48 + 16)e^1 = 76e$$

Therefore, putting $K = 76e$, $a = 0$, $b = 1$, and $n = 10$ in (7), we see that the error is at most

$$\frac{76e(1)^5}{180(10)^4} \approx 0.000115$$

(Compare this with Example 3.) Thus, correct to three decimal places, we have

$$\int_0^1 e^{x^2}\,dx \approx 1.463$$

■

Recall that it is quite possible for y to be a function of x even if no explicit formula is known for y in terms of x. If a scientific experiment establishes values for y corresponding to certain equally spaced values of x, and if there is evidence that the values are not changing rapidly, then the Trapezoidal Rule or Simpson's Rule can still be used to find an approximate value for $\int_a^b y\,dx$, the integral of y with respect to x.

EXAMPLE 7 Suppose the following data were obtained from an experiment:

x	3.0	3.25	3.5	3.75	4.0	4.25	4.5	4.75	5.0
y	6.7	7.4	8.2	9.2	10.4	11.6	12.5	13.3	14.0

Use Simpson's Rule to approximate $\int_3^5 y\,dx$.

SOLUTION We have $n = 8$ intervals and the interval length is $\Delta x = 0.25$, so Simpson's Rule gives

$$\int_3^5 y\,dx \approx \frac{0.25}{3}[6.7 + 4(7.4) + 2(8.2) + 4(9.2) + 2(10.4)$$
$$+ 4(11.6) + 2(12.5) + 4(13.3) + 14.0]$$
$$\approx 20.7$$

■

EXERCISES 8.8

1. Let $I = \int_0^4 f(x)\,dx$, where f is the function whose graph is shown.

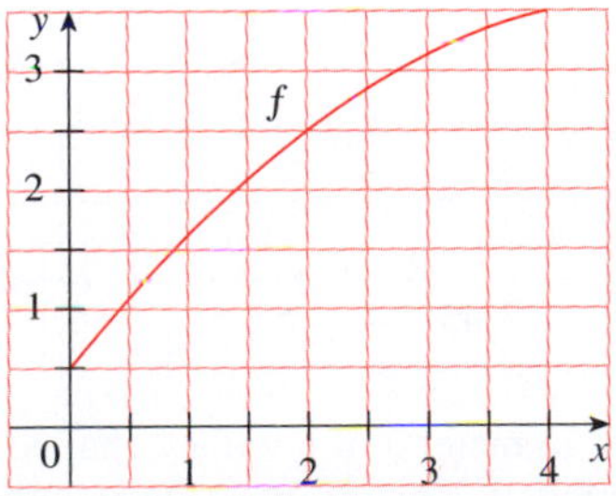

(a) Use the graph to find L_2, R_2, and M_2.
(b) Are these underestimates or overestimates of I?
(c) Use the graph to find T_2. How does it compare with I?
(d) For any value of n, list the numbers L_n, R_n, M_n, T_n, and I in increasing order.

2. Draw the graph of $f(x) = \sin(x^2/2)$ in the viewing rectangle $[0, 1]$ by $[0, 0.5]$ and let $I = \int_0^1 f(x)\,dx$.
(a) Use the graph to decide whether L_2, R_2, M_2, and T_2 underestimate or overestimate I.
(b) For any value of n, list the numbers L_n, R_n, M_n, T_n, and I in increasing order.
(c) Compute L_5, R_5, M_5, and T_5. From the graph, which do you think gives the best estimate of I?

3–6 ■ Use (a) the Trapezoidal Rule and (b) Simpson's Rule to approximate the given integral with the specified value of n. (Round your answers to six decimal places.)

3. $\int_{-1}^{1} \sqrt{1 + x^3}\, dx$, $n = 8$

4. $\int_{0}^{1} \cos(x^2)\, dx$, $n = 4$

5. $\int_{\pi/2}^{\pi} \frac{\sin x}{x}\, dx$, $n = 6$

6. $\int_{0}^{\pi/4} x \tan x\, dx$, $n = 6$

7–16 ■ Use (a) the Trapezoidal Rule, (b) the Midpoint Rule, and (c) Simpson's Rule to approximate the given integral with the specified value of n. (Round your answers to six decimal places.)

7. $\int_{0}^{1} e^{-x^2}\, dx$, $n = 10$

8. $\int_{0}^{2} \frac{1}{\sqrt{1 + x^3}}\, dx$, $n = 10$

9. $\int_{0}^{1/2} \cos(e^x)\, dx$, $n = 8$

10. $\int_{2}^{3} \frac{1}{\ln x}\, dx$, $n = 10$

11. $\int_{0}^{1} x^5 e^x\, dx$, $n = 10$

12. $\int_{0}^{1} \ln(1 + e^x)\, dx$, $n = 8$

13. $\int_{1}^{2} e^{1/x}\, dx$, $n = 4$

14. $\int_{0}^{4} \sqrt{x} \sin x\, dx$, $n = 8$

15. $\int_{0}^{3} \frac{1}{1 + x^4}\, dx$, $n = 6$

16. $\int_{2}^{4} \frac{e^x}{x}\, dx$, $n = 10$

17. (a) Find the approximations T_{10} and M_{10} for the integral $\int_0^2 e^{-x^2}\, dx$.
(b) Estimate the errors in the approximations of part (a).

18. (a) Find the approximations T_4, T_8, M_4, and M_8 for $\int_0^1 \cos(x^2)\, dx$.
(b) Estimate the errors involved in these approximations.

19. (a) Find the approximations T_{10} and S_{10} for $\int_0^1 e^x\, dx$ and the corresponding errors E_T and E_S.
(b) Compare the actual errors in part (a) with the error estimates given by (5) and (7).

20. How large do we have to choose n so that the approximations T_n, M_n, and S_n to the integral $\int_0^1 e^x\, dx$ are accurate to within 0.00001?

21. How large do we have to choose n so that the approximations T_n and M_n to the integral $\int_0^1 e^{-x^2}\, dx$ are accurate to within 0.00001?

22. How large should n be to guarantee that the Simpson's Rule approximation to $\int_0^1 e^{x^2}\, dx$ is accurate to within 0.00001?

CAS **23.** The trouble with the error estimates is that it is often very difficult to compute four derivatives and obtain a good upper bound K for $|f^{(4)}(x)|$ by hand. But computer algebra systems have no problem computing $f^{(4)}$ and graphing it, so we can easily find a value for K from a machine graph. This exercise deals with approximations to the integral $I = \int_0^{2\pi} f(x)\, dx$, where $f(x) = e^{\cos x}$.
(a) Use a graph to get a good upper bound for $|f''(x)|$.
(b) Use M_{10} to approximate I.
(c) Use part (a) to estimate the error in part (b).
(d) Use the built-in numerical integration capability of your CAS to approximate I.
(e) How does the actual error compare with the error estimate in part (c)?
(f) Use a graph to get a good upper bound for $|f^{(4)}(x)|$.
(g) Use S_{10} to approximate I.
(h) Use part (f) to estimate the error in part (g).
(i) How does the actual error compare with the error estimate in part (h)?
(j) How large should n be to guarantee that the size of the error in using S_n is less than 0.0001?

CAS **24.** Repeat Exercise 23 for the integral $\int_{-1}^{1} \sqrt{4 - x^3}\, dx$.

25–26 ■ Find the approximations L_n, R_n, T_n, and M_n for $n = 4, 8$, and 16. Then compute the corresponding errors E_L, E_R, E_T, and E_M. (Round your answers to six decimal places.) What observations can you make?

25. $\int_0^1 x^3\, dx$

26. $\int_0^2 e^x\, dx$

27–28 ■ Find the approximations T_n, M_n, and S_n for $n = 6$ and 12. Then compute the corresponding errors E_T, E_M, and E_S. (Round your answers to six decimal places.) What observations can you make? In particular, what happens to the errors when n is doubled?

27. $\int_1^4 \sqrt{x}\, dx$

28. $\int_{-1}^{2} xe^x\, dx$

29. Use the Trapezoidal Rule and the following data to estimate the value of the integral $\int_1^{3.2} y\, dx$.

x	1.0	1.2	1.4	1.6	1.8	2.0	2.2	2.4	2.6	2.8	3.0	3.2
y	4.9	5.4	5.8	6.2	6.7	7.0	7.3	7.5	8.0	8.2	8.3	8.3

30. Use Simpson's Rule and the following data to estimate the value of the integral $\int_2^6 y\, dx$.

x	2.0	2.5	3.0	3.5	4.0	4.5	5.0	5.5	6.0
y	9.22	9.01	8.76	8.30	7.52	6.83	7.32	7.69	7.91

31. The speedometer reading (v) on a car was observed at 1-minute intervals and recorded in the following chart. Use Simpson's Rule to estimate the distance traveled by the car.

t (min)	0	1	2	3	4	5	6	7	8	9	10
v (mi/h)	40	42	45	49	52	54	56	57	57	55	56

32. The widths (in meters) of a kidney-shaped swimming pool were measured at 2-meter intervals as indicated in the figure. Use Simpson's Rule to estimate the area of the pool.

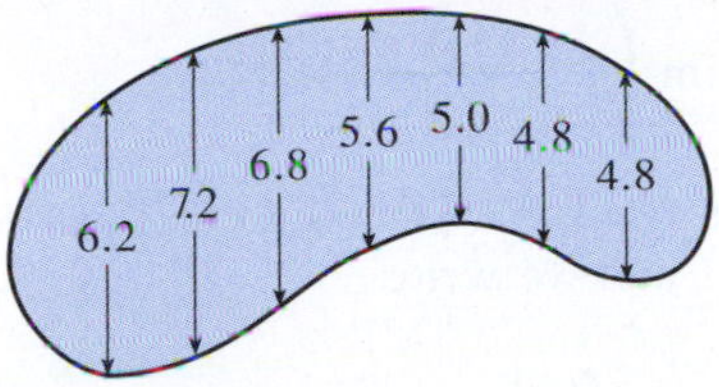

33. Estimate the area under the graph in the figure by using (a) the Trapezoidal Rule, (b) the Midpoint Rule, and (c) Simpson's Rule, each with $n = 4$.

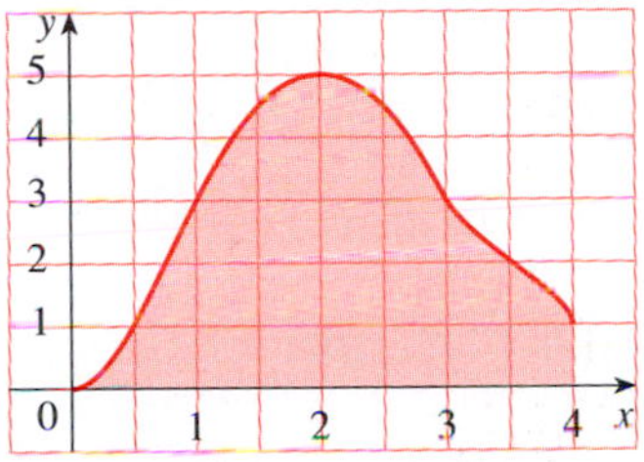

34. A log 10 meters long is cut at 1-meter intervals and its cross-sectional areas A (at a distance x from the end of the log) are listed in the following table. Use Simpson's Rule to estimate the volume of the log.

x (m)	0	1	2	3	4	5	6	7	8	9	10
A (m^2)	0.68	0.65	0.64	0.61	0.58	0.59	0.53	0.55	0.52	0.50	0.48

35. The region bounded by the curves $y = \sqrt[3]{1 + x^3}$, $y = 0$, $x = 0$, and $x = 2$ is rotated about the x-axis. Use Simpson's Rule with $n = 10$ to estimate the volume of the resulting solid.

36. The figure shows a pendulum with length L that makes a maximum angle θ_0 with the vertical. Using Newton's Second Law it can be shown that the period T (the time for one complete swing) is given by

$$T = 4\sqrt{\frac{L}{g}} \int_0^{\pi/2} \frac{dx}{\sqrt{1 - k^2 \sin^2 x}}$$

where $k = \sin(\frac{1}{2}\theta_0)$ and g is the acceleration due to gravity. If $L = 1$ m and $\theta_0 = 42°$, use Simpson's Rule with $n = 10$ to find the period.

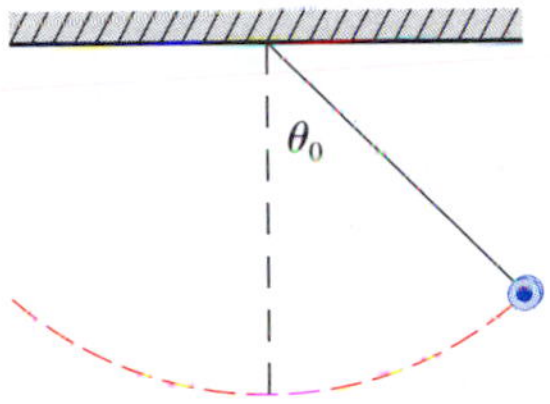

37. If f is a positive function and $f''(x) < 0$ for $a \leqslant x \leqslant b$, show that

$$T_n < \int_a^b f(x)\,dx < M_n$$

38. Show that if f is a polynomial of degree 3 or lower, then Simpson's Rule gives the exact value of $\int_a^b f(x)\,dx$.

39. Show that $\frac{1}{2}(T_n + M_n) = T_{2n}$.

40. Show that $\frac{1}{3}T_n + \frac{2}{3}M_n = S_{2n}$.

8.9 IMPROPER INTEGRALS

In defining a definite integral $\int_a^b f(x)\,dx$ we dealt with a function f defined on a finite interval $[a, b]$ and we observed that if the integral exists then f is a bounded function (see Section 6.3). In this section we extend the concept of a definite integral to the case where the interval is infinite and also to the case where f is unbounded (for instance, where f has an infinite discontinuity in $[a, b]$). In either case the integral is called an *improper* integral.

TYPE 1: INFINITE INTERVALS

Consider the infinite region S that lies under the curve $y = 1/x^2$, above the x-axis, and to the right of the line $x = 1$. You might think that, since S is infinite in extent, its area must be infinite, but let's take a closer look. The area of the part of S that lies to the left of the line $x = t$ (shaded in Figure 1) is

$$A(t) = \int_1^t \frac{1}{x^2}\,dx = -\frac{1}{x}\bigg]_1^t = 1 - \frac{1}{t}$$

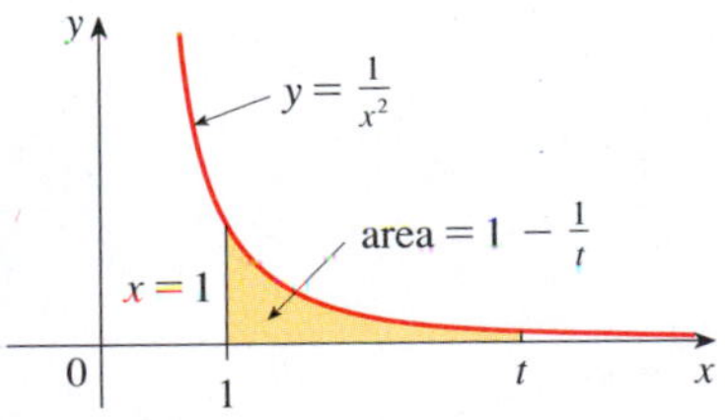

FIGURE 1

Notice that $A(t) < 1$ no matter how large t is chosen.

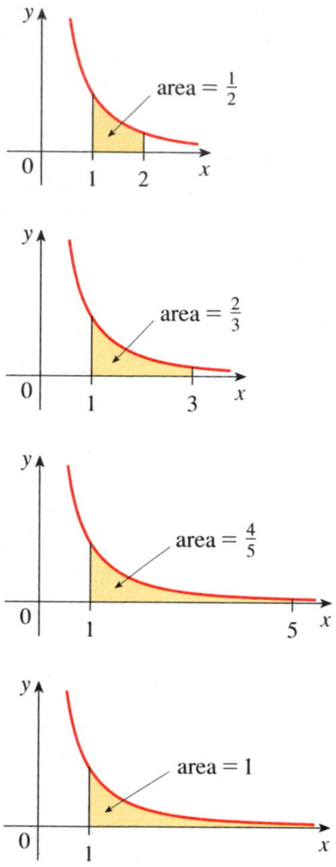

FIGURE 2

We also observe that

$$\lim_{t\to\infty} A(t) = \lim_{t\to\infty}\left(1 - \frac{1}{t}\right) = 1$$

The area of the shaded region approaches 1 as $t \to \infty$ (see Figure 2), so we say that the area of the infinite region S is equal to 1 and we write

$$\int_1^\infty \frac{1}{x^2}\,dx = \lim_{t\to\infty}\int_1^t \frac{1}{x^2}\,dx = 1$$

Using this example as a guide, we define the integral of f (not necessarily a positive function) over an infinite interval as the limit of integrals over finite intervals.

(1) DEFINITION OF AN IMPROPER INTEGRAL OF TYPE 1

(a) If $\int_a^t f(x)\,dx$ exists for every number $t \geqslant a$, then

$$\int_a^\infty f(x)\,dx = \lim_{t\to\infty}\int_a^t f(x)\,dx$$

provided this limit exists (as a finite number).

(b) If $\int_t^b f(x)\,dx$ exists for every number $t \leqslant b$, then

$$\int_{-\infty}^b f(x)\,dx = \lim_{t\to-\infty}\int_t^b f(x)\,dx$$

provided this limit exists (as a finite number).

The improper integrals in (a) and (b) are called **convergent** if the limit exists and **divergent** if the limit does not exist.

(c) If both $\int_a^\infty f(x)\,dx$ and $\int_{-\infty}^a f(x)\,dx$ are convergent, then we define

$$\int_{-\infty}^\infty f(x)\,dx = \int_{-\infty}^a f(x)\,dx + \int_a^\infty f(x)\,dx$$

In part (c) any real number a can be used (see Exercise 62).

Any of the improper integrals in Definition 1 can be interpreted as an area provided f is a positive function. For instance, in case (a) if $f(x) \geqslant 0$ and the integral $\int_a^\infty f(x)\,dx$ is convergent, then we define the area of the region $S = \{(x, y) \mid x \geqslant a,\ 0 \leqslant y \leqslant f(x)\}$ in Figure 3 to be

$$A(S) = \int_a^\infty f(x)\,dx$$

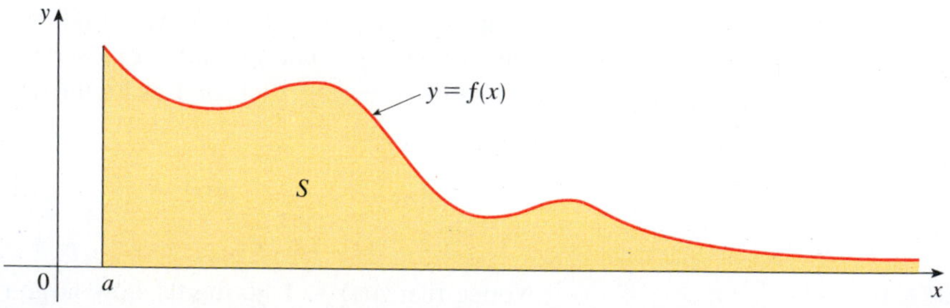

FIGURE 3

This is appropriate because $\int_a^\infty f(x)\,dx$ is the limit as $t \to \infty$ of the area under the graph of f from a to t.

EXAMPLE 1 Determine whether the integral $\int_1^\infty (1/x)\,dx$ is convergent or divergent.

SOLUTION According to part (a) of Definition 1, we have

$$\int_1^\infty \frac{1}{x}\,dx = \lim_{t\to\infty}\int_1^t \frac{1}{x}\,dx = \lim_{t\to\infty} \ln|x|\Big]_1^t$$
$$= \lim_{t\to\infty}(\ln t - \ln 1) = \lim_{t\to\infty}\ln t = \infty$$

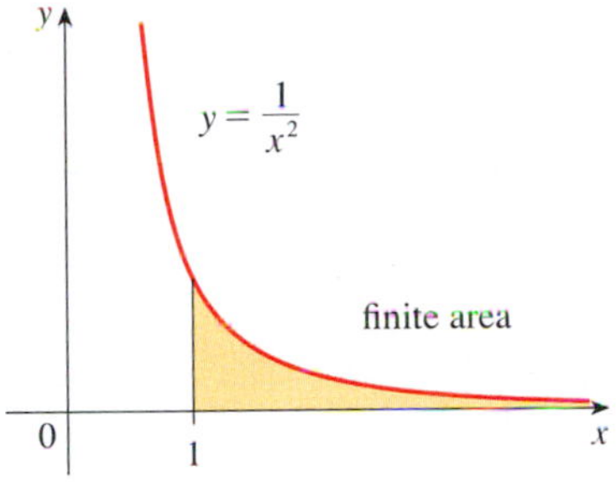

FIGURE 4

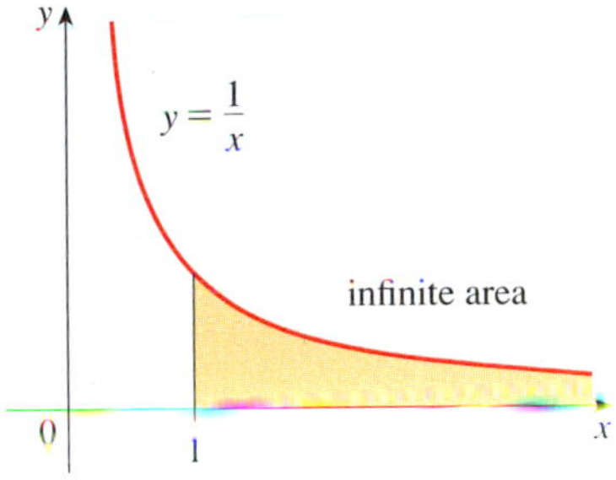

FIGURE 5

The limit does not exist as a finite number and so the improper integral $\int_1^\infty (1/x)\,dx$ is divergent. ■

Let us compare the result of Example 1 with the example at the beginning of this section:

$$\int_1^\infty \frac{1}{x^2}\,dx \text{ converges} \qquad \int_1^\infty \frac{1}{x}\,dx \text{ diverges}$$

Geometrically, this says that although the curves $y = 1/x^2$ and $y = 1/x$ look very similar for $x > 0$, the region under $y = 1/x^2$ to the right of $x = 1$ (the shaded region in Figure 4) has finite area whereas the corresponding region under $y = 1/x$ (in Figure 5) has infinite area. Note that both $1/x^2$ and $1/x$ approach 0 as $x \to \infty$ but $1/x^2$ approaches 0 faster than $1/x$.

EXAMPLE 2 Evaluate $\int_{-\infty}^0 xe^x\,dx$.

SOLUTION Using part (b) of Definition 1, we have

$$\int_{-\infty}^0 xe^x\,dx = \lim_{t\to-\infty}\int_t^0 xe^x\,dx$$

We integrate by parts with $u = x$, $dv = e^x\,dx$ so that $du = dx$, $v = e^x$:

$$\int_t^0 xe^x\,dx = xe^x\Big]_t^0 - \int_t^0 e^x\,dx$$
$$= -te^t - 1 + e^t$$

We know that $e^t \to 0$ as $t \to -\infty$, and by l'Hospital's Rule we have

$$\lim_{t\to-\infty} te^t = \lim_{t\to-\infty}\frac{t}{e^{-t}} = \lim_{t\to-\infty}\frac{1}{-e^{-t}}$$
$$= \lim_{t\to-\infty}(-e^t) = 0$$

Therefore

$$\int_{-\infty}^0 xe^x\,dx = \lim_{t\to-\infty}(-te^t - 1 + e^t)$$
$$= -0 - 1 + 0 = -1$$ ■

EXAMPLE 3 Evaluate $\displaystyle\int_{-\infty}^{\infty} \frac{1}{1 + x^2}\,dx$.

SOLUTION It is convenient to choose $a = 0$ in Definition 1(c):

$$\int_{-\infty}^{\infty} \frac{1}{1 + x^2}\,dx = \int_{-\infty}^{0} \frac{1}{1 + x^2}\,dx + \int_{0}^{\infty} \frac{1}{1 + x^2}\,dx$$

We must now evaluate the integrals on the right side separately:

$$\int_{0}^{\infty} \frac{1}{1 + x^2}\,dx = \lim_{t\to\infty} \int_{0}^{t} \frac{dx}{1 + x^2} = \lim_{t\to\infty} \tan^{-1}x\Big]_0^t$$

$$= \lim_{t\to\infty} (\tan^{-1}t - \tan^{-1}0) = \lim_{t\to\infty} \tan^{-1}t = \frac{\pi}{2}$$

$$\int_{-\infty}^{0} \frac{1}{1 + x^2}\,dx = \lim_{t\to-\infty} \int_{t}^{0} \frac{dx}{1 + x^2} = \lim_{t\to-\infty} \tan^{-1}x\Big]_t^0$$

$$= \lim_{t\to-\infty} (\tan^{-1}0 - \tan^{-1}t)$$

$$= 0 - \left(-\frac{\pi}{2}\right) = \frac{\pi}{2}$$

Since both of these integrals are convergent, the given integral is convergent and

$$\int_{-\infty}^{\infty} \frac{1}{1 + x^2}\,dx = \frac{\pi}{2} + \frac{\pi}{2} = \pi$$

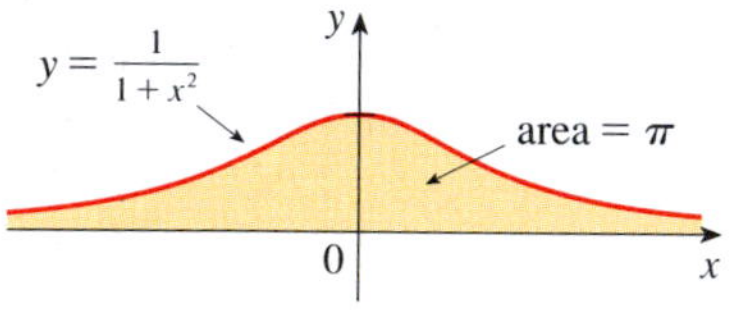

FIGURE 6

Since $1/(1 + x^2) > 0$, the given improper integral can be interpreted as the area of the infinite region that lies under the curve $y = 1/(1 + x^2)$ and above the x-axis (see Figure 6). ■

EXAMPLE 4 For what values of p is the integral

$$\int_{1}^{\infty} \frac{1}{x^p}\,dx$$

convergent?

SOLUTION We know from Example 1 that if $p = 1$, then the integral is divergent, so let's assume that $p \neq 1$. Then

$$\int_{1}^{\infty} \frac{1}{x^p}\,dx = \lim_{t\to\infty} \int_{1}^{t} \frac{1}{x^p}\,dx$$

$$= \lim_{t\to\infty} \frac{x^{-p+1}}{-p+1}\Bigg]_{x=1}^{x=t}$$

$$= \lim_{t\to\infty} \frac{1}{1 - p}\left[\frac{1}{t^{p-1}} - 1\right]$$

If $p > 1$, then $p - 1 > 0$ and so $1/t^{p-1} \to 0$ as $t \to \infty$. Therefore

$$\int_{1}^{\infty} \frac{1}{x^p}\,dx = \frac{1}{p - 1} \qquad \text{if } p > 1$$

But if $p < 1$, then $p - 1 < 0$ and so

$$\frac{1}{t^{p-1}} = t^{1-p} \to \infty \qquad \text{as } t \to \infty$$

and the integral diverges. ■

We summarize the result of Example 4 for future reference:

(2) $$\int_1^\infty \frac{1}{x^p}\,dx \text{ is convergent if } p > 1 \text{ and divergent if } p \leq 1.$$

TYPE 2: DISCONTINUOUS INTEGRANDS

Suppose that f is a positive continuous function defined on a finite interval $[a, b)$ but has a vertical asymptote at b. Let S be the unbounded region under the graph of f and above the x-axis between a and b. (For Type 1 integrals, the regions extended indefinitely in a horizontal direction. Here the region is infinite in a vertical direction.) The area of the part of S between a and t (the shaded region in Figure 7) is

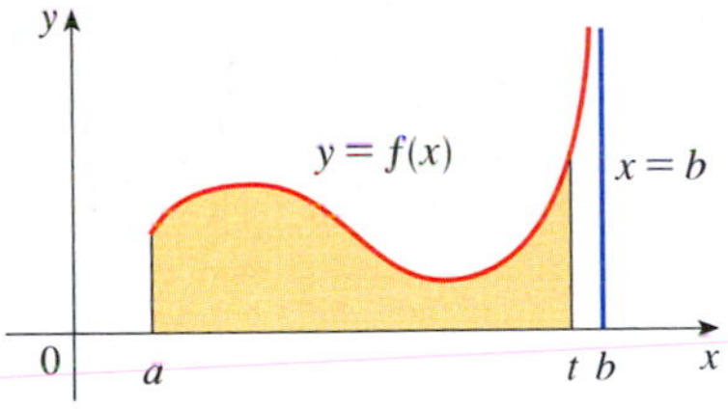

FIGURE 7

$$A(t) = \int_a^t f(x)\,dx$$

If it happens that $A(t)$ approaches a definite number A as $t \to b^-$, then we say that the area of the region S is A and we write

$$\int_a^b f(x)\,dx = \lim_{t \to b^-} \int_a^t f(x)\,dx$$

We use this equation to define an improper integral of Type 2 even when f is not a positive function, no matter what type of discontinuity f has at b.

(3) DEFINITION OF AN IMPROPER INTEGRAL OF TYPE 2

(a) If f is continuous on $[a, b)$ and is discontinuous at b, then

$$\int_a^b f(x)\,dx = \lim_{t \to b^-} \int_a^t f(x)\,dx$$

if this limit exists (as a finite number).

(b) If f is continuous on $(a, b]$ and is discontinuous at a, then

$$\int_a^b f(x)\,dx = \lim_{t \to a^+} \int_t^b f(x)\,dx$$

if this limit exists (as a finite number).

The improper integrals in (a) and (b) are called **convergent** if the limit exists and **divergent** if the limit does not exist.

(c) If f has a discontinuity at c, where $a < c < b$, and both $\int_a^c f(x)\,dx$ and $\int_c^b f(x)\,dx$ are convergent, then we define

$$\int_a^b f(x)\,dx = \int_a^c f(x)\,dx + \int_c^b f(x)\,dx$$

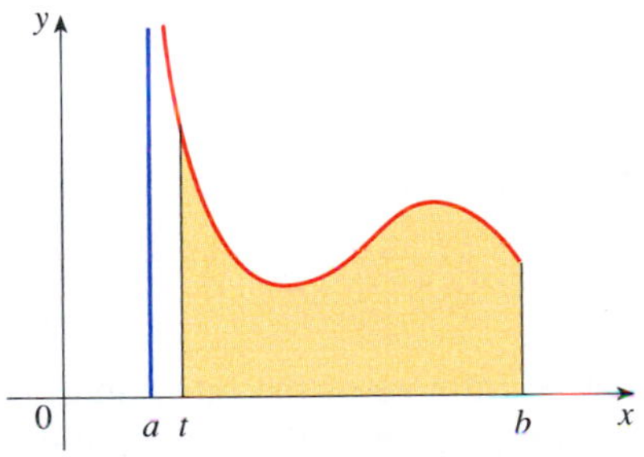

FIGURE 8

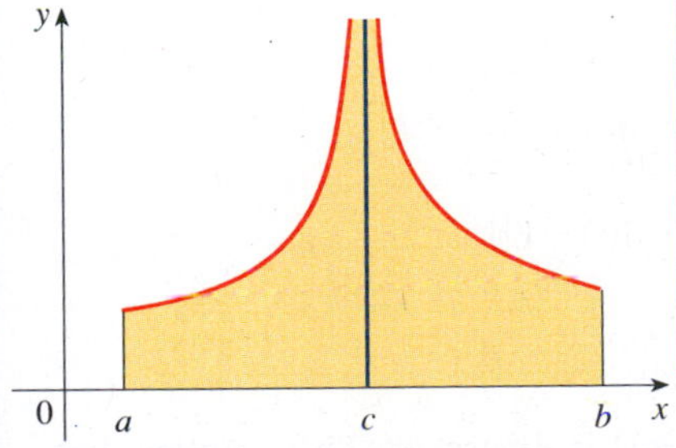

FIGURE 9

Parts (b) and (c) of Definition 3 are illustrated in Figures 8 and 9 for the case where $f(x) \geq 0$ and f has vertical asymptotes at a and c, respectively.

EXAMPLE 5 Find $\displaystyle\int_2^5 \frac{1}{\sqrt{x-2}}\,dx$.

SOLUTION We note first of all that the given integral is improper because $f(x) = 1/\sqrt{x-2}$ has the vertical asymptote $x = 2$. Since the infinite discontinuity occurs at the left endpoint of $[2, 5]$, we use part (b) of Definition 3:

$$\begin{aligned}\int_2^5 \frac{dx}{\sqrt{x-2}} &= \lim_{t\to 2^+}\int_t^5 \frac{dx}{\sqrt{x-2}}\\ &= \lim_{t\to 2^+} 2\sqrt{x-2}\,\Big]_t^5\\ &= \lim_{t\to 2^+} 2(\sqrt{3} - \sqrt{t-2})\\ &= 2\sqrt{3}\end{aligned}$$

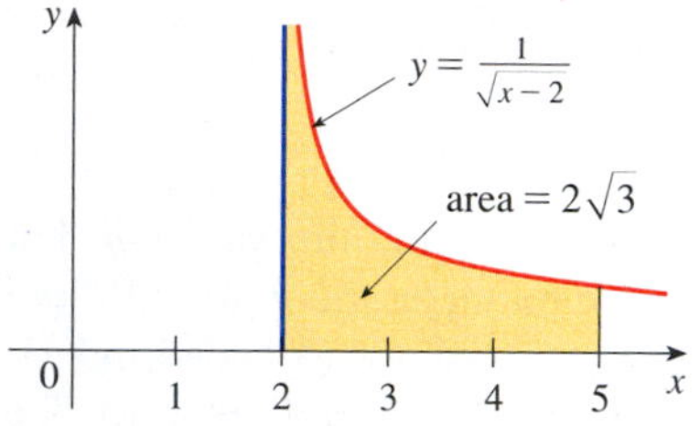

FIGURE 10

Thus, the given improper integral is convergent and, since the integrand is positive, we can interpret the value of the integral as the area of the shaded region in Figure 10. ■

EXAMPLE 6 Determine whether $\displaystyle\int_0^{\pi/2} \sec x\,dx$ converges or diverges.

SOLUTION Note that the given integral is improper because $\lim_{x\to(\pi/2)^-} \sec x = \infty$. Using part (a) of Definition 3, we have

$$\begin{aligned}\int_0^{\pi/2} \sec x\,dx &= \lim_{t\to(\pi/2)^-}\int_0^t \sec x\,dx\\ &= \lim_{t\to(\pi/2)^-} \ln|\sec x + \tan x|\,\Big]_0^t\\ &= \lim_{t\to(\pi/2)^-} [\ln(\sec t + \tan t) - \ln 1]\\ &= \infty\end{aligned}$$

because $\sec t \to \infty$ and $\tan t \to \infty$ as $t \to (\pi/2)^-$. Thus, the given improper integral is divergent. ■

EXAMPLE 7 Evaluate $\displaystyle\int_0^3 \frac{dx}{x-1}$ if possible.

SOLUTION Observe that the line $x = 1$ is a vertical asymptote of the integrand. Since it occurs in the middle of the interval $[0, 3]$, we must use part (c) of Definition 3 with $c = 1$:

$$\int_0^3 \frac{dx}{x-1} = \int_0^1 \frac{dx}{x-1} + \int_1^3 \frac{dx}{x-1}$$

where

$$\begin{aligned}\int_0^1 \frac{dx}{x-1} &= \lim_{t\to 1^-}\int_0^t \frac{dx}{x-1} = \lim_{t\to 1^-} \ln|x-1|\,\Big]_0^t\\ &= \lim_{t\to 1^-} (\ln|t-1| - \ln|-1|)\\ &= \lim_{t\to 1^-} \ln(1-t) = -\infty\end{aligned}$$

because $1 - t \to 0^+$ as $t \to 1^-$. Thus, $\int_0^1 dx/(x-1)$ is divergent. This implies that $\int_0^3 dx/(x-1)$ is divergent. [We do not need to evaluate $\int_1^3 dx/(x-1)$.] ■

Warning: If we had not noticed the asymptote $x = 1$ in Example 7 and had instead confused the integral with an ordinary integral, then we might have made the following erroneous calculation:

$$\int_0^3 \frac{dx}{x-1} = \ln|x-1|\Big]_0^3 = \ln 2 - \ln 1 = \ln 2$$

This is wrong because the integral is improper and must be calculated in terms of limits. (Compare this with Example 11 in Section 6.4.)

From now on, whenever you meet the symbol $\int_a^b f(x)\,dx$ you must decide, by looking at the function f on $[a, b]$, whether it is an ordinary definite integral or an improper integral.

EXAMPLE 8 Evaluate $\int_0^1 \ln x\,dx$.

SOLUTION We know that the function $f(x) = \ln x$ has a vertical asymptote at 0 since $\lim_{x\to 0^+} \ln x = -\infty$. Thus, the given integral is improper and we have

$$\int_0^1 \ln x\,dx = \lim_{t\to 0^+} \int_t^1 \ln x\,dx$$

Now we integrate by parts with $u = \ln x$, $dv = dx$, $du = dx/x$, and $v = x$:

$$\begin{aligned}\int_t^1 \ln x\,dx &= x\ln x\Big]_t^1 - \int_t^1 dx \\ &= 1\ln 1 - t\ln t - (1 - t) \\ &= -t\ln t - 1 + t\end{aligned}$$

To find the limit of the first term we use l'Hospital's Rule:

$$\begin{aligned}\lim_{t\to 0^+} t\ln t &= \lim_{t\to 0^+} \frac{\ln t}{1/t} = \lim_{t\to 0^+} \frac{1/t}{-1/t^2} \\ &= \lim_{t\to 0^+} (-t) = 0\end{aligned}$$

Therefore

$$\begin{aligned}\int_0^1 \ln x\,dx &= \lim_{t\to 0^+} (-t\ln t - 1 + t) \\ &= -0 - 1 + 0 = -1\end{aligned}$$

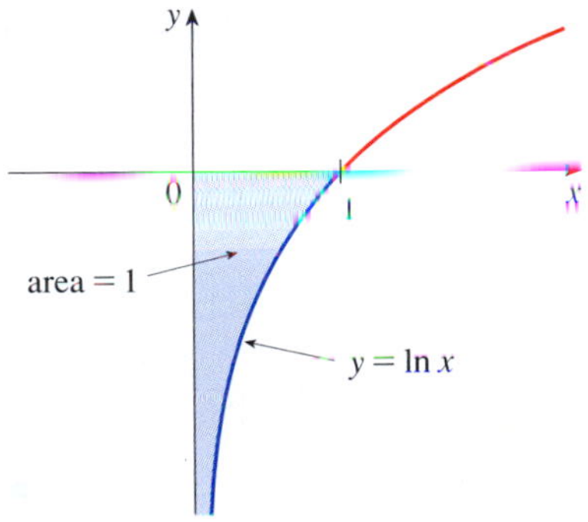

FIGURE 11

Figure 11 shows the geometric interpretation of this result. The area of the shaded region above $y = \ln x$ and below the x-axis is 1. ■

A COMPARISON TEST FOR IMPROPER INTEGRALS

Sometimes it is impossible to find the exact value of an improper integral and yet it is important to know whether it is convergent or divergent. In such cases the following theorem is useful. Although we state it for Type 1 integrals, a similar theorem is true for Type 2 integrals.

(4) COMPARISON THEOREM Suppose that f and g are continuous functions with $f(x) \geqslant g(x) \geqslant 0$ for $x \geqslant a$.

(a) If $\int_a^\infty f(x)\,dx$ is convergent, then $\int_a^\infty g(x)\,dx$ is convergent.

(b) If $\int_a^\infty g(x)\,dx$ is divergent, then $\int_a^\infty f(x)\,dx$ is divergent.

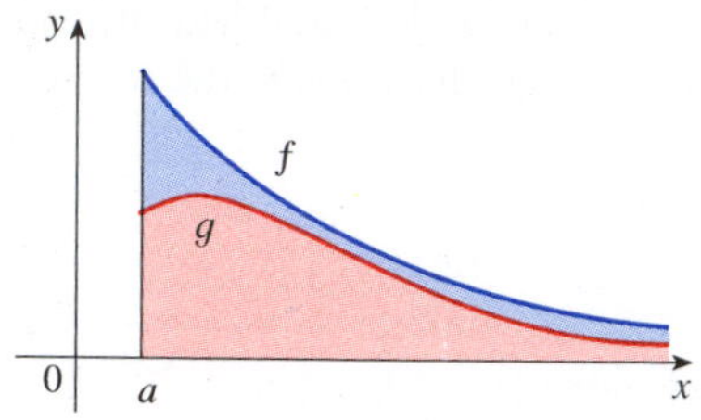

FIGURE 12

We omit the proof of the Comparison Theorem, but Figure 12 makes it seem plausible. If the area under the top curve $y = f(x)$ is finite, then so is the area under the bottom curve $y = g(x)$. And if the area under $y = g(x)$ is infinite, then so is the area under $y = f(x)$.

EXAMPLE 9 Show that $\int_0^\infty e^{-x^2}\,dx$ is convergent.

SOLUTION We cannot evaluate the integral directly because the antiderivative of e^{-x^2} is not an elementary function (as explained in Section 8.6). We write

$$\int_0^\infty e^{-x^2}\,dx = \int_0^1 e^{-x^2}\,dx + \int_1^\infty e^{-x^2}\,dx$$

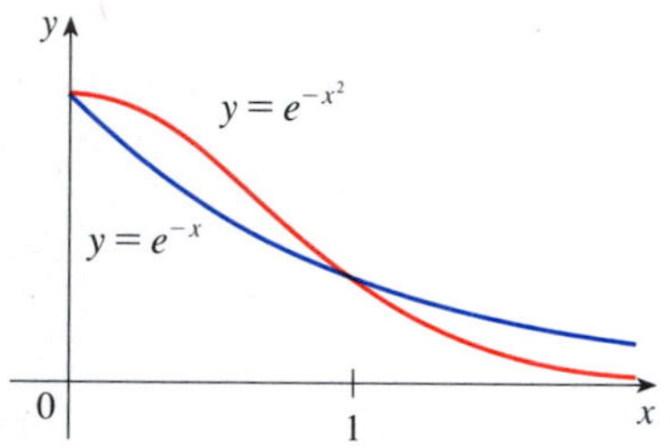

FIGURE 13

and observe that the first integral is just an ordinary definite integral. In the second integral we use the fact that for $x \geqslant 1$ we have $x^2 \geqslant x$, so $-x^2 \leqslant -x$ and therefore $e^{-x^2} \leqslant e^{-x}$ (see Figure 13). The integral of e^{-x} is easy to evaluate:

$$\int_1^\infty e^{-x}\,dx = \lim_{t\to\infty} \int_1^t e^{-x}\,dx = \lim_{t\to\infty} (e^{-1} - e^{-t}) = e^{-1}$$

Thus, taking $f(x) = e^{-x}$ and $g(x) = e^{-x^2}$ in the Comparison Theorem, we see that $\int_1^\infty e^{-x^2}\,dx$ is convergent. It follows that $\int_0^\infty e^{-x^2}\,dx$ is convergent. ■

In Example 9 we showed that $\int_0^\infty e^{-x^2}\,dx$ is convergent without computing its value. In Exercise 70 we indicate how to show that its value is approximately 0.89. In probability theory it is important to know the exact value of this improper integral, and in Section 13.5 indirect methods are introduced that give the exact value as $\sqrt{\pi}/2$.

EXAMPLE 10 The integral $\displaystyle\int_1^\infty \frac{1 + e^{-x}}{x}\,dx$ is divergent by the Comparison Theorem because

$$\frac{1 + e^{-x}}{x} > \frac{1}{x}$$

and $\int_1^\infty (1/x)\,dx$ is divergent by Example 1 [or by (2) with $p = 1$]. ■

EXERCISES 8.9

1. Find the area under the curve $y = 1/x^3$ from $x = 1$ to $x = t$ and evaluate it for $t = 10$, 100, and 1000. Then find the total area under this curve for $x \geqslant 1$.

2. (a) Graph the functions $f(x) = 1/x^{1.1}$ and $g(x) = 1/x^{0.9}$ in the viewing rectangles $[0, 10]$ by $[0, 1]$ and $[0, 100]$ by $[0, 1]$.
   (b) Find the areas under the graphs of f and g from $x = 1$ to $x = t$ and evaluate for $t = 10$, 100, 10^4, 10^6, 10^{10}, and 10^{20}.
   (c) Find the total area under each curve for $x \geqslant 1$, if it exists.

3–42 ■ Determine whether each integral is convergent or divergent. Evaluate those that are convergent.

3. $\displaystyle\int_2^\infty \frac{1}{\sqrt{x + 3}}\,dx$

4. $\displaystyle\int_2^\infty \frac{1}{(x + 3)^{3/2}}\,dx$

5. $\displaystyle\int_{-\infty}^1 \frac{1}{(2x - 3)^2}\,dx$

6. $\displaystyle\int_{-\infty}^{-1} \frac{1}{\sqrt[3]{x - 1}}\,dx$

7. $\displaystyle\int_{-\infty}^\infty x\,dx$

8. $\displaystyle\int_{-\infty}^\infty (2x^2 - x + 3)\,dx$

9. $\int_0^\infty e^{-x}\,dx$

10. $\int_{-\infty}^0 e^{3x}\,dx$

11. $\int_{-\infty}^\infty xe^{-x^2}\,dx$

12. $\int_{-\infty}^\infty x^2e^{-x^3}\,dx$

13. $\int_0^\infty \frac{1}{(x+2)(x+3)}\,dx$

14. $\int_0^\infty \frac{x}{(x+2)(x+3)}\,dx$

15. $\int_0^\infty \cos x\,dx$

16. $\int_1^\infty \sin \pi x\,dx$

17. $\int_0^\infty \frac{5}{2x+3}\,dx$

18. $\int_{-\infty}^3 \frac{1}{x^2+9}\,dx$

19. $\int_{-\infty}^1 xe^{2x}\,dx$

20. $\int_0^\infty xe^{-x}\,dx$

21. $\int_1^\infty \frac{\ln x}{x}\,dx$

22. $\int_e^\infty \frac{1}{x(\ln x)^2}\,dx$

23. $\int_{-\infty}^\infty \frac{x}{1+x^2}\,dx$

24. $\int_{-\infty}^\infty e^{-|x|}\,dx$

25. $\int_1^\infty \frac{\ln x}{x^2}\,dx$

26. $\int_1^\infty \frac{\ln x}{x^3}\,dx$

27. $\int_0^3 \frac{1}{\sqrt{x}}\,dx$

28. $\int_0^3 \frac{1}{x\sqrt{x}}\,dx$

29. $\int_{-1}^0 \frac{1}{x^2}\,dx$

30. $\int_1^9 \frac{1}{\sqrt[3]{x-9}}\,dx$

31. $\int_{-2}^3 \frac{1}{x^4}\,dx$

32. $\int_0^2 \frac{1}{4x-5}\,dx$

33. $\int_4^5 \frac{1}{(5-x)^{2/5}}\,dx$

34. $\int_{\pi/4}^{\pi/2} \sec^2 x\,dx$

35. $\int_{\pi/4}^{\pi/2} \tan^2 x\,dx$

36. $\int_0^{\pi/4} \frac{\cos x}{\sqrt{\sin x}}\,dx$

37. $\int_0^\pi \sec x\,dx$

38. $\int_0^4 \frac{1}{x^2+x-6}\,dx$

39. $\int_{-2}^2 \frac{1}{x^2-1}\,dx$

40. $\int_0^2 \frac{x-3}{2x-3}\,dx$

41. $\int_0^1 x\ln x\,dx$

42. $\int_0^1 \frac{\ln x}{\sqrt{x}}\,dx$

43–48 ■ Sketch the region and find its area (if the area is finite).

43. $S = \{(x, y) \mid x \leqslant 1,\ 0 \leqslant y \leqslant e^x\}$

44. $S = \{(x, y) \mid x \geqslant -2,\ 0 \leqslant y \leqslant e^{-x/2}\}$

45. $S = \left\{(x, y) \mid 0 \leqslant y \leqslant \frac{1}{x^2 - 2x + 5}\right\}$

46. $S = \left\{(x, y) \mid x \geqslant 0,\ 0 \leqslant y \leqslant \frac{1}{\sqrt{x+1}}\right\}$

47. $S = \{(x, y) \mid 0 \leqslant x \leqslant \pi,\ 0 \leqslant y \leqslant \tan^2 x \sec^2 x\}$

48. $S = \left\{(x, y) \mid 3 < x \leqslant 7,\ 0 \leqslant y \leqslant \frac{1}{\sqrt{x-3}}\right\}$

49–54 ■ Use the Comparison Theorem to determine whether the integral is convergent or divergent.

49. $\int_1^\infty \frac{\sin^2 x}{x^2}\,dx$

50. $\int_1^\infty \frac{\sqrt{1+\sqrt{x}}}{\sqrt{x}}\,dx$

51. $\int_1^\infty \frac{dx}{x + e^{2x}}$

52. $\int_1^\infty \frac{1}{\sqrt{x^3+1}}\,dx$

53. $\int_0^{\pi/2} \frac{dx}{x \sin x}$

54. $\int_0^1 \frac{e^{-x}}{\sqrt{x}}\,dx$

55. The integral

$$\int_0^\infty \frac{1}{\sqrt{x}(1+x)}\,dx$$

is improper for two reasons: the interval $[0, \infty)$ is infinite and the integrand is unbounded near 0. Evaluate it by expressing it as a sum of improper integrals of Type 2 and Type 1 as follows:

$$\int_0^\infty \frac{1}{\sqrt{x}(1+x)}\,dx = \int_0^1 \frac{1}{\sqrt{x}\,(1+x)}\,dx + \int_1^\infty \frac{1}{\sqrt{x}(1+x)}\,dx$$

56. Evaluate

$$\int_2^\infty \frac{1}{x\sqrt{x^2-4}}\,dx$$

by the same method as in Exercise 55.

57–59 ■ Find the values of p for which the integral converges and evaluate the integral for those values of p.

57. $\int_0^1 \frac{1}{x^p}\,dx$

58. $\int_e^\infty \frac{1}{x(\ln x)^p}\,dx$

59. $\int_0^1 x^p \ln x\,dx$

60. (a) Evaluate the integral $\int_0^\infty x^n e^{-x}\,dx$ for $n = 0, 1, 2$, and 3.
(b) Guess the value of $\int_0^\infty x^n e^{-x}\,dx$ when n is an arbitrary positive integer.
(c) Prove your guess using mathematical induction.

61. (a) Show that $\int_{-\infty}^\infty x\,dx$ is divergent.
(b) Show that

$$\lim_{t \to \infty} \int_{-t}^t x\,dx = 0$$

This shows that we cannot define

$$\int_{-\infty}^\infty f(x)\,dx = \lim_{t \to \infty} \int_{-t}^t f(x)\,dx$$

62. If $\int_{-\infty}^\infty f(x)\,dx$ is convergent and a and b are real numbers, show that

$$\int_{-\infty}^a f(x)\,dx + \int_a^\infty f(x)\,dx = \int_{-\infty}^b f(x)\,dx + \int_b^\infty f(x)\,dx$$

63. We know from Example 1 that the region $\mathcal{R} = \{(x, y) \mid x \geq 1, 0 \leq y \leq 1/x\}$ has infinite area. Show that by rotating $\mathcal{R}$ about the x-axis we obtain a solid with finite volume.

64. Use the information and data in Exercises 25 and 26 of Section 7.4 to find the work required to propel a 1000-kg satellite out of Earth's gravitational field.

65. Find the *escape velocity* v_0 that is needed to propel a rocket of mass m out of the gravitational field of a planet with mass M and radius R. Use Newton's Law of Gravitation (see Exercise 25 in Section 7.4) and the fact that the initial kinetic energy of $\frac{1}{2}mv_0^2$ supplies the needed work.

66. A positive function f is called a *probability density function* if $\int_{-\infty}^{\infty} f(x)\,dx = 1$.

(a) Show that if $c > 0$, the function f defined by $f(x) = ce^{-cx}$ for $x \geq 0$ and $f(x) = 0$ for $x < 0$ is a probability density function.

(b) Calculate the *mean:*

$$\mu = \int_{-\infty}^{\infty} xf(x)\,dx$$

(c) Calculate the *standard deviation:*

$$\sigma = \left[\int_{-\infty}^{\infty} (x - \mu)^2 f(x)\,dx\right]^{1/2}$$

67. If $f(t)$ is continuous for $t \geq 0$, the *Laplace transform* of f is the function F defined by

$$F(s) = \int_0^{\infty} f(t)e^{-st}\,dt$$

and the domain of F is the set consisting of all numbers s for which the integral converges. Find the Laplace transforms of the following functions.

(a) $f(t) = 1$ (b) $f(t) = e^t$ (c) $f(t) = t$

68. Show that if $0 \leq f(t) \leq Me^{at}$ for $t \geq 0$, where M and a are constants, then the Laplace transform $F(s)$ exists for $s > a$.

69. Suppose that $0 \leq f(t) \leq Me^{at}$ and $0 \leq f'(t) \leq Ke^{at}$ for $t \geq 0$, where f' is continuous. If the Laplace transform of $f(t)$ is $F(s)$ and the Laplace transform of $f'(t)$ is $G(s)$, show that

$$G(s) = sF(s) - f(0) \qquad s > a$$

70. Estimate the numerical value of $\int_0^{\infty} e^{-x^2}\,dx$ by writing it as the sum of $\int_0^4 e^{-x^2}\,dx$ and $\int_4^{\infty} e^{-x^2}\,dx$. Approximate the first integral by using Simpson's Rule with $n = 8$ and show that the second integral is smaller than $\int_4^{\infty} e^{-4x}\,dx$, which is less than 0.0000001.

71. Show that $\int_0^{\infty} x^2e^{-x^2}\,dx = \frac{1}{2}\int_0^{\infty} e^{-x^2}\,dx$.

72. Show that $\int_0^{\infty} e^{-x^2}\,dx = \int_0^1 \sqrt{-\ln y}\,dy$ by interpreting the integrals as areas.

73. Find the value of the constant C for which the integral

$$\int_0^{\infty} \left(\frac{1}{\sqrt{x^2 + 4}} - \frac{C}{x + 2}\right) dx$$

converges. Evaluate the integral for this value of C.

74. Find the value of the constant C for which the integral

$$\int_0^{\infty} \left(\frac{x}{x^2 + 1} - \frac{C}{3x + 1}\right) dx$$

converges. Evaluate the integral for this value of C.

75. Use integration by parts to show that, for all $x > 0$,

$$0 < \int_0^{\infty} \frac{\sin t}{\ln(1 + x + t)}\,dt < \frac{2}{\ln(1 + x)}$$

8 REVIEW

KEY TOPICS ■ Define, state, or discuss the following.

1. Integration by parts
2. Trigonometric integrals
3. Trigonometric substitution
4. Partial fractions
5. Rationalizing substitutions
6. Strategy for integration
7. The Midpoint Rule
8. The Trapezoidal Rule
9. Simpson's Rule
10. Improper integrals
11. Comparison Theorem

EXERCISES

Note: Additional practice in techniques of integration is provided in Exercises 8.6.

1–8 ■ Determine whether the statement is true or false.

1. $\dfrac{x(x^2+4)}{x^2-4}$ can be put in the form $\dfrac{A}{x+2}+\dfrac{B}{x-2}$.

2. $\dfrac{x^2+4}{x(x^2-4)}$ can be put in the form $\dfrac{A}{x}+\dfrac{B}{x+2}+\dfrac{C}{x-2}$.

3. $\dfrac{x^2+4}{x^2(x-4)}$ can be put in the form $\dfrac{A}{x^2}+\dfrac{B}{x-4}$.

4. $\dfrac{x^2-4}{x(x^2+4)}$ can be put in the form $\dfrac{A}{x}+\dfrac{B}{x^2+4}$.

5. $\displaystyle\int_0^4 \frac{x}{x^2-1}\,dx = \tfrac{1}{2}\ln 15$

6. $\displaystyle\int_1^\infty \frac{1}{x^{\sqrt{2}}}\,dx$ is convergent.

7. If f is continuous, then $\int_{-\infty}^{\infty} f(x)\,dx = \lim_{t\to\infty}\int_{-t}^{t} f(x)\,dx$.

8. The Midpoint Rule is always more accurate than the Trapezoidal Rule.

9–42 ■ Evaluate the integral.

9. $\displaystyle\int \frac{x-1}{x+1}\,dx$

10. $\displaystyle\int \frac{\sin^3 x}{\cos x}\,dx$

11. $\displaystyle\int \frac{(\arctan x)^5}{1+x^2}\,dx$

12. $\displaystyle\int x^2 e^{-3x}\,dx$

13. $\displaystyle\int \frac{\cos x}{e^{\sin x}}\,dx$

14. $\displaystyle\int \frac{x^2+1}{x-1}\,dx$

15. $\displaystyle\int x^4 \ln x\,dx$

16. $\displaystyle\int \frac{\sec^2\theta}{1-\tan\theta}\,d\theta$

17. $\displaystyle\int x\sin(x^2)\,dx$

18. $\displaystyle\int x\sin^2 x\,dx$

19. $\displaystyle\int \frac{dx}{2x^2-5x+2}$

20. $\displaystyle\int \frac{dt}{\sin^2 t+\cos 2t}$

21. $\displaystyle\int \tan^7 x\sec^3 x\,dx$

22. $\displaystyle\int \frac{dx}{\sqrt{8+2x-x^2}}$

23. $\displaystyle\int \frac{dx}{\sqrt{1+2x}+3}$

24. $\displaystyle\int x(\tan^{-1}x)^2\,dx$

25. $\displaystyle\int \frac{e^{\sqrt{x}}}{\sqrt{x}}\,dx$

26. $\displaystyle\int \frac{dx}{x^3-2x^2+x}$

27. $\displaystyle\int \frac{dx}{(x^2-1)^{3/2}}$

28. $\displaystyle\int \frac{dx}{x^2\sqrt{1+x^2}}$

29. $\displaystyle\int \frac{dx}{x^3+x}$

30. $\displaystyle\int \frac{dx}{1+e^x}$

31. $\displaystyle\int \cot^2 x\,dx$

32. $\displaystyle\int \frac{dx}{5-3\cos x}$

33. $\displaystyle\int \frac{2x^2+3x+11}{x^3+x^2+3x-5}\,dx$

34. $\displaystyle\int \frac{x^3}{(x+1)^{10}}\,dx$

35. $\displaystyle\int \csc^4 4x\,dx$

36. $\displaystyle\int (\arcsin x)^2\,dx$

37. $\displaystyle\int \frac{\ln(\ln x)}{x}\,dx$

38. $\displaystyle\int \frac{\sin x}{1+\sin x}\,dx$

39. $\displaystyle\int \frac{1}{\sqrt{4x^2+4x+5}}\,dx$

40. $\displaystyle\int e^{-x}\sinh x\,dx$

41. $\displaystyle\int (\cos x+\sin x)^2\cos 2x\,dx$

42. $\displaystyle\int \frac{e^{2x}}{e^{4x}-7}\,dx$

43–60 ■ Evaluate the integral or show that it is divergent.

43. $\displaystyle\int_0^{\pi/2} \cos^3 x\sin 2x\,dx$

44. $\displaystyle\int_{-1}^{1} \frac{1}{2x+1}\,dx$

45. $\displaystyle\int_0^3 \frac{dx}{x^2-x-2}$

46. $\displaystyle\int_0^{\pi/4} \cos^5(2\theta)\,d\theta$

47. $\displaystyle\int_0^1 \frac{t^2-1}{t^2+1}\,dt$

48. $\displaystyle\int_2^6 \frac{y}{\sqrt{y-2}}\,dy$

49. $\displaystyle\int_0^\infty \frac{1}{(x+2)^4}\,dx$

50. $\displaystyle\int_1^4 \frac{e^{1/x}}{x^2}\,dx$

51. $\displaystyle\int_1^e \frac{dx}{x\sqrt{\ln x}}$

52. $\displaystyle\int_0^\infty \frac{dx}{(x+1)^2(x+2)}$

53. $\displaystyle\int_1^4 \frac{\sqrt{x}}{\sqrt{x}+2}\,dx$

54. $\displaystyle\int_{-3}^{3} x\sqrt{1+x^4}\,dx$

55. $\displaystyle\int_{-\infty}^{\infty} \frac{dx}{4x^2+4x+5}$

56. $\displaystyle\int_1^e \frac{dx}{x[1+(\ln x)^2]}$

57. $\displaystyle\int_1^2 \frac{\sqrt{x^2-1}}{x}\,dx$

58. $\displaystyle\int_{-1}^{1} \frac{x+1}{\sqrt[3]{x^4}}\,dx$

59. $\displaystyle\int_0^\infty e^{ax}\cos bx\,dx$

60. $\displaystyle\int_1^\infty \frac{\tan^{-1}x}{x^2}\,dx$

61. Evaluate $\int \ln(x^2+2x+2)\,dx$ and graph both the integrand and the indefinite integral (with $C=0$) to check that your answer is reasonable.

62. Graph the function $f(x)=\cos^2 x\sin^3 x$ and use the graph to guess the value of the integral $\int_0^{2\pi} f(x)\,dx$. Then evaluate the integral to confirm your guess.

63–66 ■ Use the Table of Integrals in Appendix K to evaluate the integral.

63. $\int e^x \sqrt{1 - e^{2x}}\, dx$

64. $\int \tan^5 x\, dx$

65. $\int \sqrt{x^2 + x + 1}\, dx$

66. $\int \frac{\cot x}{\sqrt{1 + 2\sin x}}\, dx$

67. Verify Formula 33 in the Table of Integrals (a) by differentiation and (b) by using a trigonometric substitution.

68. Verify Formula 62 in the Table of Integrals.

69–70 ■ Use (a) the Trapezoidal Rule, (b) the Midpoint Rule, and (c) Simpson's Rule with $n = 10$ to approximate the given integral. Round your answers to six decimal places.

69. $\int_0^1 \sqrt{1 + x^4}\, dx$

70. $\int_0^{\pi/2} \sqrt{\sin x}\, dx$

71. Estimate the errors involved in Exercise 69, parts (a) and (b).

72. Use Simpson's Rule with $n = 6$ to estimate the area under the curve $y = e^x/x$ from $x = 1$ to $x = 4$.

CAS **73.** (a) If $f(x) = \sin(\sin x)$, use a graph to find an upper bound for $|f^{(4)}(x)|$.
(b) Use Simpson's Rule with $n = 10$ to approximate $\int_0^{\pi} f(x)\, dx$ and use part (a) to estimate the error.
(c) How large should n be to guarantee that the size of the error in using S_n is less than 0.00001?

CAS **74.** (a) How would you evaluate $\int x^5 e^{-2x}\, dx$ by hand? (Don't actually carry out the integration.)
(b) How would you evaluate $\int x^5 e^{-2x}\, dx$ using tables? (Don't actually do it.)
(c) Use a CAS to evaluate $\int x^5 e^{-2x}\, dx$.
(d) Graph the integrand and the indefinite integral on the same screen.

75. Use the Comparison Theorem to determine whether the integral

$$\int_1^{\infty} \frac{x^3}{x^5 + 2}\, dx$$

is convergent or divergent.

76. Find the area of the region bounded by the hyperbola $y^2 - x^2 = 1$ and the line $y = 3$.

77. Find the area bounded by the curves $y = \cos x$ and $y = \cos^2 x$ between $x = 0$ and $x = \pi$.

78. Find the area of the region bounded by the curves $y = 1/(2 + \sqrt{x})$, $y = 1/(2 - \sqrt{x})$, and $x = 1$.

79. The region under the curve $y = \cos^2 x$, $0 \leqslant x \leqslant \pi/2$, is brotated about the x-axis. Find the volume of the resulting solid.

80. The region in Exercise 79 is rotated about the y-axis. Find the volume of the resulting solid.

81. Is it possible to find a number n such that $\int_0^{\infty} x^n\, dx$ is convergent?

82. If n is a positive integer, prove that

$$\int_0^1 (\ln x)^n\, dx = (-1)^n n!$$

83. If f' is continuous on $[0, \infty)$ and $\lim_{x\to\infty} f(x) = 0$, show that

$$\int_0^{\infty} f'(x)\, dx = -f(0)$$

84. The magnitude of the repulsive force between two point charges with the same sign, one of size 1 and the other of size q, is

$$F = \frac{q}{4\pi\varepsilon_0 r^2}$$

where r is the distance between the charges and ε_0 is a constant. The *potential* V at a point P due to the charge q is defined to be the work expended in bringing a unit charge to P from infinity along the straight line that joins q and P. Find a formula for V.

85. Use the substitution $u = 1/x$ to show that

$$\int_0^{\infty} \frac{\ln x}{1 + x^2}\, dx = 0$$

APPLICATIONS PLUS

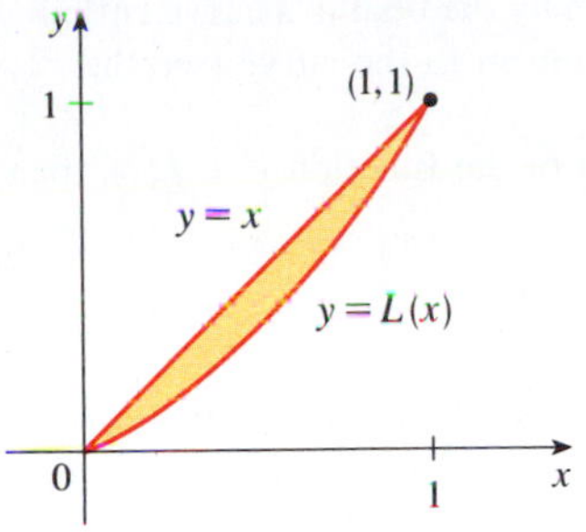

FIGURE FOR PROBLEM 1

1. Economists use a cumulative distribution called a *Lorenz curve* to describe the distribution of income between households in a given country. Typically, a Lorenz curve is defined on $[0, 1]$, passes through $(0, 0)$ and $(1, 1)$, and is continuous, increasing, and concave upward. The points on this curve are determined by ranking all households by income and then computing the percentage of households whose income is less than or equal to a given percentage of the total income of the country. For example, the point $(a/100, b/100)$ is on the Lorenz curve if the bottom $a\%$ of the households receive less than or equal to $b\%$ of the total income. *Absolute equality* of income distribution would occur if the bottom $a\%$ of the households receive $a\%$ of the income, in which case the Lorenz curve would be the line $y = x$. The area between the Lorenz curve and the line $y = x$ measures how much the income distribution differs from absolute equality. The *coefficient of inequality* is the ratio of the area between the Lorenz curve and the line $y = x$ to the area under $y = x$.
   (a) Show that the coefficient of inequality is twice the area between the Lorenz curve and the line $y = x$, that is, show that

$$\text{coefficient of inequality} = 2 \int_0^1 [x - L(x)]\, dx$$

   (b) The income distribution for a certain country is represented by the Lorenz curve defined by the equation

$$L(x) = \tfrac{5}{12}x^2 + \tfrac{7}{12}x$$

   What is the percentage of total income received by the bottom 50% of the households? Find the coefficient of inequality.
   (c) Find the coefficient of inequality for the Lorenz curve defined by the equation

$$L(x) = \frac{5x^3}{4 + x^2}$$

2. A cylindrical glass of radius r and height L is filled with water and then tilted until the water remaining in the glass exactly covers its base.
   (a) Determine a way to "slice" the water into parallel rectangular cross-sections and then *set up* a definite integral for the volume of the water in the glass.
   (b) Determine a way to "slice" the water into parallel cross-sections that are trapezoids and then *set up* a definite integral for the volume of the water.
   (c) Find the volume of water in the glass by evaluating one of the integrals in part (a) or part (b).
   (d) Find the volume of the water in the glass from purely geometric considerations.
   (e) Suppose the glass is tilted until the water exactly covers half the base. In what direction can you "slice" the water into triangular cross-sections? Rectangular cross-sections? Cross-sections that are segments of circles? Find the volume of water in the glass.

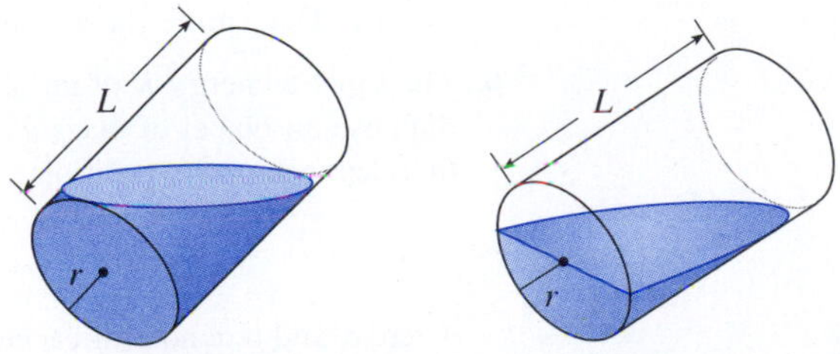

FIGURE FOR PROBLEM 2

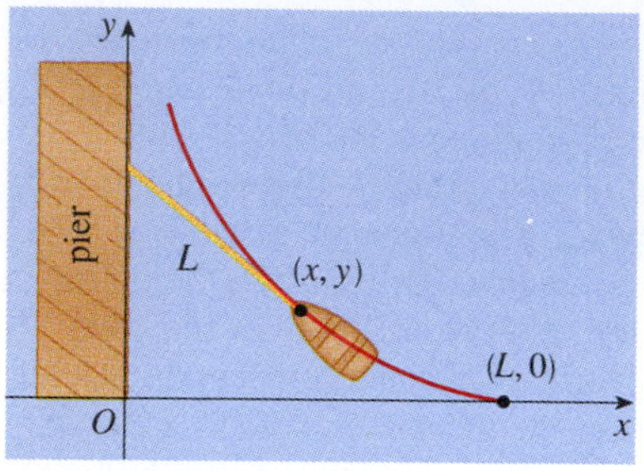

FIGURE FOR PROBLEM 3

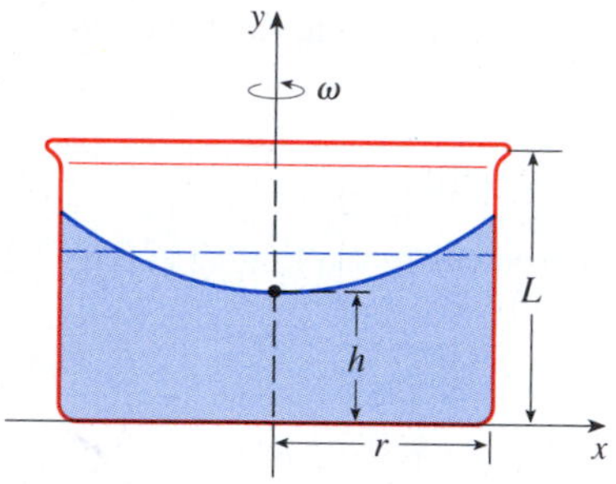

FIGURE FOR PROBLEM 4

3. A man initially at the point O walks along a pier pulling a rowboat by a rope of length L. The man keeps the rope straight and taut. The path followed by the boat is a curve called a *tractrix* and it has the property that the rope is always tangent to the curve (see the figure).

(a) Show that if the path followed by the boat is the graph of the function $y = f(x)$, then

$$f'(x) = \frac{dy}{dx} = \frac{-\sqrt{L^2 - x^2}}{x}$$

(b) Determine the function $y = f(x)$.

4. A cylindrical container of radius r and height L is partially filled with a liquid whose volume is V. If the container is rotated about its axis of symmetry with constant angular speed ω, then the container will induce a rotational motion in the liquid around the same axis. Eventually, the liquid will be rotating at the same angular speed as the container. The surface of the liquid will be convex, as indicated in the figure, because the centrifugal force on the liquid particles increases with the distance from the axis of the container. It can be shown that the surface of the liquid is a paraboloid of revolution generated by rotating the parabola

$$y = h + \frac{\omega^2 x^2}{2g}$$

about the y-axis, where g is the acceleration due to gravity.

(a) Determine h as a function of ω.

(b) At what angular speed will the surface of the liquid touch the bottom? At what speed will it spill over the top?

(c) Suppose the radius of the container is 2 ft, the height is 7 ft, and the container and liquid are rotating at the same constant angular speed. The surface of the liquid is 5 ft below the top of the tank at the central axis and 4 ft below the top of the tank 1 ft out from the central axis.

(i) Determine the angular speed of the container and the volume of the fluid.

(ii) How far below the top of the tank is the liquid at the wall of the container?

5. The momentum p of an object is the product of its mass m and its velocity v, that is, $p = mv$. Suppose an object, moving along a straight line, is acted on by a force $F = F(t)$ that is a continuous function of time.

(a) Show that the change in momentum over a time interval $[t_0, t_1]$ is equal to the integral of F from t_0 to t_1; that is, show that

$$p(t_1) - p(t_0) = \int_{t_0}^{t_1} F(t)\,dt$$

This integral is called the *impulse* of the force over time.

(b) A baseball pitcher throws a 90-mi/h fastball to a batter, who hits a line drive directly back to the pitcher. Suppose the ball is in contact with the bat for 0.01 s and leaves the bat with velocity 110 mi/h. A baseball weighs 5 oz.

(i) Determine the change in the ball's momentum.

(ii) Determine the average force on the bat.

6. The kinetic energy K of an object of mass m and velocity v is given by $K = \frac{1}{2}mv^2$. Suppose an object of mass m, moving in a straight line, is acted on by a force $F = F(s)$ that depends on its position s. According to Newton's Second Law

$$F(s) = ma = m\frac{dv}{dt}$$

where a and v denote the acceleration and velocity of the object, respectively.

(a) Show that the work done in moving the object from a position s_0 to a position s_1 is equal to the change in the object's kinetic energy; that is, show that

$$W = \int_{s_0}^{s_1} F(s)\,ds = \tfrac{1}{2}mv_1^2 - \tfrac{1}{2}mv_0^2$$

where $v_0 = v(s_0)$ and $v_1 = v(s_1)$ are the velocities of the object at the positions s_0 and s_1, respectively. *Hint:* By the Chain Rule,

$$m\frac{dv}{dt} = m\frac{dv}{ds}\frac{ds}{dt} = mv\frac{dv}{ds}$$

(b) How many foot-pounds of work does it take to throw a baseball at a speed of 90 mi/h? A baseball weighs 5 oz and $m = w/g$ where $g = 32\ \text{ft/s}^2$.

7. (a) An outfielder fields a baseball 280 ft away from home plate and throws it directly to the catcher with an initial velocity of 100 ft/s. Assume that the velocity $v(t)$ of the ball after t seconds satisfies the differential equation $dv/dt = -v/10$ because of air resistance. How long does it take for the ball to reach home plate? (Ignore any vertical motion of the ball.)

(b) The manager of the team wonders whether the ball will reach home plate sooner if it is relayed by an infielder. The shortstop can position himself directly between the outfielder and home plate, catch the ball thrown by the outfielder, turn, and throw the ball to the catcher with an initial velocity of 105 ft/s. The manager clocks the relay time of the shortstop (catching, turning, throwing) at half a second. How far from home plate should the shortstop position himself to minimize the total time for the ball to reach home plate? Should the manager encourage a direct throw or a relayed throw? What if the shortstop can throw at 115 ft/s?

(c) For what throwing velocity of the shortstop does a relayed throw take the same time as a direct throw?

8. Any object emits radiation when heated. A *blackbody* is a system that absorbs all the radiation that falls on it. For instance, a mat black surface or a large cavity with a small hole in its wall (like a blastfurnace) is a blackbody and emits blackbody radiation. Even the radiation from the Sun is close to being blackbody radiation. In 1900 the German physicist Max Planck found a function that models blackbody radiation very well. He expressed the radiant energy density of wavelength λ as

$$f(\lambda) = \frac{8\pi hc\lambda^{-5}}{e^{hc/(\lambda kT)} - 1}$$

where λ is measured in meters, T is the temperature (in kelvins), and

$$h = \text{Planck's constant} = 6.6262 \times 10^{-34}\ \text{J·s}$$

$$c = \text{speed of light} = 2.997925 \times 10^{8}\ \text{m/s}$$

$$k = \text{Boltzmann's constant} = 1.3807 \times 10^{-23}\ \text{J/K}$$

This equation is known as Planck's Law of Radiation. If we change from meters to the more convenient unit of micrometers (μm) and combine the constants, we get

$$f(\lambda) = \frac{a}{\lambda^5(e^{b/(\lambda T)} - 1)}$$

where $a = 4.99258 \times 10^{-18}$ and $b = 14{,}387.5$.

(a) Show that

$$\lim_{\lambda\to 0^+} f(\lambda) = 0 \qquad \text{and} \qquad \lim_{\lambda\to\infty} f(\lambda) = 0$$

(b) The temperature at the surface of the Sun is $T = 5700$ K. Graph f for the Sun and use the graph to estimate the value of λ for which $f(\lambda)$ is a maximum. (Notice that the constant a is so small that it's hard to find the graph. It's best to graph with $a = 1$ and rescale.)

(c) Use Newton's method to solve the equation $f'(\lambda) = 0$ correct to five decimal places. [Use your estimate from part (b) as the starting point in Newton's method.]

(d) Investigate how the graph of f changes as T varies. In particular, graph f for the stars Betelgeuse ($T = 3400$ K), Procyon ($T = 6400$ K), and Sirius ($T = 9200$ K) as well as the Sun. How does the total radiation emitted (the area under the curve) vary with T? Use the graph to comment on why Sirius is known as a blue star and Betelgeuse as a red star.

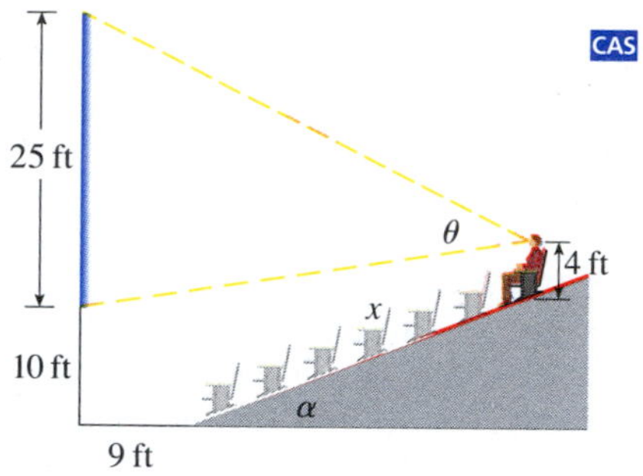

FIGURE FOR PROBLEM 9

CAS **9.** A movie theater has a screen that is positioned 10 ft off the floor and is 25 ft high. The first row of seats is 9 ft from the screen and the rows are 3 ft apart. The floor of the seating area is inclined at an angle of $\alpha = 20°$ above the horizontal and the distance up the incline that you sit is x. The theater has 21 rows of seats, so $0 \leqslant x \leqslant 60$. Suppose you decide that the best place to sit is where the angle θ subtended by the screen at your eyes is a maximum. Let's also suppose that your eyes are 4 ft above the floor, as in the figure. (In Exercise 52 in Section 5.5 we looked at a simpler version of this problem, where the floor is horizontal, but this is a more complicated situation and requires technology.)

(a) Show that

$$\theta = \arccos\left(\frac{a^2 + b^2 - 625}{2ab}\right)$$

where $$a^2 = (9 + x\cos\alpha)^2 + (31 - x\sin\alpha)^2$$

and $$b^2 = (9 + x\cos\alpha)^2 + (x\sin\alpha - 6)^2$$

(b) Use a graph of θ as a function of x to estimate the value of x that maximizes θ. In which row should you sit? What is the viewing angle θ in this row?

(c) Use your CAS to differentiate θ and find a numerical value for the root of the equation $d\theta/dx = 0$. Does this value confirm your result in part (b)?

(d) Use the graph of θ to estimate the average value of θ over the interval $0 \leqslant x \leqslant 60$. Compare with the maximum and minimum values of θ.

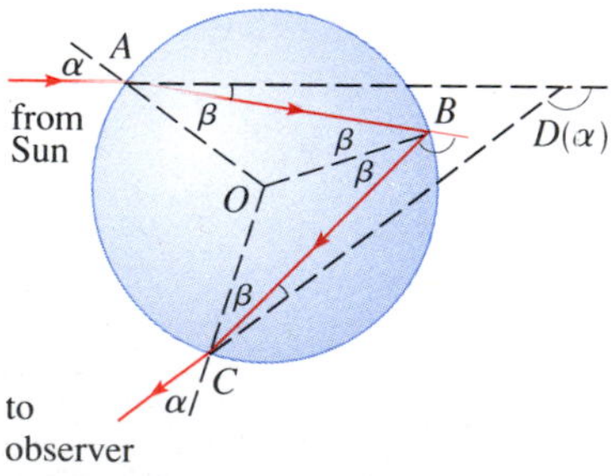

Formation of the primary rainbow

FIGURE FOR PROBLEM 10

10. Rainbows are created when raindrops scatter sunlight. They have fascinated mankind since ancient times and have inspired attempts at scientific explanation since the time of Aristotle. In this problem we use the ideas of Descartes and Newton to explain the shape, location, and colors of rainbows.

(a) The figure shows a ray of sunlight entering a spherical raindrop at A. Some of the light is reflected, but the line AB shows the path of the part that enters the drop. Notice that the light is refracted toward the normal line AO and in fact we know from Exercise 45 in Section 5.5 (Snell's Law) that $\sin\alpha = k\sin\beta$, where α is the angle of incidence, β is the angle of refraction, and $k \approx \frac{4}{3}$ is the index of refraction for water. At B some of the light passes through the drop and is refracted into the air, but the line BC shows the part that is reflected. (The angle of incidence equals the angle of reflection.) When the ray reaches C part of it is reflected, but for the time being we are more interested in the part that leaves the raindrop at C. (Notice that it is refracted away from the normal line.) The *angle of deviation* $D(\alpha)$ is the amount of clockwise rotation that the ray has undergone during this three-stage process.

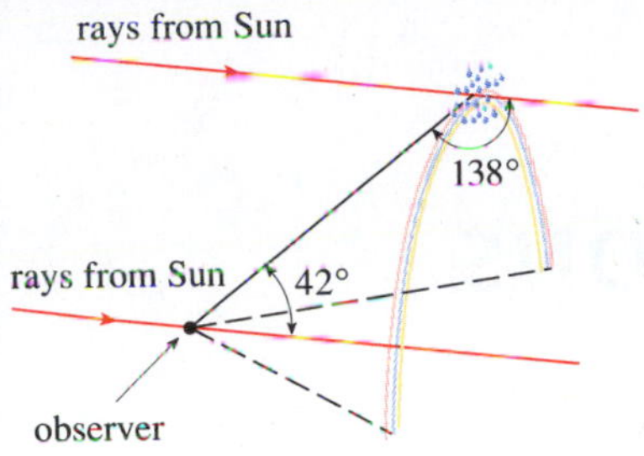

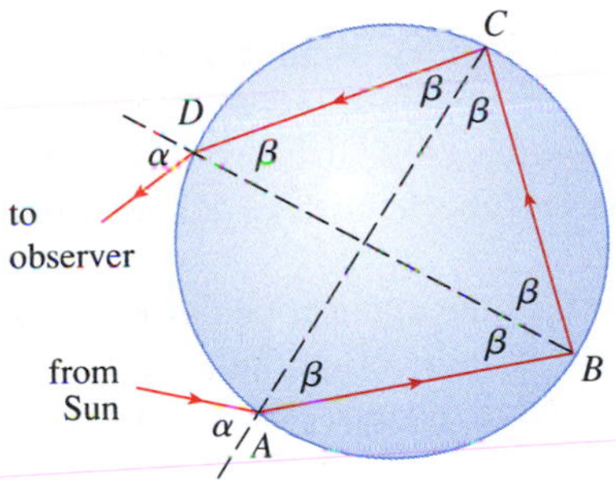

Formation of the secondary rainbow

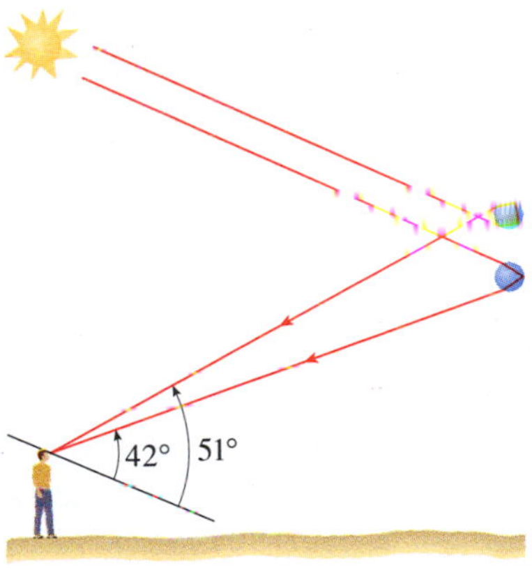

Thus

$$D(\alpha) = (\alpha - \beta) + (\pi - 2\beta) + (\alpha - \beta) = \pi + 2\alpha - 4\beta$$

Show that the minimum value of the deviation is $D(\alpha) \approx 138°$ and occurs when $\alpha \approx 59.4°$.

The significance of the minimum deviation is that when $\alpha \approx 59.4°$ we have $D'(\alpha) \approx 0$, so $\Delta D/\Delta\alpha \approx 0$. This means that many rays with $\alpha \approx 59.4°$ become deviated by approximately the same amount. It is the *concentration* of rays coming from near the direction of minimum deviation that creates the brightness of the primary rainbow. The figure shows that the angle of elevation from the observer up to the highest point on the rainbow is $180° - 138° = 42°$. (This angle is called the *rainbow angle.*)

(b) Part (a) explains the location of the primary rainbow but how do we explain the colors? Sunlight comprises a range of wavelengths, from the red range through orange, yellow, green, blue, indigo, and violet. As Newton discovered in his prism experiments of 1666, the index of refraction is different for each color. (The effect is called *dispersion*.) For red light the refractive index is $k \approx 1.3318$ whereas for violet light it is $k \approx 1.3435$. By repeating the calculation of part (a) for these values of k, show that the rainbow angle is about 42.3° for the red bow and 40.6° for the violet bow. So the rainbow really consists of seven individual bows corresponding to the seven colors.

(c) Perhaps you have seen a fainter secondary rainbow above the primary bow. That results from the part of a ray that enters a raindrop and is refracted at A, reflected twice (at B and C), and refracted as it leaves the drop at D (see the figure). This time the deviation angle $D(\alpha)$ is the total amount of counterclockwise rotation that the ray undergoes in this four-stage process. Show that

$$D(\alpha) = 2\alpha - 6\beta + 2\pi$$

and $D(\alpha)$ has a minimum value when

$$\cos\alpha = \sqrt{\frac{k^2 - 1}{8}}$$

Taking $k = \frac{4}{3}$, show that the minimum deviation is about 129° and so the rainbow angle for the secondary rainbow is about 51°, as shown in the figure.

(d) Show that the colors in the secondary rainbow appear in the opposite order from those in the primary rainbow.

9

FURTHER APPLICATIONS OF INTEGRATION

> Mathematics takes us still further from what is human, into the region of absolute necessity, to which not only the actual world, but every possible world, must conform.
>
> BERTRAND RUSSELL

We looked at some applications of integrals in Chapter 7: areas, volumes, work, and average values. Here we investigate some of the many other applications of integration to geometry and physics—the length of a curve, the area of a surface, hydrostatic force, centers of mass—as well as quantities of interest in chemistry, biology, and economics.

We start with what is perhaps the most important of all the applications of integration: differential equations. When a scientist uses calculus, more often than not it is to solve a differential equation that has arisen in the description of some physical process.

9.1 DIFFERENTIAL EQUATIONS

A **differential equation** is an equation that contains an unknown function and some of its derivatives. Here are some examples:

$$y' = xy \tag{1}$$

$$y'' + 2y' + y = 0 \tag{2}$$

$$\frac{d^3y}{dx^3} + x\frac{d^2y}{dx^2} + \frac{dy}{dx} - 2y = e^{-x} \tag{3}$$

In each of these differential equations y is an unknown function of x. The importance of differential equations lies in the fact that when a scientist or engineer formulates a physical law in mathematical terms, it frequently turns out to be a differential equation.

The **order** of a differential equation is the order of the highest derivative that occurs in the equation. Thus, Equations 1, 2, and 3 are of order 1, 2, and 3, respectively.

A function f is called a **solution** of a differential equation if the equation is satisfied when $y = f(x)$ and its derivatives are substituted into the equation. Thus, f is a solution of Equation 1 if

$$f'(x) = xf(x)$$

for all values of x in some interval.

You can easily verify that both $f(x) = \sin x$ and $g(x) = \cos x$ are solutions of the differential equation

$$y'' + y = 0 \tag{4}$$

But when we are asked to *solve* a differential equation we are expected to find all possible solutions of the equation. In Section 15.1 we will show that any solution of Equation 4 is of the form

$$y = A \sin x + B \cos x \tag{5}$$

where A and B are constants. So (5) is called the **general solution** of the differential equation, and particular solutions are obtained by substituting values for the arbitrary constants A and B.

We have already solved some particularly simple differential equations, namely, those of the form

$$y' = f(x)$$

For instance, we know that the general solution of the differential equation

$$y' = x^3$$

is given by

$$y = \frac{x^4}{4} + C$$

where C is an arbitrary constant.

But, in general, solving a differential equation is not an easy matter. There is no systematic technique that enables us to solve all differential equations. In this section we learn how to solve a certain type of differential equation called a separable equation. (Other types of equations will be discussed in Section 9.2 and in Chapter 15.) At the end of the section, however, we will see how to sketch a rough graph of a solution of a first-order differential equation, even when it's impossible to find a formula for the solution.

A **separable equation** is a first-order differential equation that can be written in the form

$$\frac{dy}{dx} = g(x)f(y)$$

The name *separable* comes from the fact that the expression on the right side can be "separated" into a function of x and a function of y. Equivalently, we could write

$$\frac{dy}{dx} = \frac{g(x)}{h(y)} \tag{6}$$

To solve this equation we rewrite it in the differential form

$$h(y)\,dy = g(x)\,dx$$

so that all y's are on one side of the equation and all x's are on the other side. Then we integrate both sides of the equation:

$$\int h(y)\,dy = \int g(x)\,dx \tag{7}$$

Equation 7 defines y implicitly as a function of x. In some cases we may be able to solve for y in terms of x.

The technique for solving separable differential equations was first used by James Bernoulli (in 1690) in solving a problem about pendulums and by Leibniz (in a letter to Huygens in 1691). John Bernoulli explained the general method in a paper published in 1694.

The justification for the step in Equation 7 comes from the Substitution Rule:

$$\int h(y)\,dy = \int h(y(x))\,\frac{dy}{dx}\,dx$$

$$= \int h(y(x))\,\frac{g(x)}{h(y(x))}\,dx \qquad \text{(from Equation 6)}$$

$$= \int g(x)\,dx$$

EXAMPLE 1 Solve the differential equation $\dfrac{dy}{dx} = \dfrac{6x^2}{2y + \cos y}$.

Some computer algebra systems can plot curves defined by implicit equations. Figure 1 shows the graphs of several members of the family of solutions of the differential equation in Example 1. As we look at the curves from left to right, the values of C are 3, 2, 1, 0, -1, -2, and -3.

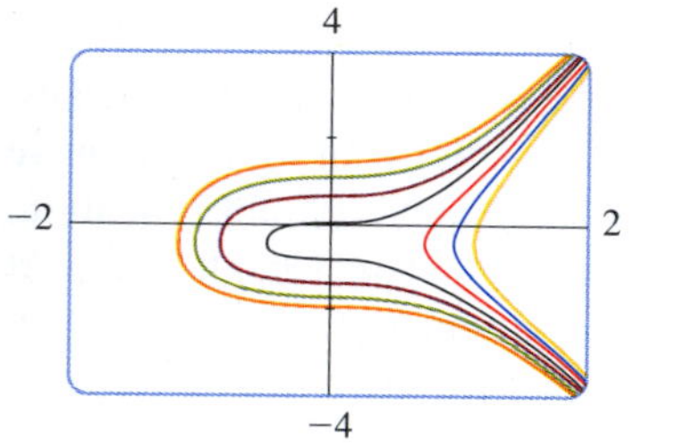

FIGURE 1

SOLUTION Writing the equation in differential form and integrating both sides, we have

$$(2y + \cos y)\,dy = 6x^2\,dx$$

$$\int (2y + \cos y)\,dy = \int 6x^2\,dx$$

$$y^2 + \sin y = 2x^3 + C \tag{8}$$

where C is an arbitrary constant. (We could have used a constant C_1 on the left side and another constant C_2 on the right side. But then we could combine these constants by writing $C = C_2 - C_1$.)

Equation 8 gives the general solution implicitly. In this case it is impossible to solve the equation to express y explicitly as a function of x. ■

EXAMPLE 2 Solve the equation $y' = x^2y$.

SOLUTION First we rewrite the equation using Leibniz notation:

$$\frac{dy}{dx} = x^2y$$

If $y \neq 0$, we can rewrite it in differential notation and integrate:

$$\frac{dy}{y} = x^2\,dx \qquad y \neq 0$$

$$\int \frac{dy}{y} = \int x^2\,dx$$

$$\ln|y| = \frac{x^3}{3} + C$$

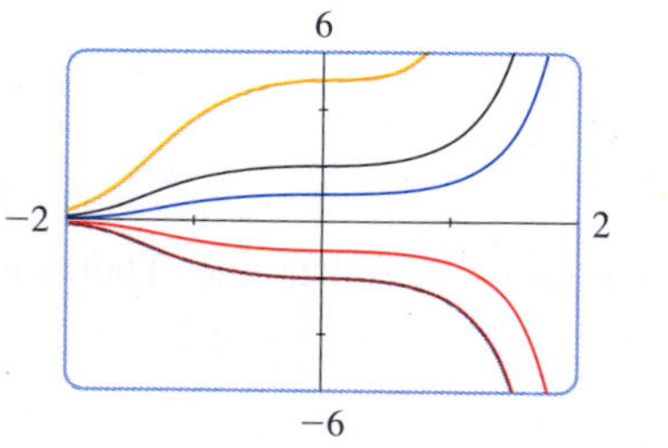

FIGURE 2

Several solutions of the differential equation in Example 2 are graphed in Figure 2. The values of A are the same as the y-intercepts.

This defines y implicitly as a function of x. But in this case we can solve explicitly for y as follows:

$$|y| = e^{\ln|y|} = e^{(x^3/3)+C} = e^C e^{x^3/3}$$

$$y = \pm e^C e^{x^3/3}$$

We note that the function $y = 0$ is also a solution of the given differential equation. So we can write the general solution in the form

$$y = Ae^{x^3/3}$$

where A is an arbitrary constant ($A = e^C$, or $A = -e^C$, or $A = 0$). ■

In Examples 1 and 2 we found the general solution of the given differential equation. But in many physical problems we need to find the particular solution that satisfies a condition of the form $y(x_0) = y_0$. This is called an **initial condition**, and the problem of finding a solution of the differential equation that satisfies the initial condition is called an **initial-value problem**.

EXAMPLE 3 Solve the initial-value problem $xy' = -y$, $x > 0$, $y(4) = 2$.

SOLUTION We write the differential equation as

$$x\frac{dy}{dx} = -y \qquad \text{or} \qquad \frac{dy}{y} = -\frac{dx}{x}$$

Therefore

$$\int \frac{dy}{y} = -\int \frac{dx}{x}$$

$$\ln|y| = -\ln|x| + C$$

$$|y| = \frac{1}{|x|}e^C$$

$$y = \frac{K}{x}$$

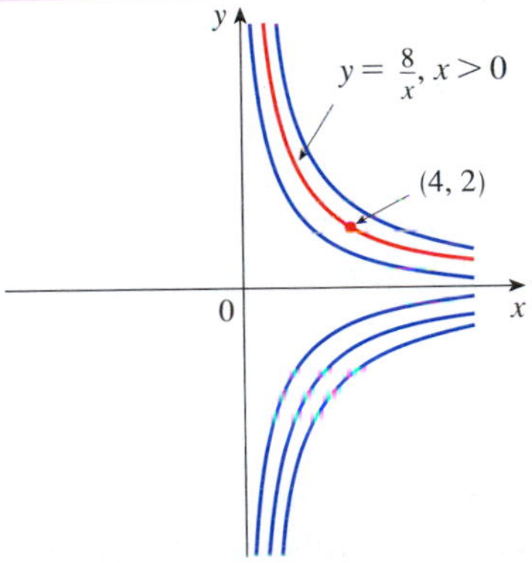

FIGURE 3

where $K = \pm e^C$ is a constant. To determine K we put $x = 4$ and $y = 2$ in this equation:

$$2 = \frac{K}{4} \qquad K = 8$$

The solution of the initial-value problem is

$$y = \frac{8}{x} \qquad x > 0$$

Figure 3 shows the family of solutions $xy = K$ for several values of K (equilateral hyperbolas) and, in particular, the solution that satisfies $y(4) = 2$ [the hyperbola that passes through the point $(4, 2)$]. ■

The graph of the solution of the initial-value problem in Example 4 is shown in Figure 4. Compare with Figure 1.

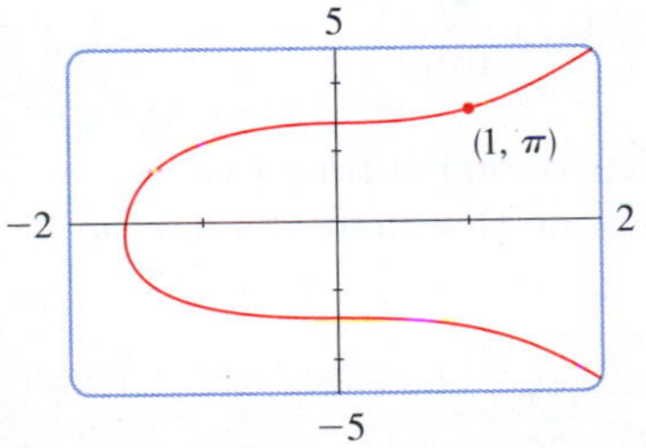

FIGURE 4

EXAMPLE 4 Find the solution of $\dfrac{dy}{dx} = \dfrac{6x^2}{2y + \cos y}$ that satisfies $y(1) = \pi$.

SOLUTION From Example 1 we know that the general solution is

$$y^2 + \sin y = 2x^3 + C$$

We are given that $y(1) = \pi$, so we substitute $x = 1$ and $y = \pi$ in this equation:

$$\pi^2 + \sin \pi = 2(1)^3 + C$$

$$C = \pi^2 - 2$$

Therefore, the solution is given implicitly by

$$y^2 + \sin y = 2x^3 + \pi^2 - 2$$ ■

EXAMPLE 5 Solve $y' = 1 + y^2 - 2x - 2xy^2$, $y(0) = 0$, and graph the solution.

SOLUTION At first glance this does not look like a separable equation, but notice that it is possible to factor the right side as the product of a function of x and a function of y as follows:

$$\frac{dy}{dx} = 1 + y^2 - 2x(1 + y^2) = (1 - 2x)(1 + y^2)$$

$$\int \frac{dy}{1 + y^2} = \int (1 - 2x)\,dx$$

$$\tan^{-1}y = x - x^2 + C$$

Putting $x = 0$ and $y = 0$, we get $C = \tan^{-1}0 = 0$, so

$$\tan^{-1}y = x - x^2$$

To graph this equation we notice that it is equivalent to

$$y = \tan(x - x^2)$$

provided that $-\pi/2 < x - x^2 < \pi/2$. Solving these inequalities using the quadratic formula, we find that

$$\tfrac{1}{2}(1 - \sqrt{1 + 2\pi}\,) < x < \tfrac{1}{2}(1 + \sqrt{1 + 2\pi}\,)$$

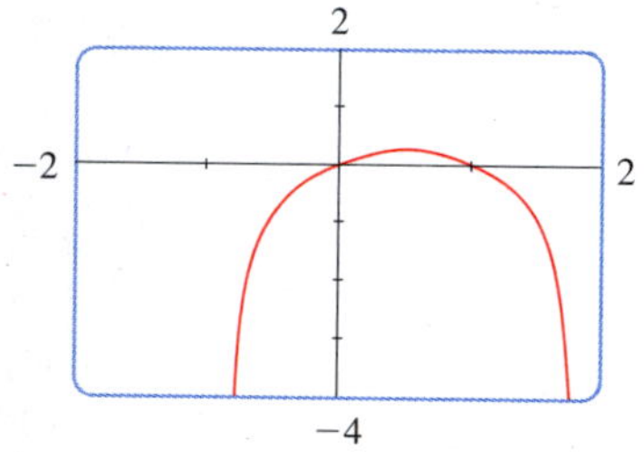

FIGURE 5

This enables us to graph the solution in Figure 5. ■

EXAMPLE 6 A tank contains 20 kg of salt dissolved in 5000 L of water. Brine that contains 0.03 kg of salt per liter of water enters the tank at a rate of 25 L/min. The solution is kept thoroughly mixed and drains from the tank at the same rate. How much salt remains in the tank after half an hour?

SOLUTION Let $y(t)$ be the amount of salt (in kilograms) after t minutes. We are given that $y(0) = 20$ and we want to find $y(30)$. We do this by finding a differential equation satisfied by $y(t)$. Note that dy/dt is the rate of change of the amount of salt, so

$$\frac{dy}{dt} = (\text{rate in}) - (\text{rate out}) \tag{9}$$

where (rate in) is the rate at which salt enters the tank and (rate out) is the rate at which salt leaves the tank. We have

$$\text{rate in} = \left(0.03\,\frac{\text{kg}}{\text{L}}\right)\left(25\,\frac{\text{L}}{\text{min}}\right) = 0.75\,\frac{\text{kg}}{\text{min}}$$

The tank always contains 5000 L of liquid, so the concentration at time t is $y(t)/5000$ (measured in kilograms per liter). Since the brine flows out at a rate of 25 L/min, we have

$$\text{rate out} = \left(\frac{y(t)}{5000}\,\frac{\text{kg}}{\text{L}}\right)\left(25\,\frac{\text{L}}{\text{min}}\right) = \frac{y(t)}{200}\,\frac{\text{kg}}{\text{min}}$$

Thus, from Equation 9 we get

$$\frac{dy}{dt} = 0.75 - \frac{y(t)}{200} = \frac{150 - y(t)}{200}$$

Solving this separable differential equation, we obtain

$$\int \frac{dy}{150 - y} = \int \frac{dt}{200}$$

$$-\ln|150 - y| = \frac{t}{200} + C$$

Figure 6 shows the graph of the function $y(t)$ of Example 6. Notice that, as time goes on, the amount of salt approaches 150 kg.

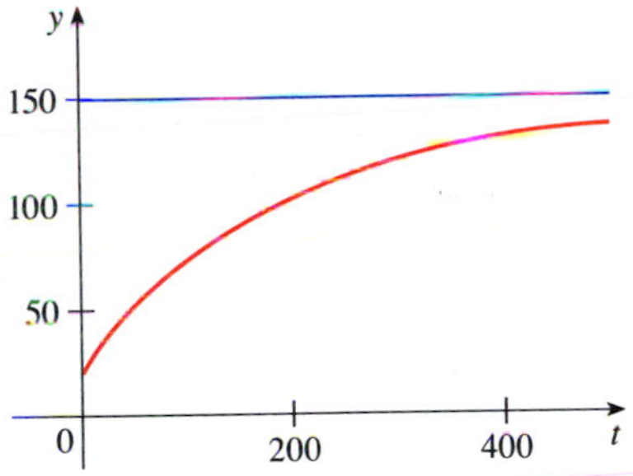

FIGURE 6

Since $y(0) = 20$, we have $-\ln 130 = C$, so

$$-\ln|150 - y| = \frac{t}{200} - \ln 130$$

Therefore $$|150 - y| = 130e^{-t/200}$$

Since $y(t)$ is continuous and $y(0) = 20$ and the right side is never 0, we deduce that $150 - y(t)$ is always positive. Thus, $|150 - y| = 150 - y$ and

$$y(t) = 150 - 130e^{-t/200}$$

The amount of salt after 30 min is

$$y(30) = 150 - 130e^{-30/200} \approx 38.1 \text{ kg}$$ ■

EXAMPLE 7 Solve the equation $\dfrac{dy}{dt} = ky$.

SOLUTION This differential equation was studied in Section 4.5, where it was called the law of natural growth (or decay). Since it is a separable equation, we can solve it by the methods of this section as follows:

$$\int \frac{dy}{y} = \int k\,dt \qquad y \neq 0$$

$$\ln|y| = kt + C$$

$$|y| = e^{kt+C} = e^C e^{kt}$$

$$y = Ae^{kt}$$

where A ($= \pm e^C$ or 0) is an arbitrary constant. ■

LOGISTIC GROWTH

The differential equation of Example 7 is appropriate for modeling population growth ($y' = ky$ says that the rate of growth is proportional to the size of the population) under conditions of unlimited environment and food supply. However, in a restricted environment and with limited food supply, the population cannot exceed a maximal size M at which it consumes its entire food supply. If we make the assumption that the rate of growth of population is jointly proportional to the size of the population (y) and the amount by which y falls short of the maximal size ($M - y$), then we have the equation

$$\frac{dy}{dt} = ky(M - y) \qquad (10)$$

where k is a constant. Equation 10 is called the **logistic differential equation** and was used by the Dutch mathematical biologist Verhulst in the 1840s to model world population growth.

The logistic equation is separable, so we write it in the form

$$\int \frac{dy}{y(M - y)} = \int k\,dt$$

Using partial fractions, we have

$$\frac{1}{y(M - y)} = \frac{1}{M}\left[\frac{1}{y} + \frac{1}{M - y}\right]$$

and so $$\frac{1}{M}\left[\int \frac{dy}{y} + \int \frac{dy}{M - y}\right] = \int k\,dt = kt + C$$

$$\frac{1}{M}(\ln|y| - \ln|M - y|) = kt + C$$

But $|y| = y$ and $|M - y| = M - y$, since $0 < y < M$, so we have

$$\ln \frac{y}{M - y} = M(kt + C)$$

$$\frac{y}{M - y} = Ae^{kMt} \qquad (A = e^{MC})$$

If the population at time $t = 0$ is $y(0) = y_0$, then $A = y_0/(M - y_0)$, so

$$\frac{y}{M - y} = \frac{y_0}{M - y_0}e^{kMt}$$

If we solve this equation for y, we get

(11) $$y = \frac{y_0 M e^{kMt}}{M - y_0 + y_0 e^{kMt}} = \frac{y_0 M}{y_0 + (M - y_0)e^{-kMt}}$$

Using the right-hand expression for y, we see that

$$\lim_{t \to \infty} y(t) = M$$

which is to be expected.

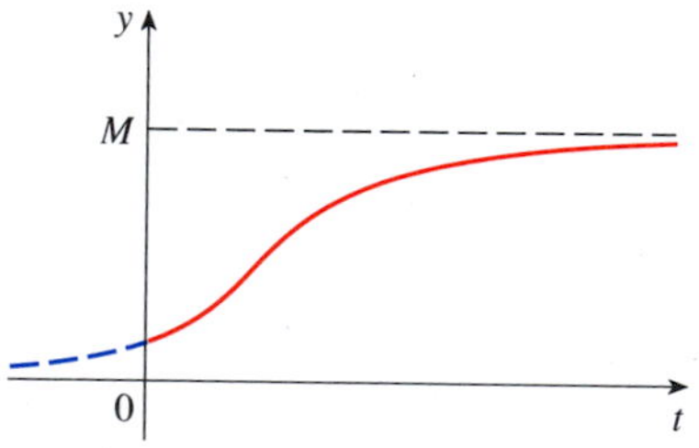

FIGURE 7
Logistic growth function

The graph of the logistic growth function is shown in Figure 7. At first the graph is concave upward and the growth curve appears to be almost exponential, but then it becomes concave downward and approaches the limiting population M.

DIRECTION FIELDS

Suppose we are given a first-order differential equation of the form

$$y' = F(x, y)$$

where $F(x, y)$ is some expression in x and y. [Recall that a separable equation is the special case in which $F(x, y)$ can be factored as a function of x times a function of y.] Even if it's impossible to find a formula for the solution, we can still visualize the solution curves by means of a direction field. If a solution curve passes through a point (x, y), then its slope at that point is y', which is equal to $F(x, y)$. If we draw short line segments with slope $F(x, y)$ at several points (x, y), the result is called a **direction field** (or **slope field**). These line segments indicate the direction in which a soution curve is heading, so the direction field helps us visualize the general shape of these curves.

EXAMPLE 8

(a) Sketch the direction field for the differential equation $y' = x^2 + y^2 - 1$.
(b) Use part (a) to sketch the solution curve that passes through the origin.

SOLUTION

(a) We start by computing the slope at several points in the following chart:

x	-2	-1	0	1	2	-2	-1	0	1	2	...
y	0	0	0	0	0	1	1	1	1	1	...
$y' = x^2 + y^2 - 1$	3	0	-1	0	3	4	1	0	1	4	...

Now we draw short line segments with these slopes at these points. The result is the direction field shown in Figure 8.

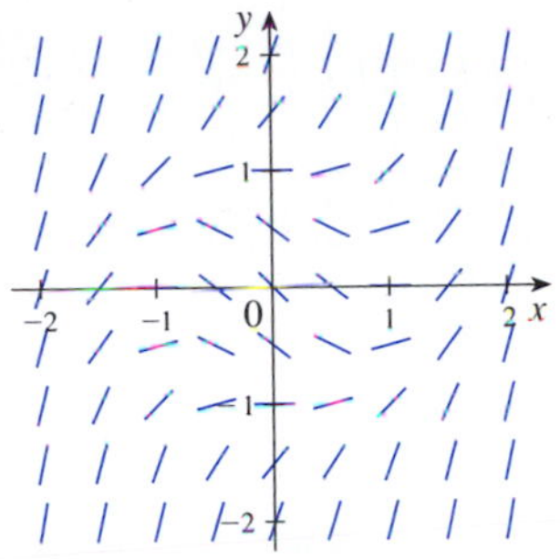

FIGURE 8

(b) We start at the origin and move to the right in the direction of the line segment (which has slope -1). We continue to draw the solution curve so that it moves parallel to the nearby line segments. The resulting solution curve is shown in Figure 9. Returning to the origin, we draw the solution curve to the left as well. ■

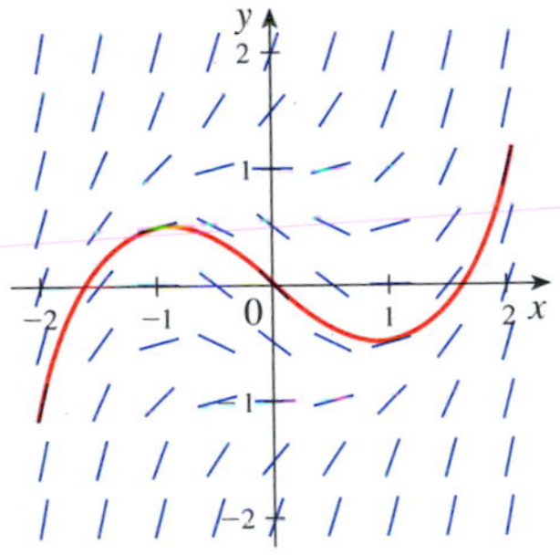

FIGURE 9

The more line segments we draw in a direction field, the clearer the picture becomes. Of course, it's tedious to compute slopes and draw line segments for a huge number of points by hand, but computers are well suited for this task. Figure 10 shows a more detailed, computer-drawn direction field for the differential equation in Example 8. It enables us to draw, with reasonable accuracy, the solution curves shown in Figure 11 with y-intercepts -2, -1, 0, 1, and 2.

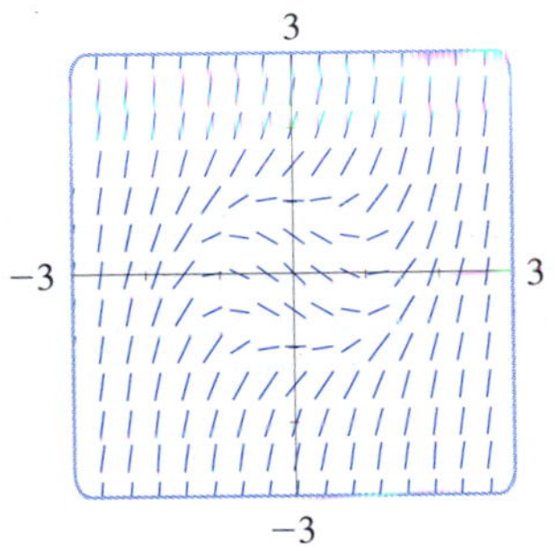

FIGURE 10

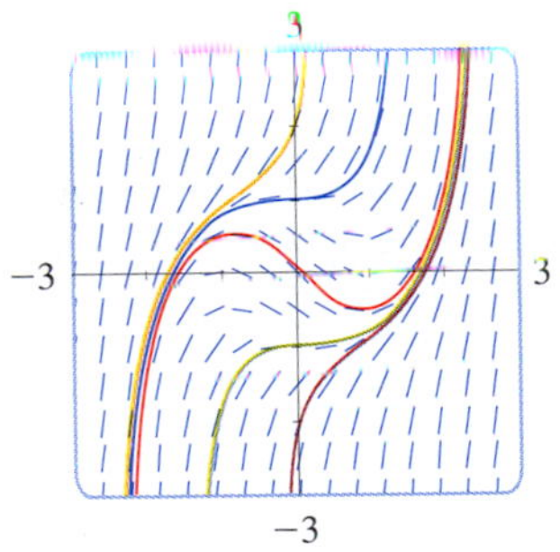

FIGURE 11

EXERCISES 9.1

1–8 ■ Solve the differential equation.

1. $\dfrac{dy}{dx} = y^2$

2. $\dfrac{dy}{dx} = \dfrac{x + \sin x}{3y^2}$

3. $yy' = x$

4. $y' = xy$

5. $x^2y' + y = 0$

6. $y' = \dfrac{\ln x}{xy + xy^3}$

7. $\dfrac{du}{dt} = e^{u+2t}$

8. $\dfrac{dx}{dt} = 1 + t - x - tx$

9–14 ■ Find the solution to the differential equation that satisfies the given initial condition.

9. $e^y y' = \dfrac{3x^2}{1 + y}$, $\quad y(2) = 0$

10. $\dfrac{dy}{dx} = \dfrac{1 + x}{xy}, \quad x > 0, \quad y(1) = -4$

11. $xe^{-t}\dfrac{dx}{dt} = t, \quad x(0) = 1$

12. $x\,dx + 2y\sqrt{x^2 + 1}\,dy = 0, \quad y(0) = 1$

13. $\dfrac{du}{dt} = \dfrac{2t + 1}{2(u - 1)}, \quad u(0) = -1$

14. $\dfrac{dy}{dt} = \dfrac{ty + 3t}{t^2 + 1}, \quad y(2) = 2$

15. Find a function f such that $f'(x) = x^3 f(x)$ and $f(0) = 1$.

16. Find a function g such that $g'(x) = g(x)(1 + g(x))$ and $g(0) = 1$.

17. Find the equation of the curve that satisfies $dy/dx = 4x^3y$ and whose y-intercept is 7.

18. Find an equation of the curve that passes through the point $(1, 1)$ and whose slope at (x, y) is y^2/x^3.

19. Solve the initial-value problem $y' = e^{x-y}$, $y(0) = 1$, and graph the solution.

20. Solve the equation $e^{-y}y' + \cos x = 0$ and graph several members of the family of solutions. How does the solution curve change as the constant C varies?

CAS **21.** Solve the initial-value problem $y' = (\sin x)/\sin y$, $y(0) = \pi/2$, and graph the solution (if your CAS does implicit plots).

CAS **22.** Solve the equation $y' = x\sqrt{x^2 + 1}/(ye^y)$ and graph several members of the family of solutions (if your CAS does implicit plots). How does the solution curve change as the constant C varies?

23. A direction field for the differential equation $y' = y - e^{-x}$ is shown. Sketch the graphs of the solutions that satisfy the given initial conditions.
(a) $y(0) = 0$
(b) $y(0) = 1$
(c) $y(0) = -1$

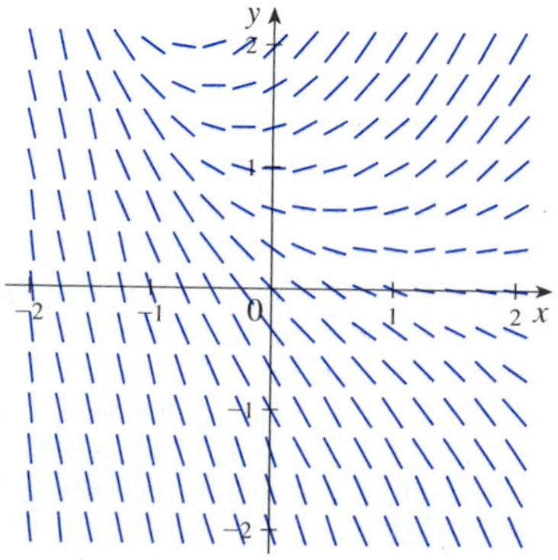

24. A direction field for the differential equation $y' = 2y(y - 2)$ is shown. Sketch the graphs of the solutions that satisfy the given initial conditions.
(a) $y(0) = 1$
(b) $y(0) = 2.5$
(c) $y(0) = -1$

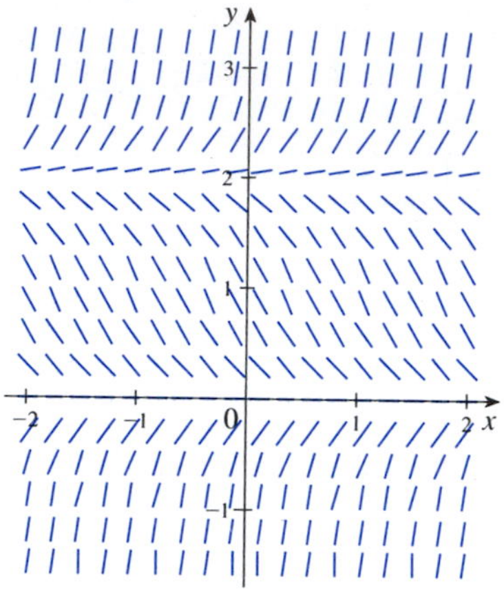

25–26 ■ Sketch a direction field for the differential equation. Then use it to sketch three solution curves.

25. $y' = x - y$ **26.** $y' = xy + y^2$

CAS **27–28** ■ Use a computer algebra system to draw a direction field for the given differential equation. Get a printout and sketch on it the solution curve that passes through $(0, 1)$. Then use the CAS to draw the solution curve and compare it with your sketch.

27. $y' = y\sin 2x$ **28.** $y' = \sin(x + y)$

29. A tank contains 1000 L of brine with 15 kg of dissolved salt. Pure water enters the tank at a rate of 10 L/min. The solution is kept thoroughly mixed and drains from the tank at the same rate. How much salt is in the tank (a) after t minutes and (b) after 20 minutes?

30. A tank contains 1000 L of pure water. Brine that contains 0.05 kg of salt per liter of water enters the tank at a rate of 5 L/min. Brine that contains 0.04 kg of salt per liter of water enters the tank at a rate of 10 L/min. The solution is kept thoroughly mixed and drains from the tank at a rate of 15 L/min. How much salt is in the tank (a) after t minutes and (b) after one hour?

31. In an elementary chemical reaction, single molecules of two reactants A and B form a molecule of the product C: $A + B \rightarrow C$. The law of mass action states that the rate of reaction is proportional to the product of the concentrations of A and B:

$$\frac{d[C]}{dt} = k[A][B]$$

(See Example 4 in Section 3.3.) Thus, if the initial concentrations are [A] = a moles/L and [B] = b moles/L and we write x = [C], then we have

$$\frac{dx}{dt} = k(a - x)(b - x)$$

Assuming that $a \neq b$, find x as a function of t.

32. (a) Find $x(t)$ in Exercise 31 assuming that $b = a$.
(b) How does this expression for $x(t)$ simplify if it is known that [C] = $a/2$ after 20 seconds?

33. (a) The population of the world was about 5 billion in 1986. Using the exponential model for population growth (see Example 7 or Section 4.5) with the recently observed rate of increase of population of 2% per year in 1986, find an expression for the population of the world in the year t.
(b) Use this model to predict the population of the world in the years (i) 2000, (ii) 2100, and (iii) 2500.
(c) The total land surface area of this planet is about 1.8×10^{15} ft^2. Under the exponential model, how many square feet of land per person will there be in each of the years considered in part (b)?

34. For a more realistic picture of long-term growth than in Exercise 33, we consider the logistic model for world population growth with y_0 = 5 billion in 1986 and an assumed maximum population of M = 100 billion. To determine the value of k, use Equation 10 and the fact that the population was increasing at a rate of 2% per year when the population was 5 billion. Use this model to predict the population of the world in the years (a) 2000, (b) 2100, and (c) 2500. Compare with the results of Exercise 33.

35. One model for the spread of a rumor is that the rate of spread is proportional to the product of the fraction y of the population who have heard the rumor and the fraction who have not heard the rumor.
(a) Write a differential equation that is satisfied by y.
(b) Solve the differential equation.
(c) A small town has 1000 inhabitants. At 8 A.M. 80 people have heard a rumor. By noon half the town has heard it. At what time will 90% of the population have heard the rumor?

36. Biologists stocked a lake with 400 fish and estimated the carrying capacity (the maximal population for the fish of that species in that lake) to be 10,000. The number of fish tripled in the first year.
(a) Assuming that the size of the fish population satisfies the logistic equation, find an expression for the size of the population after t years.
(b) How long will it take for the population to increase to 5000?

37. Show that the solution to the logistic Equation 10 increases most rapidly when $y = M/2$.

38. For a fixed value of M (say $M = 10$), the family of logistic functions given by Equation 11 depends on the initial value y_0 and the proportionality constant k. Graph several members of this family. How does the graph change when y_0 varies? How does it change when k varies?

39. Let's compare the logistic model for population growth (Equation 11) with the corresponding exponential model. If $y_0 = 1$ and $M = 10$, find the value of k for which the logistic model has the same rate of growth at $t = 0$ as the exponential model $y = e^{0.1t}$. For this value of k, graph both functions on the same screen and comment on how these models compare for small and large values of t.

40. Another model for a growth function for a limited population is given by the *Gompertz function*, which is a solution of the differential equation

$$\frac{dy}{dt} = c \ln\left(\frac{M}{y}\right) y$$

where c is a constant and M is the maximum size of the population.
(a) Solve this differential equation.
(b) Compute $\lim_{t\to\infty} y(t)$.
(c) Sketch the graph of the Gompertz growth function.

41. Let $A(t)$ be the area of a tissue culture at time t and let M be the final area of the tissue when growth is complete. Most cell divisions occur on the periphery of the tissue and the number of cells on the periphery is proportional to $\sqrt{A(t)}$. So a reasonable model for the growth of tissue is that the rate of growth of the area is jointly proportional to $\sqrt{A(t)}$ and $M - A(t)$.
(a) Formulate a differential equation and use it to show that the tissue grows fastest when $A(t) = M/3$.
(b) Solve the differential equation to find an expression for $A(t)$.

42. When a raindrop falls it increases in size, so its mass at time t is a function of t, $m(t)$. The rate of growth of the mass is $km(t)$ for some positive constant k. When we apply Newton's Law of Motion to the raindrop, we get $(mv)' = gm$, where v is the velocity of the raindrop (directed downward) and g is the acceleration due to gravity. The *terminal velocity* of the raindrop is $\lim_{t\to\infty} v(t)$. Find an expression for the terminal velocity in terms of g and k.

43. In the simple electric circuit shown in the figure, a battery supplies a constant voltage V and produces a current of $I(t)$ amperes at time t. The circuit also contains a resistor with a resistance of R ohms and an inductor with an inductance of L henries. Ohm's Law gives the voltage drop across the resistor as RI; the voltage drop due to the inductor is

$L(dI/dt)$. One of Kirchhoff's laws says that the sum of these voltage drops is the applied voltage V, so

$$RI + LI'(t) = V$$

If the switch is closed when $t = 0$, so that $I(0) = 0$, solve this equation to show that

$$I(t) = \frac{V}{R}(1 - e^{-Rt/L})$$

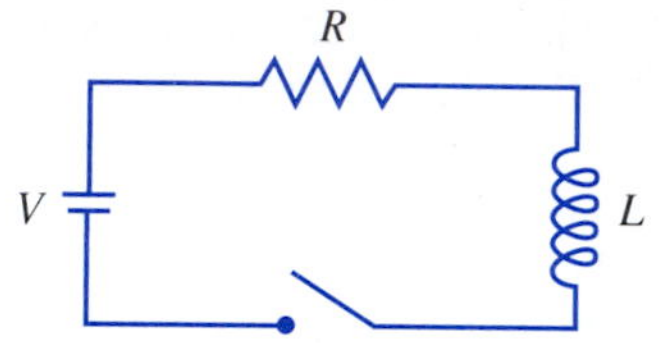

44. According to Newton's Law of Universal Gravitation, the gravitational force on an object of mass m that has been projected vertically upward from Earth's surface is

$$F = \frac{mgR^2}{(x + R)^2}$$

where $x = x(t)$ is the object's distance above the surface at time t, R is the radius of Earth, and g is the acceleration due to gravity. Also, by Newton's Second Law, $F = ma = m(dv/dt)$ and so

$$m\frac{dv}{dt} = -\frac{mgR^2}{(x + R)^2}$$

(a) Suppose a rocket is fired vertically upward with an initial velocity v_0. Let h be the maximum height above the surface reached by the object. Show that

$$v_0 = \sqrt{\frac{2gRh}{R + h}}$$

[*Hint:* By the Chain Rule, $m(dv/dt) = mv(dv/dx)$.]

(b) Calculate $v_e = \lim_{h \to \infty} v_0$. This limit is called the *escape velocity* for Earth.

(c) Use $R = 3960$ mi and $g = 32$ ft/s^2 to calculate v_e in feet per second and in miles per second.

9.2 FIRST-ORDER LINEAR EQUATIONS

A first-order **linear** differential equation is one that can be put into the form

(1) $$\frac{dy}{dx} + P(x)y = Q(x)$$

where P and Q are continuous functions on a given interval. This type of equation occurs frequently in various sciences, as we will see.

The standard method for solving Equation 1 is to multiply both sides of the equation by a suitable function $I(x)$ called an *integrating factor.* We try to find I so that the left side of Equation 1, when multiplied by $I(x)$, becomes the derivative of the product $I(x)y$:

(2) $$I(x)(y' + P(x)y) = (I(x)y)'$$

If we can find such a function I, then Equation 1 becomes

$$(I(x)y)' = I(x)Q(x)$$

Integrating both sides, we would have

$$I(x)y = \int I(x)Q(x)\,dx + C$$

so the solution would be

(3) $$y(x) = \frac{1}{I(x)}\left[\int I(x)Q(x)\,dx + C\right]$$

To find such an I we expand Equation 2 and cancel terms:

$$I(x)y' + I(x)P(x)y = (I(x)y)' = I'(x)y + I(x)y'$$

$$I(x)P(x) = I'(x)$$

This is a separable differential equation for I, which we solve as follows:

$$\int \frac{dI}{I} = \int P(x)\,dx$$

$$\ln|I| = \int P(x)\,dx$$

$$I = Ae^{\int P(x)\,dx}$$

where $A = \pm e^C$. We are looking for a particular integrating factor, not the most general one, so we take $A = 1$ and use

(4) $$I(x) = e^{\int P(x)\,dx}$$

Thus, a formula for the general solution to Equation 1 is provided by Equation 3, where I is given by Equation 4. Instead of memorizing this formula, however, we just remember the form of the integrating factor.

> To solve the linear differential equation $y' + P(x)y = Q(x)$, multiply both sides by the **integrating factor** $I(x) = e^{\int P(x)\,dx}$ and integrate both sides.

EXAMPLE 1 Solve the differential equation $\dfrac{dy}{dx} + 3x^2y = 6x^2$.

SOLUTION The given equation is linear since it has the form of Equation 1 with $P(x) = 3x^2$ and $Q(x) = 6x^2$. An integrating factor is

$$I(x) = e^{\int 3x^2\,dx} = e^{x^3}$$

Figure 1 shows the graphs of several members of the family of solutions in Example 1.

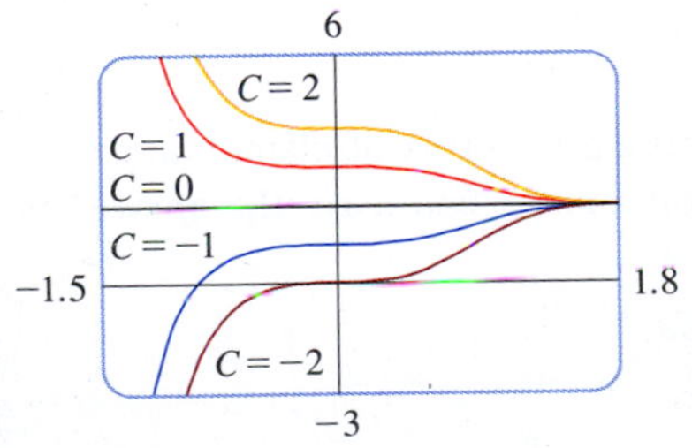

FIGURE 1

Multiplying both sides of the differential equation by e^{x^3}, we get

$$e^{x^3}\frac{dy}{dx} + 3x^2e^{x^3}y = 6x^2e^{x^3}$$

or

$$\frac{d}{dx}(e^{x^3}y) = 6x^2e^{x^3}$$

Integrating both sides, we have

$$e^{x^3}y = \int 6x^2e^{x^3}\,dx = 2e^{x^3} + C$$

$$y = 2 + Ce^{-x^3}$$ ■

EXAMPLE 2 Find the solution of the initial-value problem

$$x^2y' + xy = 1 \qquad x > 0 \qquad y(1) = 2$$

SOLUTION We must first divide both sides by the coefficient of y' to put the differential equation into standard form:

(5) $$y' + \frac{1}{x}y = \frac{1}{x^2} \qquad x > 0$$

The integrating factor is

$$I(x) = e^{\int (1/x)\,dx} = e^{\ln x} = x$$

Multiplication of Equation 5 by x gives

$$xy' + y = \frac{1}{x} \qquad \text{or} \qquad (xy)' = \frac{1}{x}$$

Then

$$xy = \int \frac{1}{x}\,dx = \ln x + C$$

and so

$$y = \frac{\ln x + C}{x}$$

Since $y(1) = 2$, we have

$$2 = \frac{\ln 1 + C}{1} = C$$

Therefore, the solution to the initial-value problem is

$$y = \frac{\ln x + 2}{x}$$

■

The solution of the initial-value problem in Example 2 is shown in Figure 2.

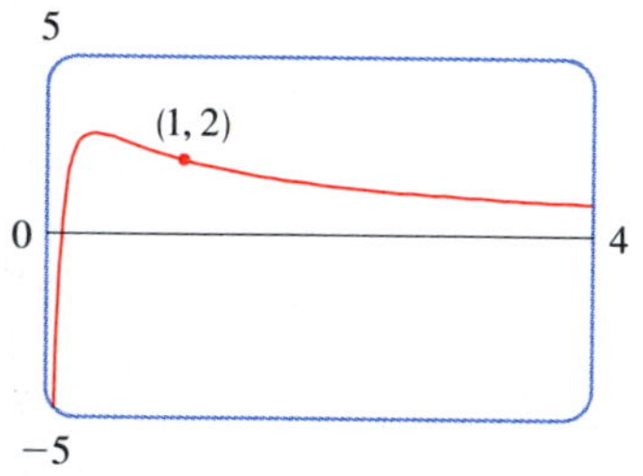

FIGURE 2

EXAMPLE 3 Solve $y' + 2xy = 1$.

SOLUTION The given equation is in the standard form for a linear equation. Multiplying by the integrating factor

$$e^{\int 2x\,dx} = e^{x^2}$$

we get

$$e^{x^2}y' + 2xe^{x^2}y = e^{x^2}$$

or

$$(e^{x^2}y)' = e^{x^2}$$

Therefore

$$e^{x^2}y = \int e^{x^2}\,dx + C$$

Recall from Section 8.6 that $\int e^{x^2}\,dx$ cannot be expressed in terms of elementary functions. Nonetheless, it is a perfectly good function and we can leave the answer as

$$y = e^{-x^2}\int e^{x^2}\,dx + Ce^{-x^2}$$

Another way of writing the solution is

$$y = e^{-x^2}\int_0^x e^{t^2}\,dt + Ce^{-x^2}$$

(Any number can be chosen for the lower limit of integration.) ■

Even though the solutions of the differential equation in Example 3 are expressed in terms of an integral, they can still be graphed by a computer algebra system (Figure 3).

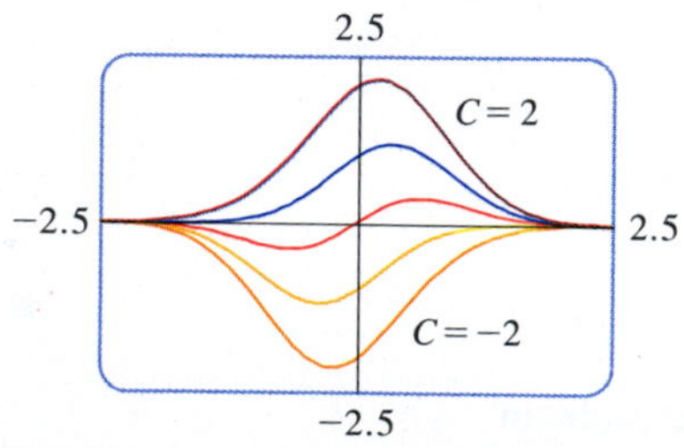

FIGURE 3

APPLICATION TO ELECTRIC CIRCUITS

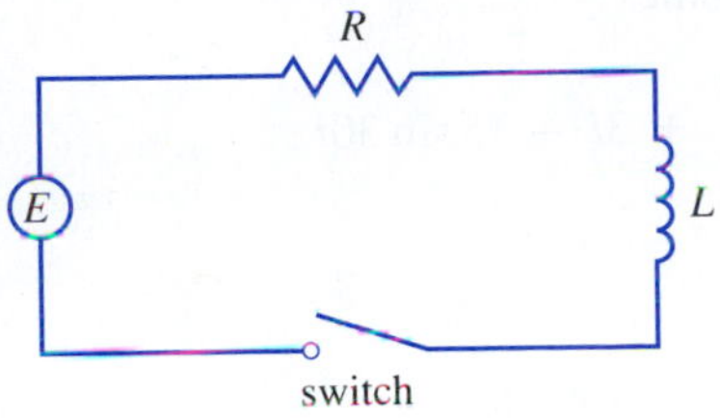

FIGURE 4

The simple electric circuit shown in Figure 4 contains an electromotive force (usually a battery or generator) that produces a voltage of $E(t)$ volts (V) and a current of $I(t)$ amperes (A) at time t. The circuit also contains a resistor with a resistance of R ohms (Ω) and an inductor with an inductance of L henries (H).

Ohm's Law gives the drop in voltage due to the resistor as RI. The voltage drop due to the inductor is $L(dI/dt)$. One of Kirchhoff's laws says that the sum of the voltage drops is equal to the supplied voltage $E(t)$. Thus, we have

$$L\frac{dI}{dt} + RI = E(t) \tag{6}$$

which is a first-order linear differential equation. The solution gives the current I at time t.

EXAMPLE 4 Suppose that in the simple circuit of Figure 4 the resistance is 12 Ω and the inductance is 4 H. If a battery gives a constant voltage of 60 V and the switch is closed when $t = 0$ so the current starts with $I(0) = 0$, find (a) $I(t)$, (b) the current after 1 s, and (c) the limiting value of the current.

SOLUTION

(a) If we put $L = 4$, $R = 12$, and $E(t) = 60$ in Equation 6, we obtain the initial-value problem

$$4\frac{dI}{dt} + 12I = 60 \qquad I(0) = 0$$

or

$$\frac{dI}{dt} + 3I = 15 \qquad I(0) = 0$$

Multiplying by the integrating factor $e^{\int 3\,dt} = e^{3t}$, we get

$$e^{3t}\frac{dI}{dt} + 3e^{3t}I = 15e^{3t}$$

$$\frac{d}{dt}(e^{3t}I) = 15e^{3t}$$

$$e^{3t}I = \int 15e^{3t}\,dt = 5e^{3t} + C$$

$$I(t) = 5 + Ce^{-3t}$$

Figure 5 shows how the current in Example 4 approaches its limiting value.

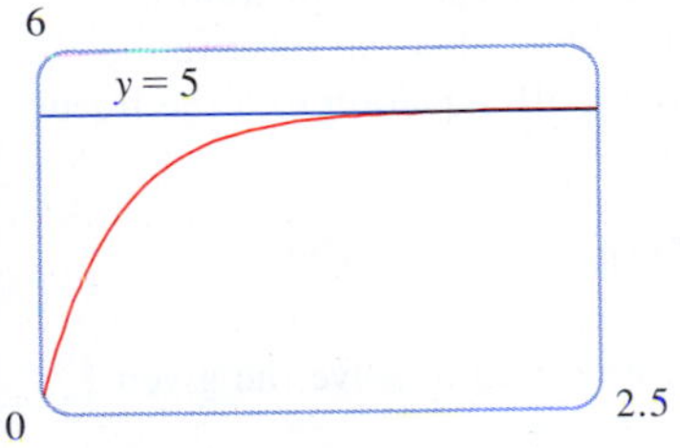

FIGURE 5

Since $I(0) = 0$, we have $5 + C = 0$, so $C = -5$ and

$$I(t) = 5(1 - e^{-3t})$$

(b) After 1 s the current is

$$I(1) = 5(1 - e^{-3}) \approx 4.75 \text{ A}$$

(c)

$$\begin{aligned}\lim_{t\to\infty} I(t) &= \lim_{t\to\infty} 5(1 - e^{-3t}) \\ &= 5 - 5\lim_{t\to\infty} e^{-3t} \\ &= 5 - 0 = 5\end{aligned}$$

■

EXAMPLE 5 Suppose that the resistance and inductance remain as in Example 4 but, instead of the battery, we use a generator that produces a variable voltage of $E(t) = 60\sin 30t$ volts. Find $I(t)$.

SOLUTION This time the differential equation becomes

$$4\frac{dI}{dt} + 12I = 60\sin 30t \qquad \text{or} \qquad \frac{dI}{dt} + 3I = 15\sin 30t$$

The same integrating factor e^{3t} gives

$$\frac{d}{dt}(e^{3t}I) = e^{3t}\frac{dI}{dt} + 3e^{3t}I = 15e^{3t}\sin 30t$$

Figure 6 shows the graph of the current when the battery is replaced by a generator.

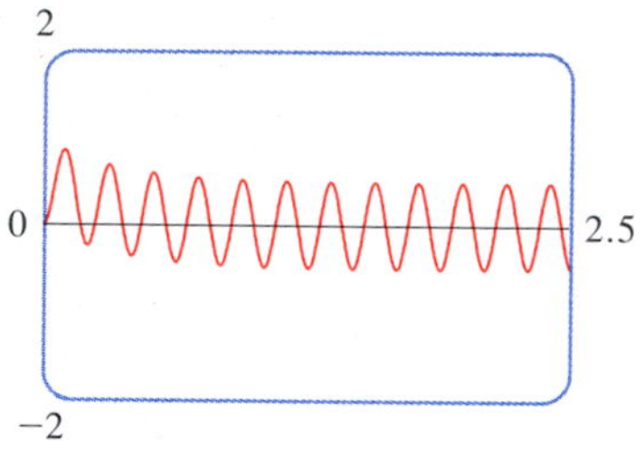

FIGURE 6

Using Formula 98 in the Table of Integrals, we have

$$e^{3t}I = \int 15e^{3t}\sin 30t\,dt = 15\frac{e^{3t}}{909}(3\sin 30t - 30\cos 30t) + C$$

$$I = \tfrac{5}{101}(\sin 30t - 10\cos 30t) + Ce^{-3t}$$

Since $I(0) = 0$, we get

$$-\tfrac{50}{101} + C = 0$$

so

$$I(t) = \tfrac{5}{101}(\sin 30t - 10\cos 30t) + \tfrac{50}{101}e^{-3t}$$

■

EXERCISES 9.2

1–4 ■ Determine whether the differential equation is linear.

1. $y' + x^2y = y^2$

2. $x^2y' - y + x = 0$

3. $xy' = x - y$

4. $yy' = \sin x$

5–14 ■ Solve the differential equation.

5. $y' - 3y = e^x$

6. $y' + 4y = x$

7. $y' - 2xy = x$

8. $xy' + 2y = e^{x^2}$

9. $y'\cos x = y\sin x + \sin 2x, \quad -\pi/2 < x < \pi/2$

10. $1 + xy = xy'$

11. $\dfrac{dy}{dx} + 2xy = x^2$

12. $\dfrac{dy}{dx} = x\sin 2x + y\tan x, \quad -\pi/2 < x < \pi/2$

13. $\dfrac{dy}{d\theta} - y\tan\theta = 1, \quad -\pi/2 < \theta < \pi/2$

14. $xy' + xy + y = e^{-x}, \quad x > 0$

15–20 ■ Solve the initial-value problem.

15. $y' + y = x + e^x, \quad y(0) = 0$

16. $xy' - 3y = x^2, \quad x > 0, \quad y(1) = 0$

17. $y' - 2xy = 2xe^{x^2}, \quad y(0) = 3$

18. $(1 + x^2)y' + 2xy = 3\sqrt{x}, \quad y(0) = 2$

19. $x^2\dfrac{dy}{dx} + 2xy = \cos x, \quad y(\pi) = 0$

20. $x\dfrac{dy}{dx} - \dfrac{y}{x+1} = x, \quad y(1) = 0, \quad x > 0$

21–22 ■ Solve the differential equation and use a graphing calculator or computer to graph several members of the family of solutions. How does the solution curve change as C varies?

21. $xy' + y = x\cos x, \ x > 0$

22. $y' + (\cos x)y = \cos x$

23. A **Bernoulli differential equation** [named after James Bernoulli (1654–1705)] is of the form

$$\frac{dy}{dx} + P(x)y = Q(x)y^n$$

Observe that, if $n = 0$ or 1, the Bernoulli equation is linear. For other values of n, show that the substitution $u = y^{1-n}$ transforms the Bernoulli equation into the linear equation

$$\frac{du}{dx} + (1 - n)P(x)u = (1 - n)Q(x)$$

24–26 ■ Use the method of Exercise 23 to solve the given differential equation.

24. $xy' + y = -xy^2$

25. $y' + \dfrac{2}{x}y = \dfrac{y^3}{x^2}$

26. $y' + y = xy^3$

27. In the circuit shown in Figure 4 a battery supplies a constant voltage of 40 V, the inductance is 2 H, the resistance is 10 Ω, and $I(0) = 0$.
(a) Find $I(t)$.
(b) Find the current after 0.1 s.

28. In the circuit shown in Figure 4 a generator supplies a voltage of $E(t) = 40 \sin 60t$ volts, the inductance is 1 H, the resistance is 20 Ω, and $I(0) = 1$ A.
(a) Find $I(t)$.
(b) Find the current after 0.1 s.
(c) Use a graphing device to draw the graph of the current function.

29. The figure shows a circuit containing an electromotive force, a capacitor with a capacitance of C farads (F), and a resistor with a resistance of R ohms (Ω). The voltage drop across the capacitor is Q/C, where Q is the charge (in coulombs), so in this case Kirchhoff's Law gives

$$RI + \frac{Q}{C} = E(t)$$

But $I = dQ/dt$ (see Example 3 in Section 3.3), so we have

$$R\frac{dQ}{dt} + \frac{1}{C}Q = E(t)$$

Suppose the resistance is 5 Ω, the capacitance is 0.05 F, a battery gives a constant voltage of 60 V, and the initial charge is $Q(0) = 0$ C. Find the charge and the current at time t.

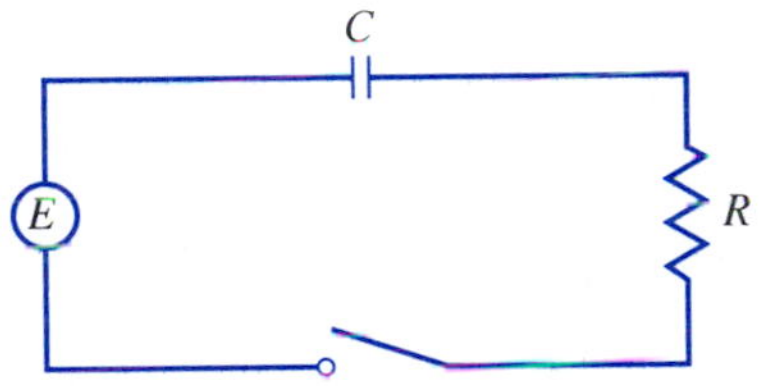

30. In the circuit of Exercise 29, $R = 2$ Ω, $C = 0.01$ F, $Q(0) = 0$, and $E(t) = 10 \sin 60t$. Find the charge and the current at time t.

31. Psychologists interested in learning theory study **learning curves**. A learning curve is the graph of a function $P(t)$, the performance of someone learning a skill as a function of the training time t. The derivative dP/dt represents the rate at which performance improves. If M is the maximum level of performance of which the learner is capable, it is reasonable to assume that dP/dt is proportional to $M - P(t)$. (At first, learning is rapid. Then, as performance increases and approaches its maximal value, the rate of learning decreases.) Thus

$$\frac{dP}{dt} = k(M - P(t))$$

where k is a positive constant. Solve this linear differential equation and sketch the learning curve.

32. Two new workers were hired for an assembly line. Jim processed 25 units during the first hour and 45 units during the second hour. Mark processed 35 units during the first hour and 50 units the second hour. Using the model of Exercise 31 and assuming that $P(0) = 0$, estimate the maximum number of units per hour that each worker is capable of processing.

33. An object with mass m is dropped from rest and we assume that the air resistance is proportional to the speed of the object. If $s(t)$ is the distance dropped after t seconds, then the speed is $v = s'(t)$ and the acceleration is $a = v'(t)$. If g is the acceleration due to gravity, then the downward force on the object is $mg - cv$, where c is a positive constant, and Newton's Second Law gives

$$m\frac{dv}{dt} = mg - cv$$

(a) Solve this differential equation to find the velocity at time t.
(b) What is the limiting velocity?
(c) Find the distance the object has fallen after t seconds.

9.3 ARC LENGTH

FIGURE 1

What do we mean by the length of a curve? We might think of fitting a piece of string to the curve in Figure 1 and then measuring the string against a ruler. But that might be difficult to do with much accuracy if we have a complicated curve.

It's easy to find the length of a polygon; we just add the lengths of the line segments that form the polygon. We are going to define the length of a curve by first approximating it by a polygon and then taking a limit. This process is familiar for the case of a circle, where the circumference is the limit of lengths of inscribed polygons (see Figure 2).

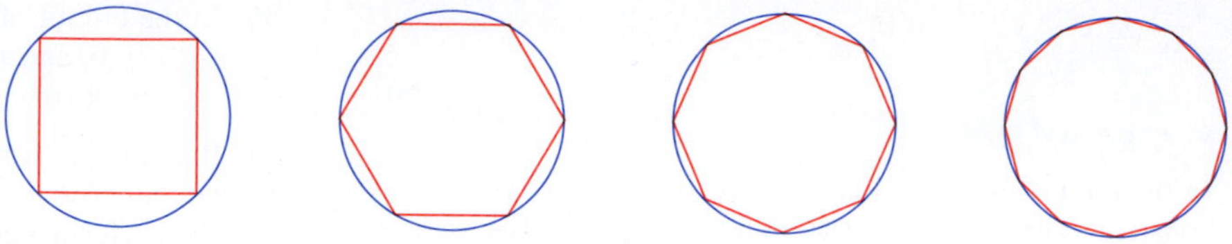

FIGURE 2

Suppose that a curve C is described by the parametric equations

$$x = f(t) \qquad y = g(t) \qquad a \leqslant t \leqslant b$$

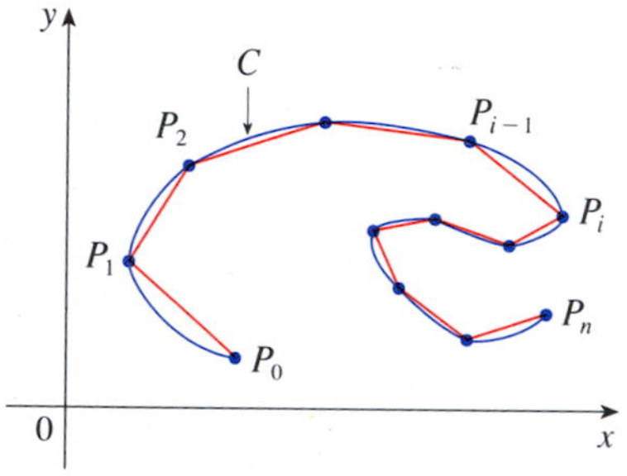

FIGURE 3

Let's assume that C is **smooth** in the sense that the derivatives $f'(t)$ and $g'(t)$ are continuous and not simultaneously zero for $a < t < b$. (This ensures that C has no sudden change in direction.) We take a partition P of the parameter interval $[a, b]$ into n subintervals of equal width Δt. If $t_0, t_1, t_2, \ldots, t_n$ are the endpoints of these subintervals, then $x_i = f(t_i)$ and $y_i = g(t_i)$ are the coordinates of points $P_i(x_i, y_i)$ that lie on C and the polygon with vertices $P_0, P_1, \ldots, P_n$ approximates C (see Figure 3). The length L of C is approximately the length of this polygon and the approximation gets better as we let $\|P\| \to 0$. (see Figure 4). Therefore, we define the **length** of C to be the limit of the lengths of these inscribed polygons:

$$L = \lim_{\|P\| \to 0} \sum_{i=1}^{n} |P_{i-1} P_i|$$

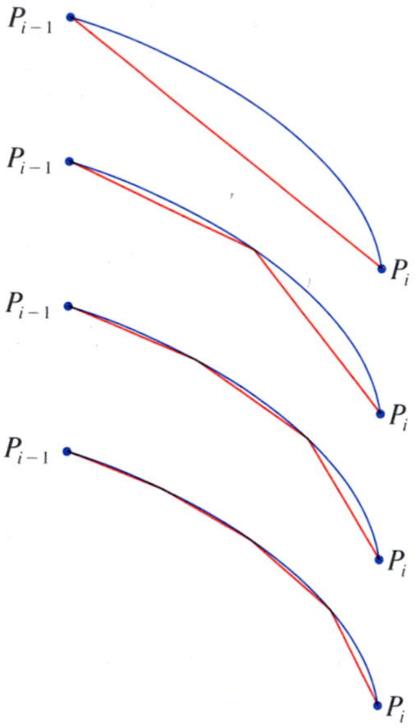

FIGURE 4

Notice that the procedure for defining arc length is very similar to the procedure we used for defining area and volume. We divided the curve into a large number of small parts. We then found the approximate lengths of the small parts and added them. Finally, we took the limit as $\|P\| \to 0$.

For computational purposes we need a more convenient expression for L. If we let $\Delta x_i = x_i - x_{i-1}$ and $\Delta y_i = y_i - y_{i-1}$, then the length of the ith line segment of the polygon is

$$|P_{i-1} P_i| = \sqrt{(\Delta x_i)^2 + (\Delta y_i)^2}$$

But from the definition of a derivative we know that

$$f'(t_i) \approx \frac{\Delta x_i}{\Delta t}$$

if Δt is small. (We could have used any sample point t_i^* in place of t_i.) Therefore

$$\Delta x_i \approx f'(t_i)\,\Delta t \qquad \Delta y_i \approx g'(t_i)\,\Delta t$$

and so

$$\begin{aligned} |P_{i-1} P_i| &= \sqrt{(\Delta x_i)^2 + (\Delta y_i)^2} \\ &\approx \sqrt{[f'(t_i)\,\Delta t]^2 + [g'(t_i)\,\Delta t]^2} \\ &= \sqrt{[f'(t_i)]^2 + [g'(t_i)]^2}\,\Delta t \end{aligned}$$

Thus

$$L \approx \sum_{i=1}^{n} \sqrt{[f'(t_i)]^2 + [g'(t_i)]^2}\,\Delta t$$

This is a Riemann sum for the function $\sqrt{[f'(t)]^2 + [g'(t)]^2}$ and so our argument suggests that

$$L = \int_a^b \sqrt{[f'(t)]^2 + [g'(t)]^2}\,dt$$

In fact, our reasoning can be made precise; this formula is correct, provided that we rule out situations where a portion of the curve is traced out more than once.

(1) ARC LENGTH FORMULA If a smooth curve with parametric equations $x = f(t)$, $y = g(t)$, $a \leqslant t \leqslant b$, is traversed exactly once as t increases from a to b, then its length is

$$L = \int_a^b \sqrt{\left(\frac{dx}{dt}\right)^2 + \left(\frac{dy}{dt}\right)^2}\, dt$$

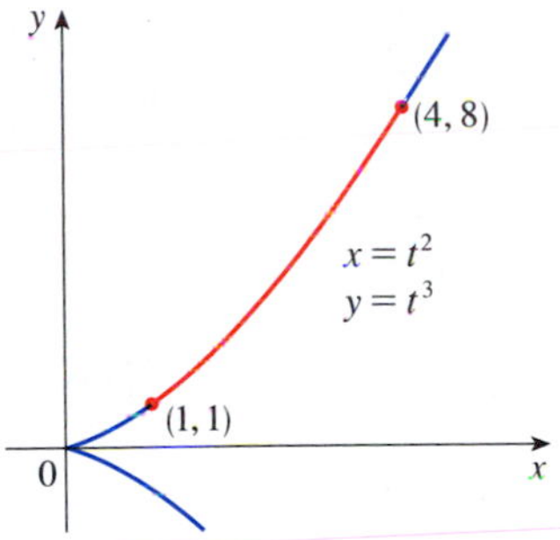

FIGURE 5

As a check on our answer to Example 1, notice from Figure 5 that it ought to be slightly larger than the distance from $(1, 1)$ to $(4, 8)$, which is

$$\sqrt{58} \approx 7.615733$$

According to our calculation in Example 1, we have

$$L = \frac{80\sqrt{10} - 13\sqrt{13}}{27} \approx 7.633705$$

Sure enough, this is a bit greater than the length of the line segment.

EXAMPLE 1 Find the length of the arc of the curve $x = t^2$, $y = t^3$ that lies between the points $(1, 1)$ and $(4, 8)$ (see Figure 5).

SOLUTION First we notice from the equations $x = t^2$ and $y = t^3$ that the portion of the curve between $(1, 1)$ and $(4, 8)$ corresponds to the parameter interval $1 \leqslant t \leqslant 2$. So the arc length formula (1) gives

$$L = \int_1^2 \sqrt{\left(\frac{dx}{dt}\right)^2 + \left(\frac{dy}{dt}\right)^2}\, dt = \int_1^2 \sqrt{(2t)^2 + (3t^2)^2}\, dt$$

$$= \int_1^2 \sqrt{4t^2 + 9t^4}\, dt = \int_1^2 t\sqrt{4 + 9t^2}\, dt$$

If we substitute $u = 4 + 9t^2$, then $du = 18t\, dt$. When $t = 1$, $u = 13$; when $t = 2$, $u = 40$. Therefore

$$L = \frac{1}{18}\int_{13}^{40} \sqrt{u}\, du = \frac{1}{18} \cdot \frac{2}{3}u^{3/2}\Big]_{13}^{40}$$

$$= \frac{1}{27}\left[40^{3/2} - 13^{3/2}\right] = \frac{1}{27}\left(80\sqrt{10} - 13\sqrt{13}\right)$$ ■

If we are given a curve with equation $y = f(x)$, $a \leqslant x \leqslant b$, then we can regard x as a parameter. Then parametric equations are $x = x$, $y = f(x)$, and Formula 1 becomes

(2)
$$L = \int_a^b \sqrt{1 + \left(\frac{dy}{dx}\right)^2}\, dx$$

Similarly, if a curve has the equation $x = f(y)$, $a \leqslant y \leqslant b$, we regard y as the parameter and the length is

(3)
$$L = \int_a^b \sqrt{\left(\frac{dx}{dy}\right)^2 + 1}\, dy$$

Because of the presence of the square root sign in Formulas 1, 2, and 3, the calculation of an arc length often leads to an integral that is very difficult or even impossible to evaluate explicitly. Thus, we often have to be content with finding an approximation to the length of a curve as in the following example.

EXAMPLE 2 Estimate the length of the portion of the hyperbola $xy = 1$ from the point $(1, 1)$ to the point $(2, \frac{1}{2})$.

SOLUTION We have

$$y = \frac{1}{x} \qquad \frac{dy}{dx} = -\frac{1}{x^2}$$

and so, from Formula 2, the length is

$$L = \int_1^2 \sqrt{1 + \left(\frac{dy}{dx}\right)^2}\, dx = \int_1^2 \sqrt{1 + \frac{1}{x^4}}\, dx$$

It is impossible to evaluate this integral exactly, so let's use Simpson's Rule (see Section 8.8) with $a = 1$, $b = 2$, $n = 10$, $\Delta x = 0.1$, and $f(x) = \sqrt{1 + 1/x^4}$. Thus

$$L = \int_1^2 \sqrt{1 + \frac{1}{x^4}}\, dx$$

$$\approx \frac{\Delta x}{3}[f(1) + 4f(1.1) + 2f(1.2) + 4f(1.3) + \cdots + 2f(1.8) + 4f(1.9) + f(2)]$$

$$\approx 1.1321$$

Checking the value of the definite integral with a more accurate approximation produced by a computer algebra system, we see that the approximation using Simpson's Rule is accurate to four decimal places. ■

EXAMPLE 3 Find the length of the arc of the parabola $y^2 = x$ from $(0, 0)$ to $(1, 1)$.

SOLUTION Since $x = y^2$, we have $dx/dy = 2y$, and Formula 3 gives

$$L = \int_0^1 \sqrt{\left(\frac{dx}{dy}\right)^2 + 1}\, dy = \int_0^1 \sqrt{4y^2 + 1}\, dy$$

Using either a computer algebra system or the Table of Integrals (use Formula 21 after substituting $u = 2y$), we find that

$$L = \frac{\sqrt{5}}{2} + \frac{\ln(\sqrt{5} + 2)}{4}$$ ■

Figure 6 shows the arc of the parabola whose length is computed in Example 2, together with polygonal approximations having $n = 1$ and $n = 2$ line segments. For $n = 1$ the approximate length is $L_1 = \sqrt{2}$, the diagonal of a square. The table shows the approximations L_n that we get by dividing $[0, 1]$ into n equal subintervals. Notice that each time we double the number of sides of the polygon, we get closer to the exact length, which is

$$L = \frac{\sqrt{5}}{2} + \frac{\ln(\sqrt{5} + 2)}{4} \approx 1.478943$$

n	L_n
1	1.414
2	1.445
4	1.464
8	1.472
16	1.476
32	1.478
64	1.479

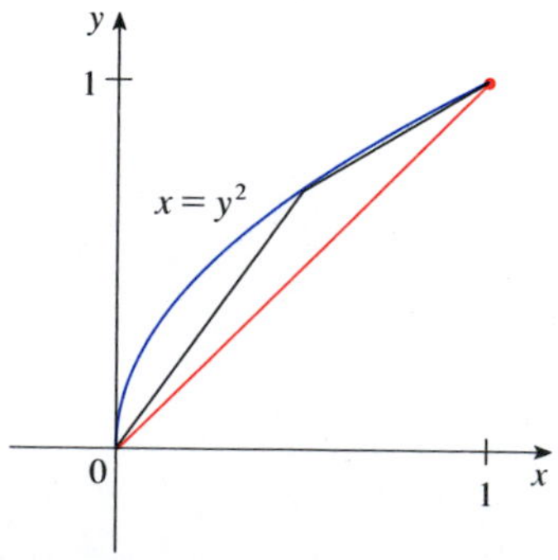

FIGURE 6

EXAMPLE 4 Find the length of one arch of the cycloid $x = r(\theta - \sin\theta)$, $y = r(1 - \cos\theta)$.

SOLUTION From Example 3 in Section 3.9 we see that one arch is described by the parameter interval $0 \leqslant \theta \leqslant 2\pi$. Since

$$\frac{dx}{d\theta} = r(1 - \cos\theta) \qquad \text{and} \qquad \frac{dy}{d\theta} = r\sin\theta$$

we have

$$L = \int_0^{2\pi} \sqrt{\left(\frac{dx}{d\theta}\right)^2 + \left(\frac{dy}{d\theta}\right)^2}\, d\theta = \int_0^{2\pi} \sqrt{r^2(1 - \cos\theta)^2 + r^2\sin^2\theta}\, d\theta$$

$$= \int_0^{2\pi} \sqrt{r^2(1 - 2\cos\theta + \cos^2\theta + \sin^2\theta)}\, d\theta = r\int_0^{2\pi} \sqrt{2(1 - \cos\theta)}\, d\theta$$

This integral could be evaluated after using further trigonometric identities. Instead we use a computer algebra system:

$$L = r\int_0^{2\pi} \sqrt{2(1 - \cos\theta)}\, d\theta = 8r$$

■

The result of Example 4 says that the length of one arch of a cycloid is eight times the radius of the generating circle (see Figure 7). This was first proved in 1658 by Sir Christopher Wren, who later became the architect of St. Paul's Cathedral in London.

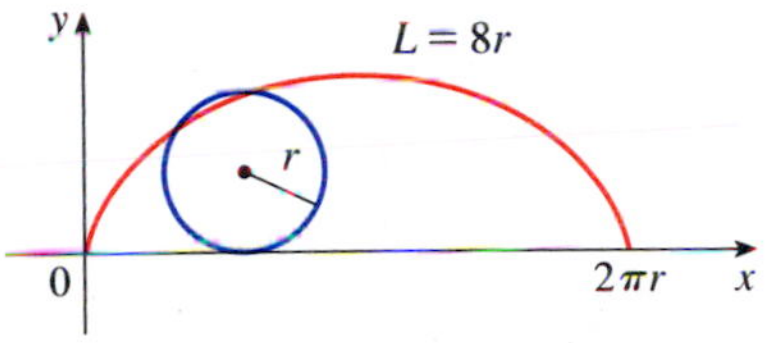

FIGURE 7

THE ARC LENGTH FUNCTION

If a smooth curve C has the equation $y = f(x)$, $a \leqslant x \leqslant b$, let $s(x)$ be the distance along C from the initial point $P_0(a, f(a))$ to the point $Q(x, f(x))$. Then s is a function, called the **arc length function**, and, by Formula 2,

$$s(x) = \int_a^x \sqrt{1 + [f'(t)]^2}\, dt \tag{4}$$

(We have replaced the dummy variable of integration by t so that x does not have two meanings.) We can use Part 1 of the Fundamental Theorem of Calculus to differentiate Equation 4 (since the integrand is continuous):

$$\frac{ds}{dx} = \sqrt{1 + [f'(x)]^2} = \sqrt{1 + \left(\frac{dy}{dx}\right)^2} \tag{5}$$

Equation 5 shows that the rate of change of s with respect to x is always at least 1 and is equal to 1 when $f'(x)$, the slope of the curve, is 0. The differential of arc length is

$$ds = \sqrt{1 + \left(\frac{dy}{dx}\right)^2}\, dx \tag{6}$$

and this equation is sometimes written in the symmetric form

$$(ds)^2 = (dx)^2 + (dy)^2 \tag{7}$$

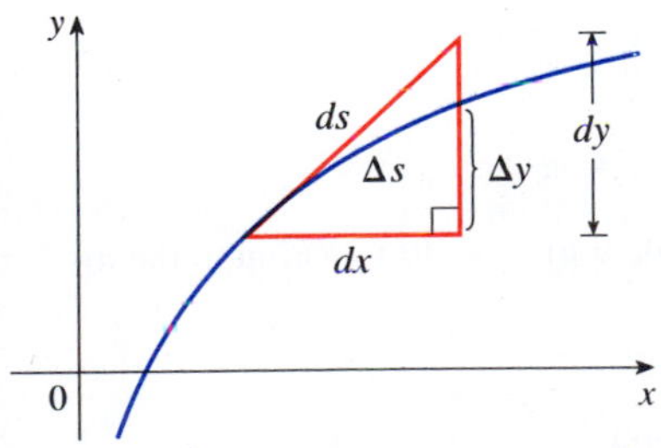

FIGURE 8

The geometric interpretation of Equation 7 is shown in Figure 8. It can be used as a mnemonic device for remembering Formula 1 (and Formulas 2 and 3). If we write $L = \int ds$, then from Equation 7 either we can solve to get (6), which gives (2), or we can solve to get

$$ds = \sqrt{1 + \left(\frac{dx}{dy}\right)^2}\, dy$$

which gives (3).

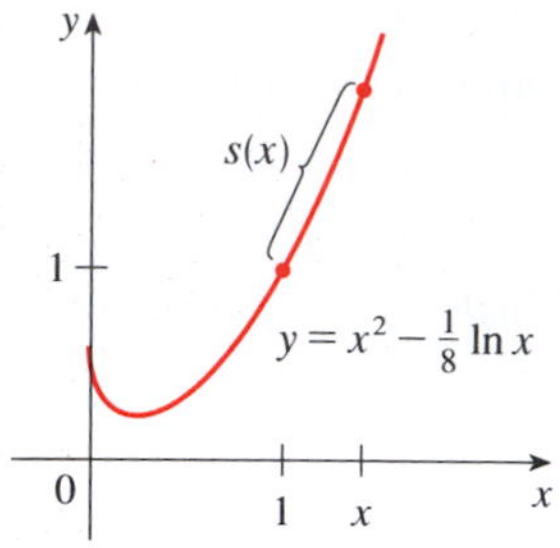

FIGURE 9

Figure 9 shows the interpretation of the arc length function in Example 5. Figure 10 shows the graph of this arc length function.

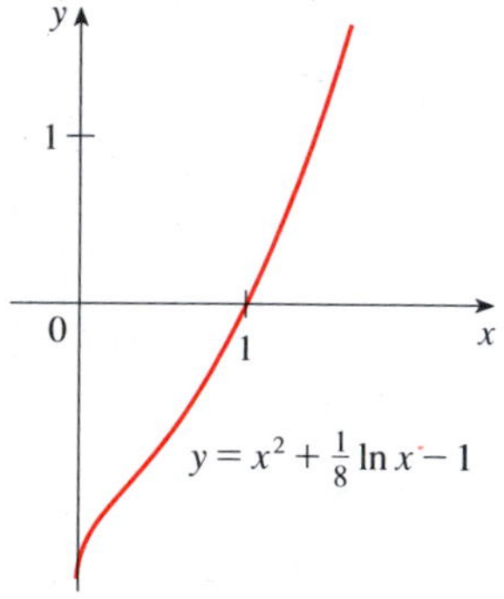

FIGURE 10

EXAMPLE 5 Find the arc length function for the curve $y = x^2 - (\ln x)/8$ taking $P_0(1, 1)$ as the starting point.

SOLUTION If $f(x) = x^2 - (\ln x)/8$, then

$$f'(x) = 2x - \frac{1}{8x}$$

$$1 + [f'(x)]^2 = 1 + \left(2x - \frac{1}{8x}\right)^2 = 1 + 4x^2 - \frac{1}{2} + \frac{1}{64x^2}$$

$$= 4x^2 + \frac{1}{2} + \frac{1}{64x^2} = \left(2x + \frac{1}{8x}\right)^2$$

$$\sqrt{1 + [f'(x)]^2} = 2x + \frac{1}{8x}$$

Thus, the arc length function is given by

$$s(x) = \int_1^x \sqrt{1 + [f'(t)]^2}\, dt$$

$$= \int_1^x \left(2t - \frac{1}{8t}\right) dt = t^2 + \tfrac{1}{8}\ln t\,\Big]_1^x$$

$$= x^2 + \tfrac{1}{8}\ln x - 1$$

■

EXERCISES 9.3

1. Use the arc length formula (2) to find the length of the curve $y = 2x + 1$, $-1 \leq x \leq 3$. Check your answer by noting that the curve is a line segment and calculating its length by the distance formula.

2. (a) In Section 1.3 we showed that the parametric equations $x = \cos t$, $y = \sin t$, $0 \leq t \leq 2\pi$, represent the unit circle. Use these equations to show that the length of the unit circle has the expected value.
   (b) In Section 1.3 we showed that the equations $x = \sin 2t$, $y = \cos 2t$, $0 \leq t \leq 2\pi$, also represent the unit circle. What value does the integral in Formula 1 give? How do you explain the discrepancy?

3–6 ■ Graph the curve and find its exact length.

3. $x = e^t \cos t$, $y = e^t \sin t$, $0 \leq t \leq \pi$

4. $x = 3t - t^3$, $y = 3t^2$, $0 \leq t \leq 2$

5. $x = y^{3/2}$, $0 \leq y \leq 1$

6. $x = a(\cos\theta + \theta\sin\theta)$, $y = a(\sin\theta - \theta\cos\theta)$, $0 \leq \theta \leq \pi$

7–10 ■ Find the length of the curve.

7. $y = \frac{1}{3}(x^2 + 2)^{3/2}$, $0 \leq x \leq 1$

8. $y = \dfrac{x^3}{6} + \dfrac{1}{2x}$, $1 \leq x \leq 2$

9. $y = \dfrac{x^4}{4} + \dfrac{1}{8x^2}$, $1 \leq x \leq 3$

10. $y = \ln(\sin x)$, $\pi/6 \leq x \leq \pi/3$

11–13 ■ Use Simpson's Rule with $n = 10$ to estimate the arc length of the curve.

11. $y = x^3$, $0 \leq x \leq 1$

12. $y = \tan x$, $0 \leq x \leq \pi/4$

13. $y = \sin x$, $0 \leq x \leq \pi$

14. The curve given by the parametric equations $x = 2a \cot\theta$, $y = 2a \sin^2\theta$ is called the witch of Maria Agnesi. Graph this curve and use Simpson's Rule with $n = 4$ to estimate the length of the arc of this curve given by the parameter interval $\pi/4 \leqslant \theta \leqslant \pi/2$.

15. (a) Graph the curve $y = x\sqrt[3]{4 - x}$, $0 \leqslant x \leqslant 4$.
(b) Compute the lengths of inscribed polygons with $n = 1$, 2, and 4 sides. (Divide the interval into equal subintervals.) Illustrate by sketching these polygons (as in Figure 6).
(c) Set up an integral for the length of the curve.
(d) If your calculator (or CAS) evaluates definite integrals, use it to find the length of the curve to four decimal places. If not, use Simpson's Rule. Compare with the approximations in part (b).

16. Repeat Exercise 15 for the curve $y = x + \sin x$, $0 \leqslant x \leqslant 2\pi$.

CAS **17–19** ■ Use either a CAS or a table of integrals to find the exact length of the curve.

17. $x = t^3$, $y = t^4$, $0 \leqslant t \leqslant 1$

18. $x = \ln(1 - y^2)$, $0 \leqslant y \leqslant \frac{1}{2}$

19. $y = \ln(\cos x)$, $0 \leqslant x \leqslant \pi/4$

20. Find the arc length function for the curve $y = 2x^{3/2}$ with starting point $P_0(1, 2)$.

21. (a) Graph the curve $y = x^3/3 + 1/(4x)$, $x > 0$.
(b) Find the arc length function for this curve with starting point $P_0(1, \frac{7}{12})$.
(c) Graph the arc length function.

22. If a bomb is dropped from an aircraft flying at 200 m/s at an altitude of 4500 m, then the parabolic trajectory of the bomb is described by the equation

$$y = 4500 - \frac{x^2}{8000}$$

until it hits the ground, where y is its height above the ground and x is the horizontal distance traveled in meters. Calculate the distance traveled by the bomb from the time it is dropped to the time it hits the ground. Express your answer correct to the nearest meter.

23. (a) The figure shows a telephone wire hanging between two poles at $x = -b$ and $x = b$. It takes the shape of a catenary with equation $y = a\cosh(x/a)$. Find the length of the wire.

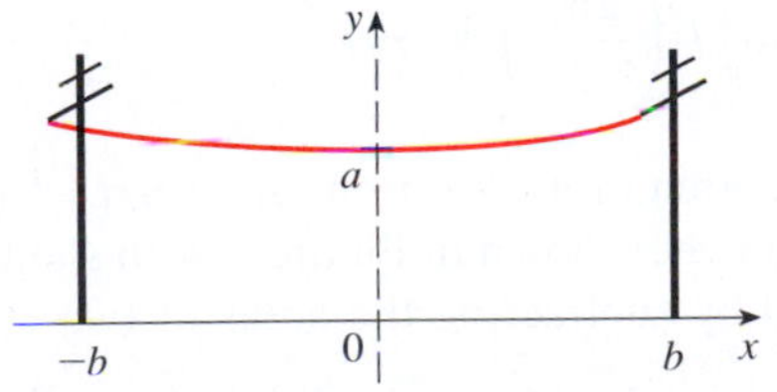

(b) Suppose two telephone poles are 50 ft apart and the length of the wire between the poles is 56 ft. How far above the ground is the lowest point of the wire?

CAS **24.** Use either a computer algebra system or a table of integrals to find the exact length of the arc of the exponential curve $y = e^x$ that lies between the points $(0, 1)$ and $(1, e)$. If your CAS has trouble evaluating the integral, make a substitution that changes the integral into one that the CAS can evaluate.

25. A manufacturer of corrugated metal roofing wants to produce panels that are 28 in. wide and 2 in. thick by processing flat sheets of metal as shown in the figure. The profile of the roofing takes the shape of a sine wave. Verify that the sine curve has equation $y = \sin(\pi x/7)$ and find the width w of a flat metal sheet that is needed to make a 28-inch panel. (If your calculator or CAS evaluates definite integrals, use it. Otherwise, use Simpson's Rule.)

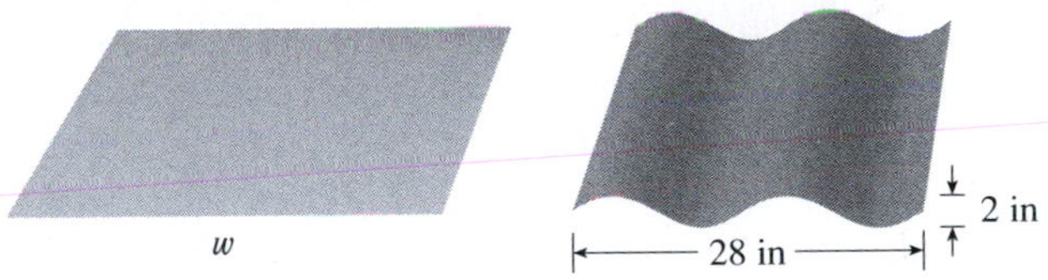

26. Find the total length of the astroid

$$x = a\cos^3\theta \qquad y = a\sin^3\theta$$

27. Show that the total length of the ellipse $x = a\sin\theta$, $y = b\cos\theta$, $a > b > 0$, is

$$L = 4a\int_0^{\pi/2} \sqrt{1 - e^2\sin^2\theta}\, d\theta$$

where e is the eccentricity of the ellipse ($e = c/a$, where $c = \sqrt{a^2 - b^2}$).

28. The curves with equations $x^n + y^n = 1$, $n = 4, 6, 8, \ldots$, are called **fat circles**. Graph the curves with $n = 2, 4, 6$, 8, and 10 to see why. Set up an integral for the length L_{2k} of the fat circle with $n = 2k$. Without attempting to evaluate this integral, state the value of

$$\lim_{k \to \infty} L_{2k}$$

CAS **29.** (a) Graph the epitrochoid with equations

$$x = 11\cos t - 4\cos(11t/2)$$

$$y = 11\sin t - 4\sin(11t/2)$$

What parameter interval gives the complete curve?
(b) Use your CAS to find the approximate length of this curve.

CAS **30.** A curve called **Cornu's spiral** is defined by the parametric equations

$$x = C(t) = \int_0^t \cos\left(\frac{\pi u^2}{2}\right) du$$

$$y = S(t) = \int_0^t \sin\left(\frac{\pi u^2}{2}\right) du$$

where C and S are the Fresnel functions that were introduced in Section 6.4.

(a) Graph this curve. What happens as $t \to \infty$ and as $t \to -\infty$?

(b) Find the length of Cornu's spiral from the origin to the point with parameter value t.

9.4 AREA OF A SURFACE OF REVOLUTION

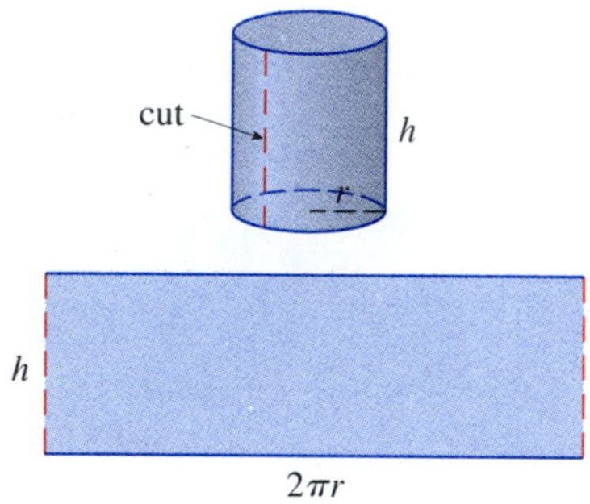

FIGURE 1

A surface of revolution is formed when a curve is rotated about a line. Such a surface is the lateral boundary of a solid of revolution of the type discussed in Sections 7.2 and 7.3.

We want to define the area of a surface of revolution in such a way that it corresponds to our intuition. We can think of peeling away a very thin outer layer of the solid of revolution and laying it out flat so that we can measure its area. Or, if the surface area is A, we can imagine that painting the surface would require the same amount of paint as does a flat region with area A.

Let's start with some simple surfaces. The lateral surface area of a circular cylinder with radius r and height h is taken to be $A = 2\pi rh$ because we can imagine cutting the cylinder and unrolling it (as in Figure 1) to obtain a rectangle with dimensions $2\pi r$ and h.

Likewise, we can take a circular cone with base radius r and slant height l, cut it along the broken line in Figure 2, and flatten it to form a sector of a circle with radius l and central angle $\theta = 2\pi r/l$.

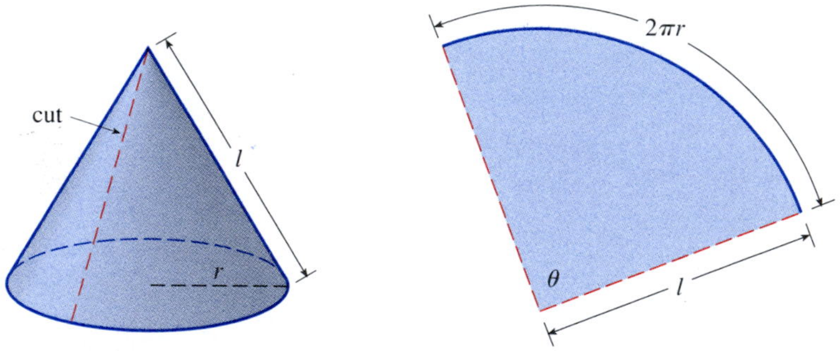

FIGURE 2

We know that, in general, the area of a sector of a circle with radius l and angle θ is $\frac{1}{2}l^2\theta$ (see Exercise 31 in Section 8.3) and so in this case it is

$$A = \tfrac{1}{2}l^2\theta = \tfrac{1}{2}l^2\left(\frac{2\pi r}{l}\right) = \pi rl$$

Therefore, we define the lateral surface area of a cone to be $A = \pi rl$.

The area of the band (or frustum of a cone) shown in Figure 3 with slant height l and upper and lower radii r_1 and r_2 is found by subtracting the areas of two cones:

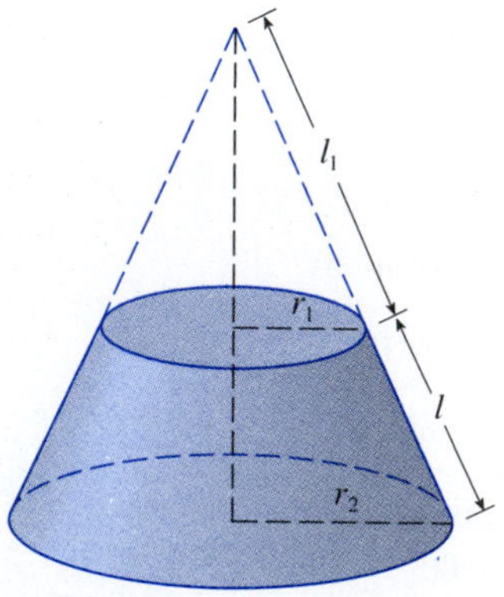

FIGURE 3

(1) $$A = \pi r_2(l_1 + l) - \pi r_1 l_1 = \pi[(r_2 - r_1)l_1 + r_2 l]$$

From similar triangles we have

$$\frac{l_1}{r_1} = \frac{l_1 + l}{r_2}$$

which gives

$$r_2 l_1 = r_1 l_1 + r_1 l \qquad \text{or} \qquad (r_2 - r_1) l_1 = r_1 l$$

Putting this in Equation 1, we get

$$A = \pi(r_1 l + r_2 l)$$

or

(2) $$A = 2\pi r l$$

where $r = \frac{1}{2}(r_1 + r_2)$ is the average radius of the band.

Now we consider the surface shown in Figure 4, which is obtained by rotating the curve $y = f(x)$, $a \leqslant x \leqslant b$, about the x-axis, where f is positive and has a continuous derivative. In order to define its surface area, we use a method similar to the one for arc length. We take a partition P of $[a, b]$ by points $a = x_0, x_1, \ldots, x_n = b$, and let $y_i = f(x_i)$ so that the point $P_i(x_i, y_i)$ lies on the curve. The part of the surface between x_{i-1} and x_i is approximated by taking the line segment $P_{i-1}P_i$ and rotating it about the x-axis. The result is a band (a frustum of a cone) with slant height $l = |P_{i-1}P_i|$ and average radius $r = \frac{1}{2}(y_{i-1} + y_i)$ so, by Formula 2, its surface area is

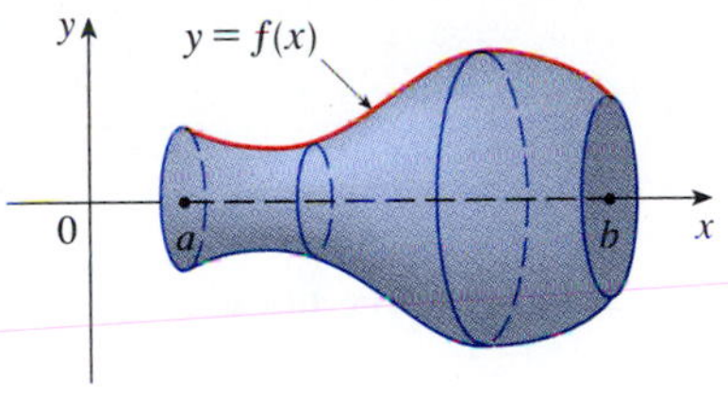

(a) Surface of revolution

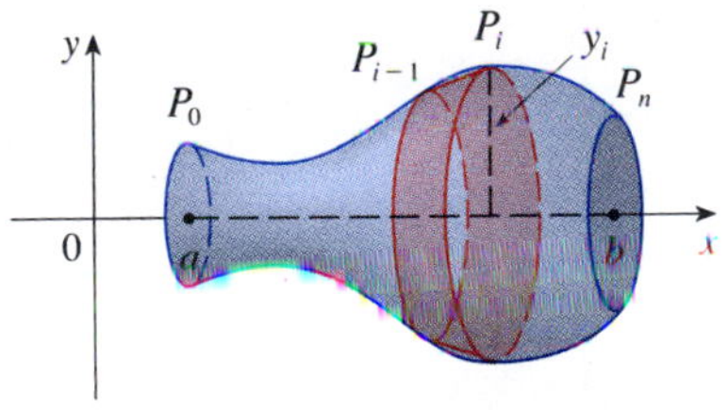

(b) Approximating band

FIGURE 4

$$2\pi \frac{y_{i-1} + y_i}{2} |P_{i-1}P_i|$$

As in our discussion of arc length in Section 9.3, we have

$$\begin{aligned} |P_{i-1}P_i| &= \sqrt{(\Delta x_i)^2 + (\Delta y_i)^2} \\ &\approx \sqrt{(\Delta x_i)^2 + [f'(x_i^*)\,\Delta x_i]^2} \\ &= \sqrt{1 + [f'(x_i^*)]^2}\;\Delta x_i \end{aligned}$$

where $x_i^* \in [x_{i-1}, x_i]$. When Δx_i is small we have $y_i = f(x_i) \approx f(x_i^*)$ and also $y_{i-1} = f(x_{i-1}) \approx f(x_i^*)$, since f is continuous. Therefore

$$2\pi \frac{y_{i-1} + y_i}{2} |P_{i-1}P_i| \approx 2\pi f(x_i^*) \sqrt{1 + [f'(x_i^*)]^2}\;\Delta x_i$$

and so an approximation to what we think of as the area of the complete surface of revolution is

(3) $$\sum_{i=1}^{n} 2\pi f(x_i^*) \sqrt{1 + [f'(x_i^*)]^2}\;\Delta x_i$$

This approximation appears to become better as $\|P\| \to 0$ and, recognizing (3) as a Riemann sum for the function $g(x) = 2\pi f(x)\sqrt{1 + [f'(x)]^2}$, we have

$$\lim_{\|P\| \to 0} \sum_{i=1}^{n} 2\pi f(x_i^*) \sqrt{1 + [f'(x_i^*)]^2}\;\Delta x_i = \int_a^b 2\pi f(x) \sqrt{1 + [f'(x)]^2}\,dx$$

Therefore, in the case where f is positive and has a continuous derivative, we define the **surface area** of the surface obtained by rotating the curve $y = f(x)$, $a \leqslant x \leqslant b$, about the x-axis as

(4)
$$S = \int_a^b 2\pi f(x)\sqrt{1 + [f'(x)]^2}\,dx$$

With the Leibniz notation for derivatives, this formula becomes

(5)
$$S = \int_a^b 2\pi y \sqrt{1 + \left(\frac{dy}{dx}\right)^2}\,dx$$

If the curve is described as $x = g(y)$, $c \leqslant y \leqslant d$, then the formula for surface area becomes

(6)
$$S = \int_c^d 2\pi y \sqrt{1 + \left(\frac{dx}{dy}\right)^2}\,dy$$

and both Formulas 5 and 6 can be summarized symbolically, using the notation for arc length given in Section 9.3, as

(7)
$$S = \int 2\pi y\,ds$$

For rotation about the y-axis, the surface area formula becomes

(8)
$$S = \int 2\pi x\,ds$$

where, as before, we can use either

$$ds = \sqrt{1 + \left(\frac{dy}{dx}\right)^2}\,dx \qquad \text{or} \qquad ds = \sqrt{1 + \left(\frac{dx}{dy}\right)^2}\,dy$$

These formulas can be remembered by thinking of $2\pi y$ or $2\pi x$ as the circumference of a circle traced out by the point (x, y) on the curve as it is rotated about the x-axis or y-axis, respectively (see Figure 5).

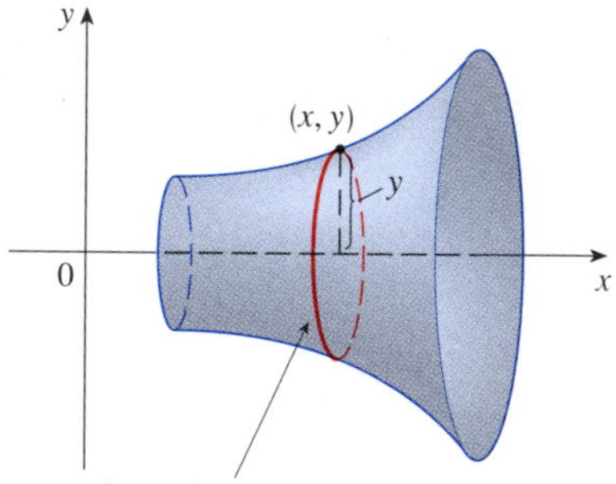

(a) Rotation about x-axis: $S = \int 2\pi y\,ds$

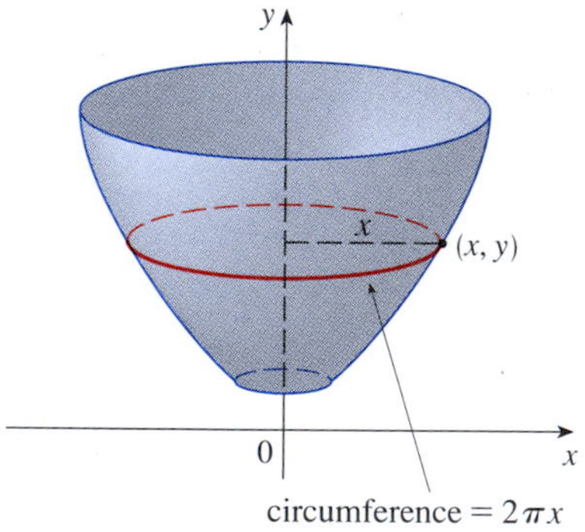

(b) Rotation about y-axis: $S = \int 2\pi x\,ds$

FIGURE 5

EXAMPLE 1 The arc of the parabola $y = x^2$ from $(1, 1)$ to $(2, 4)$ is rotated about the y-axis. Find the area of the resulting surface.

SOLUTION 1 Using

$$y = x^2 \qquad \text{and} \qquad \frac{dy}{dx} = 2x$$

Figure 6 shows the surface of revolution whose area is computed in Example 1.

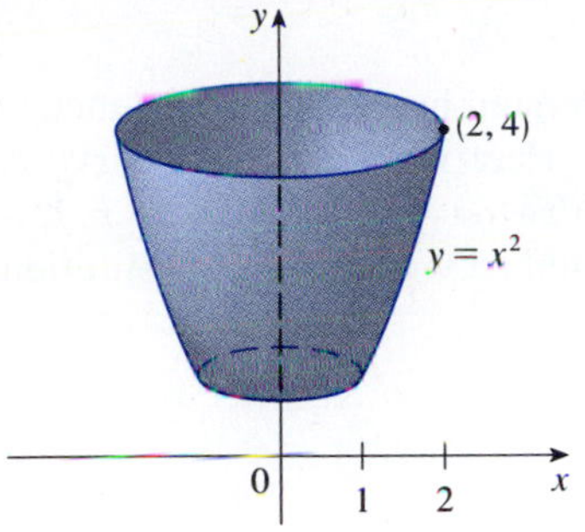

FIGURE 6

we have, from Formula 8,

$$S = \int 2\pi x\,ds = \int_1^2 2\pi x \sqrt{1 + \left(\frac{dy}{dx}\right)^2}\,dx$$

$$= 2\pi \int_1^2 x\sqrt{1 + 4x^2}\,dx$$

Substituting $u = 1 + 4x^2$, we have $du = 8x\,dx$. Remembering to change the limits of integration, we have

$$S = \frac{\pi}{4}\int_5^{17} \sqrt{u}\,du = \frac{\pi}{4}\left[\tfrac{2}{3}u^{3/2}\right]_5^{17}$$

$$= \frac{\pi}{6}(17\sqrt{17} - 5\sqrt{5})$$

SOLUTION 2 Using

$$x = \sqrt{y} \qquad \text{and} \qquad \frac{dx}{dy} = \frac{1}{2\sqrt{y}}$$

we have

$$S = \int 2\pi x\,ds = \int_1^4 2\pi x\sqrt{1 + \left(\frac{dx}{dy}\right)^2}\,dy$$

$$= 2\pi \int_1^4 \sqrt{y}\sqrt{1 + \frac{1}{4y}}\,dy = \pi \int_1^4 \sqrt{4y + 1}\,dy$$

$$= \frac{\pi}{4}\int_5^{17} \sqrt{u}\,du \qquad \text{(where } u = 1 + 4y\text{)}$$

$$= \frac{\pi}{6}(17\sqrt{17} - 5\sqrt{5}) \qquad \text{(as in Solution 1)}$$

■

EXAMPLE 2 Find the area of the surface generated by rotating the curve $y = e^x$, $0 \leqslant x \leqslant 1$, about the x-axis.

Another method: Use Formula 6 with $x = \ln y$.

SOLUTION Using Formula 5 with

$$y = e^x \qquad \text{and} \qquad \frac{dy}{dx} = e^x$$

we have

$$S = \int_0^1 2\pi y\sqrt{1 + \left(\frac{dy}{dx}\right)^2}\,dx = 2\pi \int_0^1 e^x\sqrt{1 + e^{2x}}\,dx$$

$$= 2\pi \int_1^e \sqrt{1 + u^2}\,du \qquad \text{(where } u = e^x\text{)}$$

$$= 2\pi \int_{\pi/4}^{\alpha} \sec^3\theta\,d\theta \qquad \text{(where } u = \tan\theta \text{ and } \alpha = \tan^{-1}e\text{)}$$

Or use Formula 21 in the Table of Integrals.

$$= 2\pi \cdot \tfrac{1}{2}\left[\sec\theta\tan\theta + \ln|\sec\theta + \tan\theta|\right]_{\pi/4}^{\alpha} \qquad \text{(by Example 9 in Section 8.2)}$$

$$= \pi\left[\sec\alpha\tan\alpha + \ln(\sec\alpha + \tan\alpha) - \sqrt{2} - \ln(\sqrt{2} + 1)\right]$$

Since $\tan\alpha = e$, we have $\sec^2\alpha = 1 + \tan^2\alpha = 1 + e^2$ and

$$S = \pi\left[e\sqrt{1 + e^2} + \ln\left(e + \sqrt{1 + e^2}\right) - \sqrt{2} - \ln\left(\sqrt{2} + 1\right)\right]$$ ■

Formula 5 can be adapted to the case of a surface generated by rotating a parametric curve $x = x(t)$, $y = y(t)$, $\alpha \leqslant t \leqslant \beta$, about the x-axis. Here we assume the curve is traversed from left to right as t increases from α to β with $x(\alpha) = a$ and $x(\beta) = b$. Using (5) together with $dy/dx = (dy/dt)/(dx/dt)$ (from Formula 3.9.1) and the Substitution Rule, we have

$$S = \int_a^b 2\pi y \sqrt{1 + \left(\frac{dy}{dx}\right)^2}\, dx \qquad \text{[from (5)]}$$

$$= \int_\alpha^\beta 2\pi y(t) \sqrt{1 + \frac{(dy/dt)^2}{(dx/dt)^2}}\, \frac{dx}{dt}\, dt$$

or

(9) $$S = \int_\alpha^\beta 2\pi y(t) \sqrt{\left(\frac{dx}{dt}\right)^2 + \left(\frac{dy}{dt}\right)^2}\, dt$$

In a similar manner, if this curve is rotated about the y-axis, then

(10) $$S = \int_\alpha^\beta 2\pi x(t) \sqrt{\left(\frac{dx}{dt}\right)^2 + \left(\frac{dy}{dt}\right)^2}\, dt$$

Formulas 9 and 10 can be summarized as $S = \int 2\pi y\, ds$ and $S = \int 2\pi x\, ds$ where

$$ds = \sqrt{\left(\frac{dx}{dt}\right)^2 + \left(\frac{dy}{dt}\right)^2}\, dt$$

Although these formulas have been derived from Formula 5, they hold whenever a parametric curve is traversed exactly once as t increases from α to β.

EXAMPLE 3 Show that the surface area of a sphere of radius r is $4\pi r^2$.

SOLUTION The sphere is obtained by rotating the semicircle

$$x = r\cos t \qquad y = r\sin t \qquad 0 \leqslant t \leqslant \pi$$

about the x-axis. Therefore, from Formula 9, we get

$$S = \int_0^\pi 2\pi r\sin t\sqrt{(-r\sin t)^2 + (r\cos t)^2}\, dt$$

$$= 2\pi \int_0^\pi r\sin t\sqrt{r^2(\sin^2 t + \cos^2 t)}\, dt$$

$$= 2\pi r^2 \int_0^\pi \sin t\, dt = 2\pi r^2(-\cos t)\Big]_0^\pi = 4\pi r^2$$ ■

EXAMPLE 4 Find the area of the surface generated by rotating one arch of the cycloid $x = r(\theta - \sin\theta)$, $y = r(1 - \cos\theta)$ about the x-axis.

Figure 7 shows the surface whose area is computed in Example 4. This problem was solved by Blaise Pascal (1623–1662) without using the Fundamental Theorem of Calculus.

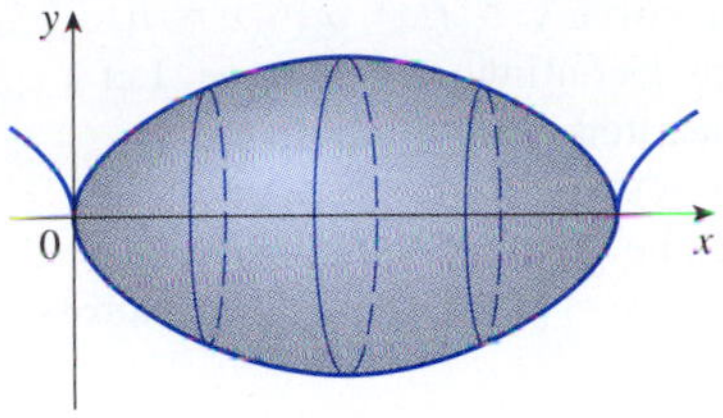

FIGURE 7

SOLUTION Using Formula 9, we have

$$S = \int_0^{2\pi} 2\pi y \sqrt{\left(\frac{dx}{d\theta}\right)^2 + \left(\frac{dy}{d\theta}\right)^2}\, d\theta$$

$$= \int_0^{2\pi} 2\pi r(1 - \cos\theta)\sqrt{r^2(1 - \cos\theta)^2 + r^2\sin^2\theta}\, d\theta$$

$$= 2\pi r^2 \int_0^{2\pi} (1 - \cos\theta)\sqrt{2(1 - \cos\theta)}\, d\theta$$

This integral can be computed using trigonometric identities or a computer algebra system to obtain

$$S = \frac{64\pi r^2}{3}$$

■

EXERCISES 9.4

1–10 ■ Find the area of the surface obtained by rotating the curve about the x-axis.

1. $y = \sqrt{x},\ 4 \leqslant x \leqslant 9$

2. $y^2 = 4x + 4,\ 0 \leqslant x \leqslant 8$

3. $y = x^3,\ 0 \leqslant x \leqslant 2$

4. $y = \dfrac{x^2}{4} - \dfrac{\ln x}{2},\ 1 \leqslant x \leqslant 4$

5. $y = \sin x,\ 0 \leqslant x \leqslant \pi$

6. $y = \cos x,\ 0 \leqslant x \leqslant \pi/3$

7. $y = \cosh x,\ 0 \leqslant x \leqslant 1$

8. $2y = 3x^{2/3},\ 1 \leqslant x \leqslant 8$

9. $x = \frac{1}{3}(y^2 + 2)^{3/2},\ 1 \leqslant y \leqslant 2$

10. $x = 1 + 2y^2,\ 1 \leqslant y \leqslant 2$

11–16 ■ The given curve is rotated about the y-axis. Find the area of the resulting surface.

11. $y = \sqrt[3]{x},\quad 1 \leqslant y \leqslant 2$

12. $x = \sqrt{2y - y^2},\quad 0 \leqslant y \leqslant 1$

13. $x = e^{2y},\quad 0 \leqslant y \leqslant \frac{1}{2}$

14. $y = 1 - x^2,\quad 0 \leqslant x \leqslant 1$

15. $x = \dfrac{1}{2\sqrt{2}}(y^2 - \ln y),\quad 1 \leqslant y \leqslant 2$

16. $x = a\cosh(y/a),\ -a \leqslant y \leqslant a$

17–18 ■ Use Simpson's Rule with $n = 10$ to find the area of the surface obtained by rotating the curve about the x-axis.

17. $y = x^4,\quad 0 \leqslant x \leqslant 1$

18. $y = \tan x,\quad 0 \leqslant x \leqslant \pi/4$

19. Find the surface area generated by rotating a loop of the curve $8y^2 = x^2(1 - x^2)$ about the x-axis.

20. If the infinite curve $y = e^{-x}$, $x \geqslant 0$, is rotated about the x-axis, find the area of the resulting surface.

21. If the region $\mathcal{R} = \{(x, y) \mid x \geqslant 1,\ 0 \leqslant y \leqslant 1/x\}$ is rotated about the x-axis, the volume of the resulting solid is finite (see Exercise 63 in Section 8.9). Show that the surface area is infinite. (The surface is shown in the figure and is known as **Gabriel's horn**.)

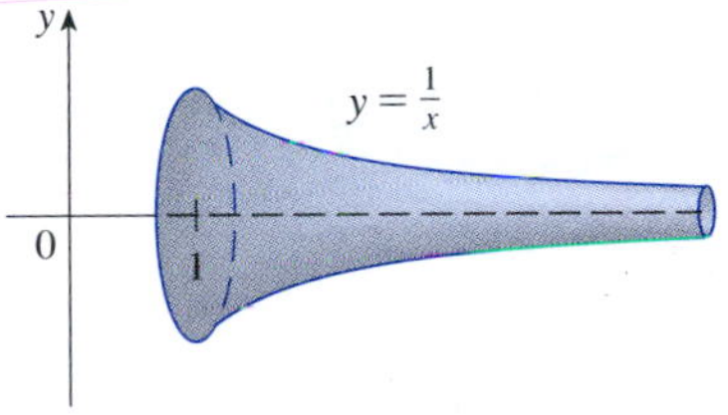

22. Find the surface area of the torus in Exercise 63 in Section 7.2.

23–25 ■ Find the area of the surface obtained by rotating the curve about the x-axis.

23. $x = t^3,\quad y = t^2,\quad 0 \leqslant t \leqslant 1$

24. $x = 3t - t^3,\quad y = 3t^2,\quad 0 \leqslant t \leqslant 1$

25. $x = a\cos^3\theta,\quad y = a\sin^3\theta,\quad 0 \leqslant \theta \leqslant \pi/2$

26. Graph the curve

$$x = 2\cos\theta - \cos 2\theta \qquad y = 2\sin\theta - \sin 2\theta$$

If this curve is rotated about the x-axis, find the area of the resulting surface. (Use your graph to help find the correct parameter interval.)

27–28 ■ Find the surface area generated by rotating the curve about the y-axis.

27. $x = 3t^2,\quad y = 2t^3,\quad 0 \leqslant t \leqslant 5$

28. $x = e^t - t,\quad y = 4e^{t/2},\quad 0 \leqslant t \leqslant 1$

29. Find the surface area of the ellipsoid obtained by rotating the ellipse $x = a\cos\theta$, $y = b\sin\theta$ $(a > b)$ about (a) the x-axis and (b) the y-axis.

30. Show that the surface area of a zone of a sphere that lies between two parallel planes is $S = \pi dh$, where d is the diameter of the sphere and h is the distance between the planes. (Notice that S depends only on the distance between the planes and not on their location, provided that both planes intersect the sphere.)

31. Formula 4 is valid only when $f(x) \geqslant 0$. Show that when $f(x)$ is not necessarily positive, the formula for surface area becomes

$$S = \int_a^b 2\pi |f(x)| \sqrt{1 + [f'(x)]^2}\, dx$$

32. If the curve $y = f(x)$, $a \leqslant x \leqslant b$, is rotated about the horizontal line $y = c$, where $f(x) \leqslant c$, find a formula for the area of the resulting surface.

33. Find the area of the surface obtained by rotating the circle $x^2 + y^2 = r^2$ about the line $y = r$.

34. Let L be the length of the curve $y = f(x)$, $a \leqslant x \leqslant b$, where f is positive and has a continuous derivative. Let S_f be the surface area generated by rotating the curve about the x-axis. If c is a positive constant, define $g(x) = f(x) + c$ and let S_g be the corresponding surface area generated by the curve $y = g(x)$, $a \leqslant x \leqslant b$. Express S_g in terms of S_f and L.

9.5 MOMENTS AND CENTERS OF MASS

The main object of this section is to find the point P on which a thin plate of any given shape balances horizontally as in Figure 1. This point is called the **center of mass** (or center of gravity) of the plate.

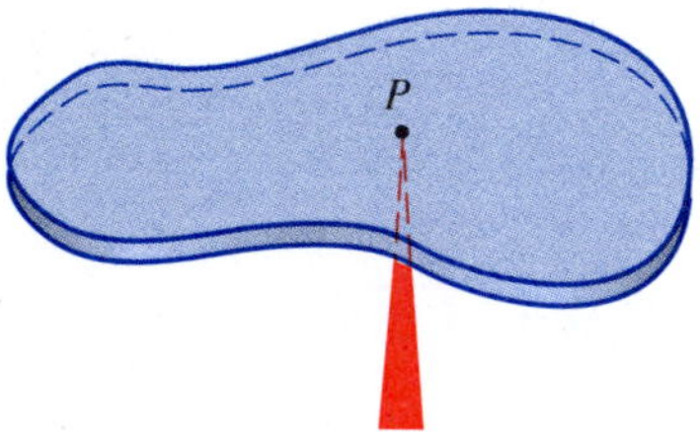

FIGURE 1

We first consider the simpler situation illustrated in Figure 2, where two masses m_1 and m_2 are attached to a rod of negligible mass on opposite sides of a fulcrum and at distances d_1 and d_2 from the fulcrum. The rod will balance if

$$m_1 d_1 = m_2 d_2 \tag{1}$$

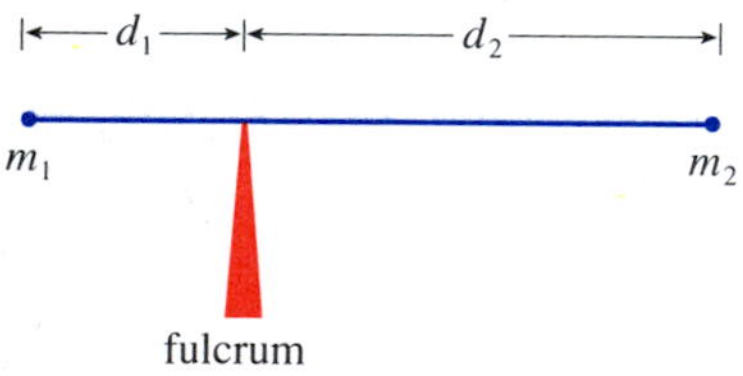

FIGURE 2

This is an experimental fact discovered by Archimedes and called the Law of the Lever. (Think of a lighter person balancing a heavier one on a seesaw by sitting farther away from the center.)

Now suppose that the rod lies along the x-axis with m_1 at x_1 and m_2 at x_2 and the center of mass at $\bar{x}$. If we compare Figures 2 and 3, we see that $d_1 = \bar{x} - x_1$ and $d_2 = x_2 - \bar{x}$ and so Equation 1 gives

$$m_1(\bar{x} - x_1) = m_2(x_2 - \bar{x})$$

$$m_1\bar{x} + m_2\bar{x} = m_1 x_1 + m_2 x_2$$

$$\bar{x} = \frac{m_1 x_1 + m_2 x_2}{m_1 + m_2} \tag{2}$$

The numbers $m_1 x_1$ and $m_2 x_2$ are called the **moments** of the masses m_1 and m_2 (with respect to the origin), and Equation 2 says that the center of mass $\bar{x}$ is obtained by adding the moments of the masses and dividing by the total mass $m = m_1 + m_2$.

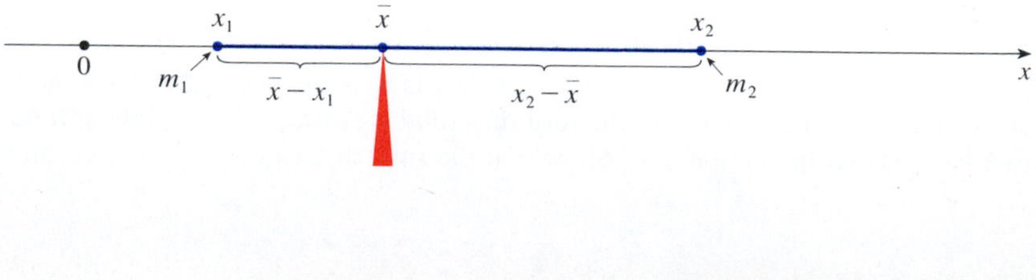

FIGURE 3

In general, if we have a system of n particles with masses $m_1, m_2, \ldots, m_n$ located at the points $x_1, x_2, \ldots, x_n$ on the x-axis, it can be shown similarly that the center of mass of the system is located at

$$\bar{x} = \frac{\sum_{i=1}^{n} m_i x_i}{\sum_{i=1}^{n} m_i} = \frac{\sum_{i=1}^{n} m_i x_i}{m} \tag{3}$$

where $m = \sum m_i$ is the total mass of the system, and the sum of the individual moments

$$M = \sum_{i=1}^{n} m_i x_i \tag{4}$$

is called the moment of the system with respect to the origin. Then Equation 3 could be rewritten as $m\bar{x} = M$, which says that if the total mass were considered as being concentrated at the center of mass $\bar{x}$, then its moment would be the same as the moment of the system.

EXAMPLE 1 Find the center of mass of a system of four objects with masses 10 g, 45 g, 32 g, and 24 g that are located at the points −4, 1, 3, and 8, respectively, on the x-axis.

SOLUTION Using Equation 3, we have

$$\bar{x} = \frac{10(-4) + 45(1) + 32(3) + 24(8)}{10 + 45 + 32 + 24}$$

$$= \tfrac{293}{111}$$ ■

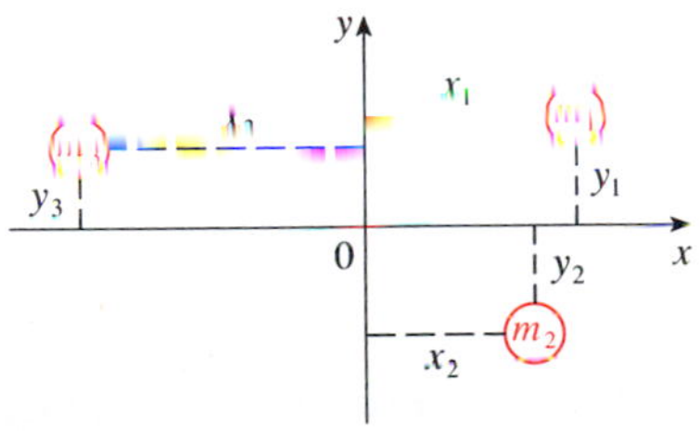

FIGURE 4

Now we consider a system of n particles with masses $m_1, m_2, \ldots, m_n$ located at the points $(x_1, y_1), (x_2, y_2), \ldots, (x_n, y_n)$ in the xy-plane as shown in Figure 4. By analogy with the one-dimensional case, we define the **moment of the system about the y-axis** to be

$$M_y = \sum_{i=1}^{n} m_i x_i \tag{5}$$

and the **moment of the system about the x-axis** as

$$M_x = \sum_{i=1}^{n} m_i y_i \tag{6}$$

Then M_y measures the tendency to rotate about the y-axis and M_x measures the tendency to rotate about the x-axis.

As in the one-dimensional case, the coordinates $(\bar{x}, \bar{y})$ of the center of mass are given in terms of the moments by the formulas

$$\bar{x} = \frac{M_y}{m} \qquad \bar{y} = \frac{M_x}{m} \tag{7}$$

where $m = \sum m_i$ is the total mass. Since $m\bar{x} = M_y$ and $m\bar{y} = M_x$, the center of mass $(\bar{x}, \bar{y})$ is the point where a single particle of mass m would have the same moments as the system.

EXAMPLE 2 Find the moments and center of mass of the system of objects that have masses 3, 4, and 8 at the points $(-1, 1)$, $(2, -1)$, and $(3, 2)$.

SOLUTION We use Equations 5 and 6 to compute the moments:

$$M_y = 3(-1) + 4(2) + 8(3) = 29$$

$$M_x = 3(1) + 4(-1) + 8(2) = 15$$

Since $m = 3 + 4 + 8 = 15$, we use Equations 7 to obtain

$$\bar{x} = \frac{M_y}{m} = \frac{29}{15} \qquad \bar{y} = \frac{M_x}{m} = \frac{15}{15} = 1$$

Thus, the center of mass is $(1\frac{14}{15}, 1)$ (see Figure 5). ■

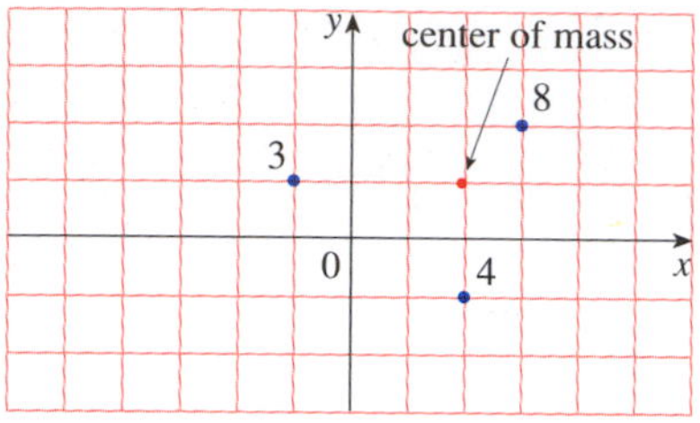

FIGURE 5

Next we consider a flat plate (called a *lamina*) with uniform density ρ that occupies a region $\mathcal{R}$ of the plane. We wish to locate the center of mass of the plate, which is called the **centroid** of $\mathcal{R}$. In doing so we use the following physical principles: The **symmetry principle** says that if $\mathcal{R}$ is symmetric about a line l, then the centroid of $\mathcal{R}$ lies on l. (If $\mathcal{R}$ is reflected about l, then $\mathcal{R}$ remains the same so its centroid remains fixed. But the only fixed points lie on l.) Thus, the centroid of a rectangle is its center. Moments should be defined so that if the entire mass of a region is concentrated at the center of mass, then its moments remain unchanged. Also, the moment of the union of two nonoverlapping regions should be the sum of the moments of the individual regions.

First we suppose that the region $\mathcal{R}$ is of the type shown in Figure 6(a); that is, $\mathcal{R}$ lies between the lines $x = a$ and $x = b$, above the x-axis, and beneath the graph of f, where f is a continuous function. We take a partition P by points x_i with $a = x_0 < x_1 < \cdots < x_n = b$ and choose x_i^* to be the midpoint of the ith subinterval, that is, $x_i^* = (x_{i-1} + x_i)/2$. This determines the polygonal approximation to $\mathcal{R}$ shown in Figure 6(b). The centroid of the ith approximating rectangle R_i is its center $C_i(x_i^*, \frac{1}{2}f(x_i^*))$. Its area is $f(x_i^*)\,\Delta x_i$ so its mass is

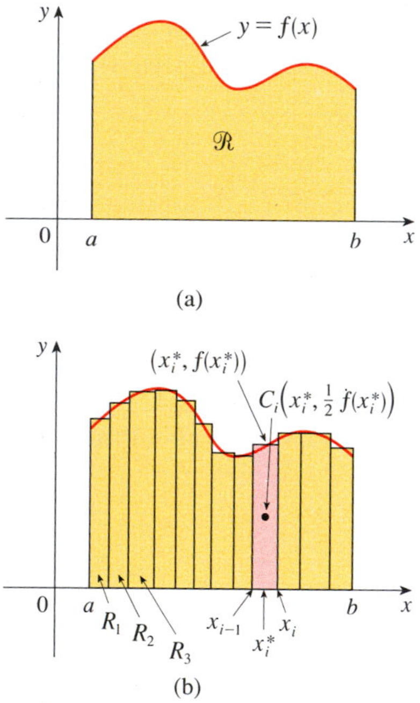

FIGURE 6

$$\rho f(x_i^*)\,\Delta x_i$$

The moment of R_i about the y-axis is the product of its mass and the distance from C_i to the y-axis, which is x_i^*. Thus

$$M_y(R_i) = [\rho f(x_i^*)\,\Delta x_i]x_i^* = \rho x_i^* f(x_i^*)\,\Delta x_i$$

Adding these moments, we obtain the moment of the polygonal approximation to $\mathcal{R}$, and then by taking the limit as $\|P\| \to 0$ we obtain the moment of $\mathcal{R}$ itself about the y-axis:

(8)
$$M_y = \lim_{\|P\| \to 0} \sum_{i=1}^{n} \rho x_i^* f(x_i^*)\,\Delta x_i = \rho \int_a^b x f(x)\,dx$$

In a similar fashion we compute the moment of R_i about the x-axis as the product of its mass and the distance from C_i to the x-axis:

$$M_x(R_i) = [\rho f(x_i^*)\,\Delta x_i]\tfrac{1}{2}f(x_i^*) = \rho \cdot \tfrac{1}{2}[f(x_i^*)]^2\,\Delta x_i$$

Again we add these moments and take the limit to obtain the moment of $\mathcal{R}$ about the x-axis:

(9)
$$M_x = \lim_{\|P\|\to 0} \sum_{i=1}^{n} \rho \cdot \tfrac{1}{2}[f(x_i^*)]^2 \Delta x_i = \rho \int_a^b \tfrac{1}{2}[f(x)]^2\, dx$$

Just as for systems of particles, the center of mass of the plate is defined so that $m\bar{x} = M_y$ and $m\bar{y} = M_x$. But the mass of the plate is the product of its density and its area:

$$m = \rho A = \rho \int_a^b f(x)\, dx$$

and so

$$\bar{x} = \frac{M_y}{m} = \frac{\rho \int_a^b xf(x)\, dx}{\rho \int_a^b f(x)\, dx} = \frac{\int_a^b xf(x)\, dx}{\int_a^b f(x)\, dx}$$

$$\bar{y} = \frac{M_x}{m} = \frac{\rho \int_a^b \frac{1}{2}[f(x)]^2\, dx}{\rho \int_a^b f(x)\, dx} = \frac{\int_a^b \frac{1}{2}[f(x)]^2\, dx}{\int_a^b f(x)\, dx}$$

Notice the cancellation of the ρ's. The location of the center of mass is independent of the density.

In summary, the center of mass of the plate (or the centroid of $\mathcal{R}$) is located at the point $(\bar{x}, \bar{y})$, where

(10)
$$\bar{x} = \frac{1}{A}\int_a^b xf(x)\, dx \qquad \bar{y} = \frac{1}{A}\int_a^b \tfrac{1}{2}[f(x)]^2\, dx$$

EXAMPLE 3 Find the center of mass of a semicircular plate of radius r.

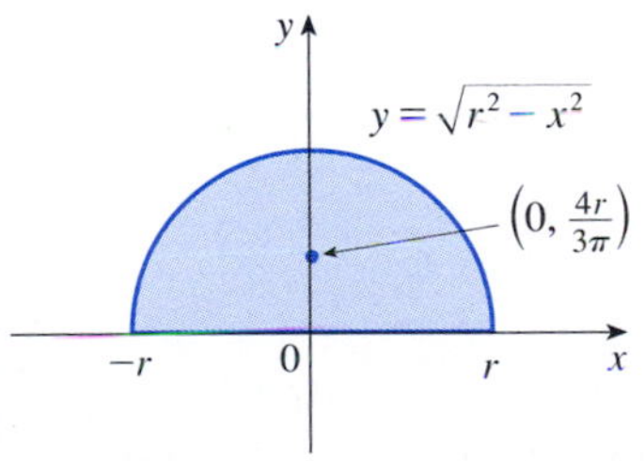

FIGURE 7

SOLUTION In order to use (10) we place the semicircle as in Figure 7 so that $f(x) = \sqrt{r^2 - x^2}$ and $a = -r$, $b = r$. Here there is no need to use the formula to calculate $\bar{x}$ because, by the symmetry principle, the center of mass must lie on the y-axis, so $\bar{x} = 0$. The area of the semicircle is $A = \pi r^2/2$, so

$$\begin{aligned}
\bar{y} &= \frac{1}{A}\int_{-r}^{r} \tfrac{1}{2}[f(x)]^2\, dx \\
&= \frac{1}{\pi r^2/2} \cdot \frac{1}{2}\int_{-r}^{r} \left(\sqrt{r^2 - x^2}\right)^2 dx \\
&= \frac{1}{\pi r^2}\int_{-r}^{r} (r^2 - x^2)\, dx = \frac{1}{\pi r^2}\left[r^2 x - \frac{x^3}{3}\right]_{-r}^{r} \\
&= \frac{1}{\pi r^2}\,\frac{4r^3}{3} = \frac{4r}{3\pi}
\end{aligned}$$

The center of mass is located at the point $(0, 4r/(3\pi))$. ■

EXAMPLE 4 Find the centroid of the region bounded by the curves $y = \cos x$, $y = 0$, $x = 0$, and $x = \pi/2$.

SOLUTION The area of the region is

$$A = \int_0^{\pi/2} \cos x\,dx = \sin x\Big]_0^{\pi/2} = 1$$

so Formulas 10 give

$$\bar{x} = \frac{1}{A}\int_0^{\pi/2} xf(x)\,dx = \int_0^{\pi/2} x\cos x\,dx$$

$$= x\sin x\Big]_0^{\pi/2} - \int_0^{\pi/2} \sin x\,dx \qquad \text{(by integration by parts)}$$

$$= \frac{\pi}{2} - 1$$

$$\bar{y} = \frac{1}{A}\int_0^{\pi/2} \tfrac{1}{2}[f(x)]^2\,dx = \tfrac{1}{2}\int_0^{\pi/2} \cos^2 x\,dx$$

$$= \tfrac{1}{4}\int_0^{\pi/2} (1 + \cos 2x)\,dx = \tfrac{1}{4}\Big[x + \tfrac{1}{2}\sin 2x\Big]_0^{\pi/2}$$

$$= \frac{\pi}{8}$$

The centroid is $((\pi/2) - 1, \pi/8)$ and is shown in Figure 8. ■

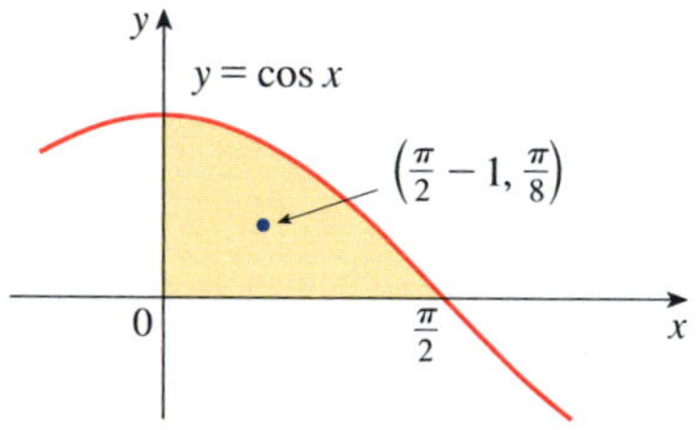

FIGURE 8

If the region $\mathcal{R}$ lies between two curves $y = f(x)$ and $y = g(x)$, where $f(x) \geqslant g(x)$, as illustrated in Figure 9, then the same sort of argument that led to Formulas 10 can be used to show that the centroid of $\mathcal{R}$ is $(\bar{x}, \bar{y})$, where

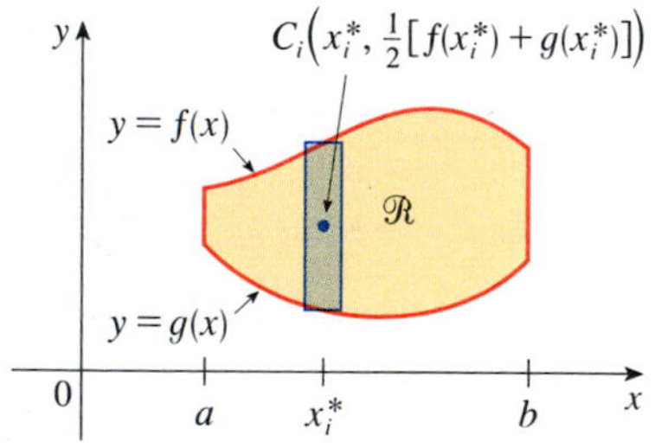

FIGURE 9

(11)

$$\bar{x} = \frac{1}{A}\int_a^b x[f(x) - g(x)]\,dx$$

$$\bar{y} = \frac{1}{A}\int_a^b \tfrac{1}{2}\{[f(x)]^2 - [g(x)]^2\}\,dx$$

(See Exercise 29.)

EXAMPLE 5 Find the centroid of the region bounded by the line $y = x$ and the parabola $y = x^2$.

SOLUTION The region is sketched in Figure 10. We take $f(x) = x$, $g(x) = x^2$, $a = 0$, and $b = 1$ in Formulas 11. First we note that the area of the region is

$$A = \int_0^1 (x - x^2)\,dx = \frac{x^2}{2} - \frac{x^3}{3}\Big]_0^1 = \frac{1}{6}$$

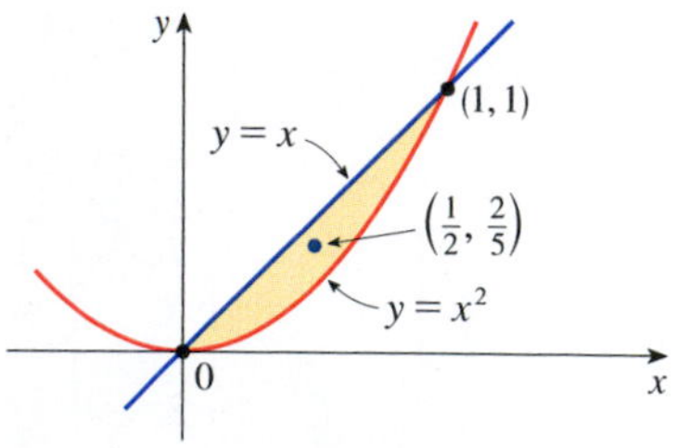

FIGURE 10

Therefore

$$\bar{x} = \frac{1}{A}\int_0^1 x[f(x) - g(x)]\,dx = \frac{1}{\frac{1}{6}}\int_0^1 x(x - x^2)\,dx$$

$$= 6\int_0^1 (x^2 - x^3)\,dx = 6\left[\frac{x^3}{3} - \frac{x^4}{4}\right]_0^1 = \frac{1}{2}$$

$$\bar{y} = \frac{1}{A}\int_0^1 \tfrac{1}{2}\{[f(x)]^2 - [g(x)]^2\}\,dx = \frac{1}{\frac{1}{6}}\int_0^1 \tfrac{1}{2}(x^2 - x^4)\,dx$$

$$= 3\left[\frac{x^3}{3} - \frac{x^5}{5}\right]_0^1 = \frac{2}{5}$$

The centroid is $(\frac{1}{2}, \frac{2}{5})$. ■

We end this section by showing how centroids can be used in finding volumes of revolution.

This theorem is named after the Greek mathematician Pappus of Alexandria, who lived in the fourth century A.D.

THEOREM OF PAPPUS Let $\mathcal{R}$ be a plane region that lies entirely on one side of a line l in the plane. If $\mathcal{R}$ is rotated about l, then the volume of the resulting solid is the product of the area A of $\mathcal{R}$ and the distance d traveled by the centroid of $\mathcal{R}$.

PROOF We give the proof for the special case in which the region lies between $y = f(x)$ and $y = g(x)$ as in Figure 9 and the line l is the y-axis. Using the method of cylindrical shells (see Section 7.3), we have

$$V = \int_a^b 2\pi x[f(x) - g(x)]\,dx$$

$$= 2\pi \int_a^b x[f(x) - g(x)]\,dx$$

$$= 2\pi(\bar{x}A) \qquad \text{(by Formulas 11)}$$

$$= (2\pi\bar{x})A = Ad$$

where $d = 2\pi\bar{x}$ is the distance traveled by the centroid during one rotation about the y-axis. □

EXAMPLE 6 A torus is formed by rotating a circle of radius r about a line in the plane of the circle that is a distance R ($> r$) from the center of the circle. Find the volume of the torus.

SOLUTION The circle has area $A = \pi r^2$. By the symmetry principle, its centroid is its center and so the distance traveled by the centroid during a rotation is $d = 2\pi R$. Therefore, by the Theorem of Pappus, the volume of the torus is

$$V = Ad = (\pi r^2)(2\pi R) = 2\pi^2 r^2 R$$

■

The method of Example 6 should be compared with the method of Exercise 63 in Section 7.2.

EXERCISES 9.5

1–4 ■ The masses m_i are located at the points P_i. Find the moments M_x and M_y and the center of mass of the system.

1. $m_1 = 4$, $m_2 = 8$; $P_1(-1, 2)$, $P_2(2, 4)$

2. $m_1 = 2$, $m_2 = 3$, $m_3 = 1$; $P_1(5, 1)$, $P_2(3, -2)$, $P_3(-2, 4)$

3. $m_1 = 4$, $m_2 = 2$, $m_3 = 5$;
$P_1(-1, -2)$, $P_2(-2, 4)$, $P_3(5, -3)$

4. $m_1 = 3$, $m_2 = 3$, $m_3 = 8$, $m_4 = 6$;
$P_1(0, 0)$, $P_2(1, 8)$, $P_3(3, -4)$, $P_4(-6, -5)$

5–16 ■ Find the centroid of the region bounded by the given curves.

5. $y = x^2$, $y = 0$, $x = 2$

6. $y = 1 - x^2$, $y = 0$

7. $y = 3x + 5$, $y = 0$, $x = -1$, $x = 2$

8. $y = \sqrt{x}$, $y = 0$, $x = 4$

9. $y = \cos 2x$, $y = 0$, $x = -\pi/4$, $x = \pi/4$

10. $y = \sin x$, $y = 0$, $x = 0$, $x = \pi/2$

11. $y = e^x$, $y = 0$, $x = 0$, $x = 1$

12. $y = \ln x$, $y = 0$, $x = e$

13. $y = \sqrt{x}$, $y = x$

14. $y = x^2$, $y = 8 - x^2$

15. $y = \sin x$, $y = \cos x$, $x = 0$, $x = \pi/4$

16. $y = x$, $y = 0$, $y = 1/x$, $x = 2$

17–20 ■ Calculate the moments M_x and M_y and the center of mass of a lamina with the given density and shape.

17. $\rho = 1$

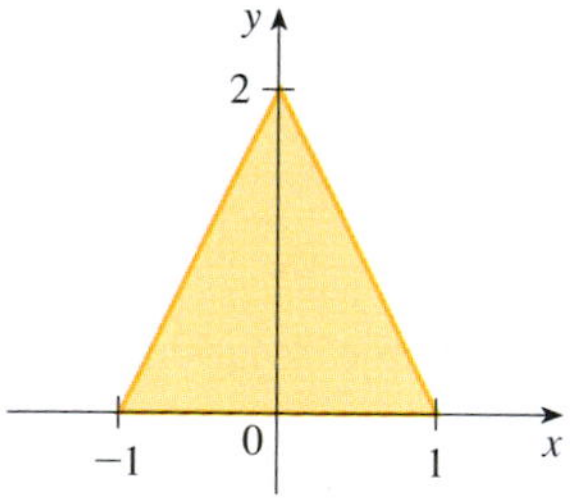

18. $\rho = 2$

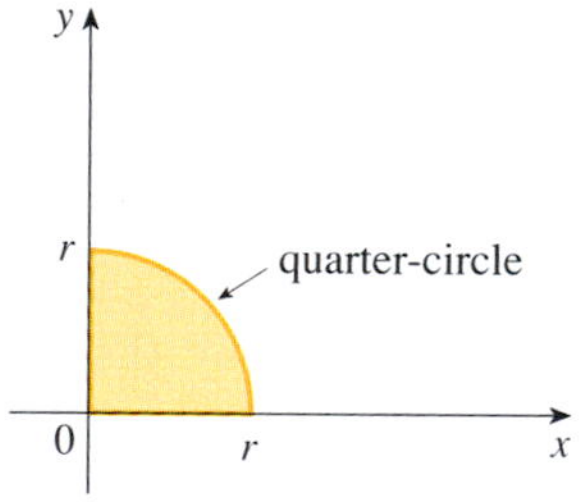

19. $\rho = 4$

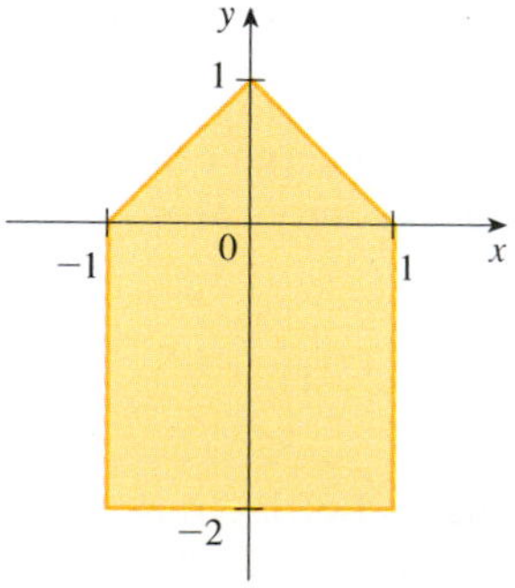

20. $\rho = 5$

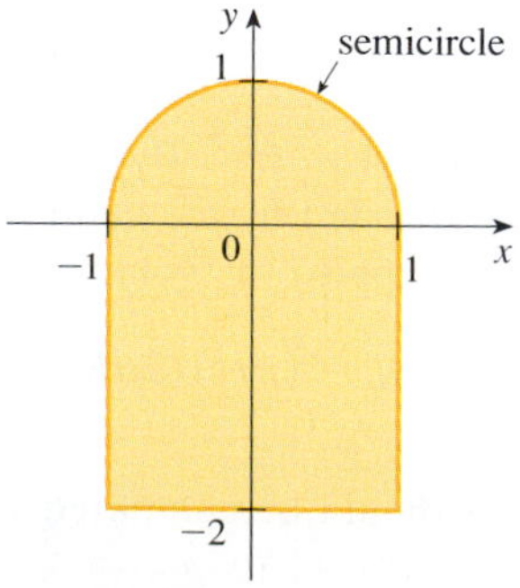

21. Prove that the centroid of any triangle is located at the point of intersection of the medians. [*Hints:* Place the axes so that the vertices are $(a, 0)$, $(0, b)$, and $(c, 0)$. Recall that a median is a line segment from a vertex to the midpoint of the opposite side. Recall also that the medians intersect at a point two-thirds of the way from each vertex (along the median) to the opposite side.]

22–25 ■ Find the centroid of the region shown, not by integration, but by locating the centroids of the rectangles and triangles (from Exercise 21) and using additivity of moments.

22.

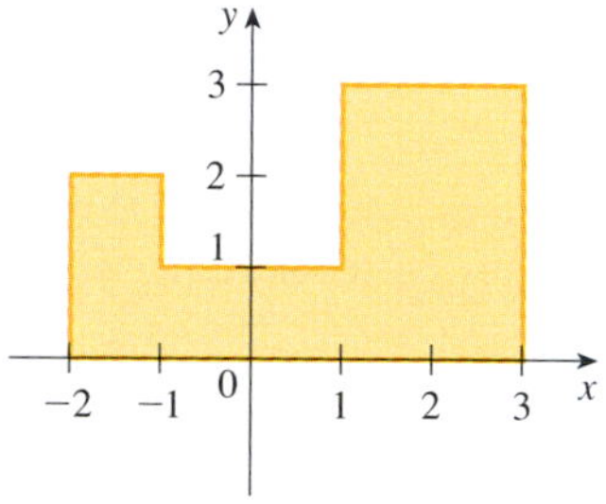

23.

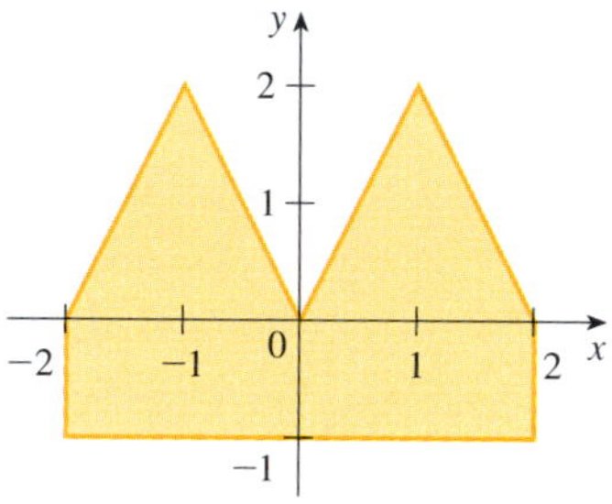

24.

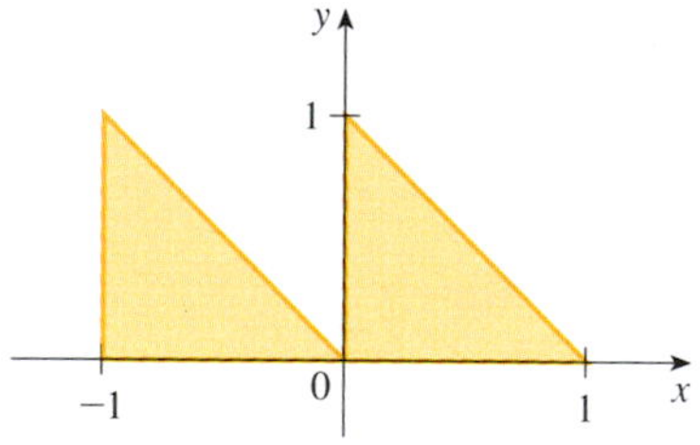

25.

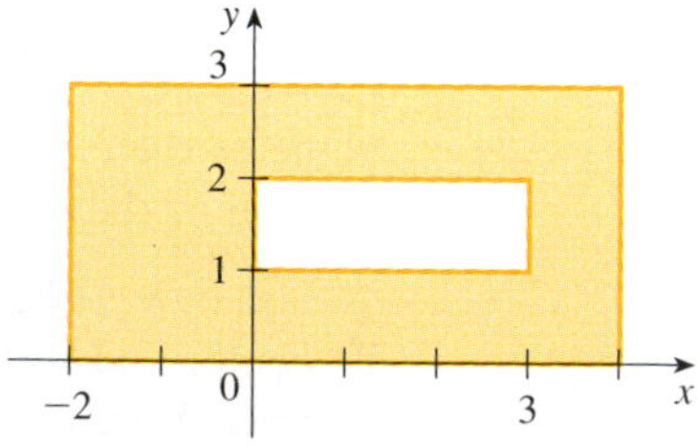

26–28 ■ Use the Theorem of Pappus to find the volume of the given solid.

26. A sphere of radius r (Use Example 3.)

27. A cone with height h and base radius r

28. The solid obtained by rotating the quadrilateral with vertices $(0, 0)$, $(1, 4)$, $(7, 4)$, and $(6, 0)$ about the y-axis

29. Prove Formulas 11.

30. Let $\mathcal{R}$ be the region that lies between the curves $y = x^m$ and $y = x^n$, $0 \leqslant x \leqslant 1$, where m and n are integers with $0 \leqslant n < m$.

(a) Sketch the region $\mathcal{R}$.

(b) Find the coordinates of the centroid of $\mathcal{R}$.

(c) Try to find values of m and n such that the centroid lies *outside* $\mathcal{R}$.

9.6 HYDROSTATIC PRESSURE AND FORCE

Divers realize that water pressure increases as they dive deeper. This is because the weight of the water above them increases.

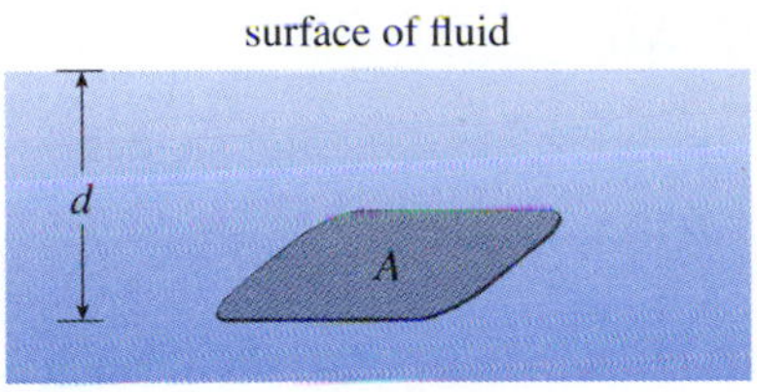

FIGURE 1

In general, suppose that a thin horizontal plate with area A square meters is submerged in a fluid of density ρ kilograms per cubic meter at a depth d meters below the surface of the fluid as in Figure 1. The fluid directly above the plate has volume $V = Ad$ so its mass is $m = \rho V = \rho A d$. The force exerted by the fluid on the plate is therefore

$$F = mg = \rho g A d$$

where g is the acceleration due to gravity. The pressure P on the plate is defined to be the force per unit area:

$$P = \frac{F}{A} = \rho g d$$

The SI unit for measuring pressure is newtons per square meter, which is called a pascal (abbreviation: $1 \text{ N/m}^2 = 1 \text{ Pa}$). Since this is a small unit, the kilopascal (kPa) is often used. For instance, because the density of water is $\rho = 1000 \text{ kg/m}^3$, the pressure at the bottom of a swimming pool 2 m deep is

$$P = \rho g d = 1000 \text{ kg/m}^3 \times 9.8 \text{ m/s}^2 \times 2 \text{ m}$$
$$= 19{,}600 \text{ Pa} = 19.6 \text{ kPa}$$

When using British units, we write $P = \rho g d = \delta d$, where $\delta = \rho g$ is the weight density (as opposed to ρ, which is the mass density). For instance, the weight density of water is $\delta = 62.5 \text{ lb/ft}^3$.

An important principle of fluid pressure is the experimentally verified fact that *at any point in a liquid the pressure is the same in all directions.* (A diver feels the same pressure on both ears and nose.) Thus, the pressure in *any* direction at a depth d in a fluid with mass density ρ is given by

$$P = \rho g d = \delta d \qquad \textbf{(1)}$$

This helps us determine the hydrostatic force against a *vertical* plate or wall or dam in a fluid. This is not a straightforward problem, because the pressure is not constant but increases as the depth increases.

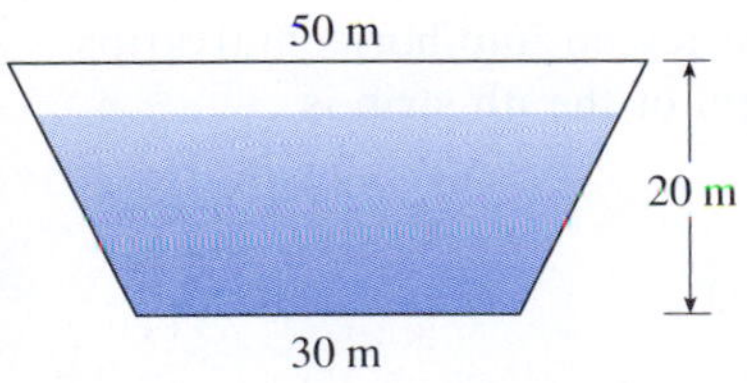

FIGURE 2

EXAMPLE 1 A dam has the shape of the trapezoid shown in Figure 2. The height is 20 m and the width is 50 m at the top and 30 m at the bottom. Find the force on the dam due to hydrostatic pressure if the water level is 4 m from the top of the dam.

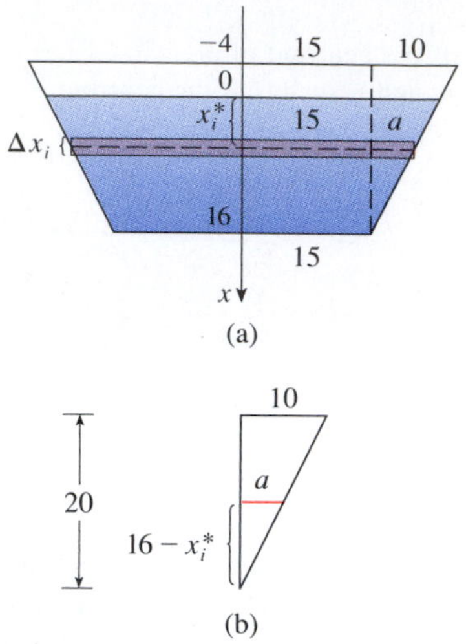

FIGURE 3

SOLUTION We choose a vertical x-axis with origin at the surface of the water as in Figure 3(a). The depth of the water is 16 m so we consider a partition P of the interval $[0, 16]$ by points x_i and we choose $x_i^* \in [x_{i-1}, x_i]$. The ith horizontal strip of the dam is approximated by a rectangle with height Δx_i and width w_i, where, from similar triangles in Figure 3(b),

$$\frac{a}{16 - x_i^*} = \frac{10}{20} \qquad a = \frac{16 - x_i^*}{2} = 8 - \frac{x_i^*}{2}$$

and so
$$w_i = 2(15 + a) = 2\left(15 + 8 - \frac{x_i^*}{2}\right) = 46 - x_i^*$$

If A_i is the area of the ith strip, then

$$A_i \approx w_i \, \Delta x_i = (46 - x_i^*) \, \Delta x_i$$

If Δx_i is small, then the pressure P_i on the ith strip is almost constant and we can use Equation 1 to write

$$P_i \approx 1000 g x_i^*$$

The hydrostatic force F_i acting on the ith strip is the product of the pressure and the area:

$$F_i = P_i A_i \approx 1000 g x_i^* (46 - x_i^*) \, \Delta x_i$$

Adding these forces and taking the limit as $\|P\| \to 0$, we obtain the total hydrostatic force on the dam:

$$\begin{aligned} F &= \lim_{\|P\| \to 0} \sum_{i=1}^{n} 1000 g x_i^* (46 - x_i^*) \, \Delta x_i \\ &= \int_0^{16} 1000 g x (46 - x) \, dx \\ &= 1000(9.8) \int_0^{16} (46x - x^2) \, dx \\ &= 9800 \left[23x^2 - \frac{x^3}{3} \right]_0^{16} \\ &\approx 4.43 \times 10^7 \text{ N} \end{aligned}$$

■

EXAMPLE 2 Find the hydrostatic force on one end of a cylindrical drum with radius 3 ft if the drum is submerged in water 10 ft deep.

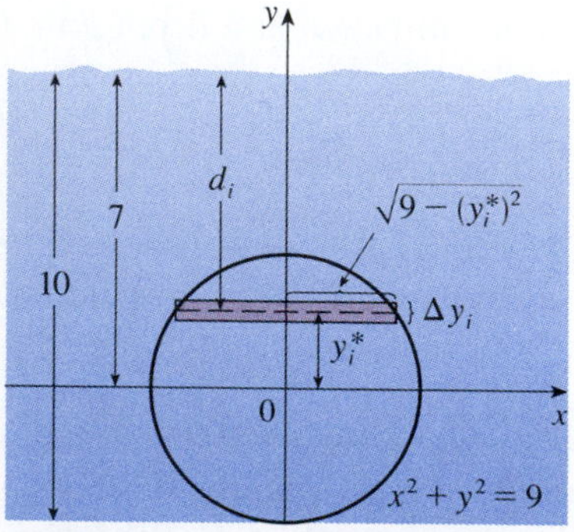

FIGURE 4

SOLUTION In this example it is convenient to choose the axes as in Figure 4 so that the origin is placed at the center of the drum. Then the circle has a simple equation, $x^2 + y^2 = 9$. As in Example 1 we divide the circular region into horizontal strips. From the equation of the circle, we see that the length of the ith strip is $2\sqrt{9 - (y_i^*)^2}$ and so its area is

$$A_i = 2\sqrt{9 - (y_i^*)^2} \, \Delta y_i$$

The pressure on this strip is approximately

$$\delta d_i = 62.5(7 - y_i^*)$$

and so the force on the strip is approximately

$$\delta d_i A_i = 62.5(7 - y_i^*)2\sqrt{9 - (y_i^*)^2}\ \Delta y_i$$

The total force is obtained by adding the forces on all the strips and taking the limit:

$$\begin{aligned} F &= \lim_{\|P\|\to 0} \sum_{i=1}^{n} 62.5(7 - y_i^*)2\sqrt{9 - (y_i^*)^2}\ \Delta y_i \\ &= 125 \int_{-3}^{3} (7 - y)\sqrt{9 - y^2}\, dy \\ &= 125 \cdot 7 \int_{-3}^{3} \sqrt{9 - y^2}\, dy - 125 \int_{-3}^{3} y\sqrt{9 - y^2}\, dy \end{aligned}$$

The second integral is 0 because the integrand is an odd function (see Theorem 6.5.7). The first integral can be evaluated using the trigonometric substitution $y = 3\sin\theta$, but it's simpler to observe that it is the area of a semicircular disk with radius 3. Thus

$$\begin{aligned} F &= 875 \int_{-3}^{3} \sqrt{9 - y^2}\, dy = 875 \cdot \tfrac{1}{2}\pi(3)^2 \\ &= \frac{7875\pi}{2} \approx 12{,}370 \text{ lb} \end{aligned}$$

■

EXERCISES 9.6

1. An aquarium 2 m long, 1 m wide, and 1 m deep is full of water. Find (a) the hydrostatic pressure on the bottom of the aquarium, (b) the hydrostatic force on the bottom, and (c) the hydrostatic force on one end of the aquarium.

2. A swimming pool 5 m wide, 10 m long, and 3 m deep is filled with seawater of density 1030 kg/m^3 to a depth of 2.5 m. Find (a) the hydrostatic pressure at the bottom of the pool, (b) the hydrostatic force on the bottom, and (c) the hydrostatic force on one end of the pool.

3–12 ■ A tank contains water. The end of the tank is vertical and has the indicated shape. Find the hydrostatic force against the end of the tank.

3.
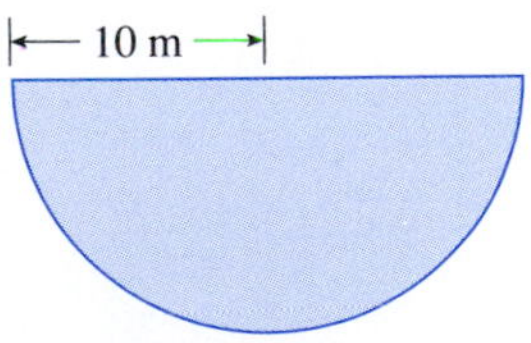

4.
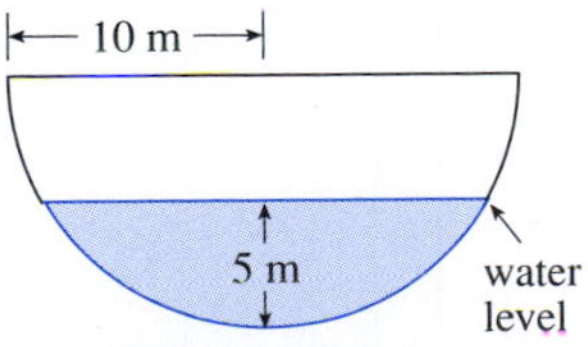

5.
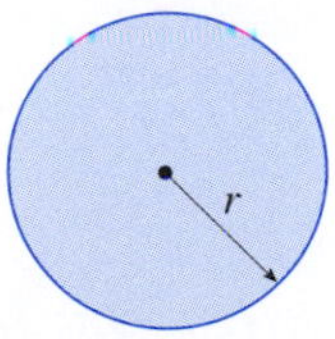

6.
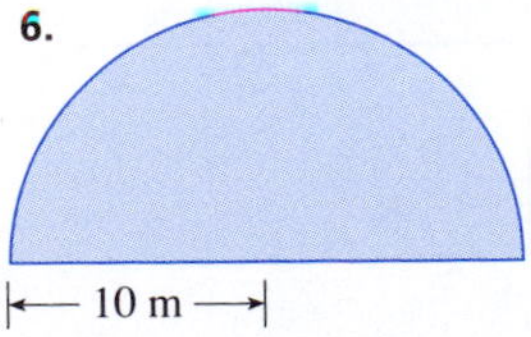

7.
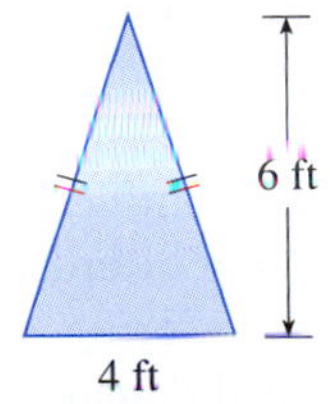

8.
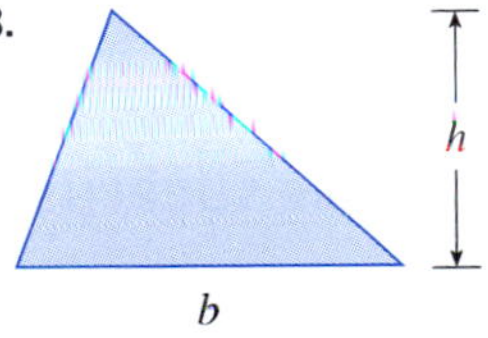

9.
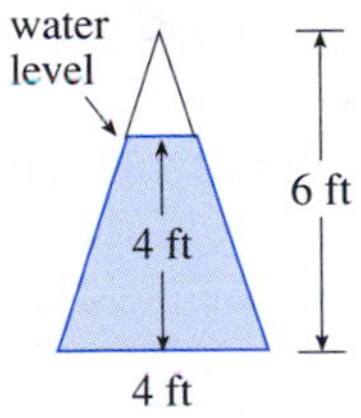

10.
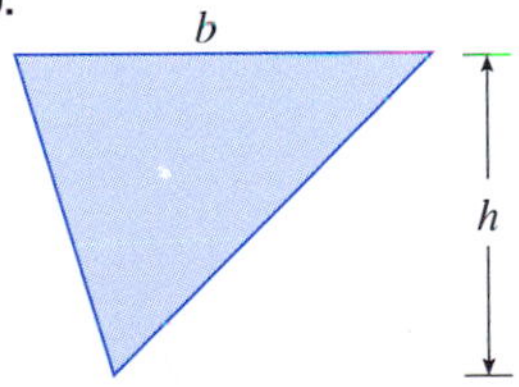

11.
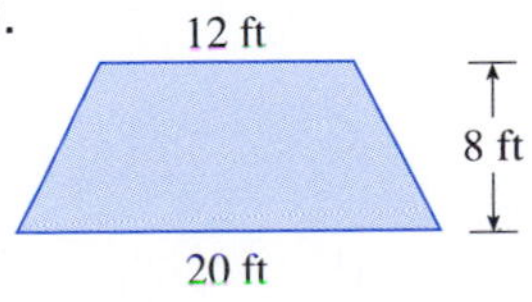

12.
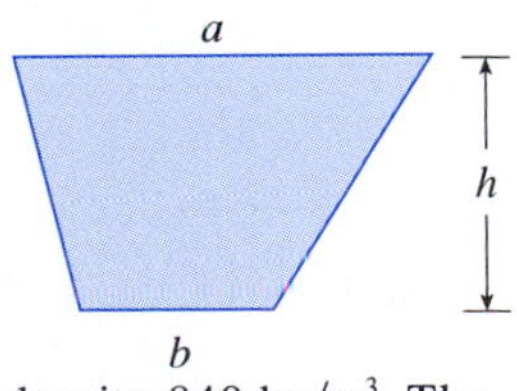

13. A trough is filled with a liquid of density 840 kg/m^3. The ends of the trough are equilateral triangles with sides 8 m long and vertex at the bottom. Find the hydrostatic force on one end of the trough.

14. Work Exercise 13 if the height of the liquid is 4 m.

15. A cube with 20-cm-long sides is sitting on the bottom of an aquarium in which the water is one meter deep. Find the hydrostatic force on (a) the top of the cube and (b) one of the sides of the cube.

16. A vertical dam has a semicircular gate as shown in the figure. Find the hydrostatic force against the gate.

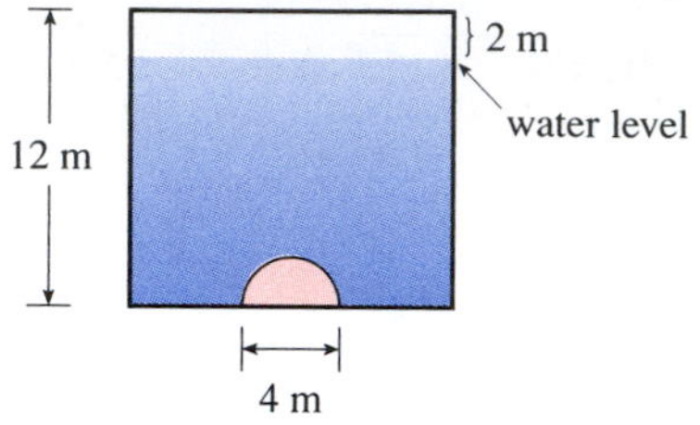

17. A trough 2 m high is filled with water. Its ends are vertical but its sides are 3 m wide and are inclined at an angle of 45° from the vertical. Find the hydrostatic force on one of the sides of the trough.

18. A dam is inclined at an angle of 30° from the vertical and has the shape of an isosceles trapezoid 100 ft wide at the top and 50 ft wide at the bottom and with a slant height of 70 ft. Find the hydrostatic force on the dam when it is full of water.

19. A swimming pool is 20 ft wide and 40 ft long and its bottom is an inclined plane, the shallow end having a depth of 3 ft and the deep end 9 ft. Find the hydrostatic force on (a) the shallow end, (b) the deep end, (c) one of the sides, and (d) the bottom of the pool.

20. Suppose that a plate is immersed vertically in a fluid with density ρ and the width of the plate is $w(x)$ at a depth of x meters beneath the surface of the fluid. If the top of the plate is at depth a and the bottom is at depth b, show that the hydrostatic force on one side of the plate is

$$F = \int_a^b \rho g x w(x)\, dx$$

21. Use the formula of Exercise 20 to show that

$$F = (\rho g \bar{x})A$$

where $\bar{x}$ is the x-coordinate of the centroid of the plate and A is its area. This equation shows that the hydrostatic force against a vertical plane region is the same as if the region were horizontal at the depth of the centroid of the region.

22. Use the result of Exercise 21 to give another solution to Exercise 5.

9.7 APPLICATIONS TO ECONOMICS AND BIOLOGY

In this section we consider some applications of integration to economics (consumer surplus, present value of future income) and biology (blood flow, cardiac output). Others are found in the exercises.

CONSUMER SURPLUS

Recall from Section 5.6 that the demand function $p(x)$ is the price that a company has to charge in order to sell x units of a commodity. Usually, selling larger quantities requires lowering prices, so the demand function is a decreasing function. The graph of a typical demand function, called a **demand curve**, is shown in Figure 1. If X is the amount of the commodity that is currently available, then $P = p(X)$ is the current selling price.

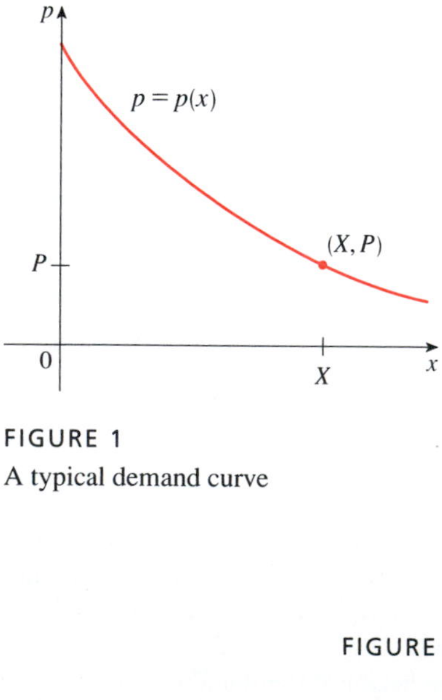

FIGURE 1
A typical demand curve

We partition the interval $[0, X]$ into n subintervals, each of length $\Delta x = X/n$, and let $x_i^* = x_i$ be the right endpoint of the ith subinterval, as in Figure 2. If, after the first x_{i-1} units were sold, only a total of x_i units had been available and the price per unit had

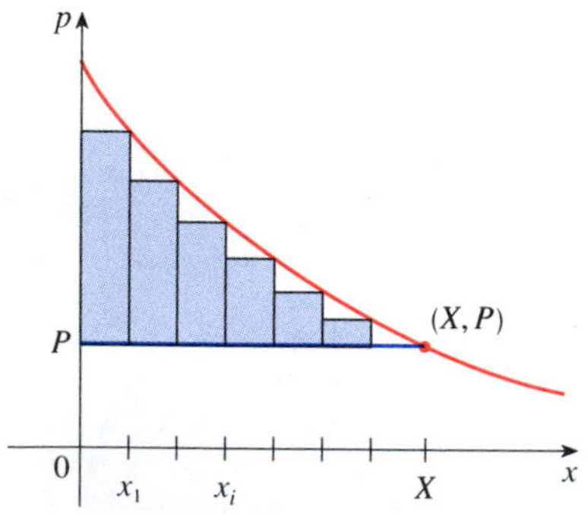

FIGURE 2

been set at $p(x_i)$ dollars, then the additional Δx units could have been sold (but no more). The consumers who would have paid $p(x_i)$ dollars placed a high value on the product; they would have paid what it was worth to them. So in paying only P dollars they have saved an amount of

$$(\text{savings per unit})(\text{number of units}) = [p(x_i) - P]\Delta x$$

Considering similar groups of willing consumers for each of the subintervals and adding the savings, we get the total savings:

$$\sum_{i=1}^{n} [p(x_i) - P]\Delta x$$

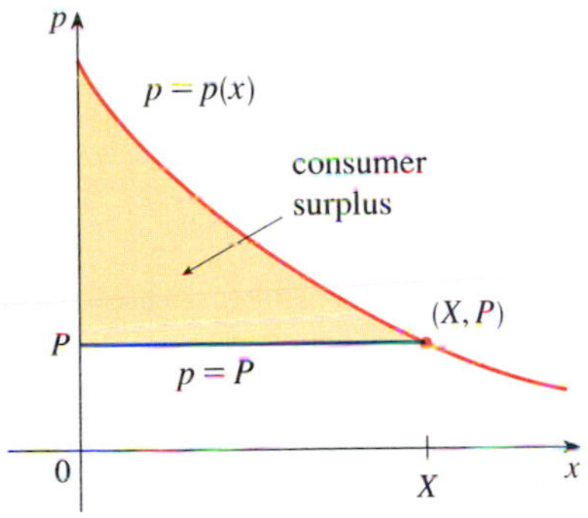

FIGURE 3

If we let $n \to \infty$, this Riemann sum approaches the integral

(1) $$\int_0^X [p(x) - P]\,dx$$

which economists call the **consumer surplus** for the commodity.

The consumer surplus represents the amount of money saved by consumers in purchasing the commodity at price P, corresponding to an amount demanded of X. Figure 3 shows the interpretation of the consumer surplus as the area under the demand curve and above the line $p = P$.

EXAMPLE 1 The demand for a product, in dollars, is $p = 1200 - 0.2x - 0.0001x^2$. Find the consumer surplus when the sales level is 500.

SOLUTION Since the number of products sold is $X = 500$, the corresponding price is

$$P = 1200 - (0.2)(500) - (0.0001)(500)^2 = 1075$$

Therefore, from the definition in (1), the consumer surplus is

$$\int_0^{500} [p(x) - P]\,dx = \int_0^{500} (1200 - 0.2x - 0.0001x^2 - 1075)\,dx$$

$$= \int_0^{500} (125 - 0.2x - 0.0001x^2)\,dx$$

$$= 125x - 0.1x^2 - (0.0001)\left(\frac{x^3}{3}\right)\Bigg]_0^{500}$$

$$= (125)(500) - (0.1)(500)^2 - \frac{(0.0001)(500)^3}{3}$$

$$= \$33{,}333.33$$ ■

PRESENT VALUE OF AN INCOME STREAM

In Example 4 in Section 4.5 we discussed continuous compounding of interest and found that the value of a dollar after t years, at an annual interest rate r compounded continuously, is e^{rt}. It follows that if A is the amount that will grow to one dollar in t years, then $Ae^{rt} = 1$ and so $A = e^{-rt}$; this is called the *present value* of one dollar. For instance, if the interest rate is 9%, the present value of a dollar to be paid 5 years from now is $e^{-(0.09)(5)} = e^{-0.45} \approx 64$ cents.

We now suppose that income will be received, not in one lump sum, but over a period of time from $t = a$ to $t = b$ at a rate of $f(t)$ dollars per year at time t. This is referred to as an **income stream**. To find the total present value of this income we partition the interval $[a, b]$ into n subintervals of equal length Δt. From time $t = t_{i-1}$ to time $t = t_i$ the income received will be about $f(t_i)\,\Delta t$ dollars, with a present value of

$$e^{-rt_i} f(t_i)\,\Delta t$$

So an approximation to the present value of the total income is

$$\sum_{i=1}^{n} e^{-rt_i} f(t_i)\,\Delta t$$

If we let $n \to \infty$, this Riemann sum approaches the integral

$$\textbf{(2)} \qquad \int_a^b e^{-rt} f(t)\,dt$$

which is called the **present value of the income stream** $f(t)$ at an interest rate r over the time period from $t = a$ to $t = b$.

EXAMPLE 2 A trust fund pays $8000 a year for 10 years, starting 5 years from now, at a rate of 10% per year compounded continuously.
(a) Find the present value of the trust fund.
(b) Find the value 3 years from now.

SOLUTION
(a) Here the income stream is constant: $f(t) = 8000$. Using (2) with $a = 5$, $b = 15$, and $r = 0.1$, we find that the present value of the trust fund is

$$\begin{aligned}\int_5^{15} e^{-(0.1)t}(8000)\,dt &= (8000)\frac{e^{-(0.1)t}}{-0.1}\bigg]_5^{15} \\ &= 80{,}000(e^{-0.5} - e^{-1.5}) \\ &= \$30{,}672.04\end{aligned}$$

(b) The value 3 years from now is

$$(30{,}672.04)e^{(0.1)3} = \$41{,}402.92$$

■

If we let $b \to \infty$ in Formula 2, we get the improper integral

$$\textbf{(3)} \qquad \int_a^\infty e^{-rt} f(t)\,dt$$

This represents the present value of a **perpetuity**, or perpetual annuity, under which income will be received at a rate of $f(t)$ dollars per year forever. (See Exercises 13 and 14.)

BLOOD FLOW

In Example 7 in Section 3.3 we discussed the law of laminar flow:

$$v(r) = \frac{P}{4\eta l}(R^2 - r^2)$$

which gives the velocity v of blood that flows along a blood vessel with radius R and length l at a distance r from the central axis, where P is the pressure difference between the ends of the vessel and η is the viscosity of the blood. Now in order to compute the flux (volume per unit time) we consider radii r_i, where

$$0 = r_0 < r_1 < r_2 < \cdots < r_n = R$$

The approximate area of the annulus with inner radius r_{i-1} and outer radius r_i is

$$2\pi r_i \,\Delta r_i \qquad \text{where } \Delta r_i = r_i - r_{i-1}$$

If Δr_i is small, then the velocity is almost constant throughout this annulus and can be approximated by $v(r_i)$. Thus, the volume of blood per unit time that flows across the annulus is approximately

$$(2\pi r_i \,\Delta r_i)v(r_i) = 2\pi r_i v(r_i)\,\Delta r_i$$

and the total volume of blood that flows across a cross-section per unit time is approximately

$$\sum_{i=1}^{n} 2\pi r_i v(r_i)\,\Delta r_i$$

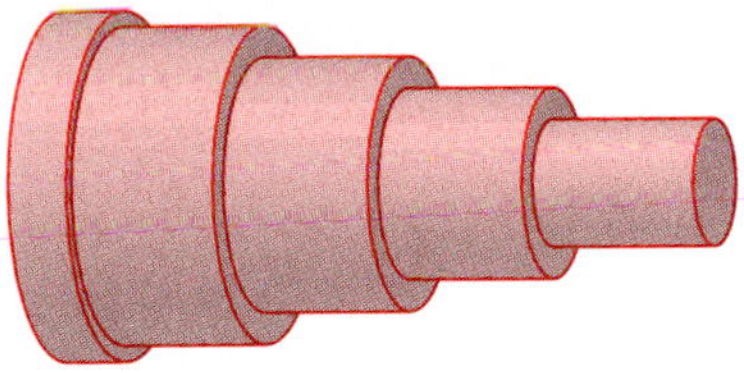

FIGURE 4

This approximation is illustrated in Figure 4. Notice that the velocity (and hence the volume per unit time) increases toward the center of the blood vessel. The approximation gets better as we take finer subdivisions, that is, as the norm $\|P\| \to 0$. When we take the limit we get the exact value of the *flux* (or *discharge*), which is the volume of blood that passes a cross-section per unit time:

$$\begin{aligned}
F &= \lim_{\|P\|\to 0} \sum_{i=1}^{n} 2\pi r_i v(r_i)\,\Delta r_i \\
&= \int_0^R 2\pi r v(r)\,dr \\
&= \int_0^R 2\pi r \frac{P}{4\eta l}(R^2 - r^2)\,dr \\
&= \frac{\pi P}{2\eta l}\int_0^R (R^2 r - r^3)\,dr = \frac{\pi P}{2\eta l}\left[R^2\frac{r^2}{2} - \frac{r^4}{4}\right]_{r=0}^{r=R} \\
&= \frac{\pi P}{2\eta l}\left[\frac{R^4}{2} - \frac{R^4}{4}\right] = \frac{\pi P R^4}{8\eta l}
\end{aligned}$$

The resulting equation

(4)
$$F = \frac{\pi P R^4}{8\eta l}$$

is called **Poiseuille's Law** and shows that the flux is proportional to the fourth power of the radius of the blood vessel.

CARDIAC OUTPUT

In the human cardiovascular system, blood returns from the body through the veins, enters the right atrium of the heart, and is pumped to the lungs through the pulmonary arteries for oxygenation, back into the left atrium through the pulmonary veins, and then out to the rest of the body through the aorta (see Figure 5). The **cardiac output** of the heart is the volume of blood pumped by the heart per unit time, that is, the rate of flow into the aorta.

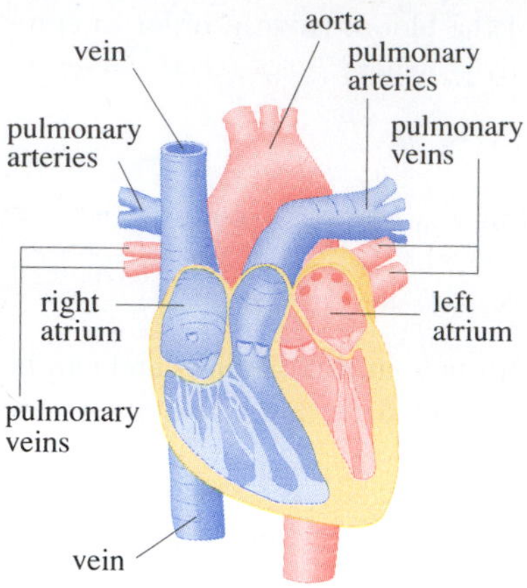

FIGURE 5

The *dye dilution method* is used to measure the cardiac output. Dye is injected into the right atrium and flows through the heart into the aorta. A probe inserted into the aorta measures the concentration of the dye leaving the heart at equally spaced times over a time interval $[0, T]$ until the dye has cleared. Let $c(t)$ be the concentration of the dye at time t. If we partition $[0, T]$ into subintervals of equal length Δt, then the amount of dye that flows past the measuring point during the subinterval from $t = t_{i-1}$ to $t = t_i$ is approximately

$$\text{(concentration) (volume)} = c(t_i)(F\,\Delta t)$$

where F is the rate of flow that we are trying to determine. Thus, the total amount of dye is approximately

$$\sum_{i=1}^{n} c(t_i)F\Delta t = F \sum_{i=1}^{n} c(t_i)\,\Delta t$$

and, letting $n \to \infty$, we find that the amount of dye is

$$A = F \int_0^T c(t)\,dt$$

Thus, the cardiac output is given by

(5) $$F = \frac{A}{\int_0^T c(t)\,dt}$$

where the amount of dye A is known and the integral can be approximated from the concentration readings.

EXAMPLE 3 A 5-mg bolus of dye is injected into a right atrium. The concentration of the dye (in milligrams per liter) is measured in the aorta at one-second intervals as shown in the chart. Estimate the cardiac output.

t	0	1	2	3	4	5	6	7	8	9	10
$c(t)$	0	0.4	2.8	6.5	9.8	8.9	6.1	4.0	2.3	1.1	0

SOLUTION Here $A = 5$, $\Delta t = 1$, and $T = 10$. We use Simpson's Rule to approximate the integral of the concentration:

$$\begin{aligned} \int_0^{10} c(t)\,dt &\approx \tfrac{1}{3}[0 + 4(0.4) + 2(2.8) + 4(6.5) + 2(9.8) + 4(8.9) \\ &\qquad + 2(6.1) + 4(4.0) + 2(2.3) + 4(1.1) + 0] \\ &\approx 41.87 \end{aligned}$$

Thus, Formula 5 gives the cardiac output to be

$$\begin{aligned} F &= \frac{A}{\int_0^{10} c(t)\,dt} \approx \frac{5}{41.87} \\ &\approx 0.12 \text{ L/s} \approx 7.17 \text{ L/min} \end{aligned}$$

■

EXERCISES 9.7

1. The marginal cost function $C'(x)$ was defined to be the derivative of the cost function. (See Sections 3.3 and 5.6.) If the marginal cost of manufacturing x units of a product is $C'(x) = 0.006x^2 - 1.5x + 8$ (measured in dollars per unit) and the fixed start-up cost is $C(0) = \$1,500,000$, use the Fundamental Theorem of Calculus to find the cost of producing the first 2000 units.

2. The marginal revenue from selling x items is $90 - 0.02x$. The revenue from the sale of the first 100 items is \$8800. What is the revenue from the sale of the first 200 items?

3. The marginal cost of producing x units of a certain product is $140 - 0.5x + 0.012x^2$ (in dollars per unit). Find the increase in cost if the production level is raised from 3000 units to 5000 units.

4. The demand function for a certain commodity is $p = 5 - x/10$. Find the consumer surplus when the sales level is 30. Illustrate by drawing the demand curve and identifying the consumer surplus as an area.

5. A demand curve is given by $p = 1000/(x + 20)$. Find the consumer surplus when the selling price is \$20.

6. The **supply function** $p_S(x)$ for a commodity gives the relation between the selling price and the number of units that manufacturers will produce at that price. For a higher price, manufacturers will produce more units, so p_S is an increasing function of x. Let X be the amount of the commodity currently produced and let $P = p_S(X)$ be the current price. Some producers would be willing to make and sell the commodity for a lower selling price and are therefore receiving more than their minimal price. The excess is called the **producer surplus**. An argument similar to that for consumer surplus shows that the surplus is given by the integral

$$\int_0^X [P - p_S(x)]\,dx$$

Calculate the producer surplus for the supply function $p_S(x) = 3 + 0.01x^2$ at the sales level $X = 10$. Illustrate by drawing the supply curve and identifying the producer surplus as an area.

7. A supply curve is given by $p = 5 + \frac{1}{10}\sqrt{x}$. Find the producer surplus when the selling price is \$10.

8. For a given commodity and pure competition, the number of units produced and the price per unit are determined as the coordinates of the point of intersection of the supply and demand curves. Given the demand curve $p = 50 - x/20$ and the supply curve $p = 20 + x/10$, find the consumer surplus and the producer surplus. Illustrate by sketching the supply and demand curves and identifying the surpluses as areas.

9. A manufacturer has been selling 1000 television sets a week at \$450 each. A market survey indicates that for every \$10 that the price is reduced, the number of sets sold will increase by 100 a week. Find the demand function and calculate the consumer surplus when the selling price is set at \$400.

10. A trust fund pays \$2000 a year for 5 years, starting immediately. The interest rate is 7% per year compounded continuously. Find the present value of the trust fund.

11. A trust fund starts 10 years from now and pays \$12,000 a year for 15 years. The interest rate is 6% per year compounded continuously. Find the value of the trust fund 5 years from now.

12. A baseball player signs a salary contract whereby he receives a sum that increases continuously and linearly from a starting salary of \$1,000,000 a year and reaches \$3,000,000 a year after 4 years. Thus, his salary after t years (in millions of dollars) is $f(t) = 1 + \frac{1}{2}t$. Find the present value of the contract assuming an interest rate of 6% per year compounded continuously.

13. (a) Use (3) to show that the present value of a perpetuity under which you and your heirs forever receive an amount A annually is A/r, where r is the annual interest rate, compounded continuously.
(b) Assuming an interest rate of 8% per year, what is the present value of a perpetuity that pays an annual amount of \$5000?

14. Suppose that a perpetuity pays an amount of $5000 + 1000t$ dollars t years from now. Find the present value.

15. If the amount of capital that a company has at time t is $f(t)$, then the derivative, $f'(t)$, is called the *net investment flow*. Suppose that the net investment flow is $\sqrt{t}$ million dollars per year (where t is measured in years). Find the increase in capital (the capital formation) from the fourth year to the eighth year.

16. An animal population is increasing at a rate of $200 + 50t$ per year (where t is measured in years). By how much does the animal population increase between the fourth and tenth years?

17. Use Poiseuille's Law to calculate the rate of flow in a typical human artery where we can take $\eta = 0.027$, $R = 0.008$ cm, $l = 2$ cm, and $P = 4000$ dynes/cm^2.

18. High blood pressure results from constriction of the arteries. To maintain the same flow rate (flux) the heart has to pump harder, thus increasing the blood pressure. Use Poiseuille's Law to show that if R_0 and P_0 are normal values of the radius and pressure in an artery and the constricted values are R and P, then for the flux to remain constant, P and R are related by the equation

$$\frac{P}{P_0} = \left(\frac{R_0}{R}\right)^4$$

Deduce that if the radius of an artery is reduced to three-fourths of its former value, then the pressure is more than tripled.

19. The dye dilution method is used to measure cardiac output with 8 mg of dye. The dye concentrations, in mg/L, are given by $c(t) = \frac{1}{4}t(12 - t)$, $0 \leqslant t \leqslant 12$, where t is measured in seconds. Find the cardiac output.

20. After a 6-mg injection of dye, the readings of dye concentrations at two-second intervals are as shown in the table. Use Simpson's Rule to estimate the cardiac output.

t	0	2	4	6	8	10	12	14	16	18	20
$c(t)$	0	2.1	4.5	7.3	5.8	3.6	2.8	1.4	0.6	0.2	0

9 REVIEW

KEY TOPICS ■ Define, state, or discuss the following.

1. Differential equation
2. Separable equation
3. Direction field
4. Logistic growth
5. First-order linear differential equation
6. Arc length
7. Area of a surface of revolution
8. Moments and centroid of a plane region
9. Theorem of Pappus
10. Pressure and force exerted by a fluid
11. Consumer surplus
12. Present value of an income stream
13. Rate of blood flow
14. Cardiac output

EXERCISES

1–3 ■ Solve the differential equation.

1. $y^2 \dfrac{dy}{dx} = x + \sin x$ **2.** $\dfrac{dy}{dx} = \dfrac{y^2 + 1}{xy}$, $x > 0$

3. $y' = \dfrac{1}{x^2y - 2x^2 + y - 2}$

4. Solve the initial-value problem $2yy' = xe^x$, $y(0) = 1$, and graph the solution.

5. Solve: $xyy' = \ln x$, $y(1) = 2$

6. Sketch a direction field for the equation $y' = x/y$. Then use it to sketch four solutions.

7–9 ■ Solve the differential equation.

7. $y' + 2xy = 2x^3$ **8.** $xy' - 2y = x^3$

9. $y' = 2 + 2x^2 + y + x^2y$

10. Solve the equation $y' = y + e^{2x}$ and graph several members of the family of solutions. How does the solution curve change as the constant C varies?

11. Solve the initial-value problem: $y' + y = \sqrt{x}e^{-x}$, $y(0) = 3$

12. Consider a population $P = P(t)$ with constant birth and death rates α and β, respectively, and a constant emigration rate m, where α, β, and m are positive constants. Assume that $\alpha > \beta$. Then the rate of change of the population at time t is given by

$$\frac{dP}{dt} = kP - m \qquad k = \alpha - \beta$$

(a) Find the solution of this differential equation that satisfies the initial condition $P(0) = P_0$.
(b) What conditions on m will lead to an exponential expansion of the population?
(c) What conditions on m will result in a constant population? A population decline?

13–14 ■ Find the length of the curve.

13. $3x = 2(y - 1)^{3/2}$, $2 \leqslant y \leqslant 5$

14. $y = \sqrt{x - x^2} + \sin^{-1}\sqrt{x}$

15. (a) Find the length of the curve

$$y = \frac{x^3}{6} + \frac{1}{2x} \qquad 1 \leqslant x \leqslant 2$$

(b) Find the area of the surface obtained by rotating the curve in part (a) about the x-axis.

16. (a) The curve $y = x^2$, $0 \leqslant x \leqslant 1$, is rotated about the y-axis. Find the area of the resulting surface.
(b) Find the area of the surface obtained by rotating the curve in part (a) about the x-axis.

17. Use Simpson's Rule with $n = 10$ to estimate the length of the arc of the curve $y = 1/x^2$ from $(1, 1)$ to $(2, \frac{1}{4})$.

18. Use Simpson's Rule with $n = 10$ to estimate the area of the surface obtained by rotating the arc of the curve $y = 1/x^2$ from $(1, 1)$ to $(2, \frac{1}{4})$ about the x-axis.

19. Find the surface area generated by rotating the loop of the curve $9ay^2 = x(3a - x)^2$ about the y-axis.

20. If the loop in Exercise 19 is rotated about the x-axis, find the area of the resulting surface.

21–22 ■ Find the centroid of the region bounded by the given curves.

21. $y = 4 - x^2$, $y = x + 2$

22. $y = 4x - x^2$, $y = 0$

23–24 ■ Find the centroid of the region shown.

23.

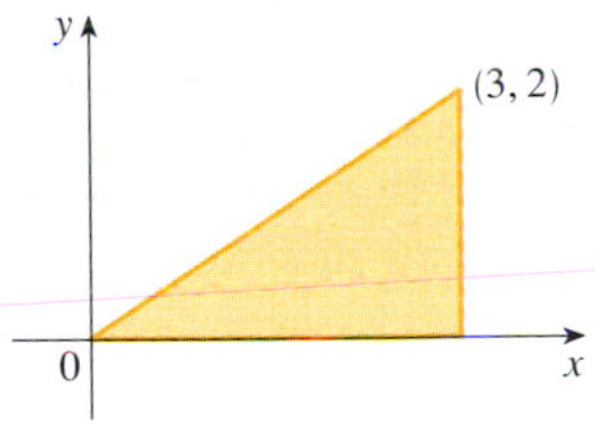

24.

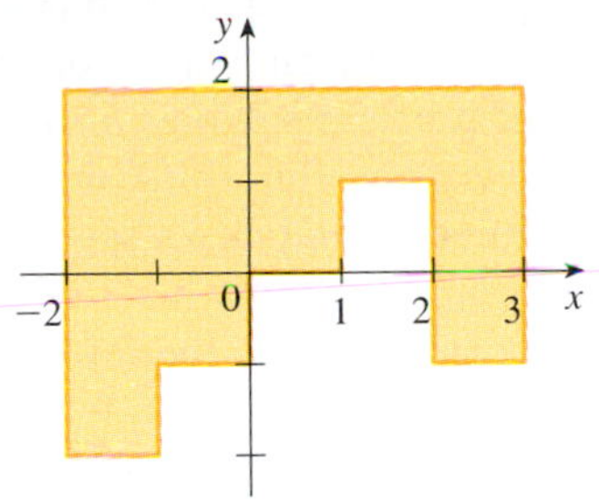

25. Find the volume obtained when the circle of radius 1 with center $(1, 0)$ is rotated about the y-axis.

26. Use the Theorem of Pappus and the fact that the volume of a sphere of radius r is $\frac{4}{3}\pi r^3$ to find the centroid of the semicircular region bounded by the curve $y = \sqrt{r^2 - x^2}$ and the x-axis.

27. A gate in an irrigation canal is in the form of a trapezoid 3 ft wide at the bottom, 5 ft wide at the top, and 2 ft high. It is placed vertically in the canal, with the water extending to its top. Find the hydrostatic force on one side of the gate.

28. A trough is filled with water and its vertical ends have the shape of the parabolic region in the figure. Find the hydrostatic force on one end of the trough.

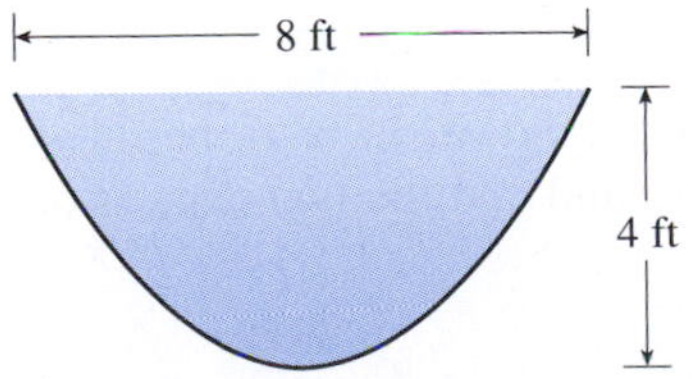

29. The demand function for a commodity is given by $p = 2000 - 0.1x - 0.01x^2$. Find the consumer surplus when the sales level is 100.

30. A trust fund starts 8 years from now and pays \$10,000 a year for 7 years. The interest rate is 8% per year compounded continuously.
(a) Find the present value of the trust fund.
(b) Find the value of the trust fund 5 years from now.
(c) Find the present value if the fund lasts not just 7 years but forever.

31. The von Bertalanffy growth model is used to predict the length $L(t)$ of a fish over a period of time. If L_∞ is the largest length for a species, then the hypothesis is that the rate of growth in length is proportional to $L_\infty - L$, the length yet to be achieved.
(a) Formulate and solve a differential equation to find an expression for $L(t)$.
(b) For the North Sea haddock it has been determined that $L_\infty = 53$ cm, $L(0) = 10$ cm, and the constant of proportionality is 0.2. What does the expression for $L(t)$ become with these data?

32. A tank contains 100 L of pure water. Brine that contains 0.1 kg of salt per liter enters the tank at a rate of 10 L/min. The solution is kept thoroughly mixed and drains from the tank at the same rate. How much salt is in the tank after 6 min?

33. One model for the spread of an epidemic is that the rate of spread is jointly proportional to the number of infected people and the number of uninfected people. In an isolated town of 5000 inhabitants, 160 people have a disease at the beginning of the week and 1200 have it at the end of the week. How long does it take for 80% of the population to be infected?

34. Barbara weighs 60 kg and is on a diet of 1600 calories per day of which 850 are used automatically by basal metabolism. She spends about 15 cal/kg/day times her weight doing exercise. If 1 kg of fat contains 10,000 cal and we assume that the storage of calories in the form of fat is 100% efficient, formulate a differential equation and solve it to find her weight as a function of time. Does her weight ultimately approach an equilibrium weight?

35. A planning engineer for a new alum plant must present some estimates to his company regarding the capacity of a silo designed to contain bauxite ore until it is processed into alum. The ore resembles pink talcum powder and is poured from a conveyor at the top of the silo. The silo is a cylinder 100 ft high with a radius of 200 ft. The conveyor carries $60{,}000\pi$ ft^3/h and the ore maintains a conical shape whose radius is 1.5 times its height.
(a) If, at a certain time t, the pile is 60 ft high, how long will it take for the pile to reach the top of the silo?
(b) Management wants to know how much room will be left in the floor area of the silo when the pile is 60 ft high. How fast is the floor area of the pile growing at that height?
(c) Suppose a loader starts removing the ore at the rate of $20{,}000\pi$ ft^3/h when the height of the pile reaches 90 ft. Suppose, also, that the pile continues to maintain its shape. How long will it take for the pile to reach the top of the silo under these conditions?

PROBLEMS PLUS

Cover up the solution to the example and try it yourself first.

EXAMPLE

(a) Prove that if f is a continuous function, then

$$\int_0^a f(x)\,dx = \int_0^a f(a-x)\,dx$$

(b) Use part (a) to show that

$$\int_0^{\pi/2} \frac{\sin^n x}{\sin^n x + \cos^n x}\,dx = \frac{\pi}{4}$$

for all positive numbers n.

SOLUTION

The principles of problem solving are discussed in Section 4 of Review and Preview.

(a) At first sight, the given equation may appear somewhat baffling. How is it possible to connect the left side to the right side? Connections can often be made through one of the principles of problem solving: *Introduce something extra.* Here the extra ingredient is a new variable. We often think of introducing a new variable when we use the Substitution Rule to integrate a specific function. But that technique is still useful in the present circumstance in which we have a general function f.

Once we think of making a substitution, the form of the right side suggests that it should be $u = a - x$. Then $du = -dx$. When $x = 0$, $u = a$; when $x = a$, $u = 0$. So

$$\int_0^a f(a-x)\,dx = -\int_a^0 f(u)\,du = \int_0^a f(u)\,du$$

But this integral on the right side is just another way of writing $\int_0^a f(x)\,dx$. So the given equation is proved.

(b) If we let the given integral be I and apply part (a) with $a = \pi/2$, we get

$$I = \int_0^{\pi/2} \frac{\sin^n x}{\sin^n x + \cos^n x}\,dx = \int_0^{\pi/2} \frac{\sin^n(\pi/2 - x)}{\sin^n(\pi/2 - x) + \cos^n(\pi/2 - x)}\,dx$$

The computer graphs in Figure 1 make it seem plausible that all of the integrals in the example have the same value. The graph of each integrand is labeled with the corresponding value of n.

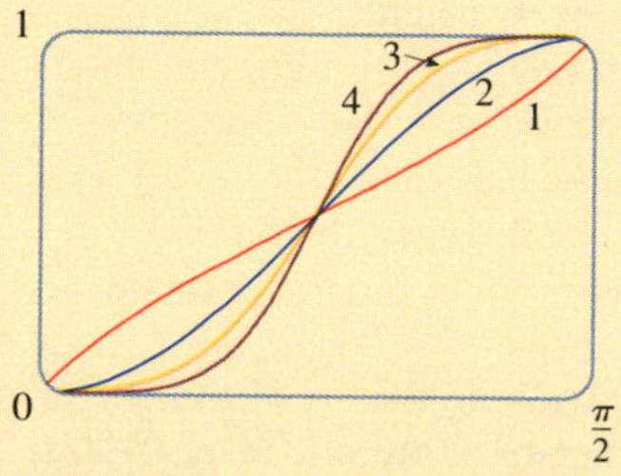

FIGURE 1

A well-known trigonometric identity tells us that $\sin(\pi/2 - x) = \cos x$ and $\cos(\pi/2 - x) = \sin x$, so we get

$$I = \int_0^{\pi/2} \frac{\cos^n x}{\cos^n x + \sin^n x}\,dx$$

Notice that the two expressions for I are very similar. In fact, the integrands have the same denominator. This suggests that we should add the two expressions. If we do so, we get

$$2I = \int_0^{\pi/2} \frac{\sin^n x + \cos^n x}{\sin^n x + \cos^n x}\,dx = \int_0^{\pi/2} 1\,dx = \frac{\pi}{2}$$

Therefore, $I = \pi/4$. ■

PROBLEMS

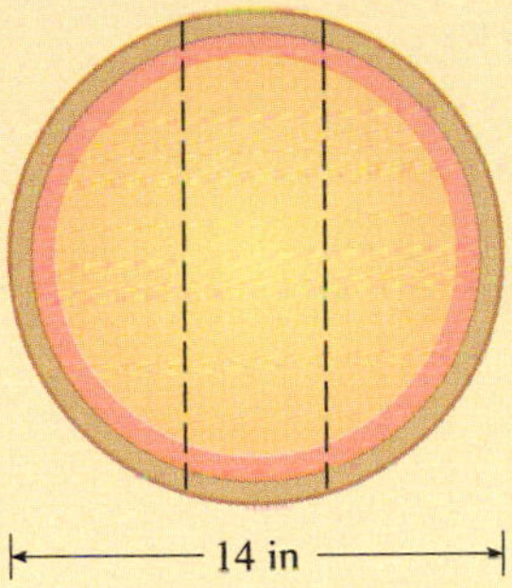

FIGURE FOR PROBLEM 1

1. Three mathematics students have ordered a 14-inch pizza. Instead of slicing it in the traditional way, they decide to slice it by parallel cuts, as in the figure. Being mathematics majors, they are able to determine where to slice so that each gets the same amount of pizza. Where are the cuts made?

2. A student forgot the Product Rule for differentiation and made the mistake of thinking that $(fg)' = f'g'$. However, he was lucky and got the correct answer. The function f that he used was $f(x) = e^{x^2}$ and the domain of his problem was the interval $(\frac{1}{2}, \infty)$. What was the function g?

3. Let f be a function with the property that $f(0) = 1$, $f'(0) = 1$, and $f(a + b) = f(a)f(b)$ for all real numbers a and b. Show that $f'(x) = f(x)$ for all x and deduce that $f(x) = e^x$.

4. Find the centroid of the region enclosed by the loop of the curve $y^2 = x^3 - x^4$.

5. Let a and b be positive numbers. Show that not both of the numbers $a(1 - b)$ and $b(1 - a)$ can be greater than $\frac{1}{4}$.

6. Evaluate $\displaystyle\int \frac{1}{x^7 - x}\,dx$.

   The straightforward approach would be to start with partial fractions, but that would be unthinkably arduous. Try a substitution.

7. Let f be a continuous function on $[a, b]$. Prove that there exists a number x in $[a, b]$ such that

$$\int_a^x f(t)\,dt = \int_x^b f(t)\,dt$$

8. (a) Show that an observer at height H above the north pole of a sphere of radius r can see a part of the sphere that has area

$$\frac{2\pi r^2 H}{r + H}$$

   (b) Two spheres with radii r and R are placed so that the distance between their centers is d, where $d > r + R$. Where should a light be placed on the line joining the centers of the spheres in order to illuminate the largest total surface?

9. Evaluate $\int_0^1 \left(\sqrt[3]{1 - x^7} - \sqrt[7]{1 - x^3}\right) dx$.

10. Snow began to fall during the morning of February 2 and continued steadily into the afternoon. A snowplow began to clear a street at noon, traveling at a constant rate. The plow traveled 6 km from noon to 1 P.M. but only 3 km from 1 P.M. to 2 P.M. When did the snow begin to fall? [*Hints:* To get started, let t be the time measured in hours after noon; let $x(t)$ be the distance traveled by the plow at time t; then the speed of the plow is dx/dt. Let b be the number of hours before noon that it began to snow. Find an expression for the height of the snow at time t. Then use the given information that the rate of removal R (in m^3/h) is constant.]

11. A function f is defined by $f(x) = 3x - 2x^3$ with the restriction that its domain is $\{x \mid 5x^2 \geqslant x^4 + 4\}$. Find the maximum value of f.

12. Find all functions f that satisfy the equation

$$\left(\int f(x)\,dx\right)\left(\int \frac{1}{f(x)}\,dx\right) = -1$$

13. A function f is defined by

$$f(x) = \int_0^{\pi} \cos t \cos(x - t)\, dt \qquad 0 \leqslant x \leqslant 2\pi$$

Find the minimum value of f.

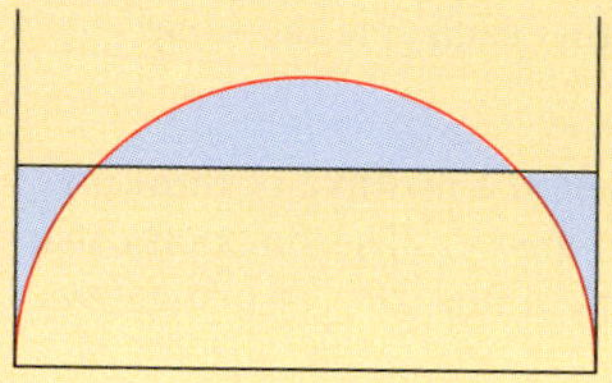

FIGURE FOR PROBLEM 14

14. The figure shows a semicircle with radius 1, horizontal diameter PQ, and tangent lines at P and Q. At what height above the diameter should the horizontal line be placed so as to minimize the shaded area?

15. If $0 < a < b$, find $\lim_{t \to 0} \left\{ \int_0^1 [bx + a(1 - x)]^t\, dx \right\}^{1/t}$.

16. Graph $f(x) = \sin(e^x)$ and use the graph to estimate the value of t such that $\int_t^{t+1} f(x)\, dx$ is a maximum. Then find the exact value of t that maximizes this integral.

17. Show that

$$\int_0^1 (1 - x^2)^n\, dx = \frac{2^{2n}(n!)^2}{(2n + 1)!}$$

Hint: Start by showing that if I_n denotes the integral, then

$$I_{k+1} = \frac{2k + 2}{2k + 3} I_k$$

18. Suppose that f is a positive function such that f' is continuous.
   (a) How is the graph of $y = f(x) \sin nx$ related to the graph of $y = f(x)$? What happens as $n \to \infty$?
   (b) Make a guess as to the value of the limit

$$\lim_{n \to \infty} \int_0^1 f(x) \sin nx\, dx$$

   based on graphs of the integrand.
   (c) Using integration by parts, confirm the guess that you made in part (b). [Use the fact that, since f' is continuous, there is a constant M such that $|f'(x)| \leqslant M$ for $0 \leqslant x \leqslant 1$.]

19. Find all functions f such that f' is continuous and

$$[f(x)]^2 = 100 + \int_0^x \{[f(t)]^2 + [f'(t)]^2\}\, dt \qquad \text{for all real } x$$

20. The **Chebyshev polynomials** T_n are defined by $T_n(x) = \cos(n \arccos x)$ for $n = 0, 1, 2, 3, \ldots$.
   (a) What are the domain and range of these functions?
   (b) We know that $T_0(x) = 1$ and $T_1(x) = x$. Express T_2 explicitly as a quadratic polynomial and T_3 as a cubic polynomial.
   (c) Show that, for $n \geqslant 1$,

$$T_{n+1}(x) = 2xT_n(x) - T_{n-1}(x)$$

   (d) Use part (c) to show that T_n is a polynomial of degree n.
   (e) Use parts (b) and (c) to express T_4, T_5, T_6, and T_7 explicitly as polynomials.
   (f) What are the zeros of T_n? At what numbers does T_n have local maximum and minimum values?

(g) Graph T_2, T_3, T_4, and T_5 on a common screen.

(h) Graph T_5, T_6, and T_7 on a common screen.

(i) Based on your observations from parts (g) and (h), how are the zeros of T_n related to the zeros of T_{n+1}? What about the x-coordinates of the maximum and minimum values?

(j) Based on your graphs in parts (g) and (h), what can you say about $\int_{-1}^{1} T_n(x)\,dx$ when n is odd and when n is even?

(k) Use the substitution $u = \arccos x$ to evaluate the integral in part (j).

(l) The family of functions $f(x) = \cos(c \arccos x)$ are defined even when c is not an integer (but then f is not a polynomial). Describe how the graph of f changes as c increases.

21. Let P be a pyramid with a square base of side $2b$ and suppose that S is a sphere with its center on the base of P and is tangent to all eight edges of P. Find the height of P. Then find the volume of the intersection of S and P.

22. Let C be the arc of the curve $y = f(x)$ between the points $P(p, f(p))$ and $Q(q, f(q))$ and let $\mathcal{R}$ be the region bounded by C, by the line $y = mx + b$ (which lies entirely on one side of C), and by the perpendiculars to the line from P and Q.

(a) Show that the area of $\mathcal{R}$ is

$$\frac{1}{1 + m^2} \int_p^q [f(x) - mx - b][1 + mf'(x)]\,dx$$

(b) Find a formula similar to the one in part (a) for the volume of the solid obtained by rotating $\mathcal{R}$ about the line $y = mx + b$.

(c) Find a formula for the area of the surface obtained by rotating C about the line $y = mx + b$.

[*Hint:* The formula in part (a) can be verified by subtracting areas, but it is more instructive to derive it by first approximating the area using rectangles perpendicular to the line, as shown in the figure. This will also help in finding the formulas for parts (b) and (c). Use the figure to help express Δu in terms of Δx.]

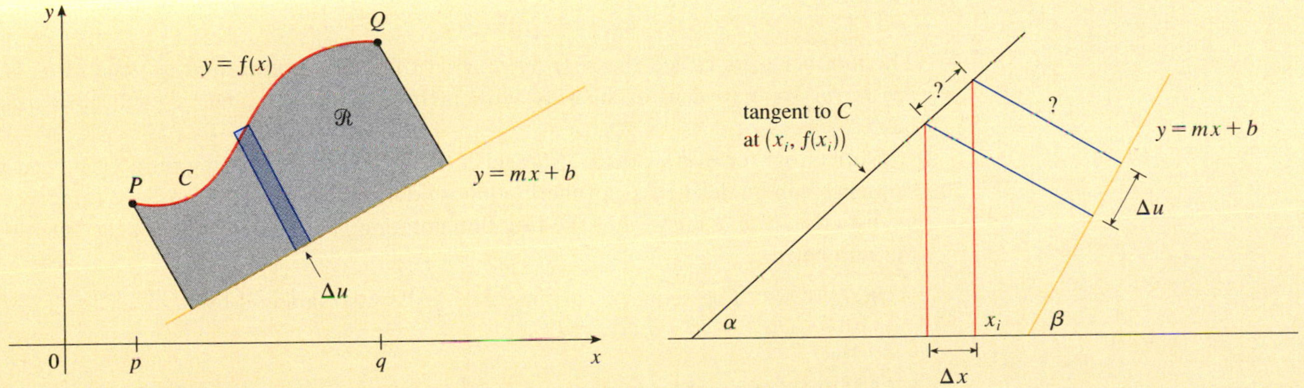

FIGURE FOR PROBLEM 22

10

INFINITE SEQUENCES AND SERIES

> Our minds are finite, and yet even in those circumstances of finitude, we are surrounded by possibilities that are infinite, and the purpose of human life is to grasp as much as we can out of that infinitude.
>
> ALFRED NORTH WHITEHEAD

Infinite sequences and series were introduced briefly in the *Preview of Calculus* (Section 5 in Review and Preview) in connection with Zeno's paradoxes and the decimal representation of numbers. Their importance in calculus stems from Newton's idea of representing functions as sums of infinite series. For instance, in finding areas he often integrated a function by first expressing it as a series and then integrating each term of the series. We will pursue his idea in Section 10.7 in order to integrate such functions as e^{-x^2}. (Recall that we have previously been unable to do this). Many of the functions that arise in mathematical physics and chemistry, such as Bessel functions, are defined as sums of series, so it is important to be familiar with the basic concepts of convergence of infinite sequences and series.

Physicists also use series in another way, as we will see in Section 10.9. In studying fields as diverse as optics, special relativity, and electromagnetism, they analyze phenomena by replacing a function with the first few terms in the series that represents it.

10.1 SEQUENCES

A **sequence** can be thought of as a list of numbers written in a definite order:

$$a_1, \quad a_2, \quad a_3, \quad a_4, \quad \ldots, \quad a_n, \quad \ldots$$

The number a_1 is called the *first term*, a_2 is the *second term*, and in general a_n is the *nth term*. We will deal exclusively with infinite sequences and so each term a_n will have a successor a_{n+1}.

Notice that for every positive integer n there is a corresponding number a_n and so a sequence can be defined as a function whose domain is the set of positive integers. But we usually write a_n instead of the function notation $f(n)$ for the value of the function at the number n.

NOTATION: The sequence $\{a_1, a_2, a_3, \ldots\}$ is also denoted by

$$\{a_n\} \qquad \text{or} \qquad \{a_n\}_{n=1}^{\infty}$$

EXAMPLE 1 Some sequences can be defined by giving a formula for the nth term. In the following examples we give three descriptions of the sequence: one by using the preceding notation, another by using the defining formula, and a third by writing out the terms of the sequence. Notice that n doesn't have to start at 1.

(a) $\left\{\dfrac{n}{n+1}\right\}_{n=1}^{\infty} \qquad a_n = \dfrac{n}{n+1} \qquad \left\{\dfrac{1}{2}, \dfrac{2}{3}, \dfrac{3}{4}, \dfrac{4}{5}, \ldots, \dfrac{n}{n+1}, \ldots\right\}$

(b) $\left\{\dfrac{(-1)^n(n+1)}{3^n}\right\}$ $\quad a_n = \dfrac{(-1)^n(n+1)}{3^n}$ $\quad \left\{-\dfrac{2}{3}, \dfrac{3}{9}, -\dfrac{4}{27}, \dfrac{5}{81}, \ldots, \dfrac{(-1)^n(n+1)}{3^n}, \ldots\right\}$

(c) $\{\sqrt{n-3}\}_{n=3}^{\infty}$ $\quad a_n = \sqrt{n-3},\ n \geqslant 3$ $\quad \{0, 1, \sqrt{2}, \sqrt{3}, \ldots, \sqrt{n-3}, \ldots\}$

(d) $\left\{\cos\dfrac{n\pi}{6}\right\}_{n=0}^{\infty}$ $\quad a_n = \cos\dfrac{n\pi}{6},\ n \geqslant 0$ $\quad \left\{1, \dfrac{\sqrt{3}}{2}, \dfrac{1}{2}, 0, \ldots, \cos\dfrac{n\pi}{6}, \ldots\right\}$ ■

EXAMPLE 2 Here are some sequences that don't have a simple defining equation.

(a) The sequence $\{p_n\}$, where p_n is the population of the world as of January 1 in the year n.

(b) If we let a_n be the digit in the nth decimal place of the number e, then $\{a_n\}$ is a well-defined sequence whose first few terms are

$$\{7, 1, 8, 2, 8, 1, 8, 2, 8, 4, 5, \ldots\}$$

(c) The **Fibonacci sequence** $\{f_n\}$ is defined recursively by the conditions

$$f_1 = 1 \qquad f_2 = 1 \qquad f_n = f_{n-1} + f_{n-2} \qquad n \geqslant 3$$

Each term is the sum of the two preceding terms. The first few terms are

$$\{1, 1, 2, 3, 5, 8, 13, 21, \ldots\}$$

This sequence arose when the 13th-century Italian mathematician known as Fibonacci solved a problem concerning the breeding of rabbits (see Exercise 35). ■

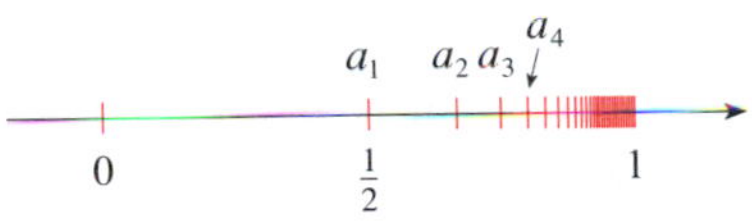

FIGURE 1

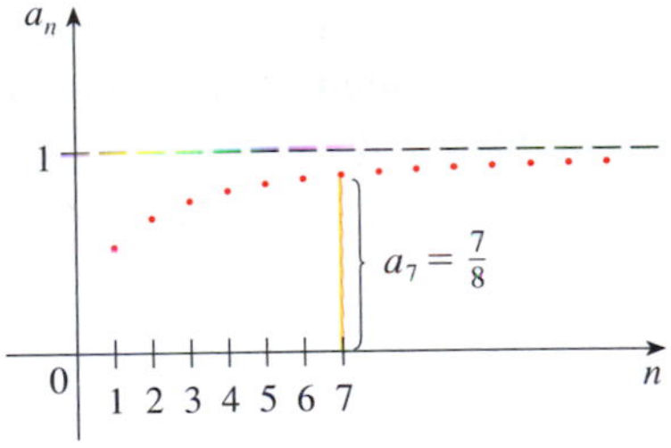

FIGURE 2

A sequence such as the one in Example 1(a), $a_n = n/(n+1)$, can be pictured either by plotting its terms on a number line, as in Figure 1, or by plotting its graph, as in Figure 2. Note that, since a sequence is a function whose domain is the set of positive integers, its graph consists of isolated points with coordinates

$$(1, a_1) \qquad (2, a_2) \qquad (3, a_3) \qquad \ldots \qquad (n, a_n) \qquad \ldots$$

From Figure 1 or 2 it appears that the terms of the sequence $a_n = n/(n+1)$ are approaching 1 as n becomes large. In fact, the difference

$$1 - \frac{n}{n+1} = \frac{1}{n+1}$$

can be made as small as we like by taking n sufficiently large. We indicate this by writing

$$\lim_{n\to\infty} \frac{n}{n+1} = 1$$

In general, the notation

$$\lim_{n\to\infty} a_n = L$$

means that the terms of the sequence $\{a_n\}$ approach L as n becomes large. Notice that the following definition of the limit of a sequence is very similar to the definition of a limit of a function at infinity given in Section 2.6.

(1) DEFINITION A sequence $\{a_n\}$ has the **limit** L and we write

$$\lim_{n\to\infty} a_n = L \qquad \text{or} \qquad a_n \to L \text{ as } n \to \infty$$

if we can make the terms a_n as close to L as we like by taking n sufficiently large. If $\lim_{n\to\infty} a_n$ exists, we say the sequence **converges** (or is **convergent**). Otherwise, we say the sequence **diverges** (or is **divergent**).

Figure 3 illustrates Definition 1 by showing the graphs of two sequences that have the limit L.

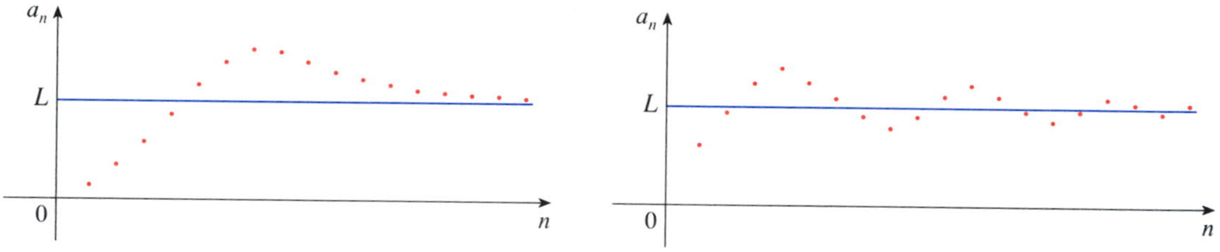

FIGURE 3
Graphs of two sequences with $\lim_{n\to\infty} a_n = L$

Here is a more precise way of stating Definition 1:

(2) DEFINITION If $\{a_n\}$ is a sequence, then

$$\lim_{n\to\infty} a_n = L$$

means that for every $\varepsilon > 0$ there is a corresponding integer N such that

$$|a_n - L| < \varepsilon \qquad \text{whenever} \qquad n > N$$

Figure 4 illustrates Definition 2. The points on the graph of $\{a_n\}$ must lie between the horizontal lines $y = L + \varepsilon$ and $y = L - \varepsilon$ if $n > N$. This picture must be valid no matter how small ε is chosen, but usually a smaller ε requires a larger N.

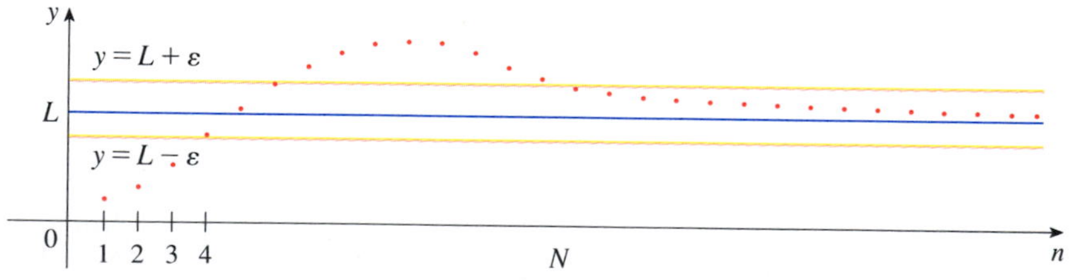

FIGURE 4

If you compare Definition 2 with Definition 5 in Section 2.6 you will see that the only difference between $\lim_{n\to\infty} a_n = L$ and $\lim_{x\to\infty} f(x) = L$ is that n is required to be an integer. Thus, we have the following theorem, which is illustrated by Figure 5.

(3) THEOREM If $\lim_{x\to\infty} f(x) = L$ and $f(n) = a_n$ when n is an integer, then $\lim_{n\to\infty} a_n = L$.

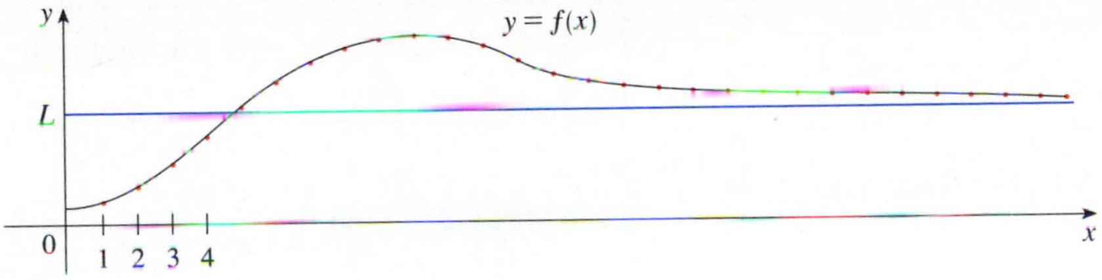

FIGURE 5

In particular, since we know from Section 2.6 that $\lim_{x\to\infty}(1/x^r) = 0$ when $r > 0$, we have

(4) $$\lim_{n\to\infty} \frac{1}{n^r} = 0 \qquad \text{if } r > 0$$

If a_n becomes large as n becomes large, we use the notation

$$\lim_{n\to\infty} a_n = \infty$$

In this case the sequence $\{a_n\}$ is divergent, but in a special way. We say that $\{a_n\}$ diverges to ∞.

The Limit Laws given in Section 2.3 also hold for the limits of sequences and their proofs are similar.

Limit Laws for Convergent Sequences

If $\{a_n\}$ and $\{b_n\}$ are convergent sequences and c is a constant, then

$$\lim_{n\to\infty}(a_n + b_n) = \lim_{n\to\infty} a_n + \lim_{n\to\infty} b_n$$

$$\lim_{n\to\infty}(a_n - b_n) = \lim_{n\to\infty} a_n - \lim_{n\to\infty} b_n$$

$$\lim_{n\to\infty} ca_n = c \lim_{n\to\infty} a_n$$

$$\lim_{n\to\infty}(a_n b_n) = \lim_{n\to\infty} a_n \cdot \lim_{n\to\infty} b_n$$

$$\lim_{n\to\infty} \frac{a_n}{b_n} = \frac{\lim_{n\to\infty} a_n}{\lim_{n\to\infty} b_n} \quad \text{if } \lim_{n\to\infty} b_n \neq 0$$

$$\lim_{n\to\infty} c = c$$

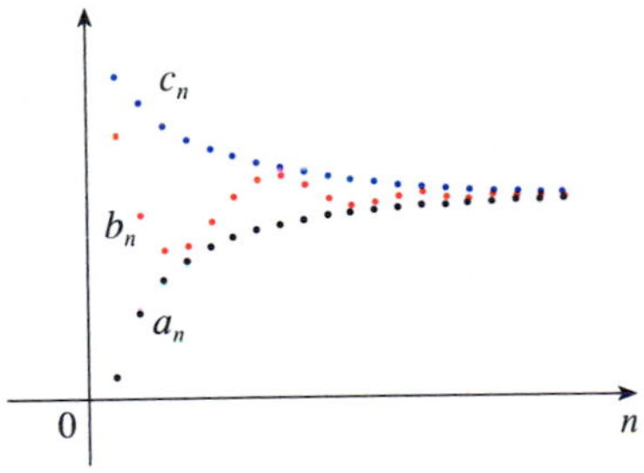

FIGURE 6
The sequence $\{b_n\}$ is squeezed between the sequences $\{a_n\}$ and $\{c_n\}$.

The Squeeze Theorem can also be adapted for sequences as follows. (See Figure 6.)

Squeeze Theorem for Sequences

If $a_n \leqslant b_n \leqslant c_n$ for $n \geqslant n_0$ and $\lim_{n\to\infty} a_n = \lim_{n\to\infty} c_n = L$, then $\lim_{n\to\infty} b_n = L$.

Another useful fact about limits of sequences is given by the following theorem, which follows from the Squeeze Theorem because $-|a_n| \leqslant a_n \leqslant |a_n|$.

(5) THEOREM If $\lim_{n\to\infty} |a_n| = 0$, then $\lim_{n\to\infty} a_n = 0$.

EXAMPLE 3 Find $\lim_{n\to\infty} \dfrac{n}{n+1}$.

SOLUTION The method is similar to the one we used in Section 2.6: Divide numerator and denominator by the highest power of n and then use the Limit Laws.

This shows that the guess we made earlier from Figures 1 and 2 was correct.

$$\lim_{n\to\infty} \frac{n}{n+1} = \lim_{n\to\infty} \frac{1}{1+\dfrac{1}{n}} = \frac{\lim_{n\to\infty} 1}{\lim_{n\to\infty} 1 + \lim_{n\to\infty} \dfrac{1}{n}}$$

$$= \frac{1}{1+0} = 1$$

Here we used Equation 4 with $r = 1$. ■

EXAMPLE 4 Calculate $\lim_{n\to\infty} \dfrac{\ln n}{n}$.

SOLUTION Notice that both numerator and denominator approach infinity as $n \to \infty$. We can't apply l'Hospital's Rule directly because it applies not to sequences but only to functions of a real variable. However, we can apply l'Hospital's Rule to the related function $f(x) = (\ln x)/x$ and obtain

$$\lim_{x\to\infty} \frac{\ln x}{x} = \lim_{x\to\infty} \frac{1/x}{1} = 0$$

Therefore, by Theorem 3 we have

$$\lim_{n\to\infty} \frac{\ln n}{n} = 0$$

■

EXAMPLE 5 Determine whether the sequence $a_n = (-1)^n$ is convergent or divergent.

SOLUTION If we write out the terms of the sequence, we obtain

$$\{-1, 1, -1, 1, -1, 1, -1, \ldots\}$$

The graph of this sequence is shown in Figure 7. Since the terms oscillate between 1 and -1 infinitely often, a_n does not approach any number. Thus, $\lim_{n\to\infty} (-1)^n$ does not exist; that is, the sequence $\{(-1)^n\}$ is divergent.

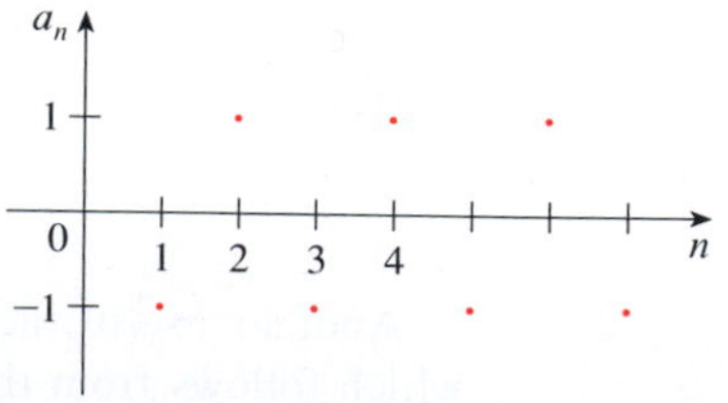

FIGURE 7

■

The graph of the sequence in Example 6 is shown in Figure 8 and supports the answer.

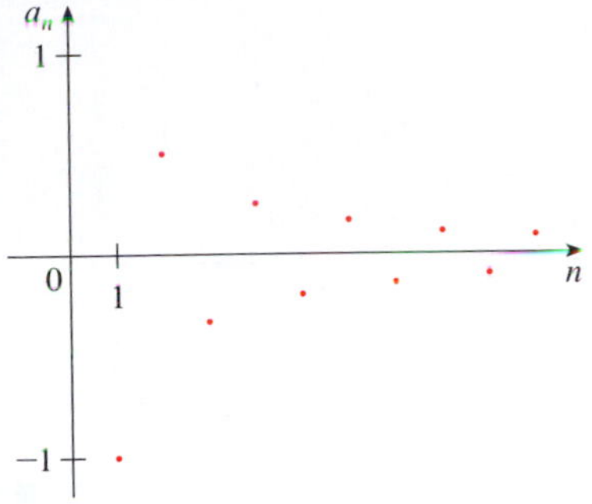

FIGURE 8

EXAMPLE 6 Evaluate $\lim_{n\to\infty} \frac{(-1)^n}{n}$ if it exists.

SOLUTION

$$\lim_{n\to\infty} \left| \frac{(-1)^n}{n} \right| = \lim_{n\to\infty} \frac{1}{n} = 0$$

Therefore, by Theorem 5,

$$\lim_{n\to\infty} \frac{(-1)^n}{n} = 0$$

■

EXAMPLE 7 Discuss the convergence of the sequence $a_n = n!/n^n$, where $n! = 1 \cdot 2 \cdot 3 \cdot \cdots \cdot n$.

SOLUTION Both numerator and denominator approach infinity as $n \to \infty$ but here we have no corresponding function for use with l'Hospital's Rule ($x!$ is not defined when x is not an integer). Let's write out a few terms to get a feeling for what happens to a_n as n gets large:

$$a_1 = 1 \qquad a_2 = \frac{1 \cdot 2}{2 \cdot 2} \qquad a_3 = \frac{1 \cdot 2 \cdot 3}{3 \cdot 3 \cdot 3}$$

$$a_n = \frac{1 \cdot 2 \cdot 3 \cdot \cdots \cdot n}{n \cdot n \cdot n \cdot \cdots \cdot n} \tag{6}$$

It appears from these expressions and the graph in Figure 9 that the terms are decreasing and perhaps approach 0. To confirm this, observe from Equation 6 that

$$a_n = \frac{1}{n}\left(\frac{2 \cdot 3 \cdot \cdots \cdot n}{n \cdot n \cdot \cdots \cdot n}\right)$$

so

$$0 < a_n \leqslant \frac{1}{n}$$

We know that $1/n \to 0$ as $n \to \infty$. Therefore, $a_n \to 0$ as $n \to \infty$ by the Squeeze Theorem. ■

CREATING GRAPHS OF SEQUENCES

Some computer algebra systems have special commands that enable us to create sequences and graph them directly. With most graphing calculators, however, sequences can be graphed by using parametric equations. For instance, the sequence in Example 7 can be graphed by entering the parametric equations

$$x = t \qquad y = t!/t^t$$

and graphing in dot mode, starting with $t = 1$ and setting the t-step equal to 1. The result is shown in Figure 9.

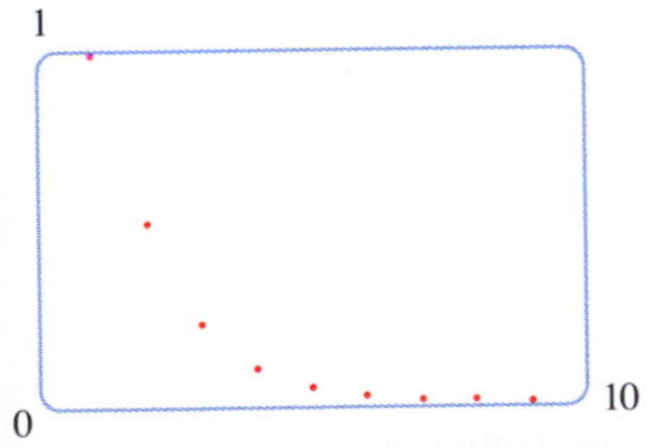

FIGURE 9

EXAMPLE 8 For what values of r is the sequence $\{r^n\}$ convergent?

SOLUTION We know from Section 2.6 and the graphs of the exponential functions in Section 4.1 that $\lim_{x\to\infty} a^x = \infty$ for $a > 1$ and $\lim_{x\to\infty} a^x = 0$ for $0 < a < 1$. Therefore, putting $a = r$ and using Theorem 3, we have

$$\lim_{n\to\infty} r^n = \begin{cases} \infty & \text{if } r > 1 \\ 0 & \text{if } 0 < r < 1 \end{cases}$$

For the cases $r = 1$ and $r = 0$ we have

$$\lim_{n\to\infty} 1^n = \lim_{n\to\infty} 1 = 1 \qquad \text{and} \qquad \lim_{n\to\infty} 0^n = \lim_{n\to\infty} 0 = 0$$

If $-1 < r < 0$, then $0 < |r| < 1$, so

$$\lim_{n\to\infty} |r^n| = \lim_{n\to\infty} |r|^n = 0$$

and, therefore, $\lim_{n\to\infty} r^n = 0$ by Theorem 5. If $r \leqslant -1$, then $\{r^n\}$ diverges as in Example 5. Figure 10 shows the graphs for various values of r. (The case $r = -1$ is shown in Figure 7.)

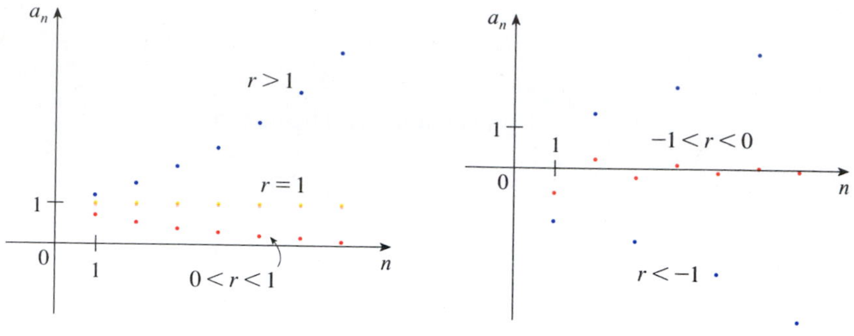

FIGURE 10
The sequence $a_n = r^n$

The results of Example 8 are summarized for future use as follows.

(7) The sequence $\{r^n\}$ is convergent if $-1 < r \leqslant 1$ and divergent for all other values of r.

$$\lim_{n\to\infty} r^n = \begin{cases} 0 & \text{if } -1 < r < 1 \\ 1 & \text{if } r = 1 \end{cases}$$

DEFINITION A sequence $\{a_n\}$ is called **increasing** if $a_n < a_{n+1}$ for all $n \geqslant 1$, that is, $a_1 < a_2 < a_3 < \cdots$. It is called **decreasing** if $a_n > a_{n+1}$ for all $n \geqslant 1$. A sequence is **monotonic** if it is either increasing or decreasing.

EXAMPLE 9 The sequence $\left\{\dfrac{3}{n+5}\right\}$ is decreasing because

$$\frac{3}{n+5} > \frac{3}{n+6}$$

for all $n \geqslant 1$. (The right side is smaller because it has a larger denominator.) ■

EXAMPLE 10 Show that the sequence $a_n = \dfrac{n}{n^2+1}$ is decreasing.

SOLUTION 1 We must show that $a_{n+1} < a_n$, that is,

$$\frac{n+1}{(n+1)^2+1} < \frac{n}{n^2+1}$$

This inequality is equivalent to the one we get by cross-multiplication:

$$\frac{n+1}{(n+1)^2+1} < \frac{n}{n^2+1} \iff (n+1)(n^2+1) < n[(n+1)^2+1]$$

$$\iff n^3 + n^2 + n + 1 < n^3 + 2n^2 + 2n$$

$$\iff 1 < n^2 + n$$

Since $n \geqslant 1$, we know that the inequality $n^2 + n > 1$ is true. Therefore, $a_{n+1} < a_n$ and so $\{a_n\}$ is decreasing.

SOLUTION 2 Consider the function $f(x) = \dfrac{x}{x^2 + 1}$:

$$f'(x) = \frac{x^2 + 1 - 2x^2}{(x^2 + 1)^2} = \frac{1 - x^2}{(x^2 + 1)^2} < 0 \qquad \text{whenever } x^2 > 1$$

Thus, f is decreasing on $(1, \infty)$ and so $f(n) > f(n + 1)$. Therefore, $\{a_n\}$ is decreasing. ■

DEFINTION A sequence $\{a_n\}$ is **bounded above** if there is a number M such that

$$a_n \leqslant M \qquad \text{for all } n \geqslant 1$$

It is **bounded below** if there is a number m such that

$$m \leqslant a_n \qquad \text{for all } n \geqslant 1$$

If it is bounded above and below, then $\{a_n\}$ is a **bounded sequence**.

For instance, the sequence $a_n = n$ is bounded below $(a_n > 0)$ but not above. The sequence $a_n = n/(n + 1)$ is bounded because $0 < a_n < 1$ for all n.

We know that not every bounded sequence is convergent [$a_n = (-1)^n$ satisfies $-1 \leqslant a_n \leqslant 1$ but is divergent, from Example 5] and not every monotonic sequence is convergent ($a_n = n \to \infty$). But if a sequence is both bounded *and* monotonic, then it must be convergent. This fact is stated as Theorem 8 and proved in Appendix F, but intuitively you can understand why it is true by looking at Figure 11. If $\{a_n\}$ is increasing and $a_n \leqslant M$ for all n, then the terms are forced to crowd together and approach some number L.

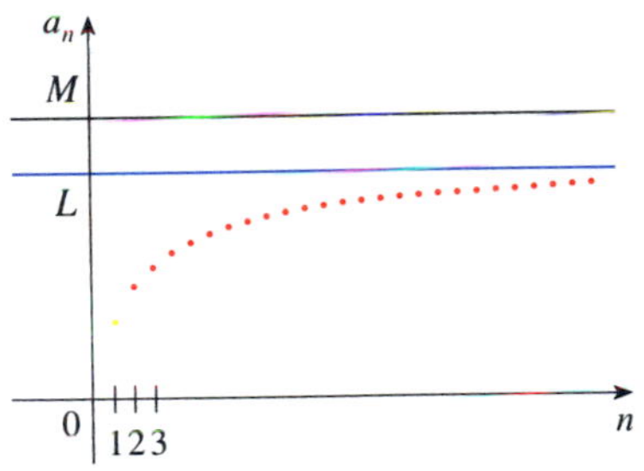

FIGURE 11

(8) MONOTONIC SEQUENCE THEOREM Every bounded, monotonic sequence is convergent.

EXAMPLE 11 Investigate the sequence $\{a_n\}$ defined by the recurrence relation

$$a_1 = 2 \qquad a_{n+1} = \tfrac{1}{2}(a_n + 6) \qquad \text{for } n = 1, 2, 3, \ldots$$

SOLUTION We begin by computing the first few terms:

$$a_1 = 2 \qquad a_2 = \tfrac{1}{2}(2 + 6) = 4 \qquad a_3 = \tfrac{1}{2}(4 + 6) = 5 \qquad a_4 = \tfrac{1}{2}(5 + 6) = 5.5$$

$$a_5 = 5.75 \qquad a_6 = 5.875 \qquad a_7 = 5.9375 \qquad a_8 = 5.96875$$

Mathematical induction is often used in dealing with recursive sequences. See Appendix E for a discussion of the Principle of Mathematical Induction.

These initial terms suggest that the sequence is increasing and the terms are approaching 6. To confirm that the sequence is increasing, we use mathematical induction to show that $a_{n+1} > a_n$ for all $n \geqslant 1$. This is true for $n = 1$ because $a_2 = 4 > a_1$. If we assume that it is true for $n = k$, then we have

$$a_{k+1} > a_k$$

so $$a_{k+1} + 6 > a_k + 6$$

and $$\tfrac{1}{2}(a_{k+1} + 6) > \tfrac{1}{2}(a_k + 6)$$

Thus $$a_{k+2} > a_{k+1}$$

We have deduced that $a_{n+1} > a_n$ is true for $n = k + 1$. Therefore, the inequality is true for all n by induction.

Next we verify that $\{a_n\}$ is bounded by showing that $a_n < 6$ for all n. (Since the sequence is increasing, we already know that it has a lower bound: $a_n \geqslant a_1 = 2$ for all n.) We know that $a_1 < 6$, so the assertion is true for $n = 1$. Suppose it is true for $n = k$. Then

$$a_k < 6$$

so $$a_k + 6 < 12$$

and $$\tfrac{1}{2}(a_k + 6) < \tfrac{1}{2}(12)$$

Thus $$a_{k+1} < 6$$

This shows, by mathematical induction, that $a_n < 6$ for all n.

Since the sequence $\{a_n\}$ is increasing and bounded, the Monotonic Sequence Theorem guarantees that it has a limit. The theorem doesn't tell us what the value of the limit is. But now that we know $L = \lim_{n\to\infty} a_n$ exists, we can use the given recurrence relation to write

$$\lim_{n\to\infty} a_{n+1} = \lim_{n\to\infty} \tfrac{1}{2}(a_n + 6) = \tfrac{1}{2}\left(\lim_{n\to\infty} a_n + 6\right) = \tfrac{1}{2}(L + 6)$$

Since $a_n \to L$, it follows that $a_{n+1} \to L$, too (as $n \to \infty$, $n + 1 \to \infty$ also). So we have

$$L = \tfrac{1}{2}(L + 6)$$

Solving this equation for L, we get $L = 6$, as we predicted. ■

EXERCISES 10.1

1. (a) What is a sequence?
(b) What does it mean to say that $\lim_{n\to\infty} a_n = 8$?
(c) What does it mean to say that $\lim_{n\to\infty} a_n = \infty$?

2. (a) What is a convergent sequence? Give two examples.
(b) What is a divergent sequence? Give two examples.

3. List the first six terms of the sequence defined by

$$a_n = \frac{n}{2n + 1}$$

Does the sequence appear to have a limit? If so, find it.

4. List the first eight terms of the sequence $\{\sin(n\pi/2)\}$. Does this sequence appear to have a limit? If so, find it. If not, explain why.

5–8 ■ Find a formula for the general term a_n of the sequence, assuming that the pattern of the first few terms continues.

5. $\left\{\frac{1}{2}, \frac{1}{4}, \frac{1}{8}, \frac{1}{16}, \ldots\right\}$

6. $\left\{\frac{1}{2}, \frac{1}{4}, \frac{1}{6}, \frac{1}{8}, \ldots\right\}$

7. $\left\{\frac{3}{16}, \frac{4}{25}, \frac{5}{36}, \frac{6}{49}, \ldots\right\}$

8. $\{0, 2, 0, 2, 0, 2, \ldots\}$

9–26 ■ Determine whether the sequence converges or diverges. If it converges, find the limit.

9. $a_n = \dfrac{1}{5^n}$

10. $a_n = 4\sqrt{n}$

11. $a_n = \dfrac{n^2 - 1}{n^2 + 1}$

12. $a_n = \dfrac{4n - 3}{3n + 4}$

13. $a_n = \dfrac{n^2}{n + 1}$

14. $\{\arctan 2n\}$

15. $a_n = \cos(n\pi/2)$

16. $a_n = (-1)^n \dfrac{n^2}{1 + n^3}$

17. $\left\{\dfrac{\pi^n}{3^n}\right\}$

18. $\left\{\dfrac{3 + (-1)^n}{n^2}\right\}$

19. $\left\{\dfrac{\ln(n^2)}{n}\right\}$

20. $\{(-1)^n \sin(1/n)\}$

21. $\{\sqrt{n+2} - \sqrt{n}\}$

22. $\left\{\dfrac{\ln(2 + e^n)}{3n}\right\}$

23. $a_n = n2^{-n}$

24. $a_n = \ln(n + 1) - \ln n$

25. $a_n = \dfrac{\cos^2 n}{2^n}$

26. $a_n = \dfrac{(-3)^n}{n!}$

27–32 ■ Use a graph of the sequence to decide whether the sequence is convergent or divergent. If the sequence is convergent, guess the value of the limit from the graph and then prove your guess. (See the margin note on page 581 for advice on graphing sequences.)

27. $a_n = (-1)^n \dfrac{n+1}{n}$

28. $a_n = 2 + (-2/\pi)^n$

29. $\left\{\arctan\left(\dfrac{2n}{2n+1}\right)\right\}$

30. $\left\{\dfrac{\sin n}{\sqrt{n}}\right\}$

31. $a_n = \dfrac{n^3}{n!}$

32. $a_n = \dfrac{1 \cdot 3 \cdot 5 \cdot \cdots \cdot (2n-1)}{(2n)^n}$

33. (a) Determine whether the sequence defined as follows is convergent or divergent:

$$a_1 = 1 \qquad a_{n+1} = 4 - a_n \quad \text{for } n \geqslant 1$$

(b) What happens if the first term is $a_1 = 2$?

34. (a) If $\lim_{n\to\infty} a_n = L$, what is the value of $\lim_{n\to\infty} a_{n+1}$?

(b) A sequence $\{a_n\}$ is defined by

$$a_1 = 1 \qquad a_{n+1} = 1/(1 + a_n) \quad \text{for } n \geqslant 1$$

Find the first ten terms of the sequence correct to five decimal places. Does it appear that the sequence is convergent? If so, estimate the value of the limit to three decimal places.

(c) Assuming that the sequence in part (b) has a limit, use part (a) to find its exact value. Compare with your estimate from part (b).

35. (a) Fibonacci posed the following problem: Suppose that rabbits live forever and that every month each pair produces a new pair, which becomes productive at age 2 months. If we start with one newborn pair, how many pairs of rabbits will we have in the nth month? Show that the answer is f_n, where $\{f_n\}$ is the Fibonacci sequence defined in Example 2(c).

(b) Let $a_n = f_{n+1}/f_n$ and show that $a_{n-1} = 1 + 1/a_{n-2}$. Assuming that $\{a_n\}$ is convergent, find its limit.

36. Find the limit of the sequence

$$\{\sqrt{2}, \sqrt{2\sqrt{2}}, \sqrt{2\sqrt{2\sqrt{2}}}, \ldots\}$$

37–40 ■ Determine whether the sequence is increasing, decreasing, or not monotonic.

37. $a_n = \dfrac{1}{3n+5}$

38. $a_n = 3 + (-1)^n/n$

39. $a_n = \dfrac{n-2}{n+2}$

40. $a_n = \dfrac{\sqrt{n+1}}{5n+3}$

41. Suppose you know that $\{a_n\}$ is a decreasing sequence and all its terms lie between the numbers 5 and 8. Explain why the sequence has a limit. What can you say about the value of the limit?

42. A sequence $\{a_n\}$ is given by $a_1 = \sqrt{2}$, $a_{n+1} = \sqrt{2 + a_n}$.

(a) By induction or otherwise, show that $\{a_n\}$ is increasing and bounded above by 3. Deduce that $\lim_{n\to\infty} a_n$ exists.

(b) Find $\lim_{n\to\infty} a_n$.

43. Show that the sequence defined by

$$a_1 = 1 \qquad a_{n+1} = 3 - 1/a_n$$

is increasing and $a_n < 3$ for all n. Deduce that $\{a_n\}$ is convergent, and find its limit.

44. Show that the sequence defined by

$$a_1 = 2 \qquad a_{n+1} = 1/(3 - a_n)$$

satisfies $0 < a_n \leqslant 2$ and is decreasing. Deduce that the sequence is convergent, and find its limit.

45. We know that $\lim_{n\to\infty} (0.8)^n = 0$ [from (7) with $r = 0.8$]. Use logarithms to determine how large n has to be so that $(0.8)^n < 0.000001$.

46. (a) Let $a_1 = a$, $a_2 = f(a)$, $a_3 = f(a_2) = f(f(a))$, $\ldots$, $a_{n+1} = f(a_n)$, where f is a continuous function. If $\lim_{n\to\infty} a_n = L$, show that $f(L) = L$.

(b) Illustrate part (a) by taking $f(x) = \cos x$, $a = 1$, and estimating the value of L to five decimal places.

47. Let a and b be positive numbers with $a > b$. Let a_1 be their arithmetic mean and b_1 their geometric mean:

$$a_1 = \frac{a+b}{2} \qquad b_1 = \sqrt{ab}$$

Repeat this process so that, in general,

$$a_{n+1} = \frac{a_n + b_n}{2} \qquad b_{n+1} = \sqrt{a_n b_n}$$

(a) Use mathematical induction to show that

$$a_n > a_{n+1} > b_{n+1} > b_n$$

(b) Deduce that both $\{a_n\}$ and $\{b_n\}$ are convergent.

(c) Show that $\lim_{n\to\infty} a_n = \lim_{n\to\infty} b_n$. Gauss called the common value of these limits the **arithmetic-geometric mean** of the numbers a and b.

48. A sequence is defined recursively by

$$a_1 = 1 \qquad a_{n+1} = 1 + \frac{1}{1 + a_n}$$

Find the first eight terms of the sequence $\{a_n\}$. What do you notice about the odd terms and the even terms? By considering the odd and even terms separately, show that $\{a_n\}$ is convergent and deduce that

$$\lim_{n \to \infty} a_n = \sqrt{2}$$

This gives the **continued fraction expansion**

$$\sqrt{2} = 1 + \cfrac{1}{2 + \cfrac{1}{2 + \cdots}}$$

10.2 SERIES

If we try to add the terms of an infinite sequence $\{a_n\}_{n=1}^{\infty}$ we get an expression of the form

$$a_1 + a_2 + a_3 + \cdots + a_n + \cdots \tag{1}$$

which is called an **infinite series** (or just a **series**) and is denoted, for short, by the symbol

$$\sum_{n=1}^{\infty} a_n \qquad \text{or} \qquad \sum a_n$$

But does it make sense to talk about the sum of infinitely many terms?

It would be impossible to find a finite sum for the series

$$1 + 2 + 3 + 4 + 5 + \cdots + n + \cdots$$

because if we start adding the terms we get the cumulative sums 1, 3, 6, 10, 15, 21, . . . and, after the nth term, $n(n + 1)/2$, which becomes very large as n increases.

However, if we start to add the terms of the series

$$\frac{1}{2} + \frac{1}{4} + \frac{1}{8} + \frac{1}{16} + \frac{1}{32} + \frac{1}{64} + \cdots + \frac{1}{2^n} + \cdots$$

n	Sum of first n terms
1	0.50000000
2	0.75000000
3	0.87500000
4	0.93750000
5	0.96875000
6	0.98437500
7	0.99218750
10	0.99902344
15	0.99996948
20	0.99999905
25	0.99999997

we get $\frac{1}{2}, \frac{3}{4}, \frac{7}{8}, \frac{15}{16}, \frac{31}{32}, \frac{63}{64}, \ldots, 1 - 1/2^n, \ldots$. The table shows that as we add more and more terms, these partial sums become closer and closer to 1. (See also Figure 11 on page 45.) In fact, by adding sufficiently many terms of the series we can make the partial sums as close as we like to 1. So it seems reasonable to say that the sum of this infinite series is 1 and to write

$$\sum_{n=1}^{\infty} \frac{1}{2^n} = \frac{1}{2} + \frac{1}{4} + \frac{1}{8} + \frac{1}{16} + \cdots + \frac{1}{2^n} + \cdots = 1$$

We use a similar idea to determine whether or not a general series (1) has a sum. We consider the **partial sums**

$$s_1 = a_1$$

$$s_2 = a_1 + a_2$$

$$s_3 = a_1 + a_2 + a_3$$

$$s_4 = a_1 + a_2 + a_3 + a_4$$

and, in general,

$$s_n = a_1 + a_2 + a_3 + \cdots + a_n = \sum_{i=1}^{n} a_i$$

These partial sums form a new sequence $\{s_n\}$, which may or may not have a limit. If $\lim_{n\to\infty} s_n = s$ exists (as a finite number), then, as in the preceding example, we call it the sum of the infinite series $\Sigma\, a_n$.

(2) DEFINITION Given a series $\sum_{n=1}^{\infty} a_n = a_1 + a_2 + a_3 + \cdots$, let s_n denote its nth partial sum:

$$s_n = \sum_{i=1}^{n} a_i = a_1 + a_2 + \cdots + a_n$$

If the sequence $\{s_n\}$ is convergent and $\lim_{n\to\infty} s_n = s$ exists as a real number, then the series $\Sigma\, a_n$ is called **convergent** and we write

$$a_1 + a_2 + \cdots + a_n + \cdots = s \qquad \text{or} \qquad \sum_{n=1}^{\infty} a_n = s$$

The number s is called the **sum** of the series. Otherwise, the series is called **divergent**.

Thus, when we write $\sum_{n=1}^{\infty} a_n = s$ we mean that by adding sufficiently many terms of the series we can get as close as we like to the number s. Notice that

$$\sum_{n=1}^{\infty} a_n = \lim_{n\to\infty} \sum_{i=1}^{n} a_i$$

EXAMPLE 1 An important example of an infinite series is the **geometric series**

$$a + ar + ar^2 + ar^3 + \cdots + ar^{n-1} + \cdots = \sum_{n=1}^{\infty} ar^{n-1} \qquad a \neq 0$$

Each term is obtained from the preceding one by multiplying it by the common ratio r. (We have already considered the special case where $a = \frac{1}{2}$ and $r = \frac{1}{2}$.)

If $r = 1$, then $s_n = a + a + \cdots + a = na \to \pm\infty$. Since $\lim_{n\to\infty} s_n$ doesn't exist, the geometric series diverges in this case.

If $r \neq 1$, we have

$$s_n = a + ar + ar^2 + \cdots + ar^{n-1}$$

and

$$rs_n = \quad ar + ar^2 + \cdots + ar^{n-1} + ar^n$$

Subtracting these equations, we get

$$s_n - rs_n = a - ar^n$$

(3)

$$s_n = \frac{a(1 - r^n)}{1 - r}$$

If $-1 < r < 1$, we know from (7) in Section 10.1 that $r^n \to 0$ as $n \to \infty$, so

$$\lim_{n\to\infty} s_n = \lim_{n\to\infty} \frac{a(1-r^n)}{1-r} = \frac{a}{1-r} - \frac{a}{1-r}\lim_{n\to\infty} r^n = \frac{a}{1-r}$$

Thus, when $|r| < 1$ the geometric series is convergent and its sum is $a/(1-r)$.

If $r \leqslant -1$ or $r > 1$, the sequence $\{r^n\}$ is divergent and so, by Equation 3, $\lim_{n\to\infty} s_n$ does not exist. Therefore, the geometric series diverges in those cases. ■

We summarize the results of Example 1 as follows.

(4) The geometric series

$$\sum_{n=1}^{\infty} ar^{n-1} = a + ar + ar^2 + \cdots$$

is convergent if $|r| < 1$ and its sum is

$$\sum_{n=1}^{\infty} ar^{n-1} = \frac{a}{1-r} \qquad |r| < 1$$

If $|r| \geqslant 1$, the geometric series is divergent.

EXAMPLE 2 Find the sum of the geometric series

$$5 - \tfrac{10}{3} + \tfrac{20}{9} - \tfrac{40}{27} + \cdots$$

SOLUTION The first term is $a = 5$ and the common ratio is $r = -\frac{2}{3}$. Since $|r| = \frac{2}{3} < 1$, the series is convergent by (4) and its sum is

$$5 - \frac{10}{3} + \frac{20}{9} - \frac{40}{27} + \cdots = \frac{5}{1-\left(-\frac{2}{3}\right)} = \frac{5}{\frac{5}{3}} = 3$$

■

What do we really mean when we say that the sum of the series in Example 2 is 3? Of course, we can't literally add an infinite number of terms, one by one. But, according to Definition 2, the total sum is the *limit* of the sequence of partial sums. So, by taking the sum of sufficiently many terms, we can get as close as we like to the number 3. The table shows the first ten partial sums s_n and the graph in Figure 1 shows how the sequence of partial sums approaches 3.

n	s_n
1	5.000000
2	1.666667
3	3.888889
4	2.407407
5	3.395062
6	2.736626
7	3.175583
8	2.882945
9	3.078037
10	2.947975

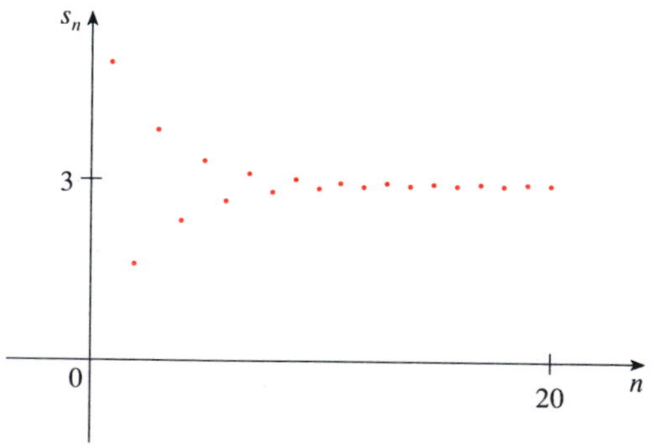

FIGURE 1

EXAMPLE 3 Is the series $\sum_{n=1}^{\infty} 2^{2n}3^{1-n}$ convergent or divergent?

SOLUTION Let's rewrite the nth term of the series in the form ar^{n-1}:

$$\sum_{n=1}^{\infty} 2^{2n}3^{1-n} = \sum_{n=1}^{\infty} \frac{4^n}{3^{n-1}} = \sum_{n=1}^{\infty} 4\left(\tfrac{4}{3}\right)^{n-1}$$

Another way to identify a and r is to write out the first few terms:

$$4 + \tfrac{16}{3} + \tfrac{64}{9} + \cdots$$

We recognize this series as a geometric series with $a = 4$ and $r = \frac{4}{3}$. Since $r > 1$, the series diverges by (4). ■

EXAMPLE 4 Write the number $2.3\overline{17} = 2.3171717\ldots$ as a ratio of integers.

SOLUTION
$$2.3171717\ldots = 2.3 + \frac{17}{10^3} + \frac{17}{10^5} + \frac{17}{10^7} + \cdots$$

After the first term we have a geometric series with $a = 17/10^3$ and $r = 1/10^2$. Therefore

$$2.3\overline{17} = 2.3 + \frac{\dfrac{17}{10^3}}{1 - \dfrac{1}{10^2}} = 2.3 + \frac{\dfrac{17}{1000}}{\dfrac{99}{100}}$$

$$= \frac{23}{10} + \frac{17}{990} = \frac{1147}{495}$$ ■

EXAMPLE 5 Find the sum of the series $\sum_{n=0}^{\infty} x^n$, where $|x| < 1$.

SOLUTION Notice that this series starts with $n = 0$ and so the first term is $x^0 = 1$. (With series, we adopt the convention that $x^0 = 1$ even when $x = 0$.) Thus

$$\sum_{n=0}^{\infty} x^n = 1 + x + x^2 + x^3 + x^4 + \cdots$$

This is a geometric series with $a = 1$ and $r = x$. Since $|r| = |x| < 1$, it converges and (4) gives

$$\sum_{n=0}^{\infty} x^n = \frac{1}{1 - x} \qquad (5)$$ ■

EXAMPLE 6 Show that the series $\sum_{n=1}^{\infty} \frac{1}{n(n+1)}$ is convergent, and find its sum.

SOLUTION This is not a geometric series, so we go back to the definition of a convergent series and compute the partial sums.

$$s_n = \sum_{i=1}^{n} \frac{1}{i(i+1)} = \frac{1}{1 \cdot 2} + \frac{1}{2 \cdot 3} + \frac{1}{3 \cdot 4} + \cdots + \frac{1}{n(n+1)}$$

Partial fractions are discussed in Section 8.4.

We can simplify this expression if we use the partial fraction decomposition

$$\frac{1}{i(i+1)} = \frac{1}{i} - \frac{1}{i+1}$$

Thus, we have

Notice that the terms cancel in pairs. This is an example of a **telescoping sum**: Because of all the cancellations, the sum collapses (like an old-fashioned collapsing telescope) into just two terms.

$$s_n = \sum_{i=1}^{n} \frac{1}{i(i+1)} = \sum_{i=1}^{n} \left(\frac{1}{i} - \frac{1}{i+1} \right)$$

$$= \left(1 - \frac{1}{2}\right) + \left(\frac{1}{2} - \frac{1}{3}\right) + \left(\frac{1}{3} - \frac{1}{4}\right) + \cdots + \left(\frac{1}{n} - \frac{1}{n+1}\right)$$

$$= 1 - \frac{1}{n+1}$$

Figure 2 illustrates Example 6 by showing the graphs of the sequence of terms $a_n = 1/[n(n+1)]$ and the sequence $\{s_n\}$ of partial sums. Notice that $a_n \to 0$ and $s_n \to 1$. See Exercises 44 and 45 for two geometric interpretations of Example 6.

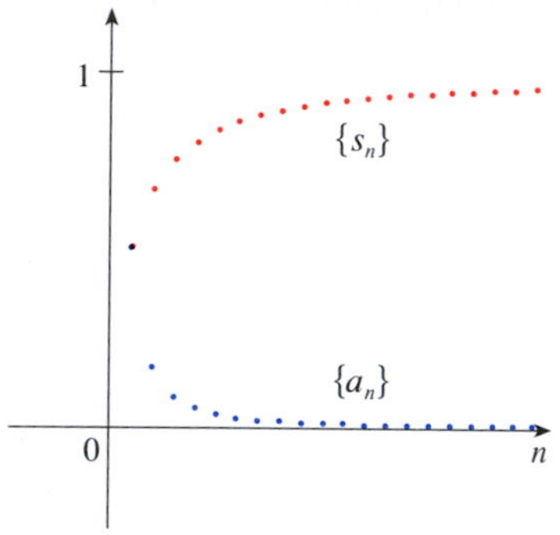

FIGURE 2

and so
$$\lim_{n\to\infty} s_n = \lim_{n\to\infty}\left(1 - \frac{1}{n+1}\right) = 1 - 0 = 1$$

Therefore, the given series is convergent and
$$\sum_{n=1}^{\infty} \frac{1}{n(n+1)} = 1$$

EXAMPLE 7 Show that the **harmonic series**
$$\sum_{n=1}^{\infty} \frac{1}{n} = 1 + \frac{1}{2} + \frac{1}{3} + \frac{1}{4} + \cdots$$
is divergent.

SOLUTION
$$s_1 = 1$$
$$s_2 = 1 + \tfrac{1}{2}$$
$$s_4 = 1 + \tfrac{1}{2} + \left(\tfrac{1}{3} + \tfrac{1}{4}\right) > 1 + \tfrac{1}{2} + \left(\tfrac{1}{4} + \tfrac{1}{4}\right) = 1 + \tfrac{2}{2}$$
$$\begin{aligned} s_8 &= 1 + \tfrac{1}{2} + \left(\tfrac{1}{3} + \tfrac{1}{4}\right) + \left(\tfrac{1}{5} + \tfrac{1}{6} + \tfrac{1}{7} + \tfrac{1}{8}\right) \\ &> 1 + \tfrac{1}{2} + \left(\tfrac{1}{4} + \tfrac{1}{4}\right) + \left(\tfrac{1}{8} + \tfrac{1}{8} + \tfrac{1}{8} + \tfrac{1}{8}\right) \\ &= 1 + \tfrac{1}{2} + \tfrac{1}{2} + \tfrac{1}{2} = 1 + \tfrac{3}{2} \end{aligned}$$
$$\begin{aligned} s_{16} &= 1 + \tfrac{1}{2} + \left(\tfrac{1}{3} + \tfrac{1}{4}\right) + \left(\tfrac{1}{5} + \cdots + \tfrac{1}{8}\right) + \left(\tfrac{1}{9} + \cdots + \tfrac{1}{16}\right) \\ &> 1 + \tfrac{1}{2} + \left(\tfrac{1}{4} + \tfrac{1}{4}\right) + \left(\tfrac{1}{8} + \cdots + \tfrac{1}{8}\right) + \left(\tfrac{1}{16} + \cdots + \tfrac{1}{16}\right) \\ &= 1 + \tfrac{1}{2} + \tfrac{1}{2} + \tfrac{1}{2} + \tfrac{1}{2} = 1 + \tfrac{4}{2} \end{aligned}$$

Similarly, $s_{32} > 1 + \tfrac{5}{2}$, $s_{64} > 1 + \tfrac{6}{2}$, and in general
$$s_{2^n} > 1 + \frac{n}{2}$$

The method used in Example 7 for showing that the harmonic series diverges is due to the French scholar Nicole Oresme (1323–1382).

This shows that $s_{2^n} \to \infty$ as $n \to \infty$ and so $\{s_n\}$ is divergent. Therefore, the harmonic series diverges. ■

(6) THEOREM If the series $\sum_{n=1}^{\infty} a_n$ is convergent, then $\lim_{n\to\infty} a_n = 0$.

PROOF Let $s_n = a_1 + a_2 + \cdots + a_n$. Then $a_n = s_n - s_{n-1}$. Since $\sum a_n$ is convergent, the sequence $\{s_n\}$ is convergent. Let $\lim_{n\to\infty} s_n = s$. Since $n - 1 \to \infty$ as $n \to \infty$, we also have $\lim_{n\to\infty} s_{n-1} = s$. Therefore
$$\begin{aligned} \lim_{n\to\infty} a_n &= \lim_{n\to\infty} (s_n - s_{n-1}) = \lim_{n\to\infty} s_n - \lim_{n\to\infty} s_{n-1} \\ &= s - s = 0 \end{aligned}$$
□

NOTE 1: With any *series* $\Sigma\, a_n$ we associate two *sequences:* the sequence $\{s_n\}$ of its partial sums and the sequence $\{a_n\}$ of its terms. If $\Sigma\, a_n$ is convergent, then the limit of the sequence $\{s_n\}$ is s (the sum of the series) and, as Theorem 6 asserts, the limit of the sequence $\{a_n\}$ is 0.

NOTE 2: The converse of Theorem 6 is not true in general. If $\lim_{n\to\infty} a_n = 0$, we cannot conclude that $\Sigma\, a_n$ is convergent. Observe that for the harmonic series $\Sigma\, 1/n$ we have $a_n = 1/n \to 0$ as $n \to \infty$, but we showed in Example 7 that $\Sigma\, 1/n$ is divergent.

(7) THE TEST FOR DIVERGENCE If $\lim\limits_{n\to\infty} a_n$ does not exist or if $\lim\limits_{n\to\infty} a_n \neq 0$, then the series $\sum\limits_{n=1}^{\infty} a_n$ is divergent.

The Test for Divergence follows from Theorem 6 because, if the series is not divergent, then it is convergent, and so $\lim_{n\to\infty} a_n = 0$.

EXAMPLE 8 Show that the series $\sum\limits_{n=1}^{\infty} \dfrac{n^2}{5n^2 + 4}$ diverges.

SOLUTION

$$\lim_{n\to\infty} a_n = \lim_{n\to\infty} \frac{n^2}{5n^2 + 4} = \lim_{n\to\infty} \frac{1}{5 + 4/n^2} = \frac{1}{5} \neq 0$$

So the series diverges by the Test for Divergence. ■

NOTE 3: If we find that $\lim_{n\to\infty} a_n \neq 0$, we know that $\Sigma\, a_n$ is divergent. If we find that $\lim_{n\to\infty} a_n = 0$, we know *nothing* about the convergence or divergence of $\Sigma\, a_n$. Remember the warning in Note 2: If $\lim_{n\to\infty} a_n = 0$, the series $\Sigma\, a_n$ might converge or it might diverge.

(8) THEOREM If $\Sigma\, a_n$ and $\Sigma\, b_n$ are convergent series, then so are the series $\Sigma\, ca_n$ (where c is a constant), $\Sigma\, (a_n + b_n)$, and $\Sigma\, (a_n - b_n)$, and

(i) $\displaystyle\sum_{n=1}^{\infty} ca_n = c \sum_{n=1}^{\infty} a_n$ (ii) $\displaystyle\sum_{n=1}^{\infty} (a_n + b_n) = \sum_{n=1}^{\infty} a_n + \sum_{n=1}^{\infty} b_n$

(iii) $\displaystyle\sum_{n=1}^{\infty} (a_n - b_n) = \sum_{n=1}^{\infty} a_n - \sum_{n=1}^{\infty} b_n$

These properties of convergent series follow from the corresponding Limit Laws for Convergent Sequences in Section 10.1. For instance, here is how part (ii) of Theorem 8 is proved:

Let

$$s_n = \sum_{i=1}^{n} a_i \qquad s = \sum_{n=1}^{\infty} a_n \qquad t_n = \sum_{i=1}^{n} b_i \qquad t = \sum_{n=1}^{\infty} b_n$$

The nth partial sum for the series $\Sigma\, (a_n + b_n)$ is

$$u_n = \sum_{i=1}^{n} (a_i + b_i)$$

and, using Theorem 2(b) in Section 6.1, we have

$$\lim_{n\to\infty} u_n = \lim_{n\to\infty} \sum_{i=1}^{n} (a_i + b_i) = \lim_{n\to\infty} \left(\sum_{i=1}^{n} a_i + \sum_{i=1}^{n} b_i \right)$$

$$= \lim_{n\to\infty} \sum_{i=1}^{n} a_i + \lim_{n\to\infty} \sum_{i=1}^{n} b_i$$

$$= \lim_{n\to\infty} s_n + \lim_{n\to\infty} t_n = s + t$$

Therefore, $\Sigma\,(a_n + b_n)$ is convergent and its sum is

$$\sum_{n=1}^{\infty} (a_n + b_n) = s + t = \sum_{n=1}^{\infty} a_n + \sum_{n=1}^{\infty} b_n$$

□

EXAMPLE 9 Find the sum of the series $\displaystyle\sum_{n=1}^{\infty} \left(\frac{3}{n(n+1)} + \frac{1}{2^n} \right)$.

SOLUTION The series $\Sigma\, 1/2^n$ is a geometric series with $a = \frac{1}{2}$ and $r = \frac{1}{2}$, so

$$\sum_{n=1}^{\infty} \frac{1}{2^n} = \frac{\frac{1}{2}}{1 - \frac{1}{2}} = 1$$

In Example 6 we found that

$$\sum_{n=1}^{\infty} \frac{1}{n(n+1)} = 1$$

So, by Theorem 8, the given series is convergent and

$$\sum_{n=1}^{\infty} \left(\frac{3}{n(n+1)} + \frac{1}{2^n} \right) = 3 \sum_{n=1}^{\infty} \frac{1}{n(n+1)} + \sum_{n=1}^{\infty} \frac{1}{2^n}$$

$$= 3 \cdot 1 + 1 = 4$$

■

NOTE 4: A finite number of terms can't affect the convergence of a series. For instance, suppose that we were able to show that the series

$$\sum_{n=4}^{\infty} \frac{n}{n^3 + 1}$$

is convergent. Since

$$\sum_{n=1}^{\infty} \frac{n}{n^3 + 1} = \frac{1}{2} + \frac{2}{9} + \frac{3}{28} + \sum_{n=4}^{\infty} \frac{n}{n^3 + 1}$$

it follows that the entire series $\Sigma_{n=1}^{\infty}\, n/(n^3 + 1)$ is convergent. Similarly, if it is known that the series $\Sigma_{n=N+1}^{\infty}\, a_n$ converges, then the full series

$$\sum_{n=1}^{\infty} a_n = \sum_{n=1}^{N} a_n + \sum_{n=N+1}^{\infty} a_n$$

is also convergent.

EXERCISES 10.2

1. (a) What is the difference between a sequence and a series?
   (b) What is a convergent series? What is a divergent series?

2. Explain what it means to say that $\sum_{n=1}^{\infty} a_n = 5$.

3–8 ■ Find at least 10 partial sums of the series. Graph both the sequence of terms and the sequence of partial sums on the same screen. Does it appear that the series is convergent or divergent? If it is convergent, find the sum. If it is divergent, explain why.

3. $\displaystyle\sum_{n=1}^{\infty} \frac{10}{3^n}$

4. $\displaystyle\sum_{n=1}^{\infty} \sin n$

5. $\displaystyle\sum_{n=1}^{\infty} \frac{n}{n+1}$

6. $\displaystyle\sum_{n=4}^{\infty} \frac{3}{n(n-1)}$

7. $\displaystyle\sum_{n=1}^{\infty} \left(\frac{1}{n^{1.5}} - \frac{1}{(n+1)^{1.5}}\right)$

8. $\displaystyle\sum_{n=1}^{\infty} \left(-\frac{2}{7}\right)^{n-1}$

9. Let $a_n = \dfrac{2n}{3n+1}$.
   (a) Determine whether $\{a_n\}$ is convergent.
   (b) Determine whether $\sum_{n=1}^{\infty} a_n$ is convergent.

10. (a) Explain the difference between
$$\sum_{i=1}^{n} a_i \quad \text{and} \quad \sum_{j=1}^{n} a_j$$
   (b) Explain the difference between
$$\sum_{i=1}^{n} a_i \quad \text{and} \quad \sum_{i=1}^{n} a_j$$

11–28 ■ Determine whether the series is convergent or divergent. If it is convergent, find its sum.

11. $4 + \frac{8}{5} + \frac{16}{25} + \frac{32}{125} + \cdots$

12. $1 - \frac{3}{2} + \frac{9}{4} - \frac{27}{8} + \cdots$

13. $\displaystyle\sum_{n=1}^{\infty} \frac{(-3)^{n-1}}{4^n}$

14. $\displaystyle\sum_{n=1}^{\infty} \frac{1}{e^{2n}}$

15. $\displaystyle\sum_{n=1}^{\infty} 3^{-n}8^{n+1}$

16. $\displaystyle\sum_{n=0}^{\infty} \frac{4^{n+1}}{5^n}$

17. $\displaystyle\sum_{n=1}^{\infty} [2(0.1)^n + (0.2)^n]$

18. $\displaystyle\sum_{n=1}^{\infty} \frac{n^2}{3(n+1)(n+2)}$

19. $\displaystyle\sum_{n=1}^{\infty} \frac{n}{\sqrt{1+n^2}}$

20. $\displaystyle\sum_{n=1}^{\infty} \left(\frac{1}{2^{n-1}} + \frac{2}{3^{n-1}}\right)$

21. $\displaystyle\sum_{n=1}^{\infty} \frac{1}{n(n+2)}$

22. $\displaystyle\sum_{n=1}^{\infty} \frac{1}{4n^2-1}$

23. $\displaystyle\sum_{n=1}^{\infty} \frac{3^n + 2^n}{6^n}$

24. $\displaystyle\sum_{n=1}^{\infty} \ln\left(\frac{n}{2n+5}\right)$

25. $\displaystyle\sum_{n=1}^{\infty} \left[\sin\left(\frac{1}{n}\right) - \sin\left(\frac{1}{n+1}\right)\right]$

26. $\displaystyle\sum_{n=1}^{\infty} \frac{1}{5 + 2^{-n}}$

27. $\displaystyle\sum_{n=1}^{\infty} \arctan n$

28. $\displaystyle\sum_{n=1}^{\infty} \ln \frac{n}{n+1}$

29–32 ■ Express the number as a ratio of integers.

29. $0.\overline{5} = 0.5555\ldots$

30. $0.\overline{15} = 0.15151515\ldots$

31. $0.\overline{307} = 0.307307307307\ldots$

32. $4.\overline{1570} = 4.157015701570\ldots$

33–36 ■ Find the values of x for which the series converges. Find the sum of the series for those values of x.

33. $\displaystyle\sum_{n=0}^{\infty} (x-3)^n$

34. $\displaystyle\sum_{n=0}^{\infty} 3^n x^n$

35. $\displaystyle\sum_{n=0}^{\infty} \frac{1}{x^n}$

36. $\displaystyle\sum_{n=0}^{\infty} \tan^n x$

CAS **37–38** ■ Use the partial fraction command on your CAS to find a convenient expression for the partial sum, and then use this expression to find the sum of the series. Check your answer by using the CAS to sum the series directly.

37. $\displaystyle\sum_{n=1}^{\infty} \frac{1}{(4n+1)(4n-3)}$

38. $\displaystyle\sum_{n=1}^{\infty} \frac{n^2+3n+1}{(n^2+n)^2}$

39. If the nth partial sum of a series $\sum_{n=1}^{\infty} a_n$ is
$$s_n = \frac{n-1}{n+1}$$
find a_n and $\sum_{n=1}^{\infty} a_n$.

40. If the nth partial sum of a series $\sum_{n=1}^{\infty} a_n$ is
$$s_n = 3 - n2^{-n}$$
find a_n and $\sum_{n=1}^{\infty} a_n$.

41. When money is spent on goods and services, those that receive the money also spend some of it. The people receiving some of the twice-spent money will spend some of that, and so on. Economists call this chain reaction the *multiplier effect*. In a hypothetical isolated community, the local government begins the process by spending D dollars. Suppose that each recipient of spent money spends $100c\%$ and saves $100s\%$ of the money that he or she receives. The values c and s are called the *marginal propensity to consume* and the *marginal propensity to save* and, of course, $c + s = 1$.
   (a) Let S_n be the total spending that has been generated after n transactions. Find an equation for S_n.

(b) Show that $\lim_{n\to\infty} S_n = kD$, where $k = 1/s$. The number k is called the *multiplier*. What is the multiplier if the marginal propensity to consume is 80%?

Note: The federal government uses this principle to justify deficit spending. Banks use this principle to justify lending a large percentage of the money that they receive in deposits.

42. A certain ball has the property that each time it falls from a height h onto a hard, level surface, it rebounds to a height rh, where $0 < r < 1$. Suppose that the ball is dropped from an initial height of H meters.

(a) Assuming that the ball continues to bounce indefinitely, find the total distance that it travels.

(b) Calculate the total time that the ball travels.

(c) Suppose that each time the ball strikes the surface with velocity v it rebounds with velocity $-kv$, where $0 < k < 1$. How long will it take for the ball to come to rest?

43. What is the value of c if $\sum_{n=2}^{\infty} (1 + c)^{-n} = 2$?

44. Graph the curves $y = x^n$, $0 \leqslant x \leqslant 1$, for $n = 0, 1, 2, 3, 4, \ldots$ on a common screen. By finding the areas between successive curves, give a geometric demonstration of the fact, shown in Example 6, that

$$\sum_{n=1}^{\infty} \frac{1}{n(n+1)} = 1$$

45. The figure shows two circles C and D of radius 1 that touch at P. T is a common tangent line; C_1 is the circle that touches C, D, and T; C_2 is the circle that touches C, D, and C_1; C_3 is the circle that touches C, D, and C_2. This procedure can be continued indefinitely and produces an infinite sequence of circles $\{C_n\}$. Find an expression for the diameter of C_n and thus provide another geometric demonstration of Example 6.

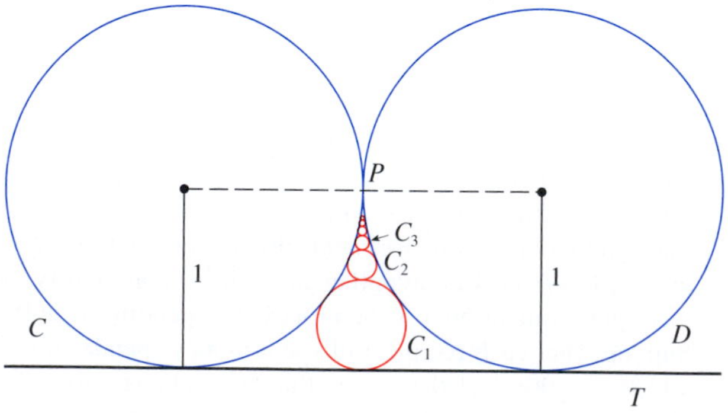

46. A right triangle ABC is given with $\angle A = \theta$ and $|AC| = b$. CD is drawn perpendicular to AB, DE is drawn perpendicular to BC, $EF \perp AB$, and this process is continued indefinitely as in the figure. Find the total length of all the perpendiculars

$$|CD| + |DE| + |EF| + |FG| + \cdots$$

in terms of b and θ.

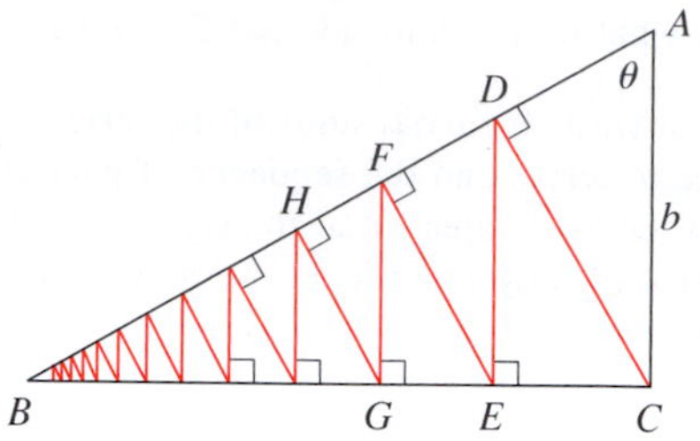

47. What is wrong with the following calculation?

$$\begin{aligned}0 &= 0 + 0 + 0 + \cdots \\ &= (1-1) + (1-1) + (1-1) + \cdots \\ &= 1 - 1 + 1 - 1 + 1 - 1 + \cdots \\ &= 1 + (-1+1) + (-1+1) + (-1+1) + \cdots \\ &= 1 + 0 + 0 + 0 + \cdots = 1\end{aligned}$$

(Guido Ubaldus thought that this proved the existence of God because "something has been created out of nothing.")

48. Suppose that $\sum_{n=1}^{\infty} a_n$ $(a_n \neq 0)$ is known to be a convergent series. Prove that $\sum_{n=1}^{\infty} 1/a_n$ is a divergent series.

49. If $\sum a_n$ is convergent and $\sum b_n$ is divergent, show that the series $\sum (a_n + b_n)$ is divergent. [*Hint:* Argue by contradiction.]

50. If $\sum a_n$ and $\sum b_n$ are both divergent, is $\sum (a_n + b_n)$ necessarily divergent?

51. Suppose that a series $\sum a_n$ has positive terms and its partial sums s_n satisfy the inequality $s_n \leqslant 1000$ for all n. Explain why $\sum a_n$ must be convergent.

52. The Fibonacci sequence was defined in Section 10.1 by the equations

$$f_1 = 1, \quad f_2 = 1, \quad f_n = f_{n-1} + f_{n-2} \qquad n \geqslant 3$$

Show that each of the following statements is true.

(a) $\dfrac{1}{f_{n-1} f_{n+1}} = \dfrac{1}{f_{n-1} f_n} - \dfrac{1}{f_n f_{n+1}}$

(b) $\displaystyle\sum_{n=2}^{\infty} \frac{1}{f_{n-1} f_{n+1}} = 1$

(c) $\displaystyle\sum_{n=2}^{\infty} \frac{f_n}{f_{n-1} f_{n+1}} = 2$

53. The **Cantor set**, named after the German mathematician Georg Cantor (1845–1918), is constructed as follows. We start with the closed interval $[0, 1]$ and remove the open interval $(\frac{1}{3}, \frac{2}{3})$. That leaves the two intervals $[0, \frac{1}{3}]$ and $[\frac{2}{3}, 1]$ and we remove the open middle third of each. Four intervals remain and again we remove the open middle third of each of them. We continue this procedure indefinitely, at each step removing the open middle third of every interval that remains from the preceding step. The Cantor set consists of the numbers that remain in $[0, 1]$ after all those intervals have been removed.

(a) Show that the total length of all the intervals that are removed is 1. Despite that, the Cantor set contains infinitely many numbers. Give examples of some numbers in the Cantor set.

(b) The **Sierpinski carpet** is a two-dimensional analogue of the Cantor set. It is constructed by removing the center one-ninth of a square of side 1, then removing the centers of the eight smaller remaining squares, and so on. (The figure shows the first three steps of the construction.) Show that the sum of the areas of the removed squares is 1. This implies that the Sierpinski carpet has area 0.

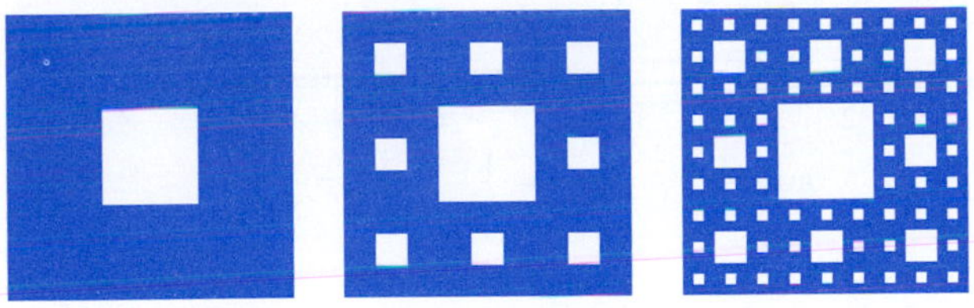

54. (a) A sequence $\{a_n\}$ is defined recursively by the equation $a_n = \frac{1}{2}(a_{n-1} + a_{n-2})$ for $n \geqslant 3$, where a_1 and a_2 can be any real numbers. Experiment with various values of a_1 and a_2 and use your calculator to guess the limit of the sequence.

(b) Find $\lim_{n\to\infty} a_n$ in terms of a_1 and a_2 by expressing $a_{n+1} - a_n$ in terms of $a_2 - a_1$ and summing a series.

55. Consider the series

$$\sum_{n=1}^{\infty} \frac{n}{(n+1)!}$$

(a) Find the partial sums s_1, s_2, s_3, and s_4. Do you recognize the denominators? Use the pattern to guess a formula for s_n.

(b) Use mathematical induction to prove your guess.

(c) Show that the given infinite series is convergent, and find its sum.

56. In the figure there are infinitely many circles approaching the vertices of an equilateral triangle, each circle touching other circles and sides of the triangle. If the triangle has sides of length 1, find the total area occupied by the circles.

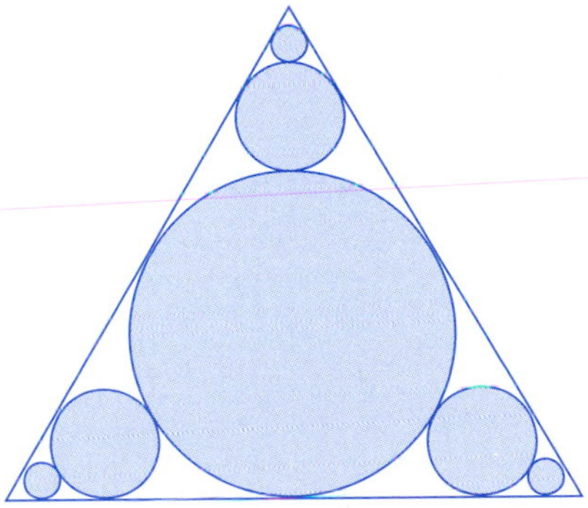

10.3 THE INTEGRAL AND COMPARISON TESTS; ESTIMATING SUMS

In general, it is difficult to find the exact sum of a series. We were able to accomplish this for geometric series and for the series $\Sigma\, 1/[n(n+1)]$ because in each of those cases we could find a simple formula for the nth partial sum s_n. But usually it is not easy to compute $\lim_{n\to\infty} s_n$. Therefore, in this section and the next we develop tests that enable us to determine whether a series is convergent or divergent without explicitly finding its sum. In some cases, however, our methods will enable us to find good estimates of the sum.

In this section we deal only with series with positive terms, so the partial sums are increasing. In view of the Monotonic Sequence Theorem, to decide whether a series is convergent or divergent, we need to determine whether the partial sums are bounded or not.

TESTING WITH AN INTEGRAL

Let's investigate the series whose terms are the reciprocals of the squares of the positive integers:

$$\sum_{n=1}^{\infty} \frac{1}{n^2} = \frac{1}{1^2} + \frac{1}{2^2} + \frac{1}{3^2} + \frac{1}{4^2} + \frac{1}{5^2} + \cdots$$

n	$s_n = \sum_{i=1}^{n} \frac{1}{i^2}$
5	1.4636
10	1.5498
50	1.6251
100	1.6350
500	1.6429
1000	1.6439
5000	1.6447

There's no simple formula for the sum s_n of the first n terms, but the computer-generated table of values given in the margin suggests that the partial sums are approaching a number near 1.64 as $n \to \infty$ and so it looks as if the series is convergent.

We can confirm this impression with a geometric argument. Figure 1 shows the curve $y = 1/x^2$ and rectangles that lie below the curve. The base of each rectangle is an interval of length 1; the height is equal to the value of the function $y = 1/x^2$ at the right endpoint of the interval. So the sum of the areas of the rectangles is

$$\frac{1}{1^2} + \frac{1}{2^2} + \frac{1}{3^2} + \frac{1}{4^2} + \frac{1}{5^2} + \cdots = \sum_{n=1}^{\infty} \frac{1}{n^2}$$

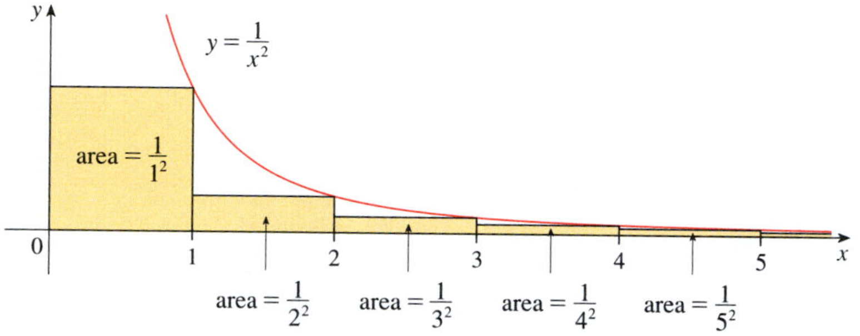

FIGURE 1

If we exclude the first rectangle, the total area of the remaining rectangles is smaller than the area under the curve $y = 1/x^2$ for $x \geqslant 1$, which is the value of the integral $\int_1^\infty (1/x^2)\,dx$. In Section 8.9 we discovered that this improper integral is convergent and has value 1. So the picture shows that all the partial sums are less than

$$\frac{1}{1^2} + \int_1^\infty \frac{1}{x^2}\,dx = 2$$

Thus, the partial sums are bounded and the series converges. The sum of the series (the limit of the partial sums) is also less than 2:

$$\sum_{n=1}^{\infty} \frac{1}{n^2} = \frac{1}{1^2} + \frac{1}{2^2} + \frac{1}{3^2} + \frac{1}{4^2} + \cdots < 2$$

[The exact sum of this series was found by the Swiss mathematician Leonhard Euler (1707–1783) to be $\pi^2/6$, but the proof of this fact is beyond the scope of this book.]

Now let's look at the series

$$\sum_{n=1}^{\infty} \frac{1}{\sqrt{n}} = \frac{1}{\sqrt{1}} + \frac{1}{\sqrt{2}} + \frac{1}{\sqrt{3}} + \frac{1}{\sqrt{4}} + \frac{1}{\sqrt{5}} + \cdots$$

n	$s_n = \sum_{i=1}^{n} \frac{1}{\sqrt{i}}$
5	3.2317
10	5.0210
50	12.7524
100	18.5896
500	43.2834
1000	61.8010
5000	139.9681

The table of values of s_n suggests that the partial sums aren't approaching a finite number, so we suspect that the given series may be divergent. Again we use a picture for confirmation. Figure 2 shows the curve $y = 1/\sqrt{x}$, but this time we use rectangles whose tops lie *above* the curve.

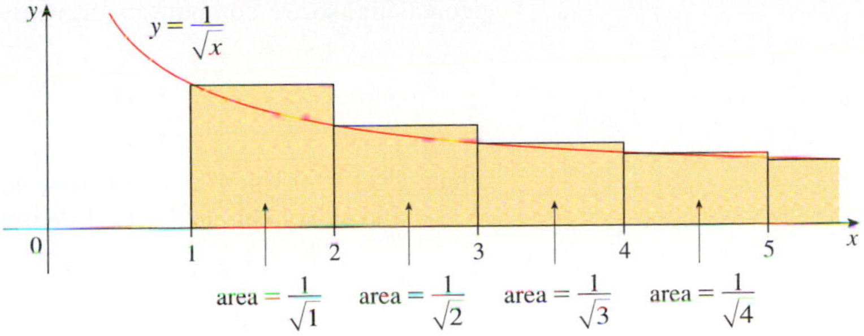

FIGURE 2

The base of each rectangle is an interval of length 1. The height is equal to the value of the function $y = 1/\sqrt{x}$ at the *left* endpoint of the interval. So the sum of the areas of all the rectangles is

$$\frac{1}{\sqrt{1}} + \frac{1}{\sqrt{2}} + \frac{1}{\sqrt{3}} + \frac{1}{\sqrt{4}} + \frac{1}{\sqrt{5}} + \cdots = \sum_{n=1}^{\infty} \frac{1}{\sqrt{n}}$$

This total area is greater than the area under the curve $y = 1/\sqrt{x}$ for $x \geqslant 1$, which is equal to the integral $\int_1^\infty (1/\sqrt{x})\,dx$. But we know from Section 8.9 that this improper integral is divergent. In other words, the area under the curve is infinite. So the sum of the series must be infinite, that is; the series is divergent.

The same sort of geometric reasoning that we used for these two series can be used to prove the following test. (See Appendix F.)

THE INTEGRAL TEST Suppose f is a continuous, positive, decreasing function on $[1, \infty)$ and let $a_n = f(n)$. Then the series $\sum_{n=1}^\infty a_n$ is convergent if and only if the improper integral $\int_1^\infty f(x)\,dx$ is convergent. In other words:

(a) If $\displaystyle\int_1^\infty f(x)\,dx$ is convergent, then $\displaystyle\sum_{n=1}^\infty a_n$ is convergent.

(b) If $\displaystyle\int_1^\infty f(x)\,dx$ is divergent, then $\displaystyle\sum_{n=1}^\infty a_n$ is divergent.

NOTE: When we use the Integral Test it is not necessary to start the series or the integral at $n = 1$. For instance, in testing the series

$$\sum_{n=4}^{\infty} \frac{1}{(n-3)^2} \qquad \text{we use} \qquad \int_4^\infty \frac{1}{(x-3)^2}\,dx$$

Also, it is not necessary that f be always decreasing. What is important is that f be *ultimately* decreasing, that is, decreasing for x larger than some number N. Then $\sum_{n=N}^\infty a_n$ is convergent, so $\sum_{n=1}^\infty a_n$ is convergent by Note 4 of Section 10.2.

EXAMPLE 1 Determine whether the series $\displaystyle\sum_{n=1}^{\infty} \frac{\ln n}{n}$ converges or diverges.

SOLUTION The function $f(x) = (\ln x)/x$ is positive and continuous for $x > 1$ because the logarithm function is continuous. But it is not obvious whether or not f is decreasing, so we compute its derivative:

$$f'(x) = \frac{x(1/x) - \ln x}{x^2} = \frac{1 - \ln x}{x^2}$$

Thus, $f'(x) < 0$ when $\ln x > 1$, that is, $x > e$. It follows that f is decreasing when $x > e$ and so we can apply the Integral Test:

$$\int_1^\infty \frac{\ln x}{x}\,dx = \lim_{t\to\infty} \int_1^t \frac{\ln x}{x}\,dx = \lim_{t\to\infty} \frac{(\ln x)^2}{2}\bigg]_1^t$$

$$= \lim_{t\to\infty} \frac{(\ln t)^2}{2} = \infty$$

Since this improper integral is divergent, the series $\Sigma\,(\ln n)/n$ is also divergent by the Integral Test. ■

EXAMPLE 2 For what values of p is the series $\displaystyle\sum_{n=1}^{\infty} \frac{1}{n^p}$ convergent?

SOLUTION If $p < 0$, then $\lim_{n\to\infty} (1/n^p) = \infty$. If $p = 0$, then $\lim_{n\to\infty} (1/n^p) = 1$. In either case $\lim_{n\to\infty} (1/n^p) \neq 0$, so the given series diverges by the Test for Divergence [see (7) in Section 10.2].

If $p > 0$, then the function $f(x) = 1/x^p$ is clearly continuous, positive, and decreasing on $[1, \infty)$. We found in Chapter 8 [see (8.9.2)] that

$$\int_1^\infty \frac{1}{x^p}\,dx \text{ converges if } p > 1 \text{ and diverges if } p \leqslant 1$$

It follows from the Integral Test that the series $\Sigma\,1/n^p$ converges if $p > 1$ and diverges if $0 < p \leqslant 1$. (For $p = 1$, this series is the harmonic series discussed in Example 7 in Section 10.2.) ■

The series in Example 2 is called the ***p*-series**. It is important in the rest of this chapter, so we summarize the results of Example 2 for future reference as follows.

(1) The p-series $\displaystyle\sum_{n=1}^{\infty} \frac{1}{n^p}$ is convergent if $p > 1$ and divergent if $p \leqslant 1$.

TESTING BY COMPARING

The series

(2) $$\sum_{n=1}^{\infty} \frac{1}{2^n + 1}$$

reminds us of the series $\Sigma_{n=1}^{\infty} 1/2^n$, which is a geometric series with $a = 1$ and $r = \frac{1}{2}$ and is therefore convergent. Because the series (2) is so similar to a convergent series, we have the feeling that it too must be convergent. Indeed, it is. The inequality

$$\frac{1}{2^n + 1} < \frac{1}{2^n}$$

shows that our given series (2) has smaller terms than those of the geometric series and therefore all its partial sums are also smaller than 1 (the sum of the geometric series). This means that its partial sums form a bounded increasing sequence, which is convergent. It also follows that the sum of the series is less than the sum of the geometric series:

$$\sum_{n=1}^{\infty} \frac{1}{2^n + 1} < 2$$

Similar reasoning is used in Appendix F to prove the following test, which applies only to series whose terms are positive. The first part says that if we have a series whose terms are *smaller* than those of a known *convergent* series, then our series is also convergent. The second part says that if we start with a series whose terms are *larger* than those of a known *divergent* series, then it too is divergent.

THE COMPARISON TEST Suppose that $\Sigma\, a_n$ and $\Sigma\, b_n$ are series with positive terms.

(a) If $\Sigma\, b_n$ is convergent and $a_n \leqslant b_n$ for all n, then $\Sigma\, a_n$ is also convergent.

(b) If $\Sigma\, b_n$ is divergent and $a_n \geqslant b_n$ for all n, then $\Sigma\, a_n$ is also divergent.

Standard series for use with the Comparison Test

In using the Comparison Test we must, of course, have some known series $\Sigma\, b_n$ for the purpose of comparison. Most of the time we use either a p-series [$\Sigma\, 1/n^p$ converges if $p > 1$ and diverges if $p \leqslant 1$; see (1)] or a geometric series [$\Sigma\, ar^{n-1}$ converges if $|r| < 1$ and diverges if $|r| \geqslant 1$; see (10.2.4)].

EXAMPLE 3 Determine whether the series $\displaystyle\sum_{n=1}^{\infty} \frac{5}{2n^2 + 4n + 3}$ converges or diverges.

SOLUTION For large n the dominant term in the denominator is $2n^2$, so we compare the given series with the series $\Sigma\, 5/(2n^2)$. Observe that

$$\frac{5}{2n^2 + 4n + 3} < \frac{5}{2n^2}$$

because the left side has a bigger denominator. (In the notation of the Comparison Test, a_n is the left side and b_n is the right side.) We know that

$$\sum_{n=1}^{\infty} \frac{5}{2n^2} = \frac{5}{2} \sum_{n=1}^{\infty} \frac{1}{n^2}$$

is convergent (p-series with $p = 2 > 1$). Therefore

$$\sum_{n=1}^{\infty} \frac{5}{2n^2 + 4n + 3}$$

is convergent by part (a) of the Comparison Test. ■

Although the condition $a_n \leqslant b_n$ or $a_n \geqslant b_n$ in the Comparison Test is given for all n, we need verify only that it holds for $n \geqslant N$, where N is some fixed integer, because the convergence of a series is not affected by a finite number of terms. This is illustrated in the next example.

EXAMPLE 4 Test the series $\sum_{n=1}^{\infty} \frac{\ln n}{n}$ for convergence or divergence.

SOLUTION We used the Integral Test to test this series in Example 1, but we can also test it by comparing it with the harmonic series. Observe that $\ln n > 1$ for $n \geqslant 3$ and so

$$\frac{\ln n}{n} > \frac{1}{n} \qquad n \geqslant 3$$

We know that $\sum 1/n$ is divergent (p-series with $p = 1$). Thus, the given series is divergent by the Comparison Test. ■

NOTE: The terms of the series being tested must be smaller than those of a convergent series or larger than those of a divergent series. If the terms are larger than the terms of a convergent series or smaller than those of a divergent series, then the Comparison Test doesn't apply. Consider, for instance, the series

$$\sum_{n=1}^{\infty} \frac{1}{2^n - 1}$$

The inequality

$$\frac{1}{2^n - 1} > \frac{1}{2^n}$$

is useless as far as the Comparison Test is concerned because $\sum b_n = \sum \left(\frac{1}{2}\right)^n$ is convergent and $a_n > b_n$. Nonetheless, we have the feeling that $\sum 1/(2^n - 1)$ ought to be convergent because it is very similar to the convergent geometric series $\sum \left(\frac{1}{2}\right)^n$. In such cases the following test can be used.

THE LIMIT COMPARISON TEST Suppose that $\sum a_n$ and $\sum b_n$ are series with positive terms. If

$$\lim_{n \to \infty} \frac{a_n}{b_n} = c$$

where c is a finite number and $c > 0$, then either both series converge or both diverge.

Although we won't prove the Limit Comparison Test, it seems reasonable because for large n, $a_n \approx cb_n$.

EXAMPLE 5 Test the series $\sum_{n=1}^{\infty} \frac{1}{2^n - 1}$ for convergence or divergence.

SOLUTION We use the Limit Comparison Test with

$$a_n = \frac{1}{2^n - 1} \qquad b_n = \frac{1}{2^n}$$

and obtain
$$\lim_{n\to\infty}\frac{a_n}{b_n}=\lim_{n\to\infty}\frac{2^n}{2^n-1}=\lim_{n\to\infty}\frac{1}{1-1/2^n}=1$$

Since this limit exists and $\Sigma\, 1/2^n$ is a convergent geometric series, the given series converges by the Limit Comparison Test. ■

ESTIMATING THE SUM OF A SERIES

Suppose we have been able to use the Integral Test to show that a series $\Sigma\, a_n$ is convergent and we now want to find an approximation to the sum s of the series. Of course, any partial sum s_n is an approximation to s because $\lim_{n\to\infty} s_n = s$. But how good is such an approximation? To find out, we need to estimate the size of the **remainder**

$$R_n = s - s_n = a_{n+1} + a_{n+2} + a_{n+3} + \cdots$$

The remainder R_n is the error made when s_n, the sum of the first n terms, is used as an approximation to the total sum.

We use the same notation and ideas as in the Integral Test. Comparing the areas of the rectangles with the area under $y = f(x)$ for $x > n$ in Figure 3, we see that

$$R_n = a_{n+1} + a_{n+2} + \cdots \leqslant \int_n^\infty f(x)\,dx$$

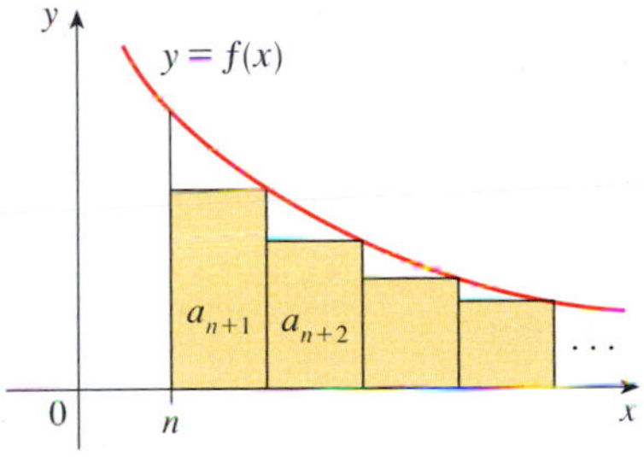

FIGURE 3

Similarly, we see from Figure 4 that

$$R_n = a_{n+1} + a_{n+2} + \cdots \geqslant \int_{n+1}^\infty f(x)\,dx$$

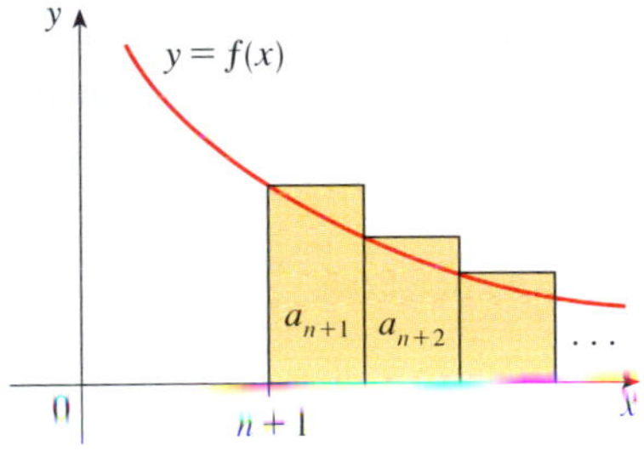

FIGURE 4

So we have proved the following error estimate.

(3) REMAINDER ESTIMATE FOR THE INTEGRAL TEST If $\Sigma\, a_n$ converges by the Integral Test and $R_n = s - s_n$, then

$$\int_{n+1}^\infty f(x)\,dx \leqslant R_n \leqslant \int_n^\infty f(x)\,dx$$

EXAMPLE 6

(a) Approximate the sum of the series $\Sigma\, 1/n^3$ by using the sum of the first 10 terms. Estimate the error involved in this approximation.

(b) How many terms are required to ensure that the sum is accurate to within 0.0005?

SOLUTION In both parts (a) and (b) we need to know $\int_n^\infty f(x)\,dx$. With $f(x) = 1/x^3$, we have

$$\int_n^\infty \frac{1}{x^3}\,dx = \lim_{t\to\infty}\left[-\frac{1}{2x^2}\right]_n^t = \lim_{t\to\infty}\left(-\frac{1}{2t^2}+\frac{1}{2n^2}\right) = \frac{1}{2n^2}$$

(a)
$$\sum_{n=1}^\infty \frac{1}{n^3} \approx s_{10} = \frac{1}{1^3}+\frac{1}{2^3}+\frac{1}{3^3}+\cdots+\frac{1}{10^3} \approx 1.1975$$

According to the remainder estimate in (3), we have

$$R_{10} \leqslant \int_{10}^\infty \frac{1}{x^3}\,dx = \frac{1}{2(10)^2} = \frac{1}{200}$$

So the size of the error is at most 0.005.

(b) Accuracy to within 0.0005 means that we have to find a value of n such that $R_n \leqslant 0.0005$. Since

$$R_n \leqslant \int_n^\infty \frac{1}{x^3}\,dx = \frac{1}{2n^2}$$

we want

$$\frac{1}{2n^2} < 0.0005$$

Solving this inequality, we get

$$n^2 > \frac{1}{0.001} = 1000 \qquad \text{or} \qquad n > \sqrt{1000} \approx 31.6$$

We need 32 terms to ensure accuracy to within 0.0005. ■

If we add s_n to each side of the inequalities in (3), we get

(4)
$$s_n + \int_{n+1}^\infty f(x)\,dx \leqslant s \leqslant s_n + \int_n^\infty f(x)\,dx$$

because $s_n + R_n = s$. The inequalities in (4) give a lower bound and an upper bound for s. They provide a more accurate approximation to the sum of the series than the partial sum s_n does.

EXAMPLE 7 Use (4) with $n = 10$ to estimate the sum of the series $\displaystyle\sum_{n=1}^\infty \frac{1}{n^3}$.

SOLUTION The inequalities in (4) become

$$s_{10} + \int_{11}^\infty \frac{1}{x^3}\,dx \leqslant s \leqslant s_{10} + \int_{10}^\infty \frac{1}{x^3}\,dx$$

From Example 6 we know that

$$\int_n^\infty \frac{1}{x^3}\,dx = \frac{1}{2n^2}$$

so

$$s_{10} + \frac{1}{2(11)^2} \leqslant s \leqslant s_{10} + \frac{1}{2(10)^2}$$

Using $s_{10} \approx 1.197532$, we get

$$1.201664 \leqslant s \leqslant 1.202532$$

If we approximate s by the midpoint of this interval, then the error is at most half the length of the interval. So

$$\sum_{n=1}^\infty \frac{1}{n^3} \approx 1.2021 \qquad \text{with error} < 0.0005$$

■

If we compare Example 7 with Example 6, we see that the improved estimate in (4) can be much better than the estimate $s \approx s_n$. To make the error smaller than 0.0005 we had to use 32 terms in Example 6 but only 10 terms in Example 7.

If we have used the Comparison Test to show that a series $\Sigma\, a_n$ converges by comparison with a series $\Sigma\, b_n$, then we may be able to estimate the sum $\Sigma\, a_n$ by comparing remainders, as the following example shows.

EXAMPLE 8 Use the sum of the first 100 terms to approximate the sum of the series $\Sigma\, 1/(n^3 + 1)$. Estimate the error involved in this approximation.

SOLUTION Since

$$\frac{1}{n^3 + 1} < \frac{1}{n^3}$$

the given series is convergent by the Comparison Test. The remainder T_n for the comparison series $\Sigma\, 1/n^3$ was estimated in Example 6. There we found that

$$T_n \leqslant \int_n^\infty \frac{1}{x^3}\, dx = \frac{1}{2n^2}$$

Therefore, the remainder R_n for the given series satisfies

$$R_n \leqslant T_n \leqslant \frac{1}{2n^2}$$

With $n = 100$ we have

$$R_{100} \leqslant \frac{1}{2(100)^2} = 0.00005$$

Using a programmable calculator or a computer, we find that

$$\sum_{n=1}^{\infty} \frac{1}{n^3 + 1} \approx \sum_{n=1}^{100} \frac{1}{n^3 + 1} \approx 0.6864538$$

with error less than 0.00005. ■

EXERCISES 10.3

1. Draw a picture to show that

$$\sum_{n=2}^{\infty} \frac{1}{n^{1.3}} < \int_1^\infty \frac{1}{x^{1.3}}\, dx$$

What can you conclude about the series?

2. Suppose f is a continuous positive decreasing function for $x \geqslant 1$ and $a_n = f(n)$. By drawing a picture, rank the following three quantities in increasing order:

$$\int_1^n f(x)\, dx \qquad \sum_{i=1}^{n-1} a_i \qquad \sum_{i=2}^{n} a_i$$

3. Suppose $\Sigma\, a_n$ and $\Sigma\, b_n$ are series with positive terms and $\Sigma\, b_n$ is known to be convergent.

(a) If $a_n > b_n$ for all n, what can you say about $\Sigma\, a_n$? Why?

(b) If $a_n < b_n$ for all n, what can you say about $\Sigma\, a_n$? Why?

4. Suppose $\Sigma\, a_n$ and $\Sigma\, b_n$ are series with positive terms and $\Sigma\, b_n$ is known to be divergent.

(a) If $a_n > b_n$ for all n, what can you say about $\Sigma\, a_n$? Why?

(b) If $a_n < b_n$ for all n, what can you say about Σa_n? Why?

5. It's important to distinguish between

$$\sum_{n=1}^{\infty} n^b \quad \text{and} \quad \sum_{n=1}^{\infty} b^n$$

What name is given to the first series? To the second? For what values of b does the first series converge? For what values of b does the second series converge?

6–22 ■ Determine whether the series is convergent or divergent.

6. $\displaystyle\sum_{n=1}^{\infty} \left(\frac{2}{n\sqrt{n}} + \frac{3}{n^3} \right)$

7. $\displaystyle\sum_{n=1}^{\infty} ne^{-n^2}$

8. $\displaystyle\sum_{n=1}^{\infty} \frac{1}{n^2 + 1}$

9. $\displaystyle\sum_{n=1}^{\infty} \frac{n}{n^2 + 1}$

10. $\displaystyle\sum_{n=1}^{\infty} \frac{\ln n}{n^2}$

11. $\displaystyle\sum_{n=2}^{\infty} \frac{1}{n \ln n}$

12. $\displaystyle\sum_{n=2}^{\infty} \frac{1}{\sqrt{n} - 1}$

13. $\displaystyle\sum_{n=1}^{\infty} \frac{1}{n^3 + n^2}$

14. $\displaystyle\sum_{n=1}^{\infty} \frac{3}{4^n + 5}$

15. $\displaystyle\sum_{n=0}^{\infty} \frac{1 + 5^n}{4^n}$

16. $\displaystyle\sum_{n=1}^{\infty} \frac{\sin^2 n}{n\sqrt{n}}$

17. $\displaystyle\sum_{n=1}^{\infty} \frac{3}{n(n + 3)}$

18. $\displaystyle\sum_{n=1}^{\infty} \frac{1 + 2^n}{1 + 3^n}$

19. $\displaystyle\sum_{n=1}^{\infty} \frac{1}{1 + \sqrt{n}}$

20. $\displaystyle\sum_{n=3}^{\infty} \frac{1}{n^2 - 4}$

21. $\displaystyle\sum_{n=1}^{\infty} \sin\left(\frac{1}{n}\right)$

22. $\displaystyle\sum_{n=1}^{\infty} \frac{n^2 - 3n}{\sqrt[3]{n^{10} - 4n^2}}$

23. Find the values of p for which the following series is convergent:

$$\sum_{n=2}^{\infty} \frac{1}{n(\ln n)^p}$$

24. (a) Find the partial sum s_{10} of the series $\Sigma_{n=1}^{\infty} 1/n^4$. Estimate the error in using s_{10} as an approximation to the sum of the series.

(b) Use (4) with $n = 10$ to give an improved estimate of the sum.

(c) Find a value of n so that s_n is within 0.00001 of the sum.

25. (a) Use the sum of the first 10 terms to estimate the sum of the series

$$\sum_{n=1}^{\infty} \frac{1}{n^2}$$

How good is this estimate?

(b) Improve this estimate using (4) with $n = 10$.

(c) Find a value of n that will ensure that the error in the approximation $s \approx s_n$ is less than 0.001.

26. Find the sum of the series $\Sigma_{n=1}^{\infty} 1/n^5$ correct to three decimal places.

27. Estimate $\Sigma_{n=1}^{\infty} n^{-3/2}$ to within 0.01.

28. How many terms of the series

$$\sum_{n=2}^{\infty} \frac{1}{n(\ln n)^2}$$

would you need to add to find its sum to within 0.01?

29–30 ■ Use the sum of the first 10 terms to approximate the sum of the series. Estimate the error.

29. $\displaystyle\sum_{n=1}^{\infty} \frac{1}{n^4 + n^2}$

30. $\displaystyle\sum_{n=1}^{\infty} \frac{n}{(n + 1)3^n}$

31. (a) Use a graph of $y = 1/x$ to show that if s_n is the nth partial sum of the harmonic series, then

$$s_n \leqslant 1 + \ln n$$

(b) The harmonic series diverges but very slowly. Use part (a) to show that the sum of the first million terms is less than 15 and the sum of the first billion terms is less than 22.

CAS **32.** (a) Show that the series

$$\sum_{n=1}^{\infty} \frac{(\ln n)^2}{n^2}$$

is convergent.

(b) Find an upper bound for the error in the approximation $s \approx s_n$.

(c) What is the smallest value of n such that this upper bound is less than 0.05?

(d) Find s_n for this value of n.

33. The meaning of the decimal representation of a number $0.d_1d_2d_3\ldots$ (where the digit d_i is one of the numbers 0, 1, 2, ..., 9) is that

$$0.d_1d_2d_3d_4\ldots = \frac{d_1}{10} + \frac{d_2}{10^2} + \frac{d_3}{10^3} + \frac{d_4}{10^4} + \cdots$$

Show that this series always converges.

34. Find all positive values of b for which the series $\Sigma_{n=1}^{\infty} b^{\ln n}$ converges.

35. If Σa_n is a convergent series with positive terms, is it true that $\Sigma \sin(a_n)$ is also convergent?

36. Show that if $a_n > 0$ and Σa_n is convergent, then $\Sigma \ln(1 + a_n)$ is convergent.

10.4 OTHER CONVERGENCE TESTS

The convergence tests that we have looked at so far apply only to series with positive terms. In this section we learn how to deal with series whose terms are not necessarily positive.

ALTERNATING SERIES

An **alternating series** is a series whose terms are alternately positive and negative. Here are two examples:

$$1 - \frac{1}{2} + \frac{1}{3} - \frac{1}{4} + \frac{1}{5} - \frac{1}{6} + \cdots = \sum_{n=1}^{\infty} (-1)^{n-1} \frac{1}{n}$$

$$-\frac{1}{2} + \frac{2}{3} - \frac{3}{4} + \frac{4}{5} - \frac{5}{6} + \frac{6}{7} - \cdots = \sum_{n=1}^{\infty} (-1)^{n} \frac{n}{n+1}$$

We see from these examples that the nth term of an alternating series is of the form

$$a_n = (-1)^{n-1} b_n \qquad \text{or} \qquad a_n = (-1)^n b_n$$

where b_n is a positive number. (In fact, $b_n = |a_n|$.)

The following test says that if the terms of an alternating series decrease to 0 in absolute value, then the series converges.

THE ALTERNATING SERIES TEST If the alternating series

$$\sum_{n=1}^{\infty} (-1)^{n-1} b_n = b_1 - b_2 + b_3 - b_4 + b_5 - b_6 + \cdots \qquad b_n > 0$$

satisfies

(a) $b_{n+1} \leqslant b_n$ for all n

(b) $\lim_{n \to \infty} b_n = 0$

then the series is convergent.

A formal proof of this test is given in Appendix F, but Figure 1 gives a picture of the idea behind the proof. We first plot $s_1 = b_1$ on a number line. To find s_2 we subtract b_2, so s_2 is to the left of s_1. Then to find s_3 we add b_3, so s_3 is to the right of s_2. But, since $b_3 < b_2$, s_3 is to the left of s_1. Continuing in this manner, we see that the partial sums oscillate back and forth. Since $b_n \to 0$, the successive steps are becoming smaller and smaller. The even partial sums $s_2, s_4, s_6, \ldots$ are increasing and the odd partial sums s_1,

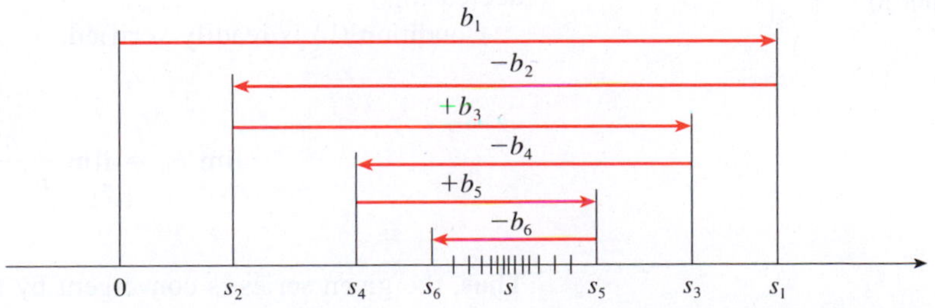

FIGURE 1

$s_3, s_5, \ldots$ are decreasing. Thus, it seems plausible that both are converging to some number s, which is the sum of the series.

Figure 2 illustrates Example 1 by showing the graphs of the terms $a_n = (-1)^{n-1}/n$ and the partial sums s_n. Notice how the values of s_n zigzag across the limiting value, which appears to be about 0.7. In fact, it can be proved that the exact sum of the series is $\ln 2 \approx 0.693$.

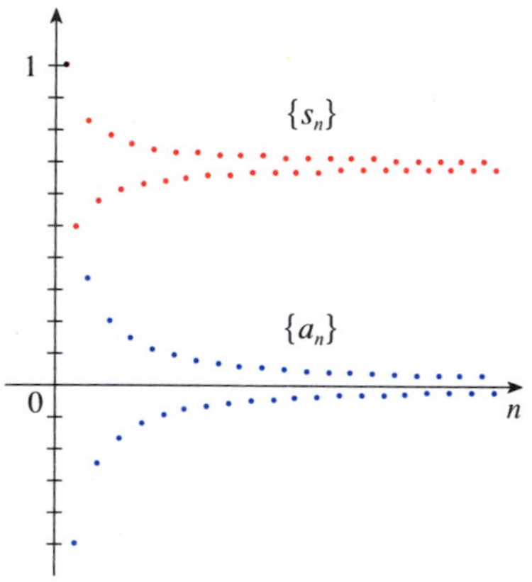

FIGURE 2

EXAMPLE 1 The alternating harmonic series

$$1 - \frac{1}{2} + \frac{1}{3} - \frac{1}{4} + \cdots = \sum_{n=1}^{\infty} \frac{(-1)^{n-1}}{n}$$

satisfies

(a) $b_{n+1} < b_n$ because $\displaystyle \frac{1}{n+1} < \frac{1}{n}$

(b) $\displaystyle \lim_{n\to\infty} b_n = \lim_{n\to\infty} \frac{1}{n} = 0$

so the series is convergent by the Alternating Series Test. ■

EXAMPLE 2 The series $\displaystyle \sum_{n=1}^{\infty} \frac{(-1)^n 3n}{4n-1}$ is alternating, but

$$\lim_{n\to\infty} b_n = \lim_{n\to\infty} \frac{3n}{4n-1} = \lim_{n\to\infty} \frac{3}{4 - \dfrac{1}{n}} = \frac{3}{4}$$

so condition (b) is not satisfied. Instead, we look at the limit of the nth term of the series:

$$\lim_{n\to\infty} a_n = \lim_{n\to\infty} \frac{(-1)^n 3n}{4n-1}$$

This limit does not exist, so the series diverges by the Test for Divergence. ■

EXAMPLE 3 Test the series $\displaystyle \sum_{n=1}^{\infty} (-1)^{n+1} \frac{n^2}{n^3+1}$ for convergence or divergence.

SOLUTION The given series is alternating so we try to verify conditions (a) and (b) of the Alternating Series Test.

Unlike the situation in Example 1, it is not obvious that the sequence given by $b_n = n^2/(n^3+1)$ is decreasing. However, if we consider the related function $f(x) = x^2/(x^3+1)$, we find that

$$f'(x) = \frac{x(2-x^3)}{(x^3+1)^2}$$

Instead of verifying condition (a) of the Alternating Series Test by computing a derivative, we could verify that $b_{n+1} < b_n$ directly by using the technique of Solution 1 of Example 10 in Section 10.1.

Since we are considering only positive x, we see that $f'(x) < 0$ if $2 - x^3 < 0$, that is, $x > \sqrt[3]{2}$. Thus, f is decreasing on the interval $(\sqrt[3]{2}, \infty)$. This means that $f(n+1) < f(n)$ and therefore $b_{n+1} < b_n$ when $n \geqslant 2$. (The inequality $b_2 < b_1$ can be verified directly but all that really matters is that the sequence $\{b_n\}$ is eventually decreasing.)

Condition (b) is readily verified:

$$\lim_{n\to\infty} b_n = \lim_{n\to\infty} \frac{n^2}{n^3+1} = \lim_{n\to\infty} \frac{\dfrac{1}{n}}{1 + \dfrac{1}{n^3}} = 0$$

Thus, the given series is convergent by the Alternating Series Test. ■

The error involved in using the partial sum s_n as an approximation to the total sum s is the remainder $R_n = s - s_n$. The next theorem says that for series that satisfy the conditions of the Alternating Series Test, the size of the error is smaller than b_{n+1}, which is the absolute value of the first neglected term.

ALTERNATING SERIES ESTIMATION THEOREM If $s = \Sigma\,(-1)^{n-1}b_n$ is the sum of an alternating series that satisfies

$$\text{(a) } 0 < b_{n+1} \leqslant b_n \qquad \text{and} \qquad \text{(b) } \lim_{n\to\infty} b_n = 0$$

then

$$|R_n| = |s - s_n| \leqslant b_{n+1}$$

You can see geometrically why this is true by looking at Figure 1. Notice that $s - s_4 < b_5$, $|s - s_5| < b_6$, and so on. A formal proof is given in Appendix F.

EXAMPLE 4 Find the sum of the series $\displaystyle\sum_{n=0}^{\infty} \frac{(-1)^n}{n!}$ correct to three decimal places. (By definition, $0! = 1$.)

SOLUTION We first observe that the series is convergent by the Alternating Series Test because

$$\text{(a)} \quad \frac{1}{(n+1)!} = \frac{1}{n!(n+1)} < \frac{1}{n!}$$

$$\text{(b)} \quad 0 < \frac{1}{n!} < \frac{1}{n} \to 0 \quad \text{so} \quad \frac{1}{n!} \to 0 \text{ as } n \to \infty$$

To give us a feel for how many terms we need to use in our approximation, let's write out the first few terms of the series:

$$s = \frac{1}{0!} - \frac{1}{1!} + \frac{1}{2!} - \frac{1}{3!} + \frac{1}{4!} - \frac{1}{5!} + \frac{1}{6!} - \frac{1}{7!} + \cdots$$

$$= 1 - 1 + \tfrac{1}{2} - \tfrac{1}{6} + \tfrac{1}{24} - \tfrac{1}{120} + \tfrac{1}{720} - \tfrac{1}{5040} + \cdots$$

Notice that $\qquad b_7 = \tfrac{1}{5040} < \tfrac{1}{5000} = 0.0002$

and $\qquad s_6 = 1 - 1 + \tfrac{1}{2} - \tfrac{1}{6} + \tfrac{1}{24} - \tfrac{1}{120} + \tfrac{1}{720} \approx 0.368056$

By the Alternating Series Estimation Theorem we know that

$$|s - s_6| \leqslant b_7 < 0.0002$$

This error of less than 0.0002 does not affect the third decimal place, so we have

$$s \approx 0.368$$

correct to three decimal places.

In Section 10.7 we will prove that $e^x = \sum_{n=0}^{\infty} x^n/n!$ for all x, so what we have obtained in this example is actually an approximation to the number e^{-1}. ■

⊘ NOTE: The rule that the error (in using s_n to approximate s) is smaller than the first neglected term is, in general, valid only for alternating series that satisfy the conditions of the Alternating Series Estimation Theorem. The rule does not apply to other types of series.

ABSOLUTE CONVERGENCE

Given any series $\Sigma\, a_n$, we can consider the corresponding series

$$\sum_{n=1}^{\infty} |a_n| = |a_1| + |a_2| + |a_3| + \cdots$$

whose terms are the absolute values of the terms of the original series.

We have convergence tests for series with positive terms and for alternating series. But what if the signs of the terms switch back and forth irregularly? We will see in Example 7 that the idea of absolute convergence sometimes helps in such cases.

DEFINITION A series $\Sigma\, a_n$ is called **absolutely convergent** if the series of absolute values $\Sigma\, |a_n|$ is convergent.

Notice that if $\Sigma\, a_n$ is a series with positive terms, then $|a_n| = a_n$ and so absolute convergence is the same as convergence.

EXAMPLE 5 The series

$$\sum_{n=1}^{\infty} \frac{(-1)^{n-1}}{n^2} = 1 - \frac{1}{2^2} + \frac{1}{3^2} - \frac{1}{4^2} + \cdots$$

is absolutely convergent because

$$\sum_{n=1}^{\infty} \left| \frac{(-1)^{n-1}}{n^2} \right| = \sum_{n=1}^{\infty} \frac{1}{n^2} = 1 + \frac{1}{2^2} + \frac{1}{3^2} + \frac{1}{4^2} + \cdots$$

is a convergent p-series ($p = 2$). ■

EXAMPLE 6 We know that the alternating harmonic series

$$\sum_{n=1}^{\infty} \frac{(-1)^{n-1}}{n} = 1 - \frac{1}{2} + \frac{1}{3} - \frac{1}{4} + \cdots$$

is convergent (see Example 1), but it is not absolutely convergent because the corresponding series of absolute values is

$$\sum_{n=1}^{\infty} \left| \frac{(-1)^{n-1}}{n} \right| = \sum_{n=1}^{\infty} \frac{1}{n} = 1 + \frac{1}{2} + \frac{1}{3} + \frac{1}{4} + \cdots$$

which is the harmonic series (p-series with $p = 1$) and is therefore divergent. ■

Example 6 shows it is possible for a series to be convergent but not absolutely convergent. However, the following theorem shows that absolute convergence implies convergence.

(1) THEOREM If a series $\Sigma\, a_n$ is absolutely convergent, then it is convergent.

To see why Theorem 1 is true, observe that the inequality

$$0 \leqslant a_n + |a_n| \leqslant 2|a_n|$$

is true because $|a_n|$ is either a_n or $-a_n$. If $\sum a_n$ is absolutely convergent, then $\sum |a_n|$ is convergent, so $\sum 2|a_n|$ is convergent. Therefore, by the Comparison Test, $\sum (a_n + |a_n|)$ is convergent. Then

$$\sum a_n = \sum (a_n + |a_n|) - \sum |a_n|$$

is the difference of two convergent series and is therefore convergent. □

Figure 3 shows the graphs of the terms a_n and partial sums s_n of the series in Example 7. Notice that the series is not alternating but has positive and negative terms.

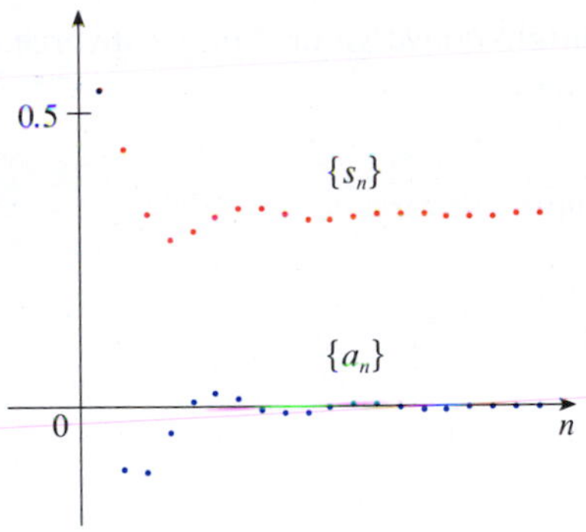

FIGURE 3

EXAMPLE 7 Determine whether the series

$$\sum_{n=1}^{\infty} \frac{\cos n}{n^2} = \frac{\cos 1}{1^2} + \frac{\cos 2}{2^2} + \frac{\cos 3}{3^2} + \cdots$$

is convergent or divergent.

SOLUTION This series has both positive and negative terms, but it is not alternating. (The first term is positive, the next three are negative, and the following three are positive. The signs change irregularly.) We can apply the Comparison Test to the series of absolute values

$$\sum_{n=1}^{\infty} \left| \frac{\cos n}{n^2} \right| = \sum_{n=1}^{\infty} \frac{|\cos n|}{n^2}$$

Since $|\cos n| \leqslant 1$ for all n, we have

$$\frac{|\cos n|}{n^2} \leqslant \frac{1}{n^2}$$

We know that $\sum 1/n^2$ is convergent (p-series with $p = 2$) and, therefore, $\sum |\cos n|/n^2$ is convergent by the Comparison Test. Thus, the given series $\sum (\cos n)/n^2$ is absolutely convergent and therefore convergent by Theorem 1. ■

THE RATIO TEST

The following test is very useful in determining whether a given series is absolutely convergent.

THE RATIO TEST

(a) If $\lim_{n \to \infty} \left| \frac{a_{n+1}}{a_n} \right| = L < 1$, then the series $\sum_{n=1}^{\infty} a_n$ is absolutely convergent (and therefore convergent).

(b) If $\lim_{n \to \infty} \left| \frac{a_{n+1}}{a_n} \right| = L > 1$ or $\lim_{n \to \infty} \left| \frac{a_{n+1}}{a_n} \right| = \infty$, then the series $\sum_{n=1}^{\infty} a_n$ is divergent.

The Ratio Test can be proved by comparing the given series to a geometric series. (See Appendix F.) It's understandable that geometric series are involved because, for those series, the ratio r of consecutive terms is constant and the series converges if $|r| < 1$. In part (a) of the Ratio Test, the ratio of consecutive terms isn't constant but $|a_{n+1}/a_n| \to L$ so, for large n, $|a_{n+1}/a_n|$ is almost constant and the series converges if $L < 1$.

NOTE: If $\lim_{n\to\infty} |a_{n+1}/a_n| = 1$, the Ratio Test gives no information. For instance, for the convergent series $\sum 1/n^2$ we have

$$\left|\frac{a_{n+1}}{a_n}\right| = \frac{\dfrac{1}{(n+1)^2}}{\dfrac{1}{n^2}} = \frac{n^2}{(n+1)^2} = \frac{1}{\left(1+\dfrac{1}{n}\right)^2} \to 1 \qquad \text{as } n \to \infty$$

whereas for the divergent series $\sum 1/n$ we have

$$\left|\frac{a_{n+1}}{a_n}\right| = \frac{\dfrac{1}{n+1}}{\dfrac{1}{n}} = \frac{n}{n+1} = \frac{1}{1+\dfrac{1}{n}} \to 1 \qquad \text{as } n \to \infty$$

Therefore, if $\lim_{n\to\infty} |a_{n+1}/a_n| = 1$, the series $\sum a_n$ might converge or it might diverge. In this case the Ratio Test fails and we must use some other test.

EXAMPLE 8 Test the series $\displaystyle\sum_{n=1}^{\infty} (-1)^n \frac{n^3}{3^n}$ for absolute convergence.

ESTIMATING SUMS

We have used various methods for estimating the sum of a series—the method depended on which test was used to prove convergence. What about series for which the Ratio Test works? There are two possibilities: If the series happens to be an alternating series, as in Example 8, then it is best to use the Alternating Series Estimation Theorem. If the terms are all positive, then use the special methods explained in Exercise 34.

SOLUTION We use the Ratio Test with $a_n = (-1)^n n^3/3^n$:

$$\left|\frac{a_{n+1}}{a_n}\right| = \left|\frac{\dfrac{(-1)^{n+1}(n+1)^3}{3^{n+1}}}{\dfrac{(-1)^n n^3}{3^n}}\right| = \frac{(n+1)^3}{3^{n+1}} \cdot \frac{3^n}{n^3}$$

$$= \frac{1}{3}\left(\frac{n+1}{n}\right)^3 = \frac{1}{3}\left(1+\frac{1}{n}\right)^3 \to \frac{1}{3} < 1$$

Thus, by the Ratio Test, the given series is absolutely convergent and therefore convergent. ■

EXAMPLE 9 Test the convergence of the series $\displaystyle\sum_{n=1}^{\infty} \frac{n^n}{n!}$.

SOLUTION Since the terms $a_n = n^n/n!$ are positive, we don't need the absolute value signs.

$$\frac{a_{n+1}}{a_n} = \frac{(n+1)^{n+1}}{(n+1)!} \cdot \frac{n!}{n^n} = \frac{(n+1)(n+1)^n}{(n+1)n!} \cdot \frac{n!}{n^n}$$

$$= \left(\frac{n+1}{n}\right)^n = \left(1+\frac{1}{n}\right)^n \to e \qquad \text{as } n \to \infty$$

(see Equation 4.4.7). Since $e > 1$, the given series is divergent by the Ratio Test. ■

NOTE: Although the Ratio Test works in Example 9, an easier method is to use the Test for Divergence. Since

$$a_n = \frac{n^n}{n!} = \frac{n \cdot n \cdot n \cdot \cdots \cdot n}{1 \cdot 2 \cdot 3 \cdot \cdots \cdot n} \geqslant n$$

it follows that a_n does not approach 0 as $n \to \infty$. Therefore, the given series is divergent by the Test for Divergence.

EXERCISES 10.4

1. (a) What is an alternating series?
   (b) Under what conditions does an alternating series converge?
   (c) If these conditions are satisfied, what can you say about the remainder after n terms?

2. What can you say about the series $\Sigma\, a_n$ in each of the following cases?

   (a) $\displaystyle\lim_{n\to\infty}\left|\frac{a_{n+1}}{a_n}\right| = 8$

   (b) $\displaystyle\lim_{n\to\infty}\left|\frac{a_{n+1}}{a_n}\right| = 0.8$

   (c) $\displaystyle\lim_{n\to\infty}\left|\frac{a_{n+1}}{a_n}\right| = 1$

3–8 ■ Test the series for convergence or divergence.

3. $\frac{3}{5} - \frac{3}{6} + \frac{3}{7} - \frac{3}{8} + \frac{3}{9} - \cdots$

4. $\displaystyle\frac{1}{\ln 2} - \frac{1}{\ln 3} + \frac{1}{\ln 4} - \frac{1}{\ln 5} + \frac{1}{\ln 6} - \cdots$

5. $\displaystyle\sum_{n=1}^{\infty} (-1)^{n+1}\frac{n}{5n+1}$

6. $\displaystyle\sum_{n=1}^{\infty} (-1)^{n}\frac{n^2}{n^2+1}$

7. $\displaystyle\sum_{n=1}^{\infty} (-1)^{n}\frac{n}{n^2+1}$

8. $\displaystyle\sum_{n=1}^{\infty} (-1)^{n-1}\frac{\ln n}{n}$

9. Is the 50th partial sum s_{50} of the alternating series $\sum_{n=1}^{\infty} (-1)^{n-1}/n$ an overestimate or an underestimate of the total sum? Explain.

10. Calculate the first 10 partial sums of the series

$$\sum_{n=1}^{\infty} \frac{(-1)^{n-1}}{n^3}$$

and graph both the sequence of terms and the sequence of partial sums on the same screen. Estimate the error in using the 10th partial sum to approximate the total sum.

11. For what values of p is the following series convergent?

$$\sum_{n=1}^{\infty} \frac{(-1)^{n-1}}{n^p}$$

12–14 ■ How many terms of the series do we need to add in order to find the sum to the indicated accuracy?

12. $\displaystyle\sum_{n=1}^{\infty} \frac{(-1)^{n+1}}{n^4}$ (error < 0.001)

13. $\displaystyle\sum_{n=1}^{\infty} \frac{(-2)^{n}}{n!}$ (error < 0.01)

14. $\displaystyle\sum_{n=1}^{\infty} \frac{(-1)^{n}n}{4^n}$ (error < 0.002)

15–16 ■ Graph both the sequence of terms and the sequence of partial sums on the same screen. Use the graph to make a rough estimate of the sum of the series. Then use the Alternating Series Estimation Theorem to estimate the sum correct to four decimal places.

15. $\displaystyle\sum_{n=1}^{\infty} \frac{(-1)^{n-1}}{(2n-1)!}$

16. $\displaystyle\sum_{n=0}^{\infty} \frac{(-1)^{n}}{(2n)!}$

17–18 ■ Approximate the sum of the series to the indicated accuracy.

17. $\displaystyle\sum_{n=0}^{\infty} \frac{(-1)^{n}}{2^n n!}$ (four decimal places)

18. $\displaystyle\sum_{n=1}^{\infty} \frac{(-1)^{n-1}}{n^6}$ (five decimal places)

19–28 ■ Determine whether the series is absolutely convergent.

19. $\displaystyle\sum_{n=1}^{\infty} \frac{(-3)^{n}}{n^3}$

20. $\displaystyle\sum_{n=0}^{\infty} \frac{(-3)^{n}}{n!}$

21. $\displaystyle\sum_{n=1}^{\infty} \frac{(-1)^{n+1}}{2n+1}$

22. $\displaystyle\sum_{n=1}^{\infty} (-1)^{n-1}\frac{\sqrt{n}}{n+1}$

23. $\displaystyle\sum_{n=1}^{\infty} \frac{\sin 2n}{n^2}$

24. $\displaystyle\sum_{n=1}^{\infty} \frac{(-1)^{n+1}5^{n-1}}{(n+1)^2 4^{n+2}}$

25. $\displaystyle\sum_{n=1}^{\infty} \frac{(n+1)5^n}{n 3^{2n}}$

26. $\displaystyle\sum_{n=1}^{\infty} \frac{\cos(n\pi/6)}{n\sqrt{n}}$

27. $\displaystyle\sum_{n=1}^{\infty} \frac{(n+2)!}{n!\,10^n}$

28. $\displaystyle\frac{1}{3} + \frac{1\cdot 4}{3\cdot 5} + \frac{1\cdot 4\cdot 7}{3\cdot 5\cdot 7} + \frac{1\cdot 4\cdot 7\cdot 10}{3\cdot 5\cdot 7\cdot 9} + \cdots + \frac{1\cdot 4\cdot 7\cdot \cdots \cdot (3n-2)}{3\cdot 5\cdot 7\cdot \cdots \cdot (2n+1)} + \cdots$

29. The terms of a series are defined recursively by the equations

$$a_1 = 2 \qquad a_{n+1} = \frac{5n+1}{4n+3}\,a_n$$

Determine whether $\Sigma\, a_n$ converges or diverges.

30. A series $\Sigma\, a_n$ is defined by the equations

$$a_1 = 1 \qquad a_{n+1} = \frac{2+\cos n}{\sqrt{n}}\,a_n$$

Determine whether $\Sigma\, a_n$ converges or diverges.

31. For which of the following series is the Ratio Test inconclusive (that is, it fails to give a definite answer)?

   (a) $\displaystyle\sum_{n=1}^{\infty} \frac{1}{n^3}$
   (b) $\displaystyle\sum_{n=1}^{\infty} \frac{n}{2^n}$
   (c) $\displaystyle\sum_{n=1}^{\infty} \frac{(-3)^{n-1}}{\sqrt{n}}$
   (d) $\displaystyle\sum_{n=1}^{\infty} \frac{\sqrt{n}}{1+n^2}$

32. For which positive integers k is the series

$$\sum_{n=1}^{\infty} \frac{(n!)^2}{(kn)!}$$

convergent?

33. (a) Show that $\sum_{n=0}^{\infty} x^n/n!$ converges for all x.
(b) Deduce that $\lim_{n\to\infty} x^n/n! = 0$ for all x.

34. Let $\sum a_n$ be a series with positive terms and let $r_n = a_{n+1}/a_n$. Suppose that $\lim_{n\to\infty} r_n = L < 1$, so $\sum a_n$ converges by the Ratio Test. As usual, we let R_n be the remainder after n terms, that is,

$$R_n = a_{n+1} + a_{n+2} + a_{n+3} + \cdots$$

(a) If $\{r_n\}$ is a decreasing sequence and $r_{n+1} < 1$, show, by summing a geometric series, that

$$R_n \leqslant \frac{a_{n+1}}{1 - r_{n+1}}$$

(b) If $\{r_n\}$ is an increasing sequence, show that

$$R_n \leqslant \frac{a_{n+1}}{1 - L}$$

35. (a) Find the partial sum s_5 of the series

$$\sum_{n=1}^{\infty} \frac{1}{n2^n}$$

Use Exercise 34 to estimate the error in using s_5 as an approximation to the sum of the series.
(b) Find a value of n so that s_n is within 0.00005 of the sum. Use this value of n to approximate the sum of the series.

36. Use the sum of the first 10 terms to approximate the sum of the series

$$\sum_{n=1}^{\infty} \frac{n}{2^n}$$

Use Exercise 34 to estimate the error.

10.5 POWER SERIES

A **power series** is a series of the form

(1)
$$\sum_{n=0}^{\infty} c_n x^n = c_0 + c_1 x + c_2 x^2 + c_3 x^3 + \cdots$$

where x is a variable and the c_n's are constants called the **coefficients** of the series. For each fixed x, the series (1) is a series of constants that we can test for convergence or divergence. A power series may converge for some values of x and diverge for other values of x. The sum of the series is a function

$$f(x) = c_0 + c_1 x + c_2 x^2 + \cdots + c_n x^n + \cdots$$

whose domain is the set of all x for which the series converges. Notice that f resembles a polynomial. The only difference is that f has infinitely many terms.

For instance, if we take $c_n = 1$ for all n, the power series becomes the geometric series

$$\sum_{n=0}^{\infty} x^n = 1 + x + x^2 + \cdots + x^n + \cdots = \frac{1}{1 - x}$$

which converges when $-1 < x < 1$ and diverges when $|x| \geqslant 1$ (see Equation 10.2.5).

More generally, a series of the form

(2)
$$\sum_{n=0}^{\infty} c_n(x - a)^n = c_0 + c_1(x - a) + c_2(x - a)^2 + \cdots$$

is called a **power series in $(x - a)$** or a **power series centered at a** or a **power series about a**. Notice that in writing out the term corresponding to $n = 0$ in Equations 1 and

2 we have adopted the convention that $(x - a)^0 = 1$ even when $x = a$. Notice also that when $x = a$ all of the terms are 0 for $n \geqslant 1$ and so the power series (2) always converges when $x = a$.

EXAMPLE 1 For what values of x is the series $\sum_{n=0}^{\infty} n!x^n$ convergent?

SOLUTION We use the Ratio Test. If we let a_n, as usual, denote the nth term of the series, then $a_n = n!x^n$. If $x \neq 0$, we have

$$\lim_{n\to\infty} \left| \frac{a_{n+1}}{a_n} \right| = \lim_{n\to\infty} \left| \frac{(n+1)!x^{n+1}}{n!x^n} \right|$$

$$= \lim_{n\to\infty} (n+1)|x| = \infty$$

By the Ratio Test, the series diverges when $x \neq 0$. Thus, the given series converges only when $x = 0$. ■

EXAMPLE 2 For what values of x does the series $\sum_{n=1}^{\infty} \frac{(x-3)^n}{n}$ converge?

SOLUTION Let $a_n = (x - 3)^n/n$. Then

$$\left| \frac{a_{n+1}}{a_n} \right| = \left| \frac{(x-3)^{n+1}}{n+1} \cdot \frac{n}{(x-3)^n} \right|$$

$$= \frac{1}{1 + \frac{1}{n}} |x - 3| \to |x - 3| \qquad \text{as } n \to \infty$$

By the Ratio Test, the given series is absolutely convergent, and therefore convergent, when $|x - 3| < 1$ and divergent when $|x - 3| > 1$. Now

$$|x - 3| < 1 \iff -1 < x - 3 < 1 \iff 2 < x < 4$$

so the series converges when $2 < x < 4$ and diverges when $x < 2$ or $x > 4$.

The Ratio Test gives no information when $|x - 3| = 1$ so we must consider $x = 2$ and $x = 4$ separately. If we put $x = 4$ in the series, it becomes $\Sigma\, 1/n$, the harmonic series, which is divergent. If $x = 2$, the series is $\Sigma\, (-1)^n/n$, which converges by the Alternating Series Test. Thus, the given power series converges for $2 \leqslant x < 4$. ■

We will see that the main use of a power series is that it provides a way to represent some of the most important functions that arise in mathematics, physics, and chemistry. In particular, the sum of the power series in the next example is called a **Bessel function**, after the German astronomer Friedrich Bessel (1784–1846), and the function given in Exercise 21 is another example of a Bessel function. In fact, these functions first arose when Bessel solved Kepler's equation for describing planetary motion. Since that time, these functions have been applied in many different physical situations, including the temperature distribution in a circular plate and the shape of a vibrating drumhead.

EXAMPLE 3 Find the domain of the Bessel function of order 0 defined by

$$J_0(x) = \sum_{n=0}^{\infty} \frac{(-1)^n x^{2n}}{2^{2n}(n!)^2}$$

SOLUTION Let $a_n = (-1)^n x^{2n}/[2^{2n}(n!)^2]$. Then

$$\left|\frac{a_{n+1}}{a_n}\right| = \left|\frac{(-1)^{n+1}x^{2(n+1)}}{2^{2(n+1)}[(n+1)!]^2} \cdot \frac{2^{2n}(n!)^2}{(-1)^n x^{2n}}\right|$$

$$= \frac{x^{2n+2}}{2^{2n+2}(n+1)^2(n!)^2} \cdot \frac{2^{2n}(n!)^2}{x^{2n}}$$

$$= \frac{x^2}{4(n+1)^2} \to 0 < 1 \qquad \text{for all } x$$

Thus, by the Ratio Test, the given series converges for all values of x. In other words, the domain of the Bessel function J_0 is $(-\infty, \infty) = \mathbb{R}$. ■

Recall that the sum of a series is equal to the limit of the sequence of partial sums. So when we define the Bessel function in Example 3 as the sum of a series we mean that, for every real number x,

$$J_0(x) = \lim_{n\to\infty} s_n(x) \qquad \text{where} \qquad s_{2n}(x) = \sum_{i=0}^{n} \frac{(-1)^i x^{2i}}{2^{2i}(i!)^2}$$

The first few partial sums are

$$s_0(x) = 1 \qquad s_2(x) = 1 - \frac{x^2}{4} \qquad s_4(x) = 1 - \frac{x^2}{4} + \frac{x^4}{64}$$

$$s_6(x) = 1 - \frac{x^2}{4} + \frac{x^4}{64} - \frac{x^6}{2304} \qquad s_8(x) = 1 - \frac{x^2}{4} + \frac{x^4}{64} - \frac{x^6}{2304} + \frac{x^8}{147{,}456}$$

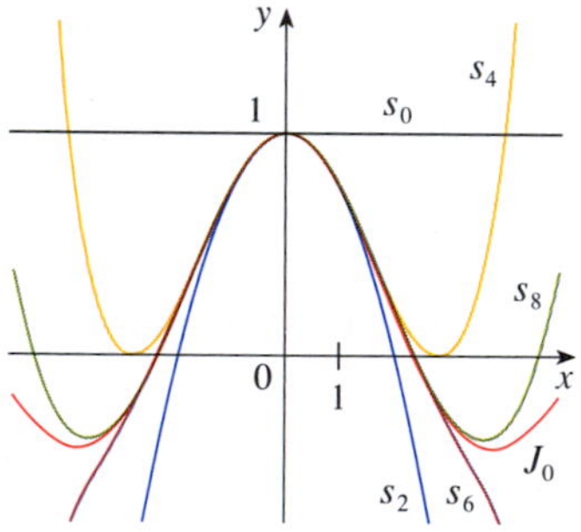

FIGURE 1
Partial sums of the Bessel function J_0

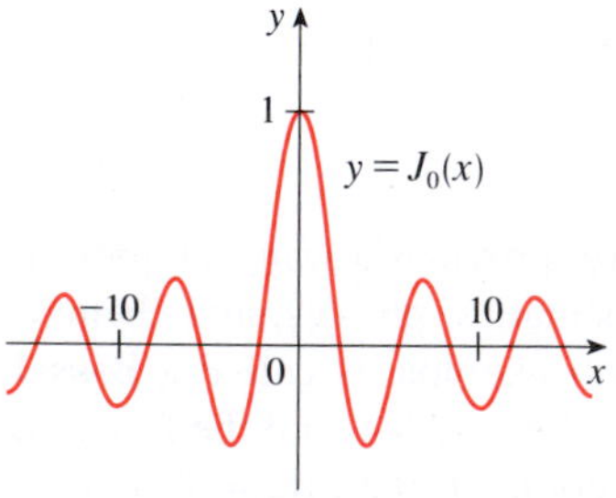

FIGURE 2

Figure 1 shows the graphs of these partial sums, which are polynomials. They are all approximations to the function J_0, but notice that the approximations become better when more terms are included. Figure 2 shows a more complete graph of the Bessel function.

For the power series that we have looked at so far, the set of values of x for which the series is convergent has always turned out to be an interval [a finite interval for the geometric series and the series in Example 2, the infinite interval $(-\infty, \infty)$ in Example 3, and a collapsed interval $[0, 0] = \{0\}$ in Example 1]. The following theorem, which we won't prove, says that this is true in general.

(3) THEOREM For a given power series $\sum_{n=0}^{\infty} c_n(x-a)^n$ there are only three possibilities:

(i) The series converges only when $x = a$.

(ii) The series converges for all x.

(iii) There is a positive number R such that the series converges if $|x - a| < R$ and diverges if $|x - a| > R$.

The number R in case (iii) is called the **radius of convergence** of the power series. By convention, the radius of convergence is $R = 0$ in case (i) and $R = \infty$ in case (ii). The **interval of convergence** of a power series is the interval that consists of all values of x for which the series converges. In case (i) the interval consists of just a single point a. In case (ii) the interval is $(-\infty, \infty)$. In case (iii) note that the inequality $|x - a| < R$ can be rewritten as $a - R < x < a + R$. When x is an *endpoint* of the interval, that is, $x = a \pm R$, anything can happen—the series might converge at one or both endpoints or it might diverge at both endpoints. Thus, in case (iii) there are four possibilities for the interval of convergence:

$$(a - R, a + R) \qquad (a - R, a + R] \qquad [a - R, a + R) \qquad [a - R, a + R]$$

The situation is illustrated in Figure 3.

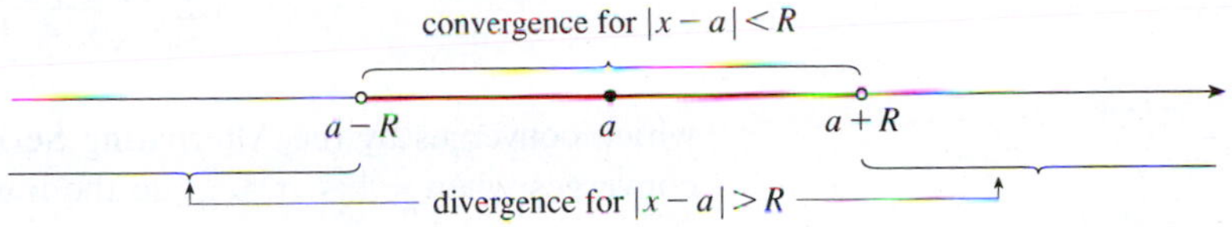

FIGURE 3

We summarize here the radius and interval of convergence for each of the examples already considered in this section.

	Series	Radius of convergence	Interval of convergence
Geometric series	$\sum_{n=0}^{\infty} x^n$	$R = 1$	$(-1, 1)$
Example 1	$\sum_{n=0}^{\infty} n!x^n$	$R = 0$	$\{0\}$
Example 2	$\sum_{n=1}^{\infty} \frac{(x-3)^n}{n}$	$R = 1$	$[2, 4)$
Example 3	$\sum_{n=0}^{\infty} \frac{(-1)^n x^{2n}}{2^{2n}(n!)^2}$	$R = \infty$	$(-\infty, \infty)$

The Ratio Test can be used to determine the radius of convergence R in most cases. The Ratio Test always fails when x is an endpoint of the interval of convergence, so the endpoints must be checked with some other test.

EXAMPLE 4 Find the radius of convergence and interval of convergence of the series

$$\sum_{n=0}^{\infty} \frac{(-3)^n x^n}{\sqrt{n+1}}$$

SOLUTION Let $a_n = (-3)^n x^n / \sqrt{n+1}$. Then

$$\left| \frac{a_{n+1}}{a_n} \right| = \left| \frac{(-3)^{n+1} x^{n+1}}{\sqrt{n+2}} \cdot \frac{\sqrt{n+1}}{(-3)^n x^n} \right|$$

$$= 3\sqrt{\frac{1 + (1/n)}{1 + (2/n)}}\, |x| \to 3|x| \quad \text{as } n \to \infty$$

By the Ratio Test, the given series converges if $3|x| < 1$ and diverges if $3|x| > 1$. Thus, it converges if $|x| < \frac{1}{3}$ and diverges if $|x| > \frac{1}{3}$. This means that the radius of convergence is $R = \frac{1}{3}$.

We know the series converges in the interval $(-\frac{1}{3}, \frac{1}{3})$, but we must now test for convergence at the endpoints of this interval. If $x = -\frac{1}{3}$, the series becomes

$$\sum_{n=0}^{\infty} \frac{(-3)^n(-\frac{1}{3})^n}{\sqrt{n+1}} = \sum_{n=0}^{\infty} \frac{1}{\sqrt{n+1}} = \frac{1}{\sqrt{1}} + \frac{1}{\sqrt{2}} + \frac{1}{\sqrt{3}} + \frac{1}{\sqrt{4}} + \cdots$$

which diverges. (Use the Integral Test or simply observe that it is a p-series with $p = \frac{1}{2} < 1$.) If $x = \frac{1}{3}$, the series is

$$\sum_{n=0}^{\infty} \frac{(-3)^n(\frac{1}{3})^n}{\sqrt{n+1}} = \sum_{n=0}^{\infty} \frac{(-1)^n}{\sqrt{n+1}}$$

which converges by the Alternating Series Test. Therefore, the given power series converges when $-\frac{1}{3} < x \leqslant \frac{1}{3}$, so the interval of convergence is $(-\frac{1}{3}, \frac{1}{3}]$. ■

EXAMPLE 5 Find the radius of convergence and interval of convergence of the series

$$\sum_{n=0}^{\infty} \frac{n(x+2)^n}{3^{n+1}}$$

SOLUTION If $a_n = n(x+2)^n/3^{n+1}$, then

$$\left|\frac{a_{n+1}}{a_n}\right| = \left|\frac{(n+1)(x+2)^{n+1}}{3^{n+2}} \cdot \frac{3^{n+1}}{n(x+2)^n}\right|$$

$$= \left(1 + \frac{1}{n}\right)\frac{|x+2|}{3} \to \frac{|x+2|}{3} \qquad \text{as } n \to \infty$$

Using the Ratio Test, we see that the series converges if $|x+2|/3 < 1$ and it diverges if $|x+2|/3 > 1$. So it converges if $|x+2| < 3$ and diverges if $|x+2| > 3$. Thus, the radius of convergence is $R = 3$.

The inequality $|x+2| < 3$ can be written as $-5 < x < 1$, so we test the series at the endpoints -5 and 1. When $x = -5$, the series is

$$\sum_{n=0}^{\infty} \frac{n(-3)^n}{3^{n+1}} = \frac{1}{3}\sum_{n=0}^{\infty} (-1)^n n$$

which diverges by the Test for Divergence [$(-1)^n n$ doesn't converge to 0]. When $x = 1$, the series is

$$\sum_{n=0}^{\infty} \frac{n(3)^n}{3^{n+1}} = \frac{1}{3}\sum_{n=0}^{\infty} n$$

which also diverges by the Test for Divergence. Thus, the series converges only when $-5 < x < 1$, so the interval of convergence is $(-5, 1)$. ■

EXERCISES 10.5

1. What is a power series?

2. (a) What is the radius of convergence of a power series? How do you find it?
   (b) What is the interval of convergence of a power series? How do you find it?

3. If $\sum_{n=0}^{\infty} c_n 4^n$ is convergent, does it follow that the following series are convergent?
   (a) $\sum_{n=0}^{\infty} c_n(-2)^n$ (b) $\sum_{n=0}^{\infty} c_n(-4)^n$

4. Suppose that $\sum_{n=0}^{\infty} c_n x^n$ converges when $x = -4$ and diverges when $x = 6$. What can be said about the convergence or divergence of the following series?
   (a) $\sum_{n=0}^{\infty} c_n$ (b) $\sum_{n=0}^{\infty} c_n 8^n$
   (c) $\sum_{n=0}^{\infty} c_n(-3)^n$ (d) $\sum_{n=0}^{\infty} (-1)^n c_n 9^n$

5–18 ■ Find the radius of convergence and interval of convergence of the series.

5. $\displaystyle\sum_{n=0}^{\infty} \frac{x^n}{n+2}$

6. $\displaystyle\sum_{n=1}^{\infty} \frac{(-1)^n x^n}{\sqrt[3]{n}}$

7. $\displaystyle\sum_{n=0}^{\infty} \frac{x^n}{n!}$

8. $\displaystyle\sum_{n=1}^{\infty} \frac{x^n}{n^2}$

9. $\displaystyle\sum_{n=1}^{\infty} \frac{(-1)^n x^n}{n2^n}$

10. $\displaystyle\sum_{n=0}^{\infty} \frac{n^2 x^n}{10^n}$

11. $\displaystyle\sum_{n=0}^{\infty} \frac{n}{4^n}(2x-1)^n$

12. $\displaystyle\sum_{n=1}^{\infty} \frac{(-1)^n x^{2n-1}}{(2n-1)!}$

13. $\displaystyle\sum_{n=1}^{\infty} (-1)^n \frac{(x-1)^n}{\sqrt{n}}$

14. $\displaystyle\sum_{n=1}^{\infty} \frac{(x-4)^n}{n5^n}$

15. $\displaystyle\sum_{n=0}^{\infty} \frac{2^n(x-3)^n}{n+3}$

16. $\displaystyle\sum_{n=1}^{\infty} \frac{(x+1)^n}{n(n+1)}$

17. $\displaystyle\sum_{n=1}^{\infty} n!(2x-1)^n$

18. $\displaystyle\sum_{n=1}^{\infty} \frac{nx^n}{1 \cdot 3 \cdot 5 \cdot \cdots \cdot (2n-1)}$

19. If k is a positive integer, find the radius of convergence of the series

$$\sum_{n=0}^{\infty} \frac{(n!)^k}{(kn)!} x^n$$

20. Graph the first several partial sums $s_n(x)$ of the series $\sum_{n=0}^{\infty} x^n$, together with the sum function $f(x) = 1/(1-x)$, on a common screen. On what interval do these partial sums appear to be converging to $f(x)$?

21. The function J_1 defined by

$$J_1(x) = \sum_{n=0}^{\infty} \frac{(-1)^n x^{2n+1}}{n!(n+1)!2^{2n+1}}$$

is called the *Bessel function of order 1*.
   (a) Find its domain.
   (b) Graph the first several partial sums on a common screen.
   CAS (c) If your CAS has built-in Bessel functions, graph J_1 on the same screen as the partial sums in part (b) and observe how the partial sums approximate J_1.

22. The function A defined by

$$A(x) = 1 + \frac{x^3}{2 \cdot 3} + \frac{x^6}{2 \cdot 3 \cdot 5 \cdot 6} + \frac{x^9}{2 \cdot 3 \cdot 5 \cdot 6 \cdot 8 \cdot 9} + \cdots$$

is called the *Airy function* after the English mathematician and astronomer Sir George Airy (1801–1892).
   (a) Find the domain of the Airy function.
   (b) Graph the first several partial sums $s_n(x)$ on a common screen.
   CAS (c) If your CAS has built-in Airy functions, graph A on the same screen as the partial sums in part (b) and observe how the partial sums approximate A.

23. A function f is defined by

$$f(x) = 1 + 2x + x^2 + 2x^3 + x^4 + \cdots$$

that is, its coefficients are $c_{2n} = 1$ and $c_{2n+1} = 2$ for all $n \geqslant 0$. Find the interval of convergence of the series and find an explicit formula for $f(x)$.

24. If $f(x) = \sum_{n=0}^{\infty} c_n x^n$, where $c_{n+4} = c_n$ for all $n \geqslant 0$, find the interval of convergence of the series and a formula for $f(x)$.

25. Suppose the series $\sum c_n x^n$ has radius of convergence 2 and the series $\sum d_n x^n$ has radius of convergence 3. What is the radius of convergence of the series $\sum (c_n + d_n)x^n$? Explain.

26. Suppose that the radius of convergence of the power series $\sum c_n x^n$ is R. What is the radius of convergence of the power series $\sum c_n x^{2n}$?

10.6 REPRESENTATIONS OF FUNCTIONS AS POWER SERIES

In this section we learn how to represent certain types of functions as sums of power series by manipulating geometric series or by differentiating or integrating such a series. You might wonder why we would ever want to express a known function as a sum of infinitely many terms. We will see later that this strategy is useful for integrating functions that don't have elementary antiderivatives, for solving differential equations, and for approximating functions by polynomials. (Scientists do this to simplify the expressions they deal with; computer scientists do this to represent functions on calculators and computers.)

We start with an equation that we have seen before:

A geometric illustration of Equation 1 is shown in Figure 1. Because the sum of a series is the limit of the sequence of partial sums, we have

$$\frac{1}{1-x} = \lim_{n\to\infty} s_n(x)$$

where

$$s_n(x) = 1 + x + x^2 + \cdots + x^n$$

is the nth partial sum. Notice that as n increases, $s_n(x)$ becomes a better approximation to $f(x)$ for $-1 < x < 1$.

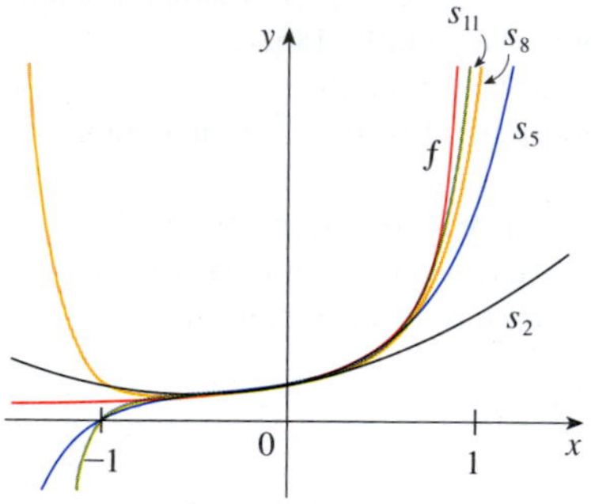

FIGURE 1

$f(x) = \dfrac{1}{1-x}$ and some partial sums

$$\textbf{(1)} \qquad \frac{1}{1-x} = 1 + x + x^2 + x^3 + \cdots = \sum_{n=0}^{\infty} x^n \qquad |x| < 1$$

We first encountered this equation in Example 5 in Section 10.2, where we obtained it by observing that it is a geometric series with $a = 1$ and $r = x$. But here our point of view is different. We now regard Equation 1 as expressing the function $f(x) = 1/(1 - x)$ as a sum of a power series.

EXAMPLE 1 Express $1/(1 + x^2)$ as the sum of a power series and find the interval of convergence.

SOLUTION Replacing x by $-x^2$ in Equation 1, we have

$$\frac{1}{1+x^2} = \frac{1}{1-(-x^2)} = \sum_{n=0}^{\infty} (-x^2)^n$$

$$= \sum_{n=0}^{\infty} (-1)^n x^{2n} = 1 - x^2 + x^4 - x^6 + x^8 - \cdots$$

Because this is a geometric series, it converges when $|-x^2| < 1$, that is, $x^2 < 1$, or $|x| < 1$. Therefore, the interval of convergence is $(-1, 1)$. (Of course, we could have determined the radius of convergence by applying the Ratio Test, but that much work is unnecessary here.) ■

EXAMPLE 2 Find a power series representation for $1/(x + 2)$.

SOLUTION In order to put this function in the form of the left side of Equation 1 we first factor a 2 from the denominator:

$$\frac{1}{2+x} = \frac{1}{2\left(1 + \dfrac{x}{2}\right)} = \frac{1}{2\left[1 - \left(-\dfrac{x}{2}\right)\right]}$$

$$= \frac{1}{2}\sum_{n=0}^{\infty}\left(-\frac{x}{2}\right)^n = \sum_{n=0}^{\infty} \frac{(-1)^n}{2^{n+1}} x^n$$

This series converges when $|-x/2| < 1$, that is, $|x| < 2$. So the interval of convergence is $(-2, 2)$. ■

EXAMPLE 3 Find a power series representation of $x^3/(x + 2)$.

SOLUTION Since this function is just x^3 times the function in Example 2, all we have to do is to multiply that series by x^3:

It's legitimate to move x^3 across the sigma sign because it doesn't depend on n. [Use Theorem 10.2.8(i) with $c = x^3$.]

$$\frac{x^3}{x+2} = x^3 \sum_{n=0}^{\infty} \frac{(-1)^n}{2^{n+1}} x^n = \sum_{n=0}^{\infty} \frac{(-1)^n}{2^{n+1}} x^{n+3}$$

$$= \tfrac{1}{2}x^3 - \tfrac{1}{4}x^4 + \tfrac{1}{8}x^5 - \tfrac{1}{16}x^6 + \cdots$$

Another way of writing this series is as follows:

$$\frac{x^3}{x+2} = \sum_{n=3}^{\infty} \frac{(-1)^{n-1}}{2^{n-2}} x^n$$

As in Example 2, the interval of convergence is $(-2, 2)$. ■

DIFFERENTIATION AND INTEGRATION OF POWER SERIES

The sum of a power series is a function $f(x) = \sum_{n=0}^{\infty} c_n(x - a)^n$ whose domain is the interval of convergence of the series. We would like to be able to differentiate and integrate such functions, and the following theorem (which we won't prove) says that we can do so by differentiating or integrating each individual term in the series, just as we would for a polynomial. This is called **term-by-term differentiation and integration**.

(2) THEOREM If the power series $\sum c_n(x - a)^n$ has radius of convergence $R > 0$, then the function f defined by

$$f(x) = c_0 + c_1(x - a) + c_2(x - a)^2 + \cdots = \sum_{n=0}^{\infty} c_n(x - a)^n$$

is differentiable (and therefore continuous) on the interval $(a - R, a + R)$ and

(a) $$f'(x) = c_1 + 2c_2(x - a) + 3c_3(x - a)^2 + \cdots = \sum_{n=1}^{\infty} nc_n(x - a)^{n-1}$$

(b) $$\int f(x)\,dx = C + c_0(x - a) + c_1\frac{(x - a)^2}{2} + c_2\frac{(x - a)^3}{3} + \cdots$$

$$= C + \sum_{n=0}^{\infty} c_n \frac{(x - a)^{n+1}}{n + 1}$$

The radii of convergence of the power series in Equations (a) and (b) are both R.

In part (b), $\int c_0\,dx = c_0x + C_1$ is written as $c_0(x - a) + C$, where $C = C_1 + ac_0$, so all the terms of the series have the same form.

NOTE 1: Equations (a) and (b) in Theorem 2 can be rewritten in the form

(c) $$\frac{d}{dx}\left[\sum_{n=0}^{\infty} c_n(x - a)^n\right] = \sum_{n=0}^{\infty} \frac{d}{dx}[c_n(x - a)^n]$$

(d) $$\int \left[\sum_{n=0}^{\infty} c_n(x - a)^n\right] dx = \sum_{n=0}^{\infty} \int c_n(x - a)^n\,dx$$

We know that, for finite sums, the derivative of a sum is the sum of the derivatives and the integral of a sum is the sum of the integrals. Equations (c) and (d) assert that the same is true for infinite sums, provided we are dealing with *power series.* (For other types of series of functions the situation is not as simple; see Exercise 32.)

NOTE 2: Although Theorem 2 says that the radius of convergence remains the same when a power series is differentiated or integrated, this does not mean that the *interval* of convergence remains the same. It may happen that the original series converges at an endpoint, whereas the differentiated series diverges there. (See Exercise 33.)

NOTE 3: The idea of differentiating a power series term by term is the basis for a powerful method for solving differential equations. We will discuss this method in Section 15.4.

EXAMPLE 4 In Example 3 in Section 10.5 we saw that the Bessel function

$$J_0(x) = \sum_{n=0}^{\infty} \frac{(-1)^n x^{2n}}{2^{2n}(n!)^2}$$

is defined for all x. Thus, by Theorem 2, J_0 is differentiable for all x and its derivative is found by term-by-term differentiation as follows:

$$J_0'(x) = \sum_{n=0}^{\infty} \frac{d}{dx} \frac{(-1)^n x^{2n}}{2^{2n}(n!)^2} = \sum_{n=1}^{\infty} \frac{(-1)^n 2n x^{2n-1}}{2^{2n}(n!)^2}$$ ■

EXAMPLE 5 Express $1/(1-x)^2$ as a power series by differentiating Equation 1. What is the radius of convergence?

SOLUTION Differentiating each side of the equation

$$\frac{1}{1-x} = 1 + x + x^2 + x^3 + \cdots = \sum_{n=0}^{\infty} x^n$$

we get

$$\frac{1}{(1-x)^2} = 1 + 2x + 3x^2 + \cdots = \sum_{n=1}^{\infty} nx^{n-1}$$

If we wish, we can replace n by $n+1$ and write the answer as

$$\frac{1}{(1-x)^2} = \sum_{n=0}^{\infty} (n+1)x^n$$

According to Theorem 2, the radius of convergence of the differentiated series is the same as the radius of convergence of the original series, namely, $R = 1$. ■

EXAMPLE 6 Find a power series representation for $\ln(1-x)$ and its radius of convergence.

SOLUTION We notice that, except for a factor of -1, the derivative of this function is $1/(1-x)$. So we integrate both sides of Equation 1:

$$-\ln(1-x) = \int \frac{1}{1-x}\,dx = C + x + \frac{x^2}{2} + \frac{x^3}{3} + \cdots$$

$$= C + \sum_{n=0}^{\infty} \frac{x^{n+1}}{n+1} = C + \sum_{n=1}^{\infty} \frac{x^n}{n} \qquad |x| < 1$$

To determine the value of C we put $x = 0$ in this equation and obtain $-\ln(1 - 0) = C$. Thus, $C = 0$ and

$$\ln(1 - x) = -x - \frac{x^2}{2} - \frac{x^3}{3} - \cdots = -\sum_{n=1}^{\infty} \frac{x^n}{n} \qquad |x| < 1$$

The radius of convergence is the same as for the original series: $R = 1$. ■

Notice what happens if we put $x = \frac{1}{2}$ in the result of Example 6. Since $\ln \frac{1}{2} = -\ln 2$, we see that

$$\ln 2 = \frac{1}{2} + \frac{1}{8} + \frac{1}{24} + \frac{1}{64} + \cdots = \sum_{n=1}^{\infty} \frac{1}{n2^n}$$

EXAMPLE 7 Find a power series representation for $f(x) = \tan^{-1}x$.

SOLUTION We observe that $f'(x) = 1/(1 + x^2)$ and find the required series by integrating the power series for $1/(1 + x^2)$ found in Example 1.

The power series for $\tan^{-1}x$ obtained in Example 7 is called *Gregory's series* after the Scottish mathematician James Gregory (1638–1675), who had anticipated some of Newton's discoveries. We have shown that Gregory's series is valid when $-1 < x < 1$, but it turns out (although it isn't easy to prove) that it is also valid when $x = \pm 1$. Notice that when $x = 1$ the series becomes

$$\frac{\pi}{4} = 1 - \frac{1}{3} + \frac{1}{5} - \frac{1}{7} + \cdots$$

This beautiful result is known as the Leibniz formula for π.

$$\tan^{-1}x = \int \frac{1}{1 + x^2}\,dx = \int (1 - x^2 + x^4 - x^6 + \cdots)\,dx$$

$$= C + x - \frac{x^3}{3} + \frac{x^5}{5} - \frac{x^7}{7} + \cdots$$

To find C we put $x = 0$ and obtain $C = \tan^{-1}0 = 0$. Therefore

$$\tan^{-1}x = x - \frac{x^3}{3} + \frac{x^5}{5} - \frac{x^7}{7} + \cdots = \sum_{n=0}^{\infty} (-1)^n \frac{x^{2n+1}}{2n + 1}$$

Since the radius of convergence of the series for $1/(1 + x^2)$ is 1, the radius of convergence of this series for $\tan^{-1}x$ is also 1. ■

EXAMPLE 8

(a) Evaluate $\int [1/(1 + x^7)]\,dx$ as a power series.

(b) Use part (a) to approximate $\int_0^{0.5} [1/(1 + x^7)]\,dx$ correct to within 10^{-7}.

SOLUTION

(a) The first step is to express the integrand, $1/(1 + x^7)$, as the sum of a power series. As in Example 1, we start with Equation 1 and replace x by $-x^7$:

$$\frac{1}{1 + x^7} = \frac{1}{1 - (-x^7)} = \sum_{n=0}^{\infty} (-x^7)^n = \sum_{n=0}^{\infty} (-1)^n x^{7n} = 1 - x^7 + x^{14} - \cdots$$

This example demonstrates one way in which power series representations are useful. Integrating $1/(1 + x^7)$ by hand is incredibly difficult. Different computer algebra systems return different forms of the answer, but they are all extremely complicated. (If you have a CAS, try it yourself.) The infinite series answer that we obtain in Example 8(a) is actually much easier to deal with than the finite answer provided by a CAS.

Now we integrate term by term:

$$\int \frac{1}{1 + x^7}\,dx = \int \sum_{n=0}^{\infty} (-1)^n x^{7n}\,dx = C + \sum_{n=0}^{\infty} (-1)^n \frac{x^{7n+1}}{7n + 1}$$

$$= C + x - \frac{x^8}{8} + \frac{x^{15}}{15} - \frac{x^{22}}{22} + \cdots$$

This series converges for $|-x^7| < 1$, that is, for $|x| < 1$.

(b) In applying the Fundamental Theorem of Calculus it doesn't matter which antiderivative we use, so let's use the antiderivative from part (a) with $C = 0$:

$$\int_0^{0.5} \frac{1}{1 + x^7}\,dx = \left[x - \frac{x^8}{8} + \frac{x^{15}}{15} - \frac{x^{22}}{22} + \cdots \right]_0^{1/2}$$

$$= \frac{1}{2} - \frac{1}{8 \cdot 2^8} + \frac{1}{15 \cdot 2^{15}} - \frac{1}{22 \cdot 2^{22}} + \cdots + \frac{(-1)^n}{(7n + 1)2^{7n+1}} + \cdots$$

This infinite series is the exact value of the definite integral, but since it is an alternating series, we can approximate the sum using the Alternating Series Estimation Theorem. If we stop adding after the term with $n = 3$, the error is smaller than the term with $n = 4$:

$$\frac{1}{29 \cdot 2^{29}} \approx 6.4 \times 10^{-11}$$

So we have

$$\int_0^{0.5} \frac{1}{1 + x^7}\,dx \approx \frac{1}{2} - \frac{1}{8 \cdot 2^8} + \frac{1}{15 \cdot 2^{15}} - \frac{1}{22 \cdot 2^{22}} \approx 0.49951374$$

■

EXERCISES 10.6

1. If the radius of convergence of the power series $\sum_{n=0}^{\infty} c_n x^n$ is 10, what is the radius of convergence of the series $\sum_{n=1}^{\infty} n c_n x^{n-1}$? Why?

2. Suppose you know that the series $\sum_{n=0}^{\infty} b_n x^n$ converges for $|x| < 2$. What can you say about the following series? Why?

$$\sum_{n=0}^{\infty} \frac{b_n}{n + 1} x^{n+1}$$

3–8 ■ Find a power series representation for the function and determine the interval of convergence.

3. $f(x) = \dfrac{1}{1 + x}$

4. $f(x) = \dfrac{x}{1 - x}$

5. $f(x) = \dfrac{1}{1 + 4x^2}$

6. $f(x) = \dfrac{1}{x^4 + 16}$

7. $f(x) = \dfrac{x}{x - 3}$

8. $f(x) = \dfrac{1 + x^2}{1 - x^2}$

9–14 ■ Find a power series representation for the function and determine the radius of convergence.

9. $f(x) = \dfrac{1}{(1 + x)^2}$

10. $f(x) = \ln(1 + x)$

11. $f(x) = \dfrac{1}{(1 + x)^3}$

12. $f(x) = x \ln(1 + x)$

13. $f(x) = \ln(5 - x)$

14. $f(x) = \dfrac{x^2}{(1 - 2x)^2}$

15–18 ■ Find a power series representation for f, and graph f and several partial sums $s_n(x)$ on the same screen. What happens as n increases?

15. $f(x) = \ln(3 + x)$

16. $f(x) = \dfrac{1}{x^2 + 25}$

17. $f(x) = \ln\left(\dfrac{1 + x}{1 - x}\right)$

18. $f(x) = \tan^{-1}(2x)$

19–22 ■ Evaluate the indefinite integral as a power series.

19. $\displaystyle\int \frac{1}{1 + x^4}\,dx$

20. $\displaystyle\int \frac{x}{1 + x^5}\,dx$

21. $\displaystyle\int \frac{\arctan x}{x}\,dx$

22. $\displaystyle\int \tan^{-1}(x^2)\,dx$

23–26 ■ Use a power series to approximate the definite integral to six decimal places.

23. $\displaystyle\int_0^{0.2} \frac{1}{1 + x^4}\,dx$

24. $\displaystyle\int_0^{1/2} \tan^{-1}(x^2)\,dx$

25. $\displaystyle\int_0^{1/3} x^2 \tan^{-1}(x^4)\,dx$

26. $\displaystyle\int_0^{0.5} \frac{dx}{1 + x^6}$

27. Use the result of Example 6 to compute $\ln 1.1$ correct to five decimal places.

28. Show that the function

$$f(x) = \sum_{n=0}^{\infty} \frac{(-1)^n x^{2n}}{(2n)!}$$

is a solution of the differential equation

$$f''(x) + f(x) = 0$$

29. (a) Show that J_0 (the Bessel function of order 0 given in Example 4) satisfies the differential equation

$$x^2 J_0''(x) + x J_0'(x) + x^2 J_0(x) = 0$$

(b) Evaluate $\int_0^1 J_0(x)\,dx$ correct to three decimal places.

30. The Bessel function of order 1 is defined by

$$J_1(x) = \sum_{n=0}^{\infty} \frac{(-1)^n x^{2n+1}}{n!(n+1)!2^{2n+1}}$$

(a) Show that J_1 satisfies the differential equation

$$x^2 J_1''(x) + x J_1'(x) + (x^2 - 1) J_1(x) = 0$$

(b) Show that $J_0'(x) = -J_1(x)$.

31. (a) Show that the function

$$f(x) = \sum_{n=0}^{\infty} \frac{x^n}{n!}$$

is a solution of the differential equation

$$f'(x) = f(x)$$

(b) Show that $f(x) = e^x$.

32. Let $f_n(x) = (\sin nx)/n^2$. Show that the series $\Sigma\, f_n(x)$ converges for all values of x but the series of derivatives $\Sigma\, f_n'(x)$ diverges when $x = 2n\pi$, n an integer. For what values of x does the series $\Sigma\, f_n''(x)$ converge?

33. Let

$$f(x) = \sum_{n=1}^{\infty} \frac{x^n}{n^2}$$

Find the intervals of convergence for f, f', and f''.

34. (a) Starting with the geometric series $\Sigma_{n=0}^{\infty} x^n$, find the sum of the series

$$\sum_{n=1}^{\infty} nx^{n-1} \qquad |x| < 1$$

(b) Find the sum of each of the following series.

(i) $\displaystyle\sum_{n=1}^{\infty} nx^n$, $|x| < 1$ (ii) $\displaystyle\sum_{n=1}^{\infty} \frac{n}{2^n}$

(c) Find the sum of each of the following series.

(i) $\displaystyle\sum_{n=2}^{\infty} n(n-1)x^n$, $|x| < 1$

(ii) $\displaystyle\sum_{n=2}^{\infty} \frac{n^2 - n}{2^n}$

(iii) $\displaystyle\sum_{n=1}^{\infty} \frac{n^2}{2^n}$

10.7 TAYLOR AND MACLAURIN SERIES

In the preceding section we were able to find power series representations for a certain restricted class of functions. Here we investigate more general problems: Which functions have power series representations? How can we find such representations?

We start by supposing that f is any function that can be represented by a power series

(1) $$f(x) = c_0 + c_1(x-a) + c_2(x-a)^2 + c_3(x-a)^3 + c_4(x-a)^4 + \cdots \qquad |x - a| < R$$

Let's try to determine what the coefficients c_n must be in terms of f. To begin, notice that if we put $x = a$ in Equation 1, then all terms after the first one are 0 and we get

$$f(a) = c_0$$

By Theorem 10.6.2, we can differentiate the series in Equation 1 term by term:

(2) $$f'(x) = c_1 + 2c_2(x-a) + 3c_3(x-a)^2 + 4c_4(x-a)^3 + \cdots \qquad |x - a| < R$$

and substitution of $x = a$ in Equation 2 gives

$$f'(a) = c_1$$

Now we differentiate both sides of Equation 2 and obtain

(3) $$f''(x) = 2c_2 + 2 \cdot 3c_3(x - a) + 3 \cdot 4c_4(x - a)^2 + \cdots \qquad |x - a| < R$$

Again we put $x = a$ in Equation 3. The result is

$$f''(a) = 2c_2$$

Let's apply the procedure one more time. Differentiation of the series in Equation 3 gives

(4) $$f'''(x) = 2 \cdot 3c_3 + 2 \cdot 3 \cdot 4c_4(x - a) + 3 \cdot 4 \cdot 5c_5(x - a)^2 + \cdots \qquad |x - a| < R$$

and substitution of $x = a$ in Equation 4 gives

$$f'''(a) = 2 \cdot 3c_3 = 3!c_3$$

By now you can see the pattern. If we continue to differentiate and substitute $x = a$, we obtain

$$f^{(n)}(a) = 2 \cdot 3 \cdot 4 \cdot \cdots \cdot nc_n = n!c_n$$

Solving this equation for the nth coefficient c_n, we get

$$c_n = \frac{f^{(n)}(a)}{n!}$$

This formula remains valid even for $n = 0$ if we adopt the conventions that $0! = 1$ and $f^{(0)} = f$. Thus, we have proved the following theorem.

(5) THEOREM If f has a power series representation (expansion) at a, that is, if

$$f(x) = \sum_{n=0}^{\infty} c_n(x - a)^n \qquad |x - a| < R$$

then its coefficients are given by the formula

$$c_n = \frac{f^{(n)}(a)}{n!}$$

Substituting this formula for c_n back into the series, we see that *if* f *has a power series expansion at* a, *then it must be of the following form:*

(6) $$f(x) = \sum_{n=0}^{\infty} \frac{f^{(n)}(a)}{n!}(x - a)^n$$

$$= f(a) + \frac{f'(a)}{1!}(x - a) + \frac{f''(a)}{2!}(x - a)^2 + \frac{f'''(a)}{3!}(x - a)^3 + \cdots$$

The Taylor series is named after the English mathematician Brook Taylor (1685–1731) and the Maclaurin series is named in honor of the Scottish mathematician Colin Maclaurin (1698–1746) despite the fact that the Maclaurin series is really just a special case of the Taylor series. But the idea of representing particular functions as sums of power series goes back to Newton, and the general Taylor series was known to the Scottish mathematician James Gregory in 1668 and to the Swiss mathematician John Bernoulli in the 1690s. Taylor was apparently unaware of the work of Gregory and Bernoulli when he published his discoveries on series in 1715 in his book *Methodus incrementorum directa et inversa*. Maclaurin series are named after Colin Maclaurin because he popularized them in his calculus textbook *Treatise of Fluxions* published in 1742.

The series in Equation 6 is called the **Taylor series of the function f at a** (or **about a** or **centered at a**). For the special case $a = 0$ the Taylor series becomes

$$f(x) = \sum_{n=0}^{\infty} \frac{f^{(n)}(0)}{n!} x^n = f(0) + \frac{f'(0)}{1!} x + \frac{f''(0)}{2!} x^2 + \cdots \tag{7}$$

This case arises frequently enough that it is given the special name **Maclaurin series**.

NOTE: We have shown that *if* f can be represented as a power series about a, then f is equal to the sum of its Taylor series. But there exist functions that are not equal to the sum of their Taylor series. An example of such a function is given in Exercise 50.

EXAMPLE 1 Find the Maclaurin series of the function $f(x) = e^x$ and its radius of convergence.

SOLUTION If $f(x) = e^x$, then $f^{(n)}(x) = e^x$, so $f^{(n)}(0) = e^0 = 1$ for all n. Therefore, the Taylor series for f at 0 (that is, the Maclaurin series) is

$$\sum_{n=0}^{\infty} \frac{f^{(n)}(0)}{n!} x^n = \sum_{n=0}^{\infty} \frac{x^n}{n!} = 1 + \frac{x}{1!} + \frac{x^2}{2!} + \frac{x^3}{3!} + \cdots$$

To find the radius of convergence we let $a_n = x^n/n!$. Then

$$\left| \frac{a_{n+1}}{a_n} \right| = \left| \frac{x^{n+1}}{(n+1)!} \cdot \frac{n!}{x^n} \right| = \frac{|x|}{n+1} \to 0 < 1$$

so, by the Ratio Test, the series converges for all x and the radius of convergence is $R = \infty$. ■

The conclusion we can draw from Theorem 5 and Example 1 is that *if* e^x has a power series expansion at 0, then

$$e^x = \sum_{n=0}^{\infty} \frac{x^n}{n!}$$

So how can we determine whether e^x *does* have a power series representation?

Let's investigate the more general question: Under what circumstances is a function $f(x)$ equal to the sum of its Taylor series? In other words, if f has derivatives of all orders, when is it true that

$$f(x) = \sum_{n=0}^{\infty} \frac{f^{(n)}(a)}{n!} (x - a)^n$$

As with any convergent series, this means that $f(x)$ is the limit of the sequence of partial sums. In the case of the Taylor series, the partial sums are

$$T_n(x) = \sum_{i=0}^{n} \frac{f^{(i)}(a)}{i!} (x - a)^i$$

$$= f(a) + \frac{f'(a)}{1!}(x - a) + \frac{f''(a)}{2!}(x - a)^2 + \cdots + \frac{f^{(n)}(a)}{n!}(x - a)^n$$

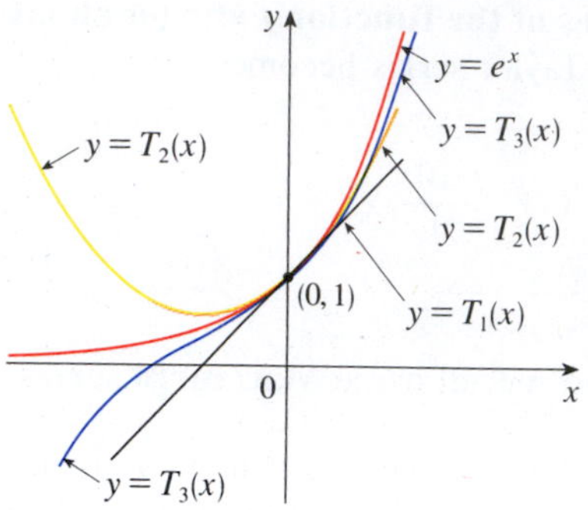

FIGURE 1

As n increases, $T_n(x)$ appears to approach e^x in Figure 1. This suggests that e^x is equal to the sum of its Taylor series.

Notice that T_n is a polynomial of degree n called the **nth-degree Taylor polynomial of f at a**. For instance, for the exponential function $f(x) = e^x$, the result of Example 1 shows that the Taylor polynomials at 0 (or Maclaurin polynomials) with $n = 1$, 2, and 3 are

$$T_1(x) = 1 + x \qquad T_2(x) = 1 + x + \frac{x^2}{2!} \qquad T_3(x) = 1 + x + \frac{x^2}{2!} + \frac{x^3}{3!}$$

The graphs of the exponential function and these three Taylor polynomials are drawn in Figure 1.

In general, $f(x)$ is the sum of its Taylor series if

$$f(x) = \lim_{n\to\infty} T_n(x)$$

If we let $R_n(x)$ be the remainder of the series, then

$$R_n(x) = f(x) - T_n(x) \qquad \text{so} \qquad f(x) = T_n(x) + R_n(x)$$

If we can somehow show that $\lim_{n\to\infty} R_n(x) = 0$, then it follows that

$$\lim_{n\to\infty} T_n(x) = \lim_{n\to\infty} [f(x) - R_n(x)] = f(x) - \lim_{n\to\infty} R_n(x) = f(x)$$

We have therefore proved the following theorem.

(8) THEOREM If $f(x) = T_n(x) + R_n(x)$, where T_n is the nth-degree Taylor polynomial of f at a and

$$\lim_{n\to\infty} R_n(x) = 0$$

for $|x - a| < R$, then f is equal to the sum of its Taylor series on the interval $|x - a| < R$.

When trying to show that $\lim_{n\to\infty} R_n(x) = 0$ for a specific function f, we usually use the following fact.

(9) TAYLOR'S INEQUALITY If $|f^{(n+1)}(x)| \leq M$ for $|x - a| < R$, then the remainder $R_n(x)$ of the Taylor series satisfies the inequality

$$|R_n(x)| \leq \frac{M}{(n+1)!}|x - a|^{n+1} \qquad \text{for } |x - a| < R$$

To see why this is true for $n = 1$, we assume that $|f''(x)| \leq M$. In particular, we have $f''(x) \leq M$, so

$$\int_a^x f''(t)\,dt \leq \int_a^x M\,dt$$

An antiderivative of f'' is f', so by the Evaluation Theorem, we have

$$f'(x) - f'(a) \leq M(x - a) \qquad \text{or} \qquad f'(x) \leq f'(a) + M(x - a)$$

Thus
$$\int_a^x f'(t)\,dt \leqslant \int_a^x [f'(a) + M(t-a)]\,dt$$

$$f(x) - f(a) \leqslant f'(a)(x-a) + M\frac{(x-a)^2}{2}$$

$$f(x) - f(a) - f'(a)(x-a) \leqslant \frac{M}{2}(x-a)^2$$

But $R_1(x) = f(x) - T_1(x) = f(x) - f(a) - f'(a)(x-a)$. So

$$R_1(x) \leqslant \frac{M}{2}(x-a)^2$$

A similar argument, using $f''(x) \geqslant -M$, shows that

$$R_1(x) \geqslant -\frac{M}{2}(x-a)^2$$

So
$$|R_1(x)| \leqslant \frac{M}{2}|x-a|^2$$

This proves Taylor's Inequality for the case where $n = 1$. The result for any n is proved in a similar way by integrating $n + 1$ times.

NOTE: In Section 10.9 we will explore the use of Taylor's Inequality in approximating functions. Our immediate use of it is in conjunction with Theorem 8.

In applying Theorems 8 and 9 it is often helpful to make use of the following fact:

(10)
$$\lim_{n\to\infty} \frac{x^n}{n!} = 0 \qquad \text{for every real number } x$$

This is true because we know from Example 1 that the series $\sum x^n/n!$ converges for all x and so its nth term approaches 0.

EXAMPLE 2 Prove that e^x is equal to the sum of its Taylor series.

SOLUTION If $f(x) = e^x$, then $f^{(n+1)}(x) = e^x$ for all n. So, for any fixed number x, we can take $M = e^x$ in Taylor's Inequality (with $a = 0$) for every value of n:

$$|R_n(x)| \leqslant \frac{e^x}{(n+1)!}|x|^{n+1}$$

But, from Equation 10, we have

$$\lim_{n\to\infty} \frac{e^x}{(n+1)!}|x|^{n+1} = e^x \lim_{n\to\infty} \frac{|x|^{n+1}}{(n+1)!} = 0$$

It follows from the Squeeze Theorem that $\lim_{n\to\infty} |R_n(x)| = 0$ and therefore $\lim_{n\to\infty} R_n(x) = 0$. By Theorem 8, e^x is equal to the sum of its Taylor series, that is,

(11)
$$e^x = \sum_{n=0}^{\infty} \frac{x^n}{n!} \qquad \text{for all } x$$

■

In particular, if we put $x = 1$ in Equation 11, we obtain the following expression for the number e as a sum of an infinite series:

(12) $$e = \sum_{n=0}^{\infty} \frac{1}{n!} = 1 + \frac{1}{1!} + \frac{1}{2!} + \frac{1}{3!} + \cdots$$

EXAMPLE 3 Find the Taylor series for $f(x) = e^x$ at $a = 2$.

SOLUTION We have $f^{(n)}(2) = e^2$ and so, putting $a = 2$ in the definition of a Taylor series (6), we get

$$\sum_{n=0}^{\infty} \frac{f^{(n)}(2)}{n!}(x-2)^n = \sum_{n=0}^{\infty} \frac{e^2}{n!}(x-2)^n$$

Again it can be verified, as in Example 1, that the radius of convergence is $R = \infty$. As in Example 2 we can verify that $\lim_{n\to\infty} R_n(x) = 0$, so

(13) $$e^x = \sum_{n=0}^{\infty} \frac{e^2}{n!}(x-2)^n \qquad \text{for all } x$$

We have two power series expansions for e^x, the Maclaurin series in Equation 11 and the Taylor series in Equation 13. The first is better if we are interested in values of x near 0 and the second is better if x is near 2.

EXAMPLE 4 Find the Maclaurin series for $\sin x$ and prove that it represents $\sin x$ for all x.

Figure 2 shows the graph of $\sin x$ together with its Taylor (or Maclaurin) polynomials

$$T_1(x) = x$$

$$T_3(x) = x - \frac{x^3}{3!}$$

$$T_5(x) = x - \frac{x^3}{3!} + \frac{x^5}{5!}$$

Notice that, as n increases, $T_n(x)$ becomes a better approximation to $\sin x$.

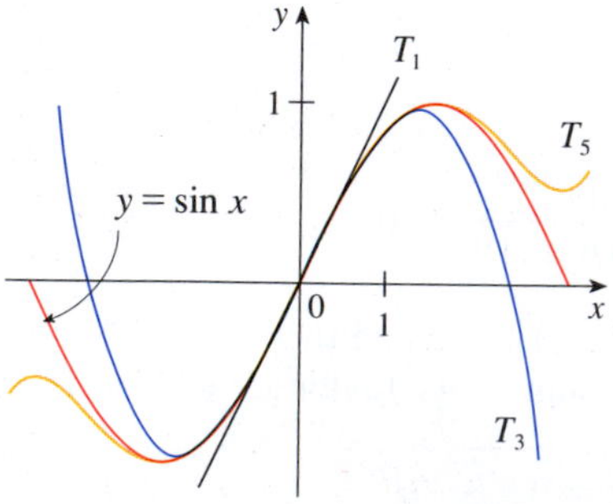

FIGURE 2

SOLUTION We arrange our computation in two columns as follows:

$$\begin{aligned} f(x) &= \sin x & f(0) &= 0 \\ f'(x) &= \cos x & f'(0) &= 1 \\ f''(x) &= -\sin x & f''(0) &= 0 \\ f'''(x) &= -\cos x & f'''(0) &= -1 \\ f^{(4)}(x) &= \sin x & f^{(4)}(0) &= 0 \end{aligned}$$

Since the derivatives repeat in a cycle of four, we can write the Maclaurin series as follows:

$$f(0) + \frac{f'(0)}{1!}x + \frac{f''(0)}{2!}x^2 + \frac{f'''(0)}{3!}x^3 + \cdots$$

$$= x - \frac{x^3}{3!} + \frac{x^5}{5!} - \frac{x^7}{7!} + \cdots = \sum_{n=0}^{\infty} (-1)^n \frac{x^{2n+1}}{(2n+1)!}$$

Since $f^{(n+1)}(x)$ is $\pm\sin x$ or $\pm\cos x$, we know that $|f^{(n+1)}(x)| \leqslant 1$ for all x. So we can take $M = 1$ in Taylor's Inequality:

(14) $$|R_n(x)| \leqslant \frac{M}{(n+1)!}|x^{n+1}| \leqslant \frac{|x|^{n+1}}{(n+1)!}$$

By Equation 10 the right side of this inequality approaches 0 as $n \to \infty$, so $|R_n(x)| \to 0$ by the Squeeze Theorem. It follows that $R_n(x) \to 0$ as $n \to \infty$, so $\sin x$ is equal to the sum of its Maclaurin series by Theorem 8. ■

We state the result of Example 4 for future reference:

(15)
$$\sin x = x - \frac{x^3}{3!} + \frac{x^5}{5!} - \frac{x^7}{7!} + \cdots$$
$$= \sum_{n=0}^{\infty} (-1)^n \frac{x^{2n+1}}{(2n+1)!} \qquad \text{for all } x$$

EXAMPLE 5 Find the Maclaurin series for $\cos x$.

SOLUTION We could proceed directly as in Example 4 but it is easier to differentiate the Maclaurin series for $\sin x$ given by Equation 15:

$$\cos x = \frac{d}{dx}(\sin x) = \frac{d}{dx}\left(x - \frac{x^3}{3!} + \frac{x^5}{5!} - \frac{x^7}{7!} + \cdots\right)$$
$$= 1 - \frac{3x^2}{3!} + \frac{5x^4}{5!} - \frac{7x^6}{7!} + \cdots = 1 - \frac{x^2}{2!} + \frac{x^4}{4!} - \frac{x^6}{6!} + \cdots$$

The Maclaurin series for e^x, $\sin x$, and $\cos x$ that we found in Examples 2, 4, and 5 were first discovered, using different methods, by Newton. These equations are remarkable because they say we know everything about each of these functions if we know all its derivatives at the single number 0.

Since the Maclaurin series for $\sin x$ converges for all x, Theorem 10.6.2 tells us that the differentiated series for $\cos x$ also converges for all x. Thus

(16)
$$\cos x = 1 - \frac{x^2}{2!} + \frac{x^4}{4!} - \frac{x^6}{6!} + \cdots$$
$$= \sum_{n=0}^{\infty} (-1)^n \frac{x^{2n}}{(2n)!} \qquad \text{for all } x$$

■

We have obtained two different series representations for $\sin x$, the Maclaurin series in Example 4 and the Taylor series in Example 7. It is best to use the Maclaurin series for values of x near 0 and the Taylor series for x near $\pi/3$. Notice that the third Taylor polynomial T_3 in Figure 3 is a good approximation to $\sin x$ near $\pi/3$ but not as good near 0. Compare it with the third Maclaurin polynomial T_3 in Figure 2, where the opposite is true.

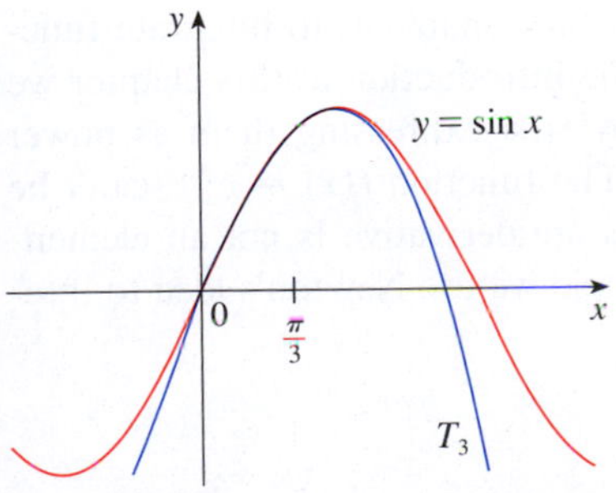

FIGURE 3

EXAMPLE 6 Find the Maclaurin series for the function $f(x) = x\cos x$.

SOLUTION Instead of computing derivatives and substituting in Equation 7, it is easier to multiply the series for $\cos x$ (Equation 16) by x:

$$x\cos x = x\sum_{n=0}^{\infty} (-1)^n \frac{x^{2n}}{(2n)!} = \sum_{n=0}^{\infty} (-1)^n \frac{x^{2n+1}}{(2n)!}$$ ■

EXAMPLE 7 Represent $f(x) = \sin x$ as the sum of its Taylor series centered at $\pi/3$.

SOLUTION Arranging our work in columns, we have

$$f(x) = \sin x \qquad f(\pi/3) = \sqrt{3}/2$$
$$f'(x) = \cos x \qquad f'(\pi/3) = \tfrac{1}{2}$$
$$f''(x) = -\sin x \qquad f''(\pi/3) = -\sqrt{3}/2$$
$$f'''(x) = -\cos x \qquad f'''(\pi/3) = -\tfrac{1}{2}$$

and this pattern repeats indefinitely. Therefore, the Taylor series at $\pi/3$ is

$$f\left(\frac{\pi}{3}\right) + \frac{f'\left(\frac{\pi}{3}\right)}{1!}\left(x - \frac{\pi}{3}\right) + \frac{f''\left(\frac{\pi}{3}\right)}{2!}\left(x - \frac{\pi}{3}\right)^2 + \frac{f'''\left(\frac{\pi}{3}\right)}{3!}\left(x - \frac{\pi}{3}\right)^3 + \cdots$$

$$= \frac{\sqrt{3}}{2} + \frac{1}{2 \cdot 1!}\left(x - \frac{\pi}{3}\right) - \frac{\sqrt{3}}{2 \cdot 2!}\left(x - \frac{\pi}{3}\right)^2 - \frac{1}{2 \cdot 3!}\left(x - \frac{\pi}{3}\right)^3 + \cdots$$

The proof that this series represents $\sin x$ for all x is very similar to that in Example 4. [Just replace x by $x - \pi/3$ in (14).] We can write the series in sigma notation if we separate the terms that contain $\sqrt{3}$:

$$\sin x = \sum_{n=0}^{\infty} \frac{(-1)^n \sqrt{3}}{2(2n)!}\left(x - \frac{\pi}{3}\right)^{2n} + \sum_{n=0}^{\infty} \frac{(-1)^n}{2(2n+1)!}\left(x - \frac{\pi}{3}\right)^{2n+1}$$ ■

The power series that we obtained by indirect methods in Examples 5 and 6 and in Section 10.6 are indeed the Taylor or Maclaurin series of the given functions because Theorem 5 asserts that, no matter how a power series representation $f(x) = \Sigma c_n(x - a)^n$ is obtained, it is always true that $c_n = f^{(n)}(a)/n!$. In other words, the coefficients are uniquely determined.

We collect in the following table, for future reference, some important Maclaurin series that we have derived in this section and the preceding one.

Important Maclaurin series and their intervals of convergence

$\frac{1}{1-x} = \sum_{n=0}^{\infty} x^n = 1 + x + x^2 + x^3 + \cdots$	$(-1, 1)$
$e^x = \sum_{n=0}^{\infty} \frac{x^n}{n!} = 1 + \frac{x}{1!} + \frac{x^2}{2!} + \frac{x^3}{3!} + \cdots$	$(-\infty, \infty)$
$\sin x = \sum_{n=0}^{\infty} (-1)^n \frac{x^{2n+1}}{(2n+1)!} = x - \frac{x^3}{3!} + \frac{x^5}{5!} - \frac{x^7}{7!} + \cdots$	$(-\infty, \infty)$
$\cos x = \sum_{n=0}^{\infty} (-1)^n \frac{x^{2n}}{(2n)!} = 1 - \frac{x^2}{2!} + \frac{x^4}{4!} - \frac{x^6}{6!} + \cdots$	$(-\infty, \infty)$
$\tan^{-1} x = \sum_{n=0}^{\infty} (-1)^n \frac{x^{2n+1}}{2n+1} = x - \frac{x^3}{3} + \frac{x^5}{5} - \frac{x^7}{7} + \cdots$	$[-1, 1]$

One reason that Taylor series are important is that they enable us to integrate functions that we couldn't previously handle. In fact, in the introduction to this chapter we mentioned that Newton often integrated functions by first expressing them as power series and then integrating the series term by term. The function $f(x) = e^{-x^2}$ can't be integrated by techniques discussed so far because its antiderivative is not an elementary function (see Section 8.6). In the following example we use Newton's idea to integrate this function.

EXAMPLE 8

(a) Evaluate $\int e^{-x^2}\,dx$ as an infinite series.

(b) Evaluate $\int_0^1 e^{-x^2}\,dx$ correct to within an error of 0.001.

SOLUTION

(a) First we find the Maclaurin series for $f(x) = e^{-x^2}$. Although it's possible to use the direct method, let's find it simply by replacing x with $-x^2$ in the series for e^x given in the table of Maclaurin series. Thus, for all values of x,

$$e^{-x^2} = \sum_{n=0}^{\infty} \frac{(-x^2)^n}{n!} = \sum_{n=0}^{\infty} (-1)^n \frac{x^{2n}}{n!} = 1 - \frac{x^2}{1!} + \frac{x^4}{2!} - \frac{x^6}{3!} + \cdots$$

Now we integrate term by term:

$$\int e^{-x^2}\,dx = \int \left(1 - \frac{x^2}{1!} + \frac{x^4}{2!} - \frac{x^6}{3!} + \cdots + (-1)^n \frac{x^{2n}}{n!} + \cdots\right) dx$$

$$= C + x - \frac{x^3}{3 \cdot 1!} + \frac{x^5}{5 \cdot 2!} - \frac{x^7}{7 \cdot 3!} + \cdots + (-1)^n \frac{x^{2n+1}}{(2n+1)n!} + \cdots$$

This series converges for all x because the original series for e^{-x^2} converges for all x.

(b) The Evaluation Theorem gives

We can take $C = 0$ in the antiderivative in part (a).

$$\int_0^1 e^{-x^2}\,dx = \left[x - \frac{x^3}{3 \cdot 1!} + \frac{x^5}{5 \cdot 2!} - \frac{x^7}{7 \cdot 3!} + \frac{x^9}{9 \cdot 4!} - \cdots\right]_0^1$$

$$= 1 - \tfrac{1}{3} + \tfrac{1}{10} - \tfrac{1}{42} + \tfrac{1}{216} - \cdots$$

$$\approx 1 - \tfrac{1}{3} + \tfrac{1}{10} - \tfrac{1}{42} + \tfrac{1}{216} \approx 0.7475$$

The Alternating Series Estimation Theorem shows that the error involved in this approximation is less than

$$\frac{1}{11 \cdot 5!} = \frac{1}{1320} < 0.001$$

■

Another use of Taylor series is illustrated in the next example. The limit could be found with l'Hospital's Rule, but instead we use a series.

EXAMPLE 9 Evaluate $\displaystyle\lim_{x \to 0} \frac{e^x - 1 - x}{x^2}$.

SOLUTION Using the Maclaurin series for e^x, we have

Some computer algebra systems compute limits in this way.

$$\lim_{x \to 0} \frac{e^x - 1 - x}{x^2} = \lim_{x \to 0} \frac{\left(1 + \frac{x}{1!} + \frac{x^2}{2!} + \frac{x^3}{3!} + \cdots\right) - 1 - x}{x^2}$$

$$= \lim_{x \to 0} \frac{\frac{x^2}{2!} + \frac{x^3}{3!} + \frac{x^4}{4!} + \cdots}{x^2}$$

$$= \lim_{x \to 0} \left(\frac{1}{2} + \frac{x}{3!} + \frac{x^2}{4!} + \frac{x^3}{5!} + \cdots\right)$$

$$= \frac{1}{2}$$

because power series are continuous functions.

■

MULTIPLICATION AND DIVISION OF POWER SERIES

If power series are added or subtracted, they behave like polynomials (Theorem 10.2.8 shows this). In fact, as the following example illustrates, they can also be multiplied and divided like polynomials. We find only the first few terms because the calculations for the later terms become tedious and the initial terms are the most important ones.

EXAMPLE 10 Find the first three nonzero terms in the Maclaurin series for (a) $e^x \sin x$ and (b) $\tan x$.

SOLUTION

(a) Using the Maclaurin series for e^x and $\sin x$ in the table, we have

$$e^x \sin x = \left(1 + \frac{x}{1!} + \frac{x^2}{2!} + \frac{x^3}{3!} + \cdots\right)\left(x - \frac{x^3}{3!} + \cdots\right)$$

We multiply these expressions, collecting like terms just as for polynomials:

$$\begin{array}{r}
1 + x + \frac{1}{2}x^2 + \frac{1}{6}x^3 + \cdots \\
x \qquad\qquad - \frac{1}{6}x^3 + \cdots \\
\hline
x + \ x^2 + \frac{1}{2}x^3 + \frac{1}{6}x^4 + \cdots \\
- \frac{1}{6}x^3 - \frac{1}{6}x^4 - \cdots \\
\hline
x + \ x^2 + \frac{1}{3}x^3 + \cdots \quad
\end{array}$$

Thus

$$e^x \sin x = x + x^2 + \tfrac{1}{3}x^3 + \cdots$$

(b) Using the Maclaurin series in the table, we have

$$\tan x = \frac{\sin x}{\cos x} = \frac{x - \dfrac{x^3}{3!} + \dfrac{x^5}{5!} - \cdots}{1 - \dfrac{x^2}{2!} + \dfrac{x^4}{4!} - \cdots}$$

We use a procedure like long division:

$$\begin{array}{r|l}
 & x + \frac{1}{3}x^3 + \frac{2}{15}x^5 + \cdots \\
\hline
1 - \frac{1}{2}x^2 + \frac{1}{24}x^4 - \cdots & x - \frac{1}{6}x^3 + \frac{1}{120}x^5 - \cdots \\
 & x - \frac{1}{2}x^3 + \frac{1}{24}x^5 - \cdots \\
\cline{2-2}
 & \qquad \frac{1}{3}x^3 - \frac{1}{30}x^5 + \cdots \\
 & \qquad \frac{1}{3}x^3 - \frac{1}{6}x^5 + \cdots \\
\cline{2-2}
 & \qquad\qquad \frac{2}{15}x^5 + \cdots
\end{array}$$

Thus

$$\tan x = x + \tfrac{1}{3}x^3 + \tfrac{2}{15}x^5 + \cdots$$

■

Although we have not attempted to justify the formal manipulations used in Example 10, they are legitimate. There is a theorem which states that if both $f(x) = \Sigma\, c_n x^n$ and $g(x) = \Sigma\, b_n x^n$ converge for $|x| < R$ and the series are multiplied as if they were polynomials, then the resulting series also converges for $|x| < R$ and represents $f(x)g(x)$. For division we require $b_0 \neq 0$; the resulting series converges for sufficiently small $|x|$.

EXERCISES 10.7

1. If $f(x) = \sum_{n=0}^{\infty} b_n(x - 5)^n$ for all x, write a formula for b_8.

2. The graph of f is shown. Explain why the series

$$2 - 0.8(x - 1) + 0.4(x - 1)^2 - 0.1(x - 1)^3 + \cdots$$

is *not* the Taylor series of f centered at 1.

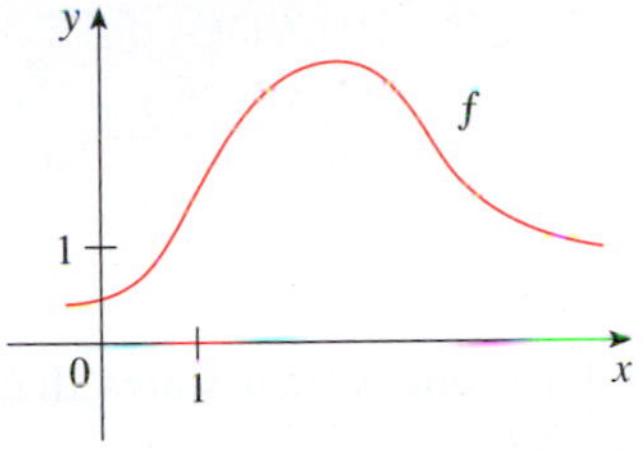

3–6 ■ Find the Maclaurin series for $f(x)$ using the definition of a Maclaurin series. [Assume that f has a power series expansion. Do not show that $R_n(x) \to 0$.] Also find the associated radius of convergence.

3. $f(x) = \cos x$

4. $f(x) = \sin 2x$

5. $f(x) = \dfrac{1}{(1 + x)^2}$

6. $f(x) = \dfrac{x}{1 - x}$

7–12 ■ Find the Taylor series for $f(x)$ at the given value of a. [Assume that f has a power series expansion. Do not show that $R_n(x) \to 0$.]

7. $f(x) = e^x$, $a = 3$

8. $f(x) = \ln x$, $a = 2$

9. $f(x) = 1/x$, $a = 1$

10. $f(x) = \sqrt{x}$, $a = 4$

11. $f(x) = \sin x$, $a = \pi/4$

12. $f(x) = \cos x$, $a = -\pi/4$

13. Prove that the series obtained in Exercise 3 represents $\cos x$ for all x.

14. Prove that the series obtained in Exercise 11 represents $\sin x$ for all x.

15–22 ■ Use a Maclaurin series derived in this section to obtain the Maclaurin series for the given function.

15. $f(x) = e^{3x}$

16. $f(x) = \sin 2x$

17. $f(x) = x^2 \cos x$

18. $f(x) = \cos(x^3)$

19. $f(x) = x \sin(x/2)$

20. $f(x) = xe^{-x}$

21. $f(x) = \sin^2 x$ [*Hint*: Use $\sin^2 x = \frac{1}{2}(1 - \cos 2x)$.]

22. $f(x) = \begin{cases} \dfrac{\sin x}{x} & \text{if } x \neq 0 \\ 1 & \text{if } x = 0 \end{cases}$

23–26 ■ Find the Maclaurin series of f (by any method) and its radius of convergence. Graph f and its first few Taylor polynomials on the same screen.

23. $f(x) = \sqrt{1 + x}$

24. $f(x) = 1/\sqrt{1 + 2x}$

25. $f(x) = (1 + x)^{-3}$

26. $f(x) = 2^x$

27. Find the Maclaurin series for $\ln(1 + x)$ and use it to calculate $\ln 1.1$ correct to five decimal places.

28. Use the Maclaurin series for $\sin x$ to compute $\sin 3^\circ$ correct to five decimal places.

29–32 ■ Evaluate the indefinite integral as an infinite series.

29. $\displaystyle\int \sin(x^2)\, dx$

30. $\displaystyle\int \frac{\sin x}{x}\, dx$

31. $\displaystyle\int \sqrt{x^3 + 1}\, dx$

32. $\displaystyle\int e^{x^3}\, dx$

33–36 ■ Use series to approximate the definite integral to within the indicated accuracy.

33. $\displaystyle\int_0^1 \sin(x^2)\, dx$ (three decimal places)

34. $\displaystyle\int_0^{0.5} \cos(x^2)\, dx$ (three decimal places)

35. $\displaystyle\int_0^{0.1} \frac{dx}{\sqrt{1 + x^3}}$ (error $< 10^{-8}$)

36. $\displaystyle\int_0^{0.5} x^2 e^{-x^2}\, dx$ (error < 0.001)

37–39 ■ Use series to evaluate the limit.

37. $\displaystyle\lim_{x \to 0} \frac{x - \tan^{-1} x}{x^3}$

38. $\displaystyle\lim_{x \to 0} \frac{1 - \cos x}{1 + x - e^x}$

39. $\displaystyle\lim_{x \to 0} \frac{\sin x - x + \frac{1}{6}x^3}{x^5}$

40. Use the series in Example 10(b) to evaluate

$$\lim_{x \to 0} \frac{\tan x - x}{x^3}$$

We found this limit in Example 4 in Section 4.8 using l'Hospital's Rule three times. Which method do you prefer?

41–44 ■ Use multiplication or division of power series to find the first three nonzero terms in the Maclaurin series for each function.

41. $y = e^{-x^2} \cos x$

42. $y = \sec x$

43. $y = \dfrac{\ln(1 - x)}{e^x}$

44. $y = e^x \ln(1 - x)$

45–49 ■ Find the sum of the series.

45. $\displaystyle\sum_{n=0}^{\infty} (-1)^n \frac{x^{4n}}{n!}$

46. $\displaystyle\sum_{n=0}^{\infty} \frac{(-1)^n \pi^{2n}}{6^{2n}(2n)!}$

47. $\displaystyle\sum_{n=0}^{\infty} \frac{(-1)^n \pi^{2n+1}}{4^{2n+1}(2n+1)!}$

48. $\displaystyle\sum_{n=2}^{\infty} \frac{x^{3n+1}}{n!}$

49. $\displaystyle\sum_{n=0}^{\infty} \frac{x^{n+1}}{(n+1)!}$

50. (a) Show that the function defined by

$$f(x) = \begin{cases} e^{-1/x^2} & \text{if } x \neq 0 \\ 0 & \text{if } x = 0 \end{cases}$$

is not equal to its Maclaurin series.

(b) Graph the function in part (a) and comment on its behavior near the origin.

10.8 THE BINOMIAL SERIES

You may be acquainted with the Binomial Theorem, which states that if a and b are any real numbers and k is a positive integer, then

$$(a+b)^k = a^k + ka^{k-1}b + \frac{k(k-1)}{2!}a^{k-2}b^2 + \frac{k(k-1)(k-2)}{3!}a^{k-3}b^3$$
$$+ \cdots + \frac{k(k-1)(k-2)\cdots(k-n+1)}{n!}a^{k-n}b^n$$
$$+ \cdots + kab^{k-1} + b^k$$

The traditional notation for the binomial coefficients is

$$\binom{k}{0} = 1 \qquad \binom{k}{n} = \frac{k(k-1)(k-2)\cdots(k-n+1)}{n!} \qquad n = 1, 2, \ldots, k$$

which enables us to write the Binomial Theorem in the abbreviated form

$$(a+b)^k = \sum_{n=0}^{k} \binom{k}{n} a^{k-n}b^n$$

In particular, if we put $a = 1$ and $b = x$, we get

(1) $$(1+x)^k = \sum_{n=0}^{k} \binom{k}{n} x^n$$

One of Newton's accomplishments was to extend the Binomial Theorem (Equation 1) to the case in which k is no longer a positive integer. In this case the expression for $(1+x)^k$ is no longer a finite sum; it becomes an infinite series. To find this series we compute the Maclaurin series of $(1+x)^k$ in the usual way:

$$\begin{aligned} f(x) &= (1+x)^k & f(0) &= 1 \\ f'(x) &= k(1+x)^{k-1} & f'(0) &= k \\ f''(x) &= k(k-1)(1+x)^{k-2} & f''(0) &= k(k-1) \\ f'''(x) &= k(k-1)(k-2)(1+x)^{k-3} & f'''(0) &= k(k-1)(k-2) \\ &\vdots & &\vdots \\ f^{(n)}(x) &= k(k-1)\cdots(k-n+1)(1+x)^{k-n} & f^{(n)}(0) &= k(k-1)\cdots(k-n+1) \end{aligned}$$

Therefore, the Maclaurin series of $f(x) = (1 + x)^k$ is

$$\sum_{n=0}^{\infty} \frac{f^{(n)}(0)}{n!} x^n = \sum_{n=0}^{\infty} \frac{k(k-1)\cdots(k-n+1)}{n!} x^n$$

This series is called the **binomial series**. If its nth term is a_n, then

$$\left| \frac{a_{n+1}}{a_n} \right| = \left| \frac{k(k-1)\cdots(k-n+1)(k-n)x^{n+1}}{(n+1)!} \cdot \frac{n!}{k(k-1)\cdots(k-n+1)x^n} \right|$$

$$= \frac{|k-n|}{n+1}|x| = \frac{\left|1 - \dfrac{k}{n}\right|}{1 + \dfrac{1}{n}}|x| \to |x| \qquad \text{as } n \to \infty$$

Thus, by the Ratio Test, the binomial series converges if $|x| < 1$ and diverges if $|x| > 1$.

The following theorem states that $(1 + x)^k$ is equal to the sum of its Maclaurin series. It is possible to prove this by showing that the remainder term $R_n(x)$ approaches 0, but that turns out to be quite difficult. The proof outlined in Exercise 15 is much easier.

(2) THE BINOMIAL SERIES If k is any real number and $|x| < 1$, then

$$(1 + x)^k = 1 + kx + \frac{k(k-1)}{2!}x^2 + \frac{k(k-1)(k-2)}{3!}x^3 + \cdots$$

$$= \sum_{n=0}^{\infty} \binom{k}{n} x^n$$

where $\displaystyle \binom{k}{n} = \frac{k(k-1)\cdots(k-n+1)}{n!} \quad (n \geqslant 1)$ and $\displaystyle \binom{k}{0} = 1$

Although the binomial series always converges when $|x| < 1$, the question of whether or not it converges at the endpoints, ± 1, depends on the value of k. It turns out that the series converges at 1 if $-1 < k \leqslant 0$ and at both endpoints if $k \geqslant 0$. Notice that if k is a positive integer and $n > k$, then the expression for $\binom{k}{n}$ contains a factor $(k - k)$, so $\binom{k}{n} = 0$ for $n > k$. This means that the series terminates and reduces to the ordinary Binomial Theorem (Equation 1) when k is a positive integer.

As we have seen, the binomial series is just a special case of the Maclaurin series; it occurs so frequently that it is worth remembering.

EXAMPLE 1 Expand $\dfrac{1}{(1 + x)^2}$ as a power series.

SOLUTION We use the binomial series with $k = -2$. The binomial coefficient is

$$\binom{-2}{n} = \frac{(-2)(-3)(-4)\cdots(-2-n+1)}{n!}$$

$$= \frac{(-1)^n 2 \cdot 3 \cdot 4 \cdot \cdots \cdot n(n+1)}{n!} = (-1)^n(n+1)$$

and so, when $|x| < 1$,

$$\frac{1}{(1+x)^2} = (1+x)^{-2} = \sum_{n=0}^{\infty} \binom{-2}{n} x^n$$
$$= \sum_{n=0}^{\infty} (-1)^n (n+1) x^n$$

■

EXAMPLE 2 Find the Maclaurin series for the function $f(x) = \dfrac{1}{\sqrt{4-x}}$ and its radius of convergence.

SOLUTION As given, $f(x)$ is not quite of the form $(1 + x)^k$ so we rewrite it as follows:

$$\frac{1}{\sqrt{4-x}} = \frac{1}{\sqrt{4\left(1-\dfrac{x}{4}\right)}} = \frac{1}{2\sqrt{1-\dfrac{x}{4}}} = \frac{1}{2}\left(1-\frac{x}{4}\right)^{-1/2}$$

Using the binomial series with $k = -\frac{1}{2}$ and with x replaced by $-x/4$, we have

$$\frac{1}{\sqrt{4-x}} = \frac{1}{2}\left(1-\frac{x}{4}\right)^{-1/2} = \frac{1}{2}\sum_{n=0}^{\infty}\binom{-\frac{1}{2}}{n}\left(-\frac{x}{4}\right)^n$$
$$= \frac{1}{2}\left[1 + \left(-\frac{1}{2}\right)\left(-\frac{x}{4}\right) + \frac{\left(-\frac{1}{2}\right)\left(-\frac{3}{2}\right)}{2!}\left(-\frac{x}{4}\right)^2 + \frac{\left(-\frac{1}{2}\right)\left(-\frac{3}{2}\right)\left(-\frac{5}{2}\right)}{3!}\left(-\frac{x}{4}\right)^3\right.$$
$$\left. + \cdots + \frac{\left(-\frac{1}{2}\right)\left(-\frac{3}{2}\right)\left(-\frac{5}{2}\right)\cdots\left(-\frac{1}{2}-n+1\right)}{n!}\left(-\frac{x}{4}\right)^n + \cdots\right]$$
$$= \frac{1}{2}\left[1 + \frac{1}{8}x + \frac{1\cdot 3}{2!\,8^2}x^2 + \frac{1\cdot 3\cdot 5}{3!\,8^3}x^3 + \cdots + \frac{1\cdot 3\cdot 5\cdot\cdots\cdot(2n-1)}{n!\,8^n}x^n + \cdots\right]$$

We know from (2) that this series converges when $|-x/4| < 1$, that is, $|x| < 4$, so the radius of convergence is $R = 4$. ■

A binomial series is a special case of a Taylor series. Figure 1 shows the graphs of the first three Taylor polynomials computed from the answer to Example 2.

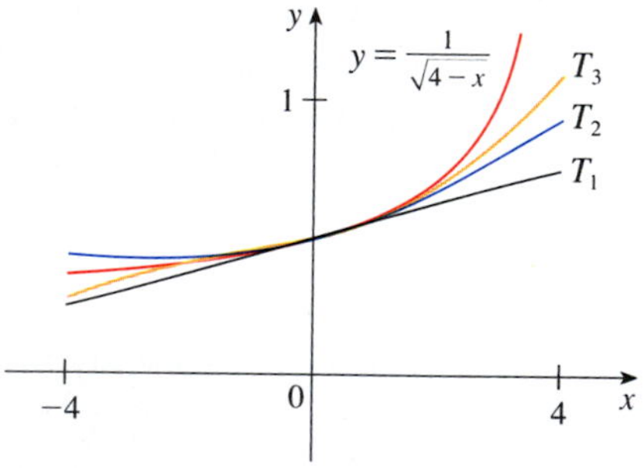

FIGURE 1

EXERCISES 10.8

1–6 ■ Use the binomial series to expand the function as a power series. State the radius of convergence.

1. $\sqrt{1+x}$

2. $\dfrac{1}{(1+x)^3}$

3. $\dfrac{1}{(1+2x)^4}$

4. $\sqrt[3]{1+x^2}$

5. $\sqrt[4]{1-x^4}$

6. $\dfrac{x^2}{\sqrt{2+x}}$

7–8 ■ Use the binomial series to expand the function as a Maclaurin series and to find the first three Taylor polynomials T_1, T_2, and T_3. Graph the function and these Taylor polynomials in the interval of convergence.

7. $\dfrac{1}{\sqrt[3]{8+x}}$ **8.** $(4+x)^{3/2}$

9. (a) Use the binomial series to expand $1/\sqrt{1-x^2}$.
(b) Use part (a) to find the Maclaurin series for $\sin^{-1}x$.

10. (a) Expand $\sqrt[3]{8+x}$ as a power series.
(b) Use part (a) to estimate $\sqrt[3]{8.2}$ correct to four decimal places.

11. (a) Expand $f(x) = x/(1-x)^2$ as a power series.
(b) Use part (a) to find the sum of the series

$$\sum_{n=1}^{\infty} \frac{n}{2^n}$$

12. (a) Expand $f(x) = (x+x^2)/(1-x)^3$ as a power series.
(b) Use part (a) to find the sum of the series

$$\sum_{n=1}^{\infty} \frac{n^2}{2^n}$$

13. (a) Use the binomial series to find the Maclaurin series of $f(x) = \sqrt{1+x^2}$.
(b) Use part (a) to evaluate $f^{(10)}(0)$.

14. (a) Use the binomial series to find the Maclaurin series of $f(x) = 1/\sqrt{1+x^3}$.
(b) Use part (a) to evaluate $f^{(9)}(0)$.

15. Use the following steps to prove (2).
(a) Let $g(x) = \sum_{n=0}^{\infty} \binom{k}{n} x^n$. Differentiate this series to show that

$$g'(x) = \frac{kg(x)}{1+x} \qquad -1 < x < 1$$

(b) Let $h(x) = (1+x)^{-k}g(x)$ and show that $h'(x) = 0$.
(c) Deduce that $g(x) = (1+x)^k$.

16. The period of a pendulum with length L that makes a maximum angle θ_0 with the vertical is

$$T = 4\sqrt{\frac{L}{g}} \int_0^{\pi/2} \frac{dx}{\sqrt{1-k^2\sin^2 x}}$$

where $k = \sin(\frac{1}{2}\theta_0)$ and g is the acceleration due to gravity. (In Exercise 36 in Section 8.8 we approximated this integral using Simpson's Rule.)
(a) Expand the integrand as a binomial series and use the result of Exercise 40 in Section 8.1 to show that

$$T = 2\pi\sqrt{\frac{L}{g}}\left[1 + \frac{1^2}{2^2}k^2 + \frac{1^2 3^2}{2^2 4^2}k^4 + \frac{1^2 3^2 5^2}{2^2 4^2 6^2}k^6 + \cdots\right]$$

If θ_0 is not too large, the approximation $T \approx 2\pi\sqrt{L/g}$, obtained by using only the first term in the series, is often used. A better approximation is obtained by using two terms:

$$T \approx 2\pi\sqrt{\frac{L}{g}}\,(1 + \tfrac{1}{4}k^2)$$

(b) Notice that all the terms in the series after the first one have coefficients that are at most $\frac{1}{4}$. Use this fact to compare this series with a geometric series and show that

$$2\pi\sqrt{\frac{L}{g}}\,(1 + \tfrac{1}{4}k^2) \leqslant T \leqslant 2\pi\sqrt{\frac{L}{g}}\,\frac{4-3k^2}{4-4k^2}$$

(c) Use the inequalities in part (b) to estimate the period of a pendulum with $L = 1$ meter and $\theta_0 = 10°$. How does it compare with the estimate $T \approx 2\pi\sqrt{L/g}$? What if $\theta_0 = 42°$?

10.9 APPLICATIONS OF TAYLOR POLYNOMIALS

Suppose that $f(x)$ is equal to the sum of its Taylor series at a:

$$f(x) = \sum_{n=0}^{\infty} \frac{f^{(n)}(a)}{n!}(x-a)^n$$

In Section 10.7 we introduced the notation $T_n(x)$ for the nth partial sum of this series and called it the nth-degree Taylor polynomial of f at a. Thus

$$\begin{aligned} T_n(x) &= \sum_{i=0}^{n} \frac{f^{(i)}(a)}{i!}(x-a)^i \\ &= f(a) + \frac{f'(a)}{1!}(x-a) + \frac{f''(a)}{2!}(x-a)^2 + \cdots + \frac{f^{(n)}(a)}{n!}(x-a)^n \end{aligned}$$

Since f is the sum of its Taylor series, we know that $T_n(x) \to f(x)$ as $n \to \infty$ and so T_n can be used as an approximation to f: $f(x) \approx T_n(x)$. It is useful to be able to approximate a function by a polynomial because polynomials are the simplest of functions. In this section we explore the use of such approximations by physical scientists and computer scientists.

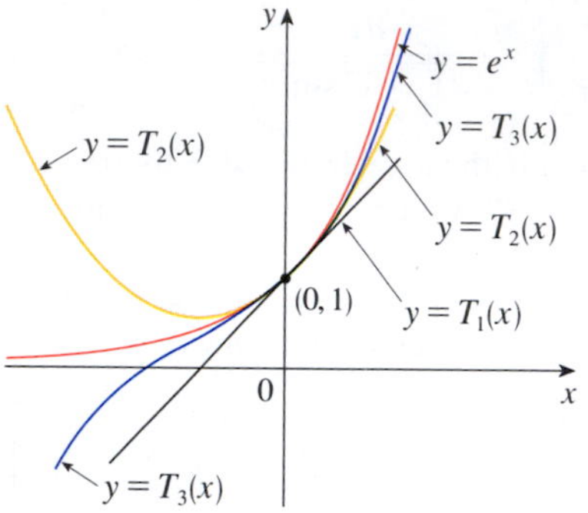

FIGURE 1

	$x = 0.2$	$x = 3.0$
$T_2(x)$	1.220000	8.500000
$T_4(x)$	1.221400	16.375000
$T_6(x)$	1.221403	19.412500
$T_8(x)$	1.221403	20.009152
$T_{10}(x)$	1.221403	20.079665
e^x	1.221403	20.085537

Notice that the first-degree Taylor polynomial

$$T_1(x) = f(a) + f'(a)(x - a)$$

is the same as the linearization of f at a that we discussed in Section 3.11. Notice also that T_1 and its derivative have the same values at a that f and f' have. In general, it can be shown that the derivatives of T_n at a agree with those of f up to and including derivatives of order n.

To illustrate these ideas let's take another look at the graphs of $y = e^x$ and its first few Taylor polynomials, as shown in Figure 1. The graph of T_1 is the tangent line to $y = e^x$ at $(0, 1)$; this tangent line is the best linear approximation to e^x near $(0, 1)$. The graph of T_2 is the parabola $y = 1 + x + x^2/2$, and the graph of T_3 is the cubic curve $y = 1 + x + x^2/2 + x^3/6$, which is a closer fit to the exponential curve $y = e^x$ than T_2. The next Taylor polynomial T_4 would be an even better approximation, and so on.

The values in the table give a numerical demonstration of the convergence of the Taylor polynomials $T_n(x)$ to the function $y = e^x$. We see that when $x = 0.2$ the convergence is very rapid, but when $x = 3$ it is somewhat slower. In fact, the farther x is from 0, the slower T_n converges to x.

When using a Taylor polynomial T_n to approximate a function f, we have to ask the questions: How good an approximation is it? How large should we take n to be in order to achieve a desired accuracy? To answer these questions we need to look at the absolute value of the remainder:

$$|R_n(x)| = |f(x) - T_n(x)|$$

There are three possible methods for estimating the size of the error:

1. If a graphing device is available, we can use it to graph $|R_n(x)|$ and thereby estimate the error.
2. If the series happens to be an alternating series, we can use the Alternating Series Estimation Theorem.
3. In all cases we can use Taylor's Inequality (Theorem 10.7.9), which says that if $|f^{(n+1)}(x)| \leqslant M$, then

$$|R_n(x)| \leqslant \frac{M}{(n+1)!}|x - a|^{n+1}$$

EXAMPLE 1

(a) Approximate the function $f(x) = \sqrt[3]{x}$ by a Taylor polynomial of degree 2 at $a = 8$.

(b) How accurate is this approximation when $7 \leqslant x \leqslant 9$?

SOLUTION

(a)

$$f(x) = \sqrt[3]{x} = x^{1/3} \qquad f(8) = 2$$
$$f'(x) = \tfrac{1}{3}x^{-2/3} \qquad f'(8) = \tfrac{1}{12}$$
$$f''(x) = -\tfrac{2}{9}x^{-5/3} \qquad f''(8) = -\tfrac{1}{144}$$
$$f'''(x) = \tfrac{10}{27}x^{-8/3}$$

Thus, the second-degree Taylor polynomial is

$$T_2(x) = f(8) + \frac{f'(8)}{1!}(x-8) + \frac{f''(8)}{2!}(x-8)^2$$
$$= 2 + \tfrac{1}{12}(x-8) - \tfrac{1}{288}(x-8)^2$$

The desired approximation is

$$\sqrt[3]{x} \approx T_2(x) = 2 + \tfrac{1}{12}(x-8) - \tfrac{1}{288}(x-8)^2$$

(b) The Taylor series is not alternating when $x < 8$, so we can't use the Alternating Series Estimation Theorem in this example. But we can use Taylor's Inequality with $n = 2$ and $a = 8$:

$$|R_2(x)| \leqslant \frac{M}{3!}|x-8|^3$$

where $|f'''(x)| \leqslant M$. Because $x \geqslant 7$, we have $x^{8/3} \geqslant 7^{8/3}$ and so

$$f'''(x) = \frac{10}{27} \cdot \frac{1}{x^{8/3}} \leqslant \frac{10}{27} \cdot \frac{1}{7^{8/3}} < 0.0021$$

Therefore, we can take $M = 0.0021$. Also $7 \leqslant x \leqslant 9$, so $-1 \leqslant x - 8 \leqslant 1$ and $|x - 8| \leqslant 1$. Then Taylor's Inequality gives

$$|R_2(x)| \leqslant \frac{0.0021}{3!} \cdot 1^3 = \frac{0.0021}{6} < 0.0004$$

Thus, if $7 \leqslant x \leqslant 9$, the approximation in part (a) is accurate to within 0.0004. ■

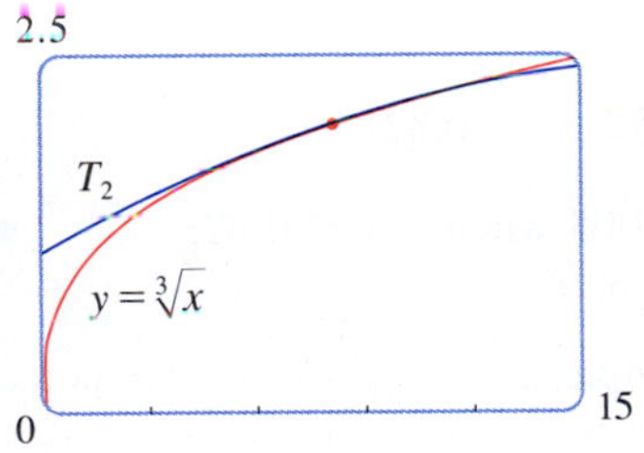

FIGURE 2

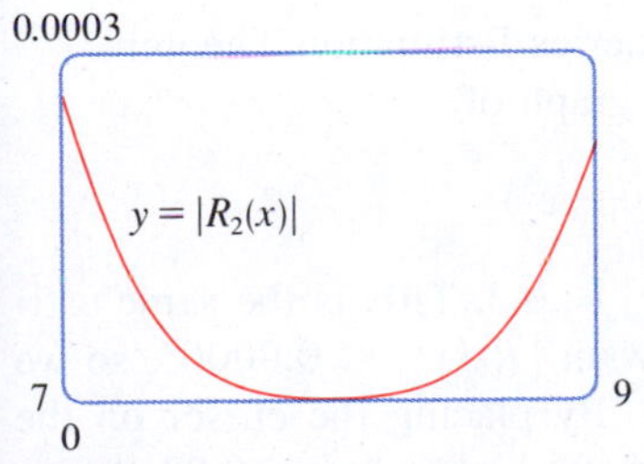

FIGURE 3

Let's use a graphing device to check the calculation in Example 1. Figure 2 shows that the graphs of $y = \sqrt[3]{x}$ and $y = T_2(x)$ are very close to each other when x is near 8. Figure 3 shows the graph of $|R_2(x)|$ computed from the expression

$$|R_2(x)| = |\sqrt[3]{x} - T_2(x)|$$

We see from the graph that

$$|R_2(x)| < 0.0003$$

when $7 \leqslant x \leqslant 9$. Thus, the error estimate from graphical methods is slightly better than the error estimate from Taylor's Inequality in this case.

EXAMPLE 2
(a) What is the maximum error possible in using the approximation

$$\sin x \approx x - \frac{x^3}{3!} + \frac{x^5}{5!}$$

when $-0.3 \leqslant x \leqslant 0.3$? Use this approximation to find sin 12° correct to six decimal places.
(b) For what values of x is this approximation accurate to within 0.00005?

SOLUTION

(a) Notice that the Maclaurin series

$$\sin x = x - \frac{x^3}{3!} + \frac{x^5}{5!} - \frac{x^7}{7!} + \cdots$$

is alternating for all nonzero values of x, so we can use the Alternating Series Estimation Theorem. The error in approximating $\sin x$ by the first three terms of its Maclaurin series is at most

$$\left| \frac{x^7}{7!} \right| = \frac{|x|^7}{5040}$$

If $-0.3 \leqslant x \leqslant 0.3$, then $|x| \leqslant 0.3$, so the error is smaller than

$$\frac{(0.3)^7}{5040} \approx 4.3 \times 10^{-8}$$

To find $\sin 12°$ we first convert to radian measure.

$$\sin 12° = \sin\left(\frac{12\pi}{180}\right) = \sin\left(\frac{\pi}{15}\right)$$

$$\approx \frac{\pi}{15} - \left(\frac{\pi}{15}\right)^3 \frac{1}{3!} + \left(\frac{\pi}{15}\right)^5 \frac{1}{5!}$$

$$\approx 0.20791169$$

Thus, correct to six decimal places, $\sin 12° \approx 0.207912$.

(b) The error will be smaller than 0.00005 if

$$\frac{|x|^7}{5040} < 0.00005$$

Solving this inequality for x, we get

$$|x|^7 < 0.252 \qquad \text{or} \qquad |x| < (0.252)^{1/7} \approx 0.821$$

So the given approximation is accurate to within 0.00005 when $|x| < 0.82$. ■

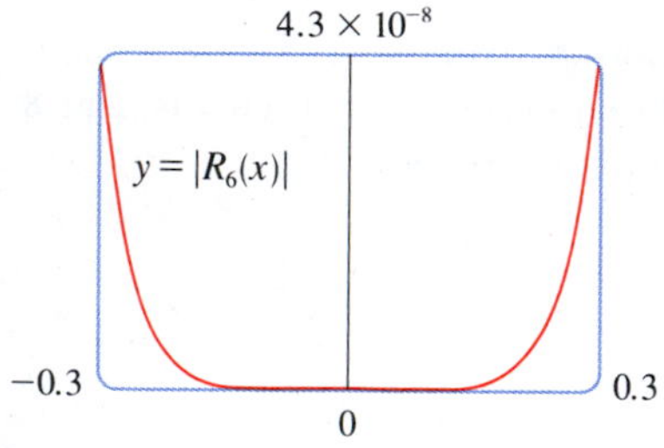

FIGURE 4

What if we use Taylor's Inequality to solve Example 2? Since $f^{(7)}(x) = -\cos x$, we have $|f^{(7)}(x)| \leqslant 1$ and so

$$|R_6(x)| \leqslant \frac{1}{7!}|x|^7$$

So we get the same estimates as with the Alternating Series Estimation Theorem.

What about graphical methods? Figure 4 shows the graph of

$$|R_6(x)| = \left|\sin x - \left(x - \tfrac{1}{6}x^3 + \tfrac{1}{120}x^5\right)\right|$$

and we see from it that $|R_6(x)| < 4.3 \times 10^{-8}$ when $|x| \leqslant 0.3$. This is the same estimate that we obtained in Example 2. For part (b) we want $|R_6(x)| < 0.00005$, so we graph both $y = |R_6(x)|$ and $y = 0.00005$ in Figure 5. By placing the cursor on the right intersection point we find that the inequality is satisfied when $|x| < 0.82$. Again this is the same estimate that we obtained in the solution to Example 2.

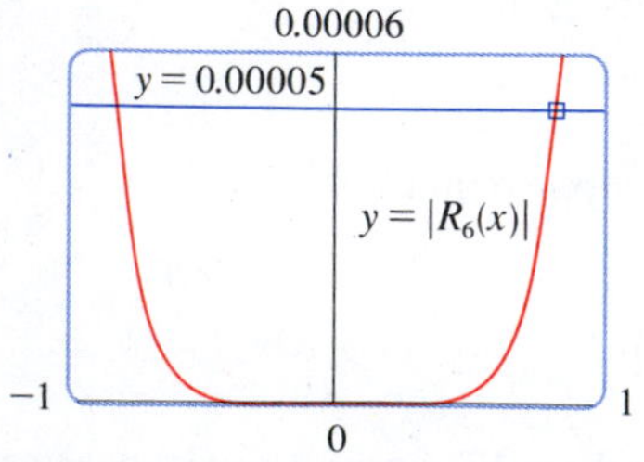

FIGURE 5

If we had been asked to approximate $\sin 72°$ instead of $\sin 12°$ in Example 2, it would have been wise to use the Taylor polynomials at $a = \pi/3$ (instead of $a = 0$) because they are better approximations to $\sin x$ for values of x close to $\pi/3$. Notice that $72°$ is close to $60°$ (or $\pi/3$ radians) and the derivatives of $\sin x$ are easy to compute at $\pi/3$.

Figure 6 shows the graphs of the Taylor polynomial approximations

$$T_1(x) = x \qquad T_3(x) = x - \frac{x^3}{3!}$$

$$T_5(x) = x - \frac{x^3}{3!} + \frac{x^5}{5!} \qquad T_7(x) = x - \frac{x^3}{3!} + \frac{x^5}{5!} - \frac{x^7}{7!}$$

to the sine curve. You can see that as n increases, $T_n(x)$ is a good approximation to $\sin x$ on a larger and larger interval.

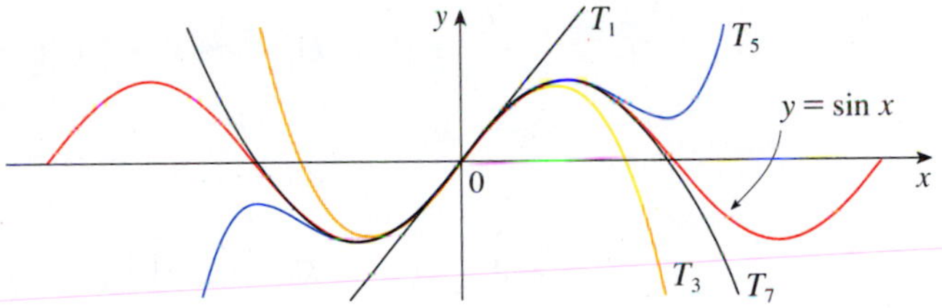

FIGURE 6

One use of the type of calculation done in Examples 1 and 2 occurs in calculators and computers. For instance, when you press the sin or e^x key on your calculator, or when a computer programmer uses a subroutine for a trigonometric or exponential or Bessel function, in many machines a polynomial approximation is calculated. The polynomial is often a Taylor polynomial that has been modified so that the error is spread more evenly throughout an interval.

APPLICATIONS TO PHYSICS

Taylor polynomials are also used frequently in physics. In order to gain insight into an equation, a physicist often simplifies a function by considering only the first two or three terms in its Taylor series. In other words, the physicist uses a Taylor polynomial as an approximation to the function. Taylor's Inequality can then be used to gauge the accuracy of the approximation. The following example shows one way in which this idea is used in special relativity.

EXAMPLE 3 In Einstein's theory of special relativity the mass of an object moving with velocity v is

$$m = \frac{m_0}{\sqrt{1 - v^2/c^2}}$$

where m_0 is the mass of the object when at rest and c is the speed of light. The kinetic energy of the object is the difference between its total energy and its energy at rest:

$$K = mc^2 - m_0c^2$$

(a) Show that when v is very small compared with c, this expression for K agrees with classical Newtonian physics: $K = \frac{1}{2}m_0v^2$.

(b) Use Taylor's Inequality to estimate the difference in these expressions for K when $|v| \leqslant 100$ m/s.

SOLUTION

(a) Using the expressions given for K and m, we get

$$K = mc^2 - m_0c^2 = \frac{m_0c^2}{\sqrt{1 - v^2/c^2}} - m_0c^2$$

$$= m_0c^2\left[\left(1 - \frac{v^2}{c^2}\right)^{-1/2} - 1\right]$$

With $x = -v^2/c^2$, the Maclaurin series for $(1 + x)^{-1/2}$ is most easily computed as a binomial series with $k = -\frac{1}{2}$. (Notice that $|x| < 1$ because $v < c$.) Therefore, we have

$$(1 + x)^{-1/2} = 1 - \tfrac{1}{2}x + \frac{(-\frac{1}{2})(-\frac{3}{2})}{2!}x^2 + \frac{(-\frac{1}{2})(-\frac{3}{2})(-\frac{5}{2})}{3!}x^3 + \cdots$$

$$= 1 - \tfrac{1}{2}x + \tfrac{3}{8}x^2 - \tfrac{5}{16}x^3 + \cdots$$

and

$$K = m_0c^2\left[\left(1 + \frac{1}{2}\frac{v^2}{c^2} + \frac{3}{8}\frac{v^4}{c^4} + \frac{5}{16}\frac{v^6}{c^6} + \cdots\right) - 1\right]$$

$$= m_0c^2\left(\frac{1}{2}\frac{v^2}{c^2} + \frac{3}{8}\frac{v^4}{c^4} + \frac{5}{16}\frac{v^6}{c^6} + \cdots\right)$$

If v is much smaller than c, then all terms after the first are very small when compared with the first term. If we omit them, we get

$$K \approx m_0c^2\left(\frac{1}{2}\frac{v^2}{c^2}\right) = \tfrac{1}{2}m_0v^2$$

(b) If $x = -v^2/c^2$, $f(x) = m_0c^2[(1 + x)^{-1/2} - 1]$, and M is a number such that $|f''(x)| \leqslant M$, then we can use Taylor's Inequality to write

$$|R_1(x)| \leqslant \frac{M}{2!}x^2$$

We have $f''(x) = \frac{3}{4}m_0c^2(1 + x)^{-5/2}$ and we are given that $|v| \leqslant 100$ m/s, so

$$|f''(x)| = \frac{3m_0c^2}{4(1 - v^2/c^2)^{5/2}} \leqslant \frac{3m_0c^2}{4(1 - 100^2/c^2)^{5/2}} \quad (= M)$$

Thus, with $c = 3 \times 10^8$ m/s,

$$|R_1(x)| \leqslant \frac{1}{2} \cdot \frac{3m_0c^2}{4(1 - 100^2/c^2)^{5/2}} \cdot \frac{100^4}{c^4} < (4.17 \times 10^{-10})m_0$$

So when $|v| \leqslant 100$ m/s, the magnitude of the error in using the Newtonian expression for kinetic energy is at most $(4.2 \times 10^{-10})m_0$. ■

Another application to physics occurs in optics. Figure 7 is adapted from a book by Eugene Hecht: *Optics,* 2d ed. (Reading, MA: Addison-Wesley, 1987), page 133. It depicts a wave from the point source S meeting a spherical interface of radius R centered at C. The ray SA is refracted toward P.

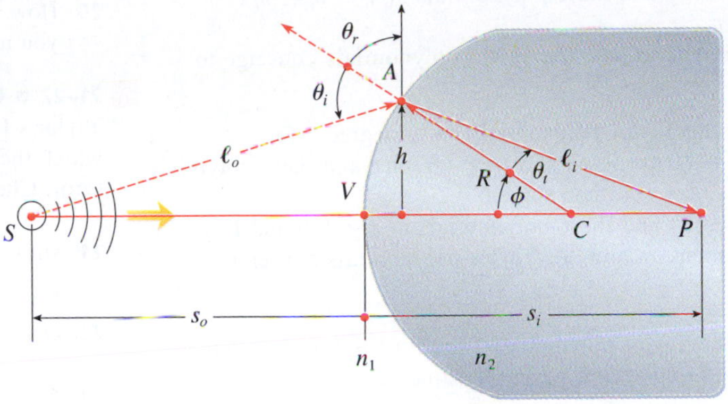

FIGURE 7
Refraction at a spherical interface

Using Fermat's principle that light travels so as to minimize the time taken, Hecht derives the equation

$$\frac{n_1}{\ell_o} + \frac{n_2}{\ell_i} = \frac{1}{R}\left(\frac{n_2 s_i}{\ell_i} - \frac{n_1 s_o}{\ell_o}\right) \tag{1}$$

where n_1 and n_2 are indexes of refraction and ℓ_o, ℓ_i, s_o, and s_i are the distances indicated in Figure 7. By the Law of Cosines, applied to triangles ACS and ACP, we have

$$\ell_o = \sqrt{R^2 + (s_o + R)^2 - 2R(s_o + R)\cos\phi} \tag{2}$$

$$\ell_i = \sqrt{R^2 + (s_i - R)^2 + 2R(s_i - R)\cos\phi}$$

Here we use the identity
$\cos(\pi - \phi) = -\cos\phi$

Because Equation 1 is cumbersome to work with, Gauss, in 1841, simplified it by using the linear approximation $\cos\phi \approx 1$ for small values of ϕ. (This amounts to using the Taylor polynomial of degree 1.) Then Equation 1 becomes the following simpler equation [as you are asked to show in Exercise 24(a)]:

$$\frac{n_1}{s_o} + \frac{n_2}{s_i} = \frac{n_2 - n_1}{R} \tag{3}$$

The resulting optical theory is known as *Gaussian optics,* or *first-order optics,* and has become the basic theoretical tool used to design lenses.

A more accurate theory is obtained by approximating $\cos\phi$ by its Taylor polynomial of degree 3 (which is the same as the Taylor polynomial of degree 2). This takes into account rays for which ϕ is not so small, that is, rays that strike the surface at greater distances h above the axis. In Exercise 24(b) you are asked to use this approximation to derive the more accurate equation

$$\frac{n_1}{s_o} + \frac{n_2}{s_i} = \frac{n_2 - n_1}{R} + h^2\left[\frac{n_1}{2s_o}\left(\frac{1}{s_o} + \frac{1}{R}\right)^2 + \frac{n_2}{2s_i}\left(\frac{1}{R} - \frac{1}{s_i}\right)^2\right] \tag{4}$$

The resulting optical theory is known as *third-order optics.*

Other applications of Taylor polynomials to physics are explored in Exercises 25–28.

EXERCISES 10.9

1. (a) Find the Taylor polynomials up to degree 6 for $f(x) = \cos x$ centered at $a = 0$. Graph f and these polynomials on a common screen.
(b) Evaluate f and these polynomials at $x = \pi/4$, $\pi/2$, and π.
(c) Comment on how the Taylor polynomials converge to $f(x)$.

2. (a) Find the Taylor polynomials up to degree 3 for $f(x) = 1/x$ centered at $a = 1$. Graph f and these polynomials on a common screen.
(b) Evaluate f and these polynomials at $x = 0.9$ and 1.3.
(c) Comment on how the Taylor polynomials converge to $f(x)$.

3–8 ■ Find the Taylor polynomial $T_n(x)$ for the function f at the number a. Graph f and T_n on the same screen.

3. $f(x) = \sin x, \quad a = \pi/6, \quad n = 3$

4. $f(x) = \cos x, \quad a = 2\pi/3, \quad n = 4$

5. $f(x) = \tan x, \quad a = 0, \quad n = 4$

6. $f(x) = \tan x, \quad a = \pi/4, \quad n = 4$

7. $f(x) = \sec x, \quad a = \pi/3, \quad n = 3$

8. $f(x) = \sqrt{x}, \quad a = 9, \quad n = 3$

CAS **9–10** ■ Use a computer algebra system to find the Taylor polynomials T_n at $a = 0$ for the given values of n. Then graph these polynomials and f on the same screen.

9. $f(x) = \sec x, \quad n = 2, 4, 6, 8$

10. $f(x) = \tan x, \quad n = 1, 3, 5, 7, 9$

11–16 ■
(a) Approximate f by a Taylor polynomial with degree n at the number a.
(b) Use Taylor's Inequality to estimate the accuracy of the approximation $f(x) \approx T_n(x)$ when x lies in the given interval.
(c) Check your result in part (b) by graphing $|R_n(x)|$.

11. $f(x) = \sqrt{1 + x}, \quad a = 0, \quad n = 1, \quad 0 \leqslant x \leqslant 0.1$

12. $f(x) = 1/x, \quad a = 1, \quad n = 3, \quad 0.8 \leqslant x \leqslant 1.2$

13. $f(x) = \sin x, \quad a = \pi/4, \quad n = 5, \quad 0 \leqslant x \leqslant \pi/2$

14. $f(x) = \cos x, \quad a = \pi/3, \quad n = 4, \quad 0 \leqslant x \leqslant 2\pi/3$

15. $f(x) = e^{x^2}, \quad a = 0, \quad n = 3, \quad 0 \leqslant x \leqslant 0.1$

16. $f(x) = \ln x, \quad a = 4, \quad n = 3, \quad 3 \leqslant x \leqslant 5$

17. Use the information from Exercise 3 to estimate $\sin 35°$ correct to five decimal places.

18. Use the information from Exercise 14 to estimate $\cos 69°$ correct to five decimal places.

19. Use Taylor's Inequality to determine the number of terms of the Maclaurin series for e^x that should be used to estimate $e^{0.1}$ to within 0.00001.

20. How many terms of the Maclaurin series for $\ln(1 + x)$ do you need to use to estimate $\ln 1.4$ to within 0.001?

21–22 ■ Use the Alternating Series Estimation Theorem or Taylor's Inequality to estimate the range of values of x for which the given approximation is accurate to within the stated error. Check your answer graphically.

21. $\sin x \approx x - \dfrac{x^3}{6}, \quad \text{error} < 0.01$

22. $\cos x \approx 1 - \dfrac{x^2}{2} + \dfrac{x^4}{24}, \quad \text{error} < 0.005$

23. A car is moving with speed 20 m/s and acceleration 2 m/s^2 at a given instant. Using a second-degree Taylor polynomial, estimate how far the car moves in the next second. Would it be reasonable to use this polynomial to estimate the distance traveled during the next minute?

24. (a) Derive Equation 3 for Gaussian optics from Equation 1 by approximating $\cos\phi$ in Equation 2 by its first-degree Taylor polynomial.
(b) Show that if $\cos\phi$ is replaced by its third-degree Taylor polynomial in Equation 2, then Equation 1 becomes Equation 4 for third-order optics. [*Hint:* Use the first two terms in the binomial series for ℓ_o^{-1} and ℓ_i^{-1}. Also, use $\phi \approx \sin\phi$.]

25. An electric dipole consists of two electric charges of equal magnitude and opposite signs. If the charges are q and $-q$ and are located at a distance d from each other, then the electric field E at the point P in the figure is

$$E = \frac{q}{D^2} - \frac{q}{(D + d)^2}$$

By expanding this expression for E as a series in powers of d/D, show that E is approximately proportional to $1/D^3$ when P is far away from the dipole.

26. The resistivity ρ of a conducting wire is the reciprocal of the conductivity and is measured in units of ohm-meters (Ω·m). The resistivity of a given metal depends on the temperature according to the equation

$$\rho(t) = \rho_{20} e^{\alpha(t-20)}$$

where t is the temperature in °C. There are tables that list the values of α (called the temperature coefficient) and ρ_{20} (the resistivity at 20 °C) for various metals. Except at very low temperatures, the resistivity varies almost linearly with temperature and so it is common to approximate the expression for $\rho(t)$ by its first- or second-degree Taylor polynomial at $t = 20$.

(a) Find expressions for these linear and quadratic approximations.

(b) For copper, the tables give $\alpha = 0.0039/°\text{C}$ and $\rho_{20} = 1.7 \times 10^{-8}\ \Omega\text{-m}$. Graph the resistivity of copper and the linear and quadratic approximations for $-250\,°\text{C} \leq t \leq 1000\,°\text{C}$.

(c) For what values of t does the linear approximation agree with the exponential expression to within one percent?

27. If a water wave with length L moves with velocity v across a body of water with depth d, as in the figure, then

$$v^2 = \frac{gL}{2\pi} \tanh \frac{2\pi d}{L}$$

(a) If the water is deep, show that $v \approx \sqrt{gL/(2\pi)}$.

(b) If the water is shallow, use the Maclaurin series for tanh to show that $v \approx \sqrt{gd}$. (Thus, in shallow water the velocity of a wave tends to be independent of the length of the wave.)

(c) Use the Alternating Series Estimation Theorem to show that if $L > 10d$, then the estimate $v^2 \approx gd$ is accurate to within $0.014gL$.

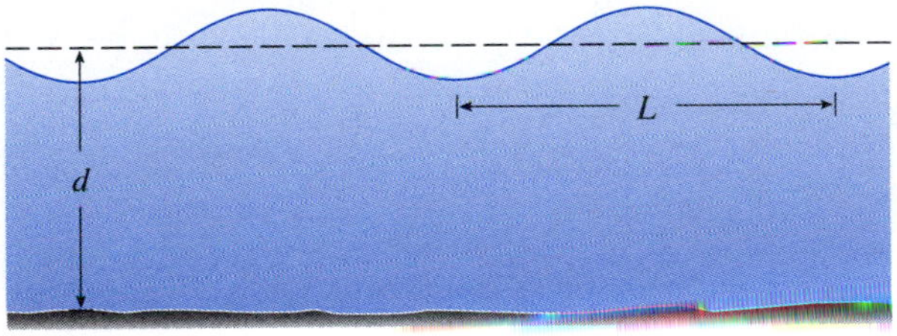

28. In the late 19th century, the Rayleigh-Jeans Law expressed the energy density of blackbody radiation of wavelength λ as

$$f(\lambda) = \frac{8\pi kT}{\lambda^4}$$

where λ is measured in meters, T is the temperature in kelvins, and k is Boltzmann's constant. (See Problem 8 on page 523 for background information on blackbody radiation.) The Rayleigh-Jeans Law agrees with experimental measurements for long wavelengths but disagrees drastically for short wavelengths. [The law predicts that $f(\lambda) \to \infty$ as $\lambda \to 0^+$ but experiments have shown that $f(\lambda) \to 0$.] This fact is known as the *ultraviolet catastrophe*.

In 1900 Max Planck found a better model (known now as Planck's Law) for blackbody radiation:

$$f(\lambda) = \frac{8\pi hc\lambda^{-5}}{e^{hc/(\lambda kT)} - 1}$$

where h is Planck's constant and c is the speed of light.

(a) Use a Taylor polynomial to show that, for large wavelengths, Planck's Law gives approximately the same values as the Rayleigh-Jeans Law.

(b) Graph f as given by both laws on the same screen and comment on the similarities and differences. Use $T = 5700$ K (the temperature of the Sun), $h = 6.6262 \times 10^{-34}$ J-s, $c = 2.997925 \times 10^8$ m/s, and $k = 1.3807 \times 10^{-23}$ J/K. (You may want to change from meters to the more convenient unit of micrometers: $1\ \mu\text{m} = 10^{-6}$ m.)

29. In Section 3.12 we considered Newton's method for approximating a root r of the equation $f(x) = 0$, and from an initial approximation x_1 we obtained successive approximations $x_2, x_3, \ldots$, where

$$x_{n+1} = x_n - \frac{f(x_n)}{f'(x_n)}$$

Use Taylor's Inequality with $n = 1$, $a = x_n$, and $x = r$ to show that if $f''(x)$ exists on an interval I containing r, x_n, and x_{n+1}, and $|f''(x)| \leq M$, $|f'(x)| \geq K$ for all $x \in I$, then

$$|x_{n+1} - r| \leq \frac{M}{2K} |x_n - r|^2$$

[This means that if x_n is accurate to d decimal places, then x_{n+1} is accurate to about $2d$ decimal places. More precisely, if the error at stage n is at most 10^{-m}, then the error at stage $n + 1$ is at most $(M/2K)10^{-2m}$.]

10 REVIEW

KEY TOPICS ■ Define, state, or discuss the following.

1. Sequence
2. Limit of a sequence
3. Convergent sequence; divergent sequence
4. Increasing, decreasing, and monotonic sequences
5. Bounded sequence
6. Convergence of bounded, monotonic sequences
7. Series
8. Partial sums
9. Convergent series; divergent series
10. Sum of a series
11. Geometric series
12. Harmonic series
13. Test for Divergence
14. Integral Test
15. Convergence of a p-series
16. Remainder Estimate for the Integral Test
17. Comparison Test
18. Limit Comparison Test
19. Alternating Series Test
20. Alternating Series Estimation Theorem
21. Absolute convergence
22. Conditional convergence
23. Relation between convergence and absolute convergence
24. Ratio Test
25. Power series
26. Radius of convergence
27. Interval of convergence
28. The power series for $1/(1 - x)$
29. Differentiation and integration of power series
30. Taylor series
31. Maclaurin series
32. Taylor polynomial
33. Taylor's Inequality
34. Maclaurin series for e^x, $\sin x$, $\cos x$
35. Binomial series

EXERCISES

1–16 ■ Determine whether the statement is true or false.

1. If $\lim_{n\to\infty} a_n = 0$, then $\Sigma\, a_n$ is convergent.
2. If $\Sigma\, c_n 6^n$ is convergent, then $\Sigma\, c_n(-2)^n$ is convergent.
3. If $\Sigma\, c_n 6^n$ is convergent, then $\Sigma\, c_n(-6)^n$ is convergent.
4. If $\Sigma\, c_n x^n$ diverges when $x = 6$, then it diverges when $x = 10$.
5. The Ratio Test can be used to determine whether $\Sigma\, 1/n^3$ converges.
6. The Ratio Test can be used to determine whether $\Sigma\, 1/n!$ converges.
7. If $0 \leqslant a_n \leqslant b_n$ and $\Sigma\, b_n$ diverges, then $\Sigma\, a_n$ diverges.
8. $\displaystyle\sum_{n=0}^{\infty} \frac{(-1)^n}{n!} = \frac{1}{e}$
9. If $-1 < \alpha < 1$, then $\lim_{n\to\infty} \alpha^n = 0$.
10. If $\Sigma\, a_n$ is divergent, then $\Sigma\, |a_n|$ is divergent.
11. If $f(x) = 2x - x^2 + \frac{1}{3}x^3 - \cdots$ converges for all x, then $f'''(0) = 2$.
12. If $\{a_n\}$ and $\{b_n\}$ are divergent, then $\{a_n + b_n\}$ is divergent.
13. If $\{a_n\}$ and $\{b_n\}$ are divergent, then $\{a_n b_n\}$ is divergent.
14. If $\{a_n\}$ is decreasing and $a_n > 0$ for all n, then $\{a_n\}$ is convergent.
15. If $a_n > 0$ and $\Sigma\, a_n$ converges, then $\Sigma\, (-1)^n a_n$ converges.
16. If $a_n > 0$ and $\lim_{n\to\infty} (a_{n+1}/a_n) < 1$, then $\lim_{n\to\infty} a_n = 0$.

17–23 ■ Determine whether the sequence is convergent or divergent. If it is convergent, find its limit.

17. $a_n = \dfrac{n}{2n + 5}$
18. $a_n = 5 - (0.9)^n$
19. $a_n = 2n + 5$
20. $a_n = n/\ln n$
21. $a_n = \sin n$
22. $a_n = (\sin n)/n$
23. $\{(1 + 3/n)^{4n}\}$

24. A sequence is defined recursively by the equations $a_1 = 1$, $a_{n+1} = \frac{1}{3}(a_n + 4)$. Show that $\{a_n\}$ is increasing and $a_n < 2$ for all n. Deduce that $\{a_n\}$ is convergent and find its limit.

25–34 ■ Determine whether the series is convergent or divergent.

25. $\displaystyle\sum_{n=1}^{\infty} \frac{n^2}{n^3 + 1}$

26. $\displaystyle\sum_{n=1}^{\infty} \frac{n + n^2}{n + n^4}$

27. $\displaystyle\sum_{n=1}^{\infty} \frac{(-1)^n}{\sqrt[4]{n}}$

28. $\displaystyle\sum_{n=1}^{\infty} \frac{n^2}{3^n}$

29. $\displaystyle\sum_{n=1}^{\infty} \frac{4^n}{n3^n}$

30. $\displaystyle\sum_{n=1}^{\infty} \sqrt{\frac{n-1}{n}}$

31. $\displaystyle\sum_{n=1}^{\infty} \frac{\sin n}{1 + n^2}$

32. $\displaystyle\sum_{n=2}^{\infty} \frac{1}{n(\ln n)^2}$

33. $\displaystyle\sum_{n=1}^{\infty} \frac{1 \cdot 3 \cdot 5 \cdot \cdots \cdot (2n-1)}{5^n n!}$

34. $\displaystyle\sum_{n=1}^{\infty} (-1)^{n+1} \frac{\ln n}{\sqrt{n}}$

35–38 ■ Find the sum of the series.

35. $\displaystyle\sum_{n=1}^{\infty} \frac{2^{2n+1}}{5^n}$

36. $\displaystyle\sum_{n=1}^{\infty} \frac{1}{n(n+3)}$

37. $\displaystyle\sum_{n=1}^{\infty} [\tan^{-1}(n+1) - \tan^{-1} n]$

38. $\displaystyle\sum_{n=0}^{\infty} \frac{(-1)^n x^n}{2^{2n} n!}$

39. Express the repeating decimal 1.2345345345 . . . as a fraction.

40. For what values of x does the series $\sum_{n=1}^{\infty} (\ln x)^n$ converge?

41. Find the sum of the series $\displaystyle\sum_{n=1}^{\infty} \frac{(-1)^{n+1}}{n^5}$ correct to four decimal places.

42. (a) Find the partial sum s_5 of the series $\sum_{n=1}^{\infty} 1/n^6$ and estimate the error in using it as an approximation to the sum of the series.
(b) Find the sum of this series correct to five decimal places.

43. Use the sum of the first eight terms to approximate the sum of the series $\sum_{n=1}^{\infty} (2 + 5^n)^{-1}$. Estimate the error involved in this approximation.

44. (a) Show that the series $\displaystyle\sum_{n=1}^{\infty} \frac{n^n}{(2n)!}$ is convergent.
(b) Deduce that $\displaystyle\lim_{n \to \infty} \frac{n^n}{(2n)!} = 0$.

45. Prove that if the series $\sum_{n=1}^{\infty} a_n$ is absolutely convergent, then the series

$$\sum_{n=1}^{\infty} \left(\frac{n+1}{n}\right) a_n$$

is also absolutely convergent.

46–49 ■ Find the radius of convergence and interval of convergence of the series.

46. $\displaystyle\sum_{n=0}^{\infty} \frac{(-3)^n x^{2n}}{n+1}$

47. $\displaystyle\sum_{n=1}^{\infty} \frac{x^n}{3^n n^3}$

48. $\displaystyle\sum_{n=1}^{\infty} \frac{(x+1)^n}{(n+1)!}$

49. $\displaystyle\sum_{n=0}^{\infty} \frac{2^n (x-3)^n}{\sqrt{n+3}}$

50. Find the radius of convergence of the series

$$\sum_{n=1}^{\infty} \frac{(2n)!}{(n!)^2} x^n$$

51. Find the Taylor series of $f(x) = \sin x$ at $a = \pi/6$.

52. Find the Taylor series of $f(x) = \cos x$ at $a = \pi/3$.

53–60 ■ Find the Maclaurin series for f and its radius of convergence. You may use either the direct method (definition of a Maclaurin series) or known series such as geometric series, binomial series, or the Maclaurin series for e^x and $\sin x$.

53. $f(x) = \dfrac{x^2}{1 + x}$

54. $f(x) = \sqrt{1 - x^2}$

55. $f(x) = \ln(1 - x)$

56. $f(x) = xe^{2x}$

57. $f(x) = \sin(x^4)$

58. $f(x) = 10^x$

59. $f(x) = 1/\sqrt[4]{16 - x}$

60. $f(x) = (1 - 3x)^{-5}$

61. Evaluate $\displaystyle\int \frac{e^x}{x}\,dx$ as an infinite series.

62. Use series to approximate $\int_0^1 \sqrt{1 + x^4}\,dx$ correct to two decimal places.

63–64 ■

(a) Approximate f by a Taylor polynomial with degree n at the number a.
(b) Graph f and T_n on a common screen.
(c) Use Taylor's Inequality to estimate the accuracy of the approximation $f(x) \approx T_n(x)$ when x lies in the given interval.
(d) Check your result in part (c) by graphing $|R_n(x)|$.

63. $f(x) = \sqrt{x}$, $a = 1$, $n = 3$, $0.9 \leqslant x \leqslant 1.1$

64. $f(x) = \sec x$, $a = 0$, $n = 2$, $0 \leqslant x \leqslant \pi/6$

65. Use series to evaluate $\lim_{x \to \infty} x^2(1 - e^{-1/x^2})$.

66. The force due to gravity on an object with mass m at a height h above the surface of Earth is

$$F = \frac{mgR^2}{(R + h)^2}$$

where R is the radius of Earth and g is the acceleration due to gravity.
(a) Express F as a series in powers of h/R.
(b) Observe that if we approximate F by the first term in the series, we get the expression $F \approx mg$ that is usually used when h is much smaller than R. Use the Alternating Series Estimation Theorem to estimate the range of values of h for which the approximation $F \approx mg$ is accurate to within 1%. (Use $R = 6400$ km.)

PROBLEMS PLUS

Before you look at the solution of the example, cover it up and try to solve the problem yourself.

EXAMPLE Find the sum of the series $\sum_{n=0}^{\infty} \frac{(x+2)^n}{(n+3)!}$.

SOLUTION The problem-solving principle that is relevant here is *recognizing something familiar*. Does the given series look anything like a series that we already know? Well, it does have some ingredients in common with the Maclaurin series for the exponential function:

$$e^x = \sum_{n=0}^{\infty} \frac{x^n}{n!} = 1 + x + \frac{x^2}{2!} + \frac{x^3}{3!} + \cdots$$

We can make this series look more like our given series by replacing x by $x + 2$:

$$e^{x+2} = \sum_{n=0}^{\infty} \frac{(x+2)^n}{n!} = 1 + (x+2) + \frac{(x+2)^2}{2!} + \frac{(x+2)^3}{3!} + \cdots$$

But here the exponent in the numerator matches the number in the denominator whose factorial is taken. To make that happen in the given series, let's multiply and divide by $(x+2)^3$:

$$\sum_{n=0}^{\infty} \frac{(x+2)^n}{(n+3)!} = \frac{1}{(x+2)^3} \sum_{n=0}^{\infty} \frac{(x+2)^{n+3}}{(n+3)!}$$

$$= (x+2)^{-3}\left[\frac{(x+2)^3}{3!} + \frac{(x+2)^4}{4!} + \cdots\right]$$

We see that the series between brackets is just the series for e^{x+2} with the first three terms missing. So

$$\sum_{n=0}^{\infty} \frac{(x+2)^n}{(n+3)!} = (x+2)^{-3}\left[e^{x+2} - 1 - (x+2) - \frac{(x+2)^2}{2!}\right]$$ ■

PROBLEMS

1. If $f(x) = \sin(x^3)$, find $f^{(15)}(0)$.

2. Let $\{P_n\}$ be a sequence of points determined as in the figure. Thus $|AP_1| = 1$, $|P_n P_{n+1}| = 2^{n-1}$, and angle $AP_n P_{n+1}$ is a right angle. Find $\lim_{n\to\infty} \angle P_n A P_{n+1}$.

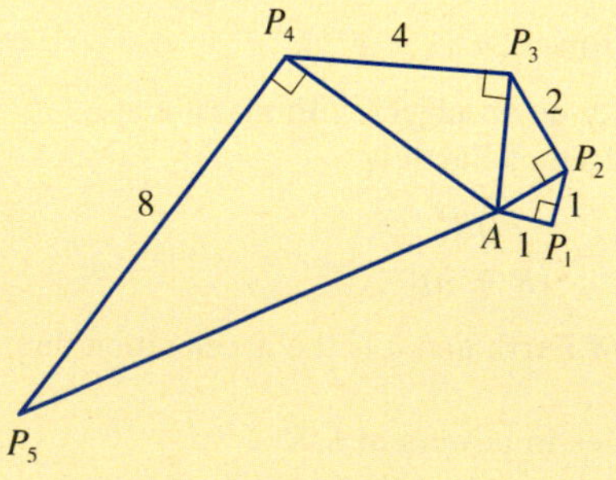

FIGURE FOR PROBLEM 2

3. (a) Show that $\tan \frac{1}{2}x = \cot \frac{1}{2}x - 2\cot x$.
(b) Find the sum of the series

$$\sum_{n=1}^{\infty} \frac{1}{2^n} \tan \frac{x}{2^n}$$

4. A function f is defined by

$$f(x) = \lim_{n\to\infty} \frac{x^{2n} - 1}{x^{2n} + 1}$$

Where is f continuous?

5. To construct the **snowflake curve**, start with an equilateral triangle with sides of length 1. Step 1 in the construction is to divide each side into three equal parts, construct an equilateral triangle on the middle part, and then delete the middle part (see the figure). Step 2 is to repeat step 1 for each side of the resulting polygon. This process is repeated at each succeeding step. The snowflake curve is the curve that results from repeating this process indefinitely.
   (a) Let s_n, l_n, and p_n represent the number of sides, the length of a side, and the total length of the nth approximating curve (the curve obtained after step n of the construction), respectively. Find formulas for s_n, l_n, and p_n.
   (b) Show that $p_n \to \infty$ as $n \to \infty$.
   (c) Sum an infinite series to find the area enclosed by the snowflake curve. Parts (b) and (c) show that the snowflake curve is infinitely long but encloses only a finite area.

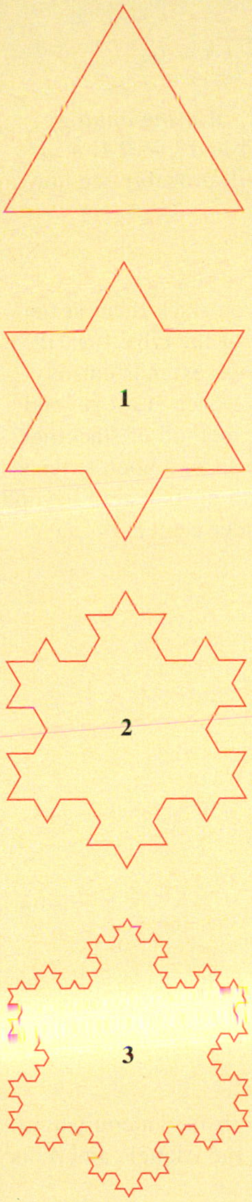

FIGURE FOR PROBLEM 5

6. Find the sum of the series

$$1 + \frac{1}{2} + \frac{1}{3} + \frac{1}{4} + \frac{1}{6} + \frac{1}{8} + \frac{1}{9} + \frac{1}{12} + \cdots$$

where the terms are the reciprocals of the positive integers whose only prime factors are 2s and 3s.

7. (a) Show that for $xy \neq -1$,

$$\arctan x - \arctan y = \arctan \frac{x - y}{1 + xy}$$

if the left side lies between $-\pi/2$ and $\pi/2$.
   (b) Show that

$$\arctan \tfrac{120}{119} - \arctan \tfrac{1}{239} = \frac{\pi}{4}$$

   (c) Deduce the following formula of John Machin (1680–1751):

$$4 \arctan \tfrac{1}{5} - \arctan \tfrac{1}{239} = \frac{\pi}{4}$$

   (d) Use the Maclaurin series for arctan to show that

$$0.197395560 < \arctan \tfrac{1}{5} < 0.197395562$$

   (e) Show that

$$0.004184075 < \arctan \tfrac{1}{239} < 0.004184077$$

   (f) Deduce that, correct to seven decimal places,

$$\pi \approx 3.1415927$$

Machin used this method in 1706 to find π correct to 100 decimal places. In this century, with the aid of computers, the value of π has been computed to increasingly greater accuracy. In 1995 Jonathan and Peter Borwein of Simon Fraser University and Yasumasa Kanada of the University of Tokyo calculated the value of π to 4,294,967,268 decimal places!

8. If $a_0 + a_1 + a_2 + \cdots + a_k = 0$, show that

$$\lim_{n \to \infty} \left(a_0\sqrt{n} + a_1\sqrt{n+1} + a_2\sqrt{n+2} + \cdots + a_k\sqrt{n+k}\right) = 0$$

If you don't see how to prove this, try the problem-solving strategy of *using analogy* (see Section 4 of Review and Preview). Try the special cases $k = 1$ and $k = 2$ first. If you can see how to prove the assertion for these cases, then you will probably see how to prove it in general.

9. Find the interval of convergence of $\sum_{n=1}^{\infty} n^3x^n$ and find its sum.

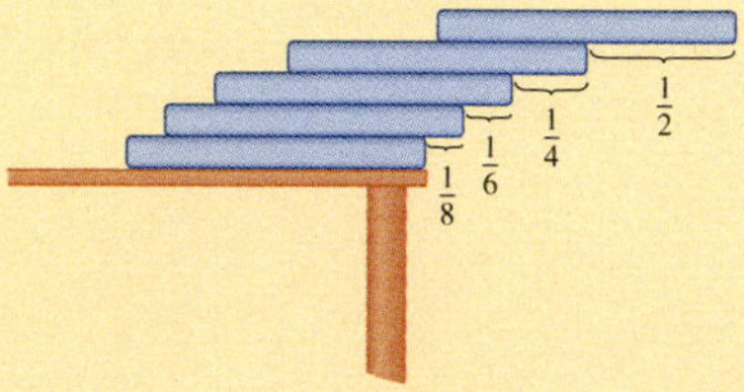

FIGURE FOR PROBLEM 10

10. Suppose you have a large supply of books, all the same size, and you stack them at the edge of a table, with each book extending farther beyond the edge of the table than the one beneath it. Show that it is possible to do this so that the top book extends entirely beyond the table. In fact, show that the top book can extend any distance at all beyond the edge of the table if the stack is high enough. Use the following method of stacking: The top book extends half its length beyond the second book. The second book extends a quarter of its length beyond the third. The third extends one-sixth of its length beyond the fourth, and so on. (Try it yourself with a deck of cards.) Consider centers of mass.

11. Let

$$u = 1 + \frac{x^3}{3!} + \frac{x^6}{6!} + \frac{x^9}{9!} + \cdots$$

$$v = x + \frac{x^4}{4!} + \frac{x^7}{7!} + \frac{x^{10}}{10!} + \cdots$$

$$w = \frac{x^2}{2!} + \frac{x^5}{5!} + \frac{x^8}{8!} + \cdots$$

Show that $u^3 + v^3 + w^3 - 3uvw = 1$.

12. If $p > 1$, evaluate the expression

$$\frac{1 + \dfrac{1}{2^p} + \dfrac{1}{3^p} + \dfrac{1}{4^p} + \cdots}{1 - \dfrac{1}{2^p} + \dfrac{1}{3^p} - \dfrac{1}{4^p} + \cdots}$$

13. Suppose that circles of equal diameter are packed tightly in n rows inside an equilateral triangle. (The figure illustrates the case $n = 4$.) If A is the area of the triangle and A_n is the total area occupied by the n rows of circles, show that

$$\lim_{n \to \infty} \frac{A_n}{A} = \frac{\pi}{2\sqrt{3}}$$

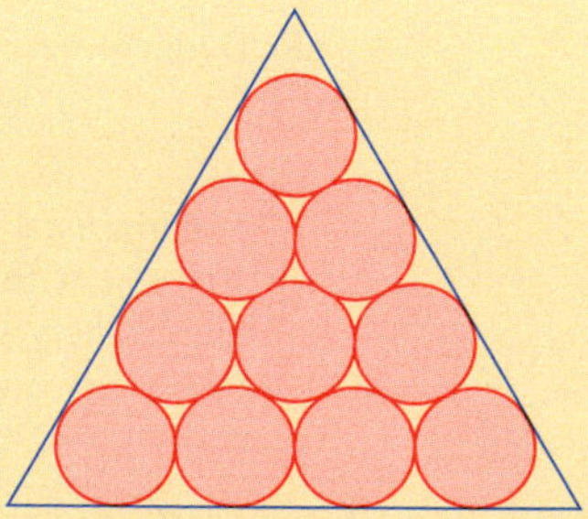

FIGURE FOR PROBLEM 13

14. Starting with the vertices $P_1(0, 1)$, $P_2(1, 1)$, $P_3(1, 0)$, $P_4(0, 0)$ of a square, we construct further points as shown in the figure: P_5 is the midpoint of P_1P_2, P_6 is the midpoint of P_2P_3, P_7 is the midpoint of P_3P_4, and so on. The polygonal spiral path $P_1P_2P_3P_4P_5P_6P_7\ldots$ approaches a point P inside the square.

(a) If the coordinates of P_n are (x_n, y_n), show that

$$\tfrac{1}{2}x_n + x_{n+1} + x_{n+2} + x_{n+3} = 2$$

and find a similar equation for the y-coordinates.

(b) Find the coordinates of P.

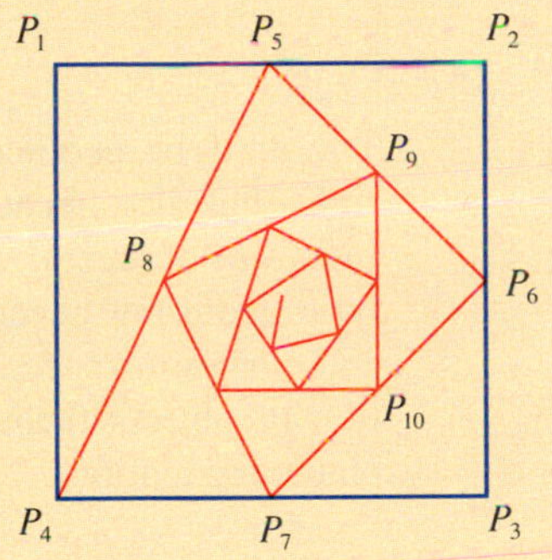

FIGURE FOR PROBLEM 14

15. Consider the series whose terms are the reciprocals of the positive integers that can be written in base 10 notation without using the digit 0. Show that this series is convergent and the sum is less than 90.

CAS **16.** A sequence that arises in ecology as a model for population growth is defined by the **logistic difference equation**

$$p_{n+1} = kp_n(1 - p_n)$$

where p_n measures the size of the population of the nth generation of a single species. To keep the numbers manageable, p_n is a fraction of the maximal size of the population, so $0 \leq p_n \leq 1$. Notice that the form of this equation is similar to the logistic differential equation in Section 9.1. The discrete model—with sequences instead of continuous functions—is preferable for modeling insect populations, where mating and death occur in a periodic fashion.

An ecologist is interested in predicting the size of the population as time goes on, and asks the questions: Will it stabilize at a limiting value? Will it change in a cyclical fashion? Or will it exhibit random behavior?

Write a program to compute the first n terms of this sequence starting with an initial population p_0, where $0 < p_0 < 1$. Use this program to do the following.

(a) Calculate 20 or 30 terms of the sequence for $p_0 = \frac{1}{2}$ and for two values of k such that $1 < k < 3$. Graph the sequences. Do they appear to converge? Repeat for a different value of p_0 between 0 and 1. Does the limit depend on the choice of p_0? Does it depend on the choice of k?

(b) Calculate terms of the sequence for a value of k between 3 and 3.4 and plot them. What do you notice about the behavior of the terms?

(c) Experiment with values of k between 3.4 and 3.5. What happens to the terms?

(d) For values of k between 3.6 and 4, compute and plot at least 100 terms and comment on the behavior of the sequence. What happens if you change p_0 by 0.001? This type of behavior is called *chaotic* and is exhibited by insect populations under certain conditions.

11

THREE-DIMENSIONAL ANALYTIC GEOMETRY AND VECTORS

I now demonstrate the frame of the System of the World.
SIR ISAAC NEWTON

The analytic geometry of three-dimensional space is important not only in its own right, but also because it will be needed in the next chapter to study the calculus of functions of several variables. We use vectors in our treatment of three-dimensional analytic geometry because vectors give particularly simple descriptions of lines, planes, and curves in space. We also see that vector functions can be used to describe the motion of objects through space. In particular, we use them to derive Kepler's laws of planetary motion.

11.1 THREE-DIMENSIONAL COORDINATE SYSTEMS

To locate a point in a plane, two numbers are necessary. We know that any point in the plane can be represented as an ordered pair (a, b) of real numbers, where a is the x-coordinate and b is the y-coordinate. For this reason, a plane is called two-dimensional. To locate a point in space, three numbers are required. We represent any point in space by an ordered triple (a, b, c) of real numbers.

In order to represent points in space, we first choose a fixed point O (the origin) and three directed lines through O that are perpendicular to each other, called the **coordinate axes** and labeled the x-axis, y-axis, and z-axis. Usually we think of the x- and y-axes as being horizontal and the z-axis as being vertical, and we draw the orientation of the axes as in Figure 1.

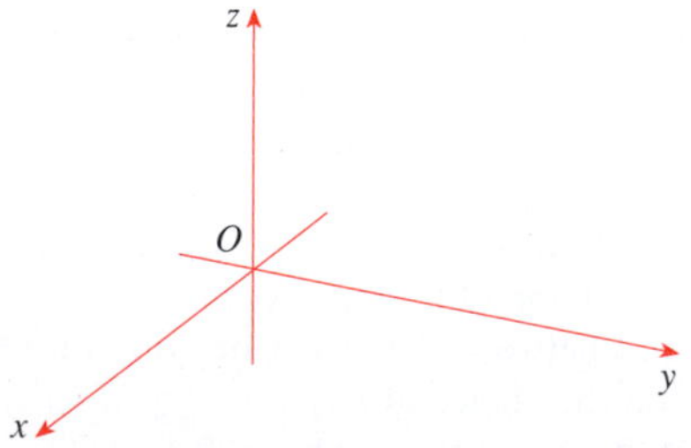

FIGURE 1
Coordinate axes

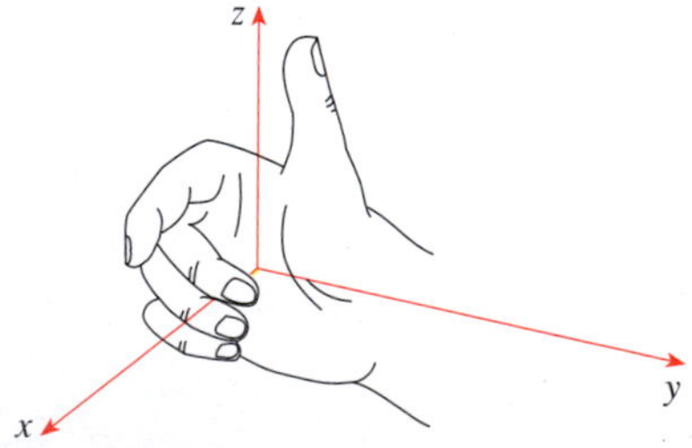

FIGURE 2
Right-hand rule

In looking at Figure 2 you can think of the y- and z-axes as lying in the plane of the paper and the x-axis as coming out of the paper toward you. The direction of the z-axis is determined by the **right-hand rule**: If you curl the fingers of your right hand around the z-axis in the direction of a 90° counterclockwise rotation from the positive x-axis to the positive y-axis, then your thumb points in the positive direction of the z-axis.

The three coordinate axes determine the three **coordinate planes** illustrated in Figure 3(a). The xy-plane is the plane that contains the x- and y-axes; the yz-plane contains the y- and z-axes; the xz-plane contains the x- and z-axes. These three coordinate planes divide space into eight parts, called **octants**. The **first octant**, in the foreground, is determined by the positive axes.

Because many people have some difficulty visualizing diagrams of three-dimensional figures, you may find it helpful to do the following [see Figure 3(b)]. Look at any bottom corner of a room and call the corner the origin. The wall on your left is in the xz-plane, the wall on your right is in the yz-plane, and the floor is in the xy-plane. The x-axis runs along the intersection of the floor and the left wall. The y-axis runs along the intersection of the floor and the right wall. The z-axis runs up from the floor toward the ceiling along the intersection of the two walls. You are situated in the first octant, and you can now imagine seven other rooms situated in the other seven octants (three on the same floor and four on the floor below), all connected by the common corner point O.

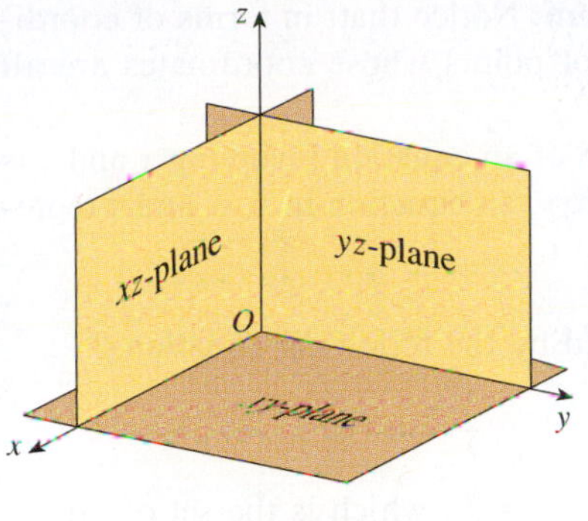

(a) Coordinate planes

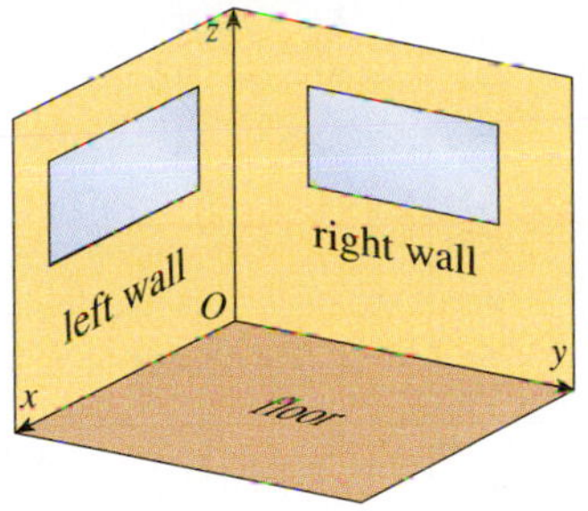

(b)

FIGURE 3

Now if P is any point in space, let a be the (directed) distance from the yz-plane to P, let b be the distance from the xz-plane to P, and let c be the distance from the xy-plane to P. We represent the point P by the ordered triple (a, b, c) of real numbers and we call a, b, and c the **coordinates** of P; a is the x-coordinate, b is the y-coordinate, and c is the z-coordinate. Thus, to locate the point (a, b, c) we can start at the origin O and move a units along the x-axis, then b units parallel to the y-axis, and then c units parallel to the z-axis as in Figure 4.

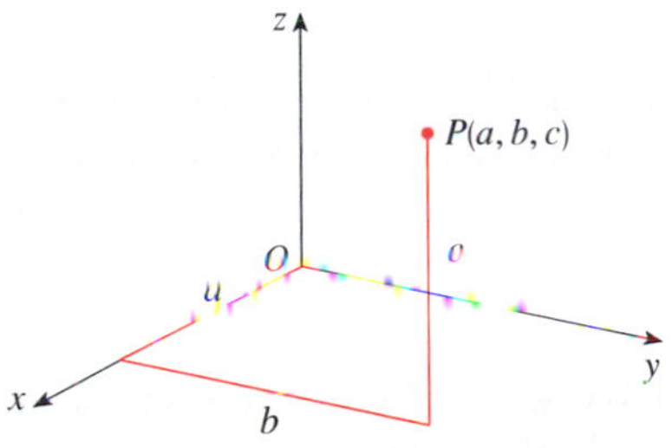

FIGURE 4

The point $P(a, b, c)$ determines a rectangular box as in Figure 5. If we drop a perpendicular from P to the xy-plane, we get a point Q with coordinates $(a, b, 0)$ called the **projection** of P on the xy-plane. Similarly, $R(0, b, c)$ and $S(a, 0, c)$ are the projections of P on the yz-plane and xz-plane, respectively.

As numerical illustrations, the points $(-4, 3, -5)$ and $(3, -2, -6)$ are plotted in Figure 6.

The Cartesian product $\mathbb{R} \times \mathbb{R} \times \mathbb{R} = \{(x, y, z) \mid x, y, z \in \mathbb{R}\}$ is the set of all ordered triples of real numbers and is denoted by $\mathbb{R}^3$. We have given a one-to-one correspondence between points P in space and ordered triples (a, b, c) in $\mathbb{R}^3$. It is called a

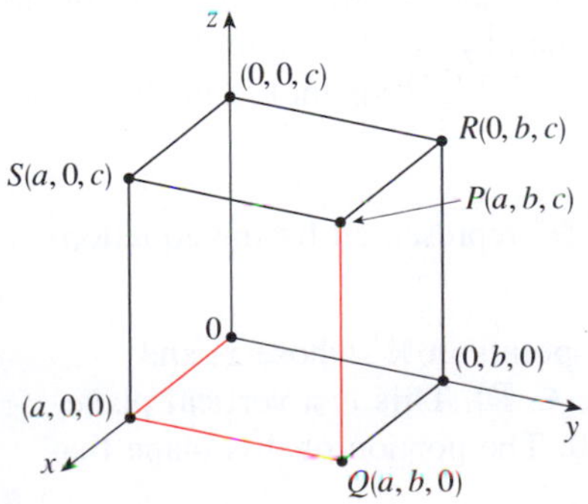

FIGURE 5

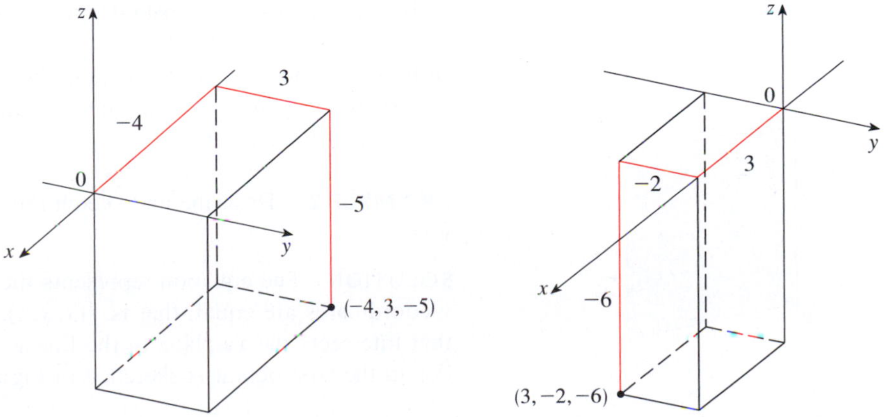

FIGURE 6

three-dimensional rectangular coordinate system. Notice that, in terms of coordinates, the first octant can be described as the set of points whose coordinates are all positive.

In two-dimensional analytic geometry, the graph of an equation involving x and y is a curve in $\mathbb{R}^2$. In three-dimensional analytic geometry, an equation in x, y, and z represents a **surface** in $\mathbb{R}^3$.

EXAMPLE 1 What surfaces in $\mathbb{R}^3$ are represented by the following equations?
(a) $z = 3$ (b) $y = 5$

SOLUTION
(a) The equation $z = 3$ represents the set $\{(x, y, z) \mid z = 3\}$, which is the set of all points in $\mathbb{R}^3$ whose z-coordinate is 3. This is the horizontal plane that is parallel to the xy-plane and three units above it as in Figure 7(a).

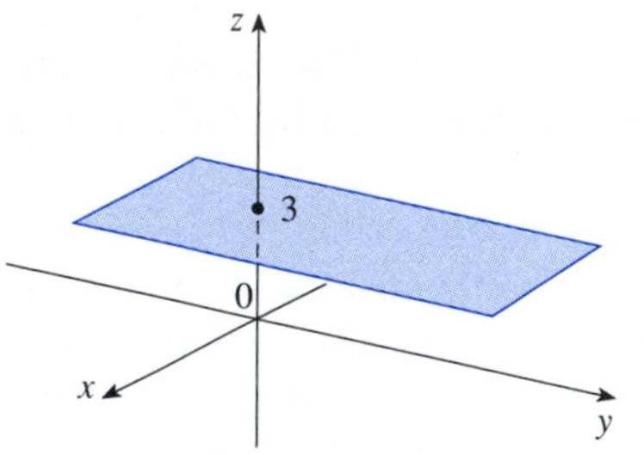

(a) $z = 3$, a plane in $\mathbb{R}^3$

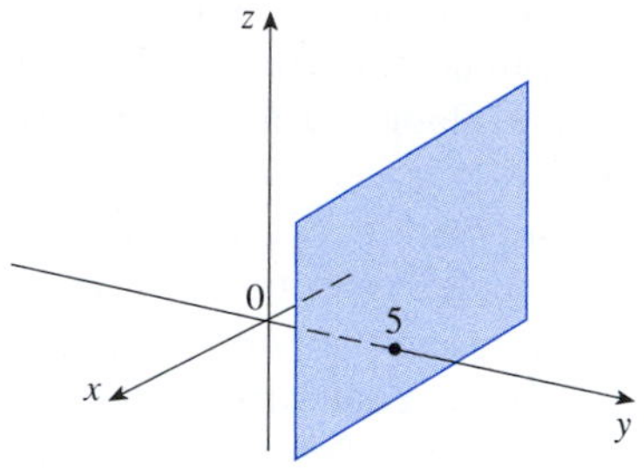

(b) $y = 5$, a plane in $\mathbb{R}^3$

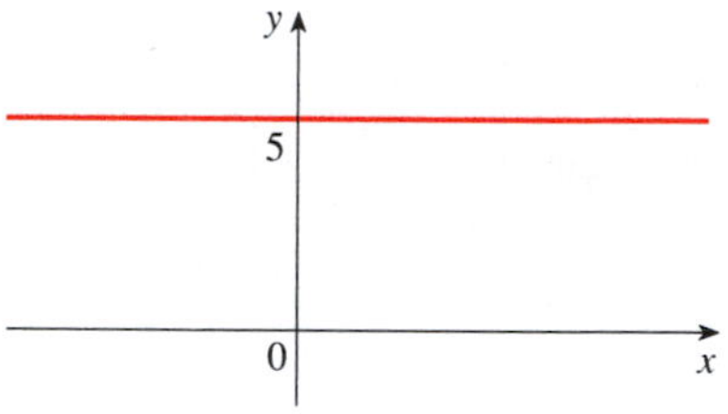

(c) $y = 5$, a line in $\mathbb{R}^2$

FIGURE 7

(b) The equation $y = 5$ represents the set of all points in $\mathbb{R}^3$ whose y-coordinate is 5. This is the vertical plane that is parallel to the xz-plane and five units to the right of it as in Figure 7(b). ■

NOTE: When an equation is given, it must be understood from the context whether it represents a curve in $\mathbb{R}^2$ or a surface in $\mathbb{R}^3$. In Example 1, $y = 5$ represents a plane in $\mathbb{R}^3$, but of course $y = 5$ can also represent a line in $\mathbb{R}^2$ if we are dealing with two-dimensional analytic geometry. See Figure 7(b) and (c).

In general, if k is a constant, then $x = k$ represents a plane parallel to the yz-plane, $y = k$ is a plane parallel to the xz-plane, and $z = k$ is a plane parallel to the xy-plane. In Figure 5, the faces of the rectangular box are formed by the three coordinate planes $x = 0$ (the yz-plane), $y = 0$ (the xz-plane), and $z = 0$ (the xy-plane), and the planes $x = a$, $y = b$, and $z = c$.

EXAMPLE 2 Describe and sketch the surface in $\mathbb{R}^3$ represented by the equation $y = x$.

SOLUTION The equation represents the set of all points in $\mathbb{R}^3$ whose x- and y-coordinates are equal, that is, $\{(x, x, z) \mid x \in \mathbb{R}, z \in \mathbb{R}\}$. This is a vertical plane that intersects the xy-plane in the line $y = x$, $z = 0$. The portion of this plane that lies in the first octant is sketched in Figure 8. ■

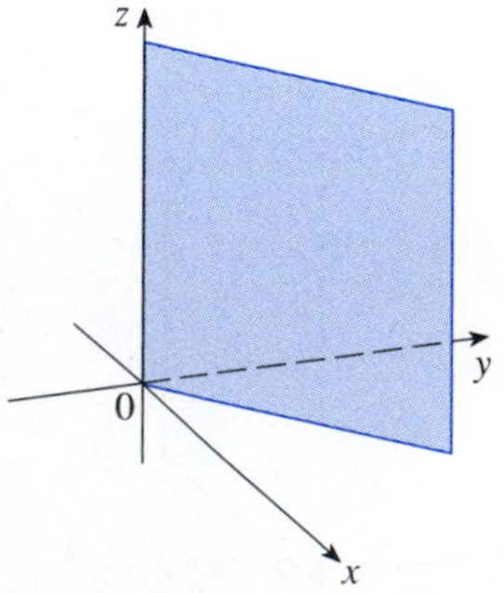

FIGURE 8
The plane $y = x$

The familiar formula for the distance between two points in a plane is easily extended to the following three-dimensional formula.

DISTANCE FORMULA IN THREE DIMENSIONS The distance $|P_1P_2|$ between the points $P_1(x_1, y_1, z_1)$ and $P_2(x_2, y_2, z_2)$ is

$$|P_1P_2| = \sqrt{(x_2 - x_1)^2 + (y_2 - y_1)^2 + (z_2 - z_1)^2}$$

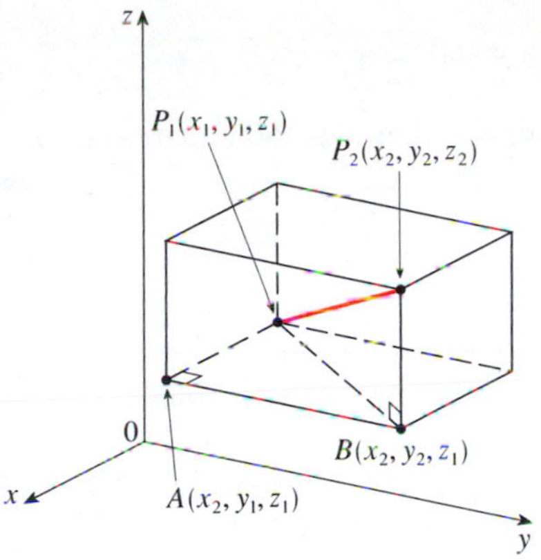

FIGURE 9

PROOF Construct a rectangular box as in Figure 9, where P_1 and P_2 are opposite vertices and the faces of the box are parallel to the coordinate planes. If $A(x_2, y_1, z_1)$ and $B(x_2, y_2, z_1)$ are the vertices of the box indicated in the figure, then

$$|P_1A| = |x_2 - x_1| \qquad |AB| = |y_2 - y_1| \qquad |BP_2| = |z_2 - z_1|$$

Because triangles P_1BP_2 and P_1AB are both right-angled, two applications of the Pythagorean Theorem give

$$|P_1P_2|^2 = |P_1B|^2 + |BP_2|^2 \qquad \text{and} \qquad |P_1B|^2 = |P_1A|^2 + |AB|^2$$

Combining these equations, we get

$$\begin{aligned} |P_1P_2|^2 &= |P_1A|^2 + |AB|^2 + |BP_2|^2 \\ &= |x_2 - x_1|^2 + |y_2 - y_1|^2 + |z_2 - z_1|^2 \\ &= (x_2 - x_1)^2 + (y_2 - y_1)^2 + (z_2 - z_1)^2 \end{aligned}$$

Therefore $\quad |P_1P_2| = \sqrt{(x_2 - x_1)^2 + (y_2 - y_1)^2 + (z_2 - z_1)^2}$ □

EXAMPLE 3 The distance from the point $P(2, -1, 7)$ to the point $Q(1, -3, 5)$ is

$$|PQ| = \sqrt{(1 - 2)^2 + (-3 + 1)^2 + (5 - 7)^2}$$
$$= \sqrt{1 + 4 + 4} = 3$$ ■

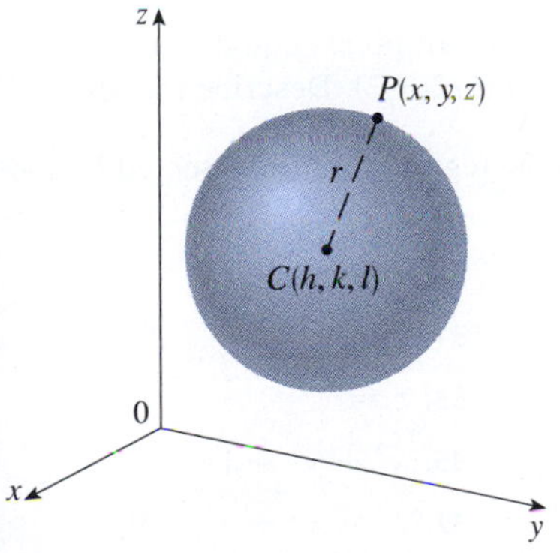

FIGURE 10

EXAMPLE 4 Find an equation of a sphere with radius r and center $C(h, k, l)$.

SOLUTION By definition, a sphere is the set of all points $P(x, y, z)$ whose distance from C is r (see Figure 10). Thus, P is on the sphere if and only if $|PC| = r$. Squaring both sides, we have $|PC|^2 = r^2$ or

$$(x - h)^2 + (y - k)^2 + (z - l)^2 = r^2$$ ■

This result is worth remembering:

EQUATION OF A SPHERE An equation of a sphere with center $C(h, k, l)$ and radius r is

$$(x - h)^2 + (y - k)^2 + (z - l)^2 = r^2$$

In particular, if the center is the origin O, then an equation of the sphere is

$$x^2 + y^2 + z^2 = r^2$$

EXAMPLE 5 Show that $x^2 + y^2 + z^2 + 4x - 6y + 2z + 6 = 0$ is the equation of a sphere, and find its center and radius.

SOLUTION We can rewrite the given equation in the form of an equation of a sphere if we complete squares:

$$(x^2 + 4x + 4) + (y^2 - 6y + 9) + (z^2 + 2z + 1) = -6 + 4 + 9 + 1$$

$$(x + 2)^2 + (y - 3)^2 + (z + 1)^2 = 8$$

Comparing this equation with the standard form, we see that it is the equation of a sphere with center $(-2, 3, -1)$ and radius $\sqrt{8} = 2\sqrt{2}$. ■

EXERCISES 11.1

1–4 ■ Draw a rectangular box that has P and Q as opposite vertices and has its faces parallel to the coordinate planes. Then find (a) the coordinates of the other six vertices of the box and (b) the length of the diagonal of the box.

1. $P(0, 0, 0)$, $Q(2, 3, 5)$
2. $P(0, 0, 0)$, $Q(-4, -1, 2)$
3. $P(1, 1, 2)$, $Q(3, 4, 5)$
4. $P(4, 3, 0)$, $Q(1, 6, -4)$

5–8 ■ Find the lengths of the sides of the triangle ABC and determine whether the triangle is isosceles, a right triangle, both, or neither.

5. $A(2, 1, 0)$, $B(3, 3, 4)$, $C(5, 4, 3)$
6. $A(5, 5, 1)$, $B(3, 3, 2)$, $C(1, 4, 4)$
7. $A(-2, 6, 1)$, $B(5, 4, -3)$, $C(2, -6, 4)$
8. $A(3, -4, 1)$, $B(5, -3, 0)$, $C(6, -7, 4)$

9–10 ■ Determine whether the given points are collinear.

9. $P(1, 2, 3)$, $Q(0, 3, 7)$, $R(3, 5, 11)$
10. $K(0, 3, -4)$, $L(1, 2, -2)$, $M(3, 0, 1)$

11–14 ■ Find the equation of the sphere with center C and radius r.

11. $C(0, 1, -1)$, $r = 4$
12. $C(-1, 2, 4)$, $r = \frac{1}{2}$
13. $C(-6, -1, 2)$, $r = 2\sqrt{3}$
14. $C(1, 2, -3)$, $r = 7$

15–20 ■ Show that the given equation represents a sphere, and find its center and radius.

15. $x^2 + y^2 + z^2 + 2x + 8y - 4z = 28$
16. $x^2 + y^2 + z^2 = 6x + 4y + 10z$
17. $x^2 + y^2 + z^2 + x - 2y + 6z - 2 = 0$
18. $2x^2 + 2y^2 + 2z^2 + 4y - 2z = 1$
19. $x^2 + y^2 + z^2 = x$
20. $x^2 + y^2 + z^2 + ax + by + cz + d = 0$, where $a^2 + b^2 + c^2 > 4d$

21. Prove that the midpoint of the line segment from $P_1(x_1, y_1, z_1)$ to $P_2(x_2, y_2, z_2)$ is

$$\left(\frac{x_1 + x_2}{2}, \frac{y_1 + y_2}{2}, \frac{z_1 + z_2}{2}\right)$$

22. Find an equation of a sphere if one of its diameters has endpoints $(2, 1, 4)$ and $(4, 3, 10)$.

23. Find the lengths of the medians of the triangle with vertices $A(1, 2, 3)$, $B(-2, 0, 5)$, and $C(4, 1, 5)$.

24. Find an equation of the sphere that has center $(1, 2, 3)$ and passes through the point $(-1, 1, -2)$.

25. Find equations of the spheres with center $(2, -3, 6)$ that touch (a) the xy-plane, (b) the yz-plane, (c) the xz-plane.

26. Consider the points P such that the distance from P to $A(-1, 5, 3)$ is twice the distance from P to $B(6, 2, -2)$. Show that the set of all such points is a sphere, and find its center and radius.

27. Find an equation of the set of all points equidistant from the points $A(-1, 5, 3)$ and $B(6, 2, -2)$. Describe the set.

28–43 ■ Describe in words the region of $\mathbb{R}^3$ represented by the equation or inequality.

28. $x = 0$
29. $x = 9$
30. $z = -8$
31. $y > 2$
32. $z \leqslant 0$
33. $z = x$
34. $y = z$
35. $x^2 + y^2 = 1$
36. $y^2 + z^2 \leqslant 4$
37. $x^2 + y^2 + z^2 > 1$
38. $1 \leqslant x^2 + y^2 + z^2 \leqslant 25$
39. $x^2 + y^2 + z^2 - 2z < 3$
40. $xy = 0$
41. $xy = 1$
42. $xyz = 0$
43. $|z| \leqslant 2$

44–47 ■ Write inequalities to describe the given region.

44. The solid rectangular box in the first octant bounded by the planes $x = 1$, $y = 2$, and $z = 3$

45. The half-space consisting of all points to the left of the xz-plane

46. The solid upper hemisphere of the sphere of radius 2 centered at the origin

47. The region consisting of all points between (but not on) the spheres of radius r and R centered at the origin, where $r < R$

48. Use a computer with three-dimensional graphing software to graph the sphere with center the origin and radius 1. (You may need to graph the upper and lower hemispheres separately.)

49. The figure shows a line L_1 in space and a second line L_2, which is the projection of L_1 on the xy-plane. (In other words, the points on L_2 are directly beneath, or above, the points on L_1.)

(a) Find the coordinates of the point P.

(b) Locate on the diagram the points A, B, and C, where the line L_1 intersects the xy-plane, the yz-plane, and the xz-plane, respectively.

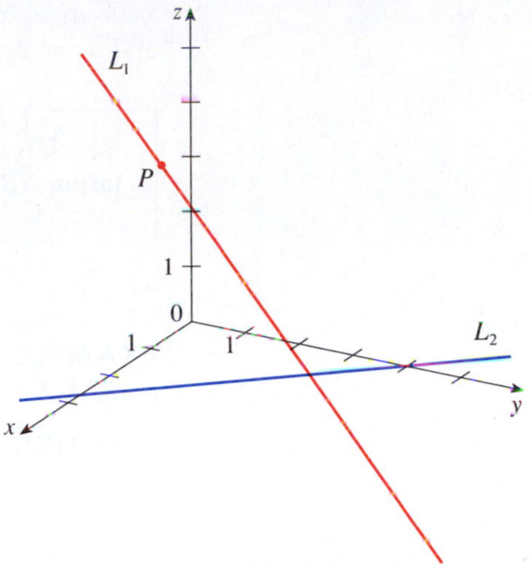

50. Find the volume of the intersection of the spheres $x^2 + y^2 + z^2 + 4x - 2y + 4z + 5 = 0$ and $x^2 + y^2 + z^2 = 4$.

11.2 VECTORS AND THE DOT PRODUCT IN THREE DIMENSIONS

In Section 1.1 we defined the idea of a two-dimensional vector. In this section we extend this concept to three dimensions.

Geometrically, a three-dimensional vector can be considered as an arrow with both a length and a direction. An arrow is a directed line segment with a starting point and an ending point. Algebraically, a three-dimensional vector is an ordered triple of numbers.

(1) DEFINITION A **three-dimensional vector** is an ordered triple $\mathbf{a} = \langle a_1, a_2, a_3 \rangle$ of real numbers. The numbers a_1, a_2, and a_3 are called the **components** of $\mathbf{a}$.

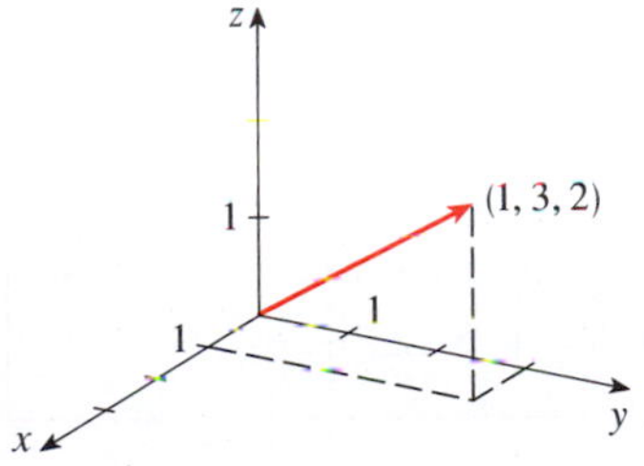

FIGURE 1

For example, the vector $\mathbf{a} = \langle 1, 3, 2 \rangle$ is the vector with components 1, 3, and 2. If the starting point of this vector is the origin $(0, 0, 0)$, then the vector $\mathbf{a} = \langle 1, 3, 2 \rangle$ is the arrow that stretches from the origin to the point $(1, 3, 2)$, as shown in Figure 1.

The starting point of a vector does not have to be the origin. If the vector $\mathbf{a} = \langle a_1, a_2, a_3 \rangle$ starts at the point $A(x, y, z)$, then this vector can be identified as the directed line segment from $A(x, y, z)$ to the point $B(x + a_1, y + a_2, z + a_3)$, as illustrated in Figure 2. This line segment is called a **representation** of the vector $\mathbf{a}$. A particular representation of $\mathbf{a}$ is the directed line segment $\overrightarrow{OP}$ from the origin O to the point $P(a_1, a_2, a_3)$. In this case $\langle a_1, a_2, a_3 \rangle$ is called the **position vector** of the point $P(a_1, a_2, a_3)$ (see Figure 2).

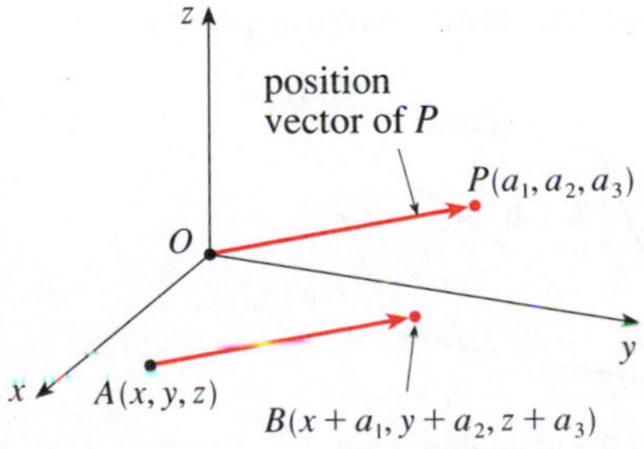

FIGURE 2
Representations of $\mathbf{a} = \langle a_1, a_2, a_3 \rangle$

For example, the position vector of the point $(1, 3, 2)$ is the vector $\langle 1, 3, 2 \rangle$, which is viewed as the directed line segment from $(0, 0, 0)$ to the point $(1, 3, 2)$. If the starting point of the vector $\mathbf{a} = \langle 1, 3, 2 \rangle$ is $(-1, 2, 4)$, then this vector is the directed line segment from $(-1, 2, 4)$ to $(0, 5, 6)$.

Observe that if $\mathbf{a} = \langle a_1, a_2, a_3 \rangle$ is a vector that has the representation $\overrightarrow{AB}$, where the initial point is $A(x_1, y_1, z_1)$ and the terminal point is $B(x_2, y_2, z_2)$, then we must have

$x_1 + a_1 = x_2$, $y_1 + a_2 = y_2$, and $z_1 + a_3 = z_2$ and so $a_1 = x_2 - x_1$, $a_2 = y_2 - y_1$, and $a_3 = z_2 - z_1$. Thus, we have the following:

(2) Given the points $A(x_1, y_1, z_1)$ and $B(x_2, y_2, z_2)$, the vector **a** with representation $\overrightarrow{AB}$ is

$$\mathbf{a} = \langle x_2 - x_1, y_2 - y_1, z_2 - z_1 \rangle$$

EXAMPLE 1 Find the vector represented by the directed line segment with initial point $A(2, -3, 4)$ and terminal point $B(-2, 1, 1)$.

SOLUTION By (2), the vector corresponding to $\overrightarrow{AB}$ is

$$\mathbf{a} = \langle -2 - 2, 1 - (-3), 1 - 4 \rangle = \langle -4, 4, -3 \rangle$$

■

The **magnitude** or **length** of the vector **v** is the length of any of its representations and is denoted by the symbol $|\mathbf{v}|$ or $\|\mathbf{v}\|$. By using the distance formula to compute the length of a segment $\overrightarrow{OP}$, we obtain the following:

(3) The length of the three-dimensional vector $\mathbf{a} = \langle a_1, a_2, a_3 \rangle$ is

$$|\mathbf{a}| = \sqrt{a_1^2 + a_2^2 + a_3^2}$$

The only vector with length 0 is the **zero vector** $\mathbf{0} = \langle 0, 0, 0 \rangle$. This vector is also the only vector with no specific direction.

According to the following definition, we add vectors by adding the corresponding components of two vectors.

(4) VECTOR ADDITION If $\mathbf{a} = \langle a_1, a_2, a_3 \rangle$ and $\mathbf{b} = \langle b_1, b_2, b_3 \rangle$, then the vector $\mathbf{a} + \mathbf{b}$ is defined by

$$\mathbf{a} + \mathbf{b} = \langle a_1, a_2, a_3 \rangle + \langle b_1, b_2, b_3 \rangle = \langle a_1 + b_1, a_2 + b_2, a_3 + b_3 \rangle$$

Vector addition is illustrated geometrically in Figure 3. As with vector addition in two dimensions, the vector $\mathbf{a} + \mathbf{b}$ can be constructed either as the third side of a triangle with sides **a** and **b** (if **b** starts where **a** ends) or as the diagonal of the parallelogram formed by the vectors **a** and **b** (if **a** and **b** have the same starting point).

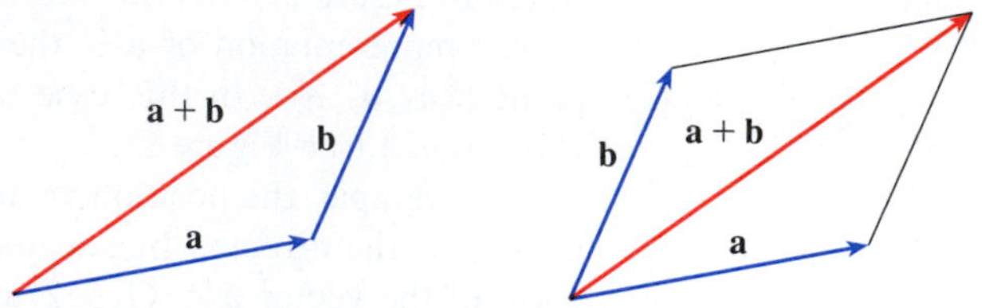

FIGURE 3 (a) Triangle Law (b) Parallelogram Law

Scalar multiplication is defined as follows.

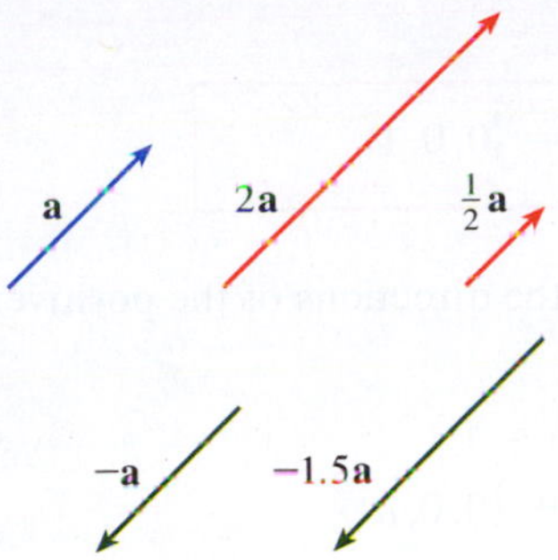

FIGURE 4
Scalar multiples of **a**

(5) MULTIPLICATION OF A VECTOR BY A SCALAR If c is a scalar and $\mathbf{a} = \langle a_1, a_2, a_3 \rangle$, then the vector $c\mathbf{a}$ is defined by

$$c\langle a_1, a_2, a_3 \rangle = \langle ca_1, ca_2, ca_3 \rangle$$

As with the two-dimensional case, scalar multiplication of the vector **a** by a scalar c has the geometric effect of stretching the vector **a** by the factor $|c|$. If $c > 0$, then the vectors **a** and $c\mathbf{a}$ point in the same direction. If $c < 0$, then **a** and $c\mathbf{a}$ point in opposite directions. Figure 4 illustrates a vector multiplied by the scalars 2, $\frac{1}{2}$, -1, and -1.5. Two vectors **a** and **b** are said to be parallel if $\mathbf{b} = c\mathbf{a}$ for some scalar c.

The **difference**, $\mathbf{a} - \mathbf{b}$, of two vectors is defined as

$$\mathbf{a} - \mathbf{b} = \mathbf{a} + (-\mathbf{b})$$

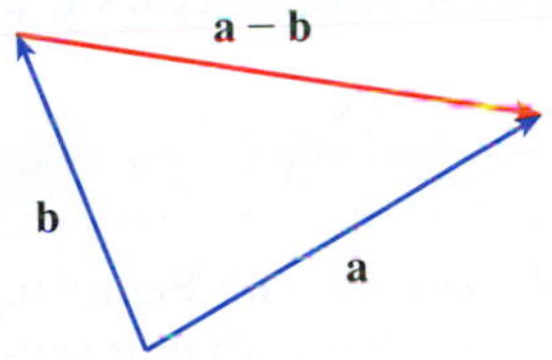

FIGURE 5

If $\mathbf{a} = \langle a_1, a_2, a_3 \rangle$ and $\mathbf{b} = \langle b_1, b_2, b_3 \rangle$, then

$$\mathbf{a} - \mathbf{b} = \langle a_1 - b_1, a_2 - b_2, a_3 - b_3 \rangle$$

As with the two-dimensional case, the vector $\mathbf{a} - \mathbf{b}$ can be geometrically described by the directed line segment starting at the endpoint of **b** and ending at the endpoint of **a**, as illustrated in Figure 5.

EXAMPLE 2 If $\mathbf{a} = \langle 4, 0, 3 \rangle$ and $\mathbf{b} = \langle -2, 1, 5 \rangle$, find $|\mathbf{a}|$ and the vectors $\mathbf{a} + \mathbf{b}$, $\mathbf{a} - \mathbf{b}$, $3\mathbf{b}$, and $2\mathbf{a} + 5\mathbf{b}$.

SOLUTION

$$|\mathbf{a}| = \sqrt{4^2 + 0^2 + 3^2} = \sqrt{25} = 5$$

$$\begin{aligned} \mathbf{a} + \mathbf{b} &= \langle 4, 0, 3 \rangle + \langle -2, 1, 5 \rangle \\ &= \langle 4 - 2, 0 + 1, 3 + 5 \rangle = \langle 2, 1, 8 \rangle \end{aligned}$$

$$\begin{aligned} \mathbf{a} - \mathbf{b} &= \langle 4, 0, 3 \rangle - \langle -2, 1, 5 \rangle \\ &= \langle 4 - (-2), 0 - 1, 3 - 5 \rangle = \langle 6, -1, -2 \rangle \end{aligned}$$

$$3\mathbf{b} = 3\langle -2, 1, 5 \rangle = \langle 3(-2), 3(1), 3(5) \rangle = \langle -6, 3, 15 \rangle$$

$$\begin{aligned} 2\mathbf{a} + 5\mathbf{b} &= 2\langle 4, 0, 3 \rangle + 5\langle -2, 1, 5 \rangle \\ &= \langle 8, 0, 6 \rangle + \langle -10, 5, 25 \rangle = \langle -2, 5, 31 \rangle \end{aligned}$$ ■

We denote by V_2 the set of all two-dimensional vectors and by V_3 the set of all three-dimensional vectors. More generally, we will later need to consider the set V_n of all n-dimensional vectors. An n-dimensional vector is an ordered n-tuple:

$$\mathbf{a} = \langle a_1, a_2, \ldots, a_n \rangle$$

where $a_1, a_2, \ldots, a_n$ are real numbers that are called the components of **a**. Addition and scalar multiplication are defined in terms of components just as for the cases $n = 2$ and $n = 3$. All the properties of vectors in Section 1.1 apply to vectors in V_n.

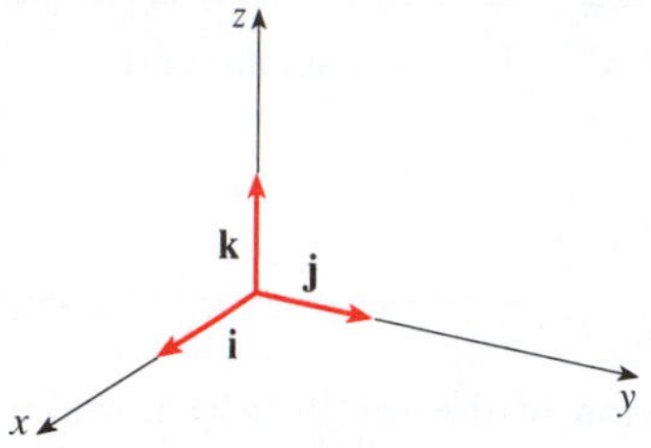

FIGURE 6
Standard basis vectors in V_3

Three vectors in V_3 play a special role. Let

$$\mathbf{i} = \langle 1, 0, 0 \rangle \qquad \mathbf{j} = \langle 0, 1, 0 \rangle \qquad \mathbf{k} = \langle 0, 0, 1 \rangle$$

Then **i**, **j**, and **k** are vectors that have length 1 and point in the directions of the positive x-, y-, and z-axes (see Figure 6).

If $\mathbf{a} = \langle a_1, a_2, a_3 \rangle$, then we can write

$$\mathbf{a} = \langle a_1, a_2, a_3 \rangle = \langle a_1, 0, 0 \rangle + \langle 0, a_2, 0 \rangle + \langle 0, 0, a_3 \rangle$$
$$= a_1\langle 1, 0, 0 \rangle + a_2\langle 0, 1, 0 \rangle + a_3\langle 0, 0, 1 \rangle$$

(6) $$\mathbf{a} = a_1\mathbf{i} + a_2\mathbf{j} + a_3\mathbf{k}$$

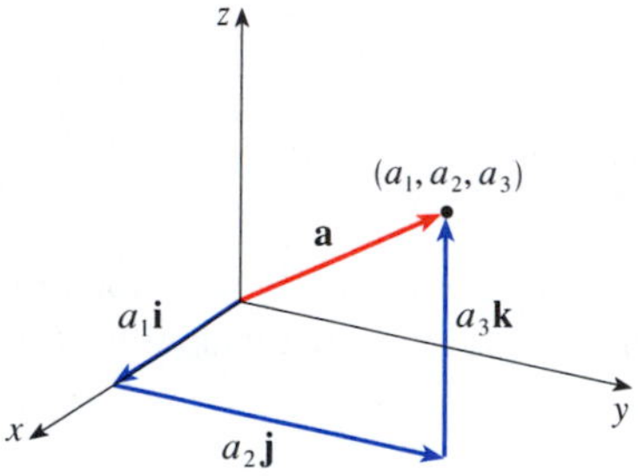

FIGURE 7
$\mathbf{a} = a_1\mathbf{i} + a_2\mathbf{j} + a_3\mathbf{k}$

Thus, any vector in V_3 can be expressed in terms of the **standard basis vectors i, j,** and **k**. For instance,

$$\langle 1, -2, 6 \rangle = \mathbf{i} - 2\mathbf{j} + 6\mathbf{k}$$

See Figure 7 for the geometric interpretation of Equation 6 and compare with Figure 6.

A **unit vector** is a vector whose length is 1. For instance, **i**, **j**, and **k** are all unit vectors. In general, if $\mathbf{a} \neq \mathbf{0}$, then the unit vector that has the same direction as **a** is

(7) $$\mathbf{u} = \frac{1}{|\mathbf{a}|}\mathbf{a} = \frac{\mathbf{a}}{|\mathbf{a}|}$$

In order to verify this, we let $c = 1/|\mathbf{a}|$. Then $\mathbf{u} = c\mathbf{a}$ and c is a positive scalar, so **u** has the same direction as **a**. Also

$$|\mathbf{u}| = |c\mathbf{a}| = |c||\mathbf{a}| = \frac{1}{|\mathbf{a}|}|\mathbf{a}| = 1$$

EXAMPLE 3 Find the unit vector in the direction of the vector $2\mathbf{i} - \mathbf{j} - 2\mathbf{k}$.

SOLUTION The given vector has length

$$|2\mathbf{i} - \mathbf{j} - 2\mathbf{k}| = \sqrt{2^2 + (-1)^2 + (-2)^2} = \sqrt{9} = 3$$

so, by Equation 7, the unit vector with the same direction is

$$\tfrac{1}{3}(2\mathbf{i} - \mathbf{j} - 2\mathbf{k}) = \tfrac{2}{3}\mathbf{i} - \tfrac{1}{3}\mathbf{j} - \tfrac{2}{3}\mathbf{k}$$ ■

THE DOT PRODUCT

In Section 1.2 we defined the dot product for two-dimensional vectors. This definition naturally generalizes to three dimensions as follows.

(8) DEFINITION The **dot product** of two nonzero vectors **a** and **b** is the number

$$\mathbf{a} \cdot \mathbf{b} = |\mathbf{a}||\mathbf{b}|\cos\theta$$

where θ is the angle between the vectors **a** and **b**, $0 \leq \theta \leq \pi$. If either **a** or **b** is **0**, then we define $\mathbf{a} \cdot \mathbf{b} = 0$.

The component formula for the dot product, given in Section 1.2, also generalizes to three dimensions. Its derivation, based on the Law of Cosines, carries over to three dimensions without change.

(9) The **dot product** of $\mathbf{a} = \langle a_1, a_2, a_3 \rangle$ and $\mathbf{b} = \langle b_1, b_2, b_3 \rangle$ is

$$\mathbf{a} \cdot \mathbf{b} = a_1 b_1 + a_2 b_2 + a_3 b_3$$

EXAMPLE 4

$$\langle -1, 7, 4 \rangle \cdot \langle 6, 2, -\tfrac{1}{2} \rangle = (-1)(6) + 7(2) + 4(-\tfrac{1}{2}) = 6$$

$$(\mathbf{i} + 2\mathbf{j} - 3\mathbf{k}) \cdot (2\mathbf{j} - \mathbf{k}) = 1(0) + 2(2) + (-3)(-1) = 7$$ ■

As a special case, the length of the vector $\mathbf{a} = \langle a_1, a_2, a_3 \rangle$ can be computed by the component formula for the dot product:

$$|\mathbf{a}| = \sqrt{a_1^2 + a_2^2 + a_3^2} = \sqrt{\mathbf{a} \cdot \mathbf{a}}$$

All the other properties of the dot product listed in Section 1.2 are valid in three dimensions.

As an immediate consequence of the definition of the dot product, we have the following formula, which is used to compute the angle between two vectors.

(10) If θ is the angle between two nonzero vectors **a** and **b**, then

$$\cos\theta = \frac{\mathbf{a} \cdot \mathbf{b}}{|\mathbf{a}||\mathbf{b}|}$$

EXAMPLE 5 Find the angle between the vectors $\mathbf{a} = \langle 2, 2, -1 \rangle$ and $\mathbf{b} = \langle 5, -3, 2 \rangle$.

SOLUTION Since

$$|\mathbf{a}| = \sqrt{2^2 + 2^2 + (-1)^2} = 3 \qquad \text{and} \qquad |\mathbf{b}| = \sqrt{5^2 + (-3)^2 + 2^2} = \sqrt{38}$$

and since $\mathbf{a} \cdot \mathbf{b} = 2(5) + 2(-3) + (-1)(2) = 2$

we have, from (10),

$$\cos\theta = \frac{\mathbf{a} \cdot \mathbf{b}}{|\mathbf{a}||\mathbf{b}|} = \frac{2}{3\sqrt{38}}$$

So the angle between **a** and **b** is

$$\theta = \cos^{-1}\left(\frac{2}{3\sqrt{38}}\right) \approx 1.46 \quad (\text{or } 84°)$$ ■

Two nonzero vectors **a** and **b** are called **perpendicular** or **orthogonal** if the angle between them is $\theta = \pi/2$. Then Definition 8 gives

$$\mathbf{a} \cdot \mathbf{b} = |\mathbf{a}||\mathbf{b}|\cos(\pi/2) = 0$$

and conversely if $\mathbf{a} \cdot \mathbf{b} = 0$, then $\cos\theta = 0$, so $\theta = \pi/2$. The zero vector $\mathbf{0}$ is considered to be perpendicular to all vectors. Therefore

(11) Two vectors $\mathbf{a}$ and $\mathbf{b}$ are orthogonal if and only if $\mathbf{a} \cdot \mathbf{b} = 0$.

EXAMPLE 6 Show that $2\mathbf{i} + 2\mathbf{j} - \mathbf{k}$ is perpendicular to $5\mathbf{i} - 4\mathbf{j} + 2\mathbf{k}$.

SOLUTION Since

$$(2\mathbf{i} + 2\mathbf{j} - \mathbf{k}) \cdot (5\mathbf{i} - 4\mathbf{j} + 2\mathbf{k}) = 2(5) + 2(-4) + (-1)(2) = 0$$

these vectors are perpendicular by (11). ■

DIRECTION ANGLES AND DIRECTION COSINES

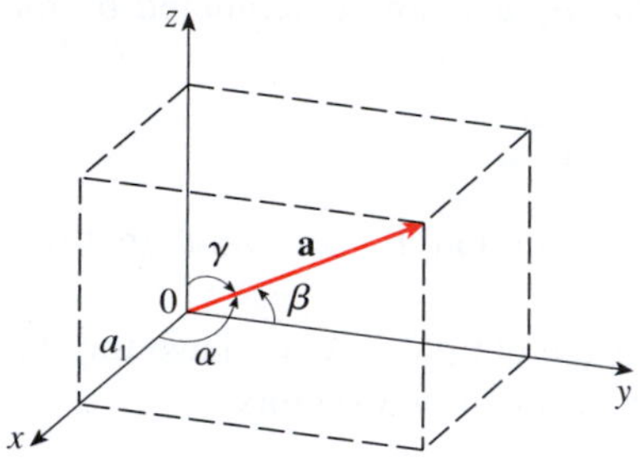

FIGURE 8

The **direction angles** of a nonzero vector $\mathbf{a}$ are the angles α, β, and γ in the interval $[0, \pi]$ that $\mathbf{a}$ makes with the positive x-, y-, and z-axes (see Figure 8).

The cosines of these direction angles, $\cos\alpha$, $\cos\beta$, and $\cos\gamma$, are called the **direction cosines** of the vector $\mathbf{a}$. Using (10) with $\mathbf{b}$ replaced by $\mathbf{i}$, we obtain

(12) $$\cos\alpha = \frac{\mathbf{a} \cdot \mathbf{i}}{|\mathbf{a}||\mathbf{i}|} = \frac{a_1}{|\mathbf{a}|}$$

(This can also be seen directly from Figure 8.) Similarly, we also have

(13) $$\cos\beta = \frac{a_2}{|\mathbf{a}|} \qquad \cos\gamma = \frac{a_3}{|\mathbf{a}|}$$

By squaring the expressions in Equations 12 and 13 and adding, we see that

(14) $$\cos^2\alpha + \cos^2\beta + \cos^2\gamma = 1$$

We can also use Equations 12 and 13 to write

$$\begin{aligned} \mathbf{a} &= \langle a_1, a_2, a_3 \rangle = \langle |\mathbf{a}|\cos\alpha, |\mathbf{a}|\cos\beta, |\mathbf{a}|\cos\gamma \rangle \\ &= |\mathbf{a}|\langle \cos\alpha, \cos\beta, \cos\gamma \rangle \end{aligned}$$

Therefore

(15) $$\frac{1}{|\mathbf{a}|}\mathbf{a} = \langle \cos\alpha, \cos\beta, \cos\gamma \rangle$$

which says that the direction cosines of $\mathbf{a}$ are the components of the unit vector in the direction of $\mathbf{a}$.

EXAMPLE 7 Find the direction angles of the vector $\mathbf{a} = \langle 1, 2, 3 \rangle$.

SOLUTION Since $|\mathbf{a}| = \sqrt{1^2 + 2^2 + 3^2} = \sqrt{14}$, Equations 12 and 13 give

$$\cos\alpha = \frac{1}{\sqrt{14}} \qquad \cos\beta = \frac{2}{\sqrt{14}} \qquad \cos\gamma = \frac{3}{\sqrt{14}}$$

and so

$$\alpha = \cos^{-1}\left(\frac{1}{\sqrt{14}}\right) \approx 74^\circ \qquad \beta = \cos^{-1}\left(\frac{2}{\sqrt{14}}\right) \approx 58^\circ \qquad \gamma = \cos^{-1}\left(\frac{3}{\sqrt{14}}\right) \approx 37^\circ$$

■

PROJECTIONS

Figure 9 shows representations $\overrightarrow{PQ}$ and $\overrightarrow{PR}$ of two vectors **a** and **b** with the same initial point P. If S is the foot of the perpendicular from R to the line containing $\overrightarrow{PQ}$, then the vector with representation $\overrightarrow{PS}$ is called the **vector projection** of **b** onto **a** and is denoted by $\text{proj}_{\mathbf{a}}\,\mathbf{b}$. The **scalar projection** of **b** onto **a** (also called the **component of b along a**) is defined to be the number $|\mathbf{b}|\cos\theta$, where θ is the angle between **a** and **b** (see Figure 10). This is denoted by $\text{comp}_{\mathbf{a}}\,\mathbf{b}$. Observe that it is negative if $\pi/2 < \theta \leqslant \pi$.

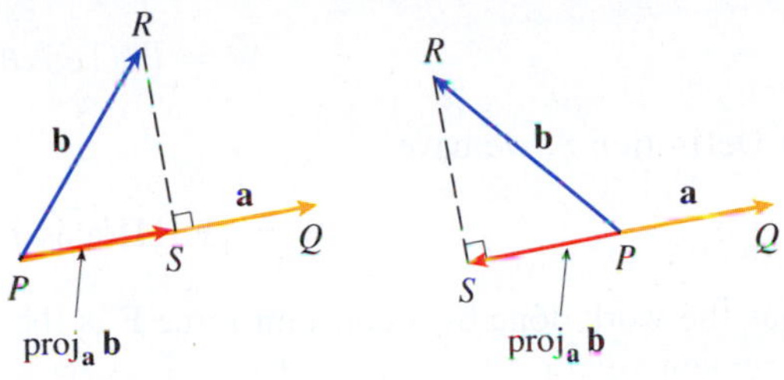

FIGURE 9
Vector projections

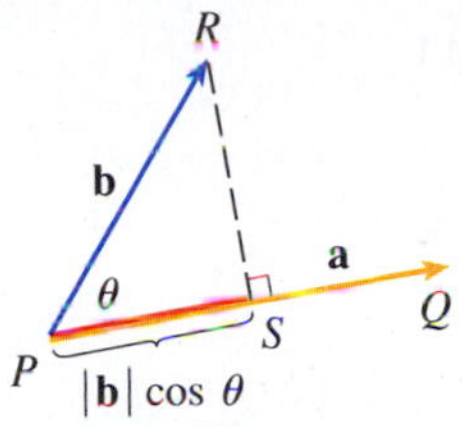

FIGURE 10
Scalar projection

The equation

$$\mathbf{a}\cdot\mathbf{b} = |\mathbf{a}||\mathbf{b}|\cos\theta = |\mathbf{a}|(|\mathbf{b}|\cos\theta)$$

shows that the dot product of **a** and **b** can be interpreted as the length of **a** times the scalar projection of **b** onto **a**. Since

$$|\mathbf{b}|\cos\theta = \frac{\mathbf{a}\cdot\mathbf{b}}{|\mathbf{a}|} = \frac{\mathbf{a}}{|\mathbf{a}|}\cdot\mathbf{b}$$

the component of **b** along **a** can be computed by taking the dot product of **b** with the unit vector in the direction of **a**. To summarize:

Scalar projection of **b** onto **a**: $\quad \text{comp}_{\mathbf{a}}\,\mathbf{b} = \dfrac{\mathbf{a}\cdot\mathbf{b}}{|\mathbf{a}|}$

Vector projection of **b** onto **a**: $\quad \text{proj}_{\mathbf{a}}\,\mathbf{b} = \left(\dfrac{\mathbf{a}\cdot\mathbf{b}}{|\mathbf{a}|}\right)\dfrac{\mathbf{a}}{|\mathbf{a}|} = \dfrac{\mathbf{a}\cdot\mathbf{b}}{|\mathbf{a}|^2}\,\mathbf{a}$

EXAMPLE 8 Find the scalar projection and vector projection of $\mathbf{b} = \langle 1, 1, 2\rangle$ onto $\mathbf{a} = \langle -2, 3, 1\rangle$.

SOLUTION Since $|\mathbf{a}| = \sqrt{(-2)^2 + 3^2 + 1^2} = \sqrt{14}$, the scalar projection of **b** onto **a** is

$$\text{comp}_{\mathbf{a}}\,\mathbf{b} = \frac{\mathbf{a}\cdot\mathbf{b}}{|\mathbf{a}|} = \frac{(-2)(1) + 3(1) + 1(2)}{\sqrt{14}} = \frac{3}{\sqrt{14}}$$

The vector projection is this scalar projection times the unit vector in the direction of **a**:

$$\text{proj}_{\mathbf{a}}\,\mathbf{b} = \frac{3}{\sqrt{14}}\,\frac{\mathbf{a}}{|\mathbf{a}|} = \frac{3}{14}\,\mathbf{a} = \left\langle -\frac{3}{7}, \frac{9}{14}, \frac{3}{14}\right\rangle$$ ■

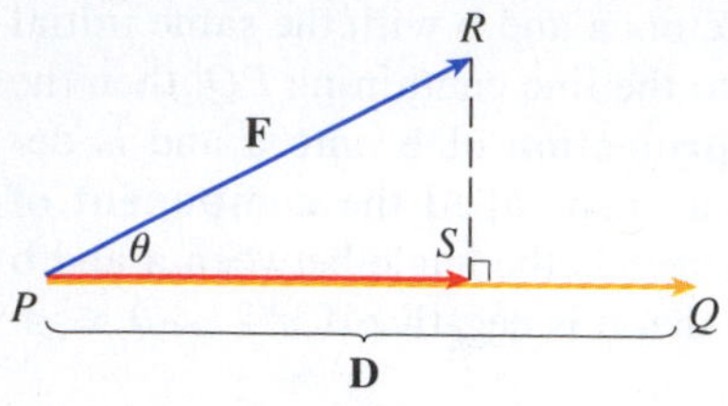

FIGURE 11

One use of projections occurs in physics in calculating work. The work done by a constant force F in moving an object through a distance d is $W = Fd$, but this applies only when the force is directed along the line of motion of the object. Suppose, however, that the constant force is a vector $\mathbf{F} = \overrightarrow{PR}$ pointing in some other direction as in Figure 11. If the force moves the object from P to Q, then the **displacement vector** is $\mathbf{D} = \overrightarrow{PQ}$. As in the two-dimensional case, the work done by this force is defined to be the product of the component of the force along $\mathbf{D}$ and the distance moved:

$$W = (|\mathbf{F}|\cos\theta)|\mathbf{D}|$$

By Definition 8, we have

$$W = |\mathbf{F}||\mathbf{D}|\cos\theta = \mathbf{F}\cdot\mathbf{D} \tag{16}$$

Thus the work done by a constant force $\mathbf{F}$ is the dot product $\mathbf{F}\cdot\mathbf{D}$, where $\mathbf{D}$ is the displacement vector.

EXAMPLE 9 A force is given by a vector $\mathbf{F} = 3\mathbf{i} + 4\mathbf{j} + 5\mathbf{k}$ and moves a particle from the point $P(2, 1, 0)$ to the point $Q(4, 6, 2)$. Find the work done.

SOLUTION The displacement vector is $\mathbf{D} = \overrightarrow{PQ} = \langle 2, 5, 2\rangle$, so by Equation 16, the work done is

$$\begin{aligned} W &= \mathbf{F}\cdot\mathbf{D} = \langle 3, 4, 5\rangle \cdot \langle 2, 5, 2\rangle \\ &= 6 + 20 + 10 = 36 \end{aligned}$$

If the unit of length is meters and the magnitude of the force is measured in newtons, then the work done is 36 joules. ■

EXERCISES 11.2

1. Find the vector represented by the directed line segment with initial point $P(1, 3, 2)$ and terminal point $Q(6, 5, 0)$.

2. Find the vector represented by the directed line segment that starts at the point $(2, -1, 6)$ and ends at $(4, 1, -8)$.

3–4 ■ Find a vector $\mathbf{a}$ with representation given by the directed line segment $\overrightarrow{AB}$. Draw $\overrightarrow{AB}$ and the equivalent representation starting at the origin.

3. $A(0, 3, 1)$, $B(2, 3, -1)$
4. $A(1, -2, 0)$, $B(1, -2, 3)$

5–6 ■ Find the sum of the vectors and illustrate geometrically.

5. $\langle 1, 0, 1\rangle$, $\langle 0, 0, 1\rangle$
6. $\langle 0, 3, 2\rangle$, $\langle 1, 0, -3\rangle$

7–10 ■ Find $|\mathbf{a}|$, $\mathbf{a} + \mathbf{b}$, $\mathbf{a} - \mathbf{b}$, $2\mathbf{a}$, and $3\mathbf{a} + 4\mathbf{b}$.

7. $\mathbf{a} = \langle 2, -3, 6\rangle$, $\mathbf{b} = \langle 1, 1, 4\rangle$
8. $\mathbf{a} = \langle 3, 2, -1\rangle$, $\mathbf{b} = \langle 0, 6, 7\rangle$
9. $\mathbf{a} = \mathbf{i} + \mathbf{j} + \mathbf{k}$, $\mathbf{b} = 2\mathbf{i} - \mathbf{j} + 3\mathbf{k}$
10. $\mathbf{a} = 6\mathbf{i} + \mathbf{k}$, $\mathbf{b} = \mathbf{i} - 2\mathbf{j} + 7\mathbf{k}$

11–14 ■ Find a unit vector that has the same direction as the given vector.

11. $\langle -2, 4, 3\rangle$
12. $\langle 1, -4, 8\rangle$
13. $\mathbf{i} + \mathbf{j}$
14. $2\mathbf{i} - 4\mathbf{j} + 7\mathbf{k}$

15–20 ■ Find $\mathbf{a}\cdot\mathbf{b}$.

15. $|\mathbf{a}| = 2$, $|\mathbf{b}| = 3$, the angle between $\mathbf{a}$ and $\mathbf{b}$ is $\pi/3$
16. $|\mathbf{a}| = 6$, $|\mathbf{b}| = \frac{1}{3}$, the angle between $\mathbf{a}$ and $\mathbf{b}$ is $\pi/4$
17. $\mathbf{a} = \langle 4, 7, -1\rangle$, $\mathbf{b} = \langle -2, 1, 4\rangle$
18. $\mathbf{a} = \langle -1, -2, -3\rangle$, $\mathbf{b} = \langle 2, 8, -6\rangle$
19. $\mathbf{a} = 2\mathbf{i} + 3\mathbf{j} - 4\mathbf{k}$, $\mathbf{b} = \mathbf{i} - 3\mathbf{j} + \mathbf{k}$
20. $\mathbf{a} = \mathbf{i} - \mathbf{k}$, $\mathbf{b} = \mathbf{i} + 2\mathbf{j}$
21. If $\mathbf{a} = \langle a_1, a_2, a_3\rangle$, show that $\mathbf{a}\cdot\mathbf{i} = a_1$, $\mathbf{a}\cdot\mathbf{j} = a_2$, and $\mathbf{a}\cdot\mathbf{k} = a_3$.
22. (a) Show that $\mathbf{i}\cdot\mathbf{j} = \mathbf{j}\cdot\mathbf{k} = \mathbf{k}\cdot\mathbf{i} = 0$.
    (b) Show that $\mathbf{i}\cdot\mathbf{i} = \mathbf{j}\cdot\mathbf{j} = \mathbf{k}\cdot\mathbf{k} = 1$.

23–26 ▪ Find the angle between the vectors. (First find an exact expression and then approximate to the nearest degree.)

23. $\mathbf{a} = \langle 1, 2, 2 \rangle$, $\mathbf{b} = \langle 3, 4, 0 \rangle$

24. $\mathbf{a} = \langle 6, 0, 2 \rangle$, $\mathbf{b} = \langle 5, 3, -2 \rangle$

25. $\mathbf{a} = 6\,\mathbf{i} - 2\,\mathbf{j} - 3\,\mathbf{k}$, $\mathbf{b} = \mathbf{i} + \mathbf{j} + \mathbf{k}$

26. $\mathbf{a} = \mathbf{i} + \mathbf{j} + 2\,\mathbf{k}$, $\mathbf{b} = 2\,\mathbf{j} - 3\,\mathbf{k}$

27–28 ▪ Find, correct to the nearest degree, the three angles of the triangle with the given vertices.

27. $A(1, 2, 3)$, $B(6, 1, 5)$, $C(-1, -2, 0)$

28. $P(0, -1, 6)$, $Q(2, 1, -3)$, $R(5, 4, 2)$

29–32 ▪ Determine whether the given vectors are orthogonal, parallel, or neither.

29. $\mathbf{a} = \langle 2, 8, -3 \rangle$, $\mathbf{b} = \langle -1, 2, 5 \rangle$

30. $\mathbf{a} = \langle -1, 5, 2 \rangle$, $\mathbf{b} = \langle 4, 2, -3 \rangle$

31. $\mathbf{a} = 3\,\mathbf{i} + \mathbf{j} - \mathbf{k}$, $\mathbf{b} = \mathbf{i} - \mathbf{j} + 2\,\mathbf{k}$

32. $\mathbf{a} = 2\,\mathbf{i} + 6\,\mathbf{j} - 4\,\mathbf{k}$, $\mathbf{b} = -3\,\mathbf{i} - 9\,\mathbf{j} + 6\,\mathbf{k}$

33–34 ▪ Find the values of x such that the given vectors are orthogonal.

33. $\langle x, 1, 2 \rangle$, $\langle 3, 4, x \rangle$

34. $\langle x, x, -1 \rangle$, $\langle 1, x, 6 \rangle$

35. Find a unit vector that is orthogonal to both $\mathbf{i} + \mathbf{j}$ and $\mathbf{i} + \mathbf{k}$.

36. For what values of c is the angle between the vectors $\langle 1, 2, 1 \rangle$ and $\langle 1, 0, c \rangle$ equal to $60°$?

37–41 ▪ Find the direction cosines and direction angles of the vector. (Give the direction angles correct to the nearest degree.)

37. $\langle 1, 2, 2 \rangle$

38. $\langle -4, -1, 2 \rangle$

39. $-8\,\mathbf{i} + 3\,\mathbf{j} + 2\,\mathbf{k}$

40. $3\,\mathbf{i} + 5\,\mathbf{j} - 4\,\mathbf{k}$

41. $\langle 2, 1.2, 0.8 \rangle$

42. If a vector has direction angles $\alpha = \pi/4$ and $\beta = \pi/3$, find the third direction angle γ.

43–48 ▪ Find the scalar and vector projections of **b** onto **a**.

43. $\mathbf{a} = \langle 2, 3 \rangle$, $\mathbf{b} = \langle 4, 1 \rangle$

44. $\mathbf{a} = \langle 3, -1 \rangle$, $\mathbf{b} = \langle 2, 3 \rangle$

45. $\mathbf{a} = \langle 4, 2, 0 \rangle$, $\mathbf{b} = \langle 1, 1, 1 \rangle$

46. $\mathbf{a} = \langle -1, -2, 2 \rangle$, $\mathbf{b} = \langle 3, 3, 4 \rangle$

47. $\mathbf{a} = \mathbf{i} + \mathbf{k}$, $\mathbf{b} = \mathbf{i} - \mathbf{j}$

48. $\mathbf{a} = 2\,\mathbf{i} - 3\,\mathbf{j} + \mathbf{k}$, $\mathbf{b} = \mathbf{i} + 6\,\mathbf{j} - 2\,\mathbf{k}$

49. Show that the vector

$$\text{orth}_{\mathbf{a}}\,\mathbf{b} = \mathbf{b} - \text{proj}_{\mathbf{a}}\,\mathbf{b}$$

is orthogonal to **a**. (It is called an **orthogonal projection** of **b**.)

50. For the vectors in Exercise 44, find $\text{orth}_{\mathbf{a}}\,\mathbf{b}$ and illustrate by drawing the vectors **a**, **b**, $\text{proj}_{\mathbf{a}}\,\mathbf{b}$, and $\text{orth}_{\mathbf{a}}\,\mathbf{b}$.

51. If $\mathbf{a} = \langle 3, 0, -1 \rangle$, find a vector **b** such that $\text{comp}_{\mathbf{a}}\,\mathbf{b} = 2$.

52. Suppose that **a** and **b** are nonzero vectors.
(a) Under what circumstances is $\text{comp}_{\mathbf{a}}\,\mathbf{b} = \text{comp}_{\mathbf{b}}\,\mathbf{a}$?
(b) Under what circumstances is $\text{proj}_{\mathbf{a}}\,\mathbf{b} = \text{proj}_{\mathbf{b}}\,\mathbf{a}$?

53. A constant force with vector representation $\mathbf{F} = 10\,\mathbf{i} + 18\,\mathbf{j} - 6\,\mathbf{k}$ moves an object along a straight line from the point $(2, 3, 0)$ to the point $(4, 9, 15)$. Find the work done if the distance is measured in meters and the magnitude of the force is measured in newtons.

54. Find the work done by a force of 20 lb acting in the direction N50°W in moving an object 4 ft due west.

55. A woman exerts a horizontal force of 25 lb on a crate as she pushes it up a ramp that is 10 ft long and inclined at an angle of 20° above the horizontal. Find the work done on the box.

56. A wagon is pulled a distance of 100 m along a horizontal path by a constant force of 50 N. The handle of the wagon is at an angle of 30° above the horizontal. How much work is done?

57. If $\mathbf{r} = \langle x, y, z \rangle$ and $\mathbf{r}_0 = \langle x_0, y_0, z_0 \rangle$, describe the set of all points (x, y, z) such that $|\mathbf{r} - \mathbf{r}_0| = 1$.

58. If $\mathbf{r} = \langle x, y, z \rangle$, $\mathbf{a} = \langle a_1, a_2, a_3 \rangle$, and $\mathbf{b} = \langle b_1, b_2, b_3 \rangle$, show that the vector equation $(\mathbf{r} - \mathbf{a}) \cdot (\mathbf{r} - \mathbf{b}) = 0$ represents a sphere, and find its center and radius.

59. Find the angle between a diagonal of a cube and one of its edges.

60. Find the angle between a diagonal of a cube and a diagonal of one of its faces.

61. A molecule of methane, CH_4, is structured with the four hydrogen atoms at the vertices of a regular tetrahedron and the carbon atom at the centroid. The *bond angle* is the angle formed by the H—C—H combination; it is the angle between the lines that join the carbon atom to two of the hydrogen atoms. Show that the bond angle is about 109.5°. [*Hint:* Take the vertices of the tetrahedron to be the points $(1, 0, 0)$, $(0, 1, 0)$, $(0, 0, 1)$, and $(1, 1, 1)$ as shown in the figure. Then the centroid is $(\frac{1}{2}, \frac{1}{2}, \frac{1}{2})$.]

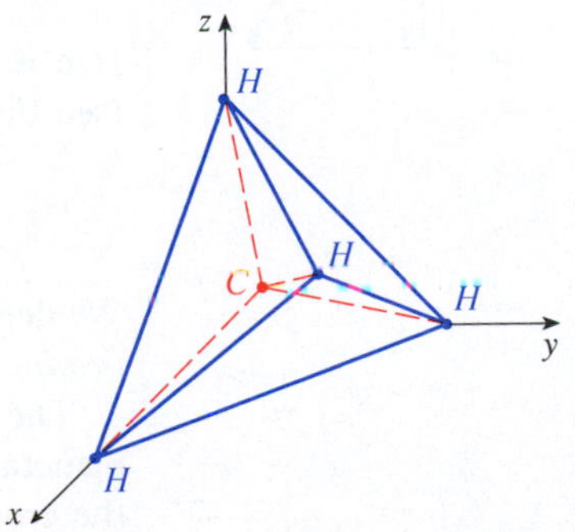

62. Suppose the three coordinate planes are all mirrored and a light ray given by the vector $\mathbf{a} = \langle a_1, a_2, a_3 \rangle$ first strikes the xz-plane, as shown in the figure. Use the fact that the angle of incidence equals the angle of reflection to show that the direction of the reflected ray is given by $\mathbf{b} = \langle a_1, -a_2, a_3 \rangle$. Deduce that, after being reflected by all three mutually perpendicular mirrors, the resulting ray is parallel to the initial ray. (American space scientists used this principle, together with laser beams and an array of corner mirrors on the Moon, to calculate very precisely the distance from Earth to the Moon.)

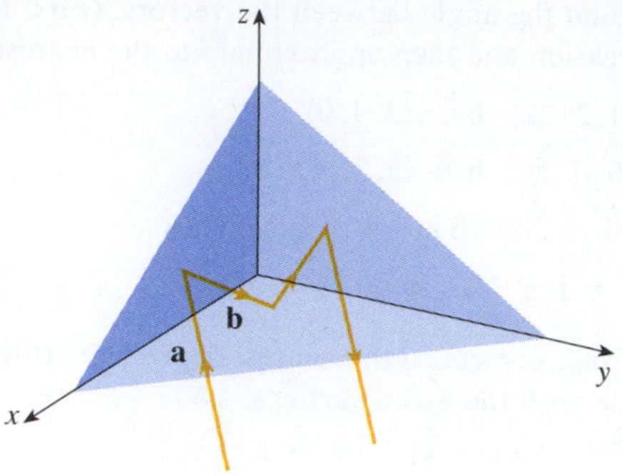

11.3 THE CROSS PRODUCT

The **cross product** $\mathbf{a} \times \mathbf{b}$ of two vectors $\mathbf{a}$ and $\mathbf{b}$, unlike the dot product, is a vector. For this reason it is also called the **vector product**. We will see that $\mathbf{a} \times \mathbf{b}$ is useful in geometry because it is perpendicular to both $\mathbf{a}$ and $\mathbf{b}$. But we introduce this product by looking at a situation where it arises in physics and engineering.

TORQUE AND THE CROSS PRODUCT

If we tighten a bolt by applying a force to a wrench as in Figure 1, we produce a turning effect called a *torque* $\boldsymbol{\tau}$ that acts along the axis of the bolt to move it forward. The magnitude of the torque depends on two things:

- The distance from the axis of the bolt to the point where the force is applied. This is $|\mathbf{r}|$, the length of the position vector $\mathbf{r}$.
- The scalar component of the force $\mathbf{F}$ in the direction perpendicular to $\mathbf{r}$. This is the only component that can cause a rotation and, from Figure 2, we see that it is

$$|\mathbf{F}| \sin \theta$$

where θ is the angle between the vectors $\mathbf{r}$ and $\mathbf{F}$.

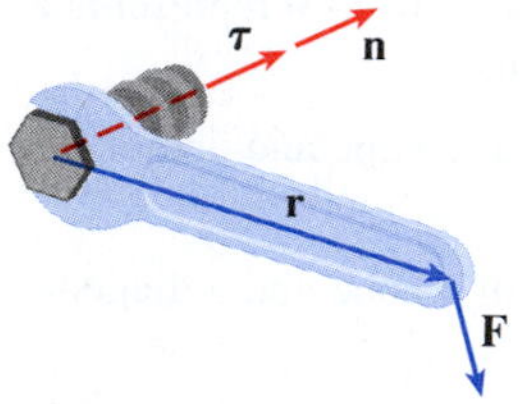

FIGURE 1

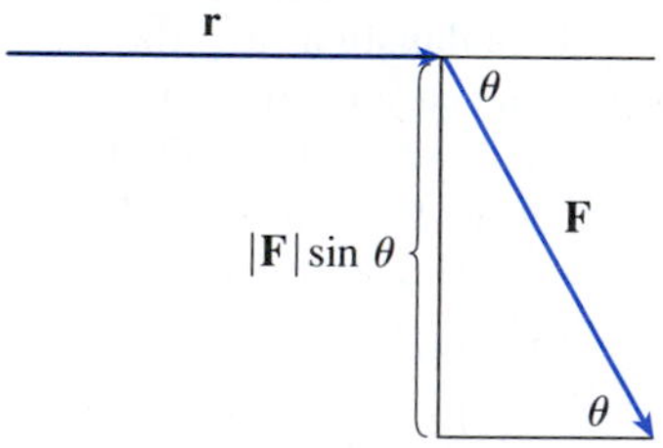

FIGURE 2

We define the magnitude of the torque vector to be the product of these two factors:

$$|\boldsymbol{\tau}| = |\mathbf{r}||\mathbf{F}| \sin \theta$$

If $\mathbf{n}$ is a unit vector that points in the direction in which a right-threaded bolt moves (see Figure 1), we define the **torque** to be the vector

$$\boldsymbol{\tau} = (|\mathbf{r}||\mathbf{F}| \sin \theta)\mathbf{n} \qquad (1)$$

We denote this torque vector by $\boldsymbol{\tau} = \mathbf{r} \times \mathbf{F}$ and we call it the *cross product* or *vector product* of $\mathbf{r}$ and $\mathbf{F}$.

The type of expression in Equation 1 occurs so frequently in the study of fluid flow, planetary motion, and other areas of physics and engineering, that we define and study the cross product of *any* pair of three-dimensional vectors $\mathbf{a}$ and $\mathbf{b}$.

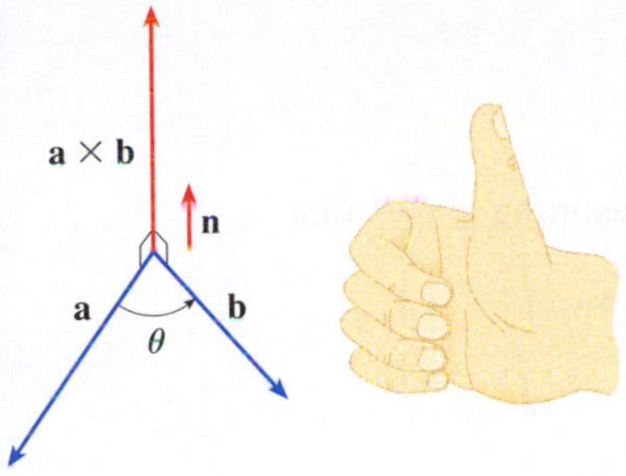

FIGURE 3
The right-hand rule gives the direction of $\mathbf{a} \times \mathbf{b}$.

DEFINITION If $\mathbf{a}$ and $\mathbf{b}$ are nonzero three-dimensional vectors, the **cross product** of $\mathbf{a}$ and $\mathbf{b}$ is the vector

$$\mathbf{a} \times \mathbf{b} = (|\mathbf{a}||\mathbf{b}|\sin\theta)\mathbf{n}$$

where θ is the angle between $\mathbf{a}$ and $\mathbf{b}$ and $\mathbf{n}$ is a unit vector perpendicular to both $\mathbf{a}$ and $\mathbf{b}$ and whose direction is given by the **right-hand rule**: If the fingers of your right hand curl through the angle θ from $\mathbf{a}$ to $\mathbf{b}$, then your thumb points in the direction of $\mathbf{n}$. (See Figure 3.)

If either $\mathbf{a}$ or $\mathbf{b}$ is $\mathbf{0}$, then we define $\mathbf{a} \times \mathbf{b}$ to be $\mathbf{0}$.
Because $\mathbf{a} \times \mathbf{b}$ is a scalar multiple of $\mathbf{n}$, it has the same direction as $\mathbf{n}$ and so

$\mathbf{a} \times \mathbf{b}$ is orthogonal to both $\mathbf{a}$ and $\mathbf{b}$.

Notice that two nonzero vectors $\mathbf{a}$ and $\mathbf{b}$ are parallel if and only if the angle between them is 0 or π. In either case, $\sin\theta = 0$ and so $\mathbf{a} \times \mathbf{b} = \mathbf{0}$.

In particular, any vector $\mathbf{a}$ is parallel to itself, so
$$\mathbf{a} \times \mathbf{a} = \mathbf{0}$$

Two nonzero vectors $\mathbf{a}$ and $\mathbf{b}$ are parallel if and only if $\mathbf{a} \times \mathbf{b} = \mathbf{0}$.

This makes sense in the torque interpretation: If we pull or push the wrench in the direction of its handle (so $\mathbf{F}$ is parallel to $\mathbf{r}$), we produce no torque.

EXAMPLE 1 A bolt is tightened by applying a 40-N force to a 0.25-m wrench as shown in Figure 4. Find the magnitude of the torque about the center of the bolt.

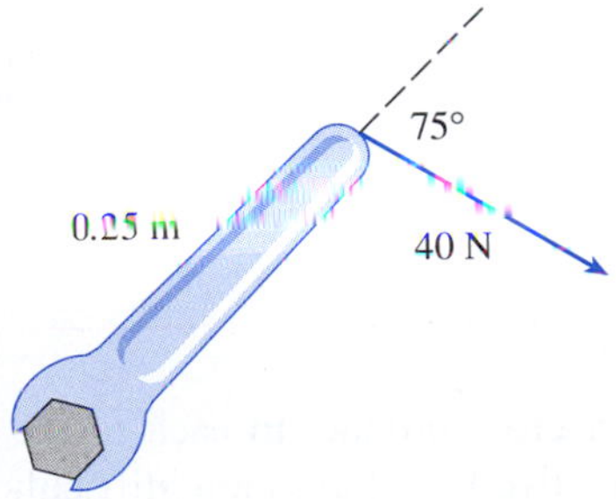

FIGURE 4

SOLUTION The magnitude of the torque vector is

$$|\boldsymbol{\tau}| = |\mathbf{r} \times \mathbf{F}| = |\mathbf{r}||\mathbf{F}|\sin 75°|\mathbf{n}| = (0.25)(40)\sin 75°$$
$$= 10\sin 75° \approx 9.66 \text{ N·m} = 9.66 \text{ J}$$

If the bolt is right-threaded, then the torque vector itself is

$$\boldsymbol{\tau} = |\boldsymbol{\tau}|\mathbf{n} \approx 9.66\mathbf{n}$$

where $\mathbf{n}$ is a unit vector directed down into the page. ■

EXAMPLE 2 Find $\mathbf{i} \times \mathbf{j}$ and $\mathbf{j} \times \mathbf{i}$.

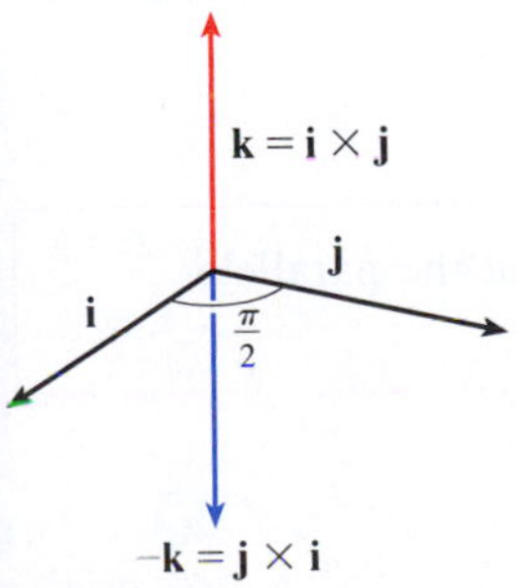

FIGURE 5

SOLUTION The standard basis vectors $\mathbf{i}$ and $\mathbf{j}$ both have length 1 and the angle between them is $\pi/2$. By the right-hand rule, the unit vector perpendicular to $\mathbf{i}$ and $\mathbf{j}$ is $\mathbf{n} = \mathbf{k}$ (see Figure 5), so

$$\mathbf{i} \times \mathbf{j} = (|\mathbf{i}||\mathbf{j}|\sin(\pi/2))\mathbf{k} = \mathbf{k}$$

But if we apply the right-hand rule to the vectors $\mathbf{j}$ and $\mathbf{i}$ (in that order), we see that $\mathbf{n}$ points downward and so $\mathbf{n} = -\mathbf{k}$. Thus

$$\mathbf{j} \times \mathbf{i} = -\mathbf{k}$$ ■

From Example 2 we see that

$$\mathbf{i} \times \mathbf{j} \neq \mathbf{j} \times \mathbf{i}$$

so the cross product is not commutative. Similar reasoning shows that

$$\mathbf{j} \times \mathbf{k} = \mathbf{i} \qquad \mathbf{k} \times \mathbf{j} = -\mathbf{i}$$

$$\mathbf{k} \times \mathbf{i} = \mathbf{j} \qquad \mathbf{i} \times \mathbf{k} = -\mathbf{j}$$

In general, the right-hand rule shows that

$$\mathbf{b} \times \mathbf{a} = -\mathbf{a} \times \mathbf{b}$$

Another algebraic law that fails for the cross product is the associative law for multiplication; that is, in general,

$$(\mathbf{a} \times \mathbf{b}) \times \mathbf{c} \neq \mathbf{a} \times (\mathbf{b} \times \mathbf{c})$$

For instance, if $\mathbf{a} = \mathbf{i}$, $\mathbf{b} = \mathbf{i}$, and $\mathbf{c} = \mathbf{j}$, then

$$(\mathbf{i} \times \mathbf{i}) \times \mathbf{j} = \mathbf{0} \times \mathbf{j} = \mathbf{0}$$

whereas

$$\mathbf{i} \times (\mathbf{i} \times \mathbf{j}) = \mathbf{i} \times \mathbf{k} = -\mathbf{j}$$

However, some of the usual laws of algebra do hold for cross products:

PROPERTIES OF THE CROSS PRODUCT If $\mathbf{a}$, $\mathbf{b}$, and $\mathbf{c}$ are vectors and c is a scalar, then

1. $\mathbf{a} \times \mathbf{b} = -\mathbf{b} \times \mathbf{a}$
2. $(c\mathbf{a}) \times \mathbf{b} = c(\mathbf{a} \times \mathbf{b}) = \mathbf{a} \times (c\mathbf{b})$
3. $\mathbf{a} \times (\mathbf{b} + \mathbf{c}) = \mathbf{a} \times \mathbf{b} + \mathbf{a} \times \mathbf{c}$
4. $(\mathbf{a} + \mathbf{b}) \times \mathbf{c} = \mathbf{a} \times \mathbf{c} + \mathbf{b} \times \mathbf{c}$

Property 2 is proved by applying the definition of a cross product to each of the three expressions. Properties 3 and 4 (the Vector Distributive Laws) are more difficult to establish; we won't do so here.

A geometric interpretation of the length of the cross product can be seen by looking at Figure 6. If $\mathbf{a}$ and $\mathbf{b}$ are represented by directed line segments with the same initial point, then they determine a parallelogram with base $|\mathbf{a}|$, altitude $|\mathbf{b}|\sin\theta$, and area

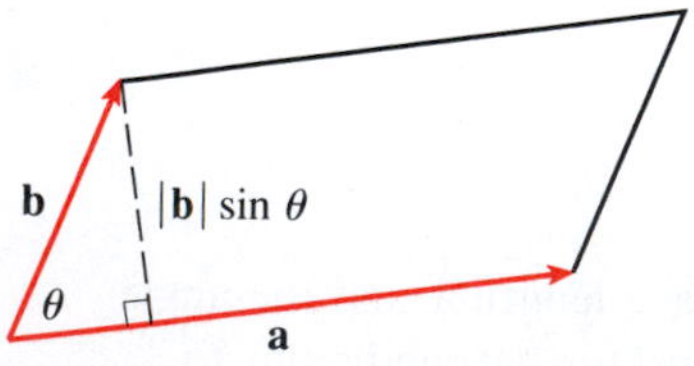

FIGURE 6

$$A = |\mathbf{a}|(|\mathbf{b}|\sin\theta) = |\mathbf{a} \times \mathbf{b}|$$

The length of the cross product $\mathbf{a} \times \mathbf{b}$ is equal to the area of the parallelogram determined by $\mathbf{a}$ and $\mathbf{b}$.

THE CROSS PRODUCT IN COMPONENT FORM

Suppose $\mathbf{a}$ and $\mathbf{b}$ are given in component form:

$$\mathbf{a} = a_1\mathbf{i} + a_2\mathbf{j} + a_3\mathbf{k} \qquad \mathbf{b} = b_1\mathbf{i} + b_2\mathbf{j} + b_3\mathbf{k}$$

We can express $\mathbf{a} \times \mathbf{b}$ in component form by using the Vector Distributive Laws together with the results from Example 2:

$$\begin{aligned} \mathbf{a} \times \mathbf{b} &= (a_1\mathbf{i} + a_2\mathbf{j} + a_3\mathbf{k}) \times (b_1\mathbf{i} + b_2\mathbf{j} + b_3\mathbf{k}) \\ &= a_1b_1\mathbf{i} \times \mathbf{i} + a_1b_2\mathbf{i} \times \mathbf{j} + a_1b_3\mathbf{i} \times \mathbf{k} \\ &\quad + a_2b_1\mathbf{j} \times \mathbf{i} + a_2b_2\mathbf{j} \times \mathbf{j} + a_2b_3\mathbf{j} \times \mathbf{k} \\ &\quad + a_3b_1\mathbf{k} \times \mathbf{i} + a_3b_2\mathbf{k} \times \mathbf{j} + a_3b_3\mathbf{k} \times \mathbf{k} \\ &= a_1b_2\mathbf{k} + a_1b_3(-\mathbf{j}) + a_2b_1(-\mathbf{k}) + a_2b_3\mathbf{i} + a_3b_1\mathbf{j} + a_3b_2(-\mathbf{i}) \\ &= (a_2b_3 - a_3b_2)\mathbf{i} + (a_3b_1 - a_1b_3)\mathbf{j} + (a_1b_2 - a_2b_1)\mathbf{k} \end{aligned}$$

Note that

$\mathbf{i} \times \mathbf{i} = \mathbf{0}$ $\quad \mathbf{j} \times \mathbf{j} = \mathbf{0}$ $\quad \mathbf{k} \times \mathbf{k} = \mathbf{0}$

(2) If $\mathbf{a} = \langle a_1, a_2, a_3 \rangle$ and $\mathbf{b} = \langle b_1, b_2, b_3 \rangle$, then

$$\mathbf{a} \times \mathbf{b} = \langle a_2b_3 - a_3b_2, a_3b_1 - a_1b_3, a_1b_2 - a_2b_1 \rangle$$

In order to make this expresion for $\mathbf{a} \times \mathbf{b}$ easier to remember, we use the notation of determinants. A **determinant of order 2** is defined by

$$\begin{vmatrix} a & b \\ c & d \end{vmatrix} = ad - bc$$

For example,

$$\begin{vmatrix} 2 & 1 \\ -6 & 4 \end{vmatrix} = 2(4) - 1(-6) = 14$$

A **determinant of order 3** can be defined in terms of second-order determinants as follows:

$$\begin{vmatrix} a_1 & a_2 & a_3 \\ b_1 & b_2 & b_3 \\ c_1 & c_2 & c_3 \end{vmatrix} = a_1\begin{vmatrix} b_2 & b_3 \\ c_2 & c_3 \end{vmatrix} - a_2\begin{vmatrix} b_1 & b_3 \\ c_1 & c_3 \end{vmatrix} + a_3\begin{vmatrix} b_1 & b_2 \\ c_1 & c_2 \end{vmatrix} \qquad (3)$$

Observe that each term on the right side of Equation 3 involves a number a_i in the first row of the determinant, and a_i is multiplied by the second-order determinant obtained from the left side by deleting the row and column in which a_i appears. Notice also the minus sign in the second term. For example,

$$\begin{aligned} \begin{vmatrix} 1 & 2 & -1 \\ 3 & 0 & 1 \\ -5 & 4 & 2 \end{vmatrix} &= 1\begin{vmatrix} 0 & 1 \\ 4 & 2 \end{vmatrix} - 2\begin{vmatrix} 3 & 1 \\ -5 & 2 \end{vmatrix} + (-1)\begin{vmatrix} 3 & 0 \\ -5 & 4 \end{vmatrix} \\ &= 1(0 - 4) - 2(6 + 5) + (-1)(12 - 0) = -38 \end{aligned}$$

If we now rewrite (2) using second-order determinants and the standard basis vectors $\mathbf{i}$, $\mathbf{j}$, and $\mathbf{k}$, we see that the cross product of $\mathbf{a} = a_1\mathbf{i} + a_2\mathbf{j} + a_3\mathbf{k}$ and $\mathbf{b} = b_1\mathbf{i} + b_2\mathbf{j} + b_3\mathbf{k}$ is

$$\mathbf{a} \times \mathbf{b} = \begin{vmatrix} a_2 & a_3 \\ b_2 & b_3 \end{vmatrix}\mathbf{i} - \begin{vmatrix} a_1 & a_3 \\ b_1 & b_3 \end{vmatrix}\mathbf{j} + \begin{vmatrix} a_1 & a_2 \\ b_1 & b_2 \end{vmatrix}\mathbf{k} \qquad (4)$$

In view of the similarity between Equations 3 and 4, we often write

$$\textbf{(5)} \qquad \mathbf{a} \times \mathbf{b} = \begin{vmatrix} \mathbf{i} & \mathbf{j} & \mathbf{k} \\ a_1 & a_2 & a_3 \\ b_1 & b_2 & b_3 \end{vmatrix}$$

Although the first row of the symbolic determinant in Equation 5 consists of vectors, if we expand it as if it were an ordinary determinant using the rule in Equation 3, we obtain Equation 4. The symbolic formula in Equation 5 is probably the easiest way of remembering and computing cross products.

EXAMPLE 3 If $\mathbf{a} = \langle 1, 3, 4 \rangle$ and $\mathbf{b} = \langle 2, 7, -5 \rangle$, then

$$\mathbf{a} \times \mathbf{b} = \begin{vmatrix} \mathbf{i} & \mathbf{j} & \mathbf{k} \\ 1 & 3 & 4 \\ 2 & 7 & -5 \end{vmatrix}$$

$$= \begin{vmatrix} 3 & 4 \\ 7 & -5 \end{vmatrix} \mathbf{i} - \begin{vmatrix} 1 & 4 \\ 2 & -5 \end{vmatrix} \mathbf{j} + \begin{vmatrix} 1 & 3 \\ 2 & 7 \end{vmatrix} \mathbf{k}$$

$$= (-15 - 28)\,\mathbf{i} - (-5 - 8)\,\mathbf{j} + (7 - 6)\,\mathbf{k} = -43\,\mathbf{i} + 13\,\mathbf{j} + \mathbf{k}$$ ■

EXAMPLE 4 Find a vector perpendicular to the plane that passes through the points $P(1, 4, 6)$, $Q(-2, 5, -1)$, and $R(1, -1, 1)$.

SOLUTION The vector $\overrightarrow{PQ} \times \overrightarrow{PR}$ is perpendicular to both $\overrightarrow{PQ}$ and $\overrightarrow{PR}$ and is therefore perpendicular to the plane through P, Q, and R. We know from (2) in Section 11.2 that

$$\overrightarrow{PQ} = (-2 - 1)\mathbf{i} + (5 - 4)\mathbf{j} + (-1 - 6)\mathbf{k} = -3\,\mathbf{i} + \mathbf{j} - 7\,\mathbf{k}$$

$$\overrightarrow{PR} = (1 - 1)\mathbf{i} + (-1 - 4)\mathbf{j} + (1 - 6)\mathbf{k} = -5\,\mathbf{j} - 5\,\mathbf{k}$$

We compute the cross product of these vectors:

$$\overrightarrow{PQ} \times \overrightarrow{PR} = \begin{vmatrix} \mathbf{i} & \mathbf{j} & \mathbf{k} \\ -3 & 1 & -7 \\ 0 & -5 & -5 \end{vmatrix}$$

$$= (-5 - 35)\,\mathbf{i} - (15 - 0)\,\mathbf{j} + (15 - 0)\,\mathbf{k} = -40\,\mathbf{i} - 15\,\mathbf{j} + 15\,\mathbf{k}$$

So the vector $\langle -40, -15, 15 \rangle$ is perpendicular to the given plane. Any nonzero scalar multiple of this vector, such as $\langle -8, -3, 3 \rangle$, would also work. ■

EXAMPLE 5 Find the area of the triangle with vertices $P(1, 4, 6)$, $Q(-2, 5, -1)$, and $R(1, -1, 1)$.

SOLUTION In Example 4 we computed that $\overrightarrow{PQ} \times \overrightarrow{PR} = \langle -40, -15, 15 \rangle$. The area of the parallelogram with adjacent sides PQ and PR is the length of this cross product:

$$|\overrightarrow{PQ} \times \overrightarrow{PR}| = \sqrt{(-40)^2 + (-15)^2 + 15^2} = 5\sqrt{82}$$

The area A of triangle PQR is half the area of this parallelogram, that is, $\frac{5}{2}\sqrt{82}$. ■

TRIPLE PRODUCTS

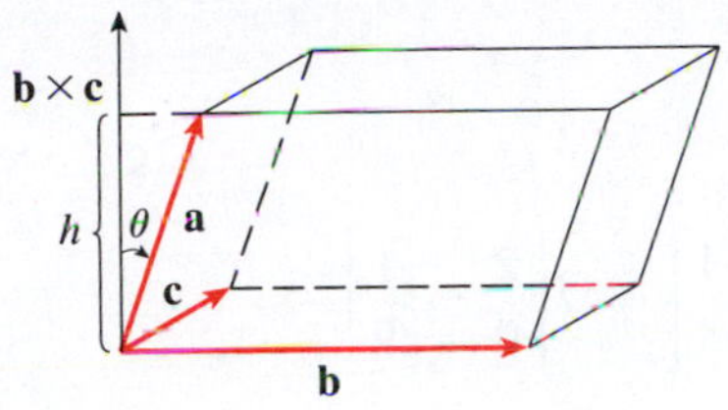

FIGURE 7

The product $\mathbf{a} \cdot (\mathbf{b} \times \mathbf{c})$ is called the **scalar triple product** of the vectors $\mathbf{a}$, $\mathbf{b}$, and $\mathbf{c}$. Its geometric significance can be seen by considering the parallelepiped determined by the vectors $\mathbf{a}$, $\mathbf{b}$, and $\mathbf{c}$ (Figure 7). The area of the base parallelogram is $A = |\mathbf{b} \times \mathbf{c}|$. If θ is the angle between $\mathbf{a}$ and $\mathbf{b} \times \mathbf{c}$, then the height h of the parallelepiped is $h = |\mathbf{a}||\cos\theta|$. (We must use $|\cos\theta|$ instead of $\cos\theta$ in case $\theta > \pi/2$.) Thus, the volume of the parallelepiped is

$$V = Ah = |\mathbf{b} \times \mathbf{c}||\mathbf{a}||\cos\theta| = |\mathbf{a} \cdot (\mathbf{b} \times \mathbf{c})|$$

Therefore, we have proved the following:

> The volume of the parallelepiped determined by the vectors $\mathbf{a}$, $\mathbf{b}$, and $\mathbf{c}$ is the magnitude of their scalar triple product:
>
> $$V = |\mathbf{a} \cdot (\mathbf{b} \times \mathbf{c})|$$

Instead of thinking of the parallelepiped as having its base parallelogram determined by $\mathbf{b}$ and $\mathbf{c}$, we can think of it with base parallelogram determined by $\mathbf{a}$ and $\mathbf{b}$. In this way, we see that

$$\mathbf{a} \cdot (\mathbf{b} \times \mathbf{c}) = \mathbf{c} \cdot (\mathbf{a} \times \mathbf{b})$$

But the dot product is commutative, so we can write

(6) $$\mathbf{a} \cdot (\mathbf{b} \times \mathbf{c}) = (\mathbf{a} \times \mathbf{b}) \cdot \mathbf{c}$$

Suppose that $\mathbf{a}$, $\mathbf{b}$, and $\mathbf{c}$ are given in component form:

$$\mathbf{a} = a_1\mathbf{i} + a_2\mathbf{j} + a_3\mathbf{k} \qquad \mathbf{b} = b_1\mathbf{i} + b_2\mathbf{j} + b_3\mathbf{k} \qquad \mathbf{c} = c_1\mathbf{i} + c_2\mathbf{j} + c_3\mathbf{k}$$

Then

$$\mathbf{a} \cdot (\mathbf{b} \times \mathbf{c}) = \mathbf{a} \cdot \left[\begin{vmatrix} b_2 & b_3 \\ c_2 & c_3 \end{vmatrix} \mathbf{i} - \begin{vmatrix} b_1 & b_3 \\ c_1 & c_3 \end{vmatrix} \mathbf{j} + \begin{vmatrix} b_1 & b_2 \\ c_1 & c_2 \end{vmatrix} \mathbf{k} \right]$$

$$= a_1 \begin{vmatrix} b_2 & b_3 \\ c_2 & c_3 \end{vmatrix} - a_2 \begin{vmatrix} b_1 & b_3 \\ c_1 & c_3 \end{vmatrix} + a_3 \begin{vmatrix} b_1 & b_2 \\ c_1 & c_2 \end{vmatrix}$$

This shows that we can write the scalar triple product of $\mathbf{a}$, $\mathbf{b}$, and $\mathbf{c}$ as the determinant whose rows are the components of these vectors:

(7) $$\mathbf{a} \cdot (\mathbf{b} \times \mathbf{c}) = \begin{vmatrix} a_1 & a_2 & a_3 \\ b_1 & b_2 & b_3 \\ c_1 & c_2 & c_3 \end{vmatrix}$$

EXAMPLE 6 Use the scalar triple product to show that the vectors $\mathbf{a} = \langle 1, 4, -7 \rangle$, $\mathbf{b} = \langle 2, -1, 4 \rangle$, and $\mathbf{c} = \langle 0, -9, 18 \rangle$ are coplanar; that is, they lie in the same plane.

SOLUTION We use Equation 7 to compute their scalar triple product:

$$\mathbf{a}\cdot(\mathbf{b}\times\mathbf{c}) = \begin{vmatrix} 1 & 4 & -7 \\ 2 & -1 & 4 \\ 0 & -9 & 18 \end{vmatrix}$$

$$= 1\begin{vmatrix} -1 & 4 \\ -9 & 18 \end{vmatrix} - 4\begin{vmatrix} 2 & 4 \\ 0 & 18 \end{vmatrix} - 7\begin{vmatrix} 2 & -1 \\ 0 & -9 \end{vmatrix}$$

$$= 1(18) - 4(36) - 7(-18) = 0$$

Therefore, the volume of the parallelepiped determined by **a**, **b**, and **c** is 0. This means that **a**, **b**, and **c** are coplanar. ■

The product $\mathbf{a}\times(\mathbf{b}\times\mathbf{c})$ is called the **vector triple product** of **a**, **b**, and **c**. The proof of the following formula for the vector triple product is left as Exercise 28.

(8)
$$\mathbf{a}\times(\mathbf{b}\times\mathbf{c}) = (\mathbf{a}\cdot\mathbf{c})\mathbf{b} - (\mathbf{a}\cdot\mathbf{b})\mathbf{c}$$

Formula 8 will be used to derive Kepler's First Law of planetary motion in Section 11.8.

EXERCISES 11.3

1. State whether each expression is meaningful. If not, explain why. If so, state whether it is a vector or a scalar.
(a) $\mathbf{a}\cdot(\mathbf{b}\times\mathbf{c})$ (b) $\mathbf{a}\times(\mathbf{b}\cdot\mathbf{c})$
(c) $\mathbf{a}\times(\mathbf{b}\times\mathbf{c})$ (d) $(\mathbf{a}\cdot\mathbf{b})\times\mathbf{c}$
(e) $(\mathbf{a}\cdot\mathbf{b})\times(\mathbf{c}\cdot\mathbf{d})$ (f) $(\mathbf{a}\times\mathbf{b})\cdot(\mathbf{c}\times\mathbf{d})$

2–3 ■ Find $|\mathbf{u}\times\mathbf{v}|$ and determine whether $\mathbf{u}\times\mathbf{v}$ is directed into the page or out of the page.

2.

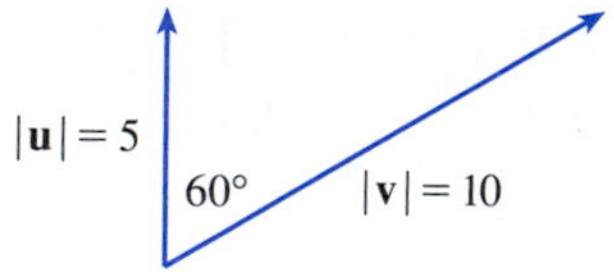

3.

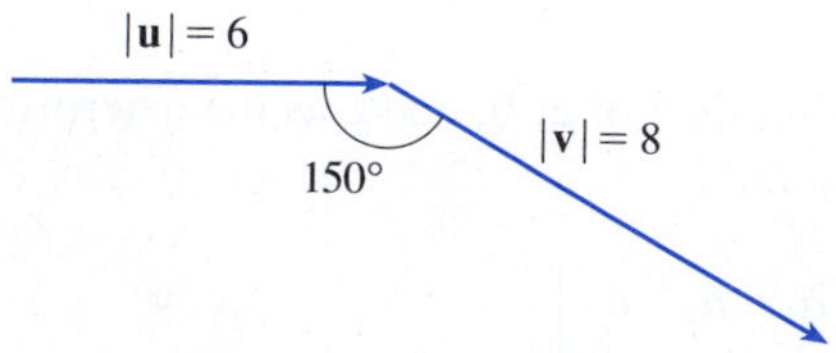

4. The figure shows a vector **a** in the xy-plane and a vector **b** in the direction of **k**. Their lengths are $|\mathbf{a}| = 3$ and $|\mathbf{b}| = 2$.
(a) Find $|\mathbf{a}\times\mathbf{b}|$.
(b) Use the right-hand rule to decide whether the components of $\mathbf{a}\times\mathbf{b}$ are positive, negative, or 0.

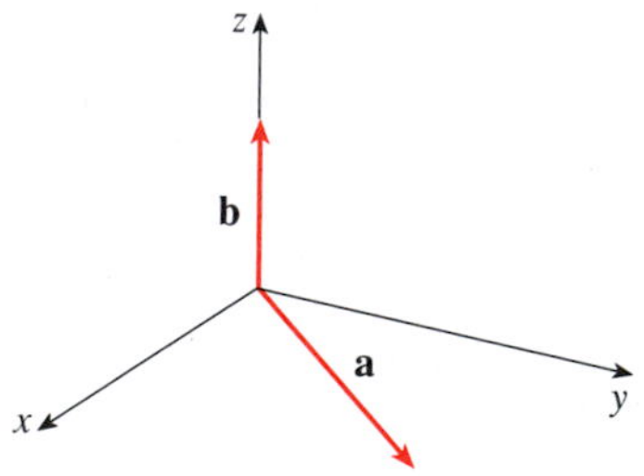

5. A bicycle pedal is pushed by a foot with a 60-N force as shown. The shaft of the pedal is 18 cm long. Find the magnitude of the torque about P.

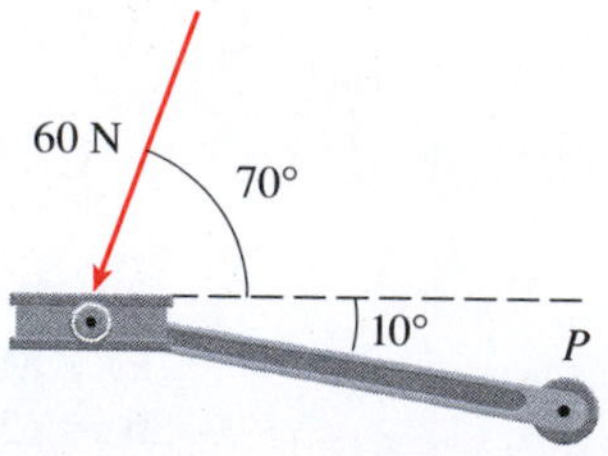

6. Find the magnitude of the torque about P if a 36-lb force is applied as shown.

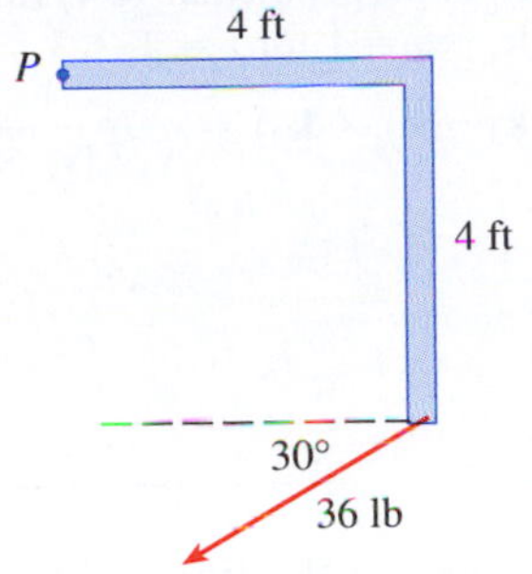

7–9 ■ Find the cross product $\mathbf{a} \times \mathbf{b}$.

7. $\mathbf{a} = \langle -2, 3, 4 \rangle$, $\mathbf{b} = \langle 3, 0, 1 \rangle$

8. $\mathbf{a} = \langle 1, 2, -3 \rangle$, $\mathbf{b} = \langle 5, -1, -2 \rangle$

9. $\mathbf{a} = \mathbf{i} + \mathbf{j} + \mathbf{k}$, $\mathbf{b} = \mathbf{i} + \mathbf{j} - \mathbf{k}$

10. If $\mathbf{a} = \mathbf{i} - 2\mathbf{k}$ and $\mathbf{b} = \mathbf{j} + \mathbf{k}$, find $\mathbf{a} \times \mathbf{b}$. Sketch $\mathbf{a}$, $\mathbf{b}$, and $\mathbf{a} \times \mathbf{b}$ as vectors starting at the origin.

11. Find two unit vectors orthogonal to both $\langle 1, -1, 1 \rangle$ and $\langle 0, 4, 4 \rangle$.

12. Find two unit vectors orthogonal to both $\mathbf{i} + \mathbf{j}$ and $\mathbf{i} - \mathbf{j} + \mathbf{k}$.

13. Find the area of the parallelogram with vertices $A(0, 1)$, $B(3, 0)$, $C(5, -2)$, and $D(2, -1)$.

14. Find the area of the parallelogram with vertices $P(0, 0, 0)$, $Q(5, 0, 0)$, $R(2, 6, 6)$, and $S(7, 6, 6)$.

15–16 ■ (a) Find a vector orthogonal to the plane through the points P, Q, and R, and (b) find the area of triangle PQR.

15. $P(1, 0, 0)$, $Q(0, 2, 0)$, $R(0, 0, 3)$

16. $P(1, 0, -1)$, $Q(2, 4, 5)$, $R(3, 1, 7)$

17. A wrench 30 cm long lies along the positive y-axis and grips a bolt at the origin. A force is applied in the direction $\langle 0, 3, -4 \rangle$ at the end of the wrench. Find the magnitude of the force needed to supply 100 J of torque to the bolt.

18. Let $\mathbf{v} = 5\mathbf{j}$ and let $\mathbf{u}$ be a vector with length 3 that starts at the origin and rotates in the xy-plane. Find the maximum and minimum values of the length of the vector $\mathbf{u} \times \mathbf{v}$. In what direction does $\mathbf{u} \times \mathbf{v}$ point?

19–20 ■ Find the volume of the parallelepiped determined by the vectors $\mathbf{a}$, $\mathbf{b}$, and $\mathbf{c}$.

19. $\mathbf{a} = \langle 1, 0, 6 \rangle$, $\mathbf{b} = \langle 2, 3, -8 \rangle$, $\mathbf{c} = \langle 8, -5, 6 \rangle$

20. $\mathbf{a} = 2\mathbf{i} + 3\mathbf{j} - 2\mathbf{k}$, $\mathbf{b} = \mathbf{i} - \mathbf{j}$, $\mathbf{c} = 2\mathbf{i} + 3\mathbf{k}$

21–22 ■ Find the volume of the parallelepiped with adjacent edges PQ, PR, and PS.

21. $P(1, 1, 1)$, $Q(2, 0, 3)$, $R(4, 1, 7)$, $S(3, -1, -2)$

22. $P(0, 1, 2)$, $Q(2, 4, 5)$, $R(-1, 0, 1)$, $S(6, -1, 4)$

23. Use the scalar triple product to verify that the vectors $\mathbf{a} = 2\mathbf{i} + 3\mathbf{j} + \mathbf{k}$, $\mathbf{b} = \mathbf{i} - \mathbf{j}$, and $\mathbf{c} = 7\mathbf{i} + 3\mathbf{j} + 2\mathbf{k}$ are coplanar.

24. Use the scalar triple product to determine whether the points $P(1, 0, 1)$, $Q(2, 4, 6)$, $R(3, -1, 2)$, and $S(6, 2, 8)$ lie in the same plane.

25. (a) Let P be a point not on the line L that passes through the points Q and R. Show that the distance d from the point P to the line L is

$$d = \frac{|\mathbf{a} \times \mathbf{b}|}{|\mathbf{a}|}$$

where $\mathbf{a} = \overrightarrow{QR}$ and $\mathbf{b} = \overrightarrow{QP}$.

(b) Use the formula in part (a) to find the distance from the point $P(1, 1, 1)$ to the line through $Q(0, 6, 8)$ and $R(-1, 4, 7)$.

26. (a) Let P be a point not on the plane that passes through the points Q, R, and S. Show that the distance d from P to the plane is

$$d = \frac{|\mathbf{a} \cdot (\mathbf{b} \times \mathbf{c})|}{|\mathbf{a} \times \mathbf{b}|}$$

where $\mathbf{a} = \overrightarrow{QR}$, $\mathbf{b} = \overrightarrow{QS}$, and $\mathbf{c} = \overrightarrow{QP}$.

(b) Use the formula in part (a) to find the distance from the point $P(2, 1, 4)$ to the plane through the points $Q(1, 0, 0)$, $R(0, 2, 0)$, and $S(0, 0, 3)$.

27. Prove that $(\mathbf{a} - \mathbf{b}) \times (\mathbf{a} + \mathbf{b}) = 2(\mathbf{a} \times \mathbf{b})$.

28. Prove the following formula for the vector triple product:

$$\mathbf{a} \times (\mathbf{b} \times \mathbf{c}) = (\mathbf{a} \cdot \mathbf{c})\mathbf{b} - (\mathbf{a} \cdot \mathbf{b})\mathbf{c}$$

29. Use Exercise 28 to prove that

$$\mathbf{a} \times (\mathbf{b} \times \mathbf{c}) + \mathbf{b} \times (\mathbf{c} \times \mathbf{a}) + \mathbf{c} \times (\mathbf{a} \times \mathbf{b}) = \mathbf{0}$$

30. Prove that

$$(\mathbf{a} \times \mathbf{b}) \cdot (\mathbf{c} \times \mathbf{d}) = \begin{vmatrix} \mathbf{a} \cdot \mathbf{c} & \mathbf{b} \cdot \mathbf{c} \\ \mathbf{a} \cdot \mathbf{d} & \mathbf{b} \cdot \mathbf{d} \end{vmatrix}$$

31. Suppose that $\mathbf{a} \neq \mathbf{0}$.
(a) If $\mathbf{a} \cdot \mathbf{b} = \mathbf{a} \cdot \mathbf{c}$, does it follow that $\mathbf{b} = \mathbf{c}$?
(b) If $\mathbf{a} \times \mathbf{b} = \mathbf{a} \times \mathbf{c}$, does it follow that $\mathbf{b} = \mathbf{c}$?
(c) If $\mathbf{a} \cdot \mathbf{b} = \mathbf{a} \cdot \mathbf{c}$ and $\mathbf{a} \times \mathbf{b} = \mathbf{a} \times \mathbf{c}$, does it follow that $\mathbf{b} = \mathbf{c}$?

32. If $\mathbf{v}_1$, $\mathbf{v}_2$, and $\mathbf{v}_3$ are noncoplanar vectors, let

$$\mathbf{k}_1 = \frac{\mathbf{v}_2 \times \mathbf{v}_3}{\mathbf{v}_1 \cdot (\mathbf{v}_2 \times \mathbf{v}_3)}$$

$$\mathbf{k}_2 = \frac{\mathbf{v}_3 \times \mathbf{v}_1}{\mathbf{v}_1 \cdot (\mathbf{v}_2 \times \mathbf{v}_3)}$$

$$\mathbf{k}_3 = \frac{\mathbf{v}_1 \times \mathbf{v}_2}{\mathbf{v}_1 \cdot (\mathbf{v}_2 \times \mathbf{v}_3)}$$

(These vectors occur in the study of crystallography. Vectors of the form $n_1\mathbf{v}_1 + n_2\mathbf{v}_2 + n_3\mathbf{v}_3$, where each n_i is an integer, form a *lattice* for a crystal. Vectors written similarly in terms of $\mathbf{k}_1$, $\mathbf{k}_2$, and $\mathbf{k}_3$ form the *reciprocal lattice*.)

(a) Show that $\mathbf{k}_i$ is perpendicular to $\mathbf{v}_j$ if $i \neq j$.

(b) Show that $\mathbf{k}_i \cdot \mathbf{v}_i = 1$ for $i = 1, 2, 3$.

(c) Show that $\mathbf{k}_1 \cdot (\mathbf{k}_2 \times \mathbf{k}_3) = \dfrac{1}{\mathbf{v}_1 \cdot (\mathbf{v}_2 \times \mathbf{v}_3)}$.

11.4 EQUATIONS OF LINES AND PLANES

A line in the xy-plane is determined when a point on the line and the direction of the line (its slope or angle of inclination) are given. The equation of the line can then be written using the point-slope form.

Likewise, a line L in three-dimensional space is determined when we know a point $P_0(x_0, y_0, z_0)$ on L and the direction of L. Our discussion here parallels the discussion of parametric equations of lines in the plane given in Section 1.3. In three dimensions the direction of a line is conveniently described by a vector, so we let $\mathbf{v}$ be a vector parallel to L. Let $P(x, y, z)$ be an arbitrary point on L and let $\mathbf{r}_0$ and $\mathbf{r}$ be the position vectors of P_0 and P (that is, they have representations $\overrightarrow{OP_0}$ and $\overrightarrow{OP}$). If $\mathbf{a}$ is the vector with representation $\overrightarrow{P_0P}$, as shown in Figure 1, then the Triangle Law for vector addition gives $\mathbf{r} = \mathbf{r}_0 + \mathbf{a}$. But, since $\mathbf{a}$ and $\mathbf{v}$ are parallel vectors, there is a scalar t such that $\mathbf{a} = t\mathbf{v}$. Thus

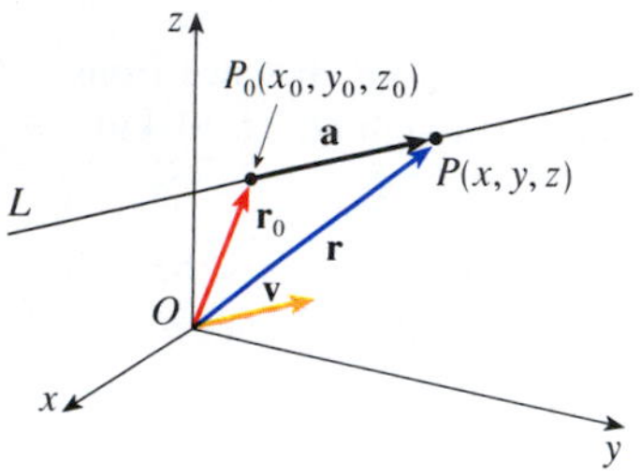

FIGURE 1

(1) $$\mathbf{r} = \mathbf{r}_0 + t\mathbf{v}$$

which is a **vector equation** of L. Each value of the **parameter** t gives the position vector $\mathbf{r}$ of a point on L. As Figure 2 indicates, positive values of t correspond to points on L that lie on one side of P_0, whereas negative values of t correspond to points that lie on the other side of P_0.

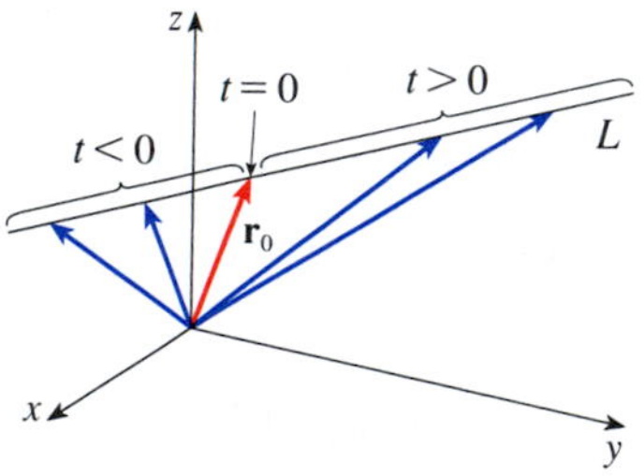

FIGURE 2

If the vector $\mathbf{v}$ that gives the direction of L is written in component form as $\mathbf{v} = \langle a, b, c\rangle$, then we have $t\mathbf{v} = \langle ta, tb, tc\rangle$. We can also write $\mathbf{r} = \langle x, y, z\rangle$ and $\mathbf{r}_0 = \langle x_0, y_0, z_0\rangle$, so the vector equation (1) becomes

$$\langle x, y, z\rangle = \langle x_0 + ta,\ y_0 + tb,\ z_0 + tc\rangle$$

Two vectors are equal if and only if corresponding components are equal. Therefore, we have the three scalar equations:

(2) $$x = x_0 + at \qquad y = y_0 + bt \qquad z = z_0 + ct$$

where $t \in \mathbb{R}$. These equations are called **parametric equations** of the line L through the point $P_0(x_0, y_0, z_0)$ and parallel to the vector $\mathbf{v} = \langle a, b, c\rangle$. Each value of the parameter t gives a point (x, y, z) on L.

EXAMPLE 1

(a) Find a vector equation and parametric equations for the line that passes through the point $(5, 1, 3)$ and is parallel to the vector $\mathbf{i} + 4\mathbf{j} - 2\mathbf{k}$.

(b) Find two other points on the line.

Figure 3 shows the line L in Example 1 and its relation to the given point and to the vector that gives its direction.

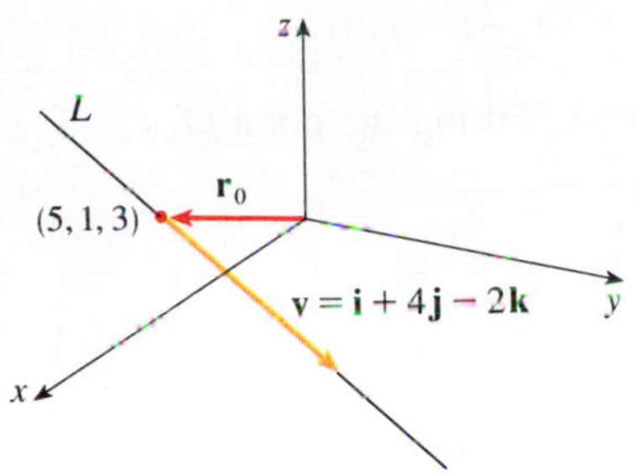

FIGURE 3

SOLUTION

(a) Here $\mathbf{r}_0 = \langle 5, 1, 3 \rangle = 5\,\mathbf{i} + \mathbf{j} + 3\,\mathbf{k}$ and $\mathbf{v} = \mathbf{i} + 4\,\mathbf{j} - 2\,\mathbf{k}$, so the vector equation (1) becomes

$$\mathbf{r} = (5\,\mathbf{i} + \mathbf{j} + 3\,\mathbf{k}) + t(\mathbf{i} + 4\,\mathbf{j} - 2\,\mathbf{k})$$

or

$$\mathbf{r} = (5 + t)\mathbf{i} + (1 + 4t)\mathbf{j} + (3 - 2t)\mathbf{k}$$

Parametric equations are

$$x = 5 + t \qquad y = 1 + 4t \qquad z = 3 - 2t$$

(b) Choosing the parameter value $t = 1$ gives $x = 6$, $y = 5$, and $z = 1$, so $(6, 5, 1)$ is a point on the line. Similarly, $t = -1$ gives the point $(4, -3, 5)$. ■

The vector equation and parametric equations of a line are not unique. If we change the point or the parameter or choose a different parallel vector, then the equations change. For instance, if, instead of $(5, 1, 3)$, we choose the point $(6, 5, 1)$ in Example 1, then the parametric equations of the line become

$$x = 6 + t \qquad y = 5 + 4t \qquad z = 1 - 2t$$

Or, if we stay with the point $(5, 1, 3)$ but choose the parallel vector $2\,\mathbf{i} + 8\,\mathbf{j} - 4\,\mathbf{k}$, we arrive at the equations

$$x = 5 + 2t \qquad y = 1 + 8t \qquad z = 3 - 4t$$

In general, if a vector $\mathbf{v} = \langle a, b, c \rangle$ is used to describe the direction of a line L, then the numbers a, b, and c are called **direction numbers** of L. Since any vector parallel to $\mathbf{v}$ could also be used, we see that any three numbers proportional to a, b, and c could also be used as a set of direction numbers for L.

Another way of describing a line L is to eliminate the parameter t from Equations 2. If none of a, b, or c is 0, we can solve each of these equations for t, equate the results, and obtain

(3)
$$\frac{x - x_0}{a} = \frac{y - y_0}{b} = \frac{z - z_0}{c}$$

These equations are called **symmetric equations** of L. Notice that the numbers a, b, and c that appear in the denominators of Equations 3 are direction numbers of L, that is, components of a vector parallel to L. If one of a, b, or c is 0, we can still eliminate t. For instance, if $a = 0$, we could write the equations of L as

$$x = x_0 \qquad \frac{y - y_0}{b} = \frac{z - z_0}{c}$$

This means that L lies in the vertical plane $x = x_0$.

EXAMPLE 2

(a) Find parametric equations and symmetric equations of the line that passes through the points $A(2, 4, -3)$ and $B(3, -1, 1)$.

(b) At what point does this line intersect the xy-plane?

SOLUTION

(a) We are not explicitly given a vector parallel to the line, but observe that the vector $\mathbf{v}$ with representation $\overrightarrow{AB}$ is parallel to the line and

$$\mathbf{v} = \langle 3 - 2, -1 - 4, 1 - (-3) \rangle = \langle 1, -5, 4 \rangle$$

Figure 4 shows the line L in Example 2 and the point P where it intersects the xy-plane.

Thus, direction numbers are $a = 1$, $b = -5$, and $c = 4$. Taking the point $(2, 4, -3)$ as P_0, we see that parametric equations (2) are

$$x = 2 + t \qquad y = 4 - 5t \qquad z = -3 + 4t$$

and symmetric equations (3) are

$$\frac{x - 2}{1} = \frac{y - 4}{-5} = \frac{z + 3}{4}$$

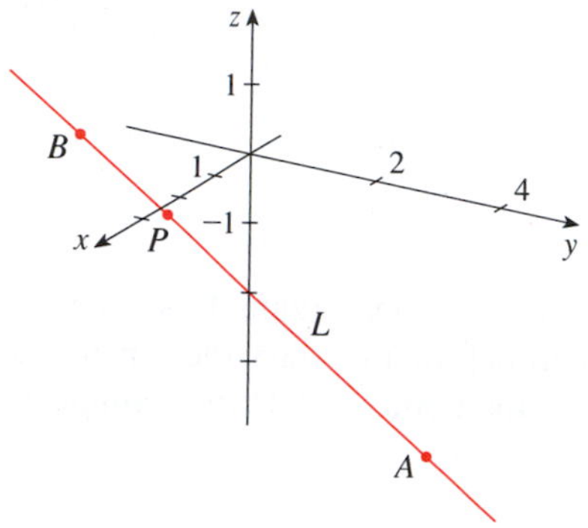

FIGURE 4

(b) The line intersects the xy-plane when $z = 0$, so we put $z = 0$ in the symmetric equations and obtain

$$\frac{x - 2}{1} = \frac{y - 4}{-5} = \frac{3}{4}$$

This gives $x = \frac{11}{4}$ and $y = \frac{1}{4}$, so the line intersects the xy-plane at the point $\left(\frac{11}{4}, \frac{1}{4}, 0\right)$. ■

In general, the procedure of Example 2 shows that direction numbers of the line L through the points $P_0(x_0, y_0, z_0)$ and $P_1(x_1, y_1, z_1)$ are $x_1 - x_0$, $y_1 - y_0$, and $z_1 - z_0$ and so symmetric equations of L are

$$\frac{x - x_0}{x_1 - x_0} = \frac{y - y_0}{y_1 - y_0} = \frac{z - z_0}{z_1 - z_0}$$

The lines L_1 and L_2 in Example 3, shown in Figure 5, are skew lines.

EXAMPLE 3 Show that the lines L_1 and L_2 with parametric equations

$$x = 1 + t \qquad y = -2 + 3t \qquad z = 4 - t$$

$$x = 2s \qquad y = 3 + s \qquad z = -3 + 4s$$

are **skew lines**; that is, they do not intersect and are not parallel (and therefore do not lie in the same plane).

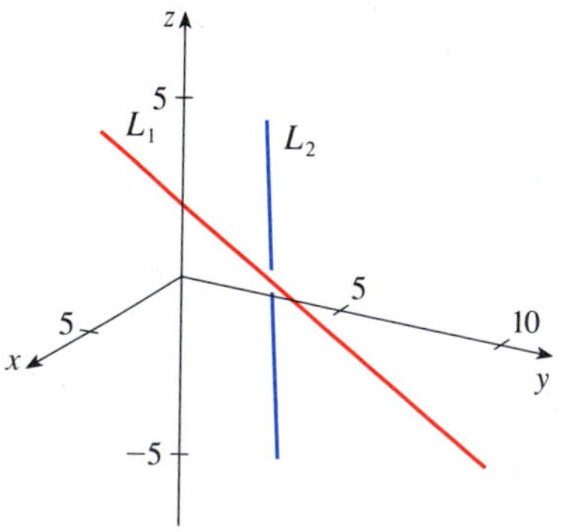

FIGURE 5

SOLUTION The lines are not parallel because the corresponding vectors $\langle 1, 3, -1 \rangle$ and $\langle 2, 1, 4 \rangle$ are not parallel. (Their components are not proportional.) If L_1 and L_2 had a point of intersection, there would be values of t and s such that

$$1 + t = 2s$$
$$-2 + 3t = 3 + s$$
$$4 - t = -3 + 4s$$

But if we solve the first two equations, we get $t = \frac{11}{5}$ and $s = \frac{8}{5}$, and these values do not satisfy the third equation. Therefore, there are no values of t and s that satisfy the three equations. Thus, L_1 and L_2 do not intersect. Hence L_1 and L_2 are skew lines. ■

PLANES

A plane in space is determined by a point $P_0(x_0, y_0, z_0)$ in the plane and a vector $\mathbf{n}$ that is orthogonal to the plane. This orthogonal vector $\mathbf{n}$ is called a **normal vector**. Let $P(x, y, z)$ be an arbitrary point in the plane and let $\mathbf{r}_0$ and $\mathbf{r}$ be the position vectors of P_0 and P. Then the vector $\mathbf{r} - \mathbf{r}_0$ is represented by $\overrightarrow{P_0P}$ (see Figure 6). The normal vector $\mathbf{n}$ is orthogonal to every vector in the given plane. In particular, $\mathbf{n}$ is orthogonal to $\mathbf{r} - \mathbf{r}_0$ and so we have

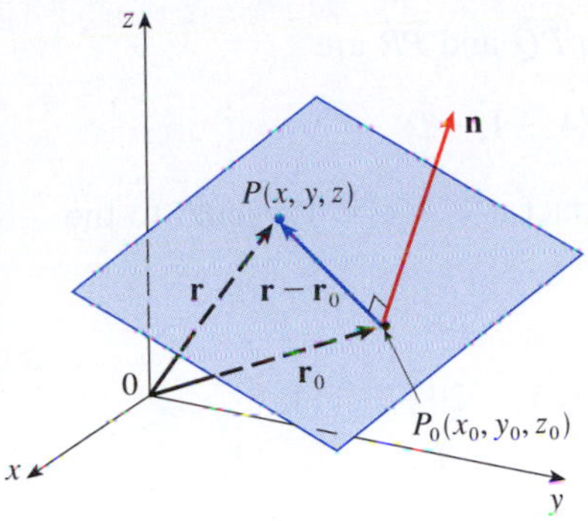

FIGURE 6

(4) $$\mathbf{n} \cdot (\mathbf{r} - \mathbf{r}_0) = 0$$

which can be rewritten as

(5) $$\mathbf{n} \cdot \mathbf{r} = \mathbf{n} \cdot \mathbf{r}_0$$

Either Equation 4 or Equation 5 is called a **vector equation of the plane**.

To obtain a scalar equation for the plane, we write $\mathbf{n} = \langle a, b, c \rangle$, $\mathbf{r} = \langle x, y, z \rangle$, and $\mathbf{r}_0 = \langle x_0, y_0, z_0 \rangle$. Then the vector equation (4) becomes

$$\langle a, b, c \rangle \cdot \langle x - x_0, y - y_0, z - z_0 \rangle = 0$$

or

(6) $$a(x - x_0) + b(y - y_0) + c(z - z_0) = 0$$

Equation 6 is the **scalar equation of the plane through** $P_0(x_0, y_0, z_0)$ **with normal vector** $\mathbf{n} = \langle a, b, c \rangle$.

EXAMPLE 4 Find an equation of the plane through the point $(2, 4, -1)$ with normal vector $\mathbf{n} = \langle 2, 3, 4 \rangle$. Find the intercepts and sketch the plane.

SOLUTION Putting $a = 2$, $b = 3$, $c = 4$, $x_0 = 2$, $y_0 = 4$, and $z_0 = -1$ in Equation 6, we see that an equation of the plane is

$$2(x - 2) + 3(y - 4) + 4(z + 1) = 0$$

or $$2x + 3y + 4z = 12$$

To find the x-intercept we set $y = z = 0$ in this equation and obtain $x = 6$. Similarly, the y-intercept is 4 and the z-intercept is 3. This enables us to sketch the portion of the plane that lies in the first octant (see Figure 7). ■

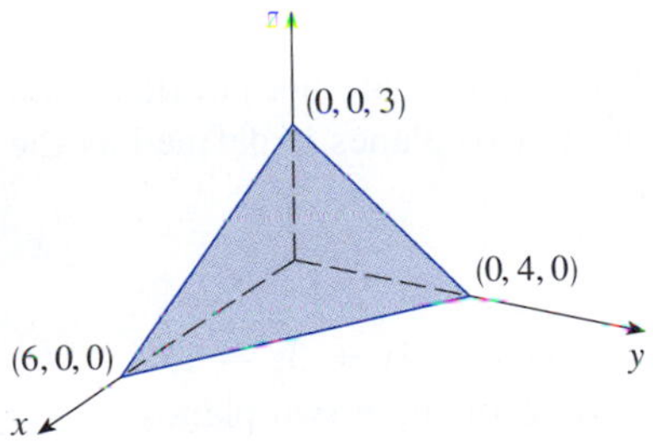

FIGURE 7

By collecting terms in Equation 6 as we did in Example 4, we can rewrite the equation of a plane as

(7) $$ax + by + cz = d$$

where $d = ax_0 + by_0 + cz_0$. Equation 7 is called a **linear equation** in x, y, and z. Conversely, it can be shown that if a, b, and c are not all 0, then the linear equation (7) represents a plane with normal vector $\langle a, b, c \rangle$. (See Exercise 71.)

Figure 8 shows the portion of the plane in Example 5 that is enclosed by triangle PQR.

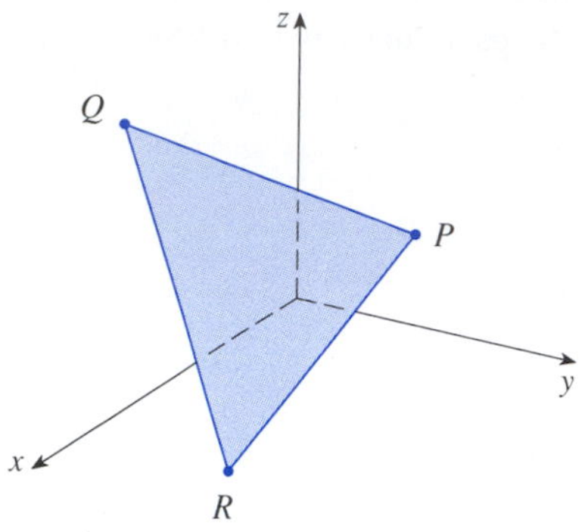

FIGURE 8

EXAMPLE 5 Find an equation of the plane that passes through the points $P(1, 3, 2)$, $Q(3, -1, 6)$, and $R(5, 2, 0)$.

SOLUTION The vectors $\mathbf{a}$ and $\mathbf{b}$ corresponding to $\overrightarrow{PQ}$ and $\overrightarrow{PR}$ are

$$\mathbf{a} = \langle 2, -4, 4 \rangle \qquad \mathbf{b} = \langle 4, -1, -2 \rangle$$

Since both $\mathbf{a}$ and $\mathbf{b}$ lie in the plane, their cross product $\mathbf{a} \times \mathbf{b}$ is orthogonal to the plane and can be taken as the normal vector. Thus

$$\mathbf{n} = \mathbf{a} \times \mathbf{b} = \begin{vmatrix} \mathbf{i} & \mathbf{j} & \mathbf{k} \\ 2 & -4 & 4 \\ 4 & -1 & -2 \end{vmatrix} = 12\,\mathbf{i} + 20\,\mathbf{j} + 14\,\mathbf{k}$$

and an equation of the plane is

$$12(x - 1) + 20(y - 3) + 14(z - 2) = 0$$

or

$$6x + 10y + 7z = 50$$

■

EXAMPLE 6 Find the point at which the line with parametric equations $x = 2 + 3t$, $y = -4t$, $z = 5 + t$ intersects the plane $4x + 5y - 2z = 18$.

SOLUTION We substitute the expressions for x, y, and z from the parametric equations into the equation of the plane:

$$4(2 + 3t) + 5(-4t) - 2(5 + t) = 18$$

This simplifies to $-10t = 20$, so $t = -2$. Therefore, the point of intersection occurs when the parameter value is $t = -2$. Then $x = 2 + 3(-2) = -4$, $y = -4(-2) = 8$, $z = 5 - 2 = 3$, and so the point of intersection is $(-4, 8, 3)$. ■

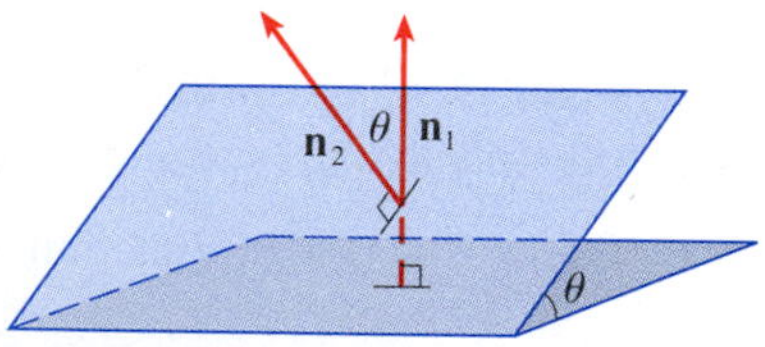

FIGURE 9

Two planes are **parallel** if their normal vectors are parallel. For instance, the planes $x + 2y - 3z = 4$ and $2x + 4y - 6z = 3$ are parallel because their normal vectors are $\mathbf{n}_1 = \langle 1, 2, -3 \rangle$ and $\mathbf{n}_2 = \langle 2, 4, -6 \rangle$ and $\mathbf{n}_2 = 2\mathbf{n}_1$. If two planes are not parallel, then they intersect in a straight line and the angle between the two planes is defined as the acute angle between their normal vectors (see Figure 9).

Figure 10 shows the planes in Example 7 and their line of intersection L.

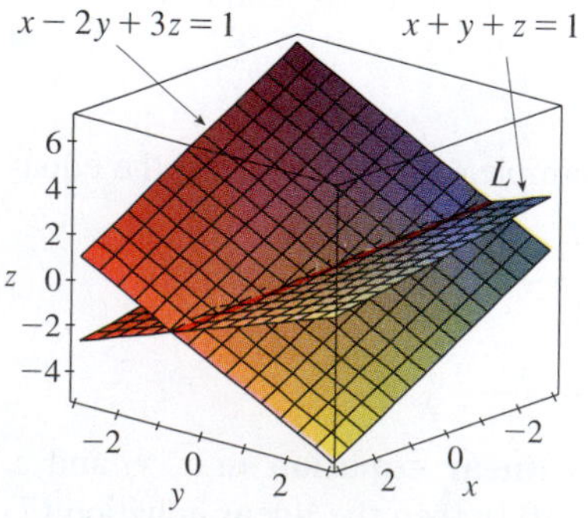

FIGURE 10

EXAMPLE 7

(a) Find the angle between the planes $x + y + z = 1$ and $x - 2y + 3z = 1$.

(b) Find symmetric equations for the line of intersection L of these two planes.

SOLUTION

(a) The normal vectors of these planes are

$$\mathbf{n}_1 = \langle 1, 1, 1 \rangle \qquad \mathbf{n}_2 = \langle 1, -2, 3 \rangle$$

and so if θ is the angle between the planes, (11.2.10) gives

$$\cos\theta = \frac{\mathbf{n}_1 \cdot \mathbf{n}_2}{|\mathbf{n}_1|\,|\mathbf{n}_2|} = \frac{1(1) + 1(-2) + 1(3)}{\sqrt{1 + 1 + 1}\,\sqrt{1 + 4 + 9}} = \frac{2}{\sqrt{42}}$$

$$\theta = \cos^{-1}\left(\frac{2}{\sqrt{42}}\right) \approx 72^\circ$$

(b) We first need to find a point on L. For instance, we can find the point where the line intersects the xy-plane by setting $z = 0$ in the equations of both planes. This gives the equations $x + y = 1$ and $x - 2y = 1$, whose solution is $x = 1$, $y = 0$. So the point $(1, 0, 0)$ lies on L.

Another way to find the line of intersection is to solve the equations of the planes for two of the variables in terms of the third, which can be taken as the parameter.

Now we observe that, since L lies in both planes, it is perpendicular to both of the normal vectors. Thus, a vector $\mathbf{v}$ parallel to L is given by the cross product

$$\mathbf{v} = \mathbf{n}_1 \times \mathbf{n}_2 = \begin{vmatrix} \mathbf{i} & \mathbf{j} & \mathbf{k} \\ 1 & 1 & 1 \\ 1 & -2 & 3 \end{vmatrix} = 5\,\mathbf{i} - 2\,\mathbf{j} - 3\,\mathbf{k}$$

and so the symmetric equations of L can be written as

$$\frac{x-1}{5} = \frac{y}{-2} = \frac{z}{-3}$$

■

NOTE: Since a linear equation in x, y, and z represents a plane and two nonparallel planes intersect in a line, it follows that two linear equations can represent a line. The points (x, y, z) that satisfy both $a_1x + b_1y + c_1z = d_1$ and $a_2x + b_2y + c_2z = d_2$ lie on both of these planes and so the pair of linear equations represents the line of intersection of the planes (if they are not parallel). For instance, in Example 7 the line L was given as the line of intersection of the planes $x + y + z = 1$ and $x - 2y + 3z = 1$. The symmetric equations that we found for L could be written as

$$\frac{x-1}{5} = \frac{y}{-2} \qquad \text{and} \qquad \frac{y}{-2} = \frac{z}{-3}$$

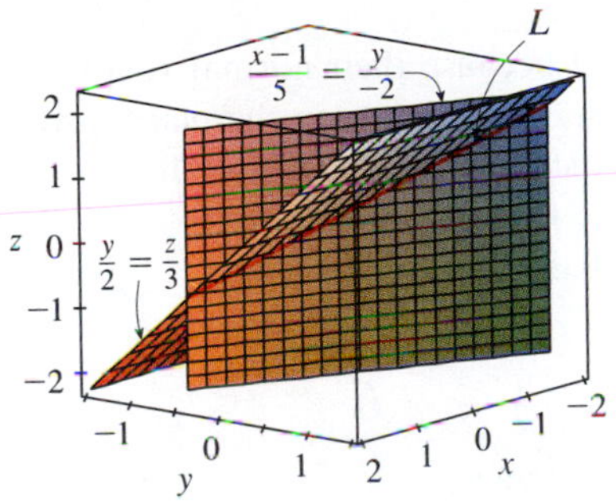

FIGURE 11

Figure 11 shows how the line L in Example 7 can also be regarded as the line of intersection of planes derived from its symmetric equations.

which is again a pair of linear equations. They exhibit L as the line of intersection of the planes $(x - 1)/5 = y/(-2)$ and $y/(-2) = z/(-3)$. (See Figure 11.)

In general, when we write the equations of a line in the symmetric form

$$\frac{x - x_0}{a} = \frac{y - y_0}{b} = \frac{z - z_0}{c}$$

we can regard the line as the line of intersection of the two planes

$$\frac{x - x_0}{a} = \frac{y - y_0}{b} \qquad \text{and} \qquad \frac{y - y_0}{b} = \frac{z - z_0}{c}$$

EXAMPLE 8 Find a formula for the distance D from a point $P_1(x_1, y_1, z_1)$ to the plane $ax + by + cz + d = 0$.

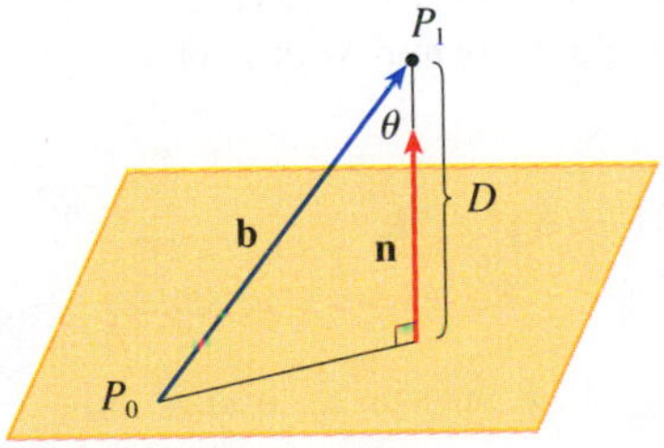

FIGURE 12

SOLUTION Let $P_0(x_0, y_0, z_0)$ be any point in the given plane and let $\mathbf{b}$ be the vector corresponding to $\overrightarrow{P_0P_1}$. Then

$$\mathbf{b} = \langle x_1 - x_0,\, y_1 - y_0,\, z_1 - z_0 \rangle$$

From Figure 12 you can see that the distance D from P_1 to the plane is equal to the absolute value of the scalar projection of $\mathbf{b}$ onto the normal vector $\mathbf{n} = \langle a, b, c \rangle$. (See

Section 11.2.) Thus

$$D = \text{comp}_{\mathbf{n}}\,\mathbf{b} = \frac{|\mathbf{n}\cdot\mathbf{b}|}{|\mathbf{n}|}$$

$$= \frac{|a(x_1 - x_0) + b(y_1 - y_0) + c(z_1 - z_0)|}{\sqrt{a^2 + b^2 + c^2}}$$

$$= \frac{|(ax_1 + by_1 + cz_1) - (ax_0 + by_0 + cz_0)|}{\sqrt{a^2 + b^2 + c^2}}$$

Since P_0 lies in the plane, its coordinates satisfy the equation of the plane and so we have $ax_0 + by_0 + cz_0 + d = 0$. Thus, the formula for D can be written as

(8)
$$D = \frac{|ax_1 + by_1 + cz_1 + d|}{\sqrt{a^2 + b^2 + c^2}}$$

■

EXAMPLE 9 Find the distance between the parallel planes $10x + 2y - 2z = 5$ and $5x + y - z = 1$.

SOLUTION First we note that the planes are parallel because their normal vectors $\langle 10, 2, -2\rangle$ and $\langle 5, 1, -1\rangle$ are parallel. To find the distance D between the planes, we choose any point on one plane and calculate its distance to the other plane. In particular, if we put $y = z = 0$ in the equation of the first plane, we get $10x = 5$ and so $(\frac{1}{2}, 0, 0)$ is a point in this plane. By Formula 8, the distance between $(\frac{1}{2}, 0, 0)$ and the plane $5x + y - z - 1 = 0$ is

$$D = \frac{|5(\frac{1}{2}) + 1(0) - 1(0) - 1|}{\sqrt{5^2 + 1^2 + (-1)^2}} = \frac{\frac{3}{2}}{3\sqrt{3}} = \frac{\sqrt{3}}{6}$$

So the distance between the planes is $\sqrt{3}/6$. ■

EXAMPLE 10 In Example 3 we showed that the lines

$$L_1:\quad x = 1 + t \qquad y = -2 + 3t \qquad z = 4 - t$$

$$L_2:\quad x = 2s \qquad y = 3 + s \qquad z = -3 + 4s$$

are skew. Find the distance between them.

SOLUTION Since the two lines L_1 and L_2 are skew, they can be viewed as lying on two parallel planes P_1 and P_2. The distance between L_1 and L_2 is the same as the distance between P_1 and P_2, which can be computed as in Example 9. The common normal vector to both planes must be orthogonal to both $\mathbf{v}_1 = \langle 1, 3, -1\rangle$ (the direction of L_1) and $\mathbf{v}_2 = \langle 2, 1, 4\rangle$ (the direction of L_2). So a normal vector is

$$\mathbf{n} = \mathbf{v}_1 \times \mathbf{v}_2 = \begin{vmatrix} \mathbf{i} & \mathbf{j} & \mathbf{k} \\ 1 & 3 & -1 \\ 2 & 1 & 4 \end{vmatrix} = 13\mathbf{i} - 6\mathbf{j} - 5\mathbf{k}$$

If we put $s = 0$ in the equations of L_2, we get the point $(0, 3, -3)$ on L_2 and so an equation for P_2 is

$$13(x - 0) - 6(y - 3) - 5(z + 3) = 0 \qquad \text{or} \qquad 13x - 6y - 5z + 3 = 0$$

If we now set $t = 0$ in the equations for L_1, we get the point $(1, -2, 4)$ on P_1. So the distance between L_1 and L_2 is the same as the distance from $(1, -2, 4)$ to $13x - 6y - 5z + 3 = 0$. By Formula 8, this distance is

$$D = \frac{|13(1) - 6(-2) - 5(4) + 3|}{\sqrt{13^2 + (-6)^2 + (-5)^2}} = \frac{8}{\sqrt{230}} \approx 0.53$$

EXERCISES 11.4

1–4 ■ Find the vector equation and parametric equations for the line passing through the given point and parallel to the vector **a**.

1. $(3, -1, 8)$, $\mathbf{a} = \langle 2, 3, 5 \rangle$
2. $(-2, 4, 5)$, $\mathbf{a} = \langle 3, -1, 6 \rangle$
3. $(0, 1, 2)$, $\mathbf{a} = 6\,\mathbf{i} + 3\,\mathbf{j} + 2\,\mathbf{k}$
4. $(1, -1, -2)$, $\mathbf{a} = 2\,\mathbf{i} - 7\,\mathbf{k}$

5–10 ■ Find parametric equations and symmetric equations for the line passing through the given points.

5. $(2, 1, 8)$, $(6, 0, 3)$
6. $(-1, 0, 5)$, $(4, -3, 3)$
7. $(3, 1, -1)$, $(3, 2, -6)$
8. $(3, 1, \frac{1}{2})$, $(-1, 4, 1)$
9. $(-\frac{1}{3}, 1, 1)$, $(0, 5, -8)$
10. $(2, -7, 5)$, $(-4, 2, 5)$
11. Show that the line through the points $(2, -1, -5)$ and $(8, 8, 7)$ is parallel to the line through the points $(4, 2, -6)$ and $(8, 8, 2)$.
12. Show that the line through the points $(0, 1, 1)$ and $(1, -1, 6)$ is perpendicular to the line through the points $(-4, 2, 1)$ and $(-1, 6, 2)$.
13. (a) Find symmetric equations for the line that passes through the point $(0, 2, -1)$ and is parallel to the line with parametric equations $x = 1 + 2t$, $y = 3t$, and $z = 5 - 7t$.
    (b) Find the points in which the required line in part (a) intersects the coordinate planes.
14. (a) Find parametric equations for the line through $(5, 1, 0)$ that is perpendicular to the plane $2x - y + z = 1$.
    (b) In what points does this line intersect the coordinate planes?

15–18 ■ Determine whether the lines L_1 and L_2 are parallel, skew, or intersecting. If they intersect, find the point of intersection.

15. L_1: $\dfrac{x-4}{2} = \dfrac{y+5}{4} = \dfrac{z-1}{-3}$, L_2: $\dfrac{x-2}{1} = \dfrac{y+1}{3} = \dfrac{z}{2}$
16. L_1: $\dfrac{x-1}{2} = \dfrac{y}{1} = \dfrac{z-1}{4}$, L_2: $\dfrac{x}{1} = \dfrac{y+2}{2} = \dfrac{z+2}{3}$
17. L_1: $x = -6t$, $y = 1 + 9t$, $z = -3t$
    L_2: $x = 1 + 2s$, $y = 4 - 3s$, $z = s$
18. L_1: $x = 1 + t$, $y = 2 - t$, $z = 3t$
    L_2: $x = 2 - s$, $y = 1 + 2s$, $z = 4 + s$

19–22 ■ Find an equation of the plane through the given point and with the specified normal vector.

19. $(1, 4, 5)$, $\mathbf{n} = \langle 7, 1, 4 \rangle$
20. $(-5, 1, 2)$, $\mathbf{n} = \langle 3, -5, 2 \rangle$
21. $(1, 2, 3)$, $\mathbf{n} = 15\,\mathbf{i} + 9\,\mathbf{j} - 12\,\mathbf{k}$
22. $(-1, -6, -4)$, $\mathbf{n} = -5\,\mathbf{i} + 2\,\mathbf{j} - 2\,\mathbf{k}$

23–26 ■ Find an equation of the plane passing through the given point and parallel to the indicated plane.

23. $(6, 5, -2)$, $x + y - z + 1 = 0$
24. $(3, 0, 8)$, $2x + 5y + 8z = 17$
25. $(-1, 3, -8)$, $3x - 4y - 6z = 9$
26. $(2, -4, 5)$, $z = 2x + 3y$

27–30 ■ Find an equation of the plane passing through the three given points.

27. $(0, 0, 0)$, $(1, 1, 1)$, $(1, 2, 3)$
28. $(-1, 1, -1)$, $(1, -1, 2)$, $(4, 0, 3)$
29. $(1, 0, -3)$, $(0, -2, -4)$, $(4, 1, 6)$
30. $(2, 1, -3)$, $(5, -1, 4)$, $(2, -2, 4)$

31–34 ■ Find an equation of the plane that passes through the given point and contains the indicated line.

31. $(1, 6, -4)$; $x = 1 + 2t$, $y = 2 - 3t$, $z = 3 - t$
32. $(-1, -3, 2)$; $x = -1 - 2t$, $y = 4t$, $z = 2 + t$
33. $(0, 1, 2)$; $x = y = z$
34. $(-1, 0, 1)$; $x = 5t$, $y = 1 + t$, $z = -t$

35–38 ■ Find the point at which the given line intersects the specified plane.

35. $x = 1 + t$, $y = 2t$, $z = 3t$; $x + y + z = 1$
36. $x = 5$, $y = 4 - t$, $z = 2t$; $2x - y + z = 5$
37. $x = 1 + 2t$, $y = -1$, $z = t$; $2x + y - z + 5 = 0$
38. $x = 1 - t$, $y = t$, $z = 1 + t$; $z = 1 - 2x + y$

39. Find direction numbers for the line of intersection of the planes $x + y + z = 1$ and $x + z = 0$.

40. Find the cosine of the angle between the planes $x + y + z = 0$ and $x + 2y + 3z = 1$.

41–46 ▪ Determine whether the planes are parallel, perpendicular, or neither. If neither, find the angle between them.

41. $x + z = 1, \quad y + z = 1$

42. $-8x - 6y + 2z = 1, \quad z = 4x + 3y$

43. $x + 4y - 3z = 1, \quad -3x + 6y + 7z = 0$

44. $2x + 2y - z = 4, \quad 6x - 3y + 2z = 5$

45. $2x + 4y - 2z = 1, \quad -3x - 6y + 3z = 10$

46. $2x - 5y + z = 3, \quad 4x + 2y + 2z = 1$

47–48 ▪ (a) Find symmetric equations for the line of intersection of the planes and (b) find the angle between the planes.

47. $x + y - z = 2, \quad 3x - 4y + 5z = 6$

48. $x - 2y + z = 1, \quad 2x + y + z = 1$

49–50 ▪ Find parametric equations for the line of intersection of the planes.

49. $z = x + y, \quad 2x - 5y - z = 1$

50. $2x + 5z + 3 = 0, \quad x - 3y + z + 2 = 0$

51. Find an equation for the plane consisting of all points that are equidistant from the points $(1, 1, 0)$ and $(0, 1, 1)$.

52. Find an equation for the plane consisting of all points that are equidistant from the points $(-4, 2, 1)$ and $(2, -4, 3)$.

53. Find an equation of the plane that passes through the line of intersection of the planes $x + y - z = 2$ and $2x - y + 3z = 1$ and passes through the point $(-1, 2, 1)$.

54. Find an equation of the plane that passes through the line of intersection of the planes $x - z = 1$ and $y + 2z = 3$ and is perpendicular to the plane $x + y - 2z = 1$.

55. Find an equation of the plane with x-intercept a, y-intercept b, and z-intercept c.

56. (a) Find the point at which the lines $\mathbf{r} = \langle 1, 1, 0 \rangle + t\langle 1, -1, 2 \rangle$ and $\mathbf{r} = \langle 2, 0, 2 \rangle + s\langle -1, 1, 0 \rangle$ intersect.
(b) Find an equation of the plane that contains these lines.

57. Find parametric equations for the line through the point $(0, 1, 2)$ that is parallel to the plane $x + y + z = 2$ and perpendicular to the line $x = 1 + t$, $y = 1 - t$, $z = 2t$.

58. Find parametric equations for the line through the point $(0, 1, 2)$ that is perpendicular to the line $x = 1 + t$, $y = 1 - t$, $z = 2t$ and intersects this line.

59. Which of the following four planes are parallel? Are any of them identical?

P_1: $4x - 2y + 6z = 3$ $\qquad$ P_2: $4x - 2y - 2z = 6$

P_3: $-6x + 3y - 9z = 5$ $\qquad$ P_4: $z = 2x - y - 3$

60. Which of the following four lines are parallel? Are any of them identical?

L_1: $x = 1 + t, \quad y = t, \quad z = 2 - 5t$

L_2: $x + 1 = y - 2 = 1 - z$

L_3: $x = 1 + t, \quad y = 4 + t, \quad z = 1 - t$

L_4: $\mathbf{r} = \langle 2, 1, -3 \rangle + t\langle 2, 2, -10 \rangle$

61–62 ▪ Use the formula in Exercise 25 in Section 11.3 to find the distance from the point to the given line.

61. $(1, 2, 3); \quad x = 2 + t, \ y = 2 - 3t, \ z = 5t$

62. $(1, 0, -1); \quad x = 5 - t, \ y = 3t, \ z = 1 + 2t$

63–64 ▪ Find the distance from the point to the given plane.

63. $(2, 8, 5), \quad x - 2y - 2z = 1$

64. $(3, -2, 7), \quad 4x - 6y + z = 5$

65–66 ▪ Find the distance between the given parallel planes.

65. $z = x + 2y + 1, \quad 3x + 6y - 3z = 4$

66. $3x + 6y - 9z = 4, \quad x + 2y - 3z = 1$

67. Show that the distance between the parallel planes $ax + by + cz = d_1$ and $ax + by + cz = d_2$ is

$$D = \frac{|d_1 - d_2|}{\sqrt{a^2 + b^2 + c^2}}$$

68. Find equations of the planes that are parallel to the plane $x + 2y - 2z = 1$ and two units away from it.

69. Show that the lines with symmetric equations $x = y = z$ and $x + 1 = y/2 = z/3$ are skew lines, and find the distance between these lines.

70. Find the distance between the skew lines with parametric equations $x = 1 + t$, $y = 1 + 6t$, $z = 2t$ and $x = 1 + 2s$, $y = 5 + 15s$, $z = -2 + 6s$.

71. If a, b, and c are not all 0, show that the equation $ax + by + cz = d$ represents a plane and $\langle a, b, c \rangle$ is a normal vector to the plane. *Hint:* Suppose $a \neq 0$ and rewrite the equation in the form

$$a\left(x - \frac{d}{a}\right) + b(y - 0) + c(z - 0) = 0$$

72. Give a geometric description of each family of planes.
(a) $x + y + z = c$
(b) $x + y + cz = 1$
(c) $y \cos\theta + z \sin\theta = 1$

11.5 QUADRIC SURFACES

A **quadric surface** is the graph of a second-degree equation in three variables x, y, and z. The most general such equation is

$$Ax^2 + By^2 + Cz^2 + Dxy + Eyz + Fxz + Gx + Hy + Iz + J = 0$$

where $A, B, C, \ldots, J$ are constants, but by translation and rotation it can be brought into one of the two standard forms

$$Ax^2 + By^2 + Cz^2 + J = 0 \qquad \text{or} \qquad Ax^2 + By^2 + Iz = 0$$

Quadric surfaces are the analogues in three dimensions of the conic sections in the plane. (See Appendix I for a review of conic sections.)

In order to sketch the graph of a quadric surface (or any surface), it is useful to determine the curves of intersection of the surface with planes parallel to the coordinate planes. These curves are called **traces** (or cross-sections) of the surface.

ELLIPSOIDS The quadric surface with equation

$$\textbf{(1)} \qquad \frac{x^2}{a^2} + \frac{y^2}{b^2} + \frac{z^2}{c^2} = 1$$

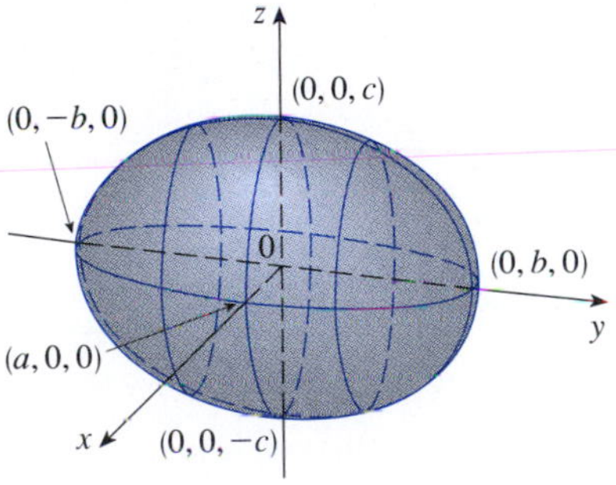

FIGURE 1

The ellipsoid $x^2 + \frac{y^2}{9} + \frac{z^2}{4} = 1$

is called an **ellipsoid** because its traces are ellipses. For instance, the horizontal plane $z = k$ (where $-c < k < c$) intersects the surface in the ellipse

$$\frac{x^2}{a^2} + \frac{y^2}{b^2} = 1 - \frac{k^2}{c^2} \qquad z = k$$

and, in particular, the trace in the xy-plane is just the ellipse $x^2/a^2 + y^2/b^2 = 1$, $z = 0$. Similarly, the traces in the other coordinate planes are the ellipses with equations $y^2/b^2 + z^2/c^2 = 1$, $x = 0$, and $x^2/a^2 + z^2/c^2 = 1$, $y = 0$. Figure 1 shows how drawing some traces indicates the shape of the ellipsoid.

The six intercepts of the ellipsoid are $(\pm a, 0, 0)$, $(0, \pm b, 0)$, and $(0, 0, \pm c)$ and the ellipsoid lies in the box

$$|x| \leqslant a \qquad |y| \leqslant b \qquad |z| \leqslant c$$

Since the equation involves only even powers of x, y, and z, the ellipsoid is symmetric with respect to each coordinate plane.

If two of the three semiaxes a, b, and c are equal, then the ellipsoid is a surface of revolution. For instance, if $c = a$, then the ellipsoid could be obtained by revolving the ellipse $x^2/a^2 + y^2/b^2 = 1$, $z = 0$, around the y-axis. If $a = b = c$, the ellipsoid is a sphere.

HYPERBOLOIDS The quadric surface

$$\textbf{(2)} \qquad \frac{x^2}{a^2} + \frac{y^2}{b^2} - \frac{z^2}{c^2} = 1$$

is also symmetric with respect to the coordinate planes. The trace in any horizontal

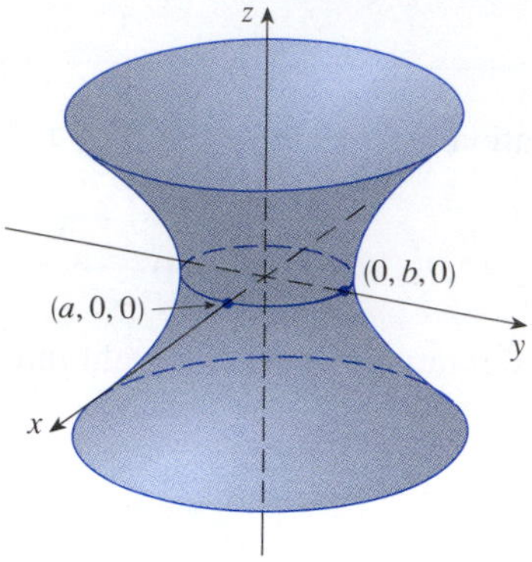

FIGURE 2
Hyperboloid of one sheet
$\frac{x^2}{a^2} + \frac{y^2}{b^2} - \frac{z^2}{c^2} = 1$

plane $z = k$ is the ellipse

$$\frac{x^2}{a^2} + \frac{y^2}{b^2} = 1 + \frac{k^2}{c^2} \qquad z = k$$

but the traces in the xz- and yz-planes are the hyperbolas

$$\frac{x^2}{a^2} - \frac{z^2}{c^2} = 1, \quad y = 0 \qquad \text{and} \qquad \frac{y^2}{b^2} - \frac{z^2}{c^2} = 1, \quad x = 0$$

Furthermore, in contrast to the next example, this surface consists of just one piece, so it is called a **hyperboloid of one sheet** (see Figure 2). The z-axis is called the **axis** of this hyperboloid. (If the minus sign in Equation 2 occurs in front of the first or second term instead of the third term, then the axis is the x- or y-axis, respectively.)

If $a = b$ in Equation 2, the surface is a hyperboloid of revolution and is obtained by rotating a hyperbola about the z-axis.

Now consider the surface

$$-\frac{x^2}{a^2} - \frac{y^2}{b^2} + \frac{z^2}{c^2} = 1 \tag{3}$$

Traces in the xz- and yz-planes are the hyperbolas

$$-\frac{x^2}{a^2} + \frac{z^2}{c^2} = 1, \quad y = 0 \qquad \text{and} \qquad -\frac{y^2}{b^2} + \frac{z^2}{c^2} = 1, \quad x = 0$$

If $|k| > c$, the horizontal plane $z = k$ intersects the surface in the ellipse

$$\frac{x^2}{a^2} + \frac{y^2}{b^2} = \frac{k^2}{c^2} - 1 \qquad z = k$$

whereas if $|k| < c$, the plane $z = k$ does not intersect the surface at all. Thus, the surface consists of two parts, one above the plane $z = c$ and one below the plane $z = -c$, and is called a **hyperboloid of two sheets** whose axis is the z-axis. (See Figure 3.)

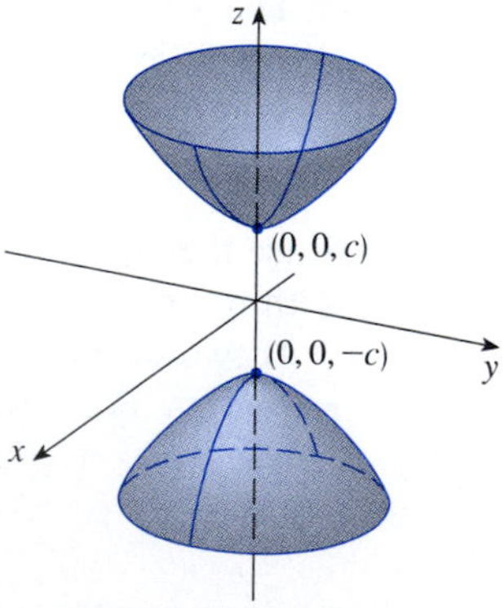

FIGURE 3
Hyperboloid of two sheets
$-\frac{x^2}{a^2} - \frac{y^2}{b^2} + \frac{z^2}{c^2} = 1$

Notice, in comparing Equations 2 and 3, that the number of minus signs in the equation indicates the number of sheets of the hyperboloid.

CONES If we replace the right side of Equation 2 or 3 by 0, we get the surface

$$\frac{z^2}{c^2} = \frac{x^2}{a^2} + \frac{y^2}{b^2} \tag{4}$$

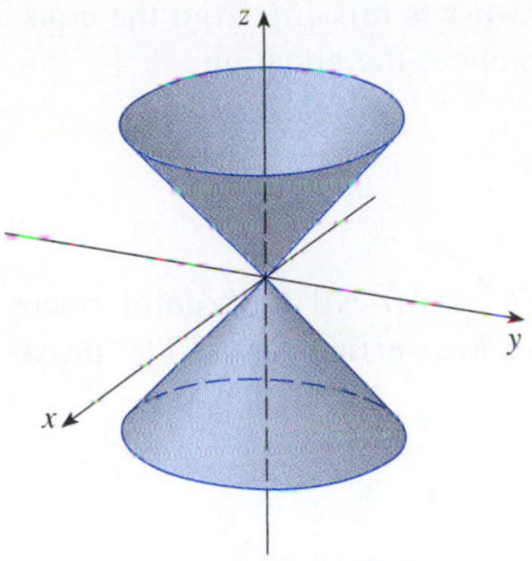

FIGURE 4

The cone $\frac{z^2}{c^2} = \frac{x^2}{a^2} + \frac{y^2}{b^2}$

which is a **cone**. This surface has the property that if P is any point on the cone, then the line OP lies entirely on the cone.

You can verify that traces in horizontal planes $z = k$ are ellipses and traces in vertical planes $x = k$ or $y = k$ are hyperbolas if $k \neq 0$ but are pairs of lines if $k = 0$.

The cone given by Equation 4 is asymptotic to both of the hyperboloids given by Equations 2 and 3. (Compare Figures 2, 3, and 4.)

PARABOLOIDS The surface

$$\frac{z}{c} = \frac{x^2}{a^2} + \frac{y^2}{b^2} \tag{5}$$

is called an **elliptic paraboloid** because its traces in horizontal planes $z = k$ are ellipses, whereas its traces in vertical planes $x = k$ or $y = k$ are parabolas. For instance, its trace in the yz-plane is the parabola

$$z = \frac{c}{b^2} y^2 \qquad x = 0$$

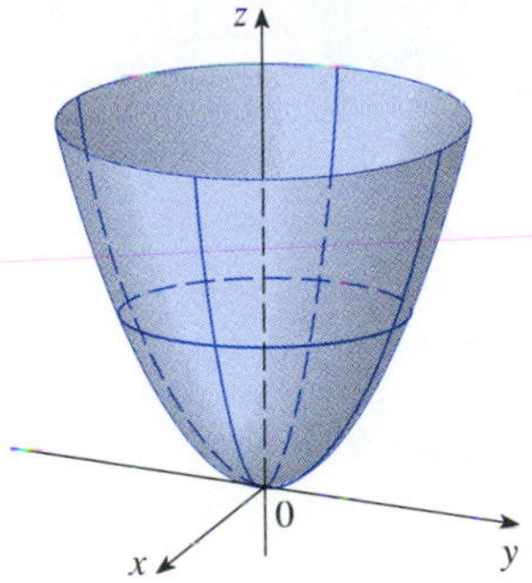

FIGURE 5

Elliptic paraboloid

$\frac{z}{c} = \frac{x^2}{a^2} + \frac{y^2}{b^2}$, $c > 0$

The **axis** of the paraboloid given by Equation 5 is the z-axis and its **vertex** is the origin. The case where $c > 0$ is illustrated in Figure 5. If $a = b$, the surface is a **circular paraboloid**, also called a paraboloid of revolution.

The **hyperbolic paraboloid**

$$\frac{z}{c} = \frac{x^2}{a^2} - \frac{y^2}{b^2} \tag{6}$$

also has parabolas as its vertical traces, but it has hyperbolas as its horizontal traces. The case where $c < 0$ is illustrated in Figure 6. Notice that the shape of the surface near the origin resembles that of a saddle. This surface will be investigated further in Section 12.7 when we discuss saddle points.

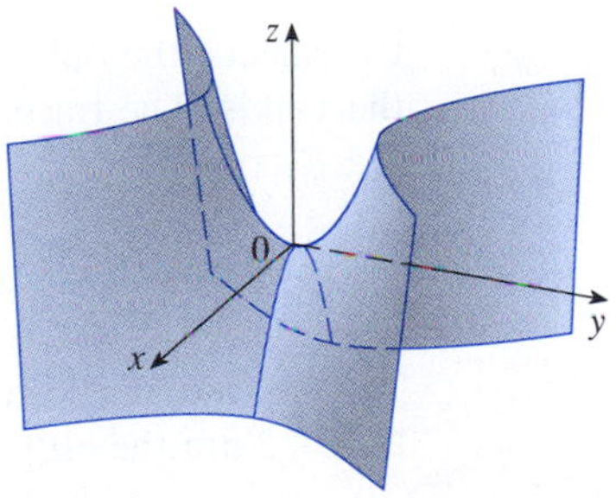

FIGURE 6

Hyperbolic paraboloid

$\frac{z}{c} = \frac{x^2}{a^2} - \frac{y^2}{b^2}$, $c < 0$

The idea of using traces to draw a surface is employed in three-dimensional graphing software for computers. Figure 7 shows two paraboloids drawn by a computer using vertical traces.

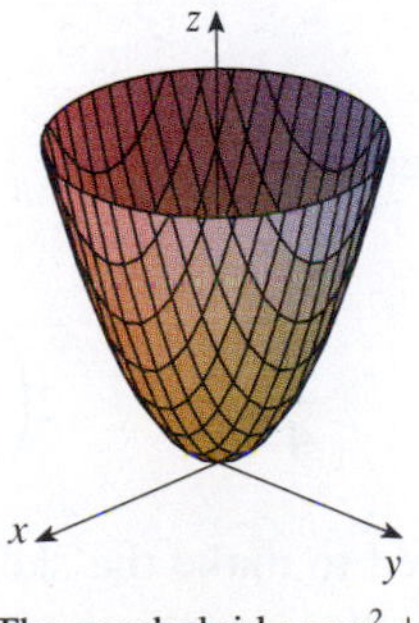

(a) The paraboloid $z = x^2 + y^2$

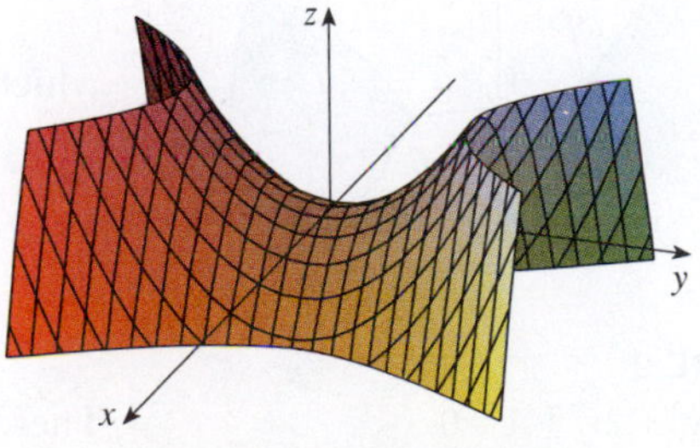

(b) The paraboloid $z = y^2 - x^2$

FIGURE 7

QUADRIC CYLINDERS When one of the variables x, y, or z is missing from the equation of a surface, then the surface is a cylinder. For instance, the equation

$$\frac{x^2}{a^2} + \frac{y^2}{b^2} = 1 \tag{7}$$

represents the **elliptic cylinder** $\{(x, y, z) \mid x^2/a^2 + y^2/b^2 = 1\}$. All horizontal traces are congruent ellipses and the generators of the cylinder are vertical lines. The **parabolic cylinder** $y = ax^2$ is illustrated in Figure 8.

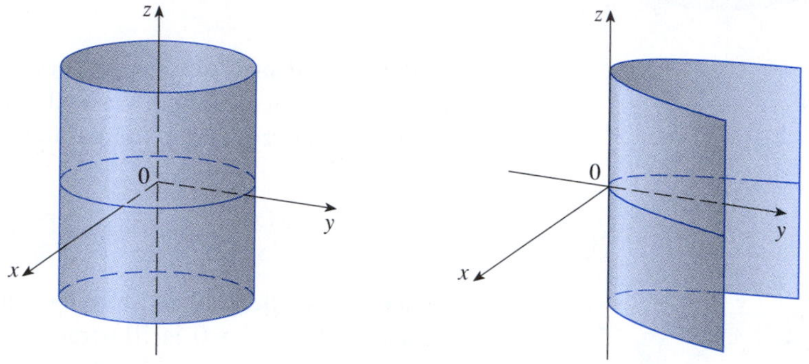

FIGURE 8

(a) Elliptic cylinder $\frac{x^2}{a^2} + \frac{y^2}{b^2} = 1$

(b) Parabolic cylinder $y = ax^2$

EXAMPLE 1 Identify and sketch the surface $4x^2 - y^2 + 2z^2 + 4 = 0$.

SOLUTION Dividing by -4, we first put the equation in standard form:

$$-x^2 + \frac{y^2}{4} - \frac{z^2}{2} = 1$$

Comparing this equation with Equation 3, we see that it represents a hyperboloid of two sheets, the only difference being that in this case the axis of the hyperboloid is the y-axis. The traces in the xy- and yz-planes are the hyperbolas

$$-x^2 + \frac{y^2}{4} = 1, \quad z = 0 \qquad \text{and} \qquad \frac{y^2}{4} - \frac{z^2}{2} = 1, \quad x = 0$$

The surface has no trace in the xz-plane, but traces in the vertical planes $y = k$ for $|k| > 2$ are the ellipses

$$x^2 + \frac{z^2}{2} = \frac{k^2}{4} - 1 \qquad y = k$$

which can be written as

$$\frac{x^2}{\dfrac{k^2}{4} - 1} + \frac{z^2}{2\left(\dfrac{k^2}{4} - 1\right)} = 1 \qquad y = k$$

These traces are used to make the sketch in Figure 9. ■

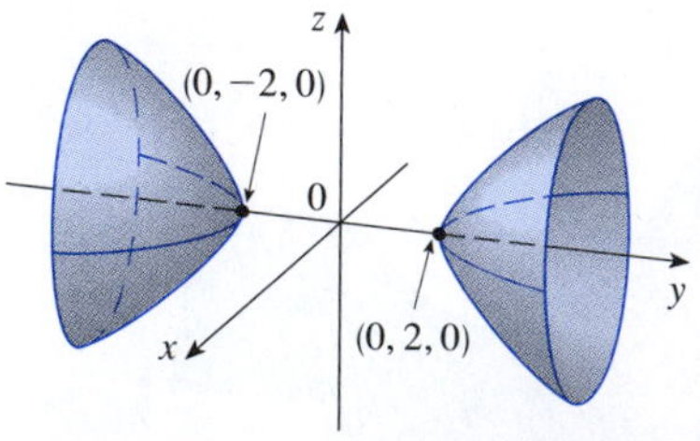

FIGURE 9
$4x^2 - y^2 + 2z^2 + 4 = 0$

EXAMPLE 2 Classify the quadric surface $x^2 + 2z^2 - 6x - y + 10 = 0$.

SOLUTION By completing the square we rewrite the equation as

$$y - 1 = (x - 3)^2 + 2z^2$$

Comparing this equation with Equation 5, we see that it represents an elliptic paraboloid. Here, however, the axis of the paraboloid is parallel to the y-axis and its vertex is the point $(3, 1, 0)$. The traces in the plane $y = k$ $(k > 1)$ are the ellipses $(x - 3)^2 + 2z^2 = k - 1$, $y = k$. The trace in the xy-plane is the parabola with equation $y = 1 + (x - 3)^2$, $z = 0$. The paraboloid is sketched in Figure 10. ■

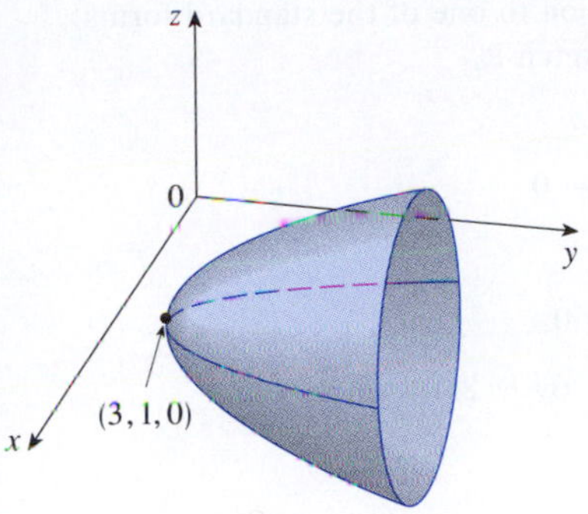

FIGURE 10
$x^2 + 2z^2 - 6x - y + 10 = 0$

EXAMPLE 3 Identify and sketch the surfaces:
(a) $x^2 + y^2 = 1$ (b) $x^2 + z^2 = 1$

SOLUTION
(a) Since z is missing and the equations $x^2 + y^2 = 1$, $z = k$, represent a circle with radius 1 in the plane $z = k$, the surface $x^2 + y^2 = 1$ is a circular cylinder whose axis is the z-axis (see Figure 11).

(b) In this case y is missing and the surface is a circular cylinder whose axis is the y-axis (see Figure 12). It is obtained by taking the circle $x^2 + z^2 = 1$, $y = 0$, in the xz-plane and moving it parallel to the y-axis.

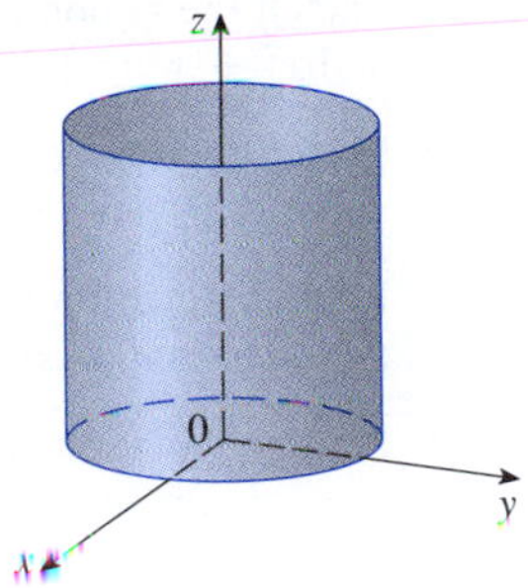

FIGURE 11
$x^2 + y^2 = 1$

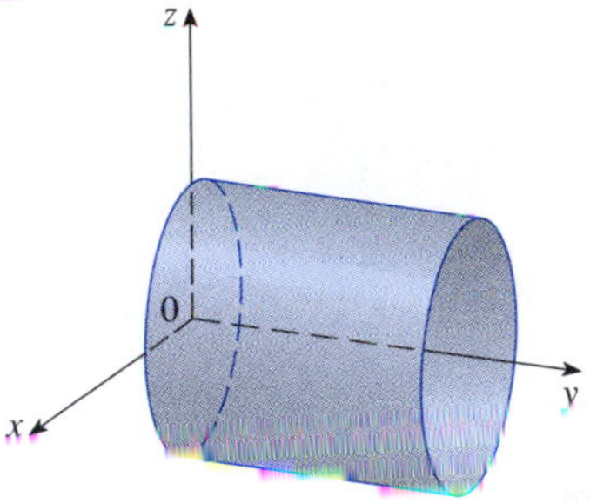

FIGURE 12
$x^2 + z^2 = 1$ ■

NOTE: When you are dealing with surfaces, it is important to recognize that an equation like $x^2 + y^2 = 1$ represents a cylinder and not a circle. The trace of the cylinder $x^2 + y^2 = 1$ in the xy-plane is the circle with equations $x^2 + y^2 = 1$, $z = 0$.

EXERCISES 11.5

1–16 ■ Find the traces of the given surface in the planes $x = k$, $y = k$, $z = k$. Then identify the surface and sketch it.

1. $x^2 - y^2 + z^2 = 1$

2. $x = y^2 + z^2$

3. $4x^2 + 9y^2 + 36z^2 = 36$

4. $2x^2 + z^2 = 4$

5. $4z^2 - x^2 - y^2 = 1$

6. $z = x^2 - y^2$

7. $z = y^2$

8. $25y^2 + z^2 = 100 + 4x^2$

9. $y^2 = x^2 + z^2$

10. $9x^2 - y^2 - z^2 = 9$

11. $x^2 + 4z^2 - y = 0$

12. $x^2 - y^2 = 1$

13. $y^2 + 9z^2 = 9$

14. $x^2 + 4y^2 + z^2 = 4$

15. $y = z^2 - x^2$

16. $16x^2 = y^2 + 4z^2$

17–24 ■ Match the equation with its graph (labeled I–VIII). Give reasons for your choices.

17. $x^2 + 4y^2 + 9z^2 = 1$

18. $9x^2 + 4y^2 + z^2 = 1$

19. $x^2 - y^2 + z^2 = 1$

20. $-x^2 + y^2 - z^2 = 1$

21. $y = 2x^2 + z^2$

22. $y^2 = x^2 + 2z^2$

23. $x^2 + 2z^2 = 1$

24. $y = x^2 - z^2$

I

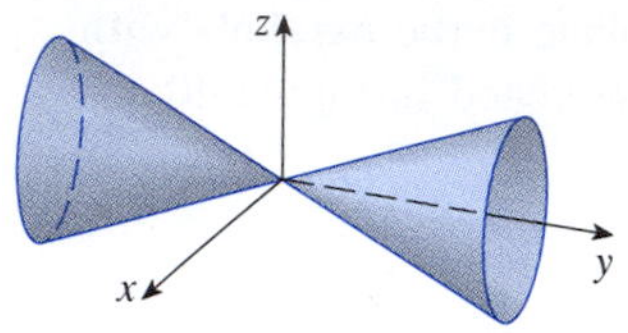

II

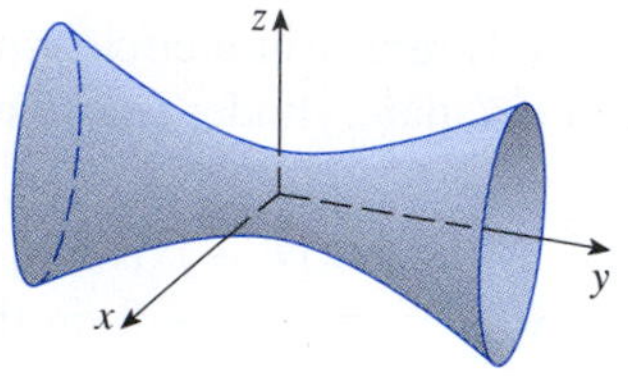

III

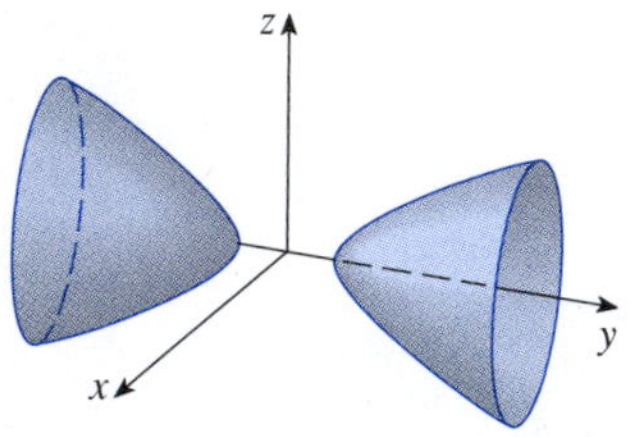

IV

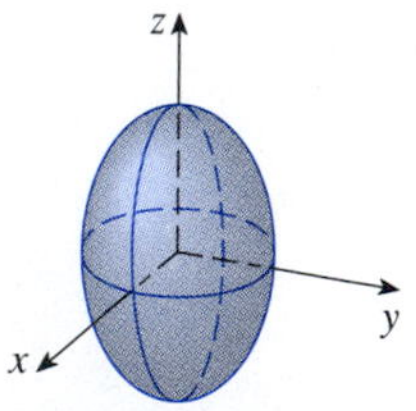

V

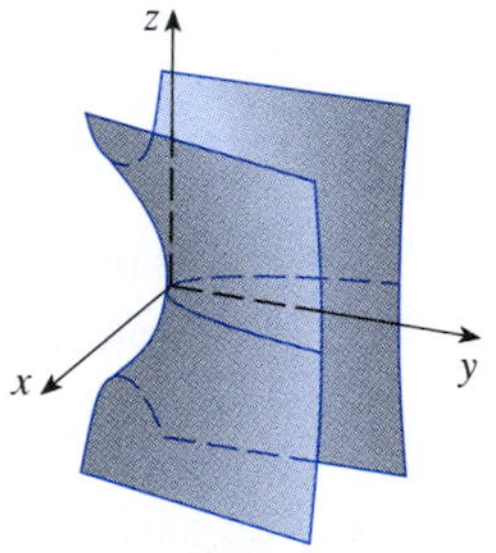

VI

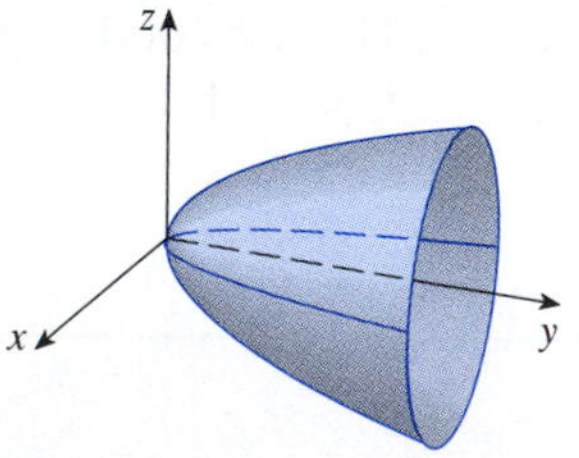

VII

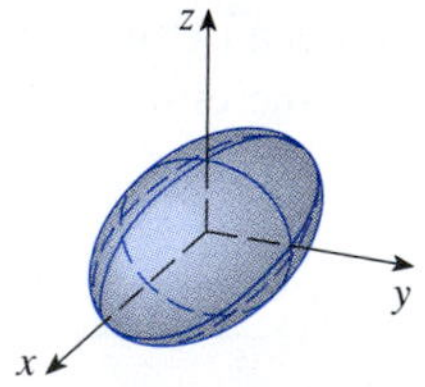

VIII

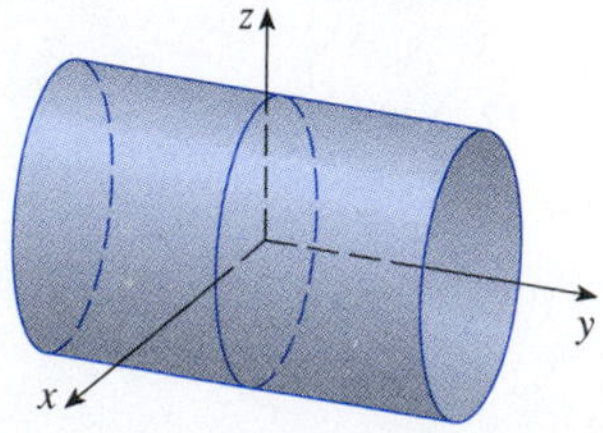

25–34 ■ Reduce the equation to one of the standard forms, classify the surface, and sketch it.

25. $z^2 = 3x^2 + 4y^2 - 12$

26. $4x^2 - 9y^2 + z^2 + 36 = 0$

27. $z = x^2 + y^2 + 1$

28. $x^2 + 4y^2 + z^2 - 2x = 0$

29. $x^2 + y^2 - 4z^2 + 4x - 6y - 8z = 13$

30. $4x = y^2 - 2z^2$

31. $x^2 + 4y^2 = 100$

32. $9x^2 + y^2 - z^2 - 2y + 2z = 0$

33. $x^2 - y^2 + 4y + z = 4$

34. $4x^2 - y^2 + z^2 + 8x + 8z + 24 = 0$

35–38 ■ Use a computer with three-dimensional graphing software to graph the surface.

35. $z = 3x^2 - 5y^2$

36. $8x^2 + 15y^2 + 5z^2 = 100$

37. $z^2 = x^2 + 4y^2$

38. $z = y^2 + xy$

39. Sketch the region bounded by the surfaces $z = \sqrt{x^2 + y^2}$ and $x^2 + y^2 = 1$ for $1 \leqslant z \leqslant 2$.

40. Sketch the region bounded by the paraboloids $z = x^2 + y^2$ and $z = 2 - x^2 - y^2$.

41. Find an equation for the surface obtained by rotating the parabola $y = x^2$ about the y-axis.

42. Find an equation for the surface obtained by rotating the line $x = 3y$ about the x-axis.

43. Find an equation for the surface consisting of all points that are equidistant from the point $(-1, 0, 0)$ and the plane $x = 1$. Identify the surface.

44. Find an equation for the surface consisting of all points P for which the distance from P to the x-axis is twice the distance from P to the yz-plane. Identify the surface.

45. Show that if the point (a, b, c) lies on the hyperbolic paraboloid $z = y^2 - x^2$, then the lines with parametric equations

$$x = a + t \qquad y = b + t \qquad z = c + 2(b - a)t$$

and

$$x = a + t \qquad y = b - t \qquad z = c - 2(b + a)t$$

both lie entirely on this paraboloid. (This shows that the hyperbolic paraboloid is what is called a **ruled surface**; that is, it can be generated by the motion of a straight line. In fact, this exercise shows that through each point on the hyperbolic paraboloid there are two generating lines. The only other quadric surfaces that are ruled surfaces are cylinders, cones, and hyperboloids of one sheet.)

46. Show that the curve of intersection of the surfaces $x^2 + 2y^2 - z^2 + 3x = 1$ and $2x^2 + 4y^2 - 2z^2 - 5y = 0$ lies in a plane.

47. Graph the surfaces $z = x^2 + y^2$ and $z = 1 - y^2$ on a common screen using the domain $|x| \leq 1.2$, $|y| \leq 1.2$ and observe the curve of intersection of these surfaces. Show that the projection of this curve onto the xy-plane is an ellipse.

48. Investigate the family of surfaces $z = x^2 + y^2 + cxy$. In particular, you should determine the transitional values of c for which the surface changes from one type of quadric surface to another.

11.6 VECTOR FUNCTIONS AND SPACE CURVES

The functions that we have used so far have been either real-valued functions or vector-valued functions in two dimensions (see Section 1.3). We now study functions whose values are three-dimensional vectors because such functions are needed to describe curves in space and the motion of particles in space.

Let $\mathbf{r}$ be a **vector function** whose range is a set of three-dimensional vectors. This means that for every number t in the domain of $\mathbf{r}$ there is a unique vector in V_3 denoted by $\mathbf{r}(t)$. If $f(t)$, $g(t)$, and $h(t)$ are the components of the vector $\mathbf{r}(t)$, then f, g, and h are real-valued functions called the **component functions** of $\mathbf{r}$ and we can write

$$\mathbf{r}(t) = \langle f(t), g(t), h(t)\rangle = f(t)\,\mathbf{i} + g(t)\,\mathbf{j} + h(t)\,\mathbf{k}$$

We use the letter t to denote the independent variable because it represents time in most applications of vector functions.

EXAMPLE 1 If

$$\mathbf{r}(t) = \langle t^3, \ln(3 - t), \sqrt{t}\,\rangle$$

then the component functions are

$$f(t) = t^3 \qquad g(t) = \ln(3 - t) \qquad h(t) = \sqrt{t}$$

By our usual convention, the domain of $\mathbf{r}$ consists of all values of t for which the expression for $\mathbf{r}(t)$ is defined. The expressions t^3, $\ln(3 - t)$, and $\sqrt{t}$ are all defined when $3 - t > 0$ and $t \geq 0$. Therefore, the domain of $\mathbf{r}$ is the interval $[0, 3)$. ■

The **limit** of a vector function $\mathbf{r}$ is defined by taking the limits of its component functions as follows:

(1) If $\mathbf{r}(t) = \langle f(t), g(t), h(t)\rangle$, then

$$\lim_{t\to a} \mathbf{r}(t) = \left\langle \lim_{t\to a} f(t), \lim_{t\to a} g(t), \lim_{t\to a} h(t)\right\rangle$$

provided the limits of the component functions exist.

Equivalently, we could have used an ε, δ definition (see Exercise 64). Limits of vector functions obey the same rules as limits of real-valued functions. (See Exercise 63.)

EXAMPLE 2 Find $\lim_{t\to 0} \mathbf{r}(t)$ where $\mathbf{r}(t) = (1 + t^3)\,\mathbf{i} + te^{-t}\mathbf{j} + \dfrac{\sin t}{t}\,\mathbf{k}$.

SOLUTION

$$\begin{aligned}\lim_{t\to 0} \mathbf{r}(t) &= \Big[\lim_{t\to 0} (1 + t^3)\Big]\mathbf{i} + \Big[\lim_{t\to 0} te^{-t}\Big]\mathbf{j} + \Big[\lim_{t\to 0} \frac{\sin t}{t}\Big]\mathbf{k} \\ &= \mathbf{i} + \mathbf{k} \qquad \text{[by (3.4.4)]}\end{aligned}$$

■

A vector function $\mathbf{r}$ is **continuous at *a*** if

$$\lim_{t\to a} \mathbf{r}(t) = \mathbf{r}(a)$$

In view of Definition 1, we see that $\mathbf{r}$ is continuous at a if and only if its component functions f, g, and h are continuous at a.

There is a close connection between continuous vector functions and space curves. Suppose that f, g, and h are continuous real-valued functions on an interval I. Then the set C of all points (x, y, z) in space, where

(2) $$x = f(t) \qquad y = g(t) \qquad z = h(t)$$

and t varies throughout the interval I, is called a **space curve**. The equations in (2) are called **parametric equations of *C*** and t is called a **parameter**. We can think of C as being traced out by a moving particle whose position at time t is $(f(t), g(t), h(t))$. If we now consider the vector function $\mathbf{r}(t) = \langle f(t), g(t), h(t)\rangle$, then $\mathbf{r}(t)$ is the position vector of the point $P(f(t), g(t), h(t))$ on C. Thus, any continuous vector function $\mathbf{r}$ defines a space curve C that is traced out by the tip of the moving vector $\mathbf{r}(t)$, as shown in Figure 1.

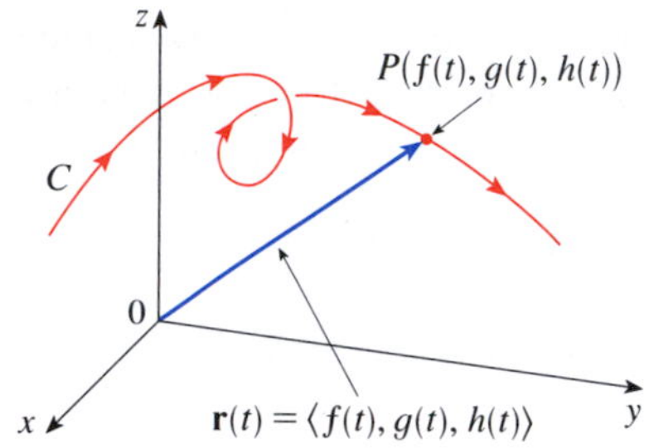

FIGURE 1
C is traced out by the tip of a moving position vector $\mathbf{r}(t)$.

EXAMPLE 3 Describe the curve defined by the vector function

$$\mathbf{r}(t) = \langle 1 + t, 2 + 5t, -1 + 6t\rangle$$

SOLUTION The corresponding parametric equations are

$$x = 1 + t \qquad y = 2 + 5t \qquad z = -1 + 6t$$

which we recognize from Equations 11.4.2 as parametric equations of a line passing through the point $(1, 2, -1)$ and parallel to the vector $\langle 1, 5, 6\rangle$. Alternatively, we could observe that the function can be written as $\mathbf{r} = \mathbf{r}_0 + t\mathbf{v}$, where $\mathbf{r}_0 = \langle 1, 2, -1\rangle$ and $\mathbf{v} = \langle 1, 5, 6\rangle$, and this is the vector equation of a line as given by Equation 11.4.1. ■

EXAMPLE 4 Sketch the curve whose vector equation is

$$\mathbf{r}(t) = 2\cos t\,\mathbf{i} + \sin t\,\mathbf{j} + t\,\mathbf{k}$$

SOLUTION The parametric equations for this curve are

$$x = 2\cos t \qquad y = \sin t \qquad z = t$$

Since $(x/2)^2 + y^2 = \cos^2 t + \sin^2 t = 1$, the curve must lie on the elliptical cylinder $x^2/4 + y^2 = 1$. Since $z = t$, the curve spirals upward around the cylinder as t increases. The curve, shown in Figure 2, is called a **helix**. ■

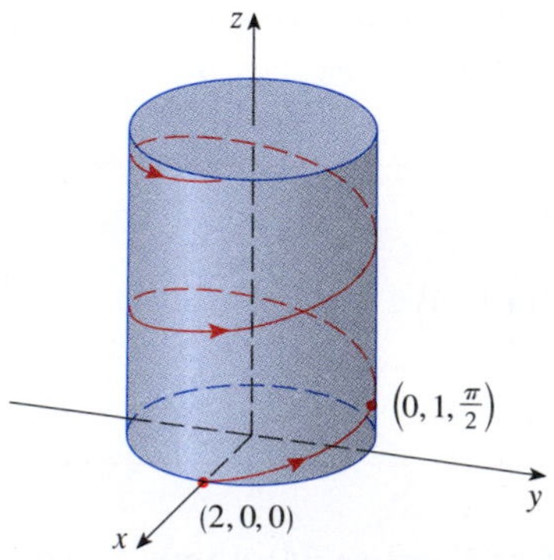

FIGURE 2

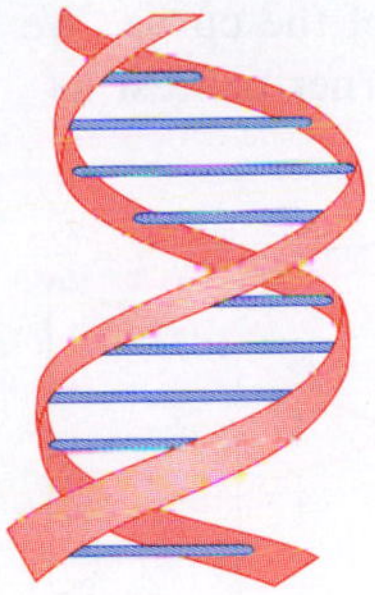

FIGURE 3

The corkscrew shape of the helix in Example 4 is familiar from its occurrence in coiled springs. It also occurs in the model of DNA (deoxyribonucleic acid, the genetic material of living cells). In 1953 James Watson and Francis Crick showed that the structure of the DNA molecule is that of two linked, parallel helices that are intertwined as in Figure 3.

As shown in Section 1.3, plane curves can also be represented in vector notation. For instance, the curve given by the parametric equations $x = \sin t$, $y = \sin^2 t$ could also be described by the vector equation

$$\mathbf{r}(t) = \langle \sin t, \sin^2 t \rangle = \sin t\,\mathbf{i} + \sin^2 t\,\mathbf{j}$$

where $\mathbf{i} = \langle 1, 0 \rangle$ and $\mathbf{j} = \langle 0, 1 \rangle$.

USING COMPUTERS TO DRAW SPACE CURVES

Space curves are inherently more difficult to draw by hand than plane curves; for an accurate representation we need to use technology. Although most graphing calculators are presently unable to graph space curves, computer graphing programs with this capability are now widely available. For instance, Figure 4 shows a computer-generated graph of the curve with parametric equations

$$x = (4 + \sin 20t)\cos t \qquad y = (4 + \sin 20t)\sin t \qquad z = \cos 20t$$

It is called a **toroidal spiral** because it lies on a torus. Another interesting curve, the **trefoil knot**, with equations

$$x = (2 + \cos 1.5t)\cos t \qquad y = (2 + \cos 1.5t)\sin t \qquad z = \sin 1.5t$$

is graphed in Figure 5. It would not be easy to plot either of these curves by hand.

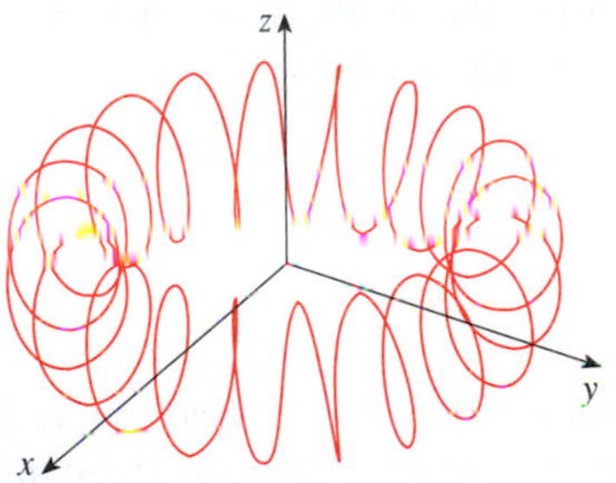

FIGURE 4
A toroidal spiral

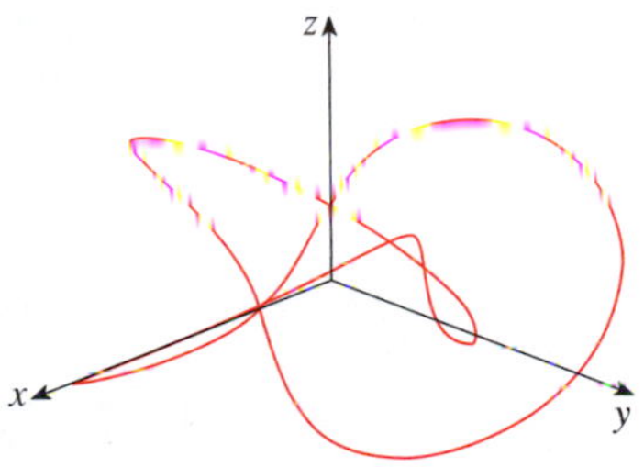

FIGURE 5
A trefoil knot

Even when a computer is used to draw a space curve, optical illusions make it difficult to get a good impression of what the curve really looks like. (This is especially true in Figure 5.) The next example shows how to cope with this problem.

EXAMPLE 5 Use a computer to sketch the curve with vector equation $\mathbf{r}(t) = \langle t, t^2, t^3 \rangle$. This curve is called a **twisted cubic**.

SOLUTION We start by using the computer to plot the curve with parametric equations $x = t$, $y = t^2$, $z = t^3$ for $-2 \leqslant t \leqslant 2$. The result is shown in Figure 6(a), but it's hard to see the true nature of the curve from that graph alone. Most three-dimensional computer graphing programs allow the user to enclose a curve or surface in a box instead of displaying the coordinate axes. When we look at the same curve in a box in Figure 6(b), we have a much clearer picture of the curve. We can see that it climbs from a lower corner of the box to the upper corner nearest us, and it twists as it climbs.

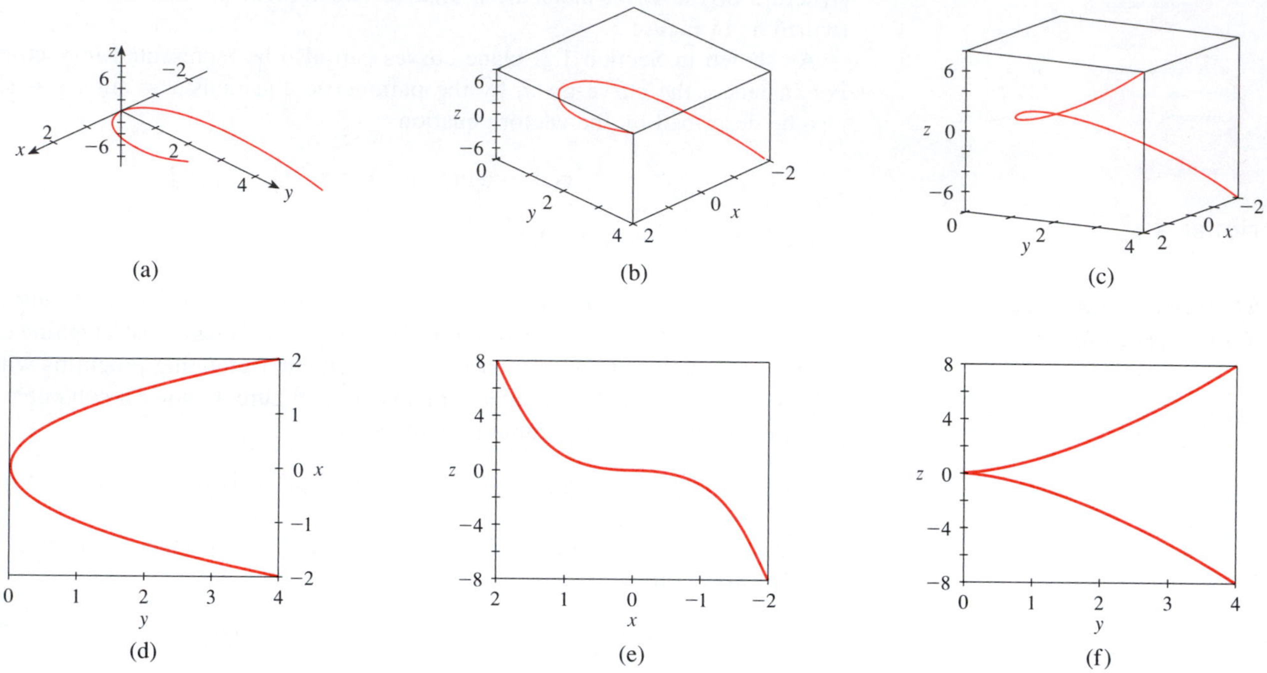

FIGURE 6
Views of the twisted cubic

We get an even better idea of the curve when we view it from different vantage points. Part (c) shows the result of rotating the box to give another viewpoint. Parts (d), (e), and (f) show the views we get when we look directly at a face of the box. In particular, part (d) shows the view from directly above the box. It is the projection of the curve on the xy-plane, namely, the parabola $y = x^2$. Part (e) shows the projection on the xz-plane, the cubic curve $z = x^3$. It is now obvious why the given curve is called a twisted cubic. ■

Another method of visualizing a space curve is to draw it on a surface. For instance, the twisted cubic in Example 5 lies on the parabolic cylinder $y = x^2$. (Eliminate the parameter from the first two parametric equations, $x = t$ and $y = t^2$.) Figure 7 shows both the cylinder and the twisted cubic, and we see that the curve moves upward from the origin along the surface of the cylinder.

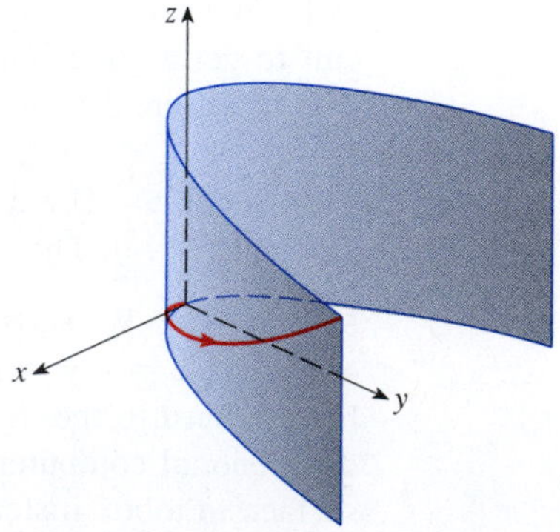

FIGURE 7

A third method for visualizing the twisted cubic is to realize that it also lies on the cylinder $z = x^3$. So it can be viewed as the curve of intersection of the cylinders $y = x^2$ and $z = x^3$.

DERIVATIVES AND INTEGRALS

The **derivative** $\mathbf{r}'$ of a vector function $\mathbf{r}$ is defined just as for a real-valued function:

(3)
$$\frac{d\mathbf{r}}{dt} = \mathbf{r}'(t) = \lim_{h \to 0} \frac{\mathbf{r}(t + h) - \mathbf{r}(t)}{h}$$

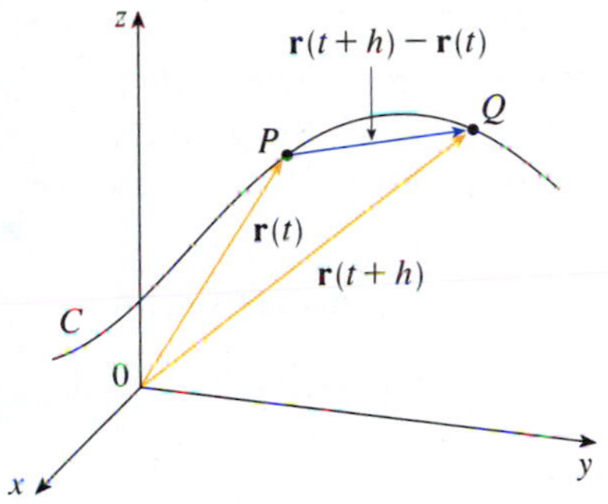

(a) The secant vector

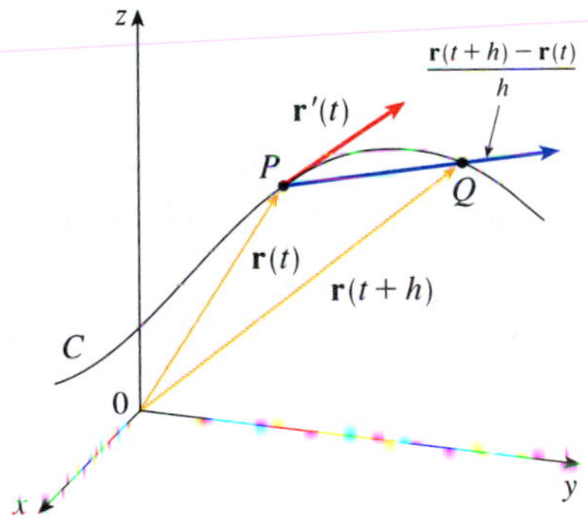

(b) The tangent vector

FIGURE 8

if this limit exists. The geometric significance of this definition is shown in Figure 8. If P and Q have position vectors $\mathbf{r}(t)$ and $\mathbf{r}(t + h)$, then $\overrightarrow{PQ}$ represents the vector $\mathbf{r}(t + h) - \mathbf{r}(t)$, which can therefore be regarded as a secant vector. If $h > 0$, the scalar multiple $(1/h)(\mathbf{r}(t + h) - \mathbf{r}(t))$ has the same direction as $\mathbf{r}(t + h) - \mathbf{r}(t)$. As $h \to 0$, it appears that this vector approaches a vector that lies on the tangent line. For this reason, the vector $\mathbf{r}'(t)$ is called the **tangent vector** to the curve defined by $\mathbf{r}$ at the point P, provided that $\mathbf{r}'(t)$ exists and $\mathbf{r}'(t) \neq \mathbf{0}$. The **tangent line** to C at P is defined to be the line through P parallel to the tangent vector $\mathbf{r}'(t)$. We will also have occasion to consider the **unit tangent vector**, which is

$$\mathbf{T}(t) = \frac{\mathbf{r}'(t)}{|\mathbf{r}'(t)|}$$

The following theorem gives us a convenient method for computing the derivative of a vector function $\mathbf{r}$; just differentiate each component of $\mathbf{r}$.

(4) THEOREM If $\mathbf{r}(t) = \langle f(t), g(t), h(t) \rangle = f(t)\,\mathbf{i} + g(t)\,\mathbf{j} + h(t)\,\mathbf{k}$, where f, g, and h are differentiable functions, then

$$\mathbf{r}'(t) = \langle f'(t), g'(t), h'(t) \rangle = f'(t)\,\mathbf{i} + g'(t)\,\mathbf{j} + h'(t)\,\mathbf{k}$$

PROOF

$$\mathbf{r}'(t) = \lim_{\Delta t \to 0} \frac{1}{\Delta t}[\mathbf{r}(t + \Delta t) - \mathbf{r}(t)]$$

$$= \lim_{\Delta t \to 0} \frac{1}{\Delta t}[\langle f(t + \Delta t), g(t + \Delta t), h(t + \Delta t) \rangle - \langle f(t), g(t), h(t) \rangle]$$

$$= \lim_{\Delta t \to 0} \left\langle \frac{f(t + \Delta t) - f(t)}{\Delta t}, \frac{g(t + \Delta t) - g(t)}{\Delta t}, \frac{h(t + \Delta t) - h(t)}{\Delta t} \right\rangle$$

$$= \left\langle \lim_{\Delta t \to 0} \frac{f(t + \Delta t) - f(t)}{\Delta t}, \lim_{\Delta t \to 0} \frac{g(t + \Delta t) - g(t)}{\Delta t}, \lim_{\Delta t \to 0} \frac{h(t + \Delta t) - h(t)}{\Delta t} \right\rangle$$

$$= \langle f'(t), g'(t), h'(t) \rangle$$

□

EXAMPLE 6

(a) Find the derivative of $\mathbf{r}(t) = (1 + t^3)\,\mathbf{i} + te^{-t}\,\mathbf{j} + \sin 2t\,\mathbf{k}$.

(b) Find the unit tangent vector at the point where $t = 0$.

SOLUTION

(a) Theorem 4 gives

$$\mathbf{r}'(t) = 3t^2\,\mathbf{i} + (1 - t)e^{-t}\,\mathbf{j} + 2\cos 2t\,\mathbf{k}$$

(b) Since $\mathbf{r}(0) = \mathbf{i}$ and $\mathbf{r}'(0) = \mathbf{j} + 2\,\mathbf{k}$, the unit tangent vector at the point $(1, 0, 0)$ is

$$\mathbf{T}(0) = \frac{\mathbf{r}'(0)}{|\mathbf{r}'(0)|} = \frac{\mathbf{j} + 2\,\mathbf{k}}{\sqrt{1+4}} = \frac{1}{\sqrt{5}}\,\mathbf{j} + \frac{2}{\sqrt{5}}\,\mathbf{k}$$

■

EXAMPLE 7 For the curve $\mathbf{r}(t) = \sqrt{t}\,\mathbf{i} + (2 - t)\mathbf{j}$, find $\mathbf{r}'(t)$ and sketch the position vector $\mathbf{r}(1)$ and the tangent vector $\mathbf{r}'(1)$.

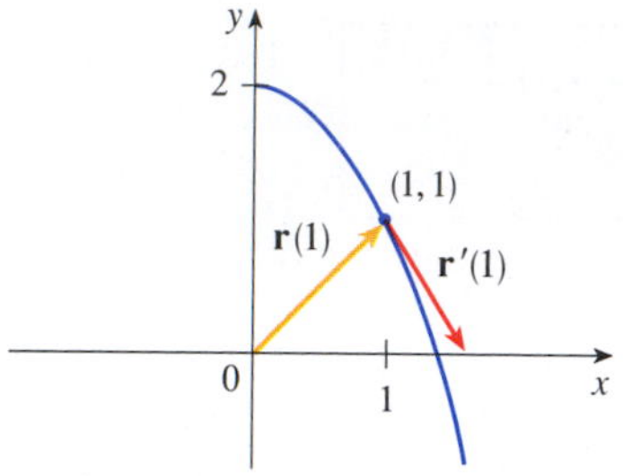

FIGURE 9

SOLUTION We have

$$\mathbf{r}'(t) = \frac{1}{2\sqrt{t}}\,\mathbf{i} - \mathbf{j} \qquad \text{and} \qquad \mathbf{r}'(1) = \frac{1}{2}\,\mathbf{i} - \mathbf{j}$$

The curve is a plane curve and elimination of the parameter from the equations $x = \sqrt{t}$, $y = 2 - t$ gives $y = 2 - x^2$, $x \geqslant 0$. In Figure 9 we draw the position vector $\mathbf{r}(1) = \mathbf{i} + \mathbf{j}$ starting at the origin and the tangent vector $\mathbf{r}'(1)$ starting at the corresponding point $(1, 1)$. ■

The helix and the tangent line in Example 8 are shown in Figure 10.

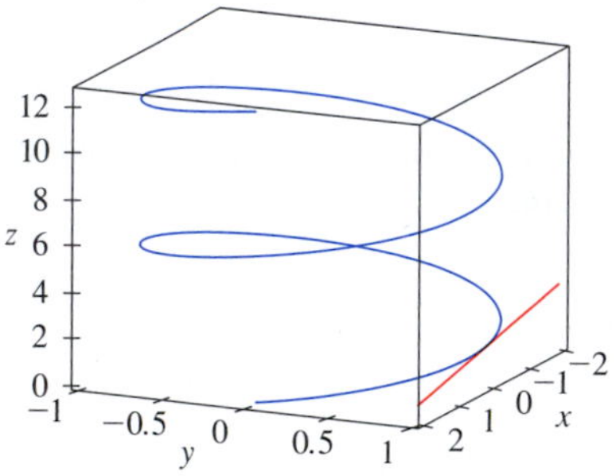

FIGURE 10

EXAMPLE 8 Find parametric equations for the tangent line to the helix with parametric equations

$$x = 2\cos t \qquad y = \sin t \qquad z = t$$

at the point $(0, 1, \pi/2)$.

SOLUTION The vector equation of the helix is $\mathbf{r}(t) = \langle 2\cos t, \sin t, t\rangle$, so

$$\mathbf{r}'(t) = \langle -2\sin t, \cos t, 1\rangle$$

The parameter value corresponding to the point $(0, 1, \pi/2)$ is $t = \pi/2$, so the tangent vector there is $\mathbf{r}'(\pi/2) = \langle -2, 0, 1\rangle$. The tangent line is the line through $(0, 1, \pi/2)$ parallel to the vector $\langle -2, 0, 1\rangle$, so by Equations 11.4.2 its parametric equations are

$$x = -2t \qquad y = 1 \qquad z = \frac{\pi}{2} + t$$

■

The next theorem shows that the differentiation formulas for real-valued functions have their counterparts for vector-valued functions.

(5) THEOREM Suppose $\mathbf{u}$ and $\mathbf{v}$ are differentiable vector functions, c is a scalar, and f is a real-valued function. Then

1. $\dfrac{d}{dt}[\mathbf{u}(t) + \mathbf{v}(t)] = \mathbf{u}'(t) + \mathbf{v}'(t)$
2. $\dfrac{d}{dt}[c\mathbf{u}(t)] = c\mathbf{u}'(t)$
3. $\dfrac{d}{dt}[f(t)\mathbf{u}(t)] = f'(t)\mathbf{u}(t) + f(t)\mathbf{u}'(t)$
4. $\dfrac{d}{dt}[\mathbf{u}(t) \cdot \mathbf{v}(t)] = \mathbf{u}'(t) \cdot \mathbf{v}(t) + \mathbf{u}(t) \cdot \mathbf{v}'(t)$
5. $\dfrac{d}{dt}[\mathbf{u}(t) \times \mathbf{v}(t)] = \mathbf{u}'(t) \times \mathbf{v}(t) + \mathbf{u}(t) \times \mathbf{v}'(t)$
6. $\dfrac{d}{dt}[\mathbf{u}(f(t))] = f'(t)\mathbf{u}'(f(t))$ (Chain Rule)

This theorem can be proved either directly from Definition 3 or by using Theorem 4 and the corresponding differentiation formulas for real-valued functions. The proofs of Formulas 1, 2, 3, 5, and 6 are left as exercises.

PROOF OF FORMULA 4 Let

$$\mathbf{u}(t) = \langle f_1(t), f_2(t), f_3(t) \rangle \qquad \mathbf{v}(t) = \langle g_1(t), g_2(t), g_3(t) \rangle$$

Then $$\mathbf{u}(t) \cdot \mathbf{v}(t) = f_1(t)g_1(t) + f_2(t)g_2(t) + f_3(t)g_3(t) = \sum_{i=1}^{3} f_i(t)g_i(t)$$

so the ordinary Product Rule gives

$$\begin{aligned}
\frac{d}{dt}[\mathbf{u}(t) \cdot \mathbf{v}(t)] &= \frac{d}{dt} \sum_{i=1}^{3} f_i(t)g_i(t) = \sum_{i=1}^{3} \frac{d}{dt}[f_i(t)g_i(t)] \\
&= \sum_{i=1}^{3} [f_i'(t)g_i(t) + f_i(t)g_i'(t)] \\
&= \sum_{i=1}^{3} f_i'(t)g_i(t) + \sum_{i=1}^{3} f_i(t)g_i'(t) \\
&= \mathbf{u}'(t) \cdot \mathbf{v}(t) + \mathbf{u}(t) \cdot \mathbf{v}'(t)
\end{aligned}$$ □

EXAMPLE 9 Show that if $|\mathbf{r}(t)| = c$ (a constant), then $\mathbf{r}'(t)$ is orthogonal to $\mathbf{r}(t)$ for all t.

SOLUTION Since

$$\mathbf{r}(t) \cdot \mathbf{r}(t) = |\mathbf{r}(t)|^2 = c^2$$

and c^2 is a constant, Formula 4 of Theorem 5 gives

$$0 = \frac{d}{dt}[\mathbf{r}(t) \cdot \mathbf{r}(t)] = \mathbf{r}'(t) \cdot \mathbf{r}(t) + \mathbf{r}(t) \cdot \mathbf{r}'(t) = 2\mathbf{r}'(t) \cdot \mathbf{r}(t)$$

Thus, $\mathbf{r}'(t) \cdot \mathbf{r}(t) = 0$, which says that $\mathbf{r}'(t)$ is orthogonal to $\mathbf{r}(t)$.

Geometrically, this result says that if a curve lies on a sphere, then the tangent vector $\mathbf{r}'(t)$ is always perpendicular to the position vector $\mathbf{r}(t)$. ■

The **definite integral** of a continuous vector function $\mathbf{r}(t)$ can be defined in much the same way as for real-valued functions except that the integral is a vector. But then we can express the integral of $\mathbf{r}$ in terms of the integrals of its component functions f, g, and h as follows. (We use the notation of Chapter 6.)

$$\begin{aligned}
\int_a^b \mathbf{r}(t)\,dt &= \lim_{\|P\|\to 0} \sum_{i=1}^{n} \mathbf{r}(t_i^*)\,\Delta t_i \\
&= \lim_{\|P\|\to 0} \left[\left(\sum_{i=1}^{n} f(t_i^*)\,\Delta t_i \right) \mathbf{i} + \left(\sum_{i=1}^{n} g(t_i^*)\,\Delta t_i \right) \mathbf{j} + \left(\sum_{i=1}^{n} h(t_i^*)\,\Delta t_i \right) \mathbf{k} \right]
\end{aligned}$$

and so

$$\int_a^b \mathbf{r}(t)\,dt = \left(\int_a^b f(t)\,dt \right) \mathbf{i} + \left(\int_a^b g(t)\,dt \right) \mathbf{j} + \left(\int_a^b h(t)\,dt \right) \mathbf{k}$$

This means that we can evaluate an integral of a vector function by integrating each component function.

The Fundamental Theorem of Calculus for continuous vector functions says that

$$\int_a^b \mathbf{r}(t)\,dt = \mathbf{R}(t)\Big]_a^b = \mathbf{R}(b) - \mathbf{R}(a)$$

where $\mathbf{R}$ is an antiderivative of $\mathbf{r}$, that is, $\mathbf{R}'(t) = \mathbf{r}(t)$. We use the notation $\int \mathbf{r}(t)\,dt$ for indefinite integrals (antiderivatives).

EXAMPLE 10 If $\mathbf{r}(t) = 2\cos t\,\mathbf{i} + \sin t\,\mathbf{j} + 2t\,\mathbf{k}$, then

$$\int \mathbf{r}(t)\,dt = \left(\int 2\cos t\,dt\right)\mathbf{i} + \left(\int \sin t\,dt\right)\mathbf{j} + \left(\int 2t\,dt\right)\mathbf{k}$$

$$= 2\sin t\,\mathbf{i} - \cos t\,\mathbf{j} + t^2\mathbf{k} + \mathbf{C}$$

where $\mathbf{C}$ is a vector constant of integration, and

$$\int_0^{\pi/2} \mathbf{r}(t)\,dt = \left[2\sin t\,\mathbf{i} - \cos t\,\mathbf{j} + t^2\mathbf{k}\right]_0^{\pi/2} = 2\,\mathbf{i} + \mathbf{j} + \frac{\pi^2}{4}\,\mathbf{k}$$ ■

EXERCISES 11.6

1–6 ■ Match the parametric equations with the graphs (labeled I–VI). Give reasons for your choices.

1. $x = \cos 4t, \quad y = t, \quad z = \sin 4t$

2. $x = t^2 - 2, \quad y = t^3, \quad z = t^4 + 1$

3. $x = t, \quad y = 1/(1 + t^2), \quad z = t^2$

4. $x = \sin 3t\cos t, \quad y = \sin 3t\sin t, \quad z = t$

5. $x = \cos t, \quad y = \sin t, \quad z = \sin 5t$

6. $x = \cos t, \quad y = \sin t, \quad z = \ln t$

I

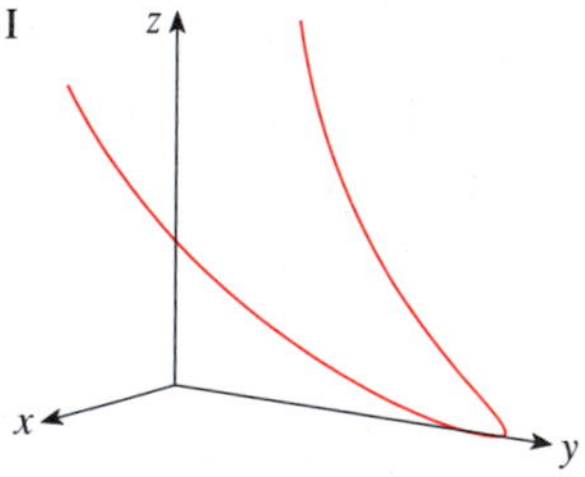

II

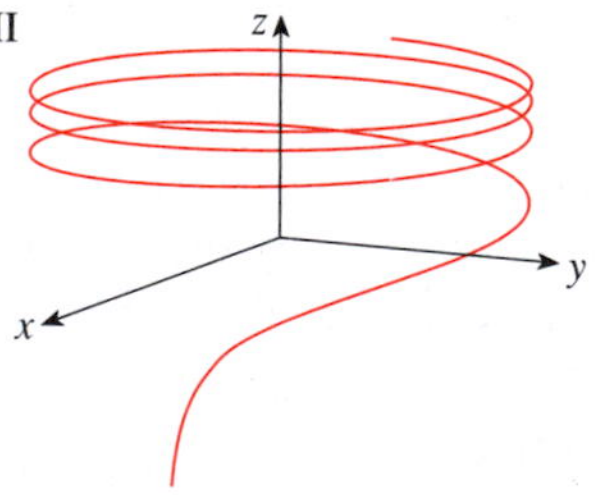

III

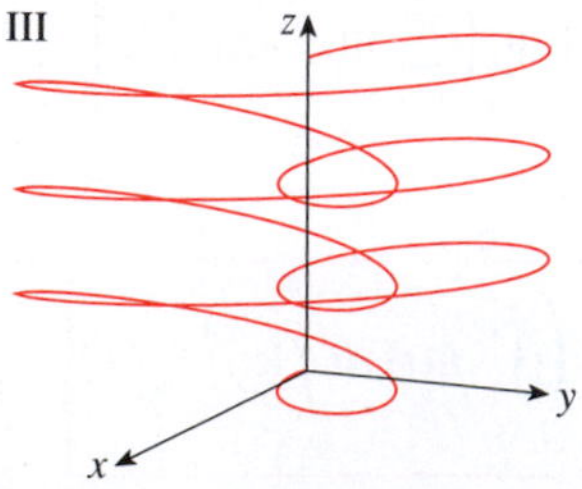

IV

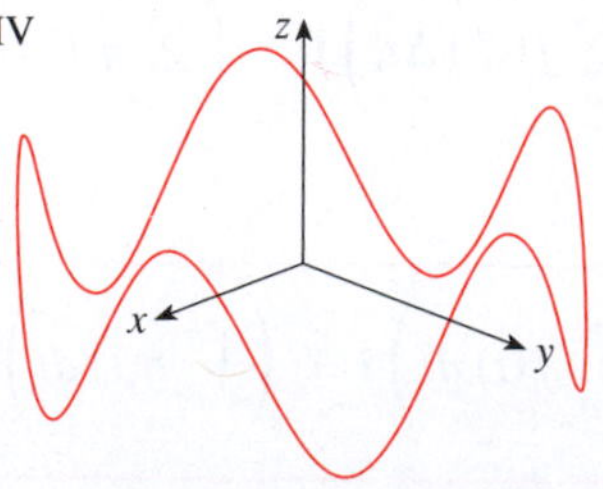

V

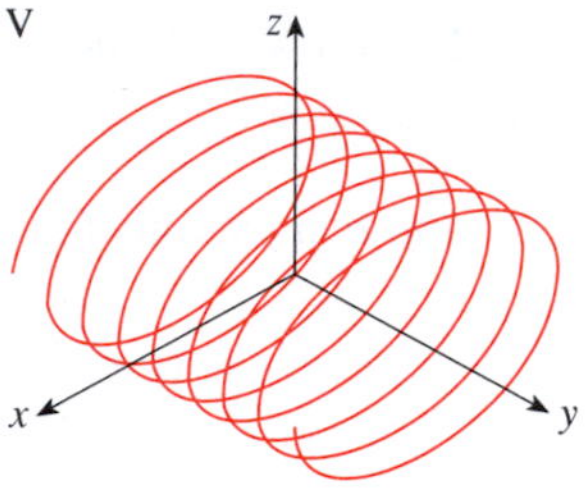

VI

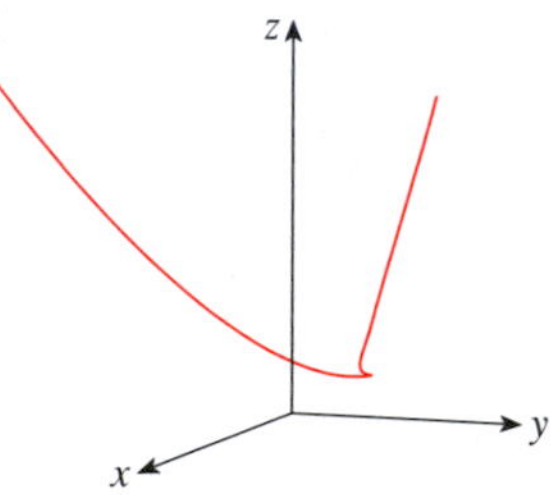

7–14 ■ Sketch the curve with the given vector equation. Indicate with an arrow the direction in which t increases.

7. $\mathbf{r}(t) = \langle t, -t, 2t\rangle$

8. $\mathbf{r}(t) = \langle t^2, t, 2\rangle$

9. $\mathbf{r}(t) = \langle \sin t, 3, \cos t\rangle$

10. $\mathbf{r}(t) = \langle \sin t, t, \cos t\rangle$

11. $\mathbf{r}(t) = (t^4 + 1)\,\mathbf{i} + t\,\mathbf{k}$

12. $\mathbf{r}(t) = t\,\mathbf{i} + t\,\mathbf{j} + \cos t\,\mathbf{k}$

13. $\mathbf{r}(t) = t^2\mathbf{i} + t^4\mathbf{j} + t^6\mathbf{k}$

14. $\mathbf{r}(t) = \sin t\,\mathbf{i} + \sin t\,\mathbf{j} + \sqrt{2}\cos t\,\mathbf{k}$

15. Show that the curve with parametric equations

$$x = t\cos t \qquad y = t\sin t \qquad z = t$$

lies on the cone $z^2 = x^2 + y^2$, and use this fact to help sketch the curve.

16. Show that the curve with parametric equations

$$x = \sin t \qquad y = \cos t \qquad z = \sin^2 t$$

is the curve of intersection of the surfaces $z = x^2$ and $x^2 + y^2 = 1$. Use this fact to help sketch the curve.

17–20 ■ Use a computer to graph the curve with the given vector equation. Make sure you choose a parameter domain and viewpoints that reveal the true nature of the curve.

17. $\mathbf{r}(t) = \langle \sin t, \cos t, t^2 \rangle$

18. $\mathbf{r}(t) = \langle t^2, t^3 - t, t \rangle$

19. $\mathbf{r}(t) = \langle \sqrt{t}, t, t^2 - 2 \rangle$

20. $\mathbf{r}(t) = \langle \sin t, \sin 2t, \sin 3t \rangle$

21. Graph the curve with parametric equations $x = (1 + \cos 16t)\cos t$, $y = (1 + \cos 16t)\sin t$, $z = 1 + \cos 16t$. Explain the appearance of the graph by showing that it lies on a cone.

22. Graph the curve with parametric equations $x = \sqrt{1 - 0.25\cos^2 10t}\cos t$, $y = \sqrt{1 - 0.25\cos^2 10t}\sin t$, $z = 0.5\cos 10t$. Explain the appearance of the graph by showing that it lies on a sphere.

23–26 ■ Find the limit.

23. $\lim_{t \to 0} \langle t, \cos t, 2 \rangle$

24. $\lim_{t \to 0} \left\langle \dfrac{1 - \cos t}{t}, t^3, e^{-1/t^2} \right\rangle$

25. $\lim_{t \to 1} \left(\sqrt{t + 3}\,\mathbf{i} + \dfrac{t - 1}{t^2 - 1}\,\mathbf{j} + \dfrac{\tan t}{t}\,\mathbf{k} \right)$

26. $\lim_{t \to \infty} \left(e^{-t}\,\mathbf{i} + \dfrac{t - 1}{t + 1}\,\mathbf{j} + \tan^{-1}t\,\mathbf{k} \right)$

27–34 ■ Find the domain and derivative of the vector function.

27. $\mathbf{r}(t) = \langle t, t^2, t^3 \rangle$

28. $\mathbf{r}(t) = \langle t^2 - 4, \sqrt{t - 4}, \sqrt{6 - t} \rangle$

29. $\mathbf{r}(t) = \mathbf{i} + \tan t\,\mathbf{j} + \sec t\,\mathbf{k}$

30. $\mathbf{r}(t) = te^{2t}\,\mathbf{i} + \dfrac{t - 1}{t + 1}\,\mathbf{j} + \tan^{-1}t\,\mathbf{k}$

31. $\mathbf{r}(t) = \ln(4 - t^2)\,\mathbf{i} + \sqrt{1 + t}\,\mathbf{j} - 4e^{3t}\,\mathbf{k}$

32. $\mathbf{r}(t) = e^{-t}\cos t\,\mathbf{i} + e^{-t}\sin t\,\mathbf{j} + \ln|t|\,\mathbf{k}$

33. $\mathbf{r}(t) = \mathbf{a} + t\,\mathbf{b} + t^2\,\mathbf{c}$

34. $\mathbf{r}(t) = t\,\mathbf{a} \times (\mathbf{b} + t\,\mathbf{c})$

35–40 ■
(a) Sketch the plane curve with the given vector equation.
(b) Find $\mathbf{r}'(t)$.
(c) Sketch the position vector $\mathbf{r}(t)$ and the tangent vector $\mathbf{r}'(t)$ for the given value of t.

35. $\mathbf{r}(t) = \langle \cos t, \sin t \rangle$, $t = \pi/4$

36. $\mathbf{r}(t) = \langle t^3, t^2 \rangle$, $t = 1$

37. $\mathbf{r}(t) = (1 + t)\,\mathbf{i} + t^2\,\mathbf{j}$, $t = 1$

38. $\mathbf{r}(t) = 2\sin t\,\mathbf{i} + 3\cos t\,\mathbf{j}$, $t = \pi/3$

39. $\mathbf{r}(t) = e^t\,\mathbf{i} + e^{-2t}\,\mathbf{j}$, $t = 0$

40. $\mathbf{r}(t) = \sec t\,\mathbf{i} + \tan t\,\mathbf{j}$, $t = \pi/4$

41–44 ■ Find the unit tangent vector $\mathbf{T}(t)$ at the point with the given value of the parameter t.

41. $\mathbf{r}(t) = \langle 2t, 3t^2, 4t^3 \rangle$, $t = 1$

42. $\mathbf{r}(t) = \langle e^{2t}, e^{-2t}, te^{2t} \rangle$, $t = 0$

43. $\mathbf{r}(t) = t\,\mathbf{i} + 2\sin t\,\mathbf{j} + 3\cos t\,\mathbf{k}$, $t = \pi/6$

44. $\mathbf{r}(t) = e^{2t}\cos t\,\mathbf{i} + e^{2t}\sin t\,\mathbf{j} + e^{2t}\,\mathbf{k}$, $t = \pi/2$

45–48 ■ Find parametric equations for the tangent line to the curve with the given parametric equations at the specified point.

45. $x = t$, $y = t^2$, $z = t^3$; $(1, 1, 1)$

46. $x = 1 + 2t$, $y = 1 + t - t^2$, $z = 1 - t + t^2 - t^3$; $(1, 1, 1)$

47. $x = t\cos 2\pi t$, $y = t\sin 2\pi t$, $z = 4t$; $(0, \frac{1}{4}, 1)$

48. $x = \sin \pi t$, $y = \sqrt{t}$, $z = \cos \pi t$; $(0, 1, -1)$

49–50 ■ Find parametric equations for the tangent line to the curve with the given parametric equations at the specified point. Illustrate by graphing both the curve and the tangent line on a common screen.

49. $x = t$, $y = \sqrt{2}\cos t$, $z = \sqrt{2}\sin t$; $(\pi/4, 1, 1)$

50. $x = \cos t$, $y = 3e^{2t}$, $z = 3e^{-2t}$; $(1, 3, 3)$

51. The curves $\mathbf{r}_1(t) = \langle t, t^2, t^3 \rangle$ and $\mathbf{r}_2(t) = \langle \sin t, \sin 2t, t \rangle$ intersect at the origin. Find their angle of intersection correct to the nearest degree.

52. At what point do the curves $\mathbf{r}_1(t) = \langle t, 1 - t, 3 + t^2 \rangle$ and $\mathbf{r}_2(s) = \langle 3 - s, s - 2, s^2 \rangle$ intersect? Find their angle of intersection correct to the nearest degree.

53. Show that the curve with parametric equations $x = t^2$, $y = 1 - 3t$, $z = 1 + t^3$ passes through the points $(1, 4, 0)$ and $(9, -8, 28)$ but not the point $(4, 7, -6)$.

54. (a) Find the point of intersection of the tangent lines to the curve $\mathbf{r}(t) = \langle \sin \pi t, 2\sin \pi t, \cos \pi t \rangle$ at the points where $t = 0$ and $t = 0.5$.
(b) Illustrate by graphing the curve and both tangent lines.

55. Try to sketch by hand the curve of intersection of the circular cylinder $x^2 + y^2 = 4$ and the parabolic cylinder $z = x^2$. Then find parametric equations for this curve using the parameter t, where $x = 2\cos t$, and use these equations and a computer to graph the curve.

56. Try to sketch by hand the curve of intersection of the parabolic cylinder $y = x^2$ and the top half of the ellipsoid $x^2 + 4y^2 + 4z^2 = 16$. Then find parametric equations for this curve using the parameter t, where $x = t$, and use these equations and a computer to graph the curve.

57–60 ■ Evaluate the integral.

57. $\displaystyle\int_0^1 (t\,\mathbf{i} + t^2\,\mathbf{j} + t^3\,\mathbf{k})\,dt$

58. $\displaystyle\int_1^2 [(1 + t^2)\,\mathbf{i} - 4t^4\,\mathbf{j} - (t^2 - 1)\,\mathbf{k}]\,dt$

59. $\int_0^{\pi/4} (\cos 2t\,\mathbf{i} + \sin 2t\,\mathbf{j} + t\sin t\,\mathbf{k})\,dt$

60. $\int_1^4 \left(\sqrt{t}\,\mathbf{i} + te^{-t}\mathbf{j} + \frac{1}{t^2}\mathbf{k}\right) dt$

61. Find $\mathbf{r}(t)$ if $\mathbf{r}'(t) = t^2\mathbf{i} + 4t^3\mathbf{j} - t^2\mathbf{k}$ and $\mathbf{r}(0) = \mathbf{j}$.

62. Find $\mathbf{r}(t)$ if $\mathbf{r}'(t) = \sin t\,\mathbf{i} - \cos t\,\mathbf{j} + 2t\,\mathbf{k}$ and $\mathbf{r}(0) = \mathbf{i} + \mathbf{j} + 2\mathbf{k}$.

63. Suppose $\mathbf{u}$ and $\mathbf{v}$ are vector functions that possess limits as $t \to a$ and let c be a constant. Prove the following properties of limits.

(a) $\lim_{t\to a} [\mathbf{u}(t) + \mathbf{v}(t)] = \lim_{t\to a} \mathbf{u}(t) + \lim_{t\to a} \mathbf{v}(t)$

(b) $\lim_{t\to a} c\mathbf{u}(t) = c \lim_{t\to a} \mathbf{u}(t)$

(c) $\lim_{t\to a} [\mathbf{u}(t) \cdot \mathbf{v}(t)] = \lim_{t\to a} \mathbf{u}(t) \cdot \lim_{t\to a} \mathbf{v}(t)$

(d) $\lim_{t\to a} [\mathbf{u}(t) \times \mathbf{v}(t)] = \lim_{t\to a} \mathbf{u}(t) \times \lim_{t\to a} \mathbf{v}(t)$

64. Show that $\lim_{t\to a} \mathbf{r}(t) = \mathbf{b}$ if and only if for every $\varepsilon > 0$ there is a number $\delta > 0$ such that $|\mathbf{r}(t) - \mathbf{b}| < \varepsilon$ whenever $0 < |t - a| < \delta$.

65. Prove Formula 1 of Theorem 5.

66. Prove Formula 3 of Theorem 5.

67. Prove Formula 5 of Theorem 5.

68. Prove Formula 6 of Theorem 5.

69. If $\mathbf{u}(t) = \mathbf{i} - 2t^2\mathbf{j} + 3t^3\mathbf{k}$ and $\mathbf{v}(t) = t\,\mathbf{i} + \cos t\,\mathbf{j} + \sin t\,\mathbf{k}$, find $D_t[\mathbf{u}(t) \cdot \mathbf{v}(t)]$.

70. If $\mathbf{u}$ and $\mathbf{v}$ are the vector functions in Exercise 69, find $D_t[\mathbf{u}(t) \times \mathbf{v}(t)]$.

71. Show that if $\mathbf{r}$ is a vector function such that $\mathbf{r}'' = (\mathbf{r}')'$ exists, then

$$\frac{d}{dt}[\mathbf{r}(t) \times \mathbf{r}'(t)] = \mathbf{r}(t) \times \mathbf{r}''(t)$$

72. Find an expression for $\frac{d}{dt}[\mathbf{u}(t) \cdot (\mathbf{v}(t) \times \mathbf{w}(t))]$.

73. If $\mathbf{r}(t) \neq \mathbf{0}$, show that $\frac{d}{dt}|\mathbf{r}(t)| = \frac{1}{|\mathbf{r}(t)|}\mathbf{r}(t) \cdot \mathbf{r}'(t)$.

74. If a curve has the property that the position vector $\mathbf{r}(t)$ is always perpendicular to the tangent vector $\mathbf{r}'(t)$, show that the curve lies on a sphere with center the origin.

75. If $\mathbf{u}(t) = \mathbf{r}(t) \cdot [\mathbf{r}'(t) \times \mathbf{r}''(t)]$, show that

$$\mathbf{u}'(t) = \mathbf{r}(t) \cdot [\mathbf{r}'(t) \times \mathbf{r}'''(t)]$$

11.7 ARC LENGTH AND CURVATURE

We already know how to find the length of a two-dimensional curve with parametric equations $x = f(t)$, $y = g(t)$, $a \leqslant t \leqslant b$. Formula 9.3.1 says that this length is given by

$$L = \int_a^b \sqrt{\left(\frac{dx}{dt}\right)^2 + \left(\frac{dy}{dt}\right)^2}\, dt$$

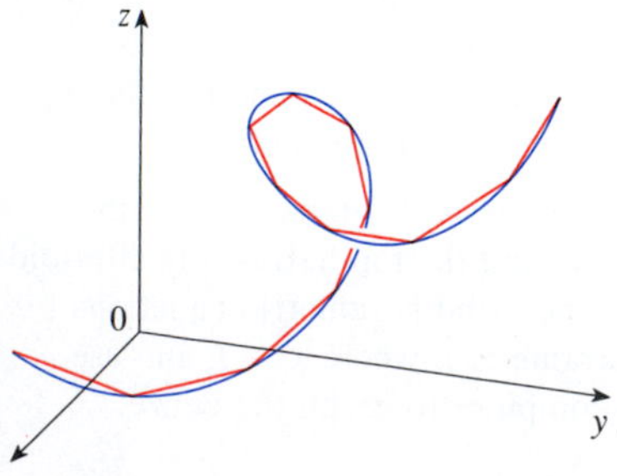

FIGURE 1
The length of a space curve is the limit of lengths of inscribed polygons.

The length of a space curve is defined in the same way as for plane curves (see Figure 1). Suppose that the curve has the vector equation $\mathbf{r}(t) = \langle f(t), g(t), h(t)\rangle$, $a \leqslant t \leqslant b$, or, equivalently, parametric equations $x = f(t)$, $y = g(t)$, $z = h(t)$, where f', g', and h' are continuous. If the curve is traversed exactly once as t increases from a to b, then it can be shown that its length is

(1)
$$L = \int_a^b \sqrt{[f'(t)]^2 + [g'(t)]^2 + [h'(t)]^2}\, dt$$
$$= \int_a^b \sqrt{\left(\frac{dx}{dt}\right)^2 + \left(\frac{dy}{dt}\right)^2 + \left(\frac{dz}{dt}\right)^2}\, dt$$

using a derivation similar to that given for two-dimensional curves. Notice that the arc length formula (1) can be put into the more compact form

(2) $$L = \int_a^b |\mathbf{r}'(t)|\,dt$$

because, for plane curves $\mathbf{r}(t) = f(t)\,\mathbf{i} + g(t)\,\mathbf{j}$,

$$|\mathbf{r}'(t)| = |f'(t)\,\mathbf{i} + g'(t)\,\mathbf{j}| = \sqrt{[f'(t)]^2 + [g'(t)]^2}$$

whereas, for space curves $\mathbf{r}(t) = f(t)\,\mathbf{i} + g(t)\,\mathbf{j} + h(t)\,\mathbf{k}$,

$$|\mathbf{r}'(t)| = |f'(t)\,\mathbf{i} + g'(t)\,\mathbf{j} + h'(t)\,\mathbf{k}| = \sqrt{[f'(t)]^2 + [g'(t)]^2 + [h'(t)]^2}$$

Figure 2 shows the arc of the helix whose length is computed in Example 1.

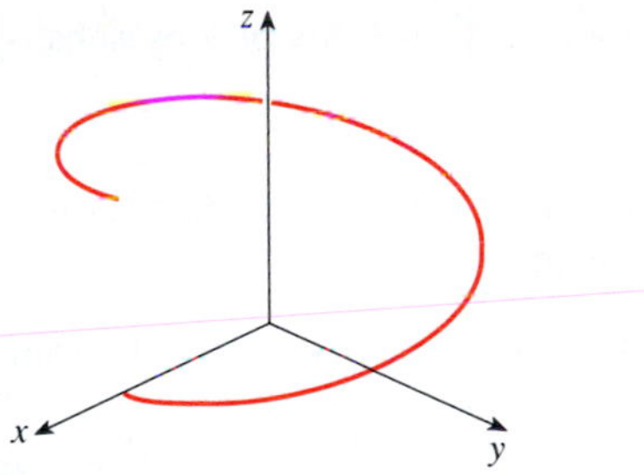

FIGURE 2

EXAMPLE 1 Find the length of the arc of the circular helix with vector equation $\mathbf{r}(t) = \cos t\,\mathbf{i} + \sin t\,\mathbf{j} + t\,\mathbf{k}$ from the point $(1, 0, 0)$ to the point $(1, 0, 2\pi)$.

SOLUTION Since $\mathbf{r}'(t) = -\sin t\,\mathbf{i} + \cos t\,\mathbf{j} + \mathbf{k}$, we have

$$|\mathbf{r}'(t)| = \sqrt{(-\sin t)^2 + \cos^2 t + 1} = \sqrt{2}$$

The arc from $(1, 0, 0)$ to $(1, 0, 2\pi)$ is described by the parameter interval $0 \leqslant t \leqslant 2\pi$ and so, from Formula 2, we have

$$L = \int_0^{2\pi} |\mathbf{r}'(t)|\,dt = \int_0^{2\pi} \sqrt{2}\,dt = 2\sqrt{2}\,\pi$$ ■

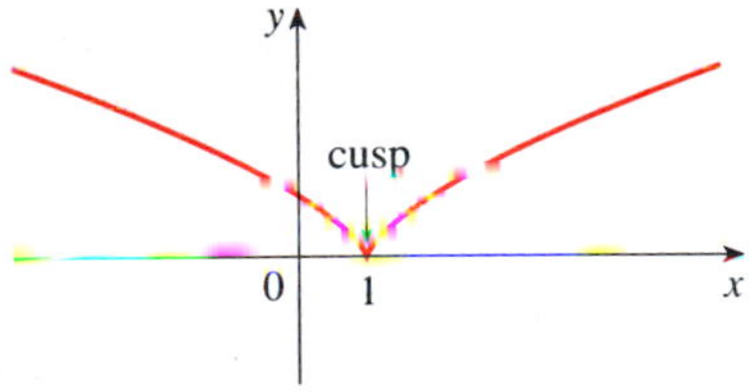

FIGURE 3
The curve $\mathbf{r}(t) = \langle 1 + t^3, t^2 \rangle$ is not smooth.

A curve given by a vector function $\mathbf{r}(t)$ on an interval I is called **smooth** if $\mathbf{r}'$ is continuous and $\mathbf{r}'(t) \neq \mathbf{0}$ (except possibly at any endpoints of I). The significance of the condition $\mathbf{r}'(t) \neq \mathbf{0}$ is illustrated by the graph of the semicubical parabola $\mathbf{r}(t) = \langle 1 + t^3, t^2 \rangle$ that is shown in Figure 3. Since $\mathbf{r}'(t) = \langle 3t^2, 2t \rangle$, we have $\mathbf{r}'(0) = \langle 0, 0 \rangle = \mathbf{0}$. The point that corresponds to $t = 0$ is $(1, 0)$ and here there is a sharp corner that is called a **cusp**. We can think of a smooth curve as a curve with no such cusp. A curve, such as the semicubical parabola, that is made up of a finite number of smooth pieces is called **piecewise smooth**. The arc length formula (2) holds for piecewise-smooth curves.

A single curve C can be represented by more than one vector function. For instance, the twisted cubic

(3) $$\mathbf{r}_1(t) = \langle t, t^2, t^3 \rangle \qquad 1 \leqslant t \leqslant 2$$

could also be represented by the function

(4) $$\mathbf{r}_2(u) = \langle e^u, e^{2u}, e^{3u} \rangle \qquad 0 \leqslant u \leqslant \ln 2$$

where the connection between the parameters t and u is given by $t = e^u$. We say that Equations 3 and 4 are **parametrizations** of the curve C. If we were to compute the length of C using Equation 1, we would get the same answer. In general, it can be shown (see Exercise 43) that when Equation 2 is used to compute the length of any piecewise-smooth curve, the arc length is independent of the parametrization that is used.

Now we suppose that C is a piecewise-smooth curve given by a vector function $\mathbf{r}(t) = f(t)\,\mathbf{i} + g(t)\,\mathbf{j} + h(t)\,\mathbf{k}$, $a \leqslant t \leqslant b$, and at least one of f, g, h is one-to-one on (a, b). As in Section 9.3, we define its **arc length function** s by

$$s(t) = \int_a^t |\mathbf{r}'(u)|\,du = \int_a^t \sqrt{\left(\frac{dx}{du}\right)^2 + \left(\frac{dy}{du}\right)^2 + \left(\frac{dz}{du}\right)^2}\,du \tag{5}$$

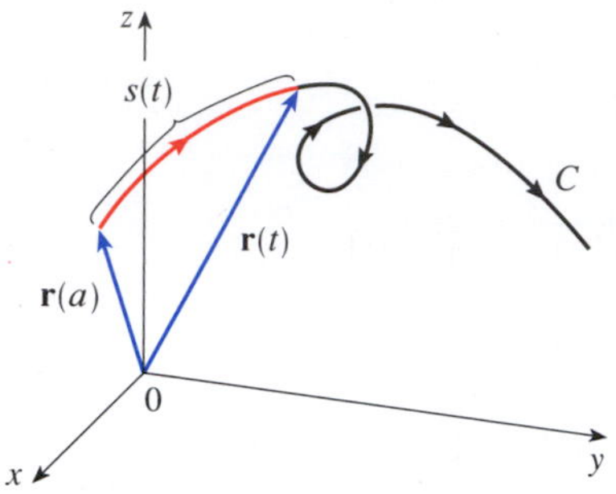

FIGURE 4

Thus, $s(t)$ is the length of the part of C between $\mathbf{r}(a)$ and $\mathbf{r}(t)$ (see Figure 4). If we differentiate both sides of Equation 5 using Part 1 of the Fundamental Theorem of Calculus, we obtain

$$\frac{ds}{dt} = |\mathbf{r}'(t)| \tag{6}$$

It is often useful to **parametrize a curve with respect to arc length** because arc length arises naturally from the shape of the curve and does not depend on a particular coordinate system. If a curve $\mathbf{r}(t)$ is already given in terms of a parameter t and $s(t)$ is the arc length function given by Equation 5, then we may be able to solve for t as a function of s: $t = t(s)$. Then the curve can be reparametrized in terms of s by substituting for t: $\mathbf{r} = \mathbf{r}(t(s))$.

EXAMPLE 2 Reparametrize the helix $\mathbf{r}(t) = \cos t\,\mathbf{i} + \sin t\,\mathbf{j} + t\,\mathbf{k}$ with respect to arc length measured from $(1, 0, 0)$ in the direction of increasing t.

SOLUTION The initial point $(1, 0, 0)$ corresponds to the parameter value $t = 0$. From Example 1 we have

$$\frac{ds}{dt} = |\mathbf{r}'(t)| = \sqrt{2}$$

and so
$$s = s(t) = \int_0^t |\mathbf{r}'(u)|\,du = \int_0^t \sqrt{2}\,du = \sqrt{2}\,t$$

Therefore, $t = s/\sqrt{2}$ and the required reparametrization is obtained by substituting for t:

$$\mathbf{r}(t(s)) = \cos(s/\sqrt{2}\,)\,\mathbf{i} + \sin(s/\sqrt{2}\,)\,\mathbf{j} + (s/\sqrt{2}\,)\,\mathbf{k}$$

■

CURVATURE

If C is a smooth curve defined by the vector function $\mathbf{r}$, then $\mathbf{r}'(t) \neq \mathbf{0}$. Recall that the unit tangent vector $\mathbf{T}(t)$ is given by

$$\mathbf{T}(t) = \frac{\mathbf{r}'(t)}{|\mathbf{r}'(t)|}$$

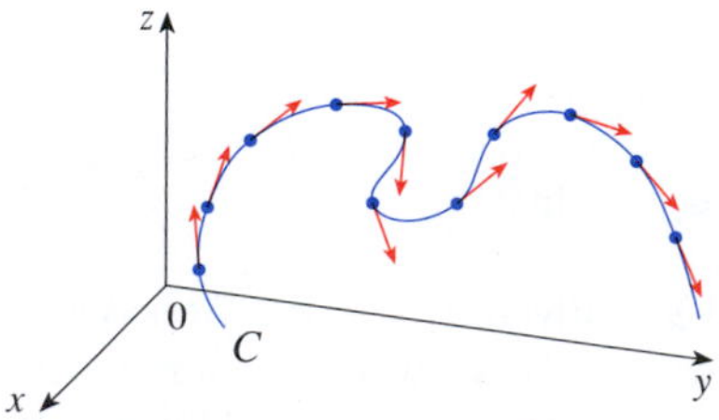

FIGURE 5
Unit tangent vectors at equally spaced points on C

and indicates the direction of the curve. From Figure 5 you can see that $\mathbf{T}(t)$ changes direction very slowly when C is fairly straight, but it changes direction more quickly when C bends or twists more sharply.

The curvature of C at a given point is a measure of how quickly the curve changes direction at that point. Specifically, we define it to be the magnitude of the rate of change of the unit tangent vector with respect to arc length. (We use arc length so that the curvature will be independent of the parametrization.)

(7) DEFINITION The **curvature** of a curve is

$$\kappa = \left| \frac{d\mathbf{T}}{ds} \right|$$

where $\mathbf{T}$ is the unit tangent vector.

The curvature is easier to compute if it is expressed in terms of the parameter t instead of s, so we use the Chain Rule (Theorem 11.6.5, Formula 6) to write

$$\frac{d\mathbf{T}}{dt} = \frac{d\mathbf{T}}{ds}\frac{ds}{dt} \qquad \text{and} \qquad \kappa = \left| \frac{d\mathbf{T}}{ds} \right| = \left| \frac{d\mathbf{T}/dt}{ds/dt} \right|$$

But $ds/dt = |\mathbf{r}'(t)|$ from Equation 6, so

(8)

$$\kappa(t) = \frac{|\mathbf{T}'(t)|}{|\mathbf{r}'(t)|}$$

EXAMPLE 3 Show that the curvature of a circle of radius a is $1/a$.

SOLUTION We can take the circle to have center the origin, and then a parametrization is

$$\mathbf{r}(t) = a\cos t\,\mathbf{i} + a\sin t\,\mathbf{j}$$

Therefore $\qquad \mathbf{r}'(t) = -a\sin t\,\mathbf{i} + a\cos t\,\mathbf{j} \qquad$ and $\qquad |\mathbf{r}'(t)| = a$

so

$$\mathbf{T}(t) = \frac{\mathbf{r}'(t)}{|\mathbf{r}'(t)|} = -\sin t\,\mathbf{i} + \cos t\,\mathbf{j}$$

and

$$\mathbf{T}'(t) = -\cos t\,\mathbf{i} - \sin t\,\mathbf{j}$$

This gives $|\mathbf{T}'(t)| = 1$, so using Equation 8, we have

$$\kappa(t) = \frac{|\mathbf{T}'(t)|}{|\mathbf{r}'(t)|} = \frac{1}{a}$$

■

The result of Example 3 shows that small circles have large curvature and large circles have small curvature, in accordance with our intuition. We can see directly from the definition of curvature that the curvature of a straight line is always 0 because the tangent vector is constant.

Although Formula 8 can be used in all cases to compute the curvature, the formula given by the following theorem is often more convenient to apply.

(9) THEOREM The curvature of the curve given by the vector function $\mathbf{r}$ is

$$\kappa(t) = \frac{|\mathbf{r}'(t) \times \mathbf{r}''(t)|}{|\mathbf{r}'(t)|^3}$$

PROOF Since $\mathbf{T} = \mathbf{r}'/|\mathbf{r}'|$ and $|\mathbf{r}'| = ds/dt$, we have

$$\mathbf{r}' = |\mathbf{r}'|\mathbf{T} = \frac{ds}{dt}\mathbf{T}$$

so the Product Rule (Theorem 11.6.5, Formula 3) gives

$$\mathbf{r}'' = \frac{d^2s}{dt^2}\mathbf{T} + \frac{ds}{dt}\mathbf{T}'$$

Using the fact that $\mathbf{T} \times \mathbf{T} = \mathbf{0}$ (Section 11.3), we have

$$\mathbf{r}' \times \mathbf{r}'' = \left(\frac{ds}{dt}\right)^2 (\mathbf{T} \times \mathbf{T}')$$

Now $|\mathbf{T}(t)| = 1$ for all t, so $\mathbf{T}$ and $\mathbf{T}'$ are orthogonal by Example 9 in Section 11.6. Therefore, by the definition of the cross product in Section 11.3,

$$|\mathbf{r}' \times \mathbf{r}''| = \left(\frac{ds}{dt}\right)^2 |\mathbf{T} \times \mathbf{T}'| = \left(\frac{ds}{dt}\right)^2 |\mathbf{T}||\mathbf{T}'| = \left(\frac{ds}{dt}\right)^2 |\mathbf{T}'|$$

Thus
$$|\mathbf{T}'| = \frac{|\mathbf{r}' \times \mathbf{r}''|}{(ds/dt)^2} = \frac{|\mathbf{r}' \times \mathbf{r}''|}{|\mathbf{r}'|^2}$$

and
$$\kappa = \frac{|\mathbf{T}'|}{|\mathbf{r}'|} = \frac{|\mathbf{r}' \times \mathbf{r}''|}{|\mathbf{r}'|^3}$$
□

EXAMPLE 4 Find the curvature of the twisted cubic $\mathbf{r}(t) = \langle t, t^2, t^3 \rangle$ at a general point and at $(0, 0, 0)$.

SOLUTION We first compute the required ingredients:

$$\mathbf{r}'(t) = \langle 1, 2t, 3t^2 \rangle$$

$$\mathbf{r}''(t) = \langle 0, 2, 6t \rangle$$

$$|\mathbf{r}'(t)| = \sqrt{1 + 4t^2 + 9t^4}$$

$$\mathbf{r}'(t) \times \mathbf{r}''(t) = \begin{vmatrix} \mathbf{i} & \mathbf{j} & \mathbf{k} \\ 1 & 2t & 3t^2 \\ 0 & 2 & 6t \end{vmatrix} = 6t^2\mathbf{i} - 6t\mathbf{j} + 2\mathbf{k}$$

$$|\mathbf{r}'(t) \times \mathbf{r}''(t)| = \sqrt{36t^4 + 36t^2 + 4} = 2\sqrt{9t^4 + 9t^2 + 1}$$

Theorem 9 then gives

$$\kappa(t) = \frac{|\mathbf{r}'(t) \times \mathbf{r}''(t)|}{|\mathbf{r}'(t)|^3} = \frac{2\sqrt{1 + 9t^2 + 9t^4}}{(1 + 4t^2 + 9t^4)^{3/2}}$$

At the origin the curvature is $\kappa(0) = 2$. ■

For the special case of a plane curve with equation $y = f(x)$, we can choose x as the parameter and write $\mathbf{r}(x) = x\,\mathbf{i} + f(x)\,\mathbf{j}$. Then $\mathbf{r}'(x) = \mathbf{i} + f'(x)\,\mathbf{j}$, $\mathbf{r}''(x) = f''(x)\,\mathbf{j}$, and, since $\mathbf{i} \times \mathbf{j} = \mathbf{k}$ and $\mathbf{j} \times \mathbf{j} = \mathbf{0}$, we have $\mathbf{r}'(x) \times \mathbf{r}''(x) = f''(x)\,\mathbf{k}$. We also have

$|\mathbf{r}'(x)| = \sqrt{1 + [f'(x)]^2}$, so, by Theorem 9,

(10) $$\kappa(x) = \frac{|f''(x)|}{[1 + (f'(x))^2]^{3/2}}$$

EXAMPLE 5 Find the curvature of the parabola $y = x^2$ at the points $(0, 0)$, $(1, 1)$, and $(2, 4)$.

SOLUTION Since $y' = 2x$ and $y'' = 2$, Formula 10 gives

$$\kappa(x) = \frac{|y''|}{[1 + (y')^2]^{3/2}} = \frac{2}{(1 + 4x^2)^{3/2}}$$

The curvature at $(0, 0)$ is $\kappa(0) = 2$. At $(1, 1)$ it is $\kappa(1) = 2/5^{3/2} \approx 0.18$. At $(2, 4)$ it is $\kappa(2) = 2/17^{3/2} \approx 0.03$. Observe that $\kappa(x) \to 0$ as $x \to \pm\infty$. This corresponds to the fact that the parabola appears to become flatter as $x \to \pm\infty$. ■

THE NORMAL AND BINORMAL VECTORS

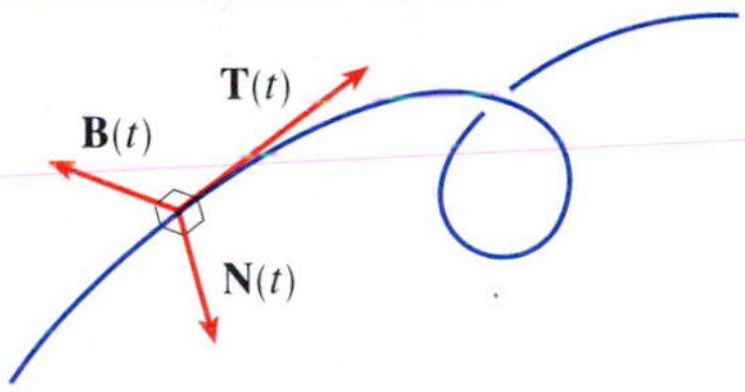

FIGURE 6

At a given point on a smooth space curve $\mathbf{r}(t)$, there are many vectors that are orthogonal to the unit tangent vector $\mathbf{T}(t)$. We single out one by observing that, since $|\mathbf{T}(t)| = 1$ for all t, we have $\mathbf{T}(t) \cdot \mathbf{T}'(t) = 0$ by Example 9 in Section 11.6, so $\mathbf{T}'(t)$ is orthogonal to $\mathbf{T}(t)$. Note that $\mathbf{T}'(t)$ is itself not a unit vector. But if $\mathbf{r}'$ is also smooth, we can define the **principal unit normal vector** $\mathbf{N}(t)$ (or **unit normal** for short) as

$$\mathbf{N}(t) = \frac{\mathbf{T}'(t)}{|\mathbf{T}'(t)|}$$

The vector $\mathbf{B}(t) = \mathbf{T}(t) \times \mathbf{N}(t)$ is called the **binormal vector**. It is perpendicular to both $\mathbf{T}$ and $\mathbf{N}$ and is also a unit vector. (See Figure 6.)

Figure 7 illustrates Example 6 by showing the vectors **T**, **N**, and **B** at two locations on the helix. In general, the vectors **T**, **N**, and **B**, starting at the various points on a curve, form a set of orthogonal vectors, called the **TNB** frame, that moves along the curve as t varies. This **TNB** frame plays an important role in the branch of mathematics known as differential geometry and in its applications to the motion of spacecraft.

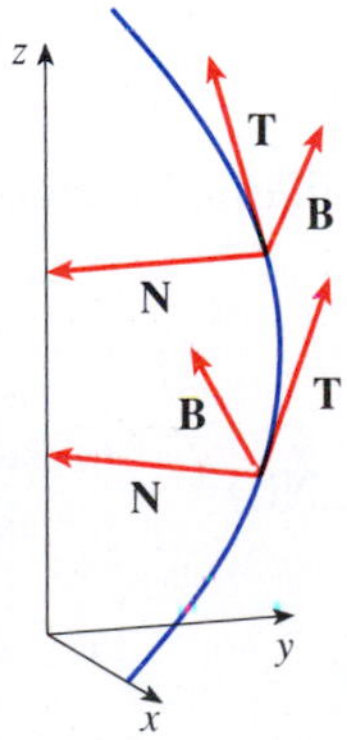

FIGURE 7

EXAMPLE 6 Find the unit normal and binormal vectors for the circular helix

$$\mathbf{r}(t) = \cos t\,\mathbf{i} + \sin t\,\mathbf{j} + t\,\mathbf{k}$$

SOLUTION $\mathbf{r}'(t) = -\sin t\,\mathbf{i} + \cos t\,\mathbf{j} + \mathbf{k} \qquad |\mathbf{r}'(t)| = \sqrt{2}$

$$\mathbf{T}(t) = \frac{\mathbf{r}'(t)}{|\mathbf{r}'(t)|} = \frac{1}{\sqrt{2}}(-\sin t\,\mathbf{i} + \cos t\,\mathbf{j} + \mathbf{k})$$

$$\mathbf{T}'(t) = \frac{1}{\sqrt{2}}(-\cos t\,\mathbf{i} - \sin t\,\mathbf{j}) \qquad |\mathbf{T}'(t)| = \frac{1}{\sqrt{2}}$$

$$\mathbf{N}(t) = \frac{\mathbf{T}'(t)}{|\mathbf{T}'(t)|} = -\cos t\,\mathbf{i} - \sin t\,\mathbf{j} = \langle -\cos t, -\sin t, 0 \rangle$$

This shows that the normal vector at a point on the helix is horizontal and points toward the z-axis. The binormal vector is

$$\mathbf{B}(t) = \mathbf{T}(t) \times \mathbf{N}(t) = \frac{1}{\sqrt{2}} \begin{vmatrix} \mathbf{i} & \mathbf{j} & \mathbf{k} \\ -\sin t & \cos t & 1 \\ -\cos t & -\sin t & 0 \end{vmatrix} = \frac{1}{\sqrt{2}} \langle \sin t, -\cos t, 1 \rangle$$ ■

The plane determined by the normal and binormal vectors **N** and **B** at a point P on a curve C is called the **normal plane** of C at P. It consists of all lines that are orthogonal to the tangent vector **T**. The plane determined by the vectors **T** and **N** is called the **osculating plane** of C at P. The name comes from the Latin *osculum*, meaning "kiss." It is the plane that comes closest to containing the part of the curve near P. (For a plane curve, the osculating plane is simply the plane that contains the curve.)

The circle that lies in the osculating plane of C at P, has the same tangent as C at P, lies on the concave side of C (toward which **N** points), and has radius $\rho = 1/\kappa$ (the reciprocal of the curvature) is called the **osculating circle** (or the **circle of curvature**) of C at P. It is the circle that best describes how C behaves near P; it shares the same tangent, normal, and curvature at P.

Figure 8 shows the helix and the osculating plane in Example 7.

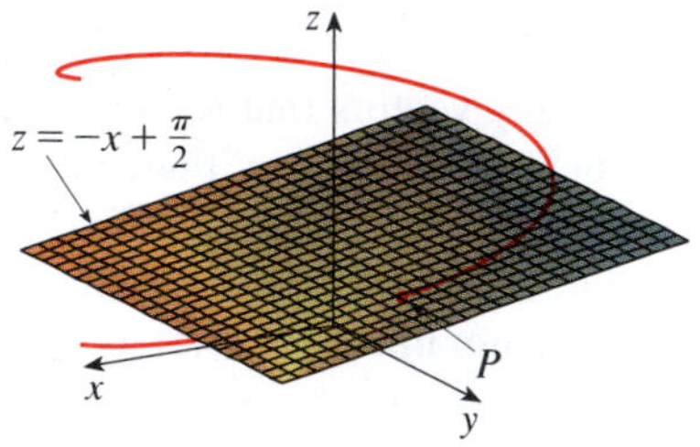

FIGURE 8

EXAMPLE 7 Find the equations of the normal plane and osculating plane of the helix in Example 6 at the point $P(0, 1, \pi/2)$.

SOLUTION The normal plane at P has normal vector $\mathbf{r}'(\pi/2) = \langle -1, 0, 1 \rangle$, so an equation is

$$-1(x-0) + 0(y-1) + 1\left(z - \frac{\pi}{2}\right) = 0 \qquad \text{or} \qquad z = x + \frac{\pi}{2}$$

The osculating plane at P contains the vectors **T** and **N**, so its normal vector is $\mathbf{T} \times \mathbf{N} = \mathbf{B}$. From Example 6 we have

$$\mathbf{B}(t) = \frac{1}{\sqrt{2}}\langle \sin t, -\cos t, 1 \rangle \qquad \mathbf{B}\left(\frac{\pi}{2}\right) = \left\langle \frac{1}{\sqrt{2}}, 0, \frac{1}{\sqrt{2}} \right\rangle$$

A simpler normal vector is $\langle 1, 0, 1 \rangle$, so an equation of the osculating plane is

$$1(x-0) + 0(y-1) + 1\left(z - \frac{\pi}{2}\right) = 0 \qquad \text{or} \qquad z = -x + \frac{\pi}{2}$$ ■

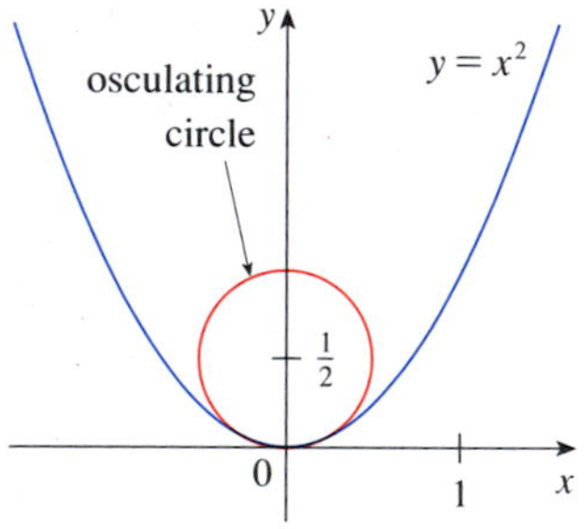

FIGURE 9

EXAMPLE 8 Find and graph the osculating circle of the parabola $y = x^2$ at the origin.

SOLUTION From Example 5 the curvature of the parabola at the origin is $\kappa(0) = 2$. So the radius of the osculating circle at the origin is $1/\kappa = \frac{1}{2}$ and its center is $(0, \frac{1}{2})$. Its equation is therefore

$$x^2 + \left(y - \tfrac{1}{2}\right)^2 = \tfrac{1}{4}$$

For the graph in Figure 9 we use parametric equations of this circle:

$$x = \tfrac{1}{2}\cos t \qquad y = \tfrac{1}{2} + \tfrac{1}{2}\sin t$$ ■

We summarize here the formulas for unit tangent, unit normal and binormal vectors, and curvature.

$$\mathbf{T}(t) = \frac{\mathbf{r}'(t)}{|\mathbf{r}'(t)|} \qquad \mathbf{N}(t) = \frac{\mathbf{T}'(t)}{|\mathbf{T}'(t)|} \qquad \mathbf{B}(t) = \mathbf{T}(t) \times \mathbf{N}(t)$$

$$\kappa = \left|\frac{d\mathbf{T}}{ds}\right| = \frac{|\mathbf{T}'(t)|}{|\mathbf{r}'(t)|} = \frac{|\mathbf{r}'(t) \times \mathbf{r}''(t)|}{|\mathbf{r}'(t)|^3}$$

EXERCISES 11.7

1–4 ■ Find the length of the given curve.

1. $\mathbf{r}(t) = \langle 2t, 3\sin t, 3\cos t\rangle$, $a \leqslant t \leqslant b$

2. $\mathbf{r}(t) = \langle e^t, e^t\sin t, e^t\cos t\rangle$, $0 \leqslant t \leqslant 2\pi$

3. $\mathbf{r}(t) = 6t\,\mathbf{i} + 3\sqrt{2}t^2\,\mathbf{j} + 2t^3\,\mathbf{k}$, $0 \leqslant t \leqslant 1$

4. $\mathbf{r}(t) = t^2\,\mathbf{i} + 2t\,\mathbf{j} + \ln t\,\mathbf{k}$, $1 \leqslant t \leqslant e$

5. Use Simpson's Rule with $n = 10$ to estimate the length of the arc of the twisted cubic $x = t$, $y = t^2$, $z = t^3$ from the origin to the point $(2, 4, 8)$.

6. Use a computer to graph the curve with parametric equations $x = \cos t$, $y = \sin 3t$, $z = \sin t$. If your computer (or calculator) evaluates definite integrals, use it to find the total length of this curve correct to four decimal places. Otherwise, use Simpson's Rule.

7–10 ■ Reparametrize the curve with respect to arc length measured from the point where $t = 0$ in the direction of increasing t.

7. $\mathbf{r}(t) = e^t\sin t\,\mathbf{i} + e^t\cos t\,\mathbf{j}$

8. $\mathbf{r}(t) = (1 + 2t)\,\mathbf{i} + (3 + t)\,\mathbf{j} - 5t\,\mathbf{k}$

9. $\mathbf{r}(t) = 3\sin t\,\mathbf{i} + 4t\,\mathbf{j} + 3\cos t\,\mathbf{k}$

10. $\mathbf{r}(t) = \cos^3 t\,\mathbf{i} + \sin^3 t\,\mathbf{j} + \cos 2t\,\mathbf{k}$, $0 \leqslant t \leqslant \pi/2$

11–14 ■
(a) Find the unit tangent and unit normal vectors $\mathbf{T}(t)$ and $\mathbf{N}(t)$.
(b) Use Formula 8 to find the curvature.

11. $\mathbf{r}(t) = \langle \sin 4t, 3t, \cos 4t\rangle$

12. $\mathbf{r}(t) = \langle 6t, 3\sqrt{2}\,t^2, 2t^3\rangle$

13. $\mathbf{r}(t) = \langle \sqrt{2}\cos t, \sin t, \sin t\rangle$

14. $\mathbf{r}(t) = \langle t^2, 2t, \ln t\rangle$

15–19 ■ Use Theorem 9 to find the curvature.

15. $\mathbf{r}(t) = \mathbf{i} + t\,\mathbf{j} + t^2\,\mathbf{k}$

16. $\mathbf{r}(t) = (1 + t)\,\mathbf{i} + (1 - t)\,\mathbf{j} + 3t^2\,\mathbf{k}$

17. $\mathbf{r}(t) = 2t^3\,\mathbf{i} - 3t^2\,\mathbf{j} + 6t\,\mathbf{k}$

18. $\mathbf{r}(t) = (t^2 + 2)\,\mathbf{i} + (t^2 - 4t)\,\mathbf{j} + 2t\,\mathbf{k}$

19. $\mathbf{r}(t) = \sin t\,\mathbf{i} + \cos t\,\mathbf{j} + \sin t\,\mathbf{k}$

20. Graph the curve with parametric equations $x = t$, $y = 4t^{3/2}$, $z = -t^2$, and find the curvature at the point $(1, 4, -1)$.

21–24 ■ Use Formula 10 to find the curvature.

21. $y = x^3$

22. $y = \sqrt{x}$

23. $y = \sin x$

24. $y = \ln x$

25. At what point does the curve $y = e^x$ have maximum curvature?

26–27 ■ Use a graphing calculator or computer to graph both the curve and its curvature function $\kappa(x)$ on the same screen. Is the graph of κ what you would expect?

26. $y = xe^{-x}$

27. $y = x^4$

28. Use Theorem 9 to show that the curvature of a plane parametric curve $x = f(t)$, $y = g(t)$ is

$$\kappa = \frac{|\dot{x}\ddot{y} - \dot{y}\ddot{x}|}{[\dot{x}^2 + \dot{y}^2]^{3/2}}$$

where the dots indicate derivatives with respect to t.

29–30 ■ Use the formula in Exercise 28 to find the curvature.

29. $x = t^3$, $y = t^2$

30. $x = t\sin t$, $y = t\cos t$

31–32 ■ Find the vectors **T**, **N**, and **B** at the given point.

31. $\mathbf{r}(t) = \langle t^2, \frac{2}{3}t^3, t\rangle$, $(1, \frac{2}{3}, 1)$

32. $\mathbf{r}(t) = \langle e^t, e^t\sin t, e^t\cos t\rangle$, $(1, 0, 1)$

33–34 ■ Find equations of the normal plane and osculating plane of the curve at the given point.

33. $x = 2\sin 3t$, $y = t$, $z = 2\cos 3t$; $(0, \pi, -2)$

34. $x = t$, $y = t^2$, $z = t^3$; $(1, 1, 1)$

35. Find the equations of the osculating circles of the ellipse $9x^2 + 4y^2 = 36$ at the points $(2, 0)$ and $(0, 3)$. Use a graphing calculator or computer to graph the ellipse and both osculating circles on the same screen.

36. Find the equations of the osculating circles of the parabola $y = \frac{1}{2}x^2$ at the points $(0, 0)$ and $(1, \frac{1}{2})$. Graph both osculating circles and the parabola.

37. At what point on the curve $x = t^3$, $y = 3t$, $z = t^4$ is the normal plane parallel to the plane $6x + 6y - 8z = 1$?

CAS 38. Is there a point on the curve in Exercise 37 where the osculating plane is parallel to the plane $x + y + z = 1$? (*Note:* You will need a CAS for differentiating, for simplifying, and for computing a cross product.)

39. Show that the curvature κ is related to the tangent and normal vectors by the equation

$$\frac{d\mathbf{T}}{ds} = \kappa\mathbf{N}$$

40. Show that the curvature of a plane curve is $\kappa = |d\phi/ds|$, where ϕ is the angle between **T** and **i**; that is, ϕ is the angle of inclination of the tangent line. (This shows that the definition of curvature is consistent with the definition for plane curves.)

41. (a) Show that $d\mathbf{B}/ds$ is perpendicular to $\mathbf{B}$.
(b) Show that $d\mathbf{B}/ds$ is perpendicular to $\mathbf{T}$.
(c) Deduce from parts (a) and (b) that $d\mathbf{B}/ds = -\tau(s)\mathbf{N}$ for some number $\tau(s)$ called the **torsion** of the curve. (The torsion measures the degree of twisting of a curve.)

42. Show that for a plane curve the torsion is $\tau(s) = 0$.

43. Show that arc length is independent of parametrization. [*Hint*: Suppose C is given by $\mathbf{r}_1(t)$, $a \leqslant t \leqslant b$, and also by $\mathbf{r}_2(u)$, $\alpha \leqslant u \leqslant \beta$, where $t = g(u)$ and $g'(u) > 0$. If $L_1 = \int_a^b |\mathbf{r}_1'(t)|\,dt$ and $L_2 = \int_\alpha^\beta |\mathbf{r}_2'(u)|\,du$, show that $L_1 = L_2$.]

44. The following formulas, called the **Frenet-Serret formulas**, are of fundamental importance in differential geometry:
1. $d\mathbf{T}/ds = \kappa\mathbf{N}$
2. $d\mathbf{N}/ds = -\kappa\mathbf{T} + \tau\mathbf{B}$
3. $d\mathbf{B}/ds = -\tau\mathbf{N}$

(Formula 1 comes from Exercise 39 and Formula 3 comes from Exercise 41.) Use the fact that $\mathbf{N} = \mathbf{B} \times \mathbf{T}$ to deduce Formula 2 from Formulas 1 and 3.

45. Use the Frenet-Serret formulas to prove each of the following. (Primes denote derivatives with respect to t. Start as in the proof of Theorem 9.)
(a) $\mathbf{r}'' = s''\mathbf{T} + \kappa(s')^2\mathbf{N}$
(b) $\mathbf{r}' \times \mathbf{r}'' = \kappa(s')^3\mathbf{B}$
(c) $\mathbf{r}''' = [s''' - \kappa^2(s')^3]\mathbf{T} + [3\kappa s's'' + \kappa'(s')^2]\mathbf{N} + \kappa\tau(s')^3\mathbf{B}$
(d) $\tau = \dfrac{(\mathbf{r}' \times \mathbf{r}'') \cdot \mathbf{r}'''}{|\mathbf{r}' \times \mathbf{r}''|^2}$

46–47 ■ Use the formula in Exercise 45(d) to find the torsion of the curve.

46. $\mathbf{r}(t) = \langle \cos t, \sin t, t \rangle$

47. $\mathbf{r}(t) = \langle t, \frac{1}{2}t^2, \frac{1}{3}t^3 \rangle$

48. Find the curvature and torsion of the curve $x = \sinh t$, $y = \cosh t$, $z = t$ at the point $(0, 1, 0)$.

49. The DNA molecule has the shape of a double helix (see Figure 3 on page 691). The radius of each helix is about 10 angstroms (1 angstrom $= 10^{-8}$ cm). Each helix rises about 34 angstroms during each complete turn and there are about 2.9×10^8 complete turns. Estimate the length of each helix.

50. Let's consider the problem of designing a railroad track to make a smooth transition between sections of straight track. Existing track along the negative x-axis is to be joined smoothly to a track along the line $y = 1$ for $x \geqslant 1$.
(a) Find a polynomial $P = P(x)$ of degree 5 such that the function F defined by

$$F(x) = \begin{cases} 0 & \text{if } x \leqslant 0 \\ P(x) & \text{if } 0 < x < 1 \\ 1 & \text{if } x \geqslant 1 \end{cases}$$

is continuous and has continuous slope and continuous curvature.
(b) Use a graphing calculator or computer to draw the graph of F.

11.8 MOTION IN SPACE: VELOCITY AND ACCELERATION

In this section we show how the ideas of tangent and normal vectors and curvature can be used in physics to study the motion of an object, including its velocity and acceleration, along a space curve. In particular, we follow in the footsteps of Newton by using these methods to derive Kepler's laws of planetary motion.

Suppose a particle moves through space so that its position vector at time t is $\mathbf{r}(t)$. As shown in Section 3.7, for small values of h the vector

$$\frac{\mathbf{r}(t+h) - \mathbf{r}(t)}{h} \tag{1}$$

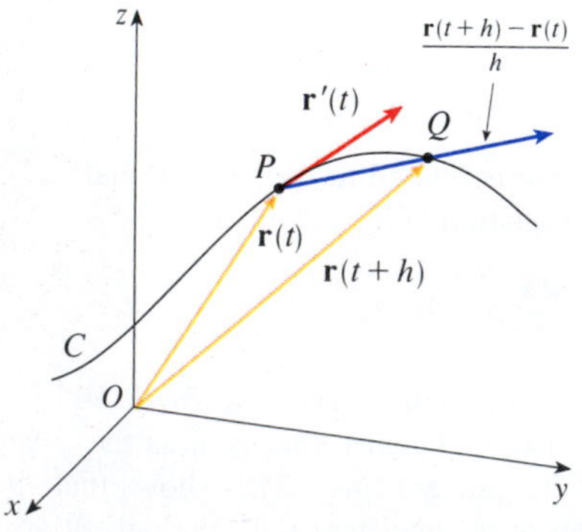

FIGURE 1

approximates the direction of the particle moving along the curve $\mathbf{r}(t)$ (see Figure 1). Its magnitude measures the size of the displacement vector per unit time. The vector (1) gives the average velocity over a time interval of length h and its limit is the **velocity vector** $\mathbf{v}(t)$ at time t:

$$\mathbf{v}(t) = \lim_{h \to 0} \frac{\mathbf{r}(t+h) - \mathbf{r}(t)}{h} = \mathbf{r}'(t) \tag{2}$$

Thus, the velocity vector points in the direction of the unit tangent vector.

The **speed** of the particle at time t is the magnitude of the velocity vector, that is, $|\mathbf{v}(t)|$. This is appropriate because, from Equations 2 and 11.7.6, we have

$$|\mathbf{v}(t)| = |\mathbf{r}'(t)| = \frac{ds}{dt} = \text{rate of change of distance with respect to time}$$

As in the case of one-dimensional motion, the **acceleration** of the particle is defined as the derivative of the velocity:

$$\mathbf{a}(t) = \mathbf{v}'(t) = \mathbf{r}''(t)$$

EXAMPLE 1 The position vector of an object moving in a plane is given by $\mathbf{r}(t) = t^3\,\mathbf{i} + t^2\,\mathbf{j}$, $t \geqslant 0$. Find its velocity, speed, and acceleration when $t = 1$ and illustrate geometrically.

SOLUTION The velocity and acceleration at time t are

$$\mathbf{v}(t) = \mathbf{r}'(t) = 3t^2\,\mathbf{i} + 2t\,\mathbf{j}$$

$$\mathbf{a}(t) = \mathbf{r}''(t) = 6t\,\mathbf{i} + 2\,\mathbf{j}$$

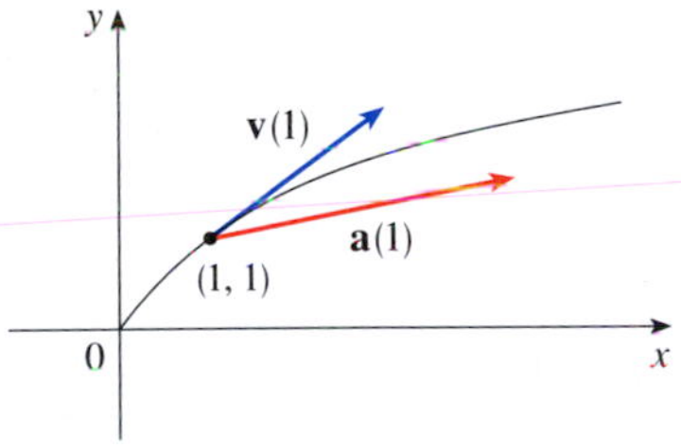

FIGURE 2

and the speed is

$$|\mathbf{v}(t)| = \sqrt{(3t^2)^2 + (2t)^2} = \sqrt{9t^4 + 4t^2}$$

When $t = 1$, we have

$$\mathbf{v}(1) = 3\,\mathbf{i} + 2\,\mathbf{j} \qquad \mathbf{a}(1) = 6\,\mathbf{i} + 2\,\mathbf{j} \qquad |\mathbf{v}(1)| = \sqrt{13}$$

These velocity and acceleration vectors are shown in Figure 2. ■

EXAMPLE 2 Find the velocity, acceleration, and speed of a particle with position vector $\mathbf{r}(t) = \langle t^2, e^t, te^t \rangle$.

SOLUTION

$$\mathbf{v}(t) = \mathbf{r}'(t) = \langle 2t, e^t, (1 + t)e^t \rangle$$

$$\mathbf{a}(t) = \mathbf{v}'(t) = \langle 2, e^t, (2 + t)e^t \rangle$$

$$|\mathbf{v}(t)| = \sqrt{4t^2 + e^{2t} + (1 + t)^2 e^{2t}}$$ ■

Figure 3 shows the path of the particle in Example 2 with the velocity and acceleration vectors when $t = 1$.

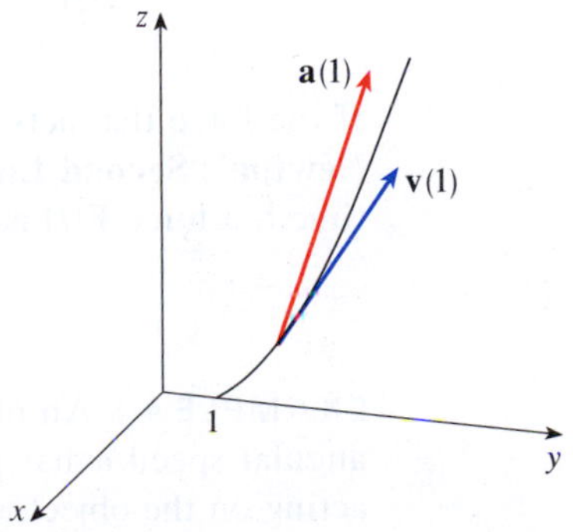

FIGURE 3

The vector integrals that were introduced in Section 11.6 can be used to find position vectors when velocity or acceleration vectors are known, as in the following example.

EXAMPLE 3 A moving particle starts at an initial position $\mathbf{r}(0) = \langle 1, 0, 0 \rangle$ with initial velocity $\mathbf{v}(0) = \mathbf{i} - \mathbf{j} + \mathbf{k}$. Its acceleration is $\mathbf{a}(t) = 4t\,\mathbf{i} + 6t\,\mathbf{j} + \mathbf{k}$. Find its velocity and position at time t.

SOLUTION Since $\mathbf{a}(t) = \mathbf{v}'(t)$, we have

$$\mathbf{v}(t) = \int \mathbf{a}(t)\,dt = \int (4t\,\mathbf{i} + 6t\,\mathbf{j} + \mathbf{k})\,dt$$

$$= 2t^2\mathbf{i} + 3t^2\mathbf{j} + t\,\mathbf{k} + \mathbf{C}$$

To determine the value of the constant vector $\mathbf{C}$, we use the fact that $\mathbf{v}(0) = \mathbf{i} - \mathbf{j} + \mathbf{k}$. The preceding equation gives $\mathbf{v}(0) = \mathbf{C}$, so $\mathbf{C} = \mathbf{i} - \mathbf{j} + \mathbf{k}$ and

$$\mathbf{v}(t) = 2t^2\mathbf{i} + 3t^2\mathbf{j} + t\,\mathbf{k} + \mathbf{i} - \mathbf{j} + \mathbf{k}$$

$$= (2t^2 + 1)\,\mathbf{i} + (3t^2 - 1)\,\mathbf{j} + (t + 1)\,\mathbf{k}$$

The expression for $\mathbf{r}(t)$ that we obtained in Example 3 was used to plot the path of the particle in Figure 4 for $0 \leq t \leq 3$.

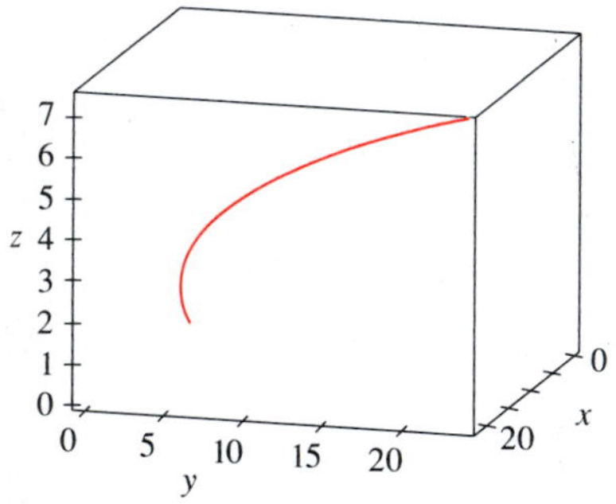

FIGURE 4

Since $\mathbf{v}(t) = \mathbf{r}'(t)$, we have

$$\mathbf{r}(t) = \int \mathbf{v}(t)\,dt$$

$$= \int [(2t^2 + 1)\,\mathbf{i} + (3t^2 - 1)\,\mathbf{j} + (t + 1)\,\mathbf{k}]\,dt$$

$$= \left(\frac{2t^3}{3} + t\right)\mathbf{i} + (t^3 - t)\,\mathbf{j} + \left(\frac{t^2}{2} + t\right)\mathbf{k} + \mathbf{D}$$

Putting $t = 0$, we find that $\mathbf{D} = \mathbf{r}(0) = \mathbf{i}$, so

$$\mathbf{r}(t) = \left(\frac{2t^3}{3} + t + 1\right)\mathbf{i} + (t^3 - t)\,\mathbf{j} + \left(\frac{t^2}{2} + t\right)\mathbf{k}$$ ■

In general, vector integrals allow us to recover velocity when acceleration is known and position when velocity is known:

$$\mathbf{v}(t) = \mathbf{v}(t_0) + \int_{t_0}^{t} \mathbf{a}(u)\,du \qquad \mathbf{r}(t) = \mathbf{r}(t_0) + \int_{t_0}^{t} \mathbf{v}(u)\,du$$

If the force that acts on a particle is known, then the acceleration can be found from **Newton's Second Law of Motion**. The vector version of this law states that if, at any time t, a force $\mathbf{F}(t)$ acts on an object of mass m producing an acceleration $\mathbf{a}(t)$, then

$$\mathbf{F}(t) = m\mathbf{a}(t)$$

EXAMPLE 4 An object with mass m that moves in an elliptical path with constant angular speed ω has position vector $\mathbf{r}(t) = a\cos\omega t\,\mathbf{i} + b\sin\omega t\,\mathbf{j}$. Find the force acting on the object and show that it is directed toward the origin.

The angular speed of the object moving with position P is $\omega = d\theta/dt$, where θ is the angle shown in Figure 5.

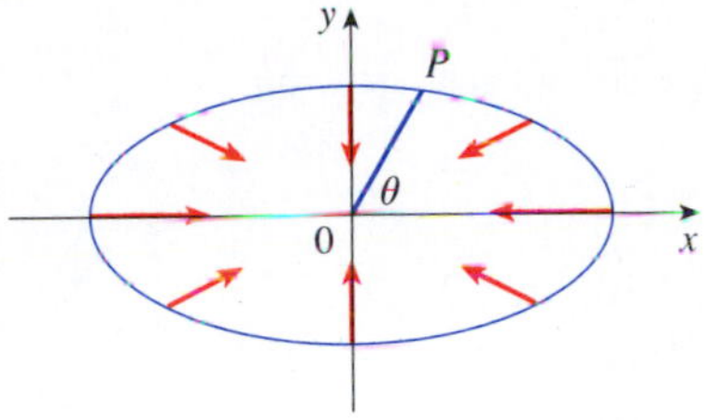

FIGURE 5

SOLUTION

$$\mathbf{v}(t) = \mathbf{r}'(t) = -a\omega \sin \omega t\, \mathbf{i} + b\omega \cos \omega t\, \mathbf{j}$$

$$\mathbf{a}(t) = \mathbf{v}'(t) = -a\omega^2 \cos \omega t\, \mathbf{i} - b\omega^2 \sin \omega t\, \mathbf{j}$$

Therefore, Newton's Second Law gives the force as

$$\mathbf{F}(t) = m\mathbf{a}(t) = -m\omega^2(a \cos \omega t\, \mathbf{i} + b \sin \omega t\, \mathbf{j})$$

Notice that $\mathbf{F}(t) = -m\omega^2\mathbf{r}(t)$. This shows that the force acts in the direction opposite to the radius vector $\mathbf{r}(t)$ and therefore points toward the origin (see Figure 5). Such a force is called a *centripetal* (center-seeking) force. ■

TANGENTIAL AND NORMAL COMPONENTS OF ACCELERATION

When we study the motion of a particle, it is often useful to resolve the acceleration into two components, one in the direction of the tangent and the other in the direction of the normal. If we write $v = |\mathbf{v}|$ for the speed of the particle, then

$$\mathbf{T}(t) = \frac{\mathbf{r}'(t)}{|\mathbf{r}'(t)|} = \frac{\mathbf{v}(t)}{|\mathbf{v}(t)|} = \frac{\mathbf{v}}{v}$$

and so

$$\mathbf{v} = v\mathbf{T}$$

If we differentiate both sides of this equation with respect to t, we get

(3) $$\mathbf{a} = \mathbf{v}' = v'\mathbf{T} + v\mathbf{T}'$$

If we use the expression for the curvature given by Equation 11.7.8, then we have

(4) $$\kappa = \frac{|\mathbf{T}'|}{|\mathbf{r}'|} = \frac{|\mathbf{T}'|}{v} \qquad \text{so} \qquad |\mathbf{T}'| = \kappa v$$

The unit normal vector was defined in the preceding section as $\mathbf{N} = \mathbf{T}'/|\mathbf{T}'|$, so (4) gives

$$\mathbf{T}' = |\mathbf{T}'|\mathbf{N} = \kappa v \mathbf{N}$$

and Equation 3 becomes

(5) $$\boxed{\mathbf{a} = v'\mathbf{T} + \kappa v^2 \mathbf{N}}$$

Writing a_T and a_N for the tangential and normal components of acceleration, we have

$$\mathbf{a} = a_T \mathbf{T} + a_N \mathbf{N}$$

where

(6) $$a_T = v' \qquad \text{and} \qquad a_N = \kappa v^2$$

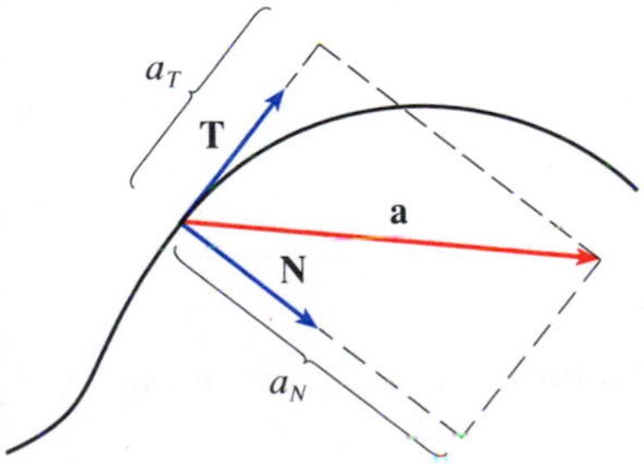

FIGURE 6

This resolution is illustrated in Figure 6.

Although we have expressions for the tangential and normal components of acceleration in Equations 6, it is desirable to have expressions that depend only on $\mathbf{r}$, $\mathbf{r}'$, and $\mathbf{r}''$.

To this end we take the dot product of $\mathbf{v} = v\mathbf{T}$ with $\mathbf{a}$ as given by Equation 5:

$$\begin{aligned} \mathbf{v} \cdot \mathbf{a} &= v\mathbf{T} \cdot (v'\mathbf{T} + \kappa v^2 \mathbf{N}) \\ &= vv'\mathbf{T} \cdot \mathbf{T} + \kappa v^3 \mathbf{T} \cdot \mathbf{N} \\ &= vv' \qquad \text{(since } \mathbf{T} \cdot \mathbf{T} = 1 \text{ and } \mathbf{T} \cdot \mathbf{N} = 0) \end{aligned}$$

Therefore

(7) $$a_T = v' = \frac{\mathbf{v} \cdot \mathbf{a}}{v} = \frac{\mathbf{r}'(t) \cdot \mathbf{r}''(t)}{|\mathbf{r}'(t)|}$$

Using the formula for curvature given by Theorem 11.7.9, we have

(8) $$a_N = \kappa v^2 = \frac{|\mathbf{r}'(t) \times \mathbf{r}''(t)|}{|\mathbf{r}'(t)|^3} |\mathbf{r}'(t)|^2 = \frac{|\mathbf{r}'(t) \times \mathbf{r}''(t)|}{|\mathbf{r}'(t)|}$$

EXAMPLE 6 A particle moves with position function $\mathbf{r}(t) = \langle t^2, t^2, t^3 \rangle$. Find the tangential and normal components of acceleration.

SOLUTION

$$\begin{aligned} \mathbf{r}(t) &= t^2\,\mathbf{i} + t^2\,\mathbf{j} + t^3\,\mathbf{k} \\ \mathbf{r}'(t) &= 2t\,\mathbf{i} + 2t\,\mathbf{j} + 3t^2\,\mathbf{k} \\ \mathbf{r}''(t) &= 2\,\mathbf{i} + 2\,\mathbf{j} + 6t\,\mathbf{k} \\ |\mathbf{r}'(t)| &= \sqrt{8t^2 + 9t^4} \end{aligned}$$

Therefore, Equation 7 gives the tangential component as

$$a_T = \frac{\mathbf{r}'(t) \cdot \mathbf{r}''(t)}{|\mathbf{r}'(t)|} = \frac{8t + 18t^3}{\sqrt{8t^2 + 9t^4}}$$

Since

$$\mathbf{r}'(t) \times \mathbf{r}''(t) = \begin{vmatrix} \mathbf{i} & \mathbf{j} & \mathbf{k} \\ 2t & 2t & 3t^2 \\ 2 & 2 & 6t \end{vmatrix} = 6t^2\,\mathbf{i} - 6t^2\,\mathbf{j}$$

Equation 8 gives the normal component as

$$a_N = \frac{|\mathbf{r}'(t) \times \mathbf{r}''(t)|}{|\mathbf{r}'(t)|} = \frac{6\sqrt{2}\,t^2}{\sqrt{8t^2 + 9t^4}}$$

■

KEPLER'S LAWS OF PLANETARY MOTION

We now describe one of the great accomplishments of calculus by showing how the material of this chapter can be used to prove Kepler's laws of planetary motion. After 20 years of studying the astronomical observations of the Danish astronomer Tycho Brahe, the German mathematician and astronomer Johannes Kepler (1571–1630) formulated the following three laws.

KEPLER'S LAWS

1. A planet revolves around the Sun in an elliptical orbit with the Sun at one focus.
2. The line joining the Sun to a planet sweeps out equal areas in equal times.
3. The square of the period of revolution of a planet is proportional to the cube of the length of the major axis of its orbit.

In his book *Principia Mathematica* of 1687, Sir Isaac Newton was able to show that these three laws are consequences of two of his own laws, the Second Law of Motion and the Law of Universal Gravitation. In what follows we prove Kepler's First Law. The remaining laws are left as exercises (with hints).

Since the gravitational force of the Sun on a planet is so much larger than the forces exerted by other celestial bodies, we can safely ignore all bodies in the universe except for the Sun and one planet revolving about it. We use a coordinate system with the Sun at the origin and we let $\mathbf{r} = \mathbf{r}(t)$ be the position vector of the planet. (Equally well, $\mathbf{r}$ could be the position vector of the Moon or a satellite moving around Earth or a comet moving around a star.) The velocity vector is $\mathbf{v} = \mathbf{r}'$ and the acceleration vector is $\mathbf{a} = \mathbf{r}''$. We use the following laws of Newton:

$$\text{Second Law of Motion:} \quad \mathbf{F} = m\mathbf{a}$$

$$\text{Law of Gravitation:} \quad \mathbf{F} = -\frac{GMm}{r^3}\mathbf{r} = -\frac{GMm}{r^2}\mathbf{u}$$

where $\mathbf{F}$ is the gravitational force on the planet, m and M are the masses of the planet and the Sun, G is the gravitational constant, $r = |\mathbf{r}|$, and $\mathbf{u} = (1/r)\mathbf{r}$ is the unit vector in the direction of $\mathbf{r}$.

We first show that the planet moves in one plane. By equating the expressions for $\mathbf{F}$ in Newton's two laws we find that

$$\mathbf{a} = -\frac{GM}{r^3}\mathbf{r}$$

and so $\mathbf{a}$ is parallel to $\mathbf{r}$. It follows that $\mathbf{r} \times \mathbf{a} = \mathbf{0}$. We use Formula 5 in Theorem 11.6.5 to write

$$\begin{aligned} \frac{d}{dt}(\mathbf{r} \times \mathbf{v}) &= \mathbf{r}' \times \mathbf{v} + \mathbf{r} \times \mathbf{v}' \\ &= \mathbf{v} \times \mathbf{v} + \mathbf{r} \times \mathbf{a} = \mathbf{0} + \mathbf{0} = \mathbf{0} \end{aligned}$$

Therefore

$$\mathbf{r} \times \mathbf{v} = \mathbf{h}$$

where $\mathbf{h}$ is a constant vector. (We may assume that $\mathbf{h} \neq \mathbf{0}$; that is, $\mathbf{r}$ and $\mathbf{v}$ are not parallel.) This means that the vector $\mathbf{r} = \mathbf{r}(t)$ is perpendicular to $\mathbf{h}$ for all values of t and so the planet always lies in the plane through the origin perpendicular to $\mathbf{h}$. Thus, the orbit of the planet is a plane curve.

To prove Kepler's First Law we rewrite the vector $\mathbf{h}$ as follows:

$$\begin{aligned} \mathbf{h} &= \mathbf{r} \times \mathbf{v} = \mathbf{r} \times \mathbf{r}' = r\mathbf{u} \times (r\mathbf{u})' \\ &= r\mathbf{u} \times (r\mathbf{u}' + r'\mathbf{u}) = r^2(\mathbf{u} \times \mathbf{u}') + rr'(\mathbf{u} \times \mathbf{u}) \\ &= r^2(\mathbf{u} \times \mathbf{u}') \end{aligned}$$

Then

$$\begin{aligned} \mathbf{a} \times \mathbf{h} &= \frac{-GM}{r^2}\mathbf{u} \times (r^2\mathbf{u} \times \mathbf{u}') = -GM\,\mathbf{u} \times (\mathbf{u} \times \mathbf{u}') \\ &= -GM[(\mathbf{u} \cdot \mathbf{u}')\,\mathbf{u} - (\mathbf{u} \cdot \mathbf{u})\,\mathbf{u}'] \qquad \text{(by Formula 11.3.8)} \end{aligned}$$

But $\mathbf{u} \cdot \mathbf{u} = |\mathbf{u}|^2 = 1$ and, since $|\mathbf{u}(t)| = 1$, it follows from Example 9 in Section 11.6

that $\mathbf{u} \cdot \mathbf{u}' = 0$. Therefore

$$\mathbf{a} \times \mathbf{h} = GM\mathbf{u}'$$

and so

$$(\mathbf{v} \times \mathbf{h})' = \mathbf{v}' \times \mathbf{h} = \mathbf{a} \times \mathbf{h} = GM\mathbf{u}'$$

Integrating both sides of this equation, we get

(9)
$$\mathbf{v} \times \mathbf{h} = GM\mathbf{u} + \mathbf{c}$$

where $\mathbf{c}$ is a constant vector.

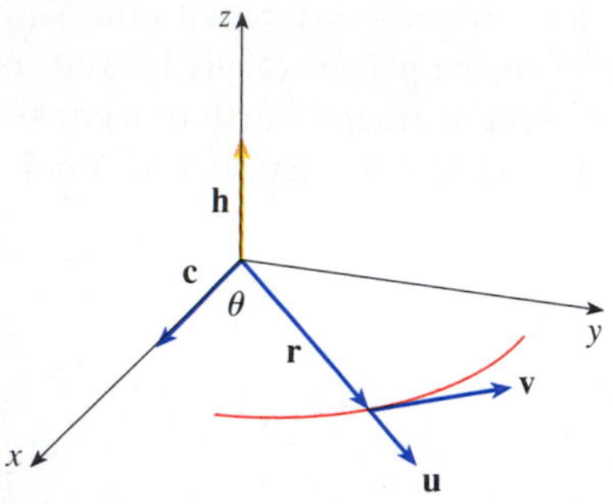

FIGURE 7

At this point it is convenient to choose the coordinate axes so that the standard basis vector $\mathbf{k}$ points in the direction of the vector $\mathbf{h}$. Then the planet moves in the xy-plane. Since both $\mathbf{v} \times \mathbf{h}$ and $\mathbf{u}$ are perpendicular to $\mathbf{h}$, Equation 9 shows that $\mathbf{c}$ lies in the xy-plane. This means that we can choose the x- and y-axes so that the vector $\mathbf{i}$ lies in the direction of $\mathbf{c}$, as shown in Figure 7.

If θ is the angle between $\mathbf{c}$ and $\mathbf{r}$, then (r, θ) are polar coordinates of the planet. (Polar coordinates are covered in Section 13.4 and Appendix J.) From Equation 9 we have

$$\begin{aligned} \mathbf{r} \cdot (\mathbf{v} \times \mathbf{h}) &= \mathbf{r} \cdot (GM\mathbf{u} + \mathbf{c}) = GM\mathbf{r} \cdot \mathbf{u} + \mathbf{r} \cdot \mathbf{c} \\ &= GMr\,\mathbf{u} \cdot \mathbf{u} + |\mathbf{r}||\mathbf{c}|\cos\theta = GMr + rc\cos\theta \end{aligned}$$

where $c = |\mathbf{c}|$. Then

$$r = \frac{\mathbf{r} \cdot (\mathbf{v} \times \mathbf{h})}{GM + c\cos\theta} = \frac{1}{GM}\,\frac{\mathbf{r} \cdot (\mathbf{v} \times \mathbf{h})}{1 + e\cos\theta}$$

where $e = c/(GM)$. But

$$\mathbf{r} \cdot (\mathbf{v} \times \mathbf{h}) = (\mathbf{r} \times \mathbf{v}) \cdot \mathbf{h} = \mathbf{h} \cdot \mathbf{h} = |\mathbf{h}|^2 = h^2$$

where $h = |\mathbf{h}|$. So

$$r = \frac{h^2/(GM)}{1 + e\cos\theta} = \frac{eh^2/c}{1 + e\cos\theta}$$

Writing $d = h^2/c$, we obtain the equation

(10)
$$r = \frac{ed}{1 + e\cos\theta}$$

In Appendix J it is shown that Equation 10 is the polar equation of a conic section with focus at the origin and eccentricity e. We know that the orbit of a planet is a closed curve and so the conic must be an ellipse.

This completes the derivation of Kepler's First Law. We will guide you through the derivation of the second and third laws in Exercises 29 and 30. The proofs of these three laws show that the methods of this chapter provide a powerful tool for describing some of the laws of nature.

EXERCISES 11.8

1–6 ■ Find the velocity, acceleration, and speed of a particle with the given position function. Sketch the path of the particle and draw the velocity and acceleration vectors for the given value of t.

1. $\mathbf{r}(t) = \langle t^2 - 1, t \rangle, \quad t = 1$

2. $\mathbf{r}(t) = \langle \sqrt{t}, 1 - t \rangle, \quad t = 1$

3. $\mathbf{r}(t) = e^t \mathbf{i} + e^{-t} \mathbf{j}, \quad t = 0$

4. $\mathbf{r}(t) = \sin t \, \mathbf{i} + 2 \cos t \, \mathbf{j}, \quad t = \pi/6$

5. $\mathbf{r}(t) = \sin t \, \mathbf{i} + t \, \mathbf{j} + \cos t \, \mathbf{k}, \quad t = 0$

6. $\mathbf{r}(t) = t \, \mathbf{i} + t^2 \mathbf{j} + t^3 \mathbf{k}, \quad t = 1$

7–12 ■ Find the velocity, acceleration, and speed of a particle with the given position function.

7. $\mathbf{r}(t) = \langle t^3, t^2 + 1, t^3 - 1 \rangle$

8. $\mathbf{r}(t) = \langle \sqrt{t}, t, t\sqrt{t} \rangle$

9. $\mathbf{r}(t) = (1/t) \, \mathbf{i} + \mathbf{j} + t^2 \mathbf{k}$

10. $\mathbf{r}(t) = e^t \mathbf{i} + 2t \, \mathbf{j} + e^{-t} \mathbf{k}$

11. $\mathbf{r}(t) = e^t (\cos t \, \mathbf{i} + \sin t \, \mathbf{j} + t \, \mathbf{k})$

12. $\mathbf{r}(t) = \cosh t \, \mathbf{i} + \sinh t \, \mathbf{j} + t \, \mathbf{k}$

13–14 ■ Find the velocity and position vectors of a particle that has the given acceleration and the specified initial velocity and position.

13. $\mathbf{a}(t) = \mathbf{k}, \quad \mathbf{v}(0) = \mathbf{i} - \mathbf{j}, \quad \mathbf{r}(0) = \mathbf{0}$

14. $\mathbf{a}(t) = -10 \, \mathbf{k}, \quad \mathbf{v}(0) = \mathbf{i} + \mathbf{j} - \mathbf{k}, \quad \mathbf{r}(0) = 2 \, \mathbf{i} + 3 \, \mathbf{j}$

15–16 ■

(a) Find the position vector of a particle that has the given acceleration and the specified initial velocity and position.

(b) Use a computer to graph the path of the particle.

15. $\mathbf{a}(t) = \mathbf{i} + 2 \, \mathbf{j} + 2t \, \mathbf{k}, \quad \mathbf{v}(0) = \mathbf{0}, \quad \mathbf{r}(0) = \mathbf{i} + \mathbf{k}$

16. $\mathbf{a}(t) = t \, \mathbf{i} + t^2 \mathbf{j} + \cos 2t \, \mathbf{k}, \quad \mathbf{v}(0) = \mathbf{i} + \mathbf{k}, \quad \mathbf{r}(0) = \mathbf{j}$

17. The position function of a particle is given by $\mathbf{r}(t) = \langle t^2, 5t, t^2 - 16t \rangle$. When is the speed a minimum?

18. What force is required so that a particle of mass m has the position function $\mathbf{r}(t) = t^3 \mathbf{i} + t^2 \mathbf{j} + t^3 \mathbf{k}$?

19. A force with magnitude 20 N acts directly upward from the xy-plane on an object with mass 4 kg. The object starts at the origin with initial velocity $\mathbf{v}(0) = \mathbf{i} - \mathbf{j}$. Find its position function and its speed at time t.

20. Show that if a particle moves with constant speed, then the velocity and acceleration vectors are orthogonal.

21–25 ■ Find the tangential and normal components of the acceleration vector.

21. $\mathbf{r}(t) = (t - \sin t) \, \mathbf{i} + (1 - \cos t) \, \mathbf{j}$

22. $\mathbf{r}(t) = t \, \mathbf{i} + 4 \sin t \, \mathbf{j} + 4 \cos t \, \mathbf{k}$

23. $\mathbf{r}(t) = t^3 \mathbf{i} + t^2 \mathbf{j} + t \, \mathbf{k}$

24. $\mathbf{r}(t) = t \, \mathbf{i} + \cos^2 t \, \mathbf{j} + \sin^2 t \, \mathbf{k}$

25. $\mathbf{r}(t) = e^t \mathbf{i} + \sqrt{2} \, t \, \mathbf{j} + e^{-t} \mathbf{k}$

26. If a particle with mass m moves with position vector $\mathbf{r}(t)$, its **angular momentum** is defined as $\mathbf{L}(t) = m\mathbf{r}(t) \times \mathbf{v}(t)$ and its **torque** as $\boldsymbol{\tau}(t) = m\mathbf{r}(t) \times \mathbf{a}(t)$. Show that $\mathbf{L}'(t) = \boldsymbol{\tau}(t)$. Deduce that if $\boldsymbol{\tau}(t) = \mathbf{0}$ for all t, then $\mathbf{L}(t)$ is constant. (This is the law of conservation of angular momentum.)

27. The position function of a spaceship is

$$\mathbf{r}(t) = (3 + t) \, \mathbf{i} + (2 + \ln t) \, \mathbf{j} + \left(7 - \frac{4}{t^2 + 1}\right) \mathbf{k}$$

and the coordinates of a space station are $(6, 4, 9)$. The captain wants the spaceship to coast into the space station. When should the engines be turned off?

28. A rocket burning its onboard fuel while moving through space has velocity $\mathbf{v}(t)$ and mass $m(t)$ at time t. If the exhaust gases escape with velocity $\mathbf{v}_e$ relative to the rocket, it can be deduced from Newton's Second Law of Motion that

$$m \frac{d\mathbf{v}}{dt} = \frac{dm}{dt} \mathbf{v}_e$$

(a) Show that $\mathbf{v}(t) = \mathbf{v}(0) - \ln \dfrac{m(0)}{m(t)} \mathbf{v}_e$.

(b) For the rocket to accelerate in a straight line from rest to twice the speed of its own exhaust gases, what fraction of its initial mass would the rocket have to burn as fuel?

29. Use the following steps to prove Kepler's Second Law. The notation is the same as in the proof of the first law. In particular, use polar coordinates so that $\mathbf{r} = (r \cos \theta) \, \mathbf{i} + (r \sin \theta) \, \mathbf{j}$.

(a) Show that $\mathbf{h} = r^2 \dfrac{d\theta}{dt} \mathbf{k}$.

(b) Deduce that $r^2 \dfrac{d\theta}{dt} = h$.

(c) If $A = A(t)$ is the area swept out by the radius vector $\mathbf{r} = \mathbf{r}(t)$ in the time interval $[t_0, t]$, show that

$$\frac{dA}{dt} = \frac{1}{2} r^2 \frac{d\theta}{dt}$$

(d) Deduce that

$$\frac{dA}{dt} = \tfrac{1}{2}h = \text{constant}$$

This says that the rate at which A is swept out is constant and proves Kepler's Second Law.

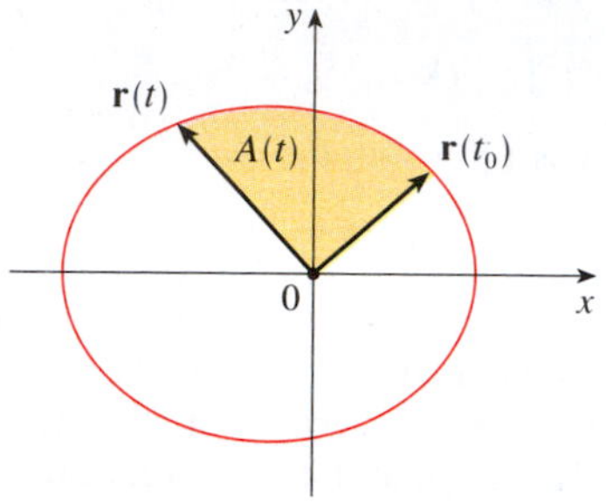

30. Let T be the period of a planet about the Sun; that is, T is the time required for it to travel once around its elliptical orbit. Suppose that the lengths of the major and minor axes of the ellipse are $2a$ and $2b$.

(a) Use Exercise 29 part (d) to show that $T = 2\pi ab/h$.

(b) Show that $\dfrac{h^2}{GM} = ed = \dfrac{b^2}{a}$.

(c) Use parts (a) and (b) to show that $T^2 = \dfrac{4\pi^2}{GM}a^3$.

This proves Kepler's Third Law. [Notice that the proportionality constant $4\pi^2/(GM)$ is independent of the planet.]

31. The period of Earth's orbit is approximately 365.25 days. Use this fact and Exercise 30 to find the length of the major axis of Earth's orbit. You will need the mass of the Sun, $M = 1.99 \times 10^{30}$ kg, and the gravitational constant, $G = 6.67 \times 10^{-11}$ N·m²/kg².

11 REVIEW

KEY TOPICS ■ Define, state, or discuss the following.

1. $\mathbb{R}^3$
2. Distance formula in $\mathbb{R}^3$
3. Equation of a sphere
4. Vectors in V_3
5. Position vector
6. Length of a vector
7. Addition of vectors
8. Multiplication of a vector by a scalar
9. Unit vector
10. Dot product and its properties
11. Angle between two vectors
12. Orthogonal vectors
13. Parallel vectors
14. Direction cosines
15. Vector and scalar projections
16. Work done by a constant force
17. Cross product and its properties
18. Scalar triple product
19. Vector, parametric, and symmetric equations of a line
20. Vector and scalar equations of a plane
21. Distance from a point to a plane
22. Standard equations of ellipsoids, hyperboloids, paraboloids, and cones
23. Vector functions and component functions
24. Limit of a vector function
25. Continuity of a vector function
26. Derivative of a vector function
27. Space curve
28. Tangent vector and tangent line
29. Differentiation formulas for vector functions
30. Integrals of vector functions
31. Length of a curve
32. Smooth curve
33. Arc length function
34. Curvature
35. Unit normal vector
36. Binormal vector
37. Normal plane
38. Osculating plane; osculating circle
39. Velocity and acceleration along a curve
40. Tangential and normal components of acceleration
41. Kepler's Laws

EXERCISES

1–14 ■ Determine whether the statement is true or false.

1. For any vectors **u** and **v** in V_3, $\mathbf{u} \cdot \mathbf{v} = \mathbf{v} \cdot \mathbf{u}$.
2. For any vectors **u** and **v** in V_3, $\mathbf{u} \times \mathbf{v} = \mathbf{v} \times \mathbf{u}$.
3. For any vectors **u** and **v** in V_3, $|\mathbf{u} \times \mathbf{v}| = |\mathbf{v} \times \mathbf{u}|$.
4. For any vectors **u** and **v** in V_3 and any scalar k, $k(\mathbf{u} \cdot \mathbf{v}) = (k\mathbf{u}) \cdot \mathbf{v}$.
5. For any vectors **u** and **v** in V_3 and any scalar k, $k(\mathbf{u} \times \mathbf{v}) = (k\mathbf{u}) \times \mathbf{v}$.
6. For any vectors **u**, **v**, and **w** in V_3, $(\mathbf{u} + \mathbf{v}) \times \mathbf{w} = \mathbf{u} \times \mathbf{w} + \mathbf{v} \times \mathbf{w}$.
7. For any vectors **u**, **v**, and **w** in V_3, $\mathbf{u} \cdot (\mathbf{v} \times \mathbf{w}) = (\mathbf{u} \times \mathbf{v}) \cdot \mathbf{w}$.
8. For any vectors **u**, **v**, and **w** in V_3, $\mathbf{u} \times (\mathbf{v} \times \mathbf{w}) = (\mathbf{u} \times \mathbf{v}) \times \mathbf{w}$.
9. For any vectors **u** and **v** in V_3, $(\mathbf{u} \times \mathbf{v}) \cdot \mathbf{u} = 0$.
10. For any vectors **u** and **v** in V_3, $(\mathbf{u} + \mathbf{v}) \times \mathbf{v} = \mathbf{u} \times \mathbf{v}$.
11. The cross product of two unit vectors is a unit vector.
12. A linear equation $Ax + By + Cz + D = 0$ represents a line in space.
13. The set of points $\{(x, y, z) \mid x^2 + y^2 = 1\}$ is a circle.
14. The curve with vector equation $\mathbf{r}(t) = t^3\mathbf{i} + 2t^3\mathbf{j} + 3t^3\mathbf{k}$ is a line.

15. Find the lengths of the sides of the triangle with vertices $A(2, 6, -4)$, $B(-1, 2, 8)$, and $C(0, 1, 2)$.

16. Find an equation of the sphere with center $(1, -1, 2)$ and radius 3.

17. Find the center and radius of the sphere

$$x^2 + y^2 + z^2 + 4x + 6y - 10z + 2 = 0$$

18–30 ■ Calculate the given quantity if

$$\mathbf{a} = \mathbf{i} + \mathbf{j} - 2\mathbf{k} \qquad \mathbf{b} = 3\mathbf{i} - 2\mathbf{j} + \mathbf{k} \qquad \mathbf{c} = \mathbf{j} - 5\mathbf{k}$$

18. $2\mathbf{a} + 3\mathbf{b}$
19. $6\mathbf{a} - 5\mathbf{c}$
20. $|\mathbf{b}|$
21. $\mathbf{a} \cdot \mathbf{b}$
22. $\mathbf{a} \times \mathbf{b}$
23. $|\mathbf{b} \times \mathbf{c}|$
24. $\mathbf{a} \cdot (\mathbf{b} \times \mathbf{c})$
25. $\mathbf{c} \times \mathbf{c}$
26. $\mathbf{a} \times (\mathbf{b} \times \mathbf{c})$
27. The angle between **a** and **b** (correct to the nearest degree)
28. The direction cosines of **b**
29. The scalar projection of **b** onto **a**
30. The vector projection of **b** onto **a**

31. Find the value of x such that the vectors $\langle 2, x, 4\rangle$ and $\langle 2x, 3, -7\rangle$ are orthogonal.

32. Find two unit vectors orthogonal to both $\langle 1, 0, 1\rangle$ and $\langle 2, 3, 4\rangle$.

33. Suppose that $\mathbf{u} \cdot (\mathbf{v} \times \mathbf{w}) = 2$. Find
    (a) $(\mathbf{u} \times \mathbf{v}) \cdot \mathbf{w}$ (b) $\mathbf{u} \cdot (\mathbf{w} \times \mathbf{v})$
    (c) $\mathbf{v} \cdot (\mathbf{u} \times \mathbf{w})$ (d) $(\mathbf{u} \times \mathbf{v}) \cdot \mathbf{v}$

34. Show that if **a**, **b**, and **c** are in V_3, then

$$(\mathbf{a} \times \mathbf{b}) \cdot [(\mathbf{b} \times \mathbf{c}) \times (\mathbf{c} \times \mathbf{a})] = [\mathbf{a} \cdot (\mathbf{b} \times \mathbf{c})]^2$$

35. Describe a method for determining whether three points P, Q, and R lie on the same line.

36. Describe a method for determining whether four points P, Q, R, and S lie in the same plane.

37. Find the acute angle between two diagonals of a cube.

38. Given the points $A(1, 0, 1)$, $B(2, 3, 0)$, $C(-1, 1, 4)$, and $D(0, 3, 2)$, find the volume of the parallelepiped with adjacent edges AB, AC, and AD.

39. (a) Find a vector perpendicular to the plane through the points $A(1, 0, 0)$, $B(2, 0, -1)$, and $C(1, 4, 3)$.
    (b) Find the area of triangle ABC.

40. A constant force $\mathbf{F} = 3\mathbf{i} + 5\mathbf{j} + 10\mathbf{k}$ moves an object along the line segment from $(1, 0, 2)$ to $(5, 3, 8)$. Find the work done if the distance is measured in meters and the force in newtons.

41. A boat is pulled onto shore using two ropes, as shown in the diagram. If a force of 255 N is needed to pull the boat, find the magnitude of the force in each rope.

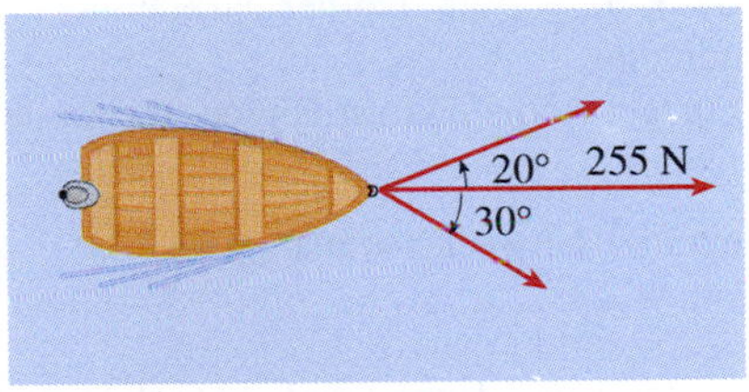

42. Find the magnitude of the torque about P if a 50-N force is applied as shown.

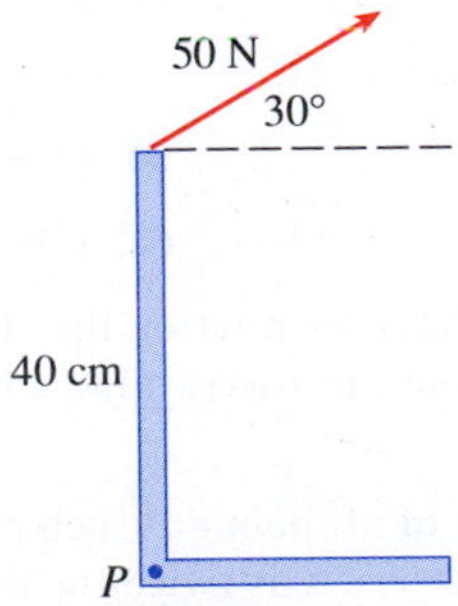

43–45 ■ Find parametric equations for the line that satisfies the given conditions.

43. Passing through $(1, 2, 4)$ and in the direction of $\mathbf{v} = 2\mathbf{i} - \mathbf{j} + 3\mathbf{k}$

44. Passing through $(-6, -1, 0)$ and $(2, -3, 5)$

45. Passing through $(1, 0, 1)$ and parallel to the line with parametric equations $x = 4t$, $y = 1 - 3t$, $z = 2 + 5t$

46–48 ■ Find an equation of the plane that satisfies the given conditions.

46. Passing through $(4, -1, -1)$ and with normal vector $\langle 2, 6, -3 \rangle$

47. Passing through $(-4, 1, 2)$ and parallel to the plane $x + 2y + 5z = 3$

48. Passing through $(-1, 2, 0)$, $(2, 0, 1)$, and $(-5, 3, 1)$

49. Find the point in which the line with parametric equations $x = 2 - t$, $y = 1 + 3t$, $z = 4t$ intersects the plane $2x - y + z = 2$.

50. Find the distance from the origin to the line $x = 1 + t$, $y = 2 - t$, $z = -1 + 2t$.

51. Determine whether the lines given by the symmetric equations

$$\frac{x-1}{2} = \frac{y-2}{3} = \frac{z-3}{4}$$

and

$$\frac{x+1}{6} = \frac{y-3}{-1} = \frac{z+5}{2}$$

are parallel, skew, or intersecting.

52. (a) Show that the planes $x + y - z = 1$ and $2x - 3y + 4z = 5$ are neither parallel nor perpendicular.

(b) Find, correct to the nearest degree, the angle between these planes.

53. Find the distance between the planes $3x + y - 4z = 2$ and $3x + y - 4z = 24$.

54–62 ■ Identify and sketch the graph of each surface.

54. $x = 6$

55. $y = z$

56. $6x + 4y + 3z = 12$

57. $y = x^2 + z^2$

58. $225x^2 + 9y^2 + 25z^2 = 225$

59. $x^2 = y^2 + 4z^2$

60. $y^2 + z^2 = 1 + x^2$

61. $-4x^2 + y^2 - 4z^2 = 4$

62. $y = z^2$

63. An ellipsoid is created by rotating the ellipse $4x^2 + y^2 = 16$ about the x-axis. Find an equation of the ellipsoid.

64. A surface consists of all points P such that the distance from P to the plane $y = 1$ is twice the distance from P to the point $(0, -1, 0)$. Find an equation for this surface and identify it.

65. (a) Sketch the curve with vector function

$$\mathbf{r}(t) = 2\mathbf{i} + \sin t\,\mathbf{j} + \cos t\,\mathbf{k}$$

(b) Find $\mathbf{r}'(t)$ and $\mathbf{r}''(t)$.

66. Find parametric equations for the tangent line to the curve $x = t^2$, $y = t^4$, $z = t^3$ at the point $(1, 1, 1)$. Graph the curve and the tangent line on a common screen.

67. If $\mathbf{r}(t) = (t + t^2)\,\mathbf{i} + (2 + t^3)\,\mathbf{j} + t^4\mathbf{k}$, evaluate $\int_0^1 \mathbf{r}(t)\,dt$.

68. Let C be the curve with equations $x = 2 - t^3$, $y = 2t - 1$, $z = \ln t$. Find (a) the point where C intersects the xz-plane, (b) parametric equations of the tangent line at $(1, 1, 0)$, and (c) an equation of the normal plane to C at $(1, 1, 0)$.

69. Use Simpson's Rule with $n = 4$ to estimate the length of the arc of the curve with equations $x = \sqrt{t}$, $y = 4/t$, $z = t^2 + 1$ from $(1, 4, 2)$ to $(2, 1, 17)$.

70. Find the length of the curve $\mathbf{r}(t) = \langle 2t^{3/2}, \cos 2t, \sin 2t \rangle$, $0 \leqslant t \leqslant 1$.

71. The helix $\mathbf{r}_1(t) = \cos t\,\mathbf{i} + \sin t\,\mathbf{j} + t\,\mathbf{k}$ intersects the curve $\mathbf{r}_2(t) = (1 + t)\,\mathbf{i} + t^2\mathbf{j} + t^3\mathbf{k}$ at the point $(1, 0, 0)$. Find the angle of intersection of these curves.

72. Reparametrize the curve $\mathbf{r}(t) = e^t\mathbf{i} + e^t \sin t\,\mathbf{j} + e^t \cos t\,\mathbf{k}$ with respect to arc length measured from the point $(1, 0, 1)$ in the direction of increasing t.

73. For the curve given by $\mathbf{r}(t) = \langle t^3/3, t^2/2, t \rangle$, find (a) the unit tangent vector, (b) the unit normal vector, and (c) the curvature.

74. Find the curvature of the ellipse $x = 3\cos t$, $y = 4\sin t$ at the points $(3, 0)$ and $(0, 4)$.

75. Find the curvature of the curve $y = x^4$ at the point $(1, 1)$.

76. Find an equation of the osculating circle of the curve $y = x^4 - x^2$ at the origin. Graph both the curve and its osculating circle.

77. Find an equation of the osculating plane of the curve $x = \sin 2t$, $y = t$, $z = \cos 2t$ at the point $(0, \pi, 1)$.

78. A particle moves with position function $\mathbf{r}(t) = 2\sqrt{2}\,t\,\mathbf{i} + e^{2t}\mathbf{j} + e^{-2t}\mathbf{k}$. Find the velocity, speed, and acceleration of the particle.

79. A particle starts at the origin with initial velocity $\mathbf{i} + 2\mathbf{j} + \mathbf{k}$. Its acceleration is $\mathbf{a}(t) = t\,\mathbf{i} + \mathbf{j} + t^2\mathbf{k}$. Find its position function.

80. Find the tangential and normal components of the acceleration vector of a particle with position function $\mathbf{r}(t) = t\,\mathbf{i} + 2t\,\mathbf{j} + t^2\mathbf{k}$.

81. A disk of radius 1 is rotating in the counterclockwise direction at a constant angular speed ω. A particle starts at the center of the disk and moves toward the edge along a fixed radius so that its position at time t, $t \geqslant 0$, is given by $\mathbf{r}(t) = t\mathbf{R}(t)$, where

$$\mathbf{R}(t) = \cos \omega t\,\mathbf{i} + \sin \omega t\,\mathbf{j}$$

(a) Show that the velocity $\mathbf{v}$ of the particle is

$$\mathbf{v} = \cos \omega t\,\mathbf{i} + \sin \omega t\,\mathbf{j} + t\mathbf{v}_d$$

where $\mathbf{v}_d = \mathbf{R}'(t)$ is the velocity of a point on the edge of the disk.

(b) Show that the acceleration $\mathbf{a}$ of the particle is

$$\mathbf{a} = 2\mathbf{v}_d + t\mathbf{a}_d$$

where $\mathbf{a}_d = \mathbf{R}''(t)$ is the acceleration of a point on the rim of the disk. The extra term $2\mathbf{v}_d$ is called the *Coriolis acceleration;* it is the result of the interaction of the rotation of the disk and the motion of the particle. One can obtain a physical demonstration of this acceleration by walking toward the edge of a moving merry-go-round.

(c) Determine the Coriolis acceleration of a particle that moves on a rotating disk according to

$$\mathbf{r}(t) = e^{-t}\cos \omega t\,\mathbf{i} + e^{-t}\sin \omega t\,\mathbf{j}$$

APPLICATIONS PLUS

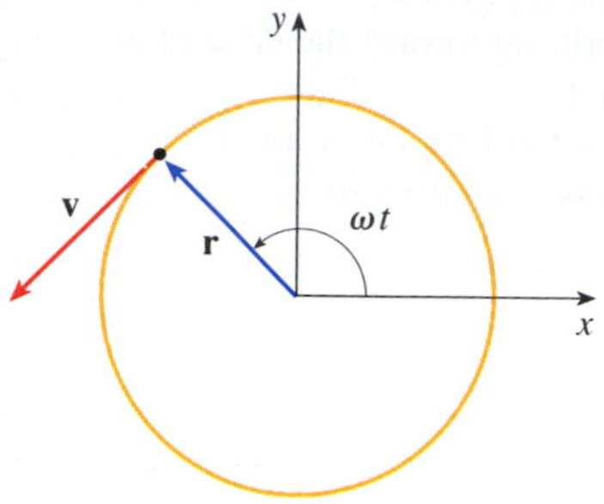

FIGURE FOR PROBLEM 1

1. A particle P moves with constant angular speed ω around a circle whose center is at the origin and whose radius is R. The particle is said to be in *uniform circular motion.* Assume that the motion is counterclockwise and that the particle is at the point $(R, 0)$ when $t = 0$. The position vector at time $t \geq 0$ is

$$\mathbf{r}(t) = R\cos\omega t\,\mathbf{i} + R\sin\omega t\,\mathbf{j}$$

(a) Find the velocity vector $\mathbf{v}$ and show that $\mathbf{v}\cdot\mathbf{r} = 0$. Conclude that $\mathbf{v}$ is tangent to the circle and points in the direction of the motion.

(b) Show that the speed $|\mathbf{v}|$ of the particle is the constant ωR. The *period* T of the particle is the time required for one complete revolution. Conclude that

$$T = \frac{2\pi R}{|\mathbf{v}|} = \frac{2\pi}{\omega}$$

(c) Find the acceleration vector $\mathbf{a}$. Show that it is proportional to $\mathbf{r}$ and that it points toward the origin. An acceleration with this property is called a *centripetal acceleration.* Show that the magnitude of the acceleration vector is $|\mathbf{a}| = R\omega^2$.

(d) Suppose that the particle has mass m. Show that the magnitude of the force $\mathbf{F}$ that is required to produce this motion, called a *centripetal force*, is

$$|\mathbf{F}| = \frac{m|\mathbf{v}|^2}{R}$$

2. A circular curve of radius R on a highway is banked at an angle θ so that a car can safely traverse the curve without skidding when there is no friction between the road and the tires. The loss of friction could occur, for example, if the road is covered with a film of water or ice. The rated speed v_R of the curve is the maximum speed that a car can attain without skidding. Suppose a car of mass m is traversing the curve at the rated speed v_R. Two forces are acting on the car: the vertical force, mg, due to the weight of the car, and a force $\mathbf{F}$ exerted by, and normal to, the road. (See the figure.)

The vertical component of $\mathbf{F}$ balances the weight of the car, so that $|\mathbf{F}|\cos\theta = mg$. The horizontal component of $\mathbf{F}$ produces a centripetal force on the car so that, by Newton's Second Law and part (d) of Problem 1,

$$|\mathbf{F}|\sin\theta = \frac{mv_R^2}{R}$$

(a) Show that $v_R^2 = Rg\tan\theta$.

(b) Find the rated speed of a circular curve with radius 400 ft that is banked at an angle of 12°.

(c) Suppose the design engineers want to keep the banking at 12°, but wish to increase the rated speed by 50%. What should the radius of the curve be?

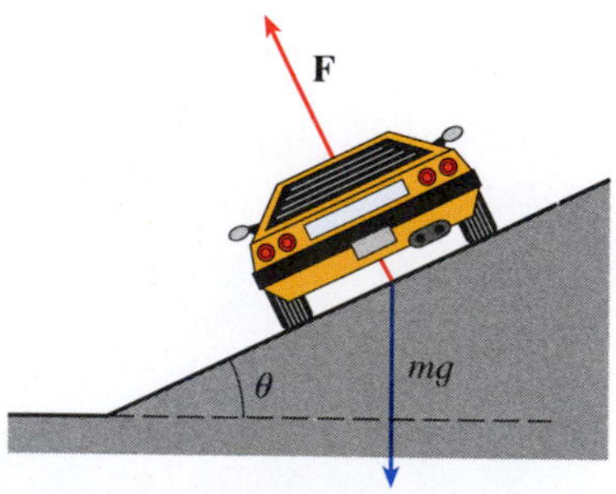

FIGURE FOR PROBLEM 2

3. A projectile is fired from the origin with angle of elevation α and initial speed v_0. Assuming that air resistance is negligible and that the only force acting on the projectile is gravity, g, we showed in Example 11 in Section 5.7 that the position vector of the projectile is

$$\mathbf{r}(t) = (v_0 \cos \alpha)t\,\mathbf{i} + \left[(v_0 \sin \alpha)t - \tfrac{1}{2}gt^2\right]\mathbf{j}$$

We also showed that the maximum horizontal distance of the projectile is achieved when $\alpha = 45°$ and in this case the range is $R = v_0^2/g$.

(a) At what angle should the projectile be fired to achieve maximum height and what is the maximum height?

(b) Fix the initial speed v_0 and consider the parabola $x^2 + 2Ry - R^2 = 0$, whose graph is shown in the figure. Show that the projectile can hit any target inside or on the boundary of the region bounded by the parabola and the x-axis, and that it cannot hit any target outside this region.

(c) Suppose that the gun is elevated to an angle of inclination α in order to aim at a target that is suspended at a height h directly over a point D units downrange. The target is released at the instant the gun is fired. Show that the projectile always hits the target, regardless of the value v_0, provided the projectile does not hit the ground "before" D.

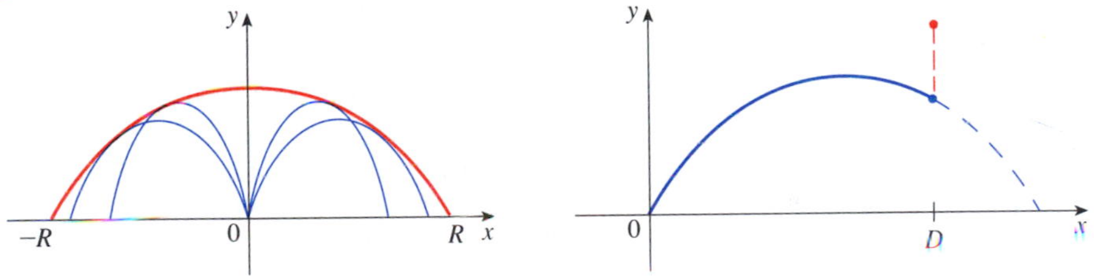

FIGURE FOR PROBLEM 3

4. A projectile is fired from the origin down an inclined plane that makes an angle θ with the horizontal. The angle of elevation of the gun and the initial velocity of the projectile are α and v_0, respectively.

(a) Find the position vector of the projectile and the parametric equations of the path of the projectile as functions of the time t. (Ignore air resistance.)

(b) Determine the angle of elevation α that will maximize the downhill range.

(c) Suppose the projectile is fired up an inclined plane whose angle of inclination is θ. At what angle should the projectile be fired to maximize the (uphill) range?

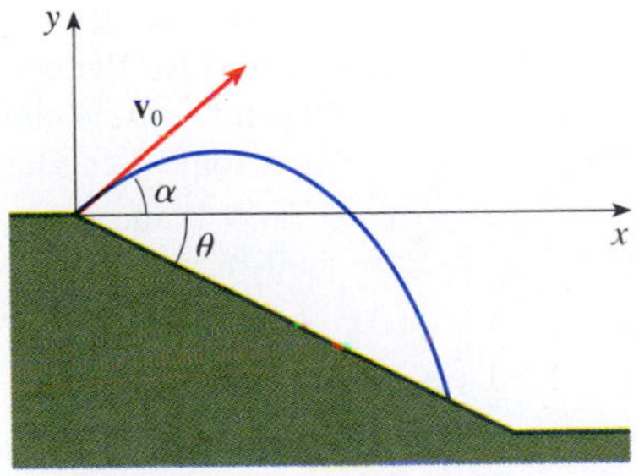

FIGURE FOR PROBLEM 4

5. A projectile of mass m is fired from the origin at an angle of elevation α. In addition to gravity, assume that air resistance provides a force that is proportional to the velocity and that opposes the motion. Then, by Newton's Second Law, the total force acting on the projectile satisfies the equation

$$m\frac{d^2\mathbf{R}}{dt^2} = -mg\,\mathbf{j} - k\frac{d\mathbf{R}}{dt} \tag{1}$$

where $\mathbf{R}$ is the position vector and $k > 0$ is the constant of proportionality.

(a) Show that Equation 1 can be integrated to obtain the equation

$$\frac{d\mathbf{R}}{dt} + \frac{k}{m}\mathbf{R} = \mathbf{v}_0 - gt\,\mathbf{j}$$

where $\mathbf{v}_0 = \mathbf{v}(0) = \dfrac{d\mathbf{R}}{dt}(0)$.

(b) Multiply both sides of the equation in part (a) by $e^{(k/m)t}$ and show that the left-hand side of the resulting equation is the derivative of the product $e^{(k/m)t}\mathbf{R}(t)$. Then integrate to find an expression for the position vector $\mathbf{R}(t)$.

6. Based on clinical evidence, it can be assumed that the rate of decrease of the concentration C of a drug in the bloodstream is proportional to the concentration itself; that is, $dC/dt = -kC$, where k is a positive constant. Thus, the concentration at time t is given by $C(t) = C_0e^{-kt}$, where C_0 is the initial dosage. Now suppose that the drug is to be administered in equal dosages C_0 at equally spaced time intervals of length T. Assume that when the drug is administered it is diffused immediately throughout the bloodstream; that is, assume that there is an instantaneous rise in the concentration at each administration of the drug.

(a) Let C_{n-1} be the concentration of the drug immediately after the nth injection and let R_n be the concentration immediately before the $(n+1)$th injection. Thus, C_{n-1} and R_n are the concentrations at the beginning and end, respectively, of the nth interval. Find expressions for C_{n-1} and R_n.

(b) Calculate $\lim_{n\to\infty} C_{n-1}$ and $\lim_{n\to\infty} R_n$.

(c) Let $C = C(t)$ denote the concentration of the drug in the bloodstream at any time t. Sketch the graph of C.

(d) Suppose that the drug is ineffective below the concentration level C_L and is harmful above the concentration level C_H. Find a dosage C_0 and a time interval T so that, eventually

$$C_L \leqslant C(t) \leqslant C_H$$

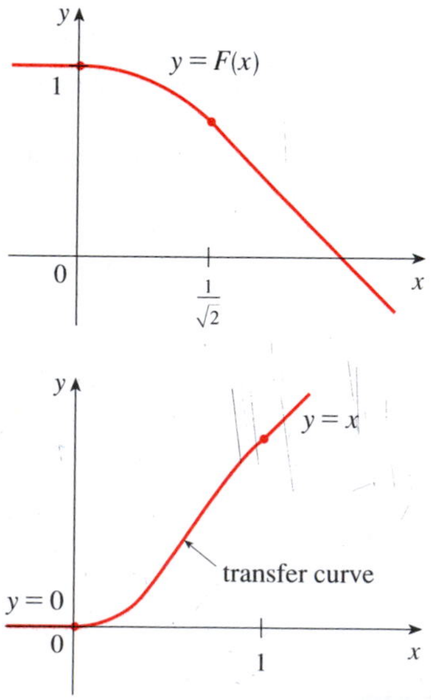

FIGURE FOR PROBLEM 7

7. In designing *transfer curves* to connect sections of straight railroad tracks, it is important to realize that the acceleration of the train should be continuous so that the reactive force exerted by the train on the track is also continuous. Because of the formulas for the components of acceleration in Section 11.8, this will be the case if the curvature varies continuously.

(a) A logical candidate for a transfer curve to join existing tracks given by $y = 1$ for $x \leqslant 0$ and $y = \sqrt{2} - x$ for $x \geqslant 1/\sqrt{2}$ might be the function $f(x) = \sqrt{1 - x^2}$, $0 < x < 1/\sqrt{2}$, whose graph is the arc of the circle shown in the figure. It looks reasonable at first glance. Show that the function

$$F(x) = \begin{cases} 1 & \text{if } x \leqslant 0 \\ \sqrt{1 - x^2} & \text{if } 0 < x < 1/\sqrt{2} \\ \sqrt{2} - x & \text{if } x \geqslant 1/\sqrt{2} \end{cases}$$

is continuous and has continuous slope, but does not have continuous curvature. Therefore, f is not an appropriate transfer curve.

(b) Find a fifth-degree polynomial to serve as a transfer curve between the following straight line segments: $y = 0$ for $x \leqslant 0$ and $y = x$ for $x \geqslant 1$. Could this be done with a fourth-degree polynomial? Use a graphing calculator or computer to sketch the graph of the "connected" function and check to see that it looks like the one in the figure.

FIGURE FOR PROBLEM 8

8. A ball rolls off a table with a speed of 2 ft/s. The table is 3.5 ft high.

(a) Determine the point at which the ball hits the floor and find its speed at the instant of impact.

(b) Find the angle θ between the path of the ball and the vertical line drawn through the point of impact. (See the figure.)

(c) Suppose the ball rebounds from the floor at the same angle with which it hits the floor, but loses 20% of its speed due to energy absorbed by the ball on impact. Where does the ball strike the floor on the second bounce?

12

PARTIAL DERIVATIVES

Strange as it may seem, the power of mathematics rests on its evasion of all unnecessary thought and on its wonderful savings of mental operations.

ERNST MACH

So far we have dealt with the calculus of functions of a single variable. But, in the real world, physical quantities usually depend on two or more variables, so in this chapter we turn our attention to functions of several variables and extend the basic ideas of differential calculus to such functions.

12.1 FUNCTIONS OF SEVERAL VARIABLES

The temperature T at a point on the surface of the earth at any given time depends on the longitude x and latitude y of the point. We can think of T as being a function of the two variables x and y, or as a function of the pair (x, y). We indicate this functional dependence by writing $T = f(x, y)$.

The volume V of a circular cylinder depends on its radius r and height h. In fact, we know from Chapter 7 that $V = \pi r^2 h$. We say that V is a function of r and h, and we write $V(r, h) = \pi r^2 h$.

(1) DEFINITION Let $D \subset \mathbb{R}^2$. A **function f of two variables** is a rule that assigns to each ordered pair (x, y) in D a unique real number denoted by $f(x, y)$. The set D is the **domain** of f and its **range** is the set of values that f takes on, that is, $\{f(x, y) \mid (x, y) \in D\}$.

We often write $z = f(x, y)$ to make explicit the value taken on by f at the general point (x, y). The variables x and y are **independent variables** and z is the **dependent variable**. [Compare this with the notation $y = f(x)$ for functions of a single variable.]

The situation described in Definition 1 is indicated by the notation $f: D \to \mathbb{R}$. A function of two variables is just a special case of the general idea of a function $f: X \to Y$ (see Section 1 of Review and Preview), where $X = D \subset \mathbb{R}^2$ and $Y \subset \mathbb{R}$. One way of visualizing such a function is by means of an arrow diagram (see Figure 1), where the domain D is represented as a subset of the xy-plane.

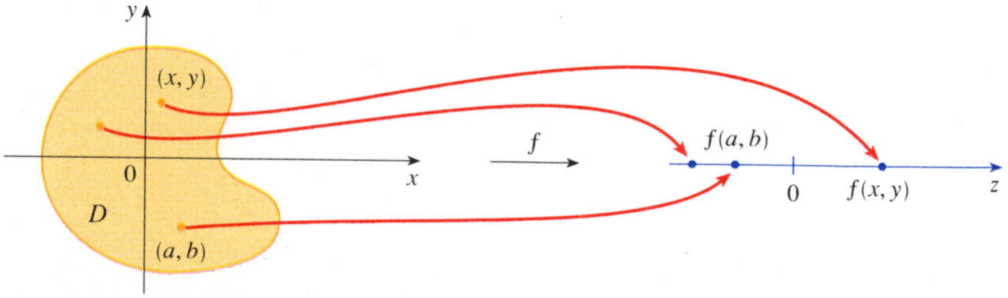

FIGURE 1

If a function f is given by a formula and no domain is specified, then the domain of f is understood to be the set of all pairs (x, y) for which the given expression is a well-defined real number.

EXAMPLE 1 Find the domain of each of the following functions and evaluate $f(3, 2)$.

(a) $f(x, y) = \dfrac{\sqrt{x + y + 1}}{x - 1}$ (b) $f(x, y) = x\ln(y^2 - x)$

SOLUTION

(a) $$f(3, 2) = \frac{\sqrt{3 + 2 + 1}}{3 - 1} = \frac{\sqrt{6}}{2}$$

The expression for f makes sense if the denominator is not 0 and the quantity under the square root sign is nonnegative. So the domain of f is

$$D = \{(x, y) \mid x + y + 1 \geqslant 0,\ x \neq 1\}$$

The inequality $x + y + 1 \geqslant 0$, or $y \geqslant -x - 1$, describes the points that lie on or above the line $y = -x - 1$, while $x \neq 1$ means that the points on the line $x = 1$ must be excluded from the domain. (See Figure 2.)

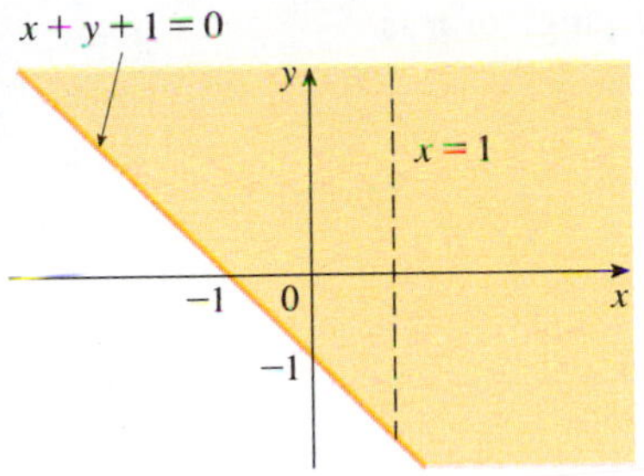

FIGURE 2

Domain of $f(x, y) = \dfrac{\sqrt{x + y + 1}}{x - 1}$

(b) $$f(3, 2) = 3\ln(2^2 - 3) = 3\ln 1 = 0$$

Since $\ln(y^2 - x)$ is defined only when $y^2 - x > 0$, that is, $x < y^2$, the domain of f is $D = \{(x, y) \mid x < y^2\}$. This is the set of points to the left of the parabola $x = y^2$. (See Figure 3.) ■

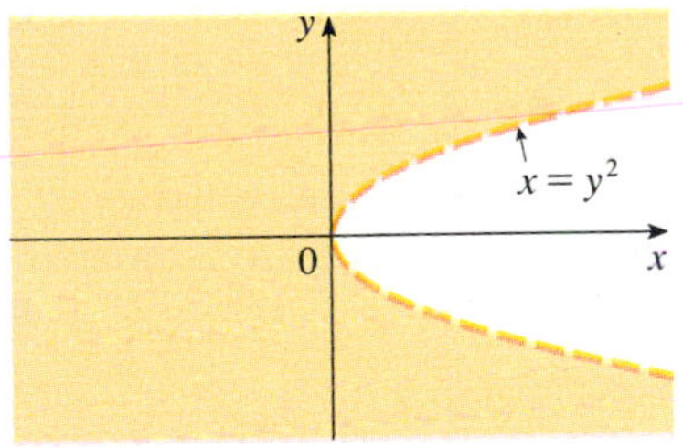

FIGURE 3

Domain of $f(x, y) = x\ln(y^2 - x)$

(2) DEFINITION If f is a function of two variables with domain D, the **graph** of f is the set

$$S = \{(x, y, z) \in \mathbb{R}^3 \mid z = f(x, y),\ (x, y) \in D\}$$

Just as the graph of a function f of one variable is a curve C with equation $y = f(x)$, so the graph of a function f of two variables is a surface S with equation $z = f(x, y)$. We can visualize the graph S of f as lying directly above or below its domain D in the xy-plane (see Figure 4).

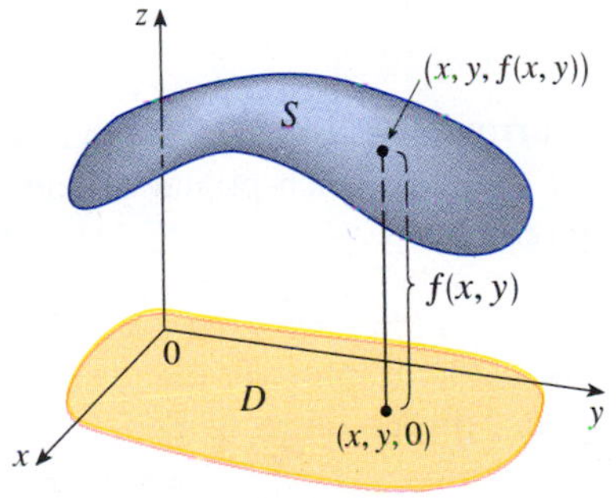

FIGURE 4

EXAMPLE 2 Sketch the graph of the function $f(x, y) = 6 - 3x - 2y$.

SOLUTION The graph of f has the equation $z = 6 - 3x - 2y$, or $3x + 2y + z = 6$, which is a plane. The portion of this graph that lies in the first octant is sketched in Figure 5.

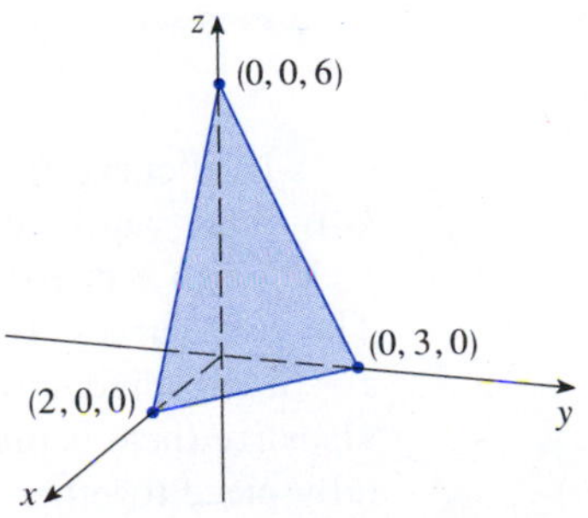

FIGURE 5

■

EXAMPLE 3 Find the domain and range and sketch the graph of

$$g(x, y) = \sqrt{9 - x^2 - y^2}$$

SOLUTION The domain of g is

$$D = \{(x, y) \mid 9 - x^2 - y^2 \geqslant 0\} = \{(x, y) \mid x^2 + y^2 \leqslant 9\}$$

which is the disk with center $(0, 0)$ and radius 3. The range of g is

$$\{z \mid z = \sqrt{9 - x^2 - y^2},\ (x, y) \in D\}$$

Since z is a positive square root, $z \geqslant 0$. Also

$$9 - x^2 - y^2 \leqslant 9 \quad \Rightarrow \quad \sqrt{9 - x^2 - y^2} \leqslant 3$$

So the range is

$$\{z \mid 0 \leqslant z \leqslant 3\} = [0, 3]$$

The graph has equation $z = \sqrt{9 - x^2 - y^2}$. We square both sides of this equation to obtain $z^2 = 9 - x^2 - y^2$, or $x^2 + y^2 + z^2 = 9$, which we recognize as an equation of the sphere with center the origin and radius 3. But, since $z \geqslant 0$, the graph of g is just the top half of this sphere (see Figure 6). ■

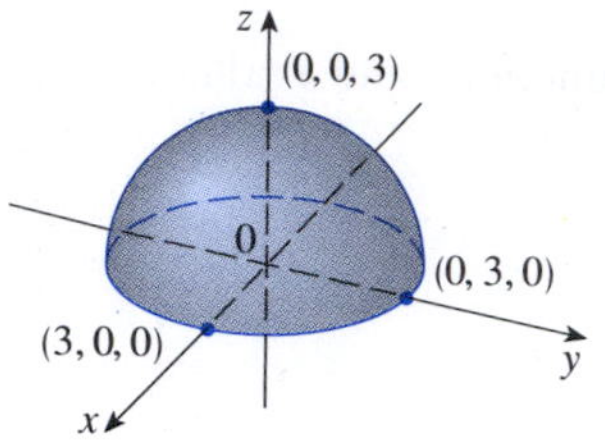

FIGURE 6
Graph of $g(x, y) = \sqrt{9 - x^2 - y^2}$

EXAMPLE 4 Sketch the graph of the function $h(x, y) = 4x^2 + y^2$.

SOLUTION The graph of h has the equation $z = 4x^2 + y^2$, which is an elliptic paraboloid (see Equation 11.5.5). Traces in the horizontal planes $z = k$ are ellipses if $k > 0$. Vertical traces are parabolas. (See Figure 7.) ■

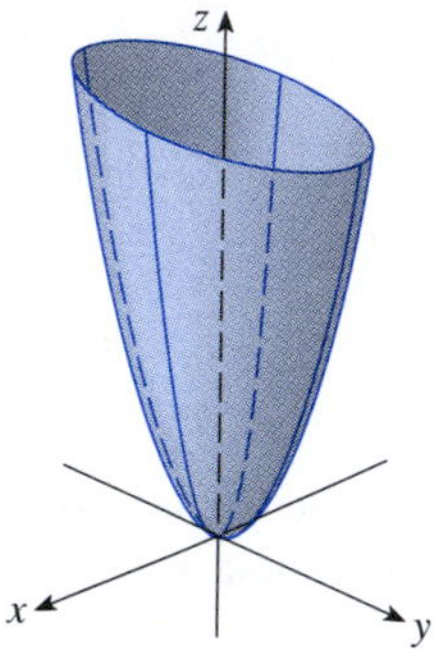

FIGURE 7
Graph of $h(x, y) = 4x^2 + y^2$

Computer programs are now readily available for graphing functions of two variables. In most such programs, traces in the vertical planes $x = k$ and $y = k$ are drawn for equally spaced values of k and parts of the graph are eliminated using hidden line removal. Figure 8 shows computer-generated graphs of several functions. Notice that we get an especially good picture of a function when rotation is used to give views from different vantage points.

LEVEL CURVES

So far we have two methods for visualizing functions: arrow diagrams and graphs. A third method, borrowed from mapmakers, is a contour map on which points of constant elevation are joined to form *contour curves,* or *level curves.*

> **(3) DEFINITION** The **level curves** of a function f of two variables are the curves with equations $f(x, y) = k$, where k is a constant (in the range of f).

A level curve $f(x, y) = k$ is the locus of all points at which f takes on a given value k. In other words, it shows where the graph of f has height k.

You can see from Figure 9 the relation between level curves and horizontal traces. The level curves $f(x, y) = k$ are just the traces of the graph of f in the horizontal plane $z = k$ projected down to the xy-plane. So if you draw the level curves of a function and visualize them being lifted up to the surface at the indicated height, then you can mentally piece together a picture of the graph. The surface is steep where the level curves are close together. It is somewhat flatter where they are farther apart.

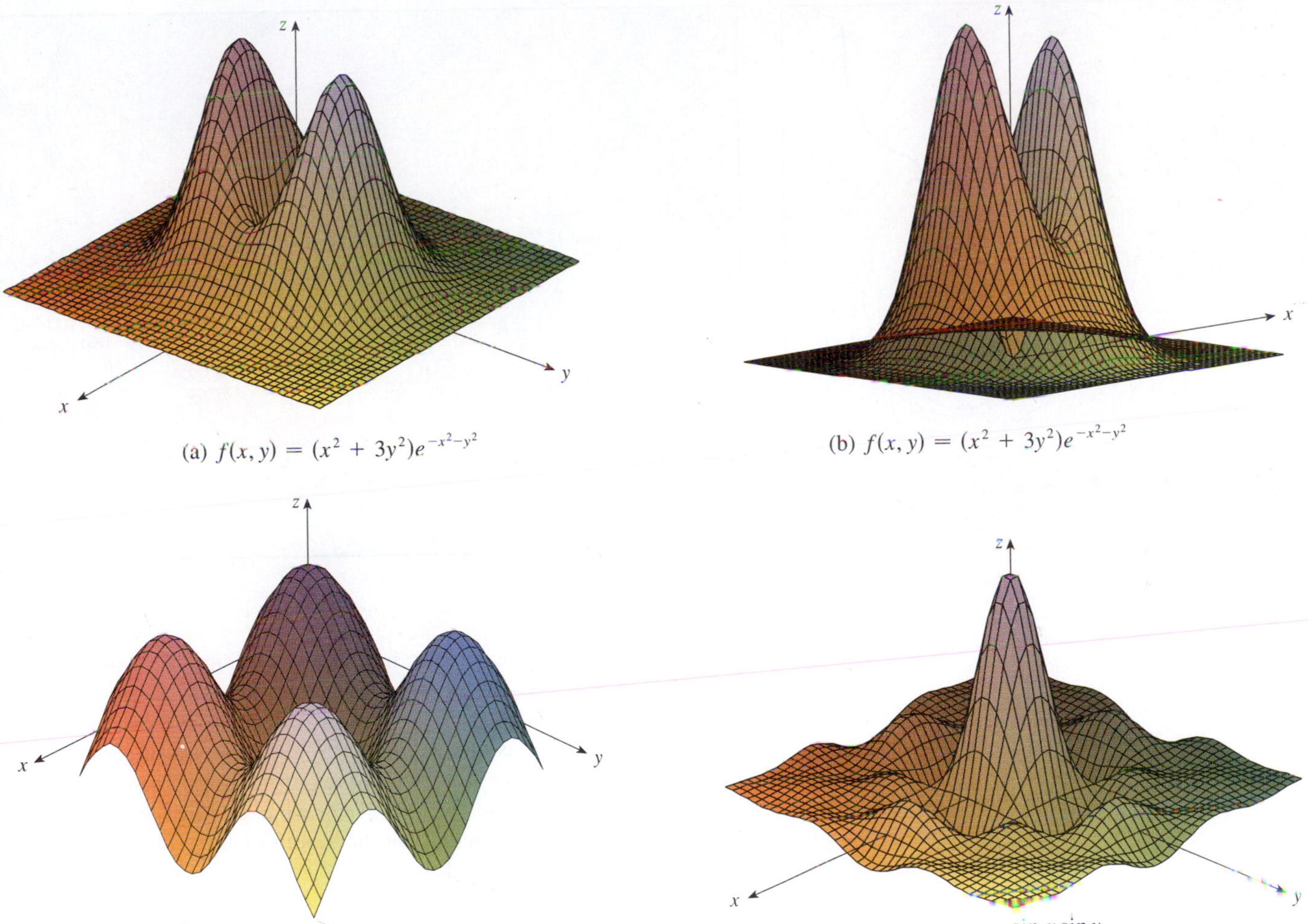

(a) $f(x, y) = (x^2 + 3y^2)e^{-x^2-y^2}$

(b) $f(x, y) = (x^2 + 3y^2)e^{-x^2-y^2}$

(c) $f(x, y) = \sin x + \sin y$

(d) $f(x, y) = \dfrac{\sin x \sin y}{xy}$

FIGURE 8

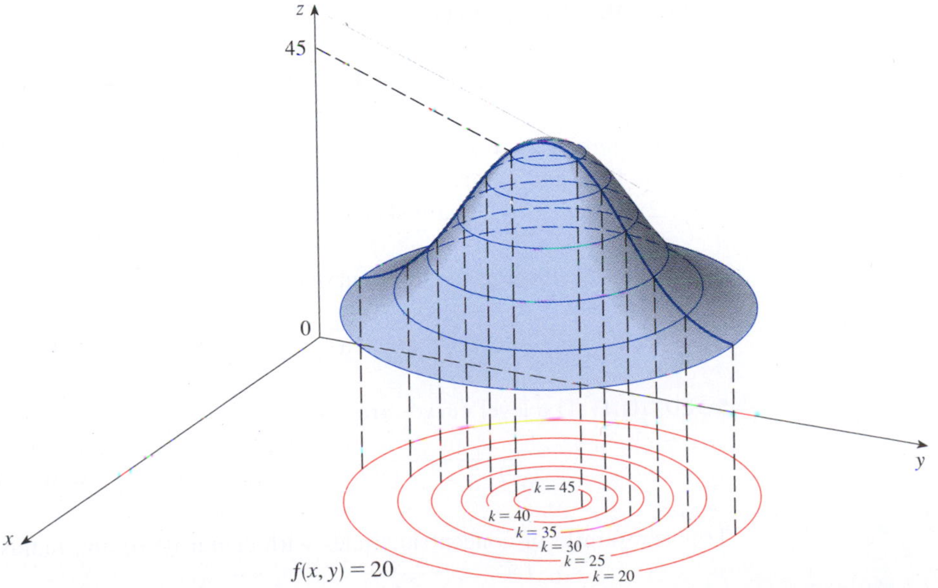

FIGURE 9

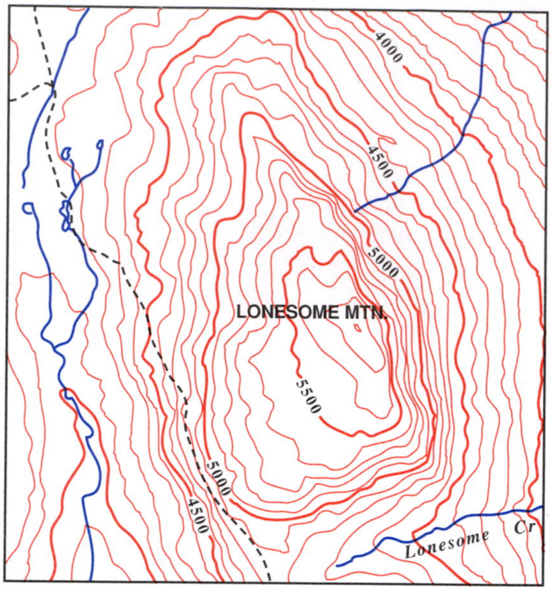

FIGURE 10

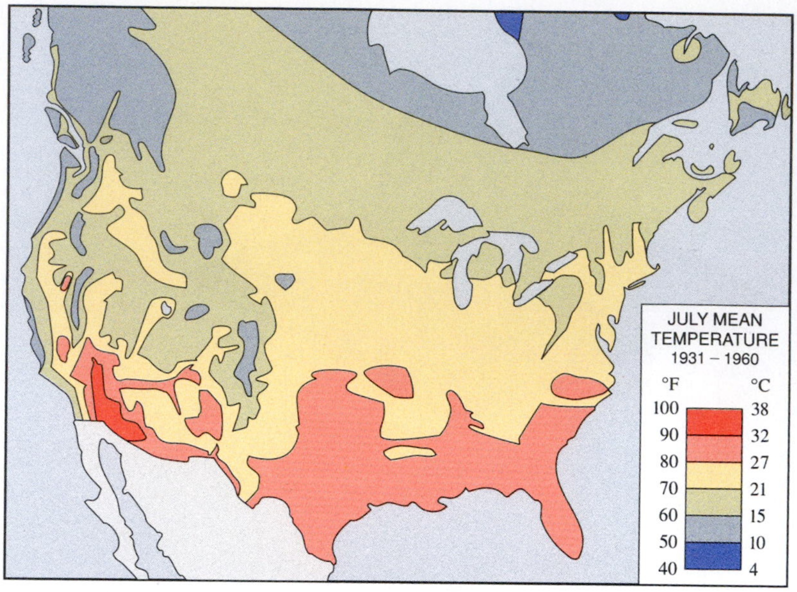

FIGURE 11

One common example of level curves occurs in topographic maps of mountainous regions, such as the map in Figure 10. The level curves are curves of constant elevation above sea level. If you walk along one of these contour lines you neither ascend nor descend. Another common example is the temperature function introduced in the opening paragraph of this section. Here the level curves are called **isothermals** and join locations with the same temperature. Figure 11 shows a weather map of North America indicating the average temperature during the month of July. The isothermals are the curves that separate the colored bands.

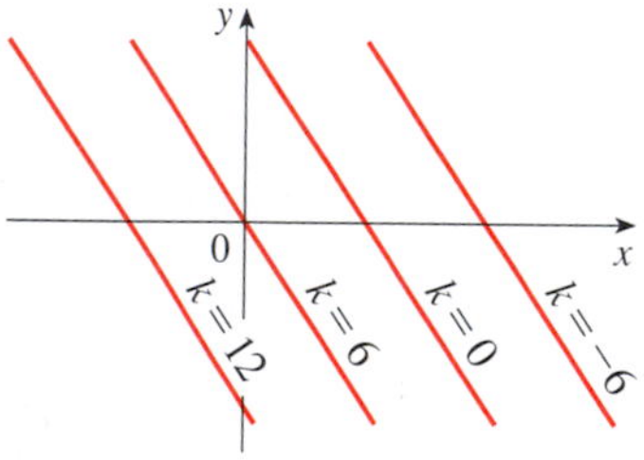

FIGURE 12
Contour map of $f(x, y) = 6 - 3x - 2y$

EXAMPLE 5 Sketch the level curves of the function $f(x, y) = 6 - 3x - 2y$ for the values $k = -6, 0, 6, 12$.

SOLUTION The level curves are

$$6 - 3x - 2y = k \qquad \text{or} \qquad 3x + 2y + (k - 6) = 0$$

This is a family of lines with slope $-\frac{3}{2}$. The four particular level curves with $k = -6, 0, 6$, and 12 are $3x + 2y - 12 = 0$, $3x + 2y - 6 = 0$, $3x + 2y = 0$, and $3x + 2y + 6 = 0$. They are sketched in Figure 12. ■

EXAMPLE 6 Sketch the level curves of the function

$$g(x, y) = \sqrt{9 - x^2 - y^2} \qquad \text{for} \qquad k = 0, 1, 2, 3$$

SOLUTION The level curves are

$$\sqrt{9 - x^2 - y^2} = k \qquad \text{or} \qquad x^2 + y^2 = 9 - k^2$$

This is a family of concentric circles with center $(0, 0)$ and radius $\sqrt{9 - k^2}$. The cases $k = 0, 1, 2, 3$ are shown in Figure 13. Try to visualize these level curves lifted up to form a surface and compare with the graph of g in Figure 6.

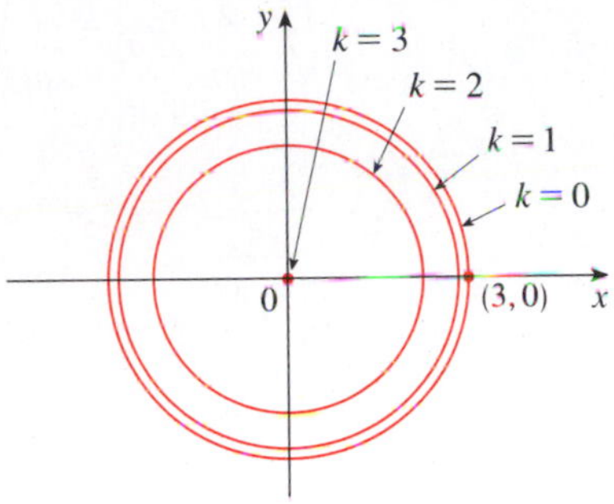

FIGURE 13
Contour map of $g(x, y) = \sqrt{9 - x^2 - y^2}$

EXAMPLE 7 Sketch some level curves of the function $h(x, y) = 4x^2 + y^2$.

SOLUTION The level curves are

$$4x^2 + y^2 = k \qquad \text{or} \qquad \frac{x^2}{k/4} + \frac{y^2}{k} = 1$$

which, for $k > 0$, describes a family of ellipses with semiaxes $\sqrt{k}/2$ and $\sqrt{k}$. Compare the sketches of these level curves in Figure 14 with the graph of h in Figure 7. ■

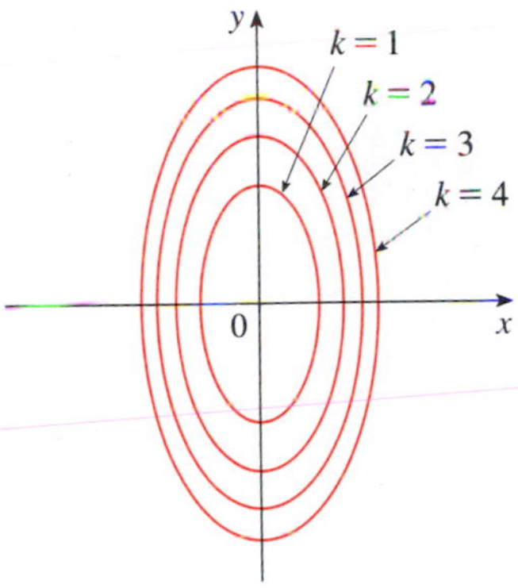

FIGURE 14
Contour map of $h(x, y) = 4x^2 + y^2$

EXAMPLE 8 Sketch the level curves of $f(x, y) = x + y^2$ for $k = -1, 0, 1, 2, 3$.

SOLUTION The level curves are $x + y^2 = k$. These form a family of parabolas with x-intercept k [see Figure 15(a)]. We see from Figure 15(b) how the graph $z = x + y^2$ is put together from the level curves.

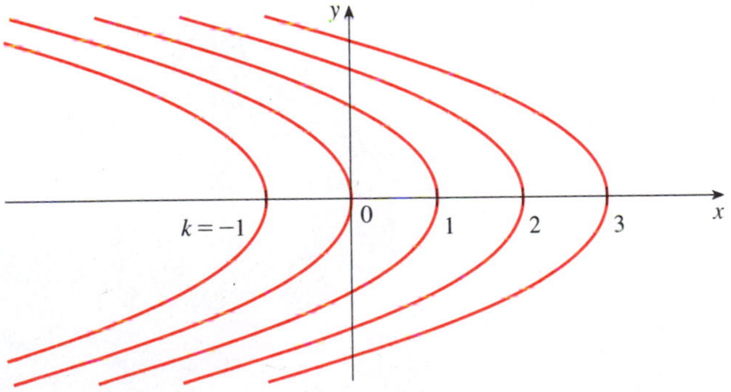

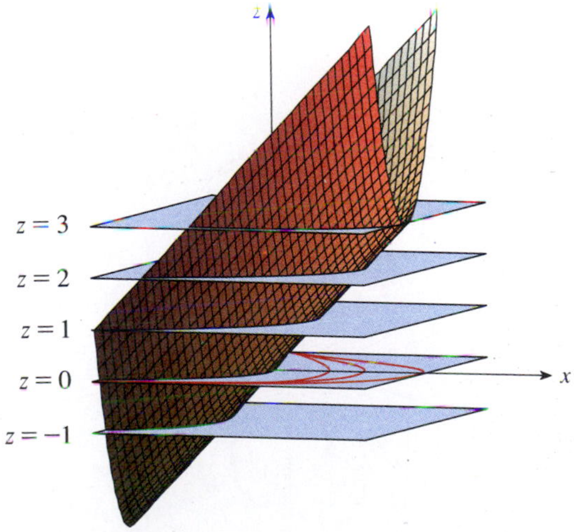

FIGURE 15 (a) Contour map of $f(x, y) = x + y^2$ (b) Graph of $f(x, y) = x + y^2$ ■

Figure 16 on page 728 shows some computer-generated level curves together with the corresponding computer-generated graphs.

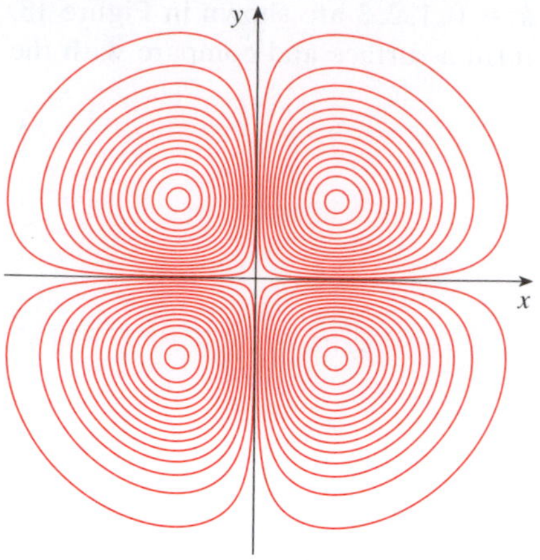

(a) Level curves of $f(x, y) = -xye^{-x^2-y^2}$

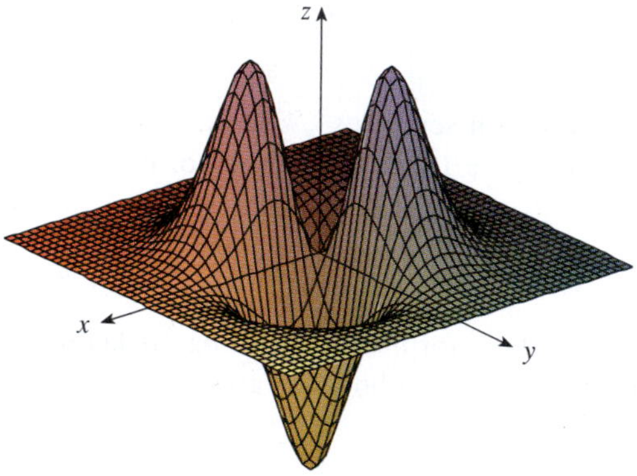

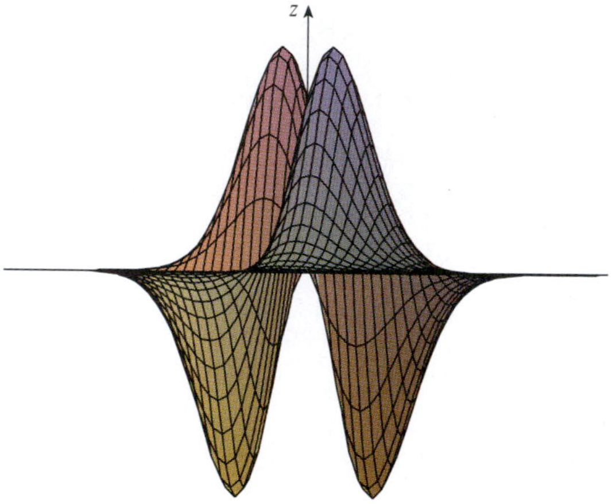

(b) Two views of $f(x, y) = -xye^{-x^2-y^2}$

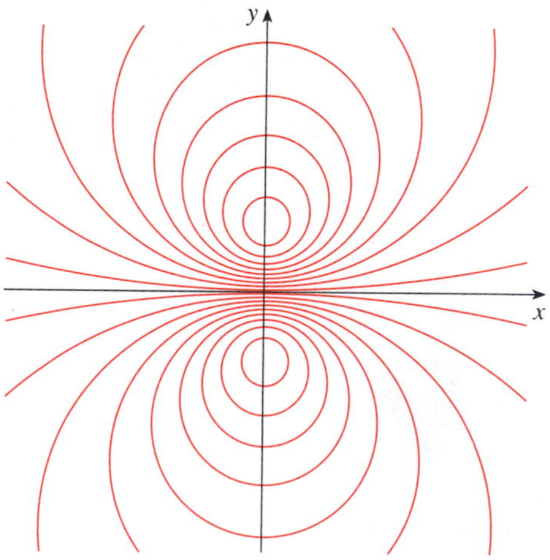

(c) Level curves of $f(x, y) = \dfrac{-3y}{x^2 + y^2 + 1}$

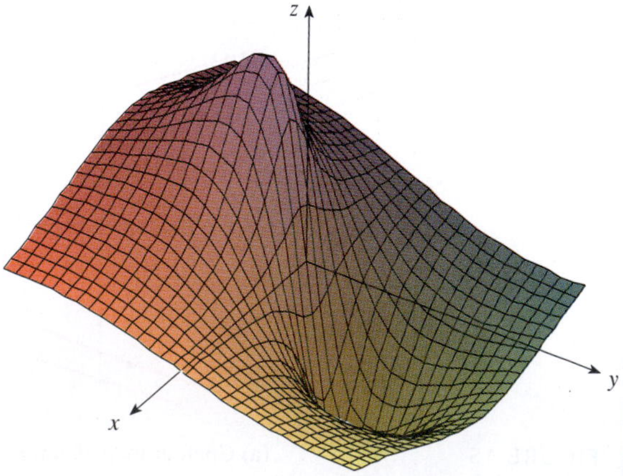

(d) $f(x, y) = \dfrac{-3y}{x^2 + y^2 + 1}$

FIGURE 16

FUNCTIONS OF THREE OR MORE VARIABLES

A **function of three variables**, f, is a rule that assigns to each ordered triple (x, y, z) in a domain $D \subset \mathbb{R}^3$ a unique real number denoted by $f(x, y, z)$. For instance, in the illustration at the beginning of this section the temperature T depends on the time t as well as on the longitude x and latitude y, so we could write $T = f(x, y, t)$.

It is possible to visualize a function f of three variables by an arrow diagram but not by its graph, since that would lie in a four-dimensional space. However we do gain some insight into f by examining its **level surfaces**, which are the surfaces with equations $f(x, y, z) = k$, where k is a constant. If the point (x, y, z) moves along a level surface, the value of $f(x, y, z)$ remains fixed.

EXAMPLE 9 Find the domain of f if $f(x, y, z) = \ln(z - y) + xy \sin z$.

SOLUTION The expression for $f(x, y, z)$ is defined as long as $z - y > 0$, so the domain of f is

$$D = \{(x, y, z) \in \mathbb{R}^3 \mid z > y\}$$

This is a **half-space** consisting of all points that lie above the plane $z = y$. ■

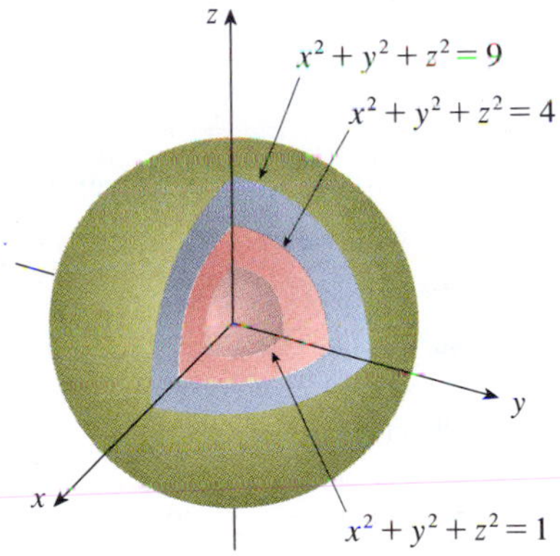

FIGURE 17

EXAMPLE 10 Find the level surfaces of the function

$$f(x, y, z) = x^2 + y^2 + z^2$$

SOLUTION The level surfaces are $x^2 + y^2 + z^2 = k$, where $k \geq 0$. These form a family of concentric spheres with radius $\sqrt{k}$ (see Figure 17). Thus, as (x, y, z) varies over any sphere with center O, the value of $f(x, y, z)$ remains fixed. ■

Functions of any number of variables can also be considered. A **function of n variables** is a rule that assigns a number $z = f(x_1, x_2, \ldots, x_n)$ to an n-tuple $(x_1, x_2, \ldots, x_n)$ of real numbers. For example, if a company uses n different ingredients in making a food product, c_i is the cost per unit of the ith ingredient, and x_i units of the ith ingredient are used, then the total cost C of the ingredients is a function of the n variables $x_1, x_2, \ldots, x_n$:

(4) $$C = f(x_1, x_2, \ldots, x_n) = c_1x_1 + c_2x_2 + \cdots + c_nx_n$$

The notation $$f : D \subset \mathbb{R}^n \to \mathbb{R}$$

is used to signify that f is a real-valued function whose domain D is a subset of $\mathbb{R}^n$. Sometimes we will use vector notation to write functions more compactly: If $\mathbf{x} = \langle x_1, x_2, \ldots, x_n \rangle$, we often write $f(\mathbf{x})$ in place of $f(x_1, x_2, \ldots, x_n)$. With this notation we can rewrite the function defined in Equation 4 as

$$f(\mathbf{x}) = \mathbf{c} \cdot \mathbf{x}$$

where $\mathbf{c} = \langle c_1, c_2, \ldots, c_n \rangle$.

In view of the one-to-one correspondence between points $(x_1, x_2, \ldots, x_n)$ in $\mathbb{R}^n$ and their position vectors $\mathbf{x} = \langle x_1, x_2, \ldots, x_n \rangle$ in V_n, we have three ways of looking at a function $f : \mathbb{R}^n \to \mathbb{R}$:

1. As a function of n real variables $x_1, x_2, \ldots, x_n$
2. As a function of a single point variable $(x_1, x_2, \ldots, x_n)$
3. As a function of a single vector variable $\mathbf{x} = \langle x_1, x_2, \ldots, x_n \rangle$

We will see that all three points of view are useful.

EXERCISES 12.1

1. If $f(x, y) = x^2 - y^2 + 4xy - 7x + 10$, find
(a) $f(2, 1)$ (b) $f(-3, 5)$
(c) $f(x + h, y)$ (d) $f(x, y + k)$
(e) $f(x, x)$

2. If $g(x, y) = \ln(xy + y - 1)$, find
(a) $g(1, 1)$ (b) $g(e, 1)$
(c) $g(x, 1)$ (d) $g(x + h, y)$
(e) $g(x, y + k)$

3. If $F(x, y) = 3xy/(x^2 + 2y^2)$, find
(a) $F(1, 1)$ (b) $F(-1, 2)$
(c) $F(t, 1)$ (d) $F(-1, y)$
(e) $F(x, x^2)$

4. If $G(x, y, z) = x \sin y \cos z$, find
(a) $G(2, \pi/6, \pi/3)$ (b) $G(4, \pi/4, 0)$
(c) $G(t, t, t)$ (d) $G(u, v, 0)$
(e) $G(x, x + y, x)$

5–14 ■ Find the domain and range of the function.

5. $f(x, y) = x + 2y - 5$
6. $f(x, y) = \sqrt{x - y}$
7. $f(x, y) = 2/(x + y)$
8. $f(x, y) = \tan^{-1}(y/x)$
9. $f(x, y) = e^{x^2 - y}$
10. $f(x, y) = \sqrt{36 - 9x^2 - 4y^2}$
11. $f(x, y, z) = x^2 \ln(x - y + z)$
12. $f(x, y, z) = x/(yz)$
13. $f(x, y, z) = x \sin(y + z)$
14. $f(x, y, z) = 1/\sqrt{x^2 + y^2 + z^2 - 1}$

15–30 ■ Find and sketch the domain of the function.

15. $f(x, y) = \sqrt[4]{y - 2x}$
16. $f(x, y) = \sqrt{x} + \sqrt{y}$
17. $f(x, y) = \dfrac{\sqrt{9 - x^2 - y^2}}{x + 2y}$
18. $f(x, y) = \dfrac{x^2 + y^2}{x^2 - y^2}$
19. $f(x, y) = xy\sqrt{x^2 + y}$
20. $f(x, y) = \tan(x - y)$
21. $f(x, y) = \ln(xy - 1)$
22. $f(x, y) = \ln(x^2 - y^2)$
23. $f(x, y) = x^2 \sec y$
24. $f(x, y) = \sqrt{x^2 + y^2 - 1} + \ln(4 - x^2 - y^2)$
25. $f(x, y) = \sin^{-1}(x + y)$
26. $f(x, y) = \sqrt{4 - 2x^2 - y^2}$
27. $f(x, y) = \ln x + \ln \sin y$
28. $f(x, y) = \sqrt{y - x} \ln(y + x)$
29. $f(x, y, z) = \sqrt{1 - x^2 - y^2 - z^2}$
30. $f(x, y, z) = \ln(16 - 4x^2 - 4y^2 - z^2)$

31–42 ■ Sketch the graph of the function.

31. $f(x, y) = 3$
32. $f(x, y) = x$
33. $f(x, y) = 1 - x - y$
34. $f(x, y) = y^2$
35. $f(x, y) = x^2 + 9y^2$
36. $f(x, y) = 3 - x^2 - y^2$
37. $f(x, y) = \sqrt{x^2 + y^2}$
38. $f(x, y) = \sqrt{16 - x^2 - 16y^2}$
39. $f(x, y) = y^2 - x^2$
40. $f(x, y) = \sin y$
41. $f(x, y) = 1 - x^2$
42. $f(x, y) = x^2 + y^2 - 4x - 2y + 5$

43–52 ■ Draw a contour map of the function showing several level curves.

43. $f(x, y) = xy$
44. $f(x, y) = x^2 - y^2$
45. $f(x, y) = x^2 + 9y^2$
46. $f(x, y) = e^{xy}$
47. $f(x, y) = \dfrac{x}{y}$
48. $f(x, y) = \dfrac{x + y}{x - y}$
49. $f(x, y) = \sqrt{x + y}$
50. $f(x, y) = y - \cos x$
51. $f(x, y) = x - y^2$
52. $f(x, y) = e^{1/(x^2 + y^2)}$

53–56 ■ Describe the level surfaces of the function.

53. $f(x, y, z) = x + 3y + 5z$
54. $f(x, y, z) = x^2 + 3y^2 + 5z^2$
55. $f(x, y, z) = x^2 - y^2 + z^2$
56. $f(x, y, z) = x^2 - y^2$

57. A thin metal plate, located in the xy-plane, has temperature $T(x, y)$ at the point (x, y). The level curves of T are called *isothermals* because at all points on an isothermal the temperature is the same. Sketch some isothermals if the temperature function is given by

$$T(x, y) = 100/(1 + x^2 + 2y^2)$$

58. If $V(x, y)$ is the electric potential at a point (x, y) in the xy-plane, then the level curves of V are called *equipotential curves* because at all points on such a curve the electric potential is the same. Sketch some equipotential curves if $V(x, y) = c/\sqrt{r^2 - x^2 - y^2}$, where c is a positive constant.

59–64 ■ Match the function (a) with its graph (labeled A–F) and (b) with its contour map (labeled I–VI). Give reasons for your choices.

59. $z = \sin\sqrt{x^2 + y^2}$
60. $z = x^2y^2e^{-x^2-y^2}$
61. $z = \dfrac{1}{x^2 + 4y^2}$
62. $z = x^3 - 3xy^2$
63. $z = \sin x \sin y$
64. $z = \sin^2 x + \frac{1}{4}y^2$

A

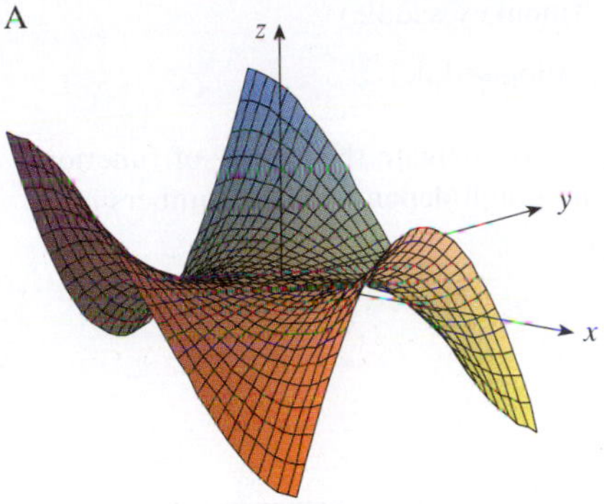

B

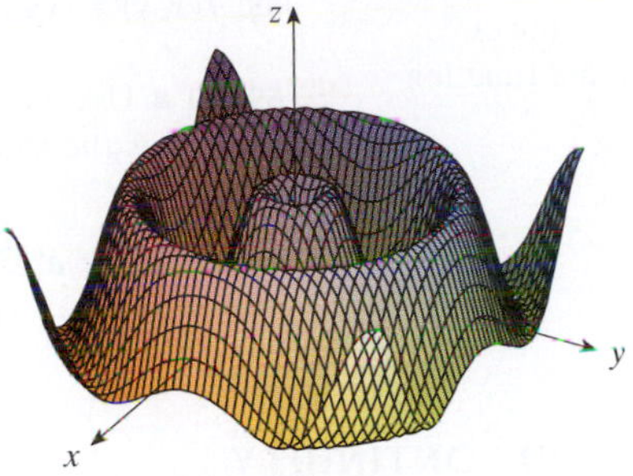

C

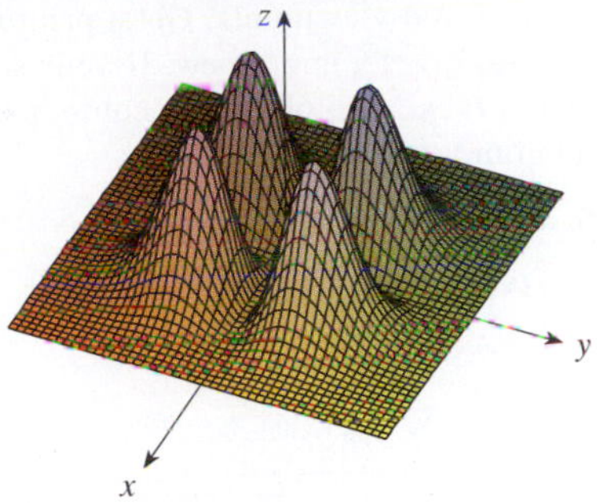

D

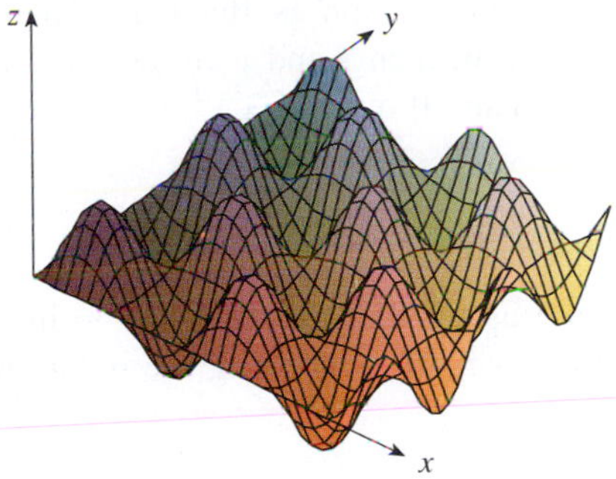

E

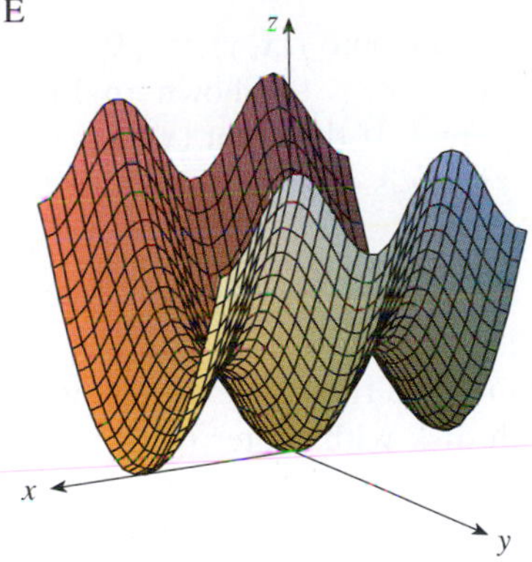

F

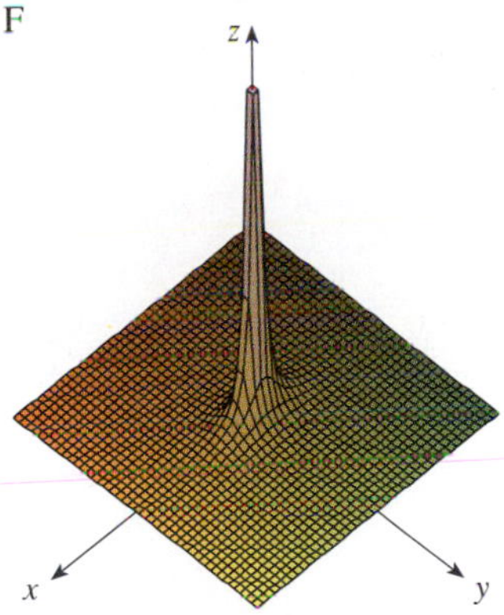

I

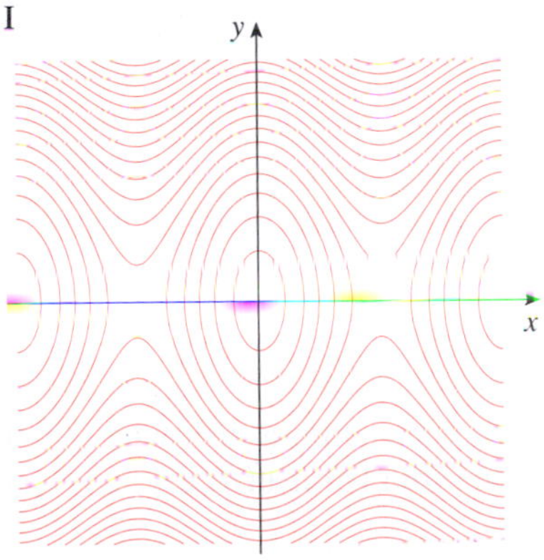

II

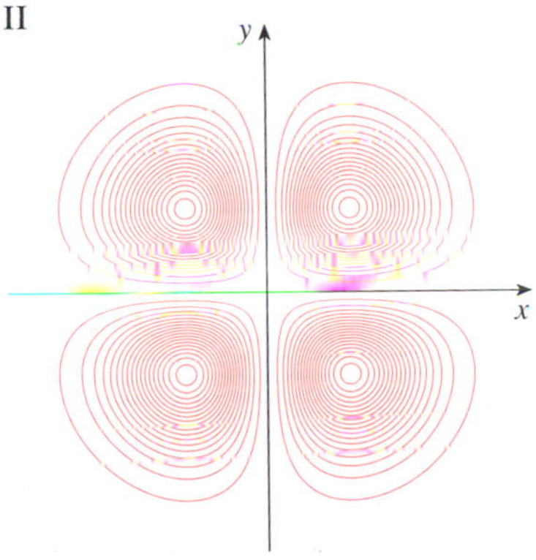

III

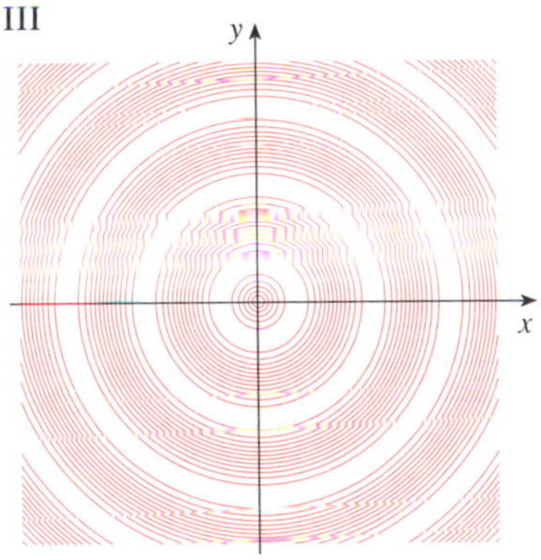

IV

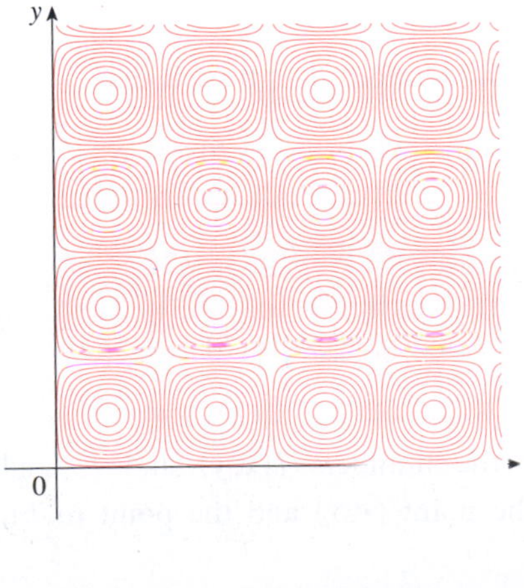

V

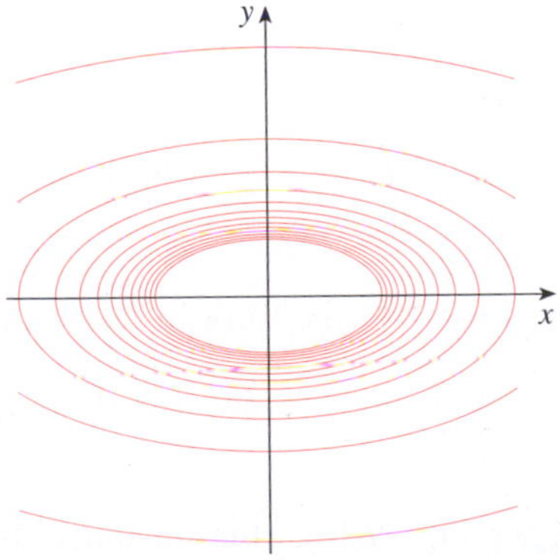

VI

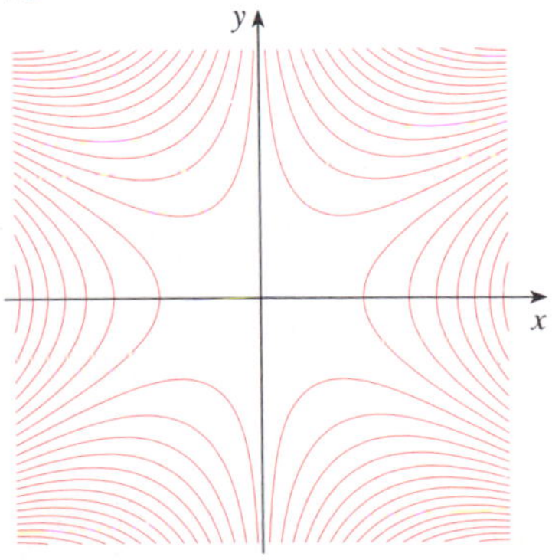

65–68 ■ Use a computer to graph the function using various domains and viewpoints. Get a printout of one that, in your opinion, gives a good view. If your software also produces level curves, then plot some contour lines of the same function and compare with the graph.

65. $f(x, y) = x^3 + y^3$

66. $f(x, y) = \sin(ye^{-x})$

67. $f(x, y) = xy^2 - x^3$ (monkey saddle)

68. $f(x, y) = xy^3 - yx^3$ (dog saddle)

69–70 ■ Use a computer to investigate the family of functions. How does the shape of the graph depend on the numbers a and b?

69. $f(x, y) = e^{ax^2+by^2}$

70. $f(x, y) = (ax^2 + by^2)e^{-x^2-y^2}$

12.2 LIMITS AND CONTINUITY

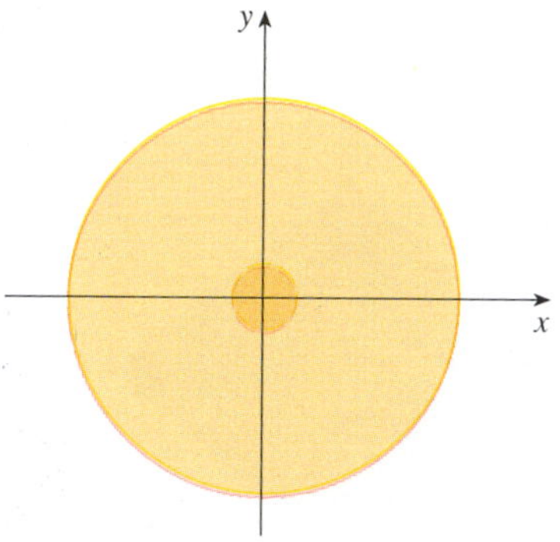

FIGURE 1
Domain of f

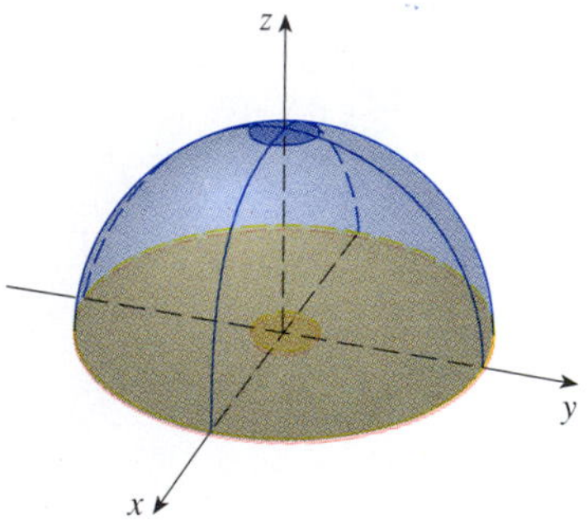

FIGURE 2
Graph of f

Consider the function $f(x, y) = \sqrt{9 - x^2 - y^2}$ whose domain is the closed disk $D = \{(x, y) \mid x^2 + y^2 \leq 9\}$ shown in Figure 1 and whose graph is the hemisphere shown in Figure 2. If the point (x, y) is close to the origin, then x and y are both close to 0, and so $f(x, y)$ is close to 3. In fact, if (x, y) lies in a small open disk $x^2 + y^2 < \delta^2$, then

$$f(x, y) = \sqrt{9 - (x^2 + y^2)} > \sqrt{9 - \delta^2}$$

Thus, we can make the values of $f(x, y)$ as close to 3 as we like by taking (x, y) in a small enough disk with center $(0, 0)$. We describe this situation by using the notation

$$\lim_{(x, y)\to(0, 0)} \sqrt{9 - x^2 - y^2} = 3$$

In general, the notation

$$\lim_{(x, y)\to(a, b)} f(x, y) = L$$

means that the values of $f(x, y)$ can be made as close as we wish to the number L by taking the point (x, y) close enough to the point (a, b). A more precise definition follows.

(1) DEFINITION Let f be a function of two variables defined on a disk with center (a, b), except possibly at (a, b). Then we say that the **limit of $f(x, y)$ as (x, y) approaches (a, b)** is L and we write

$$\lim_{(x, y)\to(a, b)} f(x, y) = L$$

if for every number $\varepsilon > 0$ there is a corresponding number $\delta > 0$ such that

$$|f(x, y) - L| < \varepsilon \qquad \text{whenever} \qquad 0 < \sqrt{(x - a)^2 + (y - b)^2} < \delta$$

Other notations for the limit in Definition 1 are

$$\lim_{\substack{x\to a\\ y\to b}} f(x, y) = L \qquad \text{and} \qquad f(x, y) \to L \text{ as } (x, y) \to (a, b)$$

Since $|f(x, y) - L|$ is the distance between the numbers $f(x, y)$ and L, and $\sqrt{(x - a)^2 + (y - b)^2}$ is the distance between the point (x, y) and the point (a, b),

Definition 1 says that the distance between $f(x, y)$ and L can be made arbitrarily small by making the distance from (x, y) to (a, b) sufficiently small (but not 0). Figure 3 illustrates Definition 1 by means of an arrow diagram. If any small interval $(L - \varepsilon, L + \varepsilon)$ is given around L, then we can find a disk D_δ with center (a, b) and radius $\delta > 0$ such that f maps all the points in D_δ [except possibly (a, b)] into the interval $(L - \varepsilon, L + \varepsilon)$.

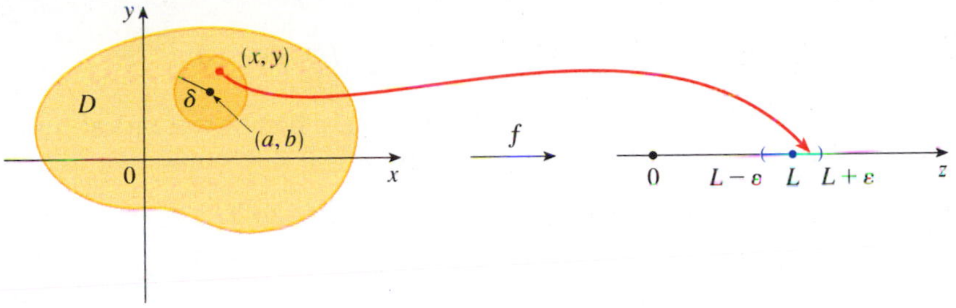

FIGURE 3

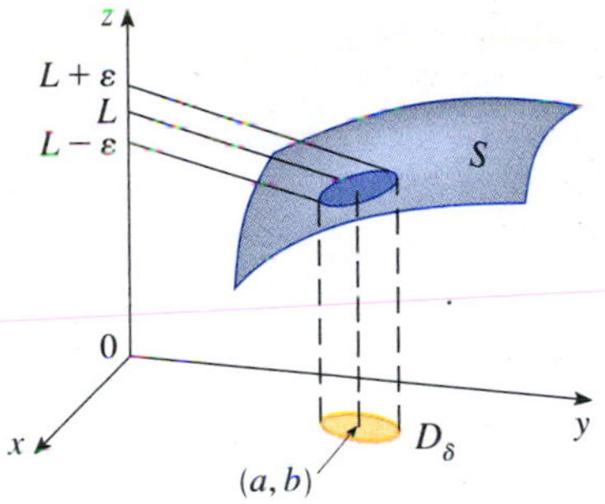

FIGURE 4

Another illustration of Definition 1 is given in Figure 4 where the surface S is the graph of f. If $\varepsilon > 0$ is given, we can find $\delta > 0$ such that if (x, y) is restricted to lie in the disk D_δ and $(x, y) \neq (a, b)$, then the corresponding part of S lies between the horizontal planes $z = L - \varepsilon$ and $z = L + \varepsilon$.

For functions of a single variable, when we let x approach a, there are only two possible directions of approach, from the left or from the right. We recall from Chapter 2 that if $\lim_{x \to a^-} f(x) \neq \lim_{x \to a^+} f(x)$, then $\lim_{x \to a} f(x)$ does not exist.

For functions of two variables the situation is not as simple because we can let (x, y) approach (a, b) from an infinite number of directions in any manner whatsoever (see Figure 5).

Definition 1 refers only to the *distance* between (x, y) and (a, b). It does not refer to the direction of approach. Therefore, if the limit exists, then $f(x, y)$ must approach the same limit no matter how (x, y) approaches (a, b). Thus, if we can find two different paths of approach along which $f(x, y)$ has different limits, then it follows that $\lim_{(x, y) \to (a, b)} f(x, y)$ does not exist.

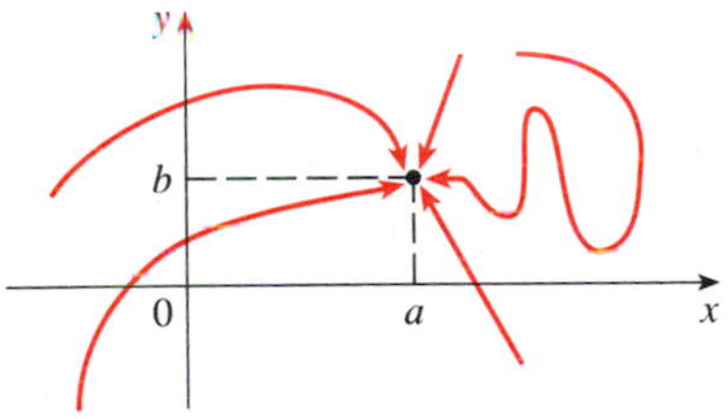

FIGURE 5

> If $f(x, y) \to L_1$ as $(x, y) \to (a, b)$ along a path C_1 and $f(x, y) \to L_2$ as $(x, y) \to (a, b)$ along a path C_2, where $L_1 \neq L_2$, then $\lim_{(x, y) \to (a, b)} f(x, y)$ does not exist.

EXAMPLE 1 Find $\displaystyle\lim_{(x, y) \to (0, 0)} \frac{x^2 - y^2}{x^2 + y^2}$ if it exists.

SOLUTION Let $f(x, y) = (x^2 - y^2)/(x^2 + y^2)$. First let's approach $(0, 0)$ along the x-axis. Then $y = 0$ gives $f(x, 0) = x^2/x^2 = 1$ for all $x \neq 0$, so

$$f(x, y) \to 1 \qquad \text{as} \qquad (x, y) \to (0, 0) \text{ along the } x\text{-axis}$$

We now approach along the y-axis by putting $x = 0$. Then $f(0, y) = -y^2/y^2 = -1$ for all $y \neq 0$, so

$$f(x, y) \to -1 \qquad \text{as} \qquad (x, y) \to (0, 0) \text{ along the } y\text{-axis}$$

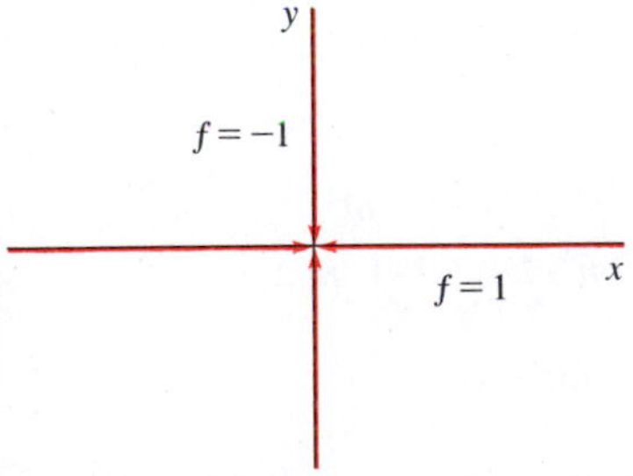

FIGURE 6

(See Figure 6.) Since f has two different limits along two different lines, the given limit does not exist. ■

EXAMPLE 2 If $f(x, y) = xy/(x^2 + y^2)$, does $\lim_{(x, y) \to (0, 0)} f(x, y)$ exist?

SOLUTION If $y = 0$, then $f(x, 0) = 0/x^2 = 0$. Therefore

$$f(x, y) \to 0 \quad \text{as} \quad (x, y) \to (0, 0) \text{ along the } x\text{-axis}$$

If $x = 0$, then $f(0, y) = 0/y^2 = 0$, so

$$f(x, y) \to 0 \quad \text{as} \quad (x, y) \to (0, 0) \text{ along the } y\text{-axis}$$

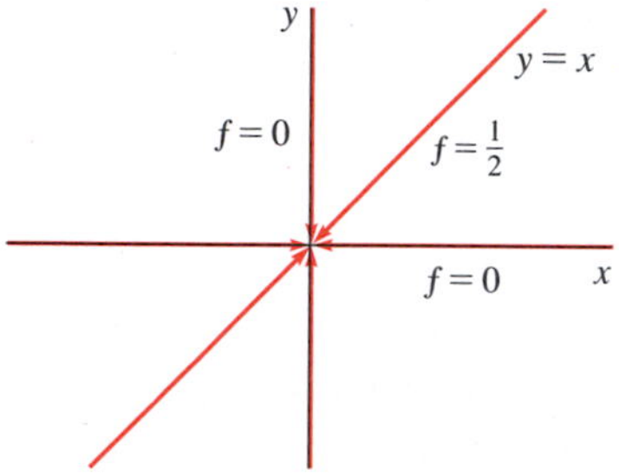

FIGURE 7

Although we have obtained identical limits along the axes, that does not show that the given limit is 0. Let's now approach $(0, 0)$ along another line, say $y = x$. For all $x \neq 0$,

$$f(x, x) = \frac{x^2}{x^2 + x^2} = \frac{1}{2}$$

Therefore $\qquad f(x, y) \to \frac{1}{2} \quad \text{as} \quad (x, y) \to (0, 0) \text{ along } y = x$

(See Figure 7.) Since we obtained different limits along different paths, the given limit does not exist. ■

Figure 8 sheds some light on Example 2. The ridge that occurs above the line $y = x$ corresponds to the fact that $f(x, y) = \frac{1}{2}$ for all points (x, y) on that line except the origin.

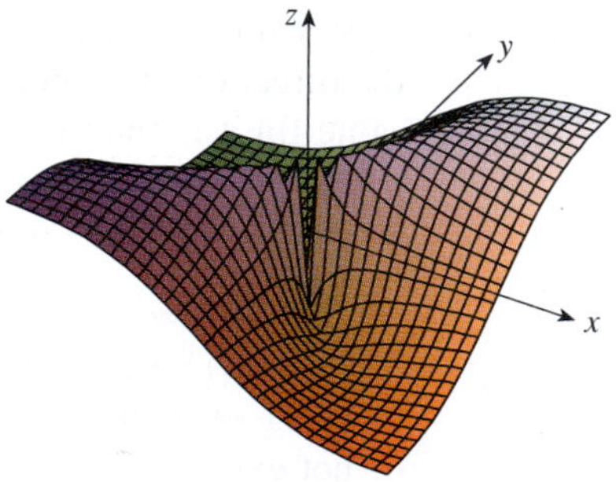

FIGURE 8
$f(x, y) = \dfrac{xy}{x^2 + y^2}$

EXAMPLE 3 If $f(x, y) = \dfrac{xy^2}{x^2 + y^4}$, does $\lim_{(x, y) \to (0, 0)} f(x, y)$ exist?

SOLUTION With the solution of Example 2 in mind, let's try to save time by letting $(x, y) \to (0, 0)$ along any line through the origin. Then $y = mx$, where m is the slope, and if $m \neq 0$,

$$f(x, y) = f(x, mx) = \frac{x(mx)^2}{x^2 + (mx)^4} = \frac{m^2x^3}{x^2 + m^4x^4} = \frac{m^2x}{1 + m^4x^2}$$

So $\qquad f(x, y) \to 0 \quad \text{as} \quad (x, y) \to (0, 0) \text{ along } y = mx$

Thus, f has the same limiting value along every line through the origin. But that does not show that the given limit is 0, for if we now let $(x, y) \to (0, 0)$ along the parabola

Figure 9 shows the graph of the function in Example 3. Notice the ridge above the parabola $x = y^2$.

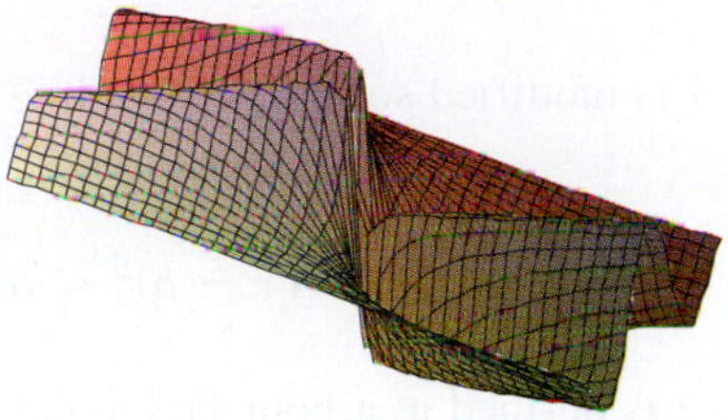

FIGURE 9

$x = y^2$, we have

$$f(x, y) = f(y^2, y) = \frac{y^2 \cdot y^2}{(y^2)^2 + y^4} = \frac{y^4}{2y^4} = \frac{1}{2}$$

so $$f(x, y) \to \tfrac{1}{2} \quad \text{as} \quad (x, y) \to (0, 0) \text{ along } x = y^2$$

Since different paths lead to different limiting values, the given limit does not exist. ■

EXAMPLE 4 Find $\displaystyle\lim_{(x, y)\to(0,0)} \frac{3x^2y}{x^2 + y^2}$ if it exists.

SOLUTION As in Example 3, one can show that the limit along any line through the origin is 0. This does not prove that the given limit is 0, but the limits along the parabolas $y = x^2$ and $x = y^2$ also turn out to be 0, so we begin to suspect that the limit does exist.

Let $\varepsilon > 0$. We want to find $\delta > 0$ such that

$$\left| \frac{3x^2y}{x^2 + y^2} - 0 \right| < \varepsilon \quad \text{whenever} \quad 0 < \sqrt{x^2 + y^2} < \delta$$

that is, $$\frac{3x^2|y|}{x^2 + y^2} < \varepsilon \quad \text{whenever} \quad 0 < \sqrt{x^2 + y^2} < \delta$$

But $x^2 \leqslant x^2 + y^2$ since $y^2 \geqslant 0$, so

(2) $$\frac{3x^2|y|}{x^2 + y^2} \leqslant 3|y| = 3\sqrt{y^2} \leqslant 3\sqrt{x^2 + y^2}$$

Thus, if we choose $\delta = \varepsilon/3$ and let $0 < \sqrt{x^2 + y^2} < \delta$, then

$$\left| \frac{3x^2y}{x^2 + y^2} - 0 \right| \leqslant 3\sqrt{x^2 + y^2} < 3\delta = 3\left(\frac{\varepsilon}{3}\right) = \varepsilon$$

Another way to do Example 4 is to use the Squeeze Theorem instead of Definition 1. From (4) it follows that

$$\lim_{(x, y)\to(0,0)} 3|y| = 0$$

and so the first inequality in (2) shows that the given limit is 0.

Hence, by Definition 1,

$$\lim_{(x, y)\to(0,0)} \frac{3x^2y}{x^2 + y^2} = 0$$ ■

Just as for functions of one variable, the calculation of limits can be greatly simplified by the use of properties of limits and by the use of continuity.

The properties of limits listed in Section 2.3 can be extended to functions of two variables. The limit of a sum is the sum of the limits, the limit of a product is the product of the limits, and so on.

Recall that evaluating limits of *continuous* functions of a single variable is easy. It can be accomplished by direct substitution because the defining property of a continuous function is $\lim_{x\to a} f(x) = f(a)$. Continuous functions of two variables are also defined by the direct substitution property.

(3) DEFINITION Let f be a function of two variables defined on a disk with center (a, b). Then f is called **continuous at** (a, b) if

$$\lim_{(x, y)\to(a, b)} f(x, y) = f(a, b)$$

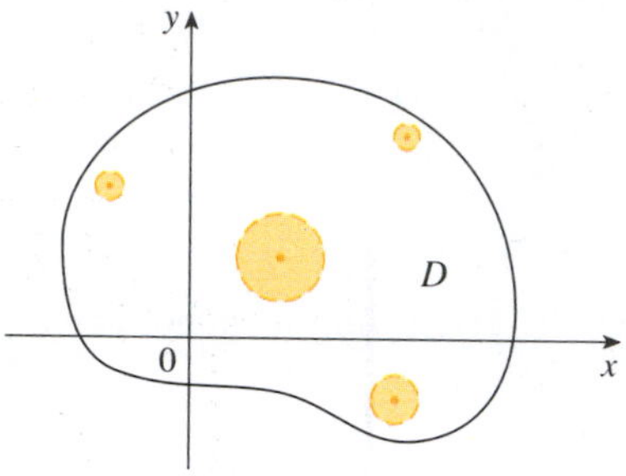

FIGURE 10
Interior points of D

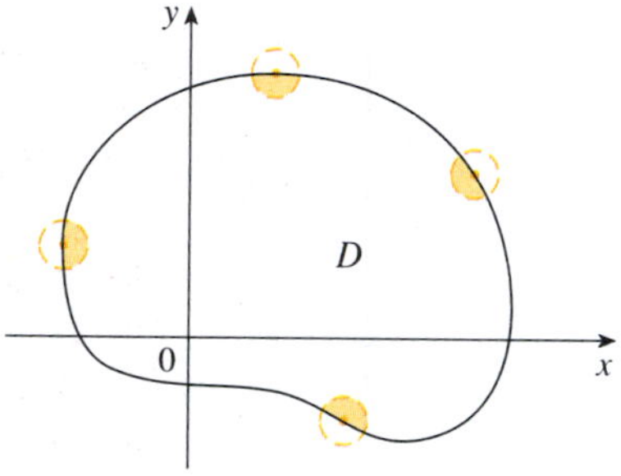

FIGURE 11
Boundary points of D

If the domain of f is a set $D \subset \mathbb{R}^2$, then Definition 3 defines the continuity of f at an **interior point** (a, b) of D, that is, a point that is contained in a disk $D_\delta \subset D$ (see Figure 10). But D may also contain a **boundary point**, that is, a point (a, b) such that every disk with center (a, b) contains points in D and also points not in D (see Figure 11).

If (a, b) is a boundary point of D, then Definition 1 is modified so that the last line reads

$$|f(x, y) - L| < \varepsilon \quad \text{whenever} \quad (x, y) \in D \quad \text{and} \quad 0 < \sqrt{(x - a)^2 + (y - b)^2} < \delta$$

With this convention, Definition 3 also applies when f is defined at a boundary point (a, b) of D.

Finally, we say f is **continuous on** D if f is continuous at every point (a, b) in D.

The intuitive meaning of continuity is that if the point (x, y) changes by a small amount, then the value of $f(x, y)$ changes by a small amount. This means that a surface that is the graph of a continuous function has no hole or break.

Using the properties of limits, you can see that sums, differences, products, and quotients of continuous functions are continuous on their domains. Let's use this fact to give examples of continuous functions.

A **polynomial function of two variables** (or polynomial, for short) is a sum of terms of the form $cx^m y^n$, where c is a constant and m and n are nonnegative integers. A **rational function** is a ratio of polynomials. For instance,

$$f(x, y) = x^4 + 5x^3y^2 + 6xy^4 - 7y + 6$$

is a polynomial, whereas

$$g(x, y) = \frac{2xy + 1}{x^2 + y^2}$$

is a rational function.

From Definition 1 it can be shown (see Exercise 47) that

$$\textbf{(4)} \qquad \lim_{(x, y) \to (a, b)} x = a \qquad \lim_{(x, y) \to (a, b)} y = b \qquad \lim_{(x, y) \to (a, b)} c = c$$

These limits show that the functions $f(x, y) = x$, $g(x, y) = y$, and $h(x, y) = c$ are continuous. Since any polynomial can be built up out of the simple functions f, g, and h by multiplication and addition, it follows that *all polynomials are continuous on* $\mathbb{R}^2$. Likewise, any rational function is continuous on its domain since it is a quotient of continuous functions.

EXAMPLE 5 Evaluate $\displaystyle\lim_{(x, y) \to (1, 2)} (x^2y^3 - x^3y^2 + 3x + 2y)$.

SOLUTION Since $f(x, y) = x^2y^3 - x^3y^2 + 3x + 2y$ is a polynomial, it is continuous everywhere, so the limit can be found by direct substitution:

$$\lim_{(x, y) \to (1, 2)} (x^2y^3 - x^3y^2 + 3x + 2y) = 1^2 \cdot 2^3 - 1^3 \cdot 2^2 + 3 \cdot 1 + 2 \cdot 2 = 11$$

■

EXAMPLE 6 Where is the function $f(x, y) = \dfrac{x^2 - y^2}{x^2 + y^2}$ continuous?

SOLUTION The function f is discontinuous at $(0, 0)$ because it is not defined there. Since f is a rational function, it is continuous on its domain, which is the set $D = \{(x, y) \mid (x, y) \neq (0, 0)\}$. ■

EXAMPLE 7 Let

$$g(x, y) = \begin{cases} \dfrac{x^2 - y^2}{x^2 + y^2} & \text{if } (x, y) \neq (0, 0) \\ 0 & \text{if } (x, y) = (0, 0) \end{cases}$$

Here g is defined at $(0, 0)$ but g is still discontinuous at 0 because $\lim_{(x, y) \to (0, 0)} g(x, y)$ does not exist (see Example 1). ■

Figure 12 shows the graph of the continuous function in Example 8.

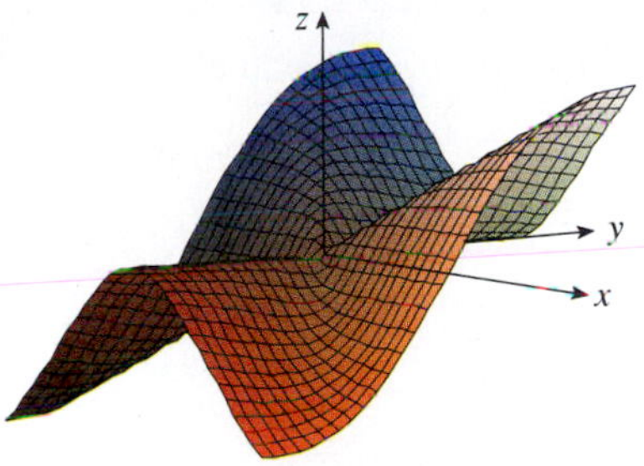

FIGURE 12

EXAMPLE 8 Let

$$f(x, y) = \begin{cases} \dfrac{3x^2 y}{x^2 + y^2} & \text{if } (x, y) \neq (0, 0) \\ 0 & \text{if } (x, y) = (0, 0) \end{cases}$$

We know f is continuous for $(x, y) \neq (0, 0)$ since it is equal to a rational function there. Also, from Example 4, we have

$$\lim_{(x, y) \to (0, 0)} f(x, y) = \lim_{(x, y) \to (0, 0)} \frac{3x^2 y}{x^2 + y^2} = 0 = f(0, 0)$$

Therefore, f is continuous at $(0, 0)$, and so it is continuous on $\mathbb{R}^2$. ■

EXAMPLE 9 Let

$$g(x, y) = \begin{cases} \dfrac{3x^2 y}{x^2 + y^2} & \text{if } (x, y) \neq (0, 0) \\ 17 & \text{if } (x, y) = (0, 0) \end{cases}$$

Again from Example 4, we have

$$\lim_{(x, y) \to (0, 0)} g(x, y) = \lim_{(x, y) \to (0, 0)} \frac{3x^2 y}{x^2 + y^2} = 0 \neq 17 = g(0, 0)$$

and this shows that g is discontinuous at $(0, 0)$. However, g is continuous on the set $S = \{(x, y) \mid (x, y) \neq (0, 0)\}$ since it is equal to a rational function on S. ■

Composition is another way of combining two continuous functions to get a third. The proof of the following theorem is similar to that of Theorem 2.5.8.

> **(5) THEOREM** If f is continuous at (a, b) and g is a function of a single variable that is continuous at $f(a, b)$, then the composite function $h = g \circ f$ defined by $h(x, y) = g(f(x, y))$ is continuous at (a, b).

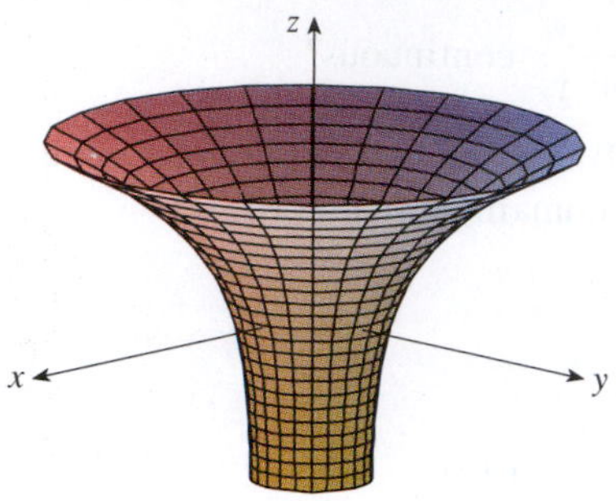

FIGURE 13

EXAMPLE 10 On what set is the function $h(x, y) = \ln(x^2 + y^2 - 1)$ continuous?

SOLUTION Let $f(x, y) = x^2 + y^2 - 1$ and $g(t) = \ln t$. Then

$$g(f(x, y)) = \ln(x^2 + y^2 - 1) = h(x, y)$$

so $h = g \circ f$. Now f is continuous everywhere since it is a polynomial, and g is continuous on its domain $\{t \mid t > 0\}$. Thus, by Theorem 5, h is continuous on its domain

$$D = \{(x, y) \mid x^2 + y^2 - 1 > 0\} = \{(x, y) \mid x^2 + y^2 > 1\}$$

which consists of all points outside the circle $x^2 + y^2 = 1$. A computer-drawn graph of h is shown in Figure 13. ■

Everything that we have done in this section can be extended to functions of three or more variables. The distance between two points (x, y, z) and (a, b, c) in $\mathbb{R}^3$ is given by $\sqrt{(x - a)^2 + (y - b)^2 + (z - c)^2}$, so the definitions of limit and continuity of a function of three variables are as follows.

(6) DEFINITION Let f be a function of three variables defined inside a sphere with center (a, b, c), except possibly at (a, b, c).

(a)
$$\lim_{(x, y, z) \to (a, b, c)} f(x, y, z) = L$$

means that for every number $\varepsilon > 0$ there is a corresponding number $\delta > 0$ such that

$$|f(x, y, z) - L| < \varepsilon \quad \text{whenever} \quad 0 < \sqrt{(x - a)^2 + (y - b)^2 + (z - c)^2} < \delta$$

(b) f is **continuous** at (a, b, c) if

$$\lim_{(x, y, z) \to (a, b, c)} f(x, y, z) = f(a, b, c)$$

If we use the vector notation introduced at the end of Section 12.1, then we can write the definitions of a limit for functions of two or three variables in a single compact form as follows.

(7) If $f: D \subset \mathbb{R}^n \to \mathbb{R}$, then $\lim_{\mathbf{x} \to \mathbf{a}} f(\mathbf{x}) = L$ means that for every number $\varepsilon > 0$ there is a corresponding number $\delta > 0$ such that

$$|f(\mathbf{x}) - L| < \varepsilon \quad \text{whenever} \quad 0 < |\mathbf{x} - \mathbf{a}| < \delta$$

Notice that if $n = 1$, then $\mathbf{x} = x$ and $\mathbf{a} = a$, and (7) is just the definition of a limit for functions of a single variable. For the case $n = 2$, we have $\mathbf{x} = \langle x, y \rangle$, $\mathbf{a} = \langle a, b \rangle$, and $|\mathbf{x} - \mathbf{a}| = \sqrt{(x - a)^2 + (y - b)^2}$, so (7) becomes Definition 1. If $n = 3$, then $\mathbf{x} = \langle x, y, z \rangle$, $\mathbf{a} = \langle a, b, c \rangle$, and (7) becomes part (a) of Definition 6. In each case the definition of continuity can be written as

$$\lim_{\mathbf{x} \to \mathbf{a}} f(\mathbf{x}) = f(\mathbf{a})$$

EXERCISES 12.2

1–26 ■ Find the limit, if it exists, or show that the limit does not exist.

1. $\lim_{(x,y)\to(2,3)} (x^2y^2 - 2xy^5 + 3y)$

2. $\lim_{(x,y)\to(-3,4)} (x^3 + 3x^2y^2 - 5y^3 + 1)$

3. $\lim_{(x,y)\to(0,0)} \dfrac{x^2y^3 + x^3y^2 - 5}{2 - xy}$

4. $\lim_{(x,y)\to(-2,1)} \dfrac{x^2 + xy + y^2}{x^2 - y^2}$

5. $\lim_{(x,y)\to(\pi,\pi)} x\sin\left(\dfrac{x+y}{4}\right)$

6. $\lim_{(x,y)\to(1,4)} e^{\sqrt{x+2y}}$

7. $\lim_{(x,y)\to(0,0)} \dfrac{x - y}{x^2 + y^2}$

8. $\lim_{(x,y)\to(0,0)} \dfrac{x^2}{x^2 + y^2}$

9. $\lim_{(x,y)\to(0,0)} \dfrac{8x^2y^2}{x^4 + y^4}$

10. $\lim_{(x,y)\to(0,0)} \dfrac{x^3 + xy^2}{x^2 + y^2}$

11. $\lim_{(x,y)\to(0,0)} \dfrac{2xy}{x^2 + 2y^2}$

12. $\lim_{(x,y)\to(0,0)} \dfrac{(x+y)^2}{x^2 + y^2}$

13. $\lim_{(x,y)\to(0,0)} \dfrac{xy}{\sqrt{x^2 + y^2}}$

14. $\lim_{(x,y)\to(0,0)} \dfrac{xy + 1}{x^2 + y^2 + 1}$

15. $\lim_{(x,y)\to(0,0)} \dfrac{2x^2y}{x^4 + y^2}$

16. $\lim_{(x,y)\to(0,0)} \dfrac{x^3y^2}{x^2 + y^2}$

17. $\lim_{(x,y)\to(0,0)} \dfrac{x^2 + y^2}{\sqrt{x^2 + y^2 + 1} - 1}$

18. $\lim_{(x,y)\to(0,0)} \dfrac{\sqrt{x^2y^2 + 1} - 1}{x^2 + y^2}$

19. $\lim_{(x,y)\to(0,1)} \dfrac{xy - x}{x^2 + y^2 - 2y + 1}$

20. $\lim_{(x,y)\to(1,-1)} \dfrac{x^2 + y^2 - 2x - 2y}{x^2 + y^2 - 2x + 2y + 2}$

21. $\lim_{(x,y,z)\to(1,2,3)} \dfrac{xz^2 - y^2z}{xyz - 1}$

22. $\lim_{(x,y,z)\to(2,3,0)} [xe^z + \ln(2x - y)]$

23. $\lim_{(x,y,z)\to(0,0,0)} \dfrac{x^2 - y^2 - z^2}{x^2 + y^2 + z^2}$

24. $\lim_{(x,y,z)\to(0,0,0)} \dfrac{xy + yz + zx}{x^2 + y^2 + z^2}$

25. $\lim_{(x,y,z)\to(0,0,0)} \dfrac{xy + yz^2 + xz^2}{x^2 + y^2 + z^4}$

26. $\lim_{(x,y,z)\to(0,0,0)} \dfrac{x^2y^2z^2}{x^2 + y^2 + z^2}$

27–28 ■ Use a computer graph of the function to explain why the limit does not exist.

27. $\lim_{(x,y)\to(0,0)} \dfrac{2x^2 + 3xy + 4y^2}{3x^2 + 5y^2}$

28. $\lim_{(x,y)\to(0,0)} \dfrac{xy^3}{x^2 + y^6}$

29–32 ■ Find $h(x, y) = g(f(x, y))$ and the set on which h is continuous.

29. $g(t) = e^{-t}\cos t, \quad f(x, y) = x^4 + x^2y^2 + y^4$

30. $g(t) = \dfrac{\sqrt{t} - 1}{\sqrt{t} + 1}, \quad f(x, y) = x^2 - y$

31. $g(t) = t^2 + \sqrt{t}, \quad f(x, y) = 2x + 3y - 6$

32. $g(z) = \sin z, \quad f(x, y) = y\ln x$

33–46 ■ Determine the largest set on which the function is continuous.

33. $F(x, y) = \dfrac{x^2 + y^2 + 1}{x^2 + y^2 - 1}$

34. $F(x, y) = \dfrac{x^6 + x^3y^3 + y^6}{x^3 + y^3}$

35. $F(x, y) = \tan(x^4 - y^4)$

36. $F(x, y) = \ln(2x + 3y)$

37. $G(x, y) = e^{xy}\sin(x + y)$

38. $G(x, y) = \sin^{-1}(x^2 + y^2)$

39. $G(x, y) = \sqrt{x + y} - \sqrt{x - y}$

40. $G(x, y) = 2^{x\tan y}$

41. $f(x, y, z) = x\ln(yz)$

42. $f(x, y, z) = x + y\sqrt{x + z}$

43. $f(x, y) = \begin{cases} \dfrac{2x^2 - y^2}{2x^2 + y^2} & \text{if } (x, y) \neq (0, 0) \\ 0 & \text{if } (x, y) = (0, 0) \end{cases}$

44. $f(x, y) = \begin{cases} \dfrac{x^2y^3}{2x^2 + y^2} & \text{if } (x, y) \neq (0, 0) \\ 0 & \text{if } (x, y) = (0, 0) \end{cases}$

45. $f(x, y) = \begin{cases} \dfrac{x^2y^3}{2x^2 + y^2} & \text{if } (x, y) \neq (0, 0) \\ 1 & \text{if } (x, y) = (0, 0) \end{cases}$

46. $f(x, y) = \begin{cases} \dfrac{xy}{x^2 + xy + y^2} & \text{if } (x, y) \neq (0, 0) \\ 0 & \text{if } (x, y) = (0, 0) \end{cases}$

47. Prove, using Definition 1, that

(a) $\displaystyle\lim_{(x,y)\to(a,b)} x = a$ (b) $\displaystyle\lim_{(x,y)\to(a,b)} y = b$

(c) $\displaystyle\lim_{(x,y)\to(a,b)} c = c$

48. For what values of the number r is the function

$$f(x, y, z) = \begin{cases} \dfrac{(x + y + z)^r}{x^2 + y^2 + z^2} & \text{if } (x, y, z) \neq (0, 0, 0) \\ 0 & \text{if } (x, y, z) = (0, 0, 0) \end{cases}$$

continuous on $\mathbb{R}^3$?

49. Show that the function $f : \mathbb{R}^n \to \mathbb{R}$ given by $f(\mathbf{x}) = |\mathbf{x}|$ is continuous on $\mathbb{R}^n$.
[*Hint*: Consider $|\mathbf{x} - \mathbf{a}|^2 = (\mathbf{x} - \mathbf{a}) \cdot (\mathbf{x} - \mathbf{a})$.]

50. If $\mathbf{c} \in V_n$, show that the function $f : \mathbb{R}^n \to \mathbb{R}$ given by $f(\mathbf{x}) = \mathbf{c} \cdot \mathbf{x}$ is continuous on $\mathbb{R}^n$.

12.3 PARTIAL DERIVATIVES

If f is a function of two variables x and y, suppose we let only x vary while keeping y fixed, say $y = b$, where b is a constant. Then we are really considering a function of a single variable x, namely, $g(x) = f(x, b)$. If g has a derivative at a, then we call it the **partial derivative of f with respect to x at (a, b)** and denote it by $f_x(a, b)$. Thus

(1) $$f_x(a, b) = g'(a) \quad \text{where} \quad g(x) = f(x, b)$$

By Definition 3.1.2 we know that

$$g'(a) = \lim_{h\to 0} \frac{g(a + h) - g(a)}{h}$$

and so Equation 1 becomes

(2) $$f_x(a, b) = \lim_{h\to 0} \frac{f(a + h, b) - f(a, b)}{h}$$

Similarly, the **partial derivative of f with respect to y at (a, b)**, denoted by $f_y(a, b)$, is obtained by keeping x fixed ($x = a$) and finding the ordinary derivative at b of the function $G(y) = f(a, y)$:

(3) $$f_y(a, b) = \lim_{h\to 0} \frac{f(a, b + h) - f(a, b)}{h}$$

If we now let the point (a, b) vary, f_x and f_y become functions of two variables.

(4) If f is a function of two variables, its **partial derivatives** are the functions f_x and f_y defined by

$$f_x(x, y) = \lim_{h\to 0} \frac{f(x + h, y) - f(x, y)}{h}$$

$$f_y(x, y) = \lim_{h\to 0} \frac{f(x, y + h) - f(x, y)}{h}$$

There are many alternative notations for partial derivatives. For instance, instead of f_x we can write f_1 or $D_1 f$ (to indicate differentiation with respect to the *first* variable) or $\partial f/\partial x$. But here $\partial f/\partial x$ cannot be interpreted as a ratio of differentials.

NOTATIONS FOR PARTIAL DERIVATIVES If $z = f(x, y)$, we write

$$f_x(x, y) = f_x = \frac{\partial f}{\partial x} = \frac{\partial}{\partial x} f(x, y) = \frac{\partial z}{\partial x} = f_1 = D_1 f = D_x f$$

$$f_y(x, y) = f_y = \frac{\partial f}{\partial y} = \frac{\partial}{\partial y} f(x, y) = \frac{\partial z}{\partial y} = f_2 = D_2 f = D_y f$$

To compute partial derivatives, all we have to do is remember from Equation 1 that the partial derivative with respect to x is just the *ordinary* derivative of the function g of a single variable that we get by keeping y fixed. Thus, we have the following rule:

RULE FOR FINDING PARTIAL DERIVATIVES OF $z = f(x, y)$

1. To find f_x, regard y as a constant and differentiate $f(x, y)$ with respect to x.
2. To find f_y, regard x as a constant and differentiate $f(x, y)$ with respect to y.

EXAMPLE 1 If $f(x, y) = x^3 + x^2y^3 - 2y^2$, find $f_x(2, 1)$ and $f_y(2, 1)$.

SOLUTION Holding y constant and differentiating with respect to x, we get

$$f_x(x, y) = 3x^2 + 2xy^3$$

and so

$$f_x(2, 1) = 3 \cdot 2^2 + 2 \cdot 2 \cdot 1^3 = 16$$

Holding x constant and differentiating with respect to y, we get

$$f_y(x, y) = 3x^2y^2 - 4y$$

$$f_y(2, 1) = 3 \cdot 2^2 \cdot 1^2 - 4 \cdot 1 = 8$$

■

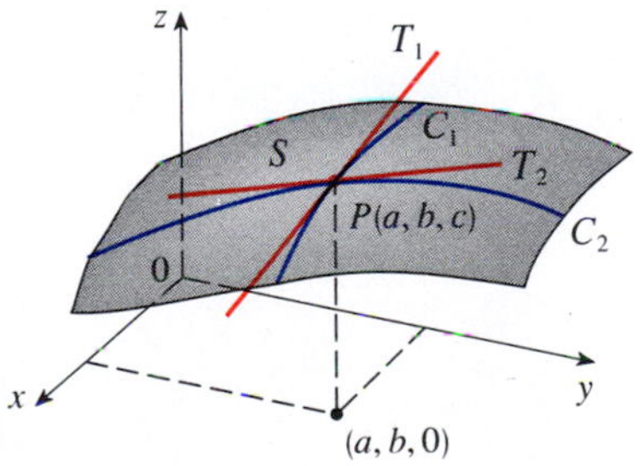

FIGURE 1
The partial derivatives of f at (a, b) are the slopes of the tangents to C_1 and C_2.

To give a geometric interpretation of partial derivatives, we recall that the equation $z = f(x, y)$ represents a surface S (the graph of f). If $f(a, b) = c$, then the point $P(a, b, c)$ lies on S. The vertical plane $y = b$ intersects S in a curve C_1. (In other words, C_1 is the trace of S in the plane $y = b$.) Likewise, the vertical plane $x = a$ intersects S in a curve C_2. Both of the curves C_1 and C_2 pass through the point P (see Figure 1).

Notice that the curve C_1 is the graph of the function $g(x) = f(x, b)$, so the slope of its tangent T_1 at P is $g'(a) = f_x(a, b)$. The curve C_2 is the graph of the function $G(y) = f(a, y)$, so the slope of its tangent T_2 at P is $G'(b) = f_y(a, b)$.

Thus, the partial derivatives $f_x(a, b)$ and $f_y(a, b)$ can be interpreted geometrically as the slopes of the tangent lines at $P(a, b, c)$ to the traces C_1 and C_2 of S in the planes $y = b$ and $x = a$.

Partial derivatives can also be interpreted as rates of change. If $z = f(x, y)$, then $\partial z/\partial x$ represents the rate of change of z with respect to x when y is fixed. Similarly, $\partial z/\partial y$ represents the rate of change of z with respect to y when x is fixed. For instance, referring to the temperature function T at the beginning of Section 12.1, we see that $\partial T/\partial x$ is the rate at which the temperature changes in the east-west direction and $\partial T/\partial y$ is the rate at which it changes in the north-south direction.

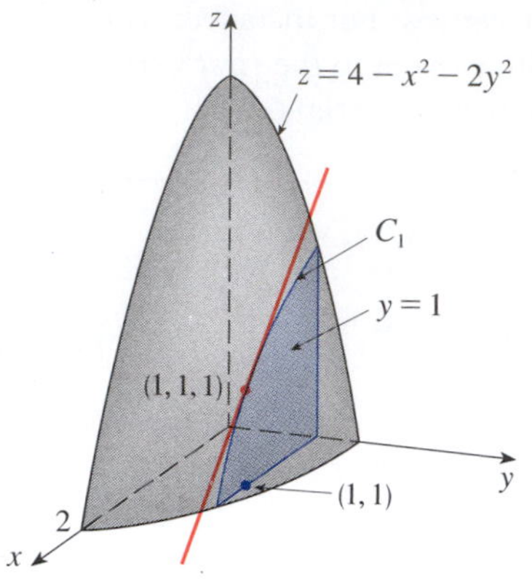

FIGURE 2

EXAMPLE 2 If $f(x, y) = 4 - x^2 - 2y^2$, find $f_x(1, 1)$ and $f_y(1, 1)$ and interpret these numbers as slopes.

SOLUTION We have

$$f_x(x, y) = -2x \qquad f_y(x, y) = -4y$$

$$f_x(1, 1) = -2 \qquad f_y(1, 1) = -4$$

The graph of f is the paraboloid $z = 4 - x^2 - 2y^2$ and the vertical plane $y = 1$ intersects it in the parabola $z = 2 - x^2$, $y = 1$. (As in the preceding discussion, we label it C_1 in Figure 2.) The slope of the tangent line to this parabola at the point $(1, 1, 1)$ is $f_x(1, 1) = -2$. Similarly, the curve C_2 in which the plane $x = 1$ intersects the paraboloid is the parabola $z = 3 - 2y^2$, $x = 1$ and the slope of the tangent line at $(1, 1, 1)$ is $f_y(1, 1) = -4$ (see Figure 3). ■

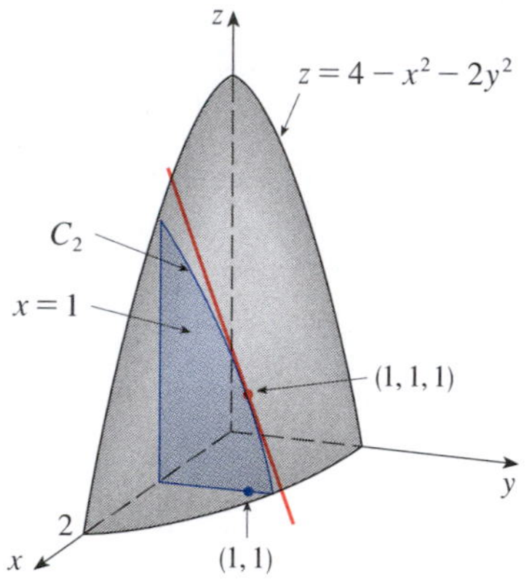

FIGURE 3

EXAMPLE 3 If $f(x, y) = \sin\left(\dfrac{x}{1 + y}\right)$, calculate $\dfrac{\partial f}{\partial x}$ and $\dfrac{\partial f}{\partial y}$.

SOLUTION Using the Chain Rule for functions of one variable, we have

$$\frac{\partial f}{\partial x} = \cos\left(\frac{x}{1 + y}\right) \cdot \frac{\partial}{\partial x}\left(\frac{x}{1 + y}\right) = \cos\left(\frac{x}{1 + y}\right) \cdot \frac{1}{1 + y}$$

$$\frac{\partial f}{\partial y} = \cos\left(\frac{x}{1 + y}\right) \cdot \frac{\partial}{\partial y}\left(\frac{x}{1 + y}\right) = -\cos\left(\frac{x}{1 + y}\right) \cdot \frac{x}{(1 + y)^2}$$ ■

EXAMPLE 4 Find $\partial z/\partial x$ and $\partial z/\partial y$ if z is defined implicitly as a function of x and y by the equation

$$x^3 + y^3 + z^3 + 6xyz = 1$$

Some computer algebra systems can plot surfaces defined by implicit equations in three variables. Figure 4 shows such a plot of the surface defined by the equation in Example 4.

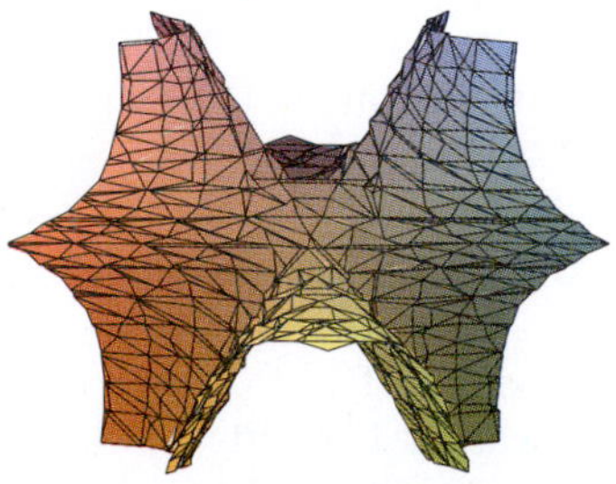

FIGURE 4

SOLUTION To find $\partial z/\partial x$, we differentiate implicitly with respect to x, being careful to treat y as a constant:

$$3x^2 + 3z^2 \frac{\partial z}{\partial x} + 6yz + 6xy \frac{\partial z}{\partial x} = 0$$

Solving this equation for $\partial z/\partial x$, we obtain

$$\frac{\partial z}{\partial x} = -\frac{x^2 + 2yz}{z^2 + 2xy}$$

Similarly, implicit differentiation with respect to y gives

$$\frac{\partial z}{\partial y} = -\frac{y^2 + 2xz}{z^2 + 2xy}$$ ■

FUNCTIONS OF MORE THAN TWO VARIABLES

Partial derivatives can also be defined for functions of three or more variables. For example, if f is a function of three variables x, y, and z, then its partial derivative with respect to x is defined as

$$f_x(x, y, z) = \lim_{h \to 0} \frac{f(x + h, y, z) - f(x, y, z)}{h}$$

and it is found by regarding y and z as constants and differentiating $f(x, y, z)$ with respect to x.

In general, if u is a function of n variables, $u = f(x_1, x_2, \ldots, x_n)$, its partial derivative with respect to the ith variable x_i is

$$\frac{\partial u}{\partial x_i} = \lim_{h \to 0} \frac{f(x_1, \ldots, x_{i-1}, x_i + h, x_{i+1}, \ldots, x_n) - f(x_1, \ldots, x_i, \ldots, x_n)}{h}$$

and we also write

$$\frac{\partial u}{\partial x_i} = \frac{\partial f}{\partial x_i} = f_{x_i} = f_i = D_i f$$

EXAMPLE 5 Find f_x, f_y, and f_z if $f(x, y, z) = e^{xy} \ln z$.

SOLUTION Holding y and z constant and differentiating with respect to x, we have

$$f_x = ye^{xy} \ln z$$

Similarly, $f_y = xe^{xy} \ln z$ and $f_z = \dfrac{e^{xy}}{z}$ ■

HIGHER DERIVATIVES

If f is a function of two variables, then its partial derivatives f_x and f_y are also functions of two variables, so we can consider their partial derivatives $(f_x)_x$, $(f_x)_y$, $(f_y)_x$, and $(f_y)_y$, which are called the **second partial derivatives** of f. If $z = f(x, y)$, we use the following notation:

$$(f_x)_x = f_{xx} = f_{11} = \frac{\partial}{\partial x}\left(\frac{\partial f}{\partial x}\right) = \frac{\partial^2 f}{\partial x^2} = \frac{\partial^2 z}{\partial x^2}$$

$$(f_x)_y = f_{xy} = f_{12} = \frac{\partial}{\partial y}\left(\frac{\partial f}{\partial x}\right) = \frac{\partial^2 f}{\partial y\, \partial x} = \frac{\partial^2 z}{\partial y\, \partial x}$$

$$(f_y)_x = f_{yx} = f_{21} = \frac{\partial}{\partial x}\left(\frac{\partial f}{\partial y}\right) = \frac{\partial^2 f}{\partial x\, \partial y} = \frac{\partial^2 z}{\partial x\, \partial y}$$

$$(f_y)_y = f_{yy} = f_{22} = \frac{\partial}{\partial y}\left(\frac{\partial f}{\partial y}\right) = \frac{\partial^2 f}{\partial y^2} = \frac{\partial^2 z}{\partial y^2}$$

Thus, the notation f_{xy} (or $\partial^2 f/\partial y\, \partial x$) means that we first differentiate with respect to x and then with respect to y, whereas in computing f_{yx} the order is reversed.

EXAMPLE 6 Find the second partial derivatives of

$$f(x, y) = x^3 + x^2y^3 - 2y^2$$

SOLUTION In Example 1 we found that

$$f_x(x, y) = 3x^2 + 2xy^3 \qquad f_y(x, y) = 3x^2y^2 - 4y$$

Therefore

$$f_{xx} = \frac{\partial}{\partial x}(3x^2 + 2xy^3) = 6x + 2y^3$$

$$f_{xy} = \frac{\partial}{\partial y}(3x^2 + 2xy^3) = 6xy^2$$

$$f_{yx} = \frac{\partial}{\partial x}(3x^2y^2 - 4y) = 6xy^2$$

$$f_{yy} = \frac{\partial}{\partial y}(3x^2y^2 - 4y) = 6x^2y - 4$$

Figure 5 shows the graph of the function f in Example 6 and the graphs of its first- and second-order partial derivatives for $-2 \leqslant x \leqslant 2$, $-2 \leqslant y \leqslant 2$. Notice that these graphs are consistent with our interpretations of f_x and f_y as slopes of tangent lines to traces of the graph of f. For instance, the graph of f decreases if we start at $(0, -2)$ and move in the positive x-direction. This is reflected in the negative values of f_x. You should compare the graphs of f_{yx} and f_{yy} with the graph of f_y to see the relationships.

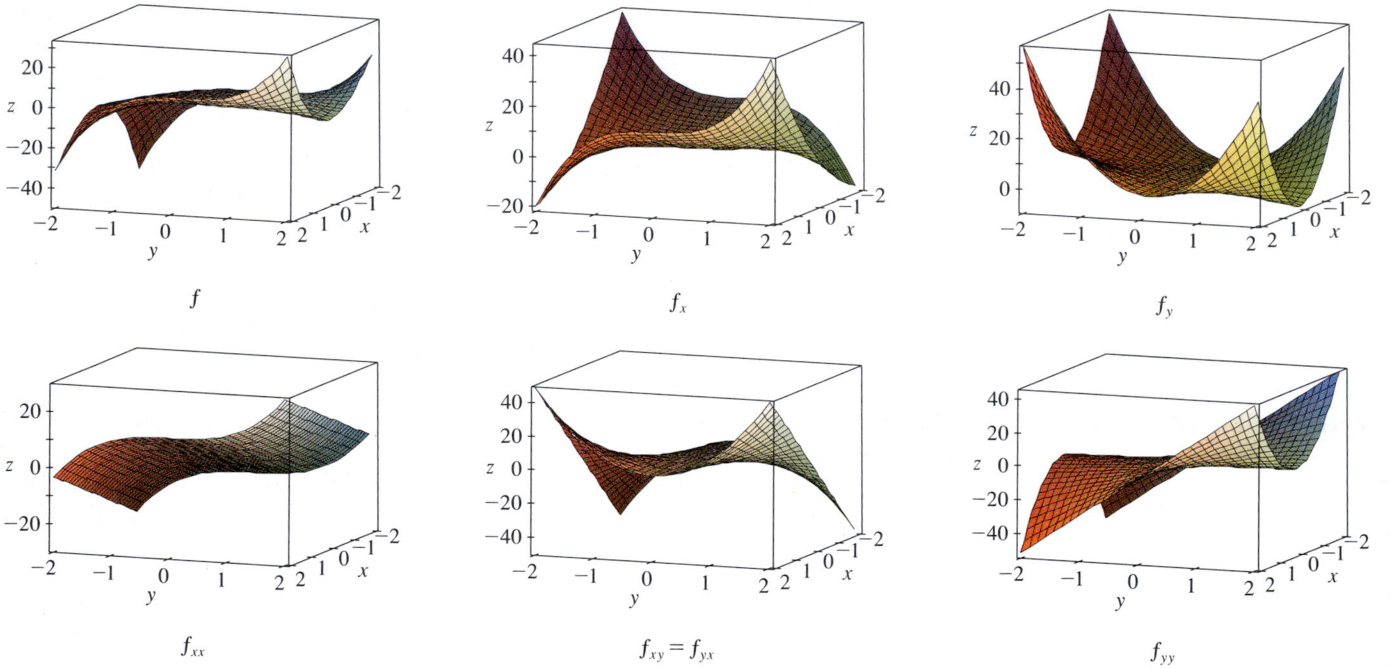

FIGURE 5

Notice that $f_{xy} = f_{yx}$ in Example 6. This is not just a coincidence. It turns out that the mixed partial derivatives f_{xy} and f_{yx} are equal for most functions that one meets in practice. The following theorem, discovered by the French mathematician Alexis Clairaut (1713–1765), gives conditions under which we can assert that $f_{xy} = f_{yx}$. The proof is given in Appendix F.

(5) CLAIRAUT'S THEOREM Suppose f is defined on a disk D that contains the point (a, b). If the functions f_{xy} and f_{yx} are both continuous on D, then

$$f_{xy}(a, b) = f_{yx}(a, b)$$

Alexis Clairaut was a child prodigy in mathematics, having read l'Hospital's textbook on calculus when he was ten and presented a paper on geometry to the French Academy of Sciences when he was 13. At the age of 18, Clairaut published *Recherches sur les courbes à double courbure*, which was the first systematic treatise on three-dimensional analytic geometry and included the calculus of space curves.

Partial derivatives of order 3 or higher can also be defined. For instance,

$$f_{xyy} = (f_{xy})_y = \frac{\partial}{\partial y}\left(\frac{\partial^2 z}{\partial y\,\partial x}\right) = \frac{\partial^3 z}{\partial y^2\,\partial x}$$

and using Clairaut's Theorem it can be shown that $f_{xyy} = f_{yxy} = f_{yyx}$ if these functions are continuous.

EXAMPLE 7 Calculate f_{xxyz} if $f(x, y, z) = \sin(3x + yz)$.

SOLUTION

$$f_x = 3\cos(3x + yz)$$

$$f_{xx} = -9\sin(3x + yz)$$

$$f_{xxy} = -9z\cos(3x + yz)$$

$$f_{xxyz} = -9\cos(3x + yz) + 9yz\sin(3x + yz)$$

■

PARTIAL DIFFERENTIAL EQUATIONS

Partial derivatives occur in *partial differential equations* that express certain physical laws. For instance, the partial differential equation

$$\frac{\partial^2 u}{\partial x^2} + \frac{\partial^2 u}{\partial y^2} = 0$$

is called **Laplace's equation** after Pierre Laplace (1749–1827). Solutions of this equation are called **harmonic functions** and play a role in problems of heat conduction, fluid flow, and electric potential.

EXAMPLE 8 Show that the function $u(x, y) = e^x \sin y$ is a solution of Laplace's equation.

SOLUTION

$$u_x = e^x \sin y \qquad u_y = e^x \cos y$$

$$u_{xx} = e^x \sin y \qquad u_{yy} = -e^x \sin y$$

$$u_{xx} + u_{yy} = e^x \sin y - e^x \sin y = 0$$

Therefore, u satisfies Laplace's equation. ■

The **wave equation**

$$\frac{\partial^2 u}{\partial t^2} = a^2 \frac{\partial^2 u}{\partial x^2}$$

describes the motion of a waveform, which could be an ocean wave, a sound wave, a light wave, or a wave traveling along a vibrating string. For instance, if $u(x, t)$ represents the displacement of a vibrating violin string at time t and at a distance x from one end of the string (as in Figure 6), then $u(x, t)$ satisfies the wave equation. Here the constant a depends on the density of the string and on the tension in the string.

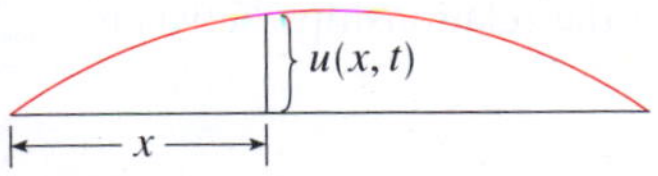

FIGURE 6

EXAMPLE 9 Verify that the function $u(x, t) = \sin(x - at)$ satisfies the wave equation.

SOLUTION

$$u_x = \cos(x - at) \qquad u_{xx} = -\sin(x - at)$$

$$u_t = -a\cos(x - at) \qquad u_{tt} = -a^2\sin(x - at) = a^2 u_{xx}$$

So u satisfies the wave equation. ■

EXERCISES 12.3

1. If $f(x, y) = 16 - 4x^2 - y^2$, find $f_x(1, 2)$ and $f_y(1, 2)$ and interpret these numbers as slopes. Illustrate with sketches.

2. If $f(x, y) = \sqrt{4 - x^2 - 4y^2}$, find $f_x(1, 0)$ and $f_y(1, 0)$ and interpret these numbers as slopes. Illustrate with sketches.

3–16 ■ Find the indicated partial derivatives.

3. $f(x, y) = x^3y^5$; $f_x(3, -1)$

4. $f(x, y) = \sqrt{2x + 3y}$; $f_y(2, 4)$

5. $f(x, y) = xe^{-y} + 3y$; $\dfrac{\partial f}{\partial y}(1, 0)$

6. $f(x, y) = \sin(y - x)$; $\dfrac{\partial f}{\partial x}(3, 3)$

7. $z = \dfrac{x^3 + y^3}{x^2 + y^2}$; $\dfrac{\partial z}{\partial x}, \dfrac{\partial z}{\partial y}$

8. $z = x\sqrt{y} - \dfrac{y}{\sqrt{x}}$; $\dfrac{\partial z}{\partial x}, \dfrac{\partial z}{\partial y}$

9. $xy + yz = xz$; $\dfrac{\partial z}{\partial x}, \dfrac{\partial z}{\partial y}$

10. $xyz = \cos(x + y + z)$; $\dfrac{\partial z}{\partial x}, \dfrac{\partial z}{\partial y}$

11. $x^2 + y^2 - z^2 = 2x(y + z)$; $\dfrac{\partial z}{\partial x}, \dfrac{\partial z}{\partial y}$

12. $xy^2z^3 + x^3y^2z = x + y + z$; $\dfrac{\partial z}{\partial x}, \dfrac{\partial z}{\partial y}$

13. $f(x, y, z) = xyz$; $f_y(0, 1, 2)$

14. $f(x, y, z) = \sqrt{x^2 + y^2 + z^2}$; $f_z(0, 3, 4)$

15. $u = xy + yz + zx$; u_x, u_y, u_z

16. $u = x^2y^3t^4$; u_x, u_y, u_t

17–44 ■ Find the first partial derivatives of the function.

17. $f(x, y) = x^3y^5 - 2x^2y + x$

18. $f(x, y) = x^2y^2(x^4 + y^4)$

19. $f(x, y) = x^4 + x^2y^2 + y^4$

20. $f(x, y) = \ln(x^2 + y^2)$

21. $f(x, y) = \dfrac{x - y}{x + y}$

22. $f(x, y) = x^y$

23. $f(x, y) = e^x\tan(x - y)$

24. $f(s, t) = s/\sqrt{s^2 + t^2}$

25. $f(u, v) = \tan^{-1}(u/v)$

26. $f(x, t) = e^{\sin(t/x)}$

27. $g(x, y) = y\tan(x^2y^3)$

28. $g(x, y) = \ln(x + \ln y)$

29. $z = \ln(x + \sqrt{x^2 + y^2})$

30. $z = x^{x^y}$

31. $f(x, y) = \displaystyle\int_x^y e^{t^2}\,dt$

32. $f(x, y) = \displaystyle\int_y^x \frac{e^t}{t}\,dt$

33. $f(x, y, z) = x^2yz^3 + xy - z$

34. $f(x, y, z) = x\sqrt{yz}$

35. $f(x, y, z) = x^{yz}$

36. $f(x, y, z) = xe^y + ye^z + ze^x$

37. $u = z\sin\dfrac{y}{x + z}$

38. $u = x^{y/z}$

39. $u = xy^2z^3\ln(x + 2y + 3z)$

40. $u = x^{y^z}$

41. $f(x, y, z, t) = \dfrac{x - y}{z - t}$

42. $f(x, y, z, t) = xy^2z^3t^4$

43. $u = \sqrt{x_1^2 + x_2^2 + \cdots + x_n^2}$

44. $u = \sin(x_1 + 2x_2 + \cdots + nx_n)$

45–46 ■ Use the definition of partial derivatives as limits (4) to find $f_x(x, y)$ and $f_y(x, y)$.

45. $f(x, y) = x^2 - xy + 2y^2$

46. $f(x, y) = \sqrt{3x - y}$

47–48 ■ Find f_x and f_y and graph f, f_x, and f_y with domains and viewpoints that enable you to see the relationships between them.

47. $f(x, y) = x^2 + y^2 + x^2y$

48. $f(x, y) = xe^{-x^2-y^2}$

49. The following surfaces, labeled a, b, and c, are graphs of a function f and its partial derivatives f_x and f_y. Identify each surface and give reasons for your choices.

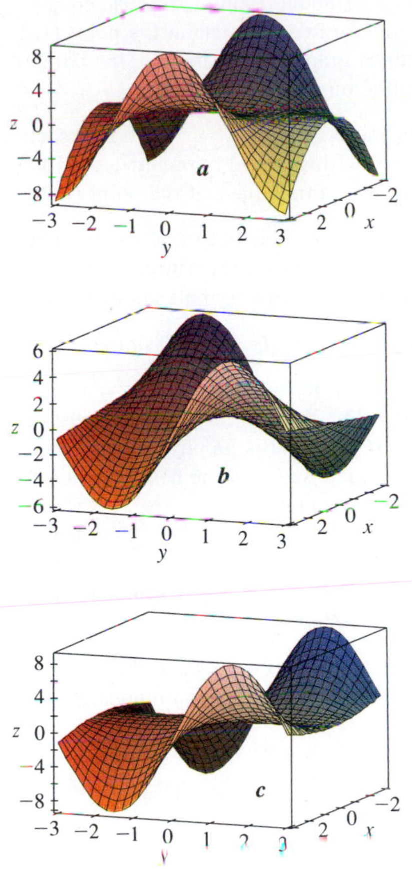

50. A contour map is given for a function f. Use it to estimate $f_x(2, 1)$ and $f_y(2, 1)$.

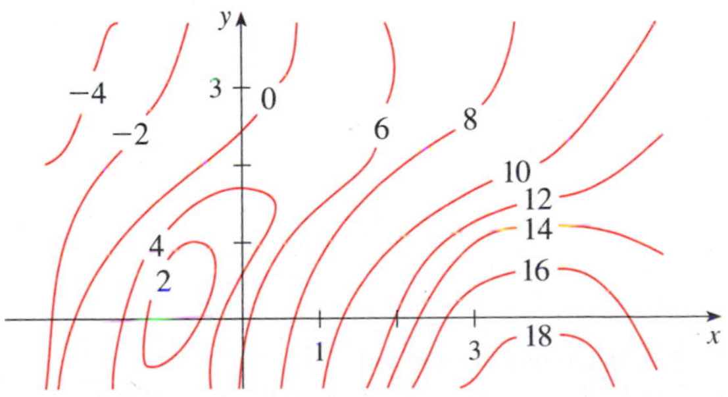

51–56 ■ Find $\partial z/\partial x$ and $\partial z/\partial y$.

51. $z = f(x) + g(y)$

52. $z = f(x)g(y)$

53. $z = f(x + y)$

54. $z = f(xy)$

55. $z = f(x/y)$

56. $z = f(ax + by)$

57–62 ■ Find all the second partial derivatives.

57. $f(x, y) = x^2y + x\sqrt{y}$

58. $f(x, y) = \sin(x + y) + \cos(x - y)$

59. $z = (x^2 + y^2)^{3/2}$

60. $z = \cos^2(5x + 2y)$

61. $z = t\sin^{-1}\sqrt{x}$

62. $z = x^{\ln t}$

63–66 ■ Verify that the conclusion of Clairaut's Theorem holds, that is, $u_{xy} = u_{yx}$.

63. $u = x^5y^4 - 3x^2y^3 + 2x^2$

64. $u = \sin^2 x\cos y$

65. $u = \sin^{-1}(xy^2)$

66. $u = x^2y^3z^4$

67–74 ■ Find the indicated partial derivative.

67. $f(x, y) = x^2y^3 - 2x^4y$; f_{xxx}

68. $f(x, y) = e^{xy^2}$; f_{xxy}

69. $f(x, y, z) = x^5 + x^4y^4z^3 + yz^2$; f_{xyz}

70. $f(x, y, z) = e^{xyz}$; f_{yzy}

71. $z = x\sin y$; $\dfrac{\partial^3 z}{\partial y^2\,\partial x}$

72. $z = \ln\sin(x - y)$; $\dfrac{\partial^3 z}{\partial y\,\partial x^2}$

73. $u = \ln(x + 2y^2 + 3z^3)$; $\dfrac{\partial^3 u}{\partial x\,\partial y\,\partial z}$

74. $u = x^a y^b z^c$; $\dfrac{\partial^6 u}{\partial x\,\partial y^2\,\partial z^3}$

75. Verify that the function $u = e^{-\alpha^2k^2t}\sin kx$ is a solution of the heat conduction equation $u_t = \alpha^2u_{xx}$.

76. Determine whether each of the following functions is a solution of Laplace's equation $u_{xx} + u_{yy} = 0$.
(a) $u = x^2 + y^2$
(b) $u = x^2 - y^2$
(c) $u = x^3 + 3xy^2$
(d) $u = \ln\sqrt{x^2 + y^2}$
(e) $u = \sin x\cosh y + \cos x\sinh y$
(f) $u = e^{-x}\cos y - e^{-y}\cos x$

77. Verify that the function $u = 1/\sqrt{x^2 + y^2 + z^2}$ is a solution of the three-dimensional Laplace equation $u_{xx} + u_{yy} + u_{zz} = 0$.

78. Show that each of the following functions is a solution of the wave equation $u_{tt} = a^2u_{xx}$.
(a) $u = \sin(kx)\sin(akt)$
(b) $u = t/(a^2t^2 - x^2)$
(c) $u = (x - at)^6 + (x + at)^6$
(d) $u = \sin(x - at) + \ln(x + at)$

79. If f and g are twice differentiable functions of a single variable, show that the function

$$u(x, t) = f(x + at) + g(x - at)$$

is a solution of the wave equation given in Exercise 78.

80. If f and g are twice differentiable functions of a single variable, show that the function

$$u(x, y) = xf(x + y) + yg(x + y)$$

satisfies the equation $u_{xx} - 2u_{xy} + u_{yy} = 0$.

81. Show that the function $z = xe^y + ye^x$ is a solution of the equation

$$\frac{\partial^3 z}{\partial x^3} + \frac{\partial^3 z}{\partial y^3} = x\frac{\partial^3 z}{\partial x\, \partial y^2} + y\frac{\partial^3 z}{\partial x^2\, \partial y}$$

82. If $u = e^{a_1x_1 + a_2x_2 + \cdots + a_nx_n}$, where $a_1^2 + a_2^2 + \cdots + a_n^2 = 1$, show that

$$\frac{\partial^2 u}{\partial x_1^2} + \frac{\partial^2 u}{\partial x_2^2} + \cdots + \frac{\partial^2 u}{\partial x_n^2} = u$$

83. Show that the function

$$f(x_1, \ldots, x_n) = (x_1^2 + \cdots + x_n^2)^{(2-n)/2}$$

satisfies the equation

$$\frac{\partial^2 f}{\partial x_1^2} + \cdots + \frac{\partial^2 f}{\partial x_n^2} = 0$$

84. The temperature at a point (x, y) on a flat metal plate is given by $T(x, y) = 60/(1 + x^2 + y^2)$, where T is measured in °C and x, y in meters. Find the rate of change of temperature with respect to distance at the point $(2, 1)$ in (a) the x-direction and (b) the y-direction.

85. The total resistance R produced by three conductors with resistances R_1, R_2, R_3 connected in a parallel electrical circuit is given by the formula

$$\frac{1}{R} = \frac{1}{R_1} + \frac{1}{R_2} + \frac{1}{R_3}$$

Find $\partial R/\partial R_1$.

86. The gas law for a fixed mass m of an ideal gas at absolute temperature T, pressure P, and volume V is $PV = mRT$, where R is the gas constant. Show that

$$\frac{\partial P}{\partial V}\frac{\partial V}{\partial T}\frac{\partial T}{\partial P} = -1$$

87. The kinetic energy of a body with mass m and velocity v is $K = \frac{1}{2}mv^2$. Show that

$$\frac{\partial K}{\partial m}\frac{\partial^2 K}{\partial v^2} = K$$

88. If a, b, c are the sides of a triangle and A, B, C are the opposite angles, find $\partial A/\partial a$, $\partial A/\partial b$, $\partial A/\partial c$ by implicit differentiation of the Law of Cosines.

89. You are told that there is a function $f(x, y)$ whose partial derivatives are $f_x(x, y) = x + 4y$ and $f_y(x, y) = 3x - y$. Should you believe it?

90. The paraboloid $z = 6 - x - x^2 - 2y^2$ intersects the plane $x = 1$ in a parabola. Find parametric equations for the tangent line to this parabola at the point $(1, 2, -4)$. Use a computer to graph the paraboloid, the parabola, and the tangent line on the same screen.

91. The ellipsoid $4x^2 + 2y^2 + z^2 = 16$ intersects the plane $y = 2$ in an ellipse. Find parametric equations for the tangent line to this ellipse at the point $(1, 2, 2)$.

92. In a study of frost penetration it was found that the temperature T at time t (measured in days) at a depth x (measured in feet) can be modeled by the function

$$T(x, t) = T_0 + T_1 e^{-\lambda x}\sin(\omega t - \lambda x)$$

where $\omega = 2\pi/365$ and λ is a positive constant.

(a) Find $\partial T/\partial x$. What is its physical significance?

(b) Find $\partial T/\partial t$. What is its physical significance?

(c) Show that T satisfies the heat equation $T_t = kT_{xx}$ for a certain constant k.

(d) If $\lambda = 0.2$, $T_0 = 0$, and $T_1 = 10$, use a computer to graph $T(x, t)$.

(e) What is the physical significance of the term $-\lambda x$ in the expression $\sin(\omega t - \lambda x)$?

93. Use Clairaut's Theorem to show that if the third-order partial derivatives of f are continuous, then

$$f_{xyy} = f_{yxy} = f_{yyx}$$

94. (a) How many nth-order partial derivatives does a function of two variables have?

(b) If these partial derivatives are all continuous, how many of them can be distinct?

(c) Answer the question in part (a) for a function of three variables.

95. If $f(x, y) = x(x^2 + y^2)^{-3/2}e^{\sin(x^2y)}$, find $f_x(1, 0)$. [*Hint:* Instead of finding $f_x(x, y)$ first, note that it is easier to use Equation 1 or Equation 2.]

96. If $f(x, y) = \sqrt[3]{x^3 + y^3}$, find $f_x(0, 0)$.

97. Let

$$f(x, y) = \begin{cases} \dfrac{x^3y - xy^3}{x^2 + y^2} & \text{if } (x, y) \neq (0, 0) \\ 0 & \text{if } (x, y) = (0, 0) \end{cases}$$

(a) Use a computer to graph f.

(b) Find $f_x(x, y)$ and $f_y(x, y)$ when $(x, y) \neq (0, 0)$.

(c) Find $f_x(0, 0)$ and $f_y(0, 0)$ using Equations 2 and 3.

(d) Show that $f_{xy}(0, 0) = -1$ and $f_{yx}(0, 0) = 1$.

CAS (e) Does the result of part (d) contradict Clairaut's Theorem? Use graphs of f_{xy} and f_{yx} to illustrate your answer.

12.4 TANGENT PLANES AND DIFFERENTIALS

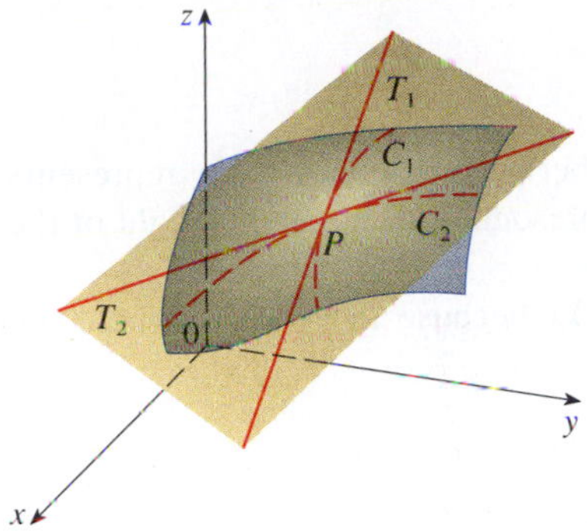

FIGURE 1

Suppose a surface S has equation $z = f(x, y)$, where f has continuous first partial derivatives, and let $P(x_0, y_0, z_0)$ be a point on S. As in the preceding section, let C_1 and C_2 be the curves obtained by intersecting the vertical planes $y = y_0$ and $x = x_0$ with the surface S. Then the point P lies on both C_1 and C_2. Let T_1 and T_2 be the tangent lines to the curves C_1 and C_2 at the point P. Then the **tangent plane** to the surface S at the point P is defined to be the plane that contains both of the tangent lines T_1 and T_2 (see Figure 1).

We will see in Section 12.6 that if C is any other curve that lies on the surface S and passes through P, then its tangent line at P also lies in the tangent plane. Therefore, you can think of the tangent plane to S at P as consisting of all possible tangent lines at P to curves that lie on S and pass through P. The tangent plane at P is the plane that most closely approximates the surface S near the point P.

We know from Equation 11.4.6 that any plane passing through $P(x_0, y_0, z_0)$ has an equation of the form

$$A(x - x_0) + B(y - y_0) + C(z - z_0) = 0$$

By dividing this equation by C and letting $a = -A/C$ and $b = -B/C$, we can write it in the form

$$z - z_0 = a(x - x_0) + b(y - y_0) \tag{1}$$

If Equation 1 represents the tangent plane at P, then its intersection with the plane $y = y_0$ must be the tangent line T_1. Setting $y = y_0$ in Equation 1 gives

$$z - z_0 = a(x - x_0) \qquad y = y_0$$

and we recognize these as the equations (in slope-point form) of a line with slope a. But from Section 12.3 we know that the slope of T_1 is $f_x(x_0, y_0)$. Therefore, we have $a = f_x(x_0, y_0)$.

Similarly, putting $x = x_0$ in Equation 1, we get $z - z_0 = b(y - y_0)$, which must represent the tangent line T_2, so $b = f_y(x_0, y_0)$.

Note the similarity between the equation of a tangent plane and the equation of a tangent line:

$$y - y_0 = f'(x_0)(x - x_0)$$

> **(2)** An equation of the tangent plane to the surface $z = f(x, y)$ at the point $P(x_0, y_0, z_0)$ is
>
> $$z - z_0 = f_x(x_0, y_0)(x - x_0) + f_y(x_0, y_0)(y - y_0)$$

EXAMPLE 1 Find the tangent plane to the elliptic paraboloid $z = 2x^2 + y^2$ at the point $(1, 1, 3)$.

SOLUTION Let $f(x, y) = 2x^2 + y^2$. Then

$$f_x(x, y) = 4x \qquad f_y(x, y) = 2y$$

$$f_x(1, 1) = 4 \qquad f_y(1, 1) = 2$$

Then (2) gives the equation of the tangent plane at $(1, 1, 3)$ as

$$z - 3 = 4(x - 1) + 2(y - 1)$$

or

$$4x + 2y - z = 3 \qquad \blacksquare$$

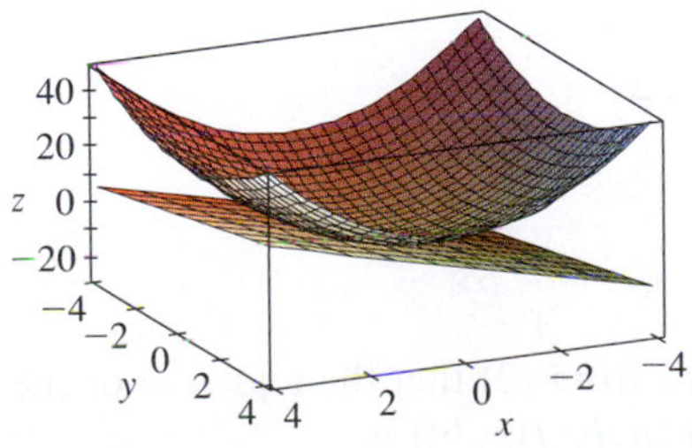

FIGURE 2

Figure 2 shows the elliptic paraboloid and the tangent plane in Example 1.

DIFFERENTIALS

Recall that for a function of one variable, $y = f(x)$, we defined the increment of y as

$$\Delta y = f(x + \Delta x) - f(x)$$

and the differential of y as

$$dy = f'(x)\,dx \tag{3}$$

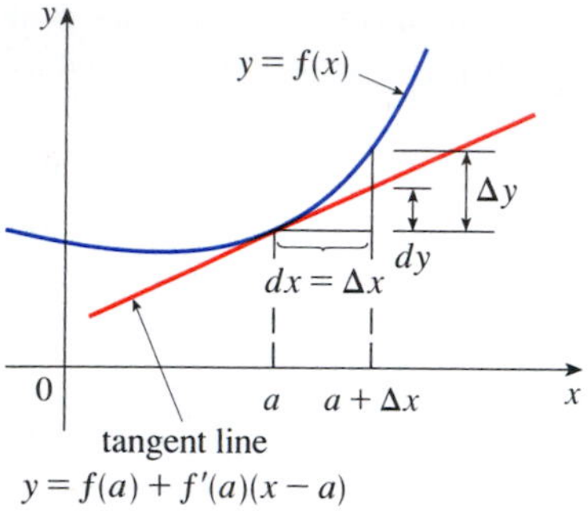

FIGURE 3

(See Section 3.11.) Figure 3 shows the relationship between Δy and dy: Δy represents the change in height of the curve $y = f(x)$ and dy represents the change in height of the tangent line when x changes by an amount $dx = \Delta x$.

We note that $\Delta y - dy$ approaches 0 faster than Δx because if we let ε denote the ratio of these quantities, then

$$\begin{aligned}\varepsilon = \frac{\Delta y - dy}{\Delta x} &= \frac{f(x + \Delta x) - f(x) - f'(x)\,\Delta x}{\Delta x}\\ &= \frac{f(x + \Delta x) - f(x)}{\Delta x} - f'(x)\\ &\to f'(x) - f'(x) = 0 \qquad \text{as } \Delta x \to 0\end{aligned}$$

Therefore, we have

$$\Delta y = dy + \varepsilon\,\Delta x \qquad \text{where } \varepsilon \to 0 \text{ as } \Delta x \to 0 \tag{4}$$

Now consider a function of two variables, $z = f(x, y)$. If x and y are given increments Δx and Δy, then the corresponding **increment** of z is

$$\Delta z = f(x + \Delta x, y + \Delta y) - f(x, y) \tag{5}$$

Thus, the increment Δz represents the change in the value of f when (x, y) changes to $(x + \Delta x, y + \Delta y)$.

The **differentials** dx and dy are independent variables; that is, they can be given any values. Then the **differential** dz, also called the **total differential**, is defined by

$$dz = f_x(x, y)\,dx + f_y(x, y)\,dy = \frac{\partial z}{\partial x}\,dx + \frac{\partial z}{\partial y}\,dy \tag{6}$$

(Compare with Equation 3.) Sometimes the notation df is used in place of dz.

If we take

$$dx = \Delta x = x - a \qquad dy = \Delta y = y - b$$

in Equation 6, then the differential of z is

$$dz = f_x(a, b)\,(x - a) + f_y(a, b)\,(y - b) \tag{7}$$

On the other hand, if f_x and f_y are continuous, we see from (2) that the equation of the tangent plane to the surface $z = f(x, y)$ at the point $(a, b, f(a, b))$ is

$$z - f(a, b) = f_x(a, b)\,(x - a) + f_y(a, b)\,(y - b) \tag{8}$$

Comparing Equations 7 and 8, we see that dz represents the change in height of the tangent plane, whereas Δz represents the change in height of the surface $z = f(x, y)$ when

(x, y) changes from (a, b) to $(a + \Delta x, b + \Delta y)$. (See Figure 4 and compare it with Figure 3.)

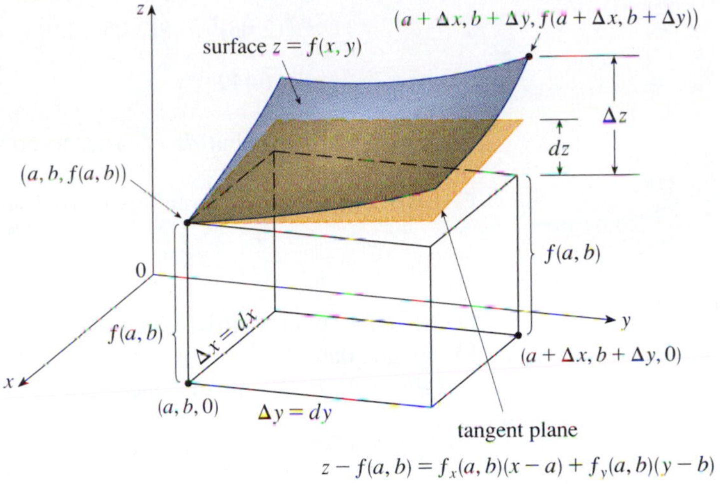

FIGURE 4

It is proved later in this section (Theorem 10) that if f_x and f_y are continuous, then

$$\Delta z - dz = \varepsilon_1 \Delta x + \varepsilon_2 \Delta y$$

where ε_1 and ε_2 are functions of Δx and Δy that approach 0 as Δx and Δy approach 0. This means that $\Delta z - dz \approx 0$ and so

$$\Delta z \approx dz$$

which says that the actual change in z is approximately equal to the differential dz when Δx and Δy are small. This allows us to estimate the value of $f(a + \Delta x, b + \Delta y)$ when $f(a, b)$ is known:

(9) $$f(a + \Delta x, b + \Delta y) \approx f(a, b) + dz$$

When we use the approximation in (9) we are using the tangent plane at $P(a, b, f(a, b))$ as an approximation to the surface $z = f(x, y)$ when (x, y) is close to (a, b).

EXAMPLE 2

(a) If $z = f(x, y) = x^2 + 3xy - y^2$, find the differential dz.
(b) If x changes from 2 to 2.05 and y changes from 3 to 2.96, compare the values of Δz and dz.

SOLUTION

(a) Definition 6 gives

$$dz = \frac{\partial z}{\partial x}\,dx + \frac{\partial z}{\partial y}\,dy = (2x + 3y)\,dx + (3x - 2y)\,dy$$

(b) Putting $x = 2$, $dx = \Delta x = 0.05$, $y = 3$, and $dy = \Delta y = -0.04$, we get

$$\begin{aligned} dz &= [2(2) + 3(3)]0.05 + [3(2) - 2(3)](-0.04) \\ &= 0.65 \end{aligned}$$

In Example 2, dz is close to Δz because the tangent plane is a good approximation to the surface $z = x^2 + 3xy - y^2$ near $(2, 3, 13)$. (See Figure 5.)

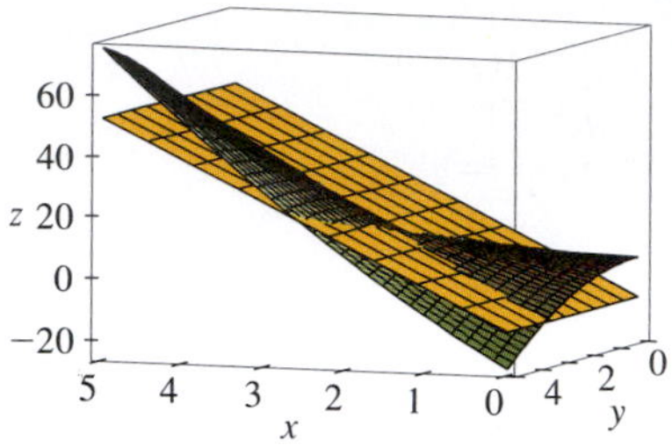

FIGURE 5

The increment of z is

$$\begin{aligned}\Delta z &= f(2.05, 2.96) - f(2, 3)\\ &= [(2.05)^2 + 3(2.05)(2.96) - (2.96)^2] - [2^2 + 3(2)(3) - 3^2]\\ &= 0.6449\end{aligned}$$

Notice that $\Delta z \approx dz$ but dz is easier to compute. ■

EXAMPLE 3 Use differentials to find an approximate value for

$$\sqrt{9(1.95)^2 + (8.1)^2}$$

SOLUTION Consider the function $z = f(x, y) = \sqrt{9x^2 + y^2}$ and observe that we can easily calculate $f(2, 8) = 10$. Therefore, we take $a = 2$, $b = 8$, $dx = \Delta x = -0.05$, and $dy = \Delta y = 0.1$ in (9). Since

$$f_x(x, y) = \frac{9x}{\sqrt{9x^2 + y^2}} \qquad \text{and} \qquad f_y(x, y) = \frac{y}{\sqrt{9x^2 + y^2}}$$

we have

$$\begin{aligned}\sqrt{9(1.95)^2 + (8.1)^2} &= f(1.95, 8.1) \approx f(2, 8) + dz\\ &= f(2, 8) + f_x(2, 8)\,dx + f_y(2, 8)\,dy\\ &= 10 + \tfrac{18}{10}(-0.05) + \tfrac{8}{10}(0.1)\\ &= 9.99\end{aligned}$$

You can check with a calculator that this approximation is accurate to two decimal places.

■

EXAMPLE 4 The base radius and height of a right circular cone are measured as 10 cm and 25 cm, respectively, with a possible error in measurement of as much as 0.1 cm in each. Use differentials to estimate the maximum error in the calculated volume of the cone.

SOLUTION The volume V of a cone with base radius r and height h is $V = \pi r^2 h/3$. So the differential of V is

$$dV = \frac{\partial V}{\partial r}\,dr + \frac{\partial V}{\partial h}\,dh = \frac{2\pi rh}{3}\,dr + \frac{\pi r^2}{3}\,dh$$

Since the errors are at most 0.1 cm, we have $|\Delta r| \leqslant 0.1$, $|\Delta h| \leqslant 0.1$. To find the largest error in the volume we take the largest error in the measurement of r and h. Therefore, we take $dr = 0.1$ and $dh = 0.1$ along with $r = 10$, $h = 25$. This gives

$$dV = \frac{500\pi}{3}(0.1) + \frac{100\pi}{3}(0.1) = 20\pi$$

Thus, the maximum error in the calculated volume is about $20\pi\ \text{cm}^3 \approx 63\ \text{cm}^3$. ■

The following theorem says that dz is a good approximation to Δz when Δx and Δy are small, provided that f_x and f_y are both continuous.

(10) THEOREM Suppose that f_x and f_y exist on a rectangular region R with sides parallel to the axes and containing the points (a, b) and $(a + \Delta x, b + \Delta y)$. Suppose that f_x and f_y are continuous at the point (a, b) and let

$$\Delta z = f(a + \Delta x, b + \Delta y) - f(a, b)$$

Then $$\Delta z = f_x(a, b)\,\Delta x + f_y(a, b)\,\Delta y + \varepsilon_1\,\Delta x + \varepsilon_2\,\Delta y$$

where ε_1 and ε_2 are functions of Δx and Δy that approach 0 as $(\Delta x, \Delta y) \to (0, 0)$.

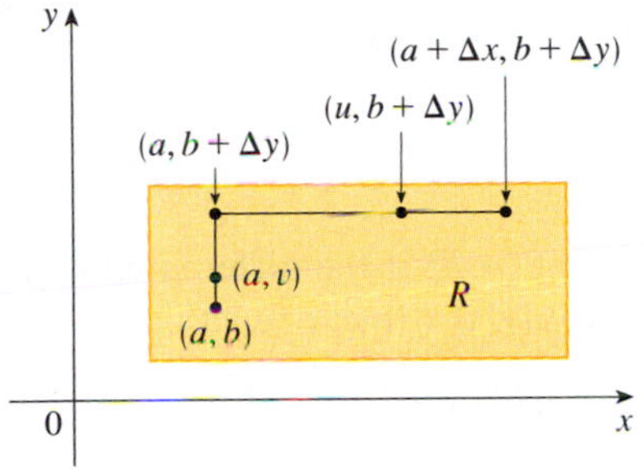

FIGURE 6

PROOF Referring to Figure 6, we write

$$\textbf{(11)}\quad \Delta z = [f(a + \Delta x, b + \Delta y) - f(a, b + \Delta y)] + [f(a, b + \Delta y) - f(a, b)]$$

Observe that the function of a single variable

$$g(x) = f(x, b + \Delta y)$$

is defined on the interval $[a, a + \Delta x]$ and $g'(x) = f_x(x, b + \Delta y)$. If we apply the Mean Value Theorem to g, we get

$$g(a + \Delta x) - g(a) = g'(u)\,\Delta x$$

where u is some number between a and $a + \Delta x$. In terms of f, this equation becomes

$$f(a + \Delta x, b + \Delta y) - f(a, b + \Delta y) = f_x(u, b + \Delta y)\,\Delta x$$

This gives us an expression for the first part of the right side of Equation 11. For the second part we let $h(y) = f(a, y)$. Then h is a function of a single variable defined on the interval $[b, b + \Delta y]$ and $h'(y) = f_y(a, y)$. A second application of the Mean Value Theorem then gives

$$h(b + \Delta y) - h(b) = h'(v)\,\Delta y$$

where v is some number between b and $b + \Delta y$. In terms of f, this becomes

$$f(a, b + \Delta y) - f(a, b) = f_y(a, v)\,\Delta y$$

We now substitute these expressions into Equation 11 and obtain

$$\begin{aligned}\Delta z &= f_x(u, b + \Delta y)\,\Delta x + f_y(a, v)\,\Delta y \\ &= f_x(a, b)\,\Delta x + [f_x(u, b + \Delta y) - f_x(a, b)]\,\Delta x + f_y(a, b)\,\Delta y \\ &\qquad + [f_y(a, v) - f_y(a, b)]\,\Delta y \\ &= f_x(a, b)\,\Delta x + f_y(a, b)\,\Delta y + \varepsilon_1\,\Delta x + \varepsilon_2\,\Delta y\end{aligned}$$

where $$\varepsilon_1 = f_x(u, b + \Delta y) - f_x(a, b)$$

$$\varepsilon_2 = f_y(a, v) - f_y(a, b)$$

Since $(u, b + \Delta y) \to (a, b)$ and $(a, v) \to (a, b)$ as $(\Delta x, \Delta y) \to (0, 0)$ and since f_x and f_y are continuous at (a, b), we see that $\varepsilon_1 \to 0$ and $\varepsilon_2 \to 0$ as $(\Delta x, \Delta y) \to (0, 0)$. □

The conclusion of Theorem 10 can be written as

$$\Delta z = dz + \varepsilon_1 \Delta x + \varepsilon_2 \Delta y$$

where ε_1 and $\varepsilon_2 \to 0$ as $(\Delta x, \Delta y) \to (0, 0)$. This is the two-dimensional version of Equation 4, which is equivalent to the differentiability of a function of a single variable. Therefore, we use the following definition for the differentiability of a function of two variables.

(12) DEFINITION If $z = f(x, y)$, then f is **differentiable** at (a, b) if Δz can be expressed in the form

$$\Delta z = f_x(a, b)\,\Delta x + f_y(a, b)\,\Delta y + \varepsilon_1 \Delta x + \varepsilon_2 \Delta y$$

where ε_1 and $\varepsilon_2 \to 0$ as $(\Delta x, \Delta y) \to (0, 0)$.

Definition 12 says that a function f is differentiable at (a, b) if the differential dz is a good approximation to the increment Δz. In other words, the linear function

$$z = f(a, b) + f_x(a, b)(x - a) + f_y(a, b)(y - b)$$

is a good approximation to the function f near (a, b).

Thus, Theorem 10 says that if f_x and f_y exist near (a, b) and are continuous at (a, b), then f is differentiable at (a, b) and the tangent plane is a good approximation to the graph of f near (a, b). In particular, polynomials and rational functions are differentiable on their domains because their partial derivatives are continuous.

It can be proved that, just as in single-variable calculus, all differentiable functions are continuous (see Exercise 39). But, unlike the situation in one-dimensional calculus, Exercise 40 gives an example of a function whose partial derivatives exist but that is not differentiable.

FUNCTIONS OF THREE OR MORE VARIABLES

Differentials and differentiability can be defined in a similar manner for functions of more than two variables. For instance, if $w = f(x, y, z)$, then the **increment** of w is

$$\Delta w = f(x + \Delta x, y + \Delta y, z + \Delta z) - f(x, y, z)$$

The **differential** dw is defined in terms of the differentials dx, dy, and dz of the independent variables by

$$dw = \frac{\partial w}{\partial x}\,dx + \frac{\partial w}{\partial y}\,dy + \frac{\partial w}{\partial z}\,dz$$

If $dx = \Delta x$, $dy = \Delta y$, and $dz = \Delta z$ are all small and f has continuous partial derivatives, then dw can be used to approximate Δw.

Differentiability can be defined by an expression similar to the one in Definition 12.

EXAMPLE 5 The dimensions of a rectangular box are measured to be 75 cm, 60 cm, and 40 cm, and each measurement is correct to within 0.2 cm. Use differentials to estimate the largest possible error when the volume of the box is calculated from these measurements.

SOLUTION If the dimensions of the box are x, y, and z, its volume is $V = xyz$ and so

$$dV = \frac{\partial V}{\partial x}\,dx + \frac{\partial V}{\partial y}\,dy + \frac{\partial V}{\partial z}\,dz = yz\,dx + xz\,dy + xy\,dz$$

We are given that $|\Delta x| \leqslant 0.2$, $|\Delta y| \leqslant 0.2$, and $|\Delta z| \leqslant 0.2$. To find the largest error in the volume, we therefore use $dx = 0.2$, $dy = 0.2$, and $dz = 0.2$ together with $x = 75$, $y = 60$, and $z = 40$:

$$\Delta V \approx dV = (60)(40)(0.2) + (75)(40)(0.2) + (75)(60)(0.2)$$
$$= 1980$$

Thus, an error of only 0.2 cm in measuring each dimension could lead to an error of as much as 1980 cm^3 in the calculated volume! ■

EXERCISES 12.4

1–6 ■ Find an equation of the tangent plane to the given surface at the specified point.

1. $z = x^2 + 4y^2$, $(2, 1, 8)$

2. $z = x^2 - y^2$, $(3, -2, 5)$

3. $z = 5 + (x - 1)^2 + (y + 2)^2$, $(2, 0, 10)$

4. $z = \sin(x + y)$, $(1, -1, 0)$

5. $z = \ln(2x + y)$, $(-1, 3, 0)$

6. $z = e^x \ln y$, $(3, 1, 0)$

7–8 ■ Use a computer to graph the surface and the tangent plane at the given point. Choose the domain and viewpoint so that you get a good view of both the surface and the tangent plane.

7. $z = xy$, $(-1, 2, -2)$ **8.** $z = \sqrt{x - y}$, $(5, 1, 2)$

CAS **9–10** ■ Draw the graph of f and its tangent plane at the given point. Use your computer algebra system both to compute the partial derivatives and to graph the surface and its tangent plane.

9. $f(x, y) = e^{-(x^2+y^2)/15}(\sin^2 x + \cos^2 y)$, $(2, 3, f(2, 3))$

10. $f(x, y) = \dfrac{\sqrt{1 + 4x^2 + 4y^2}}{1 + x^4 + y^4}$, $(1, 1, 1)$

11–18 ■ Find the differential of the function.

11. $z = x^2 y^3$ **12.** $z = ye^{xy}$

13. $u = e^x \cos xy$ **14.** $v = \ln(2x - 3y)$

15. $w = x^2 y + y^2 z$ **16.** $w = x \sin yz$

17. $w = \ln\sqrt{x^2 + y^2 + z^2}$ **18.** $w = \dfrac{x + y}{y + z}$

19. If $z = 5x^2 + y^2$ and (x, y) changes from $(1, 2)$ to $(1.05, 2.1)$, compare the values of Δz and dz.

20. If $z = x^2 - xy + 3y^2$ and (x, y) changes from $(3, -1)$ to $(2.96, -0.95)$, compare the values of Δz and dz.

21–24 ■ Use differentials to approximate the value of f at the given point.

21. $f(x, y) = \sqrt{20 - x^2 - 7y^2}$, $(1.95, 1.08)$

22. $f(x, y) = \ln(x - 3y)$, $(6.9, 2.06)$

23. $f(x, y, z) = x^2 y^3 z^4$, $(1.05, 0.9, 3.01)$

24. $f(x, y, z) = xy^2 \sin \pi z$, $(3.99, 4.98, 4.03)$

25–28 ■ Use differentials to approximate the number.

25. $8.94\sqrt{9.99 - (1.01)^3}$ **26.** $(\sqrt{99} + \sqrt[3]{124})^4$

27. $\sqrt{0.99}\, e^{0.02}$ **28.** $\sqrt{(3.02)^2 + (1.97)^2 + (5.99)^2}$

29. The length and width of a rectangle are measured as 30 cm and 24 cm, respectively, with an error in measurement of at most 0.1 cm in each. Use differentials to estimate the maximum error in the calculated area of the rectangle.

30. The dimensions of a closed rectangular box are measured as 80 cm, 60 cm, and 50 cm, respectively, with a possible error of 0.2 cm in each dimension. Use differentials to estimate the maximum error in calculating the surface area of the box.

31. Use differentials to estimate the amount of tin in a closed tin can with diameter 8 cm and height 12 cm if the tin is 0.04 cm thick.

32. Use differentials to estimate the amount of metal in a closed cylindrical can that is 10 cm high and 4 cm in diameter if the metal in the wall is 0.05 cm thick and the metal in the top and bottom is 0.1 cm thick.

33. A boundary stripe 3 in. wide is painted around a rectangle whose dimensions are 100 ft by 200 ft. Use differentials to approximate the number of square feet of paint in the stripe.

34. The pressure, volume, and temperature of a mole of an ideal gas are related by the equation $PV = 8.31T$, where P is measured in kilopascals, V in liters, and T in kelvins. Use differentials to find the approximate change in the pressure if the volume increases from 12 L to 12.3 L and the temperature decreases from 310 K to 305 K.

35. If R is the total resistance of three resistors, connected in parallel, with resistances R_1, R_2, R_3, then

$$\frac{1}{R} = \frac{1}{R_1} + \frac{1}{R_2} + \frac{1}{R_3}$$

If the resistances are measured as $R_1 = 25$ ohms, $R_2 = 40$ ohms, and $R_3 = 50$ ohms, with possible errors of 0.5% in each case, estimate the maximum error in the calculated value of R.

36. Four positive numbers, each less than 50, are rounded to the first decimal place and then multiplied together. Use differentials to estimate the maximum possible error in the computed product that might result from the rounding.

37–38 ■ Show that the function is differentiable by finding values of ε_1 and ε_2 that satisfy Definition 12.

37. $f(x, y) = x^2 + y^2$

38. $f(x, y) = xy - 5y^2$

39. Prove that if f is a function of two variables that is differentiable at (a, b), then f is continuous at (a, b). *Hint:* Show that

$$\lim_{(\Delta x, \Delta y) \to (0, 0)} f(a + \Delta x, b + \Delta y) = f(a, b)$$

40. The function

$$f(x, y) = \begin{cases} \dfrac{xy}{x^2 + y^2} & \text{if } (x, y) \neq (0, 0) \\ 0 & \text{if } (x, y) = (0, 0) \end{cases}$$

was graphed in Figure 8 in Section 12.2. Show that $f_x(0, 0)$ and $f_y(0, 0)$ both exist but f is not differentiable at $(0, 0)$. [*Hint:* Use the result of Exercise 39.]

12.5 THE CHAIN RULE

We recall that the Chain Rule for functions of a single variable gives the rule for differentiating a composite function: If $y = f(x)$ and $x = g(t)$, where f and g are differentiable functions, then y is indirectly a differentiable function of t and

(1)
$$\frac{dy}{dt} = \frac{dy}{dx} \frac{dx}{dt}$$

For functions of more than one variable, the Chain Rule has several versions, each of them giving a rule for differentiating a composite function. The first version (Theorem 2) deals with the case where $z = f(x, y)$ and each of the variables x and y is in turn a function of a variable t. This means that z is indirectly a function of t $[z = f(g(t), h(t))]$ and the Chain Rule gives a formula for differentiating z as a function of t. We assume that f is differentiable (Definition 12.4.12). Recall from Theorem 12.4.10 that this is the case when f_x and f_y are continuous.

(2) THE CHAIN RULE (CASE 1) Suppose that $z = f(x, y)$ is a differentiable function of x and y, where $x = g(t)$ and $y = h(t)$ are both differentiable functions of t. Then z is a differentiable function of t and

$$\frac{dz}{dt} = \frac{\partial f}{\partial x} \frac{dx}{dt} + \frac{\partial f}{\partial y} \frac{dy}{dt}$$

PROOF A change of Δt in t produces changes of Δx in x and Δy in y. These, in turn, produce a change of Δz in z, and from Definition 12.4.12 we have

$$\Delta z = \frac{\partial f}{\partial x} \Delta x + \frac{\partial f}{\partial y} \Delta y + \varepsilon_1 \Delta x + \varepsilon_2 \Delta y$$

where $\varepsilon_1 \to 0$ and $\varepsilon_2 \to 0$ as $(\Delta x, \Delta y) \to (0, 0)$. [If the functions ε_1 and ε_2 are not defined at $(0, 0)$, we can define them to be 0 there.] Dividing both sides of this equation by Δt, we have

$$\frac{\Delta z}{\Delta t} = \frac{\partial f}{\partial x} \frac{\Delta x}{\Delta t} + \frac{\partial f}{\partial y} \frac{\Delta y}{\Delta t} + \varepsilon_1 \frac{\Delta x}{\Delta t} + \varepsilon_2 \frac{\Delta y}{\Delta t}$$

If we now let $\Delta t \to 0$, then $\Delta x = g(t + \Delta t) - g(t) \to 0$ because g is differentiable and therefore continuous. Similarly, $\Delta y \to 0$. This, in turn, means that $\varepsilon_1 \to 0$ and $\varepsilon_2 \to 0$, so

$$\begin{aligned}
\frac{dz}{dt} &= \lim_{\Delta t \to 0} \frac{\Delta z}{\Delta t} \\
&= \frac{\partial f}{\partial x} \lim_{\Delta t \to 0} \frac{\Delta x}{\Delta t} + \frac{\partial f}{\partial y} \lim_{\Delta t \to 0} \frac{\Delta y}{\Delta t} + \lim_{\Delta t \to 0} \varepsilon_1 \lim_{\Delta t \to 0} \frac{\Delta x}{\Delta t} + \lim_{\Delta t \to 0} \varepsilon_2 \lim_{\Delta t \to 0} \frac{\Delta y}{\Delta t} \\
&= \frac{\partial f}{\partial x} \frac{dx}{dt} + \frac{\partial f}{\partial y} \frac{dy}{dt} + 0 \cdot \frac{dx}{dt} + 0 \cdot \frac{dy}{dt} \\
&= \frac{\partial f}{\partial x} \frac{dx}{dt} + \frac{\partial f}{\partial y} \frac{dy}{dt}
\end{aligned}$$

□

Since we often write $\partial z/\partial x$ in place of $\partial f/\partial x$, we can rewrite the Chain Rule in the form

Notice the similarity to the definition of the differential:

$$dz = \frac{\partial z}{\partial x} dx + \frac{\partial z}{\partial y} dy$$

$$\frac{dz}{dt} = \frac{\partial z}{\partial x} \frac{dx}{dt} + \frac{\partial z}{\partial y} \frac{dy}{dt}$$

EXAMPLE 1 If $z = x^2y + 3xy^4$, where $x = e^t$ and $y = \sin t$, find dz/dt.

SOLUTION The Chain Rule gives

$$\begin{aligned}
\frac{dz}{dt} &= \frac{\partial z}{\partial x} \frac{dx}{dt} + \frac{\partial z}{\partial y} \frac{dy}{dt} \\
&= (2xy + 3y^4)e^t + (x^2 + 12xy^3)\cos t \\
&= (2e^t \sin t + 3\sin^4 t)e^t + (e^{2t} + 12e^t \sin^3 t)\cos t
\end{aligned}$$

■

NOTE 1: Although we have expressed the answer to Example 1 totally in terms of t, the expression in terms of x, y, and t is adequate for some purposes. For instance, if we had been asked to find the value of dz/dt when $t = 0$, we could simply observe that $x = 1$ and $y = 0$ when $t = 0$ and so

$$\left.\frac{dz}{dt}\right|_{t=0} = 0e^0 + 1\cos 0 = 1$$

NOTE 2: If $T(x, y)$ represents the temperature at a point (x, y) and $x = f(t)$ and $y = g(t)$ are the parametric equations of a curve C, then the composite function $z = T(f(t), g(t))$ represents the temperature at points on C and the derivative dz/dt represents the rate of change of temperature along the curve.

EXAMPLE 2 The pressure P (in kilopascals), volume V (in liters), and temperature T (in kelvins) of a mole of an ideal gas are related by the equation $PV = 8.31T$. Find the rate at which the pressure is changing when the temperature is 300 K and increasing at a rate of 0.1 K/s and the volume is 100 L and increasing at a rate of 0.2 L/s.

SOLUTION If t represents the time elapsed in seconds, then at the given instant we have $T = 300$, $dT/dt = 0.1$, $V = 100$, $dV/dt = 0.2$. Since

$$P = 8.31 \frac{T}{V}$$

the Chain Rule gives

$$\frac{dP}{dt} = \frac{\partial P}{\partial T}\frac{dT}{dt} + \frac{\partial P}{\partial V}\frac{dV}{dt} = \frac{8.31}{V}\frac{dT}{dt} - \frac{8.31T}{V^2}\frac{dV}{dt}$$

$$= \frac{8.31}{100}(0.1) - \frac{8.31(300)}{100^2}(0.2)$$

$$= -0.04155$$

The pressure is decreasing at a rate of about 0.042 kPa/s. ■

We now consider the situation where $z = f(x, y)$ but each of x and y is a function of two variables s and t: $x = g(s, t)$, $y = h(s, t)$. Then z is indirectly a function of s and t and we wish to find $\partial z/\partial s$ and $\partial z/\partial t$. Recall that in computing $\partial z/\partial t$ we hold s fixed and compute the ordinary derivative of z with respect to t. Therefore, we can apply Theorem 2 to obtain

$$\frac{\partial z}{\partial t} = \frac{\partial z}{\partial x}\frac{\partial x}{\partial t} + \frac{\partial z}{\partial y}\frac{\partial y}{\partial t}$$

A similar argument holds for $\partial z/\partial s$ and so we have proved the following version of the Chain Rule.

(3) THE CHAIN RULE (CASE 2) Suppose that $z = f(x, y)$ is a differentiable function of x and y, where $x = g(s, t)$, $y = h(s, t)$, and the partial derivatives g_s, g_t, h_s, and h_t exist. Then

$$\frac{\partial z}{\partial s} = \frac{\partial z}{\partial x}\frac{\partial x}{\partial s} + \frac{\partial z}{\partial y}\frac{\partial y}{\partial s}$$

$$\frac{\partial z}{\partial t} = \frac{\partial z}{\partial x}\frac{\partial x}{\partial t} + \frac{\partial z}{\partial y}\frac{\partial y}{\partial t}$$

EXAMPLE 3 If $z = e^x \sin y$, where $x = st^2$ and $y = s^2t$, find $\partial z/\partial s$ and $\partial z/\partial t$.

SOLUTION Applying Case 2 of the Chain Rule, we get

$$\frac{\partial z}{\partial s} = \frac{\partial z}{\partial x}\frac{\partial x}{\partial s} + \frac{\partial z}{\partial y}\frac{\partial y}{\partial s} = (e^x \sin y)(t^2) + (e^x \cos y)(2st)$$

$$= t^2 e^{st^2}\sin(s^2t) + 2ste^{st^2}\cos(s^2t)$$

$$\frac{\partial z}{\partial t} = \frac{\partial z}{\partial x}\frac{\partial x}{\partial t} + \frac{\partial z}{\partial y}\frac{\partial y}{\partial t} = (e^x \sin y)(2st) + (e^x \cos y)(s^2)$$

$$= 2ste^{st^2}\sin(s^2t) + s^2 e^{st^2}\cos(s^2t)$$ ■

Case 2 of the Chain Rule contains three types of variables: s and t are **independent** variables, x and y are called **intermediate** variables, and z is the **dependent** variable. Notice that Theorem 3 has one term for each intermediate variable and each of these terms resembles the one-dimensional Chain Rule in Equation 1.

To remember the Chain Rule it is helpful to draw the **tree diagram** in Figure 1. We draw branches from the dependent variable z to the intermediate variables x and y to indicate that z is a function of x and y. Then we draw branches from x and y to the independent variables s and t. On each branch we write the corresponding partial derivative. To find $\partial z/\partial s$ we find the product of the partial derivatives along each path from z to s and then add these products:

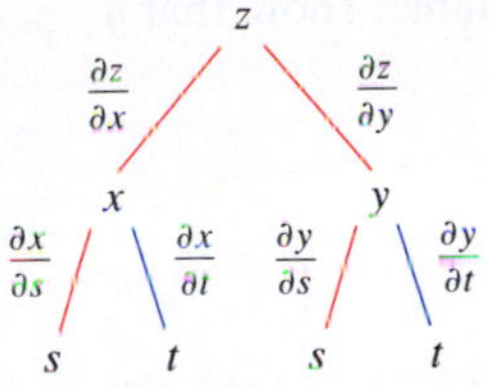

FIGURE 1

$$\frac{\partial z}{\partial s} = \frac{\partial z}{\partial x}\frac{\partial x}{\partial s} + \frac{\partial z}{\partial y}\frac{\partial y}{\partial s}$$

Similarly, we find $\partial z/\partial t$ by using the paths from z to t.

Now we consider the general situation in which a dependent variable u is a function of n intermediate variables $x_1, \ldots, x_n$, each of which is, in turn, a function of m independent variables $t_1, \ldots, t_m$. Notice that there are n terms, one for each intermediate variable. The proof is similar to that of Case 1.

(4) THE CHAIN RULE (GENERAL VERSION) Suppose that u is a differentiable function of the n variables $x_1, x_2, \ldots, x_n$ and each x_j is a function of the m variables $t_1, t_2, \ldots, t_m$ such that all the partial derivatives $\partial x_j/\partial t_i$ exist ($j = 1, 2, \ldots, n$; $i = 1, 2, \ldots, m$). Then u is a function of $t_1, t_2, \ldots, t_m$ and

$$\frac{\partial u}{\partial t_i} = \frac{\partial u}{\partial x_1}\frac{\partial x_1}{\partial t_i} + \frac{\partial u}{\partial x_2}\frac{\partial x_2}{\partial t_i} + \cdots + \frac{\partial u}{\partial x_n}\frac{\partial x_n}{\partial t_i}$$

for each $i = 1, 2, \ldots, m$.

EXAMPLE 4 Write out the Chain Rule for the case where $w = f(x, y, z, t)$ and $x = x(u, v)$, $y = y(u, v)$, $z = z(u, v)$, and $t = t(u, v)$.

SOLUTION We apply Theorem 4 with $n = 4$ and $m = 2$. Figure 2 shows the tree diagram. Although we have not written the derivatives on the branches, it is understood that if a branch leads from y to u, then the partial derivative for that branch is $\partial y/\partial u$. With the aid of the tree diagram we can now write the required expressions:

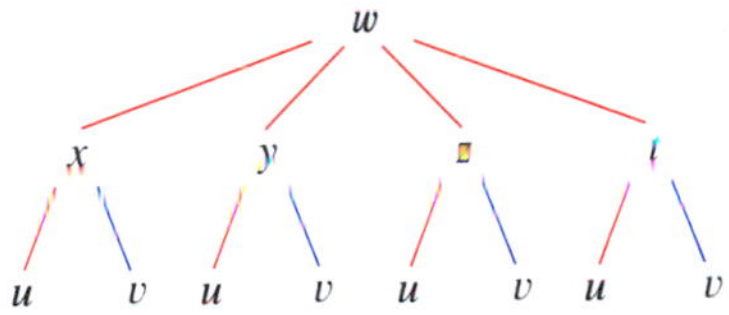

FIGURE 2

$$\frac{\partial w}{\partial u} = \frac{\partial w}{\partial x}\frac{\partial x}{\partial u} + \frac{\partial w}{\partial y}\frac{\partial y}{\partial u} + \frac{\partial w}{\partial z}\frac{\partial z}{\partial u} + \frac{\partial w}{\partial t}\frac{\partial t}{\partial u}$$

$$\frac{\partial w}{\partial v} = \frac{\partial w}{\partial x}\frac{\partial x}{\partial v} + \frac{\partial w}{\partial y}\frac{\partial y}{\partial v} + \frac{\partial w}{\partial z}\frac{\partial z}{\partial v} + \frac{\partial w}{\partial t}\frac{\partial t}{\partial v}$$ ■

EXAMPLE 5 If $u = x^4y + y^2z^3$, where $x = rse^t$, $y = rs^2e^{-t}$, and $z = r^2s\sin t$, find the value of $\partial u/\partial s$ when $r = 2$, $s = 1$, $t = 0$.

SOLUTION With the help of the tree diagram in Figure 3, we have

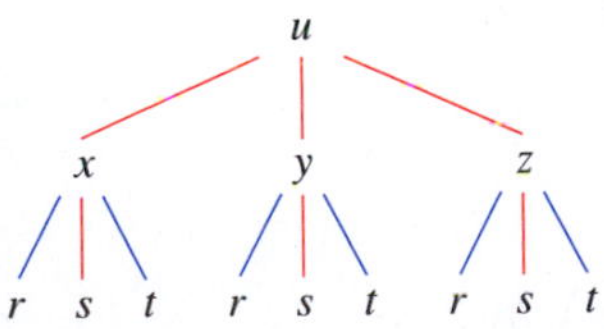

FIGURE 3

$$\frac{\partial u}{\partial s} = \frac{\partial u}{\partial x}\frac{\partial x}{\partial s} + \frac{\partial u}{\partial y}\frac{\partial y}{\partial s} + \frac{\partial u}{\partial z}\frac{\partial z}{\partial s}$$

$$= (4x^3y)(re^t) + (x^4 + 2yz^3)(2rse^{-t}) + (3y^2z^2)(r^2\sin t)$$

When $r = 2$, $s = 1$, and $t = 0$, we have $x = 2$, $y = 2$, and $z = 0$, so

$$\frac{\partial u}{\partial s} = (64)(2) + (16)(4) + (0)(0) = 192$$ ■

EXAMPLE 6 If $g(s,t) = f(s^2 - t^2, t^2 - s^2)$ and f is differentiable, show that g satisfies the equation

$$t\frac{\partial g}{\partial s} + s\frac{\partial g}{\partial t} = 0$$

SOLUTION Let $x = s^2 - t^2$ and $y = t^2 - s^2$. Then $g(s,t) = f(x,y)$ and the Chain Rule gives

$$\frac{\partial g}{\partial s} = \frac{\partial f}{\partial x}\frac{\partial x}{\partial s} + \frac{\partial f}{\partial y}\frac{\partial y}{\partial s} = \frac{\partial f}{\partial x}(2s) + \frac{\partial f}{\partial y}(-2s)$$

$$\frac{\partial g}{\partial t} = \frac{\partial f}{\partial x}\frac{\partial x}{\partial t} + \frac{\partial f}{\partial y}\frac{\partial y}{\partial t} = \frac{\partial f}{\partial x}(-2t) + \frac{\partial f}{\partial y}(2t)$$

Therefore

$$t\frac{\partial g}{\partial s} + s\frac{\partial g}{\partial t} = \left(2st\frac{\partial f}{\partial x} - 2st\frac{\partial f}{\partial y}\right) + \left(-2st\frac{\partial f}{\partial x} + 2st\frac{\partial f}{\partial y}\right) = 0$$ ■

EXAMPLE 7 If $z = f(x,y)$ has continuous second-order partial derivatives and $x = r^2 + s^2$ and $y = 2rs$, find (a) $\partial z/\partial r$ and (b) $\partial^2 z/\partial r^2$.

SOLUTION

(a) The Chain Rule gives

$$\frac{\partial z}{\partial r} = \frac{\partial z}{\partial x}\frac{\partial x}{\partial r} + \frac{\partial z}{\partial y}\frac{\partial y}{\partial r} = \frac{\partial z}{\partial x}(2r) + \frac{\partial z}{\partial y}(2s)$$

(b) Applying the Product Rule to the expression in part (a), we get

(5)
$$\frac{\partial^2 z}{\partial r^2} = \frac{\partial}{\partial r}\left(2r\frac{\partial z}{\partial x} + 2s\frac{\partial z}{\partial y}\right)$$
$$= 2\frac{\partial z}{\partial x} + 2r\frac{\partial}{\partial r}\left(\frac{\partial z}{\partial x}\right) + 2s\frac{\partial}{\partial r}\left(\frac{\partial z}{\partial y}\right)$$

But, using the Chain Rule again, we have

$$\frac{\partial}{\partial r}\left(\frac{\partial z}{\partial x}\right) = \frac{\partial}{\partial x}\left(\frac{\partial z}{\partial x}\right)\frac{\partial x}{\partial r} + \frac{\partial}{\partial y}\left(\frac{\partial z}{\partial x}\right)\frac{\partial y}{\partial r}$$
$$= \frac{\partial^2 z}{\partial x^2}(2r) + \frac{\partial^2 z}{\partial y\,\partial x}(2s)$$

$$\frac{\partial}{\partial r}\left(\frac{\partial z}{\partial y}\right) = \frac{\partial}{\partial x}\left(\frac{\partial z}{\partial y}\right)\frac{\partial x}{\partial r} + \frac{\partial}{\partial y}\left(\frac{\partial z}{\partial y}\right)\frac{\partial y}{\partial r}$$
$$= \frac{\partial^2 z}{\partial x\,\partial y}(2r) + \frac{\partial^2 z}{\partial y^2}(2s)$$

Putting these expressions into Equation 5 and using the equality of the mixed second-order derivatives, we obtain

$$\frac{\partial^2 z}{\partial r^2} = 2\frac{\partial z}{\partial x} + 2r\left(2r\frac{\partial^2 z}{\partial x^2} + 2s\frac{\partial^2 z}{\partial y\,\partial x}\right) + 2s\left(2r\frac{\partial^2 z}{\partial x\,\partial y} + 2s\frac{\partial^2 z}{\partial y^2}\right)$$

$$= 2\frac{\partial z}{\partial x} + 4r^2\frac{\partial^2 z}{\partial x^2} + 8rs\frac{\partial^2 z}{\partial x\,\partial y} + 4s^2\frac{\partial^2 z}{\partial y^2}$$ ■

IMPLICIT DIFFERENTIATION

The Chain Rule can be used to give a more complete description of the process of implicit differentiation that was introduced in Sections 3.6 and 12.3. We suppose that an equation of the form $F(x, y) = 0$ defines y implicitly as a differentiable function of x, that is, $y = f(x)$, where $F(x, f(x)) = 0$ for all x in the domain of f. If F is differentiable, we can apply Case 1 of the Chain Rule to differentiate both sides of the equation $F(x, y) = 0$ with respect to x. Since both x and y are functions of x, we obtain

$$\frac{\partial F}{\partial x}\frac{dx}{dx} + \frac{\partial F}{\partial y}\frac{dy}{dx} = 0$$

But $dx/dx = 1$, so if $\partial F/\partial y \neq 0$ we solve for dy/dx and obtain

(6)
$$\frac{dy}{dx} = -\frac{\dfrac{\partial F}{\partial x}}{\dfrac{\partial F}{\partial y}} = -\frac{F_x}{F_y}$$

To derive this equation we assumed that $F(x, y) = 0$ defines y implicitly as a function of x. The **Implicit Function Theorem**, proved in advanced calculus, gives conditions under which this assumption is valid. It states that if F is defined on an open disk containing (a, b), where $F(a, b) = 0$, $F_y(a, b) \neq 0$, and F_x and F_y are continuous on the disk, then the equation $F(x, y) = 0$ defines y as a function of x near the point (a, b) and the derivative of this function is given by Equation 6.

EXAMPLE 8 Find y' if $x^3 + y^3 = 6xy$.

SOLUTION The given equation can be written as

$$F(x, y) = x^3 + y^3 - 6xy = 0$$

so Equation 6 gives

$$\frac{dy}{dx} = -\frac{F_x}{F_y} = -\frac{3x^2 - 6y}{3y^2 - 6x} = -\frac{x^2 - 2y}{y^2 - 2x}$$ ■

The solution to Example 8 should be compared to the one in Example 2 in Section 3.6.

Now we suppose that z is given implicitly as a function $z = f(x, y)$ by an equation of the form $F(x, y, z) = 0$. This means that $F(x, y, f(x, y)) = 0$ for all (x, y) in the domain of f. If F is differentiable and f_x and f_y exist, then we can use the Chain Rule to differentiate the equation $F(x, y, z) = 0$ as follows:

$$\frac{\partial F}{\partial x}\frac{\partial x}{\partial x} + \frac{\partial F}{\partial y}\frac{\partial y}{\partial x} + \frac{\partial F}{\partial z}\frac{\partial z}{\partial x} = 0$$

But $\dfrac{\partial}{\partial x}(x) = 1$ and $\dfrac{\partial}{\partial x}(y) = 0$

so this equation becomes

$$\frac{\partial F}{\partial x} + \frac{\partial F}{\partial z}\frac{\partial z}{\partial x} = 0$$

If $\partial F/\partial z \neq 0$, we solve for $\partial z/\partial x$ and obtain the first formula in Equations 7. The formula for $\partial z/\partial y$ is obtained in a similar manner.

(7)
$$\frac{\partial z}{\partial x} = -\frac{\dfrac{\partial F}{\partial x}}{\dfrac{\partial F}{\partial z}} \qquad \frac{\partial z}{\partial y} = -\frac{\dfrac{\partial F}{\partial y}}{\dfrac{\partial F}{\partial z}}$$

Again, a version of the **Implicit Function Theorem** gives conditions under which our assumption is valid. If F is defined within a sphere containing (a, b, c), where $F(a, b, c) = 0$, $F_z(a, b, c) \neq 0$, and F_x, F_y, and F_z are continuous inside the sphere, then the equation $F(x, y, z) = 0$ defines z as a function of x and y near the point (a, b, c) and the partial derivatives of this function are given by (7).

EXAMPLE 9 Find $\dfrac{\partial z}{\partial x}$ and $\dfrac{\partial z}{\partial y}$ if $x^3 + y^3 + z^3 + 6xyz = 1$.

SOLUTION Let $F(x, y, z) = x^3 + y^3 + z^3 + 6xyz - 1$. Then, from Equations 7, we have

$$\frac{\partial z}{\partial x} = -\frac{F_x}{F_z} = -\frac{3x^2 + 6yz}{3z^2 + 6xy} = -\frac{x^2 + 2yz}{z^2 + 2xy}$$

$$\frac{\partial z}{\partial y} = -\frac{F_y}{F_z} = -\frac{3y^2 + 6xz}{3z^2 + 6xy} = -\frac{y^2 + 2xz}{z^2 + 2xy}$$ ■

The solution to Example 9 should be compared to the one in Example 4 in Section 12.3.

EXERCISES 12.5

1–6 ■ Use the Chain Rule to find dz/dt or dw/dt.

1. $z = x^2 + y^2$, $x = t^3$, $y = 1 + t^2$

2. $z = x^2y^3$, $x = 1 + \sqrt{t}$, $y = 1 - \sqrt{t}$

3. $z = \ln(x + y^2)$, $x = \sqrt{1 + t}$, $y = 1 + \sqrt{t}$

4. $z = xe^{x/y}$, $x = \cos t$, $y = e^{2t}$

5. $w = xy^2z^3$, $x = \sin t$, $y = \cos t$, $z = 1 + e^{2t}$

6. $w = \dfrac{x}{y} + \dfrac{y}{z}$, $x = \sqrt{t}$, $y = \cos 2t$, $z = e^{-3t}$

7–12 ■ Use the Chain Rule to find $\partial z/\partial s$ and $\partial z/\partial t$.

7. $z = x^2 \sin y$, $x = s^2 + t^2$, $y = 2st$

8. $z = \sin x \cos y$, $x = (s - t)^2$, $y = s^2 - t^2$

9. $z = x^2 - 3x^2y^3$, $x = se^t$, $y = se^{-t}$

10. $z = x \tan^{-1}(xy)$, $x = t^2$, $y = se^t$

11. $z = 2^{x-3y}$, $x = s^2t$, $y = st^2$

12. $z = xe^y + ye^{-x}$, $x = e^t$, $y = st^2$

13–16 ■ Write out the Chain Rule for the given case (as in Example 4).

13. $u = f(x, y)$, where $x = x(r, s, t)$, $y = y(r, s, t)$

14. $w = f(x, y, z)$, where $x = x(t, u)$, $y = y(t, u)$, $z = z(t, u)$

15. $v = f(p, q, r)$, where $p = p(x, y, z)$, $q = q(x, y, z)$, $r = r(x, y, z)$

16. $u = f(s, t)$, where $s = s(w, x, y, z)$, $t = t(w, x, y, z)$

17–22 ■ Use the Chain Rule to find the indicated partial derivatives.

17. $w = x^2 + y^2 + z^2$, $x = st$, $y = s\cos t$, $z = s\sin t$;
$\dfrac{\partial w}{\partial s}, \dfrac{\partial w}{\partial t}$ when $s = 1$, $t = 0$

18. $u = xy + yz + zx$, $x = st$, $y = e^{st}$, $z = t^2$;
$\dfrac{\partial u}{\partial s}, \dfrac{\partial u}{\partial t}$ when $s = 0$, $t = 1$

19. $z = y^2 \tan x$, $x = t^2uv$, $y = u + tv^2$;
$\dfrac{\partial z}{\partial t}, \dfrac{\partial z}{\partial u}, \dfrac{\partial z}{\partial v}$ when $t = 2$, $u = 1$, $v = 0$

20. $z = \dfrac{x}{y}$, $x = re^{st}$, $y = rse^t$;
$\dfrac{\partial z}{\partial r}, \dfrac{\partial z}{\partial s}, \dfrac{\partial z}{\partial t}$ when $r = 1$, $s = 2$, $t = 0$

21. $u = \dfrac{x + y}{y + z}$, $x = p + r + t$, $y = p - r + t$,
$z = p + r - t$; $\dfrac{\partial u}{\partial p}, \dfrac{\partial u}{\partial r}, \dfrac{\partial u}{\partial t}$

22. $t = z\sec(xy)$, $x = uv$, $y = vw$, $z = wu$; $\dfrac{\partial t}{\partial u}, \dfrac{\partial t}{\partial v}, \dfrac{\partial t}{\partial w}$

23–26 ■ Use Equation 6 to find dy/dx.

23. $x^2 - xy + y^3 = 8$

24. $y^5 + 3x^2y^2 + 5x^4 = 12$

25. $2y^2 + \sqrt[3]{xy} = 3x^2 + 17$

26. $x\cos y + y\cos x = 1$

27–32 ■ Use Equations 7 to find $\partial z/\partial x$ and $\partial z/\partial y$.

27. $xy + yz - xz = 0$

28. $xyz = \cos(x + y + z)$

29. $x^2 + y^2 - z^2 = 2x(y + z)$

30. $xy^2z^3 + x^3y^2z = x + y + z$

31. $xe^y + yz + ze^x = 0$

32. $y^2ze^{x+y} - \sin(xyz) = 0$

33. The radius of a right circular cylinder is decreasing at a rate of 1.2 cm/s while its height is increasing at a rate of 3 cm/s. At what rate is the volume of the cylinder changing when the radius is 80 cm and the height is 150 cm?

34. The radius of a right circular cone is increasing at a rate of 1.8 in/s while its height is decreasing at a rate of 2.5 in/s. At what rate is the volume of the cone changing when the radius is 120 in. and the height is 140 in.?

35. The length ℓ, width w, and height h of a box change with time. At a certain instant the dimensions are $\ell = 1$ m and $w = h = 2$ m, and ℓ and w are increasing at a rate of 2 m/s while h is decreasing at a rate of 3 m/s. At that instant find the rates at which the following quantities are changing.
(a) The volume
(b) The surface area
(c) The length of a diagonal

36. The voltage V in a simple electrical circuit is slowly decreasing as the battery wears out. The resistance R is slowly increasing as the resistor heats up. Use Ohm's Law, $V = IR$, to find how the current I is changing at the moment when $R = 400\,\Omega$, $I = 0.08$ A, $dV/dt = -0.01$ V/s, and $dR/dt = 0.03\,\Omega$/s.

37. The pressure of 1 mole of an ideal gas is increasing at a rate of 0.05 kPa/s and the temperature is increasing at a rate of 0.15 K/s. Use the equation in Example 2 to find the rate of change of the volume when the pressure is 20 kPa and the temperature is 320 K.

38. Car A is traveling north on Highway 16 at 90 km/h. Car B is traveling west on Highway 83 at 80 km/h. Each car is approaching the intersection of these highways. How fast is the distance between the cars changing when car A is 0.3 km from the intersection and car B is 0.4 km from the intersection?

39–42 ■ Assume that all the given functions are differentiable.

39. If $z = f(x, y)$, where $x = r\cos\theta$ and $y = r\sin\theta$, (a) find $\partial z/\partial r$ and $\partial z/\partial\theta$ and (b) show that

$$\left(\frac{\partial z}{\partial x}\right)^2 + \left(\frac{\partial z}{\partial y}\right)^2 = \left(\frac{\partial z}{\partial r}\right)^2 + \frac{1}{r^2}\left(\frac{\partial z}{\partial \theta}\right)^2$$

40. If $u = f(x, y)$, where $x = e^s\cos t$ and $y = e^s\sin t$, show that

$$\left(\frac{\partial u}{\partial x}\right)^2 + \left(\frac{\partial u}{\partial y}\right)^2 = e^{-2s}\left[\left(\frac{\partial u}{\partial s}\right)^2 + \left(\frac{\partial u}{\partial t}\right)^2\right]$$

41. If $z = f(x - y)$, show that $\dfrac{\partial z}{\partial x} + \dfrac{\partial z}{\partial y} = 0$.

42. If $z = f(x, y)$, where $x = s + t$ and $y = s - t$, show that

$$\left(\frac{\partial z}{\partial x}\right)^2 - \left(\frac{\partial z}{\partial y}\right)^2 = \frac{\partial z}{\partial s}\frac{\partial z}{\partial t}$$

43–48 ■ Assume that all the given functions have continuous second-order partial derivatives.

43. Show that any function of the form

$$z = f(x + at) + g(x - at)$$

is a solution of the wave equation

$$\frac{\partial^2 z}{\partial t^2} = a^2\frac{\partial^2 z}{\partial x^2}$$

[*Hint:* Let $u = x + at$, $v = x - at$.]

44. If $u = f(x, y)$, where $x = e^s\cos t$ and $y = e^s\sin t$, show that

$$\frac{\partial^2 u}{\partial x^2} + \frac{\partial^2 u}{\partial y^2} = e^{-2s}\left[\frac{\partial^2 u}{\partial s^2} + \frac{\partial^2 u}{\partial t^2}\right]$$

45. If $z = f(x, y)$, where $x = r^2 + s^2$, $y = 2rs$, find $\partial^2 z/\partial r\,\partial s$. (Compare with Example 7.)

46. If $z = f(x, y)$, where $x = r\cos\theta$, $y = r\sin\theta$, find (a) $\partial z/\partial r$, (b) $\partial z/\partial\theta$, and (c) $\partial^2 z/\partial r\,\partial\theta$.

47. If $z = f(x, y)$, where $x = r\cos\theta$, $y = r\sin\theta$, show that

$$\frac{\partial^2 z}{\partial x^2} + \frac{\partial^2 z}{\partial y^2} = \frac{\partial^2 z}{\partial r^2} + \frac{1}{r^2}\frac{\partial^2 z}{\partial\theta^2} + \frac{1}{r}\frac{\partial z}{\partial r}$$

48. Suppose $z = f(x, y)$, where $x = g(s, t)$ and $y = h(s, t)$.

(a) Show that

$$\frac{\partial^2 z}{\partial t^2} = \frac{\partial^2 z}{\partial x^2}\left(\frac{\partial x}{\partial t}\right)^2 + 2\frac{\partial^2 z}{\partial x\,\partial y}\frac{\partial x}{\partial t}\frac{\partial y}{\partial t} + \frac{\partial^2 z}{\partial y^2}\left(\frac{\partial y}{\partial t}\right)^2 + \frac{\partial z}{\partial x}\frac{\partial^2 x}{\partial t^2} + \frac{\partial z}{\partial y}\frac{\partial^2 y}{\partial t^2}$$

(b) Find a similar formula for $\partial^2 z/\partial s\,\partial t$.

49–51 ■ Each function f is **homogeneous of degree n**; that is, f satisfies the equation $f(tx, ty) = t^n f(x, y)$ for all t, where n is a positive integer and f has continuous second-order partial derivatives. Verify that f satisfies the equation.

49. $x\dfrac{\partial f}{\partial x} + y\dfrac{\partial f}{\partial y} = nf(x, y)$

50. $x^2\dfrac{\partial^2 f}{\partial x^2} + 2xy\dfrac{\partial^2 f}{\partial x\,\partial y} + y^2\dfrac{\partial^2 f}{\partial y^2} = n(n-1)f(x, y)$

51. $f_x(tx, ty) = t^{n-1}f_x(x, y)$

52. Suppose that the equation $F(x, y, z) = 0$ implicitly defines each of the three variables x, y, and z as functions of the other two: $z = f(x, y)$, $y = g(x, z)$, $x = h(y, z)$. If F is differentiable and F_x, F_y, and F_z are all nonzero, show that

$$\frac{\partial z}{\partial x}\frac{\partial x}{\partial y}\frac{\partial y}{\partial z} = -1$$

12.6 DIRECTIONAL DERIVATIVES AND THE GRADIENT VECTOR

Recall that if $z = f(x, y)$, then the partial derivatives f_x and f_y are defined as

(1)
$$f_x(x_0, y_0) = \lim_{h\to 0}\frac{f(x_0 + h, y_0) - f(x_0, y_0)}{h}$$

$$f_y(x_0, y_0) = \lim_{h\to 0}\frac{f(x_0, y_0 + h) - f(x_0, y_0)}{h}$$

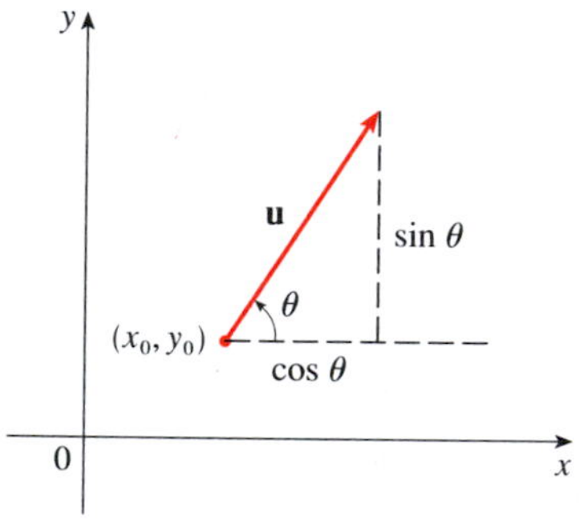

FIGURE 1
A unit vector $\mathbf{u} = \langle a, b\rangle = \langle\cos\theta, \sin\theta\rangle$

and represent the rates of change of z in the x- and y-directions, that is, in the directions of the unit vectors **i** and **j**.

Suppose that we now wish to find the rate of change of z at (x_0, y_0) in the direction of an arbitrary unit vector $\mathbf{u} = \langle a, b\rangle$ (see Figure 1). To do this we consider the surface S with equation $z = f(x, y)$ (the graph of f) and we let $z_0 = f(x_0, y_0)$. Then the point $P(x_0, y_0, z_0)$ lies on S. The vertical plane that passes through P in the direction **u** intersects S in a curve C (see Figure 2). The slope of the tangent line T to C at P is the rate of change of z in the direction of **u**.

If $Q(x, y, z)$ is another point on C and P', Q' are the projections of P, Q on the xy-plane, then the vector $\overrightarrow{P'Q'}$ is parallel to **u** and so

$$\overrightarrow{P'Q'} = h\mathbf{u} = \langle ha, hb\rangle$$

for some scalar h. Therefore, $x - x_0 = ha$, $y - y_0 = hb$, so $x = x_0 + ha$, $y = y_0 + hb$, and

$$\frac{\Delta z}{h} = \frac{z - z_0}{h} = \frac{f(x_0 + ha, y_0 + hb) - f(x_0, y_0)}{h}$$

If we take the limit as $h \to 0$, we obtain the rate of change of z (with respect to distance) in the direction of **u**, which is called the directional derivative of f in the direction of **u**.

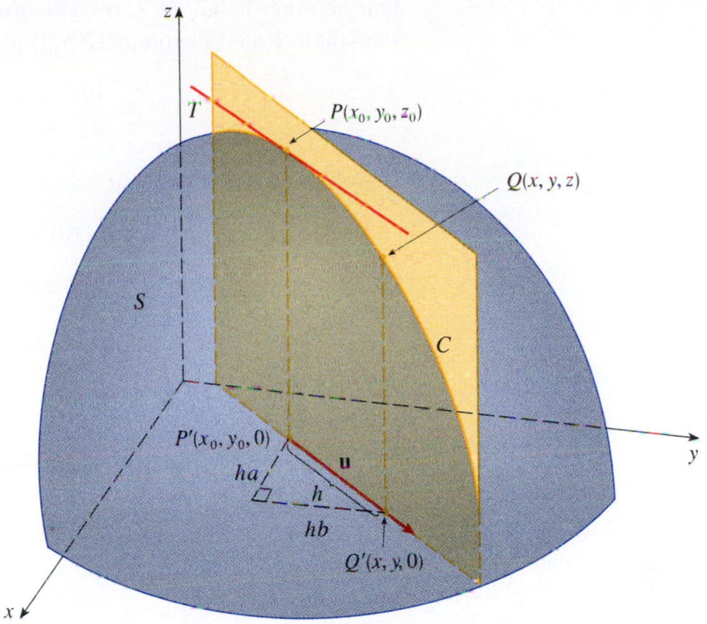

FIGURE 2

(2) DEFINITION The **directional derivative** of f at (x_0, y_0) in the direction of a unit vector $\mathbf{u} = \langle a, b \rangle$ is

$$D_{\mathbf{u}} f(x_0, y_0) = \lim_{h \to 0} \frac{f(x_0 + ha,\, y_0 + hb) - f(x_0, y_0)}{h}$$

if this limit exists.

By comparing Definition 2 with Equations 1, we see that if $\mathbf{u} = \mathbf{i} = \langle 1, 0 \rangle$, then $D_{\mathbf{i}} f = f_x$ and if $\mathbf{u} = \mathbf{j} = \langle 0, 1 \rangle$, then $D_{\mathbf{j}} f = f_y$. In other words, the partial derivatives of f with respect to x and y are just special cases of the directional derivative.

For computational purposes we generally use the formula given by the following theorem.

(3) THEOREM If f is a differentiable function of x and y, then f has a directional derivative in the direction of any unit vector $\mathbf{u} = \langle a, b \rangle$ and

$$D_{\mathbf{u}} f(x, y) = f_x(x, y)a + f_y(x, y)b$$

PROOF If we define a function g of the single variable h by

$$g(h) = f(x_0 + ha,\, y_0 + hb)$$

then by the definition of a derivative we have

$$\textbf{(4)} \qquad g'(0) = \lim_{h \to 0} \frac{g(h) - g(0)}{h} = \lim_{h \to 0} \frac{f(x_0 + ha,\, y_0 + hb) - f(x_0, y_0)}{h}$$

$$= D_{\mathbf{u}} f(x_0, y_0)$$

On the other hand, we can write $g(h) = f(x, y)$, where $x = x_0 + ha$, $y = y_0 + hb$, so the Chain Rule (Theorem 12.5.2) gives

$$g'(h) = \frac{\partial f}{\partial x}\frac{dx}{dh} + \frac{\partial f}{\partial y}\frac{dy}{dh} = f_x(x, y)a + f_y(x, y)b$$

If we now put $h = 0$, then $x = x_0$, $y = y_0$, and

(5) $$g'(0) = f_x(x_0, y_0)a + f_y(x_0, y_0)b$$

Comparing Equations 4 and 5, we see that

$$D_{\mathbf{u}} f(x_0, y_0) = f_x(x_0, y_0)a + f_y(x_0, y_0)b$$ □

If the unit vector **u** makes an angle θ with the positive x-axis (as in Figure 1), then we can write $\mathbf{u} = \langle \cos\theta, \sin\theta \rangle$ and the formula in Theorem 3 becomes

(6) $$D_{\mathbf{u}} f(x, y) = f_x(x, y)\cos\theta + f_y(x, y)\sin\theta$$

The directional derivative $D_{\mathbf{u}} f(1, 2)$ in Example 1 represents the rate of change of z in the direction of **u**. This is the slope of the tangent line to the curve of intersection of the surface $z = x^3 - 3xy + 4y^2$ and the vertical plane through $(1, 2, 0)$ in the direction of **u** shown in Figure 3.

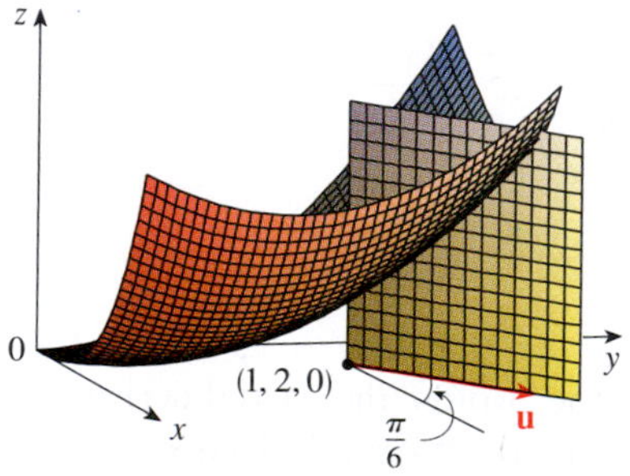

FIGURE 3

EXAMPLE 1 Find the directional derivative $D_{\mathbf{u}} f(x, y)$ if

$$f(x, y) = x^3 - 3xy + 4y^2$$

and **u** is the unit vector given by angle $\theta = \pi/6$. What is $D_{\mathbf{u}} f(1, 2)$?

SOLUTION Formula 6 gives

$$\begin{aligned} D_{\mathbf{u}} f(x, y) &= f_x(x, y)\cos\frac{\pi}{6} + f_y(x, y)\sin\frac{\pi}{6} \\ &= (3x^2 - 3y)\frac{\sqrt{3}}{2} + (-3x + 8y)\tfrac{1}{2} \\ &= \tfrac{1}{2}\left[3\sqrt{3}\,x^2 - 3x + (8 - 3\sqrt{3}\,)y\right] \end{aligned}$$

Therefore

$$D_{\mathbf{u}} f(1, 2) = \tfrac{1}{2}\left[3\sqrt{3}(1)^2 - 3(1) + (8 - 3\sqrt{3}\,)(2)\right] = \frac{13 - 3\sqrt{3}}{2}$$ ■

Notice from Theorem 3 that the directional derivative can be written as the dot product of two vectors:

(7) $$\begin{aligned} D_{\mathbf{u}} f(x, y) &= f_x(x, y)a + f_y(x, y)b \\ &= \langle f_x(x, y), f_y(x, y) \rangle \cdot \langle a, b \rangle \\ &= \langle f_x(x, y), f_y(x, y) \rangle \cdot \mathbf{u} \end{aligned}$$

The first vector in this dot product occurs not only in computing directional derivatives but in many other contexts as well. So we give it a special name (the *gradient* of f) and a special notation (**grad** f or ∇f, which is read "del f").

(8) DEFINITION If f is a function of two variables x and y, then the **gradient** of f is the vector function ∇f defined by

$$\nabla f(x, y) = \langle f_x(x, y), f_y(x, y) \rangle = \frac{\partial f}{\partial x}\mathbf{i} + \frac{\partial f}{\partial y}\mathbf{j}$$

EXAMPLE 2 If $f(x, y) = \sin x + e^{xy}$, then

$$\nabla f(x, y) = \langle \cos x + ye^{xy}, xe^{xy} \rangle$$

and
$$\nabla f(0, 1) = \langle 2, 0 \rangle$$

■

With this notation for the gradient vector, we can rewrite the expression (7) for the directional derivative as

(9)
$$D_{\mathbf{u}} f(x, y) = \nabla f(x, y) \cdot \mathbf{u}$$

This expresses the directional derivative in the direction of **u** as the scalar projection of the gradient vector onto **u**.

EXAMPLE 3 Find the directional derivative of the function $f(x, y) = x^2y^3 - 4y$ at the point $(2, -1)$ in the direction of the vector $\mathbf{v} = 2\,\mathbf{i} + 5\,\mathbf{j}$.

The gradient vector $\nabla f(2, -1)$ in Example 3 is shown in Figure 4 with initial point $(2, -1)$. Also shown is the vector **v** that gives the direction of the directional derivative. Both of these vectors are superimposed on a contour plot of the graph of f.

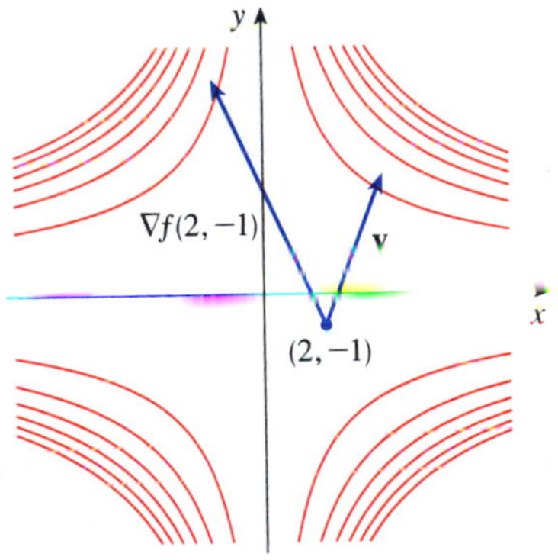

FIGURE 4

SOLUTION We first compute the gradient vector at $(2, -1)$:

$$\nabla f(x, y) = 2xy^3\,\mathbf{i} + (3x^2y^2 - 4)\,\mathbf{j}$$

$$\nabla f(2, -1) = -4\,\mathbf{i} + 8\,\mathbf{j}$$

Note that **v** is not a unit vector, but since $|\mathbf{v}| = \sqrt{29}$, the unit vector in the direction of **v** is

$$\mathbf{u} = \frac{\mathbf{v}}{|\mathbf{v}|} = \frac{2}{\sqrt{29}}\,\mathbf{i} + \frac{5}{\sqrt{29}}\,\mathbf{j}$$

Therefore, by Equation 9, we have

$$D_{\mathbf{u}} f(2, -1) = \nabla f(2, -1) \cdot \mathbf{u} = (-4\,\mathbf{i} + 8\,\mathbf{j}) \cdot \left(\frac{2}{\sqrt{29}}\,\mathbf{i} + \frac{5}{\sqrt{29}}\,\mathbf{j} \right)$$

$$= \frac{-4 \cdot 2 + 8 \cdot 5}{\sqrt{29}} = \frac{32}{\sqrt{29}}$$

■

For functions of three variables we can define directional derivatives in a similar manner. Again $D_{\mathbf{u}} f(x, y, z)$ can be interpreted as the rate of change of the function in the direction of a unit vector **u**.

(10) DEFINITION The **directional derivative** of f at (x_0, y_0, z_0) in the direction of a unit vector $\mathbf{u} = \langle a, b, c \rangle$ is

$$D_{\mathbf{u}} f(x_0, y_0, z_0) = \lim_{h \to 0} \frac{f(x_0 + ha, y_0 + hb, z_0 + hc) - f(x_0, y_0, z_0)}{h}$$

if this limit exists.

If we use vector notation, then we can write both definitions (2 and 10) of the directional derivative in the compact form

(11) $$D_{\mathbf{u}} f(\mathbf{x}_0) = \lim_{h \to 0} \frac{f(\mathbf{x}_0 + h\mathbf{u}) - f(\mathbf{x}_0)}{h}$$

where $\mathbf{x}_0 = \langle x_0, y_0 \rangle$ if $n = 2$ and $\mathbf{x}_0 = \langle x_0, y_0, z_0 \rangle$ if $n = 3$. This is reasonable because the vector equation of the line through $\mathbf{x}_0$ in the direction of the vector $\mathbf{u}$ is given by $\mathbf{x} = \mathbf{x}_0 + t\mathbf{u}$ (Equation 11.4.1) and so $f(\mathbf{x}_0 + h\mathbf{u})$ represents the value of f at a point on this line.

If $f(x, y, z)$ is differentiable and $\mathbf{u} = \langle a, b, c \rangle$, then the same method that was used to prove Theorem 3 can be used to show that

(12) $$D_{\mathbf{u}} f(x, y, z) = f_x(x, y, z)a + f_y(x, y, z)b + f_z(x, y, z)c$$

For a function f of three variables, the **gradient vector**, denoted by ∇f or **grad** f, is

$$\nabla f(x, y, z) = \langle f_x(x, y, z), f_y(x, y, z), f_z(x, y, z) \rangle$$

or, for short,

(13) $$\nabla f = \langle f_x, f_y, f_z \rangle = \frac{\partial f}{\partial x}\mathbf{i} + \frac{\partial f}{\partial y}\mathbf{j} + \frac{\partial f}{\partial z}\mathbf{k}$$

Then, just as with functions of two variables, Formula 12 for the directional derivative can be rewritten as

(14) $$D_{\mathbf{u}} f(x, y, z) = \nabla f(x, y, z) \cdot \mathbf{u}$$

EXAMPLE 4 If $f(x, y, z) = x \sin yz$, (a) find the gradient of f and (b) find the directional derivative of f at $(1, 3, 0)$ in the direction of $\mathbf{v} = \mathbf{i} + 2\mathbf{j} - \mathbf{k}$.

SOLUTION

(a) The gradient of f is

$$\begin{aligned}\nabla f(x, y, z) &= \langle f_x(x, y, z), f_y(x, y, z), f_z(x, y, z) \rangle \\ &= \langle \sin yz, xz \cos yz, xy \cos yz \rangle\end{aligned}$$

(b) At $(1, 3, 0)$ we have $\nabla f(1, 3, 0) = \langle 0, 0, 3 \rangle$. The unit vector in the direction of $\mathbf{v} = \mathbf{i} + 2\mathbf{j} - \mathbf{k}$ is

$$\mathbf{u} = \frac{1}{\sqrt{6}}\mathbf{i} + \frac{2}{\sqrt{6}}\mathbf{j} - \frac{1}{\sqrt{6}}\mathbf{k}$$

Therefore, Equation 14 gives

$$\begin{aligned}D_{\mathbf{u}} f(1, 3, 0) &= \nabla f(1, 3, 0) \cdot \mathbf{u} \\ &= 3\mathbf{k} \cdot \left(\frac{1}{\sqrt{6}}\mathbf{i} + \frac{2}{\sqrt{6}}\mathbf{j} - \frac{1}{\sqrt{6}}\mathbf{k}\right) \\ &= 3\left(-\frac{1}{\sqrt{6}}\right) = -\sqrt{\frac{3}{2}}\end{aligned}$$

■

Suppose we have a function f of two or three variables and we consider all possible directional derivatives of f at a given point. These give the rates of change of f in all possible directions. We can then ask the questions: In which of these directions does f change fastest and what is the maximum rate of change? The answers are provided by the following theorem.

(15) THEOREM Suppose f is a differentiable function of two or three variables. The maximum value of the directional derivative $D_{\mathbf{u}} f(\mathbf{x})$ is $|\nabla f(\mathbf{x})|$ and it occurs when $\mathbf{u}$ has the same direction as the gradient vector $\nabla f(\mathbf{x})$.

PROOF From Equation 9 or 14 we have

$$
\begin{aligned}
D_{\mathbf{u}} f &= \nabla f \cdot \mathbf{u} \\
&= |\nabla f||\mathbf{u}|\cos\theta \qquad \text{(from Definition 11.2.8)} \\
&= |\nabla f|\cos\theta
\end{aligned}
$$

where θ is the angle between ∇f and $\mathbf{u}$. The maximum value of $\cos\theta$ is 1 and this occurs when $\theta = 0$. Therefore, the maximum value of $D_{\mathbf{u}} f$ is $|\nabla f|$ and it occurs when $\theta = 0$, that is, when $\mathbf{u}$ has the same direction as ∇f. □

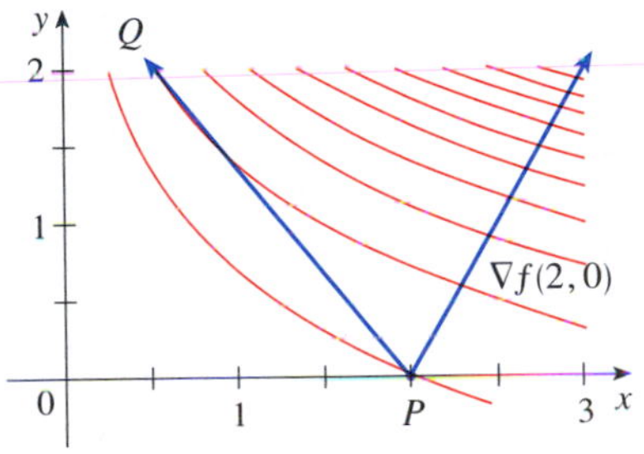

FIGURE 5

At $(2, 0)$ the function in Example 5 increases fastest in the direction of the gradient vector $\nabla f(2, 0) = \langle 1, 2 \rangle$. Notice from Figure 5 that this vector appears to be perpendicular to the level curve through $(2, 0)$. Figure 6 shows the graph of f and the gradient vector.

EXAMPLE 5
(a) If $f(x, y) = xe^y$, find the rate of change of f at the point $P(2, 0)$ in the direction from P to $Q(\frac{1}{2}, 2)$.
(b) In what direction does f have the maximum rate of change? What is this maximum rate of change?

SOLUTION
(a) We first compute the gradient vector:

$$\nabla f(x, y) = \langle f_x, f_y \rangle = \langle e^y, xe^y \rangle$$

$$\nabla f(2, 0) = \langle 1, 2 \rangle$$

The unit vector in the direction of $\overrightarrow{PQ} = \langle -1.5, 2 \rangle$ is $\mathbf{u} = \langle -\frac{3}{5}, \frac{4}{5} \rangle$, so the rate of change of f in the direction from P to Q is

$$
\begin{aligned}
D_{\mathbf{u}} f(2, 0) &= \nabla f(2, 0) \cdot \mathbf{u} = \langle 1, 2 \rangle \cdot \langle -\tfrac{3}{5}, \tfrac{4}{5} \rangle \\
&= 1(-\tfrac{3}{5}) + 2(\tfrac{4}{5}) = 1
\end{aligned}
$$

(b) According to Theorem 15, f increases fastest in the direction of the gradient vector $\nabla f(2, 0) = \langle 1, 2 \rangle$. The maximum rate of change is

$$|\nabla f(2, 0)| = |\langle 1, 2 \rangle| = \sqrt{5}$$ ■

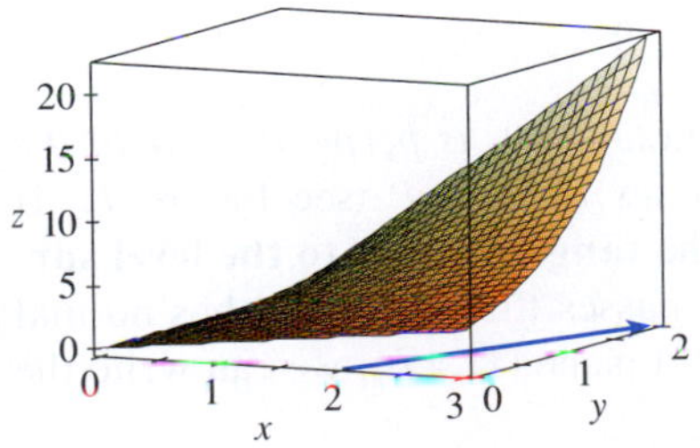

FIGURE 6

EXAMPLE 6 Suppose that the temperature at a point (x, y, z) in space is given by $T(x, y, z) = 80/(1 + x^2 + 2y^2 + 3z^2)$, where T is measured in °C and x, y, z in meters. In which direction does the temperature increase fastest at the point $(1, 1, -2)$? What is the maximum rate of increase?

SOLUTION The gradient of T is

$$
\begin{aligned}
\nabla T &= \frac{\partial T}{\partial x}\mathbf{i} + \frac{\partial T}{\partial y}\mathbf{j} + \frac{\partial T}{\partial z}\mathbf{k} \\
&= -\frac{160x}{(1 + x^2 + 2y^2 + 3z^2)^2}\mathbf{i} - \frac{320y}{(1 + x^2 + 2y^2 + 3z^2)^2}\mathbf{j} - \frac{480z}{(1 + x^2 + 2y^2 + 3z^2)^2}\mathbf{k}
\end{aligned}
$$

$$= \frac{160}{(1 + x^2 + 2y^2 + 3z^2)^2}(-x\,\mathbf{i} - 2y\,\mathbf{j} - 3z\,\mathbf{k})$$

At the point $(1, 1, -2)$ the gradient vector is

$$\nabla T(1, 1, -2) = \tfrac{160}{256}(-\mathbf{i} - 2\,\mathbf{j} + 6\,\mathbf{k}) = \tfrac{5}{8}(-\mathbf{i} - 2\,\mathbf{j} + 6\,\mathbf{k})$$

By Theorem 15 the temperature increases fastest in the direction of the gradient vector $\nabla T(1, 1, -2) = \tfrac{5}{8}(-\mathbf{i} - 2\,\mathbf{j} + 6\,\mathbf{k})$ or, equivalently, in the direction of $-\mathbf{i} - 2\,\mathbf{j} + 6\,\mathbf{k}$ or the unit vector $(-\mathbf{i} - 2\,\mathbf{j} + 6\,\mathbf{k})/\sqrt{41}$. The maximum rate of increase is the length of the gradient vector:

$$|\nabla T(1, 1, -2)| = \tfrac{5}{8}|-\mathbf{i} - 2\,\mathbf{j} + 6\,\mathbf{k}| = \frac{5\sqrt{41}}{8}$$

Therefore, the maximum rate of increase of temperature is $5\sqrt{41}/8 \approx 4°\text{C/m}$. ■

TANGENT PLANES TO LEVEL SURFACES

Suppose S is a surface with equation $F(x, y, z) = k$, that is, it is a level surface of a function F of three variables, and let $P(x_0, y_0, z_0)$ be a point on S. Let C be any curve that lies on the surface S and passes through the point P. Recall from Section 11.6 that C is described by a continuous vector function $\mathbf{r}(t) = \langle x(t), y(t), z(t)\rangle$. Let t_0 be the parameter value corresponding to P; that is, $\mathbf{r}(t_0) = \langle x_0, y_0, z_0\rangle$. Since C lies on S, any point $(x(t), y(t), z(t))$ must satisfy the equation of S, that is,

(16) $$F(x(t), y(t), z(t)) = k$$

If x, y, and z are differentiable functions of t and F is also differentiable, then we can use the Chain Rule to differentiate both sides of Equation 16 as follows:

(17) $$\frac{\partial F}{\partial x}\frac{dx}{dt} + \frac{\partial F}{\partial y}\frac{dy}{dt} + \frac{\partial F}{\partial z}\frac{dz}{dt} = 0$$

But, since $\nabla F = \langle F_x, F_y, F_z\rangle$ and $\mathbf{r}'(t) = \langle x'(t), y'(t), z'(t)\rangle$, Equation 17 can be written in terms of a dot product as

$$\nabla F \cdot \mathbf{r}'(t) = 0$$

In particular, when $t = t_0$ we have $\mathbf{r}(t_0) = \langle x_0, y_0, z_0\rangle$, so

(18) $$\nabla F(x_0, y_0, z_0) \cdot \mathbf{r}'(t_0) = 0$$

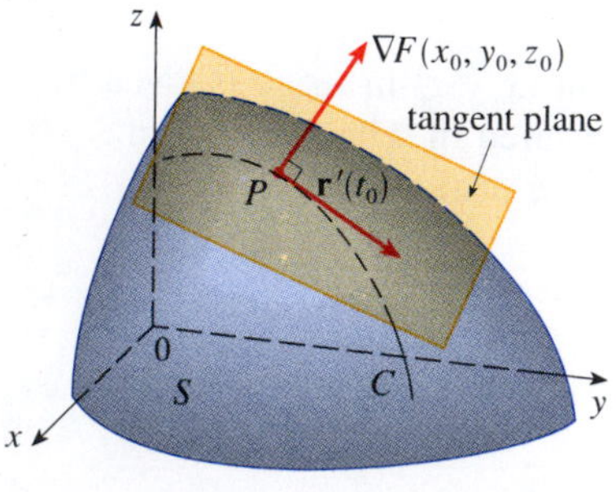

FIGURE 7

Equation 18 says that *the gradient vector at P, $\nabla F(x_0, y_0, z_0)$, is perpendicular to the tangent vector $\mathbf{r}'(t_0)$ to any curve C on S that passes through P* (see Figure 7). If $\nabla F(x_0, y_0, z_0) \neq \mathbf{0}$, it is therefore natural to define the **tangent plane to the level surface $F(x, y, z) = k$ at $P(x_0, y_0, z_0)$** as the plane that passes through P and has normal vector $\nabla F(x_0, y_0, z_0)$. Using the standard equation of a plane (11.4.6), we can write the equation of this tangent plane as

(19) $$F_x(x_0, y_0, z_0)(x - x_0) + F_y(x_0, y_0, z_0)(y - y_0) + F_z(x_0, y_0, z_0)(z - z_0) = 0$$

The **normal line** to S at P is the line passing through P and perpendicular to the tangent plane. Its direction is therefore given by the gradient vector $\nabla F(x_0, y_0, z_0)$ and so, by Equation 11.4.3, its symmetric equations are

$$\textbf{(20)} \qquad \frac{x - x_0}{F_x(x_0, y_0, z_0)} = \frac{y - y_0}{F_y(x_0, y_0, z_0)} = \frac{z - z_0}{F_z(x_0, y_0, z_0)}$$

In the special case in which the equation of a surface S is of the form $z = f(x, y)$ (that is, S is the graph of a function f of two variables), we can rewrite it as

$$F(x, y, z) = f(x, y) - z = 0$$

and regard S as a level surface (with $k = 0$) of F. Then

$$F_x(x_0, y_0, z_0) = f_x(x_0, y_0) \qquad F_y(x_0, y_0, z_0) = f_y(x_0, y_0) \qquad F_z(x_0, y_0, z_0) = -1$$

so Equation 19 becomes

$$f_x(x_0, y_0)(x - x_0) + f_y(x_0, y_0)(y - y_0) - (z - z_0) = 0$$

which is equivalent to (12.4.2). Thus our new, more general definition of a tangent plane is consistent with the definition that was given for the special case of Section 12.4.

EXAMPLE 7 Find the equations of the tangent plane and normal line at the point $(-2, 1, -3)$ to the ellipsoid

$$\frac{x^2}{4} + y^2 + \frac{z^2}{9} = 3$$

Figure 8 shows the ellipsoid, tangent plane, and normal line in Example 7.

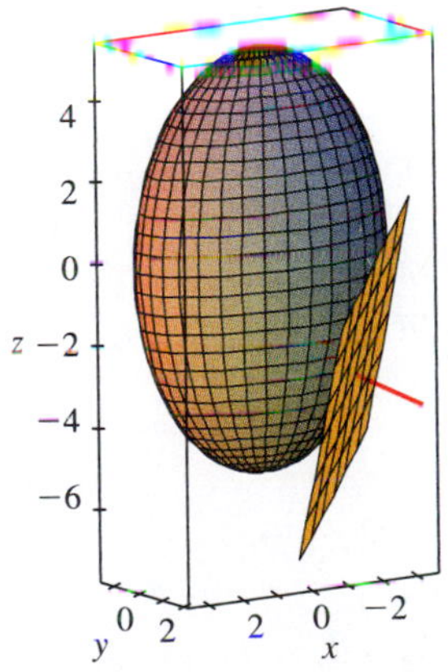

FIGURE 8

SOLUTION The ellipsoid is the level surface (with $k = 3$) of the function

$$F(x, y, z) = \frac{x^2}{4} + y^2 + \frac{z^2}{9}$$

Therefore, we have

$$F_x(x, y, z) = \frac{x}{2} \qquad F_y(x, y, z) = 2y \qquad F_z(x, y, z) = \frac{2z}{9}$$

$$F_x(-2, 1, -3) = -1 \qquad F_y(-2, 1, -3) = 2 \qquad F_z(-2, 1, -3) = -\tfrac{2}{3}$$

Then Equation 19 gives the equation of the tangent plane at $(-2, 1, -3)$ as

$$-1(x + 2) + 2(y - 1) - \tfrac{2}{3}(z + 3) = 0$$

which simplifies to $3x - 6y + 2z + 18 = 0$.

By Equation 20, symmetric equations of the normal line are

$$\frac{x + 2}{-1} = \frac{y - 1}{2} = \frac{z + 3}{-\frac{2}{3}}$$

■

SIGNIFICANCE OF THE GRADIENT VECTOR

Let's summarize the ways in which the gradient vector is significant. We first consider a function f of three variables and a point $P(x_0, y_0, z_0)$ in its domain. On the one hand, we know from Theorem 15 that the gradient vector $\nabla f(x_0, y_0, z_0)$ gives the direction of fastest increase of f. On the other hand, we know that $\nabla f(x_0, y_0, z_0)$ is orthogonal to the level surface S of f through P. (Refer to Figure 7.) These two properties are quite compatible intuitively because as we move away from P on the level surface S, the value of f does not change at all. So it seems reasonable that if we move in the perpendicular direction, we get the maximum increase.

In like manner we consider a function f of two variables and a point $P(x_0, y_0)$ in its domain. Again the gradient vector $\nabla f(x_0, y_0)$ gives the direction of fastest increase of f. Also, by considerations similar to our discussion of tangent planes, it can be shown that $\nabla f(x_0, y_0)$ is perpendicular to the level curve $f(x, y) = k$ that passes through P. Again this is intuitively plausible because the values of f remain constant as we move along the curve. (See Figure 9.)

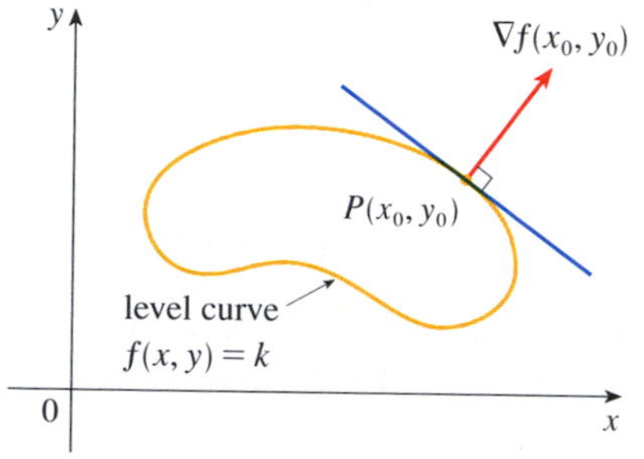

FIGURE 9

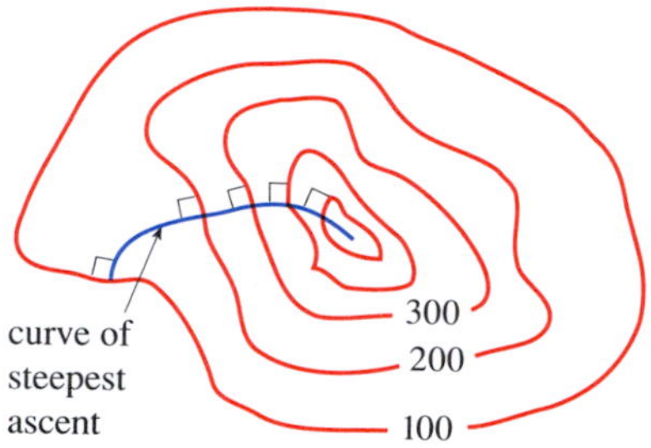

FIGURE 10

If we consider a topographical map of a hill and let $f(x, y)$ represent the height above sea level at a point with coordinates (x, y), then a curve of steepest ascent can be drawn as in Figure 10 by making it perpendicular to all of the contour lines. This phenomenon can also be noticed in Figure 10 in Section 12.1, where Lonesome Creek follows a curve of steepest descent.

EXERCISES 12.6

1–4 ■ Find the directional derivative of f at the given point in the direction indicated by the angle θ.

1. $f(x, y) = x^2y^3 + 2x^4y$, $(1, -2)$, $\theta = \pi/3$

2. $f(x, y) = (x^2 - y)^3$, $(3, 1)$, $\theta = 3\pi/4$

3. $f(x, y) = y^x$, $(1, 2)$, $\theta = \pi/2$

4. $f(x, y) = \sin(x + 2y)$, $(4, -2)$, $\theta = -2\pi/3$

5–8 ■

(a) Find the gradient of f.

(b) Evaluate the gradient at the point P.

(c) Find the rate of change of f at P in the direction of the vector $\mathbf{u}$.

5. $f(x, y) = x^3 - 4x^2y + y^2$, $P(0, -1)$, $\mathbf{u} = \langle \frac{3}{5}, \frac{4}{5} \rangle$

6. $f(x, y) = e^x \sin y$, $P(1, \pi/4)$, $\mathbf{u} = \left\langle \dfrac{-1}{\sqrt{5}}, \dfrac{2}{\sqrt{5}} \right\rangle$

7. $f(x, y, z) = xy^2z^3$, $P(1, -2, 1)$, $\mathbf{u} = \left\langle \dfrac{1}{\sqrt{3}}, \dfrac{-1}{\sqrt{3}}, \dfrac{1}{\sqrt{3}} \right\rangle$

8. $f(x, y, z) = xy + yz^2 + xz^3$, $P(2, 0, 3)$, $\mathbf{u} = \langle -\frac{2}{3}, -\frac{1}{3}, \frac{2}{3} \rangle$

9–16 ■ Find the directional derivative of the function at the given point in the direction of the vector $\mathbf{v}$.

9. $f(x, y) = \sqrt{x - y}$, $(5, 1)$, $\mathbf{v} = \langle 12, 5 \rangle$

10. $f(x, y) = x/y$, $(6, -2)$, $\mathbf{v} = \langle -1, 3 \rangle$

11. $g(x, y) = xe^{xy}$, $(-3, 0)$, $\mathbf{v} = 2\,\mathbf{i} + 3\,\mathbf{j}$

12. $g(x, y) = e^x \cos y$, $(1, \pi/6)$, $\mathbf{v} = \mathbf{i} - \mathbf{j}$

13. $f(x, y, z) = \sqrt{xyz}$, $(2, 4, 2)$, $\mathbf{v} = \langle 4, 2, -4 \rangle$

14. $f(x, y, z) = z^3 - x^2y$, $(1, 6, 2)$, $\mathbf{v} = \langle 3, 4, 12 \rangle$

15. $g(x, y, z) = x \tan^{-1}(y/z)$, $(1, 2, -2)$, $\mathbf{v} = \mathbf{i} + \mathbf{j} - \mathbf{k}$

16. $g(x, y, z) = xe^{yz} + xye^z$, $(-2, 1, 1)$, $\mathbf{v} = \mathbf{i} - 2\,\mathbf{j} + 3\,\mathbf{k}$

17–22 ■ Find the maximum rate of change of f at the given point and the direction in which it occurs.

17. $f(x, y) = xe^{-y} + 3y$, $(1, 0)$

18. $f(x, y) = \ln(x^2 + y^2)$, $(1, 2)$

19. $f(x, y) = \sqrt{x^2 + 2y}$, $(4, 10)$

20. $f(x, y, z) = x + y/z$, $(4, 3, -1)$

21. $f(x, y) = \cos(3x + 2y)$, $(\pi/6, -\pi/8)$

22. $f(x, y, z) = \dfrac{x}{y} + \dfrac{y}{z}$, $(4, 2, 1)$

23. Show that a differentiable function f decreases most rapidly at $\mathbf{x}$ in the direction opposite to the gradient vector, that is, in the direction of $-\nabla f(\mathbf{x})$.

24. Use the result of Exercise 23 to find the direction in which the function $f(x, y) = x^4y - x^2y^3$ decreases fastest at the point $(2, -3)$.

25. The temperature T in a metal ball is inversely proportional to the distance from the center of the ball, which we take to be the origin. The temperature at the point $(1, 2, 2)$ is $120°$.
(a) Find the rate of change of T at $(1, 2, 2)$ in the direction toward the point $(2, 1, 3)$.
(b) Show that at any point in the ball the direction of greatest increase in temperature is given by a vector that points toward the origin.

26. The temperature at a point (x, y, z) is given by

$$T(x, y, z) = 200e^{-x^2-3y^2-9z^2}$$

where T is measured in °C and x, y, z in meters.
(a) Find the rate of change of temperature at the point $P(2, -1, 2)$ in the direction toward the point $(3, -3, 3)$.
(b) In which direction does the temperature increase fastest at P?
(c) Find the maximum rate of increase at P.

27. Suppose that over a certain region of space the electrical potential V is given by $V(x, y, z) = 5x^2 - 3xy + xyz$.
(a) Find the rate of change of the potential at $P(3, 4, 5)$ in the direction of the vector $\mathbf{v} = \mathbf{i} + \mathbf{j} - \mathbf{k}$.
(b) In which direction does V change most rapidly at P?
(c) What is the maximum rate of change at P?

28. Suppose that you are climbing a hill whose shape is given by the equation $z = 1000 - 0.01x^2 - 0.02y^2$ and you are standing at a point with coordinates $(60, 100, 764)$.
(a) In which direction should you proceed initially in order to reach the top of the hill fastest?
(b) If you climb in that direction, at what angle above the horizontal will you be climbing initially?

29. Let f be a function of two variables that has continuous partial derivatives and consider the points $A(1, 3)$, $B(3, 3)$, $C(1, 7)$, and $D(6, 15)$. The directional derivative of f at A in the direction of the vector $\overrightarrow{AB}$ is 3 and the directional derivative at A in the direction of $\overrightarrow{AC}$ is 26. Find the directional derivative of f at A in the direction of the vector $\overrightarrow{AD}$.

30. For the given contour map draw the curves of steepest ascent starting at P and at Q.

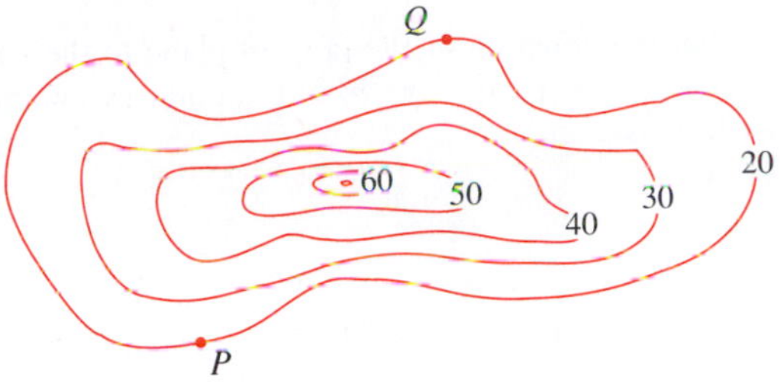

31–34 ■ Show that the operation of taking the gradient of a function has the given property. Assume that u and v are differentiable functions of x and y and a, b are constants.

31. $\nabla(au + bv) = a\nabla u + b\nabla v$ **32.** $\nabla(uv) = u\nabla v + v\nabla u$

33. $\nabla\left(\dfrac{u}{v}\right) = \dfrac{v\nabla u - u\nabla v}{v^2}$ **34.** $\nabla u^n = nu^{n-1}\nabla u$

35–40 ■ Find equations of (a) the tangent plane and (b) the normal line to the given surface at the specified point.

35. $4x^2 + y^2 + z^2 = 24$, $(2, 2, 2)$

36. $x^2 - 2y^2 + z^2 = 3$, $(-1, 1, -2)$

37. $x^2 + y^2 - z^2 - 2xy + 4xz = 4$, $(1, 0, 1)$

38. $x^2 - 2y^2 - 3z^2 + xyz = 4$, $(3, -2, -1)$

39. $z + 1 = xe^y\cos z$, $(1, 0, 0)$

40. $xe^{yz} = 1$, $(1, 0, 5)$

41–42 ■ Use a computer to graph the surface, the tangent plane, and the normal line on the same screen. Choose the domain carefully so that you avoid extraneous vertical planes. Choose the viewpoint so that you get a good view of all three objects.

41. $xy + yz + zx = 3$, $(1, 1, 1)$

42. $xyz = 6$, $(1, 2, 3)$

43. If $f(x, y) = x^2 + 4y^2$, find the gradient vector $\nabla f(2, 1)$ and use it to find the tangent line to the level curve $f(x, y) = 8$ at the point $(2, 1)$. Sketch the level curve, the tangent line, and the gradient vector.

44. If $g(x, y) = x - y^2$, find the gradient vector $\nabla g(3, -1)$ and use it to find the tangent line to the level curve $g(x, y) = 2$ at the point $(3, -1)$. Sketch the level curve, the tangent line, and the gradient vector.

45. Show that the equation of the tangent plane to the ellipsoid $x^2/a^2 + y^2/b^2 + z^2/c^2 = 1$ at the point (x_0, y_0, z_0) can be written as

$$\frac{xx_0}{a^2} + \frac{yy_0}{b^2} + \frac{zz_0}{c^2} = 1$$

46. Find the equation of the tangent plane to the hyperboloid $x^2/a^2 + y^2/b^2 - z^2/c^2 = 1$ at (x_0, y_0, z_0) and express it in a form similar to the one in Exercise 45.

47. Show that the equation of the tangent plane to the elliptic paraboloid $z/c = x^2/a^2 + y^2/b^2$ at the point (x_0, y_0, z_0) can be written as

$$\frac{2xx_0}{a^2} + \frac{2yy_0}{b^2} = \frac{z + z_0}{c}$$

48. Find the points on the ellipsoid $x^2 + 2y^2 + 3z^2 = 1$ where the tangent plane is parallel to the plane $3x - y + 3z = 1$.

49. Find the points on the hyperboloid $x^2 - y^2 + 2z^2 = 1$ where the normal line is parallel to the line that joins the points $(3, -1, 0)$ and $(5, 3, 6)$.

50. Show that the ellipsoid $3x^2 + 2y^2 + z^2 = 9$ and the sphere $x^2 + y^2 + z^2 - 8x - 6y - 8z + 24 = 0$ are tangent to each other at the point $(1, 1, 2)$. (This means that they have a common tangent plane at the point.)

51. Show that every plane that is tangent to the cone $x^2 + y^2 = z^2$ passes through the origin.

52. Show that every normal line to the sphere $x^2 + y^2 + z^2 = r^2$ passes through the center of the sphere.

53. Show that the sum of the x-, y-, and z-intercepts of any tangent plane to the surface $\sqrt{x} + \sqrt{y} + \sqrt{z} = \sqrt{c}$ is a constant.

54. Show that the product of the x-, y-, and z-intercepts of any tangent plane to the surface $xyz = c^3$ is a constant.

55. Find parametric equations for the tangent line to the curve of intersection of the paraboloid $z = x^2 + y^2$ and the ellipsoid $4x^2 + y^2 + z^2 = 9$ at the point $(-1, 1, 2)$.

56. (a) The plane $y + z = 3$ intersects the cylinder $x^2 + y^2 = 5$ in an ellipse. Find parametric equations for the tangent line to this ellipse at the point $(1, 2, 1)$.

(b) Use a computer to graph the cylinder, the plane, and the tangent line on the same screen.

57. (a) Two surfaces are called **orthogonal** at a point of intersection if their normal lines are perpendicular at that point. Show that surfaces with equations $F(x, y, z) = 0$ and $G(x, y, z) = 0$ are orthogonal at a point P where $\nabla F \neq \mathbf{0}$ and $\nabla G \neq \mathbf{0}$ if and only if

$$F_x G_x + F_y G_y + F_z G_z = 0$$

at P.

(b) Use part (a) to show that the surfaces $z^2 = x^2 + y^2$ and $x^2 + y^2 + z^2 = r^2$ are orthogonal at every point of intersection. Can you see why this is true without using calculus?

58. (a) Show that the function $f(x, y) = \sqrt[3]{xy}$ is continuous and the partial derivatives f_x and f_y exist at the origin but the directional derivatives in all other directions do not exist.

(b) Use a computer to graph f near the origin and comment on how the graph confirms part (a).

59. Suppose that the directional derivatives of $f(x, y)$ are known at a given point in two nonparallel directions given by unit vectors $\mathbf{u}$ and $\mathbf{v}$. Is it possible to find ∇f at this point? If so, how would you do it?

60. Show that if $z = f(x, y)$ is differentiable at $\mathbf{x}_0 = \langle x_0, y_0 \rangle$, then

$$\lim_{\mathbf{x} \to \mathbf{x}_0} \frac{f(\mathbf{x}) - f(\mathbf{x}_0) - \nabla f(\mathbf{x}_0) \cdot (\mathbf{x} - \mathbf{x}_0)}{|\mathbf{x} - \mathbf{x}_0|} = 0$$

[*Hint:* Use Definition 12.4.12 directly.]

12.7 MAXIMUM AND MINIMUM VALUES

As we saw in Chapter 5, one of the main uses of ordinary derivatives is in finding maximum and minimum values. In this section we see how to use partial derivatives to locate maxima and minima of functions of two variables.

(1) DEFINITION A function of two variables has a **local maximum** at (a, b) if $f(x, y) \leqslant f(a, b)$ for all points (x, y) in some disk with center (a, b). The number $f(a, b)$ is called a **local maximum value**. If $f(x, y) \geqslant f(a, b)$ for all (x, y) in such a disk, $f(a, b)$ is a **local minimum value**.

If the inequalities in Definition 1 hold for *all* points (x, y) in the domain of f, then f has an **absolute maximum** (or **absolute minimum**) at (a, b).

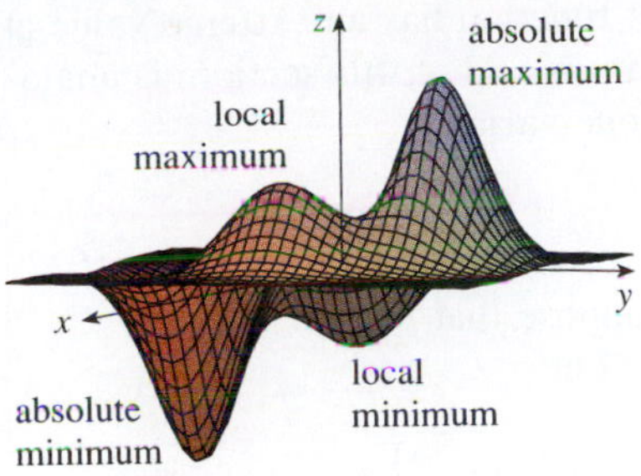

FIGURE 1

The graph of a function with several maxima and minima is shown in Figure 1. You can think of the local maxima as mountain peaks and the local minima as valley bottoms.

> **(2) THEOREM** If f has a local extremum (that is, a local maximum or minimum) at (a, b) and the first-order partial derivatives of f exist there, then $f_x(a, b) = 0$ and $f_y(a, b) = 0$.

PROOF Let $g(x) = f(x, b)$. If f has a local extremum at (a, b), then g has a local extremum at a, so $g'(a) = 0$ by Fermat's Theorem (5.2.4). But, by Equation 12.3.1, $g'(a) = f_x(a, b)$ and so $f_x(a, b) = 0$. Similarly, by applying Fermat's Theorem to the function $G(y) = f(a, y)$, we obtain $f_y(a, b) = 0$. □

If we put $f_x(a, b) = 0$ and $f_y(a, b) = 0$ in the equation of a tangent plane (12.4.2), we get $z = z_0$. Thus, the geometric interpretation of Theorem 2 is that if the graph of f has a tangent plane at a local extremum, then the tangent plane must be horizontal.

A point (a, b) such that $f_x(a, b) = 0$ and $f_y(a, b) = 0$, or one of these partial derivatives does not exist, is called a **critical point** (or *stationary point*) of f. Theorem 2 says that if f has a local extremum at (a, b), then (a, b) is a critical point of f. However, as in single-variable calculus, not all critical points give rise to extrema. At a critical point, a function could have a local maximum or a local minimum or neither.

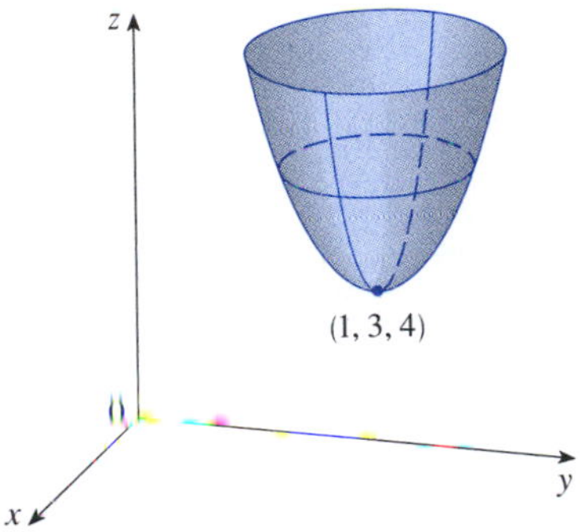

FIGURE 2
$z = x^2 + y^2 - 2x - 6y + 14$

EXAMPLE 1 Let $f(x, y) = x^2 + y^2 - 2x - 6y + 14$. Then

$$f_x(x, y) = 2x - 2 \qquad f_y(x, y) = 2y - 6$$

These partial derivatives are equal to 0 when $x = 1$ and $y = 3$, so the only critical point is $(1, 3)$. By completing the square, we find that

$$f(x, y) = 4 + (x - 1)^2 + (y - 3)^2$$

Since $(x - 1)^2 \geqslant 0$ and $(y - 3)^2 \geqslant 0$, we have $f(x, y) \geqslant 4$ for all values of x and y. Therefore, $f(1, 3) = 4$ is a local minimum, and in fact it is the absolute minimum of f. This can be confirmed geometrically from the graph of f, which is the elliptic paraboloid with vertex $(1, 3, 4)$ shown in Figure 2. ■

EXAMPLE 2 Find the extreme values of $f(x, y) = y^2 - x^2$.

SOLUTION Since $f_x = -2x$ and $f_y = 2y$, the only critical point is $(0, 0)$. Notice that for points on the x-axis we have $y = 0$, so $f(x, y) = -x^2 < 0$ (if $x \neq 0$). However, for points on the y-axis we have $x = 0$, so $f(x, y) = y^2 > 0$ (if $y \neq 0$). Thus, every disk with center $(0, 0)$ contains points where f takes positive values as well as points where f takes negative values. Therefore, $f(0, 0) = 0$ cannot be an extreme value for f, so f has no extreme values. ■

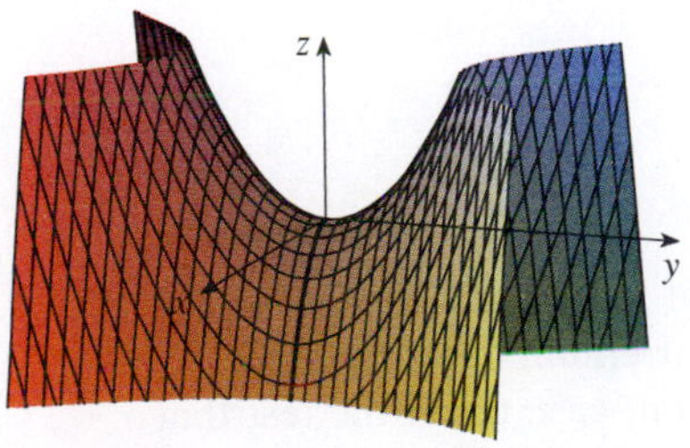

FIGURE 3
$z = y^2 - x^2$

Example 2 illustrates the fact that a function need not have a maximum or minimum value at a critical point. Figure 3 shows how this is possible. The graph of f is the hyperbolic paraboloid $z = y^2 - x^2$, which has a horizontal tangent plane ($z = 0$) at the origin. You can see that $f(0, 0) = 0$ is a maximum in the direction of the x-axis but a minimum in the direction of the y-axis. Near the origin the graph has the shape of a saddle and so $(0, 0)$ is called a *saddle point* of f.

We need to be able to determine whether or not a function has an extreme value at a critical point. The following test, which is proved at the end of this section, is analogous to the Second Derivative Test for functions of one variable.

(3) SECOND DERIVATIVES TEST Suppose the second partial derivatives of f are continuous in a disk with center (a, b), and suppose that $f_x(a, b) = 0$ and $f_y(a, b) = 0$ [that is, (a, b) is a critical point of f]. Let

$$D = D(a, b) = f_{xx}(a, b)f_{yy}(a, b) - [f_{xy}(a, b)]^2$$

(a) If $D > 0$ and $f_{xx}(a, b) > 0$, then $f(a, b)$ is a local minimum.
(b) If $D > 0$ and $f_{xx}(a, b) < 0$, then $f(a, b)$ is a local maximum.
(c) If $D < 0$, then $f(a, b)$ is not a local extremum.

NOTE 1: In case (c) the point (a, b) is called a **saddle point** of f and the graph of f crosses its tangent plane at (a, b).

NOTE 2: If $D = 0$, the test gives no information: f could have a local maximum or local minimum at (a, b), or (a, b) could be a saddle point of f.

NOTE 3: To remember the formula for D it is helpful to write it as a determinant:

$$D = \begin{vmatrix} f_{xx} & f_{xy} \\ f_{yx} & f_{yy} \end{vmatrix} = f_{xx}f_{yy} - (f_{xy})^2$$

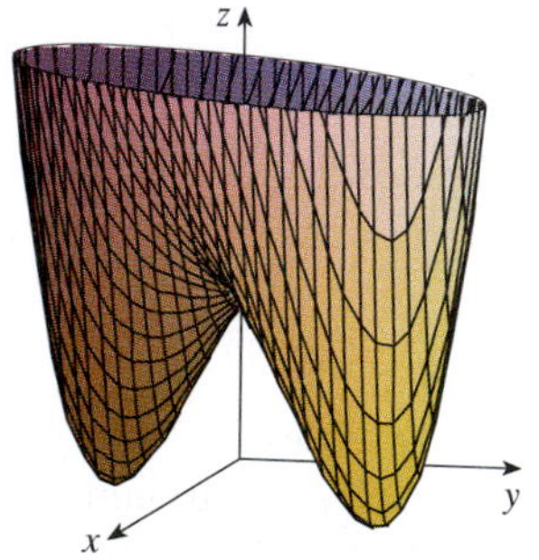

FIGURE 4
$z = x^4 + y^4 - 4xy + 1$

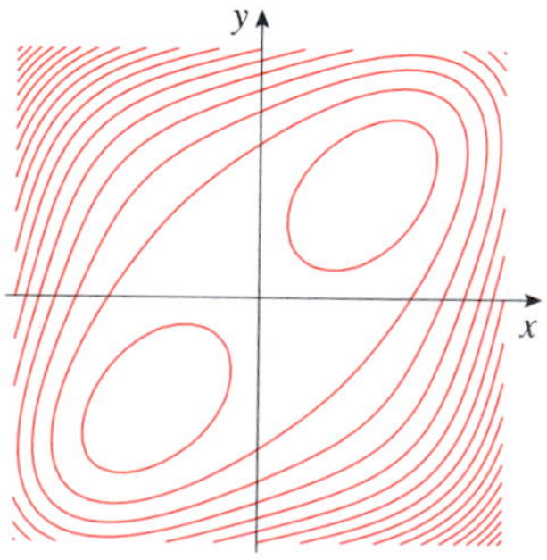

FIGURE 5
Contour map of
$z = x^4 + y^4 - 4xy + 1$

EXAMPLE 3 Find the local extrema of $f(x, y) = x^4 + y^4 - 4xy + 1$.

SOLUTION We first locate the critical points:

$$f_x = 4x^3 - 4y \qquad f_y = 4y^3 - 4x$$

Setting these partial derivatives equal to 0, we obtain the equations

$$x^3 - y = 0 \quad \text{and} \quad y^3 - x = 0$$

To solve these equations we substitute $y = x^3$ from the first equation into the second one. This gives

$$0 = x^9 - x = x(x^8 - 1) = x(x^4 - 1)(x^4 + 1) = x(x^2 - 1)(x^2 + 1)(x^4 + 1)$$

so there are three real roots: $x = 0, 1, -1$. The three critical points are $(0, 0)$, $(1, 1)$, and $(-1, -1)$.

Next we calculate the second partial derivatives and $D(x, y)$:

$$f_{xx} = 12x^2 \qquad f_{xy} = -4 \qquad f_{yy} = 12y^2$$

$$D(x, y) = f_{xx}f_{yy} - (f_{xy})^2 = 144x^2y^2 - 16$$

Since $D(0, 0) = -16 < 0$, it follows from case (c) of the Second Derivatives Test that the origin is a saddle point; that is, f has no local extremum at $(0, 0)$. Since $D(1, 1) = 128 > 0$ and $f_{xx}(1, 1) = 12 > 0$, we see from case (a) of the test that $f(1, 1) = -1$ is a local minimum. Similarly, we have $D(-1, -1) = 128 > 0$ and $f_{xx}(-1, -1) = 12 > 0$, so $f(-1, -1) = -1$ is also a local minimum.

The graph of f and its contour map are shown in Figures 4 and 5. ■

EXAMPLE 4 Find and classify the critical points of the function

$$f(x, y) = 10x^2y - 5x^2 - 4y^2 - x^4 - 2y^4$$

Also find the highest point on the graph of f.

SOLUTION The first-order partial derivatives are

$$f_x = 20xy - 10x - 4x^3 \qquad f_y = 10x^2 - 8y - 8y^3$$

So to find the critical points we need to solve the equations

$$2x(10y - 5 - 2x^2) = 0 \tag{4}$$

$$5x^2 - 4y - 4y^3 = 0 \tag{5}$$

From Equation 4 we see that either

$$x = 0 \qquad \text{or} \qquad 10y - 5 - 2x^2 = 0$$

In the first case ($x = 0$), Equation 5 becomes $-4y(1 + y^2) = 0$, so $y = 0$ and we have the critical point $(0, 0)$.

In the second case ($10y - 5 - 2x^2 = 0$), we get

$$x^2 = 5y - 2.5 \tag{6}$$

and, putting this in Equation 5, we have $25y - 12.5 - 4y - 4y^3 = 0$. So we have to solve the cubic equation

$$4y^3 - 21y + 12.5 = 0 \tag{7}$$

Using a graphing calculator or computer to graph the function

$$g(y) = 4y^3 - 21y + 12.5$$

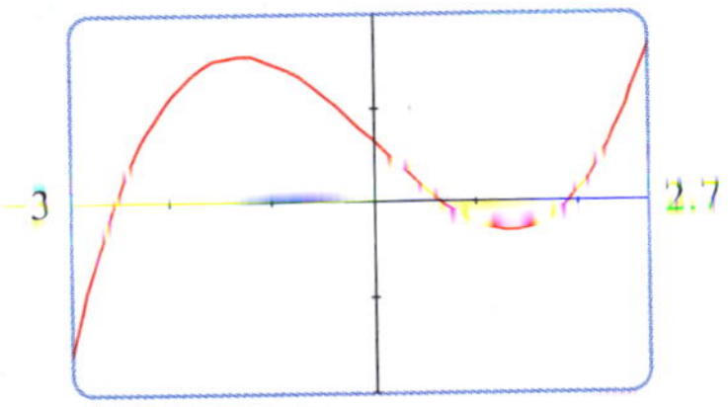

FIGURE 6

as in Figure 6, we see that Equation 7 has three real roots. By zooming in, we can find the roots to four decimal places:

$$y \approx -2.5452 \qquad y \approx 0.6468 \qquad y \approx 1.8984$$

(Alternatively, we could have used Newton's method or a computer algebra system to locate these roots.) From Equation 6, the corresponding x-values are given by

$$x = \pm\sqrt{5y - 2.5}$$

If $y \approx -2.5452$, then there is no corresponding real value of x. If $y \approx 0.6468$, then $x \approx \pm 0.8567$. If $y \approx 1.8984$, then $x \approx \pm 2.6442$. So we have a total of five critical points, which are analyzed in the following chart. All quantities are rounded to two decimal places.

Critical point	Value of f	f_{xx}	D	Conclusion
$(0, 0)$	0.00	−10.00	80.00	local maximum
$(\pm 2.64, 1.90)$	8.50	−55.93	2488.71	local maximum
$(\pm 0.86, 0.65)$	−1.48	−5.87	−187.64	saddle point

The five critical points are shown in blue in Figure 7. Notice that the level curves have shapes similar to ellipses near the maximum points and similar to hyperbolas near the saddle points.

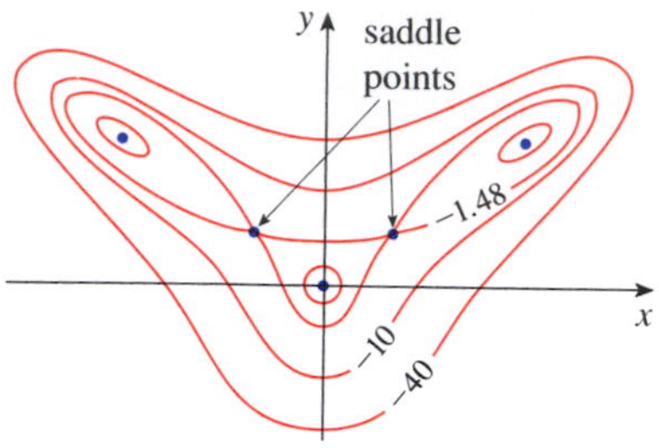

FIGURE 7

Figure 7 shows the location of the critical points on a contour map of f. Figures 8 and 9 give two views of the graph of f and we see that the surface opens downward. [This can also be seen from the expression for $f(x, y)$: the dominant terms are $-x^2 - 2y^4$ when $|x|$ and $|y|$ are large.] Comparing the values of f at its local maximum points, we see that the absolute maximum value of f is $f(\pm 2.64, 1.90) \approx 8.50$. In other words, the highest points on the graph of f are $(\pm 2.64, 1.90, 8.50)$.

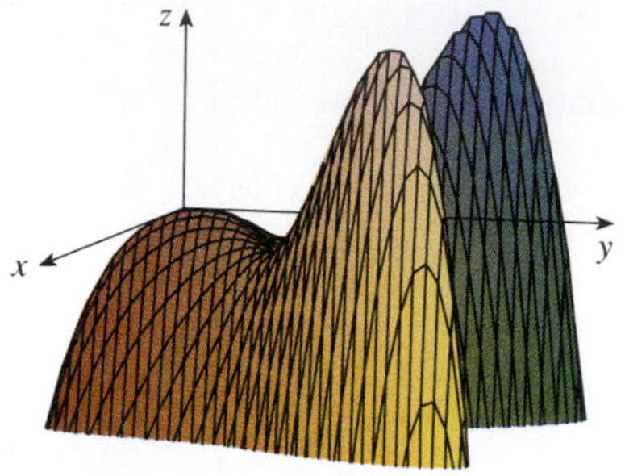

FIGURE 8

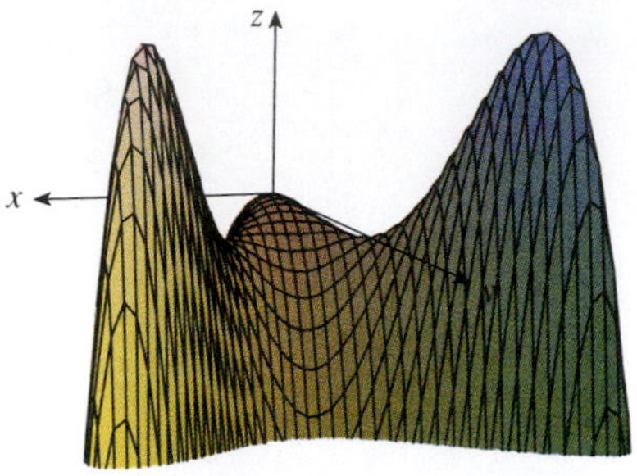

FIGURE 9

■

EXAMPLE 5 Find the shortest distance from the point $(1, 0, -2)$ to the plane $x + 2y + z = 4$.

SOLUTION The distance from any point (x, y, z) to the point $(1, 0, -2)$ is

$$d = \sqrt{(x - 1)^2 + y^2 + (z + 2)^2}$$

but if (x, y, z) lies on the plane $x + 2y + z = 4$, then $z = 4 - x - 2y$ and so we have $d = \sqrt{(x - 1)^2 + y^2 + (6 - x - 2y)^2}$. We can minimize d by minimizing the simpler expression

$$d^2 = f(x, y) = (x - 1)^2 + y^2 + (6 - x - 2y)^2$$

By solving the equations

$$f_x = 2(x - 1) - 2(6 - x - 2y) = 4x + 4y - 14 = 0$$

$$f_y = 2y - 4(6 - x - 2y) = 4x + 10y - 24 = 0$$

we find that the only critical point is $(\frac{11}{6}, \frac{5}{3})$. Since $f_{xx} = 4$, $f_{xy} = 4$, and $f_{yy} = 10$, we have $D(x, y) = f_{xx} f_{yy} - (f_{xy})^2 = 24 > 0$ and $f_{xx} > 0$, so by the Second Derivatives Test f has a local minimum at $(\frac{11}{6}, \frac{5}{3})$. Intuitively, we can see that this local minimum is actually an absolute minimum because there must be a point on the given plane that is closest to $(1, 0, -2)$. If $x = \frac{11}{6}$ and $y = \frac{5}{3}$, then

$$d = \sqrt{(x - 1)^2 + y^2 + (6 - x - 2y)^2} = \sqrt{(\tfrac{5}{6})^2 + (\tfrac{5}{3})^2 + (\tfrac{5}{6})^2} = \frac{5\sqrt{6}}{6}$$

Example 5 could also be solved using vectors. Compare with the methods of Section 11.4.

The shortest distance from $(1, 0, -2)$ to the plane $x + 2y + z = 4$ is $5\sqrt{6}/6$. ■

EXAMPLE 6 A rectangular box without a lid is to be made from 12 m^2 of cardboard. Find the maximum volume of such a box.

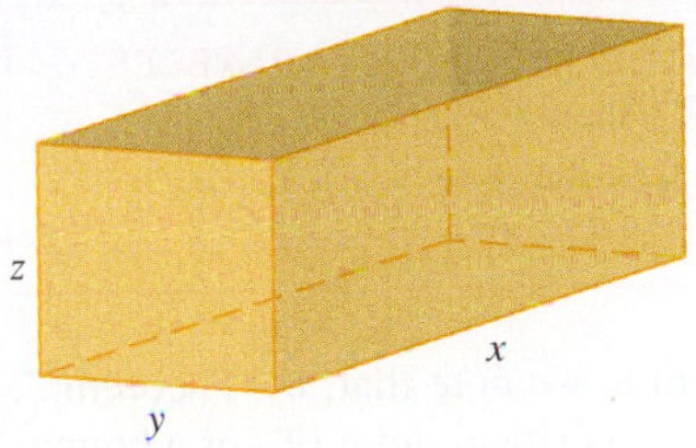

FIGURE 10

SOLUTION Let the length, width, and height of the box (in meters) be x, y, and z, as shown in Figure 10. Then the volume of the box is

$$V = xyz$$

We can express V as a function of just two variables x and y by using the fact that the area of the four sides and the bottom of the box is

$$2xz + 2yz + xy = 12$$

Solving this equation for z, we get $z = (12 - xy)/[2(x + y)]$, so the expression for V becomes

$$V = xy\frac{12 - xy}{2(x + y)} = \frac{12xy - x^2y^2}{2(x + y)}$$

We compute the partial derivatives:

$$\frac{\partial V}{\partial x} = \frac{y^2(12 - 2xy - x^2)}{2(x + y)^2} \qquad \frac{\partial V}{\partial y} = \frac{x^2(12 - 2xy - y^2)}{2(x + y)^2}$$

If V is a maximum, then $\partial V/\partial x = \partial V/\partial y = 0$, but $x = 0$ or $y = 0$ gives $V = 0$, so we must solve the equations

$$12 - 2xy - x^2 = 0 \qquad 12 - 2xy - y^2 = 0$$

These imply that $x^2 = y^2$ and so $x = y$. (Note that x and y must both be positive in this problem.) If we put $x = y$ in either equation we get $12 - 3x^2 = 0$, which gives $x = 2$, $y = 2$, and $z = (12 - 2 \cdot 2)/[2(2 + 2)] = 1$.

We could use the Second Derivatives Test to show that this gives a local maximum of V, or we could simply argue from the physical nature of this problem that there must be an absolute maximum volume, which has to occur at a critical point of V, so it must occur when $x = 2$, $y = 2$, $z = 1$. Then $V = 2 \cdot 2 \cdot 1 = 4$, so the maximum volume of the box is 4 m^3. ■

ABSOLUTE MAXIMUM AND MINIMUM VALUES

For a function f of one variable the Extreme Value Theorem says that if f is continuous on a closed interval $[a, b]$, then f has an absolute minimum value and an absolute maximum value. According to the Closed Interval Method (5.2.7) we found these by evaluating f not only at the critical numbers but also at the endpoints a and b.

There is a similar situation for functions of two variables. Just as a closed interval contains its endpoints, a **closed set** in $\mathbb{R}^2$ is one that contains all its boundary points. (See the discussion following Definition 12.2.3.) For instance, the disk

$$D = \{(x, y) \mid x^2 + y^2 \leq 1\}$$

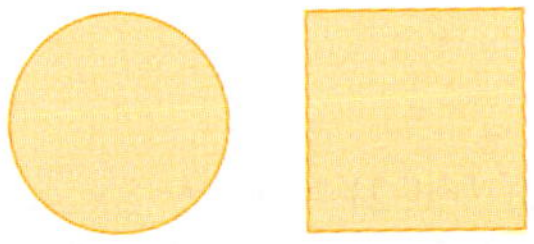
(a) Closed sets

(b) Sets that are not closed

FIGURE 11

consisting of all points on and inside the circle $x^2 + y^2 = 1$ is a closed set because it contains all of its boundary points (which are the points on the circle $x^2 + y^2 = 1$). But if even one point on the boundary curve were omitted, the set would not be closed. (See Figure 11.)

A **bounded set** in $\mathbb{R}^2$ is one that is contained within some disk. In other words, it is finite in extent. Then, in terms of closed and bounded sets, we can state the following analogue of the Extreme Value Theorem in two dimensions.

(8) EXTREME VALUE THEOREM FOR FUNCTIONS OF TWO VARIABLES If f is continuous on a closed, bounded set D in $\mathbb{R}^2$, then f attains an absolute maximum value $f(x_1, y_1)$ and an absolute minimum value $f(x_2, y_2)$ at some points (x_1, y_1) and (x_2, y_2) in D.

To find the extreme values guaranteed by Theorem 8, we note that, by Theorem 2, if f has an extreme value at (x_1, y_1), then (x_1, y_1) is either a critical point of f or a boundary point of D. Thus, we have the following analogue of the Closed Interval Method (5.2.7) for finding absolute extreme values of a function of one variable.

(9) To find the absolute maximum and minimum values of a continuous function f on a closed, bounded set D:

1. Find the values of f at the critical points of f in D.
2. Find the extreme values of f on the boundary of D.
3. The largest of the values from steps 1 and 2 is the absolute maximum value; the smallest of these values is the absolute minimum value.

EXAMPLE 7 Find the absolute maximum and minimum values of the function $f(x, y) = x^2 - 2xy + 2y$ on the rectangle $D = \{(x, y) \mid 0 \leq x \leq 3, 0 \leq y \leq 2\}$.

SOLUTION Since f is a polynomial, it is continuous on the closed, bounded rectangle D, so Theorem 8 tells us there is both an absolute maximum and an absolute minimum. According to step 1 we first find the critical points. These occur when

$$f_x = 2x - 2y = 0 \qquad f_y = -2x + 2 = 0$$

so the only critical point is $(1, 1)$, and the value of f there is $f(1, 1) = 1$.

In step 2 we look at the values of f on the boundary of D, which consists of the four line segments L_1, L_2, L_3, L_4 shown in Figure 12. On L_1 we have $y = 0$ and

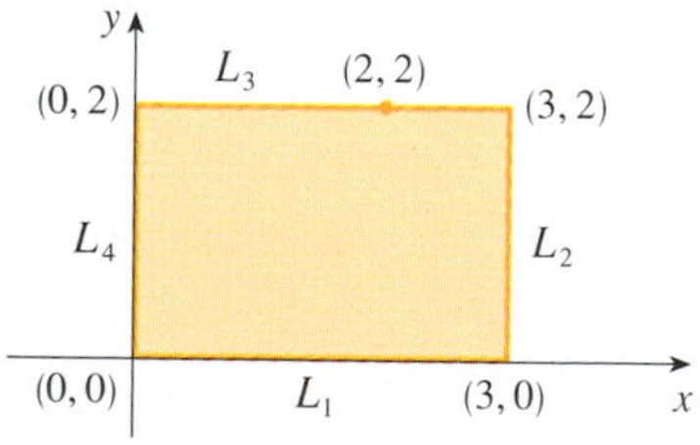

FIGURE 12

$$f(x, 0) = x^2 \qquad 0 \leq x \leq 3$$

This is an increasing function of x, so its minimum value is $f(0, 0) = 0$ and its maximum value is $f(3, 0) = 9$. On L_2 we have $x = 3$ and

$$f(3, y) = 9 - 4y \qquad 0 \leq y \leq 2$$

This is a decreasing function of y, so its maximum value is $f(3, 0) = 9$ and its minimum value is $f(3, 2) = 1$. On L_3 we have $y = 2$ and

$$f(x, 2) = x^2 - 4x + 4 \qquad 0 \leq x \leq 3$$

By the methods of Chapter 5, or simply by observing that $f(x, 2) = (x - 2)^2$, we see that the minimum value of this function is $f(2, 2) = 0$ and the maximum value is $f(0, 2) = 4$. Finally, on L_4 we have $x = 0$ and

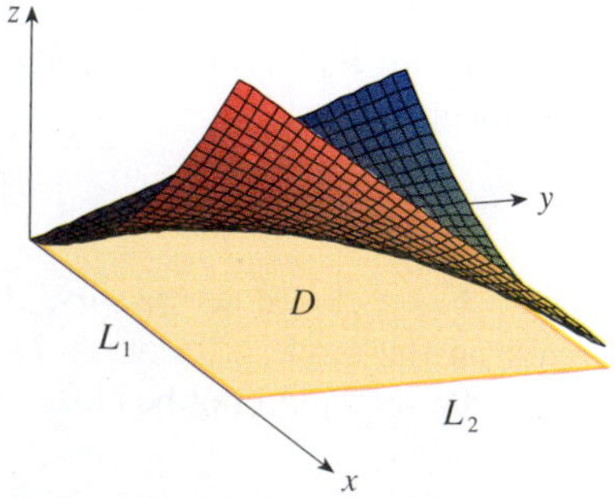

FIGURE 13
$f(x, y) = x^2 - 2xy + 2y$

$$f(0, y) = 2y \qquad 0 \leq y \leq 2$$

with maximum value $f(0, 2) = 4$ and minimum value $f(0, 0) = 0$. Thus, on the boundary, the minimum value of f is 0 and the maximum is 9.

In step 3 we compare these values with the value $f(1, 1) = 1$ at the critical point and conclude that the absolute maximum value of f on D is $f(3, 0) = 9$ and the absolute minimum value is $f(0, 0) = f(2, 2) = 0$. Figure 13 shows the graph of f. ■

We close this section by giving a proof of the first part of the Second Derivatives Test. Notice the similarity to the proof of the Second Derivative Test for functions of one variable. Parts (b) and (c) have similar proofs.

PROOF OF THEOREM 3, PART (a) We compute the second-order directional derivative of f in the direction of $\mathbf{u} = \langle h, k \rangle$. The first-order derivative is given by Theorem 12.6.3:

$$D_{\mathbf{u}} f = f_x h + f_y k$$

Applying this theorem a second time, we have

$$\begin{aligned} D_{\mathbf{u}}^2 f = D_{\mathbf{u}}(D_{\mathbf{u}} f) &= \frac{\partial}{\partial x}(D_{\mathbf{u}} f)h + \frac{\partial}{\partial y}(D_{\mathbf{u}} f)k \\ &= (f_{xx}h + f_{yx}k)h + (f_{xy}h + f_{yy}k)k \\ &= f_{xx}h^2 + 2f_{xy}hk + f_{yy}k^2 \qquad \text{(by Clairaut's Theorem)} \end{aligned}$$

If we complete the square in this expression, we obtain

$$D_{\mathbf{u}}^2 f = f_{xx}\left(h + \frac{f_{xy}}{f_{xx}}k\right)^2 + \frac{k^2}{f_{xx}}(f_{xx}f_{yy} - f_{xy}^2) \qquad (10)$$

We are given that $f_{xx}(a, b) > 0$ and $D(a, b) > 0$. But f_{xx} and $D = f_{xx}f_{yy} - f_{xy}^2$ are continuous functions, so there is a disk B with center (a, b) and radius $\delta > 0$ such that $f_{xx}(x, y) > 0$ and $D(x, y) > 0$ whenever (x, y) is in B. Therefore, by looking at Equation 10, we see that $D_{\mathbf{u}}^2 f(x, y) > 0$ whenever (x, y) is in B. This means that if C is the curve obtained by intersecting the graph of f with the vertical plane through $P(a, b, f(a, b))$ in the direction of $\mathbf{u}$, then C is concave upward on an interval of length 2δ. This is true in the direction of every vector $\mathbf{u}$, so if we restrict (x, y) to lie in B, the graph of f lies above its horizontal tangent plane at P. Thus $f(x, y) \geqslant f(a, b)$ whenever (x, y) is in B. This shows that $f(a, b)$ is a local minimum. □

EXERCISES 12.7

1–16 ■ Find the local maximum and minimum values and saddle points of the function. If you have three-dimensional graphing software, use a computer to graph the function with a domain and viewpoint that reveal all the important aspects of the function.

1. $f(x, y) = x^2 + y^2 + 4x - 6y$
2. $f(x, y) = 4x^2 + y^2 - 4x + 2y$
3. $f(x, y) = 2x^2 + y^2 + 2xy + 2x + 2y$
4. $f(x, y) = 1 + 2xy - x^2 - y^2$
5. $f(x, y) = x^2 + y^2 + x^2y + 4$
6. $f(x, y) = 2x^3 + xy^2 + 5x^2 + y^2$
7. $f(x, y) = x^3 - 3xy + y^3$
8. $f(x, y) = y\sqrt{x} - y^2 - x + 6y$
9. $f(x, y) = xy - 2x - y$
10. $f(x, y) = xy(1 - x - y)$
11. $f(x, y) = \dfrac{x^2y^2 - 8x + y}{xy}$
12. $f(x, y) = x^2 + y^2 + \dfrac{1}{x^2y^2}$
13. $f(x, y) = e^x \cos y$
14. $f(x, y) = (2x - x^2)(2y - y^2)$
15. $f(x, y) = x \sin y$
16. $f(x, y) = \dfrac{(x + y + 1)^2}{x^2 + y^2 + 1}$

17–20 ■ Use a graph and/or level curves to estimate the local maximum and minimum values and saddle points of the function. Then use calculus to find these values precisely.

17. $f(x, y) = 3x^2y + y^3 - 3x^2 - 3y^2 + 2$
18. $f(x, y) = xye^{-x^2-y^2}$

19. $f(x, y) = \sin x + \sin y + \sin(x + y)$,
$0 \leq x \leq 2\pi$, $0 \leq y \leq 2\pi$

20. $f(x, y) = \sin x + \sin y + \cos(x + y)$,
$0 \leq x \leq \pi/4$, $0 \leq y \leq \pi/4$

21–24 ■ Use a graphing calculator or computer as in Example 4 (or Newton's method or a CAS) to find the critical points of f correct to three decimal places. Then classify the critical points and find the highest or lowest points on the graph.

21. $f(x, y) = x^4 - 5x^2 + y^2 + 3x + 2$

22. $f(x, y) = 5 - 10xy - 4x^2 + 3y - y^4$

23. $f(x, y) = 2x + 4x^2 - y^2 + 2xy^2 - x^4 - y^4$

24. $f(x, y) = e^x + y^4 - x^3 + 4\cos y$

25–32 ■ Find the absolute maximum and minimum values of f on the set D.

25. $f(x, y) = 5 - 3x + 4y$, D is the closed triangular region with vertices $(0, 0)$, $(4, 0)$, and $(4, 5)$

26. $f(x, y) = x^2 + 2xy + 3y^2$, D is the closed triangular region with vertices $(-1, 1)$, $(2, 1)$, and $(-1, -2)$

27. $f(x, y) = x^2 + y^2 + x^2y + 4$, $D = \{(x, y) \mid |x| \leq 1, |y| \leq 1\}$

28. $f(x, y) = y\sqrt{x} - y^2 - x + 6y$,
$D = \{(x, y) \mid 0 \leq x \leq 9, 0 \leq y \leq 5\}$

29. $f(x, y) = 1 + xy - x - y$, D is the region bounded by the parabola $y = x^2$ and the line $y = 4$

30. $f(x, y) = 2x^2 + x + y^2 - 2$, $D = \{(x, y) \mid x^2 + y^2 \leq 4\}$

31. $f(x, y) = 2x^3 + y^4$, $D = \{(x, y) \mid x^2 + y^2 \leq 1\}$

32. $f(x, y) = x^3 - 3x - y^3 + 12y$, D is the quadrilateral whose vertices are $(-2, 3)$, $(2, 3)$, $(2, 2)$, and $(-2, -2)$.

33. It is impossible for a continuous function of one variable to have two local maxima and no local minimum. But there do exist functions of *two* variables with this property. Show that the function

$$f(x, y) = -(x^2 - 1)^2 - (x^2y - x - 1)^2$$

has only two critical points, but has local maxima at both of them. Then use a computer to produce a graph with a carefully chosen domain and viewpoint to see how this is possible.

34. If a function of one variable is continuous on an interval and has only one critical number, then a local maximum has to be an absolute maximum. But this is not true for functions of two variables. Show that the function

$$f(x, y) = 3xe^y - x^3 - e^{3y}$$

has exactly one critical point, and that f has a local maximum there that is not an absolute maximum. Then use a computer to produce a graph with a carefully chosen domain and viewpoint to see how this is possible.

35. Find the point on the plane $x + 2y + 3z = 4$ that is closest to the origin.

36. Find the point on the plane $2x - y + z = 1$ that is closest to the point $(-4, 1, 3)$.

37. Find the shortest distance from the point $(2, -2, 3)$ to the plane $6x + 4y - 3z = 2$.

38. Find the shortest distance from the point (x_0, y_0, z_0) to the plane $Ax + By + Cz + D = 0$.

39. Find the points on the surface $z^2 = xy + 1$ that are closest to the origin.

40. Find the points on the surface $x^2y^2z = 1$ that are closest to the origin.

41. Find three positive numbers whose sum is 100 and whose product is a maximum.

42. Find three positive numbers x, y, and z whose sum is 100 such that $x^ay^bz^c$ is a maximum.

43. Find the volume of the largest rectangular box with edges parallel to the axes that can be inscribed in the ellipsoid $9x^2 + 36y^2 + 4z^2 = 36$.

44. Solve the problem in Exercise 43 for a general ellipsoid $x^2/a^2 + y^2/b^2 + z^2/c^2 = 1$.

45. Find the volume of the largest rectangular box in the first octant with three faces in the coordinate planes and one vertex in the plane $x + 2y + 3z = 6$.

46. Solve the problem in Exercise 45 for a general plane $x/a + y/b + z/c = 1$, where a, b, c are positive numbers.

47. Find the dimensions of a rectangular box of maximum volume such that the sum of the lengths of its 12 edges is a constant c.

48. Find the dimensions of the rectangular box with largest volume if the total surface area is given as 64 cm^2.

49. A cardboard box without a lid is to have a volume of 32,000 cm^3. Find the dimensions that minimize the amount of cardboard used.

50. The base of an aquarium with given volume V is made of slate and the sides are made of glass. If slate costs five times as much (per unit area) as glass, find the dimensions of the aquarium that minimize the cost of the materials.

51. Suppose that a scientist has reason to believe that two quantities x and y are related linearly, that is, $y = mx + b$, at least approximately, for some values of m and b. The

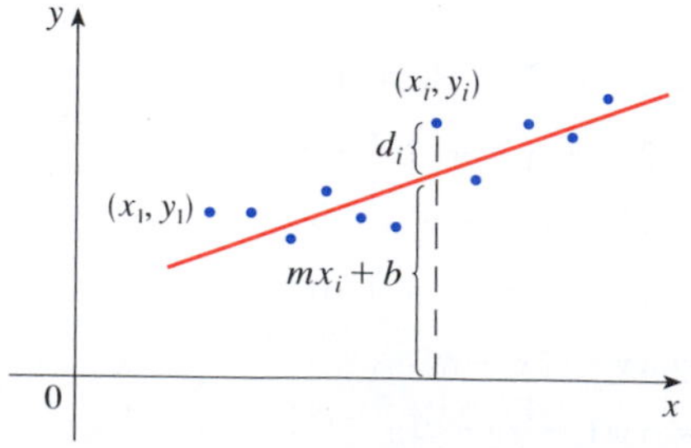

scientist performs an experiment and collects data in the form of points (x_1, y_1), (x_2, y_2), $\ldots$, (x_n, y_n), and then plots these points. The points do not lie exactly on a straight line, so the scientist wants to find constants m and b so that the line $y = mx + b$ "fits" the points as well as possible. (See the figure.) Let $d_i = y_i - (mx_i + b)$ be the vertical deviation of the point (x_i, y_i) from the line. The **method of least squares** determines m and b so as to minimize $\sum_{i=1}^{n} d_i^2$, the sum of the squares of these deviations. Show that, according to this method, the line of best fit is obtained when

$$m \sum_{i=1}^{n} x_i + bn = \sum_{i=1}^{n} y_i$$

$$m \sum_{i=1}^{n} x_i^2 + b \sum_{i=1}^{n} x_i = \sum_{i=1}^{n} x_i y_i$$

Thus, the line is found by solving these two equations in the two unknowns m and b.

52. (a) Use the method of least squares (Exercise 51) to find the straight line $y = mx + b$ that best fits the data points $(1, 5.1)$, $(2, 6.8)$, $(3, 9.4)$, and $(4, 10.5)$.

(b) As a check on your calculations in part (a), plot the four points and graph the line.

53. The following data give the height x (in inches) and weight y (in pounds) of ten 18-year-old boys. Use the method of least squares (Exercise 51) to fit these data to a straight line. Then use it to predict the weight of an 18-year-old boy who is 6 ft tall.

x	69	65	71	73	68	63	70	67	69	70
y	138	127	178	185	141	122	158	135	145	162

54. Find an equation of the plane that passes through the point $(1, 2, 3)$ and cuts off the smallest volume in the first octant.

12.8 LAGRANGE MULTIPLIERS

In Example 6 in Section 12.7 we maximized a volume function $V = xyz$ subject to the constraint $2xz + 2yz + xy = 12$, which expressed the side condition that the surface area was $12\ \text{m}^2$. In this section we present Lagrange's method for maximizing or minimizing a general function $f(x, y, z)$ subject to a constraint (or side condition) of the form $g(x, y, z) = k$.

It's easier to explain the geometric basis of Lagrange's method for functions of two variables. So we start by trying to find the extreme values of $f(x, y)$ subject to a constraint of the form $g(x, y) = k$. In other words, we seek the extreme values of $f(x, y)$ when the point (x, y) is restricted to lie on the level curve $g(x, y) = k$. Figure 1 shows this curve together with several level curves of f. These have the equations $f(x, y) = c$, where $c = 7, 8, 9, 10, 11$. To maximize $f(x, y)$ subject to $g(x, y) = k$ is to find the largest value of c such that the level curve $f(x, y) = c$ intersects $g(x, y) = k$. It appears from Figure 1 that this happens when these curves just touch each other, that is, when they have a common tangent line. This means that the normal lines at the point (x_0, y_0) where they touch are identical. Therefore, the gradient vectors are parallel; that is, $\nabla f(x_0, y_0) = \lambda \nabla g(x_0, y_0)$ for some scalar λ.

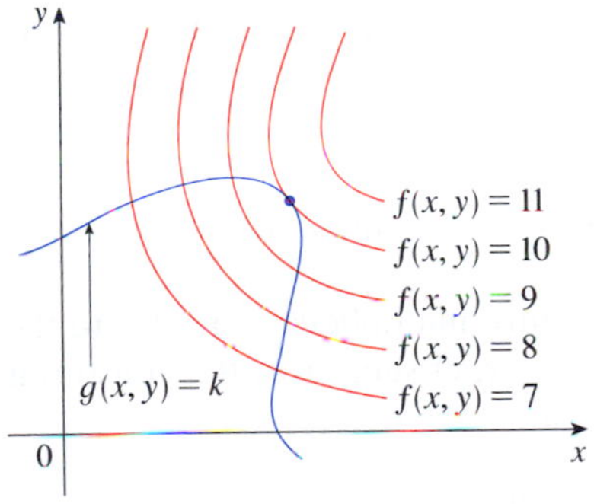

FIGURE 1

This kind of argument also applies to the problem of finding the extreme values of $f(x, y, z)$ subject to the constraint $g(x, y, z) = k$. Thus, the point (x, y, z) is restricted to lie on the level surface S with equation $g(x, y, z) = k$. Instead of the level curves in Figure 1, we consider the level surfaces $f(x, y, z) = c$ and argue that if the maximum value of f is $f(x_0, y_0, z_0) = c$, then the level surface $f(x, y, z) = c$ is tangent to the level surface $g(x, y, z) = k$ and so the corresponding gradient vectors are parallel.

This intuitive argument can be made precise as follows. Suppose that f has an extreme value at a point $P(x_0, y_0, z_0)$ on S and let C be a curve with vector equation $\mathbf{r}(t) = \langle x(t), y(t), z(t) \rangle$ that lies on S and passes through P. If t_0 is the parameter value corresponding to the point P, then $\mathbf{r}(t_0) = \langle x_0, y_0, z_0 \rangle$. The composite function $h(t) = f(x(t), y(t), z(t))$ represents the values that f takes on the curve C. Since f has an extreme value at (x_0, y_0, z_0), it follows that h has an extreme value at t_0, so $h'(t_0) = 0$.

But if f is differentiable, we can use the Chain Rule to write

$$\begin{aligned} 0 = h'(t_0) &= f_x(x_0, y_0, z_0)x'(t_0) + f_y(x_0, y_0, z_0)y'(t_0) + f_z(x_0, y_0, z_0)z'(t_0) \\ &= \nabla f(x_0, y_0, z_0) \cdot \mathbf{r}'(t_0) \end{aligned}$$

This shows that the gradient vector $\nabla f(x_0, y_0, z_0)$ is orthogonal to the tangent vector $\mathbf{r}'(t_0)$ to every such curve C. But we already know from Section 12.6 that the gradient vector of g, $\nabla g(x_0, y_0, z_0)$, is also orthogonal to $\mathbf{r}'(t_0)$ (see Equation 12.6.18). This means that the gradient vectors $\nabla f(x_0, y_0, z_0)$ and $\nabla g(x_0, y_0, z_0)$ must be parallel. Therefore, if $\nabla g(x_0, y_0, z_0) \neq \mathbf{0}$, there is a number λ such that

(1) $$\nabla f(x_0, y_0, z_0) = \lambda \nabla g(x_0, y_0, z_0)$$

Lagrange multipliers are named after the French-Italian mathematician Joseph-Louis Lagrange (1736–1813). See page 315 for a biographical sketch of Lagrange.

The number λ in Equation 1 is called a **Lagrange multiplier**. The procedure based on Equation 1 is called the **method of Lagrange multipliers** and is as follows.

> To find the maximum and minimum values of $f(x, y, z)$ subject to the constraint $g(x, y, z) = k$ (assuming that these extreme values exist):
>
> (a) Find all values of x, y, z, and λ such that
>
> $$\nabla f(x, y, z) = \lambda \nabla g(x, y, z)$$
>
> and $$g(x, y, z) = k$$
>
> (b) Evaluate f at all the points (x, y, z) that arise from step (a). The largest of these values is the maximum value of f; the smallest is the minimum value of f.

If we write the vector equation $\nabla f = \lambda \nabla g$ in terms of its components, then the equations in step (a) become

$$f_x = \lambda g_x \qquad f_y = \lambda g_y \qquad f_z = \lambda g_z \qquad g(x, y, z) = k$$

This is a system of four equations in the four unknowns x, y, z, and λ, but it is not necessary to find explicit values for λ.

For functions of two variables the method of Lagrange multipliers is similar to the method just described. To find the extreme values of $f(x, y)$ subject to the constraint $g(x, y) = k$, we look for values of x, y, and λ such that

$$\nabla f(x, y) = \lambda \nabla g(x, y) \qquad \text{and} \qquad g(x, y) = k$$

This amounts to solving three equations in three unknowns:

$$f_x = \lambda g_x \qquad f_y = \lambda g_y \qquad g(x, y) = k$$

Our first illustration of Lagrange's method is to reconsider the problem given in Example 6 in Section 12.7.

EXAMPLE 1 A rectangular box without a lid is to be made from 12 m^2 of cardboard. Find the maximum volume of such a box.

SOLUTION As in Example 6 in Section 12.7 we let x, y, and z be the length, width, and height, respectively, of the box in meters. Then we wish to maximize

$$V = xyz$$

subject to the constraint

$$g(x, y, z) = 2xz + 2yz + xy = 12$$

Using the method of Lagrange multipliers, we look for values of x, y, z, and λ such that $\nabla V = \lambda \nabla g$ and $g(x, y, z) = 12$. This gives the equations

$$V_x = \lambda g_x \qquad V_y = \lambda g_y \qquad V_z = \lambda g_z \qquad 2xz + 2yz + xy = 12$$

which become

$$yz = \lambda(2z + y) \tag{2}$$

$$xz = \lambda(2z + x) \tag{3}$$

$$xy = \lambda(2x + 2y) \tag{4}$$

$$2xz + 2yz + xy = 12 \tag{5}$$

There are no general rules for solving systems of equations. Sometimes some ingenuity is required. In the present example you might notice that if we multiply (2) by x, (3) by y, and (4) by z, then the left sides of these equations will be identical. Doing this, we have

$$xyz = \lambda(2xz + xy) \tag{6}$$

$$xyz = \lambda(2yz + xy) \tag{7}$$

$$xyz = \lambda(2xz + 2yz) \tag{8}$$

We observe that $\lambda \neq 0$ because $\lambda = 0$ would imply $yz = xz = xy = 0$ from (2), (3), and (4) and this would contradict (5). Therefore, from (6) and (7) we have

$$2xz + xy = 2yz + xy$$

which gives $xz = yz$. But $z \neq 0$ (since $z = 0$ would give $V = 0$), so $x = y$. From (7) and (8) we have

$$2yz + xy = 2xz + 2yz$$

which gives $2xz = xy$ and so (since $x \neq 0$) $y = 2z$. If we now put $x = y = 2z$ in (5), we get

$$4z^2 + 4z^2 + 4z^2 = 12$$

Since x, y, and z are all positive, we therefore have $z = 1$, $x = 2$, and $y = 2$ as before. ■

EXAMPLE 2 Find the extreme values of the function $f(x, y) = x^2 + 2y^2$ on the circle $x^2 + y^2 = 1$.

SOLUTION We are asked for the extreme values of f subject to the constraint $g(x, y) = x^2 + y^2 = 1$. Using Lagrange multipliers, we solve the equations $\nabla f = \lambda \nabla g$, $g(x, y) = 1$, which can be written as

$$f_x = \lambda g_x \qquad f_y = \lambda g_y \qquad g(x, y) = 1$$

In geometric terms, Example 2 asks for the highest and lowest points on the curve C in Figure 2 that lies on the paraboloid $z = x^2 + 2y^2$ and directly above the constraint circle $x^2 + y^2 = 1$.

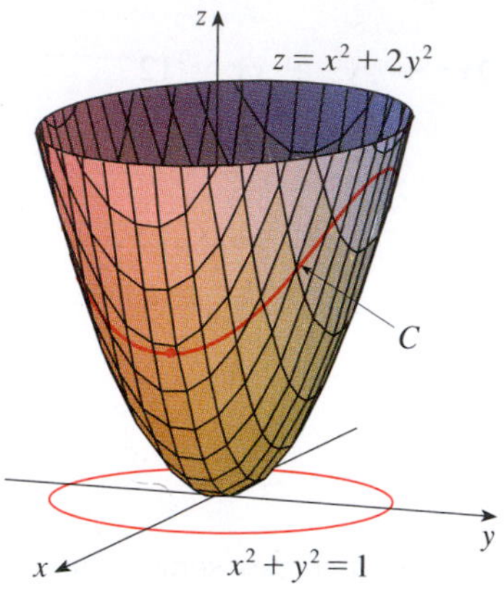

FIGURE 2

The geometry behind the use of Lagrange multipliers in Example 2 is shown in Figure 3. The extreme values of $f(x, y) = x^2 + 2y^2$ correspond to the level curves that touch the circle $x^2 + y^2 = 1$.

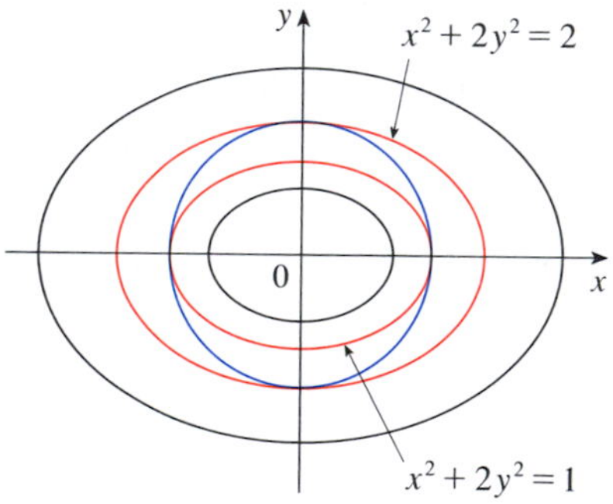

FIGURE 3

or as

(9) $$2x = 2x\lambda$$

(10) $$4y = 2y\lambda$$

(11) $$x^2 + y^2 = 1$$

From (9) we have $x = 0$ or $\lambda = 1$. If $x = 0$, then (11) gives $y = \pm 1$. If $\lambda = 1$, then $y = 0$ from (10), so then (11) gives $x = \pm 1$. Therefore, f has possible extreme values at the points $(0, 1)$, $(0, -1)$, $(1, 0)$, and $(-1, 0)$. Evaluating f at these four points, we find that

$$f(0, 1) = 2 \qquad f(0, -1) = 2 \qquad f(1, 0) = 1 \qquad f(-1, 0) = 1$$

Therefore, the maximum value of f on the circle $x^2 + y^2 = 1$ is $f(0, \pm 1) = 2$ and the minimum value is $f(\pm 1, 0) = 1$. Checking with Figure 2, we see that these values look reasonable. ■

EXAMPLE 3 Find the extreme values of $f(x, y) = x^2 + 2y^2$ on the disk $x^2 + y^2 \leqslant 1$.

SOLUTION According to the procedure in (12.7.9), we compare the values of f at the critical points with those at the points on the boundary. Since $f_x = 2x$ and $f_y = 4y$, the only critical point is $(0, 0)$. We compare the value of f at that point with the extreme values on the boundary from Example 2:

$$f(0, 0) = 0 \qquad f(\pm 1, 0) = 1 \qquad f(0, \pm 1) = 2$$

Therefore, the maximum value of f on the disk $x^2 + y^2 \leqslant 1$ is $f(0, \pm 1) = 2$ and the minimum value is $f(0, 0) = 0$. ■

EXAMPLE 4 Find the points on the sphere $x^2 + y^2 + z^2 = 4$ that are closest to and farthest from the point $(3, 1, -1)$.

SOLUTION The distance from a point (x, y, z) to the point $(3, 1, -1)$ is

$$d = \sqrt{(x - 3)^2 + (y - 1)^2 + (z + 1)^2}$$

but the algebra is simpler if we instead maximize and minimize the square of the distance:

$$d^2 = f(x, y, z) = (x - 3)^2 + (y - 1)^2 + (z + 1)^2$$

The constraint is that the point (x, y, z) lies on the sphere, that is,

$$g(x, y, z) = x^2 + y^2 + z^2 = 4$$

According to the method of Lagrange multipliers, we solve $\nabla f = \lambda \nabla g$, $g = 4$. This gives

(12) $$2(x - 3) = 2x\lambda$$

(13) $$2(y - 1) = 2y\lambda$$

(14) $$2(z + 1) = 2z\lambda$$

(15) $$x^2 + y^2 + z^2 = 4$$

The simplest way to solve these equations is to solve for x, y, and z in terms of λ from (12), (13), and (14), and then substitute these values into (15). From (12) we have

$$x - 3 = x\lambda \quad \text{or} \quad x(1 - \lambda) = 3 \quad \text{or} \quad x = \frac{3}{1 - \lambda}$$

[Note that $1 - \lambda \neq 0$ because $\lambda = 1$ is impossible from (12).] Similarly, (13) and (14) give

$$y = \frac{1}{1 - \lambda} \qquad z = -\frac{1}{1 - \lambda}$$

Figure 4 shows the sphere and the nearest point P in Example 4. Can you see how to find the coordinates of P without using calculus?

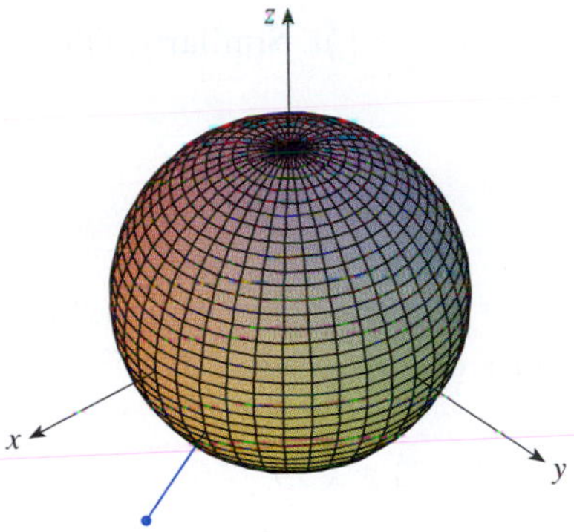

FIGURE 4

Therefore, from (15) we have

$$\frac{3^2}{(1 - \lambda)^2} + \frac{1^2}{(1 - \lambda)^2} + \frac{(-1)^2}{(1 - \lambda)^2} = 4$$

which gives $(1 - \lambda)^2 = \frac{11}{4}$, $1 - \lambda = \pm\sqrt{11}/2$, so

$$\lambda = 1 \pm \frac{\sqrt{11}}{2}$$

These values of λ then give the corresponding points (x, y, z):

$$\left(\frac{6}{\sqrt{11}}, \frac{2}{\sqrt{11}}, -\frac{2}{\sqrt{11}}\right) \quad \text{and} \quad \left(-\frac{6}{\sqrt{11}}, -\frac{2}{\sqrt{11}}, \frac{2}{\sqrt{11}}\right)$$

It is easy to see that f has a smaller value at the first of these points, so the closest point is $(6/\sqrt{11}, 2/\sqrt{11}, -2/\sqrt{11})$ and the farthest is $(-6/\sqrt{11}, -2/\sqrt{11}, 2/\sqrt{11})$. ■

TWO CONSTRAINTS

Suppose now that we want to find the maximum and minimum values of $f(x, y, z)$ subject to two constraints (side conditions) of the form $g(x, y, z) = k$ and $h(x, y, z) = c$. Geometrically, this means that we are looking for the extreme values of f when (x, y, z) is restricted to lie on the curve of intersection of the level surfaces $g(x, y, z) = k$ and $h(x, y, z) = c$. It can be shown that if an extreme value occurs at (x_0, y_0, z_0), then the gradient vector $\nabla f(x_0, y_0, z_0)$ is in the plane determined by $\nabla g(x_0, y_0, z_0)$ and $\nabla h(x_0, y_0, z_0)$. (We assume that these gradient vectors are not parallel.) So there are numbers λ and μ (called Lagrange multipliers) such that

(16) $$\nabla f(x_0, y_0, z_0) = \lambda \nabla g(x_0, y_0, z_0) + \mu \nabla h(x_0, y_0, z_0)$$

In this case Lagrange's method is to look for extreme values by solving five equations in the five unknowns x, y, z, λ, and μ. These equations are obtained by writing Equation 16 in terms of its components and using the constraint equations:

$$f_x = \lambda g_x + \mu h_x$$
$$f_y = \lambda g_y + \mu h_y$$
$$f_z = \lambda g_z + \mu h_z$$
$$g(x, y, z) = k$$
$$h(x, y, z) = c$$

The cylinder $x^2 + y^2 = 1$ intersects the plane $x - y + z = 1$ in an ellipse (Figure 5). Example 5 asks for the maximum value of f when (x, y, z) is restricted to lie on the ellipse.

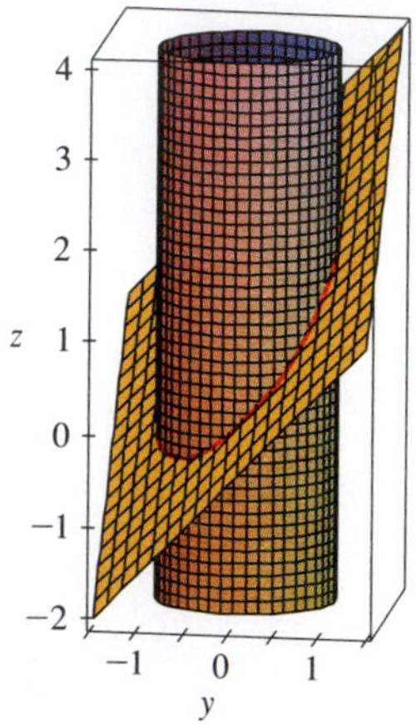

FIGURE 5

EXAMPLE 5 Find the maximum value of the function $f(x, y, z) = x + 2y + 3z$ on the curve of intersection of the plane $x - y + z = 1$ and the cylinder $x^2 + y^2 = 1$.

SOLUTION We maximize the function $f(x, y, z) = x + 2y + 3z$ subject to the constraints $g(x, y, z) = x - y + z = 1$ and $h(x, y, z) = x^2 + y^2 = 1$. The Lagrange condition is $\nabla f = \lambda \nabla g + \mu \nabla h$, so we solve the equations

$$1 = \lambda + 2x\mu \tag{17}$$

$$2 = -\lambda + 2y\mu \tag{18}$$

$$3 = \lambda \tag{19}$$

$$x - y + z = 1 \tag{20}$$

$$x^2 + y^2 = 1 \tag{21}$$

Putting $\lambda = 3$ [from (19)] in (17), we get $2x\mu = -2$, so $x = -1/\mu$. Similarly, (18) gives $y = 5/(2\mu)$. Substitution in (21) then gives

$$\frac{1}{\mu^2} + \frac{25}{4\mu^2} = 1$$

and so $\mu^2 = \frac{29}{4}$, $\mu = \pm\sqrt{29}/2$. Then $x = \mp 2/\sqrt{29}$, $y = \pm 5/\sqrt{29}$, and, from (20), $z = 1 - x + y = 1 \pm 7/\sqrt{29}$. The corresponding values of f are

$$\mp\frac{2}{\sqrt{29}} + 2\left(\pm\frac{5}{\sqrt{29}}\right) + 3\left(1 \pm \frac{7}{\sqrt{29}}\right) = 3 \pm \sqrt{29}$$

Therefore, the maximum value of f on the given curve is $3 + \sqrt{29}$. ■

EXERCISES 12.8

1–15 ■ Use Lagrange multipliers to find the maximum and minimum values of the function subject to the given constraint or constraints.

1. $f(x, y) = x^2 - y^2$; $x^2 + y^2 = 1$

2. $f(x, y) = 2x + y$; $x^2 + 4y^2 = 1$

3. $f(x, y) = xy$; $9x^2 + y^2 = 4$

4. $f(x, y) = x^2 + y^2$; $x^4 + y^4 = 1$

5. $f(x, y, z) = x + 3y + 5z$; $x^2 + y^2 + z^2 = 1$

6. $f(x, y, z) = x - y + 3z$; $x^2 + y^2 + 4z^2 = 4$

7. $f(x, y, z) = xyz$; $x^2 + 2y^2 + 3z^2 = 6$

8. $f(x, y, z) = x^2y^2z^2$; $x^2 + y^2 + z^2 = 1$

9. $f(x, y, z) = x^2 + y^2 + z^2$; $x^4 + y^4 + z^4 = 1$

10. $f(x, y, z) = x^4 + y^4 + z^4$; $x^2 + y^2 + z^2 = 1$

11. $f(x, y, z, t) = x + y + z + t$; $x^2 + y^2 + z^2 + t^2 = 1$

12. $f(x_1, x_2, \ldots, x_n) = x_1 + x_2 + \cdots + x_n$; $x_1^2 + x_2^2 + \cdots + x_n^2 = 1$

13. $f(x, y, z) = x + 2y$; $x + y + z = 1$, $y^2 + z^2 = 4$

14. $f(x, y, z) = 3x - y - 3z$; $x + y - z = 0$, $x^2 + 2z^2 = 1$

15. $f(x, y, z) = yz + xy$; $xy = 1$, $y^2 + z^2 = 1$

16–17 ■ Find the extreme values of f on the region described by the inequality.

16. $f(x, y) = 2x^2 + 3y^2 - 4x - 5$, $x^2 + y^2 \leq 16$

17. $f(x, y) = e^{-xy}$, $x^2 + 4y^2 \leq 1$

18. (a) Use a graphing calculator or computer to graph the circle $x^2 + y^2 = 1$. On the same screen graph several curves of the form $x^2 + y = c$ until you find two that just touch the circle. What is the significance of the values of c for these two curves?

(b) Use Lagrange multipliers to find the extreme values of $f(x, y) = x^2 + y$ subject to the constraint $x^2 + y^2 = 1$. Compare your answers with those in part (a).

CAS **19.** (a) If your computer algebra system plots implicitly defined curves, use it to estimate the minimum and maximum values of $f(x, y) = x^3 + y^3 + 3xy$ subject to the constraint $(x - 3)^2 + (y - 3)^2 = 9$ by graphical methods.

(b) Solve the problem in part (a) with the aid of Lagrange multipliers. Use your CAS to solve the equations numerically. Compare your answers with those in part (a).

20. Find the maximum and minimum volumes of a rectangular box whose surface area is 1500 cm^2 and whose total edge length is 200 cm.

21. A manufacturer produces a quantity Q of a certain product and Q depends on the amount x of labor used and the amount y of capital. A simple model that is sometimes used to express Q explicitly as a function of x and y is the Cobb-Douglas production function: $Q = Kx^{\alpha}y^{1-\alpha}$, where K and α are constants with $K > 0$ and $0 < \alpha < 1$. If the cost of a unit of labor is m and the cost of a unit of capital is n, and the company can spend only p dollars as its total budget, then maximizing the production Q is subject to the constraint $mx + ny = p$. Show that the maximum production occurs when

$$x = \frac{\alpha p}{m} \qquad \text{and} \qquad y = \frac{(1-\alpha)p}{n}$$

22. Referring to Exercise 21, we now suppose that the production is fixed at $Kx^{\alpha}y^{1-\alpha} = Q$, where Q is a constant. What values of x and y minimize the cost function $C(x, y) = mx + ny$?

23. Use Lagrange multipliers to prove that the rectangle with maximum area that has a given perimeter p is a square.

24. Use Lagrange multipliers to prove that the triangle with maximum area that has a given perimeter p is equilateral. [*Hint:* Use Heron's formula for the area: $A = \sqrt{s(s-x)(s-y)(s-z)}$, where $s = p/2$ and x, y, z are the lengths of the sides.]

25–40 ■ Use Lagrange multipliers to give an alternate solution to the indicated exercise in Section 12.7.

25. Exercise 35
26. Exercise 36
27. Exercise 37
28. Exercise 38
29. Exercise 39
30. Exercise 40
31. Exercise 41
32. Exercise 42
33. Exercise 43
34. Exercise 44
35. Exercise 45
36. Exercise 46
37. Exercise 47
38. Exercise 48
39. Exercise 49
40. Exercise 50

41. The plane $x + y + 2z = 2$ intersects the paraboloid $z = x^2 + y^2$ in an ellipse. Find the points on this ellipse that are nearest to and farthest from the origin.

42. (a) Maximize $\Sigma_{i=1}^{n} x_i y_i$ subject to the constraints $\Sigma_{i=1}^{n} x_i^2 = 1$ and $\Sigma_{i=1}^{n} y_i^2 = 1$.

(b) Put

$$x_i = \frac{a_i}{\sqrt{\Sigma a_i^2}} \qquad \text{and} \qquad y_i = \frac{b_i}{\sqrt{\Sigma b_i^2}}$$

to show that

$$\sum a_i b_i \leq \sqrt{\Sigma a_i^2}\,\sqrt{\Sigma b_i^2}$$

for any numbers $a_1, \ldots, a_n, b_1, \ldots, b_n$. This inequality is known as the Cauchy-Schwarz Inequality.

12 REVIEW

KEY TOPICS ■ Define, state, or discuss the following.

1. Functions of two variables and their graphs
2. Level curves
3. Functions of three variables
4. Level surfaces
5. Limit of a function of two or three variables
6. Continuity of a function of two or three variables
7. Partial derivatives
8. Clairaut's Theorem
9. Equation of a tangent plane to the surface $z = f(x, y)$
10. Equation of a tangent plane to the surface $F(x, y, z) = k$
11. Differential
12. Differentiable function
13. The Chain Rule
14. Implicit differentiation
15. Directional derivative
16. Gradient and its significance
17. Local and absolute maximum and minimum values
18. Critical point
19. Saddle point
20. Second Derivatives Test
21. Extreme Value Theorem for functions of two variables
22. Lagrange multipliers

EXERCISES

1–12 ■ Determine whether the statement is true or false.

1. $f_y(a, b) = \lim_{y \to b} \dfrac{f(a, y) - f(a, b)}{y - b}$

2. There exists a function f whose partial derivatives are $f_x(x, y) = x + y^2$ and $f_y(x, y) = x - y^2$.

3. $f_{xy} = \dfrac{\partial^2 f}{\partial x\,\partial y}$

4. $D_{\mathbf{k}}\, f(x, y, z) = f_z(x, y, z)$

5. If $f(x, y) \to L$ as $(x, y) \to (a, b)$ along every straight line through (a, b), then $\lim_{(x,y)\to(a,b)} f(x, y) = L$.

6. If $f_x(a, b)$ and $f_y(a, b)$ both exist, then f is differentiable at (a, b).

7. If f has a local minimum at (a, b) and f is differentiable at (a, b), then $\nabla f(a, b) = \mathbf{0}$.

8. $\displaystyle\lim_{(x,y)\to(1,1)} \frac{x - y}{x^2 - y^2} = \lim_{(x,y)\to(1,1)} \frac{1}{x + y} = \frac{1}{2}$

9. If $f(x, y) = \ln y$, then $\nabla f(x, y) = 1/y$.

10. If (a, b) is a critical point of f and $f_{xx}(a, b) f_{yy}(a, b) < [f_{xy}(a, b)]^2$, then f has a saddle point at (a, b).

11. If $f(x, y) = \sin x + \sin y$, then $-\sqrt{2} \leqslant D_{\mathbf{u}}\, f(x, y) \leqslant \sqrt{2}$.

12. If $f(x, y)$ has two local maxima, then f must have a local minimum.

13–16 ■ Find and sketch the domain of the function.

13. $f(x, y) = \dfrac{\ln(x + y + 1)}{x - 1}$

14. $f(x, y) = \sqrt{\sin \pi(x^2 + y^2)}$

15. $f(x, y) = \cos^{-1}x + \tan^{-1}y$

16. $f(x, y, z) = \sqrt{z - x^2 - y^2}$

17–18 ■ Sketch the graph of the function.

17. $f(x, y) = 1 - x^2 - y^2$

18. $f(x, y) = \sqrt{x^2 + y^2 - 1}$

19–20 ■ Sketch several level curves of the function.

19. $f(x, y) = e^{-(x^2+y^2)}$

20. $f(x, y) = x^2 + 4y$

21–22 ■ Evaluate the limit or show that it does not exist.

21. $\displaystyle\lim_{(x,y)\to(0,0)} \frac{x^2y^2}{x^2 + 2y^2}$

22. $\displaystyle\lim_{(x,y)\to(0,0)} \frac{2xy}{x^2 + 2y^2}$

23–28 ■ Find the first partial derivatives.

23. $f(x, y) = 3x^4 - x\sqrt{y}$

24. $g(x, y) = \dfrac{x}{\sqrt{x + 2y}}$

25. $f(s, t) = e^{2s}\cos \pi t$

26. $g(r, s) = r \sin\sqrt{r^2 + s^2}$

27. $f(x, y, z) = xy^z$

28. $g(u, v, w) = w^2 e^{u/v}$

29–32 ■ Find all second partial derivatives of f.

29. $f(x, y) = x^2y^3 - 2x^4 + y^2$

30. $f(x, y) = x^3 \ln(x - y)$

31. $f(x, y, z) = xy^2z^3$

32. $f(x, y, z) = xe^y \cos z$

33. If $u = x^y$, show that $\dfrac{x}{y}\dfrac{\partial u}{\partial x} + \dfrac{1}{\ln x}\dfrac{\partial u}{\partial y} = 2u$.

34. If $\rho = \sqrt{x^2 + y^2 + z^2}$, show that

$$\frac{\partial^2 \rho}{\partial x^2} + \frac{\partial^2 \rho}{\partial y^2} + \frac{\partial^2 \rho}{\partial z^2} = \frac{2}{\rho}$$

35–39 ■ Find an equation of the tangent plane to the given surface at the specified point.

35. $z = x^2 + y^2 + 4y, \quad (0, 1, 5)$

36. $z = xe^y, \quad (1, 0, 1)$

37. $xy^2z^3 = 12, \quad (3, 2, 1)$

38. $x^2 + y^2 + z^2 = 29, \quad (2, 3, 4)$

39. $x^2 + 2y^2 - 3z^2 = 14, \quad (3, 2, -1)$

40. Use a computer to graph the surface $z = x^3 + 2xy$ and its tangent plane and normal line at $(1, 2, 5)$ on the same screen. Choose the domain and viewpoint so that you get a good view of all three objects.

41. Find the points on the sphere $x^2 + y^2 + z^2 = 1$ where the tangent plane is parallel to the plane $2x + y - 3z = 2$.

42. Find dz if $z = x^2 \tan^{-1}y$.

43. Use differentials to approximate the number

$$(1.98)^3\sqrt{(3.01)^2 + (3.97)^2}$$

44. The two legs of a right triangle are measured as 5 m and 12 m, respectively, with a possible error in measurement of at most 0.2 cm in each. Use differentials to estimate the maximum error in the calculated value of (a) the area of the triangle and (b) the length of the hypotenuse.

45. If $w = \sqrt{x} + y^2/z$, where $x = e^{2t}$, $y = t^3 + 4t$, and $z = t^2 - 4$, use the Chain Rule to find dw/dt.

46. If $z = \cos xy + y \cos x$, where $x = u^2 + v$ and $y = u - v^2$, use the Chain Rule to find $\partial z/\partial u$ and $\partial z/\partial v$.

47. If $z = y + f(x^2 - y^2)$, where f is differentiable, show that

$$y\frac{\partial z}{\partial x} + x\frac{\partial z}{\partial y} = x$$

48. The length x of a side of a triangle is increasing at a rate of 3 in/s, the length y of another side is decreasing at a rate of 2 in/s, and the contained angle θ is increasing at a rate of 0.05 radian/s. How fast is the area of the triangle changing when $x = 40$ in., $y = 50$ in., and $\theta = \pi/6$?

49. If $z = f(u, v)$, where $u = xy$, $v = y/x$, and f has continuous second partial derivatives, show that

$$x^2 \frac{\partial^2 z}{\partial x^2} - y^2 \frac{\partial^2 z}{\partial y^2} = -4uv \frac{\partial^2 z}{\partial u\, \partial v} + 2v \frac{\partial z}{\partial v}$$

50. If $yz^4 + x^2z^3 = e^{xyz}$, find $\dfrac{\partial z}{\partial x}$ and $\dfrac{\partial z}{\partial y}$.

51. Find the gradient of the function $f(x, y, z) = z^2 e^{x\sqrt{y}}$.

52. (a) How are the directional derivative and the gradient of $f(x, y)$ related?
(b) When is the directional derivative of f a maximum?
(c) When is it a minimum?
(d) When is it 0?
(e) When is it half of its maximum value?

53–54 ▪ Find the directional derivative of f at the given point in the indicated direction.

53. $f(x, y) = 2\sqrt{x} - y^2$, $(1, 5)$, in the direction toward the point $(4, 1)$

54. $f(x, y, z) = x^2y + x\sqrt{1 + z}$, $(1, 2, 3)$, in the direction of $\mathbf{v} = 2\,\mathbf{i} + \mathbf{j} - 2\,\mathbf{k}$

55. Find the maximum rate of change of $f(x, y) = x^2y + \sqrt{y}$ at the point $(2, 1)$. In which direction does it occur?

56. Find the direction in which $f(x, y, z) = ze^{xy}$ increases most rapidly at the point $(0, 1, 2)$. What is the maximum rate of increase?

57. Evaluate $\displaystyle\lim_{(x, y)\to(1, 0)} \tan^{-1}\left(\frac{x^2 + y^2}{(x - 1)^2 + y^2}\right)$.

58. Find parametric equations of the tangent line at the point $(-2, 2, 4)$ to the curve of intersection of the surface $z = 2x^2 - y^2$ and the plane $z = 4$.

59–62 ▪ Find the local maximum and minimum values and saddle points of the function. If you have three-dimensional graphing software, use a computer to graph the function with a domain and viewpoint that reveal all the important aspects of the function.

59. $f(x, y) = x^2 - xy + y^2 + 9x - 6y + 10$

60. $f(x, y) = x^3 - 6xy + 8y^3$

61. $f(x, y) = 3xy - x^2y - xy^2$

62. $f(x, y) = (x^2 + y)e^{y/2}$

63–64 ▪ Find the absolute maximum and minimum values of f on the set D.

63. $f(x, y) = 4xy^2 - x^2y^2 - xy^3$; D is the closed triangular region in the xy-plane with vertices $(0, 0)$, $(0, 6)$, and $(6, 0)$

64. $f(x, y) = e^{-x^2-y^2}(x^2 + 2y^2)$; D is the disk $x^2 + y^2 \leq 4$

65. Use a graph and/or level curves to estimate the local maximum and minimum values and saddle points of $f(x, y) = x^3 - 3x + y^4 - 2y^2$. Then use calculus to find these values precisely.

66. Use a graphing calculator or computer (or Newton's method or a computer algebra system) to find the critical points of $f(x, y) = 12 + 10y - 2x^2 - 8xy - y^4$ correct to three decimal places. Then classify the critical points and find the highest point on the graph.

67–70 ▪ Use Lagrange multipliers to find the maximum and minimum values of f subject to the given constraints.

67. $f(x, y) = x^2y$; $x^2 + y^2 = 1$

68. $f(x, y) = \dfrac{1}{x} + \dfrac{1}{y}$; $\dfrac{1}{x^2} + \dfrac{1}{y^2} = 1$

69. $f(x, y, z) = x + y + z$; $\dfrac{1}{x} + \dfrac{1}{y} + \dfrac{1}{z} = 1$

70. $f(x, y, z) = x^2 + 2y^2 + 3z^2$; $x + y + z = 1$, $x - y + 2z = 2$

71. Find the points on the surface $xy^2z^3 = 2$ that are closest to the origin.

72. A package in the shape of a rectangular box can be mailed parcel post if the sum of its length and girth (the perimeter of a cross-section perpendicular to the length) is at most 84 in. Find the dimensions of the package with largest volume that can be mailed parcel post.

73. A pentagon is formed by placing an isosceles triangle on a rectangle, as shown in the figure. If the pentagon has fixed perimeter P, find the lengths of the sides of the pentagon that maximize the area of the pentagon.

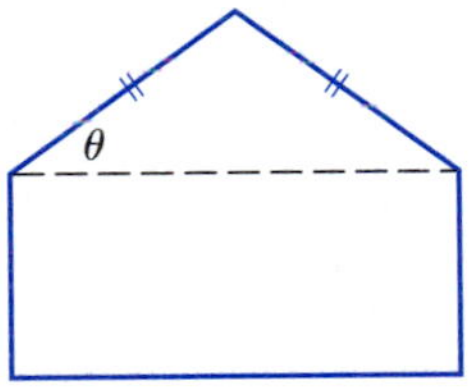

74. A particle of mass m moves on the surface $z = f(x, y)$. Let $x = x(t)$, $y = y(t)$ be the x- and y-coordinates of the particle at time t.
(a) Find the velocity vector $\mathbf{v}$ and the kinetic energy $K = \frac{1}{2}m|\mathbf{v}|^2$ of the particle.
(b) Determine the acceleration vector $\mathbf{a}$.
(c) Let $z = x^2 + y^2$ and $x(t) = t\cos t$, $y(t) = t\sin t$. Find the velocity vector, the kinetic energy, and the acceleration vector.

PROBLEMS PLUS

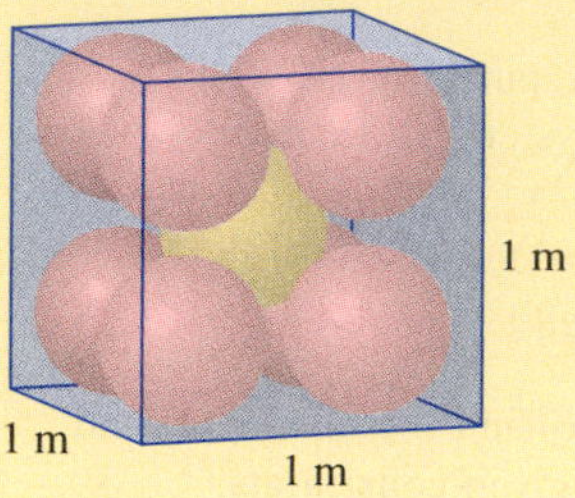

FIGURE FOR PROBLEM 1

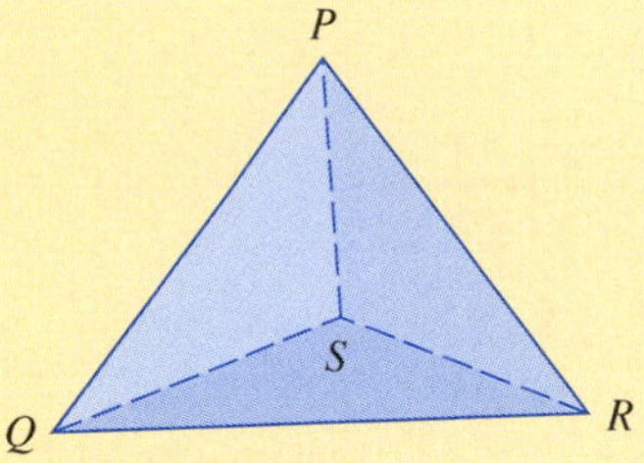

FIGURE FOR PROBLEM 2

1. Each edge of a cubical box has length 1 m. The box contains nine spherical balls with the same radius r. The center of one ball is at the center of the cube and it touches the other eight balls. Each of the other eight balls touches three sides of the box. Thus the balls are tightly packed in the box. Find r. (If you have trouble with this problem, read about the problem-solving strategy entitled *Use analogy* in Section 4 of Review and Preview.)

2. A tetrahedron is a solid with four vertices, P, Q, R, and S, and four triangular faces. (See the figure.)
   (a) Let $\mathbf{v}_1$, $\mathbf{v}_2$, $\mathbf{v}_3$, and $\mathbf{v}_4$ be vectors with magnitudes equal to the areas of the faces opposite the vertices P, Q, R, and S, respectively, and directions parallel to the outward normals of the respective faces. Show that
   $$\mathbf{v}_1 + \mathbf{v}_2 + \mathbf{v}_3 + \mathbf{v}_4 = \mathbf{0}$$
   (b) The volume V of a tetrahedron is one-third the distance from a vertex to the opposite face, times the area of that face. Find a formula for the volume of the tetrahedron with vertices P, Q, R, and S.
   (c) Determine the volume of the tetrahedron whose vertices are $P(1, 1, 1)$, $Q(1, 2, 3)$, $R(1, 1, 2)$, and $S(3, -1, 2)$.

3. Suppose the tetrahedron in Problem 2 has a trirectangular vertex S. (This means that the three angles at S are all right angles.) Let A, B, and C be the areas of the three faces that meet at S and let D be the area of the opposite face PQR. Show that
   $$D^2 = A^2 + B^2 + C^2$$
   (This is a three-dimensional version of the Pythagorean Theorem.)

4. Let B be a solid box with length L, width W, and height H. Let S be the set of all points that are a distance at most 1 from some point of B. Express the volume of S in terms of L, W, and H.

5. A rectangle with length L and width W is cut into four smaller rectangles by two lines parallel to the sides. Find the maximum and minimum values of the sum of the squares of the areas of the smaller rectangles.

6. Find the curvature of the curve with parametric equations
   $$x = \int_0^t \sin \tfrac{1}{2}\pi\theta^2 \, d\theta \qquad y = \int_0^t \cos \tfrac{1}{2}\pi\theta^2 \, d\theta$$

7. Suppose f is a differentiable function of one variable. Show that all tangent planes to the surface $z = xf(y/x)$ intersect in a common point.

8. The plane $x + y + z = 24$ intersects the paraboloid $z = x^2 + y^2$ in an ellipse. Find the highest and lowest points on the ellipse.

9. Newton's method for approximating a root of an equation $f(x) = 0$ (see Section 3.12) can be adapted to approximating a solution of a system of equations $f(x, y) = 0$ and $g(x, y) = 0$. The surfaces $z = f(x, y)$ and $z = g(x, y)$ intersect in a curve that intersects the xy-plane at the point (r, s), which is the solution of the system. If an initial approxi-

mation (x_1, y_1) is close to this point, then the tangent planes to the surfaces at (x_1, y_1) intersect in a straight line that intersects the xy-plane in a point (x_2, y_2), which should be closer to (r, s). (Compare with Figure 1 in Section 3.12.) Show that

$$x_2 = x_1 - \frac{fg_y - f_y g}{f_x g_y - f_y g_x} \quad \text{and} \quad y_2 = y_1 - \frac{f_x g - fg_x}{f_x g_y - f_y g_x}$$

where f, g, and their partial derivatives are evaluated at (x_1, y_1). If we continue this procedure, we obtain successive approximations (x_n, y_n).

10. Use Newton's method (see Problem 9) to find the points of intersection of the curves $x^2 + y^2 = 4$ and $\sin y = x$ correct to four decimal places.

11. The temperature at a point (x, y) on a metal plate is given by $T(x, y) = 100 - x^2 - 2y^2$. A heat-seeking particle starts at the point $(4, 2)$ and moves at each instant in the direction of maximum temperature increase.
(a) In what direction does the particle move initially?
(b) Draw a contour map of T and sketch the path followed by the particle.
(c) Find an equation of this path by setting up and solving two differential equations.

12. A cable has radius r and length L and is wound around a spool with radius R without overlapping. What is the shortest length along the spool that is covered by the cable?

13. If the ellipse $x^2/a^2 + y^2/b^2 = 1$ is to contain the circle $x^2 + y^2 = 2y$ in its interior, what values of a and b minimize the area of the ellipse?

13

MULTIPLE INTEGRALS

■ Mathematics is the tool specially suited to dealing with abstract concepts of any kind and there is no limit to its power in this field. For this reason a book on the new physics, if not purely descriptive of experimental work, must be essentially mathematical.

P. A. M. DIRAC

In this chapter we extend the idea of a definite integral to double and triple integrals of functions of two or three variables. These ideas are then used to compute volumes, masses, and centroids of more general regions than we were able to consider in Chapters 7 and 9.

13.1 DOUBLE INTEGRALS OVER RECTANGLES

As a guide to defining the double integral of a function of two variables, let's first recall the basic facts concerning definite integrals of functions of a single variable. If f is defined on a closed interval $[a, b]$, we start by taking a partition P of $[a, b]$ into subintervals $[x_{i-1}, x_i]$ where

$$a = x_0 < x_1 < \cdots < x_{n-1} < x_n = b$$

We choose points x_i^* in $[x_{i-1}, x_i]$ and let $\Delta x_i = x_i - x_{i-1}$ and $\|P\| = \max\{\Delta x_i\}$. Then we form the Riemann sum

(1) $$\sum_{i=1}^{n} f(x_i^*)\,\Delta x_i$$

and take the limit of such sums as $\|P\| \to 0$ to obtain the definite integral of f from a to b:

(2) $$\int_a^b f(x)\,dx = \lim_{\|P\|\to 0} \sum_{i=1}^{n} f(x_i^*)\,\Delta x_i$$

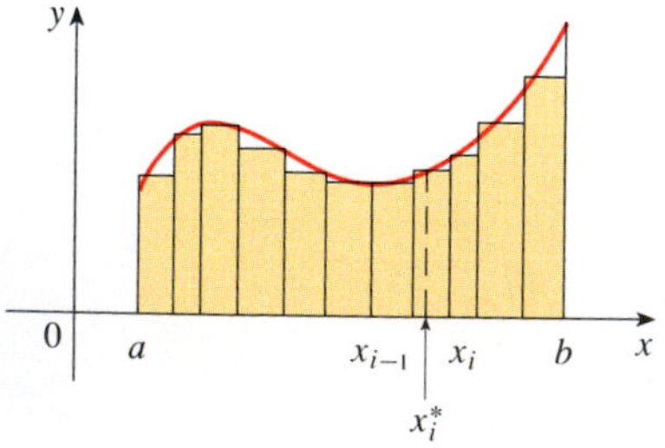

FIGURE 1

In the special case where $f(x) \geqslant 0$, the Riemann sum can be interpreted as the sum of the areas of the approximating rectangles in Figure 1, and $\int_a^b f(x)\,dx$ represents the area under the curve $y = f(x)$ from a to b.

In a similar manner we define the double integral of a function f of two variables that is defined on a closed rectangle

$$R = [a, b] \times [c, d] = \{(x, y) \in \mathbb{R}^2 \mid a \leqslant x \leqslant b,\ c \leqslant y \leqslant d\}$$

The first step is to take a partition P of R into subrectangles. This is accomplished by partitioning the intervals $[a, b]$ and $[c, d]$ as follows:

$$a = x_0 < x_1 < \cdots < x_{i-1} < x_i < \cdots < x_m = b$$

$$c = y_0 < y_1 < \cdots < y_{j-1} < y_j < \cdots < y_n = d$$

By drawing lines parallel to the coordinate axes through these partition points as in Figure 2, we form the subrectangles

$$R_{ij} = [x_{i-1}, x_i] \times [y_{j-1}, y_j] = \{(x, y) \mid x_{i-1} \leqslant x \leqslant x_i, y_{j-1} \leqslant y \leqslant y_j\}$$

for $i = 1, \ldots, m$ and $j = 1, \ldots, n$. There are mn of these subrectangles, and they cover R. If we let

$$\Delta x_i = x_i - x_{i-1} \qquad \Delta y_j = y_j - y_{j-1}$$

then the area of R_{ij} is $\qquad \Delta A_{ij} = \Delta x_i \, \Delta y_j$

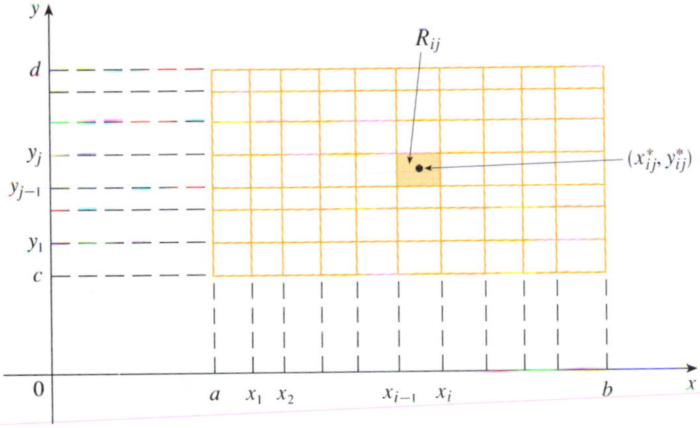

FIGURE 2
Partition of a rectangle

Next we choose a point (x_{ij}^*, y_{ij}^*) in R_{ij} and, by analogy with the Riemann sum (1), we form the **double Riemann sum**

$$\sum_{i=1}^{m} \sum_{j=1}^{n} f(x_{ij}^*, y_{ij}^*) \, \Delta A_{ij} \tag{3}$$

This double sum means that for each subrectangle we evaluate f at the chosen point and multiply by the area of the subrectangle, and then we add the results. When written out in full, the double Riemann sum (3) becomes

$$\begin{aligned} & f(x_{11}^*, y_{11}^*) \, \Delta A_{11} + f(x_{12}^*, y_{12}^*) \, \Delta A_{12} + \cdots + f(x_{1n}^*, y_{1n}^*) \, \Delta A_{1n} \\ & + f(x_{21}^*, y_{21}^*) \, \Delta A_{21} + f(x_{22}^*, y_{22}^*) \, \Delta A_{22} + \cdots + f(x_{2n}^*, y_{2n}^*) \, \Delta A_{2n} \\ & + \cdots \\ & + f(x_{m1}^*, y_{m1}^*) \, \Delta A_{m1} + f(x_{m2}^*, y_{m2}^*) \, \Delta A_{m2} + \cdots + f(x_{mn}^*, y_{mn}^*) \, \Delta A_{mn} \end{aligned}$$

The final ingredient that we need for the definition of a double integral is the **norm** of the partition, which is the length of the longest diagonal of all the subrectangles R_{ij} and is denoted by $\|P\|$. Note that if we let $\|P\| \to 0$, then the partition becomes finer. By analogy with the defining equation for a single integral (2), we make the following definition.

(4) DEFINITION The **double integral** of f over the rectangle R is

$$\iint_R f(x, y) \, dA = \lim_{\|P\| \to 0} \sum_{i=1}^{m} \sum_{j=1}^{n} f(x_{ij}^*, y_{ij}^*) \, \Delta A_{ij}$$

if this limit exists.

NOTE 1: The precise meaning of the limit that defines the double integral is as follows:

$$\iint_R f(x, y)\,dA = I$$

means that for every $\varepsilon > 0$ there is a corresponding number $\delta > 0$ such that

$$\left| I - \sum_{i=1}^{m} \sum_{j=1}^{n} f(x_{ij}^*, y_{ij}^*)\,\Delta A_{ij} \right| < \varepsilon$$

for all partitions P of R with $\|P\| < \delta$ and for all possible choices of (x_{ij}^*, y_{ij}^*) in R_{ij}. In other words, Riemann sums can be made arbitrarily close to I by taking sufficiently fine partitions.

NOTE 2: In view of the fact that $\Delta A_{ij} = \Delta x_i\,\Delta y_j$, another notation that is sometimes used for the double integral is

$$\iint_R f(x, y)\,dA = \iint_R f(x, y)\,dx\,dy$$

NOTE 3: A function f is called **integrable** if the limit in Definition 4 exists. It is shown in courses on advanced calculus that all continuous functions are integrable. In fact, the double integral of f exists provided that f is "not too discontinuous." In particular, if f is bounded on R and is continuous there, except on a finite number of smooth curves, then f is integrable over R.

EXAMPLE 1 Find an approximate value for the integral $\iint_R (x - 3y^2)\,dA$, where $R = \{(x, y) \mid 0 \leqslant x \leqslant 2,\ 1 \leqslant y \leqslant 2\}$, by computing the double Riemann sum with partition lines $x = 1$ and $y = \frac{3}{2}$ and taking (x_{ij}^*, y_{ij}^*) to be the center of each rectangle.

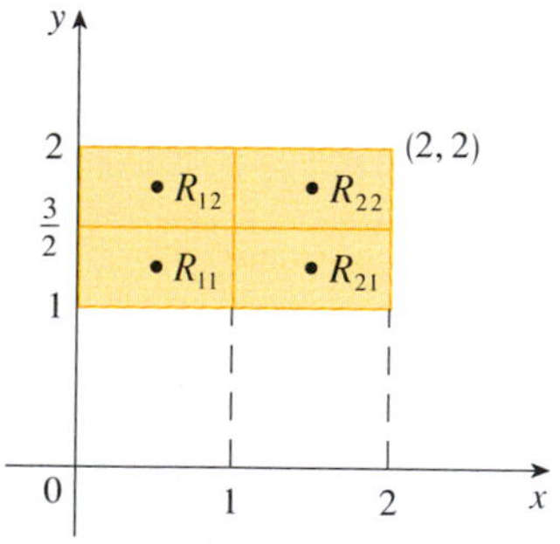

FIGURE 3

SOLUTION The partition is shown in Figure 3. The area of each subrectangle is $\Delta A_{ij} = \frac{1}{2}$, (x_{ij}^*, y_{ij}^*) is the center of R_{ij}, and $f(x, y) = x - 3y^2$. So the corresponding Riemann sum is

$$\begin{aligned}
\sum_{i=1}^{2} \sum_{j=1}^{2} f(x_{ij}^*, y_{ij}^*)\,\Delta A_{ij}
&= f(x_{11}^*, y_{11}^*)\,\Delta A_{11} + f(x_{12}^*, y_{12}^*)\,\Delta A_{12} + f(x_{21}^*, y_{21}^*)\,\Delta A_{21} + f(x_{22}^*, y_{22}^*)\,\Delta A_{22} \\
&= f\left(\tfrac{1}{2}, \tfrac{5}{4}\right)\Delta A_{11} + f\left(\tfrac{1}{2}, \tfrac{7}{4}\right)\Delta A_{12} + f\left(\tfrac{3}{2}, \tfrac{5}{4}\right)\Delta A_{21} + f\left(\tfrac{3}{2}, \tfrac{7}{4}\right)\Delta A_{22} \\
&= \left(-\tfrac{67}{16}\right)\tfrac{1}{2} + \left(-\tfrac{139}{16}\right)\tfrac{1}{2} + \left(-\tfrac{51}{16}\right)\tfrac{1}{2} + \left(-\tfrac{123}{16}\right)\tfrac{1}{2} \\
&= -\tfrac{95}{8} = -11.875
\end{aligned}$$

Thus, we have

$$\iint_R (x - 3y^2)\,dA \approx -11.875$$

■

NOTE 4: It is very difficult to calculate double integrals directly from the definition. In the next section we will develop an efficient method for computing double integrals and then we will see that the exact value of the double integral in Example 1 is -12.

Number of subrectangles	Midpoint Rule approximations
1	−11.5000
4	−11.8750
16	−11.9687
64	−11.9922
256	−11.9980
1024	−11.9995

NOTE 5: The methods that we used for approximating single integrals (the Midpoint Rule, the Trapezoidal Rule, Simpson's Rule) all have analogues for double integrals. In the **Midpoint Rule for double integrals** we use a double Riemann sum to approximate the double integral, where the point (x_{ij}^*, y_{ij}^*) in R_{ij} is chosen to be the center $(\bar{x}_i, \bar{y}_j)$ of R_{ij}. In other words, $\bar{x}_i$ is the midpoint of $[x_{i-1}, x_i]$ and $\bar{y}_j$ is the midpoint of $[y_{j-1}, y_j]$. In effect, the approximation we got in Example 1 is the result of using the Midpoint Rule with four subrectangles. If we keep dividing each subrectangle into four smaller ones with similar shape, we get the Midpoint Rule approximations displayed in the chart in the margin. Notice how these approximations approach the exact value of the double integral, -12.

INTERPRETATION OF DOUBLE INTEGRALS AS VOLUMES

Just as single integrals of positive functions can be interpreted as areas, double integrals of positive functions can be interpreted as volumes in the following way. Suppose that $f(x, y) \geqslant 0$ and f is defined on the rectangle $R = [a, b] \times [c, d]$. The graph of f is a surface with equation $z = f(x, y)$. Let S be the solid that lies above R and under the graph of f, that is,

$$S = \{(x, y, z) \in \mathbb{R}^3 \mid 0 \leqslant z \leqslant f(x, y),\ (x, y) \in R\}$$

If we partition R into subrectangles R_{ij} and choose (x_{ij}^*, y_{ij}^*) in R_{ij}, then we can approximate the part of S that lies above R_{ij} by a thin rectangular box (or "column") with base R_{ij} and height $f(x_{ij}^*, y_{ij}^*)$ as shown in Figure 4. (Compare with Figure 1.) The volume of this box is

$$V_{ij} = f(x_{ij}^*, y_{ij}^*)\,\Delta A_{ij}$$

If we follow this procedure for all the rectangles and add the volumes of the corresponding boxes, we get an approximation to the total volume of S:

$$V \approx \sum_{i=1}^{m} \sum_{j=1}^{n} f(x_{ij}^*, y_{ij}^*)\,\Delta A_{ij} \tag{5}$$

(See Figure 5.) Notice that this double sum is just the Riemann sum (3). Our intuition tells us that the approximation given in (5) becomes better if we use a finer partition P

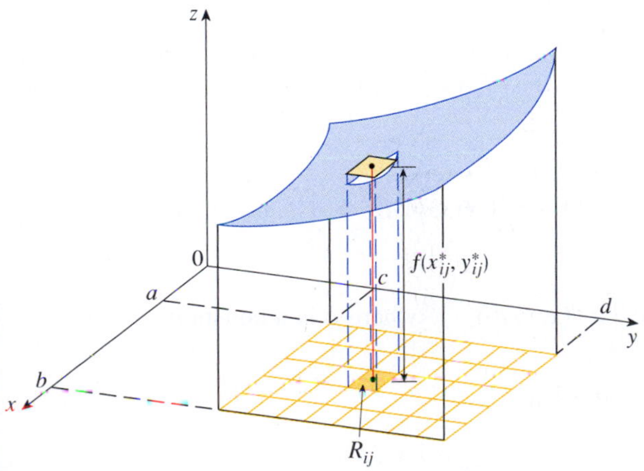

FIGURE 4

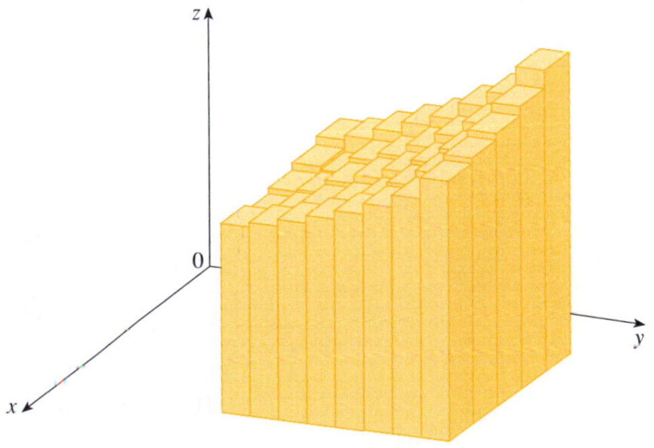

FIGURE 5

and so we would expect that

$$V = \lim_{\|P\|\to 0} \sum_{i=1}^{m} \sum_{j=1}^{n} f(x_{ij}^*, y_{ij}^*)\,\Delta A_{ij} = \iint_R f(x, y)\,dA$$

(the latter equality being true by the definition of a double integral). It can be shown, using our previous definition of volume (7.2.1), that this expectation is correct.

(6) THEOREM If $f(x, y) \geqslant 0$ and f is continuous on the rectangle R, then the volume V of the solid that lies above R and under the surface $z = f(x, y)$ is

$$V = \iint_R f(x, y)\,dA$$

It should be remembered that the interpretation of a double integral as a volume is valid only when the integrand f is a *positive* function. The integrand in Example 1 is not a positive function, so its integral is not a volume.

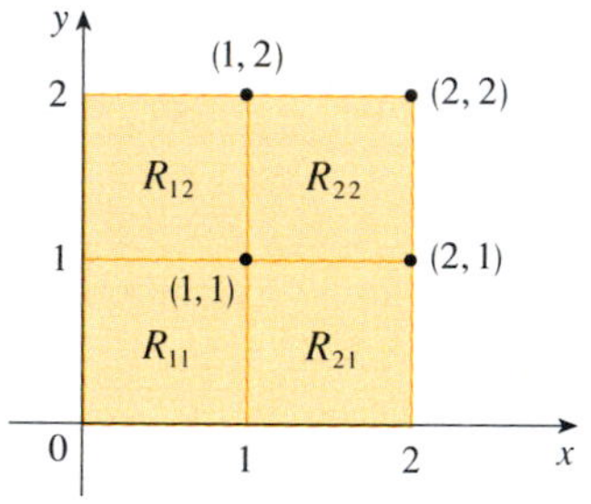

FIGURE 6

EXAMPLE 2 Estimate the volume of the solid that lies above the square $R = [0, 2] \times [0, 2]$ and below the elliptic paraboloid $z = 16 - x^2 - 2y^2$. Use the partition of R into four equal squares and choose (x_{ij}^*, y_{ij}^*) to be the upper right corner of R_{ij}. Sketch the solid and the approximating rectangular boxes.

SOLUTION The partition is shown in Figure 6. The paraboloid is the graph of $f(x, y) = 16 - x^2 - 2y^2$ and the area of each square is 1. Approximating the volume by the Riemann sum, we have

$$\begin{aligned} V &\approx f(1, 1)\,\Delta A_{11} + f(1, 2)\,\Delta A_{12} + f(2, 1)\,\Delta A_{21} + f(2, 2)\,\Delta A_{22} \\ &= 13(1) + 7(1) + 10(1) + 4(1) = 34 \end{aligned}$$

This is the volume of the approximating rectangular boxes shown in Figure 7. ■

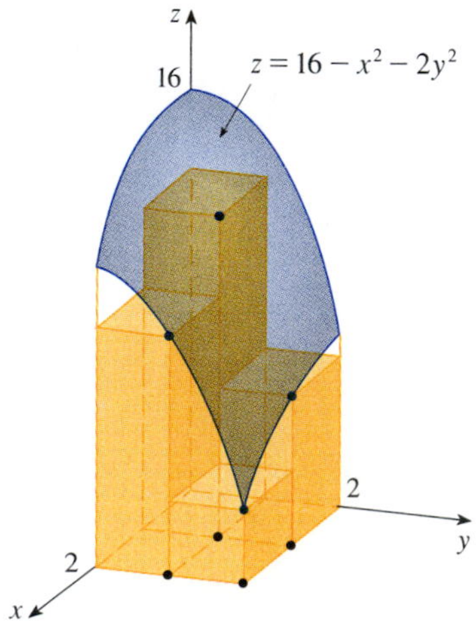

FIGURE 7

We list here three properties of double integrals that can be proved in the same manner as in Section 6.3. We assume that all of the integrals exist. Properties 7 and 8 are referred to as the *linearity* of the integral.

$$\text{(7)} \qquad \iint_R [f(x, y) + g(x, y)]\,dA = \iint_R f(x, y)\,dA + \iint_R g(x, y)\,dA$$

$$\text{(8)} \qquad \iint_R c f(x, y)\,dA = c \iint_R f(x, y)\,dA \qquad \text{where } c \text{ is a constant}$$

If $f(x, y) \geqslant g(x, y)$ for all (x, y) in R, then

$$\text{(9)} \qquad \iint_R f(x, y)\,dA \geqslant \iint_R g(x, y)\,dA$$

EXERCISES 13.1

1. Find approximations to $\iint_R (x - 3y^2)\, dA$ using the same partition as in Example 1 but choosing (x_{ij}^*, y_{ij}^*) to be the (a) upper left corner, (b) upper right corner, (c) lower left corner, (d) lower right corner of R_{ij}.

2. Find the approximation to the volume in Example 2 if (x_{ij}^*, y_{ij}^*) is chosen to be the center of each square.

3–8 ■ Calculate the double Riemann sum of f for the partition of R given by the indicated lines and the given choice of (x_{ij}^*, y_{ij}^*). Also calculate the norm of the partition.

3. $f(x, y) = x^2 + 4y$, $R = \{(x, y) \mid 0 \leq x \leq 2, 0 \leq y \leq 3\}$, $x = 1, y = 1, y = 2$; $(x_{ij}^*, y_{ij}^*) =$ upper right corner of R_{ij}

4. $f(x, y) = x^2 + 4y$, $R = \{(x, y) \mid 0 \leq x \leq 2, 0 \leq y \leq 3\}$, $x = 1, y = 1, y = 2$; $(x_{ij}^*, y_{ij}^*) =$ center of R_{ij}

5. $f(x, y) = xy - y^2$, $R = \{(x, y) \mid 0 \leq x \leq 5, 0 \leq y \leq 4\}$, $x = 1, x = 2, x = 3, x = 4, y = 2$; $(x_{ij}^*, y_{ij}^*) =$ center of R_{ij}

6. $f(x, y) = 2x + x^2y$, $R = \{(x, y) \mid -2 \leq x \leq 2, -1 \leq y \leq 1\}$, $x = -1, x = 0, x = 1, y = -\frac{1}{2}, y = 0, y = \frac{1}{2}$; $(x_{ij}^*, y_{ij}^*) =$ lower left corner of R_{ij}

7. $f(x, y) = x^2 - y^2$, $R = [0, 5] \times [0, 2]$, $x = 1, x = 3, x = 4, y = \frac{1}{2}, y = 1$; $(x_{ij}^*, y_{ij}^*) =$ upper left corner of R_{ij}

8. $f(x, y) = 5xy^2$, $R = [1, 3] \times [1, 4]$, $x = 1.8, x = 2.5, y = 2, y = 3$; $(x_{ij}^*, y_{ij}^*) =$ lower right corner of R_{ij}

9. Let V be the volume of the solid that lies under the graph of $f(x, y) = \sqrt{52 - x^2 - y^2}$ and above the rectangle given by $2 \leq x \leq 4, 2 \leq y \leq 6$. For the partition of R defined by the lines $x = 3$ and $y = 4$, let L and U be the Riemann sums computed using lower left corners and upper right corners, respectively. Without calculating the numbers V, L, and U, arrange them in increasing order and explain your reasoning.

10. The figure shows level curves of a function f in the square $R = [0, 1] \times [0, 1]$. Use them to estimate $\iint_R f(x, y)\, dA$ to the nearest integer.

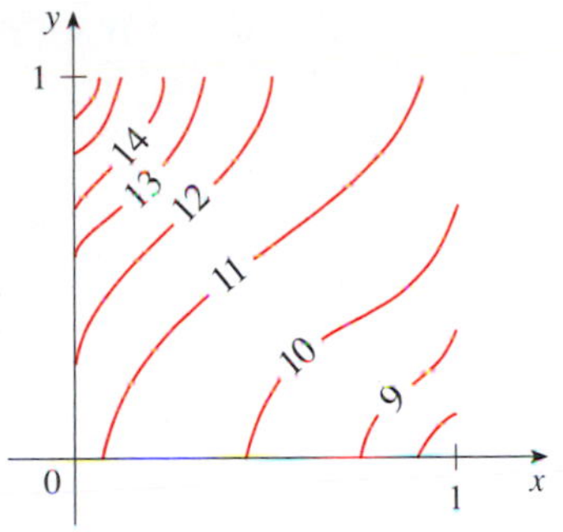

11. Use a programmable calculator or computer (or the sum command on a CAS) to estimate $\iint_R e^{-x^2-y^2}\, dA$, where $R = [0, 1] \times [0, 1]$. Use the Midpoint Rule with the following numbers of squares of equal size: 1, 4, 16, 64, 256, and 1024.

12. Repeat Exercise 11 for the integral $\iint_R \cos(x^4 + y^4)\, dA$.

13. Evaluate the integral $\iint_R (4 - 2y)\, dA$, where $R = [0, 1] \times [0, 1]$, by first identifying it as the volume of a solid.

14. The integral $\iint_R \sqrt{9 - y^2}\, dA$, where $R = [0, 4] \times [0, 2]$, represents the volume of a solid. Sketch the solid.

15. If f is a constant function, $f(x, y) = k$, and $R = [a, b] \times [c, d]$, show that

$$\iint_R k\, dA = k(b - a)(d - c)$$

16. If $R = [0, 1] \times [0, 1]$, show that

$$0 \leq \iint_R \sin(x + y)\, dA \leq 1$$

13.2 ITERATED INTEGRALS

Recall that it is usually difficult to evaluate single integrals from first principles, but the Fundamental Theorem of Calculus provides a much easier method. The evaluation of double integrals from first principles is even more difficult, but in this section we see how to express a double integral as an iterated integral, which can then be evaluated by calculating two single integrals.

Suppose f is a function of two variables that is integrable over the rectangle $R = [a, b] \times [c, d]$. We use the notation $\int_c^d f(x, y)\, dy$ to mean that x is held fixed and $f(x, y)$ is integrated with respect to y from $y = c$ to $y = d$. This procedure is called *partial integration with respect to y*. (Notice its similarity to partial differentiation.)

Now $\int_c^d f(x, y)\,dy$ is a number that depends on the value of x, so it defines a function of x:

$$A(x) = \int_c^d f(x, y)\,dy$$

If we now integrate the function A with respect to x from $x = a$ to $x = b$, we get

$$\int_a^b A(x)\,dx = \int_a^b \left[\int_c^d f(x, y)\,dy \right] dx \tag{1}$$

The integral on the right side of Equation 1 is called an **iterated integral**. Usually the brackets are omitted. Thus

$$\int_a^b \int_c^d f(x, y)\,dy\,dx = \int_a^b \left[\int_c^d f(x, y)\,dy \right] dx \tag{2}$$

means that we first integrate with respect to y from c to d and then with respect to x from a to b.

Similarly, the iterated integral

$$\int_c^d \int_a^b f(x, y)\,dx\,dy = \int_c^d \left[\int_a^b f(x, y)\,dx \right] dy \tag{3}$$

means that we first integrate with respect to x (holding y fixed) from $x = a$ to $x = b$ and then we integrate the resulting function of y with respect to y from $y = c$ to $y = d$. Notice that in both Equations 2 and 3 we work from the inside out.

EXAMPLE 1 Evaluate the iterated integrals:

(a) $\displaystyle\int_0^3 \int_1^2 x^2 y\,dy\,dx$ (b) $\displaystyle\int_1^2 \int_0^3 x^2 y\,dx\,dy$

SOLUTION

(a) Regarding x as a constant, we obtain

$$\int_1^2 x^2 y\,dy = \left[x^2 \frac{y^2}{2} \right]_{y=1}^{y=2}$$
$$= x^2\left(\frac{2^2}{2}\right) - x^2\left(\frac{1^2}{2}\right) = \tfrac{3}{2}x^2$$

Thus, the function A in the preceding discussion is given by $A(x) = \frac{3}{2}x^2$ in this example. We now integrate this function of x from 0 to 3:

$$\int_0^3 \int_1^2 x^2 y\,dy\,dx = \int_0^3 \left[\int_1^2 x^2 y\,dy \right] dx$$
$$= \int_0^3 \tfrac{3}{2}x^2\,dx = \frac{x^3}{2}\bigg]_0^3 = \frac{27}{2}$$

(b) Here we first integrate with respect to x:

$$\int_1^2 \int_0^3 x^2 y\,dx\,dy = \int_1^2 \left[\int_0^3 x^2 y\,dx \right] dy = \int_1^2 \left[\frac{x^3}{3} y \right]_{x=0}^{x=3} dy$$
$$= \int_1^2 9y\,dy = 9\frac{y^2}{2}\bigg]_1^2 = \frac{27}{2}$$

■

Notice that in Example 1 we obtained the same answer whether we integrated with respect to y or x first. In general, it turns out (see Theorem 4) that the two iterated integrals in Equations 2 and 3 are always equal; that is, the order of integration does not matter. (This is similar to Clairaut's Theorem on the equality of the mixed partial derivatives.)

The following theorem gives a practical method for evaluating a double integral by expressing it as an iterated integral (in either order).

Theorem 4 is named after the Italian mathematician Guido Fubini (1879–1943), who proved a very general version of this theorem in 1907. But the version for continuous functions was known to the French mathematician Augustin-Louis Cauchy almost a century earlier.

(4) FUBINI'S THEOREM If f is continuous on the rectangle $R = \{(x, y) \mid a \leqslant x \leqslant b,\ c \leqslant y \leqslant d\}$, then

$$\iint_R f(x, y)\,dA = \int_a^b \int_c^d f(x, y)\,dy\,dx = \int_c^d \int_a^b f(x, y)\,dx\,dy$$

More generally, this is true if we assume that f is bounded on R, f is discontinuous only on a finite number of smooth curves, and the iterated integrals exist.

The proof of Fubini's Theorem is too difficult to include in this book, but we can at least give an intuitive indication of why it is true for the case where $f(x, y) \geqslant 0$. Recall that if f is positive, then we can interpret the double integral $\iint_R f(x, y)\,dA$ as the volume V of the solid S that lies above R and under the surface $z = f(x, y)$. But we have another formula that we used for volume in Chapter 7, namely,

$$V = \int_a^b A(x)\,dx$$

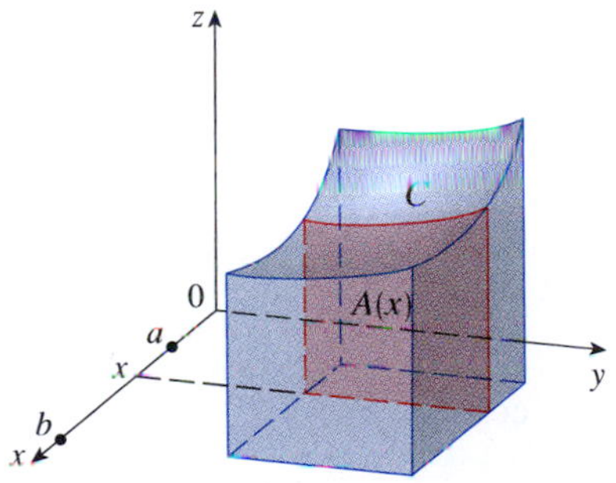

FIGURE 1

where $A(x)$ is the area of a cross-section of S in the plane through x perpendicular to the x-axis. From Figure 1 you can see that $A(x)$ is the area under the curve C whose equation is $z = f(x, y)$, where x is held constant and $c \leqslant y \leqslant d$. Therefore

$$A(x) = \int_c^d f(x, y)\,dy$$

and we have

$$\iint_R f(x, y)\,dA = V = \int_a^b A(x)\,dx = \int_a^b \int_c^d f(x, y)\,dy\,dx$$

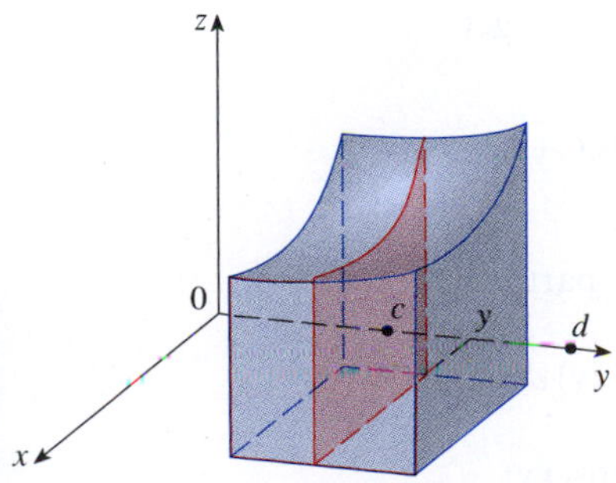

FIGURE 2

A similar argument, using cross-sections perpendicular to the y-axis as in Figure 2, shows that

$$\iint_R f(x, y)\,dA = \int_c^d \int_a^b f(x, y)\,dx\,dy$$

EXAMPLE 2 Evaluate the double integral $\iint_R (x - 3y^2)\,dA$, where $R = \{(x, y) \mid 0 \leqslant x \leqslant 2,\ 1 \leqslant y \leqslant 2\}$. (Compare with Example 1 in Section 13.1.)

Notice the negative answer in Example 2; nothing is wrong with that. The function f in that example is not a positive function, so its integral does not represent a volume. From Figure 3 we see that f is always negative on R, so the value of the integral is the *negative* of the volume that lies *above* the graph of f and *below* R.

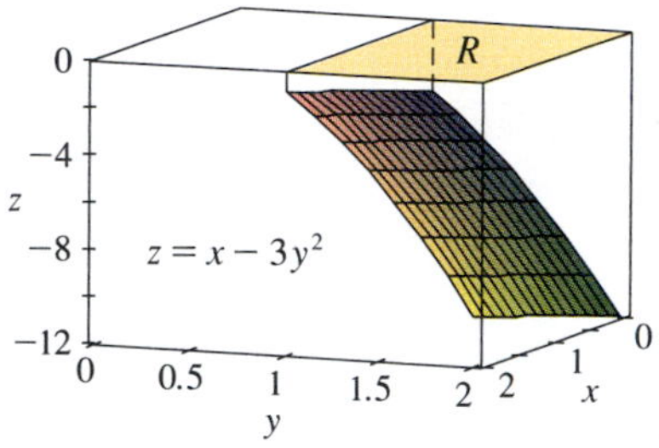

FIGURE 3

SOLUTION 1 Fubini's Theorem gives

$$\begin{aligned}\iint_R (x - 3y^2)\,dA &= \int_0^2 \int_1^2 (x - 3y^2)\,dy\,dx \\ &= \int_0^2 \Big[xy - y^3\Big]_{y=1}^{y=2} dx \\ &= \int_0^2 (x - 7)\,dx = \frac{x^2}{2} - 7x\Big]_0^2 = -12\end{aligned}$$

SOLUTION 2 Again applying Fubini's Theorem, but this time integrating with respect to x first, we have

$$\begin{aligned}\iint_R (x - 3y^2)\,dA &= \int_1^2 \int_0^2 (x - 3y^2)\,dx\,dy \\ &= \int_1^2 \left[\frac{x^2}{2} - 3xy^2\right]_{x=0}^{x=2} dy \\ &= \int_1^2 (2 - 6y^2)\,dy = 2y - 2y^3\Big]_1^2 = -12\end{aligned}$$

■

EXAMPLE 3 Evaluate $\iint_R y\sin(xy)\,dA$, where $R = [1, 2] \times [0, \pi]$.

SOLUTION 1 If we first integrate with respect to x, we get

$$\begin{aligned}\iint_R y\sin(xy)\,dA &= \int_0^\pi \int_1^2 y\sin(xy)\,dx\,dy \\ &= \int_0^\pi \Big[-\cos(xy)\Big]_{x=1}^{x=2} dy \\ &= \int_0^\pi (-\cos 2y + \cos y)\,dy \\ &= -\tfrac{1}{2}\sin 2y + \sin y\Big]_0^\pi = 0\end{aligned}$$

For a function f that takes on both positive and negative values, $\iint_R f(x, y)\,dA$ is a difference of volumes: $V_1 - V_2$, where V_1 is the volume above R and below the graph of f and V_2 is the volume below R and above the graph. The fact that the integral in Example 3 is 0 means that these two volumes V_1 and V_2 are equal. (See Figure 4.)

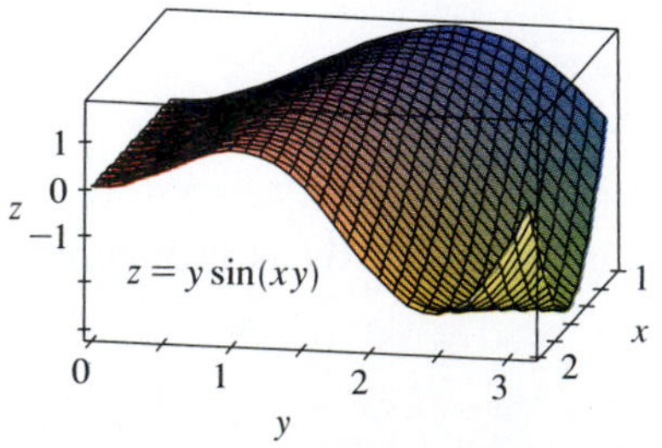

FIGURE 4

SOLUTION 2 If we reverse the order of integration, we get

$$\iint_R y\sin(xy)\,dA = \int_1^2 \int_0^\pi y\sin(xy)\,dy\,dx$$

To evaluate the inner integral we use integration by parts with

$$u = y \qquad dv = \sin(xy)\,dy$$

$$du = dy \qquad v = -\frac{\cos(xy)}{x}$$

and so
$$\int_0^\pi y\sin(xy)\,dy = -\frac{y\cos(xy)}{x}\bigg]_{y=0}^{y=\pi} + \frac{1}{x}\int_0^\pi \cos(xy)\,dy$$
$$= -\frac{\pi\cos\pi x}{x} + \frac{1}{x^2}\big[\sin(xy)\big]_{y=0}^{y=\pi}$$
$$= -\frac{\pi\cos\pi x}{x} + \frac{\sin\pi x}{x^2}$$

If we now integrate the first term by parts with $u = -1/x$ and $dv = \pi\cos\pi x\,dx$, we get $du = dx/x^2$, $v = \sin\pi x$, and

$$\int\left(-\frac{\pi\cos\pi x}{x}\right)dx = -\frac{\sin\pi x}{x} - \int\frac{\sin\pi x}{x^2}\,dx$$

Therefore
$$\int\left(-\frac{\pi\cos\pi x}{x} + \frac{\sin\pi x}{x^2}\right)dx = -\frac{\sin\pi x}{x}$$

and so
$$\int_1^2\int_0^\pi y\sin(xy)\,dy\,dx = \left[-\frac{\sin\pi x}{x}\right]_1^2$$
$$= -\frac{\sin 2\pi}{2} + \sin\pi = 0$$

In Example 2, Solutions 1 and 2 are equally straightforward, but in Example 3 the first solution is much easier than the second one. Therefore, when we evaluate double integrals it is wise to choose the order of integration that gives simpler integrals.

EXAMPLE 4 Find the volume of the solid S that is bounded by the elliptic paraboloid $x^2 + 2y^2 + z = 16$, the planes $x = 2$ and $y = 2$, and the three coordinate planes.

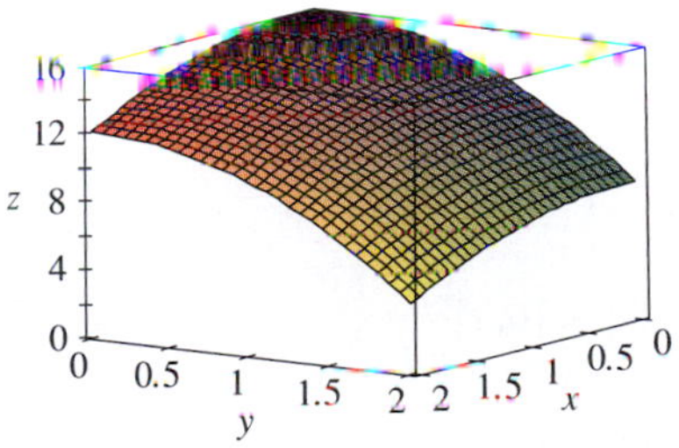

FIGURE 5

SOLUTION We first observe that S is the solid that lies under the surface $z = 16 - x^2 - 2y^2$ and above the square $R = [0, 2] \times [0, 2]$ (see Figure 5). This solid was considered in Example 2 in Section 13.1, but we are now in a position to evaluate the double integral using Fubini's Theorem. Therefore

$$V = \iint_R (16 - x^2 - 2y^2)\,dA = \int_0^2\int_0^2 (16 - x^2 - 2y^2)\,dx\,dy$$
$$= \int_0^2 \left[16x - \frac{x^3}{3} - 2y^2x\right]_{x=0}^{x=2} dy$$
$$= \int_0^2 \left(\tfrac{88}{3} - 4y^2\right)dy = \left[\tfrac{88}{3}y - \tfrac{4}{3}y^3\right]_0^2 = 48$$

In the special case where $f(x, y)$ can be factored as the product of a function of x only and a function of y only, the double integral of f can be written in a particularly simple form. To be specific, suppose that $f(x, y) = g(x)h(y)$ and $R = [a, b] \times [c, d]$. Then Fubini's Theorem gives

$$\iint_R f(x, y)\,dA = \int_c^d\int_a^b g(x)h(y)\,dx\,dy = \int_c^d\left[\int_a^b g(x)h(y)\,dx\right]dy$$

In the inner integral y is a constant, so $h(y)$ is a constant and we can write

$$\int_c^d \left[\int_a^b g(x)h(y)\,dx\right]dy = \int_c^d \left[h(y)\left(\int_a^b g(x)\,dx\right)\right]dy$$
$$= \int_a^b g(x)\,dx \int_c^d h(y)\,dy$$

since $\int_a^b g(x)\,dx$ is a constant. Therefore, in this case, the double integral of f can be written as the product of two single integrals:

$$\iint_R g(x)h(y)\,dA = \int_a^b g(x)\,dx \int_c^d h(y)\,dy \qquad \text{where } R = [a, b] \times [c, d]$$

The function $f(x, y) = \sin x \cos y$ in Example 5 is positive on R, so the integral represents the volume of the solid that lies above R and below the graph of f shown in Figure 6.

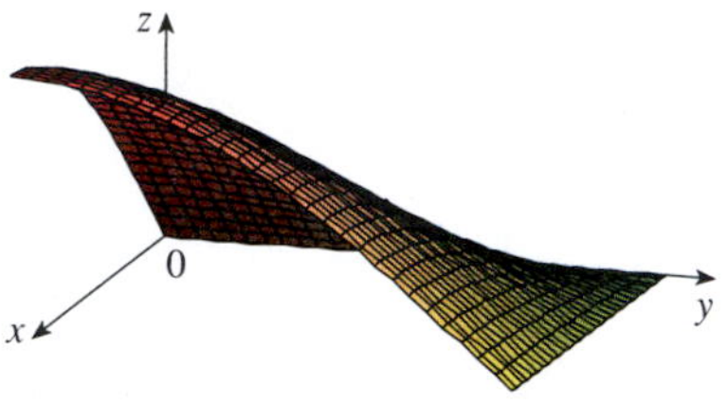

FIGURE 6

EXAMPLE 5 If $R = [0, \pi/2] \times [0, \pi/2]$, then

$$\iint_R \sin x \cos y\,dA = \int_0^{\pi/2} \sin x\,dx \int_0^{\pi/2} \cos y\,dy$$
$$= \left[-\cos x\right]_0^{\pi/2}\left[\sin y\right]_0^{\pi/2}$$
$$= 1 \cdot 1 = 1$$

■

EXERCISES 13.2

1–4 ■ Find $\int_0^2 f(x, y)\,dy$ and $\int_0^1 f(x, y)\,dx$.

1. $f(x, y) = x^2y^3$
2. $f(x, y) = 2xy - 3x^2$
3. $f(x, y) = xe^{x+y}$
4. $f(x, y) = \dfrac{x}{y^2 + 1}$

5–14 ■ Calculate the iterated integral.

5. $\int_0^4 \int_0^2 x\sqrt{y}\,dx\,dy$
6. $\int_0^2 \int_0^3 e^{x-y}\,dy\,dx$
7. $\int_{-1}^1 \int_0^1 (x^3y^3 + 3xy^2)\,dy\,dx$
8. $\int_0^1 \int_1^2 (x^4 - y^2)\,dx\,dy$
9. $\int_0^3 \int_0^1 \sqrt{x + y}\,dx\,dy$
10. $\int_0^{\pi/2} \int_0^{\pi/2} \sin(x + y)\,dy\,dx$
11. $\int_0^{\pi/4} \int_0^3 \sin x\,dy\,dx$
12. $\int_1^2 \int_0^1 (x + y)^{-2}\,dx\,dy$
13. $\int_0^{\ln 2} \int_0^{\ln 5} e^{2x-y}\,dx\,dy$
14. $\int_0^1 \int_0^1 \dfrac{xy}{\sqrt{x^2 + y^2 + 1}}\,dy\,dx$

15–22 ■ Calculate the double integral.

15. $\iint_R (2y^2 - 3xy^3)\,dA$, $R = \{(x, y) \mid 1 \le x \le 2,\ 0 \le y \le 3\}$
16. $\iint_R \left(xy^2 + \dfrac{y}{x}\right)dA$, $R = \{(x, y) \mid 2 \le x \le 3,\ -1 \le y \le 0\}$
17. $\iint_R x \sin y\,dA$, $R = \{(x, y) \mid 1 \le x \le 4,\ 0 \le y \le \pi/6\}$
18. $\iint_R \dfrac{1 + x}{1 + y}\,dA$, $R = \{(x, y) \mid -1 \le x \le 2,\ 0 \le y \le 1\}$
19. $\iint_R x \sin(x + y)\,dA$, $R = [0, \pi/6] \times [0, \pi/3]$
20. $\iint_R xe^{xy}\,dA$, $R = [0, 1] \times [0, 1]$
21. $\iint_R \dfrac{1}{x + y}\,dA$, $R = [1, 2] \times [0, 1]$
22. $\iint_R xye^{xy^2}\,dA$, $R = [0, 1] \times [0, 1]$

23–24 ■ Sketch the solid whose volume is given by the iterated integral.

23. $\int_0^1 \int_0^1 (4 - x - 2y)\,dx\,dy$
24. $\int_0^1 \int_0^1 (2 - x^2 - y^2)\,dy\,dx$

25. Find the volume of the solid lying under the plane $z = 2x + 5y + 1$ and above the rectangle $\{(x, y) \mid -1 \le x \le 0,\ 1 \le y \le 4\}$.

26. Find the volume of the solid lying under the circular paraboloid $z = x^2 + y^2$ and above the rectangle $R = [-2, 2] \times [-3, 3]$.

27. Find the volume of the solid lying under the elliptic paraboloid $x^2/4 + y^2/9 + z = 1$ and above the rectangle $R = [-1, 1] \times [-2, 2]$.

28. Find the volume of the solid lying under the hyperbolic paraboloid $z = y^2 - x^2$ and above the square $R = [-1, 1] \times [1, 3]$.

29. Find the volume of the solid bounded by the surface $z = x\sqrt{x^2 + y}$ and the planes $x = 0$, $x = 1$, $y = 0$, $y = 1$, and $z = 0$.

30. Find the volume of the solid bounded by the elliptic paraboloid $z = 1 + (x - 1)^2 + 4y^2$, the planes $x = 3$ and $y = 2$, and the coordinate planes.

31. Find the volume of the solid in the first octant bounded by the cylinder $z = 9 - y^2$ and the plane $x = 2$.

32. (a) Find the volume of the solid bounded by the surface $z = 6 - xy$ and the planes $x = 2$, $x = -2$, $y = 0$, $y = 3$, and $z = 0$.
(b) Use a computer to draw the solid.

CAS **33.** Use a computer algebra system to find the exact value of the integral $\iint_R x^5y^3e^{xy}\,dA$, where $R = [0, 1] \times [0, 1]$. Then use the CAS to draw the solid whose volume is given by the integral.

CAS **34.** Graph the solid that lies between the surfaces $z = e^{-x^2}\cos(x^2 + y^2)$ and $z = 2 - x^2 - y^2$ for $|x| \leq 1$, $|y| \leq 1$. Use a computer algebra system to approximate the volume of this solid correct to four decimal places.

35–36 ■ The **average value** of a function $f(x, y)$ over a rectangle R is defined to be

$$f_{\text{ave}} = \frac{1}{A(R)} \iint_R f(x, y)\,dA$$

where $A(R)$ is the area of R. (Compare with the definition for functions of one variable in Equation 7.5.1.) Find the average value of f over the given rectangle.

35. $f(x, y) = x^2y$, R has vertices $(-1, 0)$, $(-1, 5)$, $(1, 5)$, $(1, 0)$

36. $f(x, y) = x\sin xy$, $R = [0, \pi/2] \times [0, 1]$

CAS **37.** Use your CAS to compute the iterated integrals

$$\int_0^1\int_0^1 \frac{x - y}{(x + y)^3}\,dy\,dx \quad \text{and} \quad \int_0^1\int_0^1 \frac{x - y}{(x + y)^3}\,dx\,dy$$

Do the answers contradict Fubini's Theorem? Explain what is happening.

38. (a) In what way are the theorems of Fubini and Clairaut similar?
(b) If $f(x, y)$ is continuous on $[a, b] \times [c, d]$ and

$$g(x, y) = \int_a^x\int_c^y f(s, t)\,dt\,ds \qquad \text{for } a < x < b,\ c < y < d$$

show that $g_{xy} = g_{yx} = f(x, y)$.

13.3 DOUBLE INTEGRALS OVER GENERAL REGIONS

For single integrals, the region over which we integrate is always an interval. But, for double integrals, we want to be able to integrate not just over rectangles but also over regions D of more general shape, such as the one illustrated in Figure 1. We suppose that D is a bounded region, which means that D can be enclosed in a rectangular region R as in Figure 2. Then we define a new function F with domain R by

$$\textbf{(1)} \qquad F(x, y) = \begin{cases} f(x, y) & \text{if } (x, y) \text{ is in } D \\ 0 & \text{if } (x, y) \text{ is in } R \text{ but not in } D \end{cases}$$

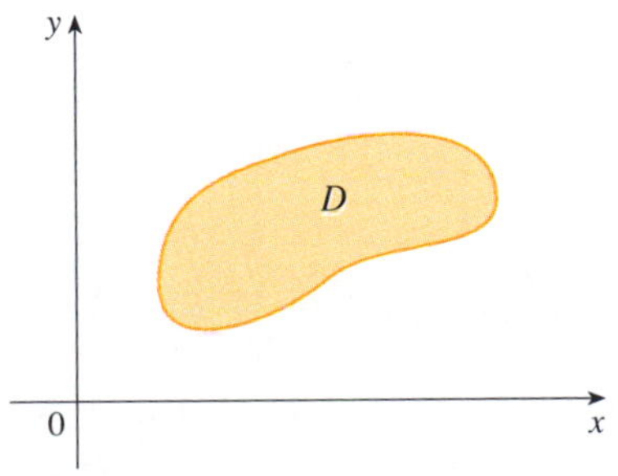

FIGURE 1

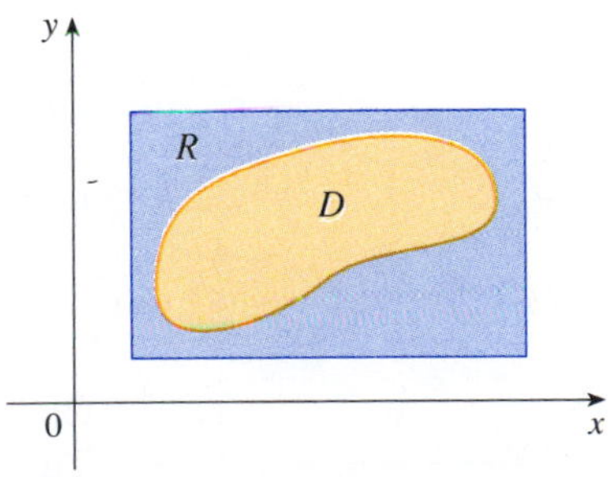

FIGURE 2

If F is integrable over R, then we say f is **integrable** over D and we define the **double integral of f over D** by

(2) $$\iint_D f(x, y)\, dA = \iint_R F(x, y)\, dA \qquad \text{where } F \text{ is given by Equation 1.}$$

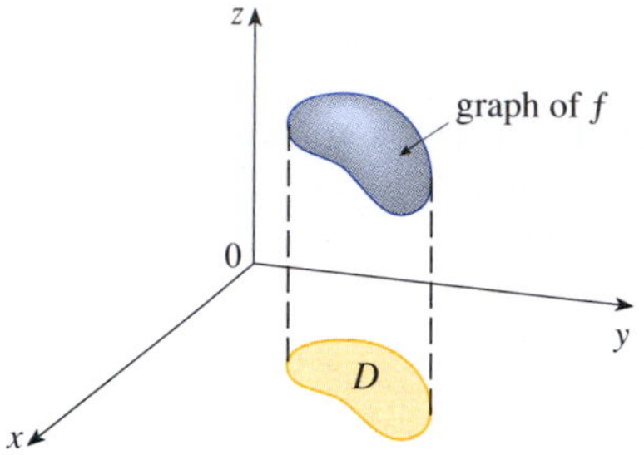

FIGURE 3

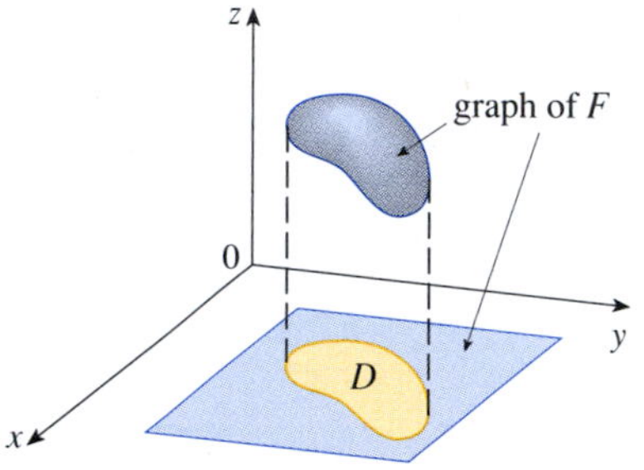

FIGURE 4

Definition 2 makes sense because R is a rectangle and so $\iint_R F(x, y)\, dA$ has been previously defined in Section 13.1. The procedure that we have used is reasonable because the values of $F(x, y)$ are 0 when (x, y) lies outside D and so they contribute nothing to the integral. This means that it doesn't matter what rectangle R we use as long as it contains D.

In the case where $f(x, y) \geqslant 0$ we can still interpret $\iint_D f(x, y)\, dA$ as the volume of the solid that lies above D and under the surface $z = f(x, y)$ (the graph of f). You can see that this is reasonable by comparing the graphs of f and F in Figures 3 and 4 and remembering that $\iint_R F(x, y)\, dA$ is the volume under the graph of F.

Figure 4 also shows that F is likely to have discontinuities at the boundary points of D. Nonetheless if f is continuous on D and the boundary curve of D is "well behaved" (in a sense outside the scope of this book), then it can be shown that F is integrable over R and therefore f is integrable over D. In particular, this is the case for the following types of regions.

A plane region D is said to be of **type I** if it lies between the graphs of two continuous functions of x, that is,

$$D = \{(x, y) \mid a \leqslant x \leqslant b,\ g_1(x) \leqslant y \leqslant g_2(x)\}$$

where g_1 and g_2 are continuous on $[a, b]$. Some examples of type I regions are shown in Figure 5.

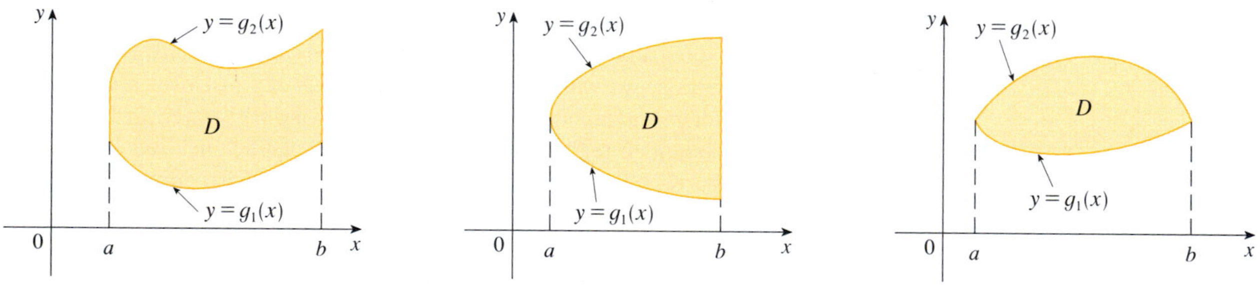

FIGURE 5 Some type I regions

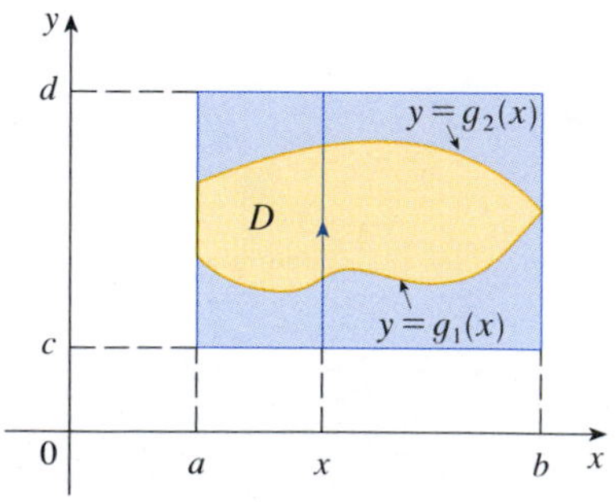

FIGURE 6

In order to evaluate $\iint_D f(x, y)\, dA$ when D is a region of type I, we choose a rectangle $R = [a, b] \times [c, d]$ that contains D, as in Figure 6, and we let F be the function given by Equation 1; that is, F agrees with f on D and F is 0 outside D. Then, by Fubini's Theorem,

$$\iint_D f(x, y)\, dA = \iint_R F(x, y)\, dA = \int_a^b \int_c^d F(x, y)\, dy\, dx$$

Observe that $F(x, y) = 0$ if $y < g_1(x)$ or $y > g_2(x)$ since (x, y) then lies outside D. Therefore

$$\int_c^d F(x, y)\, dy = \int_{g_1(x)}^{g_2(x)} F(x, y)\, dy = \int_{g_1(x)}^{g_2(x)} f(x, y)\, dy$$

because $F(x, y) = f(x, y)$ when $g_1(x) \leqslant y \leqslant g_2(x)$. Thus, we have the following formula that enables us to evaluate the double integral as an iterated integral.

(3) If f is continuous on a type I region D such that

$$D = \{(x, y) \mid a \leqslant x \leqslant b,\ g_1(x) \leqslant y \leqslant g_2(x)\}$$

then
$$\iint_D f(x, y)\, dA = \int_a^b \int_{g_1(x)}^{g_2(x)} f(x, y)\, dy\, dx$$

The integral on the right side of (3) is an iterated integral that is similar to the ones we considered in the preceding section, except that in the inner integral we regard x as being constant not only in $f(x, y)$ but also in the limits of integration, $g_1(x)$ and $g_2(x)$.

We also consider plane regions of **type II**, which can be expressed as

(4)
$$D = \{(x, y) \mid c \leqslant y \leqslant d,\ h_1(y) \leqslant x \leqslant h_2(y)\}$$

where h_1 and h_2 are continuous. Two such regions are illustrated in Figure 7.

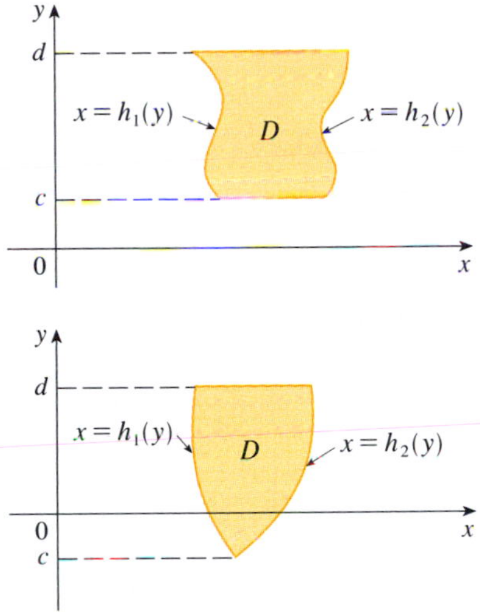

FIGURE 7
Some type II regions

Using the same methods that were used in establishing (4), we can show that

(5)
$$\iint_D f(x, y)\, dA = \int_c^d \int_{h_1(y)}^{h_2(y)} f(x, y)\, dx\, dy$$

where D is a type II region given by Equation 4.

EXAMPLE 1 Evaluate $\iint_D (x + 2y)\, dA$, where D is the region bounded by the parabolas $y = 2x^2$ and $y = 1 + x^2$.

SOLUTION The parabolas intersect when $2x^2 = 1 + x^2$, that is, $x^2 = 1$, so $x = \pm 1$. We note that the region D, sketched in Figure 8, is a type I region but not a type II region and we can write

$$D = \{(x, y) \mid -1 \leqslant x \leqslant 1,\ 2x^2 \leqslant y \leqslant 1 + x^2\}$$

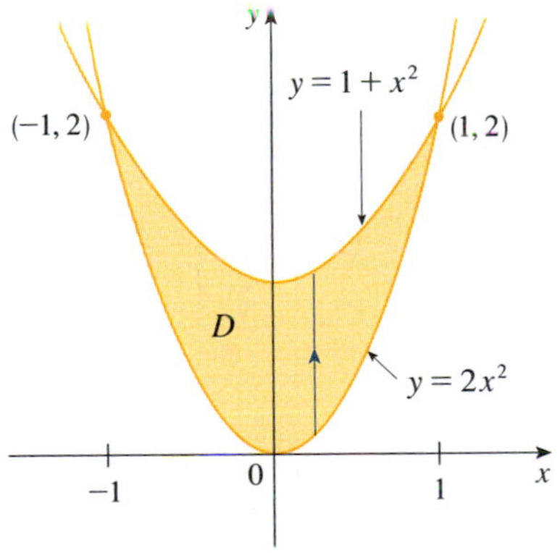

FIGURE 8

Since the lower boundary is $y = 2x^2$ and the upper boundary is $y = 1 + x^2$, Equation 3 gives

$$\iint_D (x + 2y)\, dA = \int_{-1}^{1} \int_{2x^2}^{1+x^2} (x + 2y)\, dy\, dx$$

$$= \int_{-1}^{1} \Big[xy + y^2\Big]_{y=2x^2}^{y=1+x^2} dx$$

$$= \int_{-1}^{1} [x(1 + x^2) + (1 + x^2)^2 - x(2x^2) - (2x^2)^2]\, dx$$

$$= \int_{-1}^{1} (-3x^4 - x^3 + 2x^2 + x + 1)\, dx$$

$$= -3\frac{x^5}{5} - \frac{x^4}{4} + 2\frac{x^3}{3} + \frac{x^2}{2} + x\Big]_{-1}^{1} = \frac{32}{15}$$ ■

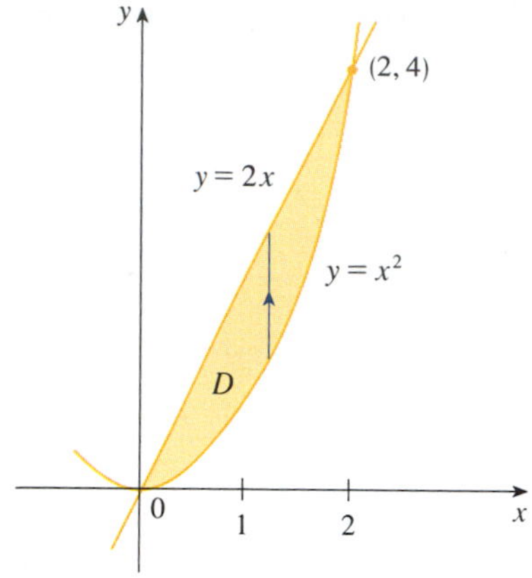

FIGURE 9
D as a type I region

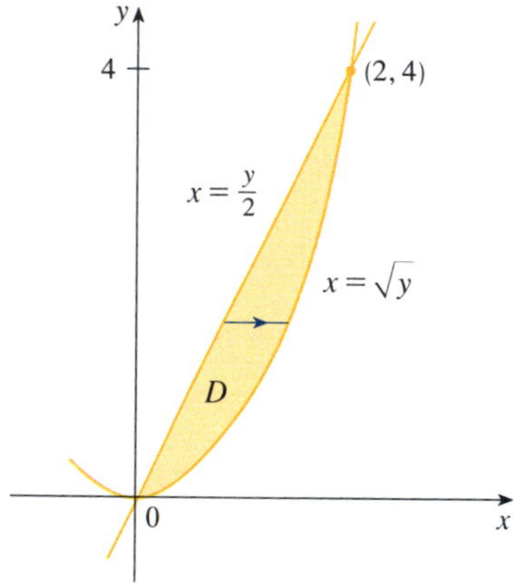

FIGURE 10
D as a type II region

Figure 11 shows the solid whose volume is calculated in Example 2. It lies above the xy-plane, below the paraboloid $z = x^2 + y^2$, and between the plane $y = 2x$ and the parabolic cylinder $y = x^2$.

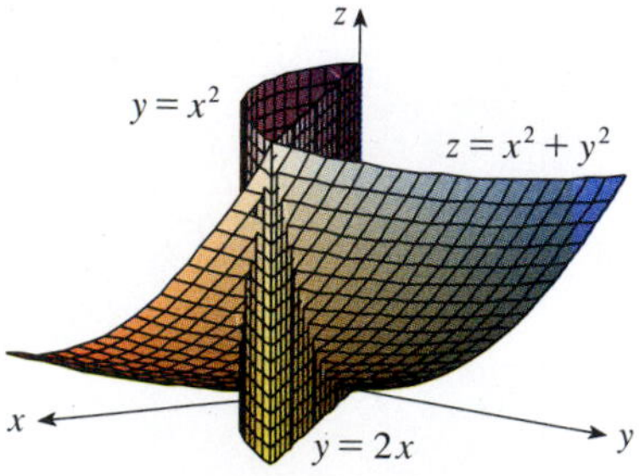

FIGURE 11

NOTE: When we set up a double integral as in Example 1, it is essential to draw a diagram. Often it is helpful to draw a vertical arrow as in Figure 8. Then the limits of integration for the *inner* integral can be read from the diagram as follows: The arrow starts at the lower boundary $y = g_1(x)$, which gives the lower limit in the integral, and the arrow ends at the upper boundary $y = g_2(x)$, which gives the upper limit of integration. For a type II region the arrow is horizontal and goes from the left boundary to the right boundary.

EXAMPLE 2 Find the volume of the solid that lies under the paraboloid $z = x^2 + y^2$ and above the region D in the xy-plane bounded by the line $y = 2x$ and the parabola $y = x^2$.

SOLUTION 1 From Figure 9 we see that D is a type I region and

$$D = \{(x, y) \mid 0 \leqslant x \leqslant 2,\ x^2 \leqslant y \leqslant 2x\}$$

Therefore, the volume under $z = x^2 + y^2$ and above D is

$$V = \iint_D (x^2 + y^2)\,dA = \int_0^2 \int_{x^2}^{2x} (x^2 + y^2)\,dy\,dx$$

$$= \int_0^2 \left[x^2y + \frac{y^3}{3}\right]_{y=x^2}^{y=2x} dx = \int_0^2 \left[x^2(2x) + \frac{(2x)^3}{3} - x^2x^2 - \frac{(x^2)^3}{3}\right] dx$$

$$= \int_0^2 \left(-\frac{x^6}{3} - x^4 + \frac{14x^3}{3}\right) dx = -\frac{x^7}{21} - \frac{x^5}{5} + \frac{7x^4}{6}\Bigg]_0^2 = \frac{216}{35}$$

SOLUTION 2 From Figure 10 we see that D can also be written as a type II region:

$$D = \left\{(x, y) \,\middle|\, 0 \leqslant y \leqslant 4,\ \frac{y}{2} \leqslant x \leqslant \sqrt{y}\right\}$$

Therefore, another expression for V is

$$V = \iint_D (x^2 + y^2)\,dA = \int_0^4 \int_{y/2}^{\sqrt{y}} (x^2 + y^2)\,dx\,dy$$

$$= \int_0^4 \left[\frac{x^3}{3} + y^2x\right]_{x=y/2}^{x=\sqrt{y}} dy = \int_0^4 \left(\frac{y^{3/2}}{3} + y^{5/2} - \frac{y^3}{24} - \frac{y^3}{2}\right) dy$$

$$= \tfrac{2}{15}y^{5/2} + \tfrac{2}{7}y^{7/2} - \tfrac{13}{96}y^4\Big]_0^4 = \tfrac{216}{35}$$ ■

EXAMPLE 3 Evaluate $\iint_D xy\,dA$ where D is the region bounded by the line $y = x - 1$ and the parabola $y^2 = 2x + 6$.

SOLUTION The region D is shown in Figure 12. Again D is both type I and type II, but the description of D as a type I region is more complicated because the lower boundary consists of two parts. Therefore, we prefer to express D as a type II region:

$$D = \left\{(x, y) \,\middle|\, -2 \leqslant y \leqslant 4,\ \frac{y^2}{2} - 3 \leqslant x \leqslant y + 1\right\}$$

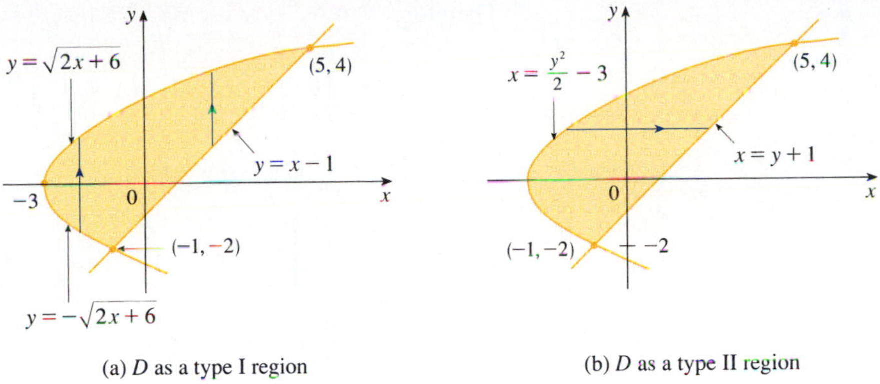

FIGURE 12

Then (5) gives

$$\iint_D xy\,dA = \int_{-2}^{4}\int_{y^2/2-3}^{y+1} xy\,dx\,dy = \int_{-2}^{4}\left[\frac{x^2}{2}y\right]_{x=y^2/2-3}^{x=y+1} dy$$

$$= \frac{1}{2}\int_{-2}^{4} y\left[(y+1)^2 - \left(\frac{y^2}{2}-3\right)^2\right] dy$$

$$= \frac{1}{2}\int_{-2}^{4}\left(-\frac{y^5}{4} + 4y^3 + 2y^2 - 8y\right) dy$$

$$= \frac{1}{2}\left[-\frac{y^6}{24} + y^4 + 2\frac{y^3}{3} - 4y^2\right]_{-2}^{4} = 36$$

If we had expressed D as a type I region using Figure 12(a), then we would have obtained

$$\iint_D xy\,dA = \int_{-3}^{-1}\int_{-\sqrt{2x+6}}^{\sqrt{2x+6}} xy\,dy\,dx + \int_{-1}^{5}\int_{x-1}^{\sqrt{2x+6}} xy\,dy\,dx$$

but this would have involved more work than the above method. ■

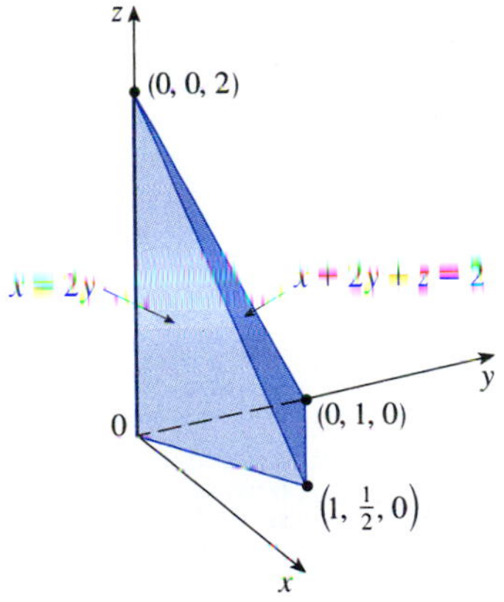

FIGURE 13

EXAMPLE 4 Find the volume of the tetrahedron bounded by the planes $x + 2y + z = 2$, $x = 2y$, $x = 0$, and $z = 0$.

SOLUTION In a question such as this, it is wise to draw two diagrams: one of the three-dimensional solid and another of the plane region D over which it lies. Figure 13 shows the tetrahedron T bounded by the coordinate planes $x = 0$, $z = 0$, the vertical plane $x = 2y$, and the plane $x + 2y + z = 2$. Since the plane $x + 2y + z = 2$ intersects the xy-plane (whose equation is $z = 0$) in the line $x + 2y = 2$, we see that T lies above the triangular region D in the xy-plane bounded by the lines $x = 2y$, $x + 2y = 2$, and $x = 0$ (see Figure 14).

The plane $x + 2y + z = 2$ can be written as $z = 2 - x - 2y$, so the required volume lies under the graph of the function $z = 2 - x - 2y$ and above

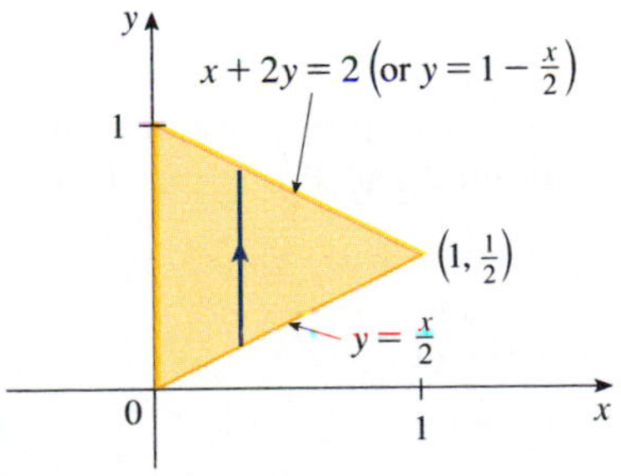

FIGURE 14

$$D = \left\{(x, y) \,\middle|\, 0 \leqslant x \leqslant 1,\ \frac{x}{2} \leqslant y \leqslant 1 - \frac{x}{2}\right\}$$

Therefore

$$V = \iint_D (2 - x - 2y)\,dA = \int_0^1 \int_{x/2}^{1-x/2} (2 - x - 2y)\,dy\,dx$$

$$= \int_0^1 \left[2y - xy - y^2\right]_{y=x/2}^{y=1-x/2} dx$$

$$= \int_0^1 \left[2 - x - x\left(1 - \frac{x}{2}\right) - \left(1 - \frac{x}{2}\right)^2 - x + \frac{x^2}{2} + \frac{x^2}{4}\right] dx$$

$$= \int_0^1 (x^2 - 2x + 1)\,dx = \frac{x^3}{3} - x^2 + x\Big]_0^1 = \frac{1}{3}$$ ■

EXAMPLE 5 Evaluate the iterated integral $\int_0^1 \int_x^1 \sin(y^2)\,dy\,dx$.

SOLUTION If we try to evaluate the integral as it stands, we are faced with the task of first evaluating $\int \sin(y^2)\,dy$. But it is impossible to do so in finite terms since $\int \sin(y^2)\,dy$ is not an elementary function. (See the end of Section 8.6.) So we must change the order of integration. This is accomplished by first expressing the given iterated integral as a double integral. Using (3) backward, we have

$$\int_0^1 \int_x^1 \sin(y^2)\,dy\,dx = \iint_D \sin(y^2)\,dA$$

where $D = \{(x, y) \mid 0 \leqslant x \leqslant 1,\ x \leqslant y \leqslant 1\}$

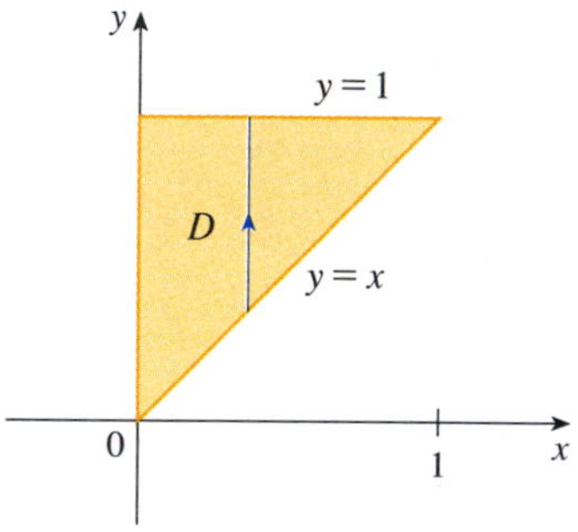

FIGURE 15
D as a type I region

We sketch this region D in Figure 15. Then from Figure 16 we see that an alternative description of D is

$$D = \{(x, y) \mid 0 \leqslant y \leqslant 1,\ 0 \leqslant x \leqslant y\}$$

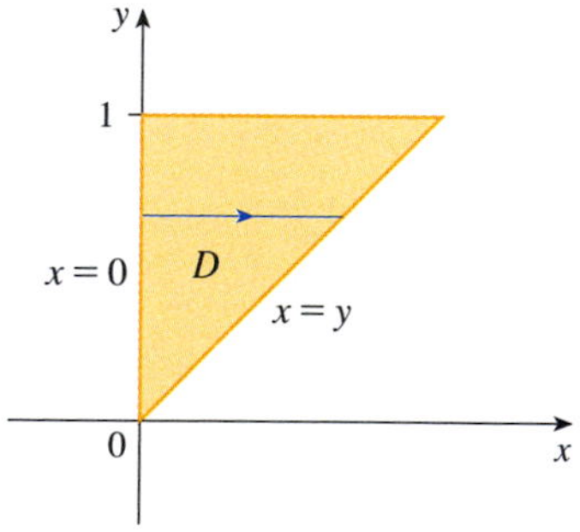

FIGURE 16
D as a type II region

This enables us to use (5) to express the double integral as an iterated integral in the reverse order:

$$\int_0^1 \int_x^1 \sin(y^2)\,dy\,dx = \iint_D \sin(y^2)\,dA$$

$$= \int_0^1 \int_0^y \sin(y^2)\,dx\,dy = \int_0^1 \left[x \sin(y^2)\right]_{x=0}^{x=y} dy$$

$$= \int_0^1 y \sin(y^2)\,dy = -\tfrac{1}{2} \cos(y^2)\Big]_0^1$$

$$= \tfrac{1}{2}(1 - \cos 1)$$ ■

PROPERTIES OF DOUBLE INTEGRALS

We assume that all of the following integrals exist. The first three properties of double integrals over a region D follow immediately from Definition 2 and Properties 13.1.7, 13.1.8, and 13.1.9.

(6) $$\iint_D [f(x, y) + g(x, y)]\,dA = \iint_D f(x, y)\,dA + \iint_D g(x, y)\,dA$$

(7) $$\iint_D c f(x, y)\,dA = c \iint_D f(x, y)\,dA$$

If $f(x, y) \geqslant g(x, y)$ for all (x, y) in D, then

(8)
$$\iint_D f(x, y)\, dA \geqslant \iint_D g(x, y)\, dA$$

The next property of double integrals is the analogue of the property of single integrals given by the equation $\int_a^b f(x)\, dx = \int_a^c f(x)\, dx + \int_c^b f(x)\, dx$.

If $D = D_1 \cup D_2$, where D_1 and D_2 do not overlap except perhaps on their boundaries (see Figure 17), then

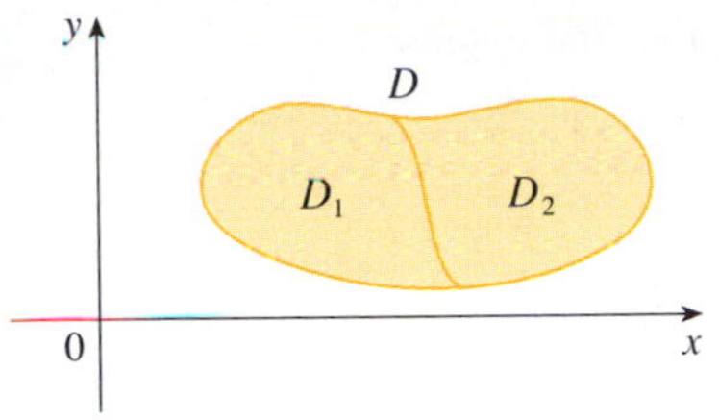

FIGURE 17

(9)
$$\iint_D f(x, y)\, dA = \iint_{D_1} f(x, y)\, dA + \iint_{D_2} f(x, y)\, dA$$

Property 9 can be used to evaluate double integrals over regions D that are neither type I nor type II but can be expressed as a union of type I or II regions. Figure 18 illustrates this procedure. (See Exercises 45 and 46.)

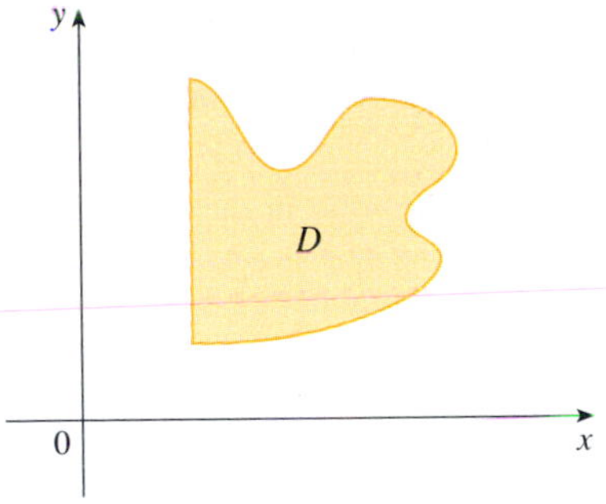

(a) D is neither type I nor type II.

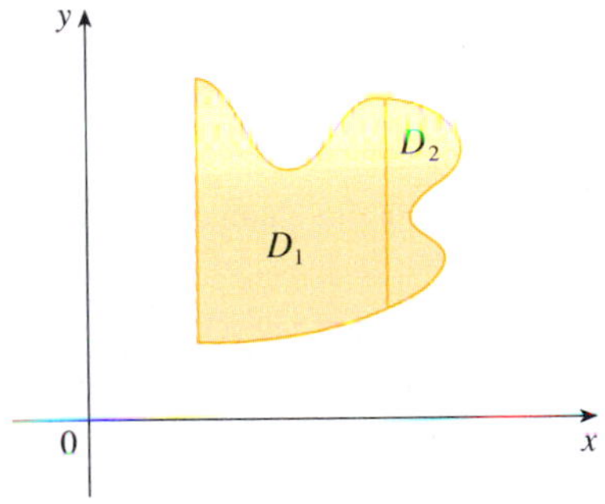

(b) $D = D_1 \cup D_2$, D_1 is type I, D_2 is type II.

FIGURE 18

The next property of integrals says that if we integrate the constant function $f(x, y) = 1$ over a region D, we get the area of D:

(10)
$$\iint_D 1\, dA = A(D)$$

For instance, if D is a type I region and we put $f(x, y) = 1$ in Formula 3, we get

$$\iint_D 1\, dA = \int_a^b \int_{g_1(x)}^{g_2(x)} 1\, dy\, dx = \int_a^b [g_2(x) - g_1(x)]\, dx = A(D)$$

by Equation 7.1.2.

Finally, we obtain an analogue of Property 8 of single integrals (see Section 6.3) by combining Properties 7, 8, and 10. (See Exercise 49.)

(11) If $m \leqslant f(x, y) \leqslant M$ for all (x, y) in D, then

$$mA(D) \leqslant \iint_D f(x, y)\, dA \leqslant MA(D)$$

EXAMPLE 6 Use Property 11 to estimate the integral $\iint_D e^{\sin x \cos y}\, dA$, where D is the disk with center the origin and radius 2.

SOLUTION Since $-1 \leqslant \sin x \leqslant 1$ and $-1 \leqslant \cos y \leqslant 1$, we have $-1 \leqslant \sin x \cos y \leqslant 1$ and therefore

$$e^{-1} \leqslant e^{\sin x \cos y} \leqslant e^1 = e$$

Thus, using $m = e^{-1} = 1/e$, $M = e$, and $A(D) = \pi(2)^2$ in Property 11, we obtain

$$\frac{4\pi}{e} \leqslant \iint_D e^{\sin x \cos y}\, dA \leqslant 4\pi e$$ ■

EXERCISES 13.3

1–6 ■ Evaluate the iterated integral.

1. $\int_0^1 \int_0^y x\,dx\,dy$

2. $\int_0^1 \int_0^y y\,dx\,dy$

3. $\int_0^2 \int_{\sqrt{x}}^3 (x^2 + y)\,dy\,dx$

4. $\int_0^1 \int_{1-x}^{1+x} (2x - 3y^2)\,dy\,dx$

5. $\int_0^1 \int_0^x \sin(x^2)\,dy\,dx$

6. $\int_0^1 \int_{x-1}^0 \dfrac{2y}{x+1}\,dy\,dx$

7–18 ■ Evaluate the double integral.

7. $\iint_D xy\,dA, \quad D = \{(x, y) \mid 0 \leqslant x \leqslant 1,\ x^2 \leqslant y \leqslant \sqrt{x}\}$

8. $\iint_D (x - 2y)\,dA,$
$D = \{(x, y) \mid 1 \leqslant x \leqslant 3,\ 1 + x \leqslant y \leqslant 2x\}$

9. $\iint_D (x^2 - 2xy)\,dA,$
$D = \{(x, y) \mid 0 \leqslant x \leqslant 1,\ \sqrt{x} \leqslant y \leqslant 2 - x\}$

10. $\iint_D x \sin y\,dA, \quad D = \{(x, y) \mid 0 \leqslant y \leqslant \pi/2,\ 0 \leqslant x \leqslant \cos y\}$

11. $\iint_D e^{x/y}\,dA, \quad D = \{(x, y) \mid 1 \leqslant y \leqslant 2,\ y \leqslant x \leqslant y^3\}$

12. $\iint_D \dfrac{1}{x}\,dA, \quad D = \{(x, y) \mid 1 \leqslant y \leqslant e,\ y^2 \leqslant x \leqslant y^4\}$

13. $\iint_D x \cos y\,dA$, D is bounded by $y = 0$, $y = x^2$, $x = 1$

14. $\iint_D e^{x+y}\,dA$, D is bounded by $y = 0$, $y = x$, $x = 1$

15. $\iint_D 4y^3\,dA$, D is bounded by $y = x - 6$, $y^2 = x$

16. $\iint_D (y^2 - x)\,dA$, D is bounded by $x = y^2$, $x = 3 - 2y^2$

17. $\iint_D xy\,dA$, D is the first-quadrant part of the disk with center $(0, 0)$ and radius 1

18. $\iint_D ye^x\,dA$, D is the triangular region with vertices $(0, 0)$, $(2, 4)$, and $(6, 0)$

19–28 ■ Find the volume of the given solid.

19. Under the paraboloid $z = x^2 + y^2$ and above the region bounded by $y = x^2$ and $x = y^2$

20. Under the paraboloid $z = 3x^2 + y^2$ and above the region bounded by $y = x$ and $x = y^2 - y$

21. Under the surface $z = xy$ and above the triangle with vertices $(1, 1)$, $(4, 1)$, and $(1, 2)$

22. Bounded by the paraboloid $z = x^2 + y^2 + 4$ and the planes $x = 0$, $y = 0$, $z = 0$, $x + y = 1$

23. Bounded by the cylinder $x^2 + z^2 = 9$ and the planes $x = 0$, $y = 0$, $z = 0$, $x + 2y = 2$ in the first octant

24. Bounded by the cylinder $y^2 + z^2 = 4$ and the planes $x = 2y$, $x = 0$, $z = 0$ in the first octant

25. Bounded by the planes $x = 0$, $y = 0$, $z = 0$, and $x + y + z = 1$

26. Bounded by the planes $y = 0$, $z = 0$, $y = x$, and $6x + 2y + 3z = 6$

27. Bounded by the cylinder $x^2 + y^2 = 1$ and the planes $y = z$, $x = 0$, $z = 0$ in the first octant

28. Bounded by the cylinders $x^2 + y^2 = r^2$ and $y^2 + z^2 = r^2$

29. Use a graphing calculator or computer to estimate the x-coordinates of the points of intersection of the curves $y = x^4$ and $y = 3x - x^2$. If D is the region bounded by these curves, estimate $\iint_D x\,dA$.

30. Find the approximate volume of the solid in the first octant that is bounded by the planes $y = x$, $z = 0$, and $z = x$ and the cylinder $y = \cos x$. (Use a graphing device to estimate the points of intersection.)

CAS **31–32** ■ Use a computer algebra system to find the exact volume of the solid.

31. Under the surface $z = x^3y^4 + xy^2$ and above the region bounded by the curves $y = x^3 - x$ and $y = x^2 + x$ for $x \geqslant 0$

32. Between the paraboloids $z = 2x^2 + y^2$ and $z = 8 - x^2 - 2y^2$ and inside the cylinder $x^2 + y^2 = 1$

33–38 ■ Sketch the region of integration and change the order of integration.

33. $\int_0^1 \int_0^x f(x, y)\,dy\,dx$

34. $\int_0^{\pi/2} \int_0^{\sin x} f(x, y)\,dy\,dx$

35. $\int_1^2 \int_0^{\ln x} f(x, y)\,dy\,dx$

36. $\int_0^1 \int_{y^2}^{2-y} f(x, y)\,dx\,dy$

37. $\int_0^4 \int_{y/2}^2 f(x, y)\,dx\,dy$

38. $\int_0^1 \int_{\arctan x}^{\pi/4} f(x, y)\,dy\,dx$

39–44 ■ Evaluate the integral by reversing the order of integration.

39. $\int_0^1 \int_{3y}^3 e^{x^2}\,dx\,dy$

40. $\int_0^1 \int_{\sqrt{y}}^1 \sqrt{x^3 + 1}\,dx\,dy$

41. $\int_0^3 \int_{y^2}^9 y \cos(x^2)\,dx\,dy$

42. $\int_0^1 \int_{x^2}^1 x^3 \sin(y^3)\,dy\,dx$

43. $\int_0^1 \int_{\arcsin y}^{\pi/2} \cos x \sqrt{1 + \cos^2 x}\,dx\,dy$

44. $\int_0^8 \int_{\sqrt[3]{y}}^2 e^{x^4}\,dx\,dy$

45–46 ■ Express D as a union of regions of type I or type II and evaluate the integral.

45. $\iint_D x^2\,dA$

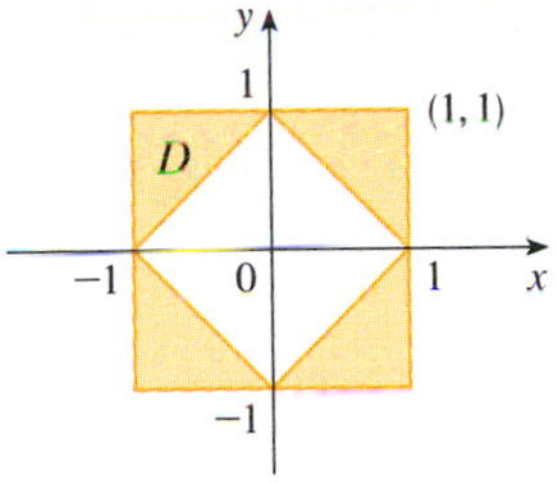

46. $\iint_D xy\,dA$

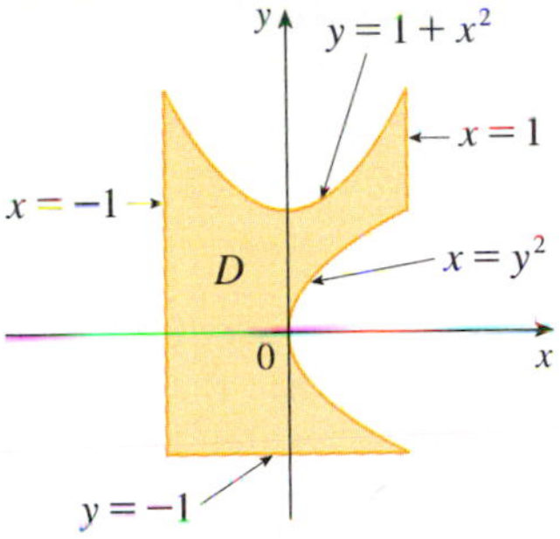

47–48 ■ Use Property 11 to estimate the value of the integral.

47. $\iint_D \sqrt{x^3 + y^3}\,dA, \quad D = [0, 1] \times [0, 1]$

48. $\iint_D e^{x^2+y^2}\,dA$, D is the disk with center the origin and radius $\frac{1}{2}$

49. Prove Property 11.

50. In evaluating a double integral over a region D, a sum of iterated integrals was obtained as follows:

$$\iint_D f(x, y)\,dA = \int_0^1 \int_0^{2y} f(x, y)\,dx\,dy + \int_1^3 \int_0^{3-y} f(x, y)\,dx\,dy$$

Sketch the region D and express the double integral as an iterated integral with reversed order of integration.

51. Evaluate $\iint_D (x^2 \tan x + y^3 + 4)\,dA$, where $D = \{(x, y) \mid x^2 + y^2 \leq 2\}$.
[*Hint:* Exploit the fact that D is symmetric with respect to both axes.]

CAS 52. Graph the solid bounded by the plane $x + y + z = 1$ and the paraboloid $z = 4 - x^2 - y^2$ and find its exact volume. (Use your CAS to do the graphing, to find the equations of the boundary curves of the region of integration, and to evaluate the double integral.)

13.4 POLAR COORDINATES

A coordinate system represents a point in the plane by an ordered pair of numbers called coordinates. So far we have been using Cartesian coordinates, which are directed distances from two perpendicular axes. In this section we describe a coordinate system introduced by Newton, called the **polar coordinate system**, which is more convenient for many purposes.

We choose a point in the plane that is called the **pole** (or origin) and labeled O. Then we draw a ray (half-line) starting at O called the **polar axis**. This axis is usually drawn horizontally to the right and corresponds to the positive x-axis in Cartesian coordinates.

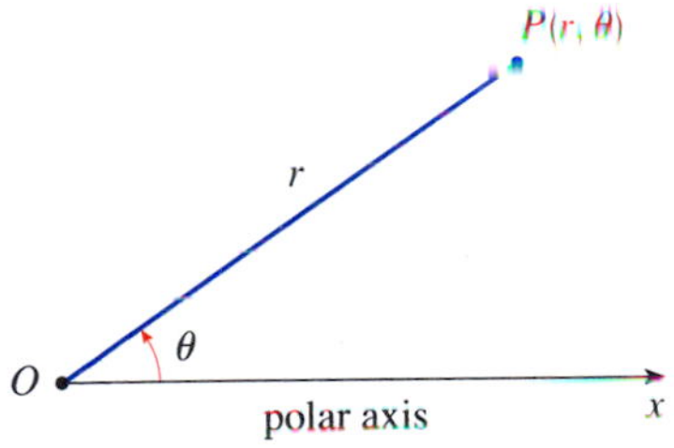

FIGURE 1

If P is any other point in the plane, let r be the distance from O to P and let θ be the angle (usually measured in radians) between the polar axis and the line OP as in Figure 1. Then the point P is represented by the ordered pair (r, θ) and r, θ are called **polar coordinates** of P. We use the convention that an angle is positive if measured in the counterclockwise direction from the polar axis and negative in the clockwise direction. If $P = O$, then $r = 0$ and we agree that $(0, \theta)$ represents the pole for any value of θ.

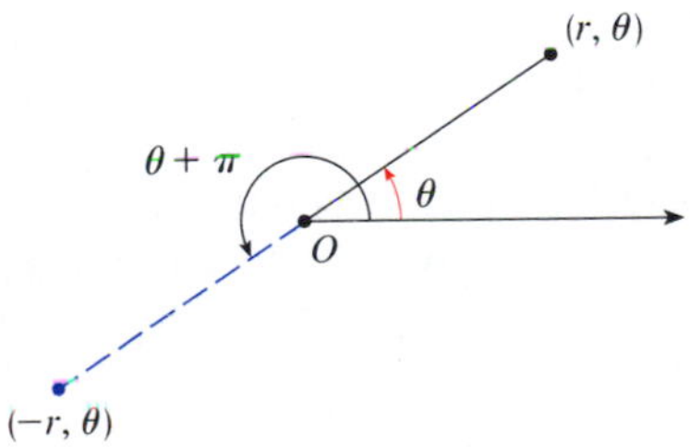

FIGURE 2

We extend the meaning of polar coordinates (r, θ) to the case in which r is negative by agreeing that, as in Figure 2, the points $(-r, \theta)$ and (r, θ) lie on the same line through O and at the same distance $|r|$ from O, but on opposite sides of O. If $r > 0$, the point (r, θ) lies in the same quadrant as θ; if $r < 0$, it lies in the quadrant on the opposite side of the pole. Notice that $(-r, \theta)$ represents the same point as $(r, \theta + \pi)$.

EXAMPLE 1 Plot the points whose polar coordinates are given:
(a) $(1, 5\pi/4)$ (b) $(2, 3\pi)$
(c) $(2, -2\pi/3)$ (d) $(-3, 3\pi/4)$

SOLUTION The points are plotted in Figure 3. In part (d) the point $(-3, 3\pi/4)$ is located three units from the pole in the fourth quadrant because the angle $3\pi/4$ is in the second quadrant and $r = -3$ is negative. ■

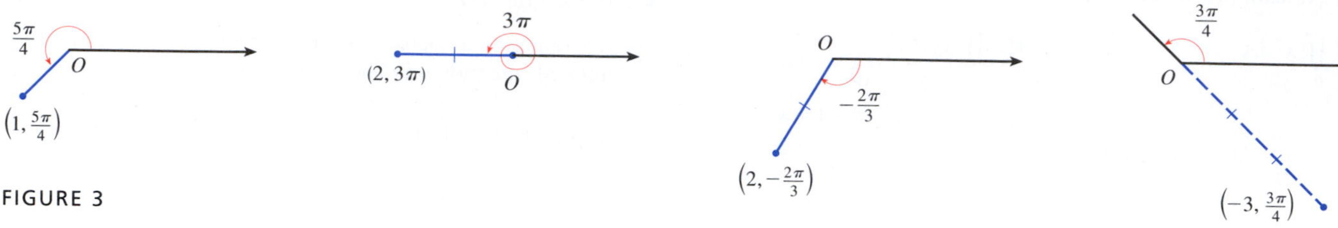

FIGURE 3

In the Cartesian coordinate system every point has only one representation, but in the polar coordinate system each point has many representations. For instance, the point $(1, 5\pi/4)$ in Example 1(a) could be written as $(1, -3\pi/4)$ or $(1, 13\pi/4)$ or $(-1, \pi/4)$ (see Figure 4).

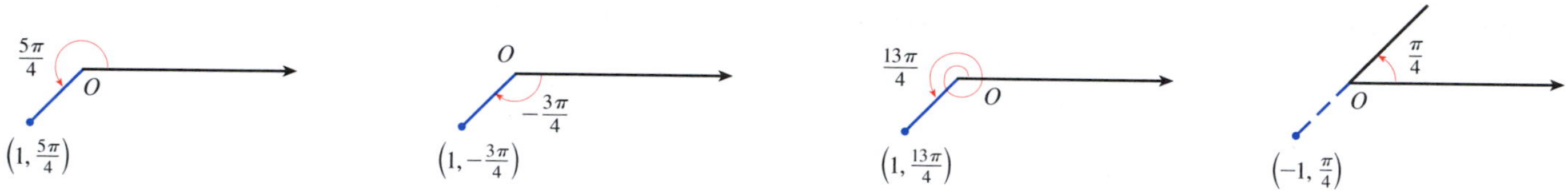

FIGURE 4

In fact, since a complete counterclockwise rotation is given by an angle 2π, the point represented by polar coordinates (r, θ) is also represented by

$$(r, \theta + 2n\pi) \qquad \text{and} \qquad (-r, \theta + (2n + 1)\pi)$$

where n is any integer.

The connection between polar and Cartesian coordinates can be seen from Figure 5, in which the pole corresponds to the origin and the polar axis coincides with the positive x-axis. If the point P has Cartesian coordinates (x, y) and polar coordinates (r, θ), then, from the figure, we have

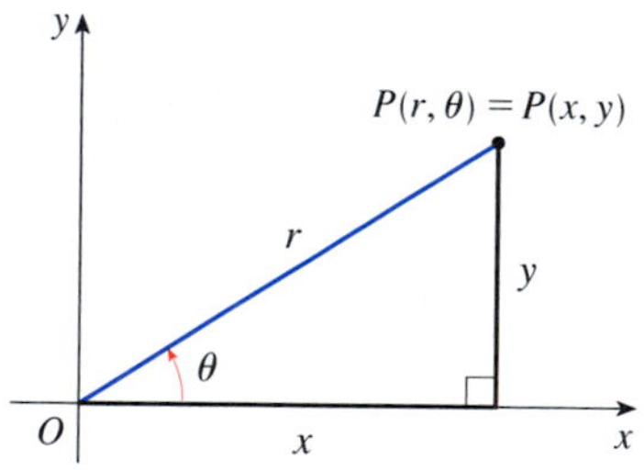

FIGURE 5

$$\cos\theta = \frac{x}{r} \qquad \sin\theta = \frac{y}{r}$$

and so

(1) $$x = r\cos\theta \qquad y = r\sin\theta$$

Although Equations 1 were deduced from Figure 5, which illustrates the case where $r > 0$ and $0 < \theta < \pi/2$, these equations are valid for all values of r and θ. (See the general definition of $\sin\theta$ and $\cos\theta$ in Appendix D.)

Equations 1 allow us to find the Cartesian coordinates of a point when the polar coordinates are known. To find r and θ when x and y are known, we use the equations

(2) $$r^2 = x^2 + y^2 \qquad \tan\theta = \frac{y}{x}$$

which can be deduced from Equations 1 or simply read from Figure 5.

EXAMPLE 2 Convert the point $(2, \pi/3)$ from polar to Cartesian coordinates.

SOLUTION Since $r = 2$ and $\theta = \pi/3$, Equations 1 give

$$x = r\cos\theta = 2\cos\frac{\pi}{3} = 2 \cdot \frac{1}{2} = 1$$

$$y = r\sin\theta = 2\sin\frac{\pi}{3} = 2 \cdot \frac{\sqrt{3}}{2} = \sqrt{3}$$

Therefore, the point is $(1, \sqrt{3})$ in Cartesian coordinates. ■

EXAMPLE 3 Represent the point with Cartesian coordinates $(1, -1)$ in terms of polar coordinates.

SOLUTION If we choose r to be positive, then Equations 2 give

$$r = \sqrt{x^2 + y^2} = \sqrt{1^2 + (-1)^2} = \sqrt{2}$$

$$\tan\theta = \frac{y}{x} = -1$$

Since the point $(1, -1)$ lies in the fourth quadrant, we can choose $\theta = -\pi/4$ or $\theta = 7\pi/4$. Thus, one possible answer is $(\sqrt{2}, -\pi/4)$. Another is $(\sqrt{2}, 7\pi/4)$. ■

NOTE: Equations 2 do not uniquely determine θ when x and y are given because, as θ increases through the interval $0 \leqslant \theta < 2\pi$, each value of $\tan\theta$ occurs twice. Therefore, in converting from Cartesian to polar coordinates, it is not good enough just to find r and θ that satisfy Equations 2. As in Example 3, we must choose θ so that the point (r, θ) lies in the correct quadrant.

The **graph of a polar equation** $r = f(\theta)$, or more generally $F(r, \theta) = 0$, consists of all points P that have at least one polar representation (r, θ) whose coordinates satisfy the equation.

EXAMPLE 4 What curve is represented by the polar equation $r = 2$?

SOLUTION The curve consists of all points (r, θ) with $r = 2$. Since r represents the distance from the point to the pole, the curve $r = 2$ represents the circle with center O and radius 2. In general, the equation $r = a$ represents a circle with center O and radius $|a|$ (see Figure 6).

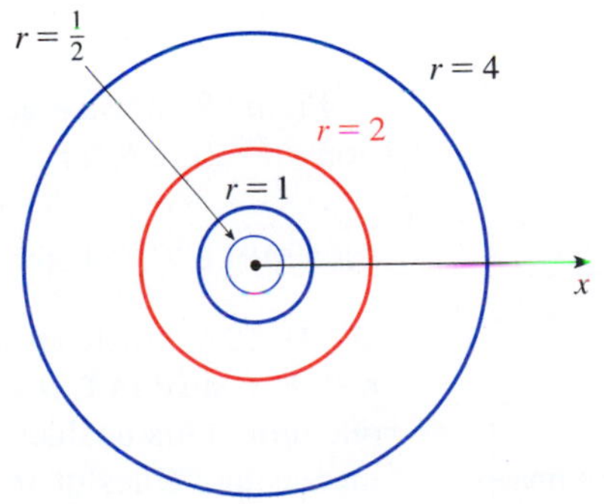

FIGURE 6 ■

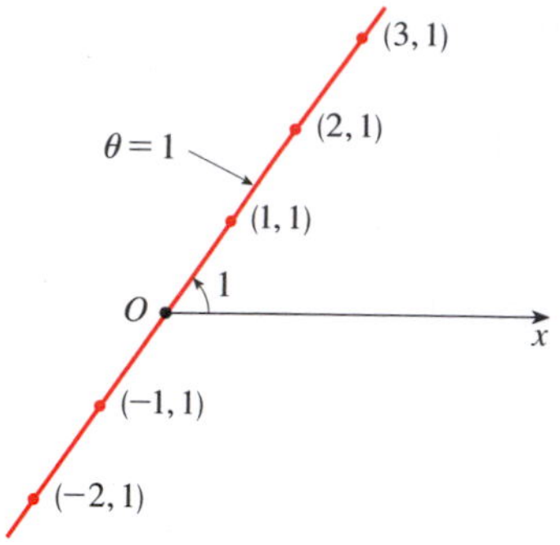

FIGURE 7

EXAMPLE 5 Sketch the polar curve $\theta = 1$.

SOLUTION This curve consists of all points (r, θ) such that the polar angle θ is 1 radian. It is the straight line that passes through O and makes an angle of 1 radian with the polar axis (see Figure 7). Notice that the points $(r, 1)$ on the line with $r > 0$ are in the first quadrant, whereas those with $r < 0$ are in the third quadrant. ■

EXAMPLE 6

(a) Sketch the curve with polar equation $r = 2\cos\theta$.
(b) Find a Cartesian equation for this curve.

SOLUTION

(a) In Figure 8 we find the values of r for some convenient values of θ and plot the corresponding points (r, θ). Then we join these points to sketch the curve, which appears to be a circle. We have used only values of θ between 0 and π, since if we let θ increase beyond π, we obtain the same points again.

θ	$r = 2\cos\theta$
0	2
$\pi/6$	$\sqrt{3}$
$\pi/4$	$\sqrt{2}$
$\pi/3$	1
$\pi/2$	0
$2\pi/3$	-1
$3\pi/4$	$-\sqrt{2}$
$5\pi/6$	$-\sqrt{3}$
π	-2

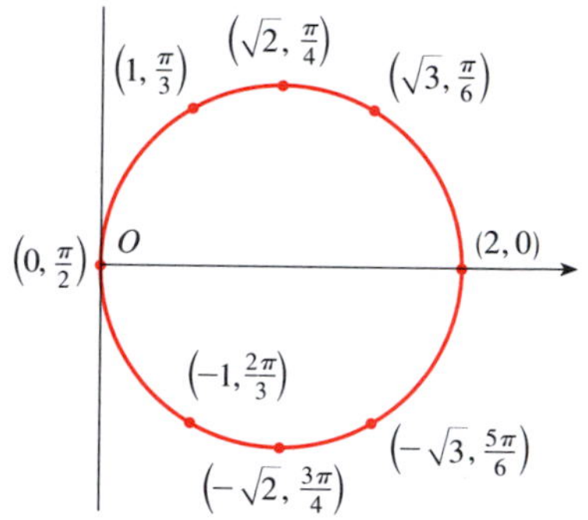

FIGURE 8 $r = 2\cos\theta$

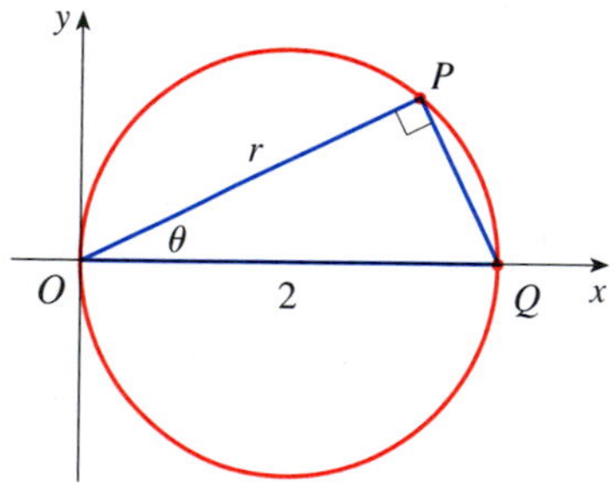

FIGURE 9

(b) To convert the given equation into a Cartesian equation we use Equations 1 and 2. From $x = r\cos\theta$ we have $\cos\theta = x/r$, so the equation $r = 2\cos\theta$ becomes $r = 2x/r$, which gives

$$2x = r^2 = x^2 + y^2 \qquad \text{or} \qquad x^2 + y^2 - 2x = 0$$

Completing the square, we obtain

$$(x - 1)^2 + y^2 = 1$$

which is the equation of a circle with center $(1, 0)$ and radius 1. ■

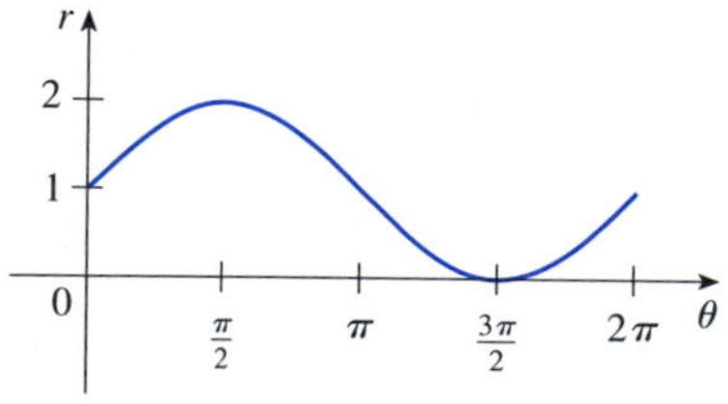

FIGURE 10
$r = 1 + \sin\theta$ in Cartesian coordinates, $0 \leq \theta \leq 2\pi$

Figure 9 shows a geometrical illustration that the circle in Example 6 has the equation $r = 2\cos\theta$. The angle OPQ is a right angle (why?) and so $r/2 = \cos\theta$.

EXAMPLE 7 Sketch the curve $r = 1 + \sin\theta$.

SOLUTION Instead of plotting points as in Example 6, we first sketch the graph of $r = 1 + \sin\theta$ in *Cartesian* coordinates in Figure 10 by shifting the sine curve up one unit. This enables us to read at a glance the values of r that correspond to increasing values of θ. For instance, we see that as θ increases from 0 to $\pi/2$, r (the distance from O) increases from 1 to 2, so we sketch the corresponding part of the

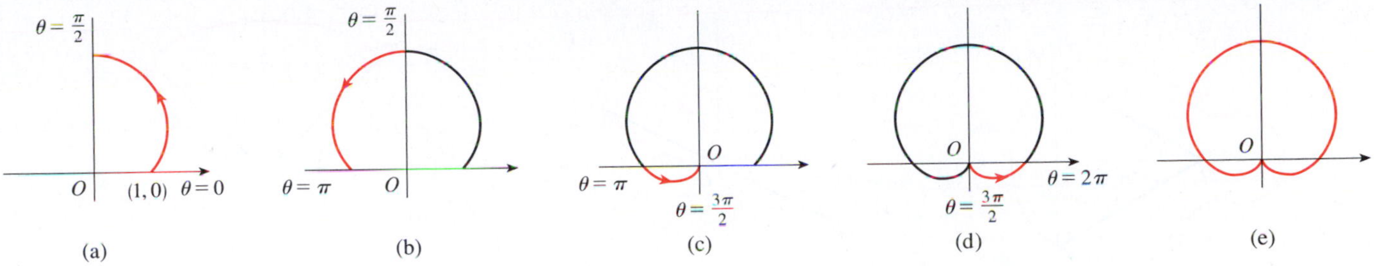

FIGURE 11
Stages in sketching the cardioid $r = 1 + \sin\theta$

polar curve in Figure 11(a). As θ increases from $\pi/2$ to π, Figure 10 shows that r decreases from 2 to 1, so we sketch the next part of the curve as in Figure 11(b). As θ increases from π to $3\pi/2$, r decreases from 1 to 0 as shown in part (c). Finally, as θ increases from $3\pi/2$ to 2π, r increases from 0 to 1 as shown in part (d). If we let θ increase beyond 2π or decrease beyond 0, we would simply retrace our path. Putting together the parts of the curve from Figure 11(a)–(d), we sketch the complete curve in Figure 11(e). It is called a **cardioid** because it is shaped like a heart. ■

EXAMPLE 8 Sketch the curve $r = \cos 2\theta$.

SOLUTION As in Example 7, we first sketch $r = \cos 2\theta$, $0 \leqslant \theta \leqslant 2\pi$, in Cartesian coordinates in Figure 12. As θ increases from 0 to $\pi/4$, Figure 12 shows that r decreases from 1 to 0 and so we draw the corresponding portion of the polar curve in Figure 13 (indicated by a single arrow). As θ increases from $\pi/4$ to $\pi/2$, r goes from 0 to -1. This means that the distance from O increases from 0 to 1, but instead of being in the first quadrant this portion of the polar curve (indicated by a double arrow) lies on the opposite side of the pole in the third quadrant. The remainder of the curve is drawn in a similar fashion, with the arrows and numbers indicating the order in which the portions are traced out. The resulting curve has four loops and is called a **four-leaved rose**.

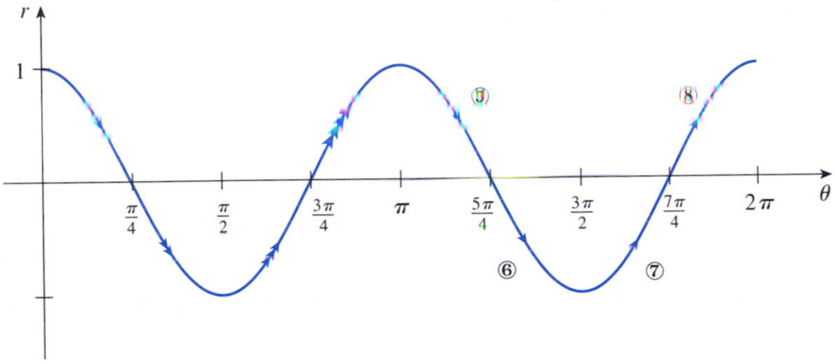

FIGURE 12
$r = \cos 2\theta$ in Cartesian coordinates

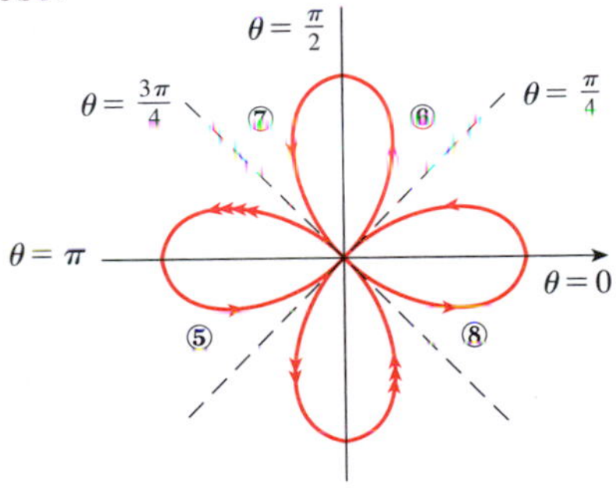

FIGURE 13
Four-leaved rose $r = \cos 2\theta$ ■

When we sketch polar curves it is sometimes helpful to take advantage of symmetry. The following three rules are explained by Figure 14.

(a) If a polar equation is unchanged when θ is replaced by $-\theta$, the curve is symmetric about the polar axis.

(b) If the equation is unchanged when r is replaced by $-r$, the curve is symmetric about the pole.

(c) If the equation is unchanged when θ is replaced by $\pi - \theta$, the curve is symmetric about the vertical line $\theta = \pi/2$.

The curves sketched in Examples 6 and 8 are symmetric about the polar axis, since $\cos(-\theta) = \cos\theta$. The curves in Examples 7 and 8 are symmetric about $\theta = \pi/2$

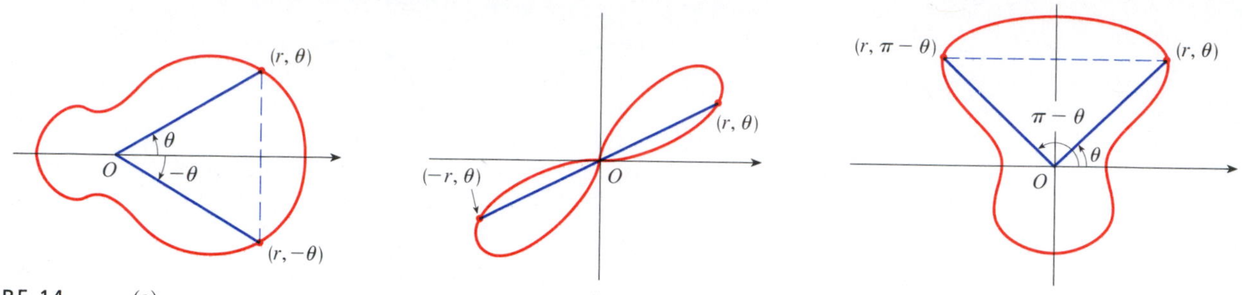

FIGURE 14 (a) (b) (c)

because $\sin(\pi - \theta) = \sin\theta$ and $\cos 2(\pi - \theta) = \cos 2\theta$. The four-leaved rose is also symmetric about the pole. These symmetry properties could have been used in sketching the curves. For instance, in Example 6 we need only have plotted points for $0 \leqslant \theta \leqslant \pi/2$ and then reflected about the polar axis to obtain the complete circle.

GRAPHING POLAR CURVES WITH GRAPHING DEVICES

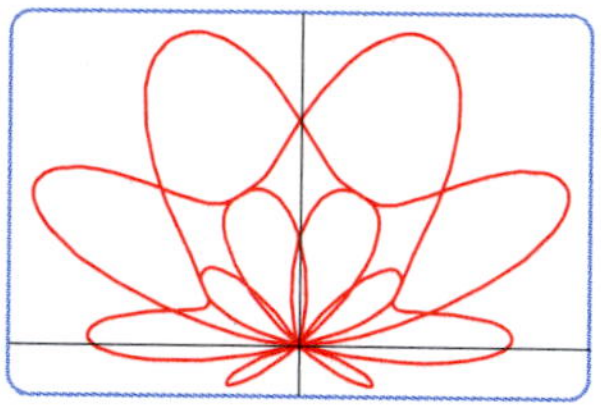

FIGURE 15
$r = \sin\theta + \sin^3(5\theta/2)$

Although it is useful to be able to sketch simple polar curves by hand, we need to use a graphing calculator or computer when we are faced with a curve as complicated as the one shown in Figure 15.

Some graphing devices have commands that enable us to graph polar curves directly. With other machines it is necessary to convert to parametric equations first. In the latter case we take the polar equation $r = f(\theta)$ and write its parametric equations as

$$x = r\cos\theta = f(\theta)\cos\theta \qquad y = r\sin\theta = f(\theta)\sin\theta$$

Some machines require that the parameter be called t rather than θ.

EXAMPLE 9 Graph the curve $r = \sin(8\theta/5)$.

SOLUTION Let's assume that our graphing device doesn't have a built-in polar graphing command. In this case we need to work with the corresponding parametric equations, which are

$$x = r\cos\theta = \sin(8\theta/5)\cos\theta \qquad y = r\sin\theta = \sin(8\theta/5)\sin\theta$$

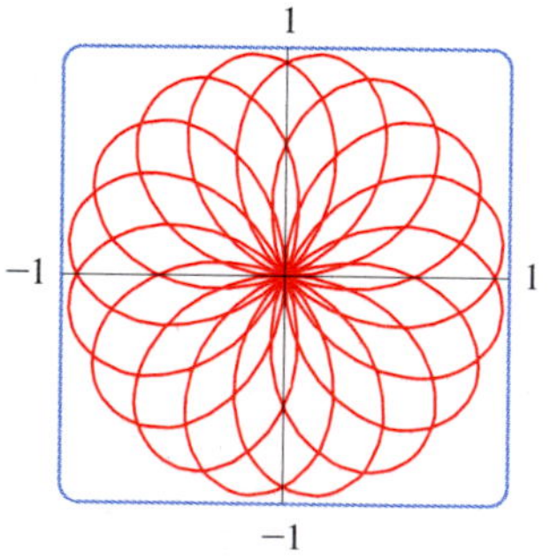

FIGURE 16
$r = \sin(8\theta/5)$

In any case we need to determine the domain for θ. So we ask ourselves: How many complete rotations are required until the curve starts to repeat itself? If the answer is n, then

$$\sin\frac{8(\theta + 2n\pi)}{5} = \sin\frac{8\theta}{5}$$

and so we require that $16n\pi/5$ be an even multiple of π. This will first occur when $n = 5$. Therefore, we will graph the entire curve if we specify that $0 \leqslant \theta \leqslant 10\pi$. Switching from θ to t, we have the equations

$$x = \sin(8t/5)\cos t \qquad y = \sin(8t/5)\sin t \qquad 0 \leqslant t \leqslant 10\pi$$

and Figure 16 shows the resulting curve. Notice that this rose has 16 loops. ■

EXAMPLE 10 Investigate the family of polar curves given by $r = 1 + c\sin\theta$. How does the shape change as c changes? (These curves are called **limaçons**, after a French word for snail, because of the shape of the curves for certain values of c.)

In Exercise 63 you are asked to prove analytically what we have discovered from the graphs.

SOLUTION Figure 17 shows computer-drawn graphs for various values of c. For $c > 1$ there is a loop that decreases in size as c decreases. When $c = 1$ the loop disappears and the curve becomes the cardioid that we sketched in Example 7. For c between 1 and $\frac{1}{2}$ the cardioid's cusp is smoothed out and becomes a "dimple." When c decreases from $\frac{1}{2}$ to 0, the limaçon is shaped like an oval. This oval becomes more circular as $c \to 0$, and when $c = 0$ the curve is just the circle $r = 1$.

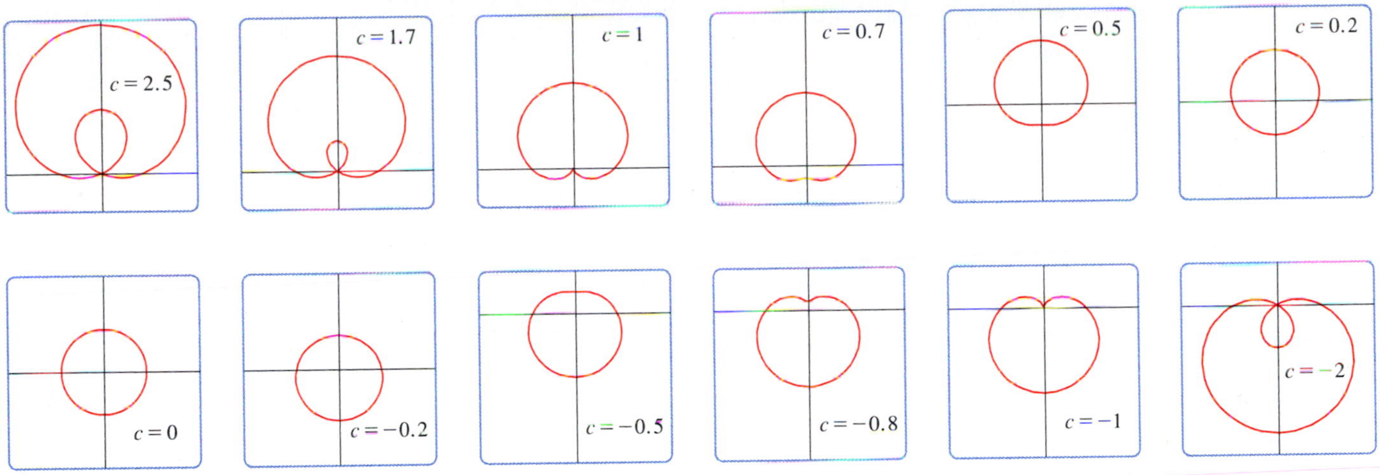

FIGURE 17
Members of the family of limaçons $r = 1 + c\sin\theta$

The remaining parts of Figure 17 show that as c becomes negative, the shapes change in reverse order. In fact, these curves are reflections about the horizontal axis of the corresponding curves with positive c. ■

EXERCISES 13.4

1–6 ■ Plot the point whose polar coordinates are given. Then find two other pairs of polar coordinates of this point, one with $r > 0$ and one with $r < 0$.

1. $(1, \pi/2)$ **2.** $(3, 0)$ **3.** $(-1, \pi/5)$

4. $(2, -\pi/7)$ **5.** $(3, 2)$ **6.** $(-1, \pi)$

7–12 ■ Plot the point whose polar coordinates are given. Then find the Cartesian coordinates of the point.

7. $(\sqrt{2}, \pi/4)$ **8.** $(2, 2\pi/3)$ **9.** $(1.5, 3\pi/2)$

10. $(4, 3\pi)$ **11.** $(-1, \pi/3)$ **12.** $(-2, -5\pi/6)$

13–16 ■ The Cartesian coordinates of a point are given. Find the polar coordinates (r, θ) of the point, where $r > 0$ and $0 \leqslant \theta < 2\pi$.

13. $(-1, 1)$ **14.** $(-1, -\sqrt{3})$

15. $(2\sqrt{3}, -2)$ **16.** $(3, 4)$

17–22 ■ Sketch the region in the plane consisting of points whose polar coordinates satisfy the given conditions.

17. $r > 1$ **18.** $0 \leqslant \theta \leqslant \pi/3$

19. $0 \leqslant r \leqslant 2,\ \pi/2 \leqslant \theta \leqslant \pi$

20. $1 \leqslant r < 3,\ -\pi/4 \leqslant \theta \leqslant \pi/4$

21. $3 < r < 4,\ -\pi/2 \leqslant \theta \leqslant \pi$

22. $-1 \leqslant r \leqslant 1,\ \pi/4 \leqslant \theta \leqslant 3\pi/4$

23. Find the distance between the points with polar coordinates $(1, \pi/6)$ and $(3, 3\pi/4)$.

24. Find a formula for the distance between the points with polar coordinates (r_1, θ_1) and (r_2, θ_2).

25–30 ■ Find a Cartesian equation for the curve described by the given polar equation.

25. $r\sin\theta = 2$ **26.** $r = 2\sin\theta$

27. $r = \dfrac{1}{1 - \cos\theta}$ **28.** $r = \dfrac{5}{3 - 4\sin\theta}$

29. $r^2 = \sin 2\theta$ **30.** $r^2 = \theta$

31–36 ■ Find a polar equation for the curve represented by the given Cartesian equation.

31. $y = 5$ **32.** $y = x + 1$

33. $x^2 + y^2 = 25$ **34.** $x^2 = 4y$

35. $2xy = 1$ **36.** $x^2 - y^2 = 1$

37–58 ■ Sketch the curve of each polar equation.

37. $r = 5$

38. $\theta = 3\pi/4$

39. $r = 2\sin\theta$

40. $r = -4\sin\theta$

41. $r = -\cos\theta$

42. $r = 2\sin\theta + 2\cos\theta$

43. $r = 3(1 - \cos\theta)$

44. $r = 1 + \cos\theta$

45. $r = \theta,\ \theta \geqslant 0$ (spiral)

46. $r = \theta/2,\ -4\pi \leqslant \theta \leqslant 4\pi$

47. $r = 1/\theta$

48. $r = e^{\theta}$

49. $r = 1 - 2\cos\theta$

50. $r = 2 + \cos\theta$

51. $r = \sin 2\theta$

52. $r = 2\cos 3\theta$

53. $r = 2\cos 4\theta$

54. $r = \sin 5\theta$ (five-leaved rose)

55. $r^2 = 4\cos 2\theta$

56. $r^2 = \sin 2\theta$ (lemniscate)

57. $r = 2\cos(3\theta/2)$

58. $r^2\theta = 1$ (lituus)

59. Show that the polar curve $r = 4 + 2\sec\theta$ (called a **conchoid**) has the line $x = 2$ as a vertical asymptote by showing that $\lim_{r\to\pm\infty} x = 2$. Use this fact to help sketch the conchoid.

60. Show that the curve $r = 2 - \csc\theta$ (also a conchoid) has the line $y = -1$ as a horizontal asymptote by showing that $\lim_{r\to\pm\infty} y = -1$. Use this fact to help sketch the conchoid.

61. Show that the curve $r = \sin\theta\tan\theta$ (called a **cissoid of Diocles**) has the line $x = 1$ as a vertical asymptote. Show also that the curve lies entirely within the vertical strip $0 \leqslant x < 1$. Use these facts to help sketch the cissoid.

62. Sketch the curve $(x^2 + y^2)^3 = 4x^2y^2$.

63. (a) In Example 10 the graphs suggest that the limaçon $r = 1 + c\sin\theta$ has an inner loop when $|c| > 1$. Prove that this is true, and find the values of θ that correspond to the inner loop.
(b) From Figure 17 it appears that the limaçon loses its dimple when $c = \frac{1}{2}$. Prove this.

64. Match the polar equations with the graphs labeled I–VI. Give reasons for your choices.
(a) $r = \sin(\theta/2)$ (b) $r = \sin(\theta/4)$
(c) $r = \sec(3\theta)$ (d) $r = \theta\sin\theta$
(e) $r = 1 + 4\cos 5\theta$ (f) $r = 1/\sqrt{\theta}$

I

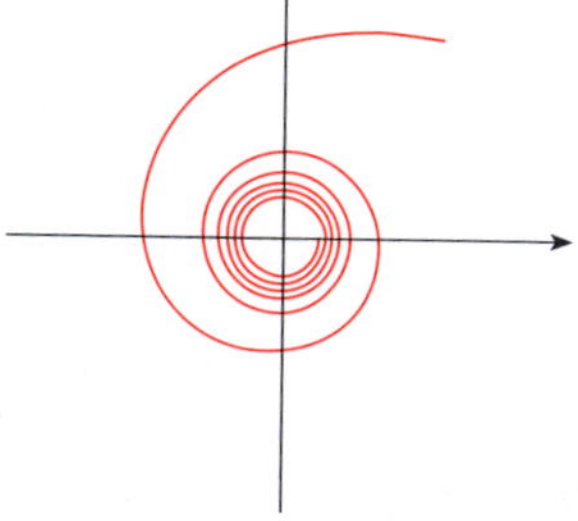

II

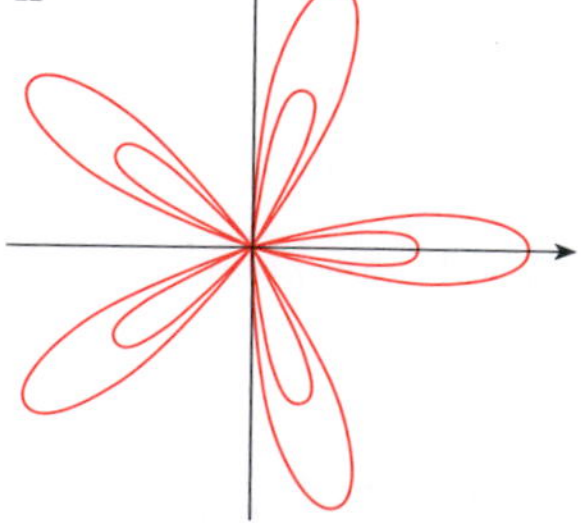

III

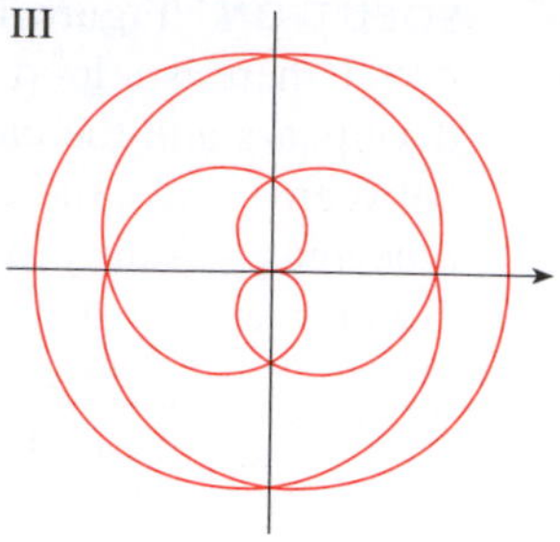

IV

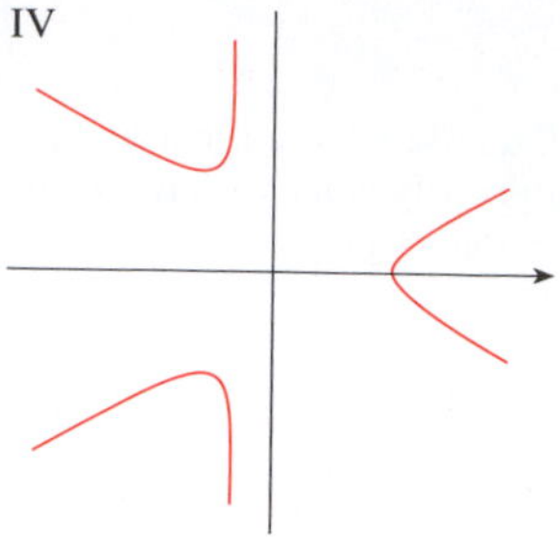

V

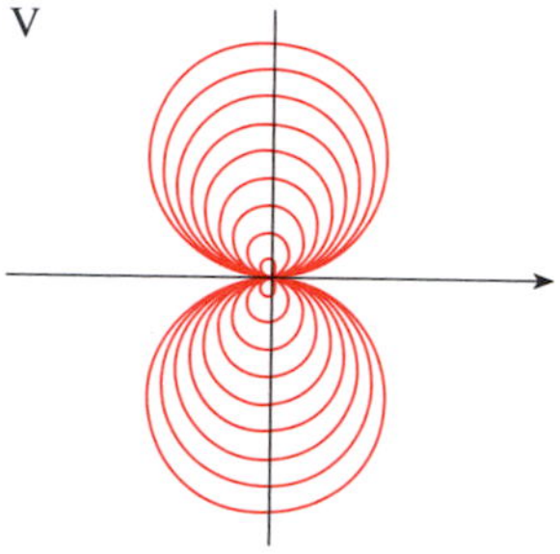

VI

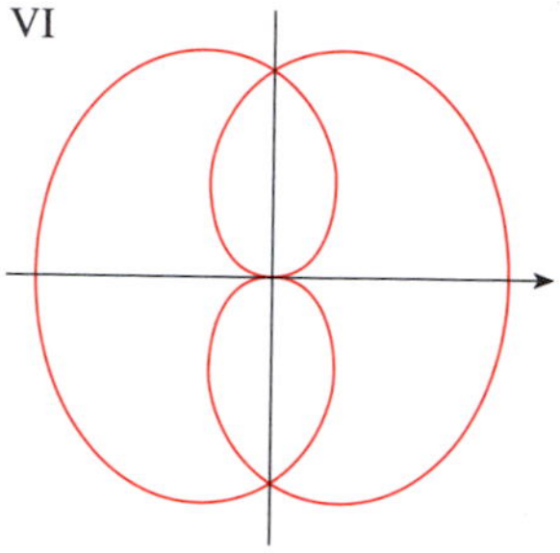

65–68 ■ Use a graphing device to graph the polar curve. Choose the parameter interval to make sure that you produce the entire curve.

65. $r = 1 + 2\sin(\theta/2)$ (nephroid of Freeth)

66. $r = \sqrt{1 - 0.8\sin^2\theta}$ (hippopede)

67. $r = \sin(9\theta/4)$

68. $r = 1 + 4\cos(\theta/3)$

69. How are the graphs of $r = 1 + \sin(\theta - \pi/6)$ and $r = 1 + \sin(\theta - \pi/3)$ related to the graph of $r = 1 + \sin\theta$? In general, how is the graph of $r = f(\theta - \alpha)$ related to the graph of $r = f(\theta)$?

70. Use a graph to estimate the y-coordinate of the highest points on the curve $r = \sin 2\theta$. Then use calculus to find the exact value.

71. (a) Investigate the family of curves defined by the polar equations $r = \sin n\theta$, where n is a positive integer. How is the number of loops related to n?
(b) What happens if the equation in part (a) is replaced by $r = |\sin n\theta|$?

72. A family of curves is given by the equations $r = 1 + c\sin n\theta$, where c is a real number and n is a positive integer. How does the graph change as n increases? How does it change as c changes? Illustrate by graphing enough members of the family to support your conclusions.

73. A family of curves has polar equations

$$r = \frac{1 - a\cos\theta}{1 + a\cos\theta}$$

Investigate how the graph changes as the number a changes. In particular, you should identify the transitional values of a for which the basic shape of the curve changes.

74. The astronomer Giovanni Cassini (1625–1712) studied the family of curves with polar equations

$$r^4 - 2c^2r^2\cos 2\theta + c^4 - a^4 = 0$$

where a and c are positive real numbers. These curves are called the **ovals of Cassini** even though they are oval-shaped only for certain values of a and c. (Cassini thought that these curves might represent planetary orbits better than Kepler's ellipses.) Investigate the variety of shapes that these curves may have. In particular, how are a and c related to each other when the curve splits into two parts?

13.5 DOUBLE INTEGRALS IN POLAR COORDINATES

Suppose that we want to evaluate a double integral $\iint_R f(x, y)\,dA$, where R is one of the regions shown in Figure 1. In either case the description of R in terms of rectangular coordinates is rather complicated but R is easily described using polar coordinates.

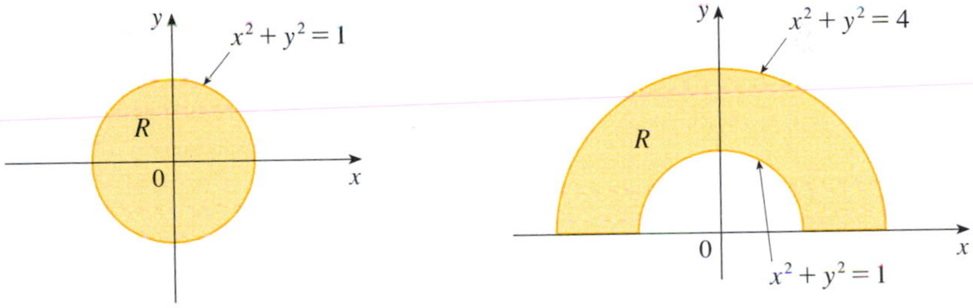

FIGURE 1

(a) $R = \{(r, \theta) \mid 0 \leqslant r \leqslant 1, 0 \leqslant \theta \leqslant 2\pi\}$ (b) $R = \{(r, \theta) \mid 1 \leqslant r \leqslant 2, 0 \leqslant \theta \leqslant \pi\}$

Recall from the preceding section that the polar coordinates (r, θ) of a point are related to the rectangular coordinates by the equations

$$r^2 = x^2 + y^2 \qquad x = r\cos\theta \qquad y = r\sin\theta$$

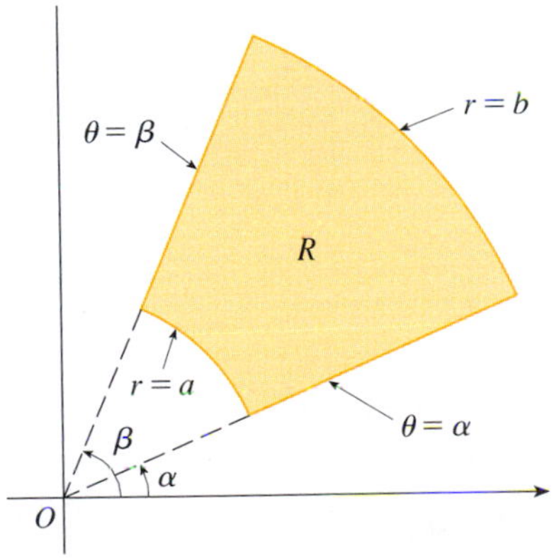

FIGURE 2
Polar rectangle

The regions in Figure 1 are special cases of a **polar rectangle**

$$R = \{(r, \theta) \mid a \leqslant r \leqslant b,\ \alpha \leqslant \theta \leqslant \beta\}$$

which is shown in Figure 2. In order to compute $\iint_R f(x, y)\,dA$, where R is a polar rectangle, we start with a partition of $[a, b]$ into m subintervals:

$$a = r_0 < r_1 < r_2 < \cdots < r_{i-1} < r_i < \cdots < r_m = b$$

and a partition of $[\alpha, \beta]$ into n subintervals:

$$\alpha = \theta_0 < \theta_1 < \theta_2 < \cdots < \theta_{j-1} < \theta_j < \cdots < \theta_n = \beta$$

Then the circles $r = r_i$ and the rays $\theta = \theta_j$ determine a **polar partition** P of R into the

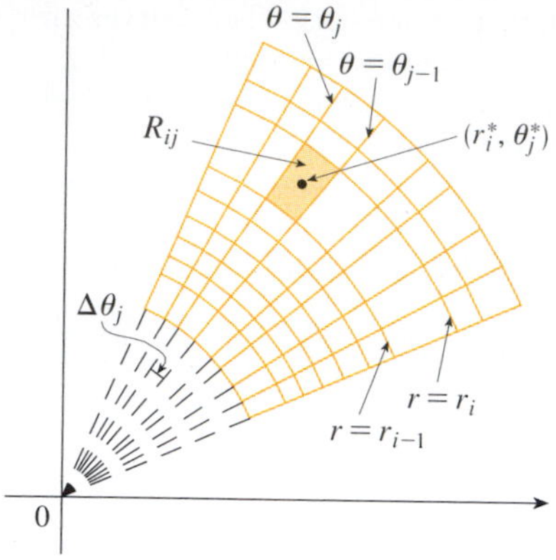

FIGURE 3
Polar partition

small polar rectangles shown in Figure 3. The norm $\|P\|$ of the polar partition is the length of the longest diagonal of all the polar subrectangles.

The "center" of the polar subrectangle

$$R_{ij} = \{(r, \theta) \mid r_{i-1} \leqslant r \leqslant r_i,\ \theta_{j-1} \leqslant \theta \leqslant \theta_j\}$$

has polar coordinates

$$r_i^* = \tfrac{1}{2}(r_{i-1} + r_i) \qquad \theta_j^* = \tfrac{1}{2}(\theta_{j-1} + \theta_j)$$

We compute the area of R_{ij} using the fact that the area of a sector of a circle with radius r and central angle θ is $\frac{1}{2}r^2\theta$. Subtracting the areas of two such sectors, each of which has central angle $\Delta\theta_j = \theta_j - \theta_{j-1}$, we find that the area of R_{ij} is

$$\begin{aligned}\Delta A_{ij} &= \tfrac{1}{2}r_i^2\,\Delta\theta_j - \tfrac{1}{2}r_{i-1}^2\,\Delta\theta_j \\ &= \tfrac{1}{2}(r_i^2 - r_{i-1}^2)\,\Delta\theta_j \\ &= \tfrac{1}{2}(r_i + r_{i-1})(r_i - r_{i-1})\,\Delta\theta_j \\ &= r_i^*\,\Delta r_i\,\Delta\theta_j\end{aligned}$$

where $\Delta r_i = r_i - r_{i-1}$.

Although we have defined the double integral $\iint_R f(x, y)\,dA$ in terms of rectangular partitions, it can be shown that, for continuous functions f, we always obtain the same answer using polar partitions. The rectangular coordinates of the center of R_{ij} are $(r_i^* \cos\theta_j^*, r_i^* \sin\theta_j^*)$, so a typical Riemann sum is

(1) $$\sum_{i=1}^{m}\sum_{j=1}^{n} f(r_i^*\cos\theta_j^*, r_i^*\sin\theta_j^*)\,\Delta A_{ij} = \sum_{i=1}^{m}\sum_{j=1}^{n} f(r_i^*\cos\theta_j^*, r_i^*\sin\theta_j^*)\,r_i^*\,\Delta r_i\,\Delta\theta_j$$

If we write $g(r, \theta) = rf(r\cos\theta, r\sin\theta)$, then the Riemann sum in Equation 1 can be written as

$$\sum_{i=1}^{m}\sum_{j=1}^{n} g(r_i^*, \theta_j^*)\,\Delta r_i\,\Delta\theta_j$$

which is a Riemann sum for the double integral

$$\int_\alpha^\beta \int_a^b g(r, \theta)\,dr\,d\theta$$

Therefore, we have

$$\begin{aligned}\iint_R f(x, y)\,dA &= \lim_{\|P\|\to 0} \sum_{i=1}^{m}\sum_{j=1}^{n} f(r_i^*\cos\theta_j^*, r_i^*\sin\theta_j^*)\,\Delta A_{ij} \\ &= \lim_{\|P\|\to 0} \sum_{i=1}^{m}\sum_{j=1}^{n} g(r_i^*, \theta_j^*)\,\Delta r_i\,\Delta\theta_j \\ &= \int_\alpha^\beta \int_a^b g(r, \theta)\,dr\,d\theta \\ &= \int_\alpha^\beta \int_a^b f(r\cos\theta, r\sin\theta)\,r\,dr\,d\theta\end{aligned}$$

(2) CHANGE TO POLAR COORDINATES IN A DOUBLE INTEGRAL If f is continuous on a polar rectangle R given by $0 \leqslant a \leqslant r \leqslant b$, $\alpha \leqslant \theta \leqslant \beta$, where $0 \leqslant \beta - \alpha \leqslant 2\pi$, then

$$\iint_R f(x, y)\, dA = \int_\alpha^\beta \int_a^b f(r\cos\theta, r\sin\theta)\, r\, dr\, d\theta$$

The formula in (2) says that we convert from rectangular to polar coordinates in a double integral by writing $x = r\cos\theta$ and $y = r\sin\theta$, using the appropriate limits of integration for r and θ, and replacing dA by $r\,dr\,d\theta$. Be careful not to forget the additional factor r on the right side of Formula 2. A classical method for remembering this is shown in Figure 4, where the "infinitesimal" polar rectangle can be thought of as an ordinary rectangle with dimensions $r\,d\theta$ and dr and therefore has "area" $dA = r\,dr\,d\theta$.

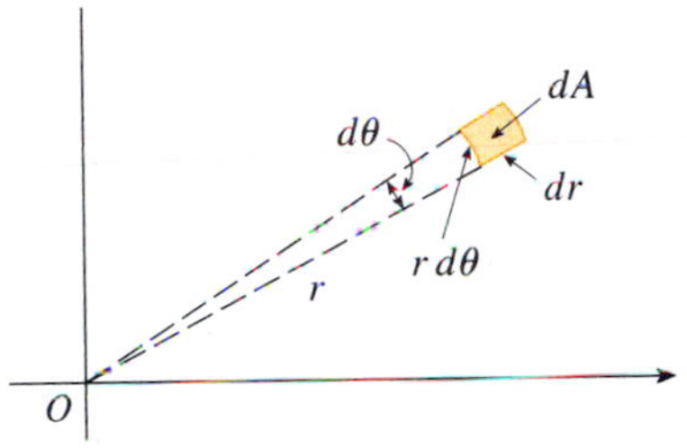

FIGURE 4

EXAMPLE 1 Evaluate $\iint_R (3x + 4y^2)\, dA$, where R is the region in the upper half-plane bounded by the circles $x^2 + y^2 = 1$ and $x^2 + y^2 = 4$.

SOLUTION The region R can be described as

$$R = \{(x, y) \mid y \geqslant 0,\ 1 \leqslant x^2 + y^2 \leqslant 4\}$$

It is the half-ring shown in Figure 1(b), and in polar coordinates it is given by $1 \leqslant r \leqslant 2$, $0 \leqslant \theta \leqslant \pi$. Therefore, by Formula 2,

$$\begin{aligned}
\iint_R (3x + 4y^2)\, dA &= \int_0^\pi \int_1^2 (3r\cos\theta + 4r^2\sin^2\theta)\, r\, dr\, d\theta \\
&= \int_0^\pi \int_1^2 (3r^2\cos\theta + 4r^3\sin^2\theta)\, dr\, d\theta \\
&= \int_0^\pi \left[r^3\cos\theta + r^4\sin^2\theta\right]_{r=1}^{r=2} d\theta \\
&= \int_0^\pi (7\cos\theta + 15\sin^2\theta)\, d\theta \\
&= \int_0^\pi \left[7\cos\theta + \tfrac{15}{2}(1 - \cos 2\theta)\right] d\theta \\
&= 7\sin\theta + \frac{15\theta}{2} - \frac{15}{4}\sin 2\theta\Big]_0^\pi = \frac{15\pi}{2}
\end{aligned}$$

EXAMPLE 2 Find the volume of the solid bounded by the plane $z = 0$ and the paraboloid $z = 1 - x^2 - y^2$.

SOLUTION If we put $z = 0$ in the equation of the paraboloid, we get $x^2 + y^2 = 1$. This means that the plane intersects the paraboloid in the circle $x^2 + y^2 = 1$, so the solid lies under the paraboloid and above the circular disk D given by $x^2 + y^2 \leqslant 1$

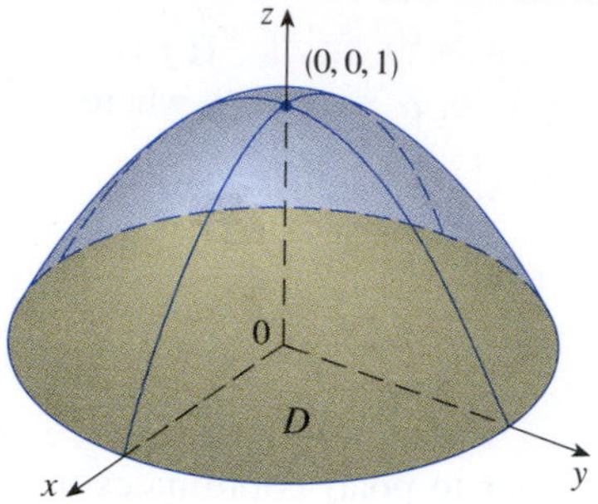

FIGURE 5

[see Figures 5 and 1(a)]. In polar coordinates D is given by $0 \leqslant r \leqslant 1$, $0 \leqslant \theta \leqslant 2\pi$. Since $1 - x^2 - y^2 = 1 - r^2$, the volume is

$$V = \iint_D (1 - x^2 - y^2)\, dA = \int_0^{2\pi} \int_0^1 (1 - r^2)\, r\, dr\, d\theta$$

$$= \int_0^{2\pi} d\theta \int_0^1 (r - r^3)\, dr = 2\pi \left[\frac{r^2}{2} - \frac{r^4}{4} \right]_0^1 = \frac{\pi}{2}$$

If we had used rectangular coordinates instead of polar coordinates, then we would have obtained

$$V = \iint_D (1 - x^2 - y^2)\, dA = \int_{-1}^{1} \int_{-\sqrt{1-x^2}}^{\sqrt{1-x^2}} (1 - x^2 - y^2)\, dy\, dx$$

which is not easy to evaluate because it involves finding the following integrals:

$$\int \sqrt{1 - x^2}\, dx \qquad \int x^2 \sqrt{1 - x^2}\, dx \qquad \int (1 - x^2)^{3/2}\, dx$$

■

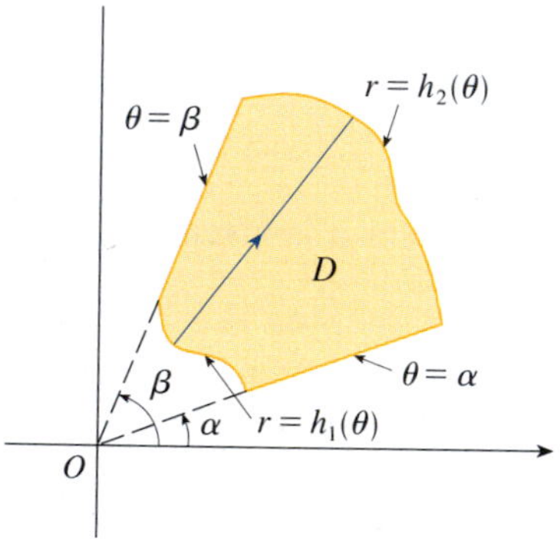

FIGURE 6
Type II polar region,
$D = \{(r, \theta) \mid \alpha \leqslant \theta \leqslant \beta, h_1(\theta) \leqslant r \leqslant h_2(\theta)\}$

What we have done so far can be extended to the more complicated type of region shown in Figure 6. We could call it a type II polar region because it is analogous to the type II rectangular regions considered in Section 13.3. In fact, by combining Formulas 2 and 13.3.5, we obtain the following formula:

(3) If f is continuous on a polar region of the form

$$D = \{(r, \theta) \mid \alpha \leqslant \theta \leqslant \beta,\ h_1(\theta) \leqslant r \leqslant h_2(\theta)\}$$

then
$$\iint_D f(x, y)\, dA = \int_\alpha^\beta \int_{h_1(\theta)}^{h_2(\theta)} f(r\cos\theta, r\sin\theta)\, r\, dr\, d\theta$$

In particular, taking $f(x, y) = 1$, $h_1(\theta) = 0$, and $h_2(\theta) = h(\theta)$ in this formula, we see that the area of the region D bounded by $\theta = \alpha$, $\theta = \beta$, and $r = h(\theta)$ is

$$A(D) = \iint_D 1\, dA = \int_\alpha^\beta \int_0^{h(\theta)} r\, dr\, d\theta$$

$$= \int_\alpha^\beta \left[\frac{r^2}{2} \right]_0^{h(\theta)} d\theta = \int_\alpha^\beta \tfrac{1}{2}[h(\theta)]^2\, d\theta$$

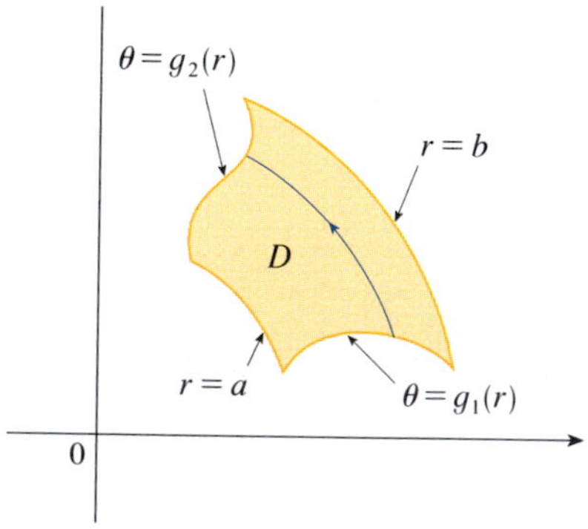

FIGURE 7
Type I polar region,
$D = \{(r, \theta) \mid a \leqslant r \leqslant b, g_1(r) \leqslant \theta \leqslant g_2(r)\}$

Similarly, the analogue of a type I rectangular region is a polar region of the form

$$D = \{(r, \theta) \mid a \leqslant r \leqslant b,\ g_1(r) \leqslant \theta \leqslant g_2(r)\}$$

(see Figure 7), and Formula 13.3.3, together with Formula 2, shows that

(4)
$$\iint_D f(x, y)\, dA = \int_a^b \int_{g_1(r)}^{g_2(r)} f(r\cos\theta, r\sin\theta)\, r\, d\theta\, dr$$

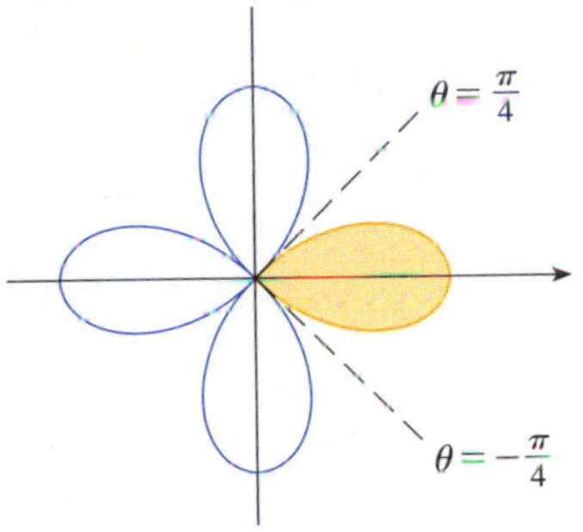

FIGURE 8

EXAMPLE 3 Use a double integral to find the area enclosed by one loop of the four-leaved rose $r = \cos 2\theta$.

SOLUTION From the sketch of the curve in Figure 8 we see that a loop is given by the region

$$D = \left\{(r, \theta) \;\middle|\; -\frac{\pi}{4} \leqslant \theta \leqslant \frac{\pi}{4},\; 0 \leqslant r \leqslant \cos 2\theta\right\}$$

So the area is

$$A(D) = \iint_D dA = \int_{-\pi/4}^{\pi/4} \int_0^{\cos 2\theta} r\,dr\,d\theta$$

$$= \int_{-\pi/4}^{\pi/4} \left[\tfrac{1}{2}r^2\right]_0^{\cos 2\theta} d\theta = \tfrac{1}{2}\int_{-\pi/4}^{\pi/4} \cos^2 2\theta\,d\theta$$

$$= \tfrac{1}{4}\int_{-\pi/4}^{\pi/4} (1 + \cos 4\theta)\,d\theta = \tfrac{1}{4}\left[\theta + \tfrac{1}{4}\sin 4\theta\right]_{-\pi/4}^{\pi/4} = \frac{\pi}{8}$$ ■

EXAMPLE 4 Find the volume of the solid that lies under the paraboloid $z = x^2 + y^2$, above the xy-plane, and inside the cylinder $x^2 + y^2 = 2x$.

SOLUTION The solid lies above the disk D whose boundary circle has equation $x^2 + y^2 = 2x$ or

$$(x - 1)^2 + y^2 = 1$$

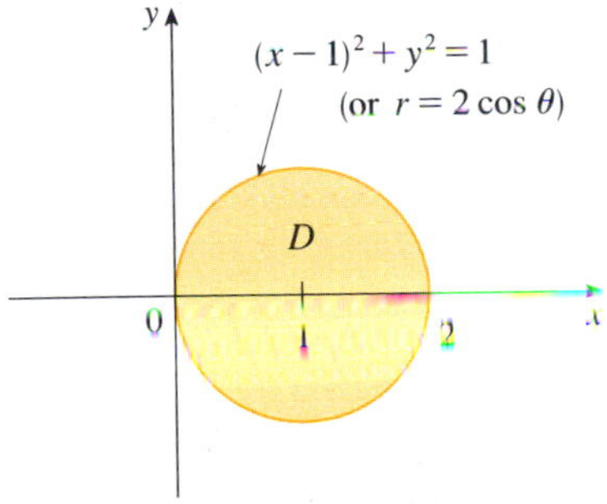

FIGURE 9

(see Figures 9 and 10). In polar coordinates we have $x^2 + y^2 = r^2$ and $x = r\cos\theta$, so the boundary circle becomes $r^2 = 2r\cos\theta$ or $r = 2\cos\theta$. Thus, the disk D is given by

$$D = \left\{(r, \theta) \;\middle|\; -\frac{\pi}{2} \leqslant \theta \leqslant \frac{\pi}{2},\; 0 \leqslant r \leqslant 2\cos\theta\right\}$$

and, by Formula 3, we have

$$V = \iint_D (x^2 + y^2)\,dA = \int_{-\pi/2}^{\pi/2} \int_0^{2\cos\theta} r^2 r\,dr\,d\theta$$

$$= \int_{-\pi/2}^{\pi/2} \left[\frac{r^4}{4}\right]_0^{2\cos\theta} d\theta = 4\int_{-\pi/2}^{\pi/2} \cos^4\theta\,d\theta$$

$$= 8\int_0^{\pi/2} \cos^4\theta\,d\theta = 8\int_0^{\pi/2} \left(\frac{1 + \cos 2\theta}{2}\right)^2 d\theta$$

$$= 2\int_0^{\pi/2} \left[1 + 2\cos 2\theta + \tfrac{1}{2}(1 + \cos 4\theta)\right] d\theta$$

$$= 2\left[\tfrac{3}{2}\theta + \sin 2\theta + \tfrac{1}{8}\sin 4\theta\right]_0^{\pi/2} = 2\left(\frac{3}{2}\right)\left(\frac{\pi}{2}\right) = \frac{3\pi}{2}$$ ■

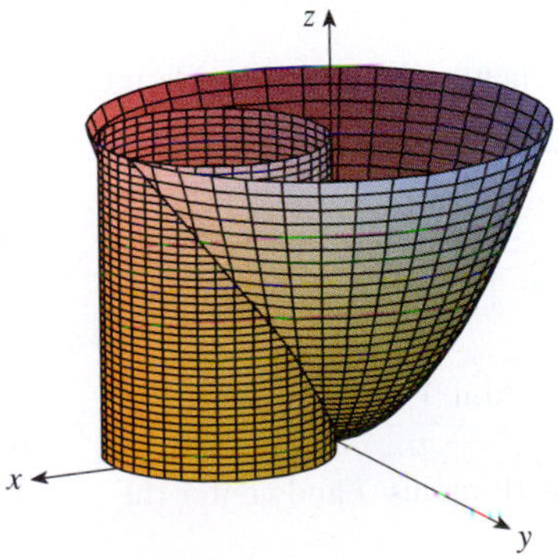

FIGURE 10

EXERCISES 13.5

1–8 ■ Evaluate the given integral by changing to polar coordinates.

1. $\iint_R x\,dA$, where R is the disk with center the origin and radius 5

2. $\iint_R y\,dA$, where R is the region in the first quadrant bounded by the circle $x^2 + y^2 = 9$ and the lines $y = x$ and $y = 0$

3. $\iint_R xy\,dA$, where R is the region in the first quadrant that lies between the circles $x^2 + y^2 = 4$ and $x^2 + y^2 = 25$

4. $\iint_R \sin(x^2 + y^2)\,dA$, where R is the annular region $1 \leq x^2 + y^2 \leq 16$

5. $\iint_D 1/\sqrt{x^2 + y^2}\,dA$, where D is the region that lies inside the cardioid $r = 1 + \sin\theta$ and outside the circle $r = 1$

6. $\iint_D \sqrt{x^2 + y^2}\,dA$, where D is the region bounded by the cardioid $r = 1 + \cos\theta$

7. $\iint_D (x^2 + y^2)\,dA$, where D is the region bounded by the spirals $r = \theta$ and $r = 2\theta$ for $0 \leq \theta \leq 2\pi$

8. $\iint_D x\,dA$, where D is the region in the first quadrant that lies between the circles $x^2 + y^2 = 4$ and $x^2 + y^2 = 2x$

9–14 ■ Use a double integral to find the area of the region.

9. One loop of the rose $r = \cos 3\theta$

10. The region enclosed by the cardioid $r = 1 - \sin\theta$

11. The region enclosed by the lemniscate $r^2 = 4\cos 2\theta$

12. The region inside the circle $r = 4\sin\theta$ and outside the circle $r = 2$

13. The region inside the circle $r = 3\cos\theta$ and outside the cardioid $r = 1 + \cos\theta$

14. The smaller region bounded by the spiral $r\theta = 1$, the circles $r = 1$ and $r = 3$, and the polar axis

15–23 ■ Use polar coordinates to find the volume of the given solid.

15. Under the paraboloid $z = x^2 + y^2$ and above the disk $x^2 + y^2 \leq 9$

16. Under the cone $z = \sqrt{x^2 + y^2}$ and above the ring $4 \leq x^2 + y^2 \leq 25$

17. Under the plane $6x + 4y + z = 12$ and above the disk with boundary circle $x^2 + y^2 = y$

18. Bounded by the paraboloid $z = 10 - 3x^2 - 3y^2$ and the plane $z = 4$

19. Above the cone $z = \sqrt{x^2 + y^2}$ and below the sphere $x^2 + y^2 + z^2 = 1$

20. Bounded by the paraboloids $z = 3x^2 + 3y^2$ and $z = 4 - x^2 - y^2$

21. Inside both the cylinder $x^2 + y^2 = 4$ and the ellipsoid $4x^2 + 4y^2 + z^2 = 64$

22. Inside the sphere $x^2 + y^2 + z^2 = 4a^2$ and outside the cylinder $x^2 + y^2 = 2ax$

23. A sphere of radius a

24. (a) A cylindrical drill with radius r_1 is used to bore a hole through the center of a sphere of radius r_2. Find the volume of the ring-shaped solid that remains.
(b) Express the volume in part (a) in terms of the height h of the ring. Notice that the volume depends only on h, not on r_1 or r_2.

25–28 ■ Evaluate the iterated integral by converting to polar coordinates.

25. $\displaystyle\int_0^1 \int_0^{\sqrt{1-x^2}} e^{x^2+y^2}\,dy\,dx$

26. $\displaystyle\int_{-a}^a \int_0^{\sqrt{a^2-y^2}} (x^2 + y^2)^{3/2}\,dx\,dy$

27. $\displaystyle\int_0^2 \int_{-\sqrt{4-y^2}}^{\sqrt{4-y^2}} x^2y^2\,dx\,dy$

28. $\displaystyle\int_0^2 \int_0^{\sqrt{2x-x^2}} \sqrt{x^2 + y^2}\,dy\,dx$

29. Convert the integral $\int_0^1 \int_0^{\sqrt{1-x^2}} e^{-(x^2+y^2)^2}\,dy\,dx$ to polar coordinates. If your calculator or CAS evaluates definite integrals numerically, use it to find the integral correct to three decimal places. If not, use Simpson's Rule.

30. An agricultural sprinkler distributes water in a circular pattern of radius 100 ft. It supplies water to a depth of e^{-r} feet per hour at a distance of r feet from the sprinkler.
(a) What is the total amount of water supplied per hour to the region inside the circle of radius R centered at the sprinkler?
(b) Determine an expression for the average amount of water per hour per square foot supplied to the region inside the circle of radius R.

31. Use polar coordinates to combine the sum

$$\int_{1/\sqrt{2}}^1 \int_{\sqrt{1-x^2}}^x xy\,dy\,dx + \int_1^{\sqrt{2}} \int_0^x xy\,dy\,dx + \int_{\sqrt{2}}^2 \int_0^{\sqrt{4-x^2}} xy\,dy\,dx$$

into one double integral. Then evaluate the double integral.

32. (a) We define the improper integral (over the entire plane $\mathbb{R}^2$)

$$I = \iint_{\mathbb{R}^2} e^{-(x^2+y^2)}\,dA = \int_{-\infty}^\infty \int_{-\infty}^\infty e^{-(x^2+y^2)}\,dy\,dx$$

$$= \lim_{a\to\infty} \iint_{D_a} e^{-(x^2+y^2)}\,dA$$

where D_a is the disk with radius a and center the origin. Show that

$$\int_{-\infty}^\infty \int_{-\infty}^\infty e^{-(x^2+y^2)}\,dA = \pi$$

(b) An equivalent definition of the improper integral in part (a) is

$$\iint_{\mathbb{R}^2} e^{-(x^2+y^2)}\,dA = \lim_{a\to\infty} \iint_{S_a} e^{-(x^2+y^2)}\,dA$$

where S_a is the square with vertices $(\pm a, \pm a)$. Use this to show that

$$\int_{-\infty}^{\infty} e^{-x^2}\,dx \int_{-\infty}^{\infty} e^{-y^2}\,dy = \pi$$

(c) Deduce that

$$\int_{-\infty}^{\infty} e^{-x^2}\,dx = \sqrt{\pi}$$

(d) By making the change of variable $t = \sqrt{2}x$, show that

$$\int_{-\infty}^{\infty} e^{-x^2/2}\,dx = \sqrt{2\pi}$$

(This is a fundamental result for probability and statistics.)

33. Use the result of Exercise 32 part (c) to evaluate the following integrals.

(a) $\int_0^{\infty} x^2 e^{-x^2}\,dx$ (b) $\int_0^{\infty} \sqrt{x}\,e^{-x}\,dx$

13.6 APPLICATIONS OF DOUBLE INTEGRALS

We have already seen one application of double integrals: computing volumes. Another geometric application is finding areas of surfaces and this will be done in the next section. In this section we explore physical applications such as computing mass, electric charge, center of mass, and moment of inertia.

In Chapter 9 we were able to use single integrals to compute moments and the center of mass of a thin plate or lamina with constant density. But now, equipped with the double integral, we can consider a lamina with variable density. Suppose the lamina occupies a region D of the xy-plane and its density (in units of mass per unit area) at a point (x, y) in D is given by $\rho(x, y)$, where ρ is a continuous function on D. This means that

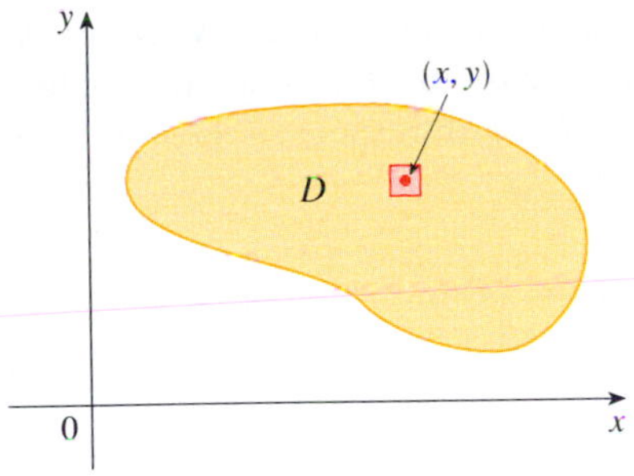

FIGURE 1

$$\rho(x, y) = \lim \frac{\Delta m}{\Delta A}$$

where Δm and ΔA are the mass and area of a small rectangle that contains (x, y) and the limit is taken as the dimensions of the rectangle approach 0 (see Figure 1).

To find the total mass m of the lamina we partition a rectangle R containing D into subrectangles R_{ij} (as in Figure 2) and consider $\rho(x, y)$ to be 0 outside D. If we choose a point (x_{ij}^*, y_{ij}^*) in R_{ij}, then the mass of the part of the lamina that occupies R_{ij} is approximately $\rho(x_{ij}^*, y_{ij}^*)\,\Delta A_{ij}$, where ΔA_{ij} is the area of R_{ij}. If we add all such masses, we get an approximation to the total mass:

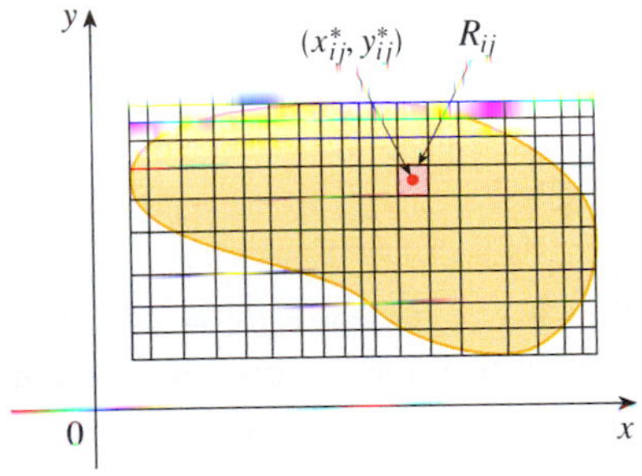

FIGURE 2

$$m \approx \sum_{i=1}^{m} \sum_{j=1}^{n} \rho(x_{ij}^*, y_{ij}^*)\,\Delta A_{ij}$$

If we now take finer partitions, we obtain the total mass m of the lamina as the limiting value of the approximations:

(1) $$m = \lim_{\|P\|\to 0} \sum_{i=1}^{m} \sum_{j=1}^{n} \rho(x_{ij}^*, y_{ij}^*)\,\Delta A_{ij} = \iint_D \rho(x, y)\,dA$$

Physicists also consider other types of density that can be treated in the same manner. For example, if an electric charge is distributed over a region D and the charge density (in units of charge per unit area) is given by $\sigma(x, y)$ at a point (x, y) in D, then the total charge Q is given by

(2) $$Q = \iint_D \sigma(x, y)\,dA$$

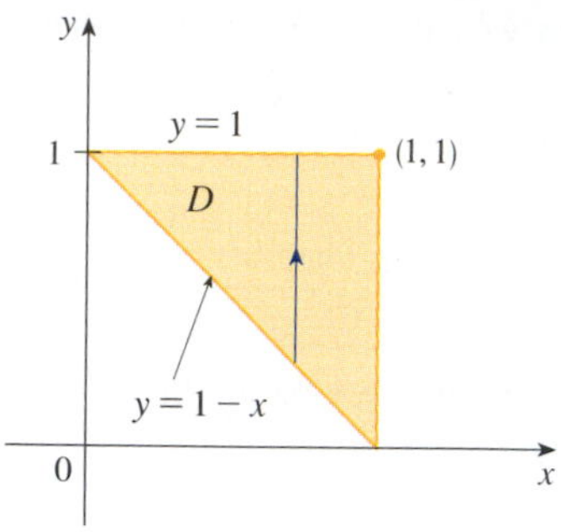

FIGURE 3

EXAMPLE 1 Charge is distributed over the triangular region D in Figure 3 so that the charge density at (x, y) is $\sigma(x, y) = xy$, measured in coulombs per square meter (C/m^2). Find the total charge.

SOLUTION From Equation 2 and Figure 3 we have

$$Q = \iint_D \sigma(x, y)\, dA = \int_0^1 \int_{1-x}^1 xy\, dy\, dx$$

$$= \int_0^1 \left[x\frac{y^2}{2} \right]_{y=1-x}^{y=1} dx = \int_0^1 \frac{x}{2}[1^2 - (1 - x)^2]\, dx$$

$$= \frac{1}{2} \int_0^1 (2x^2 - x^3)\, dx = \frac{1}{2}\left[\frac{2x^3}{3} - \frac{x^4}{4} \right]_0^1 = \frac{5}{24}$$

Thus, the total charge is $\frac{5}{24}$ C. ■

Now let's find the center of mass of a lamina with density function $\rho(x, y)$ that occupies a region D. Recall that in Section 9.5 we defined the moment of a particle about an axis as the product of its mass and its directed distance from the axis. We partition D into small rectangles as in Figure 2. Then the mass of R_{ij} is approximately $\rho(x_{ij}^*, y_{ij}^*)\, \Delta A_{ij}$, so we can approximate the moment of R_{ij} with respect to the x-axis by

$$[\rho(x_{ij}^*, y_{ij}^*)\, \Delta A_{ij}]\, y_{ij}^*$$

If we now add these quantities and take the limit as the norm $\|P\|$ approaches 0, we obtain the **moment** of the entire lamina **about the *x*-axis**:

(3) $$M_x = \lim_{\|P\| \to 0} \sum_{i=1}^{m} \sum_{j=1}^{n} y_{ij}^* \rho(x_{ij}^*, y_{ij}^*)\, \Delta A_{ij} = \iint_D y\rho(x, y)\, dA$$

Similarly, the **moment about the *y*-axis** is

(4) $$M_y = \lim_{\|P\| \to 0} \sum_{i=1}^{m} \sum_{j=1}^{n} x_{ij}^* \rho(x_{ij}^*, y_{ij}^*)\, \Delta A_{ij} = \iint_D x\rho(x, y)\, dA$$

As before, we define the center of mass $(\bar{x}, \bar{y})$ so that $m\bar{x} = M_y$ and $m\bar{y} = M_x$. The physical significance is that the lamina behaves as if its entire mass is concentrated at its center of mass. Thus, the lamina balances horizontally when supported at its center of mass. (See Figure 1 in Section 9.5.)

(5) The coordinates $(\bar{x}, \bar{y})$ of the center of mass of a lamina occupying the region D and having density function $\rho(x, y)$ are

$$\bar{x} = \frac{M_y}{m} = \frac{1}{m} \iint_D x\rho(x, y)\, dA \qquad \bar{y} = \frac{M_x}{m} = \frac{1}{m} \iint_D y\rho(x, y)\, dA$$

where the mass m is given by

$$m = \iint_D \rho(x, y)\, dA$$

EXAMPLE 2 Find the mass and center of mass of a triangular lamina with vertices $(0, 0)$, $(1, 0)$, and $(0, 2)$ if the density function is $\rho(x, y) = 1 + 3x + y$.

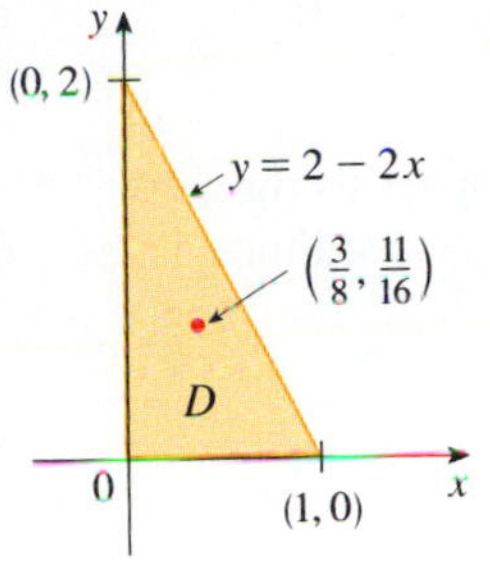

FIGURE 4

SOLUTION The triangle is shown in Figure 4. (Note that the equation of the upper boundary is $y = 2 - 2x$.) The mass of the lamina is

$$\begin{aligned} m &= \iint_D \rho(x, y)\, dA = \int_0^1 \int_0^{2-2x} (1 + 3x + y)\, dy\, dx \\ &= \int_0^1 \left[y + 3xy + \frac{y^2}{2} \right]_{y=0}^{y=2-2x} dx \\ &= 4 \int_0^1 (1 - x^2)\, dx = 4\left[x - \frac{x^3}{3} \right]_0^1 = \frac{8}{3} \end{aligned}$$

Then the formulas in (5) give

$$\begin{aligned} \bar{x} &= \frac{1}{m} \iint_D x\rho(x, y)\, dA = \tfrac{3}{8} \int_0^1 \int_0^{2-2x} (x + 3x^2 + xy)\, dy\, dx \\ &= \tfrac{3}{8} \int_0^1 \left[xy + 3x^2y + x\frac{y^2}{2} \right]_{y=0}^{y=2-2x} dx \\ &= \tfrac{3}{2} \int_0^1 (x - x^3)\, dx = \left[\frac{x^2}{2} - \frac{x^4}{4} \right]_0^1 = \frac{3}{8} \end{aligned}$$

$$\begin{aligned} \bar{y} &= \frac{1}{m} \iint_D y\rho(x, y)\, dA = \tfrac{3}{8} \int_0^1 \int_0^{2-2x} (y + 3xy + y^2)\, dy\, dx \\ &= \tfrac{3}{8} \int_0^1 \left[\frac{y^2}{2} + 3x\frac{y^2}{2} + \frac{y^3}{3} \right]_{y=0}^{y=2-2x} dx \\ &= \tfrac{1}{4} \int_0^1 (7 - 9x - 3x^2 + 5x^3)\, dx \\ &= \frac{1}{4}\left[7x - 9\frac{x^2}{2} - x^3 + 5\frac{x^4}{4} \right]_0^1 = \frac{11}{16} \end{aligned}$$

The center of mass is at the point $\left(\frac{3}{8}, \frac{11}{16}\right)$. ■

EXAMPLE 3 The density at any point on a semicircular lamina is proportional to the distance from the center of the circle. Find the center of mass of the lamina.

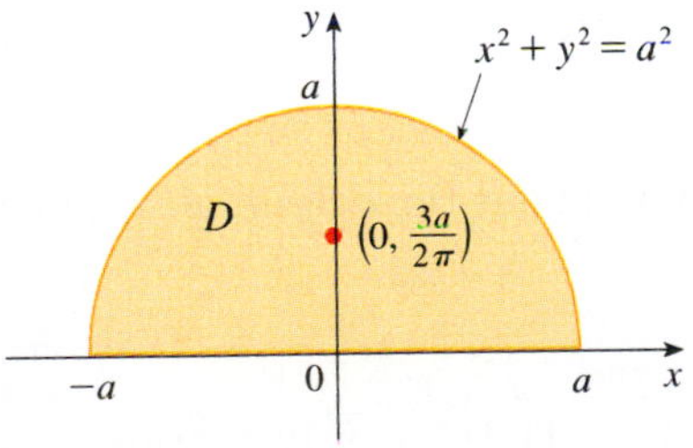

FIGURE 5

SOLUTION Let's place the lamina as the upper half of the circle $x^2 + y^2 = a^2$ (see Figure 5). Then the distance from a point (x, y) to the center of the circle (the origin) is $\sqrt{x^2 + y^2}$. Therefore, the density function is

$$\rho(x, y) = K\sqrt{x^2 + y^2}$$

where K is some constant. Both the density function and the shape of the lamina suggest that we convert to polar coordinates. Then $\sqrt{x^2 + y^2} = r$ and the region D is given by $0 \leqslant r \leqslant a$, $0 \leqslant \theta \leqslant \pi$. Thus, the mass of the lamina is

$$m = \iint_D \rho(x, y)\, dA = \iint_D K\sqrt{x^2 + y^2}\, dA$$

$$= \int_0^{\pi} \int_0^a (Kr) r\,dr\,d\theta = K \int_0^{\pi} d\theta \int_0^a r^2\,dr$$

$$= K\pi \frac{r^3}{3}\bigg]_0^a = \frac{K\pi a^3}{3}$$

Both the lamina and the density function are symmetric with respect to the y-axis, so the center of mass must lie on the y-axis, that is, $\bar{x} = 0$. The y-coordinate is given by

$$\bar{y} = \frac{1}{m} \iint_D y\rho(x, y)\,dA$$

$$= \frac{3}{K\pi a^3} \int_0^{\pi} \int_0^a r\sin\theta\,(Kr) r\,dr\,d\theta$$

$$= \frac{3}{\pi a^3} \int_0^{\pi} \sin\theta\,d\theta \int_0^a r^3\,dr$$

$$= \frac{3}{\pi a^3} \big[-\cos\theta\big]_0^{\pi} \left[\frac{r^4}{4}\right]_0^a$$

$$= \frac{3}{\pi a^3} \frac{2a^4}{4} = \frac{3a}{2\pi}$$

Therefore, the center of mass is at the point $(0, 3a/(2\pi))$. [Compare this with Example 3 in Section 9.5 where we found that the center of mass of a lamina with the same shape but uniform density is at the point $(0, 4a/(3\pi))$.] ■

MOMENT OF INERTIA

The **moment of inertia** (also called the **second moment**) of a particle of mass m about an axis is defined to be mr^2, where r is the distance from the particle to the axis. We extend this concept to a lamina with density function $\rho(x, y)$ and occupying a region D by proceeding as we did for ordinary moments. We partition D into small rectangles, approximate the moment of inertia of each subrectangle about the x-axis, and take the limit of the sum as $\|P\|$ approaches 0. The result is the **moment of inertia** of the lamina **about the x-axis**:

(6) $$I_x = \lim_{\|P\| \to 0} \sum_{i=1}^{m} \sum_{j=1}^{n} (y_{ij}^*)^2 \rho(x_{ij}^*, y_{ij}^*)\,\Delta A_{ij} = \iint_D y^2 \rho(x, y)\,dA$$

Similarly, the **moment of inertia about the y-axis** is

(7) $$I_y = \lim_{\|P\| \to 0} \sum_{i=1}^{m} \sum_{j=1}^{n} (x_{ij}^*)^2 \rho(x_{ij}^*, y_{ij}^*)\,\Delta A_{ij} = \iint_D x^2 \rho(x, y)\,dA$$

It is also of interest to consider the **moment of inertia about the origin**, also called the **polar moment of inertia**:

(8) $$I_0 = \lim_{\|P\| \to 0} \sum_{i=1}^{m} \sum_{j=1}^{n} [(x_{ij}^*)^2 + (y_{ij}^*)^2] \rho(x_{ij}^*, y_{ij}^*)\,\Delta A_{ij} = \iint_D (x^2 + y^2) \rho(x, y)\,dA$$

Note that $I_0 = I_x + I_y$.

EXAMPLE 4 Find the moments of inertia I_x, I_y, and I_0 of a homogeneous disk D with density $\rho(x, y) = \rho$, center the origin, and radius a.

SOLUTION The boundary of D is the circle $x^2 + y^2 = a^2$ and in polar coordinates D is described by $0 \leqslant \theta \leqslant 2\pi$, $0 \leqslant r \leqslant a$. Let's compute I_0 first:

$$I_0 = \iint_D (x^2 + y^2)\rho \, dA = \rho \int_0^{2\pi} \int_0^a r^2 r \, dr \, d\theta$$

$$= \rho \int_0^{2\pi} d\theta \int_0^a r^3 \, dr = 2\pi\rho \left[\frac{r^4}{4} \right]_0^a = \frac{\pi \rho a^4}{2}$$

Instead of computing I_x and I_y directly, we use the facts that $I_x + I_y = I_0$ and $I_x = I_y$ (from the symmetry of the problem). Thus

$$I_x = I_y = \frac{I_0}{2} = \frac{\pi \rho a^4}{4}$$

■

In Example 4 notice that the mass of the disk is

$$m = \text{density} \times \text{area} = \rho(\pi a^2)$$

so the moment of inertia of the disk about the origin (like a wheel about its axle) can be written as

$$I_0 = \tfrac{1}{2} m a^2$$

Thus, if we increase the mass or the radius of the disk, we thereby increase the moment of inertia. In general, the moment of inertia plays much the same role in rotational motion that mass plays in linear motion. The moment of inertia of a wheel is what makes it difficult to start or stop the rotation of the wheel, just as the mass of a car is what makes it difficult to start or stop the motion of the car.

The **radius of gyration of a lamina about an axis** is the number R such that

$$mR^2 = I \tag{9}$$

where m is the mass of the lamina and I is the moment of inertia about the given axis. Equation 9 says that if the mass of the lamina were concentrated at a distance R from the axis, then the moment of inertia of this "point mass" would be the same as the moment of inertia of the lamina.

In particular, the radius of gyration $\bar{\bar{y}}$ with respect to the x-axis and the radius of gyration $\bar{\bar{x}}$ with respect to the y-axis are given by the equations

$$m\bar{\bar{y}}^2 = I_x \qquad m\bar{\bar{x}}^2 = I_y \tag{10}$$

Thus $(\bar{\bar{x}}, \bar{\bar{y}})$ is the point at which the mass of the lamina can be concentrated without changing the moments of inertia with respect to the coordinate axes. (Note the analogy with the center of mass.)

EXAMPLE 5 Find the radius of gyration about the x-axis of the disk in Example 4.

SOLUTION As noted, the mass of the disk is $m = \rho\pi a^2$, so from Equations 10 we have

$$\bar{\bar{y}}^2 = \frac{I_x}{m} = \frac{\frac{1}{4}\pi\rho a^4}{\rho\pi a^2} = \frac{a^2}{4}$$

Therefore, the radius of gyration about the x-axis is

$$\bar{\bar{y}} = \frac{a}{2}$$

which is half the radius of the disk. ■

EXERCISES 13.6

1. Electric charge is distributed over the rectangle $0 \leq x \leq 2$, $1 \leq y \leq 2$ so that the charge density at (x, y) is $\sigma(x, y) = x^2 + 3y^2$ (measured in coulombs per square meter). Find the total charge on the rectangle.

2. Electric charge is distributed over the unit disk $x^2 + y^2 \leq 1$ so that the charge density at (x, y) is $\sigma(x, y) = 1 + x^2 + y^2$ (measured in coulombs per square meter). Find the total charge on the disk.

3–12 ■ Find the mass and center of mass of the lamina that occupies the region D and has the given density function ρ.

3. $D = \{(x, y) \mid -1 \leq x \leq 1,\ 0 \leq y \leq 1\}$; $\rho(x, y) = x^2$

4. $D = \{(x, y) \mid 0 \leq x \leq 2,\ 0 \leq y \leq 3\}$; $\rho(x, y) = y$

5. D is the triangular region with vertices $(0, 0)$, $(2, 1)$, $(0, 3)$; $\rho(x, y) = x + y$

6. D is the triangular region with vertices $(0, 0)$, $(1, 1)$, $(4, 0)$; $\rho(x, y) = x$

7. D is the region in the first quadrant bounded by the parabola $y = x^2$ and the line $y = 1$; $\rho(x, y) = xy$

8. D is bounded by the parabola $y = 9 - x^2$ and the x-axis; $\rho(x, y) = y$

9. D is bounded by the parabola $x = y^2$ and the line $y = x - 2$; $\rho(x, y) = 3$

CAS 10. D is bounded by the cardioid $r = 1 + \sin\theta$; $\rho(x, y) = 2$

11. $D = \{(x, y) \mid 0 \leq y \leq \sin x,\ 0 \leq x \leq \pi\}$; $\rho(x, y) = y$

12. $D = \{(x, y) \mid 0 \leq y \leq \cos x,\ 0 \leq x \leq \pi/2\}$; $\rho(x, y) = x$

13. A lamina occupies the part of the disk $x^2 + y^2 \leq 1$ in the first quadrant. Find its center of mass if the density at any point is proportional to its distance from the x-axis.

14. Find the center of mass of the lamina in Exercise 13 if the density at any point is proportional to the square of its distance from the origin.

15. Find the center of mass of a lamina in the shape of an isosceles right triangle with equal sides of length a if the density at any point is proportional to the square of the distance from the vertex opposite the hypotenuse.

16. A lamina occupies the region inside the circle $x^2 + y^2 = 2y$ but outside the circle $x^2 + y^2 = 1$. Find the center of mass if the density at any point is inversely proportional to its distance from the origin.

17. Find the moments of inertia I_x, I_y, I_0 for the lamina of Exercise 7.

18. Find the moments of inertia I_x, I_y, I_0 for the lamina of Exercise 4.

19. Find the moments of inertia I_x, I_y, I_0 for the lamina of Exercise 9.

20. Find the moments of inertia I_x, I_y, I_0 for the lamina of Exercise 14.

21. A lamina with constant density $\rho(x, y) = \rho$ occupies a square with vertices $(0, 0)$, $(a, 0)$, (a, a), and $(0, a)$. Find the moments of inertia I_x and I_y and the radii of gyration $\bar{\bar{x}}$ and $\bar{\bar{y}}$.

22. A lamina with constant density $\rho(x, y) = \rho$ occupies the region under the curve $y = \sin x$ from $x = 0$ to $x = \pi$. Find the moments of inertia I_x and I_y and the radii of gyration $\bar{\bar{x}}$ and $\bar{\bar{y}}$.

23. Show that the formulas for $\bar{x}$ and $\bar{y}$ in (5) become Equations 9.5.11 when $\rho(x, y) = \rho$ (the density is constant).

24. (a) A lamina has constant density ρ and takes the shape of a disk with center the origin and radius R. Use Newton's Law of Gravitation (see Section 11.8) to show that the magnitude of the force of attraction that the lamina exerts on a body with mass m located at the point $(0, 0, d)$ on the positive z-axis is

$$F = 2\pi G m \rho d\left(\frac{1}{d} - \frac{1}{\sqrt{R^2 + d^2}}\right)$$

[*Hint:* Consider a polar partition as in Figure 3 in Section 13.5 and first compute the vertical component of the force exerted by the polar subrectangle R_{ij}.]

(b) Show that the magnitude of the force of attraction of a lamina with density ρ that occupies an entire plane on an object with mass m located at a distance d from the plane is

$$F = 2\pi G m \rho$$

Notice that this expression does not depend on d.

13.7 SURFACE AREA

In Section 14.6 we will deal with areas of more general surfaces, called parametric surfaces, and so it is possible to omit this section if that later section will be covered.

In this section we apply double integrals to the problem of computing the area of a surface. In Section 9.4 we found the area of a very special type of surface—a surface of revolution—by the methods of single-variable calculus. Here we compute the area of a surface with equation $z = f(x, y)$, the graph of a function of two variables.

Let S be a surface with equation $z = f(x, y)$, where f has continuous partial derivatives. For simplicity in deriving the surface area formula, we assume that $f(x, y) \geq 0$ and the domain D of f is a rectangle. We consider a partition P of D into small rectangles R_{ij} with areas $\Delta A_{ij} = \Delta x_i \, \Delta y_j$. If (x_i, y_j) is the corner of R_{ij} closest to the origin, let $P_{ij}(x_i, y_j, f(x_i, y_j))$ be the point on S directly above it (see Figure 1). The tangent plane to S at P_{ij} is an approximation to S near P_{ij}. So the area ΔT_{ij} of the part of this tangent plane (a parallelogram) that lies directly above R_{ij} is an approximation to the area ΔS_{ij} of the part of S that lies directly above R_{ij}. Thus, the sum $\Sigma\Sigma\, \Delta T_{ij}$ is an approximation to the total area of S, and this approximation appears to improve as $\|P\| \to 0$. Therefore, we define the **surface area** of S to be

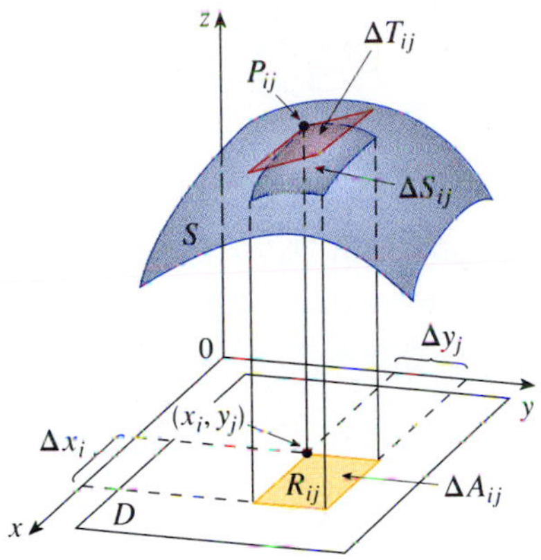

FIGURE 1

(1)
$$A(S) = \lim_{\|P\| \to 0} \sum_{i=1}^{m} \sum_{j=1}^{n} \Delta T_{ij}$$

To find a formula that is more convenient than Equation 1 for computational purposes, we let **a** and **b** be the vectors that start at P_{ij} and lie along the sides of the parallelogram with area ΔT_{ij} (see Figure 2). Then $\Delta T_{ij} = |\mathbf{a} \times \mathbf{b}|$. Recall from Section 12.3 that $f_x(x_i, y_j)$ and $f_y(x_i, y_j)$ are the slopes of the tangent lines through P_{ij} in the directions of **a** and **b**. Therefore

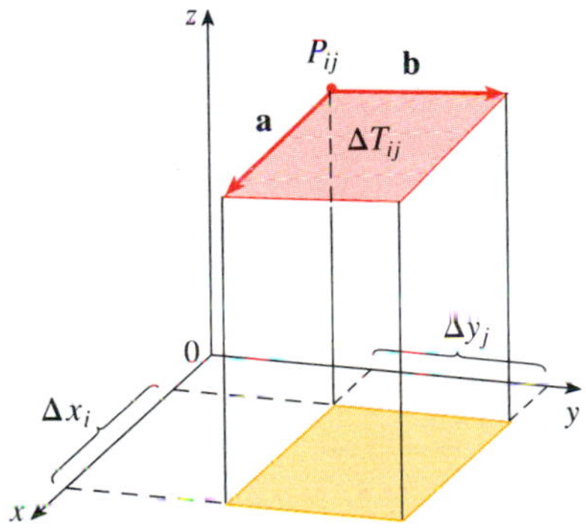

FIGURE 2

$$\mathbf{a} = \Delta x_i \, \mathbf{i} + f_x(x_i, y_j)\, \Delta x_i \, \mathbf{k}$$

$$\mathbf{b} = \Delta y_j \, \mathbf{j} + f_y(x_i, y_j)\, \Delta y_j \, \mathbf{k}$$

and

$$\mathbf{a} \times \mathbf{b} = \begin{vmatrix} \mathbf{i} & \mathbf{j} & \mathbf{k} \\ \Delta x_i & 0 & f_x(x_i, y_j)\, \Delta x_i \\ 0 & \Delta y_j & f_y(x_i, y_j)\, \Delta y_j \end{vmatrix}$$

$$= -f_x(x_i, y_j)\, \Delta x_i \, \Delta y_j \, \mathbf{i} - f_y(x_i, y_j)\, \Delta x_i \, \Delta y_j \, \mathbf{j} + \Delta x_i \, \Delta y_j \, \mathbf{k}$$

$$= [-f_x(x_i, y_j)\, \mathbf{i} - f_y(x_i, y_j)\, \mathbf{j} + \mathbf{k}]\, \Delta A_{ij}$$

Thus
$$\Delta T_{ij} = |\mathbf{a} \times \mathbf{b}| = \sqrt{[f_x(x_i, y_j)]^2 + [f_y(x_i, y_j)]^2 + 1}\; \Delta A_{ij}$$

From Definition 1 we then have

$$A(S) = \lim_{\|P\| \to 0} \sum_{i=1}^{m} \sum_{j=1}^{n} \Delta T_{ij}$$

$$= \lim_{\|P\| \to 0} \sum_{i=1}^{m} \sum_{j=1}^{n} \sqrt{[f_x(x_i, y_j)]^2 + [f_y(x_i, y_j)]^2 + 1}\; \Delta A_{ij}$$

and by the definition of a double integral we get the following formula.

> **(2)** The area of the surface with equation $z = f(x, y)$, $(x, y) \in D$, where f_x and f_y are continuous, is
>
> $$A(S) = \iint_D \sqrt{[f_x(x, y)]^2 + [f_y(x, y)]^2 + 1}\, dA$$

We will verify in Section 14.6 that this formula is consistent with our previous formula for the area of a surface of revolution. Notice the similarity of Formula 2 to the arc length formula

$$L = \int_a^b \sqrt{[f'(x)]^2 + 1}\, dx$$

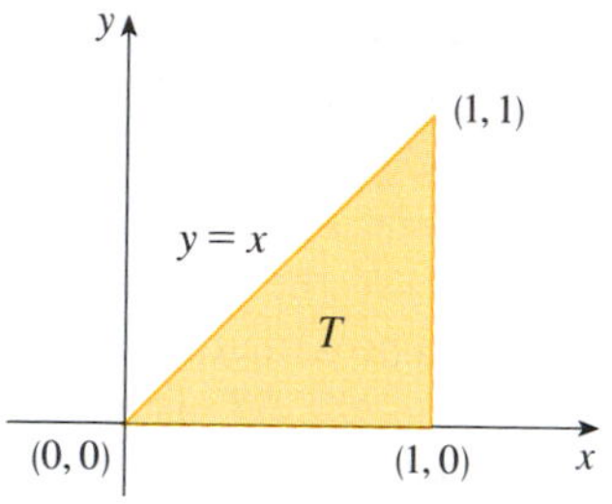

FIGURE 3

EXAMPLE 1 Find the surface area of the part of the surface $z = x^2 + 2y$ that lies above the triangular region T in the xy-plane with vertices $(0, 0)$, $(1, 0)$, and $(1, 1)$.

SOLUTION The region T is shown in Figure 3 and is described by

$$T = \{(x, y) \mid 0 \leqslant x \leqslant 1,\ 0 \leqslant y \leqslant x\}$$

Using Formula 2 with $f(x, y) = x^2 + 2y$, we get

$$A = \iint_T \sqrt{(2x)^2 + (2)^2 + 1}\, dA = \int_0^1 \int_0^x \sqrt{4x^2 + 5}\, dy\, dx$$

$$= \int_0^1 x\sqrt{4x^2 + 5}\, dx = \tfrac{1}{8} \cdot \tfrac{2}{3}(4x^2 + 5)^{3/2}\Big]_0^1 = \tfrac{1}{12}(27 - 5\sqrt{5})$$

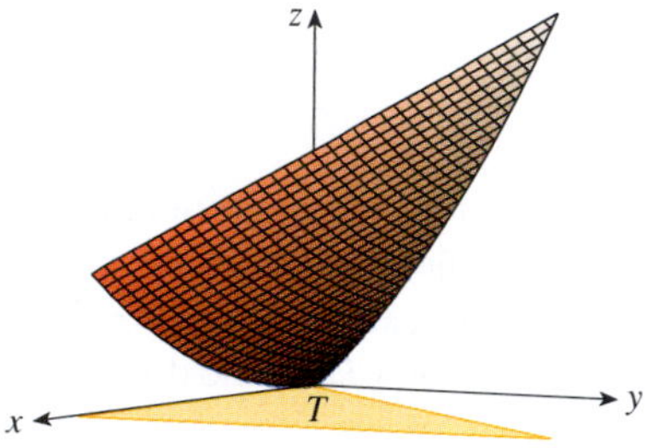

FIGURE 4

Figure 4 shows the portion of the surface whose area we have just computed. ■

EXAMPLE 2 Find the area of the part of the paraboloid $z = x^2 + y^2$ that lies under the plane $z = 9$.

SOLUTION The plane intersects the paraboloid in the circle $x^2 + y^2 = 9$, $z = 9$. Therefore, the given surface lies above the disk D with center the origin and radius 3 (see Figure 5). Using Formula 2, we have

$$A = \iint_D \sqrt{(2x)^2 + (2y)^2 + 1}\, dA = \iint_D \sqrt{4(x^2 + y^2) + 1}\, dA$$

Converting to polar coordinates, we obtain

$$A = \int_0^{2\pi} \int_0^3 \sqrt{4r^2 + 1}\, r\, dr\, d\theta = \int_0^{2\pi} d\theta \int_0^3 \tfrac{1}{8}\sqrt{4r^2 + 1}\,(8r)\, dr$$

$$= 2\pi\left(\tfrac{1}{8}\right)\tfrac{2}{3}(4r^2 + 1)^{3/2}\Big]_0^3 = \frac{\pi}{6}(37\sqrt{37} - 1)$$

■

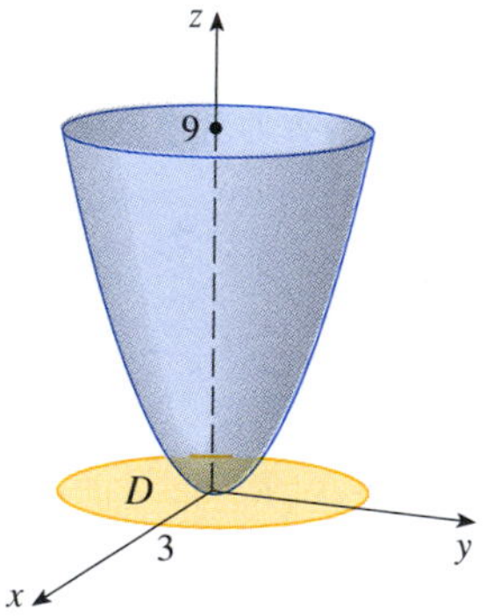

FIGURE 5

EXERCISES 13.7

1–10 ■ Find the area of the surface.

1. The part of the plane $x + 2y + z = 4$ that lies inside the cylinder $x^2 + y^2 = 4$

2. The part of the plane $2x + 3y - z + 1 = 0$ that lies above the rectangle $[1, 4] \times [2, 4]$

3. The part of the cylinder $y^2 + z^2 = 9$ that lies above the rectangle with vertices $(0, 0)$, $(4, 0)$, $(0, 2)$, and $(4, 2)$

4. The part of the surface $z = x + y^2$ that lies above the triangle with vertices $(0, 0)$, $(1, 1)$, and $(0, 1)$

5. The part of the hyperbolic paraboloid $z = y^2 - x^2$ that lies between the cylinders $x^2 + y^2 = 1$ and $x^2 + y^2 = 4$

6. The part of the paraboloid $z = 4 - x^2 - y^2$ that lies above the xy-plane

7. The part of the surface $z = xy$ that lies within the cylinder $x^2 + y^2 = 1$

8. The part of the sphere $x^2 + y^2 + z^2 = 4$ that lies above the plane $z = 1$

9. The part of the sphere $x^2 + y^2 + z^2 = a^2$ that lies within the cylinder $x^2 + y^2 = ax$ and above the xy-plane

10. The part of the sphere $x^2 + y^2 + z^2 = 4z$ that lies inside the paraboloid $z = x^2 + y^2$

11. Use the Midpoint Rule for double integrals (see Section 13.1) with four squares to estimate the surface area of the portion of the paraboloid $z = x^2 + y^2$ that lies above the square $[0, 1] \times [0, 1]$.

CAS **12.** Use a computer algebra system to approximate the surface area in Exercise 11 to four decimal places. Compare with the answer to that exercise.

CAS **13.** Find the exact area of the surface $z = x^2 + 2y$, $0 \le x \le 1$, $0 \le y \le 1$.

CAS **14.** Find the exact area of the surface

$$z = 1 + x + y + x^2, \quad -2 \le x \le 1, \; -1 \le y \le 1$$

Illustrate by graphing the surface.

CAS **15.** Find, to four decimal places, the area of the part of the surface $z = 1 + x^2y^2$ that lies above the disk $x^2 + y^2 \le 1$.

CAS **16.** Find, to four decimal places, the area of the part of the surface $z = (1 + x^2)/(1 + y^2)$ that lies above the square $|x| + |y| \le 1$. Illustrate by graphing this part of the surface.

17. Show that the area of the part of the plane $z = ax + by + c$ that projects onto a region D in the xy-plane with area $A(D)$ is $\sqrt{a^2 + b^2 + 1}\, A(D)$.

18. If you attempt to use Formula 2 to find the area of the top half of the sphere $x^2 + y^2 + z^2 = a^2$, you have a slight problem because the double integral is improper. In fact, the integrand has an infinite discontinuity at every point of the boundary circle $x^2 + y^2 = a^2$. However, the integral can be computed as the limit of the integral over the disk $x^2 + y^2 \le t^2$ as $t \to a^-$. Use this method to show that the area of a sphere of radius a is $4\pi a^2$.

13.8 TRIPLE INTEGRALS

Just as we defined single integrals for functions of one variable and double integrals for functions of two variables, so we can define triple integrals for functions of three variables. Let's first deal with the simplest case where f is defined on a rectangular box:

$$B = \{(x, y, z) \mid a \le x \le b,\; c \le y \le d,\; r \le z \le s\} \tag{1}$$

The first step is to partition the intervals $[a, b]$, $[c, d]$, and $[r, s]$ as follows:

$$a = x_0 < x_1 < \cdots < x_{i-1} < x_i < \cdots < x_l = b$$

$$c = y_0 < y_1 < \cdots < y_{j-1} < y_j < \cdots < y_m = d$$

$$r = z_0 < z_1 < \cdots < z_{k-1} < z_k < \cdots < z_n = s$$

The planes through these partition points parallel to the coordinate planes divide the box B into lmn sub-boxes

$$B_{ijk} = [x_{i-1}, x_i] \times [y_{j-1}, y_j] \times [z_{k-1}, z_k]$$

which are shown in Figure 1.

The volume of B_{ijk} is

$$\Delta V_{ijk} = \Delta x_i \, \Delta y_j \, \Delta z_k$$

where $\quad \Delta x_i = x_i - x_{i-1} \qquad \Delta y_j = y_j - y_{j-1} \qquad \Delta z_k = z_k - z_{k-1}$

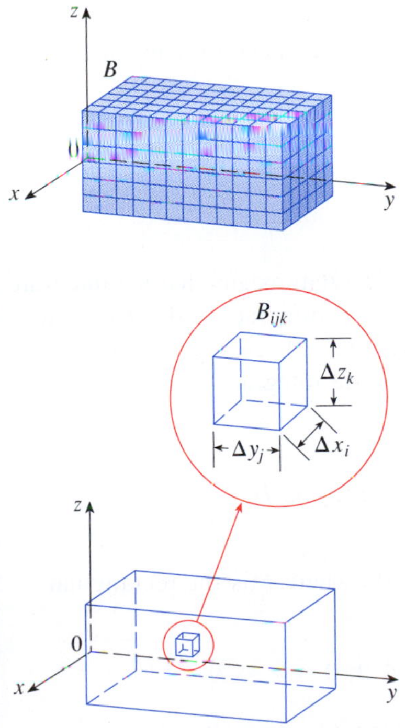

FIGURE 1

Then we form the **triple Riemann sum**

(2)
$$\sum_{i=1}^{l} \sum_{j=1}^{m} \sum_{k=1}^{n} f(x_{ijk}^*, y_{ijk}^*, z_{ijk}^*)\, \Delta V_{ijk}$$

where $(x_{ijk}^*, y_{ijk}^*, z_{ijk}^*)$ is in B_{ijk}. We define the **norm** of the partition P to be the length of the longest diagonal of all the boxes B_{ijk} and we denote the norm by $\|P\|$. Then, by analogy with the definition of a double integral (13.1.4), we define the triple integral as the limit of the triple Riemann sums in (2).

(3) DEFINITION The **triple integral** of f over the box B is

$$\iiint_B f(x, y, z)\, dV = \lim_{\|P\| \to 0} \sum_{i=1}^{l} \sum_{j=1}^{m} \sum_{k=1}^{n} f(x_{ijk}^*, y_{ijk}^*, z_{ijk}^*)\, \Delta V_{ijk}$$

if this limit exists.

The precise meaning of the limit in Definition 3 is similar to that given in Note 1 after Definition 13.1.4. Again, the triple integral always exists if f is continuous. An alternative notation is

$$\iiint_B f(x, y, z)\, dV = \iiint_B f(x, y, z)\, dx\, dy\, dz$$

Just as for double integrals, the practical method for evaluating triple integrals is to express them as iterated integrals as follows:

(4) FUBINI'S THEOREM FOR TRIPLE INTEGRALS If f is continuous on the rectangular box $B = [a, b] \times [c, d] \times [r, s]$, then

$$\iiint_B f(x, y, z)\, dV = \int_r^s \int_c^d \int_a^b f(x, y, z)\, dx\, dy\, dz$$

The iterated integral on the right side of Fubini's Theorem means that we integrate first with respect to x (keeping y and z fixed), then we integrate with respect to y (keeping z fixed), and finally we integrate with respect to z. There are five other possible orders in which we can integrate. For instance, if we integrate with respect to y, then z, and then x, we have

$$\iiint_B f(x, y, z)\, dV = \int_a^b \int_r^s \int_c^d f(x, y, z)\, dy\, dz\, dx$$

EXAMPLE 1 Evaluate the triple integral $\iiint_B xyz^2\, dV$, where B is the rectangular box given by

$$B = \{(x, y, z) \mid 0 \leqslant x \leqslant 1,\ -1 \leqslant y \leqslant 2,\ 0 \leqslant z \leqslant 3\}$$

SOLUTION We could use any of the six possible orders of integration. If we choose to integrate with respect to x, then y, and then z, we obtain

$$\iiint_B xyz^2\,dV = \int_0^3\int_{-1}^2\int_0^1 xyz^2\,dx\,dy\,dz = \int_0^3\int_{-1}^2\left[\frac{x^2yz^2}{2}\right]_{x=0}^{x=1} dy\,dz$$

$$= \int_0^3\int_{-1}^2 \frac{yz^2}{2}\,dy\,dz = \int_0^3\left[\frac{y^2z^2}{4}\right]_{y=-1}^{y=2} dz$$

$$= \int_0^3 \frac{3z^2}{4}\,dz = \frac{z^3}{4}\Bigg]_0^3 = \frac{27}{4}$$ ■

Now we define the **triple integral over a general bounded region E** in three-dimensional space (a solid) by the same procedure that we used for double integrals (13.3.2). We enclose E in a box B of the type given by Equation 1. Then we define F so that it agrees with f on E but is 0 for points in B that are outside E. By definition,

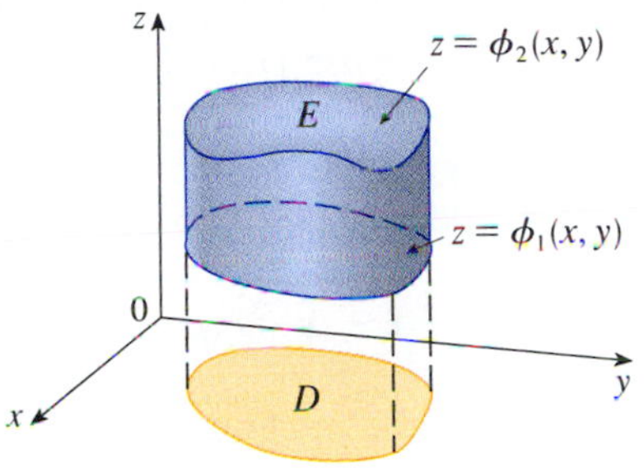

FIGURE 2
A type 1 solid region

$$\iiint_E f(x, y, z)\,dV = \iiint_B F(x, y, z)\,dV$$

This integral exists if f is continuous and the boundary of E is "reasonably smooth." The triple integral has essentially the same properties as the double integral (13.3.6–13.3.9).

We restrict our attention to continuous functions f and to certain simple types of regions. A solid region E is said to be of **type 1** if it lies between the graphs of two continuous functions of x and y, that is,

(5) $$E = \{(x, y, z) \mid (x, y) \in D,\ \phi_1(x, y) \leqslant z \leqslant \phi_2(x, y)\}$$

where D is the projection of E onto the xy-plane as shown in Figure 2. Notice that the upper boundary of the solid E is the surface with equation $z = \phi_2(x, y)$, while the lower boundary is the surface $z = \phi_1(x, y)$.

By the same sort of argument that led to (13.3.3), it can be shown that, if E is a type 1 region given by Equation 5, then

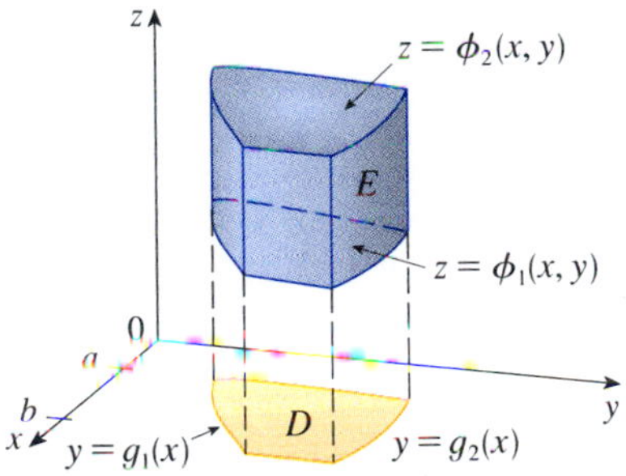

FIGURE 3
A type 1 solid region

(6) $$\iiint_E f(x, y, z)\,dV = \iint_D\left[\int_{\phi_1(x, y)}^{\phi_2(x, y)} f(x, y, z)\,dz\right] dA$$

The meaning of the inner integral on the right side of Equation 6 is that x and y are held fixed, and therefore $\phi_1(x, y)$ and $\phi_2(x, y)$ are regarded as constants, while $f(x, y, z)$ is integrated with respect to z.

In particular, if the projection D of E onto the xy-plane is a type I plane region (as in Figure 3), then

$$E = \{(x, y, z) \mid a \leqslant x \leqslant b,\ g_1(x) \leqslant y \leqslant g_2(x),\ \phi_1(x, y) \leqslant z \leqslant \phi_2(x, y)\}$$

and Equation 6 becomes

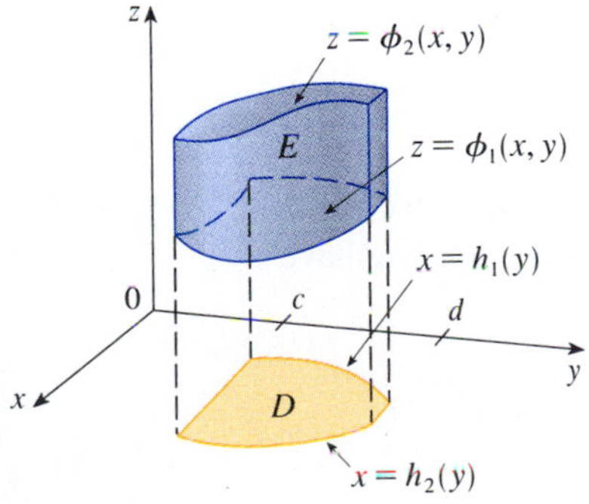

FIGURE 4
Another type 1 solid region

(7) $$\iiint_E f(x, y, z)\,dV = \int_a^b\int_{g_1(x)}^{g_2(x)}\int_{\phi_1(x, y)}^{\phi_2(x, y)} f(x, y, z)\,dz\,dy\,dx$$

If, on the other hand, D is a type II plane region (as in Figure 4), then

$$E = \{(x, y, z) \mid c \leqslant y \leqslant d,\ h_1(y) \leqslant x \leqslant h_2(y),\ \phi_1(x, y) \leqslant z \leqslant \phi_2(x, y)\}$$

and Equation 6 becomes

(8)
$$\iiint_E f(x, y, z)\, dV = \int_c^d \int_{h_1(y)}^{h_2(y)} \int_{\phi_1(x, y)}^{\phi_2(x, y)} f(x, y, z)\, dz\, dx\, dy$$

EXAMPLE 2 Evaluate $\iiint_E z\, dV$, where E is the solid tetrahedron bounded by the four planes $x = 0$, $y = 0$, $z = 0$, and $x + y + z = 1$.

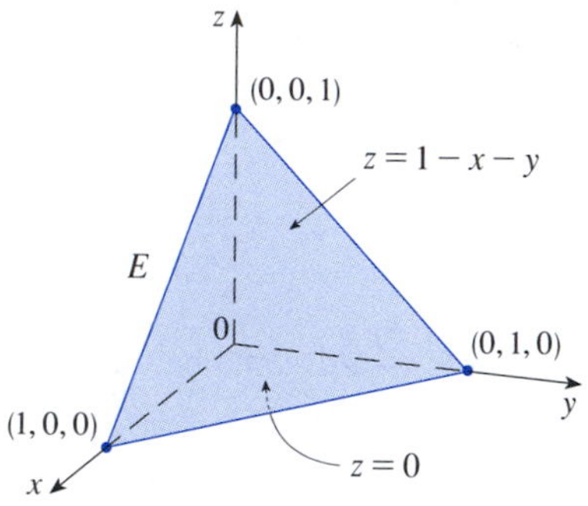

FIGURE 5

SOLUTION When we set up a triple integral it is wise to draw *two* diagrams: one of the solid region E (see Figure 5) and one of its projection D on the xy-plane (see Figure 6). The lower boundary of the tetrahedron is the plane $z = 0$ and the upper boundary is the plane $x + y + z = 1$ (or $z = 1 - x - y$), so we use $\phi_1(x, y) = 0$ and $\phi_2(x, y) = 1 - x - y$ in Formula 7. Notice that the planes $x + y + z = 1$ and $z = 0$ intersect in the line $x + y = 1$ (or $y = 1 - x$) in the xy-plane. So the projection of E is the triangular region shown in Figure 6, and we have

(9)
$$E = \{(x, y, z) \mid 0 \leq x \leq 1,\ 0 \leq y \leq 1 - x,\ 0 \leq z \leq 1 - x - y\}$$

This description of E as a type 1 region enables us to evaluate the integral as follows:

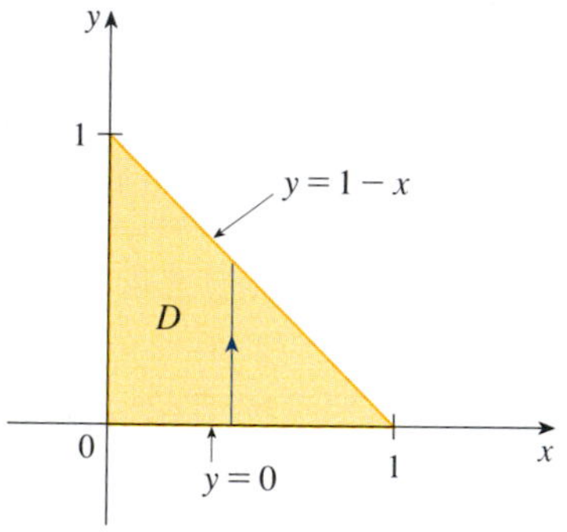

FIGURE 6

$$\begin{aligned}
\iiint_E z\, dV &= \int_0^1 \int_0^{1-x} \int_0^{1-x-y} z\, dz\, dy\, dx \\
&= \int_0^1 \int_0^{1-x} \left[\frac{z^2}{2} \right]_{z=0}^{z=1-x-y} dy\, dx \\
&= \tfrac{1}{2} \int_0^1 \int_0^{1-x} (1 - x - y)^2\, dy\, dx \\
&= \tfrac{1}{2} \int_0^1 \left[-\frac{(1 - x - y)^3}{3} \right]_{y=0}^{y=1-x} dx \\
&= \tfrac{1}{6} \int_0^1 (1 - x)^3\, dx = \tfrac{1}{6} \left[-\frac{(1 - x)^4}{4} \right]_0^1 = \frac{1}{24}
\end{aligned}$$
■

A solid region E is of **type 2** if it is of the form

$$E = \{(x, y, z) \mid (y, z) \in D,\ \phi_1(y, z) \leq x \leq \phi_2(y, z)\}$$

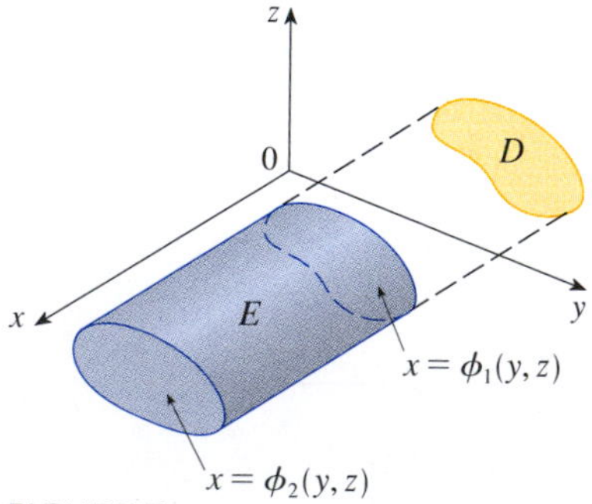

FIGURE 7
A type 2 region

where, this time, D is the projection of E onto the yz-plane (see Figure 7). The back surface is $x = \phi_1(y, z)$, the front surface is $x = \phi_2(y, z)$, and we have

(10)
$$\iiint_E f(x, y, z)\, dV = \iint_D \left[\int_{\phi_1(y, z)}^{\phi_2(y, z)} f(x, y, z)\, dx \right] dA$$

Finally, a **type 3** region is of the form

$$E = \{(x, y, z) \mid (x, z) \in D,\ \phi_1(x, z) \leq y \leq \phi_2(x, z)\}$$

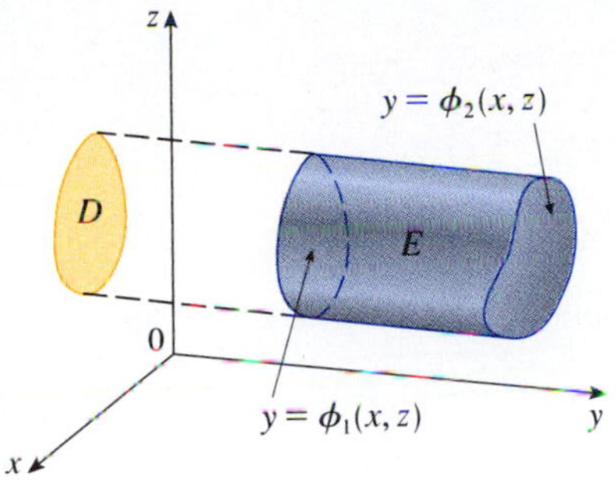

FIGURE 8
A type 3 region

where D is the projection of E onto the xz-plane, $y = \phi_1(x, z)$ is the left surface, and $y = \phi_2(x, z)$ is the right surface (see Figure 8). For this type of region we have

(11) $$\iiint_E f(x, y, z)\, dV = \iint_D \left[\int_{\phi_1(x,z)}^{\phi_2(x,z)} f(x, y, z)\, dy \right] dA$$

In each of Equations 10 and 11 there may be two possible expressions for the integral depending on whether D is a type I or II plane region (and corresponding to Equations 7 and 8).

EXAMPLE 3 Evaluate $\iiint_E \sqrt{x^2 + z^2}\, dV$, where E is the region bounded by the paraboloid $y = x^2 + z^2$ and the plane $y = 4$.

SOLUTION The solid E is shown in Figure 9. If we regard it as a type 1 region, then we need to consider its projection D_1 onto the xy-plane, which is the parabolic region in Figure 10. (The trace of $y = x^2 + z^2$ in the plane $z = 0$ is the parabola $y = x^2$.)

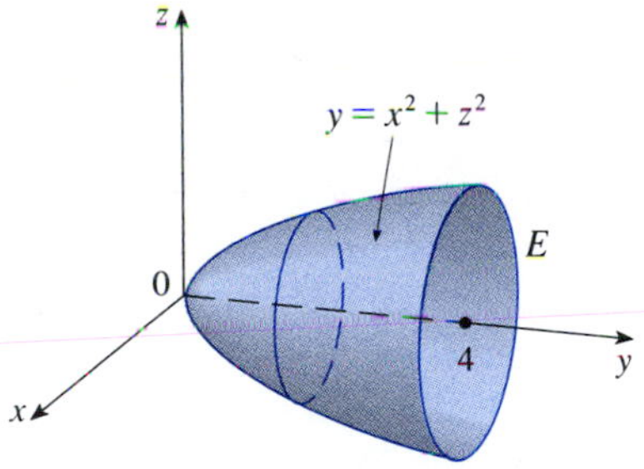

FIGURE 9
Region of integration

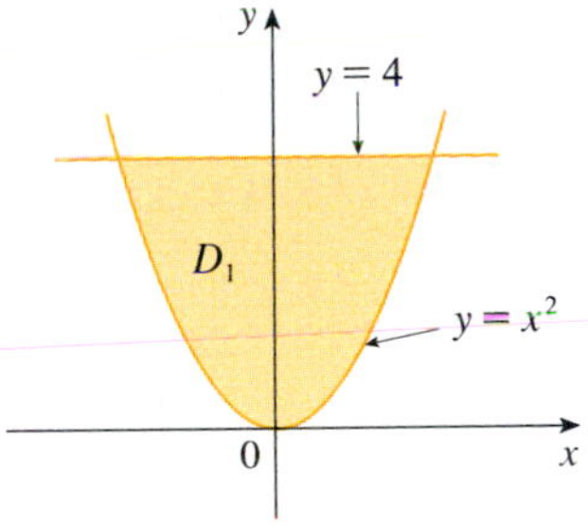

FIGURE 10
Projection on xy-plane

From $y = x^2 + z^2$ we obtain $z = \pm\sqrt{y - x^2}$, so the lower boundary surface of E is $z = -\sqrt{y - x^2}$ and the upper surface is $z = \sqrt{y - x^2}$. Therefore, the description of E as a type 1 region is

$$E = \left\{(x, y, z) \mid -2 \leqslant x \leqslant 2,\ x^2 \leqslant y \leqslant 4,\ -\sqrt{y - x^2} \leqslant z \leqslant \sqrt{y - x^2}\right\}$$

and so we obtain

$$\iiint_E \sqrt{x^2 + z^2}\, dV = \int_{-2}^{2} \int_{x^2}^{4} \int_{-\sqrt{y-x^2}}^{\sqrt{y-x^2}} \sqrt{x^2 + z^2}\, dz\, dy\, dx$$

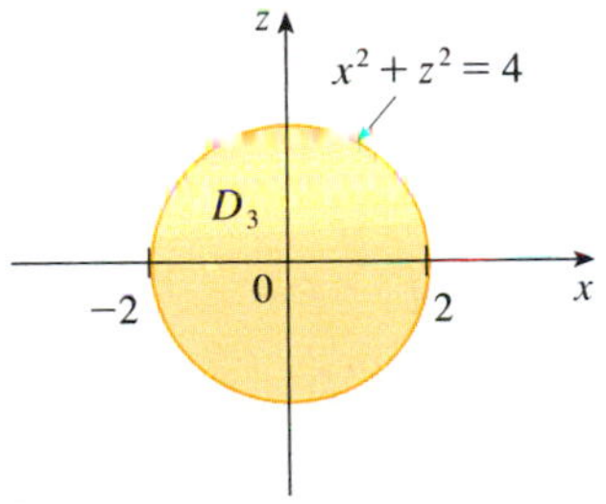

FIGURE 11
Projection on xz-plane

The most difficult step in evaluating a triple integral is setting up an expression for the region of integration (such as Equation 9 in Example 2). Remember that the limits of integration in the inner integral contain at most two variables, the limits of integration in the middle integral contain at most one variable, and the limits of integration in the outer integral must be constants.

Although this expression is correct, it is extremely difficult to evaluate. So let's instead consider E as a type 3 region. As such, its projection D_3 onto the xz-plane is the disk $x^2 + z^2 \leqslant 4$ shown in Figure 11.

Then the left boundary of E is the paraboloid $y = x^2 + z^2$ and the right boundary is the plane $y = 4$, so taking $\phi_1(x, z) = x^2 + z^2$ and $\phi_2(x, z) = 4$ in Equation 11, we have

$$\iiint_E \sqrt{x^2 + z^2}\, dV = \iint_{D_3} \left[\int_{x^2+z^2}^{4} \sqrt{x^2 + z^2}\, dy \right] dA$$

$$= \iint_{D_3} (4 - x^2 - z^2)\sqrt{x^2 + z^2}\, dA$$

Although this integral could be written as

$$\int_{-2}^{2}\int_{-\sqrt{4-x^2}}^{\sqrt{4-x^2}} (4 - x^2 - z^2)\sqrt{x^2 + z^2}\, dz\, dx$$

it's easier to convert to polar coordinates in the xz-plane: $x = r\cos\theta$, $z = r\sin\theta$. This gives

$$\iiint_E \sqrt{x^2 + z^2}\, dV = \iint_{D_3} (4 - x^2 - z^2)\sqrt{x^2 + z^2}\, dA$$

$$= \int_0^{2\pi}\int_0^2 (4 - r^2) r\, r\, dr\, d\theta = \int_0^{2\pi} d\theta \int_0^2 (4r^2 - r^4)\, dr$$

$$= 2\pi\left[\frac{4r^3}{3} - \frac{r^5}{5}\right]_0^2 = \frac{128\pi}{15}$$

■

APPLICATIONS OF TRIPLE INTEGRALS

Recall that if $f(x) \geqslant 0$, then the single integral $\int_a^b f(x)\, dx$ represents the area under the curve $y = f(x)$ from a to b, and if $f(x, y) \geqslant 0$, then the double integral $\iint_D f(x, y)\, dA$ represents the volume under the surface $z = f(x, y)$ and above D. The corresponding interpretation of a triple integral $\iiint_E f(x, y, z)\, dV$, where $f(x, y, z) \geqslant 0$, is not very useful because it would be the "hypervolume" of a four-dimensional object and, of course, that is impossible to visualize. (Remember that E is just the *domain* of the function f; the graph of f lies in four-dimensional space.) Nonetheless, the triple integral $\iiint_E f(x, y, z)\, dV$ can be interpreted in different ways in different physical situations, depending on the physical interpretations of x, y, z and $f(x, y, z)$.

Let's begin with the special case where $f(x, y, z) = 1$ for all points in E. Then the triple integral does represent the volume of E:

(12)
$$V(E) = \iiint_E dV$$

For example, you can see this in the case of a type 1 region by putting $f(x, y, z) = 1$ in Formula 6:

$$\iiint_E 1\, dV = \iint_D \left[\int_{\phi_1(x, y)}^{\phi_2(x, y)} dz\right] dA = \iint_D [\phi_2(x, y) - \phi_1(x, y)]\, dA$$

and from Section 13.3 we know this represents the volume that lies between the surfaces $z = \phi_1(x, y)$ and $z = \phi_2(x, y)$.

EXAMPLE 4 Use a triple integral to find the volume of the tetrahedron T bounded by the planes $x + 2y + z = 2$, $x = 2y$, $x = 0$, and $z = 0$.

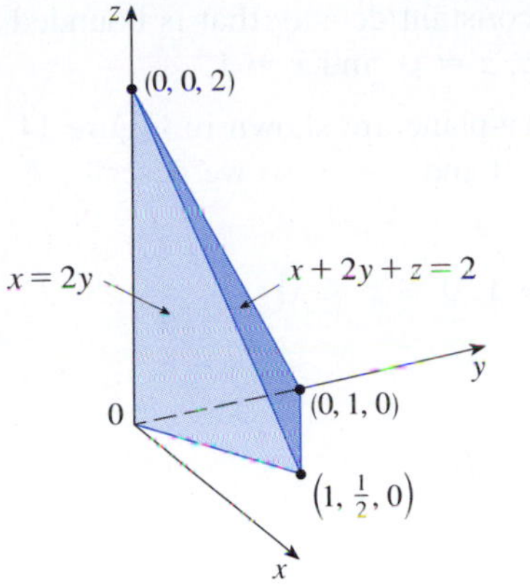

FIGURE 12

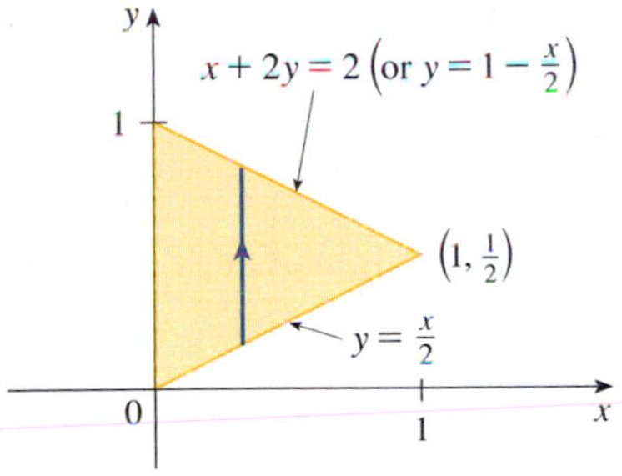

FIGURE 13

SOLUTION The tetrahedron T and its projection D on the xy-plane are shown in Figures 12 and 13. The lower boundary of T is the plane $z = 0$ and the upper boundary is the plane $x + 2y + z = 2$, that is, $z = 2 - x - 2y$. Therefore, we have

$$V(T) = \iiint_T dV = \int_0^1 \int_{x/2}^{1-x/2} \int_0^{2-x-2y} dz\,dy\,dx$$

$$= \int_0^1 \int_{x/2}^{1-x/2} (2 - x - 2y)\,dy\,dx = \tfrac{1}{3}$$

by the same calculation as in Example 4 in Section 13.3.

(Notice that it is not necessary to use triple integrals to compute volumes. They simply give an alternative method for setting up the calculation.) ■

All the applications of double integrals in Section 13.6 can be immediately extended to triple integrals. For example, if the density function of a solid object that occupies the region E is $\rho(x, y, z)$, in units of mass per unit volume, at any given point (x, y, z), then its **mass** is

$$m = \iiint_E \rho(x, y, z)\,dV \tag{13}$$

and its **moments** about the three coordinate planes are

$$M_{yz} = \iiint_E x\rho(x, y, z)\,dV \qquad M_{xz} = \iiint_E y\rho(x, y, z)\,dV \tag{14}$$

$$M_{xy} = \iiint_E z\rho(x, y, z)\,dV$$

The **center of mass** is located at the point $(\bar{x}, \bar{y}, \bar{z})$ where

$$\bar{x} = \frac{M_{yz}}{m} \qquad \bar{y} = \frac{M_{xz}}{m} \qquad \bar{z} = \frac{M_{xy}}{m} \tag{15}$$

If the density is constant, the center of mass of the solid is called the **centroid** of E. The **moments of inertia** about the three coordinate axes are

$$I_x = \iiint_E (y^2 + z^2)\rho(x, y, z)\,dV \qquad I_y = \iiint_E (x^2 + z^2)\rho(x, y, z)\,dV \tag{16}$$

$$I_z = \iiint_E (x^2 + y^2)\rho(x, y, z)\,dV$$

Also, as in Section 13.6, the total *electric charge* on a solid object occupying a region E and having charge density $\sigma(x, y, z)$ is

$$Q = \iiint_E \sigma(x, y, z)\,dV$$

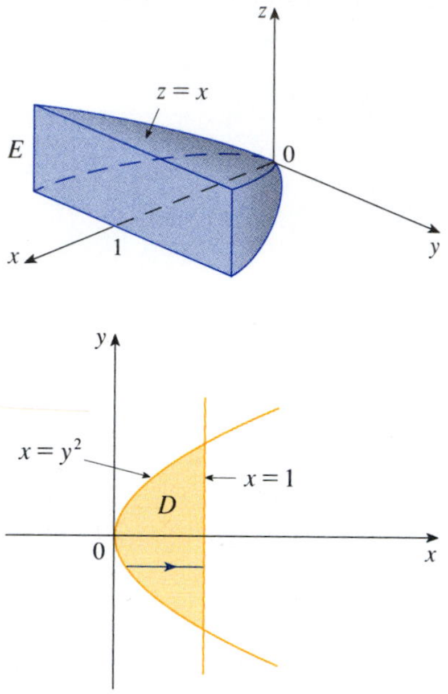

FIGURE 14

EXAMPLE 5 Find the center of mass of a solid of constant density that is bounded by the parabolic cylinder $x = y^2$ and the planes $x = z$, $z = 0$, and $x = 1$.

SOLUTION The solid E and its projection onto the xy-plane are shown in Figure 14. The lower and upper surfaces of E are the planes $z = 0$ and $z = x$, so we describe E as a type 1 region:

$$E = \{(x, y, z) \mid -1 \leqslant y \leqslant 1, y^2 \leqslant x \leqslant 1, 0 \leqslant z \leqslant x\}$$

Then, if the density is $\rho(x, y, z) = \rho$, the mass is

$$\begin{aligned} m &= \iiint_E \rho\, dV = \int_{-1}^{1} \int_{y^2}^{1} \int_{0}^{x} \rho\, dz\, dx\, dy \\ &= \rho \int_{-1}^{1} \int_{y^2}^{1} x\, dx\, dy = \rho \int_{-1}^{1} \left[\frac{x^2}{2} \right]_{x=y^2}^{x=1} dy \\ &= \frac{\rho}{2} \int_{-1}^{1} (1 - y^4)\, dy = \rho \int_{0}^{1} (1 - y^4)\, dy \\ &= \rho \left[y - \frac{y^5}{5} \right]_0^1 = \frac{4\rho}{5} \end{aligned}$$

Because of the symmetry of E and ρ about the xz-plane, we can immediately say that $M_{xz} = 0$ and, therefore, $\bar{y} = 0$. The other moments are

$$\begin{aligned} M_{yz} &= \iiint_E x\rho\, dV = \int_{-1}^{1} \int_{y^2}^{1} \int_{0}^{x} x\rho\, dz\, dx\, dy \\ &= \rho \int_{-1}^{1} \int_{y^2}^{1} x^2\, dx\, dy = \rho \int_{-1}^{1} \left[\frac{x^3}{3} \right]_{x=y^2}^{x=1} dy \\ &= \frac{2\rho}{3} \int_{0}^{1} (1 - y^6)\, dy = \frac{2\rho}{3} \left[y - \frac{y^7}{7} \right]_0^1 = \frac{4\rho}{7} \end{aligned}$$

$$\begin{aligned} M_{xy} &= \iiint_E z\rho\, dV = \int_{-1}^{1} \int_{y^2}^{1} \int_{0}^{x} z\rho\, dz\, dx\, dy \\ &= \rho \int_{-1}^{1} \int_{y^2}^{1} \left[\frac{z^2}{2} \right]_{z=0}^{z=x} dx\, dy = \frac{\rho}{2} \int_{-1}^{1} \int_{y^2}^{1} x^2\, dx\, dy \\ &= \frac{\rho}{3} \int_{0}^{1} (1 - y^6)\, dy = \frac{2\rho}{7} \end{aligned}$$

Therefore, the center of mass is

$$(\bar{x}, \bar{y}, \bar{z}) = \left(\frac{M_{yz}}{m}, \frac{M_{xz}}{m}, \frac{M_{xy}}{m} \right) = \left(\tfrac{5}{7}, 0, \tfrac{5}{14}\right)$$

■

EXERCISES 13.8

1. Evaluate the integral in Example 1, integrating first with respect to z, then x, and then y.

2. Evaluate the integral $\iiint_E (x^2 + yz)\, dV$, where

$$E = \{(x, y, z) \mid 0 \leqslant x \leqslant 2, -3 \leqslant y \leqslant 0, -1 \leqslant z \leqslant 1\}$$

using three different orders of integration.

3–6 ■ Evaluate the iterated integral.

3. $\displaystyle\int_0^1 \int_0^z \int_0^y xyz\, dx\, dy\, dz$

4. $\displaystyle\int_0^1 \int_x^{2x} \int_0^{x+y} 2xy\, dz\, dy\, dx$

5. $\displaystyle\int_0^{\pi} \int_0^2 \int_0^{\sqrt{4-z^2}} z \sin y\, dx\, dz\, dy$

6. $\displaystyle\int_0^3 \int_0^{\sqrt{9-x^2}} \int_0^x yz\, dy\, dz\, dx$

7–16 ■ Evaluate the triple integral.

7. $\iiint_E yz\,dV$, where $E = \{(x, y, z) \mid 0 \leqslant z \leqslant 1,\ 0 \leqslant y \leqslant 2z,\ 0 \leqslant x \leqslant z + 2\}$

8. $\iiint_E e^x\,dV$, where $E = \{(x, y, z) \mid 0 \leqslant y \leqslant 1,\ 0 \leqslant x \leqslant y,\ 0 \leqslant z \leqslant x + y\}$

9. $\iiint_E y\,dV$, where E lies under the plane $z = x + 2y$ and above the region in the xy-plane bounded by the curves $y = x^2$, $y = 0$, and $x = 1$

10. $\iiint_E x\,dV$, where E is bounded by the planes $x = 0$, $y = 0$, $z = 0$, and $3x + 2y + z = 6$

11. $\iiint_E xy\,dV$, where E is the solid tetrahedron with vertices $(0, 0, 0)$, $(1, 0, 0)$, $(0, 2, 0)$, and $(0, 0, 3)$

12. $\iiint_E xz\,dV$, where E is the solid tetrahedron with vertices $(0, 0, 0)$, $(0, 1, 0)$, $(1, 1, 0)$, and $(0, 1, 1)$

13. $\iiint_E z\,dV$, where E is bounded by the planes $x = 0$, $y = 0$, $z = 0$, $y + z = 1$, and $x + z = 1$

14. $\iiint_E (x + 2y)\,dV$, where E is bounded by the parabolic cylinder $y = x^2$ and the planes $x = z$, $x = y$, and $z = 0$

15. $\iiint_E x\,dV$, where E is bounded by the paraboloid $x = 4y^2 + 4z^2$ and the plane $x = 4$

16. $\iiint_E z\,dV$, where E is bounded by the cylinder $y^2 + z^2 = 9$ and the planes $x = 0$, $y = 3x$, and $z = 0$ in the first octant

17–20 ■ Use a triple integral to find the volume of the given solid.

17. The tetrahedron bounded by the coordinate planes and the plane $2x + 3y + 6z = 12$

18. The solid bounded by the elliptic cylinder $4x^2 + z^2 = 4$ and the planes $y = 0$ and $y = z + 2$

19. The solid bounded by the cylinder $x = y^2$ and the planes $z = 0$ and $x + z = 1$

20. The solid enclosed by the paraboloids $z = x^2 + y^2$ and $z = 18 - x^2 - y^2$

21. (a) Express the volume of the wedge in the first octant that is cut from the cylinder $y^2 + z^2 = 1$ by the planes $y = x$ and $x = 1$ as a triple integral.

CAS (b) Use either the Table of Integrals (in Appendix K) or a computer algebra system to find the exact value of the triple integral in part (a).

CAS 22. Evaluate the triple integral exactly:

$$\int_0^{\pi}\int_0^{\pi}\int_0^{\pi} \ldots$$

23. In the **Midpoint Rule for triple integrals** we use a triple Riemann sum to approximate a triple integral over a box B, where $f(x, y, z)$ is evaluated at the center $(\bar{x}_i, \bar{y}_j, \bar{z}_k)$ of the box B_{ijk}. Use the Midpoint Rule to estimate $\iiint_B e^{-x^2-y^2-z^2}\,dV$, where B is the cube defined by $0 \leqslant x \leqslant 1$, $0 \leqslant y \leqslant 1$, $0 \leqslant z \leqslant 1$. Use the partition of B into eight cubes of equal size. What is the norm of this partition?

CAS 24. Use a computer algebra system to approximate the integral in Exercise 23 to two decimal places. Compare with the answer to that exercise.

25–26 ■ Sketch the solid whose volume is given by the iterated integral.

25. $\displaystyle\int_0^1\int_0^{1-x}\int_0^{2-2z} dy\,dz\,dx$ 26. $\displaystyle\int_0^2\int_0^{2-y}\int_0^{4-y^2} dx\,dz\,dy$

27–30 ■ Express the integral $\iiint_E f(x, y, z)\,dV$ as an iterated integral in six different ways, where E is the solid bounded by the given surfaces.

27. $x^2 + z^2 = 4$, $y = 0$, $y = 6$

28. $z = 0$, $x = 0$, $y = 2$, $z = y - 2x$

29. $z = 0$, $z = y$, $x^2 = 1 - y$

30. $9x^2 + 4y^2 + z^2 = 1$

31. The figure shows the region of integration for the integral

$$\int_0^1\int_{\sqrt{x}}^1\int_0^{1-y} f(x, y, z)\,dz\,dy\,dx$$

Rewrite this integral as an equivalent iterated integral in the five other orders.

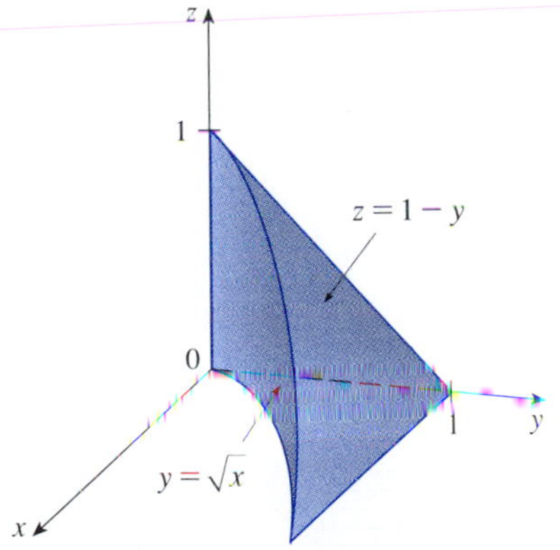

32. The figure shows the region of integration for the integral

$$\int_0^1\int_0^{1-x^2}\int_0^{1-x} f(x, y, z)\,dy\,dz\,dx$$

Rewrite this integral as an equivalent iterated integral in the five other orders.

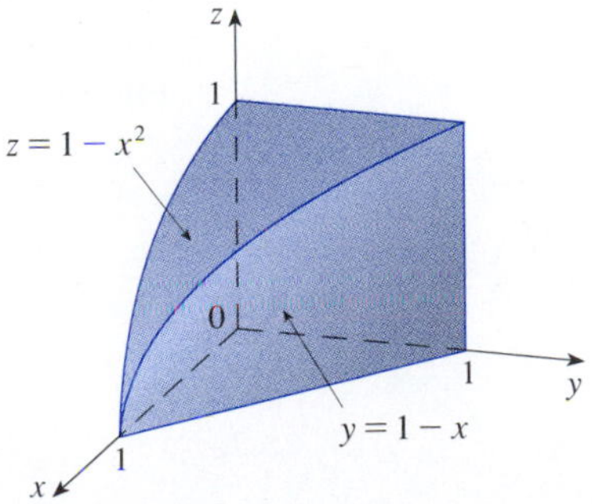

33–34 ■ Write five other iterated integrals that are equal to the given iterated integral.

33. $\int_0^1 \int_y^1 \int_0^y f(x, y, z)\, dz\, dx\, dy$

34. $\int_0^1 \int_0^{x^2} \int_0^y f(x, y, z)\, dz\, dy\, dx$

35–38 ■ Find the mass and center of mass of the given solid E with the given density function ρ.

35. E is the solid of Exercise 9; $\rho(x, y, z) = 2$

36. E is bounded by the parabolic cylinder $z = 1 - y^2$ and the planes $x + z = 1$, $x = 0$, and $z = 0$; $\rho(x, y, z) = 4$

37. E is the cube given by $0 \leqslant x \leqslant a$, $0 \leqslant y \leqslant a$, $0 \leqslant z \leqslant a$; $\rho(x, y, z) = x^2 + y^2 + z^2$

38. E is the tetrahedron bounded by the planes $x = 0$, $y = 0$, $z = 0$, $x + y + z = 1$; $\rho(x, y, z) = y$

39–42 ■ Set up, but do not evaluate, integral expressions for (a) the mass, (b) the center of mass, and (c) the moment of inertia about the z-axis.

39. The solid in the first octant bounded by the cylinder $x^2 + y^2 = 1$ and the planes $y = z$, $x = 0$, and $z = 0$; $\rho(x, y, z) = 1 + x + y + z$

40. The hemisphere $x^2 + y^2 + z^2 = 1$, $z \geqslant 0$; $\rho(x, y, z) = \sqrt{x^2 + y^2 + z^2}$

41. The solid of Exercise 15; $\rho(x, y, z) = x^2 + y^2 + z^2$

42. The solid of Exercise 16; $\rho(x, y, z) = x^2 + y^2$

CAS **43.** Evaluate the quantities in Exercise 39 exactly.

CAS **44.** Evaluate the quantities in Exercise 42 correct to three decimal places.

45. Find the moments of inertia for a cube of constant density k and side length L if one vertex is at the origin and three edges lie along the coordinate axes.

46. Find the moments of inertia for a rectangular brick with dimensions a, b, and c, mass M, and constant density if the center of the brick is at the origin and the edges are parallel to the coordinate axes.

47–48 ■ The average value of a function $f(x, y, z)$ over a solid region E is defined to be

$$f_{\text{ave}} = \frac{1}{V(E)} \iiint_E f(x, y, z)\, dV$$

where $V(E)$ is the volume of E. For instance, if ρ is a density function, then ρ_{ave} is the average density of E.

47. Find the average value of the function $f(x, y, z) = xyz$ over the cube with side length L that lies in the first octant with one vertex at the origin and edges parallel to the coordinate axes.

48. Find the average value of the function $f(x, y, z) = x + y + z$ over the tetrahedron with vertices $(0, 0, 0)$, $(1, 0, 0)$, $(0, 1, 0)$, and $(0, 0, 1)$.

49. Find the region E for which the triple integral

$$\iiint_E (1 - x^2 - 2y^2 - 3z^2)\, dV$$

is a maximum.

13.9 CYLINDRICAL AND SPHERICAL COORDINATES

Recall that in Section 13.4 we introduced the polar coordinate system in order to give a more convenient description of certain curves and regions. In three dimensions there are two coordinate systems that are similar to polar coordinates and give convenient descriptions of some commonly occurring surfaces and solids. They will be especially useful in Section 13.10 when we use them to compute volumes and triple integrals.

In the **cylindrical coordinate system**, a point P in three-dimensional space is represented by the ordered triple (r, θ, z), where r and θ are polar coordinates of the projection of P onto the xy-plane and z is the directed distance from the xy-plane to P (see Figure 1).

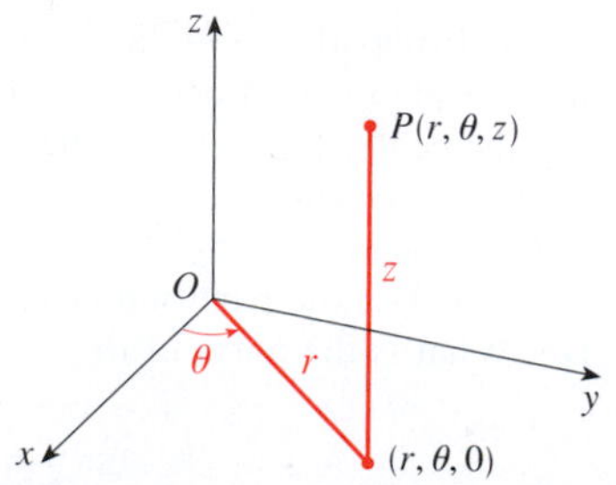

FIGURE 1

To convert from cylindrical to rectangular coordinates we use the equations

(1) $$x = r\cos\theta \qquad y = r\sin\theta \qquad z = z$$

whereas to convert from rectangular to cylindrical coordinates we use

(2) $$r^2 = x^2 + y^2 \qquad \tan\theta = \frac{y}{x} \qquad z = z$$

These equations follow from Equations 13.4.1 and 13.4.2.

EXAMPLE 1

(a) Plot the point with cylindrical coordinates $(2, 2\pi/3, 1)$ and find its rectangular coordinates.

(b) Find the cylindrical coordinates of the point with rectangular coordinates $(3, -3, -7)$.

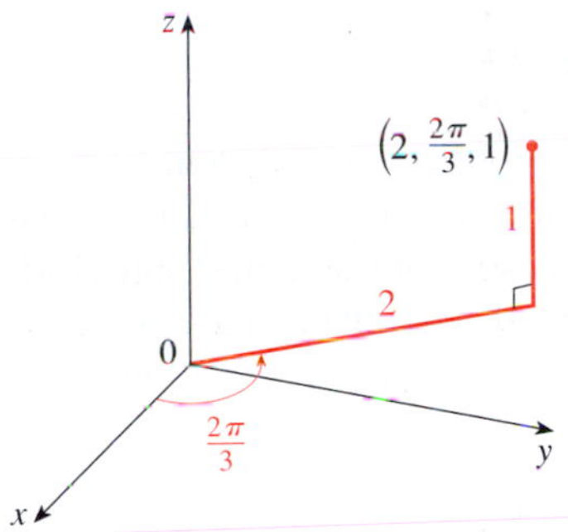

FIGURE 2

SOLUTION

(a) The point with cylindrical coordinates $(2, 2\pi/3, 1)$ is plotted in Figure 2. From Equations 1, its rectangular coordinates are

$$x = 2\cos\frac{2\pi}{3} = 2\left(-\frac{1}{2}\right) = -1 \qquad y = 2\sin\frac{2\pi}{3} = 2\left(\frac{\sqrt{3}}{2}\right) = \sqrt{3} \qquad z = 1$$

Thus, the point is $(-1, \sqrt{3}, 1)$ in rectangular coordinates.

(b) From Equations 2 we have

$$r = \sqrt{3^2 + (-3)^2} = 3\sqrt{2}$$

$$\tan\theta = \frac{-3}{3} = -1 \qquad \text{so} \qquad \theta = \frac{7\pi}{4} + 2n\pi$$

$$z = -7$$

Therefore, one set of cylindrical coordinates is $(3\sqrt{2}, 7\pi/4, -7)$. Another is $(3\sqrt{2}, -\pi/4, -7)$. As with polar coordinates, there are infinitely many choices. ■

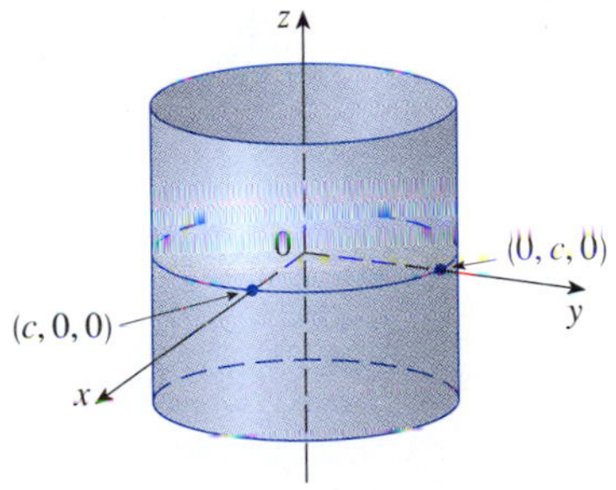

FIGURE 3

$r = c$, a cylinder

Cylindrical coordinates are useful in problems that involve symmetry about the z-axis. For instance, the axis of the circular cylinder with Cartesian equation $x^2 + y^2 = c^2$ is the z-axis. In cylindrical coordinates this cylinder has the very simple equation $r = c$ (see Figure 3). This is the reason for the name "cylindrical" coordinates.

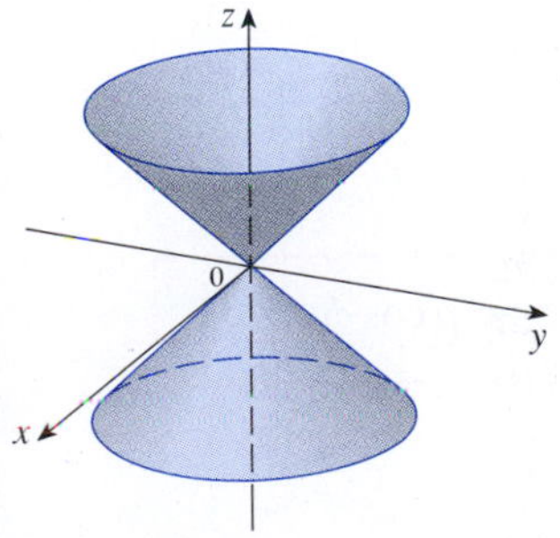

FIGURE 4

$z = r$, a cone

EXAMPLE 2 Describe the surface whose equation in cylindrical coordinates is $z = r$.

SOLUTION We first convert to an equation in rectangular coordinates. From the first equation in (2) we have

$$z^2 = r^2 = x^2 + y^2$$

We recognize the equation $z^2 = x^2 + y^2$ (by comparison with Equation 11.5.4) as being a circular cone whose axis is the z-axis (see Figure 4). ■

EXAMPLE 3 Find an equation in cylindrical coordinates for the ellipsoid $4x^2 + 4y^2 + z^2 = 1$.

SOLUTION Since $r^2 = x^2 + y^2$ from Equations 2, we have

$$z^2 = 1 - 4(x^2 + y^2) = 1 - 4r^2$$

So the equation of the ellipsoid in cylindrical coordinates is $z^2 = 1 - 4r^2$. ■

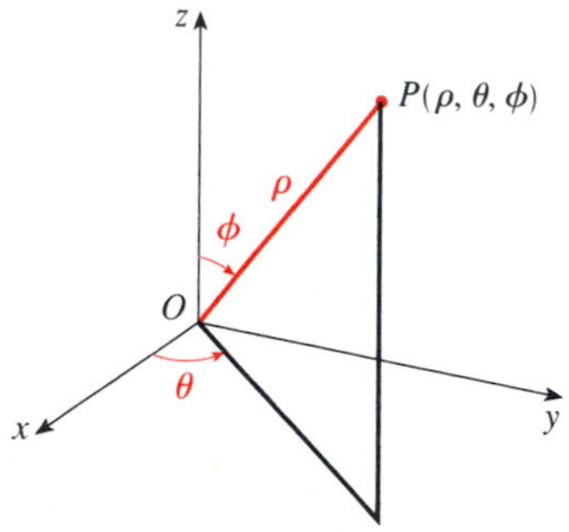

FIGURE 5

The **spherical coordinates** (ρ, θ, ϕ) of a point P in space are shown in Figure 5, where $\rho = |OP|$ is the distance from the origin to P, θ is the same angle as in cylindrical coordinates, and ϕ is the angle between the positive z-axis and the line segment OP. Note that

$$\rho \geqslant 0 \qquad 0 \leqslant \phi \leqslant \pi$$

The spherical coordinate system is especially useful in problems where there is symmetry about the origin. For example, the sphere with center the origin and radius c has the simple equation $\rho = c$ (see Figure 6) and this is the reason for the name "spherical" coordinates. The graph of the equation $\theta = c$ is a vertical half-plane (see Figure 7), and the equation $\phi = c$ represents a half-cone with the z-axis as its axis (see Figure 8).

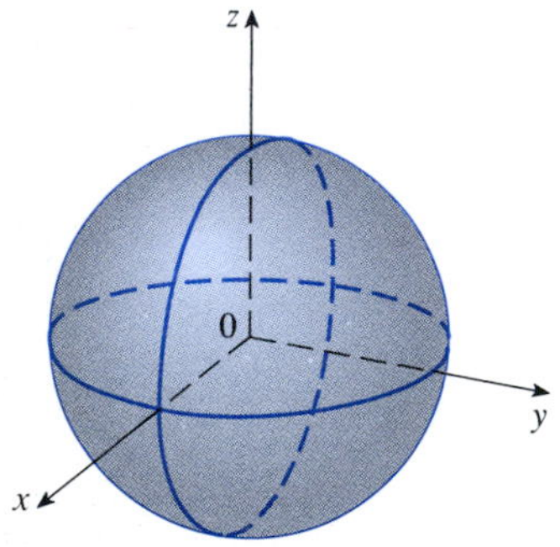

FIGURE 6
$r = c$, a sphere

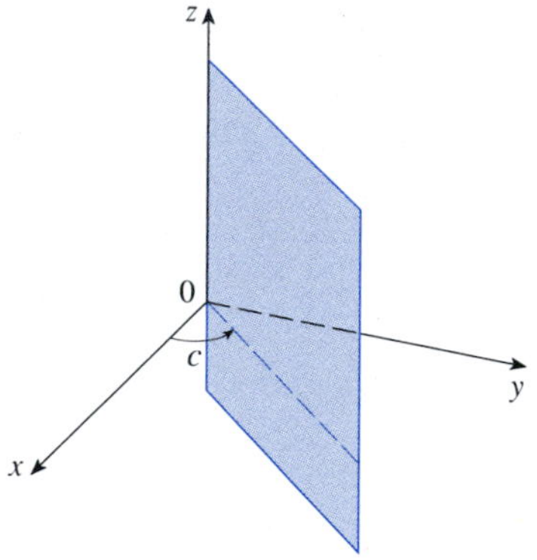

FIGURE 7
$\theta = c$, a half-plane

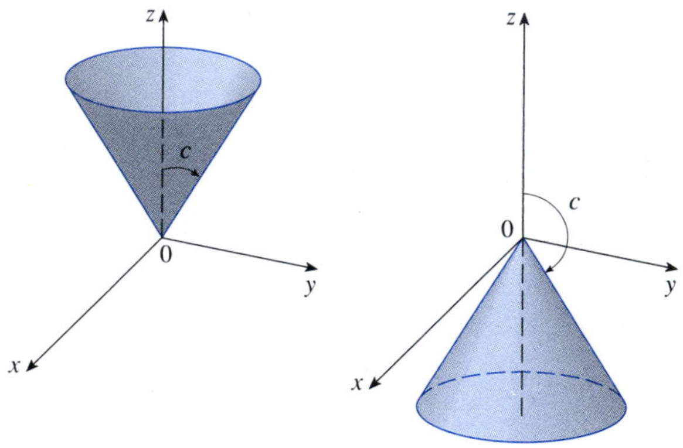

$0 < c < \pi/2$

$\pi/2 < c < \pi$

FIGURE 8
$\phi = c$, a half-cone

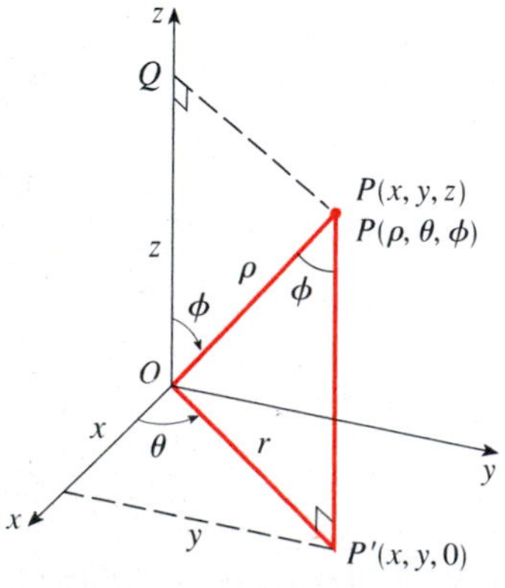

FIGURE 9

The relationship between rectangular and spherical coordinates can be seen from Figure 9. From triangles OPQ and OPP' we have

$$z = \rho \cos \phi \qquad r = \rho \sin \phi$$

But $x = r \cos \theta$ and $y = r \sin \theta$, so

(3) $$x = \rho \sin \phi \cos \theta \qquad y = \rho \sin \phi \sin \theta \qquad z = \rho \cos \phi$$

Also, the distance formula shows that

(4) $$\rho^2 = x^2 + y^2 + z^2$$

EXAMPLE 4 The point $(2, \pi/4, \pi/3)$ is given in spherical coordinates. Find its rectangular coordinates.

SOLUTION From Equations 3 we have

$$x = \rho \sin\phi \cos\theta = 2 \sin\frac{\pi}{3} \cos\frac{\pi}{4} = 2\left(\frac{\sqrt{3}}{2}\right)\left(\frac{1}{\sqrt{2}}\right) = \sqrt{\frac{3}{2}}$$

$$y = \rho \sin\phi \sin\theta = 2 \sin\frac{\pi}{3} \sin\frac{\pi}{4} = 2\left(\frac{\sqrt{3}}{2}\right)\left(\frac{1}{\sqrt{2}}\right) = \sqrt{\frac{3}{2}}$$

$$z = \rho \cos\phi = 2 \cos\frac{\pi}{3} = 2\left(\tfrac{1}{2}\right) = 1$$

Thus, the point $(2, \pi/4, \pi/3)$ is $(\sqrt{3/2}, \sqrt{3/2}, 1)$ in rectangular coordinates. ■

EXAMPLE 5 The point $(0, 2\sqrt{3}, -2)$ is given in rectangular coordinates. Find its spherical coordinates.

SOLUTION From Equation 4 we have

$$\rho = \sqrt{x^2 + y^2 + z^2} = \sqrt{0 + 12 + 4} = 4$$

and so Equations 3 give

$$\cos\phi = \frac{z}{\rho} = \frac{-2}{4} = -\frac{1}{2} \qquad \phi = \frac{2\pi}{3}$$

$$\cos\theta = \frac{x}{\rho \sin\phi} = 0 \qquad \theta = \frac{\pi}{2}$$

(Note that $\theta \neq 3\pi/2$ because $y = 2\sqrt{3} > 0$.) Therefore, the spherical coordinates of the given point are $(4, \pi/2, 2\pi/3)$. ■

EXAMPLE 6 Find an equation in spherical coordinates for the hyperboloid of two sheets with equation $x^2 - y^2 - z^2 = 1$.

SOLUTION Substituting the expressions in Equations 3 into the given equation, we have

$$\rho^2 \sin^2\phi \cos^2\theta - \rho^2 \sin^2\phi \sin^2\theta - \rho^2 \cos^2\phi = 1$$

$$\rho^2[\sin^2\phi\,(\cos^2\theta - \sin^2\theta) - \cos^2\phi] = 1$$

or

$$\rho^2(\sin^2\phi \cos 2\theta - \cos^2\phi) = 1$$ ■

EXAMPLE 7 Find a rectangular equation for the surface whose spherical equation is $\rho = \sin\theta \sin\phi$.

SOLUTION From Equations 4 and 3 we have

$$x^2 + y^2 + z^2 = \rho^2 = \rho \sin\theta \sin\phi = y$$

or

$$x^2 + (y - \tfrac{1}{2})^2 + z^2 = \tfrac{1}{4}$$

which is the equation of a sphere with center $(0, \tfrac{1}{2}, 0)$ and radius $\tfrac{1}{2}$. ■

EXAMPLE 8 Use a computer to draw a picture of the solid that remains when a hole of radius 3 is drilled through the center of a sphere of radius 4.

SOLUTION To keep the equations simple, let's choose the coordinate system so that the center of the sphere is at the origin and the axis of the cylinder that forms the hole is the z-axis. We could use either cylindrical or spherical coordinates to describe the solid, but the description is somewhat simpler if we use cylindrical coordinates. Then the equation of the cylinder is $r = 3$ and the equation of the sphere is $x^2 + y^2 + z^2 = 16$ or $r^2 + z^2 = 16$. The points in the solid lie outside the cylinder and inside the sphere, so they satisfy the inequalities

$$3 \leqslant r \leqslant \sqrt{16 - z^2}$$

To ensure that the computer graphs only the appropriate parts of these surfaces, we find where they intersect by solving the equations $r = 3$ and $r = \sqrt{16 - z^2}$:

$$\sqrt{16 - z^2} = 3 \quad \Rightarrow \quad 16 - z^2 = 9 \quad \Rightarrow \quad z^2 = 7 \quad \Rightarrow \quad z = \pm\sqrt{7}$$

The solid lies between $z = -\sqrt{7}$ and $z = \sqrt{7}$, so we use the computer to graph the surfaces with the following equations and domains:

$$r = 3 \qquad 0 \leqslant \theta \leqslant 2\pi \qquad -\sqrt{7} \leqslant z \leqslant \sqrt{7}$$

$$r = \sqrt{16 - z^2} \qquad 0 \leqslant \theta \leqslant 2\pi \qquad -\sqrt{7} \leqslant z \leqslant \sqrt{7}$$

The resulting picture, shown in Figure 10, is exactly what we want. ■

Most three-dimensional graphing programs can graph surfaces whose equations are given in cylindrical or spherical coordinates. As Example 8 demonstrates, this is often the most convenient way of drawing a solid.

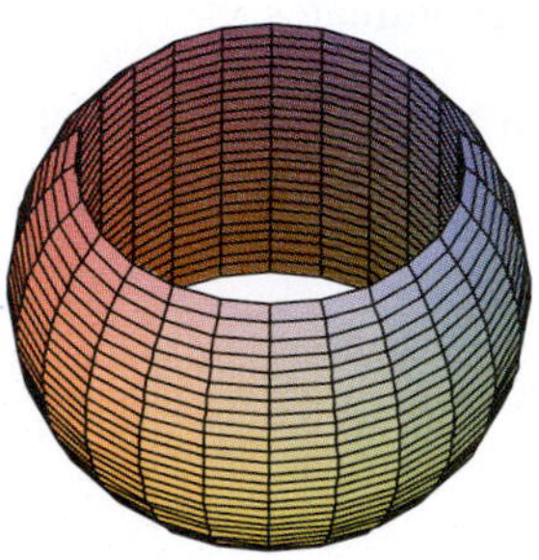

FIGURE 10

EXERCISES 13.9

1–6 ■ Change from cylindrical to rectangular coordinates.

1. $(3, \pi/2, 1)$ **2.** $(\sqrt{2}, \pi/4, \sqrt{2})$

3. $(2, 4\pi/3, 8)$ **4.** $(6, -\pi/6, 5)$

5. $(3, 0, -6)$ **6.** $(1, \pi, e)$

7–12 ■ Change from rectangular to cylindrical coordinates.

7. $(-1, 0, 0)$ **8.** $(1, 1, 1)$

9. $(\sqrt{3}, 1, 4)$ **10.** $(-\sqrt{2}, \sqrt{2}, 0)$

11. $(4, 4, 4)$ **12.** $(-1, \sqrt{3}, 2)$

13–18 ■ Change from spherical to rectangular coordinates.

13. $(1, 0, 0)$ **14.** $(3, 0, \pi)$

15. $(1, \pi/6, \pi/6)$ **16.** $(2, \pi/2, 3\pi/4)$

17. $(4, \pi/4, \pi/6)$ **18.** $(2, \pi/4, \pi/4)$

19–24 ■ Change from rectangular to spherical coordinates.

19. $(-3, 0, 0)$ **20.** $(1, 1, \sqrt{2})$

21. $(\sqrt{3}, 0, 1)$ **22.** $(-\sqrt{3}, -3, -2)$

23. $(1, -1, -\sqrt{2})$ **24.** $(\sqrt{3}, 1, 2\sqrt{3})$

25–28 ■ Change from cylindrical to spherical coordinates.

25. $(\sqrt{2}, \pi/4, 0)$ **26.** $(1, \pi/2, 1)$

27. $(4, \pi/3, 4)$ **28.** $(12, \pi, 5)$

29–32 ■ Change from spherical to cylindrical coordinates.

29. $(2, 0, 0)$ **30.** $(2\sqrt{2}, 3\pi/2, \pi/2)$

31. $(8, \pi/6, \pi/2)$ **32.** $(4, \pi/4, \pi/3)$

33–50 ■ Identify the surface whose equation is given.

33. $r = 3$ **34.** $\rho = 3$

35. $\phi = \pi/3$ **36.** $\theta = \pi/3$

37. $z = r^2$ **38.** $r = 4\sin\theta$

39. $\rho\cos\phi = 2$ **40.** $\rho\sin\phi = 2$

41. $\phi = 0$ **42.** $\phi = \pi/2$

43. $r = 2\cos\theta$ **44.** $\rho = 2\cos\phi$

45. $r^2 + z^2 = 25$ **46.** $r^2 - 2z^2 = 4$

47. $\rho^2(\sin^2\phi\cos^2\theta + \cos^2\phi) = 4$

48. $\rho^2(\sin^2\phi - 4\cos^2\phi) = 1$

49. $r^2 = r$ **50.** $\rho^2 - 6\rho + 8 = 0$

51–58 ■ Write the equation (a) in cylindrical coordinates and (b) in spherical coordinates.

51. $x^2 + y^2 + z^2 = 16$

52. $x^2 + y^2 - z^2 = 16$

53. $x + 2y + 3z = 6$

54. $x^2 + y^2 = 2z$

55. $x^2 - y^2 - 2z^2 = 4$

56. $y^2 + z^2 = 1$

57. $x^2 + y^2 = 2y$

58. $z = x^2 - y^2$

59–62 ■ Sketch the solid described by the given inequalities.

59. $r^2 \leqslant z \leqslant 2 - r^2$

60. $0 \leqslant \theta \leqslant \pi/2, \quad r \leqslant z \leqslant 2$

61. $-\pi/2 \leqslant \theta \leqslant \pi/2, \quad 0 \leqslant \phi \leqslant \pi/6, \quad 0 \leqslant \rho \leqslant \sec\phi$

62. $0 \leqslant \phi \leqslant \pi/3, \quad \rho \leqslant 2$

63. A solid lies above the cone $z = \sqrt{x^2 + y^2}$ and below the sphere $x^2 + y^2 + z^2 = z$. Write a description of the solid in terms of inequalities involving spherical coordinates.

64. Use a computer to draw the solid enclosed by the paraboloids $z = x^2 + y^2$ and $z = 5 - x^2 - y^2$.

65. Use a computer to draw a silo consisting of a cylinder with radius 3 and height 10 surmounted by a hemisphere.

66. The latitude and longitude of a point P in the Northern Hemisphere are related to spherical coordinates ρ, θ, ϕ as follows. We take the origin to be the center of the earth and the positive z-axis to pass through the North Pole. The positive x-axis passes through the point where the prime meridian (the meridian through Greenwich, England) intersects the equator. Then the latitude of P is $\alpha = 90° - \phi°$ and the longitude is $\beta = 360° - \theta°$. Find the great-circle distance from Los Angeles (latitude 34.06° north, longitude 118.25° west) to Montreal (latitude 45.50° north, longitude 73.60° west). Take the radius of Earth to be 3960 mi. (A *great circle* is the circle of intersection of a sphere and a plane through the center of the sphere.)

13.10 TRIPLE INTEGRALS IN CYLINDRICAL AND SPHERICAL COORDINATES

We saw in Section 13.5 that some double integrals are easier to evaluate using polar coordinates. In this section we see that some triple integrals are easier to evaluate using cylindrical or spherical coordinates.

CYLINDRICAL COORDINATES

Recall from the preceding section that the cylindrical coordinates of a point P are (r, θ, z), where r, θ, and z are shown in Figure 1. Suppose that E is a type 1 region whose projection D on the xy-plane is conveniently described in polar coordinates (see Figure 2). In particular, suppose that f is continuous and

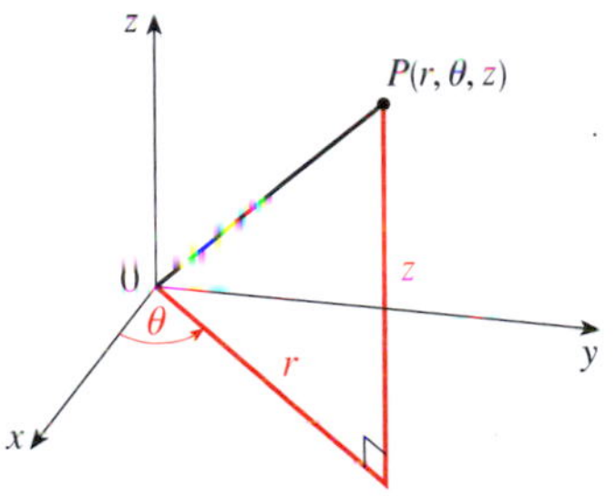

FIGURE 1

$$E = \{(x, y, z) \mid (x, y) \in D, \ \phi_1(x, y) \leqslant z \leqslant \phi_2(x, y)\}$$

where D is given in polar coordinates by

$$D = \{(r, \theta) \mid \alpha \leqslant \theta \leqslant \beta, \ h_1(\theta) \leqslant r \leqslant h_2(\theta)\}$$

We know from Equations 13.8.6 that

$$\iiint_E f(x, y, z)\, dV = \iint_D \left[\int_{\phi_1(x, y)}^{\phi_2(x, y)} f(x, y, z)\, dz \right] dA \tag{1}$$

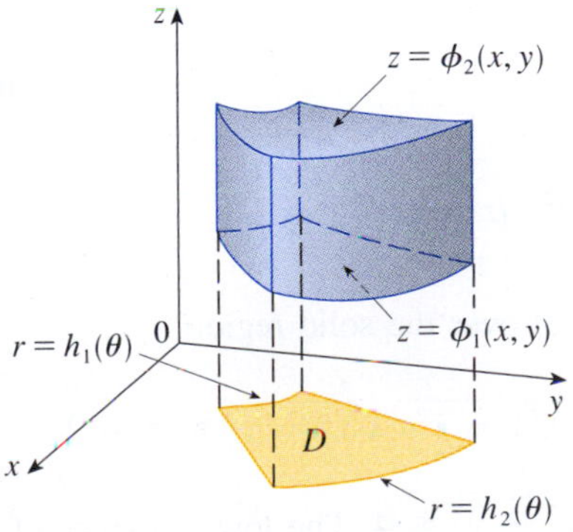

FIGURE 2

But we also know how to evaluate double integrals in polar coordinates. In fact, combining Equation 1 with (13.5.3), we obtain

$$\iiint_E f(x, y, z)\, dV = \int_\alpha^\beta \int_{h_1(\theta)}^{h_2(\theta)} \int_{\phi_1(r\cos\theta,\, r\sin\theta)}^{\phi_2(r\cos\theta,\, r\sin\theta)} f(r\cos\theta, r\sin\theta, z)\, r\, dz\, dr\, d\theta \tag{2}$$

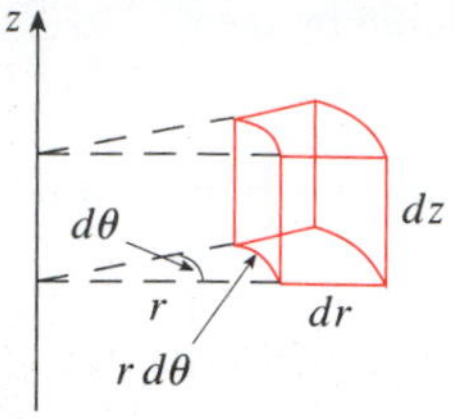

FIGURE 3
Volume element in cylindrical coordinates: $dV = r\,dz\,dr\,d\theta$

Formula 2 is the **formula for triple integration in cylindrical coordinates.** It says that we convert a triple integral from rectangular to cylindrical coordinates by writing $x = r\cos\theta$, $y = r\sin\theta$, leaving z as it is, using the appropriate limits of integration for z, r, and θ, and replacing dV by $r\,dz\,dr\,d\theta$. (Figure 3 shows how to remember this.) It is worthwhile to use this formula when E is a solid region easily described in cylindrical coordinates, and especially when the function $f(x, y, z)$ involves the expression $x^2 + y^2$.

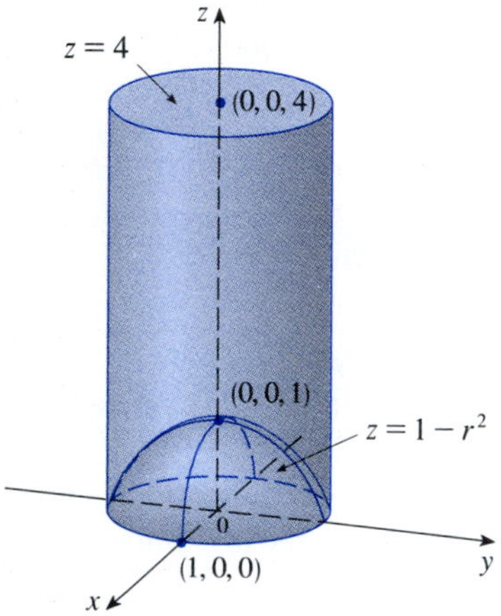

FIGURE 4

EXAMPLE 1 A solid E lies within the cylinder $x^2 + y^2 = 1$, below the plane $z = 4$, and above the paraboloid $z = 1 - x^2 - y^2$ (see Figure 4). The density at any point is proportional to its distance from the axis of the cylinder. Find the mass of E.

SOLUTION In cylindrical coordinates the cylinder is $r = 1$ and the paraboloid is $z = 1 - r^2$, so we can write

$$E = \{(r, \theta, z) \mid 0 \leq \theta \leq 2\pi,\ 0 \leq r \leq 1,\ 1 - r^2 \leq z \leq 4\}$$

Since the density at (x, y, z) is proportional to the distance from the z-axis, the density function is

$$f(x, y, z) = K\sqrt{x^2 + y^2} = Kr$$

where K is the proportionality constant. Therefore, from Formula 13.8.13, the mass of E is

$$\begin{aligned} m &= \iiint_E K\sqrt{x^2 + y^2}\,dV \\ &= \int_0^{2\pi}\int_0^1\int_{1-r^2}^4 (Kr)\,r\,dz\,dr\,d\theta \\ &= \int_0^{2\pi} d\theta \int_0^1 Kr^2[4 - (1 - r^2)]\,dr \\ &= 2\pi K \int_0^1 (3r^2 + r^4)\,dr \\ &= 2\pi K\left[r^3 + \frac{r^5}{5}\right]_0^1 = \frac{12\pi K}{5} \end{aligned}$$

■

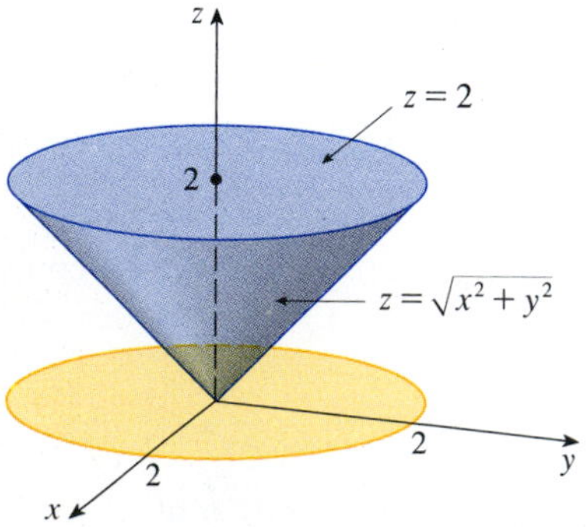

FIGURE 5

EXAMPLE 2 Evaluate $\displaystyle\int_{-2}^2 \int_{-\sqrt{4-x^2}}^{\sqrt{4-x^2}} \int_{\sqrt{x^2+y^2}}^2 (x^2 + y^2)\,dz\,dy\,dx$.

SOLUTION This iterated integral is a triple integral over the solid region

$$E = \{(x, y, z) \mid -2 \leq x \leq 2,\ -\sqrt{4 - x^2} \leq y \leq \sqrt{4 - x^2},\ \sqrt{x^2 + y^2} \leq z \leq 2\}$$

and the projection of E in the xy-plane is the disk $x^2 + y^2 \leq 4$. The lower surface of E is the cone $z = \sqrt{x^2 + y^2}$ and its upper surface is the plane $z = 2$ (see Figure 5).

This region has a much simpler description in cylindrical coordinates:

$$E = \{(r, \theta, z) \mid 0 \leq \theta \leq 2\pi,\ 0 \leq r \leq 2,\ r \leq z \leq 2\}$$

Therefore, we have

$$\int_{-2}^{2} \int_{-\sqrt{4-x^2}}^{\sqrt{4-x^2}} \int_{\sqrt{x^2+y^2}}^{2} (x^2 + y^2)\, dz\, dy\, dx = \iiint_E (x^2 + y^2)\, dV$$

$$= \int_0^{2\pi} \int_0^2 \int_r^2 r^2\, r\, dz\, dr\, d\theta$$

$$= \int_0^{2\pi} d\theta \int_0^2 r^3(2 - r)\, dr$$

$$= 2\pi\left[\tfrac{1}{2}r^4 - \tfrac{1}{5}r^5\right]_0^2 = \frac{16\pi}{5}$$

SPHERICAL COORDINATES

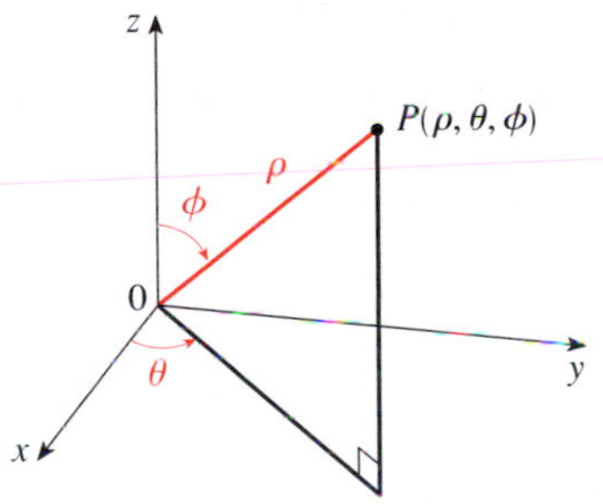

FIGURE 6
Spherical coordinates of P

In the preceding section we defined the spherical coordinates (ρ, θ, ϕ) of a point (see Figure 6) and we demonstrated the following relationships between rectangular coordinates and spherical coordinates:

(3) $\quad x = \rho \sin\phi \cos\theta \qquad y = \rho \sin\phi \sin\theta \qquad z = \rho \cos\phi$

In this coordinate system the analogue of a rectangular box is a **spherical wedge**

$$E = \{(\rho, \theta, \phi) \mid a \leq \rho \leq b,\ \alpha \leq \theta \leq \beta,\ c \leq \phi \leq d\}$$

where $a \geq 0$, $\beta - \alpha \leq 2\pi$, and $d - c \leq \pi$. To integrate over such a region we consider a spherical partition P of E into smaller spherical wedges E_{ijk} by means of spheres $\rho = \rho_i$, half-planes $\theta = \theta_j$, and half-cones $\phi = \phi_k$. The norm of P, $\|P\|$, is the length of the longest diagonal of these wedges. If $\|P\|$ is small, then Figure 7 shows that E_{ijk} is approximately a rectangular box with dimensions $\Delta\rho_i$, $\rho_i \Delta\phi_k$ (arc of a circle with radius ρ_i, angle $\Delta\phi_k$), and $\rho_i \sin\phi_k \Delta\theta_j$ (arc of a circle with radius $\rho_i \sin\phi_k$, angle $\Delta\theta_j$). So an approximation to the volume of E_{ijk} is given by

$$\Delta V_{ijk} \approx \rho_i^2 \sin\phi_k\, \Delta\rho_i\, \Delta\theta_j\, \Delta\phi_k$$

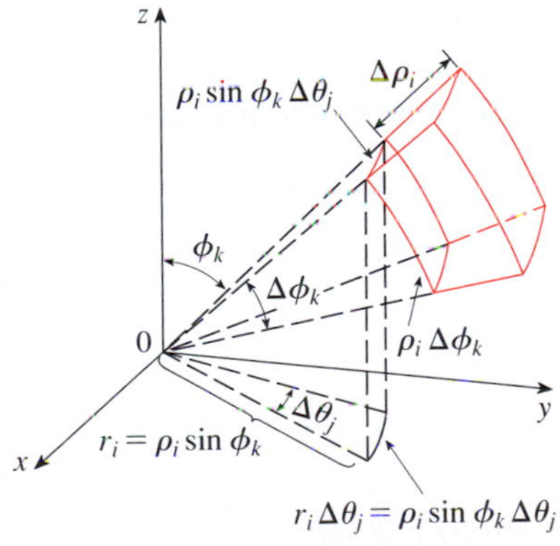

FIGURE 7

In fact, it can be shown, with the aid of the Mean Value Theorem (Exercise 37), that the volume of E_{ijk} is given exactly by

$$\Delta V_{ijk} = \tilde{\rho}_i^2 \sin\tilde{\phi}_k\, \Delta\rho_i\, \Delta\theta_j\, \Delta\phi_k$$

where $(\tilde{\rho}_i, \tilde{\theta}_j, \tilde{\phi}_k)$ is some point in E_{ijk}. Let $(x_{ijk}^*, y_{ijk}^*, z_{ijk}^*)$ be the rectangular coordinates of this point. Although triple integrals were defined using rectangular partitions, it can be shown that spherical partitions could be used instead. Therefore

$$\iiint_E f(x, y, z)\, dV = \lim_{\|P\| \to 0} \sum_{i=1}^{l} \sum_{j=1}^{m} \sum_{k=1}^{n} f(x_{ijk}^*, y_{ijk}^*, z_{ijk}^*)\, \Delta V_{ijk}$$

$$= \lim_{\|P\| \to 0} \sum_{i=1}^{l} \sum_{j=1}^{m} \sum_{k=1}^{n} f(\tilde{\rho}_i \sin\tilde{\phi}_k \cos\tilde{\theta}_j, \tilde{\rho}_i \sin\tilde{\phi}_k \sin\tilde{\theta}_j, \tilde{\rho}_i \cos\tilde{\phi}_k)\, \tilde{\rho}_i^2 \sin\tilde{\phi}_k\, \Delta\rho_i\, \Delta\theta_j\, \Delta\phi_k$$

But this sum is a Riemann sum for the function

$$F(\rho, \theta, \phi) = \rho^2 \sin\phi f(\rho\sin\phi\cos\theta, \rho\sin\phi\sin\theta, \rho\cos\phi)$$

Consequently, we have the following **formula for triple integration in spherical coordinates**:

(4) $$\iiint_E f(x, y, z)\, dV = \int_c^d \int_\alpha^\beta \int_a^b f(\rho\sin\phi\cos\theta, \rho\sin\phi\sin\theta, \rho\cos\phi)\, \rho^2 \sin\phi\, d\rho\, d\theta\, d\phi$$

where E is a spherical wedge given by

$$E = \{(\rho, \theta, \phi) \mid a \leqslant \rho \leqslant b,\ \alpha \leqslant \theta \leqslant \beta,\ c \leqslant \phi \leqslant d\}$$

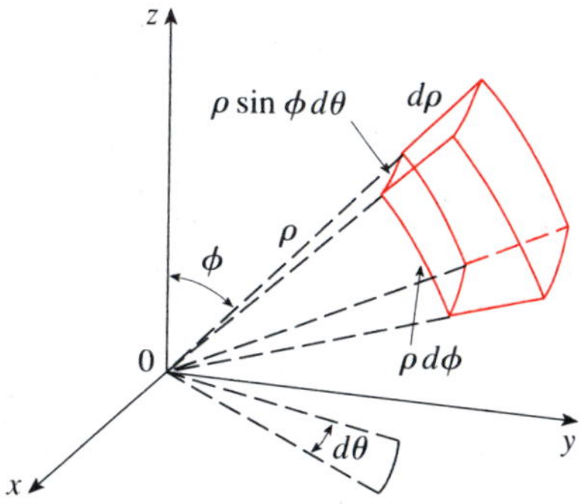

FIGURE 8
Volume element in spherical coordinates: $dV = \rho^2 \sin\phi\, d\rho\, d\theta\, d\phi$

Formula 4 says that we convert a triple integral from rectangular coordinates to spherical coordinates by writing

$$x = \rho\sin\phi\cos\theta \qquad y = \rho\sin\phi\sin\theta \qquad z = \rho\cos\phi$$

using the appropriate limits of integration, and replacing dV by $\rho^2 \sin\phi\, d\rho\, d\theta\, d\phi$. This is illustrated in Figure 8.

This formula can be extended to include more general spherical regions such as

$$E = \{(\rho, \theta, \phi) \mid \alpha \leqslant \theta \leqslant \beta,\ c \leqslant \phi \leqslant d,\ g_1(\theta, \phi) \leqslant \rho \leqslant g_2(\theta, \phi)\}$$

In this case the formula is the same as in (4) except that the limits of integration for ρ are $g_1(\theta, \phi)$ and $g_2(\theta, \phi)$.

Usually spherical coordinates are used in triple integrals when surfaces such as cones and spheres form the boundary of the region of integration.

EXAMPLE 3 Evaluate $\iiint_B e^{(x^2+y^2+z^2)^{3/2}}\, dV$, where B is the unit ball:

$$B = \{(x, y, z) \mid x^2 + y^2 + z^2 \leqslant 1\}$$

SOLUTION Since the boundary of B is a sphere, we use spherical coordinates:

$$B = \{(\rho, \theta, \phi) \mid 0 \leqslant \rho \leqslant 1,\ 0 \leqslant \theta \leqslant 2\pi,\ 0 \leqslant \phi \leqslant \pi\}$$

In addition, spherical coordinates are appropriate because

$$x^2 + y^2 + z^2 = \rho^2$$

Thus, (4) gives

$$\iiint_B e^{(x^2+y^2+z^2)^{3/2}}\, dV = \int_0^\pi \int_0^{2\pi} \int_0^1 e^{(\rho^2)^{3/2}} \rho^2 \sin\phi\, d\rho\, d\theta\, d\phi$$

$$= \int_0^\pi \sin\phi\, d\phi \int_0^{2\pi} d\theta \int_0^1 \rho^2 e^{\rho^3}\, d\rho$$

$$= \left[-\cos\phi\right]_0^\pi (2\pi) \left[\tfrac{1}{3} e^{\rho^3}\right]_0^1 = \frac{4\pi}{3}(e - 1)$$

■

NOTE: It would have been extremely awkward to evaluate the integral in Example 3 without spherical coordinates. In rectangular coordinates the iterated integral would have been

$$\int_{-1}^{1}\int_{-\sqrt{1-x^2}}^{\sqrt{1-x^2}}\int_{-\sqrt{1-x^2-y^2}}^{\sqrt{1-x^2-y^2}} e^{(x^2+y^2+z^2)^{3/2}}\,dz\,dy\,dx$$

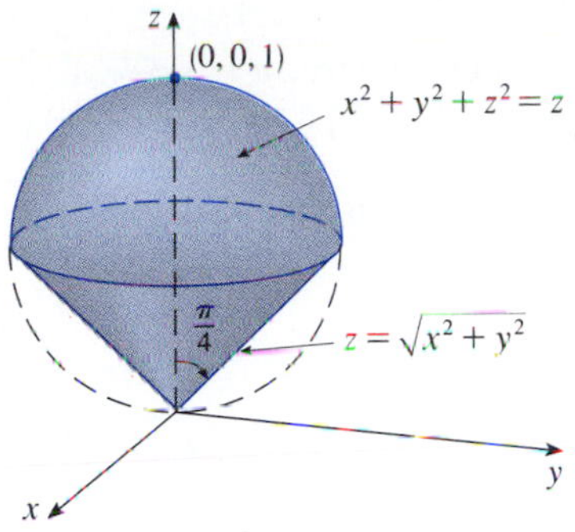

FIGURE 9

Figure 10 gives another look (this time drawn by Maple) at the solid of Example 4.

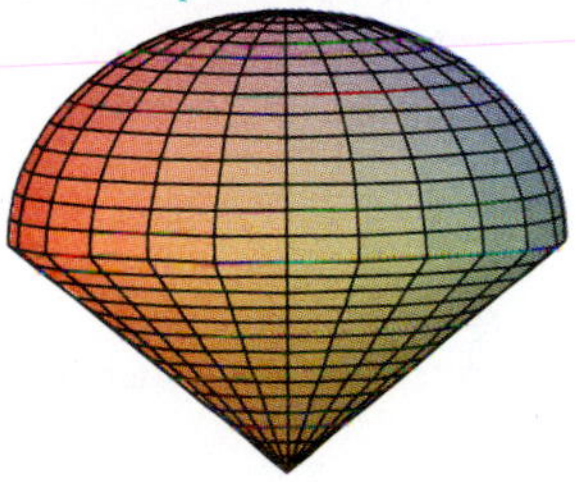

FIGURE 10

EXAMPLE 4 Use spherical coordinates to find the volume of the solid that lies above the cone $z = \sqrt{x^2 + y^2}$ and below the sphere $x^2 + y^2 + z^2 = z$ (see Figure 9).

SOLUTION Notice that the sphere passes through the origin and has center $(0, 0, \frac{1}{2})$. We write the equation of the sphere in spherical coordinates as

$$\rho^2 = \rho\cos\phi \qquad \text{or} \qquad \rho = \cos\phi$$

The cone can be written as

$$\rho\cos\phi = \sqrt{\rho^2\sin^2\phi\cos^2\theta + \rho^2\sin^2\phi\sin^2\theta} = \rho\sin\phi$$

This gives $\sin\phi = \cos\phi$ or $\phi = \pi/4$. Therefore, the description of the solid E in spherical coordinates is

$$E = \{(\rho, \theta, \phi) \mid 0 \leqslant \theta \leqslant 2\pi,\ 0 \leqslant \phi \leqslant \pi/4,\ 0 \leqslant \rho \leqslant \cos\phi\}$$

and its volume is

$$V(E) = \iiint_E dV = \int_0^{2\pi}\int_0^{\pi/4}\int_0^{\cos\phi} \rho^2\sin\phi\,d\rho\,d\phi\,d\theta$$

$$= \int_0^{2\pi} d\theta \int_0^{\pi/4} \sin\phi\left[\frac{\rho^3}{3}\right]_{\rho=0}^{\rho=\cos\phi} d\phi$$

$$= \frac{2\pi}{3}\int_0^{\pi/4} \sin\phi\cos^3\phi\,d\phi = \frac{2\pi}{3}\left[-\frac{\cos^4\phi}{4}\right]_0^{\pi/4} = \frac{\pi}{8}$$ ■

EXERCISES 13.10

1–4 ■ Sketch the solid whose volume is given by the integral and evaluate the integral.

1. $\displaystyle\int_0^{2\pi}\int_0^2\int_0^{4-r^2} r\,dz\,dr\,d\theta$

2. $\displaystyle\int_1^3\int_0^{\pi/2}\int_r^3 r\,dz\,d\theta\,dr$

3. $\displaystyle\int_0^{\pi/2}\int_0^{\pi/2}\int_0^1 \rho^2\sin\phi\,d\rho\,d\theta\,d\phi$

4. $\displaystyle\int_0^{\pi/3}\int_0^{2\pi}\int_0^{\sec\phi} \rho^2\sin\phi\,d\rho\,d\theta\,d\phi$

5–14 ■ Use cylindrical coordinates.

5. Evaluate $\iiint_E (x^2 + y^2)\,dV$, where E is the region bounded by the cylinder $x^2 + y^2 = 4$ and the planes $z = -1$ and $z = 2$.

6. Evaluate $\iiint_E \sqrt{x^2 + y^2}\,dV$, where E is the solid bounded by the paraboloid $z = 9 - x^2 - y^2$ and the xy-plane.

7. Evaluate $\iiint_E y\,dV$, where E is the solid that lies between the cylinders $x^2 + y^2 = 1$ and $x^2 + y^2 = 4$, above the xy-plane, and below the plane $z = x + 2$.

8. Evaluate $\iiint_E xz\,dV$, where E is bounded by the planes $z = 0$, $z = y$, and the cylinder $x^2 + y^2 = 1$ in the half-space $y \geqslant 0$.

9. Evaluate $\iiint_E x^2\,dV$, where E is the solid that lies within the cylinder $x^2 + y^2 = 1$, above the plane $z = 0$, and below the cone $z^2 = 4x^2 + 4y^2$.

10. (a) Find the volume of the solid that the cylinder $r = a\cos\theta$ cuts out of the sphere of radius a centered at the origin.
(b) Illustrate the solid of part (a) by graphing the sphere and the cylinder on the same screen.

11. Find the volume of the region E bounded by the paraboloids $z = x^2 + y^2$ and $z = 36 - 3x^2 - 3y^2$.

12. Find the centroid of the region E of Exercise 11.

13. Find the mass and center of mass of the solid S bounded by the paraboloid $z = 4x^2 + 4y^2$ and the plane $z = a$ $(a > 0)$ if S has constant density K.

14. Find the mass of a ball B given by $x^2 + y^2 + z^2 \leqslant a^2$ if the density at any point is proportional to its distance from the z-axis.

15–28 ■ Use spherical coordinates.

15. Evaluate $\iiint_B (x^2 + y^2 + z^2)\,dV$, where B is the unit ball $x^2 + y^2 + z^2 \leqslant 1$.

16. Evaluate $\iiint_H (x^2 + y^2)\,dV$, where H is the hemispherical region that lies above the xy-plane and below the sphere $x^2 + y^2 + z^2 = 1$.

17. Evaluate $\iiint_E y^2\,dV$, where E is the part of the unit ball $x^2 + y^2 + z^2 \leqslant 1$ that lies in the first octant.

18. Evaluate $\iiint_E xe^{(x^2+y^2+z^2)^2}\,dV$, where E is the solid that lies between the spheres $x^2 + y^2 + z^2 = 1$ and $x^2 + y^2 + z^2 = 4$ in the first octant.

19. Evaluate $\iiint_E \sqrt{x^2 + y^2 + z^2}\,dV$, where E is bounded below by the cone $\phi = \pi/6$ and above by the sphere $\rho = 2$.

20. Evaluate $\iiint_E x^2\,dV$, where E lies between the spheres $\rho = 1$ and $\rho = 3$ and above the cone $\phi = \pi/4$.

21. Find the volume of the solid that lies above the cone $\phi = \pi/3$ and below the sphere $\rho = 4\cos\phi$.

22. Find the centroid of the solid in Exercise 21.

23. Find the mass of a solid hemisphere H of radius a if the density at any point is proportional to its distance from the center of the base.

24. Find the center of mass of the solid H in Exercise 23.

25. Find the moment of inertia of the solid H in Exercise 23 about its axis.

26. Find the centroid of a solid homogeneous hemisphere of radius a.

27. Find the moment of inertia about a diameter of the base of a solid homogeneous hemisphere of radius a.

28. Find the mass and center of mass of a solid hemisphere of radius a if the density at any point is proportional to its distance from the base.

29–32 ■ Use cylindrical or spherical coordinates, whichever seems more appropriate.

29. Find the volume and centroid of the solid E that lies above the cone $z = \sqrt{x^2 + y^2}$ and below the sphere $x^2 + y^2 + z^2 = 1$.

30. Find the volume of the smaller wedge cut from a sphere of radius a by two planes that intersect along a diameter at an angle of $\pi/6$.

CAS **31.** Evaluate $\iiint_E z\,dV$, where E lies above the paraboloid $z = x^2 + y^2$ and below the plane $z = 2y$. Use either the Table of Integrals (in Appendix K) or a computer algebra system to evaluate the integral.

32. (a) Find the volume enclosed by the torus $\rho = \sin\phi$.
(b) Use a computer to draw the torus.

33–34 ■ Evaluate the integral by changing to cylindrical coordinates.

33. $\displaystyle\int_{-1}^{1}\int_{-\sqrt{1-x^2}}^{\sqrt{1-x^2}}\int_{x^2+y^2}^{2-x^2-y^2} (x^2 + y^2)^{3/2}\,dz\,dy\,dx$

34. $\displaystyle\int_0^1\int_0^{\sqrt{1-y^2}}\int_{x^2+y^2}^{\sqrt{x^2+y^2}} xyz\,dz\,dx\,dy$

35–36 ■ Evaluate the integral by changing to spherical coordinates.

35. $\displaystyle\int_{-3}^{3}\int_{-\sqrt{9-x^2}}^{\sqrt{9-x^2}}\int_0^{\sqrt{9-x^2-y^2}} z\sqrt{x^2 + y^2 + z^2}\,dz\,dy\,dx$

36. $\displaystyle\int_0^3\int_0^{\sqrt{9-y^2}}\int_{\sqrt{x^2+y^2}}^{\sqrt{18-x^2-y^2}} (x^2 + y^2 + z^2)\,dz\,dx\,dy$

37. (a) Use cylindrical coordinates to show that the volume of the solid bounded above by the sphere $r^2 + z^2 = a^2$ and below by the cone $z = r\cot\phi_0$ (or $\phi = \phi_0$), where $0 < \phi_0 < \pi/2$, is

$$V = \frac{2\pi a^3}{3}(1 - \cos\phi_0)$$

(b) Deduce that the volume of the spherical wedge given by $\rho_1 \leqslant \rho \leqslant \rho_2$, $\theta_1 \leqslant \theta \leqslant \theta_2$, $\phi_1 \leqslant \phi \leqslant \phi_2$ is

$$\Delta V = \frac{\rho_2^3 - \rho_1^3}{3}(\cos\phi_1 - \cos\phi_2)(\theta_2 - \theta_1)$$

(c) Use the Mean Value Theorem to show that the volume in part (b) can be written as

$$\Delta V = \tilde{\rho}^2 \sin\tilde{\phi}\,\Delta\rho\,\Delta\theta\,\Delta\phi$$

where $\tilde{\rho}$ lies between ρ_1 and ρ_2, $\tilde{\phi}$ lies between ϕ_1 and ϕ_2, $\Delta\rho = \rho_2 - \rho_1$, $\Delta\theta = \theta_2 - \theta_1$, and $\Delta\phi = \phi_2 - \phi_1$.

38. Show that

$$\int_{-\infty}^{\infty}\int_{-\infty}^{\infty}\int_{-\infty}^{\infty} \sqrt{x^2 + y^2 + z^2}\, e^{-(x^2+y^2+z^2)}\,dx\,dy\,dz = 2\pi$$

(The improper triple integral is defined as the limit of a triple integral over a solid sphere as the radius of the sphere increases indefinitely.)

13.11 CHANGE OF VARIABLES IN MULTIPLE INTEGRALS

In one-dimensional calculus we often use a change of variable (a substitution) to simplify an integral. By reversing the roles of x and u, we can write the Substitution Rule (6.5.6) as

(1) $$\int_a^b f(x)\,dx = \int_c^d f(g(u))g'(u)\,du$$

where $x = g(u)$ and $a = g(c)$, $b = g(d)$. Another way of writing Formula 1 is as follows:

(2) $$\int_a^b f(x)\,dx = \int_c^d f(x(u))\,\frac{dx}{du}\,du$$

A change of variables can also be useful in double integrals. We have already seen one example of this: conversion to polar coordinates. The new variables r and θ are related to the old variables x and y by the equations

$$x = r\cos\theta \qquad y = r\sin\theta$$

and the change of variables formula (13.5.2) can be written as

$$\iint_R f(x, y)\,dA = \iint_S f(r\cos\theta, r\sin\theta)\,r\,dA$$

where S is the region in the $r\theta$-plane that corresponds to the region R in the xy-plane.

More generally, we consider a change of variables that is given by a **transformation** T from the uv-plane to the xy-plane:

$$T(u, v) = (x, y)$$

where x and y are related to u and v by the equations

(3) $$x = g(u, v) \qquad y = h(u, v)$$

or, as we sometimes write,

$$x = x(u, v) \qquad y = y(u, v)$$

We usually assume that T is a C^1 **transformation**, which means that g and h have continuous first-order partial derivatives.

A transformation T is really just a function whose domain and range are both subsets of $\mathbb{R}^2$. If $T(u_1, v_1) = (x_1, y_1)$, then the point (x_1, y_1) is called the **image** of the point (u_1, v_1). If no two points have the same image, T is called **one-to-one**. Figure 1 shows

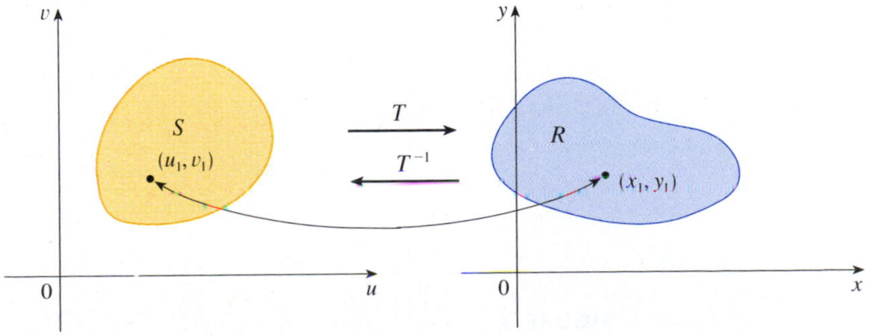

FIGURE 1

the effect of a transformation T on a region S in the uv-plane. T transforms S into a region R in the xy-plane called the **image of S**, consisting of the images of all points in S.

If T is a one-to-one transformation, then it has an **inverse transformation** T^{-1} from the xy-plane to the uv-plane and it may be possible to solve Equations 3 for u and v in terms of x and y:

$$u = G(x, y) \qquad v = H(x, y)$$

EXAMPLE 1 A transformation is defined by the equations

$$x = u^2 - v^2 \qquad y = 2uv$$

Find the image of the square $S = \{(u, v) \mid 0 \leqslant u \leqslant 1, 0 \leqslant v \leqslant 1\}$.

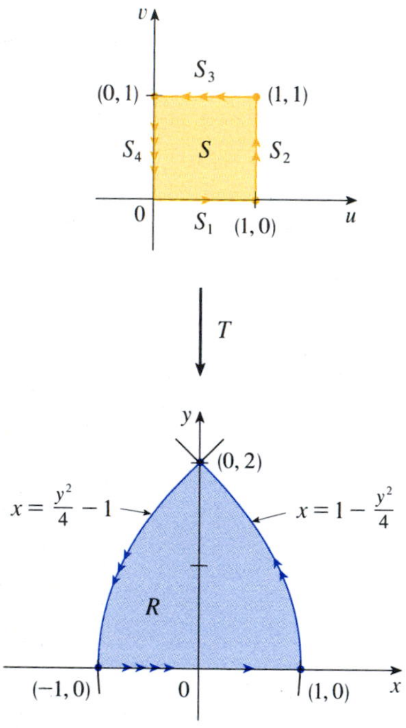

FIGURE 2

SOLUTION First let's find the images of the sides of S. The first side, S_1, is given by $v = 0$ $(0 \leqslant u \leqslant 1)$ (see Figure 2). From the given equations we have $x = u^2$, $y = 0$, and so $0 \leqslant x \leqslant 1$. Thus, S_1 is mapped into the line segment from $(0, 0)$ to $(1, 0)$ in the xy-plane. The second side, S_2, is $u = 1$ $(0 \leqslant v \leqslant 1)$ and, putting $u = 1$ in the given equations, we get

$$x = 1 - v^2 \qquad y = 2v$$

Eliminating v, we obtain

$$x = 1 - \frac{y^2}{4} \qquad 0 \leqslant x \leqslant 1 \tag{4}$$

which is part of a parabola. Similarly, S_3 is given by $v = 1$ $(0 \leqslant u \leqslant 1)$, whose image is the parabolic arc

$$x = \frac{y^2}{4} - 1 \qquad -1 \leqslant x \leqslant 0 \tag{5}$$

Finally, S_4 is given by $u = 0$ $(0 \leqslant v \leqslant 1)$ whose image is $x = -v^2$, $y = 0$, that is, $-1 \leqslant x \leqslant 0$. (Notice that as we move around the square in the counterclockwise direction, we also move around the parabolic region in the counterclockwise direction.) The image of S is the region R (shown in Figure 2) bounded by the x-axis and the parabolas given by Equations 4 and 5. ■

Now let's see how a change of variables affects a double integral. We start with a small rectangle S in the uv-plane whose lower left corner is the point (u_0, v_0) and whose dimensions are Δu and Δv (see Figure 3). Its image is a region R in the xy-plane, one of

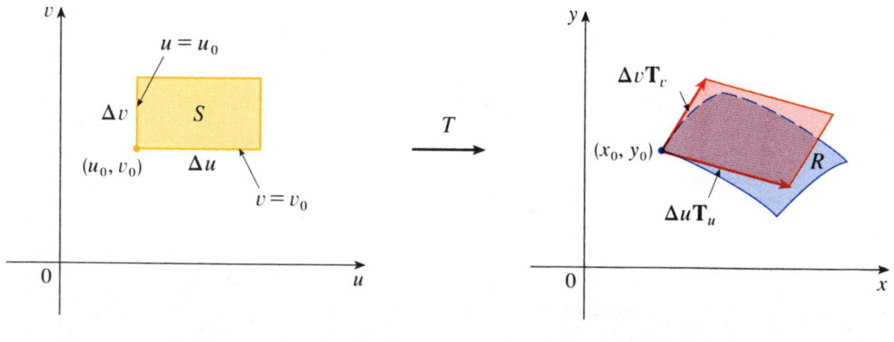

FIGURE 3

whose boundary points is $(x_0, y_0) = T(u_0, v_0)$. The equation of the lower side of S is $v = v_0$, whose image curve is given by $x = g(u, v_0)$, $y = h(u, v_0)$, or, in vector form,

$$g(u, v_0)\,\mathbf{i} + h(u, v_0)\,\mathbf{j}$$

The tangent vector at (x_0, y_0) to this image curve is

$$\mathbf{T}_u = g_u(u_0, v_0)\,\mathbf{i} + h_u(u_0, v_0)\,\mathbf{j} = \frac{\partial x}{\partial u}\,\mathbf{i} + \frac{\partial y}{\partial u}\,\mathbf{j}$$

Similarly, the tangent vector at (x_0, y_0) to the image curve of the left side of S (namely, $u = u_0$) is

$$\mathbf{T}_v = g_v(u_0, v_0)\,\mathbf{i} + h_v(u_0, v_0)\,\mathbf{j} = \frac{\partial x}{\partial v}\,\mathbf{i} + \frac{\partial y}{\partial v}\,\mathbf{j}$$

This means that we can approximate the image region $R = T(S)$ by a parallelogram determined by the vectors $\Delta u\,\mathbf{T}_u$ and $\Delta v\,\mathbf{T}_v$ (see Figure 3). Therefore, we can approximate the area of R by the area of this parallelogram, which, from Section 11.3, is

(6) $$|(\Delta u\,\mathbf{T}_u) \times (\Delta v\,\mathbf{T}_v)| = |\mathbf{T}_u \times \mathbf{T}_v|\,\Delta u\,\Delta v$$

Computing the cross product, we obtain

$$\mathbf{T}_u \times \mathbf{T}_v = \begin{vmatrix} \mathbf{i} & \mathbf{j} & \mathbf{k} \\ \dfrac{\partial x}{\partial u} & \dfrac{\partial y}{\partial u} & 0 \\ \dfrac{\partial x}{\partial v} & \dfrac{\partial y}{\partial v} & 0 \end{vmatrix} = \begin{vmatrix} \dfrac{\partial x}{\partial u} & \dfrac{\partial y}{\partial u} \\ \dfrac{\partial x}{\partial v} & \dfrac{\partial y}{\partial v} \end{vmatrix} \mathbf{k} = \begin{vmatrix} \dfrac{\partial x}{\partial u} & \dfrac{\partial x}{\partial v} \\ \dfrac{\partial y}{\partial u} & \dfrac{\partial y}{\partial v} \end{vmatrix} \mathbf{k}$$

The Jacobian is named after the German mathematician Carl Gustav Jacob Jacobi (1804–1851). Although the French mathematician Cauchy first used these special determinants involving partial derivatives, Jacobi developed them into a method for evaluating multiple integrals.

The determinant that arises in this calculation is called the *Jacobian* of the transformation and is given a special notation.

(7) DEFINITION The **Jacobian** of the transformation T given by $x = g(u, v)$ and $y = h(u, v)$ is

$$\frac{\partial(x, y)}{\partial(u, v)} = \begin{vmatrix} \dfrac{\partial x}{\partial u} & \dfrac{\partial x}{\partial v} \\ \dfrac{\partial y}{\partial u} & \dfrac{\partial y}{\partial v} \end{vmatrix} = \frac{\partial x}{\partial u}\frac{\partial y}{\partial v} - \frac{\partial x}{\partial v}\frac{\partial y}{\partial u}$$

With this notation we can use Equation 6 to give an approximation to the area ΔA of R:

(8) $$\Delta A \approx \left|\frac{\partial(x, y)}{\partial(u, v)}\right| \Delta u\,\Delta v$$

where the Jacobian is evaluated at (u_0, v_0).

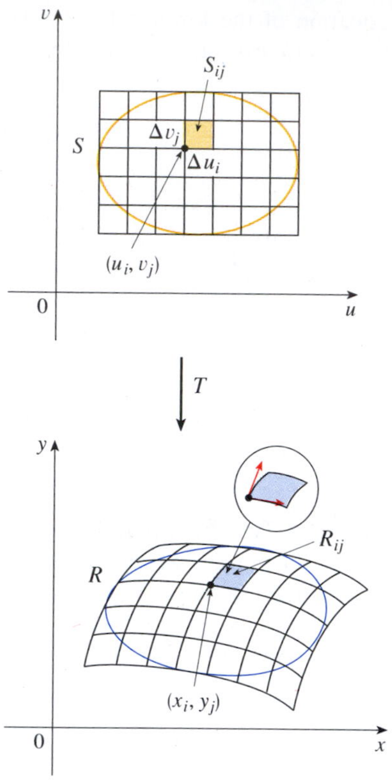

FIGURE 4

Next we partition a region S in the uv-plane into rectangles S_{ij} and call their images in the xy-plane R_{ij} (see Figure 4). Applying the approximation (8) to each R_{ij}, we approximate the double integral of f over R as follows:

$$\iint_R f(x, y)\, dA \approx \sum_{i=1}^{m} \sum_{j=1}^{n} f(x_i, y_j)\, \Delta A_{ij}$$

$$\approx \sum_{i=1}^{m} \sum_{j=1}^{n} f(g(u_i, v_j),\, h(u_i, v_j)) \left| \frac{\partial(x, y)}{\partial(u, v)} \right| \Delta u_i\, \Delta v_j$$

where the Jacobian is evaluated at (u_i, v_j). Notice that this double sum is a Riemann sum for the integral

$$\iint_S f(g(u, v),\, h(u, v)) \left| \frac{\partial(x, y)}{\partial(u, v)} \right| du\, dv$$

The foregoing argument suggests that the following theorem is true. (A full proof is given in books on advanced calculus.)

(9) CHANGE OF VARIABLES IN A DOUBLE INTEGRAL Suppose that T is a one-to-one C^1 transformation whose Jacobian is nonzero and that maps a region S in the uv-plane onto a region R in the xy-plane. Suppose that f is continuous on R and that R and S are type I or type II plane regions. Then

$$\iint_R f(x, y)\, dx\, dy = \iint_S f(x(u, v), y(u, v)) \left| \frac{\partial(x, y)}{\partial(u, v)} \right| du\, dv$$

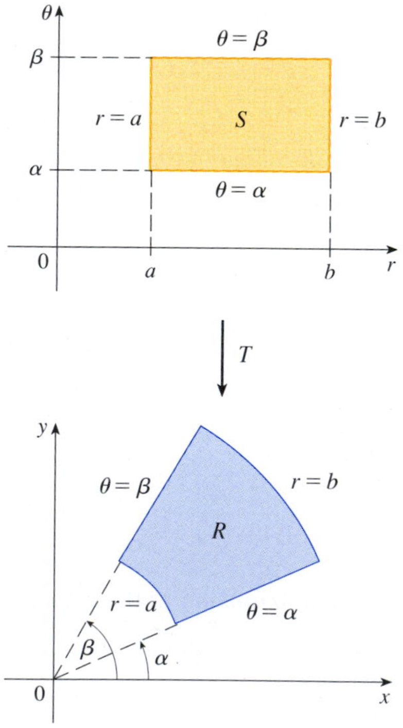

FIGURE 5
The polar coordinate transformation

Theorem 9 says that we change from an integral in x and y to an integral in u and v by expressing x and y in terms of u and v and writing

$$dA = \left| \frac{\partial(x, y)}{\partial(u, v)} \right| du\, dv$$

Notice the similarity between Theorem 9 and the one-dimensional formula in Equation 2. Instead of the derivative dx/du, we have the absolute value of the Jacobian: $|\partial(x, y)/\partial(u, v)|$.

As a first illustration of Theorem 9, let's show that the formula for integration in polar coordinates is just a special case. Here the transformation T from the $r\theta$-plane to the xy-plane is given by

$$x = g(r, \theta) = r\cos\theta \qquad y = h(r, \theta) = r\sin\theta$$

and the geometry of the transformation is shown in Figure 5. T maps an ordinary rectangle in the $r\theta$-plane to a polar rectangle in the xy-plane. The Jacobian of T is

$$\frac{\partial(x, y)}{\partial(r, \theta)} = \begin{vmatrix} \dfrac{\partial x}{\partial r} & \dfrac{\partial x}{\partial \theta} \\ \dfrac{\partial y}{\partial r} & \dfrac{\partial y}{\partial \theta} \end{vmatrix} = \begin{vmatrix} \cos\theta & -r\sin\theta \\ \sin\theta & r\cos\theta \end{vmatrix} = r\cos^2\theta + r\sin^2\theta = r > 0$$

Thus, Theorem 9 gives

$$\iint_R f(x, y)\,dx\,dy = \iint_S f(r\cos\theta, r\sin\theta)\left|\frac{\partial(x, y)}{\partial(r, \theta)}\right| dr\,d\theta$$

$$= \int_\alpha^\beta \int_a^b f(r\cos\theta, r\sin\theta)\,r\,dr\,d\theta$$

which is the same as (13.5.2).

EXAMPLE 2 Use the change of variables $x = u^2 - v^2$, $y = 2uv$ to evaluate the integral $\iint_R y\,dA$, where R is the region bounded by the x-axis and the parabolas $y^2 = 4 - 4x$ and $y^2 = 4 + 4x$.

SOLUTION The region R is pictured in Figure 2. In Example 1 we discovered that $T(S) = R$, where S is the square $[0, 1] \times [0, 1]$. Indeed, the reason for making the change of variables to evaluate the integral is that S is a much simpler region than R. First we need to compute the Jacobian:

$$\frac{\partial(x, y)}{\partial(u, v)} = \begin{vmatrix} \dfrac{\partial x}{\partial u} & \dfrac{\partial x}{\partial v} \\ \dfrac{\partial y}{\partial u} & \dfrac{\partial y}{\partial v} \end{vmatrix} = \begin{vmatrix} 2u & -2v \\ 2v & 2u \end{vmatrix} = 4u^2 + 4v^2 > 0$$

Therefore, by Theorem 9,

$$\iint_R y\,dA = \iint_S 2uv\left|\frac{\partial(x, y)}{\partial(u, v)}\right| dA$$

$$= \int_0^1 \int_0^1 (2uv)4(u^2 + v^2)\,du\,dv$$

$$= 8\int_0^1 \int_0^1 (u^3v + uv^3)\,du\,dv = 8\int_0^1 \left[\frac{u^4}{4}v + \frac{u^2}{2}v^3\right]_{u=0}^{u=1} dv$$

$$= \int_0^1 (2v + 4v^3)\,dv = \left[v^2 + v^4\right]_0^1 = 2$$

■

NOTE: Example 2 was not a very difficult problem to solve because we were given a suitable change of variables. If we are not supplied with a transformation, then the first step is to think of an appropriate change of variables. If $f(x, y)$ is difficult to integrate, then the form of $f(x, y)$ may suggest a transformation. If the region of integration R is awkward, then the transformation should be chosen so that the corresponding region S in the uv-plane has a convenient description.

EXAMPLE 3 Evaluate the integral $\iint_R e^{(x+y)/(x-y)}\,dA$, where R is the trapezoidal region with vertices $(1, 0)$, $(2, 0)$, $(0, -2)$, and $(0, -1)$.

SOLUTION Since it is not easy to integrate $e^{(x+y)/(x-y)}$, we make a change of variables suggested by the form of this function:

(10) $$u = x + y \qquad v = x - y$$

These equations define a transformation T^{-1} from the xy-plane to the uv-plane. Theorem 9 talks about a transformation T from the uv-plane to the xy-plane. It is obtained by solving Equations 10 for x and y:

(11) $$x = \tfrac{1}{2}(u + v) \qquad y = \tfrac{1}{2}(u - v)$$

The Jacobian of T is

$$\frac{\partial(x, y)}{\partial(u, v)} = \begin{vmatrix} \dfrac{\partial x}{\partial u} & \dfrac{\partial x}{\partial v} \\ \dfrac{\partial y}{\partial u} & \dfrac{\partial y}{\partial v} \end{vmatrix} = \begin{vmatrix} \frac{1}{2} & \frac{1}{2} \\ \frac{1}{2} & -\frac{1}{2} \end{vmatrix} = -\tfrac{1}{2}$$

To find the region S in the uv-plane corresponding to R, we note that the sides of R lie on the lines

$$y = 0 \qquad x - y = 2 \qquad x = 0 \qquad x - y = 1$$

and, from either Equations 10 or Equations 11, the image lines in the uv-plane are

$$u = v \qquad v = 2 \qquad u = -v \qquad v = 1$$

Thus, the region S is the trapezoidal region with vertices $(1, 1)$, $(2, 2)$, $(-2, 2)$, and $(-1, 1)$ shown in Figure 6. Since

$$S = \{(u, v) \mid 1 \leqslant v \leqslant 2,\ -v \leqslant u \leqslant v\}$$

Theorem 9 gives

$$\iint_R e^{(x+y)/(x-y)}\, dA = \iint_S e^{u/v} \left| \frac{\partial(x, y)}{\partial(u, v)} \right| du\, dv$$

$$= \int_1^2 \int_{-v}^{v} e^{u/v}\left(\tfrac{1}{2}\right) du\, dv = \tfrac{1}{2} \int_1^2 \left[ve^{u/v}\right]_{u=-v}^{u=v} dv$$

$$= \tfrac{1}{2} \int_1^2 (e - e^{-1})v\, dv = \tfrac{3}{4}(e - e^{-1})$$

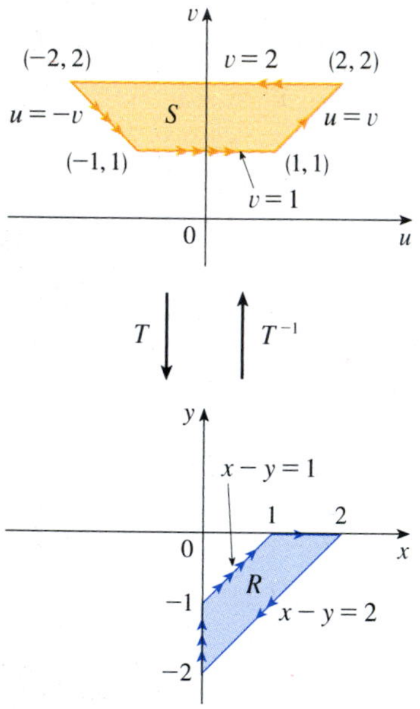

FIGURE 6

There is a similar change of variables formula for triple integrals. Let T be a transformation that maps a region S in uvw-space onto a region R in xyz-space by means of the equations

$$x = g(u, v, w) \qquad y = h(u, v, w) \qquad z = k(u, v, w)$$

The **Jacobian** of T is the following 3×3 determinant:

(12) $$\frac{\partial(x, y, z)}{\partial(u, v, w)} = \begin{vmatrix} \dfrac{\partial x}{\partial u} & \dfrac{\partial x}{\partial v} & \dfrac{\partial x}{\partial w} \\ \dfrac{\partial y}{\partial u} & \dfrac{\partial y}{\partial v} & \dfrac{\partial y}{\partial w} \\ \dfrac{\partial z}{\partial u} & \dfrac{\partial z}{\partial v} & \dfrac{\partial z}{\partial w} \end{vmatrix}$$

Under hypotheses similar to those in Theorem 9, we have the following formula for triple integrals:

(13)
$$\iiint_R f(x, y, z)\, dx\, dy\, dz = \iiint_S f(x(u, v, w), y(u, v, w), z(u, v, w)) \left| \frac{\partial(x, y, z)}{\partial(u, v, w)} \right| du\, dv\, dw$$

EXAMPLE 4 Use Formula 13 to derive the formula for triple integration in spherical coordinates.

SOLUTION Here the change of variables is given by

$$x = \rho \sin\phi \cos\theta \qquad y = \rho \sin\phi \sin\theta \qquad z = \rho\cos\phi$$

We compute the Jacobian as follows:

$$\frac{\partial(x, y, z)}{\partial(\rho, \theta, \phi)} = \begin{vmatrix} \sin\phi\cos\theta & -\rho\sin\phi\sin\theta & \rho\cos\phi\cos\theta \\ \sin\phi\sin\theta & \rho\sin\phi\cos\theta & \rho\cos\phi\sin\theta \\ \cos\phi & 0 & -\rho\sin\phi \end{vmatrix}$$

$$= \cos\phi \begin{vmatrix} -\rho\sin\phi\sin\theta & \rho\cos\phi\cos\theta \\ \rho\sin\phi\cos\theta & \rho\cos\phi\sin\theta \end{vmatrix} - \rho\sin\phi \begin{vmatrix} \sin\phi\cos\theta & -\rho\sin\phi\sin\theta \\ \sin\phi\sin\theta & \rho\sin\phi\cos\theta \end{vmatrix}$$

$$= \cos\phi\,(-\rho^2\sin\phi\cos\phi\sin^2\theta - \rho^2\sin\phi\cos\phi\cos^2\theta)$$

$$- \rho\sin\phi\,(\rho\sin^2\phi\cos^2\theta + \rho\sin^2\phi\sin^2\theta)$$

$$= -\rho^2\sin\phi\cos^2\phi - \rho^2\sin\phi\sin^2\phi = -\rho^2\sin\phi$$

Since $0 \leqslant \phi \leqslant \pi$, we have $\sin\phi \geqslant 0$. Therefore

$$\left| \frac{\partial(x, y, z)}{\partial(\rho, \theta, \phi)} \right| = |-\rho^2\sin\phi| = \rho^2\sin\phi$$

and Formula 13 gives

$$\iiint_R f(x, y, z)\, dx\, dy\, dz = \iiint_S f(\rho\sin\phi\cos\theta, \rho\sin\phi\sin\theta, \rho\cos\phi)\rho^2\sin\phi\, d\rho\, d\theta\, d\phi$$

which is equivalent to Formula 13.10.4. ■

EXERCISES 13.11

1–6 ■ Find the Jacobian of the transformation.

1. $x = u - 2v, \quad y = 2u - v$

2. $x = u - v^2, \quad y = u + v^2$

3. $x = e^{2u}\cos v, \quad y = e^{2u}\sin v$

4. $x = se^t, \quad y = se^{-t}$

5. $x = u + v + w, \quad y = u + v - w, \quad z = u - v + w$

6. $x = 2u, \quad y = 3v^2, \quad z = 4w^3$

7–10 ■ Find the image of the set S under the given transformation.

7. $S = \{(u, v) \mid 0 \leqslant u \leqslant 2, 0 \leqslant v \leqslant 1\}$;
$x = u - 2v, y = 2u - v$

8. $S = \{(u, v) \mid 0 \leqslant u \leqslant 1, u \leqslant v \leqslant 1\}$; $x = u^2, y = v$

9. S is the triangular region with vertices $(0, 0)$, $(1, 0)$, $(0, 1)$; $x = 4u + 3v$, $y = 4v$.

10. S is the disk given by $u^2 + v^2 \leqslant 1$; $x = au$, $y = bv$.

11–18 ■ Use the given transformation to evaluate the integral.

11. $\iint_R (3x + 4y)\, dA$, where R is the region bounded by the lines $y = x$, $y = x - 2$, $y = -2x$, and $y = 3 - 2x$; $x = \frac{1}{3}(u + v)$, $y = \frac{1}{3}(v - 2u)$

12. $\iint_R (x + y)\, dA$, where R is the square with vertices $(0, 0)$, $(2, 3)$, $(5, 1)$, and $(3, -2)$; $x = 2u + 3v$, $y = 3u - 2v$

13. $\iint_R x^2\, dA$, where R is the region bounded by the ellipse $9x^2 + 4y^2 = 36$; $x = 2u$, $y = 3v$

14. $\iint_R (x^2 - xy + y^2)\, dA$, where R is the region bounded by the ellipse $x^2 - xy + y^2 = 2$; $x = \sqrt{2}\, u - \sqrt{2/3}\, v$, $y = \sqrt{2}\, u + \sqrt{2/3}\, v$

15. $\iint_R xy\, dA$, where R is the region in the first quadrant bounded by the lines $y = x$ and $y = 3x$ and the hyperbolas $xy = 1$, $xy = 3$; $x = u/v$, $y = v$

16. $\iint_R y^2\, dA$, where R is the region bounded by the curves $xy = 1$, $xy = 2$, $xy^2 = 1$, $xy^2 = 2$; $u = xy$, $v = xy^2$. Illustrate by using a graphing calculator or computer to draw R.

17. $\iiint_E dV$, where E is the solid enclosed by the ellipsoid $x^2/a^2 + y^2/b^2 + z^2/c^2 = 1$; $x = au$, $y = bv$, $z = cw$ (This integral gives the volume of the ellipsoid.)

18. $\iiint_E x^2y\, dV$, where E is the solid of Exercise 17

19–23 ■ Evaluate the integral by making an appropriate change of variables.

19. $\iint_R xy\, dA$, where R is the region bounded by the lines $2x - y = 1$, $2x - y = -3$, $3x + y = 1$, and $3x + y = -2$

20. $\displaystyle\iint_R \frac{x + 2y}{\cos(x - y)}\, dA$, where R is the parallelogram bounded by the lines $y = x$, $y = x - 1$, $x + 2y = 0$, and $x + 2y = 2$

21. $\displaystyle\iint_R \cos\left(\frac{y - x}{y + x}\right) dA$, where R is the trapezoidal region with vertices $(1, 0)$, $(2, 0)$, $(0, 2)$, and $(0, 1)$

22. $\iint_R \sin(9x^2 + 4y^2)\, dA$, where R is the region in the first quadrant bounded by the ellipse $9x^2 + 4y^2 = 1$

23. $\iint_R e^{x+y}\, dA$, where R is given by the inequality $|x| + |y| \leqslant 1$

24. Let f be continuous on $[0, 1]$ and let R be the triangular region with vertices $(0, 0)$, $(1, 0)$, and $(0, 1)$. Show that

$$\iint_R f(x + y)\, dA = \int_0^1 u f(u)\, du$$

13 REVIEW

KEY TOPICS ■ Define, state, or discuss the following.

1. Double Riemann sum
2. Double integral over a rectangle
3. Iterated integral
4. Fubini's Theorem
5. Double integral over a general region
6. Regions of type I and type II
7. Evaluating double integrals
8. Volume
9. Properties of double integrals
10. Polar coordinates
11. Conversion from rectangular to polar coordinates in a double integral
12. Electric charge
13. Mass and center of mass of a lamina
14. Moment of inertia
15. Radius of gyration
16. Triple integral
17. Evaluation of triple integrals
18. Mass, center of mass, and moments of inertia of a solid region
19. Surface area
20. Cylindrical and spherical coordinates
21. Triple integrals in cylindrical coordinates

22. Triple integrals in spherical coordinates

23. Jacobian of a transformation

24. Formula for change of variables in a double integral

25. Formula for change of variables in a triple integral

EXERCISES

1–6 ■ Determine whether the statement is true or false.

1. $\int_{-1}^{2}\int_{0}^{6} x^2\sin(x-y)\,dx\,dy = \int_{0}^{6}\int_{-1}^{2} x^2\sin(x-y)\,dy\,dx$

2. $\int_{-1}^{1}\int_{0}^{1} e^{x^2+y^2}\sin y\,dx\,dy = 0$

3. If D is the disk given by $x^2+y^2 \leqslant 4$, then
$$\iint_D \sqrt{4-x^2-y^2}\,dA = \frac{16\pi}{3}$$

4. $\int_{1}^{4}\int_{0}^{1} (x^2+\sqrt{y})\sin(x^2y^2)\,dx\,dy \leqslant 9$

5. The integral
$$\int_0^{2\pi}\int_0^2\int_r^2 dz\,dr\,d\theta$$
represents the volume enclosed by the cone $z=\sqrt{x^2+y^2}$ and the plane $z=2$.

6. The integral $\iiint_E kr^3\,dz\,dr\,d\theta$ represents the moment of inertia about the z-axis of a solid E with constant density k.

7–12 ■ Calculate the iterated integral.

7. $\int_{-2}^{2}\int_{0}^{4} (4x^3+3xy^2)\,dx\,dy$

8. $\int_{0}^{\pi}\int_{0}^{1} x\cos(xy)\,dy\,dx$

9. $\int_{1}^{2}\int_{0}^{x^2} \frac{1}{x+y}\,dy\,dx$

10. $\int_{0}^{1}\int_{0}^{y^2} e^{y^3}\,dx\,dy$

11. $\int_{0}^{1}\int_{0}^{x^2}\int_{0}^{y} y^2z\,dz\,dy\,dx$

12. $\int_{0}^{1}\int_{\sqrt{y}}^{1}\int_{0}^{y} xy\,dz\,dx\,dy$

13. Describe the region whose area is given by the integral
$$\int_0^{\pi}\int_1^{1+\sin\theta} r\,dr\,d\theta$$

14. Describe the solid whose volume is given by the integral
$$\int_0^{2\pi}\int_0^{\pi/6}\int_1^3 \rho^2\sin\phi\,d\rho\,d\phi\,d\theta$$
and evaluate the integral.

15–16 ■ Calculate the iterated integral by first reversing the order of integration.

15. $\int_{0}^{1}\int_{x}^{1} e^{x/y}\,dy\,dx$

16. $\int_{0}^{1}\int_{y^2}^{1} y\sin(x^2)\,dx\,dy$

17–30 ■ Calculate the value of the multiple integral.

17. $\iint_R \frac{1}{(x-y)^2}\,dA$, where $R=\{(x,y)\,|\,0\leqslant x\leqslant 1,\ 2\leqslant y\leqslant 4\}$

18. $\iint_D x^3\,dA$, where $D=\{(x,y)\,|\,-1\leqslant x\leqslant 1,\ x^2-1\leqslant y\leqslant x+1\}$

19. $\iint_D xy\,dA$, where D is bounded by $y^2=x^3$ and $y=x$

20. $\iint_D xe^y\,dA$, where D is bounded by $y=0$, $y=x^2$, $x=1$

21. $\iint_D (xy+2x+3y)\,dA$, where D is the region in the first quadrant bounded by $x=1-y^2$, $y=0$, $x=0$

22. $\iint_D y\,dA$, where D is the region in the first quadrant that lies above the hyperbola $xy=1$ and the line $y=x$ and below the line $y=2$

23. $\iint_D (x^2+y^2)^{3/2}\,dA$, where D is the region in the first quadrant bounded by the lines $y=0$ and $y=\sqrt{3}\,x$ and the circle $x^2+y^2=9$

24. $\iint_D \sqrt{x^2+y^2}\,dA$, where D is the closed disk with radius 1 and center $(0,1)$

25. $\iiint_E x^2z\,dV$, where $E=\{(x,y,z)\,|\,0\leqslant x\leqslant 2,\ 0\leqslant y\leqslant 2x,\ 0\leqslant z\leqslant x\}$

26. $\iiint_T y\,dV$, where T is the tetrahedron bounded by the planes $x=0$, $y=0$, $z=0$, and $2x+y+z=2$

27. $\iiint_E y^2z^2\,dV$, where E is bounded by the paraboloid $x=1-y^2-z^2$ and the plane $x=0$

28. $\iiint_E z\,dV$, where E is bounded by the planes $y=0$, $z=0$, $x+y=2$ and the cylinder $y^2+z^2=1$ in the first octant

29. $\iiint_E yz\,dV$, where E lies above the plane $z=0$, below the plane $z=y$, and inside the cylinder $x^2+y^2=4$

30. $\iiint_H z^3\sqrt{x^2+y^2+z^2}\,dV$, where H is the solid hemisphere with center the origin, radius 1, that lies above the xy-plane

31–36 ■ Find the volume of the given solid.

31. Under the paraboloid $z=x^2+4y^2$ and above the rectangle $R=[0,2]\times[1,4]$

32. Under the surface $z=x^2y$ and above the triangle in the xy-plane with vertices $(1,0)$, $(2,1)$, and $(4,0)$

33. The solid tetrahedron with vertices $(0,0,0)$, $(0,0,1)$, $(0,2,0)$, and $(2,2,0)$

34. Bounded by the cylinder $x^2 + y^2 = 4$ and the planes $z = 0$ and $y + z = 3$

35. One of the wedges cut from the cylinder $x^2 + 9y^2 = a^2$ by the planes $z = 0$ and $z = mx$

36. Above the paraboloid $z = x^2 + y^2$ and below the half-cone $z = \sqrt{x^2 + y^2}$

37. Find the mass and center of mass of a lamina that occupies the region D bounded by the parabola $x = 1 - y^2$ and the coordinate axes in the first quadrant if the density function is $\rho(x, y) = y$.

38. Find the moments of inertia and radii of gyration about the x- and y-axes for the lamina of Exercise 37.

39. A lamina occupies the part of the disk $x^2 + y^2 \leqslant a^2$ that lies in the first quadrant. Find the centroid of the lamina.

40. Find the center of mass of the lamina in Exercise 39 if the density function is $\rho(x, y) = xy^2$.

41. Find the centroid of a right circular cone with height h and base radius a. (Place the cone so that its base is in the xy-plane with center the origin and its axis along the positive z-axis.)

42. Find the moment of inertia of the cone of Exercise 41 about its axis (the z-axis).

43. Find the area of the part of the plane $3x + 2y + 6z = 6$ in the first octant.

44. Find the area of the part of the cone $z^2 = a^2(x^2 + y^2)$ between the planes $z = 1$ and $z = 2$.

45–52 ■ Sketch the polar curve.

45. $r = 1 + 3\cos\theta$

46. $r = 3 - \sin\theta$

47. $r^2 = \sec 2\theta$

48. $r = \tan\theta$

49. $r = 2\cos^2(\theta/2)$

50. $r = 2\cos(\theta/2)$

51. $r = \dfrac{1}{1 + \cos\theta}$

52. $r = \dfrac{5}{1 - 3\sin\theta}$

53–54 ■ Find a polar equation for the curve represented by the given Cartesian equation.

53. $x^2 + y^2 = 4x$

54. $x + y^2 = 0$

55. The cylindrical coordinates of a point are $(2, \pi/6, 2)$. Find the rectangular and spherical coordinates of the point.

56. The rectangular coordinates of a point are $(2, 2, -1)$. Find the cylindrical and spherical coordinates of the point.

57. The spherical coordinates of a point are $(4, \pi/3, \pi/6)$. Find the rectangular and cylindrical coordinates of the point.

58–61 ■ Identify the surface whose equation is given.

58. $\phi = \pi/4$

59. $\theta = \pi/4$

60. $r = \cos\theta$

61. $\rho = 3\sec\phi$

62–64 ■ Write the equation in cylindrical coordinates and in spherical coordinates.

62. $x^2 + y^2 = 4$

63. $x^2 + y^2 + z^2 = 4$

64. $x^2 + y^2 + z^2 = 2x$

65. The parabola $z = 4y^2$, $x = 0$ is rotated about the z-axis. Write an equation of the resulting surface in cylindrical coordinates.

66. Sketch the solid consisting of all points with spherical coordinates (ρ, θ, ϕ) such that $0 \leqslant \theta \leqslant \pi/2$, $0 \leqslant \phi \leqslant \pi/6$, and $0 \leqslant \rho \leqslant 2\cos\phi$.

67. Use polar coordinates to evaluate

$$\int_0^{\sqrt{2}} \int_y^{\sqrt{4-y^2}} \frac{1}{1 + x^2 + y^2}\, dx\, dy$$

68. Use spherical coordinates to evaluate

$$\int_0^1 \int_0^{\sqrt{1-x^2}} \int_0^{\sqrt{1-x^2-y^2}} (x^2 + y^2 + z^2)^2\, dz\, dy\, dx$$

69. If D is the region bounded by the curves $y = 1 - x^2$ and $y = e^x$, find the approximate value of the integral $\iint_D y^2\, dA$. (Use a graphing device to estimate the points of intersection of the curves.)

CAS **70.** Graph the surface $z = x\sin y$, $-3 \leqslant x \leqslant 3$, $-\pi \leqslant y \leqslant \pi$, and find its surface area correct to four decimal places.

CAS **71.** Find the center of mass of the solid tetrahedron with vertices $(0, 0, 0)$, $(1, 0, 0)$, $(0, 2, 0)$, $(0, 0, 3)$ and density function $\rho(x, y, z) = x^2 + y^2 + z^2$.

CAS **72.** Find the exact value of the triple integral:

$$\int_0^1 \int_0^{e^y} \int_x^{y^2} (x^2 y + z\sin 4y)\, dz\, dx\, dy$$

73. Rewrite the integral

$$\int_{-1}^1 \int_{x^2}^1 \int_0^{1-y} f(x, y, z)\, dz\, dy\, dx$$

as an iterated integral in the order $dx\, dy\, dz$.

74. Give five other iterated integrals that are equal to

$$\int_0^2 \int_0^{y^3} \int_0^{y^2} f(x, y, z)\, dz\, dx\, dy$$

75. Use the transformation $u = x - y$, $v = x + y$ to evaluate $\iint_R (x - y)/(x + y)\, dA$, where R is the square with vertices $(0, 2)$, $(1, 1)$, $(2, 2)$, and $(1, 3)$.

76. Use the transformation $x = u^2$, $y = v^2$, $z = w^2$ to find the volume of the region bounded by the surface $\sqrt{x} + \sqrt{y} + \sqrt{z} = 1$ and the coordinate planes.

77. Use the change of variables formula and an appropriate transformation to evaluate $\iint_R xy\, dA$, where R is the square with vertices $(0, 0)$, $(1, 1)$, $(2, 0)$, and $(1, -1)$.

78. The **Mean Value Theorem for double integrals** says that if f is a continuous function on a plane region D that is of type I or II, then there exists a point (x_0, y_0) in D such that

$$\iint_D f(x, y)\, dA = f(x_0, y_0)\, A(D)$$

Use the Extreme Value Theorem (12.7.8) and Property 13.3.11 of integrals to prove this theorem. (Use the proof of the single-variable version in Exercise 19 of Section 7.5 as a guide.)

79. Suppose that f is continuous on a domain D in the plane and let (a, b) be an interior point of D. Let D_r be the closed disk with center (a, b) and radius r. Use the Mean Value Theorem for double integrals (see Exercise 78) to show that

$$\lim_{r \to 0} \frac{1}{\pi r^2} \iint_{D_r} f(x, y)\, dA = f(a, b)$$

80. (a) Evaluate

$$\iint_D \frac{1}{(x^2 + y^2)^{n/2}}\, dA$$

where n is an integer and D is the region bounded by the circles with center the origin and radii r and R, $0 < r < R$.

(b) For what values of n does the integral in part (a) have a limit as $r \to 0^+$?

(c) Find

$$\iiint_E \frac{1}{(x^2 + y^2 + z^2)^{n/2}}\, dV$$

where E is the region bounded by the spheres with center the origin and radii r and R, $0 < r < R$.

(d) For what values of n does the integral in part (c) have a limit as $r \to 0^+$?

APPLICATIONS PLUS

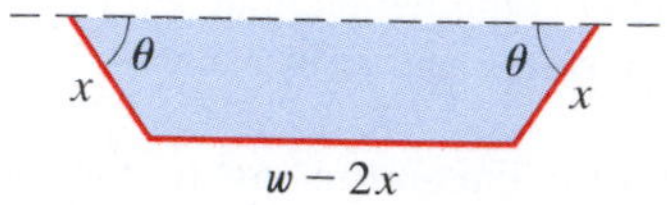

FIGURE FOR PROBLEM 1

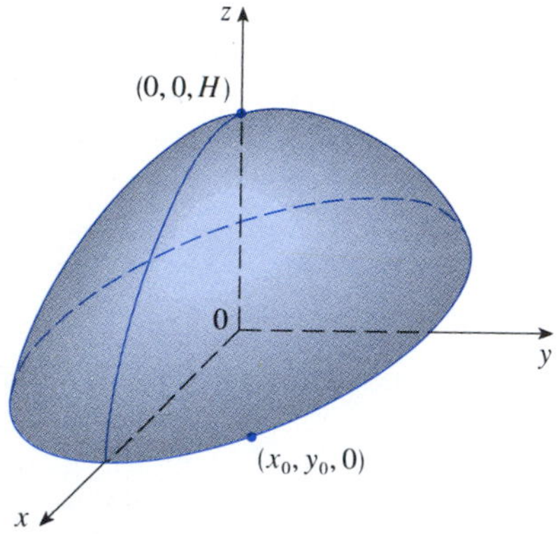

FIGURE FOR PROBLEM 2

1. A long piece of galvanized sheet metal w inches wide is to be bent into a symmetric form with three straight sides to make a rain gutter. The cross-section is shown in the figure.

(a) Determine the dimensions that allow the maximum possible flow; that is, find the dimensions that give the maximum possible cross-sectional area.

(b) Would it be better to bend the metal into a gutter with a semicircular cross-section than a three-sided cross-section?

2. A "sugarloaf" mountain has the equation

$$z = f(x, y) = H - \alpha(x^2 + \mu y^2) \qquad \alpha > 0,\ \mu > 1$$

Starting at a point $(x_0, y_0, 0)$ at the mountain base, a climber wants to ascend the mountain by means of the steepest ascent curve.

(a) Identify the level curves of the surface $z = f(x, y)$ and sketch a few members of the family.

(b) Calculate the gradient vector for the surface at an arbitrary point (x, y) in the domain of f and determine the differential equation for the steepest ascent curve.

(c) Find the solution of the differential equation in part (b) that satisfies the initial condition $y(x_0) = y_0$.

(d) Find parametric equations $x = x(t)$, $y = y(t)$, $z = z(t)$ for the steepest ascent curve on the mountain surface.

3. When studying the spread of an epidemic, we assume that the probability that an infected individual will spread the disease to an uninfected individual is a function of the distance between them. Consider a circular city of radius 10 mi in which the population is uniformly distributed. For an uninfected individual at a fixed point $A(x_0, y_0)$, assume that the probability function is given by

$$f(P) = \tfrac{1}{20}[20 - d(P, A)]$$

where $d(P, A)$ denotes the distance between P and A.

(a) Suppose the exposure of a person to the disease is the sum of the probabilities of catching the disease from all members of the population. Assume that the infected people are uniformly distributed throughout the city, with k infected individuals per square mile. Find a double integral that represents the exposure of a person residing at A.

(b) Evaluate the integral for the case in which A is the center of the city and for the case in which A is located on the edge of the city. Where would you prefer to live?

4. Consider a one-parameter family of curves $F(x, y, C) = 0$. A curve Γ is an *envelope* of the given family if for each curve $F(x, y, C) = 0$ in the family, Γ intersects and is tangent to $F(x, y, C) = 0$ at their point of intersection. It can be shown that the equation of an envelope can be obtained by eliminating the parameter α between the two equations

$$F(x, y, \alpha) = 0 \quad \text{and} \quad F_\alpha(x, y, \alpha) = 0$$

(a) Identify the family of curves defined by

$$\frac{x}{\cos C} + \frac{y}{\sin C} = 1$$

and determine the envelope for this family.

(b) Sketch the members of the family and the envelope in the same coordinate system.

(c) Neglecting air resistance, the path of a projectile fired from the origin at an angle of elevation α and initial velocity v_0 is the parabola whose equation is

$$y = (\tan\alpha)x - \left(\frac{g}{2v_0^2\cos^2\alpha}\right)x^2$$

Fix the initial velocity v_0 and regard this equation as a one-parameter family of curves determined by the angle of elevation α. Find the envelope of this family. (See Applications Plus Problem 3, page 719.)

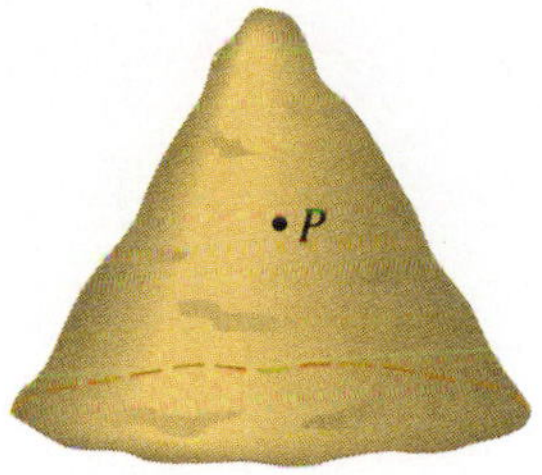

FIGURE FOR PROBLEM 5

5. When studying the formation of mountain ranges, geologists estimate the amount of work required to lift a mountain from sea level. Consider a mountain that is essentially in the shape of a right circular cone. Suppose that the weight density of the material in the vicinity of a point P is $g(P)$ and the height is $h(P)$.
(a) Find a definite integral that represents the total work done in forming the mountain.
(b) Assume that Mount Fuji in Japan is in the shape of a right circular cone with radius 62,000 ft, height 12,400 ft, and density a constant 200 lb/ft^3. How much work was done in forming Mount Fuji if the land was initially at sea level?

6. Marine biologists have determined that when a shark detects the presence of blood in the water, it will swim in the direction in which the concentration of the blood increases most rapidly. Based on certain tests in seawater, the concentration of blood (in parts per million) at a point $P(x, y)$ on the surface is approximated by

$$C(x, y) = e^{-(x^2+2y^2)/10^4}$$

where x and y are measured in meters in a rectangular coordinate system with the blood source at the origin.
(a) Identify the level curves of the concentration function and sketch several members of this family.
(b) Suppose a shark is at the point (x_0, y_0) when it first detects the presence of blood in the water. Determine the path that the shark will follow to the source (the origin).

FIGURE FOR PROBLEM 7

7. Suppose that a solid ball (a marble), a hollow ball (a squash ball), a solid cylinder (a steel bar), and a hollow cylinder (a lead pipe) roll down a slope. Which of these objects reaches the bottom first? (Make a guess before proceeding.)

To answer this question we consider a ball or cylinder with mass m, radius r, and moment of inertia I (about the axis of rotation). If the vertical drop is h, then the potential energy at the top is mgh. Suppose the object reaches the bottom with velocity v and angular velocity ω, so $v = \omega r$. The kinetic energy at the bottom consists of two parts: $\frac{1}{2}mv^2$ from translation (moving down the slope) and $\frac{1}{2}I\omega^2$ from rotation. If we assume that energy loss from rolling friction is negligible, then conservation of energy gives

$$mgh = \tfrac{1}{2}mv^2 + \tfrac{1}{2}I\omega^2$$

(a) Show that $\qquad v^2 = \dfrac{2gh}{1 + I^*} \qquad$ where $I^* = \dfrac{I}{mr^2}$

(b) If $y(t)$ is the vertical distance traveled at time t, then the same reasoning as used in part (a) shows that $v^2 = 2gy/(1 + I^*)$ at any time t. Use this result to show that y satisfies the differential equation

$$\frac{dy}{dt} = \sqrt{\frac{2g}{1 + I^*}}\,(\sin\alpha)\sqrt{y}$$

where α is the angle of inclination of the plane.

(c) By solving the differential equation in part (b), show that the total travel time is

$$T = \sqrt{\frac{2h(1 + I^*)}{g\sin^2\alpha}}$$

This shows that the object with the smallest value of I^* wins the race.

(d) Show that $I^* = \frac{1}{2}$ for a solid cylinder and $I^* = 1$ for a hollow cylinder.
(e) Show that $I^* = \frac{2}{5}$ for a solid ball and $I^* = \frac{2}{3}$ for a hollow ball. Thus, the objects finish in the following order: solid ball, solid cylinder, hollow ball, hollow cylinder.
(f) Calculate I^* for a partly hollow ball with inner radius a and outer radius r. Express your answer in terms of $b = a/r$. What happens as $a \to 0$ and as $a \to r$?

14

VECTOR CALCULUS

■ It is mathematics that offers the exact mathematical sciences a certain measure of security which, without mathematics, they could not obtain.

ALBERT EINSTEIN

In this chapter we study the calculus of vector fields. (These are functions that assign vectors to points in space.) In particular we define line integrals (which can be used to find the work done by a force field in moving an object along a curve). Then we define surface integrals (which can be used to find the rate of fluid flow across a surface). The connections between these new types of integrals and the single, double, and triple integrals that we have already met are given by the higher-dimensional analogues of the Fundamental Theorem of Calculus: Green's Theorem, Stokes' Theorem, and the Divergence Theorem.

14.1 VECTOR FIELDS

In Sections 1.3 and 11.6 we studied vector functions whose domains were sets of real numbers and whose ranges were sets of vectors. Now we look at a type of function, called a vector field, whose domain is a set of points in $\mathbb{R}^2$ (or $\mathbb{R}^3$) and whose range is a set of vectors in V_2 (or V_3). We also see how vector fields arise in various areas of physics.

(1) DEFINITION Let D be a set in $\mathbb{R}^2$ (a plane region). A **vector field on** $\mathbb{R}^2$ is a function $\mathbf{F}$ that assigns to each point (x, y) in D a two-dimensional vector $\mathbf{F}(x, y)$.

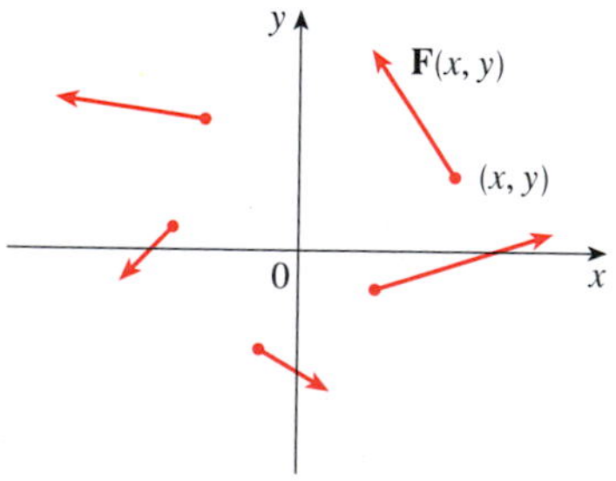

FIGURE 1
Vector field on $\mathbb{R}^2$

The best way to picture a vector field is to draw the arrow representing the vector $\mathbf{F}(x, y)$ starting at the point (x, y). Of course, it is impossible to do this for all points (x, y), but we can gain a reasonable impression of $\mathbf{F}$ by doing it for a few representative points in D as in Figure 1. Since $\mathbf{F}(x, y)$ is a two-dimensional vector, we can write it in terms of its **component functions** P and Q as follows:

$$\mathbf{F}(x, y) = P(x, y)\,\mathbf{i} + Q(x, y)\,\mathbf{j} = \langle P(x, y), Q(x, y)\rangle$$

or, for short,

$$\mathbf{F} = P\,\mathbf{i} + Q\,\mathbf{j}$$

Notice that P and Q are scalar functions of two variables and are sometimes called **scalar fields** to distinguish them from vector fields.

(2) DEFINITION Let E be a subset of $\mathbb{R}^3$. A **vector field on** $\mathbb{R}^3$ is a function $\mathbf{F}$ that assigns to each point (x, y, z) in E a three-dimensional vector $\mathbf{F}(x, y, z)$.

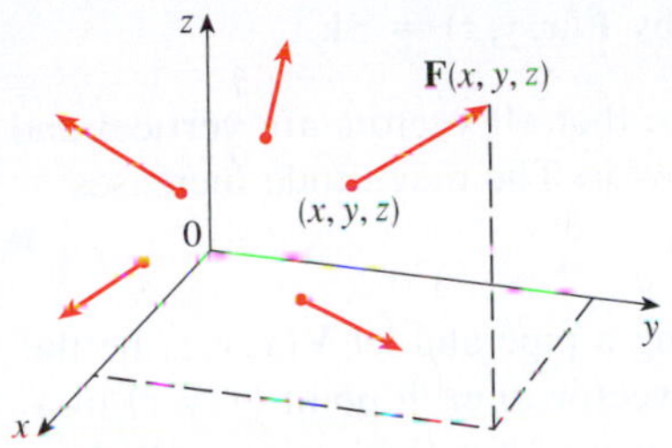

FIGURE 2
Vector field on $\mathbb{R}^3$

A vector field $\mathbf{F}$ on $\mathbb{R}^3$ is pictured in Figure 2. We can express it in terms of its component functions P, Q, and R as

$$\mathbf{F}(x, y, z) = P(x, y, z)\,\mathbf{i} + Q(x, y, z)\,\mathbf{j} + R(x, y, z)\,\mathbf{k}$$

As with the vector functions in Section 11.6, we can define continuity of vector fields and show that $\mathbf{F}$ is continuous if and only if its component functions P, Q, and R are continuous.

We sometimes identify a point (x, y, z) with its position vector $\mathbf{x} = \langle x, y, z\rangle$ and write $\mathbf{F}(\mathbf{x})$ instead of $\mathbf{F}(x, y, z)$. Then $\mathbf{F}$ becomes a function that assigns a vector $\mathbf{F}(\mathbf{x})$ to a vector $\mathbf{x}$.

EXAMPLE 1 A vector field on $\mathbb{R}^2$ is defined by

$$\mathbf{F}(x, y) = -y\,\mathbf{i} + x\,\mathbf{j}$$

Describe $\mathbf{F}$ by sketching some of the vectors $\mathbf{F}(x, y)$ as in Figure 1.

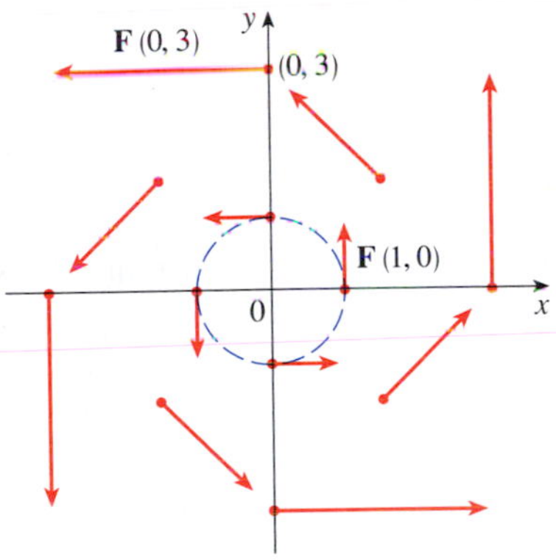

FIGURE 3
$\mathbf{F}(x, y) = -y\,\mathbf{i} + x\,\mathbf{j}$

SOLUTION Since $\mathbf{F}(1, 0) = \mathbf{j}$, we draw the vector $\mathbf{j} = \langle 0, 1\rangle$ starting at the point $(1, 0)$ in Figure 3. Since $\mathbf{F}(0, 1) = -\mathbf{i}$, we draw the vector $\langle -1, 0\rangle$ with starting point $(0, 1)$. Continuing in this way, we draw a number of representative vectors to represent the vector field in Figure 3.

It appears that each arrow is tangent to a circle with center the origin. To confirm this, we take the dot product of the position vector $\mathbf{x} = x\,\mathbf{i} + y\,\mathbf{j}$ with the vector $\mathbf{F}(\mathbf{x}) = \mathbf{F}(x, y)$:

$$\begin{aligned} \mathbf{x} \cdot \mathbf{F}(\mathbf{x}) &= (x\,\mathbf{i} + y\,\mathbf{j}) \cdot (-y\,\mathbf{i} + x\,\mathbf{j}) \\ &= -xy + yx = 0 \end{aligned}$$

This shows that $\mathbf{F}(x, y)$ is perpendicular to the position vector $\langle x, y\rangle$ and is therefore tangent to a circle with center the origin and radius $|\mathbf{x}| = \sqrt{x^2 + y^2}$. Notice also that

$$|\mathbf{F}(x, y)| = \sqrt{(-y)^2 + x^2} = \sqrt{x^2 + y^2} = |\mathbf{x}|$$

so the magnitude of the vector $\mathbf{F}(x, y)$ is equal to the radius of the circle. ■

CAS Some computer algebra systems are capable of plotting vector fields in two or three dimensions. They give a better impression of the vector field than is possible by hand because the computer can plot a large number of representative vectors. Figure 4 shows a computer plot of the vector field in Example 1; Figures 5 and 6 show two other vector fields. Notice that the computer scales the lengths of the vectors so they are not too long and yet are proportional to their true lengths.

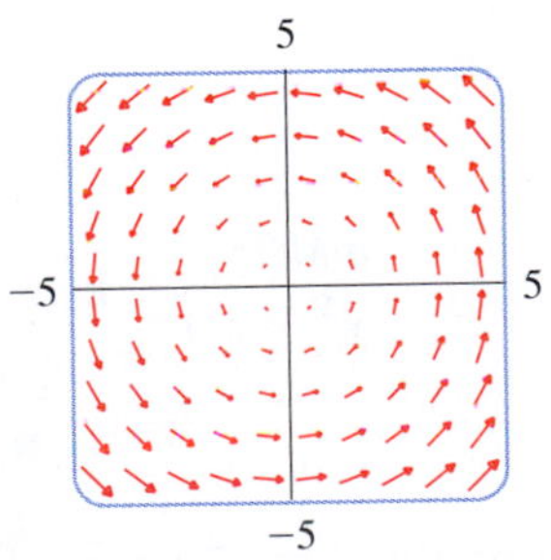

FIGURE 4
$\mathbf{F}(x, y) = \langle -y, x\rangle$

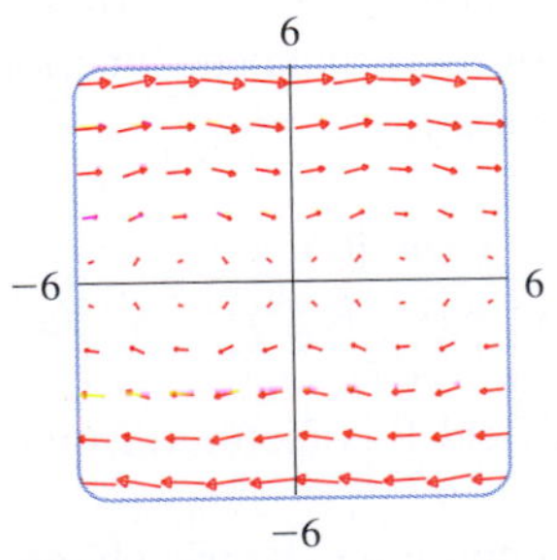

FIGURE 5
$\mathbf{F}(x, y) = \langle y, \sin x\rangle$

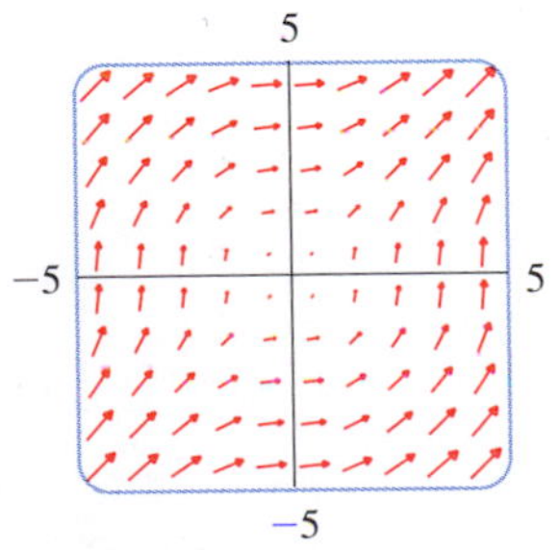

FIGURE 6
$\mathbf{F}(x, y) = \langle \ln(1 + y^2), \ln(1 + x^2)\rangle$

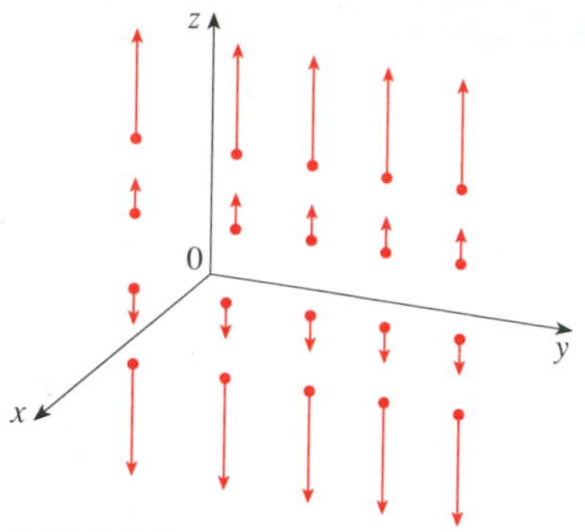

FIGURE 7
$\mathbf{F}(x, y, z) = z\,\mathbf{k}$

EXAMPLE 2 Sketch the vector field on $\mathbb{R}^3$ given by $\mathbf{F}(x, y, z) = z\,\mathbf{k}$.

SOLUTION The sketch is shown in Figure 7. Notice that all vectors are vertical and point upward above the xy-plane and downward below it. The magnitude increases with the distance from the xy-plane. ■

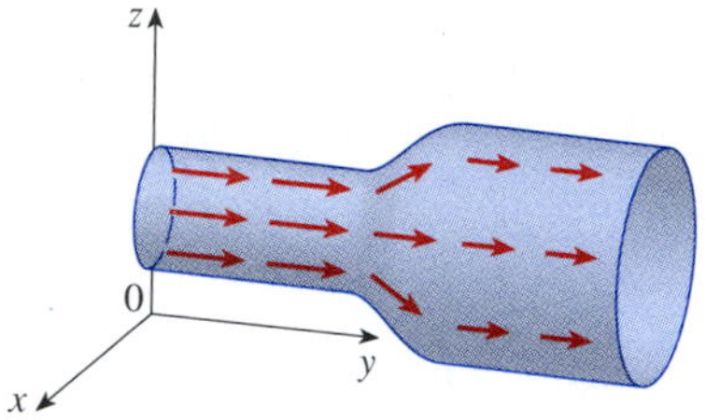

FIGURE 8
Velocity field in fluid flow

EXAMPLE 3 Imagine a fluid flowing steadily along a pipe and let $\mathbf{V}(x, y, z)$ be the velocity vector at a point (x, y, z). Then $\mathbf{V}$ assigns a vector to each point (x, y, z) in a certain domain E (the interior of the pipe) and so $\mathbf{V}$ is a vector field on $\mathbb{R}^3$ called a **velocity field**. A possible velocity field is illustrated in Figure 8. The speed at any given point is indicated by the length of the arrow.

Velocity fields also occur in other areas of physics. For instance, the vector field in Example 1 could be used as the velocity field describing the counterclockwise rotation of a wheel. ■

EXAMPLE 4 Newton's Law of Gravitation states that the magnitude of the gravitational force between two objects with masses m and M is

$$|\mathbf{F}| = \frac{mMG}{r^2}$$

where r is the distance between the objects and G is the gravitational constant. (This is an example of an inverse square law.) Let's assume that the object with mass M is located at the origin in $\mathbb{R}^3$. (For instance, M could be the mass of Earth and the origin would be at the center of Earth.) Let the position vector of the object with mass m be $\mathbf{x} = \langle x, y, z\rangle$. Then $r = |\mathbf{x}|$, so $r^2 = |\mathbf{x}|^2$. The gravitational force exerted on this second object acts toward the origin, and the unit vector in this direction is

$$-\frac{\mathbf{x}}{|\mathbf{x}|}$$

Therefore, the gravitational force acting on the object at $\mathbf{x} = \langle x, y, z\rangle$ is

$$\mathbf{F}(\mathbf{x}) = -\frac{mMG}{|\mathbf{x}|^3}\,\mathbf{x} \tag{3}$$

[Physicists often use the notation $\mathbf{r}$ instead of $\mathbf{x}$ for the position vector, so you may see Formula 3 written in the form $\mathbf{F} = -(mMG/r^3)\mathbf{r}$.] The function given by Equation 3 is an example of a vector field, called the **gravitational field**, because it associates a vector [the force $\mathbf{F}(\mathbf{x})$] with every point $\mathbf{x}$ in space.

Formula 3 is a compact way of writing the gravitational field, but we can also write it in terms of its component functions by using the facts that $\mathbf{x} = x\,\mathbf{i} + y\,\mathbf{j} + z\,\mathbf{k}$ and $|\mathbf{x}| = \sqrt{x^2 + y^2 + z^2}$:

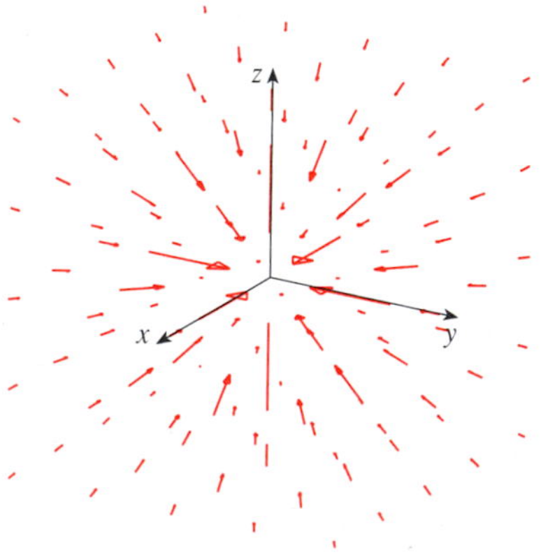

FIGURE 9
Gravitational force field

$$\mathbf{F}(x, y, z) = \frac{-mMGx}{(x^2 + y^2 + z^2)^{3/2}}\,\mathbf{i} + \frac{-mMGy}{(x^2 + y^2 + z^2)^{3/2}}\,\mathbf{j} + \frac{-mMGz}{(x^2 + y^2 + z^2)^{3/2}}\,\mathbf{k}$$

The gravitational field $\mathbf{F}$ is pictured in Figure 9. ■

EXAMPLE 5 Suppose an electric charge Q is located at the origin. According to Coulomb's Law, the electric force $\mathbf{F}(\mathbf{x})$ exerted by this charge on a charge q located at a point (x, y, z) with position vector $\mathbf{x} = \langle x, y, z \rangle$ is

(4) $$\mathbf{F}(\mathbf{x}) = \frac{\varepsilon q Q}{|\mathbf{x}|^3}\mathbf{x}$$

where ε is a constant (which depends on the units that are used). For like charges, we have $qQ > 0$ and the force is repulsive; for unlike charges, we have $qQ < 0$ and the force is attractive. Notice the similarity between Formulas 3 and 4. Both vector fields are examples of **force fields**.

Instead of considering the electric force $\mathbf{F}$, physicists often consider the force per unit charge:

$$\mathbf{E}(\mathbf{x}) = \frac{1}{q}\mathbf{F}(\mathbf{x}) = \frac{\varepsilon Q}{|\mathbf{x}|^3}\mathbf{x}$$

Then $\mathbf{E}$ is a vector field on $\mathbb{R}^3$ called the **electric field** of Q. ■

Figure 10 shows the gradient vector field

$$\nabla f(x, y) = 2xy\,\mathbf{i} + (x^2 - 3y^2)\,\mathbf{j}$$

of the function $f(x, y) = x^2y - y^3$ in Example 6.

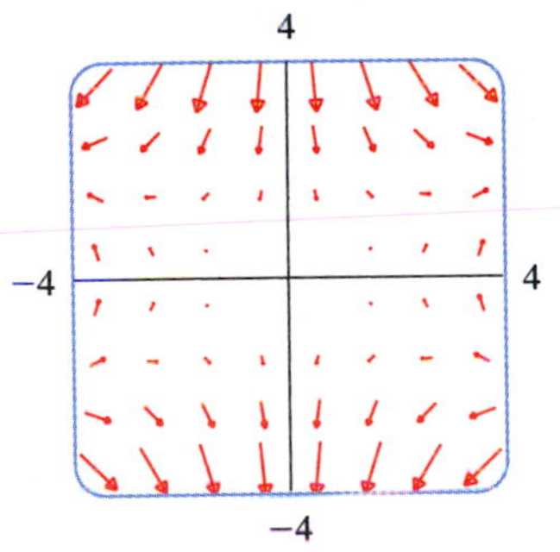

FIGURE 10

EXAMPLE 6 If f is a scalar function of two variables, recall from Section 12.6 that its gradient ∇f (or grad f) is defined by

$$\nabla f(x, y) = f_x(x, y)\,\mathbf{i} + f_y(x, y)\,\mathbf{j}$$

Therefore, ∇f is really a vector field on $\mathbb{R}^2$ and is called a **gradient vector field**. For instance, if

$$f(x, y) = x^2y - y^3$$

then its gradient vector field is given by

$$\nabla f(x, y) = 2xy\,\mathbf{i} + (x^2 - 3y^2)\,\mathbf{j}$$

Likewise, if f is a scalar function of three variables, its gradient is a vector field on $\mathbb{R}^3$ given by

$$\nabla f(x, y, z) = f_x(x, y, z)\,\mathbf{i} + f_y(x, y, z)\,\mathbf{j} + f_z(x, y, z)\,\mathbf{k}$$ ■

A vector field $\mathbf{F}$ is called a **conservative vector field** if it is the gradient of some scalar function, that is, if there exists a function f such that $\mathbf{F} = \nabla f$. In this situation f is called a **potential function** for $\mathbf{F}$.

Not all vector fields are conservative, but such fields do arise frequently in physics. For example, the gravitational field $\mathbf{F}$ in Example 4 is conservative because if we define

$$f(x, y, z) = \frac{mMG}{\sqrt{x^2 + y^2 + z^2}}$$

then

$$\begin{aligned}\nabla f(x, y, z) &= \frac{\partial f}{\partial x}\mathbf{i} + \frac{\partial f}{\partial y}\mathbf{j} + \frac{\partial f}{\partial z}\mathbf{k} \\ &= \frac{-mMGx}{(x^2 + y^2 + z^2)^{3/2}}\mathbf{i} + \frac{-mMGy}{(x^2 + y^2 + z^2)^{3/2}}\mathbf{j} + \frac{-mMGz}{(x^2 + y^2 + z^2)^{3/2}}\mathbf{k} \\ &= \mathbf{F}(x, y, z)\end{aligned}$$

Later in this chapter (see Theorems 14.3.6 and 14.5.4) we learn how to tell whether or not a given vector field is conservative.

EXERCISES 14.1

1–10 ■ Sketch the vector field $\mathbf{F}$ by drawing a diagram like Figure 3 or Figure 7.

1. $\mathbf{F}(x, y) = x\,\mathbf{i} + y\,\mathbf{j}$

2. $\mathbf{F}(x, y) = x\,\mathbf{i} - y\,\mathbf{j}$

3. $\mathbf{F}(x, y) = y\,\mathbf{i} + \mathbf{j}$

4. $\mathbf{F}(x, y) = -x\,\mathbf{i} + 2y\,\mathbf{j}$

5. $\mathbf{F}(x, y) = \dfrac{y\,\mathbf{i} + x\,\mathbf{j}}{\sqrt{x^2 + y^2}}$

6. $\mathbf{F}(x, y) = \dfrac{y\,\mathbf{i} - x\,\mathbf{j}}{\sqrt{x^2 + y^2}}$

7. $\mathbf{F}(x, y, z) = \mathbf{j}$

8. $\mathbf{F}(x, y, z) = z\,\mathbf{j}$

9. $\mathbf{F}(x, y, z) = y\,\mathbf{j}$

10. $\mathbf{F}(x, y, z) = \mathbf{j} + \mathbf{k}$

11–14 ■ Match the vector fields with the plots labeled I–IV. Give reasons for your choices.

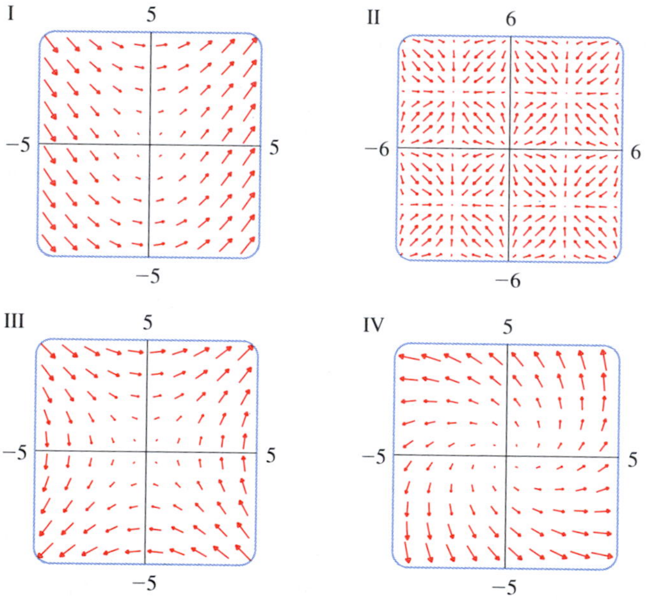

11. $\mathbf{F}(x, y) = \langle y, x \rangle$

12. $\mathbf{F}(x, y) = \langle 2x - 3y, 2x + 3y \rangle$

13. $\mathbf{F}(x, y) = \langle \sin x, \sin y \rangle$

14. $\mathbf{F}(x, y) = \langle \ln(1 + x^2 + y^2), x \rangle$

CAS **15.** If you have a CAS that plots vector fields (the command is fieldplot in Maple and PlotVectorField in Mathematica), use it to plot $\mathbf{F}(x, y) = (y^2 - 2xy)\,\mathbf{i} + (3xy - 6x^2)\,\mathbf{j}$. Explain the appearance by finding the set of points (x, y) such that $\mathbf{F}(x, y) = \mathbf{0}$.

CAS **16.** Let $\mathbf{F}(\mathbf{x}) = (r^2 - 2r)\mathbf{x}$, where $\mathbf{x} = \langle x, y \rangle$ and $r = |\mathbf{x}|$. Use a CAS to plot this vector field in various domains until you can see what is happening. Describe the appearance of the plot and explain it by finding the points where $\mathbf{F}(\mathbf{x}) = \mathbf{0}$.

17–22 ■ Find the gradient vector field of f.

17. $f(x, y) = x^5 - 4x^2y^3$

18. $f(x, y) = \sin(2x + 3y)$

19. $f(x, y) = e^{3x}\cos 4y$

20. $f(x, y, z) = xyz$

21. $f(x, y, z) = xy^2 - yz^3$

22. $f(x, y, z) = x\ln(y - z)$

23–24 ■ Find the gradient vector field ∇f of f and sketch it.

23. $f(x, y) = x^2 - \frac{1}{2}y^2$

24. $f(x, y) = \ln\sqrt{x^2 + y^2}$

25. The **flow lines** (or **streamlines**) of a vector field are the paths followed by a particle whose velocity field is the given vector field. Thus, the vectors in a vector field are tangent to the flow lines.

(a) Sketch some flow lines for the vector field $\mathbf{F}(x, y) = x\,\mathbf{i} - y\,\mathbf{j}$. From your sketches, can you guess the equations of the flow lines?

(b) Find an equation of the flow line that passes through the point $(1, 1)$ by solving the differential equations $dx/dt = x$ and $dy/dt = -y$.

14.2 LINE INTEGRALS

In this section we define an integral that is similar to a single integral except that instead of integrating over an interval $[a, b]$, we integrate over a curve C. Such integrals are called *line integrals*, although "curve integrals" would be better terminology.

We start with a plane curve C given by the parametric equations

$$x = x(t) \qquad y = y(t) \qquad a \leqslant t \leqslant b \tag{1}$$

or, equivalently, by the vector equation $\mathbf{r}(t) = x(t)\,\mathbf{i} + y(t)\,\mathbf{j}$, and we assume that C is a smooth curve. [This means that $\mathbf{r}'$ is continuous and $\mathbf{r}'(t) \neq \mathbf{0}$. See Section 11.7.] A partition of the parameter interval $[a, b]$ by points t_i with

$$a = t_0 < t_1 < t_2 < \cdots < t_n = b$$

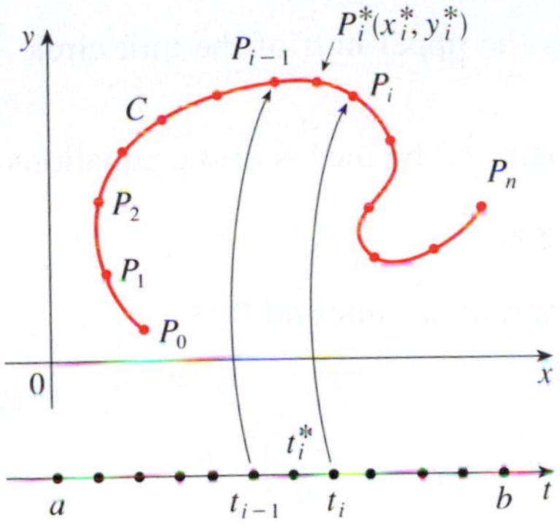

FIGURE 1
A partition of $[a, b]$ determines a partition of C

determines a partition P of the curve by points $P_i(x_i, y_i)$, where $x_i = x(t_i)$ and $y_i = y(t_i)$ (see Figure 1). These points P_i divide C into n subarcs with lengths $\Delta s_1, \Delta s_2, \ldots, \Delta s_n$. The **norm** $\|P\|$ of the partition is the longest of these lengths. We choose any point $P_i^*(x_i^*, y_i^*)$ in the ith subarc. (This corresponds to a point t_i^* in $[t_{i-1}, t_i]$.) Now if f is any function of two variables whose domain includes the curve C, we evaluate f at the point (x_i^*, y_i^*), multiply by the length Δs_i of the subarc, and form the sum

$$\sum_{i=1}^{n} f(x_i^*, y_i^*)\,\Delta s_i$$

which is similar to a Riemann sum. Then we take the limit of these sums and make the following definition by analogy with a single integral.

(2) DEFINITION If f is defined on a smooth curve C given by Equations 1, then the **line integral of f along C** is

$$\int_C f(x, y)\,ds = \lim_{\|P\| \to 0} \sum_{i=1}^{n} f(x_i^*, y_i^*)\,\Delta s_i$$

if this limit exists.

It can be shown that if f is a continuous function, then the limit in Definition 2 always exists and the following formula can be used to evaluate the line integral:

$$\text{(3)} \qquad \int_C f(x, y)\,ds = \int_a^b f(x(t), y(t)) \sqrt{\left(\frac{dx}{dt}\right)^2 + \left(\frac{dy}{dt}\right)^2}\,dt$$

The value of the line integral does not depend on the parametrization of the curve, provided that the curve is traversed exactly once as t increases from a to b.

The way to remember Formula 3 is to express everything in terms of the parameter t. Use the parametric equations to express x and y in terms of t and recall the arc length formula from Section 9.4:

$$ds = \sqrt{\left(\frac{dx}{dt}\right)^2 + \left(\frac{dy}{dt}\right)^2}\,dt$$

In the special case where C is the line segment that joins $(a, 0)$ to $(b, 0)$, using x as the parameter, we can write the parametric equations of C as follows: $x = x$, $y = 0$, $a \leqslant x \leqslant b$. Formula 3 then becomes

$$\int_C f(x, y)\,ds = \int_a^b f(x, 0)\,dx$$

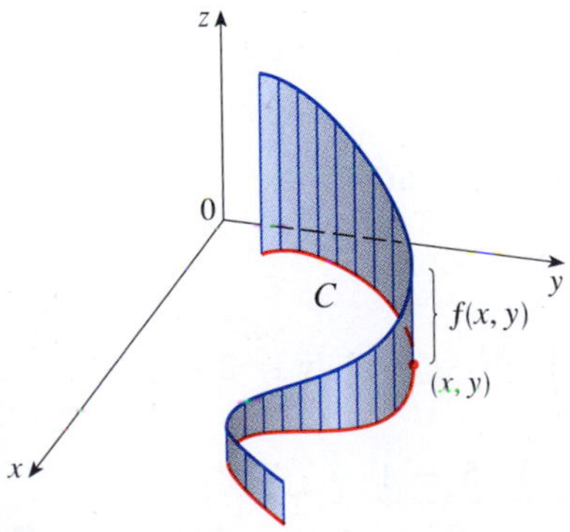

FIGURE 2

and so the line integral reduces to an ordinary single integral in this case.

Just as for an ordinary single integral, we can interpret the line integral of a *positive* function as an area. In fact, if $f(x, y) \geqslant 0$, $\int_C f(x, y)\,ds$ represents the area of one side of the "fence" or "curtain" in Figure 2 whose base is C and whose height above the point (x, y) is $f(x, y)$.

EXAMPLE 1 Evaluate $\int_C (2 + x^2y)\,ds$, where C is the upper half of the unit circle $x^2 + y^2 = 1$.

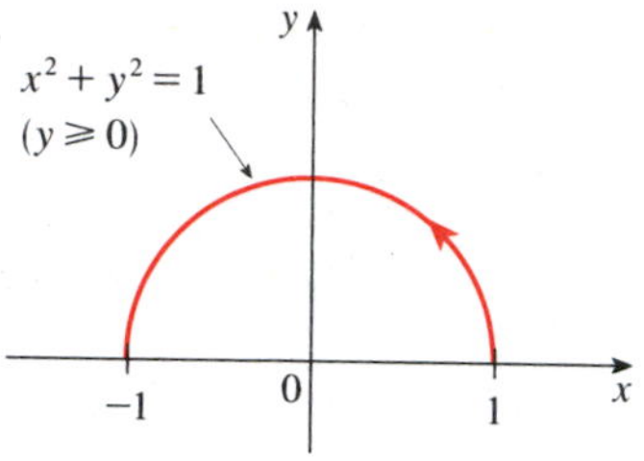

FIGURE 3

SOLUTION Recall that the unit circle can be parametrized by means of the equations

$$x = \cos t \qquad y = \sin t$$

and the upper half of the circle is described by the parameter interval $0 \leq t \leq \pi$ (see Figure 3). Therefore, Formula 3 gives

$$\begin{aligned}\int_C (2 + x^2y)\,ds &= \int_0^\pi (2 + \cos^2 t \sin t)\sqrt{\left(\frac{dx}{dt}\right)^2 + \left(\frac{dy}{dt}\right)^2}\,dt \\ &= \int_0^\pi (2 + \cos^2 t \sin t)\sqrt{\sin^2 t + \cos^2 t}\,dt \\ &= \int_0^\pi (2 + \cos^2 t \sin t)\,dt = \left[2t - \frac{\cos^3 t}{3}\right]_0^\pi \\ &= 2\pi + \tfrac{2}{3}\end{aligned}$$

■

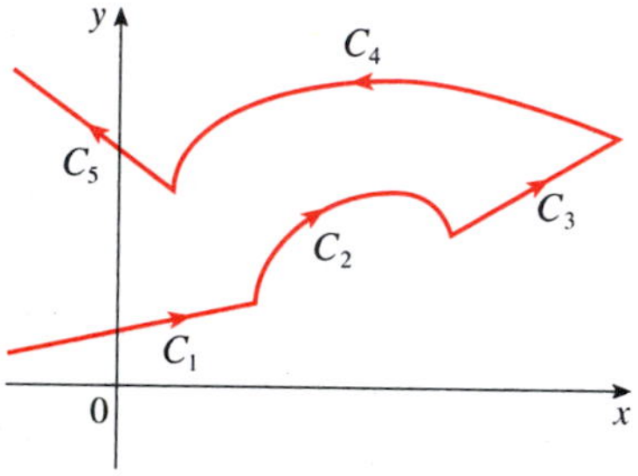

FIGURE 4
A piecewise-smooth curve

Suppose now that C is a **piecewise-smooth curve**; that is, C is a union of a finite number of smooth curves $C_1, C_2, \ldots, C_n$, where, as illustrated in Figure 4, the initial point of C_{i+1} is the terminal point of C_i. Then we define the integral of f along C as the sum of the integrals of f along each of the smooth pieces of C:

$$\int_C f(x, y)\,ds = \int_{C_1} f(x, y)\,ds + \int_{C_2} f(x, y)\,ds + \cdots + \int_{C_n} f(x, y)\,ds$$

EXAMPLE 2 Evaluate $\int_C 2x\,ds$, where C consists of the arc C_1 of the parabola $y = x^2$ from $(0, 0)$ to $(1, 1)$ followed by the vertical line segment C_2 from $(1, 1)$ to $(1, 2)$.

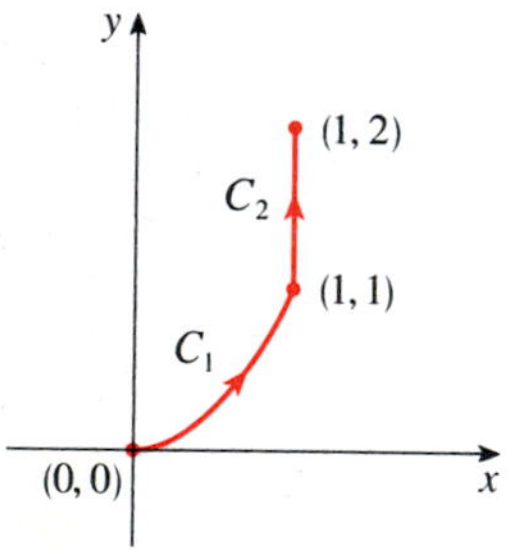

FIGURE 5
$C = C_1 \cup C_2$

SOLUTION The curve C is shown in Figure 5. C_1 is the graph of a function of x, so we can choose x as the parameter and the equations for C_1 become

$$x = x \qquad y = x^2 \qquad 0 \leq x \leq 1$$

Therefore

$$\begin{aligned}\int_{C_1} 2x\,ds &= \int_0^1 2x\sqrt{\left(\frac{dx}{dx}\right)^2 + \left(\frac{dy}{dx}\right)^2}\,dx \\ &= \int_0^1 2x\sqrt{1 + 4x^2}\,dx = \tfrac{1}{4} \cdot \tfrac{2}{3}(1 + 4x^2)^{3/2}\Big]_0^1 = \frac{5\sqrt{5} - 1}{6}\end{aligned}$$

On C_2 we choose y as the parameter, so the equations of C_2 are

$$x = 1 \qquad y = y \qquad 1 \leq y \leq 2$$

and
$$\int_{C_2} 2x\,ds = \int_1^2 2(1)\sqrt{\left(\frac{dx}{dy}\right)^2 + \left(\frac{dy}{dy}\right)^2}\,dy = \int_1^2 2\,dy = 2$$

Thus
$$\int_C 2x\,ds = \int_{C_1} 2x\,ds + \int_{C_2} 2x\,ds = \frac{5\sqrt{5} - 1}{6} + 2$$

■

Any physical interpretation of a line integral $\int_C f(x, y)\, ds$ depends on the physical interpretation of the function f. Suppose that $\rho(x, y)$ represents the linear density at a point (x, y) of a thin wire shaped like a curve C. Then the mass of the part of the wire from P_{i-1} to P_i in Figure 1 is approximately $\rho(x_i^*, y_i^*)\, \Delta s_i$ and so the total mass of the wire is approximately $\Sigma\, \rho(x_i^*, y_i^*)\, \Delta s_i$. By taking finer partitions of the curve, we obtain the **mass** m of the wire as the limiting value of these approximations:

$$m = \lim_{\|P\| \to 0} \sum_{i=1}^{n} \rho(x_i^*, y_i^*)\, \Delta s_i = \int_C \rho(x, y)\, ds$$

[For example, if $f(x, y) = 2 + x^2 y$ represents the density of a semicircular wire, then the integral in Example 1 would represent the mass of the wire.] The **center of mass** of the wire with density function ρ is at the point $(\bar{x}, \bar{y})$, where

$$\bar{x} = \frac{1}{m} \int_C x\rho(x, y)\, ds \qquad \bar{y} = \frac{1}{m} \int_C y\rho(x, y)\, ds \tag{4}$$

Other physical interpretations of line integrals will be discussed later in this chapter.

EXAMPLE 3 A wire takes the shape of the semicircle $x^2 + y^2 = 1$, $y \geqslant 0$, and is thicker near its base than near the top. Find the center of mass of the wire if the linear density at any point is proportional to its distance from the line $y = 1$.

SOLUTION As in Example 1 we use the parametrization $x = \cos t$, $y = \sin t$, $0 \leqslant t \leqslant \pi$, and find that $ds = dt$. The linear density is

$$\rho(x, y) = k(1 - y)$$

where k is a constant, and so the mass of the wire is

$$m = \int_C k(1 - y)\, ds = \int_0^\pi k(1 - \sin t)\, dt$$

$$= k\Big[t + \cos t\Big]_0^\pi = k(\pi - 2)$$

From Equations 4 we have

$$\bar{y} = \frac{1}{m} \int_C y\rho(x, y)\, ds = \frac{1}{k(\pi - 2)} \int_C yk(1 - y)\, ds$$

$$= \frac{1}{\pi - 2} \int_0^\pi (\sin t - \sin^2 t)\, dt = \frac{1}{\pi - 2}\Big[-\cos t - \tfrac{1}{2}t + \tfrac{1}{4}\sin 2t\Big]_0^\pi$$

$$= \frac{4 - \pi}{2(\pi - 2)}$$

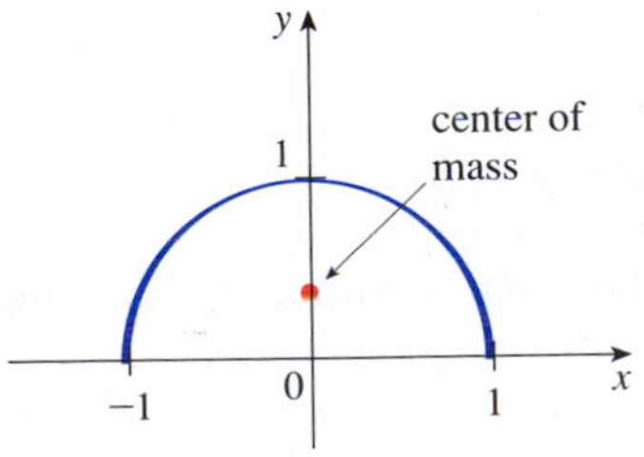

FIGURE 6

By symmetry we see that $\bar{x} = 0$, so the center of mass is at

$$\left(0, \frac{4 - \pi}{2(\pi - 2)}\right) \approx (0, 0.38)$$

See Figure 6. ■

Two other line integrals are obtained by replacing Δs_i by $\Delta x_i = x_i - x_{i-1}$ or $\Delta y_i = y_i - y_{i-1}$ in Definition 2. They are called the **line integrals of f along C with respect to x and y**:

(5) $$\int_C f(x, y)\, dx = \lim_{\|P\| \to 0} \sum_{i=1}^{n} f(x_i^*, y_i^*)\, \Delta x_i$$

(6) $$\int_C f(x, y)\, dy = \lim_{\|P\| \to 0} \sum_{i=1}^{n} f(x_i^*, y_i^*)\, \Delta y_i$$

When we want to distinguish the original line integral $\int_C f(x, y)\, ds$ from those in Equations 5 and 6, we call it the **line integral with respect to arc length**.

The following formulas say that line integrals with respect to x and y can also be evaluated by expressing everything in terms of t: $x = x(t)$, $y = y(t)$, $dx = x'(t)\, dt$, $dy = y'(t)\, dt$.

(7) $$\int_C f(x, y)\, dx = \int_a^b f(x(t), y(t))\, x'(t)\, dt$$

$$\int_C f(x, y)\, dy = \int_a^b f(x(t), y(t))\, y'(t)\, dt$$

It frequently happens that line integrals with respect to x and y occur together. When this happens, it is customary to abbreviate by writing

$$\int_C P(x, y)\, dx + \int_C Q(x, y)\, dy = \int_C P(x, y)\, dx + Q(x, y)\, dy$$

When you are setting up a line integral, sometimes the most difficult thing is to think of a parametric representation for a curve whose geometric description is given. It is useful to remember that a vector representation of the line segment that starts at $\mathbf{r}_0$ and ends at $\mathbf{r}_1$ is given by

(8) $$\mathbf{r}(t) = (1 - t)\,\mathbf{r}_0 + t\,\mathbf{r}_1 \qquad 0 \leqslant t \leqslant 1$$

(See Equation 11.4.1 with $\mathbf{v} = \mathbf{r}_1 - \mathbf{r}_0$.)

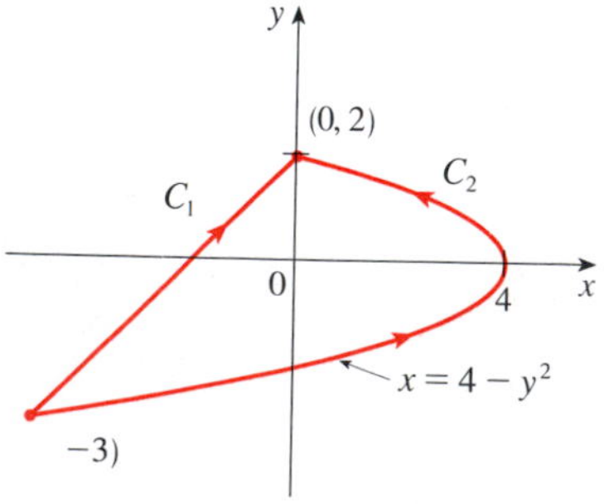

FIGURE 7

EXAMPLE 4 Evaluate $\int_C y^2\, dx + x\, dy$, where (a) $C = C_1$ is the line segment from $(-5, -3)$ to $(0, 2)$ and (b) $C = C_2$ is the arc of the parabola $x = 4 - y^2$ from $(-5, -3)$ to $(0, 2)$ (see Figure 7).

SOLUTION

(a) A parametric representation for the line segment is

$$x = 5t - 5 \qquad y = 5t - 3 \qquad 0 \leqslant t \leqslant 1$$

(Use Equation 8 with $\mathbf{r}_0 = \langle -5, -3 \rangle$ and $\mathbf{r}_1 = \langle 0, 2 \rangle$.) Then $dx = 5\, dt$, $dy = 5\, dt$, and Formula 7 gives

$$\int_{C_1} y^2\, dx + x\, dy = \int_0^1 (5t - 3)^2 (5\, dt) + (5t - 5)(5\, dt)$$

$$= 5 \int_0^1 (25t^2 - 25t + 4)\, dt$$

$$= 5\left[\frac{25t^3}{3} - \frac{25t^2}{2} + 4t\right]_0^1 = -\frac{5}{6}$$

(b) Since the parabola is given as a function of y, let's take y as the parameter and write C_2 as

$$x = 4 - y^2 \qquad y = y \qquad -3 \leq y \leq 2$$

Then $dx = -2y\,dy$ and by Formula 7 we have

$$\int_{C_2} y^2\,dx + x\,dy = \int_{-3}^{2} y^2(-2y)\,dy + (4 - y^2)\,dy$$

$$= \int_{-3}^{2} (-2y^3 - y^2 + 4)\,dy$$

$$= \left[-\frac{y^4}{2} - \frac{y^3}{3} + 4y\right]_{-3}^{2} = 40\tfrac{5}{6}$$ ■

Notice that we got different answers in parts (a) and (b) of Example 4 even though the two curves had the same endpoints. Thus, in general, the value of a line integral depends not just on the endpoints of the curve but also on the path. (But see Section 14.3 for conditions under which the integral is independent of the path.)

Notice also that the answers in Example 4 depend on the direction, or orientation, of the curve. If $-C_1$ denotes the line segment from $(0, 2)$ to $(-5, -3)$, you can verify, using the parametrization

$$x = -5t \qquad y = 2 - 5t \qquad 0 \leq t \leq 1$$

that

$$\int_{-C_1} y^2\,dx + x\,dy = \tfrac{5}{6}$$

In general, a given parametrization $x = x(t)$, $y = y(t)$, $a \leq t \leq b$, determines an **orientation** of a curve C, with the positive direction corresponding to increasing values of the parameter t. (See Figure 8, where the initial point A corresponds to the parameter value a and the terminal point B corresponds to $t = b$.)

If $-C$ denotes the curve consisting of the same points as C but with the opposite orientation (from initial point B to terminal point A in Figure 8), then we have

$$\int_{-C} f(x, y)\,dx = -\int_C f(x, y)\,dx \qquad \int_{-C} f(x, y)\,dy = -\int_C f(x, y)\,dy$$

But if we integrate with respect to arc length, the value of the line integral does *not* change when we reverse the orientation of the curve:

$$\int_{-C} f(x, y)\,ds = \int_C f(x, y)\,ds$$

(This is because Δs_i is always positive, whereas Δx_i and Δy_i change sign when we reverse the orientation of C.)

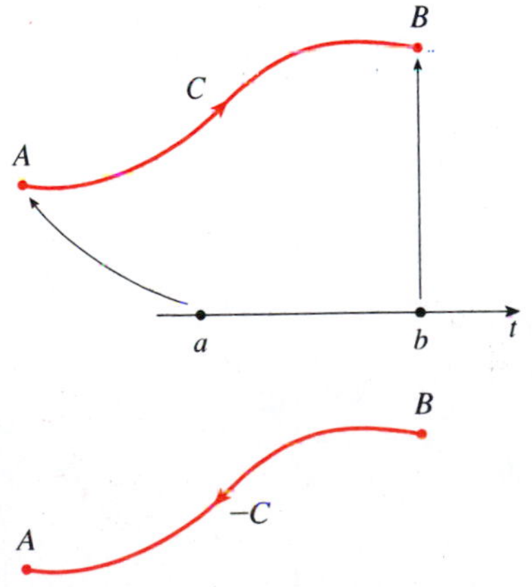

FIGURE 8

LINE INTEGRALS IN SPACE

We now suppose that C is a smooth space curve given by the parametric equations

$$x = x(t) \qquad y = y(t) \qquad z = z(t) \qquad a \leqslant t \leqslant b$$

or by a vector equation $\mathbf{r}(t) = x(t)\,\mathbf{i} + y(t)\,\mathbf{j} + z(t)\,\mathbf{k}$. If f is a function of three variables that is continuous on some region containing C, then we define the **line integral of f along C** (with respect to arc length) in a manner similar to that for plane curves:

$$\int_C f(x, y, z)\, ds = \lim_{\|P\| \to 0} \sum_{i=1}^{n} f(x_i^*, y_i^*, z_i^*)\, \Delta s_i$$

We evaluate it using a formula similar to Formula 3:

$$\textbf{(9)} \qquad \int_C f(x, y, z)\, ds = \int_a^b f(x(t), y(t), z(t)) \sqrt{\left(\frac{dx}{dt}\right)^2 + \left(\frac{dy}{dt}\right)^2 + \left(\frac{dz}{dt}\right)^2}\, dt$$

Observe that the integrals in both Formulas 3 and 9 can be written in the more compact vector notation

$$\int_a^b f(\mathbf{r}(t))\,|\mathbf{r}'(t)|\, dt$$

For the special case $f(x, y, z) = 1$, we get

$$\int_C ds = \int_a^b |\mathbf{r}'(t)|\, dt = L$$

where L is the length of the curve C (see Formula 11.7.3).

Line integrals along C with respect to x, y, and z can also be defined. For example,

$$\int_C f(x, y, z)\, dz = \lim_{\|P\| \to 0} \sum_{i=1}^{n} f(x_i^*, y_i^*, z_i^*)\, \Delta z_i$$

$$= \int_a^b f(x(t), y(t), z(t))\, z'(t)\, dt$$

Therefore, as with line integrals in the plane, we evaluate integrals of the form

$$\textbf{(10)} \qquad \int_C P(x, y, z)\, dx + Q(x, y, z)\, dy + R(x, y, z)\, dz$$

by expressing everything (x, y, z, dx, dy, dz) in terms of the parameter t.

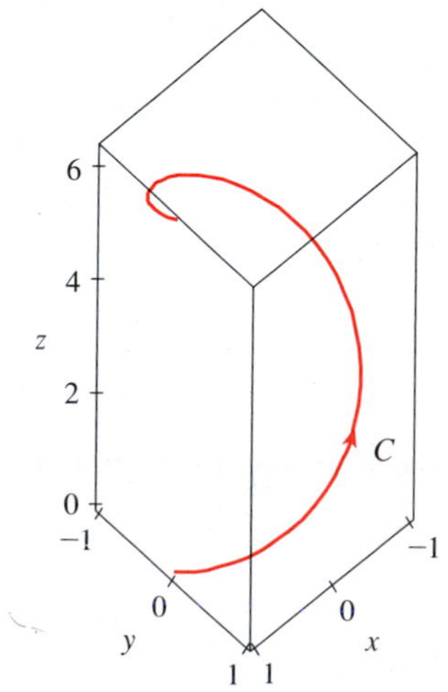

FIGURE 9

EXAMPLE 5 Evaluate $\int_C y \sin z\, ds$, where C is the circular helix given by the equations $x = \cos t$, $y = \sin t$, $z = t$, $0 \leqslant t \leqslant 2\pi$ (see Figure 9).

SOLUTION Formula 9 gives

$$\int_C y \sin z\, ds = \int_0^{2\pi} (\sin t) \sin t \sqrt{\left(\frac{dx}{dt}\right)^2 + \left(\frac{dy}{dt}\right)^2 + \left(\frac{dz}{dt}\right)^2}\, dt$$

$$= \int_0^{2\pi} \sin^2 t \sqrt{\sin^2 t + \cos^2 t + 1}\, dt$$

$$= \sqrt{2} \int_0^{2\pi} \tfrac{1}{2}(1 - \cos 2t)\, dt = \frac{\sqrt{2}}{2}\Big[t - \tfrac{1}{2}\sin 2t\Big]_0^{2\pi} = \sqrt{2}\,\pi$$ ■

EXAMPLE 6 Evaluate $\int_C y\,dx + z\,dy + x\,dz$, where C consists of the line segment C_1 from $(2, 0, 0)$ to $(3, 4, 5)$ followed by the vertical line segment C_2 from $(3, 4, 5)$ to $(3, 4, 0)$.

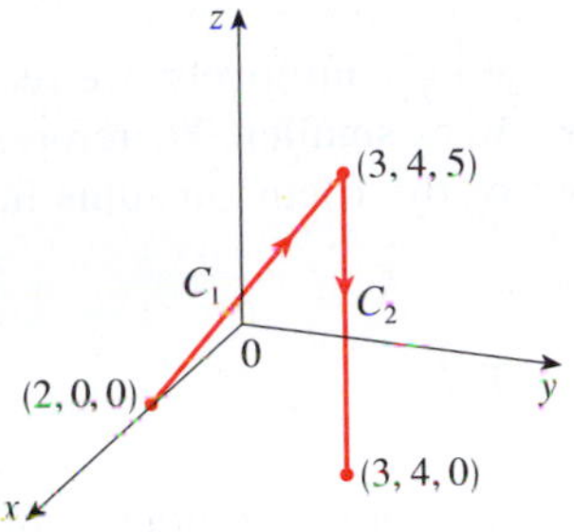

FIGURE 10

SOLUTION The curve C is shown in Figure 10. Using Equation 8, we write C_1 as

$$\mathbf{r}(t) = (1 - t)\langle 2, 0, 0\rangle + t\langle 3, 4, 5\rangle = \langle 2 + t, 4t, 5t\rangle$$

or, in parametric form, as

$$x = 2 + t \qquad y = 4t \qquad z = 5t \qquad 0 \leqslant t \leqslant 1$$

Thus

$$\int_{C_1} y\,dx + z\,dy + x\,dz = \int_0^1 (4t)\,dt + (5t)4\,dt + (2 + t)5\,dt$$

$$= \int_0^1 (10 + 29t)\,dt = 10t + 29\frac{t^2}{2}\bigg]_0^1 = 24.5$$

Likewise, C_2 can be written in the form

$$\mathbf{r}(t) = (1 - t)\langle 3, 4, 5\rangle + t\langle 3, 4, 0\rangle = \langle 3, 4, 5 - 5t\rangle$$

or

$$x = 3 \qquad y = 4 \qquad z = 5 - 5t \qquad 0 \leqslant t \leqslant 1$$

Then $dx = 0 = dy$, so

$$\int_{C_2} y\,dx + z\,dy + x\,dz = \int_0^1 3(-5)\,dt = -15$$

Adding the values of these integrals together, we obtain

$$\int_C y\,dx + z\,dy + x\,dz = 24.5 - 15 = 9.5$$

■

LINE INTEGRALS OF VECTOR FIELDS

Recall from Section 7.4 that the work done by a variable force $f(x)$ in moving a particle from a to b along the x-axis is $W = \int_a^b f(x)\,dx$. Then in Section 11.2 we found that the work done by a constant force $\mathbf{F}$ in moving an object from a point P to another point Q in space is $W = \mathbf{F} \cdot \mathbf{D}$, where $\mathbf{D} = \overrightarrow{PQ}$ is the displacement vector.

Now suppose that $\mathbf{F} = P\,\mathbf{i} + Q\,\mathbf{j} + R\,\mathbf{k}$ is a continuous force field on $\mathbb{R}^3$, such as the gravitational field of Example 4 in Section 14.1 or the electric force field of Example 5 in Section 14.1. (A force field on $\mathbb{R}^2$ could be regarded as a special case where $R = 0$ and P and Q depend only on x and y.) We wish to compute the work done by this force in moving a particle along a smooth curve C.

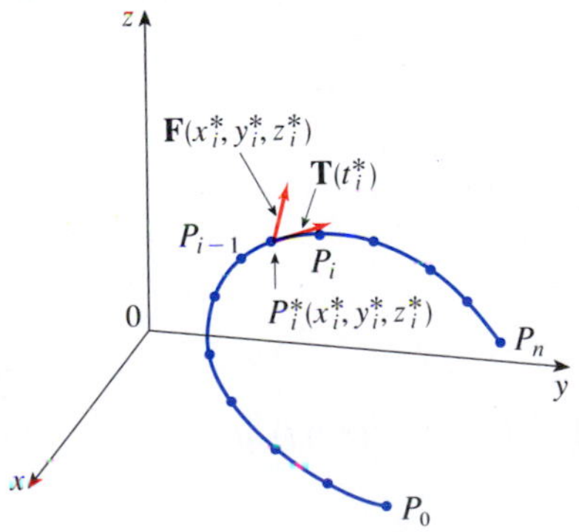

FIGURE 11

We partition C into subarcs $P_{i-1}P_i$ with lengths Δs_i by means of a partition of the parameter interval $[a, b]$. (See Figure 1 for the two-dimensional case or Figure 11 for the three-dimensional case.) Choose a point $P_i^*(x_i^*, y_i^*, z_i^*)$ on the ith subarc corresponding to the parameter value t_i^*. If Δs_i is small, then as the particle moves from P_{i-1} to P_i along the curve, it proceeds approximately in the direction of $\mathbf{T}(t_i^*)$, the unit tangent vector at P_i^*. Therefore, the work done by the force $\mathbf{F}$ in moving the particle from P_{i-1} to P_i is approximately

$$\mathbf{F}(x_i^*, y_i^*, z_i^*) \cdot [\Delta s_i\,\mathbf{T}(t_i^*)] = [\mathbf{F}(x_i^*, y_i^*, z_i^*) \cdot \mathbf{T}(t_i^*)]\,\Delta s_i$$

and the total work done in moving the particle along C is approximately

(11) $$\sum_{i=1}^{n} [\mathbf{F}(x_i^*, y_i^*, z_i^*) \cdot \mathbf{T}(x_i^*, y_i^*, z_i^*)] \Delta s_i$$

where $\mathbf{T}(x, y, z)$ is the unit tangent vector at the point (x, y, z) on C. Intuitively, we see that these approximations ought to become better as $\|P\|$ becomes smaller. Therefore, we define the **work** W done by the force field $\mathbf{F}$ as the limit of the Riemann sums in (11), namely,

(12) $$W = \int_C \mathbf{F}(x, y, z) \cdot \mathbf{T}(x, y, z)\, ds = \int_C \mathbf{F} \cdot \mathbf{T}\, ds$$

Equation 12 says that *work is the integral with respect to arc length of the tangential component of the force.*

If the curve C is given by the vector equation $\mathbf{r}(t) = x(t)\,\mathbf{i} + y(t)\,\mathbf{j} + z(t)\,\mathbf{k}$, then $\mathbf{T}(t) = \mathbf{r}'(t)/|\mathbf{r}'(t)|$, so using Equation 9 we can rewrite Equation 12 in the form

$$W = \int_a^b \left[\mathbf{F}(\mathbf{r}(t)) \cdot \frac{\mathbf{r}'(t)}{|\mathbf{r}'(t)|}\right] |\mathbf{r}'(t)|\, dt$$

$$= \int_a^b \mathbf{F}(\mathbf{r}(t)) \cdot \mathbf{r}'(t)\, dt$$

This latter integral is often abbreviated as $\int_C \mathbf{F} \cdot d\mathbf{r}$ and occurs in other areas of physics as well. Therefore, we make the following definition for the line integral of *any* continuous vector field.

(13) DEFINITION Let $\mathbf{F}$ be a continuous vector field defined on a smooth curve C given by a vector function $\mathbf{r}(t)$, $a \leqslant t \leqslant b$. Then the **line integral of F along C** is

$$\int_C \mathbf{F} \cdot d\mathbf{r} = \int_a^b \mathbf{F}(\mathbf{r}(t)) \cdot \mathbf{r}'(t)\, dt = \int_C \mathbf{F} \cdot \mathbf{T}\, ds$$

When you use Definition 13, remember that $\mathbf{F}(\mathbf{r}(t))$ is just short for $\mathbf{F}(x(t), y(t), z(t))$, so we simply evaluate $\mathbf{F}(\mathbf{r}(t))$ by putting $x = x(t)$, $y = y(t)$, and $z = z(t)$ in the expression for $\mathbf{F}(x, y, z)$. Notice also that we can formally write $d\mathbf{r} = \mathbf{r}'(t)\, dt$.

EXAMPLE 7 Find the work done by the force field $\mathbf{F}(x, y) = x^2\,\mathbf{i} - xy\,\mathbf{j}$ in moving a particle along the quarter-circle $\mathbf{r}(t) = \cos t\,\mathbf{i} + \sin t\,\mathbf{j}$, $0 \leqslant t \leqslant \pi/2$.

SOLUTION Since $x = \cos t$ and $y = \sin t$, we have

$$\mathbf{F}(\mathbf{r}(t)) = \cos^2 t\,\mathbf{i} - \cos t \sin t\,\mathbf{j}$$

and $$\mathbf{r}'(t) = -\sin t\,\mathbf{i} + \cos t\,\mathbf{j}$$

Therefore, the work done is

$$\int_C \mathbf{F} \cdot d\mathbf{r} = \int_0^{\pi/2} \mathbf{F}(\mathbf{r}(t)) \cdot \mathbf{r}'(t)\, dt = \int_0^{\pi/2} (-2\cos^2 t \sin t)\, dt$$

$$= 2\frac{\cos^3 t}{3}\Bigg]_0^{\pi/2} = -\frac{2}{3}$$ ■

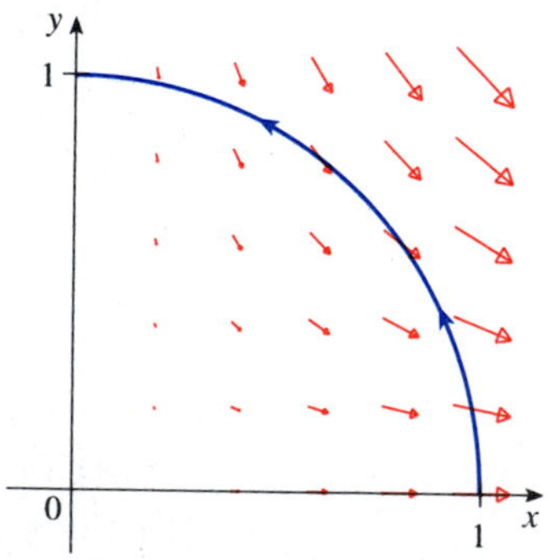

FIGURE 12

Figure 12 shows the force field and the curve in Example 7. The work done is negative because the field impedes movement along the curve.

NOTE: Even though $\int_C \mathbf{F} \cdot d\mathbf{r} = \int_C \mathbf{F} \cdot \mathbf{T}\, ds$ and integrals with respect to arc length are unchanged when orientation is reversed, it is still true that

$$\int_{-C} \mathbf{F} \cdot d\mathbf{r} = -\int_C \mathbf{F} \cdot d\mathbf{r}$$

because the unit tangent vector $\mathbf{T}$ is replaced by its negative when C is replaced by $-C$.

Figure 13 shows the twisted cubic C in Example 8 and some typical vectors acting at three points on C.

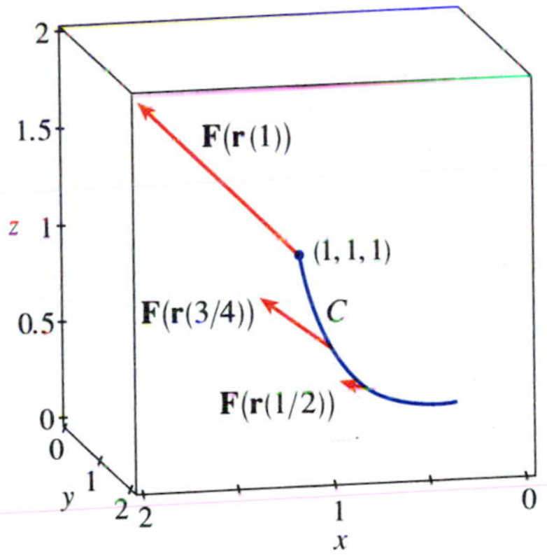

FIGURE 13

EXAMPLE 8 Evaluate $\int_C \mathbf{F} \cdot d\mathbf{r}$, where $\mathbf{F}(x, y, z) = xy\,\mathbf{i} + yz\,\mathbf{j} + zx\,\mathbf{k}$ and C is the twisted cubic given by

$$x = t \qquad y = t^2 \qquad z = t^3 \qquad 0 \leqslant t \leqslant 1$$

SOLUTION We have

$$\mathbf{r}(t) = t\,\mathbf{i} + t^2\,\mathbf{j} + t^3\,\mathbf{k}$$

$$\mathbf{r}'(t) = \mathbf{i} + 2t\,\mathbf{j} + 3t^2\,\mathbf{k}$$

$$\mathbf{F}(\mathbf{r}(t)) = t^3\,\mathbf{i} + t^5\,\mathbf{j} + t^4\,\mathbf{k}$$

Thus

$$\int_C \mathbf{F} \cdot d\mathbf{r} = \int_0^1 \mathbf{F}(\mathbf{r}(t)) \cdot \mathbf{r}'(t)\, dt$$

$$= \int_0^1 (t^3 + 5t^6)\, dt = \frac{t^4}{4} + \frac{5t^7}{7}\Big]_0^1 = \frac{27}{28}$$ ■

Finally, we note the connection between line integrals of vector fields and line integrals of scalar fields. Suppose the vector field $\mathbf{F}$ on $\mathbb{R}^3$ is given in component form by the equation $\mathbf{F} = P\,\mathbf{i} + Q\,\mathbf{j} + R\,\mathbf{k}$. We use Definition 13 to compute its line integral along C:

$$\int_C \mathbf{F} \cdot d\mathbf{r} = \int_a^b \mathbf{F}(\mathbf{r}(t)) \cdot \mathbf{r}'(t)\, dt$$

$$= \int_a^b (P\,\mathbf{i} + Q\,\mathbf{j} + R\,\mathbf{k}) \cdot (x'(t)\,\mathbf{i} + y'(t)\,\mathbf{j} + z'(t)\,\mathbf{k})\, dt$$

$$= \int_a^b [P(x(t), y(t), z(t))\, x'(t) + Q(x(t), y(t), z(t))\, y'(t) + R(x(t), y(t), z(t))\, z'(t)]\, dt$$

But this last integral is precisely the line integral in (10). Therefore, we have

$$\int_C \mathbf{F} \cdot d\mathbf{r} = \int_C P\,dx + Q\,dy + R\,dz \qquad \text{where } \mathbf{F} = P\,\mathbf{i} + Q\,\mathbf{j} + R\,\mathbf{k}$$

For example, the integral $\int_C y\,dx + z\,dy + x\,dz$ in Example 6 could be expressed as $\int_C \mathbf{F} \cdot d\mathbf{r}$ where

$$\mathbf{F}(x, y, z) = y\,\mathbf{i} + z\,\mathbf{j} + x\,\mathbf{k}$$

EXERCISES 14.2

1–16 ■ Evaluate the line integral, where C is the given curve.

1. $\int_C x\,ds$, C: $x = t^3$, $y = t$, $0 \leqslant t \leqslant 1$
2. $\int_C y\,ds$, C: $x = t^3$, $y = t^2$, $0 \leqslant t \leqslant 1$
3. $\int_C xy^4\,ds$, C is the right half of the circle $x^2 + y^2 = 16$
4. $\int_C xy\,ds$, C is the line segment joining $(-1, 1)$ to $(2, 3)$
5. $\int_C (x - 2y^2)\,dy$, C is the arc of the parabola $y = x^2$ from $(-2, 4)$ to $(1, 1)$
6. $\int_C \sin x\,dx$, C is the arc of the curve $x = y^4$ from $(1, -1)$ to $(1, 1)$
7. $\int_C xy\,dx + (x - y)\,dy$, C consists of line segments from $(0, 0)$ to $(2, 0)$ and from $(2, 0)$ to $(3, 2)$
8. $\int_C x\sqrt{y}\,dx + 2y\sqrt{x}\,dy$, C consists of the arc of the circle $x^2 + y^2 = 1$ from $(1, 0)$ to $(0, 1)$ and the line segment from $(0, 1)$ to $(4, 3)$
9. $\int_C xyz\,ds$, C: $x = 2t$, $y = 3\sin t$, $z = 3\cos t$, $0 \leqslant t \leqslant \pi/2$
10. $\int_C x^2z\,ds$, C: $x = \sin 2t$, $y = 3t$, $z = \cos 2t$, $0 \leqslant t \leqslant \pi/4$
11. $\int_C xy^2z\,ds$, C is the line segment from $(1, 0, 1)$ to $(0, 3, 6)$
12. $\int_C xz\,ds$, C: $x = 6t$, $y = 3\sqrt{2}t^2$, $z = 2t^3$, $0 \leqslant t \leqslant 1$
13. $\int_C x^3y^2z\,dz$, C: $x = 2t$, $y = t^2$, $z = t^2$, $0 \leqslant t \leqslant 1$
14. $\int_C yz\,dy + xy\,dz$, C: $x = \sqrt{t}$, $y = t$, $z = t^2$, $0 \leqslant t \leqslant 1$
15. $\int_C z^2\,dx - z\,dy + 2y\,dz$, C consists of line segments from $(0, 0, 0)$ to $(0, 1, 1)$, from $(0, 1, 1)$ to $(1, 2, 3)$, and from $(1, 2, 3)$ to $(1, 2, 4)$
16. $\int_C yz\,dx + xz\,dy + xy\,dz$, C consists of line segments from $(0, 0, 0)$ to $(2, 0, 0)$, from $(2, 0, 0)$ to $(1, 3, -1)$, and from $(1, 3, -1)$ to $(1, 3, 0)$

17–20 ■ Evaluate the line integral $\int_C \mathbf{F} \cdot d\mathbf{r}$, where C is given by the vector function $\mathbf{r}(t)$.

17. $\mathbf{F}(x, y) = x^2y\,\mathbf{i} - xy\,\mathbf{j}$, $\mathbf{r}(t) = t^3\,\mathbf{i} + t^4\,\mathbf{j}$, $0 \leqslant t \leqslant 1$
18. $\mathbf{F}(x, y, z) = (y + z)\,\mathbf{i} - x^2\,\mathbf{j} - 4y^2\,\mathbf{k}$, $\mathbf{r}(t) = t\,\mathbf{i} + t^2\,\mathbf{j} + t^4\,\mathbf{k}$, $0 \leqslant t \leqslant 1$
19. $\mathbf{F}(x, y, z) = \sin x\,\mathbf{i} + \cos y\,\mathbf{j} + xz\,\mathbf{k}$, $\mathbf{r}(t) = t^3\,\mathbf{i} - t^2\,\mathbf{j} + t\,\mathbf{k}$, $0 \leqslant t \leqslant 1$
20. $\mathbf{F}(x, y, z) = x^2\,\mathbf{i} + xy\,\mathbf{j} + z^2\,\mathbf{k}$, $\mathbf{r}(t) = \sin t\,\mathbf{i} + \cos t\,\mathbf{j} + t^2\,\mathbf{k}$, $0 \leqslant t \leqslant \pi/2$
21. (a) Evaluate the line integral $\int_C \mathbf{F} \cdot d\mathbf{r}$, where $\mathbf{F}(x, y) = e^{x-1}\,\mathbf{i} + xy\,\mathbf{j}$ and C is given by $\mathbf{r}(t) = t^2\,\mathbf{i} + t^3\,\mathbf{j}$, $0 \leqslant t \leqslant 1$.
    (b) Illustrate part (a) by using a graphing calculator or computer to graph C and the vectors from the vector field corresponding to $t = 0$, $1/\sqrt{2}$, and 1 (as in Figure 13).
22. (a) Evaluate the line integral $\int_C \mathbf{F} \cdot d\mathbf{r}$, where $\mathbf{F}(x, y, z) = x\,\mathbf{i} - z\,\mathbf{j} + y\,\mathbf{k}$ and C is given by $\mathbf{r}(t) = 2t\,\mathbf{i} + 3t\,\mathbf{j} - t^2\,\mathbf{k}$, $-1 \leqslant t \leqslant 1$.
    (b) Illustrate part (a) by using a computer to graph C and the vectors from the vector field corresponding to $t = \pm 1$ and $\pm\frac{1}{2}$ (as in Figure 13).

23–24 ■ If your calculator or CAS evaluates integrals numerically, use it to find the integral to three decimal places. Otherwise, use the Midpoint Rule with $n = 4$ to approximate the integral.

23. $\int x\sin y\,ds$, C: $x = \ln t$, $y = e^{-t}$, $1 \leqslant t \leqslant 2$
24. $\int z^2\ln(1 + x^2 + y^2)\,ds$, C: $x = t$, $y = t^2$, $z = t^3$, $0 \leqslant t \leqslant 1$
25. Find the exact value of $\int_C x^3y^5\,ds$, where C is the part of the astroid $x = \cos^3 t$, $y = \sin^3 t$ in the first quadrant.
26. Find the exact value of $\int_C \mathbf{F} \cdot d\mathbf{r}$, where $\mathbf{F}(x, y, z) = x^4e^y\,\mathbf{i} + \ln z\,\mathbf{j} + \sqrt{y^2 + z^2}\,\mathbf{k}$ and C is the line segment from $(1, 2, 1)$ to $(6, 4, 5)$.
27. Let $\mathbf{F}$ be the vector field shown in the figure.
    (a) If C_1 is the vertical line segment from $(-3, -3)$ to $(-3, 3)$, determine whether $\int_{C_1} \mathbf{F} \cdot d\mathbf{r}$ is positive, negative, or zero.
    (b) If C_2 is the counterclockwise-oriented circle with radius 3 and center the origin, determine whether $\int_{C_2} \mathbf{F} \cdot d\mathbf{r}$ is positive, negative, or zero.

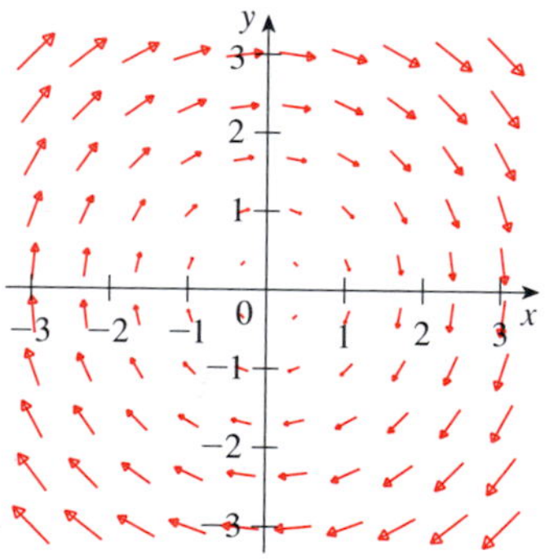

28. (a) Find the work done by the force field $\mathbf{F}(x, y) = x^2\,\mathbf{i} + xy\,\mathbf{j}$ on a particle that moves once around the circle $x^2 + y^2 = 4$ oriented in the counterclockwise direction.
    (b) Use a computer algebra system to graph the force field and circle on the same screen. Use the graph to explain your answer to part (a).
29. A thin wire is bent into the shape of a semicircle $x^2 + y^2 = 4$, $x \geqslant 0$. If the linear density is a constant k, find the mass and center of mass of the wire.
30. Find the mass and center of mass of a thin wire in the shape of a quarter-circle $x^2 + y^2 = r^2$, $x \geqslant 0$, $y \geqslant 0$, if the density function is $\rho(x, y) = x + y$.

31. (a) Write the formulas analogous to Equations 4 for the center of mass $(\bar{x}, \bar{y}, \bar{z})$ of a thin wire with density function $\rho(x, y, z)$ in the shape of a space curve C.
(b) Find the center of mass of a wire in the shape of the helix $x = 2\sin t$, $y = 2\cos t$, $z = 3t$, $0 \leqslant t \leqslant 2\pi$, if the density is a constant k.

32. Find the mass and center of mass of a wire in the shape of the helix $x = t$, $y = \cos t$, $z = \sin t$, $0 \leqslant t \leqslant 2\pi$, if the density at any point is equal to the square of the distance from the origin.

33. If a wire with linear density $\rho(x, y)$ lies along a plane curve C, its **moments of inertia** about the x- and y-axes are defined as

$$I_x = \int_C y^2 \rho(x, y)\, ds \qquad I_y = \int_C x^2 \rho(x, y)\, ds$$

Find the moments of inertia for the wire in Example 3.

34. If a wire with linear density $\rho(x, y, z)$ lies along a space curve C, its **moments of inertia** about the x-, y-, and z-axes are defined as

$$I_x = \int_C (y^2 + z^2)\rho(x, y, z)\, ds$$

$$I_y = \int_C (x^2 + z^2)\rho(x, y, z)\, ds$$

$$I_z = \int_C (x^2 + y^2)\rho(x, y, z)\, ds$$

Find the moments of inertia for the wire in Example 3.

35. Find the work done by the force field $\mathbf{F}(x, y) = x\,\mathbf{i} + (y + 2)\,\mathbf{j}$ in moving an object along an arch of the cycloid $\mathbf{r}(t) = (t - \sin t)\,\mathbf{i} + (1 - \cos t)\,\mathbf{j}$, $0 \leqslant t \leqslant 2\pi$.

36. Find the work done by the force field $\mathbf{F}(x, y) = x \sin y\,\mathbf{i} + y\,\mathbf{j}$ on a particle that moves along the parabola $y = x^2$ from $(-1, 1)$ to $(2, 4)$.

37. Find the work done by the force field $\mathbf{F}(x, y, z) = xz\,\mathbf{i} + yx\,\mathbf{j} + zy\,\mathbf{k}$ on a particle that moves along the curve $\mathbf{r}(t) = t^2\,\mathbf{i} - t^3\,\mathbf{j} + t^4\,\mathbf{k}$, $0 \leqslant t \leqslant 1$.

38. The force exerted by an electric charge at the origin on a charged particle at a point (x, y, z) with position vector $\mathbf{r} = \langle x, y, z\rangle$ is $\mathbf{F}(\mathbf{r}) = K\mathbf{r}/|\mathbf{r}|^3$ where K is a constant. (See Example 5 in Section 14.1.) Find the work done as the particle moves along a straight line from $(2, 0, 0)$ to $(2, 1, 5)$.

39. A 160-lb man carries a 25-lb can of paint up a helical staircase that encircles a silo with a radius of 20 ft. If the silo is 90 ft high and the man makes exactly three complete revolutions, how much work is done by the man against gravity in climbing to the top?

40. Suppose there is a hole in the can of paint in Exercise 39 and 9 lb of paint leak steadily out of the can during the man's ascent. How much work is done?

41. Experiments show that a steady current I in a long wire produces a magnetic field $\mathbf{B}$ that is tangent to any circle that lies in the plane perpendicular to the wire and whose center is the axis of the wire (as in the figure). *Ampère's Law* relates the electric current to its magnetic effects and states that

$$\int_C \mathbf{B} \cdot d\mathbf{r} = \mu_0 I$$

where I is the net current that passes through any surface bounded by a closed curve C and μ_0 is a constant called the permeability of free space. By taking C to be a circle with radius r, show that the magnitude $B = |\mathbf{B}|$ of the magnetic field at a distance r from the center of the wire is

$$B = \frac{\mu_0 I}{2\pi r}$$

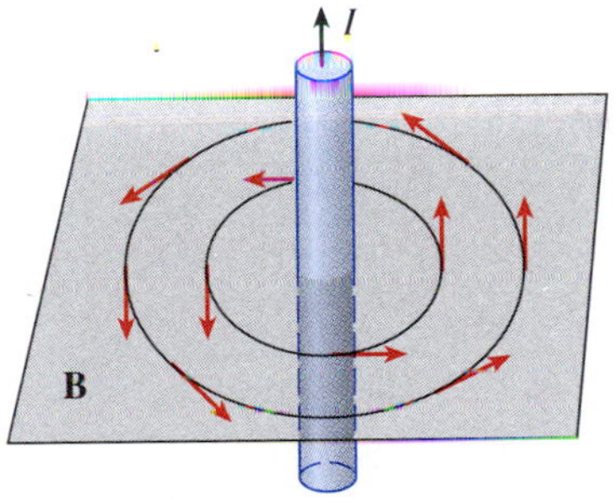

14.3 THE FUNDAMENTAL THEOREM FOR LINE INTEGRALS

Recall from Section 6.4 that Part 2 of the Fundamental Theorem of Calculus can be written as

$$\textbf{(1)} \qquad \int_a^b F'(x)\, dx = F(b) - F(a)$$

where F' is continuous on $[a, b]$. If we think of the gradient vector ∇f of a function f of two or three variables as a sort of derivative of f, then the following theorem can be regarded as a version of the fundamental theorem for line integrals.

(2) THEOREM Let C be a smooth curve given by the vector function $\mathbf{r}(t)$, $a \leqslant t \leqslant b$. Let f be a differentiable function of two or three variables whose gradient vector ∇f is continuous on C. Then

$$\int_C \nabla f \cdot d\mathbf{r} = f(\mathbf{r}(b)) - f(\mathbf{r}(a))$$

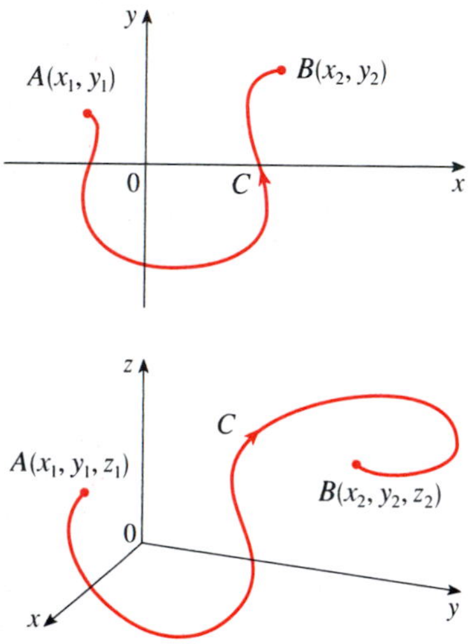

FIGURE 1

NOTE: Theorem 2 says that we can evaluate the line integral of a conservative vector field (the gradient vector field of the potential function f) simply by knowing the value of f at the endpoints of C. If f is a function of two variables and C is a plane curve with initial point $A(x_1, y_1)$ and terminal point $B(x_2, y_2)$, as in Figure 1, then Theorem 2 becomes

$$\int_C \nabla f \cdot d\mathbf{r} = f(x_2, y_2) - f(x_1, y_1)$$

If f is a function of three variables and C is a space curve joining $A(x_1, y_1, z_1)$ to $B(x_2, y_2, z_2)$, then we have

$$\int_C \nabla f \cdot d\mathbf{r} = f(x_2, y_2, z_2) - f(x_1, y_1, z_1)$$

Let's prove Theorem 2 for this last case.

PROOF OF THEOREM 2 Using Definition 14.2.13, we have

$$\begin{aligned}
\int_C \nabla f \cdot d\mathbf{r} &= \int_a^b \nabla f(\mathbf{r}(t)) \cdot \mathbf{r}'(t)\, dt \\
&= \int_a^b \left(\frac{\partial f}{\partial x}\frac{dx}{dt} + \frac{\partial f}{\partial y}\frac{dy}{dt} + \frac{\partial f}{\partial z}\frac{dz}{dt} \right) dt \\
&= \int_a^b \frac{d}{dt} f(\mathbf{r}(t))\, dt && \text{(by the Chain Rule)} \\
&= f(\mathbf{r}(b)) - f(\mathbf{r}(a))
\end{aligned}$$

The last step follows from the Fundamental Theorem of Calculus (Equation 1). □

Although we have proved Theorem 2 for smooth curves, it is also true for piecewise-smooth curves. This can be seen by subdividing C into a finite number of smooth curves and adding the resulting integrals.

EXAMPLE 1 Find the work done by the gravitational field

$$\mathbf{F}(\mathbf{x}) = -\frac{mMG}{|\mathbf{x}|^3}\mathbf{x}$$

in moving a particle with mass m from the point $(3, 4, 12)$ to the point $(2, 2, 0)$ along a piecewise-smooth curve C. (See Example 4 in Section 14.1.)

SOLUTION From Section 14.1 we know that $\mathbf{F}$ is a conservative vector field and, in fact, $\mathbf{F} = \nabla f$, where

$$f(x, y, z) = \frac{mMG}{\sqrt{x^2 + y^2 + z^2}}$$

Therefore, by Theorem 2, the work done is

$$\begin{aligned} W &= \int_C \mathbf{F} \cdot d\mathbf{r} = \int_C \nabla f \cdot d\mathbf{r} \\ &= f(2, 2, 0) - f(3, 4, 12) \\ &= \frac{mMG}{\sqrt{2^2 + 2^2}} - \frac{mMG}{\sqrt{3^2 + 4^2 + 12^2}} = mMG\left(\frac{1}{2\sqrt{2}} - \frac{1}{13}\right) \end{aligned}$$

■

INDEPENDENCE OF PATH

Suppose C_1 and C_2 are two piecewise-smooth curves (which are called **paths**) that have the same initial point A and terminal point B. We know from Example 4 in Section 14.2 that, in general, $\int_{C_1} \mathbf{F} \cdot d\mathbf{r} \neq \int_{C_2} \mathbf{F} \cdot d\mathbf{r}$. But one implication of Theorem 2 is that $\int_{C_1} \nabla f \cdot d\mathbf{r} = \int_{C_2} \nabla f \cdot d\mathbf{r}$ whenever ∇f is continuous. In other words, the line integral of a conservative vector field depends only on the initial point and terminal point of a curve.

In general, if $\mathbf{F}$ is a continuous vector field with domain D, we say that the line integral $\int_C \mathbf{F} \cdot d\mathbf{r}$ is **independent of path** if $\int_{C_1} \mathbf{F} \cdot d\mathbf{r} = \int_{C_2} \mathbf{F} \cdot d\mathbf{r}$ for any two paths C_1 and C_2 in D that have the same initial and terminal points. With this terminology we can say that **line integrals of conservative vector fields are independent of path**.

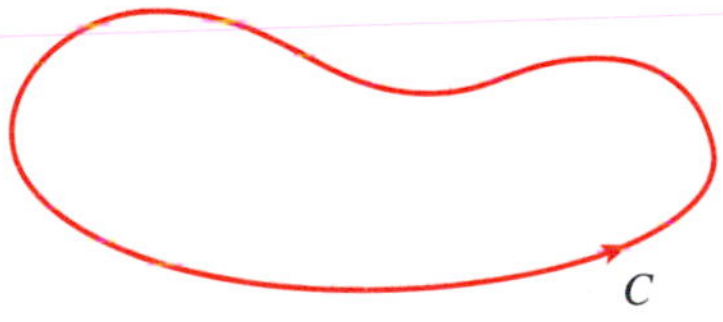

FIGURE 2
A closed curve

A curve is called **closed** if its terminal point coincides with its initial point, that is, $\mathbf{r}(b) = \mathbf{r}(a)$ (see Figure 2). If $\int_C \mathbf{F} \cdot d\mathbf{r}$ is independent of path in D and C is any closed path in D, we can choose any two points A and B on C and regard C as being composed of the path C_1 from A to B followed by the path C_2 from B to A (see Figure 3). Then

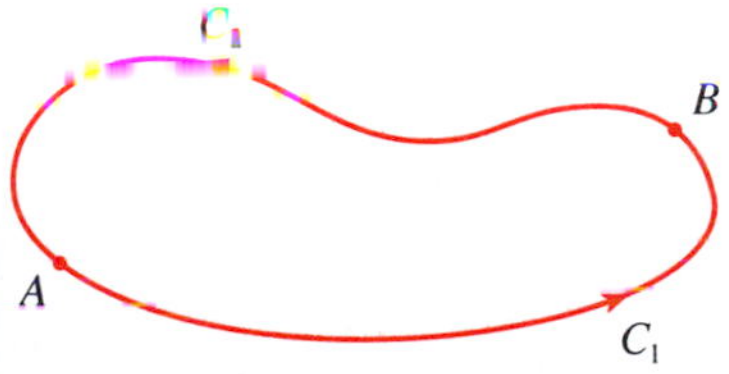

FIGURE 3

$$\int_C \mathbf{F} \cdot d\mathbf{r} = \int_{C_1} \mathbf{F} \cdot d\mathbf{r} + \int_{C_2} \mathbf{F} \cdot d\mathbf{r} = \int_{C_1} \mathbf{F} \cdot d\mathbf{r} - \int_{-C_2} \mathbf{F} \cdot d\mathbf{r} = 0$$

since C_1 and $-C_2$ have the same initial and terminal points.

Conversely, if it is true that $\int_C \mathbf{F} \cdot d\mathbf{r} = 0$ whenever C is a closed path in D, then we demonstrate independence of path as follows. Take any two paths C_1 and C_2 from A to B in D and define C to be the curve consisting of C_1 followed by $-C_2$. Then

$$0 = \int_C \mathbf{F} \cdot d\mathbf{r} = \int_{C_1} \mathbf{F} \cdot d\mathbf{r} + \int_{-C_2} \mathbf{F} \cdot d\mathbf{r} = \int_{C_1} \mathbf{F} \cdot d\mathbf{r} - \int_{C_2} \mathbf{F} \cdot d\mathbf{r}$$

and so $\int_{C_1} \mathbf{F} \cdot d\mathbf{r} = \int_{C_2} \mathbf{F} \cdot d\mathbf{r}$. Thus, we have proved the following theorem:

(3) THEOREM $\int_C \mathbf{F} \cdot d\mathbf{r}$ is independent of path in D if and only if $\int_C \mathbf{F} \cdot d\mathbf{r} = 0$ for every closed path C in D.

Since we know that the line integral of any conservative vector field $\mathbf{F}$ is independent of path, it follows that $\int_C \mathbf{F} \cdot d\mathbf{r} = 0$ for closed paths. The physical interpretation is that the work done by a conservative force field (such as the gravitational or electric field in Section 14.1) as it moves an object around a closed path is 0.

The following theorem says that the *only* vector fields that are independent of path are conservative. It is stated and proved for plane curves, but there is a similar version for space curves. We assume that D is **open,** which means that for every point P in D there is a disk with center P that lies entirely in D. (Thus, every point in D is an interior point in the sense of Section 12.2.) In addition, we assume that D is **connected.** This means that any two points in D can be joined by a path that lies in D.

(4) THEOREM Suppose $\mathbf{F}$ is a vector field that is continuous on an open connected region D. If $\int_C \mathbf{F} \cdot d\mathbf{r}$ is independent of path in D, then $\mathbf{F}$ is a conservative vector field on D; that is, there exists a function f such that $\nabla f = \mathbf{F}$.

PROOF Let $A(a, b)$ be a fixed point in D. We construct the desired potential function f by defining

$$f(x, y) = \int_{(a,b)}^{(x,y)} \mathbf{F} \cdot d\mathbf{r}$$

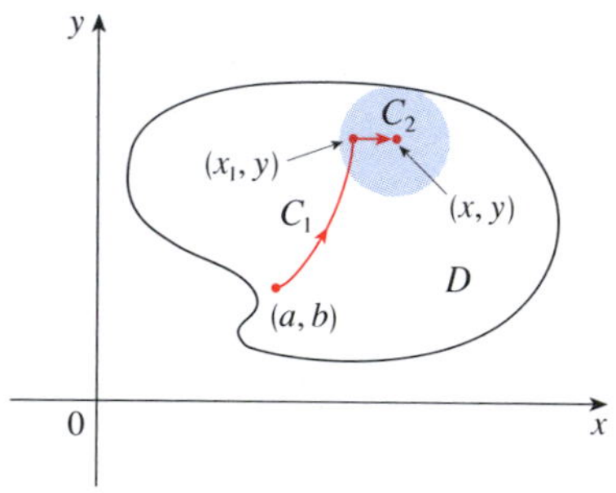

FIGURE 4

for any point (x, y) in D. Since $\int_C \mathbf{F} \cdot d\mathbf{r}$ is independent of path, it does not matter which path C from (a, b) to (x, y) is used to evaluate $f(x, y)$. Since D is open, there exists a disk contained in D with center (x, y). Choose any point (x_1, y) in the disk with $x_1 < x$ and let C consist of any path C_1 from (a, b) to (x_1, y) followed by the horizontal line segment C_2 from (x_1, y) to (x, y) (see Figure 4). Then

$$f(x, y) = \int_{C_1} \mathbf{F} \cdot d\mathbf{r} + \int_{C_2} \mathbf{F} \cdot d\mathbf{r} = \int_{(a,b)}^{(x_1,y)} \mathbf{F} \cdot d\mathbf{r} + \int_{C_2} \mathbf{F} \cdot d\mathbf{r}$$

Notice that the first of these integrals does not depend on x, so

$$\frac{\partial}{\partial x} f(x, y) = 0 + \frac{\partial}{\partial x} \int_{C_2} \mathbf{F} \cdot d\mathbf{r}$$

If we write $\mathbf{F} = P\,\mathbf{i} + Q\,\mathbf{j}$, then

$$\int_{C_2} \mathbf{F} \cdot d\mathbf{r} = \int_{C_2} P\,dx + Q\,dy$$

On C_2, y is constant, so $dy = 0$. Using t as the parameter, where $x_1 \leqslant t \leqslant x$, we have

$$\frac{\partial}{\partial x} f(x, y) = \frac{\partial}{\partial x} \int_{C_2} P\,dx + Q\,dy$$

$$= \frac{\partial}{\partial x} \int_{x_1}^{x} P(t, y)\,dt = P(x, y)$$

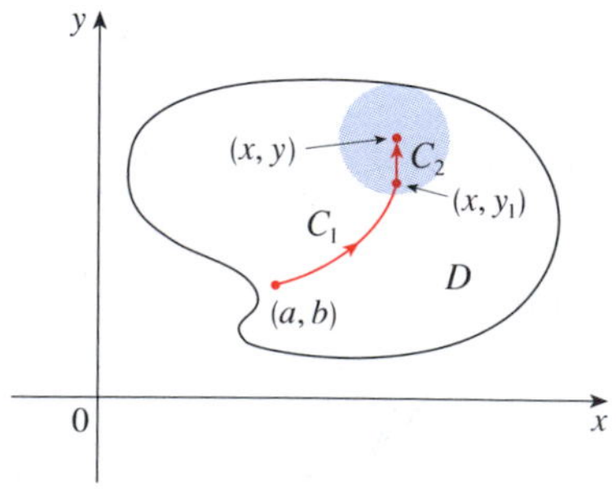

FIGURE 5

by Part 1 of the Fundamental Theorem of Calculus (6.4.2). A similar argument, using a vertical line segment (see Figure 5), shows that

$$\frac{\partial}{\partial y} f(x, y) = \frac{\partial}{\partial y} \int_{C_2} P\,dx + Q\,dy = \frac{\partial}{\partial y} \int_{y_1}^{y} Q(x, t)\,dt = Q(x, y)$$

Thus
$$\mathbf{F} = P\,\mathbf{i} + Q\,\mathbf{j} = \frac{\partial f}{\partial x}\,\mathbf{i} + \frac{\partial f}{\partial y}\,\mathbf{j} = \nabla f$$

which says that **F** is conservative.

The question remains: How is it possible to determine whether or not a vector field **F** is conservative? Suppose it is known that $\mathbf{F} = P\,\mathbf{i} + Q\,\mathbf{j}$ is conservative, where P and Q have continuous first-order partial derivatives. Then there is a function f such that $\mathbf{F} = \nabla f$, that is,

$$P = \frac{\partial f}{\partial x} \quad \text{and} \quad Q = \frac{\partial f}{\partial y}$$

Therefore, by Clairaut's Theorem,

$$\frac{\partial P}{\partial y} = \frac{\partial^2 f}{\partial y\,\partial x} = \frac{\partial^2 f}{\partial x\,\partial y} = \frac{\partial Q}{\partial x}$$

(5) THEOREM If $\mathbf{F}(x, y) = P(x, y)\,\mathbf{i} + Q(x, y)\,\mathbf{j}$ is a conservative vector field, where P and Q have continuous first-order partial derivatives on a domain D, then throughout D we have

$$\frac{\partial P}{\partial y} = \frac{\partial Q}{\partial x}$$

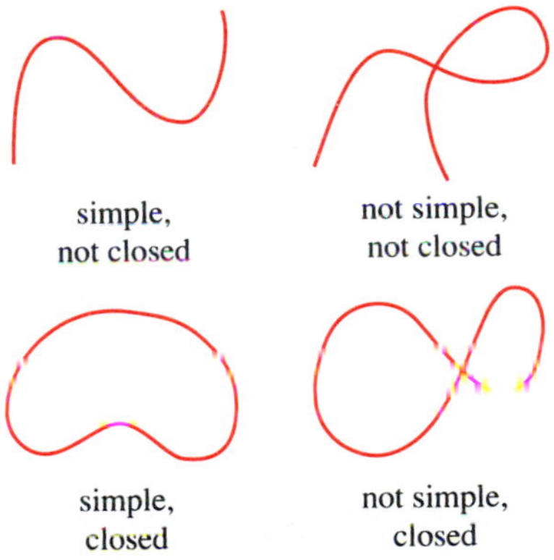

FIGURE 6
Types of curves

The converse of Theorem 5 is true only for a special type of region. To explain this, we first need the concept of a **simple curve**, which is a curve that does not intersect itself anywhere between its endpoints. [See Figure 6; $\mathbf{r}(a) = \mathbf{r}(b)$ for a simple closed curve, but $\mathbf{r}(t_1) \neq \mathbf{r}(t_2)$ when $a < t_1 < t_2 < b$.]

In Theorem 4 we needed an open connected region. For the next theorem we need a stronger condition. A **simply-connected region** in the plane is a connected region D such that every simple closed curve in D encloses only points that are in D. Notice from Figure 7 that, intuitively speaking, a simply-connected region contains no hole and cannot consist of two separate pieces.

In terms of simply-connected regions we can now state a partial converse to Theorem 5 that gives a convenient method for verifying that a vector field on $\mathbb{R}^2$ is conservative. The proof will be sketched in the next section as a consequence of Green's Theorem.

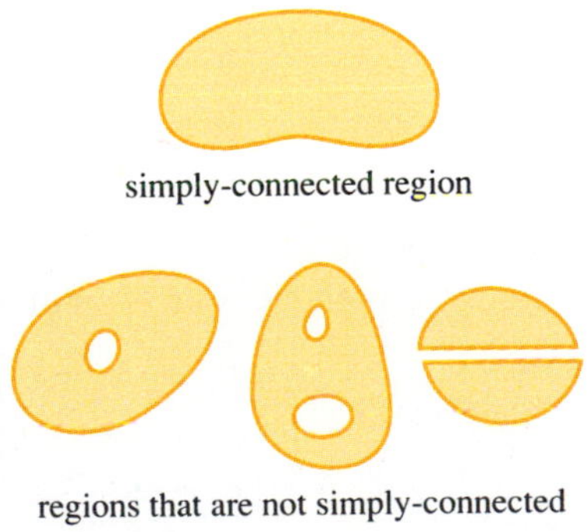

FIGURE 7

(6) THEOREM Let $\mathbf{F} = P\,\mathbf{i} + Q\,\mathbf{j}$ be a vector field on an open simply-connected region D. Suppose that P and Q have continuous first-order derivatives and

$$\frac{\partial P}{\partial y} = \frac{\partial Q}{\partial x} \quad \text{throughout } D$$

Then **F** is conservative.

EXAMPLE 2 Determine whether or not the vector field

$$\mathbf{F}(x, y) = (x - y)\,\mathbf{i} + (x - 2)\,\mathbf{j}$$

is conservative.

SOLUTION Let $P(x, y) = x - y$ and $Q(x, y) = x - 2$. Then

$$\frac{\partial P}{\partial y} = -1 \qquad \frac{\partial Q}{\partial x} = 1$$

Since $\partial P/\partial y \neq \partial Q/\partial x$, $\mathbf{F}$ is not conservative by Theorem 5. ■

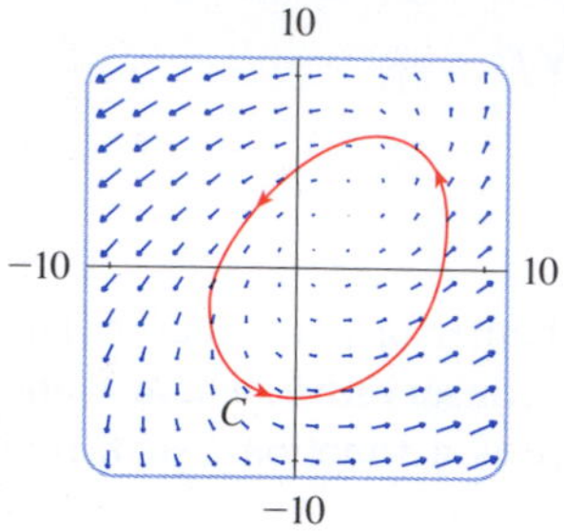

FIGURE 8

Figures 8 and 9 show the vector fields in Examples 2 and 3, respectively. The vectors in Figure 8 that start on the closed curve C all appear to point in roughly the same direction as C. So it looks as if $\int_C \mathbf{F} \cdot d\mathbf{r} > 0$ and therefore $\mathbf{F}$ is not conservative. The calculation in Example 2 confirms this impression. Some of the vectors near the curves C_1 and C_2 in Figure 9 point in approximately the same direction as the curves, whereas others point in the opposite direction. So it appears plausible that line integrals around all closed paths are 0. Example 3 shows that $\mathbf{F}$ is indeed conservative.

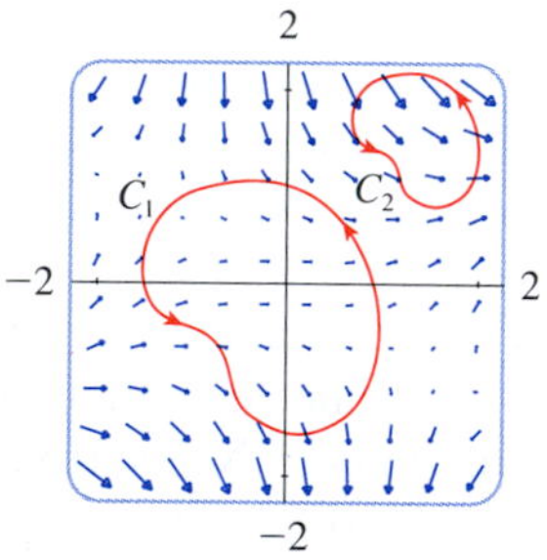

FIGURE 9

EXAMPLE 3 Determine whether or not the vector field

$$\mathbf{F}(x, y) = (3 + 2xy)\,\mathbf{i} + (x^2 - 3y^2)\,\mathbf{j}$$

is conservative.

SOLUTION Let $P(x, y) = 3 + 2xy$ and $Q(x, y) = x^2 - 3y^2$. Then

$$\frac{\partial P}{\partial y} = 2x = \frac{\partial Q}{\partial x}$$

Also the domain of $\mathbf{F}$ is the entire plane ($D = \mathbb{R}^2$), which is open and simply-connected. Therefore, we can apply Theorem 6 and conclude that $\mathbf{F}$ is conservative. ■

In Example 3, Theorem 6 told us that $\mathbf{F}$ is conservative, but it did not tell us how to find the (potential) function f such that $\mathbf{F} = \nabla f$. The proof of Theorem 4 gives us a clue as to how to find f. We use "partial integration" as in the following example.

EXAMPLE 4
(a) If $\mathbf{F}(x, y) = (3 + 2xy)\,\mathbf{i} + (x^2 - 3y^2)\,\mathbf{j}$, find a function f such that $\mathbf{F} = \nabla f$.
(b) Evaluate the line integral $\int_C \mathbf{F} \cdot d\mathbf{r}$, where C is the curve given by $\mathbf{r}(t) = e^t \sin t\,\mathbf{i} + e^t \cos t\,\mathbf{j}$, $0 \leqslant t \leqslant \pi$.

SOLUTION
(a) From Example 3 we know that $\mathbf{F}$ is conservative and so there exists a function f with $\nabla f = \mathbf{F}$, that is,

$$f_x(x, y) = 3 + 2xy \tag{7}$$

$$f_y(x, y) = x^2 - 3y^2 \tag{8}$$

Integrating (7) with respect to x, we obtain

$$f(x, y) = 3x + x^2 y + g(y) \tag{9}$$

Notice that the constant of integration is a constant with respect to x, that is, a function of y, which we have called $g(y)$. Next we differentiate both sides of (9) with respect to y:

$$f_y(x, y) = x^2 + g'(y) \tag{10}$$

Comparing (8) and (10), we see that

$$g'(y) = -3y^2$$

Integrating with respect to y, we have

$$g(y) = -y^3 + K$$

where K is a constant. Putting this in (9), we have

$$f(x, y) = 3x + x^2y - y^3 + K$$

as the desired potential function.

(b) To use Theorem 2 all we have to know are the initial and terminal points of C, namely, $\mathbf{r}(0) = (0, 1)$ and $\mathbf{r}(\pi) = (0, -e^{\pi})$. In the expression for $f(x, y)$ in part (a), any value of the constant K will do, so let's choose $K = 0$. Then we have

$$\int_C \mathbf{F} \cdot d\mathbf{r} = \int_C \nabla f \cdot d\mathbf{r} = f(0, -e^{\pi}) - f(0, 1)$$

$$= e^{3\pi} - (-1) = e^{3\pi} + 1$$

This method is much shorter than the straightforward method for evaluating line integrals that we learned in Section 14.2. ■

A criterion for determining whether or not a vector field $\mathbf{F}$ on $\mathbb{R}^3$ is conservative is given in Section 14.5. Meanwhile, the next example shows that the technique for finding the potential function is much the same as for vector fields on $\mathbb{R}^2$.

EXAMPLE 5 If $\mathbf{F}(x, y, z) = y^2\,\mathbf{i} + (2xy + e^{3z})\,\mathbf{j} + 3ye^{3z}\,\mathbf{k}$, find a function f such that $\nabla f = \mathbf{F}$.

SOLUTION If there is such a function f, then

$$f_x(x, y, z) = y^2 \tag{11}$$

$$f_y(x, y, z) = 2xy + e^{3z} \tag{12}$$

$$f_z(x, y, z) = 3ye^{3z} \tag{13}$$

Integrating (11) with respect to x, we get

$$f(x, y, z) = xy^2 + g(y, z) \tag{14}$$

where $g(y, z)$ is a constant with respect to x. Then differentiating (14) with respect to y, we have

$$f_y(x, y, z) = 2xy + g_y(y, z)$$

and comparison with (12) gives

$$g_y(y, z) = e^{3z}$$

Thus $g(y, z) = ye^{3z} + h(z)$ and we rewrite (14) as

$$f(x, y, z) = xy^2 + ye^{3z} + h(z)$$

Finally, differentiating with respect to z and comparing with (13), we obtain $h'(z) = 0$ and, therefore, $h(z) = K$, a constant. The desired function is

$$f(x, y, z) = xy^2 + ye^{3z} + K$$

It is easily verified that $\nabla f = \mathbf{F}$. ■

CONSERVATION OF ENERGY

Let's apply the ideas of this chapter to a continuous force field $\mathbf{F}$ that moves an object along a path C given by $\mathbf{r}(t)$, $a \leqslant t \leqslant b$, where $\mathbf{r}(a) = A$ is the initial point and $\mathbf{r}(b) = B$ is the terminal point of C. According to Newton's Second Law of Motion (see Section 11.8), the force $\mathbf{F}(\mathbf{r}(t))$ at a point on C is related to the acceleration $\mathbf{a}(t) = \mathbf{r}''(t)$ by the equation

$$\mathbf{F}(\mathbf{r}(t)) = m\mathbf{r}''(t)$$

So the work done by the force on the object is

$$\begin{aligned} W &= \int_C \mathbf{F} \cdot d\mathbf{r} = \int_a^b \mathbf{F}(\mathbf{r}(t)) \cdot \mathbf{r}'(t)\,dt \\ &= \int_a^b m\mathbf{r}''(t) \cdot \mathbf{r}'(t)\,dt \\ &= \frac{m}{2}\int_a^b \frac{d}{dt}[\mathbf{r}'(t) \cdot \mathbf{r}'(t)]\,dt \qquad \text{(Theorem 11.6.5, Formula 4)} \\ &= \frac{m}{2}\int_a^b \frac{d}{dt}|\mathbf{r}'(t)|^2\,dt \\ &= \frac{m}{2}\Big[|\mathbf{r}'(t)|^2\Big]_a^b \qquad \text{(Fundamental Theorem of Calculus)} \\ &= \frac{m}{2}(|\mathbf{r}'(b)|^2 - |\mathbf{r}'(a)|^2) \end{aligned}$$

Therefore

$$W = \tfrac{1}{2}m|\mathbf{v}(b)|^2 - \tfrac{1}{2}m|\mathbf{v}(a)|^2 \qquad (15)$$

where $\mathbf{v} = \mathbf{r}'$ is the velocity.

The quantity $\frac{1}{2}m|\mathbf{v}(t)|^2$, that is, half the mass times the square of the speed, is called the **kinetic energy** of the object. Therefore, we can rewrite Equation 15 as

$$W = K(B) - K(A) \qquad (16)$$

which says that the work done by the force field along C is equal to the change in kinetic energy at the endpoints of C.

Now let us further assume that $\mathbf{F}$ is a conservative force field; that is, we can write $\mathbf{F} = \nabla f$. In physics, the **potential energy** of an object at the point (x, y, z) is defined as $P(x, y, z) = -f(x, y, z)$, so we have $\mathbf{F} = -\nabla P$. Then by Theorem 2 we have

$$\begin{aligned} W &= \int_C \mathbf{F} \cdot d\mathbf{r} = -\int_C \nabla P \cdot d\mathbf{r} \\ &= -[P(\mathbf{r}(b)) - P(\mathbf{r}(a))] \\ &= P(A) - P(B) \end{aligned}$$

Comparing this equation with Equation 16, we see that

$$P(A) + K(A) = P(B) + K(B)$$

which says that if an object moves from one point A to another point B under the influence of a conservative force field, then the sum of its potential energy and its kinetic energy remains constant. This is called the **Law of Conservation of Energy** and it is the reason the vector field is called **conservative**.

EXERCISES 14.3

1–10 ■ Determine whether or not $\mathbf{F}$ is a conservative vector field. If it is, find a function f such that $\mathbf{F} = \nabla f$.

1. $\mathbf{F}(x, y) = (2x - 3y)\,\mathbf{i} + (2y - 3x)\,\mathbf{j}$

2. $\mathbf{F}(x, y) = (3x^2 - 4y)\,\mathbf{i} + (4y^2 - 2x)\,\mathbf{j}$

3. $\mathbf{F}(x, y) = (x^2 + y)\,\mathbf{i} + x^2\mathbf{j}$

4. $\mathbf{F}(x, y) = (x^2 + y)\,\mathbf{i} + (y^2 + x)\,\mathbf{j}$

5. $\mathbf{F}(x, y) = (1 + 4x^3y^3)\,\mathbf{i} + 3x^4y^2\,\mathbf{j}$

6. $\mathbf{F}(x, y) = (y\cos x - \cos y)\,\mathbf{i} + (\sin x + x\sin y)\,\mathbf{j}$

7. $\mathbf{F}(x, y) = (e^{2x} + x\sin y)\,\mathbf{i} + x^2\cos y\,\mathbf{j}$

8. $\mathbf{F}(x, y) = (ye^{xy} + 4x^3y)\,\mathbf{i} + (xe^{xy} + x^4)\,\mathbf{j}$

9. $\mathbf{F}(x, y) = (ye^x + \sin y)\,\mathbf{i} + (e^x + x\cos y)\,\mathbf{j}$

10. $\mathbf{F}(x, y) = (x + y^2)\,\mathbf{i} + (2xy + y^2)\,\mathbf{j}$

11–18 ■ (a) Find a function f such that $\mathbf{F} = \nabla f$ and (b) use part (a) to evaluate $\int_C \mathbf{F} \cdot d\mathbf{r}$ along the given curve C.

11. $\mathbf{F}(x, y) = x\,\mathbf{i} + y\,\mathbf{j}$,
C is the arc of the parabola $y = x^2$ from $(-1, 1)$ to $(3, 9)$

12. $\mathbf{F}(x, y) = y\,\mathbf{i} + x\,\mathbf{j}$,
C is the arc of the curve $y = x^4 - x^3$ from $(1, 0)$ to $(2, 8)$

13. $\mathbf{F}(x, y) = 2xy^3\,\mathbf{i} + 3x^2y^2\,\mathbf{j}$,
C: $\mathbf{r}(t) = \sin t\,\mathbf{i} + (t^2 + 1)\,\mathbf{j}$, $0 \leqslant t \leqslant \pi/2$

14. $\mathbf{F}(x, y) = e^{2y}\,\mathbf{i} + (1 + 2xe^{2y})\,\mathbf{j}$,
C: $\mathbf{r}(t) = te^t\,\mathbf{i} + (1 + t)\,\mathbf{j}$, $0 \leqslant t \leqslant 1$

15. $\mathbf{F}(x, y, z) = y\,\mathbf{i} + (x + z)\,\mathbf{j} + y\,\mathbf{k}$,
C is the line segment from $(2, 1, 4)$ to $(8, 3, -1)$

16. $\mathbf{F}(x, y, z) = 2xy^3z^4\,\mathbf{i} + 3x^2y^2z^4\,\mathbf{j} + 4x^2y^3z^3\,\mathbf{k}$,
C: $x = t$, $y = t^2$, $z = t^3$, $0 \leqslant t \leqslant 2$

17. $\mathbf{F}(x, y, z) = (2xz + \sin y)\,\mathbf{i} + x\cos y\,\mathbf{j} + x^2\,\mathbf{k}$,
C: $\mathbf{r}(t) = \cos t\,\mathbf{i} + \sin t\,\mathbf{j} + t\,\mathbf{k}$, $0 \leqslant t \leqslant 2\pi$

18. $\mathbf{F}(x, y, z) = 4xe^z\,\mathbf{i} + \cos y\,\mathbf{j} + 2x^2e^z\,\mathbf{k}$,
C: $\mathbf{r}(t) = t\,\mathbf{i} + t^2\,\mathbf{j} + t^4\,\mathbf{k}$, $0 \leqslant t \leqslant 1$

19–20 ■ Show that the line integral is independent of path and evaluate the integral.

19. $\int_C 2x\sin y\,dx + (x^2\cos y - 3y^2)\,dy$,
C is any path from $(-1, 0)$ to $(5, 1)$

20. $\int_C (2y^2 - 12x^3y^3)\,dx + (4xy - 9x^4y^2)\,dy$,
C is any path from $(1, 1)$ to $(3, 2)$

21–22 ■ Find the work done by the force field $\mathbf{F}$ in moving an object from P to Q.

21. $\mathbf{F}(x, y) = x^2y^3\,\mathbf{i} + x^3y^2\,\mathbf{j}$; $P(0, 0)$, $Q(2, 1)$

22. $\mathbf{F}(x, y) = (y^2/x^2)\,\mathbf{i} - (2y/x)\,\mathbf{j}$; $P(1, 1)$, $Q(4, -2)$

23. Is the vector field shown in the figure conservative? Explain.

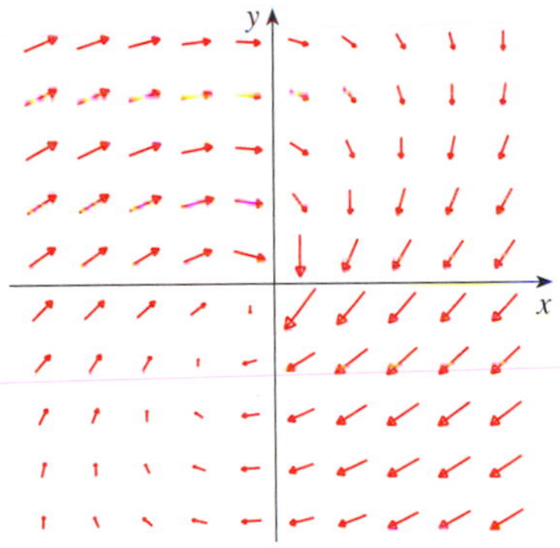

CAS **24–25** ■ From a plot of $\mathbf{F}$ guess whether it is conservative. Then determine whether your guess is correct.

24. $\mathbf{F}(x, y) = (2xy + \sin y)\,\mathbf{i} + (x^2 + x\cos y)\,\mathbf{j}$

25. $\mathbf{F}(x, y) = \dfrac{(x - 2y)\,\mathbf{i} + (x - 2)\,\mathbf{j}}{\sqrt{1 + x^2 + y^2}}$

26. Let $\mathbf{F} = \nabla f$, where $f(x, y) = \sin(x - 2y)$. Find curves C_1 and C_2 that are not closed and satisfy the equation.

(a) $\int_{C_1} \mathbf{F} \cdot d\mathbf{r} = 0$ (b) $\int_{C_2} \mathbf{F} \cdot d\mathbf{r} = 1$

27. Show that if the vector field $\mathbf{F} = P\,\mathbf{i} + Q\,\mathbf{j} + R\,\mathbf{k}$ is conservative and P, Q, R have continuous first-order partial derivatives, then

$$\frac{\partial P}{\partial y} = \frac{\partial Q}{\partial x} \qquad \frac{\partial P}{\partial z} = \frac{\partial R}{\partial x} \qquad \frac{\partial Q}{\partial z} = \frac{\partial R}{\partial y}$$

28. Use Exercise 27 to show that the line integral $\int_C y\,dx + x\,dy + xyz\,dz$ is not independent of path.

29–32 ■ Determine whether or not the given set is (a) open, (b) connected, and (c) simply-connected.

29. $\{(x, y) \mid x > 0, y > 0\}$

30. $\{(x, y) \mid x \neq 0\}$

31. $\{(x, y) \mid 1 < x^2 + y^2 < 4\}$

32. $\{(x, y) \mid x^2 + y^2 \leq 1 \text{ or } 4 \leq x^2 + y^2 \leq 9\}$

33. Let $\mathbf{F}(x, y) = \dfrac{-y\,\mathbf{i} + x\,\mathbf{j}}{x^2 + y^2}$.

(a) Show that $\partial P/\partial y = \partial Q/\partial x$.

(b) Show that $\int_C \mathbf{F} \cdot d\mathbf{r}$ is not independent of path. [*Hint:* Compute $\int_{C_1} \mathbf{F} \cdot d\mathbf{r}$ and $\int_{C_2} \mathbf{F} \cdot d\mathbf{r}$, where C_1 and C_2 are the upper and lower halves of the circle $x^2 + y^2 = 1$ from $(1, 0)$ to $(-1, 0)$.] Does this contradict Theorem 6?

34. (a) Suppose that $\mathbf{F}$ is an inverse square force field, that is,

$$\mathbf{F}(\mathbf{r}) = \frac{c\mathbf{r}}{|\mathbf{r}|^3}$$

for some constant c, where $\mathbf{r} = x\,\mathbf{i} + y\,\mathbf{j} + z\,\mathbf{k}$. Find the work done by $\mathbf{F}$ in moving an object from a point P_1 along a path to a point P_2 in terms of the distances d_1 and d_2 from these points to the origin.

(b) An example of an inverse square field is the gravitational field $\mathbf{F} = -(mMG)\mathbf{r}/|\mathbf{r}|^3$ discussed in Example 4 in Section 14.1. Use part (a) to find the work done by the gravitational field when the earth moves from aphelion (at a maximum distance of 1.52×10^8 km from the sun) to perihelion (at a minimum distance of 1.47×10^8 km). (Use the values $m = 5.97 \times 10^{24}$ kg, $M = 1.99 \times 10^{30}$ kg, and $G = 6.67 \times 10^{-11}$ N·m^2/kg^2.)

(c) Another example of an inverse square field is the electric field $\mathbf{E} = \varepsilon q Q \mathbf{r}/|\mathbf{r}|^3$ discussed in Example 5 in Section 14.1. Suppose that an electron with a charge of -1.6×10^{-19} C is at the origin. A positive unit charge is at a distance 10^{-12} m from the electron and moves to a position half that distance from the electron. Use part (a) to find the work done by the electric field. (Use the value $\varepsilon = 8.985 \times 10^{10}$.)

14.4 GREEN'S THEOREM

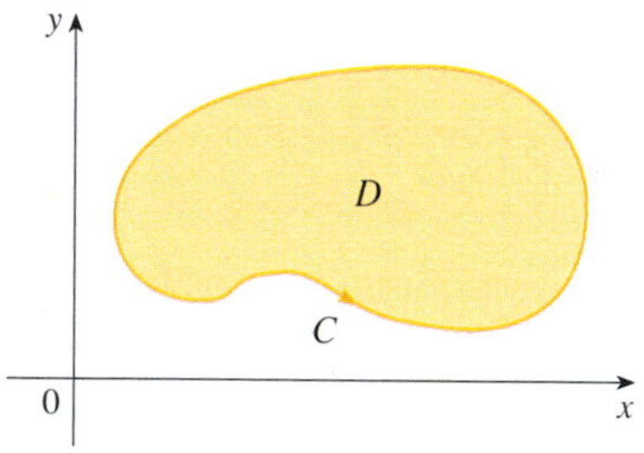

FIGURE 1

Green's Theorem gives the relationship between a line integral around a simple closed curve C and a double integral over the plane region D bounded by C. (See Figure 1. We assume that D consists of all points inside C as well as all points on C.) In stating Green's Theorem we use the convention that the **positive orientation** of a simple closed curve C refers to a single *counterclockwise* traversal of C. Thus, if C is given by the vector function $\mathbf{r}(t)$, $a \leq t \leq b$, then the region D is always on the left as the point $\mathbf{r}(t)$ traverses C (see Figure 2).

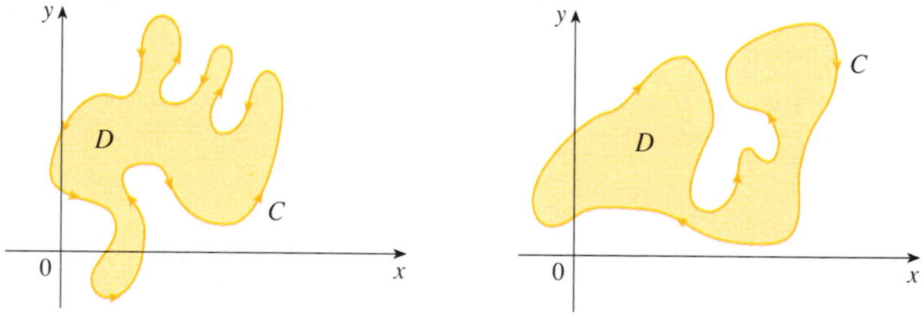

FIGURE 2 (a) Positive orientation (b) Negative orientation

> **(1) GREEN'S THEOREM** Let C be a positively oriented, piecewise-smooth, simple closed curve in the plane and let D be the region bounded by C. If P and Q have continuous partial derivatives on an open region that contains D, then
>
> $$\int_C P\,dx + Q\,dy = \iint_D \left(\frac{\partial Q}{\partial x} - \frac{\partial P}{\partial y} \right) dA$$

NOTE: The notation

$$\oint_C P\,dx + Q\,dy \qquad \text{or} \qquad \oint_C P\,dx + Q\,dy$$

is sometimes used to indicate that the line integral is calculated using the positive orientation of the closed curve C. Another notation for the positively oriented boundary curve of D is ∂D, so the equation in Green's Theorem can be written as

(2) $$\iint_D \left(\frac{\partial Q}{\partial x} - \frac{\partial P}{\partial y}\right) dA = \int_{\partial D} P\,dx + Q\,dy$$

Green's Theorem should be regarded as the analogue of the Fundamental Theorem of Calculus for double integrals. Compare Equation 2 with the statement of the Fundamental Theorem of Calculus, Part 2, in the following equation:

$$\int_a^b F'(x)\,dx = F(b) - F(a)$$

In both cases there is an integral involving derivatives (F', $\partial Q/\partial x$, and $\partial P/\partial y$) on the left side of the equation. And in both cases the right side involves the values of the original functions (F, Q, and P) only on the *boundary* of the domain. (In the one-dimensional case, the domain is an interval $[a, b]$ whose boundary consists of just two points, a and b.)

Green's Theorem is named after the self-taught English scientist George Green (1793–1841). He worked in his father's bakery from an early age and taught himself mathematics from library books. In 1828 he published privately *An Essay on the Application of Mathematical Analysis to the Theories of Electricity and Magnetism*, but only 100 copies were printed and most of those went to friends. What we know as Green's Theorem was in this book, but it did not become widely known at that time. Finally, at age 40, Green entered Cambridge University as an undergraduate but died four years after graduation. In 1846 William Thomson (Lord Kelvin) located a copy of Green's essay, realized its significance, and had it reprinted. Green was the first person to try to formulate a mathematical theory of electricity and magnetism. His work was the basis for the subsequent electromagnetic theories of Thomson, Stokes, Rayleigh, and Maxwell.

Green's Theorem is not easy to prove in the generality stated in Theorem 1, but we can give a proof for the special case where the region is both of type I and of type II (see Section 13.3). Let's call such regions **simple regions**.

PROOF OF GREEN'S THEOREM FOR THE CASE IN WHICH D IS A SIMPLE REGION

Notice that Green's Theorem will be proved if we can show that

(3) $$\int_C P\,dx = -\iint_D \frac{\partial P}{\partial y}\,dA$$

and

(4) $$\int_C Q\,dy = \iint_D \frac{\partial Q}{\partial x}\,dA$$

We prove Equation 3 by expressing D as a type I region:

$$D = \{(x, y) \mid a \leqslant x \leqslant b,\ g_1(x) \leqslant y \leqslant g_2(x)\}$$

where g_1 and g_2 are continuous functions. This enables us to compute the double integral on the right side of Equation 3 as follows:

(5) $$\iint_D \frac{\partial P}{\partial y}\,dA = \int_a^b \int_{g_1(x)}^{g_2(x)} \frac{\partial P}{\partial y}(x, y)\,dy\,dx = \int_a^b [P(x, g_2(x)) - P(x, g_1(x))]\,dx$$

where the last step follows from the Fundamental Theorem of Calculus.

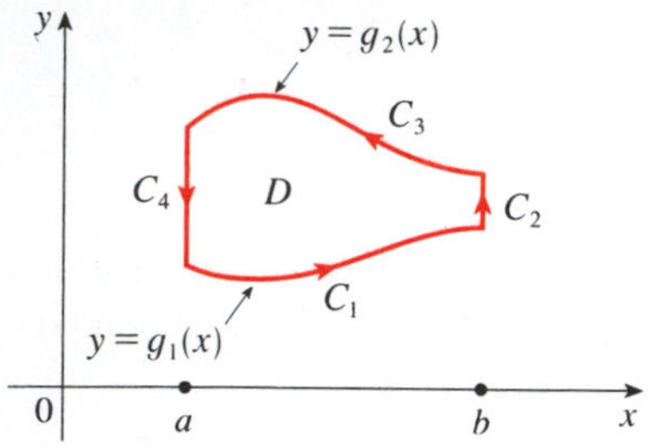

FIGURE 3

Now we compute the left side of Equation 3 by breaking up C as the union of the four curves C_1, C_2, C_3, and C_4 shown in Figure 3. On C_1 we take x as the parameter and write the parametric equations as $x = x$, $y = g_1(x)$, $a \leqslant x \leqslant b$. Thus

$$\int_{C_1} P(x, y)\, dx = \int_a^b P(x, g_1(x))\, dx$$

Observe that C_3 goes from right to left but $-C_3$ goes from left to right, so we can write the parametric equations of $-C_3$ as $x = x$, $y = g_2(x)$, $a \leqslant x \leqslant b$. Therefore

$$\int_{C_3} P(x, y)\, dx = -\int_{-C_3} P(x, y)\, dx = -\int_a^b P(x, g_2(x))\, dx$$

On C_2 or C_4 (either of which might reduce to just a single point), x is constant, so $dx = 0$ and

$$\int_{C_2} P(x, y)\, dx = 0 = \int_{C_4} P(x, y)\, dx$$

Hence

$$\int_C P(x, y)\, dx = \int_{C_1} P(x, y)\, dx + \int_{C_2} P(x, y)\, dx + \int_{C_3} P(x, y)\, dx + \int_{C_4} P(x, y)\, dx$$

$$= \int_a^b P(x, g_1(x))\, dx - \int_a^b P(x, g_2(x))\, dx$$

Comparing this expression with the one in Equation 5, we see that

$$\int_C P(x, y)\, dx = -\iint_D \frac{\partial P}{\partial y}\, dA$$

Equation 4 can be proved in much the same way by expressing D as a type II region (see Exercise 30). Then, by adding Equations 3 and 4, we obtain Green's Theorem. □

EXAMPLE 1 Evaluate $\int_C x^4\, dx + xy\, dy$, where C is the triangular curve consisting of the line segments from $(0, 0)$ to $(1, 0)$, from $(1, 0)$ to $(0, 1)$, and from $(0, 1)$ to $(0, 0)$.

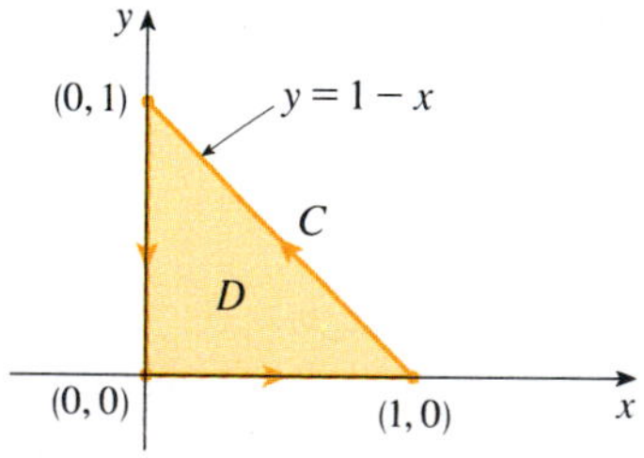

FIGURE 4

SOLUTION Although the given line integral could be evaluated as usual by the methods of Section 14.2, that would involve setting up three separate integrals along the three sides of the triangle, so let's instead use Green's Theorem. Notice that the region D enclosed by C is simple and C has positive orientation (see Figure 4). If we let $P(x, y) = x^4$ and $Q(x, y) = xy$, then we have

$$\int_C x^4\, dx + xy\, dy = \iint_D \left(\frac{\partial Q}{\partial x} - \frac{\partial P}{\partial y} \right) dA = \int_0^1 \int_0^{1-x} (y - 0)\, dy\, dx$$

$$= \int_0^1 \left[\frac{y^2}{2} \right]_{y=0}^{y=1-x} dx = \frac{1}{2} \int_0^1 (1 - x)^2\, dx$$

$$= -\frac{1}{6}(1 - x)^3 \Big]_0^1 = \frac{1}{6}$$

■

EXAMPLE 2 Evaluate $\oint_C (3y - e^{\sin x})\, dx + (7x + \sqrt{y^4 + 1}\,)\, dy$, where C is the circle $x^2 + y^2 = 9$.

SOLUTION The region D bounded by C is the disk $x^2 + y^2 \leq 9$, so let's change to polar coordinates after applying Green's Theorem:

$$\oint_C (3y - e^{\sin x})\,dx + (7x + \sqrt{y^4 + 1})\,dy$$

$$= \iint_D \left[\frac{\partial}{\partial x}(7x + \sqrt{y^4+1}) - \frac{\partial}{\partial y}(3y - e^{\sin x})\right] dA$$

$$= \int_0^{2\pi} \int_0^3 (7 - 3)\,r\,dr\,d\theta$$

$$= 4\int_0^{2\pi} d\theta \int_0^3 r\,dr = 36\pi$$

Instead of using polar coordinates, we could simply use the fact that D is a disk of radius 3 and write

$$\iint_D 4\,dA = 4 \cdot \pi(3)^2 = 36\pi$$

In Examples 1 and 2 we found that the double integral was easier to evaluate than the line integral. (Try setting up the line integral in Example 2 and you will be convinced!) But sometimes it is easier to evaluate the line integral, and Green's Theorem is used in the reverse direction. For instance, if it is known that $P(x, y) = Q(x, y) = 0$ on the curve C, then Green's Theorem gives

$$\iint_D \left(\frac{\partial Q}{\partial x} - \frac{\partial P}{\partial y}\right) dA = \int_C P\,dx + Q\,dy = 0$$

no matter what values P and Q assume in the region D.

Another application of the reverse direction of Green's Theorem is in computing areas. Since the area of D is $\iint_D 1\,dA$, we wish to choose P and Q so that

$$\frac{\partial Q}{\partial x} - \frac{\partial P}{\partial y} = 1$$

There are several possibilities: $P(x, y) = 0$ and $Q(x, y) = x$; $P(x, y) = -y$ and $Q(x, y) = 0$; or $P(x, y) = -y/2$ and $Q(x, y) = x/2$. Then Green's Theorem gives the following formulas for the area of D:

(6) $$A = \oint_C x\,dy = -\oint_C y\,dx = \tfrac{1}{2}\oint_C x\,dy - y\,dx$$

EXAMPLE 3 Find the area enclosed by the ellipse $\dfrac{x^2}{a^2} + \dfrac{y^2}{b^2} = 1$.

SOLUTION The ellipse has parametric equations $x = a\cos t$ and $y = b\sin t$, where $0 \leq t \leq 2\pi$. Using the third formula in Equation 6, we have

$$A = \tfrac{1}{2}\int_C x\,dy - y\,dx$$

$$= \tfrac{1}{2}\int_0^{2\pi} (a\cos t)(b\cos t)\,dt - (b\sin t)(-a\sin t)\,dt$$

$$= \frac{ab}{2}\int_0^{2\pi} dt = \pi ab$$

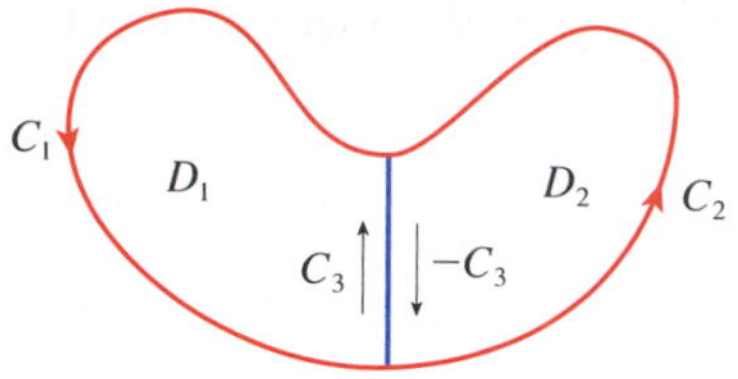

FIGURE 5

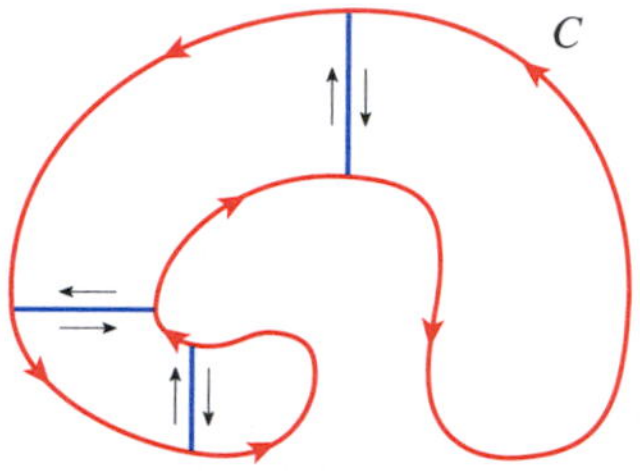

FIGURE 6

Although we have proved Green's Theorem only for the case where D is simple, we can now extend it to the case where D is a finite union of simple regions. For example, if D is the region shown in Figure 5, then we can write $D = D_1 \cup D_2$, where D_1 and D_2 are both simple. The boundary of D_1 is $C_1 \cup C_3$ and the boundary of D_2 is $C_2 \cup (-C_3)$ so, applying Green's Theorem to D_1 and D_2 separately, we get

$$\int_{C_1 \cup C_3} P\,dx + Q\,dy = \iint_{D_1} \left(\frac{\partial Q}{\partial x} - \frac{\partial P}{\partial y} \right) dA$$

$$\int_{C_2 \cup (-C_3)} P\,dx + Q\,dy = \iint_{D_2} \left(\frac{\partial Q}{\partial x} - \frac{\partial P}{\partial y} \right) dA$$

If we add these two equations, the line integrals along C_3 and $-C_3$ cancel, so we get

$$\int_{C_1 \cup C_2} P\,dx + Q\,dy = \iint_{D} \left(\frac{\partial Q}{\partial x} - \frac{\partial P}{\partial y} \right) dA$$

which is Green's Theorem for $D = D_1 \cup D_2$, since its boundary is $C = C_1 \cup C_2$.

The same sort of argument allows us to establish Green's Theorem for any finite union of simple regions (see Figure 6).

EXAMPLE 4 Evaluate $\oint_C y^2\,dx + 3xy\,dy$, where C is the boundary of the semiannular region D in the upper half-plane between the circles $x^2 + y^2 = 1$ and $x^2 + y^2 = 4$.

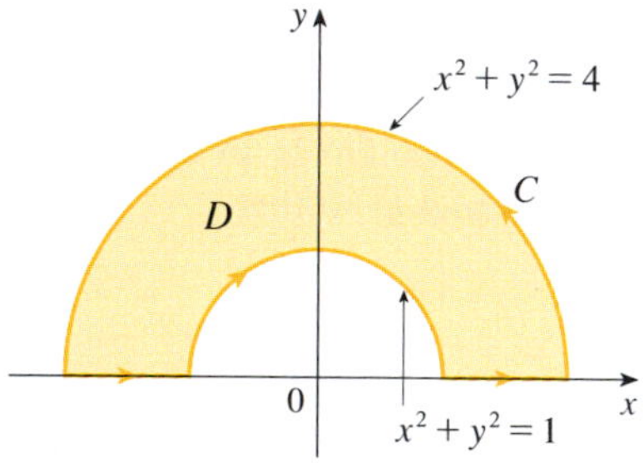

FIGURE 7

SOLUTION Notice that although D is not simple, the y-axis divides it into two simple regions (see Figure 7). In polar coordinates we can write

$$D = \{(r, \theta) \mid 1 \leqslant r \leqslant 2,\ 0 \leqslant \theta \leqslant \pi\}$$

Therefore, Green's Theorem gives

$$\begin{aligned}
\int_C y^2\,dx + 3xy\,dy &= \iint_D \left[\frac{\partial}{\partial x}(3xy) - \frac{\partial}{\partial y}(y^2) \right] dA \\
&= \iint_D y\,dA = \int_0^{\pi} \int_1^2 (r \sin\theta) r\,dr\,d\theta \\
&= \int_0^{\pi} \sin\theta\,d\theta \int_1^2 r^2\,dr = [-\cos\theta]_0^{\pi} \left[\frac{r^3}{3} \right]_1^2 \\
&= \frac{14}{3}
\end{aligned}$$

■

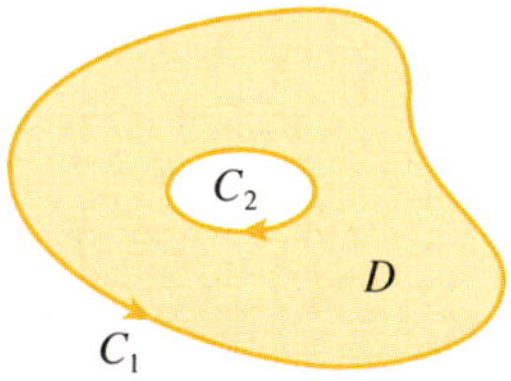

FIGURE 8

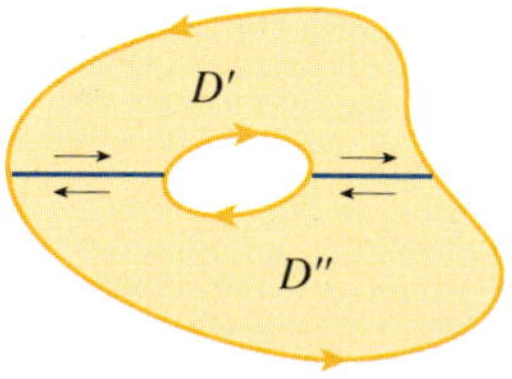

FIGURE 9

Green's Theorem can be extended to apply to regions with holes, that is, regions that are not simply-connected. Observe that the boundary C of the region D in Figure 8 consists of two simple closed curves C_1 and C_2. We assume that these boundary curves are oriented so that the region D is always on the left as the curve C is traversed. Thus the positive direction is counterclockwise for the outer curve C_1 but clockwise for the inner curve C_2. If we divide D into two regions D' and D'' by means of the lines shown in Figure 9 and then apply Green's Theorem to each of D' and D'', we get

$$\begin{aligned}
\iint_D \left(\frac{\partial Q}{\partial x} - \frac{\partial P}{\partial y} \right) dA &= \iint_{D'} \left(\frac{\partial Q}{\partial x} - \frac{\partial P}{\partial y} \right) dA + \iint_{D''} \left(\frac{\partial Q}{\partial x} - \frac{\partial P}{\partial y} \right) dA \\
&= \int_{\partial D'} P\,dx + Q\,dy + \int_{\partial D''} P\,dx + Q\,dy
\end{aligned}$$

Since the line integrals along the common boundary lines are in opposite directions, they cancel and we get

$$\iint_D \left(\frac{\partial Q}{\partial x} - \frac{\partial P}{\partial y} \right) dA = \int_{C_1} P\,dx + Q\,dy + \int_{C_2} P\,dx + Q\,dy = \int_C P\,dx + Q\,dy$$

which is Green's Theorem for the region D.

EXAMPLE 5 If $\mathbf{F}(x, y) = (-y\,\mathbf{i} + x\,\mathbf{j})/(x^2 + y^2)$, show that $\int_C \mathbf{F} \cdot d\mathbf{r} = 2\pi$ for every positively oriented simple closed path that encloses the origin.

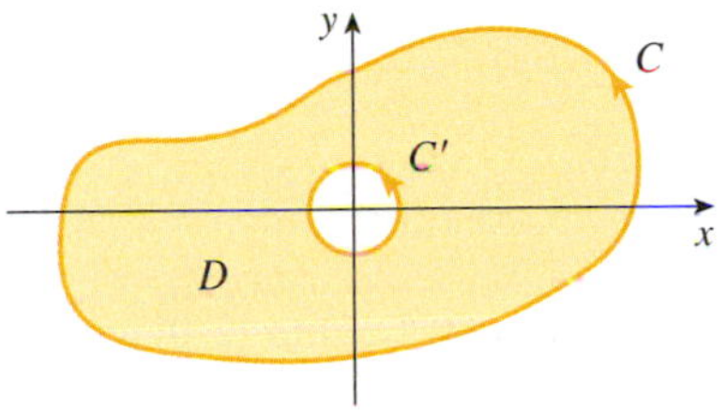

FIGURE 10

SOLUTION Since C is an *arbitrary* closed path that encloses the origin, it is difficult to compute the given integral directly. So let's consider a counterclockwise-oriented circle C' with center the origin and radius a, where a is chosen to be small enough that C' lies inside C (see Figure 10). Let D be the region bounded by C and C'. Then its positively oriented boundary is $C \cup (-C')$ and so the general version of Green's Theorem gives

$$\begin{aligned} \int_C P\,dx + Q\,dy + \int_{-C'} P\,dx + Q\,dy &= \iint_D \left(\frac{\partial Q}{\partial x} - \frac{\partial P}{\partial y} \right) dA \\ &= \iint_D \left[\frac{y^2 - x^2}{(x^2 + y^2)^2} - \frac{y^2 - x^2}{(x^2 + y^2)^2} \right] dA \\ &= 0 \end{aligned}$$

Therefore
$$\int_C P\,dx + Q\,dy = \int_{C'} P\,dx + Q\,dy$$

that is,
$$\int_C \mathbf{F} \cdot d\mathbf{r} = \int_{C'} \mathbf{F} \cdot d\mathbf{r}$$

We now easily compute the latter integral using the parametrization given by $\mathbf{r}(t) = a \cos t\,\mathbf{i} + a \sin t\,\mathbf{j}$, $0 \leqslant t \leqslant 2\pi$. Thus

$$\begin{aligned} \int_C \mathbf{F} \cdot d\mathbf{r} = \int_{C'} \mathbf{F} \cdot d\mathbf{r} &= \int_0^{2\pi} \mathbf{F}(\mathbf{r}(t)) \cdot \mathbf{r}'(t)\,dt \\ &= \int_0^{2\pi} \frac{(-a \sin t)(-a \sin t) + (a \cos t)(a \cos t)}{a^2 \cos^2 t + a^2 \sin^2 t}\,dt \\ &= \int_0^{2\pi} dt = 2\pi \end{aligned}$$

■

We end this section by using Green's Theorem to discuss a result that was stated in the preceding section.

SKETCH OF PROOF OF THEOREM 14.3.6 We are assuming that $\mathbf{F} = P\,\mathbf{i} + Q\,\mathbf{j}$ is a vector field on an open simply-connected region D, that P and Q have continuous first-order partial derivatives, and that

$$\frac{\partial P}{\partial y} = \frac{\partial Q}{\partial x} \qquad \text{throughout } D$$

If C is any simple closed path in D and R is the region that C encloses, then Green's Theorem gives

$$\oint_C \mathbf{F} \cdot d\mathbf{r} = \oint_C P\,dx + Q\,dy = \iint_R \left(\frac{\partial Q}{\partial x} - \frac{\partial P}{\partial y} \right) dA = \iint_R 0\,dA = 0$$

A curve that is not simple crosses itself at one or more points and can be broken up into a number of simple curves. We have shown that the line integrals of $\mathbf{F}$ around these simple curves are all 0 and, adding these integrals, we see that $\int_C \mathbf{F} \cdot d\mathbf{r} = 0$ for any closed curve C. Therefore, $\int_C \mathbf{F} \cdot d\mathbf{r}$ is independent of path in D by Theorem 14.3.3. It follows that $\mathbf{F}$ is a conservative vector field by Theorem 14.3.4. □

EXERCISES 14.4

1–4 ■ Evaluate the line integral by two methods: (a) directly and (b) using Green's Theorem.

1. $\oint_C x^2y\,dx + xy^3\,dy$,
C is the square with vertices $(0,0)$, $(1,0)$, $(1,1)$, and $(0,1)$

2. $\oint_C x\,dx - x^2y^2\,dy$,
C is the triangle with vertices $(0,0)$, $(1,1)$, and $(0,1)$

3. $\oint_C (x + 2y)\,dx + (x - 2y)\,dy$,
C consists of the arc of the parabola $y = x^2$ from $(0,0)$ to $(1,1)$ followed by the line segment from $(1,1)$ to $(0,0)$

4. $\oint_C (x^2 + y^2)\,dx + 2xy\,dy$, C consists of the arc of the parabola $y = x^2$ from $(0,0)$ to $(2,4)$ and the line segments from $(2,4)$ to $(0,4)$ and from $(0,4)$ to $(0,0)$

CAS **5–6** ■ Verify Green's Theorem by using a computer algebra system to evaluate both the line integral and the double integral.

5. $P(x,y) = x^4y^5$, $Q(x,y) = -x^7y^6$; C: $x^2 + y^2 = 1$

6. $P(x,y) = y^2 \sin x$, $Q(x,y) = x^2 \sin y$;
C is the curve in Exercise 3

7–18 ■ Use Green's Theorem to evaluate the line integral along the given positively oriented curve.

7. $\int_C xy\,dx + y^5\,dy$,
C is the triangle with vertices $(0,0)$, $(2,0)$, and $(2,1)$

8. $\int_C x^2y\,dx + xy^5\,dy$,
C is the square with vertices $(\pm1, \pm1)$

9. $\int_C \left(y + e^{\sqrt{x}}\right) dx + (2x + \cos y^2)\,dy$,
C is the boundary of the region enclosed by the parabolas $y = x^2$ and $x = y^2$

10. $\int_C (y^2 - \tan^{-1}x)\,dx + (3x + \sin y)\,dy$,
C is the boundary of the region enclosed by the parabola $y = x^2$ and the line $y = 4$

11. $\int_C x^2\,dx + y^2\,dy$, C is the curve $x^6 + y^6 = 1$

12. $\int_C x^2y\,dx - 3y^2\,dy$, C is the circle $x^2 + y^2 = 1$

13. $\int_C xy\,dx + 2x^2\,dy$,
C consists of the line segment from $(-2,0)$ to $(2,0)$ and the top half of the circle $x^2 + y^2 = 4$

14. $\int_C 2xy\,dx + x^2\,dy$,
C is the cardioid $r = 1 + \cos\theta$

15. $\int_C (xy + e^{x^2})\,dx + (x^2 - \ln(1 + y))\,dy$,
C consists of the line segment from $(0,0)$ to $(\pi,0)$ and the curve $y = \sin x$, $0 \leqslant x \leqslant \pi$

16. $\int_C (x^3 - y^3)\,dx + (x^3 + y^3)\,dy$,
C is the boundary of the region between the circles $x^2 + y^2 = 1$ and $x^2 + y^2 = 9$

17. $\int_C \mathbf{F} \cdot d\mathbf{r}$, where $\mathbf{F}(x,y) = (y^2 - x^2y)\,\mathbf{i} + xy^2\,\mathbf{j}$ and C consists of the circle $x^2 + y^2 = 4$ from $(2,0)$ to $(\sqrt{2}, \sqrt{2})$ and the line segments from $(\sqrt{2}, \sqrt{2})$ to $(0,0)$ and from $(0,0)$ to $(2,0)$

18. $\int_C \mathbf{F} \cdot d\mathbf{r}$, where $\mathbf{F}(x,y) = x^3y\,\mathbf{i} + x^4\,\mathbf{j}$ and C is the curve $x^4 + y^4 = 1$

19. Use Green's Theorem to find the work done by the force $\mathbf{F}(x,y) = x(x + y)\,\mathbf{i} + xy^2\,\mathbf{j}$ in moving a particle from the origin along the x-axis to $(1,0)$, then along the line segment to $(0,1)$, and then back to the origin along the y-axis.

20. A particle starts at the point $(-2,0)$, moves along the x-axis to $(2,0)$, and then along the semicircle $y = \sqrt{4 - x^2}$ to the starting point. Use Green's Theorem to find the work done on this particle by the force field $\mathbf{F}(x,y) = \langle x, x^3 + 3xy^2 \rangle$.

21–22 ■ Find the area of the given region using one of the formulas in Equations 6.

21. The region bounded by the hypocycloid with vector equation $\mathbf{r}(t) = \cos^3 t\,\mathbf{i} + \sin^3 t\,\mathbf{j}$, $0 \leqslant t \leqslant 2\pi$

22. The region bounded by the curve with vector equation $\mathbf{r}(t) = \cos t\,\mathbf{i} + \sin^3 t\,\mathbf{j}$, $0 \leqslant t \leqslant 2\pi$

23. (a) If C is the line segment connecting the point (x_1, y_1) to the point (x_2, y_2), show that

$$\int_C x\,dy - y\,dx = x_1y_2 - x_2y_1$$

(b) If the vertices of a polygon, in counterclockwise order, are (x_1, y_1), (x_2, y_2), $\ldots$, (x_n, y_n), show that the area of the polygon is

$$A = \tfrac{1}{2}[(x_1y_2 - x_2y_1) + (x_2y_3 - x_3y_2) + \cdots + (x_{n-1}y_n - x_ny_{n-1}) + (x_ny_1 - x_1y_n)]$$

(c) Find the area of the pentagon with vertices $(0, 0)$, $(2, 1)$, $(1, 3)$, $(0, 2)$, and $(-1, 1)$.

24. Let D be a region bounded by a simple closed path C in the xy-plane. Use Green's Theorem to prove that the coordinates of the centroid $(\bar{x}, \bar{y})$ of D are

$$\bar{x} = \frac{1}{2A}\oint_C x^2\,dy \qquad \bar{y} = -\frac{1}{2A}\oint_C y^2\,dx$$

where A is the area of D.

25. Use Exercise 24 to find the centroid of the triangle with vertices $(0, 0)$, $(1, 0)$, and $(0, 1)$.

26. Use Exercise 24 to find the centroid of a semicircular region of radius a.

27. A plane lamina with constant density $\rho(x, y) = \rho$ occupies a region in the xy-plane bounded by a simple closed path C. Show that its moments of inertia about the axes are

$$I_x = -\frac{\rho}{3}\oint_C y^3\,dx$$

$$I_y = \frac{\rho}{3}\oint_C x^3\,dy$$

28. Use Exercise 27 to find the moment of inertia of a circular disk of radius a with constant density ρ about a diameter. (Compare with Example 4 in Section 13.6.)

29. If $\mathbf{F}$ is the vector field of Example 5, show that $\int_C \mathbf{F} \cdot d\mathbf{r} = 0$ for every simple closed path that does not pass through or enclose the origin.

30. Complete the proof of the special case of Green's Theorem by proving Equation 4.

31. Use Green's Theorem to prove the change of variables formula for a double integral (13.11.9) for the case where $f(x, y) = 1$:

$$\iint_R dx\,dy = \iint_S \left|\frac{\partial(x, y)}{\partial(u, v)}\right| du\,dv$$

Here R is the region in the xy-plane that corresponds to the region S in the uv-plane under the transformation given by $x = g(u, v)$, $y = h(u, v)$. [*Hint:* Note that the left side is $A(R)$ and apply the first part of Equation 6. Convert the line integral over ∂R to a line integral over ∂S and apply Green's Theorem in the uv-plane.]

14.5 CURL AND DIVERGENCE

In this section we define two operations that can be performed on vector fields and that play a basic role in the applications of vector calculus. Each operation resembles differentiation, but one produces a vector field whereas the other produces a scalar field.

CURL

If $\mathbf{F} = P\,\mathbf{i} + Q\,\mathbf{j} + R\,\mathbf{k}$ is a vector field on $\mathbb{R}^3$ and the partial derivatives of P, Q, and R all exist, then the **curl** of $\mathbf{F}$ is the vector field on $\mathbb{R}^3$ defined by

$$\text{(1)} \qquad \operatorname{curl} \mathbf{F} = \left(\frac{\partial R}{\partial y} - \frac{\partial Q}{\partial z}\right)\mathbf{i} + \left(\frac{\partial P}{\partial z} - \frac{\partial R}{\partial x}\right)\mathbf{j} + \left(\frac{\partial Q}{\partial x} - \frac{\partial P}{\partial y}\right)\mathbf{k}$$

As an aid to the memory, let us rewrite Equation 1 using operator notation. We introduce the vector differential operator ∇ ("del") as

$$\nabla = \mathbf{i}\frac{\partial}{\partial x} + \mathbf{j}\frac{\partial}{\partial y} + \mathbf{k}\frac{\partial}{\partial z}$$

It has meaning when it operates on a scalar function to produce the gradient of f:

$$\nabla f = \mathbf{i}\frac{\partial f}{\partial x} + \mathbf{j}\frac{\partial f}{\partial y} + \mathbf{k}\frac{\partial f}{\partial z} = \frac{\partial f}{\partial x}\mathbf{i} + \frac{\partial f}{\partial y}\mathbf{j} + \frac{\partial f}{\partial z}\mathbf{k}$$

If we think of ∇ as a vector with components $\partial/\partial x$, $\partial/\partial y$, and $\partial/\partial z$, we can also consider the formal cross product of ∇ with the vector field $\mathbf{F}$ as follows:

$$\begin{aligned}\nabla \times \mathbf{F} &= \begin{vmatrix} \mathbf{i} & \mathbf{j} & \mathbf{k} \\ \dfrac{\partial}{\partial x} & \dfrac{\partial}{\partial y} & \dfrac{\partial}{\partial z} \\ P & Q & R \end{vmatrix} \\ &= \left(\frac{\partial R}{\partial y} - \frac{\partial Q}{\partial z}\right)\mathbf{i} + \left(\frac{\partial P}{\partial z} - \frac{\partial R}{\partial x}\right)\mathbf{j} + \left(\frac{\partial Q}{\partial x} - \frac{\partial P}{\partial y}\right)\mathbf{k} \\ &= \text{curl } \mathbf{F}\end{aligned}$$

Thus, the easiest way to remember Definition 1 is by means of the symbolic expression

(2)
$$\text{curl } \mathbf{F} = \nabla \times \mathbf{F}$$

EXAMPLE 1 If $\mathbf{F}(x, y, z) = xz\,\mathbf{i} + xyz\,\mathbf{j} - y^2\,\mathbf{k}$, find curl $\mathbf{F}$.

SOLUTION Using Equation 2, we have

$$\begin{aligned}\text{curl } \mathbf{F} = \nabla \times \mathbf{F} &= \begin{vmatrix} \mathbf{i} & \mathbf{j} & \mathbf{k} \\ \dfrac{\partial}{\partial x} & \dfrac{\partial}{\partial y} & \dfrac{\partial}{\partial z} \\ xz & xyz & -y^2 \end{vmatrix} \\ &= \left[\frac{\partial}{\partial y}(-y^2) - \frac{\partial}{\partial z}(xyz)\right]\mathbf{i} - \left[\frac{\partial}{\partial x}(-y^2) - \frac{\partial}{\partial z}(xz)\right]\mathbf{j} \\ &\quad + \left[\frac{\partial}{\partial x}(xyz) - \frac{\partial}{\partial y}(xz)\right]\mathbf{k} \\ &= (-2y - xy)\,\mathbf{i} - (0 - x)\,\mathbf{j} + (yz - 0)\,\mathbf{k} \\ &= -y(2 + x)\,\mathbf{i} + x\,\mathbf{j} + yz\,\mathbf{k}\end{aligned}$$

■

Most computer algebra systems have commands that compute the curl and divergence of vector fields. If you have access to a CAS, use these commands to check the answers to the examples and exercises in this section.

Recall that the gradient of a function f of three variables is a vector field on $\mathbb{R}^3$ and so we can compute its curl. The following theorem says that the curl of a gradient vector field is $\mathbf{0}$. Notice the similarity to what we know from Section 11.3: $\mathbf{a} \times \mathbf{a} = \mathbf{0}$ for every three-dimensional vector $\mathbf{a}$.

(3) THEOREM If f is a function of three variables that has continuous second-order partial derivatives, then

$$\text{curl}(\nabla f) = \mathbf{0}$$

PROOF We have

$$\text{curl}(\nabla f) = \nabla \times (\nabla f) = \begin{vmatrix} \mathbf{i} & \mathbf{j} & \mathbf{k} \\ \dfrac{\partial}{\partial x} & \dfrac{\partial}{\partial y} & \dfrac{\partial}{\partial z} \\ \dfrac{\partial f}{\partial x} & \dfrac{\partial f}{\partial y} & \dfrac{\partial f}{\partial z} \end{vmatrix}$$

$$= \left(\frac{\partial^2 f}{\partial y\, \partial z} - \frac{\partial^2 f}{\partial z\, \partial y} \right) \mathbf{i} + \left(\frac{\partial^2 f}{\partial z\, \partial x} - \frac{\partial^2 f}{\partial x\, \partial z} \right) \mathbf{j} + \left(\frac{\partial^2 f}{\partial x\, \partial y} - \frac{\partial^2 f}{\partial y\, \partial x} \right) \mathbf{k}$$

$$= 0\,\mathbf{i} + 0\,\mathbf{j} + 0\,\mathbf{k} = \mathbf{0}$$

by Clairaut's Theorem.

Since a conservative vector field is one for which $\mathbf{F} = \nabla f$, Theorem 3 can be rephrased as saying that if $\mathbf{F}$ is conservative, then curl $\mathbf{F} = \mathbf{0}$. (Compare this with Exercise 27 in Section 14.3.) This gives us a way of verifying that a vector field is not conservative.

EXAMPLE 2 Show that the vector field $\mathbf{F}(x, y, z) = xz\,\mathbf{i} + xyz\,\mathbf{j} - y^2\mathbf{k}$ is not conservative.

SOLUTION In Example 1 we showed that

$$\text{curl } \mathbf{F} = -y(2 + x)\,\mathbf{i} + x\,\mathbf{j} + yz\,\mathbf{k}$$

This shows that curl $\mathbf{F} \neq \mathbf{0}$ and so, by Theorem 3, $\mathbf{F}$ is not conservative. ■

The converse of Theorem 3 is not true in general, but the following theorem says the converse is true if $\mathbf{F}$ is defined everywhere. (More generally it is true if the domain is simply-connected, that is, "has no hole.") Theorem 4 is the three-dimensional analogue of Theorem 14.3.6. Its proof requires Stokes' Theorem and is sketched at the end of Section 14.8.

(4) THEOREM If $\mathbf{F}$ is a vector field defined on all of $\mathbb{R}^3$ whose component functions have continuous partial derivatives and curl $\mathbf{F} = \mathbf{0}$, then $\mathbf{F}$ is a conservative vector field.

EXAMPLE 3

(a) Show that $\mathbf{F}(x, y, z) = y^2z^3\,\mathbf{i} + 2xyz^3\,\mathbf{j} + 3xy^2z^2\,\mathbf{k}$ is a conservative vector field.

(b) Find a function f such that $\mathbf{F} = \nabla f$.

SOLUTION

(a) We compute the curl of $\mathbf{F}$:

$$\text{curl } \mathbf{F} = \nabla \times \mathbf{F} = \begin{vmatrix} \mathbf{i} & \mathbf{j} & \mathbf{k} \\ \dfrac{\partial}{\partial x} & \dfrac{\partial}{\partial y} & \dfrac{\partial}{\partial z} \\ y^2z^3 & 2xyz^3 & 3xy^2z^2 \end{vmatrix}$$

$$= (6xyz^2 - 6xyz^2)\,\mathbf{i} - (3y^2z^2 - 3y^2z^2)\,\mathbf{j} + (2yz^3 - 2yz^3)\,\mathbf{k}$$

$$= \mathbf{0}$$

Since curl $\mathbf{F} = \mathbf{0}$ and the domain of $\mathbf{F}$ is $\mathbb{R}^3$, $\mathbf{F}$ is a conservative vector field by Theorem 4.

(b) The technique for finding f was given in Section 14.3. We have

(5) $$f_x(x, y, z) = y^2z^3$$

(6) $$f_y(x, y, z) = 2xyz^3$$

(7) $$f_z(x, y, z) = 3xy^2z^2$$

Integrating (5) with respect to x, we obtain

(8) $$f(x, y, z) = xy^2z^3 + g(y, z)$$

Differentiating (8) with respect to y, we get $f_y(x, y, z) = 2xyz^3 + g_y(y, z)$, so comparison with (6) gives $g_y(y, z) = 0$. Thus, $g(y, z) = h(z)$ and

$$f_z(x, y, z) = 3xy^2z^2 + h'(z)$$

Then (7) gives $h'(z) = 0$. Therefore

$$f(x, y, z) = xy^2z^3 + K$$ ■

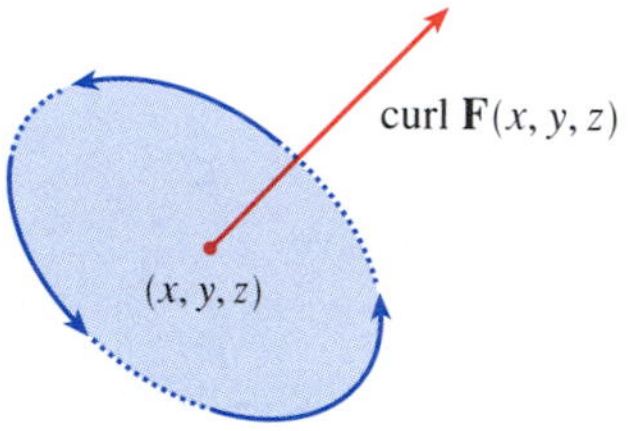

FIGURE 1

The reason for the name *curl* is that the curl vector is associated with rotations. One connection is explained in Exercise 43. Another occurs when **F** represents the velocity field in fluid flow (see Example 3 in Section 14.1). Particles near (x, y, z) in the fluid tend to rotate about the axis that points in the direction of curl $\mathbf{F}(x, y, z)$ and the length of this curl vector is a measure of how quickly the particles move around the axis (see Figure 1). If curl $\mathbf{F} = \mathbf{0}$ at a point P, then the fluid is free from rotations at P and **F** is called **irrotational** at P. In other words, there is no whirlpool or eddy. If curl $\mathbf{F} = \mathbf{0}$, then a tiny paddle wheel moves with the fluid but does not rotate about its axis. If curl $\mathbf{F} \neq \mathbf{0}$, the paddle wheel rotates about its axis. We give a more detailed explanation in Section 14.8 as a consequence of Stokes' Theorem.

DIVERGENCE

If $\mathbf{F} = P\,\mathbf{i} + Q\,\mathbf{j} + R\,\mathbf{k}$ is a vector field on $\mathbb{R}^3$ and $\partial P/\partial x$, $\partial Q/\partial y$, and $\partial R/\partial z$ exist, then the **divergence of F** is the function of three variables defined by

(9) $$\operatorname{div} \mathbf{F} = \frac{\partial P}{\partial x} + \frac{\partial Q}{\partial y} + \frac{\partial R}{\partial z}$$

Observe that curl **F** is a vector field but div **F** is a scalar field. In terms of the gradient operator $\nabla = (\partial/\partial x)\,\mathbf{i} + (\partial/\partial y)\,\mathbf{j} + (\partial/\partial z)\,\mathbf{k}$, the divergence of **F** can be written symbolically as the dot product of ∇ and **F**:

(10) $$\operatorname{div} \mathbf{F} = \nabla \cdot \mathbf{F}$$

EXAMPLE 4 If $\mathbf{F}(x, y, z) = xz\,\mathbf{i} + xyz\,\mathbf{j} - y^2\,\mathbf{k}$, find div **F**.

SOLUTION By the definition of divergence (Equation 9 or 10) we have

$$\operatorname{div} \mathbf{F} = \nabla \cdot \mathbf{F} = \frac{\partial}{\partial x}(xz) + \frac{\partial}{\partial y}(xyz) + \frac{\partial}{\partial z}(-y^2)$$

$$= z + xz$$ ■

If **F** is a vector field on $\mathbb{R}^3$, then curl **F** is also a vector field on $\mathbb{R}^3$. As such, we can compute its divergence. The next theorem shows that the result is 0. [Note the analogy with the scalar triple product: $\mathbf{a} \cdot (\mathbf{a} \times \mathbf{b}) = 0$.]

(11) THEOREM If $\mathbf{F} = P\,\mathbf{i} + Q\,\mathbf{j} + R\,\mathbf{k}$ is a vector field on $\mathbb{R}^3$ and P, Q, and R have continuous second-order partial derivatives, then

$$\operatorname{div}\operatorname{curl}\mathbf{F} = 0$$

PROOF Using the definitions of divergence and curl, we have

$$\begin{aligned}\operatorname{div}\operatorname{curl}\mathbf{F} &= \nabla\cdot(\nabla\times\mathbf{F})\\ &= \frac{\partial}{\partial x}\left(\frac{\partial R}{\partial y}-\frac{\partial Q}{\partial z}\right)+\frac{\partial}{\partial y}\left(\frac{\partial P}{\partial z}-\frac{\partial R}{\partial x}\right)+\frac{\partial}{\partial z}\left(\frac{\partial Q}{\partial x}-\frac{\partial P}{\partial y}\right)\\ &= \frac{\partial^2 R}{\partial x\,\partial y}-\frac{\partial^2 Q}{\partial x\,\partial z}+\frac{\partial^2 P}{\partial y\,\partial z}-\frac{\partial^2 R}{\partial y\,\partial x}+\frac{\partial^2 Q}{\partial z\,\partial x}-\frac{\partial^2 P}{\partial z\,\partial y}\\ &= 0\end{aligned}$$

because the terms cancel in pairs by Clairaut's Theorem. □

EXAMPLE 5 Show that the vector field $\mathbf{F}(x, y, z) = xz\,\mathbf{i} + xyz\,\mathbf{j} - y^2\,\mathbf{k}$ cannot be written as the curl of another vector field, that is, $\mathbf{F} \neq \operatorname{curl}\mathbf{G}$.

SOLUTION In Example 4 we showed that

$$\operatorname{div}\mathbf{F} = z + xz$$

and therefore $\operatorname{div}\mathbf{F} \neq 0$. If it were true that $\mathbf{F} = \operatorname{curl}\mathbf{G}$, then Theorem 11 would give

$$\operatorname{div}\mathbf{F} = \operatorname{div}\operatorname{curl}\mathbf{G} = 0$$

which contradicts $\operatorname{div}\mathbf{F} \neq 0$. Therefore, $\mathbf{F}$ is not the curl of another vector field. ■

The reason for this interpretation of div **F** will be explained at the end of Section 14.9 as a consequence of the Divergence Theorem.

Again, the reason for the name *divergence* can be understood in the context of fluid flow. If $\mathbf{F}(x, y, z)$ is the velocity of a fluid (or gas), then $\operatorname{div}\mathbf{F}(x, y, z)$ represents the net rate of change (with respect to time) of the mass of fluid (or gas) flowing from the point (x, y, z) per unit volume. In other words, $\operatorname{div}\mathbf{F}(x, y, z)$ measures the tendency of the fluid to diverge from the point (x, y, z). If $\operatorname{div}\mathbf{F} = 0$, then $\mathbf{F}$ is said to be **incompressible**.

Another differential operator occurs when we compute the divergence of a gradient vector field ∇f. If f is a function of three variables, we have

$$\operatorname{div}(\nabla f) = \nabla\cdot(\nabla f) = \frac{\partial^2 f}{\partial x^2}+\frac{\partial^2 f}{\partial y^2}+\frac{\partial^2 f}{\partial z^2}$$

and this expression occurs so often that we abbreviate it as $\nabla^2 f$. The operator

$$\nabla^2 = \nabla\cdot\nabla$$

is called the **Laplace operator** because of its relation to **Laplace's equation**

$$\nabla^2 f = \frac{\partial^2 f}{\partial x^2}+\frac{\partial^2 f}{\partial y^2}+\frac{\partial^2 f}{\partial z^2} = 0$$

We can also apply the Laplace operator ∇^2 to a vector field

$$\mathbf{F} = P\,\mathbf{i} + Q\,\mathbf{j} + R\,\mathbf{k}$$

in terms of its components:

$$\nabla^2\mathbf{F} = \nabla^2 P\,\mathbf{i} + \nabla^2 Q\,\mathbf{j} + \nabla^2 R\,\mathbf{k}$$

VECTOR FORMS OF GREEN'S THEOREM

The curl and divergence operators allow us to rewrite Green's Theorem in versions that will be useful in our later work. We suppose that the plane region D, its boundary curve C, and the functions P and Q satisfy the hypotheses of Green's Theorem (14.4.1). Then we consider the vector field $\mathbf{F} = P\,\mathbf{i} + Q\,\mathbf{j}$. Its line integral is

$$\oint_C \mathbf{F} \cdot d\mathbf{r} = \oint_C P\,dx + Q\,dy$$

and its curl is

$$\operatorname{curl} \mathbf{F} = \begin{vmatrix} \mathbf{i} & \mathbf{j} & \mathbf{k} \\ \dfrac{\partial}{\partial x} & \dfrac{\partial}{\partial y} & \dfrac{\partial}{\partial z} \\ P(x, y) & Q(x, y) & 0 \end{vmatrix} = \left(\frac{\partial Q}{\partial x} - \frac{\partial P}{\partial y}\right)\mathbf{k}$$

Therefore
$$(\operatorname{curl} \mathbf{F}) \cdot \mathbf{k} = \left(\frac{\partial Q}{\partial x} - \frac{\partial P}{\partial y}\right)\mathbf{k} \cdot \mathbf{k} = \frac{\partial Q}{\partial x} - \frac{\partial P}{\partial y}$$

and we can now rewrite the equation in Green's Theorem in the vector form

(12)
$$\oint_C \mathbf{F} \cdot d\mathbf{r} = \iint_D (\operatorname{curl} \mathbf{F}) \cdot \mathbf{k}\,dA$$

Equation 12 expresses the line integral of the tangential component of $\mathbf{F}$ along C as the double integral of $(\operatorname{curl} \mathbf{F}) \cdot \mathbf{k}$ over the region D enclosed by C. We now derive a similar formula involving the *normal* component of $\mathbf{F}$.

If C is given by the vector equation

$$\mathbf{r}(t) = x(t)\,\mathbf{i} + y(t)\,\mathbf{j} \qquad a \leqslant t \leqslant b$$

then the unit tangent vector (see Section 11.6) is

$$\mathbf{T}(t) = \frac{x'(t)}{|\mathbf{r}'(t)|}\,\mathbf{i} + \frac{y'(t)}{|\mathbf{r}'(t)|}\,\mathbf{j}$$

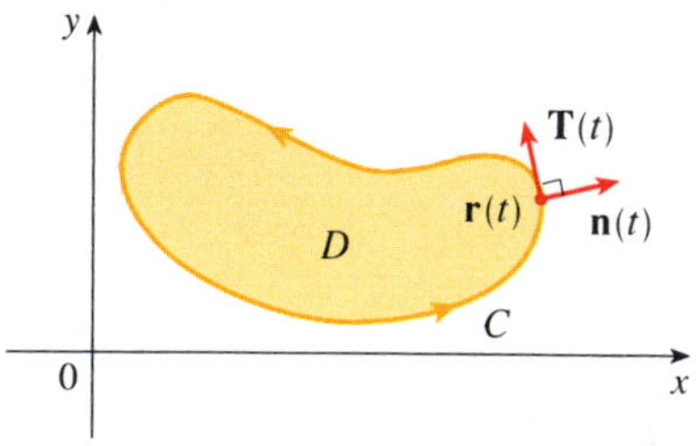

FIGURE 2

You can verify that the outward unit normal vector to C is given by

$$\mathbf{n}(t) = \frac{y'(t)}{|\mathbf{r}'(t)|}\,\mathbf{i} - \frac{x'(t)}{|\mathbf{r}'(t)|}\,\mathbf{j}$$

(See Figure 2.) Then, from Equation 14.2.3, we have

$$\oint_C \mathbf{F} \cdot \mathbf{n}\,ds = \int_a^b (\mathbf{F} \cdot \mathbf{n})(t)\,|\mathbf{r}'(t)|\,dt$$

$$= \int_a^b \left[\frac{P(x(t), y(t))\, y'(t)}{|\mathbf{r}'(t)|} - \frac{Q(x(t), y(t))\, x'(t)}{|\mathbf{r}'(t)|} \right] |\mathbf{r}'(t)|\, dt$$

$$= \int_a^b P(x(t), y(t))\, y'(t)\, dt - Q(x(t), y(t))\, x'(t)\, dt$$

$$= \int_C P\, dy - Q\, dx$$

$$= \iint_D \left(\frac{\partial P}{\partial x} + \frac{\partial Q}{\partial y} \right) dA$$

by Green's Theorem. But the integrand in this double integral is just the divergence of **F**. So we have a second vector form of Green's Theorem.

(13) $$\oint_C \mathbf{F} \cdot \mathbf{n}\, ds = \iint_D \operatorname{div} \mathbf{F}(x, y)\, dA$$

This version says that the line integral of the normal component of **F** along C is equal to the double integral of the divergence of **F** over the region D enclosed by C.

EXERCISES 14.5

1–10 ■ Find (a) the curl and (b) the divergence of the vector field.

1. $\mathbf{F}(x, y, z) = x\,\mathbf{i} + y\,\mathbf{j} + z\,\mathbf{k}$

2. $\mathbf{F}(x, y, z) = x^2y\,\mathbf{i} + yz^2\,\mathbf{j} + zx^2\,\mathbf{k}$

3. $\mathbf{F}(x, y, z) = yz\,\mathbf{i} + xz\,\mathbf{j} + xy\,\mathbf{k}$

4. $\mathbf{F}(x, y, z) = y^2z\,\mathbf{i} - x^2yz\,\mathbf{k}$

5. $\mathbf{F}(x, y, z) = xy\,\mathbf{j} + xyz\,\mathbf{k}$

6. $\mathbf{F}(x, y, z) = \sin x\,\mathbf{i} + \cos x\,\mathbf{j} + z^2\,\mathbf{k}$

7. $\mathbf{F}(x, y, z) = e^{xz}\,\mathbf{i} - 2e^{yz}\,\mathbf{j} + 3xe^y\,\mathbf{k}$

8. $\mathbf{F}(x, y, z) = (x + 3y - 5z)\,\mathbf{i} + (z - 3y)\,\mathbf{j} + (5x + 6y - z)\,\mathbf{k}$

9. $\mathbf{F}(x, y, z) = xe^y\,\mathbf{i} - ze^{-y}\,\mathbf{j} + y\ln z\,\mathbf{k}$

10. $\mathbf{F}(x, y, z) = e^{xyz}\,\mathbf{i} + \sin(x - y)\,\mathbf{j} - \dfrac{xy}{z}\,\mathbf{k}$

11–18 ■ Determine whether or not the given vector field is conservative. If it is conservative, find a function f such that $\mathbf{F} = \nabla f$.

11. $\mathbf{F}(x, y, z) = y\,\mathbf{i} + x\,\mathbf{j} + \mathbf{k}$

12. $\mathbf{F}(x, y, z) = x\,\mathbf{i} + y\,\mathbf{j} + x\,\mathbf{k}$

13. $\mathbf{F}(x, y, z) = yz\,\mathbf{i} - z^2\,\mathbf{j} + x^2\,\mathbf{k}$

14. $\mathbf{F}(x, y, z) = z\,\mathbf{i} + 2yz\,\mathbf{j} + (x + y^2)\,\mathbf{k}$

15. $\mathbf{F}(x, y, z) = \cos y\,\mathbf{i} + \sin x\,\mathbf{j} + \tan z\,\mathbf{k}$

16. $\mathbf{F}(x, y, z) = x\,\mathbf{i} + e^y \sin z\,\mathbf{j} + e^y \cos z\,\mathbf{k}$

17. $\mathbf{F}(x, y, z) = yz\,\mathbf{i} + (y^2 + xz)\,\mathbf{j} + xy\,\mathbf{k}$

18. $\mathbf{F}(x, y, z) = zx\,\mathbf{i} + xy\,\mathbf{j} + yz\,\mathbf{k}$

19. Is there a vector field **G** on $\mathbb{R}^3$ such that curl $\mathbf{G} = xy^2\,\mathbf{i} + yz^2\,\mathbf{j} + zx^2\,\mathbf{k}$? Explain.

20. Is there a vector field **G** on $\mathbb{R}^3$ such that curl $\mathbf{G} = yz\,\mathbf{i} + xyz\,\mathbf{j} + xy\,\mathbf{k}$? Explain.

21. Show that any vector field of the form $\mathbf{F}(x, y, z) = f(x)\,\mathbf{i} + g(y)\,\mathbf{j} + h(z)\,\mathbf{k}$, where f, g, h are differentiable functions, is irrotational.

22. Show that any vector field of the form $\mathbf{F}(x, y, z) = f(y, z)\,\mathbf{i} + g(x, z)\,\mathbf{j} + h(x, y)\,\mathbf{k}$ is incompressible.

23–30 ■ Prove the identity, assuming that the appropriate partial derivatives exist and are continuous. If f is a scalar field and **F**, **G** are vector fields, then $f\mathbf{F}$, $\mathbf{F} \cdot \mathbf{G}$, and $\mathbf{F} \times \mathbf{G}$ are vector fields defined by

$$(f\mathbf{F})(x, y, z) = f(x, y, z)\mathbf{F}(x, y, z)$$
$$(\mathbf{F} \cdot \mathbf{G})(x, y, z) = \mathbf{F}(x, y, z) \cdot \mathbf{G}(x, y, z)$$
$$(\mathbf{F} \times \mathbf{G})(x, y, z) = \mathbf{F}(x, y, z) \times \mathbf{G}(x, y, z)$$

23. $\operatorname{div}(\mathbf{F} + \mathbf{G}) = \operatorname{div} \mathbf{F} + \operatorname{div} \mathbf{G}$

24. $\operatorname{curl}(\mathbf{F} + \mathbf{G}) = \operatorname{curl} \mathbf{F} + \operatorname{curl} \mathbf{G}$

25. $\operatorname{div}(f\mathbf{F}) = f \operatorname{div} \mathbf{F} + \mathbf{F} \cdot \nabla f$

26. $\text{curl}(f\,\mathbf{F}) = f\,\text{curl}\,\mathbf{F} + (\nabla f) \times \mathbf{F}$

27. $\text{div}(\mathbf{F} \times \mathbf{G}) = \mathbf{G} \cdot \text{curl}\,\mathbf{F} - \mathbf{F} \cdot \text{curl}\,\mathbf{G}$

28. $\text{div}(\nabla f \times \nabla g) = 0$

29. $\text{curl curl}\,\mathbf{F} = \text{grad div}\,\mathbf{F} - \nabla^2\mathbf{F}$

30. $\nabla(\mathbf{F} \cdot \mathbf{G}) = (\mathbf{F} \cdot \nabla)\mathbf{G} + (\mathbf{G} \cdot \nabla)\mathbf{F} + \mathbf{F} \times \text{curl}\,\mathbf{G} + \mathbf{G} \times \text{curl}\,\mathbf{F}$

31. Let f be a scalar field and $\mathbf{F}$ a vector field. State whether each expression is a scalar field, a vector field, or meaningless.

(a) curl f (b) grad f
(c) div $\mathbf{F}$ (d) curl(grad f)
(e) grad $\mathbf{F}$ (f) grad(div $\mathbf{F}$)
(g) div(grad f) (h) grad(div f)
(i) curl(curl $\mathbf{F}$) (j) div(div $\mathbf{F}$)
(k) (grad f) × (div $\mathbf{F}$) (l) div(curl(grad f))

32–38 ■ Verify the identity, where $\mathbf{r} = x\,\mathbf{i} + y\,\mathbf{j} + z\,\mathbf{k}$ and $r = |\mathbf{r}|$.

32. $\nabla r = \mathbf{r}/r$

33. $\nabla \cdot \mathbf{r} = 3$

34. $\nabla \times \mathbf{r} = \mathbf{0}$

35. $\nabla(1/r) = -\mathbf{r}/r^3$

36. $\nabla \cdot (r\mathbf{r}) = 4r$

37. $\nabla \ln r = \mathbf{r}/r^2$

38. $\nabla^2 r^3 = 12r$

39–40 ■ Is the divergence of the vector field positive, negative, or zero? Explain.

39.

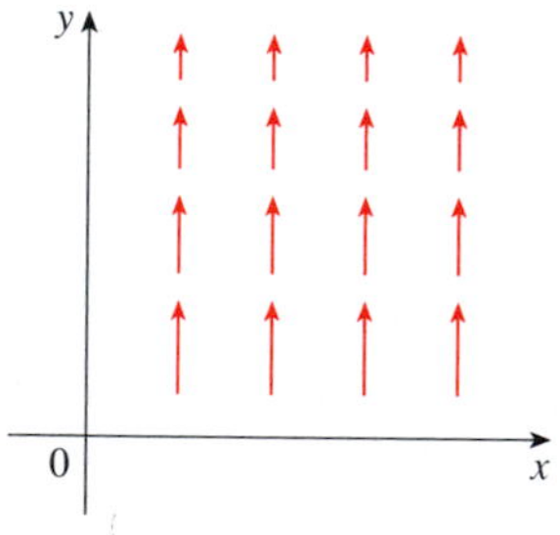

40.

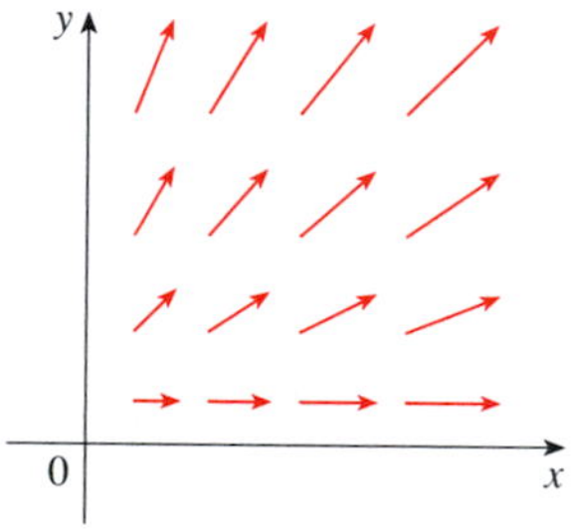

41. Use Green's Theorem in the form of Equation 13 to prove **Green's first identity**:

$$\iint_D f\nabla^2 g\,dA = \oint_C f(\nabla g) \cdot \mathbf{n}\,ds - \iint_D \nabla f \cdot \nabla g\,dA$$

where D and C satisfy the hypotheses of Green's Theorem and the appropriate partial derivatives of f and g exist and are continuous. (The quantity $\nabla g \cdot \mathbf{n} = D_{\mathbf{n}}g$ occurs in the line integral. This is the directional derivative in the direction of the normal vector $\mathbf{n}$ and is called the **normal derivative** of g.)

42. Use Green's first identity (Exercise 41) to prove **Green's second identity**:

$$\iint_D (f\nabla^2 g - g\nabla^2 f)\,dA = \oint_C (f\nabla g - g\nabla f) \cdot \mathbf{n}\,ds$$

where D and C satisfy the hypotheses of Green's Theorem and the appropriate partial derivatives of f and g exist and are continuous.

43. This exercise demonstrates a connection between the curl vector and rotations. Let B be a rigid body rotating about the z-axis. The rotation can be described by the vector $\mathbf{w} = \omega\mathbf{k}$, where ω is the angular speed of B, that is, the tangential speed of any point P in B divided by the distance d from the axis of rotation. Let $\mathbf{r} = \langle x, y, z \rangle$ be the position vector of P.

(a) By considering the angle θ in the figure, show that the velocity field of B is given by $\mathbf{v} = \mathbf{w} \times \mathbf{r}$.

(b) Show that $\mathbf{v} = -\omega y\,\mathbf{i} + \omega x\,\mathbf{j}$.

(c) Show that curl $\mathbf{v} = 2\mathbf{w}$.

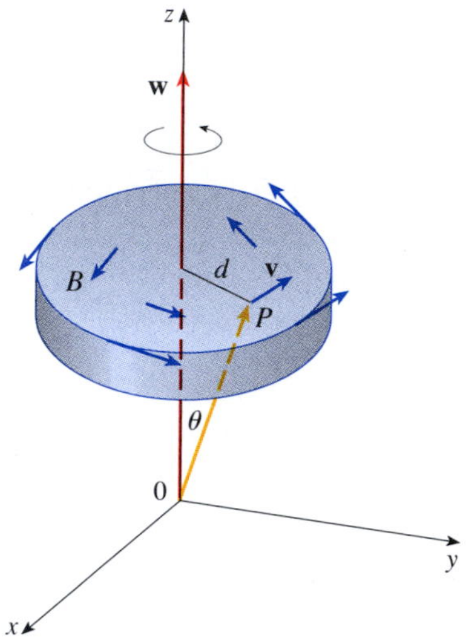

44. Maxwell's equations relating the electric field $\mathbf{E}$ and magnetic field $\mathbf{H}$ as they vary with time in a region containing no charge and no current can be stated as follows:

$$\text{div}\,\mathbf{E} = 0 \qquad \text{div}\,\mathbf{H} = 0$$

$$\text{curl}\,\mathbf{E} = -\frac{1}{c}\frac{\partial \mathbf{H}}{\partial t} \qquad \text{curl}\,\mathbf{H} = \frac{1}{c}\frac{\partial \mathbf{E}}{\partial t}$$

where c is the speed of light. Use these equations to prove the following:

(a) $\nabla \times (\nabla \times \mathbf{E}) = -\dfrac{1}{c^2}\dfrac{\partial^2 \mathbf{E}}{\partial t^2}$

(b) $\nabla \times (\nabla \times \mathbf{H}) = -\dfrac{1}{c^2}\dfrac{\partial^2 \mathbf{H}}{\partial t^2}$

(c) $\nabla^2 \mathbf{E} = \dfrac{1}{c^2}\dfrac{\partial^2 \mathbf{E}}{\partial t^2}$ [*Hint:* Use Exercise 29.]

(d) $\nabla^2 \mathbf{H} = \dfrac{1}{c^2}\dfrac{\partial^2 \mathbf{H}}{\partial t^2}$

14.6 PARAMETRIC SURFACES AND THEIR AREAS

In Section 9.4 we found the area of a surface of revolution and in Section 13.7 we found the area of a surface with the equation $z = f(x, y)$. Here we discuss more general surfaces, called parametric surfaces, and compute their areas.

In much the same way that we describe a space curve by a vector function $\mathbf{r}(t)$ of a single parameter t, we can describe a surface by a vector function $\mathbf{r}(u, v)$ of two parameters u and v. We suppose that

$$\mathbf{r}(u, v) = x(u, v)\,\mathbf{i} + y(u, v)\,\mathbf{j} + z(u, v)\,\mathbf{k} \tag{1}$$

is a vector-valued function defined on a region D in the uv-plane and the partial derivatives of x, y, and z with respect to u and v are all continuous. The set of all points (x, y, z) in $\mathbb{R}^3$, such that

$$x = x(u, v) \qquad y = y(u, v) \qquad z = z(u, v) \tag{2}$$

and (u, v) varies throughout D, is called a **parametric surface** S and Equations 2 are called **parametric equations** of S. In other words, the surface S is traced out by the position vector $\mathbf{r}(u, v)$ as (u, v) moves throughout the region D (see Figure 1).

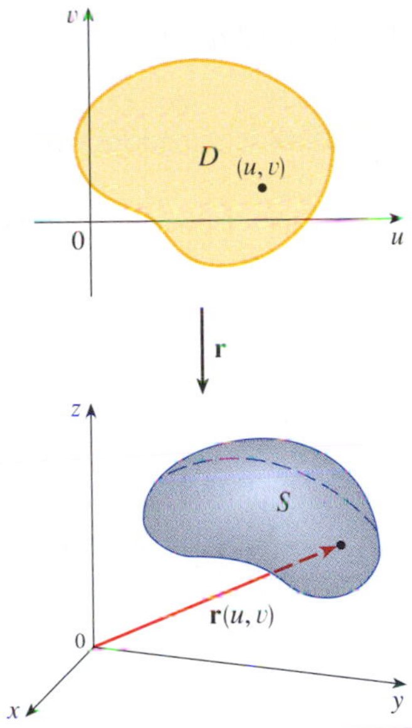

FIGURE 1
A parametric surface

EXAMPLE 1 Find a parametric representation of the sphere

$$x^2 + y^2 + z^2 = a^2$$

SOLUTION The sphere has a simple representation $\rho = a$ in spherical coordinates, so let's choose the angles ϕ and θ in spherical coordinates as the parameters (see Section 13.9). Then, putting $\rho = a$ in the equations for conversion from spherical to rectangular coordinates (Equations 13.9.3), we obtain

$$x = a \sin\phi \cos\theta \qquad y = a \sin\phi \sin\theta \qquad z = a \cos\phi$$

as the parametric equations of the sphere. The corresponding vector equation is

$$\mathbf{r}(\phi, \theta) = a \sin\phi \cos\theta\,\mathbf{i} + a \sin\phi \sin\theta\,\mathbf{j} + a \cos\phi\,\mathbf{k}$$

We have $0 \leqslant \phi \leqslant \pi$ and $0 \leqslant \theta \leqslant 2\pi$, so the parameter domain is the rectangle $D = [0, \pi] \times [0, 2\pi]$. ■

One of the uses of parametric surfaces is in computer graphics. Figure 2 shows the result of trying to graph the sphere $x^2 + y^2 + z^2 = 1$ by solving the equation for z and graphing the top and bottom hemispheres separately. Part of the sphere appears to be missing because of the rectangular grid system used by the computer. The much better picture in Figure 3 was produced by a computer using the parametric equations found in Example 1.

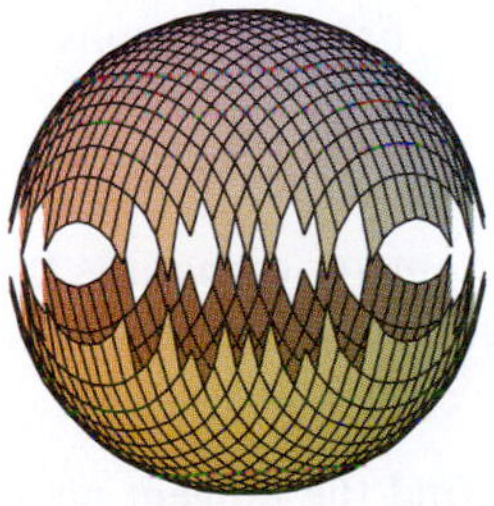

FIGURE 2

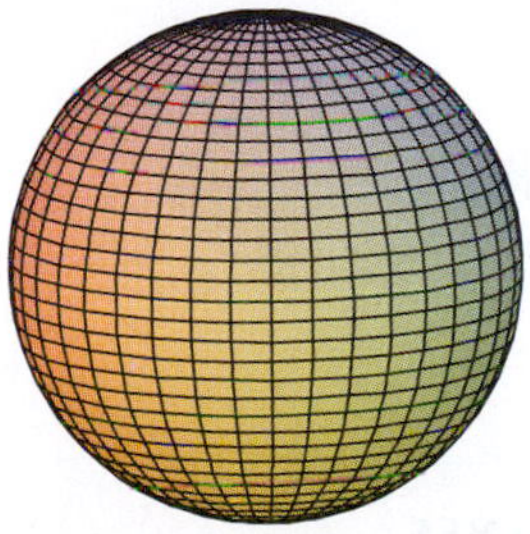

FIGURE 3

EXAMPLE 2 Find a parametric representation for the cylinder

$$x^2 + y^2 = 4 \qquad 0 \leqslant z \leqslant 1$$

SOLUTION The cylinder has a simple representation $r = 2$ in cylindrical coordinates, so we choose as parameters θ and z in cylindrical coordinates. Then the parametric equations of the cylinder are

$$x = 2\cos\theta \qquad y = 2\sin\theta \qquad z = z$$

where $0 \leqslant \theta \leqslant 2\pi$ and $0 \leqslant z \leqslant 1$. ■

EXAMPLE 3 Find a parametric representation for the elliptic paraboloid $z = x^2 + 2y^2$.

SOLUTION If we regard x and y as parameters, then the parametric equations are simply

$$x = x \qquad y = y \qquad z = x^2 + 2y^2$$

and the vector equation is

$$\mathbf{r}(x, y) = x\,\mathbf{i} + y\,\mathbf{j} + (x^2 + 2y^2)\,\mathbf{k}$$ ■

In general, a surface given as the graph of a function of x and y, that is, with an equation of the form $z = f(x, y)$, can always be regarded as a parametric surface by taking x and y as parameters and writing the parametric equations as

$$x = x \qquad y = y \qquad z = f(x, y)$$

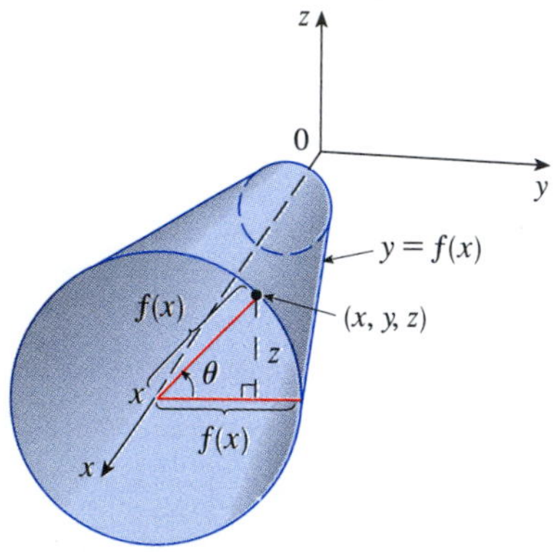

FIGURE 4

Surfaces of revolution can be represented parametrically and thus graphed using a computer. For instance, let us consider the surface S obtained by rotating the curve $y = f(x)$, $a \leqslant x \leqslant b$, about the x-axis, where $f(x) \geqslant 0$ and f' is continuous. Let θ be the angle of rotation as shown in Figure 4. If (x, y, z) is a point on S, then

(3) $$x = x \qquad y = f(x)\cos\theta \qquad z = f(x)\sin\theta$$

Therefore, we take x and θ as parameters and regard Equations 3 as parametric equations of S. The parameter domain is given by $a \leqslant x \leqslant b$, $0 \leqslant \theta \leqslant 2\pi$.

EXAMPLE 4 Find parametric equations for the surface generated by rotating the curve $y = \sin x$, $0 \leqslant x \leqslant 2\pi$, about the x-axis. Use these equations to graph the surface of revolution.

SOLUTION From Equations 3, the parametric equations are

$$x = x \qquad y = \sin x \cos\theta \qquad z = \sin x \sin\theta$$

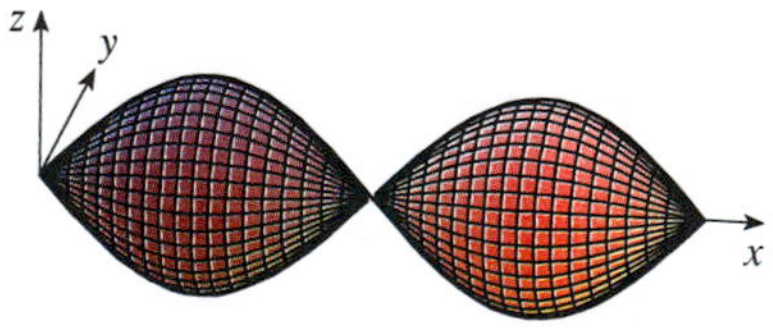

FIGURE 5

and the parameter domain is $0 \leqslant x \leqslant 2\pi$, $0 \leqslant \theta \leqslant 2\pi$. Using a computer to plot these equations and rotate the image, we obtain the graph in Figure 5. ■

TANGENT PLANES

Let's now find the tangent plane to a parametric surface S given by a vector function $\mathbf{r}(u, v)$ at a point P_0 with position vector $\mathbf{r}(u_0, v_0)$. If we keep u constant by putting $u = u_0$, then $\mathbf{r}(u_0, v)$ becomes a vector function of the single parameter v and defines a curve C_1 lying on S (see Figure 6). The tangent vector to C_1 at P_0 is obtained by taking the partial derivative of $\mathbf{r}$ with respect to v:

(4) $$\mathbf{r}_v = \frac{\partial x}{\partial v}(u_0, v_0)\,\mathbf{i} + \frac{\partial y}{\partial v}(u_0, v_0)\,\mathbf{j} + \frac{\partial z}{\partial v}(u_0, v_0)\,\mathbf{k}$$

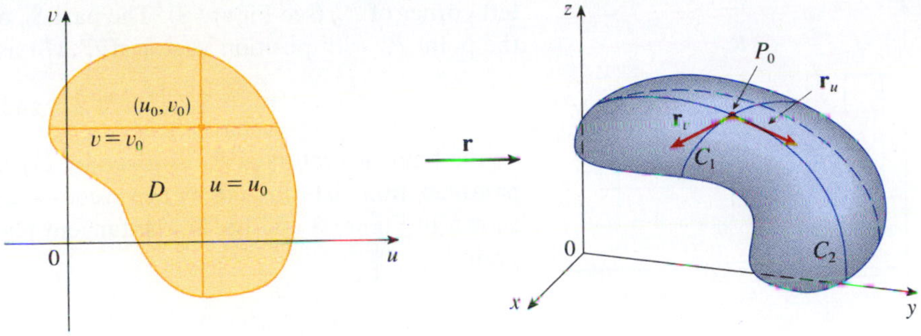

FIGURE 6

Similarly, if we keep v constant by putting $v = v_0$, we get a curve C_2 given by $\mathbf{r}(u, v_0)$ that lies on S, and its tangent vector at P_0 is

(5) $$\mathbf{r}_u = \frac{\partial x}{\partial u}(u_0, v_0)\,\mathbf{i} + \frac{\partial y}{\partial u}(u_0, v_0)\,\mathbf{j} + \frac{\partial z}{\partial u}(u_0, v_0)\,\mathbf{k}$$

If the **normal vector** $\mathbf{r}_u \times \mathbf{r}_v$ is not $\mathbf{0}$, then the surface S is called **smooth**. (It has no "corner.") In this case the tangent plane to S at P_0 exists and can be found as usual using the normal vector.

EXAMPLE 5 Find the tangent plane to the surface with parametric equations $x = u^2$, $y = v^2$, $z = u + 2v$ at the point $(1, 1, 3)$.

SOLUTION We first compute the tangent vectors:

$$\mathbf{r}_u = \frac{\partial x}{\partial u}\,\mathbf{i} + \frac{\partial y}{\partial u}\,\mathbf{j} + \frac{\partial z}{\partial u}\,\mathbf{k} = 2u\,\mathbf{i} + \mathbf{k}$$

$$\mathbf{r}_v = \frac{\partial x}{\partial v}\,\mathbf{i} + \frac{\partial y}{\partial v}\,\mathbf{j} + \frac{\partial z}{\partial v}\,\mathbf{k} = 2v\,\mathbf{j} + 2\,\mathbf{k}$$

Figure 7 shows the self-intersecting surface in Example 5 and its tangent plane at $(1, 1, 3)$.

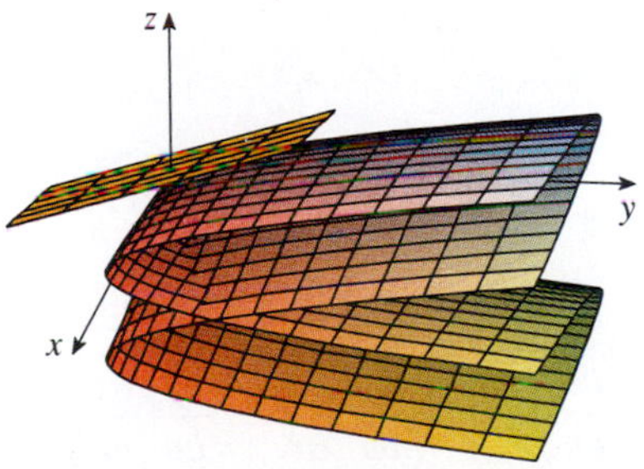

FIGURE 7

Thus, the normal vector is

$$\mathbf{r}_u \times \mathbf{r}_v = \begin{vmatrix} \mathbf{i} & \mathbf{j} & \mathbf{k} \\ 2u & 0 & 1 \\ 0 & 2v & 2 \end{vmatrix} = -2v\,\mathbf{i} - 4u\,\mathbf{j} + 4uv\,\mathbf{k}$$

Notice that the point $(1, 1, 3)$ corresponds to the parameter values $u = 1$ and $v = 1$, so the normal vector there is

$$-2\,\mathbf{i} - 4\,\mathbf{j} + 4\,\mathbf{k}$$

Therefore, the tangent plane at $(1, 1, 3)$ is

$$-2(x - 1) - 4(y - 1) + 4(z - 3) = 0$$

or

$$x + 2y - 2z + 3 = 0$$ ■

SURFACE AREA

Now we define the surface area of a general parametric surface given by Equation 1. For simplicity we start by considering a surface whose parameter domain D is a rectangle, and we partition it into subrectangles R_{ij}. Let's choose (u_i^*, v_j^*) to be the lower

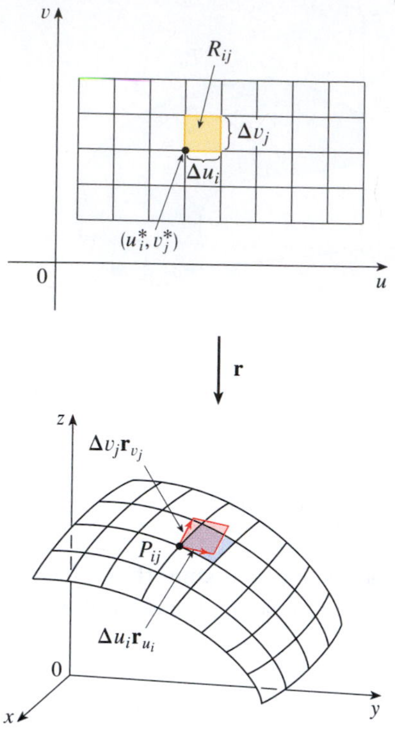

FIGURE 8

left corner of R_{ij} (see Figure 8). The part S_{ij} of the surface S that corresponds to R_{ij} has the point P_{ij} with position vector $\mathbf{r}(u_i^*, v_j^*)$ as one of its corners. Let

$$\mathbf{r}_{u_i} = \mathbf{r}_u(u_i^*, v_j^*) \qquad \text{and} \qquad \mathbf{r}_{v_j} = \mathbf{r}_v(u_i^*, v_j^*)$$

be the tangent vectors at P_{ij} as given by Equations 5 and 4. We approximate S_{ij} by the parallelogram determined by the vectors $\Delta u_i \mathbf{r}_{u_i}$ and $\Delta v_j \mathbf{r}_{v_j}$. (This parallelogram is shown in Figure 8 and lies in the tangent plane to S at P_{ij}.) The area of this parallelogram is

$$|(\Delta u_i \mathbf{r}_{u_i}) \times (\Delta v_j \mathbf{r}_{v_j})| = |\mathbf{r}_{u_i} \times \mathbf{r}_{v_j}| \Delta u_i \Delta v_j$$

and so an approximation to the area of S is

$$\sum_{i=1}^{m} \sum_{j=1}^{n} |\mathbf{r}_{u_i} \times \mathbf{r}_{v_j}| \Delta u_i \Delta v_j$$

Our intuition tells us that this approximation gets better as $\|P\| \to 0$, and we recognize the double sum as a Riemann sum for the double integral $\iint_D |\mathbf{r}_u \times \mathbf{r}_v| \, du \, dv$. This motivates the following definition:

(6) DEFINITION If a smooth parametric surface S is given by the equation

$$\mathbf{r}(u, v) = x(u, v)\,\mathbf{i} + y(u, v)\,\mathbf{j} + z(u, v)\,\mathbf{k} \qquad (u, v) \in D$$

and S is covered just once as (u, v) ranges throughout the parameter domain D, then the **surface area** of S is

$$A(S) = \iint_D |\mathbf{r}_u \times \mathbf{r}_v| \, dA$$

where $\mathbf{r}_u = \dfrac{\partial x}{\partial u}\mathbf{i} + \dfrac{\partial y}{\partial u}\mathbf{j} + \dfrac{\partial z}{\partial u}\mathbf{k} \qquad \mathbf{r}_v = \dfrac{\partial x}{\partial v}\mathbf{i} + \dfrac{\partial y}{\partial v}\mathbf{j} + \dfrac{\partial z}{\partial v}\mathbf{k}$

Computing the cross product of the tangent vectors, we get

$$\mathbf{r}_u \times \mathbf{r}_v = \begin{vmatrix} \mathbf{i} & \mathbf{j} & \mathbf{k} \\ \dfrac{\partial x}{\partial u} & \dfrac{\partial y}{\partial u} & \dfrac{\partial z}{\partial u} \\ \dfrac{\partial x}{\partial v} & \dfrac{\partial y}{\partial v} & \dfrac{\partial z}{\partial v} \end{vmatrix}$$

$$= \left(\frac{\partial y}{\partial u}\frac{\partial z}{\partial v} - \frac{\partial z}{\partial u}\frac{\partial y}{\partial v} \right)\mathbf{i} + \left(\frac{\partial z}{\partial u}\frac{\partial x}{\partial v} - \frac{\partial x}{\partial u}\frac{\partial z}{\partial v} \right)\mathbf{j} + \left(\frac{\partial x}{\partial u}\frac{\partial y}{\partial v} - \frac{\partial y}{\partial u}\frac{\partial x}{\partial v} \right)\mathbf{k}$$

$$= \frac{\partial(y, z)}{\partial(u, v)}\mathbf{i} + \frac{\partial(z, x)}{\partial(u, v)}\mathbf{j} + \frac{\partial(x, y)}{\partial(u, v)}\mathbf{k}$$

where we have used the notation for Jacobians given in Definition 13.11.7. Thus, the formula for surface area in Definition 6 becomes

Notice the similarity between the surface area formula in Equation 7 and the arc length formula in Equation 11.7.1:

$$L = \int_a^b \sqrt{\left(\frac{dx}{dt}\right)^2 + \left(\frac{dy}{dt}\right)^2 + \left(\frac{dz}{dt}\right)^2} \, dt$$

(7) $$A(S) = \iint_D \sqrt{\left[\frac{\partial(x, y)}{\partial(u, v)}\right]^2 + \left[\frac{\partial(y, z)}{\partial(u, v)}\right]^2 + \left[\frac{\partial(x, z)}{\partial(u, v)}\right]^2} \, dA$$

EXAMPLE 6 Find the surface area of a sphere of radius a.

SOLUTION In Example 1 we found the parametric representation

$$x = a \sin\phi \cos\theta \qquad y = a \sin\phi \sin\theta \qquad z = a \cos\phi$$

where the parameter domain is

$$D = \{(\phi, \theta) \mid 0 \leqslant \phi \leqslant \pi,\ 0 \leqslant \theta \leqslant 2\pi\}$$

We first compute the cross product of the tangent vectors:

$$\mathbf{r}_\phi \times \mathbf{r}_\theta = \begin{vmatrix} \mathbf{i} & \mathbf{j} & \mathbf{k} \\ \dfrac{\partial x}{\partial \phi} & \dfrac{\partial y}{\partial \phi} & \dfrac{\partial z}{\partial \phi} \\ \dfrac{\partial x}{\partial \theta} & \dfrac{\partial y}{\partial \theta} & \dfrac{\partial z}{\partial \theta} \end{vmatrix} = \begin{vmatrix} \mathbf{i} & \mathbf{j} & \mathbf{k} \\ a\cos\phi\cos\theta & a\cos\phi\sin\theta & -a\sin\phi \\ -a\sin\phi\sin\theta & a\sin\phi\cos\theta & 0 \end{vmatrix}$$

$$= a^2 \sin^2\phi \cos\theta\,\mathbf{i} + a^2 \sin^2\phi \sin\theta\,\mathbf{j} + a^2 \sin\phi\cos\phi\,\mathbf{k}$$

Thus

$$|\mathbf{r}_\phi \times \mathbf{r}_\theta| = \sqrt{a^4 \sin^4\phi \cos^2\theta + a^4 \sin^4\phi \sin^2\theta + a^4 \sin^2\phi \cos^2\phi}$$

$$= \sqrt{a^4 \sin^4\phi + a^4 \sin^2\phi\cos^2\phi} = a^2\sqrt{\sin^2\phi} = a^2 \sin\phi$$

since $\sin\phi \geqslant 0$ for $0 \leqslant \phi \leqslant \pi$. Therefore, the area of the sphere is

$$A = \iint_D |\mathbf{r}_\phi \times \mathbf{r}_\theta|\, dA = \int_0^{2\pi}\int_0^{\pi} a^2 \sin\phi\, d\phi\, d\theta$$

$$= a^2 \int_0^{2\pi} d\theta \int_0^{\pi} \sin\phi\, d\phi = a^2(2\pi)2 = 4\pi a^2$$

For the special case of a surface S with equation $z = f(x, y)$, where (x, y) lies in D and f has continuous partial derivatives, we take x and y as parameters. The parametric equations are

$$x = x \qquad y = y \qquad z = f(x, y)$$

so

$$\mathbf{r}_x = \mathbf{i} + \left(\frac{\partial f}{\partial x}\right)\mathbf{k}$$

$$\mathbf{r}_y = \mathbf{j} + \left(\frac{\partial f}{\partial y}\right)\mathbf{k}$$

and

(8)
$$\mathbf{r}_x \times \mathbf{r}_y = \begin{vmatrix} \mathbf{i} & \mathbf{j} & \mathbf{k} \\ 1 & 0 & \dfrac{\partial f}{\partial x} \\ 0 & 1 & \dfrac{\partial f}{\partial y} \end{vmatrix} = -\frac{\partial f}{\partial x}\mathbf{i} - \frac{\partial f}{\partial y}\mathbf{j} + \mathbf{k}$$

Thus, the surface area formula in Definition 6 becomes

(9)
$$A(S) = \iint_D \sqrt{1 + \left(\frac{\partial z}{\partial x}\right)^2 + \left(\frac{\partial z}{\partial y}\right)^2}\, dA$$

which agrees with Formula 13.7.2. Equation 9 should be compared with the arc length formula

$$L = \int_a^b \sqrt{1 + \left(\frac{dy}{dx}\right)^2}\, dx$$

EXAMPLE 7 Find the area of the part of the paraboloid $z = x^2 + y^2$ that lies under the plane $z = 9$.

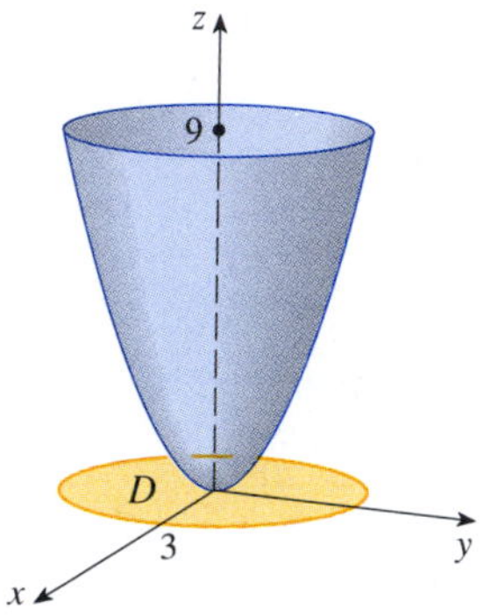

FIGURE 9

SOLUTION The plane intersects the paraboloid in the circle $x^2 + y^2 = 9$, $z = 9$. Therefore, the given surface lies above the disk D with center the origin and radius 3 (see Figure 9). Using Formula 9, we have

$$A = \iint_D \sqrt{1 + \left(\frac{\partial z}{\partial x}\right)^2 + \left(\frac{\partial z}{\partial y}\right)^2}\, dA = \iint_D \sqrt{1 + (2x)^2 + (2y)^2}\, dA$$
$$= \iint_D \sqrt{1 + 4(x^2 + y^2)}\, dA$$

Converting to polar coordinates, we obtain

$$A = \int_0^{2\pi}\int_0^3 \sqrt{1 + 4r^2}\, r\,dr\,d\theta = \int_0^{2\pi} d\theta \int_0^3 r\sqrt{1 + 4r^2}\, dr$$
$$= 2\pi\left(\frac{1}{8}\right)\frac{2}{3}(1 + 4r^2)^{3/2}\Big]_0^3 = \frac{\pi}{6}(37\sqrt{37} - 1)$$ ■

The question remains whether our definition of surface area (6) is consistent with the surface area formula from single-variable calculus (9.4.4).

We consider the surface S obtained by rotating the curve $y = f(x)$, $a \leqslant x \leqslant b$, about the x-axis, where $f(x) \geqslant 0$ and f' is continuous. From Equations 3 we know that parametric equations of S are

$$x = x \qquad y = f(x)\cos\theta \qquad z = f(x)\sin\theta \qquad a \leqslant x \leqslant b \qquad 0 \leqslant \theta \leqslant 2\pi$$

To compute the surface area of S we need the tangent vectors

$$\mathbf{r}_x = \mathbf{i} + f'(x)\cos\theta\,\mathbf{j} + f'(x)\sin\theta\,\mathbf{k}$$
$$\mathbf{r}_\theta = -f(x)\sin\theta\,\mathbf{j} + f(x)\cos\theta\,\mathbf{k}$$

Thus

$$\mathbf{r}_x \times \mathbf{r}_\theta = \begin{vmatrix} \mathbf{i} & \mathbf{j} & \mathbf{k} \\ 1 & f'(x)\cos\theta & f'(x)\sin\theta \\ 0 & -f(x)\sin\theta & f(x)\cos\theta \end{vmatrix}$$
$$= f(x)f'(x)\,\mathbf{i} - f(x)\cos\theta\,\mathbf{j} - f(x)\sin\theta\,\mathbf{k}$$

and so $$|\mathbf{r}_x \times \mathbf{r}_\theta| = \sqrt{[f(x)]^2[f'(x)]^2 + [f(x)]^2\cos^2\theta + [f(x)]^2\sin^2\theta}$$
$$= \sqrt{[f(x)]^2[1 + (f'(x))^2]} = f(x)\sqrt{1 + [f'(x)]^2}$$

because $f(x) \geqslant 0$. Therefore, the area of S is

$$A = \iint_D |\mathbf{r}_x \times \mathbf{r}_\theta|\, dA = \int_0^{2\pi}\int_a^b f(x)\sqrt{1 + [f'(x)]^2}\, dx\, d\theta$$
$$= 2\pi \int_a^b f(x)\sqrt{1 + [f'(x)]^2}\, dx$$

This is precisely the formula that was used to define the area of a surface of revolution in single-variable calculus (9.4.4).

EXERCISES 14.6

1–8 ■ Find a parametric representation for the surface.

1. The upper half of the ellipsoid $3x^2 + 2y^2 + z^2 = 1$
2. The part of the hyperboloid $-x^2 - y^2 + z^2 = 1$ that lies below the rectangle $[-1, 1] \times [-3, 3]$
3. The part of the elliptic paraboloid $y = 6 - 3x^2 - 2z^2$ that lies to the right of the xz-plane
4. The part of the elliptic paraboloid $x + y^2 + 2z^2 = 4$ that lies in front of the plane $x = 0$
5. The part of the sphere $x^2 + y^2 + z^2 = 4$ that lies above the cone $z = \sqrt{x^2 + y^2}$
6. The part of the cylinder $x^2 + z^2 = 1$ that lies between the planes $y = -1$ and $y = 3$
7. The part of the plane $z = 5$ that lies inside the cylinder $x^2 + y^2 = 16$
8. The part of the plane $z = x + 3$ that lies inside the cylinder $x^2 + y^2 = 1$

9–12 ■ Match the equations with the graphs (labeled I–IV). Give reasons for your answer.

9. $\mathbf{r}(u, v) = u\cos v\,\mathbf{i} + u\sin v\,\mathbf{j} + v\,\mathbf{k}$
10. $\mathbf{r}(u, v) = u\cos v\,\mathbf{i} + u\sin v\,\mathbf{j} + u\,\mathbf{k}$
11. $x = (u - \sin u)\cos v,\quad y = (1 - \cos u)\sin v,\quad z = u$
12. $x = (1 - u)(3 + \cos v)\cos 4\pi u,$
$y = (1 - u)(3 + \cos v)\sin 4\pi u,\quad z = 3u + (1 - u)\sin v$

I

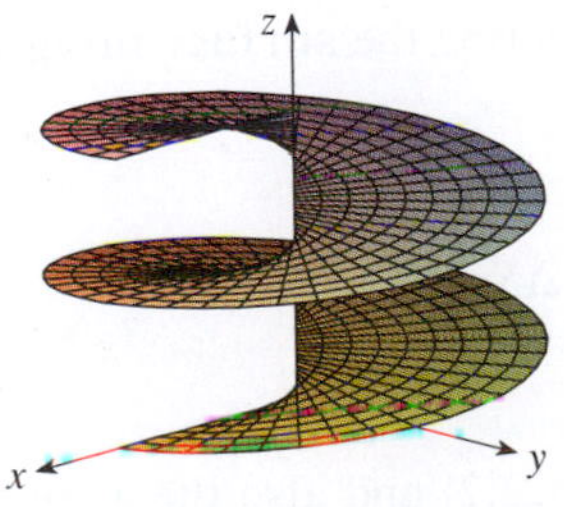

II

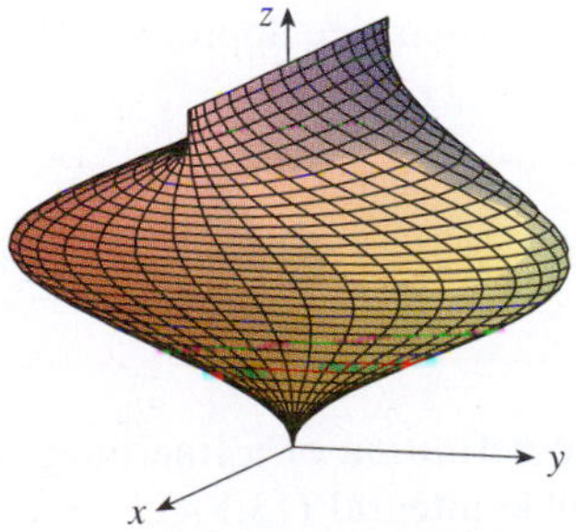

III

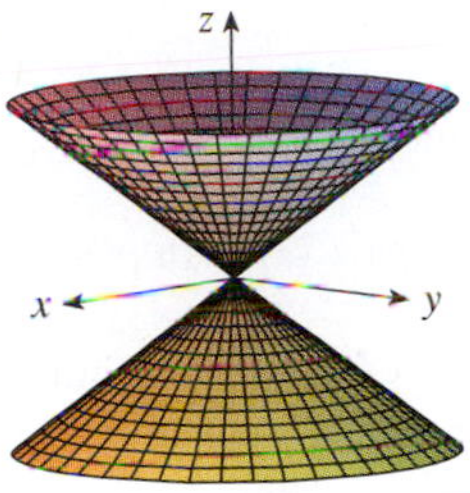

IV

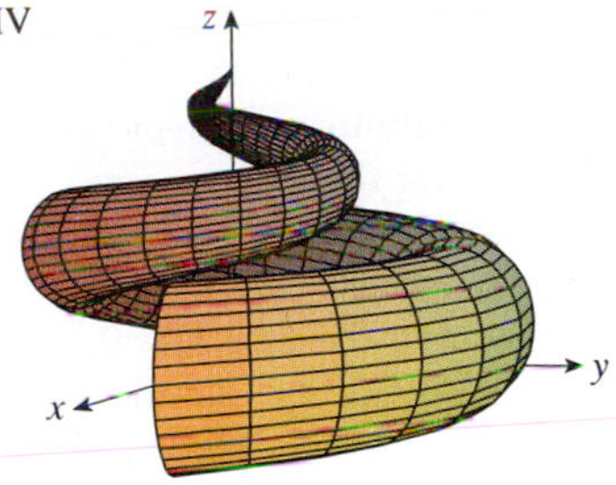

13. Find parametric equations for the surface obtained by rotating the curve $y = e^{-x}$, $0 \leqslant x \leqslant 3$, about the x-axis and use them to graph the surface.
14. Find parametric equations for the surface obtained by rotating the curve $x = 4y^2 - y^4$, $-2 \leqslant y \leqslant 2$, about the y-axis and use them to graph the surface.

15–18 ■ Find an equation of the tangent plane to the given parametric surface at the specified point. If you have software that graphs parametric surfaces, use a computer to graph the surface and the tangent plane.

15. $x = u + v,\quad y = 3u^2,\quad z = u - v;\quad (2, 3, 0)$
16. $x = u^2,\quad y = u - v^2,\quad z = v^2;\quad (1, 0, 1)$
17. $\mathbf{r}(u, v) = uv\,\mathbf{i} + ue^v\,\mathbf{j} + ve^u\,\mathbf{k};\quad (0, 0, 0)$
18. $\mathbf{r}(u, v) = (u + v)\,\mathbf{i} + u\cos v\,\mathbf{j} + v\sin u\,\mathbf{k};\quad (1, 1, 0)$

19–27 ■ Find the area of the given surface.

19. The part of the plane $x + 2y + z = 4$ that lies inside the cylinder $x^2 + y^2 = 4$
20. The part of the surface $z = x + y^2$ that lies above the triangle with vertices $(0, 0)$, $(1, 1)$, and $(0, 1)$
21. The part of the hyperbolic paraboloid $z = y^2 - x^2$ that lies between the cylinders $x^2 + y^2 = 1$ and $x^2 + y^2 = 4$
22. The part of the paraboloid $x = y^2 + z^2$ that lies inside the cylinder $y^2 + z^2 = 9$

23. The part of the surface $y = 4x + z^2$ that lies between the planes $x = 0$, $x = 1$, $z = 0$, and $z = 1$

24. The part of the cylinder $x^2 + z^2 = a^2$ that lies inside the cylinder $x^2 + y^2 = a^2$

25. The part of the sphere $x^2 + y^2 + z^2 = a^2$ that lies inside the cylinder $x^2 + y^2 = ax$

26. The helicoid (or spiral ramp) with vector equation $\mathbf{r}(u, v) = u \cos v\, \mathbf{i} + u \sin v\, \mathbf{j} + v\, \mathbf{k}$, $0 \leqslant u \leqslant 1$, $0 \leqslant v \leqslant \pi$

27. The surface with parametric equations $x = uv$, $y = u + v$, $z = u - v$, $u^2 + v^2 \leqslant 1$

28. (a) Set up, but do not evaluate, a double integral for the area of the surface with parametric equations $x = au \cos v$, $y = bu \sin v$, $z = u^2$, $0 \leqslant u \leqslant 2$, $0 \leqslant v \leqslant 2\pi$.
(b) Eliminate the parameters to show that the surface is an elliptic paraboloid and set up another double integral for the surface area.
(c) Use the parametric equations in part (a) with $a = 2$ and $b = 3$ to graph the surface.
CAS (d) For the case $a = 2$, $b = 3$, use a computer algebra system to find the surface area correct to four decimal places.

29. (a) Show that the parametric equations $x = a \sin u \cos v$, $y = b \sin u \sin v$, $z = c \cos u$, $0 \leqslant u \leqslant \pi$, $0 \leqslant v \leqslant 2\pi$, represent an ellipsoid.
(b) Use the parametric equations in part (a) to graph the ellipsoid for the case $a = 1$, $b = 2$, $c = 3$.
(c) Set up, but do not evaluate, a double integral for the surface area of the ellipsoid in part (b).

30. (a) Show that the parametric equations $x = a \cosh u \cos v$, $y = b \cosh u \sin v$, $z = c \sinh u$, represent a hyperboloid of one sheet.
(b) Use the parametric equations in part (a) to graph the hyperboloid for the case $a = 1$, $b = 2$, $c = 3$.
(c) Set up, but do not evaluate, a double integral for the surface area of the part of the hyperboloid in part (b) that lies between the planes $z = -3$ and $z = 3$.

31. Find a parametric representation for the torus obtained by rotating about the z-axis the circle in the xz-plane with center $(b, 0, 0)$ and radius $a < b$. [*Hint:* Take as parameters the angles θ and α shown in the figure.]

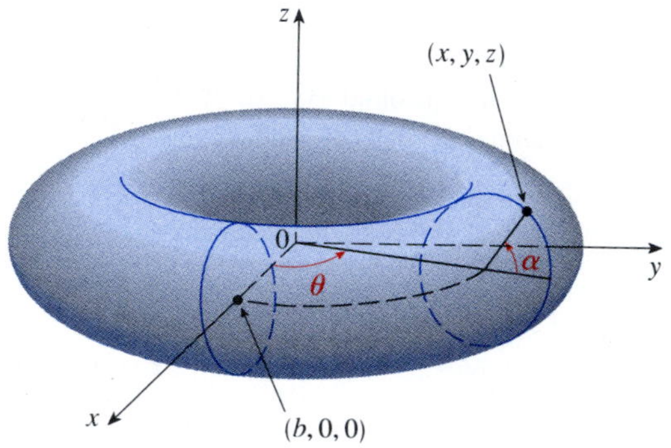

32. Use the parametric representation from Exercise 31 to find the surface area of the torus.

14.7 SURFACE INTEGRALS

The relationship between surface integrals and surface area is much the same as the relationship between line integrals and arc length. Suppose f is a function of three variables whose domain includes a surface S. We divide S into patches S_{ij} with area ΔS_{ij}. We evaluate f at a point P_{ij}^* in each patch, multiply by the area ΔS_{ij}, and form the sum

$$\sum_{i=1}^{m} \sum_{j=1}^{n} f(P_{ij}^*)\, \Delta S_{ij}$$

Then we take the limit as the patch size approaches 0 and define the **surface integral of f over the surface S** as

(1) $$\iint_S f(x, y, z)\, dS = \lim_{\|P\| \to 0} \sum_{i=1}^{m} \sum_{j=1}^{n} f(P_{ij}^*)\, \Delta S_{ij}$$

Notice the analogy with the definition of a line integral (14.2.2) and also the analogy with the definition of a double integral (13.1.4).

To evaluate the surface integral in Equation 1 we approximate the patch area ΔS_{ij} by the area ΔT_{ij} of an approximating parallelogram in the tangent plane, and the limit becomes a double integral. We now explain the details for two types of surfaces, graphs and parametric surfaces.

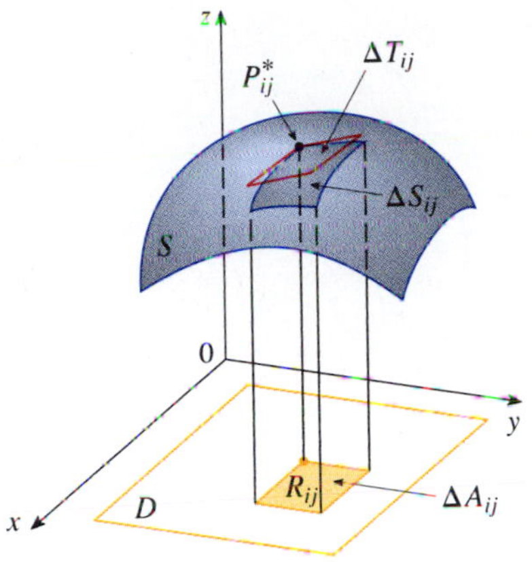

FIGURE 1

GRAPHS If the surface S is a graph of a function of two variables, then it has an equation of the form $z = g(x, y)$, $(x, y) \in D$, and is illustrated in Figure 1. The patch S_{ij} lies directly above a rectangle R_{ij} in a partition P of D and the point P^*_{ij} in S_{ij} is of the form $(x_i^*, y_j^*, g(x_i^*, y_j^*))$. As with surface area in Section 13.7, we use the approximation

$$\Delta S_{ij} \approx \Delta T_{ij} = \sqrt{[g_x(x_i, y_j)]^2 + [g_y(x_i, y_j)]^2 + 1}\ \Delta A_{ij}$$

It can be shown that, if f is continuous on S and g has continuous derivatives, then Definition 1 becomes

$$\iint_S f(x, y, z)\, dS$$

$$= \lim_{\|P\|\to 0} \sum_{i=1}^{m} \sum_{j=1}^{n} f(x_i^*, y_j^*, g(x_i^*, y_j^*))\sqrt{[g_x(x_i, y_j)]^2 + [g_y(x_i, y_j)]^2 + 1}\ \Delta A_{ij}$$

$$= \iint_D f(x, y, g(x, y))\sqrt{[g_x(x, y)]^2 + [g_y(x, y)]^2 + 1}\ dA$$

This formula is usually written in the following form:

(2) $$\iint_S f(x, y, z)\, dS = \iint_D f(x, y, g(x, y))\sqrt{\left(\frac{\partial z}{\partial x}\right)^2 + \left(\frac{\partial z}{\partial y}\right)^2 + 1}\ dA$$

Similar formulas apply when it is more convenient to project S onto the yz-plane or xz-plane. For instance, if S is a surface with equation $y = h(x, z)$ and D is its projection on the xz-plane, then

$$\iint_S f(x, y, z)\, dS = \iint_D f(x, h(x, z), z)\sqrt{\left(\frac{\partial y}{\partial x}\right)^2 + \left(\frac{\partial y}{\partial z}\right)^2 + 1}\ dA$$

Surface integrals have applications similar to those for the integrals we have previously considered. For example, if a thin sheet (say, of aluminum foil) has the shape of a surface S and the density (mass per unit area) at the point (x, y, z) is $\rho(x, y, z)$, then the total **mass** of the sheet is

$$m = \iint_S \rho(x, y, z)\, dS$$

and the **center of mass** is $(\bar{x}, \bar{y}, \bar{z})$, where

$$\bar{x} = \frac{1}{m}\iint_S x\rho(x, y, z)\, dS \qquad \bar{y} = \frac{1}{m}\iint_S y\rho(x, y, z)\, dS \qquad \bar{z} = \frac{1}{m}\iint_S z\rho(x, y, z)\, dS$$

Moments of inertia can also be defined as before (see Exercise 33).

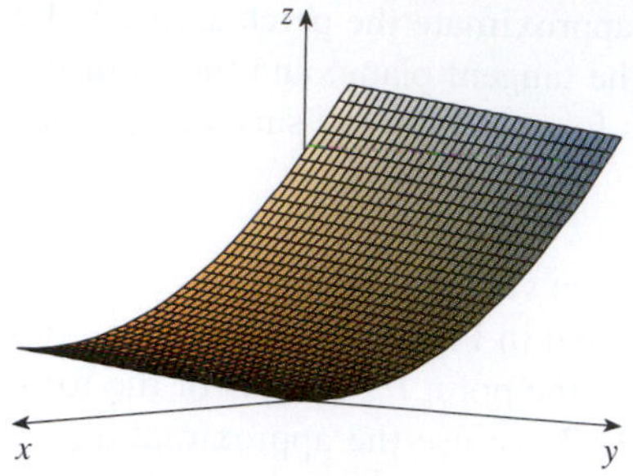

FIGURE 2

EXAMPLE 1 Evaluate $\iint_S y\,dS$, where S is the surface $z = x + y^2$, $0 \leqslant x \leqslant 1$, $0 \leqslant y \leqslant 2$ (see Figure 2).

SOLUTION Since

$$\frac{\partial z}{\partial x} = 1 \qquad \text{and} \qquad \frac{\partial z}{\partial y} = 2y$$

Formula 2 gives

$$\begin{aligned}\iint_S y\,dS &= \iint_D y\sqrt{1 + \left(\frac{\partial z}{\partial x}\right)^2 + \left(\frac{\partial z}{\partial y}\right)^2}\,dA \\ &= \int_0^1 \int_0^2 y\sqrt{1 + 1 + 4y^2}\,dy\,dx \\ &= \int_0^1 dx\,\sqrt{2}\int_0^2 y\sqrt{1 + 2y^2}\,dy \\ &= \sqrt{2}\left(\tfrac{1}{4}\right)\tfrac{2}{3}(1 + 2y^2)^{3/2}\Big]_0^2 = \frac{13\sqrt{2}}{3}\end{aligned}$$

■

PARAMETRIC SURFACES Suppose that a surface S has a vector equation

$$\mathbf{r}(u, v) = x(u, v)\,\mathbf{i} + y(u, v)\,\mathbf{j} + z(u, v)\,\mathbf{k} \qquad (u, v) \in D$$

We first assume that the parameter domain D is a rectangle. A partition of D into subrectangles R_{ij} with dimensions Δu_i and Δv_j determines a partition of S into curvilinear regions S_{ij} that lie on S (see Figure 3). A point (u_i^*, v_j^*) in R_{ij} determines a point on S_{ij} with position vector $\mathbf{r}(u_i^*, v_j^*)$. As in Section 14.6 we approximate S_{ij} by a parallelogram with area

$$\Delta T_{ij} = |\mathbf{r}_{u_i} \times \mathbf{r}_{v_j}|\,\Delta u_i\,\Delta v_j$$

where $\mathbf{r}_u = \dfrac{\partial x}{\partial u}\,\mathbf{i} + \dfrac{\partial y}{\partial u}\,\mathbf{j} + \dfrac{\partial z}{\partial u}\,\mathbf{k} \qquad \mathbf{r}_v = \dfrac{\partial x}{\partial v}\,\mathbf{i} + \dfrac{\partial y}{\partial v}\,\mathbf{j} + \dfrac{\partial z}{\partial v}\,\mathbf{k}$

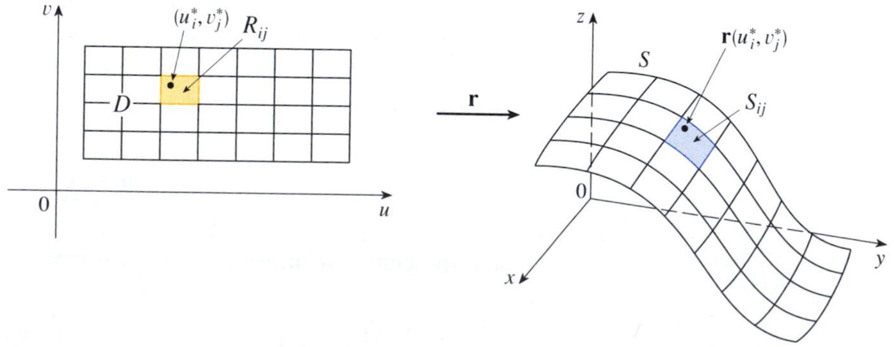

FIGURE 3

If the components are continuous and $\mathbf{r}_u$ and $\mathbf{r}_v$ are nonzero and nonparallel in the interior of D, it can be shown from Definition 1, even when D is not a rectangle, that

We assume that the surface is covered only once as (u, v) ranges throughout D. The value of the surface integral does not depend on the parametrization that is used.

(3)
$$\iint_S f(x, y, z)\, dS = \iint_D f(\mathbf{r}(u, v))\,|\mathbf{r}_u \times \mathbf{r}_v|\, dA$$

This should be compared with the formula for a line integral:

$$\int_C f(x, y, z)\, ds = \int_a^b f(\mathbf{r}(t))\,|\mathbf{r}'(t)|\, dt$$

Formula 3 allows us to compute a surface integral by converting it into a double integral over the parameter domain D. When using this formula, remember that $f(\mathbf{r}(u, v))$ is evaluated by writing $x = x(u, v)$, $y = y(u, v)$, and $z = z(u, v)$ in the formula for $f(x, y, z)$. In fact, by using the same calculation that led to Equation 14.6.7, we can write the right side of Equation 3 as

$$\iint_D f(x(u, v), y(u, v), z(u, v)) \sqrt{\left[\frac{\partial(x, y)}{\partial(u, v)}\right]^2 + \left[\frac{\partial(y, z)}{\partial(u, v)}\right]^2 + \left[\frac{\partial(x, z)}{\partial(u, v)}\right]^2}\, dA$$

which is the analogue of Equation 14.2.9. Observe that $\iint_S 1\, dS = A(S)$.

Any surface S with equation $z = g(x, y)$ can be regarded as a parametric surface with parametric equations

$$x = x \qquad y = y \qquad z = g(x, y)$$

and from Equation 14.6.8 we have

$$|\mathbf{r}_x \times \mathbf{r}_y| = \sqrt{1 + \left(\frac{\partial z}{\partial x}\right)^2 + \left(\frac{\partial z}{\partial y}\right)^2}$$

Therefore, in this case, Formula 3 becomes Formula 2.

EXAMPLE 2 Compute the surface integral $\iint_S x^2\, dS$, where S is the unit sphere $x^2 + y^2 + z^2 = 1$.

SOLUTION As in Example 1 in Section 14.6, we use the parametric representation

$$x = \sin\phi\cos\theta \qquad y = \sin\phi\sin\theta \qquad z = \cos\phi \qquad 0 \leqslant \phi \leqslant \pi \qquad 0 \leqslant \theta \leqslant 2\pi$$

that is, $\qquad \mathbf{r}(\phi, \theta) = \sin\phi\cos\theta\,\mathbf{i} + \sin\phi\sin\theta\,\mathbf{j} + \cos\phi\,\mathbf{k}$

As in Example 6 in Section 14.6, we can compute that

$$|\mathbf{r}_\phi \times \mathbf{r}_\theta| = \sin\phi$$

Therefore, by Formula 3,

$$\begin{aligned}\iint_S x^2\, dS &= \iint_D (\sin\phi\cos\theta)^2\,|\mathbf{r}_\phi \times \mathbf{r}_\theta|\, dA \\ &= \int_0^{2\pi}\int_0^{\pi} \sin^2\phi\cos^2\theta\sin\phi\, d\phi\, d\theta = \int_0^{2\pi}\cos^2\theta\, d\theta \int_0^{\pi}\sin^3\phi\, d\phi \\ &= \int_0^{2\pi} \tfrac{1}{2}(1 + \cos 2\theta)\, d\theta \int_0^{\pi} (\sin\phi - \sin\phi\cos^2\phi)\, d\phi\end{aligned}$$

$$= \tfrac{1}{2}\left[\theta + \tfrac{1}{2}\sin 2\theta\right]_0^{2\pi}\left[-\cos\phi + \tfrac{1}{3}\cos^3\phi\right]_0^{\pi} = \frac{4\pi}{3}$$ ■

If S is a piecewise-smooth surface, that is, a finite union of smooth surfaces $S_1, S_2, \ldots, S_n$ that intersect only along their boundaries, then the surface integral of f over S is defined by

$$\iint_S f(x, y, z)\,dS = \iint_{S_1} f(x, y, z)\,dS + \cdots + \iint_{S_n} f(x, y, z)\,dS$$

EXAMPLE 3 Evaluate $\iint_S z\,dS$, where S is the surface whose sides S_1 are given by the cylinder $x^2 + y^2 = 1$, whose bottom S_2 is the disk $x^2 + y^2 \leqslant 1$ in the plane $z = 0$, and whose top S_3 is the part of the plane $z = x + 1$ that lies above S_2.

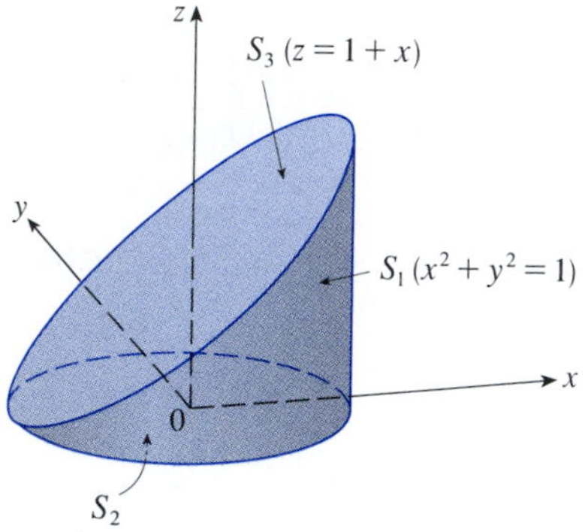

FIGURE 4

SOLUTION The surface S is shown in Figure 4. (We have changed the usual position of the axes to get a better look at S.) For S_1 we use θ and z as parameters (see Example 2 in Section 14.6) and write its parametric equations as

$$x = \cos\theta \qquad y = \sin\theta \qquad z = z$$

where $\qquad 0 \leqslant \theta \leqslant 2\pi \qquad$ and $\qquad 0 \leqslant z \leqslant 1 + x = 1 + \cos\theta$

Therefore

$$\mathbf{r}_\theta \times \mathbf{r}_z = \begin{vmatrix} \mathbf{i} & \mathbf{j} & \mathbf{k} \\ -\sin\theta & \cos\theta & 0 \\ 0 & 0 & 1 \end{vmatrix} = \cos\theta\,\mathbf{i} + \sin\theta\,\mathbf{j}$$

and $$|\mathbf{r}_\theta \times \mathbf{r}_z| = \sqrt{\cos^2\theta + \sin^2\theta} = 1$$

Thus, the surface integral over S_1 is

$$\iint_{S_1} z\,dS = \iint_D z\,|\mathbf{r}_\theta \times \mathbf{r}_z|\,dA$$

$$= \int_0^{2\pi}\int_0^{1+\cos\theta} z\,dz\,d\theta = \int_0^{2\pi} \tfrac{1}{2}(1 + \cos\theta)^2\,d\theta$$

$$= \tfrac{1}{2}\int_0^{2\pi} \left[1 + 2\cos\theta + \tfrac{1}{2}(1 + \cos 2\theta)\right] d\theta$$

$$= \tfrac{1}{2}\left[\tfrac{3}{2}\theta + 2\sin\theta + \tfrac{1}{4}\sin 2\theta\right]_0^{2\pi} = \frac{3\pi}{2}$$

Since S_2 lies in the plane $z = 0$, we have

$$\iint_{S_2} z\,dS = \iint_{S_2} 0\,dS = 0$$

The top surface S_3 lies above the unit disk D and is part of the plane $z = 1 + x$. So, taking $g(x, y) = 1 + x$ in Formula 2 and converting to polar coordinates, we have

$$\iint_{S_3} z\,dS = \iint_D (1 + x)\sqrt{1 + \left(\frac{\partial z}{\partial x}\right)^2 + \left(\frac{\partial z}{\partial y}\right)^2}\,dA$$

$$= \int_0^{2\pi} \int_0^1 (1 + r\cos\theta)\sqrt{1 + 1 + 0}\; r\,dr\,d\theta$$

$$= \sqrt{2} \int_0^{2\pi} \int_0^1 (r + r^2\cos\theta)\,dr\,d\theta$$

$$= \sqrt{2} \int_0^{2\pi} \left(\tfrac{1}{2} + \tfrac{1}{3}\cos\theta\right) d\theta = \sqrt{2}\left[\frac{\theta}{2} + \frac{\sin\theta}{3}\right]_0^{2\pi} = \sqrt{2}\,\pi$$

Therefore
$$\iint_S z\,dS = \iint_{S_1} z\,dS + \iint_{S_2} z\,dS + \iint_{S_3} z\,dS$$

$$= \frac{3\pi}{2} + 0 + \sqrt{2}\,\pi = \left(\tfrac{3}{2} + \sqrt{2}\right)\pi$$ ■

ORIENTED SURFACES

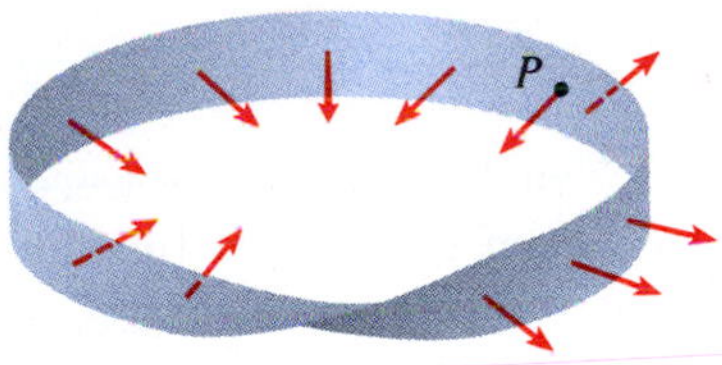

FIGURE 5
A Möbius strip

In order to define surface integrals of vector fields, we need to rule out nonorientable surfaces such as the *Möbius strip* shown in Figure 5. [It is named after the German geometer August Möbius (1790–1868).] You can construct one for yourself by taking a long rectangular strip of paper, giving it a half-twist, and taping the short edges together as in Figure 6. If an ant were to crawl along the Möbius strip starting at a point P, it would end up on the "other side" of the strip (that is, with its upper side pointing in the opposite direction). Then, if the ant continued to crawl in the same direction, it would end up back at the same point P without ever having crossed an edge. (If you have constructed a Möbius strip, try drawing a pencil line down the middle.) Therefore, a Möbius strip really has only one side.

FIGURE 6
Constructing a Möbius strip

From now on we consider only orientable (two-sided) surfaces. We start with a surface S that has a tangent plane at every point (x, y, z) on S (except at any boundary points). There are two unit normal vectors $\mathbf{n}_1$ and $\mathbf{n}_2 = -\mathbf{n}_1$ at (x, y, z) (see Figure 7). If it is possible to choose a unit normal vector $\mathbf{n}$ at every such point (x, y, z) so that $\mathbf{n}$ varies continuously over S, then S is called an **oriented surface** and the given choice of

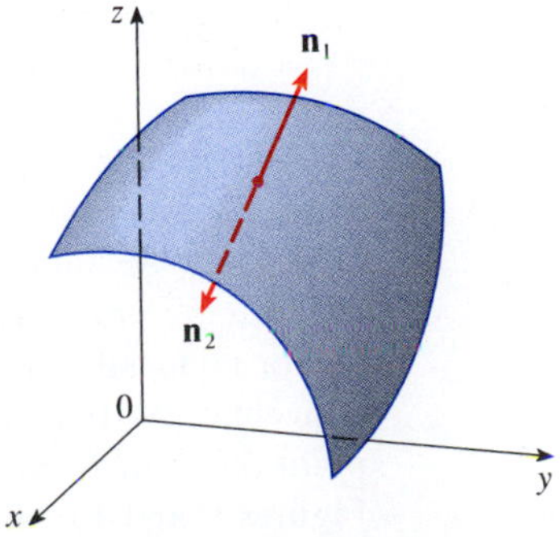

FIGURE 7

n provides S with an **orientation**. There are two possible orientations for any orientable surface (see Figure 8).

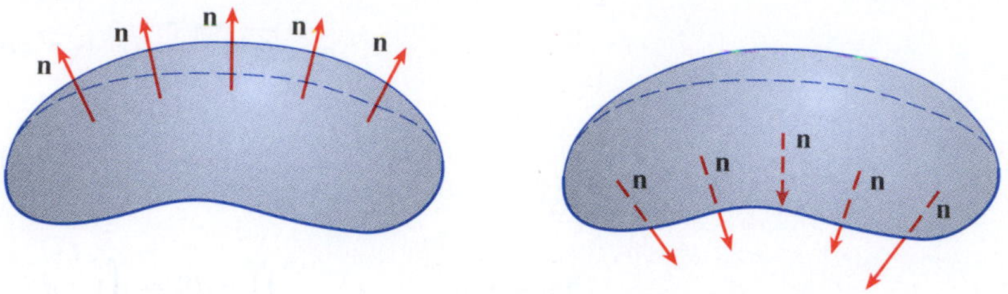

FIGURE 8
The two orientations of an orientable surface

For a surface $z = g(x, y)$ given as the graph of g, we use Equation 14.6.8 and see that the induced orientation is given by the unit normal vector

(4)
$$\mathbf{n} = \frac{-\dfrac{\partial g}{\partial x}\mathbf{i} - \dfrac{\partial g}{\partial y}\mathbf{j} + \mathbf{k}}{\sqrt{1 + \left(\dfrac{\partial g}{\partial x}\right)^2 + \left(\dfrac{\partial g}{\partial y}\right)^2}}$$

Since the **k**-component is positive, this gives the *upward* orientation of the surface.

If S is a smooth orientable surface given in parametric form by a vector function $\mathbf{r}(u, v)$, then it is automatically supplied with the orientation of the unit normal vector

(5)
$$\mathbf{n} = \frac{\mathbf{r}_u \times \mathbf{r}_v}{|\mathbf{r}_u \times \mathbf{r}_v|}$$

and the opposite orientation is given by $-\mathbf{n}$. For instance, in Example 1 in Section 14.6 we found the parametric representation

$$\mathbf{r}(\phi, \theta) = a \sin\phi \cos\theta\, \mathbf{i} + a \sin\phi \sin\theta\, \mathbf{j} + a \cos\phi\, \mathbf{k}$$

for the sphere $x^2 + y^2 + z^2 = a^2$. Then in Example 6 in Section 14.6 we found that

$$\mathbf{r}_\phi \times \mathbf{r}_\theta = a^2 \sin^2\phi \cos\theta\, \mathbf{i} + a^2 \sin^2\phi \sin\theta\, \mathbf{j} + a^2 \sin\phi \cos\phi\, \mathbf{k}$$

and
$$|\mathbf{r}_\phi \times \mathbf{r}_\theta| = a^2 \sin\phi$$

So the orientation induced by $\mathbf{r}(\phi, \theta)$ is defined by the unit normal vector

$$\mathbf{n} = \frac{\mathbf{r}_\phi \times \mathbf{r}_\theta}{|\mathbf{r}_\phi \times \mathbf{r}_\theta|} = \sin\phi \cos\theta\, \mathbf{i} + \sin\phi \sin\theta\, \mathbf{j} + \cos\phi\, \mathbf{k} = \frac{1}{a}\mathbf{r}(\phi, \theta)$$

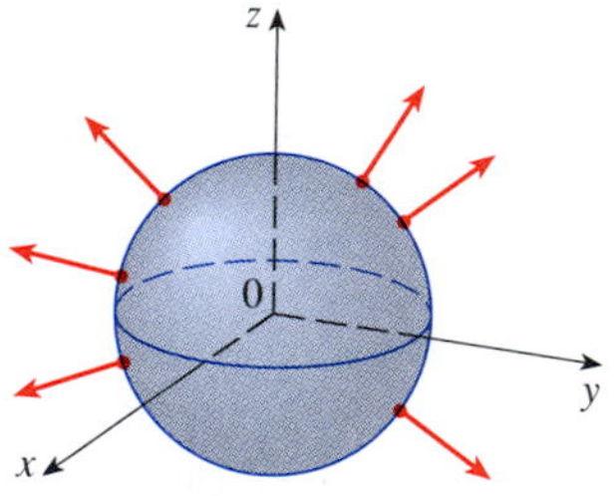

FIGURE 9
Positive orientation

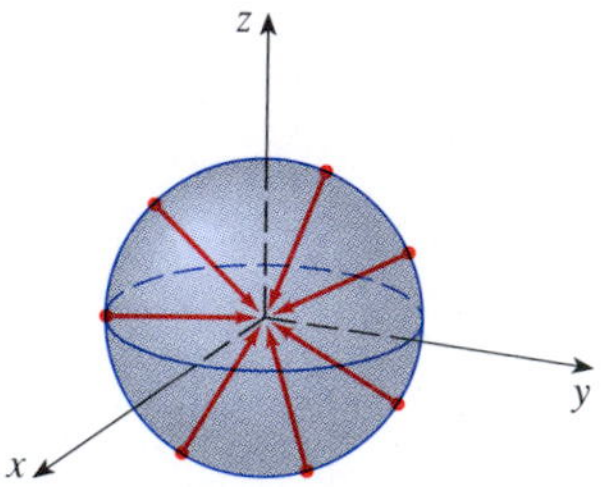

FIGURE 10
Negative orientation

Observe that **n** points in the same direction as the position vector, that is, outward from the sphere (see Figure 9). The opposite (inward) orientation would have been obtained (see Figure 10) if we had reversed the order of the parameters because $\mathbf{r}_\theta \times \mathbf{r}_\phi = -\mathbf{r}_\phi \times \mathbf{r}_\theta$.

For a **closed surface**, that is, a surface that is the boundary of a solid region E, the convention is that the **positive orientation** is the one for which the normal vectors point *outward* from E, and inward-pointing normals give the negative orientation (see Figures 9 and 10).

SURFACE INTEGRALS OF VECTOR FIELDS

Suppose that S is an oriented surface with unit normal vector $\mathbf{n}$, and imagine a fluid with density $\rho(x, y, z)$ and velocity field $\mathbf{v}(x, y, z)$ flowing through S. (Think of S as an imaginary surface so that it does not impede the fluid flow.) Then the rate of flow (mass per unit time) per unit area is $\rho\mathbf{v}$. If we partition S into small parts S_{ij}, as in Figure 11 (compare with Figures 1 and 3), then S_{ij} is nearly planar and so we can approximate the mass of fluid crossing S_{ij} in the direction of the normal $\mathbf{n}$ per unit time by the quantity

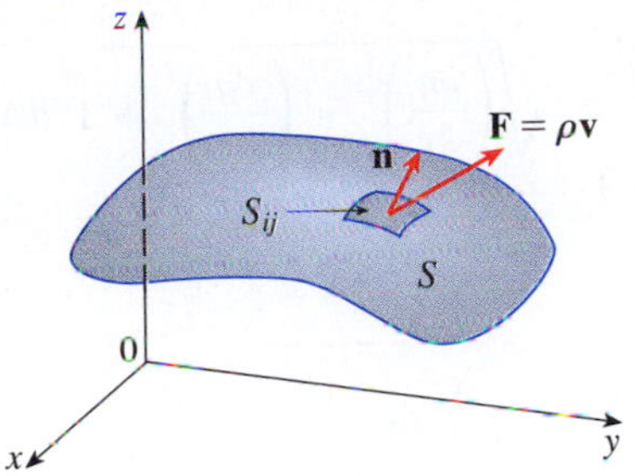

FIGURE 11

$$\rho\mathbf{v} \cdot \mathbf{n}A(S_{ij})$$

where ρ, $\mathbf{v}$, and $\mathbf{n}$ are evaluated at some point on S_{ij}. (Recall that the component of the vector $\rho\mathbf{v}$ in the direction of the unit vector $\mathbf{n}$ is $\rho\mathbf{v} \cdot \mathbf{n}$.) By summing these quantities and taking the limit we get, according to Definition 1, the surface integral of the function $\rho\mathbf{v} \cdot \mathbf{n}$ over S:

$$\textbf{(6)} \qquad \iint_S \rho\mathbf{v} \cdot \mathbf{n}\, dS = \iint_S \rho(x, y, z)\mathbf{v}(x, y, z) \cdot \mathbf{n}(x, y, z)\, dS$$

and this is interpreted physically as the rate of flow through S.

If we write $\mathbf{F} = \rho\mathbf{v}$, then $\mathbf{F}$ is also a vector field on $\mathbb{R}^3$ and the integral in Equation 6 becomes

$$\iint_S \mathbf{F} \cdot \mathbf{n}\, dS$$

A surface integral of this form occurs frequently in physics, even when $\mathbf{F}$ is not $\rho\mathbf{v}$, and is called the *surface integral* (or *flux integral*) of $\mathbf{F}$ over S.

(7) DEFINITION If $\mathbf{F}$ is a continuous vector field defined on an oriented surface S with unit normal vector $\mathbf{n}$, then the **surface integral of F over S** is

$$\iint_S \mathbf{F} \cdot d\mathbf{S} = \iint_S \mathbf{F} \cdot \mathbf{n}\, dS$$

This integral is also called the **flux** of $\mathbf{F}$ across S.

In words, Definition 7 says that the surface integral of a vector field over S is equal to the surface integral of its normal component over S (as previously defined).

GRAPHS In the case of a surface S given by a graph $z = g(x, y)$, we can find $\mathbf{n}$ by noting that S is also the level surface $f(x, y, z) = z - g(x, y) = 0$. We know that the gradient $\nabla f(x, y, z)$ is normal to this surface at (x, y, z) and so a unit normal vector is

$$\mathbf{n} = \frac{\nabla f(x, y, z)}{|\nabla f(x, y, z)|} = \frac{-g_x(x, y)\,\mathbf{i} - g_y(x, y)\,\mathbf{j} + \mathbf{k}}{\sqrt{[g_x(x, y)]^2 + [g_y(x, y)]^2 + 1}}$$

Since the $\mathbf{k}$-component is positive, this is the upward unit normal. If we now use Formula 2 to evaluate the surface integral (7) with

$$\mathbf{F}(x, y, z) = P(x, y, z)\,\mathbf{i} + Q(x, y, z)\,\mathbf{j} + R(x, y, z)\,\mathbf{k}$$

we get

$$\iint_S \mathbf{F} \cdot d\mathbf{S} = \iint_S \mathbf{F} \cdot \mathbf{n}\, dS$$

$$= \iint_D (P\,\mathbf{i} + Q\,\mathbf{j} + R\,\mathbf{k}) \cdot \frac{-\dfrac{\partial g}{\partial x}\,\mathbf{i} - \dfrac{\partial g}{\partial y}\,\mathbf{j} + \mathbf{k}}{\sqrt{\left(\dfrac{\partial g}{\partial x}\right)^2 + \left(\dfrac{\partial g}{\partial y}\right)^2 + 1}} \sqrt{\left(\frac{\partial g}{\partial x}\right)^2 + \left(\frac{\partial g}{\partial y}\right)^2 + 1}\, dA$$

or

(8) $$\iint_S \mathbf{F} \cdot d\mathbf{S} = \iint_D \left(-P\frac{\partial g}{\partial x} - Q\frac{\partial g}{\partial y} + R\right) dA$$

For a downward unit normal we multiply by -1. Similar formulas can be worked out if S is given by $y = h(x, z)$ or $x = k(y, z)$. (See Exercises 29 and 30.)

EXAMPLE 4 Evaluate $\iint_S \mathbf{F} \cdot d\mathbf{S}$, where $\mathbf{F}(x, y, z) = y\,\mathbf{i} + x\,\mathbf{j} + z\,\mathbf{k}$ and S is the boundary of the solid region E enclosed by the paraboloid $z = 1 - x^2 - y^2$ and the plane $z = 0$.

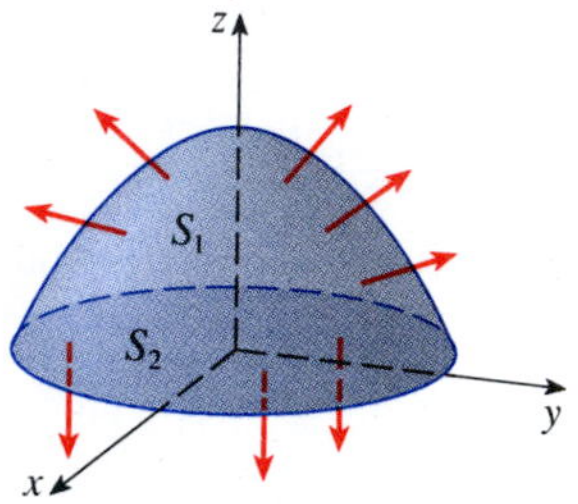

FIGURE 12

SOLUTION S consists of a parabolic top surface S_1 and a circular bottom surface S_2 (see Figure 12). Since S is a closed surface, we use the convention of positive (outward) orientation. This means that S_1 is oriented upward and we can use Equation 8 with D being the projection of S_1 on the xy-plane, namely, the disk $x^2 + y^2 \leq 1$. Since

$$P(x, y, z) = y \qquad Q(x, y, z) = x \qquad R(x, y, z) = z = 1 - x^2 - y^2$$

on S_1 and

$$\frac{\partial g}{\partial x} = -2x \qquad \frac{\partial g}{\partial y} = -2y$$

we have

$$\iint_{S_1} \mathbf{F} \cdot d\mathbf{S} = \iint_D \left(-P\frac{\partial g}{\partial x} - Q\frac{\partial g}{\partial y} + R\right) dA$$

$$= \iint_D [-y(-2x) - x(-2y) + 1 - x^2 - y^2]\, dA$$

$$= \iint_D (1 + 4xy - x^2 - y^2)\, dA$$

$$= \int_0^{2\pi}\int_0^1 (1 + 4r^2\cos\theta\sin\theta - r^2)r\, dr\, d\theta$$

$$= \int_0^{2\pi}\int_0^1 (r - r^3 + 4r^3\cos\theta\sin\theta)\, dr\, d\theta$$

$$= \int_0^{2\pi} \left(\tfrac{1}{4} + \cos\theta\sin\theta\right) d\theta = \tfrac{1}{4}(2\pi) + 0 = \frac{\pi}{2}$$

The disk S_2 is oriented downward, so its unit normal vector is $\mathbf{n} = -\mathbf{k}$ and we have

$$\iint_{S_2} \mathbf{F} \cdot d\mathbf{S} = \iint_{S_2} \mathbf{F} \cdot (-\mathbf{k})\, dS = \iint_D (-z)\, dA = \iint_D 0\, dA = 0$$

since $z = 0$ on S_2. Finally, we compute, by definition, $\iint_S \mathbf{F} \cdot d\mathbf{S}$ as the sum of the surface integrals of $\mathbf{F}$ over the pieces S_1 and S_2:

$$\iint_S \mathbf{F} \cdot d\mathbf{S} = \iint_{S_1} \mathbf{F} \cdot d\mathbf{S} + \iint_{S_2} \mathbf{F} \cdot d\mathbf{S} = \frac{\pi}{2} + 0 = \frac{\pi}{2}$$ ■

PARAMETRIC SURFACES If S is given by a vector function $\mathbf{r}(u, v)$, then $\mathbf{n}$ is given by Equation 5, and from Definition 7 and Equation 3 we have

$$\iint_S \mathbf{F} \cdot d\mathbf{S} = \iint_S \mathbf{F} \cdot \frac{\mathbf{r}_u \times \mathbf{r}_v}{|\mathbf{r}_u \times \mathbf{r}_v|}\, dS$$

$$= \iint_D \left[\mathbf{F}(\mathbf{r}(u, v)) \cdot \frac{\mathbf{r}_u \times \mathbf{r}_v}{|\mathbf{r}_u \times \mathbf{r}_v|} \right] |\mathbf{r}_u \times \mathbf{r}_v|\, dA$$

where D is the parameter domain. Thus, we have

Compare Equation 9 to the similar expression for evaluating line integrals of vector fields in Definition 14.2.13:

$$\int_C \mathbf{F} \cdot d\mathbf{r} = \int_a^b \mathbf{F}(\mathbf{r}(t)) \cdot \mathbf{r}'(t)\, dt$$

(9)
$$\iint_S \mathbf{F} \cdot d\mathbf{S} = \iint_D \mathbf{F} \cdot (\mathbf{r}_u \times \mathbf{r}_v)\, dA$$

Notice that, in view of Equation 14.6.8, we have

$$\mathbf{r}_x \times \mathbf{r}_y = -\frac{\partial g}{\partial x}\mathbf{i} - \frac{\partial g}{\partial y}\mathbf{j} + \mathbf{k}$$

and so Formula 8 is just a special case of Formula 9.

Figure 13 shows the vector field $\mathbf{F}$ in Example 5 at points on the unit sphere.

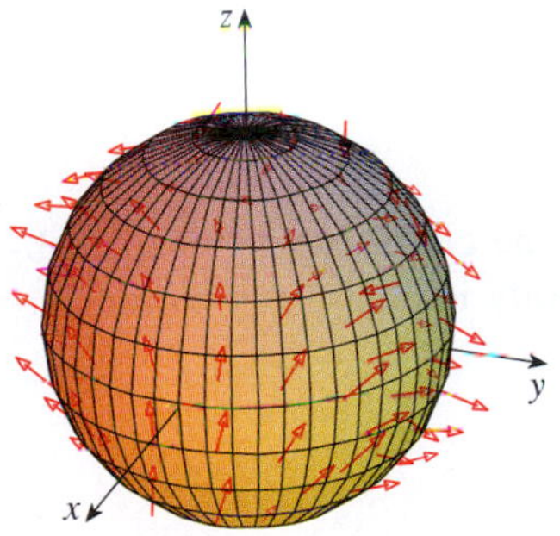

FIGURE 13

EXAMPLE 5 Find the flux of the vector field $\mathbf{F}(x, y, z) = z\,\mathbf{i} + y\,\mathbf{j} + x\,\mathbf{k}$ across the unit sphere $x^2 + y^2 + z^2 = 1$.

SOLUTION Using the parametric representation

$$\mathbf{r}(\phi, \theta) = \sin\phi\cos\theta\,\mathbf{i} + \sin\phi\sin\theta\,\mathbf{j} + \cos\phi\,\mathbf{k} \qquad 0 \le \phi \le \pi \qquad 0 \le \theta \le 2\pi$$

we have

$$\mathbf{F}(\mathbf{r}(\phi, \theta)) = \cos\phi\,\mathbf{i} + \sin\phi\sin\theta\,\mathbf{j} + \sin\phi\cos\theta\,\mathbf{k}$$

and, from Example 6 in Section 14.6,

$$\mathbf{r}_\phi \times \mathbf{r}_\theta = \sin^2\phi\cos\theta\,\mathbf{i} + \sin^2\phi\sin\theta\,\mathbf{j} + \sin\phi\cos\phi\,\mathbf{k}$$

Therefore

$$\mathbf{F}(\mathbf{r}(\phi, \theta)) \cdot (\mathbf{r}_\phi \times \mathbf{r}_\theta) = \cos\phi\sin^2\phi\cos\theta + \sin^3\phi\sin^2\theta + \sin^2\phi\cos\phi\cos\theta$$

and, by Formula 9, the flux is

$$\iint_S \mathbf{F} \cdot d\mathbf{S} = \iint_D \mathbf{F} \cdot (\mathbf{r}_\phi \times \mathbf{r}_\theta)\, dA$$

$$= \int_0^{2\pi}\int_0^{\pi} (2\sin^2\phi\cos\phi\cos\theta + \sin^3\phi\sin^2\theta)\, d\phi\, d\theta$$

$$= 2 \int_0^{\pi} \sin^2\phi \cos\phi \, d\phi \int_0^{2\pi} \cos\theta \, d\theta + \int_0^{\pi} \sin^3\phi \, d\phi \int_0^{2\pi} \sin^2\theta \, d\theta$$

$$= 0 + \int_0^{\pi} \sin^3\phi \, d\phi \int_0^{2\pi} \sin^2\theta \, d\theta \qquad \left(\text{since } \int_0^{2\pi} \cos\theta \, d\theta = 0\right)$$

$$= \frac{4\pi}{3}$$

by the same calculation as in Example 2. ■

If, for instance, the vector field in Example 5 is a velocity field describing the flow of a fluid with density 1, then the answer, $4\pi/3$, represents the rate of flow through the unit sphere in units of mass per unit time.

Although we motivated the surface integral of a vector field using the example of fluid flow, this concept also arises in other physical situations. For instance, if $\mathbf{E}$ is an electric field (see Example 5 in Section 14.1), then the surface integral

$$\iint_S \mathbf{E} \cdot d\mathbf{S}$$

is called the **electric flux** of $\mathbf{E}$ through the surface S. One of the important laws of electrostatics is **Gauss's Law**, which says that the net charge enclosed by a closed surface S is

$$Q = \varepsilon_0 \iint_S \mathbf{E} \cdot d\mathbf{S} \tag{10}$$

where ε_0 is a constant (called the permittivity of free space) that depends on the units used. (In the SI system, $\varepsilon_0 \approx 8.8542 \times 10^{-12}\ \mathrm{C^2/N{\cdot}m^2}$.) Thus, if the vector field $\mathbf{F}$ in Example 5 represents an electric field, we can conclude that the charge enclosed by S is $Q = 4\pi\varepsilon_0/3$.

Another application of surface integrals occurs in the study of heat flow. Suppose the temperature at a point (x, y, z) in a body is $u(x, y, z)$. Then the **heat flow** is defined as the vector field

$$\mathbf{F} = -K \nabla u$$

where K is an experimentally determined constant called the **conductivity** of the substance. The rate of heat flow across the surface S in the body is then given by the surface integral

$$\iint_S \mathbf{F} \cdot d\mathbf{S} = -K \iint_S \nabla u \cdot d\mathbf{S}$$

EXAMPLE 6 The temperature u in a metal ball is proportional to the square of the distance from the center of the ball. Find the rate of heat flow across a sphere S of radius a with center at the center of the ball.

SOLUTION Taking the center of the ball to be at the origin, we have

$$u(x, y, z) = C(x^2 + y^2 + z^2)$$

where C is the proportionality constant. Then the heat flow is

$$\mathbf{F}(x, y, z) = -K \nabla u = -KC(2x\,\mathbf{i} + 2y\,\mathbf{j} + 2z\,\mathbf{k})$$

where K is the conductivity of the metal. Instead of using the usual parametrization of the sphere as in Example 5, we observe that the outward unit normal to the sphere $x^2 + y^2 + z^2 = a^2$ at the point (x, y, z) is

$$\mathbf{n} = \frac{1}{a}(x\,\mathbf{i} + y\,\mathbf{j} + z\,\mathbf{k})$$

and so

$$\mathbf{F} \cdot \mathbf{n} = -\frac{2KC}{a}(x^2 + y^2 + z^2)$$

But on S we have $x^2 + y^2 + z^2 = a^2$, so $\mathbf{F} \cdot \mathbf{n} = -2aKC$. Therefore, the rate of heat flow across S is

$$\begin{aligned}\iint_S \mathbf{F} \cdot d\mathbf{S} &= \iint_S \mathbf{F} \cdot \mathbf{n}\, dS \\ &= -2aKC \iint_S dS = -2aKCA(S) \\ &= -2aKC(4\pi a^2) = -8KC\pi a^3\end{aligned}$$

■

EXERCISES 14.7

1. Let S be the cube with vertices $(\pm1, \pm1, \pm1)$. Approximate $\iint_S \sqrt{x^2 + 2y^2 + 3z^2}\, dS$ by using a Riemann sum as in Definition 1, taking the patches S_{ij} to be the squares that are the faces of the cube and the points P_{ij}^* to be the centers of the squares.

2. Suppose that $f(x, y, z) = g(\sqrt{x^2 + y^2 + z^2})$, where g is a function of one variable such that $g(2) = -5$. Evaluate $\iint_S f(x, y, z)\, dS$, where S is the sphere $x^2 + y^2 + z^2 = 4$.

3–14 ■ Evaluate the given surface integral.

3. $\iint_S y\, dS$, S is the part of the plane $3x + 2y + z = 6$ that lies in the first octant

4. $\iint_S xz\, dS$,
S is the triangle with vertices $(1, 0, 0)$, $(0, 1, 0)$, and $(0, 0, 1)$

5. $\iint_S x\, dS$,
S is the surface $y = x^2 + 4z$, $0 \leq x \leq 2$, $0 \leq z \leq 2$

6. $\iint_S (y^2 + z^2)\, dS$,
S is the part of the paraboloid $x = 4 - y^2 - z^2$ that lies in front of the plane $x = 0$

7. $\iint_S yz\, dS$, S is the part of the plane $z = y + 3$ that lies inside the cylinder $x^2 + y^2 = 1$

8. $\iint_S xy\, dS$,
S is the boundary of the region enclosed by the cylinder $x^2 + z^2 = 1$ and the planes $y = 0$ and $x + y = 2$

9. $\iint_S (x^2 z + y^2 z)\, dS$,
S is the hemisphere $x^2 + y^2 + z^2 = 4$, $z \geq 0$

10. $\iint_S xyz\, dS$, S is the part of the sphere $x^2 + y^2 + z^2 = 1$ that lies above the cone $z = \sqrt{x^2 + y^2}$

11. $\iint_S (x^2 y + z^2)\, dS$, S is the part of the cylinder $x^2 + y^2 = 9$ between the planes $z = 0$ and $z = 2$

12. $\iint_S (x^2 + y^2 + z^2)\, dS$, S consists of the cylinder in Exercise 11 together with its top and bottom disks

13. $\iint_S yz\, dS$, S is the surface with parametric equations $x = uv$, $y = u + v$, $z = u - v$, $u^2 + v^2 \leq 1$

14. $\iint_S \sqrt{1 + x^2 + y^2}\, dS$, S is the helicoid with vector equation $\mathbf{r}(u, v) = u\cos v\,\mathbf{i} + u\sin v\,\mathbf{j} + v\,\mathbf{k}$, $0 \leq u \leq 1$, $0 \leq v \leq \pi$

15–24 ■ Evaluate the surface integral $\iint_S \mathbf{F} \cdot d\mathbf{S}$ for the given vector field $\mathbf{F}$ and the given oriented surface S. In other words, find the flux of $\mathbf{F}$ across S. For closed surfaces, use the positive (outward) orientation.

15. $\mathbf{F}(x, y, z) = e^y\,\mathbf{i} + ye^x\,\mathbf{j} + x^2 y\,\mathbf{k}$, S is the part of the paraboloid $z = x^2 + y^2$ that lies above the square $0 \leq x \leq 1$, $0 \leq y \leq 1$ and has upward orientation

16. $\mathbf{F}(x, y, z) = x^2 y\,\mathbf{i} - 3xy^2\,\mathbf{j} + 4y^3\,\mathbf{k}$,
S is the part of the elliptic paraboloid $z = x^2 + y^2 - 9$ that lies below the rectangle $0 \leq x \leq 2$, $0 \leq y \leq 1$ and has downward orientation

17. $\mathbf{F}(x, y, z) = x\,\mathbf{i} + xy\,\mathbf{j} + xz\,\mathbf{k}$,
S is the surface of Exercise 3 with upward orientation

18. $\mathbf{F}(x, y, z) = -x\,\mathbf{i} - y\,\mathbf{j} + z^2\,\mathbf{k}$,
S is the part of the cone $z = \sqrt{x^2 + y^2}$ between the planes $z = 1$ and $z = 2$ with upward orientation

19. $\mathbf{F}(x, y, z) = x\,\mathbf{i} + y\,\mathbf{j} + z\,\mathbf{k}$,
S is the sphere $x^2 + y^2 + z^2 = 9$

20. $\mathbf{F}(x, y, z) = -y\,\mathbf{i} + x\,\mathbf{j} + 3z\,\mathbf{k}$, S is the hemisphere $z = \sqrt{16 - x^2 - y^2}$ with upward orientation

21. $\mathbf{F}(x, y, z) = y\,\mathbf{j} - z\,\mathbf{k}$, S consists of the paraboloid $y = x^2 + z^2$, $0 \leqslant y \leqslant 1$, and the disk $x^2 + z^2 \leqslant 1$, $y = 1$

22. $\mathbf{F}(x, y, z) = x\,\mathbf{i} + y\,\mathbf{j} + 5\,\mathbf{k}$, S is the surface of Exercise 8

23. $\mathbf{F}(x, y, z) = x\,\mathbf{i} + 2y\,\mathbf{j} + 3z\,\mathbf{k}$, S is the cube with vertices $(\pm1, \pm1, \pm1)$

24. $\mathbf{F}(x, y, z) = y\,\mathbf{i} + x\,\mathbf{j} + z^2\mathbf{k}$, S is the helicoid of Exercise 14

CAS **25.** Evaluate $\iint_S xyz\,dS$ correct to four decimal places, where S is the surface $z = xy$, $0 \leqslant x \leqslant 1$, $0 \leqslant y \leqslant 1$.

CAS **26.** Find the exact value of $\iint_S x^2yz\,dS$, where S is the surface in Exercise 25.

CAS **27.** Find the value of $\iint_S x^2y^2z^2\,dS$ correct to four decimal places, where S is the part of the paraboloid $z = 3 - 2x^2 - y^2$ that lies above the xy-plane.

CAS **28.** Find the flux of $\mathbf{F}(x, y, z) = \sin(xyz)\,\mathbf{i} + x^2y\,\mathbf{j} + z^2e^{x/5}\,\mathbf{k}$ across the part of the cylinder $4y^2 + z^2 = 4$ that lies above the xy-plane and between the planes $x = -2$ and $x = 2$ with upward orientation. Illustrate by using a computer algebra system to draw the cylinder and the vector field on the same screen.

29. Find a formula for $\iint_S \mathbf{F} \cdot d\mathbf{S}$ similar to Formula 8 for the case where S is given by $y = h(x, z)$ and $\mathbf{n}$ is the unit normal that points toward the left.

30. Find a formula for $\iint_S \mathbf{F} \cdot d\mathbf{S}$ similar to Formula 8 for the case where S is given by $x = k(y, z)$ and $\mathbf{n}$ is the unit normal that points forward (that is, toward the viewer when the axes are drawn in the usual way).

31. Find the center of mass of the hemisphere $x^2 + y^2 + z^2 = a^2$, $z \geqslant 0$, if it has constant density.

32. Find the mass of a thin funnel in the shape of a cone $z = \sqrt{x^2 + y^2}$, $1 \leqslant z \leqslant 4$ if its density function is $\rho(x, y, z) = 10 - z$.

33. (a) Give an integral expression for the moment of inertia I_z about the z-axis of a thin sheet in the shape of a surface S if the density function is ρ.
(b) Find the moment of inertia about the z-axis of the funnel in Exercise 32.

34. The conical surface $z^2 = x^2 + y^2$, $0 \leqslant z \leqslant a$, has constant density k. Find (a) the center of mass and (b) the moment of inertia about the z-axis.

35. A fluid with density 1200 flows with velocity

$$\mathbf{v} = y\,\mathbf{i} + \mathbf{j} + z\,\mathbf{k}$$

Find the rate of flow upward through the paraboloid $z = 9 - (x^2 + y^2)/4$, $x^2 + y^2 \leqslant 36$.

36. A fluid has density 1500 and velocity field

$$\mathbf{v} = -y\,\mathbf{i} + x\,\mathbf{j} + 2z\,\mathbf{k}$$

Find the rate of flow outward through the sphere $x^2 + y^2 + z^2 = 25$.

37. Use Gauss's Law to find the charge contained in the solid hemisphere

$$x^2 + y^2 + z^2 \leqslant a^2 \qquad z \geqslant 0$$

if the electric field is $\mathbf{E}(x, y, z) = x\,\mathbf{i} + y\,\mathbf{j} + 2z\,\mathbf{k}$.

38. Use Gauss's Law to find the charge enclosed by the cube with vertices $(\pm1, \pm1, \pm1)$ if the electric field is $\mathbf{E}(x, y, z) = x\,\mathbf{i} + y\,\mathbf{j} + z\,\mathbf{k}$.

39. The temperature at the point (x, y, z) in a substance with conductivity $K = 6.5$ is $u(x, y, z) = 2y^2 + 2z^2$. Find the rate of heat flow inward across the cylindrical surface

$$y^2 + z^2 = 6 \qquad 0 \leqslant x \leqslant 4$$

40. The temperature at a point in a ball with conductivity K is inversely proportional to the distance from the center of the ball. Find the rate of heat flow across a sphere S of radius a with center at the center of the ball.

14.8 STOKES' THEOREM

Stokes' Theorem can be regarded as a higher-dimensional version of Green's Theorem. Whereas Green's Theorem relates a double integral over a plane region D to a line integral around its plane boundary curve, Stokes' Theorem relates a surface integral over a surface S to a line integral around the boundary curve of S (which is a space curve). Figure 1 shows an oriented surface with unit normal vector $\mathbf{n}$. The orientation of S induces the **positive orientation of the boundary curve C** shown in the figure. This means that if you walk in the positive direction around C with your head pointing in the direction of $\mathbf{n}$, then the surface will always be on your left.

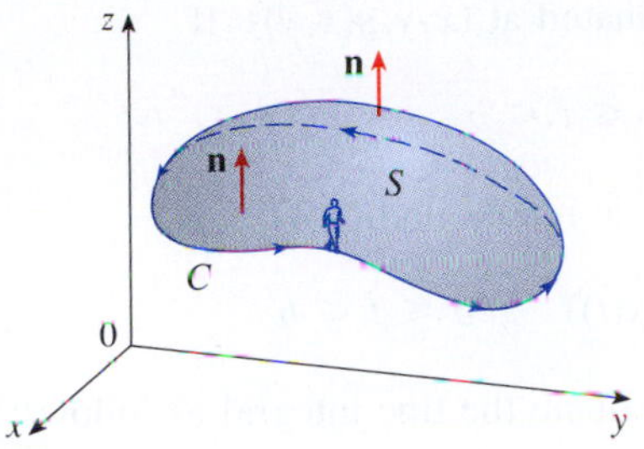

FIGURE 1

(1) STOKES' THEOREM Let S be an oriented piecewise-smooth surface that is bounded by a simple, closed, piecewise-smooth boundary curve C with positive orientation. Let $\mathbf{F}$ be a vector field whose components have continuous partial derivatives on an open region in $\mathbb{R}^3$ that contains S. Then

$$\int_C \mathbf{F} \cdot d\mathbf{r} = \iint_S \operatorname{curl} \mathbf{F} \cdot d\mathbf{S}$$

Since

$$\int_C \mathbf{F} \cdot d\mathbf{r} = \int_C \mathbf{F} \cdot \mathbf{T}\, ds \qquad \text{and} \qquad \iint_S \operatorname{curl} \mathbf{F} \cdot d\mathbf{S} = \iint_S \operatorname{curl} \mathbf{F} \cdot \mathbf{n}\, dS$$

Stokes' Theorem says that the line integral around the boundary curve of S of the tangential component of $\mathbf{F}$ is equal to the surface integral of the normal component of the curl of $\mathbf{F}$.

Stokes' Theorem is named after the Irish mathematical physicist Sir George Stokes (1819–1903). Stokes was a professor at Cambridge University (in fact he held the same position as Newton, Lucasian Professor of Mathematics) and was especially noted for his studies of fluid flow and light. What we call Stokes' Theorem was actually discovered by the Scottish physicist Sir William Thomson (1824–1907, known as Lord Kelvin). Stokes learned of this theorem in a letter from Thomson in 1850 and asked students to prove it on an examination at Cambridge University in 1854. We don't know if any of those students was able to do so.

The positively oriented boundary curve of the oriented surface S is often written as ∂S, so Stokes' Theorem can be expressed as

$$\iint_S \operatorname{curl} \mathbf{F} \cdot d\mathbf{S} = \int_{\partial S} \mathbf{F} \cdot d\mathbf{r} \tag{2}$$

There is an analogy among Stokes' Theorem, Green's Theorem, and the Fundamental Theorem of Calculus. As before, there is an integral involving derivatives on the left side of Equation 2 (recall that curl $\mathbf{F}$ is a sort of derivative of $\mathbf{F}$) and the right side involves the values of $\mathbf{F}$ only on the *boundary* of S.

In fact, in the special case where the surface S is flat and lies in the xy-plane with upward orientation, the unit normal is $\mathbf{k}$, the surface integral becomes a double integral, and Stokes' Theorem becomes

$$\int_C \mathbf{F} \cdot d\mathbf{r} = \iint_S \operatorname{curl} \mathbf{F} \cdot d\mathbf{S} = \iint_S (\operatorname{curl} \mathbf{F}) \cdot \mathbf{k}\, dA$$

This is precisely the vector form of Green's Theorem given in Equation 14.5.12. Thus we see that Green's Theorem is really a special case of Stokes' Theorem.

Although Stokes' Theorem is too difficult for us to prove in its full generality, we can give a proof when S is a graph and $\mathbf{F}$, S, and C are well behaved.

PROOF OF A SPECIAL CASE OF STOKES' THEOREM We assume that the equation of S is $z = g(x, y)$, $(x, y) \in D$, where g has continuous second-order partial derivatives and D is a simple plane region whose boundary curve C_1 corresponds to C. If the orientation of S is upward, then the positive orientation of C corresponds to the positive orientation of C_1 (see Figure 2). We are also assuming that $\mathbf{F} = P\,\mathbf{i} + Q\,\mathbf{j} + R\,\mathbf{k}$, where the partial derivatives of P, Q, and R are continuous.

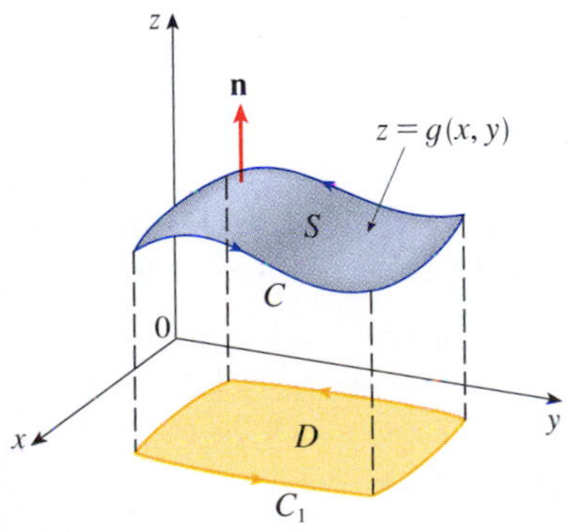

FIGURE 2

Since S is a graph, we can apply Formula 14.7.8 with $\mathbf{F}$ replaced by curl $\mathbf{F}$. The result is

$$\iint_S \operatorname{curl} \mathbf{F} \cdot d\mathbf{S} = \iint_D \left[-\left(\frac{\partial R}{\partial y} - \frac{\partial Q}{\partial z} \right) \frac{\partial z}{\partial x} - \left(\frac{\partial P}{\partial z} - \frac{\partial R}{\partial x} \right) \frac{\partial z}{\partial y} + \left(\frac{\partial Q}{\partial x} - \frac{\partial P}{\partial y} \right) \right] dA \tag{3}$$

where the partial derivatives of P, Q, and R are evaluated at $(x, y, g(x, y))$. If

$$x = x(t) \qquad y = y(t) \qquad a \leqslant t \leqslant b$$

is a parametric representation of C_1, then a parametric representation of C is

$$x = x(t) \qquad y = y(t) \qquad z = g(x(t), y(t)) \qquad a \leqslant t \leqslant b$$

This allows us, with the aid of the Chain Rule, to evaluate the line integral as follows:

$$\begin{aligned}
\int_C \mathbf{F} \cdot d\mathbf{r} &= \int_a^b \left(P \frac{dx}{dt} + Q \frac{dy}{dt} + R \frac{dz}{dt} \right) dt \\
&= \int_a^b \left[P \frac{dx}{dt} + Q \frac{dy}{dt} + R \left(\frac{\partial z}{\partial x} \frac{dx}{dt} + \frac{\partial z}{\partial y} \frac{dy}{dt} \right) \right] dt \\
&= \int_a^b \left[\left(P + R \frac{\partial z}{\partial x} \right) \frac{dx}{dt} + \left(Q + R \frac{\partial z}{\partial y} \right) \frac{dy}{dt} \right] dt \\
&= \int_{C_1} \left(P + R \frac{\partial z}{\partial x} \right) dx + \left(Q + R \frac{\partial z}{\partial y} \right) dy \\
&= \iint_D \left[\frac{\partial}{\partial x} \left(Q + R \frac{\partial z}{\partial y} \right) - \frac{\partial}{\partial y} \left(P + R \frac{\partial z}{\partial x} \right) \right] dA
\end{aligned}$$

where we have used Green's Theorem in the last step. Then, using the Chain Rule again and remembering that P, Q, and R are functions of x, y, and z and that z is itself a function of x and y, we get

$$\begin{aligned}
\int_C \mathbf{F} \cdot d\mathbf{r} = \iint_D \Bigg[&\left(\frac{\partial Q}{\partial x} + \frac{\partial Q}{\partial z} \frac{\partial z}{\partial x} + \frac{\partial R}{\partial x} \frac{\partial z}{\partial y} + \frac{\partial R}{\partial z} \frac{\partial z}{\partial x} \frac{\partial z}{\partial y} + R \frac{\partial^2 z}{\partial x \, \partial y} \right) \\
&- \left(\frac{\partial P}{\partial y} + \frac{\partial P}{\partial z} \frac{\partial z}{\partial y} + \frac{\partial R}{\partial y} \frac{\partial z}{\partial x} + \frac{\partial R}{\partial z} \frac{\partial z}{\partial y} \frac{\partial z}{\partial x} + R \frac{\partial^2 z}{\partial y \, \partial x} \right) \Bigg] dA
\end{aligned}$$

Four of the terms in this double integral cancel and the remaining six terms can be arranged to coincide with the right side of Equation 3. Therefore

$$\int_C \mathbf{F} \cdot d\mathbf{r} = \iint_S \operatorname{curl} \mathbf{F} \cdot d\mathbf{S}$$

□

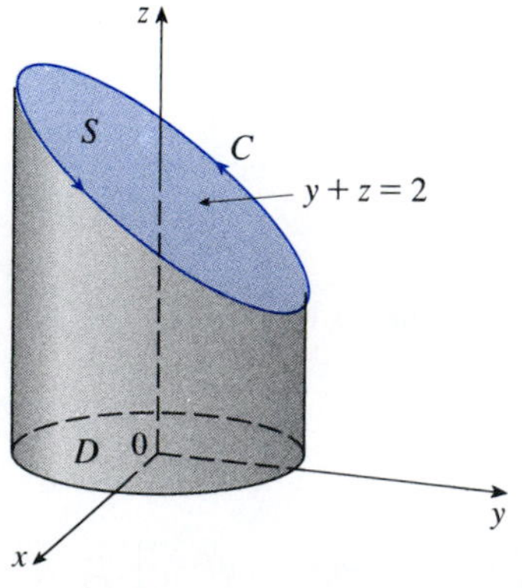

FIGURE 3

EXAMPLE 1 Evaluate $\int_C \mathbf{F} \cdot d\mathbf{r}$, where $\mathbf{F}(x, y, z) = -y^2\mathbf{i} + x\mathbf{j} + z^2\mathbf{k}$ and C is the curve of intersection of the plane $y + z = 2$ and the cylinder $x^2 + y^2 = 1$. (Orient C to be counterclockwise when viewed from above.)

SOLUTION The curve C (an ellipse) is shown in Figure 3. Although $\int_C \mathbf{F} \cdot d\mathbf{r}$ could be evaluated directly, it is not easy to parametrize C, so let us use Stokes' Theorem instead. We first compute

$$\operatorname{curl} \mathbf{F} = \begin{vmatrix} \mathbf{i} & \mathbf{j} & \mathbf{k} \\ \dfrac{\partial}{\partial x} & \dfrac{\partial}{\partial y} & \dfrac{\partial}{\partial z} \\ -y^2 & x & z^2 \end{vmatrix} = (1 + 2y)\,\mathbf{k}$$

Let's choose S to be the elliptical region in the plane $y + z = 2$ that is bounded by C. If we orient S upward, then C has the induced positive orientation. The projection D of S on the xy-plane is the disk $x^2 + y^2 \leq 1$ and so using Equation 14.7.8 with $z = g(x, y) = 2 - y$, we have

$$\begin{aligned}\int_C \mathbf{F} \cdot d\mathbf{r} &= \iint_S \operatorname{curl} \mathbf{F} \cdot d\mathbf{S} = \iint_D (1 + 2y)\, dA \\ &= \int_0^{2\pi} \int_0^1 (1 + 2r \sin\theta) r\, dr\, d\theta \\ &= \int_0^{2\pi} \left[\frac{r^2}{2} + 2\frac{r^3}{3} \sin\theta \right]_0^1 d\theta = \int_0^{2\pi} \left(\tfrac{1}{2} + \tfrac{2}{3}\sin\theta\right) d\theta \\ &= \tfrac{1}{2}(2\pi) + 0 = \pi\end{aligned}$$

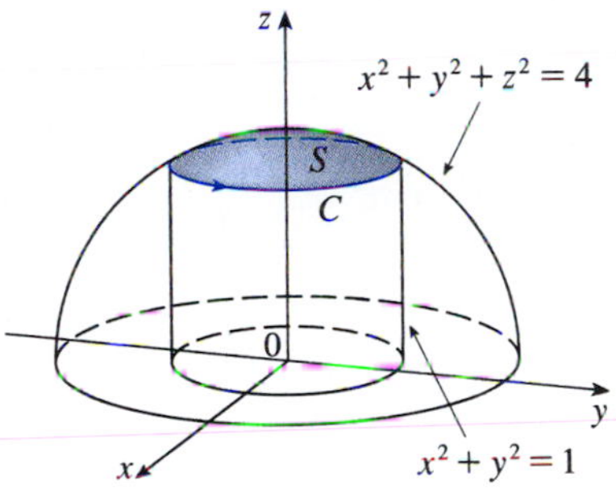

FIGURE 4

EXAMPLE 2 Use Stokes' Theorem to compute the integral $\iint_S \operatorname{curl} \mathbf{F} \cdot d\mathbf{S}$, where $\mathbf{F}(x, y, z) = yz\,\mathbf{i} + xz\,\mathbf{j} + xy\,\mathbf{k}$ and S is the part of the sphere $x^2 + y^2 + z^2 = 4$ that lies inside the cylinder $x^2 + y^2 = 1$ and above the xy-plane (see Figure 4).

SOLUTION To find the boundary curve C we solve the equations $x^2 + y^2 + z^2 = 4$ and $x^2 + y^2 = 1$. Subtracting, we get $z^2 = 3$ and so $z = \sqrt{3}$ (since $z > 0$). Thus, C is the circle given by the equations $x^2 + y^2 = 1$, $z = \sqrt{3}$. The vector equation of C is

$$\mathbf{r}(t) = \cos t\,\mathbf{i} + \sin t\,\mathbf{j} + \sqrt{3}\,\mathbf{k} \qquad 0 \leq t \leq 2\pi$$

so

$$\mathbf{r}'(t) = -\sin t\,\mathbf{i} + \cos t\,\mathbf{j}$$

Also, we have

$$\mathbf{F}(\mathbf{r}(t)) = \sqrt{3}\sin t\,\mathbf{i} + \sqrt{3}\cos t\,\mathbf{j} + \cos t \sin t\,\mathbf{k}$$

Therefore, by Stokes' Theorem,

$$\begin{aligned}\iint_S \operatorname{curl} \mathbf{F} \cdot d\mathbf{S} &= \int_C \mathbf{F} \cdot d\mathbf{r} = \int_0^{2\pi} \mathbf{F}(\mathbf{r}(t)) \cdot \mathbf{r}'(t)\, dt \\ &= \int_0^{2\pi} (-\sqrt{3}\sin^2 t + \sqrt{3}\cos^2 t)\, dt \\ &= \sqrt{3} \int_0^{2\pi} \cos 2t\, dt = 0\end{aligned}$$

Note that in Example 2 we computed a surface integral simply by knowing the values of $\mathbf{F}$ on the boundary curve C. This means that if we have a different oriented surface with the same boundary curve C, then we get exactly the same value for the surface integral!

In general, if S_1 and S_2 are oriented surfaces with the same oriented boundary curve C and both satisfy the hypotheses of Stokes' Theorem, then

$$\textbf{(4)} \qquad \iint_{S_1} \operatorname{curl} \mathbf{F} \cdot d\mathbf{S} = \int_C \mathbf{F} \cdot d\mathbf{r} = \iint_{S_2} \operatorname{curl} \mathbf{F} \cdot d\mathbf{S}$$

This fact is useful when it is difficult to integrate over one surface but easy to integrate over the other.

Let's now use Stokes' Theorem to throw some light on the meaning of the curl vector. Suppose that C is an oriented closed curve and $\mathbf{v}$ represents the velocity field in fluid flow. Consider the line integral

$$\int_C \mathbf{v} \cdot d\mathbf{r} = \int_C \mathbf{v} \cdot \mathbf{T}\, ds$$

and recall that $\mathbf{v} \cdot \mathbf{T}$ is the component of $\mathbf{v}$ in the direction of the unit tangent vector $\mathbf{T}$. This means that the closer the direction of $\mathbf{v}$ is to the direction of $\mathbf{T}$, the larger the value of $\mathbf{v} \cdot \mathbf{T}$. Thus, $\int_C \mathbf{v} \cdot d\mathbf{r}$ is a measure of the tendency of the fluid to move around C and is called the **circulation** of $\mathbf{v}$ around C (see Figure 5).

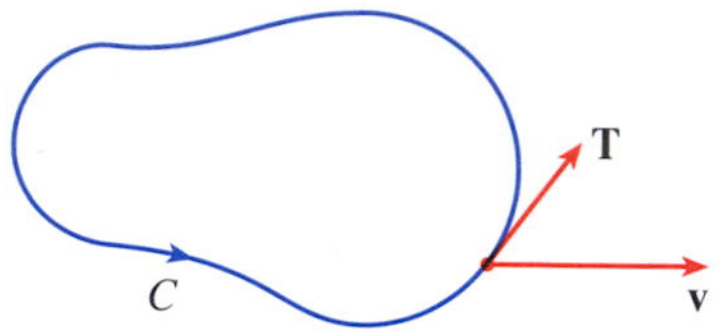

(a) $\int_C \mathbf{v} \cdot d\mathbf{r} > 0$, positive circulation

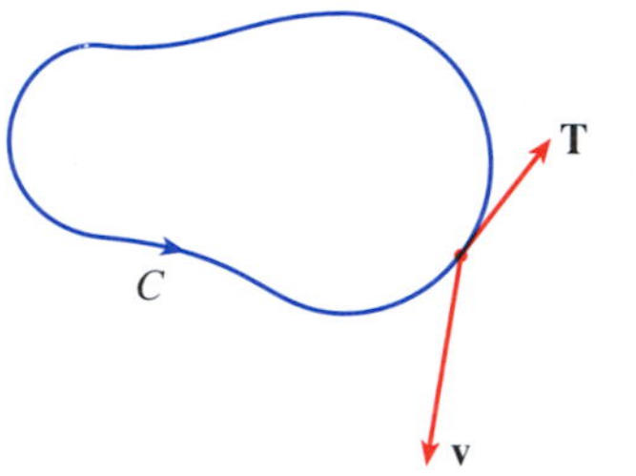

(b) $\int_C \mathbf{v} \cdot d\mathbf{r} < 0$, negative circulation

FIGURE 5

Now let $P_0(x_0, y_0, z_0)$ be a point in the fluid and let S_a be a small disk with radius a and center P_0. Then $(\text{curl}\,\mathbf{F})(P) \approx (\text{curl}\,\mathbf{F})(P_0)$ for all points P on S_a because curl $\mathbf{F}$ is continuous. Thus, by Stokes' Theorem, we get the following approximation to the circulation around the boundary circle C_a:

$$\begin{aligned}\int_{C_a} \mathbf{v} \cdot d\mathbf{r} &= \iint_{S_a} \text{curl}\,\mathbf{v} \cdot d\mathbf{S} = \iint_{S_a} \text{curl}\,\mathbf{v} \cdot \mathbf{n}\, dS \\ &\approx \iint_{S_a} \text{curl}\,\mathbf{v}(P_0) \cdot \mathbf{n}(P_0)\, dS \\ &= \text{curl}\,\mathbf{v}(P_0) \cdot \mathbf{n}(P_0)\pi a^2\end{aligned}$$

Imagine a tiny paddle wheel placed in the fluid at a point P, as in Figure 6; the paddle wheel rotates fastest when its axis is parallel to curl $\mathbf{v}$.

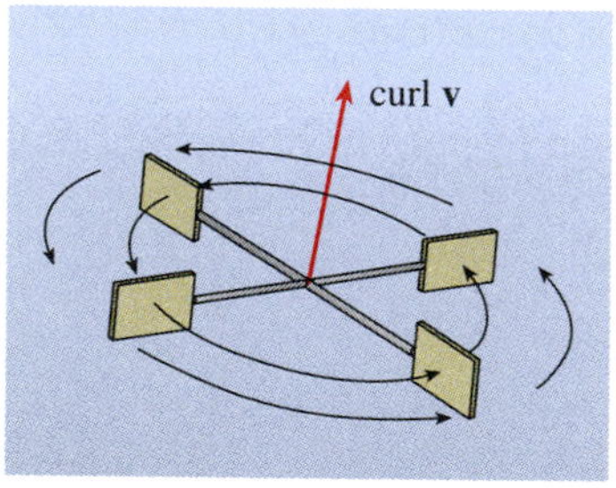

FIGURE 6

This approximation becomes better as $a \to 0$ and we have

$$\textbf{(5)} \qquad \text{curl}\,\mathbf{v}(P_0) \cdot \mathbf{n}(P_0) = \lim_{a \to 0} \frac{1}{\pi a^2} \int_{C_a} \mathbf{v} \cdot d\mathbf{r}$$

Equation 5 gives the relationship between the curl and the circulation. It shows that curl $\mathbf{v} \cdot \mathbf{n}$ is a measure of the rotating effect of the fluid about the axis $\mathbf{n}$. The curling effect is greatest about the axis parallel to curl $\mathbf{v}$.

Finally, we mention that Stokes' Theorem can be used to prove Theorem 14.5.4 (which states that if curl $\mathbf{F} = \mathbf{0}$ on all of $\mathbb{R}^3$, then $\mathbf{F}$ is conservative). From our previous work (Theorems 14.3.3 and 14.3.4), we know that $\mathbf{F}$ is conservative if $\int_C \mathbf{F} \cdot d\mathbf{r} = 0$ for every closed path C. Given C, suppose we can find an orientable surface S whose boundary is C. (This can be done, but the proof requires advanced techniques.) Then Stokes' Theorem gives

$$\int_C \mathbf{F} \cdot d\mathbf{r} = \iint_S \text{curl}\,\mathbf{F} \cdot d\mathbf{S} = \iint_S \mathbf{0} \cdot d\mathbf{S} = 0$$

A curve that is not simple can be broken into a number of simple curves, and the integrals around these simple curves are all 0. Adding these integrals, we obtain $\int_C \mathbf{F} \cdot d\mathbf{r} = 0$ for any closed curve C.

EXERCISES 14.8

1–6 ■ Use Stokes' Theorem to evaluate $\iint_S \operatorname{curl} \mathbf{F} \cdot d\mathbf{S}$.

1. $\mathbf{F}(x, y, z) = xyz\,\mathbf{i} + x\,\mathbf{j} + e^{xy}\cos z\,\mathbf{k}$, S is the hemisphere $x^2 + y^2 + z^2 = 1$, $z \geqslant 0$, oriented upward

2. $\mathbf{F}(x, y, z) = y^2z\,\mathbf{i} + xz\,\mathbf{j} + x^2y^2\,\mathbf{k}$, S is the part of the paraboloid $z = x^2 + y^2$ that lies inside the cylinder $x^2 + y^2 = 1$, oriented upward

3. $\mathbf{F}(x, y, z) = yz^3\,\mathbf{i} + \sin(xyz)\,\mathbf{j} + x^3\,\mathbf{k}$, S is the part of the paraboloid $y = 1 - x^2 - z^2$ that lies to the right of the xz-plane, oriented toward the xz-plane

4. $\mathbf{F}(x, y, z) = (x + \tan^{-1}yz)\,\mathbf{i} + y^2z\,\mathbf{j} + z\,\mathbf{k}$, S is the part of the hemisphere $x = \sqrt{9 - y^2 - z^2}$ that lies inside the cylinder $y^2 + z^2 = 4$, oriented in the direction of the positive x-axis

5. $\mathbf{F}(x, y, z) = xyz\,\mathbf{i} + xy\,\mathbf{j} + x^2yz\,\mathbf{k}$, S consists of the top and the four sides (but not the bottom) of the cube with vertices $(\pm1, \pm1, \pm1)$, oriented outward [*Hint:* Use Equation 4.]

6. $\mathbf{F}(x, y, z) = xy\,\mathbf{i} + e^z\,\mathbf{j} + xy^2\,\mathbf{k}$, S consists of the four sides of the pyramid with vertices $(0, 0, 0)$, $(1, 0, 0)$, $(0, 0, 1)$, $(1, 0, 1)$, and $(0, 1, 0)$ that lie to the right of the xz-plane, oriented in the direction of the positive y-axis [*Hint:* Use Equation 4.]

7–10 ■ Use Stokes' Theorem to evaluate $\int_C \mathbf{F} \cdot d\mathbf{r}$. In each case C is oriented counterclockwise as viewed from above.

7. $\mathbf{F}(x, y, z) = xz\,\mathbf{i} + 2xy\,\mathbf{j} + 3xy\,\mathbf{k}$, C is the boundary of the part of the plane $3x + y + z = 3$ in the first octant

8. $\mathbf{F}(x, y, z) = z^2\,\mathbf{i} + y^2\,\mathbf{j} + xy\,\mathbf{k}$, C is the triangle with vertices $(1, 0, 0)$, $(0, 1, 0)$, and $(0, 0, 2)$

9. $\mathbf{F}(x, y, z) = 2z\,\mathbf{i} + 4x\,\mathbf{j} + 5y\,\mathbf{k}$, C is the curve of intersection of the plane $z = x + 4$ and the cylinder $x^2 + y^2 = 4$

10. $\mathbf{F}(x, y, z) = x\,\mathbf{i} + y\,\mathbf{j} + (x^2 + y^2)\,\mathbf{k}$, C is the boundary of the part of the paraboloid $z = 1 - x^2 - y^2$ in the first octant

11. (a) Use Stokes' Theorem to evaluate $\int_C \mathbf{F} \cdot d\mathbf{r}$, where

$$\mathbf{F}(x, y, z) = x^2z\,\mathbf{i} + xy^2\,\mathbf{j} + z^2\,\mathbf{k}$$

and C is the curve of intersection of the plane $x + y + z = 1$ and the cylinder $x^2 + y^2 = 9$ oriented counterclockwise as viewed from above.

(b) Graph both the plane and the cylinder with domains chosen so that you can see the curve C and the surface that you used in part (a).

(c) Find parametric equations for C and use them to graph C.

12. (a) Use Stokes' Theorem to evaluate $\int_C \mathbf{F} \cdot d\mathbf{r}$, where $\mathbf{F}(x, y, z) = x^2y\,\mathbf{i} + \frac{1}{3}x^3\,\mathbf{j} + xy\,\mathbf{k}$ and C is the curve of intersection of the hyperbolic paraboloid $z = y^2 - x^2$ and the cylinder $x^2 + y^2 = 1$ oriented counterclockwise as viewed from above.

(b) Graph both the hyperbolic paraboloid and the cylinder with domains chosen so that you can see the curve C and the surface that you used in part (a).

(c) Find parametric equations for C and use them to graph C.

13–16 ■ Verify that Stokes' Theorem is true for the given vector field $\mathbf{F}$ and surface S.

13. $\mathbf{F}(x, y, z) = 3y\,\mathbf{i} + 4z\,\mathbf{j} - 6x\,\mathbf{k}$, S is the part of the paraboloid $z = 9 - x^2 - y^2$ that lies above the xy-plane, oriented upward

14. $\mathbf{F}(x, y, z) = xy\,\mathbf{i} + yz\,\mathbf{j} + xz\,\mathbf{k}$, S is the hemisphere $z = \sqrt{a^2 - x^2 - y^2}$, oriented upward

15. $\mathbf{F}(x, y, z) = y\,\mathbf{i} + z\,\mathbf{j} + x\,\mathbf{k}$, S is the part of the plane $x + y + z = 1$ that lies in the first octant, oriented upward

16. $\mathbf{F}(x, y, z) = y\,\mathbf{i} + z\,\mathbf{j} + x\,\mathbf{k}$, S is the helicoid of Exercise 26 in Section 14.6

17. Calculate the work done by the force field

$$\mathbf{F}(x, y, z) = (x^x + z^2)\,\mathbf{i} + (y^y + x^2)\,\mathbf{j} + (z^z + y^2)\,\mathbf{k}$$

when a particle moves under its influence around the edge of the part of the sphere $x^2 + y^2 + z^2 = 4$ that lies in the first octant, in a counterclockwise direction as viewed from above.

18. Evaluate $\int_C (y + \sin x)\,dx + (z^2 + \cos y)\,dy + x^3\,dz$, where C is the curve $\mathbf{r}(t) = \langle \sin t, \cos t, \sin 2t \rangle$, $0 \leqslant t \leqslant 2\pi$. [*Hint:* Observe that C lies on the surface $z = 2xy$.]

19. If S is a sphere and $\mathbf{F}$ satisfies the hypotheses of Stokes' Theorem, show that $\iint_S \operatorname{curl} \mathbf{F} \cdot d\mathbf{S} = 0$.

20. If S and C satisfy the hypotheses of Stokes' Theorem and f, g have continuous second-order partial derivatives, show that

(a) $\int_C (f\nabla g) \cdot d\mathbf{r} = \iint_S (\nabla f \times \nabla g) \cdot d\mathbf{S}$

(b) $\int_C (f\nabla f) \cdot d\mathbf{r} = 0$

(c) $\int_C (f\nabla g + g\nabla f) \cdot d\mathbf{r} = 0$

14.9 THE DIVERGENCE THEOREM

In Section 14.5 we rewrote Green's Theorem in a vector version as

$$\int_C \mathbf{F} \cdot \mathbf{n}\, ds = \iint_D \operatorname{div} \mathbf{F}(x, y)\, dA$$

where C is the positively oriented boundary curve of the plane region D. If we were seeking to extend this theorem to vector fields on $\mathbb{R}^3$, we might make the guess that

(1) $$\iint_S \mathbf{F} \cdot \mathbf{n}\, dS = \iiint_E \operatorname{div} \mathbf{F}(x, y, z)\, dV$$

where S is the boundary surface of the solid region E. It turns out that Equation 1 is true, under appropriate hypotheses, and is called the Divergence Theorem. Notice its similarity to Green's Theorem and Stokes' Theorem in that it relates the integral of a derivative of a function (div $\mathbf{F}$ in this case) over a region to the integral of the original function $\mathbf{F}$ over the boundary of the region.

The Divergence Theorem is sometimes called Gauss's Theorem after the great German mathematician Karl Friedrich Gauss (1777–1855) who discovered this theorem during his investigation of electrostatics. In Eastern Europe the Divergence Theorem is known as Ostrogradsky's Theorem after the Russian mathematician Mikhail Ostrogradsky (1801–1862), who published this result in 1826.

At this stage you may wish to review the various types of regions over which we were able to evaluate triple integrals in Section 13.8. We state and prove the Divergence Theorem for regions E that are simultaneously of types 1, 2, and 3 and we call such regions **simple solid regions**. (For instance, regions bounded by ellipsoids or rectangular boxes are simple solid regions.) The boundary of E is a closed surface, and we use the convention, introduced in Section 14.7, that the positive orientation is outward; that is, the unit normal vector $\mathbf{n}$ is directed outward from E.

(2) THE DIVERGENCE THEOREM Let E be a simple solid region whose boundary surface S has positive (outward) orientation. Let $\mathbf{F}$ be a vector field whose component functions have continuous partial derivatives on an open region that contains E. Then

$$\iint_S \mathbf{F} \cdot d\mathbf{S} = \iiint_E \operatorname{div} \mathbf{F}\, dV$$

PROOF Let $\mathbf{F} = P\,\mathbf{i} + Q\,\mathbf{j} + R\,\mathbf{k}$. Then

$$\operatorname{div} \mathbf{F} = \frac{\partial P}{\partial x} + \frac{\partial Q}{\partial y} + \frac{\partial R}{\partial z}$$

so

$$\iiint_E \operatorname{div} \mathbf{F}\, dV = \iiint_E \frac{\partial P}{\partial x}\, dV + \iiint_E \frac{\partial Q}{\partial y}\, dV + \iiint_E \frac{\partial R}{\partial z}\, dV$$

If $\mathbf{n}$ is the unit outward normal of S, then the surface integral on the left side of the Divergence Theorem is

$$\iint_S \mathbf{F} \cdot d\mathbf{S} = \iint_S \mathbf{F} \cdot \mathbf{n}\, dS = \iint_S (P\,\mathbf{i} + Q\,\mathbf{j} + R\,\mathbf{k}) \cdot \mathbf{n}\, dS$$

$$= \iint_S P\,\mathbf{i} \cdot \mathbf{n}\, dS + \iint_S Q\,\mathbf{j} \cdot \mathbf{n}\, dS + \iint_S R\,\mathbf{k} \cdot \mathbf{n}\, dS$$

Therefore, to prove the Divergence Theorem, it suffices to prove the three equations:

(3) $$\iint_S P\,\mathbf{i} \cdot \mathbf{n}\,dS = \iiint_E \frac{\partial P}{\partial x}\,dV$$

(4) $$\iint_S Q\,\mathbf{j} \cdot \mathbf{n}\,dS = \iiint_E \frac{\partial Q}{\partial y}\,dV$$

(5) $$\iint_S R\,\mathbf{k} \cdot \mathbf{n}\,dS = \iiint_E \frac{\partial R}{\partial z}\,dV$$

To prove Equation 5 we use the fact that E is a type 1 region:

$$E = \{(x, y, z) \mid (x, y) \in D,\ \phi_1(x, y) \leqslant z \leqslant \phi_2(x, y)\}$$

where D is the projection of E onto the xy-plane. By Equation 13.8.6, we have

$$\iiint_E \frac{\partial R}{\partial z}\,dV = \iint_D \left[\int_{\phi_1(x,y)}^{\phi_2(x,y)} \frac{\partial R}{\partial z}(x, y, z)\,dz \right] dA$$

and, therefore, by the Fundamental Theorem of Calculus,

(6) $$\iiint_E \frac{\partial R}{\partial z}\,dV = \iint_D [R(x, y, \phi_2(x, y)) - R(x, y, \phi_1(x, y))]\,dA$$

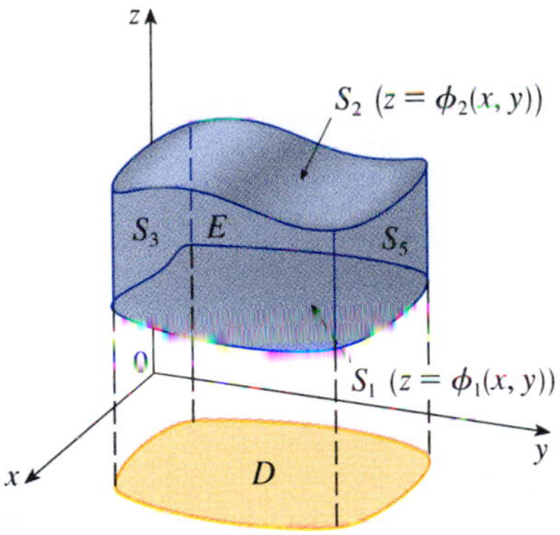

FIGURE 1

The boundary surface S consists of six pieces: the bottom surface S_1, the top surface S_2, and the four vertical sides S_3, S_4, S_5, and S_6 (see Figure 1). Notice that on each of the vertical sides we have $\mathbf{k} \cdot \mathbf{n} = 0$ and so

$$\iint_{S_i} R\,\mathbf{k} \cdot \mathbf{n}\,dS = \iint_{S_i} 0\,dS = 0 \qquad i = 3, 4, 5, 6$$

This gives

(7) $$\iint_S R\,\mathbf{k} \cdot \mathbf{n}\,dS = \iint_{S_1} R\,\mathbf{k} \cdot \mathbf{n}\,dS + \iint_{S_2} R\,\mathbf{k} \cdot \mathbf{n}\,dS$$

(It may happen that some of the vertical sides do not appear, as in the case of an ellipsoid, but Equation 7 is still true.)

The equation of S_2 is $z = \phi_2(x, y)$, $(x, y) \in D$, and the outward normal $\mathbf{n}$ points upward, so from Equation 14.7.8 (with $\mathbf{F}$ replaced by $R\,\mathbf{k}$) we have

$$\iint_{S_2} R\,\mathbf{k} \cdot \mathbf{n}\,dS = \iint_D R(x, y, \phi_2(x, y))\,dA$$

On S_1 we have $z = \phi_1(x, y)$, but here the outward normal $\mathbf{n}$ points downward, so

$$\mathbf{n} = \frac{1}{\sqrt{\left(\frac{\partial \phi_1}{\partial x}\right)^2 + \left(\frac{\partial \phi_1}{\partial y}\right)^2 + 1}} \left(\frac{\partial \phi_1}{\partial x}\,\mathbf{i} + \frac{\partial \phi_1}{\partial y}\,\mathbf{j} - \mathbf{k} \right)$$

and

$$\iint_{S_1} R\,\mathbf{k}\cdot\mathbf{n}\,dS$$

$$= \iint_D R(x, y, \phi_1(x, y))\,\frac{(-1)}{\sqrt{\left(\dfrac{\partial \phi_1}{\partial x}\right)^2 + \left(\dfrac{\partial \phi_1}{\partial y}\right)^2 + 1}}\sqrt{1 + \left(\frac{\partial \phi_1}{\partial x}\right)^2 + \left(\frac{\partial \phi_1}{\partial y}\right)^2}\,dA$$

$$= -\iint_D R(x, y, \phi_1(x, y))\,dA$$

Therefore, Equation 7 gives

$$\iint_S R\,\mathbf{k}\cdot\mathbf{n}\,dS = \iint_D [R(x, y, \phi_2(x, y)) - R(x, y, \phi_1(x, y))]\,dA$$

Comparison with Equation 6 shows that

$$\iint_S R\,\mathbf{k}\cdot\mathbf{n}\,dS = \iiint_E \frac{\partial R}{\partial z}\,dV$$

Notice that the method of proof of the Divergence Theorem is very similar to that of Green's Theorem.

Equations 3 and 4 are proved in a similar manner using the expressions for E as a type 2 or type 3 region. □

EXAMPLE 1 Find the flux of the vector field $\mathbf{F}(x, y, z) = z\,\mathbf{i} + y\,\mathbf{j} + x\,\mathbf{k}$ over the unit sphere $x^2 + y^2 + z^2 = 1$.

SOLUTION First we compute the divergence of $\mathbf{F}$:

$$\text{div}\,\mathbf{F} = \frac{\partial}{\partial x}(z) + \frac{\partial}{\partial y}(y) + \frac{\partial}{\partial z}(x) = 1$$

The unit sphere S is the boundary of the unit ball B given by $x^2 + y^2 + z^2 \leqslant 1$. Thus, the Divergence Theorem gives the flux as

$$\iint_S \mathbf{F}\cdot d\mathbf{S} = \iiint_B \text{div}\,\mathbf{F}\,dV = \iiint_B 1\,dV$$

$$= V(B) = \tfrac{4}{3}\pi(1)^3 = \frac{4\pi}{3}$$ ■

The solution in Example 1 should be compared with the solution in Example 5 in Section 14.7.

EXAMPLE 2 Evaluate $\iint_S \mathbf{F}\cdot d\mathbf{S}$, where

$$\mathbf{F}(x, y, z) = xy\,\mathbf{i} + (y^2 + e^{xz^2})\,\mathbf{j} + \sin(xy)\,\mathbf{k}$$

and S is the surface of the region E bounded by the parabolic cylinder $z = 1 - x^2$ and the planes $z = 0$, $y = 0$, and $y + z = 2$ (see Figure 2).

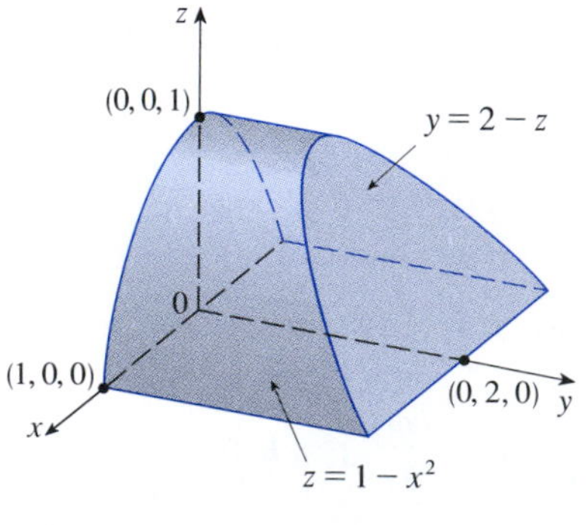

FIGURE 2

SOLUTION It would be extremely difficult to evaluate the given surface integral directly. (We would have to evaluate four surface integrals corresponding to the four pieces of S.) Furthermore, the divergence of $\mathbf{F}$ is much less complicated than $\mathbf{F}$ itself:

$$\text{div}\,\mathbf{F} = \frac{\partial}{\partial x}(xy) + \frac{\partial}{\partial y}(y^2 + e^{xz^2}) + \frac{\partial}{\partial z}(\sin xy)$$

$$= y + 2y = 3y$$

Therefore, we use the Divergence Theorem to transform the given surface integral into a triple integral. The easiest way to evaluate the triple integral is to express E as a type 3 region:

$$E = \{(x, y, z) \mid -1 \leqslant x \leqslant 1, 0 \leqslant z \leqslant 1 - x^2, 0 \leqslant y \leqslant 2 - z\}$$

Then we have

$$\begin{aligned}
\iint_S \mathbf{F} \cdot d\mathbf{S} &= \iiint_E \operatorname{div} \mathbf{F}\, dV = \iiint_E 3y\, dV \\
&= 3 \int_{-1}^{1} \int_0^{1-x^2} \int_0^{2-z} y\, dy\, dz\, dx \\
&= 3 \int_{-1}^{1} \int_0^{1-x^2} \frac{(2-z)^2}{2}\, dz\, dx \\
&= \frac{3}{2} \int_{-1}^{1} \left[-\frac{(2-z)^3}{3} \right]_0^{1-x^2} dx \\
&= -\tfrac{1}{2} \int_{-1}^{1} [(x^2+1)^3 - 8]\, dx \\
&= -\int_0^1 (x^6 + 3x^4 + 3x^2 - 7)\, dx = \frac{184}{35}
\end{aligned}$$

■

Although we have proved the Divergence Theorem only for simple solid regions, it can be proved for regions that are finite unions of simple solid regions. (The procedure is analogous to the one we used in Section 14.4 to extend Green's Theorem.)

For example, let's consider the region E that lies between the closed surfaces S_1 and S_2, where S_1 lies inside S_2. Let $\mathbf{n}_1$ and $\mathbf{n}_2$ be outward normals of S_1 and S_2. Then the boundary surface of E is $S = S_1 \cup S_2$ and its normal $\mathbf{n}$ is given by $\mathbf{n} = -\mathbf{n}_1$ on S_1 and $\mathbf{n} = \mathbf{n}_2$ on S_2 (see Figure 3). Applying the Divergence Theorem to S, we get

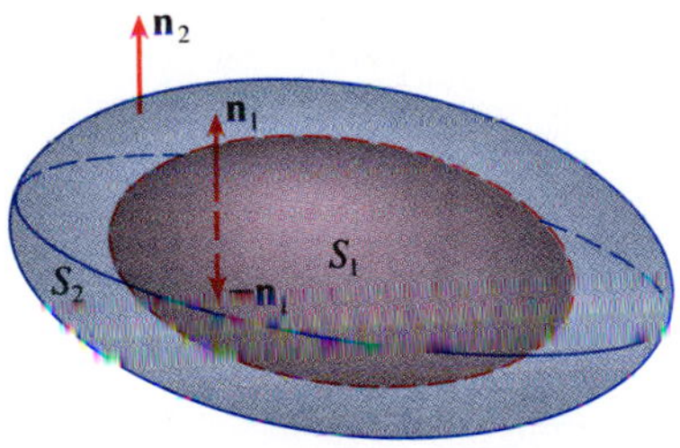

FIGURE 3

$$\begin{aligned}
\iiint_E \operatorname{div} \mathbf{F}\, dV &= \iint_S \mathbf{F} \cdot d\mathbf{S} = \iint_S \mathbf{F} \cdot \mathbf{n}\, dS \\
&= \iint_{S_1} \mathbf{F} \cdot (-\mathbf{n}_1)\, dS + \iint_{S_2} \mathbf{F} \cdot \mathbf{n}_2\, dS \\
&= -\iint_{S_1} \mathbf{F} \cdot d\mathbf{S} + \iint_{S_2} \mathbf{F} \cdot d\mathbf{S}
\end{aligned} \tag{8}$$

Let's apply this to the electric field (see Example 5 in Section 14.1):

$$\mathbf{E}(\mathbf{x}) = \frac{\varepsilon Q}{|\mathbf{x}|^3} \mathbf{x}$$

where S_1 is a small sphere with radius a and center the origin. You can verify that $\operatorname{div} \mathbf{E} = 0$ (see Exercise 19). Therefore, Equation 8 gives

$$\begin{aligned}
\iint_{S_2} \mathbf{E} \cdot d\mathbf{S} &= \iint_{S_1} \mathbf{E} \cdot d\mathbf{S} + \iiint_E \operatorname{div} \mathbf{E}\, dV \\
&= \iint_{S_1} \mathbf{E} \cdot d\mathbf{S} = \iint_{S_1} \mathbf{E} \cdot \mathbf{n}\, dS
\end{aligned}$$

The point of this calculation is that we can compute the surface integral over S_1 because S_1 is a sphere. The normal vector at $\mathbf{x}$ is $\mathbf{x}/|\mathbf{x}|$. Therefore

$$\mathbf{E} \cdot \mathbf{n} = \frac{\varepsilon Q}{|\mathbf{x}|^3}\mathbf{x} \cdot \left(\frac{\mathbf{x}}{|\mathbf{x}|}\right) = \frac{\varepsilon Q}{|\mathbf{x}|^4}\mathbf{x} \cdot \mathbf{x}$$

$$= \frac{\varepsilon Q}{|\mathbf{x}|^2} = \frac{\varepsilon Q}{a^2}$$

since the equation of S_1 is $|\mathbf{x}| = a$. Thus, we have

$$\iint_{S_2} \mathbf{E} \cdot d\mathbf{S} = \iint_{S_1} \mathbf{E} \cdot \mathbf{n}\, dS$$

$$= \frac{\varepsilon Q}{a^2}\iint_{S_1} dS = \frac{\varepsilon Q}{a^2}A(S_1)$$

$$= \frac{\varepsilon Q}{a^2}4\pi a^2 = 4\pi\varepsilon Q$$

This shows that the electric flux of $\mathbf{E}$ is $4\pi\varepsilon Q$ through *any* closed surface S_2 that contains the origin. [This is a special case of Gauss's Law (Equation 14.7.10) for a single charge. The relationship between ε and ε_0 is $\varepsilon = 1/(4\pi\varepsilon_0)$.]

Another application of the Divergence Theorem occurs in fluid flow. Let $\mathbf{v}(x, y, z)$ be the velocity field of a fluid with constant density ρ. Then $\mathbf{F} = \rho\mathbf{v}$ is the rate of flow per unit area. If $P_0(x_0, y_0, z_0)$ is a point in the fluid and B_a is a ball with center P_0 and very small radius a, then $\operatorname{div} \mathbf{F}(P) \approx \operatorname{div} \mathbf{F}(P_0)$ for all points in B_a since $\operatorname{div} \mathbf{F}$ is continuous. We approximate the flux over the boundary sphere S_a as follows:

$$\iint_{S_a} \mathbf{F} \cdot d\mathbf{S} = \iiint_{B_a} \operatorname{div} \mathbf{F}\, dV$$

$$\approx \iiint_{B_a} \operatorname{div} \mathbf{F}(P_0)\, dV$$

$$= \operatorname{div} \mathbf{F}(P_0)V(B_a)$$

This approximation becomes better as $a \to 0$ and suggests that

(9) $$\operatorname{div} \mathbf{F}(P_0) = \lim_{a \to 0} \frac{1}{V(B_a)} \iint_{S_a} \mathbf{F} \cdot d\mathbf{S}$$

Equation 9 says that $\operatorname{div} \mathbf{F}(P_0)$ is the net rate of outward flux per unit volume at P_0. (This is the reason for the name *divergence*.) If $\operatorname{div} \mathbf{F}(P) > 0$, the net flow is outward near P and P is called a **source**. If $\operatorname{div} \mathbf{F}(P) < 0$, the net flow is inward near P and P is called a **sink**.

EXERCISES 14.9

1–2 ■ Verify that the Divergence Theorem is true for the vector field $\mathbf{F}$ on the region E.

1. $\mathbf{F}(x, y, z) = 3x\,\mathbf{i} + xy\,\mathbf{j} + 2xz\,\mathbf{k}$; E is the cube bounded by the planes $x = 0$, $x = 1$, $y = 0$, $y = 1$, $z = 0$, and $z = 1$

2. $\mathbf{F}(x, y, z) = xz\,\mathbf{i} + yz\,\mathbf{j} + 3z^2\mathbf{k}$; E is the solid bounded by the paraboloid $z = x^2 + y^2$ and the plane $z = 1$

3–15 ■ Use the Divergence Theorem to calculate the surface integral $\iint_S \mathbf{F} \cdot d\mathbf{S}$; that is, calculate the flux of $\mathbf{F}$ across S.

3. $\mathbf{F}(x, y, z) = 3y^2z^3\,\mathbf{i} + 9x^2yz^2\,\mathbf{j} - 4xy^2\,\mathbf{k}$,
S is the surface of the cube with vertices $(\pm1, \pm1, \pm1)$

4. $\mathbf{F}(x, y, z) = x^2y\,\mathbf{i} - x^2z\,\mathbf{j} + z^2y\,\mathbf{k}$,
S is the surface of the rectangular box bounded by the planes $x = 0$, $x = 3$, $y = 0$, $y = 2$, $z = 0$, and $z = 1$

5. $\mathbf{F}(x, y, z) = -xz\,\mathbf{i} - yz\,\mathbf{j} + z^2\,\mathbf{k}$,
S is the ellipsoid $x^2/a^2 + y^2/b^2 + z^2/c^2 = 1$

6. $\mathbf{F}(x, y, z) = 3xy\,\mathbf{i} + y^2\,\mathbf{j} - x^2y^4\,\mathbf{k}$,
S is the surface of the tetrahedron with vertices $(0, 0, 0)$, $(1, 0, 0)$, $(0, 1, 0)$, and $(0, 0, 1)$

7. $\mathbf{F}(x, y, z) = z\cos y\,\mathbf{i} + x\sin z\,\mathbf{j} + xz\,\mathbf{k}$,
S is the surface of the tetrahedron bounded by the planes $x = 0$, $y = 0$, $z = 0$, and $2x + y + z = 2$

8. $\mathbf{F}(x, y, z) = (x + e^{y\tan z})\,\mathbf{i} + 3xe^{xz}\,\mathbf{j} + (\cos y - z)\,\mathbf{k}$,
S is the surface with equation $x^4 + y^4 + z^4 = 1$

9. $\mathbf{F}(x, y, z) = x^3\,\mathbf{i} + y^3\,\mathbf{j} + z^3\,\mathbf{k}$,
S is the sphere $x^2 + y^2 + z^2 = 1$

10. $\mathbf{F}(x, y, z) = x^3\,\mathbf{i} + 2xz^2\,\mathbf{j} + 3y^2z\,\mathbf{k}$,
S is the surface of the solid bounded by the paraboloid $z = 4 - x^2 - y^2$ and the xy-plane

11. $\mathbf{F}(x, y, z) = ye^{z^2}\,\mathbf{i} + y^2\,\mathbf{j} + e^{xy}\,\mathbf{k}$,
S is the surface of the solid bounded by the cylinder $x^2 + y^2 = 9$ and the planes $z = 0$ and $z = y - 3$

12. $\mathbf{F}(x, y, z) = (x^3 + y\sin z)\,\mathbf{i} + (y^3 + z\sin x)\,\mathbf{j} + 3z\,\mathbf{k}$,
S is the surface of the solid bounded by the hemispheres $z = \sqrt{4 - x^2 - y^2}$, $z = \sqrt{1 - x^2 - y^2}$ and the plane $z = 0$

13. $\mathbf{F}(x, y, z) = xy^2\,\mathbf{i} + yz\,\mathbf{j} + zx^2\,\mathbf{k}$, S is the surface of the solid that lies between the cylinders $x^2 + y^2 = 1$ and $x^2 + y^2 = 4$ and between the planes $z = 1$ and $z = 3$

14. $\mathbf{F}(x, y, z) = (x^3 + yz)\,\mathbf{i} + x^2y\,\mathbf{j} + xy^2\,\mathbf{k}$,
S is the surface of the solid bounded by the spheres $x^2 + y^2 + z^2 = 4$ and $x^2 + y^2 + z^2 = 9$

CAS **15.** $\mathbf{F}(x, y, z) = e^y\tan z\,\mathbf{i} + y\sqrt{3 - x^2}\,\mathbf{j} + x\sin y\,\mathbf{k}$, S is the surface of the solid that lies above the xy-plane and below the surface $z = 2 - x^4 - y^4$, $-1 \leqslant x \leqslant 1$, $-1 \leqslant y \leqslant 1$

CAS **16.** Use a computer algebra system to plot the vector field $\mathbf{F}(x, y, z) = \sin x\cos^2 y\,\mathbf{i} + \sin^3 y\cos^4 z\,\mathbf{j} + \sin^5 z\cos^6 x\,\mathbf{k}$ in the cube cut from the first octant by the planes $x = \pi/2$, $y = \pi/2$, and $z = \pi/2$. Then compute the flux across the surface of the cube.

17. Use the Divergence Theorem to evaluate $\iint_S \mathbf{F} \cdot d\mathbf{S}$, where $\mathbf{F}(x, y, z) = z^2x\,\mathbf{i} + (\frac{1}{3}y^3 + \tan z)\,\mathbf{j} + (x^2z + y^2)\,\mathbf{k}$ and S is the top half of the sphere $x^2 + y^2 + z^2 = 1$. [*Hint:* Note that S is not a closed surface. First compute integrals over S_1 and S_2, where S_1 is the disk $x^2 + y^2 \leqslant 1$ oriented downward and $S_2 = S \cup S_1$.]

18. Let $\mathbf{F}(x, y, z) = z\tan^{-1}(y^2)\,\mathbf{i} + z^3\ln(x^2 + 1)\,\mathbf{j} + z\,\mathbf{k}$. Find the flux of $\mathbf{F}$ across the part of the paraboloid $x^2 + y^2 + z = 2$ that lies above the plane $z = 1$ and is oriented upward.

19. Verify that $\operatorname{div}\mathbf{E} = 0$ for the electric field $\mathbf{E}(\mathbf{x}) = \varepsilon Q\mathbf{x}/|\mathbf{x}|^3$.

20. Use the Divergence Theorem to evaluate

$$\iint_S (2x + 2y + z^2)\,dS$$

where S is the sphere $x^2 + y^2 + z^2 = 1$.

21–26 ■ Prove each identity, assuming that S and E satisfy the conditions of the Divergence Theorem and the scalar functions and components of the vector fields have continuous second-order partial derivatives.

21. $\displaystyle\iint_S \mathbf{a} \cdot \mathbf{n}\,dS = 0$, where $\mathbf{a}$ is a constant vector

22. $V(E) = \frac{1}{3}\displaystyle\iint_S \mathbf{F} \cdot d\mathbf{S}$, where $\mathbf{F}(x, y, z) = x\,\mathbf{i} + y\,\mathbf{j} + z\,\mathbf{k}$

23. $\displaystyle\iint_S \operatorname{curl}\mathbf{F} \cdot d\mathbf{S} = 0$

24. $\displaystyle\iint_S D_{\mathbf{n}} f\,dS = \iiint_E \nabla^2 f\,dV$

25. $\displaystyle\iint_S (f\nabla g) \cdot \mathbf{n}\,dS = \iiint_E (f\nabla^2 g + \nabla f \cdot \nabla g)\,dV$

26. $\displaystyle\iint_S (f\nabla g - g\nabla f) \cdot \mathbf{n}\,dS = \iiint_E (f\nabla^2 g - g\nabla^2 f)\,dV$

14.10 SUMMARY

The main results of this chapter are all higher-dimensional analogues of the Fundamental Theorem of Calculus. To help you remember them, we collect them together here (without hypotheses) so that you can see more easily their essential similarity. Notice that in each case we have an integral of a "derivative" over a region on the left side, and the right side involves the values of the original function only on the *boundary* of the region.

Fundamental Theorem of Calculus

$$\int_a^b F'(x)\,dx = F(b) - F(a)$$

Fundamental Theorem for Line Integrals

$$\int_C \nabla f \cdot d\mathbf{r} = f(\mathbf{r}(b)) - f(\mathbf{r}(a))$$

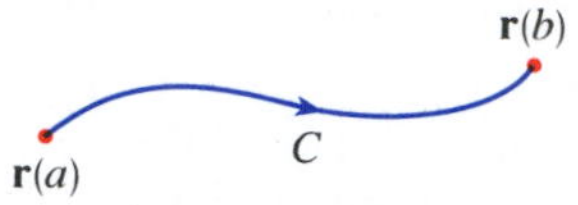

Green's Theorem

$$\iint_D \left(\frac{\partial Q}{\partial x} - \frac{\partial P}{\partial y} \right) dA = \int_C P\,dx + Q\,dy$$

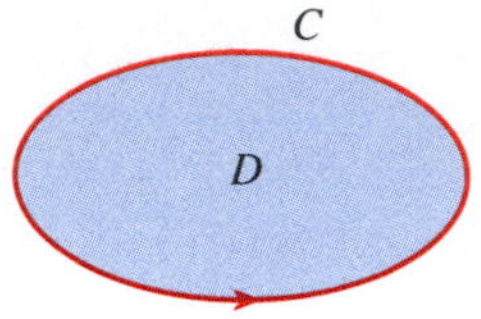

Stokes' Theorem

$$\iint_S \operatorname{curl} \mathbf{F} \cdot d\mathbf{S} = \int_C \mathbf{F} \cdot d\mathbf{r}$$

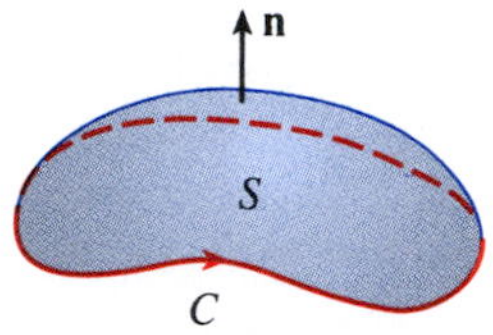

Divergence Theorem

$$\iiint_E \operatorname{div} \mathbf{F}\,dV = \iint_S \mathbf{F} \cdot d\mathbf{S}$$

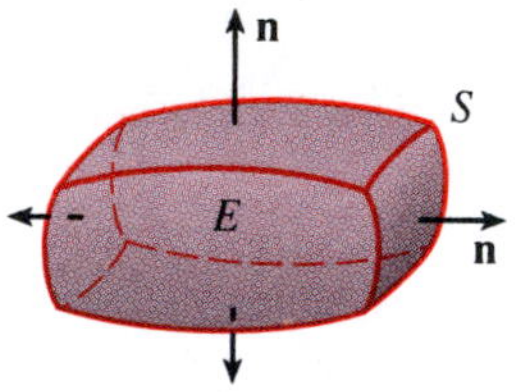

14 REVIEW

KEY TOPICS ■ Define, state, or discuss the following.

1. Vector field
2. Conservative vector field
3. Potential function
4. Line integral of a scalar function with respect to arc length
5. Line integral of a scalar function with respect to x, y, and z
6. Line integral of a vector field
7. Work done by a force field
8. Fundamental theorem for line integrals
9. Independence of path
10. Green's Theorem
11. Curl
12. Divergence
13. Vector forms of Green's Theorem
14. Parametric surface
15. Normal vector to a parametric surface
16. Smooth surface
17. Surface area of a parametric surface
18. Formula for the surface area of a graph $z = f(x, y)$
19. Surface integral of a scalar function
20. Oriented surface
21. Surface integral of a vector field
22. Stokes' Theorem
23. The Divergence Theorem

EXERCISES

1–8 ■ Determine whether the statement is true or false.

1. If $\mathbf{F}$ is a vector field, then div $\mathbf{F}$ is a vector field.
2. If $\mathbf{F}$ is a vector field, then curl $\mathbf{F}$ is a vector field.
3. If f has continuous partial derivatives of all orders on $\mathbb{R}^3$, then $\operatorname{div}(\operatorname{curl} \nabla f) = 0$.
4. If f has continuous partial derivatives on $\mathbb{R}^3$ and C is any circle, then $\int_C \nabla f \cdot d\mathbf{r} = 0$.
5. If $\mathbf{F} = P\,\mathbf{i} + Q\,\mathbf{j}$ and $P_y = Q_x$ in an open region D, then $\mathbf{F}$ is conservative.
6. $\int_{-C} f(x, y)\, ds = -\int_C f(x, y)\, ds$
7. If S is a sphere and $\mathbf{F}$ is a constant vector field, then $\iint_S \mathbf{F} \cdot d\mathbf{S} = 0$.
8. There is a vector field $\mathbf{F}$ such that curl $\mathbf{F} = x\,\mathbf{i} + y\,\mathbf{j} + z\,\mathbf{k}$.

9–17 ■ Evaluate the given line integral.

9. $\int_C y\, ds$, C is the arc of the parabola $y^2 = 2x$ from $(0, 0)$ to $(2, 2)$
10. $\int_C yz^2\, ds$, C is the line segment from $(-1, 1, 3)$ to $(0, 3, 5)$
11. $\int_C x^3 z\, ds$, C: $x = 2\sin t$, $y = t$, $z = 2\cos t$, $0 \leqslant t \leqslant \pi/2$
12. $\int_C xy\, dx + y\, dy$, C is the sine curve $y = \sin x$, $0 \leqslant x \leqslant \pi/2$
13. $\int_C x^3 y\, dx - x\, dy$, C is the circle $x^2 + y^2 = 1$ with counterclockwise orientation
14. $\int_C x \sin y\, dx + xyz\, dz$, C is given by $\mathbf{r}(t) = t\,\mathbf{i} + t^2\,\mathbf{j} + t^3\,\mathbf{k}$, $0 \leqslant t \leqslant 1$
15. $\int_C y\, dx + z\, dy + x\, dz$, C consists of the line segments from $(0, 0, 0)$ to $(1, 1, 2)$ and from $(1, 1, 2)$ to $(3, 1, 4)$
16. $\int_C \mathbf{F} \cdot d\mathbf{r}$, where $\mathbf{F}(x, y) = x^2 y\,\mathbf{i} + e^y\,\mathbf{j}$ and C is given by $\mathbf{r}(t) = t^2\,\mathbf{i} - t^3\,\mathbf{j}$, $0 \leqslant t \leqslant 1$
17. $\int_C \mathbf{F} \cdot d\mathbf{r}$, where $\mathbf{F}(x, y, z) = (x + y)\,\mathbf{i} + z\,\mathbf{j} + x^2 y\,\mathbf{k}$ and C is given by $\mathbf{r}(t) = 2t\,\mathbf{i} + t^2\,\mathbf{j} + t^4\,\mathbf{k}$, $0 \leqslant t \leqslant 1$
18. Find the work done by the force field $\mathbf{F}(x, y, z) = z\,\mathbf{i} + x\,\mathbf{j} + y\,\mathbf{k}$ in moving a particle from the point $(3, 0, 0)$ to the point $(0, \pi/2, 3)$ (a) along a straight line and (b) along the helix $x = 3\cos t$, $y = t$, $z = 3\sin t$.

19–20 ■ Show that $\mathbf{F}$ is a conservative vector field. Then find a function f such that $\mathbf{F} = \nabla f$.

19. $\mathbf{F}(x, y) = \sin y\,\mathbf{i} + (x\cos y + \sin y)\,\mathbf{j}$
20. $\mathbf{F}(x, y, z) = (2xy^3 + z^2)\,\mathbf{i} + (3x^2y^2 + 2yz)\,\mathbf{j} + (y^2 + 2xz)\,\mathbf{k}$

21–22 ■ Show that $\mathbf{F}$ is conservative and use this fact to evaluate $\int_C \mathbf{F} \cdot d\mathbf{r}$ along the given curve.

21. $\mathbf{F}(x, y) = (2x + y^2 + 3x^2y)\,\mathbf{i} + (2xy + x^3 + 3y^2)\,\mathbf{j}$, C is the arc of the curve $y = x\sin x$ from $(0, 0)$ to $(\pi, 0)$
22. $\mathbf{F}(x, y, z) = yz(2x + y)\,\mathbf{i} + xz(x + 2y)\,\mathbf{j} + xy(x + y)\,\mathbf{k}$, C is given by $\mathbf{r}(t) = (1 + t)\,\mathbf{i} + (1 + 2t^2)\,\mathbf{j} + (1 + 3t^3)\,\mathbf{k}$, $0 \leqslant t \leqslant 1$
23. Verify that Green's Theorem is true for the line integral $\int_C xy\, dx + x^2\, dy$, where C is the triangle with vertices $(0, 0)$, $(1, 0)$, and $(1, 2)$.
24. Use Green's Theorem to evaluate
$$\int_C (1 + \tan y)\, dx + (x^2 + e^y)\, dy$$
where C is the positively oriented boundary of the region enclosed by the curves $y = \sqrt{x}$, $x = 1$, and $y = 0$.
25. Use Green's Theorem to evaluate
$$\int_C x^2 y\, dx - xy^2\, dy$$
where C is the circle $x^2 + y^2 = 4$ with counterclockwise orientation.
26. Find curl $\mathbf{F}$ and div $\mathbf{F}$ if
$$\mathbf{F}(x, y, z) = x^2 z\,\mathbf{i} + 2x\sin y\,\mathbf{j} + 2z\cos y\,\mathbf{k}$$
27. Show that there is no vector field $\mathbf{G}$ such that
$$\operatorname{curl} \mathbf{G} = 2x\,\mathbf{i} + 3yz\,\mathbf{j} - xz^2\,\mathbf{k}$$
28. Show that, under conditions to be stated on the vector fields $\mathbf{F}$ and $\mathbf{G}$,
$$\operatorname{curl}(\mathbf{F} \times \mathbf{G}) = \mathbf{F} \operatorname{div} \mathbf{G} - \mathbf{G} \operatorname{div} \mathbf{F} + (\mathbf{G} \cdot \nabla)\mathbf{F} - (\mathbf{F} \cdot \nabla)\mathbf{G}$$
29. If C is any piecewise-smooth simple closed plane curve and f and g are differentiable functions, show that
$$\int_C f(x)\, dx + g(y)\, dy = 0$$
30. If f and g are twice differentiable functions, show that
$$\nabla^2(fg) = f\nabla^2 g + g\nabla^2 f + 2\nabla f \cdot \nabla g$$
31. If f is a harmonic function, that is, $\nabla^2 f = 0$, show that the line integral $\int f_y\, dx - f_x\, dy$ is independent of path in any simple region D.
32. (a) Sketch the curve C with parametric equations
$$x = \cos t \qquad y = \sin t \qquad z = \sin t \qquad 0 \leqslant t \leqslant 2\pi$$
(b) Find $\int_C 2xe^{2y}\, dx + (2x^2e^{2y} + 2y\cot z)\, dy - y^2\csc^2 z\, dz$.
33. Find the area of the part of the surface $z = x^2 + 2y$ that lies above the triangle with vertices $(0, 0)$, $(1, 0)$, and $(1, 2)$.

34. (a) Find an equation of the tangent plane at the point $(4, -2, 1)$ to the parametric surface S given by $\mathbf{r}(u, v) = v^2\,\mathbf{i} - uv\,\mathbf{j} + u^2\,\mathbf{k}$, $0 \leqslant u \leqslant 3$, $-3 \leqslant v \leqslant 3$.

(b) Use a computer to graph the surface S and the tangent plane found in part (a).

(c) Set up, but do not evaluate, an integral for the surface area of S.

(d) If

$$\mathbf{F}(x, y, z) = \frac{z^2}{1 + x^2}\,\mathbf{i} + \frac{x^2}{1 + y^2}\,\mathbf{j} + \frac{y^2}{1 + z^2}\,\mathbf{k}$$

find $\iint_S \mathbf{F} \cdot d\mathbf{S}$ correct to four decimal places.

35–38 ■ Evaluate the surface integral.

35. $\iint_S z\,dS$, where S is the part of the paraboloid $z = x^2 + y^2$ that lies under the plane $z = 4$

36. $\iint_S (x^2z + y^2z)\,dS$, where S is the part of the plane $z = 4 + x + y$ that lies inside the cylinder $x^2 + y^2 = 4$

37. $\iint_S \mathbf{F} \cdot d\mathbf{S}$, where $\mathbf{F}(x, y, z) = xz\,\mathbf{i} - 2y\,\mathbf{j} + 3x\,\mathbf{k}$ and S is the sphere $x^2 + y^2 + z^2 = 4$ with outward orientation

38. $\iint_S \mathbf{F} \cdot d\mathbf{S}$, where $\mathbf{F}(x, y, z) = x^2\,\mathbf{i} + xy\,\mathbf{j} + z\,\mathbf{k}$ and S is the part of the paraboloid $z = x^2 + y^2$ below the plane $z = 1$ with upward orientation

39. Verify that Stokes' Theorem is true for the vector field $\mathbf{F}(x, y, z) = x^2\,\mathbf{i} + y^2\,\mathbf{j} + z^2\,\mathbf{k}$, where S is the part of the paraboloid $z = 1 - x^2 - y^2$ that lies above the xy-plane, and S has upward orientation.

40. Use Stokes' Theorem to evaluate $\iint_S \operatorname{curl} \mathbf{F} \cdot d\mathbf{S}$, where $\mathbf{F}(x, y, z) = x^2yz\,\mathbf{i} + yz^2\,\mathbf{j} + z^3e^{xy}\,\mathbf{k}$; S is the part of the sphere $x^2 + y^2 + z^2 = 5$ that lies above the plane $z = 1$, and S is oriented upward.

41. Use Stokes' Theorem to evaluate $\int_C \mathbf{F} \cdot d\mathbf{r}$, where $\mathbf{F}(x, y, z) = xy\,\mathbf{i} + yz\,\mathbf{j} + zx\,\mathbf{k}$ and C is the triangle with vertices $(1, 0, 0)$, $(0, 1, 0)$, and $(0, 0, 1)$, oriented counterclockwise as viewed from above.

42. Use the Divergence Theorem to calculate the surface integral $\iint_S \mathbf{F} \cdot d\mathbf{S}$, where $\mathbf{F}(x, y, z) = x^3\,\mathbf{i} + y^3\,\mathbf{j} + z^3\,\mathbf{k}$ and S is the surface of the solid bounded by the cylinder $x^2 + y^2 = 1$ and the planes $z = 0$ and $z = 2$.

43. Verify that the Divergence Theorem is true for the vector field $\mathbf{F}(x, y, z) = x\,\mathbf{i} + y\,\mathbf{j} + z\,\mathbf{k}$, where E is the unit ball $x^2 + y^2 + z^2 \leqslant 1$.

44. Compute the outward flux of

$$\mathbf{F}(x, y, z) = \frac{x\,\mathbf{i} + y\,\mathbf{j} + z\,\mathbf{k}}{(x^2 + y^2 + z^2)^{3/2}}$$

through the ellipsoid $4x^2 + 9y^2 + 6z^2 = 36$.

45. Let

$$\mathbf{F}(x, y, z) = (3x^2yz - 3y)\,\mathbf{i} + (x^3z - 3x)\,\mathbf{j} + (x^3y + 2z)\,\mathbf{k}$$

Evaluate $\int_C \mathbf{F} \cdot d\mathbf{r}$, where C is the curve with initial point $(0, 0, 2)$ and terminal point $(0, 3, 0)$ shown in the figure.

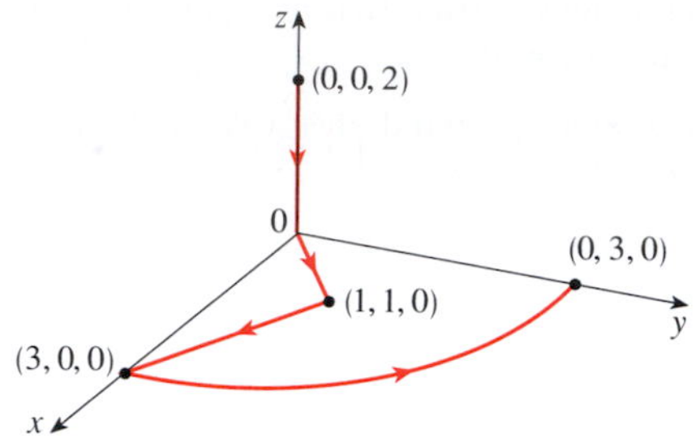

46. Let

$$\mathbf{F}(x, y) = \frac{(2x^3 + 2xy^2 - 2y)\,\mathbf{i} + (2y^3 + 2x^2y + 2x)\,\mathbf{j}}{x^2 + y^2}$$

Evaluate $\oint_C \mathbf{F} \cdot d\mathbf{r}$, where C is shown in the figure.

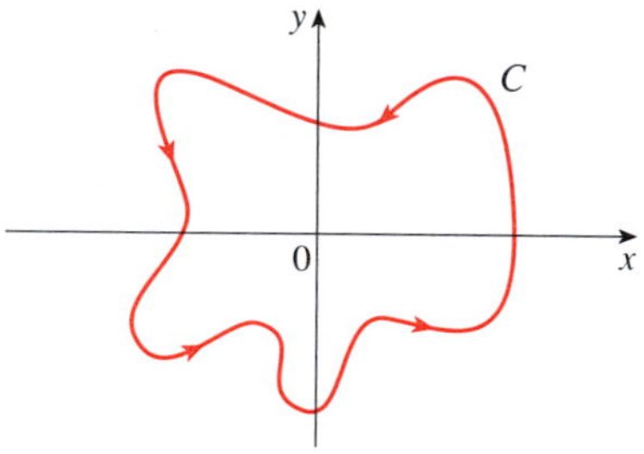

47. Find $\iint_S \mathbf{F} \cdot \mathbf{n}\,dS$, where $\mathbf{F}(x, y, z) = x\,\mathbf{i} + y\,\mathbf{j} + z\,\mathbf{k}$ and S is the outwardly oriented surface shown in the figure (the boundary surface of a cube with a unit corner cube removed).

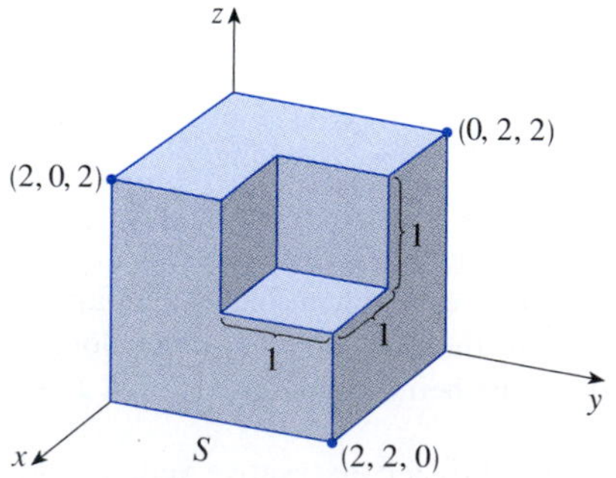

48. If the components of $\mathbf{F}$ have continuous second partial derivatives and S is the boundary surface of a simple solid region, show that $\iint_S \operatorname{curl} \mathbf{F} \cdot d\mathbf{S} = 0$.

PROBLEMS PLUS

1. The double integral $\int_0^1 \int_0^1 \frac{1}{1 - xy}\, dx\, dy$ is an improper integral and could be defined as the limit of double integrals over the rectangle $[0, t] \times [0, t]$ as $t \to 1^-$. But if we expand the integrand as a geometric series, we can express the integral as the sum of an infinite series. Show that

$$\int_0^1 \int_0^1 \frac{1}{1 - xy}\, dx\, dy = \sum_{n=1}^{\infty} \frac{1}{n^2}$$

2. Leonhard Euler was able to find the exact sum of the series in Problem 1. In 1736 he proved that

$$\sum_{n=1}^{\infty} \frac{1}{n^2} = \frac{\pi^2}{6}$$

In this problem we ask you to prove this fact by evaluating the double integral in Problem 1. Start by making the change of variables

$$x = \frac{u - v}{\sqrt{2}} \qquad y = \frac{u + v}{\sqrt{2}}$$

This gives a rotation about the origin through the angle $\pi/4$. You will need to sketch the corresponding region in the uv-plane.

[*Hint*: If, in evaluating the integral, you encounter either of the expressions $(1 - \sin\theta)/\cos\theta$ or $(\cos\theta)/(1 + \sin\theta)$, you might like to use the identity $\cos\theta = \sin((\pi/2) - \theta)$ and the corresponding identity for $\sin\theta$.]

3. (a) Show that

$$\int_0^1 \int_0^1 \int_0^1 \frac{1}{1 - xyz}\, dx\, dy\, dz = \sum_{n=1}^{\infty} \frac{1}{n^3}$$

(Nobody has ever been able to find the exact value of the sum of this series.)

(b) Show that

$$\int_0^1 \int_0^1 \int_0^1 \frac{1}{1 + xyz}\, dx\, dy\, dz = \sum_{n=1}^{\infty} \frac{(-1)^{n-1}}{n^3}$$

and use this equation to evaluate the triple integral correct to two decimal places.

4. Find the simple closed curve C for which the value of the line integral

$$\int_C (y^3 - y)\, dx - 2x^3\, dy$$

is a maximum.

5. If $[\![x]\!]$ denotes the greatest integer in x, evaluate the integral

$$\iint_R [\![x + y]\!]\, dA$$

where $R = \{(x, y) \mid 1 \leq x \leq 3,\ 2 \leq y \leq 5\}$.

6. Evaluate the integral

$$\int_0^1 \int_0^1 e^{\max\{x^2, y^2\}}\, dy\, dx$$

7. Find the average value of the function $f(x) = \int_x^1 \cos(t^2)\, dt$ on the interval $[0, 1]$.

8. Among all planes that are tangent to the surface $xy^2z^2 = 1$, find the ones that are farthest from the origin.

9. If f is continuous, show that

$$\int_0^x \int_0^y \int_0^z f(t)\, dt\, dz\, dy = \frac{1}{2} \int_0^x (x - t)^2 f(t)\, dt$$

10. Show that

$$\int_0^\infty \frac{\arctan \pi x - \arctan x}{x}\, dx = \frac{\pi}{2} \ln \pi$$

by first expressing the integral as an iterated integral.

11. If **a**, **b**, and **c** are constant vectors, **r** is the position vector $x\,\mathbf{i} + y\,\mathbf{j} + z\,\mathbf{k}$, and E is given by the inequalities $0 \leqslant \mathbf{a} \cdot \mathbf{r} \leqslant \alpha$, $0 \leqslant \mathbf{b} \cdot \mathbf{r} \leqslant \beta$, $0 \leqslant \mathbf{c} \cdot \mathbf{r} \leqslant \gamma$, show that

$$\iiint_E (\mathbf{a} \cdot \mathbf{r})(\mathbf{b} \cdot \mathbf{r})(\mathbf{c} \cdot \mathbf{r})\, dV = \frac{(\alpha\beta\gamma)^2}{8\,|\mathbf{a} \cdot (\mathbf{b} \times \mathbf{c})|}$$

12. In this problem we find formulas for the volume enclosed by a hypersphere in n-dimensional space.
    (a) Use a double integral and trigonometric substitution to find the area of a circle with radius r.
    (b) Use a triple integral and trigonometric substitution to find the volume of a sphere with radius r.
    (c) Use a quadruple integral to find the hypervolume enclosed by the hypersphere $x^2 + y^2 + z^2 + w^2 = r^2$ in $\mathbb{R}^4$. (Use only trigonometric substitution and the reduction formulas for $\int \sin^n x\, dx$ or $\int \cos^n x\, dx$.)
    (d) Use an n-tuple integral to find the volume enclosed by a hypersphere of radius r in n-dimensional space $\mathbb{R}^n$. [*Hint*: The formulas for n even and for n odd are different.]

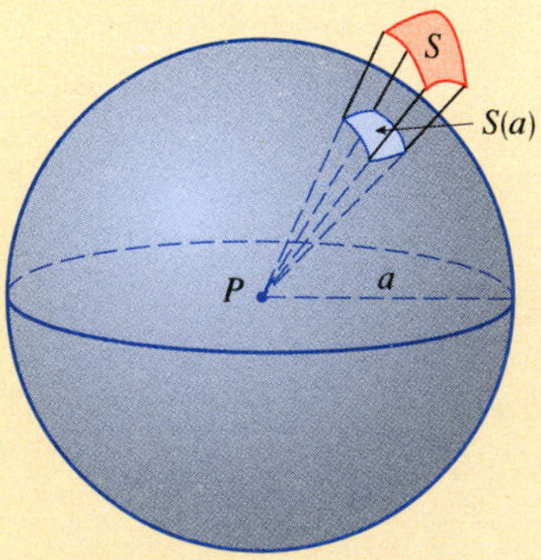

FIGURE FOR PROBLEM 13

13. Let S be a smooth parametric surface and let P be a point such that each line that starts at P intersects S at most once. The **solid angle** $\Omega(S)$ subtended by S at P is the set of lines starting at P and passing through S. Let $S(a)$ be the intersection of $\Omega(S)$ with the surface of the sphere with center P and radius a. Then the measure of the solid angle (in steradians) is defined to be

$$|\Omega(S)| = \frac{\text{area of } S(a)}{a^2}$$

Apply the Divergence Theorem to the part of $\Omega(S)$ between $S(a)$ and S to show that

$$|\Omega(S)| = \iint_S \frac{\mathbf{r} \cdot \mathbf{n}}{r^3}\, dS$$

where $\mathbf{r}$ is the radius vector from P to any point on S, $r = |\mathbf{r}|$, and the unit normal vector $\mathbf{n}$ is directed away from P.

This shows that the definition of the measure of a solid angle is independent of the radius a of the sphere. Thus the measure of the solid angle is equal to the area subtended on a *unit* sphere. (Note the analogy with the definition of radian measure.) The total solid angle subtended by a sphere at its center is thus 4π steradians.

14. (a) Investigate the solid enclosed by the three cylinders $x^2 + y^2 = 1$, $x^2 + z^2 = 1$, and $y^2 + z^2 = 1$. Describe the solid and sketch it carefully. Then compute its volume.

CAS (b) Use a computer algebra system to draw the edges of the solid.

(c) What happens to the solid in part (a) if the radius of the first cylinder is different from 1?

15

SECOND-ORDER DIFFERENTIAL EQUATIONS

Among all the mathematical disciplines the theory of differential equations is the most important. It furnishes the explanation of all those elementary manifestations of nature which involve time.

SOPHUS LIE

The basic ideas of differential equations were explained in Section 9.1 and we learned how to solve separable equations and first-order linear equations in Sections 9.1 and 9.2. In this chapter we study second-order linear differential equations and learn how they can be applied to solve problems concerning the vibrations of springs and electric circuits. We will also see how infinite series can be used to solve differential equations.

15.1 SECOND-ORDER LINEAR EQUATIONS

A **second-order linear differential equation** has the form

(1) $$P(x)\frac{d^2y}{dx^2} + Q(x)\frac{dy}{dx} + R(x)y = G(x)$$

where P, Q, R, and G are continuous functions. We will see in Section 15.3 that equations of this type arise in the study of the motion of a spring as well as in electric circuits.

In this section we study the case where $G(x) = 0$, for all x, in Equation 1. Such equations are called **homogeneous** linear equations. Thus, the form of a second-order linear homogeneous differential equation is

(2) $$P(x)\frac{d^2y}{dx^2} + Q(x)\frac{dy}{dx} + R(x)y = 0$$

If $G(x) \neq 0$ for some x, Equation 1 is **nonhomogeneous** and is discussed in Section 15.2.

Two basic facts enable us to solve homogeneous linear equations. The first of these says that if we know two solutions y_1 and y_2 of such an equation, then the **linear combination** $y = c_1y_1 + c_2y_2$ is also a solution.

(3) THEOREM If $y_1(x)$ and $y_2(x)$ are both solutions of the linear homogeneous equation (2) and c_1 and c_2 are any constants, then the function

$$y(x) = c_1y_1(x) + c_2y_2(x)$$

is also a solution of Equation 2.

PROOF Since y_1 and y_2 are solutions of Equation 2, we have

$$P(x)y_1'' + Q(x)y_1' + R(x)y_1 = 0$$

and

$$P(x)y_2'' + Q(x)y_2' + R(x)y_2 = 0$$

Therefore, using the basic rules for differentiation, we have

$$\begin{aligned} P(x)y'' + Q(x)y' + R(x)y &\\ &= P(x)(c_1y_1 + c_2y_2)'' + Q(x)(c_1y_1 + c_2y_2)' + R(x)(c_1y_1 + c_2y_2) \\ &= P(x)(c_1y_1'' + c_2y_2'') + Q(x)(c_1y_1' + c_2y_2') + R(x)(c_1y_1 + c_2y_2) \\ &= c_1[P(x)y_1'' + Q(x)y_1' + R(x)y_1] + c_2[P(x)y_2'' + Q(x)y_2' + R(x)y_2] \\ &= c_1(0) + c_2(0) = 0 \end{aligned}$$

Thus, $y = c_1y_1 + c_2y_2$ is a solution of Equation 2. □

The other fact we need is given by the following theorem, which is proved in more advanced courses. It says that the general solution is a linear combination of two **linearly independent** solutions y_1 and y_2. This means that neither y_1 nor y_2 is a constant multiple of the other. For instance, the functions $f(x) = x^2$ and $g(x) = 5x^2$ are linearly dependent, but $f(x) = e^x$ and $g(x) = xe^x$ are linearly independent.

(4) THEOREM If y_1 and y_2 are linearly independent solutions of Equation 2, then the general solution is given by

$$y(x) = c_1y_1(x) + c_2y_2(x)$$

where c_1 and c_2 are arbitrary constants.

Theorem 4 is very useful because it says that if we know *two* particular linearly independent solutions, then we know *every* solution.

In general, it is not easy to discover particular solutions to a second-order linear equation. But it is always possible to do so if the coefficient functions P, Q, and R are constant functions, that is, if the differential equation has the form

$$ay'' + by' + cy = 0 \tag{5}$$

where a, b, and c are constants.

It's not hard to think of some likely candidates for particular solutions of Equation 5 if we state the equation verbally. We are looking for a function y such that a constant times its second derivative y'' plus another constant times y' plus a third constant times y is equal to 0. We know that the exponential function $y = e^{rx}$ (where r is a constant) has the property that its derivative is a constant multiple of itself: $y' = re^{rx}$. Furthermore, $y'' = r^2e^{rx}$. If we substitute these expressions into Equation 5, we see that $y = e^{rx}$ is a solution if

$$ar^2e^{rx} + bre^{rx} + ce^{rx} = 0$$

or

$$(ar^2 + br + c)e^{rx} = 0$$

But e^{rx} is never 0. Therefore, $y = e^{rx}$ is a solution of Equation 5 if r is a root of the equation

(6) $$ar^2 + br + c = 0$$

Equation 6 is called the **auxiliary equation** (or **characteristic equation**) of the differential equation $ay'' + by' + cy = 0$. Notice that it is an algebraic equation that is obtained from the differential equation by replacing y'' by r^2, y' by r, and y by 1.

Sometimes the roots r_1 and r_2 of the auxiliary equation can be found by factoring. In other cases they are found by using the quadratic formula:

(7) $$r_1 = \frac{-b + \sqrt{b^2 - 4ac}}{2a} \qquad r_2 = \frac{-b - \sqrt{b^2 - 4ac}}{2a}$$

We distinguish three cases according to the sign of the discriminant $b^2 - 4ac$.

CASE I $b^2 - 4ac > 0$

In this case the roots r_1 and r_2 of the auxiliary equation are real and distinct, so $y_1 = e^{r_1 x}$ and $y_2 = e^{r_2 x}$ are two linearly independent solutions of Equation 5. (Note that $e^{r_2 x}$ is not a constant multiple of $e^{r_1 x}$.) Thus, by Theorem 4, we have the following:

(8) If the roots r_1 and r_2 of the auxiliary equation $ar^2 + br + c = 0$ are real and unequal, then the general solution of $ay'' + by' + cy = 0$ is

$$y = c_1 e^{r_1 x} + c_2 e^{r_2 x}$$

EXAMPLE 1 Solve the equation $y'' + y' - 6y = 0$.

In Figure 1 the graphs of the basic solutions $f(x) = e^{2x}$ and $g(x) = e^{-3x}$ of the differential equation in Example 1 are shown in black and red, respectively. Some of the other solutions, linear combinations of f and g, are shown in blue.

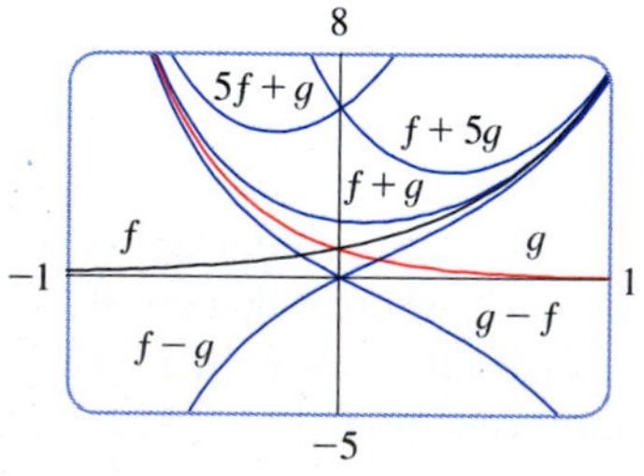

FIGURE 1

SOLUTION The auxiliary equation is

$$r^2 + r - 6 = (r - 2)(r + 3) = 0$$

whose roots are $r = 2, -3$. Therefore, by (8) the general solution of the given differential equation is

$$y = c_1 e^{2x} + c_2 e^{-3x}$$ ■

EXAMPLE 2 Solve $3\dfrac{d^2y}{dx^2} + \dfrac{dy}{dx} - y = 0$.

SOLUTION To solve the auxiliary equation $3r^2 + r - 1 = 0$ we use the quadratic formula:

$$r = \frac{-1 \pm \sqrt{13}}{6}$$

Since the roots are real and distinct, the general solution is

$$y = c_1 e^{(-1+\sqrt{13})x/6} + c_2 e^{(-1-\sqrt{13})x/6}$$ ■

CASE II $b^2 - 4ac = 0$

In this case $r_1 = r_2$; that is, the roots of the auxiliary equation are real and equal. Let's denote by r the common value of r_1 and r_2. Then, from Equations 7, we have

$$r = -\frac{b}{2a} \quad \text{so} \quad 2ar + b = 0 \tag{9}$$

We know that $y_1 = e^{rx}$ is one solution of Equation 5. We now verify that $y_2 = xe^{rx}$ is also a solution:

$$\begin{aligned} ay_2'' + by_2' + cy_2 &= a(2re^{rx} + r^2xe^{rx}) + b(e^{rx} + rxe^{rx}) + cxe^{rx} \\ &= (2ar + b)e^{rx} + (ar^2 + br + c)xe^{rx} \\ &= 0(e^{rx}) + 0(xe^{rx}) = 0 \end{aligned}$$

The first term is 0 by Equations 9; the second term is 0 because r is a root of the auxiliary equation. Since $y_1 = e^{rx}$ and $y_2 = xe^{rx}$ are linearly independent solutions, Theorem 4 provides us with the general solution.

(10) If the auxiliary equation $ar^2 + br + c = 0$ has only one real root r, then the general solution of $ay'' + by' + cy = 0$ is

$$y = c_1e^{rx} + c_2xe^{rx}$$

EXAMPLE 3 Solve the equation $4y'' + 12y' + 9y = 0$.

Figure 2 shows the basic solutions $f(x) = e^{-3x/2}$ and $g(x) = xe^{-3x/2}$ in Example 3 and some other members of the family of solutions. Notice that all of them approach 0 as $x \to \infty$.

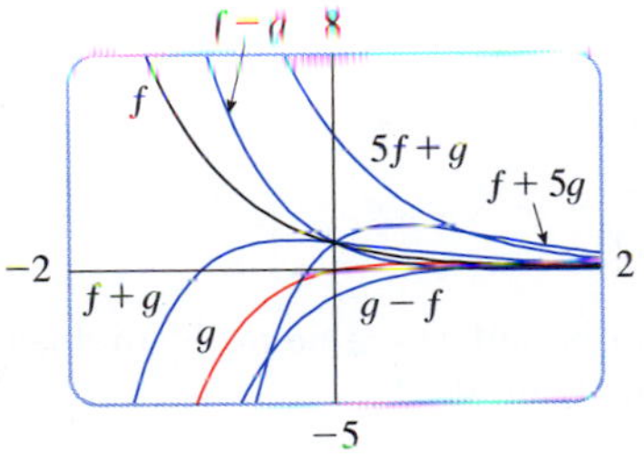

FIGURE 2

SOLUTION The auxiliary equation $4r^2 + 12r + 9 = 0$ can be factored as

$$(2r + 3)^2 = 0$$

so the only root is $r = -\frac{3}{2}$. By (10) the general solution is

$$y = c_1e^{-3x/2} + c_2xe^{-3x/2}$$

■

CASE III $b^2 - 4ac < 0$

In this case the roots r_1 and r_2 of the auxiliary equation are complex numbers. (See Appendix H for information about complex numbers.) We can write

$$r_1 = \alpha + i\beta \qquad r_2 = \alpha - i\beta$$

where α and β are real numbers. [In fact, $\alpha = -b/(2a)$, $\beta = \sqrt{4ac - b^2}/(2a)$.] Then, using Euler's equation

$$e^{i\theta} = \cos\theta + i\sin\theta$$

from Appendix H, we write the solution of the differential equation as

$$\begin{aligned} y &= C_1e^{r_1x} + C_2e^{r_2x} \\ &= C_1e^{(\alpha+i\beta)x} + C_2e^{(\alpha-i\beta)x} \\ &= C_1e^{\alpha x}(\cos\beta x + i\sin\beta x) + C_2e^{\alpha x}(\cos\beta x - i\sin\beta x) \\ &= e^{\alpha x}[(C_1 + C_2)\cos\beta x + i(C_1 - C_2)\sin\beta x] \\ &= e^{\alpha x}(c_1\cos\beta x + c_2\sin\beta x) \end{aligned}$$

where $c_1 = C_1 + C_2$, $c_2 = i(C_1 - C_2)$. This gives all solutions (real or complex) of the differential equation. The solutions are real when the constants c_1 and c_2 are real. We summarize the discussion as follows:

(11) If the roots of the auxiliary equation $ar^2 + br + c = 0$ are the complex numbers $r_1 = \alpha + i\beta$, $r_2 = \alpha - i\beta$, then the general solution of $ay'' + by' + cy = 0$ is

$$y = e^{\alpha x}(c_1 \cos \beta x + c_2 \sin \beta x)$$

Figure 3 shows the graphs of the solutions in Example 4, $f(x) = e^{3x} \cos 2x$ and $g(x) = e^{3x} \sin 2x$, together with some linear combinations. All solutions approach 0 as $x \to -\infty$.

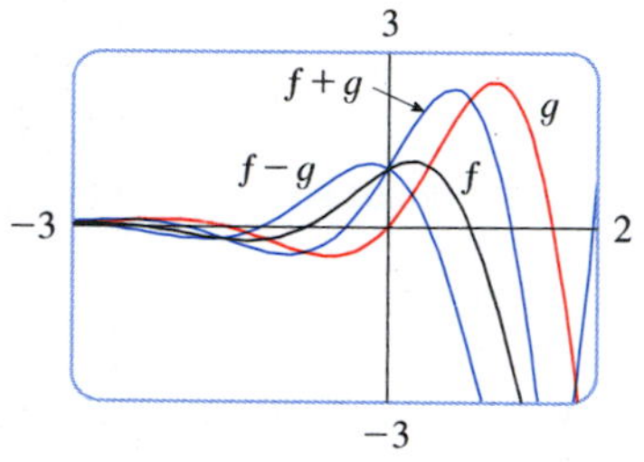

FIGURE 3

EXAMPLE 4 Solve the equation $y'' - 6y' + 13y = 0$.

SOLUTION The auxiliary equation is $r^2 - 6r + 13 = 0$. By the quadratic formula, the roots are

$$r = \frac{6 \pm \sqrt{36 - 52}}{2} = \frac{6 \pm \sqrt{-16}}{2} = 3 \pm 2i$$

By (11) the general solution of the differential equation is

$$y = e^{3x}(c_1 \cos 2x + c_2 \sin 2x)$$ ■

INITIAL-VALUE AND BOUNDARY-VALUE PROBLEMS

An **initial-value problem** for the second-order Equation 1 or 2 consists of finding a solution y of the differential equation that also satisfies initial conditions of the form

$$y(x_0) = y_0 \qquad y'(x_0) = y_1$$

where y_0 and y_1 are given constants. If P, Q, R, and G are continuous on an interval and $P(x) \neq 0$ there, then a theorem found in more advanced books guarantees the existence and uniqueness of a solution to this initial-value problem. Examples 5 and 6 illustrate the technique for solving such a problem.

A **boundary-value problem** for Equation 1 consists of finding a solution y of the differential equation that also satisfies boundary conditions of the form

$$y(x_0) = y_0 \qquad y(x_1) = y_1$$

In contrast with the situation for initial-value problems, a boundary-value problem does not always have a solution. The method is illustrated in Example 7.

EXAMPLE 5 Solve the initial-value problem

$$y'' + y' - 6y = 0 \qquad y(0) = 1 \qquad y'(0) = 0$$

SOLUTION From Example 1 we know that the general solution of the differential equation is

$$y(x) = c_1 e^{2x} + c_2 e^{-3x}$$

Differentiating this solution, we get

$$y'(x) = 2c_1 e^{2x} - 3c_2 e^{-3x}$$

Figure 4 shows the graph of the solution of the initial-value problem in Example 5. Compare with Figure 1.

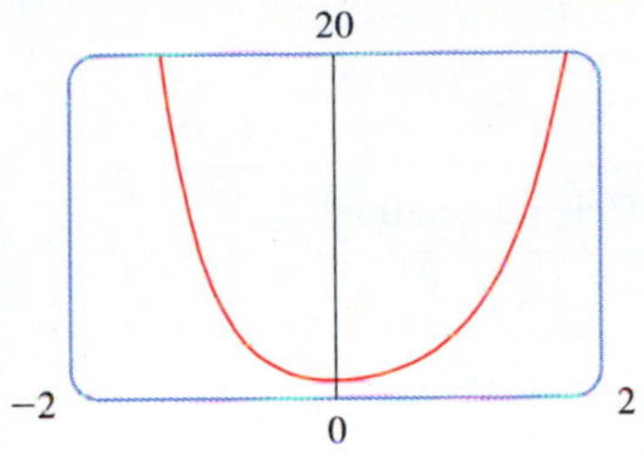

FIGURE 4

To satisfy the initial conditions we require that

$$y(0) = c_1 + c_2 = 1 \qquad (12)$$

$$y'(0) = 2c_1 - 3c_2 = 0 \qquad (13)$$

From (13) we have $c_2 = \frac{2}{3}c_1$ and so (12) gives

$$c_1 + \tfrac{2}{3}c_1 = 1 \qquad c_1 = \tfrac{3}{5} \qquad c_2 = \tfrac{2}{5}$$

Thus, the required solution of the initial-value problem is

$$y = \tfrac{3}{5}e^{2x} + \tfrac{2}{5}e^{-3x}$$ ■

EXAMPLE 6 Solve the initial-value problem

$$y'' + y = 0 \qquad y(0) = 2 \qquad y'(0) = 3$$

SOLUTION The auxiliary equation is $r^2 + 1 = 0$, or $r^2 = -1$, whose roots are $\pm i$. Thus $\alpha = 0$, $\beta = 1$, and since $e^{0x} = 1$, the general solution is

$$y(x) = c_1 \cos x + c_2 \sin x$$

Since

$$y'(x) = -c_1 \sin x + c_2 \cos x$$

the initial conditions become

$$y(0) = c_1 = 2 \qquad y'(0) = c_2 = 3$$

Therefore, the solution of the initial-value problem is

$$y(x) = 2\cos x + 3\sin x$$ ■

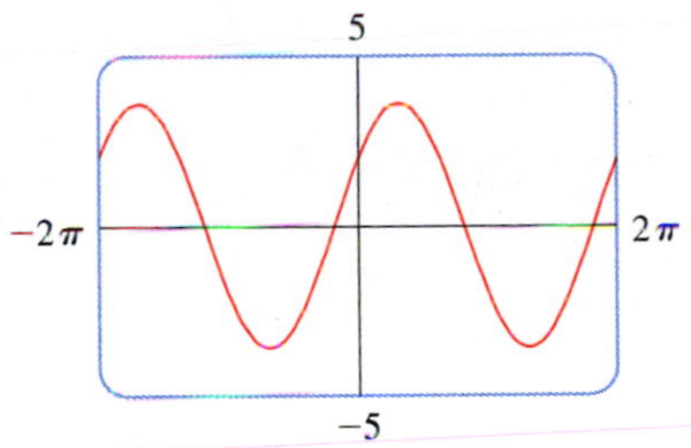

FIGURE 5

The solution to Example 6 is graphed in Figure 5. It appears to be a shifted sine curve and, indeed, you can verify that another way of writing the solution is $y = \sqrt{13}\sin(x + \phi)$ where $\tan\phi = \frac{2}{3}$

EXAMPLE 7 Solve the boundary-value problem

$$y'' + 2y' + y = 0 \qquad y(0) = 1 \qquad y(1) = 3$$

SOLUTION The auxiliary equation is

$$r^2 + 2r + 1 = 0 \qquad \text{or} \qquad (r + 1)^2 = 0$$

whose only root is $r = -1$. Therefore, the general solution is

$$y(x) = c_1 e^{-x} + c_2 x e^{-x}$$

Figure 6 shows the graph of the solution of the boundary-value problem in Example 7.

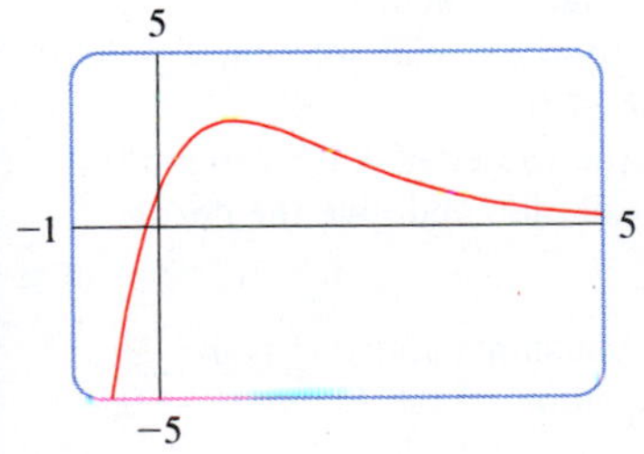

FIGURE 6

The boundary conditions are satisfied if

$$y(0) = c_1 = 1$$

$$y(1) = c_1 e^{-1} + c_2 e^{-1} = 3$$

The first condition gives $c_1 = 1$, so the second condition becomes

$$e^{-1} + c_2 e^{-1} = 3$$

Solving this equation for c_2 by first multiplying through by e, we get

$$1 + c_2 = 3e \qquad c_2 = 3e - 1$$

Thus, the solution of the boundary-value problem is

$$y = e^{-x} + (3e - 1)xe^{-x}$$ ■

SUMMARY: SOLUTIONS OF $ay'' + by' + c = 0$

Roots of $ar^2 + br + c = 0$	General solution
r_1, r_2 real and distinct	$y = c_1e^{r_1x} + c_2e^{r_2x}$
$r_1 = r_2 = r$	$y = c_1e^{rx} + c_2xe^{rx}$
r_1, r_2 complex: $\alpha \pm i\beta$	$y = e^{\alpha x}(c_1 \cos \beta x + c_2 \sin \beta x)$

EXERCISES 15.1

1–17 ■ Solve the differential equation.

1. $y'' - 3y' + 2y = 0$

2. $y'' - y' = 0$

3. $3y'' - 8y' - 3y = 0$

4. $y'' + 9y' + 20y = 0$

5. $y'' + 2y' + 10y = 0$

6. $y'' + 10y' + 41y = 0$

7. $y'' = y$

8. $9y'' - 30y' + 25y = 0$

9. $y'' + 25y = 0$

10. $y'' - 4y' + 13y = 0$

11. $2y'' + y' = 0$

12. $y'' - 2y' - 4y = 0$

13. $y'' - y' + 2y = 0$

14. $y'' = -5y$

15. $\dfrac{d^2y}{dx^2} + 2\dfrac{dy}{dx} - y = 0$

16. $2\dfrac{d^2y}{dx^2} + 5\dfrac{dy}{dx} + y = 0$

17. $2\dfrac{d^2y}{dx^2} + \dfrac{dy}{dx} + 3y = 0$

18–20 ■ Graph the two basic solutions of the differential equation and several other solutions. What features do the solutions have in common?

18. $6\dfrac{d^2y}{dx^2} - \dfrac{dy}{dx} - 2y = 0$

19. $\dfrac{d^2y}{dx^2} - 8\dfrac{dy}{dx} + 16y = 0$

20. $\dfrac{d^2y}{dx^2} - 2\dfrac{dy}{dx} + 5y = 0$

21–28 ■ Solve the initial-value problem.

21. $y'' + 3y' - 4y = 0, \quad y(0) = 2, \quad y'(0) = -3$

22. $y'' - 4y = 0, \quad y(0) = 1, \quad y'(0) = 0$

23. $y'' - 2y' + 2y = 0, \quad y(0) = 1, \quad y'(0) = 2$

24. $y'' + 4y' + 6y = 0, \quad y(0) = 2, \quad y'(0) = 4$

25. $y'' - 2y' - 3y = 0, \quad y(1) = 3, \quad y'(1) = 1$

26. $y'' - 2y' + y = 0, \quad y(2) = 0, \quad y'(2) = 1$

27. $y'' + 9y = 0, \quad y(\pi/3) = 0, \quad y'(\pi/3) = 1$

28. $y'' + 4y = 0, \quad y(\pi/6) = 1, \quad y'(\pi/6) = 0$

29–36 ■ Solve each boundary-value problem if possible.

29. $y'' + 4y' + 4y = 0, \quad y(0) = 0, \quad y(1) = 3$

30. $y'' + 5y' - 6y = 0, \quad y(0) = 0, \quad y(2) = 1$

31. $y'' + y = 0, \quad y(0) = 1, \quad y(\pi) = 0$

32. $y'' + 9y = 0, \quad y(0) = 1, \quad y(\pi/2) = 0$

33. $y'' - y' - 2y = 0, \quad y(-1) = 1, \quad y(1) = 0$

34. $y'' + 4y' + 3y = 0, \quad y(1) = 0, \quad y(3) = 2$

35. $y'' + 4y' + 13y = 0, \quad y(0) = 2, \quad y(\pi/2) = 1$

36. $y'' + 2y' + 5y = 0, \quad y(0) = 1, \quad y(\pi) = 2$

37. (a) Show that the boundary-value problem $y'' + \lambda y = 0$, $y(0) = 0$, $y(L) = 0$ has only the trivial solution $y = 0$ for the cases $\lambda = 0$ and $\lambda < 0$.

(b) For the case $\lambda > 0$, find the values of λ for which this problem has a nontrivial solution and give the corresponding solution.

38. If a, b, and c are all positive constants and $y(x)$ is a solution of the differential equation $ay'' + by' + cy = 0$, show that $\lim_{x \to \infty} y(x) = 0$.

15.2 NONHOMOGENEOUS LINEAR EQUATIONS

In this section we learn how to solve second-order nonhomogeneous linear differential equations with constant coefficients, that is, equations of the form

$$ay'' + by' + cy = G(x) \tag{1}$$

where a, b, c are constants and G is a continuous function. The related homogeneous equation

$$ay'' + by' + cy = 0 \tag{2}$$

is called the **complementary equation** and plays an important role in the solution of the original nonhomogeneous equation (1).

(3) THEOREM The general solution of the nonhomogeneous differential equation (1) can be written as

$$y(x) = y_p(x) + y_c(x)$$

where y_p is a particular solution of Equation 1 and y_c is the general solution of the complementary Equation 2.

PROOF All we have to do is verify that if y is any solution of Equation 1, then $y - y_p$ is a solution of the complementary Equation 2. Indeed

$$\begin{aligned} a(y - y_p)'' + b(y - y_p)' + c(y - y_p) &= ay'' - ay_p'' + by' - by_p' + cy - cy_p \\ &= (ay'' + by' + cy) - (ay_p'' + by_p' + cy_p) \\ &= g(x) - g(x) = 0 \end{aligned}$$

□

We know from Section 15.1 how to solve the complementary equation. (Recall that the solution is $y_c = c_1 y_1 + c_2 y_2$, where y_1 and y_2 are linearly independent solutions of Equation 2.) Therefore, Theorem 3 says that we know the general solution of the nonhomogeneous equation as soon as we know a particular solution y_p. There are two methods for finding a particular solution. The method of undetermined coefficients is straightforward but works only for a restricted class of functions G. The method of variation of parameters works for every function G but is usually more difficult to apply in practice.

THE METHOD OF UNDETERMINED COEFFICIENTS

We first illustrate the method of undetermined coefficients for the equation

$$ay'' + by' + cy = G(x)$$

where $G(x)$ is a polynomial. It is reasonable to guess that there is a particular solution y_p that is a polynomial of the same degree as G because if y is a polynomial, then $ay'' + by' + cy$ is also a polynomial. We therefore substitute $y_p(x) =$ a polynomial (of the same degree as G) into the differential equation and determine the coefficients.

EXAMPLE 1 Solve the equation $y'' + y' - 2y = x^2$.

SOLUTION The auxiliary equation of $y'' + y' - 2y = 0$ is

$$r^2 + r - 2 = (r - 1)(r + 2) = 0$$

with roots $r = 1, -2$. So the solution of the complementary equation is

$$y_c = c_1 e^x + c_2 e^{-2x}$$

Since $G(x) = x^2$ is a polynomial of degree 2, we seek a particular solution of the form

$$y_p(x) = Ax^2 + Bx + C$$

Figure 1 shows four solutions of the differential equation in Example 1 in terms of the particular solution y_p and the functions $f(x) = e^x$ and $g(x) = e^{-2x}$.

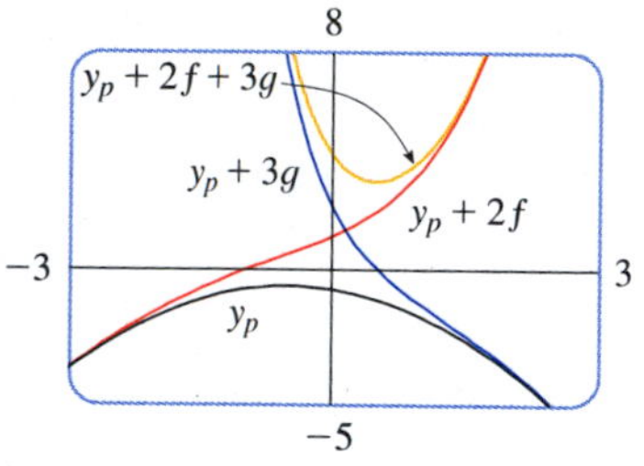

FIGURE 1

Then $y_p' = 2Ax + B$ and $y_p'' = 2A$ so, substituting into the given differential equation, we have

$$(2A) + (2Ax + B) - 2(Ax^2 + Bx + C) = x^2$$

or

$$-2Ax^2 + (2A - 2B)x + (2A + B - 2C) = x^2$$

Polynomials are equal when their coefficients are equal. Thus

$$-2A = 1 \qquad 2A - 2B = 0 \qquad 2A + B - 2C = 0$$

The solution of this system of equations is

$$A = -\tfrac{1}{2} \qquad B = -\tfrac{1}{2} \qquad C = -\tfrac{3}{4}$$

A particular solution is therefore

$$y_p(x) = -\tfrac{1}{2}x^2 - \tfrac{1}{2}x - \tfrac{3}{4}$$

and, by Theorem 3, the general solution is

$$y = y_c + y_p = c_1 e^x + c_2 e^{-2x} - \tfrac{1}{2}x^2 - \tfrac{1}{2}x - \tfrac{3}{4}$$ ■

If $G(x)$ (the right side of Equation 1) is of the form Ce^{kx}, where C and k are constants, then we take as a trial solution a function of the same form, $y_p(x) = Ae^{kx}$, because the derivatives of e^{kx} are constant multiples of e^{kx}.

Figure 2 shows solutions of the differential equation in Example 2 in terms of y_p and the functions $f(x) = \cos 2x$ and $g(x) = \sin 2x$. Notice that all solutions approach ∞ as $x \to \infty$ and all solutions resemble sine functions when x is negative.

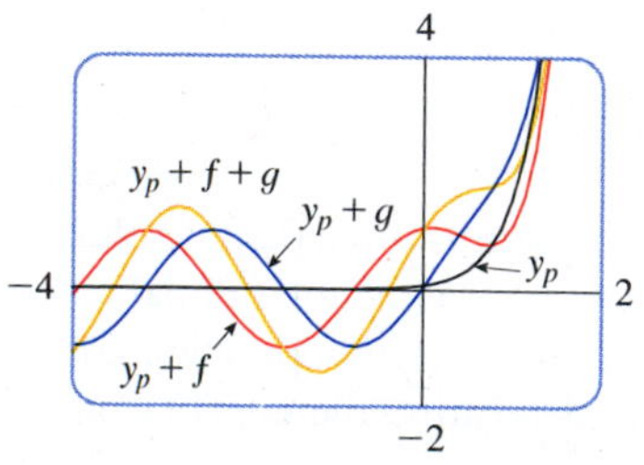

FIGURE 2

EXAMPLE 2 Solve $y'' + 4y = e^{3x}$.

SOLUTION The auxiliary equation is $r^2 + 4 = 0$ with roots $\pm 2i$, so the solution of the complementary equation is

$$y_c(x) = c_1 \cos 2x + c_2 \sin 2x$$

For a particular solution we try $y_p(x) = Ae^{3x}$. Then $y_p' = 3Ae^{3x}$ and $y_p'' = 9Ae^{3x}$. Substituting into the differential equation, we have

$$9Ae^{3x} + 4(Ae^{3x}) = e^{3x}$$

so $13Ae^{3x} = e^{3x}$ and $A = \frac{1}{13}$. Thus, a particular solution is

$$y_p(x) = \tfrac{1}{13}e^{3x}$$

and the general solution is

$$y(x) = c_1 \cos 2x + c_2 \sin 2x + \tfrac{1}{13}e^{3x}$$ ■

If $G(x)$ is either $C\cos kx$ or $C\sin kx$, then, because of the rules for differentiating the sine and cosine functions, we take as a trial particular solution a function of the form

$$y_p(x) = A\cos kx + B\sin kx$$

EXAMPLE 3 Solve $y'' + y' - 2y = \sin x$.

SOLUTION We try a particular solution

$$y_p(x) = A\cos x + B\sin x$$

Then $\qquad y_p' = -A\sin x + B\cos x \qquad y_p'' = -A\cos x - B\sin x$

so substitution in the differential equation gives

$$(-A\cos x - B\sin x) + (-A\sin x + B\cos x) - 2(A\cos x + B\sin x) = \sin x$$

or

$$(-3A + B)\cos x + (-A - 3B)\sin x = \sin x$$

This is true if

$$-3A + B = 0 \qquad \text{and} \qquad -A - 3B = 1$$

The solution of this system is

$$A = -\tfrac{1}{10} \qquad B = -\tfrac{3}{10}$$

so a particular solution is

$$y_p(x) = -\tfrac{1}{10}\cos x - \tfrac{3}{10}\sin x$$

In Example 1 we determined that the solution of the complementary equation is $y_c = c_1e^x + c_2e^{-2x}$. Thus, the general solution of the given equation is

$$y(x) = c_1e^x + c_2e^{-2x} - \tfrac{1}{10}(\cos x + 3\sin x)$$ ■

If $G(x)$ is a product of functions of the preceding types, then we take the trial solution to be a product of functions of the same type. For instance, in solving the differential equation

$$y'' + 2y' + 4y = x\cos 3x$$

we would try

$$y_p(x) = (Ax + B)\cos 3x + (Cx + D)\sin 3x$$

If $G(x)$ is a sum of functions of these types, we use the easily verified *principle of superposition*, which says that if y_{p_1} and y_{p_2} are solutions of

$$ay'' + by' + cy = G_1(x) \qquad ay'' + by' + cy = G_2(x)$$

respectively, then $y_{p_1} + y_{p_2}$ is a solution of

$$ay'' + by' + cy = G_1(x) + G_2(x)$$

EXAMPLE 4 Solve $y'' - 4y = xe^x + \cos 2x$.

SOLUTION The auxiliary equation is $r^2 - 4 = 0$ with roots ± 2, so the solution of the complementary equation is $y_c(x) = c_1 e^{2x} + c_2 e^{-2x}$. For the equation $y'' - 4y = xe^x$ we try

$$y_{p_1}(x) = (Ax + B)e^x$$

Then $y'_{p_1} = (Ax + A + B)e^x$, $y''_{p_1} = (Ax + 2A + B)e^x$, so substitution in the equation gives

$$(Ax + 2A + B)e^x - 4(Ax + B)e^x = xe^x$$

or

$$(-3Ax + 2A - 3B)e^x = xe^x$$

Thus, $-3A = 1$ and $2A - 3B = 0$, so $A = -\frac{1}{3}$, $B = -\frac{2}{9}$, and

$$y_{p_1}(x) = \left(-\tfrac{1}{3}x - \tfrac{2}{9}\right)e^x$$

In Figure 3 we show the particular solution y_p of the differential equation in Example 4. The other solutions are given in terms of $f(x) = e^{2x}$ and $g(x) = e^{-2x}$.

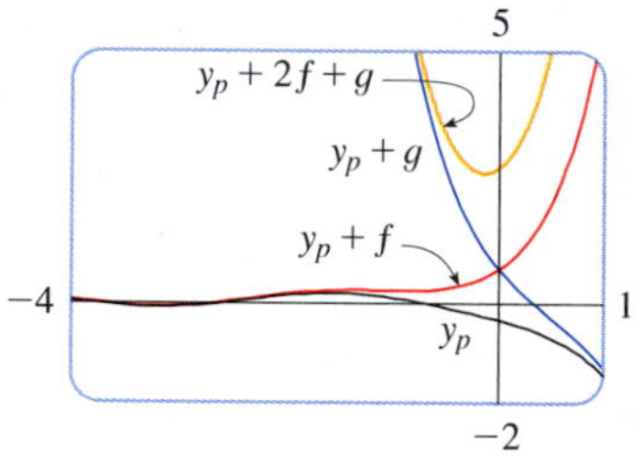

FIGURE 3

For the equation $y'' - 4y = \cos 2x$, we try

$$y_{p_2}(x) = C\cos 2x + D\sin 2x$$

Substitution gives

$$-4C\cos 2x - 4D\sin 2x - 4(C\cos 2x + D\sin 2x) = \cos 2x$$

or

$$-8C\cos 2x - 8D\sin 2x = \cos 2x$$

Therefore, $-8C = 1$, $-8D = 0$, and

$$y_{p_2}(x) = -\tfrac{1}{8}\cos 2x$$

By the superposition principle, the general solution is

$$y = y_c + y_{p_1} + y_{p_2} = c_1 e^{2x} + c_2 e^{-2x} - \left(\tfrac{1}{3}x + \tfrac{2}{9}\right)e^x - \tfrac{1}{8}\cos 2x \qquad \blacksquare$$

Finally we note that the recommended trial solution y_p sometimes turns out to be a solution of the complementary equation and therefore cannot be a solution of the nonhomogeneous equation. In such cases we multiply the recommended trial solution by x (or by x^2 if necessary) so that no term in $y_p(x)$ is a solution of the complementary equation.

EXAMPLE 5 Solve $y'' + y = \sin x$.

SOLUTION The auxiliary equation is $r^2 + 1 = 0$ with roots $\pm i$, so the solution of the complementary equation is

$$y_c(x) = c_1\cos x + c_2\sin x$$

Ordinarily, we would use the trial solution

$$y_p(x) = A\cos x + B\sin x$$

but we observe that it is a solution of the complementary equation, so instead we try

$$y_p(x) = Ax\cos x + Bx\sin x$$

The graphs of four solutions of the differential equation in Example 5 are shown in Figure 4.

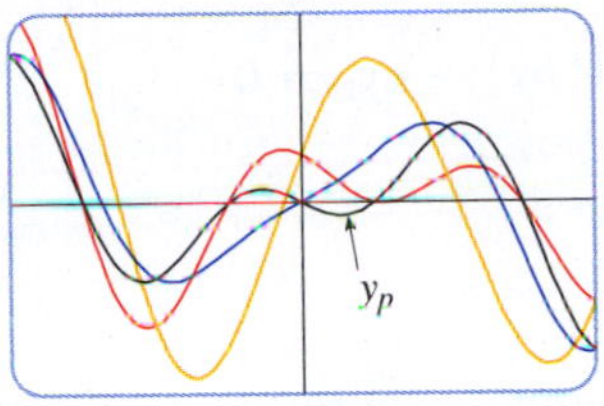

FIGURE 4

Then $$y_p'(x) = A\cos x - Ax\sin x + B\sin x + Bx\cos x$$

$$y_p''(x) = -2A\sin x - Ax\cos x + 2B\cos x - Bx\sin x$$

Substitution in the differential equation gives

$$y_p'' + y_p = -2A\sin x + 2B\cos x = \sin x$$

so $A = -\frac{1}{2}$, $B = 0$, and

$$y_p(x) = -\tfrac{1}{2}x\cos x$$

The general solution is

$$y(x) = c_1\cos x + c_2\sin x - \tfrac{1}{2}x\cos x$$

■

SUMMARY OF UNDETERMINED COEFFICIENTS

$G(x) =$	First try $y_p =$
$C_1x^n + \cdots + C_1x + C_0$ Ce^{kx} $C\cos kx + D\sin kx$	$A_1x^n + \cdots + A_1x + A_0$ Ae^{kx} $A\cos kx + B\sin kx$
Modification: If any term of y_p is a solution of the complementary equation, multiply y_p by x (or by x^2 if necessary).	

THE METHOD OF VARIATION OF PARAMETERS

Suppose we have already solved the homogeneous equation $ay'' + by' + cy = 0$ and written the solution as

(4) $$y(x) = c_1y_1(x) + c_2y_2(x)$$

where y_1 and y_2 are linearly independent solutions. Let's replace the constants (or parameters) c_1 and c_2 in Equation 4 by arbitrary functions $u_1(x)$ and $u_2(x)$. We look for a particular solution of the nonhomogeneous equation $ay'' + by' + cy = G(x)$ of the form

(5) $$y_p(x) = u_1(x)y_1(x) + u_2(x)y_2(x)$$

(This method is called **variation of parameters** because we have varied the parameters c_1 and c_2 to make them functions.) Differentiating Equation 5, we get

(6) $$y_p' = (u_1'y_1 + u_2'y_2) + (u_1y_1' + u_2y_2')$$

Since u_1 and u_2 are arbitrary functions, we can impose two conditions on them. One condition is that y_p is a solution of the differential equation; we can choose the other condition so as to simplify our calculations. In view of the expression in Equation 6, let us impose the condition that

(7) $$u_1'y_1 + u_2'y_2 = 0$$

Then $$y_p'' = u_1'y_1' + u_2'y_2' + u_1y_1'' + u_2y_2''$$

Substituting in the differential equation, we get

$$a(u_1'y_1' + u_2'y_2' + u_1y_1'' + u_2y_2'') + b(u_1y_1' + u_2y_2') + c(u_1y_1 + u_2y_2) = G$$

or

(8) $$u_1(ay_1'' + by_1' + cy_1) + u_2(ay_2'' + by_2' + cy_2) + a(u_1' y_1' + u_2' y_2') = G$$

But y_1 and y_2 are solutions of the complementary equation, so

$$ay_1'' + by_1' + cy_1 = 0 \qquad \text{and} \qquad ay_2'' + by_2' + cy_2 = 0$$

and Equation 8 simplifies to

(9) $$a(u_1' y_1' + u_2' y_2') = G$$

Equations 7 and 9 form a system of two equations in the unknown functions u_1' and u_2'. After solving this system we may be able to integrate to find u_1 and u_2 and then the particular solution is given by Equation 5.

EXAMPLE 6 Solve the equation $y'' + y = \tan x$, $0 < x < \pi/2$.

SOLUTION The auxiliary equation is $r^2 + 1 = 0$ with roots $\pm i$, so the solution of $y'' + y = 0$ is $c_1 \sin x + c_2 \cos x$. Using variation of parameters, we seek a solution of the form

$$y_p(x) = u_1(x) \sin x + u_2(x) \cos x$$

Then $$y_p' = (u_1' \sin x + u_2' \cos x) + (u_1 \cos x - u_2 \sin x)$$

Set

(10) $$u_1' \sin x + u_2' \cos x = 0$$

Then $$y_p'' = u_1' \cos x - u_2' \sin x - u_1 \sin x - u_2 \cos x$$

For y_p to be a solution we must have

(11) $$y_p'' + y_p = u_1' \cos x - u_2' \sin x = \tan x$$

Solving Equations 10 and 11, we get

$$u_1'(\sin^2 x + \cos^2 x) = \cos x \tan x$$

$$u_1' = \sin x \qquad u_1(x) = -\cos x$$

(We seek a particular solution, so we do not need a constant of integration here.) Then, from Equation 10, we obtain

Figure 5 shows four solutions of the differential equation in Example 6.

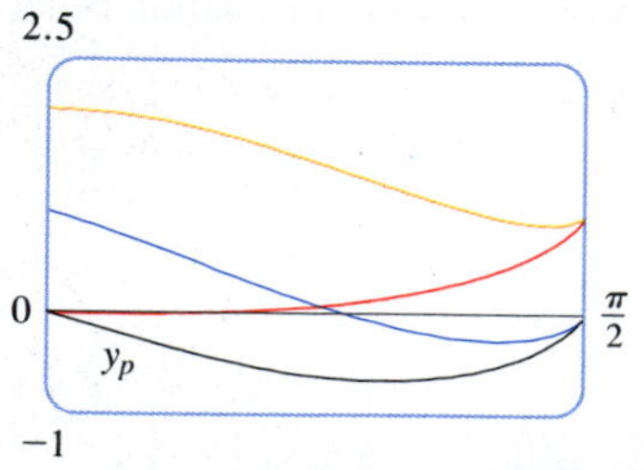

FIGURE 5

$$u_2' = -\frac{\sin x}{\cos x} u_1' = -\frac{\sin^2 x}{\cos x} = \frac{\cos^2 x - 1}{\cos x} = \cos x - \sec x$$

So $$u_2(x) = \sin x - \ln(\sec x + \tan x)$$

(Note that $\sec x + \tan x > 0$ for $0 < x < \pi/2$.) Therefore

$$y_p(x) = -\cos x \sin x + (\sin x - \ln(\sec x + \tan x)) \cos x$$
$$= -\cos x \ln(\sec x + \tan x)$$

and the general solution is

$$y(x) = c_1 \sin x + c_2 \cos x - \cos x \ln(\sec x + \tan x)$$ ■

EXERCISES 15.2

1–10 ■ Solve the differential equation or initial-value problem using the method of undetermined coefficients.

1. $y'' - y' - 6y = \cos 3x$
2. $y'' + 2y' + 2y = x^3 - 1$
3. $y'' - 4y' + 4y = e^{-x}$
4. $y'' - 7y' + 12y = \sin x - \cos x$
5. $y'' + 36y = 2x^2 - x$
6. $y'' + 2y' + y = xe^{-x}$
7. $y'' - 2y' + 5y = x + \sin 3x, \quad y(0) = 1, \quad y'(0) = 2$
8. $y'' + 2y = e^x \sin x, \quad y(0) = 1, \quad y'(0) = 1$
9. $y'' - y = xe^{3x}, \quad y(0) = 0, \quad y'(0) = 1$
10. $y'' - 3y' + 2y = 2x - e^{-2x}, \quad y(0) = 1, \quad y'(0) = 0$

11–12 ■ Graph the particular solution and several other solutions. What characteristics do these solutions have in common?

11. $4y'' + 5y' + y = e^x$
12. $2y'' + 3y' + y = 1 + \cos 2x$

13–16 ■ Write a trial solution for the method of undetermined coefficients. Do not determine the coefficients.

13. $y'' + 2y' + 6y = x^4 e^{2x}$
14. $y'' + 6y' + 2y = x^3 + e^x \sin 2x$
15. $y'' - 2y' + 2y = e^x \cos x$
16. $y'' + 3y' = 1 + xe^{-3x}$

17–20 ■ Solve the differential equation using (a) undetermined coefficients and (b) variation of parameters.

17. $y'' + 4y = x$
18. $y'' - 3y' + 2y = \sin x$
19. $y'' - 2y' + y = e^{2x}$
20. $y'' - y' = e^x$

21–26 ■ Solve the differential equation using the method of variation of parameters.

21. $y'' + y = \sec x, \ 0 < x < \pi/2$
22. $y'' + y = \cot x, \ 0 < x < \pi/2$
23. $y'' - 3y' + 2y = \dfrac{1}{1 + e^{-x}}$
24. $y'' + 3y' + 2y = \sin(e^x)$
25. $y'' - y = 1/x$
26. $y'' + 4y' + 4y = \dfrac{e^{-2x}}{x^3}$

15.3 APPLICATIONS OF SECOND-ORDER DIFFERENTIAL EQUATIONS

Second-order linear differential equations have a variety of applications in science and engineering. In this section we explore two of them: the vibration of springs and electric circuits.

VIBRATING SPRINGS

We consider the motion of an object with mass m at the end of a spring that is either vertical (as in Figure 1) or horizontal on a level surface (as in Figure 2).

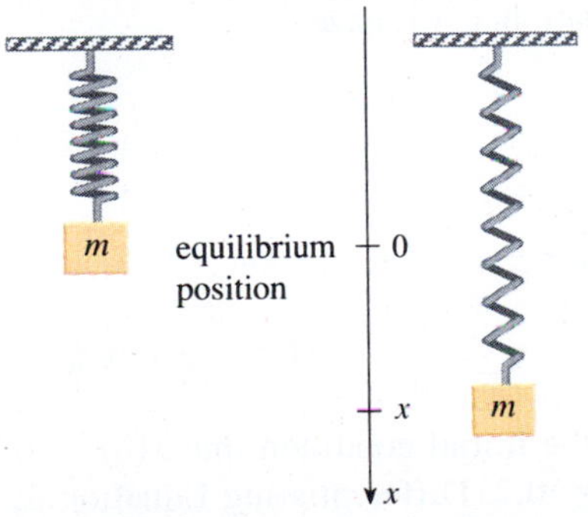

FIGURE 1

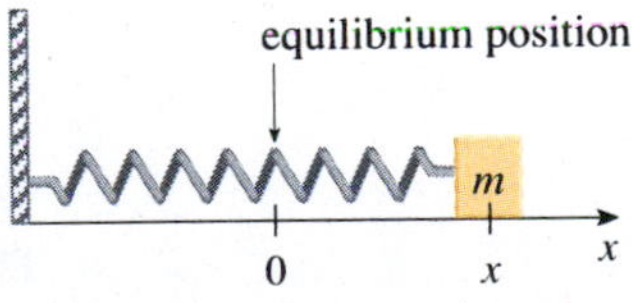

FIGURE 2

In Section 7.4 we discussed Hooke's Law, which says that if the spring is stretched (or compressed) x units from its natural length, then it exerts a force that is proportional to x:

$$\text{restoring force} = -kx$$

where k is a positive constant (called the *spring constant*). If we ignore any external resisting forces (due to air resistance or friction) then, by Newton's Second Law (force equals mass times acceleration), we have

(1) $$m\frac{d^2x}{dt^2} = -kx \qquad \text{or} \qquad m\frac{d^2x}{dt^2} + kx = 0$$

This is a second-order linear differential equation. Its auxiliary equation is $mr^2 + k = 0$ with roots $r = \pm\omega i$, where $\omega = \sqrt{k/m}$. Thus, the general solution is

$$x(t) = c_1 \cos \omega t + c_2 \sin \omega t$$

which can also be written as

$$x(t) = A\cos(\omega t + \delta)$$

where $$\omega = \sqrt{\frac{k}{m}} \qquad \text{(frequency)}$$

$$A = \sqrt{c_1^2 + c_2^2} \qquad \text{(amplitude)}$$

$$\cos\delta = \frac{c_1}{A} \qquad \sin\delta = -\frac{c_2}{A} \qquad (\delta \text{ is the phase angle})$$

(see Exercise 13). This type of motion is called **simple harmonic motion**.

EXAMPLE 1 A spring with a mass of 2 kg has natural length 0.5 m. A force of 25.6 N is required to maintain it stretched to a length of 0.7 m. If the spring is stretched to a length of 0.7 m and then released with initial velocity 0, find the position of the mass at any time t.

SOLUTION From Hooke's Law, the force required to stretch the spring is

$$k(0.2) = 25.6$$

so $k = 25.6/0.2 = 128$. Using this value of the spring constant k, together with $m = 2$ in Equation 1, we have

$$2\frac{d^2x}{dt^2} + 128x = 0$$

As in the earlier general discussion, the solution of this equation is

(2) $$x(t) = c_1 \cos 8t + c_2 \sin 8t$$

We are given the initial condition that $x(0) = 0.2$. But, from Equation 2, $x(0) = c_1$. Therefore, $c_1 = 0.2$. Differentiating Equation 2, we get

$$x'(t) = -8c_1 \sin 8t + 8c_2 \cos 8t$$

Since the initial velocity is given as $x'(0) = 0$, we have $c_2 = 0$ and so the solution is

$$x(t) = \tfrac{1}{5}\cos 8t$$ ■

DAMPED VIBRATIONS

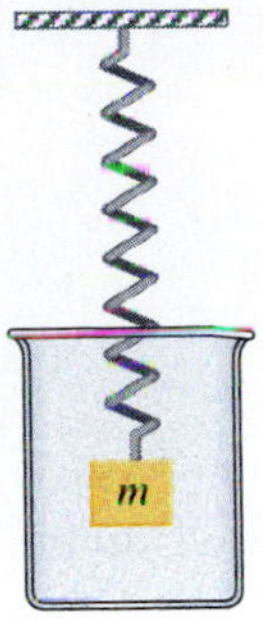

FIGURE 3

We next consider the motion of a spring that is subject to a frictional force (in the case of the horizontal spring of Figure 2) or a damping force (in the case where a vertical spring moves through a fluid as in Figure 3). An example is the damping force supplied by a shock absorber in a car.

We assume that the damping force is proportional to the velocity of the mass and acts in the direction opposite to the motion. (This has been confirmed, at least approximately, by some physical experiments.) Thus

$$\text{damping force} = -c\,\frac{dx}{dt}$$

where c is a positive constant, called the **damping constant**. Thus, in this case, Newton's Second Law gives

$$m\,\frac{d^2x}{dt^2} = \text{restoring force} + \text{damping force}$$

$$= -kx - c\,\frac{dx}{dt}$$

or

$$m\,\frac{d^2x}{dt^2} + c\,\frac{dx}{dt} + kx = 0 \qquad (3)$$

This is a second-order linear differential equation and its auxiliary equation is $mr^2 + cr + k = 0$. The roots are

$$r_1 = \frac{-c + \sqrt{c^2 - 4mk}}{2m} \qquad r_2 = \frac{-c - \sqrt{c^2 - 4mk}}{2m} \qquad (4)$$

According to Section 15.1 we need to discuss three cases.

CASE I $\mathbf{c^2 - 4mk > 0}$ **(overdamping)**
In this case r_1 and r_2 are distinct real roots and

$$x = c_1e^{r_1t} + c_2e^{r_2t}$$

Since c, m, and k are all positive, we have $\sqrt{c^2 - 4mk} < c$, so the roots r_1 and r_2 given by Equations 4 must both be negative. This shows that $x \to 0$ as $t \to \infty$. Typical graphs of x as a function of t are shown in Figure 4. Notice that oscillations do not occur. This is because $c^2 > 4mk$ means that there is a strong damping force (high-viscosity oil or grease) compared with a weak spring or small mass.

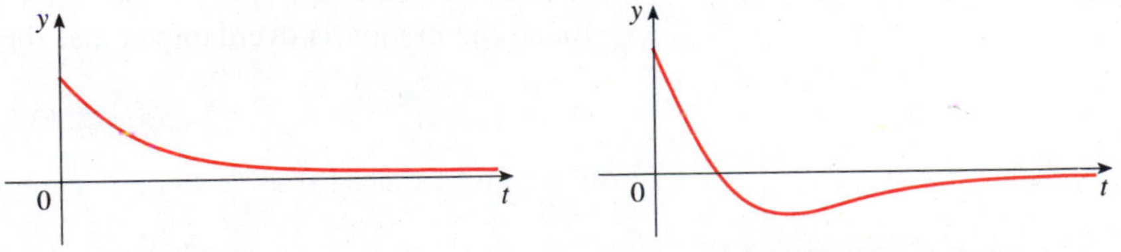

(a) c_1 and c_2 are positive

(b) c_1 and c_2 have opposite signs

FIGURE 4
Overdamping

CASE II $c^2 - 4mk = 0$ **(critical damping)**
This case corresponds to equal roots

$$r_1 = r_2 = -\frac{c}{2m}$$

The solution is given by

$$x = (c_1 + c_2 t)e^{-(c/2m)t}$$

and a typical graph is shown in Figure 5. It is similar to Case I, but the damping is just sufficient to suppress vibrations. Any decrease in the viscosity of the fluid leads to the vibrations of the following case.

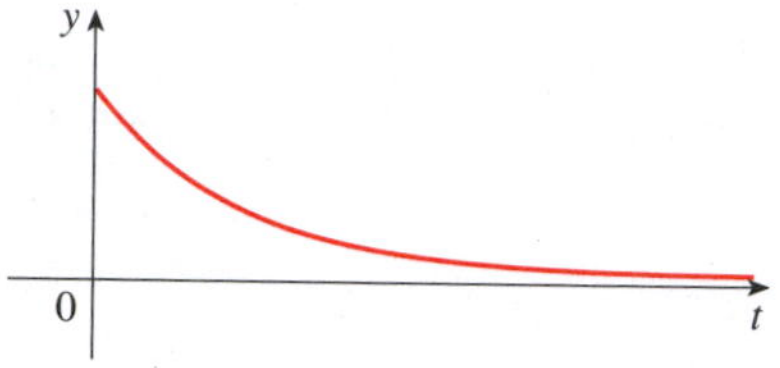

FIGURE 5
Critical damping

CASE III $c^2 - 4mk < 0$ **(underdamping)**
Here the roots are complex:

$$\left.\begin{matrix} r_1 \\ r_2 \end{matrix}\right\} = -\frac{c}{2m} \pm \omega i$$

where

$$\omega = \frac{\sqrt{4mk - c^2}}{2m}$$

The solution is given by

$$x = e^{-(c/2m)t}(c_1 \cos \omega t + c_2 \sin \omega t)$$

We see that there are oscillations that are damped by the factor $e^{-(c/2m)t}$. Since $c > 0$ and $m > 0$, we have $-(c/2m) < 0$ so $e^{-(c/2m)t} \to 0$ as $t \to \infty$. This implies that $x \to 0$ as $t \to \infty$; that is, the motion decays to 0 as time increases. A typical graph is shown in Figure 6.

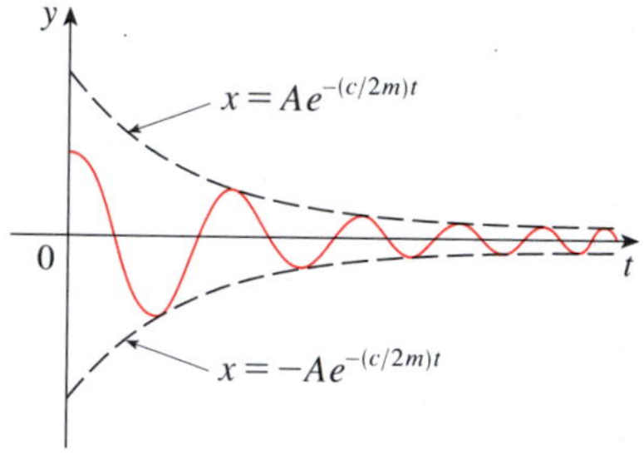

FIGURE 6
Underdamping

EXAMPLE 2 Suppose that the spring of Example 1 is immersed in a fluid with damping constant $c = 40$. Find the position of the mass at any time t if it starts from the equilibrium position and is given a push to start it with an initial velocity of 0.6 m/s.

SOLUTION From Example 1 the mass is $m = 2$ and the spring constant is $k = 128$, so the differential equation (3) becomes

$$2\frac{d^2x}{dt^2} + 40\frac{dx}{dt} + 128x = 0$$

or

$$\frac{d^2x}{dt^2} + 20\frac{dx}{dt} + 64x = 0$$

Figure 7 shows the graph of the position function for the overdamped motion in Example 2.

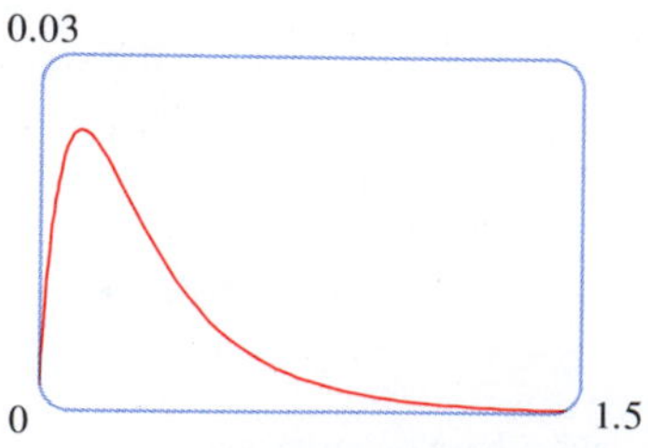

FIGURE 7

The auxiliary equation is $r^2 + 20r + 64 = (r + 4)(r + 16) = 0$ with roots -4 and -16, so the motion is overdamped and the solution is

$$x(t) = c_1 e^{-4t} + c_2 e^{-16t}$$

We are given that $x(0) = 0$, so $c_1 + c_2 = 0$. Differentiating, we get

$$x'(t) = -4c_1 e^{-4t} - 16c_2 e^{-16t}$$

so

$$x'(0) = -4c_1 - 16c_2 = 0.6$$

Since $c_2 = -c_1$, this gives $12c_1 = 0.6$ or $c_1 = 0.05$. Therefore

$$x = 0.05(e^{-4t} - e^{-16t})$$ ■

FORCED VIBRATIONS

Suppose that, in addition to the restoring force and the damping force, the motion of the spring is affected by an external force $F(t)$. Then Newton's Second Law gives

$$m\frac{d^2x}{dt^2} = \text{restoring force} + \text{damping force} + \text{external force}$$

$$= -kx - c\frac{dx}{dt} + F(t)$$

Thus, instead of the homogeneous equation (3), the motion of the spring is now governed by the following nonhomogeneous differential equation:

$$m\frac{d^2x}{dt^2} + c\frac{dx}{dt} + kx = F(t) \qquad (5)$$

The motion of the spring can be determined by the methods of Section 15.2.

A commonly occurring type of external force is a periodic force function

$$F(t) = F_0 \cos \omega_0 t \qquad \text{where } \omega_0 \neq \omega = \sqrt{\frac{k}{m}}$$

In this case, and in the absence of a damping force ($c = 0$), you are asked in Exercise 7 to use the method of undetermined coefficients to show that

$$x(t) = c_1 \cos \omega t + c_2 \sin \omega t + \frac{F_0}{m(\omega^2 - \omega_0^2)} \cos \omega_0 t \qquad (6)$$

If $\omega_0 = \omega$, then the applied frequency reinforces the natural frequency and the result is vibrations of large amplitude. This is the phenomenon of **resonance** (see Exercise 8).

ELECTRIC CIRCUITS

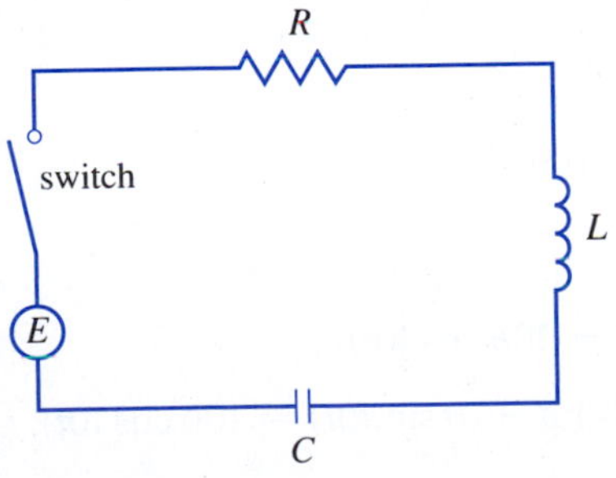

FIGURE 8

In Section 9.2 we were able to use first-order linear equations to analyze electric circuits that contain a resistor and inductor (see Figure 4 in Section 9.2) or a resistor and capacitor (see Exercise 29 in Section 9.2). Now that we know how to solve second-order linear equations, we are in a position to analyze the circuit shown in Figure 8. It contains an electromotive force E (supplied by a battery or generator), a resistor R, an inductor L, and a capacitor C, in series. If the charge on the capacitor at time t is $Q = Q(t)$, then the current is the rate of change of Q with respect to t: $I = dQ/dt$. As in Section 9.2, it is known from physics that the voltage drops across the resistor, inductor, and capacitor are

$$RI \qquad L\frac{dI}{dt} \qquad \frac{Q}{C}$$

respectively. Kirchhoff's voltage law says that the sum of these voltage drops is equal to the supplied voltage:

$$L\frac{dI}{dt} + RI + \frac{Q}{C} = E(t)$$

Since $I = dQ/dt$, this equation becomes

$$L\frac{d^2Q}{dt^2} + R\frac{dQ}{dt} + \frac{1}{C}Q = E(t) \tag{7}$$

which is a second-order linear differential equation with constant coefficients. If the charge Q_0 and the current I_0 are known at time 0, then we have the initial conditions

$$Q(0) = Q_0 \qquad Q'(0) = I(0) = I_0$$

and the initial-value problem can be solved by the methods of Section 15.2.

A differential equation for the current can be obtained by differentiating Equation 7 with respect to t and remembering that $I = dQ/dt$:

$$L\frac{d^2I}{dt^2} + R\frac{dI}{dt} + \frac{1}{C}I = E'(t)$$

EXAMPLE 3 Find the charge and current at time t in the circuit of Figure 8 if $R = 40\ \Omega$, $L = 1$ H, $C = 16 \times 10^{-4}$ F, $E(t) = 100\cos 10t$, and the initial charge and current are both 0.

SOLUTION With the given values of L, R, C, and $E(t)$, Equation 7 becomes

$$\frac{d^2Q}{dt^2} + 40\frac{dQ}{dt} + 625Q = 100\cos 10t \tag{8}$$

The auxiliary equation is $r^2 + 40r + 625 = 0$ with roots

$$r = \frac{-40 \pm \sqrt{-900}}{2} = -20 \pm 15i$$

so the solution of the complementary equation is

$$Q_c(t) = e^{-20t}(c_1\cos 15t + c_2\sin 15t)$$

For the method of undetermined coefficients we try the particular solution

$$Q_p(t) = A\cos 10t + B\sin 10t$$

Then

$$Q_p'(t) = -10A\sin 10t + 10B\cos 10t$$

$$Q_p''(t) = -100A\cos 10t - 100B\sin 10t$$

Substituting into Equation 8, we have

$$(-100A\cos 10t - 100B\sin 10t) + 40(-10A\sin 10t + 10B\cos 10t)$$
$$+ 625(A\cos 10t + B\sin 10t) = 100\cos 10t$$

or

$$(525A + 400B)\cos 10t + (-400A + 525B)\sin 10t = 100\cos 10t$$

Equating coefficients, we have

$$\begin{aligned} 525A + 400B &= 100 \\ -400A + 525B &= 0 \end{aligned} \qquad \text{or} \qquad \begin{aligned} 21A + 16B &= 4 \\ -16A + 21B &= 0 \end{aligned}$$

The solution of this system is $A = \frac{84}{697}$ and $B = \frac{64}{697}$, so a particular solution is

$$Q_p(t) = \tfrac{1}{697}(84 \cos 10t + 64 \sin 10t)$$

and the general solution is

$$\begin{aligned} Q(t) &= Q_c(t) + Q_p(t) \\ &= e^{-20t}(c_1 \cos 15t + c_2 \sin 15t) + \tfrac{4}{697}(21 \cos 10t + 16 \sin 10t) \end{aligned}$$

Imposing the initial condition $Q(0) = 0$, we get

$$Q(0) = c_1 + \tfrac{84}{697} = 0 \qquad c_1 = -\frac{84}{697}$$

To impose the other initial condition we first differentiate to find the current:

$$I = \frac{dQ}{dt} = e^{-20t}[(-20c_1 + 15c_2) \cos 15t + (-15c_1 - 20c_2) \sin 15t]$$
$$+ \tfrac{40}{697}(-21 \sin 10t + 16 \cos 10t)$$

$$I(0) = -20c_1 + 15c_2 + \tfrac{640}{697} = 0 \qquad c_2 = -\frac{464}{2091}$$

Thus, the formula for the charge is

$$Q(t) = \frac{4}{697}\left[\frac{e^{-20t}}{3}(-63 \cos 15t - 116 \sin 15t) + (21 \cos 10t + 16 \sin 10t)\right]$$

and the expression for the current is

$$I(t) = \tfrac{1}{2091}[e^{-20t}(-1920 \cos 15t + 13{,}060 \sin 15t) + 120(-21 \sin 10t + 16 \cos 10t)]$$

■

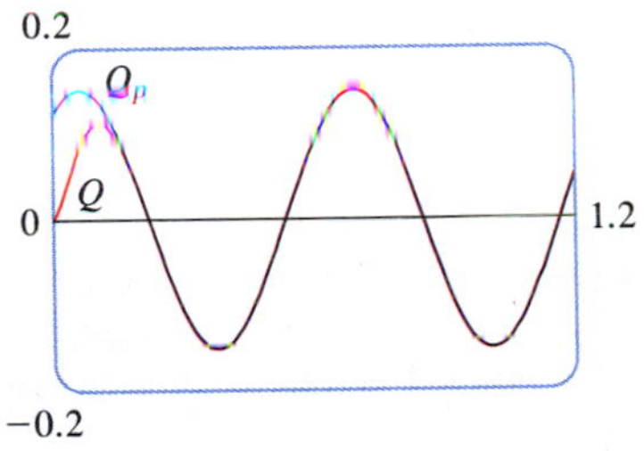

FIGURE 9

NOTE 1: In Example 3 the solution for $Q(t)$ consists of two parts. Since $e^{-20t} \to 0$ as $t \to \infty$ and both $\cos 15t$ and $\sin 15t$ are bounded functions,

$$Q_c(t) = \tfrac{4}{2091} e^{-20t}(-63 \cos 15t - 116 \sin 15t) \to 0 \qquad \text{as } t \to \infty$$

So, for large values of t,

$$Q(t) \approx Q_p(t) = \tfrac{4}{697}(21 \cos 10t + 16 \sin 10t)$$

and, for this reason, $Q_p(t)$ is called the **steady state solution**. Figure 9 shows how the graph of the steady state solution compares with the graph of Q in this case.

$$m \frac{d^2x}{dt^2} + c \frac{dx}{dt} + kx = F(t)$$

$$L \frac{d^2Q}{dt^2} + R \frac{dQ}{dt} + \frac{1}{C} Q = E(t)$$

NOTE 2: Comparing Equations 5 and 7, we see that mathematically they are identical. This suggests the analogies given in the following chart between physical situations that, at first glance, are very different.

Spring system		Electric circuit	
x	displacement	Q	charge
dx/dt	velocity	$I = dQ/dt$	current
m	mass	L	inductance
c	damping constant	R	resistance
k	spring constant	$1/C$	elastance
$F(t)$	external force	$E(t)$	electromotive force

We can also transfer other ideas from one situation to the other. For instance, the steady state solution discussed in Note 1 makes sense in the spring system. And the phenomenon of resonance in the spring system can be usefully carried over to electric circuits as electrical resonance.

EXERCISES 15.3

1. A spring with a 3-kg mass is held stretched 0.6 m beyond its natural length by a force of 20 N. If the spring begins at its equilibrium position but a push gives it an initial velocity of 1.2 m/s, find the position of the mass after t seconds.

2. A spring with a 4-kg mass has natural length 1 m and is maintained stretched to a length of 1.3 m by a force of 24.3 N. If the spring is compressed to a length of 0.8 m and then released with zero velocity, find the position of the mass at any time t.

3. A spring with a mass of 2 kg has damping constant 14, and a force of 6 N is required to keep the spring stretched 0.5 m beyond its natural length. The spring is stretched 1 m beyond its natural length and then released with zero velocity. Find the position of the mass at any time t.

4. A spring with a mass of 3 kg has damping constant 30 and spring constant 123.
(a) Find the position of the mass at time t if it starts at the equilibrium position with a velocity of 2 m/s.
(b) Graph the position function of the mass.

5. For the spring in Exercise 3, find the mass that would produce critical damping.

6. For the spring in Exercise 4, find the damping constant that would produce critical damping.

7. Suppose a spring has mass m and spring constant k and let $\omega = \sqrt{k/m}$. Suppose that the damping constant is so small that the damping force is negligible. If an external force $F(t) = F_0 \cos \omega_0 t$ is applied, where $\omega_0 \neq \omega$, use the method of undetermined coefficients to show that the motion of the mass is described by Equation 6.

8. As in Exercise 7, consider a spring with mass m, spring constant k, and damping constant $c = 0$, and let $\omega = \sqrt{k/m}$. If an external force $F(t) = F_0 \cos \omega t$ is applied (the applied frequency equals the natural frequency), use the method of undetermined coefficients to show that the motion of the mass is given by

$$x(t) = c_1 \cos \omega t + c_2 \sin \omega t + \frac{F_0}{2m\omega} t \sin \omega t$$

9. A series circuit consists of a resistor with $R = 20\ \Omega$, an inductor with $L = 1$ H, a capacitor with $C = 0.002$ F, and a 12-V battery. If the initial charge and current are both 0, find the charge and current at time t.

10. A series circuit contains a resistor with $R = 24\ \Omega$, an inductor with $L = 2$ H, a capacitor with $C = 0.005$ F, and a 12-V battery. The initial charge is $Q = 0.001$ C and the initial current is 0.
(a) Find the charge and current at time t.
(b) Graph the charge and current functions.

11. The battery in Exercise 9 is replaced by a generator producing a voltage of $E(t) = 12 \sin 10t$. Find the charge at time t.

12. The battery in Exercise 10 is replaced by a generator producing a voltage of $E(t) = 12 \sin 10t$.
(a) Find the charge at time t.
(b) Graph the charge function.

13. Verify that the solution to Equation 1 can be written in the form $x(t) = A \cos(\omega t + \delta)$.

15.4 USING SERIES TO SOLVE DIFFERENTIAL EQUATIONS

Many differential equations can't be solved explicitly in terms of finite combinations of simple familiar functions. This is true even for a simple-looking equation like

$$y'' - 2xy' + y = 0 \tag{1}$$

But it is important to be able to solve equations such as Equation 1 because they arise from physical problems and, in particular, in connection with the Schrödinger equation in quantum mechanics. In such a case we use the method of power series; that is, we look for a solution of the form

$$y = f(x) = \sum_{n=0}^{\infty} c_n x^n = c_0 + c_1 x + c_2 x^2 + c_3 x^3 + \cdots$$

The method is to substitute this expression into the differential equation and determine the values of the coefficients $c_0, c_1, c_2, \ldots$.

Before using power series to solve Equation 1, we illustrate the method on the simpler equation $y'' + y = 0$ in Example 1.

EXAMPLE 1 Use power series to solve the equation $y'' + y = 0$.

SOLUTION We assume there is a solution of the form

$$y = c_0 + c_1 x + c_2 x^2 + c_3 x^3 + \cdots = \sum_{n=0}^{\infty} c_n x^n \tag{2}$$

We can differentiate power series term by term, so

$$y' = c_1 + 2c_2 x + 3c_3 x^2 + \cdots = \sum_{n=1}^{\infty} n c_n x^{n-1}$$

$$y'' = 2c_2 + 2 \cdot 3c_3 x + \cdots = \sum_{n=2}^{\infty} n(n-1)c_n x^{n-2} \tag{3}$$

In order to compare the expressions for y and y'' more easily, we rewrite y'' as follows:

By writing out the first few terms of (4), you can see that it is the same as (3). To obtain (4) we replaced n by $n + 2$ and began the summation at 0 instead of 2.

$$y'' = \sum_{n=0}^{\infty} (n+2)(n+1)c_{n+2} x^n \tag{4}$$

Substituting the expressions in Equations 2 and 4 into the differential equation, we obtain

$$\sum_{n=0}^{\infty} (n+2)(n+1)c_{n+2} x^n + \sum_{n=0}^{\infty} c_n x^n = 0$$

or

$$\sum_{n=0}^{\infty} [(n+2)(n+1)c_{n+2} + c_n]x^n = 0 \tag{5}$$

If two power series are equal, then the corresponding coefficients must be equal. Therefore, the coefficients of x^n in Equation 5 must be 0:

$$(n+2)(n+1)c_{n+2} + c_n = 0$$

$$c_{n+2} = -\frac{c_n}{(n+1)(n+2)} \qquad n = 0, 1, 2, 3, \ldots \tag{6}$$

Equation 6 is called a recursion relation. If c_0 and c_1 are known, this equation allows us to determine the remaining coefficients recursively by putting $n = 0, 1, 2, 3, \dots$ in succession.

$$\text{Put } n = 0: \qquad c_2 = -\frac{c_0}{1 \cdot 2}$$

$$\text{Put } n = 1: \qquad c_3 = -\frac{c_1}{2 \cdot 3}$$

$$\text{Put } n = 2: \qquad c_4 = -\frac{c_2}{3 \cdot 4} = \frac{c_0}{1 \cdot 2 \cdot 3 \cdot 4} = \frac{c_0}{4!}$$

$$\text{Put } n = 3: \qquad c_5 = -\frac{c_3}{4 \cdot 5} = \frac{c_1}{2 \cdot 3 \cdot 4 \cdot 5} = \frac{c_1}{5!}$$

$$\text{Put } n = 4: \qquad c_6 = -\frac{c_4}{5 \cdot 6} = -\frac{c_0}{4!\,5 \cdot 6} = -\frac{c_0}{6!}$$

$$\text{Put } n = 5: \qquad c_7 = -\frac{c_5}{6 \cdot 7} = -\frac{c_1}{5!\,6 \cdot 7} = -\frac{c_1}{7!}$$

By now we see the pattern:

$$\text{For the even coefficients, } c_{2n} = (-1)^n \frac{c_0}{(2n)!}$$

$$\text{For the odd coefficients, } c_{2n+1} = (-1)^n \frac{c_1}{(2n+1)!}$$

Putting these values back into Equation 2, we write the solution as

$$\begin{aligned}
y &= c_0 + c_1 x + c_2 x^2 + c_3 x^3 + c_4 x^4 + c_5 x^5 + \cdots \\
&= c_0\left(1 - \frac{x^2}{2!} + \frac{x^4}{4!} - \frac{x^6}{6!} + \cdots + (-1)^n \frac{x^{2n}}{(2n)!} + \cdots\right) \\
&\quad + c_1\left(x - \frac{x^3}{3!} + \frac{x^5}{5!} - \frac{x^7}{7!} + \cdots + (-1)^n \frac{x^{2n+1}}{(2n+1)!} + \cdots\right) \\
&= c_0 \sum_{n=0}^{\infty} (-1)^n \frac{x^{2n}}{(2n)!} + c_1 \sum_{n=0}^{\infty} (-1)^n \frac{x^{2n+1}}{(2n+1)!}
\end{aligned}$$

Notice that there are two arbitrary constants, c_0 and c_1. ■

NOTE 1: We recognize the series obtained in Example 1 as being the Maclaurin series for $\cos x$ and $\sin x$. Therefore, we could write the solution as

$$y(x) = c_0 \cos x + c_1 \sin x$$

But we are not usually able to express power series solutions of differential equations in terms of known functions.

EXAMPLE 2 Solve $y'' - 2xy' + y = 0$.

SOLUTION We assume there is a solution of the form

$$y = \sum_{n=0}^{\infty} c_n x^n$$

Then
$$y' = \sum_{n=1}^{\infty} n c_n x^{n-1}$$

and
$$y'' = \sum_{n=2}^{\infty} n(n-1)c_n x^{n-2} = \sum_{n=0}^{\infty} (n+2)(n+1)c_{n+2}x^n$$

as in Example 1. Substituting in the differential equation, we get

$$\sum_{n=0}^{\infty} (n+2)(n+1)c_{n+2}x^n - 2x \sum_{n=1}^{\infty} n c_n x^{n-1} + \sum_{n=0}^{\infty} c_n x^n = 0$$

$$\sum_{n=1}^{\infty} 2nc_n x^n = \sum_{n=0}^{\infty} 2nc_n x^n$$

$$\sum_{n=0}^{\infty} (n+2)(n+1)c_{n+2}x^n - \sum_{n=1}^{\infty} 2nc_n x^n + \sum_{n=0}^{\infty} c_n x^n = 0$$

$$\sum_{n=0}^{\infty} [(n+2)(n+1)c_{n+2} - (2n-1)c_n]x^n = 0$$

This equation is true if the coefficient of x^n is 0:

$$(n+2)(n+1)c_{n+2} - (2n-1)c_n = 0$$

(7)
$$c_{n+2} = \frac{2n-1}{(n+1)(n+2)} c_n \qquad n = 0, 1, 2, 3, \ldots$$

We solve this recursion relation by putting $n = 0, 1, 2, 3, \ldots$ successively in Equation 7:

Put $n = 0$: $$c_2 = \frac{-1}{1 \cdot 2} c_0$$

Put $n = 1$: $$c_3 = \frac{1}{2 \cdot 3} c_1$$

Put $n = 2$: $$c_4 = \frac{3}{3 \cdot 4} c_2 = -\frac{3}{1 \cdot 2 \cdot 3 \cdot 4} c_0 = -\frac{3}{4!} c_0$$

Put $n = 3$: $$c_5 = \frac{5}{4 \cdot 5} c_3 = \frac{1 \cdot 5}{2 \cdot 3 \cdot 4 \cdot 5} c_1 = \frac{1 \cdot 5}{5!} c_1$$

Put $n = 4$: $$c_6 = \frac{7}{5 \cdot 6} c_4 = -\frac{3 \cdot 7}{4!\,5 \cdot 6} c_0 = -\frac{3 \cdot 7}{6!} c_0$$

Put $n = 5$: $$c_7 = \frac{9}{6 \cdot 7} c_5 = \frac{1 \cdot 5 \cdot 9}{5!\,6 \cdot 7} c_1 = \frac{1 \cdot 5 \cdot 9}{7!} c_1$$

Put $n = 6$: $\quad c_8 = \dfrac{11}{7 \cdot 8} c_6 = -\dfrac{3 \cdot 7 \cdot 11}{8!} c_0$

Put $n = 7$: $\quad c_9 = \dfrac{13}{8 \cdot 9} c_7 = \dfrac{1 \cdot 5 \cdot 9 \cdot 13}{9!} c_1$

In general, the even coefficients are given by

$$c_{2n} = -\frac{3 \cdot 7 \cdot 11 \cdot \cdots \cdot (4n-5)}{(2n)!} c_0$$

and the odd coefficients are given by

$$c_{2n+1} = \frac{1 \cdot 5 \cdot 9 \cdot \cdots \cdot (4n-3)}{(2n+1)!} c_1$$

The solution is

$$\begin{aligned} y &= c_0 + c_1 x + c_2 x^2 + c_3 x^3 + c_4 x^4 + \cdots \\ &= c_0\left(1 - \frac{1}{2!}x^2 - \frac{3}{4!}x^4 - \frac{3 \cdot 7}{6!}x^6 - \frac{3 \cdot 7 \cdot 11}{8!}x^8 - \cdots\right) \\ &\quad + c_1\left(x + \frac{1}{3!}x^3 + \frac{1 \cdot 5}{5!}x^5 + \frac{1 \cdot 5 \cdot 9}{7!}x^7 + \frac{1 \cdot 5 \cdot 9 \cdot 13}{9!}x^9 + \cdots\right) \end{aligned}$$

or

(8)
$$\begin{aligned} y &= c_0\left(1 - \frac{1}{2!}x^2 - \sum_{n=2}^{\infty} \frac{3 \cdot 7 \cdot \cdots \cdot (4n-5)}{(2n)!} x^{2n}\right) \\ &\quad + c_1\left(x + \sum_{n=1}^{\infty} \frac{1 \cdot 5 \cdot 9 \cdot \cdots \cdot (4n-3)}{(2n+1)!} x^{2n+1}\right) \end{aligned}$$

■

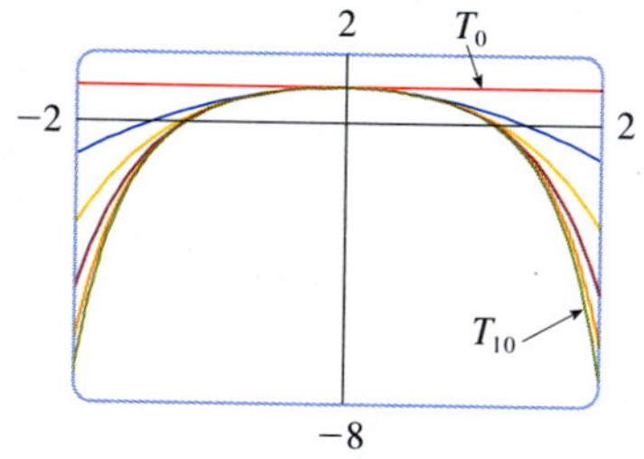

FIGURE 1

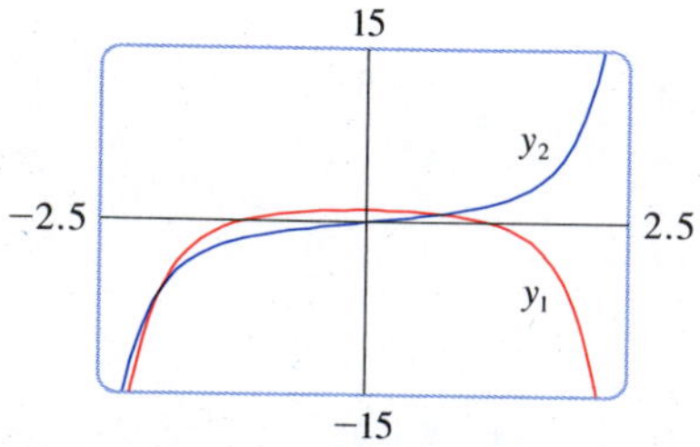

FIGURE 2

NOTE 2: In Example 2 we had to assume that the differential equation had a series solution. But now we could verify directly that the function given by Equation 8 is indeed a solution.

NOTE 3: Unlike the situation of Example 1, the power series that arise in the solution of Example 2 do not define elementary functions. The functions

$$y_1(x) = 1 - \frac{1}{2!}x^2 - \sum_{n=2}^{\infty} \frac{3 \cdot 7 \cdot \cdots \cdot (4n-5)}{(2n)!} x^{2n}$$

and

$$y_2(x) = x + \sum_{n=1}^{\infty} \frac{1 \cdot 5 \cdot 9 \cdot \cdots \cdot (4n-3)}{(2n+1)!} x^{2n+1}$$

are perfectly good functions but they can't be expressed in terms of familiar functions. We can use these power series expressions for y_1 and y_2 to compute approximate values of the functions and even to graph them. Figure 1 shows the first few partial sums $T_0, T_2, T_4, \ldots$ (Taylor polynomials) for $y_1(x)$, and we see how they converge to y_1. In this way we can graph both y_1 and y_2 in Figure 2.

NOTE 4: If we were asked to solve the initial-value problem

$$y'' - 2xy' + y = 0 \qquad y(0) = 0 \qquad y'(0) = 1$$

we would observe that

$$c_0 = y(0) = 0 \qquad c_1 = y'(0) = 1$$

This would simplify the calculations in Example 2, since all of the even coefficients would be 0. The solution to the initial-value problem is

$$y(x) = x + \sum_{n=1}^{\infty} \frac{1 \cdot 5 \cdot 9 \cdot \cdots \cdot (4n - 3)}{(2n + 1)!} x^{2n+1}$$

EXERCISES 15.4

1–9 ■ Use power series to solve the differential equation.

1. $y' = 6y$
2. $y' = xy$
3. $y' = x^2y$
4. $y'' = y$
5. $y'' + 3xy' + 3y = 0$
6. $y'' = xy$
7. $y'' - xy' - y = 0$, $y(0) = 1$, $y'(0) = 0$
8. $y'' + x^2y = 0$, $y(0) = 1$, $y'(0) = 0$
9. $y'' + x^2y' + xy = 0$, $y(0) = 0$, $y'(0) = 1$

10. The solution of the initial-value problem

$$x^2y'' + xy' + x^2y = 0 \qquad y(0) = 1 \qquad y'(0) = 0$$

is called a Bessel function of order 0.

(a) Solve the initial-value problem to find a power series expansion for the Bessel function.

(b) Graph several Taylor polynomials until you reach one that looks like a good approximation to the Bessel function on the interval $[-5, 5]$.

15 REVIEW

KEY TOPICS ■

Write the general form of each type of second-order linear differential equation with constant coefficients:

1. Homogeneous
2. Nonhomogeneous

Then discuss the method of solution.

EXERCISES

1–4 ■ Determine whether the statement is true or false.

1. If y_1 and y_2 are solutions of $y'' + y = 0$, then $y_1 + y_2$ is also a solution of the equation.
2. If y_1 and y_2 are solutions of $y'' + 6y' + 5y = x$, then $c_1y_1 + c_2y_2$ is also a solution of the equation.
3. The general solution of $y'' - y = 0$ can be written as $y = c_1 \cosh x + c_2 \sinh x$.
4. The equation $y'' - y = e^x$ has a particular solution of the form $y_p = Ae^x$.

5–14 ■ Solve the differential equation.

5. $y'' - 6y' + 34y = 0$

6. $4y'' - 20y' + 25y = 0$

7. $2y'' + y' = y$

8. $3y'' + 2y' + y = 0$

9. $\dfrac{d^2y}{dx^2} + 2\dfrac{dy}{dx} + y = \sin 3x$

10. $\dfrac{d^2y}{dx^2} - 2\dfrac{dy}{dx} - 3y = \cos 4x$

11. $4\dfrac{d^2y}{dx^2} + 9y = 2x^2 - 3$

12. $\dfrac{d^2y}{dx^2} - 4\dfrac{dy}{dx} + 20y = xe^x$

13. $\dfrac{d^2y}{dx^2} - 3\dfrac{dy}{dx} + 2y = e^{2x}$

14. $\dfrac{d^2y}{dx^2} + y = \csc x, \quad 0 < x < \pi/2$

15–18 ■ Solve the initial-value problem.

15. $y'' + 6y' = 0, \quad y(1) = 3, \quad y'(1) = 12$

16. $y'' - 6y' + 25y = 0, \quad y(0) = 2, \quad y'(0) = 1$

17. $y'' - 5y' + 4y = 0, \quad y(0) = 0, \quad y'(0) = 1$

18. $9y'' + y = 3x + e^{-x}, \quad y(0) = 1, \quad y'(0) = 2$

19. Use power series to solve the initial-value problem

$$y'' + xy' + y = 0 \qquad y(0) = 0 \qquad y'(0) = 1$$

20. Use power series to solve the equation

$$y'' - xy' - 2y = 0$$

21. A series circuit contains a resistor with $R = 40\ \Omega$, an inductor with $L = 2$ H, a capacitor with $C = 0.0025$ F, and a 12-V battery. The initial charge is $Q = 0.01$ C and the initial current is 0. Find the charge at time t.

22. A spring with a mass of 2 kg has damping constant 16, and a force of 12.8 N keeps the spring stretched 0.2 m beyond its natural length. Find the position of the mass at time t if it starts at the equilibrium position with a velocity of 2.4 m/s.

23. Assume that Earth is a solid sphere of uniform density with mass M and radius $R = 3960$ mi. For a particle of mass m within the Earth at a distance r from Earth's center, the gravitational force attracting the particle to the center is

$$F_r = \frac{-GM_r m}{r^2}$$

where G is the gravitational constant and M_r is the mass of Earth within the sphere of radius r.

(a) Show that $F_r = \dfrac{-GMm}{R^3}r$.

(b) Suppose a hole is drilled through Earth along a diameter. Show that if a particle of mass m is dropped from rest, at the surface, into the hole, then the distance $y = y(t)$ of the particle from the center of Earth at time t is given by

$$y''(t) = -k^2 y(t)$$

where $k^2 = GM/R^3 = g/R$.

(c) Conclude from part (b) that the particle undergoes simple harmonic motion. Find the period T.

(d) With what speed does the particle pass through the center of Earth?

APPENDIXES

A NUMBERS, INEQUALITIES, AND ABSOLUTE VALUES

Calculus is based on the real number system. We start with the **integers**:

$$\ldots, -3, -2, -1, 0, 1, 2, 3, 4, \ldots$$

Then we construct the **rational numbers**, which are ratios of integers. Thus, any rational number r can be expressed as

$$r = \frac{m}{n} \qquad \text{where } m \text{ and } n \text{ are integers and } n \neq 0$$

Examples are

$$\tfrac{1}{2} \qquad -\tfrac{3}{7} \qquad 46 = \tfrac{46}{1} \qquad 0.17 = \tfrac{17}{100}$$

(Recall that division by 0 is always ruled out, so expressions like $\frac{3}{0}$ and $\frac{0}{0}$ are undefined.) Some real numbers, such as $\sqrt{2}$, cannot be expressed as a ratio of integers and are therefore called **irrational numbers**. It can be shown, with varying degrees of difficulty, that the following are also irrational numbers:

$$\sqrt{3} \qquad \sqrt{5} \qquad \sqrt[3]{2} \qquad \pi \qquad \sin 1^\circ \qquad \log_{10} 2$$

The set of all real numbers is usually denoted by the symbol $\mathbb{R}$. When we use the word *number* without qualification, we mean "real number."

Every number has a decimal representation. If the number is rational, then the corresponding decimal is repeating. For example,

$$\tfrac{1}{2} = 0.5000\ldots = 0.5\overline{0} \qquad \tfrac{2}{3} = 0.66666\ldots = 0.\overline{6}$$

$$\tfrac{157}{495} = 0.3171717\ldots = 0.3\overline{17} \qquad \tfrac{9}{7} = 1.285714285714\ldots = 1.\overline{285714}$$

(The bar indicates that the sequence of digits repeats forever.) On the other hand, if the number is irrational, the decimal is nonrepeating:

$$\sqrt{2} = 1.414213562373095\ldots \qquad \pi = 3.141592653589793\ldots$$

If we stop the decimal expansion of any number at a certain place, we get an approximation to the number. For instance, we can write

$$\pi \approx 3.14159265$$

where the symbol $\approx$ is read "is approximately equal to." The more decimal places we retain, the better the approximation we get.

The real numbers can be represented by points on a line as in Figure 1. The positive direction (to the right) is indicated by an arrow. We choose an arbitrary reference point O, called the **origin**, which corresponds to the real number 0. Given any convenient unit of measurement, each positive number x is represented by the point on the line a distance of x units to the right of the origin, and each negative number $-x$ is represented by the point x units to the left of the origin. Thus, every real number is represented by a point on the line, and every point P on the line corresponds to exactly one real number. The number associated with the point P is called the **coordinate** of P and the line is then called a **coordinate line**, or a **real number line**, or simply a **real line**. Often we identify the point with its coordinate and think of a number as being a point on the real line.

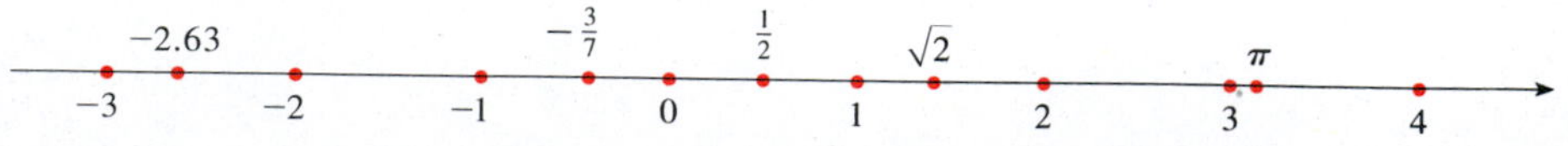

FIGURE 1

The real numbers are ordered. We say *a is less than b* and write $a < b$ if $b - a$ is a positive number. Geometrically, this means that a lies to the left of b on the number line. (Equivalently, we say *b is greater than a* and write $b > a$.) The symbol $a \leqslant b$ (or $b \geqslant a$) means that either $a < b$ or $a = b$ and is read "a is less than or equal to b." For instance, the following are true inequalities:

$$7 < 7.4 < 7.5 \qquad -3 > -\pi \qquad \sqrt{2} < 2 \qquad \sqrt{2} \leqslant 2 \qquad 2 \leqslant 2$$

In what follows we need to use set notation. A **set** is a collection of objects, and these objects are called the **elements** of the set. If S is a set, the notation $a \in S$ means that a is an element of S, and $a \notin S$ means that a is not an element of S. For example, if Z represents the set of integers, then $-3 \in Z$ but $\pi \notin Z$. If S and T are sets, then their **union** $S \cup T$ is the set consisting of all elements that are in S or T (or in both S and T). The **intersection** of S and T is the set $S \cap T$ consisting of all elements that are in both S and T. In other words, $S \cap T$ is the common part of S and T. The empty set, denoted by $\varnothing$, is the set that contains no element.

Some sets can be described by listing their elements between braces. For instance, the set A consisting of all positive integers less than 7 can be written as

$$A = \{1, 2, 3, 4, 5, 6\}$$

We could also write A in set-builder notation as

$$A = \{x \mid x \text{ is an integer and } 0 < x < 7\}$$

which is read "A is the set of x such that x is an integer and $0 < x < 7$."

Certain sets of real numbers, called **intervals**, occur frequently in calculus and correspond geometrically to line segments. For example, if $a < b$, the **open interval** from a to b consists of all numbers between a and b and is denoted by the symbol (a, b). Using set-builder notation, we can write

$$(a, b) = \{x \mid a < x < b\}$$

FIGURE 2
Open interval (a, b)

Notice that the endpoints of the interval—namely, a and b—are excluded. This is indicated by the round brackets ( ) and by the open dots in Figure 2. The **closed interval** from a to b is the set

$$[a, b] = \{x \mid a \leqslant x \leqslant b\}$$

FIGURE 3
Closed interval $[a, b]$

Here the endpoints of the interval are included. This is indicated by the square brackets [ ] and by the solid dots in Figure 3. It is also possible to include only one endpoint in an interval, as shown in Table 1.

(1) TABLE OF INTERVALS

Notation	Set description	Picture
(a, b)	$\{x \mid a < x < b\}$	
$[a, b]$	$\{x \mid a \leqslant x \leqslant b\}$	
$[a, b)$	$\{x \mid a \leqslant x < b\}$	
$(a, b]$	$\{x \mid a < x \leqslant b\}$	
(a, ∞)	$\{x \mid x > a\}$	
$[a, \infty)$	$\{x \mid x \geqslant a\}$	
$(-\infty, b)$	$\{x \mid x < b\}$	
$(-\infty, b]$	$\{x \mid x \leqslant b\}$	
$(-\infty, \infty)$	$\mathbb{R}$ (set of all real numbers)	

Table 1 lists the nine possible types of intervals. When these intervals are discussed, it is always assumed that $a < b$.

We also need to consider infinite intervals such as

$$(a, \infty) = \{x \mid x > a\}$$

This does not mean that ∞ ("infinity") is a number. The notation (a, ∞) stands for the set of all numbers that are greater than a, so the symbol ∞ simply indicates that the interval extends indefinitely far in the positive direction.

When working with inequalities, note the following rules:

(2) RULES FOR INEQUALITIES

1. If $a < b$, then $a + c < b + c$.
2. If $a < b$ and $c < d$, then $a + c < b + d$.
3. If $a < b$ and $c > 0$, then $ac < bc$.
4. If $a < b$ and $c < 0$, then $ac > bc$.
5. If $0 < a < b$, then $1/a > 1/b$.

Rule 1 says that we can add any number to both sides of an inequality, and Rule 2 says that two inequalities can be added. However, we have to be careful with multiplication. Rule 3 says that we can multiply both sides of an inequality by a *positive* number, but Rule 4 says that *if we multiply both sides of an inequality by a negative number, then we reverse the direction of the inequality.* For example, if we take the inequality $3 < 5$ and multiply by 2, we get $6 < 10$, but if we multiply by -2, we get $-6 > -10$. Finally, Rule 5 says that if we take reciprocals, then we reverse the direction of an inequality (provided the numbers are positive).

EXAMPLE 1 Solve the inequality $1 + x < 7x + 5$.

SOLUTION The given inequality is satisfied by some values of x but not by others. To *solve* an inequality means to determine the set of numbers x for which the inequality is true. This is called the *solution set*.

First we subtract 1 from each side of the inequality (using Rule 1 with $c = -1$):

$$x < 7x + 4$$

Then we subtract $7x$ from both sides (Rule 1 with $c = -7x$):

$$-6x < 4$$

Now we divide both sides by -6 (Rule 4 with $c = -\frac{1}{6}$):

$$x > -\tfrac{4}{6} = -\tfrac{2}{3}$$

These steps can all be reversed, so the solution set consists of all numbers greater than $-\frac{2}{3}$. In other words, the solution of the inequality is the interval $(-\frac{2}{3}, \infty)$. ■

EXAMPLE 2 Solve the inequalities $4 \leqslant 3x - 2 < 13$.

SOLUTION Here the solution set consists of all values of x that satisfy both inequalities. Using the rules given in (2), we see that the following inequalities are equivalent:

$$4 \leqslant 3x - 2 < 13$$

$$6 \leqslant 3x < 15 \qquad \text{(add 2)}$$

$$2 \leqslant x < 5 \qquad \text{(divide by 3)}$$

Therefore, the solution set is $[2, 5)$. ■

EXAMPLE 3 Solve $2x + 1 \leq 4x - 3 \leq x + 7$.

SOLUTION This time we first solve the inequalities separately:

$$2x + 1 \leq 4x - 3 \qquad 4x - 3 \leq x + 7$$

$$4 \leq 2x \qquad 3x \leq 10$$

$$2 \leq x \qquad x \leq \tfrac{10}{3}$$

Since x must satisfy both inequalities, we have

$$2 \leq x \leq \tfrac{10}{3}$$

Thus, the solution set is the closed interval $[2, \frac{10}{3}]$. ■

EXAMPLE 4 Solve the inequality $x^2 - 5x + 6 \leq 0$.

SOLUTION First we factor the left side:

$$(x - 2)(x - 3) \leq 0$$

We know that the corresponding equation $(x - 2)(x - 3) = 0$ has the solutions 2 and 3. The numbers 2 and 3 divide the real line into three intervals:

$$(-\infty, 2) \qquad (2, 3) \qquad (3, \infty)$$

On each of these intervals we determine the signs of the factors. For instance,

$$x \in (-\infty, 2) \quad \Rightarrow \quad x < 2 \quad \Rightarrow \quad x - 2 < 0$$

Then we record these signs in the following chart:

Interval	$x - 2$	$x - 3$	$(x - 2)(x - 3)$
$x < 2$	$-$	$-$	$+$
$2 < x < 3$	$+$	$-$	$-$
$x > 3$	$+$	$+$	$+$

Another method for obtaining the information in the chart is to use *test values*. For instance, if we use the test value $x = 1$ for the interval $(-\infty, 2)$, then substitution in $x^2 - 5x + 6$ gives

$$1^2 - 5(1) + 6 = 2$$

The polynomial $x^2 - 5x + 6$ does not change sign inside any of the three intervals, so we conclude that it is positive on $(-\infty, 2)$.

Then we read from the chart that $(x - 2)(x - 3)$ is negative when $2 < x < 3$. Thus, the solution of the inequality $(x - 2)(x - 3) \leq 0$ is

$$\{x \mid 2 \leq x \leq 3\} = [2, 3]$$

Notice that we have included the endpoints 2 and 3 because we seek values of x such that the product is either negative or zero. The solution is illustrated in Figure 4.

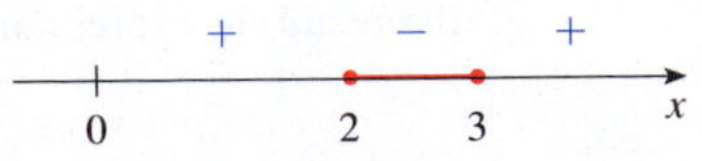

FIGURE 4

■

EXAMPLE 5 Solve $\dfrac{1+x}{1-x} > 1$.

SOLUTION 1 One method is to take all nonzero terms to the left side and use a common denominator:

$$\frac{1+x}{1-x} > 1$$

$$\frac{1+x}{1-x} - 1 > 0$$

$$\frac{1+x-1+x}{1-x} > 0$$

$$\frac{2x}{1-x} > 0$$

The numerator is zero when $x = 0$ and the denominator is zero when $x = 1$. As before, we can set up a chart to determine the sign on each of the intervals $(-\infty, 0)$, $(0, 1)$, and $(1, \infty)$.

Interval	$2x$	$1-x$	$\dfrac{2x}{1-x}$
$x < 0$	$-$	$+$	$-$
$0 < x < 1$	$+$	$+$	$+$
$x > 1$	$+$	$-$	$-$

From the chart we see that the solution set is $\{x \mid 0 < x < 1\} = (0, 1)$.

SOLUTION 2 Another method is to multiply both sides by $1 - x$, but in view of Rules 3 and 4 we must consider separately the cases in which $1 - x$ is positive and negative.

CASE I

If $1 - x > 0$, that is, $x < 1$, then multiplying the given inequality by $1 - x$ gives

$$1 + x > 1 - x$$

which becomes $2x > 0$, that is, $x > 0$. So we have

$$0 < x < 1$$

CASE II

If $1 - x < 0$, that is, $x > 1$, then multiplying the given inequality by $1 - x$ gives

$$1 + x < 1 - x$$

which becomes $2x < 0$, that is, $x < 0$. But the conditions $x > 1$ and $x < 0$ are incompatible, so there is no solution in Case II.

Therefore, the solution set is the open interval $(0, 1)$. ■

EXAMPLE 6 Solve $x^3 + 3x^2 > 4x$.

SOLUTION First we take all nonzero terms to one side of the inequality sign and factor the resulting expression.

$$x^3 + 3x^2 - 4x > 0 \qquad \text{or} \qquad x(x-1)(x+4) > 0$$

As in Example 4 we solve the corresponding equation $x(x - 1)(x + 4) = 0$ and use the solutions $x = -4$, $x = 0$, and $x = 1$ to divide the real line into four intervals $(-\infty, -4)$, $(-4, 0)$, $(0, 1)$, and $(1, \infty)$. On each interval the product keeps a constant sign as shown in the following chart:

Interval	x	$x - 1$	$x + 4$	$x(x - 1)(x + 4)$
$x < -4$	$-$	$-$	$-$	$-$
$-4 < x < 0$	$-$	$-$	$+$	$+$
$0 < x < 1$	$+$	$-$	$+$	$-$
$x > 1$	$+$	$+$	$+$	$+$

Then we read from the chart that the solution set is

$$\{x \mid -4 < x < 0 \text{ or } x > 1\} = (-4, 0) \cup (1, \infty)$$

FIGURE 5

The solution is illustrated in Figure 5. ■

ABSOLUTE VALUE

The **absolute value** of a number a, denoted by $|a|$, is the distance from a to 0 on the real number line. Distances are always positive or 0, so we have

$$|a| \geqslant 0 \qquad \text{for every number } a$$

For example,

$$|3| = 3 \qquad |-3| = 3 \qquad |0| = 0 \qquad |\sqrt{2} - 1| = \sqrt{2} - 1 \qquad |3 - \pi| = \pi - 3$$

In general, we have

(3)
$$|a| = a \qquad \text{if } a \geqslant 0$$
$$|a| = -a \qquad \text{if } a < 0$$

(Remember that if a is negative, then $-a$ is positive.)

EXAMPLE 7 Express $|3x - 2|$ without using the absolute value symbol.

SOLUTION

$$|3x - 2| = \begin{cases} 3x - 2 & \text{if } 3x - 2 \geqslant 0 \\ -(3x - 2) & \text{if } 3x - 2 < 0 \end{cases}$$

$$= \begin{cases} 3x - 2 & \text{if } x \geqslant \frac{2}{3} \\ 2 - 3x & \text{if } x < \frac{2}{3} \end{cases}$$ ■

Let's recall that the symbol $\sqrt{\ }$ means "the positive square root of." Thus, $\sqrt{r} = s$ means $s^2 = r$ and $s \geqslant 0$. Therefore, the equation $\sqrt{a^2} = a$ is not always true. It is true only when $a \geqslant 0$. If $a < 0$, then $-a > 0$, so we have $\sqrt{a^2} = -a$. In view of (3), we then have the equation

(4)
$$\sqrt{a^2} = |a|$$

which is true for all values of a.

Hints for the proofs of the following properties are given in the exercises.

(5) PROPERTIES OF ABSOLUTE VALUES Suppose a and b are any real numbers and n is an integer. Then

1. $|ab| = |a||b|$ **2.** $\left|\dfrac{a}{b}\right| = \dfrac{|a|}{|b|}$ $(b \neq 0)$ **3.** $|a^n| = |a|^n$

For solving equations or inequalities involving absolute values, it is often very helpful to use the following statements:

(6) Suppose $a > 0$. Then

4. $|x| = a$ if and only if $x = \pm a$

5. $|x| < a$ if and only if $-a < x < a$

6. $|x| > a$ if and only if $x > a$ or $x < -a$

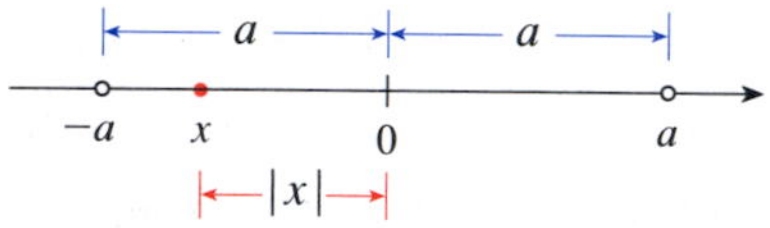

FIGURE 6

For instance, the inequality $|x| < a$ says that the distance from x to the origin is less than a, and you can see from Figure 6 that this is true if and only if x lies between $-a$ and a.

If a and b are any real numbers, then the distance between a and b is the absolute value of the difference, namely, $|a - b|$, which is also equal to $|b - a|$ (see Figure 7).

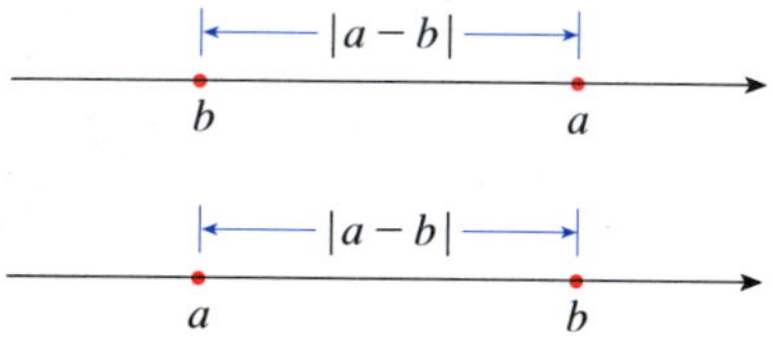

FIGURE 7
Length of a line segment = $|a - b|$

EXAMPLE 8 Solve $|2x - 5| = 3$.

SOLUTION By Property 4 of (6), $|2x - 5| = 3$ is equivalent to

$$2x - 5 = 3 \quad \text{or} \quad 2x - 5 = -3$$

So $2x = 8$, or $2x = 2$. Thus $x = 4$, or $x = 1$. ■

EXAMPLE 9 Solve $|x - 5| < 2$.

SOLUTION 1 By Property 5 of (6), $|x - 5| < 2$ is equivalent to

$$-2 < x - 5 < 2$$

Therefore, adding 5 to each side, we have

$$3 < x < 7$$

and the solution set is the open interval $(3, 7)$.

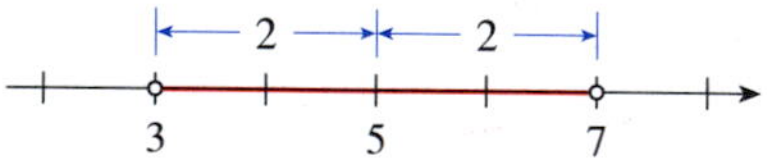

FIGURE 8

SOLUTION 2 Geometrically, the solution set consists of all numbers x whose distance from 5 is less than 2. From Figure 8 we see that this is the interval $(3, 7)$. ■

EXAMPLE 10 Solve $|3x + 2| \geqslant 4$.

SOLUTION By Properties 4 and 6 of (6), $|3x + 2| \geqslant 4$ is equivalent to

$$3x + 2 \geqslant 4 \quad \text{or} \quad 3x + 2 \leqslant -4$$

In the first case $3x \geqslant 2$, which gives $x \geqslant \frac{2}{3}$. In the second case $3x \leqslant -6$, which gives $x \leqslant -2$. So the solution set is

$$\{x \mid x \leqslant -2 \text{ or } x \geqslant \tfrac{2}{3}\} = (-\infty, -2] \cup [\tfrac{2}{3}, \infty)$$

■

Another important property of absolute value, called the Triangle Inequality, is used frequently not only in calculus but throughout mathematics in general.

(7) THE TRIANGLE INEQUALITY If a and b are any real numbers, then

$$|a + b| \leq |a| + |b|$$

Observe that if the numbers a and b are both positive or both negative, then the two sides in the Triangle Inequality are actually equal. But if a and b have opposite signs, the left side involves a subtraction and the right side does not. This makes the Triangle Inequality seem reasonable, but we can prove it as follows.

Notice that

$$-|a| \leq a \leq |a|$$

is always true because a equals either $|a|$ or $-|a|$. The corresponding statement for b is

$$-|b| \leq b \leq |b|$$

Adding these inequalities, we get

$$-(|a| + |b|) \leq a + b \leq |a| + |b|$$

If we now apply Properties 4 and 5 (with x replaced by $a + b$ and a by $|a| + |b|$), we obtain

$$|a + b| \leq |a| + |b|$$

which is what we wanted to show.

EXAMPLE 11 If $|x - 4| < 0.1$ and $|y - 7| < 0.2$, use the Triangle Inequality to estimate $|(x + y) - 11|$.

SOLUTION In order to use the given information, we use the Triangle Inequality with $a = x - 4$ and $b = y - 7$:

$$\begin{aligned} |(x + y) - 11| &= |(x - 4) + (y - 7)| \\ &\leq |x - 4| + |y - 7| \\ &< 0.1 + 0.2 = 0.3 \end{aligned}$$

Thus $$|(x + y) - 11| < 0.3$$ ■

EXERCISES A

1–12 ■ Rewrite the expression without using the absolute value symbol.

1. $|5 - 23|$

2. $|5| - |-23|$

3. $|-\pi|$

4. $|\pi - 2|$

5. $|\sqrt{5} - 5|$

6. $||-2| - |-3||$

7. $|x - 2|$ if $x < 2$

8. $|x - 2|$ if $x > 2$

9. $|x + 1|$

10. $|2x - 1|$

11. $|x^2 + 1|$

12. $|1 - 2x^2|$

13–46 ■ Solve the inequality in terms of intervals and illustrate the solution set on the real number line.

13. $2x + 7 > 3$

14. $3x - 11 < 4$

15. $1 - x \leq 2$

16. $4 - 3x \geq 6$

17. $2x + 1 < 5x - 8$

18. $1 + 5x > 5 - 3x$

19. $-1 < 2x - 5 < 7$

20. $1 < 3x + 4 \leqslant 16$

21. $0 \leqslant 1 - x < 1$

22. $-5 \leqslant 3 - 2x \leqslant 9$

23. $4x < 2x + 1 \leqslant 3x + 2$

24. $2x - 3 < x + 4 < 3x - 2$

25. $1 - x \geqslant 3 - 2x \geqslant x - 6$

26. $x > 1 - x \geqslant 3 + 2x$

27. $(x - 1)(x - 2) > 0$

28. $(2x + 3)(x - 1) \geqslant 0$

29. $2x^2 + x \leqslant 1$

30. $x^2 < 2x + 8$

31. $x^2 + x + 1 > 0$

32. $x^2 + x > 1$

33. $x^2 < 3$

34. $x^2 \geqslant 5$

35. $x^3 - x^2 \leqslant 0$

36. $(x + 1)(x - 2)(x + 3) \geqslant 0$

37. $x^3 > x$

38. $x^3 + 3x < 4x^2$

39. $\dfrac{1}{x} < 4$

40. $-3 < \dfrac{1}{x} \leqslant 1$

41. $\dfrac{4}{x} < x$

42. $\dfrac{x}{x + 1} > 3$

43. $\dfrac{2x + 1}{x - 5} < 3$

44. $\dfrac{2 + x}{3 - x} \leqslant 1$

45. $\dfrac{x^2 - 1}{x^2 + 1} \geqslant 0$

46. $\dfrac{x^2 - 2x}{x^2 - 2} > 0$

47. The relationship between the Celsius and Fahrenheit temperature scales is given by $C = \frac{5}{9}(F - 32)$, where C is the temperature in degrees Celsius and F is the temperature in degrees Fahrenheit. What interval on the Celsius scale corresponds to the temperature range $50 \leqslant F \leqslant 95$?

48. Use the relationship between C and F given in Exercise 47 to find the interval on the Fahrenheit scale corresponding to the temperature range $20 \leqslant C \leqslant 30$.

49. As dry air moves upward, it expands and in so doing cools at a rate of about 1°C for each 100-m rise, up to about 12 km.
(a) If the ground temperature is 20°C, write a formula for the temperature at height h.
(b) What range of temperature can be expected if a plane takes off and reaches a maximum height of 5 km?

50. If a ball is thrown upward from the top of a building 128 ft high with an initial velocity of 16 ft/s, then the height h above the ground t seconds later will be

$$h = 128 + 16t - 16t^2$$

During what time interval will the ball be at least 32 ft above the ground?

51–54 ■ Solve the equation for x.

51. $|2x| = 3$

52. $|3x + 5| = 1$

53. $|x + 3| = |2x + 1|$

54. $\left|\dfrac{2x - 1}{x + 1}\right| = 3$

55–68 ■ Solve the inequality.

55. $|x| < 3$

56. $|x| \geqslant 3$

57. $|x - 4| < 1$

58. $|x - 6| < 0.1$

59. $|x + 5| \geqslant 2$

60. $|x + 1| \geqslant 3$

61. $|2x - 3| \leqslant 0.4$

62. $|5x - 2| < 6$

63. $1 \leqslant |x| \leqslant 4$

64. $0 < |x - 5| < \frac{1}{2}$

65. $|x| > |x - 1|$

66. $|2x - 5| \leqslant |x + 4|$

67. $\left|\dfrac{x}{2 + x}\right| < 1$

68. $\left|\dfrac{2 - 3x}{1 + 2x}\right| \leqslant 4$

69–70 ■ Solve for x, assuming a, b, and c are positive constants.

69. $a(bx - c) \geqslant bc$

70. $a \leqslant bx + c < 2a$

71–72 ■ Solve for x, assuming a, b, and c are negative constants.

71. $ax + b < c$

72. $\dfrac{ax + b}{c} \leqslant b$

73. Suppose that $|x - 2| < 0.01$ and $|y - 3| < 0.04$. Use the Triangle Inequality to show that $|(x + y) - 5| < 0.05$.

74. Show that if $|x + 3| < \frac{1}{2}$, then $|4x + 13| < 3$.

75. Show that if $a < b$, then $a < \dfrac{a + b}{2} < b$.

76. Use Rule 3 to prove Rule 5 of (2).

77. Prove that $|ab| = |a||b|$. [*Hint:* Use Equation 4.]

78. Prove that $\left|\dfrac{a}{b}\right| = \dfrac{|a|}{|b|}$.

79. Show that if $0 < a < b$, then $a^2 < b^2$.

80. Prove that $|x - y| \geqslant |x| - |y|$. [*Hint:* Use the Triangle Inequality with $a = x - y$ and $b = y$.]

81. Show that the sum, difference, and product of rational numbers are rational numbers.

82. (a) Is the sum of two irrational numbers always an irrational number?
(b) Is the product of two irrational numbers always an irrational number?

B COORDINATE GEOMETRY AND LINES

Just as the points on a line can be identified with real numbers by assigning them coordinates, as described in Appendix A, so the points in a plane can be identified with ordered pairs of real numbers. We start by drawing two perpendicular coordinate lines that intersect at the origin O on each line. Usually one line is horizontal with positive direction to the right and is called the x-axis; the other line is vertical with positive direction upward and is called the y-axis.

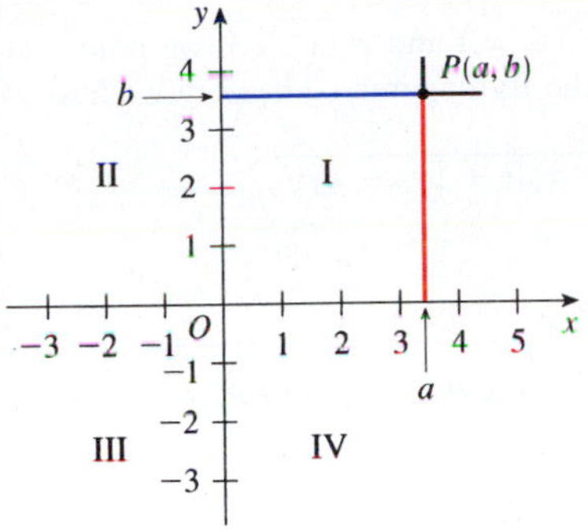

FIGURE 1

Any point P in the plane can be located by a unique ordered pair of numbers as follows. Draw lines through P perpendicular to the x- and y-axes. These lines intersect the axes in points with coordinates a and b as shown in Figure 1. Then the point P is assigned the ordered pair (a, b). The first number a is called the **x-coordinate** (or **abscissa**) of P; the second number b is called the **y-coordinate** (or **ordinate**) of P. We say that P is the point with coordinates (a, b), and we denote the point by the symbol $P(a, b)$. Several points are labeled with their coordinates in Figure 2.

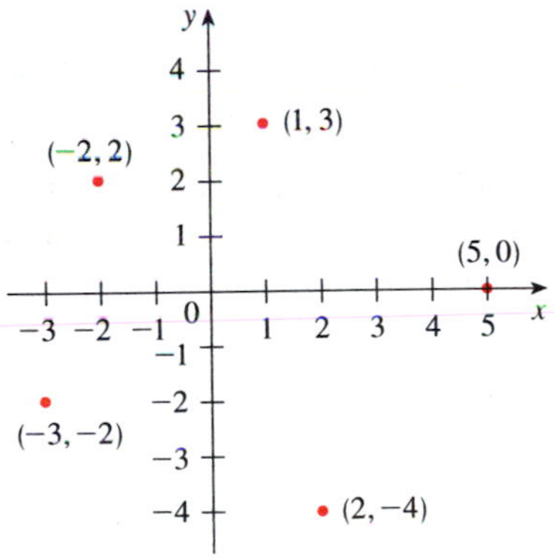

FIGURE 2

By reversing the preceding process we can start with an ordered pair (a, b) and arrive at the corresponding point P. Often we identify the point P with the ordered pair (a, b) and refer to "the point (a, b)." [Although the notation used for an open interval (a, b) is the same as the notation used for a point (a, b), you will be able to tell from the context which meaning is intended.]

This coordinate system is called the **rectangular coordinate system** or the **Cartesian coordinate system** in honor of the French mathematician René Descartes (1596–1650), even though another Frenchman, Pierre Fermat (1601–1665), invented the principles of analytic geometry at about the same time as Descartes. The plane supplied with this coordinate system is called the **coordinate plane** or the **Cartesian plane** and is denoted by $\mathbb{R}^2$.

The x- and y-axes are called the **coordinate axes** and divide the Cartesian plane into four quadrants, which are labeled I, II, III, and IV in Figure 1. Notice that the first quadrant consists of those points whose x- and y-coordinates are both positive.

EXAMPLE 1 Describe and sketch the regions given by the following sets:

(a) $\{(x, y) \mid x \geqslant 0\}$ (b) $\{(x, y) \mid y = 1\}$ (c) $\{(x, y) \mid |y| < 1\}$

SOLUTION

(a) The points whose x-coordinates are 0 or positive lie on the y-axis or to the right of it [see Figure 3(a)].

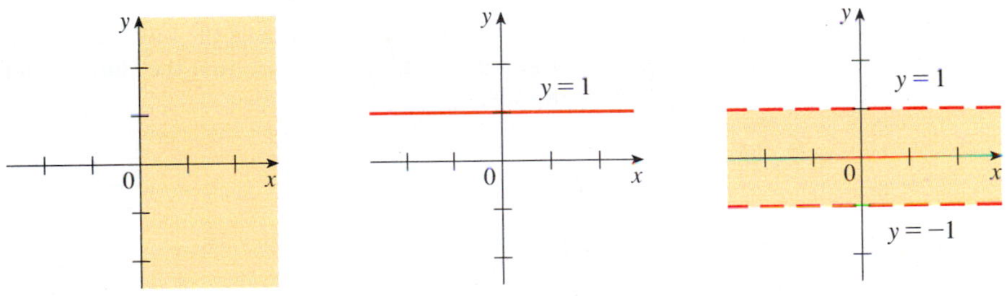

FIGURE 3

(a) $x \geqslant 0$ (b) $y = 1$ (c) $|y| < 1$

(b) The set of all points with y-coordinate 1 is a horizontal line one unit above the x-axis [see Figure 3(b)].

(c) Recall from Appendix A that

$$|y| < 1 \quad \text{if and only if} \quad -1 < y < 1$$

The given region consists of those points in the plane whose y-coordinates lie between -1 and 1. Thus, the region consists of all points that lie between (but not on) the horizontal lines $y = 1$ and $y = -1$. [These lines are shown as broken lines in Figure 3(c) to indicate that the points on these lines do not lie in the set.] ■

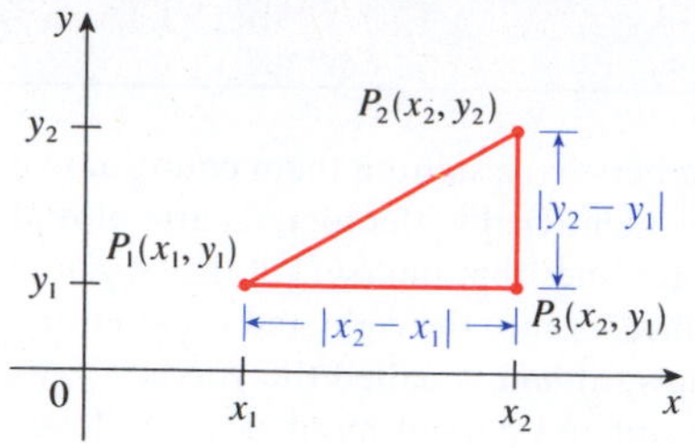

FIGURE 4

Recall from Appendix A that the distance between points a and b on a number line is $|a - b| = |b - a|$. Thus, the distance between points $P_1(x_1, y_1)$ and $P_3(x_2, y_1)$ on a horizontal line must be $|x_2 - x_1|$ and the distance between $P_2(x_2, y_2)$ and $P_3(x_2, y_1)$ on a vertical line must be $|y_2 - y_1|$ (see Figure 4).

To find the distance $|P_1P_2|$ between any two points $P_1(x_1, y_1)$ and $P_2(x_2, y_2)$, we note that triangle $P_1P_2P_3$ in Figure 4 is a right triangle, and so by the Pythagorean Theorem we have

$$|P_1P_2| = \sqrt{|P_1P_3|^2 + |P_2P_3|^2} = \sqrt{|x_2 - x_1|^2 + |y_2 - y_1|^2} = \sqrt{(x_2 - x_1)^2 + (y_2 - y_1)^2}$$

(1) DISTANCE FORMULA The distance between the points $P_1(x_1, y_1)$ and $P_2(x_2, y_2)$ is

$$|P_1P_2| = \sqrt{(x_2 - x_1)^2 + (y_2 - y_1)^2}$$

EXAMPLE 2 The distance between $(1, -2)$ and $(5, 3)$ is

$$\sqrt{(5 - 1)^2 + [3 - (-2)]^2} = \sqrt{4^2 + 5^2} = \sqrt{41}$$ ■

LINES

We want to find an equation of a given line L; such an equation is satisfied by the coordinates of the points on L and by no other points. To find the equation of L we use its *slope*, which is a measure of the steepness of the line.

(2) DEFINITION The **slope** of a nonvertical line that passes through the points $P_1(x_1, y_1)$ and $P_2(x_2, y_2)$ is

$$m = \frac{y_2 - y_1}{x_2 - x_1}$$

The slope of a vertical line is not defined.

Thus, the slope of a line is the ratio of the change in y to the change in x. From the similar triangles in Figure 5 we see that the slope is independent of which two points are chosen on the line:

$$\frac{y_2 - y_1}{x_2 - x_1} = \frac{y_2' - y_1'}{x_2' - x_1'}$$

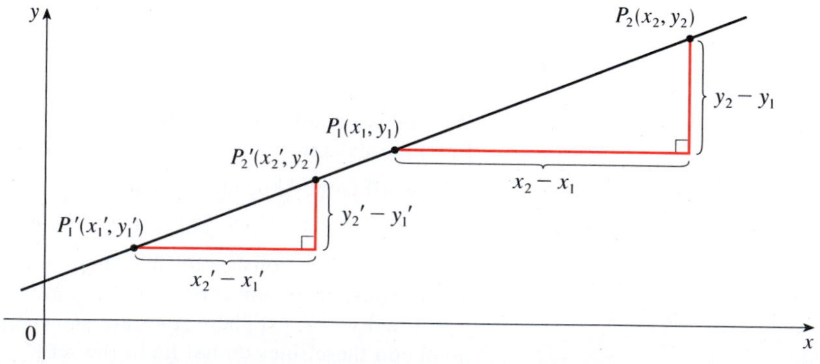

FIGURE 5

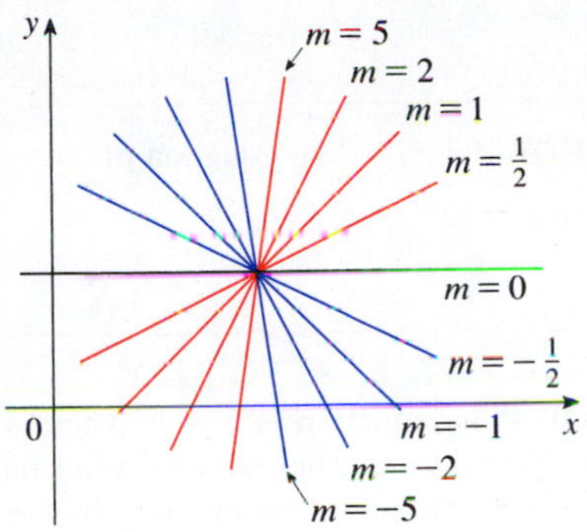

FIGURE 6

Figure 6 shows several lines labeled with their slopes. Notice that lines with positive slope slant upward to the right, whereas lines with negative slope slant downward to the right. Notice also that the steepest lines are the ones for which the absolute value of the slope is largest, and a horizontal line has slope 0.

Now let's find the equation of the line that passes through a given point $P_1(x_1, y_1)$ and has slope m. A point $P(x, y)$ with $x \neq x_1$ lies on this line if and only if the slope of the line through P_1 and P is equal to m; that is,

$$\frac{y - y_1}{x - x_1} = m$$

This equation can be rewritten in the form

$$y - y_1 = m(x - x_1)$$

and we observe that this equation is also satisfied when $x = x_1$ and $y = y_1$. Therefore, it is an equation of the given line.

(3) POINT-SLOPE FORM OF THE EQUATION OF A LINE An equation of the line passing through the point $P_1(x_1, y_1)$ and having slope m is

$$y - y_1 = m(x - x_1)$$

EXAMPLE 3 Find an equation of the line through $(1, -7)$ with slope $-\frac{1}{2}$.

SOLUTION Using (3) with $m = -\frac{1}{2}$, $x_1 = 1$, and $y_1 = -7$, we obtain an equation of the line as

$$y + 7 = -\tfrac{1}{2}(x - 1)$$

which we can rewrite as

$$2y + 14 = -x + 1 \qquad \text{or} \qquad x + 2y + 13 = 0$$

EXAMPLE 4 Find an equation of the line through the points $(-1, 2)$ and $(3, -4)$.

SOLUTION By Definition 2 the slope of the line is

$$m = \frac{-4 - 2}{3 - (-1)} = -\frac{3}{2}$$

Using the point-slope form with $x_1 = -1$ and $y_1 = 2$, we obtain

$$y - 2 = -\tfrac{3}{2}(x + 1)$$

which simplifies to

$$3x + 2y = 1$$

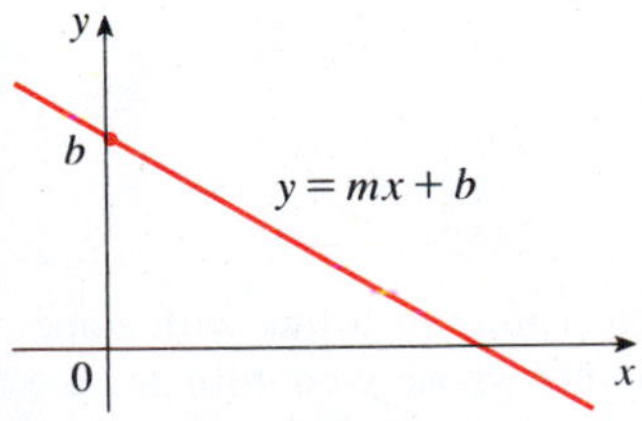

FIGURE 7

Suppose a nonvertical line has slope m and y-intercept b (see Figure 7). This means it intersects the y-axis at the point $(0, b)$, so the point-slope form of the equation of the line, with $x_1 = 0$ and $y_1 = b$, becomes

$$y - b = m(x - 0)$$

This simplifies as follows:

(4) SLOPE-INTERCEPT FORM OF THE EQUATION OF A LINE An equation of the line with slope m and y-intercept b is

$$y = mx + b$$

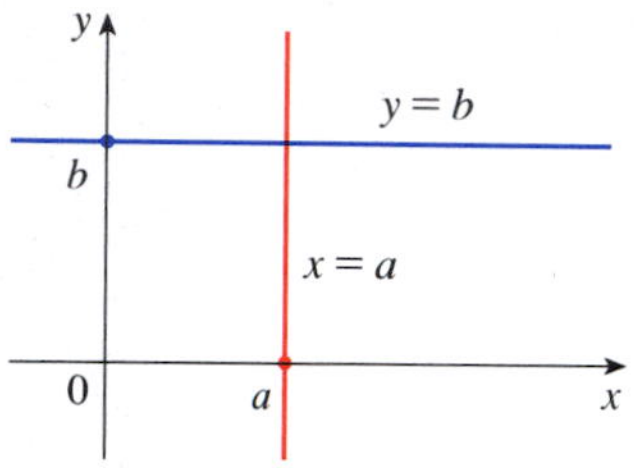

FIGURE 8

In particular, if a line is horizontal, its slope is $m = 0$, so its equation is $y = b$, where b is the y-intercept (see Figure 8). A vertical line does not have a slope, but we can write its equation as $x = a$, where a is the x-intercept, because the x-coordinate of every point on the line is a.

Observe that the equation of every line can be written in the form

$$Ax + By + C = 0 \qquad (5)$$

because a vertical line has the equation $x = a$ or $x - a = 0$ ($A = 1$, $B = 0$, $C = -a$) and a nonvertical line has the equation $y = mx + b$ or $-mx + y - b = 0$ ($A = -m$, $B = 1$, $C = -b$). Conversely, if we start with a general first-degree equation, that is, an equation of the form (5), where A, B, and C are constants and A and B are not both 0, then we can show that it is the equation of a line. If $B = 0$, it becomes $Ax + C = 0$ or $x = -C/A$, which represents a vertical line with x-intercept $-C/A$. If $B \neq 0$, the equation can be rewritten by solving for y:

$$y = -\frac{A}{B}x - \frac{C}{B}$$

and we recognize this as being the slope-intercept form of the equation of a line ($m = -A/B$, $b = -C/B$). Therefore, an equation of the form (5) is called a **linear equation** or the **general equation of a line**. For brevity, we often refer to "the line $Ax + By + C = 0$" instead of "the line whose equation is $Ax + By + C = 0$."

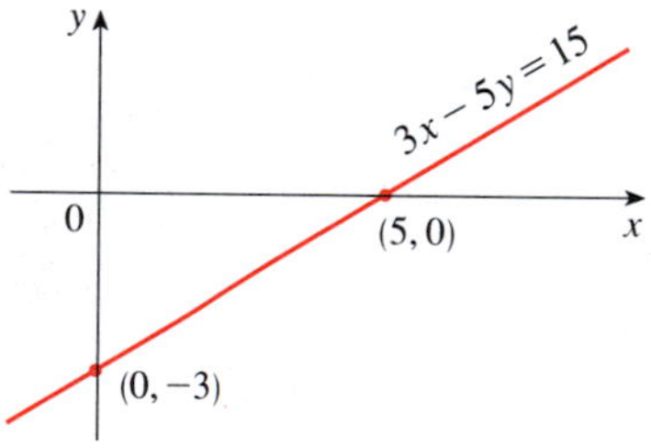

FIGURE 9

EXAMPLE 5 Sketch the graph of the equation $3x - 5y = 15$.

SOLUTION Since the equation is linear, its graph is a line. To draw the graph, we can simply find two points on the line. It is easiest to find the intercepts. Substituting $y = 0$ (the equation of the x-axis) in the given equation, we get $3x = 15$, so $x = 5$ is the x-intercept. Substituting $x = 0$ in the equation, we see that the y-intercept is -3. This allows us to sketch the graph as in Figure 9. ■

EXAMPLE 6 Graph the inequality $x + 2y > 5$.

SOLUTION We are asked to sketch the graph of the set $\{(x, y) \mid x + 2y > 5\}$ and we do so by solving the inequality for y:

$$x + 2y > 5$$

$$2y > -x + 5$$

$$y > -\tfrac{1}{2}x + \tfrac{5}{2}$$

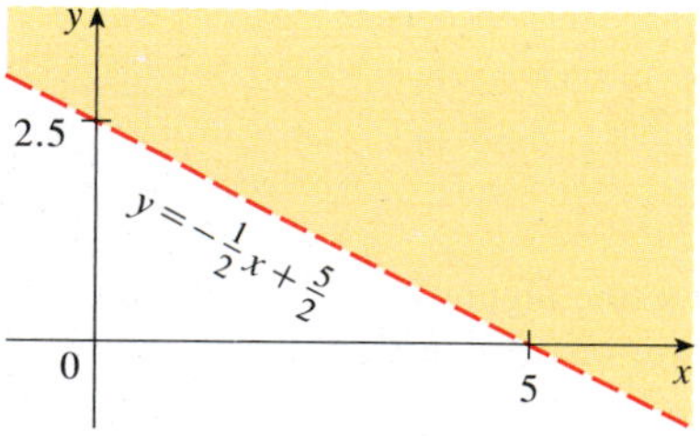

FIGURE 10

Compare this inequality with the equation $y = -\frac{1}{2}x + \frac{5}{2}$, which represents a line with slope $-\frac{1}{2}$ and y-intercept $\frac{5}{2}$. We see that the given graph consists of points whose y-coordinates are *larger* than those on the line $y = -\frac{1}{2}x + \frac{5}{2}$. Thus, the graph is the region that lies *above* the line, as illustrated in Figure 10. ■

PARALLEL AND PERPENDICULAR LINES

Slopes can be used to show that lines are parallel or perpendicular. The following facts are proved, for instance, in *Precalculus: Mathematics for Calculus, Third Edition* by Stewart, Redlin, and Watson (Brooks/Cole Publishing Co., Pacific Grove, CA, 1998).

(6) PARALLEL AND PERPENDICULAR LINES

1. Two nonvertical lines are parallel if and only if they have the same slope.
2. Two lines with slopes m_1 and m_2 are perpendicular if and only if $m_1 m_2 = -1$; that is, their slopes are negative reciprocals:

$$m_2 = -\frac{1}{m_1}$$

EXAMPLE 7 Find an equation of the line through the point $(5, 2)$ that is parallel to the line $4x + 6y + 5 = 0$.

SOLUTION The given line can be written in the form

$$y = -\tfrac{2}{3}x - \tfrac{5}{6}$$

which is in slope-intercept form with $m = -\frac{2}{3}$. Parallel lines have the same slope, so the required line has slope $-\frac{2}{3}$ and its equation in point-slope form is

$$y - 2 = -\tfrac{2}{3}(x - 5)$$

This simplifies to $2x + 3y = 16$. ■

EXAMPLE 8 Show that the lines $2x + 3y = 1$ and $6x - 4y - 1 = 0$ are perpendicular.

SOLUTION The equations can be written as

$$y = -\tfrac{2}{3}x + \tfrac{1}{3} \qquad \text{and} \qquad y = \tfrac{3}{2}x - \tfrac{1}{4}$$

from which we see that the slopes are

$$m_1 = -\tfrac{2}{3} \qquad \text{and} \qquad m_2 = \tfrac{3}{2}$$

Since $m_1 m_2 = -1$, the lines are perpendicular. ■

EXERCISES B

1–6 ■ Find the distance between the given points.

1. $(1, 1)$, $(4, 5)$
2. $(1, -3)$, $(5, 7)$
3. $(6, -2)$, $(-1, 3)$
4. $(1, -6)$, $(-1, -3)$
5. $(2, 5)$, $(4, -7)$
6. (a, b), (b, a)

7–10 ■ Find the slope of the line through P and Q.

7. $P(1, 5)$, $Q(4, 11)$
8. $P(-1, 6)$, $Q(4, -3)$
9. $P(-3, 3)$, $Q(-1, -6)$
10. $P(-1, -4)$, $Q(6, 0)$
11. Show that the triangle with vertices $A(0, 2)$, $B(-3, -1)$, and $C(-4, 3)$ is isosceles.
12. (a) Show that the triangle with vertices $A(6, -7)$, $B(11, -3)$, and $C(2, -2)$ is a right triangle using the converse of the Pythagorean Theorem.
    (b) Use slopes to show that ABC is a right triangle.
    (c) Find the area of the triangle.
13. Show that the points $(-2, 9)$, $(4, 6)$, $(1, 0)$, and $(-5, 3)$ are the vertices of a square.
14. (a) Show that the points $A(-1, 3)$, $B(3, 11)$, and $C(5, 15)$ are collinear by showing that $|AB| + |BC| = |AC|$.
    (b) Use slopes to show that A, B, and C are collinear.
15. Show that $A(1, 1)$, $B(7, 4)$, $C(5, 10)$, and $D(-1, 7)$ are vertices of a parallelogram.

16. Show that $A(1, 1)$, $B(11, 3)$, $C(10, 8)$, and $D(0, 6)$ are vertices of a rectangle.

17–20 ■ Sketch the graph of the equation.

17. $x = 3$

18. $y = -2$

19. $xy = 0$

20. $|y| = 1$

21–36 ■ Find an equation of the line that satisfies the given conditions.

21. Through $(2, -3)$, slope 6

22. Through $(-1, 4)$, slope -3

23. Through $(1, 7)$, slope $\frac{2}{3}$

24. Through $(-3, -5)$, slope $-\frac{7}{2}$

25. Through $(2, 1)$ and $(1, 6)$

26. Through $(-1, -2)$ and $(4, 3)$

27. Slope 3, y-intercept -2

28. Slope $\frac{2}{5}$, y-intercept 4

29. x-intercept 1, y-intercept -3

30. x-intercept -8, y-intercept 6

31. Through $(4, 5)$, parallel to the x-axis

32. Through $(4, 5)$, parallel to the y-axis

33. Through $(1, -6)$, parallel to the line $x + 2y = 6$

34. y-intercept 6, parallel to the line $2x + 3y + 4 = 0$

35. Through $(-1, -2)$, perpendicular to the line $2x + 5y + 8 = 0$

36. Through $\left(\frac{1}{2}, -\frac{2}{3}\right)$, perpendicular to the line $4x - 8y = 1$

37–42 ■ Find the slope and y-intercept of the line and draw its graph.

37. $x + 3y = 0$

38. $2x - 5y = 0$

39. $y = -2$

40. $2x - 3y + 6 = 0$

41. $3x - 4y = 12$

42. $4x + 5y = 10$

43–52 ■ Sketch the region in the xy-plane.

43. $\{(x, y) \mid x < 0\}$

44. $\{(x, y) \mid y > 0\}$

45. $\{(x, y) \mid xy < 0\}$

46. $\{(x, y) \mid x \geq 1 \text{ and } y < 3\}$

47. $\{(x, y) \mid |x| \leq 2\}$

48. $\{(x, y) \mid |x| < 3 \text{ and } |y| < 2\}$

49. $\{(x, y) \mid 0 \leq y \leq 4 \text{ and } x \leq 2\}$

50. $\{(x, y) \mid y > 2x - 1\}$

51. $\{(x, y) \mid 1 + x \leq y \leq 1 - 2x\}$

52. $\{(x, y) \mid -x \leq y < \frac{1}{2}(x + 3)\}$

53. Find a point on the y-axis that is equidistant from $(5, -5)$ and $(1, 1)$.

54. Show that the midpoint of the line segment from $P_1(x_1, y_1)$ to $P_2(x_2, y_2)$ is

$$\left(\frac{x_1 + x_2}{2}, \frac{y_1 + y_2}{2}\right)$$

55. Find the midpoint of the line segment joining the points
(a) $(1, 3)$ and $(7, 15)$ (b) $(-1, 6)$ and $(8, -12)$

56. Find the lengths of the medians of the triangle with vertices $A(1, 0)$, $B(3, 6)$, and $C(8, 2)$. (A median is a line segment from a vertex to the midpoint of the opposite side.)

57. Show that the lines $2x - y = 4$ and $6x - 2y = 10$ are not parallel and find their point of intersection.

58. Show that the lines $3x - 5y + 19 = 0$ and $10x + 6y - 50 = 0$ are perpendicular and find their point of intersection.

59. Find the equation of the perpendicular bisector of the line segment joining the points $A(1, 4)$ and $B(7, -2)$.

60. (a) Find equations for the sides of the triangle with vertices $P(1, 0)$, $Q(3, 4)$, and $R(-1, 6)$.
(b) Find equations for the medians of this triangle. Where do they intersect?

61. (a) Show that if the x- and y-intercepts of a line are nonzero numbers a and b, then the equation of the line can be put in the form

$$\frac{x}{a} + \frac{y}{b} = 1$$

This equation is called the **two-intercept form** of an equation of a line.
(b) Use part (a) to find an equation of the line whose x-intercept is 6 and whose y-intercept is -8.

62. A car leaves Detroit at 2:00 P.M., traveling at a constant speed west along I-90. It passes Ann Arbor, 40 mi from Detroit, at 2:50 P.M.
(a) Express the distance traveled in terms of the time elapsed.
(b) Draw the graph of the equation in part (a).
(c) What is the slope of this line? What does it represent?

C GRAPHS OF SECOND-DEGREE EQUATIONS

In Appendix B we saw that a first-degree, or linear, equation $Ax + By + C = 0$ represents a line. In this section we discuss second-degree equations such as

$$x^2 + y^2 = 1 \qquad y = x^2 + 1 \qquad \frac{x^2}{9} + \frac{y^2}{4} = 1 \qquad x^2 - y^2 = 1$$

which represent a circle, a parabola, an ellipse, and a hyperbola, respectively.

The graph of such an equation in x and y is the set of all points (x, y) that satisfy the equation; it gives a visual representation of the equation. Conversely, given a curve in the xy-plane, we may have to find an equation that represents it, that is, an equation satisfied by the coordinates of the points on the curve and by no other points. This is the other half of the basic principle of analytic geometry as formulated by Descartes and Fermat. The idea is that if a geometric curve can be represented by an algebraic equation, then the rules of algebra can be used to analyze the geometric problem.

CIRCLES

As an example of this type of problem, let's find the equation of a circle with radius r and center (h, k). By definition, the circle is the set of all points $P(x, y)$ whose distance from the center $C(h, k)$ is r (see Figure 1). Thus, P is on the circle if and only if $|PC| = r$. From the distance formula, we have

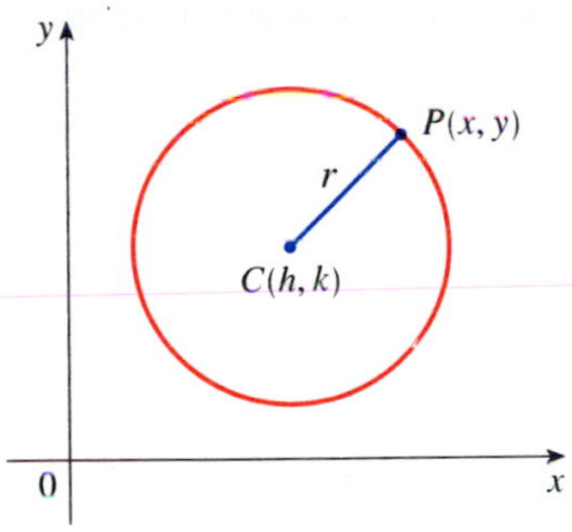

FIGURE 1

$$\sqrt{(x - h)^2 + (y - k)^2} = r$$

or equivalently, squaring both sides, we get

$$(x - h)^2 + (y - k)^2 = r^2$$

This is the desired equation.

(1) EQUATION OF A CIRCLE The equation of a circle with center (h, k) and radius r is

$$(x - h)^2 + (y - k)^2 = r^2$$

In particular, if the center is the origin $(0, 0)$, the equation is

$$x^2 + y^2 = r^2$$

EXAMPLE 1 Find an equation of the circle with radius 3 and center $(2, -5)$.

SOLUTION From Equation 1 with $r = 3$, $h = 2$, and $k = -5$, we obtain

$$(x - 2)^2 + (y + 5)^2 = 9$$ ■

EXAMPLE 2 Sketch the graph of the equation $x^2 + y^2 + 2x - 6y + 7 = 0$ by first showing that it represents a circle and then finding its center and radius.

SOLUTION We first group the x-terms and y-terms as follows:

$$(x^2 + 2x) + (y^2 - 6y) = -7$$

Then we complete the square within each grouping, adding the appropriate constants to both sides of the equation:

$$(x^2 + 2x + 1) + (y^2 - 6y + 9) = -7 + 1 + 9$$

or

$$(x + 1)^2 + (y - 3)^2 = 3$$

Comparing this equation with the standard equation of a circle (1), we see that $h = -1$, $k = 3$, and $r = \sqrt{3}$, so the given equation represents a circle with center $(-1, 3)$ and radius $\sqrt{3}$. It is sketched in Figure 2.

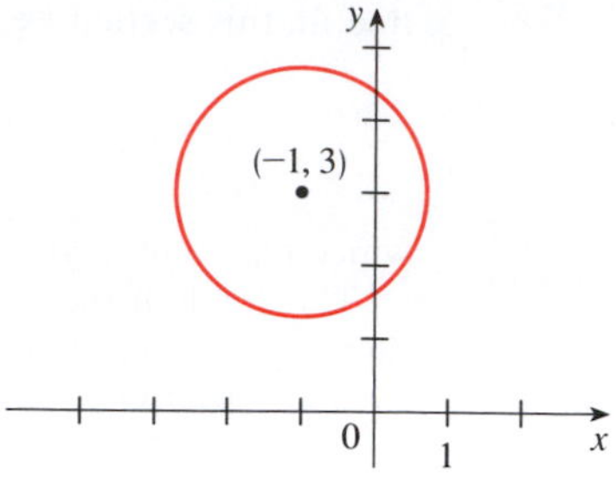

FIGURE 2
$x^2 + y^2 + 2x - 6y + 7 = 0$

PARABOLAS

The geometric properties of parabolas are reviewed in Appendix I. Here we regard a parabola as a graph of an equation of the form $y = ax^2 + bx + c$.

EXAMPLE 3 Draw the graph of the parabola $y = x^2$.

SOLUTION We set up a table of values, plot points, and join them by a smooth curve to obtain the graph in Figure 3.

x	$y = x^2$
0	0
$\pm\frac{1}{2}$	$\frac{1}{4}$
± 1	1
± 2	4
± 3	9

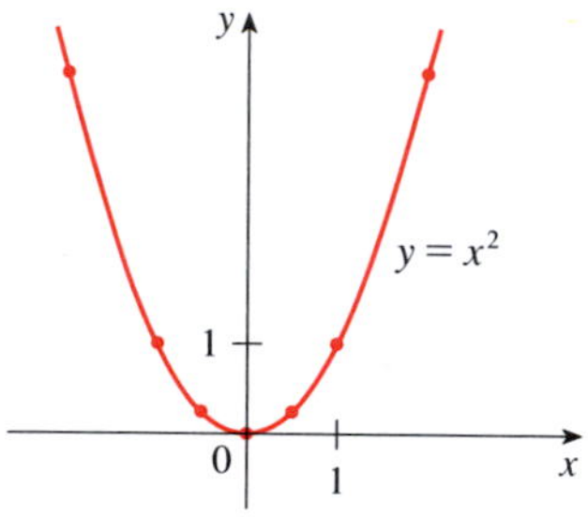

FIGURE 3

Figure 4 shows the graphs of several parabolas with equations of the form $y = ax^2$ for various values of the number a. In each case the *vertex*, the point where the parabola changes direction, is the origin. We see that the parabola $y = ax^2$ opens upward if $a > 0$ and downward if $a < 0$ (as in Figure 5).

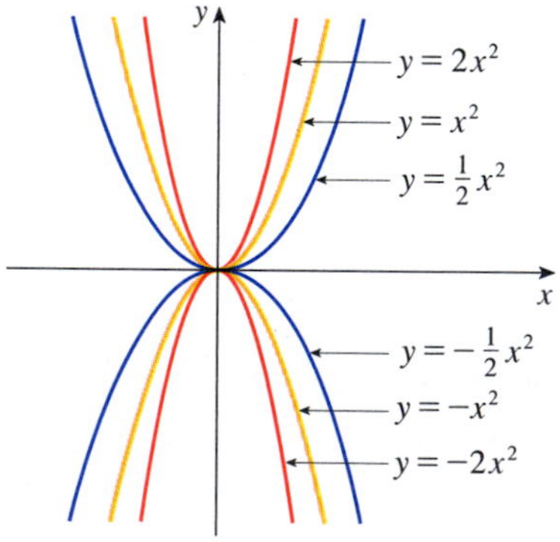

FIGURE 4

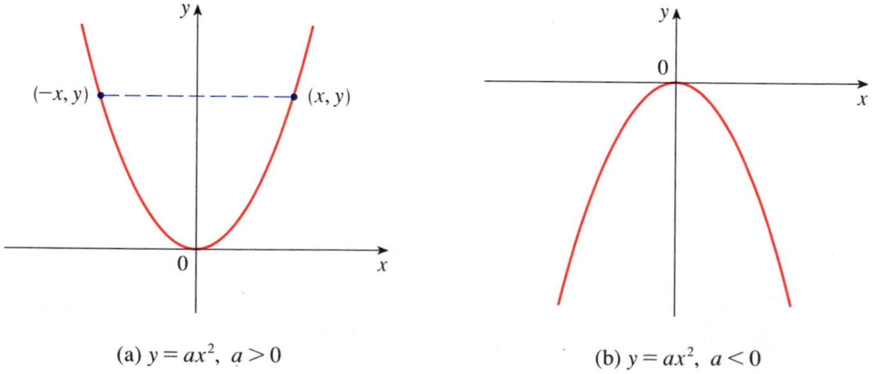

FIGURE 5

Notice that if (x, y) satisfies $y = ax^2$, then so does $(-x, y)$. This corresponds to the geometric fact that if the right half of the graph is reflected about the y-axis, then the left half of the graph is obtained. We say that the graph is **symmetric with respect to the y-axis**.

The graph of an equation is symmetric with respect to the y-axis if the equation is unchanged when x is replaced by $-x$.

If we interchange x and y in the equation $y = ax^2$, the result is $x = ay^2$, which also represents a parabola. (Interchanging x and y amounts to reflecting about the diagonal line $y = x$.) The parabola $x = ay^2$ opens to the right if $a > 0$ and to the left if $a < 0$ (see Figure 6). This time the parabola is symmetric with respect to the x-axis because if (x, y) satisfies $x = ay^2$, then so does $(x, -y)$.

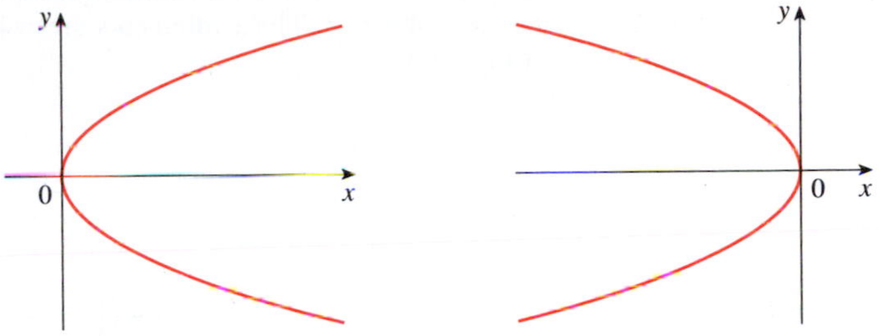

FIGURE 6 (a) $x = ay^2$, $a > 0$ (b) $x = ay^2$, $a < 0$

The graph of an equation is symmetric with respect to the x-axis if the equation is unchanged when y is replaced by $-y$.

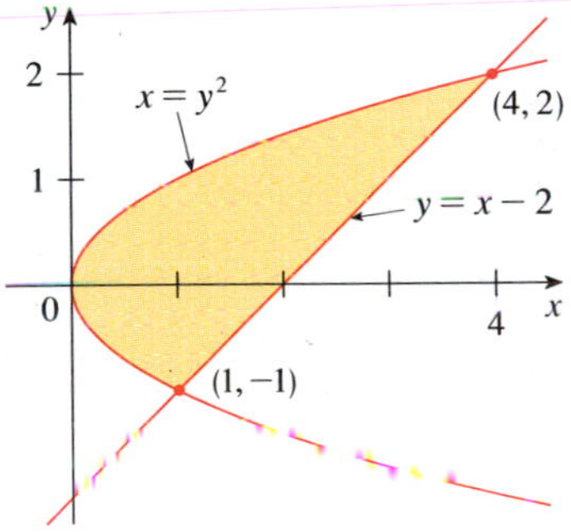

FIGURE 7

EXAMPLE 4 Sketch the region bounded by the parabola $x = y^2$ and the line $y = x - 2$.

SOLUTION First we find the points of intersection by solving the two equations. Substituting $x = y + 2$ into the equation $x = y^2$, we get $y + 2 = y^2$, which gives

$$0 = y^2 - y - 2 = (y - 2)(y + 1)$$

so $y = 2$ or -1. Thus the points of intersection are $(4, 2)$ and $(1, -1)$, and we draw the line $y = x - 2$ passing through these points. We then sketch the parabola $x = y^2$ by referring to Figure 6(a) and having the parabola pass through $(4, 2)$ and $(1, -1)$. The region bounded by $x = y^2$ and $y = x - 2$ means the finite region whose boundaries are these curves. It is sketched in Figure 7. ■

ELLIPSES

The curve with equation

(2) $$\frac{x^2}{a^2} + \frac{y^2}{b^2} = 1$$

where a and b are positive numbers, is called an **ellipse** in standard position. (Geometric properties of ellipses are discussed in Appendix I.) Observe that Equation 2 is unchanged if x is replaced by $-x$ or y is replaced by $-y$, so the ellipse is symmetric with respect to both axes. As a further aid to sketching the ellipse, we find its intercepts.

The **x-intercepts** of a graph are the x-coordinates of the points where the graph intersects the x-axis. They are found by setting $y = 0$ in the equation of the graph. The **y-intercepts** are the y-coordinates of the points where the graph intersects the y-axis. They are found by setting $x = 0$ in its equation.

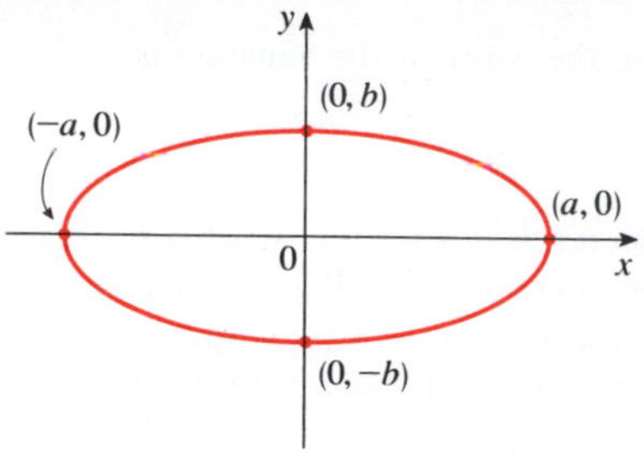

FIGURE 8
$\frac{x^2}{a^2}+\frac{y^2}{b^2}=1$

If we set $y = 0$ in Equation 2, we get $x^2 = a^2$ and so the x-intercepts are $\pm a$. Setting $x = 0$, we get $y^2 = b^2$, so the y-intercepts are $\pm b$. Using this information, together with symmetry, we sketch the ellipse in Figure 8. If $a = b$, the ellipse is a circle with radius a.

EXAMPLE 5 Sketch the graph of $9x^2 + 16y^2 = 144$.

SOLUTION We divide both sides of the equation by 144:

$$\frac{x^2}{16}+\frac{y^2}{9}=1$$

The equation is now in the standard form for an ellipse (2), so we have $a^2 = 16$, $b^2 = 9$, $a = 4$, and $b = 3$. The x-intercepts are ± 4; the y-intercepts are ± 3. The graph is sketched in Figure 9.

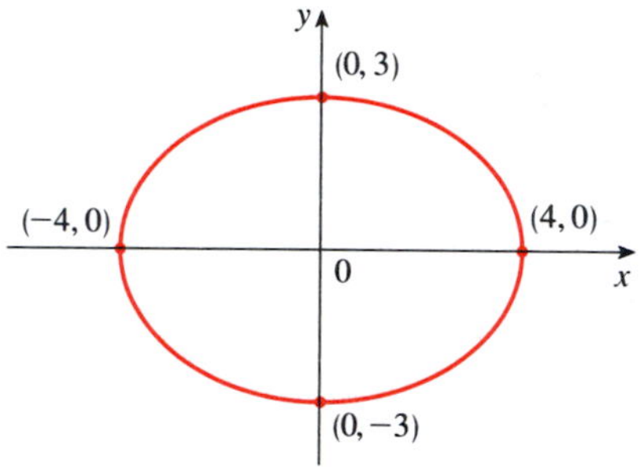

FIGURE 9
$9x^2+16y^2=144$

■

HYPERBOLAS

The curve with equation

(3)
$$\frac{x^2}{a^2}-\frac{y^2}{b^2}=1$$

is called a **hyperbola** in standard position. Again, Equation 3 is unchanged when x is replaced by $-x$ or y is replaced by $-y$, so the hyperbola is symmetric with respect to both axes. To find the x-intercepts we set $y = 0$ and obtain $x^2 = a^2$ and $x = \pm a$. However, if we put $x = 0$ in Equation 3, we get $y^2 = -b^2$, which is impossible, so there is no y-intercept. In fact, from Equation 3 we obtain

$$\frac{x^2}{a^2}=1+\frac{y^2}{b^2}\geqslant 1$$

which shows that $x^2 \geqslant a^2$ and so $|x| = \sqrt{x^2} \geqslant a$. Therefore, we have $x \geqslant a$ or $x \leqslant -a$. This means that the hyperbola consists of two parts, called its *branches*. It is sketched in Figure 10.

In drawing a hyperbola it is useful to draw first its *asymptotes,* which are the lines $y = (b/a)x$ and $y = -(b/a)x$ shown in Figure 10. Both branches of the hyperbola approach the asymptotes; that is, they come arbitrarily close to the asymptotes. This involves the idea of a limit, which is discussed in Chapter 2.

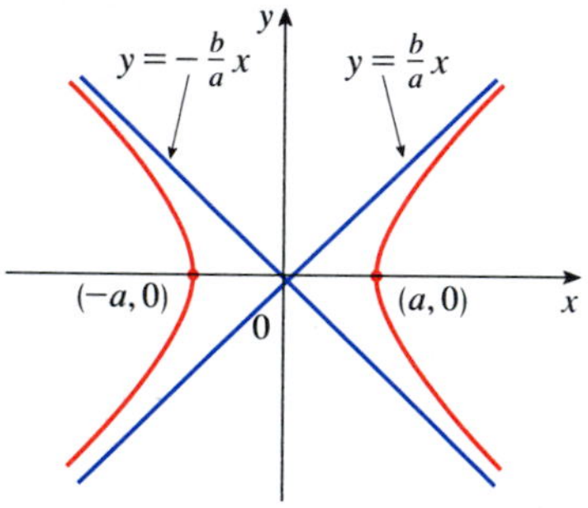

FIGURE 10
The hyperbola $\frac{x^2}{a^2}-\frac{y^2}{b^2}=1$

By interchanging the roles of x and y we get an equation of the form

$$\frac{y^2}{a^2}-\frac{x^2}{b^2}=1$$

which also represents a hyperbola and is sketched in Figure 11.

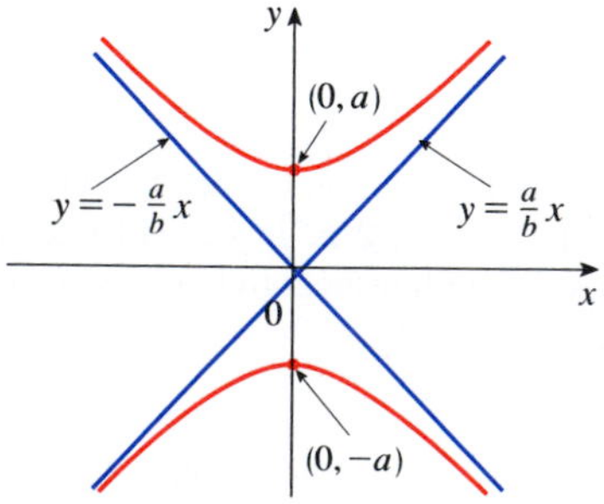

FIGURE 11
The hyperbola $\frac{y^2}{a^2}-\frac{x^2}{b^2}=1$

EXAMPLE 6 Sketch the curve $9x^2 - 4y^2 = 36$.

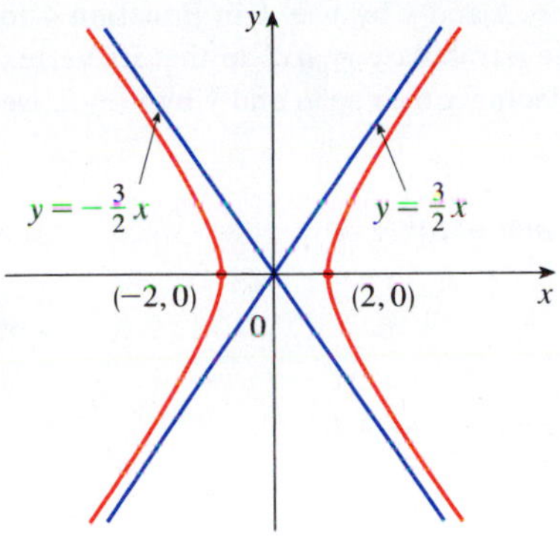

FIGURE 12
The hyperbola $9x^2 - 4y^2 = 36$

SOLUTION Dividing both sides by 36, we obtain

$$\frac{x^2}{4} - \frac{y^2}{9} = 1$$

which is the standard form of the equation of a hyperbola (Equation 3). Since $a^2 = 4$, the x-intercepts are ± 2. Since $b^2 = 9$, we have $b = 3$ and the asymptotes are $y = \pm(\frac{3}{2})x$. The hyperbola is sketched in Figure 12. ■

If $b = a$, a hyperbola has the equation $x^2 - y^2 = a^2$ (or $y^2 - x^2 = a^2$) and is called an *equilateral hyperbola* [see Figure 13(a)]. Its asymptotes are $y = \pm x$, which are perpendicular. If an equilateral hyperbola is rotated by 45°, the asymptotes become the x- and y-axes, and it can be shown that the new equation of the hyperbola is $xy = k$, where k is a constant [see Figure 13(b)].

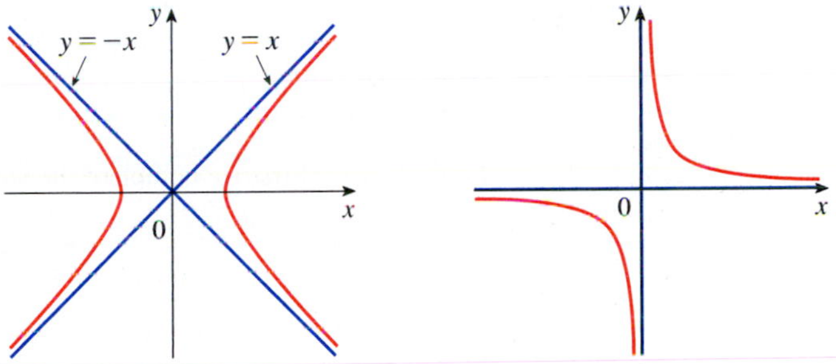

FIGURE 13
Equilateral hyperbolas

(a) $x^2 - y^2 = a^2$ (b) $xy = k$ $(k > 0)$

SHIFTED CONICS

Recall that the equation of a circle with center the origin and radius r is $x^2 + y^2 = r^2$, but if the center is the point (h, k), then the equation of the circle becomes

$$(x - h)^2 + (y - k)^2 = r^2$$

Similarly, if we take the ellipse with equation

(4) $$\frac{x^2}{a^2} + \frac{y^2}{b^2} = 1$$

and translate it (shift it) so that its center is the point (h, k), then its equation becomes

(5) $$\frac{(x - h)^2}{a^2} + \frac{(y - k)^2}{b^2} = 1$$

(See Figure 14.)

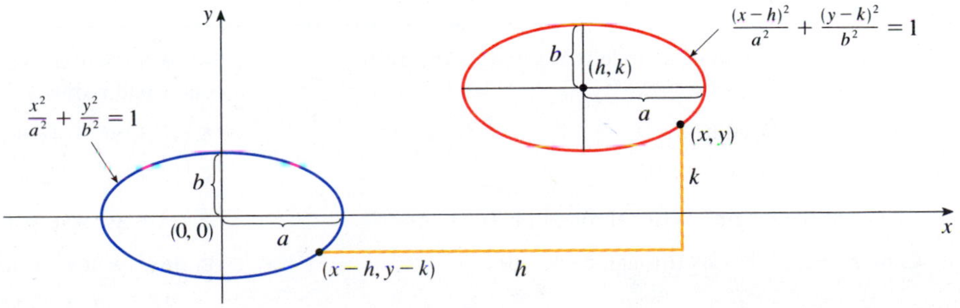

FIGURE 14

Notice that in shifting the ellipse, we replaced x by $x - h$ and y by $y - k$ in Equation 4 to obtain Equation 5. We use the same procedure to shift the parabola $y = ax^2$ so that its vertex (the origin) becomes the point (h, k) as in Figure 15. Replacing x by $x - h$ and y by $y - k$, we see that the new equation is

$$y - k = a(x - h)^2 \quad \text{or} \quad y = a(x - h)^2 + k$$

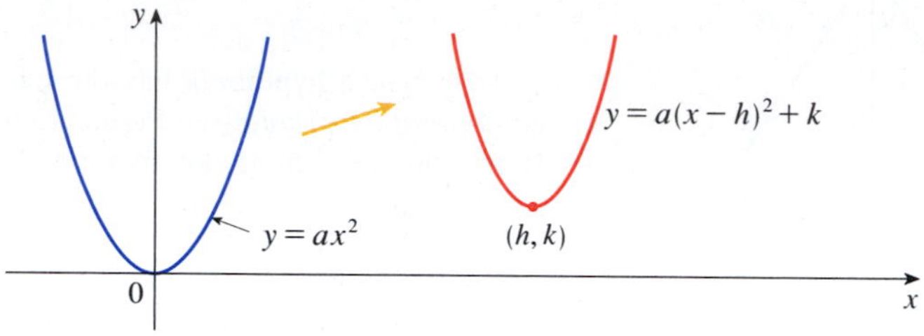

FIGURE 15

EXAMPLE 7 Sketch the graph of the equation $y = 2x^2 - 4x + 1$.

SOLUTION First we complete the square:

$$y = 2(x^2 - 2x) + 1 = 2(x - 1)^2 - 1$$

In this form we see that the equation represents the parabola obtained by shifting $y = 2x^2$ so that its vertex is at the point $(1, -1)$. The graph is sketched in Figure 16. ■

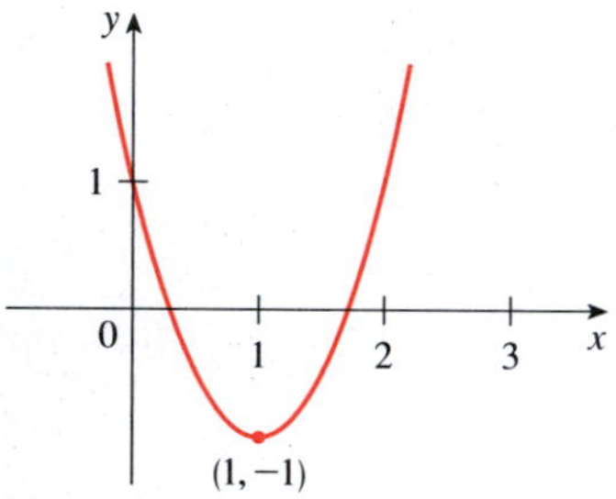

FIGURE 16
$y = 2x^2 - 4x + 1$

EXAMPLE 8 Sketch the curve $x = 1 - y^2$.

SOLUTION This time we start with the parabola $x = -y^2$ (as in Figure 6 with $a = -1$) and shift one unit to the right to get the graph of $x = 1 - y^2$ (see Figure 17).

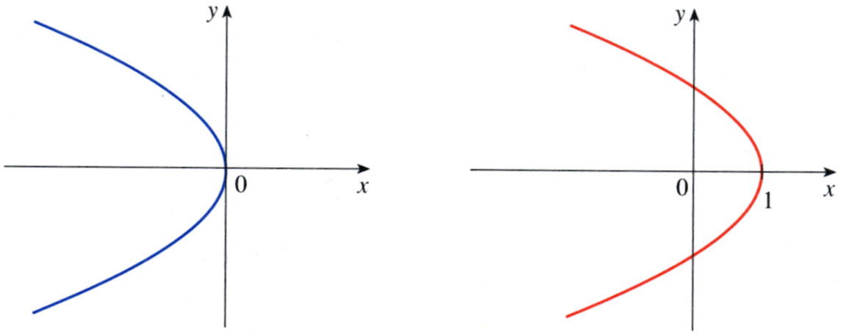

FIGURE 17 (a) $x = -y^2$ (b) $x = 1 - y^2$ ■

EXERCISES C

1–4 ■ Find an equation of a circle that satisfies the given conditions.

1. Center $(3, -1)$, radius 5
2. Center $(-2, -8)$, radius 10
3. Center at the origin, passes through $(4, 7)$
4. Center $(-1, 5)$, passes through $(-4, -6)$

5–9 ■ Show that the equation represents a circle and find the center and radius.

5. $x^2 + y^2 - 4x + 10y + 13 = 0$
6. $x^2 + y^2 + 6y + 2 = 0$
7. $x^2 + y^2 + x = 0$
8. $16x^2 + 16y^2 + 8x + 32y + 1 = 0$

9. $2x^2 + 2y^2 - x + y = 1$

10. Under what condition on the coefficients a, b, and c does the equation $x^2 + y^2 + ax + by + c = 0$ represent a circle? When that condition is satisfied, find the center and radius of the circle.

11–32 ■ Identify the type of curve and sketch the graph. Do not plot points. Just use the standard graphs given in Figures 5, 6, 8, 10, and 11 and shift if necessary.

11. $y = -x^2$

12. $y^2 - x^2 = 1$

13. $x^2 + 4y^2 = 16$

14. $x = -2y^2$

15. $16x^2 - 25y^2 = 400$

16. $25x^2 + 4y^2 = 100$

17. $4x^2 + y^2 = 1$

18. $y = x^2 + 2$

19. $x = y^2 - 1$

20. $9x^2 - 25y^2 = 225$

21. $9y^2 - x^2 = 9$

22. $2x^2 + 5y^2 = 10$

23. $xy = 4$

24. $y = x^2 + 2x$

25. $9(x - 1)^2 + 4(y - 2)^2 = 36$

26. $16x^2 + 9y^2 - 36y = 108$

27. $y = x^2 - 6x + 13$

28. $x^2 - y^2 - 4x + 3 = 0$

29. $x = 4 - y^2$

30. $y^2 - 2x + 6y + 5 = 0$

31. $x^2 + 4y^2 - 6x + 5 = 0$

32. $4x^2 + 9y^2 - 16x + 54y + 61 = 0$

33–34 ■ Sketch the region bounded by the curves.

33. $y = 3x$, $y = x^2$

34. $y = 4 - x^2$, $x - 2y = 2$

35. Find an equation of the parabola with vertex $(1, -1)$ that passes through the points $(-1, 3)$ and $(3, 3)$.

36. Find an equation of the ellipse with center at the origin that passes through the points $(1, -10\sqrt{2}/3)$ and $(-2, 5\sqrt{5}/3)$.

37–40 ■ Sketch the graph of the set.

37. $\{(x, y) \mid x^2 + y^2 \leqslant 1\}$

38. $\{(x, y) \mid x^2 + y^2 > 4\}$

39. $\{(x, y) \mid y \geqslant x^2 - 1\}$

40. $\{(x, y) \mid x^2 + 4y^2 \leqslant 4\}$

D TRIGONOMETRY

ANGLES

Angles can be measured in degrees or in radians (abbreviated as rad). The angle given by a complete revolution contains 360°, which is the same as 2π rad. Therefore

(1)
$$\pi \text{ rad} = 180°$$

and

(2)
$$1 \text{ rad} = \left(\frac{180}{\pi}\right)^\circ \approx 57.3° \qquad 1° = \frac{\pi}{180} \text{ rad} \approx 0.017 \text{ rad}$$

EXAMPLE 1

(a) Find the radian measure of 60°. (b) Express $5\pi/4$ rad in degrees.

SOLUTION

(a) From Equation 1 or 2 we see that to convert from degrees to radians we multiply by $\pi/180$. Therefore

$$60° = 60\left(\frac{\pi}{180}\right) = \frac{\pi}{3} \text{ rad}$$

(b) To convert from radians to degrees we multiply by $180/\pi$. Thus

$$\frac{5\pi}{4} \text{ rad} = \frac{5\pi}{4}\left(\frac{180}{\pi}\right) = 225°$$

■

In calculus we use radians to measure angles except when otherwise indicated. The following table gives the correspondence between degree and radian measures of some common angles.

Degrees	0°	30°	45°	60°	90°	120°	135°	150°	180°	270°	360°
Radians	0	$\frac{\pi}{6}$	$\frac{\pi}{4}$	$\frac{\pi}{3}$	$\frac{\pi}{2}$	$\frac{2\pi}{3}$	$\frac{3\pi}{4}$	$\frac{5\pi}{6}$	π	$\frac{3\pi}{2}$	2π

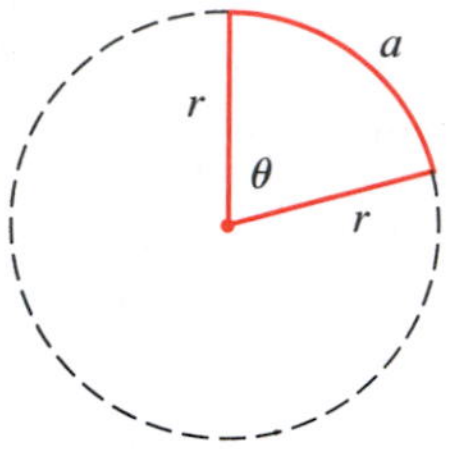

FIGURE 1

Figure 1 shows a sector of a circle with central angle θ and radius r subtending an arc with length a. Since the length of the arc is proportional to the size of the angle, and since the entire circle has circumference $2\pi r$ and central angle 2π, we have

$$\frac{\theta}{2\pi} = \frac{a}{2\pi r}$$

Solving this equation for θ and for a, we obtain

$$\textbf{(3)} \qquad \boxed{\theta = \frac{a}{r}} \qquad \boxed{a = r\theta}$$

Remember that Equations 3 are valid only when θ is measured in radians.

In particular, putting $a = r$ in Equation 3, we see that an angle of 1 rad is the angle subtended at the center of a circle by an arc equal in length to the radius of the circle (see Figure 2).

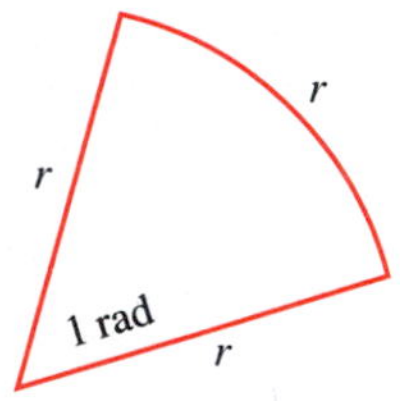

FIGURE 2

EXAMPLE 2

(a) If the radius of a circle is 5 cm, what angle is subtended by an arc of 6 cm?

(b) If a circle has radius 3 cm, what is the length of an arc subtended by a central angle of $3\pi/8$ rad?

SOLUTION

(a) Using Equation 3 with $a = 6$ and $r = 5$, we see that the angle is

$$\theta = \tfrac{6}{5} = 1.2 \text{ rad}$$

(b) With $r = 3$ cm and $\theta = 3\pi/8$ rad, the arc length is

$$a = r\theta = 3\left(\frac{3\pi}{8}\right) = \frac{9\pi}{8} \text{ cm}$$

■

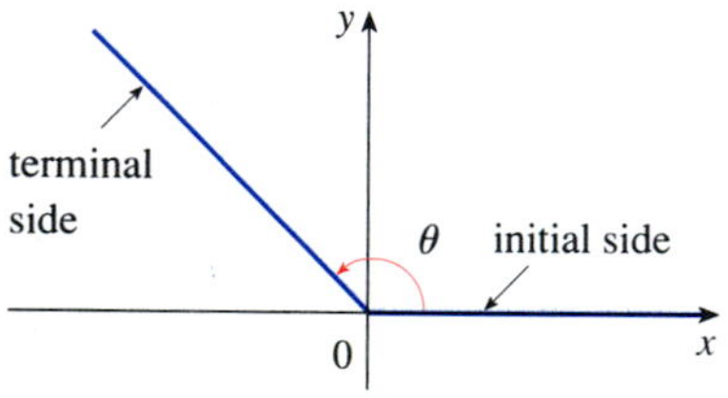

FIGURE 3

$\theta \geqslant 0$

The **standard position** of an angle occurs when we place its vertex at the origin of a coordinate system and its initial side on the positive x-axis as in Figure 3. A **positive** angle is obtained by rotating the initial side counterclockwise until it coincides with the terminal side. Likewise, **negative** angles are obtained by clockwise rotation as in Figure 4. Figure 5 shows several examples of angles in standard position. Notice that different angles can have the same terminal side. For instance, the angles $3\pi/4$, $-5\pi/4$, and $11\pi/4$ have the same initial and terminal sides because

$$\frac{3\pi}{4} - 2\pi = -\frac{5\pi}{4} \qquad \frac{3\pi}{4} + 2\pi = \frac{11\pi}{4}$$

and 2π rad represents a complete revolution.

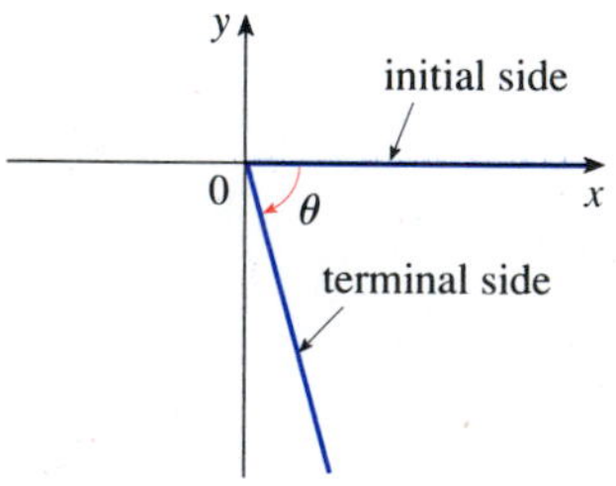

FIGURE 4

$\theta < 0$

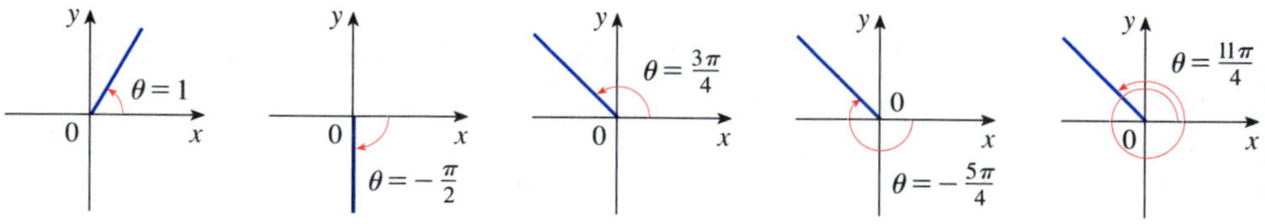

FIGURE 5

Angles in standard position

THE TRIGONOMETRIC FUNCTIONS

For an acute angle θ the six trigonometric functions are defined as ratios of lengths of sides of a right triangle as follows (see Figure 6):

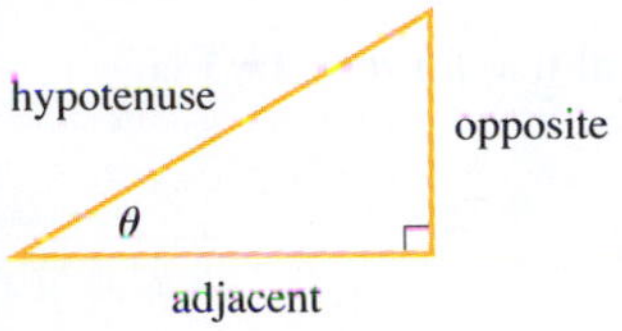

FIGURE 6

(4)

$$\sin\theta = \frac{\text{opp}}{\text{hyp}} \qquad \csc\theta = \frac{\text{hyp}}{\text{opp}}$$

$$\cos\theta = \frac{\text{adj}}{\text{hyp}} \qquad \sec\theta = \frac{\text{hyp}}{\text{adj}}$$

$$\tan\theta = \frac{\text{opp}}{\text{adj}} \qquad \cot\theta = \frac{\text{adj}}{\text{opp}}$$

This definition does not apply to obtuse or negative angles, so for a general angle θ in standard position we let $P(x, y)$ be any point on the terminal side of θ and we let r be the distance $|OP|$ as in Figure 7. Then we define

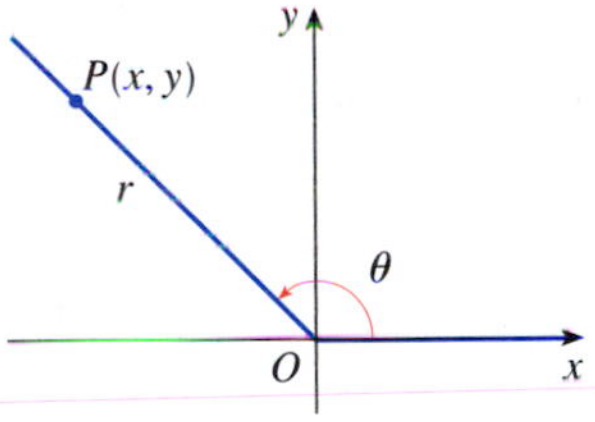

FIGURE 7

(5)

$$\sin\theta = \frac{y}{r} \qquad \csc\theta = \frac{r}{y}$$

$$\cos\theta = \frac{x}{r} \qquad \sec\theta = \frac{r}{x}$$

$$\tan\theta = \frac{y}{x} \qquad \cot\theta = \frac{x}{y}$$

Since division by 0 is not defined, $\tan\theta$ and $\sec\theta$ are undefined when $x = 0$ and $\csc\theta$ and $\cot\theta$ are undefined when $y = 0$. Notice that the definitions in (4) and (5) are consistent when θ is an acute angle.

If θ is a number, the convention is that $\sin\theta$ means the sine of the angle whose *radian* measure is θ. For example, the expression $\sin 3$ implies that we are dealing with an angle of 3 rad. When finding a calculator approximation to this number we must remember to set our calculator in radian mode, and then we obtain

$$\sin 3 \approx 0.14112$$

If we want to know the sine of the angle 3° we would write $\sin 3°$ and, with our calculator in degree mode, we find that

$$\sin 3° \approx 0.05234$$

The exact trigonometric ratios for certain angles can be read from the triangles in Figure 8. For instance,

$$\sin\frac{\pi}{4} = \frac{1}{\sqrt{2}} \qquad \sin\frac{\pi}{6} = \frac{1}{2} \qquad \sin\frac{\pi}{3} = \frac{\sqrt{3}}{2}$$

$$\cos\frac{\pi}{4} = \frac{1}{\sqrt{2}} \qquad \cos\frac{\pi}{6} = \frac{\sqrt{3}}{2} \qquad \cos\frac{\pi}{3} = \frac{1}{2}$$

$$\tan\frac{\pi}{4} = 1 \qquad \tan\frac{\pi}{6} = \frac{1}{\sqrt{3}} \qquad \tan\frac{\pi}{3} = \sqrt{3}$$

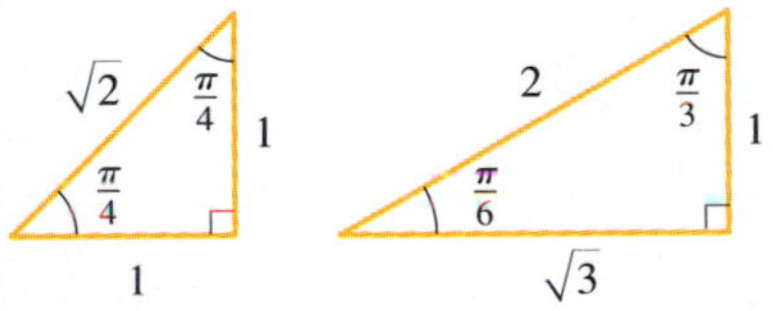

FIGURE 8

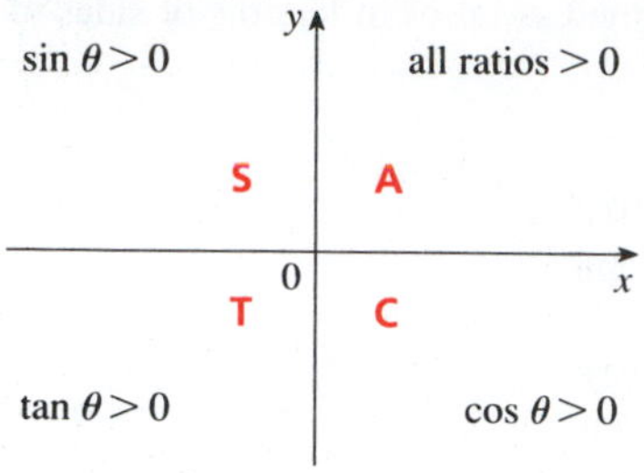

FIGURE 9

The signs of the trigonometric functions for angles in each of the four quadrants can be remembered by means of the rule "All Students Take Calculus" shown in Figure 9.

EXAMPLE 3 Find the exact trigonometric ratios for $\theta = 2\pi/3$.

SOLUTION From Figure 10 we see that a point on the terminal line for $\theta = 2\pi/3$ is $P(-1, \sqrt{3})$. Therefore, taking

$$x = -1 \qquad y = \sqrt{3} \qquad r = 2$$

in the definitions of the trigonometric ratios, we have

$$\sin\frac{2\pi}{3} = \frac{\sqrt{3}}{2} \qquad \cos\frac{2\pi}{3} = -\frac{1}{2} \qquad \tan\frac{2\pi}{3} = -\sqrt{3}$$

$$\csc\frac{2\pi}{3} = \frac{2}{\sqrt{3}} \qquad \sec\frac{2\pi}{3} = -2 \qquad \cot\frac{2\pi}{3} = -\frac{1}{\sqrt{3}}$$

■

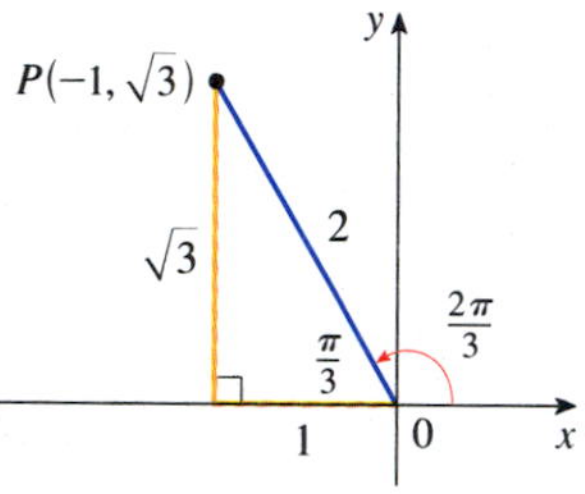

FIGURE 10

The following table gives some values of $\sin\theta$ and $\cos\theta$ found by the method of Example 3.

θ	0	$\frac{\pi}{6}$	$\frac{\pi}{4}$	$\frac{\pi}{3}$	$\frac{\pi}{2}$	$\frac{2\pi}{3}$	$\frac{3\pi}{4}$	$\frac{5\pi}{6}$	π	$\frac{3\pi}{2}$	2π
$\sin\theta$	0	$\frac{1}{2}$	$\frac{1}{\sqrt{2}}$	$\frac{\sqrt{3}}{2}$	1	$\frac{\sqrt{3}}{2}$	$\frac{1}{\sqrt{2}}$	$\frac{1}{2}$	0	-1	0
$\cos\theta$	1	$\frac{\sqrt{3}}{2}$	$\frac{1}{\sqrt{2}}$	$\frac{1}{2}$	0	$-\frac{1}{2}$	$-\frac{1}{\sqrt{2}}$	$-\frac{\sqrt{3}}{2}$	-1	0	1

EXAMPLE 4 If $\cos\theta = \frac{2}{5}$ and $0 < \theta < \pi/2$, find the other five trigonometric functions of θ.

SOLUTION Since $\cos\theta = \frac{2}{5}$, we can label the hypotenuse as having length 5 and the adjacent side as having length 2 in Figure 11. If the opposite side has length x, then the Pythagorean Theorem gives $x^2 + 4 = 25$ and so $x^2 = 21$, $x = \sqrt{21}$. We can now use the diagram to write the other five trigonometric functions:

$$\sin\theta = \frac{\sqrt{21}}{5} \qquad \tan\theta = \frac{\sqrt{21}}{2}$$

$$\csc\theta = \frac{5}{\sqrt{21}} \qquad \sec\theta = \frac{5}{2} \qquad \cot\theta = \frac{2}{\sqrt{21}}$$

■

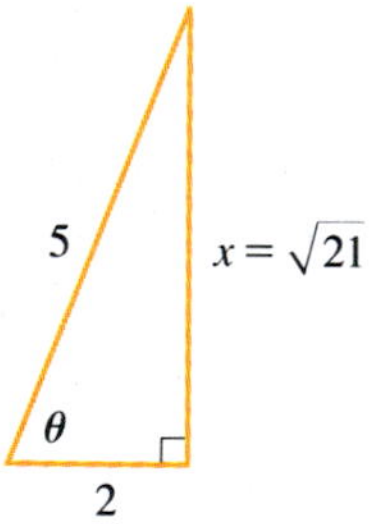

FIGURE 11

EXAMPLE 5 Use a calculator to approximate the value of x in Figure 12.

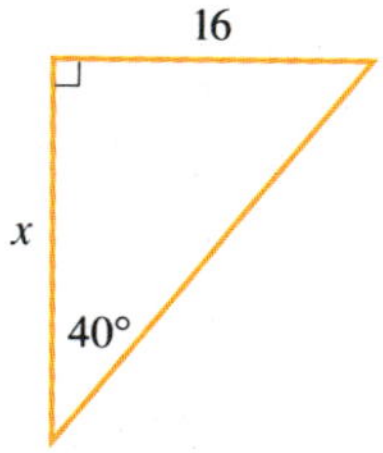

FIGURE 12

SOLUTION From the diagram we see that

$$\tan 40^\circ = \frac{16}{x}$$

Therefore
$$x = \frac{16}{\tan 40^\circ} \approx 19.07$$

■

TRIGONOMETRIC IDENTITIES

A trigonometric identity is a relationship among the trigonometric functions. The most elementary are the following, which are immediate consequences of the definitions of the trigonometric functions.

(6)
$$\csc\theta = \frac{1}{\sin\theta} \qquad \sec\theta = \frac{1}{\cos\theta} \qquad \cot\theta = \frac{1}{\tan\theta}$$
$$\tan\theta = \frac{\sin\theta}{\cos\theta} \qquad \cot\theta = \frac{\cos\theta}{\sin\theta}$$

For the next identity we refer to Figure 7. The distance formula (or, equivalently, the Pythagorean Theorem) tells us that $x^2 + y^2 = r^2$. Therefore

$$\sin^2\theta + \cos^2\theta = \frac{y^2}{r^2} + \frac{x^2}{r^2} = \frac{x^2 + y^2}{r^2} = \frac{r^2}{r^2} = 1$$

We have therefore proved one of the most useful of all trigonometric identities:

(7)
$$\sin^2\theta + \cos^2\theta = 1$$

If we now divide both sides of Equation 7 by $\cos^2\theta$ and use Equations 6, we get

(8)
$$\tan^2\theta + 1 = \sec^2\theta$$

Similarly, if we divide both sides of Equation 7 by $\sin^2\theta$, we get

(9)
$$1 + \cot^2\theta = \csc^2\theta$$

The identities

(10a)
$$\sin(-\theta) = -\sin\theta$$

(10b)
$$\cos(-\theta) = \cos\theta$$

show that sin is an odd function and cos is an even function. (See Section 1 of Review and Preview.) They are easily proved by drawing a diagram showing θ and $-\theta$ in standard position (see Exercise 39).

Since the angles θ and $\theta + 2\pi$ have the same terminal side, we have

(11)
$$\sin(\theta + 2\pi) = \sin\theta \qquad \cos(\theta + 2\pi) = \cos\theta$$

These identities show that the sine and cosine functions are periodic with period 2π.

The remaining trigonometric identities are all consequences of two basic identities called the **addition formulas**:

(12a)
$$\sin(x + y) = \sin x\cos y + \cos x\sin y$$

(12b)
$$\cos(x + y) = \cos x\cos y - \sin x\sin y$$

The proofs of these addition formulas are outlined in Exercises 85, 86, and 87.

By substituting $-y$ for y in Equations 12a and 12b and using Equations 10a and 10b, we obtain the following **subtraction formulas**:

$$\sin(x - y) = \sin x \cos y - \cos x \sin y \tag{13a}$$

$$\cos(x - y) = \cos x \cos y + \sin x \sin y \tag{13b}$$

Then, by dividing the formulas in Equations 12 or Equations 13, we obtain the corresponding formulas for $\tan(x \pm y)$:

$$\tan(x + y) = \frac{\tan x + \tan y}{1 - \tan x \tan y} \tag{14a}$$

$$\tan(x - y) = \frac{\tan x - \tan y}{1 + \tan x \tan y} \tag{14b}$$

If we put $y = x$ in the addition formulas (12), we get the **double-angle formulas**:

$$\sin 2x = 2 \sin x \cos x \tag{15a}$$

$$\cos 2x = \cos^2 x - \sin^2 x \tag{15b}$$

Then, by using the identity $\sin^2 x + \cos^2 x = 1$, we obtain the following alternate forms of the double-angle formulas for $\cos 2x$:

$$\cos 2x = 2 \cos^2 x - 1 \tag{16a}$$

$$\cos 2x = 1 - 2 \sin^2 x \tag{16b}$$

If we now solve these equations for $\cos^2 x$ and $\sin^2 x$, we get the following **half-angle formulas**, which are useful in integral calculus:

$$\cos^2 x = \frac{1 + \cos 2x}{2} \tag{17a}$$

$$\sin^2 x = \frac{1 - \cos 2x}{2} \tag{17b}$$

Finally, we state the **product formulas**, which can be deduced from Equations 12 and 13.

$$\sin x \cos y = \tfrac{1}{2}[\sin(x + y) + \sin(x - y)] \tag{18a}$$

$$\cos x \cos y = \tfrac{1}{2}[\cos(x + y) + \cos(x - y)] \tag{18b}$$

$$\sin x \sin y = \tfrac{1}{2}[\cos(x - y) - \cos(x + y)] \tag{18c}$$

There are many other trigonometric identities, but those we have stated are the ones used most often in calculus. If you forget any of them, remember that they can all be deduced from Equations 12a and 12b.

EXAMPLE 6 Find all values of x in the interval $[0, 2\pi]$ such that $\sin x = \sin 2x$.

SOLUTION Using the double-angle formula (15a), we rewrite the given equation as

$$\sin x = 2 \sin x \cos x \qquad \text{or} \qquad \sin x(1 - 2\cos x) = 0$$

Therefore, there are two possibilities for x:

$$\sin x = 0 \qquad \text{or} \qquad 1 - 2\cos x = 0$$

$$x = 0, \pi, 2\pi \qquad\qquad \cos x = \tfrac{1}{2}$$

$$x = \frac{\pi}{3}, \frac{5\pi}{3}$$

The given equation has five solutions: 0, $\pi/3$, π, $5\pi/3$, and 2π. ■

GRAPHS OF THE TRIGONOMETRIC FUNCTIONS

The graph of the function $f(x) = \sin x$, shown in Figure 13(a), is obtained by plotting points for $0 \leq x \leq 2\pi$ and then using the periodic nature of the function (from Equation 11) to complete the graph. Notice that the zeros of the sine function occur at the integer multiples of π, that is,

$$\sin x = 0 \qquad \text{whenever } x = n\pi,\ n \text{ an integer}$$

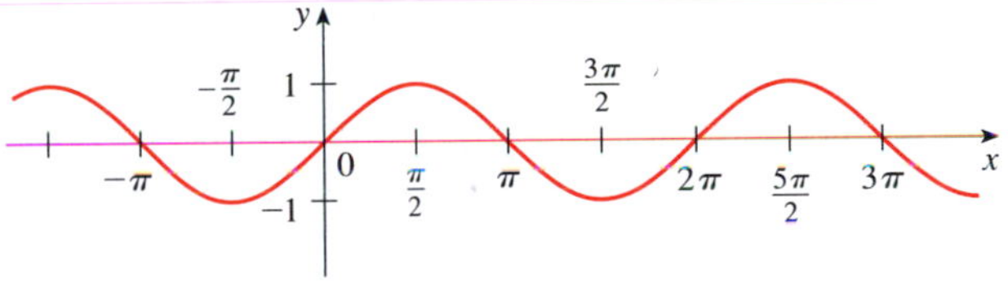

(a) $f(x) = \sin x$

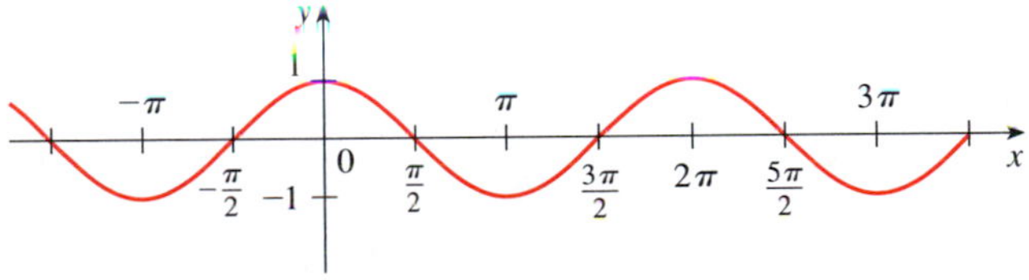

(b) $g(x) = \cos x$

FIGURE 13

Because of the identity

$$\cos x = \sin\left(x + \frac{\pi}{2}\right)$$

(which can be verified using Equation 12a), the graph of cosine is obtained by shifting the graph of sine by an amount $\pi/2$ to the left [see Figure 13(b)]. Note that for both the sine and cosine functions the domain is $(-\infty, \infty)$ and the range is the closed interval $[-1, 1]$. Thus, for all values of x, we have

$$-1 \leq \sin x \leq 1 \qquad\qquad -1 \leq \cos x \leq 1$$

The graphs of the remaining four trigonometric functions are shown in Figure 14 and their domains are indicated there. Notice that tangent and cotangent have range $(-\infty, \infty)$, whereas cosecant and secant have range $(-\infty, -1] \cup [1, \infty)$. All four functions are periodic: tangent and cotangent have period π, whereas cosecant and secant have period 2π.

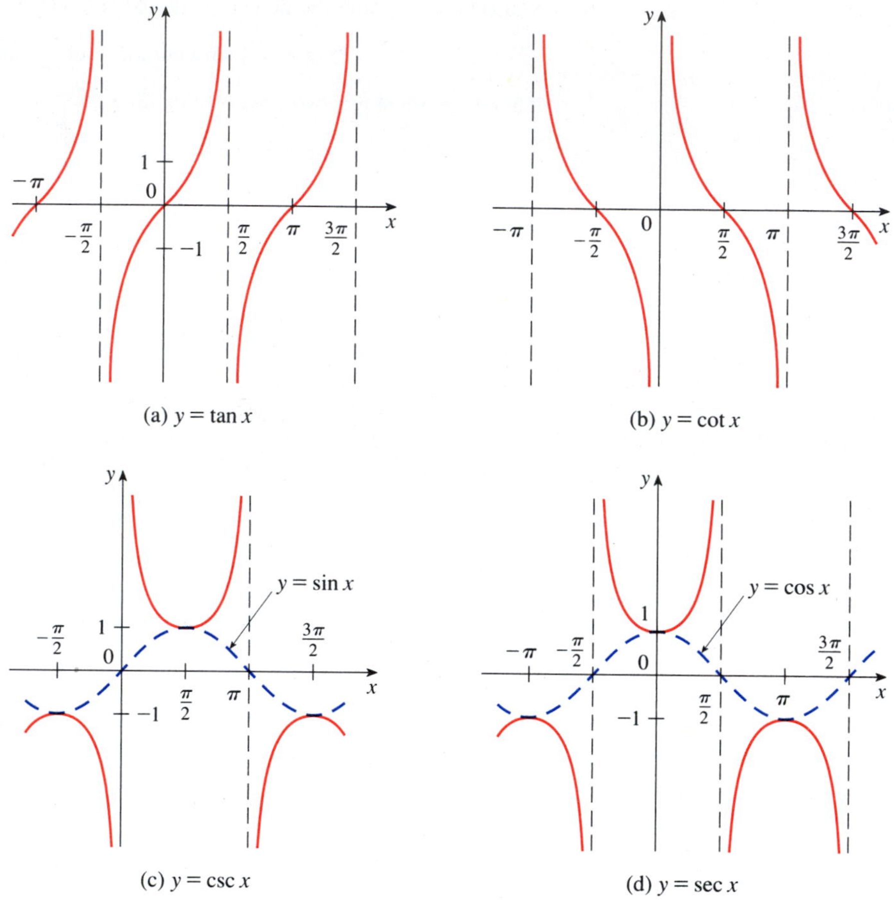

FIGURE 14

EXERCISES D

1–6 ■ Convert from degrees to radians.

1. 210° **2.** 300° **3.** 9°

4. −315° **5.** 900° **6.** 36°

7–12 ■ Convert from radians to degrees.

7. 4π **8.** $-\dfrac{7\pi}{2}$ **9.** $\dfrac{5\pi}{12}$

10. $\dfrac{8\pi}{3}$ **11.** $-\dfrac{3\pi}{8}$ **12.** 5

13. Find the length of a circular arc subtended by an angle of $\pi/12$ rad if the radius of the circle is 36 cm.

14. If a circle has radius 10 cm, what is the length of the arc subtended by a central angle of 72°?

15. A circle has radius 1.5 m. What angle is subtended at the center of the circle by an arc 1 m long?

16. Find the radius of a circular sector with angle $3\pi/4$ and arc length 6 cm.

17–22 ■ Draw, in standard position, the angle whose measure is given.

17. 315° **18.** −150° **19.** $-\dfrac{3\pi}{4}$ rad

20. $\dfrac{7\pi}{3}$ rad **21.** 2 rad **22.** −3 rad

23–28 ■ Find the exact trigonometric ratios for the angle whose radian measure is given.

23. $\frac{3\pi}{4}$ **24.** $\frac{4\pi}{3}$ **25.** $\frac{9\pi}{2}$

26. -5π **27.** $\frac{5\pi}{6}$ **28.** $\frac{11\pi}{4}$

29–34 ■ Find the remaining trigonometric ratios.

29. $\sin\theta = \frac{3}{5}, \quad 0 < \theta < \frac{\pi}{2}$

30. $\tan\alpha = 2, \quad 0 < \alpha < \frac{\pi}{2}$

31. $\sec\phi = -1.5, \quad \frac{\pi}{2} < \phi < \pi$

32. $\cos x = -\frac{1}{3}, \quad \pi < x < \frac{3\pi}{2}$

33. $\cot\beta = 3, \quad \pi < \beta < 2\pi$

34. $\csc\theta = -\frac{4}{3}, \quad \frac{3\pi}{2} < \theta < 2\pi$

35–38 ■ Find, correct to five decimal places, the length of the side labeled x.

35.

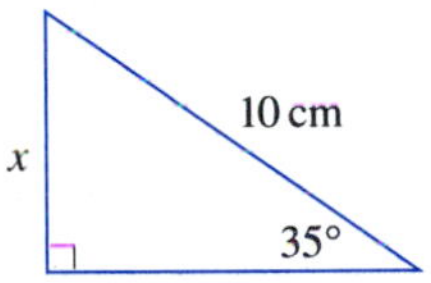

36.

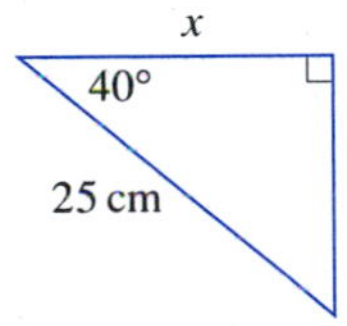

37.

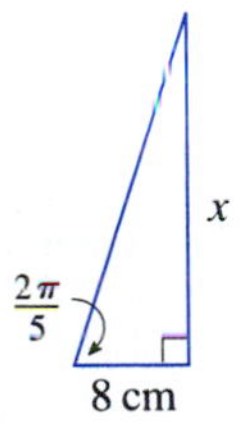

38.

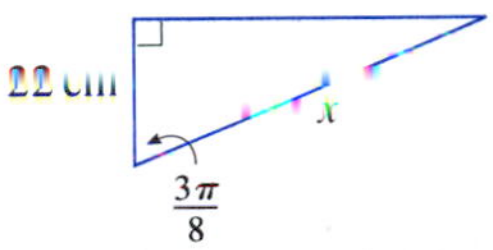

39–41 ■ Prove each equation.

39. (a) Equation 10a (b) Equation 10b

40. (a) Equation 14a (b) Equation 14b

41. (a) Equation 18a (b) Equation 18b
(c) Equation 18c

42–58 ■ Prove each identity.

42. $\cos\left(\frac{\pi}{2} - x\right) = \sin x$ **43.** $\sin\left(\frac{\pi}{2} + x\right) = \cos x$

44. $\sin(\pi - x) = \sin x$ **45.** $\sin\theta\cot\theta = \cos\theta$

46. $(\sin x + \cos x)^2 = 1 + \sin 2x$

47. $\sec y - \cos y = \tan y \sin y$

48. $\tan^2\alpha - \sin^2\alpha = \tan^2\alpha \sin^2\alpha$

49. $\cot^2\theta + \sec^2\theta = \tan^2\theta + \csc^2\theta$

50. $2\csc 2t = \sec t \csc t$

51. $\tan 2\theta = \frac{2\tan\theta}{1 - \tan^2\theta}$

52. $\frac{1}{1 - \sin\theta} + \frac{1}{1 + \sin\theta} = 2\sec^2\theta$

53. $\sin x \sin 2x + \cos x \cos 2x = \cos x$

54. $\sin^2 x - \sin^2 y = \sin(x + y)\sin(x - y)$

55. $\frac{\sin\phi}{1 - \cos\phi} = \csc\phi + \cot\phi$

56. $\tan x + \tan y = \frac{\sin(x + y)}{\cos x \cos y}$

57. $\sin 3\theta + \sin\theta = 2\sin 2\theta\cos\theta$

58. $\cos 3\theta = 4\cos^3\theta - 3\cos\theta$

59–64 ■ If $\sin x = \frac{1}{3}$ and $\sec y = \frac{5}{4}$, where x and y lie between 0 and $\pi/2$, evaluate the expression.

59. $\sin(x + y)$ **60.** $\cos(x + y)$

61. $\cos(x - y)$ **62.** $\sin(x - y)$

63. $\sin 2y$ **64.** $\cos 2y$

65–72 ■ Find all values of x in the interval $[0, 2\pi]$ that satisfy the equation.

65. $2\cos x - 1 = 0$ **66.** $3\cot^2 x = 1$

67. $2\sin^2 x = 1$ **68.** $|\tan x| = 1$

69. $\sin 2x = \cos x$ **70.** $2\cos x + \sin 2x = 0$

71. $\sin x = \tan x$ **72.** $2 + \cos 2x = 3\cos x$

73–76 ■ Find all values of x in the interval $[0, 2\pi]$ that satisfy the inequality.

73. $\sin x \leqslant \frac{1}{2}$ **74.** $2\cos x + 1 > 0$

75. $-1 < \tan x < 1$ **76.** $\sin x > \cos x$

77–82 ■ Graph the function by starting with the graphs in Figures 13 and 14 and applying the transformations of Section 2 of Review and Preview where appropriate.

77. $y = \cos\left(x - \frac{\pi}{3}\right)$ **78.** $y = \tan 2x$

79. $y = \frac{1}{3}\tan\left(x - \frac{\pi}{2}\right)$ **80.** $y = 1 + \sec x$

81. $y = |\sin x|$ **82.** $y = 2 + \sin\left(x + \frac{\pi}{4}\right)$

83. Prove the **Law of Cosines**: If a triangle has sides with lengths a, b, and c, and θ is the angle between the sides with lengths a and b, then

$$c^2 = a^2 + b^2 - 2ab\cos\theta$$

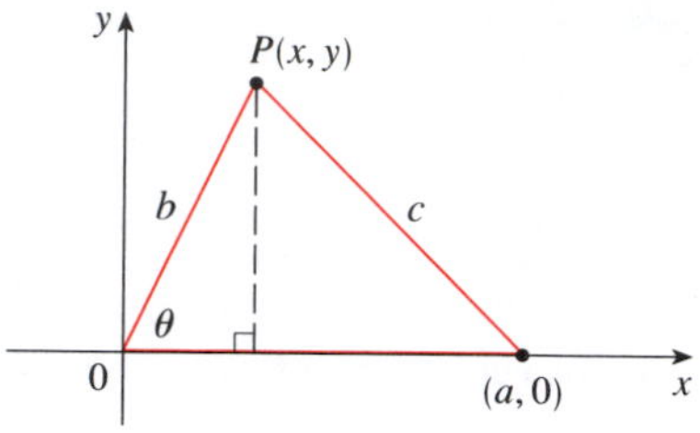

[*Hint:* Introduce a coordinate system so that θ is in standard position as in the figure. Express x and y in terms of θ and then use the distance formula to compute c.]

84. In order to find the distance $|AB|$ across a small inlet, a point C is located as in the figure and the following measurements were recorded: $\angle C = 103°$, $|AC| = 820$ m, $|BC| = 910$ m. Use the Law of Cosines from Exercise 83 to find the required distance.

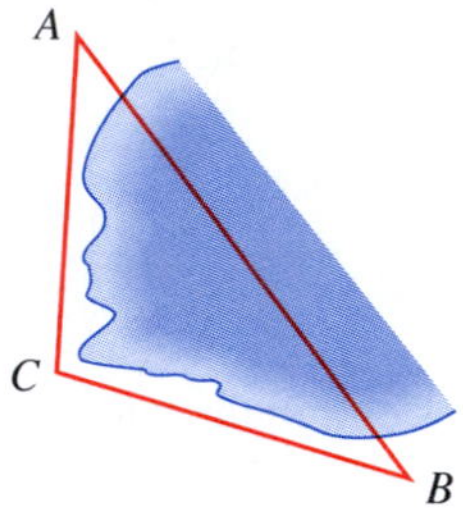

85. Use the figure to prove the subtraction formula

$$\cos(\alpha - \beta) = \cos\alpha\cos\beta + \sin\alpha\sin\beta$$

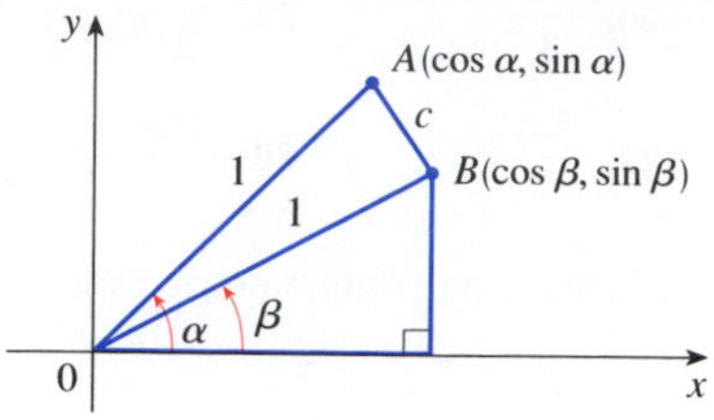

[*Hint:* Compute c^2 in two ways (using the Law of Cosines from Exercise 83 and also using the distance formula) and compare the two expressions.]

86. Use the formula in Exercise 85 to prove the addition formula for cosine (12b).

87. Use the addition formula for cosine and the identities

$$\cos\left(\frac{\pi}{2} - \theta\right) = \sin\theta \qquad \sin\left(\frac{\pi}{2} - \theta\right) = \cos\theta$$

to prove the subtraction formula for the sine function.

88. Show that the area of a triangle with sides of lengths a and b and with included angle θ is

$$A = \tfrac{1}{2}ab\sin\theta$$

89. Find the area of triangle ABC, correct to five decimal places, if $|AB| = 10$ cm, $|BC| = 3$ cm, and $\angle ABC = 107°$.

E MATHEMATICAL INDUCTION

The principle of mathematical induction is useful when we need to prove a statement S_n about the positive integer n. For instance, if S_n is the statement

$$(ab)^n = a^n b^n$$

then $\qquad S_1$ says that $ab = ab$

$\qquad S_2$ says that $(ab)^2 = a^2 b^2$

and so on.

PRINCIPLE OF MATHEMATICAL INDUCTION Let S_n be a statement about the positive integer n. Suppose that

1. S_1 is true.
2. S_{k+1} is true whenever S_k is true.

Then S_n is true for all positive integers n.

This is reasonable because, since S_1 is true, it follows from condition 2 (with $k = 1$) that S_2 is true. Then, using condition 2 with $k = 2$, we see that S_3 is true. Again using condition 2, this time with $k = 3$, we have that S_4 is true. This procedure can be followed indefinitely.

In using the principle of mathematical induction, we follow three steps.

Step 1: Prove that S_n is true when $n = 1$.

Step 2: Assume that S_n is true when $n = k$ and deduce that S_n is true when $n = k + 1$.

Step 3: Conclude that S_n is true for all n by the principle of mathematical induction.

EXAMPLE 1 If a and b are real numbers, prove that $(ab)^n = a^n b^n$ for every positive integer n.

SOLUTION Let S_n be the given statement.

1. S_1 is true because $(ab)^1 = ab = a^1 b^1$.
2. Assume that S_k is true, that is, $(ab)^k = a^k b^k$. Then

$$\begin{aligned}(ab)^{k+1} &= (ab)^k(ab) = a^k b^k ab \\ &= (a^k a)(b^k b) = a^{k+1}b^{k+1}\end{aligned}$$

   This says that S_{k+1} is true.
3. Therefore, by the principle of mathematical induction, S_n is true for all n; that is, $(ab)^n = a^n b^n$ for every positive integer n. ■

EXAMPLE 2 Prove that, for every positive integer n,

$$1 + 2 + 3 + \cdots + n = \frac{n(n+1)}{2}$$

SOLUTION Let S_n be the given statement.

1. S_1 is true because

$$1 = \frac{1(1+1)}{2}$$

2. Assume that S_k is true, that is

$$1 + 2 + \cdots + k = \frac{k(k+1)}{2}$$

Then
$$\begin{aligned}1 + 2 + \cdots + (k+1) &= (1 + 2 + \cdots + k) + (k+1) \\ &= \frac{k(k+1)}{2} + k + 1 \\ &= \frac{k(k+1) + 2(k+1)}{2} \\ &= \frac{(k+1)(k+2)}{2}\end{aligned}$$

Thus
$$1 + 2 + \cdots + (k+1) = \frac{(k+1)[(k+1)+1]}{2}$$

   which shows that S_{k+1} is true.
3. Therefore, S_n is true for all n by mathematical induction, that is,

$$1 + 2 + \cdots + (n+1) = \frac{n(n+1)}{2}$$

   for every positive integer n. ■

EXERCISES E

1–10 ■ The variable n represents a positive integer. Use mathematical induction to prove each statement.

1. $2^n > n$

2. $3^n > 2n$

3. $(1 + x)^n \geqslant 1 + nx$ (where $x \geqslant -1$)

4. If $0 \leqslant a < b$, then $a^n < b^n$.

5. $7^n - 1$ is divisible by 6.

6. $\left(\dfrac{a}{b}\right)^n = \dfrac{a^n}{b^n}$

7. $1 + 3 + 5 + \cdots + (2n - 1) = n^2$

8. $2 + 6 + 12 + \cdots + n(n + 1) = \dfrac{n(n + 1)(n + 2)}{3}$

9. $\dfrac{1}{2} + \dfrac{1}{6} + \dfrac{1}{12} + \cdots + \dfrac{1}{n(n + 1)} = \dfrac{n}{n + 1}$

10. $a + ar + ar^2 + \cdots + ar^{n-1} = \dfrac{a(1 - r^n)}{1 - r}$ $(r \neq 1)$

F PROOFS OF THEOREMS

In this appendix we present proofs of several theorems that are stated in the main body of the text. The sections in which they occur are indicated in the margin.

SECTION 2.3

LIMIT LAWS Suppose that c is a constant and the limits

$$\lim_{x \to a} f(x) = L \quad \text{and} \quad \lim_{x \to a} g(x) = M$$

exist. Then

1. $\lim_{x \to a} [f(x) + g(x)] = L + M$

2. $\lim_{x \to a} [f(x) - g(x)] = L - M$

3. $\lim_{x \to a} [cf(x)] = cL$

4. $\lim_{x \to a} [f(x)g(x)] = LM$

5. $\lim_{x \to a} \dfrac{f(x)}{g(x)} = \dfrac{L}{M}$ if $M \neq 0$

PROOF OF LAW 4 Let $\varepsilon > 0$ be given. We want to find $\delta > 0$ such that

$$|f(x)g(x) - LM| < \varepsilon \quad \text{whenever} \quad 0 < |x - a| < \delta$$

In order to get terms that contain $|f(x) - L|$ and $|g(x) - M|$, we add and subtract $Lg(x)$ as follows:

$$\begin{aligned} |f(x)g(x) - LM| &= |f(x)g(x) - Lg(x) + Lg(x) - LM| \\ &= |[f(x) - L]g(x) + L[g(x) - M]| \\ &\leqslant |[f(x) - L]g(x)| + |L[g(x) - M]| \qquad \text{(Triangle Inequality)} \\ &= |f(x) - L||g(x)| + |L||g(x) - M| \end{aligned}$$

We want to make each of these terms less than $\varepsilon/2$.

Since $\lim_{x \to a} g(x) = M$, there is a number $\delta_1 > 0$ such that

$$|g(x) - M| < \frac{\varepsilon}{2(1 + |L|)} \quad \text{whenever} \quad 0 < |x - a| < \delta_1$$

Also, there is a number $\delta_2 > 0$ such that if $0 < |x - a| < \delta_2$, then

$$|g(x) - M| < 1$$

and therefore

$$|g(x)| = |g(x) - M + M| \leqslant |g(x) - M| + |M| < 1 + |M|$$

Since $\lim_{x \to a} f(x) = L$, there is a number $\delta_3 > 0$ such that

$$|f(x) - L| < \frac{\varepsilon}{2(1 + |M|)} \qquad \text{whenever} \quad 0 < |x - a| < \delta_3$$

Let $\delta = \min\{\delta_1, \delta_2, \delta_3\}$. If $0 < |x - a| < \delta$, then we have $0 < |x - a| < \delta_1$, $0 < |x - a| < \delta_2$, and $0 < |x - a| < \delta_3$, so we can combine the inequalities to obtain

$$\begin{aligned} |f(x)g(x) - LM| &\leqslant |f(x) - L||g(x)| + |L||g(x) - M| \\ &< \frac{\varepsilon}{2(1 + |M|)}(1 + |M|) + |L|\frac{\varepsilon}{2(1 + |L|)} \\ &< \frac{\varepsilon}{2} + \frac{\varepsilon}{2} = \varepsilon \end{aligned}$$

This shows that $\lim_{x \to a} f(x)g(x) = LM$. □

PROOF OF LAW 3 If we take $g(x) = c$ in Law 4, we get

$$\begin{aligned} \lim_{x \to a} [cf(x)] &= \lim_{x \to a} [g(x)f(x)] = \lim_{x \to a} g(x) \cdot \lim_{x \to a} f(x) \\ &= \lim_{x \to a} c \cdot \lim_{x \to a} f(x) \\ &= c \lim_{x \to a} f(x) \qquad \text{(by Law 7)} \end{aligned}$$

□

PROOF OF LAW 2 Using Law 1 and Law 3 with $c = -1$, we have

$$\begin{aligned} \lim_{x \to a} [f(x) - g(x)] &= \lim_{x \to a} [f(x) + (-1)g(x)] = \lim_{x \to a} f(x) + \lim_{x \to a} (-1)g(x) \\ &= \lim_{x \to a} f(x) + (-1) \lim_{x \to a} g(x) = \lim_{x \to a} f(x) - \lim_{x \to a} g(x) \end{aligned}$$

□

PROOF OF LAW 5 First let's show that

$$\lim_{x \to a} \frac{1}{g(x)} = \frac{1}{M}$$

To do this we must show that, given $\varepsilon > 0$, there exists $\delta > 0$ such that

$$\left| \frac{1}{g(x)} - \frac{1}{M} \right| < \varepsilon \qquad \text{whenever} \quad 0 < |x - a| < \delta$$

Observe that

$$\left| \frac{1}{g(x)} - \frac{1}{M} \right| = \frac{|M - g(x)|}{|Mg(x)|}$$

We know that we can make the numerator small. But we also need to know that the denominator is not small when x is near a. Since $\lim_{x\to a} g(x) = M$, there is a number $\delta_1 > 0$ such that, whenever $0 < |x - a| < \delta_1$, we have

$$|g(x) - M| < \frac{|M|}{2}$$

and therefore

$$\begin{aligned}|M| = |M - g(x) + g(x)| &\leqslant |M - g(x)| + |g(x)| \\ &< \frac{|M|}{2} + |g(x)|\end{aligned}$$

This shows that

$$|g(x)| > \frac{|M|}{2} \qquad \text{whenever} \quad 0 < |x - a| < \delta_1$$

and so, for these values of x,

$$\frac{1}{|Mg(x)|} = \frac{1}{|M||g(x)|} < \frac{1}{|M|} \cdot \frac{2}{|M|} = \frac{2}{M^2}$$

Also, there exists $\delta_2 > 0$ such that

$$|g(x) - M| < \frac{M^2}{2}\varepsilon \qquad \text{whenever} \quad 0 < |x - a| < \delta_2$$

Let $\delta = \min\{\delta_1, \delta_2\}$. Then, for $0 < |x - a| < \delta$, we have

$$\left|\frac{1}{g(x)} - \frac{1}{M}\right| = \frac{|M - g(x)|}{|Mg(x)|} < \frac{2}{M^2}\frac{M^2}{2}\varepsilon = \varepsilon$$

It follows that $\lim_{x\to a} 1/g(x) = 1/M$. Finally, using Law 4, we obtain

$$\begin{aligned}\lim_{x\to a} \frac{f(x)}{g(x)} &= \lim_{x\to a} f(x)\left(\frac{1}{g(x)}\right) \\ &= \lim_{x\to a} f(x) \lim_{x\to a} \frac{1}{g(x)} = L \cdot \frac{1}{M} = \frac{L}{M}\end{aligned}$$

□

(2) THEOREM If $f(x) \leqslant g(x)$ for all x in an open interval that contains a (except possibly at a) and

$$\lim_{x\to a} f(x) = L \qquad \text{and} \qquad \lim_{x\to a} g(x) = M$$

then $L \leqslant M$.

PROOF We use the method of proof by contradiction. Suppose, if possible, that $L > M$. Law 2 of limits says that

$$\lim_{x\to a} [g(x) - f(x)] = M - L$$

Therefore, for any $\varepsilon > 0$, there exists $\delta > 0$ such that

$$|[g(x) - f(x)] - (M - L)| < \varepsilon \qquad \text{whenever} \quad 0 < |x - a| < \delta$$

In particular, taking $\varepsilon = L - M$ (noting that $L - M > 0$ by hypothesis), we have a number $\delta > 0$ such that

$$|[g(x) - f(x)] - (M - L)| < L - M \quad \text{whenever} \quad 0 < |x - a| < \delta$$

Since $a \leqslant |a|$ for any number a, we have

$$[g(x) - f(x)] - (M - L) < L - M \quad \text{whenever} \quad 0 < |x - a| < \delta$$

which simplifies to

$$g(x) < f(x) \quad \text{whenever} \quad 0 < |x - a| < \delta$$

But this contradicts $f(x) \leqslant g(x)$. Thus, the inequality $L > M$ must be false. Therefore, $L \leqslant M$. □

(3) THE SQUEEZE THEOREM If $f(x) \leqslant g(x) \leqslant h(x)$ for all x in an open interval that contains a (except possibly at a) and

$$\lim_{x \to a} f(x) = \lim_{x \to a} h(x) = L$$

then

$$\lim_{x \to a} g(x) = L$$

PROOF Let $\varepsilon > 0$ be given. Since $\lim_{x \to a} f(x) = L$, there is a number $\delta_1 > 0$ such that

$$|f(x) - L| < \varepsilon \quad \text{whenever} \quad 0 < |x - a| < \delta_1$$

that is, $$L - \varepsilon < f(x) < L + \varepsilon \quad \text{whenever} \quad 0 < |x - a| < \delta_1$$

Since $\lim_{x \to a} h(x) = L$, there is a number $\delta_2 > 0$ such that

$$|h(x) - L| < \varepsilon \quad \text{whenever} \quad 0 < |x - a| < \delta_2$$

that is, $$L - \varepsilon < h(x) < L + \varepsilon \quad \text{whenever} \quad 0 < |x - a| < \delta_2$$

Let $\delta = \min\{\delta_1, \delta_2\}$. If $0 < |x - a| < \delta$, then $0 < |x - a| < \delta_1$ and $0 < |x - a| < \delta_2$, so

$$L - \varepsilon < f(x) \leqslant g(x) \leqslant h(x) < L + \varepsilon$$

In particular,

$$L - \varepsilon < g(x) < L + \varepsilon$$

and so $|g(x) - L| < \varepsilon$. Therefore, $\lim_{x \to a} g(x) = L$. □

SECTION 2.5

(7) THEOREM If f is continuous at b and $\lim_{x \to a} g(x) = b$, then

$$\lim_{x \to a} f(g(x)) = f(b)$$

PROOF Let $\varepsilon > 0$ be given. We want to find a number $\delta > 0$ such that

$$|f(g(x)) - f(b)| < \varepsilon \quad \text{whenever} \quad 0 < |x - a| < \delta$$

Since f is continuous at b, we have

$$\lim_{y \to b} f(y) = f(b)$$

and so there exists $\delta_1 > 0$ such that

$$|f(y) - f(b)| < \varepsilon \quad \text{whenever} \quad 0 < |y - b| < \delta_1$$

Since $\lim_{x\to a} g(x) = b$, there exists $\delta > 0$ such that

$$|g(x) - b| < \delta_1 \quad \text{whenever} \quad 0 < |x - a| < \delta$$

Combining these two statements, we see that whenever $0 < |x - a| < \delta$ we have $|g(x) - b| < \delta_1$, which implies that $|f(g(x)) - f(b)| < \varepsilon$. Therefore, we have proved that $\lim_{x\to a} f(g(x)) = f(b)$. □

SECTION 3.4

The proof of the following result was promised in the proof of Theorem 3.4.4.

THEOREM If $0 < \theta < \frac{\pi}{2}$, then $\theta \leqslant \tan\theta$.

PROOF Figure 1 shows a sector of a circle with center O, central angle θ, and radius 1. Then

$$|AD| = |OA|\tan\theta = \tan\theta$$

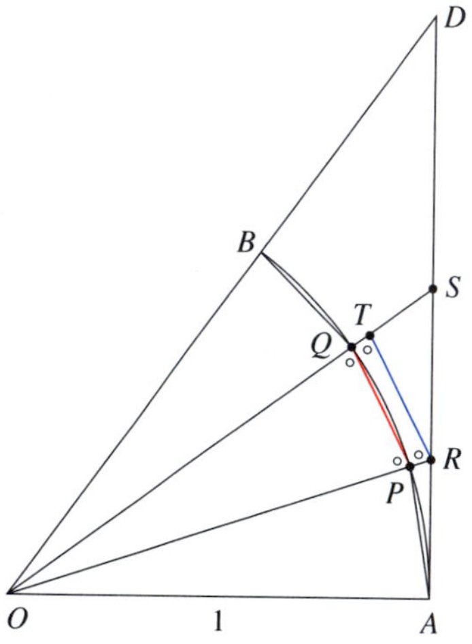

FIGURE 1

We approximate the arc AB by an inscribed polygon consisting of n equal line segments and we look at a typical segment PQ. We extend the lines OP and OQ to meet AD in the points R and S. Then we draw $RT \parallel PQ$ as in Figure 1. Observe that

$$\angle RTO = \angle PQO < 90°$$

and so $\angle RTS > 90°$. Therefore, we have

$$|PQ| < |RT| < |RS|$$

If we add n such inequalities, we get

$$L_n < |AD| = \tan\theta$$

where L_n is the length of the inscribed polygon. Thus, by Theorem 2.3.2, we have

$$\lim_{n\to\infty} L_n \leqslant \tan\theta$$

But the arc length was defined in Section 9.3 as the limit of the lengths of inscribed polygons, so

$$\theta = \lim_{n\to\infty} L_n \leqslant \tan\theta$$

□

SECTION 4.2

(6) THEOREM If f is a one-to-one continuous function defined on an interval (a, b), then its inverse function f^{-1} is also continuous.

PROOF First we show that if f is both one-to-one and continuous on (a, b), then it must be either increasing or decreasing on (a, b). If it were neither increasing nor decreasing, then there would exist numbers x_1, x_2, and x_3 in (a, b) with $x_1 < x_2 < x_3$ such that $f(x_2)$ does not lie between $f(x_1)$ and $f(x_3)$. There are two possibilities: either (1) $f(x_3)$ lies between $f(x_1)$ and $f(x_2)$ or (2) $f(x_1)$ lies between $f(x_2)$ and $f(x_3)$. (Draw a picture.) In case (1) we apply the Intermediate Value Theorem to the continuous function f to get a number c between x_1 and x_2 such that $f(c) = f(x_3)$. In case (2) the Intermediate Value Theorem gives a number c between x_2 and x_3 such that $f(c) = f(x_1)$. In either case we have contradicted the fact that f is one-to-one.

Let's assume, for the sake of definiteness, that f is increasing on (a, b). We take any number y_0 in the domain of f^{-1} and we let $f^{-1}(y_0) = x_0$; that is, x_0 is the number in (a, b) such that $f(x_0) = y_0$. To show that f^{-1} is continuous at y_0 we take any $\varepsilon > 0$ such that the interval $(x_0 - \varepsilon, x_0 + \varepsilon)$ is contained in the interval (a, b). Since f is increasing, it maps the numbers in the interval $(x_0 - \varepsilon, x_0 + \varepsilon)$ onto the numbers in the interval $(f(x_0 - \varepsilon), f(x_0 + \varepsilon))$ and f^{-1} reverses the correspondence. If we let δ denote the smaller of the numbers $\delta_1 = y_0 - f(x_0 - \varepsilon)$ and $\delta_2 = f(x_0 + \varepsilon) - y_0$, then the interval $(y_0 - \delta, y_0 + \delta)$ is contained in the interval $(f(x_0 - \varepsilon), f(x_0 + \varepsilon))$ and so is mapped into the interval $(x_0 - \varepsilon, x_0 + \varepsilon)$ by f^{-1}. (See the arrow diagram in Figure 2.) We have therefore found a number $\delta > 0$ such that

$$|f(y) - f(b)| < \varepsilon \qquad \text{whenever} \qquad 0 < |y - b| < \delta_1$$

This shows that $\lim_{y \to y_0} f^{-1}(y) = f^{-1}(y_0)$ and so f^{-1} is continuous at any number y_0 in its domain.

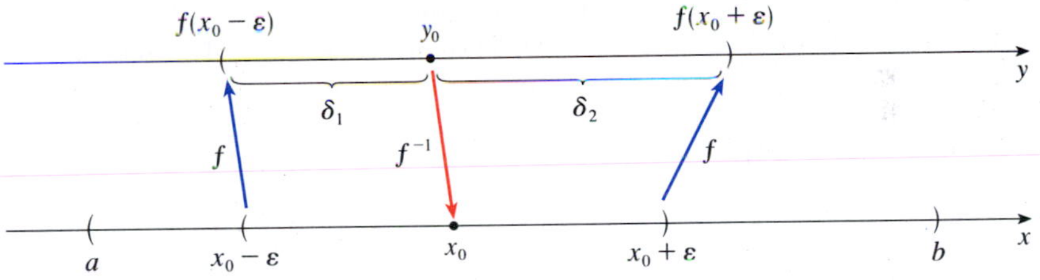

FIGURE 2

□

SECTION 5.2

FERMAT'S THEOREM If f has a local extremum (that is, maximum or minimum) at c, and if $f'(c)$ exists, then $f'(c) = 0$.

PROOF Suppose, for the sake of definiteness, that f has a local maximum at c. Then $f(c) \geqslant f(x)$ if x is sufficiently close to c. This implies that if h is sufficiently close to 0, with h being positive or negative, then

$$f(c) \geqslant f(c + h)$$

and therefore

(1) $$f(c + h) - f(c) \leqslant 0$$

We can divide both sides of an inequality by a positive number. Thus, if $h > 0$ and h is sufficiently small, we have

$$\frac{f(c + h) - f(c)}{h} \leqslant 0$$

Taking the right-hand limit of both sides of this inequality (using Theorem 2.3.2), we get

$$\lim_{h \to 0^+} \frac{f(c + h) - f(c)}{h} \leqslant \lim_{h \to 0^+} 0 = 0$$

But since $f'(c)$ exists, we have

$$f'(c) = \lim_{h \to 0} \frac{f(c+h) - f(c)}{h} = \lim_{h \to 0^+} \frac{f(c+h) - f(c)}{h}$$

and so we have shown that $f'(c) \leqslant 0$.

If $h < 0$, then the direction of the inequality (1) is reversed when we divide by h:

$$\frac{f(c+h) - f(c)}{h} \geqslant 0 \qquad h < 0$$

So, taking the left-hand limit, we have

$$f'(c) = \lim_{h \to 0} \frac{f(c+h) - f(c)}{h}$$

$$= \lim_{h \to 0^-} \frac{f(c+h) - f(c)}{h} \geqslant 0$$

We have shown that $f'(c) \geqslant 0$ and also that $f'(c) \leqslant 0$. Since both of these inequalities must be true, the only possibility is that $f'(c) = 0$.

We have proved Fermat's Theorem for the case of a local maximum. The case of a local minimum can be proved in a similar manner. □

SECTION 6.3

PROPERTY 5 OF INTEGRALS

$$\int_a^b f(x)\,dx = \int_a^c f(x)\,dx + \int_c^b f(x)\,dx$$

if all these integrals exist.

PROOF We first assume that $a < c < b$. Since we are assuming that $\int_a^b f(x)\,dx$ exists, we can compute it as a limit of Riemann sums using only partitions P that include c as one of the partition points. If P is such a partition, let P_1 be the corresponding partition of $[a, c]$ determined by those partition points of P that lie in $[a, c]$. Similarly, P_2 will denote the corresponding partition of $[c, b]$. Note that $\|P_1\| \leqslant \|P\|$ and $\|P_2\| \leqslant \|P\|$. Thus, if $\|P\| \to 0$, it follows that $\|P_1\| \to 0$ and $\|P_2\| \to 0$. If $\{x_i \mid 1 \leqslant i \leqslant n\}$ is the set of partition points for P and $n = k + m$, where k is the number of subintervals in $[a, c]$ and m is the number of subintervals in $[c, b]$, then $\{x_i \mid 1 \leqslant i \leqslant k\}$ is the set of partition points for P_1. If we write $t_j = x_{k+j}$ for the partition points to the right of c, then $\{t_j \mid 1 \leqslant j \leqslant m\}$ is the set of partition points for P_2. Thus, we have

$$a = x_0 < x_1 < \cdots < x_k < x_{k+1} < \cdots < x_n = b$$

$$c < t_1 < \cdots < t_m = b$$

Choosing $x_i^* = x_i$ and letting $\Delta t_j = t_j - t_{j-1}$, we compute $\int_a^b f(x)\,dx$ as follows:

$$\int_a^b f(x)\,dx = \lim_{\|P\| \to 0} \sum_{i=1}^{n} f(x_i)\,\Delta x_i$$

$$= \lim_{\|P\| \to 0} \left[\sum_{i=1}^{k} f(x_i)\,\Delta x_i + \sum_{i=k+1}^{n} f(x_i)\,\Delta x_i \right]$$

$$= \lim_{\|P\|\to 0} \left[\sum_{i=1}^{k} f(x_i)\,\Delta x_i + \sum_{j=1}^{m} f(t_j)\,\Delta t_j \right]$$

$$= \lim_{\|P_1\|\to 0} \sum_{i=1}^{k} f(x_i)\,\Delta x_i + \lim_{\|P_2\|\to 0} \sum_{j=1}^{m} f(t_j)\,\Delta t_j$$

$$= \int_a^c f(x)\,dx + \int_c^b f(t)\,dt$$

Now suppose that $c < a < b$. By what we have already proved, we have

$$\int_c^b f(x)\,dx = \int_c^a f(x)\,dx + \int_a^b f(x)\,dx$$

Therefore

$$\int_a^b f(x)\,dx = -\int_c^a f(x)\,dx + \int_c^b f(x)\,dx$$

$$= \int_a^c f(x)\,dx + \int_c^b f(x)\,dx$$

(See Note 6 in Section 6.3.) The proofs are similar for the remaining four orderings of a, b, and c. □

SECTION 10.1

The proof of the Monotonic Sequence Theorem is based on the **Completeness Axiom** for the set $\mathbb{R}$ of real numbers, which says that if S is a nonempty set of real numbers that has an upper bound M ($x \leqslant M$ for all x in S), then S has a **least upper bound** b. (This means that b is an upper bound for S, but if M is any other upper bound, then $b \leqslant M$.) The Completeness Axiom is an expression of the fact that there is no gap or hole in the real number line.

> **MONOTONIC SEQUENCE THEOREM** Every bounded, monotonic sequence is convergent.

PROOF Suppose $\{a_n\}$ is an increasing sequence. Since $\{a_n\}$ is bounded, the set $S = \{a_n \mid n \geqslant 1\}$ has an upper bound. By the Completeness Axiom it has a least upper bound L. Given $\varepsilon > 0$, $L - \varepsilon$ is *not* an upper bound for S (since L is the *least* upper bound). Therefore

$$a_N > L - \varepsilon \qquad \text{for some integer } N$$

But the sequence is increasing so $a_n \geqslant a_N$ for every $n > N$. Thus, if $n > N$ we have

$$a_n > L - \varepsilon$$

so

$$0 \leqslant L - a_n < \varepsilon$$

since $a_n \leqslant L$. Thus

$$|L - a_n| < \varepsilon \qquad \text{whenever} \qquad n > N$$

so $\lim_{n\to\infty} a_n = L$.

A similar proof (using the greatest lower bound) works if $\{a_n\}$ is decreasing. □

SECTION 10.3

THE INTEGRAL TEST Suppose f is a continuous, positive, decreasing function on $[1, \infty)$ and let $a_n = f(n)$. Then the series $\sum_{n=1}^{\infty} a_n$ is convergent if and only if the improper integral $\int_1^{\infty} f(x)\,dx$ is convergent. In other words:

(a) If $\int_1^{\infty} f(x)\,dx$ is convergent, then $\sum_{n=1}^{\infty} a_n$ is convergent.

(b) If $\int_1^{\infty} f(x)\,dx$ is divergent, then $\sum_{n=1}^{\infty} a_n$ is divergent.

PROOF The basic idea behind the Integral Test can be seen by looking at Figures 3 and 4. The area of the first shaded rectangle in Figure 3 is the value of f at the right endpoint of $[1, 2]$, that is, $f(2) = a_2$. So, comparing the areas of the shaded rectangles with the area under $y = f(x)$ from 1 to n, we see that

(1) $$a_2 + a_3 + \cdots + a_n \leqslant \int_1^n f(x)\,dx$$

(Notice that this inequality depends on the fact that f is decreasing.) Likewise, Figure 4 shows that

(2) $$\int_1^n f(x)\,dx \leqslant a_1 + a_2 + \cdots + a_{n-1}$$

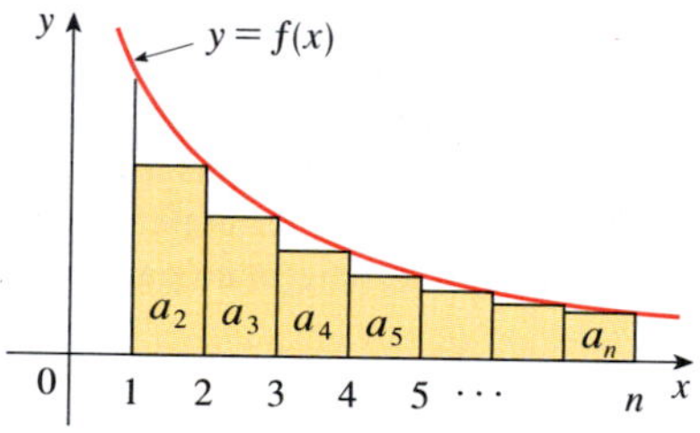

FIGURE 3

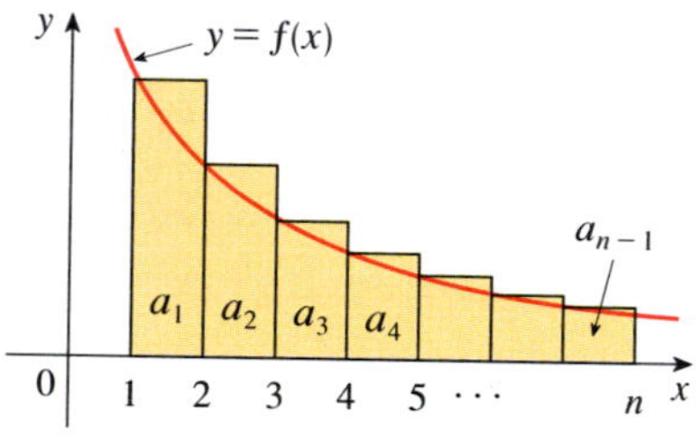

FIGURE 4

(a) If $\int_1^{\infty} f(x)\,dx$ is convergent, then (1) gives

$$\sum_{i=2}^{n} a_i \leqslant \int_1^n f(x)\,dx \leqslant \int_1^{\infty} f(x)\,dx$$

since $f(x) \geqslant 0$. Therefore

$$s_n = a_1 + \sum_{i=2}^{n} a_i \leqslant a_1 + \int_1^{\infty} f(x)\,dx = M, \text{ say}$$

Since $s_n \leqslant M$ for all n, the sequence $\{s_n\}$ is bounded above. Also

$$s_{n+1} = s_n + a_{n+1} \geqslant s_n$$

since $a_{n+1} = f(n + 1) \geqslant 0$. Thus, $\{s_n\}$ is an increasing bounded sequence and so it is convergent by Theorem 10.1.8. This means that $\sum a_n$ is convergent.

(b) If $\int_1^\infty f(x)\,dx$ is divergent, then $\int_1^n f(x)\,dx \to \infty$ as $n \to \infty$ because $f(x) \geqslant 0$. But (2) gives

$$\int_1^n f(x)\,dx \leqslant \sum_{i=1}^{n-1} a_i = s_{n-1}$$

and so $s_{n-1} \to \infty$. This implies that $s_n \to \infty$ and so $\Sigma\, a_n$ diverges. □

THE COMPARISON TEST Suppose that $\Sigma\, a_n$ and $\Sigma\, b_n$ are series with positive terms.

(a) If $\Sigma\, b_n$ is convergent and $a_n \leqslant b_n$ for all n, then $\Sigma\, a_n$ is also convergent.

(b) If $\Sigma\, b_n$ is divergent and $a_n \geqslant b_n$ for all n, then $\Sigma\, a_n$ is also divergent.

It is important to keep in mind the distinction between a sequence and a series. A sequence is a list of numbers, whereas a series is a sum. With every series $\Sigma\, a_n$ there are associated two sequences: the sequence $\{a_n\}$ of terms and the sequence $\{s_n\}$ of partial sums.

PROOF

(a) Let
$$s_n = \sum_{i=1}^{n} a_i \qquad t_n = \sum_{i=1}^{n} b_i \qquad t = \sum_{n=1}^{\infty} b_n$$

Since both series have positive terms, the sequences $\{s_n\}$ and $\{t_n\}$ are increasing $(s_{n+1} = s_n + a_{n+1} \geqslant s_n)$. Also $t_n \to t$, so $t_n \leqslant t$ for all n. Since $a_i \leqslant b_i$, we have $s_n \leqslant t_n$. Thus, $s_n \leqslant t$ for all n. This means that $\{s_n\}$ is increasing and bounded above and therefore converges by Theorem 10.1.8. Thus, $\Sigma\, a_n$ converges.

(b) If $\Sigma\, b_n$ is divergent, then $t_n \to \infty$ (since $\{t_n\}$ is increasing). But $a_i \geqslant b_i$, so $s_n \geqslant t_n$. Thus, $s_n \to \infty$. Therefore, $\Sigma\, a_n$ diverges. □

SECTION 10.4

THE ALTERNATING SERIES TEST If the alternating series

$$\sum_{n=1}^{\infty} (-1)^{n-1} b_n = b_1 - b_2 + b_3 - b_4 + b_5 - b_6 + \cdots \qquad b_n > 0$$

satisfies

(a) $b_{n+1} \leqslant b_n$ for all n

(b) $\lim_{n\to\infty} b_n = 0$

then the series is convergent.

PROOF We first consider the even partial sums:

$$s_2 = b_1 - b_2 \geqslant 0 \qquad \text{since } b_2 \leqslant b_1$$

$$s_4 = s_2 + (b_3 - b_4) \geqslant s_2 \qquad \text{since } b_4 \leqslant b_3$$

In general

$$s_{2n} = s_{2n-2} + (b_{2n-1} - b_{2n}) \geqslant s_{2n-2} \qquad \text{since } b_{2n} \leqslant b_{2n-1}$$

Thus
$$0 \leqslant s_2 \leqslant s_4 \leqslant s_6 \leqslant \cdots \leqslant s_{2n} \leqslant \cdots$$

But we can also write

$$s_{2n} = b_1 - (b_2 - b_3) - (b_4 - b_5) - \cdots - (b_{2n-2} - b_{2n-1}) - b_{2n}$$

Every term in brackets is positive, so $s_{2n} \leqslant b_1$ for all n. Therefore, the sequence $\{s_{2n}\}$ of even partial sums is increasing and bounded above. It is therefore convergent by Theorem 10.1.8. Let's call its limit s, that is,

$$\lim_{n\to\infty} s_{2n} = s$$

Now we compute the limit of the odd partial sums:

$$\begin{aligned}
\lim_{n\to\infty} s_{2n+1} &= \lim_{n\to\infty} (s_{2n} + b_{2n+1}) \\
&= \lim_{n\to\infty} s_{2n} + \lim_{n\to\infty} b_{2n+1} \\
&= s + 0 \qquad \text{[by condition (b)]} \\
&= s
\end{aligned}$$

Since both the even and odd partial sums converge to s, we have $\lim_{n\to\infty} s_n = s$ and so the series is convergent. □

ALTERNATING SERIES ESTIMATION THEOREM If $s = \Sigma\, (-1)^{n-1} b_n$ is the sum of an alternating series that satisfies

$$\text{(a) } 0 \leqslant b_{n+1} \leqslant b_n \qquad \text{and} \qquad \text{(b) } \lim_{n\to\infty} b_n = 0$$

then

$$|R_n| = |s - s_n| \leqslant b_{n+1}$$

PROOF The idea is similar to the one for the proof of the Alternating Series Test. We have

$$\begin{aligned}
s - s_n &= (-1)^n b_{n+1} + (-1)^{n+1} b_{n+2} + (-1)^{n+2} b_{n+3} + \cdots \\
&= (-1)^n [b_{n+1} - b_{n+2} + b_{n+3} - \cdots]
\end{aligned}$$

and so

$$\begin{aligned}
|s - s_n| &= (b_{n+1} - b_{n+2}) + (b_{n+3} - b_{n+4}) + \cdots \\
&= b_{n+1} - (b_{n+2} - b_{n+3}) - (b_{n+4} - b_{n+5}) - \cdots
\end{aligned}$$

Every term in brackets is positive, so $|s - s_n| \leqslant b_{n+1}$. □

THE RATIO TEST

(a) If $\lim_{n\to\infty} \left| \dfrac{a_{n+1}}{a_n} \right| = L < 1$, then the series $\sum_{n=1}^{\infty} a_n$ is absolutely convergent (and therefore convergent).

(b) If $\lim_{n\to\infty} \left| \dfrac{a_{n+1}}{a_n} \right| = L > 1$ or $\lim_{n\to\infty} \left| \dfrac{a_{n+1}}{a_n} \right| = \infty$, then the series $\sum_{n=1}^{\infty} a_n$ is divergent.

PROOF

(a) The idea is to compare the given series with a convergent geometric series. Since $L < 1$, we can choose a number r such that $L < r < 1$. Since

$$\lim_{n \to \infty} \left| \frac{a_{n+1}}{a_n} \right| = L \qquad \text{and} \qquad L < r$$

the ratio $|a_{n+1}/a_n|$ will eventually be less than r; that is, there exists an integer N such that

$$\left| \frac{a_{n+1}}{a_n} \right| < r \qquad \text{whenever} \qquad n \geqslant N$$

or, equivalently,

(1) $$|a_{n+1}| < |a_n| r \qquad \text{whenever} \qquad n \geqslant N$$

Putting n successively equal to $N, N + 1, N + 2, \ldots$ in (1), we obtain

$$|a_{N+1}| < |a_N| r$$

$$|a_{N+2}| < |a_{N+1}| r < |a_N| r^2$$

$$|a_{N+3}| < |a_{N+2}| r < |a_N| r^3$$

and, in general,

(2) $$|a_{N+k}| < |a_N| r^k \qquad \text{for all } k \geqslant 1$$

Now the series

$$\sum_{k=1}^{\infty} |a_N| r^k = |a_N| r + |a_N| r^2 + |a_N| r^3 + \cdots$$

is convergent because it is a geometric series with $0 < r < 1$. So the inequality (2), together with the Comparison Test, shows that the series

$$\sum_{n=N+1}^{\infty} |a_n| = \sum_{k=1}^{\infty} |a_{N+k}| = |a_{N+1}| + |a_{N+2}| + |a_{N+3}| + \cdots$$

is also convergent. It follows that the series $\sum_{n=1}^{\infty} |a_n|$ is convergent. (Recall that a finite number of terms cannot affect convergence.) Therefore, $\sum a_n$ is absolutely convergent.

(b) If $|a_{n+1}/a_n| \to L > 1$ or $|a_{n+1}/a_n| \to \infty$, then the ratio $|a_{n+1}/a_n|$ will eventually be greater than 1; that is, there exists an integer N such that

$$\left| \frac{a_{n+1}}{a_n} \right| > 1 \qquad \text{whenever} \qquad n \geqslant N$$

This means that $|a_{n+1}| > |a_n|$ whenever $n \geqslant N$ and so

$$\lim_{n \to \infty} a_n \neq 0$$

Therefore, $\sum a_n$ diverges by the Test for Divergence. □

SECTION 12.3

(5) CLAIRAUT'S THEOREM Suppose f is defined on a disk D containing the point (a, b). If the functions f_{xy} and f_{yx} are both continuous on D, then $f_{xy}(a, b) = f_{yx}(a, b)$.

PROOF For small values of h, $h \neq 0$, consider the difference

$$\Delta(h) = [f(a + h, b + h) - f(a + h, b)] - [f(a, b + h) - f(a, b)]$$

Notice that if we let $g(x) = f(x, b + h) - f(x, b)$, then

$$\Delta(h) = g(a + h) - g(a)$$

By the Mean Value Theorem, there is a number c between a and $a + h$ such that

$$g(a + h) - g(a) = g'(c)h = h[f_x(c, b + h) - f_x(c, b)]$$

Applying the Mean Value Theorem again, this time to f_x, we get a number d between b and $b + h$ such that

$$f_x(c, b + h) - f_x(c, b) = f_{xy}(c, d)h$$

Combining these equations, we obtain

$$\Delta(h) = h^2 f_{xy}(c, d)$$

If $h \to 0$, then $(c, d) \to (a, b)$, so the continuity of f_{xy} at (a, b) gives

$$\lim_{h \to 0} \frac{\Delta(h)}{h^2} = \lim_{(c,d) \to (a,b)} f_{xy}(c, d) = f_{xy}(a, b)$$

Similarly, by writing

$$\Delta(h) = [f(a + h, b + h) - f(a, b + h)] - [f(a + h, b) - f(a, b)]$$

and using the Mean Value Theorem twice and the continuity of f_{yx} at (a, b), we obtain

$$\lim_{h \to 0} \frac{\Delta(h)}{h^2} = f_{yx}(a, b)$$

It follows that $f_{xy}(a, b) = f_{yx}(a, b)$. □

G LIES MY CALCULATOR AND COMPUTER TOLD ME

See Section 3 of Review and Preview for a discussion of graphing calculators and computers with graphing software.

A wide variety of pocket-size calculating devices are currently marketed. Some can run programs prepared by the user; some have pre-programmed packages for frequently used calculus procedures, including the display of graphs. All have certain limitations in common: a limited range of magnitude (usually less than 10^{100} for calculators) and a bound on accuracy (typically eight to thirteen digits).

A calculator usually comes with an owner's manual. Read it! The manual will tell you about further limitations (for example, for angles when entering trigonometric functions) and perhaps how to overcome them.

Program packages for microcomputers (even the most fundamental ones, which realize arithmetical operations and elementary functions) often suffer from hidden flaws. You will be made aware of some of them in the examples in this appendix, and you are encouraged to experiment using the ideas presented here.

PRELIMINARY EXPERIMENTS WITH YOUR CALCULATOR OR COMPUTER

To have a first look at the limitations and quality of your calculator, make it compute $2 \div 3$. Of course, the answer is not a terminating decimal so it cannot be represented exactly on your calculator. If the last displayed digit is 6 rather than 7, then your calculator approximates $\frac{2}{3}$ by truncating instead of rounding, so be prepared for slightly greater loss of accuracy in longer calculations.

Now multiply the result by 3; that is, calculate $(2 \div 3) \times 3$. If the answer is 2, then subtract 2 from the result, thereby calculating $(2 \div 3) \times 3 - 2$. Instead of obtaining 0 as the answer, you might obtain a small negative number, which depends on the construction of the circuits. (The calculator keeps, in this case, a few "spare" digits that are remembered but not shown.) This is all right because, as previously mentioned, the finite number of digits makes it impossible to represent $2 \div 3$ exactly.

A similar situation occurs when you calculate $(\sqrt{6})^2 - 6$. If you do not obtain 0, the order of magnitude of the result will tell you how many digits the calculator uses internally.

Next, try to compute $(-1)^5$ using the y^x key. Many calculators will indicate an error because they are built to attempt $e^{5 \ln(-1)}$. One way to overcome this is to use the fact that $(-1)^k = \cos k\pi$ whenever k is an integer.

Calculators are usually constructed to operate in the decimal number system. In contrast, some microcomputer packages of arithmetical programs operate in a number system with base other than 10 (typically 2 or 16). Here the list of unwelcome tricks your device can play on you is even longer, since not all terminating decimal numbers are represented exactly. A recent implementation of the BASIC language shows (in double precision) examples of incorrect conversion from one number system into another, for example,

$$8 \times 0.1 \stackrel{?}{=} 0.79999\ 99999\ 99999\ 9$$

whereas

$$19 \times 0.1 \stackrel{?}{=} 1.90000\ 00000\ 00001$$

Yet another implementation, apparently free of the preceding anomalies, will not calculate standard functions in double precision. For example, the number $\pi = 4 \times \tan^{-1} 1$, whose representation with sixteen decimal digits should be 3.14159 26535 89793, appears as 3.14159 29794 31152; this is off by more than 3×10^{-7}. What is worse, the cosine function is programmed so badly that its "cos" $0 = 1 + 2^{-23}$. (Can you invent a situation when this could ruin your calculations?) These or similar defects exist in other programming languages, too.

THE PERILS OF SUBTRACTION

You might have observed that subtraction of two numbers that are close to each other is a tricky operation. The difficulty is similar to this thought exercise: Imagine that you walk blindfolded 100 steps forward and then turn around and walk 99 steps. Are you sure that you end up exactly one step from where you started?

The name of this phenomenon is "loss of significant digits." To illustrate, let's calculate

$$8721\sqrt{3} - 10{,}681\sqrt{2}$$

The approximations from my calculator are

$$8721\sqrt{3} \approx 15105.21509 \qquad \text{and} \qquad 10{,}681\sqrt{2} \approx 15105.21506$$

and so we get $8721\sqrt{3} - 10{,}681\sqrt{2} \approx 0.00003$. Even with three spare digits exposed, the difference comes out as 0.00003306. As you can see, the two ten-digit numbers agree in nine digits that, after subtraction, become zeros before the first nonzero digit. To make things worse, the formerly small errors in the square roots become more visible. In this particular example we can use rationalization to write

$$8721\sqrt{3} - 10{,}681\sqrt{2} = \frac{1}{8721\sqrt{3} + 10{,}681\sqrt{2}}$$

(work out the details!) and now the loss of significant digits does not occur:

$$\frac{1}{8721\sqrt{3} + 10{,}681\sqrt{2}} \approx 0.00003310115 \quad \text{to seven digits}$$

(It would take too much space to explain why all seven digits are reliable; the subject *numerical analysis* deals with these and similar situations.) See Exercise 7 for another instance of restoring lost digits.

Now you can see why in Exercise 28 in Section 2.2 your guess at the limit was bound to go wrong: $\tan x$ becomes so close to x that the values will eventually agree in all digits that the calculator is capable of carrying. Similarly, if you start with just about any continuous function f and try to guess the value of

$$f'(x) = \lim_{h \to 0} \frac{f(x+h) - f(x)}{h}$$

long enough using a calculator, you will end up with a zero, despite all the rules in Section 3.2!

WHERE CALCULUS IS MORE POWERFUL THAN CALCULATORS AND COMPUTERS

One of the secrets of success of calculus in overcoming the difficulties connected with subtraction is symbolic manipulation. For instance, $(a + b) - a$ is always b, although the calculated value may be different. Try it with $a = 10^7$ and $b = \sqrt{2} \times 10^{-5}$. Another powerful tool is the use of inequalities; a good example is the Squeeze Theorem (2.3.3) as demonstrated in Example 13 in Section 2.3. Yet another method for avoiding computational difficulties is provided by the Mean Value Theorem (see Section 5.3 and Exercise 5) and its consequences, such as l'Hospital's Rule (4.8.3) (which helps solve the aforementioned Exercise 28 in Section 2.2 and others) and Taylor's Inequality (10.7.9).

The limitations of calculators and computers are further illustrated by infinite series. A common misconception is that a series can be summed by adding terms until there is "practically nothing to add" and "the error is less than the first neglected term." This last statement is true for certain alternating series (see the Alternating Series Estimation Theorem in Section 10.4) but not in general; a modified version is true for another class of series (Exercise 10). As an example to refute these misconceptions, let's consider the series

$$\sum_{n=1}^{\infty} \frac{1}{n^{1.001}}$$

which is a convergent p-series ($p = 1.001 > 1$). Suppose we were to try to sum this series, correct to eight decimal places, by adding terms until they are less than 5 in the ninth decimal place. In other words, we would stop when

$$\frac{1}{n^{1.001}} < 0.00000\,0005$$

that is, when $n = N = 196{,}217{,}284$. (This would require a high-speed computer and increased precision.) After going to all this trouble, we would end up with the approximating partial sum

$$S_N = \sum_{n=1}^{N} \frac{1}{n^{1.001}} < 19.5$$

But, from the proof of the Integral Test (see Figure 4 in Appendix F), we have

$$\sum_{n=1}^{\infty} \frac{1}{n^{1.001}} > \int_1^{\infty} \frac{dx}{x^{1.001}} = 1000$$

Thus, the machine result represents less than 2% of the correct answer!

Suppose that we then wanted to add a huge number of terms of this series, say, 10^{100} terms, in order to approximate the infinite sum more closely. (This number 10^{100}, called a *googol,* is outside the range of pocket calculators and is much larger than the number of elementary particles in our solar system.) If we were to add 10^{100} terms of the above series (only in theory; a million years is less than 10^{26} microseconds), we would still obtain a sum of less than 207 compared with the true sum of more than 1000. (This estimate of 207 is obtained by using a more precise form of the Integral Test, known as the Euler-Maclaurin Formula, and only then using a calculator. The formula provides a way to accelerate the convergence of this and other series.)

If the two preceding approaches did not give the right information about the accuracy of the partial sums, what does? A suitable inequality satisfied by the remainder of the series, as you can see from Exercise 6.

Computers and calculators are not replacements for mathematical thought. They are just replacements for some kinds of mathematical labor, either numerical or symbolic. There are, and always will be, mathematical problems that cannot be solved by a calculator or computer, regardless of its size and speed. A calculator or computer does stretch the human capacity for handling numbers and symbols, but there is still considerable scope and necessity for "thinking before doing."

EXERCISES G

1. Guess the value of

$$\lim_{x \to 0} \left(\frac{1}{\sin^2 x} - \frac{1}{x^2} \right)$$

and determine when to stop guessing before the loss of significant digits destroys your results. (The answer will depend on your calculator.) Then find the precise answer using an appropriate calculus method.

2. Guess the value of

$$\lim_{h \to 0} \frac{\ln(1 + h)}{h}$$

and determine when to stop guessing before the loss of significant digits destroys your result. This time the detrimental subtraction takes place inside the machine; explain how (assuming that the Taylor series with center $c = 1$ is used to approximate $\ln x$). Then find the precise answer using an appropriate calculus method.

3. Even innocent-looking calculus problems can lead to numbers beyond the calculator range. Show that the maximum value of the function

$$f(x) = \frac{x^{25}}{(1.0001)^x}$$

is greater than 10^{124}. [*Hint:* Use logarithms.] What is the limit of $f(x)$ as $x \to \infty$?

4. What is a numerically reliable expression to replace $\sqrt{1 - \cos x}$, especially when x is a small number? Refer to the identities in Appendix D for help. (Recall that some computer packages would signal an unnecessary error condition, or even switch to complex arithmetic, when $x = 0$.)

5. Try to evaluate

$$D = \ln\ln(10^9 + 1) - \ln\ln(10^9)$$

on your calculator. These numbers are so close together that you will likely obtain 0 or just a few digits of accuracy. However, we can use the Mean Value Theorem to achieve much greater accuracy.

(a) Let $f(x) = \ln\ln x$, $a = 10^9$, and $b = 10^9 + 1$. Then the Mean Value Theorem gives

$$f(b) - f(a) = f'(c)(b - a) = f'(c)$$

where $a < c < b$. Since f' is decreasing, we have $f'(a) > f'(c) > f'(b)$. Use this to estimate the value of D.

(b) Use the Mean Value Theorem a second time to discover why the quantities $f'(a)$ and $f'(b)$ in part (a) are so close to each other.

6. For the series $\Sigma_{n=1}^{\infty} n^{-1.001}$, studied in the text, exactly how many terms do we need (in theory) to make the error less than 5 in the ninth decimal place? You can use the inequalities from the proof of the Integral Test:

$$\int_{N+1}^{\infty} f(x)\,dx < \sum_{n=N+1}^{\infty} f(n) < \int_{N}^{\infty} f(x)\,dx$$

7. Archimedes found an approximation to 2π by considering the perimeter p of a regular 96-gon inscribed in a circle of radius 1. His formula, in modern notation, is

$$p = 96\sqrt{2 - \sqrt{2 + \sqrt{2 + \sqrt{2 + \sqrt{3}}}}}$$

(a) Carry out the calculations and compare with the value of p from more accurate sources, say $p = 192\sin(\pi/96)$. How many digits did you lose?

(b) Perform rationalization to avoid subtraction of approximate numbers and count the exact digits again.

8. This exercise is related to Exercise 2. Suppose that your computing device has an excellent program for the exponential function $\exp(x) = e^x$ but a poor program for $\ln x$. Use the identity

$$\ln a = b + \ln\left(1 + \frac{a - e^b}{e^b}\right)$$

and Taylor's Inequality (10.7.9) to improve the accuracy of $\ln x$.

9. The cubic equation

$$x^3 + px + q = 0$$

where we assume for simplicity that $p > 0$, has a classical solution formula for the real root, called Cardano's formula:

$$x = \frac{1}{3}\left[\left(\frac{27q + \sqrt{729q^2 + 108p^3}}{2}\right)^{1/3} + \left(\frac{27q - \sqrt{729q^2 + 108p^3}}{2}\right)^{1/3}\right]$$

For a user of a pocket-size calculator, as well as for an inexperienced programmer, the solution presents several stumbling blocks. First, the second radicand is negative and the fractional power key or routine may not handle it. Next, even if we fix the negative radical problem, when q is small in magnitude and p is of moderate size, the small number x is the difference of two numbers close to $\sqrt{p/3}$.

(a) Show that all these troubles are avoided by the formula

$$x = \frac{-9q}{a^{2/3} + 3p + 9p^2a^{-2/3}}$$

where $a = \dfrac{27|q| + \sqrt{729q^2 + 108p^3}}{2}$

Hint: Use the factorization formula

$$A + B = \frac{A^3 + B^3}{A^2 - AB + B^2}$$

(b) Evaluate

$$u = \frac{4}{(2 + \sqrt{5})^{2/3} + 1 + (2 + \sqrt{5})^{-2/3}}$$

If the result is simple, relate it to part (a), that is, restore the cubic equation whose root is u written in this form.

10. (a) A series convergent by the Ratio Test offers an error estimate similar to the "first neglected term" rule (Alternating Series Estimation Theorem in Section 10.4). When the index N is large enough to make

$$\left|\frac{a_{n+1}}{a_n}\right| \leq r < 1 \qquad \text{for every } n \geq N$$

(and we know that it will happen if the limit of the quotient is < 1), we can use this constant r to write

$$|s_n - s| = |a_{n+1} + a_{n+2} + \cdots|$$
$$\leq |a_{n+1}| + |a_{n+2}| + \cdots \leq |a_n| \cdot \frac{r}{1 - r}$$

where s_n is the partial sum and s is the sum. [This estimate becomes especially simple when $r \leq \frac{1}{2}$, making $r/(1 - r) \leq 1$: "the error is not more than the last used term."] Find the place in the proof of the Ratio Test (on page A45) where you can draw this conclusion.

(b) Consider the power series

$$f(x) = \sum_{n=1}^{\infty} \frac{x^n}{100^n + 1}$$

It is easy to show that its radius of convergence is $r = 100$. The series will converge rather slowly at $x = 99$: find out how many terms will make the error less than 5×10^{-7}.

(c) We can speed up the convergence of the series in part (b). Show that

$$f(x) = \frac{x}{100 - x} - f\left(\frac{x}{100}\right)$$

and find the number of the terms of this transformed series that leads to an error less than 5×10^{-7}. [*Hint:* Compare with the series $\Sigma_{n=1}^{\infty} (x^n/100^n)$, whose sum you know.]

11. The positive numbers

$$a_n = \int_0^1 e^{1-x}x^n\,dx$$

can, in theory, be calculated from a reduction formula obtained by integration by parts: $a_0 = e - 1$, $a_n = na_{n-1} - 1$. Prove, using $1 \leq e^{1-x} \leq e$ and the

Squeeze Theorem, that $\lim_{n \to \infty} a_n = 0$. Then try to calculate a_{20} from the reduction formula using your calculator. What went wrong?

The initial term $a_0 = e - 1$ cannot be represented exactly in a calculator. Let's call c the approximation of $e - 1$ that we can enter. Verify from the reduction formula (by observing the pattern after a few steps) that

$$a_n = \left[c - \left(\frac{1}{1!} + \frac{1}{2!} + \cdots + \frac{1}{n!}\right)\right] n!$$

and recall from Equation 10.7.12 that

$$\frac{1}{1!} + \frac{1}{2!} + \cdots + \frac{1}{n!}$$

converges to $e - 1$ as $n \to \infty$. The expression in square brackets converges to $c - (e - 1)$, a nonzero number, which gets multiplied by a fast-growing factor $n!$. We conclude that even if all further calculations (after entering a_0) were performed without errors, the initial inaccuracy would cause the computed sequence $\{a_n\}$ to diverge.

12. (a) A consolation after the catastrophic outcome of Exercise 11: If we rewrite the reduction formula to read

$$a_{n-1} = \frac{1 + a_n}{n}$$

we can use the inequality used in the squeeze argument to obtain improvements of the approximations of a_n. Try a_{20} again using this reverse approach.

(b) We used the reversed reduction formula to calculate quantities for which we have elementary formulas. To see that the idea is even more powerful, develop it for the integrals

$$\int_0^1 x^{n-\theta} e^{1-x}\, dx$$

where θ is a constant, $0 < \theta < 1$, and $n = 0, 1, \ldots$. For such θ, the integrals are no longer elementary (not solvable in "finite terms"), but the numbers can be calculated quickly. Find the integrals for the particular choice $\theta = \frac{1}{3}$ and $n = 0, 1, \ldots, 5$ to five digits of accuracy.

13. An advanced calculator has a key for a peculiar function:

$$E(x) = \begin{cases} 1 & \text{if } x = 0 \\ \dfrac{e^x - 1}{x} & \text{if } x \neq 0 \end{cases}$$

After so many warnings about the subtraction of close numbers, you may appreciate that the definition $\sinh x = \frac{1}{2}(e^x - e^{-x})$ gives inaccurate results for small x, where $\sinh x$ is close to x. Show that the use of the accurately evaluated function $E(x)$ helps restore the accuracy of $\sinh x$ for small x.

H COMPLEX NUMBERS

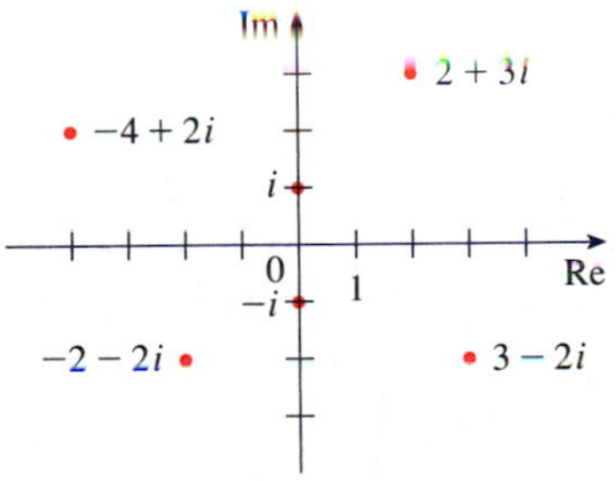

FIGURE 1

Complex numbers as points in the Argand plane

A **complex number** can be represented by an expression of the form $a + bi$, where a and b are real numbers and i is a symbol with the property that $i^2 = -1$. The complex number $a + bi$ can also be represented by the ordered pair (a, b) and plotted as a point in a plane (called the Argand plane) as in Figure 1. Thus, the complex number $i = 0 + 1 \cdot i$ is identified with the point $(0, 1)$.

The **real part** of the complex number $a + bi$ is the real number a and the **imaginary part** is the real number b. Thus, the real part of $4 - 3i$ is 4 and the imaginary part is -3. Two complex numbers $a + bi$ and $c + di$ are **equal** if $a = c$ and $b = d$, that is, their real parts are equal and their imaginary parts are equal. In the Argand plane the x-axis is called the real axis and the y-axis is called the imaginary axis.

The sum and difference of two complex numbers are defined by adding or subtracting their real parts and their imaginary parts, respectively:

$$(a + bi) + (c + di) = (a + c) + (b + d)i$$

$$(a + bi) - (c + di) = (a - c) + (b - d)i$$

For instance, $$(1 - i) + (4 + 7i) = (1 + 4) + (-1 + 7)i = 5 + 6i$$

The product of complex numbers is defined so that the usual commutative and distributive laws hold:

$$\begin{aligned}(a + bi)(c + di) &= a(c + di) + (bi)(c + di) \\ &= ac + adi + bci + bdi^2\end{aligned}$$

Since $i^2 = -1$, this becomes

$$(a + bi)(c + di) = (ac - bd) + (ad + bc)i$$

EXAMPLE 1

$$(-1 + 3i)(2 - 5i) = -(2 - 5i) + 3i(2 - 5i)$$
$$= -2 + 5i + 6i - 15(-1) = 13 + 11i$$ ■

Division of complex numbers is much like rationalizing the denominator of a rational expression. For the complex number $z = a + bi$, we define its **complex conjugate** to be $\bar{z} = a - bi$. To find the quotient of two complex numbers we multiply numerator and denominator by the complex conjugate of the denominator.

EXAMPLE 2 Express the number $\dfrac{-1 + 3i}{2 + 5i}$ in the form $a + bi$.

SOLUTION We multiply numerator and denominator by the complex conjugate of $2 + 5i$, namely $2 - 5i$, and we take advantage of the result of Example 1:

$$\frac{-1 + 3i}{2 + 5i} = \frac{-1 + 3i}{2 + 5i} \cdot \frac{2 - 5i}{2 - 5i} = \frac{13 + 11i}{2^2 + 5^2} = \frac{13}{29} + \frac{11}{29}i$$ ■

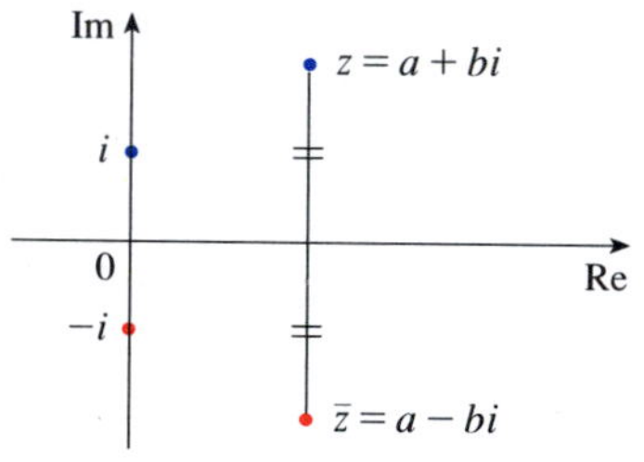

FIGURE 2

The geometric interpretation of the complex conjugate is shown in Figure 2: $\bar{z}$ is the reflection of z in the real axis. We list some of the properties of the complex conjugate in the following box. The proofs follow from the definition and are requested in Exercise 18.

PROPERTIES OF CONJUGATES

$$\overline{z + w} = \bar{z} + \bar{w} \qquad \overline{zw} = \bar{z}\,\bar{w} \qquad \overline{z^n} = \bar{z}^n$$

The **modulus**, or **absolute value**, $|z|$ of a complex number $z = a + bi$ is its distance from the origin. From Figure 3 we see that if $z = a + bi$, then

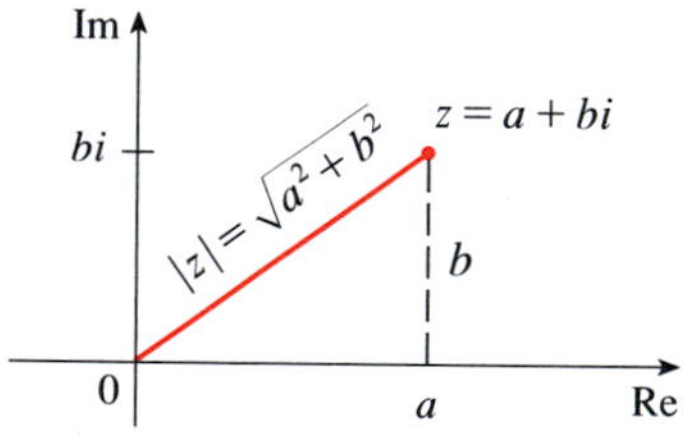

FIGURE 3

$$|z| = \sqrt{a^2 + b^2}$$

Notice that

$$z\bar{z} = (a + bi)(a - bi) = a^2 + abi - abi - b^2i^2 = a^2 + b^2$$

and so

$$z\bar{z} = |z|^2$$

This explains why the division procedure in Example 2 works in general:

$$\frac{z}{w} = \frac{z\bar{w}}{w\bar{w}} = \frac{z\bar{w}}{|w|^2}$$

Since $i^2 = -1$, we can think of i as a square root of -1. But we also have $(-i)^2 = i^2 = -1$ and so $-i$ is also a square root of -1. We say that i is the **principal square root** of -1 and write $\sqrt{-1} = i$. In general, if c is any positive number, we write

$$\sqrt{-c} = \sqrt{c}\; i$$

With this convention the usual derivation and formula for the roots of the quadratic equation

$ax^2 + bx + c = 0$ are valid even when $b^2 - 4ac < 0$:

$$x = \frac{-b \pm \sqrt{b^2 - 4ac}}{2a}$$

EXAMPLE 3 Find the roots of the equation $x^2 + x + 1 = 0$.

SOLUTION Using the quadratic formula, we have

$$x = \frac{-1 \pm \sqrt{1^2 - 4 \cdot 1}}{2} = \frac{-1 \pm \sqrt{-3}}{2} = \frac{-1 \pm \sqrt{3}\, i}{2}$$ ■

We observe that the solutions of the equation in Example 3 are complex conjugates of each other. In general, the solutions of any quadratic equation $ax^2 + bx + c = 0$ with real coefficients a, b, and c are always complex conjugates. (If z is real, $\bar{z} = z$, so z is its own conjugate.)

We have seen that if we allow complex numbers as solutions, then every quadratic equation has a solution. More generally, it is true that every polynomial equation

$$a_n x^n + a_{n-1} x^{n-1} + \cdots + a_1 x + a_0 = 0$$

of degree at least one has a solution among the complex numbers. This fact is known as the Fundamental Theorem of Algebra and was proved by Gauss.

POLAR FORM

We know that any complex number $z = a + bi$ can be considered as a point (a, b) and that any such point can be represented by polar coordinates (r, θ) with $r \geqslant 0$. In fact,

$$a = r\cos\theta \qquad b = r\sin\theta$$

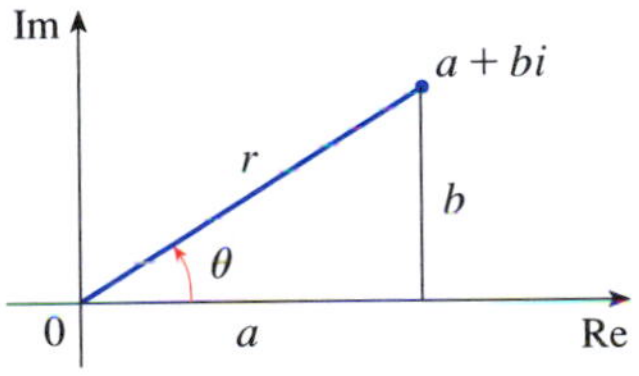

FIGURE 4

as in Figure 4. Therefore, we have

$$z = a + bi = (r\cos\theta) + (r\sin\theta)i$$

Thus, we can write any complex number z in the form

$$\boxed{z = r(\cos\theta + i\sin\theta)}$$

where $\qquad r = |z| = \sqrt{a^2 + b^2} \qquad$ and $\qquad \tan\theta = \dfrac{b}{a}$

The angle θ is called the **argument** of z and we write $\theta = \arg(z)$. Note that $\arg(z)$ is not unique; any two arguments of z differ by an integer multiple of 2π.

EXAMPLE 4 Write the following numbers in polar form:
(a) $z = 1 + i$ (b) $w = \sqrt{3} - i$

SOLUTION
(a) We have $r = |z| = \sqrt{1 + 1} = \sqrt{2}$ and $\tan\theta = 1$, so we can take $\theta = \pi/4$. Therefore, the polar form is

$$z = \sqrt{2}\left(\cos\frac{\pi}{4} + i\sin\frac{\pi}{4}\right)$$

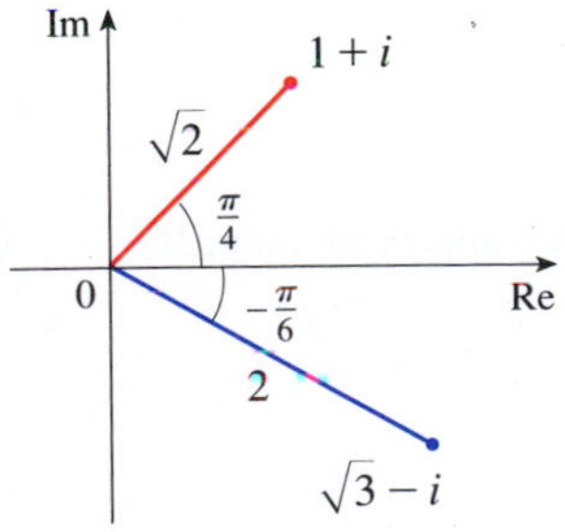

FIGURE 5

(b) Here we have $r = |w| = \sqrt{3 + 1} = 2$ and $\tan\theta = -1/\sqrt{3}$. Since w lies in the fourth quadrant, we take $\theta = -\pi/6$ and

$$w = 2\left[\cos\left(-\frac{\pi}{6}\right) + i\sin\left(-\frac{\pi}{6}\right)\right]$$

The numbers z and w are shown in Figure 5. ■

The polar form of complex numbers gives insight into multiplication and division. Let

$$z_1 = r_1(\cos\theta_1 + i\sin\theta_1) \qquad z_2 = r_2(\cos\theta_2 + i\sin\theta_2)$$

be two complex numbers written in polar form. Then

$$\begin{aligned} z_1 z_2 &= r_1 r_2(\cos\theta_1 + i\sin\theta_1)(\cos\theta_2 + i\sin\theta_2) \\ &= r_1 r_2[(\cos\theta_1\cos\theta_2 - \sin\theta_1\sin\theta_2) + i(\sin\theta_1\cos\theta_2 + \cos\theta_1\sin\theta_2)] \end{aligned}$$

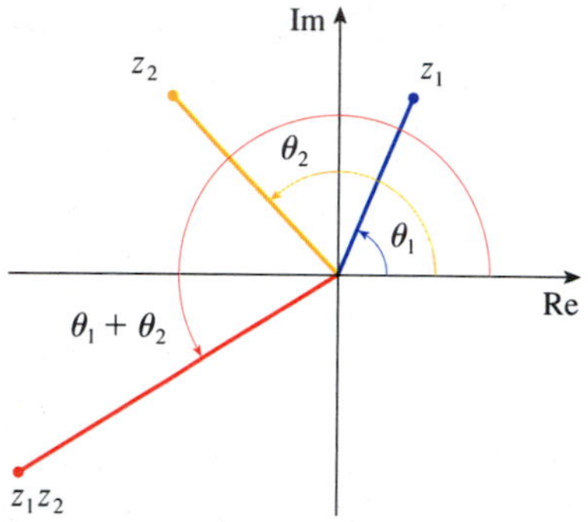

FIGURE 6

Therefore, using the addition formulas for cosine and sine, we have

(1) $$z_1 z_2 = r_1 r_2[\cos(\theta_1 + \theta_2) + i\sin(\theta_1 + \theta_2)]$$

This formula says that *to multiply two complex numbers we multiply the moduli and add the arguments.* (See Figure 6.)

A similar argument using the subtraction formulas for sine and cosine shows that *to divide two complex numbers we divide the moduli and subtract the arguments.*

$$\frac{z_1}{z_2} = \frac{r_1}{r_2}[\cos(\theta_1 - \theta_2) + i\sin(\theta_1 - \theta_2)] \qquad z_2 \neq 0$$

In particular, taking $z_1 = 1$ and $z_2 = z$, we have the following, which is illustrated in Figure 7.

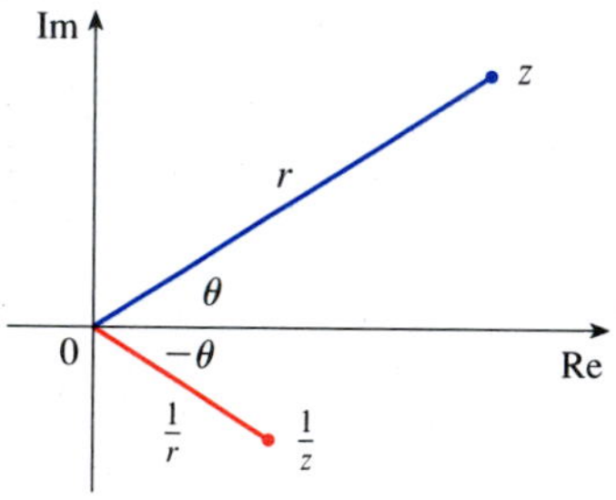

FIGURE 7

$$\text{If} \quad z = r(\cos\theta + i\sin\theta), \quad \text{then} \quad \frac{1}{z} = \frac{1}{r}(\cos\theta - i\sin\theta).$$

EXAMPLE 5 Find the product of the complex numbers $1 + i$ and $\sqrt{3} - i$ in polar form.

SOLUTION From Example 4 we have

$$1 + i = \sqrt{2}\left(\cos\frac{\pi}{4} + i\sin\frac{\pi}{4}\right)$$

and

$$\sqrt{3} - i = 2\left[\cos\left(-\frac{\pi}{6}\right) + i\sin\left(-\frac{\pi}{6}\right)\right]$$

So, by Equation 1,

$$\begin{aligned} (1 + i)(\sqrt{3} - i) &= 2\sqrt{2}\left[\cos\left(\frac{\pi}{4} - \frac{\pi}{6}\right) + i\sin\left(\frac{\pi}{4} - \frac{\pi}{6}\right)\right] \\ &= 2\sqrt{2}\left(\cos\frac{\pi}{12} + i\sin\frac{\pi}{12}\right) \end{aligned}$$

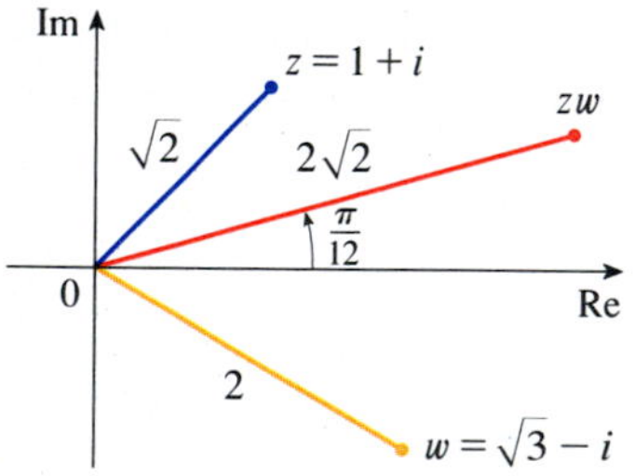

FIGURE 8

This is illustrated in Figure 8. ■

Repeated use of Formula 1 shows how to compute powers of a complex number. If

$$z = r(\cos\theta + i\sin\theta)$$

then

$$z^2 = r^2(\cos 2\theta + i\sin 2\theta)$$

and

$$z^3 = zz^2 = r^3(\cos 3\theta + i\sin 3\theta)$$

In general, we obtain the following result, which is named after the French mathematician Abraham De Moivre (1667–1754).

(2) DE MOIVRE'S THEOREM If $z = r(\cos\theta + i\sin\theta)$ and n is a positive integer, then

$$z^n = [r(\cos\theta + i\sin\theta)]^n = r^n(\cos n\theta + i\sin n\theta)$$

This says that *to take the* n*th power of a complex number we take the* n*th power of the modulus and multiply the argument by* n.

EXAMPLE 6 Find $(\frac{1}{2} + \frac{1}{2}i)^{10}$.

SOLUTION Since $\frac{1}{2} + \frac{1}{2}i = \frac{1}{2}(1 + i)$, it follows from Example 4(a) that $\frac{1}{2} + \frac{1}{2}i$ has the polar form

$$\tfrac{1}{2} + \tfrac{1}{2}i = \frac{\sqrt{2}}{2}\left(\cos\frac{\pi}{4} + i\sin\frac{\pi}{4}\right)$$

So by De Moivre's Theorem,

$$\left(\tfrac{1}{2} + \tfrac{1}{2}i\right)^{10} = \left(\frac{\sqrt{2}}{2}\right)^{10}\left(\cos\frac{10\pi}{4} + i\sin\frac{10\pi}{4}\right)$$

$$= \frac{2^5}{2^{10}}\left(\cos\frac{5\pi}{2} + i\sin\frac{5\pi}{2}\right) = \frac{1}{32}i$$

■

De Moivre's Theorem can also be used to find the nth roots of complex numbers. An nth root of the complex number z is a complex number w such that

$$w^n = z$$

Writing these two numbers in trigonometric form as

$$w = s(\cos\phi + i\sin\phi) \qquad \text{and} \qquad z = r(\cos\theta + i\sin\theta)$$

and using De Moivre's Theorem, we get

$$s^n(\cos n\phi + i\sin n\phi) = r(\cos\theta + i\sin\theta)$$

The equality of these two complex numbers shows that

$$s^n = r \qquad \text{or} \qquad s = r^{1/n}$$

and

$$\cos n\phi = \cos\theta \qquad \text{and} \qquad \sin n\phi = \sin\theta$$

From the fact that sine and cosine have period 2π it follows that

$$n\phi = \theta + 2k\pi \qquad \text{or} \qquad \phi = \frac{\theta + 2k\pi}{n}$$

Thus

$$w = r^{1/n}\left[\cos\left(\frac{\theta + 2k\pi}{n}\right) + i\sin\left(\frac{\theta + 2k\pi}{n}\right)\right]$$

Since this expression gives a different value of w for $k = 0, 1, 2, \ldots, n - 1$, we have the following:

(3) ROOTS OF A COMPLEX NUMBER Let $z = r(\cos\theta + i\sin\theta)$ and let n be a positive integer. Then z has the n distinct nth roots

$$w_k = r^{1/n}\left[\cos\left(\frac{\theta + 2k\pi}{n}\right) + i\sin\left(\frac{\theta + 2k\pi}{n}\right)\right]$$

where $k = 0, 1, 2, \ldots, n - 1$.

Notice that each of the nth roots of z has modulus $|w_k| = r^{1/n}$. Thus, all the nth roots of z lie on the circle of radius $r^{1/n}$ in the complex plane. Also, since the argument of each successive nth root exceeds the argument of the previous root by $2\pi/n$, we see that the nth roots of z are equally spaced on this circle.

EXAMPLE 7 Find the six sixth roots of $z = -8$ and graph these roots in the complex plane.

SOLUTION In trigonometric form, $z = 8(\cos\pi + i\sin\pi)$. Applying Equation 3 with $n = 6$, we get

$$w_k = 8^{1/6}\left(\cos\frac{\pi + 2k\pi}{6} + i\sin\frac{\pi + 2k\pi}{6}\right)$$

We get the six sixth roots of -8 by taking $k = 0, 1, 2, 3, 4, 5$ in this formula:

$$w_0 = 8^{1/6}\left(\cos\frac{\pi}{6} + i\sin\frac{\pi}{6}\right) = \sqrt{2}\left(\frac{\sqrt{3}}{2} + \frac{1}{2}i\right)$$

$$w_1 = 8^{1/6}\left(\cos\frac{\pi}{2} + i\sin\frac{\pi}{2}\right) = \sqrt{2}\,i$$

$$w_2 = 8^{1/6}\left(\cos\frac{5\pi}{6} + i\sin\frac{5\pi}{6}\right) = \sqrt{2}\left(-\frac{\sqrt{3}}{2} + \frac{1}{2}i\right)$$

$$w_3 = 8^{1/6}\left(\cos\frac{7\pi}{6} + i\sin\frac{7\pi}{6}\right) = \sqrt{2}\left(-\frac{\sqrt{3}}{2} - \frac{1}{2}i\right)$$

$$w_4 = 8^{1/6}\left(\cos\frac{3\pi}{2} + i\sin\frac{3\pi}{2}\right) = -\sqrt{2}\,i$$

$$w_5 = 8^{1/6}\left(\cos\frac{11\pi}{6} + i\sin\frac{11\pi}{6}\right) = \sqrt{2}\left(\frac{\sqrt{3}}{2} - \frac{1}{2}i\right)$$

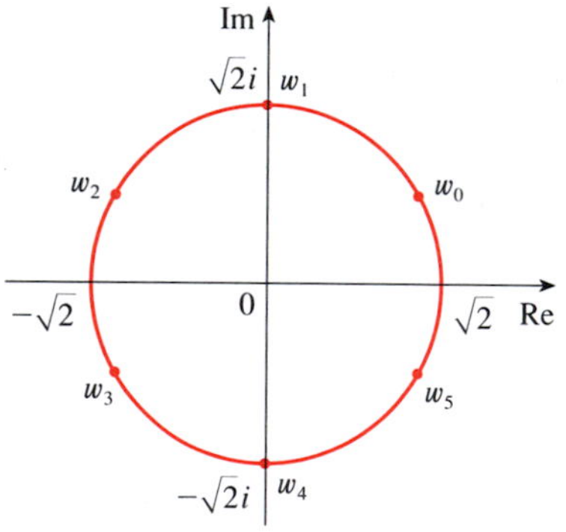

FIGURE 9
The six sixth roots of $z = -8$

All these points lie on the circle of radius $\sqrt{2}$ as shown in Figure 9. ■

COMPLEX EXPONENTIALS

We also need to give a meaning to the expression e^z when $z = x + iy$ is a complex number. The theory of infinite series as developed in Chapter 10 can be extended to the case where the terms are complex numbers. Using the Taylor series for e^x (10.7.11) as our guide, we define

$$e^z = \sum_{n=0}^{\infty} \frac{z^n}{n!} = 1 + z + \frac{z^2}{2!} + \frac{z^3}{3!} + \cdots \qquad \textbf{(4)}$$

and it turns out that this complex exponential function has the same properties as the real exponential function. In particular, it is true that

(5) $$e^{z_1+z_2} = e^{z_1}e^{z_2}$$

If we put $z = iy$, where y is a real number, in Equation 4, and use the facts that

$$i^2 = -1, \quad i^3 = i^2 i = -i, \quad i^4 = 1, \quad i^5 = i, \quad \ldots$$

we get
$$\begin{aligned} e^{iy} &= 1 + iy + \frac{(iy)^2}{2!} + \frac{(iy)^3}{3!} + \frac{(iy)^4}{4!} + \frac{(iy)^5}{5!} + \cdots \\ &= 1 + iy - \frac{y^2}{2!} - i\frac{y^3}{3!} + \frac{y^4}{4!} + i\frac{y^5}{5!} + \cdots \\ &= \left(1 - \frac{y^2}{2!} + \frac{y^4}{4!} - \frac{y^6}{6!} + \cdots\right) + i\left(y - \frac{y^3}{3!} + \frac{y^5}{5!} - \cdots\right) \\ &= \cos y + i\sin y \end{aligned}$$

Here we have used the Taylor series for $\cos y$ and $\sin y$ (Equations 10.7.15 and 10.7.14). The result is a famous formula called **Euler's formula**:

(6) $$e^{iy} = \cos y + i\sin y$$

Combining Euler's formula with Equation 5, we get

(7) $$e^{x+iy} = e^x e^{iy} = e^x(\cos y + i\sin y)$$

EXAMPLE 8 Evaluate: (a) $e^{i\pi}$ (b) $e^{-1+i\pi/2}$

SOLUTION

(a) From Euler's equation (6) we have

$$e^{i\pi} = \cos\pi + i\sin\pi = -1 + i(0) = -1$$

(b) Using Equation 7 we get

$$e^{-1+i\pi/2} = e^{-1}\left(\cos\frac{\pi}{2} + i\sin\frac{\pi}{2}\right) = \frac{1}{e}[0 + i(1)] = \frac{i}{e}$$ ■

Finally, we note that Euler's equation provides us with an easier method of proving De Moivre's Theorem:

$$[r(\cos\theta + i\sin\theta)]^n = (re^{i\theta})^n = r^n e^{in\theta} = r^n(\cos n\theta + i\sin n\theta)$$

EXERCISES H

1–14 ■ Evaluate the expression and write your answer in the form $a + bi$.

1. $(3 + 2i) + (7 - 3i)$
2. $(1 + i) - (2 - 3i)$
3. $(3 - i)(4 + i)$
4. $(4 - 7i)(1 + 3i)$
5. $\overline{12 + 7i}$
6. $\overline{2i(\frac{1}{2} - i)}$
7. $\dfrac{2 + 3i}{1 - 5i}$
8. $\dfrac{5 - i}{3 + 4i}$
9. $\dfrac{1}{1 + i}$
10. $\dfrac{3}{4 - 3i}$
11. i^3
12. i^{100}

13. $\sqrt{-25}$

14. $\sqrt{-3}\sqrt{-12}$

15–17 ■ Find the complex conjugate and the modulus of each number.

15. $3 + 4i$ **16.** $\sqrt{3} - i$ **17.** $-4i$

18. Prove the following properties of complex numbers.
(a) $\overline{z + w} = \bar{z} + \bar{w}$
(b) $\overline{zw} = \bar{z}\,\bar{w}$
(c) $\overline{z^n} = \bar{z}^n$, where n is a positive integer
[*Hint:* Write $z = a + bi$, $w = c + di$.]

19–24 ■ Find all solutions of the equation.

19. $4x^2 + 9 = 0$

20. $x^4 = 1$

21. $x^2 - 8x + 17 = 0$

22. $x^2 - 4x + 5 = 0$

23. $z^2 + z + 2 = 0$

24. $z^2 + \frac{1}{2}z + \frac{1}{4} = 0$

25–28 ■ Write the number in polar form with argument between 0 and 2π.

25. $-3 + 3i$

26. $1 - \sqrt{3}\,i$

27. $3 + 4i$

28. $8i$

29–32 ■ Find polar forms for zw, z/w, and $1/z$ by first putting z and w into polar form.

29. $z = \sqrt{3} + i, \quad w = 1 + \sqrt{3}\,i$

30. $z = 4\sqrt{3} - 4i, \quad w = 8i$

31. $z = 2\sqrt{3} - 2i, \quad w = -1 + i$

32. $z = 4(\sqrt{3} + i), \quad w = -3 - 3i$

33–36 ■ Find the indicated power using De Moivre's Theorem.

33. $(1 + i)^{20}$

34. $(1 - \sqrt{3}\,i)^5$

35. $(2\sqrt{3} + 2i)^5$

36. $(1 - i)^8$

37–40 ■ Find the indicated roots. Sketch the roots in the complex plane.

37. The eighth roots of 1

38. The fifth roots of 32

39. The cube roots of i

40. The cube roots of $1 + i$

41–46 ■ Write the number in the form $a + bi$.

41. $e^{i\pi/2}$ **42.** $e^{2\pi i}$ **43.** $e^{i3\pi/4}$

44. $e^{-i\pi}$ **45.** $e^{2+i\pi}$ **46.** e^{1+2i}

47. Use De Moivre's Theorem with $n = 3$ to express $\cos 3\theta$ and $\sin 3\theta$ in terms of $\cos\theta$ and $\sin\theta$.

48. Use Euler's formula to prove the following formulas for $\cos x$ and $\sin x$:

$$\cos x = \frac{e^{ix} + e^{-ix}}{2}$$

$$\sin x = \frac{e^{ix} - e^{-ix}}{2i}$$

49. If $u(x) = f(x) + ig(x)$ is a complex-valued function of a real variable x and the real and imaginary parts $f(x)$ and $g(x)$ are differentiable functions of x, then the derivative of u is defined to be $u'(x) = f'(x) + ig'(x)$. Use this together with Equation 7 to prove that if $F(x) = e^{rx}$, then $F'(x) = re^{rx}$ when $r = a + bi$ is a complex number.

I CONIC SECTIONS

In this section we give geometric definitions of parabolas, ellipses, and hyperbolas and derive their standard equations. They are called **conic sections**, or **conics**, because they result from intersecting a cone with a plane as shown in Figure 1.

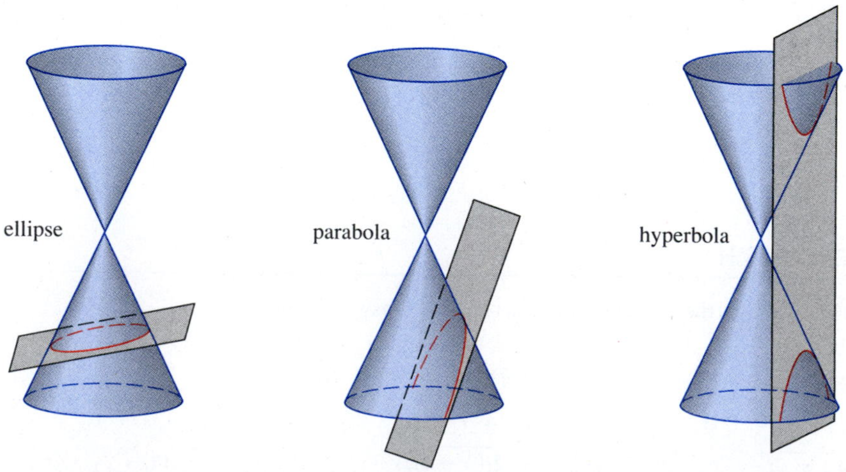

FIGURE 1
Conics

PARABOLAS

A **parabola** is the set of points in a plane that are equidistant from a fixed point F (called the **focus**) and a fixed line (called the **directrix**). This definition is illustrated by Figure 2. Notice that the point halfway between the focus and the directrix lies on the parabola; it is called the **vertex**. The line through the focus perpendicular to the directrix is called the **axis** of the parabola.

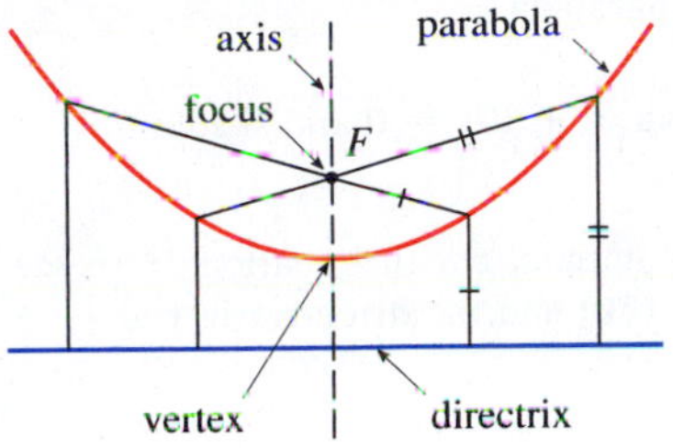

FIGURE 2

In the 16th century Galileo showed that the path of a projectile that is shot into the air at an angle to the ground is a parabola. Since then, parabolic shapes have been used in designing automobile headlights, reflecting telescopes, and suspension bridges. (See Problem 22 on page 241 for the reflection property of parabolas that makes them so useful.)

We obtain a particularly simple equation for a parabola if we place its vertex at the origin O and its directrix parallel to the x-axis as in Figure 3. If the focus is the point $(0, p)$, then the directrix has the equation $y = -p$. If $P(x, y)$ is any point on the parabola, then the distance from P to the focus is

$$|PF| = \sqrt{x^2 + (y - p)^2}$$

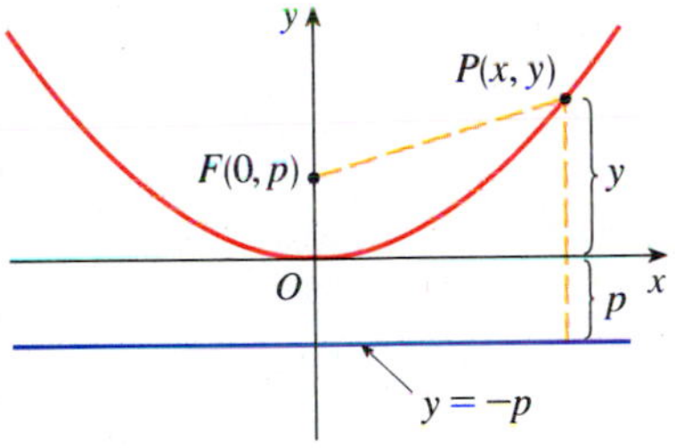

FIGURE 3

and the distance from P to the directrix is $|y + p|$. (Figure 3 illustrates the case where $p > 0$.) The defining property of a parabola is that these distances are equal:

$$\sqrt{x^2 + (y - p)^2} = |y + p|$$

We get an equivalent equation by squaring and simplifying:

$$x^2 + (y - p)^2 = |y + p|^2 = (y + p)^2$$

$$x^2 + y^2 - 2py + p^2 = y^2 + 2py + p^2$$

$$x^2 = 4py$$

(1) The equation of a parabola with focus $(0, p)$ and directrix $y = -p$ is

$$x^2 = 4py$$

If we write $a = 1/(4p)$, then the standard equation of a parabola (1) becomes $y = ax^2$. It opens upward if $p > 0$ and downward if $p < 0$ [see Figure 4, parts (a) and (b)]. The graph is symmetric with respect to the y-axis because (1) is unchanged when x is replaced by $-x$.

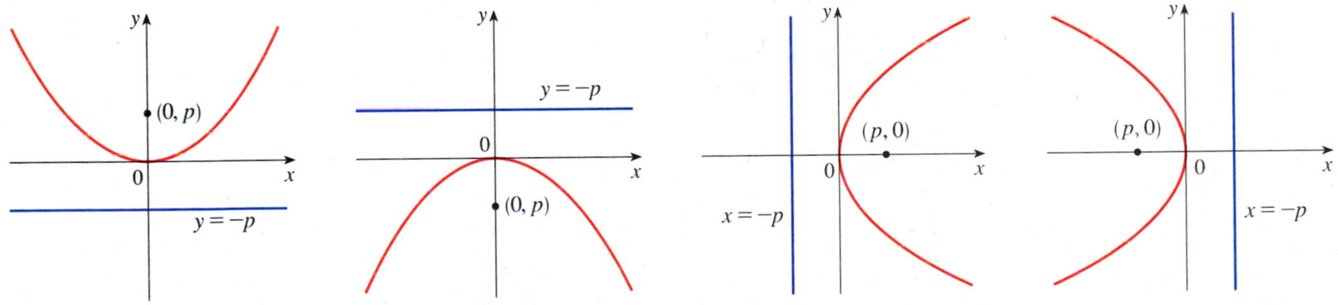

(a) $x^2 = 4py$, $p > 0$ (b) $x^2 = 4py$, $p < 0$ (c) $x^2 = 4px$, $p > 0$ (d) $x^2 = 4px$, $p < 0$

FIGURE 4

If we interchange x and y in (1), we obtain

(2) $$y^2 = 4px$$

which is the equation of a parabola with focus $(p, 0)$ and directrix $x = -p$. (Interchanging x and y amounts to reflecting about the diagonal line $y = x$.) The parabola opens to the right if $p > 0$ and to the left if $p < 0$ [see Figure 4, parts (c) and (d)]. In both cases the graph is symmetric with respect to the x-axis, which is the axis of the parabola.

EXAMPLE 1 Find the focus and directrix of the parabola $y^2 + 10x = 0$ and sketch the graph.

SOLUTION If we write the equation as $y^2 = -10x$ and compare it with Equation 2, we see that $4p = -10$, so $p = -\frac{5}{2}$. Thus, the focus is $(p, 0) = (-\frac{5}{2}, 0)$ and the directrix is $x = \frac{5}{2}$. The sketch is shown in Figure 5.

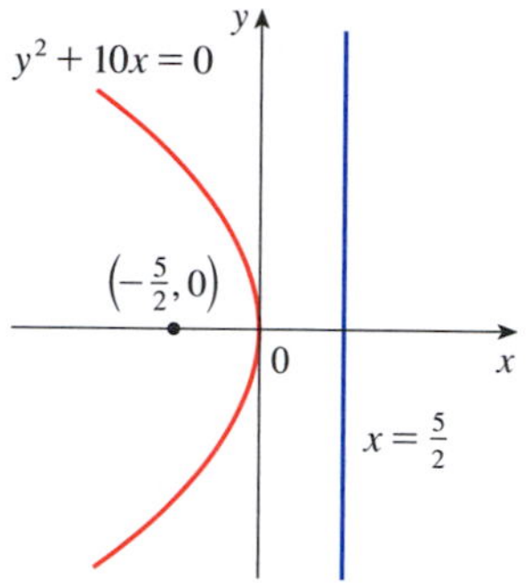

FIGURE 5

■

ELLIPSES

An **ellipse** is the set of points in a plane the sum of whose distances from two fixed points F_1 and F_2 is a constant (see Figure 6). These two fixed points are called the **foci** (plural of **focus**). One of Kepler's laws is that the orbits of the planets in the solar system are ellipses with the Sun at one focus.

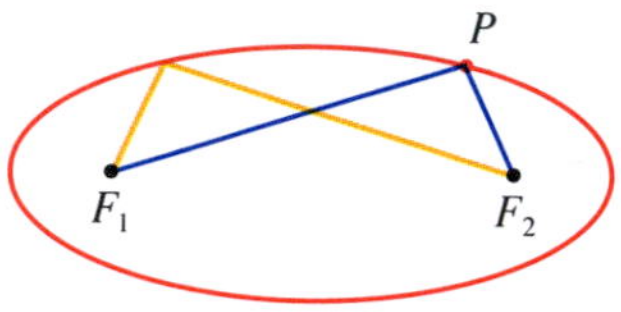

FIGURE 6

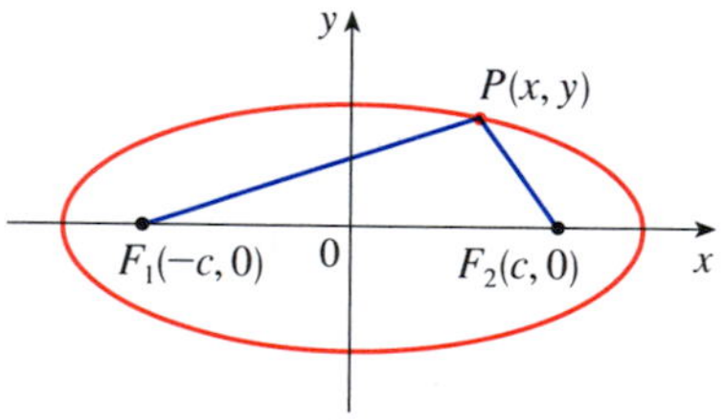

FIGURE 7

In order to obtain the simplest equation for an ellipse, we place the foci on the x-axis at the points $(-c, 0)$ and $(c, 0)$ as in Figure 7 so that the origin is halfway between the foci. Let the sum of the distances from a point on the ellipse to the foci be $2a > 0$. Then $P(x, y)$ is a point on the ellipse when

$$|PF_1| + |PF_2| = 2a$$

that is,
$$\sqrt{(x + c)^2 + y^2} + \sqrt{(x - c)^2 + y^2} = 2a$$

or
$$\sqrt{(x - c)^2 + y^2} = 2a - \sqrt{(x + c)^2 + y^2}$$

Squaring both sides, we have

$$x^2 - 2cx + c^2 + y^2 = 4a^2 - 4a\sqrt{(x + c)^2 + y^2} + x^2 + 2cx + c^2 + y^2$$

which simplifies to

$$a\sqrt{(x + c)^2 + y^2} = a^2 + cx$$

We square again:

$$a^2(x^2 + 2cx + c^2 + y^2) = a^4 + 2a^2cx + c^2x^2$$

which becomes

$$(a^2 - c^2)x^2 + a^2y^2 = a^2(a^2 - c^2)$$

From triangle F_1F_2P in Figure 7 we see that $2c < 2a$, so $c < a$ and, therefore, $a^2 - c^2 > 0$. For convenience, let $b^2 = a^2 - c^2$. Then the equation of the ellipse becomes

$b^2x^2 + a^2y^2 = a^2b^2$ or, if both sides are divided by a^2b^2,

(3) $$\frac{x^2}{a^2} + \frac{y^2}{b^2} = 1$$

Since $b^2 = a^2 - c^2 < a^2$, it follows that $b < a$. The x-intercepts are found by setting $y = 0$. Then $x^2/a^2 = 1$, or $x^2 = a^2$, so $x = \pm a$. The corresponding points $(a, 0)$ and $(-a, 0)$ are called the **vertices** of the ellipse and the line segment joining the vertices is called the **major axis**. To find the y-intercepts we set $x = 0$ and obtain $y^2 = b^2$, so $y = \pm b$. Equation 3 is unchanged if x is replaced by $-x$ or y is replaced by $-y$, so the ellipse is symmetric about both axes. Notice that if the foci coincide, then $c = 0$, so $a = b$ and the ellipse becomes a circle with radius $r = a = b$.

We summarize this discussion as follows (see also Figure 8):

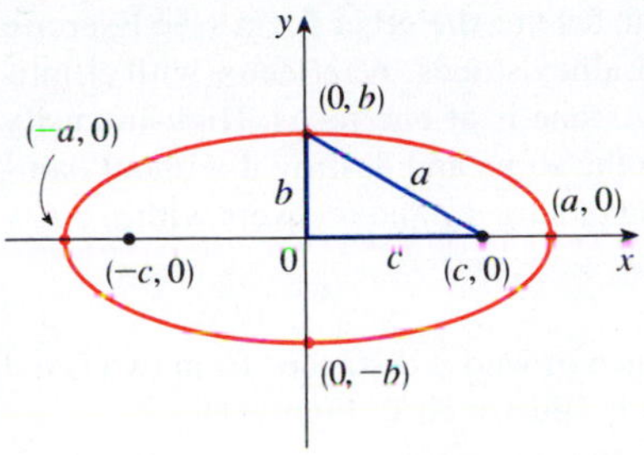

FIGURE 8
$\frac{x^2}{a^2} + \frac{y^2}{b^2} = 1$

(4) The ellipse

$$\frac{x^2}{a^2} + \frac{y^2}{b^2} = 1 \qquad a \geqslant b > 0$$

has foci $(\pm c, 0)$, where $c^2 = a^2 - b^2$, and vertices $(\pm a, 0)$.

If the foci of an ellipse are located on the y-axis at $(0, \pm c)$, then we can find its equation by interchanging x and y in (4) (see Figure 9).

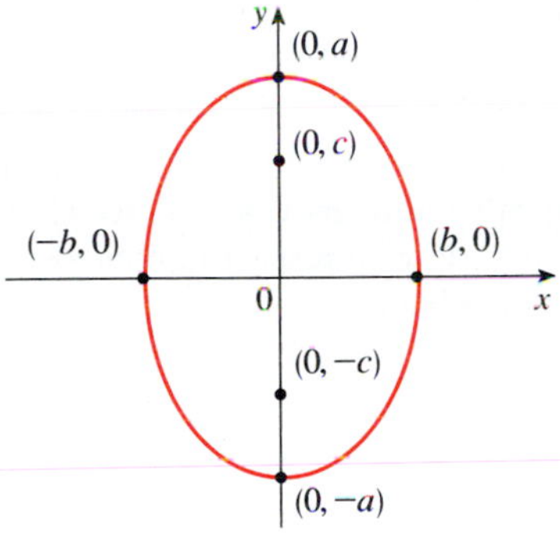

FIGURE 9
$\frac{x^2}{b^2} + \frac{y^2}{a^2} = 1,\ a \geqslant b$

(5) The ellipse

$$\frac{x^2}{b^2} + \frac{y^2}{a^2} = 1 \qquad a \geqslant b > 0$$

has foci $(0, \pm c)$, where $c^2 = a^2 - b^2$, and vertices $(0, \pm a)$.

EXAMPLE 2 Sketch the graph of $9x^2 + 16y^2 = 144$ and locate the foci.

SOLUTION Divide both sides of the equation by 144:

$$\frac{x^2}{16} + \frac{y^2}{9} = 1$$

The equation is now in the standard form for an ellipse, so we have $a^2 = 16$, $b^2 = 9$, $a = 4$, and $b = 3$. The x-intercepts are ± 4 and the y-intercepts are ± 3. Also, $c^2 = a^2 - b^2 = 7$, so $c = \sqrt{7}$ and the foci are $(\pm\sqrt{7}, 0)$. The graph is sketched in Figure 10. ■

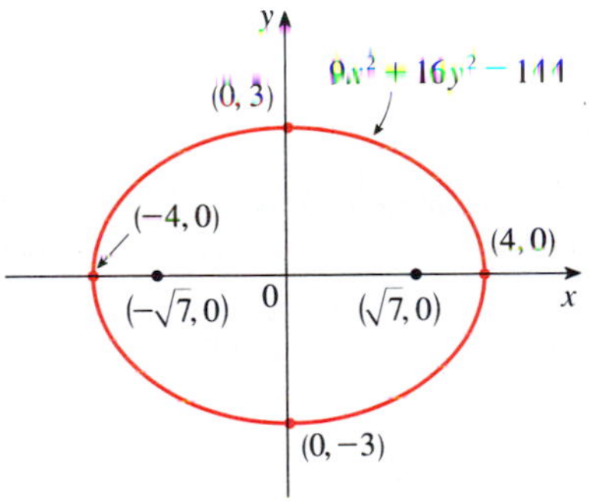

FIGURE 10

EXAMPLE 3 Find an equation of the ellipse with foci $(0, \pm 2)$ and vertices $(0, \pm 3)$.

SOLUTION Using the notation of (5), we have $c = 2$ and $a = 3$. Then we obtain $b^2 = a^2 - c^2 = 9 - 4 = 5$, so the equation of the ellipse is

$$\frac{x^2}{5} + \frac{y^2}{9} = 1$$

Another way of writing the equation is $9x^2 + 5y^2 = 45$. ■

Like parabolas, ellipses have an interesting reflection property that has practical consequences. If a source of light or sound is placed at one focus of a surface with elliptical cross-sections, then all the light or sound is reflected off the surface to the other focus (see Exercise 49). This principle is used in *lithotripsy*, a treatment for kidney stones. A reflector with elliptical cross-section is placed in such a way that the kidney stone is at one focus. High-intensity sound waves generated at the other focus are reflected to the stone and destroy it without damaging surrounding tissue. The patient is spared the trauma of surgery and recovers within a few days.

HYPERBOLAS

A **hyperbola** is the set of all points in a plane the difference of whose distances from two fixed points F_1 and F_2 (the foci) is a constant. This definition is illustrated in Figure 11.

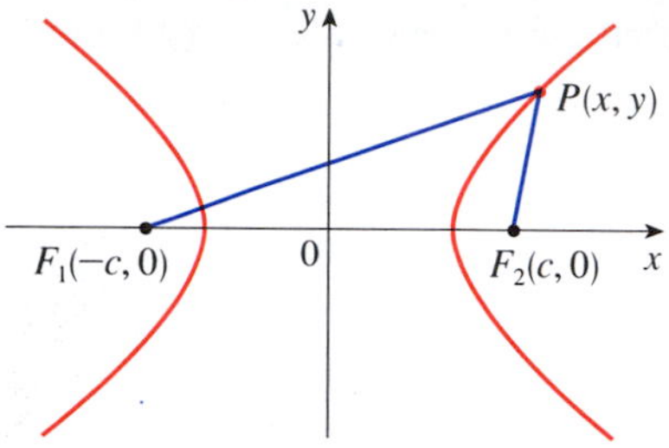

FIGURE 11

P is on the hyperbola when $|PF_1| - |PF_2| = \pm 2a$

Notice that the definition of a hyperbola is similar to that of an ellipse; the only change is that the sum of distances has become a difference of distances. In fact, the derivation of the equation of a hyperbola is also similar to the one given earlier for an ellipse. It is left as Exercise 42 to show that when the foci are on the x-axis at $(\pm c, 0)$ and the difference of distances is $|PF_1| - |PF_2| = \pm 2a$, then the equation of the hyperbola is

(6) $$\frac{x^2}{a^2} - \frac{y^2}{b^2} = 1$$

where $c^2 = a^2 + b^2$. Notice that the x-intercepts are again $\pm a$ and the points $(a, 0)$ and $(-a, 0)$ are the **vertices** of the hyperbola. But if we put $x = 0$ in Equation 6 we get $y = -b^2$, which is impossible, so there is no y-intercept. The hyperbola is symmetric with respect to both axes.

To analyze the hyperbola further, we look at Equation 6 and obtain

$$\frac{x^2}{a^2} = 1 + \frac{y^2}{b^2} \geqslant 1$$

This shows that $x^2 \geqslant a^2$, so $|x| = \sqrt{x^2} \geqslant a$. Therefore, we have $x \geqslant a$ or $x \leqslant -a$. This means that the hyperbola consists of two parts, called its *branches.*

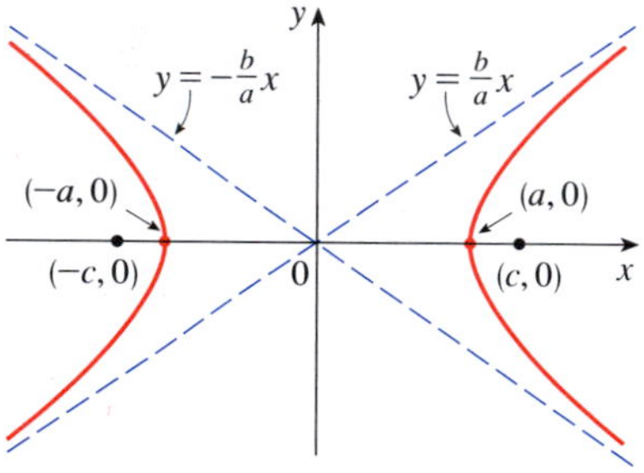

FIGURE 12

$\frac{x^2}{a^2} - \frac{y^2}{b^2} = 1$

When we draw a hyperbola it's useful to first draw its asymptotes, which are the lines $y = (b/a)x$ and $y = -(b/a)x$ shown in Figure 12. Both branches of the hyperbola approach the asymptotes; that is, they come arbitrarily close to the asymptotes.

(7) The hyperbola

$$\frac{x^2}{a^2} - \frac{y^2}{b^2} = 1$$

has foci $(\pm c, 0)$, where $c^2 = a^2 + b^2$, vertices $(\pm a, 0)$, and asymptotes $y = \pm(b/a)x$.

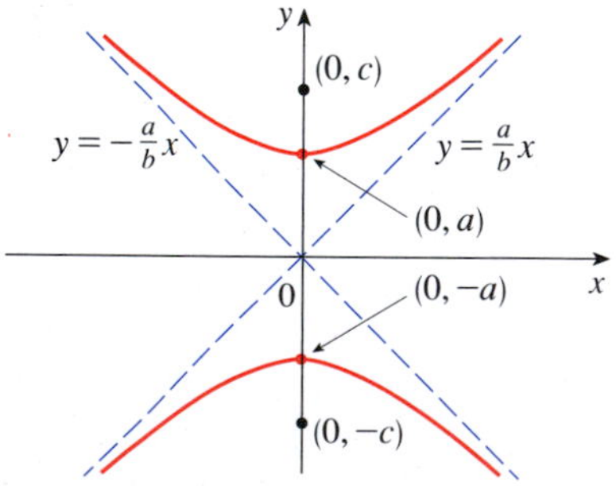

FIGURE 13

$\frac{y^2}{a^2} - \frac{x^2}{b^2} = 1$

If the foci of a hyperbola are on the y-axis, then by reversing the roles of x and y we obtain the following information, which is illustrated in Figure 13.

(8) The hyperbola

$$\frac{y^2}{a^2} - \frac{x^2}{b^2} = 1$$

has foci $(0, \pm c)$, where $c^2 = a^2 + b^2$, vertices $(0, \pm a)$, and asymptotes $y = \pm(a/b)x$.

EXAMPLE 4 Find the foci and asymptotes of the hyperbola $9x^2 - 16y^2 = 144$ and sketch its graph.

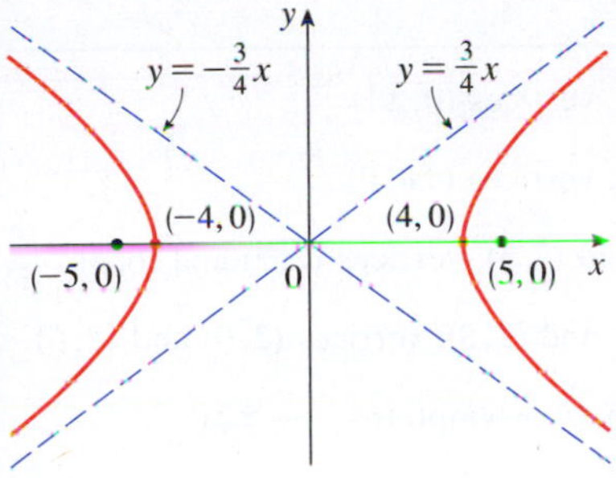

FIGURE 14
$9x^2 - 16y^2 = 144$

SOLUTION If we divide both sides of the equation by 144, it becomes

$$\frac{x^2}{16} - \frac{y^2}{9} = 1$$

which is of the form given in (7) with $a = 4$ and $b = 3$. Since $c^2 = 16 + 9 = 25$, the foci are $(\pm 5, 0)$. The asymptotes are the lines $y = \frac{3}{4}x$ and $y = -\frac{3}{4}x$. The graph is shown in Figure 14. ■

EXAMPLE 5 Find the foci and equation of the hyperbola with vertices $(0, \pm 1)$ and asymptote $y = 2x$.

SOLUTION From (8) and the given information, we see that $a = 1$ and $a/b = 2$. Thus, $b = a/2 = \frac{1}{2}$ and $c^2 = a^2 + b^2 = \frac{5}{4}$. The foci are $(0, \pm\sqrt{5}/2)$ and the equation of the hyperbola is

$$y^2 - 4x^2 = 1$$ ■

SHIFTED CONICS

As discussed in Appendix C, we shift conics by taking the standard equations (1), (2), (4), (5), (7), and (8) and replacing x and y by $x - h$ and $y - k$.

EXAMPLE 6 Find an equation of the ellipse with foci $(2, -2)$, $(4, -2)$ and vertices $(1, -2)$, $(5, -2)$.

SOLUTION The major axis is the line segment that joins the vertices $(1, -2)$, $(5, -2)$ and has length 4, so $a = 2$. The distance between the foci is 2, so $c = 1$. Thus, $b^2 = a^2 - c^2 = 3$. Since the center of the ellipse is $(3, -2)$, we replace x and y in (4) by $x - 3$ and $y + 2$ to obtain

$$\frac{(x-3)^2}{4} + \frac{(y+2)^2}{3} = 1$$

as the equation of the ellipse. ■

EXAMPLE 7 Sketch the conic

$$9x^2 - 4y^2 - 72x + 8y + 176 = 0$$

and find its foci.

SOLUTION We complete the squares as follows:

$$4(y^2 - 2y) - 9(x^2 - 8x) = 176$$

$$4(y^2 - 2y + 1) - 9(x^2 - 8x + 16) = 176 + 4 - 144$$

$$4(y-1)^2 - 9(x-4)^2 = 36$$

$$\frac{(y-1)^2}{9} - \frac{(x-4)^2}{4} = 1$$

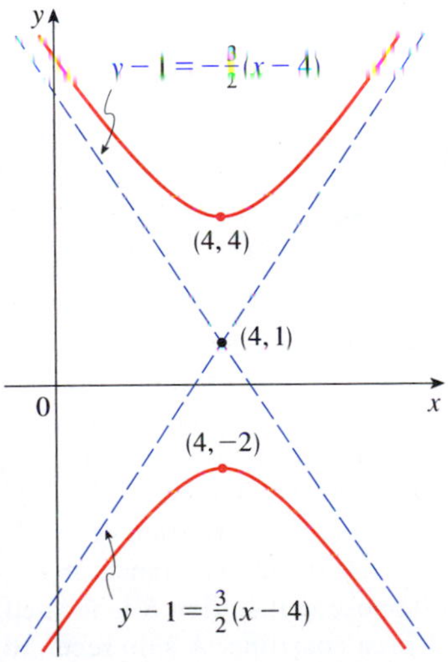

FIGURE 15
$9x^2 - 4y^2 - 72x + 8y + 176 = 0$

This is in the form (8) except that x and y are replaced by $x - 4$ and $y - 1$. Thus $a^2 = 9$, $b^2 = 4$, and $c^2 = 13$. The hyperbola is shifted four units to the right and one unit upward. The foci are $(4, 1 + \sqrt{13})$ and $(4, 1 - \sqrt{13})$ and the vertices are $(4, 4)$ and $(4, -2)$. The asymptotes are $y - 1 = \pm\frac{3}{2}(x - 4)$. The hyperbola is sketched in Figure 15. ■

EXERCISES I

1–8 ■ Find the vertex, focus, and directrix of the parabola and sketch its graph.

1. $x^2 = -8y$

2. $x = -5y^2$

3. $y^2 = x$

4. $2x^2 = y$

5. $x + 1 = 2(y - 3)^2$

6. $x^2 - 6x + 8y = 7$

7. $2x + y^2 - 8y + 12 = 0$

8. $x^2 + 12x - y + 39 = 0$

9–20 ■ Find the vertices and foci of the conic and sketch its graph. In the case of a hyperbola, find the asymptotes.

9. $\dfrac{x^2}{16} + \dfrac{y^2}{4} = 1$

10. $\dfrac{x^2}{4} + \dfrac{y^2}{25} = 1$

11. $25x^2 + 9y^2 = 225$

12. $x^2 + 4y^2 = 4$

13. $\dfrac{x^2}{144} - \dfrac{y^2}{25} = 1$

14. $\dfrac{y^2}{25} - \dfrac{x^2}{144} = 1$

15. $9y^2 - x^2 = 9$

16. $x^2 - y^2 = 1$

17. $9x^2 - 18x + 4y^2 = 27$

18. $16x^2 - 9y^2 + 64x - 90y = 305$

19. $2y^2 - 3x^2 - 4y + 12x + 8 = 0$

20. $x^2 + 2y^2 - 6x + 4y + 7 = 0$

21–38 ■ Find an equation for the conic that satisfies the given conditions.

21. Parabola, focus $(0, 3)$, directrix $y = -3$

22. Parabola, focus $(-2, 0)$, directrix $x = 2$

23. Parabola, focus $(3, 0)$, directrix $x = 1$

24. Parabola, focus $(1, -1)$, directrix $y = 5$

25. Parabola, vertex $(0,0)$, axis the x-axis, passing through $(1, -4)$

26. Parabola, vertical axis, passing through $(-2, 3)$, $(0, 3)$, and $(1, 9)$

27. Ellipse, foci $(\pm 1, 0)$, vertices $(\pm 2, 0)$

28. Ellipse, foci $(0, \pm 4)$, vertices $(0, \pm 5)$

29. Ellipse, foci $(3, \pm 1)$, vertices $(3, \pm 3)$

30. Ellipse, foci $(\pm 1, 2)$, length of major axis 6

31. Ellipse, center $(2, 2)$, focus $(0, 2)$, vertex $(5, 2)$

32. Ellipse, foci $(\pm 2, 0)$, passing through $(2, 1)$

33. Hyperbola, foci $(0, \pm 3)$, vertices $(0, \pm 1)$

34. Hyperbola, foci $(\pm 6, 0)$, vertices $(\pm 4, 0)$

35. Hyperbola, foci $(1, 3)$ and $(7, 3)$, vertices $(2, 3)$ and $(6, 3)$

36. Hyperbola, foci $(2, -2)$ and $(2, 8)$, vertices $(2, 0)$ and $(2, 6)$

37. Hyperbola, vertices $(\pm 3, 0)$, asymptotes $y = \pm 2x$

38. Hyperbola, foci $(2, 2)$ and $(6, 2)$, asymptotes $y = x - 2$ and $y = 6 - x$

39. The point in a lunar orbit nearest the surface of the moon is called *perilune* and the point farthest from the surface is called *apolune*. The Apollo 11 spacecraft was placed in an elliptical lunar orbit with perilune altitude 110 km and apolune altitude 314 km (above the moon). Find an equation of this ellipse if the radius of the moon is 1728 km and the center of the moon is at one focus.

40. A cross-section of a parabolic reflector is shown in the figure. The bulb is located at the focus and the opening at the focus is 10 cm.
(a) Find an equation of the parabola.
(b) Find the diameter of the opening $|CD|$, 11 cm from the vertex.

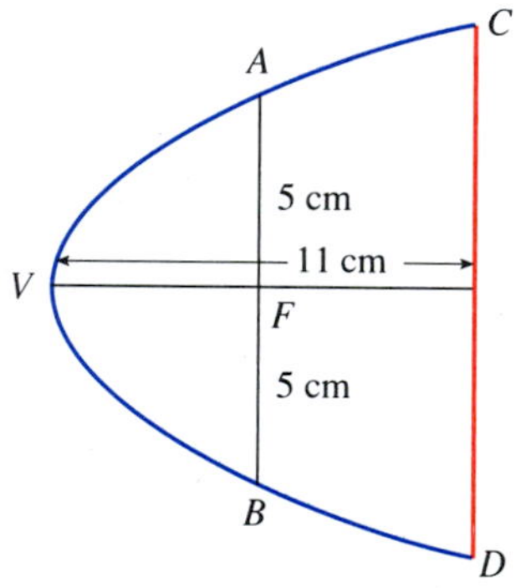

41. In the LORAN (LOng RAnge Navigation) radio navigation system, two radio stations located at A and B transmit simultaneous signals to a ship or an aircraft located at P. The onboard computer converts the time difference in receiving these signals into a distance difference $|PA| - |PB|$, and this, according to the definition of a hyperbola, locates the ship or aircraft on one branch of a hyperbola (see the figure). Suppose that station B is located 400 mi due east of station A on a coastline. A ship received the signal from B 1200 microseconds (μs) before it received the signal from A.
(a) Assuming that radio signals travel at a speed of 980 ft/μs, find the equation of the hyperbola on which the ship lies.

(b) If the ship is due north of B, how far off the coastline is the ship?

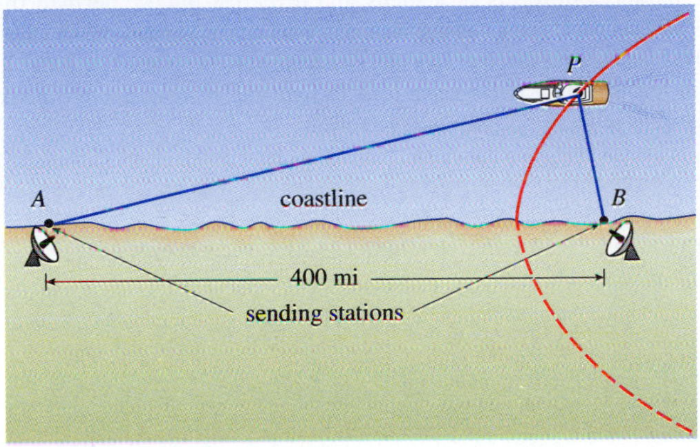

42. Use the definition of a hyperbola to derive Equation 6 for a hyperbola with foci $(\pm c, 0)$ and vertices $(\pm a, 0)$.

43. Show that the function defined by the upper branch of the hyperbola $y^2/a^2 - x^2/b^2 = 1$ is concave upward.

44. Find an equation for the ellipse with foci $(1, 1)$ and $(-1, -1)$ and major axis of length 4.

45. Determine the type of curve represented by the equation

$$\frac{x^2}{k} + \frac{y^2}{k - 16} = 1$$

in each of the following cases:

(a) $k > 16$ (b) $0 < k < 16$ (c) $k < 0$

(d) Show that all the curves in parts (a) and (b) have the same foci, no matter what the value of k is.

46. (a) Show that the equation of the tangent line to the parabola $y^2 = 4px$ at the point (x_0, y_0) can be written as

$$y_0 y = 2p(x + x_0)$$

(b) What is the x-intercept of this tangent line? Use this fact to draw the tangent line.

47. Use Simpson's Rule with $n = 10$ to estimate the length of the ellipse $x^2 + 4y^2 = 4$.

48. The planet Pluto travels in an elliptical orbit around the Sun (at one focus). The length of the major axis is 1.18×10^{10} km and the length of the minor axis is 1.14×10^{10} km. Use Simpson's Rule with $n = 10$ to estimate the distance traveled by the planet during one complete orbit around the Sun.

49. Let $P(x_1, y_1)$ be a point on the ellipse $x^2/a^2 + y^2/b^2 = 1$ with foci F_1 and F_2 and let α and β be the angles between the lines PF_1, PF_2 and the ellipse as in the figure. Prove that $\alpha = \beta$. This explains how whispering galleries and lithotripsy work. Sound coming from one focus is reflected and passes through the other focus. [*Hint:* Use the formula in Problem 21 on page 241 to show that $\tan \alpha = \tan \beta$.]

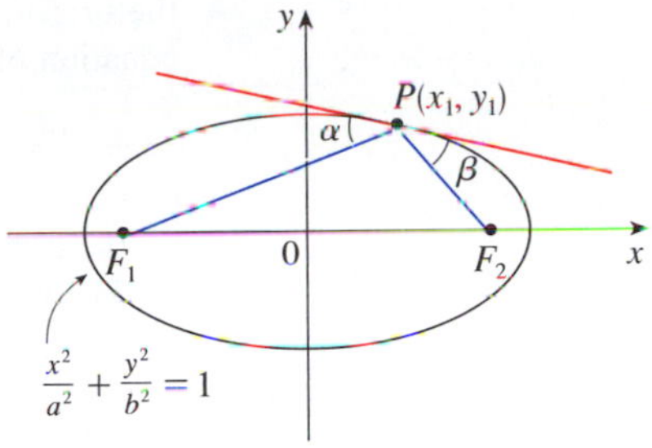

50. Let $P(x_1, y_1)$ be a point on the hyperbola $x^2/a^2 - y^2/b^2 = 1$ with foci F_1 and F_2 and let α and β be the angles between the lines PF_1, PF_2 and the hyperbola as shown in the figure. Prove that $\alpha = \beta$. (This is the "reflection property" of the hyperbola. It shows that light aimed at a focus F_2 of a hyperbolic mirror is reflected toward the other focus F_1.)

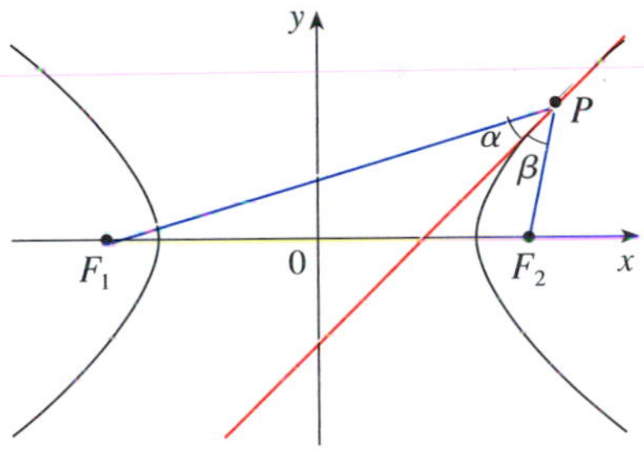

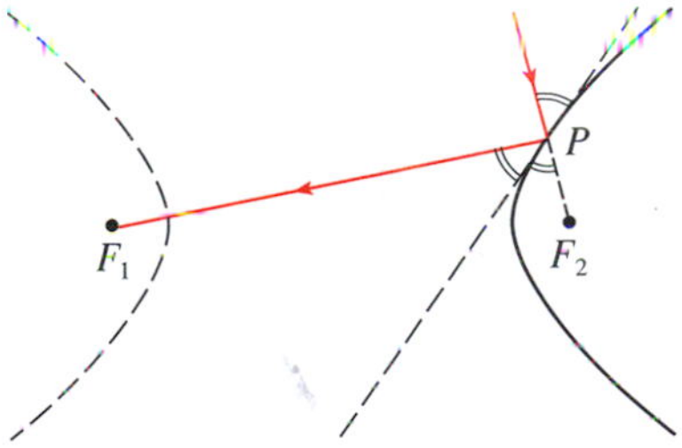

51. Show that the upper half of the hyperbola given by Equation 6 is the graph of the function

$$f(x) = \frac{b}{a}\sqrt{x^2 - a^2}$$

Then show that

$$\lim_{x \to \infty} \left[\frac{b}{a}x - f(x) \right] = 0$$

This shows that the line $y = (b/a)x$ is an asymptote of the hyperbola because the vertical distance between the line and the hyperbola approaches 0 as $x \to \infty$.

J CONIC SECTIONS IN POLAR COORDINATES

In Appendix I we defined the parabola in terms of a focus and directrix but we defined the ellipse and hyperbola in terms of two foci. Here we give a more unified treatment of all three types of conic sections in terms of a focus and directrix. Furthermore, if we place the focus at the origin, then a conic section has a simple polar equation. In Section 11.8 we use the polar equation of an ellipse to derive Kepler's laws of planetary motion.

(1) THEOREM Let F be a fixed point (called the **focus**) and l be a fixed line (called the **directrix**) in a plane. Let e be a fixed positive number (called the **eccentricity**). The set of all points P in the plane such that

$$\frac{|PF|}{|Pl|} = e$$

(that is, the ratio of the distance from F to the distance from l is the constant e) is a conic section. The conic is

(a) an ellipse if $e < 1$
(b) a parabola if $e = 1$
(c) a hyperbola if $e > 1$

PROOF Notice that if the eccentricity is $e = 1$, then $|PF| = |Pl|$ and so the given condition simply becomes the definition of a parabola as given in Appendix I.

Let's place the focus F at the origin and the directrix parallel to the y-axis and d units to the right. Thus, the directrix has equation $x = d$ and is perpendicular to the polar axis. If the point P has polar coordinates (r, θ), we see from Figure 1 that

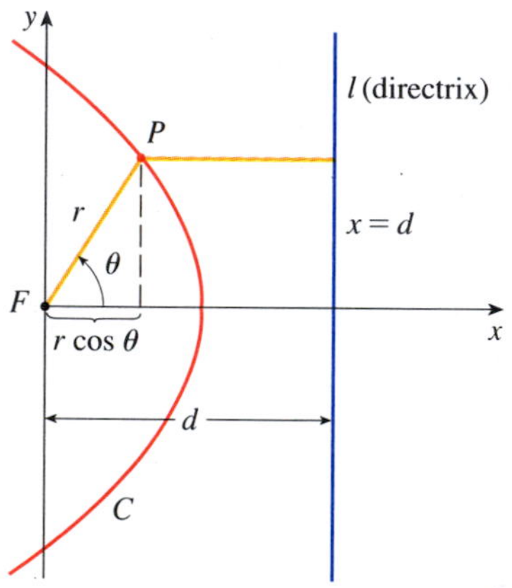

FIGURE 1

$$|PF| = r \qquad |Pl| = d - r\cos\theta$$

Thus, the condition $|PF|/|Pl| = e$, or $|PF| = e|Pl|$, becomes

$$r = e(d - r\cos\theta) \tag{2}$$

If we square both sides of this polar equation and convert to rectangular coordinates, we get

$$x^2 + y^2 = e^2(d - x)^2 = e^2(d^2 - 2dx + x^2)$$

or

$$(1 - e^2)x^2 + 2de^2x + y^2 = e^2d^2$$

After completing the square, we have

$$\left(x + \frac{e^2d}{1 - e^2}\right)^2 + \frac{y^2}{1 - e^2} = \frac{e^2d^2}{(1 - e^2)^2} \tag{3}$$

If $e < 1$, we recognize Equation 3 as the equation of an ellipse. In fact, it is of the form

$$\frac{(x - h)^2}{a^2} + \frac{y^2}{b^2} = 1$$

where

$$h = -\frac{e^2d}{1 - e^2} \qquad a^2 = \frac{e^2d^2}{(1 - e^2)^2} \qquad b^2 = \frac{e^2d^2}{1 - e^2} \tag{4}$$

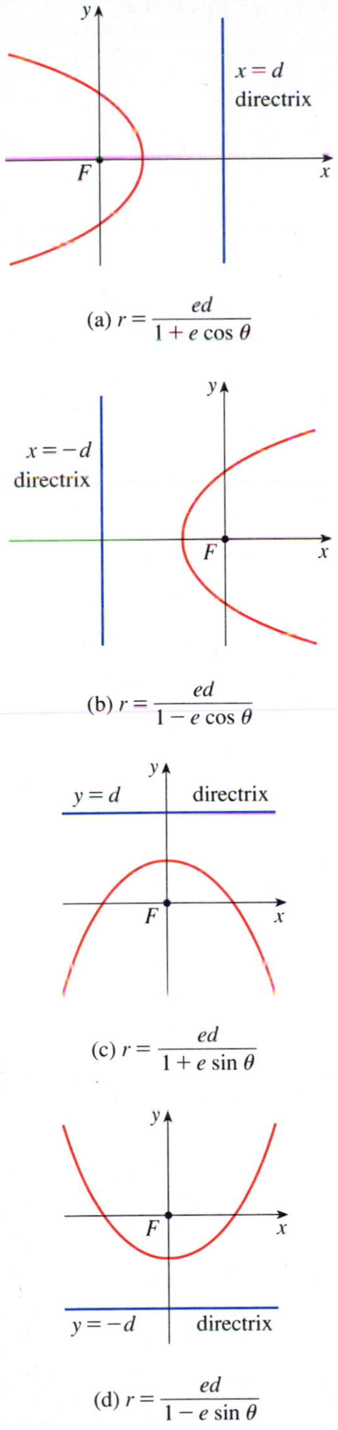

FIGURE 2
Polar equations of conics

In Appendix I we found that the foci of an ellipse are at a distance c from the center, where

(5) $$c^2 = a^2 - b^2 = \frac{e^4d^2}{(1 - e^2)^2}$$

This shows that
$$c = \frac{e^2d}{1 - e^2} = -h$$

and confirms that the focus as defined in Theorem 1 means the same as the focus defined in Appendix I. It also follows from Equations 4 and 5 that the eccentricity is given by

$$e = \frac{c}{a}$$

If $e > 1$, then $1 - e^2 < 0$ and we see that Equation 3 represents a hyperbola. Just as we did before, we could rewrite Equation 3 in the form

$$\frac{(x - h)^2}{a^2} - \frac{y^2}{b^2} = 1$$

and see that

$$e = \frac{c}{a} \qquad \text{where } c^2 = a^2 + b^2$$ □

By solving Equation 2 for r, we see that the polar equation of the conic shown in Figure 1 can be written as

$$r = \frac{ed}{1 + e\cos\theta}$$

If the directrix is chosen to be to the left of the focus as $x = -d$, or if the directrix is chosen to be parallel to the polar axis as $y = \pm d$, then the polar equation of the conic is given by the following theorem, which is illustrated by Figure 2. (See Exercises 21–23.)

(6) THEOREM A polar equation of the form

$$r = \frac{ed}{1 \pm e\cos\theta} \qquad \text{or} \qquad r = \frac{ed}{1 \pm e\sin\theta}$$

represents a conic section with eccentricity e. The conic is an ellipse if $e < 1$, a parabola if $e = 1$, or a hyperbola if $e > 1$.

EXAMPLE 1 Find a polar equation for a parabola that has its focus at the origin and whose directrix is the line $y = -6$.

SOLUTION Using Theorem 6 with $e = 1$ and $d = 6$, and using part (d) of Figure 2, we see that the equation of the parabola is

$$r = \frac{6}{1 - \sin\theta}$$ ■

EXAMPLE 2 A conic is given by the polar equation

$$r = \frac{10}{3 - 2\cos\theta}$$

Find the eccentricity, identify the conic, locate the directrix, and sketch the conic.

SOLUTION Dividing numerator and denominator by 3, we write the equation as

$$r = \frac{\frac{10}{3}}{1 - \frac{2}{3}\cos\theta}$$

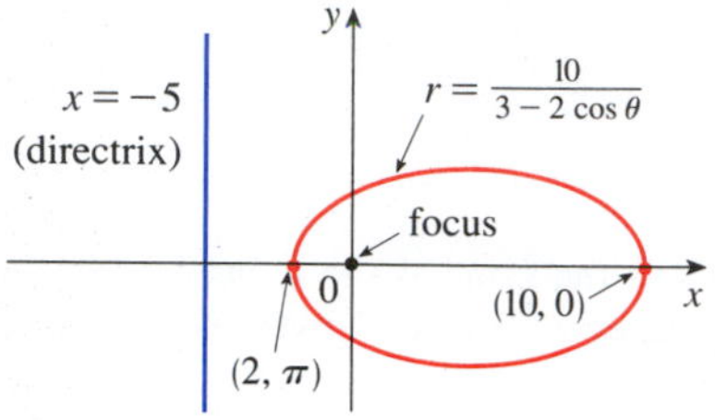

FIGURE 3

From Theorem 6 we see that this represents an ellipse with $e = \frac{2}{3}$. Since $ed = \frac{10}{3}$, we have

$$d = \frac{\frac{10}{3}}{e} = \frac{\frac{10}{3}}{\frac{2}{3}} = 5$$

so the directrix has Cartesian equation $x = -5$. When $\theta = 0$, $r = 10$; when $\theta = \pi$, $r = 2$. So the vertices have polar coordinates $(10, 0)$, $(2, \pi)$. The ellipse is sketched in Figure 3. ■

EXAMPLE 3 Sketch the conic $r = \dfrac{12}{2 + 4\sin\theta}$.

SOLUTION Writing the equation in the form

$$r = \frac{6}{1 + 2\sin\theta}$$

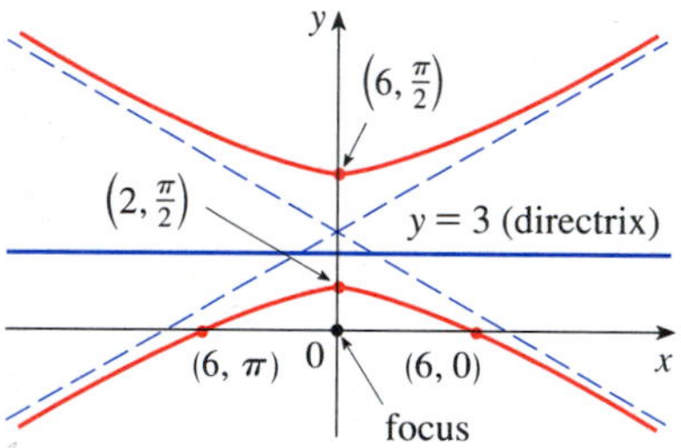

FIGURE 4

$r = \dfrac{12}{2 + 4\sin\theta}$

we see that the eccentricity is $e = 2$ and the equation therefore represents a hyperbola. Since $ed = 6$, $d = 3$ and the directrix has equation $y = 3$. The vertices occur when $\theta = \pi/2$ and $3\pi/2$, so they are $(2, \pi/2)$ and $(-6, 3\pi/2) = (6, \pi/2)$. It is also useful to plot the x-intercepts. These occur when $\theta = 0$, π and in both cases $r = 6$. For additional accuracy we could draw the asymptotes. Note that $r \to \pm\infty$ when $2 + 4\sin\theta \to 0^+$ or 0^- and $2 + 4\sin\theta = 0$ when $\sin\theta = -\frac{1}{2}$. Thus, the asymptotes are parallel to the rays $\theta = 7\pi/6$ and $\theta = 11\pi/6$. The hyperbola is sketched in Figure 4. ■

When rotating conic sections, we find it much more convenient to use polar equations than Cartesian equations. We just use the fact (see Exercise 69 in Section 13.4) that the graph of $r = f(\theta - \alpha)$ is the graph of $r = f(\theta)$ rotated counterclockwise about the origin through an angle α.

EXAMPLE 4 If the ellipse of Example 2 is rotated through an angle $\pi/4$ about the origin, find a polar equation and graph the resulting ellipse.

SOLUTION We get the equation of the rotated ellipse by replacing θ with $\theta - \pi/4$ in the equation given in Example 2. So the new equation is

$$r = \frac{10}{3 - 2\cos(\theta - \pi/4)}$$

We use this equation to graph the rotated ellipse in Figure 5. Notice that the ellipse has been rotated about its left focus.

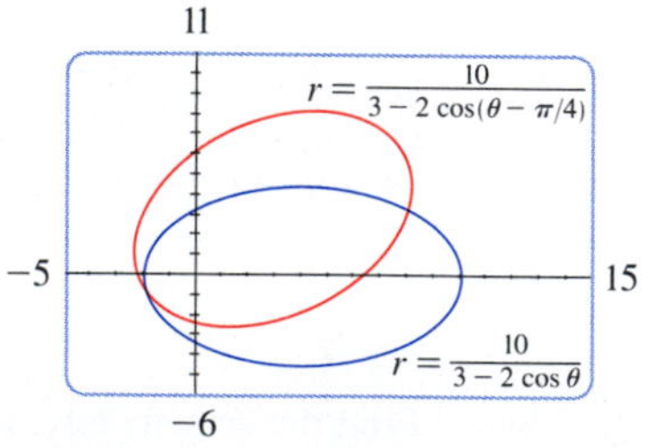

FIGURE 5

■

In Figure 6 we use a computer to sketch a number of conics to demonstrate the effect of varying the eccentricity e. Notice that when e is close to 0 the ellipse is nearly circular, whereas it becomes more elongated as $e \to 1^-$. When $e = 1$, of course, the conic is a parabola.

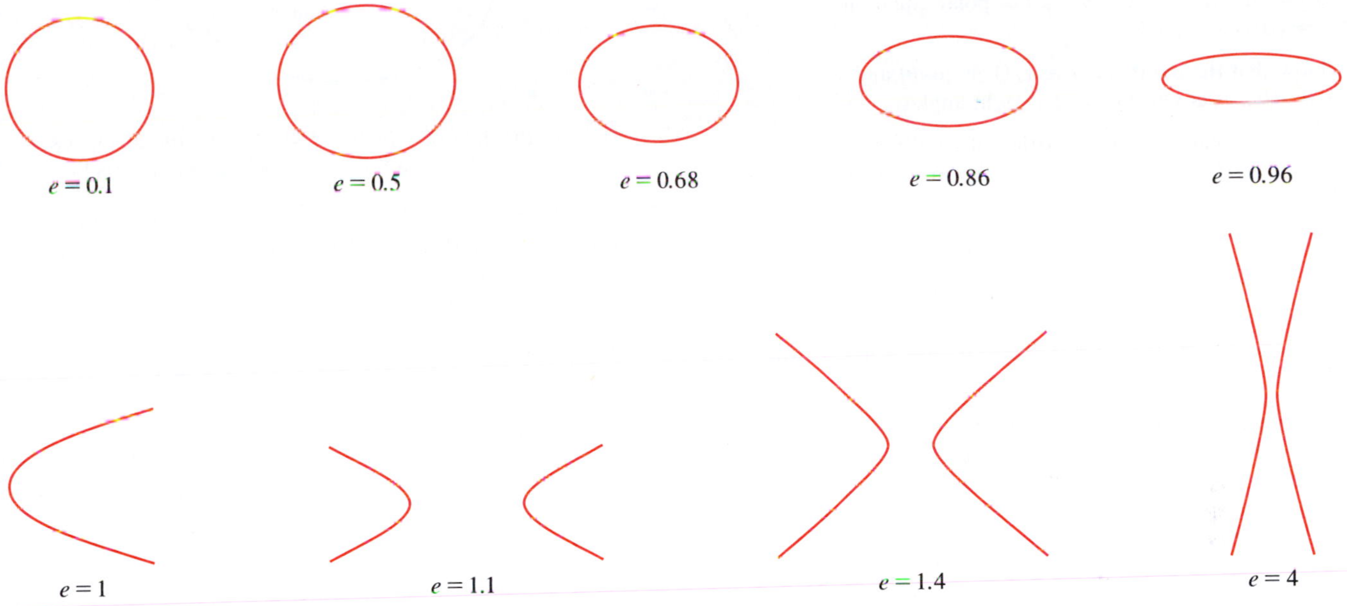

FIGURE 6

EXERCISES J

1–8 ■ Write a polar equation of a conic with the focus at the origin and the given data.

1. Ellipse, eccentricity $\frac{2}{3}$, directrix $x = 3$

2. Hyperbola, eccentricity $\frac{4}{3}$, directrix $x = -3$

3. Parabola, directrix $y = 2$

4. Ellipse, eccentricity $\frac{1}{2}$, directrix $y = -4$

5. Hyperbola, eccentricity 4, directrix $r = 5 \sec\theta$

6. Ellipse, eccentricity 0.6, directrix $r = 2 \csc\theta$

7. Parabola, vertex at $(5, \pi/2)$

8. Ellipse, eccentricity 0.4, vertex at $(2, 0)$

9–16 ■ (a) Find the eccentricity, (b) identify the conic, (c) give an equation of the directrix, and (d) sketch the conic.

9. $r = \dfrac{4}{1 + 3\cos\theta}$

10. $r = \dfrac{8}{3 + 3\cos\theta}$

11. $r = \dfrac{2}{1 - \cos\theta}$

12. $r = \dfrac{10}{3 - 2\sin\theta}$

13. $r = \dfrac{6}{2 + \sin\theta}$

14. $r = \dfrac{5}{2 - 3\sin\theta}$

15. $r = \dfrac{7}{2 - 5\sin\theta}$

16. $r = \dfrac{8}{3 + \cos\theta}$

17. (a) Find the eccentricity and directrix of the conic $r = 1/(4 - 3\cos\theta)$ and graph the conic and its directrix.
(b) If this conic is rotated counterclockwise about the origin through an angle $\pi/3$, write the resulting equation and graph its curve.

18. Graph the parabola $r = 5/(2 + 2\sin\theta)$ and its directrix. Also graph the curve obtained by rotating this parabola about its focus through an angle $\pi/6$.

19. Graph the conics $r = e/(1 - e\cos\theta)$ with $e = 0.4, 0.6, 0.8$, and 1.0 on a common screen. How does the value of e affect the shape of the curve?

20. (a) Graph the conics $r = ed/(1 + e\sin\theta)$ for $e = 1$ and various values of d. How does the value of d affect the shape of the conic?
(b) Graph these conics for $d = 1$ and various values of e. How does the value of e affect the shape of the conic?

21. Show that a conic with focus at the origin, eccentricity e, and directrix $x = -d$ has polar equation $r = ed/(1 - e\cos\theta)$.

22. Show that a conic with focus at the origin, eccentricity e, and directrix $y = d$ has polar equation $r = ed/(1 + e \sin\theta)$.

23. Show that a conic with focus at the origin, eccentricity e, and directrix $y = -d$ has polar equation $r = ed/(1 - e \sin\theta)$.

24. Show that the parabolas $r = c/(1 + \cos\theta)$ and $r = d/(1 - \cos\theta)$ intersect at right angles.

25. (a) Show that the polar equation of an ellipse with directrix $x = -d$ can be written in the form

$$r = \frac{a(1 - e^2)}{1 - e\cos\theta}$$

(b) Find an approximate polar equation for Earth's elliptical orbit around the Sun (at one focus) given that the eccentricity is about 0.017 and the length of the major axis is about 2.99×10^8 km.

26. (a) The planets move around the Sun in elliptical orbits with the Sun at one focus. The positions of a planet that are closest to and farthest from the Sun are called its *perihelion* and *aphelion*, respectively. Use Exercise 25(a) to show that the perihelion distance from a planet to the Sun is $a(1 - e)$ and the aphelion distance is $a(1 + e)$.

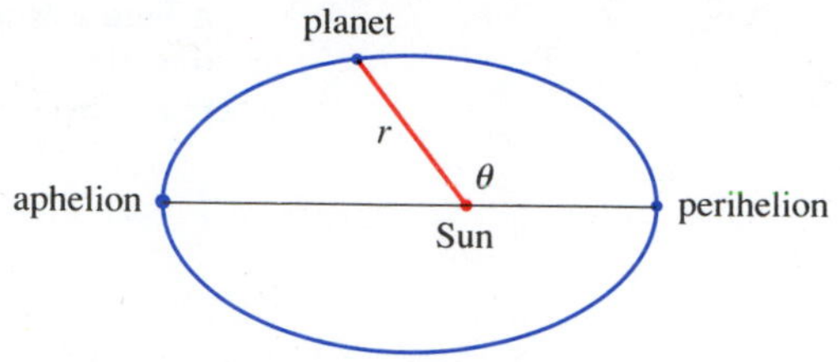

(b) Use the data of Exercise 25(b) to find the distances from Earth to the Sun at perihelion and at aphelion.

27. The planet Mercury travels in an elliptical orbit with eccentricity 0.206. Its minimum distance from the Sun is 4.6×10^7 km. Use the results of Exercise 26(a) to find its maximum distance from the Sun.

28. The distance from the planet Pluto to the Sun is 4.43×10^9 km at perihelion and 7.37×10^9 km at aphelion. Use Exercise 26 to find the eccentricity of Pluto's orbit.

29. Using the data from Exercise 27, find the distance traveled by the planet Mercury during one complete orbit around the Sun. (If your calculator or computer algebra system evaluates definite integrals, use it. Otherwise, use Simpson's Rule.)

K TABLE OF INTEGRALS

BASIC FORMS

1. $\int u\,dv = uv - \int v\,du$

2. $\int u^n\,du = \frac{1}{n+1}u^{n+1} + C, \quad n \neq -1$

3. $\int \frac{du}{u} = \ln|u| + C$

4. $\int e^u\,du = e^u + C$

5. $\int a^u\,du = \frac{1}{\ln a}a^u + C$

6. $\int \sin u\,du = -\cos u + C$

7. $\int \cos u\,du = \sin u + C$

8. $\int \sec^2 u\,du = \tan u + C$

9. $\int \csc^2 u\,du = -\cot u + C$

10. $\int \sec u \tan u\,du = \sec u + C$

11. $\int \csc u \cot u\,du = -\csc u + C$

12. $\int \tan u\,du = \ln|\sec u| + C$

13. $\int \cot u\,du = \ln|\sin u| + C$

14. $\int \sec u\,du = \ln|\sec u + \tan u| + C$

15. $\int \csc u\,du = \ln|\csc u - \cot u| + C$

16. $\int \frac{du}{\sqrt{a^2 - u^2}} = \sin^{-1}\frac{u}{a} + C$

17. $\int \frac{du}{a^2 + u^2} = \frac{1}{a}\tan^{-1}\frac{u}{a} + C$

18. $\int \frac{du}{u\sqrt{u^2 - a^2}} = \frac{1}{a}\sec^{-1}\frac{u}{a} + C$

19. $\int \frac{du}{a^2 - u^2} = \frac{1}{2a}\ln\left|\frac{u + a}{u - a}\right| + C$

20. $\int \frac{du}{u^2 - a^2} = \frac{1}{2a}\ln\left|\frac{u - a}{u + a}\right| + C$

FORMS INVOLVING $\sqrt{a^2 + u^2}$, $a > 0$

21. $\displaystyle\int \sqrt{a^2 + u^2}\, du = \frac{u}{2}\sqrt{a^2 + u^2} + \frac{a^2}{2}\ln\left(u + \sqrt{a^2 + u^2}\right) + C$

22. $\displaystyle\int u^2\sqrt{a^2 + u^2}\, du = \frac{u}{8}(a^2 + 2u^2)\sqrt{a^2 + u^2} - \frac{a^4}{8}\ln\left(u + \sqrt{a^2 + u^2}\right) + C$

23. $\displaystyle\int \frac{\sqrt{a^2 + u^2}}{u}\, du = \sqrt{a^2 + u^2} - a\ln\left|\frac{a + \sqrt{a^2 + u^2}}{u}\right| + C$

24. $\displaystyle\int \frac{\sqrt{a^2 + u^2}}{u^2}\, du = -\frac{\sqrt{a^2 + u^2}}{u} + \ln\left(u + \sqrt{a^2 + u^2}\right) + C$

25. $\displaystyle\int \frac{du}{\sqrt{a^2 + u^2}} = \ln\left(u + \sqrt{a^2 + u^2}\right) + C$

26. $\displaystyle\int \frac{u^2\, du}{\sqrt{a^2 + u^2}} = \frac{u}{2}\sqrt{a^2 + u^2} - \frac{a^2}{2}\ln\left(u + \sqrt{a^2 + u^2}\right) + C$

27. $\displaystyle\int \frac{du}{u\sqrt{a^2 + u^2}} = -\frac{1}{a}\ln\left|\frac{\sqrt{a^2 + u^2} + a}{u}\right| + C$

28. $\displaystyle\int \frac{du}{u^2\sqrt{a^2 + u^2}} = -\frac{\sqrt{a^2 + u^2}}{a^2 u} + C$

29. $\displaystyle\int \frac{du}{(a^2 + u^2)^{3/2}} = \frac{u}{a^2\sqrt{a^2 + u^2}} + C$

FORMS INVOLVING $\sqrt{a^2 - u^2}$, $a > 0$

30. $\displaystyle\int \sqrt{a^2 - u^2}\, du = \frac{u}{2}\sqrt{a^2 - u^2} + \frac{a^2}{2}\sin^{-1}\frac{u}{a} + C$

31. $\displaystyle\int u^2\sqrt{a^2 - u^2}\, du = \frac{u}{8}(2u^2 - a^2)\sqrt{a^2 - u^2} + \frac{a^4}{8}\sin^{-1}\frac{u}{a} + C$

32. $\displaystyle\int \frac{\sqrt{a^2 - u^2}}{u}\, du = \sqrt{a^2 - u^2} - a\ln\left|\frac{a + \sqrt{a^2 - u^2}}{u}\right| + C$

33. $\displaystyle\int \frac{\sqrt{a^2 - u^2}}{u^2}\, du = -\frac{1}{u}\sqrt{a^2 - u^2} - \sin^{-1}\frac{u}{a} + C$

34. $\displaystyle\int \frac{u^2\, du}{\sqrt{a^2 - u^2}} = -\frac{u}{2}\sqrt{a^2 - u^2} + \frac{a^2}{2}\sin^{-1}\frac{u}{a} + C$

35. $\displaystyle\int \frac{du}{u\sqrt{a^2 - u^2}} = -\frac{1}{a}\ln\left|\frac{a + \sqrt{a^2 - u^2}}{u}\right| + C$

36. $\displaystyle\int \frac{du}{u^2\sqrt{a^2 - u^2}} = -\frac{1}{a^2 u}\sqrt{a^2 - u^2} + C$

37. $\displaystyle\int (a^2 - u^2)^{3/2}\, du = -\frac{u}{8}(2u^2 - 5a^2)\sqrt{a^2 - u^2} + \frac{3a^4}{8}\sin^{-1}\frac{u}{a} + C$

38. $\displaystyle\int \frac{du}{(a^2 - u^2)^{3/2}} = \frac{u}{a^2\sqrt{a^2 - u^2}} + C$

FORMS INVOLVING $\sqrt{u^2 - a^2}$, $a > 0$

39. $\displaystyle\int \sqrt{u^2 - a^2}\, du = \frac{u}{2}\sqrt{u^2 - a^2} - \frac{a^2}{2}\ln\left|u + \sqrt{u^2 - a^2}\right| + C$

40. $\displaystyle\int u^2\sqrt{u^2 - a^2}\, du = \frac{u}{8}(2u^2 - a^2)\sqrt{u^2 - a^2} - \frac{a^4}{8}\ln\left|u + \sqrt{u^2 - a^2}\right| + C$

41. $\displaystyle\int \frac{\sqrt{u^2 - a^2}}{u}\,du = \sqrt{u^2 - a^2} - a\cos^{-1}\frac{a}{u} + C$

42. $\displaystyle\int \frac{\sqrt{u^2 - a^2}}{u^2}\,du = -\frac{\sqrt{u^2 - a^2}}{u} + \ln|u + \sqrt{u^2 - a^2}| + C$

43. $\displaystyle\int \frac{du}{\sqrt{u^2 - a^2}} = \ln|u + \sqrt{u^2 - a^2}| + C$

44. $\displaystyle\int \frac{u^2\,du}{\sqrt{u^2 - a^2}} = \frac{u}{2}\sqrt{u^2 - a^2} + \frac{a^2}{2}\ln|u + \sqrt{u^2 - a^2}| + C$

45. $\displaystyle\int \frac{du}{u^2\sqrt{u^2 - a^2}} = \frac{\sqrt{u^2 - a^2}}{a^2 u} + C$

46. $\displaystyle\int \frac{du}{(u^2 - a^2)^{3/2}} = -\frac{u}{a^2\sqrt{u^2 - a^2}} + C$

FORMS INVOLVING $a + bu$

47. $\displaystyle\int \frac{u\,du}{a + bu} = \frac{1}{b^2}(a + bu - a\ln|a + bu|) + C$

48. $\displaystyle\int \frac{u^2\,du}{a + bu} = \frac{1}{2b^3}[(a + bu)^2 - 4a(a + bu) + 2a^2\ln|a + bu|] + C$

49. $\displaystyle\int \frac{du}{u(a + bu)} = \frac{1}{a}\ln\left|\frac{u}{a + bu}\right| + C$

50. $\displaystyle\int \frac{du}{u^2(a + bu)} = -\frac{1}{au} + \frac{b}{a^2}\ln\left|\frac{a + bu}{u}\right| + C$

51. $\displaystyle\int \frac{u\,du}{(a + bu)^2} = \frac{a}{b^2(a + bu)} + \frac{1}{b^2}\ln|a + bu| + C$

52. $\displaystyle\int \frac{du}{u(a + bu)^2} = \frac{1}{a(a + bu)} - \frac{1}{a^2}\ln\left|\frac{a + bu}{u}\right| + C$

53. $\displaystyle\int \frac{u^2\,du}{(a + bu)^2} = \frac{1}{b^3}\left(a + bu - \frac{a^2}{a + bu} - 2a\ln|a + bu|\right) + C$

54. $\displaystyle\int u\sqrt{a + bu}\,du = \frac{2}{15b^2}(3bu - 2a)(a + bu)^{3/2} + C$

55. $\displaystyle\int \frac{u\,du}{\sqrt{a + bu}} = \frac{2}{3b^2}(bu - 2a)\sqrt{a + bu} + C$

56. $\displaystyle\int \frac{u^2\,du}{\sqrt{a + bu}} = \frac{2}{15b^3}(8a^2 + 3b^2u^2 - 4abu)\sqrt{a + bu} + C$

57. $\displaystyle\int \frac{du}{u\sqrt{a + bu}} = \frac{1}{\sqrt{a}}\ln\left|\frac{\sqrt{a + bu} - \sqrt{a}}{\sqrt{a + bu} + \sqrt{a}}\right| + C, \quad \text{if } a > 0$

$\displaystyle= \frac{2}{\sqrt{-a}}\tan^{-1}\sqrt{\frac{a + bu}{-a}} + C, \qquad \text{if } a < 0$

58. $\displaystyle\int \frac{\sqrt{a + bu}}{u}\,du = 2\sqrt{a + bu} + a\int \frac{du}{u\sqrt{a + bu}}$

59. $\displaystyle\int \frac{\sqrt{a + bu}}{u^2}\,du = -\frac{\sqrt{a + bu}}{u} + \frac{b}{2}\int \frac{du}{u\sqrt{a + bu}}$

60. $\displaystyle\int u^n\sqrt{a + bu}\,du = \frac{2}{b(2n + 3)}\left[u^n(a + bu)^{3/2} - na\int u^{n-1}\sqrt{a + bu}\,du\right]$

61. $\displaystyle\int \frac{u^n\,du}{\sqrt{a+bu}} = \frac{2u^n\sqrt{a+bu}}{b(2n+1)} - \frac{2na}{b(2n+1)}\int \frac{u^{n-1}\,du}{\sqrt{a+bu}}$

62. $\displaystyle\int \frac{du}{u^n\sqrt{a+bu}} = -\frac{\sqrt{a+bu}}{a(n-1)u^{n-1}} - \frac{b(2n-3)}{2a(n-1)}\int \frac{du}{u^{n-1}\sqrt{a+bu}}$

TRIGONOMETRIC FORMS

63. $\displaystyle\int \sin^2 u\,du = \tfrac{1}{2}u - \tfrac{1}{4}\sin 2u + C$

64. $\displaystyle\int \cos^2 u\,du = \tfrac{1}{2}u + \tfrac{1}{4}\sin 2u + C$

65. $\displaystyle\int \tan^2 u\,du = \tan u - u + C$

66. $\displaystyle\int \cot^2 u\,du = -\cot u - u + C$

67. $\displaystyle\int \sin^3 u\,du = -\tfrac{1}{3}(2 + \sin^2 u)\cos u + C$

68. $\displaystyle\int \cos^3 u\,du = \tfrac{1}{3}(2 + \cos^2 u)\sin u + C$

69. $\displaystyle\int \tan^3 u\,du = \tfrac{1}{2}\tan^2 u + \ln|\cos u| + C$

70. $\displaystyle\int \cot^3 u\,du = -\tfrac{1}{2}\cot^2 u - \ln|\sin u| + C$

71. $\displaystyle\int \sec^3 u\,du = \tfrac{1}{2}\sec u\tan u + \tfrac{1}{2}\ln|\sec u + \tan u| + C$

72. $\displaystyle\int \csc^3 u\,du = -\tfrac{1}{2}\csc u\cot u + \tfrac{1}{2}\ln|\csc u - \cot u| + C$

73. $\displaystyle\int \sin^n u\,du = -\frac{1}{n}\sin^{n-1}u\cos u + \frac{n-1}{n}\int \sin^{n-2}u\,du$

74. $\displaystyle\int \cos^n u\,du = \frac{1}{n}\cos^{n-1}u\sin u + \frac{n-1}{n}\int \cos^{n-2}u\,du$

75. $\displaystyle\int \tan^n u\,du = \frac{1}{n-1}\tan^{n-1}u - \int \tan^{n-2}u\,du$

76. $\displaystyle\int \cot^n u\,du = \frac{-1}{n-1}\cot^{n-1}u - \int \cot^{n-2}u\,du$

77. $\displaystyle\int \sec^n u\,du = \frac{1}{n-1}\tan u\sec^{n-2}u + \frac{n-2}{n-1}\int \sec^{n-2}u\,du$

78. $\displaystyle\int \csc^n u\,du = \frac{-1}{n-1}\cot u\csc^{n-2}u + \frac{n-2}{n-1}\int \csc^{n-2}u\,du$

79. $\displaystyle\int \sin au\sin bu\,du = \frac{\sin(a-b)u}{2(a-b)} - \frac{\sin(a+b)u}{2(a+b)} + C$

80. $\displaystyle\int \cos au\cos bu\,du = \frac{\sin(a-b)u}{2(a-b)} + \frac{\sin(a+b)u}{2(a+b)} + C$

81. $\displaystyle\int \sin au\cos bu\,du = -\frac{\cos(a-b)u}{2(a-b)} - \frac{\cos(a+b)u}{2(a+b)} + C$

82. $\int u \sin u \, du = \sin u - u \cos u + C$

83. $\int u \cos u \, du = \cos u + u \sin u + C$

84. $\int u^n \sin u \, du = -u^n \cos u + n \int u^{n-1} \cos u \, du$

85. $\int u^n \cos u \, du = u^n \sin u - n \int u^{n-1} \sin u \, du$

86. $$\int \sin^n u \cos^m u \, du = -\frac{\sin^{n-1} u \cos^{m+1} u}{n + m} + \frac{n - 1}{n + m} \int \sin^{n-2} u \cos^m u \, du$$
$$= \frac{\sin^{n+1} u \cos^{m-1} u}{n + m} + \frac{m - 1}{n + m} \int \sin^n u \cos^{m-2} u \, du$$

INVERSE TRIGONOMETRIC FORMS

87. $\int \sin^{-1} u \, du = u \sin^{-1} u + \sqrt{1 - u^2} + C$

88. $\int \cos^{-1} u \, du = u \cos^{-1} u - \sqrt{1 - u^2} + C$

89. $\int \tan^{-1} u \, du = u \tan^{-1} u - \frac{1}{2} \ln(1 + u^2) + C$

90. $\int u \sin^{-1} u \, du = \frac{2u^2 - 1}{4} \sin^{-1} u + \frac{u\sqrt{1 - u^2}}{4} + C$

91. $\int u \cos^{-1} u \, du = \frac{2u^2 - 1}{4} \cos^{-1} u - \frac{u\sqrt{1 - u^2}}{4} + C$

92. $\int u \tan^{-1} u \, du = \frac{u^2 + 1}{2} \tan^{-1} u - \frac{u}{2} + C$

93. $\int u^n \sin^{-1} u \, du = \frac{1}{n + 1} \left[u^{n+1} \sin^{-1} u - \int \frac{u^{n+1} \, du}{\sqrt{1 - u^2}} \right], \quad n \neq -1$

94. $\int u^n \cos^{-1} u \, du = \frac{1}{n + 1} \left[u^{n+1} \cos^{-1} u + \int \frac{u^{n+1} \, du}{\sqrt{1 - u^2}} \right], \quad n \neq -1$

95. $\int u^n \tan^{-1} u \, du = \frac{1}{n + 1} \left[u^{n+1} \tan^{-1} u - \int \frac{u^{n+1} \, du}{1 + u^2} \right], \quad n \neq -1$

EXPONENTIAL AND LOGARITHMIC FORMS

96. $\int u e^{au} \, du = \frac{1}{a^2}(au - 1)e^{au} + C$

97. $\int u^n e^{au} \, du = \frac{1}{a} u^n e^{au} - \frac{n}{a} \int u^{n-1} e^{au} \, du$

98. $\int e^{au} \sin bu \, du = \frac{e^{au}}{a^2 + b^2}(a \sin bu - b \cos bu) + C$

99. $\int e^{au} \cos bu \, du = \frac{e^{au}}{a^2 + b^2}(a \cos bu + b \sin bu) + C$

100. $\int \ln u \, du = u \ln u - u + C$

101. $\int u^n \ln u \, du = \frac{u^{n+1}}{(n + 1)^2}[(n + 1) \ln u - 1] + C$

102. $\int \frac{1}{u \ln u} \, du = \ln |\ln u| + C$

HYPERBOLIC FORMS

103. $\int \sinh u\,du = \cosh u + C$

104. $\int \cosh u\,du = \sinh u + C$

105. $\int \tanh u\,du = \ln \cosh u + C$

106. $\int \coth u\,du = \ln|\sinh u| + C$

107. $\int \operatorname{sech} u\,du = \tan^{-1}|\sinh u| + C$

108. $\int \operatorname{csch} u\,du = \ln|\tanh \tfrac{1}{2}u| + C$

109. $\int \operatorname{sech}^2 u\,du = \tanh u + C$

110. $\int \operatorname{csch}^2 u\,du = -\coth u + C$

111. $\int \operatorname{sech} u \tanh u\,du = -\operatorname{sech} u + C$

112. $\int \operatorname{csch} u \coth u\,du = -\operatorname{csch} u + C$

FORMS INVOLVING $\sqrt{2au - u^2}$

113. $\int \sqrt{2au - u^2}\,du = \dfrac{u-a}{2}\sqrt{2au - u^2} + \dfrac{a^2}{2}\cos^{-1}\left(\dfrac{a-u}{a}\right) + C$

114. $\int u\sqrt{2au - u^2}\,du = \dfrac{2u^2 - au - 3a^2}{6}\sqrt{2au - u^2} + \dfrac{a^3}{2}\cos^{-1}\left(\dfrac{a-u}{a}\right) + C$

115. $\int \dfrac{\sqrt{2au - u^2}}{u}\,du = \sqrt{2au - u^2} + a\cos^{-1}\left(\dfrac{a-u}{a}\right) + C$

116. $\int \dfrac{\sqrt{2au - u^2}}{u^2}\,du = -\dfrac{2\sqrt{2au - u^2}}{u} - \cos^{-1}\left(\dfrac{a-u}{a}\right) + C$

117. $\int \dfrac{du}{\sqrt{2au - u^2}} = \cos^{-1}\left(\dfrac{a-u}{a}\right) + C$

118. $\int \dfrac{u\,du}{\sqrt{2au - u^2}} = -\sqrt{2au - u^2} + a\cos^{-1}\left(\dfrac{a-u}{a}\right) + C$

119. $\int \dfrac{u^2\,du}{\sqrt{2au - u^2}} = -\dfrac{(u+3a)}{2}\sqrt{2au - u^2} + \dfrac{3a^2}{2}\cos^{-1}\left(\dfrac{a-u}{a}\right) + C$

120. $\int \dfrac{du}{u\sqrt{2au - u^2}} = -\dfrac{\sqrt{2au - u^2}}{au} + C$

L ANSWERS TO ODD-NUMBERED EXERCISES

REVIEW AND PREVIEW

Exercises 1 ■ page 15

1. $-4,\ 10,\ 3\sqrt{2},\ 5 + 7\sqrt{2},\ 2x^2 - 3x - 4,\ 2x^2 + 7x + 1,$
$4x^2 + 6x - 8,\ 8x^2 + 6x - 4$

3. $-(h^2 + 3h + 2),\ x + h - x^2 - 2xh - h^2,\ 1 - 2x - h$

5.

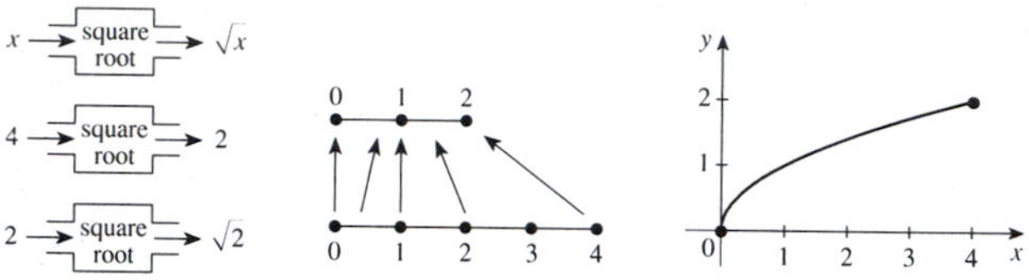

7. $\{0, 1, 2, 4\}$

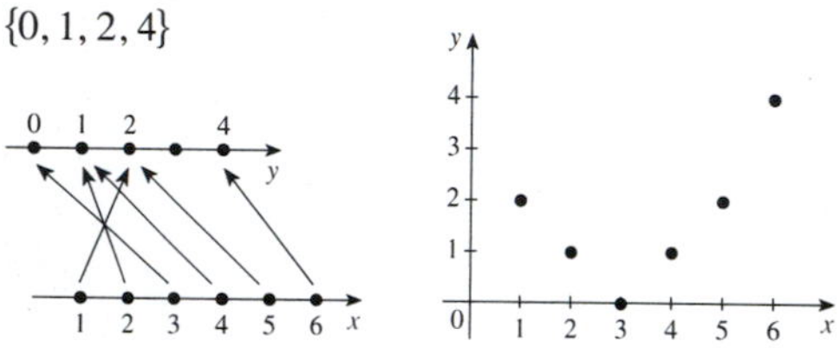

9. $[-2, 3], [-6, 14]$ **11.** $[\frac{5}{2}, \infty), [0, \infty)$

13. $\{x \mid |x| \leqslant 1\} = [-1, 1], [0, 1]$

15. $\{x \mid x \neq \pm 1\} = (-\infty, -1) \cup (-1, 1) \cup (1, \infty)$

17. $\{x \mid x \leqslant 0 \text{ or } x \geqslant 6\} = (-\infty, 0] \cup [6, \infty)$

19. $[0, \pi)$ **21.** $(-\infty, \infty)$

23. $(-\infty, \infty)$

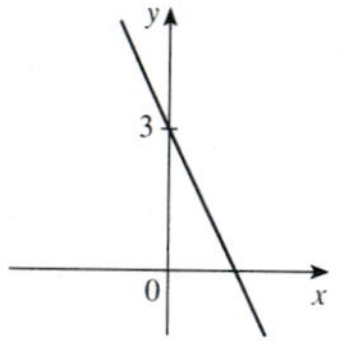

25. $(-\infty, \infty)$

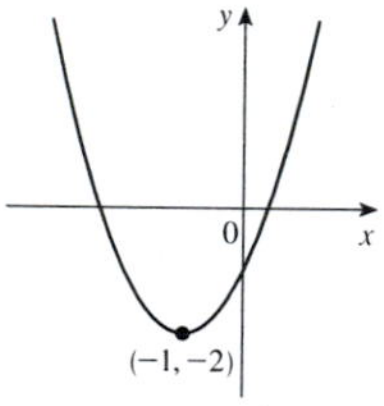

27. $(-\infty, 0]$

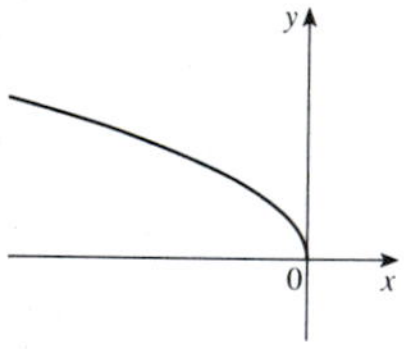

29. $[-2, 2]$

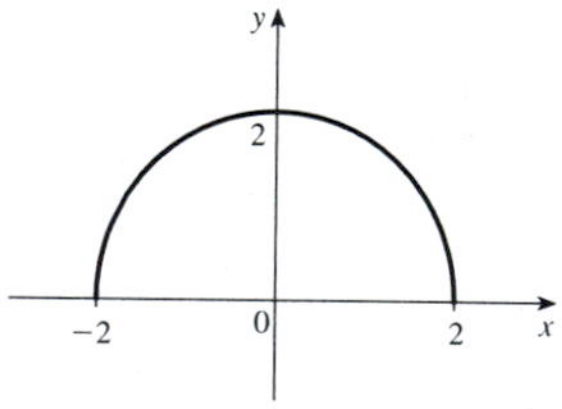

31. $\{x \mid x \neq 0\}$

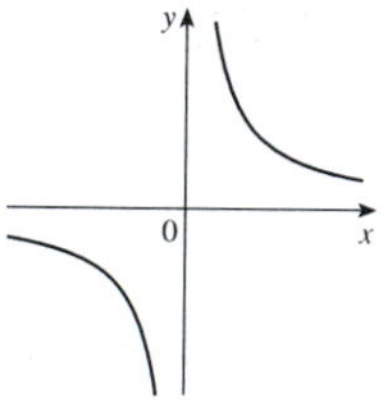

33. $(-\infty, \infty)$

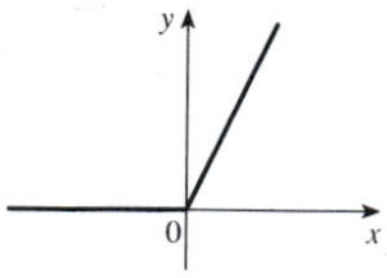

35. $(-\infty, \infty)$

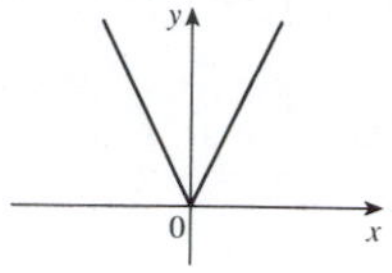

37. $(-\infty, 0) \cup (0, \infty)$

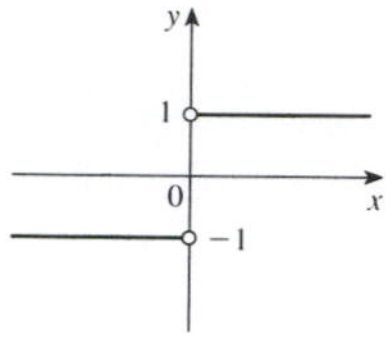

39. $(-\infty, 1) \cup (1, \infty)$

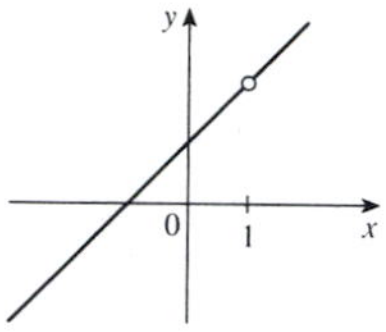

41. $(-\infty, \infty)$

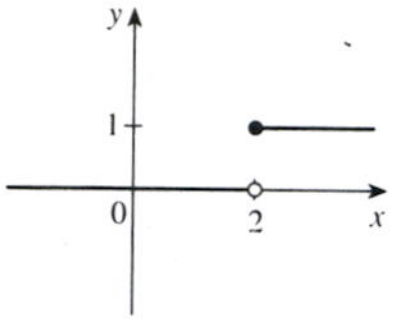

43. $(-\infty, \infty)$

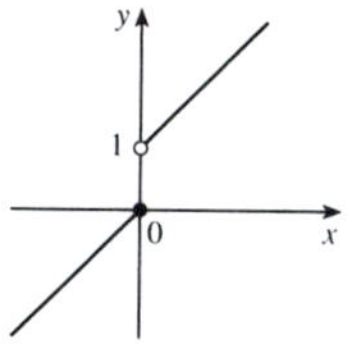

45. $(-\infty, \infty)$

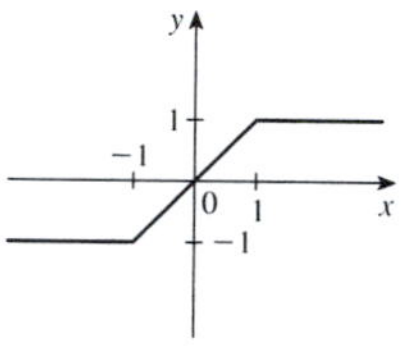

47. $(-\infty, \infty)$

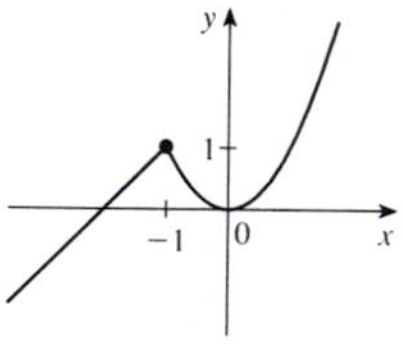

49. $(-\infty, \infty)$

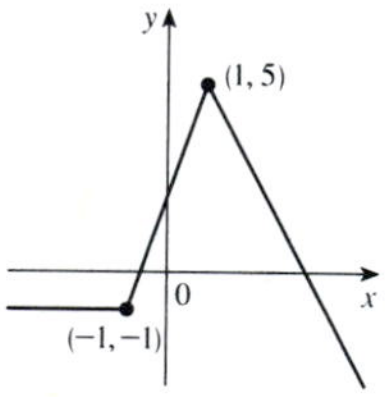

51. Yes, $[-3, 2], [-2, 2]$ **53.** No

55. $f(x) = -\frac{7}{6}x - \frac{4}{3},\ -2 \leqslant x \leqslant 4$ **57.** $f(x) = 1 - \sqrt{-x}$

59. $f(x) = \begin{cases} x + 1 & \text{if } -1 \leqslant x \leqslant 2 \\ 6 - 1.5x & \text{if } 2 < x \leqslant 4 \end{cases}$

61. $A(L) = 10L - L^2,\ 0 < L < 10$

63. $A(x) = \sqrt{3}\,x^2/4,\ x > 0$ **65.** $S(x) = x^2 + (8/x),\ x > 0$

67. $V(x) = 4x^3 - 64x^2 + 240x,\ 0 < x < 6$

69. (a) $T(h) = 20 - 10h$
(b) The rate of change of temperature with respect to height
(c) $-5\,°\text{C}$

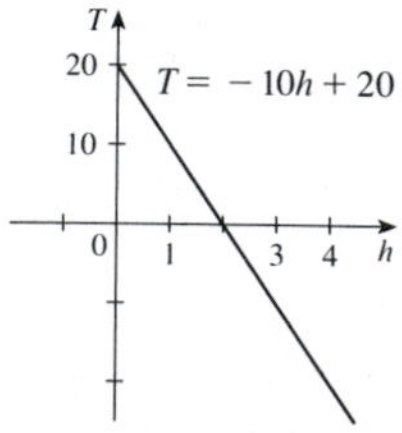

71.

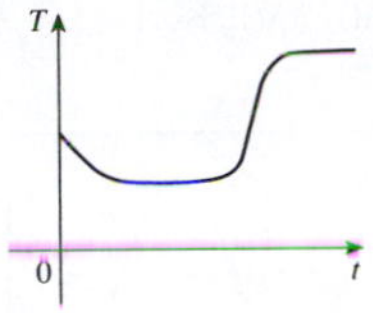

73.

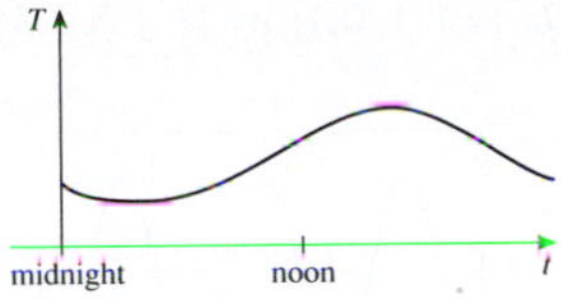

75. (a)

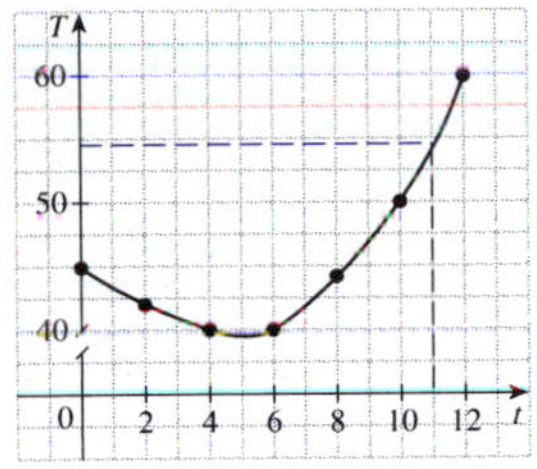

(b) 54 °F

77. Even

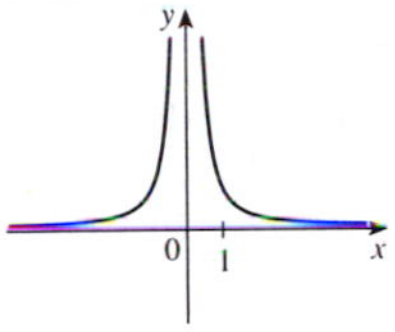

79. Neither **81.** Odd

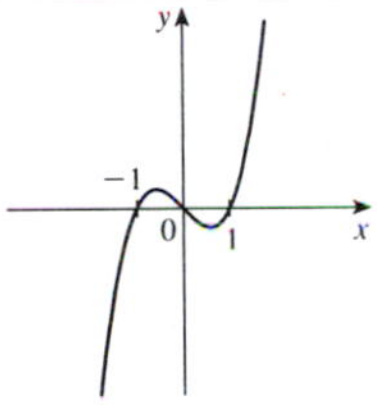

83. $(f + g)(x) = x^3 + 5x^2 - 1, (-\infty, \infty)$;
$(f - g)(x) = x^3 - x^2 + 1, (-\infty, \infty)$;
$(fg)(x) = 3x^5 + 6x^4 - x^3 - 2x^2, (-\infty, \infty)$;
$(f/g)(x) = (x^3 + 2x^2)/(3x^2 - 1), \{x \mid x \neq \pm 1/\sqrt{3}\}$

85.

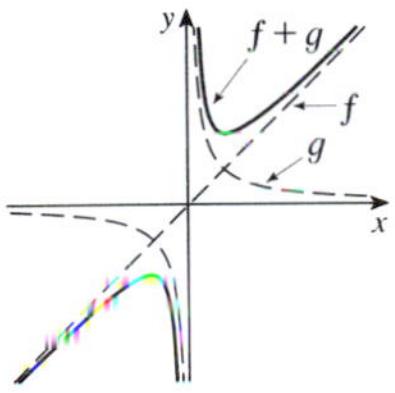

87. $(f \circ g)(x) = 3(6x^2 + 7x + 2), (-\infty, \infty)$;
$(g \circ f)(x) = 6x^2 - 3x + 2, (-\infty, \infty)$;
$(f \circ f)(x) = 8x^4 - 8x^3 + x, (-\infty, \infty)$;
$(g \circ g)(x) = 9x + 8, (-\infty, \infty)$

89. $(f \circ g)(x) = 1/(x^3 + 2x), \{x \mid x \neq 0\}$;
$(g \circ f)(x) = (1/x^3) + (2/x), \{x \mid x \neq 0\}$;
$(f \circ f)(x) = x, \{x \mid x \neq 0\}$;
$(g \circ g)(x) = x^9 + 6x^7 + 12x^5 + 10x^3 + 4x, (-\infty, \infty)$

91. $(f \circ g)(x) = \sqrt[3]{1 - \sqrt{x}}, [0, \infty)$; $(g \circ f)(x) = 1 - \sqrt[6]{x}, [0, \infty)$;
$(f \circ f)(x) = \sqrt[9]{x}, (-\infty, \infty)$; $(g \circ g)(x) = 1 - \sqrt{1 - \sqrt{x}}, [0, 1]$

93. $(f \circ g)(x) = (3x - 4)/(3x - 2), \{x \mid x \neq 2, \frac{2}{3}\}$;
$(g \circ f)(x) = -(x + 2)/(3x), \{x \mid x \neq 0, -\frac{1}{2}\}$;
$(f \circ f)(x) = (5x + 4)/(4x + 5), \{x \mid x \neq -\frac{1}{2}, -\frac{5}{4}\}$;
$(g \circ g)(x) = x/(4 - x), \{x \mid x \neq 2, 4\}$

95. $(f \circ g \circ h)(x) = \sqrt{x - 1} - 1$

97. $(f \circ g \circ h)(x) = (\sqrt{x} - 5)^4 + 1$

99. $g(x) = x - 9, f(x) = x^5$

101. $g(x) = x^2, f(x) = x/(x + 4)$

103. $h(x) = x^2, g(x) = x + 1, f(x) = 1/x$

105. $A = 3600\pi t^2$ **107.** $g(x) = x^2 + x - 1$

109. $f \circ f$ has domain $\{x \mid x \neq 0\}$

Exercises 2 ■ page 26

1. (a) Root (b) Algebraic (c) Polynomial (degree 9)
(d) Rational (e) Trigonometric (f) Logarithmic

3. (a)

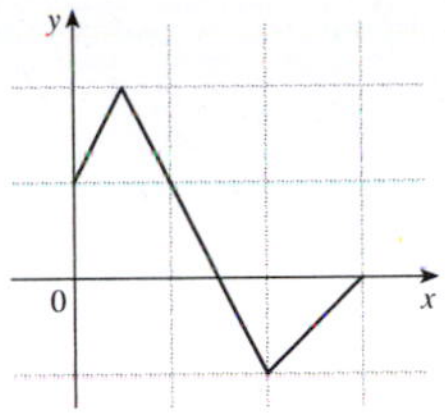

(b)

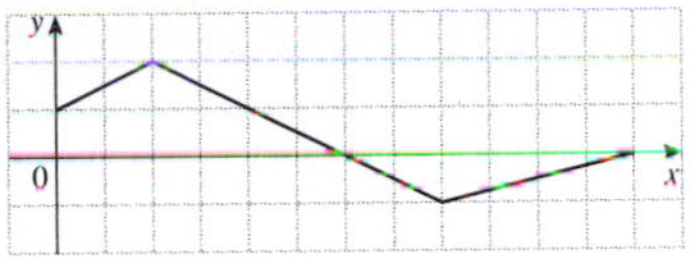

(c)

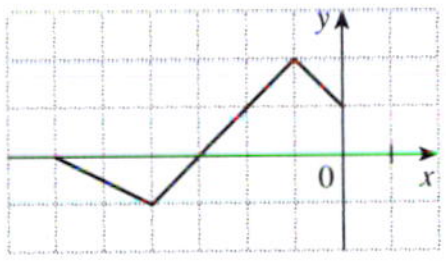

(d)

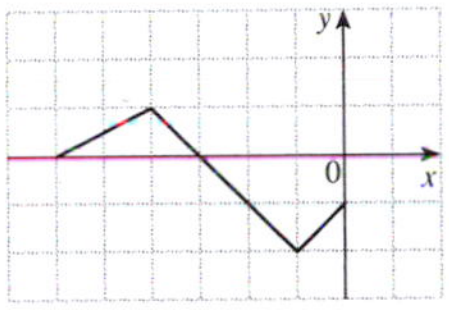

5.

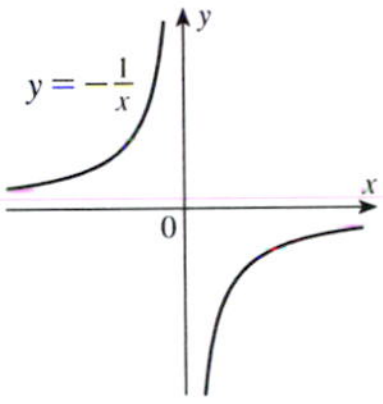

7.

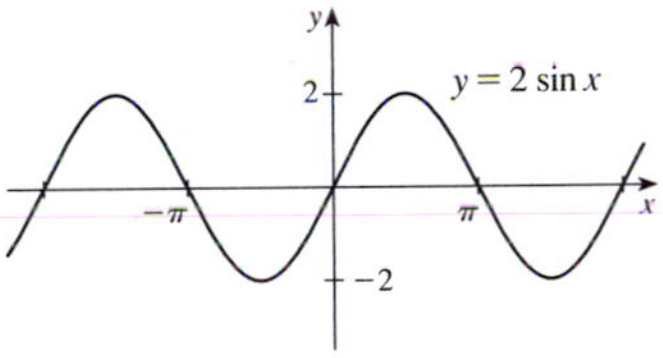

9.

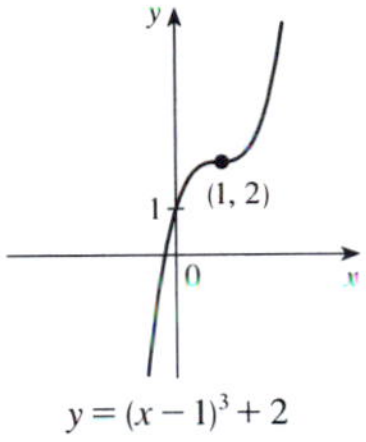

11.

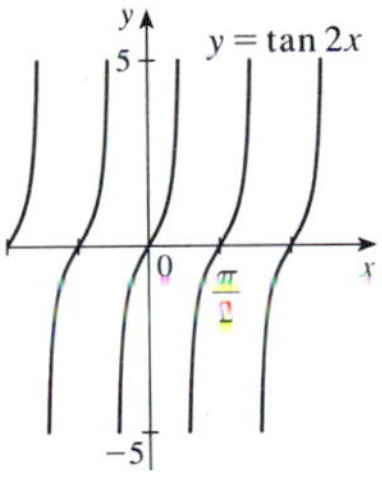

13.

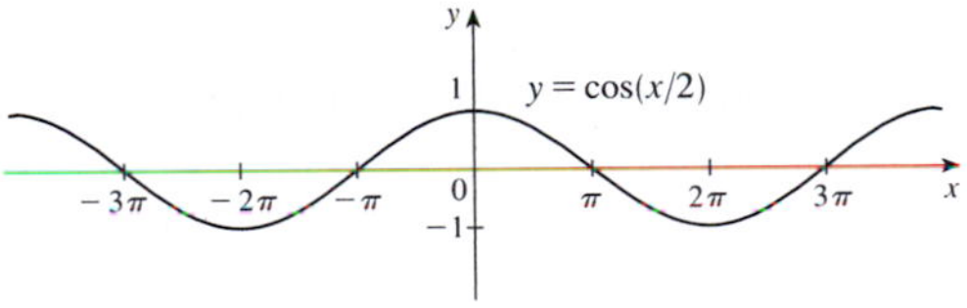

15.

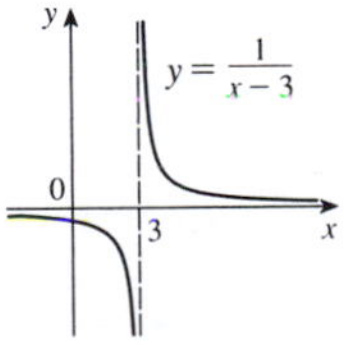

17.

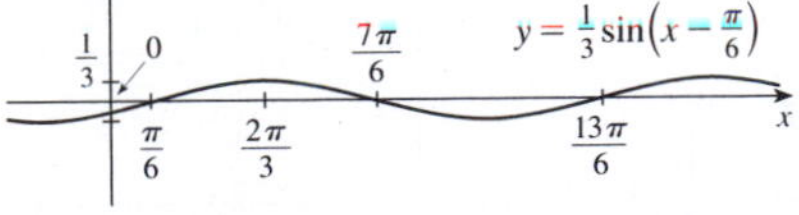

19.

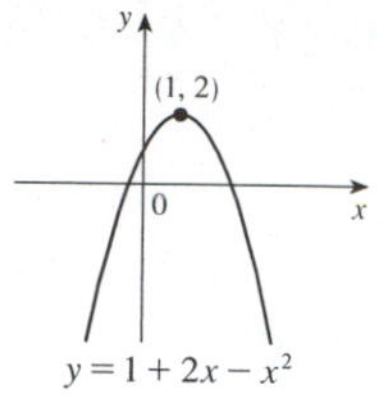

21.

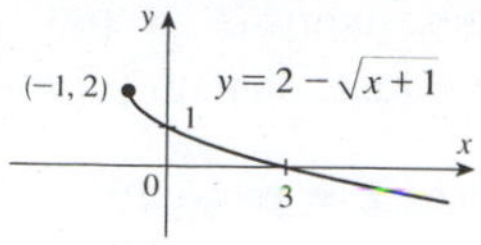

23.

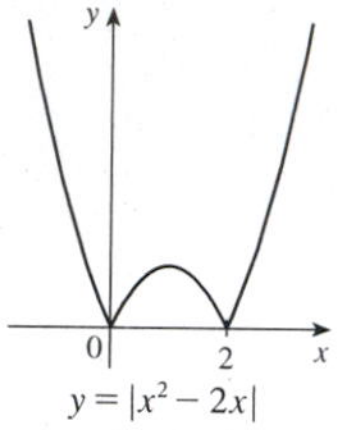

25.

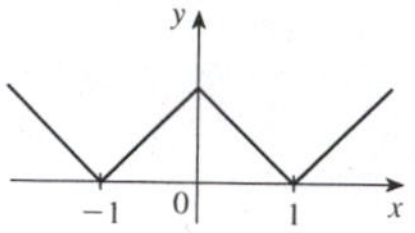

27. (a) The portion of the graph of $y = f(x)$ to the right of the y-axis is reflected in the y-axis.

(b)

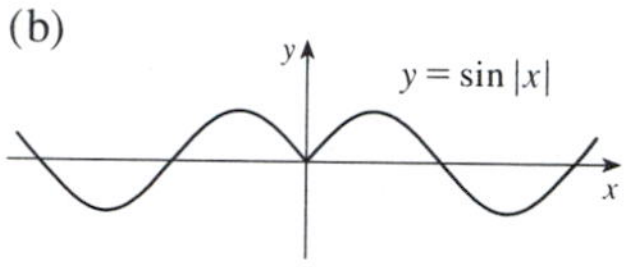

29.

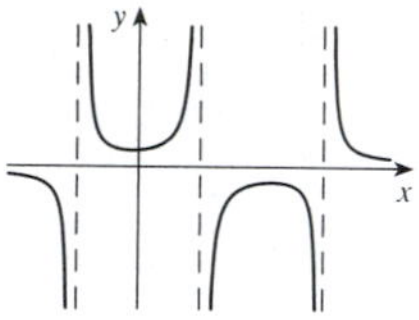

Exercises 3 ■ page 32

1. (d) **3.** (c)

5. $[-4, 10]$ by $[-10, 20]$

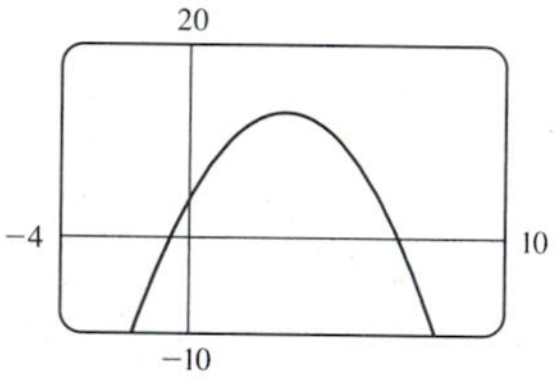

7. $[-20, 20]$ by $[-2, 6]$

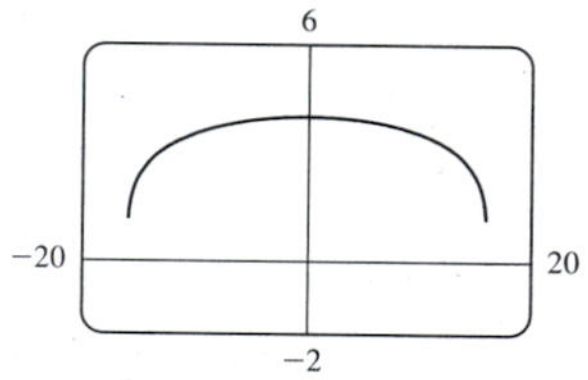

9. $[-50, 150]$ by $[-2000, 2000]$

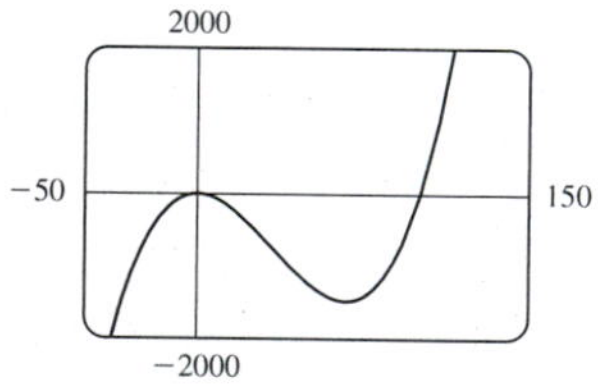

11. $[-10, 10]$ by $[-0.1, 0.1]$

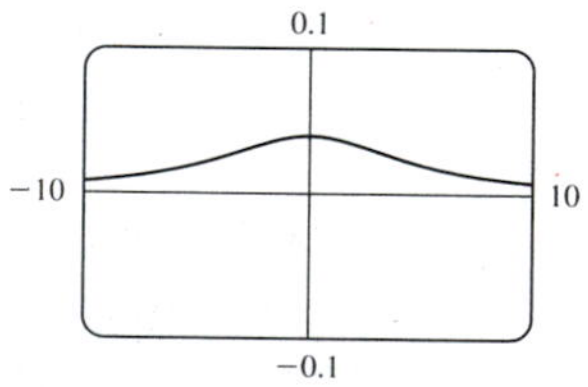

13. $[-4, 6]$ by $[-50, 100]$

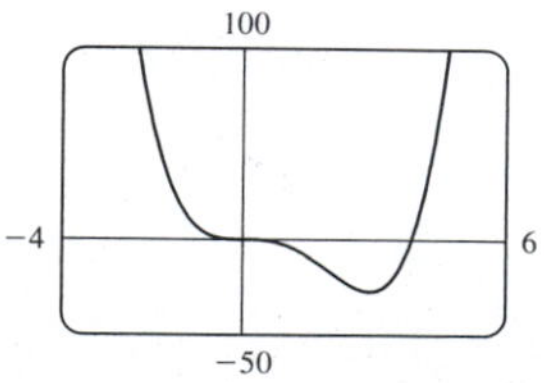

15. $[-15, 15]$ by $[-15, 15]$

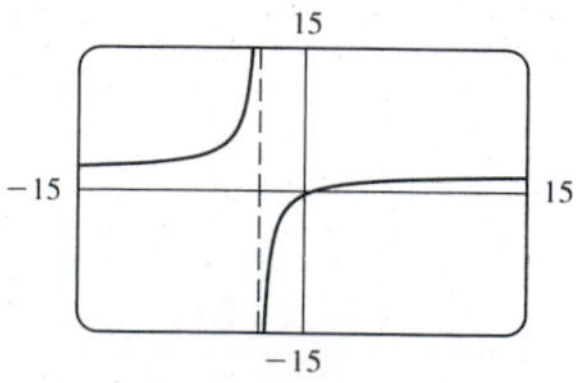

17. $[-0.1, 0.1]$ by $[-1.5, 1.5]$

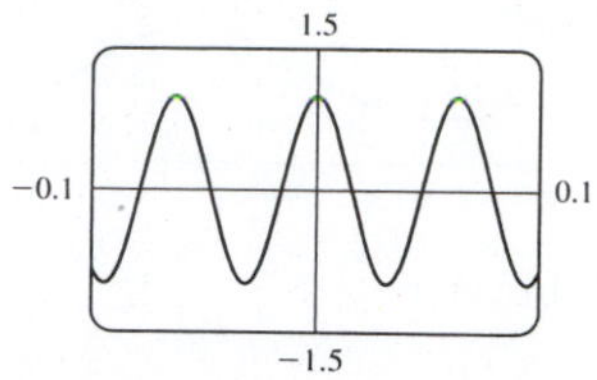

19. $[-250, 250]$ by $[-1.5, 1.5]$

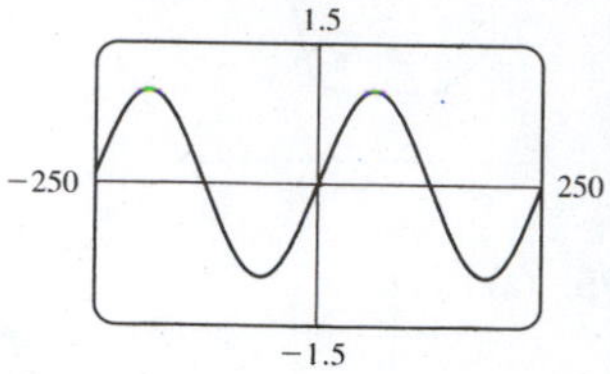

21. $[-5, 5]$ by $[0, 4]$

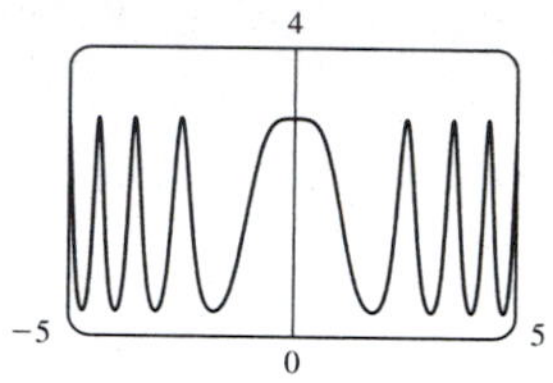

23.

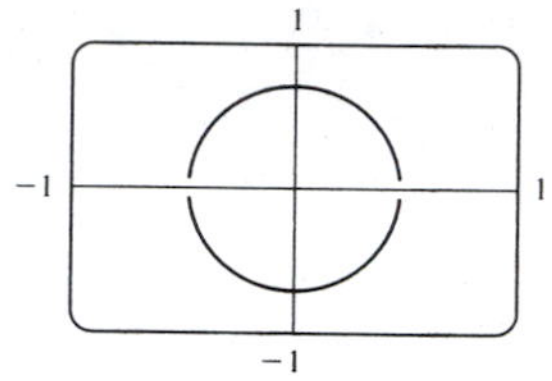

25.

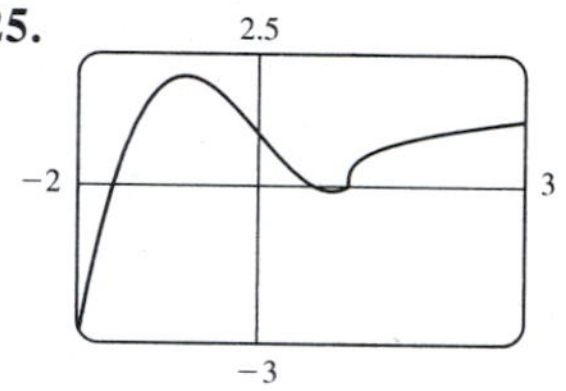

27. 0.67

29. $-1.90, 0, 1.90$

31. g

33. (a) Eventually f grows much more quickly than g
(b) 1.2, 22.4

35. $-0.85 < x < 0.85$

37. (a)

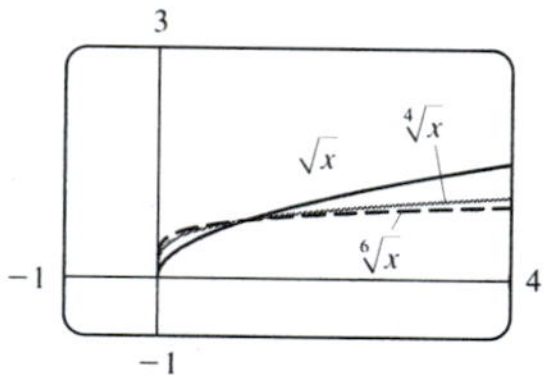

(b)

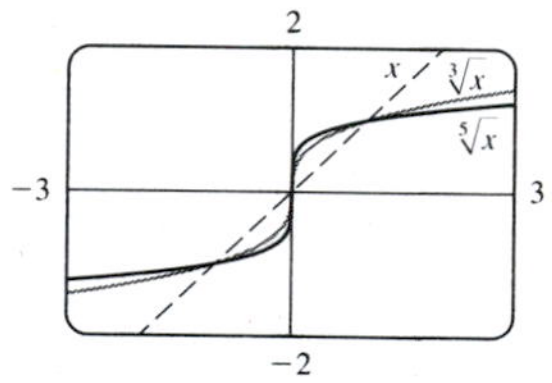

(c)

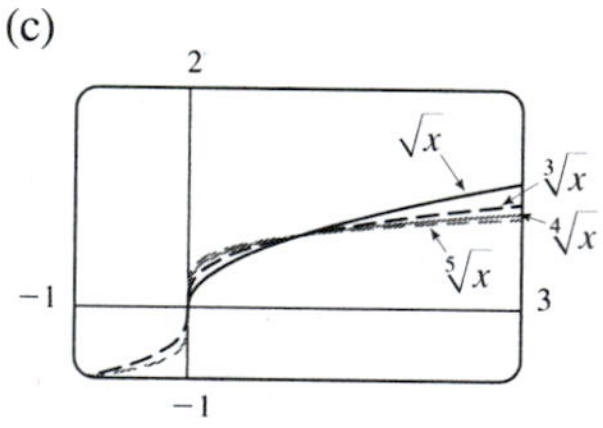

(d) Graphs of even roots are similar to $\sqrt{x}$, graphs of odd roots are similar to $\sqrt[3]{x}$. As n increases, the graph of $y = \sqrt[n]{x}$ becomes steeper near 0 and flatter for $x > 1$.

39.

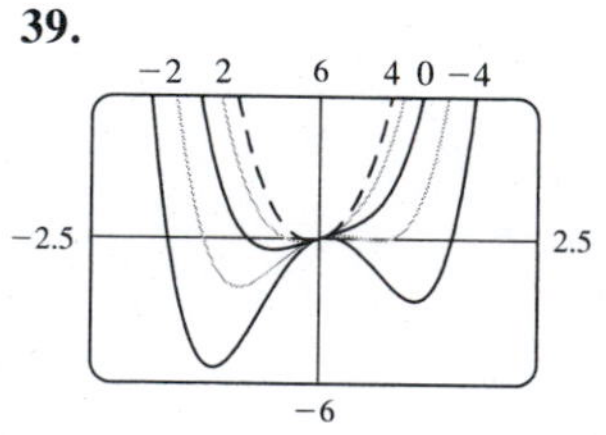

If $c < 0$, there are three humps: two minimum points and a maximum point. These humps get flatter as c increases until at $c = 0$ two of the humps disappear and there is only one minimum point. This single hump then moves to the right and approaches the origin as c increases.

41. The hump gets larger and moves to the right.

43. If $c < 0$, the loop is to the right of the origin, and if $c > 0$, the loop is to the left. The closer c is to 0, the larger the loop.

Exercises 4 ■ page 40

1. $-5 \leqslant x \leqslant 0$ **3.** $-\frac{7}{3}, 9$ **5.** 9 **7.** $f_n(x) = x^{2^{n+1}}$

9.

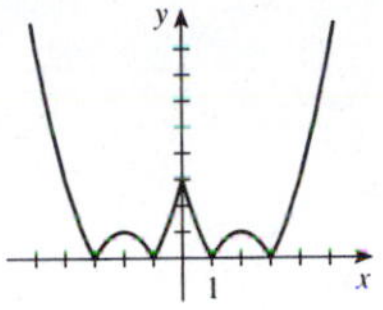

11. 2 **13.**

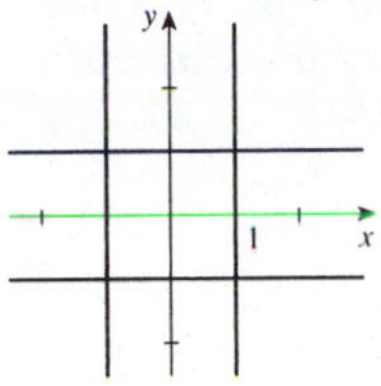

15.

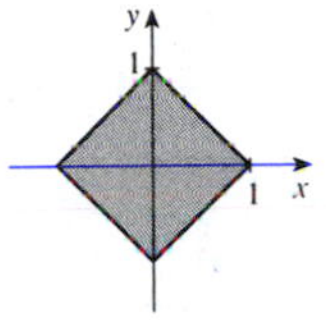

17. (a) $7920\pi \approx 24{,}880$ mi
(b) $2\pi \approx 6.3$ ft
19. 2.4 cm
23. False
25. $\frac{200}{13} \approx 15.4$ in.

CHAPTER 1

Exercises 1.1 ■ page 53

1. $\mathbf{a} = \langle 3, 1 \rangle$

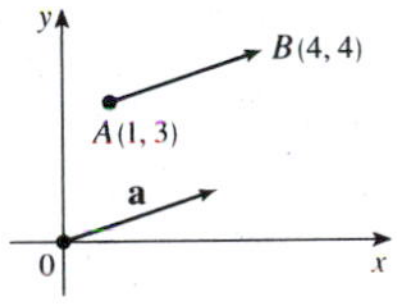

3. $\mathbf{a} = \langle 0, -2 \rangle$

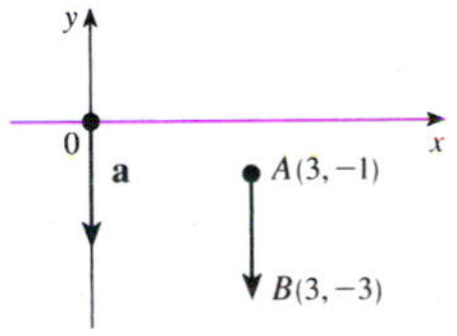

5. $\langle 5, -1 \rangle$

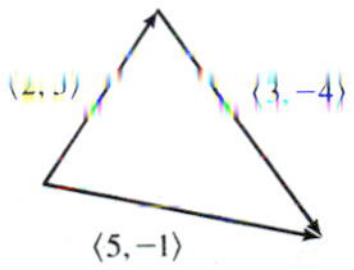

7. $\langle 1, 1 \rangle$

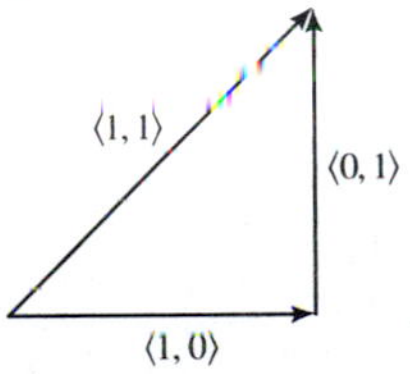

9. $13, \langle 3, -4 \rangle, \langle 7, -20 \rangle, \langle 10, -24 \rangle, \langle 7, -4 \rangle$
11. $\sqrt{13}, \langle 3, 1 \rangle, \langle 1, -7 \rangle, \langle 4, -6 \rangle, \langle 10, 7 \rangle$
13. $\sqrt{2}, 2\,\mathbf{i}, -2\,\mathbf{j}, 2\,\mathbf{i} - 2\,\mathbf{j}, 7\,\mathbf{i} + \mathbf{j}$
15. $\sqrt{2}, 3\,\mathbf{i}, -\mathbf{i} + 2\,\mathbf{j}, 2\,\mathbf{i} + 2\,\mathbf{j}, 11\,\mathbf{i} - \mathbf{j}$
17. $\langle 1/\sqrt{5}, 2/\sqrt{5} \rangle$ **19.** $(\mathbf{i} + \mathbf{j})/\sqrt{2}$
21. $\mathbf{i} = \frac{1}{5}\mathbf{a} + \frac{3}{5}\mathbf{b}, \mathbf{j} = \frac{1}{5}\mathbf{a} - \frac{2}{5}\mathbf{b}$ **23.** $\mathbf{0}$
25. (a), (b)

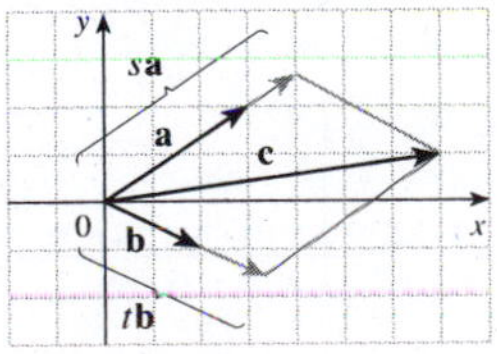

(d) $s = \frac{9}{7}, t = \frac{11}{7}$

27. $\mathbf{F} = (6\sqrt{3} - 5\sqrt{2})\,\mathbf{i} + (6 + 5\sqrt{2})\,\mathbf{j} \approx 3.32\,\mathbf{i} + 13.07\,\mathbf{j}$, $|\mathbf{F}| \approx 13.5$ lb, $\theta \approx 76°$ **29.** $\sqrt{493} \approx 22.2$ mi/h N8°W
31. A circle with radius 1, centered at (x_0, y_0)

Exercises 1.2 ■ page 60

1. (b), (c), (d) are meaningful **3.** 3 **5.** -1 **7.** -7
9. $\frac{1}{2}, -\frac{1}{2}$ **13.** $\cos^{-1}(11/(5\sqrt{5})) \approx 10°$
15. $\cos^{-1}(2/(13\sqrt{5})) \approx 86°$ **17.** $\cos^{-1}(1/\sqrt{5}) \approx 63°$
19. 105°, 35°, 40° **21.** Parallel **23.** Neither
25. Orthogonal **27.** ± 4 **29.** 0
31. $\langle 3/\sqrt{10}, -1/\sqrt{10} \rangle$ **33.** $11/\sqrt{13}, \langle \frac{22}{13}, \frac{33}{13} \rangle$
35. $3/\sqrt{5}, \langle \frac{6}{5}, \frac{3}{5} \rangle$ **37.** $1, \langle 1, 0 \rangle$ **39.** $\langle -\frac{1}{5}, \frac{2}{5} \rangle$
41. $\langle \frac{12}{13}, \frac{8}{13} \rangle$ **43.** $1/\sqrt{5}$ **45.** $\frac{13}{5}$ **47.** $6/\sqrt{5}$
49. They are equal.
51. $\langle 0, -2\sqrt{10} \rangle$ or any vector of the form $\langle s, 3s - 2\sqrt{10} \rangle$, $s \in \mathbb{R}$
53. 128 J **55.** $250 \cos 20° \approx 235$ ft-lb **57.** (b) $\frac{13}{5}$

Exercises 1.3 ■ page 68

1. (a)

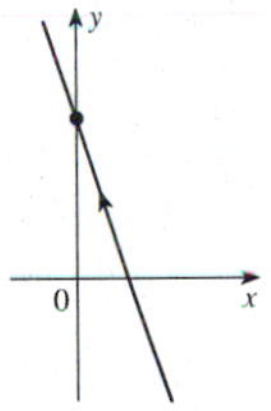

(b) $3x + y = 5$

3. (a)

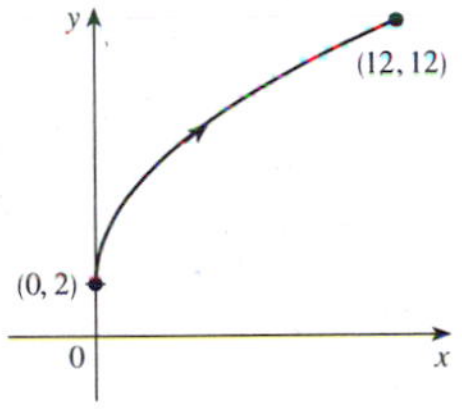

(b) $x = \frac{3}{25}(y - 2)^2, 2 \leqslant y \leqslant 12$

5. (a)

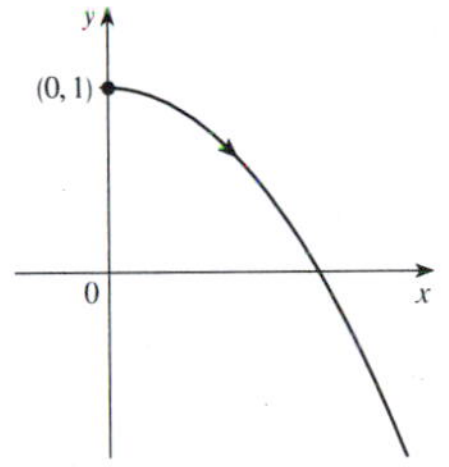

(b) $y = 1 - x^2, x \geqslant 0$

7. (a)

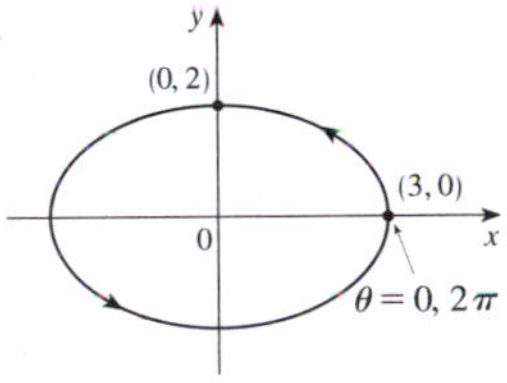

(b) $\frac{1}{9}x^2 + \frac{1}{4}y^2 = 1$

9. (a)

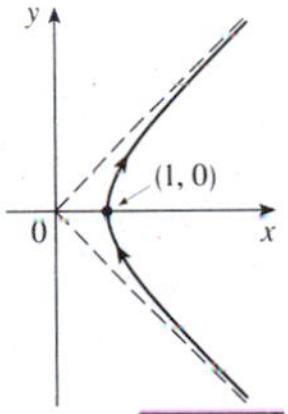

(b) $x = \sqrt{y^2 + 1}$

11. (a)

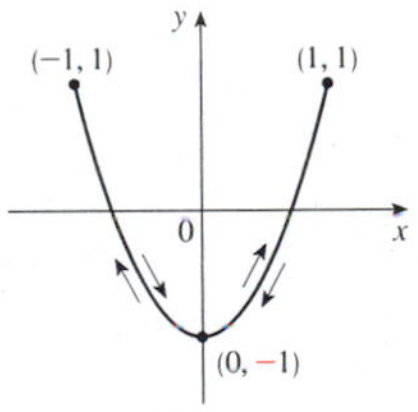

(b) $y + 1 = 2x^2, -1 \leqslant x \leqslant 1$

13. (a)

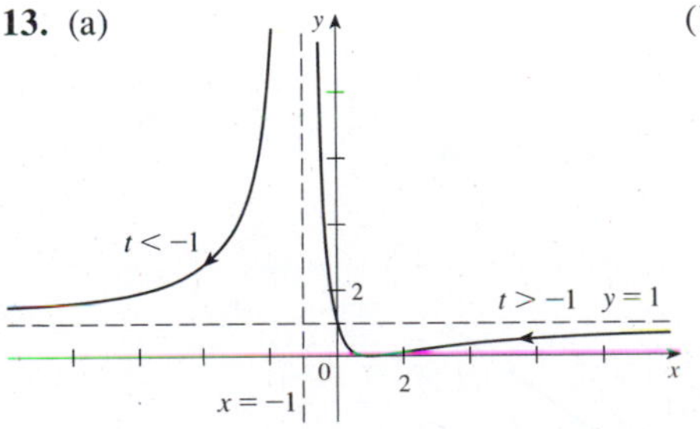

(b) $y = \dfrac{(x - 1)^2}{(x + 1)^2}$

15. Moves counterclockwise along a circle of radius 1 centered at (2, 3), starting and ending at (3, 3)

17. Moves along the parabola $x = y^2$, starting at $(1, 1)$, down to $(1, -1)$, back up to $(1, 1)$, down to $(1, -1)$, then back up to $(1, 1)$
19. Moves along the hyperbola $y = 1/x$, from $(\frac{1}{2}, 2)$ to $(\sin 1, \csc 1)$

21.

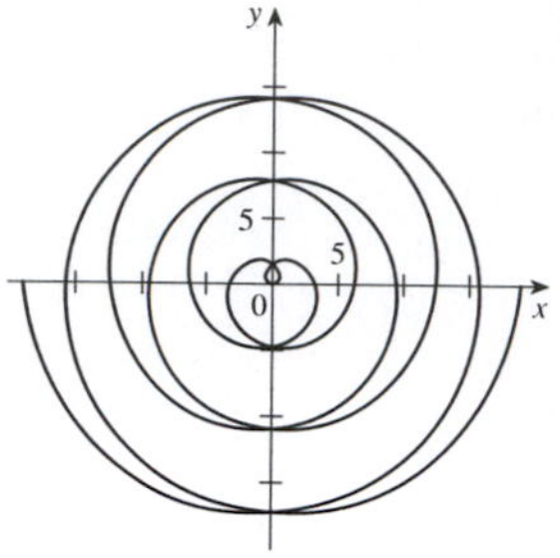

23.

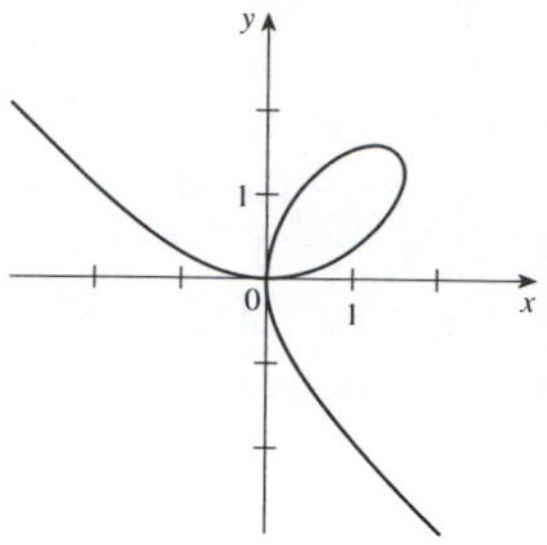

25. (a) $\mathbf{r}(t) = \langle 1, 3\rangle + t\langle -3, 4\rangle$ (b) $x = 1 - 3t$, $y = 3 + 4t$
27. (a) $\mathbf{r}(t) = \langle 4, -1\rangle + t\langle -6, 6\rangle$
(b) $x = 4 - 6t$, $y = -1 + 6t$
29. (a) $\mathbf{r}(t) = \langle -4, 5\rangle + t\langle -2, 6\rangle$
(b) $x = -4 - 2t$, $y = 5 + 6t$ (c) $y = -3x - 7$
31. Neither; $(\frac{37}{11}, \frac{9}{11})$ **33.** Perpendicular; $(\frac{33}{26}, \frac{17}{26})$
35. (a) $(8, 6)$ (b) $t = 3$ (c) No (d) $x = y^2 - 6y + 8$
37. For $c = 0$, there is a cusp; for $c > 0$, there is a loop whose size increases as c increases.

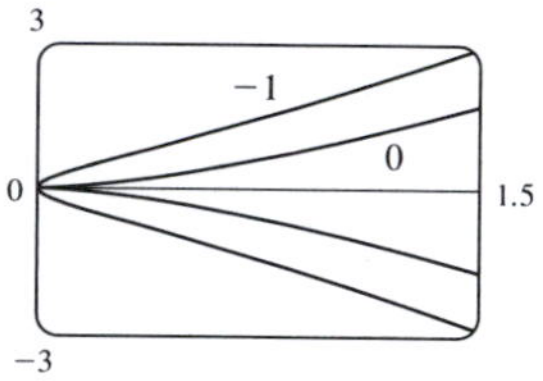

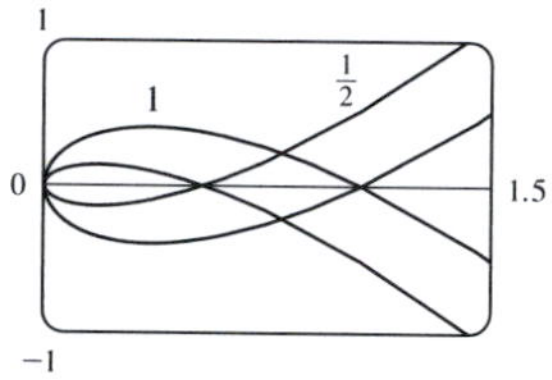

39. As n increases, the number of oscillations increases; a and b control the width and height.

Chapter 1 Review Exercises ■ page 70

1. $\mathbf{a} = \langle 1, -2\rangle$

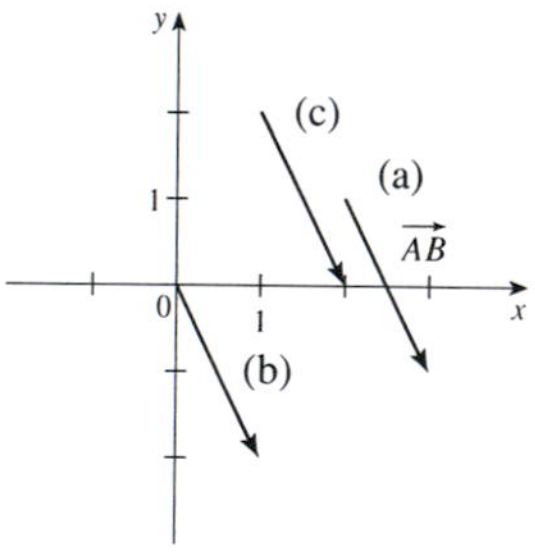

3. $-2\mathbf{i} - 2\mathbf{j}$

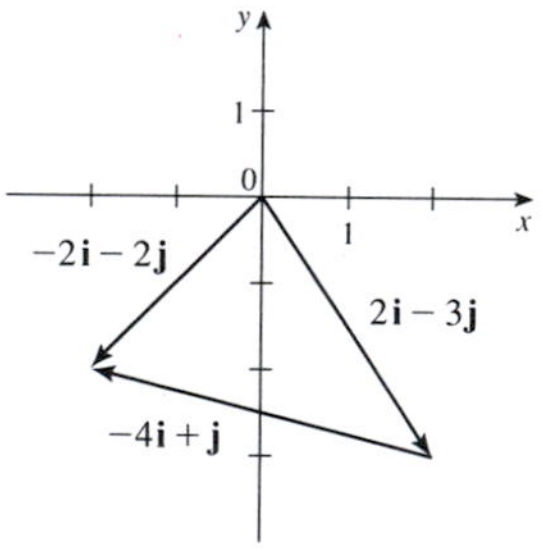

5. $\langle 3, 5\rangle$

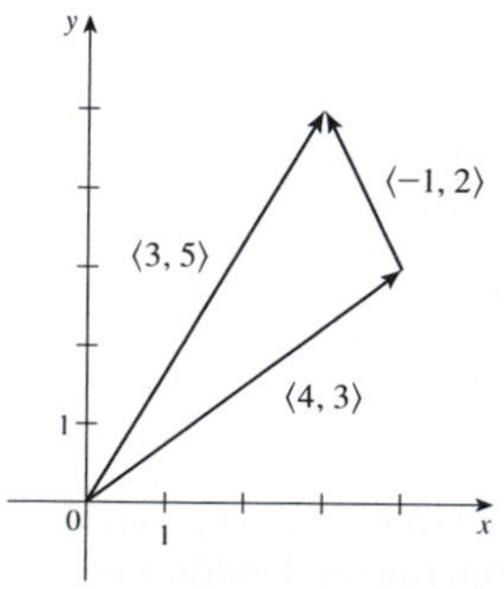

7. $(-1, 3)$ **9.** $(4, 1)$
11. (a) 5 (b) $\langle \frac{3}{5}, -\frac{4}{5}\rangle$ (c) $\langle \frac{4}{5}, \frac{3}{5}\rangle$ (d) $\langle -9, 25\rangle$
(e) $s = \frac{5}{13}$, $t = \frac{11}{13}$
13. (a) $5\sqrt{2}$ (b) $\langle 1/\sqrt{50}, -7/\sqrt{50}\rangle$ (c) $\langle 7/\sqrt{50}, 1/\sqrt{50}\rangle$
(d) $\langle 2, 37\rangle$ (e) $s = -\frac{8}{17}$, $t = \frac{4}{17}$
15. (a) 4 (b) $60°$ (c) $4/\sqrt{5}$ (d) $\langle \frac{4}{5}, \frac{8}{5}\rangle$
17. (a) -8 (b) $143°$ (c) $-\frac{8}{5}$ (d) $\langle \frac{24}{25}, -\frac{32}{25}\rangle$
19. (a) 0 (b) $90°$ (c) 0 (d) $\langle 0, 0\rangle$
21. $6/\sqrt{5}$ **23.** $1/\sqrt{5}$ **25.** 21

27.

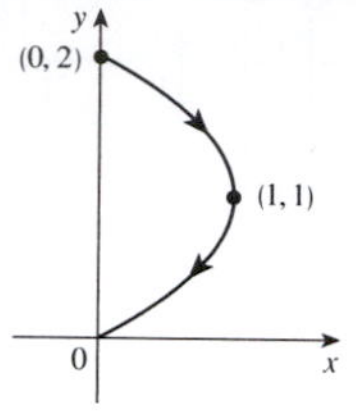

$x = 2y - y^2$, $0 \leqslant y \leqslant 2$

29.

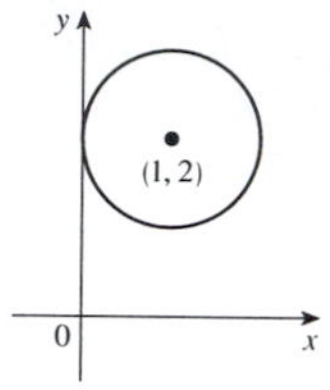

$(x - 1)^2 + (y - 2)^2 = 1$

31. (a) $\mathbf{r}(t) = \langle 1, 2\rangle + t\langle -4, 2\rangle$ (b) $x = 1 - 4t$, $y = 2 + 2$
(c) $y = -\frac{1}{2}x + \frac{5}{2}$
33. (a) $\mathbf{r}(t) = \langle 2, 3\rangle + t\langle 1, -5\rangle$ (b) $x = 2 + t$, $y = 3 - 5t$
(c) $y = -5x + 13$
35. (a) $\langle 6, -2\rangle$ (b) $t = 3$ (c) No
37. They are perpendicular.

CHAPTER 2

Exercises 2.1 ■ page 78

1. (a) -0.43, -0.35, 0.2, 0.8, 1.1 (b) 0.5
3. (a) (i) $\langle 1, 5\rangle$, $\langle 1, 4.5\rangle$, $\langle 1, 4.1\rangle$, $\langle 1, 4.01\rangle$, $\langle 1, 4.001\rangle$
(ii) $\langle 1, 3\rangle$, $\langle 1, 3.5\rangle$, $\langle 1, 3.9\rangle$, $\langle 1, 3.99\rangle$, $\langle 1, 3.999\rangle$
(b) $\langle 1, 4\rangle$ (c) $\mathbf{L}(t) = \langle 2, 3\rangle + t\langle 1, 4\rangle$
5. (a) $\langle 3, -27\rangle$ (ii) $\langle 3, -20.6\rangle$ (iii) $\langle 3, -19.8\rangle$
(iv) $\langle 3, -19.16\rangle$ (b) $\langle 3, -19\rangle$
7. (a) (i) 0.236068 (ii) 0.242641 (iii) 0.248457
(iv) 0.249844 (v) 0.249984 (vi) 0.267949
(vii) 0.258343 (viii) 0.251582 (ix) 0.250156
(x) 0.250016 (b) $\frac{1}{4}$ (c) $x - 4y + 4 = 0$
9. (a) (i) -32 ft/s (ii) -25.6 ft/s (iii) -24.8 ft/s
(iv) -24.16 ft/s (b) -24 ft/s
11. (a) (i) $\frac{13}{6}$ ft/s (ii) $\frac{7}{6}$ ft/s (iii) $\frac{19}{24}$ ft/s (iv) $\frac{331}{600}$ ft/s
(b) $\frac{1}{2}$ ft/s
(c)

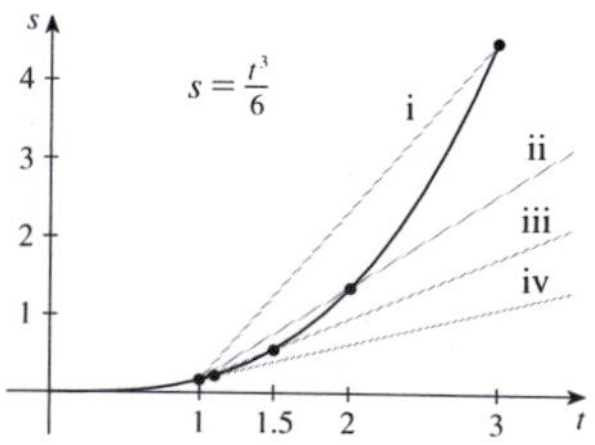

(d)

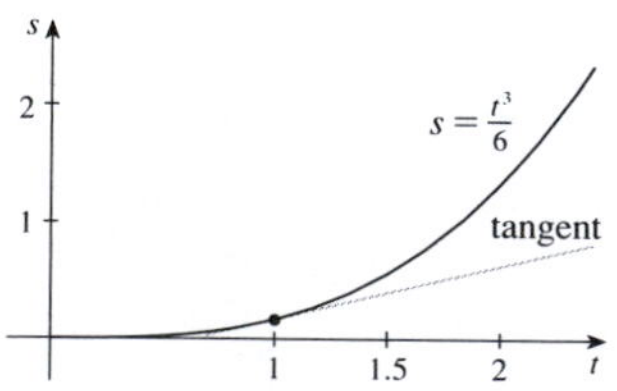

Exercises 2.2 ■ page 89

1. (a) 3 (b) 2 (c) -2 (d) Does not exist
(e) 1 (f) -1 (g) -1 (h) -1 (i) -3

3. (a) 2 (b) -1 (c) 1 (d) 1 (e) 2
(f) Does not exist
5. (a) ∞ (b) $-\infty$ (c) $-\infty$ (d) ∞ (e) $-\infty$
(f) $x = -9, x = -4, x = 3, x = 7$
7. (a)

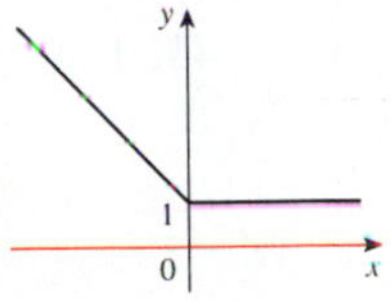

(b) (i) 1 (ii) 1 (iii) 1

9. 0.806452, 0.641026, 0.510204, 0.409836, 0.369004, 0.336689, 0.165563, 0.193798, 0.229358, 0.274725, 0.302115, 0.330022; $\frac{1}{3}$
11. -0.003884, -0.003941, -0.003988, -0.003994, -0.003999, -0.004124, -0.004061, -0.004012, -0.004006, -0.004001; -0.004
13. 0.459698, 0.489670, 0.493369, 0.496261, 0.498336, 0.499583, 0.499896, 0.499996; $\frac{1}{2}$
15. $-0.6\mathbf{i} + 1.2\mathbf{j}$, $-0.2\mathbf{i} + 1.4\mathbf{j}$, $0.2\mathbf{i} + 1.6\mathbf{j}$, $0.6\mathbf{i} + 1.8\mathbf{j}$, $0.8\mathbf{i} + 1.9\mathbf{j}$, $0.98\mathbf{i} + 1.99\mathbf{j}$; $2.6\mathbf{i} + 2.8\mathbf{j}$, $2.2\mathbf{i} + 2.6\mathbf{j}$, $1.8\mathbf{i} + 2.4\mathbf{j}$, $1.4\mathbf{i} + 2.2\mathbf{j}$, $1.2\mathbf{i} + 2.1\mathbf{j}$, $1.02\mathbf{i} + 2.01\mathbf{j}$; $\mathbf{i} + 2\mathbf{j}$
17. ∞ **19.** ∞ **21.** $-\infty$ **23.** $-\infty, \infty$
25. 2.0, 2.593742, 2.704814, 2.716924, 2.718146, 2.718268, 2.718280, 2.718282, 2.718282, 2.718282; 2.71828
27. (a) 0.998000, 0.638259, 0.358484, 0.158680, 0.038851, 0.008928, 0.001465; 0
(b) 0.000572, -0.000614, -0.000907, -0.000978, -0.000993, -0.001000; -0.001
29. 4 **31.** No matter how many times we zoom in toward the origin, the graph appears to consist of almost-vertical lines. This indicates more and more frequent oscillations as $x \to 0$.

Exercises 2.3 ▪ page 99

1. 75 **3.** 60 **5.** $\frac{1}{2}$ **7.** 2 **9.** -3 **11.** -2
13. (a) 5 (b) 9 (c) 2 (d) $-\frac{1}{3}$ (e) $-\frac{3}{8}$ (f) 0
(g) Does not exist (h) $-\frac{6}{11}$
15. Does not exist **17.** -3 **19.** 1 **21.** -10 **23.** 4
25. $-\frac{1}{5}$ **27.** 6 **29.** $-\sqrt{2}/4$ **31.** 108 **33.** $-\frac{1}{2}$
35. $\frac{2}{3}$ **37.** $\frac{1}{16}$ **39.** $-\mathbf{i} + \mathbf{j}$ **41.** $\langle 2 + \sqrt{2}, 4 \rangle$
43. $\frac{4}{3}\mathbf{i} + 4\mathbf{j}$ **47.** 1 **51.** 0 **53.** 0
55. Does not exist **57.** -2 **59.** -3 **61.** 0
63. Does not exist
65. (a)

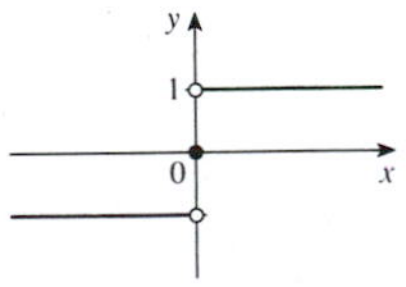

(b) (i) 1 (ii) -1
(iii) Does not exist (iv) 1

67. (a) 1, 1 (b) Yes, 1 (c)

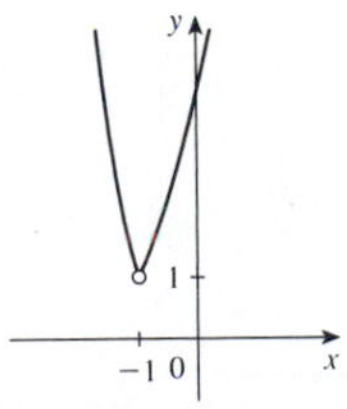

69. (a) (i) $n - 1$ (ii) n (b) a is not an integer

71. (a) (i) 2 (ii) -2 (b) No (c)

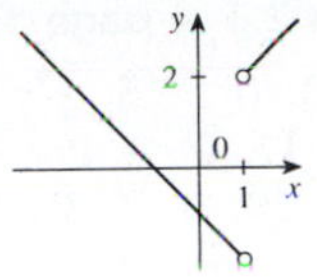

79. $c/3$ **81.** 15, -1

Exercises 2.4 ▪ page 110

1. (a) $|x - 3| < \frac{1}{60}$ (b) $|x - 3| < \frac{1}{600}$
3. $\frac{4}{7}$ (or any smaller positive number)
5. 0.6875 (or any smaller positive number)
7. 0.11, 0.012 (or smaller positive numbers)
9. 0.07 (or any smaller positive number) **35.** Within 0.1
39. For every number $\varepsilon > 0$ there is a corresponding number $\delta > 0$ such that $|\mathbf{r}(t) - \mathbf{v}| < \varepsilon$ whenever $a < t < a + \delta$.
41. (a) 0.01 (b) $0.001/\sqrt{5}$ (c) $0.0001/\sqrt{5}$

Exercises 2.5 ▪ page 120

1. (a) $-5, -3, -1, 3, 5, 8, 10$
(b) Left, left, neither, neither, neither, right, neither
13. $f(1)$ undefined **15.** $\lim_{x\to 1} f(x)$ does not exist

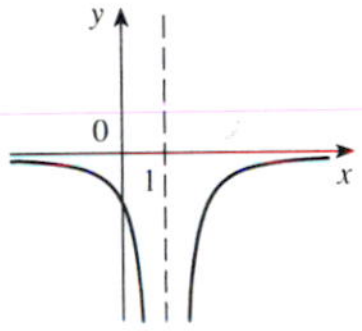

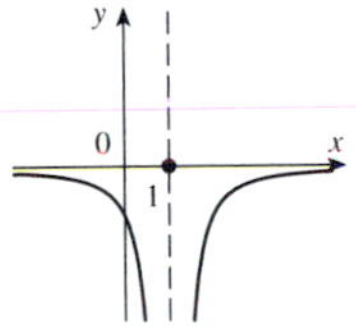

17. $\lim_{x\to -3} f(x) \neq f(-3)$

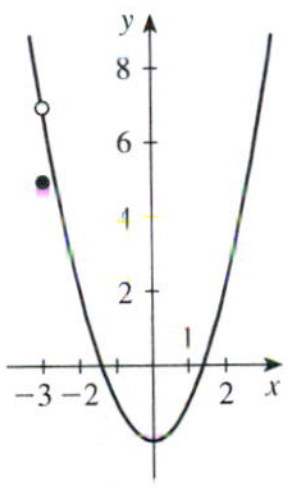

19. $\mathbb{R}$ **21.** $(-1, \infty)$ **23.** $\mathbb{R}$ **25.** $\mathbb{R}$ **27.** $\mathbb{R}$
31. 0, continuous from the right **33.** Continuous at all points

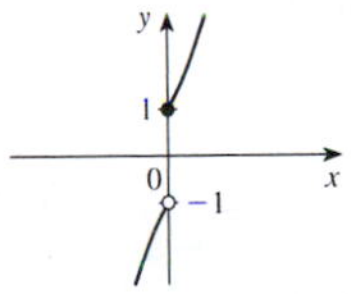

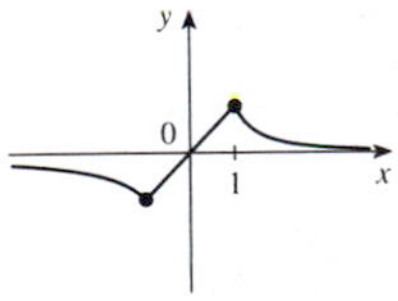

35. $\{n/2 \mid n \text{ an integer}\}$, continuous from the right

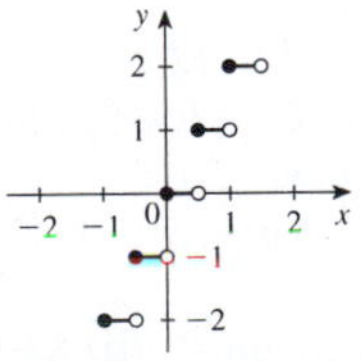

37. $\frac{1}{3}$ **39.** $c = 2, d = 0$ **49.** (b) $(-1.33, -1.32)$
51. (b) 1.434 **55.** None **57.** Yes

Exercises 2.6 ■ page 133

1. 0 **3.** 0 **5.** $\frac{1}{6}$ **7.** 0 **9.** 0 **11.** 2 **13.** -1
15. 0 **17.** 0 **19.** $-\frac{1}{2}$ **21.** ∞ **23.** ∞ **25.** $-\infty$
27. ∞ **29.** 0 **31.** 0
33. $y = 1, x = -4$ **35.** $x = 2, x = -5$

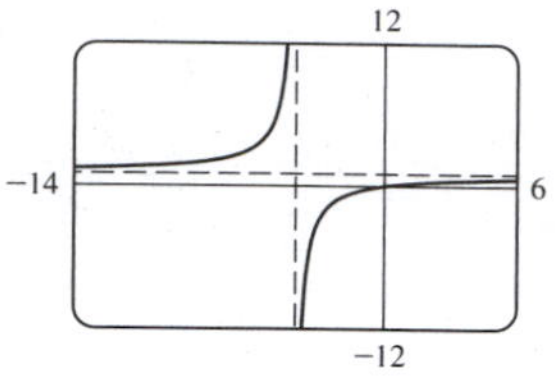

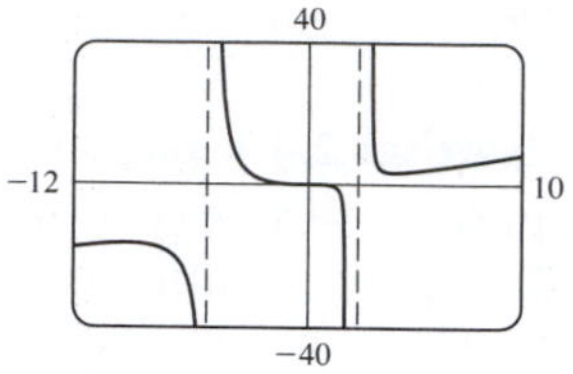

37. $y = \pm 1$

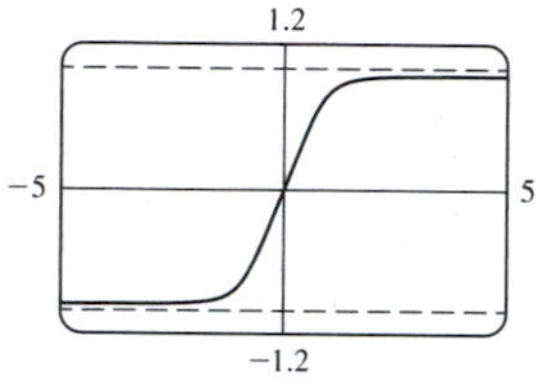

39. $-\infty, -\infty$

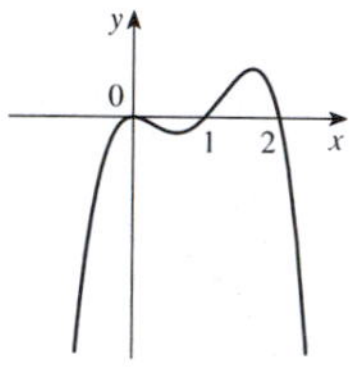

41. $\infty, -\infty$

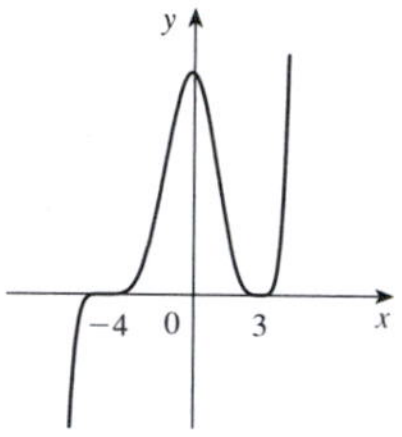

43. (a) 0 (b) ∞ or $-\infty$ **45.** 4 **47.** $N \geqslant 13$
49. $N \leqslant -6, N \leqslant -22$ **51.** (a) $x > 100$

Exercises 2.7 ■ page 143

1. (a) (i) -4 (ii) -4
(b) $4x + y + 9 = 0$
(c)

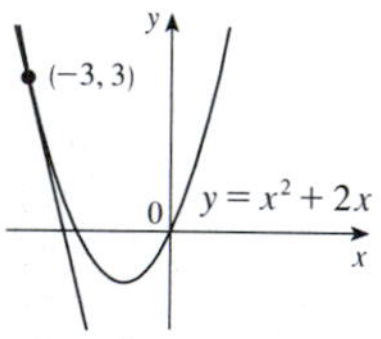

3. $10x - y + 13 = 0$
5. $x - 4y + 3 = 0$
7. (a) $-2/(a + 3)^2$
(b) (i) $-\frac{1}{2}$ (ii) $-\frac{2}{9}$ (iii) $-\frac{1}{8}$

9. (a) $3a^2 - 4$
(b) $x + y + 1 = 0, y = 8x - 15$
(c) See graph at right.

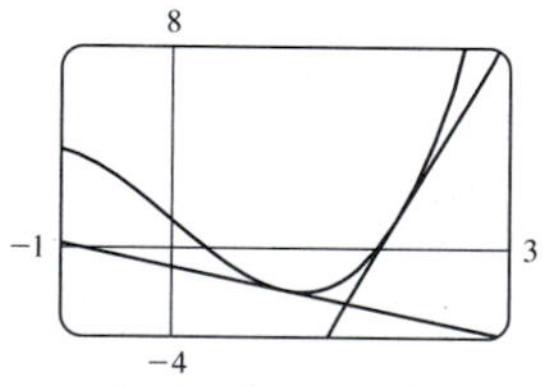

11. -24 ft/s **13.** $12a^2 + 6$, 18 m/s, 54 m/s, 114 m/s
15. (a) 0 (b) C (c) Speeding up, slowing down, neither
(d) The car stopped.
17. (a) (i) -1.2 °/h (ii) -1.25 °/h (iii) -1.3 °/h
(b) -1.6 °/h
19. (a) (i) \$20.25/unit (ii) \$20.05/unit (b) \$20/unit
21. (a) $5\mathbf{i} - \frac{1}{2}\mathbf{j}$ (b) $x = 6 + 5t, y = 1 - \frac{1}{2}t, t \neq 0$
23. (a) $10\mathbf{i} - 2\mathbf{j}$ (b) $\sqrt{104}$ ft/s

Chapter 2 Review Exercises ■ page 146

1. False **3.** True **5.** False **7.** True **9.** True
11. True **13.** $\sqrt{6}$ **15.** $\frac{1}{2}$ **17.** 2 **19.** 0 **21.** ∞
23. $-\frac{1}{8}$ **25.** -1 **27.** 0 **29.** $-\frac{1}{2}$ **31.** $\frac{1}{2}$
33. $-\infty$ **35.** $12\mathbf{i} + 9\mathbf{j}$ **37.** $\langle \frac{1}{2}, -3 \rangle$ **43.** 1
45. (a) (i) 3 (ii) 0 (iii) Does not exist (iv) 0 (v) 0 (vi) 0
(b) At 0 and 3 (c)

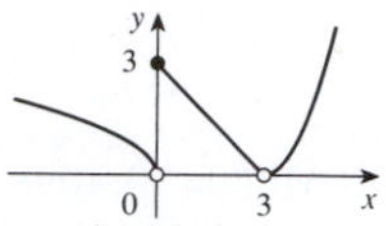

47. $\mathbb{R}$ **51.** (a) -8 (b) $8x + y = 17$
53. (a) $8\mathbf{i} + 5\mathbf{j}$ (b) $\mathbf{L}(t) = \langle 4, 12 \rangle + t\langle 8, 5 \rangle$
(c) $x = 4 + 8t, y = 12 + 5t$
55. (a) $\langle 9, 24 \rangle, \langle 8, 20 \rangle, \langle 7.5, 18 \rangle, \langle 7.1, 16.4 \rangle$ (b) $\langle 7, 16 \rangle$
57. (a) (i) 3 m/s (ii) 2.75 m/s (iii) 2.625 m/s
(iv) 2.525 m/s (b) 2.5 m/s
59. 0.09 (or any smaller positive number)
61. 0 **63.** $\frac{3}{4}$

CHAPTER 3

Exercises 3.1 ■ page 157

1. $7, 7x - y - 12 = 0$
3. (a) $-2, 2x + y + 1 = 0$ (b)

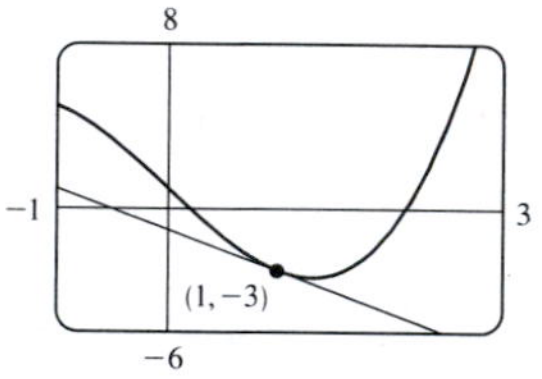

5. -2 m/s **7.** $1 - 4a$ **9.** $-1/(2a - 1)^2$
11. $1/(3 - a)^{3/2}$ **13.** $f(x) = \sqrt{x}, a = 1$
15. $f(x) = x^9, a = 1$
17. $f(x) = \sin x, a = \pi/2$ **19.** $f'(x) = 5, \mathbb{R}, \mathbb{R}$
21. $f'(x) = 3x^2 - 2x + 2, \mathbb{R}, \mathbb{R}$
23. $g'(x) = 1/\sqrt{1 + 2x}, [-\frac{1}{2}, \infty), (-\frac{1}{2}, \infty)$
25. $G'(x) = -10/(2 + x)^2, \{x \mid x \neq -2\}, \{x \mid x \neq -2\}$
27. $f'(x) = 4x^3, \mathbb{R}, \mathbb{R}$ **29.** $1, 2x, 3x^2, nx^{n-1}, 5x^4$
31. (a) $f'(x) = 1 + 2/x^2$
33. (a) -2
(b) 0.8
(c) -1
(d) -0.5

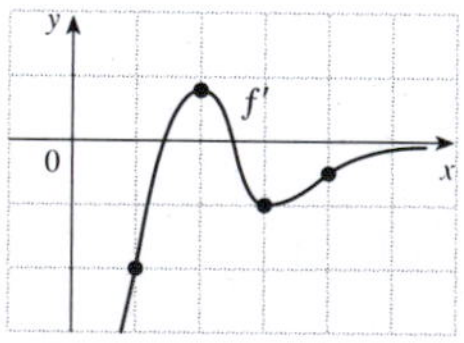

35.

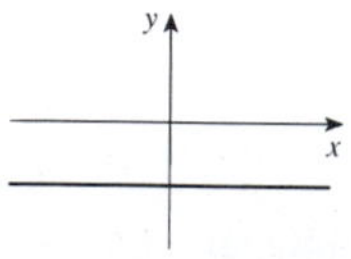

37.

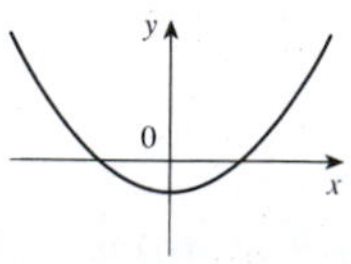

39. 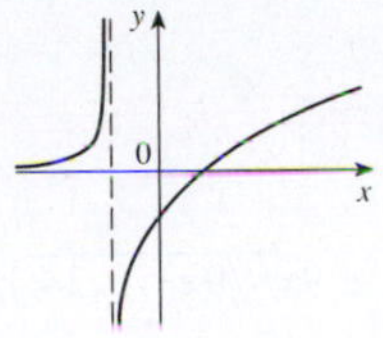

41. 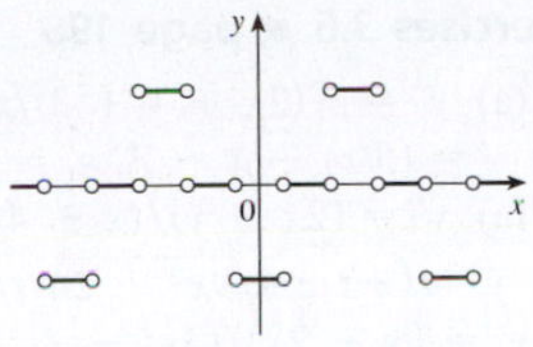

43. 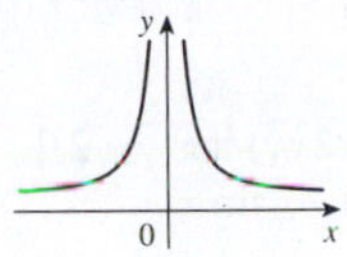

45. 3.296 **47.** −5, 4, 8, 9, 5, −0.5, −8
49. (a) $1/(3a^{2/3})$

51. −1, 11 (vertical tangents), 4 (discontinuity), 8 (corner)

53. $f'(x) = \begin{cases} -1 & \text{if } x < 6 \\ 1 & \text{if } x > 6 \end{cases}$

or $f'(x) = \dfrac{x-6}{|x-6|}$

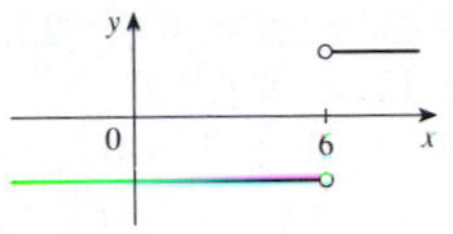

55. (a) 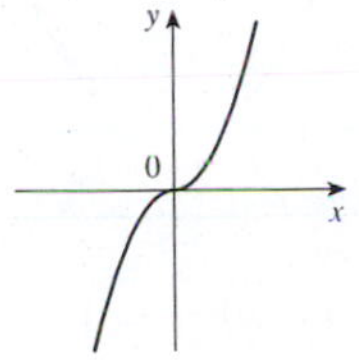

(b) All x
(c) $f'(x) = 2|x|$

57. (a) −5, 5 **59.** Does not exist **63.** 63°

Exercises 3.2 ■ page 168

1. $f'(x) = 2x - 10$ **3.** $V'(r) = 4\pi r^2$ **5.** $F'(x) = 12288x^2$
7. $Y'(t) = -54t^{-10}$ **9.** $g'(x) = 2x - (2/x^3)$
11. $h'(x) = -3/(x-1)^2$
13. $G'(s) = (2s+1)(s^2+2) + (s^2+s+1)(2s)$
$[= 4s^3 + 3s^2 + 6s + 2]$
15. $y' = \frac{3}{2}\sqrt{x} + (2/\sqrt{x}) - 3/(2x\sqrt{x})$ **17.** $y' = \sqrt{5}/(2\sqrt{x})$
19. $y' = -(4x^3 + 2x)/(x^4 + x^2 + 1)^2$ **21.** $y' = 2ax + b$
23. $y' = (-3t^2 + 14t + 23)/(t^2 + 5t - 4)^2$
25. $y' = 1 + 2/(5\sqrt[5]{x^3})$ **27.** $u' = \sqrt{2}\,x^{\sqrt{2}-1}$
29. $v' = \frac{3}{2}\sqrt{x} - 5/(2x^3\sqrt{x})$ **31.** $f'(x) = 2cx/(x^2 + c)^2$
33. $f'(x) = 2x^4(x^3 - 5)/(x^3 - 2)^2$
35. $P'(x) = na_n x^{n-1} + (n-1)a_{n-1}x^{n-2} + \cdots + 2a_2 x + a_1$
37. $y = 4$ **39.** $3x - 2y + 1 = 0$
41. (a) $x - 2y + 2 = 0$ (b)

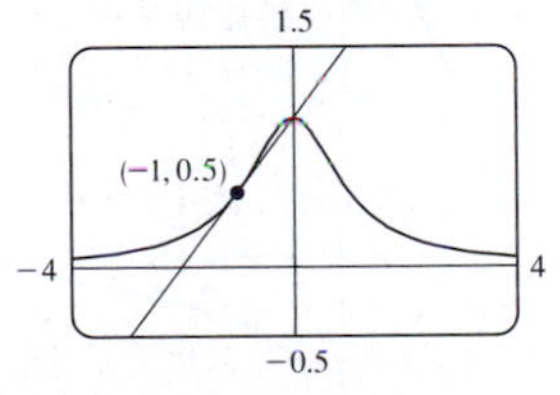

43. (a) $45x^{14} - 15x^2$ **45.** (4, 8)
47. (1, 0), $(-\frac{1}{3}, \frac{32}{27})$ **49.** 2, $(-2 \pm \sqrt{3}, (1 \mp \sqrt{3})/2)$
53. $x - 4y - 14 = 0$ **55.** $12x + y + 98 = 0$

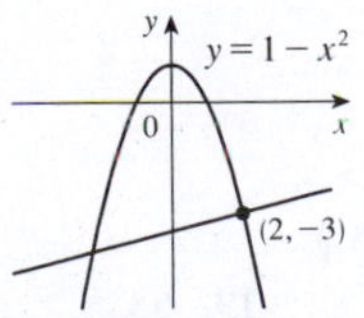

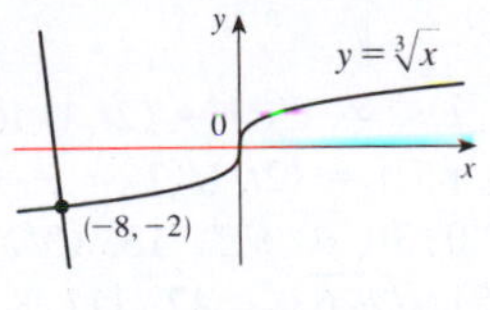

57. $(-\frac{1}{4}, \frac{1}{256})$ **59.** (a) −16 (b) $-\frac{20}{9}$ (c) 20
61. (a) 0 (b) $-\frac{2}{3}$
65. $y' = (x^4 + x + 1)(2x - 3)/(2\sqrt{x})$
$+ \sqrt{x}\,[(4x^3 + 1)(2x - 3) + 2(x^4 + x + 1)]$
67. No

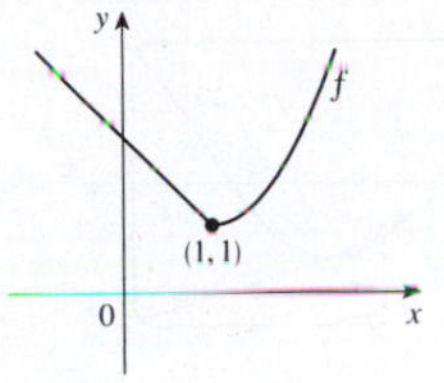

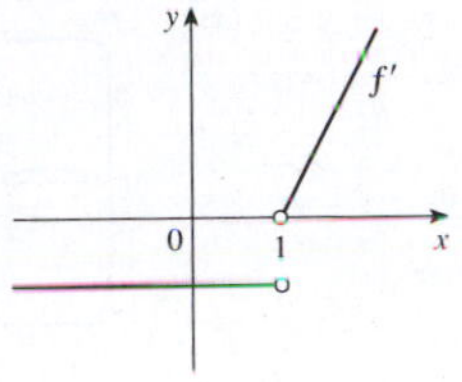

69. (a) Not differentiable at 3 or −3
$f'(x) = \begin{cases} 2x & \text{if } |x| > 3 \\ -2x & \text{if } |x| < 3 \end{cases}$
(b)

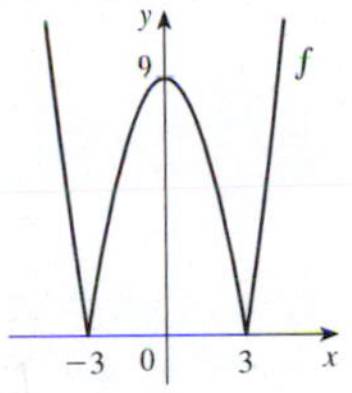

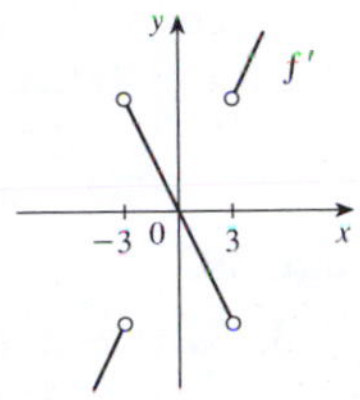

71. $a = -\frac{1}{2}$, $b = 2$ **75.** 1000

Exercises 3.3 ■ page 178

1. (a) $2t - 6$ (b) −2 ft/s (c) $t = 3$ (d) $t > 3$
(e) 10 ft (f)

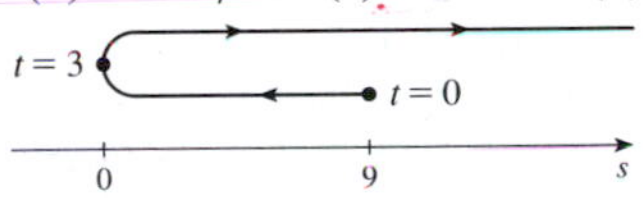

3. (a) $6t^2 - 18t + 12$ (b) 0 ft/s (c) $t = 1, 2$
(d) $0 \leqslant t < 1$, $t > 2$ (e) 34 ft
(f)

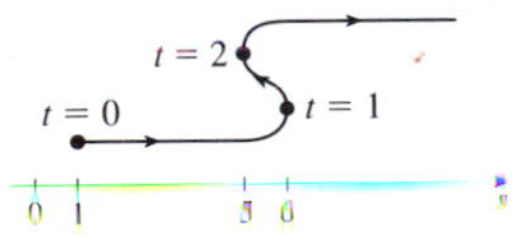

5. (a) $(1 - t^2)/(t^2 + 1)^2$ (b) $-\frac{3}{25}$ ft/s (c) $t = 1$
(d) $0 \leqslant t < 1$ (e) $\frac{13}{17}$ ft (f)

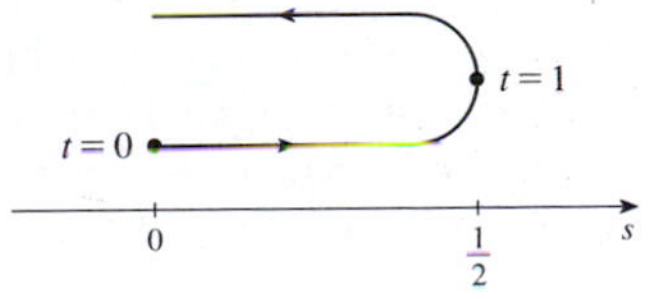

7. $t = 4$ s
9. (a) (i) 91 (ii) 76.51 (iii) 75.1501 (b) 75
11. (a) 7200π cm²/s (b) $21{,}600\pi$ cm²/s (c) $36{,}000\pi$ cm²/s
13. (a) 8π ft²/ft (b) 16π ft²/ft (c) 24π ft²/ft
15. (a) 6 kg/m (b) 12 kg/m (c) 18 kg/m
17. (a) 4.75 A (b) 5 A **19.** (a) $dV/dP = -C/P^2$
21. (a) $a^2k/(akt + 1)^2$ **23.** −0.04
25. ≈ −92.6 (cm/s)/cm
27. $C'(x) = 1.5 + 0.004x$; \$1.90, \$1.902
29. $C'(x) = 3 + 0.02x + 0.0006x^2$; \$11, \$11.07

Exercises 3.4 ■ page 185

1. 1 **3.** $(\sqrt{3} - 1)/2$ **5.** $2\sqrt{2}/(3\pi)$ **7.** 5 **9.** sin 1
11. $1/\pi$ **13.** 0 **15.** $\frac{1}{2}$ **21.** $\cos x - \sin x$

23. $-\csc x \cot^2 x - \csc^3 x$ **25.** $(x \sec^2 x - \tan x)/x^2$
27. $(\sin x + \cos x + x \sin x - x \cos x)/(1 + \sin 2x)$
29. $x^{-4} \sin x(-3 \tan x + x + x \sec^2 x)$
31. $(2x \tan x + x^2)/\sec x$ **33.** $4x - 2y = \pi - 2$
35. (a) $y = -x$ (b)

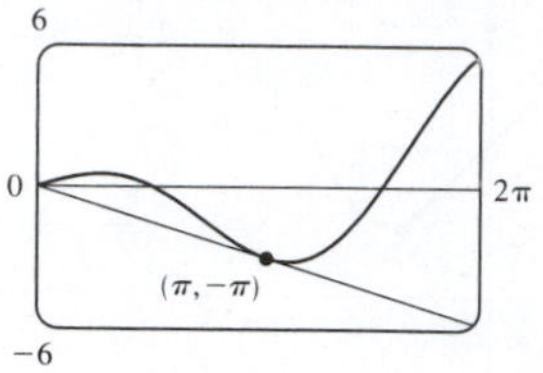

37. $(2n + 1)\pi \pm \pi/3$, n an integer **39.** 5 ft/rad
43. $\frac{1}{2}$ **45.** $\frac{1}{2}$ **47.** $\frac{1}{2}$ **49.** -1 **51.** 1 **53.** 1
55. (a) $\sec^2 x = 1/\cos^2 x$
(b) $\sec x \tan x = (\sin x)/\cos^2 x$
(c) $\cos x - \sin x = (\cot x - 1)/\csc x$
57. 1

Exercises 3.5 ■ page 192

1. $4u(x + 1)$, 48 **3.** $3u^2(1 - 1/x^2)$, 0
5. $F'(x) = 10(x^2 + 4x + 6)^4(x + 2)$
7. $G'(x) = 6(3x - 2)^9(5x^2 - x + 1)^{11}(85x^2 - 51x + 9)$
9. $f'(t) = -16(2t^2 - 6t + 1)^{-9}(2t - 3)$
11. $g'(x) = (2x - 7)/(2\sqrt{x^2 - 7x})$
13. $h'(t) = \frac{3}{2}(t - 1/t)^{1/2}(1 + 1/t^2)$
15. $F'(y) = 39(y - 6)^2/(y + 7)^4$
17. $f'(z) = -\frac{2}{5}(2z - 1)^{-6/5}$
19. $y' = 8(2x - 5)^3(8x^2 - 5)^{-4}(-4x^2 + 30x - 5)$
21. $y' = 3 \sec^2 3x$ **23.** $y' = -3x^2 \sin(x^3)$
25. $y' = -12 \cos x \sin x\,(1 + \cos^2 x)^5$
27. $y' = -\sin(\tan x) \sec^2 x$ **29.** $y' = 0$
31. $y' = -(1/3) \csc(x/3) \cot(x/3)$
33. $y' = 3 \sin x \cos x\,(\sin x - \cos x)$
35. $y' = -\cos(1/x)/x^2$ **37.** $y' = 4(\cos 2x)/(1 - \sin 2x)^2$
39. $y' = 6x^2 \tan(x^3) \sec^2(x^3)$ **41.** $y' = 0$
43. $y' = [1 + 1/(2\sqrt{x})]/(2\sqrt{x + \sqrt{x}})$
45. $f'(x) = 9[x^3 + (2x - 1)^3]^2(9x^2 - 8x + 2)$
47. $y' = \cos(\tan\sqrt{\sin x})(\sec^2\sqrt{\sin x})[1/(2\sqrt{\sin x})](\cos x)$
49. $y = 0$ **51.** $3x + 16y = 44$
53. (a) $y = \pi x - \pi + 1$
(b)

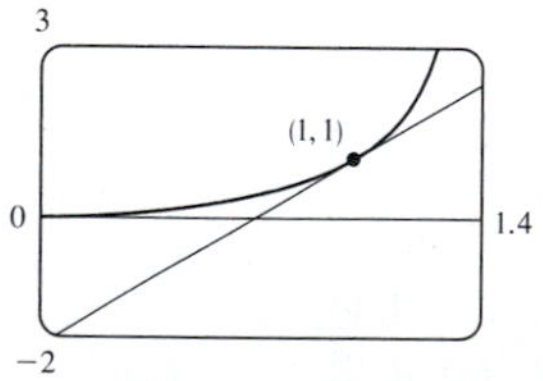

55. (a) $-1/(x^2\sqrt{1 - x^2})$
57. $((\pi/2) + 2n\pi, 3)$, $((3\pi/2) + 2n\pi, -1)$, n any integer
59. 28 **61.** $v(t) = (5\pi/2) \cos(10\pi t)$ cm/s
63. (a) $dB/dt = (7\pi/54) \cos(2\pi t/5.4)$ (b) 0.16
65. (a) On $(0, \infty)$ (b) $G'(x) = h'(\sqrt{x})/(2\sqrt{x})$
67. (a) $F'(x) = -\sin x\, f'(\cos x)$ (b) $G'(x) = -\sin(f(x)) f'(x)$
69. 0 **75.** $f'(x) = x/|x|$
77. $h'(x) = |2x - 1| + 2x(2x - 1)/|2x - 1|$

Exercises 3.6 ■ page 198

1. (a) $y' = -(2x + y + 3)/x$
(b) $y = (5/x) - x - 3$, $y' = -(5/x^2) - 1$
3. (a) $y' = (2x - y)/(x + 4y)$
(b) $y = (-x \pm \sqrt{9x^2 + 24})/4$, $y' = (-1 \pm 9x/\sqrt{9x^2 + 24})/4$
5. $y' = (y - 2x)/(3y^2 - x)$
7. $y' = (18x - x^{-2/3}y^{1/3})/(12y + x^{1/3}y^{-2/3})$
9. $y' = -x^3/y^3$
11. $y' = (y/x) + 2(x - y)^2$ [or $(3x^2 + 1 - 2xy)/(x^2 + 2)$]
13. $y' = [\sin(x - y) + y \cos x]/[\sin(x - y) - \sin x]$
15. $y' = -y/x$
17. $dx/dy = (1 - 4y^3 - 2x^2y - x^4)/(2xy^2 + 4yx^3)$ **19.** $-\frac{1}{6}$
21. $5x + 4y + 16 = 0$ **23.** $y = x$
25. $9x + 13y - 40 = 0$
27. (a) $9x - 2y - 5 = 0$ (b)

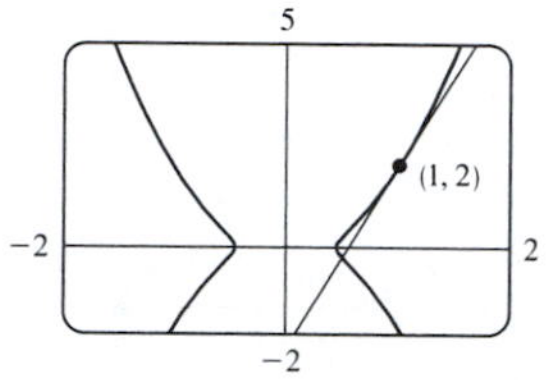

29. $(\pm 5\sqrt{3}/4, \pm 5/4)$ **31.** $(x_0x/a^2) - (y_0y/b^2) = 1$
37.

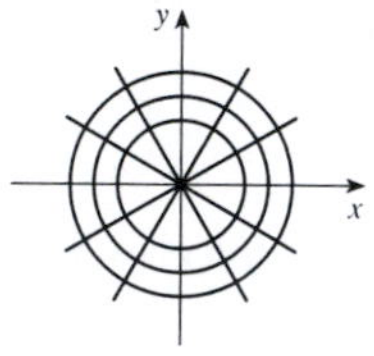

39.

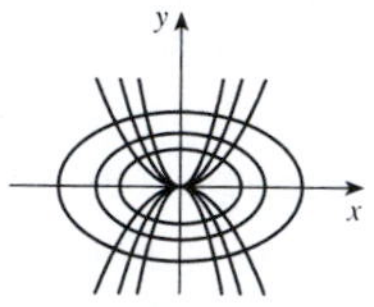

41. $(\pm\sqrt{3}, 0)$ **43.** $(-1, -1)$, $(1, 1)$ **45.** 2

Exercises 3.7 ■ page 203

1.

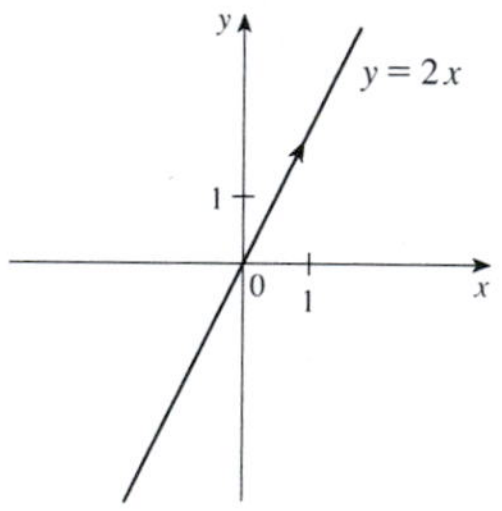

3.

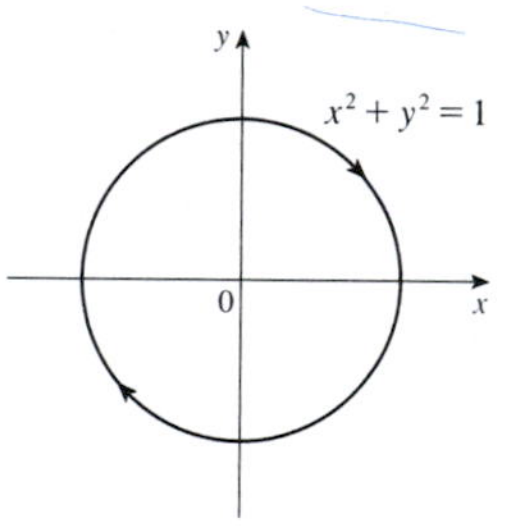

5.

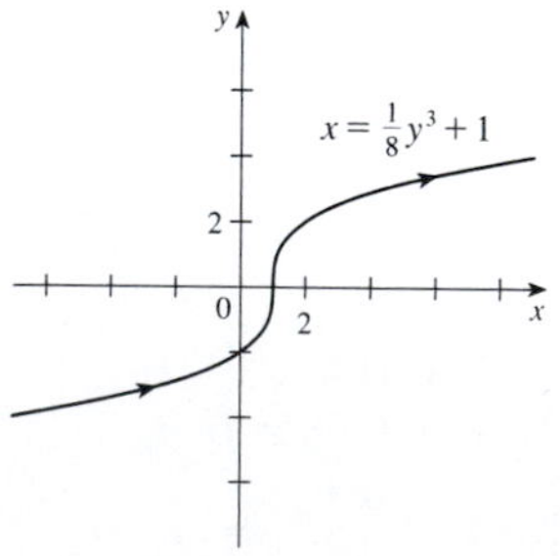

7. $-\infty < t < \infty$, $\mathbf{r}'(t) = \langle 2t, 1 \rangle$ for $t \in \mathbb{R}$
9. $t \geqslant 4$, $\mathbf{r}'(t) = \langle 2t, 1/(2\sqrt{t - 4}) \rangle$ for $t > 4$
11. $\langle 1/\sqrt{10}, 3/\sqrt{10} \rangle$ **13.** $\langle 3/\sqrt{13}, 2/\sqrt{13} \rangle$
15. $\langle 1, 15 \rangle$, $\sqrt{226}$ **17.** $11\mathbf{i} + 4\mathbf{j}$, $\sqrt{137}$ **19.** 63°

Exercises 3.8 ■ page 208

1. $a = f, b = f', c = f''$
3. $f'(x) = 4x^3 - 9x^2 + 16,\ f''(x) = 12x^2 - 18x$
5. $h'(x) = x/\sqrt{x^2 + 1},\ h''(x) = 1/(x^2 + 1)^{3/2}$
7. $F'(s) = 24(3s + 5)^7,\ F''(s) = 504(3s + 5)^6$
9. $y' = 1/(1 - x)^2,\ y'' = 2/(1 - x)^3$
11. $y' = -\frac{3}{2}x(1 - x^2)^{-1/4},\ y'' = \frac{3}{4}(1 - x^2)^{-5/4}(x^2 - 2)$
13. $H'(t) = 6\tan^2(2t - 1)\sec^2(2t - 1)$,
$H''(t) = 24\tan(2t - 1)\sec^4(2t - 1)$
$+ 24\tan^3(2t - 1)\sec^2(2t - 1)$
15. (a) $f'(x) = 2\sin x\,(\cos x - 1) = \sin 2x - 2\sin x$,
$f''(x) = 2(\cos 2x - \cos x)$
17. $\frac{375}{8}(5t - 1)^{-5/2}$
19. $1/\sqrt{2},\ 3/(4\sqrt{2}),\ 27/(16\sqrt{2}),\ 405/(64\sqrt{2})$
21. -80 **23.** $-2x/y^5$ **25.** $64/(3x + y)^3$
27. $f'(x) = 1 - 2x + 3x^2 - 4x^3 + 5x^4 - 6x^5$,
$f''(x) = -2 + 6x - 12x^2 + 20x^3 - 30x^4$,
$f'''(x) = 6 - 24x + 60x^2 - 120x^3$,
$f^{(4)}(x) = -24 + 120x - 360x^2$,
$f^{(5)}(x) = 120 - 720x,\ f^{(6)}(x) = -720$,
$f^{(n)}(x) = 0$ for $7 \leqslant n \leqslant 73$
29. $n!$ **31.** $(-1)^n(n + 2)!/(6x^{n+3})$ **33.** $-2^{50}\cos 2x$
35. (a) $v(t) = 3t^2 - 3,\ a(t) = 6t$ (b) $6\ \text{m/s}^2$
(c) $a(1) = 6\ \text{m/s}^2$
37. (a) $v(t) = 2At + B,\ a(t) = 2A$ (b) $2A\ \text{m/s}^2$
(c) $2A\ \text{m/s}^2$
39. (a) $t = 0, 2$ (b) $s(0) = 2$ m, $v(0) = 0$ m/s,
$s(2) = -14$ m, $v(2) = -16$ m/s
41. (a), (c)

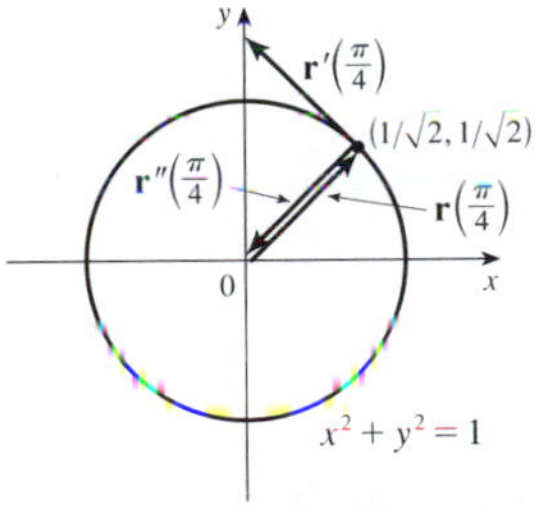

(b) $\langle -\sin t, \cos t\rangle$, $\langle -\cos t, -\sin t\rangle$

43. (a), (c)

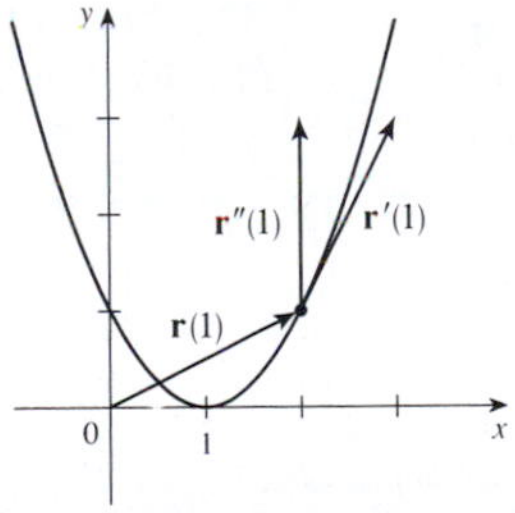

(b) $\langle 1, 2t\rangle$, $\langle 0, 2\rangle$

45. $\langle 0, -10\rangle$ **47.** $12\,\mathbf{i} + 2\,\mathbf{j}$
49. $\mathbf{v}(t) \approx \langle -27{,}030\sin 3.56t, 27{,}030\cos 3.56t\rangle$,
$|\mathbf{v}| \approx 27{,}030$ km/h,
$\mathbf{a}(t) \approx \langle -96{,}131\cos 3.56t, -96{,}131\sin 3.56t\rangle$
51. $\mathbf{v}(t) \approx \langle 500, -866\rangle,\ |\mathbf{v}| = 1000$ ft/s, $\mathbf{a}(t) = \langle 0, -32\rangle$
53. (a) $v(t) = A\omega\cos\omega t,\ a(t) = -A\omega^2\sin\omega t$
55. $P(x) = x^2 - x + 3$
59. $f''(x) = 6xg'(x^2) + 4x^3g''(x^2)$
61. $f''(x) = [\sqrt{x}\,g''(\sqrt{x}) - g'(\sqrt{x})]/(4x\sqrt{x})$
63. (a) $f'(x) = -(2x + 1)/(x^2 + x)^2$,
$f''(x) = 2(3x^2 + 3x + 1)/(x^2 + x)^3$,
$f'''(x) = -6(4x^3 + 6x^2 + 4x + 1)/(x^2 + x)^4$,
$f^{(4)}(x) = 24(5x^4 + 10x^3 + 10x^2 + 5x + 1)/(x^2 + x)^5$
(b) $f^{(n)}(x) = (-1)^n n![x^{-(n+1)} - (x + 1)^{-(n+1)}]$

Exercises 3.9 ■ page 214

1. $x + y = 0$ **3.** $y = \frac{1}{36}x + \frac{13}{9}$ **5.** $y = (1/\pi)x - \pi$
7. $y = 1$ **9.** $y = -\frac{3}{4}x + \frac{25}{4}$
11. Horizontal at $(1, 5), (1, -1)$; vertical at $(-1, 2), (3, 2)$
13. Horizontal at $(0, -9)$; vertical at $(\pm 2, -6)$
15. Horizontal at $(0, 0), (2^{1/3}, 2^{2/3})$; vertical at $(2^{2/3}, 2^{1/3}), (0, 0)$
17. $y = x,\ y = -x$

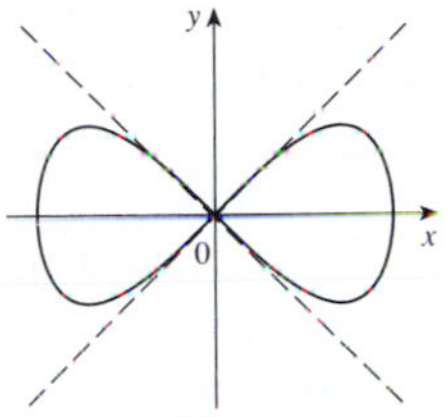

19. $(0, 0)$; $y = -\sqrt{3}\,x,\ y = \sqrt{3}\,x$ **21.** $(-5, 6), (-\frac{208}{27}, \frac{32}{3})$
25. (b)

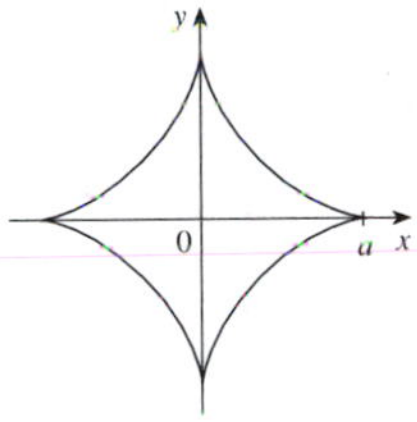

(c)

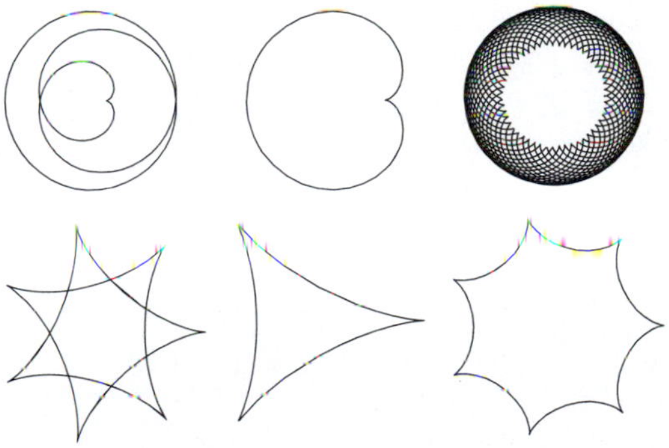

27. (a) $d\sin\theta/(r - d\cos\theta)$

Exercises 3.10 ■ page 219

1. $dV/dt = 3x^2\,dx/dt$ **3.** -1 **5.** $1/(50\pi)$ cm/min
7. (a) $\frac{25}{3}$ ft/s (b) $\frac{10}{3}$ ft/s **9.** $250\sqrt{3}$ mi/h **11.** 65 mi/h
13. $\frac{720}{13} \approx 55.4$ km/h **15.** -1.6 cm/min
17. $(10{,}000 + 800{,}000\pi/9) \approx 2.89 \times 10^5\ \text{cm}^3/\text{min}$
19. $\frac{10}{3}$ cm/min **21.** $6/(5\pi)$ ft/min **23.** $0.3\ \text{m}^2/\text{s}$
25. $80\ \text{cm}^3/\text{min}$ **27.** (a) 360 ft/s (b) 0.096 rad/s
29. $\sqrt{2}/5$ rad/s **31.** $1650/\sqrt{31} \approx 296$ km/h
33. $7\sqrt{15}/4 \approx 6.78$ m/s

Exercises 3.11 ■ page 227

1. $dy = 5x^4\,dx$ **3.** $dy = [(2x^3 + x)/\sqrt{x^4 + x^2 + 1}\,]\,dx$

5. $dy = 2\cos 2x\,dx$ **7.** (a) $dy = -2x\,dx$ (b) -5
9. (a) $dy = 6x(x^2+5)^2\,dx$ (b) 10.8
11. (a) $dy = -\sin x\,dx$ (b) -0.025
13. $\Delta y = 1.25$, $dy = 1$ **15.** $\Delta y = 1.44$, $dy = 1.6$

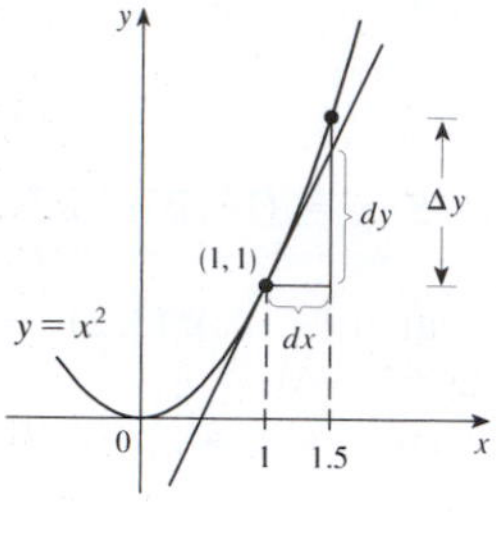

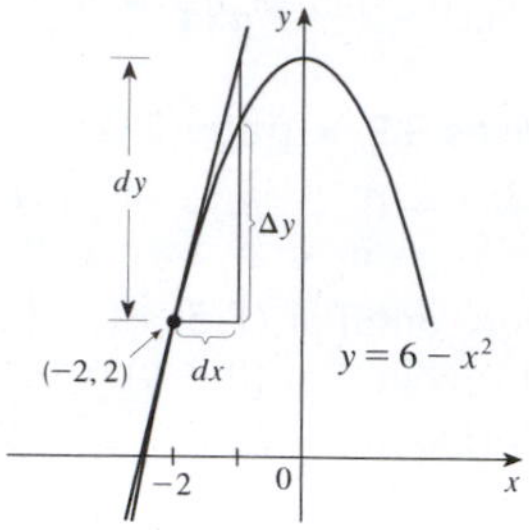

17. $\Delta y = 77, 33.25, 5.882, 0.571802$;
$dy = 57, 28.5, 5.7, 0.57$; $\Delta y - dy = 20, 4.75, 0.182, 0.001802$
19. $6 + \frac{1}{120} \approx 6.0083$ **21.** 0.099
23. 0.857 **25.** (a) 270 cm^3 (b) 36 cm^2
27. (a) $84/\pi \approx 27\text{ cm}^2$ (b) $\frac{1}{84} \approx 0.012 = 1.2\%$
29. (a) $2\pi rh\,\Delta r$ (b) $\pi(\Delta r)^2 h$
31. $L(x) = 3x - 2$ **33.** $L(x) = \frac{1}{2} - \frac{1}{16}x$
39. $\sqrt{1-x} \approx 1 - \frac{1}{2}x$,
$\sqrt{0.9} \approx 0.95$,
$\sqrt{0.99} \approx 0.995$

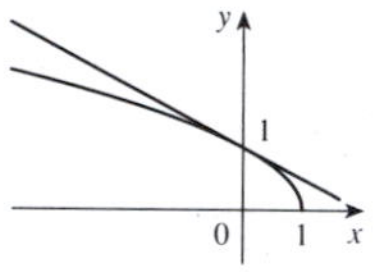

41. $-0.69 < x < 1.09$ **43.** $-0.045 < x < 0.055$
45. $1/x \approx \frac{1}{4} - \frac{1}{16}(x-4) + \frac{1}{64}(x-4)^2$ **47.** $\sec x \approx 1 + \frac{1}{2}x^2$
49. $\sqrt{x} \approx 1 + \frac{1}{2}(x-1),\ 1 + \frac{1}{2}(x-1) - \frac{1}{8}(x-1)^2$
51. (a) $\cos x \approx \sqrt{3}/2 - \frac{1}{2}(x - \pi/6)$,
$\sqrt{3}/2 - \frac{1}{2}(x-\pi/6) - (\sqrt{3}/4)(x-\pi/6)^2$
(b)

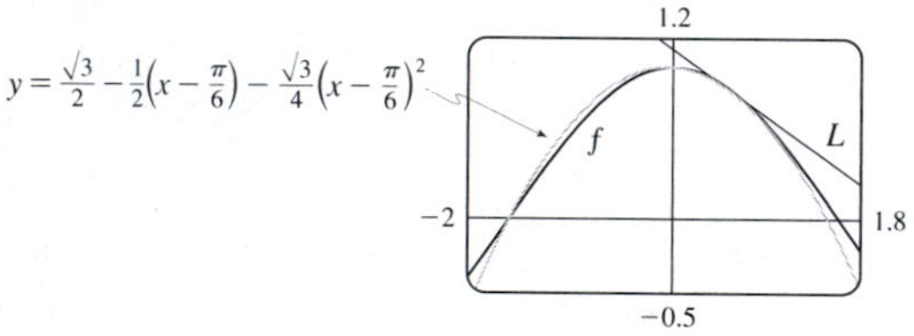

(c) $0.06 < x < 1.03$ (d) $-1.82 < x < 1.48$
55. $x - x^3/6$

Exercises 3.12 ■ page 232

1. $x_2 \approx 2.3$, $x_3 \approx 3$ **3.** -0.6860 **5.** 1.5850
7. 2.165737 **9.** 1.618034 **11.** 1.895494
13. $-2.114908, 0.254102, 1.860806$ **15.** $1, -0.569840$
17. 0, 1.109144, 3.698154 **19.** -3.20614267, 1.37506470 **21.** 0.15438500, 0.84561500
23. (b) 31.622777 **29.** 11.28 ft **31.** 0.76286%

Chapter 3 Review Exercises ■ page 234

1. False **3.** False **5.** True **7.** False **9.** True
11. False **13.** $f'(x) = 3x^2 + 5$
15. $f'(x) = -5/(2\sqrt{3-5x})$
17. $y' = 2(7x+18)(x+2)^7(x+3)^5$
19. $y' = (9-2x)/(9-4x)^{3/2}$
21. $y' = (1 - 2xy^3)/(3x^2y^2 + 6y + 4)$ **23.** $y' = \frac{7}{8}x^{-1/8}$
25. $y' = 8/(8-3x)^2$ **27.** $y' = \frac{1}{5}(x\tan x)^{-4/5}(\tan x + x\sec^2 x)$
29. $y' = 2x/(2y+1)$ **31.** $y' = 2(2x-5)/[(x-2)^2(x-3)^2]$
33. $y' = -(\sec^2\sqrt{1-x})/(2\sqrt{1-x})$
35. $y' = \cos(\tan\sqrt{1+x^3})(\sec^2\sqrt{1+x^3})(3x^2/(2\sqrt{1+x^3}))$
37. $y' = -6x\csc^2(3x^2+5)$ **39.** $y' = -\sin(2\tan x)\sec^2 x$
41. -120 **43.** $-5x^4/y^{11}$ **45.** 1 **47.** $3x + 2y - 8 = 0$
49. $4x - y + \sqrt{3} - (4\pi/3) = 0$
51. $(\pi/4, \sqrt{2}), (5\pi/4, -\sqrt{2})$ **55.** (a) 2 (b) 44
57. $a = f$, $c = f'$, $b = f''$
59. (a) $(10-3x)/(2\sqrt{5-x})$
(b) $7x - 4y + 1 = 0$, $x + y - 8 = 0$
(c)

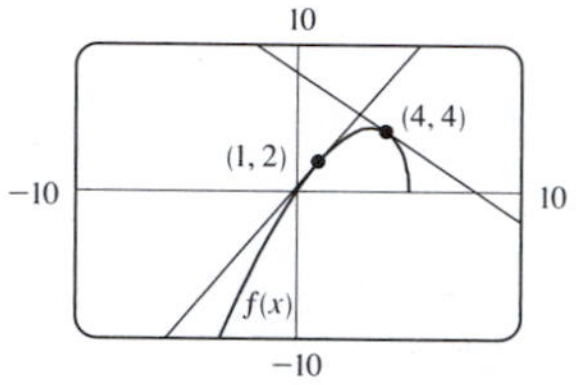

61. $f'(x) = 2xg(x) + x^2g'(x)$ **63.** $f'(x) = 2g(x)g'(x)$
65. $f'(x) = g'(g(x))\,g'(x)$
67. $h'(x) = (f'(x)[g(x)]^2 + g'(x)[f(x)]^2)/[f(x) + g(x)]^2$
69. $h'(x) = f'(g(\sin 4x))\,g'(\sin 4x)(\cos 4x)(4)$
71. (a) $v(t) = 3t^2 - 12$, $a(t) = 6t$ (b) Upward when $t > 2$, downward when $0 \le t < 2$ (c) 23
73. (a), (c)

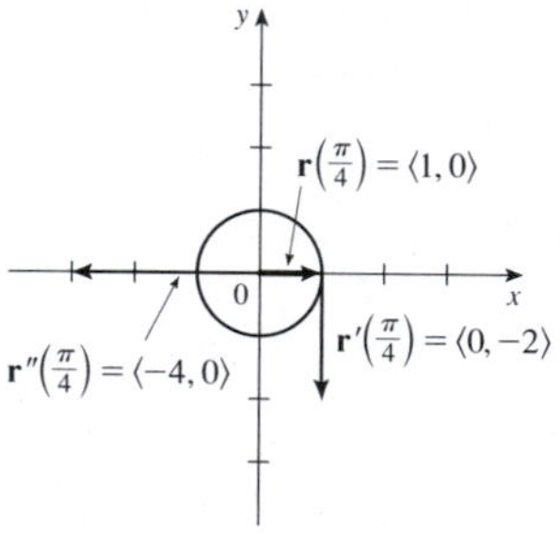

(b) $\langle 2\cos 2t, -2\sin 2t\rangle$, $\langle -4\sin 2t, -4\cos 2t\rangle$
75. (a), (c)

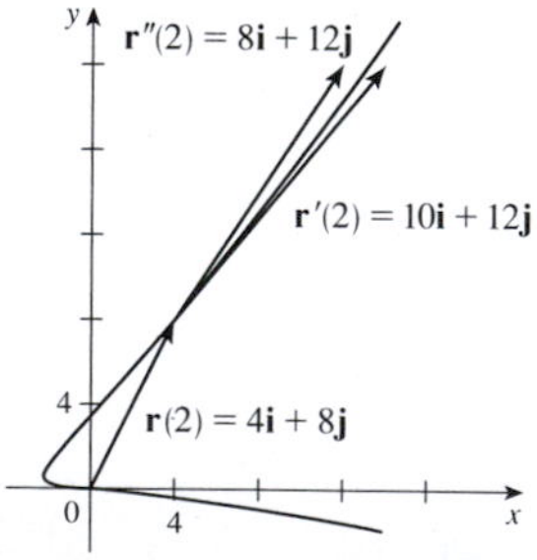

(b) $(8t-6)\,\mathbf{i} + 3t^2\,\mathbf{j}$, $8\,\mathbf{i} + 6t\,\mathbf{j}$
77. $-3\,\mathbf{i} - 4\,\mathbf{j}$, $-2\,\mathbf{i} - 2\,\mathbf{j}$, $2\,\mathbf{i} + 6\,\mathbf{j}$, $2\sqrt{2}$
79. $\langle \frac{3}{2}\sqrt{2}, -2\sqrt{2}\rangle$, $\langle -3\sqrt{2}, -4\sqrt{2}\rangle$, $\langle -6\sqrt{2}, 8\sqrt{2}\rangle$, $5\sqrt{2}$
81. $\mathbf{L}(t) = \langle 3, 0\rangle + t\langle 4, 2\rangle$
83. $$\mathbf{L}(t) = \left[\left(\frac{\sqrt{2}}{2} - \frac{\sqrt{2}\,\pi}{8}\right)t + \frac{\sqrt{2}\,\pi}{8}\right]\mathbf{i} + \left[\left(\frac{\sqrt{2}}{2} + \frac{\sqrt{2}\,\pi}{8}\right)t + \frac{\sqrt{2}\,\pi}{8}\right]\mathbf{j}$$

85. 4 kg/m **87.** $\frac{4}{3}$ cm²/min **89.** 13 ft/s
91. 400 ft/h **93.** 0.8
95. $L(x) = 1 + x$, $\sqrt[3]{1 + 3x} \approx 1 + x$, $\sqrt[3]{1.03} \approx 1.01$
97. $-0.23 < x < 0.40$ **99.** 0.724492
101. $(0.9340, -2.0634)$ **103.** 192 **105.** 27 **107.** $\frac{1}{4}$

PROBLEMS PLUS ■ page 239

1. $(\pm\sqrt{3}/2, \frac{1}{4})$ **3.** True for all x **5.** -4
9. $x_T \in (3, \infty)$, $y_T \in (2, \infty)$, $x_N \in (0, \frac{5}{3})$, $y_N \in (-\frac{5}{2}, 0)$
11. (a) $[-1, 2]$
(b) $-1/(8\sqrt{3 - x}\,\sqrt{2 - \sqrt{3 - x}}\,\sqrt{1 - \sqrt{2 - \sqrt{3 - x}}})$
13. (a)

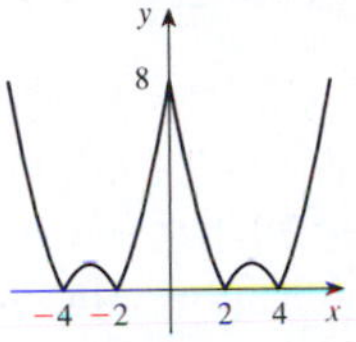

(b) 0, ±2, ±4

17. (a) 0 (b) 4 (c) $\frac{1}{2}$ **19.** $-\sin a$
21. (b) (i) 53° (or 127°) (ii) 63° (or 117°)
23. It approaches the midpoint of the radius AO.
25. $(1, -2)$, $(-1, 0)$ **27.** $\sqrt{29}/58$

CHAPTER 4

Exercises 4.1 ■ page 248

1.

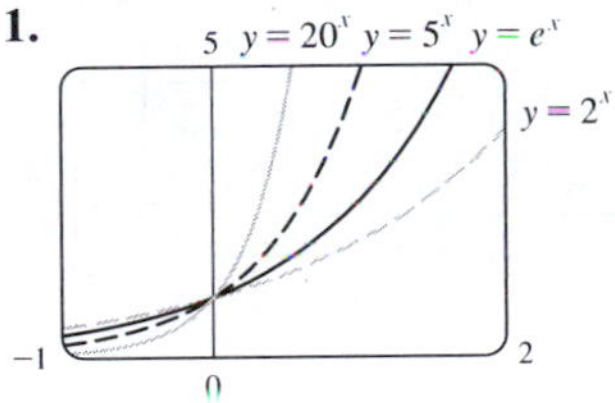

All approach 0 as $x \to -\infty$, all pass through (0, 1), and all increase. The larger the base, the faster the rate of increase for $x > 0$.

3.

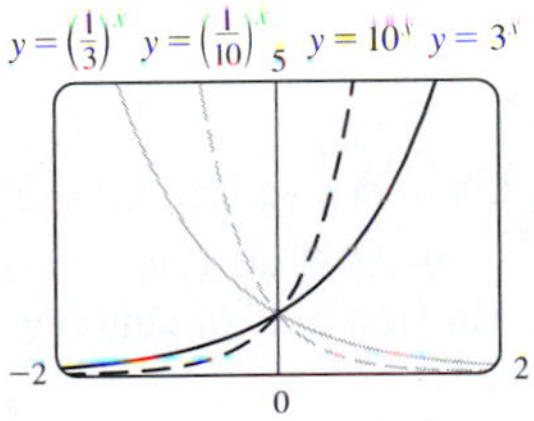

The functions with base greater than 1 are increasing and those with base less than 1 are decreasing. The latter are reflections of the former about the y-axis.

5.

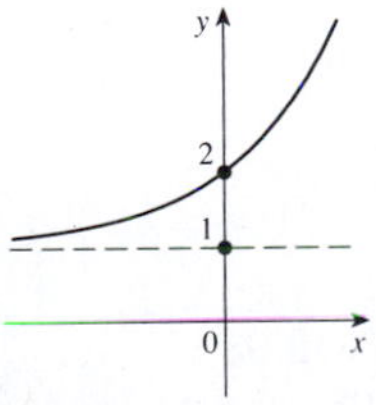

7.

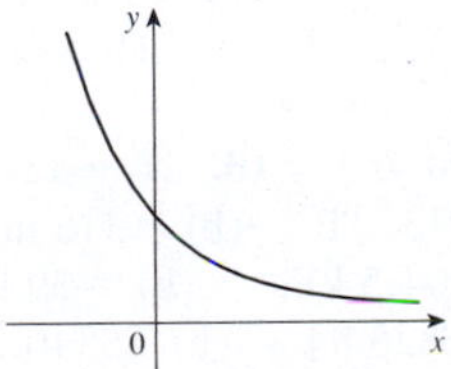

9.

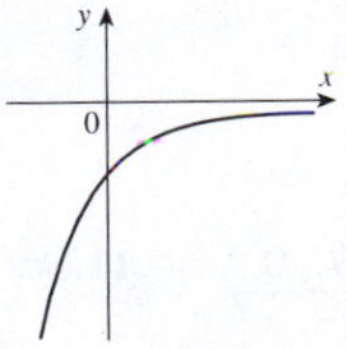

11.

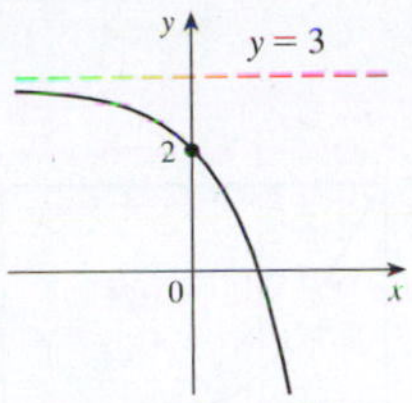

13. ∞ **15.** ∞ **17.** 1
19. 0 **21.** 0

25. (a) 1.4870, 1.3959, 1.3873, 1.3864; slope of a secant line
(b) 1.39
(c) Slope of the tangent line to the curve $y = 4^x$ at (0, 1)
27. $f'(x) = e^{\sqrt{x}}/(2\sqrt{x})$ **29.** $y' = e^{2x}(1 + 2x)$
31. $h'(t) = -e^t/(2\sqrt{1 - e^t})$
33. $y' = e^{x\cos x}(\cos x - x\sin x)$
35. $y' = e^{-1/x}/x^2$ **37.** $y' = 3e^{3x-2}\sec^2(e^{3x-2})$
39. $y' = (3e^{3x} + 2e^{4x})/(1 + e^x)^2$ **41.** $y' = ex^{e-1}$
43. $x + e^{\pi}y = \pi$ **45.** $y' = 1 + e^y(1 + x)/\sin(x - y)$
49. $r = 1, -6$ **51.** $256e^{-2x}$ **53.** (b) -0.567143
55. (a) 1 (b) $kae^{-kt}/(1 + ae^{-kt})^2$
(c)

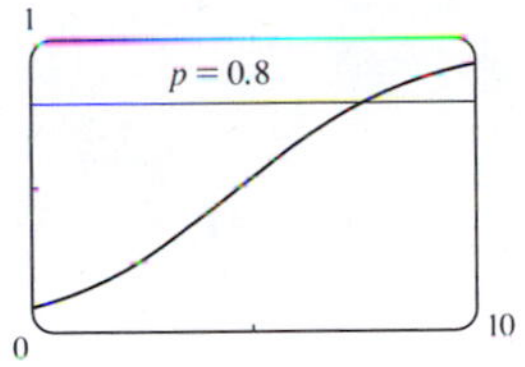

$t \approx 7.4$ h

57. -1 **59.**

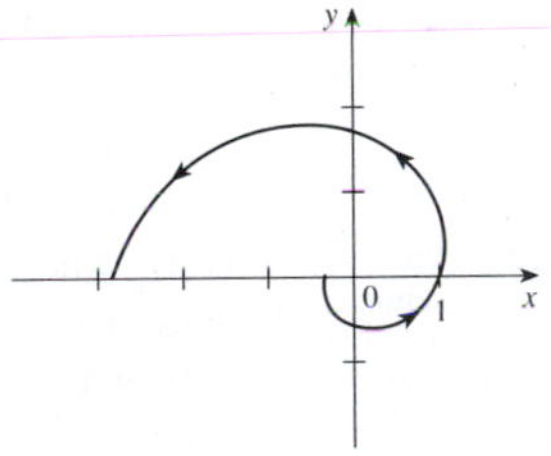

61. $-e^{3t}(1 + 2t)$

Exercises 4.2 ■ page 256

1. No **3.** Yes **5.** No **7.** Yes **9.** Yes **11.** No
13. $f^{-1}(x) = (x - 7)/4$ **15.** $f^{-1}(x) = (5x - 1)/(2x + 3)$
17. $f^{-1}(x) = (x^2 - 2)/5$, $x \geq 0$
19. (b) $\frac{1}{2}$
(c) $g(x) = (x - 1)/2$,
domain = ℝ = range
(e)

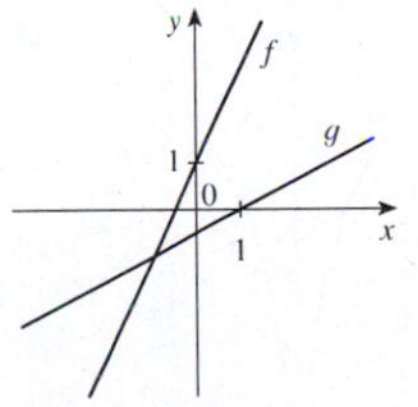

21. (b) $\frac{1}{12}$
(c) $g(x) = \sqrt[3]{x}$,
domain = ℝ = range
(e)

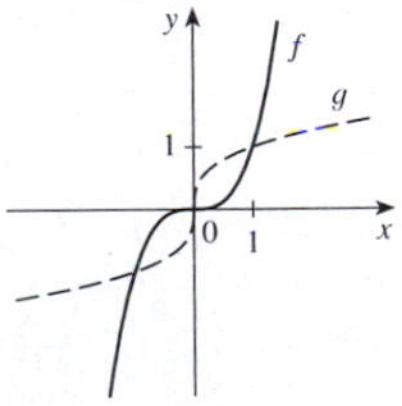

23. (b) $-\frac{1}{2}$
(c) $g(x) = \sqrt{9 - x}$,
domain = [0, 9],
range = [0, 3]
(e) See graph at right.

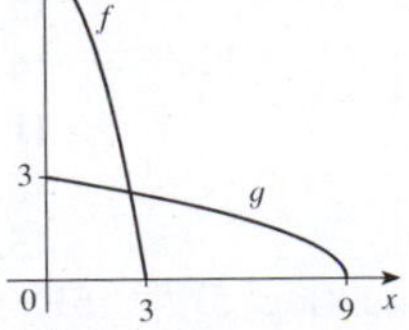

25. 1 **27.** $2/\pi$ **29.** 1 **31.** $\frac{3}{2}$

33. $f^{-1}(x) = \sqrt{2/(1-x)}$ **35.**

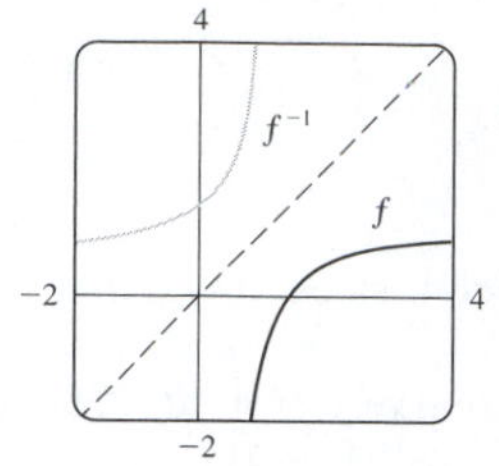

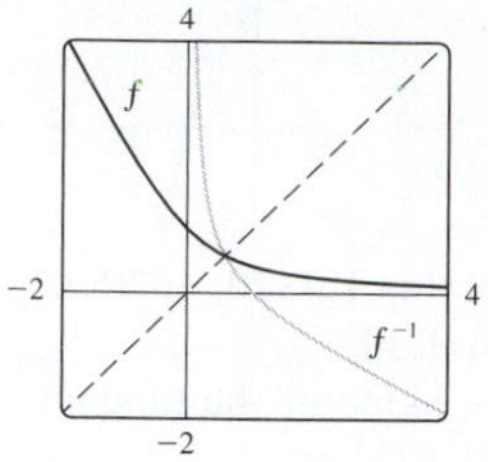

37. (a) $f^{-1}(x) = (x + \sqrt[5]{x})^5$

(b)

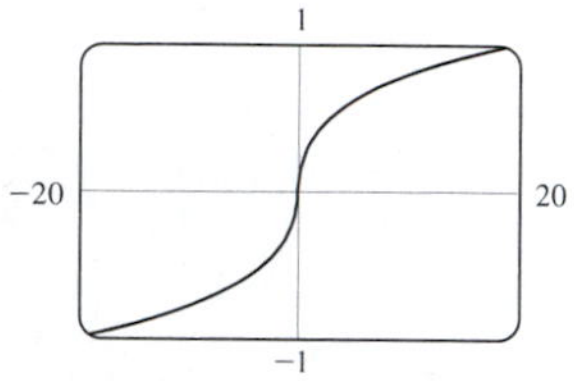

Exercises 4.3 ■ page 262

1. 6 **3.** $\frac{1}{3}$ **5.** −3 **7.** $\sqrt{2}$ **9.** 2 **11.** 1

13. 15 **15.** $\log_5(ab/c)$ **17.** $3\ln 2$

19. $\ln[\sqrt[3]{x}/(2x+3)^4]$

21. (a) 2.321928 (b) 2.025563 (c) 0.910239

(d) −7.399054

23.

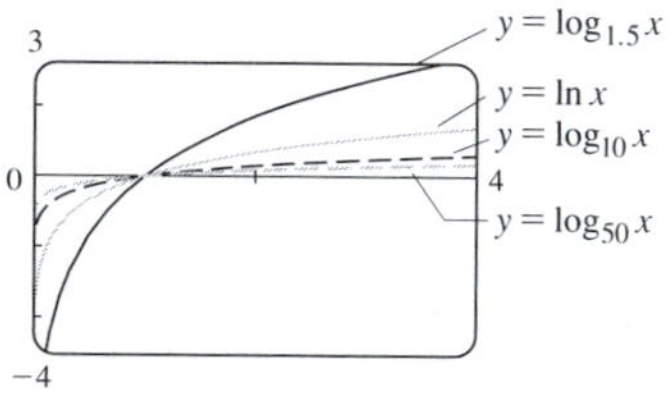

All approach $-\infty$ as $x \to 0^+$, all pass through (1, 0), and all increase. The larger the base, the slower the rate of increase for $x > 0$.

25.

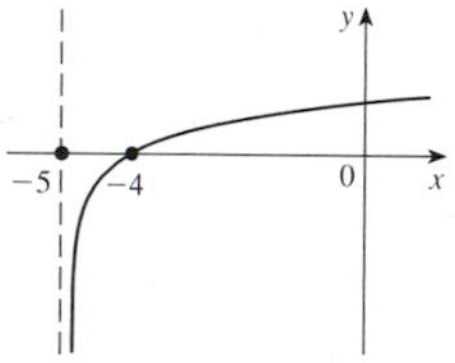

27.

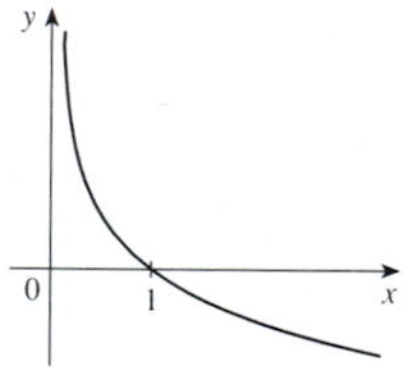

29.

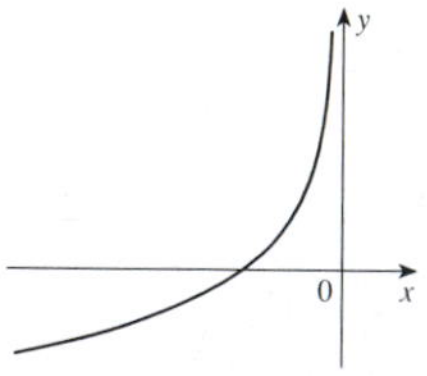

31.

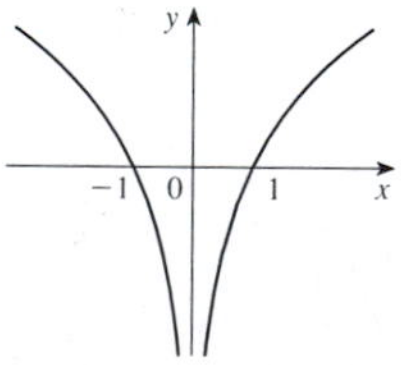

33.

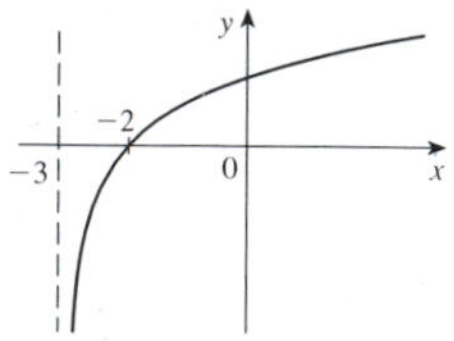

35. 8

37. $4\ln 2$

39. $(e^3+1)/2$

41. $\log_3 m - 2$

43. 40

45. 3

47. e^e **49.** $\log_3(\log_2 5)$ **51.** 4 **53.** $(\ln C)/(a-b)$

55. 25.0855 **57.** −0.3319 **59.** About 1,084,588 mi

61. 8.3 **63.** $-\infty$ **65.** ∞ **67.** $-\infty$ **69.** 0

71. $(-\infty, 1)$, $(-\infty, \infty)$ **73.** $(1, \infty)$, $(-\infty, \infty)$

75. $y = e^x - 3$ **77.** $y = (\ln x)^2$, $x \geq 1$

79. $y = \log_{10}[x/(1-x)]$

81. (b) $y = \frac{1}{2}(e^x - e^{-x})$ **83.** $\log_9 82$

87. $-1 \leq x < 1 - \sqrt{3}$ or $1 + \sqrt{3} < x \leq 3$

Exercises 4.4 ■ page 269

1. $f'(x) = 1/(x+1)$, $(-1, \infty)$, $(-1, \infty)$

3. $f'(x) = 2x\ln(1-x^2) - 2x^3/(1-x^2)$, $(-1, 1)$, $(-1, 1)$

5. $f'(x) = 2x/[(x^2-4)\ln 3]$, $|x| > 2$, $|x| > 2$

7. $y' = 1 + \ln x$, $y'' = 1/x$

9. $y' = 1/(x\ln 10)$, $y'' = -1/(x^2\ln 10)$

11. $f'(x) = (2 + \ln x)/(2\sqrt{x})$ **13.** $g'(x) = -2a/(a^2 - x^2)$

15. $F'(x) = 1/(2x)$ **17.** $f'(t) = (4t^3 - 2t)/[(t^4 - t^2 + 1)\ln 2]$

19. $g'(u) = -2/[u(1 + \ln u)^2]$ **21.** $y' = 3\cot x\,(\ln\sin x)^2$

23. $y' = [1 + x^2(1 - 2\ln x)]/[x(1+x^2)^2]$

25. $y' = (3x-2)/[x(x-1)]$ **27.** $F'(x) = e^x(\ln x + 1/x)$

29. $f'(t) = -\pi^{-t}\ln\pi$ **31.** $h'(t) = 3t^2 - 3^t\ln 3$

33. $y' = -x/(1+x)$ **35.** $y' = x^{\sin x}[\cos x\ln x + (\sin x)/x]$

37. $e^x x^{e^x}(\ln x + 1/x)$ **39.** $y' = (\ln x)^x(\ln\ln x + 1/\ln x)$

41. $y' = 0$ **43.** $y' = -\sin(x^{\sqrt{x}})x^{\sqrt{x}}[(\ln x + 2)/(2\sqrt{x})]$

45. 0 **47.** $f'(x) = \cos x + 1/x$ **49.** $x - ey = e$

51. $y = 2ex + e$ **53.** $y' = 2x/(x^2 + y^2 - 2y)$

55. $f^{(n)}(x) = (-1)^{n-1}(n-1)!/(x-1)^n$ **57.** 1.309800

59. $y' = (3x-7)^4(8x^2-1)^3[12/(3x-7) + 48x/(8x^2-1)]$

61. $y' = \dfrac{(x+1)^4(x-5)^3}{(x-3)^8}\left[\dfrac{4}{x+1} + \dfrac{3}{x-5} - \dfrac{8}{x-3}\right]$

63. $y' = \dfrac{e^x\sqrt{x^5+2}}{(x+1)^4(x^2+3)^2}\left[1 + \dfrac{5x^4}{2(x^5+2)} - \dfrac{4}{x+1} - \dfrac{4x}{x^2+3}\right]$

65. $\frac{1}{3}$

Exercises 4.5 ■ page 274

1. (a) 100×2^{3t} (b) $\approx 1.07 \times 10^{11}$

(c) $(\ln 100)/(3\ln 2) \approx 2.2$ h

3. (a) $500 \times 16^{t/3}$ (b) $\approx$20,159 (c) $(3\ln 60)/\ln 16 \approx 4.4$ h

5. (a) 1403 million, 1746 million (b) 2208 million

(c) 3667 million Explanation: Wars in first half of century, increased life expectancy in second half

7. (a) $Ce^{-0.0005t}$ (b) $-2000\ln 0.9 \approx 211$ s

9. (a) $50 \times 2^{-t/0.00014}$ (b) $\approx 1.57 \times 10^{-20}$ mg

(c) $\approx 4.5 \times 10^{-5}$ s

11. $\approx$2500 yr **13.** $21 + 12e^{-t/10}$

15. (a) $\approx$137 °F (b) $\approx$116 min

17. (a) $\approx$64.5 kPa (b) $\approx$39.9 kPa

19. (a) \$3828.84 (b) \$3840.25 (c) \$3850.08

(d) \$3851.61 (e) \$3852.01 (f) \$3852.08

21. (a) $450e^{-0.4} \approx 302$ kg (b) $\approx$30.4 min

Exercises 4.6 ■ page 281

1. π **3.** $\pi/3$ **5.** $\pi/4$ **7.** $5\pi/6$ **9.** 0.7 **11.** $\pi/3$

13. $\frac{3}{5}$ **15.** $-\pi/4$ **17.** $\frac{119}{169}$ **21.** $x/\sqrt{1+x^2}$

23.

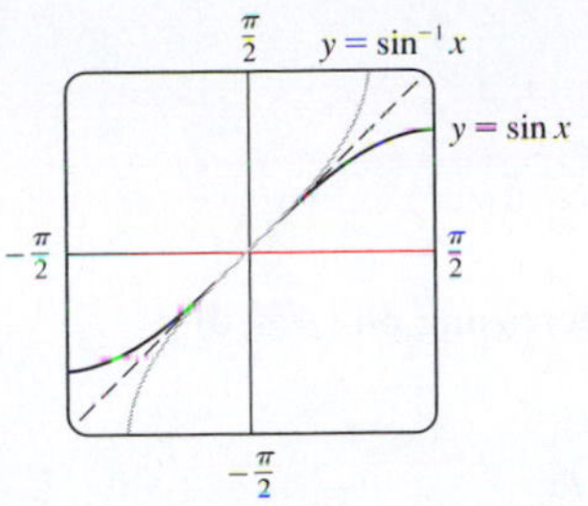

The second graph is the reflection of the first graph about the line $y = x$.

31. $g'(x) = 3x^2/(1 + x^6)$ **33.** $y' = 2x/\sqrt{1 - x^4}$
35. $H'(x) = 1 + 2x \arctan x$ **37.** $g'(t) = -4/\sqrt{t^4 - 16t^2}$
39. $G'(t) = -1/\sqrt{2(-2t^2 + 3t - 1)}$
41. $y' = x/[|x|(1 + x^2)]$ **43.** $y' = \cos x/(1 + \sin^2 x)$
45. $y' = -1/[(1 + x^2)(\tan^{-1}x)^2]$
47. $y' = 2x \cot^{-1}(3x) - 3x^2/(1 + 9x^2)$
49. $y' = \sqrt{a^2 - b^2}/(a + b \cos x)$
51. $g'(x) = 3/\sqrt{-9x^2 - 6x}$, $[-\frac{2}{3}, 0]$, $(-\frac{2}{3}, 0)$
53. $S'(x) = 1/[(1 + x^2)\sqrt{1 - (\tan^{-1}x)^2}\,]$, $[-\tan 1, \tan 1]$, $(-\tan 1, \tan 1)$
55. $U'(t) = 2^{\arctan t}(\ln 2)/(1 + t^2)$, $(-\infty, \infty)$, $(-\infty, \infty)$
57. $\pi/6$ **59.** $f'(x) = e^x - x^2/(1 + x^2) - 2x \arctan x$
61. $-\pi/2$ **63.** $\pi/2$ **65.** $\frac{1}{4}$ rad/s
67.

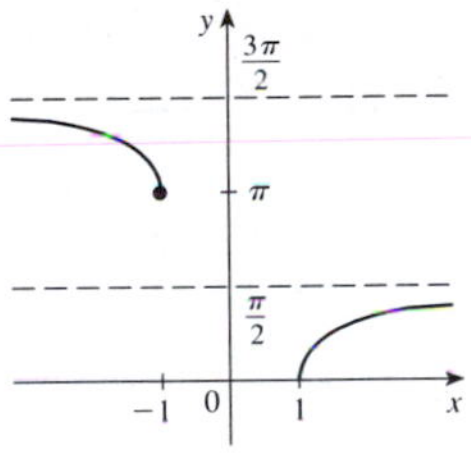

Exercises 4.7 ■ page 287

1. (a) 0 (b) 1 **3.** (a) $\frac{3}{4}$ (b) $\frac{1}{2}(e^2 - e^{-2}) \approx 3.62686$
5. (a) 1 (b) 0
21. $\coth x = \frac{5}{4}$, $\operatorname{sech} x = \frac{3}{5}$, $\cosh x = \frac{5}{3}$, $\sinh x = \frac{4}{3}$, $\operatorname{csch} x = \frac{3}{4}$
23. (a) 1 (b) -1 (c) ∞ (d) $-\infty$ (e) 0 (f) 1 (g) ∞ (h) $-\infty$ (i) 0
31. $f'(x) = 3 \operatorname{sech}^2 3x$ **33.** $h'(x) = 4x^3 \sinh(x^4)$
35. $G'(x) = 2x \operatorname{sech} x - x^2 \operatorname{sech} x \tanh x$
37. $H'(t) = e^t \operatorname{sech}^2(e^t)$
39. $y' = x^{\cosh x}[\sinh x \ln x + (\cosh x)/x]$
41. $y' = 2x/\sqrt{x^4 - 1}$ **43.** $y' = \ln(\operatorname{sech} 4x) - 4x \tanh 4x$
45. $\sinh^{-1}(x/3)$ **47.** $y' = -1/(x\sqrt{x^2 + 1})$
49. $(x^2/16) - (y^2/9) = 1$, $x \geqslant 4$ **51.** $\langle 0, 1 \rangle$
53. (b) $y = 2 \sinh 3x - 4 \cosh 3x$ **55.** $(\ln(1 + \sqrt{2}), \sqrt{2})$

Exercises 4.8 ■ page 295

1. $\frac{1}{4}$ **3.** $\frac{3}{2}$ **5.** 1 **7.** ∞ **9.** $\frac{1}{2}$ **11.** 0 **13.** ∞
15. $1/(3a^{2/3})$ **17.** $\frac{1}{2}$ **19.** 0 **21.** $\frac{1}{2}$ **23.** ∞ **25.** 0
27. $\frac{2}{3}$ **29.** α **31.** $\frac{2}{3}$ **33.** -2 **35.** 0 **37.** $\frac{1}{2}$
39. 0 **41.** 0 **43.** 0 **45.** 1 **47.** ∞ **49.** 0
51. 0 **53.** 0 **55.** 1 **57.** e^{-2} **59.** e^3 **61.** 1
63. 1 **65.** $1/e$ **67.** 1 **69.** 5 **71.** $\frac{1}{4}$ **73.** $\frac{16}{9}a$
79. (a) 0

Chapter 4 Review Exercises ■ page 297

1. False **3.** True **5.** True **7.** False **9.** False
11. False **13.** True **15.** True
17.

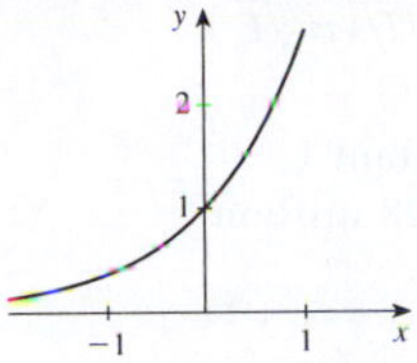

19.

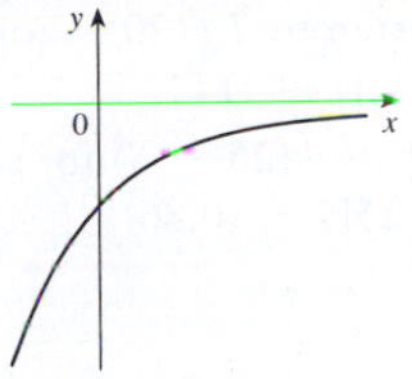

21.

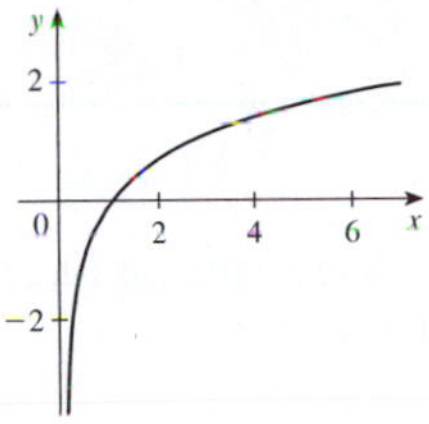

23.

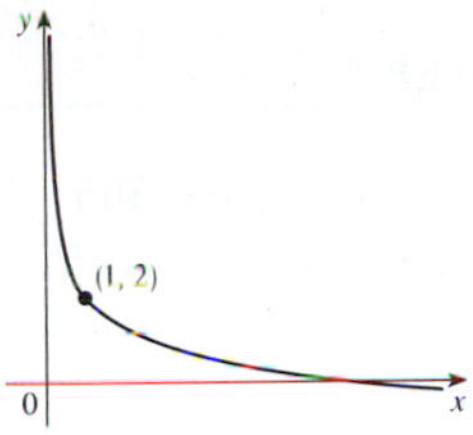

25.

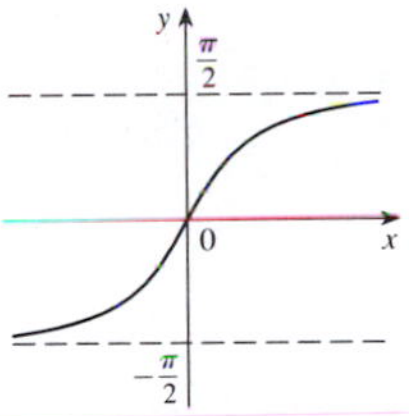

27. $\ln 5$
29. $\ln 10$
31. $e^{2/\pi}$

33. $\tan^{-1} 4 + n\pi$, n an integer **35.** $(2x - 1)/[(x^2 - x) \ln 10]$
37. $\dfrac{\sqrt{x + 1}(2 - x)^5}{(x + 3)^7}\left[\dfrac{1}{2(x + 1)} - \dfrac{5}{2 - x} - \dfrac{7}{x + 3}\right]$
39. $(1 + c^2)e^{cx} \sin x$ **41.** $2 \tan x$ **43.** $e^{-1/x}(1 + 1/x)$
45. $(\cos^{-1}x)^{\sin^{-1}x - 1}[\cos^{-1}x \ln(\cos^{-1}x) - \sin^{-1}x]/\sqrt{1 - x^2}$
47. $e^{x + e^x}$ **49.** $-1/x - 1/[x(\ln x)^2]$ **51.** $7^{\sqrt{2x}}(\ln 7)/\sqrt{2x}$
53. $3 \tanh 3x$ **55.** $(\cosh x)/\sqrt{\sinh^2 x - 1}$
57. $\cot x - \sin x \cos x$ **59.** $y = 1/[\sqrt{x}(x + 1)]$
61. $1/(x^4 + x^2 + 1)$ **63.** $2^x(\ln 2)^n$
67. $3x - 2y + \ln 4 = 0$
69.

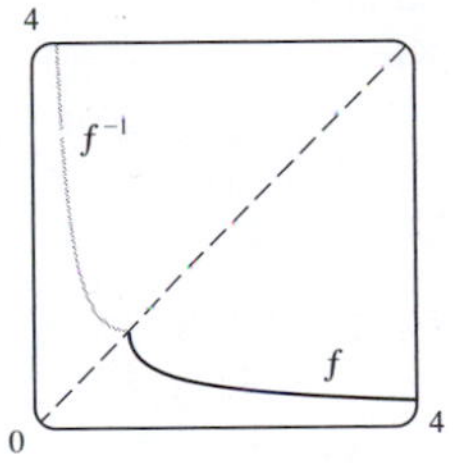

71. $(-3, 0)$
73. $y = ex$
75. ∞
77. $-\infty$
79. ∞
81. 0
83. $\pi/3$
85. $-1/(2\pi)$
87. 0 **89.** $-\frac{1}{3}$ **91.** 0 **93.** 0 **95.** 1
97. (a) $1000e^{(\ln 9)t/2} = 1000 \times 3^t$ (b) 27,000 (c) $(\ln 2)/\ln 3$ h
99. (a) \$12,624.77 (b) \$12,667.70 (c) \$12,689.86 (d) \$12,704.89 (e) \$12,712.24 (f) \$12,712.49
101. (a) $C_0 e^{-kt}$ (b) $\approx$100 h
103. $v(t) = -Ae^{-ct}[c \cos(\omega t + \delta) + \omega \sin(\omega t + \delta)]$, $a(t) = Ae^{-ct}[(c^2 - \omega^2)\cos(\omega t + \delta) + 2c\omega \sin(\omega t + \delta)]$
105. $f'(x) = e^{g(x)}g'(x)$ **107.** $f'(x) = g'(x)/g(x)$
109. $f'(x) = [g'(e^x)e^x]/g(e^x)$ **111.** $\frac{2}{3}$

APPLICATIONS PLUS ■ page 300

1. (a) $4\pi\sqrt{3}/\sqrt{11}$ rad/s (b) $40(\cos\theta + \sqrt{8 + \cos^2\theta})$ cm
(c) $-480\pi \sin\theta\,(1 + \sin\theta\cos\theta/\sqrt{8 + \cos^2\theta})$ cm/s
3. (b) Shorten by $L/120$ (c) $-g\sqrt{g}\,dT/(\pi\sqrt{L})$
5. $3/(\sqrt[3]{2} - 1) \approx 11\frac{1}{2}$ h
7. (a) $dS/dt = k(25 - S)$ for some constant k
(b) $S(t) = 25[1 - (0.86)^{t/3}]$ (c) \$8.28 million
(d) 13.8 yr

CHAPTER 5

Exercises 5.1 ■ page 305

1. (a) Increasing on $(-\infty, 0)$ and $(3, \infty)$; decreasing on $(0, 3)$
(b) Local maximum at 0, local minimum at 3
(c)

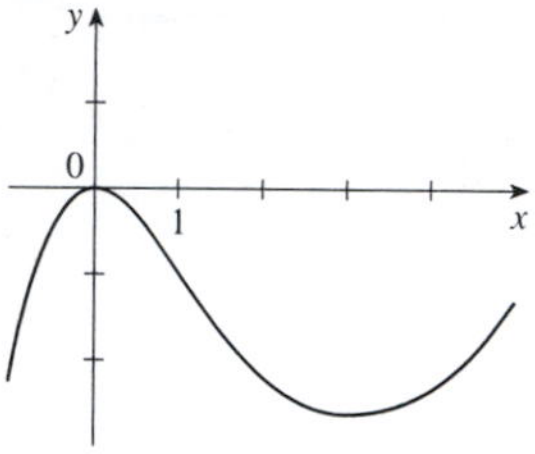

3. Increasing on $(2, 5)$; decreasing on $(-\infty, 2)$ and $(5, \infty)$
5. If $D(t)$ is the size of the deficit as a function of time, then at the time of the speech $D'(t) > 0$, but $D''(t) < 0$.
7. (a) The rate starts small, grows rapidly, levels off, then decreases and becomes negative.
(b) (1932, 2.5) and (1937, 4.3); the rate of change of population density starts to decrease in 1932 and starts to increase in 1937.
9. $K(3) - K(2)$; concave downward
11. (a) Increasing on $(0, 2)$, $(4, 6)$, and $(8, \infty)$; decreasing on $(2, 4)$ and $(6, 8)$
(b) Local maxima at $x = 2, 6$; minima at $x = 4, 8$
(c) Concave upward on $(3, 6)$ and $(6, \infty)$, concave downward on $(0, 3)$
(d) 3
(e)

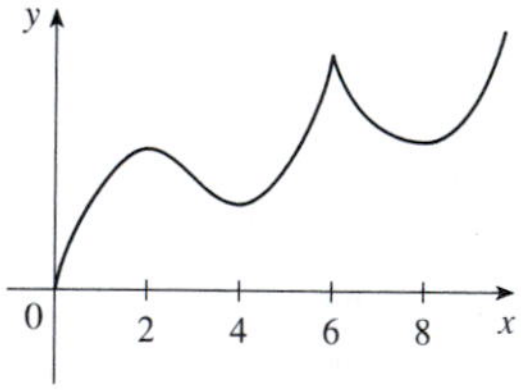

13.

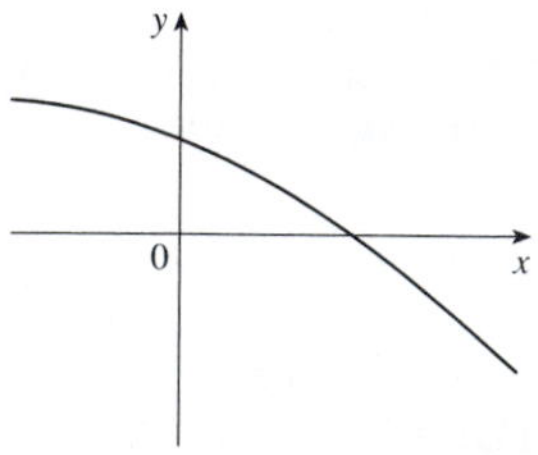

15.

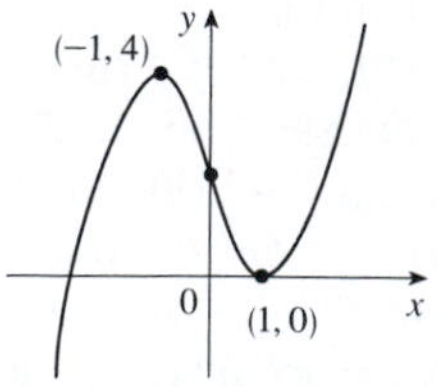

17.

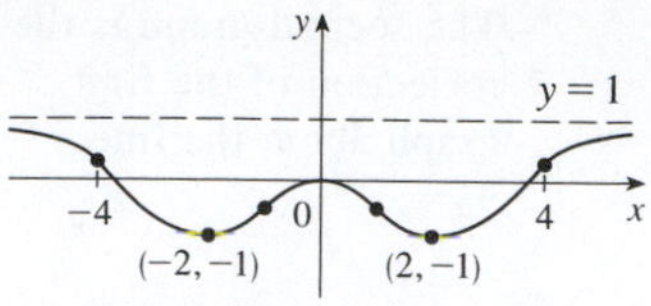

19. (a) Increasing on $(0, \infty)$; decreasing on $(-\infty, 0)$
(b) Minimum at $x = 0$
21. b

Exercises 5.2 ■ page 312

1. Absolute minimum: smallest function value on the entire domain of the function; local minimum at c: smallest function value when x is near c
3. Absolute maximum at b, local maxima at b and e, absolute minimum at d, local minima at d and s
5. Absolute maximum $f(4) = 4$; absolute minimum $f(7) = 0$; local maxima $f(4) = 4$ and $f(6) = 3$; local minima $f(2) = 1$ and $f(5) = 2$
7.

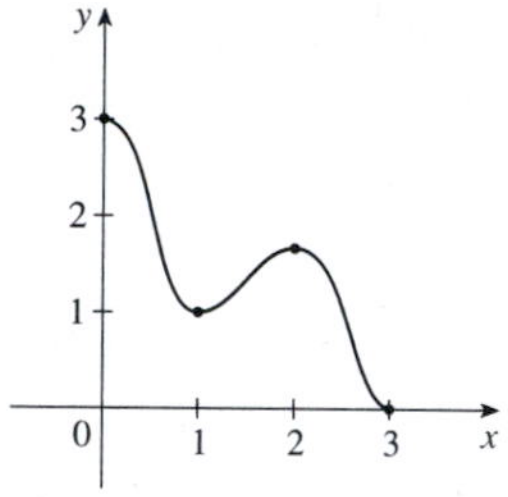

9.

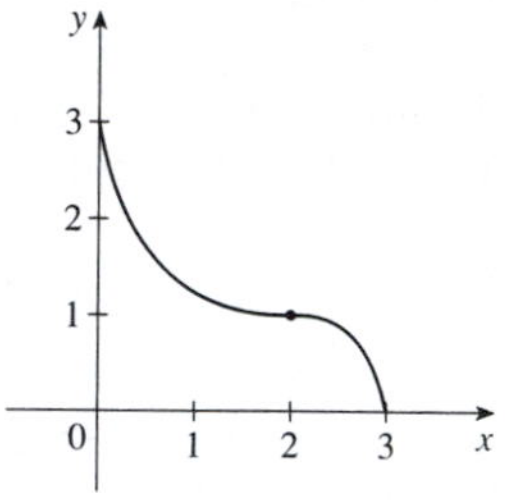

11. (a)

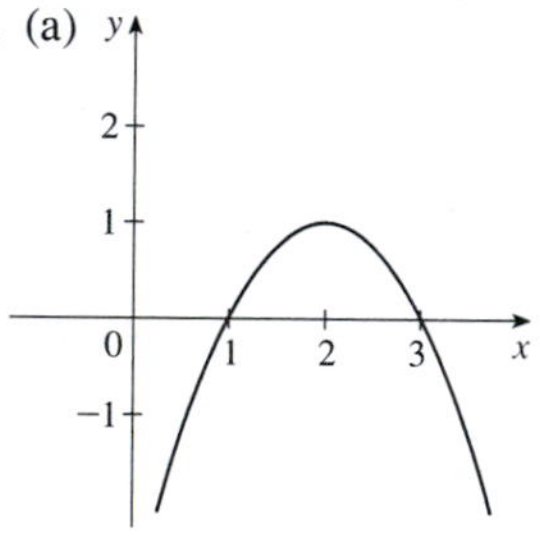

(b)

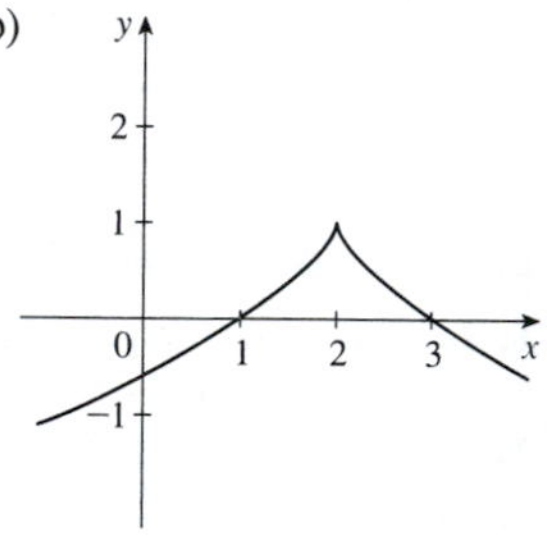

(c)

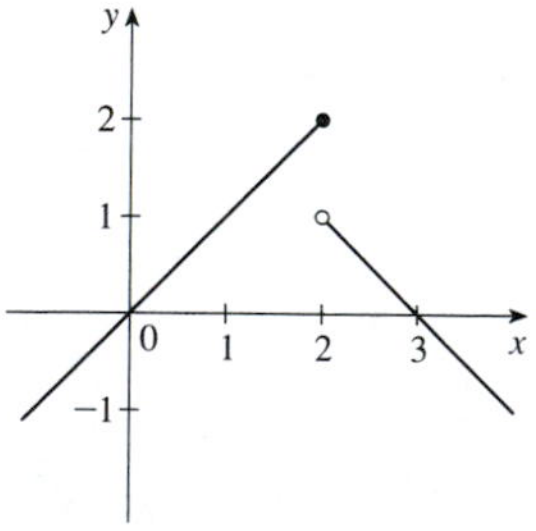

13. (a)

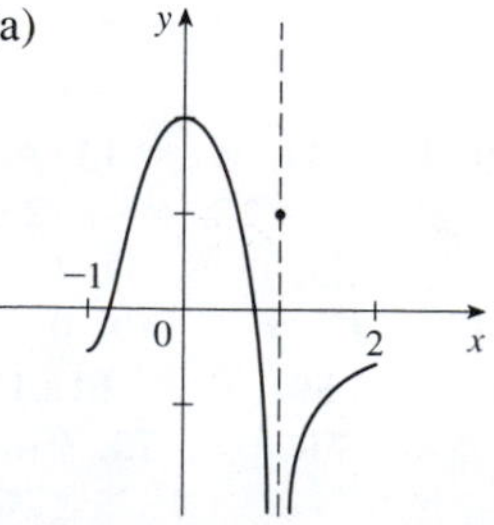

(b)

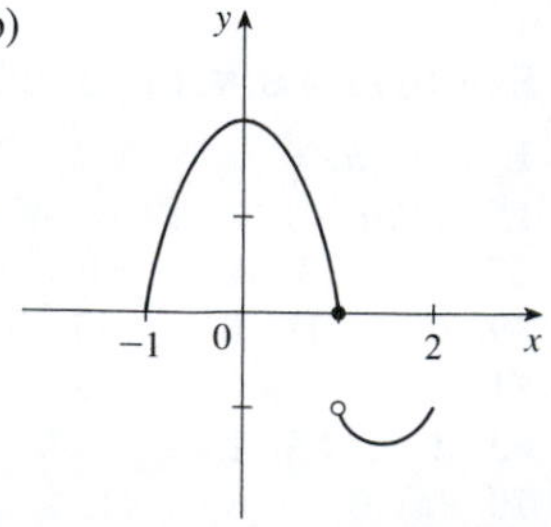

15.

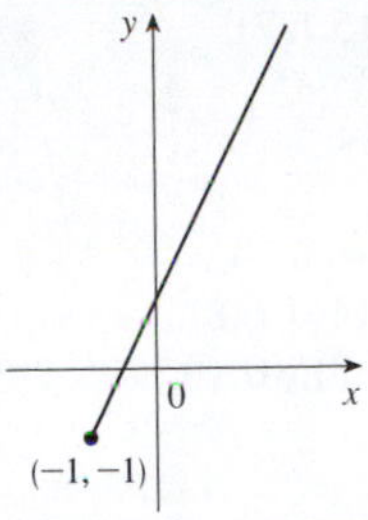

No maximum, absolute minimum $f(-1) = -1$

17.

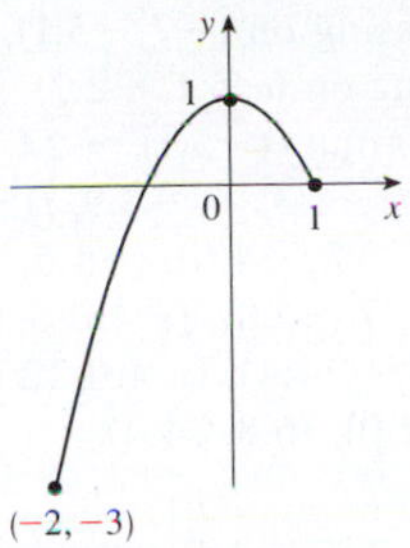

Local and absolute maximum $f(0) = 1$, absolute minimum $f(-2) = -3$

19.

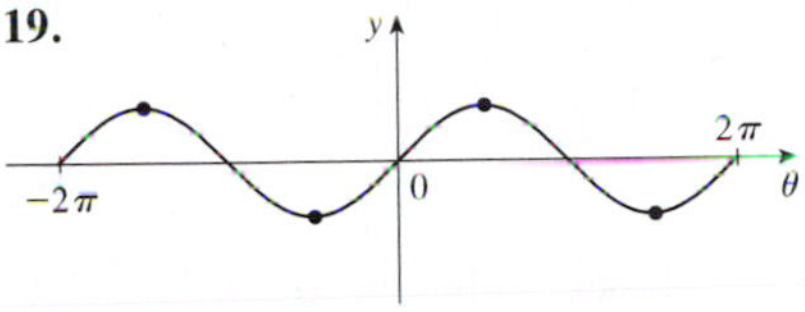

Local and absolute maxima $f(\pi/2) = f(-3\pi/2) = 1$, local and absolute minima $f(3\pi/2) = f(-\pi/2) = -1$

21.

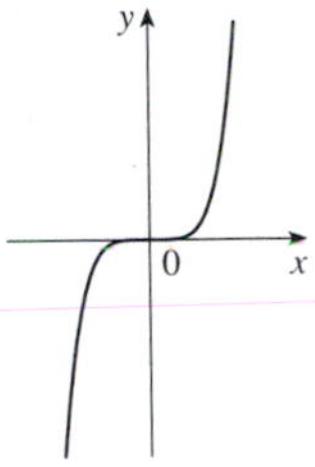

No maximum or minimum

23.

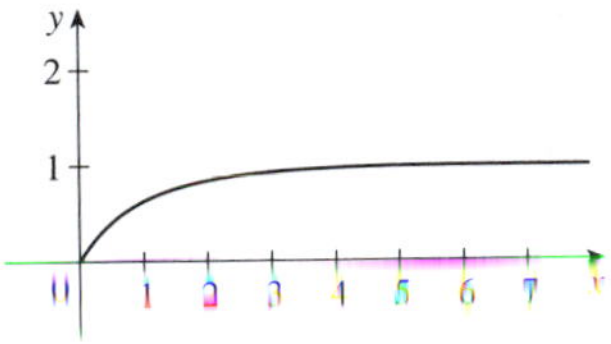

No maximum, absolute minimum $f(0) = 0$

25. $-\frac{1}{2}, 2$ **27.** $0, (-3 \pm \sqrt{5})/2$ **29.** ± 1

31. $0, \frac{8}{7}, 4$ **33.** $n\pi/4$ (n any integer) **35.** $1/e$

37. $f(3) = 5, f(1) = 1$ **39.** $f(2) = 55, f(-2) = -57$

41. $f(2) = 5, f(1) = 3$ **43.** $f(\pi/4) = \sqrt{2}, f(0) = 1$

45. $f(1) = 1/e, f(0) = 0$

47. (a) 9.71, −7.71 (b) $1 \pm 32\sqrt{6}/9$

49. (a) 0.32, 0.00 (b) $3\sqrt{3}/16, 0$ **51.** 3.9665 °C

53. $a(126) \approx 62.87 \text{ m/s}^2$, $a(23.12) \approx 21.52 \text{ m/s}^2$

55. (a) $r = \frac{2}{3}r_0$ (b) $v = \frac{4}{27}kr_0^3$

(c)

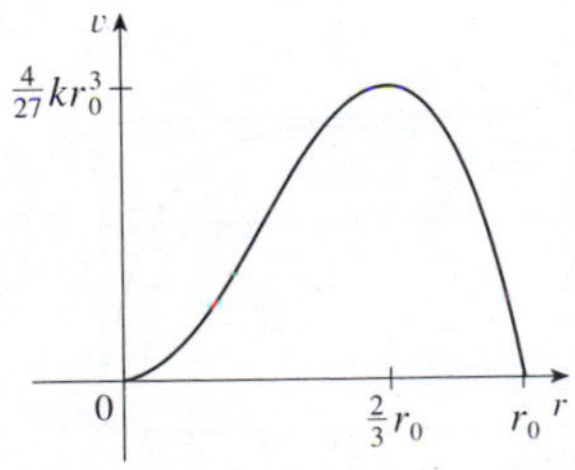

Exercises 5.3 ■ page 322

Abbreviations: CD, concave downward; CU, concave upward; HA, horizontal asymptote; IP, inflection point; VA, vertical asymptote

1. 0.8, 3.2, 4.4, 6.1

3. (a) I/D Test (b) Concavity Test
(c) Find points at which the concavity changes.

5. $x = 1, 7$

7. (a) Increasing on $(-2, \infty)$, decreasing on $(-\infty, -2)$
(b) No local maximum, local minimum $f(-2) = -303$
(c) CU on $(-\infty, \infty)$, no IP

9. (a) Increasing on $(-1, \infty)$, decreasing on $(-\infty, -1)$
(b) No local maximum, local minimum $f(-1) = -1/e$
(c) CD on $(-\infty, -2)$, CU on $(-2, \infty)$; IP $(-2, -2e^{-2})$

11. (a) Increasing on $(0, e^2)$, decreasing on (e^2, ∞)
(b) Local maximum $f(e^2) = 2/e$, no local minimum
(c) CD on $(0, e^{8/3})$, CU on $(e^{8/3}, \infty)$; IP $(e^{8/3}, \frac{8}{3}e^{-4/3})$

13. (a) Increasing on $(\frac{1}{3}, 3)$; decreasing on $(-\infty, \frac{1}{3})$, $(3, \infty)$
(b) Local maximum $f(3) = 10$, local minimum $f(\frac{1}{3}) = \frac{14}{27}$
(c) CD on $(\frac{5}{3}, \infty)$, CU on $(-\infty, \frac{5}{3})$; IP $(\frac{5}{3}, \frac{142}{27})$
(d) See graph at right.

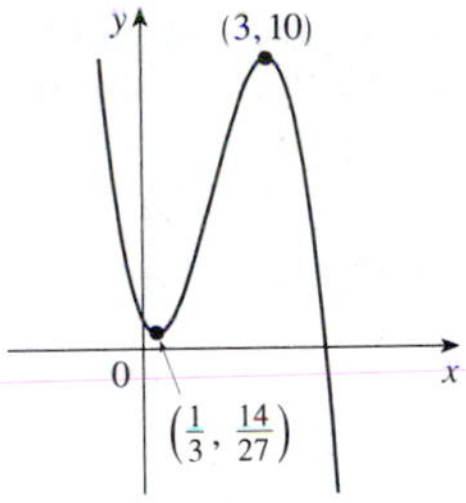

15. (a) Increasing on $(0, \infty)$, decreasing on $(-\infty, 0)$
(b) Local minimum $f(0) = -1$
(c) CD on $(-1, -1/\sqrt{5})$, $(1/\sqrt{5}, 1)$; CU on $(-\infty, -1)$, $(-1/\sqrt{5}, 1/\sqrt{5})$, $(1, \infty)$; IP $(\pm 1, 0)$, $(\pm 1/\sqrt{5}, -\frac{64}{125})$
(d) See graph at right.

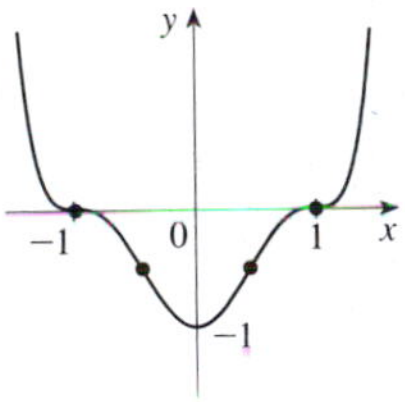

17. (a) Increasing on $(-\infty, -3)$, $(-1, \infty)$, decreasing on $(-3, -1)$
(b) Local maximum $f(-3) = 0$, local minimum $f(-1) = -\sqrt[3]{4}$
(c) CD on $(0, \infty)$, CU on $(-\infty, -3)$, $(-3, 0)$; IP $(0, 0)$
(d) See graph at right.

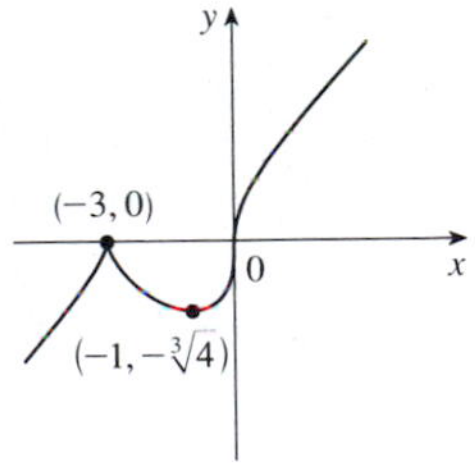

19. (a) Increasing on $((2n - 1)\pi, 2n\pi)$, decreasing on $(2n\pi, (2n + 1)\pi)$
(b) Local maxima $f(2n\pi) = 2$, local minima $f((2n + 1)\pi) = -2$
(c) CD on $(2n\pi - \frac{2}{3}\pi, 2n\pi + \frac{2}{3}\pi)$, CU on remaining intervals; IP $(2n\pi \pm \frac{2}{3}\pi, -\frac{1}{4})$
(d) See graph at right.

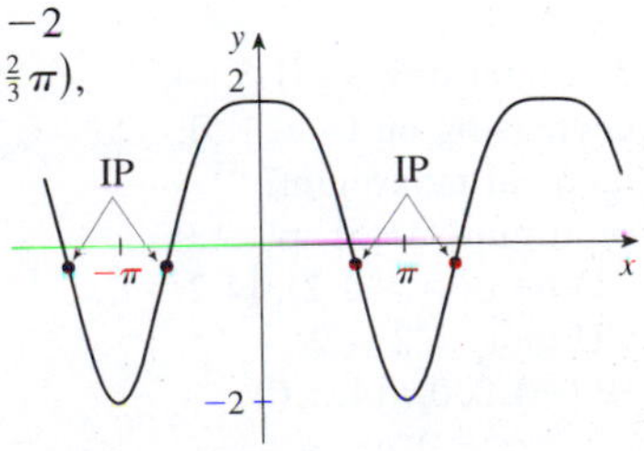

21. (a) VA $x = \pm 1$, HA $y = -1$
(b) Increasing on $(0, 1)$, $(1, \infty)$;
decreasing on $(-\infty, -1)$, $(-1, 0)$
(c) Local minimum $f(0) = 1$
(d) CD on $(-\infty, -1)$, $(1, \infty)$;
CU on $(-1, 1)$; no IP
(e) See graph at right.

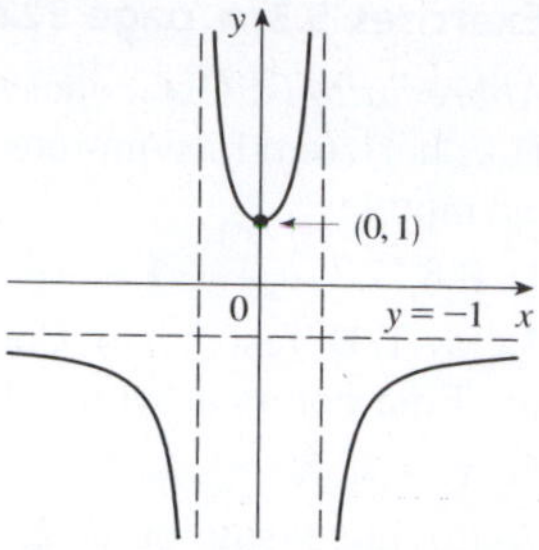

23. (a) HA $y = 0$
(b) Decreasing on $(-\infty, \infty)$
(c) No local maximum or minimum
(d) CU on $(-\infty, \infty)$; no IP
(e) See graph at right.

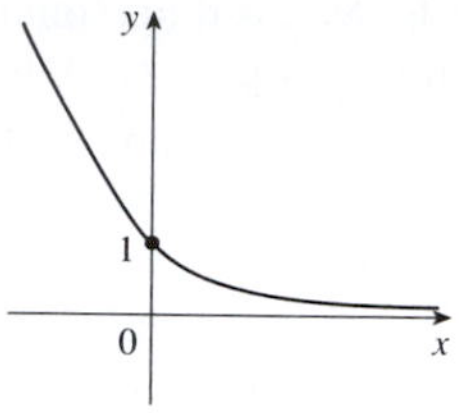

25. (a) HA $y = 1$, VA $x = -1$
(b) Increasing on $(-\infty, -1)$, $(-1, \infty)$
(c) No local maximum or minimum
(d) CD on $(-\frac{1}{2}, \infty)$;
CU on $(-\infty, -1)$, $(-1, -\frac{1}{2})$;
IP $(-\frac{1}{2}, 1/e^2)$
(e) See graph at right.

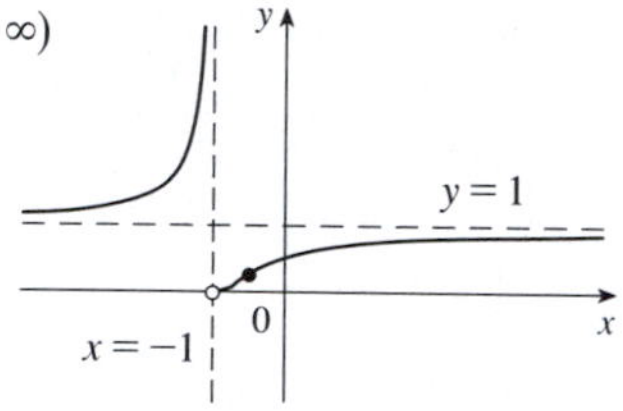

27. CD on $(-\infty, -2.1)$, $(0.25, 2)$; CU on $(-2.1, 0.25)$, $(1.9, \infty)$; IP $(-2.1, 380)$, $(0.25, 1.3)$, $(1.9, -92)$
29. (a) Local and absolute maximum $f(1) = \sqrt{2}$, no minimum (b) $(3 - \sqrt{17})/4$
31. When $t \approx 7.17$ **33.** CD on $(-\infty, 0.1)$, CU on $(0.1, \infty)$
35. $f(x) = \frac{1}{9}(2x^3 + 3x^2 - 12x + 7)$ **43.** 17

Exercises 5.4 ■ page 330

1. Increasing on $(-1.1, 0.3)$, $(0.7, \infty)$; decreasing on $(-\infty, -1.1)$, $(0.3, 0.7)$; local maximum $f(0.3) \approx 6.6$, local minima $f(-1.1) \approx -1.0$, $f(0.7) \approx 6.3$; CD on $(-0.5, 0.5)$, CU on $(-\infty, -0.5)$, $(0.5, \infty)$; IP $(-0.5, 2.5)$, $(0.5, 6.5)$

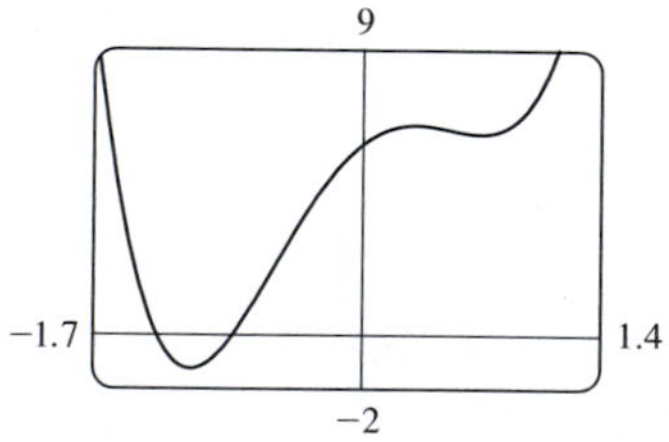

3. Increasing on $(1.5, \infty)$, decreasing on $(-\infty, 1.5)$; no local maximum, minimum $f(1.5) \approx -1.9$; CD on $(-\infty, -1.2)$, $(4.2, \infty)$; CU on $(-1.2, 4.2)$; IP $(-1.2, 0)$, $(4.2, 0)$

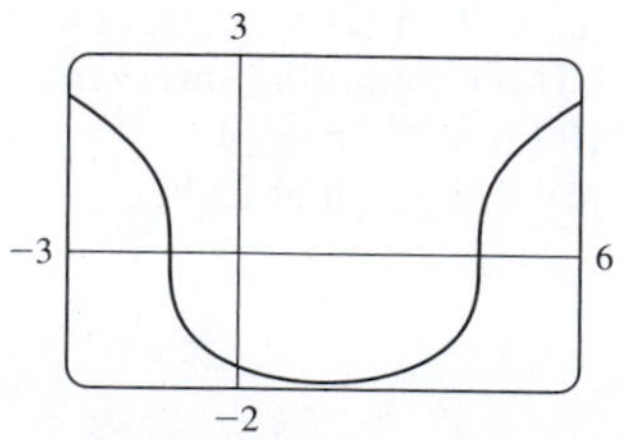

5. Increasing on $(-7, -5.1)$, $(-2.3, 2.3)$, $(5.1, 7)$; decreasing on $(-5.1, -2.3)$, $(2.3, 5.1)$; local maxima $f(-5.1) \approx 24.1$, $f(2.3) \approx 3.9$; minima $f(-2.3) \approx -3.9$, $f(5.1) \approx -24.1$; CD on $(-6.8, -4.0)$, $(-1.5, 0)$, $(1.5, 4.0)$, $(6.8, 7)$; CU on $(-7, -6.8)$, $(-4.0, -1.5)$, $(0, 1.5)$, $(4.0, 6.8)$; IP $(-6.8, -24.4)$, $(-4.0, 12.0)$, $(-1.5, -2.3)$, $(0, 0)$, $(1.5, 2.3)$, $(4.0, -12.0)$, $(6.8, 24.4)$

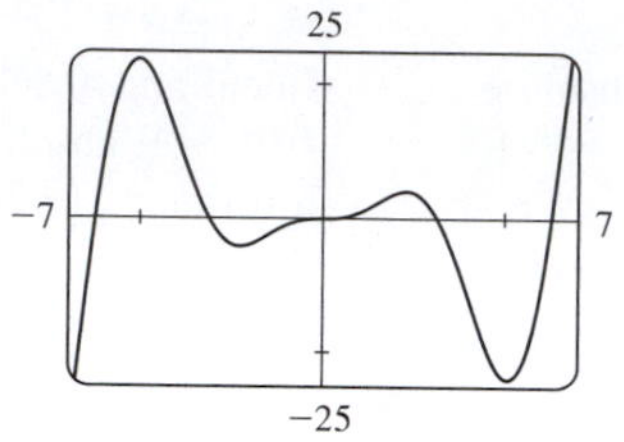

7. Increasing on $(-\infty, 0)$, $(\frac{1}{4}, \infty)$; decreasing on $(0, \frac{1}{4})$; local maximum $f(0) = -10$, minimum $f(\frac{1}{4}) = -\frac{161}{16} \approx -10.1$; CD on $(-\infty, \frac{1}{8})$; CU on $(\frac{1}{8}, \infty)$; IP $(\frac{1}{8}, -\frac{321}{32})$

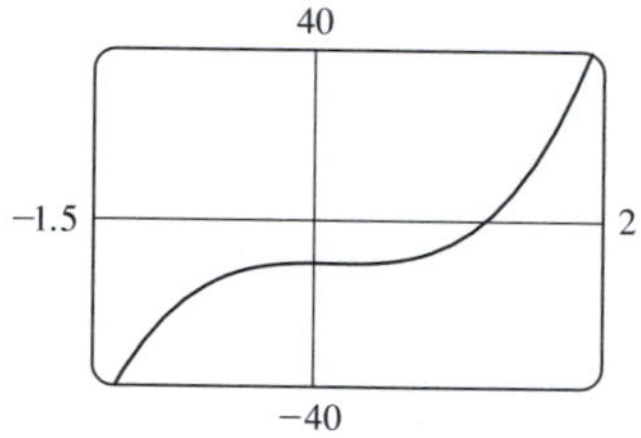

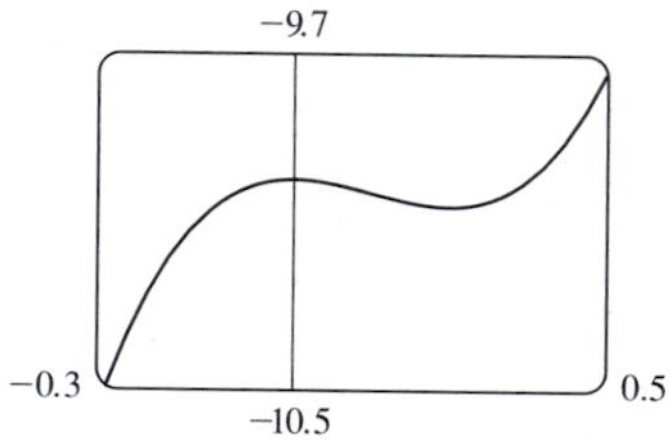

9. Local maximum $f(-1/\sqrt{3}) = e^{2\sqrt{3}/9} \approx 1.5$; local minimum $f(1/\sqrt{3}) = e^{-2\sqrt{3}/9} \approx 0.7$; IP $(-0.15, 1.15)$, $(-1.09, 0.82)$

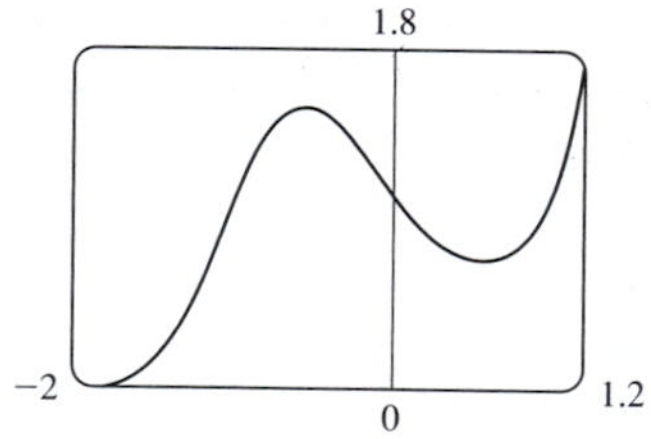

11. Local maxima $f(-5.6) \approx 0.018$, $f(0.82) \approx -281.5$, $f(5.2) \approx 0.0145$; minimum $f(3) = 0$

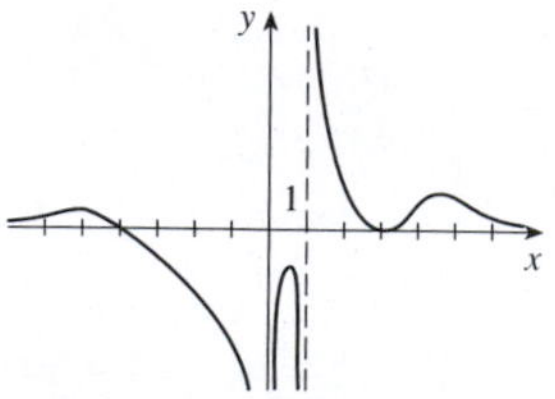

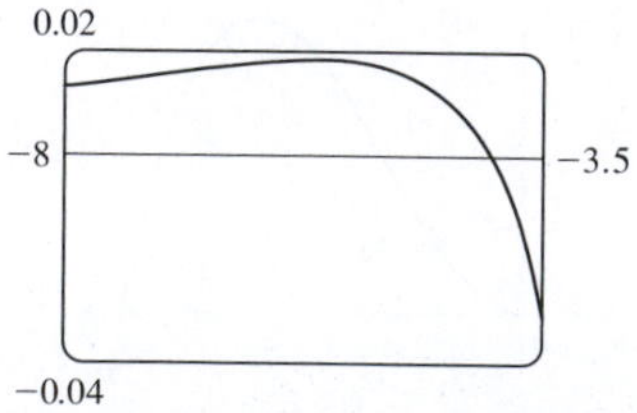

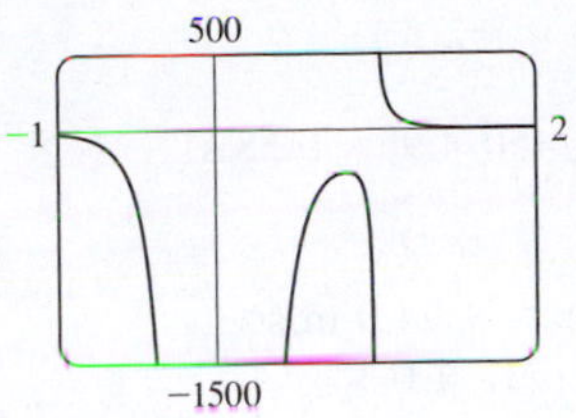

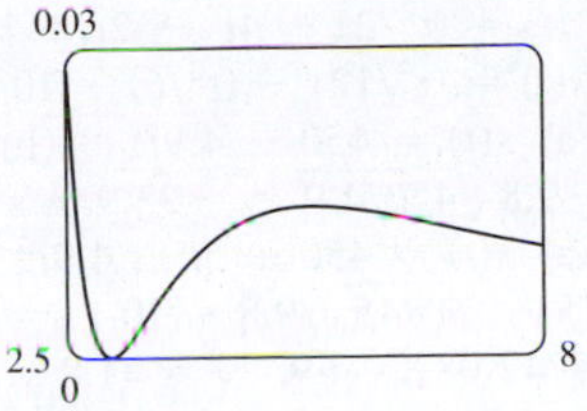

13. $f'(x) = -\dfrac{x(x+1)^2(x^3+18x^2-44x-16)}{(x-2)^3(x-4)^5}$,
$f''(x) =$
$2\dfrac{(x+1)(x^6+36x^5+6x^4-628x^3+684x^2+672x+64)}{(x-2)^4(x-4)^6}$
CU on $(-\infty,-5.0)$, $(-1,-0.5)$, $(-0.1,2)$, $(2,4)$, $(4,\infty)$;
CD on $(-5,-1)$, $(-0.5,-0.1)$; IP $(-5.0,-0.005)$, $(-1,0)$, $(-0.5,0.00001)$, $(-0.1,0.0000066)$

15. Maxima $f(0.59) \approx 1$, $f(0.68) \approx 1$, $f(1.96) \approx 1$; minima $f(0.64) \approx 0.99996$, $f(1.46) \approx 0.49$, $f(2.73) \approx -0.51$;
IP $(0.61, 0.99998)$, $(0.66, 0.99998)$, $(1.17, 0.72)$, $(1.75, 0.77)$, $(2.28, 0.34)$

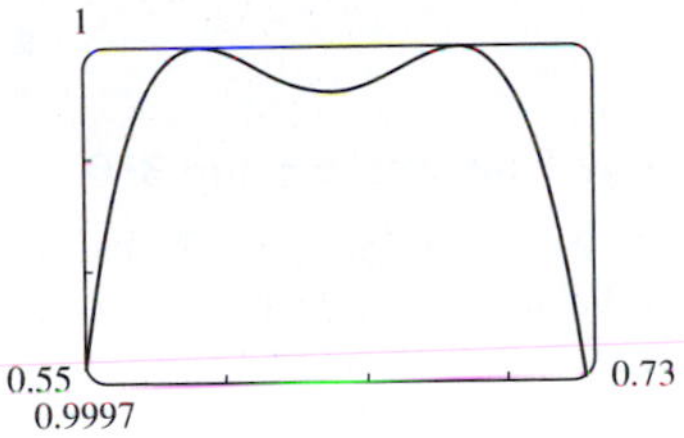

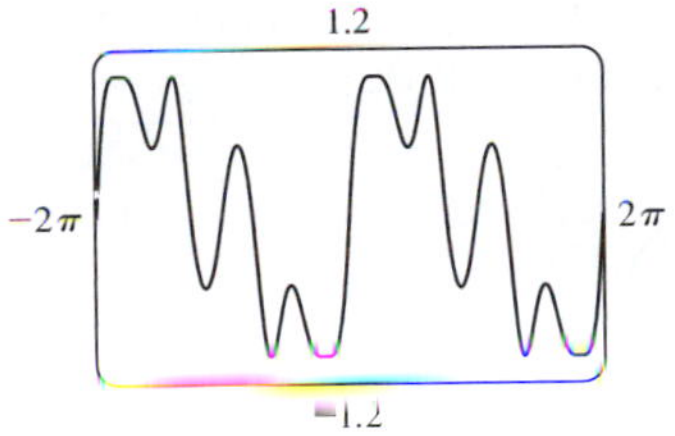

17. Vertical tangents at $(0,0)$, $(-\frac{3}{16}, \frac{3}{8})$, $(-8,6)$;
horizontal tangents at
$(-(2\sqrt{3}+5)/9, -2\sqrt{3}/9)$,
$((2\sqrt{3}-5)/9, 2\sqrt{3}/9)$

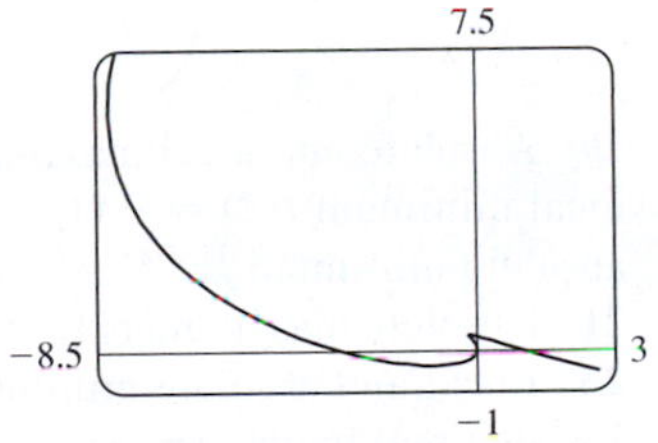

19. For $c = 0$, there is a cusp; for $c > 0$, there is a loop whose size increases as c increases and the curve intersects itself at $(0, c)$; leftmost point $(2c\sqrt{3c}/9, c/3)$, rightmost point $(-2c\sqrt{3c}/9, c/3)$

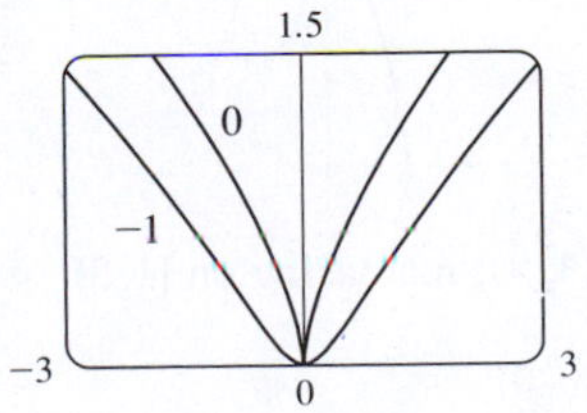

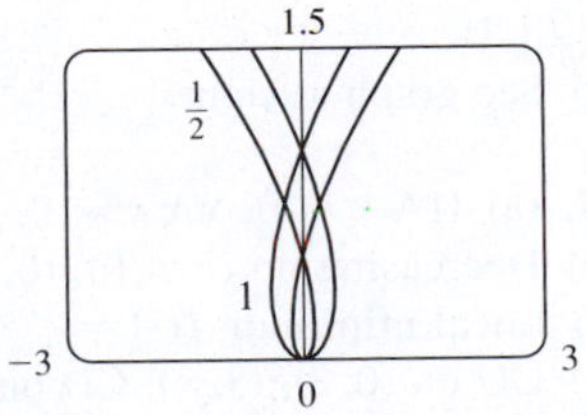

21. For $c > 0$, the maximum and minimum values are always $\pm\frac{1}{2}$, but the extreme points and IP move closer to the y-axis as c increases.
$c = 0$ is a transitional value: when c is replaced by $-c$, the curve is reflected in the x-axis.

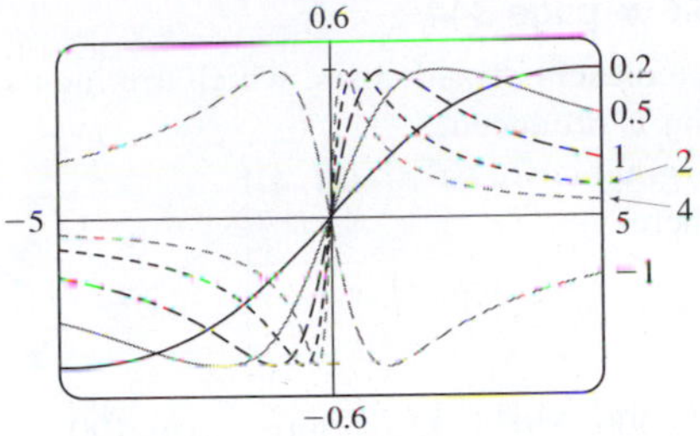

23. There is no maximum or minimum, regardless of the value of c. For $c < 0$, there is a vertical asymptote at $x = 0$, $\lim_{x\to 0} f(x) = \infty$, and $\lim_{x\to\pm\infty} f(x) = 1$.
$c = 0$ is a transitional value at which $f(x) = 1$ for $x \neq 0$.
For $c > 0$, $\lim_{x\to 0} f(x) = 0$, $\lim_{x\to\pm\infty} f(x) = 1$, and there are two IPs, which move away from the y-axis as $c \to \infty$.

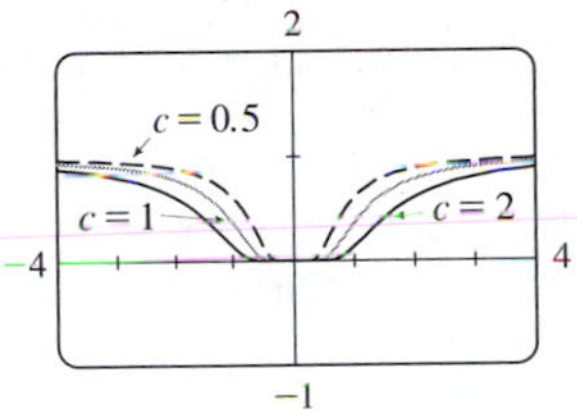

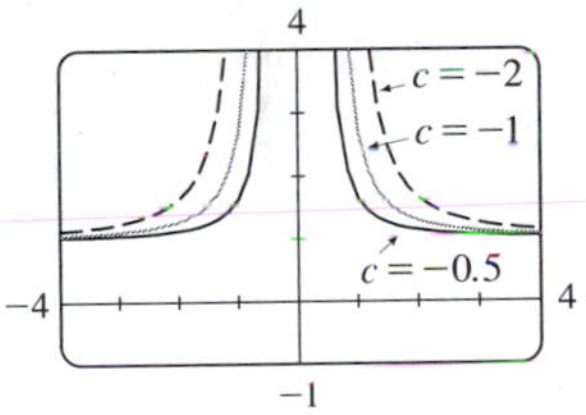

25. For $c \geq 0$, there is no IP and only one extreme point, the origin. For $c < 0$, there is a maximum point at the origin, two minimum points, and two IPs, which move downward and away from the origin as $c \to -\infty$.

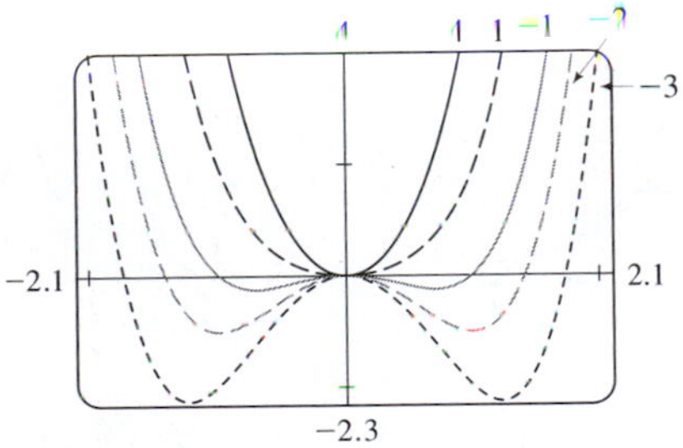

Exercises 5.5 ■ page 337

1. 50, 50 **3.** 10, 10 **7.** 14,062.5 ft² **9.** 4000 cm³
11. \$191.28 **13.** $(1.2, -0.6)$ **15.** $(1, \pm\sqrt{5})$ **17.** 2
19. Square, side $\sqrt{2}\,r$ **21.** $L/2$, $\sqrt{3}\,L/4$
23. Base $\sqrt{3}\,r$, height $3r/2$ **25.** $4\pi r^3/(3\sqrt{3})$
27. $\pi r^2(1+\sqrt{5})$ **29.** 24 cm, 36 cm
31. (a) All of the wire for the square
(b) $40\sqrt{3}/(9+4\sqrt{3})$ m for the square
33. Height = radius = $\sqrt[3]{V/\pi}$ cm **35.** $V = 2\pi R^3/(9\sqrt{3})$
37. (a) $\frac{3}{2}s^2 \csc\theta(\csc\theta - \sqrt{3}\cot\theta)$ (b) $\cos^{-1}(1/\sqrt{3}) \approx 55°$
(c) $6s[h + s/(2\sqrt{2})]$ **39.** Row directly to B

41. $10\sqrt[3]{3}/(1 + \sqrt[3]{3})$ ft from the stronger source
47. $x = 6$ in. **49.** $\pi/6$
51. At a distance $5 - 2\sqrt{5}$ from A
53. $(L + W)^2/2$ **55.** 9.35 m

Exercises 5.6 ■ page 344

1. (a) $C(0)$ represents fixed costs, which are incurred even when nothing is produced.
(b) The marginal cost is a minimum there.
(c)

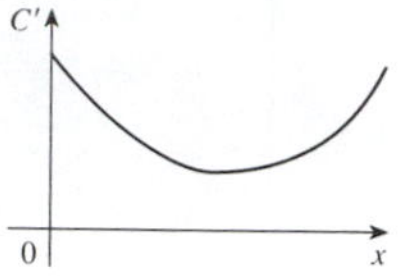

3. (a) \$1,035,000; \$1035; \$2025/unit (b) 100 (c) \$225
5. (a) \$2330.71, \$2.33, \$4.07/unit (b) 159 (c) \$1.07
7. (a) \$188.25, \$0.19, \$0.28/unit (b) 400 (c) \$0.15
9. 400 **11.** 16,667 **13.** 672 **15.** 100
17. (a) $p(x) = 19 - (x/3000)$ (b) \$9.50
19. (a) $p(x) = 550 - (x/10)$ (b) \$175 (c) \$100

Exercises 5.7 ■ page 353

1. $4x^3 + 3x^2 - 5x + C$ **3.** $(3x^{10}/5) - (x^8/2) + x^3 + x + C$
5. $(2x^{3/2}/3) + (3x^{4/3}/4) + C$
7. $-3/(2x^4) + C_1$ if $x > 0$, $-3/(2x^4) + C_2$ if $x < 0$
9. $(2t^{7/2}/7) + (4t^{5/2}/5) + C$ **11.** $-\cos x - 2\sin x + C$
13. $\tan t + (t^3/3) + C_n$, $(2n - 1)\pi/2 < t < (2n + 1)\pi/2$
15. $x^2 + 5\sin^{-1}x + C$
17. $(x^4/12) + (x^5/20) + Cx + D$ **19.** $(x^2/2) + Cx + D$
21. $x^4 + (Cx^2/2) + Dx + E$ **23.** $2x^2 + 3x - 9$
25. $2x^{3/2} - 2\sqrt{x} + 2$ **27.** $3\sin x - 5\cos x + 9$
29. $2\ln|x| + 7$ **31.** $(x^3/6) + 2x - 3$
33. $(x^4/12) - 3\cos x + 3x + 5$ **35.** $x^3 + 3x^2 - 5x + 4$
37. $f(x) = 1/(2x) + (x/4) - (3/4)$
39. $f(x) = -\ln x + (\ln 2)x - \ln 2$
41. 10 **43.** b

45.

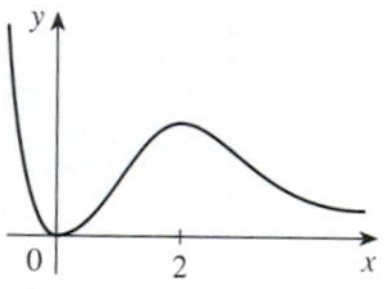

47.

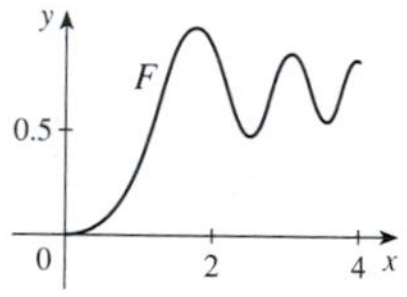

49.

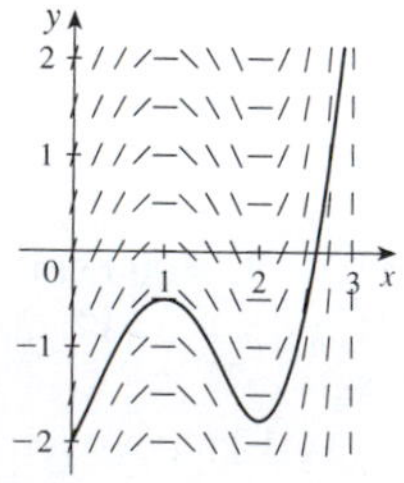

51.

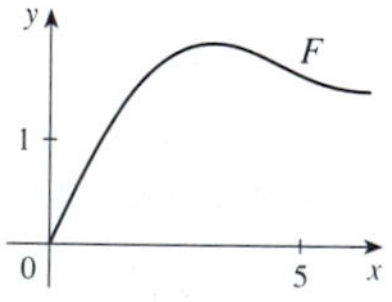

53.

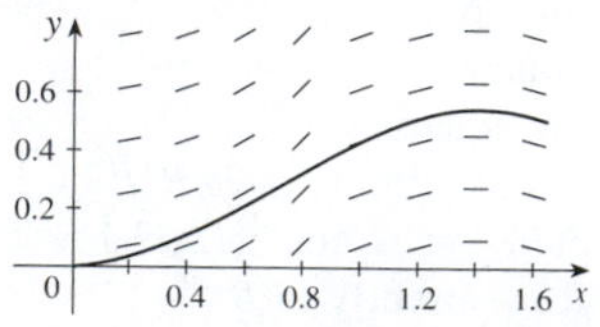

55. $s(t) = 3t - t^2 + 4$

57. $s(t) = (t^3/2) + 4t^2 - 2t + 1$
59. $s(t) = (t^4/12) - (t^3/6) - 10t$
61. (a) $s(t) = 450 - 4.9t^2$ (b) $\sqrt{450/4.9} \approx 9.58$ s
(c) $-9.8\sqrt{450/4.9} \approx -93.9$ m/s
63. (a) $s(t) = 450 + 5t - 4.9t^2$
(b) $(5 + \sqrt{8845})/9.8 \approx 10.1$ s (c) ≈ -94.0 m/s
67. \$742.08 **69.** $\frac{130}{11} \approx 11.8$ s **71.** $\frac{88}{15}$ ft/s^2
73. 225 ft
75. $\mathbf{v}(t) = \mathbf{i} + (-10t + 1)\,\mathbf{j}$, $\mathbf{r}(t) = t\,\mathbf{i} + (-5t^2 + t)\,\mathbf{j}$
77. (a) $\mathbf{v}(t) = \frac{1}{3}t^3\,\mathbf{i} + (1 + \frac{1}{2}\sin 2t)\,\mathbf{j}$;
$\mathbf{r}(t) = (1 + \frac{1}{12}t^4)\,\mathbf{i} + (t - \frac{1}{4}\cos 2t + \frac{1}{4})\,\mathbf{j}$
(b)

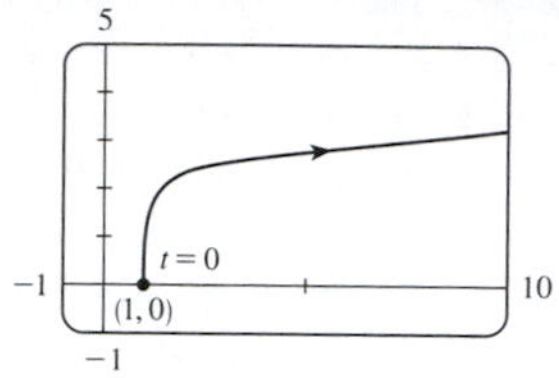

79. (a) 22,092.5 m (b) 3,188.8 m (c) 500 m/s
81. 29.7 m/s

Chapter 5 Review Exercises ■ page 356

1. False **3.** False **5.** True **7.** False **9.** True
11. True **13.** False **15.** True
17. (a) Increasing on $(-2, 0)$ and $(2, \infty)$; decreasing on $(-\infty, -2)$ and $(0, 2)$
(b) Maximum at 0; minima at -2 and 2
(c) Concave upward on $(-\infty, -1)$ and $(1, \infty)$; concave downward on $(-1, 1)$
(d)

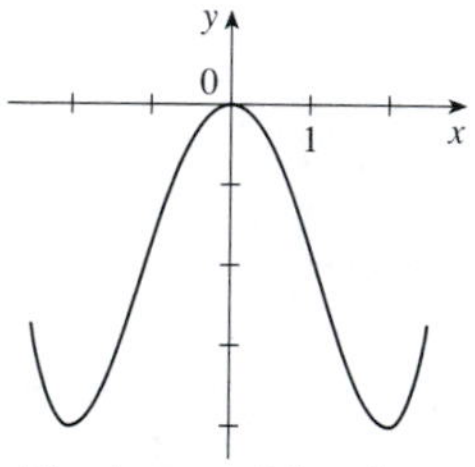

19. Absolute and local maximum $f(-2) = 21$, local minimum $f(2) = -11$, absolute minimum $f(-5) = -60$
21. Absolute maximum $f(4) = \frac{1}{3}$, absolute minimum $f(0) = -1$
23. Local and absolute minimum $f(\pi/4) = (\pi/4) - 1$, absolute maximum $f(\pi) = \pi$
25. (a) None
(b) Increasing on $(-\infty, \infty)$
(c) None
(d) CD on $(-\infty, 0)$, CU on $(0, \infty)$, IP $(0, 1)$
(e) See graph at right.

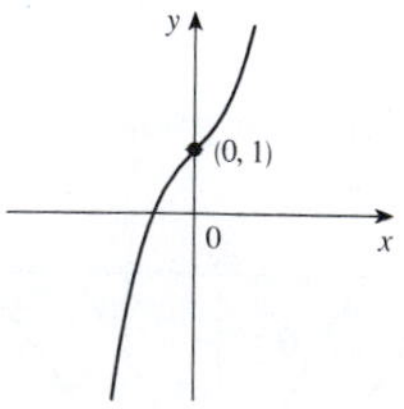

27. (a) HA $y = 0$, VA $x = 0$, $x = 3$
(b) Decreasing on $(-\infty, 0)$, $(0, 1]$, $(3, \infty)$; increasing on $[1, 3)$
(c) Local minimum $f(1) = \frac{1}{4}$
(d) CU on $(0, 3)$, $(3, \infty)$, CD on $(-\infty, 0)$

(e)

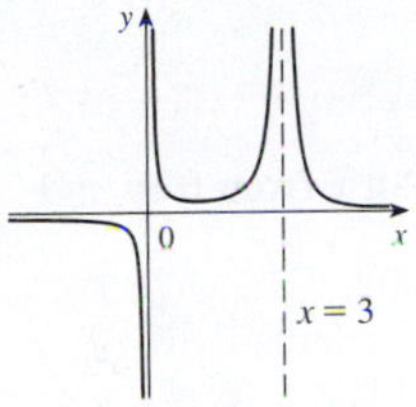

29. (a) None
(b) Increasing on $(-\infty, \frac{10}{3}]$, decreasing on $[\frac{10}{3}, 5]$
(c) Local maximum
$f(\frac{10}{3}) = 10\sqrt{5}/(3\sqrt{3})$
(d) CD on $(-\infty, 5)$
(e) See graph at right.

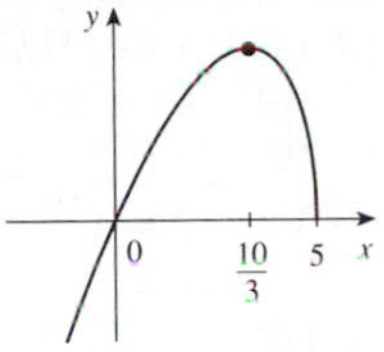

31. (a) VA $x = 0$
(b) Decreasing on $[-1, 0)$ and $(0, 1]$
(c) No local maximum or minimum
(d) CU on $(-1, -\sqrt{2/3})$ and $(0, \sqrt{2/3})$,
CD on $(-\sqrt{2/3}, 0)$ and $(\sqrt{2/3}, 1)$,
IP $(\pm\sqrt{2/3}, \pm 1/\sqrt{2})$
(e) See graph at right.

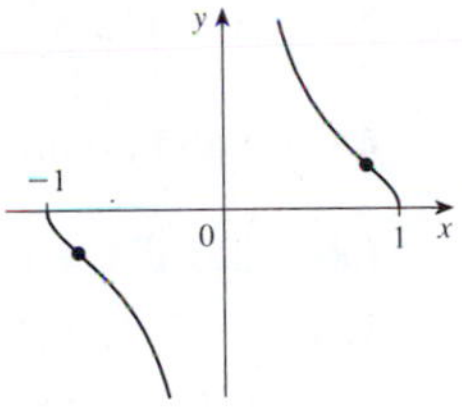

33. (a) None
(b) Decreasing on $[0, (\frac{2}{3})^6]$,
increasing on $[(\frac{2}{3})^6, \infty)$
(c) Local minimum $f((\frac{2}{3})^6) = -\frac{4}{27}$
(d) CU on $(0, (\frac{8}{9})^6)$ and
CD on $((\frac{8}{9})^6, \infty)$, IP $(\frac{8}{9}, -\frac{64}{729})$
(e) See graph at right.

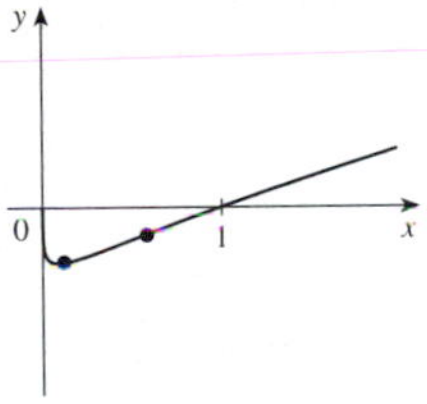

35. (a) HA $y = 0$
(b) Decreasing on $(-\infty, -1]$ and $[1, \infty)$
(c) No local extrema; absolute maximum $f(1) = \frac{\pi}{2}$,
absolute minimum $f(-1) = -\frac{\pi}{2}$
(d) CU on $(1, \infty)$ and CD on $(-\infty, -1)$
(e)

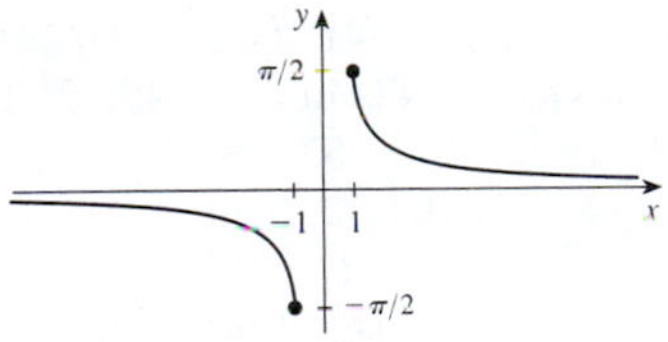

37. (a) None
(b) Decreasing on $(-\infty, \frac{1}{4}\ln 3]$,
increasing on $[\frac{1}{4}\ln 3, \infty)$
(c) Local minimum
$f(\frac{1}{4}\ln 3) = 3^{1/4} + 3^{-3/4}$
(d) CU on $(-\infty, \infty)$
(e) See graph at right.

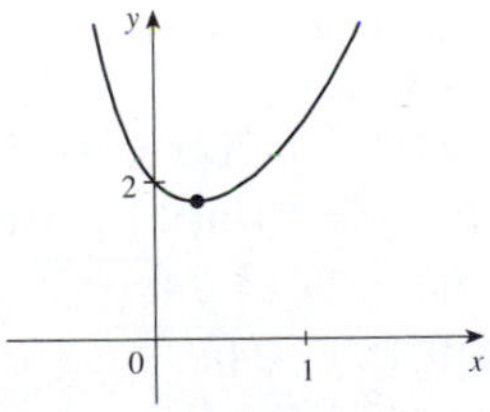

39. Increasing on $[-\sqrt{3}, 0)$ and $(0, \sqrt{3}]$,
decreasing on $(-\infty, -\sqrt{3}]$ and $[\sqrt{3}, \infty)$;
local maximum $f(\sqrt{3}) = 2\sqrt{3}/9$,
minimum $f(-\sqrt{3}) = -2\sqrt{3}/9$;
CU on $(-\sqrt{6}, 0)$ and $(\sqrt{6}, \infty)$, CD on $(-\infty, -\sqrt{6})$ and $(0, \sqrt{6})$;
IP $(\sqrt{6}, 5\sqrt{6}/36)$ and
$(-\sqrt{6}, -5\sqrt{6}/36)$

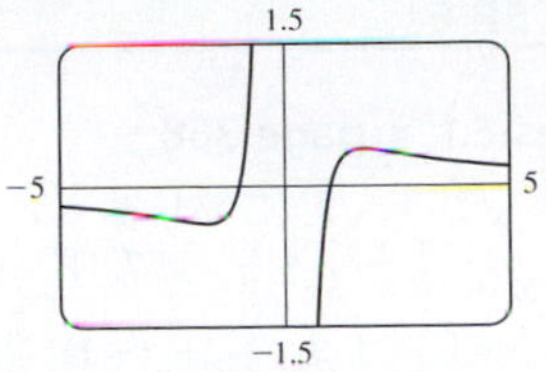

41. Increasing on $[-0.2, 0]$ and $[1.6, \infty)$,
decreasing on $(-\infty, -0.2]$ and $[0, 1.6]$;
local maximum $f(0) = 2$,
minima $f(-0.2) \approx 1.96$ and $f(1.6) \approx -19.2$;
CU on $(-\infty, -0.1)$ and $(1.2, \infty)$, CD on $(-0.1, 1.2)$;
IP $(-0.1, 2.0)$ and $(1.2, -12.1)$

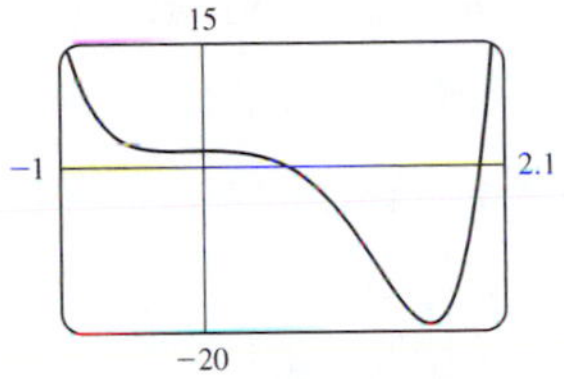

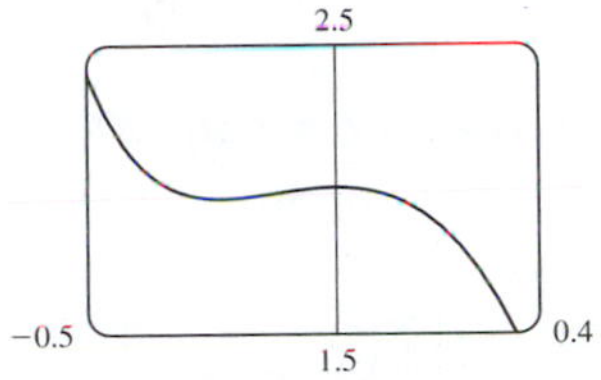

43.

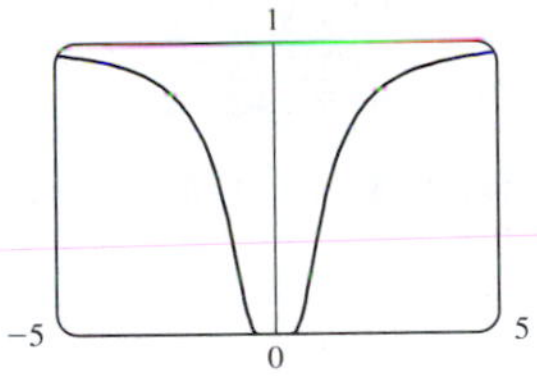

$(\pm\sqrt{2/3}, e^{-3/2})$

45. Maximum at $x = 0$, minima at $x \approx \pm 0.87$,
IP at $x \approx \pm 0.52$

51.

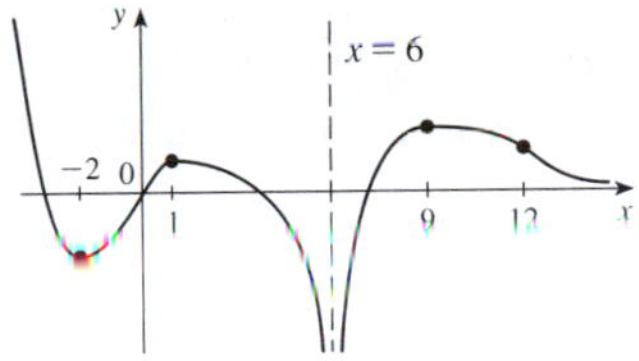

53. $a = -3, b = 7$

57. $3\sqrt{3}r^2$ **59.** $4/\sqrt{3}$ cm from D **61.** $L = C$
63. \$11.50 **65.** $(x^2/2) - (4x^{5/4}/5) + C$
67. $2\sqrt{x} + (2x^{3/2}/3) - (8/3)$
69. $(x^5/20) + (x^3/6) + x - 1$ **71.** $2\tan^{-1}x - 1$
73. $\mathbf{r}(t) = (\frac{1}{2}t^2 - 1)\,\mathbf{i} + (t^2 + t + 1)\,\mathbf{j}$
75. $\mathbf{r}(t) = (\frac{1}{3}t^3 + t + 1)\,\mathbf{i} + (\frac{3}{2}t^2 - t + 2)\,\mathbf{j}$
77. $\mathbf{r}(t) = [\frac{3}{2}t^2 + (6/\sqrt{10})t + 1]\,\mathbf{i} + [-\frac{1}{2}t^2 - (2/\sqrt{10})t + 2]\,\mathbf{j}$
79.

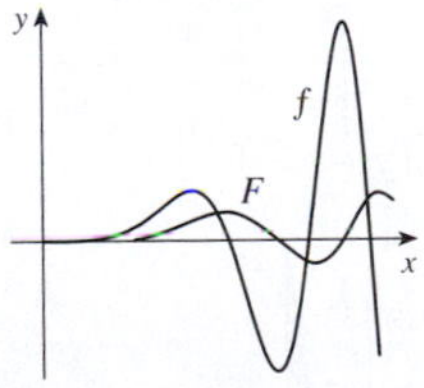

81. No
83. (b) About 8.4 in. by 2 in.
(c) $20/\sqrt{3}$ in. by $20\sqrt{2/3}$ in.

PROBLEMS PLUS ■ page 360

5. $(-2, 4), (2, -4)$ **9.** $\frac{4}{3}$ **11.** $(0, \frac{5}{4})$
17. $(m/2, m^2/4)$ **19.** 1 **21.** $-3.5 < a < -2.5$
23. (a) $x/(x^2 + 1)$ (b) $\frac{1}{2}$ **25.** $2(\sqrt{2} - 1)$
27. $a \leqslant e^{1/e}$

CHAPTER 6

Exercises 6.1 ■ page 368

1. $\sqrt{1} + \sqrt{2} + \sqrt{3} + \sqrt{4} + \sqrt{5}$ **3.** $3^4 + 3^5 + 3^6$
5. $-1 + \frac{1}{3} + \frac{3}{5} + \frac{5}{7} + \frac{7}{9}$ **7.** $1^{10} + 2^{10} + 3^{10} + \cdots + n^{10}$
9. $1 - 1 + 1 - 1 + \cdots + (-1)^{n-1}$ **11.** $\sum_{i=1}^{10} i$
13. $\sum_{i=1}^{19} \frac{i}{i+1}$ **15.** $\sum_{i=1}^{n} 2i$ **17.** $\sum_{i=0}^{5} 2^i$ **19.** $\sum_{i=1}^{n} x^i$
21. 80 **23.** 3276 **25.** 0 **27.** 61 **29.** $n(n+1)$
31. $n(n^2 + 6n + 17)/3$ **33.** $n(n^2 + 6n + 11)/3$
35. $n(n^3 + 2n^2 - n - 10)/4$
41. (a) n^4 (b) $5^{100} - 1$ (c) $\frac{97}{300}$ (d) $a_n - a_0$
43. $\frac{1}{3}$ **45.** 14 **49.** $2^{n+1} + n^2 + n - 2$ **51.** 12

Exercises 6.2 ■ page 377

1. (a) 1 (b) 50
(c)

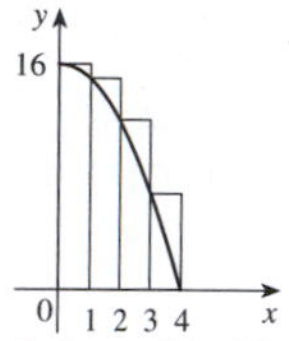

3. (a) 1 (b) 43
(c)

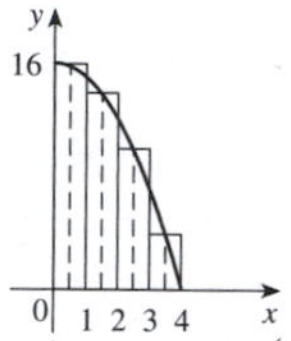

5. (a) 0.5 (b) 12.1875
(c)

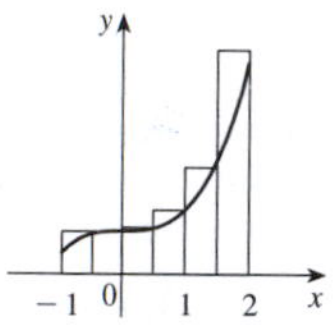

7. (a) $\pi/4$ (b) $\pi(1 + \sqrt{3})/2$
(c)

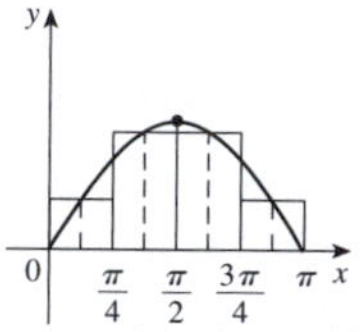

9. (a)

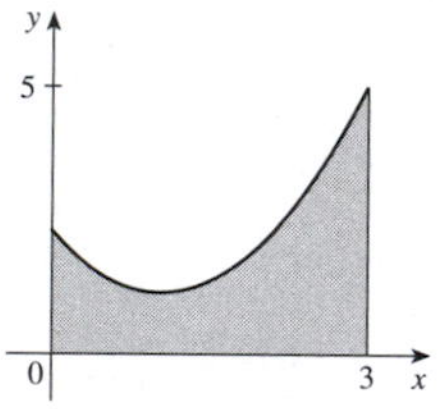

(b) $6 + 9/(2n) + 9/(2n^2)$
(c) 6.875, 6.40625, 6.1953125
(d) 6

11.

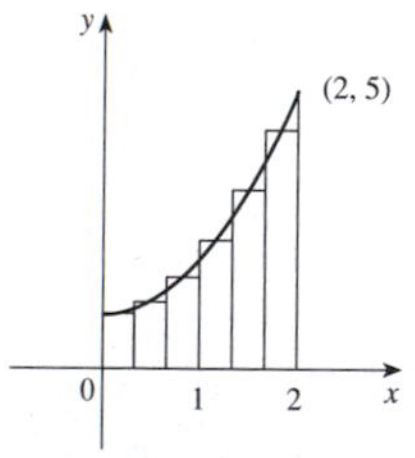

13. 30

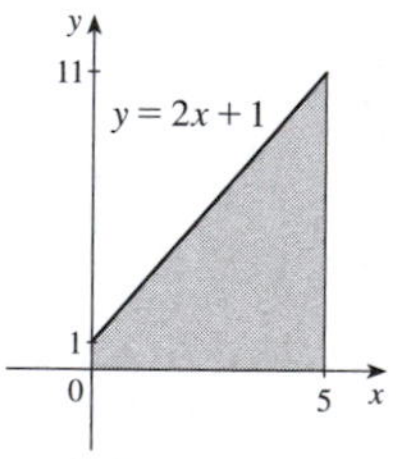

15. $\frac{175}{3}$

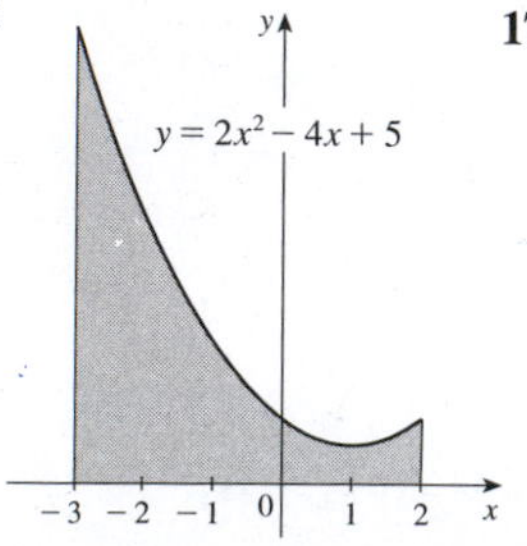

17. $\frac{17}{12}$

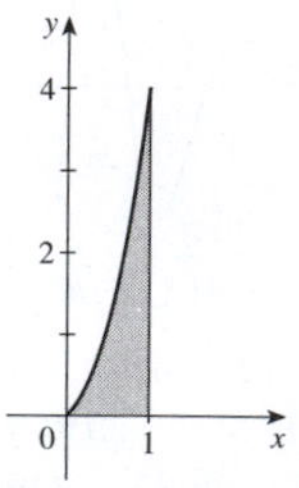

19. 1.9835, 1.9982, 1.9993; 2
21. (a) Left: 4.5148, 4.6165, 4.6366; right: 4.8148, 4.7165, 4.6966
23. The region under the graph of $y = \tan x$ from 0 to $\pi/4$
25. 2

Exercises 6.3 ■ page 386

1. (a) 1.2 (b) 4 **3.** (a) 1 (b) 2.336
5. (a) 0.5 (b) -0.028 **7.** (a) 4 (b) 6 (c) 10
9. 153.125 **11.** 1.8100
13. 1.81001414, 1.81007263, 1.81008347 **15.** $(b - a)c$
17. 15 **19.** $\frac{1}{4}b^4 + 2b^2$ **23.** 10 **25.** $3 + 9\pi/4$
27. 0 **29.** $\int_0^1 (2x^2 - 5x)\,dx$ **31.** $\int_0^\pi \cos x\,dx$
33. $\int_0^1 x^4\,dx$ **35.** $\int_1^3 (3x^5 - 6)\,dx$ **37.** $-\frac{38}{3}$
39. $3\sqrt{3}$ **41.** 22.5 **43.** 2 **45.** $\int_1^{12} f(x)\,dx$
47. $\int_7^{10} f(x)\,dx$ **55.** $\frac{1}{2} \le \int_1^2 dx/x \le 1$
57. $-3 \le \int_{-3}^{0} (x^2 + 2x)\,dx \le 9$
59. $2 \le \int_{-1}^{1} \sqrt{1 + x^4}\,dx \le 2\sqrt{2}$
69. (a) and (c) are integrable **73.** $\frac{1}{2}$

Exercises 6.4 ■ page 397

1. (a) 0, 2, 5, 7, 3
(b) [0, 3] (c) $x = 3$
(d)

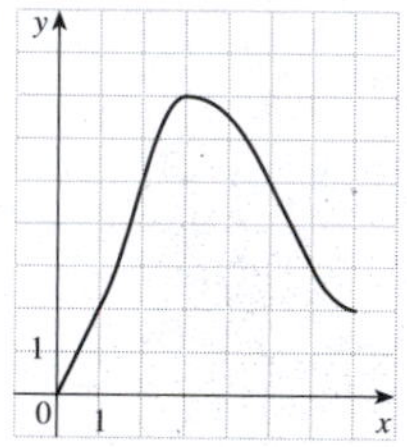

3. $g'(x) = 1 + x^2$

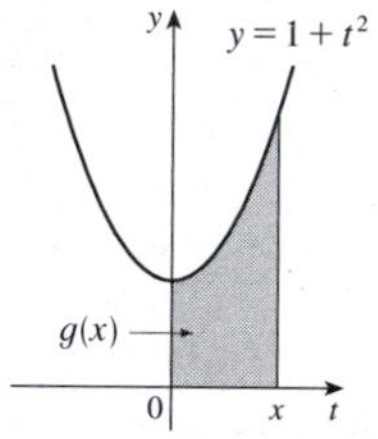

5. $g'(x) = (x^2 - 1)^{20}$ **7.** $g'(u) = 1/(1 + u^4)$
9. $F'(x) = -\cos(x^2)$ **11.** $h'(x) = -\sin^4(1/x)/x^2$
13. $y' = -\sin(\tan^4 x)\sec^2 x$ **15.** $y' = 5/(25x^2 + 10x - 4)$
17. -12 **19.** -1 **21.** 231 **23.** $\frac{16}{3}$ **25.** $\frac{28}{81}$
27. $6(3\sqrt{2} - 2)/5$ **29.** $\frac{29}{35}$ **31.** $\frac{28}{3}$ **33.** 0
35. Does not exist **37.** $\frac{2}{3}$ **39.** $\frac{1}{4}$ **41.** $(\sqrt{2} - 1)/2$
43. Does not exist **45.** $2\sqrt{3}/3$ **47.** ln 2 **49.** $2^8/\ln 2$
51. $\pi/2$ **53.** $\frac{1}{2}e^2 + e - \frac{1}{2}$ **55.** 1 **57.** -3.5
59. 10.7 **61.** $\frac{243}{4}$ **63.** 2 **65.** 0, 1.32; 0.84
71. $\frac{2}{5}x^{5/2} + C$ **73.** $4x - \frac{8}{3}x^{3/2} + \frac{1}{2}x^2 + C$
75. $x^2 + \sec x + C$ **77.** (a) $-\frac{3}{2}$ m (b) $\frac{41}{6}$ m
79. (a) $v(t) = \frac{1}{2}t^2 + 4t + 5$ (b) $\frac{1250}{3}$ m **81.** $46\frac{2}{3}$ kg
83. 1.4 mi
85. $g'(x) = -2(2x - 1)/(2x + 1) + 3(3x - 1)/(3x + 1)$
87. $y' = 3x^{7/2}\sin(x^3) - (\sin\sqrt{x})/(2\sqrt[4]{x})$ **89.** $\sqrt{257}$
91. (a) $-2\sqrt{n}$, $\sqrt{4n - 2}$, n any integer > 0
(b) $(0, 1)$, $(-\sqrt{4n - 1}, -\sqrt{4n - 3})$, and $(\sqrt{4n - 1}, \sqrt{4n + 1})$, n any integer > 0
(c) 0.7
93. (a) Local maxima at 1 and 5; minima at 3 and 7
(b) 9 (c) $(\frac{1}{2}, 2)$, $(4, 6)$, $(8, 9)$
(d) See graph at right.

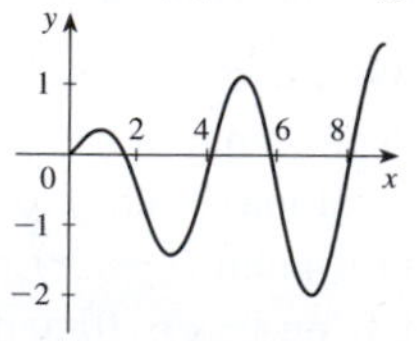

95. $\frac{1}{4}$
101. The weight gained by the child between the ages of 5 and 10
103. $f(x) = x^{3/2}$, $a = 9$

Exercises 6.5 ■ page 407

1. $(x^2 - 1)^{100}/200 + C$ **3.** $-\frac{1}{4}\cos 4x + C$
5. $-1/[2(x^2 + 6x)] + C$ **7.** $(x^2 + x + 1)^4/4 + C$
9. $\frac{2}{3}(x - 1)^{3/2} + C$ **11.** $(2 + x^4)^{3/2}/6 + C$
13. $-2/[5(t + 1)^5] + C$ **15.** $-(1 - 2y)^{2.3}/4.6 + C$
17. $\frac{1}{2}\sin 2\theta + C$ **19.** $\frac{4}{7}(x + 2)^{7/4} - \frac{8}{3}(x + 2)^{3/4} + C$
21. $-\frac{1}{2}\cos(t^2) + C$ **23.** $\frac{1}{7}(1 - x^2)^{7/2} - \frac{1}{5}(1 - x^2)^{5/2} + C$
25. $\frac{2}{3}(1 + \sec x)^{3/2} + C$ **27.** $-\frac{1}{5}\cos^5 x + C$
29. $-\frac{1}{2}\cos(2x + 3) + C$ **31.** $(\sin 3\alpha)x + \frac{1}{3}\cos 3x + C$
33. $2(b + cx^{a+1})^{3/2}/[3c(a + 1)] + C$
35. $\frac{1}{2}\ln|2x - 1| + C$ **37.** $(\ln x)^3/3 + C$
39. $(1 + e^x)^{11}/11 + C$ **41.** $\ln|\ln x| + C$
43. $x - e^{-x} + C$ **45.** $\frac{1}{2}\ln|x^2 + 2x| + C$
47. $\tan^{-1}x + \frac{1}{2}\ln(1 + x^2) + C$
49. $\dfrac{-1}{6(3x^2 - 2x + 1)^3} + C$

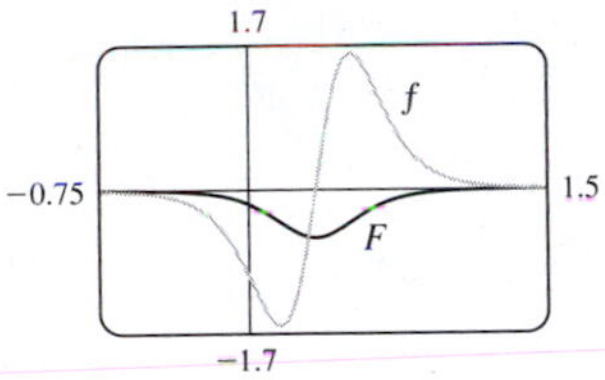

51. $\frac{1}{4}\sin^4 x + C$

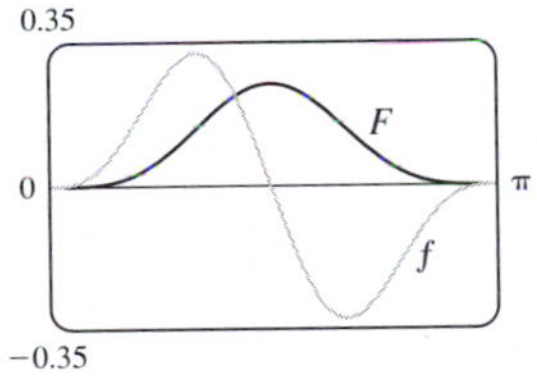

53. $\frac{1}{101}$ **55.** [illegible] **57.** $\frac{16}{15}$ **59.** 0
61. $(4\sqrt{2}/3) - (5\sqrt{5}/12)$ **63.** 1 **65.** 3
67. Does not exist **69.** $\frac{1}{3}(2\sqrt{2} - 1)a^3$ **71.** $\frac{1}{2}\ln 3$
73. 2 **75.** $\sqrt{3} - \frac{1}{3}$ **77.** 6π
79. $[5/(4\pi)][1 - \cos(2\pi t/5)]$ L **81.** 5 **87.** $\pi^2/4$

Exercises 6.6 ■ page 415

1. (b) 0.405

Chapter 6 Review Exercises ■ page 416

1. True **3.** True **5.** True **7.** False **9.** True
11. True **13.** False **15.** False
17. $f = c$, $f' = b$, $\int_0^x f(t)\,dt = a$
19. 7.75

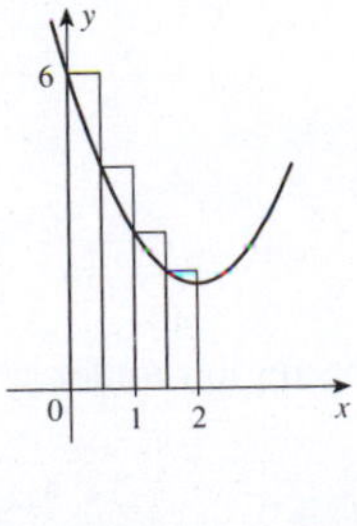

21. -18 **23.** $\frac{875}{12}$ **25.** $\frac{9}{10}$ **27.** $\frac{1209}{28}$
29. 3480 **31.** 2 **33.** Does not exist
35. $-1/[25(2 + x^5)^5] + C$ **37.** $-(\cos \pi x)/\pi + C$
39. $-\sin(1/t) + C$ **41.** $\pi/6$ **43.** $\ln 2 - \frac{7}{4}$
45. $\ln(e^x + 1) + C$ **47.** $2e^{\sqrt{x}} + C$
49. $-\frac{1}{2}[\ln(\cos x)]^2 + C$ **51.** $\frac{1}{4}\ln(1 + x^4) + C$
53. $\ln|1 + \sec\theta| + C$ **55.** $\frac{1}{3}\sinh 3t + C$
57. 4 **59.** $2\sqrt{1 + \sin x} + C$ **61.** $\frac{64}{5}$
63. $F'(x) = \sqrt{1 + x^4}$ **65.** $g'(x) = 3x^5/\sqrt{1 + x^9}$
67. $(2\cos x - \cos\sqrt{x})/(2x)$
69. $4 \leqslant \int_1^3 \sqrt{x^2 + 3}\,dx \leqslant 4\sqrt{3}$ **75.** 1.11 **77.** 8π
79. $F(x) = \int_1^x t^2\sin(t^2)\,dt$ **81.** $c \approx 1.62$ **83.** $2/e$
85. $\frac{2}{3}$ **89.** $f(x) = e^{2x}(2x + 1)/(1 - e^{-x})$

APPLICATIONS PLUS ■ page 419

1. (a) $\sqrt{800} \approx 28$ ft
(b) $dI/dt = -480k(h - 4)/[(h - 4)^2 + 1600]^{5/2}$, where k is the constant of proportionality
3. (a) 22.9125 mi (b) 21.675 mi
5. (b) Average expenditure over $[0, t]$. Minimize average expenditure.
7. (a) $2/\pi$ (b) $1/\pi$ (c) $2/(5\pi)$
9. (a) $T_1 = D/c_1$, $T_2 = (2h\sec\theta)/c_1 + (D - 2h\tan\theta)/c_2$, $T_3 = \sqrt{4h^2 + D^2}/c_1$
(c) $c_1 = 4$ km/s, $c_2 = 6$ km/s, $h = 1/\sqrt{5}$ km

CHAPTER 7

Exercises 7.1 ■ page 427

1. $\frac{20}{3}$ **3.** $\frac{8}{5}$
5. $\frac{1}{6}$

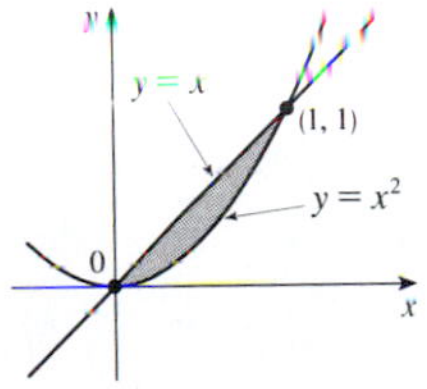

7. $\frac{1}{3}$

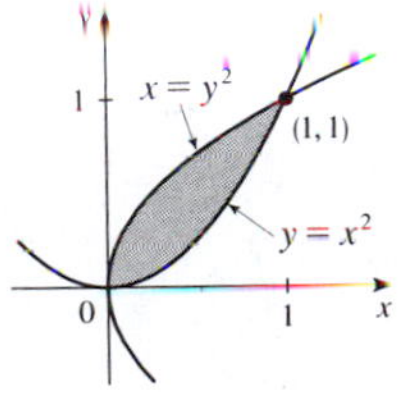

9. $\frac{4}{3}$

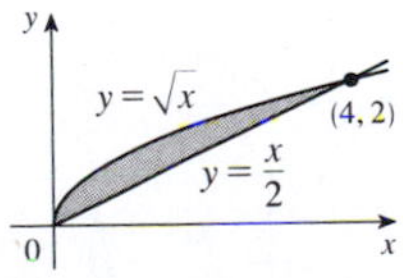

11. 4

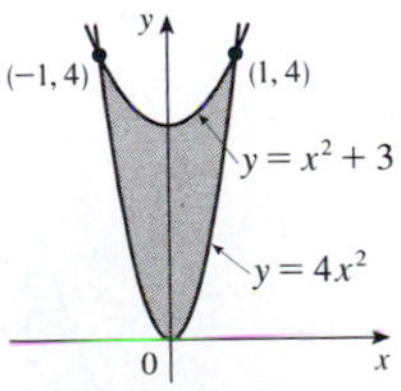

13. 36

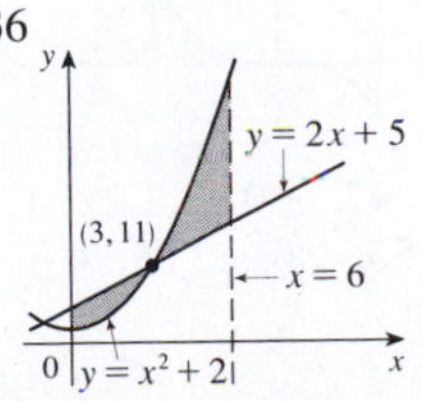

15. $\frac{32}{3}$

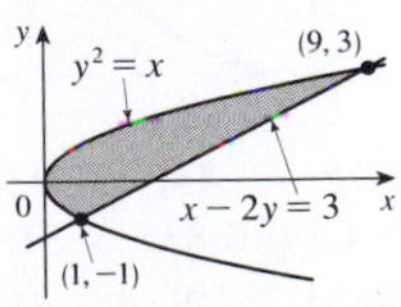

17. $\frac{8}{3}$

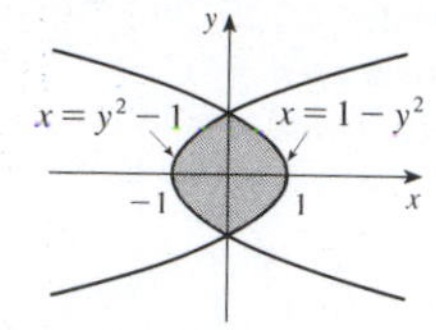

19. $\frac{5}{32}\pi^2 + (1/\sqrt{2}) - 2$

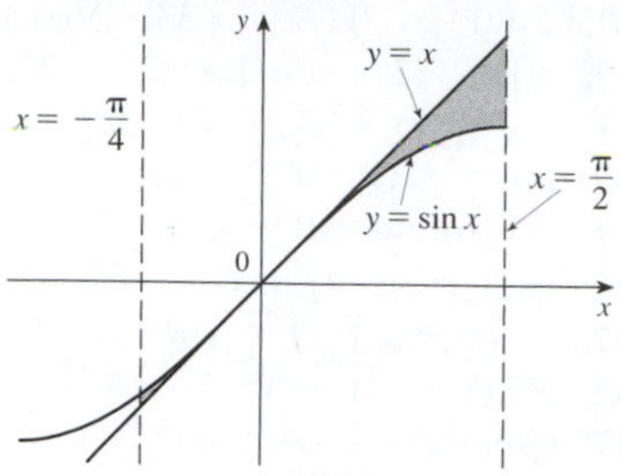

21. $\frac{1}{2}$

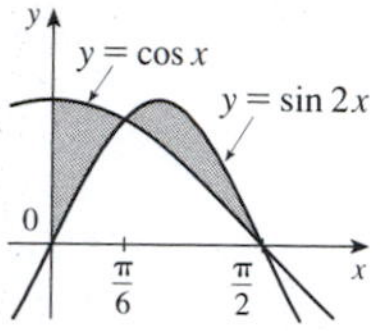

23. $\frac{1}{2}$

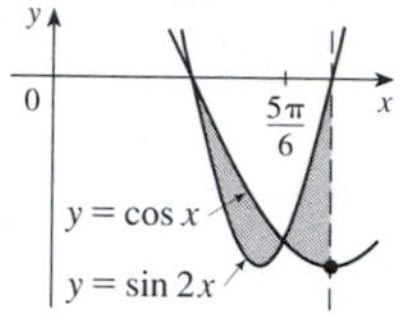

25. 34

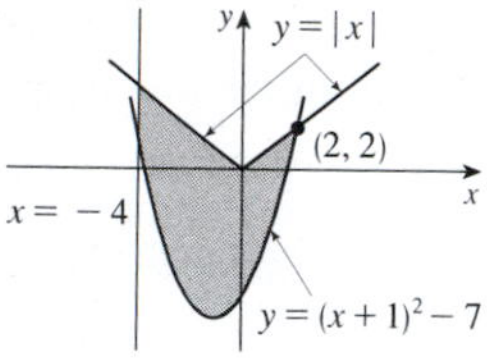

27. 12

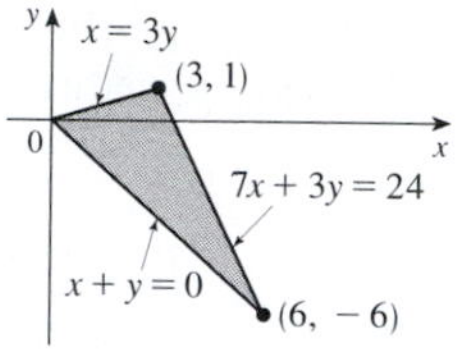

29. $\ln 2 - \frac{1}{2}$

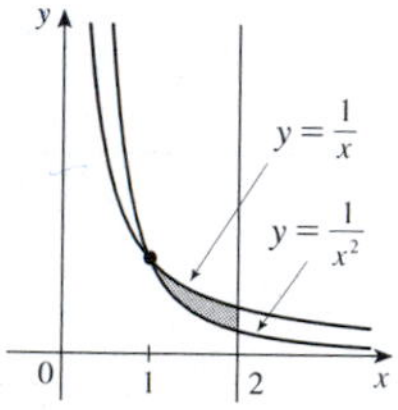

31. $\pi - \frac{2}{3}$

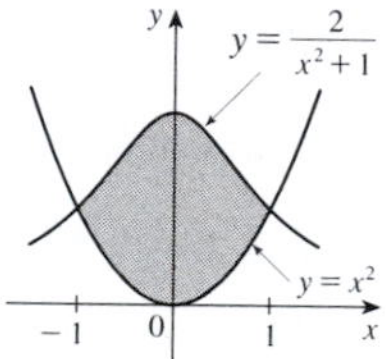

33. $\frac{1}{3}e^3 - e + \frac{2}{3}$ **35.** 9 **37.** 14.5

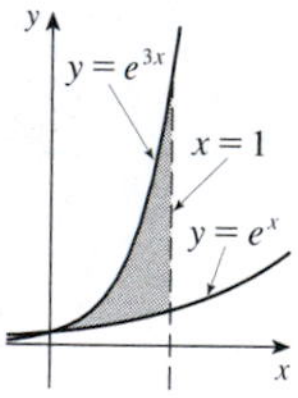

39. 1.5

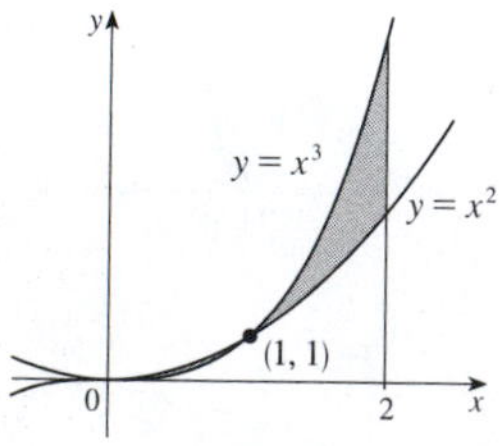

41. 3.75

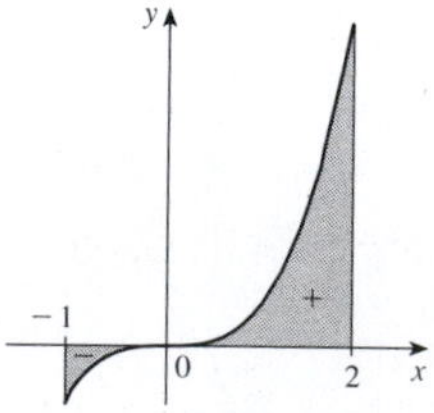

43. 3.22 **45.** $-1.02, 1.02$; 2.70 **47.** $-0.72, 1.22$; 1.38
49. 0, 1.19; 0.83 **51.** $24\sqrt{3}/5$ **53.** ± 6 **55.** $4^{2/3}$
57. (a) A (b) A (c) About 2.25
59. $0 < m < 1$; $m - \ln m - 1$

Exercises 7.2 ■ page 438

1. $\pi/5$

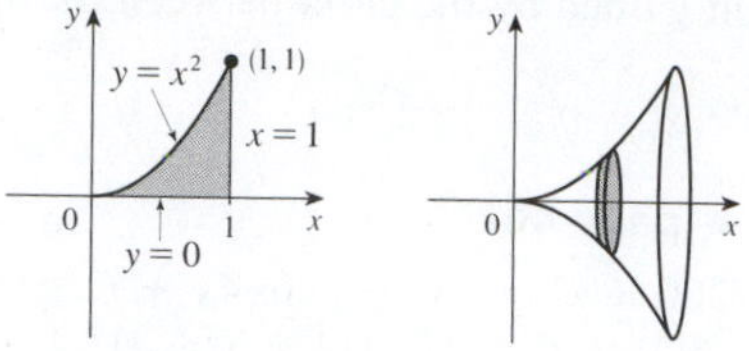

3. $\pi/3$

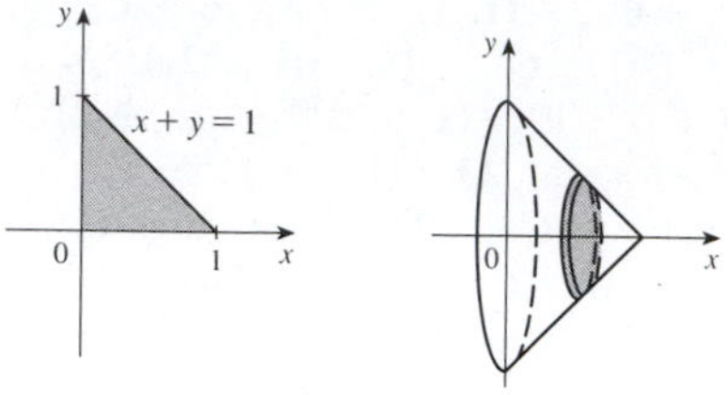

5. 8π

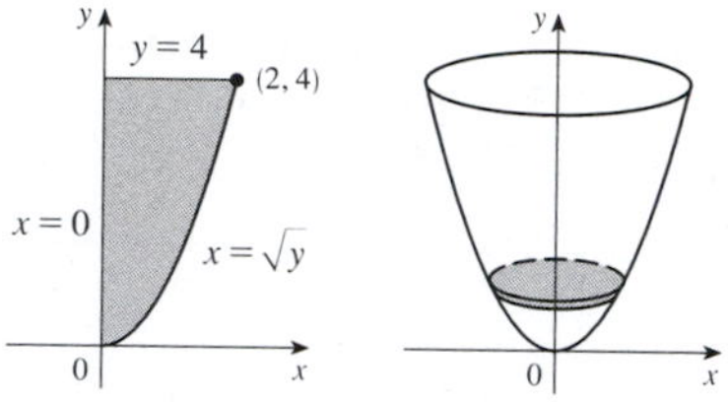

7. $3\pi/10$

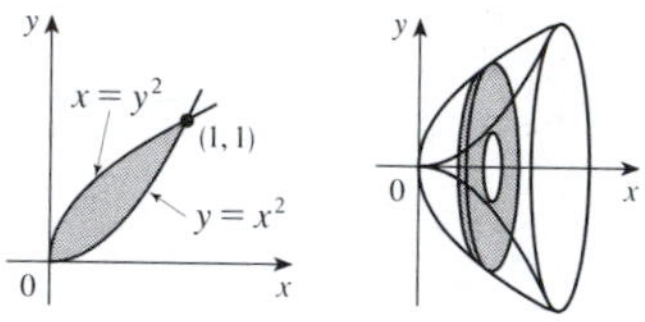

9. $64\pi/15$

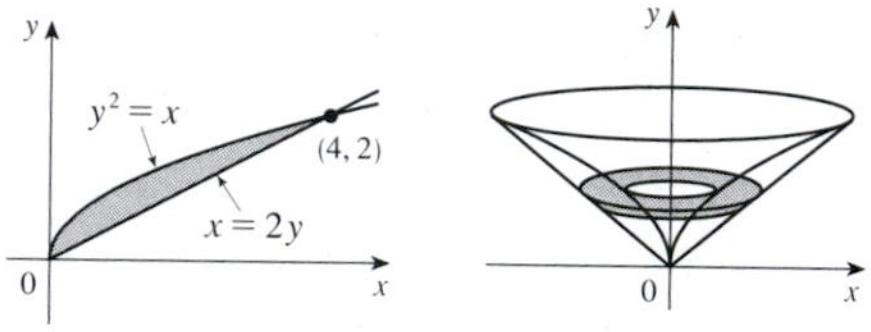

11. $208\pi/45$

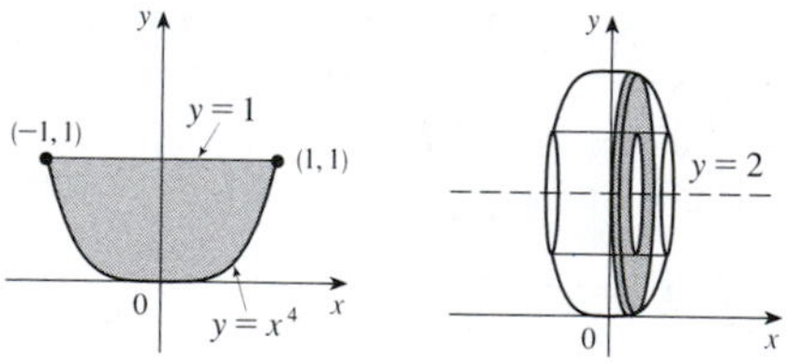

13. $32\pi/3$ **15.** $128\pi/3$ **17.** $128\pi/15$ **19.** $112\pi/15$
21. $64\pi/5$ **23.** $16\pi/5$ **25.** $46\pi/15$ **27.** $\pi(e^2 - 1)/2$
29. $2\pi(\tan 1 - 1)$ **31.** 3π **33.** $\pi \int_1^e [1 - (\ln x)^2]\,dx$
35. $\pi \int_3^6 \{[6 - (x - 4)^2]^2 - (8 - x)^2\}\,dx$
37. $\pi \int_0^{\pi/2} [(1 + \cos x)^2 - 1]\,dx$ **39.** -0.72 1.22; 5.80
41. 21π

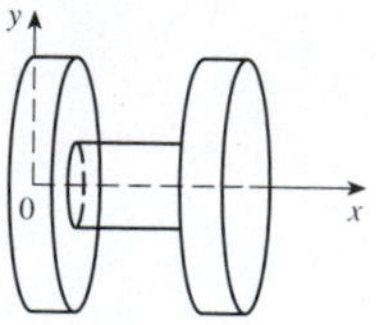

43. Solid obtained by rotating the region under $y = \tan x$ from 0 to $\pi/4$ about the x-axis

45. Solid obtained by rotating the region between $x = \sqrt{y}$ and $x = y$ about the y-axis
47. Solid obtained by rotating the region bounded by $y = 2x$ and $y = 2x^2$ about the line $y = 5$ [or the region bounded by $y = 5 - 2x$ and $y = 5 - 2x^2$ about the x-axis]
49. $\pi r^2 h/3$ **51.** $\pi h^2[r - (h/3)]$ **53.** $2b^2h/3$
55. 10 cm^3 **57.** 24 **59.** 2 **61.** $\frac{1}{3}$
63. (a) $8\pi R \int_0^r \sqrt{r^2 - y^2}\, dy$ (b) $2\pi^2 r^2 R$ **65.** $\frac{5}{12}\pi r^3$
69. $8 \int_0^r \sqrt{R^2 - y^2}\, \sqrt{r^2 - y^2}\, dy$ **71.** (b) $\pi r^2 h$

Exercises 7.3 ■ page 444

1. $15\pi/2$ **3.** $\pi(1 - 1/e)$ **5.** 8π **7.** $\pi/10$
9. $4096\pi/9$ **11.** $1944\pi/5$ **13.** $5\pi/6$
15. $124\pi/5$

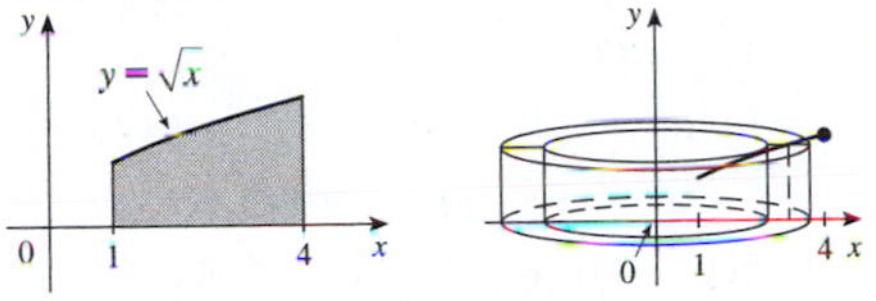

17. $17\pi/6$

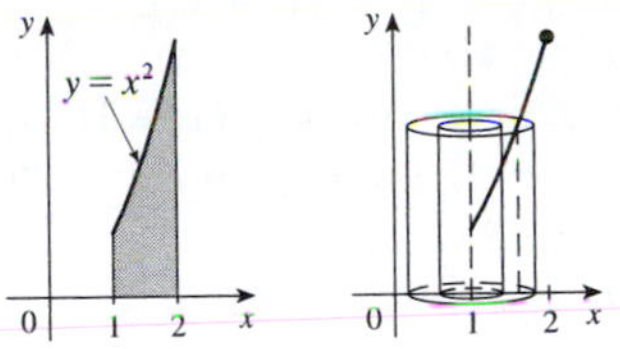

19. 24π

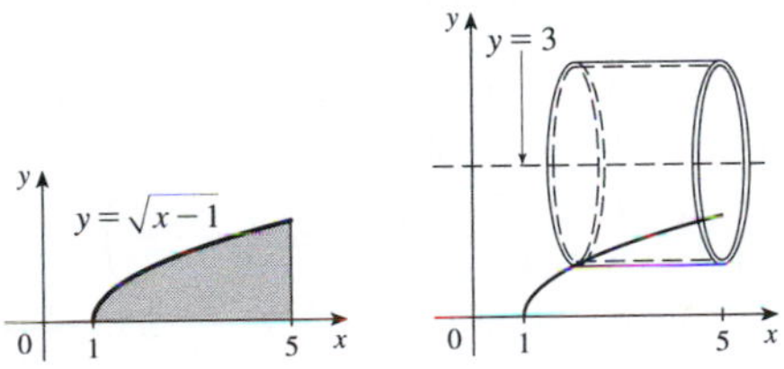

21. $\int_{2\pi}^{3\pi} 2\pi x \sin x\, dx$ **23.** $\int_0^{\pi/4} 2\pi y \cos y\, dy$
25. $\int_0^1 2\pi(x + 1)[\sin(\pi x/2) - x^4]\, dx$
27. Solid obtained by rotating the region under $y = \cos x$ from 0 to $\pi/2$ about the y-axis
29. Solid obtained by rotating the region bounded by $y = x^2$ and $y = x^6$ about the y-axis
31. 0, 1.32; 4.05 **33.** $81\pi/10$ **35.** $16\pi/15$ **37.** $4\pi/3$
39. $\frac{4}{3}\pi r^3$ **41.** $\frac{1}{3}\pi r^2 h$ **45.** 1.142

Exercises 7.4 ■ page 448

1. 7200 J **3.** $\frac{5030}{3}$ ft-lb **5.** $\frac{15}{4}$ ft-lb **7.** $\frac{25}{24} \approx 1.04$ J
9. 10.8 cm **11.** 625 ft-lb **13.** 650,000 ft-lb
15. 2.45×10^3 J **17.** $\approx 1.06 \times 10^6$ J
19. $\approx 5.8 \times 10^3$ ft-lb **21.** 2.0 m
25. $Gm_1m_2[(1/a) - (1/b)]$

Exercises 7.5 ■ page 451

1. 0 **3.** $\frac{1}{5}$ **5.** $(\sqrt{2} + 4)/(9\pi)$
7. (a) $\frac{8}{3}$ (b) $2/\sqrt{3}$
(c)

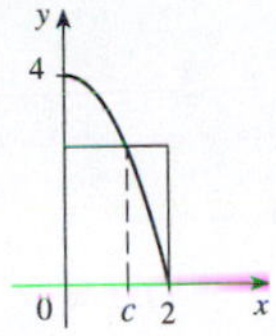

9. (a) 2 (b) ≈ 1.32 (c)

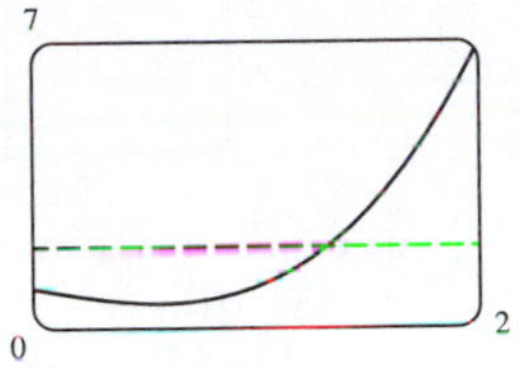

13. $(50 + 28/\pi)\,°\text{F} \approx 59\,°\text{F}$ **15.** 6 kg/m
17. $5/(4\pi) \approx 0.4$ L

Chapter 7 Review Exercises ■ page 452

1. (a) $A = \int_a^b [f(x) - g(x)]\, dx$ (b) $A = \int_c^d [u(y) - v(y)]\, dy$
(c) $V = \pi \int_a^b \{[f(x)]^2 - [g(x)]^2\}\, dx$
(d) $V = 2\pi \int_a^b x[f(x) - g(x)]\, dx$
(e) $V = 2\pi \int_c^d y[u(y) - v(y)]\, dy$
(f) $V = \pi \int_c^d \{[u(y)]^2 - [v(y)]^2\}\, dy$
3. 108 **5.** 1 **7.** $2\sqrt{2}$ **9.** 2π
11. $16\pi/3$ **13.** $4\pi(2ah + h^2)^{3/2}/3$
15. $\int_0^1 \pi[(1 - x^3)^2 - (1 - x^2)^2]\, dx$
17. (a) $2\pi/15$ (b) $\pi/6$ (c) $8\pi/15$
19. (a) 0.38 (b) 0.87
21. Solid obtained by rotating the region under $y = \sin x$ from 0 to π about the x-axis
23. Solid obtained by rotating the region in the first quadrant bounded by $x = 4 - y^2$ and the axes about the x-axis
25. 36 **27.** $125\sqrt{3}/3 \text{ m}^3$ **29.** 3.2 J
31. (a) $8000\pi/3$ ft-lb (b) 2.1 ft **33.** $f(x)$

PROBLEMS PLUS ■ page 455

1. $\pi/2$ **5.** 1 **7.** Does not exist
9. $f(x) = \sqrt{2x/\pi}$ **13.** e^{-2} **15.** $[-1, 2]$
17. $1/\sqrt{2}$ **19.** $\sqrt{1 + \sin^4 x}\, \cos x$ **23.** $c = -1; \frac{1}{2}$
25. $2x \int_0^x \sin(t^2)\, dt + x^2 \sin(x^2)$ **27.** $y = \frac{32}{9}x^2$
31. $\pi/(30\sqrt{2})$

CHAPTER 8

Exercises 8.1 ■ page 463

1. $(xe^{2x}/2) - (e^{2x}/4) + C$ **3.** $-\frac{1}{4}x \cos 4x + \frac{1}{16} \sin 4x + C$
5. $\frac{1}{3}x^2 \sin 3x + \frac{2}{9}x \cos 3x - \frac{2}{27} \sin 3x + C$
7. $x(\ln x)^2 - 2x \ln x + 2x + C$
9. $\frac{1}{8}(\sin 2\theta - 2\theta \cos 2\theta) + C$ **11.** $t^3(3 \ln t - 1)/9 + C$
13. $e^{2\theta}(2 \sin 3\theta - 3 \cos 3\theta)/13 + C$
15. $y \cosh y - \sinh y + C$ **17.** $1 - 2/e$
19. $-\frac{1}{2}$ **21.** $(\pi + 6 - 3\sqrt{3})/6$
23. $\sin x\, (\ln \sin x - 1) + C$ **25.** $(2x + 1)e^x + C$
27. $x(\cos \ln x + \sin \ln x)/2 + C$ **29.** $4 \ln 2 - \frac{3}{2}$

31. $2(\sin\sqrt{x} - \sqrt{x}\cos\sqrt{x}) + C$
33. $e^{x^2}[(x^4/2) - x^2 + 1] + C$
35. $(x\sin\pi x)/\pi + (\cos\pi x)/\pi^2 + C$

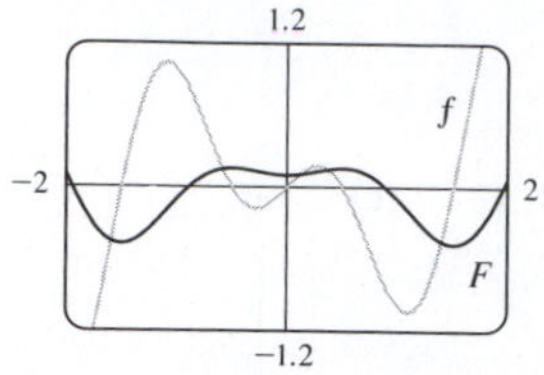

37. (b) $-\frac{1}{4}\cos x\sin^3 x + (3x/8) - \frac{3}{16}\sin 2x + C$
39. (b) $\frac{2}{3}, \frac{8}{15}$ **45.** $x[(\ln x)^3 - 3(\ln x)^2 + 6\ln x - 6] + C$
47. $(\pi + 6\sqrt{3} - 12)/12$ **49.** $0, 0.70;\ 0.08$
51. $10\pi^2$ **53.** $2\pi e$ **55.** $2 - e^{-t}(t^2 + 2t + 2)$ m
59. 1

Exercises 8.2 ■ page 470

1. $\pi/4$ **3.** $\frac{3}{8}x + \frac{1}{4}\sin 2x + \frac{1}{32}\sin 4x + C$
5. $\frac{1}{7}\cos^7 x - \frac{1}{5}\cos^5 x + C$ **7.** $(3\pi - 4)/192$
9. $(3x/2) + \cos 2x - \frac{1}{8}\sin 4x + C$
11. $\frac{1}{6}\sin^6 x - \frac{1}{4}\sin^8 x + \frac{1}{10}\sin^{10} x + C$
[or $-\frac{1}{6}\cos^6 x + \frac{1}{4}\cos^8 x - \frac{1}{10}\cos^{10} x + C_1$]
13. $[\frac{2}{7}\cos^3 x - \frac{2}{3}\cos x]\sqrt{\cos x} + C$
15. $\frac{1}{2}\cos^2 x - \ln|\cos x| + C$ **17.** $\ln(1 + \sin x) + C$
19. $\tan x - x + C$ **21.** $\tan x + \frac{1}{3}\tan^3 x + C$ **23.** $\frac{1}{5}$
25. $\frac{1}{3}\sec^3 x + C$ **27.** $\frac{1}{4}\sec^4 x - \tan^2 x + \ln|\sec x| + C$
29. $\frac{38}{15}$ **31.** $\frac{1}{2}\tan^2 x + C$ **33.** $\sqrt{3} - (\pi/3)$
35. $-\frac{1}{5}\cot^5 x - \frac{1}{7}\cot^7 x + C$ **37.** $\ln|\csc x - \cot x| + C$
39. $\frac{1}{2}[\frac{1}{3}\sin 3x - \frac{1}{7}\sin 7x] + C$ **41.** $\frac{1}{2}[\sin x + \frac{1}{7}\sin 7x] + C$
43. $\frac{1}{2}\sin 2x + C$
45. $-\frac{1}{5}\cos^5 x + \frac{2}{3}\cos^3 x - \cos x + C$

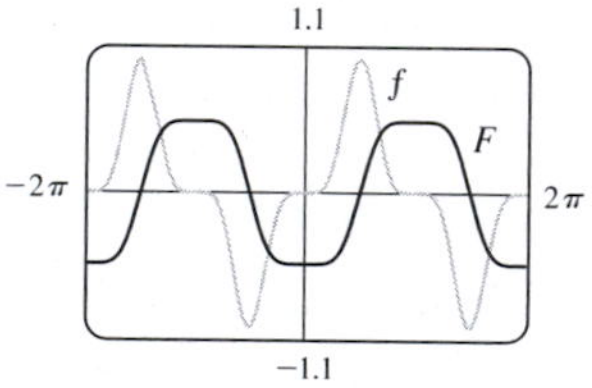

47. 0 **49.** $\frac{1}{3}$ **51.** 0 **53.** $\pi^2/4$ **55.** $2\pi + (\pi^2/4)$
57. $s = (1 - \cos^3\omega t)/(3\omega)$

Exercises 8.3 ■ page 476

1. $2/\sqrt{3}$ **3.** $-\sqrt{1 - x^2} + C$
5. $\frac{1}{4}\sin^{-1}(2x) + \frac{1}{2}x\sqrt{1 - 4x^2} + C$ **7.** $\ln(1 + \sqrt{2})$
9. $[\sqrt{x^2 - 16}/(32x^2)] + \frac{1}{128}\sec^{-1}(x/4) + C$
11. $\sqrt{9x^2 - 4} - 2\sec^{-1}(3x/2) + C$
13. $(x/\sqrt{a^2 - x^2}) - \sin^{-1}(x/a) + C$
15. $(1/\sqrt{3})\ln|(\sqrt{x^2 + 3} - \sqrt{3})/x| + C$
17. $\frac{64}{1215}$ **19.** $\frac{5}{3}(1 + x^2)^{3/2} + C$
21. $\frac{1}{2}[\sin^{-1}(x - 1) + (x - 1)\sqrt{2x - x^2}] + C$
23. $\frac{1}{3}\ln|3x + 1 + \sqrt{9x^2 + 6x - 8}| + C$
25. $\frac{1}{2}[\tan^{-1}(x + 1) + (x + 1)/(x^2 + 2x + 2)] + C$
27. $\frac{1}{2}[e^t\sqrt{9 - e^{2t}} + 9\sin^{-1}(e^t/3)] + C$ **33.** $0.81, 2;\ 2.10$

35. $r\sqrt{R^2 - r^2} + \pi r^2/2 - R^2\arcsin(r/R)$ **37.** $2\pi^2 Rr^2$

Exercises 8.4 ■ page 485

1. $\dfrac{A}{x - 1} + \dfrac{B}{x + 2}$ **3.** $\dfrac{A}{2x - 1} + \dfrac{B}{(2x - 1)^2} + \dfrac{C}{2x + 3}$
5. $\dfrac{A}{x} + \dfrac{B}{x^2} + \dfrac{C}{x^3} + \dfrac{D}{x - 1}$ **7.** $1 + \dfrac{A}{x - 1} + \dfrac{B}{x + 1}$
9. $\dfrac{A}{x} + \dfrac{Bx + C}{x^2 + 2}$ **11.** $\dfrac{Ax + B}{x^2 + 1} + \dfrac{Cx + D}{x^2 + 4} + \dfrac{Ex + F}{(x^2 + 4)^2}$
13. $\dfrac{Ax + B}{x^2 + 9} + \dfrac{Cx + D}{(x^2 + 9)^2} + \dfrac{Ex + F}{(x^2 + 9)^3}$
15. $\dfrac{A}{x} + \dfrac{B}{x^2} + \dfrac{Cx + D}{x^2 + x + 2}$ **17.** $\dfrac{x^2}{2} - x + \ln|x + 1| + C$
19. $\ln 3 + 3\ln 6 - 3\ln 4 = \ln\frac{81}{8}$
21. $3x - 7\ln|2x + 3| + C$
23. $x - \ln|x| + 2\ln|x - 1| + C = x + \ln\dfrac{(x - 1)^2}{|x|} + C$
25. $2\ln 2 + \frac{1}{2}$ **27.** $4\ln 6 - 3\ln 5$
29. $-1/[5(x - 1)] + \frac{1}{25}\ln|(x + 4)/(x - 1)| + C$
31. $2\ln|x| + 3\ln|x + 2| + (1/x) + C$
33. $\frac{1}{3}\ln|x^3 + 3x^2 + 4| + C$
35. $\ln|x + 1| + 2/(x + 1) - 1/[2(x + 1)^2] + C$
37. $(1/x) + \frac{1}{2}\ln|(x - 1)/(x + 1)| + C$ **39.** $(1 - \ln 2)/2$
41. $\ln\sqrt{3} - (\sqrt{3}\pi/18)$
43. $\ln(x - 1)^2 + \ln\sqrt{x^2 + 1} - 3\tan^{-1}x + C$
45. $\frac{1}{3}\ln|x - 1| - \frac{1}{6}\ln(x^2 + x + 1) - \dfrac{1}{\sqrt{3}}\tan^{-1}\dfrac{2x + 1}{\sqrt{3}} + C$
47. $\ln|x - 1| - \ln\sqrt{x^2 + 1} + [1/(x - 1)] + \tan^{-1}x + C$
49. $\frac{3}{2}\ln(x^2 + 1) - 3\tan^{-1}x + \sqrt{2}\tan^{-1}(x/\sqrt{2}) + C$
51. $\dfrac{-1}{2(x^2 + 2x + 4)} - \dfrac{2\sqrt{3}}{9}\tan^{-1}\left(\dfrac{x + 1}{\sqrt{3}}\right) - \dfrac{2(x + 1)}{3(x^2 + 2x + 4)} + C$
53. $\ln|\sin^2 x - 3\sin x + 2| + C$ **55.** $-\frac{1}{2}\ln 3 \approx -0.55$
59. $\frac{1}{2}\ln|(x - 2)/x| + C$
61. $\frac{1}{2}\ln|x^2 + x - 1| - \dfrac{\sqrt{5}}{10}\ln\left|\dfrac{2x + 1 - \sqrt{5}}{2x + 1 + \sqrt{5}}\right| + C$
63. $1 + 2\ln 2$ **65.** $2\pi\ln\frac{9}{8}$
67. (a) $\dfrac{24{,}110}{4879}\dfrac{1}{5x + 2} - \dfrac{668}{323}\dfrac{1}{2x + 1} - \dfrac{9438}{80{,}155}\dfrac{1}{3x - 7} + \dfrac{1}{260{,}015}\dfrac{22{,}098x + 48{,}935}{x^2 + x + 5}$
(b) $\dfrac{4822}{4879}\ln|5x + 2| - \dfrac{334}{323}\ln|2x + 1| - \dfrac{3146}{80{,}155}\ln|3x - 7| + \dfrac{11{,}049}{260{,}015}\ln(x^2 + x + 5) + \dfrac{75{,}772}{260{,}015\sqrt{19}}\tan^{-1}\dfrac{2x + 1}{\sqrt{19}} + C$
The CAS omits the absolute value signs and the constant of integration.

Exercises 8.5 ■ page 488

1. $2(1 - \ln 2)$ **3.** $2(\sqrt{x} - \tan^{-1}\sqrt{x}) + C$
5. $\frac{3}{2}\ln|x^{2/3} - 1| + C$ **7.** $\frac{1676}{15}$
9. $\frac{4}{3}(\sqrt{1 + \sqrt{x}})^3 - 4\sqrt{1 + \sqrt{x}} + C$
11. $x + 4\sqrt{x} + 4\ln|\sqrt{x} - 1| + C$
13. $\frac{3}{10}(x^2 + 1)^{5/3} - \frac{3}{4}(x^2 + 1)^{2/3} + C$

15. $2\sqrt{x} - 4\sqrt[4]{x} + 4\ln(1 + \sqrt[4]{x}) + C$
17. $\sqrt{x - x^2} - \tan^{-1}\sqrt{(1 - x)/x} + C$
19. $\ln[(e^x + 2)^2/(e^x + 1)] + C$
21. $2\sqrt{1 - e^x} + \ln[(1 - \sqrt{1 - e^x})/(1 + \sqrt{1 - e^x})] + C$
23. $-\sqrt{2}\ln(\sqrt{2} - 1) = (1/\sqrt{2})\ln(3 + 2\sqrt{2})$
25. $\frac{1}{5}\ln|(2\tan(x/2) + 1)/(\tan(x/2) - 2)| + C$
27. $\frac{1}{4}\ln|\tan(x/2)| + \frac{1}{8}\tan^2(x/2) + C$
29. $\dfrac{1}{\sqrt{a^2 + b^2}}\ln\left|\dfrac{b\tan(x/2) - a + \sqrt{a^2 + b^2}}{b\tan(x/2) - a - \sqrt{a^2 + b^2}}\right| + C$

Exercises 8.6 ■ page 494

1. $2x + 11\ln|x - 3| + C$ **3.** $\frac{1}{3}\sin^3 x - \frac{1}{5}\sin^5 x + C$
5. $1 - (\sqrt{3}/2)$ **7.** $2\sqrt{x - 2} - 4\tan^{-1}(\sqrt{x - 2}/2) + C$
9. $x\ln(1 + x^2) - 2x + 2\tan^{-1}x + C$
11. $\frac{4097}{45}$ **13.** $\frac{1}{2}\ln(x^2 - 2x + 2) + \tan^{-1}(x - 1) + C$
15. $3\ln|(3 - \sqrt{9 - x^2})/x| + \sqrt{9 - x^2} + C$
17. $(x^2 + 2)\sinh x - 2x\cosh x + C$ **19.** $\tan^{-1}(\sin x) + C$
21. $2/\pi$ **23.** $e^{3x}(5\sin 5x + 3\cos 5x)/34 + C$
25. $\frac{1}{2}[\ln|x + 1| - \frac{1}{2}\ln(x^2 + 1) + \tan^{-1}x] + C$
27. $-\frac{1}{3}(x^3 + 1)e^{-x^3} + C$
29. $\frac{1}{3}\ln|3x + 2 + \sqrt{9x^2 + 12x - 5}| + C$
31. $\frac{3}{4}x^{4/3} - \frac{6}{11}x^{11/6} + C$ **33.** $\frac{1}{6}\tan^{-1}[(x^2 + 1)/3] + C$
35. $\frac{1}{16}(x - \frac{1}{4}\sin 4x + \frac{1}{3}\sin^3 2x) + C$
37. $-\ln(1 + \sqrt{1 - x^2}) + C$
39. $\frac{1}{2}\ln|(e^x - 1)/(e^x + 1)| + C$ **41.** 0 **43.** $\frac{86}{3}$
45. $\frac{1}{2}(\ln\sin x)^2 + C$ **47.** $\frac{1}{6}\ln[(x^2 + 1)/(x^2 + 4)] + C$
49. $\frac{3}{7}(x + c)^{7/3} - \frac{3}{4}c(x + c)^{4/3} + C$
51. $3\ln(\sqrt{x + 1} + 3) - \ln(\sqrt{x + 1} + 1) + C$
53. $\frac{1}{2}(x + 2)\sin 2x - \frac{1}{4}(2x^2 + 8x - 7)\cos 2x + C$
55. $\frac{1}{2}\sin^{-1}(x^2/4) + C$ **57.** $\frac{1}{6}\csc^3 2x - \frac{1}{10}\csc^5 2x + C$
59. $e^{\arctan x} + C$ **61.** $-\frac{1}{8}e^{-2t}(4t^3 + 6t^2 + 6t + 3) + C$
63. $\frac{1}{24}\cos 6x - \frac{1}{16}\cos 4x - \frac{1}{8}\cos 2x + C$
65. $\sin^{-1}x - \sqrt{1 - x^2} + C$
67. $\ln\sqrt{x^2 + a^2} + \tan^{-1}(x/a) + C$ **69.** $\frac{1}{20}\tan^{-1}(x^5/4) + C$
71. $x\sec x - \ln|\sec x + \tan x| + C$
73. $\frac{2}{3}[(x + 1)^{3/2} - x^{3/2}] + C$
75. $2\sqrt{x}\tan^{-1}\sqrt{x} - \ln(1 + x) + C$
77. $e^{-x} + \frac{1}{2}\ln|(e^x - 1)/(e^x + 1)| + C$
79. $\frac{2}{5}\tan^{-1}(\sqrt{2x - 25}/5) + C$

Exercises 8.7 ■ page 499

1. $\frac{1}{25}e^{-3x}(-3\cos 4x + 4\sin 4x) + C$
3. $(-\sqrt{9x^2 - 1}/x) + 3\ln|3x + \sqrt{9x^2 - 1}| + C$
5. $e^{3x}(9x^2 - 6x + 2)/27 + C$
7. $\frac{1}{2}[x^2\sin^{-1}(x^2) + \sqrt{1 - x^4}] + C$ **9.** $\tan^{-1}(\sinh e^x) + C$
11. $\frac{1}{2}(x + 2)\sqrt{5 - 4x - x^2} + \frac{9}{2}\sin^{-1}[(x + 2)/3] + C$
13. $\frac{1}{4}\tan x\sec^3 x + \frac{3}{8}\tan x\sec x + \frac{3}{8}\ln|\sec x + \tan x| + C$
15. $\frac{1}{9}\sin^3 x\,[3\ln(\sin x) - 1] + C$
17. $-2\sqrt{2 + 3\cos x} - \sqrt{2}\ln\left|\dfrac{\sqrt{2 + 3\cos x} - \sqrt{2}}{\sqrt{2 + 3\cos x} + \sqrt{2}}\right| + C$
19. $\frac{8}{15}$ **21.** $\frac{1}{5}\ln|x^5 + \sqrt{x^{10} - 2}| + C$
23. $(1 + e^x)\ln(1 + e^x) - e^x + C$
25. $\sqrt{e^{2x} - 1} - \cos^{-1}(e^{-x}) + C$ **27.** $(2\pi/25)(\ln 6 - \frac{5}{6})$
31. $-\frac{1}{4}x(5 - x^2)^{3/2} + \frac{5}{8}x\sqrt{5 - x^2} + \frac{25}{8}\sin^{-1}(x/\sqrt{5}) + C$
33. $-\frac{1}{5}\sin^2 x\cos^3 x - \frac{2}{15}\cos^3 x + C$
35. $\frac{1}{10}(1 + 2x)^{5/2} - \frac{1}{6}(1 + 2x)^{3/2} + C$
37. $\frac{1}{2}\tan^2 x - \frac{1}{2}\ln(1 + \tan^2 x) + C = \frac{1}{2}\tan^2 x - \ln|\sec x| + C$
39. $F(x) = \frac{1}{2}\ln(x^2 - x + 1) - \frac{1}{2}\ln(x^2 + x + 1)$,
maximum at -1, minimum at 1,
IP at -1.7, 0, and 1.7

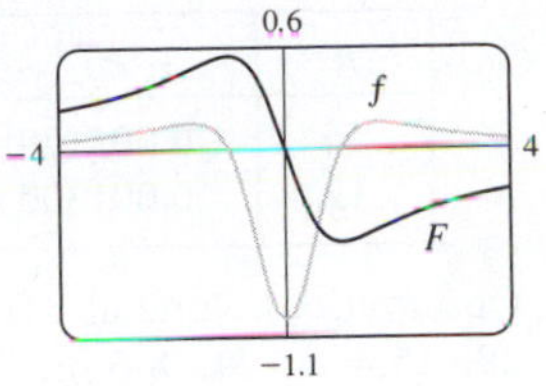

41. $F(x) = -\frac{1}{10}\sin^3 x\cos^7 x - \frac{3}{80}\sin x\cos^7 x + \frac{1}{160}\cos^5 x\sin x + \frac{1}{128}\cos^3 x\sin x + \frac{3}{256}\cos x\sin x + \frac{3}{256}x$

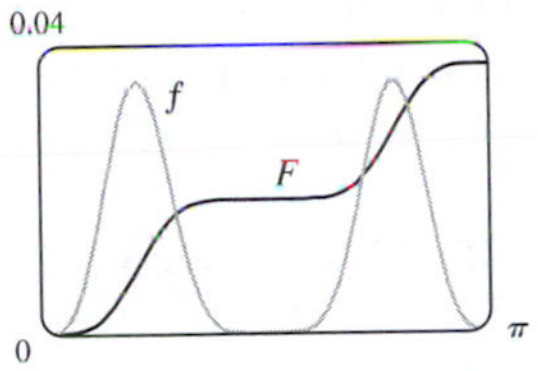

Exercises 8.8 ■ page 507

1. (a) $L_2 = 6$, $R_2 = 12$, $M_2 \approx 9.8$
(b) L_2 is an underestimate, R_2 and M_2 are overestimates.
(c) $T_2 = 9 < I$ (d) $L_n < T_n < I < M_n < R_n$
3. (a) 1.913972 (b) 1.934766
5. (a) 0.481672 (b) 0.481172
7. (a) 0.746211 (b) 0.747131 (c) 0.746825
9. (a) 0.132465 (b) 0.132857 (c) 0.132727
11. (a) 0.409140 (b) 0.388849 (c) 0.395802
13. (a) 2.031893 (b) 2.014207 (c) 2.020651
15. (a) 1.098004 (b) 1.098709 (c) 1.109031
17. (a) $T_{10} \approx 0.881839$, $M_{10} \approx 0.882202$
(b) $|E_T| \leqslant 0.013$, $|E_M| \leqslant 0.006$
19. (a) $T_{10} \approx 1.719713$, $E_T \approx -0.001432$,
$S_{10} \approx 1.718283$, $E_S \approx -0.000001$
(b) $|E_T| \leqslant 0.002266$, $|E_S| \leqslant 0.0000016$
21. $n = 130$ for T_n, $n = 92$ for M_n
23. (a) 2.8 (b) 7.954926518 (c) 0.287 (d) 7.954926521
(e) The actual error is much smaller (f) 10.9
(g) 7.953789422 (h) 0.0593
(i) The actual error is smaller (j) $n \geqslant 50$

25.

n	L_n	R_n	T_n	M_n
4	0.140625	0.390625	0.265625	0.242188
8	0.191406	0.316406	0.253906	0.248047
16	0.219727	0.282227	0.250977	0.249512

n	E_L	E_R	E_T	E_M
4	0.109375	−0.140625	−0.015625	0.007813
8	0.058594	−0.066406	−0.003906	0.001953
16	0.030273	−0.032227	−0.000977	0.000488

27.

n	T_n	M_n	S_n
6	4.661488	4.669245	4.666563
12	4.665367	4.667316	4.666659

n	E_T	E_M	E_S
6	0.005179	−0.002578	0.000104
12	0.001300	−0.000649	0.000007

Observations same as after Example 1.

29. 15.4 **31.** 8.6 mi **33.** (a) 11.5 (b) 12 (c) $11.\overline{6}$
35. 12.3251

Exercises 8.9 ■ page 516

1. $\frac{1}{2} - 1/(2t^2)$; 0.495, 0.49995, 0.4999995; 0.5 **3.** Divergent
5. $\frac{1}{2}$ **7.** Divergent **9.** 1 **11.** 0 **13.** $-\ln\frac{2}{3}$
15. Divergent **17.** Divergent **19.** $e^2/4$ **21.** Divergent
23. Divergent **25.** 1 **27.** $2\sqrt{3}$ **29.** Divergent
31. Divergent **33.** $\frac{5}{3}$ **35.** Divergent **37.** Divergent
39. Divergent **41.** $-\frac{1}{4}$
43. e

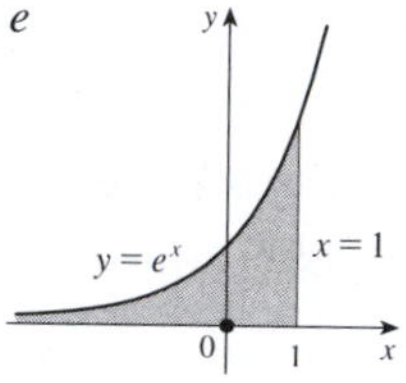

45. $\pi/2$

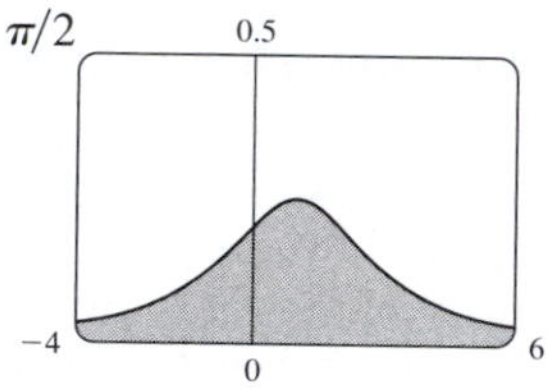

47. Infinite

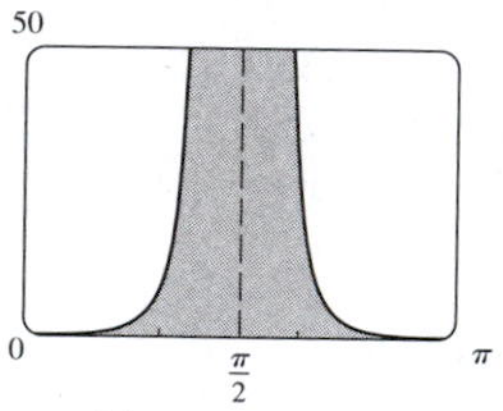

49. Convergent
51. Convergent
53. Divergent
55. π
57. $1/(1-p)$, $p < 1$
59. $-1/(p+1)^2$, $p > -1$
65. $\sqrt{2GM/R}$
67. (a) $F(s) = 1/s$, $s > 0$ (b) $F(s) = 1/(s-1)$, $s > 1$
(c) $F(s) = 1/s^2$, $s > 0$
73. $C = 1$, $\ln 2$

Chapter 8 Review Exercises ■ page 519

1. False **3.** False **5.** False **7.** False
9. $x - 2\ln|x+1| + C$ **11.** $\frac{1}{6}(\arctan x)^6 + C$
13. $-1/e^{\sin x} + C$ **15.** $\frac{1}{25}x^5(5\ln x - 1) + C$
17. $-\frac{1}{2}\cos(x^2) + C$ **19.** $\frac{1}{3}\ln|(x-2)/(2x-1)| + C$
21. $\frac{1}{9}\sec^9 x - \frac{3}{7}\sec^7 x + \frac{3}{5}\sec^5 x - \frac{1}{3}\sec^3 x + C$
23. $\sqrt{1+2x} - 3\ln(\sqrt{1+2x} + 3) + C$ **25.** $2e^{\sqrt{x}} + C$
27. $-x/\sqrt{x^2-1} + C$ **29.** $\ln|x| - \frac{1}{2}\ln(x^2+1) + C$
31. $-\cot x - x + C$
33. $2\ln|x-1| - \frac{1}{2}\tan^{-1}[(x+1)/2] + C$
35. $-\frac{1}{12}(\cot^3 4x + 3\cot 4x) + C$ **37.** $\ln x\,[\ln(\ln x) - 1] + C$
39. $\frac{1}{2}\ln(2x + 1 + \sqrt{4x^2+4x+5}) + C$
41. $\frac{1}{2}\sin 2x - \frac{1}{8}\cos 4x + C$ **43.** $\frac{2}{5}$ **45.** Divergent
47. $1 - (\pi/2)$ **49.** $\frac{1}{24}$ **51.** 2 **53.** $8\ln\frac{4}{3} - 1$
55. $\pi/4$ **57.** $\sqrt{3} - (\pi/3)$
59. $-a/(a^2+b^2)$ if $a < 0$, divergent if $a \geq 0$
61. $(x+1)\ln(x^2+2x+2) + 2\arctan(x+1) - 2x + C$
63. $\frac{1}{2}[e^x\sqrt{1-e^{2x}} + \sin^{-1}(e^x)] + C$
65. $\frac{1}{4}(2x+1)\sqrt{x^2+x+1} +$
$\frac{3}{8}\ln\left|x + \frac{1}{2} + \sqrt{x^2+x+1}\right| + C$
69. (a) 1.090608 (b) 1.088840 (c) 1.089429
71. (a) 0.0067 (b) $0.00\overline{3}$
73. (a) 3.8 (b) 1.7867, 0.000646 (c) $n \geq 30$
75. Convergent **77.** 2 **79.** $3\pi^2/16$ **81.** No

APPLICATIONS PLUS ■ page 521

1. (b) At most 40%, $\frac{5}{36}$ (c) $-4 + 20\ln 1.25 \approx 0.46$
3. (b) $f(x) = L\ln[(L + \sqrt{L^2-x^2})/x] - \sqrt{L^2-x^2}$
5. (b) (i) ≈2.86 slug-ft/s (ii) ≈286 lb
7. (a) 3.285 s (b) 165 ft, direct, relayed (c) 112.8 ft/s
9. (b) 8.3, fourth row, 48.5° (c) 8.25306209
(d) Average 36°, minimum 22°

CHAPTER 9

Exercises 9.1 ■ page 533

1. $y = -1/(x+C)$ or $y = 0$ **3.** $x^2 - y^2 = C$
5. $y = Ce^{1/x}$ **7.** $u = -\ln(C - \frac{1}{2}e^{2t})$ **9.** $ye^y = x^3 - 8$
11. $x = \sqrt{2(t-1)e^t + 3}$ **13.** $u = 1 - \sqrt{t^2+t+4}$
15. $f(x) = e^{x^4/4}$ **17.** $y = 7e^{x^4}$
19. $y = \ln(e^x + e - 1)$ **21.** $\cos y = \cos x - 1$

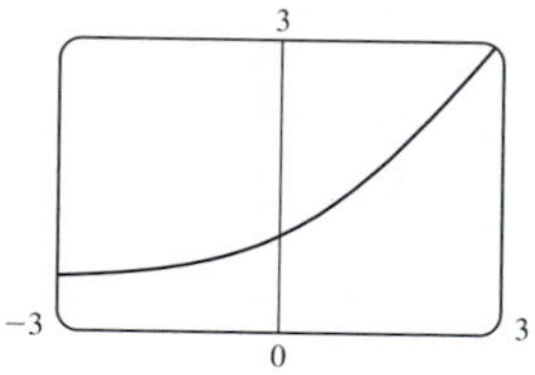

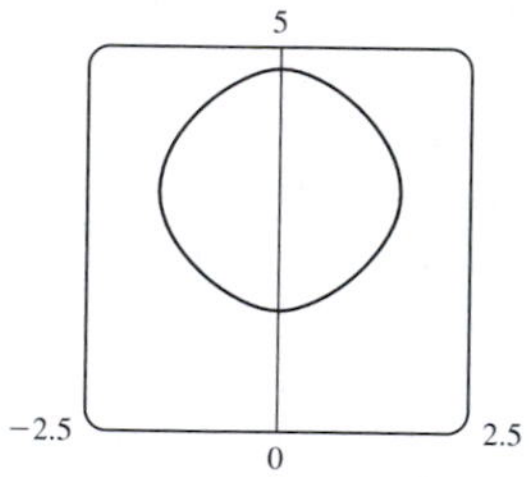

23.

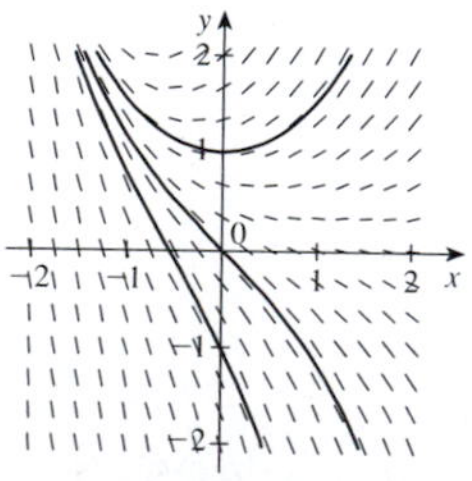

25.

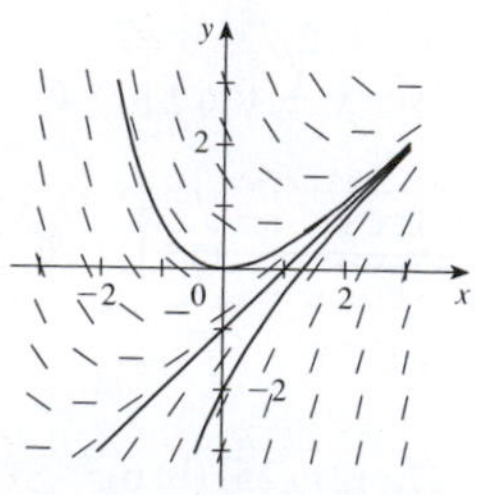

27.

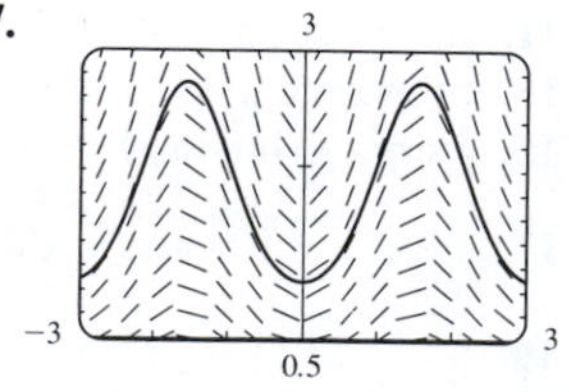

29. (a) $15e^{-t/100}$ kg (b) $15e^{-0.2} \approx 12.3$ kg
31. $x = ab(e^{(b-a)kt} - 1)/(be^{(b-a)kt} - a)$
33. (a) $5 \times 10^9 e^{0.02(t-1986)}$ (b) (i) 6.6 billion (ii) 49 billion (iii) 146 trillion (c) ≈270,000; ≈37,000; ≈12
35. (a) $dy/dt = ky(1 - y)$ (b) $y = y_0/[y_0 + (1 - y_0)e^{-kt}]$ (c) 3:36 P.M.
39. $k = 1/90$

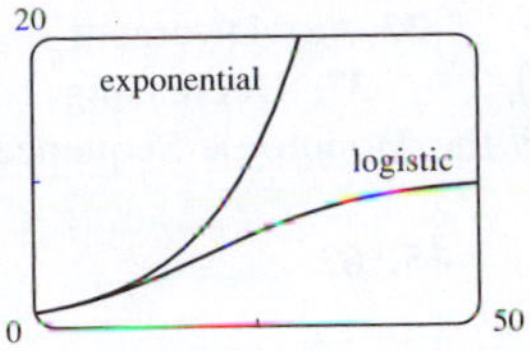

41. (a) $dA/dt = k\sqrt{A}(M - A)$
(b) $A = M[(Ce^{\sqrt{M}kt} - 1)/(Ce^{\sqrt{M}kt} + 1)]^2$, where $C = (\sqrt{M} + \sqrt{A_0})/(\sqrt{M} - \sqrt{A_0})$ and $A_0 = A(0)$. [If $A_0 = 0$, then $C = 1$.]

Exercises 9.2 ■ page 540

1. Not linear **3.** Linear **5.** $y = Ce^{3x} - \frac{1}{2}e^x$
7. $y = Ce^{x^2} - \frac{1}{2}$ **9.** $y = \frac{1}{2}\sec x - \cos x + C\sec x$
11. $y = \frac{1}{2}x + Ce^{-x^2} - \frac{1}{2}e^{-x^2}\int e^{x^2}\,dx$
13. $y = \tan\theta + C\sec\theta$
15. $y = x - 1 + \frac{1}{2}(e^x + e^{-x}) = x - 1 + \cosh x$
17. $y = (x^2 + 3)e^{x^2}$
19. $y = (\sin x)/x^2$
21. $y = \sin x + (\cos x)/x + C/x$

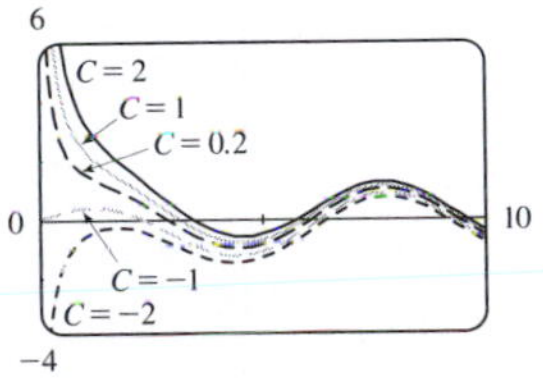

25. $y = \pm[Cx^4 + 2/(5x)]^{-1/2}$
27. (a) $I(t) = 4 - 4e^{-3t}$ (b) $4 - 4e^{-1/2} \approx 1.57$ A
29. $Q(t) = 3(1 - e^{-4t})$, $I(t) = 12e^{-4t}$
31. $P(t) = M + Ce^{-kt}$

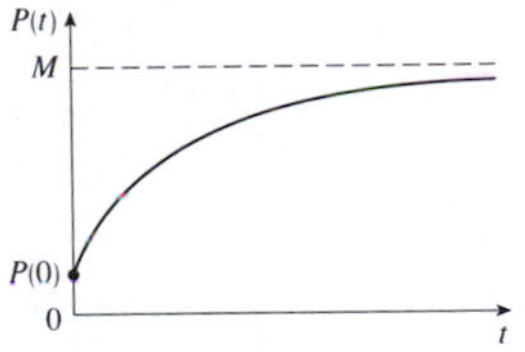

33. (a) $(mg/c)(1 - e^{-ct/m})$ (b) mg/c
(c) $(mg/c)[t + (m/c)e^{-ct/m}] - m^2g/c^2$

Exercises 9.3 ■ page 546

1. $4\sqrt{5}$ **3.** $\sqrt{2}(e^{\pi} - 1)$

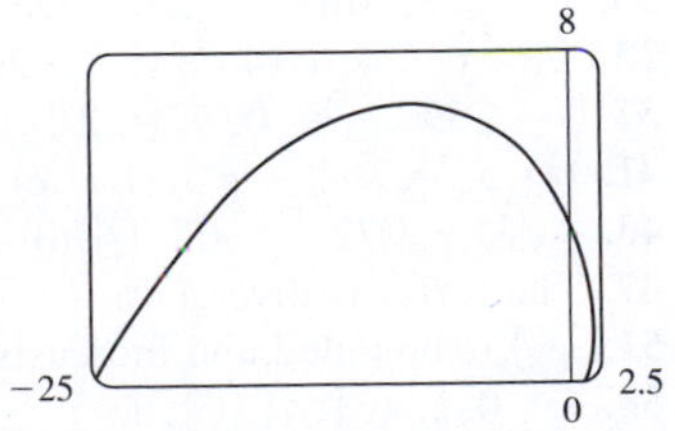

5. $(13\sqrt{13} - 8)/27$

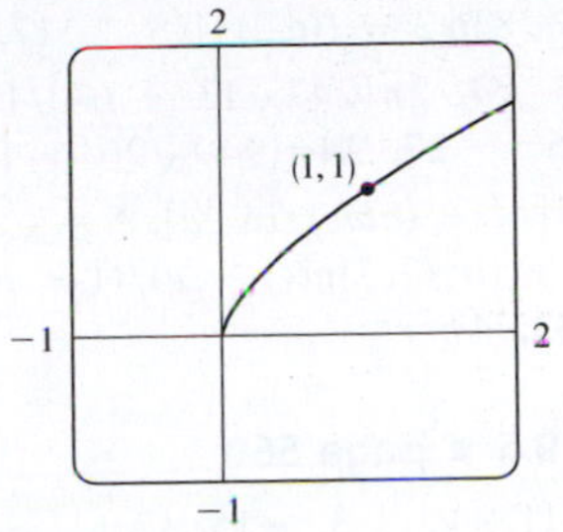

7. $\frac{4}{3}$ **9.** $\frac{181}{9}$ **11.** 1.548 **13.** 3.820
15. (a), (b)

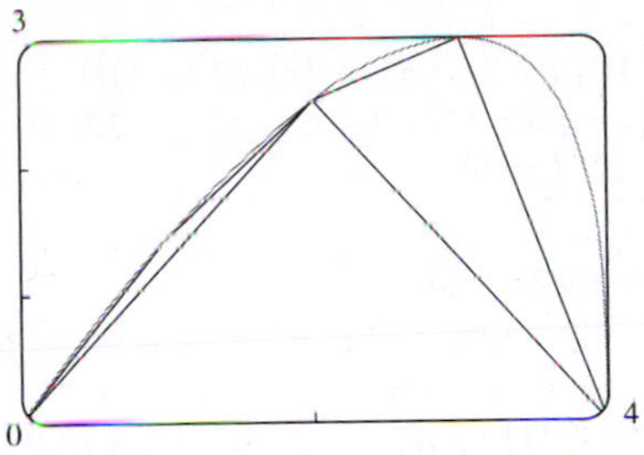

$L_1 = 4$, $L_2 \approx 6.43$, $L_4 \approx 7.50$

(c) $\int_0^4 \sqrt{1 + [4(3 - x)/(3(4 - x)^{2/3})]^2}\,dx$ (d) 7.7988
17. $(81 \ln 3)/512$ **19.** $\ln(\sqrt{2} + 1)$
21. (a)

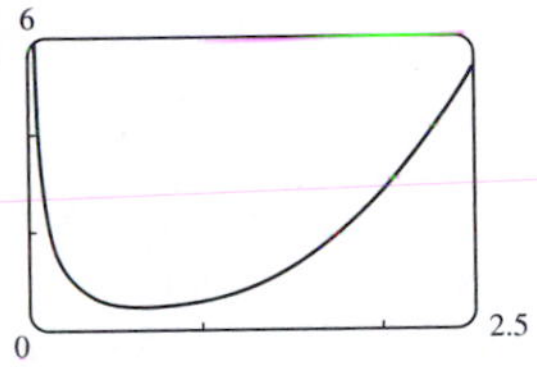

(b) $s(x) = \frac{1}{3}x^3 - 1/(4x) - \frac{1}{12}$, $x > 0$
(c)

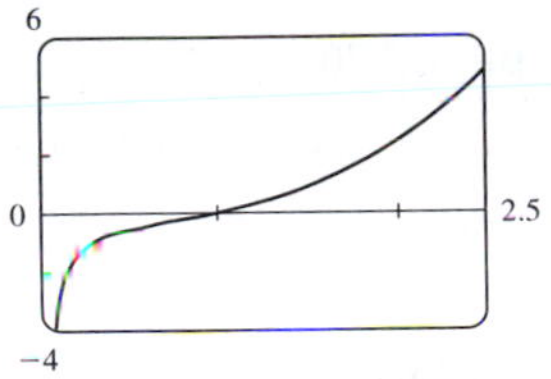

23. (a) $2a\sinh(b/a)$ (b) ≈30 ft **25.** 29.36
29. (a)

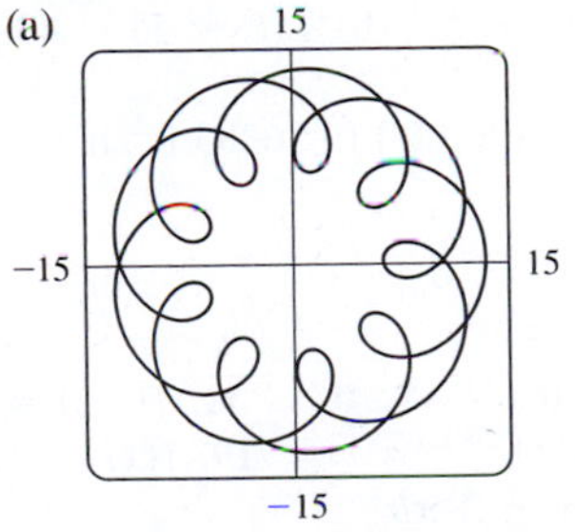

$0 \leqslant t \leqslant 4\pi$ (b) ≈294

Exercises 9.4 ■ page 553

1. $\pi(37\sqrt{37} - 17\sqrt{17})/6$ **3.** $\pi(145\sqrt{145} - 1)/27$
5. $2\pi[\sqrt{2} + \ln(\sqrt{2} + 1)]$ **7.** $\pi[1 + \frac{1}{4}(e^2 - e^{-2})]$
9. $21\pi/2$ **11.** $\pi(145\sqrt{145} - 10\sqrt{10})/27$
13. $\dfrac{\pi}{4}\left[2e\sqrt{1 + 4e^2} - 2\sqrt{5} + \ln\left[\dfrac{2e + \sqrt{1 + 4e^2}}{2 + \sqrt{5}}\right]\right]$

15. $\pi[21 - 8\ln 2 - (\ln 2)^2]/8$ **17.** ≈ 3.44
19. $\pi/4$ **23.** $2\pi(247\sqrt{13} + 64)/1215$
25. $6\pi a^2/5$ **27.** $24\pi(949\sqrt{26} + 1)/5$
29. (a) $2\pi[b^2 + (ab/e)\sin^{-1}e]$, $e = \sqrt{a^2 - b^2}/a$ = eccentricity
(b) $2\pi\{a^2 + (b^2/2e)\ln[(1 + e)/(1 - e)]\}$
31. $\frac{1}{4}$ **33.** $4\pi^2 r^2$

Exercises 9.5 ■ page 560

1. 40, 12, $(1, \frac{10}{3})$ **3.** -15, 17, $(\frac{17}{11}, -\frac{15}{11})$ **5.** $(1.5, 1.2)$
7. $(\frac{11}{13}, \frac{49}{13})$ **9.** $(0, \pi/8)$ **11.** $(1/(e - 1), (e + 1)/4)$
13. $(0.4, 0.5)$
15. $((\pi\sqrt{2} - 4)/[4(\sqrt{2} - 1)], 1/[4(\sqrt{2} - 1)])$
17. $\frac{4}{3}$, 0, $(0, \frac{2}{3})$ **19.** $-\frac{44}{3}$, 0, $(0, -\frac{11}{15})$ **23.** $(0, \frac{1}{12})$
25. $(\frac{9}{10}, \frac{3}{2})$ **27.** $\frac{1}{3}\pi r^2 h$

Exercises 9.6 ■ page 563

1. (a) 9.8 kPa (b) 1.96×10^4 N (c) 4.90×10^3 N
3. 6.5×10^6 N **5.** $1000 g\pi r^3$ N **7.** 3.00×10^3 lb
9. 1.56×10^3 lb **11.** 3.47×10^4 lb **13.** 5.27×10^5 N
15. (a) 314 N (b) 353 N **17.** 8.32×10^4 N
19. (a) 5.63×10^3 lb (b) 5.06×10^4 lb (c) 4.88×10^4 lb
(d) 3.03×10^5 lb

Exercises 9.7 ■ page 569

1. \$14,516,000 **3.** \$388,280,000 **5.** \$316.29
7. \$4166.67 **9.** \$112,500 **11.** \$87,924.80
13. (b) \$62,500 **15.** $16(2\sqrt{2} - 1)/3 \approx$ \$9.75 million
17. 1.19×10^{-4} cm^3/s **19.** $\frac{1}{9}$ L/s

Chapter 9 Review Exercises ■ page 570

1. $y = \sqrt[3]{(3x^2/2) - 3\cos x + K}$ **3.** $y = 2 \pm \sqrt{2\tan^{-1}x + C}$
5. $y = \sqrt{(\ln x)^2 + 4}$ **7.** $y = x^2 - 1 + Ce^{-x^2}$
9. $y = ke^{x + x^3/3} - 2$ **11.** $y = e^{-x}(\frac{2}{3}x^{3/2} + 3)$
13. $\frac{2}{3}(5\sqrt{5} - 2\sqrt{2})$ **15.** (a) $\frac{17}{12}$ (b) $\frac{47}{16}\pi$ **17.** 1.297
19. $56\sqrt{3}\pi a^2/5$ **21.** $(-\frac{1}{2}, \frac{12}{5})$ **23.** $(2, \frac{2}{3})$ **25.** $2\pi^2$
27. 458 lb **29.** \$7166.67
31. (a) $L(t) = L_\infty - [L_\infty - L(0)]e^{-kt}$ (b) $L(t) = 53 - 43e^{-0.2t}$
33. 15 days
35. (a) 9.8 h (b) $31{,}900\pi \approx 100{,}000$ ft^2; 6283 ft^2/h
(c) 5.1 h

PROBLEMS PLUS ■ page 573

1. About 1.85 in. from the center **9.** 0 **11.** $f(-2) = 10$
13. $f(\pi) = -\pi/2$ **15.** $(b^b a^{-a})^{1/(b-a)}e^{-1}$ **19.** $f(x) = \pm 10e^x$
21. Height $\sqrt{2}\,b$, volume $(\frac{28}{27}\sqrt{6} - 2)\pi b^3$

CHAPTER 10

Exercises 10.1 ■ page 584

1. (a) A sequence is an ordered list of numbers. It can also be defined as a function whose domain is the set of positive integers.
(b) The terms a_n approach 8 as n becomes large.
(c) The terms a_n become large as n becomes large.
3. $\frac{1}{3}, \frac{2}{5}, \frac{3}{7}, \frac{4}{9}, \frac{5}{11}, \frac{6}{13}$; yes; $\frac{1}{2}$ **5.** $a_n = 1/2^n$
7. $a_n = (n + 2)/(n + 3)^2$ **9.** 0 **11.** 1
13. Diverges (to ∞) **15.** Diverges **17.** Diverges (to ∞)
19. 0 **21.** 0 **23.** 0 **25.** 0 **27.** Diverges
29. $\pi/4$ **31.** 0 **33.** (a) Divergent (b) Convergent
35. (b) $(1 + \sqrt{5})/2$ **37.** Decreasing **39.** Increasing
41. Convergent by the Monotonic Sequence Theorem; $5 \leqslant L < 8$
43. $(3 + \sqrt{5})/2$ **45.** 62

Exercises 10.2 ■ page 593

1. (a) A sequence is an ordered list of numbers whereas a series is the *sum* of a list of numbers.
(b) A series is convergent if the sequence of partial sums is a convergent sequence. A series is divergent if it is not convergent.
3. 3.33333, 4.44444, 4.81481, 4.93827, 4.97942, 4.99314, 4.99771, 4.99924, 4.99975, 4.99992
Convergent, sum = 5

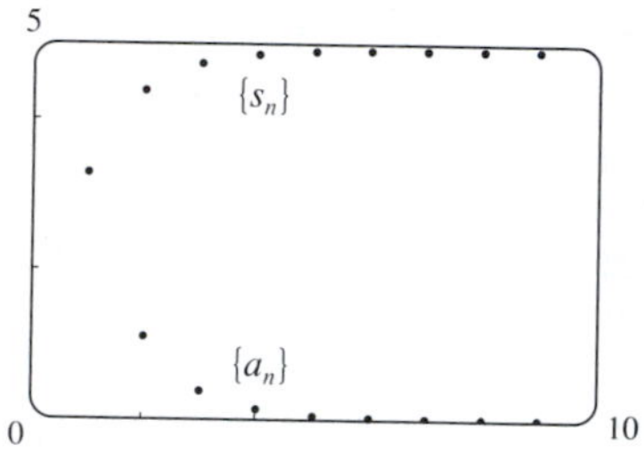

5. 0.50000, 1.16667, 1.91667, 2.71667, 3.55000, 4.40714, 5.28214, 6.17103, 7.07103, 7.98012
Divergent (terms do not approach 0)

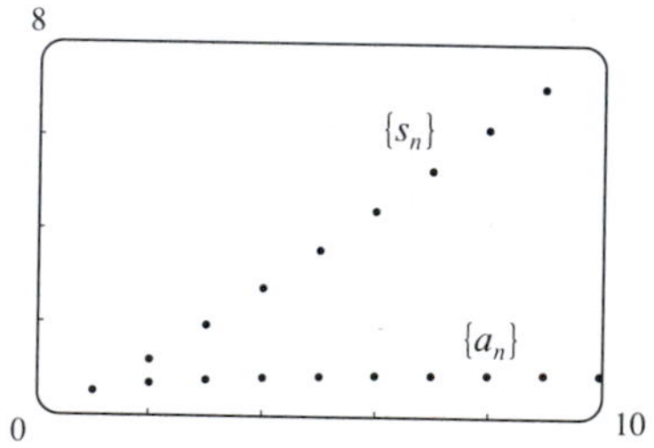

7. 0.64645, 0.80755, 0.87500, 0.91056, 0.93196, 0.94601, 0.95581, 0.96296, 0.96838, 0.97259
Convergent, sum = 1

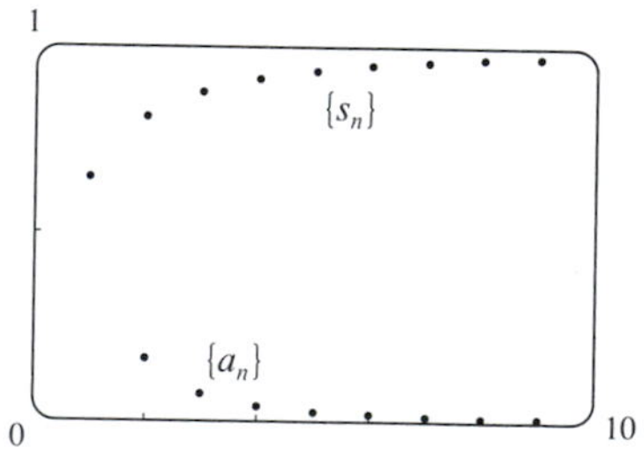

9. (a) Convergent (b) Divergent **11.** $\frac{20}{3}$ **13.** $\frac{1}{7}$
15. Divergent **17.** $\frac{17}{36}$ **19.** Divergent **21.** $\frac{3}{4}$
23. $\frac{3}{2}$ **25.** $\sin 1$ **27.** Divergent **29.** $\frac{5}{9}$ **31.** $\frac{307}{999}$
33. $2 < x < 4$, $1/(4 - x)$ **35.** $|x| > 1$, $x/(x - 1)$
37. $\frac{1}{4}$ **39.** $a_1 = 0$, $a_n = 2/[n(n + 1)]$ for $n > 1$, sum = 1
41. (a) $S_n = D(1 - c^n)/(1 - c)$ (b) 5
43. $(\sqrt{3} - 1)/2$ **45.** $1/[n(n + 1)]$
47. The series is divergent.
51. $\{s_n\}$ is bounded and increasing.
53. (a) $0, \frac{1}{9}, \frac{2}{9}, \frac{1}{3}, \frac{2}{3}, \frac{7}{9}, \frac{8}{9}, 1$

55. (a) $\frac{1}{2}, \frac{5}{6}, \frac{23}{24}, \frac{119}{120}$; $[(n + 1)! - 1]/(n + 1)!$ (c) 1

Exercises 10.3 ▪ page 603

Abbreviations: C, convergent, D, divergent

1.

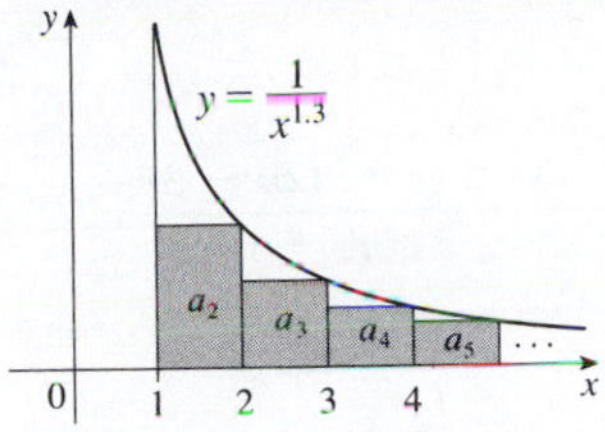

Convergent

3. (a) Nothing (b) Convergent
5. p-series; geometric series; $b < -1$; $-1 < b < 1$
7. C **9.** D **11.** D **13.** C **15.** D **17.** C
19. D **21.** D **23.** $p > 1$
25. (a) 1.54977, error $\leqslant 0.1$ (b) 1.64522, error $\leqslant 0.005$
(c) $n > 1000$
27. 2.6124 **29.** 0.567975, error $\leqslant 0.0003$ **35.** Yes

Exercises 10.4 ▪ page 611

1. (a) A series whose terms are alternately positive and negative
(b) $0 < b_{n+1} \leqslant b_n$ and $\lim_{n\to\infty} b_n = 0$, where $b_n = |a_n|$
(c) $|R_n| \leqslant b_{n+1}$
3. C **5.** D **7.** C **9.** Underestimate **11.** $p > 0$
13. 7 **15.** 0.8415

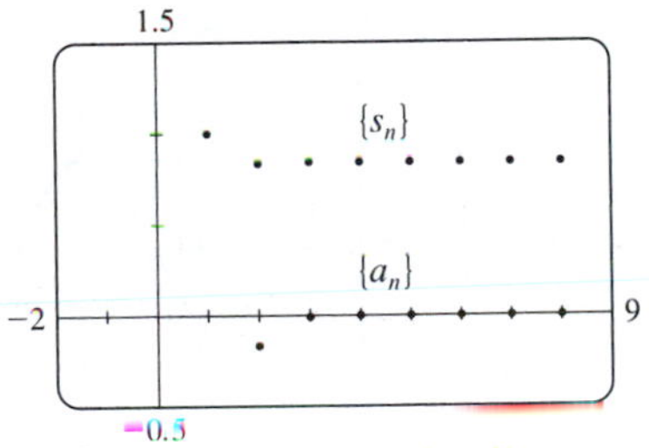

17. 0.6065 **19.** No **21.** No **23.** Yes **25.** Yes
27. Yes **29.** Diverges **31.** (a) and (d)
35. (a) $\frac{661}{960} \approx 0.68854$, error < 0.00521
(b) $n \geqslant 11$, 0.693109

Exercises 10.5 ▪ page 617

1. A series of the form $\sum_{n=0}^{\infty} c_n(x - a)^n$, where x is a variable and a and the c_n's are constants
3. (a) Yes (b) No **5.** 1, $[-1, 1)$ **7.** ∞, $(-\infty, \infty)$
9. 2, $(-2, 2]$ **11.** 2, $(-\frac{3}{2}, \frac{5}{2})$ **13.** 1, $(0, 2]$
15. 0.5, $[2.5, 3.5)$ **17.** 0, $\{\frac{1}{2}\}$ **19.** k^k
21. (a) $(-\infty, \infty)$ (b), (c)

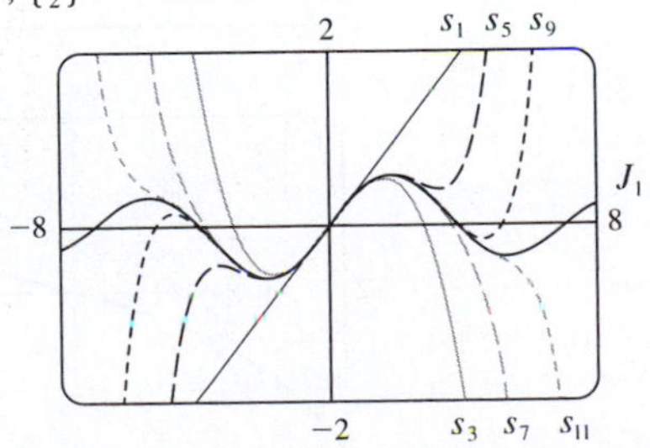

23. $(-1, 1)$, $f(x) = (1 + 2x)/(1 - x^2)$ **25.** 2

Exercises 10.6 ▪ page 622

1. 10
3. $\sum_{n=0}^{\infty} (-1)^n x^n$, $(-1, 1)$ **5.** $\sum_{n=0}^{\infty} (-1)^n 4^n x^{2n}$, $(-\frac{1}{2}, \frac{1}{2})$
7. $-\sum_{n=1}^{\infty} \left(\frac{x}{3}\right)^n$, $(-3, 3)$ **9.** $\sum_{n=0}^{\infty} (-1)^n (n + 1)x^n$, $R = 1$
11. $\frac{1}{2}\sum_{n=0}^{\infty} (-1)^n (n + 2)(n + 1)x^n$, $R = 1$
13. $\ln 5 - \sum_{n=1}^{\infty} \frac{x^n}{n5^n}$, $R = 5$
15. $\ln 3 + \sum_{n=1}^{\infty} \frac{(-1)^{n-1}}{n3^n} x^n$, $R = 3$

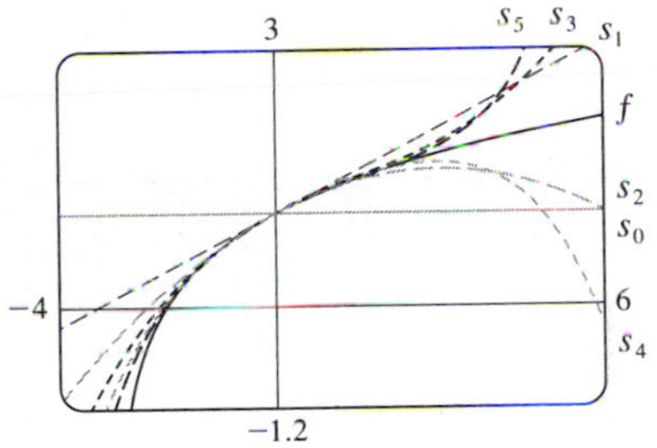

17. $\sum_{n=0}^{\infty} \frac{2x^{2n+1}}{2n + 1}$, $R = 1$

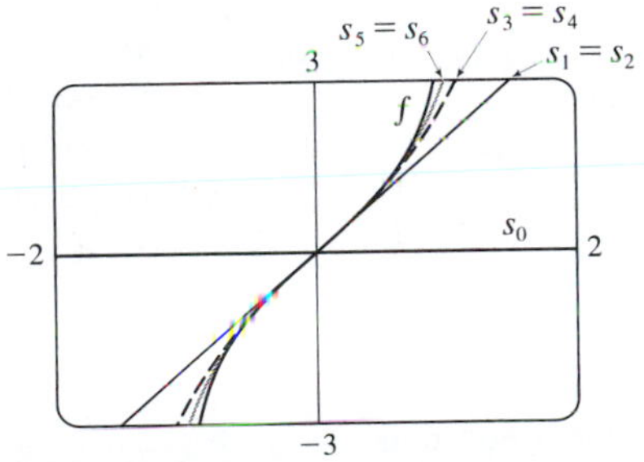

19. $C + \sum_{n=0}^{\infty} \frac{(-1)^n x^{4n+1}}{4n + 1}$ **21.** $C + \sum_{n=0}^{\infty} (-1)^n \frac{x^{2n+1}}{(2n + 1)^2}$
23. 0.199936 **25.** 0.000065 **27.** 0.09531
29. (b) 0.920 **33.** $[-1, 1]$, $[-1, 1)$, $(-1, 1)$

Exercises 10.7 ▪ page 633

1. $b_8 = f^{(8)}(5)/8!$ **3.** $\sum_{n=0}^{\infty} (-1)^n \frac{x^{2n}}{(2n)!}$, $R = \infty$
5. $\sum_{n=0}^{\infty} (-1)^n (n + 1)x^n$, $R = 1$ **7.** $\sum_{n=0}^{\infty} \frac{e^3}{n!}(x - 3)^n$, $R = \infty$
9. $\sum_{n=0}^{\infty} (-1)^n (x - 1)^n$, $R = 1$
11. $\sum_{n=0}^{\infty} \frac{(-1)^{n(n-1)/2}(x - \pi/4)^n}{\sqrt{2}\,n!}$, $R = \infty$
15. $\sum_{n=0}^{\infty} \frac{3^n x^n}{n!}$, $R = \infty$ **17.** $\sum_{n=0}^{\infty} \frac{(-1)^n x^{2n+2}}{(2n)!}$, $R = \infty$

19. $\displaystyle\sum_{n=0}^{\infty} \frac{(-1)^n x^{2n+2}}{2^{2n+1}(2n+1)!}$, $R = \infty$

21. $\displaystyle\sum_{n=1}^{\infty} \frac{(-1)^{n+1}2^{2n-1}x^{2n}}{(2n)!}$, $R = \infty$

23. $\displaystyle 1 + \frac{x}{2} + \sum_{n=2}^{\infty} (-1)^{n-1} \frac{1 \cdot 3 \cdot 5 \cdot \cdots \cdot (2n-3)}{2^n n!} x^n$, $R = 1$

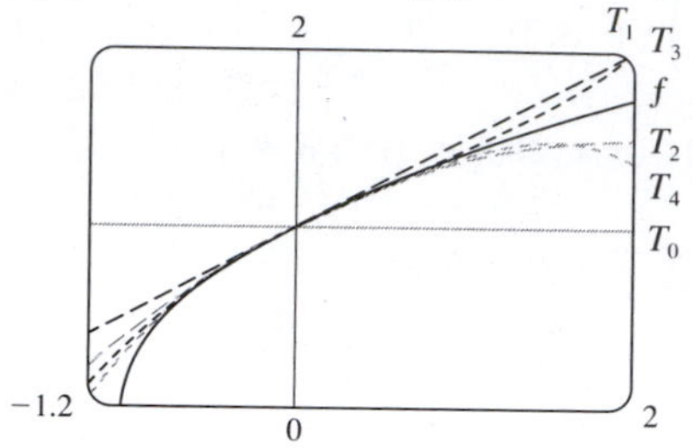

25. $\displaystyle\sum_{n=0}^{\infty} \frac{(-1)^n}{2}(n+1)(n+2)x^n$, $R = 1$

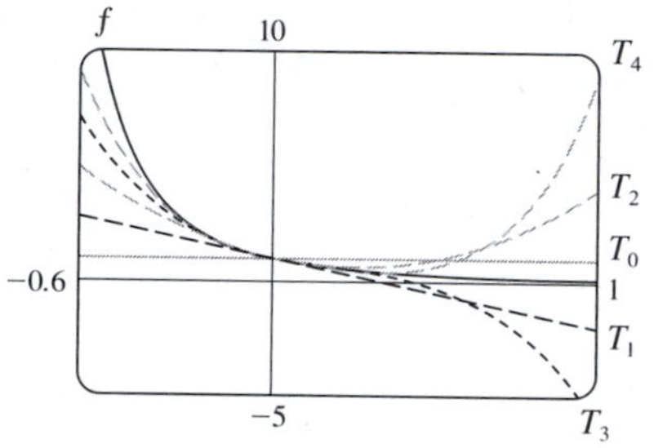

27. $\displaystyle\sum_{n=1}^{\infty} (-1)^{n-1}\frac{x^n}{n}$, 0.09531 **29.** $\displaystyle C + \sum_{n=0}^{\infty} \frac{(-1)^n x^{4n+3}}{(4n+3)(2n+1)!}$

31. $\displaystyle C + x + \frac{x^4}{8} + \sum_{n=2}^{\infty} (-1)^{n-1} \frac{1 \cdot 3 \cdot 5 \cdot \cdots \cdot (2n-3)}{2^n n!\,(3n+1)} x^{3n+1}$

33. 0.310 **35.** 0.09998750 **37.** $\frac{1}{3}$ **39.** $\frac{1}{120}$

41. $1 - \frac{3}{2}x^2 + \frac{25}{24}x^4$ **43.** $-x + \frac{1}{2}x^2 - \frac{1}{3}x^3$ **45.** e^{-x^4}

47. $1/\sqrt{2}$ **49.** $e^x - 1$

Exercises 10.8 ■ page 636

1. $\displaystyle 1 + \frac{x}{2} + \sum_{n=2}^{\infty} (-1)^{n-1} \frac{1 \cdot 3 \cdot 5 \cdot \cdots \cdot (2n-3)}{2^n n!} x^n$, $R = 1$

3. $\displaystyle\sum_{n=0}^{\infty} (-1)^n \frac{(n+1)(n+2)(n+3)2^n}{6} x^n$, $R = \frac{1}{2}$

5. $\displaystyle 1 - \frac{x^4}{4} - \sum_{n=2}^{\infty} \frac{3 \cdot 7 \cdot 11 \cdot \cdots \cdot (4n-5)}{4^n n!} x^{4n}$, $R = 1$

7. $\displaystyle \frac{1}{2} + \frac{1}{2}\sum_{n=1}^{\infty} \frac{(-1)^n 1 \cdot 4 \cdot 7 \cdot \cdots \cdot (3n-2)}{24^n n!} x^n$, $R = 8$

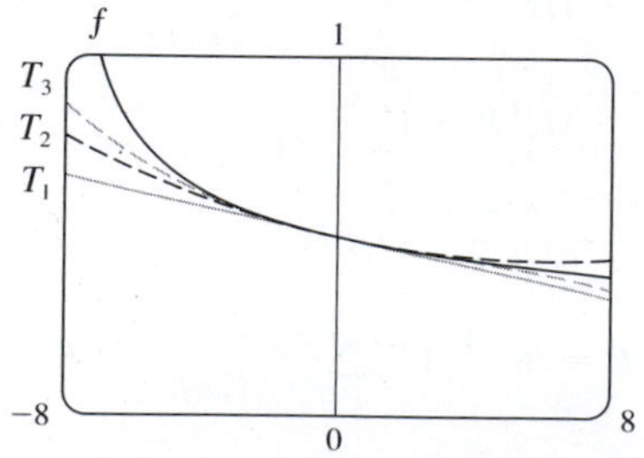

9. (a) $\displaystyle 1 + \sum_{n=1}^{\infty} \frac{1 \cdot 3 \cdot 5 \cdot \cdots \cdot (2n-1)}{2^n n!} x^{2n}$

(b) $\displaystyle x + \sum_{n=1}^{\infty} \frac{1 \cdot 3 \cdot 5 \cdot \cdots \cdot (2n-1)}{2^n n!} \frac{x^{2n+1}}{2n+1}$

11. (a) $\displaystyle\sum_{n=1}^{\infty} nx^n$ (b) 2

13. (a) $\displaystyle 1 + \frac{x^2}{2} + \sum_{n=2}^{\infty} (-1)^{n-1} \frac{1 \cdot 3 \cdot 5 \cdot \cdots \cdot (2n-3)}{2^n n!} x^{2n}$

(b) 99,225

Exercises 10.9 ■ page 644

1. (a) $T_0(x) = 1 = T_1(x)$, $T_2(x) = 1 - \frac{1}{2}x^2 = T_3(x)$,
$T_4(x) = 1 - \frac{1}{2}x^2 + \frac{1}{24}x^4 = T_5(x)$,
$T_6(x) = 1 - \frac{1}{2}x^2 + \frac{1}{24}x^4 - \frac{1}{720}x^6$

(b)

x	f	T_1	$T_2 = T_3$	$T_4 = T_5$	T_6
$\frac{\pi}{4}$	0.7071	1	0.6916	0.7074	0.7071
$\frac{\pi}{2}$	0	1	−0.2337	0.0200	−0.0009
π	−1	1	−3.9348	0.1239	−1.2114

(c) An n increases, $T_n(x)$ is a good approximation to $f(x)$ on a larger and larger interval.

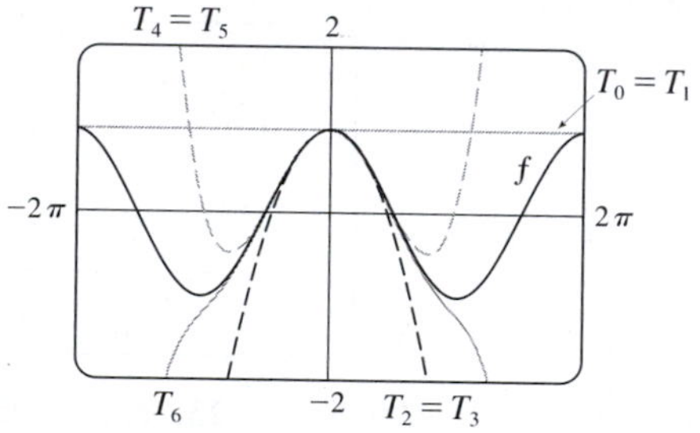

3. $\displaystyle \frac{1}{2} + \frac{\sqrt{3}}{2}\left(x - \frac{\pi}{6}\right) - \frac{1}{4}\left(x - \frac{\pi}{6}\right)^2 - \frac{\sqrt{3}}{12}\left(x - \frac{\pi}{6}\right)^3$

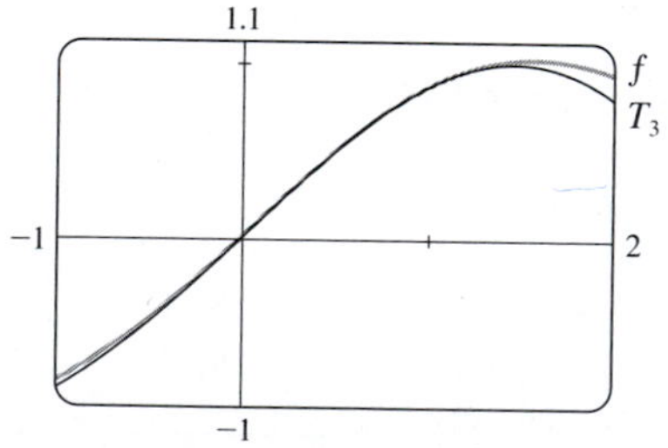

5. $x + \frac{1}{3}x^3$

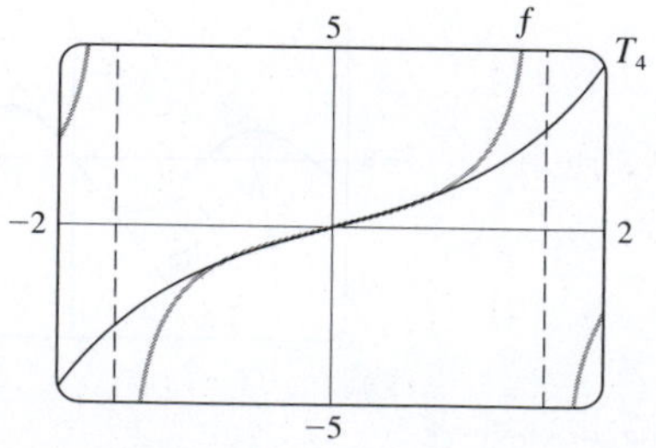

7. $2 + 2\sqrt{3}\left(x - \frac{\pi}{3}\right) + 7\left(x - \frac{\pi}{3}\right)^2 + \frac{23\sqrt{3}}{3}\left(x - \frac{\pi}{3}\right)^3$

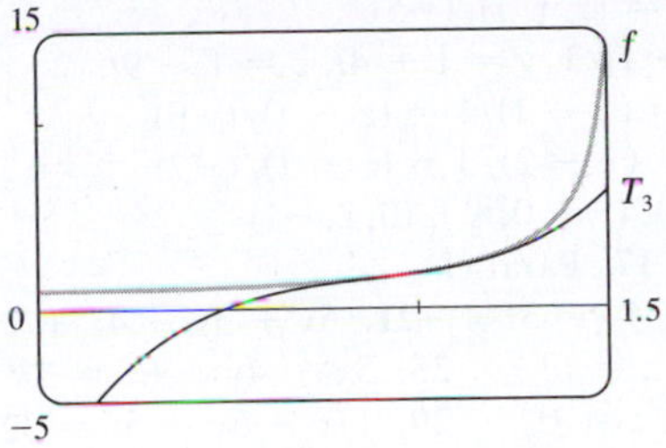

9. $T_8(x) = 1 + \frac{1}{2}x^2 + \frac{5}{24}x^4 + \frac{61}{720}x^6 + \frac{277}{8064}x^8$

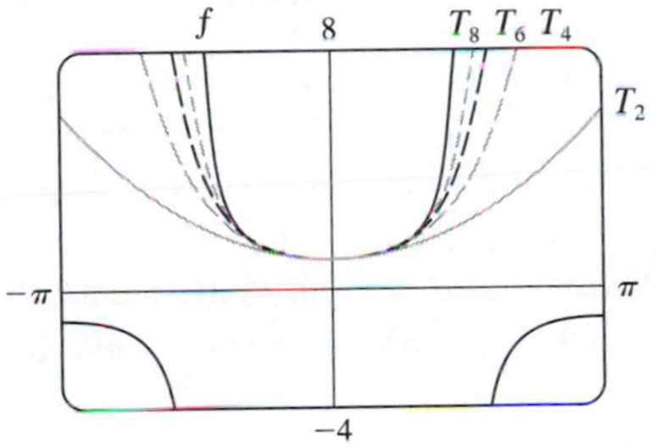

11. (a) $1 + \frac{1}{2}x$ (b) 0.00125

13. (a) $\frac{1}{\sqrt{2}} + \frac{1}{\sqrt{2}}\left(x - \frac{\pi}{4}\right) - \frac{1}{2\sqrt{2}}\left(x - \frac{\pi}{4}\right)^2 - \frac{1}{6\sqrt{2}}\left(x - \frac{\pi}{4}\right)^3 + \frac{1}{24\sqrt{2}}\left(x - \frac{\pi}{4}\right)^4 + \frac{1}{120\sqrt{2}}\left(x - \frac{\pi}{4}\right)^5$
(b) 0.00033

15. (a) $1 + x^2$ (b) 0.00006 **17.** 0.57358 **19.** Four

21. $-1.037 < x < 1.037$ **23.** 21 m, no

Chapter 10 Review Exercises ■ page 646

1. False **3.** False **5.** False **7.** False **9.** True
11. True **13.** False **15.** True **17.** C, $\frac{1}{2}$ **19.** D
21. D **23.** C, e^{12} **25.** D **27.** C **29.** D
31. C **33.** C **35.** 8 **37.** $\pi/4$ **39.** $\frac{4111}{3330}$
41. 0.9721 **43.** 0.18976224, error $< 6.4 \times 10^{-7}$
47. 3, $[-3, 3]$ **49.** 0.5, $[2.5, 3.5)$

51. $\frac{1}{2} + \frac{\sqrt{3}}{2}\left(x - \frac{\pi}{6}\right) - \frac{1}{2}\frac{1}{2!}\left(x - \frac{\pi}{6}\right)^2 - \frac{\sqrt{3}}{2}\frac{1}{3!}\left(x - \frac{\pi}{6}\right)^3 + \cdots$
$= \frac{1}{2}\sum_{n=0}^{\infty}(-1)^n\left[\frac{1}{(2n)!}\left(x - \frac{\pi}{6}\right)^{2n} + \frac{\sqrt{3}}{(2n+1)!}\left(x - \frac{\pi}{6}\right)^{2n+1}\right]$

53. $\sum_{n=0}^{\infty}(-1)^n x^{n+2}$, 1 **55.** $-\sum_{n=1}^{\infty}\frac{x^n}{n}$, 1

57. $\sum_{n=0}^{\infty}(-1)^n\frac{x^{8n+4}}{(2n+1)!}$, ∞

59. $\frac{1}{2} + \sum_{n=1}^{\infty}\frac{1 \cdot 5 \cdot 9 \cdot \cdots \cdot (4n-3)}{n!\, 2^{6n+1}}x^n$, 16

61. $\ln|x| + C + \sum_{n=1}^{\infty}\frac{x^n}{n \cdot n!}$

63. (a) $1 + \frac{1}{2}(x - 1) - \frac{1}{8}(x - 1)^2 + \frac{1}{16}(x - 1)^3$
(b)

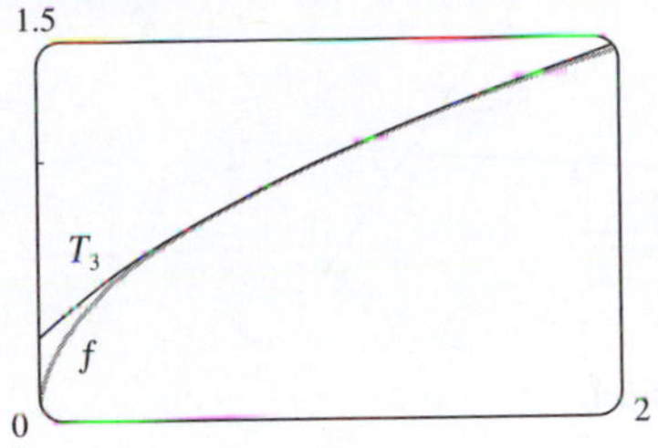

(c) 0.000006

65. 1

PROBLEMS PLUS ■ page 648

1. $15!/5! = 10{,}897{,}286{,}400$
3. (b) 0 if $x = 0$, $(1/x) - \cot x$ if $x \neq n\pi$, n an integer
5. (a) $s_n = 3 \cdot 4^n$, $l_n = 1/3^n$, $p_n = 4^n/3^{n-1}$ (c) $2\sqrt{3}/5$
9. $(-1, 1)$, $(x^3 + 4x^2 + x)/(1 - x)^4$

CHAPTER 11

Exercises 11.1 ■ page 656

1. (a)

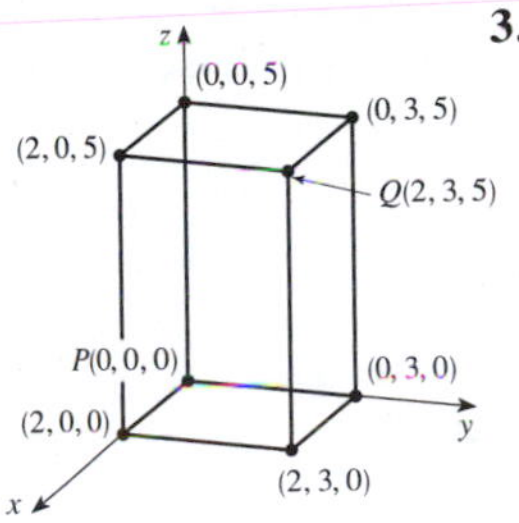

(b) $|PQ| = \sqrt{38}$

3. (a)

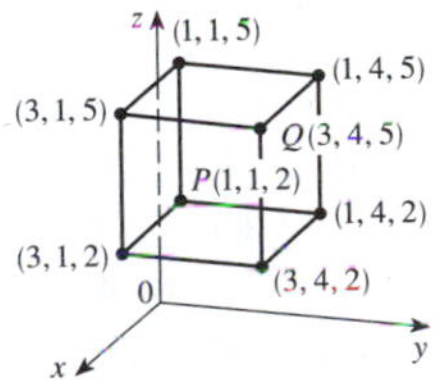

(b) $|PQ| = \sqrt{22}$

5. $|AB| = \sqrt{21}$, $|BC| = \sqrt{6}$, $|CA| = 3\sqrt{3}$; right triangle
7. $|AB| = \sqrt{69}$, $|BC| = \sqrt{158}$, $|CA| = 13$; neither
9. Not collinear **11.** $x^2 + (y - 1)^2 + (z + 1)^2 = 16$
13. $(x + 6)^2 + (y + 1)^2 + (z - 2)^2 = 12$ **15.** $(-1, -4, 2)$, 7
17. $(-\frac{1}{2}, 1, -3)$, $\frac{7}{2}$ **19.** $(\frac{1}{2}, 0, 0)$, $\frac{1}{2}$ **23.** $\sqrt{85}/2$, $\frac{5}{2}$, $\sqrt{94}/2$
25. (a) $(x - 2)^2 + (y + 3)^2 + (z - 6)^2 = 36$
(b) $(x - 2)^2 + (y + 3)^2 + (z - 6)^2 = 4$
(c) $(x - 2)^2 + (y + 3)^2 + (z - 6)^2 = 9$
27. $14x - 6y - 10z = 9$, a plane perpendicular to AB
29. A plane parallel to the yz-plane and 9 units in front of it
31. A half-space consisting of all points to the right of the plane $y = 2$
33. A plane perpendicular to the xz-plane and intersecting it in the line $x = z$, $y = 0$
35. Circular cylinder, radius 1, axis the z-axis
37. All points outside the sphere with radius 1 and center O
39. All points inside the sphere with radius 2 and center $(0, 0, 1)$ **41.** Hyperbolic cylinder
43. All points on and between the horizontal planes $z = 2$ and $z = -2$
45. $y < 0$ **47.** $r^2 < x^2 + y^2 + z^2 < R^2$

49. (a) $(2, 1, 4)$ (b)

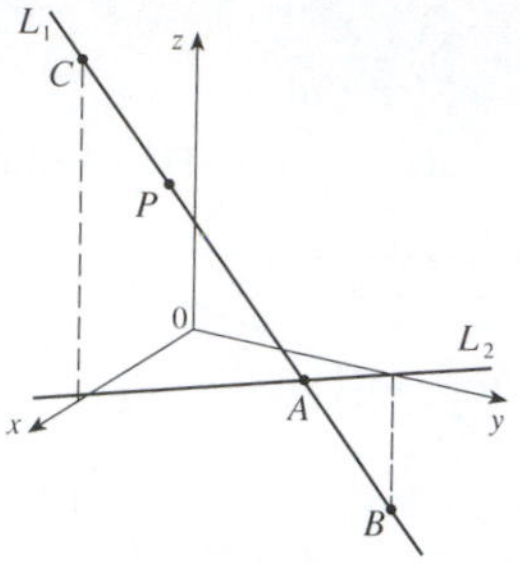

Exercises 11.2 ■ page 664

1. $\langle 5, 2, -2\rangle$
3. $\mathbf{a} = \langle 2, 0, -2\rangle$ **5.** $\langle 1, 0, 2\rangle$

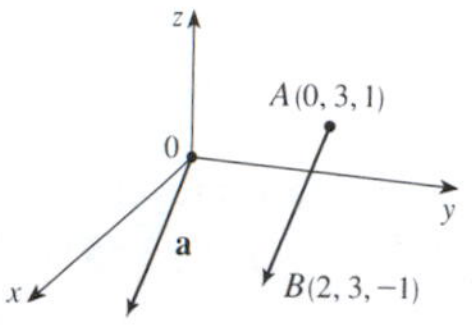

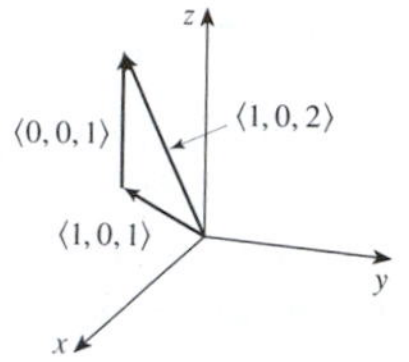

7. $7, \langle 3, -2, 10\rangle, \langle 1, -4, 2\rangle, \langle 4, -6, 12\rangle, \langle 10, -5, 34\rangle$
9. $\sqrt{3}, 3\mathbf{i} + 4\mathbf{k}, -\mathbf{i} + 2\mathbf{j} - 2\mathbf{k}, 2\mathbf{i} + 2\mathbf{j} + 2\mathbf{k}$, $11\mathbf{i} - \mathbf{j} + 15\mathbf{k}$
11. $\langle -2/\sqrt{29}, 4/\sqrt{29}, 3/\sqrt{29}\rangle$ **13.** $(\mathbf{i} + \mathbf{j})/\sqrt{2}$
15. 3 **17.** -5 **19.** -11
23. $\cos^{-1}\frac{11}{15} \approx 43°$ **25.** $\cos^{-1}(1/(7\sqrt{3})) \approx 85°$
27. $114°, 33°, 33°$ **29.** Neither
31. Orthogonal **33.** $-\frac{4}{5}$
35. $(\mathbf{i} - \mathbf{j} - \mathbf{k})/\sqrt{3}$ [or $(-\mathbf{i} + \mathbf{j} + \mathbf{k})/\sqrt{3}$]
37. $\frac{1}{3}, \frac{2}{3}, \frac{2}{3}$; $71°, 48°, 48°$
39. $-8/\sqrt{77}, 3/\sqrt{77}, 2/\sqrt{77}$; $156°, 70°, 77°$
41. $5/\sqrt{38}, 3/\sqrt{38}, 2/\sqrt{38}$; $36°, 61°, 71°$
43. $11/\sqrt{13}, \langle 22/13, 33/13\rangle$ **45.** $3/\sqrt{5}, \langle 6/5, 3/5, 0\rangle$
47. $1/\sqrt{2}, (\mathbf{i} + \mathbf{k})/2$
51. $\langle 0, 0, -2\sqrt{10}\rangle$ or any vector of the form $\langle s, t, 3s - 2\sqrt{10}\rangle$, $s, t \in \mathbb{R}$ **53.** 38 J **55.** $250 \cos 20° \approx 235$ ft-lb
57. A sphere with radius 1, centered at (x_0, y_0, z_0)
59. $\cos^{-1}(1/\sqrt{3}) \approx 55°$

Exercises 11.3 ■ page 672

1. (a) Scalar (b) Meaningless (c) Vector (d) Meaningless (e) Meaningless (f) Scalar
3. 24; into the page **5.** $10.8 \sin 100° \approx 10.6$ J
7. $\langle 3, 14, -9\rangle$ **9.** $-2\mathbf{i} + 2\mathbf{j}$
11. $\langle -2/\sqrt{6}, -1/\sqrt{6}, 1/\sqrt{6}\rangle, \langle 2/\sqrt{6}, 1/\sqrt{6}, -1/\sqrt{6}\rangle$
13. 4 **15.** (a) $\langle 6, 3, 2\rangle$ (b) $\frac{7}{2}$ **17.** ≈ 417 N
19. 226 **21.** 21 **25.** (b) $\sqrt{97/3}$
31. (a) No (b) No (c) Yes

Exercises 11.4 ■ page 681

1. $\mathbf{r} = \langle 3, -1, 8\rangle + t\langle 2, 3, 5\rangle$, $x = 3 + 2t, y = -1 + 3t, z = 8 + 5t$
3. $\mathbf{r} = (\mathbf{j} + 2\mathbf{k}) + t(6\mathbf{i} + 3\mathbf{j} + 2\mathbf{k})$, $x = 6t, y = 1 + 3t, z = 2 + 2t$
5. $x = 2 + 4t, y = 1 - t, z = 8 - 5t$; $(x - 2)/4 = (y - 1)/(-1) = (z - 8)/(-5)$
7. $x = 3, y = 1 + t, z = -1 - 5t$; $x = 3, y - 1 = (z + 1)/(-5)$
9. $x = (-1 + t)/3, y = 1 + 4t, z = 1 - 9t$; $(x + \frac{1}{3})/(\frac{1}{3}) = (y - 1)/4 = (z - 1)/(-9)$
13. (a) $x/2 = (y - 2)/3 = (z + 1)/(-7)$
(b) $(-\frac{2}{7}, \frac{11}{7}, 0), (-\frac{4}{3}, 0, \frac{11}{3}), (0, 2, -1)$
15. Skew **17.** Parallel
19. $7x + y + 4z = 31$ **21.** $5x + 3y - 4z + 1 = 0$
23. $x + y - z = 13$ **25.** $3x - 4y - 6z = 33$
27. $x - 2y + z = 0$ **29.** $17x - 6y - 5z = 32$
31. $25x + 14y + 8z = 77$ **33.** $x - 2y + z = 0$
35. $(1, 0, 0)$ **37.** $(-3, -1, -2)$ **39.** $1, 0, -1$
41. Neither, $60°$ **43.** Perpendicular **45.** Parallel
47. (a) $x - 2 = y/(-8) = z/(-7)$
(b) $\cos^{-1}(-\sqrt{6}/5) \approx 119°$ (or $61°$)
49. $x = 6t, y = -\frac{1}{6} + t, z = -\frac{1}{6} + 7t$ **51.** $x = z$
53. $x - 2y + 4z + 1 = 0$ **55.** $(x/a) + (y/b) + (z/c) = 1$
57. $x = 3t, y = 1 - t, z = 2 - 2t$
59. P_1 and P_3 are parallel, P_2 and P_4 are identical
61. $\sqrt{22/5}$ **63.** $\frac{25}{3}$ **65.** $7\sqrt{6}/18$ **69.** $1/\sqrt{6}$

Exercises 11.5 ■ page 687

1. $x = k, z^2 - y^2 = 1 - k^2$, hyperbola
$y = k, x^2 + z^2 = 1 + k^2$, circle
$z = k, x^2 - y^2 = 1 - k^2$, hyperbola
Hyperboloid of one sheet with axis the y-axis

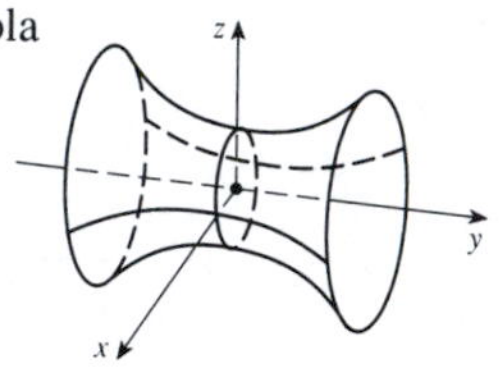

3. $x = k, y^2 + 4z^2 = 4 - 4k^2/9$, ellipse ($|k| < 3$)
$y = k, x^2 + 9z^2 = 9 - 9k^2/4$, ellipse ($|k| < 2$)
$z = k, 4x^2 + 9y^2 = 36(1 - k^2)$, ellipse ($|k| < 1$)
Ellipsoid

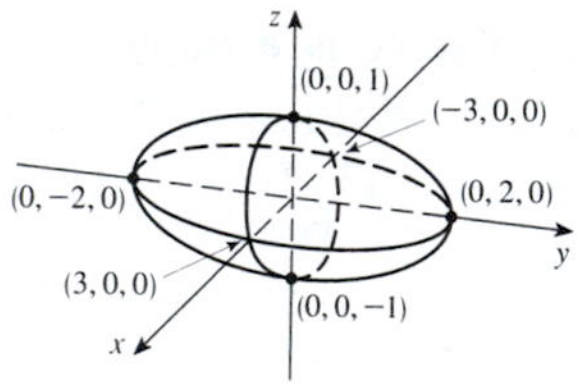

5. $x = k, 4z^2 - y^2 = 1 + k^2$, hyperbola
$y = k, 4z^2 - x^2 = 1 + k^2$, hyperbola
$z = k, x^2 + y^2 = 4k^2 - 1$, circle ($|k| > \frac{1}{2}$)
Hyperboloid of two sheets

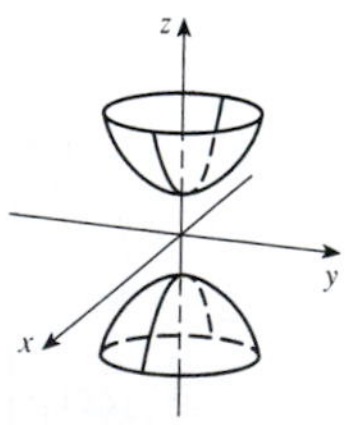

7. $x = k, z = y^2$, parabola
$y = k, z = k^2$, line
$z = k, y = \pm\sqrt{k}$, lines ($k > 0$)
Parabolic cylinder

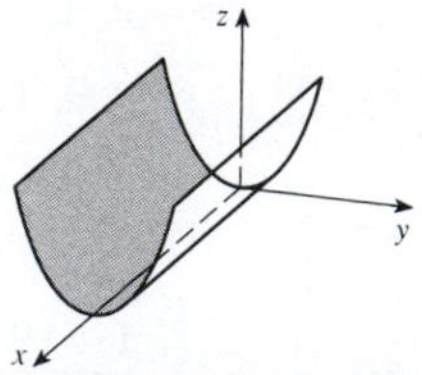

9. $x = k$, $y^2 - z^2 = k^2$, hyperbola $(k \neq 0)$
$y = k$, $x^2 + z^2 = k^2$, circle $(k \neq 0)$
$z = k$, $y^2 - x^2 = k^2$,
hyperbola $(k \neq 0)$
$x = 0$, $y = \pm z$, lines
$z = 0$, $y = \pm x$, lines
Cone

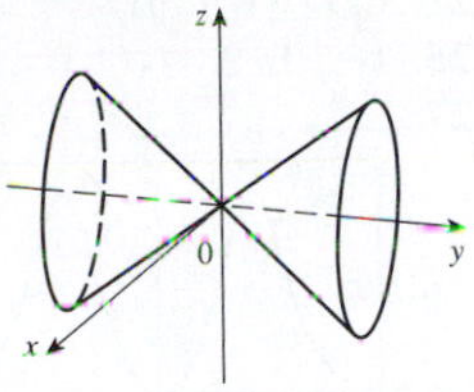

11. $x = k$, $y = 4z^2 + k^2$, parabola
$y = k$, $x^2 + 4z^2 = k$, ellipse $(k > 0)$
$z = k$, $y = x^2 + 4k^2$, parabola
Elliptic paraboloid with
axis the y-axis

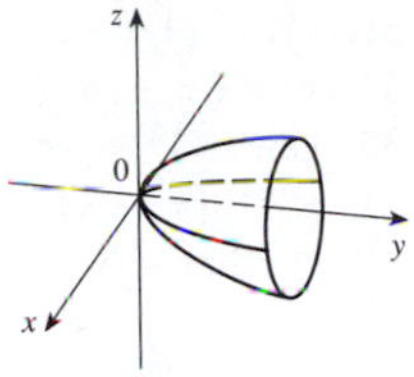

13. $x = k$, $y^2 + 9z^2 = 9$, ellipse
$y = k$, $z = \pm\sqrt{1 - (k^2/9)}$, pairs of lines $(|k| < 3)$
$z = k$, $y = \pm 3\sqrt{1 - k^2}$, pairs of lines $(|k| < 1)$
Elliptic cylinder with axis the x-axis

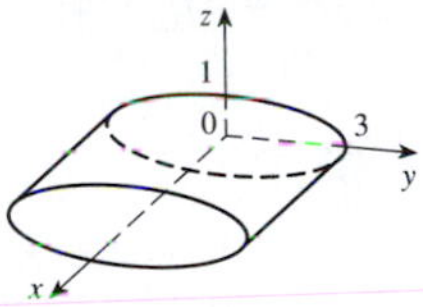

15. $x = k$, $y = z^2 - k^2$, parabola
$y = k$, $z^2 - x^2 = k$, hyperbola $(k \neq 0)$
$z = k$, $y = k^2 - x^2$, parabola
Hyperbolic paraboloid

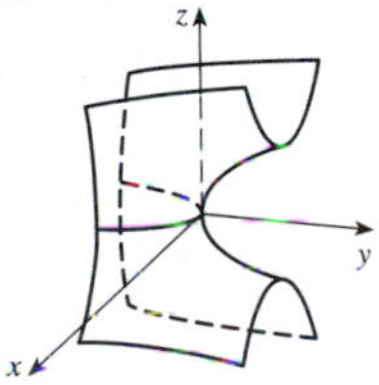

17. VII **19.** II **21.** VI **23.** VIII

25. $(x^2/4) + (y^2/3) - (z^2/12) = 1$
Hyperboloid of one sheet with
axis the z-axis

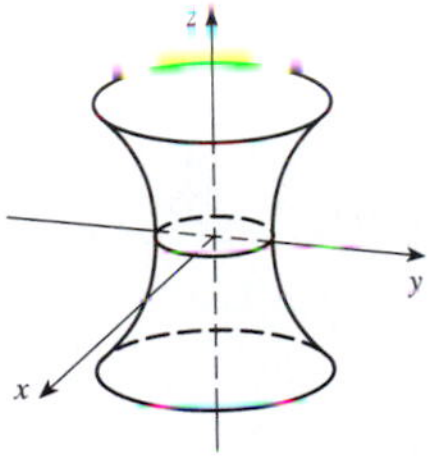

27. $z - 1 = x^2 + y^2$
Circular paraboloid with vertex $(0, 0, 1)$
and axis the z-axis

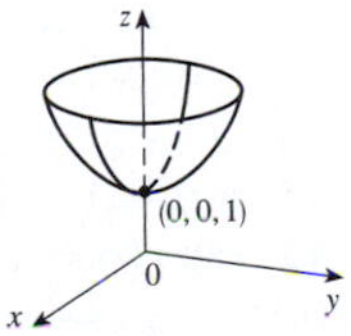

29. $(x + 2)^2 + (y - 3)^2 - 4(z + 1)^2 = 22$
Hyperboloid of one sheet
with center $(-2, 3, -1)$ and
axis parallel to the z-axis

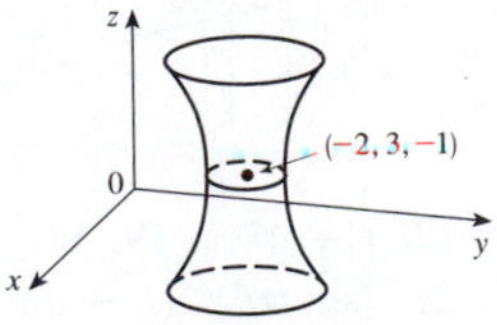

31. $(x^2/100) + (y^2/25) = 1$
Elliptic cylinder with
axis the z-axis

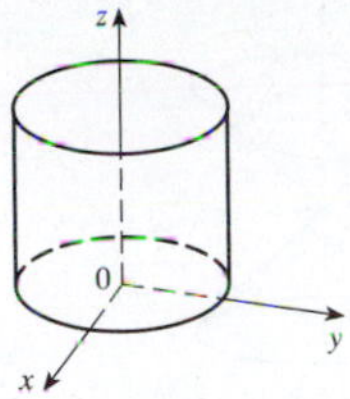

33. $z = (y - 2)^2 - x^2$
Hyperbolic paraboloid
with center $(0, 2, 0)$

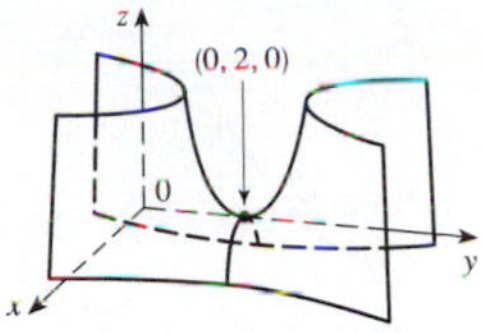

35.

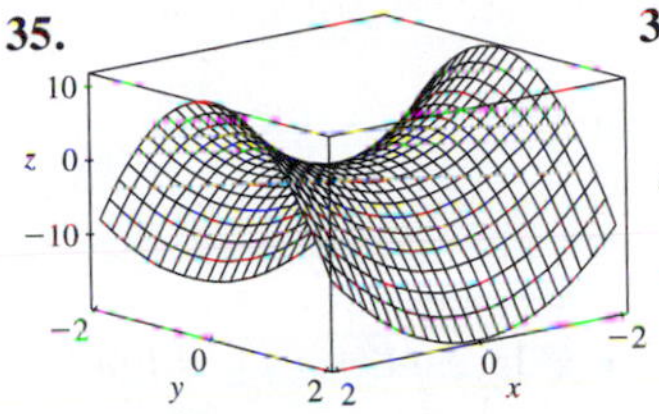

37.

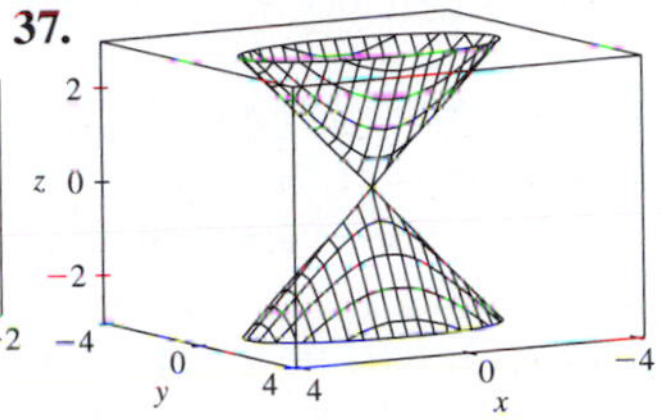

39.

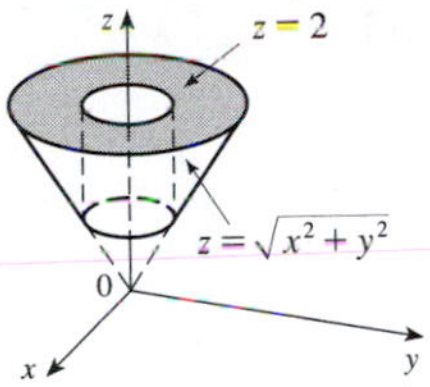

41. $y = x^2 + z^2$ **43.** $-4x = y^2 + z^2$, paraboloid

47.

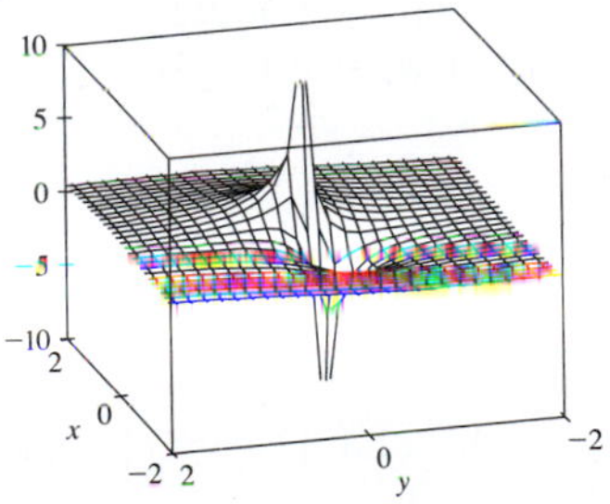

Exercises 11.6 ■ page 696

1. V **3.** I **5.** IV

7.

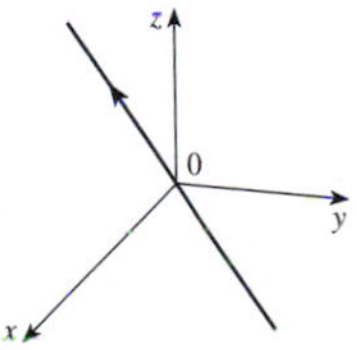

9.

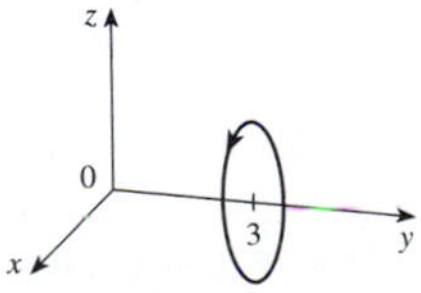

11.

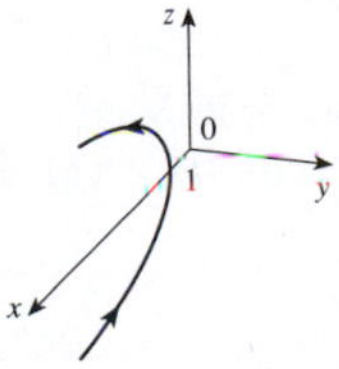

13.

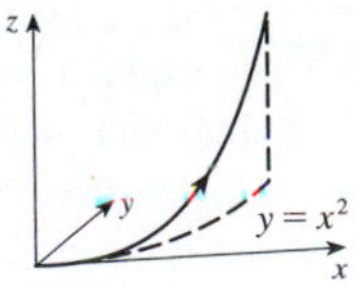

15.

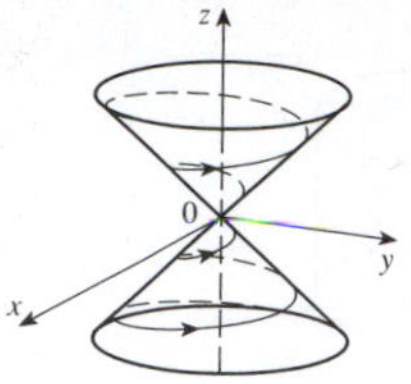

17.

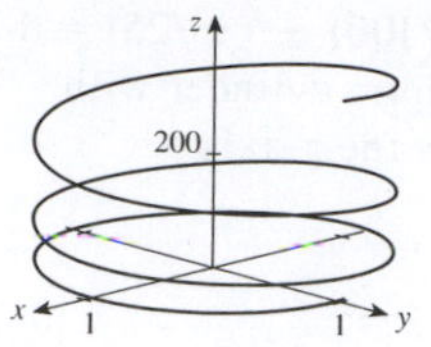

19.

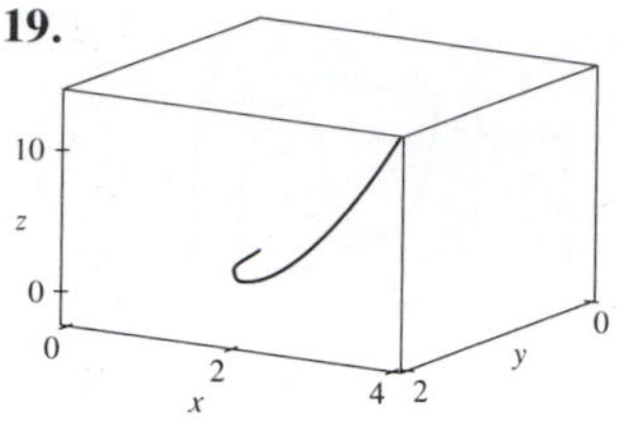

21.

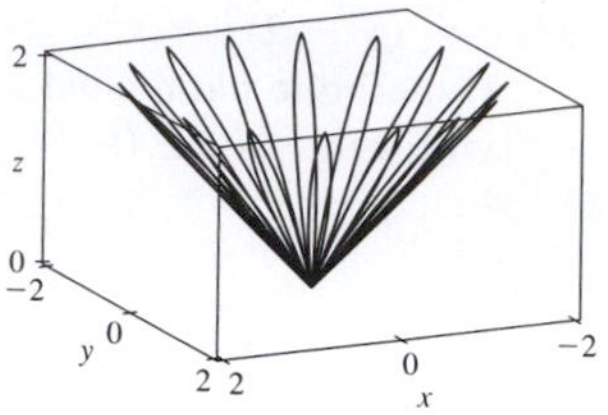

23. $\langle 0, 1, 2\rangle$ **25.** $2\,\mathbf{i} + \frac{1}{2}\mathbf{j} + (\tan 1)\,\mathbf{k}$

27. $\mathbb{R}$, $\mathbf{r}'(t) = \langle 1, 2t, 3t^2\rangle$

29. $\{t \mid t \neq (2n+1)\pi/2,\ n \text{ an integer}\}$, $\mathbf{r}'(t) = \sec^2 t\,\mathbf{j} + \sec t \tan t\,\mathbf{k}$

31. $[-1, 2)$, $\mathbf{r}'(t) = -2t/(4 - t^2)\,\mathbf{i} + 1/(2\sqrt{1+t}\,)\,\mathbf{j} - 12e^{3t}\,\mathbf{k}$

33. $\mathbb{R}$, $\mathbf{r}(t) = \mathbf{b} + 2t\mathbf{c}$

35. (a), (c)

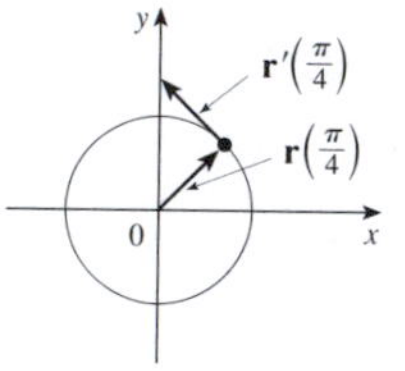

(b) $\mathbf{r}'(t) = \langle -\sin t, \cos t\rangle$

37. (a), (c)

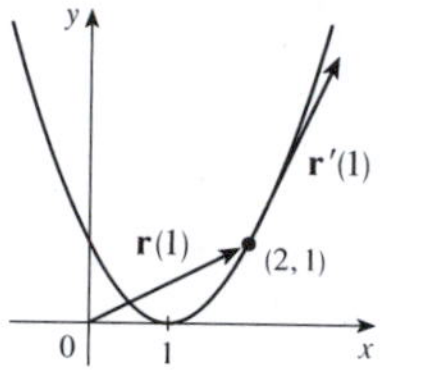

(b) $\mathbf{r}'(t) = \mathbf{i} + 2t\,\mathbf{j}$

39. (a), (c)

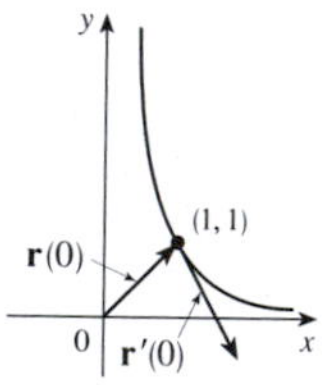

(b) $\mathbf{r}'(t) = e^t\,\mathbf{i} - 2e^{-2t}\,\mathbf{j}$

41. $\langle 1/\sqrt{46}, 3/\sqrt{46}, 6/\sqrt{46}\,\rangle$ **43.** $\frac{2}{5}\mathbf{i} + (2\sqrt{3}/5)\,\mathbf{j} - \frac{3}{5}\mathbf{k}$

45. $x = 1 + t,\ y = 1 + 2t,\ z = 1 + 3t$

47. $x = -(\pi/2)t,\ y = \frac{1}{4} + t,\ z = 1 + 4t$

49. $x = (\pi/4) + t,\ y = 1 - t,\ z = 1 + t$ **51.** $66°$

55. $x = 2\cos t,\ y = 2\sin t,\ z = 4\cos^2 t$ **57.** $\frac{1}{2}\,\mathbf{i} + \frac{1}{3}\,\mathbf{j} + \frac{1}{4}\,\mathbf{k}$

59. $\frac{1}{2}\,\mathbf{i} + \frac{1}{2}\,\mathbf{j} + (4 - \pi)/(4\sqrt{2}\,)\,\mathbf{k}$

61. $(t^3/3)\,\mathbf{i} + (t^4 + 1)\,\mathbf{j} - (t^3/3)\,\mathbf{k}$

69. $1 - 4t\cos t + 11t^2 \sin t + 3t^3 \cos t$

Exercises 11.7 ■ page 705

1. $\sqrt{13}\,(b - a)$ **3.** 8 **5.** 9.5706

7. $\mathbf{r}(t(s)) = (1 + s/\sqrt{2}\,)\sin[\ln(1 + s/\sqrt{2}\,)]\,\mathbf{i} + (1 + s/\sqrt{2}\,)\cos[\ln(1 + s/\sqrt{2}\,)]\,\mathbf{j}$

9. $\mathbf{r}(t(s)) = 3\sin(s/5)\,\mathbf{i} + (4s/5)\,\mathbf{j} + 3\cos(s/5)\,\mathbf{k}$

11. (a) $\langle \frac{4}{5}\cos 4t, \frac{3}{5}, -\frac{4}{5}\sin 4t\rangle$, $\langle -\sin 4t, 0, -\cos 4t\rangle$ (b) $\frac{16}{25}$

13. (a) $\langle -\sqrt{2}\,\sin t, \cos t, \cos t\rangle/\sqrt{2}$, $\langle -\sqrt{2}\,\cos t, -\sin t, -\sin t\rangle/\sqrt{2}$ (b) $1/\sqrt{2}$

15. $2/(4t^2 + 1)^{3/2}$

17. $\sqrt{t^4 + 4t^2 + 1}\,/[6(t^4 + t^2 + 1)^{3/2}]$ **19.** $\sqrt{2}\,/(1 + \cos^2 t)^{3/2}$

21. $6|x|/(1 + 9x^4)^{3/2}$ **23.** $|\sin x|/(1 + \cos^2 x)^{3/2}$

25. $(-\frac{1}{2}\ln 2, 1/\sqrt{2}\,)$

27.

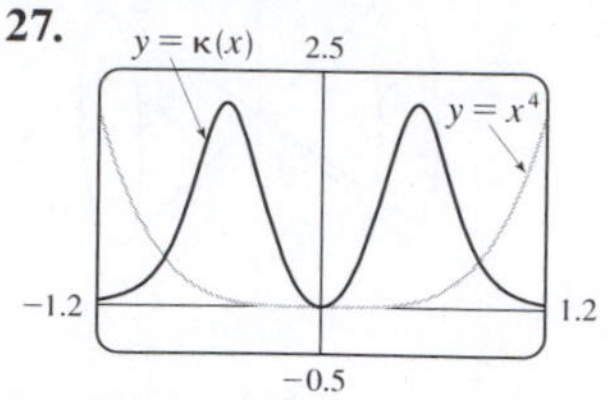

29. $6/[\,|t|(9t^2 + 4)^{3/2}]$

31. $\langle \frac{2}{3}, \frac{2}{3}, \frac{1}{3}\rangle$, $\langle -\frac{1}{3}, \frac{2}{3}, -\frac{2}{3}\rangle$, $\langle -\frac{2}{3}, \frac{1}{3}, \frac{2}{3}\rangle$

33. $y = 6x + \pi,\ x + 6y = 6\pi$

35. $(x + \frac{5}{2})^2 + y^2 = \frac{81}{4}$, $x^2 + (y - \frac{5}{3})^2 = \frac{16}{9}$

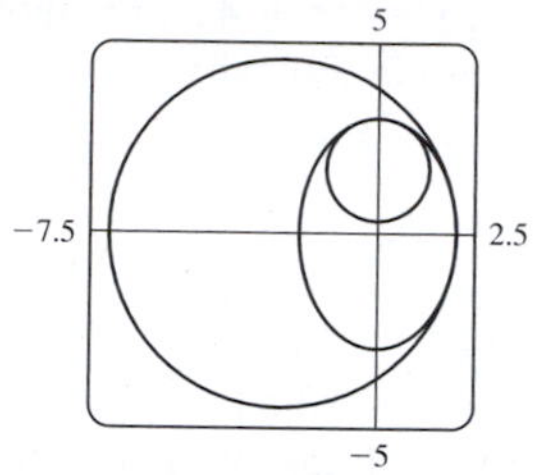

37. $(-1, -3, 1)$ **47.** $2/(t^4 + 4t^2 + 1)$ **49.** 2 m

Exercises 11.8 ■ page 713

1. $\mathbf{v}(t) = \langle 2t, 1\rangle$
$\mathbf{a}(t) = \langle 2, 0\rangle$
$|\mathbf{v}(t)| = \sqrt{4t^2 + 1}$

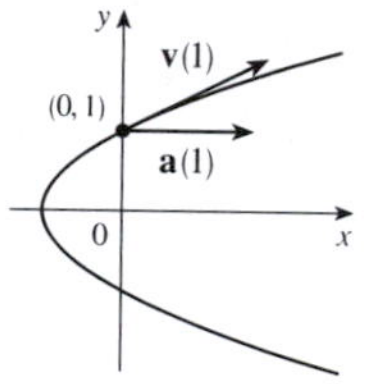

3. $\mathbf{v}(t) = e^t\,\mathbf{i} - e^{-t}\,\mathbf{j}$
$\mathbf{a}(t) = e^t\,\mathbf{i} + e^{-t}\,\mathbf{j}$
$|\mathbf{v}(t)| = \sqrt{e^{2t} + e^{-2t}}$

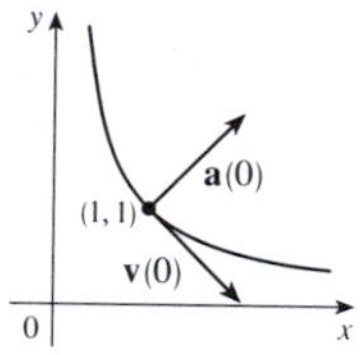

5. $\mathbf{v}(t) = \cos t\,\mathbf{i} + \mathbf{j} - \sin t\,\mathbf{k}$
$\mathbf{a}(t) = -\sin t\,\mathbf{i} - \cos t\,\mathbf{k}$
$|\mathbf{v}(t)| = \sqrt{2}$

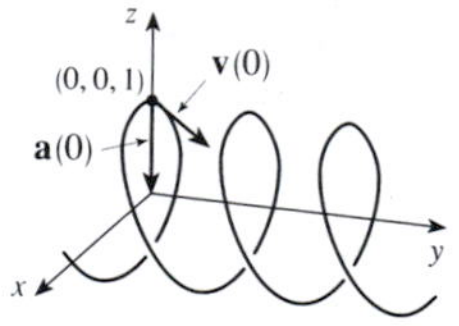

7. $\langle 3t^2, 2t, 3t^2\rangle$, $\langle 6t, 2, 6t\rangle$, $\sqrt{18t^4 + 4t^2}$

9. $-t^{-2}\,\mathbf{i} + 2t\,\mathbf{k}$, $2t^{-3}\,\mathbf{i} + 2\,\mathbf{k}$, $\sqrt{t^{-4} + 4t^2}$

11. $e^t[(\cos t - \sin t)\,\mathbf{i} + (\sin t + \cos t)\,\mathbf{j} + (t + 1)\,\mathbf{k}]$, $e^t[-2\sin t\,\mathbf{i} + 2\cos t\,\mathbf{j} + (t + 2)\,\mathbf{k}]$, $e^t\sqrt{t^2 + 2t + 3}$

13. $\mathbf{v}(t) = \mathbf{i} - \mathbf{j} + t\,\mathbf{k}$, $\mathbf{r}(t) = t\,\mathbf{i} - t\,\mathbf{j} + (t^2/2)\,\mathbf{k}$

15. (a) $\mathbf{r}(t) = (1 + t^2/2)\,\mathbf{i} + t^2\,\mathbf{j} + (1 + t^3/3)\,\mathbf{k}$
(b)

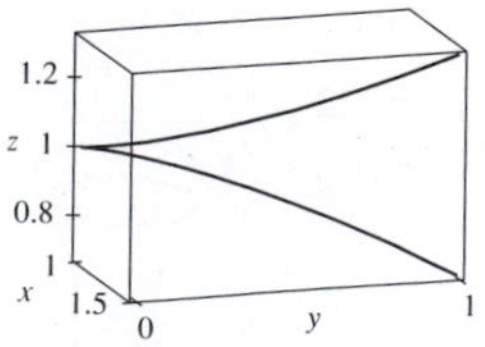

17. $t = 4$ **19.** $\mathbf{r}(t) = t\,\mathbf{i} - t\,\mathbf{j} + \frac{5}{2}t^2\mathbf{k}$, $|\mathbf{v}(t)| = \sqrt{25t^2 + 2}$

21. $(\sin t)/\sqrt{2(1 - \cos t)}$, $\sqrt{(1 - \cos t)/2}$

23. $(18t^3 + 4t)/\sqrt{9t^4 + 4t^2 + 1}$,
$2\sqrt{9t^4 + 9t^2 + 1}/\sqrt{9t^4 + 4t^2 + 1}$
25. $e^t - e^{-t}, \sqrt{2}$ **27.** $t = 1$ **31.** 2.99×10^8 km

Chapter 11 Review Exercises ■ page 715

1. True **3.** True **5.** True **7.** True **9.** True
11. False **13.** False
15. $|AB| = 13, |BC| = \sqrt{38}, |AC| = \sqrt{65}$
17. $(-2, -3, 5)$, 6 **19.** $6\,\mathbf{i} + \mathbf{j} + 13\,\mathbf{k}$ **21.** -1
23. $3\sqrt{35}$ **25.** **0** **27.** $96°$ **29.** $-1/\sqrt{6}$ **31.** 4
33. (a) 2 (b) -2 (c) -2 (d) 0
35. If $\overrightarrow{PR}$ is a scalar multiple of $\overrightarrow{PQ}$ (or if $\overrightarrow{PQ} \times \overrightarrow{PR} = \mathbf{0}$), then P, Q, R are collinear.
37. $\cos^{-1}\frac{1}{3} \approx 71°$ (or $109°$) **39.** (a) $\langle 4, -3, 4\rangle$ (b) $\sqrt{41}/2$
41. 166 N, 114 N **43.** $x = 1 + 2t, y = 2 - t, z = 4 + 3t$
45. $x = 1 + 4t, y = -3t, z = 1 + 5t$ **47.** $x + 2y + 5z = 8$
49. $(1, 4, 4)$ **51.** Skew **53.** $22/\sqrt{26}$
55. Plane

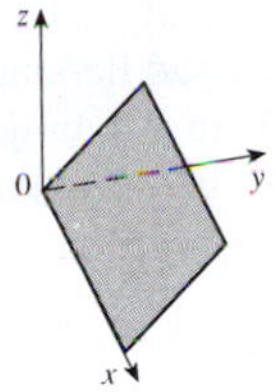

57. Circular paraboloid

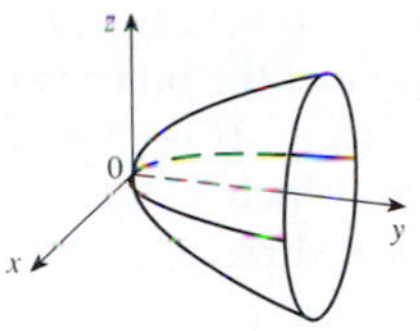

59. Cone

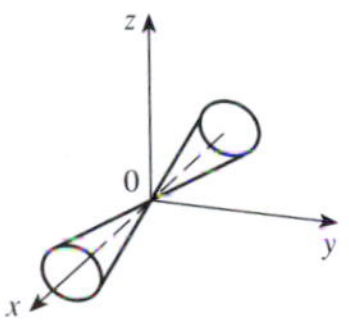

61. Hyperboloid of two sheets

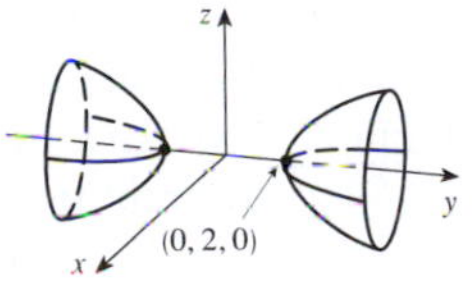

63. $4x^2 + y^2 + z^2 = 16$
65. (a)

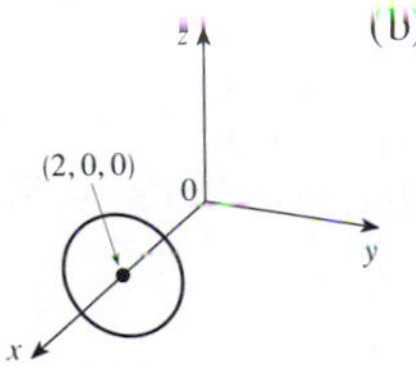

(b) $\cos t\,\mathbf{j} - \sin t\,\mathbf{k}, -\sin t\,\mathbf{j} - \cos t\,\mathbf{k}$

67. $\frac{5}{6}\mathbf{i} + \frac{9}{4}\mathbf{j} + \frac{1}{5}\mathbf{k}$ **69.** 15.9241 **71.** $\pi/2$
73. (a) $\langle t^2, t, 1\rangle/\sqrt{t^4 + t^2 + 1}$
(b) $\langle 2t, 1 - t^4, -2t^3 - t\rangle/\sqrt{t^8 + 4t^6 + 2t^4 + 5t^2}$
(c) $\sqrt{t^8 + 4t^6 + 2t^4 + 5t^2}/(t^4 + t^2 + 1)^2$ **75.** $12/17^{3/2}$
77. $x - 2y + 2\pi = 0$ **79.** $\langle t + t^3/6, 2t + t^2/2, t + t^4/12\rangle$
81. (c) $-2e^{-t}\mathbf{v}_d + e^{-t}\mathbf{R}$

APPLICATIONS PLUS ■ page 718

1. (a) $\mathbf{v} = \omega R(-\sin\omega t\,\mathbf{i} + \cos\omega t\,\mathbf{j})$ (c) $\mathbf{a} = -\omega^2\mathbf{r}$
3. (a) $90°, v_0^2/(2g)$
5. (b) $\mathbf{R}(t) = (m/k)(1 - e^{-kt/m})\mathbf{v}_0 + (gm/k)[(m/k)(1 - e^{-kt/m}) - t]\,\mathbf{j}$
7. $3x^5 - 8x^4 + 6x^3$, no

CHAPTER 12

Exercises 12.1 ■ page 730

1. (a) 7 (b) -45
(c) $x^2 + 2xh + h^2 - y^2 + 4xy + 4hy - 7x - 7h + 10$
(d) $x^2 - y^2 - 2yk - k^2 + 4xy + 4xk - 7x + 10$
(e) $4x^2 - 7x + 10$
3. (a) 1 (b) $-\frac{2}{3}$ (c) $3t/(t^2 + 2)$
(d) $-3y/(1 + 2y^2)$ (e) $3x/(1 + 2x^2)$
5. $\mathbb{R}^2; \mathbb{R}$ **7.** $\{(x, y) \mid x + y \neq 0\}; \{z \mid z \neq 0\}$
9. $\mathbb{R}^2; \{z \mid z > 0\}$ **11.** $\{(x, y, z) \mid x - y + z > 0\}; \mathbb{R}$
13. $\mathbb{R}^3; \mathbb{R}$
15. $\{(x, y) \mid y \geq 2x\}$

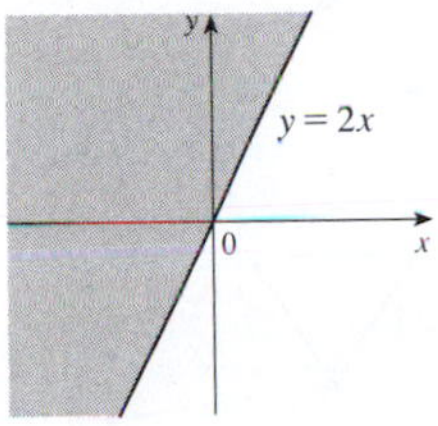

17. $\{(x, y) \mid x^2 + y^2 \leq 9, x \neq -2y\}$

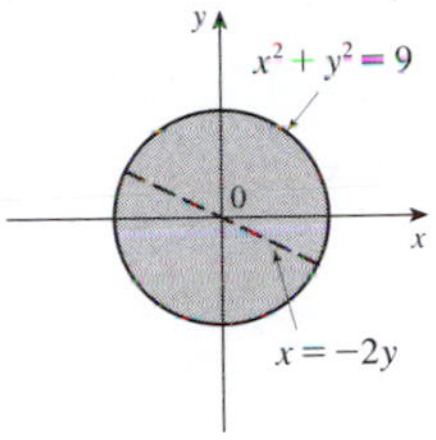

19. $\{(x, y) \mid y \geq -x^2\}$

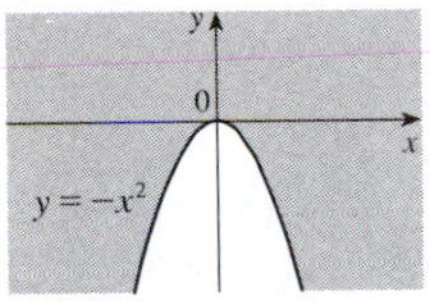

21. $\{(x, y) \mid xy > 1\}$

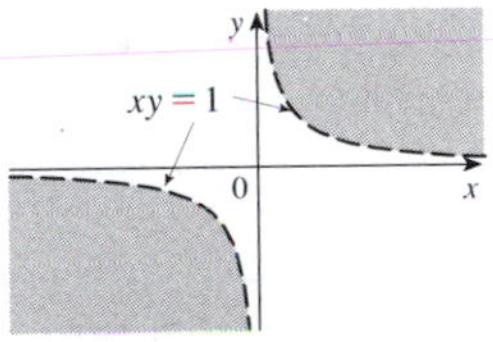

23. $\{(x, y) \mid y \neq (2n + 1)\pi/2, n$ an integer$\}$

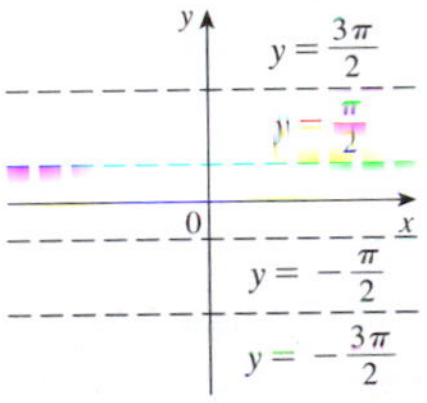

25. $\{(x, y) \mid -1 \leq x + y \leq 1\}$

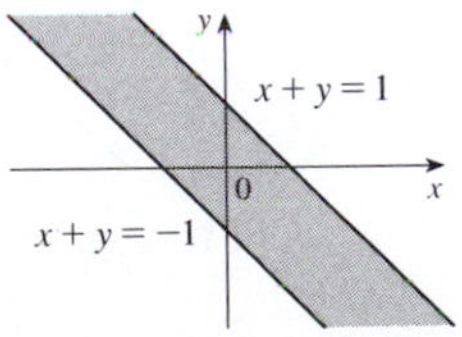

27. $\{(x, y) \mid x > 0, 2n\pi < y < (2n + 1)\pi, n$ an integer$\}$

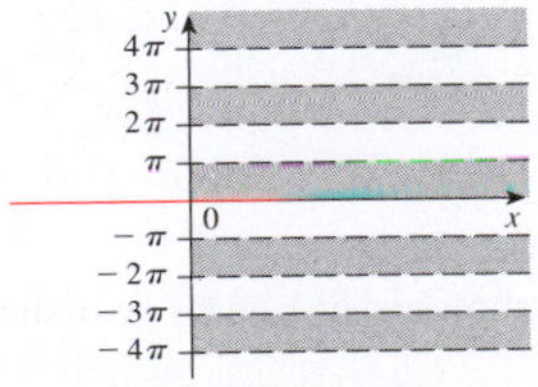

29. $\{(x, y, z) \mid x^2 + y^2 + z^2 \leqslant 1\}$

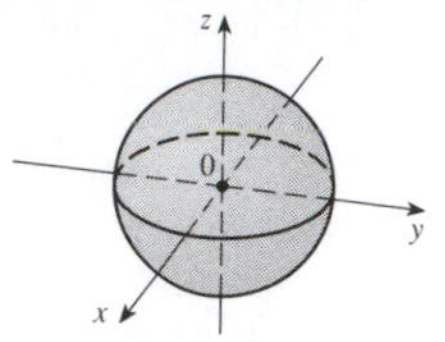

31. $z = 3$, horizontal plane

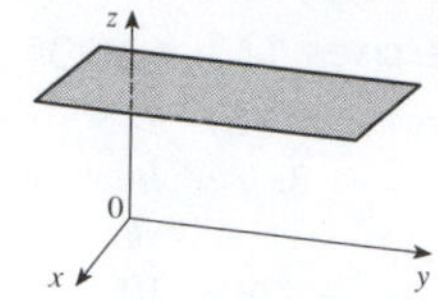

33. $x + y + z = 1$, plane

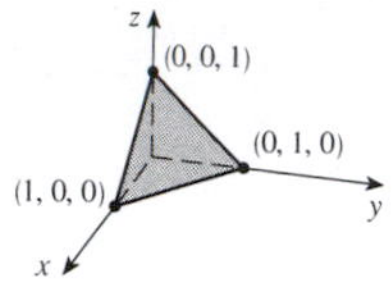

35. $z = x^2 + 9y^2$, elliptic paraboloid

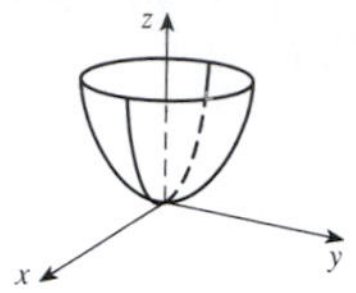

37. $z = \sqrt{x^2 + y^2}$, top half of cone

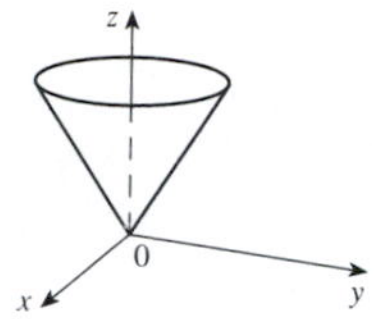

39. $z = y^2 - x^2$, hyperbolic paraboloid. See Figures 6 and 7(b) in Section 11.5.

41. $z = 1 - x^2$, parabolic cylinder

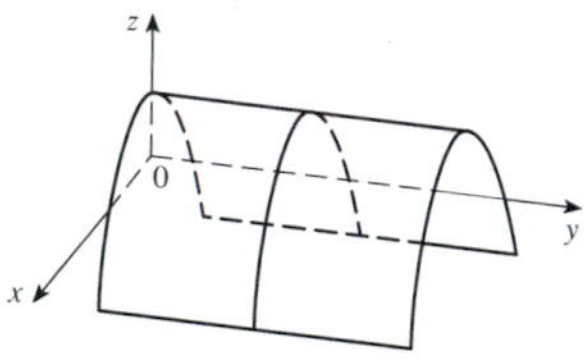

43. $xy = k$

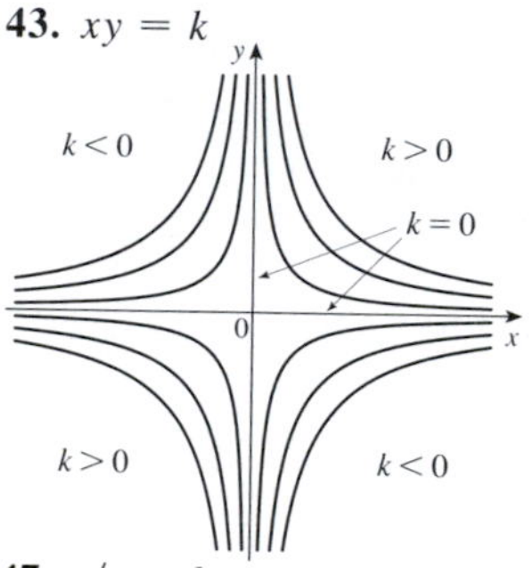

45. $x^2 + 9y^2 = k$

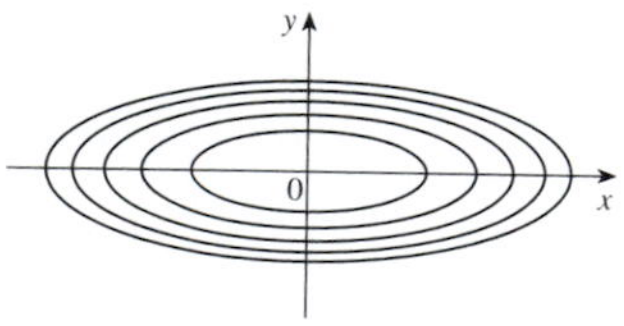

47. $x/y = k$

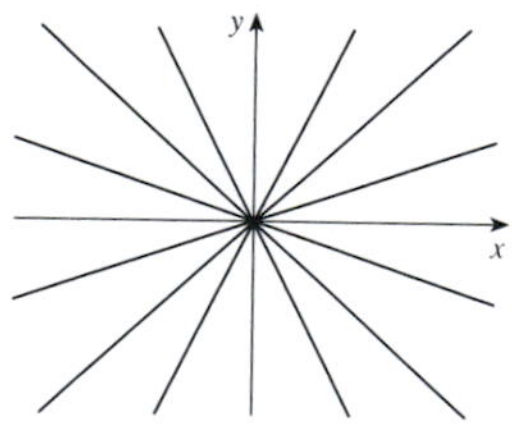

49. $\sqrt{x + y} = k$

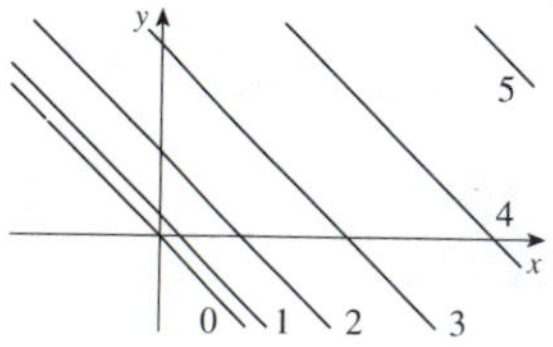

51. $x = y^2 + k$

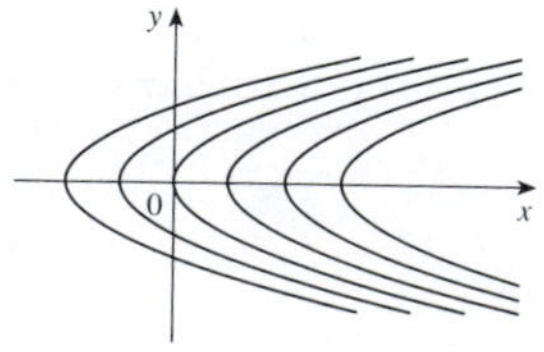

53. Family of parallel planes

55. Family of hyperboloids of one or two sheets with axis the y-axis

57.

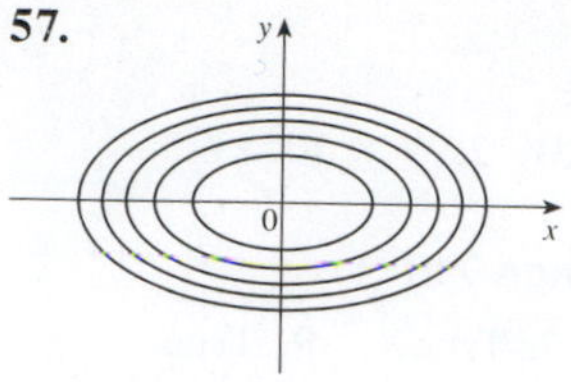

59. (a) B (b) III
61. (a) F (b) V
63. (a) D (b) IV

65.

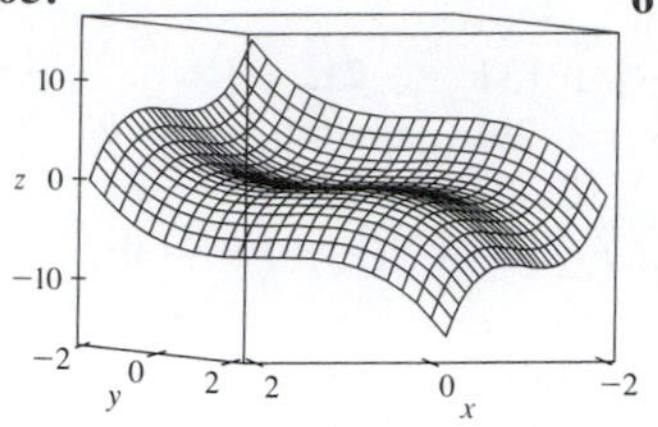

67.

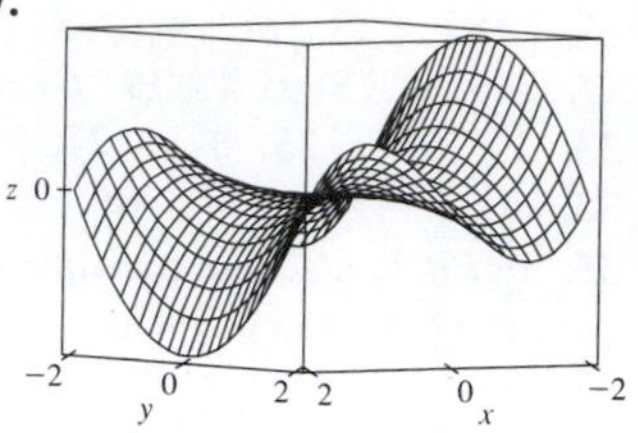

69. The level curves are elliptic if a and b have the same sign, hyperbolic if they have opposite signs. For both a and b positive, the graph increases rapidly as we leave the origin. If one of a and b is 0, the graph is a cylinder. If one of a and b is positive and the other is negative, the graph is saddle-shaped near the origin. If both a and b are negative, there is a bulge at the origin, which is elongated in the y-direction if a/b is large and in the x-direction if a/b is small.

Exercises 12.2 ■ page 739

1. -927 **3.** $-\frac{5}{2}$ **5.** π **7.** Does not exist
9. Does not exist **11.** Does not exist **13.** 0
15. Does not exist **17.** 2 **19.** Does not exist
21. $-\frac{3}{5}$ **23.** Does not exist **25.** Does not exist
27. The graph shows that the function approaches different numbers along different lines.
29. $h(x, y) = e^{-(x^4+x^2y^2+y^4)}\cos(x^4 + x^2y^2 + y^4)$; $\mathbb{R}^2$
31. $h(x, y) = 4x^2 + 9y^2 + 12xy - 24x - 36y + 36 + \sqrt{2x + 3y - 6}$; $\{(x, y) \mid 2x + 3y \geqslant 6\}$
33. $\{(x, y) \mid x^2 + y^2 \neq 1\}$
35. $\{(x, y) \mid x^4 - y^4 \neq (2n + 1)\pi/2, n \text{ an integer}\}$ **37.** $\mathbb{R}^2$
39. $\{(x, y) \mid x \geqslant |y|\}$ **41.** $\{(x, y, z) \mid yz > 0\}$
43. $\{(x, y) \mid (x, y) \neq (0, 0)\}$ **45.** $\{(x, y) \mid (x, y) \neq (0, 0)\}$

Exercises 12.3 ■ page 746

1. $f_x(1, 2) = -8 =$ slope of C_1, $f_y(1, 2) = -4 =$ slope of C_2

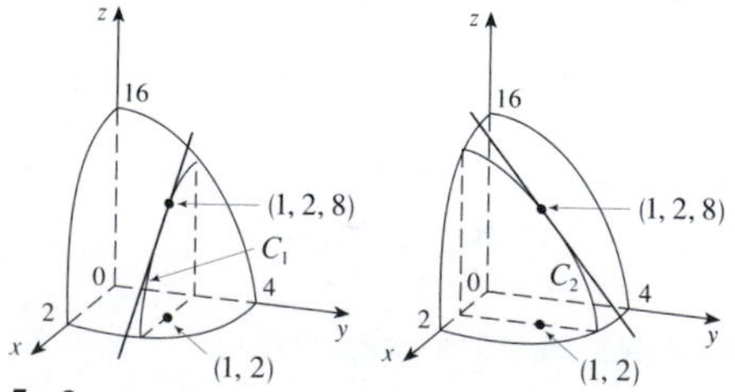

3. -27 **5.** 2
7. $(x^4 + 3x^2y^2 - 2xy^3)/(x^2 + y^2)^2$, $(y^4 + 3x^2y^2 - 2x^3y)/(x^2 + y^2)^2$
9. $(y - z)/(x - y)$, $(x + z)/(x - y)$
11. $(x - y - z)/(x + z)$, $(y - x)/(x + z)$ **13.** 0

15. $y + z, x + z, x + y$
17. $f_x(x, y) = 3x^2y^5 - 4xy + 1, f_y(x, y) = 5x^3y^4 - 2x^2$
19. $f_x(x, y) = 4x^3 + 2xy^2, f_y(x, y) = 2x^2y + 4y^3$
21. $f_x(x, y) = 2y/(x + y)^2, f_y(x, y) = -2x/(x + y)^2$
23. $f_x = e^x[\tan(x - y) + \sec^2(x - y)],\ f_y = -e^x \sec^2(x - y)$
25. $f_u = v/(u^2 + v^2), f_v = -u/(u^2 + v^2)$
27. $g_x = 2xy^4 \sec^2(x^2y^3),\ g_y = \tan(x^2y^3) + 3x^2y^3 \sec^2(x^2y^3)$
29. $\partial z/\partial x = 1/\sqrt{x^2 + y^2}, \partial z/\partial y = y/(x^2 + y^2 + x\sqrt{x^2 + y^2})$
31. $f_x = -e^{x^2}, f_y = e^{y^2}$
33. $f_x = 2xyz^3 + y, f_y = x^2z^3 + x, f_z = 3x^2yz^2 - 1$
35. $f_x = yzx^{yz-1}, f_y = zx^{yz} \ln x, f_z = yx^{yz} \ln x$
37. $u_x = -\dfrac{yz\cos(y/(x + z))}{(x + z)^2}, u_y = \dfrac{z\cos(y/(x + z))}{x + z},$
$u_z = \sin\left(\dfrac{y}{x + z}\right) - \dfrac{yz\cos(y/(x + z))}{(x + z)^2}$
39. $u_x = y^2z^3[\ln(x + 2y + 3z) + x/(x + 2y + 3z)],$
$u_y = 2xyz^3[\ln(x + 2y + 3z) + y/(x + 2y + 3z)],$
$u_z = 3xy^2z^2[\ln(x + 2y + 3z) + z/(x + 2y + 3z)]$
41. $f_x = 1/(z - t), f_y = 1/(t - z), f_z = (y - x)/(z - t)^2,$
$f_t = (x - y)/(z - t)^2$
43. $\partial u/\partial x_i = x_i/\sqrt{x_1^2 + x_2^2 + \cdots + x_n^2}$
45. $f_x(x, y) = 2x - y, f_y(x, y) = 4y - x$
47. $f_x = 2x + 2xy, f_y = 2y + x^2$

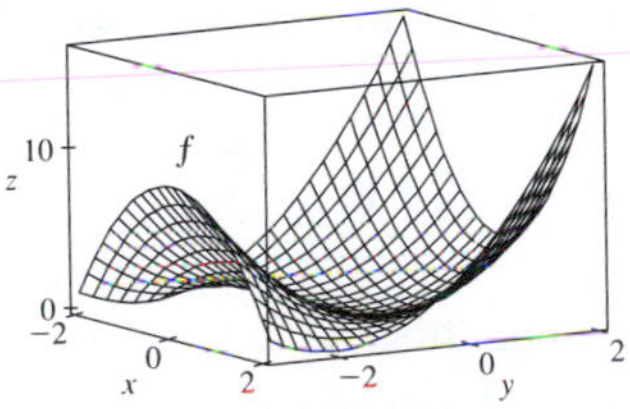

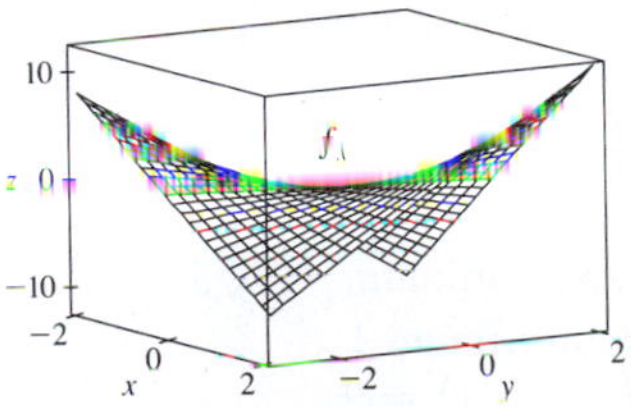

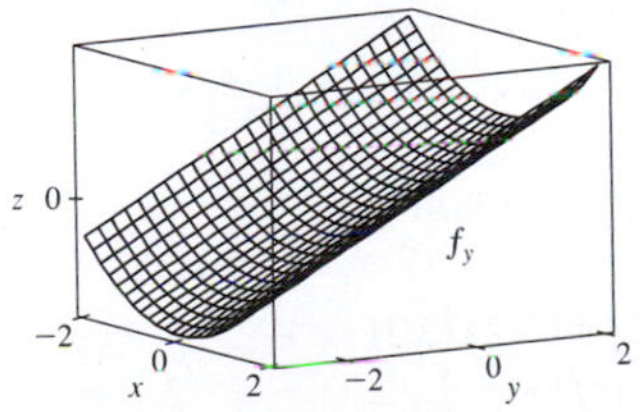

49. $c = f, b = f_x, a = f_y$ **51.** $f'(x), g'(y)$
53. $f'(x + y), f'(x + y)$ **55.** $f'(x/y)/y, -xf'(x/y)/y^2$
57. $f_{xx} = 2y, f_{xy} = 2x + 1/(2\sqrt{y}) = f_{yx}, f_{yy} = -x/(4y\sqrt{y})$
59. $z_{xx} = \dfrac{3(2x^2 + y^2)}{\sqrt{x^2 + y^2}}, z_{xy} = \dfrac{3xy}{\sqrt{x^2 + y^2}} = z_{yx},$
$z_{yy} = \dfrac{3(x^2 + 2y^2)}{\sqrt{x^2 + y^2}}$
61. $z_{tt} = 0, z_{tx} = \dfrac{1}{2\sqrt{x - x^2}} = z_{xt}, z_{xx} = \dfrac{t(2x - 1)}{4(x - x^2)^{3/2}}$
67. $-48xy$ **69.** $48x^3y^3z^2$ **71.** $-\sin y$
73. $72yz^2/(x + 2y^2 + 3z^3)^3$ **85.** R^2/R_1^2 **89.** No
91. $x = 1 + t, y = 2, z = 2 - 2t$ **95.** -2
97. (a)

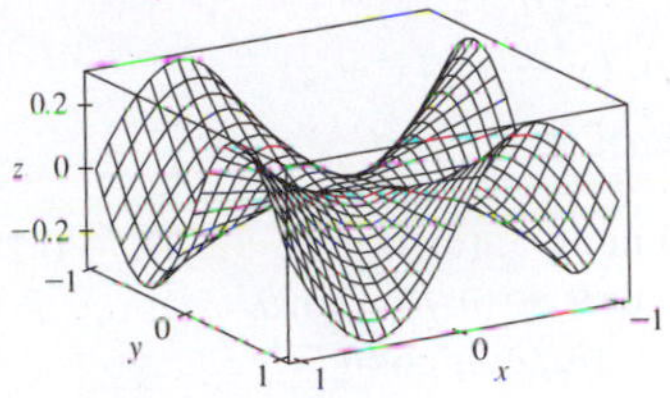

(b) $f_x(x, y) = \dfrac{x^4y + 4x^2y^3 - y^5}{(x^2 + y^2)^2}, f_y(x, y) = \dfrac{x^5 - 4x^3y^2 - xy^4}{(x^2 + y^2)^2}$
(c) 0, 0 (e) No, since f_{xy} and f_{yx} are not continuous.

Exercises 12.4 ■ page 755

1. $4x + 8y - z = 8$ **3.** $2x + 4y - z + 6 = 0$
5. $2x + y - z = 1$
7.

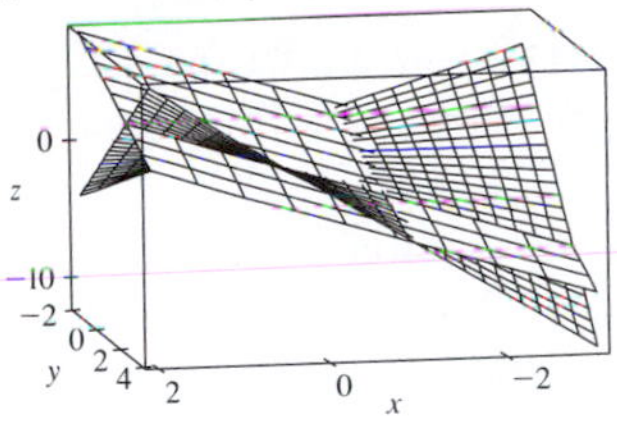

9.

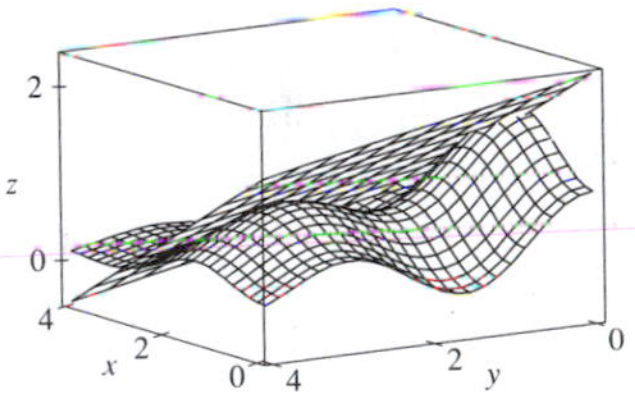

11. $dz = 2xy^3\,dx + 3x^2y^2\,dy$
13. $du = e^x(\cos xy - y \sin xy)\,dx - xe^x \sin xy\,dy$
15. $dw = 2xy\,dx + (x^2 + 2yz)\,dy + y^2\,dz$
17. $dw = (x^2 + y^2 + z^2)^{-1}(x\,dx + y\,dy + z\,dz)$
19. $\Delta z = 0.9225, dz = 0.9$ **21.** 2.846 **23.** 65.88
25. 26.76 **27.** 1.015 **29.** 5.4 cm^2 **31.** 16 cm^3
33. 150 **35.** $\frac{1}{17} \approx 0.059$ ohm **37.** $\varepsilon_1 = \Delta x, \varepsilon_2 = \Delta y$

Exercises 12.5 ■ page 762

1. $6t^5 + 4t^3 + 4t$ **3.** $(x + y^2)^{-1}(1/(2\sqrt{1 + t}) + y/\sqrt{t})$
5. $y^2z^3 \cos t - 2xyz^3 \sin t + 6xy^2z^2e^{2t}$
7. $\partial z/\partial s = 4sx \sin y + 2tx^2 \cos y, \partial z/\partial t = 4tx \sin y + 2sx^2 \cos y$
9. $\partial z/\partial s = 2x(1 - 3y^3)e^t - 9x^2y^2e^{-t},$
$\partial z/\partial t = 2x(1 - 3y^3)se^t + 9x^2y^2se^{-t}$
11. $\partial z/\partial s = 2^{x-3y}(\ln 2)(2st - 3t^2),$
$\partial z/\partial t = 2^{x-3y}(\ln 2)(s^2 - 6st)$
13. $\dfrac{\partial u}{\partial r} = \dfrac{\partial u}{\partial x}\dfrac{\partial x}{\partial r} + \dfrac{\partial u}{\partial y}\dfrac{\partial y}{\partial r}, \dfrac{\partial u}{\partial s} = \dfrac{\partial u}{\partial x}\dfrac{\partial x}{\partial s} + \dfrac{\partial u}{\partial y}\dfrac{\partial y}{\partial s},$
$\dfrac{\partial u}{\partial t} = \dfrac{\partial u}{\partial x}\dfrac{\partial x}{\partial t} + \dfrac{\partial u}{\partial y}\dfrac{\partial y}{\partial t}$
15. $\dfrac{\partial v}{\partial x} = \dfrac{\partial v}{\partial p}\dfrac{\partial p}{\partial x} + \dfrac{\partial v}{\partial q}\dfrac{\partial q}{\partial x} + \dfrac{\partial v}{\partial r}\dfrac{\partial r}{\partial x},$
$\dfrac{\partial v}{\partial y} = \dfrac{\partial v}{\partial p}\dfrac{\partial p}{\partial y} + \dfrac{\partial v}{\partial q}\dfrac{\partial q}{\partial y} + \dfrac{\partial v}{\partial r}\dfrac{\partial r}{\partial y},$
$\dfrac{\partial v}{\partial z} = \dfrac{\partial v}{\partial p}\dfrac{\partial p}{\partial z} + \dfrac{\partial v}{\partial q}\dfrac{\partial q}{\partial z} + \dfrac{\partial v}{\partial r}\dfrac{\partial r}{\partial z}$

17. 2, 0 **19.** 0, 0, 4
21. $\partial u/\partial p = 2(z - x)/(y + z)^2 = -t/p^2$, $\partial u/\partial r = 0$, $\partial u/\partial t = 2/(y + z) = 1/p$
23. $(y - 2x)/(3y^2 - x)$
25. $(18x - x^{-2/3}y^{1/3})/(12y + x^{1/3}y^{-2/3})$
27. $(y - z)/(x - y)$, $(x + z)/(x - y)$
29. $(x - y - z)/(x + z)$, $(y - x)/(x + z)$
31. $-(e^y + ze^x)/(y + e^x)$, $-(xe^y + z)/(y + e^x)$
33. $-9600\pi\ \text{cm}^3/\text{s}$
35. (a) $6\ \text{m}^3/\text{s}$ (b) $10\ \text{m}^2/\text{s}$ (c) $0\ \text{m/s}$ **37.** -0.27 L/s
39. (a) $\partial z/\partial r = (\partial z/\partial x)\cos\theta + (\partial z/\partial y)\sin\theta$, $\partial z/\partial\theta = -(\partial z/\partial x)r\sin\theta + (\partial z/\partial y)r\cos\theta$
45. $4rs\,\partial^2 z/\partial x^2 + (4r^2 + 4s^2)\,\partial^2 z/\partial x\,\partial y + 4rs\,\partial^2 z/\partial y^2 + 2\,\partial z/\partial y$

Exercises 12.6 ▪ page 772

1. $7\sqrt{3} - 16$ **3.** 1
5. (a) $\nabla f(x, y) = \langle 3x^2 - 8xy, -4x^2 + 2y\rangle$ (b) $\langle 0, -2\rangle$ (c) $-\frac{8}{5}$
7. (a) $\nabla f(x, y, z) = \langle y^2z^3, 2xyz^3, 3xy^2z^2\rangle$ (b) $\langle 4, -4, 12\rangle$ (c) $20/\sqrt{3}$
9. $\frac{7}{52}$ **11.** $29/\sqrt{13}$ **13.** $\frac{1}{6}$ **15.** $-\pi/(4\sqrt{3})$
17. $\sqrt{5}$, $\langle 1, 2\rangle$ **19.** $\sqrt{17}/6$, $\langle 4, 1\rangle$ **21.** $\sqrt{13/2}$, $\langle -3, -2\rangle$
25. (a) $-40/(3\sqrt{3})$
27. (a) $32/\sqrt{3}$ (b) $\langle 38, 6, 12\rangle$ (c) $2\sqrt{406}$ **29.** $\frac{327}{13}$
35. (a) $4x + y + z = 12$ (b) $(x - 2)/4 = y - 2 = z - 2$
37. (a) $3x - y + z = 4$ (b) $(x - 1)/3 = -y = z - 1$
39. (a) $x + y - z = 1$ (b) $x - 1 = y = -z$
41.

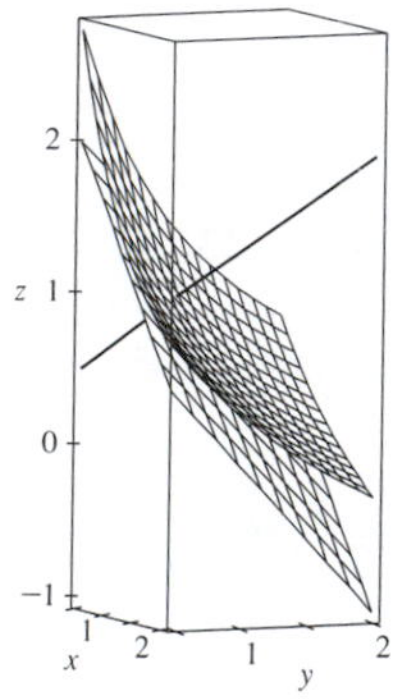

43. $\langle 4, 8\rangle$, $x + 2y = 4$

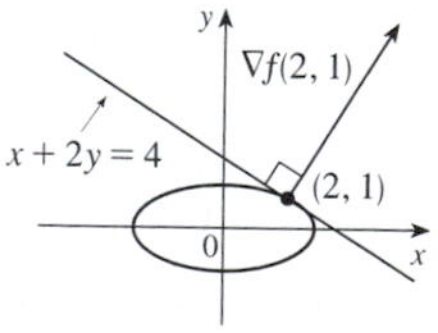

49. $(\pm\sqrt{6}/3, \mp 2\sqrt{6}/3, \pm\sqrt{6}/2)$
55. $x = -1 - 10t$, $y = 1 - 16t$, $z = 2 - 12t$
59. If $\mathbf{u} = \langle a, b\rangle$ and $\mathbf{v} = \langle c, d\rangle$, then $af_x + bf_y$ and $cf_x + df_y$ are known, so we solve linear equations for f_x and f_y.

Exercises 12.7 ▪ page 781

1. Minimum $f(-2, 3) = -13$
3. Minimum $f(0, -1) = -1$
5. Minimum $f(0, 0) = 4$, saddle points $(\pm\sqrt{2}, -1)$
7. Minimum $f(1, 1) = -1$, saddle point $(0, 0)$
9. Saddle point $(1, 2)$ **11.** Maximum $f(-\frac{1}{2}, 4) = -6$
13. None **15.** Saddle points $(0, n\pi)$, n an integer
17. Maximum $f(0, 0) = 2$, minimum $f(0, 2) = -2$, saddle points $(\pm 1, 1)$
19. Maximum $f(\pi/3, \pi/3) = 3\sqrt{3}/2$, minimum $f(5\pi/3, 5\pi/3) = -3\sqrt{3}/2$
21. Minima $f(-1.714, 0) \approx -9.200$, $f(1.402, 0) \approx 0.242$, saddle point $(0.312, 0)$, lowest point $(-1.714, 0, -9.200)$
23. Maxima $f(-1.267, 0) \approx 1.310$, $f(1.629, \pm 1.063) \approx 8.105$, saddle points $(-0.259, 0)$, $(1.526, 0)$, highest points $(1.629, \pm 1.063, 8.105)$
25. Minimum $f(4, 0) = -7$, maximum $f(4, 5) = 13$
27. Maximum $f(\pm 1, 1) = 7$, minimum $f(0, 0) = 4$
29. Maximum $f(2, 4) = 3$, minimum $f(-2, 4) = -9$
31. Maximum $f(1, 0) = 2$, minimum $f(-1, 0) = -2$
33.

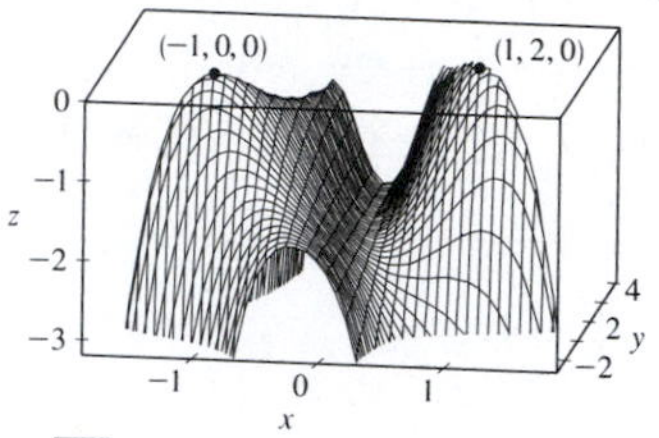

35. $(\frac{2}{7}, \frac{4}{7}, \frac{6}{7})$
37. $7/\sqrt{61}$ **39.** $(0, 0, 1)$, $(0, 0, -1)$ **41.** $\frac{100}{3}, \frac{100}{3}, \frac{100}{3}$
43. $16/\sqrt{3}$ **45.** $\frac{4}{3}$ **47.** Cube, edge length $c/12$
49. Square base of side 40 cm, height 20 cm
53. $y = 6.67x - 308$, 172.2 lb

Exercises 12.8 ▪ page 788

1. Maxima $f(\pm 1, 0) = 1$, minima $f(0, \pm 1) = -1$
3. Maxima $f(\pm\sqrt{2}/3, \pm\sqrt{2}) = \frac{2}{3}$, minima $f(\pm\sqrt{2}/3, \mp\sqrt{2}) = -\frac{2}{3}$
5. Maximum $f(1/\sqrt{35}, 3/\sqrt{35}, 5/\sqrt{35}) = \sqrt{35}$, minimum $f(-1/\sqrt{35}, -3/\sqrt{35}, -5/\sqrt{35}) = -\sqrt{35}$
7. Maximum $2/\sqrt{3}$, minimum $-2/\sqrt{3}$
9. Maximum $\sqrt{3}$, minimum 1
11. Maximum $f(\frac{1}{2}, \frac{1}{2}, \frac{1}{2}, \frac{1}{2}) = 2$, minimum $f(-\frac{1}{2}, -\frac{1}{2}, -\frac{1}{2}, -\frac{1}{2}) = -2$
13. Maximum $f(1, \sqrt{2}, -\sqrt{2}) = 1 + 2\sqrt{2}$, minimum $f(1, -\sqrt{2}, \sqrt{2}) = 1 - 2\sqrt{2}$
15. Maximum $\frac{3}{2}$, minimum $\frac{1}{2}$
17. Maxima $f(\pm 1/\sqrt{2}, \mp 1/(2\sqrt{2})) = e^{1/4}$, minima $f(\pm 1/\sqrt{2}, \pm 1/(2\sqrt{2})) = e^{-1/4}$
19. (b) Minimum $f(3 - \frac{3}{2}\sqrt{2}, 3 - \frac{3}{2}\sqrt{2}) = (351 - 243\sqrt{2})/2 \approx 3.7$, maximum $f(3 + \frac{3}{2}\sqrt{2}, 3 + \frac{3}{2}\sqrt{2}) = (351 + 243\sqrt{2})/2 \approx 347$
25.–39. See Exercises 35–49 in Section 12.7.
41. Nearest $(\frac{1}{2}, \frac{1}{2}, \frac{1}{2})$, farthest $(-1, -1, 2)$

Chapter 12 Review Exercises ▪ page 790

1. True **3.** False **5.** False **7.** True **9.** False

11. True

13. $\{(x, y) \mid y > -x - 1, x \neq 1\}$

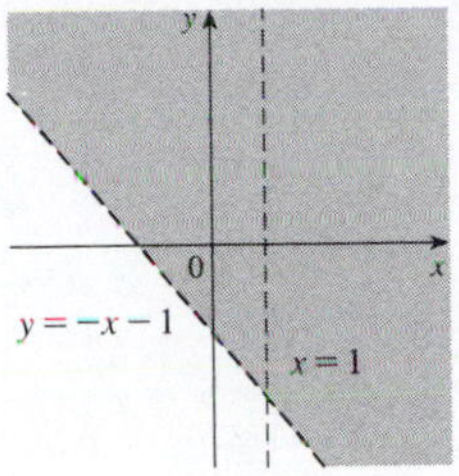

15. $\{(x, y) \mid -1 \leqslant x \leqslant 1\}$

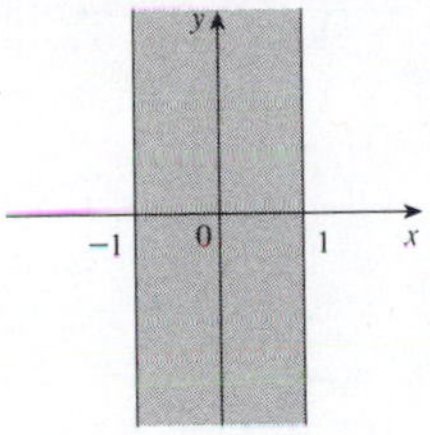

17.

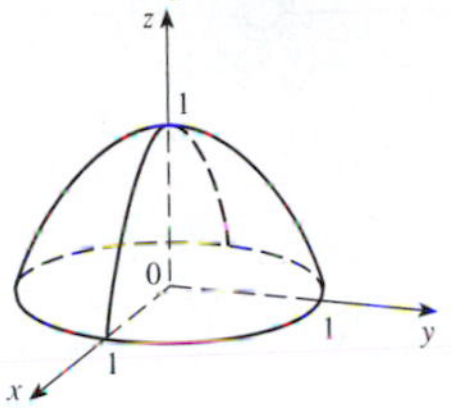

19.

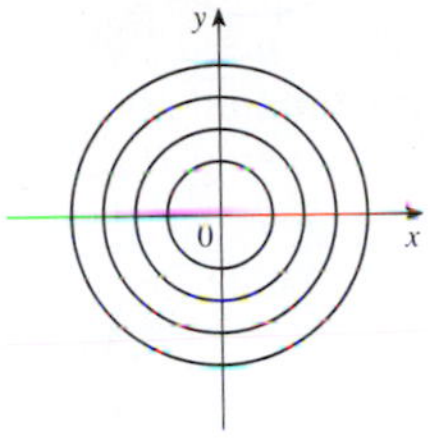

21. 0 **23.** $f_x = 12x^3 - \sqrt{y},\, f_y = -\frac{1}{2}x/\sqrt{y}$

25. $f_s = 2e^{2s}\cos \pi t,\, f_t = -\pi e^{2s}\sin \pi t$

27. $f_x = y^z,\, f_y = xzy^{z-1},\, f_z = xy^z \ln y$

29. $f_{xx} = 2y^3 - 24x^2,\, f_{xy} = 6xy^2,\, f_{yy} = 6x^2y + 2$

31. $f_{xx} = 0,\, f_{yy} = 2xz^3,\, f_{zz} = 6xy^2z,\, f_{xy} = 2yz^3,$
$f_{xz} = 3y^2z^2,\, f_{yz} = 6xyz^2$

35. $6y - z = 1$ **37.** $x + 3y + 9z = 18$

39. $3x + 4y + 3z = 14$ **41.** $(\pm\sqrt{2/7}, \pm 1/\sqrt{14}, \mp 3/\sqrt{14})$

43. 38.656 **45.** $e^t + 2(y/z)(3t^2 + 4) - 2t(y^2/z^2)$

51. $ze^{x\sqrt{y}}\langle z\sqrt{y}, xz/(2\sqrt{y}), 2\rangle$ **53.** $\frac{43}{5}$ **55.** $\sqrt{145}/2,\, \langle 4, \frac{9}{2}\rangle$

57. $\pi/2$ **59.** Minimum $f(-4, 1) = -11$

61. Maximum $f(1, 1) = 1$, saddle points $(0, 0), (0, 3), (3, 0)$

63. Maximum $f(1, 2) = 4$, minimum $f(2, 4) = -64$

65. Maximum $f(-1, 0) = 2$, minima $f(1, \pm 1) = -3$, saddle points $(-1, \pm 1), (1, 0)$

67. Maximum $f(\pm\sqrt{2/3}, 1/\sqrt{3}) = 2/(3\sqrt{3})$, minimum $f(\pm\sqrt{2/3}, -1/\sqrt{3}) = -2/(3\sqrt{3})$

69. Maximum $f(3, 3, 3) = 9$, minimum $f(1, 1, -1) = f(1, -1, 1) = f(-1, 1, 1) = 1$

71. $(\pm 3^{-1/4}, 3^{-1/4}\sqrt{2}, \pm 3^{1/4}),\; (\pm 3^{-1/4}, -3^{-1/4}\sqrt{2}, \pm 3^{1/4})$

73. $P(2 - \sqrt{3}),\, P(3 - \sqrt{3})/6,\; P(2\sqrt{3} - 3)/3$

PROBLEMS PLUS ■ page 792

1. $(\sqrt{3} - 1.5)$ m **5.** $L^2W^2, \frac{1}{4}L^2W^2$

11. (a) $\langle -8, -8\rangle$, or $\langle -1/\sqrt{2}, -1/\sqrt{2}\rangle$ as a unit vector

(b)

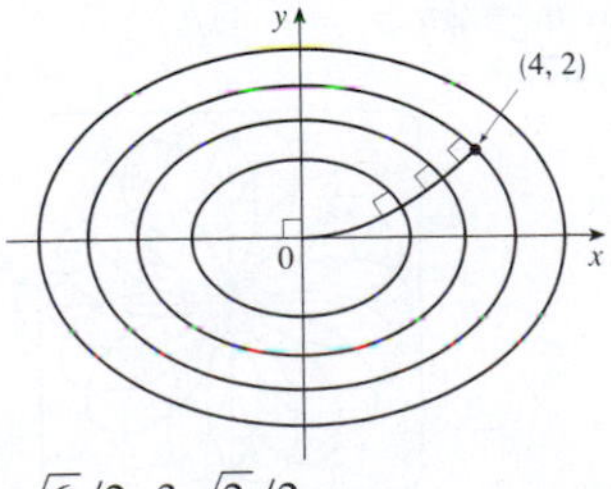

(c) Part of the parabola $y = \frac{1}{8}x^2$

13. $\sqrt{6}/2,\, 3\sqrt{2}/2$

CHAPTER 13

Exercises 13.1 ■ page 799

1. (a) -17.75 (b) -15.75 (c) -8.75 (d) -6.75

3. $63, \sqrt{2}$ **5.** $0, \sqrt{5}$ **7.** $\frac{247}{8}, \sqrt{5}$ **9.** $U < V < L$

11. 0.6065, 0.5694, 0.5606, 0.5585, 0.5579, 0.5578 **13.** 3

Exercises 13.2 ■ page 804

1. $4x^2, y^3/3$ **3.** $(e^2 - 1)xe^x, e^y$ **5.** $\frac{32}{3}$ **7.** 0

9. $\frac{4}{15}(31 - 9\sqrt{3})$ **11.** $3[1 - (1/\sqrt{2})]$ **13.** 6

15. $-\frac{585}{8}$ **17.** $\frac{15}{4}(2 - \sqrt{3})$ **19.** $[(\sqrt{3} - 1)/2] - (\pi/12)$

21. $\ln \frac{27}{16}$ **23.**

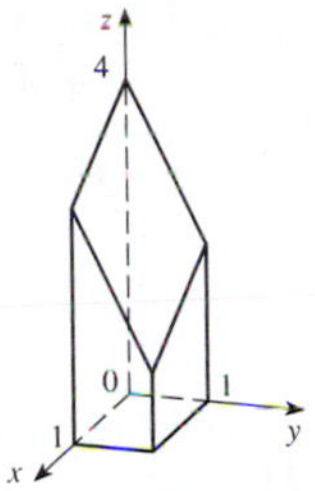

25. 37.5 **27.** $\frac{166}{27}$ **29.** $\frac{4}{15}(2\sqrt{2} - 1)$ **31.** 36

33. $21e - 57$

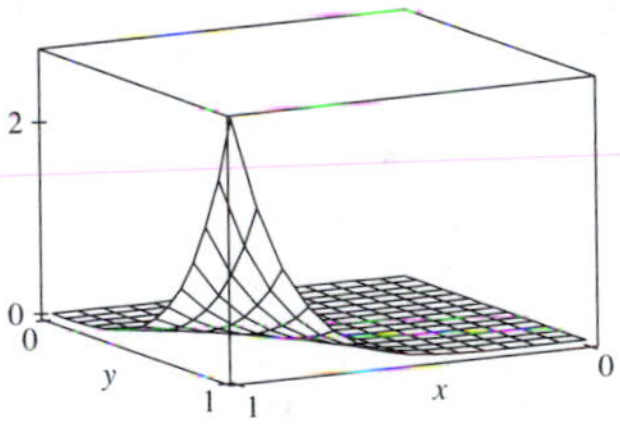

35. $\frac{5}{6}$

37. Fubini's Theorem does not apply. The integrand has an infinite discontinuity at the origin.

Exercises 13.3 ■ page 812

1. $\frac{1}{6}$ **3.** $16[1 - (\sqrt{2}/7)]$ **5.** $(1 - \cos 1)/2$ **7.** $\frac{1}{12}$

9. $-\frac{19}{42}$ **11.** $\frac{1}{2}e^4 - 2e$ **13.** $(1 - \cos 1)/2$ **15.** $\frac{500}{3}$

17. $\frac{1}{8}$ **19.** $\frac{6}{35}$ **21.** $\frac{31}{8}$ **23.** $\frac{1}{6}(11\sqrt{5} - 27) + \frac{9}{2}\sin^{-1}\frac{2}{3}$

25. $\frac{1}{6}$ **27.** $\frac{1}{3}$ **29.** 0, 1.213, 0.713

31. 13,984,735,616/14,549,535

33. $\int_0^1 \int_y^1 f(x, y)\, dx\, dy$

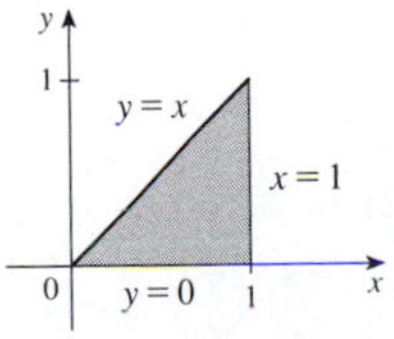

35. $\int_0^{\ln 2} \int_{e^y}^{2} f(x, y)\, dx\, dy$

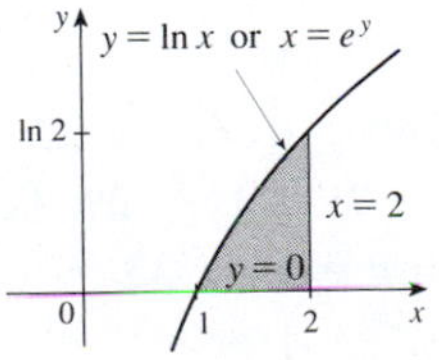

37. $\int_0^2 \int_0^{2x} f(x, y)\, dy\, dx$

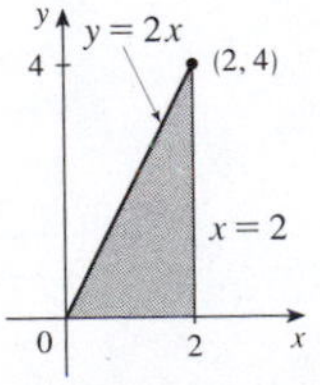

39. $(e^9 - 1)/6$ **41.** $\frac{1}{4}\sin 81$ **43.** $(2\sqrt{2} - 1)/3$ **45.** 1
47. $0 \leqslant \iint_D \sqrt{x^3 + y^3}\, dA \leqslant \sqrt{2}$ **51.** 8π

Exercises 13.4 ■ page 819

1.

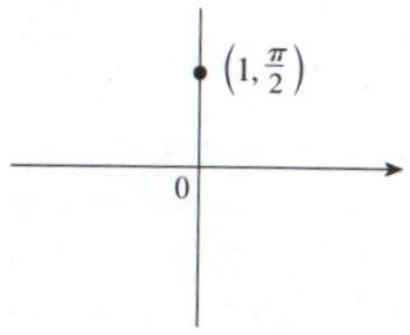

$(1, 5\pi/2)$, $(-1, 3\pi/2)$

3.

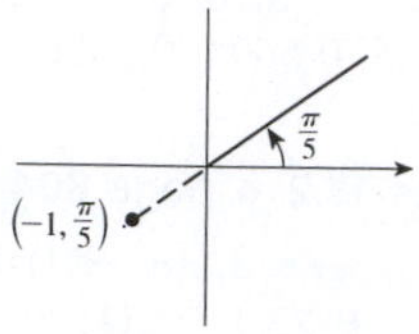

$(1, 6\pi/5)$, $(-1, 11\pi/5)$

5.

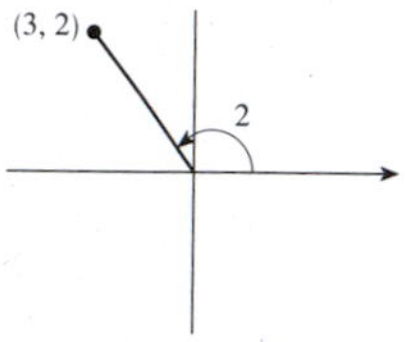

$(3, 2 + 2\pi)$, $(-3, 2 + \pi)$

7.

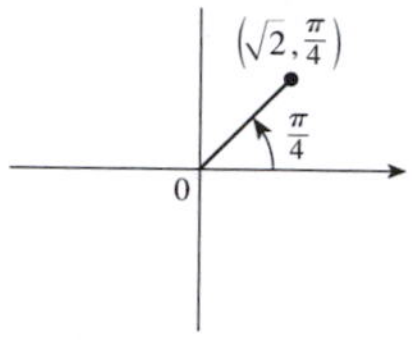

$(1, 1)$

9.

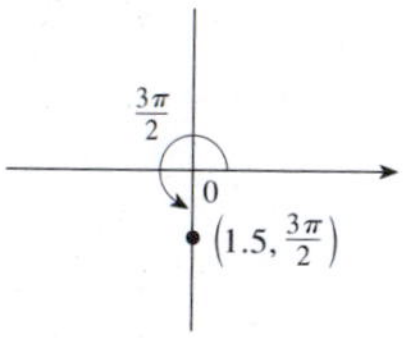

$(0, -1.5)$

11.

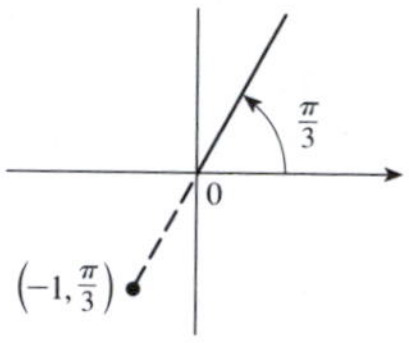

$(-\frac{1}{2}, -\sqrt{3}/2)$

13. $(\sqrt{2}, 3\pi/4)$ **15.** $(4, 11\pi/6)$

17.

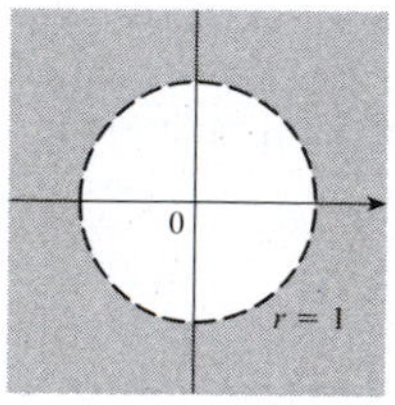

19.

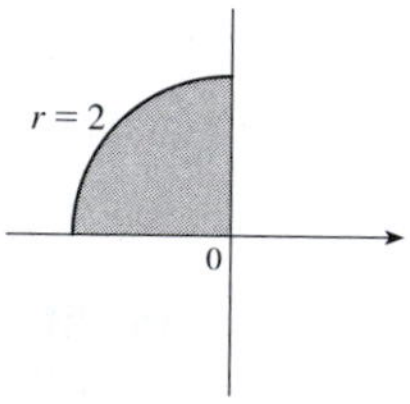

21.

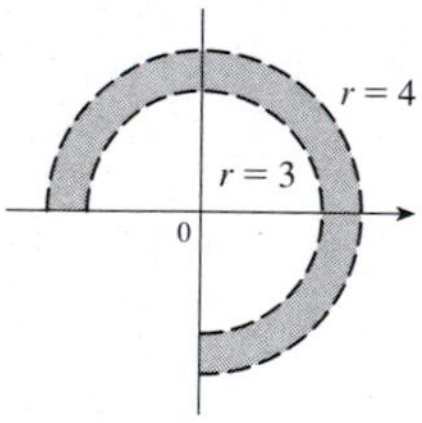

23. $\frac{1}{2}\sqrt{40 + 6\sqrt{6} - 6\sqrt{2}}$ **25.** $y = 2$
27. $y^2 = 2x + 1$ **29.** $(x^2 + y^2)^2 = 2xy$
31. $r \sin\theta = 5$ **33.** $r = 5$ **35.** $r^2 = \csc 2\theta$

37.

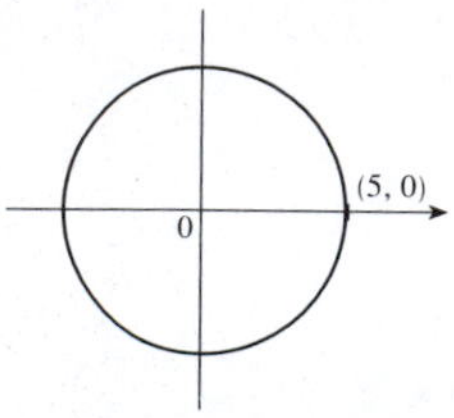

39.

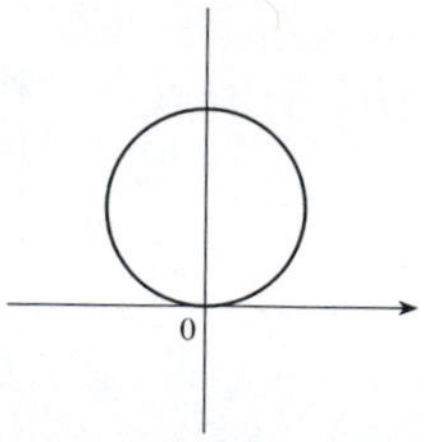

41.

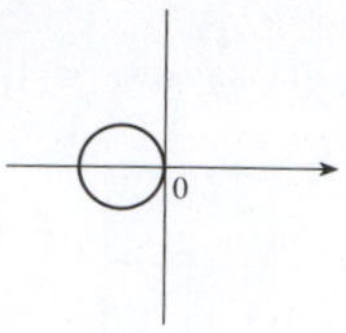

43.

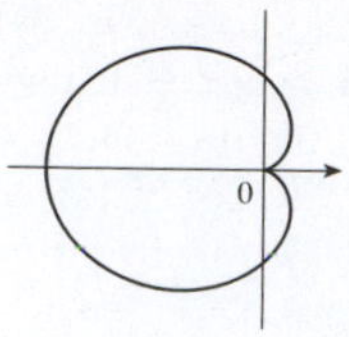

45.

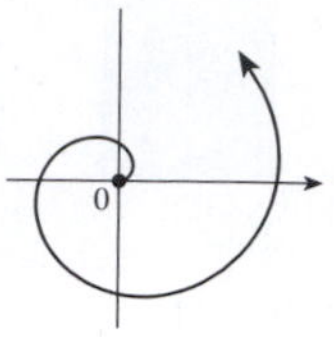

47.

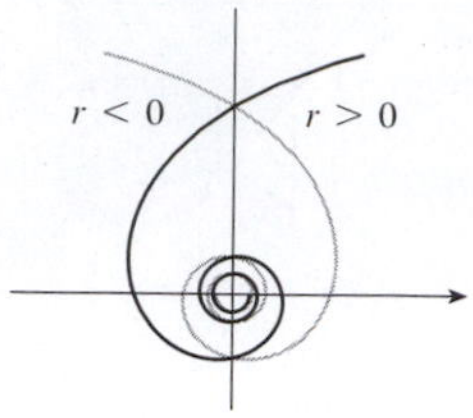

49.

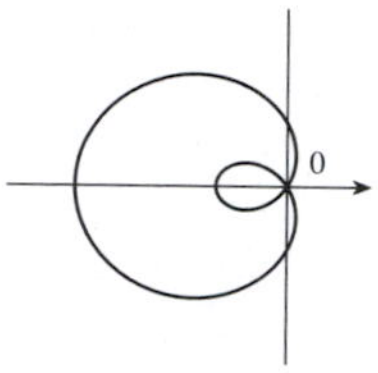

51.

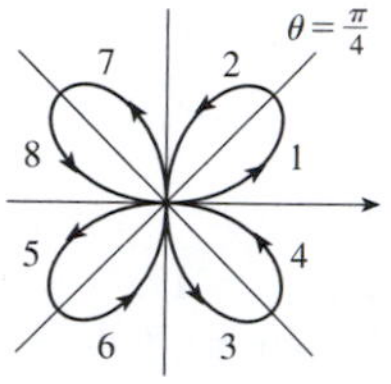

53.

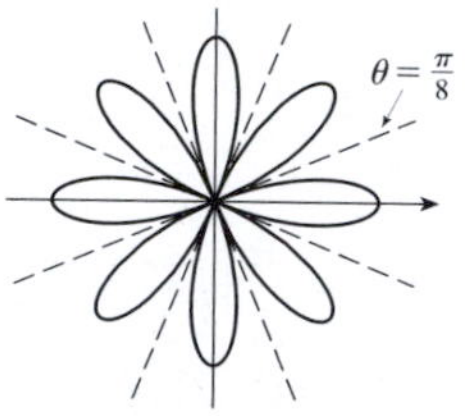

55.

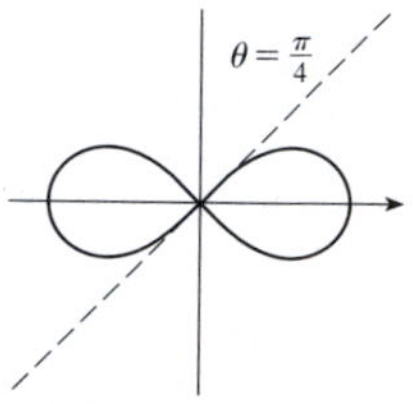

57.

59.

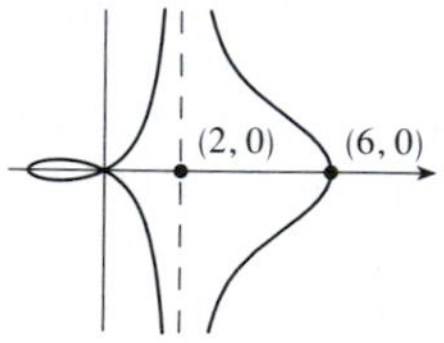

61.

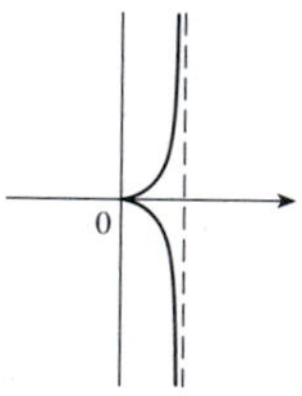

63. (a) For $c < -1$, the loop begins at $\theta = \sin^{-1}(-1/c)$ and ends at $\theta = \pi - \sin^{-1}(-1/c)$; for $c > 1$, it begins at $\theta = \pi + \sin^{-1}(1/c)$ and ends at $\theta = 2\pi - \sin^{-1}(1/c)$.

65.

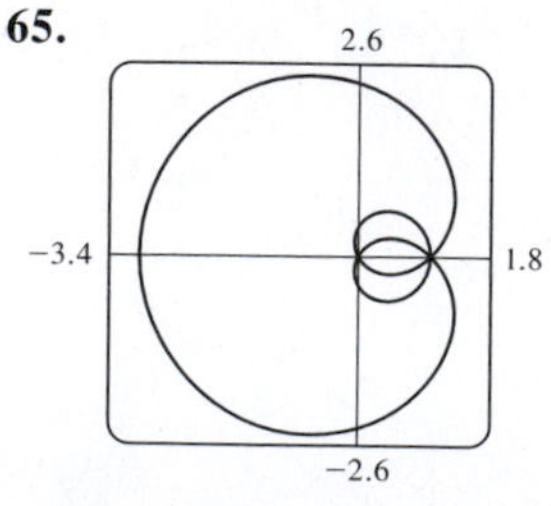

67.

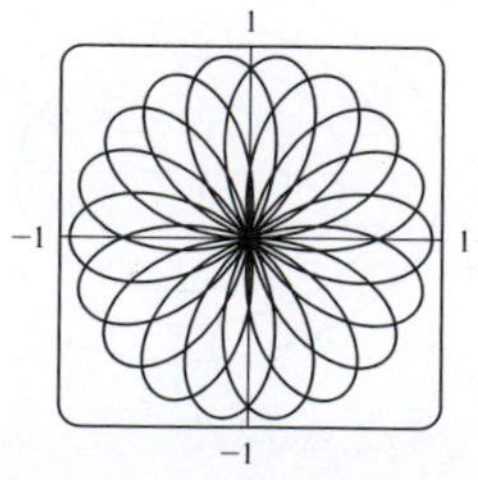

69. By counterclockwise rotation through angle $\pi/6$, $\pi/3$, or α about the origin
71. (a) A rose with n loops if n is odd and $2n$ loops if n is even (b) Number of loops is always $2n$
73. For $0 < a < 1$, the curve is an oval, which develops a dimple as $a \to 1^-$. When $a > 1$, the curve splits into two parts, one of which has a loop.

Exercises 13.5 ■ page 826

1. 0 **3.** $\frac{609}{8}$ **5.** 2 **7.** $24\pi^5$ **9.** $\pi/12$ **11.** 4
13. π **15.** $81\pi/2$ **17.** $5\pi/2$ **19.** $(2\pi/3)[1 - (1/\sqrt{2})]$
21. $(8\pi/3)(64 - 24\sqrt{3})$ **23.** $\frac{4}{3}\pi a^3$
25. $(\pi/4)(e - 1)$ **27.** $4\pi/3$ **29.** 0.587 **31.** $\frac{15}{16}$
33. (a) $\sqrt{\pi}/4$ (b) $\sqrt{\pi}/2$

Exercises 13.6 ■ page 832

1. $\frac{50}{3}$ C **3.** $\frac{2}{3}$, $(0, \frac{1}{2})$ **5.** 6, $(\frac{3}{4}, \frac{3}{2})$ **7.** $\frac{1}{6}$, $(\frac{4}{7}, \frac{3}{4})$
9. $\frac{27}{2}$, $(\frac{8}{5}, \frac{1}{2})$ **11.** $\pi/4$, $(\pi/2, 16/(9\pi))$ **13.** $(\frac{3}{8}, 3\pi/16)$
15. $(2a/5, 2a/5)$ if vertex is $(0, 0)$ and sides are along positive axes
17. $\frac{1}{10}$, $\frac{1}{16}$, $\frac{13}{80}$ **19.** $\frac{189}{20}$, $\frac{1269}{28}$, $\frac{1917}{35}$
21. $\rho a^4/3$, $\rho a^4/3$; $a/\sqrt{3}$, $a/\sqrt{3}$

Exercises 13.7 ■ page 834

1. $4\sqrt{6}\,\pi$ **3.** $12\sin^{-1}\frac{2}{3}$ **5.** $(\pi/6)(17\sqrt{17} - 5\sqrt{5})$
7. $(2\pi/3)(2\sqrt{2} - 1)$ **9.** $a^2(\pi - 2)$ **11.** 1.83
13. $\frac{3}{2} + \frac{5}{8}\ln 5$ **15.** 3.3213

Exercises 13.8 ■ page 842

3. $\frac{1}{48}$ **5.** $\frac{16}{3}$ **7.** $\frac{7}{5}$ **9.** $\frac{5}{28}$ **11.** $\frac{1}{10}$ **13.** $\frac{1}{12}$
15. $16\pi/3$ **17.** 8 **19.** $\frac{8}{15}$
21. (a) $\int_0^1 \int_0^x \int_0^{\sqrt{1-y^2}} dz\,dy\,dx$ (b) $\frac{1}{4}\pi - \frac{1}{3}$
23. 0.43, $\sqrt{3}/2$ **25.**

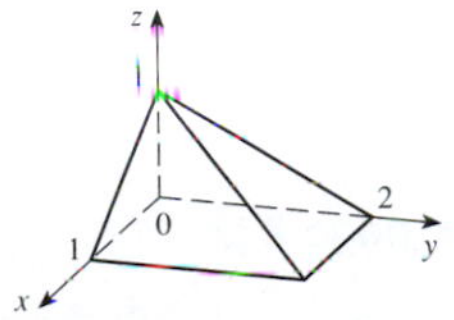

27. $\int_{-2}^2 \int_0^6 \int_{-\sqrt{4-x^2}}^{\sqrt{4-x^2}} f(x, y, z)\,dz\,dy\,dx$
$= \int_0^6 \int_{-2}^2 \int_{-\sqrt{4-x^2}}^{\sqrt{4-x^2}} f(x, y, z)\,dz\,dx\,dy$
$= \int_{-2}^2 \int_0^6 \int_{-\sqrt{4-z^2}}^{\sqrt{4-z^2}} f(x, y, z)\,dx\,dy\,dz$
$= \int_0^6 \int_{-2}^2 \int_{-\sqrt{4-z^2}}^{\sqrt{4-z^2}} f(x, y, z)\,dx\,dz\,dy$
$= \int_{-2}^2 \int_{-\sqrt{4-x^2}}^{\sqrt{4-x^2}} \int_0^6 f(x, y, z)\,dy\,dz\,dx$
$= \int_{-2}^2 \int_{-\sqrt{4-z^2}}^{\sqrt{4-z^2}} \int_0^6 f(x, y, z)\,dy\,dx\,dz$
29. $\int_{-1}^1 \int_0^{1-x^2} \int_0^y f(x, y, z)\,dz\,dy\,dx$
$= \int_0^1 \int_{-\sqrt{1-y}}^{\sqrt{1-y}} \int_0^y f(x, y, z)\,dz\,dx\,dy$
$= \int_0^1 \int_z^1 \int_{-\sqrt{1-y}}^{\sqrt{1-y}} f(x, y, z)\,dx\,dy\,dz$
$= \int_0^1 \int_0^y \int_{-\sqrt{1-y}}^{\sqrt{1-y}} f(x, y, z)\,dx\,dz\,dy$
$= \int_{-1}^1 \int_0^{1-x^2} \int_z^{1-x^2} f(x, y, z)\,dy\,dz\,dx$
$= \int_0^1 \int_{-\sqrt{1-z}}^{\sqrt{1-z}} \int_z^{1-x^2} f(x, y, z)\,dy\,dx\,dz$
31. $\int_0^1 \int_{\sqrt{x}}^1 \int_0^{1-y} f(x, y, z)\,dz\,dy\,dx = \int_0^1 \int_0^{y^2} \int_0^{1-y} f(x, y, z)\,dz\,dx\,dy$
$= \int_0^1 \int_0^{1-z} \int_0^{y^2} f(x, y, z)\,dx\,dy\,dz = \int_0^1 \int_0^{1-y} \int_0^{y^2} f(x, y, z)\,dx\,dz\,dy$
$= \int_0^1 \int_0^{1-\sqrt{x}} \int_{\sqrt{x}}^{1-z} f(x, y, z)\,dy\,dz\,dx$
$= \int_0^1 \int_0^{(1-z)^2} \int_{\sqrt{x}}^{1-z} f(x, y, z)\,dy\,dx\,dz$
33. $\int_0^1 \int_0^x \int_0^y f(x, y, z)\,dz\,dy\,dx = \int_0^1 \int_z^1 \int_y^1 f(x, y, z)\,dx\,dy\,dz$
$= \int_0^1 \int_0^y \int_y^1 f(x, y, z)\,dx\,dz\,dy = \int_0^1 \int_0^x \int_z^x f(x, y, z)\,dy\,dz\,dx$
$= \int_0^1 \int_z^1 \int_z^x f(x, y, z)\,dy\,dx\,dz$
35. $\frac{9}{10}$, $(\frac{22}{27}, \frac{25}{63}, \frac{152}{189})$
37. a^5, $(7a/12, 7a/12, 7a/12)$
39. (a) $m = \int_0^1 \int_0^{\sqrt{1-x^2}} \int_0^y (1 + x + y + z)\,dz\,dy\,dx$
(b) $(\bar{x}, \bar{y}, \bar{z})$, where
$\bar{x} = (1/m) \int_0^1 \int_0^{\sqrt{1-x^2}} \int_0^y x(1 + x + y + z)\,dz\,dy\,dx$,
$\bar{y} = (1/m) \int_0^1 \int_0^{\sqrt{1-x^2}} \int_0^y y(1 + x + y + z)\,dz\,dy\,dx$,
$\bar{z} = (1/m) \int_0^1 \int_0^{\sqrt{1-x^2}} \int_0^y z(1 + x + y + z)\,dz\,dy\,dx$
(c) $\int_0^1 \int_0^{\sqrt{1-x^2}} \int_0^y (x^2 + y^2)(1 + x + y + z)\,dz\,dy\,dx$
41. (a) $m = \int_{-1}^1 \int_{-\sqrt{1-y^2}}^{\sqrt{1-y^2}} \int_{4y^2+4z^2}^4 (x^2 + y^2 + z^2)\,dx\,dz\,dy$
(b) $(\bar{x}, \bar{y}, \bar{z})$, where
$\bar{x} = (1/m) \int_{-1}^1 \int_{-\sqrt{1-y^2}}^{\sqrt{1-y^2}} \int_{4y^2+4z^2}^4 x(x^2 + y^2 + z^2)\,dx\,dz\,dy$,
$\bar{y} = (1/m) \int_{-1}^1 \int_{-\sqrt{1-y^2}}^{\sqrt{1-y^2}} \int_{4y^2+4z^2}^4 y(x^2 + y^2 + z^2)\,dx\,dz\,dy$,
$\bar{z} = (1/m) \int_{-1}^1 \int_{-\sqrt{1-y^2}}^{\sqrt{1-y^2}} \int_{4y^2+4z^2}^4 z(x^2 + y^2 + z^2)\,dx\,dz\,dy$
(c) $\int_{-1}^1 \int_{-\sqrt{1-y^2}}^{\sqrt{1-y^2}} \int_{4y^2+4z^2}^4 (x^2 + y^2)(x^2 + y^2 + z^2)\,dx\,dz\,dy$
43. (a) $\frac{3}{32}\pi + \frac{11}{24}$ (b) $(\bar{x}, \bar{y}, \bar{z})$, where $\bar{x} = 28/(9\pi + 44)$,
$\bar{y} = 2(15\pi + 64)/[5(9\pi + 44)]$,
$\bar{z} = (45\pi + 208)/[15(9\pi + 44)]$
(c) $(68 + 15\pi)/240$
45. $I_x = I_y = I_z = \frac{2}{3}kL^5$ **47.** $L^3/8$
49. The region bounded by the ellipsoid $x^2 + 2y^2 + 3z^2 = 1$

Exercises 13.9 ■ page 848

1. $(0, 3, 1)$ **3.** $(-1, -\sqrt{3}, 8)$ **5.** $(3, 0, -6)$
7. $(1, \pi, 0)$ **9.** $(2, \pi/6, 4)$ **11.** $(4\sqrt{2}, \pi/4, 4)$
13. $(0, 0, 1)$ **15.** $(\sqrt{3}/4, \frac{1}{4}, \sqrt{3}/2)$
17. $(\sqrt{2}, \sqrt{2}, 2\sqrt{3})$ **19.** $(3, \pi, \pi/2)$
21. $(2, 0, \pi/3)$ **23.** $(2, 7\pi/4, 3\pi/4)$ **25.** $(\sqrt{2}, \pi/4, \pi/2)$
27. $(4\sqrt{2}, \pi/3, \pi/4)$ **29.** $(0, 0, 2)$ **31.** $(8, \pi/6, 0)$
33. Cylinder with radius 3 **35.** Half-cone
37. Circular paraboloid **39.** Horizontal plane
41. Positive z-axis
43. Circular cylinder, radius 1, axis parallel to the z-axis
45. Sphere, radius 5, center the origin
47. Circular cylinder, radius 2, axis the y-axis
49. Cylinder, radius 1, together with the z-axis
51. (a) $r^2 + z^2 = 16$ (b) $\rho = 4$
53. (a) $r\cos\theta + 2r\sin\theta + 3z = 6$
(b) $\rho(\sin\phi\cos\theta + 2\sin\phi\sin\theta + 3\cos\phi) = 6$
55. (a) $2z^2 = r^2\cos 2\theta - 4$
(b) $\rho^2(\sin^2\phi\cos 2\theta - 2\cos^2\phi) = 4$
57. (a) $r = 2\sin\theta$ (b) $\rho\sin\phi = 2\sin\theta$

59.

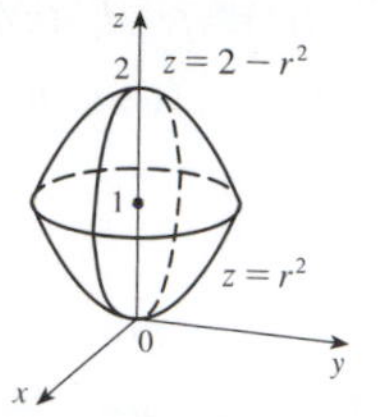

61.

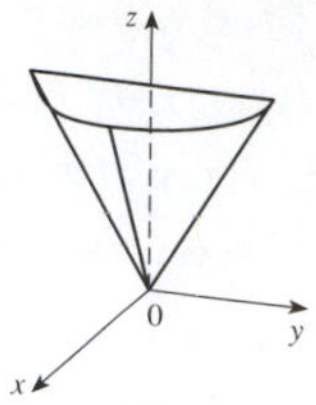

63. $0 \le \phi \le \pi/4,\ 0 \le \rho \le \cos\phi$

65.

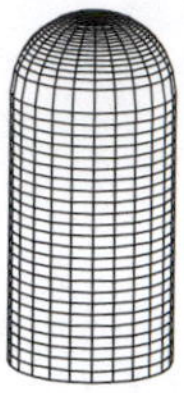

Exercises 13.10 ■ page 853

1. 8π **3.** $\pi/6$

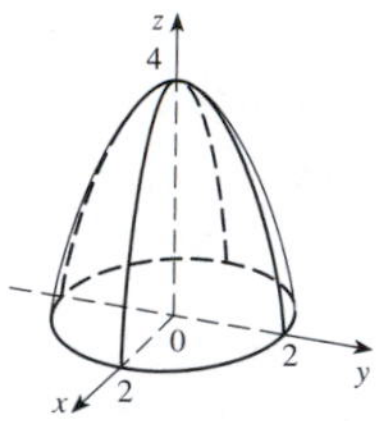

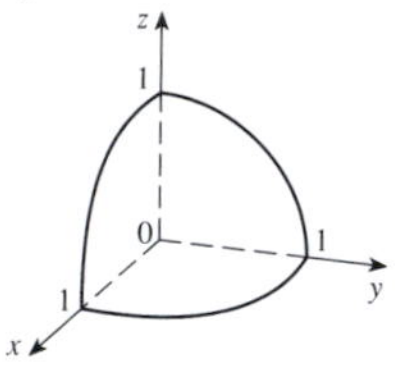

5. 24π **7.** 0 **9.** $2\pi/5$ **11.** 162π
13. $\pi K a^2/8,\ (0, 0, 2a/3)$ **15.** $4\pi/5$ **17.** $\pi/30$
19. $4\pi(2 - \sqrt{3})$ **21.** 10π
23. $K\pi a^4/2$, where K is the proportionality constant
25. $2\pi K a^6/9$ **27.** $4K\pi a^5/15$
29. $(2\pi/3)[1 - (1/\sqrt{2})],\ (0, 0, 3/[8(2 - \sqrt{2})])$ **31.** $5\pi/6$
33. $8\pi/35$ **35.** $243\pi/5$

Exercises 13.11 ■ page 861

1. 3 **3.** $2e^{4u}$ **5.** -4
7. The parallelogram bounded by the lines $y = 2x$, $y = 2x + 3$, $x = 2y$, $x = 2y - 6$
9. The triangular region with vertices $(0, 0)$, $(4, 0)$, $(3, 4)$
11. $\frac{11}{3}$ **13.** 6π **15.** $2\ln 3$ **17.** $\frac{4}{3}\pi abc$ **19.** $-\frac{66}{125}$
21. $\frac{3}{2}\sin 1$ **23.** $e - e^{-1}$

Chapter 13 Review Exercises ■ page 863

1. True **3.** True **5.** False **7.** 1152
9. $\ln(\frac{27}{4}) - 1$ **11.** $\frac{1}{110}$
13. The region outside the circle $r = 1$ and inside the cardioid $r = 1 + \sin\theta$
15. $(e - 1)/2$ **17.** $\ln\frac{3}{2}$ **19.** $\frac{1}{40}$ **21.** $\frac{41}{30}$ **23.** $81\pi/5$
25. $\frac{32}{3}$ **27.** $\pi/96$ **29.** $\frac{64}{15}$ **31.** 176 **33.** $\frac{2}{3}$
35. $2ma^3/9$ **37.** $\frac{1}{4},\ (\frac{1}{3}, \frac{8}{15})$ **39.** $(4a/(3\pi), 4a/(3\pi))$
41. $(0, 0, h/4)$ **43.** $\frac{7}{2}$

45.

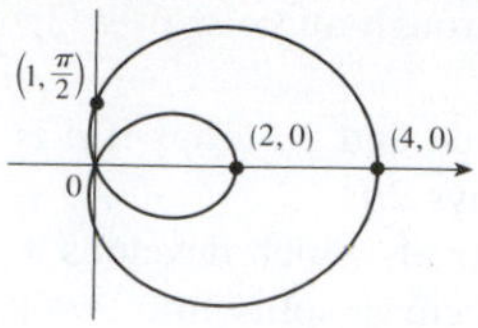

47.

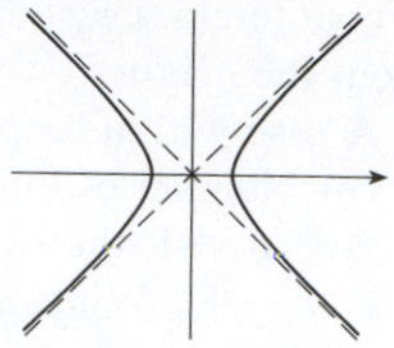

49.

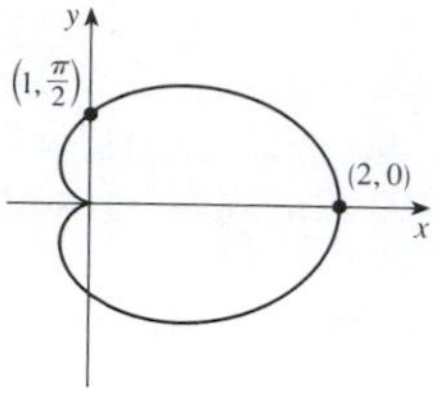

51.

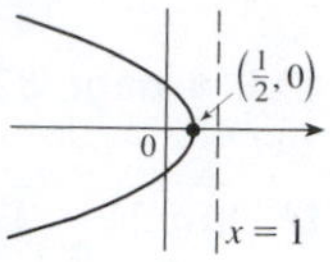

53. $r = 4\cos\theta$ **55.** $(\sqrt{3}, 1, 2), (2\sqrt{2}, \pi/6, \pi/4)$
57. $(1, \sqrt{3}, 2\sqrt{3}), (2, \pi/3, 2\sqrt{3})$ **59.** A half-plane
61. The horizontal plane $z = 3$ **63.** $r^2 + z^2 = 4, \rho = 2$
65. $z = 4r^2$ **67.** $(\pi/8)\ln 5$ **69.** 0.0512
71. $(\frac{4}{21}, \frac{11}{21}, \frac{8}{7})$ **73.** $\int_0^1 \int_0^{1-z} \int_{-\sqrt{y}}^{\sqrt{y}} f(x, y, z)\, dx\, dy\, dz$
75. $-\ln 2$ **77.** 0

APPLICATIONS PLUS ■ page 866

1. (a) $x = w/3$, base $= w/3$ (b) Yes
3. (a) $\iint_D (k/20)\left[20 - \sqrt{(x - x_0)^2 + (y - y_0)^2}\,\right] dA$, where D is the disk with radius 10 mi centered at the center of the city
(b) $200\pi k/3 \approx 209k$, $200(\pi/2 - \frac{8}{9})k \approx 136k$, on the edge
5. (a) $\iiint_C h(P)g(P)\, dV$, where C is the cone
(b) $\approx 3.1 \times 10^{19}$ ft-lb
7. (f) $\frac{2}{5}(1 - b^5)/(1 - b^3)$; this approaches $\frac{2}{5}$ as $a \to 0$ and $\frac{2}{3}$ as $a \to r$

CHAPTER 14

Exercises 14.1 ■ page 872

1.

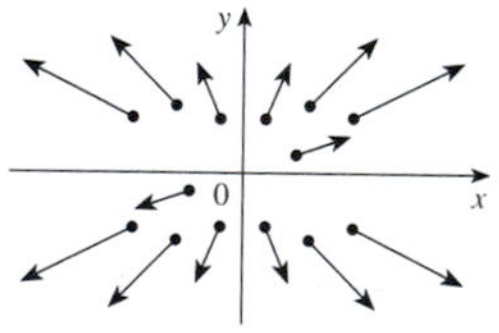

3.

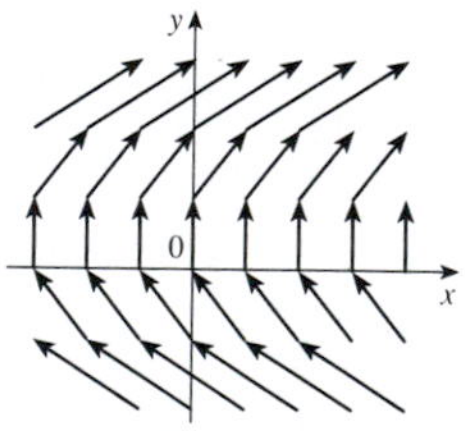

5.

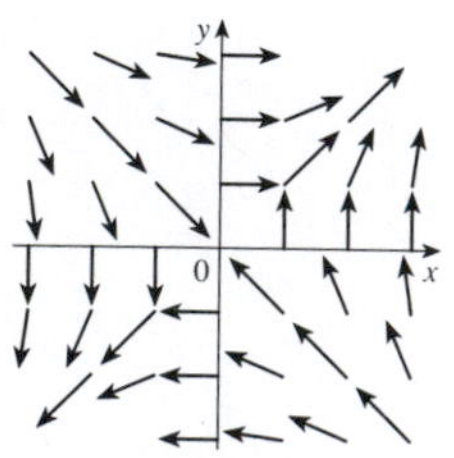

7.

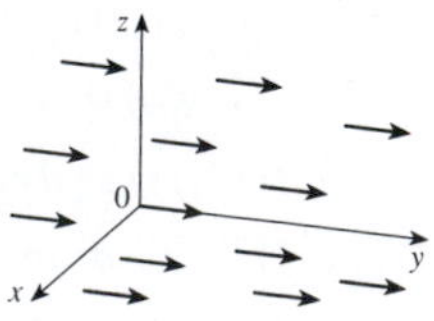

9.

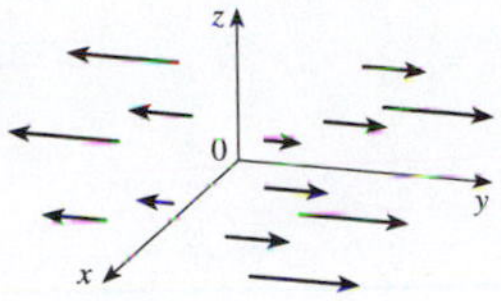

11. III **13.** II

15.

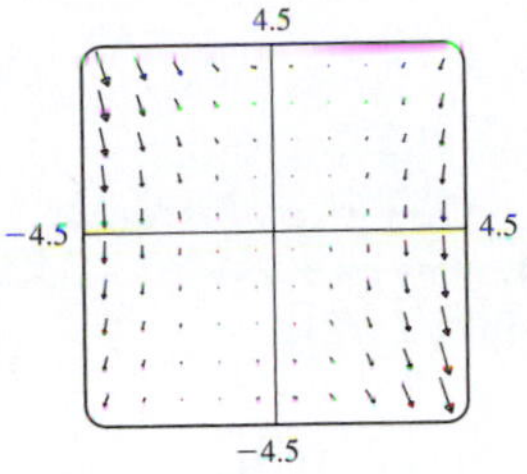

The line $y = 2x$

17. $\nabla f(x, y) = (5x^4 - 8xy^3)\,\mathbf{i} - 12x^2y^2\,\mathbf{j}$
19. $\nabla f(x, y) = 3e^{3x}\cos 4y\,\mathbf{i} - 4e^{3x}\sin 4y\,\mathbf{j}$
21. $\nabla f(x, y, z) = y^2\,\mathbf{i} + (2xy - z^3)\,\mathbf{j} - 3yz^2\,\mathbf{k}$
23. $\nabla f(x, y) = 2x\,\mathbf{i} - y\,\mathbf{j}$

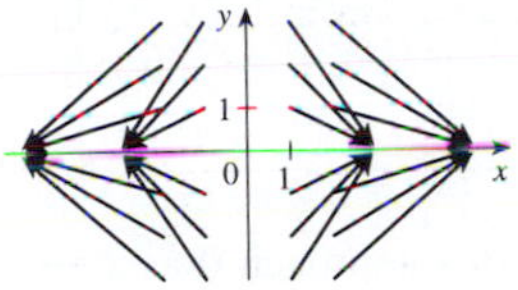

25. (a)

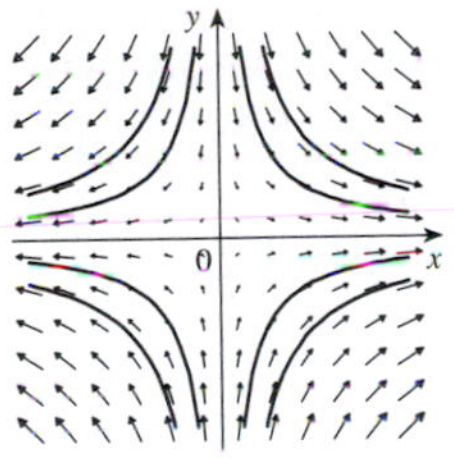

$y = C/x$

(b) $y = 1/x,\ x > 0$

Exercises 14.2 ■ page 882

1. $(10\sqrt{10} - 1)/54$ **3.** 1638.4 **5.** 48 **7.** $\frac{17}{3}$
9. $9\sqrt{13}\,\pi/4$ **11.** $3\sqrt{35}$ **13.** $\frac{10}{11}$ **15.** $\frac{77}{6}$ **17.** $\frac{10}{143}$
19. $\frac{6}{5} - \cos 1 - \sin 1$
21. (a) $\frac{11}{8} - 1/e$ (b)

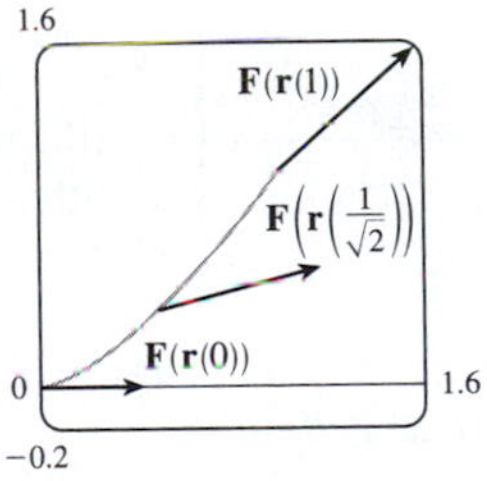

23. 0.052 **25.** $\frac{945}{16{,}777{,}216}\,\pi$
27. (a) Positive (b) Negative **29.** $2\pi k,\ (4/\pi, 0)$
31. (a) $\bar{x} = (1/m)\int_C x\rho(x, y, z)\,ds$,
$\bar{y} = (1/m)\int_C y\rho(x, y, z)\,ds$,
$\bar{z} = (1/m)\int_C z\rho(x, y, z)\,ds$, where $m = \int_C \rho(x, y, z)\,ds$
(b) $2\sqrt{13}\,k\pi,\ (0, 0, 3\pi)$
33. $I_x = k\left((\pi/2) - \frac{4}{3}\right),\ I_y = k\left((\pi/2) - \frac{2}{3}\right)$ **35.** $2\pi^2$ **37.** $\frac{23}{88}$
39. 1.67×10^4 ft-lb

Exercises 14.3 ■ page 891

1. $f(x, y) = x^2 - 3xy + y^2 + K$ **3.** Not conservative
5. $f(x, y) = x^4y^3 + x + K$ **7.** Not conservative
9. $f(x, y) = ye^x + x\sin y + K$
11. (a) $f(x, y) = (x^2 + y^2)/2$ (b) 44
13. (a) $f(x, y) = x^2y^3$ (b) $[1 + (\pi^2/4)]^3$
15. (a) $f(x, y, z) = xy + yz$ (b) 15
17. (a) $f(x, y, z) = x^2z + x\sin y$ (b) 2π
19. $25\sin 1 - 1$ **21.** $\frac{8}{3}$ **23.** No **25.** No
29. (a) Yes (b) Yes (c) Yes
31. (a) Yes (b) Yes (c) No

Exercises 14.4 ■ page 898

1. $-\frac{1}{12}$ **3.** $-\frac{1}{6}$ **7.** $-\frac{4}{3}$ **9.** $\frac{1}{3}$ **11.** 0 **13.** 0
15. π **17.** $\pi + \frac{16}{3}[(1/\sqrt{2}) - 1]$ **19.** $-\frac{1}{12}$ **21.** $3\pi/8$
23. (c) $\frac{9}{2}$ **25.** $\left(\frac{1}{3}, \frac{1}{3}\right)$

Exercises 14.5 ■ page 905

1. (a) $\mathbf{0}$ (b) 3 **3.** (a) $\mathbf{0}$ (b) 0
5. (a) $xz\,\mathbf{i} - yz\,\mathbf{j} + y\,\mathbf{k}$ (b) $x + xy$
7. (a) $(3xe^y + 2ye^{yz})\,\mathbf{i} + (xe^{xz} - 3e^y)\,\mathbf{j}$ (b) $ze^{xz} - 2ze^{yz}$
9. (a) $(\ln z + e^{-y})\,\mathbf{i} - xe^y\,\mathbf{k}$ (b) $e^y + ze^{-y} + (y/z)$
11. $f(x, y, z) = xy + z + K$ **13.** Not conservative
15. Not conservative **17.** $f(x, y, z) = xyz + (y^3/3) + K$
19. No
31. (a) Meaningless (b) Vector field (c) Scalar field
(d) Vector field (e) Meaningless (f) Vector field
(g) Scalar field (h) Meaningless (i) Vector field
(j) Meaningless (k) Meaningless (l) Scalar field
39. Negative

Exercises 14.6 ■ page 913

1. $x = x,\ y = y,\ z = \sqrt{1 - 3x^2 - 2y^2}$
[or $x = (1/\sqrt{3})\sin\phi\cos\theta,\ y = (1/\sqrt{2})\sin\phi\sin\theta,\ z = \cos\phi$,
$0 \leqslant \phi \leqslant \pi/2,\ 0 \leqslant \theta \leqslant 2\pi$]
3. $x = x,\ y = 6 - 3x^2 - 2z^2,\ z = z,\ 3x^2 + 2z^2 \leqslant 6$
5. $x = 2\sin\phi\cos\theta,\ y = 2\sin\phi\sin\theta$,
$z = 2\cos\phi,\ 0 \leqslant \phi \leqslant \pi/4,\ 0 \leqslant \theta \leqslant 2\pi$
7. $x = r\cos\theta,\ y = r\sin\theta,\ z = 5,\ 0 \leqslant r \leqslant 4,\ 0 \leqslant \theta \leqslant 2\pi$
[or $x = x,\ y = y,\ z = 5,\ x^2 + y^2 \leqslant 16$]
9. I **11.** II
13. $x = x,\ y = e^{-x}\cos\theta$,
$z = e^{-x}\sin\theta,\ 0 \leqslant x \leqslant 3$,
$0 \leqslant \theta \leqslant 2\pi$

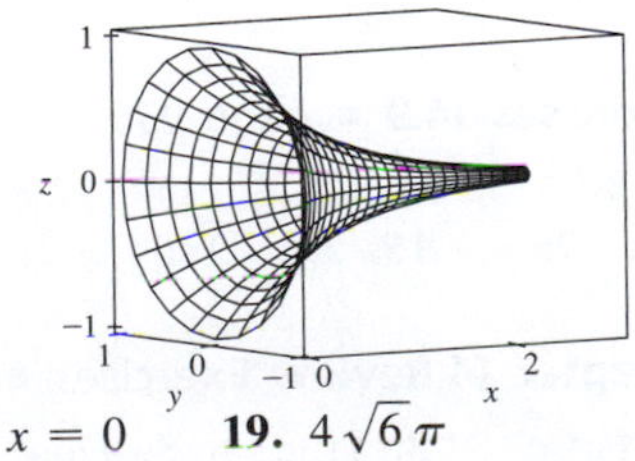

15. $3x - y + 3z = 3$ **17.** $x = 0$ **19.** $4\sqrt{6}\,\pi$
21. $(\pi/6)(17\sqrt{17} - 5\sqrt{5})$
23. $(\sqrt{21}/2) + \frac{17}{4}[\ln(2 + \sqrt{21}) - \ln\sqrt{17}]$
25. $2a^2(\pi - 2)$ **27.** $\pi(2\sqrt{6} - \frac{8}{3})$

29. (b)

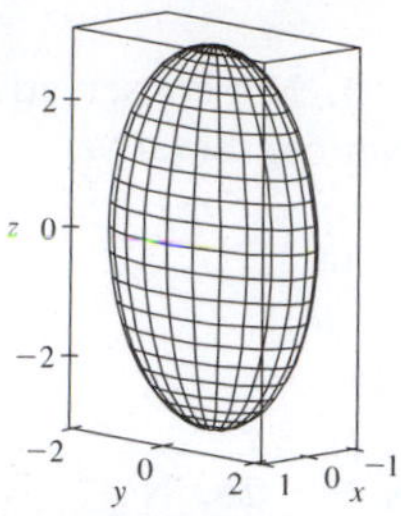

(c) $\int_0^{2\pi}\int_0^{\pi}\sqrt{36\sin^4 u\cos^2 v + 9\sin^4 u\sin^2 v + 4\cos^2 u\sin^2 u}\;du\,dv$

31. $x = (b + a\cos\alpha)\cos\theta$, $y = (b + a\cos\alpha)\sin\theta$, $z = a\sin\alpha$, $0 \leqslant \theta \leqslant 2\pi$, $0 \leqslant \alpha \leqslant 2\pi$

Exercises 14.7 ■ page 925

1. $8(1 + \sqrt{2} + \sqrt{3}) \approx 33.17$ **3.** $3\sqrt{14}$

5. $(33\sqrt{33} - 17\sqrt{17})/6$ **7.** $\pi\sqrt{2}/4$ **9.** 16π **11.** 16π

13. 0 **15.** $(11 - 10e)/6$ **17.** 12 **19.** 108π **21.** 0

23. 48 **25.** 0.1642 **27.** 3.4895

29. $\iint_S \mathbf{F}\cdot d\mathbf{S} = \iint_D [P(\partial h/\partial x) - Q + R(\partial h/\partial z)]\,dA$, where D = projection on xz-plane

31. $(0, 0, a/2)$

33. (a) $I_z = \iint_S (x^2 + y^2)\rho(x, y, z)\,dS$ (b) $4329\sqrt{2}\,\pi/5$

35. $194{,}400\pi$ **37.** $8\pi a^3\varepsilon_0/3$ **39.** 1248π

Exercises 14.8 ■ page 931

1. π **3.** $3\pi/4$ **5.** 0 **7.** 3.5 **9.** -4π

11. (a) $81\pi/2$ (b)

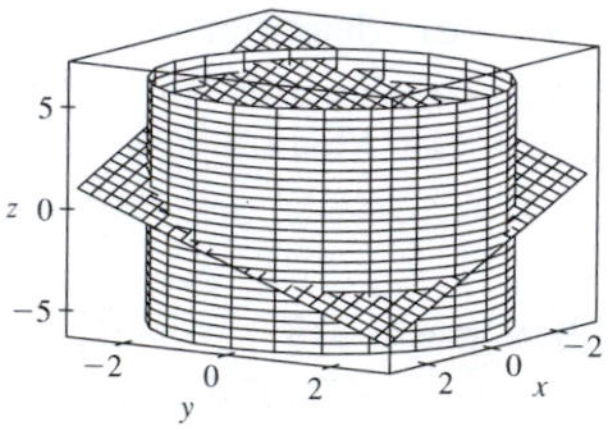

(c) $x = 3\cos t$, $y = 3\sin t$, $z = 1 - 3(\cos t + \sin t)$, $0 \leqslant t \leqslant 2\pi$

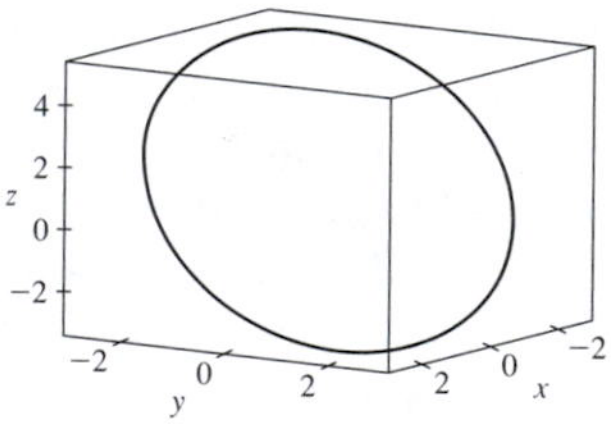

17. 16

Exercises 14.9 ■ page 936

3. 8 **5.** 0 **7.** $\frac{1}{6}$ **9.** $12\pi/5$ **11.** $-81\pi/2$

13. 27π **15.** $341\sqrt{2}/60 + \frac{81}{20}\arcsin(\sqrt{3}/3)$ **17.** $13\pi/20$

Chapter 14 Review Exercises ■ page 939

1. False **3.** True **5.** False **7.** True

9. $(5\sqrt{5} - 1)/3$ **11.** $4\sqrt{5}$ **13.** $-\pi$ **15.** $\frac{17}{2}$

17. 5 **19.** $f(x, y) = x\sin y - \cos y$ **21.** π^2 **25.** -8π

33. $\frac{1}{6}(27 - 5\sqrt{5})$ **35.** $\pi(391\sqrt{17} + 1)/60$

37. $-64\pi/3$ **41.** $-\frac{1}{2}$ **45.** -4 **47.** 21

PROBLEMS PLUS ■ page 941

3. (b) 0.90 **5.** 30 **7.** $\frac{1}{2}\sin 1$

CHAPTER 15

Exercises 15.1 ■ page 950

1. $y = c_1e^x + c_2e^{2x}$ **3.** $y = c_1e^{-x/3} + c_2e^{3x}$

5. $y = e^{-x}(c_1\cos 3x + c_2\sin 3x)$ **7.** $y = c_1e^x + c_2e^{-x}$

9. $y = c_1\cos 5x + c_2\sin 5x$ **11.** $y = c_1 + c_2e^{-x/2}$

13. $y = e^{x/2}[c_1\cos(\sqrt{7}x/2) + c_2\sin(\sqrt{7}x/2)]$

15. $y = c_1e^{(-1+\sqrt{2})x} + c_2e^{(-1-\sqrt{2})x}$

17. $y = e^{-x/4}[c_1\cos(\sqrt{23}x/4) + c_2\sin(\sqrt{23}x/4)]$

19.

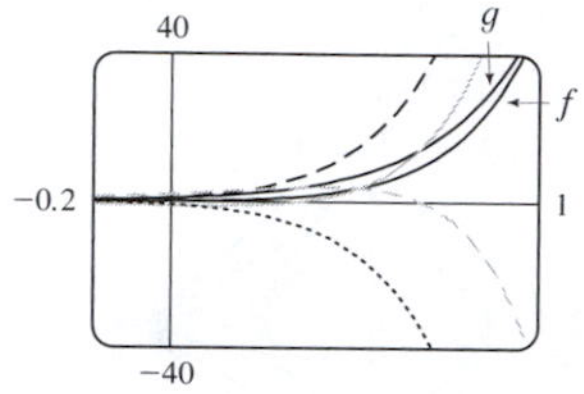

All solutions approach 0 as $x \to -\infty$ and approach $\pm\infty$ as $x \to \infty$.

21. $y = e^x + e^{-4x}$ **23.** $y = e^x(\cos x + \sin x)$

25. $y = 2e^{1-x} + e^{3(x-1)}$ **27.** $y = -\frac{1}{3}\sin 3x$

29. $y = 3xe^{-2x+2}$ **31.** No solution

37. (b) $\lambda = n^2\pi^2/L^2$, n a positive integer; $y = C\sin(n\pi x/L)$

Exercises 15.2 ■ page 957

1. $y = c_1e^{3x} + c_2e^{-2x} - \frac{5}{78}\cos 3x - \frac{1}{78}\sin 3x$

3. $y = e^{2x}(c_1x + c_2) + \frac{1}{9}e^{-x}$

5. $y = c_1\cos 6x + c_2\sin 6x + (x^2/18) - (x/36) - \frac{1}{324}$

7. $y = e^x[\frac{523}{650}\cos 2x + \frac{797}{1300}\sin 2x] + \frac{1}{5}x + \frac{2}{25} + \frac{3}{26}\cos 3x - \frac{1}{13}\sin 3x$

9. $y = \frac{5}{8}e^x - \frac{17}{32}e^{-x} + e^{3x}[\frac{1}{8}x - \frac{3}{32}]$

11.

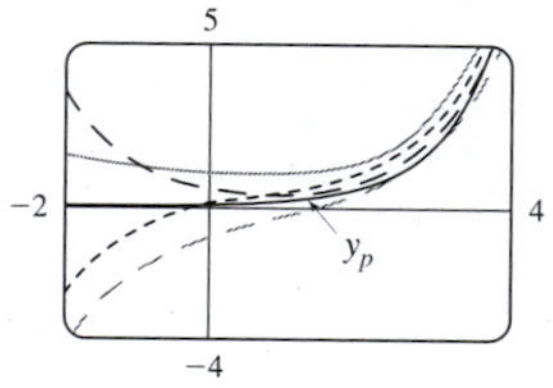

The solutions are all asymptotic to $y_p = e^x/10$ as $x \to \infty$. Except for y_p, all solutions approach either ∞ or $-\infty$ as $x \to -\infty$.

13. $y_p = (Ax^4 + Bx^3 + Cx^2 + Dx + E)e^{2x}$

15. $y_p = xe^x(A\cos x + B\sin x)$

17. $y = c_1\cos 2x + c_2\sin 2x + \frac{1}{4}x$

19. $y = c_1e^x + c_2xe^x + e^{2x}$

21. $y = (c_1 + x)\sin x + (c_2 + \ln\cos x)\cos x$

23. $y = [c_1 + \ln(1 + e^{-x})]e^x + [c_2 - e^{-x} + \ln(1 + e^{-x})]e^{2x}$

25. $y = [c_1 - \frac{1}{2}\int(e^x/x)\,dx]e^{-x} + [c_2 + \frac{1}{2}\int(e^{-x}/x)\,dx]e^x$

Exercises 15.3 ■ page 964

1. $x = 0.36\sin(10t/3)$ **3.** $x = -\frac{1}{5}e^{-6t} + \frac{6}{5}e^{-t}$ **5.** $\frac{49}{12}$ kg

9. $Q(t) = (-e^{-10t}/250)(6\cos 20t + 3\sin 20t) + \frac{3}{125}$, $I(t) = \frac{3}{5}e^{-10t}\sin 20t$

11. $Q(t) = e^{-10t}[\frac{3}{250}\cos 20t - \frac{3}{500}\sin 20t] - \frac{3}{250}\cos 10t + \frac{3}{125}\sin 10t$

Exercises 15.4 ▪ page 969

1. $\sum_{n=0}^{\infty} c_0 \frac{6^n}{n!} x^n = c_0 e^{6x}$ **3.** $\sum_{n=0}^{\infty} \frac{c_0}{3^n n!} x^{3n} = c_0 e^{x^3/3}$

5. $c_0 \sum_{n=0}^{\infty} \left(-\frac{3}{2}\right)^n \frac{1}{n!} x^{2n} + c_1 \sum_{n=0}^{\infty} \frac{(-6)^n n!}{(2n+1)!} x^{2n+1}$

7. $\sum_{n=0}^{\infty} \frac{x^{2n}}{2^n n!} = e^{x^2/2}$

9. $x + \sum_{n=1}^{\infty} \frac{(-1)^n 2^2 5^2 \cdot \cdots \cdot (3n-1)^2}{(3n+1)!} x^{3n+1}$

Chapter 15 Review Exercises ▪ page 969

1. True **3.** True

5. $y = e^{3x}(c_1 \cos 5x + c_2 \sin 5x)$ **7.** $y = c_1 e^{-x} + c_2 e^{x/2}$

9. $y = e^{-x}(c_1 + c_2 x) - \frac{3}{50}\cos 3x - \frac{2}{25}\sin 3x$

11. $y = c_1 \cos(3x/2) + c_2 \sin(3x/2) + \frac{2}{9}x^2 - \frac{43}{81}$

13. $y = c_1 e^x + c_2 e^{2x} + xe^{2x}$ **15.** $y = 5 - 2e^{-6(x-1)}$

17. $y = (e^{4x} - e^x)/3$ **19.** $\sum_{n=0}^{\infty} \frac{(-2)^n n!}{(2n+1)!} x^{2n+1}$

21. $Q(t) = -0.02e^{-10t}(\cos 10t + \sin 10t) + 0.03$

23. (c) $2\pi/k \approx 85$ min (d) $\approx$17,600 mi/h

APPENDIXES

Exercises A ▪ page A9

1. 18 **3.** π **5.** $5 - \sqrt{5}$ **7.** $2 - x$

9. $|x + 1| = \begin{cases} x + 1 & \text{for } x \geq -1 \\ -x - 1 & \text{for } x < -1 \end{cases}$ **11.** $x^2 + 1$

13. $(-2, \infty)$

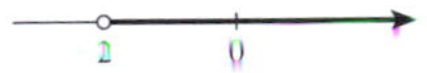

15. $[-1, \infty)$

17. $(3, \infty)$

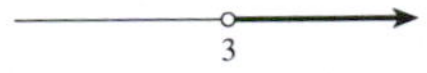

19. $(2, 6)$

21. $(0, 1]$

23. $[-1, \frac{1}{2})$

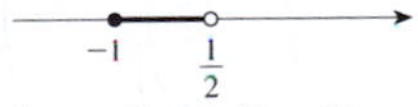

25. $[2, 3]$

27. $(-\infty, 1) \cup (2, \infty)$

29. $[-1, \frac{1}{2}]$

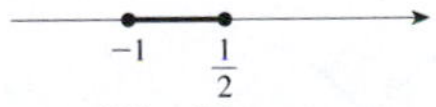

31. $(-\infty, \infty)$

33. $(-\sqrt{3}, \sqrt{3})$

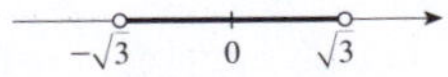

35. $(-\infty, 1]$

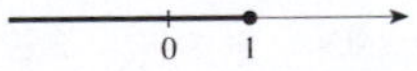

37. $(-1, 0) \cup (1, \infty)$

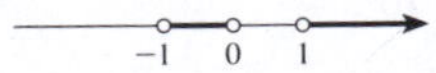

39. $(-\infty, 0) \cup (\frac{1}{4}, \infty)$

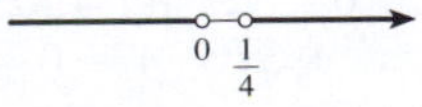

41. $(-2, 0) \cup (2, \infty)$

43. $(-\infty, 5) \cup (16, \infty)$

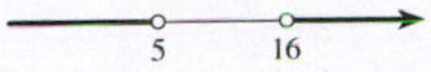

45. $(-\infty, -1] \cup [1, \infty)$

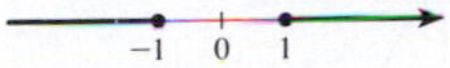

47. $10 \leq C \leq 35$ **49.** (a) $T = 20 - 10h$, $0 \leq h \leq 12$ (b) $-30\,^\circ\text{C} \leq T \leq 20\,^\circ\text{C}$ **51.** $\pm\frac{3}{2}$ **53.** $2, -\frac{4}{3}$

55. $(-3, 3)$ **57.** $(3, 5)$ **59.** $(-\infty, -7] \cup [-3, \infty)$

61. $[1.3, 1.7]$ **63.** $[-4, -1] \cup [1, 4]$ **65.** $(\frac{1}{2}, \infty)$

67. $(-1, \infty)$ **69.** $x \geq (a + b)c/(ab)$ **71.** $x > (c - b)/a$

Exercises B ▪ page A15

1. 5 **3.** $\sqrt{74}$ **5.** $2\sqrt{37}$ **7.** 2 **9.** $-\frac{9}{2}$

17.

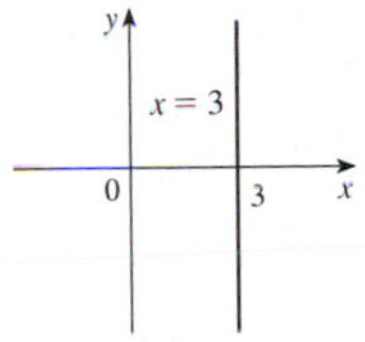

19.

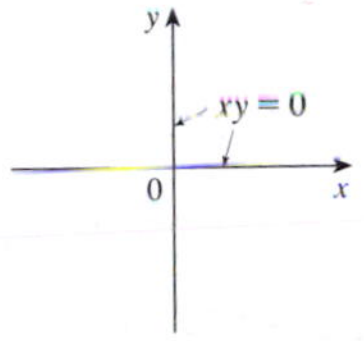

21. $y = 6x - 15$ **23.** $2x - 3y + 19 = 0$

25. $5x + y = 11$ **27.** $y = 3x - 2$ **29.** $y = 3x - 3$

31. $y = 5$ **33.** $x + 2y + 11 = 0$ **35.** $5x - 2y + 1 = 0$

37. $m = -\frac{1}{3}$, $b = 0$

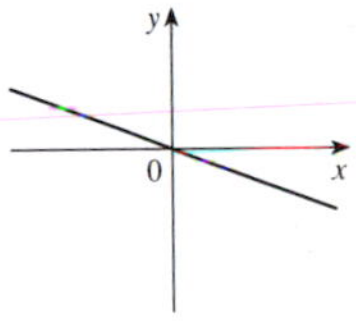

39. $m = 0$, $b = -2$

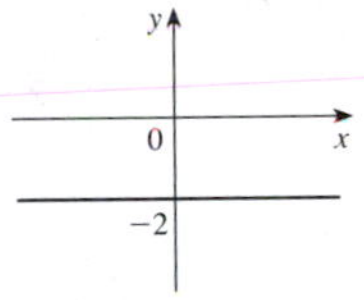

41. $m = \frac{3}{4}$, $b = -3$

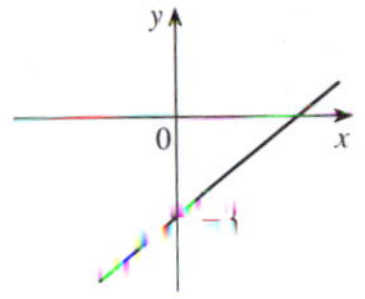

43.

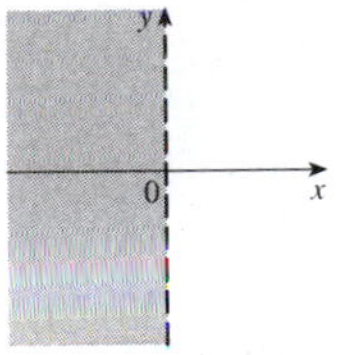

45.

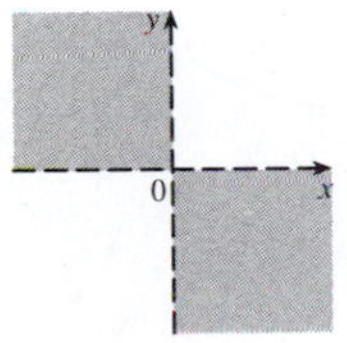

47.

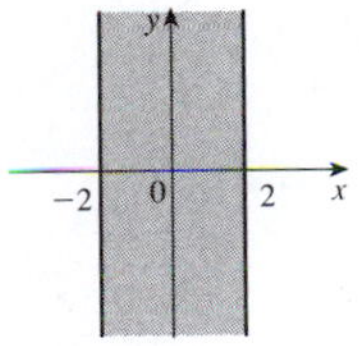

49.

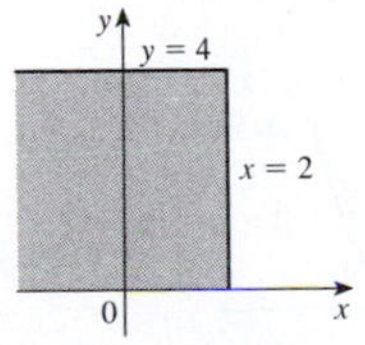

51.

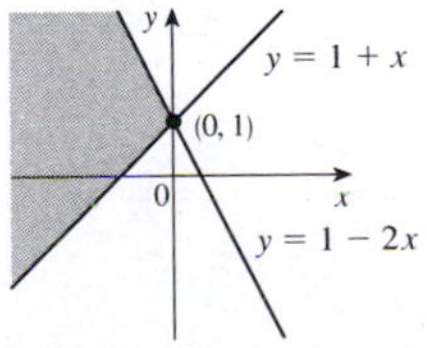

53. $(0, -4)$ **55.** (a) $(4, 9)$ (b) $(3.5, -3)$ **57.** $(1, -2)$

59. $y = x - 3$ **61.** (b) $4x - 3y - 24 = 0$

Exercises C ▪ page A22

1. $(x - 3)^2 + (y + 1)^2 = 25$ **3.** $x^2 + y^2 = 65$

5. $(2, -5)$, 4 **7.** $(-\frac{1}{2}, 0)$, $\frac{1}{2}$ **9.** $(\frac{1}{4}, -\frac{1}{4})$, $\sqrt{10}/4$

11. Parabola

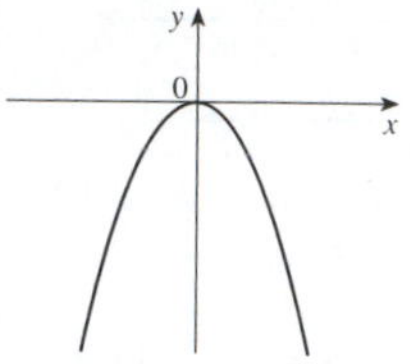

13. Ellipse

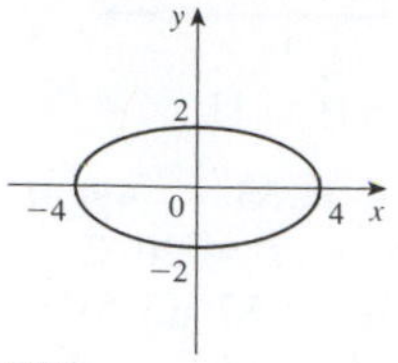

15. Hyperbola

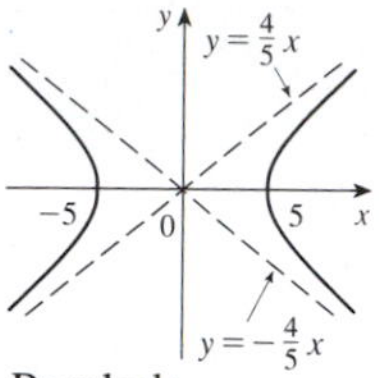

17. Ellipse

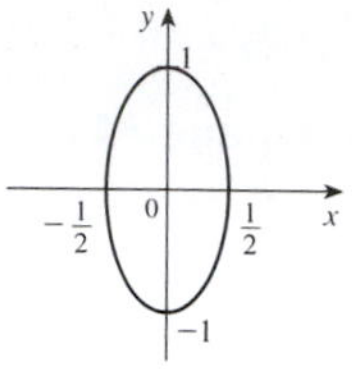

19. Parabola

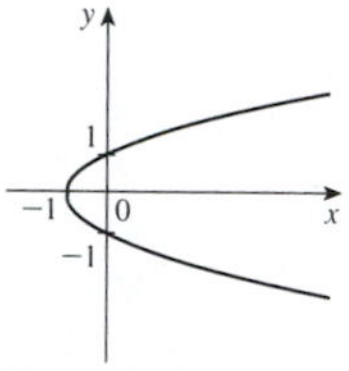

21. Hyperbola

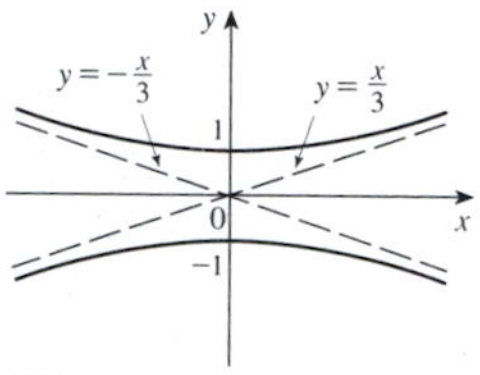

23. Hyperbola

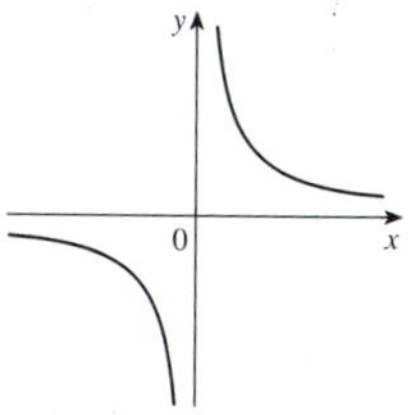

25. Ellipse

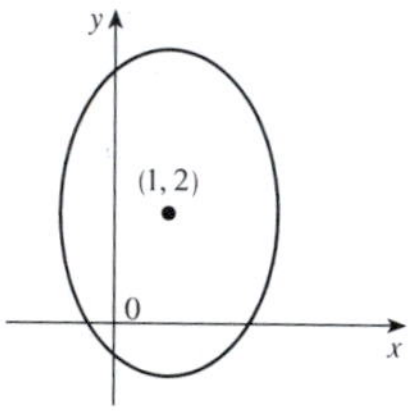

27. Parabola

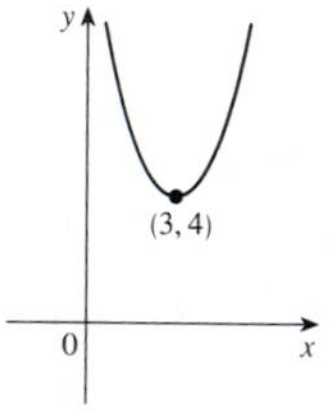

29. Parabola

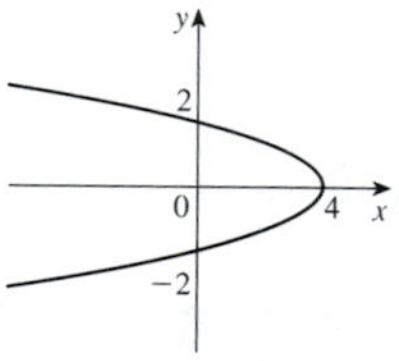

31. Ellipse

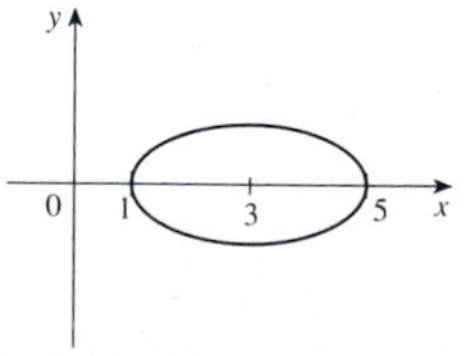

33.

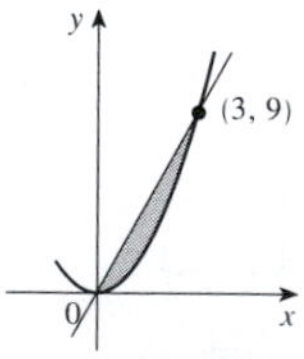

35. $y = x^2 - 2x$

37.

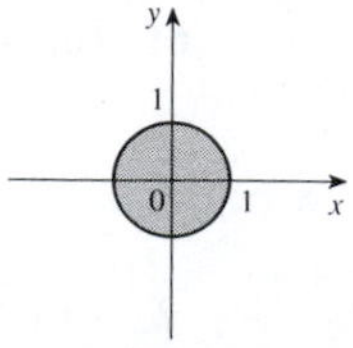

39.

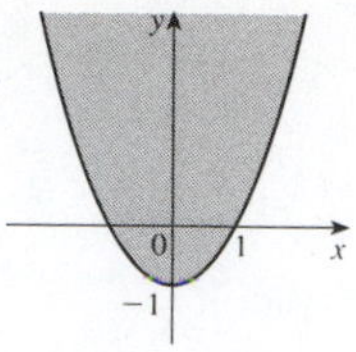

Exercises D ■ page A30

1. $7\pi/6$ **3.** $\pi/20$ **5.** 5π **7.** $720°$ **9.** $75°$
11. $-67.5°$ **13.** 3π cm **15.** $\frac{2}{3}$ rad $= (120/\pi)°$
17.

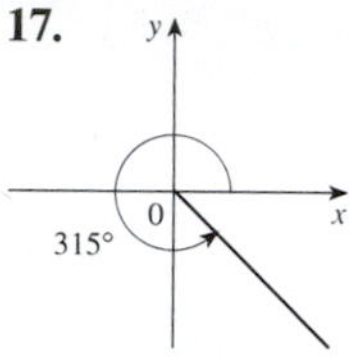

19.

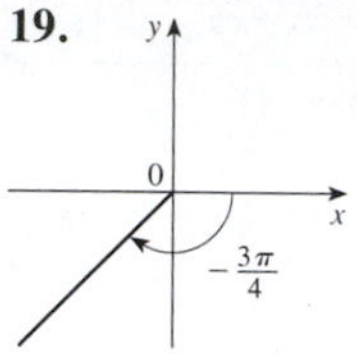

21.

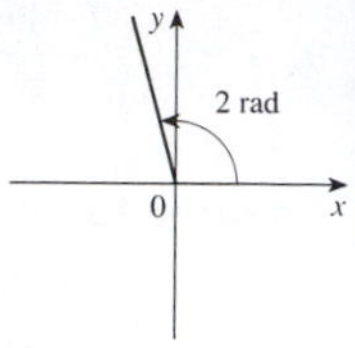

23. $\sin(3\pi/4) = 1/\sqrt{2}$, $\cos(3\pi/4) = -1/\sqrt{2}$, $\tan(3\pi/4) = -1$, $\csc(3\pi/4) = \sqrt{2}$, $\sec(3\pi/4) = -\sqrt{2}$, $\cot(3\pi/4) = -1$
25. $\sin(9\pi/2) = 1$, $\cos(9\pi/2) = 0$, $\csc(9\pi/2) = 1$, $\cot(9\pi/2) = 0$, $\tan(9\pi/2)$ and $\sec(9\pi/2)$ undefined
27. $\frac{1}{2}$, $-\sqrt{3}/2$, $-1/\sqrt{3}$, 2, $-2/\sqrt{3}$, $-\sqrt{3}$
29. $\cos\theta = \frac{4}{5}$, $\tan\theta = \frac{3}{4}$, $\csc\theta = \frac{5}{3}$, $\sec\theta = \frac{5}{4}$, $\cot\theta = \frac{4}{3}$
31. $\sin\phi = \sqrt{5}/3$, $\cos\phi = -\frac{2}{3}$, $\tan\phi = -\sqrt{5}/2$, $\csc\phi = 3/\sqrt{5}$, $\cot\phi = -2/\sqrt{5}$
33. $\sin\beta = -1/\sqrt{10}$, $\cos\beta = -3/\sqrt{10}$, $\tan\beta = \frac{1}{3}$, $\csc\beta = -\sqrt{10}$, $\sec\beta = -\sqrt{10}/3$
35. 5.73576 cm **37.** 24.62147 cm **59.** $(4 + 6\sqrt{2})/15$
61. $(3 + 8\sqrt{2})/15$ **63.** $\frac{24}{25}$ **65.** $\pi/3$, $5\pi/3$
67. $\pi/4$, $3\pi/4$, $5\pi/4$, $7\pi/4$ **69.** $\pi/6$, $\pi/2$, $5\pi/6$, $3\pi/2$
71. 0, π, 2π **73.** $0 \leqslant x \leqslant \pi/6$ and $5\pi/6 \leqslant x \leqslant 2\pi$
75. $0 \leqslant x < \pi/4$, $3\pi/4 < x < 5\pi/4$, $7\pi/4 < x \leqslant 2\pi$
77.

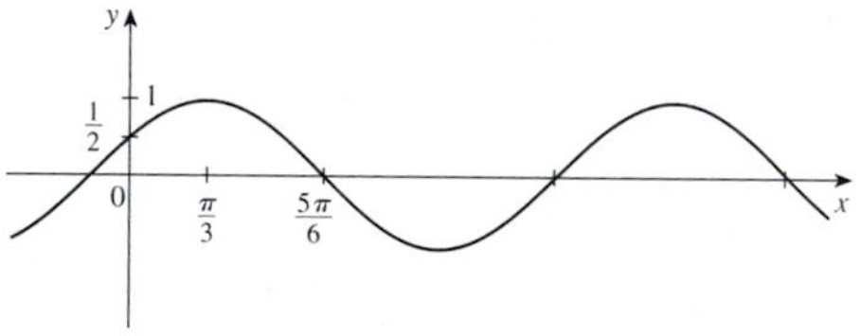

79.

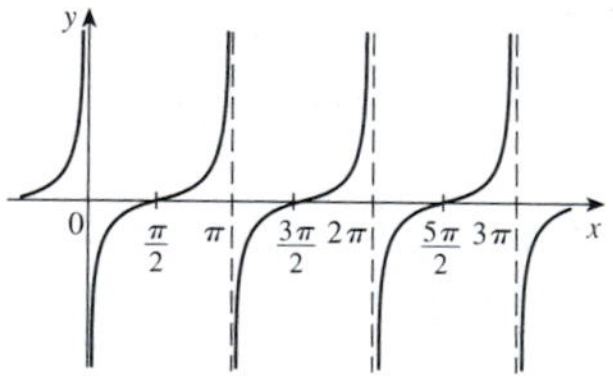

81.

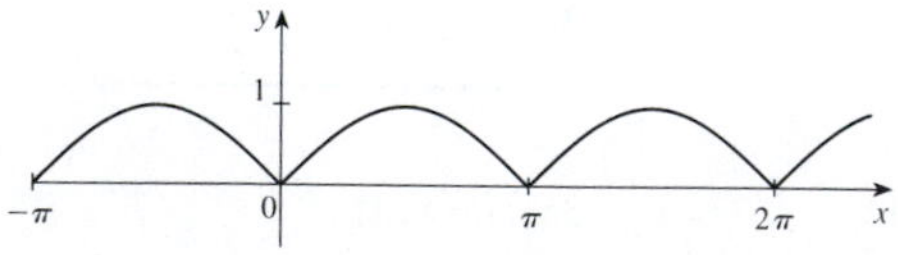

89. 14.34457

Exercises G ■ page A49

1. $\frac{1}{3}$ **3.** 0 **5.** (a) $4.82549424 \times 10^{-11}$
9. (b) 1, $x^3 + 3x - 4 = 0$

Exercises H ■ page A57

1. $10 - i$ **3.** $13 - i$ **5.** $12 - 7i$ **7.** $-\frac{1}{2} + \frac{1}{2}i$
9. $\frac{1}{2} - \frac{1}{2}i$ **11.** $-i$ **13.** $5i$ **15.** $3 - 4i$, 5 **17.** $4i$, 4
19. $\pm\frac{3}{2}i$ **21.** $4 \pm i$ **23.** $-\frac{1}{2} \pm (\sqrt{7}/2)i$
25. $3\sqrt{2}[\cos(3\pi/4) + i\sin(3\pi/4)]$
27. $5\{\cos[\tan^{-1}\frac{4}{3}] + i\sin[\tan^{-1}\frac{4}{3}]\}$
29. $4[\cos(\pi/2) + i\sin(\pi/2)]$, $\cos(-\pi/6) + i\sin(-\pi/6)$, $\frac{1}{2}[\cos(-\pi/6) + i\sin(-\pi/6)]$
31. $4\sqrt{2}[\cos(7\pi/12) + i\sin(7\pi/12)]$, $(2\sqrt{2})[\cos(13\pi/12) + i\sin(13\pi/12)]$, $\frac{1}{4}[\cos(\pi/6) + i\sin(\pi/6)]$
33. -1024 **35.** $-512\sqrt{3} + 512i$
37. ± 1, $\pm i$, $(1/\sqrt{2})(\pm 1 \pm i)$ **39.** $\pm(\sqrt{3}/2) + \frac{1}{2}i$, $-i$

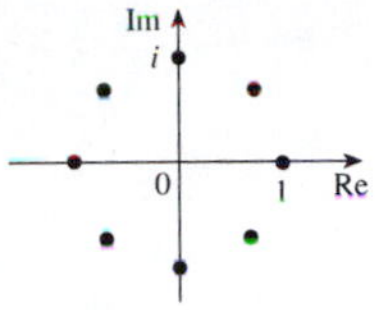

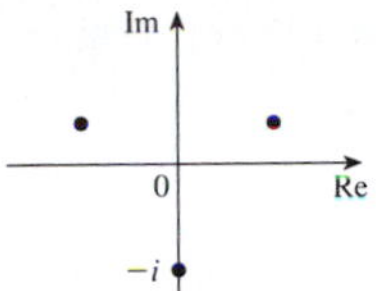

41. i **43.** $(-1/\sqrt{2}) + (1/\sqrt{2})i$ **45.** $-e^2$
47. $\cos 3\theta = \cos^3\theta - 3\cos\theta\sin^2\theta$, $\sin 3\theta = 3\cos^2\theta\sin\theta - \sin^3\theta$

Exercises I ■ page A64

1. $(0, 0)$, $(0, -2)$, $y = 2$

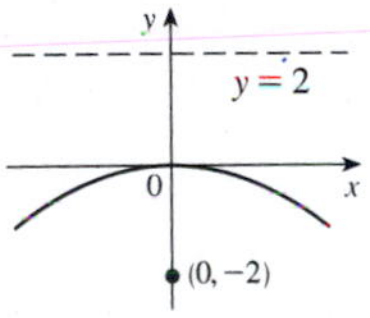

3. $(0, 0)$, $(\frac{1}{4}, 0)$, $x = -\frac{1}{4}$

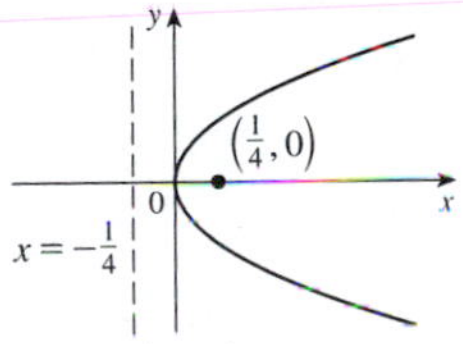

5. $(-1, 3)$, $(-\frac{7}{8}, 3)$, $x = -\frac{9}{8}$

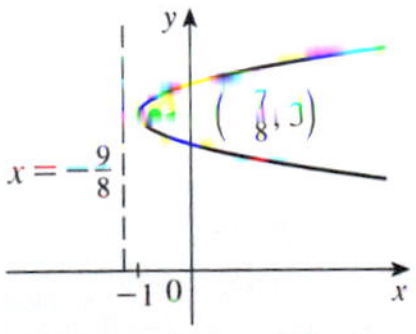

7. $(2, 4)$, $(\frac{3}{2}, 4)$, $x = \frac{5}{2}$

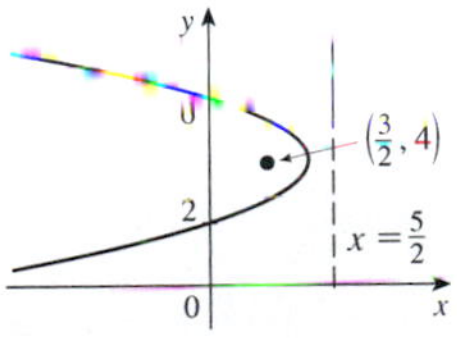

9. $(\pm 4, 0)$, $(\pm 2\sqrt{3}, 0)$

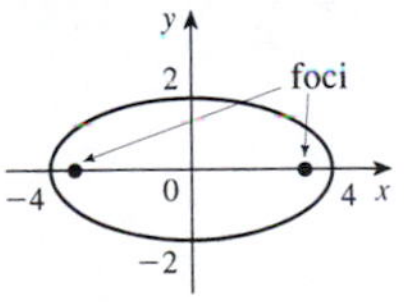

11. $(0, \pm 5)$, $(0, \pm 4)$

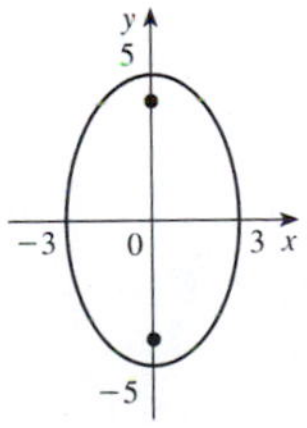

13. $(\pm 12, 0)$, $(\pm 13, 0)$, $y = \pm\frac{5}{12}x$

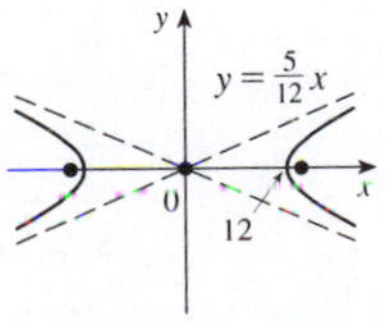

15. $(0, \pm 1)$, $(0, \pm\sqrt{10})$, $y = \pm x/3$

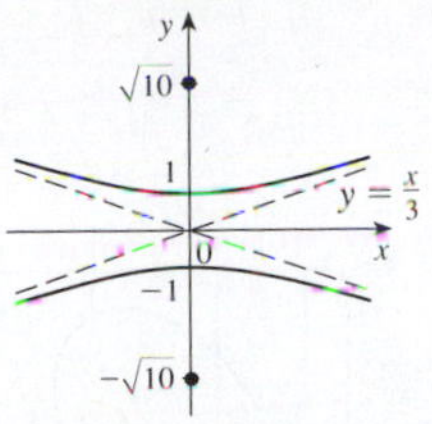

17. $(1, \pm 3)$, $(1, \pm\sqrt{5})$

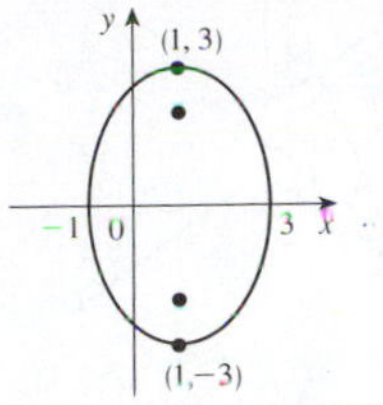

19. $(2 \pm \sqrt{6}, 1)$, $(2 \pm \sqrt{15}, 1)$, $y - 1 = \pm(\sqrt{6}/2)(x - 2)$

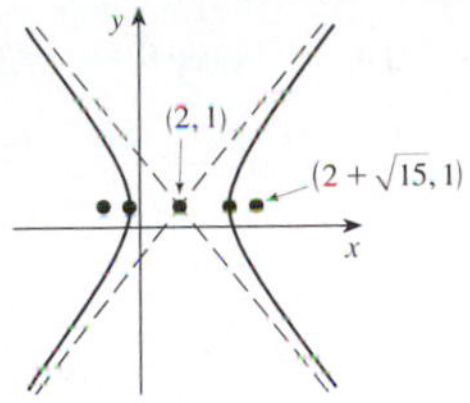

21. $x^2 = 12y$ **23.** $y^2 = 4(x - 2)$ **25.** $y^2 = 16x$
27. $(x^2/4) + (y^2/3) = 1$ **29.** $[(x - 3)^2/8] + (y^2/9) = 1$
31. $[(x - 2)^2/9] + [(y - 2)^2/5] = 1$ **33.** $y^2 - (x^2/8) = 1$
35. $[(x - 4)^2/4] - [(y - 3)^2/5] = 1$
37. $(x^2/9) - (y^2/36) = 1$
39. $(x^2/3{,}763{,}600) + (y^2/3{,}753{,}196) = 1$
41. (a) $(121x^2/1{,}500{,}625) - (121y^2/3{,}339{,}375) = 1$
(b) ≈248 mi
45. (a) Ellipse (b) Hyperbola (c) No curve
47. 9.69

Exercises J ■ page A69

1. $r = 6/(3 + 2\cos\theta)$ **3.** $r = 2/(1 + \sin\theta)$
5. $r = 20/(1 + 4\cos\theta)$ **7.** $r = 10/(1 + \sin\theta)$
9. (a) 3 (b) Hyperbola (c) $x = \frac{4}{3}$
(d)

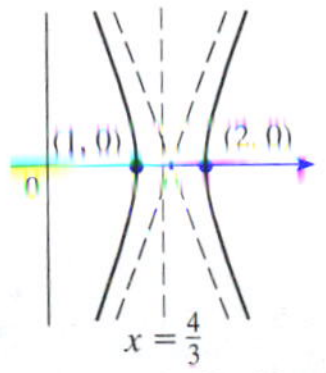

11. (a) 1 (b) Parabola (c) $x = -2$
(d)

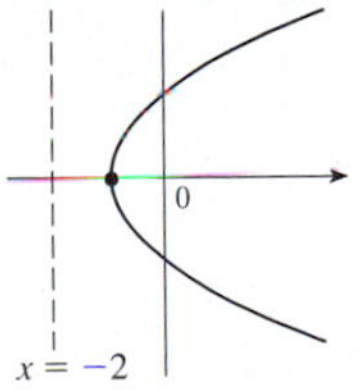

13. (a) $\frac{1}{2}$ (b) Ellipse (c) $y = 6$
(d)

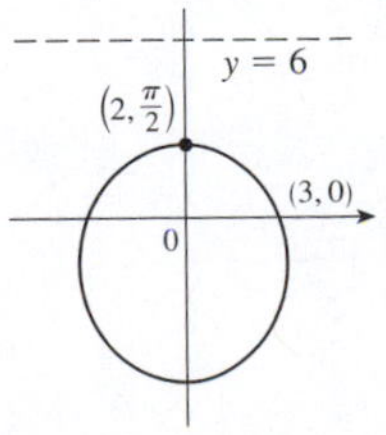

15. (a) $\frac{5}{2}$ (b) Hyperbola (c) $y = -\frac{7}{5}$
(d)

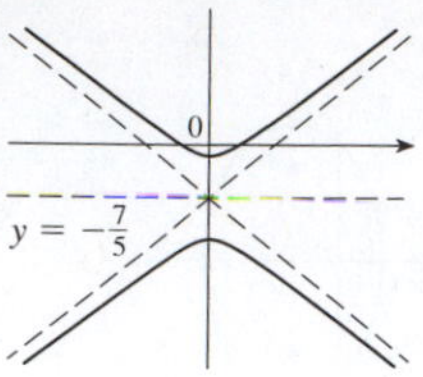

17. (a) $e = \frac{3}{4}$, directrix $x = -\frac{1}{3}$
(b) $r = 1/[4 - 3\cos(\theta - \pi/3)]$

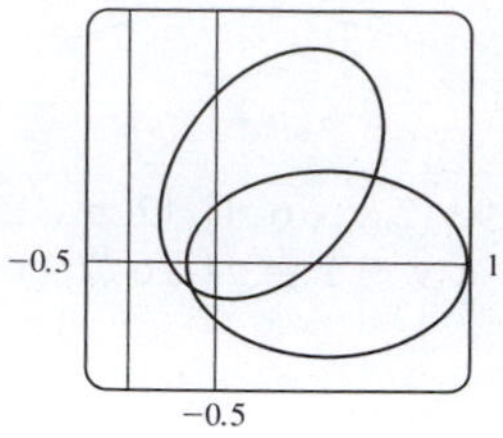

19. The ellipse is nearly circular when e is close to 0 and becomes more elongated as $e \to 1^-$. At $e = 1$, the curve becomes a parabola.

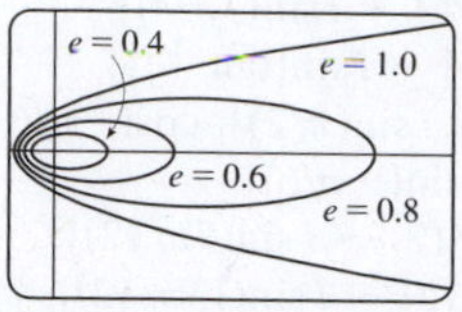

25. (b) $r = (1.49 \times 10^8)/(1 - 0.017\cos\theta)$
27. 7.0×10^7 km
29. 3.6×10^8 km

INDEX